American College of Veterinary Internal Medicine (ACVIM) Forum 2014

AA002929

In Tune with Innovation and Science

Nashville, Tennessee, USA
4-7 June 2014

ISBN: 978-1-63266-516-4

Printed from e-media with permission by:
Curran Associates, Inc.
57 Morehouse Lane
Red Hook, NY 12571

Printed by Curran Associates, Inc. (2014)

Address correspondence to:
AMERICAN COLLEGE OF VETERINARY INTERNAL MEDICINE (ACVIM)
1997 WADSWORTH BLVD.
LAKEWOOD, CO 80214-5293
www.ACVIM.org

Additional copies of this publication are available from:

Curran Associates, Inc.
57 Morehouse Lane
Red Hook, NY 12571 USA
Phone: 845-758-0400
Fax: 845-758-2634
Email: curran@proceedings.com
Web: www.proceedings.com

TABLE OF CONTENTS

RESEARCH REPORT CARDIOLOGY

RESEARCH REPORT NEUROLOGY

RESEARCH REPORT ONCOLOGY

RESEARCH REPORT SAIM

RESEARCH REPORT EQUINE

RESEARCH REPORT FAIM

CARDIOLOGY PRE-FORUM

NEUROLOGY PRE-FORUM

ONCOLOGY PRE-FORUM

SMALL ANIMAL INTERNAL MEDICINE PRE-FORUM

EQUINE PRE-FORUM

FOOD ANIMAL PRE-FORUM

ONE MEDICINE

SCIENTIFIC SESSION EQUINE CLINICAL WORKSHOP

AMERICAN ASSOCIATION OF VETERINARY NUTRITION/AMERICAN COLLEGE OF VETERINARY NUTRITION EQUINE SCIENTIFIC SESSION

AMERICAN ASSOCIATION OF VETERINARY NUTRITION/AMERICAN COLLEGE OF VETERINARY NUTRITION SMALL SCIENTIFIC SESSION

AMERICAN COLLEGE OF VETERINARY CLINICAL PHARMACOLOGY/AMERICAN OF VETERINARY PHARMACOLOGY AND THERAPEUTICS SCIENTIFIC SESSION

AMERICAN SOCIETY FOR VETERINARY CLINICAL PATHOLOGY SCIENTIFIC SESSION

AMERICAN SOCIETY OF VETERINARY NEPHROLOGY & UROLOGY SCIENTIFIC SESSION

ASSOCIATION OF VETERINARY HEMATOLOGY AND TRANSFUSION MEDICINE SCIENTIFIC SESSION

SCIENTIFIC SESSION BUSINESS MANAGEMENT

COMPANION ANIMAL PARASITE COUNCIL TRACK

SCIENTIFIC SESSION CARDIOLOGY COMPREHENSIVE REVIEW

SCIENTIFIC SESSION CARDIOLOGY CLINICAL WORKSHOP

COMPARATIVE GASTROENTEROLOGY SOCIETY SCIENTIFIC SESSION

SCIENTIFIC SESSION EQUINE COMPREHENSIVE REVIEW

SCIENTIFIC SESSION EQUINE POST-GRADUATE COURSE

ACVIM FOUNDATION SCIENTIFIC SESSION

INTERNATIONAL SOCIETY FOR COMPANION ANIMAL INFECTIOUS DISEASES SCIENTIFIC SESSION

SCIENTIFIC SESSION LIVER STUDY GROUP

SCIENTIFIC SESSION NEUROLOGY COMPREHENSIVE REVIEW

SCIENTIFIC SESSION NEUROLOGY POST-GRADUATE

SCIENTIFIC SESSION ONCOLOGY COMPREHENSIVE REVIEW

SCIENTIFIC SESSION ONCOLOGY CLINICAL WORKSHOP

SCIENTIFIC SESSION SMALL ANIMAL INTERNAL MEDICINE COMPREHENSIVE REVIEW

SCIENTIFIC SESSION SMALL ANIMAL INTERNAL MEDICINE CLINICAL WORKSHOP

SCIENTIFIC SESSION SMALL ANIMAL INTERNAL MEDICINE POST GRADUATE

SOCIETY FOR COMPARATIVE ENDOCRINOLOGY SCIENTIFIC SESSION

VETERINARY COMPARATIVE RESPIRATORY SOCIETY SCIENTIFIC SESSION

VETERINARY EMERGENCY AND CRITICAL CARE SOCIETY SCIENTIFIC SESSION

TECHNICAL PROGRAM

SEROTONIN CONCENTRATION IS ELEVATED IN PLATELET RICH PLASMA AND LEFT VENTRICULAR MYOCARDIAL AND MITRAL VALVE LEAFLET TISSUE IN DOGS WITH MYXOMATOUS MITRAL VALVE DISEASE. SE Cremer[1], GE Singletary[2], LH Olsen[1], K Wallace[3], J Häggström[4], I Ljungvall[4], K Höglund[4], CA Reynolds[2], N Pizzinat[5], MA Oyama[2]. [1]Department of Veterinary Disease Biology, University of Copenhagen, Frederiksberg, Denmark. [2]Department of Clinical Studies, School of Veterinary Medicine, University of Pennsylvania, Philadelphia PA, USA 19104. [3]Department of Pathobiology, School of Veterinary Medicine, University of Pennsylvania, Philadelphia PA, USA 19104. [4]Department of Clinical Sciences, Faculty of Veterinary Medicine and Animal Science, Swedish University of Agricultural Sciences, Uppsala, Sweden. [5]Institut National de la Santé et de la Recherche Médicale (INSERM), Institut de Médecine Moléculaire de Rangueil, Toulouse, France.

Heightened serotonin (5HT) signaling is postulated in the development and/or progression of myxomatous mitral valve disease (MMVD) in dogs. Little is known regarding platelet, myocardial, or valvular 5HT concentration ([5HT]) in affected dogs. We sought to quantify [5HT] in platelet-rich (PRP), platelet-poor plasma (PPP), left ventricular myocardium (LV), and mitral valve leaflet tissue (MV) using HPLC. PPP/PRP was prepared from 45 dogs comprising 4 different plasma groups (pG1, healthy non-Cavalier King Charles Spaniels [CKCS], n=8; pG2, healthy CKCS, n=12; pG3, MMVD CKCS, n=14; pG4, MMVD non-CKCS, n=11). PRP [5HT] was greater than PPP[5HT] (P<0.0001) indicating platelets as the primary source of circulating 5HT. Median PRP platelet [5HT] was significantly different between groups (P=0.003). PRP [5HT] was greater in pG2 (1.83femtograms/platelet[range,0.2-4.76];P=0.002), pG3 (1.58,0.70-4.03;P=0.005), and pG4 (1.72,0.85-4.44;P=0.003) vs. pG1 (0.915,0.63-1.3). There was no significant difference in median PPP [5HT] between groups. At euthanasia, LV and MV tissue was obtained from 24 dogs comprising 3 different tissue groups (tG1, MMVD, n=8; tG2, dilated cardiomyopathy [DCM]/arrhythmogenic right ventricular cardiomyopathy [ARVC], n=7; tG3, extra-cardiac disease, n=9). Median LV [5HT] was significantly different (P=0.0024) between groups. LV [5HT] was higher in tG1 (11.8ng/mg,3.96-104.83) vs. tG2 (0.87,0-10.1;P=0.011) and vs. tG3 (2.46,0-6.94;P=0.001). MV [5HT] was significantly different between groups (P=0.033) namely between tG1(32.4ng/mg,8.36-106.73) vs. tG3(3.59,0-28.26;P=0.01). In conclusion, platelet [5HT] was elevated in healthy and MMVD CKCS compared to healthy non-CKCS. LV [5HT] was elevated in MMVD compared to DCM/ARVC or non-cardiac disease. MV [5HT] was elevated in MMVD compared to non-cardiac disease.

NOVEL VASODILATOR PROTOTYPE DRUG (LASSBio 897) AND ITS ACTION ON THE CARDIOVASCULAR SYSTEM OF DOGS. P.R. Nasciutti[1], I.P. Bittar[1], A.R. Fayad[1], F.O. Carvalho[1], V. De Oliveira[1], A.B. Fraga[2], E.J. Barreiro[2], R. Oliveira Alves[1]. 1.School of Veterinary and Animal Science, Federal University of Goiás, Brazil. 2. Federal University of Rio de Janeiro, Brazil.

LASSBio 897 is a new prototype drug produced from safrole substrate, a compound extracted from the "sassafras oil", found in Brazilian plants like "canela-branca" (*Ocotea pretiosa*) which showed vasodilatory effects in *in vitro* (aorta of normotensive rats) and *in vivo* tests (rats). It was developed by Evaluation and Synthesis of Bioactive Substances Laboratory (LASSBio®) from Federal University of Rio de Janeiro. It is believed that the mechanism of action is mediated by muscarinic receptors with subsequent activation of the nitric oxide/cyclic guanosine monophosphate. Due to its pharmacological profile and the possibility of a new therapeutic strategy, an investigation of the effects on the cardiovascular system of dogs was held.

Thus, six adult Beagles, five females and one male, healthy, were used. All dogs were subjected to four treatments, with biweekly intervals, receiving, orally, specific doses of 0.5mg/kg, 1.0mg/kg and 2.0mg/kg of LASSBio 897 and 0.5mg/kg of benazepril, as positive control. The dogs underwent blood and urine sampling (before and twelve hours after oral administration of the prototype), Doppler echocardiographic and electrocardiographic examinations (before and one, two, four, six and twelve hours after oral administration) and blood pressure screening (before and fifteen, thirty, sixty and ninety minutes, two, four, six and twelve hours after administration).

The dogs showed no clinical or behavioral changes during the study. Regarding urinary, hematological and serum biochemical evaluations, although statistical differences between times were observed, the parameters remained within the reference values for the species, similar to the results obtained by Pereira et al. (2005) in a study administering benazepril. No dog had heart rhythm abnormalities. In Doppler echocardiographic evaluation, values obtained for stroke volume (SV) and cardiac output (CO) were lower than that described as normal for dogs, with the lowest values found after four (SV: 14.57mL; CO: 1.26L/min) and six (SV: 16.39mL; CO: 1.34L/min) hours of administration. In the assessment of systolic blood pressure, a decrease was observed after two hours ($p=0.009$), for all three doses of LASSBio 897 administered, similar to the results obtained with administration of benazepril, demonstrating the molecule's potential vasodilatory activity in addition to its safety, since no hypotension occurred.

Considering the results obtained and the conditions under which this experiment was conducted, it can be concluded that the oral administration of LASSBio 897 in doses of 0.5mg/kg, 1.0mg/kg, 2.0mg/kg did not cause side effects in healthy Beagles, demonstrating that the prototype drug is safe. All doses caused vasodilation after two hours of their administration, similar to what was observed with benazepril, a drug extensively used to treat hypertension in dogs.

THE INFLUENCE OF AGE AND SYSTEMIC DISEASES ON CARDIAC INFLAMMATION AND REMODELING IN CATS. Sonja Fonfara[1,2], Udo Hetzel[2,3], Shelley Hahn[2], Anja Kipar[2,3]. 1. University of Bristol, Bristol, UK. 2. University of Helsinki, Helsinki, Finland. 3. University of Zurich, Zurich, Switzerland.

Little is known about the role of myocardial inflammation and remodeling in cardiac dysfunction, in particular in association with systemic diseases and aging.

Myocardial samples (interventricular septum, right and left atrium and ventricle) of 20 cats with systemic diseases (group 1), 10 cats with FIP (group 2), 7 cats with cardiac diseases (group 3), and 8 cats without systemic or cardiac diseases (group 4) were collected. Cats from groups 1 and 4 were additionally separated into age groups (<5, 5-10, >10-15 and >15 years of age). Transcription levels of feline interleukin (IL)-1, IL-2, IL-4, IL-6, IL-18, tumour necrosis factor-α (TNF), interferon-γ (IFN), transforming growth factor-β (TGF), matrix metalloproteinase (MMP)-2, MMP-3, MMP-13, tissue inhibitors of metalloproteinase (TIMP)-1, TIMP-2, TIMP-3 were analysed in relation to the house keeping gene GAPDH, using real time RT-PCR.

Transcription of IL-4, IFN, MMP-2, MMP-3, MMP-13 was highest in group 2 ($p<0.038$), and of IL-1, IL-6, TNF, TGF, TIMP-1, TIMP-3 in groups 2 and 3 ($p<0.001$). IL-18 and TIMP-2 mRNA concentrations were not different between groups 1, 2, and 3, but higher than in group 4 ($p<0.016$). Regional differences in mRNA expression were detected in cats of groups 1, 2 and 4. Young cats showed higher transcription levels for IL-1, IL-2, IL-4, IL-6, MMP-3 ($p<0.003$), whereas TGF increased with age ($p=0.036$).

Myocardial mRNA expression of markers for inflammation and remodelling was similar in cats with systemic and cardiac diseases, which suggests an impact of systemic diseases on cardiac function. Furthermore, age might influence cardiac inflammation and remodelling.

MECHANISMS FOR SUDDEN CARDIAC DEATH IN ENDURANCE ATHLETES: SUPPORT FOR EXERCISE-INDUCED ARRHYTHMOGENESIS. Sophy A. Jesty, Sara M. Johns, Brianna Potter, Erik Berger, Irfan M. Asif. University of Tennessee, Knoxville, TN.

The purpose of this study was to determine whether intensive exercise training in dogs leads to ventricular remodeling that increases susceptibility to malignant arrhythmias. Sudden cardiac death (SCD) is the leading cause of death during exertion in people, and occurs also in animal athletes such as horses and dogs. Unfortunately, the factors explaining this phenomenon are essentially unknown. Investigation is needed to understand the mechanisms of SCD during endurance activity in order to improve outcomes in both people and animals.

This was a repeated measures design pilot study of 5 laboratory Beagle dogs serving as their own controls. Each dog underwent baseline in vivo electrophysiologic (EP) testing to assess the susceptibility to triggered ectopic ventricular activity, participated in a 16 week exercise conditioning program (treadmill running, 4 days per week, up to 90 minutes per day, 4 mph, 13% grade), and then underwent a repeat in vivo EP test to assess the effect of endurance exercise on arrhythmia inducibility.

At baseline, 0/5 (0%) of the dogs had inducible arrhythmias during EP testing. After the 16 week exercise program, 4/5 (80%) of the dogs displayed clear inducible ventricular arrhythmias during EP testing; these triggered arrhythmias appeared to be of right ventricular origin.

Although there are clear benefits to moderate intensity exercise, intensive endurance exercise training may promote adverse cardiac remodeling that acts as a substrate for malignant arrhythmias. Further understanding of SCD mechanisms may lead to enhanced prevention and resuscitation strategies.

A CROSS-SECTIONAL STUDY OF CARDIOMYOPATHY PREVALENCE IN 780 CATS IN REHOMING CENTERS (THE CATSCAN STUDY). J.R. Payne, V. Luis Fuentes, D.C. Brodbelt. Royal Veterinary College, Hatfield, Hertfordshire, United Kingdom.

The prevalence of cardiac murmurs and hypertrophic cardiomyopathy (HCM) was investigated in cats in 2 rehoming centers over a 16 month period using auscultation and 2D echocardiography. HCM was defined as a diastolic left ventricular (LV) septal or free wall thickness ≥6mm.

Cats were considered eligible if they were ≥6m of age and excluded if hypertensive, hyperthyroid, pregnant, had any clinical signs, or were too nervous/aggressive to examine. One operator (JRP) auscultated, measured blood pressure, and scanned all cats.

Of 1007 eligible cats, 780 completed the protocol. Median (range) age group (categorical) was 1-3y (range 6-12m – 17-19y); though 102 cats were ≥9y and 9 cats were ≥15y. Median (range) body weight was 3.7 (1.3-7.9) kg; 43.4% were male, and 97.2% were non-pedigree. A murmur was detected at first auscultation in 30.5% cats. A gallop was heard in 3 cats and an arrhythmia in 7 cats. HCM was diagnosed in 14.7%; 35.7% of which had systolic anterior motion of the mitral valve (SAM). Prevalence of HCM increased with age ($p<0.001$). The positive predictive values (PPV) of a murmur for detecting HCM on a single auscultation were 15.4% in juvenile cats (6-12m) and 54.2% in senior cats (≥9y); the respective negative predictive values (NPV) were 97.1% and 79.2%.

Both murmurs and HCM are extremely common in the general cat population, but many cats with a murmur have normal wall thickness, particularly when young. The PPV of a murmur to indicate HCM, and HCM prevalence, both increase with age.

DOES PIMOBENDAN THERAPY FOR CANINE SECONDARY PULMONARY ARTERIAL HYPERTENSION ALTER TRANS-SEMILUNAR VALVE SPECTRAL DOPPLER VARIABLES? D.G. Ohad, E. Gan-Mor, Y. Golani, T. Bdolah-Abram. The Koret School of Veterinary Medicine, the Hebrew University of Jerusalem, Rehovot, Israel.

Pulmonary hypertension (PH) may be present but hard to quantify or even identify in the absence of tricuspid or pulmonary insufficiency. We sought to find measurable changes in Doppler parameters of systolic blood flow through the semilunar heart valves in dogs with PH secondary to mitral valve disease (MVD), with and without pimobendan therapy.

Data were retrospectively obtained from 186 consecutive MVD dogs examined between 2009 and 2011, with Doppler evidence of PH. Dogs were sorted into 2 groups based on whether (n=96) or not (n=90) pimobendan was included in their therapy.

Only in the absence of pimobendan therapy, aortic acceleration-time (AccT), aortic ejection-time (ET), and aortic AccT/ET decreased significantly over time, potentially due to a higher systemic afterload or a decreased cardiac output due to progressive PH. In dogs receiving pimobendan, pulmonary-AccT, pulmonary-ET, and pulmonary AccT/ET decreased to a lesser extent along time, than in dogs with no pimobendan therapy.

Nevertheless, pimobendan was not unequivocally proven to reduce right ventricular afterload as per spectral Doppler imaging. The reason may be the enhancement of right ventricular stroke volume following inodilation with reduced afterload, or the progressive nature of the underlying MVD along with its complications, including progressive PH. A larger, prospective study including ECG-related systolic time intervals may establish whether the impedance to blood-flow through each of the semilunar valves progresses or decreases over time, with and without therapy.

CLINICOPATHOLOGIC FEATURES OF DOGS AND CATS PRESENTED TO A CARDIORESPIRATORY REFERRAL INSTITUTION DIAGNOSED WITH ACUTE TOXOPLASMOSIS. N. Van Israël, Animal CardioPulmonary Consultancy (ACAPULCO), Masta, Belgium.

Toxoplasma gondii infects nearly all warm-blooded animals and the consequences of its passage in the intermediate (eg. dog) and definitive host (cat) have often been underestimated. The objective of this study was to retrospectively describe the clinical and pathological features, with an emphasis on the cardiorespiratory system, in a large population with naturally occurring Toxoplasmosis. Acute Toxoplasmosis was defined as animals having positive IgM titers or a fourfold seroconvalescence of IgG titers over a 3-4 week period (as determined by IFA or ELISA techniques). Fourty-seven dogs and 5 cats fulfilled the criteria. There was no specific breed or gender predisposition. The age range was wide (cats: 5-14y median 5.5y; dogs: 4m-10y median 4.8y). Anorexia and lethargy were the main presenting signs in cats, were in dogs 61 % were exercise intolerant, 53 % experienced breathlessness, 23 % coughed and 23 % were syncopal. On electrocardiography AV-block was present in 6 dogs (3 first, 1 second, 2 third degree), V-arrhythmias in 9 dogs and supraventricular arrhythmias in 5 dogs (2 AF, 1 atrial flutter, 2 SVT). Radiographic findings ranged from pulmonary (broncho-interstitial patterns (5) and alveolar 'aspiration' infiltrates (2)) to cardiac pathology (cardiomegaly (5) and congestive heart failure (3)) but was normal in 21/35 sets of RX available. Echocardiography was normal in 25/48 animals and showed signs that might have been consistent with myocarditis in only 9/48 animals. Two dogs had developed pericardial and one pleural effusion. A neutrophilia was the most frequent hematological abnormality (15%), followed by a mildly elevated HCT (11%). Elevated liver enzymes was the most common biochemical finding (35) followed by increased CPK in 5/52. Toxoplasmosis IgM titers at initial presentation (1/40-1/640 median 1/60) were elevated in all but 1 cat that seroconverted (IgG) over a 3 week period. It could take up to 16 months for complete disappearance of the IgM titers with a median of 5 months. In two dogs the titers were re-activated over time. Troponin I levels were available in 25 dogs and elevated in 10 (range 0.34-14.5). Proteinuria at initial presentation was a feature in only 3 dogs and 1 cat. Final clinical diagnosis consisted of myocarditis in 17 animals, hepatopathy in 4, tracheobronchitis/pneumonitis/pneumonia in 7, pericarditis, chorioretinitis and PLNP in 2, laryngeal paralysis in 3 (accompanied with megaoesophagus in 2), pleuritis, polyarthritis, polymyopathy and sialodenitis respectively in 1. Nearly all animals were treated with clindamycine as first choice, S/TMP or doxycycline were the second choice. Three dogs died suddenly during their treatment and 9 were euthanized (3 cats) for complications of the disease, making this disease not as inoffensive as widely described.

COMPLICATIONS ASSOCIATED WITH TRANSCATHETER PDA OCCLUSION: RESULTS OF AN ACVIM CARDIOLOGY SURVEY. SG Gordon[1,2], AB Saunders[1,2], R Fries[1,2], MW Miller[1,2], and M Rishniw[3]. The Michael E. DeBakey Institute, College of Veterinary Medicine and Biomedical Sciences[1], Texas A&M University, College Station, TX[2], Veterinary Information Network, Davis, CA[3].

A variety of devices are used for transcatheter occlusion of patent ductus arteriosus (PDA) in veterinary medicine including various types of embolization coils, vascular plugs and occluders. Intra-operative, immediate post-operative and delayed (2 days post-operatively) embolization to the systemic or pulmonary circulation is a reported complication of transcatheter PDA occlusion, regardless of the device type used. Embolization is typically thought to occur as a result of inappropriate device sizing or operator error. Veterinary studies assessing device efficacy and complication rates are relatively small making it statistically unlikely to catch rare, but important complications. This report is based on the results of a survey designed to assess type, rate and details of complications, including embolization, associated with transcatheter PDA occlusion. In addition we will report details of preferred methods of PDA size estimation, ductal occluder selection and procedural practices such as the use of transesophageal echocardiography. An initial survey of 77 respondents indicated that a variety of devices including the canine ductal occluder (69/77), coils (31/77), vascular plugs (20/77) have been used for transcatheter occlusion Preliminary Survey Results: Six of 385 (1.6%) procedures resulted in embolization. Five of the six were intra-operative and one was after discharge from the hospital. Two of the six were in cats, and both of these devices were vascular plugs. The remaining devices were canine ductal occluders. There were two deaths. One death was associated with attempted retrieval of a vascular plug in a cat and the other a consequence of acute heart failure following delayed embolization of a ductal occluder. All respondents reported using fluoroscopy and angiography. In addition, transesophageal echocardiography was frequently used for adjunctive imaging during procedures.

DILATED CARDIOMYOPATHY IN STANDARD SCHNAUZERS WITH A HOMOZYGOUS 22 BP DELETION IN *RBM20*. SB Leach[1], GS Johnson[1], D Gilliam[1], MW Harmon[1], T Mhlanga-Mutangadura[1], L Hansen[1], JF Taylor[2], RD Schnabel[2]. [1]College of Veterinary Medicine, and [2]College of Agriculture, Food and Natural Resources, University of Missouri, Columbia, MO.

Our purpose was to characterize familial dilated cardiomyopathy (DCM) of Standard Schnauzers (SSs) and identify the causal mutation. Retrospective evaluation revealed DCM in 10 related SSs confirmed via echocardiography, necropsy, or both.

Examination of a 7-generation pedigree containing all 10 DCM confirmed cases suggested an autosomal recessive mode of inheritance. A 30-fold average coverage whole genome sequence (WGS) was generated with DNA from a male SS that died at 13.5 months of age with DCM and congestive heart failure. This WGS contained 69 homozygous sequence variants that were predicted to alter the primary structure of the gene product and absent from the WGSs of 101 canids not exhibiting DCM. A 22 bp deletion and frame shift in exon 11 of *RBM20* was the sequence variant considered most likely to be causal because mutations in human and rodent orthologs cause DCM. Eight confirmed DCM cases for which DNA was available all tested homozygous for the deletion allele (del/del). A prospective screen for the mutation identified 479 SSs that were homozygous for the wild type allele, 162 SSs that were heterozygous, and 6 additional SSs that were homozygous for the deletion allele. Two of these 6 del/del SSs have subsequently been diagnosed with occult DCM via echocardiography. Echocardiographic evaluation of 12 SSs that were heterozygous for the deletion allele, including a 15 year old SS, has revealed no echocardiographic evidence of myocardial dysfunction.

We conclude that the homozygous 22 bp deletion in *RBM20* is the likely cause of familial DCM in SSs.

COMPUTED TOMOGRAPHY ANGIOGRAPHY OF THE CORONARY ARTERIAL CIRCULATION OF BULLDOGS UNDER CONSCIOUS SEDATION IS FEASIBLE AND DEFINES ANOMALOUS ANATOMY. B.A. Scansen[1], G.A. Chiappone[2], R. Layman[2], R.D. White[2]. 1. Department of Veterinary Clinical Sciences, The Ohio State University, Columbus, OH. 2. Department of Radiology, The Ohio State University Wexner Medical Center, Columbus, OH.

Single coronary artery anomaly (SCAA) has been recognized in bulldogs characterized by a prepulmonic course that limits the ability to intervene in the setting of pulmonary valve stenosis (PS). Standard of care to define coronary arterial anatomy requires left heart catheterization under general anesthesia. We investigated the ability of a new generation CT scanner to visualize the coronary arterial circulation of bulldogs under conscious sedation.

Ten bulldogs were prospectively enrolled, one with PS. Sedation included low-dose dexmedetomidine and butorphanol to limit movement; atropine was administered as needed to normalize cycle length. The dogs were not intubated, remaining conscious during the imaging. Dogs were imaged with a dual-source/128-slice CT scanner ("Flash", Siemens Medical Systems, Erlangen, Germany); iodinated contrast (Isovue 370, 2 ml/kg) was delivered IV with the scan timed to peak aortic opacification. Structural and functional reconstructions were made of the left heart.

Visualization of the coronary circulation throughout the cardiac cycle was possible in all dogs even at heart rates of 180 per minute. SCAA with a prepulmonic left coronary artery was observed in a dog without PS. The coronary arterial anatomy of the dog with PS was clearly delineated and impacted catheterization planning.

We report a noninvasive protocol under conscious sedation for the determination of coronary arterial anatomy in a breed of dog at risk for SCAA. The existence of SCAA in an otherwise normal bulldog suggests SCAA may not be causative to the etiology of PS in bulldogs, as has been previously suggested.

CLATHRIN-MEDIATED ENDOCYTOSIS IMPACTS CARDIAC OUTFLOW TRACT MORPHOGENESIS AND A SINGLE CODON INSERTION IN PHOSPHATIDYLINOSITOL BINDING CLATHRIN ASSEMBLY PROTEIN IS ASSOCIATED WITH NEWFOUNDLAND DOG SUBVALVULAR AORTIC STENOSIS. JA Stern[1,2], Nascone-Yoder NM[3], Lehmkuhl LB[4], Meurs KM[2]. 1. Department of Medicine and Epidemiology, School of Veterinary Medicine, University of California Davis, Davis, CA. 2. Department of Veterinary Clinical Sciences, College of Veterinary Medicine, North Carolina State University, Raleigh, NC. 3. Department of Molecular Biomedical Sciences, College of Veterinary Medicine, North Carolina State University, Raleigh, NC. 4. MedVet Medical and Cancer Center for Pets, Worthington, OH.

Subvalvular aortic stenosis in Newfoundlands is described as a familial disease with an autosomal dominant pattern of inheritance, but no associated mutations have been reported.

More than 150 Newfoundland dogs were recruited for genetic studies of SAS. Dogs with left ventricular outflow tract (LVOT) velocities exceeding 2.5m/sec were considered SAS affected and compared to controls with velocities <1.8m/sec. Fresh LVOT tissue was obtained from two severely affected Newfoundlands with SAS. RNA sequence analysis (RNAseq) was performed on the SAS tissue and 6 dog samples without SAS.

A single codon insertion was identified in Phosphatidylinositol binding clathrin assembly protein (PICALM) that segregates with disease status in Newfoundlands. Significant downstream structural change is predicted in the mutant PICALM protein. Loss of function is proposed as the insert centers over a breakpoint necessary for formation of PICALM fusion complex, a transcriptional regulator. The insertion was not identified in any control dogs (180) of varied breeds. PICALM is primarily involved in clathrin-mediated endocytosis at the cellular membrane. Day 40 *Xenopus laevis* embryos were small molecule inhibited for clathrin and observed for abnormalities in outflow tract morphogenesis. Significant abnormalities in outflow tract morphogenesis were observed by immunohistochemistry.

A single codon insertion in PICALM alters the protein and is associated with SAS development in Newfoundlands. The role of PICALM in cardiac morphogenesis and pathogenesis is yet to be determined, however clathrin inhibition during embryogenesis as might be observed with a dysfunctional PICALM protein, demonstrated a previously unreported role for clathrin-mediated endocytosis during outflow tract morphogenesis.

INVESTIGATION OF DIETARY FACTORS WITH POSSIBLE ASSOCIATIONS WITH CANINE DEGENERATIVE MITRAL VALVE DISEASE. JL Sauer, LM Freeman, JE Rush. Tufts Cummings School of Veterinary Medicine, North Grafton, MA.

The pathophysiologic cause of degenerative mitral valve disease (DMVD) remains unclear. Although there are a number of ways in which diet could play a role in DMVD, including the serotonin (5-hydroxytryptamine [5HT]) pathway, dietary factors related to the development of DVMD have not been investigated. Therefore, the objective of this study was to measure dietary amino acids, choline, carnitine, serotonin, and ergovaline as possible factors that could play a role in the pathophysiology of DMVD. Thirteen commercially-available diets were selected for analysis based on a previous study comparing dogs with and without DMVD and diet histories from clinical cases. Diets were analyzed for macronutrients; amino acids; ergovaline; the indoleamines, 5HT and melatonin; choline, and free L-carnitine. There was a wide range in the concentrations of all analytes in the diets tested. No essential amino acids were below the Association of American Feed Control Officials (AAFCO) minimums. Taurine, although not an essential amino acid for dogs, was below the AAFCO Cat Food Nutrient Profile for taurine in 9 of 13 diets. Tryptophan ranged from 0.19-0.59% dry matter (median = 0.28% dry matter). All 13 samples tested had undetectable ergovaline concentrations. One sample tested positive for 5HT, and melatonin was detected in 8 diets. There also was wide variation (3-fold and >100-fold difference, respectively) in choline and free L-carnitine concentrations among diets. Additional research is needed on the effects of varying dietary intake of tryptophan and other amino acids, 5HT, choline, and carnitine on cardiac valve metabolism.

BODY SIZE IS RELATED TO METABOLIC MARKERS AND LEFT VENTRICULAR HYPERTROPHY IN CATS. JE Rush, LM Freeman, A Feugier, I van Hoek. Tufts Cummings School of Veterinary Medicine, North Grafton, MA, USA and Royal Canin SAS, Aimargues, France.

Cats with hypertrophic cardiomyopathy (HCM) are larger and have higher insulin-like growth factor-1 (IGF-1) concentrations than cats without HCM. To better understand the relationship between growth, body size, and HCM, the objective of this study was to assess adult cats to determine whether echocardiographic findings of left ventricular hypertrophy (LVH) are related to early growth patterns, body size, or metabolic markers. Adult cats from a single colony that were $\geq$ 3 years of age and for which growth curves were available were eligible for the study. Physical examination and echocardiography were performed, and body weight, body condition score (BCS), and head length and width were measured. Circulating glucose, insulin, N-terminal pro-B-type natriuretic peptide (NT pro-BNP), and IGF-1 concentrations were measured and growth data were collected. Stepwise multivariate analysis of variance and multivariate logistic regression were performed. Twenty-eight neutered adult cats (20 males, 8 females) were eligible for the study. Median age was 5.3 yrs (range, 3.2-6.8 yrs). Current body condition scores ranged from 4-9 (median, 5.5) and body weight ranged from 2.7-7.2 kg (median, 4.8 kg). Cardiac abnormalities included a cardiac murmur (n=7; 24%), gallop (n=3; 10%), and arrhythmia (n=1; 4%). Fourteen of 28 cats (50%) had echocardiographic evidence of left ventricular hypertrophy (LVH). Head width (P=0.017), body weight (P<0.001), NT pro-BNP (P=0.023), and IGF-1 (P=0.013-0.022) were significantly associated with measures of LVH. Potential associations between body size, IGF-1, LVH, and HCM warrant future prospective studies.

USEFULNESS OF PLASMA CONCENTRATION OF CHROMOGRANIN A AS AN INDICATOR OF SYMPATHETIC TONE IN DOGS WITH MYXOMATOUS MITRAL VALVE DISEASE. K. Höglund,[1] M. Stridsberg,[2] O. Höglund,[1] A. Tidholm,[3] J. Häggström,[1] I. Ljungvall[1] 1. Swedish University of Agricultural Sciences, Uppsala, Sweden. 2. Dept of Medical Sciences, Uppsala University, Uppsala, Sweden. 3. Albano Animal Hospital, Danderyd, Sweden

Chromogranin A (CgA) is stored in secretory cells of sympathetic nerve endings and the adrenal medulla and is co-released with catecholamines, while the CgA-derived peptide Catestatin acts sympathoinhibitory. The stability of these molecules in plasma makes them interesting candidates for assessment of the sympathetic nervous system. In humans, CgA is also present in ventricular myocardium, co-localized with B-type natriuretic peptide (BNP). The aim of the study was to investigate whether plasma Catestatin concentration was associated with myxomatous mitral valve disease (MMVD) severity in dogs, and to assess potential associations between plasma Catestatin concentration and dog characteristics, echocardiographic variables, heart rate (HR), systolic blood pressure (SBP) and N-terminal proBNP.

70 client-owned dogs with MMVD of varying severity were prospectively recruited. Dogs were classified according to MMVD severity (healthy, or mild, moderate or severe disease). Three dogs were in decompensated congestive heart failure (CHF). Plasma concentration of Catestatin was analyzed using radioimmunoassay.

No association between Catestatin concentration and MMVD severity groups was found. HR was higher in severely affected dogs than healthy and mildly affected, but was not associated with Catestatin concentration. Very weak negative association with left atrial and ventricular size (R^2=0.08, *P*=0.01 for both) and very weak positive association with SBP (R^2=0.07, *P*=0.02) was found for Catestatin concentration, while remaining variables were non-significant.

In the present study population, Catestatin was not associated with severity of MMVD and HR seemed to be a better indicator of sympathetic tone than Catestatin. Including more dogs in decompensated CHF might have given different results.

ASSESSMENT OF MULTIPLE ECHOCARDIOGRAPHIC PARAMETERS IN GOLDEN RETRIEVER DOGS WITH OR WITHOUT SUBAORTIC STENOSIS.

E Côté[1], MC Bélanger[2], R Vanderstichel[1]. 1. Atlantic Veterinary College, University of Prince Edward Island, Charlottetown, PE, Canada 2. Faculty of Veterinary Medicine, University of Montreal, St. Hyacinthe, QC, Canada.

Currently, a diagnosis of subaortic stenosis (SAS) in dogs principally relies on auscultation of a basilar systolic heart murmur and echocardiographic (echo) demonstration of abnormally high Doppler-derived peak left ventricular outflow tract blood velocity (LVOT Vmax). This retrospective study sought to identify associations among echo determinations of LVOT Vmax, the presence of an abnormal aortoseptal angle (AoSA), and left ventricular subendocardial hyperechogenicity (LV S-H), and to investigate a possible association between AoSA and a heart murmur of equal or greater intensity over the right heart base compared to the left.

Medical records of Golden Retrievers (GRs) seen at 1 institution[1] were reviewed if 1) a complete echo was available as cine loops, 2) no cardiac disorder other than SAS was noted, and 3) heart murmur point of maximal intensity (PMI) was noted in the record by a cardiologist at the time of echo. AoSA and S-H were determined post-hoc. GRs with LVOT Vmax <2.0 m/s were considered normal (GR-WNL). GRs with LVOT Vmax ≥ 2.5 m/s and a systolic basilar heart murmur were considered to have SAS (GR-SAS). GRs that were neither GR-WNL nor GR-SAS were considered equivocal. Regression analysis was used for investigating relationships between parameters.

Fifty-four GRs met the entry criteria: 15 GR-SAS, 26 GR-WNL, and 13 equivocal. AoSA and LVOT Vmax were inversely associated: for each 1° increase in AoSA, LVOT Vmax decreased by 0.04 m/s ($p=0.049$). Dogs with moderate LV S-H had AoSA that was on average 10.2° less than dogs without LV S-H ($p=0.03$). There was a significant association between AoSA and heart murmur PMI being equally loud or louder on the right rather than the left ($p=0.011$), with such dogs having an AoSA that was a mean of 15.7° less than dogs with murmur PMI on the left.

In this sample, associations existed among parameters and these associations may provide clinically useful information in addition to LVOT Vmax when evaluating GRs for SAS.

VALIDATION AND PROGNOSTIC VALUE OF 2 NEW HEAD TRAUMA SCORING SYSTEMS IN 84 DOGS. S. R. Platt[1], E. Beltran[2], C. May,[1] D. Keys[1], L. De Risio[2], A. Haley,[1] M. Kent,[1] M. Perez,[3] A. Shores[3] [1]College of Veterinary Medicine, University of Georgia, USA; [2]Animal Health Trust, Newmarket, England, UK; [3]College of Veterinary Medicine, Mississippi State University.

An ideal coma scoring system or scale would be reliable, valid, linear and easy to use. Scoring systems are used objectively to decide the proper course of treatment, to monitor the success of therapy and provide an estimated prognosis. The modified Glasgow coma scale (MGCS) was developed 25 years ago as a guide for management and to assist in determining the prognosis for canine and felines with head trauma (HT) Although validated and statistically relevant, the MGCS is not linear, not always easy to use, nor entirely representative of the clinical presentations seen in veterinary medicine. To address these concerns, we proposed to amend the MGCS and evaluate the validity and prognostic value of 2 derivations of this current scale. Further, we aimed to use the retrospective validation of the new scales to determine sample size necessary for a prospective multicenter treatment study.

The MGCS was amended into 2 separate scoring systems, maintaining the 3 subdivisions of brainstem reflexes, motor function and consciousness. A minimized scale (Veterinary Cranial Injury Score; VCIS) assigned each subdivision 3 points; a more expansive scale (Small Animal Coma Score; SACS) totaled 17 points; low scores indicated severe dysfunction. We retrospectively scored dogs presented to one of three referral institutions subsequent to HT. A 4-point outcome scale was developed based on the assessed quality of life (QOL) for all survivors. Data used in the study was obtained from medical records and included neurological exam, heart rate, temperature, and blood chemistry in all cases. All analyses were performed using SAS V 9.2; all tests were 2-sided and the significance level was set at 0.05. Power analyses were performed using nQuery 4.0.

Eighty-four dogs were evaluated. The mean MGCS scores and mean blood glucose values were significantly higher in dogs that survived to 72 hours. There were significant positive correlations of MGCS ($r=0.69$), SACS ($r=0.68$), VCIS ($r=0.59$), temperature ($r=0.58$) and heart rate ($r=0.36$) with QOL scores at 1 month. There were significant positive correlations of MGCS ($r=0.57$), SACS ($r=0.58$), VCIS ($r=0.59$), and temperature ($r=0.42$) with QOL scores at 3 months. MGCS presentation scores were positively correlated with SACS ($r=0.93$, $p<0.0001$) and VCIS ($r=0.82$, $p<0.0001$) scores. Sample sizes calculated for a prospective treatment study were a total of 230 animals for SACS and 157 animals for VCIS.

This study validated 2 new scoring systems as having prognostic value in dogs with HT. The new clinical outcome score can be used to assess dogs at 1 and 3 months after the trauma in terms of their QOL. Information derived from this study is being used in the initiation of a prospective multi-centered treatment trial using the power analysis results confirming feasible sample sizes.

EVALUATION OF LACTATE DEHYDROGENASE AND CREATINE KINASE ACTIVITY IN CEREBROSPINAL FLUID OF DOGS WITH NEUROLOGICAL DISEASES. AV Chen, SA Thomovsky, J Brune, N Boswell, JC Houle. Washington State University College of Veterinary Medicine, Pullman, WA.

The diagnostic significance of cerebrospinal fluid (CSF) lactate dehydrogenase (LD) and creatine kinase (CK) activity have been relatively unexplored in veterinary medicine. CSF LD and CK activity have been shown to be useful for the diagnosis and prognosis of meningitis in humans. The purpose of this study is to determine total LD and CK activity and its isoenzyme distribution in the CSF of dogs with idiopathic epilepsy (IE), primary brain tumors (PBT) and meningitis.

A total of 11 dogs were enrolled in this pilot study and divided into three groups: IE(n=4), PBT(n=4), and meningitis(n=3). CSF total LD and CK activity were measured spectrophotometrically and isoenzyme distribution was determined using gel electrophoresis and densitometric scanning. Mean total LD activity was 8, 12, 99.3U/L for IE, PBT, and meningitis, respectively. Mean LD-1 contribution was 59.7, 53.4, 11.2% while LD-3 contribution was 14.6, 11.0 and 48.1% for IE, PBT and meningitis, respectively. Mean total CK activity was 4.3, 3.8, 22.7U/L for IE, PBT, and meningitis, respectively. Mean CK-MM contribution was 69.7, 63.8, and 24.4% while CK-BB contribution was 21.5, 29.9 and 75.3% for IE, PBT and meningitis, respectively.

CSF total LD and CK activity were higher in meningitis dogs. These dogs had a suppression in LD-1 with an increase in LD-3 activity. CK-BB was also the predominate isoenzyme in meningitis dogs while CK-MM was the predominate isoenzyme in dogs with IE and PBT. LD and CK activity in canine CSF may be a useful diagnostic tool for distinguishing meningitis from other neurological diseases.

ACCURACY OF PREPOPERATIVE COMPUTED TOMOGRAPHY AND MAGNETIC RESONANCE IMAGING AS COMPARED TO SURGICAL FINDINGS IN CHONDRODYSTROPHIC DOGS WITH INTERVERTEBRAL DISC DISEASE. SA Thomovsky[1], AV Chen[1], RL Tucker[1], GR Roberts[1], JS Mattoon[1], AM Kiszonas[2], T Morimoto[1]. [1]Washington State University School of Veterinary Medicine, Pullman, WA. [2]Washington State University Department of Crop and Soil Sciences, Pullman, WA.

Intervertebral disc disease is common in chondrodystrophic dogs. Computed tomography (CT) and Magnetic Resonace (MR) Imaging are commonly used to diagnose extruded intervertebral disc material (EIVDM). The purpose of this study is to evaluate the accuracy of MR and CT in diagnosing EIVDM in chondrodystrophic dogs with thoracolumbar disc disease as compared to intraoperative findings.

Thirty-five chondrodystrophic dogs with thoracolumbar EIVDM who underwent advanced imaging and surgery were included in the study. MR, CT and hemilaminectomy surgery were performed on all dogs. Accuracy of CT and MR for location of EIVDM, identification of the disc from which herniation occurred, lateralization of EIVDM, distribution with respect to both the disc space and spinal cord, and length of EIVDM were compared to intra-operative findings. Patient temperature loss and duration of each imaging procedure were also compared.

Accuracy of site of EIVDM was 90.6% for CT and 93.8% for MR. Accuracy of disc material lateralization was 84.4% for both modalities. Accuracy of the EIVDM distribution with respect to disc space was 71.9% for CT and 68.8% for MR; while it was 59.4% for CT and 46.9% for MR with respect to the spinal cord. All accuracy data is as compared to intra-operative findings. On average, CT scanning took 14.2 minutes while MR was 64.2 minutes; temperature loss was 1.03° and 2.36° for CT and MR, respectively.

Both imaging modalities are comparable in the diagnosis and description of thoracolumbar EIVDM. Temperature loss and scan duration are significantly less in CT as compared to MR.

ANALYSIS OF CYTOKINE CONCENTRATIONS IN THE CEREBROSPINAL FLUID OF GREAT DANES WITH AND WITHOUT CLINICAL SIGNS OF CERVICAL SPONDYLOMYELOPATHY. P Martin-Vaquero[1], RC da Costa[1], SA Moore[1], AC Gross[2], TD Eubank[2]. 1. College of Veterinary Medicine, The Ohio State University, Columbus, OH. 2. Davis Heart & Lung Research Institute, The Ohio State University Medical Center, Columbus, OH.

The purpose of this study was to compare the cerebrospinal fluid (CSF) cytokine concentrations of Great Danes (GDs) with and without signs of cervical spondylomyelopathy (CSM). Associations between cytokine concentrations and administration of anti-inflammatories, disease severity, severity of spinal cord (SC) compression, and presence of SC signal changes in the CSM-affected GDs were also investigated.

Thirty client-owned GDs were prospectively enrolled, 15 clinically normal and 15 CSM-affected. All dogs underwent MRI of the cervical vertebral column and cerebellomedullary cistern CSF collection. Routine CSF analysis (total protein, cell counts) was performed. A commercially available canine multiplex immunoassay was used to measure concentrations of granulocyte macrophage colony-stimulating factor (GM-CSF), interferon-gamma (IFN-γ), inducible protein 10 (IP-10), IL-2, IL-6, IL-7, IL-8, IL-10, IL-15, IL-18, keratinocyte-derived cytokine (KC-like), monocyte chemoattractant protein-1 (MCP-1), and tumor necrosis factor-α (TNF-α). Linear regression analysis was used to compare cytokine concentrations between groups, and to test for associations with the use of anti-inflammatories, severity of signs and SC compression, and SC signal changes in the affected GDs.

There were no significant differences between normal and affected GDs in the routine CSF analysis, and concentrations of IL-6, KC-like, and IL-10. CSM-affected GDs had significantly lower concentrations of MCP-1 than normal GDs ($P = 0.028$). Moreover, affected GDs with severe ($P = 0.002$) and moderate ($P = 0.022$) SC compression on MRI had significantly lower MCP-1 concentrations than those with mild SC compression. There were no associations between cytokine concentrations and use of anti-inflammatories or disease severity in the affected GDs. Concentrations of IL-6 were significantly higher ($P < 0.001$) in GDs with SC signal changes. The remainder cytokines were below the detection limits.

Lower MCP-1 concentrations in CSM-affected GDs might interfere with monocyte attraction, slowing down the removal of myelin debris, and hindering axon regeneration and remyelination. Markedly higher CSF IL-6 concentrations in CSM-affected GDs with SC signal changes suggest more severe chronic inflammation in this subgroup of affected GDs. Histopathologic and immunohistochemical studies are needed to further evaluate the role of these cytokines in CSM.

SERUM PHENOBARBITAL LEVELS IN CATS FOLLOWING CHRONIC TRANSDERMAL ADMINISTRATION J Delamaide Gasper, HL Barnes Heller, L. Trepanier, M. Robertson. University of Wisconsin, Madison WI

The purpose of this study was to determine if chronic transdermal phenobarbital (PB) administration achieved serum phenobarbital (S-PB) concentrations between 15-45 ug/mL in healthy cats. Nineteen healthy, staff and student owned cats were divided into 3 groups. Transdermal phenobarbital in pluronic lecithin organogel (PLO) (groups 1 and 2) or PCCA-Lipoderm (Activemax®) (group 3) was applied to the pinnae for 14 days. Group 1 received 3 mg/kg q 12 hours, groups 2 and 3 received 9 mg/kg q 12 hr. Median S-PB concentration on day 15 at times 0, 2, 4, and 6 hours post drug administration were 7.5 ug/mL, 6.5 ug/mL, 6.5 ug/mL, 6.0 ug/mL for group 1; 26.0 ug/mL, 26.0 ug/mL, 25.5ug/mL, and 25.5 ug/mL group 2, and 17.0 ug/mL, 15.5 ug/mL, 15.0 ug/mL, and 15.5 ug/mL in group 3. One cat was withdrawn from group 2 on day 3 due to intolerable side effects. An additional cat was added to total 6 cats completing group 2. Side effects were mild and transient in 12 of the 18 (66%) cats completing the study. Side effects included polyuria, polydipsia, polyphagia, sedation, ataxia and mild paraparesis, pinnae erythema and increased ear grooming. The results of this study show that therapeutic serum phenobarbital concentrations can be achieved in cats, following chronic transdermal administration of PB in PLO and PCCA Lipoderm at 9 mg/kg q 12h, and that side effects were minor.

DIFFUSION TENSOR IMAGING IN DOGS WITH AND WITHOUT CERVICAL SPONDYLOMYELOPATHY. Ronaldo C da Costa[1], Juliet Armstrong[1], Seongjin Choi[2], Paula Martin-Vaquero[1], Blake Dewey[3], Seth Smith[3]. [1]Department of Veterinary Clinical Sciences, The Ohio State University, Columbus, OH, [2]Department of Radiology, The Ohio State University Medical Center, Columbus, OH, and [3]Institute of Imaging Science, School of Medicine, Vanderbilt University, Nashville, TN.

Cervical spondylomyelopathy (CSM) is the most common disease of the cervical spine of large and giant breed dogs. There is a limited understanding of the disease mechanisms and a lack of clearly defined outcome measures or prognostic indicators. Despite a large variety of surgical techniques available to treat CSM, long-term survival of affected dogs needs to be improved. Diffusion tensor imaging (DTI), an advanced magnetic resonance imaging (MRI) technique recently applied to humans, is sensitive to changes in tissue microstructure and could yield prognostic information. This study aimed at evaluating changes in fractional anisotropy (FA), median diffusivity (MD), radial diffusivity (RD) and axial diffusivity (AD) between dogs with and without CSM. We hypothesized that DTI differences would exist in CSM-affected dogs compared to unaffected dogs, furthering the understanding of the disease mechanisms.

Fourteen control and 14 CSM-affected Great Danes were enrolled for analysis with a novel DTI protocol optimized for the canine cervical vertebral column. All dogs underwent a 3.0 T MRI protocol of T1- and T2-weighted images in the sagittal and transverse planes, and a DTI protocol consisting of single-shot spin-echo echo-planar imaging with a repetition time (TR) of 24415, echo time (TE) of 42 ms, 6 diffusion-weighting directions, *b*-values: 0 and 500 s/mm^2, and slice thickness of 5 mm with no gap. Post-processing using purpose-designed proprietary software generated tensor estimation and parametric maps. Using DTI Studio's ROI Editor, FA, MD, RD, and AD were measured from 3 transverse DTI slice segments over each disc region (C2-C7). Sagittal and transverse T1 and T2 weighted images of CSM-affected dogs were also evaluated for spinal cord signal changes.

Random-effects linear regression analysis demonstrated significant differences in FA and RD ($P < 0.05$) between dogs with and without CSM, and no significant differences in MD and AD ($P > 0.05$). In CSM-affected dogs, significant differences were identified for FA, RD, and AD when comparing the sites with and without signal changes at the C5-C6 intervertebral space.

The FA and RD differences suggest that demyelination and loss of axonal integrity are present in spinal cords of CSM-affected dogs. The evaluation of spinal cord microstructure with DTI yields information that could assist in the treatment and prognostication of CSM-affected dogs.

ATLANTOAXIAL INSTABILITY WITH INCONGRUENCE: MRI FINDINGS AND SURGICAL TREATMENT IN 5 DOGS. M Dolera, L Malfassi. La Cittadina Fondazione Studi e Ricerche Veterinarie, Romanengo (CR), Italy.

Atlantoaxial instability is a surgical condition frequently described in toy breed's dogs. Various correction techniques has been described in veterinary literature. Aim of this study is to describe MRI-CT imaging findings and a novel surgical technique for stabilisation of atlantoaxial instability associated with incongruence of atlantoaxial joint in dogs.

Five dogs with a body weight of 2 kg or less with atlantoaxial instability associated with articular incongruence were considered. At clinical examination all dogs showed tetraparesis and ataxia. At CT and MRI, the articular surface of the atlas was larger then the articular surface of the axis, a laxity of all atlantoaxial ligaments was evident, so during head and neck flexion the axis was dislocated dorsally and cranially into the spinal canal. Spinal cord showed chronic compression, with focal hyperintensity in seq TSE T2W and various degrees of syringomyelia. For surgery, the anaesthetised dogs were positioned in ventral recumbency, with the head extended on the neck with slight linear traction applied at maxillary canine teeth. The reduction of atlantoaxial subluxation was verified trough fluoroscopy. A standard dorsal approach to the atlas and axis was made. Two 2.7 mm self tapping cortical screws were inserted on atlas winds on each side in a dorsal-ventral direction. Three 2.0 mm self tapping cortical screws were inserted transversally in the dorsal process of the axis. All screws were fused with polymethylmethacrylate. Serial clinical and imaging controls were made.

No intra-operative complications were observed. Functional improvement occurred in all dogs. Serial CT examinations showed a stable reduction of dorsal and cranial axis dislocation, with persistence of a residual mild spinal compression due to different size of the articular surfaces of atlas and axis.

Several means for stabilisation of atlantoaxial dislocation are described in veterinary literature. Standard treatment consist in ventral arthrodesis with cross pinning, transarticular lag screw fixation or vertebral plating. This technique requires atlantoaxial joint congruence, otherwise the means of synthesis can injure the spinal cord. Dorsal techniques utilise various non-absorbable suture materials or devices, such Kishigami Tension Band. However, cranial dislocation or impingement of the axis on spinal cord are not adequately counteracted. The technique that we describe is simple and safe and can be useful in surgical treatment in dogs with atlantoaxial dislocation and articular incongruence, a condition resembling atlantoaxial invagination described in human medicine.

CERVICAL DISC ARTHROPLASTY IN DOGS WITH DISC ASSOCIATED WOBBLER SYNDROME - LIMITATIONS AND HOW TO PREVENT POSSIBLE COMPLICATIONS.

PF Adamo[1], RC da Costa[3], R Kroll[2], C Giovannella[4], M Podell[5], P Brofman[6]. [1] East Bay Area Veterinary Specialists, Walnut Creek, CA. [2]VCA Northwest Veterinary Specialists, Clackamas, OR, [3]Department of Veterinary Clinical Sciences, The Ohio State University, Columbus, OH. . [4]Gulf Coast Veterinary Neurology & Neurosurgery, Houston, TX. [5]Chicago Veterinary Specialty Group, Chicago, IL. [6]Veterinary Specialty Care, Mt Pleasant, SC.

This study evaluated the limitations and the complications of the cervical disc arthroplasty (CDA) technique using the Adamo Spinal Disc® in dogs with disc associated wobbler syndrome (DAWS).

Thirty-three consecutive dogs with over 2 months history of DAWS diagnosed by MRI or CT-myelography were evaluated. Nineteen dogs were treated for a single level lesion, eleven dogs for a two level lesion, and three dogs for a three level lesion, for a total of a 50 treated disc sites. Median follow-up time was 12 months (range 2 weeks to 42 months). At the time of this writing, twenty-six dogs were still alive and seven dogs were deceased (five of them for non neurological diseases).

One dog (12.4 year-old, Sheltie-mix weighing 19.5lb), was euthanized 2 weeks post-surgery, because of compressive fracture of C6 with ventral displacement of the implant. Two of the three dogs presenting with over 4 months of non-ambulatory tetraparesis and severe extensor rigidity in both front legs had a poor outcome. In one dog, excessive burring of the vertebral end plate caused immediate subsidence. In two dogs, a vertebral fracture fissure occurred in surgery during distraction, this did not affect the clinical outcome. In 7 out of 50 sites (14%) subsidence was observed, this was mostly seen when narrow (side-to-side) and/or thicker (cranial to caudal) implants were used. Subsidence, except for one dog, did not affect the clinical outcome.

Correct patient selection (dog size and neurological status), size of the implant, endplate preparation, and applied distraction during surgery need to be considered as important factors that may influence the outcome.

A PILOT STUDY OF LEVETIRACETAM PULSE THERAPY FOR CANINE CLUSTER SEIZURES. R. Timothy Bentley, Melissa Logan, Andrea Sangster, George E. Moore. Purdue University College of Veterinary Medicine, West Lafayette, IN.

Canine pharamacoresistant epilepsy is common, and most pharmacoresistant dogs suffer from cluster seizures. This can lead to increased hospitalization, owner frustration and euthanasia. Levetiracetam has been proposed to be more efficacious when given intermittently than continuously. A double-blind, placebo-controlled, crossover pilot study was performed in six out-patient dogs with idiopathic epilepsy, continuing to have cluster seizures despite receiving any two maintenance anti-convulsant drugs other than levetiracetam at therapeutic serum concentrations. Dogs were randomized to a levetiracetam period (3-4 months) followed by an equal length placebo period or vice-versa, and received the same maintenance anti-convulsants throughout. The levetiracetam dose was 30mg/kg, compounded into capsules. Placebo capsules were identical in appearance. The owners started the study drug following a seizure. The owners then continued to give the study drug every 8 hours until the dog had gone 24 hours without any seizures, then stopped until the next seizure.

Dogs receiving levetiracetam pulse therapy had a median of 1 seizure per cluster (range 0-2) compared to placebo therapy (median of 4 seizures per cluster, range 1-7) ($P = 0.052$). More dogs met pre-determined treatment failure criteria during the placebo period (n=3) than the levetiracetam period (n=1). Serum concentrations of phenobarbital and potassium bromide were generally lower for the levetiracetam period than the placebo period, although this was not statistically significantly different ($P \geq 0.075$). Levetiracetam pulse therapy for cluster seizures, begun after the first seizure and continued until 24 hours without seizures, is probably effective in idiopathic epileptic dogs and warrants further investigation.

RELATIONSHIP OF DORSAL ATLANTOAXIAL JUNCTION COMPRESSIVE LESIONS (DURAL BANDS) TO CLINICAL SIGNS AND SYRINGOMYELIA IN CAVALIER KING CHARLES SPANIELS Cerda-Gonzalez S[1], Olby NJ[2], Griffith EH[2]. 1. Cornell University, Ithaca, NY. 2. North Carolina State University, Raleigh, NC.

Dorsal compressive lesions at the atlantoaxial junction (i.e. dural bands) are a well-recognized craniocervical junction anomaly, frequently seen alongside other craniocervical junction anomalies such as Chiari-like malformations. Despite their regular identification on imaging, their clinical significance remains unclear. To address this, the authors investigated their influence on clinical status and on syringomyelia in a commonly affected breed.

Thirty-six Cavalier King Charles spaniels (CKCS) were assessed for neuropathic pain and neurologic dysfunction, and were assigned a neurologic grade. Magnetic resonance imaging of the craniocervical junction was performed, with the head and neck straight and flexed. Imaging studies were then assessed for a dural band, a Chiari-like malformation (CM), and syringomyelia. Dural band and syringomyelia severity was determined using both a compression index (objective) and grading (subjective), respectively.

Of the 36 CKCS imaged, 20 demonstrated neuropathic pain. Dural bands were present in 31/36 (83.8%) dogs; 34 had a CM. Syringomyelia was found in 23 (56.8%) dogs, 21 of which also had a dural band. Dural bands were associated with both the presence and severity of clinical signs, and the presence of syringomyelia. Higher compression indices were associated with more severe syringomyelia.

Dural bands appear to play a significant clinical role. Compression indices provide a better assessment of dural band severity compared to grading.

IDENTIFICATION OF A GLIOMA SUSCEPTIBILITY LOCUS IN DOGS. P. Dickinson[1], K. Truve[2], D. York[1], A. Xiong[3], K. Jayashankar[4], K Forsberg Nilsson[3], K. Lindblad-Toh[5,6]. D. Banasch[7] 1. Department of Surgical and Radiological Sciences, School of Veterinary Medicine, Davis CA. 2. Department of Animal Breeding and Genetics, Swedish University of Agricultural Sciences Uppsala, Sweden. 3. Department of Immunology, Genetics and Pathology, Uppsala University, Sweden. 4. Ultragenyx pharmaceutical Inc., Novato, CA. 5. SciLifeLab, Department of Medical Biochemistry and Microbiology, Uppsala University, Sweden, 6. Broad Institute of Harvard and Massachusetts, MA. 7. Department of Population Health and Reproduction, School of Veterinary Medicine, Davis CA.

Gliomas are devastating primary brain tumors that are highly overrepresented in ancestral Bulldog-type breeds such as the Boxer, Bulldog and Boston Terrier, but not in other brachycephalic breeds such as the Pug. We hypothesized that across-breed genome-wide association mapping using these breeds would identify genetic susceptibility risk factors for the development of glioma.

Across-breed genome-wide association was performed for primary gliomas using the Illumina 170K canine SNP array with 40 cases and 142 controls. Targeted resequencing of control and glioma samples was performed to identify variants in glioma associated regions and putative disease causing mutations. Semi-quantitative RT-PCR, western blotting and functional analyses were performed to investigate candidate genes.

A significant association was identified on CFA 26 and appeared to be due to the oligodendroglioma cases. Across breed mapping carries with it significant population stratification but the magnitude of the effect of this association was compelling (Bonferoni $p = 2 \times 10^{-34}$). A haplotype was identified that was homozygous within breeds with high prevalence of glioma and segregated in a number of other breeds. Three candidate genes were identified with documented interactions with potential glioma pathways. Differential transcript expression was present in 2 genes comparing tumor tissue and normal brain. One candidate gene had a non-synonymous exonic single nucleotide polymorphism that results in a decrease in gene function of approximately 50%.

We have identified potential susceptibility genes for glioma that may provide insight into gliomagenesis as well as providing possible targeted therapeutic and breeding strategies for prevention and treatment of gliomas.

CANINE AWAKE AMBULATORY WIRELESS VIDEO ELECTROENCEPHALOGRAPHY. F. James[1], T. Jokinen[2], K. Tiira[2], L. Koskinen[2], E. Seppälä[2], H. Lohi[2]. 1. Ontario Veterinary College, Guelph, Ontario, Canada. 2. Research Programs Unit, Molecular Neurology, and Department of Veterinary Biosciences, University of Helsinki, and Folkhälsan Research Center, Helsinki, Finland.

Canines often present transient paroxysmal events that are hard to confirm as seizures based on description or observation alone. Electroencephalography (EEG) correlation between clinical manifestations and epileptiform activity allows this confirmation. However, sedation or general anesthesia during EEG recording may abolish the events of interest. This retrospective case series describes a novel canine awake ambulatory wireless video EEG technique.

All dogs investigated for possible seizure activity were selected from EEG cases recorded with the technique between July 2013 and March 2014, resulting in 20 cases. Subcutaneous needle or wire electrodes were affixed to the scalp by sticky bandage. If sedation was required for electrode placement, either dexmedetomidine was used and subsequently reversed with atipamezole ($n = 4$) or electrodes were placed during recovery from general anesthesia for other procedures ($n = 2$) with recording beginning after recovery and no further administration of sedatives. Up to fourteen electrodes were placed per dog, including ground and reference. Impedance was kept below 30 kOhms (ideally below 16 kOhms) where possible (replacing electrodes if necessary). Recordings were performed using a Lifelines Trackit EEG machine and reviewed using Persyst software. Recording length ranged from one to twelve hours. One dog was recorded twice, in two separate sessions. Recording took place in the clinic either as an outpatient or as an inpatient in the intensive care unit, with one dog additionally recorded at home.

Transient paroxysmal events investigated included tonic-clonic seizures ($n = 6$), altered mentation / focal twitching ($n = 6$), tail-chasing ($n = 7$), and head bobbing ($n = 1$). Paroxysmal epileptiform EEG activity confirmed three dogs as having seizures. Not all electrical seizures were accompanied by visible motor manifestations. Transient paroxysmal episodes were also recorded on video with no accompanying epileptiform EEG activity ($n = 6$). For 11 dogs, the transient episodes of interest did not occur during the recording period.

Awake ambulatory wireless video EEG can be helpful in diagnosis for 45% (9/20) of cases, similar to diagnostic yields reported in human pediatric EEG. This technique should be considered in canines presenting with frequent paroxysmal events of unknown origin.

PHARMACOKINETICS OF LEVETIRACETAM IN EPILEPTIC DOGS WHEN ADMINISTERED CONCURRENTLY WITH PHENOBARBITAL, BROMIDE, OR PHENOBARBITAL AND BROMIDE IN COMBINATION. KR Muñana, JA Nettifee-Osborne, MG Papich. North Carolina State University College of Veterinary Medicine, Raleigh, NC.

Levetiracetam (LEV) is a common add-on antiepileptic medication (AED) for dogs with refractory seizures. Concurrent phenobarbital administration alters the disposition of LEV in healthy dogs. The objective of this study was to evaluate the pharmacokinetics of LEV in epileptic dogs when administered concurrently with conventional AEDs. Eighteen client-owned dogs on maintenance therapy with LEV and phenobarbital (PB group, n=6), LEV and bromide (BR group, n=6) or LEV, phenobarbital and bromide (PB-BR group, n=6) were enrolled. All AEDs had attained steady state concentrations. Blood samples were collected at 0, 1, 2, 4 and 6 hours after a morning dose of LEV. Plasma LEV concentrations were determined by high-pressure liquid chromatography. To account for dose differences among dogs, LEV concentrations were normalized to the mean study dose (26.4 mg/kg). Pharmacokinetic analysis was performed on adjusted concentrations using a noncompartmental method, and AUC calculated using the trapezoidal method. Compared to the PB and PB-BR groups, the BR group had significantly higher C_{max} (70.6 $\pm$ 24.0 µg/mL, versus 35.5 $\pm$ 13.7 µg/mL and 25.3 $\pm$ 9.0 µg/mL, respectively) and $AUC_{0\text{-}Cn}$ (314.2 $\pm$ 114.3 h*µg/mL versus 127.0 $\pm$ 64.7 h*µg/mL and 91.0 $\pm$ 42.2 h*µg/mL, respectively), and significantly lower CL/F (68.9 $\pm$ 22.1 mL/h/kg versus 172.1 $\pm$ 81.9 mL/h/kg and 247.6 $\pm$ 126.9 mL/h/kg, respectively). Concurrent administration of phenobarbital alone or in combination with bromide significantly alters the disposition of LEV in epileptic dogs compared to concurrent administration of bromide alone. These findings warrant consideration when utilizing LEV for add-on therapy in dogs with epilepsy.

CHARACTERISATION OF GENE EXPRESSION IN NATURALLY OCCURRING FELINE DEGENERATIVE JOINT DISEASE ASSOCIATED PAIN. M. Freire[1], A. O'Nan[2], J. Benito[1], J. Hash[1], M.S. Ashwell[2], B.D.X. Lascelles[1]. 1.Comparative Pain Research Laboratory, North Carolina State University College of Veterinary Medicine, Raleigh, NC. 2. Livestock Genomics Laboratory, North Carolina State University Department of Animal Sciences, Raleigh, NC.

Degenerative joint disease (DJD) associated-pain is a clinically relevant and common condition affecting domesticated cats and other species including humans. Identification of the neurobiological signature of pain is well developed in pain models in rodents however this information is lacking from animals or humans with naturally occurring painful conditions. In this study, identification of housekeeping genes (HKG) for neuronal tissue and expression levels of genes considered associated with chronic pain in rodent models were explored in cats with naturally occurring osteoarthritic pain.

Eighteen adult cats were evaluated – 9 normal, and 9 with hind limb radiographic DJD. Expression of pain signaling genes (including *Nav1.7, Nav1.9, NGF, NK1, TNFα, TrkA, ASIC3, COX2, Nav1.8, CX3CL1, ATF3*) in lumbar spinal cord dorsal horn, lumbar dorsal root ganglia and joint capsule tissues from normal and DJD-animals were studied using quantitative RT-PCR (qPCR).

HKG identified as the most stable across all tissue samples were many of the ribosomal protein genes, such as *RPL30* and *RPS19*. qPCR results showed downregulation of *COX-2, NGF* and upregulation of *TNFα* genes in spinal cord and downregulation of *TNFα* in dorsal root ganglia collected from DJD-animals. No genes were differentially expressed in the joint capsule samples due to high levels of variability in the healthy samples.

Several genes expected to be upregulated, based on results in rodent models, were downregulated. Further work is needed to understand the neurobiology of pain in naturally occurring disease and what rodent models are predictive of these changes in naturally occurring clinical pain conditions.

CONVECTION ENHANCED DELIVERY OF CETUXIMAB CONJUGATED WITH IRON OXIDE NANOPARTICLES FOR CANINE INTRACRANIAL GLIOMAS: A PILOT STUDY.

AC Freeman[1], SR Platt[2], M Kent[2], S Holmes[2], Alexandros Bouras[3], Costas G. Hadjipanayis.[3]
[1]The Animal Health Trust, Newmarket, UK, [2]College of Veterinary Medicine, University of Georgia, Athens, GA, [3]Emory University School of Medicine, Winship Cancer Institute of Emory University, Atlanta, GA.

Convection enhanced delivery (CED) allows targeted treatment of intracranial neoplasia overcoming the restrictions of the blood brain barrier. The aim of this study was to evaluate the safety and efficacy of CED of cetuximab, an antibody against epidermal growth factor receptor (EGFR), for the adjunctive treatment of intracranial gliomas in dogs.

Nine dogs with intracranial gliomas were included. All dogs underwent surgical debulking of the tumor with intraoperative histopathologic confirmation. Two CED catheters were placed within the tumor cavity and attached to a Medtronic titanium SynchronoMed II pump containing bioconjugated cetuximab-iron oxide nanoparticles (IONPs). The pumps were secured to the head or body of the dogs during recovery and set to infuse at 0.5 µl/min over 72 hours.

The 9 dogs included 2 Boxers, 3 Boston Terriers, one Pit Bull, one French Bulldog, one English Bulldog and a mixed breed. Clinical signs included seizures (8/9), circling (3/9), and mentation changes (4/9). All tumors were supratentorial and 8/9 were oligodendrogliomas (1 astrocytoma). MRI was performed at the following time points if the patients were alive: 24 hours, 7 days, 4 weeks, 6 months, 1 year, and 2 years. No side effects of the drug delivery post-craniectomy were noted. MRI at 24 hours confirmed catheter placement and IONP presence in the tumor bed in all cases. At the time of writing, 4/9 dogs were still alive. Five dogs were euthanized because of recurrence of neurologic signs (circling, mentation changes) or seizures. Median survival time (MST) after surgery in the group of patients that were euthanized was 5.0 months (mean: 8.2 months; range 3 to 24 months) and current median survival time of those dogs which are still alive is 7.0 months (mean: 11.5 months; range 6-26 months). Overall MST including surviving and euthanized patients was 6 months (mean: 9.2 months; range 3 to 26 months).

CED of cetuximab is a safe and practical adjunctive treatment to surgical debulking in canine patients with intracranial gliomas. Based on this pilot study, CED of bioconjugated cetuximab-IONPs may be an effective treatment for canine intracranial gliomas. CED infusions can be performed in conscious patients over a 72-hour duration, which has been shown experimentally to improve volume of diffusion within the target tissue. Ultimately, a larger placebo controlled, blinded clinical trial is necessary and is underway to understand the role of this treatment for canine intracranial gliomas.

SEQUENTIAL MAGNETIC RESONANCE IMAGING OF EXPERIMENTAL INTRACEREBRAL HEMORRHAGE IN DOGS. DG An[1], J Park[2], NK Lee[1], C Lee[2], DI Jung[3], JH Kang[1], D Chang[1], MP Yang[1], BT Kang[1]. 1. Chungbuk National University College of Veterinary Medicine, Cheongju, South Korea. 2. Korea Basic Science Institute, Ochang, South Korea. 3. Gyeongsang National University College of Veterinary Medicine, Jinju, South Korea.

Intracerebral hemorrhage (ICH) is one of the most lethal stroke types. Neuroimaging techniques, particularly computed tomography and magnetic resonance imaging (MRI), have improved the accuracy of the diagnosis of intracerebral hemorrhage. In human, the MR features of five distinctive stage of ICH have already been reported, but they are not objectively verified in dogs. Therefore, the purpose of this investigation was to document the sequential MR characteristics of canine ICH according to evolving stages.

ICH was induced in 4 healthy laboratory beagle dogs by injecting 500 U of bacterial collagenase from *Clostridium histolyticum*, which was delivered into the parietal lobe over 5 min with a micro infusion pump. T1-, T2- and diffusion-weighted imaging (DWI), measurement of the apparent diffusion coefficient (ADC), and gradient-echo (GRE) imaging were performed serially at 6 different time points by using a 3T MR system: before and at 12 h, 2.5, 4.5, 9.5 and 23.5 days after ICH. The MR features of this experiment are summarized in table 1.

Table 1. The MR characteristics in 5 stages of canine ICH

Stage	Modality				
	T1-WI	T2-WI	DWI	ADC	GRE
Hyperacute (12 hours)	Hypo	Hypo	Hypo	Decreased	Hypo
Acute (2.5 days)	Iso to hypo	Mixed	Hypo	Mixed	Iso to hypo
Early subacute (4.5 days)	Hyper	Hyper	Mixed	Mixed	Mixed
Late subacute (9.5 days)	Iso to hyper	Iso to hyper	Hyper	Mixed	Mixed
Chronic (23.5 days)	Iso to hypo	Iso to hypo	Hypo	Decreased	Iso to hypo

Hypo: hypointensity; Hyper: hyperintensity; Mixed: mixed signal intensity

Most of the results were different with the previous reports of human. Individual and time variations may vary the MR findings. The present data could be used to differentiate clinical stage of ICH in canine patients.

CEREBELLAR ABIOTROPHY IN A LITTER OF BELGIAN SHEPHERDS. J. Dahan[1], MN. Lucas[1], MA. Colle[2], O. Sebille[3], I. Raymond-Letron[1], A. Diquelou[1]. 1. University of Toulouse, INP, National Veterinary School of Toulouse, Toulouse, France. 2 Oniris, Nantes, France. 3 Castres, France.

Ataxia was observed in a litter of Belgian Shepherds from phenotypically normal parents. Four puppies (out of 6) of the same litter displayed identical clinical signs. Affected puppies grew well, began to walk normally but, after the 9th week, demonstrated worsening ambulatory difficulties. Three of these puppies presented at the Teaching Hospital for evaluation. General physical examinations were normal; neurologic examinations revealed severe ataxia presumably of cerebellar origin (hypermetric ataxia, intention tremor, delayed proprioception of all limbs, no menace response). All spinal reflexes were normal. The 4th littermate, non-presented, was the most affected and unable to walk.

Results of complete biochemical and hematological examinations performed in one of the 3 presented puppies were within laboratory reference intervals, except a mild hypomagnesemia. Cerebrospinal Fluid analysis of this dog was normal, with PCR analysis negative for Distemper, Toxoplasma, Neospora and Parvovirus. Magnetic Resonance Imaging of the brain, performed on the 4 affected puppies and a normal female of the same litter, was normal in the non-affected female, but was compatible with congenital cerebellar abiotrophy in all affected puppies.

The 3 presented puppies were euthanized. Similar histopathological lesions were found, suggestive of cerebellar abiotrophy with loss of Purkinje and granule cells. To the authors 'knowledge, this affection has not been described in this breed.

OPTIMIZATION OF A METHOD FOR ISOLATION, PURIFICATION AND EXPANSION OF ADULT SCHWANN CELLS FROM DORSAL CUTANEOUS CERVICAL NERVES IN DOGS. Ji-Hey Lim, Natasha J Olby. North Carolina State University College of Veterinary Medicine, Raleigh, NC.

Cell-based therapeutic strategies are being developed for neurologic disorders. Schwann cells (SCs) support axonal regeneration and remyelination, and have the advantage of being easily retrievable from the patient, allowing for the transplantation of autologous populations of cells. Important considerations for SC preparation are to reduce contamination by fibroblasts, to obtain a large number of cells in a short period and to avoid detrimental consequences for the patient during harvesting. Here, we reported obtaining nerve biopsies from the dorsal cervical cutaneous region, from which we purified and expanded SCs. We harvested small pieces of dorsal cutaneous cervical nerves from dogs. The predegenerated tissues were cultured in conditions to enhance SC growth while reducing the number of undesired fibroblasts. Cells were immunopurified using magnetic beads and their purity was evaluated by assessing expression of cell surface antigens.

Half a million cells were generated in 10 days from less than 20mg wet weight of single nerve fascicles. More after the purification processes, the cultures contained more than 80% SCs. However, prolong cell-doubling time after purification indicated that their SCs ability to proliferate was decreased. We concluded that biopsies of the dorsal cervical cutaneous nerves are a safe and practical culture protocol that yields sufficient autologous SC population for clinical use.

CONGENITAL MYASTHENIA GRAVIS IN A FAMILY OF LABRADOR RETRIEVERS IS CAUSED BY A MUTATION IN THE *COLQ* GENE. G.D. Shelton[1], C. Rinz[2], J Levine[3], K. Minor[4], A. Starr-Moss[2], L.A. Clark[2]. 1. University of California San Diego, La Jolla, CA. 2. Clemson University, Clemson, SC. 3. Texas A&M University, College Station, TX. 4. University of Minnesota, St. Paul, MN.

Congenital myasthenic syndromes are heterogeneous disorders associated with mutations in proteins residing in the nerve terminal, the synaptic basal lamina, or the post-synaptic region of the motor endplate resulting in compromise of the safety margin of neuromuscular transmission. Congenital myasthenia gravis (CMG) has been described clinically in a few dog breeds including the Jack Russell Terrier, Smooth Fox Terrier, Long-haired Dachshund, Old Danish Pointing Dogs, and recently identified in the Labrador Retriever. To date, characterization of CMG at the molecular level has only been performed in the Old Danish Pointing Dogs (mutation in the *CHAT* gene). The purpose of this work is to identify the genetic basis of CMG in a family of Labrador Retrievers.

Causative mutations for human CMG have been identified in 18 genes. To reduce this list of candidate genes, Illumina Canine HD Infinum BeadChips were used to generate genome-wide SNP profiles for a nuclear family comprised of the sire, dam, 2 affected puppies and 5 unaffected littermates. Genotypes were analyzed for genomic inheritance patterns , with 7 chromosomes having segments consistent with an autosomal recessive mode of inheritance (i.e., heterozygosity in the parents and homozygosity in both affected puppies, with the allele absent or heterozygous in the littermates). The identified regions harbored 3 candidate genes: *CHRND, CHRNG* and *COLQ*. Sequences representing the coding regions of all 3 genes from an affected puppy, the carrier sire and an unrelated Labrador Retriever, revealed a missense mutation in *COLQ* that alters a conserved residue. Evaluation of the mutation in extended family members identified 14 carriers, but it was not found in over 250 unrelated Labrador Retrievers. Our data suggest that a *COLQ* mutation is causative for CMG in Labrador Retrievers, and that proper use of a genetic test should prevent widespread transmission in this popular breed.

DISTINCT ANATOMIC DISTRIBUTION PATTERNS IN CENTRAL NERVOUS SYSTEM LYMPHOMAS CORRELATE WITH BOTH MAGNETIC RESONANCE IMAGING PROFILES AND IMMUNOHISTOLOGICAL TYPES. S. Sisó, B.K. Sturges, P. Marco, E.R. Wisner, P.F. Moore, R.J. Higgins. School of Veterinary Medicine, University of California - Davis, Davis, CA.

Canine lymphomas comprise about 83% of all hematopoietic tumors with up to 12% causing intracranial lesions in the central nervous system (CNS). Most CNS lesions are secondary metastases as part of widely disseminated disease in other organs. In contrast, up to 4% have been defined as primary CNS lesions based on the exclusion of extraneural involvement. However there are no studies of any relationship between Magnetic Resonance Imaging (MRI) profiles, immuno- and histological phenotypes and any anatomically based patterns in either primary or secondary CNS lymphomas.

In our study of 36 lymphoma cases, which had a rigorous necropsy examination, all were classified as secondary CNS tumors. Immunophenotyping with relevant canine specific cell marker antigens identified 23 B-cell [CD79a$^+$, CD20$^+$, PAX5$^+$] and 12 T-cell [CD3$^+$, granzyme B$^+$] lymphomas with one suspected non-B/non-T cell origin. Of the seven main histologic types following the currently accepted WHO criteria, the diffuse large B-cell [n=15] and the peripheral T-cell [n=5] lymphomas were the major ones identified. Each type revealed lesions with a distinct signature of anatomical patterns comprising combinations of parenchymal, meningeal, choroid plexus, subventricular, perivascular and spinal or cranial spinal nerve involvement.

In addition, profiles of each type correlated with MRI changes. Most tumors had disseminated distribution in the spinal cord or/and brain, while the remainder were restricted to either supratentorial [33%] or infratentorial [11%] sites. The pituitary gland was involved in 68% of the cases irrespective of the immuno-histological phenotype except for the absence of the large granular T-cell lymphoma type.

EVALUATION OF CLIENT PERCEPTIONS REGARDING BARRIERS AND BENEFITS TO VETERINARY ONCOLOGY CLINICAL TRIALS PARTICIPATION. JH Burton[1], R Keiffer[2], R Livingston[2,3], CS Hardy[1], R Page[1], SE Lana[1]. 1. Colorado State University Flint Animal Cancer Center, Fort Collins, CO. 2. A-Train Marketing Communications, Inc., Fort Collins, CO. 3. Advantage Research, Loveland, CO.

Clinical trials are used in veterinary medicine to assess the utility and efficacy of new treatment protocols, surgical interventions and diagnostic techniques. While rates of enrollment into clinical trials are widely unknown and likely vary between institutions, there are assumed to be some elements of clinical research which may prevent enrollment of eligible patients into clinical trials. Possible barriers to patient enrollment such as distance from clinic, invasiveness of procedures, and risk of unknown side effects have been postulated as factors that may decrease clinical trial participation but this has not been systematically evaluated in veterinary medicine to date. The purpose of this study was to identify pet owner perceived barriers and benefits to participation in veterinary oncology clinical trials.

Clients of the Flint Animal Cancer Center (FACC) at Colorado State University from 2010 to 2012 were surveyed to evaluate barriers to clinical trial participation, motivators that would increase participation, and other factors that might influence clients' decision-making process regarding enrollment. A second population of pet owners that were not clients of the FACC from the Denver metro area was surveyed to evaluate similar criteria.

On-line surveys were completed by 465 FACC clients and 384 non-FACC pet owners. The majority of FACC clients did not report distance to the clinic as a barrier to clinical trial participation with 67% willing to drive more than 50 miles. In distinction to those clients already traveling to the FACC, 77% of non-FACC pet owners in the Denver regions were unwilling to drive greater than 49 miles to participate in a trial. Sixty-three percent of FACC clients indicated that they were referred to the FACC by their primary care veterinarian and 93% of non-FACC pet owners reported that their primary care veterinarian was their main resource for cancer treatment recommendations and referral. The reputation of the trials center and referral by the primary care veterinarian were ultimately more influential factors for trials participation than cost of services, risk of side effects, or distance to a clinical trials center for both groups. Both FACC clients and non-FACC pet owners were more likely to participate in clinical trials if they knew the research would be of benefit to other animals; the possibility that people might benefit from results of a veterinary clinical trial was not as influential.

Veterinary clinical trial recruitment should focus on outreach communication to primary care veterinarians. Availability of trials as well as information regarding benefits and risks should be made readily accessible for easy use by veterinarians. Communication to pet owners should emphasize the benefits of clinical trials specifically to animal health in order to encourage clinical trial participation.

NEW INSIGHTS INTO THE PATHOGENESIS AND TREATMENT OF CANINE THYROID CARCINOMA M Campos[1], R Ducatelle[1], M Kool[2], J Mol[2], G Rutteman[2], HS Kooistra[2], L Duchateau[1], K Peremans[1], I Van de Maele[1], H de Rooster[1], I Polis[1], S Daminet[1]. 1. Ghent University Faculty of Veterinary Medicine, Merelbeke, Belgium. 2. Utrecht University Faculty of Veterinary Medicine, Utrecht, The Netherlands.

The goals of this research were to investigate the presence of mutations and differences in relative expression of candidate genes, to explore new therapeutic targets and to investigate clinical, pathological and immunohistochemical prognostic markers in canine thyroid cancer.

The medical records of Ghent and Utrecht Universities were reviewed (1986-2013). Dogs diagnosed with histologically confirmed thyroid neoplasia were included.

Each tumor section (n=74) was reviewed and scored by a board-certified pathologist (histologic type, % necrosis, % hemorrhage, nuclear pleomorphism, mitotic index, evidence of vascular or capsular invasion) and immunohistochemistry was performed for calcitonin, Ki-67, E-cadherin, cox-2, VEGF, P-gp and p53. Differentiation between follicular cell tumors and medullary tumors was based on immunohistochemistry for calcitonin. RNA was isolated from snap-frozen tumor samples (n=59) and healthy thyroids (n=10) and sequencing was performed for mutation hotspots of RAS (K, H and N), PIK3CA, RET and BRAF genes and for the complete coding region of PTEN. Quantitative RT-PCR was performed for VEGFR1, VEGFR2, EGFR, PIK3CA, PIK3CB, PTEN, PDPK1, Akt1, Akt2, cox-2 and calcitonin.

All tumors were carcinomas; 54 were follicular thyroid carcinomas (FTC) and 20 were medullary thyroid carcinomas (MTC). In 1 FTC and 1 MTC, 2 missense mutations (G12R and E63K respectively) were found in K-RAS.

Relative expression of VEGFR1, EGFR, Akt2 and PDPK1 genes was significantly higher in FTC and MTC than in healthy controls.

80% of FTC and all MTC exhibited a very high percentage (76-100%) of neoplastic cells immunopositive for VEGF. 13% of FTC and 50% of MTC presented cox-2 expression. 7% of FTC and 70% of MTC presented P-gp expression. No tumor was immunopositive for p53.

46 dogs (30 FTC and 16 MTC; stage I-III) underwent thyroidectomy and were included in a survival analysis. Outcome was comparable between FTC and MTC. Inappetence (P=0.007), histologic vascular invasion (P=0.027) and surgical vascular invasion (P=0.005) were associated with shorter disease-free survival. Histologic vascular invasion (P=0.009), large tumor diameter (P=0.004) and high Ki-67 labeling index (P=0.004) were associated with shorter time to metastases.

Relative expression of candidate genes suggests PI3K/Akt pathway activation in both FTC and MTC. VEGF is a potential therapeutic target in both FTC and MTC, while cox-2 and P-gp seem to be attractive molecular targets in canine MTC. Intensive monitoring and adjunctive therapy might be indicated when negative prognostic factors are identified.

FRAMELESS STEREOTAXIS IN SMALL ANIMAL BRAIN & SPINE RADIOSURGERY.

L Malfassi, M Dolera. La Cittadina Fondazione Studi e Ricerche Veterinarie, Romanengo (CR), Italy.

Stereotactic radiosurgery (SRS) and hypofractionated stereotactic radiotherapy (HSRT) are advanced methods of treatment of tumours, usually performed with an head frame. The purpose of this study was to evaluated the feasibility and efficacy of frameless LINAC based SRS and HSRT in brain and spinal neoplastic disease in dogs and cats. 74 animals (64 dogs, 10 cats) with presumptive (59 cases) or histologically confirmed (15 cases) cancer of nervous system were considered. In 66 cases, RT was curative, in 8 post-surgical adjuvant. All the patients received a diagnostic high field MRI and a simulation CT. The expansion of GTV to PTV (Planning Treatment Volume) was 3 mm on each directions. For all of them, the feasibility of a frameless SRS or HSRT have been studied using a MonteCarlo statistic algorith and a CMS Monaco 3.0 treatment planning system (TPS) for LINAC 6 MV with beam modulator. To assess the treatment feasibility, treatment simulation have been performed "in air" with the Dosimetric Check software using the Iview portal ASi imager. Each animal was placed in a stereotactic cradle. Positioning verification was performed with cone beam CT. Controls were performed with serial clinical and MRI examination. All patients (37 meningiomas, 22 gliomas, 9 nerve sheath tumours, 6 pituitary tumours, 1 pineal tumour) completed the therapy. The doses delivered ranging from 16 Gy in 1 fraction to 33 Gy in 5 fractions prescribed to the 90% isodose. The RTOG toxicity show 1 G3, 3 G2, 6 G1, 64 G0. The 6 month RECIST efficacy showed 39 partial response, 19 stable disease, 11 complete response, 5 disease progression. The overall response rate was 67%. Either 89% of stable disease patients show a clinical improvement. In 93% of cases a control of disease or a clinical improvement was obtained The toxic effects were quite content. In the short and medium term the efficacy is comparable or superior to that obtainable with other fractionation schedules. SRS and HSRT are safe and effective in the control of nervous system tumours in dogs and cats.

TARGETING GANGLIOSIDES AS TUMORS ANTIGENS FOR CANCER VACCINES. RJ. Milner, N. Chimura, ME. Salute. University of Florida, College of Veterinary Medicine, Gainesville, FL.

In this research report we provide background and research information on the targeting of gangliosides as tumor antigens and provide results on prospective in-vivo and in-vitro studies. Previously we showed that GD3 (disialogangliosides) was found in high concentrations on the surface of canine melanoma cells but at lower concentrations on osteosarcoma cells. When combined with an adjuvant, the vaccine was able to elicit a humoral and cytotoxic cellular immune response in normal dogs. We report on the results from prospective IACUC approved clinical trials in dogs with melanoma.

In a Phase-1trial (n=65) dogs were entered and received 3x1-monthly vaccinations. Accrual included oral, cutaneous, pedal, nodal and anal melanoma. Melanomas were graded; low (n=13, < 3MF/10hpf), intermediate (n=8, 3-14 MF/10hpf) and high grade (n=38, >14 MF/10hpf). Immune responses were monitored using cytokines assays, antibodies, and cytotoxicity assays. No side effects were found using CBC, Chemistry, U/A, and injection site responses, only a single dog showed an allergic reaction (local swelling). Multivariate analysis found anatomic site, histological grade, and univariate analysis found IgG titers, prevaccination NK Assay, tumor stage influenced on survival. Paradoxically, cases with decreased IgG titers had prolonged survival compared to increasing and stable titers. Serum endogenous GD3 concentrations were determined in dogs with melanoma; melanoma cases had higher levels of serum GD3. Case accrual continues in phases-III of the study, and in an equine melanoma study.

RADICAL CYSTECTOMY AND BILATERAL SUBCUTANEOUS URETEROVESICULAR BYPASS (SUB) PLACEMENT FOR ADVANCED URINARY BLADDER CANCER IN DOGS. C. Weisse, A. Berent. The Animal Medical Center, New York, NY.

Canine urothelial carcinoma remains a disease with poor clinical response to therapy. Initial biological responses are non-durable and patient death typically occurs within a year or so following diagnosis. Contrary to theoretical advantages, radical surgical resections and rerouting procedure have not provided prolonged survival times; Reported complications may dissuade owners from these procedure. We hypothesized that radical cystectomies with SUB placement would permit more aggressive, wider tumor resections with few associated complications.

Three dogs with extensive bladder TCC, bilateral ureteral obstructions, and incontinence were included. Common comorbidities included severe azotemia (2/3) and previous therapies including previous surgical resection (2/3) and chemotherapy (2/3), among others. Dogs received total cystectomy, bilateral partial ureterectomies, and partial urethrectomy with placement of a bilateral SUB device (Norfolk Vet). There were no major intra-operative complications encountered and surgical times ranged from 120-160 minutes. Post-operative complications included urinary tract infections (3/3) and a kinked urethral catheter requiring an extension added to the subcutaneous port (1/3). All biopsies had tumor cells present at the urethral surgical margin only. Two dogs ultimately died or were euthanized for progressive azotemia (95 days) and a transfusion reaction with no azotemia (48 days). One dog remains alive 244 days post-operatively. SUB device flushing has confirmed patency in all three dogs. No dogs have been identified to have local tumor recurrence.

Total cystectomy with SUB placement is a potential option for dogs with extensive TCC in order to achieve wide tumor excisions. Large excisions are necessary to achieve complete tumor-free margins.

RANDOMIZED, BLINDED, PLACEBO-CONTROLLED TRIAL OF ARTESUNATE WITH DOXORUBICIN FOR B CELL LYMPHOMA OF DOGS. JN Bryan[1], D Tate[1], SM Bechtel[1], K Choy[2], L Donnelly[1], K Fitzpatrick[1], BK Flesner[1], B Fowler[1], B Gallis[3], CJ Henry[1], C Herrera[1], M Jabbes[2], KA Selting[1], T Sasaki[3]. 1. University of Missouri, Columbia, MO. 2. Washington State University, Pullman, WA. 3. University of Washington, Seattle, WA.

Artemisinin has described anti-cancer properties. Artemisinin can down-regulate VEGF, HIF-1alpha, and survivin in cancer cells, causing apoptotic death. Artesunate, a water-soluble derivative, has better oral absorption in dogs than artemisinin. The hypothesis was that oral artesunate is tolerable with doxorubicin and results in longer disease-free interval (DFI) than doxorubicin and placebo in dogs with stage 3a or 4a B-cell lymphoma.

Twenty-one qualifying dogs were randomized (blinded) to receive artesunate 20mg/kg or equivalent placebo. Dogs received study drug alone for 72 hours, then received 30mg/kg doxorubicin q.3 weeks for three doses with study drug. Thereafter, lymph node measurement, CBC, chemistry profile, and urinalysis were monitored monthly until disease progression. Differences and DFI were compared by Mann-Whitney and Kaplan-Meier log-rank tests. Significance was set at $P \leq 0.05$.

Artesunate alone did not cause clinical remission (P=0.1347). Artesunate induced no biochemical toxicosis. One dog experienced progressive disease following doxorubicin; all others experienced a complete response. Artesunate caused no greater neutropenia or thrombocytopenia following doxorubicin (P=0.6234 and P=0.2311). DFI was not different between groups (70d for placebo and 122d for artesunate, P=0.967). Dogs receiving artesunate had higher nucleated red blood cell counts (median 272nRBC/100WBC artesunate and 9nRBC/100WBC placebo, P=0.0142). Red cells displayed poikilocytosis and basophilic stippling in artesunate-treated dogs.

Artesunate did not cause remission as a single agent, nor improved DFI in dogs treated with doxorubicin. 20mg/kg artesunate caused evidence of heme stress in treated dogs. Artesunate was otherwise well tolerated with doxorubicin. Artesunate cannot be recommended as a therapy for B-cell lymphoma in dogs.

PRELIMINARY TOXICITY AND EFFICACY OF THE NOVEL ONCOLYTIC VIRUS, VSV-IFNB-NIS, IN TUMOR-BEARING DOGS. A.K. LeBlanc[1], S. Naik[2], G. Galyon[1], C. J. LeBlanc[1], S. Frazier[1], M. Federspiel[2], K. W. Peng[2], S. J. Russell[1]. 1. University of Tennessee College of Veterinary Medicine, Knoxville, TN. 2. Mayo Clinic College of Medicine, Rochester, MN.

Oncolytic virotherapy is an emerging approach to cancer therapy wherein naturally oncolytic viruses mediate tumor destruction and elicit a secondary anti-tumor immune response. VSV-IFNβ-NIS is a novel recombinant oncolytic virus with documented efficacy and safety in preclinical murine models of cancer. Safety studies in normal dogs have determined a safe dose of systemically-administered virus for clinical studies in tumor-bearing dogs.

Seven dogs with measurable malignancies were enrolled in an ongoing fixed-dose clinical trial of either VSV expressing either human Interferon-β (IFNβ) ($n = 3$) or canine IFNβ ($n = 4$) and the sodium iodide symporter (NIS). A variety of malignancies was represented, including B- and T-cell lymphoma, malignant melanoma, and multiple myeloma.

Two of 7 dogs (29%) had a demonstrable partial response to a single IV dose of virus lasting 25 and 35 days, respectively. Both dogs received VSV expressing human IFNβ and had T cell lymphoma. All other treated dogs achieved either stable ($n = 1$) or progressive ($n = 4$) disease over the 30 day study period. Clinical toxicities developed within the first 3 days of viral administration in most dogs and included mild to moderate acute, self-limiting hepatic enzymopathy; fever; and urinary tract infection. Dogs develop neutralizing antibodies within 7-10 days of viral administration, and do not appear to shed infectious virus in urine or saliva. Infectious virus was isolated from PBMCs of one dog on the day of virus administration only within the first 3 hours following systemic VSV administration

Overall the results demonstrate that measurable responses to oncolytic VSV are observed in dogs with cancer. The starting dose of 10^10 TCID50 appears well-tolerated and may allow dose escalation and/or repeated dosing of virus. Additional studies are underway to determine prognostic factors for response to therapy and a wider knowledge of response rates in dogs with hematologic and solid malignancies.

THE ENHANCED TUMORIGENICITY OF CANINE OSTEOSARCOMA CELL LINES IS ASSOCIATED WITH ENHANCED FRIZZLED-6 EXPRESSION AND SIDE POPULATION CELLS. LCS Rodrigues[1], KE Holmes[1], V Thompson[1], M Newton[2,3], TJ Stein[1,3,4]. 1. Department of Medical Sciences, 2. Department of Biostatistics, 3. Carbone Comprehensive Cancer Center, and 4. Institute for Clinical & Translational Research; University of Wisconsin-Madison, Madison, WI.

The Wnt signaling pathway aids in regulating cell proliferation, differentiation, migration, but its aberrant activity has been implicated in tumor development and tumor-initiating cells. Tumor-initiating cells have the ability to divide asymmetrically into either additional tumor-initiating cells or non-tumorigenic progeny. We performed microarray analysis comparing six canine primary osteosarcoma cell lines and identified 379 genes differentially-expressed between non-tumorigenic (n=3) and tumorigenic (n=3) cell lines. Among these genes, frizzled-6 expression was increased to the greatest extent (7.78 fold) in tumorigenic cell lines compared to non-tumorigenic cell lines. Frizzled-6 is a co-receptor for wnt ligands and has been associated inhibition of β-catenin-dependent signalling and more aggressive tumors. QPCR and Western blot analyses confirmed the increased expression of frizzled-6 in the tumorigenic cell lines. Additionally, the tumorigenic cell lines were associated with an increase in percentage of side population cells compared to non-tumorigenic cell lines (5.89% *versus* 1.58%, respectively). While additional studies are necessary to confirm, it is conceivable the increased expression of frizzdled-6 expression may be inhibiting the canonical Wnt signaling pathway, and thereby contributing to the formation of tumor-initiating cells. These newly generated canine primary osteosarcoma cell lines will be beneficial in addressing these important questions.

ACCELERATED RADIATION THERAPY AND ZOLEDRONATE FOR OSTEOSARCOMA BONE PAIN. LD Garrett, LE Selmic, RL Kamerer, TM Fan. University of Illinois Department of Veterinary Medicine, Urbana-Champaign, IL.

The purpose of this investigation was to characterize the clinical effectiveness, bone biologic activities, and biomechanical effects of an accelerated radiation protocol with adjuvant zoledronate for managing canine appendicular osteosarcoma (OS) pain. Prospectively, 54 dogs with OS were treated with ionizing radiation (10 gray x 2, Days 1 and 2), intravenous zoledronate (0.1 mg/kg, Day 2 then every 28 days), and oral analgesics (continuous). Every 28 days, clients completed a pain score assessment form, and objective data including bone biologic effects (urine NTx and DEXA) and biomechanical limb function (computerized gait analysis and videotape assessment) were obtained. Thoracic and limb radiographs were performed initially, at 6 months, and then at regular intervals.

The palliative protocol was well tolerated with no documented unacceptable radiation side effects or toxicity associated with multi-dose zoledronate administration. Fifty of the fifty-four dogs have failed, with a median of 201 days (95% CI: 131, 267) to failure. Twenty-six failed due to local disease, 15 failed due to metastasis, and 9 failed due to other causes. The median time to failure from local disease was 126.5 days (95% CI: 85, 223), while the median time to failure from metastasis was 237 days (95% CI: 119, 298). This difference approached significance ($p=0.055$). The median survival time for all dogs was 249 days (95% CI: 168, 309). On univariate analysis, elevation of ALP at diagnosis or tumor location were not found to be prognostic for time to failure or survival. Dogs achieving pain control for greater than 3 months uniformly demonstrated positive changes in bone biologic effects and biomechanical limb function.

Findings from this prospective clinical trial suggest that accelerated radiation, zoledronate, and oral analgesics are well tolerated, provide pain alleviation and exert positive bone biologic and biomechanical effects in dogs with OS.

IL-1β PROTEIN EXPRESSION IS INCREASED IN THE LAMINA PROPRIA OF DOGS WITH FOOD-RESPONSIVE CHRONIC ENTEROPATHIES AND CORRELATES WITH CLINICAL ACTIVITY INDEX. S Schmitz[1], S Priestnall[2], D Werling[2], K Allenspach[1]. [1]Department of Veterinary Clinical Sciences and Services, [2]Department of Pathology and Pathogen Biology, Royal Veterinary College (RVC), London, UK

Inflammasomes are intracellular multi-protein complexes that coordinate the maturation of Interleukin (IL)-1β and IL-18 in response to pathogens and metabolic danger signals. Recent research has shown imbalances in the mRNA expression of several signaling proteins of the inflammasome pathway in dogs with chronic enteropathies, however, IL-1β protein expression has not been assessed so far.

This study investigated IL-1β expression in the duodenal mucosa of dogs with food-responsive chronic enteropathies (FRD) before and after treatment and in healthy dogs using immunohistochemistry. Nine dogs with FRD were included in the study and treated with hydrolysed diet for 6 weeks. Biopsies from 5 of these dogs were available after treatment. Clinical scores (CCECAI, Allenspach 2007) were obtained before and after treatment. Duodenal biopsies from 11 healthy Beagles were used as controls. Immunohistochemistry was performed using a polyclonal rabbit anti-human primary antibody at 1:400 dilution (ab34837, Abcam, Cambridge, UK), followed by the ImmPress™ anti-rabbit peroxidase polymer detection kit as supplied by Vectorlabs. Slides were viewed at x40 magnification using an Olympus BX60 microscope. Cells were counted in 10 randomly selected villous regions and in 10 regions of the sub-villous lamina propria (LP) and expressed as positive cells per high power field (hpf).

The total count of IL-1β producing cells was increased in FRD dogs compared to controls, and decreased after treatment with elimination diet (median number of positive cells/hpf in healthy dogs: 9.7, range 1.5-33.2, FRD before treatment: 17.5, range 1.1-54.5; FRD after treatment 6.5, range 2.5-13; Kruskal Wallis p=0.03, paired t test FRD before *vs* after treatment p=0.01). Interestingly, this was not reflected by a quantitative change of IL-1β protein producing cells in the villi (median number of positive cells/hpf in healthy dogs: 9.6, range 1.4-29.8, FRD before treatment: 10.6, range 0.5-38; FRD after treatment 3.8, range 1-6.5; Kruskal Wallis p=0.06), but was mostly caused by an increase in positive cells in the LP, which did not entirely return to baseline values after treatment (median number of positive cells/hpf in healthy dogs: 1.2, range 0.1-5.2, FRD before treatment: 6, range 0.6-20.6; FRD after treatment 2.7, range 1.5-7.3; Kruskal Wallis p=0.002, unpaired t test healthy *vs* FRD: p= 0.0002; paired t test FRD before *vs* after treatment p=0.003). The decrease in IL-1β positive cell in the LP correlated with the CCECAI (Spearman rank r^2= 0.67). In conclusion, this study indicates that IL-1β protein expression is increased in dogs with FRD compared to healthy dogs, decreases after treatment, and correlates with CCECAI.

EVALUATION OF THE FECAL MICROBIOME IN CATS WITH DIARRHEA.

JS Suchodolski[1], ML Foster[1], M Sohail[1], C Leuchtenegger[2], E Queen[3], JM Steiner[1], SL Marks[4]. 1. Gastrointestinal Laboratory, Texas A&M University, College Station, TX; 2. IDEXX Reference Laboratories, West Sacramento, CA; 3. VCA Loomis Basin Veterinary Clinic, Loomis, CA; 4. School of Veterinary Medicine, University of California, Davis, CA.

Recent molecular studies have revealed a highly complex bacterial microbiota in the intestine of cats. There is mounting evidence that microbes play an important role in the pathogenesis of acute and chronic gastrointestinal (GI) diseases, as compositional changes of the intestinal bacterial ecosystem have been associated with intestinal inflammation in humans and dogs. The aim of this study was to evaluate the fecal microbiome in cats with diarrhea.

Fecal samples were obtained from healthy cats (n=21) and cats with diarrhea (n=48). Diseased cats were further compared based on the duration of their diarrhea: duration <7 days (n=19) vs. duration >7 days (n=29). The bacterial composition was analyzed by next generation sequencing of the 16S rRNA using the Ion-Torrent platform (Life Technologies) at a depth of 15,000 sequencing reads per sample. The 16S rRNA gene sequences were also used as input for the bioinformatics software PICRUSt to predict the functional gene content of the microbiome in each disease group. Linear discriminant analysis (LDA) effect size (LEfSe) was performed to identify bacterial groups that were significantly associated with disease and duration. Differences in bacterial groups and predicted gene content between groups were analyzed using Mann-Whitney tests and Kruskal-Wallis tests. The resulting P-values were corrected for multiple comparisons using the Benjamini & Hochberg's False Discovery Rate (adjusted P<0.05).

Significant differences in bacterial groups were detected between healthy cats and cats with diarrhea. Based on LDA, Enterobacteriaceae, *Streptococcus*, Burkholderiales, and *Collinsella* were significantly increased in diarrheic cats, while *Megamonas*, Bacteroidetes, *Helicobacter*, Campylobacterales, and *Roseburia* were significantly increased in healthy cats. Significant differences in abundances of specific bacterial groups were also observed between cats based on duration of diarrhea. Bacteroidetes were significantly decreased in cats with diarrhea >7 days, while *Lactobacillus* and Erysipelotrichi were significantly decreased in cats with diarrhea <7 days. The observed changes in bacterial groups were accompanied by significant differences in functional gene contents: metabolism of fatty acids, biosynthesis of glycosphingolipids, metabolism of biotin, metabolism of tryptophan, and ascorbate and aldarate metabolism were all significantly (P<0.001) altered in cats with diarrhea.

In conclusion, the above mentioned molecular approach revealed significant differences in the fecal microbiomes between healthy cats and cats with diarrhea. This dysbiosis was accompanied by changes in bacterial functional gene categories. Future studies are warranted to evaluate if these microbial changes correlate with changes in fecal concentrations of microbial metabolites in cats with diarrhea for the identification of potential diagnostic or therapeutic targets.

COMPARISON OF THE FECAL MICROBIOTA OF WILD WOLVES, DOGS FED COMMERCIAL DRY DIETS AND DOGS FED RAW MEAT DIETS. A Sturgeon, CM Jardine, JS Weese. University of Guelph, Guelph, ON, Canada.

The domestic dog (*Canis familiaris*) evolved from the grey wolf (*Canis lupus*) approximately 15,000 years ago. There is evidence of divergent evolution in some genes responsible for digestion, but some individuals consider the optimal diet for pet dogs to be an 'evolutionarily appropriate' diet similar to that used by wild wolves, with an emphasis on raw meat feeding. While the genome of wolves and dogs has been compared, the intestinal microbiota of wild wolves and dogs has not been compared, and there has been only limited evaluation of differences in the intestinal microbiota between dogs fed conventional diets and raw meat diets. The objectives of this study were to compare the fecal microbiota of wild wolves, dogs fed conventional dry diets (DC) and dogs fed raw meat diets (DR).

Fecal samples were collected from fifteen pet dogs (10 DC, 5 DR) and ten wild wolves (WW). Next generation sequencing of 16s rRNA gene (V4 region) PCR products was performed and data were analysed using mothur.

1,716,380 sequences passed all quality control filters and were clustered into 215,699 operational taxon units (OTUs). Singletons and doubletons represented 84% of OTUs. Sequences were subsampled to 9,736 reads for comparisons between the dog groups (DC, DR) and the wolves (WW), and to 70,293 reads for comparisons between DC and DR. A two-dimensional PCoA plot and a phylogenetic tree based on the Jaccard index demonstrated clear differences between wolves and dogs, something that was supported by analysis of molecular variance (AMOVA (<0.001)). In contrast, there was no difference in microbial population membership or structure between DC and DR groups (P=0.41). Species richness was estimated using Catchall at 76552 for dogs and 18686 for wolves.

The most prevalent phyla in the wolf scat were Firmicutes (68.3% of sequences), Proteobacteria (24.9%), Actinobacteria (3.6%), Deinococcus-Thermus (1.3%), Verrucomicrobia (0.4%), and Fusobacteria (0.4%). Unclassified bacteria represented 0.5% of sequences. Four wolves were almost completely comprised of Firmicutes while Proteobacteria dominated in six. Phylum level distribution was very similar between dogs, with main phyla being Firmicutes (38.3% of sequences), Verrucomicrobia (26.0%), Spirochaetes (2.4%), Fibrobacteres (2.0%), Proteobacteria (1.8%), Bacteroidetes (1.3%), Tenericutes (0.8%), and Fusobacteria (0.4%). In contrast to wolves, unclassified bacteria represented 26.3% of sequences in dog samples.

The fecal microbiota of wolves is highly variable, perhaps as a reflection of the variable and inconsistent diet of wild canids. In contrast, the fecal microbiota of dogs was more conserved, with no difference in population structure between raw- and conventionally-fed dogs. While diet can have an impact on the fecal microbiota of individual dogs, the fecal microbial population membership and structure is relatively conserved amongst the different feeding groups and feeding dogs raw meat does not result in development of a fecal microbiota more akin to wolves.

NUTRITIONAL IMMUNOMODULATION: SPIRULINA FOR IMMUNE & GUT HEALTH IN DOGS. Arleigh Reynolds, Ebenezer Satyaraj. Nestle Purina Research, St. Louis, MO.

Spirulina, refers to two species of blue green algae [*Arthrospira platensis*, and *A. maxima*] consumed by humans as food for centuries. Spirulina has shown immune enhancing properties in animal and human studies attributed to polysaccharides acting via TLR-dependent processes. However there are no systematic studies evaluating the immunomodulatory effect of Spirulina in dogs. The aim of this study was to evaluate the immunomodulatory effect of diet supplementation with Spirulina in dogs.

The study was conducted in two phases: Pre-test [8 wks] and Test [40 wks]. 30 dogs [mean 2.5 yrs], were randomized into two groups. In 'Pre-test' phase both groups were fed a nutritionally complete diet. At the end of 'Pre-test' phase all dogs received a rabies vaccine, and dogs in 'test group' were switched to nutritionally complete diet supplemented with dried *Arthrospira platensis* [Spirulina]. Response to rabies vaccine was evaluated by Rapid Flourescent Focus Inhibition Test (RFFIT). Gut immune response was assessed by measuring fecal IgA. Gut microflora was evaluated by Temporal Temperature Gel Electrophoresis (TTGE) methodology. Repeated measures analysis of variance was used to test for differences between groups and statistical significance considered to be $p < 0.05$.

Dogs fed diets supplemented with Spirulina demonstrated enhanced immune status by showing significantly higher vaccine response and higher levels of fecal IgA as compared to the control group. Supplementing diets with Spirulina also resulted in significantly increased gut microflora stability in the test group. In conclusion, diets supplemented with Spirulina significantly enhanced immune health and gut health in dogs.

EVALUATION OF A NOVEL TRACHEAL STENT FOR THE TREATMENT OF TRACHEAL COLLAPSE IN DOGS. D Clarke[1], E de Madron[2], and R Presley[3]. 1. Veterinary Hospital of the University of Pennsylvania, Philadelphia, PA. 2. Alta Vista Animal Hospital, Ottawa, Canada. 3. Upstate Veterinary Specialists, Greenville, SC.

The purpose of this study was to evaluate the safety, efficacy, and satisfaction associated with a novel variable diameter tracheal stent designed specifically for canine anatomy and tracheal collapse.

This was a multicenter retrospective study of 27 consecutive cases of tracheal collapse treated with the stent. Data forms requiring chart review and client interviews were distributed to the veterinarian placing each stent. Information was collected that compared symptoms pre- and post-stenting as well as a number of other parameters relating to the procedure and subsequent clinical course to assess stent performance and ease of use. Symptoms were graded on a visual line scale of 0-10. Paired t-tests or sign rank tests were used to determine changes in symptoms following stenting. A two-sided p-value of <0.05 was considered statistically significant.

Responses were obtained from 20 of 27 cases. Data was normally distributed. The mean duration of follow-up was 3 months (range >1 to 17 months). There were no procedural complications. Operative mortality was 0%. There were no stent fractures or stent related deaths; symptomatic granulation tissue developed in one case. Statistically significant improvement was seen in cough, respiratory function, quality of life, and exercise tolerance. All veterinarians stated they would use the stent again. 95% of clients would repeat the procedure and 90% expressed satisfaction with the outcome.

Tracheal stenting performed with the novel implant is both safe and subjectively effective in significantly improving clinical symptoms. Veterinarians and clients expressed a positive perception regarding outcomes associated with the stent.

TRACHEAL DIAMETER AND AREA: COMPUTED TOMOGRAPHY VERSUS FLUOROSCOPY FOR STENT SIZING IN 12 DOGS WITH TRACHEAL COLLAPSE. B.A. Scansen. Department of Veterinary Clinical Sciences, The Ohio State University, Columbus, OH.

Implant size for tracheal stenting is conventionally selected by fluoroscopic measurement of maximal dorsoventral tracheal diameter. The aim of this study was to determine if tracheal CT dimensions are comparable to fluoroscopy and also to evaluate the change in tracheal cross sectional area (CSA) between positive pressure ventilation and apnea.

Twelve dogs with tracheal collapse requiring stent implantation were prospectively enrolled. Computed tomography and fluoroscopy were performed during positive pressure and apnea with the endotracheal tube just beyond the larynx. Tracheal diameter was measured in lateral (CT) and dorsoventral (CT and fluoroscopy) planes. Tracheal CSA was calculated from fluoroscopy and directly measured from CT.

The lateral diameter was significantly ($P < 0.0001$) wider than the dorsoventral diameter in the cervical (14.0mm vs. 9.2mm) and intrathoracic (12.0mm vs. 8.4mm) trachea, indicating tracheal shape was more ellipsoid than circular. The cervical and intrathoracic dorsoventral tracheal diameters measured by CT or fluoroscopy were not different ($P = 0.69$ and $P = 0.80$, respectively). The CT CSAs of both cervical and intrathoracic trachea were different than fluoroscopy ($P < 0.02$). Tracheal CSA was significantly reduced (cervical: 33%, intrathoracic: 36%) during apnea as compared to positive pressure ($P < 0.001$).

Tracheal diameter in canine tracheal collapse is greater laterally than dorsoventrally. A stent sized to the dorsoventral fluoroscopy dimension may underestimate the true tracheal CSA, leading to a risk for implant under-sizing. Tracheal CSA in dogs with tracheal collapse is affected by positive airway pressure, with airway CSA reduced by one-third during apnea.

EFFECTS OF CYCLOSPORINE AND ASPIRIN ON CANINE PLATELET FUNCTION.

J.Thomason, T. Archer, S. Press, A. Mackin. Mississippi State University, Starkville, MS.

Immune-mediated hemolytic anemia (IMHA) has a mortality rate of around 50% in dogs, with pulmonary thromboembolism (PTE) being a common cause of death. Cyclosporine has become a popular agent for the treatment of IMHA; however, recent research in dogs has shown that immunosuppressive dosages of cyclosporine increase platelet reactivity by increasing platelet synthesis of thromboxane A_2 (TXA_2). Low-dose aspirin is often given to IMHA patients to inhibit platelet TXA_2 synthesis and prevent thromboembolism, and has been shown to decrease mortality rates in dogs with IMHA. Even though aspirin is commonly given concurrently with cyclosporine in dogs with IMHA, it has not been determined if aspirin counteracts cyclosporine-associated TXA_2 synthesis or, alternatively, if cyclosporine counteracts the anti-platelet effects of aspirin.

In a cross-over study design, seven healthy young adult dogs were given either low-dose aspirin (1 mg/kg Q24 hrs PO; Group 1), high-dose aspirin (10 mg/kg Q12 hrs PO; Group 2), cyclosporine (10 mg/kg Q12 hrs PO; Group 3), or low-dose aspirin in combination with cyclosporine (Group 4) for 7 days, with a washout period of at least one month between each cross-over. Platelet function was evaluated prior to (Day 0) and during (Days 3 & 7) therapy using turbidimetric aggregometry (collagen agonist) and PFA-100® closure time (collagen/epinephrine cartridge). Urinary levels of 11-dehydro-thromboxane B_2 (11-dTXB_2) were measured via ELISA and normalized to urine creatinine concentration. Aggregometry was performed in quadruplicate, PFA-100® performed in duplicate, and urinary 11-dTXB_2 levels were analyzed in triplicate, and results were averaged.

Compared to baseline (Day 0), low-dose aspirin decreased maximal amplitude on aggregometry by 35% & 50% on Days 3 & 7, respectively, and high-dose aspirin decreased maximal amplitude by 97% on both days. Compared to baseline, cyclosporine increased maximal amplitude by 14% in Day 3, with no change from baseline on Day 7. In Group 4, the maximal amplitude decreased from baseline by 33% and 65% on Days 3 & 7, respectively. By Day 3, PFA-100® closure times increased by 13%, 150% & 14% from baseline in Groups 1, 2 & 4, respectively; by Day 7, closure times had increased by 72%, 138% & 108%, respectively. By Day 3, urine 11-dTXB_2 levels decreased by 70% & 43% from baseline in Groups 1 & 2, respectively, but increased by 104% in Group 3 and by 24% in Group 4. By Day 7, mean urine 11-dTXB_2 levels decreased by 73%, 11% & 24% from baseline in Groups 1, 2 & 4 respectively, but increased by 168% in Group 3.

Our study confirms that low-dose aspirin inhibits platelet function, albeit not to the same extent as high-dose aspirin, and demonstrates that cyclosporine does not significantly alter the effects of low-dose aspirin on platelet function. Our study also confirms that cyclosporine markedly increases platelet thromboxane A_2 synthesis in dogs, and that when cyclosporine and aspirin are given concurrently, cyclosporine appears to significantly reduce the effectiveness of low-dose aspirin at inhibiting thromboxane synthesis. Further investigation of the effects of concurrent cyclosporine and low-dose aspirin on platelet reactivity is warranted in patients at risk of PTE.

THE EFFECT OF ORAL ACID SUPPRESSANTS ON INTRAGASTRIC pH IN CATS.

Parkinson[1], K Tolbert[1], K Messenger[2], A Odunayo[1], M Brand[1], A Reed[1], E Peters[3], G Davidson[2], MG Papich[2]. 1.Univ of Tennessee College of Veterinary Medicine, Knoxville, TN. 2.North Carolina State University, Raleigh, NC. 3.GlaxoSmithKline, Research Triangle Park, NC.

Gastrointestinal erosion and ulceration are common sequelae to gastric and non-gastric disorders in cats. Acid-suppressing drugs are a mainstay of therapy. Generic tablets containing omeprazole (OT) or famotidine are widely available and can be modified by cat owners to treat their pets. Fractionated enteric-coated OT and famotidine are among the most widely prescribed acid suppressants. However, once broken, the effect of enteric-coated OT in raising gastric pH may be significantly compromised. Moreover, studies have not been undertaken in cats to determine the efficacy of these different acid suppressants. The objectives of these studies were to compare the pharmacodynamics and pharmacokinetics of oral omeprazole formulations and famotidine in cats.

Healthy adult cats (n=6) were used in a randomized, 4-way crossover design. All cats received 0.88-1.26 mg/kg q12hr of fractionated enteric-coated omeprazole tablets, omeprazole paste (Gastrogard® diluted in cod liver oil at 10 mg/ml), famotidine tablets, and twice daily placebo (lactose capsules). Intragastric pH monitoring (Bravo system®) was used to continuously record intragastric pH for four days (treatment days 4-7) followed by a 10-day washout period. Plasma omeprazole concentrations at steady state (day 7) were determined by HPLC with ultraviolet detection. Mean percentage time (MPT) the intragastric pH was ≥ 3 and 4 were compared among groups using ANOVA. Tukey-Kramer test was used to determine which groups were different ($\alpha = 0.05$).

MPT that intragastric pH was ≥ 3 and 4 were significantly ($p < 0.0001$) greater for cats receiving omeprazole tablet and paste compared to famotidine and placebo. Famotidine was significantly different from placebo. Mean ± SD percentage of time the intragastric pH was ≥ 3 was 68.4 ±35.0% for fractionated enteric OT, 73.9 ± 23.2% for omeprazole paste, 42.8 ±18.6% for famotidine, and 16.0 ± 14.2% for placebo. Mean ± SD plasma omeprazole concentrations on day 7 (steady state) were similar in cats receiving tablet compared to paste and in a range associated with acid suppression in other studies.

These results demonstrate that both omeprazole formulations provide superior acid suppression in cats compared to famotidine or placebo. Fractionated enteric-coated OT is an effective acid suppressant despite disruption of the enteric coating.

THE DETECTION OF CALCIFYING NANOPARTICLES IN THE URINE OF CATS AND DOGS AND THEIR ROLE IN CALCIUM OXALATE UROLITHIASIS. Westropp JL[1], Mellema MM[1], Queau Y[2], Shiraki R[1], Biourge V[2]. School of Veterinary Medicine, University of California Davis[1] and Royal Canin Research Center, Aimargues, France.[2]

Calcium oxalate (CaOx) is the most common urolith submitted to our laboratory from dogs and cats. The pathophysiology of CaOx stone formation is poorly understood, but research in humans suggests a link between calcifying nanoparticles (CNP) and urolithiasis. CNPs are comprised of crystallized minerals and organic molecules; the aggregation and size of urinary CNPs is altered in lithogenic patients and reports suggest CNPs are associated with the presence of Randall's plaques Furthermore, studies in humans have reported alterations in CNP size, morphology, and aggregation that helped inhibit calculus formation in patients receiving potassium citrate, a therapeutic drug for CaOx management.

To date, there are no studies evaluating CNPs in small animals. We developed a flow cytometry protocol to detect CNPs generated *in vitro*. We confirmed these particles were indeed apatite based on transmission electron microscopy studies. Nanoparticle tracking analysis was used to evaluate canine urine samples and a uniform population of particles was detected in the exact primary CNP size range (80-90 nm). Furthermore, these particles auto-fluoresced similar to particles that contain significant amounts of mineral. Fluorescent marker studies are ongoing.

We have developed a protocol to detect CNPs in the urine of dogs. Future studies in animals will determine if 1) we are able to detect urine/plasma CNPs in cats and 2) if differences in CNP quantity, size and stability are predictive of lithogenic propensity. CNP assays may also provide a tool for identifying urinary determinants of crystal aggregation and subsequent stone formation in small animals.

METABOLIC AND GENETIC DETERMINANTS OF CALCIUM OXALATE UROLITHIASIS IN DOGS. E. Furrow, J.P. Lulich, J.R. Mickelson, P.J. Armstrong, K.M. Minor, E.E. Patterson. University of Minnesota College of Veterinary Medicine, St. Paul, MN.

The objectives of this study were to: 1) determine if hypercalciuria underlies calcium oxalate (CaOx) urolith risk in the Miniature Schnauzer (MS), Bichon Frise (BF), and Shih Tzu (ST) and 2) identify a susceptibility gene for CaOx urolithiasis in the MS using a genome-wide association study (GWAS).

MS, BF, and ST cases had a history of CaOx uroliths. Controls were ≥8 yo and had no uroliths on screening radiographs. Serum iCa^{2+} and fasting urinary calcium-to-creatinine ratios (UCa/Cr) and oxalate-to-creatinine ratios (UOx/Cr) were measured. MS dogs were genotyped on whole-genome SNP chips for the GWAS; next-generation sequencing was used to identify variants within the disease-associated locus.

The metabolic analyses included 45 MS (23 cases, 22 controls), 28 BF (14 cases, 14 controls), and 13 ST (7 cases, 6 controls). MS and BF cases had higher serum iCa^{2+} than breed-matched controls ($p<0.05$), though all values were within the reference interval. For all breeds, cases had higher UCa/Cr than breed-matched controls ($p<0.005$). UOx/Cr were not different.

The GWAS included 104 MS (62 cases, 42 controls). A strong association signal was present on chromosome 37 ($p_{raw} = 8x10^{-6}$, $p_{genome} = 0.01$). Sequence results were evaluated for 18 genes in the critical region. No coding variants were identified, but putative regulatory variants were found in a solute carrier gene, *Slc39a10*. Knockdown of the gene in a *Drosophila* CaOx model increases stone volume.

Hypercalciuria is associated with CaOx urolithiasis in 3 canine breeds. Genetic mapping in MS dogs identified *Slc39a10* as a potential CaOx susceptibility gene.

A TWO-YEAR-LONG PROSPECTIVE, RANDOMIZED, DOUBLE-MASKED STUDY OF NUTRITION ON THE RECURRENCE OF CALCIUM OXALATE UROLITHIASIS IN STONE FORMING CATS. J. Lulich[1], J. Kruger[2], J. MacLeay[3], J. Merrills[2], I. Paetau-Robinson[3], J. Brejda[4], S. Steck[2], S. Davidson[1], C. Osborne[1]. [1]University of Minnesota, St. Paul, MN., [2]Michigan State University, E. Lansing, MI., [3]Hills Pet Nutrition, Inc., Topeka, KS. [4]Alpha Statistical Consulting, Inc. Lincoln, NE.

Calcium oxalate (CaOx) uroliths account for approximately 40% of feline stone submissions to the Minnesota Urolith Center. To date, no study has followed affected cats long term to determine the rate of recurrence, confounding factors and the effect of dietary intervention.

This study was a 2- year-long, randomized, double-masked and controlled study examining 2 foods for management of CaOx urolithiasis. Compared to the Control (Ctl) food the Test food contained greater amounts of anti-oxidants and n3 fatty acids, and less Ca and Mg with a target urine pH of 6.2-6.4. The Ctl food was designed to be similar to popular grocery brands and had a target urine pH of 6.6-6.8. Foods were otherwise similar and met minimum AAFCO requirements for adult cats. Protocols were ACUC approved. Cats with a recent history of confirmed CaOx uroliths were recruited. Owners were given the choice of food form (wet, dry or combined) and then cats were randomized onto Test or Ctl foods. Cats were evaluated at 4 month intervals by physical examination, abdominal radiography, urinalyses and serum chemistries. 24-hour urine collections for chemistry and relative super saturation (RSS) calculation were conducted at 4 and 24 months or at detection of stone recurrence. Data were analyzed as a factorial combination of food, hypercalcemia (HCa), and recurrence.

36 cats were recruited and 33 (17 Test, 16 Ctl) completed the study and were included in the analysis. Of these, 12/33 (36%) had intermittent HCa. HCa was detected in 4/17 (23.5%) Test food and 8/16 (50%) Ctl food fed cats. Stone recurrence was observed in11/33 (33%) of cats overall, and was detected in 5/17 (29.4%) Test food and in 6/16 (37.5%) Ctl food fed cats. RSS CaOx and urine[Ca] were significantly higher in cats with stone recurrence, cats fed the control food, and cats with HCa. A significant ($p<0.05$) interaction between HCa and food was observed for RSS CaOx. The least square means for RSS CaOx were 5.10 (SEM 0.77) for nonHCa/Test, 4.56 (1.3) for HCa/Test, 4.15 (0.96) for nonHCa/Ctl, and 12.88 (0.96) HCa/Ctl. Urine pH was significantly different between foods, but not between cats with HCa or stone recurrence.

These results suggest that recurrence of CaOx uroliths and HCa may be lower in cats fed the nutrition of the Test food. This affect may be due to a mitigating effect of the Test food on urine calcium. The recurrence of CaOx urolithiasis is multifactorial and occult HCa may play a larger role in this disease than is typically recognized.

COMPARISON OF RELATIVE SUPER SATURATION FOR CALCIUM OXALATE TO A NEW METHOD TO DETERMINE THE PROPENSITY OF URINE TO PRECIPITATE CALCIUM OXALATE, THE CALCIUM OXALATE RISK INDEX.

J.M. MacLeay[1], H.M. Schiefelbein[1], I.V. Raymond- Loher[2], S.J Davidson[1]. [1]Hill's Pet Nutrition, Inc., Topeka, KS. [2]Aerotek Inc., Raleigh, NC.

Calcium oxalate (CaOx) urolithiasis accounts for approximately 40% of all cases of urolithiasis in cats and 33% of cats are likely to reform a stone within 2 yrs. Therefore, there is a need to develop a test that can routinely predict the risk that a cat may form or reform a stone to provide feedback that a given intervention is lowering risk. Relative super saturation (RSS) is regarded as the best method to evaluate urine. However, the test is not commercially available to the veterinary community, is labor intensive and costly. The RSS software is dated and of the 2 versions available; one, SuperSat®, is proprietary and yields results different than the other, Equil 2. Further, RSS uses a limited number of metabolites which cannot fully evaluate the complex system of urine where many promoters and inhibitors interact. As such, RSS is limited in its ability to predict the propensity of urine to form a stone. The human Bonn Risk Index uses whole urine and better reflects the risk of stone formation. The Calcium Oxalate Risk Index (CORI) is similar in principle to the human Bonn Risk Index. Briefly, whole urine is titrated with a sodium oxalate solution until precipitation occurs. The CORI value is determined by dividing the urine [Ca^{2+}] by the resultant amount of oxalate added at the point of precipitation.

A panel of 12 healthy adult cats was used in this ACUC approved study. Cats were individually housed with group exercise and social interaction in spacious rooms with natural light. Cats were fed a series of 22 different foods with varying nutritional profiles, each for 28 days, over the course of 2 years. Foods varied in form (dry or wet) and were always fed to maintain body weight. A 24-hr urine collection was performed on day 28. Equil 2 software was used to calculate RSS from urine concentrations of NH_4^+, Ca^{+2}, Cl^-, Citrate, creatinine, Mg^{+2}, oxalate, PO_4^{-3}, K^+, Na^+, SO_4^{-2} and urine pH. SAS™ version 9.2 was used for all analysis. CaOx RSS and the CORI results were moderately correlated ($r = 0.54$, $p < 0.0001$). RSS CaOx was weakly correlated to both urine pH ($r = -0.19$, $p<0.05$) and RSS Struvite ($r=-0.11$, $p = 0.05$). CORI, however, was not correlated with either urine pH ($r= 0.03$, $p = 0.62$) or RSS Struvite ($r= -0.05$, $p = 0.35$). Nutritional profiles of foods and individual urinary constituents were analyzed to determine drivers for RSS CaOx and CORI. The CORI test provides several advantages over RSS CaOx; It has the potential to be a commercially available test, it can provide veterinarians and pet owners with rapid feedback concerning the success of an intervention and, most importantly, through the use of whole urine, it takes into account the entire milieu of urinary constituents that affect the propensity to form a calcium oxalate stone. Further work will continue to determine the impact of nutrition and previous stone status on CORI.

PREVALENCE OF VARIANT ALLELES ASSOCIATED WITH PROTEIN-LOSING NEPHROPATHY IN AIREDALE TERRIERS. MP Littman, MG Raducha, PS Henthorn. University of Pennsylvania School of Veterinary Medicine, Philadelphia, PA.

Previously we showed that protein-losing nephropathy (PLN) in Soft-coated Wheaten Terriers (SCWT) is associated with variant alleles in linkage dysequilibrium in two adjacent genes, NPHS1 and KIRREL2, which encode slit diaphragm proteins nephrin and filtrin/Neph3, respectively. In SCWT, homozygous positive (2-2) dogs are at highest risk, heterozygous (1-2) dogs are at intermediate risk, and homozygous negative (1-1) dogs are at lowest risk for the development of PLN. Among 753 dogs of 114 non-SCWT breeds, only 3 dogs (2 Airedales and 1 Bloodhound) showed these alleles, including one homozygous positive Airedale terrier that died with PLN.

Cheek swab, blood, or semen samples of DNA were tested from 221 Airedale Terriers (median age 4 yrs) from widely dispersed regions of the United States and Canada to study the prevalence of the alleles in the Airedale breed and determine if breeders can safely remove carrier dogs in one generation. Results of haplotype frequencies are tabulated below. Genetic counseling would be of benefit to the Airedale community to help breeders avoid producing homozygous positive dogs.

	1-1 (%)	1-2 (%)	2-2 (%)	Variant Allele Frequency (%)
Female Airedales				
Intact, n=89	42.5	50.5	7	32
Neutered, n=46	59	41		21
Total, n=135	48	47.5	4.5	28
Male Airedales				
Intact, n=55	49	42	9	30
Neutered, n=20	65	25	10	23
Male Unknown, n=8	75	25		13
Total, n=83	55.5	36	8.5	27
Unknown Sex, n=3	33	67		33
TOTAL, n=221	51	43	6	28
YEAR OF BIRTH *includes one semen sample from a deceased dog born in 1990				
1995-2001, n=16	56	44		22
2002-2005, n=49	65	29	6	20
2006-2009, n=51	49	49	2	26
2010-2011, n=57	35	54	11	38
2012-2013, n=39	54	41	5	26
Unknown*, n=9	56	33	11	28

PLACEBO-CONTROLLED ASSESSMENT OF ALLOGENEIC MESENCHYMAL STEM CELLS FOR THE TREATMENT OF FELINE CHRONIC KIDNEY DISEASE.

JM Quimby, TL Webb, E Randall, A Marolf, A Valdes-Martinez, SW Dow. Center for Immune and Regenerative Medicine, Department of Clinical Sciences, Colorado State University, Fort Collins, CO.

Feline chronic kidney disease (CKD) is characterized by chronic tubulointerstitial nephritis, and inflammation contributes to the progression of renal fibrosis. Mesenchymal stem cells (MSC) have demonstrated anti-inflammatory and anti-fibrotic effects in rodent CKD models. The purpose of this study was to evaluate the use of allogeneic MSC for treatment of feline CKD.

MSC were isolated from the cryopreserved adipose of SPF research cats and culture-expanded. CKD cats were enrolled in a placebo-controlled, blinded one-way crossover clinical study. Four CKD cats were randomized to receive 2 x 10^6 MSC/kg intravenously at 2, 4, and 6 weeks. Three CKD cats were randomized to receive placebo, with one cat crossing over to the MSC treatment group. CBC, chemistry, and urinalysis were performed at weeks 0,2,4,6, and 8. Glomerular filtration (GFR) via nuclear scintigraphy and UPC were performed at zero and 8 weeks.

Five cats received 3 doses of fresh, culture-expanded MSC without adverse effects. No significant change in serum creatinine, GFR, UPC, or PCV was seen in cats treated with MSC. Individual changes in GFR were 12%, 8%, 8%, -13%, and -67% in treated cats compared to 16%, 36%, and 0% control cats. Discordant creatinine and GFR results were observed: the treated cat that experienced the greatest decrease in creatinine also experienced the greatest decline in GFR.

Administration of MSC expanded from cryopreserved adipose were not associated with side effects, however no discernable immediate improvement in renal function was appreciated. Long-term follow up is necessary to determine effect on progression of disease.

URINARY CLUSTERIN AND SERUM INOSINE: BIOMARKERS FOR EARLY IDENTIFICATION OF ACUTE KIDNEY INJURY IN DOGS. CA Palm[1], G Segev[2], LD Cowgill[1], J Quinn[3], Z Mirolee[3], M Yerramilli[3], E Obare[3], M Yeramelli[3]. 1. UC Davis School of Veterinary Medicine, Davis, CA. 2. Koret School of Veterinary Medicine, Hebrew University, Jerusalem, Israel. 3. IDEXX Laboratories, Inc., Westbrook, ME.

A key factor in the high mortality rate associated with acute kidney injury (AKI) in dogs is the lack of sensitive tests to identify early stages of disease. Novel urinary and serum biomarkers may identify AKI earlier than traditional markers. Urinary clusterin (uClusterin), an apolipoprotein, is overexpressed during renal tubular injury and elevations have correlated with damage severity. Serum inosine (sInosine) is a metabolite of adenosine deamination produced by proximal tubular cells in health. The purpose of this study was to evaluate the utility of uClusterin and sInosine for early AKI diagnosis in a canine model of AKI.

AKI was induced in 5 dogs by gentamicin administration (10 mg/kg SC q 8hours). Serum creatinine (sCreatinine), sInosine and uClusterin were measured twice daily during AKI induction.

AKI (defined as a 50% increase of sCreatinine above baseline) occurred on median day 16 (range, 14-19 days). Comparatively, median uClusterin increased above the cut-off value (190 ng/mL) on median day 4 (range, 4-7 days; $P = 0.008$). Baseline median uClusterin was 67 ng/mL (range, 10-71 ng/mL) and peaked at 2866 ng/mL (range, 1233-5115 ng/mL) on median day 17, a 43-fold increase above baseline. Median baseline sInosine was 241 ug/dL (range, 134-427 ug/dL) and declined steeply following gentamicin administration in all dogs (decreased by 1.7-fold on median day 2, and by 6-fold on median day 3).

In a gentamicin induced canine model of AKI, changes in uClusterin and sInosine preceded changes in sCreatinine; these novel markers show promise for early AKI detection in dogs.

EFFICACY OF CANINE ANTI-NERVE GROWTH FACTOR ANTIBODY FOR THE ALLEVIATION OF DEGENERATIVE JOINT DISEASE-ASSOCIATED PAIN IN DOGS.

BDX Lascelles[1,2], M Freire[1], D Knazovicky[1], B Case[1], K Jiamachello[1], D Gearing[3].
[1]Comparative Pain Research Laboratory, [2]Center for Comparative Medicine and Translational Research; NCSU CVM, Raleigh, NC. [3]Nexvet, Melbourne, Australia.

There is a critical need for proven drugs other than non-steroidal anti-inflammatory drugs for treatment of degenerative joint disease (DJD) pain in dogs. Antibodies against Nerve Growth Factor (NGF) are analgesic in rodent models and in humans with DJD. This pilot study aimed to evaluate the efficacy of a novel caninised anti-NGF antibody (NV-01) for the treatment of DJD pain in dogs.

In a placebo-controlled, masked clinical study, 26 dogs with DJD were randomized to NV-01 (200 mcg/kg IV) or placebo. In addition to objective accelerometry measures, owners completed clinical metrology instruments (Client-Specific Outcome Measures [CSOM] and Canine Brief Pain Inventory [CBPI]) on days 0, 14 and 28. CSOM and CBPI subscales (pain severity [PS] and pain interference [PI]) were evaluated within and between groups. Recognized success/failure criteria were applied and success compared between groups.

CBPI PS and PI scores significantly improved in the NV-01 group (PS: Day 0-14, $P = .010$ and Day 0-28, $P = .025$; PI: Day 0-14, $P = .006$ and Day 0-28, $P = .032$) but not in the placebo group. CSOM scores showed similar patterns with a significant difference between groups at day 14 and day 28 ($P = .011$ and $P = .033$, respectively), and significantly more successes at day 28 ($P = .037$). No side effects were noted. Objectively measured activity significantly increased in the NV-01 group ($P = .045$), not in the placebo group ($P = .811$).

These pilot data demonstrate a positive analgesic effect of anti-NGF antibody in dogs suffering from chronic pain. The magnitude of the effect was identical to that expected with an NSAID.

CHOROIDAL VASCULITIS DURING EXPERIMENTAL AND NATURALLY OCCURRING EHV-1 INFECTION IN HORSES. LS Goehring[1] , G Soboll Hussey[2] , DP Lunn[3] , LK Maxwell[4] , EJ Ehrhart[5]. Faculty or College of Veterinary Medicine at 1.Ludwig-Maximilians University, Munich, Germany. 2.Michigan State University, East Lansing, MI. 3.North Carolina State University, Raleigh, NC. 4.Oklahoma State University, Stillwater, OK. 5.Colorado State University, Fort Collins, CO.

Equid Herpesvirus type 1 (EHV-1) is the cause of respiratory disease, Equid Herpesvirus-associated Myelopathy (EHM), late term abortions and chorioretinopathies in horses. Following infection of the respiratory tract, a cell-associated viremia is often established leading to virus dissemination and endothelial cell (EC) infection of the spinal cord vasculature, the pregnant uterus or the eye. Previously, we showed focal or multifocal shotgun lesions in 50% or more of experimentally infected horses between 5 – 7 weeks post infection (p.i.). Based on this data our hypothesis was that eye lesions are caused by EC infection resulting in vasculitis similar to what occurs for EHM and that the incidence of EHM correlates with the incidence of ocular lesions.

To test this hypothesis, 8 mixed breed horses, aged 18 – 22 years, were infected with EHV-1 (strain OH03), and euthanized on day 14 p.i.. Clinical signs of neurological disease were evaluated between days 1-14 post infection and eyes were collected post euthanasia. In addition, eyes of a yearling horse that developed acute neurologic signs following experimental infection with EHV-1 strain Ab4 and one eye of a middle-aged gelding that developed neurological signs during the 2011 EHV Ogden, UT outbreak were collected. Neurological gait abnormalities of all horses were scored on a scale of 0 (no abnormalities) to 5 (severe abnormalities). In addition, ocular tissues were processed at Colorado State University and histological sections were collected and stained with H&E and immunohistochemistry (IHC) was performed using an anti-EHV-1 antibody. Eye lesions were described and degrees of choroiditis/vasculitis (CV) were scored according to severity and frequency on a scale from 0 (no lesions) to 3 (abundant lesions). A correlation coefficient between eye lesion score and clinical neurological gait abnormalities was determined and differences were considered significant at a p-value < 0.05.

The yearling horse and the Ogden-horse had severe, frequent CV ($\geq$grade 2). In the OH03-study, 6/8 horses were graded ≥ 1; 4/8 horses were graded ≥ 2 and 2/8 horses were graded 3. IHC was positive in the yearling and in the Ogden-horse, while horses in the OH03experiment were positive only on occasion. A clinical neurological gait abnormality score during the OH03 infection experiment did not correlate well with eye lesions ($R^2 = 0.65$).

In summary, while the pathogenesis of ocular EHV-1 and EHM might be similar and viral antigen was detected in some of the infected horses, no significant correlation between neurological disease and ocular disease was detected. This lack of viral antigen in the eyes of some of the horses exhibiting neurological disease may have been due to the time point of euthanasia and a lack of sensitivity of the assays selected.

A NOVEL STRATEGY TO BOOST ANTIBODY PRODUCTION TO EQUINE HERPESVIRUS TYPE 1 (EHV-1) IN FOALS. GA Perkins, LB Goodman, S Babasyan, H Freer, A Keggan, A Glaser, S Torsteinsdóttir, V Svansson, S Björnsdóttir, and BA Wagner. Cornell University, College of Veterinary Medicine, Ithaca, NY.

Foals respond poorly to existing vaccines that are optimized for the adult immune system. Our hypothesis is that foals can be protected against EHV-1 infection by a vaccination that activates T-cell independent B-cell memory and antibody production.

Iceland is considered free of EHV-1. Pregnant Icelandic mares (n=15) were imported to an isolation facility in February 2012. The foals born in that spring suckled colostrum that was free of maternal EHV-1 antibodies. Foals either received equine IgE at birth by nasogastric tube and 2 days later recombinant EHV-1 glycoprotein C (gC) antigen intramuscularly (IM); or received EHV-1 gC antigen IM on day 2 of life; or received no treatment. At 7-months of age, foals were weaned and infected with EHV-1. Physical examinations were done daily. EHV-1 viral DNA and anti-EHV-1 gC antibodies were measured in the blood and nasal secretions before and after infection. Repeated measures ANOVA with Dunn's post-tests to compare individual groups were done.

All foals developed a fever. At 36-h post-infection (p.i.), the vaccinated foals had a significantly lower temperature ($p<0.05$). EHV-1 DNA was detected by qPCR in nasal swabs for the first week p.i. and there was no significant difference between groups. Serum and nasal anti-EHV-1 gC antibodies increased by day 8 p.i. and this was significantly higher in magnitude in the vaccinated foals ($p<0.001$).

Our novel equine neonatal vaccination provided some protection against EHV-1 clinical disease and prepared the foal's adaptive immune system to respond more rapidly and robustly to EHV-1 infection.

COMPARISON OF INNATE IMMUNE RESPONSES IN EQUINE RESPIRATORY EPITHELIAL CELLS TO MODIFIED-LIVE EQUINE INFLUENZA VACCINE AND RELATED WILD-TYPE INFLUENZA VIRUS. HL Pecoraro[1], D. Koch[1], G Soboll Hussey[2], L Bentsen[1], GA Landolt[1]. 1. Colorado State University, Fort Collins, CO. 2. Michigan State University, East Lansing, MI.

Equine influenza virus (EIV) is an important equine respiratory pathogen that continues to cause disease and suffering in horses. In order to design effective EIV control strategies, it is crucial to understand how to invoke long-term protective immunity. While systemic IgG titers have long been used as a correlate for protection from EIV infection, systemic antibody titers are not always accurate predictors of clinical protection. The cold-adapted, modified-live virus (MLV) vaccine (FluAvert I.N.®, Merck Animal Health) induces only weak systemic antibody responses. Despite this, the vaccine has been shown to provide clinical protection for up to 12 months after a single vaccine administration. This suggests that mechanisms other than humoral responses are critically important in EIV immunity. While the adaptive immunity to EIV has been well characterized, there is a lack of information regarding the innate immune responses occurring at the airway epithelium during EIV infection.

The objective of this study was to compare the innate immune responses of primary equine respiratory epithelial cell (EREC) cultures following infection with either the MLV vaccine strain or a closely related wild-type EIV strain (A/Equine/Kentucky/1/91; Eq/KY). It was our hypothesis that infection with the MLV vaccine strain elicits similar innate immune responses in equine airway cells as seen following infection with the wild-type virus.

Equine airway epithelial cells were collected from the upper and lower respiratory tract of respiratory healthy animals and the cells were grown at the air-fluid interface until fully differentiated. Subsequently, cell cultures were inoculated with a multiplicity of infection (MOI) of 10 with either the MLV strain or Eq/KY and incubated at either 30°C or 37°C. Twelve, 24, and 48 hrs after inoculation, cell culture subnatants and cells were collected and assayed for protein and mRNA expression of IFN-α, IFN-γ, IL-4, IL-6, IL-10, IL-17, TNF-α, and TLR3 & 9.

Inoculation of ERECs with either virus resulted in up-regulation of both mRNA and cytokine expression. The degree of up-regulation was dependent on virus strain, site of cell collection (upper versus lower respiratory tract) and incubation temperature used. Moreover, the expression levels appeared to be associated with viral replication efficiency.

These results highlight the fact that immunity to influenza virus is multifaceted and involves several arms of the host's immune response. Whether innate, antibody-independent mechanisms play a role in the observed clinical protection following intranasal vaccination with the MLV vaccine remains to be investigated.

INFECTIOUS DISEASES OF WORKING EQUIDS: HAVEMEYER WORKSHOP REPORT.

N. Aklilu[1], A. Stringer[1], D.P. Lunn[2]. [1]SPANA, Ethiopia, [2]NCSU, Raleigh, NC.

A workshop was held in November, 2013 in Addis Ababa, Ethiopia, attended by 35 international delegates representing academia, NGO's, the OIE and government agencies. The goal was to identify ways to reduce the burden of infectious diseases in working equids worldwide.

The size of the working equid population is estimated to be 75% of the 112 million world domestic equid population, although this may significantly under-represent the true population. The majority of working equids live in low-income countries, and have a major socio-economic role in the livelihoods of equid-owning households. Information on the occurrence of infectious diseases in these populations is often incomplete, but there is clear evidence of wide regional distribution of high morbidity and mortality including African Horse Sickness, EIA, influenza, rabies, tetanus, glanders (Burkholderia mallei), epizootic lymphangitis, piroplasmosis, trypanosomiasis, and GI nematodes. Importantly, participatory surveys of equid owners, and observations by field workers identify both respiratory and neurological disease syndromes as having high impact and warranting further investigation. Strategies for equine disease surveillance where there is under-reporting and lack of diagnostic infrastructure were contrasted with other species in the same regions, and systematic data gathering coupled with new indicators of disease impact will be needed for progress. Understanding equine ecology was repeatedly demonstrated to be critical to disease control, and promoting control measures that can be adopted by owners through education is a top priority.

The workshop made two key sets of recommendations. The first was in regards impediments to progress, and these fell into three categories: Technical –principally lack of data including equid numbers and movement, and disease surveillance,; Behavioral – principally lack of equine management education and understanding of socio-economic impact,; Institutional: principally lack of equine status affecting all aspects of resource allocation. The second set of recommendations were in regards how to address disease challenges in three categories. For the first set of diseases the major technical barriers to progress are limited, and what is needed is advocacy to address behavioral and institutional barriers to progress; examples include African Horse Sickness and tetanus. For the second set of diseases significant knowledge and technical barriers remain in addition to behavioral and institutional barriers, and research is urgently needed to address these gaps; examples include epizootic lymphangitis, and blood borne parasites. The third set of diseases represent essentially syndromic diagnoses, and these require thorough investigation and renewed surveillance to determine diagnosis with a high degree of certainty. The need and opportunity to reduce infectious disease in working equids is enormous, and increased global and local advocacy based on equine welfare and the socio-economic impacts will be essential for success. Global networks of investigators can play a critical role in building regional capacity towards addressing these needs.

DIFFERENTIATION OF EQUINE MESENCHYMAL STROMAL CELLS INTO NEURAL CELLS: POTENTIAL FOR CLINICAL APPLICATIONS. C. Cruz Villagrán, L. Amelse, N. Neilsen, M. Dhar. University of Tennessee College of Veterinary Medicine, Knoxville, TN.

Peripheral nerve injuries are a cause of poor performance in horses. These injuries are difficult to manage and treatment mostly relies on physical therapy and anti-inflammatories; however, the long term effects are time and personnel consuming. Our objective was to test the ability of equine mesenchymal stromal cells (MSCs) to differentiate into neuronal lineage, with subsequent trans-differentiation into Schwann cells for nerve regeneration.

Bone marrow-derived MSCs were obtained from 3 young and 4 adult horses. Following demonstration of stemness, cells were neuronally induced in tissue culture plates and microscopically assessed at 3, 6, 12, 24 and 48 h. Results showed that PRIMARIA™ plates supported proliferation and differentiation. Survival was improved by high number of cells. Morphological changes were observed as early as 3 hours; at 12 h approximately 60% of the cells were neuron-like. Morphology was similar in all horses, but proliferation was variable. Protein concentration was influenced by cell number and method of protein extraction. Expression of neuronal progenitor proteins, vimentin, β3 tubulin and GFAP, was detected by immunoblot analysis; nestin was not evident.

Results demonstrate that equine MSCs can differentiate into neuronal lineage. Future work is aimed to trans-differentiate these cells into Schwann cells and to evaluate survival when co-cultured with peripheral nerve. Expression of Schwann cell specific markers, S100B, Krox20 and CD104, will be tested.

DISPOSITION OF LEVETIRACETAM IN SERUM AND CEREBROSPINAL FLUID AFTER SINGLE DOSE ORAL ADMINISTRATION OF PIONEER AND GENERIC LEVETIRACETAM IN HEALTHY ADULT HORSES. AJ Stewart, J Costello, DM Boothe, SH Duran, AD Ramos, FB Cesar. College of Veterinary Medicine, Auburn University, Auburn, AL.

Levetiracetam (LEV) is a novel anti-epileptic drug. Our purpose was to determine the disposition of pioneer (Keppra™) verses generic levetiracetam in horses.

Six healthy adult horses were administered 30mg/kg PO of pioneer and generic LEV in a cross-over design. Serum and CSF samples were collected over 48hrs for pioneer (n=5), and over 12hrs (n=3) and 24hrs (n=3) for the generic formulation. LEV concentrations were analyzed via an immunoassay. Pharmacokinetic parameters were compared using paired t-tests, with P≤0.05 considered statistically significant.

For the pioneer and generic formulations the maximum concentrations (Cmax) were similar within serum (22.1±8.5, 20.1±5.6µg/mL, P=0.4), and CSF (10.9±1.4, >11.0±3.0µg/mL, P=1.0), respectively. The predicted area under the curve to infinity (AUC) for the pioneer and generic formulations in serum were similar (364±56.6, 338±21.0 hr*µg/mL, P=0.72). AUC after pioneer administration in CSF =333.6±41.5 hr*µg/mL.

After administration of the pioneer, the Cmax, AUC, half-life ($T_{1/2}$), mean residence time (MRT), apparent volume of distribution (Vz_F) and apparent clearance (Cl) were all greater in the CSF than serum. Tmax was greater for CSF than serum.

Parameter	Pioneer				Generic
	Serum	CSF	P value		Serum
AUC (hr*µg/mL)	364±56.6	334±41.5	0.03		338±21
Cmax (µg/mL)	22.1±8.5	10.9±1.4	0.03		>20.1±5.6
Tmax (hr)	9±6.7	20.4±5.4	0.009		>9.0±5.2
T½ (hr)	6.6±1.2	9.4±1.5	0.002		7.3±0.9 (n=3)
MRT (hr)	15.5±3.4	27.0±3.6	0.00002		13.0±1.5 (n=3)
Oral Clearance (mL/hr/kg)	85.6±1.7	100±12.7	0.0009		90.3±5.7 (n=3)
VD_F (mL/kg)	835±149	1,364±198	0.0003		524±105 (n=4)

Generic and pioneer LEV formulations were considered similar, which may make the use of levetiracetam affordable for the treatment of seizures in horses.

FACTORS AFFECTING SUSCEPTIBILITY OF YOUNG ADULT HORSES TO EQUINE HERPESVIRUS MYELOENCEPHALOPATHY. LK Maxwell, TC Holbrook, LL Gilliam, GB Rezabek, R Eberle, TA Snider, D McFarlane, CA Jamieson, T Kievitsbosch, DH Black. Oklahoma State University, Stillwater, OK.

Previous work has demonstrated that aged horses are more susceptible to equine herpesvirus myeloencephalopathy (EHM) than are young horses. However, the basis for this difference has not yet been established. The objective of this work was to determine whether immunocompetency or blood brain barrier permeability contributes to susceptibility of horses to EHM. We hypothesized that: 1) decreased cell-mediated immunity, induced by cyclosporine administration, increases susceptibility to EHM, 2) increased permeability of the blood brain barrier (BBB), induced by mannitol and lipopolysaccharide administration, enhances susceptibility to EHM. Sixteen horses, mares and geldings aged 2-5 years, were randomized to 4 treatment groups: cyclosporine only, BBB disruption, cyclosporine plus BBB disruption, and untreated controls. Horses were inoculated intranasally with equine herpesvirus type 1 (EHV-1), and horses were assessed until 15 days post-inoculation (DPI). Significant findings were: 1) body temperature and nasal shedding of EHV-1 did not differ between groups ($P>0.05$), 2) viremia was higher at 5 DPI in the cyclosporine treated horses as compared with control horses ($P<0.05$), 3) viral load in neurological tissues was higher in the horses in which only cyclosporine was administered ($P<0.05$), 4) maximum ataxia and clinical scores were higher in the horses in which only cyclosporine was administered, but not significantly so ($P=0.18$, 0.11, respectively). This study suggested that immunocompetency was more important to the pathogenesis of EHM in young horses than was BBB integrity. Further studies are warranted to determine which aspects of the immune response affected by cyclosporine administration are necessary for development of EHM.

INVESTIGATION OF A UNIQUE *THEILERIA EQUI* VARIANT IN THE UNITED STATES. LN Wise, L Kappmeyer, D Knowles. Washington State University, Pullman WA.

Recent routine horse surveillance in South Texas, which included serologic testing for *Theileria equi,* identified a stray horse with inconsistent results on the commercially available competitive ELISA. Additional testing was performed and while the horse was negative for *T. equi* by cELISA, PCR and IFA, a transfer of blood into a naïve, splenectomized horse resulted in a "*T. equi* -like" infection. 18s rRNA analysis confirmed an infection with a theilerial species with almost 100% identity to isolates from horses and zebras in South Africa. Further analysis showed this isolate to be quite different from *T. equi* isolates identified to date in North America.

Genome sequencing was completed with Illumina® technology and paired bac end reads in a *de novo* assembly. Numerous differences were noted between this isolate and the genome of domestic *T. equi* but of particular interest is the *ema* (equi merozoite antigen) gene family. This gene family encodes immunodominant surface merozoite EMA proteins and antibodies against these proteins are utilized in diagnostic testing. This *T. equi* variant possesses divergent *ema* family members, which explains the inconsistent test results.

The implications of identifying this variant within the U.S. are extensive and the genetic work described here will serve as the basis for further research on this parasite.

A SYSTEMATIC REVIEW AND META-ANLYSIS OF THE EFFICACY OF FRUSEMIDE FOR EXERCISE INDUCED PULMONARY HEMORRHAGE IN THOROUGHBRED RACEHORSES. S.L. Sullivan, G.A. Anderson, T. Whittem, P.S. Morley[#], K.W. Hinchcliff
University of Melbourne, Victoria, Australia; [#]College of Veterinary Medicine, Colorado State University, Fort Collins, Colorado, USA.

Frusemide is the most commonly used medication for EIPH, however, critical evaluation of the strength of evidence for efficacy of frusemide is lacking. The objective of this systematic review and meta-analysis was to evaluate the use of frusemide to reduce the severity or frequency of detection of EIPH in Thoroughbred racehorses. Studies were identified via searches of electronic databases, relevant texts and reference lists of published articles. Studies were not restricted by language, date or publication status. Search terms included exercise induced pulmonary h*morrhage, f*semide, treatment, horse, equine. Included studies were randomised and non randomised trials, prospective observational studies and experimental studies. Data extraction was performed by one author. Three authors independently assessed each study using the GRADE recommendations of rating quality of evidence. Meta-analysis of studies was performed with a random effects model to determine if frusemide reduced the frequency of detection of EIPH (yes or no) or if frusemide reduced the severity of EIPH by at least 1 tracheobronchoscopic grade. Stata 13.1 and MedCalc 12.7.7 performed the analyses.

Seventeen studies fulfilled the inclusion criteria, and 7 provided data for meta-analysis. Frusemide reduced the risk of detecting EIPH by 17% (pooled data from 7 studies, n = 4948, RR 0.83, 95% CI 0.72 – 0.95, p = 0.006). Significant heterogeneity existed ($I^2 = 66.5\%$, $p = 0.0065$), which was not explained by a meta-regression that included if the study design was paired or unpaired (p = 0.14). When data from the two RCTs (paired designs) were used in the meta-analysis, the RR of horses treated with frusemide was 0.68 (95% CI 0.58 – 0.79, $p < 0.001$). The proportion of horses diagnosed with EIPH after a control treatment that had a reduction of at least 1 EIPH grade after frusemide was 0.69 (95% CI 0.61 – 0.76).

Meta-analysis of methodologically robust trials, using a validated outcome measure in horses from the target population, indicates frusemide reduces the frequency and severity of EIPH when compared to untreated horses or horses administered a saline placebo.

EMERGING OUTBREAKS OF CORONAVIRUS IN ADULT HORSES. N. Pusterla[1], S. Mapes[1], C. Wademan[1], A. White[1], R.L. Ball[2], K. Sapp[2], P. Burns[3], C. Ormond[4], K. Butterworth[5], J. Bartol[6], R. Steere[7], T. Hanyi[8], L. Scott[9], J. Higgins[10], L. Fielding[10], S. McIntosh[10], J. Higgins[10], K.G. Magdesian[1]. 1. School of Veterinary Medicine, University of California, Davis, CA. 2. Bracken Equine Clinic, San Antonio, TX. 3. Elkhorn Veterinary Clinic, Elkhorn, WI. 4. Oak Hill Veterinary Services, Orinda, CA. 5. SRH Veterinary Services, Ipswich, MA. 6. New England Equine Medical & Surgical Center, Dover, NH. 7. Artaurus Veterinary Clinic, Petaluma, CA. 8. South Jersey Equine Associates, Vineland, NJ. 9. Idaho Equine Hospital, Nampa, ID. 10. Loomis Basin Equine Medical Center, Penryn, CA.

Equine coronavirus (ECoV) has been identified by electron microscopy, culture and more recently by PCR in feces of foals with and without enteric disease. Two recent studies reported on the isolation of ECoV from the feces of 2- to 4-year-old horses with pyrogenic and enteric disease living in stables of a racetrack in Japan. However, little is known about ECoV, especially with regard to molecular diagnostics of field samples and the clinical significance of ECoV PCR positive fecal results. The purpose of this study was to describe clinical, hematological and fecal PCR results from 268 horses involved in outbreaks associated with ECoV.

The outbreaks happened at eight separate boarding facilities between November 2011 and November 2013 in the States of CA, TX, WI, MA, NJ and ID. The population of horses per stable ranged from 8 to 70 horses. Following the molecular detection of ECoV in the feces from the initial index cases, the remaining herdmates were closely observed for the development of clinical signs. Fecal samples were collected from sick and healthy horses for the PCR detection of ECoV. Clinical pathology from sick horses was evaluated when available.

All eight outbreaks involved primarily adult horses. Eighty horses developed clinical signs with 3 to 16 sick horses per outbreak. The main clinical signs reported were anorexia (79), lethargy (70) and fever (65). Diarrhea and colic were observed in 17 and 8 horses, respectively. Further, 4 horses develop acute onset of encephalopathy. Clinical signs generally resolved within 1-4 days with supportive care. Nine horses from six different outbreaks were euthanized or died due to rapid progression of clinical signs. The cause of death in these horses was attributed to endotoxemic shock, septicemia or metabolic derangements. Blood work was available from 22 horses and common hematological abnormalities were neutropenia and/or lymphopenia. Feces were available for ECoV testing by PCR from 61 and 173 sick and healthy horses, respectively. The overall agreement between clinical status and PCR detection of ECoV was 90%. In conclusion, ECoV infection in adult horses is generally associated with self-limiting clinical and hematological abnormalities, although compromised intestinal barrier can lead to severe secondary complications and death. ECoV infection is supported by a positive fecal PCR result.

SAFETY AND IMMUNOREACTIVITY OF A XENOGENEIC DNA PLASMID VACCINE EXPRESSING HUMAN TYROSINASE IN TUMOR-BEARING HORSES. Luis M. Lembcke[2], James T. Blackford[2], Deborah A. Grosenbaugh[3], Timothy Leard[3], Stephen A. Kania[2], Dianne J. Trent[2], Keith Prater[2], John Biggerstaff[2], Agricola Odoi[2], and Jeffrey C. Phillips[1]. 1. Lincoln Memorial University College of Veterinary Medicine, Harrogate, TN. 2. University of Tennessee, Knoxville, TN. 3. Merial Ltd., Athens, GA.

The purpose of this study was to evaluate the safety and immunologic activity of the pING-HuTyr plasmid (Oncept™) in horses diagnosed with melanoma. This vaccine has been previously reported to be effective for treating melanoma in dogs. To accomplish our goal we proposed a study wherein the vaccine was used in gradually escalating doses in horses that had been diagnosed with melanoma while temporally measuring immunologic response in these treated patients. The protocol involved treating a total of 15 horses with a series of 4-biweekly injections along with a 6-month booster. Biologic samples were collected during the protocol to measure antibody and in vitro cellular reactivity to human tyrosinase, cellular reactivity to equine tyrosinase, and finally to characterize tumor infiltrating lymphocyte populations.

No adverse reactions have been seen in any of the vaccinated patients. Objective responses were seen in 13 of 15 treated horses. Positive humoral responses were seen in all treated horses. Cellular reactivity was noted in 13 of 15 treated patients. Characterization of tumor infiltrating lymphocyte populations identified a statistically significant increase in CD8+ lymphocytes along with a decrease in CD4+/Foxp3+ regulatory T cells following vaccination. Increasing plasmid dose did not appear to be associated with an increase in either clinical activity or immunologic reactivity. This xenogenic vaccine appears to be safe and well tolerated in tumor bearing horses and appears to result in both clinical activity and a measureable immune response in treated patients.

RETROSPECTIVE ANALYSIS OF *STREPTOCOCCUS EQUI* SUBSP *EQUI* INFECTIONS IN 108 HORSES IN THE FIELD (2005-2012). L. Duffee, A. G. Boyle, R.C. Boston. University of Pennsylvania, New Bolton Center, Kennett Square, PA.

Strangles is a contagious acute infection of the upper respiratory tract caused by *Streptococcus equi* in horses. This multiple outbreak retrospective study examined the risk factors for strangles diagnosis and prevalence of disease complications.

Computerized field service medical records were examined from 2005-2012. Strangles cases (108) were selected based on the presence of classic clinical signs (fever, lymphadenopathy, and nasal discharge) in a confirmed outbreak, positive *S. equi* PCR or aerobic culture. Control cases (207) were selected by the presence of a fever (>101.5 °F) over the same time period. Data collection included signalment, clinical signs, diagnostic testing, therapies, disease complications, and hematological values and were analyzed through descriptive and regression statistical methods.

Significant risk factors ($p<0.05$) for strangles diagnosis included: 2-7 years old (OR 3.22); presence of fever, nasal discharge and external abscess (OR 96.3); hyperfibrinogenemia >700 mg/dL (OR 2.71); and anemia (OR 3.08). A leukocytosis >12,000 cells/uL trended towards strangles diagnosis (OR 2.32). Contrary to previous reports, fever and external abscesses were found in 60% and 22% of cases, respectively. 30% (14/46) of horses tested prior to treatment were PCR positive in the absence of external clinical signs.

Our study showed a higher prevalence of purpura hemorrhagica (9% [OR 2.55]), prolonged (>29 days) *S. equi* PCR positive horses (8/44 [18%]), but a lower prevalence of metastatic disease (4%) than previously reported.

Fever and abscessation may be insufficient in disease screening when used alone. Clinical signs and disease complication rates observed over multiple outbreaks differed from previously published data.

ADRENAL STEROIDS, STEROID PRECURSORS AND NEUROSTEROIDS IN HOSPITALIZED NEWBORN FOALS K.A. Dembek[1]; S.D. Hurcombe[1]; N.M. Slovis[2]; B.S. Barr[3]; R.E. Toribio[1]. [1]The Ohio State University, College of Veterinary Medicine; Columbus, OH, USA; [2]Hagyard Equine Medical Institute, Lexington, KY, USA; [3]Rood and Riddle Equine Hospital, Lexington, KY, USA.

Hypothalamic-pituitary-adrenal axis (HPAA) dysfunction (relative adrenal insufficiency; RAI), has been associated with sepsis and mortality in newborn foals. However, comprehensive information on the adrenal cortex response to sepsis in sick foals is limited. Most studies have been focused on cortisol, while other essential adrenocortical steroids, steroid precursors, and neurosteroids have been overlooked. We hypothesized that adrenal steroids, precursors, and neurosteroids concentrations will be higher in sick compared to healthy foals on admission, and will remain unchanged or increase over the next 24 h of hospitalization in septic foals. Hormone and delta steroid concentrations will be associated with severity of disease and mortality rate.

Blood samples were collected on admission (time 0) and at 24 h (time 1) from 15 septic, 15 sick non-septic (SNS) and 4 healthy foals of ≤5 days of age. Blood concentrations of corticoliberin (CRH), adrenocorticotropin (ACTH), cortisol, aldosterone, pregnenolone, progesterone, 17α-OH-progesterone, androstenedione, dehydroepiandrosterone (DHEA), and allopregnanolone were determined by immunoassays. The delta steroid was defined as steroid concentration difference between time 0 and 1.

Cortisol, ACTH and CRH concentrations were higher in septic than in healthy foals on time 0 ($p<0.05$). Septic foals had higher CRH, ACTH, cortisol, aldosterone, pregnenolone, progesterone, 17 α-OH-progesterone, androstenedione, DHEA concentrations compared to SNS and healthy foals at time 1 ($p<0.05$). There was no difference in allopregnenolone concentrations between foals groups. Delta pregnenolone and 17α-OH-progesterone were lower and delta CRH was higher in septic compared to healthy foals. Healthy foals had lower aldosterone, 17 α-OH-progesterone and pregnenolone at time 1 compared to time 0. Non-surviving hospitalized foals had higher CRH, ACTH, cortisol and progesterone concentrations compared to surviving foals at time 1. Aldosterone, pregnenolone and 17α-OH-progesterone concentrations were higher in foals that died at both time points ($p<0.05$).

The adrenocortical response to sepsis in foals is characterized by persistently increased steroid concentrations over the first 24 h of hospitalization and foals for which steroids increase after admission are more likely to die. Aldosterone, pregnenolone, progesterone, and 17α-OH-progesterone concentrations are good predictors of non-survival at admission and 24 h post-admission in hospitalized foals.

AICAR ADMINISTRATION AFFECTS GLUCOSE METABOLISM BY UPREGULATING THE NOVEL GLUCOSE TRANSPORTER, GLUT8, IN EQUINE SKELETAL MUSCLE. M.A. de Laat[1], M.A. Robinson [2], K.J. Gruntmeier[1], S. DiNello[2], L.R. Soma [2], and V.A. Lacombe[1]. [1]Center for Veterinary Health Sciences, Oklahoma State University, Stillwater, OK. [2]New Bolton Center, University of Pennsylvania, Kennett Square, PA.

Equine metabolic syndrome is characterized by obesity and insulin resistance (IR). Currently, there is no effective pharmacologic treatment for this insidious disease. Glucose uptake is mediated by a family of glucose transporters (GLUT), which is regulated by insulin-dependent and -independent pathways, including the 5-AMP-activated protein kinase (AMPK). Importantly, downregulation of AMPK has been implicated in the pathogenesis of IR. In addition, the activation of AMPK, by 5-aminoimidazole-4-carboxamide-1-D-ribofuranoside (AICAR), stimulates glucose uptake in both healthy and diabetic humans. However, whether AICAR promotes glucose uptake in horses has not been established.

Blood samples and gluteal muscle biopsies were obtained from 5 healthy horses before and up to 6 hours following an intravenous AICAR infusion. Protein expression of major GLUTs (-1, -4, -8) and (total and phosphorylated) AMPK was quantified by Western blotting in skeletal muscle. Treatment with AICAR resulted in a decrease ($P=0.001$) in blood glucose concentration and an increase ($P<0.001$) in insulin concentration. There was a trend toward an increase in skeletal muscle protein content of total and phosphorylated AMPK ($P=0.05$ and 0.06, respectively) by 1 hour after treatment. While GLUT4 and -1 protein expression remained unchanged, GLUT8 was increased ($P<0.05$) following AICAR treatment.

Upregulation of GLUT8 protein expression by AICAR suggests that this novel GLUT isoform plays an important role in equine muscle glucose transport. In addition, these data suggest that AMPK activation enhances pancreatic insulin secretion. Collectively, these data suggest that AICAR acutely promotes muscle glucose uptake in the horse and thus its therapeutic potential for managing IR requires investigation.

FACTORS ASSOCIATED WITH VARIATION IN THE EQUINE INCRETIN RESPONSE IN A COHORT OF MORGAN HORSES AND WELSH PONIES. NE Schultz[1], RJ Geor[2], RK Splan[3], ME McCue[1]. [1]College of Veterinary Medicine, University of Minnesota, St. Paul, MN, [2]College of Veterinary Medicine Michigan State University, East Lansing, MI, [3]Virginia Polytechnic Institute and State University, Blacksburg, VA.

Hyperinsulinemia and insulin resistance are components of equine metabolic syndrome (EMS), a condition associated with predisposition for laminitis development. Incretins play a key role in glucose homeostasis and the insulin secretory response. The incretin hormone glucagon-like peptide-1 (GLP-1) produced by endocrine cells of the gastrointestinal tract promotes the release of insulin in response to oral sugar administration. GLP-1 has a short half-life due to rapid inactivation by the enzyme dipeptidyl peptidase-4 (DPP-4). Our objective was to investigate variability in the equine incretin response and identify factors associated with incretin response variation. Glucose (GLU), insulin (INS), total GLP-1 (GLP1tot), and active GLP-1 (GLP1a) concentrations were determined in 56 Morgan horses and 56 Welsh Ponies from 7 farms at baseline and 15, 30, 60, 75, 90, and 120 minutes following an oral sugar challenge. Baseline DPP-4 activity, prior laminitis status, obesity status (body condition score BCS $\geq$7), age, breed, sex, and diet measurements were also collected. For repeated measures, a quadratic random regression analysis was performed with respect to time. Independent variables included time, age, breed, sex, obesity/laminitis status, baseline DPP-4 activity, and diet variables. Farm was also included as a random effect. Significant "rise-fall" curves were detected for the population's mean GLU, INS, GLP1a, and GLP1tot with peak estimates occurring at 90, 75, 60, and 60 minutes, respectively. Significant positive correlations were detected for INS/GLU, GLP1a/GLP1tot, INS/GLP1a, and INS/GLP1tot. Baseline GLU correlated positively with obesity status. Horses with a higher daily caloric intake had a lower GLU rate of change over time. Baseline INS was positively correlated with age, DPP-4 activity, obesity and prior laminitis status and was significantly lower in geldings. Baseline GLP1tot was negatively correlated with total daily caloric intake, but was associated with a significantly higher GLP1tot response over time. In contrast, starch was positively correlated with GLP1tot at baseline, but resulted in a significantly lower rate of change in GLP1tot over time. Baseline GLP1tot was significantly higher in stallions than geldings. However, the GLP1a and GLP1tot had a significantly lower rate of change over time in stallions. Baseline measurements of DPP-4 activity were significantly greater in Welsh ponies and negatively correlated with age. Findings indicate horses exhibit changes in blood GLP-1, insulin, and glucose concentrations following an oral sugar challenge characteristic of an incretin response. Significant factors associated with variation in this response include DPP-4 activity, obesity status, gender and dietary factors.

EPIDEMIOLOGY OF *SALMONELLA ENTERICA* IN A LARGE ANIMAL VETERINARY TEACHING HOSPITAL. Brandy A Burgess, Paul S Morley. Department of Clinical Sciences, College of Veterinary Medicine and Biomedical Sciences, Colorado State University, Fort Collins, CO.

Salmonella enterica is one of the most common causes of healthcare associated infections in veterinary teaching hospitals, the most common cause for equine hospital closure, and among the top three agents associated with illness in veterinary personnel. When *Salmonella* spreads among patients, environmental contamination is predictably present – whether as cause or effect. Additionally, subclinical infections and shedding in the absence of disease is much more common than clinical infections. Unfortunately, current detection methods generally lack in sensitivity; prolonging time to detection. Consequently risk recognition and the ability to rapidly identify high-risk patients and environmental contamination are critical to effective management of populations and their environments. The objectives of this project were to determine factors associated with shedding of *S. enterica* among the general inpatient population and associated environmental contamination during times of endemic disease – for the purpose of endemic and epidemic disease prevention.

From March 2002 through December 2012, fecal cultures were performed on all large animal inpatients (n=11,061) on admission and twice weekly during hospitalization; and environmental cultures were performed monthly at predetermined locations throughout the facility (n=5337). Variables characterizing the patient population and hospital were collected retrospectively from the electronic medical records database. Multivariable logistic regression was used to evaluate associations between factors of interest and both patient shedding of *S. enterica* and associated environmental contamination.

During the study period, approximately 5.9% (n=648) of patients were fecal culture positive for *S. enterica* – the majority being bovine (73.5%; 476/648) and equine (22%; 143/648). In addition, approximately 7.9% (n=423) of environmental samples were culture-positive using standard culture techniques. In general, indicators of severe disease were related to patient shedding; and environmental samples collected in the Livestock Hospital, from floor surfaces, and during times of high caseload were more likely to be positive.

Factors identified in this study not only permit refinement of existing infection control programs, but provide a better understanding of the epidemiology of *Salmonella* in veterinary hospitals – allowing for evidence based preventive measures to be implemented in infection control programs on a daily basis.

PREVALENCE OF BACTEREMIA IN DAIRY CATTLE WITH ACUTE PUERPERAL METRITIS. Brent C. Credille, Amelia R. Woolums, David J. Hurley, Michael W. Overton, Steeve Giguère. University of Georgia, College of Veterinary Medicine, Athens, GA.

The objective of this study was to determine the prevalence of bacteremia in dairy cattle with acute puerperal metritis (APM) and determine clinicopathologic factors associated with bacteremia in this population of animals.

Cows were examined at the time of identification of APM. A complete blood count, serum biochemical analysis and bacteriologic culturing of blood and lochial fluid were performed on each animal at the time of diagnosis. The same samples were collected from healthy herdmates of a similar parity and days in milk. Blood culture results and clinicopathologic parameters were compared between groups. Conditional logistic regression was used to evaluate factors associated with APM, while multivariate logistic regression was used to evaluate factors associated with bacteremia.

Bacteremia occurred in 53.1% (9/17) of cattle with APM and 53.3% (8/15) controls. *Bacillus* spp. was the organism most commonly isolated from the bloodstream in cattle of both groups. Bacteremic cattle in both groups were significantly less likely to have basophils in the peripheral circulation and more likely to have higher serum globulin concentrations.

Bacteremia is a common occurrence in post-partum dairy cattle. It is likely that bacterial colonization of the involuting uterus, peri-parturient immunosuppression, and systemic inflammation play a role in the high frequency of bacteremia. Further study is warranted to investigate the modes by which bacteria colonize the bloodstream in this population of animals and the significance of bacteremia on health and productivity of affected animals.

TRANSIT TIME IN THE LARGE INTESTINE OF COWS WITH CECAL DILATATION/DISLOCATION IN COMPARISON WITH HEALTHY CONTROL COWS AND COWS WITH LEFT DISPLACEMENT OF THE ABOMASUM.

D. Devaux[1], M. Lempen[2], E. Schelling[2], M. Meylan[1]. 1. Vetsuisse Faculty, University of Berne, Switzerland. 2. Biomedical Engineering laboratory, Faculty of Applied Sciences, University of Berne, Biel, Switzerland. 3. Swiss Tropical and Public Health Institute, University of Basel, Basel, Switzerland.

Cecal dilatation/dislocation (CDD) is a common and economically important abdominal disorder affecting mainly dairy cattle. Whereas the clinical aspects of CDD are well described, the pathogenesis of the disease remains poorly understood. Therefore, adequate recommendations for prevention of the disease remain elusive. A primary motility disturbance in the spiral colon has been suspected to be the cause of CDD. Therefore, a minimally invasive technique was developed to assess the transit time in the bovine large intestine in cows with CDD in comparison with healthy cows and cows with left-sided displacement of the abomasum (LDA) to localize the primary motility disorder leading to CDD.

Cows with spontaneous CDD (n=15), healthy control cows (n=14) and cows with LDA (n=18) were included in the study. Five water-tight polioxometalate capsules, each containing a build-in temperature sensor and electronic transmitters, were introduced in the lumen of the ileum, cecum, proximal loop of the ascending colon (PLAC) and two locations in the spiral colon to measure the transit time, defined as the time between implantation of the devices in the intestine and an abrupt drop in the measured temperature indicating excretion of the transmitters. The transit times from the different intestinal segments were compared among groups by use of the Kruskal-Wallis test.

The technique for transit time assessment in the bovine intestine was safe and reliable. A high variability in the transit times was observed in all intestinal segments and in all groups. In addition, the output sequence was irregular, i.e. the capsules implanted in the most distal intestinal segments were not always excreted first. Statistical analysis revealed a significant difference ($p=0.009$) in the transit time from the distal location in the spiral colon among the 3 groups of cows, with the longest median time in cows with CDD. A particularly marked variability and prolonged transit times were observed in the calculated transit time between the PLAC and the proximal location in the spiral colon in cows with CDD, but the difference among groups was not significant ($p>0.05$).

The high degree of variability observed will preclude the routine use of transit time measurements for indirect assessment of intestinal motility in cattle. However, the results obtained in the present study allow confirming the role of a motility disorder in the bovine colon in the pathogenesis of CDD.

THE EFFECTS OF MULTIVALENT MODIFIED-LIVE VIRAL VACCINES ON REPRODUCTIVE PERFORMANCE. P. Walz, M. Edmondson, K. Riddell, T. Braden, J. Gard, J. Bayne, K. Joiner, and M.D. Givens. College of Veterinary Medicine, Auburn University, Auburn, AL.

Prebreeding vaccination should provide fetal and abortive protection against reproductive pathogens but not impede reproduction. Our objective was to assess reproductive performance of heifers revaccinated with modified-live viral vaccine prior to synchronized natural breeding. Sixty beef heifers were randomly assigned to four treatments. Groups A and B (n=20/group) were vaccinated with Express FP at 2 days after initial detected estrus, and then revaccinated 32 days later, which corresponded to 10 days (Group A) or 31 days (Group B) before synchronized natural breeding. Groups C and D (n=10/group) served as controls and were vaccinated with Citadel VL5 at the same time points as Groups A and B, respectively. Estrous behavior was assessed using radio frequency technology. Estrus synchronization was performed, with initiation occurring at revaccination (Groups A and C) or 21 days after revaccination (Groups B and D). After synchronization, heifers were submitted to a bull breeding pasture for 45 days. At the end of breeding period, heifers were ultrasounded to assess pregnancy. Progesterone concentrations were evaluated at unsynchronized and synchronized estrus and 10 days later, at initial pregnancy check, and at study end. Vaccination with Express FP 5-VL5 did not result in significant negative reproductive impact based upon duration of interestral intervals, proportion of heifers exhibiting estrus within 5 days after synchronization, serum progesterone concentrations, pregnancy rates, and pregnancies in first five days of breeding season. Use of Express FP 5-VL5 did not impede reproduction when revaccination was performed at 10 or 31 days prior to synchronized natural breeding.

EVALUATION OF GLUCOSE RESPONSE TO THREE INSULIN TYPES AND A CONTINUOUS GLUCOSE MONITORING SYSTEM IN HEALTHY AND SICK ALPACAS. S Byers, O Beemer, A Lear, R Callan, T Holt. Colorado State University College of Veterinary Medicine and Biomedical Sciences, Fort Collins, CO.

The purpose of the study was to evaluate the effects of different insulin formulations on glucose concentrations and evaluate the use of a continuous glucose monitoring (CGM) system in alpacas. Six healthy alpacas and one hospitalized alpaca were utilized in this study. Intravenous catheters and the CGM system were placed in each healthy alpaca. Regular insulin (0.2 U/kg, IV and SC), NPH insulin (0.4 U/kg SC), insulin glargine (0.4 U/kg SC), and dextrose (300 mg/kg, IV) were administered to the healthy alpacas over a 2 week period. Following drug administration, blood samples were collected for glucose testing at set time points and the CGM recorded glucose concentrations every 5 minutes. The hospitalized alpaca also had the CGM system installed and blood samples were collected to measure glucose concentrations periodically. Data were compared using method comparison techniques, error grid plots, and ANOVA.

Blood glucose values in healthy alpacas decreased most rapidly following regular insulin administration as compared to the other formulations while the NPH insulin produced the longest glucose suppression effect. There was a lag in glucose concentrations obtained from the CGM compared to whole blood. The alpacas had no adverse reactions to the different insulin formulations. Due to the prolonged duration of effect of the NPH insulin it may be more appropriate for long term use in hyperglycemic alpacas. The CGM could be useful in monitoring glucose trends and reducing blood collection events; however it should not be the sole method for determining treatment protocols.

DETECTING EARLY BEHAVIORAL MARKERS OF RESPIRATORY DISEASE DEVELOPMENT IN YOUNG DAIRY CALVES. ME Prado[1], PD Krawczel[1], J Wilkerson[2], S McGuirk[3]. [1]Department of Animal Science and [2]Department of Biosystems Engineering, The University of Tennessee, Knoxville, TN, and [3]College of Veterinary Medicine, University of Wisconsin, Madison, WI.

Dairy calf pneumonia (DCP) is a prevalent, economically important condition affecting calves during the pre-weaning and/or post-weaning periods. The impact of DCP extends beyond the actual disease episode, having a negative effect on subsequent productivity and survivability of replacement stock. The ability to identify calves during the initial stages of respiratory disease development is difficult but objective monitoring of activity and feeding behavior may allow for earlier disease detection and intervention, potentially decreasing the consequences of DCP later on in the calf's life. The objective of this study was to identify early behavioral changes during respiratory disease development in young dairy calves subjected to different milk feeding levels (standard or high) and housing (individual or group).

Holstein calves from 3 different commercial dairy farms (standard milk/individual housing, high milk/individual housing, and standard milk/group housing) were used in this study. Overall respiratory disease incidence was approximately 20%. Social interaction/activity levels as well as feeding behavior were measured by fitting calves with collars containing sensors during the pre-weaning period. Calves were screened daily for clinical signs of disease. Deep pharyngeal swabs were obtained from sick calves for detecting pathogens associated with respiratory disease by PCR and serum samples were collected for serology.

The automated sensor system successfully recorded activity levels of calves on the different farms. Total activity levels were correlated with milk consumption showing significant increases during the time preceding the feedings. In addition, a unique time series signature that corresponds with normal specific behaviors such as sleeping, standing, walking and/or bottle feeding was established from the acceleration data (activity levels). We used these data to identify any deviations from the norm. Calves with respiratory disease had increased activity levels when compared to healthy cohorts. In conclusion, these results suggest that calf behavior may be a useful means of disease detection.

RESPONSE OF BOVINE BRONCHIAL EPITHELIAL CELLS TO CO-INFECTION WITH BOVINE RESPIRATORY SYNCYTIAL VIRUS AND MYCOPLASMA BOVIS.

A. Woolums[1], L. Berghaus[1], C. Jarrett[1], D. Vanover[2], P. Santangelo[2], T. Krunkosky[1]. 1. University of Georgia, Athens, GA. 2. Georgia Institute of Technology, Atlanta, GA.

Bovine respiratory disease (BRD) is the leading cause of bovine morbidity and mortality. Polymicrobial infection by viral and bacterial pathogens is common in BRD, leading to more severe disease than induced by single agent infection. These agents first encounter the host at the airway epithelial cell. In other species airway epithelial cells influence host immunity through release of cytokines and immune cell activation, but the role of bovine bronchial epithelial cell (BBEC) responses in BRD, and the influence of polymicrobial infection on BBEC, are poorly characterized. The objective of this study was to determine whether expression of cytokines and surface activation markers was altered in BBEC co-infected with bovine respiratory syncytial virus (BRSV) and *Mycoplasma bovis* (Mb), as compared to responses of BBEC infected with either agent alone. BBEC isolated from cattle by bronchial brushing were grown and differentiated in air-liquid interface culture. Cells were exposed to media (mock infection), 10^6 pfu BRSV, 10^8 cfu Mb, or both agents together, for 2, 4, or 24 hours. Infection was confirmed by immunofluorescence microscopy. Transepithelial resistance (TER) was measured before and after infection, and expression of mRNA for the cytokines interleukin 8 (IL-8) and RANTES and the adhesion molecule ICAM-1 was quantified by real time RT-PCR. Transepithelial resistance was decreased in all cultures at 2 hours; TER increased toward baseline in mock and BRSV cultures but not in Mb or Mb + BRSV cultures. Expression of RANTES and ICAM-1 was increased sharply at 2 hours after BRSV infection; this was suppressed by Mb. Expression of IL-8 was increased by BRSV and Mb, but was increased more by coinfection. Disease in cattle infected with BRSV + Mb may be exacerbated by impaired migration of leukocytes or excessive neutrophil influx due to alterations in expression of RANTES, ICAM-1, and IL-8.

COMPUTED TOMOGRAPHY OF TYMPANIC BULLAE IN PRE-WEANED DAIRY CALVES DIAGNOSED WITH OTITIS MEDIA. G Clausen, KP Poulsen, S Stieger-Vanegas, J Vanegas, S Nemanic. Oregon State University College of Veterinary Medicine, Corvallis, OR.

Despite short and long term costs of respiratory disease and otitis media in pre-weaned dairy replacement heifer calves, veterinarians and producers still lack the ability for effective early identification and prognostication of infected calves. The objective of this study was to use advanced imaging to improve established diagnostic techniques for calves diagnosed with otitis media.

Thirty Jersey calves were enrolled in three groups (n=10 per group) from a single farm where respiratory disease is diagnosed by farm personnel using the McGuirk Respiratory Scoring System. Calves were grouped by (1) primary disease, (2) primary treatment failure and (3) multiple treatment failure. All calves were transported to the Veterinary Medical Teaching Hospital at Oregon State University for radiographs of the thorax and skull (focusing on tympanic bullae, 35° oblique from midline) and contrast enhanced computed tomography (CT). Following imaging, calves were euthanized and submitted for diagnostic pathology. For all calves, radiographs were compared to CT images to assess lesions of the tympanic bullae using gross pathology as the gold standard. Fluid in the bullae and osteolysis were measured using a scoring system and bullae wall thickness was directly measured for comparison.

No difference was found between radiographs and CT to detect fluid in the bullae. Radiographs were better able to quantify osteolytic lesions compared to CT. CT was better able to detect bony proliferation of the bullae wall compared to radiographs. Lateral wall thickness significantly increased with treatment failure compared to ventral and medial bullae walls with thicknesses of 0.13, 0.18, 0.21cm for groups 1-3, respectively ($p<0.05$). *Mycoplasma bovis* was cultured from 8 of 10 calves in each group. Of positive calves, all nasal swabs were positive. However, in lower airway *M. bovis* cultures, 25% from groups 1 and 2 and 12.5% from group 3 were negative.

These findings suggest radiographic imaging of tympanic bullae in dairy calves with otitis media is sufficient to detect fluid in the bullae and is better than CT to quantify osteolysis. This is relevant for clinical management of otitis media where CT is not available. We also show that nasopharyngeal swabs are appropriate for *M. bovis* detection in calves with clinical signs of otitis media.

PATIENT SAFETY IN THE CATHETERIZATION LAB

Bruce W. Keene, DVM, MSc, DACVIM (Cardiology)
Raleigh, NC

Interventional cardiac catheterization is considered to be (and in retrospective studies has proven to be) a relatively safe alternative to conventional surgery for the correction or palliation of a variety of congenital heart defects, including patent ductus arteriosus (PDA), pulmonic stenosis (PS), and cor triatriatum dexter. In addition, transvenous pacemaker implantation has become the default procedure for the treatment of high grade 2nd or 3rd degree AV block or sick sinus syndrome in most dogs afflicted with these conditions.

The purpose of this presentation is to review some important factors and concepts in patient safety during interventional procedures, review the preliminary results of the ACVIM task analysis performed for 3 common interventional procedures, and to propose the careful construction and adoption of checklists to enhance patient safety during cardiac interventions, as well as the adoption of a cardiac catheterization registry, which we believe will ultimately enhance patient safety during these complex procedures.

While the dictionary definition of "safety" is clear (the condition of being safe from undergoing hurt, injury, or loss), the perception of what constitutes "safe" varies widely with activity and location. For example, according to the National Information Center on Health Services Research and Health Care Technology, the safest human cardiac catheterization laboratories over the past 5 years performed on average about 15 procedures daily (approximately 5,500 procedures yearly), and serious complications or death occurred in slightly less than 0.3% of their patients (i.e. about 15 patients per year). While such registry statistics are not currently available in veterinary medicine, it seems reasonable to speculate (based on published retrospectives and an informal survey) that even the best clinical veterinary cardiac catheterization laboratories over the past 5 years performed less than 300 procedures yearly (most perform less than 100 / year), with serious complication and death rates that vary by procedure, but average between 2 – 6 percent overall. For comparison, if commercial airline flights crashed at the rate of serious complications and deaths experienced in the very best human cardiac catheterization laboratories, there would have been 27,300 commercial airplane crashes last year.

We have been working with human factors psychologists, engineers, and interventional pediatric cardiologists, anesthesiologists, and nurses over the past 2 years to assess and improve patient safety in the cardiac catheterization lab at NC State. This effort has been directed primarily at improving operator training and patient safety in the 3 most common procedures in our clinical veterinary catheterization lab, including device closure of patent ductus arteriosus (PDA), pulmonic balloon valvuloplasty, and transvenous pacemaker implantation.

The first step in this process was performance of a task analysis – a detailed, step by step recording of all of the activities, decisions, and imaging performed on a patient including diagnosis, procedural planning, device selection, premedication, anesthetic induction, surgical preparation and patient positioning, vascular access, catheter and device manipulation, immediate results assessment, and removal of all appliances followed by anesthetic recovery.

These painstaking analyses revealed some substantive differences in how our faculty and house officers identified and addressed problems, utilized and interpreted imaging, hemodynamic, and electrocardiographic data, and how different operators communicated among themselves and with the ancillary support services (e.g. anesthesia and radiology) involved in the procedures. These detailed procedure reports, combined with extensive interviews with acknowledged experts (both experienced pediatric and veterinary interventionalists, as well as anesthesiologists, nurses and technicians) are being used to construct procedural check lists for each phase of these procedures. Complication rates will be compared before and after the implementation of these checklists.

In addition to checklist implementation, a central (ACVIM?) registry of interventional procedures would be helpful to be able to better assess the safety of various procedures, and track the efficacy of these procedures, as well as the effect of future safety enhancements.

Suggested reading

El-Said, H. G., Bratincsak, A., Foerster, S. R., et al. (2013). Safety of percutaneous patent ductus arteriosus closure: an unselected multicenter population experience. Journal of the American Heart Association, 2(6), e000424

Goodrich, K. R., Kyles, A. E., Kass, P. H., et al. (2007). Retrospective comparison of surgical ligation and transarterial catheter occlusion for treatment of PDA in two hundred and four dogs (1993-2003). Veterinary Surgery : VS, 36(1), 43–9.

Gordon, S. G., Saunders, A. B., Achen, S. E., Roland, et al. (2010). Transarterial ductal occlusion using the Amplatz Canine Duct Occluder in 40 dogs. Journal of Veterinary Cardiology 12(2), 85–92.

Henrich, E., Hildebrandt, N., Schneider, C., Hassdenteufel, E., & Schneider, M. (2010). Transvenous coil embolization of patent ductus arteriosus in small (≤3.0 kg) dogs. Journal of Veterinary Internal Medicine 25(1), 65–70.

Johnson, M. S., Martin, M., Edwards, D., French, A., & Henley, W. (2012). Pulmonic stenosis in dogs: balloon dilation improves clinical outcome. Journal of Veterinary Internal Medicine 18(5), 656–62.

Meijer, M., & Beijerink, N. J. (2012). [Patent ductus arteriosus in the dog: a retrospective study of clinical presentation, diagnostics and comparison of interventional techniques in 102 dogs (2003-2011)]. Tijdschrift Voor Diergeneeskunde, 137(6), 376–83.

Saunders, A. B., Gordon, S. G., Boggess, M. M., & Miller, M. W. (2013). Long-Term Outcome in Dogs with Patent Ductus Arteriosus: 520 Cases (1994-2009). Journal of Veterinary Internal Medicine

Oyama MA, Sisson DD, Lehmkuhl LB. Practices and outcome of artificial cardiac pacing in 154 dogs. J Vet Intern Med 2001;15:229–239.

Wess G, Thomas WP, Berger DM, Kittleson MD. Applications, complications, and outcomes of transvenous pacemaker implantation in 105 dogs (1997-2002). J Vet Intern Med 2006;20:877–84.
Johnson MS, Martin MWS, Henley W. Results of pacemaker implantation in 104 dogs. J Small Anim Pract 2007;48:4–11.
Sisson D, Thomas WP, Woodfield J, et al. Permanent transvenous pacemaker implantation in forty dogs. J Vet Intern Med 1991;5:322–31.

SHOCK AND ABLATIONS: THERAPIES FOR VENTRICULAR ARRHYTHMIAS

Roberto Santilli, DVM, PhD, DECVIM-CA (Cardiology)[a], Romain Pariaut, DVM, DACVIM, DECVIM-CA (Cardiology)[b]
[a]Samarate, Varese, Italy; [b]Baton Rouge, LA

SHOCKS FOR VENTRICULAR FIBRILLATION

High-energy electrical shocks are the only effective technique to terminate ventricular fibrillation (VF). Since their approval by the FDA in 1985, implantable cardioverter defibrillators (ICD) have been used to monitor human patients at risk of developing VF, and to deliver shocks when life-threatening rhythms are detected.[1]

ICD generators are connected to an electrode that is secured within the right heart chambers. A right ventricular defibrillation coil (RV coil) at the extremity of the electrode, and a superior vena cava coil (SVC coil) positioned 15 to 20 cm more proximally, allow to select various directions for the vector of defibrillation; high-energy shocks can be delivered between the 2 coils, between the RV coil and the generator or between the 2 coils and the generator. In dogs, ICD positioning is critical because the shock of defibrillation must travel through most of the left ventricular muscle to be effective. The generator cannot be placed in the cervical region as it is usually done with pacemakers. Defibrillation has to be programmed between the RV coil and the generator, which is placed subcutaneously on the left side of the thorax.[2,3]

Successful defibrillation therapy requires appropriate detection of the arrhythmia and interaction of the electrical shock with the myocardium. It is believed that successful defibrillation requires concurrent depolarization of 75 to 90% of the myocardium. Detection of the arrhythmia is mainly based on its rate and duration, but also onset characteristics and intracardiac electrogram morphology. Maximum programmable rate for VF detection on current devices ranges between 250 and 300 bpm. Because, sinus rhythm may reach rates approaching 250 bpm in dogs, it is challenging to avoid inappropriate shocks.[2,3,4] In addition, inappropriate therapies can occur as a result of double-counting of the T waves, that have a large amplitude in dogs.[2] One approach to prevent these complications would be to very precisely tailor the detection parameters to the characteristics of the ventricular arrhythmias in affected dogs, including ventricular tachycardia rate, duration, onset characteristics with and without antiarrhythmic therapy.

Although ICD therapy is very successful at terminating induced episodes of VF in dogs in the experimental setting, factors determining shock success or failure are numerous. What is seen as a failure to defibrillate usually corresponds to re-induction of ventricular fibrillation after termination of the treated event. In dogs with underlying structural cardiac disease, including dilated cardiomyopathy, arrhythmogenic right ventricular cardiomyopathy and myocarditis, the success of defibrillation may be lower because of myocardial remodeling. High-energy shocks also raise the concern of myocardial stunning, which sometimes require pacing for several minutes after a shock and suggest that the lowest effective energy shock should be used as first line therapy.

USE OF RADIOFREQUENCY CATHETER ABLATION TO TREAT ENGLISH BULLDOG WITH FAMILIAL RIGHT VENTRICULAR ARRHYTHMIAS

Ventricular tachycardias (VT) arising from the right ventricle usually manifest wide QRS complexes with a left bundle branch block (LBBB) configuration. The most common causes of right VT in people are arrhythmogenic right ventricular cardiomyopathy (ARVC) and idiopathic right ventricular outflow tract (RVOT) tachycardia.[5-8] Right VTs features have been described in dogs with diffuse and segmental ARVC and in a canine model of RVOT tachycardia.[9-12]

We described electrocardiographic and electrophysiologic features of right ventricular tachycardias in English Bulldogs with right ventricular outflow tract dilation with a history of syncope or cardiogenic shock.[11] In these dogs VTs are usually incessant and very rapid, present a LBBB morphology with inferior axis and a localized dyskinetic areas with diastolic bulging in the RVOT area.[11] This particular type of myocardial disorder can be a segmental phenotype of ARVC since fifty per cent of human beings with ARVC have localized right ventricular aneurysms particularly in the RVOT area, and histopathology confirmed fibro-fatty infiltration of the RVOT dilatation in a English Bulldog with right VT.[5-7,10]

Electrophysiological study with either endocardial or epicardial mapping can be performed in these familial right VTs to evaluate the inducibility of the VT, the site of VT origin and to attempt radiofrequency ablation of the tachycardia isthmus. The dogs are anesthetized as described elsewhere.[13] To perform endocardial mapping a decapolar catheter is advanced through the right external jugular vein into the coronary sinus and a quadripolar catheter is advanced through the right femoral vein and placed by the right atrium and His Bundle. A right ventricular study, including endocardial mapping and programmed ventricular stimulation, is performed using a 7-F thermocouple-tipped steerable catheter, which is subsequently use to carry out temperature-guided radiofrequency endocardial ablation. Epicardial mapping is performed through an echo-guided pericardial puncture advancing an epidural introducer needle into the pericardial space. Medium contrast is then injected to verify the exact position of the needle tip followed by the introduction of a guide wire. Finally over the wire an introducer set is placed to permit the passage of the ablation catheter into the pericardial space, to map right and left ventricular epicardium.[14] Standard 12 lead surface ECG and bipolar endocardial or epicardial electrograms were recorded on a multichannel recorder. After completion of the recordings during sinus rhythm, right ventricular programmed stimulation is performed using a train of 8 beats with cycle lengths from 350 to 300 ms delivered at the RVOT with a 2 ms pulse width at twice the late diastolic threshold. The number of extra-stimuli is limited to 2 at each cycle length from 240 to 180 ms, decreasing the pacing interval until ventricular refractoriness is reached (170 ms) or ventricular tachycardia is induced. Entrainment mapping during VT is used to characterize the presence of re-entry and to localize the tachycardia circuit isthmus. Once a suggestive diastolic electrogram is found, in order to identify the isthmus site of the VT circuit, we pace the RVOT at a cycle length of 10 to 20 ms shorter than the VT cycle length. During stimulation tachycardia should accelerate to the pacing cycle length without changes in the QRS morphology on surface leads, of ventricular activation sequence and morphology of intra-cardiac recordings. As soon as the RVOT pacing is terminated, tachycardia should return to the initial cycle length and the first post-pacing interval should be less than 30 ms shorter than VT cycle length to testify that the pacing site is within the electrical circuit and that it could be considered an optimal target for ablation. Because of the diffuse nature of the disease, the isthmus may sometimes be wide and entrainment mapping can be obtained from a rather large area. [15] In these cases, endocardial fragmented delayed potentials (FDP) during sinus rhythm can be recorded. The RVOT is mapped during sinus rhythm to detect these areas of inhomogeneous activation and slow conduction. Linear radiofrequency catheter ablation can be performed during tachycardia or, in case of very rapid ventricular rate, the ablation site can be identified during sinus rhythm searching for FDP. The endpoint of the procedure is the abolition of FDP and the induction of a bidirectional block with no re-induction of the tachycardia with programmed ventricular stimulation.

The acute success rate of RFCA of ventricular tachycardia in this disease depends on the dimension of the aneurism and the location of the circuit and range from 60 to 75 %. Recurrence rate is around 15 %, and in this case the arrhythmias can be controlled with low dose amiodarone. Complication rate in our EP laboratory is 3% with some procedure-related deaths caused by the presence of ventricular tachycardia refractory to medical treatment and causing advanced congestive heart failure.

REFERENCES

[1]Silka MJ, Bar-Cohen Y. **Heart Rhythm** 2006;3:1360-1366. [2]Nelson OL, Lahmers S, et al. **J Vet Intern Med** 2006;20:1232-1237. [3]Pariaut R, et al. **J Vet Cardiol** 2012 ;26 :389-398. [4]Pariaut R, et al. **J Vet Cardiol** 2011 ;13 :203-210. [5] Thiene G, et al. **Veterinaria** 1988;318:129. [6] Brooks R, et al. **Medicine** 1988;67:271. [7]Nakagawa M, et al. **J Cardiovasc Electrophysiol** 2002;13:633. [8]Mehta D et al. **Am J Cardiol** 1989;63:939. [9]Basso C et al. **Circulation** 2004;109:1180. [10]Santilli RA, et al. **J Vet Cardiol** 2009;11:47. [11]Santilli RA, et al. **J Small Anim Pract** 2011;52:574. [12]Zhou J, et al. **J Cardiovasc Electrophysiol** 2006;17:771. [13]Santilli RA, et al. **J Vet Cardiol** 2006;8:157. [14]Sosa E, et al. **J Cardiovasc Electrophysiol** 1996:7:531. [15]Arbelo E, et al. **J Cardiovasc Electrophysiol** 2010;21:473.

ACVIM CANDIDATE FUNCTION: ELECTROPHYSIOLOGY REVIEW

Roberto Santilli, DVM, PhD, DECVIM-CA (Cardiology)
Samarate, Varese, Italy

Electrophysiological (EP) study is considered the method of choice to recognise cardiac arrhythmias, to analyse their origin and to evaluate underlining mechanisms. It consists of a systematic analysis of dysrhythmias by recording and measuring a variety of EP events with the patient in the basal state and by evaluating the patient's response to programmed electrical stimulation.[1] Different studies reported the use to this technique in normal dogs and in dogs with supraventricular and ventricular arrhythmias. [2-11] In dog EP investigations are performed under general anesthesia. The animals are placed in dorsal recumbence. Three multipolar electrode catheters are inserted in the venous system, with a modified Seldinger technique, using fluoroscopic and intra-cardiac ECG guidance: a decapolar catheter is inserted through the right external jugular vein into the coronary sinus, two quadripolar catheter through the right femoral vein, one to record the His bundle potential and one for programmed electrical stimulation, is placed, alternatively, in high right atrium, and right ventricular apex. To map focal atrial tachycardias or macro-reentrant atrial tachycardia a second decapolar catheter is inserted into the right femoral vein and positioned along the right atrial lateral wall to record potentials from the antero-lateral aspect of the crista terminalis (HRA) with the proximal pair of electrodes, the mid-lateral right atrial (MRA) potentials with the middle pair and the low posterior-lateral right atrial (LRA) potentials with the distal pair. [7-8] To map left atrial chamber or pulmonary veins ostium, in case a patent foramen ovalis is not present, a transeptal puncture should be performed using the described technique. From the right femoral vein, under fluoroscopic guidance, a specially designed guide wire is first inserted into a transeptal Brockenbrough curved needle and pushed across the fossa ovalis into the left chamber. A jump of the tip of the needle from the thicker muscular septum to the thin wall of the fossa ovalis is used to localize fossa ovalis. Along the wire, positioned into the left atrium, the transeptal needle is pushed across the interatrial septum to enlarge the hole, and then an introducer is left in place and used to move the mapping and the ablation catheters. [7] A particular electrode named Lasso catheter is used in these cases to map pulmonary vein ostium and to perform ablation. It is composed by multiple electrodes positioned in a circular fashion.[7]

To map ventricular tachycardias either an endocardial approach, with the ablation catheter positioned in the right ventricular lumen, or a transpericardic epicardial approach can be used. With the latter technique the epicardial mapping is performed through an echo-guided pericardial puncture advancing an epidural introducer needle into the pericardial space. Medium contrast is then injected to verify the exact position of the needle tip followed by the introduction of a guide wire. Finally over the wire an introducer set is placed to permit the passage of the ablation catheter into the pericardial space, to map right and left ventricular epicardium. [9] Once all intracardiac electrodes have been placed, surface and intracardiac ECG signals are displayed on a recorder at a paper speed of 150 or 300 mm/s. Intracardiac ECGs are recorded at filter settings of 50 to 500 Hz. Pacing is performed with stimuli that are twice the diastolic threshold and 2 milliseconds in duration.

Normal conduction intervals (AH and HV), effective antegrade and retrograde refractory periods (ERP) and Wenckebach point of the normal AV node are first obtained. Normal values have been described in 15 adult dogs before and after administration of atropine and propranolol.[3] Antegrade and retrograde ERP of the AV node can be measured using programmed stimulation technique. A paced train of 8 extrastimulus (S1) are delivered respectively to ventral septal right atrium and right ventricular apex with a current of 2 mA and duration of 2 ms. S1-S1 interval or drive cycle length set is slightly minor of native sinus rhythm, after the eighth stimulus a premature extrastimulus (S2) is introduced with progressively shorter S1-S2 interval (from 300 to 180 ms) until S2 is not conducted through the AV node. ERP of the AV node is defined as the longest S1-S2 interval at which S2 failed to reach the His Bundle (antegrade) or ventral right atrium (retrograde). Similarly to human being 46 % of dogs present ventriculo-atrial (VA) conduction during the basal state, and it is enhanced by high sympathetic tone[1,3]. Antegrade and retrograde AV node Wenckebach point is obtained with incremental atrial and ventricular pacing. Pacing cycle length at which type I atrioventricular or VA block occurs is considered the Wenckebach point.[3]

During tachyarrhythmias EP study different pacing and mapping techniques can be use to define the type of tachycardia: programmed stimulation with single or double extrastimulus, rapid incremental atrial pacing, resetting and entrainment. Similarly to AV nodal ERP analysis, programmed atrial stimulation, delivered at twice the pacing threshold, is used to induce tachycardia, and to study its location, type and mechanism. Tachycardia induction is considered reproducible if initiated with the same S1-S2 interval. Reentrant tachycardia can be initiated and terminated as soon as S1-S2 interval encounter the refractory period of one of the two limb of the circuit[1-6]. Focal atrial tachycardias due to triggered activity can be initiated and terminated with rapid atrial pacing and isoproterenol infusion.[7] Automatic focal atrial tachycardia cannot be predictably and reproducibly initiated and terminated with programmed stimulation but respond to isoproterenol infusion.[7] Other tools to evaluate the location of atrial ectopic foci are mechanical interruption of the tachycardia caused by application of pressure to the area and the presence of fragmented, long duration, low amplitude electrograms. Fractioned potential consisted of three or more consecutive negative deflections with low peak-to-peak amplitude. In the dog 57% of atrial potentials of the focus area are fractionated and 31% are low-amplitude. [7] According to detailed endocardial mapping 63% of the dogs with FATs presented right-side foci (crista terminalis 41%; triangle of Koch 25%; tricuspid valve annulus 16 %; right interatrial septum 9% and right auricle 9%) and 37%

presented FATs arising from the pulmonary veins (right superior pulmonary veins 57%; left superior pulmonary veins 28%; right inferior pulmonary veins 15%). Twelve percent of dogs presented multiple ectopic foci. [7] Focal atrial tachycardia can be differentiated from reentrant tachycardia by measuring first VA interval after cessation of overdrive atrial pacing. [6-7] In latter case it is usually fixed to testify the presence of a predetermined circuit.

Anatomical distribution and electrophysiologic properties of APs can be assessed with programmed atrial and ventricular stimulation and the analysis of the sequence of activation during tachycardia. Antegrade conduction is considered present in case of ventricular pre-excitation (either abnormally short, or negative His to ventricular interval) during normal sinus rhythm or while atrial pacing at progressively shorter cycle lengths. [2,4-6,10-11] The presence of retrograde conduction along the AP is demonstrated by the evidence of eccentric VA activation during tachycardia or programmed ventricular stimulation. Retrograde atrial activation during orthodromic AVRT starts at the atrial insertion of AP. It could be defined as the point of earliest atrial activation of bipolar signal. With the exception of antero-septal bypass tracts, atrial retro-activation is eccentric and nondecremental with an all-or-none pattern of conduction. Rarely APs responsible of permanent junctional reciprocating tachycardia have decremental retrograde VA conduction. In these APs a prolongation of VA interval is observed during ventricular pacing with progressive shortening of the stimulation cycle length. [11] Concomitant unipolar recording are used to optimize the accuracy of APs localization. Sharp and negative unipolar waveform indicates the exact site of impulse spreading. [6,9-11] According to endocardial mapping in the dogs most APs are right-sided clustering near the coronary sinus ostium in the right posterior and postero-medial tricuspid valve annulus, and 73% have unidirectional retrograde activation with nondecremental properties. [10]

Resetting is the interaction of single extrastimulus with the tachycardia circuit resulting in a less than a compensatory pause (S1-S2 interval + pause = double of tachycardia cycle length). Resetting is use to prove that AP is part of supraventricular tachycardia circuit, and to differentiate atrioventricular nodal reciprocating tachycardia from orthodromic atrioventricular reciprocating tachycardia with concentric VA conduction. A single ventricular extrastimulus during tachycardia is given, while His was refractory, looking for pre-excitation of the atrium with identical activation sequence and/or termination of the tachycardia.[1-6,10-11]

The entrainment technique is defined as a stable conduction that derives from the interaction of two rhythms.[1] The degree of interaction depends on the distance between the site of stimulation and the circuit, electrophysiological properties of tissues and presence of excitable gap. Pacing the atrium with a cycle length shorter than tachycardia cycle length accelerates the tachycardia cycle with the same activation sequence and different degree of fusion depending from the distance between the site of stimulation and the circuit. In case the site of stimulation is within the circuit no fusion appears (concealed entrainment). [1] In case of apparent entrainment first beat post-interruption of pacing could be use to assess the location of the tachycardia circuit. The length of this interval is proportional to this distance. Macro-re-entrant atrial tachycardias are diagnosed according to the presence of entrainment with concealed fusion. Pacing at a site with pre-systolic activity or mid-diastolic potential with a cycle length 10-20 ms faster than the atrial tachycardia should entrain a re-entrant tachycardia to the pacing rate without changing the morphology of atrial depolarization or the intracardiac electrograms sequence. First beat post-interruption of pacing interval should result less than 30 ms. [8,13]

Entrainment technique is use to diagnose anatomical re-entry such as in case of typical and atypical atrial flutter [8,13] and ventricular tachycardia.[9] Macro-re-entrant atrial tachycardia is a re-entrant activation around a large central obstacle. Atrial flutter is a macro re-entrant atrial tachycardia characterized by an organized atrial rhythm. [14-15] Depending on the location of the protected slow conducting isthmus, atrial flutter are classified in typical and atypical. [14-15] The most common form of atrial flutter is mediated by cavo-tricuspid isthmus an area of the low posterior right atrium bounded by the inferior vena cava and the Eustachian ridge posteriorly and the tricuspid valve annulus anteriorly. [8,15] According to the rotation pattern around the tricuspid valve annulus, two forms of CTI-dependent AFL have been identified: typical and reverse typical. In the dog during typical AFL the activation wavefront turns counterclockwise around the tricuspid valve annulus, while during reverse typical AFL it turns clockwise. [8] Other two forms of atypical atrial flutter have been described in the dogs with the isthmus area located in the interatrial septum and the right atrial free wall. [13]

In veterinary medicine endocardial mapping of ventricular arrhythmias are limited to sustained ventricular tachycardias in English Bulldogs with the segmental phenotype of arrhythmogenic right ventricular cardiomyopathy (ARVC). [9] After the execution of basal measurements during sinus rhythm, right ventricular programmed stimulation is started at the endocardial or epicardial side of the RVOT using a train of 8 beats with a progressive shorter cycle length (from 300 until ventricular refractoriness is reached or ventricular tachycardia is induced) delivered with 2 ms pulse width at twice the late diastolic threshold. The number of extrastimuli is limited to 1 or 2 for each pacing cycle length with a coupling interval from 240 to 180 ms. As soon as ventricular tachycardia is induced, entrainment mapping technique is performed to characterised the presence of an anatomical re-entry and to localise the tachycardia circuit. Once a diastolic electrograms is found, in order to identify the critical isthmus area, an overdrive pacing of the tachycardia with a cycle length 10-20 ms shorter than the tachycardia cycle length should be done to obtain a concealed entrainment with a first post-pacing interval 30 ms shorter than ventricular tachycardia cycle length. In some cases, seen the diffuse damage of the endocardial and epicardial substrates, concealed entrainment can be obtained from a rather large area. In these cases the finding of fragmented delayed potentials during sinus rhythm can help to locate the critical isthmus.

Focal junctional tachycardia is a rhythm disturbance typical of Labradors retrievers, which arise from the His bundle area and it is characterised by the presence of isorhythmic atrioventricular dissociation (IAVD) in 90.9 % of the dogs affected. [16] To diagnose focal junctional tachycardia a detailed endocardial mapping of the His bundle electrogram should be performed. During focal junctional tachycardia, His potentials always precedes the ventricular deflection with a normal HV interval to testify an antegrade conduction. Atrial depolarizations are usually unrelated to His potentials, showing the classical features of IAVD, and arise from the high right atrium. AH interval can varies from positive to negative in case of IAVD with synchronization type I or can be shorter than normal and fixed in case of IAVD with synchronisation type II. In the majority of cases periods of 1:1 VA conduction can be recorded. [16]

REFERENCES

[1] Josephson ME. **Clinical Cardiac electrophysiology**. 2002.p. 322-424. [2]Atkins CE, et al. **J Vet Intern Med** 1995:9-43. [3]Wright KN, et al. **Am J Vet Res** 1996;57:1695. [4]Wright KN et al. **J Vet Intern Med** 1999;13:361. [5] Sherlag BJ et al. **J Am Vet Med Assoc** 1993;203:1147. [6]Santilli RA, et al. **J Vet Cardiol** 2006;8:157. [7] Santilli RA, et al. **J Vet Intern Med** 2010;24:539. [8] Santilli RA et al **J Vet Cardiol** 2010;12:59. [9]Santilli RA, et al. **J Small Anim Pract** 2011;52:574. [10] Santilli RA, et al. **J Am Vet Med Assoc** 2007;231:293. [11] Santilli RA, et al. **J Vet Cardiol** 2013; 15:225. [12] Sosa E, et al. **J Cardiovasc Electrophysiol** 1996:7:531. [13] Santilli RA, et al. . **J Vet Cardiol** 2014 (in press). [14] Blomstrom-Lundqvist C, et al. **J Am Coll Cardiol** 2003;42:1493. [15] Soudi N, et al. **Eur Heart J** 2001;22:1162. [16] Perego M, et al. **J Vet Intern Med** 2012:26:320.

ECHOCARDIOGRAPHIC ASSESSMENT OF LEFT VENTRICULAR SYSTOLIC FUNCTION

John D. Bonagura, DVM, DACVIM
Columbus, OH

INTRODUCTION

Quantification of left ventricular (LV) systolic function is a fundamental component of the echocardiographic examination. Deterioration in systolic performance carries prognostic implications and very often leads to therapeutic recommendations even in asymptomatic patients. LV systolic dysfunction is most often observed with idiopathic dilated cardiomyopathy in dogs and cats, but can also develop from myocarditis, drug toxicosis, taurine/L-carnitine deficiency, relentless tachycardia, or as a "cardiomyopathy" of chronic volume or pressure overload. Conversely, heart failure in the setting of normal systolic function suggests cardiac disease dominated by diastolic dysfunction, such as hypertrophic cardiomyopathy, or severe volume overload with preserved systolic function, as might be encountered in chronic mitral regurgitation (MR) of small-breed dogs or in a compensated left-to-right shunt. The prognosis and treatment of these disorders can be quite different.

The ventricles can be assessed globally – as with the determination of LV shortening fraction, LV fractional area shortening, LV ejection fraction (EF) and using systolic time intervals. Regional or segmental assessments are also common in clinical research laboratories and often involve myocardial tissue velocities and myocardial deformation (strain). The LV also wrings and untwists during the cardiac cycle: these rotations can be captured by advanced echocardiographic methods that mostly involve tissue tracking of myocardial speckles.[1-18] Both the LV and the right ventricle (RV) can be evaluated with an array M-mode, 2D, 3D, Doppler-based, and tissue-focused examinations. Increasing emphasis has been placed recently on RV function (as evidenced by a number of abstracts and presentations at Forum this year) and a number of the methods used for assessing the LV also can be applied to the right ventricle.

This lecture offers an overview of the echocardiographic methods used to assess left (and potentially right) ventricular systolic function, considering both standard and investigational forms of ultrasound imaging. Some of the current issues and practical difficulties encountered when assessing ventricular systolic function will be highlighted during the presentation. The role of 3D (4D) imaging for assessment of LV function is addressed by others at this meeting. The difficult problem of assessing the LV in the setting of primary mitral regurgitation will also be discussed (and no doubt debated!).

OVERVIEW & COMMENTARY

Modeling of the LV is generally more straightforward when compared to the complicated anatomy of the right ventricle. Most clinicians quantify LV systolic function using the underlying assumption of the ventricle as a prolate ellipsoid with a one major and two equally-sized minor dimensions. Indeed this model is applied to most of the common calculations used in clinical practice. These include calculation of shortening fraction (across one minor axis between the ventricular septum and LV free wall); fractional area shortening (using multiple cross-sectional diameters at the high papillary muscle level); and methods that calculate volumes from long axis images of the LV using Simpson's rule or an area-length method with a single or bi-plane approach. Since the heart functions as a pump, it is also possible to assess the timing of systolic phases and to estimate the actual volume ejected by the ventricles during an average beat.

However, these methods may be insensitive to mild global dysfunction or to regional abnormalities of ventricular function. Detecting mild global dysfunction is one of our greatest difficulties; whereas, most experienced examiners can "eyeball" moderate to severe ventricular dysfunction. This lack of sensitivity has fostered the development of tissue Doppler based methods, 2D myocardial speckle strain imaging, and vector imaging for the assessment of regional and segmental function. These advanced methods evaluate a small portion of the LV (or RV) and may consider longitudinal, radial, circumferential, or rotational movements or deformation as well as subendocardial and subepicardial differences in contraction. The challenges in employing these new techniques include a lack of standardization in veterinary patients of varying size and heart rate; the need to use "black-box" algorithms developed for human patients; the lack of equally sensitive gold standards (e.g., invasive or MRI-based) needed for validation; and the near absence of longitudinal studies of spontaneous disease that prove these methods are actually superior to conventional Echo variables.

Echocardiographic studies are affected by biological variation, disease, drug therapy, and technical factors; these can significantly influence the results of an evaluation and the ultimate assessment of ventricular function. Some pertinent technical issues include obtaining consistent images; employing repeatable measurement techniques; addressing the timing of systolic and diastolic measurements within the cardiac cycle; obtaining optimal endocardial border recognition; appreciating the potential for translational or rotational movements across the imaging plane; and understanding the issue of tethered regional motion (as opposed to true contractile activity or deformation).

The influence of cardiac rhythm on ventricular function cannot be underestimated. Abnormal heart rhythms (premature complexes and post-extrasystolic potentiation), varying cycle lengths, conduction disturbances (bundle branch block), and ventricular dyssynchrony can affect systolic function, sometimes underestimating actual performance. Protracted tachyarrhythmias can create transient LV systolic dysfunction that may persist for some time, even with reestablishment of sinus rhythm. Changes in ventricular preload and afterload also influence estimates of ventricular function, which may seemingly improve (or deteriorate) as a consequence of altered loading. Drug intervention may also affect heart function, by directly influencing contractility, altering heart rate, or changing loading conditions.

Inter-observer issues and day-to-day repeatability are still important concerns for many Echo measurements. Considering that day to day variation of a single examiner often approaches 10% (or more) for many Echo and Doppler variables, it may be very difficult to identify subtle changes in ventricular function within a single patient.

Additional problems occur when assessing systolic ventricular function. Methods that are touted during a laboratory study of ventricular function may be of little use when applied to a clinical population where the population variance is much greater and the important outcomes are clinically relevant, not statistically-significant, differences in cardiac function. For example, ambiguous findings of LV function are identified very often in the practice setting. These may be dismissed as insignificant when there is no clinical context to explain the abnormality, or supplemented with additional functional estimates and planned follow up examinations at suitable intervals (which is usually the most practical approach). It is quite common to perform 8 or 10 different estimates of LV function in one dog only to find that half of these variables point in one direction and the other half the opposite way. Owing to species and breed differences, as well as different methods used by examiners when measuring cardiac function, it can be difficult to apply rigid benchmarks and cutoffs for "normal" systolic function. These problems are especially common in "screening clinics", where subtleties are sought and are too often found. Decisions in these situations are very often based on multiple factors that extend beyond the echocardiographic findings (and this is probably the most prudent approach). However, when dealing with patients (as opposed to breeding stock), the importance of an "abnormal" designation cannot be underestimated, because a well-defined decrease in systolic function will prompt additional investigations, more frequent follow-up examinations, and life-long therapy. Of importance is the need for specialty working groups to standardize the interpretation of echocardiographic findings that define likelihoods for having or developing systolic dysfunction.

SPECIFIC FUNCTION INDICES

Most commonly used M-mode, 2D, and real-time 3D estimates of LV systolic function provide an estimate of LV ejection fraction (EF) and are therefore considered ejection phase indices. These are grounded in linear, area, or volumetric estimates of EF or stroke volume. Some examiners prefer to estimate global LV function based on a single diameter end-systolic diameter (or calculated end-systolic volume) index, but the author believes this index (while theoretically sound) is confounded by chamber remodeling, limitations of volume calculations based on minor dimension diameters, and the use of volumes extrapolated from human angiographic studies. Global function also can also be quantified with time intervals, such as pre-ejection period and ejection time, as well as hybrid estimates of global function like the Tei index of myocardial performance. These are calculated from M-mode, spectral Doppler, or tissue Doppler modalities. However, there are considerable overlaps between normals and affected dogs and a single point calculation can be ambiguous or misleading. In particular, the value of systolic time indices in valvular heart disease is very limited and in the author's experience, can be unduly influenced by the duration of the QRS complex and the heart rate even in dogs with suspected DCM. Additional Doppler approaches include quantifying stroke volume and cardiac output by measurement of velocity time integrals. This approach is of greatest utility when tracking trends in the intensive care setting or when applied to monitor anesthesia. Stroke volume estimates are typically insensitive for detecting mild LV or RV dysfunction.

Regional estimates of LV function are gaining more attention and include measuring tissue velocities or deformation with tissue Doppler imaging (TDI); calculating myocardial velocity gradients using color Doppler imaging (and proprietary analysis algorithms); and image processing that yields values for deformation of cardiac tissue (or strain) using 2D speckle-tracking technologies. These can be assessed regionally or combined automatically (or manually averaged) to obtain a "global" index of longitudinal, radial, or circumferential strain/strain rate. This same technology can be used for quantitation of cardiac rotation and to calculate ventricular torsion. Frequently tissue or regional assessments combine both systolic and diastolic function indices.

M-mode Estimates

One of the oldest methods for assessing LV systolic function is still among the most popular, namely, the linear estimate of shortening fraction (SF), typically measured at the high papillary muscle or chordal level. This is a minor axis approximation of global function. When the SF is recorded within a healthy reference range for species and breed, LV systolic function is generally considered "normal" and most examiners will be satisfied with the result. However, there are situations when global or regional myocardial function may be quite abnormal in the setting of normal SF. Furthermore, reference values for healthy dogs are not straightforward, and there is no single cut-off value that has been proven to have optimal sensitivity/specificity for predicting advancing heart failure across canine breeds. Many healthy dogs have SF values between 20-25% (and even lower), a factor that may be related to relatively greater shortening in other planes, especially in the apical-basilar axis. In breeds that are predisposed to dilated cardiomyopathy, the problem becomes still more difficult when the septum and free wall contract dyssynchronously due to regional wall abnormalities or conduction issues. This offset in timing has no formal adjustment in practice, aside from the eye of a practice examiner. Even in dogs with early dilated cardiomyopathy (DCM), marked dyssynchrony of the LV can produce a very low SF, while other estimates such as shortening area might show only mild to moderate LV dysfunction.

Other potential problems with SF (as well as with shortening area and EF) are the masking of myocardial failure by increased preload, decreased afterload, or compensatory LV hypertrophy. These are commonly observed in smaller breed dogs with severe mitral regurgitation where shortening fraction is invariably greater than normal (though it may be subnormal in some larger-breed dogs with chronic MR). In dogs with severe aortic stenosis and concentric hypertrophy the values for SF can be normal even when myocardial failure develops after the "supernormal" compensatory phase of hypertrophy. This phenomenon has been observed in both spontaneous and laboratory-induced cases of canine LV outflow tract obstruction. Thus, while the SF is efficient, the clinician must appreciate the limitations, particularly in healthy dogs. The M-mode situation in cats seems less confusing at least in terms of global function estimates. Most normal cats exceed a SF of 35%, though some sedatives can markedly depress these values. Horses are typically >28%, but this also can be affected by stress or sedation or dyssynchrony.

Other reported M-mode indices include percentage wall thickening, amplitude of wall excursion, E-point to septal separation (EPSS), mitral annular motion, and calculation of EF. The normal range for EPSS varies with breed and size, but this value increases in LV failure (in the absence of severe mitral regurgitation). False positives can arise in the setting of aortic regurgitation or mitral stenosis.

Mitral annular motion tracks longitudinal contraction and is recorded by an M-mode study or tissue-tracking techniques. This is one of the few M-mode methods considering apical-basilar shortening, but it is not widely measured. This can also be measured using "tissue tracking" methods available in advanced Echo systems. Subjectively watching the systolic excursion of the LV free wall and the systolic basilar to apical movement of the mitral annulus can provide insight regarding LV systolic function and are especially useful when septal motion is flat or dyssynchronous.

Most echo reporting systems automatically use M-mode study data to generate volume-estimates by diameter-cubed or Teicholz formulae; these numbers are used to calculate EF and end-systolic volume (index). There are inherent limitations to these minor axis methods, and the volumetric results may simply multiply the linear errors. While end-systolic volume index is a theoretically sound approach to identifying LV dysfunction, in the author's opinion, there are too many limitations of this approach as used currently in veterinary medicine to just accept an "increased" ESVI as proof of LV systolic dysfunction. Specific concerns include the use of a minor dimension for estimating LV volume in the setting of ventricular remodeling that also increases LV systolic dimension (regardless of functional changes); the reference standards used for any given breed; and the non-validated benchmarks used to classify various grades of dysfunction (taken from human angiographic studies). Considering most indices of global LV systolic function are hyperdynamic in the typical small breed dog with MR, and those few indices that have suggested impaired LV systolic failure have not yet been validated against a gold standard (such as MRI), one might ask if we can even assess LV systolic function in this disease using current methods? The author is simply unsure. Perhaps validated 3D methods will address this more definitively.

Two-dimensional Estimates

The 2D echocardiogram can be used to estimate global and regional myocardial function. One of the more practical area-based measurements of systolic function is the fractional area shortening of the LV. This measurement is derived from the 2DE short axis image of the left ventricle, and is calculated (like SF) as the percentage reduction in left ventricular cross-sectional area between diastole and systole (with areas traced across the bases of the papillary muscles). It has advantages over SF in that it represents wall motion across the entire short axis area. Therefore, it is less sensitive to asynchronous wall motion. However, it does not address shortening in the apical-basilar direction, and off-angle images may create inconsistent results. Values exceeding 45% appear to represent normal contraction across the minor axis plane. The gold standard for estimation of global LV function in human laboratories is calculation of LV EF using a bi-plane 2D (or 3-D reconstruction) method. Many different formulae have been used for calculation of LV volumes, but calculations based on a modified Simpson's rule seem least affected by abnormal chamber geometry. Unfortunately, left apical tomograms can yield truncated views of the LV in dogs and cats. Right parasternal long-axis views provide better length, and have been applied for single plane, EF estimation in dogs using a modified Simpson's rule or length-area method of volume estimation. Normal values in dogs for a single plane EF are typically >45 to 48%. A major concern relates the capture of true apical to basilar motion as opposed to translational movement leading to foreshortening of LV length.

Regional and global estimates of LV function also can be obtained with 2D echo, using "speckle" imaging modalities. Unique signatures of 2D images are used to calculate regional myocardial deformation or strain. This approach may have some benefits--including less angle dependency--when compared to tissue Doppler methods of strain estimation.

Intervals of Systole

Measurements of isovolumetric periods, pre-ejection intervals, and ejection times have been used for assessment of systolic or global ventricular function. The pre-ejection period and the ejection period can be recorded using either M-mode or spectral Doppler methods. Whilst systolic time intervals are influenced by loading conditions, they provide an alternative means of assessing global systolic function, provided the QRS duration is not prolonged or heart rate extreme. The ratio of PEP/LVET may be less heart rate dependent, and as myocardial dysfunction deteriorates the PEP to LVET ratio increases. Normal values from our laboratory in healthy dogs of varying sizes measured by PW Doppler have identified values of LV PEP/ET between 0.24-0.42, with small dogs generally <0.3 and large dogs generally < 0.42. For the RV PEP/ET values of 0.23 to 0.34 have been reported. Other laboratories have reported different values, demonstrating again the importance of standardizing methods or better defining reference populations. Another problem is a prolonged QRS duration. This is a major limitation in applying systolic time intervals for assessment of LV function. A less often-used function variable is the mean velocity of circumferential fiber shortening. This is an extension of SF, calculated by dividing the shortening fraction by the LV ejection time, and subject to many of the same limitations as SF. Finally, a combined index of ventricular systolic and diastolic myocardial performance has been reported by Tei using PW-Doppler {IMP = (ICT + IRT) /ET; where ICT and IRT are isovolumetric contraction and relaxation times and ET is the ejection time}. In one study of Newfoundland dogs, IMP was 0.28 +/- 0.12 in the normal dogs, and higher in dogs with DCM. We have found generally greater values for IMP in healthy dogs of different breeds, with a wide range of 0.32 to 0.50 for the left ventricle and with lower values for the RV (in every dog).

Doppler Methods for Assessment of LV Function

There are additional Doppler-derived variables used to quantify global and regional ventricular function. For example, Doppler estimates of ventricular stroke volume or cardiac output represent indices of global ventricular function. These calculations are based on the area under the aortic or pulmonic flow signal--the velocity-time integral or stroke distance--which correlate to stroke volume. The actual stroke volume (in cc) can be calculated by multiplying the stroke distance (VTI) by the calculated cross-sectional area of the aorta or pulmonary artery. Measurements of stroke volume index and cardiac index have been related favorably to invasive indicator dilution techniques.

Doppler correlations also have been demonstrated between traditional invasive indices of left ventricular function, such as the rate of LV pressure rise (LV dP/dt). LV dp/dt can be estimated non-invasively from a mitral regurgitation jet by recording the time interval in seconds between one and 3 meters per second, and dividing the value into 32 mm Hg. This is an "averaged" dp/dt over that period and is generally lower than peak dp/dt measured with a high-fidelity manometer catheter. Peak aortic velocity has been found to be abnormal in dogs with dilated cardiomyopathy, but generally, myocardial dysfunction must be very severe for peak flow velocity to be affected. Instantaneous peak aortic acceleration (dv/dt max) relates to inotropy, but is difficult to calculate (as only averaged acceleration is typically measured using conventional Doppler systems).

TDI records velocities of myocardial or annular segments in longitudinal (apical-basilar) or radial (short axis) planes. A variety of proprietary methods are used to generate these data including color M-mode imaging, color 2D imaging with velocity signals averaged from regions of interest, and single pulse wave Doppler methods. The values obtained from a specified segment are not always equivalent. The same principles apply for TDI as for conventional Doppler, including the importance of sample volume placement and alignment with movement. TDI or TDI velocity gradients have been used to assess LV systolic function by comparing epicardium to endocardium. These techniques are quite sensitive, and when standardized in an echo laboratory, can be used to identify early signs of ventricular dysfunction. Apical to basilar TDI velocities in healthy dogs are typically between 12 to 20 cm/sec near the annulus. At least for the LV, a decreasing peak tissue velocity is evident as the sample is moved from the base to the apex. Unfortunately, the author has been repeatedly disappointed by longitudinal TDI velocities, which are often relatively normal (12 to 20 cm/s) even in the setting of reduced EF.

Myocardial Strain & Twist

Calculation of regional myocardial deformation – strain and strain rate – is a recent application of TDI (as well as 2D speckle imaging). Longitudinal strain is calculated as a negative value because segments shorten during systole, while radial strain is positive (as the wall thickens). Circumferential strain and segmental rotation and ventricular twist are also available from a number of vendors.[19-21] Strain by this method is purported to overcome the TDI limitation of tethering. Values for strain obtained by TDI and speckle technology are similar, but standardization of both the methods and reference values is needed before widespread veterinary application can be recommended. Rotation of the heart at the base and apex can be done to allow estimation of ventricular twisting or torsion, yet another index of LV function. It should be noted that the values obtained are related to non-simultaneous peak twisting.

Stress Testing

Assessment of LV systolic function might be better assessed during standardized stress of exercise or perhaps stimulation with drugs such as dobutamine.[20] More data are needed with respect to methods and to safety of these approaches.

REFERENCES

[1]Sagawa K, Circ 1981; 63:1223. [2]Ross J. J Am Coll Cardiol 1985; 5:811. [3]Kittleson MD, et al. J Am Vet Med Assoc 1984;184. [4]Borgarelli M, et al. J Vet Intern Med 2007;21:61. [5]Bonagura JD et al. J Small Anim Pract 2009;50:12. [6]Spinale FG, et al. J Thorac Surg 1993;106:1147. [7]Urabe Y, et al. Circ Res 1992;70:131. [8]Edvardsen T, et al. Prog Cardiovasc Dis 2006;49:207. [9]Mazur W, et al. Curr Opin Cardiol 2001;16:246. [10]Young AA, et al. Am J Physiol 1996;271:H2689. [11]Takano H, et al. Am J Vet Res 2010;71:421. [12]Chetboul V, et al. J Vet Intern Med 2008;22:342. [13]Griffiths LG, et al. J Vet Cardiol 2011;13:115. [14]Culwell NM, et al. Am J Vet Res 2011;72:650. [15]Chetboul V, et al. J Vet Intern Med 2007;21:719. [16]Tidholm A, et al. J Vet Intern Med 2009;23:1197. [17]Smith DN, et al. J Vet Cardiol 2012, doi:10.1016/j.jvc.2011.11.002. [18]Chandra S, et al. Heart Failure Rev 2011;16:327. [19]Chetboul V, et al. J Vet Intern Med 2008; 22:342. [20]Suzuki R, et al. J Vet Intern Med 2014; [Epub ahead of print]. [21]Zois N, et al. J Vet Intern Med 2013; 27:875.

CARDIAC HEMODYNAMICS REVIEW

Mark A. Oyama, DVM, DACVIM (Cardiology)
Philadelphia PA

Various hemodynamic parameters can be calculated from data gathered during cardiac catheterization and/or echocardiography, including cardiac output and valve orifice area. These calculations yield a deeper assessment of cardiac anatomy and performance than do more simple parameters such as pressure gradient. This presentation will review several different cardiac calculations and their applicability to an example of cardiac disease.

Calculation of cardiac output using angiography or echocardiography

Cardiac output (CO) is defined as the volume of blood volume pumped by the heart over time and is commonly represented by the formula:

Cardiac output (L/min) = Stroke volume (L) x heart rate (bpm)

CO in a healthy adult human is ~5L/min. In dogs, CO varies tremendously with body size; however CO can be indexed to body surface area as cardiac index (CI). CI in dogs has been reported as 3.9 +/-0.4L/min/m^2 (Kittleson, 1998). Stroke volume can be calculated using angiography or echocardiography. The volume of the left ventricular cavity during end-diastole and end-systole is estimated using dimensions of ventricular length, width, and/or area. The most common volume formulas that have been validated in the dog include the prolate ellipse methods, cylinder-cone methods, cylinder-hemiellipsoid methods, and the modified Simpson's rule (method of discs) (Weyman, 1994). Each of these methods calculates the ventricular volume by assuming that the ventricular cavity can be defined by a particular combination of geometric shapes whose dimensions can be measured using angiography or echocardiography.

The simplest method is to assume that the ventricle is ellipsoid in nature and that its longest radius is twice the length of the two short axis radii (rz= 2 * rx).

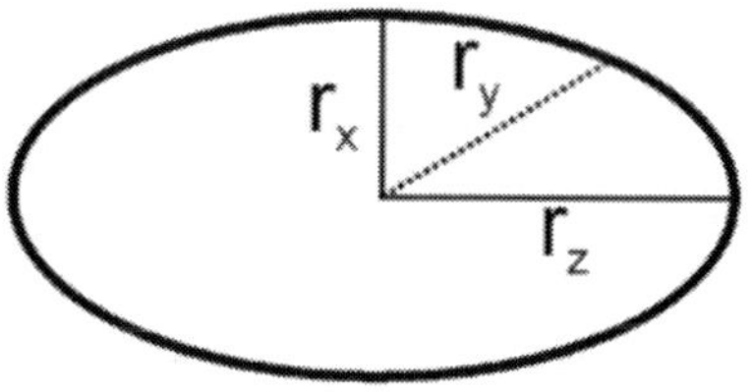

Volume of sphere

$$V=4/3 * \pi * r^3$$

Volume of prolate ellipse

$$V=4/3 * \pi * r_x * r_y * r_z$$

$$V=4/3 * \pi * \frac{d_x}{2} * \frac{d_x}{2} * \frac{2d_x}{2}$$

$$V=d_x^3$$

Thus, in a case of the prolate ellipsoid, $V=D_{shortaxis}^3$. Obviously, this is a crude approximation of the LV cavity shape but is remarkably easy to calculate. For greater accuracy, the three main axes of the ventricle can be independently measured from biplane angiography or from the left apical 4 and 2 chamber echocardiographic views.

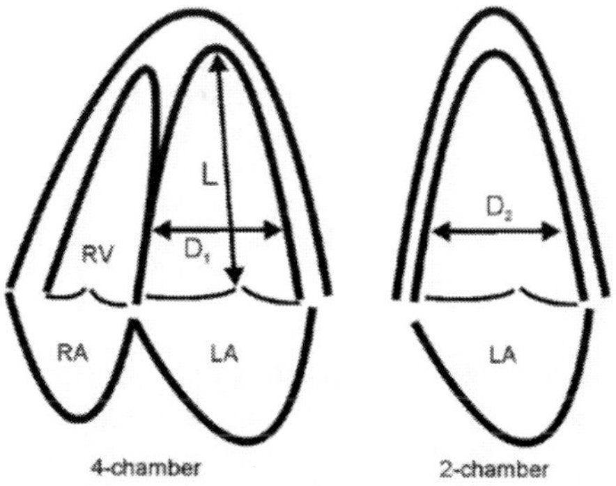

Volume of prolate ellipse

$$V=4/3 * \pi * \frac{L}{2} * \frac{D_1}{2} * \frac{D_2}{2}$$

$$V=\frac{\pi}{6} * L * D_1 * D_2$$

Going further, the shape of the ventricle can be approximated as a cylinder with either a hemiellipse or cone at one end. These formulas utilize the 2-dimensional short axis area of the LV as traced by planimetry and the LV length.

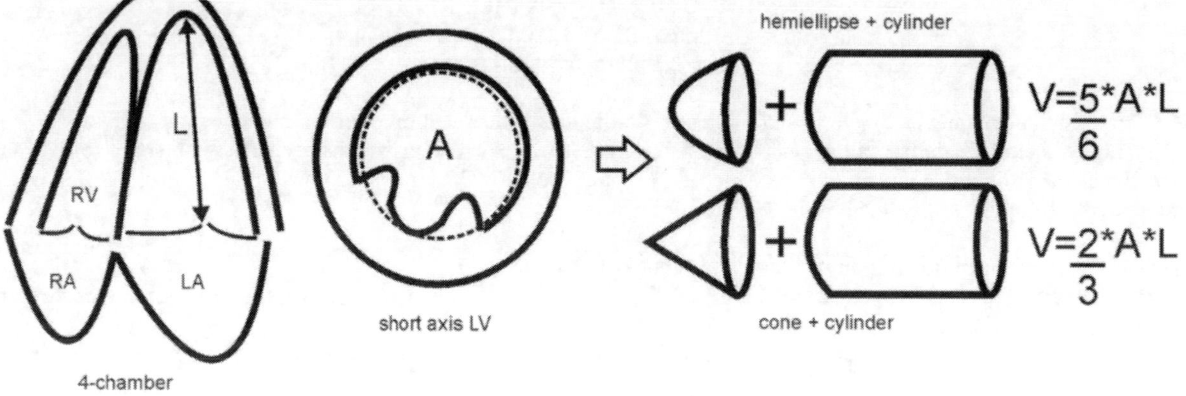

These volume calculations are performed at end-diastole and end-systole so that stroke volume can be then calculated. These techniques calculate total stroke volume of the LV and if significant mitral regurgitation is present, the forward stroke volume will be less than the total.

Calculation of valve orifice area using the Gorlin equation

Calculation of CO allows for calculation of valve orifice area. This calculation can be used to calculate the orifice area of stenotic pulmonic valves or the stenotic area of a subaortic stenosis lesion. The area of an orifice is proportional to the flow rate across the area divided by the mean pressure gradient that drives that flow. The flow rate is not equivalent to the CO but instead is the instantaneous volume of blood that moves across the orifice during the appropriate portin of the cardiac cycle (mls/s). The flow rate is the cardiac output divided by the product of the heart rate and the time during each heart beat that the flow is occurring across the valve orifice. For pulmonary artery and aortic flow, the time spent is the systolic ejection period (SEP) and for the mitral and tricuspid annulus the time spent is during the diastolic filling period (DFP)

Flow rate (mls/s) = CO (mls/min)/([SEP or DFP (sec/beat)] * HR (beats/min))

The *mean* pressure gradient (not the maximal PG) drives flow across the orifice during the entire duration of systole or diastole and is used in the denominator. The gravitational and fluid density components of the formula are captured by use of a constant of 44.3 in the denominator. In conditions of weightlessness (ie., zero gravity), this constant would not be needed. When Gorlin and Gorlin (a father and son team) compared their original calculated orifice areas to the actual areas as measured at necropsy, they found that the formula needed to be slightly modified to best approximate the necropsy gold standard. If the aortic, pulmonic, or tricuspid valves are being measured, the denominator remains unchanged by multiplying by 1, however, if the mitral valve is being measured, the denominator is multiplied by a mitral-specific constant of 0.85.

$$A_{\substack{aortic\\pulmonic}} = \frac{CO/(SEP * HR)}{1*44.3*\sqrt{PG_{mean}}} \qquad A_{mitral} = \frac{CO/(DFP * HR)}{0.85*44.3*\sqrt{PG_{mean}}}$$

A=cm^3
CO=mLs/min
HR=bpm
SEP or SFP=sec/beat
PG=mmHg

Practical use of the Gorlin formula

The success of pulmonic balloon valvuloplasty is typically assessed by measuring the pre- and post-procedure PG across the stenotic valve orifice. This can be done by direct pressure measurement using fluid-filled catheters or by Doppler echocardiography and the modified Bernoulli equation. The PG across the pulmonic orifice is used as a surrogate measure of severity, however PGs not only are affected by the stenotic orifice area (resistance) but also by the amount of flow through the orifice, as can be seen in Ohm's law (ΔP=QR). Thus, use of PGs to estimate orifice area assumes that the flow through the orifice is the same during each measurement, which might or might not be the case. For instance, the PV PG is commonly lower during anesthesia vs. the baseline measurement in the awake dog. After balloon procedure, it is reasonable to assume that flow through the stenosis is improved and the PG might not be an accurate reflection of the residual severity. Measurement of cardiac output and the resultant flow across the valve orifice allows the actual stenotic area to be calculated and this might be a better measure of disease severity and interventional success. Calculation of the valve orifice area requires calculation of cardiac output by either the aforementioned methods or by thermodilution.

In this example, CO is calculated by angiographic methods and the mean pressure gradient is calculated by either Doppler or by direct pressure tracing.

Pre-balloon

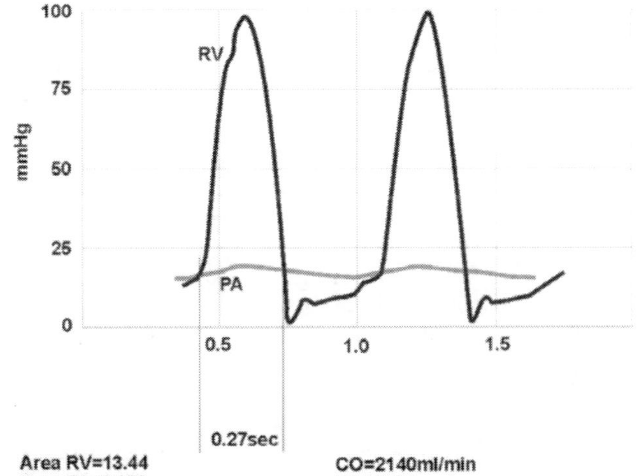

Area RV=13.44
Area PA=0.283
Area Difference=13.16
Mean Pressure Difference=13.16/0.27=48.73mmHg

CO=2140ml/min
HR=98bpm

PVA=[2140/(98*0.27)]/[44.3*SQR48]
PVA (cm^2)=0.264

Calculated PV radius=2.9mm

Post-balloon

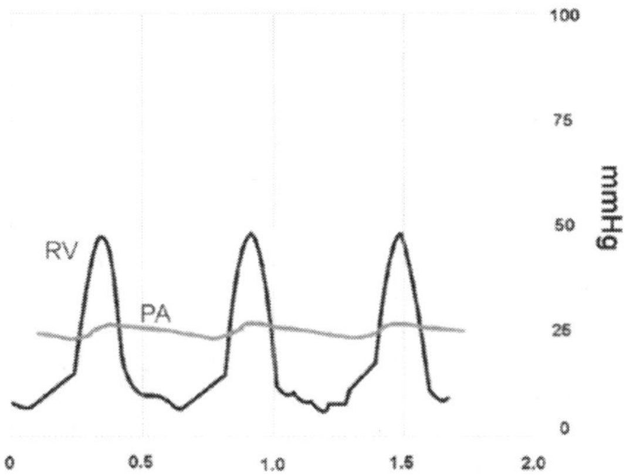

SEP=0.130sec
Area PA=0.138
Area RV=1.914
Area difference=1.776
Mean pressure difference=1.776/0.130=13.662

CO=2.38l/min
HR=114bpm

PVA=[2380/(0.130*114)]/(44.3*SQRT13.662)
PVA=0.981cm^2

Calculated PV orifice radius=5.6mm

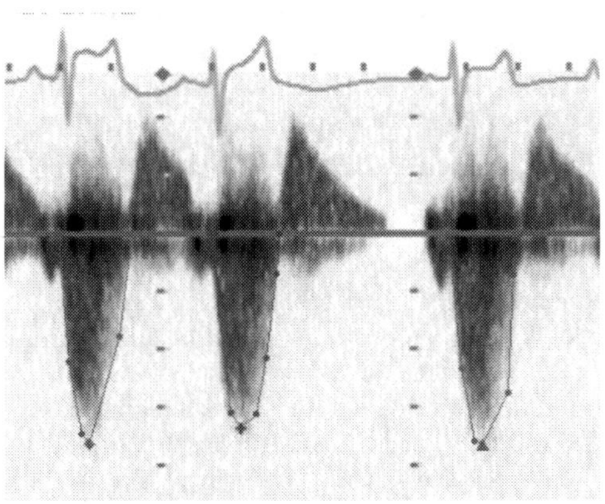

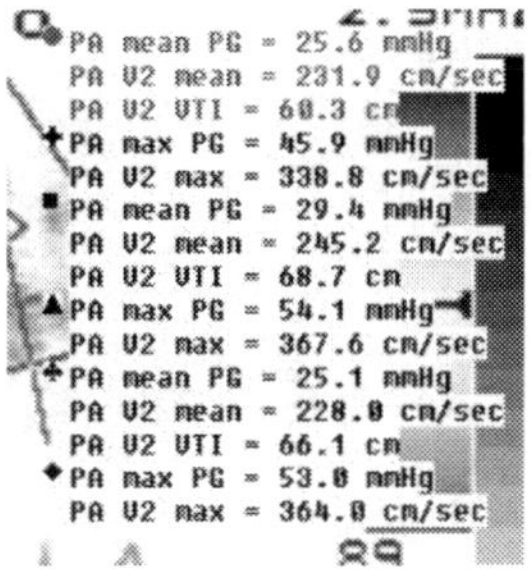

Doppler calculation of mean pressure gradient.

REFERENCES

Kittleson MD, Kienle RD. Small Animal Cardiovascular Medicine. Mosby 1998. Weyman AE. Principles and Practice of Echocardiography, 2nd ed. Williams and Wilkins, 1994.

CLINICAL TRIAL DESIGN

Nick Jeffery, BVSc PhD MSc DECVN DECVS
Ames, IA

INTRODUCTION

Clinical trials can be considered as experiments, constructed in order to answer a clinical question (analogous to a hypothesis in experimental studies). Clinical trials in human medicine are probably the most rigorously designed experiments in all fields of biology because of their stringent regulation and the considerable financial investments they represent. As a consequence the data that is derived from the most common type – the randomized controlled clinical trial (a.k.a. RCT) – are regarded as the highest form of evidence-based medicine. Although formal clinical trials are still not common in veterinary medicine, much of the methodology that underlies study design in clinical trials can be applied to observational trials and can therefore be used to improve reliability of data collection and analysis.

Clinical trials are designed with the aim of eliminating all sources of **bias**. Bias in this instance is meant to mean any deviation from the 'truth' (*i.e.* real effect of the intervention). This does not mean that all bias can be eliminated, because the play of chance can lead to imbalance between trial arms that may have an effect on the comparison of interest. However, even this possible source of bias can be ameliorated to a large extent by specific design features, as shown below.

Although there are many possible design structures that can be used for clinical trials, such as crossover, factorial, cluster, adaptive and non-inferiority types, here the focus will be on the two-arm parallel trial, but the features which minimize bias are common to all designs. Comparisons with observational trials highlight the limitation of bias achieved in RCTs. Guidelines for design and execution of clinical trials are available as the frequently updated CONSORT document[1].

SELECTION OF CASES

An obvious source of bias in observational studies is that there is an inherent difference between treatment groups in variables that might influence outcome. For instance, cases of disc herniation treated by surgery are likely to be more severe than those treated by conservative methods, because a selection process has been superimposed. In RCTs the patients are selected to receive the (two) treatments at **random**. This process alone can eliminate much of the bias associated with simple observational studies. However, there can be a great deal of resistance to the process of randomization – because many clinicians have strong prejudices about the treatments they give. The principle of ***equipoise*** is used to determine whether it would be ethical to allow randomized treatment of a patient population. For this reason a clinical trial on the efficacy of antibiotic therapy for acute pneumonia could not be conceived.

However, randomization alone does not suffice to minimize bias. It is also necessary to use **allocation concealment**. This is necessary to prevent clinicians from selectively denying entry into a trial to patients based on how well they think they may do if it could be known to which treatment arm they will be allocated. Allocation concealment means that after checking that a patient meets the inclusion criteria (and does not meet the exclusion criteria) and upon receipt of informed consent the patient is 'locked into' the treatment arm to which they are randomly allocated. The necessary process implies that the order in which patients are allocated to treatments must be truly random, meaning that devices such as relying on hospital number, or days of the week are not appropriate (because the treatment arm of allocation can be predicted for an individual patient).

SIZE OF TRIALS

An important component of reliable RCTs is their size. Although a small study may detect a 'significant' (*i.e.* $p<0.05$) difference in outcome between groups this effect may not be something that can be relied upon. Small studies are more likely to give rise to extreme (*i.e.* further from the null result) outcomes, simply because of the play of chance. Also, in a small trial, patients with unknown confounding variables might be randomized, ***purely through chance effects***, into one group. This then would lead to differences that would be inappropriately ascribed to the treatment. A secondary problem that can then arise is that, because of the prejudice of authors, reviewers and journal editors, 'positive findings' are more likely to be published. This then leads to **'publication bias'** in which the weight of published evidence can misleadingly suggest that a treatment is beneficial, whereas the truth might be that there have been many studies carried out that did not find the same result but were denied publication (because the results are not considered interesting enough).

Any clinical trials can be set up so that the known sources of bias can be controlled (through judicious use of inclusion and exclusion criteria), but a large trial also means that any ***unknown*** sources of bias can be eliminated, because any variable that might be expected to confound the outcome will, in a large group, be distributed at roughly equal proportions between the two groups. This is one reason why human clinical trials often recruit as many as tens of thousands of participants.

The size of a trial must be determined through power calculations, which can define the **power** of the study to detect a difference between groups, were it to exist. These calculations are simple to perform, but do require estimates of likely proportions or means and standard deviations of outcome measures. These can usually be derived from previous clinical or laboratory studies.

MEASURING OUTCOME

It should go without saying that the observer who measures the outcome measure of interest should be **blinded** to the patient treatment group. This is the only method to ensure that there cannot be bias in the measurements. For instance, it might erroneously be assumed that (say) force plate measurement of gait variables, cannot be influenced, whether consciously or unconsciously, by the observer because its outcomes are perceived to be objective, because they are numerical and derived from a machine. However there are still many possibilities for observer influence, such as repeated trials, rounding errors etc. The only means to reliably eliminate this potential for bias is to blind the observer. There are occasions when it can be very difficult to ensure observer blinding to some aspects of outcome (*e.g.* comparisons between surgery and non-surgical treatments) but often methods can be devised to overcome these problems (even, in human surgery, to the extent of allowing 'sham' surgery).

Measurement of outcome should always be carried out with the same instruments – which may include the same person to make the evaluations. The fact that a single observer may have a tendency to over- or under-measure a particular variable (*e.g.* on radiographs etc), so long as they are blinded to treatment allocation they will provide better data than two unblinded observers averaging (or agreeing with each other) their findings.

ANALYSIS OF OUTCOME

Reputable human clinical trials are now registered before they take place, and many trial protocols are published ***prior*** to the trial commencement. One reason for this is to ensure that the reported outcomes are those that were **pre-specified**. *Post hoc* analysis of data frequently provides highly spurious results (there are many examples of spoof analysis done to 'show' that patients that were born on specific days of the week are more likely to have poor outcomes after surgery). Therefore all analyses should be pre-specified. There is also a need to limit the number of analyses that will be performed. Multiple statistical comparisons testing will result in apparently 'significant' effects when the number of tests increases – *e.g.* with ten tests, using $p<0.05$ on each, the chances of finding a 'significant' finding is ~0.4. As an aside it is also worth considering the p values are not the be all and end all of analysis – it is the magnitude of effect that is of more interest and whether it has clinical importance.

INTERPRETATION OF OUTCOMES

There is sometimes a tendency to assume that clinical trials must be designed to test the evidence in favor of previously theorized mechanisms of effect, similar to experiments that are carried out in laboratories. However, there are two different strands of clinical trial purpose: **'explanatory'** and **'pragmatic'** and each of these serves a different purpose. Explanatory trials aim to give evidence regarding the mechanism of treatment and whether the treatment works under 'ideal conditions'. This then provides an answer about whether the therapy 'can work' if everything is in its favor. On the other hand, pragmatic trials aim to determine whether the therapy 'will work' in general clinical practice.

These two strands of clinical trial design have implication with regard to the way in which the trial will be constructed. For instance, in an explanatory trial the inclusion criteria may be very strict – *e.g.* biopsy proven brain tumor, a drug dosing schedule may mean that only certain categories of patient will be able to comply (*e.g.* because of toxicity) and the outcome measure may pertain directly to the disease process itself (*e.g.* tumor shrinkage).

In contrast, a pragmatic trial may have very loose inclusion criteria (*e.g.* MRI diagnosed 'brain lump'), the dosing schedule will allow any patient to take it (and the analysis will even include those that do not comply with the dosing protocol) and the outcome measure may be directly relevant to patients (*e.g.* survival).

In veterinary medicine there is generally a perception that explanatory trials are the only ones worth pursuing, but pragmatic trials often have substantial impact in driving clinical practice forwards. A good example of a pragmatic clinical trial is 'CRASH-2'[2], which included almost any adult patient with severe hemorrhage in which there was no contra-indication for the treatment.

REFERENCES

[1] CONSORT - http://www.consort-statement.org/ . 2. CRASH-2 trial collaborators. **Lancet** 2010; 376: 23.

BLADDER FUNCTION AND URINARY TRACT INFECTIONS IN SPINAL CORD INJURED DOGS

Natasha Olby Vet MB, PhD, DACVIM (Neurology)
Raleigh, NC

INTRODUCTION

Micturition is a complex neurologic process that combines somatic and autonomic reflexes to achieve sphincter relaxation in coordination with bladder contraction. The neuroanatomical involvement of the nervous system is widespread including cortical decision-making, the pontine micturition center and cerebellar input, white matter tracts of the spinal cord, and autonomic and somatic peripheral nerves[1]. Of these structures, disruption of spinal cord white matter is the most common cause of failure of normal urination, and urine retention and overflow incontinence are frequent in our spinal cord injured patients. The consequences of this for our patients are well recognized, the most common being a predisposition to urinary tract infections (UTIs), but also including detrusor atony, urine scald, secondary hydroureter, hydronephrosis and pyelonephritis. In chronic paraplegics, recurrent, antibiotic resistant UTIs are a major concern and they a leading cause of morbidity and shortened lifespan in humans[2-4]. This presentation will discuss the current understanding of bladder function following spinal cord injury in our patients, and describe potential strategies to effectively treat and reduce the incidence of UTIs.

RECOVERY OF MICTURITION FOLLOWING THORACOLUMBAR SPINAL CORD INJURY

There is still relatively little objective data on the recovery of micturition and full urinary continence in dogs following spinal cord injury. As a general rule paraplegic animals are unable to void following an acute injury, while dogs with some motor function, even if it is simply a voluntary tail wag are able to void. Recovery following injury tends to follow the same course, but can deviate following more severe injuries, likely as a result of a central cord syndrome. This is reflected in the fact that 32% of dogs that recovered nociception and motor function following acute disc herniations causing grade 5 injuries (paraplegia and loss of nociception) did not regain full urinary continence, even though they were able to void[5]. Closer study of the ability of dogs that suffer disc herniations to void urine has been undertaken with ultrasound (US) measurement of bladder volume pre and post voiding[6]. This study demonstrated that recovery of effective voiding occurs rapidly over the first 4 days after surgery in dogs with grades 2 and 3 injury severity, but is more prolonged and variable in dogs with grade 4 injury (paraplegia with nociception) and incomplete 3 weeks after surgery in grade 5 injuries. A subset of dogs with grade 5 injuries at the time of presentation fail to recover nociception but do recover remarkable ability to walk and wag their tails voluntarily, suggesting that they do have limited connections across the site of their injury. However, none of these dogs regain continence[5]. Many will develop reflex micturition that is extremely ineffective. Indeed, for many owners, management of UTIs and incontinence is one of the main challenges to quality of life for them and their pets[7].

URINARY TRACT INFECTIONS FOLLOWING SPINAL CORD INJURY

There are numerous risk factors for the development of UTIs, including gender (females dogs are more prone), certain systemic diseases (eg. diabetes mellitus and hyperadrenocorticism) and catheterization[8]. Myelopathy has been shown to be one of the most important causes of recurrent UTIs[8], and two studies in dogs with thoracolumbar intervertebral disc herniations described a prevalence of UTI of 21 and 27% in the 7 days after surgery[9,10]. Risk factors identified in these studies included intraoperative temperatures of <35^{0}C, duration of paraplegia, and the use of intraoperative antibiotics. One of the studies examined the method of bladder evacuation, comparing manual expression with intermittent and indwelling catheterization and did not show any difference between these techniques[10]. This was somewhat surprising given the known association between catheterization and UTIs in humans with chronic paralysis. However, it has been shown that the risk of UTI increases in dogs by 27% with each day of in-dwelling catheterization beyond the first three days[11], and so limiting the placement of indwelling catheters to 3 days or less minimizes the risk with this technique. We completed a prospective study in non-ambulatory dogs with acute thoracolumbar disc herniations, in which we looked at the prevalence of UTIs over 12 weeks post operatively[12]. In this study we found that 32% of dogs developed a UTI over the 12 week period but that the majority of UTIs occurred in weeks 1 – 6, demonstrating that this population of dogs is at risk for a longer period than anticipated. In chronically paralyzed dogs (paralyzed for more than 6 months) entering clinical trials at NCSU, all dogs have intermittent UTIs and approximately 80% of female dogs have recurrent of UTIs that generate increasing problems with antibiotic resistance.

PREVENTION STRATEGIES

In order for a urinary tract infection to establish, it is necessary for the bacteria to adhere to and colonize the bladder wall, forming a biofilm. Prevention strategies therefore focus initially on regular bladder emptying to lessen the chance of colonization. Clients have to be educated on bladder expression techniques, and the use of models to assist them in learning how much pressure to apply, and how to feel for the bladder can be extremely helpful. Drugs are frequently used in animals with increased urethral tone, typically combinations of alpha-adrenergic antagonists such as phenoxybenzamine or prazocin and a centrally acting skeletal muscle relaxant such as diazepam or methocarbamol.

Cranberries have received a lot of interest as a urinary tract prophylactic since the early 1990s, and they are widely used to improve the health of the lower urinary tract in humans[13]. Cranberries contain a mixture of organic acids (malic, quinic and citric acid) as well as fructose and glucose and phytochemicals[14,15]. *In vitro* studies have demonstrated that certain phytochemicals, such as A-type proanthocyanidins (PACs), inhibit the adhesion of type I and P-fimbriated uropathogenic bacteria, in particular *Escherichia coli* and *Klebsiella pneumoniae*, to uroepithelial cells and this is now accepted as the mechanism by which the incidence of UTIs is reduced by cranberry extract[14,15].

There are numerous, contradictory clinical trials on the efficacy of cranberry juice and extract in humans with a variety of different causes of recurrent UTI. A 2008 meta-analysis concludes that there is statistical evidence that cranberry extract reduces the recurrence of UTIs, particularly in young women[16], but an updated analysis in 2012 came to the opposite conclusion[17]. There is evidence to suggest that inhibition of bacterial adherence to bladder epithelial cells is dose-dependent and lasts for approximately 10 hours[18,19]. It is possible that a low dose or insufficient dosing frequency was used by some of the trials in which a beneficial effect was not evident. There are several studies in chronically paralyzed humans with spinal cord injury, some of which show a benefit[20,21], but some of which do not[22,23].

In order to examine whether cranberry extract was a useful prophylactic in dogs recovering from acute spinal cord injury, we designed a placebo controlled prospective clinical trial in dogs with acute disc herniations. Description of this study is limited due to submission for publication. Dogs with acute, thoracolumbar intervertebral disc herniations were recruited to the trial, were placed on active or placebo medications and were followed for six weeks with urine cultures every 2 weeks. The results underlined the importance of anti-adhesion activity in the urine for preventing *E coli* UTIs but the issue of how to reliably generate adequate levels of activity in the urine is still unresolved.

Methenamine mandelate and hippurate are also advocated as prophylactics. The mechanism of action relies on generation of formaldehyde within urine and there is some evidence that acidic urine is necessary. A recent Cochrane review in humans concluded that methanamine hippurate may be effective in populations with no renal tract abnormalities, but was otherwise ineffective, including in patients with neuropathic baldders[24]. Low doses of antibiotics given at night are also advocated as effective prophylactics and nitrofurantoin is commonly used in this way in humans. We have used this approach in chronically paraplegic female dogs with recurrent UTIs. Dogs are treated with a dose of 1mg/kg once a day orally at night with treatment starting when existing UTIs have been effectively cleared with appropriate specific antibiotics. Although not all patients respond to this therapy, and some dogs develop gastrointestinal signs associated with use of nitrofurantoin, we have had success in dogs with seemingly continuous reinfections. Newer therapies are focusing on the role of probiotics, and the most appropriate dosing regimens and routes of administration. However, given the low success rate of managing UTIs in chronic paraplegia, the use of a recently described sacral implant, with the hope that more effective voiding will eliminate or at least reduce the problem, becomes extremely attractive[25].

In conclusion, dogs with acute thoracolumbar disc herniations are predisposed to developing UTIs, with a prevalence as high as 27% reported in the first postoperative week in spite of apparent recovery of bladder function. The most severely injured patients have a higher risk of developing a UTI and have been demonstrated to have incomplete recovery of effective micturition in spite of recovery of motor function. Chronic paraplegics, especially female dogs, suffer from recurrent UTIs that can be extremely frequent. Many urinary tract prophylactics such as cranberry extract and methenamine hippurate have been found to be ineffective in a comparable human patient population, but low doses of nitrofurantoin have shown some efficacy in dogs.

REFERENCES

1. Oliver JE Jr, et al. Spinal cord representation of the micturition reflex. J Comp Neurol. 1969;137:329-46.
2. Siroky, MD. Pathogenesis of bacteriuria and infection in the spinal cord injured patient. American Journal of Medicine. 2002; 113:67S-79S.
3. Esclarín De Ruz A, et al. Epidemiology and risk factors for urinary tract infection in patients with spinal cord injury. J Urol. 2000; 164: 1285-1289.
4. Garcia Leoni ME, et al. Management of urinary tract infection in patients with spinal cord injuries. Clin Microbiol Infect. 2003; 9: 780-785.
5. Olby NJ et al. Long-term functional outcome of dogs with severe injuries of the thoracolumbar spinal cord: 87 cases (1996-2001). J Am Vet Med Assoc. 2003 Mar 15;222(6):762-9.
6. Atalan G et al. Urine volume estimations in dogs recovering from intervertebral disc prolapse surgery. Berl Munch Tierarztl Wochenschr. 2002;115: 303-305.
7. Bauer M et al. Follow-up study of owner attitudes toward home care of paraplegic dogs. J Am Vet Med Assoc. 1992;200:1809-16.
8. Seguin MA, et al. Persistent urinary tract infections and reinfections in 100 dogs (1989-1999). J Vet Intern Med. 2003; 17: 622-631
9. Stiffler KS, et al. Prevalence and characterization of urinary tract infections in dogs with surgically treated type 1 thoracolumbar intervertebral disc extrusion. Vet Surg. 2006; 35:330-336.
10. Bubenik LJ et al. Urinary tract infection in dogs with thoracolumbar intervertebral disc herniations and urinary bladder dysfunction managed by manual expression, indwelling catheterization or intermittent catheterization. Vet Surg 2008; 37:791-800.
11. Barsanti JA, et al. Urinary tract infection due to in-dwelling bladder catheters in dogs and cats. J Am Vet Med Assoc 1985;187:384–388.
12. Olby NJ et al. Incidence of urinary tract infection in dogs following surgery for thoracolumbar intervertebral disc extrusion. J Vet Intern Med, 2010;24:1106-11.
13. Guay DR. Cranberry and urinary tract infections. Drugs. 2009;69:775-807
14. Nowack R. Cranberry juice-- a well-characterized folk-remedy against bacterial urinary tract infection. Wien Med Wochenschr. 2007;157:325-30
15. Pinzón-Arango PA, et al. Role of cranberry on bacterial adhesion forces and implications for Escherichia coli-uroepithelial cell attachment. J Med Food. 2009;12:259-70.
16. Jepson RG et al. Cranberries for preventing urinary tract infections. Cochrane Database Syst Rev. 2008;(1):CD001321.
17. Jepson RG, et al. Cranberries for preventing urinary tract infections. Cochrane Database Syst Rev. 2012;(1):CD001321.
18. Santillo VM, et al. Cranberry juice for the prevention and treatment of urinary tract infections. Drugs of Today. 2007, 43: 47-54.
19. Howell, A et al. Cranberry juice and adhesion of antibiotic-resistant bacteria. JAMA. 2002;287(23):3082-3083.
20. Hess MJ, et al. Evaluation of cranberry tablets for the prevention of urinary tract infections in spinal cord injured patients with neurogenic bladder. Spinal Cord. 2008;46:622-6. Epub 2008 Apr 8.
21. Reid G, et al. Cranberry juice consumption may reduce biofilms on uroepithelial cells: pilot study in spinal cord injured patients. Spinal Cord. 2001;39:26-30
22. Lee B, et al. Spinal-injured neuropathic bladder antisepsis (SINBA) trial. Spinal Cord. 2007;45:542-50. Epub 2006 Oct 17.
23. Linsenmeyer TA, et al. Evaluation of cranberry supplement for reduction of urinary tract infections in individuals with neurogenic bladders secondary to spinal cord injury. A prospective, double-blinded, placebo-controlled, crossover study. J Spinal Cord Med. 2004;27(1):29-34.
24. Lee BS et al. Methenamine hippurate for preventing urinary tract infections. Cochrane Database Syst Rev. 2012;(10):CD003265.
25. Granger N et al. Use of an implanted sacral nerve stimulator to restore urine voiding in chronically paraplegic dogs. J Vet Intern Med. 2013 Jan-Feb;27(1):99-105.

CLINICOPATHOLOGIC STUDY OF DYSPHAGIA IN CANINE DEGENERATIVE MYELOPATHY

Joan R. Coates, DVM, MS, DACVIM (Neurology), Teresa E. Lever, PhD
Columbia, MO

INTRODUCTION

Canine degenerative myelopathy (DM) has been proposed as a naturally-occurring disease model of human amyotrophic lateral sclerosis (ALS). In both DM and familial ALS, the disease phenotype is attributed to mutations in the superoxide dismutase gene (SOD1). Clinical signs observed in DM are similar to those seen in ALS. Virtually all ALS patients develop dysphagia, which can lead to aspiration and fatal complications. Likewise, unspecified clinical signs of dysphagia have been reported in DM as the disease progresses to end-stage.[1] The nature of the dysphagia in DM has not been characterized, and the cranial nerve nuclei involved in swallowing have not been systematically investigated. Correlations between the dysphagia phenotype and histopathology in DM can then be explored and compared with human ALS. Since dysphagia occurs in latter disease of DM, there is potential that studies of swallowing function could be used as a longitudinal disease measure.

ANATOMY OF SWALLOWING

Swallowing is a complex process that requires coordinated reflex contractions of muscles to propel material from the oral pharynx to the stomach.[2] The pharynx is composed of three areas, the oropharynx (area between the soft palate and base of the tongue), the nasopharynx (area between the soft palate and the ventral aspect of the calvaria extending from the choanae [rostral] to the end of the soft palate [caudal]), and the laryngopharynx (area of the pharynx dorsal to the larynx which leads to the esophagus). The intrapharyngeal opening exists between the oropharynx and nasopharynx rostrally and the laryngopharynx caudally. At the caudal end of the laryngopharynx is the cricopharyngeus muscle which serves as the upper esophageal sphincter (UES).

Normal swallowing has been divided into three different phases, oropharyngeal, esophageal, and gastroesophageal.[2] The oropharyngeal phase can be further subdivided into oral (prehension and formation of a food bolus), pharyngeal (movement of the food bolus to the laryngopharynx), and pharyngoesophageal (movement of the food bolus into the proximal esophagus) phases. Neurologically, swallowing requires normal function of CNs V, VII, IX, X, XI, and XII [**Figure 1**]. The oral phase of swallowing requires the normal function of CNs V, VII, and XII for prehension, mastication, and formation of a food bolus. The pharyngeal and pharyngeoesophageal phase requires normal function of CNs IX, X, XI and XII. Normal function of the esophagus requires CN X function. Coordinated function of these motor nuclei is under control of the swallowing center located in the medulla between the facial and olivary nuclei. During the oral phase, food is masticated, formed into a bolus by the tongue compressing food against the palate, and moved to the base of the tongue. Once a bolus is propelled into the pharyngeal region, the swallowing reflex requires sensory input from from CNs V, VII, IX and X. During the pharyngeal phase, peristaltic waves move the food to the laryngopharynx where during the final phase, pharyngeoesophageal, the UES relaxes and is actively pulled open by hyolaryngeal excursion to permit transport of the food bolus into the cranial esophagus. These two phases proceed with rapid and well-coordinated timing. Within the esophagus, primary and secondary peristaltic waves move the bolus of food to the lower esophageal sphincter (LES), which relaxes allowing the bolus to enter the stomach.

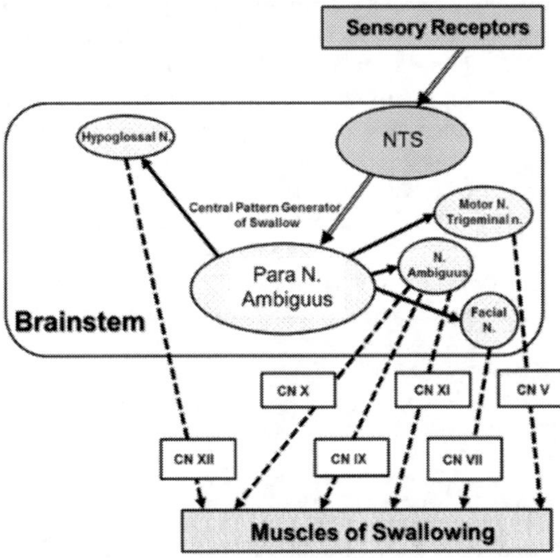

Figure 1. Schematic representing the brainstem neuronal circuit for the swallowing reflex.

The central pattern generator for swallow is responsible for establishing the motor pattern of the swallow reflex. NTS, solitary tract and nucleus; N, nucleus; CN, cranial nerve; n, nerve.

DYSPHAGIA

Traditionally, dysphagias are classified based on the anatomic location of the abnormality.[3,4] Thus, swallowing disorders are classified as problems of (1) bolus formation and passage from the mouth to the cranial esophagus (oropharyngeal dysphagia), (2) passage of the bolus from the cranial esophagus to the gastroesophageal junction, (3) disturbances in passage of the bolus from the esophagus into the stomach, and (4) mixed swallowing problems. Abnormalities in any one phase of swallowing can result in dysphagia.

Although some clinical signs are indicative of disorders involving a specific phase of swallowing, most clinical signs are nonspecific. Dogs with oral-stage deficits commonly have a reduced ability to hold food in the mouth or transport food to the base of the tongue, whereas dogs with pharyngeal and cricopharyngeal dysfunction more commonly cough, have aspiration pneumonia, and make repeated unsuccessful attempts to swallow. Other clinical signs include gagging, difficulty in drinking water or forming a solid bolus, excessive mandibular or head motion, persistent forceful ineffective swallowing efforts, foaming from the mouth, salivation, nasal regurgitation, dropping food from the mouth, reluctance to eat, coughing, malnutrition, and aspiration pneumonia.[3,4]

Dysphagia in ALS patients has been attributed solely to progressive paralysis of skeletal muscles innervated by the following cranial nerves (CN) and corresponding brainstem motor nuclei: CN V (trigeminal), CN VII (facial), CN IX, X, and XI (nucleus ambiguus, NA), and CN XII (hypoglossal).[5] Degeneration of these same neuroanatomical components of swallowing has been identified in SOD1-G93A transgenic mice,[6,7] the most widely studied, transgenic animal model of ALS. These mice develop symptoms of dysphagia which include slower lick and mastication rates, reduced water intake, and increased weight-loss compared to age-matched, nontransgenic controls. These dysphagic signs in mice resemble those of humans with ALS,[6,7] but further evaluation of dysphagia is difficult in such a small species.

In addition to motor neuron degeneration, recent research by Lever suggests that *sensory* pathology also may be contributing to dysphagia in ALS. Specifically, widespread vacuolation was identified throughout the entire rostral-caudal extent of the nucleus tractus solitarius (NTS) of SOD1-G93A transgenic mice.[7] The interstitial and intermediate subnuclei of the NTS are the main *sensory* components of the brainstem central pattern generator (CPG) for swallowing.[8] Within these NTS subnuclei are pacemaker-like neurons that orchestrate the triggering, shaping, and timing of the sequential *motor* pattern (e.g., muscle contractions) of swallowing [**Figure 1**].[9] Such pathology needs investigation in DM-affected dogs in addition to the motor nuclei involved in swallowing.

DIAGNOSTIC IMAGING

Diagnosis of dysphagia is based on owner history, observation, physical and neurologic examinations, and imaging. Dysphagia is often a dynamic abnormality that requires real-time imaging techniques such as videofluoroscopy swallow study (VFSS) to accurately localize and identify the type of dysfunction.[10] Contrast VFSSs assess for and classify oropharyngeal dysphagias and functional disorders of the esophagus and lower esophageal sphincter. Thus, VFSSs can provide qualitative and quantitative data for assessment of dysphagia. It is important to have protocol standards for body positioning and food volume and consistency/viscosity. A report described differences between sternal and lateral body positioning in the dog and found significantly delayed cervical esophageal transit times when dogs were positioned in lateral recumbency.[11]

Common qualitative assessments include tongue movement, pharyngeal activity, blockage of the oral, nasal and pharyngeal egress, cricopharyngeal activity, esophageal activity and laryngeal / tracheal aspiration.[2] Recording the study and performing frame-by-frame analysis can provide quantitative measures of swallowing function. Normal pharyngeal transit times in swallowing function have been reported.[12] Pharyngeal transit is timed from once the epiglottis closes over the larynx to until the observation of maximal constriction of the pharynx, opening of the UES and closing of the UES. Contraction of the pharynx is assessed using frame-by-frame analysis to calculate the pharyngeal constriction ratio.[13] The UES is specifically defined as the narrowest region of the esophagus throughout the study. Elevation of the pharyngeal constriction ration but normal timing for opening of the UES is the videofluoroscopic hallmark of pharyngeal weakness that is recognized in many neuromuscular diseases.[13]

The current protocol for contrast videofluoroscopic assessment of dysphagia in dogs requires physical restraint in sternal or lateral recumbency to minimize head/body movement and retain positioning within the fluoroscopy field of view.[10-13] Struggling behaviors are still common with these methods; therefore, head and body motion remains problematic. Furthermore, radiation exposure to individuals cannot be minimized because syringe and hand feeding methods are still required. Collaborations with a speech-language pathologist (TE Lever) have led to the development of a protocol for a freely-behaving VFSS that allows dogs to consume foods and liquids containing either barium or iohexol contrast agents while standing unrestrained in a custom-designed observation kennel. The kennel setting eliminates the need for physical restraints and force-feeding techniques that may result in unnatural feeding activity, which is a common complaint reported in studies of canine dysphagia. We are in the process of developing a multi-center study to further refine and standardize our self-feeding protocol for ease of use in research and clinical practice. Using our freely-behaving VFSS protocol, we hypothesize that several disease measures for dysphagia will be identified in DM-affected dogs, compared to unaffected control dogs.

REFERENCES

[1]Awano et.al **Proceed Nat Acad Sci. USA** 2009; 106:2794. [2]Watrous and Suter **Vet Radiol** 1979; 99. [3]Suter and Watrous **Vet Radiol** 1980;21:24. [4]Watrous and Suter **Vet Radiol** 1983;24:11. [5]Hillel and Miller **Head Neck** 1989;11:51. [6]Lever et.al **Dysphagia** 2009;24:180. [7]Lever et.al **Dysphagia** 2010;25:112. [8]Goyal **Am J Med** 2001;111:95S. [9]Lang **Dysphagia** 2009;24:333. [10]Pollard **ISRN Veterinary Science** 2012;15 pp. [11]Bonadio et.al **J Vet Intern Med** 2009;23:801. [12]Pollard et.al **Vet Radiol Ultrasnd** 2000;409. [13]Pollard et.al **Vet Radiol Ultrasnd** 2007;48(3):21.

REVIEW OF LARGE ANIMAL NEUROLOGY

Monica Aleman, MVZ Cert., PhD, Dipl. ACVIM (Large Animal Internal Medicine, Neurology)
Davis, CA

SUMMARY

This session will review the neurological examination, species differences, and neuroanatomical localization exemplified with clinical cases for discussion. Diagnostic work up such as CSF analysis, electrodiagnostic evaluation, neuroimaging, muscle and nerve biopsy, and histopathological review pertaining to specific cases will also be discussed.

INTRODUCTION

It is essential before performing a neurological examination to obtain relevant information of the patient to examine. Signalment (breed, gender, age), physical activity or intended use (i.e. athlete, pet, milk production, beef cattle) and a complete medical history (i.e. previous illnesses, illnesses in the farm, animals affected [age, activity]), and preventive medicine (vaccinations, deworming program). In large animals, it is important to know if a single or various animals are affected, age group, diet/supplements and water source, housing (i.e. stall, pasture, dry lot). Nutrition/toxic, infectious, and genetic disorders (in breeding farms) must be considered as possible causes of disease, if various animals are affected in the same premises. Duration (acute versus chronic), progression (non-progressive versus progressive disease), fever (especially important in large animals due to possible infectious contagious disease), and presence or absence of apparent pain are important features that will aid in the investigation of possible causes of neurologic disease. A full physical examination is essential before embarking into the neurological examination of large animals since systemic illnesses can influence the neurological status of the patient particularly in neonates. Further, metabolic, nutritional, and/or toxic disorders common in farm animals might present with neurological manifestations. There are clinical signs such as weakness that might be common to systemic illness and neurologic disease. Therefore, it is important to determine the overall health status of the patient to examine.

The basic principles of what constitutes a full neurological examination in large animals are not different to those of small animals. However, there are differences on how to perform the examination tailored to the species, domestication, size, and safety concerns. Safety must always be a priority when examining large animal species. The first goal of the neurological examination is to determine if neurological abnormalities are present or absent. The second goal is to determine neuroanatomical localization given the clinical signs localized within the major divisions of the nervous system. These two first essential steps along with the signalment, and medical history (including information of duration, progression, fever, and apparent pain) of the patient will aid in the formulation of a list of possible causes of disease. Based on this list of possible causes, a targeted diagnostic plan can be made. These steps are not different between small and large animal species. At this point, it is important to mention that the use of correct terminology is paramount and based on the Nomina Anatomica Veterinaria.[1]

NEUROLOGICAL EXAMINATION

This section will briefly review the major neuroanatomical regions for localization since this manuscript is directed to neurologists in training and other groups of interest. A major review will not be included here. It is important to emphasize that neurological examination should be tailored for the particular individual to examine (i.e. different approach in docile versus undomesticated animal, or neurological unstable animal that might risk everyone's safety). An advice is to be as thorough, complete, and consistent as possible for the examination. When performing the neurological examination, it is also important to note functional neuroanatomical differences among species. For example, the menace response (a learned response) develops earlier in prey animals (i.e. 7-10 days in horses and cattle) compared to predators (2-3 months). Prey animals such as the horse and cattle are born with developed brains and functional vision and hearing versus predator animals which continue to develop their brains (i.e. cerebellum) after birth.

Most of the neurological examination is OBSERVATION, OBSERVATION, and OBSERVATION. Become familiar with what is normal, otherwise the recognition of abnormality can be faulty. This is particularly relevant in gait differences according to breed in horses (i.e. gaited versus non-gaited breeds), for example. In large animals due to their size, the order on how the examination takes place generally starts with the head, neck, trunk, tail and limbs. The areas to evaluate consist of the following:

1. Mentation and behavior
2. Posture (head, neck, trunk, tail, limbs)
3. Cranial nerves examination
4. Palpation (i.e. muscle tone, muscle/bone asymmetry, flexion/extension of joints, apparent pain, areas of hypo- or hyperesthesia)
5. Segmental reflexes
6. Postural reactions (i.e. proprioceptive positioning)
7. Gait evaluation
8. Pain perception

The determination of mentation, behavior and posture can be done as the history is being taken but continue observing throughout the examination. Ideally, the patient should be examined in its usual environment. Cranial nerves deficits might be apparent during the initial observation before approaching the patient. The neurological examination has the static and dynamic components. The static part includes all those parts of the examination that can be done with the animal at rest, and the dynamic part includes moving the animal to assess gait, coordination, and postural reactions during movement. Animals are evaluated at the walk, trot, and canter in the case of horses if indicated and safe (tailor the exam to the patient). The gait can be evaluated in straight line, lounge line, zig-zag walk, circles (wide, tight), backing up, up and down a curb and hill, and using different surfaces to evaluate visual compensation or lack of and visualization or exacerbation of deficits. The author adds orthopedic evaluation since patients might present with concurrent orthopedic or musculoskeletal abnormalities that might complicate the neurological evaluation and its interpretation with regards to posture and gait. However, this is only done in patients that are stable and never compromise safety.

Motor tracts differ among species. For example, small ungulates have a dorsal corticospinal tract but not horses or large animal ungulates. Ungulates in general have a short ventral corticospinal tract that ends within the cervical spinal cord. This is in contrast to small animal species which have lateral (dogs and cats) and ventral (mainly cats) corticospinal tracts. The extrapyramidal motor system is the main motor system in horses and farm animals because these animals do not require a high level of fine motor function.

NEUROANATOMICAL LOCALIZATION

The clinical signs are going to depend on specific neuroanatomical localization. Understanding of applied neuroanatomy is essential when evaluating a patient with neurologic disease. The major divisions of the CNS consist of brain, spinal cord, and peripheral nervous system. Do not forget the autonomic nervous system. The major divisions and subdivisions of the brain are shown in Table 1.

Primary division	Subdivision	Structures	Lumen
Prosencephalon (Forebrain)	Telencephalon	Cerebral hemispheres Basal nuclei Limbic system Rhinencephalon	Lateral ventricle
	Diencephalon	Thalamus Hypothalamus	Most of 3rd ventricle
Mesencephalon (Midbrain)	Mesencephalon	Tectum Tegmentum	Mesencephalic aqueduct
Rhombencephalon (Hindbrain)	Metencephalon	Pons Cerebellum	Rostral 4th ventricle
	Myelencephalon	Medulla oblongata	Caudal 4th ventricle

The spinal cord segments consist of C1-C5-6, C6-T2, T3-L3, L4-S2, S1-variable by species (pelvic plexus), Cd1-variable. The following table depicts the number of cervical (C), thoracic (T), lumbar (L), sacral (S), caudal (Cd) vertebrae. In parenthesis for the cervical vertebrae is indicated the spinal cord segments. Last column shows the vertebral body at which the *conus medullaris* is located.

SPECIES	C	T	L	S	Cd	Cord ends
Dog	7 (8)	13	7	3	6-26	L: L6-7, S: L7
Cat	7 (8)	13	7	3		S1
Horse	7 (8)	18	5-6	5		S2
Cattle	7 (8)	13	6	5	18-20	Variable
Sheep	7 (8)	13	6-7	4	20-23	
Pig	7 (8)	13-17	5-7	4		
Camelid	7 (8)	12	7	4	15-20	
Human	7 (8)	12	5	5		L1-L2

The neuromuscular system is composed of motor units. Motor units consist of lower motor neurons, peripheral nerve, neuromuscular junction (presynaptic, synaptic, postsynaptic), and the myofibers they innervate. Common signs of neuromuscular disease include weakness (generalized, focal, exercise intolerance), functional deficits (i.e. paresis/paralysis, gait deficits, dysphagia, dyspnea, regurgitation), physical manifestations (muscle hypotrophy, atrophy, musculoskeletal deformities), reflex deficits (+/-).

NEUROLOGICAL DISORDERS

Due to the extent of information about neurological disorders in large animal species, examples of specific disorders pertaining to the various neuroanatomical regions of the nervous system will be provided during lecture. Differences with other species will be highlighted.

REFERENCES

[1]Waibl H, et al. **Nomina Anatomica Veterinaria**, 5th ed., 2005. [2]Lorenz MD, et al. **Handbook of Veterinary Neurology**, 5th ed. 2012. [3]Mayhew J. **Large Animal Neurology**. 2nd ed. 2009. [4]de Lahunta A, et al. **Veterinary Neuroanatomy and Clinical Neurology**. 3rd ed. 2009.

CANINE LYMPHOMA – CLINICAL TRIAL YEAR IN REVIEW

Douglas H. Thamm, VMD, DACVIM (Oncology)
Fort Collins, CO

INTRODUCTION

2013 brought a number of interesting clinical trial developments for canine lymphoma, which can be divided into 3 basic themes: (1) How do we use our existing drugs better? (2) What new drugs/protocols may be useful? (3) Are there non-"drug" therapies that may hold promise for the treatment of canine lymphoma?

HOW DO WE USE OUR EXISTING DRUGS BETTER?

Travel and time constraints can make intensive, weekly treatments challenging for some owners. Furthermore, especially in more remote areas, even every 3 week treatments can sometimes be problematic. To begin to address this concern, Higginbotham et al evaluated a discontiguous single-agent doxorubicin (DOX) protocol.[1] Briefly, the intent was to treat with 3 biweekly doses of doxorubicin, followed by regular rechecks. Single doses of doxorubicin would then be administered on an as-needed basis when relapse was noted. 18 patients, all with B-cell lymphoma, were enrolled. 14 dogs experienced a complete response (CR) (78%) and 12 dogs completed the planned 3 initial DOX doses (67%). The median first remission was 68.5 days and the median total time on doxorubicin was 80.5 days. For those dogs achieving a CR, the median time on DOX was 157.5 days, with the average dog receiving 4.5 DOX doses. The median overall survival time was 169.5 days. Stage (V) and substage (b) were negative prognostic factors, and dogs experiencing CR had improved first remission and overall survival times. When compared with historical data employing continuous DOX, these results are comparable, suggesting that this approach may be worth considering for owners with significant financial and/or travel constraints.

The use of corticosteroids in canine lymphoma is fraught with multiple questions – how much does pre-treatment with corticosteroids affect subsequent chemotherapy response? Should corticosteroids be included in single-agent DOX protocols? What should I do with my patient with lymphoma that has concurrent hyperadrenocorticism or diabetes mellitus if the owner wants to be as aggressive as possible? A new study by Zandvliet et al attempts to address this last question.[2] In the context of a randomized clinical trial, 81 dogs received a CHOP-based protocol with or without the inclusion of prednisone. Although a significantly higher percentage of dogs receiving prednisone were classified as substage b, there were otherwise no significant differences in presentation or known prognostic factors between groups. Similarly, there were no significant differences in overall response rate, %CR, first response duration, or adverse effects between groups. These results suggest that corticosteroids can be omitted from an aggressive, multiagent canine lymphoma protocol without adverse impact on outcome.

WHAT NEW DRUGS/PROTOCOLS MAY BE USEFUL?

While CHOP-based chemotherapy is an accepted standard for first-line treatment of canine lymphoma, there is no universal gold standard for the treatment of relapsed/refractory lymphoma. A multitude of single-agent and multi-agent protocols have been evaluated, and there remains a need for treatments that balance efficacy, cost, toxicity and convenience. The latest addition to the list of rescue protocols that have been evaluated is the "MOMP" protocol.[3] This protocol utilized intravenous mechlorethamine and vincristine, administered days 0 and 7, with oral melphalan and prednisone administered days 0-13, on a 28-day cycle. 88 dogs were included, 84 of which had received prior DOX. The median number of previous drugs was 4 (3-9) and the median number of previous rescue protocols was 2 (1-4). The overall response rate was 51% (12.5% CR, 38.5% PR). The overall median PFS was 28.5 days, with a PFS of 81 days for dogs experiencing CR and 49 days for those experiencing PR. Response duration was statistically longer in those dogs receiving MOMP as the first rescue protocol. There was no difference in outcome based on immunophenotype. Approximately 1/3 of dogs experienced gastrointestinal toxicity, the majority grade 1 or 2. Approximately 25% experienced hematological toxicity, again the majority being of low grade. Overall, the results obtained with this protocol are within the range reported with other protocols, and thus the use of MOMP should be a consideration for the treatment of relapsed canine lymphoma.

Lipid encapsulation, in the form of liposomes, lipid nanoemulsions, etc. have the potential to alter the pharmacokinetics and biodistribution of multiple cancer therapeutics, thereby potentially altering their efficacy spectrum and toxicities. In dogs, this is best exemplified by the changes in toxicity and efficacy observed in dogs and cats treated with Doxil® (PEGylated liposomal doxorubicin) when compared to free DOX.[4-5] A novel lipid nanoemulsion of carmustine was recently evaluated in a pilot study in dogs with lymphoma, and compared to free carmustine when utilized as part of a carmustine-vincristine-prednisone (BOP?) protocol.[6] 15 dogs with lymphoma were randomly assigned to receive either liposomal carmustine or free carmustine, at equivalent dosages, with fixed dosages of vincristine and prednisone. Objective response rate (100%/100%), %CR (71%/75%), and progression-free interval (109 days/190 days) were comparable between liposomal carmustine and free carmustine. There were also no significant differences in adverse effects, which were generally mild and self-limiting in both groups. Interestingly, the large majority of dogs treated in this study had T cell lymphomas. This is interesting preliminary data demonstrating preliminary tolerability of an equivalent dosage of carmustine when administered in a lipid nanoemulsion. However, the study was underpowered to detect any but the largest differences in outcome. Furthermore, it is possible that dose-escalation of carmustine might have been possible in the lipid arm. Finally, evaluation of the pharmacokinetics of carmustine in both arms might have shed light onto whether exposure was different between the 2 arms as a result of lipid encapsulation. It is hoped that further studies with this novel formulation are planned.

ARE THERE NON-"DRUG" THERAPIES THAT MAY HOLD PROMISE?

"Immunotherapy", in its various forms, has been somewhat of a holy grail in lymphoma therapy circles for some time. This includes monoclonal antibody-based therapies as well as vaccine strategies. 2 groups have evaluated vaccination strategies in conjunction with conventional chemotherapy in the past year, with interesting preliminary data.

In a nonrandomized, noncontrolled 2-arm study of 42 dogs, Gavazza et al evaluated COP-based chemotherapy +/- a telomerase reverse transcriptase (TERT)-based vaccine.[7] Dogs in remission after an 8-week induction COP protocol went on to a maintenance chemotherapy regimen +/- vaccination. The vaccine consisted of 2 priming vaccinations with an adenoviral TERT construct, followed by boosts with TERT plasmid-based electrovaccination. Clinical outcome as well as immune responses were measured sequentially in the 2 arms. Interestingly, while there was no difference in first remission duration, 11.4 and 11.3 weeks, between the groups, overall survival was significantly improved in the vaccinated group (76.1 vs 29.3 weeks). Most of the vaccinated dogs mounted detectable T-cell responses to the vaccine, and exploratory studies suggested a potential correlation between TERT expression in tumor and survival in vaccinated patients. While this is certainly interesting and encouraging information, caveats include lack of randomization or an active control arm, nonstandardized rescue therapy, and a lack of information comparing stage, substage or other known prognostic factors between groups. Furthermore, it is not clear whether a similar benefit would be realized in dogs treated with a more "optimal" (e.g. CHOP-based) induction chemotherapy protocol. Further studies are eagerly awaited.

A second recent study evaluated an autologous heat shock protein peptide chaperone (HSPPC)-based vaccine in conjunction with chemotherapy in dogs with B cell lymphoma in the context of a randomized, placebo-controlled, double-blinded trial.[8] 19 dogs were randomized to receive a CHOP-based chemotherapy with HSPPC vaccine or placebo. Immune response was assessed through assessment of dermal delayed-type hypersensitivity responses with visual assessment. 12 dogs were in the vaccine arm and 7 were in the placebo arm. The placebo arm contained more stage V patients and more substage b patients; there differences were not statistically significant, probably owing to the small N. The vaccinated patients had longer median first and second times to progression (304 vs 41 days and 127 vs 32 days respectively), and there was no difference in the incidence of adverse effects between the groups. "Positive" DTH responses were observed in all vaccinates but none of the placebo-treated dogs. Overall this is again very interesting preliminary data, but the very small N and unequal distribution of some known powerful prognostic factors in favor of the vaccinated group may have very likely contributed to the unexpectedly poor outcome observed in the placebo arm. Furthermore, dogs are considered a very poor species for visual DTH assessment: more sophisticated measures for assessing immune response would have been more compelling. However, this forms exciting preliminary data for a larger scale, appropriately powered study.

CONCLUSIONS

2013 brought some interesting and useful clinical results in canine lymphoma patients, some of which have immediate clinical implications. The interesting and encouraging results obtained in preliminary immunotherapy studies suggest an avenue for further exploration, which may lead to intriguing new options for canine lymphoma therapy.

REFERENCES

[1]Higginbotham ML, et. al. **J Am An Hosp Assoc** 2013;49(6):357. [2]Zandvliet M, et. al. **Vet J** 2013;197(3):656. [3]Back AR, et. al. **Vet Comp Oncol** 2013; epub ahead of print. [4]Vail DM, et.al. **Clin Cancer Res** 1998;4(6):1567 [5]Poirier VJ, et. al. **J Vet Intern Med** 2002;16(6):726. [6]Lucas SR, et. al. **Vet Comp Oncol** 2013; epub ahead of print. [7]Gavazza A, et. al. **Hum Gene Ther** 2013;24(8):728. [8]Marconato L, et. al. **Clin Cancer Res** 2014;20:668-677.

CHIMERIC ANTIGEN RECEPTORS FOR THE TREATMENT OF CANINE B CELL MALIGNANCIES

Heather Wilson-Robles, DVM, DACVIM (Oncology)
College Station, TX

INTRODUCTION

Canine B-lineage Lymphoma

Cancer is the leading cause of death for client-owned canines. The most common malignancy being Lymphoma (LSA) which accounts for up to 24% of all reported neoplasms and 85% of diagnosed hematological malignancies. Similar to humans, approximately 60-80% of canine LSA arises from malignant B cells. The most common presentation is a generalized lymphadenopathy corresponding to stages III to V using the WHO staging system, where tumor cells may be observed in the lymph nodes, spleen, liver, blood, bone marrow, and other organ systems. The current standard of care for B-lineage LSA is a combination chemotherapy regimen of cyclophosphamide, vincristine, doxorubicin, and prednisone (CHOP), which induces a temporary remission (9-11 months) in approximately 85% of dogs. With less than 50% of diagnosed dogs living one year, there is an urgent and unmet need in veterinary medicine to develop new treatment strategies for refractory. Recent promising advances in human cancer treatments have focused specifically on chemotherapies and biological therapies which harness the power of the body's own immune system to provide a durable anti-tumor response. The veterinary world is beginning to develop similar immunotherapy treatments.

Adaptive Immune System Dysfunction and LSA

A functional immune system effectively maintains the integrity of "self" by destroying and removing diseased and infected cells, while a complex network of checks and balances ensure healthy cells and tissues are not targeted. These checks and balances are provided through the collaborations between the adaptive and innate arms of the immune system, which include cell to cell interaction/cross talk, as well as, the release of soluble factors, such as cytokines. A dysfunction in the intact immune system can lead to increased risk of infections, autoimmune diseases, and the development of neoplasia, such as LSA.

The adaptive immune system has both effector and regulatory T cells which can influence tumor growth and metastases. αβ T cells encompass approximately 95% of the peripheral blood T cell pool and are primarily regarded as effector cells ($CD3^+CD8^+$) or helper cells ($CD3^+CD4^+$). $CD3^+CD4^+$ $CD25^+Foxp3^+$ T cells are regarded as regulatory cells (Tregs). The vast majority of T-cell therapies for human and dogs incorporate only αβ T cells. The remaining 1-5% of the T cell pool are regarded as γδ T cells, a rare population of T cells that do not express the αβ T-cell receptor nor are subdivided into $CD8^+$ or $CD4^+$ categories.

The T cell interacts with tumors and antigen presenting cells by interrogating the surface with their T-cell receptor (TCR). The major histocompatibility complex located on the surface of the affected cell interacts with the TCR by presenting it with the specific antigenic peptide for which the TCR is programed. The TCR is associated with a CD3 complex made up of a γ, δ, ε, and ζ signaling endodomains. When the TCR interacts with a MHC antigen complex, it triggers the CD3 complex to mediate signal which leads to pro-inflammatory cytokine production, cell mediated lysis and apoptosis[1] . Cellular death is mediated by the release of perforins from the T cell, which allows granzymes to enter into the target cells and induce apoptosis[1].

Peripheral and Central Tolerance Effect on LSA and Cellular Therapy

The primary obstacles faced by cellular based immunotherapies are over-coming both peripheral and central tolerance issues as the tumor evolves under immune pressure to become less visible. Central tolerance focuses on the deletion of self-antigen recognizing T-cells during development. In order to evade immunosurveillance and destruction by T cells, tumors routinely over-express or aberrantly express self-antigens. Conversely, peripheral tolerance is the development of a dysfunctional immune system through the formation of an immunosuppressive condition. Tumor cells commonly take advantage of this condition. One way peripheral tolerance can be overcome by is blocking inhibitory immune-modulating molecules, like CTLA-4 and PD-1 or creating a pro-inflammatory environment around the tumor. Human Phase II and Phase III trials involving monoclonal antibodies against cytotoxic T-lymphocyte antigen-4 (CTLA-4) and programmed death -1 (PD-1) alone or in combination with other therapies have resulted in potent anti-tumor responses and tolerability in melanoma, non-small cell lung cancer, and head and neck squamous cell carcinoma[2],[3]. Other tactics include the use of chemo- and radiotherapies.

Central tolerance or recognition of tumor self-antigen by T cell is more difficult to overcome. Adoptive cellular therapy (ACT) is one treatment that has shown promise against central tolerance. ACT is a biological treatment which harnesses the natural ability of the body's lymphocytes to effectively recognize and eliminate pervasive target cells while initiating and managing a systemic immune response against advanced disease. Autologous ACT involves the infusion of *ex vivo* manipulated expanded T cells post chemotherapy. The foundation for ACT was formed on two early human trials infusing (i) rapidly expanded tumor infiltrating lymphocytes (TIL) from melanoma patients and (ii) CMV virus-specific T cell clones into cancer patients with a high-risk of CMV-infection both shown to induce vigorous anti-tumor immune response[5,6]. The data from these initial ACT trials revealed three important points that would need to be taken into account for ACT to be successful: i) adoptive T-cell immunotherapy could influence endogenous anti-tumor activity; ii) future treatment strategies should incorporate the immunological contact between T-cells and tumor cells; and iii) the manufacture of clinically significant numbers of efficacious T-cell phenotypes and infusion protocols would need to be advanced and revised[7].

CANINE ADOPTIVE T CELL THERAPY FOR B-LINEAGE LSA

More recently, adoptive transfer of non-specific *ex vivo* expanded T cells has shown promise as an anti-cancer therapy in dogs especially when used in combination with chemotherapy. The phenomenon of chemotherapy's immune and tumor- modulating capabilities as well as the associated iatrogenic lymphodepletion observed post therapy prompted the testing of repeated infusions containing activated autologous non-specific T cells post CHOP to treat companion canines diagnosed with LSA. In humans, the increased rate of T-cell recovery post chemotherapy has been directly correlated with progression free and overall survival times in several malignancies[8,9]. As discussed earlier, human patients have better prognosis when there is faster lymphoid recovery post chemotherapy, increased $CD3^+CD8^+$ effector T cells trafficking to the tumor site, and circulating peripheral blood Tregs are decreased. A recent study by O'Connor et al showed a significant benefit to survival with the administration of non-specific T-cell infusions post CHOP protocol to companion canines diagnosed with LSA[10]. In this study, ex vivo expanded T cells were infused three times using an intra-patient dose escalating scheme. O'Connor et al administered non-specific autologous T cells to eight dogs with spontaneously occurring B-cell LSA in the published trial, but have infused five additional canines on this protocol since publication. The product was primarily $CD3^+CD8^+$ T cells that expressed lymph node homing molecule, CCR7, and cytolytic enzymes, granzyme B. Like humans, canines diagnosed with LSA exhibited increased numbers of $CD3^+CD4^+$ helper T cells circulating in the peripheral blood and at the sites of tumor. The adaptive immune system was further characterized as dysfunctional by the significant decreased in circulating $CD3^+$ T cells compared to healthy donors. Multiplex digital profiling using nCounter analysis system quantified T-cell specific gene expression change between peripheral blood $CD3^+$ and the expanded infusion. Gene products associated with T-cell activation (IL-21R, IL-2R), cytotoxicity (KLR, GZMH, GZMB, PRF-1, IFN-γ, CD8), proliferation (EOMES, DPP4, TNFRSF9, LCK, ZAP70), inflammation, and trafficking (CCL3, CCL4, CCL5 receptors) were upregulated in expanded T cells. mRNA sequences coding for chemokines to allow for such trafficking. LEF1, TGFBR1, FOXP3, CD4 and hypoxic related genes were upregulated in the non-expanded peripheral blood pre-infusion samples. The expanded $CD3^+CD8^+$ products secreted the pro-inflammatory cytokine protein IFN-γ which supported the mRNA findings. During cryopreservation, the expanded T cells were labeled with a red fluorescent dye, PKH-26, to allow en vivo tracking by flow cytometry and immunohistochemistry. Infused T cells persisted greater than 49 days as supported by PKH-26 detection, increased $CD3^+CD8^+$ peripheral blood population, and decreased CD4:CD8 ratios. In normal dogs, there are much higher proportions of $CD8^+$ T cells in circulation than $CD4^+$ T cells. However, in dogs with cancer, this ratio and the levels of circulating $CD4^+$ cell are increased as the immune system is dysfunctional. Infusions of $CD3^+CD8^+$ T cells were able to convert this ratio to a pre-cancer homeostatic level. In this study, dogs whose ratios significantly improved demonstrated prolonged survival over those treated with chemotherapy alone. This study demonstrated the importance of CD4:CD8 T-cell ratios as a prognostic factor or an assay to determine relapse. Within 3 hours and at 10 days post infusion, infused T cells could be observed in the lymph node. In canines with tumor burden, the percentage of lymph node $CD3^-CD21^+$ B cells decreased with the influx of $CD3^+CD8^+$ expanded T cells (data not published). Other factors that were noted to be linked with improved prognosis in this study were neutrophil:lymphocyte ratios (NLR) lower than 2.2:1 and elevated granzyme B levels of the expanded T cells before infusion. Supporting human observations, companion canines receiving two to three doses of expanded T cells had significantly longer tumor-free and overall survival compared with age and stage matched historical controls treated with CHOP chemotherapy alone. Immune reconstitution with effector T cells was able to improve the median overall survival to 392 days versus 167 days and tumor-free survival to 338 days versus 71 days. This implied that the combination therapy of CHOP and activated T cells provides more clinical benefit to canines than CHOP alone.

Based on this data, a second canine activated T-cell therapy trial has been initiated by the group.

Steriod resistant T cell infusions could be very valuable for treating dogs on concurrent steroids or in cases of graft versus host disease in humans. The development and infusion of steroid resistant T cells for these dogs could inform on the human situation. Autologous T cells have been expanded either normally or on a clinical relevant dose of dexamethasone that still yields a predominantly $CD3^+CD8^+$ product. The dexamethasone treatment does increase the percentage of $CD3^+CD4^+$ helper T cells as well as Tregs. The normal T cells are labeled with PKH-26, while the dexamethasone treated fraction is labeled with PKH-67, a green fluorescent dye. Both T-cell populations are monitored using flow cytometry of the peripheral blood and fine needle aspirates of tumor-involved lymph nodes. Currently, three patients have been infused under this protocol (O'Connor, data not published as trial is still currently enrolling at the time of publication).

The attractive benefits of adoptive polyclonal T-cell therapy range from the ability to produce long-lasting memory T cells to the reconstitution of a lympho-depleted immune system post chemo or radiotherapy. The use of chemotherapy and radiotherapy prior to infusion may allow for the expression of neo-antigens by the remaining tumor cells thus providing new TAAs to stimulated and clonally expand the infused T-cell population, which may explain the increased efficacy observed in canines treated with combination immunotherapies[10, 11].

Genetically Modified T-cell Therapy to Treat Canine LSA

Increasing efficacy of ACT has long been a goal of immunologists. New technology has allowed the detections of novel and ubiquitously expressed tumor antigens. Combined with recent advances in gene transfer and safety, T cells can be engineered to engage in target-specific killing. Stable genetic modifications may occur with the use of retroviral vectors, lentiviral vectors, TALENS, zinc fingers nucleases, and transposon/transposase systems. mRNA transfections are currently being tested as alternative method to stable DNA integration[12].

Originally, genes were transferred into T cells using γ-retroviral vector. These vectors are not widely used for transfection in clinical settings as concerns have been raised that these viruses may allow for malignant transformation in transfected primary T cells, though this is thought to be a rare occurrence in mature T cells. More recently, the HIV-1 based self-inactivating lentiviral vectors have been used to insert genes into the genome of both proliferating and quiescent T cells and has been approved for human clinical use[13]. There are four major disadvantages of lentiviral based delivery systems: i) time-consuming and expensive to prepare as high titers are required for gene therapy; ii) virus preparation may create risks of infectious agent contamination; iii) viral vectors may produce adverse cellular reactions; and iv) propagation of vectors may limit the size of the insert and require additional replication regulation sequences.

Non-viral vectors such as the *Sleeping Beauty* (*SB*) transposon system have also been investigated and have been approved for clinical use[14]. The *SB* system is comprised of two components, the transposon which carries the gene of interest between two inverted terminal repeat sequences and the *SB* transposase, which is a hyperactive enzyme that recognizes the repeat sequences. Once electroporated into the cell, the transposase cuts the gene of interest out of the transposon and inserts it randomly into the genome between TA sites[14]. The *SB* system is attractive for gene therapy use because of the low cost to generate vectors and low copy number integrations. However, the larger transposon size may inhibit genome integration.

While genetically modified (DNA) T cells can be efficacious and have potential long-term persistence, stable expression of the chimeric antigen receptor (CAR) or T-cell receptor (TCR) have led to concerns on the chronic toxic effects associated with targeting TAAs, like the B-cell antigen CD19 or epidermal growth factor receptor (EGFR), which are also expressed on normal cells. Temporary CAR expression may allow the targeting of TAAs in these categories. mRNA can be efficiently transfected into T cells to provide transient CAR or gene expression without DNA integration as mRNA is degraded by endogenous RNAses quickly in the cytoplasm and expression only last for several days. Current data shows that CAR-expression by RNA is efficacious, target specific, and may be clinically safer to use against certain targets[12].

There are currently over 30 human clinical trials investigating the application of genetically modified T cells for the treatment of cancer. Most of these trials are focused on B cell malignancies; however, trials implementing this technique for solid tumors also exist with limited success and significant side effects. The most successful trials have used a CD19-specific CAR in T cells for B cell malignancies. These CARs are second generation (i.e. they incorporate CD3ζ and CD28). The CD19-specific CAR was used in a small clinical study in 3 human patients with chemotherapy resistant B cell chronic lymphocytic leukemia. There were two clinical remissions and one long lasting partial remission. The T cells expanded out to 1000 times the original engraftment dose and persisted at these levels for six months. The only significant toxicity associated with this trial was the development of B-cell aplasia and hypoglobulinemia in these patients[13].

Because the goal of this therapy is engraftment of the modified T cells with long term survival in the host there is always the concern for toxicity. Additionally, the toxicity may become worse over time as the T cells expand in number. These cells are engineered to resist the normal signals for down regulation that cancers can secrete. For these reasons, it is important to consider a suicide gene that can allow the T cells to be eradicated in the event of a severe toxicity. One such strategy that has been employed is using herpes simplex virus thymidine kinase (HSV-tk). Incorporation of this gene allows the genetically modified T cells to be sensitive to gancyclovir or acyclovir therapy. This provides a safety net allowing for killing and inactivation of the immune response if desired. The main draw back to HSV-tk is that it can be immunogenic and lead to immune destruction of the T cells by the host. Other strategies include using a humanized caspase 9 dimer to induce apoptosis in the effector cells.

CONCLUDING REMARKS

Though the adoptive cellular treatment arm of immunotherapy is new to veterinary medicine, the successes and advances of knowledge into the intricacies of the immune system have renewed interest in clinical use for humans and companion canines alike. Comparative oncology and immunology research has better classified canine T cell and their role in tumor prevention, progression and treatment. Future trials are currently being developed to implement CAR-specific T-cell therapy in dogs with LSA and other neoplasms. Due to their ability to infiltrate and destroy solid tumor masses, non-specific and tumor-specific T-cell therapies may be able to improve survival outcomes, whether used alone or in combination with chemo-radio therapies. Additionally, using these strategies to program T cells to specifically target B cells may allow for treatment of immune mediated or allergic diseases as well. This type of approach has been used with rituximab in humans with much success for many different diseases ranging from immune mediated hemolytic anemia to asthma.

Adoptive T-cell therapy provides a new and exciting approach to cancer therapy in humans and pet animals. This new technique, though cumbersome, may provide the necessary therapeutic approach to push progression-free survival in our patients beyond one year and possibly even provide long term immune monitoring through immunologic memory in successfully treated patients.

REFERENCES

[1] Essand M, **J Intern Med.** 2013 Feb;273(2):166-81. [2] Declerck S, et al **Future Oncol**. 2014 Jan;10(1):91-105.[3] Ott PA, et al **Clin Cancer Res.** 2013 Oct 1;19(19):5300-9. [4] Greenberg PD, et al, **Ann N Y Acad Sci**. 1991 Dec 30;636:184-95. [5] Rosenberg SA, **Nat Rev Clin Oncol.** 2011 Aug 2;8(10):577-85. [7] Ruella M, et al, **Nat Rev Clin Oncol.** 2011 Aug 2;8(10):577-85. [8] Porrata LF, et al, **Blood Cancer J.** 2013 Apr 19;3:e110. [9] De Angulo G, et al, **Cancer.** 2008 Jan 15;112(2):407-15. [10] O'Connor et al, **Sci Rep**. 2012;2:249. [11] Peggs KS, et al, **Br J Haematol**. 2007 Oct;139(1):70-80. [12] Barrett KG, et al, **Drug Metab Dispos.** 2013 Aug;41(8):1505-13. [13]Porter DL, et al, **N Engl J Med.** 2011 Aug 25;365(8):725-33. [14] Singh H, **PLoS One.** 2013 May 31;8(5):e64138. [15]Cambridge G. et al, **J Autoimmun.** 2013 Dec 20. pii: S0896-8411(13)00149-2. [16] Listing J, et al, **Ann Rheum Dis.** 2013 Nov 29.

CLINICAL EVALUATION OF A CD40-ACTIVATED B CELL VACCINE TO PREVENT LYMPHOMA RELAPSE IN DOGS

Nicola J. Mason B.Vet.Med., PhD (Immunology), DACVIM (Small Animal Internal Medicine)
Philadelphia, PA

Introduction

Canine lymphoma is the most common hematopoeitic cancer in dogs with an estimated annual incidence of 30/100,000. Chemotherapy induces remission in 75-85% of dogs, however, the majority relapse with drug-resistant lymphoma within 8-10 months of diagnosis and most dogs die of their disease. Alternative approaches are required to improve overall survival. Immune therapy in the form of targeted monoclonal antibodies such as Rituximab, cell based vaccine strategies such as antigen-pulsed dendritic cells and most recently, genetically modified T cells re-directed to target universal B cell antigens such as CD19 and CD20 have all shown clinical benefit in human patients with lymphomas and leukemias. In veterinary medicine, different vaccine approaches aimed at generating anti-tumor T cell responses have been reported, some showing promising early evidence of safety and therapeutic efficacy. Here we will focus on our work with autologous RNA loaded CD40-activated B cells (CD40-B cells) that we have employed in the adjuvant setting to prevent lymphoma relapse.

Cell-based vaccine strategies

Cell-based vaccine strategies aimed at priming an effective anti-tumor cytotoxic T lymphocyte (CTL) response and generating immunological memory against multiple tumor-associated antigens have shown promise in the treatment of many different cancer types including lymphoma in humans.[1-6] Dendritic cells (DC) assume primary responsibility for T cell priming *in vivo* and as such have formed the basis of most cell-based cancer vaccines. However, their generation is technically challenging and the presence of different DC subsets, different manufacturing protocols and different vaccination strategies have made it difficult to compare trial results and determine the most beneficial DC vaccine approach.[7,8] Antigen-loaded CD40-B cells represent an alternative antigen-presenting cell (APC) to DCs with considerably less complexity and variability in their generation. Furthermore, like DCs, CD40-B cells can be loaded with tumor antigens in a variety of different ways and trigger both naïve and memory T cell responses *in vitro*.[9-12]

Dendritic cell vaccines

Dendritic cells are the most potent of the professional antigen-presenting cells (APC) and have formed the basis of most cell-based cancer vaccines aimed at stimulating effective anti-tumor immunity. The concept of this approach is that dendritic cell precursors are removed from the patient's peripheral blood, are cultured and matured *ex vivo* and then loaded with tumor-associated antigens (Fig. 1). These antigens are degraded by the proteasome and the resulting short peptides are loaded into MHC complexes within the endoplasmic reticulum and presented as peptide-MHC complexes on the surface of dendritic cells. These autologous dendritic cell vaccines are delivered intradermally or subcutaneously to the patient. Engagement of these activated, antigen-loaded dendritic cells with cognate T cells leads to tumor-specific T cell priming and activation. Activated tumor-specific T cells proliferate, produce cytokines and have potent cytotoxic capabilities that lead to tumor specific lysis. The safety and therapeutic efficacy of dendritic cell vaccines have been evaluated in multiple human clinical trials and the first dendritic cell based vaccine (for the treatment of prostatic carcinoma in humans) was licensed by the FDA in 2010 (Sipuleucel-T).[13] However, the generation of DC vaccines is technically challenging. Dendritic cells are terminally differentiated cells and do not proliferate in culture. Therefore, patients require large volume leukapheresis to procure sufficient numbers of DC precursors for vaccine generation. We have optimized protocols for large volume leukapheresis in dogs ([14] and data not shown), and although the procedure is well tolerated in healthy dogs, it is lengthy and cumbersome and not a trivial undertaking. Therefore, we and others have evaluated the use of B cells as an alternative APC to DC for use in cell-based vaccines.[15,16]

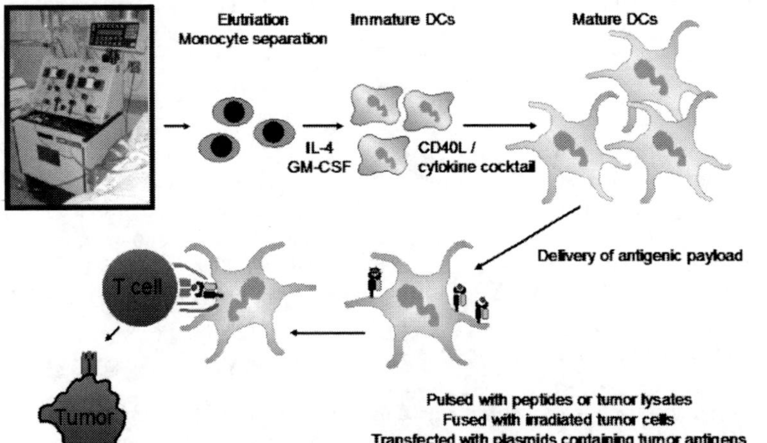

Fig. 1. Schematic of dendritic cell vaccine generation. Patients undergo large volume apheresis to isolate mononuclear cells from their circulation. Monocytes are separated by elutriation and immature dendritic cells develop in the presence of IL-4 and GM-CSF. DC maturation is achieved by engagement of surface CD40 with CD40L. Mature DCs are loaded with antigen using one of the several listed approaches and these antigen-loaded, activated DCs are administered to the patient.

CD40-B cell vaccines

Like DCs, B cells are professional APCs. However, unlike DC B cells can be readily expanded *in vitro* when cultured with IL-4 in the presence of CD40 cross-linking so that large numbers of activated CD40-B cells can be generated directly from a small volume of peripheral blood, eliminating the requirement for large volume leukapheresis.[10,17] In addition, activation of B cells via CD40 enhances their antigen-presenting function by inducing high-level surface expression of MHC and co-stimulatory molecules, stimulating the secretion of inflammatory cytokines such as IL-12 and up-regulating anti-apoptotic molecules. Like DC, CD40-B cells can be loaded with tumor RNA (the antigenic payload) and used to trigger *polyvalent* T lymphocyte responses against *multiple* TAA *in vitro*. The use of RNA as the antigenic payload allows for gene transfer without the need for viruses or vectors and allows an MHC independent, multiple antigen targeting approach. *In vitro* studies have shown that antigen-loaded CD40-B cells can prime both CD4+ and CD8+ T cells and can boost memory T cell responses.[9-12] These characteristics of CD40-B cells suggest that they represent a viable alternative to DCs in cancer vaccination protocols. We set out to explore whether this approach could be employed in the treatment of canine lymphoma.

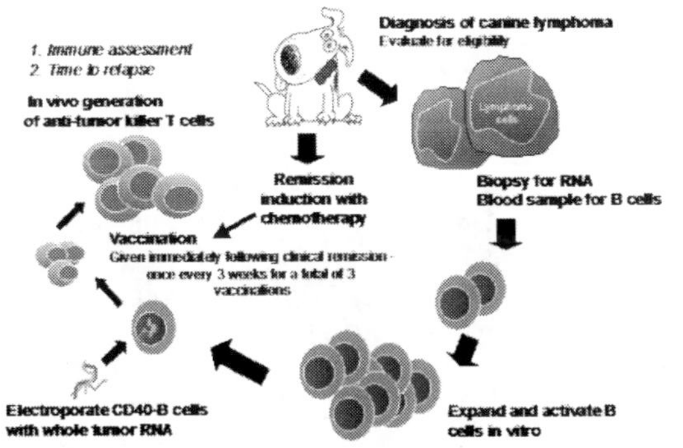

Fig. 2. Outline of the key points in CD40-B cell vaccine generation and clinical use. Following the diagnosis of lymphoma, CD40-B cells are generated from peripheral blood mononuclear cells (PBMCs) and autologous whole tumor RNA is isolated from a lymph node biopsy. When remission is confirmed, dogs are vaccinated once every 3 weeks for a total of 3 vaccines and immune and clinical responses to vaccination are determined.

Firstly, we described an *in vitro* culture method to activate and expand canine B cells from PBMCs taken from both healthy dogs and dogs with lymphoma.[15] PBMCs from healthy dogs and dogs with lymphoma were cultured with irradiated KtCD40L feeder cells (human K562 cells transfected with human CD40L) in the presence of recombinant canine (rc)IL-4 and cyclosporine. 5-10mls of peripheral blood yielded approximately 3-4 million PBMCs and approximately 2-5 x 10^5 naive B cells. After 4 weeks of culture between 40 and 80 million $CD79a^+$ CD40-B cells were generated from healthy dogs and dogs with lymphoma. We demonstrated that these CD40-B cells can be efficiently transfected with RNA. Transfection efficiencies of >80% were achieved in CD40-B cells generated from both healthy canine donors and dogs with lymphoma.[15] We then showed that CD40-B cells can promote functional, antigen specific T cell responses. By loading CD40-B cells with mRNA encoding the canine distemper virus hemagglutinin (CDV-HA) antigen and co-culturing them with autologous PBMCs for 1 week, CDV-HA-specific IFN-γ production was demonstrated by ELISpot. This result was obtained from PBMCs taken from healthy dogs, dogs with lymphoma at the time of diagnosis and dogs with lymphoma that were in clinical remission following induction chemotherapy. Taken together, our results indicated that it is feasible to generate large numbers of CD40-B cells from the peripheral blood of healthy dogs and dogs with lymphoma and that these canine CD40-B cells transfected with mRNA can induce antigen-specific, functional Th1 cell responses *in vitro*.

Next we performed a small, phase I clinical trial to evaluate the safety and therapeutic efficacy of whole tumor RNA-loaded CD40-B cells to induce anti-tumor immune responses and prolong overall survival in dogs with NHL.[16] Peripheral blood was taken from patients at the time of lymphoma diagnosis. At the same time one peripheral lymph node was removed to a) confirm the diagnosis of lymphoma, b) immunophenotype the tumor and c) provide autologous whole tumor RNA for vaccine generation. A standard 22-week L-CHOP induction chemotherapy protocol was used. Dogs were only eligible for vaccination if they were confirmed to be in clinical remission following induction chemotherapy. This was important to test the ability of the vaccine to induce anti-tumor immunity in the setting of minimal residual disease, to avoid tumor-related immunosuppressive effects on T cell priming and activation. At this time, dogs were vaccinated with autologous CD40-B cells electroporated with whole tumor RNA in one flank and electroporated with CDV-mRNA in the other flank (positive control). Immunological responses were determined by ELISpot analysis and time to relapse and overall survival were recorded. We found that 4/19 vaccinated dogs experience prolonged first remission times.[16] CD40-B vaccination did not significantly prolong time to relapse in the remaining 15 dogs. However, of 10 vaccinated dogs treated with rescue chemotherapy, 4 had significantly prolonged second remission durations and significantly improved lymphoma-specific survival compared to unvaccinated controls (Fig.3). These results suggest that our current CD40-B vaccine and vaccination protocol generates tumor-specific effector and memory T cell responses but that either this approach fails to generate *sufficient* long-lived, functional memory T cells *in vivo* or memory T cells that are generated are insufficient to prevent tumor

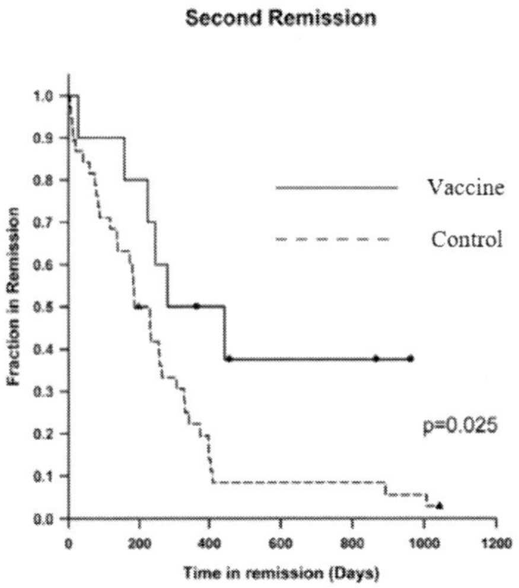

Fig. 3. Kaplan–Meier curve for lymphoma-specific survival (LSS) of relapsed dogs that received rescue chemotherapy.

recurrence. They also suggest that treatment of recurrent disease with chemotherapy in some way boosts vaccine-induced T cell responses that lead to improved overall survival (Fig.3), a previously recognized phenomenon in human cancer patients.[13,18]

CD40-B vaccines and memory T cells

Successful cancer vaccine strategies must not only act to boost the numbers of potent tumor-specific CD8+ T cells but also generate high quality memory T cell responses that mediate long-lasting anti-tumor immunity. CD4+ T cell help is required for the maintenance and secondary expansion of long-lived memory CD8+ T cells.[19] RNA loaded LPS activated B cells have been shown to stimulate antigen-specific CD4+ T cells *in vivo* and CD40-B cells have been shown to *boost* anti-viral memory T cell responses *in vitro.*[12,20] However, the ability of CD40-B cells to *generate* functional CD8+ memory T cells *in vivo* remains unclear. *In vitro* studies suggest that engagement of Toll like receptors (TLR) expressed on the surface of antigen-presenting cells (B cells, monocytes/macrophages and dendritic cells) using pathogen associated molecular patterns (e.g. bacterial CpG motifs, LPS, flagellin etc.) may augment the ability of APCs to generate memory T cell responses. Engagement of TLRs leads to the production of cytokines and chemokines that initiate and orchestrate innate and adaptive immune responses. Previous studies have demonstrated that the TLR9 agonist CpG synergizes with CD40 agonists and augments the expression of MHC I and II together with the co-stimulatory molecules CD40, CD70 and CD86 on the surface of both naïve and memory B cells.[21] Furthermore, CpG enhances CD40-B cell mediated T cell proliferation and cytokine production.[21] Interestingly, in mouse studies, the systemic administration of CD40L/CpG lead to the production of antigen-specific T cells that displayed enhanced cytokine production and cytolytic activity and generated a larger pool of functional memory T cells when compared to either agent alone.[22] Although these later findings cannot specifically be attributed to the effects of CD40 ligation and CpG on B cells, this is more likely than a direct effect on the T cells themselves. Taken together, these results suggest that the use of CpG in the maturation of CD40-B cells may enhance their ability to generate functional memory T cell responses *in vivo*. As such, the use of CpG in CD40-B cell vaccine generation may lead to a more efficacious vaccine that induces long-lived memory T cells.

To this end we are now evaluating the use of CpG in the generation of optimized CD40-B cells. We are also exploring whether repeat vaccination with RNA-loaded CD40-B cells every 2 months can maintain anti-tumor immunity and delay or prevent disease relapse. This second clinical trial is now underway. While our approach here has focused on the treatment of lymphoma, efforts to increase our understanding of how CD40-B cells may be optimized to provide long term anti-tumor immunity will hopefully have significance for other cancers given that CD40-B cells can be electroporated with RNA extracted from a wide range of tumors and employed in their treatment in the future.

REFERENCES

1. Reichardt VL, B.P. *Cytotherapy* **6**, 62-7 (2004). **2**. Banchereau, J. et al. *Cancer Res* **61**, 6451-8. (2001). **3**. Fong, L. et al. *Proc Natl Acad Sci U S A* 98, 8809-14. (2001). **4**. Hsu, F.J. et al. *Nat Med* 2, 52-8 (1996). **5**. Nestle, F.O. et al. *Nat Med* 4, 328-32 (1998). **6**. Thurner, B. et al. *J Exp Med* 190, 1669-78 (1999). **7**. Steinman, R.M. & Dhodapkar, M. *Int J Cancer* 94, 459-73. (2001). **8**. Jefford, M., et al. *Lancet Oncol* 2, 343-53 (2001). **9**. Coughlin, C.M., et al. *Blood* 103, 2046-54 (2004). **10**. Schultze, J.L. et al. *J. Clin. Invest.* 100, 2757-65 (1997). **11**. Lapointe, R. et al. *Can Res* 63, 2836-2843 (2003). **12**. von Bergwelt-Baildon, M.S. et al. *Blood* 99, 3319-25. (2002). **13**. Kantoff et al. *N. Engl J. Med.* 363, 411-422 (2010). **14**. Callan MB, et al. *Transfusion* 48, 2214-21 (2008). **15**. Mason, N.J. et al. *Gene Ther* (2008). **16**. Sorenmo KU, et al. *PLoS One* 6, 24167 (2011). **17**. Banchereau, J. et al. *Science* 251, 70-72 (1991). **18**. Antonia et al. *Clin Cancer Res* 12, 878-887 (2006). **19**. Antony, P.A. et al. *J Immunol* 174, 2591-601 (2005). **20**. Park, M.Y. et al. *Eur J Immunol* 38, 2106-17 (2008). **21**. Carpenter EL. et al. *J Transl Med* 11, 93 (2009). **22**. Ahonen, C.L. et al. *J Exp Med* 199, 775-84 (2004).

RESCUE THERAPY FOR CANINE LYMPHOMA: THE 45-DAY HURDLE

David M. Vail, DVM, DACVIM (Oncology)
Madison, WI

INTRODUCTION

When treating lymphoma, the fundamental goals of chemotherapy are to induce a complete durable (> 6 months) first remission (termed ***induction***), to reinduce a remission when the tumor recrudesces (or the patient relapses) following achievement of a remission (termed ***reinduction***), and, finally, to induce remissions when the cancer fails to respond to induction or reinduction using drugs not present in the initial protocols (termed ***rescue***). Despite the plethora of *induction* chemotherapeutic protocols for dogs with lymphoma developed over the past 15-20 years, most are modifications of 'CHOP', initially designed for human oncologic use. CHOP represents combinations of cyclophosphamide (C), doxorubicin (H, hydroxydaunorubicin), vincristine (O, Oncovin®) and prednisone (P). Conventional CHOP-based chemotherapy induces remission in approximately 80 to 95% of dogs with overall median survival times of 10 to 12 months. Eventually, the majority of dogs that achieve a remission following institution of induction will experience a relapse of their lymphoma. This usually represents the emergence of tumor clones or tumor stem cells[1] that are inherently more resistant to chemotherapy than the original tumor, so-called MDR clones that either were initially drug resistant or became so following exposure to selected chemotherapy agents. *Reinduction* and/or *rescue* therapy then becomes necessary. It should be pointed out, that most of the protocols discussed in this presentation were reported prior to the publication the Veterinary Cooperative Oncology Group (VCOG) consensus document for response evaluation criteria for peripheral nodal lymphoma in dogs (v1.0)[1] and as such, the preferred temporal endpoint measure of progression-free survival (PFS) for which efficacy comparisons between different protocols should be compared is not available. Further, few of the published protocols discussed include sufficient numbers for adequate statistical power and fewer compare treatment protocols in a randomized prospective fashion. In addition, staging, inclusion, and response criteria vary considerably between protocols presented. Therefore, evaluations of efficacy between the various protocols are subject to bias making direct comparisons difficult and indeed precarious.

REINDUCTION THERAPY

At the first recurrence of lymphoma, it is recommended that *reinduction* be attempted, first by reintroducing the *induction* protocol that was initially successful, provided the recurrence occurred temporally far enough from the conclusion of the initial protocol (e.g., ≥ 2 months) to make *reinduction* likely. Attention must be given to the cumulative dose of doxorubicin that will result from *reinduction*, and baseline cardiac assessment, the use of cardioprotectants, alternative drug choices, and client education should all be considered. In general, the length of the *reinduction* will be half that encountered in the initial therapy; however, a subset of animals will enjoy long-term *reinduction*, especially if the dog completed the initial *induction* protocol and was currently not receiving chemotherapy for several months when relapse occurred. Nearly 80 - 90% *reinduction* rates can be expected in dogs that have completed CHOP-based protocols and relapse while not currently receiving therapy.[2,3] The duration of the remission with a *reinduction* CHOP-based protocol is predicted by the duration of the interval between protocols and the duration of the first remission.[2] In their report, Flory and others found that 78% of dogs achieved a complete response (CR) to their *reinduction* CHOP protocol, with a median second remission duration of 159 days. Further, dogs having at least a 4 month remission following their *induction* CHOP experienced *reinduction* remissions approximately twice as long (214 days *versus* 98 days) as dogs having *induction* remissions less than 4 months, suggesting time off chemotherapy is an important predictor of *reinduction* success. This implies that dogs who relapse quickly after completion of their initial CHOP protocol may be better served by initiating *rescue* therapy immediately and by-passing *reindution* altogether; however, the absolute temporal cut-off for what constitutes a 'quick' relapse is currently unknown.

RESCUE THERAPY

If *reinduction* fails or the dog does not respond to the initial *induction*, the use of so-called *rescue* agents or protocols may be attempted. These are single drugs or drug combinations that are typically not found in standard CHOP protocols and are withheld for use in the drug-resistant setting. The most common *rescue* protocols used in dogs include single-agent use or a combination of actinomycin D, mitoxantrone, doxorubicin (if doxorubicin was not part of the original induction protocol), dacarbazine, temozolomide, lomustine (CCNU), L-asparaginase, mechlorethamine, vinblastine, vinorelbine, procarbazine, prednisone, and etoposide. Some *rescue* protocols are relatively easy and convenient single-agent treatments, whereas others are more complicated (and expensive) multi-agent protocols, such as MOPP (M, mechlorethamine; O, Oncovin®; P, procarbazine; P, prednisone). Overall *rescue* response rates of 40 to 90% are reported; however, responses are usually not durable with median response durations of 1.5 to 2.5 months being typical, regardless of the complexity of the protocol. A small (< 20%) subset of animals will enjoy longer *rescue* durations. ***Table 1*** provides a summary of canine *rescue* protocols with published results.[4-22] It is critical to note that current published data from *rescue* protocols do not include sufficient numbers for adequate statistical power, nor do they compare protocols in a randomized prospective fashion. Therefore, evaluations of efficacy among various protocols are subject to substantial bias, making direct comparisons difficult and indeed precarious. Therefore, at present, choice of a particular *rescue* protocol should depend on several factors, including cost, time commitment required, efficacy, adverse event profile (i.e., toxicity), and experience of the clinician with the protocols in question. As the complexity of rescue protocols does not yet appear to be associated with significant gains in rescue durability, the author tends to choose simpler and less costly protocols (e.g., LAP; ***Table 1***). That being said, the use of multiple varied rescue protocols, switching as needed based on response, continue as long as clients are comfortable with the quality of life of their dog. This sequential application of several different rescue protocols can result in several months of extended survival with acceptable quality of life.

STRATEGIES TO ENHANCE EFFECTIVENESS OF THERAPY IN DOGS WITH LYMPHOMA

Further advances in remission and survival durations await the development of new methods of delivering or targeting traditional chemotherapeutic drugs, new generations of chemotherapeutic drugs, or novel non-chemotherapeutic treatment modalities. Mechanisms of avoiding or abrogating multidrug resistance, enhancing tumor apoptosis (programmed cell death), tumor ablation (e.g., bone marrow ablation and reconstitution), and immune-system reconstitution, as well as novel immunomodulatory therapies (e.g., monoclonal antibodies), are all active areas of investigation in veterinary oncology.

Table 1: Summary of Published Response Data for Rescue Protocols in Dogs=

Protocol	Animals	Overall Response	Complete Response	Median Response Duration‡	Median Duration of Complete Response	References
Actinomycin-D	25	0%	0%	0 days	0 days	4
Actinomycin-D	49	41%	41%	129 days	129 days	5
BOPP	14	50%	28%	NR	130 days	6
DTIC	40	35%	3%	43 days	144 days	7
DTIC-anthracycline	35	72%	62%	50 days	NR	8
DMAC	54	72%	44%	61 days	112 days	9
Ifosfamide	39	3%	0%	112 days	112 days	10
Lomustine	43	27%	7%	86 days	110 days	11
Lomustine	39	60%	NR	15 days	NR	12
LAP	48	77%	65%	70 days	90 days	13
LAP	31	87%	52%	63 days	111 days	14
Lomustine-DTIC	57	35%	23%	62 days	83 days	15
LOPP	44	52%	27%	NR	112 days	6
LOPP	33	61%	36%	98 days	NR	16
Mitoxantrone	44	41%	30%	NR	127 days	17
Mitoxantrone	15	47%	47%	84 days	84 days	18
Mitoxantrone	19	21%	NR	NR	NR	19
MOMP	88	51%	12%	56 days	81 days	20
MOPP	117	65%	31%	61 days	63 days	21
MPP	41	34%	17%	56 days	238 days	22
Temozolomide-anthracycline	18	72%	50%	40 days	NR	8

= - Few of these protocols include sufficient numbers for adequate statistical power and fewer compare treatment protocols in a randomized prospective fashion. In addition, staging, inclusion, and response criteria vary considerably between protocols presented. Therefore, evaluations of efficacy between the various protocols are subject to bias making direct comparisons difficult and indeed precarious.

‡ - Various temporal response endpoints were used including disease-free interval, time to progression and progression free survival.

Abbreviations: BOPP = BCNU (carmustine), vincristine, procarbazine, prednisone; DOX = doxorubicin; DMAC = dexamethasone, melphalan, actinomycin-D, cytosine arabinoside); DTIC = decarbazine; LAP = lomustine, L-asparatinase, prednisone; LOPP = lomustine, vincristine, procarbazine, prednisone; MOMP = mechlorethamine, vincristine, melphalan, prednisone; MOPP = mechlorethamine, vincristine, procarbazine, prednisone; MPP = mechlorethamine, procarbazine, prednisone; NR = not reported

REFERENCES

1. Ito D, et al..J Vet Intern Mid 2011;25:890-896.
2. Flory AB, et al. J Am Vet Med Assoc 2011;238:501-506.
3.Garrett LD, et al. J Vet Intern Med 2002;16:704-709.
4.Moore AS, et al. J Vet Intern Med, 1994;8:343-344.
5.Bannink EO, et al. J Am Vet Med Assoc 2008;233:446-451.

6. LeBlanc AK, et al. Vet Comp Oncol 2006;4:21-32.
7.Griessmayr PB, et al. J Vet Intern Med 2009;23:1227-1231.
8.Dervisis NG, et al J Am Vet Med Assoc 2007;231:563-569.
9.Alvarez FJ, et al.. J Vet Intern Med 2006;20:1178-1183.
10. Rassnick KM, et al. J Vet Intern Med 2000;14:271-276.
11. Moore AS, et al. J Vet Intern Med, 1999;13:395-398.
12. Rusk A, et al. Clin Cancer Res 2006;12:7456-7464.
13. Saba CF, et al. J Vet Intern Med 2009;23:1058-1063.
14 .Saba CF, et al. J Vet Intern Med 2007;21:127-132.
15. Flory AB, et al. J Vet Intern Med 2008;22:164-171.
16. Fahey CE, et al. J Am Vet Med Assoc 2011;239:209-215.
17. Moore AS, et al. J Am Vet Med Assoc 1994;204:1903-1905.
18. Lucroy MD, et al. J Vet Intern Med 1998;12:325-329.
19. Ogilvie GK, et al. J Am Vet Med Assoc 1991;198:1618-1621.
20. Back AR, et al. Vet Comp Oncol 2013;E-pub ahead of print. Aug 5. Doi: 10.1111/vco.12055.
21. Rassnick KM, et al. J Vet Intern Med 2002;16:576-580.
22. Northrup NC, et al. Vet Comp Oncol 2009;7:38-44.

DIPS AND DNA: MODEL OPPORTUNITIES

Meryl P. Littman, VMD, DACVIM (SAIM)
Philadelphia, PA

INTRODUCTION

Diplomates are involved in accurate phenotypic characterizations of diseases for animals. We should all be banking DNA samples from animals with well-characterized phenotypes, because genomic technology is now available to help us determine the underlying molecular basis for a disease or trait and to discover animal models which will help all species, including man. Geneticists are skilled in defining the genotypes of the DNA samples, with tools such as genome-wide association studies and gene sequencing of candidate genes, to find variant alleles associated with the disease or trait. Diplomates are important members of the research team, to accurately identify the phenotype associated with each DNA sample, to acquire samples and pedigrees from appropriately characterized affected animals as well as properly identified controls, and to maintain important long-term relationships with owners, breeders, veterinarians, geneticists, statisticians, and the whole research team. This is a most exciting and rewarding time as genetic tools may yield findings with a relatively small population of affecteds and controls, and model opportunities must not be missed!

GETTING STARTED: DEFINE AFFECTED CASES AND CONTROLS

To get started, you need to recognize a genetic trait/disease, define the phenotype of an affected as well as a valid control case, and bank DNA samples and pedigrees of cases with well-characterized phenotypes.

Recognize a Genetic Trait

A genetic predisposition for a trait or disease must first be recognized. Consider whether the entity is genetic or acquired. Is the trait more abundant in a specific breed irrespective of its popularity? Realize that a congenital (born with) finding is not necessarily genetic or familial. A finding observed in a cluster of relatives may suggest inheritance, however consider whether the litter was exposed to an environmental cause. A genetic predisposition may have been described in the veterinary literature already, or be waiting for you and colleagues to recognize it, or perhaps breeders (or a single breeder) will bring it to your attention. For example, in 1983 in the clinics at the Veterinary Hospital of the University of Pennsylvania, I met my first soft-coated Wheaten terrier (SCWT) and found it had protein-losing enteropathy (PLE). The owner was a breeder and told me that she thought the dog's "aunt" had something similar. I said "That's interesting. Please send me the medical file." And that's how a 30 year project began!

Characterize Affected Animals

Once a phenotype is recognized, it must be further characterized. There may be many differential diagnoses that cause the phenotype and possibly many diagnostic tests recommended in order to exclude other (non-genetic or genetic) causes. Once a number of affected cases are well-documented, you will be able to ascertain breed association, age (and range) of onset, any sex predisposition, progression/prognosis, etc. Select the criteria for inclusion as well as criteria for exclusion. The definition of an excluded case is not necessarily the definition for a control case (see below). Characterization of the phenotype is VERY important. In SCWT, we soon recognized that the breed was predisposed to several diseases that mimicked each other in presentation of clinical signs and laboratory values and were getting confused with one another. The SCWT breed is predisposed to PLE (due to inflammatory bowel disease (IBD) and/or intestinal lymphangiectasia (IL), PLN (protein-losing nephropathy), Addison's disease (atypical as well as typical), food allergies and atopy, and juvenile renal disease (JRD, aka renal dysplasia).[1-3] Combinations of these, especially PLE/PLN and Addison's/PLN were also seen, because of the effects of popular sires and a small foundation stock brought from Europe. Many dogs presented with gastrointestinal signs, hypoalbuminemia, and azotemia. It was important to separate these diseases with appropriate criteria for inclusion.[3]

Characterize Proper Controls

Control (unaffected) animals will likely be the same breed and sex as your affected population, and need to be of the same overall population, based on pedigree cluster analysis, otherwise false associations may be found (eg, if European and American lines are too divergent, there will be many genetic differences found which may have nothing to do with the trait but appear to be associated with it). If the trait or disease has a very early age (with narrow range) of onset, the controls do not have to be geriatric, they just need to be well past the age when the trait/disease would be observed. In SCWT dogs with PLE and/or PLN, the age of onset was later in life (often 4-8 years old but with no age limit), so we used geriatric (>14 years of age) unaffected dogs as controls. In order to document that they were unaffected, results of CBC, Chemscreen, urinalysis, urine protein/creatinine ratio, and histopathologic examination of kidneys and small intestine after death were sought. Be sure to check the phenotype for both affecteds and controls with documentation (clinical data). Do not take the owner's, breeder's, or even veterinarian's memory or word for it; ask to see the data/proof! When choosing DNA samples for analysis, use the most thoroughly documented, classic cases of each phenotype.

Bank DNA Samples

Samples of DNA should be saved before or after death. Cheek swabs are fine once a genetic test is available, but to identify the genetic marker(s), more DNA material is better. We ask for whole blood (EDTA/purple top), semen, or after death, frozen tissue, usually kidney, spleen, liver, tongue, a toe, or skin. Some breeders save docked puppy tails for DNA extraction. Frozen tissue samples are cumbersome and expensive to ship, requiring ice packs, styrofoam box, overnight handling, and possibly papers to get through US customs. Funding from breed clubs, AKC-CHF, or other grants may be needed to help acquire such samples and to pay for confirmatory tests such as histopathology, which owners/breeders may not be able to submit otherwise. Saving samples in frost-free freezers is not recommended because of the constant freeze/thaw cycles. Freezing at -70°C is recommended.

COMMUNICATIONS WITH BREEDERS AND COLLEAGUES FOR CASES AND SAMPLES

Breeders are an important link for the recognition of cases and for further communications to acquire data. Breeders may passively or actively become informed about the health status of dogs they have bred and sold and may be able to give you contact information of the current owners and their veterinarians, and be able to supply pedigree information and historical data of ancestors of affected cases. Pedigree analysis may suggest a simple mode of inheritance, or may indicate that the inheritance is complex. Breeders may know of dogs which became sick or died prematurely. Rumors will have to be investigated, but confidentiality is paramount or you will lose the trust of the community. Anonymous health surveys by the breed club may help show prevalence of signs of illness or disease entities without pointing fingers.

Education and teamwork with breeders at their Club meetings (local and national), availability for consultations with owners'/breeders' veterinarians concerning sick animals, Club publications, websites, and listserves are important ways to encourage recognition of the problem, defray any guilt feelings, and aid in the cooperation and collaboration needed for research which requires medical records, pedigrees, and samples sent in for histopathology and DNA banking. Spread the word that you are studying this entity in the breed and that you want to know about any suspected cases as well as healthy long-lived individuals. Sharing information about suspected or affected cases may be very difficult for breeders and they are to be commended when they contact you!

After a health survey showed 10-15% SCWTs were affected with kidney and/or gastrointestinal diseases, a forward thinking Health Committee of the SCWT Club of America convinced the Board and general membership that an Open Registry[4] would be helpful. The Open Registry shared information among breeders and helped stop rumors about which dog had which disease, at what age, out of which sire/dam, how long the dog lived, and how the diagnosis was documented. It raised awareness among owners, breeders, and veterinarians about the prevalence of these problems, helped acquire funding, and encouraged other researchers to get involved in studying these problems. Do you want to get involved as an administrator of an Open Registry, documenting whether animals should be listed as affected? It will take a lot of your time doing the consultations for affected animals and suspected cases, acquiring owner/co-owner consent forms, educating owners/breeders/veterinarians about the diagnostic tests, management protocols, and monitoring for the particular individual, genetic counseling, etc, but you will get the most cases and samples for your studies.

The SCWT Open Registry, which I have been working on since 1996, lists dogs meeting the criteria for having PLN, PLE, IBD, Addison's disease, and JRD. There are more than 1100 dogs listed in the 2013 update. We can see that PLN has been the most common problem reported (722) with PLE next most common (501), and many with combined PLE/PLN (279). We found that there was no age limit for PLE/PLN (often dogs are bred before they were known to be affected). Informative families were identified and then followed throughout their lives. In the field, the mode of inheritance appeared to be autosomal recessive or complex, however Dr. Vaden's colony of "Wheagles" at NCSU (affected SCWT x Beagle) suggested there may be a dominant mode of inheritance, because some of the first generation Wheagles were affected.[2] We theorized that there may be complex inheritance with multiple genes, variable expression, incomplete penetrance, modifiers, and/or environmental triggers.

COLLABORATE WITH GENETICISTS

For many years I felt as though I had a collection of hundreds of 33-1/3 rpm albums (ie, DNA samples) but I had no record player so I didn't know what was on my records even though they were labelled (ie, with their phenotypes). With the help of geneticist Dr. Paula Henthorn, Claire A. Wiley, Michael Raducha, and Junlong Liu, and their application of the new genetic technology tools (ie, we found a record player), now we can hear the 'music' (ie, discover their genotypes)!

Breeds have extensive linkage disequilibrium and long haplotypes which make them ideal models for genetic research because relatively few cases and controls may yield statistically significant findings. The Online Mendelian Inheritance in Man (OMIM) compendium at http://www.ncbi.nlm.nih.gov/pubmed/ is a wonderful resource to find possible candidate genes known to cause certain phenotypes. The same phenotype in different breeds may be caused by identical mutations or a variety of alterations in the same candidate gene or in different genes, ultimately requiring different genetic tests to detect carrier status. Gene sequencing (fine mapping) of candidate genes may reveal additions, deletions, or single-nucleotide polymorphisms (SNPs), thereby associating variant alleles with a phenotype of interest. Finding variant alleles in candidate genes reveals the underlying molecular basis for the phenotype which is different than finding a genetic marker that happens to be linked with the trait, but is not itself causative.

For our SCWT studies, an overwhelming number of candidate genes existed, involving perhaps immunodysregulation, disorders of cell adhesion/tight junctions, abnormalities of regulatory genes for development/growth and maintenance of blood and lymphatic vasculature, etc. Pathologists were still unsure whether the main lesion of SCWT PLN was glomerulosclerosis or immune-mediated glomerulonephritis and whether the main lesion of SCWT PLE was IBD or IL. We needed to narrow the field of study so we decided to do a pilot genome-wide association study (GWAS) with 90 samples including cases of PLN, PLE, PLE/PLN, JRD, and non-affected geriatric (14-18 year old) Wheatens. The phenotypes were based on blood, urine, and histopathology findings.

The Illumina SNP-chip checks 177K SNPs in the dog genome (originally from a Boxer), looking for heterozygous SNPs among our population that might be informative, comparing DNA samples from affected cases with that of control geriatric SCWT dogs. We found >123K informative SNPs. Statistically significant SNPs mark the region of genome where mutation is present and these may be linked to variant candidate genes. These SNPS could be the actual mutation or just linked to it. Statistics comparing each pair of groups showed to our amazement that the PLN group was very different than the control group, and this was found by comparing samples from just 42 dogs with PLN with 20 geriatric controls.[5] The informative SNPs and p-values comparing PLN with control samples were mapped out for each chromosome. Many dots with low p-value scores were found on dog chromosome 1 (human chromosome 19), with the 3 lowest p-values at 10^{-8} (ie, very strong significance), which focused us to an area of interest which encompassed 321K base pairs; 64 of the lowest 100 p-values encompassed 9.7 megabytes or million base pairs on that chromosome. So the GWAS helped us focus on this area and about 65 genes which reside there. We checked OMIM to see what the genes were in this region and what their function was to see if we could narrow our focus more. There were 4 genes that might be associated with PLN in people, and the genetics team did the tedious gene sequencing of them to search for non-synonymous mutations (variant alleles).

Two of the 4 candidate genes did not show serious non-synonymous alterations but the other two did show SNPs which may alter permeability at the glomerular slit diaphragm and lead to PLN, based on PolyPhen2 assessment,[5] which checks the impact of the changes we found for the structure/function of the encoded proteins. One of these genes is NPHS1, which encodes nephrin, a major component in the slit diaphragm. There are more than 176 different mutations found in NPHS1 which cause PLN in humans, including infantile, juvenile, and adult-onset PLN. The SNP found in NPHS1 in SCWT changed a guanine → adenine in exon 22, which changed a glycine →arginine, at the extracellular edge of the fibronectin domain in nephrin. Guanine is conserved in all species that have a fibronectin domain so this may be an important change. The second candidate gene with a change that may be important is KIRREL2, which encodes for filtrin or Neph3, another slit diaphragm protein. There are no known mutations in KIRREL2 but decreased expression is associated with PLN in people. In KIRREL2, in exon 14, the found SNP changed cytosine →guanine which changes proline →arginine in an intracellular proline rich interval in filtrin/Neph3. Proline in this position is conserved all the way down to zebrafish and this may be an important change. Both nephrin and filtrin are cell adhesion molecules in the immunoglobulin superfamily. Mutations in the proteins of the slit diaphragm are called podocytopathies. There are a myriad of other slit diaphragm molecules which interact and hook up eventually to the actin cytoskeleton in the podocyte and cause the dynamic 3-dimensional changes in the shape and size of the podocyte and slit diaphragm between foot processes. Our SCWT PLN discovery is the first podocytopathy described in dogs.

Paula then developed the tests to check for these 2 specific SNPs in 145 banked SCWT samples with known phenotypes. The genes are close together on 1 chromosome, in linkage disequilibrium in all SCWTs tested so far. When one gene is normal on the chromosome, the other is normal, and when one is variant, so is the other. The 2 x 3 Fisher's test showed very strong statistical support for the association of these SNPs with the phenotype of PLN. The odds ratios were very high that a dog with 2 copies of the predisposing haplotypes ("homozygous positive") would have PLN compared with having 1 copy or no copies. The odds ratio for having 1 copy ("heterozygous") was < 1 in part of the 95% confidence interval, meaning it is sometimes no different than having no copies, however the odds ratio also includes high odds at the upper part of the confidence interval, so that dogs with 1 copy may be at intermediate risk for getting PLN. Dogs with no copies ("homozygous negative") are at low risk for developing PLN in their lifetimes.

Next we tested dogs of other breeds (non-Wheatens) to see if these mutations were found. We tested 747 dogs representing 114 breeds, including terrier breeds and breeds predisposed to PLN, including 9 dogs that died of PLN. I happened to be at Cornell speaking about our Wheaten work and Dr. Rebecca Kessler gave me a tour, including ICU where an Airedale terrier was sick with PLN. The owner gave permission for me to test its DNA from a leftover CBC tube, and it tested homozygous positive for the Wheaten PLN-associated variant alleles! Only 2 other dogs showed these variants among the 747 dogs: 1 heterozygous Airedale and a bloodhound that was heterozygous but only for the filtrin mutation.

The GWAS and sequencing of candidate genes in an area of interest helped answer why SCWT are predisposed to PLN. Structural changes causing glomerulosclerosis appear to be a primary or underlying problem. There may be more genes involved but we found a big piece of the puzzle. Perhaps a predisposition for allergy, PLE, inflammation, or increased circulating immune complexes is an additional insult on these subtly abnormal slit diaphragms. We now have a DNA-based tool[6] which can be done on cheek swabs, blood, semen, etc, to help owners know their pet's individual risk and to help breeders select the best pair to mate together to avoid producing puppies with 2 copies of the predisposing alleles, hopefully decreasing PLN in future generations. We will need to be careful and not discard all carriers lest we get new genetic problems from inbreeding, popular sires, and loss of genetic diversity. Prevalence studies for Wheaten[7] and Airedale[8] populations warn that there are many carriers of at-risk variant alleles (heterozygous and homozygous positive dogs) currently in these breeds. The first 1853 SCWT submissions showed that 44% of the dogs are heterozygous, 40% are homozygous negative, and 16% are homozygous positive. The first 217 Airedale submissions showed that 43% are heterozygous, 51% are homozygous negative, and 6% are homozygous positive. Consultations including genetic counseling will be important to help the community as a whole avoid producing dogs at risk but also avoid losing genetic diversity.

Although homozygous positive SCWT dogs are at high risk to develop PLN in their lifetimes, we don't know if that risk is (for example) 25% or 50% or 75%. Prospective studies need to be done, following dogs with known genotypes throughout their lives to assess their eventual health status (phenotypes). Additional genetic and environmental factors that may modify the phenotype in SCWT and Airedales need more study. It is possible that these 2 breeds have different risks for phenotypic expression. Future studies concerning PLE, Addison's disease, allergies/atopy, and JRD need to be done. We wish to study other breeds predisposed to PLN, and check for podocytopathies. We are very interested in retrievers with suspected Lyme nephritis; please send in data and DNA samples. With genomic technology now available, we urge YOU to save DNA samples from cases with phenotypes of YOUR interest, for future genetic studies. You may discover the underlying molecular basis for the disease process as well as a helpful genetic test!

REFERENCES

1Littman MP, et al. **J Vet Intern Med** 2000;14(1):68. 2Vaden SL, et al. **J Vet Emerg Crit Care** 2013;23(2):174. 3http://www.scwtca.org and click on "Health"; eg, http://www.scwtca.org/health/index.htm, http://www.scwtca.org/health/protocol-own.htm, http://www.scwtca.org/health/protocol-vet.htm, http://www.scwtca.org/health/healthnews.htm. 4http://www.scwtca.org/openregistry/index.htm. 5Littman MP, et al. **Mamm Genome** 2013;24(3-4):119. 6http://www.scwtca.org/health/dnatest.htm. 7Littman MP, et al. **J Vet Intern Med** 2013;27(3):736. 8Littman MP, et al. **J Vet Intern Med** 2014;28(3):TBD.

THE GENOMICS ERA IN CANINE IBD- WHAT DOES IT MEAN FOR THE PRACTITIONER?

Karin Allenspach DECVIM-CA, Royal Veterinary College, University of London, UK

INTRODUCTION

Inflammatory bowel disease (IBD) is the term applied to a group of poorly understood intestinal disorders that commonly affect animals and people. IBD in both people and animals is increasingly considered a consequence of uncontrolled intestinal inflammation in response to a combination of environmental, enteric luminal constituents (principally microbial and dietary) and immunoregulatory factors in genetically susceptible individuals[1].

In people, genetic susceptibility is linked increasingly to defects in innate immunity exemplified by mutations in the innate immune receptor NOD2/CARD15, that in the presence of the enteric microflora may lead to up-regulated mucosal cytokine production, delayed bacterial clearance and increased bacterial translocation, thereby promoting and perpetuating intestinal inflammation[2]. More recently, genome-wide association studies have shown IBD-associated polymorphisms in pathways important for intracellular bacterial clearance (autophagy-related 16-like protein 1 (ATG16L1) and in genes regulating Th_{17} cell maturation and Th_{17} cytokines, which are upregulated in the mucosa of people with IBD[3,4]. Thus, the common theme emerging in IBD genetics involves the abnormal interfacing of innate immunity with intestinal microbes, leading to abnormal inflammatory cytokine production and clinical disease.

BREED PREDISPOSITION IN CANINE IBD

In dogs, the concept of enhanced genetic susceptibility to IBD has long been supported by its' association with certain breeds. We have recently performed an epidemiological study to identify breeds that are associated with a higher risk of development of IBD in the south-eastern UK. The computer medical records at the Queen Mother Hospital for Animals, Royal Veterinary College dating from August 1, 2003 to December 31, 2009 were retrospectively searched for cases diagnosed with IBD. Five hundred and forty-six dogs with IBD were identified, representing 86 different breeds. The comparison group consisted of all dogs from these same 86 breeds without IBD admitted to the hospital during the same period that amounted to 27,463 dogs. The breeds at significantly higher risk of developing IBD compared with mixed-breed dogs consisted of weimaraner, rottweiler German shepherd dog (GSD), border collie and boxer. Further studies in other parts of the world are needed to identify more breeds that are at risk.

GENETIC BASIS OF IBD IN GSD IN THE UK: TOLL-LIKE RECEPTORS

Recent studies in the author's laboratory have identified some similarities in the molecular pathogenesis of IBD in GSD and human beings. The intestinal microbiome in GSD with IBD is altered and shows similarities to the dysbiosis found in people with IBD, with certain bacterial species, such as *E.coli*, and *Lactobacilli* being over-represented in the microbiome of GSD with IBD[5]. Furthermore, differentially high and low expression of pattern recognition receptors TLR4 and TLR5 was found in the mucosa of affected GSD, which reflects similar findings in human IBD and mouse models of IBD[6,7]. The possibility that allelic variation of pattern recognition receptors TLR2, 4 and 5 could contribute to a functionally abnormal response to the intestinal microbiota in GSD was explored using a candidate gene approach to identify single nucleotide polymorphisms (SNPs)[8]. Three non-synonymous SNPs in the TLR5 gene and 2 non-synonymous SNPs in the TLR4 gene, respectively, were identified and evaluated further in a case-control study and found to be significantly associated with IBD[8]. Moreover, the G22 mutation in the canine TLR5 (cTLR5) gene, which was associated with an 8x increased risk of developing IBD, has been shown to be functionally hyper-reactive to its natural ligand, flagellin[9]. These studies were performed *in vitro* using human embryonic kidney cells which were transfected with the mutant and wild type cTLR5, and confirmed with *ex vivo* studies in peripheral blood from dogs carrying the mutant or wild type alleles[9]. Furthermore, the TLR5 gene mutation G22 has also been shown to contribute to an increased risk of development of IBD in 38 other breeds in addition to the GSD in the UK[10]. This adds to the evidence that mutations in this particular gene may be causatively involved in the pathogenesis of the disease.

GENETIC BASIS OF IBD IN GSD IN THE UK: DLA-HAPLOTYPES

Major histocompatibility complex (MHC) class II genes polymorphisms are associated with various canine immune-mediated diseases, including some typical of GSDs such as anal furunculosis and chronic superficial keratitis. In addition, the MHC class II region is strongly associated with ulcerative colitis, a form of human IBD in people, as well as many other immune-mediated diseases in people. We therefore wanted to investigate whether MHC class II genes (DLA-DRB1, DLA-DQA1 and DLA-DQB1) are associated with IBD in GSDs. Genomic DNA was extracted from residual diagnostic EDTA blood samples of 56 GSDs with IBD and 50 geriatric, $\geq$ 8 years old, breed-matched controls without history of chronic GI signs. Dogs with previous or current history of concurrent immune-mediated or dermatological diseases were excluded from the study. MHC class II genotypes of all dogs were determined. The 'risk' haplotype DLA-DRB1*015:01/DQA1*006:01/DQB1*003:01 was significantly associated with IBD, and the haplotype DLA-DRB1*015:02/DQA1*006:01/DQB1*023:01 was found to be protective for IBD. Homozygosity for the risk haplotype further increased the risk of developing IBD suggesting a "gene-dosing" effect.

The findings of this study support the immunogenetic nature of IBD in GSD, and proves that many genetic factors likely play a role.

WHERE TO GO FROM HERE?

However, as revealed in GWAS in human IBD, it is unlikely that the mutations we found to date will account for all of the genetic susceptibility of GSD to IBD. Canine IBD is likely a polygenetic disease, and many more mutations with low penetrance could contribute to disease susceptibility. Other approaches such as genome wide association studies, as well as epidemiological studies of environmental risk factors, are warranted to try and define the pathogenesis further.

REFERENCES

1. Cavanaugh JA. IBD International Genetics Consortium: international cooperation making sense of complex disease. Inflamm Bowel Dis 2003;9:190-193. 2. Hugot JP, Chamaillard M, Zouali H, et al. Association of NOD2 leucine-rich repeat variants with susceptibility to Crohn's disease. Nature 2001;411:599-603. 3. Schmitz S, Hill S, Werling D, et al. Expression of trefoil factor genes in the duodenum and colon of dogs with inflammatory bowel disease and healthy dogs. Vet Immunol Immunopathol 2013;151:168-172. 4. Karagianni AE, Solano-Gallego L, Breitschwerdt EB, et al. Perinuclear antineutrophil cytoplasmic autoantibodies in dogs infected with various vector-borne pathogens and in dogs with immune-mediated hemolytic anemia. Am J Vet Res 2012;73:1403-1409. 5. Allenspach K, House A, Smith K, et al. Evaluation of mucosal bacteria and histopathology, clinical disease activity and expression of Toll-like receptors in German shepherd dogs with chronic enteropathies. Vet Microbiol 2010;146:326-335. 6. Cario E, Podolsky DK. Differential alteration in intestinal epithelial cell expression of toll-like receptor 3 (TLR3) and TLR4 in inflammatory bowel disease. InfectImmun 2000;68:7010-7017. 7. Rhee SH, Im E, Riegler M, et al. Pathophysiological role of Toll-like receptor 5 engagement by bacterial flagellin in colonic inflammation. Proc Natl Acad Sci U S A 2005;102:13610-13615. 8. Kathrani A, House A, Catchpole B, et al. Polymorphisms in the TLR4 and TLR5 gene are significantly associated with inflammatory bowel disease in German shepherd dogs. PLoS One 2010;5:e15740. 9. Kathrani A, Holder A, Catchpole B, et al. TLR5 Risk-Associated Haplotype for Canine Inflammatory Bowel Disease Confers Hyper-Responsiveness to Flagellin. PLoS One 2012;7:e30117. 10. Kathrani A, House A, Catchpole B, et al. Breed-independent toll-like receptor 5 polymorphisms show association with canine inflammatory bowel disease. Tissue Antigens 2011;78:94-101.

GENE THERAPY: HOW CLOSE ARE WE?

Andrew Specht, DVM, DACVIM (SAIM)
Gainesville, FL

BRIEF OVERVIEW

Gene therapy or therapeutic gene transfer involves transferring genetic material into a host to change the expression of one or more proteins in order to alleviate a disease process. The most familiar form of gene therapy to most people is "gene replacement therapy" in which a gene that encodes a functional protein is placed into a patient with genes that are lacking this functionality. However, there are actually several strategies that can be utilized for therapeutic gene transfer and even more variety in the specific techniques used to accomplish the transfer of genetic material. The primary strategies for treatment with gene therapy are commonly described as gene addition, gene correction, gene silencing, reprogramming, and cell elimination. The primary means of delivering genetic material to a host include electroporation (use of high voltage to push naked DNA or RNA into cells), liposomal packaging, viral vectors, or combining gene and cell therapy protocols.

The obvious potential for therapeutic gene transfer techniques to improve the quality of life for people suffering from genetic diseases has generated a great deal of research directed at developing techniques for safe and effective gene therapy treatments. In addition, the variety of techniques for delivery of genetic material and the numerous ways in which transferred genetic material could impact a patient has extended the potential benefits of gene transfer to include a wide variety of disease conditions beyond just genetic or heritable conditions. In almost all cases, the basic research required to develop gene transfer techniques to treat a disease condition requires research involving animal models.

The question in the title of this talk is admittedly vague and could be answered in any number of ways. The main focus of the talk, however, will be to take a closer look at what has been accomplished with therapeutic gene transfer techniques in companion animal species and to refocus the question on what would be required to start applying these techniques in clinical veterinary patients.

THERAPEUTIC GENE TRANSFER IN ANIMALS

Although the majority of experimental therapeutic gene transfer studies involving animals have been performed with mice, there has been increasing recognition that large animal (dogs and cats are typically included in this characterization) models of human disease offer significant advantages for preclinical assessment of proposed gene transfer protocols. Potential advantages compared to murine models include naturally occurring diseases with clinical presentations that might be similar to humans, less genetically homogenous populations, lifespans that allow some evaluation of long term outcomes and other effects of therapy, and larger body sizes which provides an opportunity to address issues related to scaling. Unfortunately, some of these advantages are also challenges and studies utilizing dogs or cats tend to be very costly and require more extensive clinical attention compared to laboratory rodents.

Already there have been promising results from a number of gene therapy trials utilizing large animal models with a wide variety of conditions ranging from diseases with very specific and well characterized underlying genetic abnormalities (i.e. glycogen storage disease, Leber's congenital amaurosis), to those with much more complex underlying physiology (diabetes, malignant melanoma) which have provided the basis to consider proceeding to clinical trials in people.[1-4] Veterinarians have an opportunity to apply the knowledge gained from gene transfer studies utilizing these animal models towards attempting to improve animal health. Given the success of gene therapy in treating diseases such as hemophilia, certain kinds of hereditary blindness, storage diseases, and in some cases of cancer, it seems reasonable to start evaluating whether these treatments could be used in clinical patients.

Disease Condition	**Animals treated**		**Technique**	
Leber's congenital amaurosis[5]	dogs	naturally occurring, research colony	Various AAV	Intraocular
Retinitis pigmentosa[6]	dogs	naturally occurring, research colony	Various AAV	Intraocular
Achromatopsia[7]	dogs	naturally occurring, research colony	AAV5	Intraocular
Glycogen storage diseases[1]	dogs	naturally occurring, research colony	Various AAV	IV
Lysosomal storage diseases (e.g. MPSVII)[8-10]	dogs, cats	naturally occurring, research colonies	Various AAV, RV	IV, Intracardiac
Muscular dystrophy[11]	dogs	naturally occurring, research colony	Various AAV	IM
x-linked myotubular myopathy[12]	dogs	naturally occurring, research colony	AAV8	IM
Canine leukocyte adhesion deficiency[13-14]	dogs	naturally occurring, research colony	LV or FV (w/ TSC)	IV
Hemophilia A & B[15-16]	dogs	naturally occurring, research colony	AAV, LV (w/ TSC)	local IV
Diabetes[3]	dogs	experimentally induced	AAV1	IM
Malignant melanoma[4]	dogs	naturally occurring, client owned	Adenovirus	Intratumor
Transmissible venereal tumor[17]	dogs	experimentally induced	Electroporation	Intratumor
Soft tissue sarcoma (injection site)[18]	cats	naturally occurring, client owned	Adenovirus	Intratumor

Table 1. Some disease conditions for which gene transfer techniques have been evaluated in dogs or cats. This is not a comprehensive list. (AAV = adeno-associated virus; RV = retrovirus; LV = lentivirus; TSC = transfected stem cells)

CONSIDERATIONS FOR THERAPEUTIC GENE TRANSFER IN CLINICAL VETERINARY PATIENTS

Biologic

Many of the challenges of creating a safe and effective therapeutic gene transfer protocol are present in both human and animal studies. As noted above there are a wide variety of treatment strategies and techniques and a wide variety of disease characteristics that directly affect the choice of strategy. Some genetic diseases are caused by distinct mutations in single genes while others may have a hereditary predisposition, but be caused by variations or mutations in several different genes and potentially influenced by environmental factors. Similarly, there may be a single tissue or cell type that is most affected or important for correction or there may be multiple tissues or cell types affected which all need to be corrected. When developing a treatment protocol it is critical to consider the goal of therapy, target of therapy, best means of delivery to that target, appropriate number of cells affected or level of gene product expressed, possible interactions with other intrinsic genetic sequences, and factors that might influence the duration of gene expression or cell survival, along with many other possible issues.

Just as there are several methods for using genetic material to influence protein expression, and several methods for delivering genetic material to the host, there are a number of different potential viral vectors that have been evaluated each of which has properties that might be advantageous or not depending upon the particular situation. The choice of vector influences the size of the gene(s) that can be transferred, the cell types that might be affected, the level of gene expression, the ability of the gene expression product to get to other parts of the body, the potential interaction(s) with the immune system and many other factors that could directly affect the outcome of therapy both in terms of efficacy and safety.

Whether dogs and cats are being evaluated as potential models for future gene therapy studies or are themselves the potential beneficiaries of this type of treatment, it is clear that there is still a lot to learn about how the immune system reacts to various vectors and how this might impact treatment. In a very basic sense, concerns about immune responses can be dived into three main potential problems: 1) pre-existing immunity against the vector, 2) induced immunity against the vector, and 3) immune responses directed against the transgene product.

In the case of AAV vectors, which are among the most commonly utilized, there are species differences with respect to immune responses and even in the same species there might be different reactions to different serovars or delivery methods. In humans, serologic studies indicate that most people are exposed to a wild-type AAV during childhood and the prevalence of measureable antibody titers against AAV vector serotypes in adults ranges from 20-60% in different populations and studies.[19] Two recent studies did show that research dogs had high titers for anti-AAV6 neutralizing antibodies.[20-21] Similar titers were not identified in other potential animal models including mice, rats, pigs, sheep, and rabbits.[20] There is only one known wild type canine AAV which is not closely related to AAV6 and is not associated with any recognized disease.[20] There is no feline AAV species listed by the International Committee on Taxonomy of Viruses.[22] Another related concern is whether there might be cross reactivity to other common viral infections or to vaccines used to prevent viral infections. It is important to recognize that AAV is not the same as adenovirus, of which there are a couple species that are pathogenic in canines. In fact, AAV species are much more closely related to parvoviruses. Although the wildtype canine AAV is not necessarily related to AAV serovars commonly used for gene therapy, it is possible that there could be some cross reactivity or immunity related to exposure. Similarly, in dogs and cats, exposure to other parvoviruses even in the form of vaccination could potentially result in circulating antibodies that might recognize some AAV serovars, though this does not seem to be the case for AAV6, AAV8, or AAV9.[21]

Regulatory

Due in part to the deaths of two research subjects and development of leukemia in two others, oversight of human gene therapy clinical trials has intensified in recent years.[23-24] Such trials are subject to review and oversight by multiple federal agencies and review boards/committees.[23-24] The Department of Health and Human Services (DHHS) oversees the Federal Drug Administration (FDA), Office for Human research Protection (OHRP), and National Institutes of Health (NIH). Specific regulatory functions of these agencies fall under the Center for Biologics Evaluation and Research (CBER), Office of Biotechnology Activities (OBA), and Recombinant DNA Advisory Committee (RAC). In addition to these groups, there are multiple institutional review boards or committees that also have a role in the oversight of these trials and report to those federal groups including Institutional review board (IRB), Institutional Biosafety Committee (IBC), and in some cases a Data Safety Monitoring Board (DSMB) or Data Monitoring Committee (DMC). Figure 1 is an attempt to illustrate these relationships. It is actually somewhat difficult to identify a single source of information that explains the roles of each

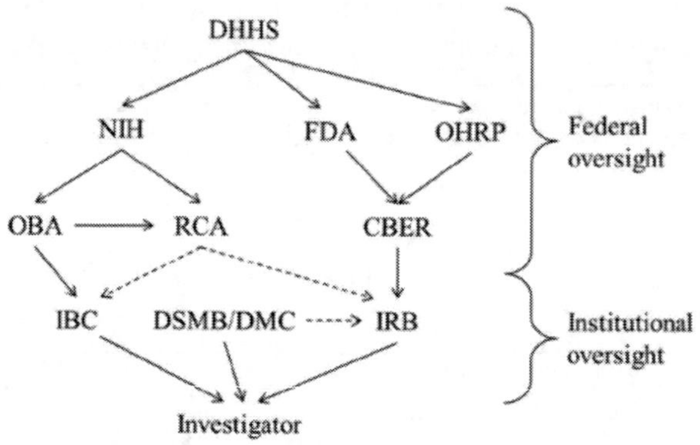

Figure 1. Schematic illustration of federal & the institutional oversight of human gene therapy trials.

of these groups or that provides specific standards. However all investigators involved in clinical trials are required to adhere to the relevant laws, regulations, and guidelines in the Code of Federal Regulations (CFR), there may be additional requirements set forth by the NIH which are updated periodically, and there are additional resources such as the International Harmonization Guidelines that help specify the responsibilities of investigators, study monitors, and sponsors.

At the current time, it is not entirely clear what federal agencies and or committees would be responsible for oversight or regulation of any purely veterinary clinical trial. A key consideration would be whether the results of a clinical trial in animal patients would even be utilized to provide supportive data for development of a potential therapy for humans.

In general, any data generated to support the potential utility of a proposed clinical vector would need to be generated under good laboratory practice (GLP).[24] Similarly, any vector administered as part of a human trial must be created under current good manufacturing practices (cGMP).[24] Although it is possible to make some changes in the vector development or production process when it is time to generate supplies for a human study, this may slow or prevent approval process and can impact the clinical relevance of the prior studies. Therefore any gene therapy trial in animals that is primarily intended as a preclinical trial for possible translation to human medicine would ideally conform to CLP, utilize cGMP, and be performed with full consideration of the regulatory oversight that will eventually be applied to any study involving human subjects.

If the goal is to generate a new therapy for dogs or cats that might be marketed, it may require federal as well as institutional oversight even in the early stages of development. On the other hand, if a gene transfer therapy was designed to treat an individual animal or to be marketed only to treat companion animals with a specific condition, the regulatory oversight structure might be very different. The FDA is likely to be the major regulatory agency for gene therapy studies in trials that involve companion animals, but there is also the potential for involvement from the USDA. There is also a question of whether animal recipients of therapeutic gene transfer could be classified as genetically engineered or genetically modified organisms. Attempts to contact the FDA's Center for Veterinary Medicine to discuss these regulatory and oversight issues had not been successful up to the point that these proceedings were being written, but more information should be available to share at the time of this talk.

Costs and other practical issues

Most therapeutic gene transfer studies are very costly and labor intensive. If there is a goal to use the data generated as evidence of safety or efficacy for potential translation to human clinical trials, the costs will likely be even higher. In most cases, development of specific gene therapy protocols directed primarily at companion animals will be prohibitively expensive. However, as technology improves our ability to safely manufacture gene constructs and vectors, these costs may decrease. When our research group looked into the cost of manufacturing vector/gene construct for an individual small dog clinical patient last year, we estimated that it would have cost us ~$10,000-12,000 to produce an adequate dose for IV administration without knowing in advance if this would be an effective therapy. Although that is very expensive compared with most treatments in veterinary medicine, if this type of on-demand treatment becomes a future reality there would likely be at least some clients willing to pay for this.

CONCLUSION

Overall, there are multiple answers to the title of talk. There are already many examples of gene therapy being applied to animals in experimental settings. There are even some therapeutic gene transfer techniques that have been used in clinical trials for client owned animals. So in these respects we are already past being "close" to using gene transfer techniques to treat animal diseases. These studies have shown enough promise to warrant continued investigation of therapeutic transfer techniques in the veterinary field. However, it will probably still be some time before there is any approved, marketed gene therapy product to treat a genetic disease in our companion animal patients.

REFERENCES

[1]Specht A. et al **J Biomed Biotech** 2011;2011:646257. [2]Stieger K, et al **ILAR J** 2009;50(2):206. [3]Callejas D, et al **Diabetes** 2013;62(5):1718. [4]Westberg S, et **al J Immunother** 2013;36(6):350. [5]Annear MJ, et al **Hum Gene Ther** 2013;24(10):883. [6]Beltran WA, et al **Proc Natl Acad Sci USA** 2012;109(6):2132. [7]Komáromy AM, et al **Hum Mol Genet** 2010;19(13):2581. [8]Ellinwood MN, et al **J Gene Med** 2004;6:481. [9]Bigg PW, et al **Mol Genet Metab** 2013;110(3):311. [10]Xing EM, at al **Mol Genet Metab** 2013;109(2):183. [11]Wang Z, et al **Mol Ther** 2007;15(6):1160-6. [12] Childers MK, et al **Sci Transl Med** 2014;6(220):220ra10. [13]Bauer TR Jr, et al **Mol Ther** 2013;21(5):964. [14]Hunter MJ, et al **Hum Gene Ther** 2011;22(6):689. [15]Du LM, et al **Nat Commun** 2013;4:2773. [16]Finn JD, et al **Blood** 2012;120(23):4521. [17]Chuang TF, et al **Int J Cancer** 2009;125(3):698. [18]Siddiqui F, et al **Mol Cancer Ther** 2007;6(1):380. [19]Mingozzi F, et al **Blood** 2013;122(1):23. [20]Rapti K, et al **Mol Ther** 2012;20(1):73. 21Shin JH, et al **Hum Gene Ther** 2012;23. [22]http://www.ictvonline.org/virusTaxonomy.asp?version=2012. [23]Cornetta **K Blood Cell Mol Dis** 2003;31:51. [24]Grilly BJ, et al **Cytotherapy** 2003;5(3):197

CAREER OPPORTUNITIES AND CONSIDERATIONS POST-RESIDENCY: THINGS YOU MAY NOT KNOW

Annie Corrigan, DVM, MS, DACIM (SAIM) and Sandra Axiak-Bechtel, DVM, DACVIM (Oncology)
Grenada, West Indies and Columbia, MO

INTRODUCTION

Perhaps one of the most difficult decisions to make post-residency is your career path. Should you choose academia or private practice? What are the benefits of each career choice? What information should I know about each path prior to making that decision? Once I've decided where to go, what are some tips that can make negotiation more clear? What is expected of me once I begin my new job? The goal of this 50 minute presentation is to provide an open discussion forum on available career paths post-residency, and to present a discussion on the unique attributes of job selection that are not usually part of the residency training programs.

CURRICULUM VITAE

Perhaps the best advice received upon finishing our residencies was to update the CV at least once monthly. Record everything you do – even those things that seem small (such as reviewing abstracts or journal articles, moderating talks, etc). You would be surprised at how much you actually accomplish! Most positions also involve yearly reviews, which is another stimulus to update your CV and make sure it is complete and accurate. Don't forget to list every research abstract with its funding source and resultant manuscript, and to note research mentees. Keep note of every course taught, number of hours, and student evaluations. Other tips on content: include your education, professional appointments and employment, publications, awards and honors, grants and fellowships, invited talks, conference activity/ participation, campus or departmental talks, teaching, research, service to the profession, departmental/ university service, community outreach, and professional memberships. You will often be able to find an example CV format at your university's tenure website.

ACADEMIA

Full-time equivalent (FTE)

FTE indicates workload with 1.0 equaling full time. In academia, it's used as a metric for measuring contribution to different areas of your job, with the big 3 being teaching, research, and service. An example standard position is 0.4 (40%) teaching, 40% research, 20% service. This differs for the practicing veterinarians because of clinical responsibilities, which can be considered teaching or service. For example, of the 52 weeks of the year, 20.8 should be spent on research, 20.8 weeks teaching, and 10.4 for service in the 40/40/20 situation (not accounting for vacation). FTEs are different at every university and for every faculty member, depending on the position negotiated. International programs can differ in their approaches to time commitment and the organization of your FTE. Knowledge of the specific requirements should be evaluated prior to, or specifically discussed during, the interview.

Tenure track versus clinical track

Many universities will offer both tenure track and clinical track positions, and it can be difficult to decide which the best route is. The definition of each and the availability for promotion for clinical track positions are variable, so make sure to carefully read the job description before interviewing. Make a list of questions in regards to position expectations to ask during your interview. Utilize any online resources available at that university and contact faculty members in each of the tracks to determine which is a better fit for you. Communication with your department head on a regular basis (such as yearly reviews) and with pre-tenure or pre-promotion committees will help focus your strengths and keep you on the right path for career success.

If you elect a tenure track position, start reading and building your tenure dossier. Every university has a different format for the dossier, and it needs to be written to explain what we do, and how your position differs from the typical university faculty member due to clinical services. Start early, and update often.

Teaching

Teaching is a large portion of time for most ACVIM diplomates in academia, be it standard didactic lectures, hands-on labs, or in-clinic teaching of veterinary students. Make sure to clarify the definition of clinical teaching and expected weeks on and off clinics. Developing and diversifying your teaching portfolio is also important regardless of tenure-track versus clinical track. Instructions for how to develop a teaching portfolio and a discussion of the topic will be initiated. For example, teaching didactic lectures to veterinary students, graduate students, and even undergraduate students are an integral part of your teaching portfolio. Student evaluations and peer evaluations will not only provide constructive feedback for improvement, but will also be included in your tenure dossier or in your CV when applying for promotion. The teaching commitments of different residency programs can differ; however most will have had ample experience in teaching from a rounds perspective and of course hands on clinical skills.

The development of a cohesive lecture based course or even just a series of lectures on a particular topic is quite a daunting task. This skill set is developed with practice and experience; however starting to develop a lesson plan; the creation of the lectures, a consideration of the method of how you will assess student performance, delivering effective feedback, and developing teaching and learning strategies are things that take time to plan. The sooner you begin to organize your thoughts will help you with these tasks as well as help you develop your basic teaching strategy or philosophy of teaching, (the beginning of your teaching portfolio) and also show potential interviewers your interest and enthusiasm for teaching. Additionally, most universities have programs for the improvement of your teaching, with seminars and discussion groups to introduce new methodologies and techniques for education. There are also some distance learning options for advancement in veterinary education.

The authors find that the teaching responsibilities are the most fulfilling aspects of academia. The creation of a list of pros and cons of teaching will be created in a discussion topic. Job satisfaction can be tied quite closely to the relationships you develop with your students. Developing an effective teaching strategy is strengthened by constructive feedback from the students. Sadly one of the cons of teaching is poor/unhelpful student reviews; specifically asking for directed feedback can help guide the students towards helpful information for you to use in the progression of your topics or courses. Using technology to make course interactive (such as the clicker system) and using case based examples to underline your teaching points are helpful to increasing student focus and understanding. Peer evaluations can be particularly helpful in providing constructive feedback on powerpoint slides and lecture style. Many universities will also offer teaching workshops, which can assist in construction of the powerpoint lecture, lecture objectives, and test question structure.

Discussion points will include the development of didactic lectures, laboratory courses and a realistic discussion of educational theory; with the purpose of assisting the development of an effective teaching style for lecture based learning. The pros and cons of teaching in the preclinical/ didactic portion of most curricula will be discussed as well as the time commitments for developing a cohesive lesson plan and strategy.

Research

The percentage of time you negotiate for research will depend on the needs of the service and your research experience. If you don't feel you have a strong research background, discuss this prior to negotiating percentages as this will be important for promotion and/or tenure. Programs are available for new clinician scientists that will discuss to aid in grant writing, how to start a project, and molecular biology skills. Participation in these programs can be negotiated into your contract. Furthermore, a start-up package may be available to assist in setting up or adding to a laboratory. Get details on what a lab has, and what you might need to start your research program when negotiating this package. Also research on-campus funding opportunities and grant writing seminars that may aid your research. Discuss potential research collaborations with established researchers to help get your program started.

Many universities have small grant programs that can assist a small project that will provide preliminary data for future, larger projects and external funding. It can at first be frustrating to apply for and achieve funding for your project, but don't give up! Have friends/ family review the grant for you, to point out things that would be confusing to reviewers and always check your formatting. Although time consuming, grant applications help organize your project prior to starting it, and reviewers often have helpful suggestions to make your project stronger, to obtain more meaningful data.

Mentorship of undergraduate, professional, and graduate student research can also help strengthen your research program and strengthen your tenure application. Often, this mentorship takes more time and effort than running the project yourself, but is an excellent opportunity for trouble-shooting and learning, and is very rewarding when projects and manuscripts are complete.

Service and Administration

Service includes committee appointments in your department, college, university, and to your profession. This also can include outreach (speaking at school career days and to general public) and continuing education. Discuss your service commitments and know what the expectations are during your interview or shortly after starting, to prevent over-commitment. The time commitment for different academic positions can vary incredibly. Discussion points will revolve around some of the basic committees' functions, responsibilities and expected duties.

Faculty and Research Mentorship

With a variety of new duties and responsibilities the development of a mentor relationship with a senior faculty member for guidance and advice is instrumental early on in your careers. We are somewhat protected during our residencies from a large amount of the red tape and bureaucracy that can exist in any department or hospital. Having an experienced mentor to help guide you through the standards and procedures for the new work environment is instrumental to your integration into new programs and departments.

PRIVATE PRACTICE

Private practice offers a high caseload with the opportunity to focus on clinical care and client communication. If you elect this route, make sure you have interviewed at multiple practices to pick the best fit for you. Additionally, locum positions can be discussed with an emphasis on state and international licensure, and expectations of time commitment and duties.

Interview at more than one practice (at least 3 if you can) before accepting a position, and take your time in making that decision. Talk to doctors and staff at that practice to get a good feel of the personalities and how you will fit in. Make a note of the equipment available – does it fit your needs? Also pay attention to the adequacy of technical staff, scheduling flexibility of appointments, and benefits offered. What is the management style and does that fit with the way you practice? Do you have other specialists nearby or in-practice that you can consult with?

RETIREMENT AND OTHER BENEFITS

It's never too early to think about your retirement! Regardless of the career path you choose, make sure you discuss whether or not a retirement plan is available and know the details prior to signing your contract. If no retirement plan is offered, how can you start planning your own retirement fund? Other benefits to consider are health insurance, paid vacation time, sick leave, maternity/ paternity leave, continuing education and funds for travel, liability insurance, licensure, etc. And have a lawyer review your contract before signing. In regards to your salary, the first chance is the best for negotiation- this applies to private practice and academia.

Know what you are worth and make sure you can live comfortably with your base salary if you have a compensation/ production package.

DISCUSSION TOPICS DEPENDING ON AUDIENCE INTEREST AND PARTICIPATION

- CV's
- FTE's
- Tenure Track considerations
- Clinical Track considerations
- Teaching Portfolio
- Teaching Pro's and Con's
- Lecture Development
- Educational Theory
- Student Assessment
- Teaching Assessment
- Faculty Mentorship
- Private Practice considerations

CONCLUSIONS

Each career choice has its share of benefits, and it's a very exciting time to contemplate where your career path may take you. Don't hesitate to contact old friends and mentors to discuss your chosen path, and gain some pearls of wisdom before making a commitment. And enjoy!

PEARLS OF GI ENDOSCOPY

Michael S. Leib, DVM, MS, DACVIM (SAIM)
Blacksburg, VA

Endoscopic examination of the gastrointestinal (GI) tract of dogs and cats is a minimally invasive method of obtaining biopsy samples of the GI mucosa. Histologic examination of mucosal biopsy samples is often necessary to obtain a diagnosis in cases with chronic GI diseases. Gross endoscopic diagnoses of gastric ulcers, pyloric hypertrophy, and intestinal lymphangiectasia can be made. In addition, endoscopy has several therapeutic indications: esophageal, gastric, and duodenal foreign body retrieval, esophageal and colonic stricture dilation, and placement of percutaneous gastrostomy tubes. This seminar will present some "Pearls of GI Endoscopy" that the author has gathered during more than 25 years of teaching GI endoscopy to internal medicine residents in the clinic and during CE courses and from hosting the ACVIM forum's Endoscopy SIG: What's your diagnosis?

GROSS ENDOSCOPIC DESCRIPTION

The use of consistent descriptive terminology is vital for accurate medical record documentation, excellent communication with pathologists and other colleagues, and publication of clinical and research findings. The gross endoscopic findings should be carefully noted in the endoscopy report. The World Small Animal Veterinary Association (WSAVA) GI standardization group has developed endoscopy report forms, which have also been published in the ACVIM Consensus Statement on guidelines for the evaluation of GI inflammation.[1] The following terms, and their definitions, are commonly used to describe the gross endoscopic appearance of the GI mucosa: hyperemia; increased redness, friability; excess hemorrhage associated with contact with the endoscope or endoscopic accessories, granularity; roughness or irregularity of the mucosal surface, erosion; superficial mucosal defect, and ulcer; deeper mucosal defect, often with raised borders. There is no way to endoscopically differentiate between a deep erosion and a superficial ulcer. Ulcer diameter and depth should be estimated and recorded. Size can be estimated by comparison with the tip of the endoscope or with the diameter of a closed or open biopsy forceps. Moderate to severe narrowing of the lumen (stricture) of the esophagus, duodenum, and colon can be detected, and the diameter and length estimated. Severe luminal dilation can rarely be observed. Foreign bodies should be grossly described and their location and size recorded in the endoscopy report. Masses should be described as sessile or pedunculated, and their size should be estimated. Excessive numbers of mucosal lymphoid follicles in the stomach or colon (lymphoid follicular hyperplasia), dilated intestinal lymphatics (lymphangiectasia)[2], and pyloric hypertrophy[3] should be documented.

ENDOSCOPIC BIOPSY

Obtaining high quality biopsy samples is a key to accurate histopathological diagnoses. Mucosal samples are best obtained with biopsy forceps oriented perpendicular to the mucosal surface. This ideal orientation is easy to obtain in the stomach, intermediate to obtain in the colon, and impossible to obtain within the small intestine. In the small bowel, adequate samples should contain at least 3 villi with subvillus lamina propria extending to the muscularis mucosa. Marginal samples contain at least 1 villus with subvillus lamina propria, but not full-thickness to the muscularis mucosa. For most small bowel diagnoses, the number of samples required for diagnosis decreases as sample quality increases.[4] Collecting a minimum of 8 duodenal samples has been suggested.[5] In addition, it has recently been shown that in some cases, sampling from the ileum results in a different histopathologic diagnosis compared to collecting biopsy samples only from the duodenum.[6-8] The ileum can be endoscopically entered in medium and large dogs (technique similar to that described for the pylorus), while "blind" biopsy samples can be obtained from small dogs and cats. To collect a "blind" mucosal sample, the closed biopsy forceps should be gently passed through the ileocolic sphincter, the forceps is opened and gently advanced until resistance is met, the forceps should then be closed, the mucosal sample broken off and the forceps withdrawn. Perforation of the ileum is a risk of this procedure, but is uncommon if careful technique is utilized. Because the ileal mucosa is thinner than duodenal mucosa, it may be easier to obtain samples of higher quality from the ilium compared to the duodenum.[6] Recently, large biopsy forceps have been shown to collect superior samples when compared to standard, oval, and pediatric forceps.[9]

In the duodenum the author prefers to collect samples at the distal duodenal flexure, where perpendicular orientation of the biopsy forceps can often be obtained. Flexures can also be sampled in the colon. When a flexure is not available to sample, or a focal lesion is present in another location, the turn and suction technique can be used.[9] When utilizing this technique the biopsy forceps should be passed into the lumen. The forceps are opened and slowly withdrawn until they are against the endoscope's tip. Using the control knobs, the endoscope's tip is turned towards the mucosal target and suction is employed. The biopsy forceps are gently advanced against the target mucosa, closed, the sample broken off, and the forceps removed. The biopsy sample should be gently removed and unfolded from the forceps using a 25g needle. Utilizing a hand lens the villus surface should be placed up on a biopsy sponge or wire cassette, prior to placement into formalin.

ENTERING THE DUODENUM – "THE FINAL FRONTEER"

Consistently entering the duodenum in dogs and cats with small bowel disease is one of the most difficult endoscopic skills for the novice to master. Entering the duodenum is easier to achieve with a smaller diameter endoscope. Minimizing gastric distention and gastric endoscopic manipulation will reduce pyloric tone. The duodenum should be evaluated and sampled prior to complete examination of the stomach. Avoiding opioid premedications in dogs, such as morphine and meperidine, will also reduce pyloric tone.[10] Overall, use of different anesthetic gases, and the administration of metoclopramide, atropine, and glucagon have not been shown to ease duodenal intubation in dogs and cats.[11,12] In cats, Hydromorphone, butorphanol, and medetomidine have been shown to not hinder duodenal intubation.[13]

To enter the duodenum, the endoscope's tip should be directed toward the center of the pyloric orifice. In most cases, the endoscope's tip will not directly pass through the sphincter, but will make contact with the mucosa surrounding the orifice, producing a "redout". Air should be insufflated and small directional turns made with the control knobs until a "black hole" can be visualized. When the "black hallway" is seen, the endoscope should be slowly advanced with gentle pressure until the mucosa of the proximal duodenum is seen. Continued air insufflation and gentle advancement of the endoscope along with a slight amount of torque to the insertion tube will deliver the endoscope into the descending duodenum, where a tunnel view should be visible. If the pyloric canal cannot be located after the "red out", the endoscope should be withdrawn, realigned with the pyloric orifice, ad the process repeated. The endoscope should be advanced as far as possible into the duodenum, as long as resistance is not met. If the pylorus cannot be successfully passed, placing the animal in dorsal recumbency can change the pyloric orientation. Additionally, air can be suctioned from the stomach if it has become distended, to help reduce pyloric tone.

ESOPHAGEAL STRICTURES

Most esophageal strictures in dogs and cats occur as an uncommon sequela to gastroesophageal reflux (GER) that occurs during general anesthesia. Most animals have a single focal stricture. Acid reflux into the esophagus has been documented in 16-80% of dogs undergoing general anesthesia. Recently, perioperative administration of metoclopramide, proton pump inhibitors, and cisapride have been shown to reduce the incidence of GER.[14-16] In a small group of dogs in which regurgitation occurred during general anesthesia, water lavage followed by administration of 4.2% bicarbonate was superior than suctioning the regurgitated material or utilizing water lavage by itself, in raising the esophageal pH.[17]

Post-dilation therapy in published studies has been too varied to reach a consensus for optimal therapy. Some clinicians feel that intralesional injection of triamcinolone (6-20mg) has resulted in fewer needed dilations. Technically, injection is easier and more effective prior to dilation.

Overall, endoscopic balloon catheter or Savary bougie dilation in dogs and cats has resulted in favorable results in 71-100% of cases reported.[18-21] The serious complication rate is approximately 7%. The mean number of reported dilations in the veterinary literature is 3/case.

Strictures secondary to medication entrapment have been reported in cats with doxycycline and clindamycin.[22,23] Dry administration of tablets and capsules to normal cats has shown slow passage through the esophagus or prolonged entrapment within the esophagus.[24,25] Doxycycline is an acidic medication that can severely damage esophageal mucosa, resulting in esophagitis and stricture formation. Medication induced esophageal strictures often occur in the cervical esophagus. A median of 4 dilations/cat have been reported in the literature.

FOREIGN BODY RETRIEVAL

Endoscopic retrieval of esophageal, gastric, and some duodenal foreign bodies is a minimally invasive and highly successful technique.[26-28] A variety of foreign body forceps are available, making removal of differing sizes and shapes of foreign bodies possible. A small or large (19mm) rat-tooth forceps will provide a very strong grasp on the foreign body. A Roth net will allow easy scooping of small foreign bodies, compared to wire snares. Large 4-wire (US Endoscopy 00711141) or 6-wire baskets (Storz 60278FK) allow retrieval of large gastric foreign bodies, such as golf balls. A technique using baby bottle liners has been described that allows capture of golf balls and other large gastric foreign bodies. Bones lodged within the esophagus are often too tightly wedged to be removed with flexible endoscopic forceps. A new semi-rigid forceps (Storz 60370 FB) placed along the side of the endoscope will facilitate removal of impacted bones. Small motile gastric foreign bodies often move to the area of the gastroesophageal junction while a dog or cat is in left lateral recumbency. Maneuvering a foreign body forceps to grasp the foreign body in the retroflexed position is often difficult. The animal should be placed in sternal recumbency and the foreign body can be easily grasped, as it will be directly in front of the endoscope within the gastric body.

Although esophagitis is a common sequela to esophageal foreign body impaction, stricture formation is relatively uncommon. However, a previously available formulation of the popular dental chew, Greenies®, resulted in a high percentage of structure formation, most likely due to severe circumferential esophagitis.[29]

PYLORIC HYPERTROPHY

Chronic hypertrophic pyloric gastropathy (CHPG) occurs most commonly in middle-aged small breeds of dogs. It causes delayed gastric emptying and vomiting of food often 8-10 hours after eating. Diagnosis can be made by observing the gross appearance of the pylorus. The pylorus for 180-360° is surrounded by thickened mucosal folds.[3] The surface of the thickened mucosal folds should be smooth. There may be retained food within the stomach, despite a 14-hour fast. The duodenum should be carefully evaluated as some proximal duodenal lesions can cause distortion of the pylorus. The thickened folds should be deeply biopsied to rule out neoplasia. The biopsy forceps should be replaced into the same location several times to obtain deep tissue to rule out neoplasia.

INTESTINAL LYMPHANGIETASIA

Intestinal lymphangiectasia (IL) is associated with protein losing enteropathy in dogs. Grossly dilated lymphatics can be visualized within the duodenum and often appear as "rice-grain nodules".[2] Biopsy of the duodenum is performed to identify a primary cause of IL; inflammatory bowel disease, diffuse lymphoma, histoplasmosis, or other infiltrative disease that can obstruct lymphatic flow through the villus. In some instances, histopathology does not identify the presence of IL. This "negative" diagnosis may be due to poor quality biopsy samples, missing the target lesion with the forceps in cases with scattered disease, or compression of the small tissue sample collapsing the thin-walled dilated lacteal.

REFERENCES

1. Washabau RJ, Day MJ, Willard MD, et al. Endoscopic, biopsy, and histopathologic guidelines for the evaluation of gastrointestinal inflammation in companion animals. J Vet Intern Med 2010;24:10-26.
2. Larson RN, Ginn JA, Bell CM, et al. Duodenal endoscopic findings and histopathologic confirmation of intestinal lymphangiectasia in dogs. Journal of veterinary internal medicine / American College of Veterinary Internal Medicine 2012;26:1087-1092.
3. Leib MS, Saunders GK, Moon ML, et al. Endoscopic diagnosis of chronic hypertrophic pyloric gastropathy in dogs. J Vet Int Med 1993;7:335-341.
4. Willard MD, Mansell J, Fosgate GT, et al. Effect of sample quality on the sensitivity of endoscopic biopsy for detecting gastric and duodenal lesions in dogs and cats. J Vet Intern Med 2008;22:1084-1089.
5. Willard M, Lovering S, Cohen N, et al. Quality of tissue specimens obtained endoscopically from the duodenum of dogs and cats. J Am Vet Med Assoc 2001;219:474-479.
6. Casamian-Sorrosal D, Willard MD, Murray JK, et al. Comparison of histopathologic findings in biopsies from the duodenum and ileum of dogs with enteropathy. J Vet Intern Med 24:80-83.
7. Procoli F, Motskula PF, Keyte SV, et al. Comparison of histopathologic findings in duodenal and ileal endoscopic biopsies in dogs with chronic small intestinal enteropathies. Journal of veterinary internal medicine / American College of Veterinary Internal Medicine 2013;27:268-274.
8. Scott KD, Zoran DL, Mansell J, et al. Utility of endoscopic biopsies of the duodenum and ileum for diagnosis of inflammatory bowel disease and small cell lymphoma in cats. J Vet Intern Med 2011;25:1253-1257.
9. Goutal-Landry CM, Mansell J, Ryan KA, et al. Effect of endoscopic forceps on quality of duodenal mucosal biopsy in healthy dogs. Journal of veterinary internal medicine / American College of Veterinary Internal Medicine 2013;27:456-461.
10. Donaldson LL, Leib MS, Boyd C, et al. Effect of preanesthetic medication on ease of endoscopic intubation of the duodenum in anesthetized dogs. Am J Vet Res 1993;54:1489-1495.
11. Matz M, Leib M, Monroe W, et al. Evaluation of atropine, glucagon, and metoclopramide for facilitation of endoscopic intubation of the duodenum in dogs. American Journal of Veterinary Medicine 1991;52:1948-1950.
12. Monroe W, Leib M, Matz M, et al. Evaulation of metoclopramide hydrochloride as an aid for passage of a flexible endoscope into the duodenum of dogs. Am J Vet Res 1992;53:149-152.
13. Smith A, Posner L, Goldstein R, et al. Evaluation of the effects of premedication on gastroduodenoscopy in cats. J Am Vet Med Assoc 2004;225:540-544.
14. Wilson DV, Evans AT, Mauer WA. Influence of metoclopramide on gastroesophageal reflux in anesthetized dogs. Am J Vet Res 2006;67:26-31.
15. Panti A, Bennett RC, Corletto F, et al. The effect of omeprazole on oesophageal pH in dogs during anaesthesia. The Journal of small animal practice 2009;50:540-544.
16. Zacuto AC, Marks SL, Osborn J, et al. The influence of esomeprazole and cisapride on gastroesophageal reflux during anesthesia in dogs. J Vet Intern Med 2012;26:518-525.
17. Wilson DV, Evans AT. The effect of topical treatment on esophageal pH during acid reflux in dogs. Vet Anaesth Analg 2007;34:339-343.
18. Bissett SA, Davis J, Subler K, et al. Risk factors and outcome of bougienage for treatment of benign esophageal strictures in dogs and cats: 28 cases (1995-2004). J Am Vet Med Assoc 2009;235:844-850.
19. Harai BH, Johnson SE, Sherding RG. Endoscopically guided balloon dilatation of benign esophageal strictures in 6 cats and 7 dogs. J Vet Int Med 1995;9:332-335.
20. Melendez L, Twedt D, Weyrauch E, et al. Conservative therapy using balloon dilation for intramural, inflammatory esophageal strictures in dogs and cats: a retrospective study of 23 cases. (1987-1997). Eur J Comp Gastroenterol 1998;3:31-36.
21. Leib M, Ward D, Reimer M, et al. Endoscopic balloon dilation of benign esophageal strictures in dogs and cats. J Vet Intern Med 2001;15:547-552.
22. German A, Cannon M, Booth M, et al. Oesophageal strictures in cats associated with doxycycline therapy. J Fel Med Surg 2005;7:33-41.
23. Beatty J, Swift N, Foster D, et al. Suspected clindamycin-associated oesophageal injury in cats: five cases. J Feline Med Surg 2006;8:412-419.
24. Westfall D, Twedt D, Steyn P, et al. Evaluation of esophageal transit of tablets and capsules in 30 cats. J Vet Intern Med 2001;15:467-470.
25. Graham J, Lipman A, Newell S, et al. Esophageal transit of capsules in clinically normal cats. Am J Vet Res 2000;61:655-657.
26. Gianella P, Pfammatter NS, Burgener IA. Oesophageal and gastric endoscopic foreign body removal: complications and follow-up of 102 dogs. J Small Anim Pract 2009;50:649-654.
27. Rousseau AD, DACVIM, DACVECC; Prittie, Jennifer DVM, DACVIM, DACVECC, Broussard, John D. DVM, DACVIM. Incidence and characterization of esophagitis following esophageal foreign body removal in dogs: 60 cases (1999–2003). Journal of Veterinary Emergency and Critical Care 2007;17:159-163.
28. Thompson HC, Cortes Y, Gannon K, et al. Esophageal foreign bodies in dogs: 34 cases (2004-2009). Journal of Veterinary Emergency and Critical Care 2012;22:253-261.
29. Leib MS, Sartor LL. Esophageal foreign body obstruction caused by a dental chew treat in 31 dogs (2000-2006). J Am Vet Med Assoc 2008;232:1021-1025.

IMMUNOSUPPRESSIVE THERAPY: WHAT'S NEW

Andrew Mackin, BSc, BVMS, MVS, DVSc, FANZCVSc, DACVIM
Starkville, MS

Several potent immunosuppressive drugs developed over the past few decades in human medicine have recently made the leap to our small animal patients, and our use of them is growing. This lecture will discuss cyclosporine, leflunomide and mycophenolate.

CYCLOSPORINE

Cyclosporine is a potent immunosuppressive drug indicated for the treatment of inflammatory and immune-mediated diseases, and for organ transplantation. Cyclosporins are cyclic polypeptide macrolides originally derived from the soil fungus *Beauveria nivea*. Cyclosporine A is the molecule developed for commercial use. The use of cyclosporine to prevent rejection of renal allografts in people was first described in the 1970s. In veterinary medicine, oral cyclosporine received FDA approval in 2003 for treatment of canine atopy, and was recently also approved for allergic skin disease in cats. Cyclosporine has been used in an extra-label fashion for many years for renal transplantation in dogs and cats[1,2], and for a range of inflammatory and immune-mediated conditions[3].

Cyclosporine's primary immunosuppressive mechanism of action is inhibition of T cell function. Cyclosporine acts to inhibit calcineurin, an intracellular protein phosphatase that activates gene transcription factors. Activation of T cells results in activation of calcineurin, which dephosphorylates inactive nuclear factor (NFAT). NFAT translocates into the nucleus, where it upregulates transcription of genes coding for several important cytokines, including IL-2, IL-4, TNF-α, and INF-γ. Calcineurin inhibitors, including cyclosporine, act by binding to intracellular cyclophilins, which are proteins that facilitate protein folding. Binding of cyclosporine to cyclophilin A creates a complex with high affinity for calcineurin. Through inhibition of calcineurin, cyclosporine specifically inhibits T cell function and thus, cell-mediated immunity, but has little immediate impact on humoral immunity. Decreased IL-2 expression in CD4+ Th1 cells associated with cyclosporine therapy leads to inhibition of proliferation and activation of both T-helper and T-cytotoxic lymphocytes, and blunting of the immune response. Cyclosporine has also been shown to have many other local anti-inflammatory and immunosuppressive effects, especially in the skin.

Cyclosporine is a large lipophilic molecule which must be solubilized prior to intestinal absorption. Commercial cyclosporine is available as two very different types of oral formulations. Cyclosporine was initially approved in humans as a vegetable-oil based preparation (Sandimmune®), but poor oral bioavailability caused marked variability in blood drug concentrations. A more recent formulation, an ultramicronized preparation approved in 1996 (Neoral®), forms a microemulsion upon contact with aqueous fluids, resulting in more predictable absorption. Oral bioavailability of the microemulsion is improved by up to 50% compared to the oil-based preparation. Because of the marked variability in bioavailability of the non-ultramicronized preparation, it is not recommended for oral use in small animals. Cyclosporine has a high binding affinity for red blood cells and plasma lipoproteins. Because up to 50% of the drug in blood is located in red cells, whole blood is recommended for therapeutic drug monitoring (TDM). Once in the circulation, cyclosporine distributes widely, accumulating in skin, liver, kidneys and fat, resulting in a large volume of distribution. Tissue levels exceed levels in serum by a factor of 3 to 14. Peak blood concentrations generally occur approximately 2 hours after oral administration. Blood concentrations then rapidly decrease over the remainder of the dosing interval. Extensive metabolism by the hepatic cytochrome P-450 system yields many different metabolites, some of which may retain therapeutic efficacy. Drugs that inhibit P-450 enzymes have been given concurrently with cyclosporine in order to decrease the dose needed to maintain adequate blood drug concentrations. Ketoconazole, in particular, has been used to decrease in oral cyclosporine dosages in dogs by as much as 75%[4,5].

The complexities of cyclosporine disposition in normal animals contribute to markedly variable blood drug concentrations both between patients and even within the same patient. Therapeutic management may therefore be facilitated by monitoring blood cyclosporine concentrations. Currently available methods for TDM include HPLC and a specific monoclonal RIA. HPLC has the advantage that the parent drug can be discriminated from metabolites. Since RIA measures metabolites as well as the parent drug, blood cyclosporine concentrations will therefore be higher by a factor of 1.5 to 1.7 compared to the same sample analyzed by HPLC. HPLC is labor intensive, and so methods such as RIA are most often employed in clinical situations, with the laboratory performing the assay typically providing recommendations regarding ideal target drug concentrations. In human medicine, trough blood levels were the initial basis for adjustment of drug dosages. Many studies in people have since suggested that area under the plasma drug concentration time curve or 2 hour peak drug concentrations are preferred. With peak cyclosporine concentrations requiring only a single sample, adjusting drug doses to attain target peak drug levels has become the single best blood concentration measurement for use during human organ transplantation. In veterinary medicine, measurement of trough cyclosporine concentrations also prevailed for many years based on initial work done in canine and feline renal transplant studies. Recommendations from laboratories offering TDM have often involved measurement of both peak and trough cyclosporine blood levels, although target peak concentrations have not been well established in small animals.

Pharmacodynamic assays investigate a drug's effect on target cells. Many pharmacodynamic markers of the immunosuppressive effects of cyclosporine have been studied in human medicine, including lymphocyte proliferation, calcineurin enzyme activity, lymphocyte surface antigen expression, and intracellular cytokine quantification. Through pharmacodynamic monitoring, human studies have shown individually distinct degrees of calcineurin inhibitor sensitivity in patients. Pharmacodynamic monitoring shows great promise for optimizing cyclosporine therapy and delivering individualized therapy. At MSU, we recently measured activated T cell expression of IL-2, IL-4, and IFN-γ via flow cytometry in dogs given two different oral cyclosporine dosages[6,7]. The dogs were first administered a high dose (10 mg/kg orally twice daily), with doses adjusted upwards to attain a target trough drug concentration greater than 600 ng/mL (HPLC), a protocol known to be sufficiently immunosuppressive for organ transplantation. With this high dose, activated T cell expression of IL-2 and IFN-γ was significantly suppressed. The dogs were then administered the dose of cyclosporine used to treat canine atopy (5 mg/kg orally once a day), a dose which has been considered to be low enough to avoid predisposing to immunosuppression and infection. Even with this low dose, however, expression of IFN-γ and IL-2 was still markedly suppressed in some dogs. Subsequent studies determined that activated T cell mRNA IL-2 and IFN-γ expression using a quantitative reverse transcriptase polymerase chain reaction (qRT-PCR) assay was comparable to flow cytometry results[8]. One advantage of the qRT-PCR assay is that it can be performed on blood samples mailed in by practitioners. Cyclosporine has been shown to have much the same effect on T cell cytokine production in cats as it does in dogs[9,10].

Cyclosporine is FDA-approved for the treatment of canine atopic dermatitis and feline allergic skin disease, and has also been used to prevent transplant rejection and to treat sebaceous adenitis, pemphigus foliaceus, anal furunculosis, feline stomatitis, inflammatory bowel disease (IBD), myasthenia gravis, non-infectious inflammatory meningoencephalitis, pure red cell aplasia, immune-mediated hemolytic anemia (IMHA), immune-mediated thrombocytopenia (IMT), and immune-mediated polyarthritis in dogs and cats. Recent pharmacodynamic research has confirmed that canine responses to the drug are comparable to the response profile that is well recognized in people: individual responses are extremely variable from dog-to-dog, both in dogs receiving the same standard oral dose, and in dogs with oral doses adjusted to attain comparable blood levels. Given this high degree of variability of individual responsiveness to cyclosporine in dogs, dosing protocols should be tailored to allow for this patient-to-patient variability. In my opinion, recommended dosing protocols in dogs with chronic, non-life-threatening inflammatory skin and gastrointestinal diseases should be quite different from the protocols used in dogs with more acute and life-threatening immune-mediated diseases.

In chronic diseases that are typically not life-threatening, such as skin conditions, anal furunculosis, and mild IBD, cyclosporine is often effective at a standard, relatively low starting dose. Cyclosporine is typically given long term, with doses adjusted upwards if needed ‘to effect’, based predominantly on clinical signs. Typically, however, starting doses do not need to be increased and, in the long-term, the drug is tapered to the lowest effective dose needed to maintain remission. Recommended starting cyclosporine doses in dogs are 5 mg/kg once daily for most skin diseases and IBD[11,12], and 5 mg/kg once to twice daily for anal furunculosis[13]. In cats with skin conditions such as allergic skin disease, eosinophilic granuloma complex and pemphigus foliaceus, a starting cyclosporine dose of around 5 mg/kg daily is recommended[14]. Blood concentrations are usually not necessary for treatment of these conditions, as remission of disease is the main criterion used to decide whether drug therapy is adequate. In fact, for many of these conditions, cyclosporine blood levels have been shown to have minimal correlation with remission, perhaps because the drug is selectively concentrated in tissues such as the skin. Even at standard low FDA-approved doses, some dogs can still develop significant suppression of T cell biomarkers of immunosuppression despite very low trough cyclosporine concentrations. This could explain the phenomenon reported by some dermatologists, that individual dogs treated for atopy can develop severe secondary infections.

In dogs suffering from more life-threatening diseases such as severe IMHA and IMT, in contrast, cyclosporine must be targeted to attain effective immunosuppression as rapidly as possible. In these patients, starting cyclosporine at a low dose and adjusting doses upwards ‘to effect’ is not recommended. Attaining effective oral doses as rapidly as possible is essential for ensuring adequate immunosuppression whilst avoiding overdosage. Currently recommended starting cyclosporine doses for life-threatening diseases range from 5 mg/kg to 10 mg/kg twice daily. Subsequent measurement of blood drug levels and/or assessment of activated T cell mRNA IL-2 and IFN-γ expression using qRT-PCR within a week of starting therapy, with dose adjustments as needed, are the best methods that are currently available to assess adequacy of therapy, and are strongly recommended in dogs with severe disease.

Side effects are uncommon with cyclosporine, with the exception of GI signs such as vomiting, diarrhea, anorexia and nausea. Giving the drug frozen and/or with food can reduce GI side effects, although such measures may also alter drug absorption profiles. Uncommonly, cyclosporine can cause an idiosyncratic hepatotoxicity, which does not seem to be dose dependent. Gingival hyperplasia and hypertrichosis have also occasionally been reported with cyclosporine. Chronic cyclosporine therapy may also predispose to neoplasia such as lymphoma. One advantage of cyclosporine is that it is not myelosuppressive. Experimentally, cyclosporine has been shown to increase some markers of platelet activation in dogs, which may be a concern in patients with IMHA, where hypercoagulability and resultant pulmonary thromboembolism can be a major contributor to patient mortality[15]. However, to date, it has not been demonstrated whether this phenomenon is clinically relevant in IMHA patients with naturally occurring disease.

Cyclosporine is an expensive drug, and clinicians are therefore tempted to explore cheaper forms of the drug. There are many approved human generic microemulsion preparations similar to Neoral®, and these have been shown to have therapeutic equivalency in people. Studies investigating the pharmacokinetics of these preparations in dogs have not been performed, and it is not safe to assume that a generic formulation is therapeutically equivalent to the approved canine product (Atopica®). Neoral® currently costs around $2 for a 25 mg capsule and $6 for a 100 mg capsule, while the generic equivalents cost around $1 and $2 for the 25 mg and 100 mg capsules accordingly. The veterinary product is priced comparably to the human proprietary products, but has the advantage of being FDA-approved and available in capsule sizes that are convenient for dosing accuracy (10 mg, 25 mg, 50 mg and 100 mg), as well as a 100 mg/ml oral suspension. Unfortunately, transdermal cyclosporine has been shown to be inadequately absorbed in cats[16].

LEFLUNOMIDE

Leflunomide is an isoxazol derivative immunosuppressive drug that was developed within the past two decades for treatment of rheumatoid arthritis and prevention of transplant rejection. Leflunomide is a prodrug for its primary active malononitriloamide metabolite, A77 1726 (teriflunomide). Malononitriloamides reversibly inhibit the mitochondrial enzyme dihydroorotate dehydrogenase, a key enzyme in pyrimidine synthesis, with resultant inhibition of the pyrimidine ribonucleotide uridine monophosphate (rUMP), and decreased DNA and RNA synthesis and G1 cell cycle arrest. Leflunomide inhibits B and T cell function, suppresses antibody production and has anti-inflammatory effects, possibly via inhibition of *de novo* pyrimidine biosynthesis and cytokine-associated and IL-2-stimulated tyrosine kinase activity.

Prior to commercial development, leflunomide was made available to the transplant research group at UC-Davis. This group described treatment of small number of canine patients with refractory inflammatory and immune-mediated diseases such as IMHA, IMT, non-infectious inflammatory meningoencephalitis, systemic histiocytosis, immune-mediated polymyositis, immune-mediated polyarthritis, and pemphigus foliaceous with leflunomide, typically with promising success rates[17]. Unfortunately, when these results were reported in the late 1990s, the drug was no longer commercially available. When leflunomide did become available as the proprietary product Arava®, drug cost limited its use in small animal clinic patients. Only recently did an affordable generic equivalent become available and, as a result, preliminary reports of leflunonide use in small animal patients are beginning to surface. There are still currently very few published reports discussing the use of leflunomide in dogs and cats[18,19]. Recently, a case series describing the use of leflunomide in 14 dogs with immune-mediated polyarthritis reported a high response rate with few side effects[20].

Leflunomide appears to be very well tolerated in dogs although, if the drug attains more common usage, it is likely that less frequent but more serious side effects will be recognized. The most common side effect seen with leflunomide in dogs is occasional inappetance, lethargy and vomiting. Serious side effects occasionally reported in people include myelosuppression, cutaneous drug reactions and hepatotoxicity. In humans, traces of teriflunomide can persist for many months after drug discontinuation, and in the instance of drug reactions, cholestyramine or activated charcoal are needed to rapidly reduce drug levels. In dogs, the terminal half-life of teriflunomide is much shorter than in humans, so the potential for persistent side effects is probably significantly less[21]. Complete blood counts and serum biochemistry (especially ALT) should be regularly monitored in small animal patients on leflunomide.

The initial recommended starting oral dose for leflunomide in dogs is 2-4 mg/kg daily, with doses adjusted to attain a plasma trough teriflunomide levels of 20 μg/ml within a few weeks of commencing therapy. Measurement of leflunomide levels is available through Auburn University. For cats with immune-mediated polyarthritis, a leflunomide dose of 10 mg (total dose) orally, once daily, combined with methotrexate, has been suggested, with dose reductions to effect[22]. Leflunomide comes in tablet sizes (10 and 20 mg) that are convenient for dosing our smaller patients. Proprietary leflunomide costs about $40 for a 10 mg or 20 mg tablet, while the generic equivalent is priced at around $1 for a 10 mg tablet and $1.50 for a 20 mg tablet. Leflunomide generics have an 'AB' rating by the FDA, meaning that the generic is 'equivalent' to Arava®. However, since 'equivalence' is often determined by pharmacokinetic data in healthy individuals, an AB rating does not guarantee identical performance in clinical patients.

MYCOPHENOLATE

Mycophenolate is the synthesized prodrug form of mycophenolic acid, a selective inhibitor of inosine monophosphate dehydrogenase, an enzyme that controls the rate of synthesis of guanine monophosphate in the *de novo* pathway of purine synthesis. Mycophenolate is a fermentation product derived from *Penicillium* fungi. Mycophenolic acid inhibits B and T cell proliferation, and decreases antibody production. Mycophenolate is primarily used in human medicine for prevention of rejection of transplanted organs, although it also used to treat immune-mediated diseases such as systemic lupus erythematosus, IMHA, IMT and pemphigus vulgaris.

The original proprietary mycophenolate mofetil, CellCept®, and the related mycophenolate sodium, Myfortic®, were expensive, and as a result the products only achieved limited usage in small animal medicine. However, recently, the availability of cheaper generic alternatives has led to increased usage of mycophenolate in small animal patients. A 250 mg CellCept® capsule currently costs around $7, whereas the equivalent generic 250 mg capsule costs less than 50c. An oral suspension version of mycophenolate mofetil (200 mg/ml) is available for more convenient dosing in smaller patients. The successful use of mycophenolate mofetil in a small animal patient with naturally-occuring disease was first described in a dog with myasthenia gravis[23]. Much of the subsequent anecdotal usage of mycophenolate for a variety of different immune-mediated diseases used dosing similar to that reported in this original paper[24-26]. Mycophenolate mofetil is also available in an injectable form, and its IV use has been described during the successful stabilization of 3 dogs with myasthenia that could not tolerate oral medications[24]. Ironically, a recent case report of 15 dogs with myasthenia gravis treated with mycophenolate mofetil reported that the drug was ineffective at attaining clinical remission[27].

A recommended starting dose for mycophenolate mofetil in dogs is 10-20 mg/kg once daily or divided twice daily, although sometimes GI signs (particularly vomiting and diarrhea) at the high end of the dose rate will necessitate dose reductions. Mycophenolate mofetil has variable oral bioavailability in dogs, so variability in response to therapy should also be expected[28]. Mycophenolate has not been used widely enough to establish the frequency of serious side effects in dogs but, in people, GI signs and, less commonly, marked myelosuppression and a rare and fatal neurologic disease (progressive multifocal leukoencephalopathy) have been reported. Complete blood counts should therefore probably be regularly monitored in dogs receiving mycophenolate. In humans, GI side effects can be reduced by replacing mycophenolate mofetil with mycophenolate sodium. Mycophenolic acid in humans is primarily excreted conjugated to glucuronide and, since cats lack the glucuronyl transferases needed for glucuronidation, the drug should used with caution in this species, although the use of mycophenolate mofetil has been described at a dose rate of 10 mg/kg twice daily, with no obvious side effects, in two cats with IMHA[30].

REFERENCES

[1]Bernsteen L, et.al. **Clinic Tech Small Anim Pract** 2000;15:40. [2]Bernsteen L, et. al. **Vet Surg** 2003;32:213. [3]Archer T. et.al. **JVIM** 2014;28:20. [4]Dahlinger J, et.al. **Vet Surg** 1998;27:64. [5]Patricelli AJ, et.al. **JAVMA** 2002;220:1009. [6]Archer TM, et.al. **JVIM** 2011;25:1391. [7]Fellman CL, et.al. **Vet Immunol Immunopathol** 2011;140:175. [8]Riggs C, et.al. **Vet Immunol Immunopathol** 2013;156:229. [9]Aronson L, et.al. **AJVR** 2011;72:541. [10]Kuga K, et.al. **J Vet Med Sc** 2008;70:1011. [11]Olivry T, et.al. **JAVMA** 2002;221:370. [12]Allenspach K, et.al. **JVIM** 2006;20:239. [13]Griffiths LG, et.al. **JSAP** 1999;40:569. [14]Wisselink M, et.al. **Vet J** 2009;180:55. [15]Thomason J, et.al. **JVIM** 2012;26:1389. [16]Miller R, et.al. **JAAHA** 2014;50:36. [17]Gregory C, et.al. **Transpl Proc** 1998;30:4143. [18]Bianco D, et.al. **JAAHA** 2009;45:147. [19]Ohno K, et.al. **J Jap Vet Med Associ** 2003;56:669. [20]Colopy S, et.al. **JAVMA** 2010;236:312. [21]Singer L, et.al. **J Vet Pharm Ther** 2011;34:609. [22]Hanna F. **VCOT**. 2005;18:94. [23]Dewey D, et.al. **JVECC** 2000;10:177. [24]Abelson A, et.al. **JVECC** 2009;19:369. [25]Ginel P, et.al. **J South Afr Vet Assoc** 2010;81:253. [26]Yuki M, et.al. **Austral Vet J** 2007;85:495. [27]Dewey C, et.al. **JAVMA** 2010;236:664. [28]Lupu M, et.al. **J Am Soc Blood Marrow Transpl** 2006;12:1352. [29]Bacek L, et.al. **JVECC** 2011;21:45.

IMMUNOSUPPRESSIVE THERAPY: BETTER USE OF 'ESTABLISHED' DRUGS

Andrew Mackin, BSc, BVMS, MVS, DVSc, FANZCVSc, DACVIM
Starkville, MS

A number of established immunosuppressive agents have been used in small animal medicine for many decades. Some have justifiably fallen out of favor whereas, for others, new and promising uses have been described in the recent veterinary literature. This lecture will discuss some 'old favorites': cyclophosphamide, chlorambucil, azathioprine, danazol and vincristine.

CYCLOPHOSPHAMIDE

Cyclophosphamide, a cell-cycle nonspecific nitrogen mustard derivative alkylating agent, was approved by the FDA over 50 years ago, and has since become well-established in human medicine as an antineoplastic drug and immunosuppressive agent. Cyclophosphamide has persisted as a core drug used in many small animal cancer chemotherapeutic protocols. In contrast, after many years as one of the most commonly immunosuppressive drugs utilized to treat immune-mediated diseases in cats and dogs, the use of cyclophosphamide as an immunosuppressive agent in small animal patients has in the past two decades essentially faded away.

Cyclophosphamide is a prodrug that is metabolized by the hepatic cytochrome P450 enzyme system to eventually form active metabolites such as 4-hydroxycyclophosphamide, 4-hydroperoxycyclophosphamide, and aldophosphamide. These metabolites can enter cell cytoplasm, where they are ultimately metabolized to phosphoramide mustard and acrolein. Phosphoramide mustard is an alkylating agent that replaces a hydrogen atom with an alkyl group on the guanine base of DNA, which interferes with nuclear DNA replication and cytoplasmic RNA transcription by forming crosslinks. Cyclophosphamide has long been reported to be a potent immunosuppressive that inhibits humoral and cell-mediated immunity, including inhibition of primary and secondary immune responses, reduction of antigen trapping in lymph nodes, and inhibition of local inflammatory responses, although in a number of experimental studies in dogs, cyclophosphamide often appears to be relatively mildly immunosuppressive compared to other drugs[1-3].

Cyclophosphamide, like most alkylating agents, can cause including GI side effects, myelosuppression, and hair loss. GI signs are relatively common, and include nausea, anorexia, vomiting and diarrhea. Myelosuppression appears to be dose dependent, and is associated with both the use of high drug doses and the use of lower doses over a sustained period of time. Myelosuppression is typically reversible with drug dose reduction or discontinuance, but can occasionally persist for weeks or even months, particularly after chronic cyclophosphamide therapy. Alopecia is most common is susceptible breeds such as poodles and old English sheepdogs. A major side effect of cyclophosphamide is sterile hemorrhagic cystitis[4,5]. Cystitis is mediated by urinary excretion of acrolein. Cystitis is often debilitating to the patient, and will not resolve until the drug is discontinued. Unfortunately, cystitis can sometimes persist for days or even weeks after drug discontinuation. The chronic local bladder inflammatory effects of cyclophosphamide have also been reported to predispose to the development of irreversible bladder wall fibrosis and transitional cell carcinoma[6].

Over the years, a number of immune-mediated and inflammatory diseases in dogs and cats have been treated with cyclophosphamide, including immune-mediated hemolytic anemia (IMHA), immune-mediated thrombocytopenia (IMT), megakaryocyte hypoplasia, pure red cell aplasia, systemic lupus erythematosus, immune-mediated polyarthritis, inflammatory bowel disease (IBD), glomerulonephritis, non-infectious inflammatory meningoencephalitis, immune-mediated vasculitis and pemphigus. Cyclophosphamide was long considered to be a 'big gun' for dogs with severe IMHA. However, about 15 years ago, a number of case studies reported that cyclophosphamide was not better than steroids alone for the treatment of IMHA and, at worst, associated with a higher mortality rate[7-10]. Given the limitations of retrospective studies, including the potential for 'case selection bias' (dogs with severe IMHA may have been given cyclophosphamide because it was perceived to be more potent), it is hard to know with any real certainty whether cyclophosphamide actually worsens prognosis in dogs with IMHA. Nevertheless, there is no doubt that, since publication of these papers, the use of cyclophosphamide to treat conditions such as canine IMHA has markedly decreased.

Cyclophosphamide is relatively cheap, with a generic 25 mg tablet costing under $2, and a 50 mg tablet costing under $4. Cyclophosphamide is difficult to dose accurately, and even more difficult to taper, especially in smaller patients. Cyclophosphamide tablets are composed of an active inner core surrounded by an inert outer shell. Because of uneven drug distribution through the tablet, cyclophosphamide tablets cannot be split without a risk of dosing inaccuracy. Without drug compounding, cyclophosphamide doses must therefore be in multiples of 25 or 50. Published immunosuppressive doses for cyclophosphamide in dogs include 50 mg/m^2 or 1.5 to 2.5 mg/kg every second day or daily on a '4 days on, 3 days off' weekly protocol. In cats and small dogs, similar total weekly doses can be used, but 'pulsed' at infrequent dosing intervals that ensure that the total weekly dose is equivalent to 7× the calculated daily dose. Complete blood counts must be performed regularly throughout the course of treatment. Cyclophosphamide is available in an IV form as well as an oral form, and a recent study in dogs confirmed that equivalent oral and IV doses of cyclophosphamide achieved comparable blood levels of the active metabolite 4-hydroxycyclophosphamide[11]. Intravenous cyclophosphamide may therefore be a viable treatment option in vomiting animals that are unable to tolerate oral immunosuppressive agents.

CHLORAMBUCIL

Chlorambucil is a nitrogen mustard derivative cell-cycle nonspecific alkylating agent that has, for many decades, been used in both human and veterinary medicine as an antineoplastic agent. Antineoplastic cytotoxicity is derived from cross-linkage of DNA and RNA by insertion of alkyl radicals on the purine base, guanine. Chlorambucil also has immunosuppressive properties, and has occasionally been used in human medicine to treat conditions such as glomerulonephritis. More than 30 years ago, some clinicians began suggesting the use of chlorambucil as an immunosuppressive agent for our small animal patients. Since then, the use of chlorambucil for the treatment of a number of feline dermatoses, such as pemphigus and eosinophilic granuloma complex, and for IMT and refractory IBD, has become well established[12-14], perhaps because of a paucity of viable alternative medications that could be accurately dosed with safety in cats. The use of chlorambucil as immunosuppressive agent in dogs has been slower to evolve, but its use has been described for the treatment of pemphigus, glomerulonephritis and, most recently, IBD[15].

Chlorambucil is metabolized predominantly in the liver, primarily to the active metabolite phenylacetic acid mustard. Compared to other alkylating agents, chlorambucil is relatively well tolerated, but does occasionally cause GI side effects such as vomiting and diarrhea, and/or myelosuppression. Alopecia and poor hair growth are sometimes reported in susceptible dog breeds, such as poodles. Chlorambucil-associated neurologic signs (including myoclonus, twitches and seizures) have been reported in cats[16].

Chlorambucil is available as a coated 2 mg tablet that cannot feasibly be divided, and dosing recommendations in smaller patients are therefore typically provided in multiples of two, and/or 'pulsed' at infrequent dosing intervals in order to avoid overdose. Chlorambucil is almost always given in combination with a glucocorticoid. In dogs, recommended starting oral immunosuppressive chlorambucil doses range from 0.1 to 0.2 mg/kg (or, alternatively, 4-6 mg/m^2) every 1-2 days. In cats (and small dogs), a starting oral chlorambucil dose of 2 mg every 2nd day, tapered to every 3rd or 4th day, is my preferred dosing immunosuppressive regime. Lower daily doses of chlorambucil, comparable to dog dosing regimes, can also be used in cats if the drug is compounded, but the effects of compounding on drug efficacy are not known. Complete blood counts must be monitored regularly and, since myelosuppression tends to be dose-dependent rather than idiosyncratic, doses can be tapered 'to effect' rather than discontinued completely.

Chlorambucil has until recently been moderately priced. Unfortunately, the patent on the only available product, Leukeran®, recently expired, leading to a change in drug company ownership, and the US price of chlorambucil has doubled as a result, to over $10 for a 2 mg tablet. There are currently no other US generic alternatives, apart from compounded products.

AZATHIOPRINE

Azathioprine has been used as an immunosuppressive agent in dogs for over 50 years. The drug was initially primarily used in studies that utilized dogs as a model for investigations of immunosuppression and organ transplantation. Within a few years, azathioprine was also being used to treat naturally occurring diseases in dogs. Despite almost half a century of clinical and research experience on the use of azathioprine in dogs, however, there have been remarkably few studies that actually elucidate the precise effects that azathioprine has on the canine immune system[17-19]. Azathioprine is a prodrug for the active metabolite 6-mercaptopurine, and the primary mechanism of action was long believed to be inhibition of the synthesis of the purines adenine and guanine by blockage of enzymes such as amidophosphoribosyltranserase, with resultant production of nonfunctional nucleic acid strands. Disruption of purine synthesis inhibits DNA and RNA synthesis, thereby inhibiting the proliferation of fast-growing cells such as lymphocytes. In the past few decades, however, multiple other mechanisms of action mediated by various azathioprine metabolites have been proposed, including blockage of T cell activation and stimulation of T cell apoptosis. Azathioprine has long been reported to be more effective against T cell function than B cell function, although strong evidence supporting this is lacking[20].

One of the key enzymes involved in azathioprine metabolism is thiopurine methyltransferase (TPMT). Individual people inherit a marked deficiency in the TPMT enzyme that renders them highly susceptible to azathioprine toxicity. Interestingly, cats also have a marked deficiency in TPMT activity[21], which may explain why azathioprine causes marked myelosuppression in cats at standard canine doses[22]. Although the use of azathioprine at a very reduced dose rates has previously been published in cats, given the narrow margin for safety it is probably wisest to recommend that azathioprine never be used in cats at any dose. Although TPMT expression in dogs is widely variable, severe deficiencies in enzyme activity of the magnitude seen in cats and some people have not been commonly reported, and TPMT deficiency does not appear to be associated with the severe drug toxicities sometimes seen in dogs[23,24].

The standard azathioprine starting dose in dogs is 2 mg/kg orally once daily. This dose is usually well tolerated and, although GI side effects such as nausea, anorexia, vomiting and diarrhea are occasionally reported, they are typically mild and self-limiting. Chronic azathioprine usage often causes a mild to moderate sub-clinical anemia. Uncommonly, azathioprine can also cause profound myelosuppression or severe hepatotoxicity in dogs[25-27]. Myelosuppression and hepatotoxicity appear to be non-dose-dependent idiosyncratic reactions, and are typically reversible if the problem is recognized early and azathioprine is discontinued. Complete blood counts and serum biochemistry (especially ALT) should therefore be monitored regularly during azathioprine therapy. Individual case reports have reported pancreatitis in dogs receiving azathioprine, but cause and effect has not been established[27-28].

Azathioprine has become well established as an 'add on' agent for the treatment of many different immune-mediated and inflammatory conditions when steroids alone are ineffective or poorly tolerated, including IMHA, IMT, IBD, chronic hepatitis, glomerulonephritis, immune-mediated polyarthritis, myasthenia gravis, non-infectious meningoencephalitis, immune-mediated skin diseases, and anal furunculosis. Despite decades of azathioprine usage, evidence supporting efficacy for many of these diseases is remarkably limited. One perceived 'limitation' of azathioprine compared to other immunosuppressive agents, that it can take many weeks or even months to exert its effects, is based on limited and dated data derived predominantly in humans. In my experience, azathioprine in a clinical setting exerts its immunosuppressive effects in dogs about as rapidly as most other comparable agents.

Compared to most other immunosuppressive agents, azathioprine is relatively inexpensive, which is an important consideration with long-term immunosuppressive therapy, especially in large dogs. While the proprietary product (Imuran® or Azasan®) typically still costs over $5 per 50 mg tablet, the generic equivalent can be obtained for less than $1 a tablet. The smallest tablet size is 50 mg (although tablet scoring permits a 25 mg dose), which can present dosing problems in small (under 20 lb) dogs.

DANAZOL

Danazol, a synthetic androgen with weak ('impeded') androgenic effects, has in the past been suggested for the treatment of canine IMHA and IMT, in combination with glucocorticoids[29]. Danazol is derived from the synthetic steroid ethisterone, a modified progestogen. Danazol's most important mechanism of action is probably to reduce macrophage Fc receptor/antibody binding affinity.

Danazol also competes with glucocorticoids for combination with steroid-binding globulin, consequently increasing the availability of active unbound steroid. Danazol therefore enables significant steroid dose reduction. Danazol may also reduce the binding of antibody and complement to cell surfaces. Side effects are uncommon, and include hepatotoxicity and masculinization of female dogs. However, although some dogs with refractory IMHA and IMT may benefit from danazol[30-32], the drug fell out of favor a few decades ago, probably because it was very expensive at the time, and response to therapy was sluggish and highly unpredictable. Reported oral danazol doses in dogs with IMHA or IMT range from 5 to 15 mg/kg daily, either given as a single dose or 2-3 divided doses. Danazol comes in 50 mg, 100 mg, and 200 mg capsules, and currently costs a little less than $2 for a 50 mg capsule.

VINCRISTINE

The vinca alkaloids are biologically-active dimeric alkaloids derived from the Madagascar periwinkle plant, *Catharanthus roseus*. Vincristine, a naturally-occurring vinca alkaloid, were characterized phytochemically more than 50 years ago. The biological effects of vincristine have traditionally been attributed to drug-induced disruption of intracellular microtubules. Microtubules are composed predominantly of complex helical polymers of the structural protein tubulin. Vincristine binds directly to tubulin, causing both inhibition of microtubule synthesis and disruption of intact microtubules. Microtubular structures susceptible to the effects of vinca-tubulin binding include the mitotic spindle in dividing cells, the neurotubules in neurons, and the cytoskeletal microtubules in platelets. Vincristine may also exert biological effects that are independent of disruption of intracellular microtubules, such as inhibition of RNA, DNA and protein synthesis, and modification of prostaglandin production.

Vincristine is a cell-cycle-specific cytotoxic agent. Vincristine disrupts microtubules within the mitotic spindle of dividing cells, thereby arresting chromosomal separation in metaphase. Vincristine at standard doses is minimally myelotoxic, and is therefore used in in veterinary cancer chemotherapy in combination with more myelosuppressive agents. The degree of immunosuppression induced by vincristine at therapeutic doses is minimal compared to other immunosuppressive agents, and vincristine is therefore not used for the treatment of most immune-mediated diseases. The one exception is IMT, where vincristine has become a mainstay of treatment.

Vincristine is usually administered IV, and inadvertent SC or IM administration causes severe local tissue irritation and necrosis. Platelets, because of their intracellular tubulin, demonstrate a remarkable ability to concentrate vincristine from plasma, and are therefore the principal circulating cellular carriers of the drug. During early clinical trials in human cancer patients, it was observed that vincristine typically caused significant but transient increases in circulating platelet numbers. A similar phenomenon has since been reported in dogs, both in research animals and in cancer patients[33-34], and appears to be due to increased megakaryocytopoiesis and thrombopoiesis, although the precise mechanisms of thrombocytosis are still uncertain. Vincristine, typically in combination with prednisone, has been reported to similarly facilitate remission of thrombocytopenia in many canine patients with IMT. Original case reports demonstrating a rapid response to vincristine in dogs with IMT[35-38] have been supported, decades later, by evidence obtained from prospective studies[39-40]. Circulating platelet numbers increase markedly within 3-5 days of vincristine administration, and the addition of vincristine to standard therapy in dogs with IMT appears to shorten hospitalization time by several days. Most authors currently recommend an intravenous vincristine bolus dose of 0.02 mg/kg for the treatment of canine IMT. Vincristine has been used in cats with IMT, although evidence of clinical efficacy is lacking. One significant advantage of vincristine compared to other therapeutic options for IMT is that vincristine is inexpensive (a 1 ml vial of 1mg/ml vincristine sulfate costs around $20).

The pathogenesis of vincristine-induced platelet responses in IMT patients is uncertain. Studies in people suggest that the main therapeutic effect of vincristine in IMT patients is not increased thrombopoiesis. Post-treatment average platelet life-spans are significantly prolonged in patients that respond to vincristine, suggesting that remission is due to reduced platelet destruction rather than increased platelet production. Since platelets are the major circulating cellular carriers of vincristine, researchers have speculated that antibody-coated platelets selectively deliver vincristine to those phagocytes within the mononuclear phagocytic system that are actively involved in platelet destruction. This proposed mechanism explains why, despite being an ineffective immunosuppressive agent for the treatment of most conditions, vincristine can still be very effective for the treatment of IMT.

Neurotoxicity, although uncommon, is the most frequent significant side-effect associated with therapeutic doses of vincristine in dogs and cats[42]. Reversible vincristine-induced neurotoxicity in the dog has been reported with chronic cancer chemotherapy, but is not likely to be an issue with the single doses used to treat IMT[43]. Other side effects such as GI disorders and alopecia, occur less frequently, and are typically mild and temporary. Vincristine at doses used for IMT typically causes minimal myelosuppression in dogs, although dogs with the ABCB1-1Δ (MDR1) gene mutation and some Border Collies have been reported to be more susceptible than other dog breeds to myelosuppression at antineoplastic vincristine doses[44-45]. An unusual transient pulmonary toxicity has been reported in a cat receiving chemotherapeutic doses of vincristine[46].

REFERENCES

[1]Legrand J, et.al. **J Immunotoxic** 2013;10:90. [2]Ogilvie G, et.al. **Vet Imm Immunopath** 1988;18:119. [3]Putnam C, et.al. **Clin Exp Imm** 1975;22:323. [4]Charney S, et.al. **JAVMA** 2003;222:1388. [5]Crow S, et.al. **JAVMA** 1977;171:259. [6]Macy DW, et.al. **JAAHA** 1983;19:965. [7]Burgess K, et.al. **JVIM** 2000;14:456. [8]Grundy S, et.al. **JAVMA** 2001;218:543. [9]Mason N, et.al. **JVIM** 2003;17:206. [10]Reimer M, et.al. **JAAHA** 1999;35:384. [11]Warry E, et.al. **JVIM** 2011;25:903. [12]Irwin K, et.al. **Vet Derm** 2012;23:403. [13]Preziosi D, et.al. **Vet Derm** 2003;14:313. [14]Wondratschek C, et.al. **JAAHA** 2010;46:12. [15]Dandrieux J, et.al. **JAVMA** 2013;242:1705. [16]Benitah N, et.al. **JAAHA** 2003;39:283. [17]Harkin K, et.al. **JAAHA** 2007;43:21. [18]Ogilvie G, et.al. **Vet Imm Immunopath** 1988;18:119. [19]Rinkardt N, et.al. **Can J Vet Res** 1999;63:18. [20]Dowling P. **Can Vet J** 1995;36:781. [21]Foster A, et.al. **JVIM** 2000;14:552. [22]Beale K, et.al. **AJVR** 1992;53:1236. [23]Kidd L, et.al. **JVIM** 2004;18:214. [24]Rodriguez D, et.al. **JVIM** 2004;18:339. [25]Rinkardt N, et.al. **Can Vet J** 1996;37:612. [26]Favrot C, et.al. **Vet Rec** 2007;160:520. [27]Houston D, et.al. **Can Vet J** 1991;32:496. [28]Moriello K, et.al. **JAVMA** 1987;191:695. [29]Miller E. **Sem Vet Med Surg** 1997;12:167. [30]Bloom J, et.al. **JAVMA** 1989;194:76. [31]Holloway S, et.al. **JAVMA** 1990;197:1045. [32]Roseler B, et.al. **J Vet Allerg Clin Imm** 1996;4:53. [33]Mackin A, et.al. **AJVR** 1995;56:100. [34]Neel J, et.al. **Vet Clin Path** 2012;41:216. [35]Greene C, et.al. **JAVMA** 1982;180:140. [36]Jans H, et.al. **JVIM** 1990;4:4. [37]Williams D, et.al. **JAVMA** 1984;185:660. [38]Wolf A. **Southwest Vet** 1983;35:209. [39]Balog K, et.al. **JVIM** 2013;27:536. [40]Rozanski E, et.al. **JAVMA** 2002;220:477. [41]Tasker S, et.al. **JSAP** 1999;40:127. [42]Cho E, et.al. **Arch Toxicol** 1983;52:83-90. [43]Hamilton T, et.al. **JAVMA** 1991;198:635. [44]Lind D, et.al. **AJVR** 2013;74:257. [45]Mealey K, et.al. **JVIM** 2008;22:996. [46]Polton G, et.al. **Vet J** 2008;177:130.

FELINE CARDIOLOGY: A REVIEW OF THE BASICS

Janice M. Bright, BSN, MS, DVM, DACVIM (Cardiology, Small Animal Internal Medicine)
Fort Collins, CO

INTRODUCTION

Diagnosis and treatment of cardiovascular disease are particularly challenging in cats because of their small body and heart size, their rapid heart rate, their often uncooperative nature, and the often abrupt onset of clinical signs, In addition, there are numerous features of clinical cardiology that are different in cats than in dogs. These "peculiarities" of feline cardiology and basic feline cardiology are reviewed in this article with emphasis placed on differences between cats and dogs.

HISTORY & CARDIOVASCULAR EXAMINATION

Dogs with heart disease often have clinical signs that are insidious in onset and slowly progressive. In contrast, cats with cardiac lesions characteristically remain asymptomatic until very late in the course of the disease at which time clinical signs are acutely manifested. For example, cats are often asymptomatic until the sudden onset of severe dyspnea from congestive heart failure (CHF) or until peracute development of pelvic limb paralysis due to arterial thromboembolism. Nor is it unusual for cats with cardiac disease to have no clinical signs yet die suddenly from hypertrophic cardiomyopathy.

Common signs of heart disease in cats include 1) respiratory distress (tachypnea & dyspnea), 2) sudden onset paralysis or paresis (usually hind limb or right forelimb), 3) syncope, and 4) anorexia, lethargy, or depression. Whereas dogs frequently cough as a result of left-sided heart disease, it is extremely rare for cats with heart disease, even severe heart disease, to cough.[1]

A stethoscope with a pediatric or neonatal size chest piece is recommended for examining cats. Since older cats are predisposed to thyrotoxic heart disease, the thyroid glands should be carefully palpated when examining the feline cardiovascular system. Ascites due to right-sided congestive heart failure (CHF) is rarely massive as in dogs and is usually accompanied by hepatomegaly and jugular distention if due to heart disease.[1] Extra diastolic sounds ("gallops") due to an audible S3 or S4 are common in cats with heart disease, and these abnormal sounds are easily missed if examination with the bell of the stethoscope is omitted. Cats often have severe pulmonary edema without audible crackles (or coughing). In fact, coughing and wheezing in cats are nearly always associated with respiratory disease rather than heart disease. It is helpful to remember that most cats with heart failure have some type of auscultable cardiac abnormality (murmur, gallop, and/or arrhythmia). Finally, it should be pointed out that functional murmurs (e.g., murmurs due to fever, anemia, volume overload, and administration of tranquilizing drugs) are common in cats.

SURVEY THORACIC RADIOGRAPHY

Normal Radiographic Findings

The overall size of the feline cardiac silhouette is relatively smaller than in dogs; in other words, cats have relatively more lung tissue on thoracic radiographs than dogs.[2] In cats the heart is farther from the chest wall and from the diaphragm than in dogs, and the normal feline heart does not contact the diaphragm on the ventrodorsal (VD) or dorsoventral (DV) views. Unlike in dogs, the main pulmonary artery (MPA) does not project beyond the edge of the cardiac silhouette in cats on the ventrodorsal (or dorsoventral) thoracic radiograph.[3] Therefore, dilation of the MPA may not create a bulge at the 1-2 o'clock position. In cats the left ventricle (LV) forms the cardiac apex and is usually slightly left of midline; however, the cardiac apex may be slightly to the right of midline on the radiograph of some normal cats, particularly if a DV view is taken.[1] The craniocaudal ventricular diameter of the cardiac silhouette on the lateral view is normally between 2 - 2.5 intercostal spaces (4th – 6th intercostal spaces),[2] and the normal vertebral heart sum in cats is 7.5.[4]

There are some unique aging changes that occur normally in the feline cardiovascular system, and these changes are reflected on survey thoracic radiographs. Specifically, as cats age, the heart becomes more horizontally positioned within the thorax and the aortic root may enlarge.[5] Consequently, in geriatric cats the cardiac silhouette on the lateral view is typically more horizontally positioned than in young cats (the sternal angle is more acute). Enlargement of the aortic root may be appreciated in older cats on either radiographic view. This change in size of the aortic root is often noted as a normal aging change in cats; however, an enlarged aortic root may also indicate systemic hypertension.

Radiographic Criteria for Cardiac Enlargement in Cats

Cardiomegaly may result in widening, elongation, altered position, or altered contour of the cardiac silhouette. It is often more difficult in cats than in dogs to determine which specific chambers of the heart are enlarged from the survey radiographs. Although a vertebral heart sum exceeding 8.0 is indicative of cardiomegaly, it may be impossible to determine this sum in cats with pleural effusion. Tracheal elevation is not always diagnostic for cardiomegaly since this finding may also occur with pleural effusion or mediastinal masses. In cats with severe left atrial enlargement the cardiac apex is often shifted toward the midline producing a "valentine-shaped" cardiac silhouette on the DV or VD view.[1] This radiographic change is often misinterpreted as biatrial enlargement. Although enlargement of the left atrium is usually easily identified on the lateral radiograph in dogs, this cardiac chamber is situated more cranially in cats, and left atrial enlargement is not always easy to appreciate on this view.[6] Whereas cardiogenic pulmonary edema in dogs is typically dorsal and perihilar in distribution, in cats the edema may also be patchy or ventral in distribution.

DOPPLER ECHOCARDIOGRAPHY

It is wise to avoid sedating feline patients with ketamine or α_2 –agonists when doing Doppler echocardiographic examination. These agents not only alter velocity and functional measurements, they may not be well tolerated by cats with cardiovascular diseases. M-mode echocardiography is the best ultrasound modality to use for measuring chamber sizes and wall thicknesses. Because most adult cats are similar in size, patient values may be compared to standard values. Diastolic thickness of the interventricular septum and diastolic thickness of the left ventricular wall are used to determine presence or absence of left ventricular hypertrophy. It may be difficult, however, to clearly distinguish normal from abnormal in cats with slightly increased wall and/or septal thickness based solely on these measurements. Two dimensional echo is more useful than M-mode for evaluating congenital lesions, regional (focal) hypertrophy, and neoplasia. Blood flow Doppler is used to evaluate diastolic function, blood flow patterns, and blood flow velocities. Tissue Doppler also provides a means of evaluating systolic and diastolic function. Interestingly, tissue Doppler has been shown to identify human patients with hypertrophic cardiomyopathy prior to the development of left ventricular hypertrophy.[7]

ELECTROCARDIOGRAPHY

Although right lateral recumbency is the standard position for ECG recording, some cats are more cooperative if allowed to remain sternal for the recording. Sternal recumbency may be used, but it is important to realize that the P and R waves are slightly taller and frontal plane axis wider in this position.[8] Because of the rapid heart rates of cats, ECG interpretation is facilitated by using a recording speed of 50 mm/s. QRS complexes may be very low voltage in some normal cats. On occasion, in fact, the P waves will be larger than the QRS complexes. It is best not to use muscle filters when obtaining a feline ECG because these filters cause significant alterations in the waveform amplitudes.[9]

A left anterior fascicular block ECG pattern is commonly associated with primary or secondary feline myocardial diseases, and the presence of this ECG abnormality should alert the clinician of possible cardiomyopathy. The ECG criteria of left anterior fascicular block include: upright (positive) QRS complexes in leads I & AVL (i.e., a qL or L pattern in these leads); abnormally deep S waves in leads II, III, and AVF; and a left axis deviation in the frontal plane (mean electrical axis 0 to -90°). Tachycardia and premature beats are also common in all forms of feline myocardial disease. Atrioventricular conduction block is not uncommonly seen in cats with restrictive cardiomyopathy.[10]

CONGESTIVE HEART FAILURE (CHF) IN CATS

Cats, unlike dogs, often have an acute or peracute onset of clinical signs of heart failure. Manifestations of left-sided CHF in cats may result from decreased cardiac output such as weakness, exercise intolerance, syncope, lethargy, hypothermia, or peripheral vasoconstriction (pale mucous membranes & prolonged capillary refill time). More commonly, however, left-sided CHF results in acute respiratory distress as a result of pulmonary edema. Therefore, clinical signs of left-sided CHF in cats include tachypnea, dyspnea, cyanosis, and increased large airway sounds. An important difference between dogs and cats with left-sided failure is that cats may develop pleural effusion as well as pulmonary edema from increased left-sided filling pressures.[1] Appropriate management of dyspnea due to left-sided heart disease may, therefore, necessitate thoracocentesis as well as diuretic administration in feline patients. It is helpful to recall that left-sided heart failure rarely causes coughing in cats as it frequently does in dogs.

Cats are relatively susceptible to development of pulmonary hypertension (with subsequent signs of right-sided failure) caused by chronic left-sided disease. It is not uncommon, therefore, for cats to have biventricular failure.

Clinical manifestations of right-sided CHF in cats may include the signs of reduced cardiac output previously described for left-sided failure. In addition, however, cats with RV failure usually have signs of increased right-sided filling pressures including jugular venous distention and pulsation, hepatojugular reflux, hepatomegaly, ascites, pleural effusion (modified transudate or chylous), and pericardial effusion.

Diagnosis of CHF is a clinical diagnosis based on history, physical examination, radiography, echocardiography, and, rarely, venous pressure measurement. Most cats with CHF will have either a systolic murmur or diastolic gallop.

CONGENITAL HEART DISEASE IN CATS

Although congenital heart disease is much less common than acquired heart disease in cats, congenital lesions occur with a prevalence of 0.02 – 0.2%.[11] Congenital heart defects in cats are reportedly more common in males than in females, and complex congenital heart disease (more than a single anomaly in a patient) occurs more commonly in cats compared to dogs. The most common congenital anomalies in cats are AV valve dysplasia, ventricular septal defects, atrioventricular septal defects (endocardial cushion defects), tetralogy of Fallot, patent ductus arteriosis, and subaortic stenosis.[11, 12] Nearly all feline congenital cardiac anomalies produce systolic murmurs, and, therefore, timing of the murmur is not as helpful diagnostically as murmur location. Because the most prevalent feline congenital heart defects are AV valve dysplasia, ventricular septal defects, and atrioventricular septal defects, and because ECG changes are not typical of any of these lesions, the ECG is usually of little help in determining the specific diagnosis. More importantly, a normal ECG does not rule out a serious congenital anomaly in a feline patient. The age at which clinical signs of congenital heart disease are present is highly variable, and all cats with murmurs present since birth are suspect for CHD, even if they are older animals. Similarly, all kittens with murmurs > 2/6 should be suspected of CHD even if asymptomatic.

FELINE MYOCARDIAL DISEASES (CARDIOMYOPATHIES)

Feline myocardial diseases comprise a heterogeneous group of disorders characterized by structural and functional abnormalities of the myocardium. Although these disorders are usually acquired, heritable forms which are not present at birth may develop at a very young age. Feline myocardial disease, like canine myocardial disease, may be *primary* (defect within the myocardium) or *secondary* to an extra cardiac disorder or non-myocardial cardiac lesion. When a modifying term (e.g., hypertensive cardiomyopathy, ischemic cardiomyopathy) is not used *primary* cardiomyopathy is implied. The World Health Organization (WHO) updated its classification for myocardial diseases in people 1996.[13] This classification, based on morphologic and pathophysiologic features, is applicable to the feline myocardial diseases as well.

Dilated Cardiomyopathy (DCM)

Feline DCM is characterized primarily by myocardial systolic failure (decreased inotropic state). Although all four cardiac chambers are usually dilated, the dilation is secondary to the decreased myocardial contractility. It has been recognized since the 1980s that feline DCM may result from reduced dietary taurine intake of decreased taurine absorption.[14] Primary feline DCM is rare.

Hypertrophic Cardiomyopathy (HCM)

Hypertrophic cardiomyopathy is characterized primarily by impaired myocardial relaxation, reduced LV diastolic volume, and unexplained myocardial hypertrophy. This disease is known to be heritable in some breeds (Maine coon, Persian, rag doll, and several lines of American shorthair). Although HCM may cause left-sided CHF, biventricular CHF, and sudden death, many cats are asymptomatic. M-mode criteria for LV hypertrophy in adult cats includes presence of septal or LV free wall thickness $\geq$5.5 mm. However, it is important to point out that in cats, as in people, hypertrophy may be present only in focal areas of the ventricle and may be missed if only standard M-mode criteria are used. Recent data from human familial HCM indicate that systolic and diastolic functional abnormalities may be recognized in people with HCM prior to the presence of recognizable hypertrophy.[7] Primary HCM must be differentiated from secondary forms of HCM, but in cats this distinction cannot be reliably made with ECG or echocardiography. It is known that longstanding feline HCM may progress to restrictive cardiomyopathy in some animals.[15, 16]

Restrictive Cardiomyopathy (RCM)

Restrictive cardiomyopathy is characterized by reduced myocardial compliance, normal LV diastolic volume, and varying degrees of LV hypertrophy. Although this disorder may be caused by previous HCM in some cats[15, 16] and to previous inflammatory disease in others[17, 18], in most cats it is idiopathic. This form of myocardial disease may cause left-sided or biventricular failure in cats. It is extremely difficult to treat and carries a very poor prognosis.

Arrhythmogenic Right Ventricular Dysplasia

Arrhythmogenic right ventricular (RV) dysplasia is much less common than HCM and RCM. This cardiomyopathy is also known as isolated right-sided cardiomyopathy; or as RV dysplasia. This myocardial disorder is characterized by progressive fibroadipose infiltration of the RV myocardium and profound RV failure.[19] It is often misdiagnosed as tricuspid valve dysplasia.

SYSTEMIC ARTERIAL THROMBOEMBOLISM (ATE)

Systemic arterial embolism is a common and serious complication of <u>any</u> form of feline heart disease! Although this problem is more prevalent in cats with severe left atrial enlargement, it is occasionally seen in cats with a normal atrium.[20] Thrombus forms in the left atrium and embolizes to a distal site in the arterial tree, most commonly to the distal aorta. Embolism to the right forelimb is also frequently noted. Cerebral or abdominal visceral embolism may also occur.

The pathogenesis of feline ATE involves stasis of blood flow (due to underlying cardiac disease), activation of platelets and/or clotting factors, and ischemic damage of distal tissues which is mediated by platelet-derived vasoactive factors.[21] Diagnosis is usually straight forward based on clinical examination, but ultrasound or angiography may be used to confirm. Serum creatine kinase, and D-dimer assays may also be helpful. Prognosis for cats with ATE is guarded at best because it is difficult to predict how much neuromuscular function will be recovered. If the affected limbs remain cyanotic and cold after 24-48 hours or if they become edematous, recovery is unlikely, and gangrenous necrosis may develop. Cats are predisposed to reoccurrence of ATE.

REFERENCES

[1]Bonagura, JD. <u>The Cat, 2nd Edition</u>, pp 819-946 (1994). [2]Rush J, et al. <u>A Practical Approach to the Radiographic Diagnosis of Heart Disease in Small Animals</u>, pp 8-49 (1996). [3]Suter, PF. Thoracic Radiography: A text Atleas of Thoracic Diseases of the Dog and Cat. 1984. [4]Litster AL, et al. JAVMA 2000;216:210. [5]Moon, ML et al. Vet Radiol Ultrasound 1993:34:315. [6]Kittleson, MD. <u>Small Animal Cardiovascular Medicine</u>, p 63 (1998). [7]Nagueh, SF et al. Circ 2001;104:128. [8]Gompf, RE et al. Am J Vet Res 1979;40:1483. [9]Schrope, DP et al. Am J Vet Res 1995;56:1534. [10]Martin, JM et al. The Veterinary ICU Book, pp 550-576 (2002). [11]Bolton, GR et al. Vet Clin North Am Small Anim Pract 1977;7:341. [12]Hunt, G et al. Aust Vet Pract 1990:20:2. [13]Richardson, P et al. Circ 1996;93:841. [14]Pion, PD et al. Science 1987;237:764. [15]Kittleson, MD et al. Proc ACVIM Forum 1996:14:747. [16]Baty, CJ et al. J Vet Intern Med 2001;15:595. [17]Stalis, IH et al. Vet Pathol 1995;32:122. [18]Meurs, KM et al. J Vet Intern Med 1998;12:201. [19]Fox, PR et al. Circ 1998:98(Suppl):297 (Abstr). [20]Laste, NJ et al. J Am Anim Hosp Assoc 1995;31:492. [21]Kittlesom, MD. <u>Small Aniimal Cardiovascular Medicine</u>, pp 540-551 (1998).

COMMON FELINE ARRRHYTHMIAS

Janice M. Bright, BSN, MS, DVM, DACVIM (Cardiology, Small Animal Internal Medicine)
Fort Collins, CO

INTRODUCTION

The arrhythmias that occur in cats are similar, in general, to those noted in dogs; however, the relative frequency and clinical significance of feline cardiac rhythm abnormalities are often different. For example, cats with complete atrioventricular (AV) block often have underlying myocardial disease whereas dogs with this arrhythmia typically do not have structural heart disease. In cats, as in dogs, arrhythmias may result from either cardiac or non-cardiac diseases, and an underlying etiology should always be sought and, if possible, specifically treated.

NORMAL FELINE HEART RATE AND RHYTHM

Normal cats in a hospital or clinic environment have a regular sinus rhythm with heart rate between 140-220 beats per minute (bpm).[1] Sinus arrhythmia and ventricular ectopy in this setting are unusual and usually abnormal. In contrast, in ambulatory or telemetric studies of heart rate and rhythm in normal cats, sinus arrhythmia is commonly identified and occasional ventricular ectopic complexes are noted.[1] The mean heart rate recorded from normal cats in the home environment using 24 hour ambulatory electrocardiographic recordings (Holter monitoring) is 132 ($\pm$ 19) bpm (STD).[1] The differences in heart rate and cardiac rhythm in home vs. hospital are most likely the result of differences in autonomic balance with sympathetic tone higher and vagal tone lower when cats are in a clinic setting.

The following are normal feline electrocardiographic (ECG) parameters obtained on standard recording in a clinical setting:[1,2] 1) heart rate 140-220 bpm, 2) regular sinus rhythm, 3) PR interval 0.05-0.09s, 4) P wave $\leq$0.2mv x <0.04s, 5) QRS complex $\leq$0.04s; R wave $\leq$0.9mv (no minimum); axis 0-180° (frontal plane), 6) QT interval 0.12-0.18s (varies with heart rate), 7) T wave positive, negative or biphasic (usually positive)

VENTRICULAR ARRHYTHMIAS

Ventricular premature complexes (VPCs) and ventricular tachycardia are the most common feline arrhythmias and, as in dogs, these arrhythmias may occur in cats from a variety of cardiac and non-cardiac causes. Ventricular premature complexes (VPCs) occur earlier than expected given the underlying rhythm, and they are ventricular in origin as discerned from their differing morphology compared to sinus origin QRS complexes. Similar to VPCs in dogs, feline VPCs are typically wide (QRS complex > 60 ms) and bizarre in shape, with a large T wave opposite in polarity from the major deflection of the QRS complex. However, the QRS morphology of feline VPCs is often not as strikingly wide and bizarre as those noted in dogs. The ectopic ventricular QRS complexes are dissociated from atrial activity. Fusion complexes may occur but are less common in cats than in dogs. Four or more consecutive VPCs constitute ventricular tachycardia (VT). As in dogs sustained, rapid VT and multifocal VT have a significant potential for deteriorating into ventricular fibrillation.

In cats ventricular tachyarrhythmias are frequently associated with some form of myocardial disease (96%), either primary or secondary. Accelerated idioventricular rhythm (AIVR) is an ectopic ventricular rhythm with a rate $\leq$ 240 bpm.[3] In other words this is a ventricular rhythm faster than a ventricular escape rhythm yet slower than a normal feline sinus rate. Thus, AIVR is a slow ventricular "tachycardia". Unlike rapid ventricular tachycardia AIRT is not usually associated with myocardial disease and is often noted in anesthetized cats and cats with gastrointestinal disease.

ATRIAL FIBRILLATION

Atrial fibrillation (AF) occurs less frequently in cats than in dogs because the relatively small size of the feline atrium in normal cats cannot readily sustain the multiple reentry circuits characteristic of this arrhythmia. Atrial enlargement secondary to structural heart disease is usually present in cats with AF, and many cats will have concurrent congestive heart failure. However, AF on rare occasion is found as an incidental finding in otherwise normal cats.[4] As in dogs with AF, this arrhythmia in cats AF is characterized by absence of P waves with irregular and chaotic baseline fluctuations (fibrillation waves or *f* waves), an irregular ventricular rhythm, and fairly normal QRS complexes. The ventricular rate is typically rapid (> 250 bpm) but may be normal or slower than normal if there is also AV conduction delay or conduction block.

ATRIOVENTRICULAR BLOCK

Atrioventricular block refers to a delay or block of impulse conduction from the atria to the ventricles at a time when the conduction system is not refractory. Conduction delay or block may arise at any site between the sinoatrial node and the ventricular myocardium (i.e., intraatrial, intranodal, or subnodal block). While certain ECG patterns of AV block more typically reflect impaired conduction in one particular segment of the cardiac conduction system, overlap occurs, and it is often impossible to determine with certainty which part of the conduction system is responsible for the block based on the surface ECG.

AV block is classified into three categories based on electrocardiographic (ECG) characteristics. First degree AV block refers to delayed AV conduction; the ECG shows a 1:1 relationship of P waves and QRS complexes with a prolonged PR interval (> 0.09 s). QRS and T waves usually are normal. First degree AV block cannot be detected on physical examination. Second degree AV block occurs when there is failure of some, but not all, P waves to be conducted to the ventricles. QRS and T waves appear normal unless there is also an intraventricular conduction abnormality. Third degree (complete) AV block is the result of complete absence of AV conduction and is characterized on ECG (in patients with sinus rhythm or other organized atrial rhythm) as a complete lack of association of P waves and QRS complexes with the ventricular rate slower than the atrial rate. In many cats with complete AV block the ventricular escape focus is near the origin of the His bundle (in the region defined as the AV junction), and in these cats the QRS complexes are narrow. The escape rate is frequently sufficiently rapid (median rate 120 bpm) to prevent clinical signs.[5]

Cats with high grade second degree or third degree AV block may have degenerative conduction system disease. However, more commonly these cats will develop AV block as a result of underlying myocardial disease (hypertrophic or restrictive cardiomyopathy).[5] Although these cats often need a permanent pacemaker to palliate clinical signs, the myocardial disease will usually require treatment and will often progress.[4]

ANTIARRHYTHMIC AGENTS

Antiarrhythmic agents are indicated in cats with life threatening cardiac rhythm abnormalities which cannot be specifically or rapidly treated and in cats with clinical signs caused by the arrhythmia. Less is known about the pharmacokinetics and adverse effects of antiarrhythmic agents in cats than in dogs; however, a summary of antiarrhythmic agents most commonly used by the author is presented below (Table 1).

Table 1. Useful Antiarrhythmic Agents for Feline Patients

Drug	Indication	Oral Dose	IV Dose
Diltiazem	atrial fibrillation & supraventricular tachycardias	1.5 – 2.0 mg/kg TID	0.1 – 0.2 mg/kg bolus 2 – 6 µg/kg/min CRI
Cardizem CD® (or generic equivalent)	" " "	10 mg/kg q24h	
Atenolol	atrial fibrillation & supraventricular tachycardias	6.26 - 12.5 mg BID	
Sotalol	ventricular tachycardia	1/8 of an 80 mg tab BID	
Lidocaine	life-threatening VT		0.25 – 0.5 mg/kg bolus
Amiodarone	VT & SVT	40 mg q24h* (*dose is anecdotal)	

REFERENCES

[1]Abbott J et al. J Feline Med Surg 2005;7:195. [2]Tilley LP. **Essentials of Canine & Feline Electrocardiography, 3rd Ed.** 1992;100. [3]Côté E. **Kirk's Current Veterinary Therapy XV** 2014;748. [4]Côté E et al. J Am Vet Med Assoc 2004;225:256. [5]Kellum HB et al. **J Vet Intern Med** 2006;20:97.

WHAT'S NEW WITH METHICILLIN-RESISTANT STAPHYLOCOCCI

J Scott Weese, DVM, DVSc, DACVIM
Guelph, Ontario, Canada

INTRODUCTION

Methicillin-resistant *Staphylococcus* spp (MRS) continue to pose significant challenges to both the veterinary and human healthcare systems. With their inherent resistance to all beta-lactam antimicrobials (including carbapenems) and frequent acquisition of other resistance genes, treatment options may be very limited. Many different staphylococcal species exist, with the coagulase-positive *S. aureus, S. pseudintermedius* and *S. schleiferi coagulans* most often involved in disease. A comprehensive review of the epidemiology, clinical relevance, diagnosis, treatment and prevention of MR-staphylococcal infections is well beyond the scope of these proceedings. However, some interesting or important new developments that may be relevant to the veterinary internist are described below.

RECENT DEVELOPMENTS AND IMPORTANT ISSUES

Rapid diagnosis

In humans, real time PCR is increasingly being used as a rapid screening tool. While not providing susceptibility data, real time PCR can indicate the presence of MRSA and allow for better initial antimicrobial selection and prompt patient isolation. A rapid test would also be desirable in animals in some situations. However, an older study indicated performance issues with a human assay in horses[1] and no further data have been published. At least one diagnostic laboratory still falsely advertises a *mecA* PCR assay as an MRSA test, despite the fact that this gene is commonly found in many different staphylococci, including minimally virulent coagulase negative staphylococci. Development of a PCR assay for MRSA or MRSP requires identification of a single target that indicates that the isolate is both the *Staphylococcus* species and is methicillin-resistant (being able to differentiate MRSA from a sample that has both methicillin-susceptible *S. aureus* and an MR-coagulase negative *Staphylococcus*). Adaptation of human MRSA assays for other species would be possible because of the similarity in strains in humans and animals, but an MRSP test would require development of the assay. This is becoming more feasible because of the recent availability of sequenced MRSP genomes, but at the time of writing, no rapid tests have been validated for animals for any MRS.

Hopefully, the coming years will be accompanied by validated real time PCR assays or other approaches such as lateral flow devices that may be practical in-clinic tools.

Susceptibility testing, part 1

The state of diagnostic laboratory susceptibility testing has improved in recent years. However, some gaps still remain and clinicians must properly scrutinize laboratory reports. Standard (e.g. CLSI) guidelines dictate reporting of all MR-staph as resistant to all beta-lactams, regardless of in vitro results. This is because false susceptible results can sometimes occur with one or more beta-lactams, yet use of those drugs would not be appropriate because treatment would be expected to fail. Yet, not all laboratories properly report MRS and automatically convert any beta-lactam "S"s to "R"s. If an MRS is reported as susceptible to any beta-lactam, the beta-lactam should not be used. Further, the erroneous reporting should be discussed with the diagnostic laboratory. Similarly, it is prudent to contact laboratories whenever there is a highly drug resistant *Staphylococcus* that is not reported as MRS. While this can happen, most often this is a false negative result for the methicillin-resistant indicator drug (cefoxitin for MRSA and oxacillin for MRSA) and retesting of the isolate would be prudent to ensure that inappropriate antimicrobials are not prescribed.

Minocycline vs doxycycline

Tetracyclines can be effective anti-staphylococcal drugs, but resistance is common. Doxycycline susceptibility is commonly reported on small animal susceptibility panels, and in vitro results are a good indicator of in vivo susceptibility to that drug. However, minocycline is another tetracycline drug that can be used, and a recent study demonstrated efficacy of minocycline in a subset of MRSP isolates that were doxycycline resistant.[2] This opens the door to another potential treatment option in some cases, something that can be very useful given the limited susceptibility of many MRSP strains. The difference between susceptibility to these two drugs is based on the tetracycline resistance gene that the isolates contain, with *tet(K)* conferring resistance to doxycycline but not minocycline and *tet(M)* conferring resistance to both drugs. The presence of these genes is strongly associated with the MRSP strain that predominates in the area. In North America, both the *tet(M)*-associated sequence type (ST) 68 and *tet(K)*-associated ST71 can be found, so minocycline susceptibility is variable. Minocycline susceptibility is more common in Europe, where ST71 is more prominent.[3] While objective study of the efficacy of minocycline is lacking, anecdotal data indicate that it can be a useful treatment option in cases where the strain is susceptible. Obtaining minocycline susceptibility data should be considered in situations where doxycycline-resistant isolates are present and where minocycline might be useful, understanding the testing limitation that is described below.

Susceptibility testing, part 2: Doxycycline and minocycline susceptibility

While these two drugs may be useful treatment options, clinical use may be complicated by ineffective interpretive breakpoints. Currently, a single breakpoint ($S \leq 4$ ug/ml, I=0.4 ug/ml, $R \geq 8$ ug/ml) is used for all members of the tetracycline family. However, recent data have suggested that doxycycline breakpoints should actually be lower ($S \leq 0.125$ ug/ml, I=0.25 ug/ml, $R \geq 0.5$ ug/ml) for *S. pseudintermedius*.[4] Thus, until interpretive breakpoints are formally changed, there may be difficulty interpreting results from commercial diagnostic laboratories. Obtaining MICs can help, but many diagnostic laboratories test a limited range of dilutions, such as only 4, 8 and 16 ug/ml. This is problematic since ≤ 4 ug/ml encompasses both the susceptible and resistant range of the proposed breakpoints. Thus, obtaining MIC data are useful but only if a method is used that will provide information that is relevant to the proposed (0.125 ug/ml) breakpoint.

Limitations in local therapy, part 1

While local therapy using impregnated materials remains a potentially useful option, a few concerns have arisen. Silver impregnated or coated materials are becoming increasingly population because of the potential antimicrobial effect of silver. Indeed, silver can be bactericidal and efficacy data are available for various materials and devices in humans. However, a recent in vitro study of silver-impregnated polymethylmethacrylate (PMMA) identified no impact of silver on MRSP biofilm formation.[5] Whether this is because of the nature of the silver compound, the concentration that was used or silver resistance is unclear, but clinicians should be aware that one cannot absolutely assume that a silver compound will be efficacious, and that proper study of veterinary patients and relevant bacterial species (i.e. MRSA and MRSP may not behave the same) is required.

Limitations in local therapy, part 2

Study of gentamicin impregnated PMMA also showed potential limitations.[5] It is often assumed that impregnated materials deliver so high concentrations that in vitro bacterial susceptibility becomes irrelevant. However, this in vitro study showed an impact of gentamicin-impregnated PMMA on MRSP biofilm formation, but only if isolates were gentamicin susceptible. Gentamicin resistant isolates were refractory to the effect of gentamicin in PMMA beads. This is presumably because the amount of gentamicin on the bead surface and in the immediate vicinity was not adequate to overcome even the resistance that was present, even though the isolates had MICs that were not much higher than the breakpoint. This study suggests that in vitro susceptibility is relevant when considering the use of impregnated PMMA beads or other materials where high concentrations at the surface and/or high elution is not necessarily present.

If we stop using antibiotics, will they go away?

There is clearly an association between antimicrobial exposure and the likelihood of MR-staph infection or colonization.[6,7] However, a lack of association between MRSP status and antimicrobial exposure has also been reported.[8] This suggests that MRSP may be so well host adapted and fit that removal of the selection pressure may not (at least rapidly) result in reduction of MRSP colonization in dogs. This would be a concerning situation, given the rapid increase in MRSP that has been encountered in recent years.

Persistence of MRS shedding

An understanding of the dynamics of MRSA and MRSP shedding is important for understanding the risk of recurrent infection, for management recommendations in the household and for managing animals admitted to a hospital with a history of MRS infection.

MRSA is poorly host adapted for companion animals and short-term colonization appears to be the norm. Colonization often persists for a short period of time after clinical infection, but this does not last long and most animals eliminate MRSA on their own within a few weeks. In contrast, MRSP is well adapted to dogs and can persist for prolonged periods of time. Some dogs may be colonized for months after clinical infection[9] and lifelong colonization cannot be ruled out. This may be part of the reason for the rapid increase in MRSP colonization in the dog population. Potential longterm colonization raises concern about ongoing risk of opportunistic infection in colonized high-risk patients, as well as the potential that any dog previously diagnosed with MRSP could be a carrier at the time of re-admission to a veterinary hospital.

Livestock-associated MRSA

The mid 2000s witnessed the emergence of a 'new' MRSA clone, sequence type (ST) 398 in pigs and people that work with pigs. It has recently been determined that methicillin-susceptible ST398 moved from humans to pigs, where it acquired methicillin-resistance, and subsequently underwent some further changes for host adaptation.[10] ST398 MRSA from livestock is now a leading cause of human MRSA infection in some European countries, and another strain, ST9 MRSA, has more recently been identified as having a livestock origin.[11] Currently, ST398 infections in companion animals are rare, with only limited reports in dogs (and most linked with direct or indirect livestock contact).[12,13] However, ST398 has emerged as an important and increasingly common cause of MRSA infection in horses in some European countries.[14,15] Livestock may represent a significant future reservoir for MRSA in pets and horses if strains that are found in livestock are able to (or evolve to be better able to) infect and colonize other species.

Zoonotic Infections from Pets

Currently, it appears that the incidence of pet-human transmission is greatly outweighed by the concern (or paranoia) about the potential role of pets in human infection. Strong evidence of pet-human transmission is lacking. It is clear that MRSA can move between pets and people, but the predominant routes, risk factors, incidence and clinical implications remain elusive. MRSA should certainly be considered a zoonotic pathogen, albeit one where the main route of transmission is human-pet, not pet-human. In the absence of better information, counseling of clients about the zoonotic risks and basic preventive measures is prudent. Screening pets for MRSA colonization in response to human infections, though, is rarely indicated. This should, at best, be reserved for very limited situations where there is ongoing transmission in people in the household, where initial control measures have failed to be effective and when there is a whole-household response that involves both people and pets. Testing pets but not addressing humans in a human-associated pathogen such as MRSA is illogical and likely ineffective.

MRSP is a zoonotic pathogen yet the incidence of clinical infection in humans appears to be very low. Concern has been expressed about the ability of human diagnostic laboratories to identify MRSP and the author is aware of a small number of MRSP infections in veterinary personnel, but the incidence of MRSP infection in people is likely truly low. This is not surprising since MRSP is no more inherently virulent than its susceptible counterpart. When one considers that most dogs are carrying methicillin susceptible *S. pseudintermedius,* that a large percentage of the human population interacts with dogs regularly and that methicillin-susceptible *S. pseudintermedius* infections are very rare in humans, it is logical that MRSP infections would be rare. However, the highly drug resistant nature of MRSP and the potential for *S. pseudintermedius* to cause serious infection in humans (albeit very rarely) make it prudent to use good general infection control and hygiene practices to reduce potential exposure by humans, particularly high-risk individuals.

Zoonotic implications for veterinary personnel

Studies continue to identify relatively high rates of MRSA colonization in veterinary personnel.[16,17] These studies have made it clear that MRSA exposure is an occupational risk in veterinary medicine. The clinical implications remain unclear. Clinical MRSA infections have been identified in veterinary personnel, yet the incidence of infection is not known, with many anecdotes but no clear data. Regardless, veterinarians should be aware of the risk of MRSA exposure from any animal, use basic infection control precautions (e.g. routine protective clothing, hand hygiene, environmental cleaning and disinfection) for all cases and enhanced precautions (e.g. enhanced protective clothing, gloves, isolation) for known or suspected MRSA cases. Additionally, considering the high rates of MRSA colonization that have been reported in veterinary personnel, veterinarians should ensure that their physicians are aware of their MRSA risk status. While this would not be expected to result in any action in most situations, there may be circumstances (e.g. prior to elective surgery, prior to undergoing immunosuppressive therapy) where knowledge of the increased risk of MRSA colonization is clinically relevant.

REFERENCES

[1] Anderson ME, et al. ***Vet Microbiol*** 2007;122:185-189.
[2] Weese JS, et al. ***Vet Microbiol*** 2012;162:968-971.
[3] Perreten V, et al. ***J Antimicrob Chemother*** 2010;65:1145-1154.
[4] Maaland MG, et al. ***J Clin Microbiol*** 2013;51:3547-3554.
[5]Singh AM, et al. Impact of gentamicin and silver on methicillin-resistant *Staphylococcus pseudintermedius* biofilm formation on polymethylmethacrylate. *3rd ASM-ESCMID conference on methicillin-resistant staphylococci*. Copenhagen, Denmark, 2013.
[6] Rota A, et al. ***Reprod Domest Anim*** 2013;48:1-6.
[7] Beck KM, et al. ***Vet Dermatol*** 2012;23:369-e367.
[8] Lehner G, et al. ***Veterinary Microbiology*** 2014;168:154-160.
[9] Laarhoven LM, et al. ***PLoS ONE*** 2011;6:e27788.
[10] Price LB, et al. ***mBio*** 2012;3:e00305-00311-e00305-00311.
[11] Cui S, et al. ***J Antimicrob Chemother*** 2009;64:680-683.
[12] Floras A, et al. ***Vet Rec*** 2010;166:826-827.
[13] Witte W, et al. ***Emerging Infect Dis*** 2007;13:255-258.
[14] Cuny C, et al. ***Microb Drug Resist*** 2008;14:307-310.
[15] Van den Eede A, et al. ***Vet Microbiol*** 2009;133:138-144.
[16] Burstiner LC, et al. ***Vet Surg*** 2010;39:150-157.
[17] Ishihara K, et al. ***Microbiol Immunol*** 2014;doi: 10.1111/1348-0421.12128.

FROM CLASSROOM TO EXAM ROOM: FACILITATING THE TRANSITION WITH POLLING TECHNOLOGY

John R. August, BVetMed, MS, MRCVS, DACVIM (SAIM)
College Station, TX

INTRODUCTION

To enhance student self-confidence, the author transformed his feline internal medicine course from a didactic format to a student-centered experience. Improved pedagogical application of an audience response system appeared to strengthen engagement and decision-making skills when applied to case studies in the flipped classroom, helping to prepare students for the transition into their hospital-based education.

THE CLASSROOM CHALLENGE

The first three years of the veterinary curriculum are an intense experience of didactic classes and laboratories, often involving 21 or more semester credit hours. By nature, veterinary medical students prefer to study independently, spend minimal time in high-level knowledge processing with their peers, and value strong direction from their instructors. Paradoxically, students state that they enjoy working in groups to solve problems with clinical relevance.[1] The fourth year is spent in the veterinary medical teaching hospital, working in small groups with faculty members, house officers, technicians, clients, and patients. The abrupt transition from the relatively passive environment of the large classroom in which students are reluctant to answer questions, to the examination room and wards where opinions and recommendations are sought constantly, can be very stressful for the learner.[2]

CRITICAL TRANSITIONS IN THE UNDERGRADUATE EXPERIENCE

Based on his observations of students at Harvard in the 1960's, William Perry identified a sequence of stages of intellectual development during the undergraduate years.[3] Depending on their individual abilities, students passed successfully through, or became trapped by, critical transition points in this nine-stage process. At the beginning of their college years, undergraduate students tended to view their world from a dualist approach; for example, considering issues as good or bad, or right or wrong. This early behavioral characteristic was reinforced by considering the instructor to be the ultimate purveyor of knowledge and the subject matter.

As his students progressed further during the early stages of their academic development, Perry noticed that some of them started to have difficulties when confronted with teachers who were not all-knowing or with situations in which they were asked to solve problems independently. By the mid-point in their development, students were no longer thinking in a dualistic way; rather, they would consider the merits of each issue in context. This mid-point appears to be one of the important transitions in student development. The later phases of academic maturity are characterized by an increasing commitment by the students to individual decision-making in an increasingly uncertain environment.

In his study, Perry observed that some students struggled with their passage through the nine phases, and when confronted with a difficult transition took one of three approaches: hesitating to take the next step, becoming detached and alienated, or retreating back to a dualistic viewpoint. Importantly, and not surprisingly, the most difficult transition for many students and their instructors was the point at which the learner must shift from an educational environment that is fact-based to one in which knowledge must be applied to unique situations and contexts. Perry reminds the reader that instructors typically praise or blame students for the responses they give in class or for their academic performance in general. Under-recognized is the courage that students must have to transition to the next phase of their development, and simple recognition of this hidden effort is an important motivational factor.

APPLYING PERRY'S MODEL TO TODAY'S VETERINARY MEDICAL CURRICULUM

At first glance, it would seem unlikely that observations made on Ivy League undergraduate students in the 1960's could be extrapolated to today's veterinary medical students. However, instructors will recognize the confusion and frustrations expressed by some students in their early clinical correlates courses as they first confront ambiguous case studies during the largely fact-based (dualistic) basic science curriculum. The trepidation felt by most third-year students as they start their clinical rotations in the veterinary medical teaching hospital, an uncertain environment in which they must assume increasing responsibility for decision-making, is particularly noticeable.

AUDIENCE RESPONSE TECHNOLOGY IN THE FLIPPED CLASSROOM

The author teaches a 16-contact-hour elective course on feline internal medicine to 80 students in the last month of the third-year curriculum. In spring semester 2010, the course was transformed from a didactic format of subject-based lectures to case-based discussions. Patient histories, laboratory data, diagnostic images, and questions were posted on the server for review before class, a concept known as flipping the classroom.[4] The change in course format was intended to provide the students with a timely opportunity to strengthen their decision-making and communication skills before starting their clinical rotations. Predictable challenges arose in engaging the students in dialogue in which they were expected to express their opinions in front of their peers.[2]

To encourage decision-making and audience engagement, a web-based audience response system (Poll Everywhere™) was incorporated in the class in 2011 and its pedagogical application was refined in 2012 and 2013. Poll Everywhere™ was chosen because the author wanted to encourage the use of smartphones, tablets, laptop computers, and e-readers in class – mobile technologies on which the students will depend increasingly for life-long learning after graduation. Also, the system allows for easy submission of narrative responses which are essential when discussing diagnostic and therapeutic strategies. Lastly, many students had developed negative opinions about the university clicker system because of its user unfriendliness and because it was linked to grades and attendance. In summary, the author wanted to create a safe, low-risk, learning environment with high expectations that would promote trust, engagement, enthusiasm, and verbal dialogue.

PEDAGOGICAL APPLICATIONS OF POLLING TECHNOLOGY

There is no consensus about the educational benefits of audience response technology in the classroom. Although its use appears to promote student attendance, engagement, self-calibration, self-improvement, and enjoyment, there is no clear evidence that it improves learning per se.[5,6] Polling technology has been used most effectively to emphasize critical concepts, assess content mastery (prior knowledge and progress), and to promote peer instruction.[7,8] Audience response activities can be used to break up, or chunk, lectures to maintain audience interest and engagement, and to provide opportunities for contingent teaching when the topic changes direction unexpectedly.[7]

The educational impact of audience response activities in the classroom depends heavily on the quality and type of questions asked.[9,10] The technology should be used sparingly, using questions of different types and increasing difficulty during the class period. Questions that prompt bimodal responses, rather than consensus, can be used to promote collaborative learning. In this scenario, the initial responses are cleared from the screen after preliminary review and the students vote again after debating their opinions with their neighbors.[10]

REFLECTIONS OF THE PRESENTER

In hindsight, the polling technology was over-used initially in 2011; the presenter was enamored with the system and was not selective with the type and number of questions asked. In subsequent years, the system has been used more sparingly to test prior knowledge, determine understanding of critical concepts, and to promote peer learning. The quality of question development and the timing of polling exercises were central to achieving positive learning outcomes. Because communication skills are an essential competency for veterinarians, the majority of questions posed in class were verbal, rather than using Poll Everywhere™. It was the presenter's anecdotal impression that the appropriate incorporation of polling technology in a low-risk format appeared to foster more relaxed verbal interaction in the classroom.

STUDENT FEEDBACK

The main goals of the change in course format and the introduction of new technology were to improve confidence in decision-making in students who were about to make the transition to their clinical year. Other goals were to provide timely opportunities for independent thinking, expression of opinions, a commitment to decisions, debate, and problem-solving.

When participating in audience-response activities, the students valued anonymity very highly. Any exercise that disrupted this trust quickly inhibited their willing expression of opinions and comments. The use of polling technology allowed students to complete their thought processes when analyzing case data – permitting them to submit, and commit to, case management recommendations without distraction. Some students commented specifically on the detrimental effect of hand-raising in their previous classes. Once a quick responder raises her/his hand in class to answer a question from the instructor, many students are triggered subconsciously by this visual cue to stop thinking about the topic, thereby losing a very important learning opportunity to complete a thought process to its end point.

Students were motivated by opportunities for peer calibration, a confidential comparison of their level of understanding of important concepts and facts with their colleagues in the class. Peer pressure may be more motivating than cajoling from the instructor for some students who are struggling to maintain interest or effort in their classes. Although answering questions correctly clearly provides encouragement and builds confidence, students commented specifically on the benefit of answering incorrectly in a safe environment. They appreciated the opportunity to choose an answer, make a decision, and be wrong. Hearing the correct answer after an incorrect response was submitted was an important motivator for self-improvement, and counter-intuitively improved understanding of concepts. In the words of one student "*when no grades are associated, I can think more clearly and learn from my mistakes*". Strategies implemented recently to improve educational outcomes based on this feedback have included a greater emphasis on metacognitive reflection of progress and closer correlation of more rigorous questions with the learning objectives of the course.

RECOMMENDATIONS

The four years of the veterinary medical curriculum are an intense period of intellectual maturation for professional students. Even the most integrated curriculum contains widely varying learning experiences for students from increasingly diverse academic and social backgrounds and with different learning and studying styles. Student stress is most obviously linked to the rigorous examination system that is necessary for documenting competency. However, educators may underestimate the stress associated with more subtle transitions in the educational continuum, as their students progress from a dualistic, fact-based environment to one of increasing ambiguity and independent decision-making. Candid conversations between academic administrators and student representatives may be useful to identify these under-recognized stress points more accurately.

Providing safe, but demanding, learning experiences in which students can complete thought processes, commit to decisions, compare themselves with their peers, and reflect on mistakes without penalty, may be helpful in facilitating confident passage through difficult transitions. Recognition of the intellectual effort that many students must expend to successfully traverse these personal challenges may significantly increase their self-confidence and motivation. The careful pedagogical application of polling technology at well-defined points in the curriculum may be particularly effective in developing skills that help students cope with an increasingly demanding and uncertain learning environment.

REFERENCES

[1]Rush BR, et al **JVME** 2010;37(2):145. [2]August JR, **Proc Lilly Conference on College and University Teaching and Learning**, Traverse City, 2013. [3]Perry WG, **Forms of Ethical and Intellectual Development in the College Years. A Scheme**. 1999, Jossey Bass. [4]Berrett D, **Chron Higher Educ** Feb 19, 2012. [5]Kay RH, et al **Computers and Education** 2009;53:819. [6]Stoddard HA, et al **Acad Med** 2010;85(10):S37. [7]Kenwright K, **TechTrends** 2009;53(1):74. [8]Kennedy GE, et al J **Comp Assist Learn** 2005;21:260. [9]Sullivan R, **J Educ Tech Syst** 2009;37(3):335. [10]Beatty ID, et al **J Sci Educ Technol** 2009;18:146.

STRATEGIES FOR USEFUL STUDENT FEEDBACK

Cynthia R. Ward, VMD, PhD, DACVIM
Athens, Georgia

The goal of teaching is to provide student learning more effectively and efficiently than students could do on their own. In many instances, learning does occur independently; however, as teachers, we hope to provide structure to that process. We hope to inspire learning and create dynamic pathways to achieve learning in our students. We hope to develop a love of learning that will continue over a lifetime.

To be effective as a teacher, learning must occur in our students as a result of what we have done. Many times, we as instructors are frustrated by evidence that students have not responded to our best teaching efforts. Exams reveal that basic concepts are not well understood and a year later in a clinical situation we might find ourselves saying: "I know you had that: I taught you that in X course." The gaps between teaching and learning can be huge and frustrating. Sometimes, by the time we realize there are large learning deficiencies, it is too late to remedy the situation or at least to affect optimal learning of the principles. So, how do we better assess learning by our students in order to become more effective teachers.

Wilbert McKeachie, a psychology teacher, handed out a manual of teaching strategies to his graduate students 1950 that has evolved into a famous guide to college teaching: McKeachie's Teaching Tips. In this book, McKeachie presents several useful assertions about teaching assessment: 1. Have learning goals for our course and do not be afraid of having goals that are difficult to measure by conventional testing; 2. Remember thaat what you value for your students to learn, you must measure; 3. Test early, often, and through a variety of means; 4. Provide feedback as quickly as possible; 5. Providing feedback is more important than assigning a grade; 5. Students need to practice self-assessment if you want them to become independent learners; 6. Non-graded evaluations are valuable. Considering feedback and assessment is a key component of integrated course design and as important as identifying learning goals and the teaching activities that support them.

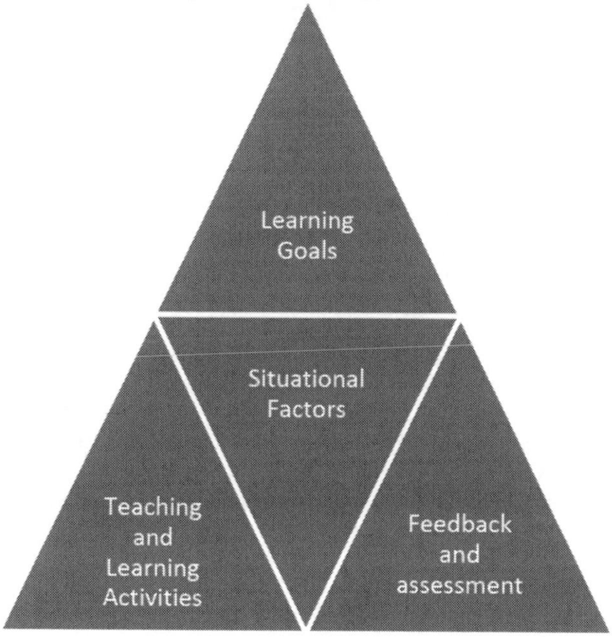

REFERENCES

McKeachie, McKeachie's teaching tips: strategies, research, and theory for college and university teachers, 13th edition.; Angelo TA, Cross, KP, Classroom Assessment Techniques: A handbook for college teachers, 2nd edition; Barkley EF, Student Engagement Techniques: A handbook for college faculty.

HOW TO MOTIVATE STUDENTS IN THE CLASSROOM AND ON THE CLINIC FLOOR

Kate E. Creevy, DVM, MS, DACVIM (SAIM)
Athens, GA

Successful education of professional students is not merely a matter of knowledge transfer or technique acquisition. We are faced with an ever-increasing rate of knowledge expansion and we must inspire and equip our students to undertake a lifelong commitment to continual expansion of their own knowledge. This practice of discovery, and the establishment of skills and habits necessary to maintain, it should have begun in primary school; for many of our students, this has not been the case. Professional students arrive with all levels of self-sufficiency as learners and must all depart with proficiency. All of them have content knowledge to uptake; for some, the uphill climb to develop the cognitive skills needed to apprehend, utilize, and continually expand that knowledge is even more daunting the massive amount of information itself. How can they possibly maintain the motivation to persist in this endeavor? How can we assist them in this pursuit?

In this interactive presentation, recent literature on this topic will be reviewed, and the results of a preliminary study on student motivation from the author's institution will be presented. The presentation is intended to initiate discussion among session attendants so that we all may benefit from the collective imagination and expertise of the room.

"Grit" - the latest buzzword in student motivation

Unless you avoid the popular press and social media altogether, you are aware of the recent interest in an idea popularized by MacArthur Fellow Angela Duckworth and colleagues,[1] that "grit," an attribute defined as "perseverance and passion for long-term goals," predicts success in all levels of education. In some studies, it appears more strongly associated with educational achievement than other classic markers, such as IQ or prior educational background. News outlets from NPR to MSNBC to USA Today have explored this notion in recent months, and school systems from kindergarten to the Ivy League are implementing strategies they believe will help their students become "grittier."

It seems self-evident that professional students who enjoy hard work, rise to challenges, and find intellectual struggle intrinsically rewarding are more likely to succeed in their academic careers, and to sustain success throughout their professional careers. Perhaps the grittiest students are also the most strongly motivated; if so, would it be possible to teach grit to those students whose energy for their studies begins to wane? Critics argue that surveying self-described perseverance to predict those who will persevere is circular and that personality traits can't be taught. Furthermore, levels of grit associate with age nearly as strongly as with educational success. Perhaps this means that early in life, it is adaptive to give up on ideas or goals that meet with failure and move on to new projects that may be more successful, while, later in life it becomes more shrewd to stick with a long-term project or path despite adversity, rather than give up on it and lose what has already been invested. In that case, so-called "grit" may simply reflect a cost-benefit analysis of the individual's current circumstances and options. The surging popularity of this educational theory is likely to create diverse classroom experiments in which to assess the integrity of the concept, the persistence of its value and the ability to teach it.

What do we know about veterinary students' motivation?

Several groups have reported that, because veterinary students are willing to work long hours to become veterinarians, they therefore must be highly motivated. As a global statement, this is likely true, but work by Mikkonen and colleagues[2] attempted to identify nuance within this presumed high level of motivation, and the impacts on student motivation over time and circumstance, by use of free-text surveys of students at the beginning and the end of the foundational courses of the veterinary program. They found that students who ultimately performed most successfully in the curriculum had tended to describe themselves as "dogged," "diligent," good students at the beginning of the program, while those who did not complete all courses on time, or who had average performance had tended to think of themselves as "slow," or easily discouraged from the outset. Similarly, most successful students had described well-defined study systems at the beginning of veterinary school, while less successful students self-reported that they had often made study plans but failed to follow them. At the end of their foundational coursework, students described changes in their enthusiasm and motivation. Common themes among the most successful students included a need to lower their standards from perfectionism to something more attainable given the heavy workload, and also that, while they had initially been interested in all aspects of school, their fatigue made it more difficult for them to maintain interest in everything. Meanwhile less successful students described difficulty in distinguishing key concepts for study among the large amount of material presented, and also confirmed that they felt their study skills were not adequate, but they lacked the time to try to develop new habits. For these reasons, they felt they had low motivation to actually learn the material, and were merely trying to memorize what they could to pass exams.

These descriptions likely sound familiar to many of us - we have seen (or perhaps have been) that student paralyzed by what seems an insurmountable workload, with no sense of how to survive, never mind how to deeply grasp core content. Mikkonen's work seems to support the notion that preparedness, whether through prior academic experiences, or through learned or innate character attributes (perhaps including "grit?") is fundamental to the ability of veterinary students to maintain motivation and to succeed; those students who could name skills they would employ to manage the curriculum were largely correct in anticipating their success. And yet, even these students described a loss of interest in their studies from time to time, "...Every now and then my interest has given out because of my enormous tiredness..." Overall, there were declines and fluctuations in the self-reported motivation of most student subjects over the course of this investigation. Average and less successful students linked their motivation to their level of interest in the topic at hand, while the most successful students pointed out that they could not maintain interest when they were overwhelmed by work. Other studies have identified a similar linkage between interest and motivation.[3-5] Together, these observations suggest that promoting learning within areas of interest, and through an appropriate workload, may be a way to help students maintain long-term motivation. Suggestions from Mikkonen's work and from others[6,7] for how to achieve this include integration of material across disciplines (students can learn about less-interesting areas by comparison to more-interesting areas), use of peer-group learning sessions (by teaching peers or by recognizing that their peers are struggling with the same material students can build confidence, which renews motivation), and allowing some degree of choice in the curriculum (so that students can refresh their levels of motivation by working on topics of particular personal interest). Of course, the devil is in the details of precisely how much integration, and how much choice, is needed, as well as where best to apply it. A final interesting point raised by Mikkonen and colleagues in their Discussion was the fact that merely asking veterinary students to reflect on their motivation, study skills, and contribution to their own learning is an unusual undertaking, and that the process of participating in the study may have impacted these students' study behaviors. Encouraging students to think about their own learning process is something that many of us could immediately, even if casually, begin.

Strategies to enhance or recover motivation

One strategy that has been reported to enhance motivation, and which is increasingly used in a myriad of academic and continuing education settings, is the audience response system. Also known as "clickers," these systems comprise software and polling devices through which students respond to embedded or spontaneous questions posed by the teacher. Thus, all students - even those who seldom speak in class - can answer the instructor's question simultaneously, and the level of understanding of a large group can be readily assessed. This enables the instructor to shift time and emphasis away from areas the group has mastered, and towards confusing or challenging content. Use of polling devices morphs the formal large-group lecture setting into an atmosphere more similar to the dynamic, small-group setting of clinical rounds. Doucet and colleagues[8] found that students self-reported improved motivation to prepare for class (by solitary and group effort), and improved ability to evaluate strengths and weaknesses in their content knowledge, when an audience response system was used. For those of us who teach primarily in a clinical setting, clickers would seldom be practical, but hopefully the nature of probing questions in case-based discussion is having the same effect.

On the other end of the spectrum, Dale and colleagues[4] investigated the factors that motivate (or discourage) engagement with lifelong learning among practicing veterinarians. It is clear that attendance of continuing education (CE) events is not synonymous with learning; in fact, prior work had identified that "extrinsic motivation" to learn, such as that imposed by licensure CE requirements, is associated with only superficial learning or rote memorization, while "intrinsic motivation" is required for deep learning and conceptual understanding to occur. Thus, Dale and colleagues surveyed 800 practitioners from 4 decades of graduating classes to identify attitudes and behaviors related to learning preferences, motivating factors, and perception of barriers to continued learning. They found that practitioners who self-reported a preference for deep learning and application of knowledge also tended to be most strongly motivated to attend CE, and tended not to identify significant barriers to their participation in CE. Interestingly, these practitioners were both intrinsically and extrinsically motivated, as well as motivated by social factors such as networking and having fun with colleagues. Conversely, those who reported a preference for surface learning and intake of knowledge (rather than application of knowledge) also tended to identify significant barriers to CE attendance, and to self-report little intrinsic, extrinsic or social motivation for attending CE. While it is unclear how practitioners develop their learning preferences, it does seem likely that to whatever extent a preference for deep learning can be promoted, motivation could be enhanced (or vice versa).

Recently, the author and colleagues undertook a study of the factors that motivate our own students, by surveying all 4th year students on the small animal internal medicine service for one academic clinical "year" (14 months at the University of Georgia, UGA). Currently, UGA does not grade students in the clinical year, providing only pass/fail scores for each rotation. Some members of our service team felt that students might be more aware of their progress if graded on the more familiar A-F scale, and that such self-awareness might help their motivation; thus, we also randomized blocks of students to receive unofficial A-F letter grades in addition to the official pass/fail scores for the rotation and assessed survey responses stratified into Proxy Letter-Graded and Conventional Pass/Fail groups. Among all students, the most strongly motivating factors were "Patient care is important to me" and "I want to be successful in practice" suggesting strong levels of intrinsic motivation. Also near the top of the list of motivating factors were "Client relations are important to me" and "I didn't want to look unprepared for cases," both of which could be identified as social motivating factors, because of the perception of impact on another person, the client. Neither group of students identified "I want to achieve a high mark" or "I desire a letter of reference from the faculty member" as strongly motivating factors, suggesting that extrinsic forces were not particularly powerful to these students. Interestingly, when asked which aspects of the rotation were most beneficial to learning, the top two choices among all students were "Topic rounds" and "One-on-one clinician discussions." In our program, topic rounds are presented by a clinician and occur in a classroom setting; while it is a small group, it is possible for students to be quite passive in these rounds, creating a knowledge-intake environment. Conversely, one-on-one case discussions with clinicians are a knowledge-application environment, in which the clinician often escalates the intellectual challenge to the student until the student's knowledge is exhausted and a learning issue is identified. Unlike the findings by Dale and colleagues[4] that preference for intake or application of knowledge segregated among survey participants, a preference for both types of learning environments seems to co-exist among our students. Further analysis of our data is underway.

Conclusions

Character traits such as grit, interest in the topical area of study, and preferences for things like deep learning and knowledge application have all been identified as factors that provide and sustain student motivation. Motivation itself can be intrinsic, extrinsic, or social, and can wax and wane over time and educational circumstances. Cultivating motivation in its many forms, and enhancing our students' ability to maintain that effort through their academic years and beyond, is our shared objective. I look forward to the discussion with the session attendants about our diverse experiences with this challenge.

REFERENCES

1. Duckworth AL, Peterson C, Matthews MD, et al. Grit: perseverance and passion for long-term goals. *J Pers Soc Psychol* 2007;92:1087.
2. Mikkonen J, Ruohoniemi M. How do veterinary students' motivation and study practices relate to academic success? *J Vet Med Educ* 2011;38:298-304.
3. Haarala-Muhonen A, Ruohoniemi M, Lindblom-Ylänne S. Factors affecting the study pace of first-year law students: in search of study counselling tools. *Stud Higher Educ* 2011;36:911-922.
4. Dale VH, Pierce SE, May SA. The importance of cultivating a preference for complexity in veterinarians for effective lifelong learning. *J Vet Med Educ* 2010;37:165-171.
5. Mikkonen J, Ruohoniemi M, Lindblom-Ylänne S. The role of individual interest and future goals during the first years of university studies. *Stud Higher Educ* 2013;38:71-86.
6. Chigerwe M, Ilkiw JE, Boudreaux KA. Influence of a Veterinary Curriculum on the Approaches and Study Skills of Veterinary Medical Students. *J Vet Med Educ* 2011;38:384-394.
7. Raidal S, Volet S. Preclinical students' predispositions towards social forms of instruction and self-directed learning: a challenge for the development of autonomous and collaborative learners. *Higher Educ* 2009;57:577-596.
8. Doucet M, Vrins A, Harvey D. Effect of using an audience response system on learning environment, motivation and long-term retention, during case-discussions in a large group of undergraduate veterinary clinical pharmacology students. *Med Teach* 2009;31:e570-e579.

CONSULTATION RELATIONSHIPS WITH THE PRIVATE PRACTITIONER

Cynthia R. Ward, VMD, PhD, DACVIM
Athens, Georgia

A physician who is called upon to consult, should observe the most honorable and scrupulous regard for the character and standing of the practitioner in attendance: the practice of the latter, if necessary, should be justified as far as it can be, consistently with a conscientious regard for truth, and no hint or insinuation should be thrown out, which could impair the confidence reposed in him, or affect his reputation.
—1847 Code of Ethics, American Medical Association

This is designed to be a roundtable discussion with several participants on the challenges and possible solutions to specialist-referral practitioner relationships. There are many dimensions to this relationship including competition for patients, the role of organized specialty societies, the establishment and enforcement of the boundaries of professional behavior, and peer review. All of these factors can create a strained relationship between the referring veterinarian and the specialty practice. We hope to explore ways to make this relationship better and stimulate discussion by the audience.

RESEARCH COLLABORATION – ACADEMIA AND PRIVATE PRACTICE

Richard Hepburn, BVSc, MS, CertEM(IntMed), DACVIM(LAIM), MRCVS
Gloucestershire, UK

INTRODUCTION

The 2011 ACVIM-LAIM survey of recently boarded diplomates showed increasing numbers are working in private practice (PPs) instead of a college of veterinary medicine (CVMs), yet at the same time they are finding it hard to promote their skillset and differentiate themselves from general practitioners[1]. Many have to combine internal medicine with ambulatory practice, and almost 50% are reported to be professionally unfulfilled. The geographic location of new diplomates is also changing, with nearly 1/3 now located outside of the USA.

In human and veterinary medicine there is a conscientious effort to move toward evidence-based medicine and yet a significant proportion of systematic reviews cannot make clinical recommendations due to insufficient basic evidence[2,3]. Within equine medicine the lack of data is more evident, with many studies involving less than 20 horses it is hard to formulate clinical guidelines and promote our speciality. PP and CVM collaborative research offers the opportunity for vastly increased study populations obtained over a short time frame, which include representation of different breeds and uses, different geographic areas, different stable managements, and the potential for treatment comparison studies. The most depressing quote of the 2011 ACVIM-LAIM survey was "Personally I believe the ACVIM has lost the battle and is basically ending up being a dusty organization where "facelifts" and management spin is used to appear to represent consensus and advancement". There is an opportunity to use the change in LAIM diplomate demographics by increasing PP-CVM research collaboration in the field of evidence based medicine, building the enthusiasm of recent diplomates and increasing the relevance and quality of our college.

WHAT CAN PRIVATE PRACTICE OFFER?

Somewhat depressingly in the 2005 National Research Council Resources for Veterinary Research, the section on private-sector resources mentioned pharmaceutical and feed companies, and private diagnostic labs but did not mention private clinical practice[4]. As PPs with ACVIM diplomates expand in size and number they will take more routine referrals; the absence of teaching time, allied to a more commercial approach to case throughput and client care, means that the absolute numbers of these routine referrals can be larger than that which would have been seen by the CVM. Greater perceived need for procedures by clients acts to further increase caseload, which can then provide a more cross sectional study population with asymptomatic horses from "at risk" populations presented rather than just symptomatic cases. Inclusion of PP caseload not only increases sample sizes, it also makes the study group more representative of a whole population, increasing clinical and statistical relevance. As most PPs will employ a maximum of 1 or 2 diplomates their clinical approach tends to be uniform. The consistency of resident training in routine referrals should limit individual variation between PPs. Case follow up can be simpler as many horses originate from the PP's own first opinion caseload, or from local general practices with which, for commercial reasons, the PP has close links. With improved access to follow up material there is the potential for collaborative research to create meaningful evidence based clinical guidelines, which can be reassessed at pre-determined intervals. First opinion populations, often of a very large size, can also be accessed in a more reliable manner, giving the potential to gather high quality data on routine, instead of just referral conditions.

WHAT DOES PRIVATE PRACTICE WANT?

The potential benefits are improved treatment regimes, increased caseload and positive public relations/advertising, as well as increased professional satisfaction. Large-scale pragmatic research has the ability to produce *generalizable* data[7], leading to improved clinical outcomes and increased client satisfaction. In this author's experience clients are willing to *self-fund* research if it is an accepted clinical procedure, the results could be of benefit to their horse, and the cost is less than what they would normally charged. For the PP this can result in cases being examined which otherwise would not, and the knock on effect of increased client awareness and increased caseload in the post study period. Client appreciation of study participation (by social and traditional media) raises the profile of the ACVIM diplomate and their practice, creating a clinical difference between them and the general practitioner. On a wider scale this helps to enhance the ACVIM brand as well as the clinical and financial relevance of internal medicine. Sadly most private practices do not reward specialists for presenting at meetings or for peer-reviewed publication, nor do they provide time for these. This limits the ability of the PP diplomate to manage and write up studies, and so these are often not their goals from study participation. Instead access to study data and results materials produced by the lead author, which can be used for PP promotion and client education is more relevant and appreciated.

WHAT CAN ACADEMIA OFFER?

Within CVMs it has been estimated that veterinary faculty can have up to 50% of their time available for research, although most will have significantly less[4]. A wide variety of clinical, basic science and statistical expertise is available which is vital to study design, acquiring funding, data analysis and study completion. Graduate students and even undergraduates can be used for data entry, whilst comparative researchers can provide insight. The academic diplomate typically will have greater experience at data manipulation and study completion (or access to greater experience), as well as allotted time off clinics for these activities. CVMs can also provide laboratory facilities for central, standardised sample analysis, often at a cost effective rate. They have access to a wide variety of funding options, including federal and state governmental sources, industry and private sources; and are more successful in obtaining funding than DVMs outside of CVMs[4]. Ethics committees at most CVMs meet regularly allowing projects to fulfil state and federal ethical guidelines. In addition some CVMs offer a field service that can also provide first opinion population, although usually of a smaller size.

WHAT DOES ACADEMIA WANT?

To advance their careers individuals at CVMs have to demonstrate research productivity in the form of publication in refereed journals, acquisition of grant funds both from sources that evaluate proposals by quality peer review and those that do not, as well as invitations to present research at scientific meetings. Time and resources are allotted to these activities. The CVM itself is interested in its research assessment ranking as well as overall ranks such as US News CVM ranking or the QS review. Higher ranks and a proven track record in research will help future grant applications, as well increasing the reputation of the CVM. With the move toward evidence-based medicine, CVMs need to broaden their research activities from bench top research to stay relevant to modern clinical practice.

WHAT TYPE OF RESEARCH?

Probably the most appropriate forms of collaborative research projects are: retrospective descriptive studies, and large-scale epidemiologic and observational surveys including case-control, cohort and cross sectional studies[6]. The purpose of increasing study population size is to increase heterogeneity, making results more *generalizable* and hence relevant to clinical practice. The most exciting potential of PP-CVM collaboration is comparative effectiveness research, in particular pragmatic trials[4,6,8,9]. These compare two or more medical interventions that are directly relevant to the delivery of clinical care, and then assess their effectiveness in real-world practice. Broad eligibility criteria are used to recruit cases from a variety of practice settings in order to include all types of horse whose care will be influenced by the trials results. All outcomes that are important to patients are measured, such as survival data, clinical response, client satisfaction, performance status, adverse effects and cost; over a time period sufficient to adequately assess benefit and risk. In pragmatic trials medical management is consistent with the usual standard of clinical care, which means study procedures such as blinding and placebos are omitted. Because of this pragmatic trials often require large sample populations to detect small differences between active treatments in a heterogeneous population, certainly larger than is needed to detect the difference between a treatment and placebo. A lack of blinding creates the potential for bias, particularly when clients give their subjective impression of efficacy, so trials are stronger when they include objective outcome measures, particularly if these can be analyzed in a blinded fashion. Thus the challenge of pragmatic research is the balance between internal validity and generalizability.

FUNDING

Pragmatic research is performed on clinical cases, which to an extent can make the study self-funding. For instance data is collected from cases that normally present and are treated, within a defined time frame, so both the examination and treatment can be client funded. Alternatively if a specific population is to be studied owners can be incentivised with a reduced cost for examination, or if treatment is to be offered then the manufacturer can contribute to its cost. Pragmatic research does not involve placebo groups, but compares accepted/potential treatments and so patient acceptance is greater[7], and hence client willingness to pay. Clients can even be willing to pay for further diagnostics such as histopathology, if these have a potential clinical benefit. Funding is still required for data analysis and publication, although these costs are comparatively small. Central organisation of funding by the principal investigator is essential, with dispersal to secondary investigators at specific periods (this author favours after data packets are received!).

WHAT PROBLEMS MIGHT ARISE IN THE COLLABORATION?

Identification of suitable collaborators and caseload

Collaborators should be routinely working in the field of study and have sufficient caseload, but do not need to be known to the primary researcher. For pragmatic research it is best if collaborators come from many diverse settings. The two obvious routes for recruiting ACVIM diplomates are the LAIM listserv and annual CE meetings, including the forum. Identifying suitable studies that would form the backbone of a consensus statement, then discussing study design with potential stakeholders at these events is an exciting possibility.

Agreement on study design

A balance between clinical utility and degree of bias, and between quantitative and qualitative data is required. Involvement of all stakeholders in the study design process is essential, again the use of specialist should improve conformity.

Quality control of data collection

Using ACVIM diplomates should give a broadly consistent standard. Introductory sessions for collaborators, either online or in person, will enhance this. These sessions also allow the principal investigator to assess the collaborator's technical skills, equipment and caseload; as well as giving potential participants a chance to drop out. Pre-study meetings are often required for ethics approval, in particular to outline patient data storage regulations and confidentiality.

Poor compliance and data return

Pragmatic studies revolve around clinical data collection, and so should not be too onerous. The use of pre-printed questionnaires, as well as online tools (such as Survey Monkey) can simplify data entry, with the online systems also improving data return. Dispersal of funds to collaborators should only occur once study materials are complete and returned.

Lack of perceived benefit

Effort should be made to ensure that the output of the collaborative research is not just peer reviewed publications and specialist meeting presentations. Production of suitable presentation material and PR accessible to all collaborators is essential, in order that CVMs and PPs can initially benefit from improved clinical outcomes, before the information disperses to general practice. The description and implementation of clinical guidelines should be present in these specialist locations, at the same time as their findings are made public in the scientific and lay press.

WHAT NEEDS TO HAPPEN NEXT?

There are two options – firstly that individual stakeholders in both CVMs and PP can continue to organise their own collaborative studies on topics of individual interest, or those promoted by interested parties such as drug companies. This approach will rely on personal networks, although the use of listserv approaches will increase participants beyond these. The second option is a more centrally coordinated approach of the sort occurring in human medicine[10] and a national veterinary governing body[11]. ACVIM has an inherent advantage in having a preformed network of *specialist* clinicians with the combination of a heterogenous caseload, research experience and access to complimentary expertise in basic sciences. The existing CE structure could be used for pre-study recruitment and discussion meetings, with consensus statements utilising study results to create meaningful clinical recommendations. ACVIM foundation grants could be used as a source of funding. With either option diplomates in PP and CVMs need to acknowledge their strengths and weaknesses and respect each other's professional aims and goals, in order to further the welfare of our patients and ensure the future relevance of our speciality.

RECENT EXAMPLES OF EQUINE MULTICENTRE STUDIES

Study	CVM participants	PP participants
Renin-Angiotensin-Aldosterone System and Hypothalamic-Pituitary-Adrenal Axis in Hospitalized Newborn Foals	6	2
Association of admission L-lactate concentration in hospitalised equine neonates with presenting complaint, periparturient events, clinical diagnosis and outcome: a prospective multicentre study.	7	6
Renin-Angiotensin-Aldosterone System and Hypothalamic-Pituitary-Adrenal Axis in Hospitalized Newborn Foals	1	2
A Multicenter Retrospective Study of 151 Renal Biopsies in Horses	12	1
Factors associated with survival of epiploic foramen entrapment colic: A multicentre, international study	12	7
Multicentre, controlled, randomised and blinded field study comparing efficacy of suxibuzone and phenylbutazone in lame horses	1	11
Unilateral and bilateral laparoscopic ovariectomy in 157 mares: a retrospective multicenter study	3	0
Evaluation of oral administration of firocoxib for the management of musculoskeletal pain and lameness associated with osteoarthritis in horses	1	80

REFERENCES

[1]Survey of ACVIM LAIM Recently-Boarded Diplomates, 2011. [2]Sackett DL et al **BMJ** 1996 312:71–2. [3]Villas Boas, P. J. **J Eval Clin Pract** 2013 Aug;19(4):633-7. [4]Sagliocca L **J Eval Clin Pract** 2013 Aug;19(4):689-696 [5]National Research Council (US) Committee on the National Needs for Research in Veterinary Science 2005.. [6]Alemayehu D et al **J Eval Clin Pract** 2013 19(4):579-583. [7]Furr M **Eq Vet J Suppl** 2012;44(41):3-4. [8]Ware **N Engl J Med** 2011;364;18. [9]Dubois R Demystifying comparative effectiveness research **National Pharmaceutical Council** 2009.
[10]http://effectivehealthcare.ahrq.gov/index.cfm/what-is-comparative-effectiveness-research [11]http://knowledge.rcvs.org.uk/evidence-based-veterinary-medicine/

HOW TO – EQUINE CT AND MRI

Sarah M. Puchalski, DVM, DACVR
Petaluma, California

INTRODUCTION

Computed tomography (CT) and magnetic resonance imaging (MRI) are powerful, diagnostic imaging tools that are increasingly available for equine practice. Historically, their use has been limited to University practice and large referral centers because of their high cost and dependence on general anesthesia. A large component of described clinical applications of these techniques have centered on lameness diagnosis and other clinical orthopedic uses. Along with increasing use of the modalities, are increasing numbers of publications surround their use for specific diseases and disease syndromes such that a good body of literature now exists for some applications[1-6]. New applications for these modalities arise as practitioners utilize the techniques to solve different clinical / diagnostic problems. This presentation will focus on the use of CT and MRI for purposes of evaluating the equine head and neck.

MODALITY SELECTION

Generalizations can be made for both techniques regarding their inherent diagnostic strengths and weakness. Knowledge of these characteristics can aid in the selection the best diagnostic test for a given case. CT and MRI are cross sectional imaging modalities that allow for the elimination of superimposition as experienced with radiograph interpretation. Both imaging modalities can be useful for most anatomic regions however some important differences exist.

CT and MRI equipment are variable in their availability. Each technique has undergone dramatic improvements in hardware and software over the last decade. Currently, CT scanners tend to be helical, multi-slice with a push towards increasing bore size (70-85 cm) allowing for larger part scanning. At the same time, small, portable CT scanners (30cm bore size) are also available and can be useful for neonatal or foal skull and neck scanning. Helical multi-slice scanners allow for very rapid, high spatial resolution images of a large tissue volume. In some locations, standing (not require general anesthesia) CT scanning of the equine head and neck is routine. MRI scanner types are highly variable in all regards. Magnetic field strength is variable with units ranging from 0.2 to 3 Tesla being used in equine practice. Increasing magnetic field strength results in increasing image resolution and decreasing scan time. Magnet configuration is also highly variable and in large part dictated by the magnetic field strength. Open, 'U' shaped configuration to traditional, tunnel shaped bores of variable size are available. Radiofrequency (RF) coils are requisite for MRI and are also variable in size, shape and configuration. The coil, a separate piece of hardware, is generally affixed to the anatomic region of interest. Some scanners, particularly smaller bore scanners, have the coil as a built in component. MRI of the skull and neck currently requires general anesthesia, a custom table and MRI compatible anesthesia equipment.

CT is a technique capable of rapid image acquisition that is best utilized in areas with high inherent differences of tissue density. E.g. the paranasal sinuses and dentition where tissues range in density from enamel (very high) to gas (very low). Because images are made based on tissue density, data regarding tissue density as a quantitative value are contained within the images. Changes in tissue density, e.g. osteolysis can be quantified. Tissues of similar density, with differing tissue composition can be difficult or impossible to separate. This characteristic makes it difficult to utilize this technique for brain imaging as grey and white matter of brain parenchyma appear very similar on CT images. CT is particularly useful for problems related to the dentition, nasal passage and paranasal sinuses, osseous structures of the skull (skull trauma / fracture), middle ear, temporohyoid apparatus, and cervical vertebrae.

MRI is variable in acquisition time, hardware capabilities and physical constraints. The manner of image acquisition is complex and difficult to conceptualize but images reflect tissue composition. Image acquisition is manipulated in various imaging sequences to emphasize different aspects of tissue composition and its response to a magnetic field and RF pulses. For these reasons, MRI is extraordinarily useful for assessment of soft tissues including the central nervous system. Highly ordered tissues such as compact bone and collagen fibrils are difficult to manipulate with the external magnetic field and RF pulses and therefore have a characteristic appearance on MRI as black or a signal void. Evaluation of osseous structures often relies on the interpretation of other imaging findings such as changes in size, shape and margination for the identification of osteolysis or osteoproliferation. In general, MRI is highly useful for lesions of the central nervous system, laryngeal and pharyngeal soft tissues, and the vertebral column. MRI provides useful but different information for all types of lesions as listed above for CT.

In many instances, either CT or MRI is available but not both. Both modalities can provide useful information for diseases of the equine head and neck.

PATIENT PREPARATION

Most CT and MRI scanners will require a horse to be placed under general anesthesia in order to obtain images. Clinical judgment must be used to determine each patient's general health and anesthetic risk. Or, careful cost benefit analysis should be performed in order that the information gained from the imaging technique outweighs the risks of general anesthesia. This is of particular concern in horses with neurological dysfunction. The imaging protocol will provide an estimate of the total imaging time and the anesthetic agents should be chosen with this in mind. Planning for anesthetic time should always include, movement to and from the table, positioning, acquisition of scout or pilot images, acquisition of the diagnostic images, and possibly the addition of contrast media if needed. If procedures such as biopsy or surgery are planned to follow, this should also be incorporated into the estimated time. A modern, multi-slice helical CT scanner is capable of acquiring an entire skull in 1-5 minutes and the scanning time is therefore a fraction of the total anesthetic time. Similar consideration must be given to MRI scanners however; the scanning time is incredibly variable, dependent on imaging protocol and magnetic field strength. The range of scan time is very large from 20 minutes to 2+ hours.

CT scanners capable of acquiring images of the head and neck in a standing, sedated horse are present and being used in equine practices. For a standing skull CT careful consideration must be given to sedation protocol and other calming techniques such as earplugs and blinkers.

CONTRAST MEDIA

Contrast media is used to augment the interpretation of diagnostic images through the provision of improved contrast resolution within tissue. Contrast media can be administered through any number of routes including: intra-vascular (venous and arterial), intra-thecal (subarachnoid), intra-synovial, or into fistulae. The most common route of administration in people and small animals is via intra-venous routes at a total body, systemic dose. In horses, this route can be prohibitively expensive depending on the agent and the geographic region. Positive contrast media for CT scanning is usually iodine-based products and for MRI scanning is gadolinium-based products. Intra-vascular contrast agents provide improved lesion delineation at areas with increased or abnormal blood flow or increased vascular permeability. Soft tissue neoplasia is a good example of a disease process with alterations in both blood flow and tissue permeability. Many soft tissue tumors are better identified and more thoroughly characterized with the addition of systemic, intra-venous contrast media. Intra-arterial delivery of CT, iodine-based contrast media to specific anatomic regions has also been described. This technique requires arterial catheterization and an infusion of contrast media during scanning. The total dose of contrast media can be reduced while still gaining the advantages of its administration. CT myelography requires the administration of intra-thecal contrast media into the sub-arachnoid space and the utility of this technique is now being discovered in hospitals with this capability. The positive contrast surrounding the spinal cord allows for the identification of spinal cord compression and impingement in sites not visible on routine myelography.

Contrast media choice, its dose and route of administration should be carefully considered. There are critically important differences between contrast agents, particularly with iodine-based media for use in CT and radiography. Ionic iodine-based agents are very hyperosmolar and should never be used with intra-thecal or intra- synovial routes. Contrast media reactions are infrequent but reported in humans and veterinary patients. The most common reactions include transient alterations in blood pressure and heart rate, however, urticaria, renal complications, and cardiovascular collapse have been reported[7-9].

SUMMARY

Maximizing the diagnostic yield from imaging techniques begins with careful case selection, imaging modality choice, patient preparation and positioning, use of contrast media in addition to the thoughtful interpretation of images. Diagnostic accuracy is improved with the provision and consideration of a pertinent clinical history and clinical examination findings. Diagnostic imaging specialists can provide insight into details of study preparation and interpretation.

REFERENCES

[1] Dyson, S., Blunden, T. and Murray, R. (2012) Comparison between magnetic resonance imaging and histological findings in the navicular bone of horses with foot pain. *Equine veterinary journal* **44**, 692-698.

[2] Dyson, S. and Murray, R. (2007) Magnetic resonance imaging evaluation of 264 horses with foot pain: the podotrochlear apparatus, deep digital flexor tendon and collateral ligaments of the distal interphalangeal joint. *Equine Vet J* **39**, 340-343.

[3] Hilton, H., Puchalski, S.M. and Aleman, M. (2009) The computed tomographic appearance of equine temporohyoid osteoarthropathy. *Vet Radiol Ultrasound* **50**, 151-156.

[4] Huggons, N.A., Bell, R.J. and Puchalski, S.M. (2011) Radiography and computed tomography in the diagnosis of nonneoplastic equine mandibular disease. *Vet Radiol Ultrasound* **52**, 53-60.

[5] Cissell, D.D., Wisner, E.R., Textor, J., Mohr, F.C., Scrivani, P.V. and Theon, A.P. (2012) Computed tomographic appearance of equine sinonasal neoplasia. *Vet Radiol Ultrasound* **53**, 245-251.

[6] Textor, J.A., Puchalski, S.M., Affolter, V.K., MacDonald, M.H., Galuppo, L.D. and Wisner, E.R. (2012) Results of computed tomography in horses with ethmoid hematoma: 16 cases (1993-2005). *J Am Vet Med Assoc* **240**, 1338-1344.

[7] Pollard, R.E. and Puchalski, S.M. (2011) Reaction to Intraarterial Ionic Iodinated Contrast Medium Administration in Anesthetized Horses. *Vet Radiol Ultrasound.*

[8] Pollard, R.E., Puchalski, S.M. and Pascoe, P.J. (2008) Hemodynamic and serum biochemical alterations associated with intravenous administration of three types of contrast media in anesthetized cats. *Am J Vet Res* **69**, 1274-1278.

[9] Pollard, R.E., Puchalski, S.M. and Pascoe, P.J. (2008) Hemodynamic and serum biochemical alterations associated with intravenous administration of three types of contrast media in anesthetized dogs. *Am J Vet Res* **69**, 1268-1273.

HOW CHALLENGING IS A RIDING HORSE'S LIFE: INFLUENCES OF CHALLENGING SITUATIONS ON PHYSIOLOGICAL AND BEHAVIOURAL RESPONSES

Cornélie M. Westermann DVM, PhD, DECEIM, Carolien C.B.M. Munsters PhD, Marianne M. Sloet van Oldruitenborgh-Oosterbaan DVM, PhD, DECEIM
Utrecht, The Netherlands

In this presentation four studies on the influence of challenging situations on physiological and behavioural responses in horses will be discussed.

1) THE INFLUENCE OF CHALLENGING OBJECTS AND HORSE-RIDER MATCHING ON HEART RATE, HEART RATE VARIABILITY AND BEHAVIOURAL SCORE IN RIDING

A good horse-rider 'match' is important in the context of equine welfare. To quantify the influence of repetition and horse-rider matching on the stress of horses encountering challenging objects, 16 Warmblood horses were ridden in a test setting on three occasions. On each occasion the horse was ridden by a different rider and was challenged by three objects (A–C). Heart rate (HR), heart rate variability (HRV) of horse and rider, and behaviour score (BS) of the horse were obtained for each object and as a total for each test. The horse-rider interaction was evaluated with each combination and assessed as 'matching' or 'mismatching', and the horses were categorised as 'compliant', 'partly-compliant' or 'non-compliant'. Horses exhibited a decreased HR ($P = 0.015$) and a decreased BS ($P = 0.004$) within and across different tests. 'Matching' horse-rider combinations exhibited less stress as indicated by reduced HR ('match' 69 ± 10 vs. 'mismatch' 72 ± 9, $P = 0.001$) and BS ('match' 1.9 ± 1.1 vs. 'mismatch' 3.8 ± 1.4, $P = 0.017$) of the horse. 'Compliant' (68 ± 8, $P < 0.001$) and 'partly-compliant' (71 ± 9, $P = 0.002$) horses had significantly lower HR than 'non-compliant' (75 ± 9) animals. The findings of the study indicate that HR and BS measurements support a subjective 'match' diagnosis and HR measurement may be a valuable tool in assess- ing horse compliance.

2) PHYSIOLOGICAL AND BEHAVIOURAL RESPONSES OF HORSES DURING POLICE TRAINING

Mounted police horses have to cope with challenging, unpredictable situations when on duty and it is essential to gain insight into how these horses handle stress to warrant their welfare. The aim of the study was to evaluate physiological and behavioural responses of 12 (six experienced and six inexperienced) police horses during police training. Horses were evaluated during four test settings at three time points over a 7-week period: outdoor track test, street track test, indoor arena test and smoke machine test. Heart rate (HR; beats/min), HR variability (HRV; root means square of successive differences; ms), behaviour score (BS; scores 0 to 5) and standard police performance score (PPS; scores 1 to 0) were obtained per test. All data were statistically evaluated using a linear mixed model (Akaike's Information criterion; t . 2.00) or logistic regression (P , 0.05). HR of horses was increased at indoor arena test (98 6 26) and smoke machine test (107 6 25) compared with outdoor track (80 6 12, t 5 2.83 and t 5 3.91, respectively) and street track tests (81 6 14, t 5 2.48 and t 5 3.52, respectively). HRV of horses at the indoor arena test (42.4 6 50.2) was significantly lower compared with street track test (85.7 6 94.3 and t 5 2.78). BS did not show significant differences between tests and HR of horses was not always correlated with the observed moderate behavioral responses. HR, HRV, PPS and BS did not differ between repetition of tests and there were no significant differences in any of the four tests between experienced and inexperienced horses. No habituation occurred during the test weeks, and experience as a police horse does not seem to be a key factor in how these horses handle stress. All horses showed only modest behavioural responses, and HR may provide complimentary information for individual evaluation and welfare assessment of these horses. Overall, little evidence of stress was observed during these police training tests. As three of these tests (excluding the indoor arena test) reflect normal police work, it is suggested that this kind of police work is not significantly stressful for horses and will have no negative impact on the horse's welfare.

3) THE EFFECTS OF TRANSPORT, RIOT CONTROL TRAINING AND NIGHT PATROLS ON THE WORKLOAD AND STRESS OF MOUNTED POLICE HORSES

Quantifying and evaluating police work is essential to avoid behavioural problems, minimise wastage and maximise welfare of police horses. Behavioural and physiological measures were used to assess the workload and/or stress police horses experience during police work. Further, the influence of the rider on the fear response and heart rate of the horse was studied. Nine police horses were evaluated during transport, riot control training and night patrols. Outcome parameters were mean and peak heart rates (HR, beats/min), plasma lactate concentration (LA, mmol/L) and behaviour score (BS, 0–10). Data were analysed using a linear mixed model test ($P<0.05$). During transport, peak HR differed between horses that were indicated as 'stressed' (116 ± 25 beats/min) and 'non-stressed' (79 ± 16 beats/min, $P=0.032$) during transport. Riot control training workload was 348±30min at a mean HR of 80±11 beats/min and LA<0.8mmol/L and night patrols workload was 198±48min at a mean HR of 51±7 beats/min. Horses experienced fire exercise (mean HR, 104 ± 18) and individual seclusion exercise (mean HR, 151 ± 17) as the most stressful events. During individual seclusion mean HR was significantly lower when the rider could control a horse's fear responses (146 ± 14 beats/min) than when not (170 ± 16 beats/min, $P = 0.004$). BS differences were too small to discriminate between training scenarios. How- ever, individual horses differed significantly (mean HR, $P < 0.001$; mean BS, $P < 0.001$) and consistently between scenarios. The workload and stress of police horses are low compared with those of horses in equine sports and probably do not cause behavioural problems and wastage. However, the rider's control over a horse's fear response seems to be important in the stress, ridden horses experience during challenging situations.

4) HEART RATE, HEARTRATE VARIABILITY AND BEHAVIOUR OF HORSES DURING AIR TRANSPORT

Heart rate (HR), HR variability (HRV) and behaviour score (BS) of nine horses were evaluated during an eight-hour air transport between The Netherlands and New York. HR and HRV parameters were calculated every five minutes during the air transport. Compared with transit (40±3), mean HRs were higher during loading into the jet stall (67±21, P<0.001), loading into the aircraft (47±6, P=0.011), taxiing (50±8, P=0.001), and during periods of in-flight turbulence (46±7, P=0.017). During the flight, individual horses showed differences in mean HR (P=0.005) and peak HR (P<0.001). By contrast with HR data, HRV data did not differ between stages or horses. BS was highest during turbulence (3.2±0.4). However, behaviour did not always correspond with HR measurements: the least responsive horse had the highest HR. Loading into the jet stall caused the highest increase in HR and was considered the most stressful event. During transit, HR was generally comparable with resting rates. Previous studies have shown that loading and transporting by road caused more elevation in HR than during loading and transporting by air. HRV data were not found to be useful, and caution is needed when interpreting HRV data. Not every horse exhibited stress through visible (evasive) behaviour, and HR measurements may provide an additional tool to assess stress in horses.

REFERENCES

C.C.B.M. Munsters et al; Vet J., 2012;192:75. C.C.B.M. Munsters et al; Animal, 2013, 822. C.C.B.M. Munsters et al; Applied Animal Behaviour Science, 2013, 143: 52. C.C.B.M. Munsters et al; Veterinary Record, 2013,172:15

ACVIM CANDIDATE FUNCTION
RESIDENCY MENTORING: DETERMINING WHERE WE ARE, WHERE WE WANT TO BE, AND HOW TO GET THERE

Julie Cary, DVM, MS, DACVS (LAIM), Nicki Wise, DVM, MS, ACVIM (LAIM)
Pullman, WA

INTRODUCTION

Effective mentorship not only aids in development of skills and knowledge but the guidance and experience provided by a mentor shapes personal and professional development and determines career success (Sambunjak 2006). A lack of mentorship has been shown to significantly hinder professional progress (Jackson 2003). Providing quality mentorship is an acquired skill and can be learned (Kashiwagi 2013). Research supports that having a well-structured mentoring program which focuses on clear expectations, support, communication and effective feedback results in an optimal situation for both the mentor and mentee (Davis 2010). Creating this environment while balancing clinical, teaching and/or research responsibilities can be daunting. Most specialists responsible for training residents do not have the opportunity to obtain any type of formal training for this critical task. This lack of preparation is due to a number of factors including but not limited to lack of protected time, funding, or administrative support (Pfund 2009; Thorndyke 2008) and tradition. Most resident advisors learn by observing and emulating their own mentors, which may be of variable efficacy. Providing mentors with the resources and tools necessary to create an optimal trainer/trainee relationship is paramount in order for these relationships to thrive. Likewise, mentee training is just as important. Understanding how to "be mentored" while learning to navigate skill development, knowledge acquisition and growing independence is vital to a resident's professional and personal development. An ideal learning environment is created when both the mentor and mentee have the same goals and possess the tools necessary to achieve them.

WORKSHOP

Today's ACVIM residents are tomorrow's mentors. The challenges facing academic and specialty medicine will be managed by the residents being trained today. Optimizing the mentorship experience during residency is an effective way to ensure that future ACVIM diplomates continue to be leaders in veterinary medicine. It is vital that diplomates are aware that their role in resident training is not simply as an advisor or clinical role model, but instead stands to have a profound impact on the future of their residents and this college. Even more importantly, the tools and resources necessary to be an effective mentor must be provided. To that end, we will conduct a workshop with the following objectives --1) determine resident perceptions of mentorship 2) introduce the evidence base in mentoring skills and 3) provide some specific skills that may be beneficial to residents in their roles as mentors in the present and the future. The general findings from objective 1 will be used to direct future training efforts for ACVIM.

REFERENCES

Davis OC, et. al.. Acad Med. 2010; 85(6): 1060-1066.
Jackson VA, et. al. Acad Med. 2003; 78: 328-334.
Kashiwagi DT, et. al. Acad Med. 2013; 88(7): 1-9.
Pfund C, et. al.. Clin Trans Sci. 2009; 6(1): 26-33.
Sambunjak D, et. al.. JAMA. 2006; 296: 1103-1115.
Thorndyke LE, et. al.. J Contin Educ Health Prof. 2008; 28: 157-164.

FOOT DISEASES IN CATTLE

Sarel R. van Amstel BVSc, MMedVet(Med), DACVIM
Knoxville, Tennessee

The foot is defined as the part of the limb below the fetlock whereas the claw is defined as the horn capsule and structures within. Conditions/diseases of the foot are listed in Table 1. Claw zones are shown in Figure1

Table 1. Conditions/diseases of the foot and claw.

Foot and Claw Condition	Claw Zone and Diagnosis	Cause
Sole ulcer	4. Exposed/protruding corium	Overgrowth; laminitis; screw claw
Toe ulcer	5 apex. Full thickness horn defect	Laminitis
Heel ulcer	Transition 5&6 Exposed/protruding corium	Long claw diagonal; thin soles; laminitis; overgrowth
Thin sole toe ulcer	5 adjacent to zones 1 and 2. Full thickness horn defect	Excessive wear
Thin sole horn/ heel horn	4, 5, 6 flexible on thumb pressure	Excessive wear
Toe abscess	5 apex Subsolar puss accumulation	Excessive wear; screw claw
White line disease	1, 2 Separation of white line	Laminitis
White line disease	3 Separation of white line	Overgrowth; laminitis; screw claw
Subsolar/ heel abscess	4, 5, 6	Secondary to sole ulcer; white line disease; foreign body penetration
Sole/ heel hemorrhage	4, 5, 6	Trauma; laminitis
White line hemorrhage	1, 2, 3	Weight bearing; laminitis
Vertical wall crack	abaxial/axial wall	Laminitis; trace mineral deficiency
Horizontal wall crack single	abaxial/axial wall	Calving, weaning
Horizontal wall crack multiple	abaxial/axial wall	Laminitis
Heel erosion	6, 10 Partial thickness horn defect	Interdigital/digital dermatitis
Interdigital dermatitis	0 Epidermitis/epidermal ulceration	Slurry, manure, moisture
Interdigital hyperplasia (corn)	0 Epidermal hyperplasia	Chronic interdigital dermatitis
Digital dermatitis	0,10 Ulcerative/proliferative epidermitis	Infectious agent
Interdigital phlegmon (foot rot)	0, 10 Soft Symmetrical swelling of foot	*Fusobacterium*
Fescu foot	Whole foot necrosis	Ergot alkaloid
Mucosal disease	0 Ulcerative dermatitis	Pestivirus
Septic distal interphalangeal joint	Unilateral swelling of foot Discharging tract on coronary band	Ascending infection through sole/white line
Septic proximal interphalangeal joint	Swelling of affected digit	Trauma; ascending infection
Phalangeal fracture (P1 ,P2, P3)	Swelling of digit except P1 fracture	Traumatic; pathological
Septic tenosynovitis	Ascending infection heel, sole, white line. Unilateral swelling	Sole ulcer, white line, trauma
Exostoses phalanges	Bony swelling	Chronic osteitis; osteomyelitis; fluorosis; laminitis affecting P3
Superficial flexor contracture	Proximal interphalageal joint subluxation. Hock overextention	Trauma
Superficial and deep flexor contracture	Walk on toes/dorsum of foot	Congenital
Flexor rupture	Overextension of the foot	Trauma
Osteochondrosis of DIP joint	Painful gait. Walk on toes	Congenital
Toe overgrowth	Long toes	Insufficient wear; long diagonal; laminitis; screw claw

Table 2. Common postural deformities associated with conditions in the foot and claw

Postural deformity	Origin
"Camped-under" posture rear feet	Laminitis
Non-weight bearing	Fractures of the claw/foot, any joint or long bone
Tiptoe stance	Stifle injuries; foot lesions palmar or plantar region, sole and heel ulcer; osteochondrosis
Knuckling of fetlock	Peroneal and tibial neuropathies; spinal injuries; foot injuries/lesions palmar and plantar; spinal lymphoma
Knuckling of proximal interphalangeal joint	Superficial flexor tendon contracture
"Camped-out" posture in front	Laminitis

Table 3. Diagnosis of postural changes associated with lameness which can be made through observation during standing and walking.

Diagnosis	Characteristic features
Ruptured flexors/suspensory ligaments	Overextension in fetlock. May be weight bearing on the plantar/palmar aspect of the foot
Chronic digital dermatitis Back foot.	"Boxy" claw. Periodic rhythmical lifting of foot
Septic distal interphalangeal joint	Unilateral swollen foot. Sinus tract at the coronary band abaxial to the insertion of the extensor tendon
Chronic laminitis	Doral wall (toe) curvature, poor quality horn (flaky, dull), lateral deviation in growth (flare)
True (Heritable) screw claw	Young animal, abxial to axial curve of wall, rotation of toe, absence of multiple horizontal grooves
Acquired screw claw	Older animal, front and or rear feet, multiple grooves, abaxial to axial curved wall, rotation of toe
Vertical wall crack	Vertical crack in the abaxial or axial wall.
Thimble	Full thickness horizontal wall crack near toe forming loose cap
Corn/Fibroma	Chronic interdigital hyperplasia of epidermis
Contracted flexors	Over flexion of fetlock and /or knee. Knuckling of proximal interphalangeal joint

Figure 1. Claw zones

0

10

FOOT CONFORMATION AND LAMENESS

The cause of lameness is multifactorial. Some of the factors playing an important role include genetically transmissible conformation traits, rate of growth and body condition, nutrition, environmental factors and normal biomechanics of weight bearing (Figure 2).

Figure 2. Lameness and premature culling.

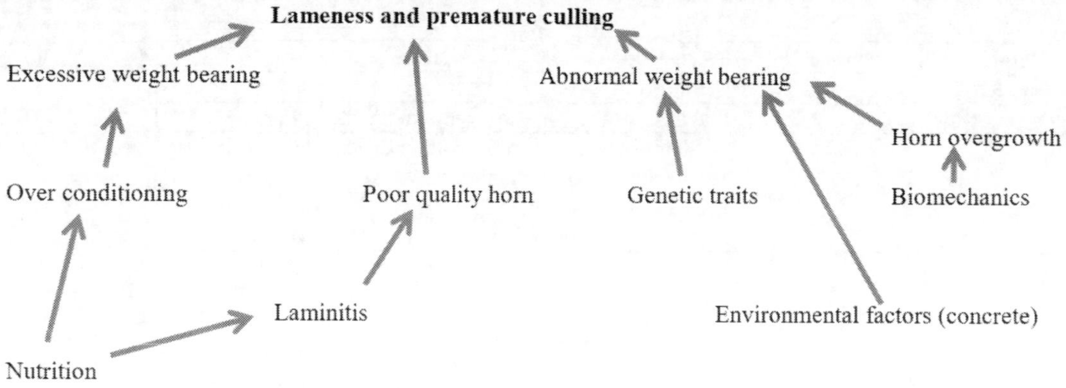

From figure 1 it can be seen that there is an overlap and interconnection between factors predisposing to lameness.

PHENOTYPIC AND GENOTYPIC TRAITS

Traits may vary depending on breed, age, parity, season, housing and management factors. Research on genetic aspects has concentrated on traits such as body conformation, claw measurements and shape, disease frequency and locomotion. A sire's breeding value can be evaluated on the basis of his daughters' expression of a trait or group of traits or potential. AI bulls may be selected on traits that can be measured and which are correlated with important characteristics in their female offspring.

Effect of environment and housing

Table 4. Effect of environment and housing on claw conformation.

Housing system	Conformation	Claw lesion/zone
Free stalls	Toe length increase; toe angle decrease	Sole ulcer(4) ; white line (3)
Concrete	Toe length decrease; toe angle increase Increased heel height	Thin sole (5 adjacent 1&2) White line (3)
Earthen surface	Toe length increase; toe angle decrease; lateral claws wider front and rear; high heels front and rear	Sole ulcer(4) ; white line (3)
Increase nutrition	Increased toe length; increased heel height	Laminitis. Sole ulcer(4) ; white line (3)

Objective claw measurements

Objective claw measurements provide an accurate means to determine the environmental and genetic effects of claws on longevity. Traits used include toe angle, toe length, claw length, claw size, heel depth and claw diagonal. Several studies showed that repeatabilities among observers are high for most measurements except heel depth. Some normal variation exists between front and rear claws. Toe length, heel height and toe:heel ratio are greater for front claws. Rear claws are longer than front claws but front claws are wider. For both front and rear claws, lateral claws are wider than medial claws. However several studies showed that measurement of one front and one rear claw give a practical and reliable estimate of claw conformation in cattle. These measurements should be taken at a young age since toe angle decreases with age whereas toe length and heel depth increase. Only the trait toe angle did not show any dependence on age

Measurements should be done before any corrective trimming and several repeat measurements should be taken to establish a baseline for an individual animal

Rear leg side view (RLSV)

Rear leg side view was found to be significantly correlated with hock and bone structure as well as foot angle. A steep foot angle was positively correlated with a straight RLSV and vice versa. Studies in Dutch black and white cows showed that the cull rate is higher for cows with a smaller angle on RLSV.

Toe angle and toe length

There is a genetic correlation between sole lesions and toe angle. A toe angle of 50 – 60^0 was found to be positively correlated with increased survival. It is reported that bulls siring offspring with steeper toe angle have daughters that live longer. Toe angle was also found to be more correlated with longevity than leg traits. Steeper toe angle and shorter toe length have been reported by several authors to be the most promising traits to ensure high claw quality thus reducing the incidence of lameness. With a shallow toe angle the toe is longer and the heels lower. This shifts more of the weight bearing towards the heel resulting in concussion and is associated with more sole ulcers and white line disease. Genetically, days open decrease as the toe angle increases.

Claw diagonal length

Diagonal length represents the distance from the apex of the toe to the skin–horn junction at the heel. This is used as a selection tool for Dutch dairy cattle and has a higher heritability score than toe length or toe angle. Based on Dutch research findings, cull rate as a result of lameness increased with increasing values for claw diagonal. Cull rates for animals with very short claw diagonals were lower than those with longer measurements for claw diagonal. A shallow toe angle is associated with a longer diagonal, which increases the probability of lameness. Claw angle was found to be the best single trait because of its high correlation toe length and toe angle to predict the likelihood of future lameness and abnormal locomotion.

Claw size

The claw size should provide a large and stable bearing surface. A large size is better able to act as shock absorber and provide a larger area to dissipate weight bearing forces thereby reducing lameness risk. One study found that heavier cattle had larger claws. However another study found that small feet in bulls were combined with fine bone and hock structure in female offspring. This was found across several breeds. In general small feet was also correlated with straight legs

For front claws, claw size is virtually equal and makes up 54% total claw volume. This results in more balanced weight bearing between claws of the front legs. For rear claws the outer claw is consistently larger and wider than inside claw so that the outer claw naturally bears more weight relative to the inside claw. This results in overgrowth which predispose to lameness

Claw shape

The most common abnormal claw shape is referred to as" screw claw" which is a common conformational defect predisposing to lameness. It is characterized by an inward curvature of the outside wall and a cork screw rotation at the toe.

Acquired screw claw

Although screw claw can be a heritable condition other non-heritable factors may predispose to a shape change which resembles screw claw. Selection of progeny tested bulls based on claw measurements and claw shape should be made at an age of 12 months because of an increase in the coefficient of variation with advancing age.

Acquired screw claw is non-heritable condition predisposing to lameness caused by several factors including nutrition, body weight and biomechanics of locomotion and weight bearing as well as environmental factors. All these factors may cause curvature of the wall and rotation of the toe. The outside claw of the rear legs is most commonly affected. The affected claw undergoes significant overgrowth, it becomes bigger than the opposite claw and as such bears more weight which predisposes to lameness because of sole ulcers and white line disease which develops as a result.

Inappropriate nutrition such as high intake of fermentable carbohydrates leads to subacute rumen acidosis which in turn causes subclinical laminitis. Subclinical laminitis results in poor quality horn due to insufficient keratinization of horn cells. The horn is soft and the wall starts to deviate and curve under. This is exacerbated if the animal is over conditioned. See Figure 3.

Figure 3. Factors playing a role in pathogenesis of acquired screw claw.

High levels of fermentable starch or lack of appropriate fiber length

Laminitis with soft horn and poor keratinization

High growth rate and body weight. Biomechanics of weight bearing. Concrete floor/overcrowding

Curvature of wall and rotation of the toe – Acquired screw claw. Outside claw of back leg mostly affected

Overgrowth with excessive weight bearing

Sole ulcers and white line disease.

Lameness

Heritable screw claw

Research in the 1950's found evidence of a misalignment between P2 and P3in cases showing typical conformational changes associated with screw claw. However heritability scores for screw claw is low. What is generally regarded as heritable screw claw is highly correlated with laminitis. A diagnosis of heritable screw claw can only be assumed if there is a family history of the problem and which manifests at an early age (12-14 months). The typical signs of curvature of the outer wall and rotation of the toe should be present. The back outer claws are usually involved. However normal biomechanics of weight bearing should be kept in mind particularly in rapid growing over-conditioned animals. Rotation of the inside claws of the front feet with normal back feet is usually the non-heritable acquired form and often results from pushing up against the feed bunk over a period of time. A presumptive diagnosis of heritable screw claw can only be made if there are no signs of laminitis. However the question whether to attribute abnormally shaped claws to inherited factors or to inadequate wear or pathological horn growth remains largely unresolved.

Foot diseases

One study showed a significant correlation between toe angle and the incidence of diseases of the foot. A high disease frequency and thin soles in bulls were correlated with a steep foot angle in female offspring. Longer toes and deeper heels were found to be affected with digital disease. Similarly a high toe to heel ratio was found to be associated with a high incidence of claw disorders

Longevity and production traits

Cows with shorter toe length, steeper toe angle were found to be less lame, had better productive efficiency, higher milk yield in later life and higher longevity than cows with longer toe lengths and smaller toe angles. Longevity was positively correlated with udder conformation, RLSV and claw diagonal. Claw diagonal is positively correlated with shorter toe length, steeper toe angle and RLSV

Conclusion

Heritabilities and genetic variation for claw measurements and claw disorders are high enough to achieve genetic improvements in these traits. Claw shape traits could therefore be measured on bulls at a young age (12 moths) during performance testing and before calving in heifers. Of the traits RSLV and claw diagonal may have a high predictive value for longevity

REFERENCES

Boelling D et al. Livestock Production Science 54(1998) 193-203; Boelling D et al. Acta Agric Scand, Sect A, Animal Sci 51 (2001) 122 – 128; Smit H et al. Livestock Production Science 15 (1986) 205-217.; Distl O et al. Livestock Production Science 25 (1990) 1-13; Huang YC et al. Livestock Production Science 44 (1995) 107 – 114; van der Waaij EH et al. J Dairy Science 88 (2005) 3672 – 3678; Vermunt JJ et al. Br Vet J 151 (1995)157 – 180; Vermunt JJ et al. Br Vet

J 152 (1996) 321 – 331; Nuss K et al The Veterinary Journal 172 (2006) 284 – 292; Bouckaert J et al. Vlaams Diergeneeskundig Tijdschriift 6 (1958)149 – 152; Van Amstel SR et al. J Dairy Sci 87 (2004) 757 – 763; Van Amstel SR The Bovine Practitioner (42 (2008)189 – 195.

WHY SHOULD WE USE THE SCIENTIFIC LITERATURE TO ANSWER CLINICAL QUESTIONS?

Bob L. Larson, DVM, PhD, ACT, ACAN, ACVPM-Epi
Manhattan, KS

INTRODUCTION

While clinical experience provides important observations about disease onset risk factors, therapeutic intervention efficacy and patient prognosis, clinical observations lack controls for bias and confounding and as such must be tempered with data from controlled experiments. Clinical experience plays a critical role in veterinary decision making and these experiences form the framework through which all literature will be interpreted. While our own experience and the experience of colleagues that is passed through communications can be useful, the human mind has significant limitations managing normal variation in results, dealing with bias, and overcoming the difficulty of interpreting the complicated biologic systems that make up the animals and populations with which we contend on a daily basis. While every research study has important limitations in either internal validity, external validity or both, well-designed studies have incorporated at least some level of control for bias and confounding and use statistical tests to distinguish treatment effects from random variation.

USING SCIENTIFIC LITERATURE TO ANSWER CLINCIAL QUESIONS

Limitations of Clinical Experience

Biologic systems are inherently more complex than other natural systems. Only biologic systems have complex homeostatic controls as well as the capability to remove insults, self-repair, and substitute for deficient factors. In addition, the hierarchy of: tissues, animals, intra-animal ecosystem, animal herds, animal populations, and ecosystems adds complexity that is absent in other natural systems. The relative simplicity of chemical, geological, and mechanical natural systems allowed for thousands of years of sustained advancements in humankind's understanding of chemistry, astronomy, and physics as evidenced by innovations in metallurgy, horology, and engineering. These advancements in other sciences were occurring even as basic understanding of disease processes was stagnant. Simple experiments with small sample size and only rudimentary controls for bias and confounding and no statistical tests to differentiate between random chance and treatment effects were used to gain great insight into the sciences of chemistry, astronomy, and physics throughout history. In contrast, until the development of specific strategies in the first half of the twentieth century to design experiments with the purpose of controlling bias and to rigorously analyze hypotheses with statistical tests, advancements in disease prevention and treatment were slow or non-existent.[1,2,3] The reasons that for hundreds of years medical sciences lagged behind mechanical and other sciences are the same reasons that veterinarians today must embrace the scientific method and the results of rigorous, well-designed studies that control for variation, minimize bias, and effectively explore interactions.

Attributes of veterinary medicine: random variation, biased and confounded observations, and complex interactions

Veterinarians continually witness the phenomena of random variation of animals and biologic systems in that every practitioner is aware of the range of behaviors, laboratory and imaging indices, and growth and reproductive performance inherent in the species with which one works. We recognize that animals treated identically exhibit a range of average daily weight gain, pain tolerance, response to anesthesia, immunologic response to vaccination, and countless other examples. Differentiating the expected variation in response between animals treated identically from the effects of an intervention requires careful collection of data and probability calculations (e.g. statistical tests) unless the magnitude of treatment effect is so large as to be obvious. However, many of the diagnostic, therapeutic, and prognostic questions that we investigate are concerning subtle differences that are impossible to identify without careful data collection and statistical tests.

In addition to accounting for the expected range of outcomes that animals exposed to the same factors will exhibit, well-designed studies also differ from clinical experience because explicit restraints are implemented to control for sources of bias. Clinical experience is particularly prone to bias because the same person provides and then evaluates interventions. Biases that occur commonly, and often inadvertently, when relying on clinical experience are typically grouped into categories of selection bias, information bias, and confounding.

Observations in clinical settings are plagued by selection bias because although this bias is a crucial flaw if one wants to compare interventions, it is perfectly appropriate and beneficial when applied to clinical case management. Essentially, selection bias occurs when animals with certain signalment, history, or physical examination findings are treated differently than animals without those case characteristics. While clinically reasonable, this practice prevents any attempt to accurately compare alternative risks or treatments.

Similarly, information bias is very common in clinical case management because we intentionally gather different types and amounts of information about different animals. For example, it is reasonable to observe some animals more closely, under different circumstances, or for longer periods of time than other animals. However, this bias can lead to incorrect associations with either disease-causation or treatment factors.

Confounding can occur when two potential causative factors are correlated but not evenly distributed and it is not clear which factor is causing the outcome of interest. Because many risk factors for disease tend to occur clustered together, confounding is a common problem when using our clinical experience to make disease causation or treatment efficacy associations. Examples of potential confounding factors in veterinary medicine include: calves experiencing dystocia and failure of passive transfer, nutrition and other management factors and farm of origin, and younger, lighter calves being transported longer distances. Without careful statistical control, it is impossible to determine which of two confounding factors is the most important factor associated with a disease outbreak or therapy success. Controlled, well-designed experiments use a number of tools including random allocation of animals, blinding (aka masking) of observers, and extensive data collection to combat bias and confounding.

Another limitation of clinical experience compared to well-designed experiments is the difficulty for the human mind to detect and quantify multiple relationships between two or more factors that affect clinically important outcomes. Unfortunately, because of the complexity of disease avoidance and the multi-faceted nature of most diseases, health outcomes are rarely influenced by a single factor. Factors that may interact with each other to affect disease onset or recovery include age, sex, breed, concurrent exposure to other infectious, toxic, or metabolic insults, nutritional status, and stress, to list a few. The interpretative challenge is that the effect of each factor may not be constant and the interrelationships may alter disease risk. Interactions usually involve many variables, making it essentially impossible to accurately determine the entire causal web involved with disease onset or recovery, and to extrapolate clinical experience in one setting to other situations.

Information Literacy

Increased information availability and widespread access to published articles at low or no cost is changing the sources of information that are available to and used by veterinary practitioners. However, the increased access to current peer-reviewed articles available on the internet; and the enhanced ability to search for pertinent articles in an efficient manner with search engines such as PubMed and CAB now allow practicing veterinarians to have access to a breadth of veterinary literature only possible within the walls of veterinary medical library a few years ago. And, with the speed of change within library science and online information availability, veterinary practitioners should be prepared for continued enhancements in their abilities to access current and valuable information and improve their 'information literacy'. The definition of information literacy from the Association of College & Research Libraries (ACRL) is "A set of abilities requiring individuals to recognize when information is needed and to have the ability to locate, evaluate, and use effectively the needed information."[4]

According to the ACRL standards, an information literate individual is able to: determine the extent of information needed, access the needed information effectively and efficiently, evaluate information and its sources critically, incorporate selected information into one's knowledge base, use information effectively to accomplish a specific purpose, and understand the economic, legal, and social issues surrounding the use of information, and access and use information ethically and legally.

Evaluating the Body of Evidence

Although all or most veterinary clinical decisions are based on evidence of some type, some evidence is very strong (rigorously tested in the target-species under natural conditions in experiments designed to prove a theory to be false) and some is very weak (not tested) and some is intermediate.[5,6,7] The hierarchy of evidence is based on the strength of evidence of causation, the ability of the study to control bias, and the similarity between the study population and the population currently being considered in a clinical setting. Sources regarded as strong evidence for therapeutic questions include randomized controlled clinical trials in the target species under a similar husbandry environment or systematic reviews of more than one trial (meta-analysis) that meet these criteria. These are followed by progressively weaker epidemiological studies: cohort studies, case-control studies, and case-series; then controlled experimental challenge trials. The weakest evidence includes opinions based on clinical experience and observations, *in vitro* research, pathophysiologic rationale, and uncontrolled observations.

The body of evidence relevant to answering clinical questions is the sum of multiple studies investigating the same area of interest. Each individual research study contributes to that body of evidence and each individual publication can be ranked on a scale from weak evidence to very strong evidence. Unfortunately, the number of studies and amount of available literature decreases as one moves from weak to strong evidence. Veterinarians considering the strength of evidence must use several perspectives to determine the reliability of research for clinical use. The first consideration is the internal validity of the research which is determined by the study method and appropriate use of controls for bias. Research reports with good internal validity provide assurance that the results represent an unbiased estimate of the true direction and magnitude of the treatment effect. The second consideration is the population used in the research and its appropriateness as a model for the population that generated the clinical question. Generally, the target species in similar housing and husbandry environments provides stronger evidence than the target species in significantly different housing and husbandry environments, related species, unrelated species, or in-vitro methods. And thirdly, the clinical relevance of the outcomes of the research should be considered with patient-/herd-oriented outcomes providing more direct evidence than disease-oriented outcome measurements. Using these considerations, the highest rating in all three dimensions would provide the highest level of evidence.

These three considerations can be used to evaluate the strength and quality of evidence in a published study, by asking the following questions: First, are the results of this individual study valid? This can be addressed by determining if the assignment of patients to treatments was randomized and if the caregivers and clinicians were blind to those assignments. It is also important that all animals in a group or treatment were evaluated (even drop-outs- for example, look for "intent-to-treat" or "deads-in and deads-out" analyses), and that all groups were treated equally except for the experimental therapy. In addition, it is important that the trial last long enough to recognize both potential positive and negative outcomes. Randomization of study subjects to treatment group is critical for study validity. Studies in which treatment is allocated by any method other than randomization tend to show larger and frequently false-positive treatment effects than do randomized trials.[8,9,10]

If it seems that the study results are valid, the second question one needs to ask is, are the valid results of this individual study important? Or asked another way, how large was the treatment effect? This question may be addressed by a report of the magnitude of treatment effect, such as relative risk reduction, absolute risk reduction, or number needed to treat. To determine if a particular magnitude of effect is important, the effect produced by the therapy in question should be compared to the magnitude of effect for other interventions and durations of therapy. The true magnitude of effect of a therapy can never be known, but a valid trial provides a point estimate around which the true effect is expected to lie. The 95% confidence interval of the point estimate of the treatment effect is commonly reported to communicate the precision of the estimate of the treatment effect, and indicates the range that includes the true treatment effect 95% of the time. The larger the sample size and the larger the number of outcome events, the greater is one's confidence that the true treatment effect is near the point estimate (narrow 95% confidence interval). To determine the importance of a reported positive treatment effect, one should look at the lower boundary of the confidence interval. If this lower boundary is important to you as a clinician, then you are confident that the treatment in question only has a 2.5% chance of failing to reach this lower boundary and the treatment can be deemed important. In contrast, if the treatment effect indicated by the lower boundary is not clinically important to you, the study cannot definitively indicate an important treatment effect even if the results are statistically significant. Similarly, to determine the risk of a reported neutral or negative treatment effect, one should look at the upper boundary of the confidence interval. If the upper boundary is clinically important, you cannot exclude the possibility that an important treatment effect is possible even though the trial indicated either no difference between treatment and controls or a negative treatment effect.

And finally, if a study appears to be valid and important, are the valid, important results of this study applicable to my client's patient or herd?[11] If the study animals or treatment setting is sufficiently different from your patient/herd that initiated the question, or if the treatment is either not feasible or not consistent with the client's/manager's values and expectations, a valid and important therapy would not be applicable in your current situation. For a reported trial to influence treatment strategies, one should ask whether all clinically important outcomes were considered. And most importantly, were the outcomes of most importance to you, the clinician, and the owner/manager reported? Trials that report substitute endpoints such as blood cell counts, blood chemistry, or body temperature are not as valuable for evaluating treatment options as indications of reduced mortality risk or improved production (weight gain, conception, etc.). Even when a report indicates one or more positive outcomes associated with a treatment, one should consider whether other negative outcomes are also associated with that treatment. This is particularly important if the negative outcomes are more important to the owner/manager than are the positive outcomes.

The role of scientific literature in critical thinking

Critical thinking is a process, and as such is never completed. Critically thinking about a clinical problem involves repeating the steps of gathering information, evaluating that information, reflecting on the information and coming to tentative conclusions. Because no single study can fully address most clinical questions and because every study has limitations either in internal and/or external validity, using scientific studies to enhance clinical decision making requires combining different pieces of evidence of varying strengths. While gaining competence in each step of the critical thinking process requires education, skill, and experience, probably the most difficult to master is to reflect deeply about what is known and what is unknown and how to tie multiple pieces of information and evidence together.

Essentially, good clinical decisions are based on good critical thinking skills. Because the quality of thinking is limited by the amount and accuracy of pertinent information available, gaining competence to access, evaluate, and integrate the scientific literature is critical to develop good critical thinking skills. However, good information literacy skills does not guarantee good critical thinking skills in that a person with poor critical thinking skills may accrue very limited benefit from gaining access to valid scientific studies.

Limitations in the use of scientific literature in clinical decision making

One of the greatest challenges to increased use of scientific literature in clinical decision making is that by knowing the strength of evidence for potential interventions, veterinary practitioners will often be less certain of his/her diagnostic, therapeutic, and prognostic pronouncements compared to client expectations for greater certainty. This puts the expectations and values of the client at odds with the expectations and values of the practitioner. It appears that many clients gravitates toward the clarity of an untested (or under-tested), but strongly stated pronouncement in preference to the skeptical, methodical, testing-rejecting-retesting pace of the scientific method.

SUMMARY

Practicing high-quality medicine by incorporating scientific literature into veterinary practices leads one to establish a problem-solving process that recognizes the important limitations that bias and complex interactions places on our ability to gain knowledge about animal health through clinical observations, and the value that well-designed studies provide to reduce bias and explore some aspects of complex interactions. However, understanding how to use the scientific literature in clinical practice also requires that one recognizes the limitations that even well-designed studies have incomplete control of bias and have limited ability to explore complex interactions. And, finally, integrating scientific literature into high-quality veterinary practice causes one to realize that gaining accurate knowledge is a very slow, iterative process with no short-cuts for the complex problems commonly encountered by veterinary practitioners.

REFERENCES

1. Fisher, RA. Statistical Methods for Research Workers, Edinburgh, Oliver and Boyd, 1925. 2. Fisher, RA. The Design of Experiments, Edinburgh and London, Oliver and Boyd, 1935. 3. Mann, HB. Analysis and Design of Experiments. New York, Dover Publications, 1949. 4. American Library Association, Association of College and Research Libraries http://www.ala.org/ala/acrl/acrlstandards/informationliteracycompetency.cfm accessed Feb 5, 2014. 5. Sackett DL. **Can J Cardiol** 1993;9:487. 6. Dans AL, et al. **J Am Med Assoc** 1998;279:545. 7. Berg AO, Dimensions of evidence. In: Geyman JP, Deyo RA, Ramsey SD, editors:. Evidence-based clinical practice: concepts and approaches, Boston, 2000, Buterworth-Heinemann. 8. Sacks HS, et. al. **Am J Med** 1982;72:233. 9. Chalmers TC, et. al. **N Eng J Med** 1983;309:1358. 10. Colditz GA, et. al. **Stat Med** 1989;8:441. 11. Sackett DL, et. al., Diagnosis and screening. In Sackett DL, Straus SE, Richardson WS, Rosenberg WA, Haynes RB, editors: Evidence-based medicine: how to practice and teach EBM, 2nd ed, Philadelphia, 2000, Churchill-Livingstone.

UROLITHIASIS IN RUMINANTS

DAVID E ANDERSON, DVM, MS, DACVS
Knoxville, Tennessee

Surgical treatment of urolithiasis in ruminants traditionally was limited to short-term salvage procedures in feedlot animals destined for early slaughter. These methods often involved resection of the penis, experienced numerous complications, and had variable success rates. More recently, an increase in the number of small ruminants kept as pets has led to the development of several other techniques for resolution of urolithiasis. The intent is to develop a procedure which does not stricture and maintains the breeding ability of intact males. Research also has led to methods for replacement of ruptured urethral tissue in order to salvage animals which would otherwise be doomed to slaughter or euthanasia. The perceived value of these animals as pets also provides the opportunity to perform procedures which would not be cost-effective in feedlot animals.

SURGICAL OPTIONS FOR ALLEVIATION OF URETHRAL OBSTRUCTION IN RUMINANTS

Urethral Process Amputation

The urethral process (vermiform process) is the single most common site of obstructive urolithiasis in sheep and goats. Thus, *urethral process amputation* is the first surgical procedure to be attempted in small ruminants. This procedure is simple to perform in intact male sheep and goats, but may not be possible without a surgical approach to the penis in males castrated at a young age (e.g. 2 weeks old). The uroliths must be present in the process itself, or in the distal penile urethra in order to pass. Examination of the urethral process should be performed in every animal with urolithiasis, and often the process is removed as a precautionary measure. The procedure is performed quickly with scissors or scalpel, and is easiest with the animal restrained in a sitting position on its rump. Unfortunately, urethral process amputation usually results in re-obstruction within a short period of time, as there are more uroliths further proximal in the urethra. Approximately half of the animals treated with amputation will have urine flow restored.[34] Amputation can be performed as an adjunct therapy for other procedures. Amputation of the urethral process is only possible in ruminants which have reached the age of puberty, at which time the prepuce separates from the penis and allows access to the distal penis.

Unlike other animals, ruminants, swine, camelids, and cervids cannot be catheterized to relieve obstructive urolithiasis, as their urethra has a diverticulum in the region of the ischial portion of the urethra.[36,37] This diverticulum results in the redirection of any catheter into the diverticulum, rather than into the bladder itself. Only with special pre-curved catheters has bladder catheterization been possible.[37] Catheterization and retrograde flushing are therefore not useful in ruminants for resolution of obstructive urolithiasis, and may result in urethral rupture if too much force is used.

Lithotrispy

Laser lithotripsy has recently been investigated as a technique for dissolution of uroliths in ruminants.[31,32] A chromium-thulium-holmium:yttrium-aluminum-garnet (Ho:YAG) laser was successfully used to rupture a urethral calculus when placed via an ischial urethrotomy in a steer.[31] In a separate study, a Ho:YAG laser was successfully used to fracture uroliths in 3 goats and 2 potbellied pigs.[32] The use of lithotripsy is dependent on the presence of a urethra sufficiently large for passage of the laser in an endoscope, and the location of the uroliths in the distal urethra. Lithotripsy, however, presents a viable option for removal of uroliths in the future, and further investigation is warranted. Advantages include reduced cost and rapid discharge from the hospital. Unfortunately, the size the urethra of small ruminants limits transurethral options. Shock wave lithotripsy offers an non-invasive method for disruption of uroliths. The position of the urolith needs to be well defined and stable for application of this modality. Economic constraints and availability of this equipment has prevented wide spread use of this option. Also, complications, such as re-obstruction, are expected to be similar to other treatment methods and have been documented in other species.

Urethrotomy

Urethrotomy can be performed over the urolith itself, or adjacent to the urolith in healthier tissue. The urolith must be located in an area which can be reached easily, determined by catheterization or palpation. Following incision at the urolith, it is removed with forceps or is crushed to allow the pieces to pass. An alternative technique involves retropulsion of a urolith from the distal urethra to an ischial urethrotomy where it can be removed. Ischial urethrotomy also allows for placement of a long urethral catheter to act as a stent for a ruptured urethra. Urethral stricture formation is a common side effect of urethrotomy, but the simplicity of the procedure makes it popular, especially in breeding animals where the penis must remain intact.[35] The urethrotomy incision may be allowed to heal without suturing, as suturing may increase the likelihood of stricture formation. Some surgeons, however, recommend suture closure of the urethrotomy as soon as possible to prevent stricture formation.

Cystotomy

Cystotomy with urolith removal combined with normograde and retrograde urethral flushing was successful in restoring urinary function in 7 of 8 small ruminants with obstructive urolithiasis.[34] The authors concluded that this technique was more successful than perineal urethrostomy in restoring urinary function. However, at the time of that study, further techniques such as tube cystostomy, urinary bladder marsupialization, and laser lithotripsy had not been described.

Tube Cystostomy

Tube cystotomy provides an excellent method of treating urolithiasis, and has been applied to both cattle and small ruminants.[22-24] In this technique, the patient is anesthetized in dorsal recumbency, and the bladder approached via a paramedian incision. A cystotomy is performed to remove any uroliths present, and to perform normograde flushing of the urethra if desired. A Foley catheter (12 to 24 Fr.) is then passed through a stab incision in the body wall of the inguinal region, placed into the bladder through a stab incision, and sutured into place with the cuff inflated. Following closure of the celiotomy incision, extra catheter is secured with sutures or bandages to the abdominal wall. The catheter is allowed to drain freely for several days, after which time it is periodically occluded with a clamp in order to test for normograde urine flow through the urethra. Once normograde flow is established, the catheter cuff is deflated and the catheter is removed. In two separate reports, urine flow was reestablished in a mean of 11.5 & 11 days.[22,23] Success of this procedure was 76 to 90% in the short term, and 86% in the long term.[22,23] Success rates have been shown to be dependent on several factors. Sheep, animals with serum potassium level >5.2mEq/L at admission, the presence of fluid in the abdomen at admission, and previous urethral process amputation have all been shown to be associated with poorer survival.[23] In a separate study, increased BUN and respiratory rate were associated with poorer survival.[25] The major drawback of surgical tube cystostomy is the cost associated with the procedure, as animals must be hospitalized during the waiting period prior to catheter occlusion. Repeat obstruction and premature tube removal have not been major problems following tube cystostomy.[23] Tube cystostomy has been used to restore breeding function in intact males, avoiding the loss of breeding function previously described.[26]

Vesicular irrigation with hemiacidrin was used in conjuction with percutaneous tube cystostomy to restore urinary function in a goat.[27] The authors concluded that irrigation resulted in rapid dissolution of calcium phosphate uroliths without the need for celiotomy and cystotomy. Vesicular irrigation could be combined with surgical or percutaneous tube cystostomy, and perhaps also with other techniques to dissolve uroliths *in situ*. The efficacy of this technique on different types of uroliths is not known.

In order to reduce the costs associated with the tube cystostomy procedure, a *percutaneous tube cystostomy* technique has been developed for placement of the cystostomy tube.[27] This technique was successful in the initial description, but subsequent reports indicated that it is 5.6 times more likely to require a second procedure to replace the tube following premature tube loss.[25] If successful, percutaneous tube cystostomy is managed similarly to the surgically-placed tubes, in that the tube is periodically occluded after a waiting period of approximately 7 days, and is removed once urination is reestablished.

Urinary Bladder Marsupialization

Urinary bladder marsupialization in small ruminants has been moderately successful in restoring urination.[28,29] It does, however, eliminate urinary continence and may, as a result, be unacceptable to some pet owners. In this technique, the apex of the bladder is marsupialized to the ventral body wall at midline during a midline celiotomy, creating a stoma which allows for free drainage of urine from the bladder. Major complications involved include bladder mucosal prolapse through the stoma, urine scald on the ventrum of the animal, and ascending urinary tract infections through the marsupialization site.[28,29] The marsupialization stoma strictured in 3 of 25 goats.[28,29] Success rates were 66% and 94% in two separate studies.[28,29] Hospitalization was significantly shorter than tube cystostomy (mean of 4 days vs. 14 days), which may reduce cost associated with the procedure.

Prepubic Urethrostomy

Prepubic urethrostomy has been described in a sheep and a goat for relief of obstructive urolithiasis.[30] In the report described, prepubic urethrostomy was used for relief of urinary obstruction following perineal urethrostomy stricture formation. The technique used is similar to that performed in dogs and cats. In dorsal recumbency, the pubic bone of the pelvis was approached, all attached muscles elevated and reflected, and the pubic bone transected with an oscillating bone saw. The urethra was then mobilized from its caudal location by blunt dissection and brought through the ventral midline incision. The pubic bones were reapposed with orthopedic wire, and all elevated muscles reapposed. The urethra was then positioned near the caudal abdomen and carefully reapposed to the skin edges. Both animals described were able to void normally after the procedure, but recurrent cystitis and further stricture formation were complications. Due to its proximal location in the urethra and the difficulty of the procedure, this technique should be reserved for revision of previously-performed techniques which stricture. The author (DEA) has performed pelvic urethra to prepuce and pelvic urethra to distal penis anastomosis with short-term success in goats. These techniques are not recommended because the retroflexion of the pelvic urethra exceeds 50 degrees and thus causes chronic partial outflow obstruction.

Urethral Mucosa Graft

Buccal mucosal grafting has recently been described for the reversal of a perineal urethrostomy in a goat.[33] In this case, the urethra was reconstructed using a strip of tissue from the buccal mucosa following unsuccessful revision of the urethrostomy site. Buccal mucosa proved to be an excellent tissue for this procedure, and healing was excellent, with good long term survival.

Perineal Urethrostomy

Perineal urethrostomy has long been the mainstay of urolithiasis treatment in ruminants. It is considered a salvage technique in most cases, though some animals have survived for long periods of time after surgery. With this technique, a permanent opening is made in the urethra through which the animal voids urine. Perineal urethrostomy does not guarantee urethral patency, as an animal can still obstruct proximal to the stoma. As a result, the urethrostomy stoma should be placed as distally as possible to allow for repeat surgery more proximally. The most common complication of urethrostomy is stricture of the stoma, which may occur rapidly (weeks to months) after surgery.[35]

The penis is approached via midline incision in a standing animal under epidural anesthesia. The ideal location for the incision is just caudal to the scrotum, as this location is usually proximal to the obstruction, but is sufficiently distal to provide penile mobility and the opportunity for more proximal urethrostomy in the future. The retractor penis muscles are located and traced to the penis itself within the thick tunica albuginea. The retractor muscles can be transected, or can be left in place at the discretion of the surgeon, though transection of the muscles helps reduce any tension which may result in breakdown of the stoma site. The penis is dissected free of the surrounding tissues and is mobilized to the incision site. A tension-free position of the penis is required to prevent breakdown of the urethrostomy. The penis is stabilized at the incision with horizontal mattress stay sutures of large monofilament non-absorbable suture material passing through the lateral aspects of the fibrous tunica albuginea and then through the skin. Once stabilized, the urethra on the dorsal aspect of the penis (under the bulbospongiosus muscle) is then opened with a single longitudinal incision. Small-gauge (0 to 2-0) nonabsorbable monofilament suture material is used to appose skin edges and urethral edges, creating a stoma.

Perineal urethrostomy can also be performed at a location much more proximal; even as high as the level of the ischium. At this location, the stoma thus created may allow catheterization of the bladder for relief of bladder distension. The proximal location is much more likely to experience urine scald on the perineal region, and does not allow for a secondary procedure to relieve any further urinary obstruction. The technique is similar to that described for the distal urethrostomy.

Penile Amputation

Penile amputation (penectomy) is similar to the technique described for perineal urethrostomy, however the penis is transected and may be completely removed rather than creating a stoma in the penis. This is a simpler procedure than the perineal urethrostomy, but has the disadvantage of increased stricture rate and speed. It is strictly a salvage procedure. The approach to the penis is identical to that described for the perineal urethrostomy, however once the penis is identified and mobilized to the incision, the retractor penis muscles are transected. The penis itself is brought to the incision and sufficiently mobilized so that the transection can be performed several inches below the ventral aspect of the skin incision to allow for tension-free positioning of the penis. The distal penis may be excised completely, in which case the dorsal vessels of the penis are ligated immediately proximal to the transection site. The distal penis may also be left in place, in which case the dorsal vessels are carefully separated from the dorsal aspect of the tunica albuginea and are protected from damage during the penile transection.

Following sharp transection of the penis at the desired location, mattress sutures of heavy non-absorbable monofilament suture material are placed at the lateral aspects to secure the penis to the skin edges. Care must be taken to avoid penetrating or ligating the urethra with these mattress sutures. The penis should be positioned such that the stump projects approximately 1 inch from the incision and is oriented such that the urine flow will be caudoventrally. Once the penis is secure, the urethra may be spatulated to create a larger opening, or may be left unspatulated. If hemorrhage is severe from the exposed corpus spongiosum and cavernosum, the cut edge of the penis may be closed with interrupted sutures of small monofilament suture.

REFERENCES

1. Fubini SL: Surgery of the vagina. InFubini SL, Ducharme NG, editors: Farm Animal Surgery, St. Louis, 2004, Saunders.
2. Hooper RN, Taylor TS: Urinary Surgery, Vet Clin North Amer Food Anim Pract 11:95-121, 1995.
3. Madison JB: Surgery of the Urinary Tract. In Wolfe DF, Moll HD, editors: Large Animal Urogenital Surgery 2nd ed., Baltimore, 1999, Williams & Wilkins.
4. StJean G, Hull BL, Robertson JT, et al: Urethral extension for correction of urovagina in cattle. A review of 14 cases, Vet Surg 17:258-262, 1988.
5. Wolfe DF, Baird AN: Female urogenital surgery in cattle, Vet Clin North Amer Food Anim Pract 9:369-388, 1993.
6. Gilbert RO, Wilson DG, Levine SA, et al: Surgical management of urovagina and associated infertility in a cow, J Am Vet Med Assoc 194:931-932, 1989.
7. Brown MP, Colahan PT, Hawkins DL: Urethral extension for treatment of urine pooling in mares, J Am Vet Med Assoc 173:1005-1007, 1978.
8. Mckinnon AO, Belden JO: A urethral extension technique to correct urine pooling (vesicovaginal reflux) in mares, J Am Vet Med Assoc 192:647-650, 1988.
9. Monin T: Vaginoplasty: a surgical treatment for urine pooling in the horse. In Proceedings 18th Annual Meet Am Assoc Equine Pract, 1973, pp. 99-102.
10. Turner AS, McIlwraith CW: Urethroplasty by caudal relocation of the transverse fold. In Turner AS, McIlwraith CW, editors: Techniques in Large Animal Surgery 2nd ed., Philadelphia, 1989, Lea & Febiger.
11. Fubini SL: Surgery of the urinary bladder and ureters. In Fubini SL, Ducharme NG, editors: Farm Animal Surgery, St. Louis, 2004, Saunders.
12. Baird AN: Surgery of the urinary bladder. In Wolfe DF, Moll HD, editors: Large Animal Urogenital Surgery 2nd ed, Baltimore, 1999, Williams & Wilkins.
13. Dart AJ, Dart CM, Hodgson DR: Surgical management of a ruptured bladder secondary to urethral obstruction in an alpaca, Aust Vet J 75:793-795, 1997.
14. Carr EA, Schott HC, Barrington GM, et al: Ruptured urinary bladder after dystocia in a cow, J Am Vet Med Assoc 202:631-632, 1993.
15. Roussel AJ, Ward DS: Ruptured urinary bladder in a heifer, J Am Vet Med Assoc 186:1310-1311, 1985.
16. Smith JA, Divers TJ, Lamp TM: Ruptured urinary bladder in a post-parturient cow, Cornell Vet 73:3-12, 1983.
17. Hooper RN, Taylor TS: Urinary surgery, Vet Clin North Amer Food Anim Pract 11:95-121, 1995.
18. Ewoldt JM, Anderson DE, Miesner MD, et al: Short- and long-term outcome and factors predicting survival after surgical tube cystostomy for treatment of obstructive urolithiasis in small ruminants, Vet Surg 35:417-422, 2006.
19. Madigan JE: Diagnosing and treating the foal with bladder rupture, Vet Med 82:1048-1052, 1987.
20. Bouré LP, Kerr CL, Pearce SG, et al: Comparison of two laparoscopic suture patterns for repair of experimentally ruptured urinary bladders in normal neonatal calves, Vet Surg 34:47-54, 2005.
21. Bouré LP, Foster RA, Palmer M, et al: Use of an endscopic suturing device for laparoscopic resection of the apex of the bladder and umbilical structures in normal neonatal calves, Vet Surg 30:319-326, 2001.
22. Rakestraw PC, Fubini SL, Gilbert RO, et al: Tube cystostomy for treatment of obstructive urolithiasis in small ruminants, Vet Surg 24:498-505, 1995.
23. Ewoldt JM, Anderson DE, Miesner MD et al: Short- and long-term outcome and factors predicting survival following surgical tube cystostomy for treatment of obstructive urolithiasis in small ruminants, Vet Surg 35:417-422, 2006.
24. Hastings DH: Retention catheters for treatment of steers with ruptured bladders, J Am Vet Med Assoc 147:1329-1330, 1965.
25. Fortier LA, Gregg AJ, Erb HN, et al: Caprine obstructive urolithiasis: requirement for 2nd surgical intervention and mortality after percutaneous tube cystostomy, surgical tube cystostomy, or urinary bladder marsupialization, Vet Surg 33:661-667, 2004.

26. Todhunter P, Baird AN, Wolfe DF. Erection failure as a sequela to obstructive urolithiasis in a male goat, J Am Vet Med Assoc 209:650-652, 1996.
27. Streeter RN, Washburn KE, McCauley CT: Percutaneous tube cystostomy and vesicular irrigation for treatment of obstructive urolithiasis in a goat, J Am Vet Med Assoc 221:546-549, 2002.
28. May KA, Moll HD, Wallace LM, et al: Urinary bladder marsupialization for treatment of obstructive urolithiasis in male goats, Vet Surg 27:583-588, 1998.
29. May KA, Moll HD, Duncan RB, et al: Experimental evaluation of urinary bladder marsupialization in male goats, Vet Surg 31:251-258, 2002.
30. Stone WC, Bjorling DE, Trostle SS, et al: Prepubic urethrostomy for relief of urethral obstruction in a sheep and a goat, J Am Vet Med Assoc 210:939-941, 1997.
31. Streeter RN, Washburn KE, Higbee RG, et al: Laser lithotripsy of a urethral calculus via ischial urethrotomy in a steer, J Am Vet Med Assoc 219:640-643, 2001.
32. Halland SK, House JK, George LW. Urethroscopy and laser lithotripsy for the diagnosis and treatment of obstructive urolithiasis in goats and pot-bellied pigs, J Am Vet Med Assoc 220:1831-1834, 2002.
33. Gill MS, Sod GA. Buccal mucosal graft urethroplasty for reversal of a perineal urethrostomy in a goat wether, Vet Surg 33:382-385, 2004.
34. Haven ML, Bowman KF, Engelbert TA, et al: Surgical management of urolithiasis in small ruminants, Cornell Vet 83:47-55, 1993.
35. Van Metre D: Urolithiasis. In Fubini SL, Ducharme, NG, editors: Farm Animal Surgery, St. Louis, 2004, Saunders.
36. Garrett PD: Urethral recess in male goats, sheep, cattle, and swine, J Am Vet Med Assoc 220:1831-1834, 1987.
37. Palmer JL, Dykes NL, Love K, et al: Contrast radiography of the lower urinary tract in the management of obstructive urolithiasis in small ruminants and swine, Vet Radiol Ultrasound 39:175-180, 2002.

Figure 2. Stay sutures holding bladder in position for surgery. Courtesy of Kimberly A May, DVM, MS, DACVS.

Figure 3. Placement of Foley catheter for tube cystostomy. The catheter enters the abdomen through the body wall (left) and enters the bladder through a stab incision. The Foley cuff has not yet been inflated. Sutured cystotomy incision for removal of uroliths is evident on bladder surface.
Figure 4. Completed tube cystostomy showing locations of abdominal incision and tube placement, with stay sutures stabilizing Foley catheter.
Figure 5. Completed bladder marsupialization surgery demonstrating placement of stoma and abdominal incision. Courtesy of Kimberly A. May, DVM, MS, DACVS.
Figure 6. Perineal urethrostomy in a steer. The penis has been mobilized and brought to the incision with good mobility, allowing for a tension-free fixation. Courtesy Bruce L. Hull, DVM, MS.
Figure 7. Perineal urethrostomy (penile amputation) in a lamb. The penis has been freed from its preputial attachments and is ready for fixation and amputation.
Figure 8. Perineal urethrostomy in a lamb showing suture placement and final appearance of penis. Note the dorsally-placed urethra visible in this photo.
Figure 9. Normal urethral process in a ram. The lumen of the process is extremely narrow and prone to obstruction with uroliths. Courtesy of Michael Rings, DVM, MS.

STANDING RESTRAINT FOR CONSCIOUS PROCEDURES

David E Anderson, DVM, MS, DACVS
Knoxville, Tennessee

Comments on Handling

While camelids are much easier to work with than many traditional livestock, these animals can be challenging at times when it becomes necessary to handle them. To help you as you begin working with these interesting animals, here are a few handling guidelines that should be used. These guidelines will make working with camelids much easier on both you and your llama or alpaca, and they will make any veterinary visits go more smoothly.

One of the most important aspects of camelid behavior to remember is their herding instinct; both llamas and alpacas are extremely herd-oriented animals. This can be both an advantage and a disadvantage as we work with them. One of the benefits of their strong herding instinct is that this can make moving groups much easier. The animals will tend to stay together and make less work for us as we move them. A disadvantage of this behavior is that it makes it difficult to single animals out for individual treatment.

Perhaps the most effective way to overcome this is to use a series of catch pens that gradually become smaller, until you are able to separate individual animals. Avoid making sudden movements that can be interpreted as threatening gestures as this may frighten the animals and make it much more difficult to catch those that you need. As you single out the individual that needs your attention, approach the animal slowly and take hold of the neck. This will allow you more control over the animal, as it will likely try to escape your grasp. For a secure hold that will be safer for you and the animal, place the head in the crook of your elbow and pull it close to your shoulder.

When properly trained, use of a halter on llamas and alpacas can be invaluable as a handling tool. Training consists of getting the animal used to handling and used to the halter from an early age. Proper halter fit is essential for its use; this is necessary due to camelid nasal anatomy. A good portion of the nose directly behind the nostrils is cartilage rather than bone. Improper halter fit can lead to pinching of the nasal passage and panic in the llama or alpaca as these animals mostly breathe through their nose. Make sure the halter is sized appropriately for a llama or alpaca and that the halter rings are almost to the corner of the eyes. This position will assure that the noseband is sitting past the cartilage on bone. Now you may use a lead rope to direct the animal, giving them an appropriate amount of slack in case something should startle the animal.

Another restraint method used for alpacas is to rest your other hand on the point of the shoulders at the base of the neck and gently but firmly press downward. This should help you to control the animal and keep it still for any treatments necessary. For an animal that is continuously swinging around and not standing still, you may use your free hand to grasp the base of the tail to help hold them in one place. You may also have another person do the same for you, especially with larger animals.

A unique camelid behavior that can be used to our advantage in handling is the "kush." This is a term for sternal recumbency, in other words, when the animal lays down with its legs up underneath its body. Depending on the animal, some llamas and alpacas will cush when you grab their tail, when they are stressed, or for no obvious reason. When the animal does go down while you are trying to handle it, allow it to do so and simply restrain the animal and apply light pressure to encourage it to maintain the posture. This handling method works well when multiple people are in the pen; one or more people can help to hold the animal while another performs the procedure. While not the least stressful handling option, it can be useful when an animal is not willing to submit to any other forms of restraint.

When additional control or restraint is required for short periods, one of the easiest methods is "earing". This technique involves bringing the hand not involved in head restraint up along the animal's neck and grabbing the outside ear. Do not use a twisting motion to control the animal, rather, simply squeeze the base of the ear firmly and hold as the animal is treated.

These are just a few handling tips that will help to make working with your new llama or alpaca go more smoothly. When you become more comfortable with these procedures and camelid behavior, your experiences as an owner and handler will be much more enjoyable. These skills will also be useful in the event that your llama or alpaca becomes ill; continuing care will be more easily accomplished when you are able to give the medicine prescribed by your veterinarian. When you have questions on animal handling, remember to ask other owners or your veterinarian. They can assist you in becoming better at working with these unique animals.

Field Anesthesia Techniques in Camelids

A regularly encountered subject during consultation with veterinarians in the field, involves recommendations for chemical restraint and analgesia to accomplish various procedures in camelids. Relative unfamiliarity with the species and potentially the drugs used, present an uneasy feeling when preparing for the task at hand. Single drugs or combinations of $\alpha 2$ agonists, opioids, ketamine with or without local anesthetics are excellent and safe methods for sedating, anesthetizing and providing analgesia in camelids. With camelids, physical restraint alone when performing painful procedures, may pose greater risks to the welfare of the patient than anesthesia or sedation. Where financial and slaughter constraints may factor into client reasoning for physical restraint to trump drug use in traditional food animal species, analgesia is an expectation of the camelid client.

Debating ethics, economics, or legalities of pain management in animals are beyond the scope of this presentation. We have little validated means of documenting and quantifying pain in animals and therefore have little scientific guidance for how analgesics should be used or if they work as we think they do. Nor does a plethora of scientific data on kinetics and dynamics of anesthetic drugs used in camelids exist. Public perception, common experience of pain, and strengthening animal welfare standards of care, warrant discussion of ways to use analgesic agents as adjunct therapeutics in food animal practice.

Patient evaluation

Administration of anesthetic agents to camelids is a source of anxiety for many veterinarians and owners. Llama and Alpaca physiology is not a mainstream topic in most veterinary schools, and, therefore, veterinarians do not feel comfortable in the administration of drugs routinely used in other livestock species. Although specific drug dosages differ between camelids and other livestock and even among llamas and alpacas, all other aspects of the "warm blooded mammal" apply. The minimum data base before using an anesthetic is rectal temperature, heart rate and rhythm, respiratory rate and character, and general demeanor of the patient. If general anesthesia is planned, a PCV and TP should be determined. If the patient is geriatric (older than 10 years for llamas and alpacas), then more exhaustive diagnostic tests may be warranted (e.g. serum biochemistry). An accurate body weight will increase the safety of the procedure, but an estimation of body weight may be obtained by evaluating the body condition score.

Local anesthesia

Local anesthetics are used sparingly in camelids. The most common indications are for suturing of wounds or other minor surgical procedures. Lidocaine 2 % HCL is used most commonly. I have used lidocaine for intravenous regional anesthesia (IVRA) in llamas. IVRA's are administered intravenously distal to a tourniquet placed on the limb. Anesthesia generally lasts as long as the tourniquet is held in place, but rapidly dissipates after removal. Although the LD_{50} dose of lidocaine has not been evaluated, toxicities have been reported. Therefore, I recommend diluting lidocaine to a 1 % solution using isotonic bicarbonate or saline. Total dose should not exceed 4.4 mg/kg body weight (1 cc of 2 % lidocaine / 5 kg body weight). Lidocaine toxicity is recognized by the onset of drowsiness, nystagmus, convulsions, and possibly death (LaRue Johnson, personal communication, 1998). The treatment of choice is diazepam (0.4 mg / kg, IV).

Epidural anesthesia

Epidural anesthesia may be used for analgesia of the perineum and tail. Lidocaine 2%, xylazine, or a combination of these may be used. Lidocaine (maximum dose 1 cc / 50 kg body weight) alone is expected to provide 2 hours of analgesia. Xylazine (0.1 mg / kg) alone is expected to provide 3 hours of analgesia. Xylazine + lidocaine combination is expected to provide 6 hours of analgesia. Epidural injection is made at the level of the most cranial palpably movable intervertebral space cranial to the base of the tail. The skin overlying this space is aseptically prepared. The needle is placed on dorsal midline with the bevel facing cranially and at a 60 degree angle to the tail head. The hub of the needle is filled with solution and advanced until the solution is seen to be pulled into the epidural space.

Sedation

The most commonly used drugs for sedation of llamas and alpacas are butorphanol, xylazine, and metdetomidine. Butorphanol (0.1 mg / kg, IV or IM) provides safe, reliable, and predictable sedation lasting approximately 20 minutes. Xylazine (0.3 mg / kg, s.c. or IM; 0.2 mg / kg, IV) provides more profound sedation and is less predictable lasting 10 to 20 minutes with maximum sedation and an additional 10 to 20 minutes of mild sedation. Metdetomidine (10 to 15 µg / kg, IV) acts similarly to xylazine. In general, alpacas are more resistant to anesthetic drugs. I find that a 20 % increase in dosage for alpacas over that used for llamas is not uncommon.

"Ketamine Stun"

Eric Abrahamsen and I have been developing an injectable chemical restraint technique to provide an enhanced level of patient cooperation and analgesia. It is called "the ketamine stun" because patients under its influence generally appear quite awake, though they are much more cooperative. We have been experimenting with various approaches to provide these benefits in both the standing and recumbent patient. These techniques are still under development and are offered here in hope that your experimentation with them will enhance all of our abilities to meet the needs of the ruminant patients in our practices. The level of analgesia varies from patient to patient and should not be counted upon for surgical procedures normally requiring a local anesthetic block.

For standing procedures in llamas and alpacas, a combination of xylazine (0.1 mg/kg), ketamine (0.2 mg/kg) and either butorphanol (0.05 mg/kg.) or morphine (0.1 mg/kg.) is administered IV or IM. Morphine is much more cost effective and we have not found its use to alter the results of the technique. Peak level of patient cooperation and analgesia will extend for approximately 15 minutes following the administration of the drug cocktail. An additional half dose of ketamine can be used to extend duration another 5-10 minutes. When recumbency is desired we increase the doses of the xylazine (0.3 mg/kg.) and ketamine (0.6 mg/kg) used in the protocol. These patients will appear awake in most instances, but will be much more cooperative than when physical restraint (alone or with xylazine) is used. As with the standing approach, additional half doses of ketamine can be used to extend the duration of patient cooperation and analgesia. Patients are often able to stand and walk upon completion of procedures using this restraint technique. We have found this technique very useful for improving both minor procedures such as casting fractures and as a supplement during more major surgeries using a local anesthetic block.

General anesthesia

Many anesthetic combinations have been used for llamas and alpacas. For field anesthesia, I prefer to use a combination of butorphanol (0.1 mg/kg, IV or IM), xylazine (0.2 mg/kg, IV or 0.4 mg/kg IM), and ketamine (2 mg/kg, IV or 4 mg/kg IM). This combination provides approximately 20 minutes of useful anesthesia for minor surgical procedures (e.g. castration). I recommend caution in the dosage of ketamine. I have seen profound respiratory depression after intravenous ketamine at 2 mg/kg. This reaction is not predictable, and I prefer to "titre" the dose based on clinical response.

Table 1 Butorphanol-Xylazine-Ketamine combined IM anesthesia for camelids. Expected duration is 30 minutes.

Alpacas	0.1 mg/kg butorphanol	0.4 mg/kg xylazine	4 mg/kg ketamine
Llamas	0.05 mg/kg butorphanol	0.3 mg/kg xylazine	3 mg/kg ketamine

Patient monitoring

Patient monitoring is critical during any period of sedation or general anesthesia. In my experience, respiratory rate is an easy and extremely useful parameter to measure to determine the status of the patient. Steady respirations are indicative of steady state sedation or anesthesia, infrequent shallow respirations indicate that the patient may be too deeply anesthetized, and rapid respirations indicate recovery from the effect of the drugs.

Selective antagonists

If the patient is determined to be too deeply anesthetized or immediate recovery is desired, selective antagonist or general stimulant drugs may be administered. Yohimbine (0.1 mg/kg, IV) is the reversal drug of choice for llamas and alpacas. This drug is safe and effective. However, caution must be observed in reversing camelids under the effects of several drugs, such as xylazine + ketamine, because the other drugs will not be antagonized. Tolazoline is a selective antagonist for xylazine, but has been associated with adverse reactions and deaths (possibly caused by profound hypotension). This drug is not recommended for use in camelids. Atapamezole (0.1 mg/kg, IV) has been used in camelids, but I have witnessed several failures to reverse metdetomidine sedation (1 alpaca actually became more sedated!).

REFERENCES

1. Baird AN, Pugh DG, Wenzel JG, et al. Comparison of two techniques for castration of llamas. JAVMA 1996;208(2):261-2.
2. Barrington GM, Meyer TF, Parish SM. Standing castration of the llama using butorphanol tartrate and local anesthesia. Equine Pract. 1993;15:35-39.
3. Fowler, ME 2000. Medical management of south american camelids. In: Selected Topics on Camelids, Gahlot TK (Ed.). The Camelid Publishers, Bikaner, India.
4. McGee Bennett, M. 2001. Medical herd management. In: The Camelid Companion. Raccoon Press.
5. Padula AM. Clinical evaluation of caudal epidural anaesthesia for neutering of alpacas. Vet Rec 2005;156:616-17.
6. Read MR, Duke T, Toews AR. Suspected tolazoline toxicosis in a llama. JAVMA 2000; 216(2):227-29.
7. Uhrig SR, Papich MG, Kukanich B, et al. Pharmacokinetics and pharmacodynamics of morphine in llamas. AJVR 2007;68(1):25-34.

MULTIDISCIPLINARYAPPROACH TO MITIGATION OF ROCKY MOUNTAIN SPOTTED FEVER, TREATMENT

Suzanne R. Todd, DVM, and the RMSF Treatment Mitigation Team
Atlanta, GA

INTRODUCTION

Successful treatment of Rocky Mountain spotted fever (RMSF), in both animal and human patients, rests on early recognition of risk factors and symptoms and correct choice of antibiotic therapy. The CDC works closely with human health care providers practicing in areas at high risk for RMSF, particularly with the Indian Health Service (IHS) in Arizona. The American Indian population of Arizona has experienced an epidemic of RMSF in the last decade (see the proceedings for "Multidisciplinary Approach to Mitigation of Rocky Mountain Spotted Fever, Prevention"). This presentation will briefly cover a clinical algorithm currently in place at several IHS facilities in Arizona. In addition, the CDC recently conducted a study examining the effects of doxycycline treatment on the color of permanent teeth when used for treating RMSF in children under the age of 8 years. Initial results of the study will be presented.

EARLY RECOGNITION: HUMAN CLINICAL ALGORITHM EXAMPLE

A clinical algorithm designed for use in areas highly endemic for RMSF has been in use since 2003 in some IHS hospitals in Arizona. This algorithm is designed to identify all possible suspect cases of RMSF and prompt the clinician to initiate immediate treatment with doxycycline, collection of blood for RMSF IFA titers, and follow-up with community public health staff.

The algorithm is integrated into the IHS electronic health record (EHR) system and is called into use as soon as a fever is reported. Entry of this data into the EHR at the start of the clinic visit initiates a series of prompts which guide the health care provider through recognition of further RMSF risk factors, collection of blood for laboratory testing and doxycycline treatment, as well as patient education and public health follow-up. Those patients fitting some, but not all, of the algorithm still receive education on RMSF warning signs and are added to the list for follow-up.

The algorithm also serves as a syndromic surveillance system aimed at early identification of all possible suspect RMSF cases. Use of the algorithm does lead to doxycycline treatment for people who do not have RMSF. However, in the remote environment in which these hospitals operate and the high risk of RMSF infection, the algorithm is a life-saving tool.

ANTIBIOTIC THERAPY: EVIDENCE THAT DOXYCYCLINE DOES NOT CAUSE TOOTH STAINING

Doxycycline is the treatment of choice for human rickettsial disease, including RMSF, regardless of the age of the patient (1, 2). Other antibiotics, including chloramphenicol and broad-spectrum antimicrobials such as fluoroquinolones, are less effective (3).

Tetracyclines bind to calcium causing intrinsic staining of permanent teeth if administered during the period when these teeth are calcifying (4). Doxycycline, like all tetracycline class antibiotics, carry a warning label stating that the drug should not be used in children under the age of 8 years due to concerns about dental staining (5). Doxycycline is a newer antibiotic in this class and recent research suggests doxycycline does not cause dental staining when used at the dose and duration recommended for rickettsial diseases. Doxycycline is thought to bind less readily to calcium and multiple studies have found no evidence that it causes visible staining when administered to children under the age of 8 years (6, 7). However, these studies examined small numbers of children.

Since 2003, 128 cases of RMSF including eight fatalities have been reported on one American Indian reservation (Reservation Y) in eastern Arizona (8). Due to intense education efforts and use of the earlier mentioned clinical algorithm at this reservation's IHS facility, healthcare providers there have routinely prescribed doxycycline to all possible suspect RMSF cases regardless of age. The result is a large population of children who have appropriately received one or more courses of doxycycline prior to the age of 8 years. The CDC undertook a study in the summer of 2013 to examine whether dental staining associated with pediatric doxycycline administration was occurring, as it should now be detectable in those children aged 8 to 16 years in this community.

Study participants were American Indian children between the ages of 8 and 16 years residing and attending schools on Reservation Y. Children who were 16 years of age at the time of the study should have had at least one crown still calcifying at the time routine use of doxycycline began on the reservation in 2003. At the other end of the age range, children 8 years of age should have had at least one anterior maxillary permanent tooth erupted at the time of the study.

In order to ascertain behaviors which might darken teeth, study participants were given a brief questionnaire pertaining to consumption of dark-colored beverages, brushing and tobacco-use habits. A dentist performed a visual examination of the anterior maxillary permanent teeth, looking for signs of tetracycline staining. A handheld spectrophotometer was used to measure the brightness shade of the same teeth. Medical charts were abstracted to determine prescription history of doxycycline and other tetracycline family antibiotics.

The results of this study were encouraging and will be published in the near future.

REFERENCES

[1]Pickering LK, ed. Red Book: 2012 Report of the Committee on Infectious Diseases. 29th ed. American Academy of Pediatrics. 2012: 620. [2]Chapman AS, et al. MMWR Recomm Rep. 2006 Mar 31; 55(RR-4):1. [3]Holman RC, et al. J Infect Dis. 2001 Dec 1; 184(11):1437. [4]Logan WHG, et al. J Am Dent Assoc 1933; 20(3):379. [5]Federal Register 66:213. 2001 Nov 02; 55679. [6]Lochary ME, et al. Pediatr Infect Dis J. 1998 May; 17(5):429. [7]Volovitz B, et al. Clin Pediatr (Phila). 2007 Mar; 46(2):121. [8]Arizona Department of Health Services unpublished data.

MULTIDISCIPLINARYAPPROACH TO MITIGATION OF ROCKY MOUNTAIN SPOTTED FEVER, PREVENTION

Suzanne R. Todd, DVM, and the RMSF Prevention Working Group
Atlanta, GA

INTRODUCTION

Rocky Mountain spotted fever (RMSF) is an established public health threat in eastern Arizona, impacting American Indian reservations (1, 2). RMSF requires an intensive "One Health" prevention strategy to stop future human RMSF cases and deaths. The Centers for Disease Control and Prevention (CDC) has been working closely with Tribe X since 2004. In early 2012, a demonstration project was proposed in collaboration with tribe, state, and federal partners. The goal of this project was to conduct a series of properly timed prevention activities in a single neighborhood in a way that controlled ticks, provided a free and easy means of spaying and neutering dogs in the community, and ultimately would prevent human RMSF cases from occurring. The 2012 project was a pilot program to see if such activities would be helpful on a larger, reservation-wide scale. At the conclusion of the project, a survey was conducted to assess knowledge, attitudes and practices of the community regarding RMSF. Data showed that within the project neighborhood there was a substantial reduction in tick vectors on dogs, and residents showed increased knowledge about tick prevention practices on dogs and in the environment, as well as a strong desire for an improved comprehensive animal control program on the reservation. The findings from the prevention project indicated that scaling up a similar program reservation-wide might provide significant benefit against tickborne diseases like RMSF.

BACKGROUND

RMSF is a severe and potentially fatal disease caused by the tickborne bacterium *Rickettsia rickettsii*. RMSF has recently emerged as a significant public health threat on some tribal lands in Arizona, including Reservation X in eastern Arizona. In this region, RMSF is transmitted by *Rhipicephalus sanguineus*, the brown dog tick, and the disease has reached epidemic proportions (1). The issue is amplified by lack of veterinary support and animal control programs, as well as dog overpopulation contributing to high tick numbers. During 2011, 77 RMSF cases were reported from eastern Arizona, the highest yearly number to date. In total, from 2003 through March 2011, at least 219 human RMSF cases have been reported from four reservations, including 16 deaths (3).

FIELD ACTIVITIES

Phase I (Summer 2012)

Previous prevention efforts undertaken by the tribe, including once a year treatment of houses and distribution of tick preventive products for dogs upon request or referral, had not been successful in controlling the epidemic, largely because the tribe was not able to apply the products at systematic intervals to achieve sustained acaracidal activity. During 2012, CDC partnered with the tribe and other state and federal agencies to implement a demonstration prevention project in one housing district. The following activities occurred as part of the phase I of the project:

1. Complete and accurate count/identification of all dogs in the community
2. Registration and consent of participating households, which agreed to:
    a. Ensure that tick products or collars placed on dogs remain in place for the duration of the project
    b. Be present for at least one of the yard treatment visits and agree to learn the procedure and continue to apply it
    c. Report results (tick counts, new puppy births) through written surveys, door-to-door surveys, or other methods for up to 3 years following the pilot project
3. Removal of unwanted/clearly unowned dogs from the community by Tribe X Animal Control officers
4. Every participating household received:
    a. Pet care items (collar, identification tag, and when requested, a tie-out stake and cable)
    b. Free spay/neuter surgery was offered for all identified owned dogs
    c. Effective tick treatment (long-acting tick collars) was provided to every dog in the neighborhood
    d. Once a month treatment of yards of all participating households
    e. Education for owner-treatment of dogs and yards

The demonstration project was designed to integrate appropriately timed environmental and veterinary tick treatments with improved access to pet care. The goal was to demonstrate a reduction in human RMSF cases through improved pet health. Within the project area, a high percentage of occupied households participated in the project. At the conclusion of phase I, a program evaluation was conducted via household survey in order to assess effectiveness of the program in decreasing tick population. The evaluation also assessed tick counts on dogs and in the environment, and evaluated behavioral changes influenced by participation in the project. Data showed substantially reduced tick vectors both on dogs and in the environment in the neighborhoods served by the prevention project, compared with homes outside the project area. The evaluation revealed strong support for development of animal control programs and the sheltering of unwanted dogs. Unfortunately, spay/neuter surgery as a method of dog population control was underutilized. At the end of phase I of the project, the CDC recommended that the project be expanded to all housing districts on the reservation.

Phase II (Summer 2013)

In 2013, a modified tick control intervention modeled after phase I of the project was implemented reservation-wide. Tribal members sprayed yards monthly in the non-project area. Free spay/neuter clinics were offered to all tribal members' dogs each month starting in January 2013. Long-acting tick collars were placed on all full-grown dogs. Within the original project area the intervention shifted to focus only on a new tick collar for dogs. Environmental pesticides were only applied where tick infestations were observed. If proven effective, this method would allow for a more cost-efficient approach to tick control over a long period of time, and would also reduce the amount of pesticides required in the environment, while still providing adequate protection from RMSF. Data showed lower tick counts in household assessments of non-project areas compared with the previous year, with sustained improvements in the project area. Additional challenges to be addressed by the tribe in finding a long-term solution to RMSF prevention are the poorly controlled dog population and securing funding for a high-functioning animal control program.

FUTURE PLANS

CDC plans to offer technical assistance to Tribe X and other Arizona tribes interested in implementing similar RMSF prevention programs. Additionally, the CDC in cooperation with Indian Health Service on Reservation X will closely monitor the incidence of febrile illness and RMSF treatment and diagnostic testing in order to determine if the incidence of possible suspected cases of RMSF have decreased following the RMSF prevention project.

REFERENCES

[1]Demma LJ, et al. N Engl J Med. 2005 Aug 11; 353(6):587-94. [2]Holman RC, et al. Am J Trop Med Hyg. 2009 Apr; 80(4):601-5. [3]Arizona Department of Health Services (http://www.azdhs.gov/phs/oids/vector/rmsf/stats.htm); accessed on 03 February 2014.

DEVELOPMENT OF ANIMAL MODELS FOR THE ENHANCEMENT OF ANIMAL AND HUMAN HEALTH (AVATARS FOR HUMANITY: CURRENT CONCEPTS AND LESSONS LEARNED)

Erica E. Carroll, DVM, PhD, DACVP and Donald G. Heppner, MD, FACP, FASTMH
Washington, DC

INTRODUCTION

An 'animal model' has been defined as "a living organism with an inherited, naturally acquired, or induced pathological process that in one or more respects closely resembles the same phenomenon occurring in man."[1] While many would agree there is no ideal animal model, scientists have identified at least five classes of models[2] useful to answer scientific questions. Considerations in choosing an animal model such as relevance to the human condition, accuracy of manifestation, predictability of outcome and species availability[3] must be coupled with thorough understanding of the entity in humans to accurately compare manifestations. Achievement of an appropriate animal model, however, has recently been called into question by more than one group of investigators, whose concerns merit contemplation, even while continuing to design and work with animal models. Species selection depends on the exact scientific questions to be asked. The species that is phylogenetically closest to humans is not necessarily the best model to answer those questions. This 'high-fidelity fallacy'[2] may lead to errors in species selection. Another error is lack of knowledge of differences in metabolic pathways between humans and potential animal models prior to species selection. Depending on the questions to be asked of the animal model, biomethods, microbiota, genetic characterization and even behavior may be important to know about potential model species as these factors may affect results of intervention.[2] Basic concepts of design of an appropriate model, including scientific, ethical and regulatory considerations are readily available on-line, including the FDA's Animal Model Qualification process and the Animal Rule. Demonstrating this process, we will offer an example of a current U.S. Government (USG) animal model qualification effort, using an induced model of a biological threat. Despite widespread use of animals in biomedical research, the predictive value of animal models has begun to be questioned, increasingly seriously, ever since a 1994 statement declaring 'virtually every medical achievement of the last century has depended directly or indirectly on research with animals.'[4] In the 21st century, innovations in safety, immunogenicity and toxicity testing of new or repurposed biological or chemical entities include *in silico* and *in vitro* testing systems which may supplement and, in the not-too-distant future, supplant certain aspects of animal model testing.

ANIMAL MODEL DESIGN

Before embarking on species selection for an animal model, it may be helpful to consider the overall purpose for which the model will be developed. Most animal models can be grouped into five general classes based on the type of investigation desired: exploratory, explanatory, predictive, spontaneous disease models and induced disease models. Induced models include genetically modified organisms such as knock-out mice.[2] An ideal animal model manifests a condition (e.g., degenerative, metabolic, oncologic or infectious) in the same way and degree as humans with that disease or condition. Identification of a suitable animal model requires, a priori, in-depth knowledge of the human condition. The world-wide body of scientific literature should be combed to identify clinical signs (both measured, and reported 'symptoms' in humans), metabolic disturbances, clinical and anatomic pathology, age of onset, duration of the illness or condition and 'behavior' of the condition (waxing/waning, chronic, acute), affected population (gender, elderly, infants), secondary effects (e.g., immunosupression, bacterial infections), geographic regions in which the condition is found (due to climate or the presence of arthropod or mammalian vectors), genetic predisposing factors (e.g., chromosome 15 abnormality, Factor V mutation, X-linked inheritance or Human Leukocyte Antigen (HLA) association). The better understood the human condition and the better defined the scientific questions to be asked [and, presumably, answered], the easier it may be to identify appropriate non-human species as a model. Although it may seem logical to find the species closest to humans in which to study a human condition or pathology, in order to answer many scientific questions, a 'simpler' model such as a species 'lower' on the phylogenetic scale may offer advantages and permit control of some variables. For example, the study of squid giant axons helped elucidate neurologic action potentials in mammals. Ethics and economics also promote use of the smallest animal possible (e.g., mice) and the one 'lowest'on the phylogenetic scale; mice have been used extensively as models for a variety of human diseases. However, mice do not always perform as humans do, especially in response to various inflammatory challenges (sepsis, trauma, burns, endotoxemia).[5] Besides selecting a species, a strain must often be chosen. Outbred mice (e.g., Swiss Webster) exhibit genetic diversity and are referred to as 'stocks' whereas inbred mice arise from more than 20 generations of brother x sister matings, resulting in mice that are almost genetically identical.[2] Inbred strains are phenotypically and genotypically much more uniform than outbred stocks, which controls for many variables and may be desirable, depending on the scientific questions asked. North Carolina State University has an on-line course which lists additional considerations below (Table 1).[2]

Besides these considerations, regulatory guidance must be followed, such as the requirement to obtain institute animal care and use committee (IACUC) approval for any research protocol involving animals. All Public Health Service-funded research, which consists of grants by the National Institutes of Health, Centers for Disease Control and Prevention, Food and Drug Administration, and others are regulated by the Office of Laboratory Animal Welfare (OLAW), which publishes the "Guide for the Care and Use of Laboratory Animals, now in its eighth edition, a veritable 'bible' of laboratory animal care. Ironically, the Animal Welfare Act, although pertaining to several common species, does NOT cover animals slaughtered for food, birds, or rats and mice used in research.

Table 1. Practical considerations for Species Selection[2]

Adequate discrimination (data extrapolatable to target species)
Adequate fidelity (necessary anatomic structures, biochemical pathways, etc.)
Good literature base; historical usage (i.e., accepted animal model)
Readily available to other researchers
Accommodation in existing animal facilities (caging, environmental controls, exercise and environmental enrichments)
Experience/training of animal care staff
Experience/training of research staff
Tractable; trainable
Low cost (purchase; maintenance)
Good reproductive performance
Minimal endogenous disease (or ability to control
Ethical considerations
Public relations implications

(Fish, RE, Unit Five – Animal models and Biomethodology, in the course, 'Humane Care and Use of Animals in Research,' from NCSU website (http://www4.ncsu.edu/~refish/HHMI/unit_5.html)

Diseased animal models

Although intuitively, one would imagine that a 'diseased animal model' would be ideal in which to test safety and toxicity of new drugs, there is strenuous objection by some, based on general lack of historical controls; heterogeneity in disease expression; limited life span in some diseased animals, thereby preventing treatment of chronic conditions; and confounding effects of the disease (e.g., altered immune function, opportunistic (secondary) infections.[6] Because of these concerns, Morgan et al (2013) state "animal models of human disease should NOT be utilized to supplant [presumably, pre-clinical safety and toxicity] testing in conventional animal models."[6] Their concerns specifically refer to pre-clinical toxicity testing of new compounds, but their concerns are factors that need to be addressed in any animal model.

ANIMAL MODEL QUALIFICATION

Animal models of infectious disease are often developed to test novel medical countermeasures (e.g., such as vaccines, post-exposure prophylaxis or therapeutics), with the ultimate goal of fulfilling regulatory requirements for FDA approval and commercial licensure. In the U.S., pre-clinical animal tests are done to fully characterize and assess the novel substance with regard to absorption, digestion, metabolism, excretion, pharmacokinetics and pharmacodynamics, minimal inhibitory concentration, maximum tolerable dose, as appropriate, prior to beginning Phase I (i.e., safety) clinical trials in people. Those that successfully complete phase I may proceed to Phase II (larger group efficacy and further safety testing) and Phase III (large study to confirm effectiveness, monitor side effects, compare to standard therapy). The FDA has issued several "Guidance for Industry" documents that include animal use in toxicity studies for 'preventive and therapeutic vaccines' for infectious disease indications; preclinical assessment of investigational 'cellular and gene therapy products' and others, which codify the rigorous requirements to achieve FDA approval of a new biologic or therapeutic entity.

The Animal Rule

For those infectious agents that cannot ethically be administered to volunteers in Phase II efficacy studies due to lack of treatment, and for those diseases for which there are too few naturally acquired cases with which to study product efficacy, a 'qualified' animal model is required before the FDA will approve a novel vaccine or therapeutic against that disease. As a Drug Development Tool, the FDA defines an animal model as "a specific combination of an animal species, challenge agent, and route of exposure that produces a disease process or pathological condition that, in multiple important aspects, corresponds to the human disease or condition of interest. The FDA's Animal Model Qualification (AMQ) program applies specifically to animal models intended for use in "adequate and well-controlled efficacy studies [i.e., under Good Laboratory Practices whenever possible] to be able to generate substantial evidence of effectiveness for drugs developed under the Animal Rule."[7] The concept is if a model is 'qualified,' which can be done through multi-agency collaborative efforts, it can then be used to provide efficacy evidence for multiple drugs without developing a model 'from scratch' for each drug. Other types of animal models, such as those used for proof-of-concept testing or for safety testing are not eligible for qualification. "(Draft) Guidance for Industry:Animal Models –Essential Elements to Address Efficacy Under the Animal Rule" (January 2009) describes the procedures to be followed for those seeking animal model qualification under the animal rule. The four phases of qualification by the FDA under the Animal Rule are Initiation, Consultation and Advice; Review; and Finalization.[8]

Unfortunately, in the first nine years since the issue of the Guidance related to the Animal Rule, only two drugs were approved: pyridostigmine bromide (a reversible inhibitor of acetylcholinesterase) for use as a pretreatment for Soman poisoning (based on studies in monkeys and guinea pigs) in 2003, and Cyanokit,® a lyophilized formulation of hydroxocobalamin for use as an antidote in treating patients with known or suspected cyanide poisoning (in 2006, based on studies in dogs).

LESSONS LEARNED AND CONCERNS

Not all species manifest infection by a given agent identically, obviously; therefore, identifying an animal model of infectious disease can present unique challenges. Investigators that choose to use mice, for ethical, economic or other reasons, sometimes must alter either the pathogen (e.g., via multiple *in vitro* passages) or the mouse strain (e.g., genetic alteration to immunosuppress or eliminate some physiologic function of the mouse) in order to elicit clinical manifestations of infection that resemble those in humans. This raises concerns that the animal model may not physiologically represent the target human population. A huge concern is predictability with regard to the preclinical testing of drugs and vaccines. Bracken (2008) stated that many cancer chemoprevention experiments do not translate well into human trials because they are poorly designed, conducted and analyzed.[9] Pound et al (2004) identified numerous methodological problems in current animal-based research such as variations in drug dosing schedules and regimens of uncertain relevance to the human condition; variability in the way animals are selected for study, methods of randomization, choice of comparison therapy (non, placebo, vehicle), and reporting of loss to follow up; small experimental groups with inadequate power, simplistic statistical analysis; variable length of follow-up that may not adequately determine disease outcome, and several others.[10] The intervention must additionally show efficacy superior to supportive care alone.

Not only are animal models often flawed due to human error in design or execution, but a growing number of scientists are reaching the conclusion that not only is there no 'ideal' animal model, where predictive ability is concerned, there may be few, if any, 'acceptable' animal models. In 1994 the US Public Health Service published a statement in *The Physiologist* titled 'The importance of Animals in Biomedical and Behavioral Research' in which they asserted, "Virtually every medical achievement of the last century has depended directly or indirectly on research with animals." Although many 'eminent' scientists have repeated similar language for years, Matthews (2008) asserts there were no citations in the literature at that time nor have there been comprehensive attempts since then to try to justify that statement.[4] On the contrary, Greek and Menache (2013), in their article on systematic reviews of animal models, observe that inadequate animal models may be an important reason why the vast majority of drug candidates do NOT achieve FDA licensure.[11] They point out that approximately 100 vaccines have been shown to be effective against an HIV-like virus in animal models but none have prevented HIV in humans; almost one-thousand drugs were described as effective for neuroprotection in animal models but none have been effective for humans. Twenty-two drugs were demonstrated to be neuroprotective in dogs with spinal cord injury; none were effective in humans. They claim that animal models involving rodents do NOT exhibit positive predictive value for new drug entities in humans and cite numerous statistics stating that the proportion of drugs that seemed efficacious in animals that achieved FDA licensure is abysmal. Additionally, pharmaceutical companies can now screen thousands of molecules to find potentially efficacious candidates, but developing non-toxic versions becomes the limiting factor, a problem which sometimes is not identified until the phase I safety trials. Murine models of inflammatory disease and septic shock are notoriously poor in predictive ability, especially studies examining genomic response. Seok, et al (2013) report, of nearly 150 clinical trials of candidate compounds intended to block the inflammatory response in critically ill patients, every single one failed.[5] They believe it is because there have been NO studies to systematically evaluate, on a molecular basis, how well the murine clinical models mimic human inflammatory diseases in patients. There is a 'striking paucity of quantitative comparative data for animal models, per Matthews (2008).[4] Matthews suggests that reasons include the commercial confidentiality of many study results, lack of published results for compounds failing to progress to clinical trials, the difficulty of establishing clear-cut end-points to permit categorization as 'true positives' or 'true negatives' (referring to sensitivity, specificity and likelihood ratios of characteristics of the models). Greek and Menache add, simplified, that the lack of predictive ability of animal models arises from the fact that mammalian species are "evolved complex systems" and that simple statistical comparisons cannot account for the myriad variables involved.[11]

One initiative, the Collaborative Cross,[12] has generated a panel of mice to reflect the genetic diversity seen in human populations, which may improve the predictive ability of some mouse models.[13] Another idea is to "raise the bar by requiring molecular detail in the animal model studies indicating whether the model mimics or fails to mimic the molecular behavior of key genes, key pathways, or the genome-wide level believed important for the relevant human disease.[5] One FDA scientist, at a 2012 animal model development workshop believed that 'performance of a vaccine or therapeutic in an animal model does not always predict success in humans, [but] failure in an animal model often <u>does</u> accurately predict failure in humans.'

TOWARDS MEDICAL COUNTERMEAUSURES AGAINST BURKHOLDERIA SPECIES

21ST CENTURY PRE-CLINICAL TESTING AND THE POTENTIAL DECLINE IN ANIMAL MODEL USE

Stemming from ethics, economics and a need for efficiency in achieving FDA approval of compounds, in the year 2000 the Interagency Coordinating Committee on the Validation of Alternative Methods (ICCVAM) was established of representatives from 15 federal agencies that generate or use toxicologic data. Their charter is:

"to establish, wherever feasible, guidelines, recommendations, and regulations that promote the regulatory acceptance of new or revised scientifically valid toxicological tests that protect human and animal health and the environment while reducing, refining, or replacing animal tests and ensuring human safety and product effectiveness."[14]

The ICCVAM, therefore, established the current nation-wide tenets of animal use: Reduce, Refine, Replace with *ex vivo, in vitro, in silico* safety, toxicity and other pre-clinical tests wherever possible. Additionally, The National Toxicology Program's Vision for the 21st Century is "to move toxicology from a predominantly observational science at the level of disease-specific [animal] models to a predominantly predictive science focused upon a broad inclusion of target-specific, mechanism-based, biological observations."[15] Since then, multiple synthetic human models by *in vitro* reconstitution of disease-related cell types or tissues have been developed which may improve current disease models, gain FDA acceptance, reduce animal use, and accelerate the licensure of new compounds to improve human and animal health. Examples include VaxDesign's MIMIC® (Modular Immune In vitro Construct), a surrogate human immune system that enables researchers to test the immune response to a vaccine, biologic, drug, environmental irritant directly in a 'human system,[16] and Wake Forest Baptist Medical Center's multi-institution project to develop 'body on a chip.'[17] Liver bioengineering and regeneration has potential for transplantation and for toxicity testing,[18] two functions with potential for reducing use of animal models in research.

REFERENCES

[1]Wessler S Introduction: What is a model?, pp.xi-xvi. IN: Animal Models of Thrombosis and Hemorrhagic Diseases. NIH, Bethesda, 1976. [2]Fish RE, Unit Five – Animal models and Biomethodology, on-line course Humane Care and Use of Animals in Research, from NCSU website.(http://www4.ncsu.edu/~refish/HHMI/unit_5.html). [3]Lake Superior State University Institute Care and Use Committee Care and Use Manual (h ttp://www.lssu.edu/faculty/jroese/AnimalCare/selecting%20animals.htm). [4]Matthews R **J R Soc Med** 2008;101:96. [5]Seok J **PNAS** 2013;110(9):3507. [6]Morgan S, et al. **Toxic Pathol** 2013;41:508. [7]Guidance for Industry and FDA Staff: Qualification Process for Drug Development Tools, 2014, U.S. Department of Health and Human Services, Food and Drug Administration, Center for Drug Evaluation and Research (CDER) (http://www.fda.gov/cder/guidance/index.htm). [8](Draft) Guidance for Industry: Animal Models –Essential Elements to Address Efficacy Under the Animal Rule'(January 2009) U.S. Department of health and Human Services, Food and Drug Administration, Center for Drug Evaluation and Research (CDER) (http://www.fda.gov/cder/guidance/index.htm). [9]Bracken M **JR Soc Med** 2008;101:120. [10]Pound, et al. **BMJ**. 2004;328:514. [11] Greek R., et al **Int J Med Sci** 2013;10:206. [12] Churchill G, et al. **Nat Gen** 2004;36:1133. [13]**Nat Med** 2013;19:4:379. [14] Public Law 106-545-DEC.19, 2000 "ICCVAM Authorization Act of 2000." [15]NTP Vision (2004) (http://ntp-server.niehs.nih.gov/). [16]VaxDesign MIMIC® Technology (http;//www.vaxdesign.com/mimic-technology). [17]Richardson K, Wake Forest Media Relations Office, 2013 (http://www.eurekalert.org/pub_releases/2013-09/wfb. [18]Booth C, et al. **World J Gastroenterol** 2012;18(47):6926.

APPLYING PRINCIPLES OF HEALTHCARE QUALITY IMPROVEMENT IN VETERINARY PRACTICe

R. Goodman, DVM, DACVIM (SAIM)
San Antonio, TX

BACKGROUND

The national debate on healthcare reform, as well as generating much emotional rhetoric, has also provided an opportunity for intellectual discussion of quality and value within the human healthcare system. In 2009, Dr. Atul Gawande also brought the concepts of quality improvement through reduction of errors to the New York Times Bestseller list with the Checklist Manifesto.[1] More recently, collaboration[2] between the *New England Journal of Medicine* and the *Harvard Business Review* provided a forum for key leaders in healthcare reform to share insights and ideas on increasing value in healthcare by improving patient outcomes and reducing costs. Quality improvement publications in veterinary medicine have been less visible, of much smaller scale, and focused on either patient outcomes or, somewhat counter to the prevailing trend in medicine, increasing patient costs. A recent episode of the news program *20/20* even questioned the quality/value of veterinary recommendations. This presentation will explore some of the key concepts and strategies in healthcare improvement in relation to veterinary medicine and describe some of the lessons learned in attempting to implement these concepts within a large multi-center practice.

KEY CONCEPTS

The Institute of Healthcare Improvement champions the Triple Aim ® - "*a framework to optimize health system performance by simultaneously focusing on three care dimensions: improving the patient experience of care, improving the health of populations, and reducing the per capita cost of health care*"[3,4]

Improving Patient Experience

The patient experience involves both the subjective feelings of the patient/client as well as the quality of care received. Subjective measures of experience may be measured actively through surveys or passively through social media. Areas to measure improvement in individual quality of care may be identified by the Institute of Medicine's Quality Chasm aims: Safe, Effective, Timely, Efficient, Equitable, and Patient-Centered.[5]

Improving Population Health

Human medicine uses various population measures of functional status, risk status, mortality, and disease burden to assess population health. The available studies of veterinary patient population health are limited to single disease[6], single corporation[7], or limited area.[8]

Reducing Per Capita Cost

While the human healthcare debate focuses on reducing cost of care through capitation, accountable care organizations, bundled payments, and insurance reform, the veterinary discussion paradoxically centers on increasing fees. [9]

IMPLEMENTATION STEPS

An October 2013 article in the *Harvard Business Review* outlined a six-part strategy to improve healthcare. [10] These six steps will frame our discussion of veterinary healthcare improvement concepts.

Organize into integrated practice units

The consolidation of human hospitals into hospital systems and the increasing specialization in human medicine have resulted in increased complexity at the patient level. Several solutions to provide patient centered care within the human medical system have been advanced. As the veterinary medical system continues the slow advance to consolidation and specialization[11], we can learn from human medicine and perhaps avoid the course corrections required to bring focus back to the patient.

Measure outcomes and costs for every patient

In order to truly assess value within healthcare delivery, the meaningful outcomes within a cycle of care must be measured in order to improve care across the system as well as provide transparent information to consumers to aid in selection of providers. Accurate recording of costs on a per patient basis not only allows internal cost management and best practice identification, but also assists in communicating value to clients.

Move to bundled payments for care cycles

Bundled payments for the entire cycle of care for predictable disease conditions or health maintenance in human medicine seeks to align provider and insurer incentives to deliver more patient value[12] and may represent a more ethical manner of provider compensation than fee for service based compensation.[13] Development of bundled payment plans within veterinary medicine is far more than a "Wellness Plan."

Integrate care across separate facilities

Coordination of care across multi-facility, multi-specialty health systems in human medicine attempts to balance the economy and improved care provided by specialized care teams with a patient-centered approach. Similar opportunities in veterinary medicine exist in the relationship between referring veterinarians, specialists, and other care providers.

Expand excellent services across geography

In human medicine, recognized local and regional expertise is slowly being scaled to new locations in the replication of best practices in a variety of business relationships. Similar opportunities may exist in veterinary medicine as corporate practices mature.[11]

Build an enabling information technology platform

Appropriately designed information platforms enable providers to effectively integrate the five elements above. These systems are centered on the patient and not the practice, utilize common data definitions, encompass all types of patient data, are accessible by all care providers, include templates and expert systems, and allow data to be extracted for both outcome and cost measurement. Emerging systems and standards in information management will impact both human and veterinary medicine.

LESSONS LEARNED

Since 2008, the United States Army has been pursuing a Global Practice Concept to integrate the 140+ clinics on military installations worldwide. The success, challenges, and opportunities experienced within the personnel, procurement, medical standardization, information technology, and leadership domains will be reviewed with a focus on applicability across a wide range of practice situations.

DISCUSSION

The discussion of quality improvement within the human healthcare system provides a number of opportunities and warnings to the veterinary profession. However, if past history is any indication, the solutions to the current challenges facing veterinary and human medicine will not be easily solved or comfortable to discuss candidly. Ultimately, health care providers of all types must answer the question addressed in the 1959 George W. Gay Lecture at Harvard Medical School, "Profession or Business?"[13]

REFERENCES

[1]Gawande, The Checklist Manifesto, 2009. [2] http://hbr.org//special-collections/insight/leading-health-care-innovation. [3] http://www.ihi.org/about/pages/innovationscontributions.aspx. [4] Triple Aim Concept Design, 2009 (www.ihi.org/engage/initiatives/TripleAim/Documents/ConceptDesign.pdf). [5] Richardson, National Academy Press 2001(http://www.nap.edu/books/0309072808/html/). [6] Brown, H et al. **Parasites & Vectors** 2012, 5:245. [7]Banfield 2013 State of Pet Health Report (http://www.stateofpethealth.com/) [8] O'Neill, DG, et al. Vet J 2012; 198:638-643. [9] Veterinary Economics Benchmarks 2013 [10] Porter, M and Lee, T **HBR** October 2013. [11]Howard, B **Vet Econ** 2007; 11:48 [12] Witkowski, M et al, (http://blogs.hbr.org/2013/10/how-to-design-a-bundled-payment-around-value/) [13]Means, JH **NEJM** 1959; 261:791-797.

THORACIC ULTRASONOGRAPHY OF THE HORSE

Michelle H. Barton DVM, PhD, DACVIM (LAIM)
Athens, GA

INTRODUCTION

Ultrasonography is a safe, affordable, and convenient technique for investigating peripheral lung diseases and pleural effusions. The greatest limitation of thoracic ultrasonography is that normal aerated lung precludes detection of deeper lesions. The lungs can be scanned with just about any transducer, but high frequency transducers (> 7.5 MHz) set to a depth of 4 to 8 cm provide the best detail.[1] When examining the cranial mediastinum, if the periphery of the lung is abnormal, or there is pleural effusion, switching to a lower frequency (≤ 5 MHz) curvilinear probe is preferable. In most situations, soaking of the hair with warm water or isopropyl alcohol over the areas to be scanned is sufficient. However in obese or heavy coated horses, clipping of the hair and use of coupling gel may be needed. Smaller footprint probes facilitate scanning in between ribs and can range from sector to straight or linear array transducers. A systematic approach should be used by scanning from the dorsal most aspect of each intercostal space to its ventral most aspect to the diaphragm, moving caudally, intercostal space to intercostal space, in the transverse (perpendicular to the spine) and slightly oblique plane as dictated by the ribs. Slowly moving the probe down each intercostal space allows the lungs to be visualized during both inspiration and expiration and may facilitate identification of subtle lesions that are only seen when the lung moves out from behind a rib or into the plane of imaging. Both sides of the thorax should be scanned from the right 3rd or the left 4th to the 15th to 16th intercostal spaces. Care must be taken to stay within the intercostal space, as the ribs will impede ultrasound penetration and will generate an acoustic shadow (anechoic area distal to the rib surface created by sound absorption).[2] The lungs may also be scanned in the dorsal plane that is parallel to the spine; however impedance of the ribs limits viewing and thus it is more tedious to perform. This presentation will use quick time virtual reality movies to illustrate normal thoracic cavity anatomy, as well as thoracic pathology. Furthermore, both normal and abnormal frozen images and video movie files of thoracic ultrasonography from equine patients will be included.

NORMAL ANATOMY and ULTRASONOGRAPHIC APPEARANCE OF THE LUNGS AND MEDIASTINUM

The lungs

The lungs of the horse are considerably simpler than most other species in that they are not clearly divided into lobes by deep interlobular fissures. However, they are paired as a left and right lung, with a small medially located accessory lobe that is associated with the right lung.[3] Both the left and right lungs are separated in terminology only into the apical or cranial lobe and the diaphragmatic or caudal lobe, as defined by the cardiac notch. On the left, the cardiac notch of the lung is adjacent to the left third to sixth intercostal spaces and on the right, from the right third to fourth intercostal spaces.[3] Thus in the areas of the cardiac notches of the lung, the pericardium is in direct contact with the parietal pleura.

Ultrasonographically, the skin, subcutaneous fat, and intercostal muscles will be seen immediately adjacent to the footprint of the transducer. They transmit sound waves well and do not generate artifacts. The interface between the pleural surface of the lung and the aerated lung deep to it will appear as a straight hyperechoic line. This hyperechoic line is compromised of the parietal and visceral pleurae and can be seen sliding back and forth as the visceral surface of the lungs move when the horse breaths. The characteristic "gliding line" is the hallmark of normal aerated lung. With the expectation that normal lung primarily contains air, which is a great reflector of ultrasound waves, normal aerated peripheral lung casts a relatively "boring" and uniform acoustic gas shadow with the additional characteristic artifact of equidistant reverberation echoes that appear as parallel hyperechoic lines that are also called "A lines."[1,2] "A lines" occur when sound waves bounce back and forth (echo) between the transducer and the lung surface. Each subsequent reverberation of sound waves takes longer to return to the probe and thus appear as a hyperechoic horizontal line deeper and deeper to the initial intercostal muscle to lung interface (Figure 1).[2] The deeper "A lines" are more distorted than the superficial ones. When over a rib, the density of the bone reflects the sound waves and scatters and blocks further transmission. A hyperechoic curved (rib surface) line will appear in the near field within the muscle, but deep to this is a dark acoustic shadow that continues to the edge of the viewing screen.

The second artifact that is created by normal aerated lung is "B lines,"[2] often referred to as "comet tails."[1,2] These are discrete vertical lines that original from the pleural surface and move with the lung. Minute areas of fluid accumulation create comet tail artifacts in normal peripheral lung (Figure 2). The ultrasound waves are transmitted through these small areas deep to the pleural surface. When the transmitted sounds reach the next interface of air in the lung, it is highly reflective and creates a discrete acoustic shadow that resembles the tail of a comet.[2] As an artifact in normal lung, comet tails are most commonly seen in the dependent lung and the acoustic shadow usually fades long before reaching the edge of the screen. In people, less than three comet tail artifacts are expected at each intercostal space.[2]

The mediastinum

Most of the mediastinum is not visible ultrasonographically in normal horses; however, the ventral most portion of the cranial mediastinum can be seen at the right 3rd intercostal space, especially if the probe is angled rostrally. Pericardial fat is commonly seen in this area and appears uniform in echogenicity, being slightly more echogenic than lung but without reverberation artifacts. When fat is present, the centrally located cranial mediastinal septum might be visible as a thin echogenic membrane. Pleural fluid is not typically seen, though small amounts of anechoic fluid can rarely be detected in the most rostral dependent areas of the thoracic cavity of normal horses.

The caudal mediastinum is not visible ultrasonographically unless there is hypoechoic pleural effusion. The caudal mediastinum reflects off the diaphragm (the diaphragmatic line of pleural reflection) asymmetrically. The accessory lung lobe displaces the caudal mediastinum to the left of the midline of the diaphragm and it extends ventral from the sternum up dorsally to the vertebral column to contain the aorta and esophagus and left phrenic nerve.[3] On the right, the caudal vena cava and phrenic nerve are not contained in the caudal mediastinum per se, but in the free edge of a special fold of the diaphragmatic parietal pleura (plica venae cavae).[3]

PULMONARY DISEASE

Consolidation occurs when fluid or infiltrate (edema, pus, mucus, blood, cancerous cells) accumulate in alveoli and ultrasonographically the affected area appears hypoechoic relative to surrounding aerated lung. Often consolidated lung is referred to as "hepatized" lung, as it appears acoustically more like liver than lung. Areas of consolidation are frequently dependent and often fairly discrete, though small areas may only be identified during expiration. Large areas of consolidation often have deep irregular hyperechoic borders that mark the interface with aerated lung. Larger areas of consolidated lung may have variable-sized scattered echogenic foci that represent focal areas of residual air within the predominantly consolidated lung. Air bronchograms or air within bronchi within areas of consolidation, appear as distinct hyperechoic linear echoes that are easiest to detect when scanning parallel to the long axis of the bronchus. Fluid bronchograms, or fluid within bronchi, appear as hypoechoic linear structures that widen with convergence toward the hilus and are flanked on each side with hyperechoic linear walls. The finding of pinpoint hyperechoic areas generating acoustic shadows within areas of consolidation is consistent with an anaerobic infection.[4] Consolidation is most commonly identified in the dependent lung fields, especially on the right side. If pulmonary necrosis is present, air bronchograms will not be clearly identifiable within the area of consolidation.[1] Pulmonary infarcts should be suspected if color flow Doppler shows abrupt disruption of blood flow in consolidated areas.[1] Common differentials for consolidated lung include pneumonia and pulmonary edema.

Atelectasis occurs when lung parenchyma is compressed by fluid (pleural effusion), air (pneumothorax), or viscera (diaphragmatic hernia). The compressed lung is devoid of air, and thus appears hypoechoic relative to adjacent aerated lung and bronchi and vessels can still be visualized in the areas of atelectasis.

Pulmonary abscesses appear as "space occupying" lesions with anechoic to mixed echogenicity. Abscesses appear similar to areas of consolidation, but lack other evidence of normal lung tissue, such as air or fluid bronchograms or vessels. Occasionally, the purulent material contained within an abscess can be seen as swirling mixed echogenic material as the lungs move. Abscesses may appear encapsulated, have loculations of compartmentalization, and free gas foci. The later is indicative of anaerobic infection and carries a guarded prognosis.[4] Occasionally, in well-encapsulated abscesses, a dorsal gas to ventral pus interface can be detected within the abscess. Only peripherally located abscesses can be detected via ultrasonography.

Comet tails that appear dorsally or extended deep to the edge of the viewing field are likely pathologic and are generated by the presence of minute peripheral areas of nonaerated lung superficial to an aerated lung interface. The sound may be transmitted past the pleural surface by fluid (edema, mucus), cellular infiltrate (leukocytes, neoplastic cells), or fibrosis. Comet tail artifacts are nonspecific to the etiology as they can be seen with many disease states.[1,2] Be mindful that mild comet tailing can be present in the dependent lung fields of normal patients.

Other space occupying lesions of the lung

Other discrete hypoechoic "space-occupying" lesions of the lung include pulmonary neoplasia, granulomas, or pulmonary fibrosis. Like pulmonary abscesses, these other space occupying lesions vary in size and echogenicity, are devoid of other evidence of lung tissue and are only detectable at the peripheral lung. Pulmonary fibrosis typically generates an increased number of comet tail artifacts and a scalloped appearance to the pleural surface.[1,2] Thoracic radiographs usually provide more global identification of the extent of space occupying pulmonary lesions, but ultrasonography will provide precise location that is more likely to facilitate guidance for biopsy.

PLEURAL CAVITY ABNORMALITIES

Pleural effusions

A frequent and easy to detect abnormality of the pleural cavity is the presence of pleural effusion. It is most commonly seen with pneumonia and neoplasia, particularly lymphosarcoma, but can also develop with cardiac failure, secondary to severe pulmonary edema, penetrating chest wounds, esophageal rupture, and diaphragmatic hernias.[1] Pleural fluid is identified as ventrally located, anechoic to mixed echogenicity fluid immediately deep to the intercostal muscles that displaces the visceral surface of the lungs away from the chest wall. The degree of displacement depends on the amount of pleural effusion, with larger amounts of fluid causing greater displacement of the lung and atelectasis. When pleural fluid is identifiable to the level of the shoulder, often 3 to 6 liters of fluid is present.[1] If the fluid is visible above the shoulder, 5 to 10 liters of fluid may be present. When pleural fluid is present in sufficient quantities, structures that are not normally seen because of interface with appropriately inflated lungs may be visualized, such as the pericardial diaphragmatic ligament, the caudal mediastinum, the vena cava, and the cranial mediastinum. The caudal mediastinum is normally displaced by the accessary lung lobe to the left of the midline of the diaphragm and when surrounded by anechoic fluid, it appears as a hyperechoic undulating string that often is misidentified as fibrin. With anechoic effusion in the right caudal hemithorax, the caudal vena cava may become visible as a "ball" (vena cava) on an undulating string (plica venae cavae).

The ultrasonographic characteristics of pleural fluid can assist with differential diagnosis and prognosis. Mostly anechoic pleural fluid is usually of low cellularity and more indicative of pleural transudate that accompanies noninfectious pleural diseases such as neoplasia, cardiac failure, and pulmonary edema. Echogenic pleural fluid, particularly in the dependent areas of the pleural cavity is likely to be more cellular and indicative of an infectious process. When the pleural fluid appears vastly different in echogenicity in each hemithorax, it indicates that either the mediastinum is complete or different degrees of pathology are present in each hemithorax. Fibrin strands can often be seen floating like "seaweed" or "fronds" in the fluid. Chronic septic pleuritis may result in lobulation and cavitation of the pleural fluid and the formation of pleural abscesses and pulmonary adhesions. The later can be detected by the appearance of restricted movement of the lung surface with each respiration. Free gas within pleural fluid will appear as hyperechoic foci and usually indicates the presence of anaerobic infection,[2] pneumothorax, iatrogenic introduction of air second to thoracocentesis or chest tube placement, or the presence of a bronchopleural fistula. Hemothorax appears as a mixed echogenic swirling fluid and can be most easily confirmed by thoracocentesis.

Pneumothorax is characterized by the presence of reverberation artifact (the equidistant horizontal "A lines") of trapped air in the chest cavity in the dorsal fields immediately adjacent to the parietal pleura.[1] When pneumothorax is present, the characteristic sliding appearance of the hyperechoic surface of lung adjacent to the intercostal muscles is obscured. Thus when looking for evidence of pneumothorax, search for the "lung point" or the interface in which the lung sliding is detected over the cycle of respiration, representing the point on the chest wall where the pneumothorax ends and the aerated lung tissue is in contact with the parietal pleura.[2] Comet tail artifacts will not be present at the parietal pleural surface in the area of pneumothorax. Pneumothorax is easiest to detect when there is concurrent pleural fluid ventrally.

Diaphragmatic hernias are not very common in the horse and direct visualization of the hernia is often not possible, especially if there is no displacement of abdominal visceral into the chest cavity.[3] However, if fluid accumulates on either side of the diaphragm, loss of continuity of the diaphragmatic musculature may be detected and the contracted free edge of the diaphragm may be visible. Herniated abdominal visceral will be seen adjacent to lung with the diaphragm and other abdominal viscera located ventrally. However, herniated abdominal viscera cannot always be identified ultrasonographically and its absence does not rule out the diagnosis.[5]

Cranial mediastinal masses (abscesses and neoplasia) may be seen at the right third intercostal space and often displace the heart caudally and the lung dorsally and cause signs of obstruction of the cranial vena cava (bilateral jugular vein distension without pulsation).[6] It is important to note that fat in the cranial ventral thorax will not displace the heart and lung.[1] Pulling the right leg forward can facilitate imaging the cranial mediastinum. In horses with heavy triceps musculature, imaging through the triceps muscle can be attempted. Masses within the cranial mediastinum sometimes can be identified by imaging at the thoracic inlet.

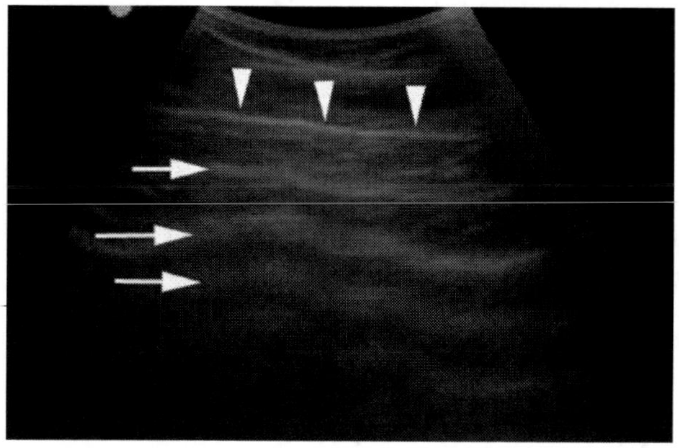

Figure 1. "A line" reverberation artifacts (horizontal arrows) created by the echoing of sound waves between the lung surface (arrowheads) and the probe.

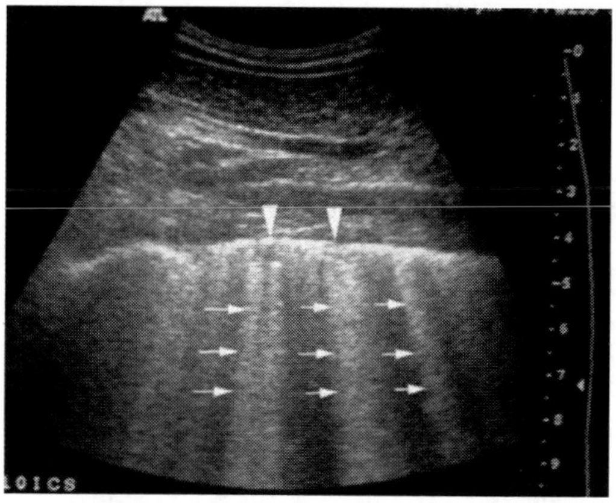

Figure 2. "B line" or comet tail artifacts (horizontal arrows) created by minute areas of nonaerated lung (arrowheads) at the lung surface.

REFERENCES

1. Reef V. Clin Techniques in Equine Practice. Elsevier. 2004. 284. 2. Stone M. Philips tutorial. http://www.healthcare.philips.com/main/education/ultrasound/emergency_medicine_cc/index.wpd 3. Sisson S: In Sisson's and Grossman's, The Anatomy of Domestic Animals, Philadelphia, PA, W. B. Saunders, 1975. 4. Reimer JM et al. J Amer Vet Med Assoc. 1989; 194: 278. 5. Hart SK et al. J Vet Emerg Crit Care. 2009; 19: 357. 6. Garber JL et al. J Amer Vet Med Assoc. 1994; 205: 1432.

THE HORSE WITH A 'LARGE' HEART: DE(HYPO)HYDRATED, ENDOTOXEMIC, HYPERTENSIVE OR ATHLETIC?

Cristobal Navas de Solis, LV, PhD, DACVIM
Bern, Switzerland

INTRODUCTION

Pseudohypertrophy, caused by dehydration or endotoxemia, and hypertensive cardiomyopathy (HC), have been recently reported in horses and added to the spectrum of conditions that create the echocardiographic appearance of hypertrophy. The purpose of this talk is to highlight the echocardiographic differences between these, and previously reported conditions such as athlete's heart. The appearance of the equine heart with concentric hypertrophy or pseudohypertrophy is typically different from the echocardiographic images obtained in classically described causes of heart disease with dilation (eccentric hypertrophy). Valvular disease, congenital heart disease or cardiomyopathies that cause chamber dilation will only be very briefly mentioned during the presentation.

DIFFERENTIAL DIAGNOSIS OF CONCENTRIC CARDIAC HYPERTROPHY IN HORSES

Pseudohypertrophy, caused by dehydration or endotoxemia, and HC have now been added to the spectrum of conditions that create the echocardiographic appearance of hypertrophy in horses.[1,2,3] In many cases the clinical presentation and non-invasive blood pressure measurements will be definitive to differentiate between these causes of abnormal echocardiographic appearance of the equine heart. However in some cases the clinical presentation may not be specific or there may be concomitant and overlapping problems. In these cases there are echocardiographic parameters that can be used to differentiate between these recently reported conditions in the horse and other previously reported conditions such as athlete's heart. Valvular disease or primary cardiomyopathies in horses will cause, almost uniformly, dilation while the rest of the above-mentioned conditions cause concentric hypertrophy. Dilation is described in some instances as eccentric hypertrophy. The appearance of the dilated heart is grossly different to the concentrically hypertrophied heart and mainly characterized by an increased internal diameter of the cardiac chambers. In humans and cats[4] hypertrophic cardiomyopathy (HCM) is a common cause of concentric left ventricular hypertrophy but to the best of our knowledge this genetic disease has not been described in horses.

Athlete's heart

Some degree of overlap in cardiac measurements and echocardiographic appearance exists between humans with an athlete's heart and those with diseases that cause myocardial hypertrophy such as HC or HCM.[4,5] Comparison of the data in a case series of equine HC or in reports of pseudohypertrophy during dehydration or endotoxemia with that collected from equine athletes[6,7] suggests that confusion caused by such overlap would be uncommon in horses. Relative wall thickness (RWT) appears to be uniformly increased in horses with HC or pseudohypertrophy and normal in equine athletes. Equine athletes develop a more proportionate left ventricular dilation with respect to the myocardial hypertrophy and therefore RWT remains normal.

Human athletes engaged in disciplines that require isometric exercises can potentially develop patterns of hypertrophy that resemble the ones observed with HC.[8] Studies in populations different than Thoroughbred and Standardbred racehorses would be necessary to determine whether other types of equine athletes develop different patterns of hypertrophy. However, the classic belief that the type of sport determines the pattern of hypertrophy has been challenged [8] and it seems unlikely that any common equestrian discipline mimics the isometric exercises that cause severe concentric hypertrophy with an increase in RWT in humans. Perhaps quarter horse racing or some non-FEI disciplines that involve pulling may be situations that could resemble sports that involve isometric exercises and in which athletes are reported to develop concentric hypertrophy but this remains speculative.

Hypertensive cardiomyopathy

In humans, HC is defined as the myocardial component of hypertensive heart disease (HHD) and is characterized by left ventricular hypertrophy and fibrosis. In a recent publication[3] a series of 5 cases of hypertensive cardiomyopathy in horses was reported. Horses had hypertension secondary to chronic laminitis or chronic renal failure. Blood pressure values (median [range]), measured noninvasively, were determined as 190 mm Hg (183 to 261 mm Hg) for systolic pressure, 126 mm Hg (100 to 190 mm Hg) for diastolic pressure, and 155 mm Hg (126 to 222 mm Hg) for mean pressure. Horses had increased relative wall thickness and ventricular mass. The interventricular septum was thickened at end diastole (n = 5) and in peak systole (4). The left ventricular internal diameter was small at end diastole (n = 4) and in peak systole (3). The left ventricular free wall was thickened at end diastole (n = 3) and in peak systole (4). All horses in this study were euthanized due to the severity of their primary disease and the hypertrophy confirmed postmortem. Interestingly fibrosis was not detected in any of the cases.

Left ventricular hypertrophy in the event of hypertension is an adaptation or maladaptation to the systemic hypertension that was easily recognizable in two-dimensional echocardiograms of horses when the myocardial hypertrophy was moderate or severe, similarly to what is described in humans with HHD[9]. Pressure overload does not appear to be the only mechanism that causes ventricular hypertrophy during hypertension. Renin angiotensin aldosterone system activation, insulin resistance, and catecholamine release play a direct role in the development of hypertensive myocardial hypertrophy in humans.[8] The cardiac response to hypertension has high individual variability and has been described as proportionate to the 'area under the lifetime blood pressure curve'.[10] The degree of left ventricular hypertrophy is not associated with blood pressure measurements in human beings with HHD [11] and therefore it is not surprising that no significant associations were found between the degree of left ventricular hypertrophy and the severity of hypertension in the horses reported in the equine cases series of HC.

The clinical findings and blood pressure measurements of horses with HC should allow their differentiation from those with pseudohypertrophy due to dehydration or endotoxemia.[1,2] The left ventricular mass (LVM) was increased in the group of horses with HC that had true hypertrophy and was reportedly normal in the study by Underwood *et al.* using an experimental model of de(hypo)hydration. The values of RWT (and mean wall thickness) in horses with HC were also higher than those reported in experimentally hypohydrated horses. Moreover, the systolic left ventricular wall measurements, which were altered in 4 horses in a series of 5 cases of HC, but within reference range in volume-depleted horses,[1] would also help in this differentiation. The data reported in a case series of equine HC suggests that systemic hypertension causes echocardiographically evident left ventricular hypertrophy in horses that can be differentiated from pseudohypertrophy by the use of the calculated values of LVM or by detection of increased wall thicknesses during systole in horses in which the clinical appearance is not definitive. As stated above differentiation with an athlete's heart could be done echocardiographically by the detection of abnormal RWT (normal in equine athletes).

De(hypo)hydration

The echocardiographic appearance of horses after experimental de(hypo)ydration was described recently by Underwood *et al.* similarly to the described in other veterinary species[12]. In this study it was observed that the left ventricular internal diameter in systole and diastole and the left atrial diameter decreased while septal wall thickness in diastole, free wall thickness in diastole and relative wall thickness increased significantly. A salient fact reported by this study is that a RWT of 0.55 would discriminate hypohydrated (≥ 0.55) and euhydrated horses (<0.55) in all cases.

This study concluded that hypohydration could mask or promote the severity of cardiac changes and advised taking this factor into account when evaluating horses with cardiac disease. Interestingly, echocardiographic changes with analogous or similar direction had been previously reported by Marr *et al.* in 1999 in a study of the effect of exercise under hot/humid conditions[13]. The effects of exercise and the different variables recorded in this earlier study do not allow for direct comparison of the results.

Endotoxemia

The echocardiographic appearance of horses after an experimental endotoxin challenge has been recently described.[2] Left ventricular internal diameter in diastole was small during the endotoxin infusion similarly to most horses with HC or dehydration. Left atrial diameter was significantly decreased during endotoxin infusion (or dehydration) and mildly so in some of the horses with HC. Left ventricular free wall thickness during systole was significantly decreased at hours 3, 4, and 5 after endotoxin infusion while it is increased or normal in horses with HC and usually normal in horses with pseudohypertrophy due to dehydration. The appearance of the left ventricular free wall in systole may be one of the keys to differentiate the echocardiographic appearance of horses with HC or dehydration from endotoxemic horses. The endotoxic challenge also decreased fractional shortening to less than 30% in some horses (normal or increased in horses with HC or dehydration) and created a marked spontaneous contrast within the left ventricle which was interpreted by the investigators as evidence of myocardial dysfunction. The study reported by Slack *et al.* is relevant for the evaluation of horses with pre-existing cardiac disease and acute disease causing endotoxemia. This may be particularly useful in the intensive care unit for the differentiation of endotoxemia and de(hypo)hydration.

SUMMARY

In summary key factors to differentiate these four conditions that can give the subjective appearance of cardiac hypertrophy are:

- The clinical presentation
- Normal RWT in horses with athlete's heart.
- Increased LVM with increased RWT in cases of HC.
- Increased thickness of the interventricular septum in systole in cases of HC.
- Decrease in the left ventricular free wall thickness in systole or decreased FS in horses with endotoxemia when compared to an increase, or normal, left ventricular free wall in cases of HC and dehydration.
- Normal to increased FS in HC, normal in cases of dehydration and decreased in some phases of the response to endotoxins.

REFERENCES

1. Underwood C, Norton JL, Nolen-Walston RD, et al. J. Echocardiographic changes in heart size in hypohydrated horses. J Vet Intern Med 2011;25:563-569. 2. B Slack J, Nolen-Walston RD, Dallap Shaer BL. Echocardiographic and ecg findings in horses with experimentally induced endotoxemia. Research Abstract Program of the 2013 ACVIM Forum. J Vet Intern Med. 2013 27 (3):649. 3. Navas de Solis C, Slack J, Boston RC, Reef VB. Hypertensive cardiomyopathy in horses: 5 cases (1995-2011). J Am Vet Med Assoc. 2013 Jul 1;243(1):126-30. 4. Maron BJ. Hypertrophic cardiomyopathy: a systematic review. JAMA 2002;287:1308-1320. 5.Lauschke J, Maisch B. Athlete's heart or hypertrophic cardiomyopathy? Clin Res Cardiol 2009;98:80-88. 6.Buhl R, Ersbøll AK, Eriksen L, Koch J .Changes over time in echocardiographic measurements in young Standardbred racehorses undergoing training and racing and association with racing performance. J Am Vet Med Assoc 2005;226(11):1881-7. 7.Young LE, Rogers K, Wood JL. Left ventricular size and systolic function in Thoroughbred racehorses and their relationships to race performance. J Appl Physiol 2005;99(4):1278-85. 8. Urhausen A, Kindermann W. Sports-specific adaptations and differentiation of the athlete's heart. Sports Med 1999;28:237-244. 9.Janardhanan R, Kramer CM. Imaging in hypertensive heart disease. Expert Rev Cardiovasc Ther 2011;9:199-209. 10. Drazner MH. The progression of hypertensive heart disease. Circulation 2011;123(3):327-34 11. Izzo JL Jr, Gradman AH. Mechanisms and management of hypertensive heart disease: from left ventricular hypertrophy to heart failure. Med Clin North Am 2004;88:1257-1271. 12.Fine DM, Durham HE Jr, Rossi NF, et al. Echocardiographic assessment of hemodynamic changes produced by two methods of inducing fluid deficit in dogs. J Vet Intern Med 2010;24:348-353. 13.Marr CM, Bright JM, Marlin DJ, Harris PA and Roberts CA. Pre and post exercise echocardiography in horses performing treadmill exercise in cool and hot/humid conditions. Equine vet. J. Suppl. 1999,30:131-136

USING THE ONTARIO DEATH REGISTRY: HISTOPATHOLOGICAL LUNG SURVEY OF ACTIVELY RACING HORSES AND MORE

Fe ter Woort, DVM, DVSc, DACVIM (LAIM)
Kennett Square, PA

INTRODUCTION

The Ontario Racing Commission (ORC) created the Death Registry in an effort to collect information about the health and welfare of racing horses in Ontario, as well as to investigate possible cases of mistreatment. Since 2003, mandatory necropsies for any horse that dies within 2 weeks of racing or qualifying, or at the discretion of the ORC for any horse that dies within 2 months of racing or qualifying. In addition to its primary goal, it has provided a golden opportunity for research, providing access to large population of actively racing horses. The ORC has given consent for a number of samples to be taken during the necropsy of these horses, and this population has been the source of 4 previously published studies, one pending publication and further sample collection is underway for a 6th study. Since the majority of these horses die following catastrophic fractures, the first 3 studies naturally focused on the musculoskeletal system, investigating the epidemiology of musculoskeletal injuries in race horses[1] and looking at subchondral bone characteristics to investigate the pathogenesis of condylar fractures[2,3].

PULMONARY ARTERY CALCIFICATION

However, this population has recently been used in a novel way, providing valuable information about the horses' pulmonary vasculature and leading to the discovery that pulmonary artery calcification is present in 82% of these horses[4]. This study further evaluated the histopathologic and ultrastructural features of the pulmonary artery and found that damaged elastic fibers of the tunica media (thinned, fragmented and calcified) may play a role in the pathogenesis of such lesion.

INVESTIGATING INFLAMMATORY AIRWAY DISEASE

Continuing this trend, this population was used to investigate Inflammatory Airway Disease. Although much has been published about this condition, many of the studies looking at the prevalence vary in their diagnostic methods. The studies evaluating larger populations of horses used the tracheal aspirate to diagnose IAD[5,6], which was later not recognized by the ACVIM consensus statement since the cytology of a tracheal wash is a poor indicator of inflammation occurring in the lower aiways[7]. Later studies using the BAL or pulmonary function testing have selected their population based on respiratory signs or poor performance[7,8]. The ACVIM consensus statement recognized the validity of a variety of diagnostic methods to diagnose the condition[9], highlighting that there is no gold standard. Histopathology has been shown to correlate well with clinical signs, BAL cytology in horses with Recurrent Airway Obstruction (RAO)[10], but the histopathologic features of IAD have not been described. We set out to investigate the presence of small airway pathology in this population of horses that died while in active racing or training. Lung samples were obtained from 95 racehorses and evaluated, using a histopathologic scoring system previously validated for RAO[10]. Small airway inflammation, smooth muscle hyperplasia and hemosiderin were a common feature in these racehorses, especially in the caudodorsal sections of the lungs. Mild to moderate inflammation was present around the airways of 82/95 horses.

In conclusion, the ORC death registry has provided a golden opportunity for researchers to sample a population of actively racing horses and has allowed us to investigate IAD in a novel way, providing valuable information about this condition. Although the number of horses submitted yearly to the ORC death registry is declining, formalin fixed-paraffin embedded samples from the past 10 years can continue to provide further research opportunities.

REFERENCES

[1] Cruz AM, et al. **Am J Vet Res**. 2007; 68:12. [2,3] Rubio-Martinez LM, et al. **Am J Vet Res**. 2008; 69: 11. [4] Arroyo LG, et al. **Vet Path**. 2008; 45:617. [5] Sweeney CR, et al. **Am J Vet Res**. 1992; 53:7 [6] Christley RM, et al, **Equine Vet J**. 2001; 33: 3 [7] Malikides N, et al. **Aust Vet J**. 2003; 81:11 [8] Hoffman AM, et al. **Am J Vet Res**. 1998; 59: 2 [9] Couetil LL, **J Vet Intern Med**. 2007; 21: 2. [10] Viel L. PhD Thesis. 1983; UoGuelph.

EQUINE GUT MICROBES IN HEALTH

Megan Shepherd, DVM, PhD, DACVN
Blacksburg, VA

INTRODUCTION

The gastrointestinal tract (gut) of horses is inhabited by a complex population of eukaryotic and prokaryotic microorganisms (microbes/microbiota), of which bacteria dominate. The number of microbial cells is reported to be 10-fold higher than cells of the host [1] and the number of genes in the gut microbiome is estimated to exceed the number of host genes by a factor of 100 [2]. Essential roles of these microbes include digestion of fibers, synthesis of vitamins, stimulation of the immune system, inhibition of pathogenic bacteria, and maintenance of colonocyte health.

FUNCTION OF THE EQUINE GUT MICROBES

The primary site of microbial fermentation in the horse is the hindgut (cecum & colon); hindgut microbes enable the horse to thrive on a forage-based diet. Forage is high in fiber and fiber (e.g. cellulose and hemicellulose) are complexes of monosaccharides linked by β-glycosidic bonds, which can not be degraded by mamallian enzymes in the small intestine. The dominant products of microbial fermentation are the volatile fatty acids acetate, propionate, and butyrate, in decreasing order of total VFAs, in the equine hindgut_ENREF_13. Once in the blood, acetate may be aerobically oxidized and directly used for energy or stored as triglyceride in adipose and skeletal tissue [3]. Most absorbed propionate is converted into glucose by the liver. Butyrate is the preferred energy substrate for colonocytes and thus plays an important role in maintenance of hindgut health. Gut microbes are important for the maintenance of host health as disturbances or shifts in the bacterial population have been associated with disease in the host.

EVALUATION OF THE GUT MICROBIOTA

The majority of gut bacteria cannot be identified by traditional culture techniques; therefore, culture-independent methods are needed to evaluate complex microbial ecosystems. The first challenge in evaluation of the gut microbiome of any species is to select appropriate samples. Fecal samples_ENREF_105 are low-cost and non-invasive method for sampling gut microbes. Serial evaluation of fecal microbial communities over time may be performed, due to the ease of sampling, without compromising the health of the host. However, the utility of fecal samples is dependent on how well fecal microbes represent gut microbes. Several reports describe the unique profile of the small intestinal microbiota compared to the hindgut, but there are several accounts of similarities in the microbiota between feces and certain regions of the hindgut [4-7]_ENREF_39_ENREF_84. Milinovich *et al.* [6] reported that feces represent cecal microbiota with an 8 hour delay, which could be explained by normal equine gastrointestinal mean transit time [8]_ENREF_52. Interestingly, differences between regions of the hindgut appear to be more prominent as Schoster *et al.* [7] reported that bacterial profiles of the colon differed from that in the cecum (and feces) and Dougal *et al.* [4] reported that bacterial profiles of the cecum differed from that in the RDC (and feces). Fecal samples are likely inadequate in the functional evaluation of the gut microbiome because RNA is an unstable molecule and RNases are abundant in any environment . Therefore, the functional profile of feces is unlikely to represent gut microbial function. Hindgut samples can be collected in horses via a large intestine cannula_ENREF_82 and surgical collection; however, both methods are invasive and impose surgical risk on the horse. Postmortem sample collection is an alternative; however, this method is limited to a one-time collection.

Extraction of microbial DNA, of suitable quality for molecular analysis, from feces and gut contents can be challenging due to the presence of host cells, plant cells, and a variety of inhibitory molecules present (i.e., polysaccharides, polyphenols, bile, etc.). Nucleic acid extraction kits have enhanced methods for removing DNA polymerase inhibitors (e.g. organic and phenolic compounds, and humic acids) . Different cell lysis methods among commercial extraction kits may result in some degree of difference in which cells are lysed as Gram-positive bacteria do not lyse with the same ease as Gram-negative bacteria, due to the thickness and density of their peptidoglycan cell wall layer . Therefore, the extraction kit used may influence results of genomic/transcriptomic-based evaluation of a microbial community [9].

The molecular target, genome (target DNA genes; identifies who is there) vs. transcriptome (expressed genes/RNA; inferring function) vs. proteome (expressed proteins; ideal target for function), for microbial analysis should be tailored to the study objective. For genomic/transcriptomic evaluation, a/the target gene(s) in the population's genome/transcriptome must be selected. Genes conserved across whole domains (i.e. 16S rRNA and *gyr* genes present in all bacteria) are commonly targeted for genomic evaluation of the microbiota as a whole, to use genes to determine what microbes are present. Functional genes (genes encoding for enzymes, sometime present in only a subset of the population) are targeted when the objective is to study microbes with a specific function. The 16S rRNA gene is common target for evaluation of microbial populations as this gene is highly conserved across bacteria, is a stable molecule, and there is a large database of known 16S rRNA gene sequences (Ribosomal Database Project [rdp.cme.msu.edu] and NCBI [http://blast.ncbi.nlm.nih.gov/Blast.cgi]). To add further complexity, there are different regions within the 16S rRNA gene sequence that are targeted for analysis.

Culture-independent molecular technologies provide a much broader description microbial communities and reduce the bias of culture-dependent techniques; however, these methods aren't free of biases and each technology introduces a different bias in the evaluation of the gut microbiota_ENREF_84. Numerous culture-independent molecular techniques are available for evaluating complex microbial communities_ENREF_49_ENREF_50_ENREF_51 like the gut microbiota. There is no single gold standard method for evaluating a complex microbial population and thus multiple methodologies are often employed. Molecular techniques currently described include Denaturing/Temperature Gradient Gel Electrophoresis (TGGE/DGGE)_ENREF_14, Terminal-Restriction Fragment Length Polymporphism (TRFLP), Ribosomal Intergenic Spacer Analysis (RISA), clone library generation, Fluorescence *in situ* Hybridization (FISH), Real-Time Quantitative PCR (qPCR), and high-throughput sequencing (e.g. pyrosequencing and Illumina). All of these differ in their utility to microbial investigation, but they are all gene-based vs. culture-based analysis. Many of these techniques have been utilized in describing the equine gut microbiota (genomics)_ENREF_21_ENREF_21.

A bacterial population may be evaluated as a whole or based on the taxonomic group(s) of interest. The number of members in a specific population (i.e. number of members within a bacterial species) represents bacterial abundance, which can be described in absolute or relative terms. A bacterial population as a whole may be described based on its richness, which represents the number of different taxonomic groups in a population (i.e. number of different species) and/or diversity, a factor of both richness and abundance. Bacterial population evenness is a measure of taxonomic distribution of a population. In order a bacteria to be detected, it must share sequence similarity with a previously identified microorganism. For example, if a sequence shares more than 97% similarity with *Clostricium cellulolyticum*, then that sequence will be classified to the bacteria domain, *Firmicutes* phyla, *Clostridia* class, *Clostridiales* order, *Clostridiacea* family, *Clostridium* genus, and *cellulolyticum* species.

While there are general themes often shared between equine gut microbe studies, there is a challenge in comparing results across studies as discrepancies exist [5,10-12] to differing microbial techniques, as described above, individual (horse), and/or diet.

EQUINE GUT/FECAL MICROBIOTA, *WHAT WE KNOW TO DATE*

The gut microbiota is established at birth and is influenced by microbes residing in the dam's rectum, vagina, cervix, mammary gland, and milk as well as other environmental sources. Development of the equine gut microbiome appears to be similar to humans in that early environmental exposure (i.e. dam) and management influences the microbial population [13]. The bacterial richness of foal feces appear to increase dramatically within the first 2 weeks of life [14,15]_ENREF_96; stabilization of the fecal microbiota is reported to occur at 4 weeks of age [14]. Interestingly, fecal microbial stabilization is described prior to weaning at 4-6 months of age when foals are fully transitioned from a liquid-based diet to a solid diet. The use of creep feeding strategies or transitioning the foal onto solid food prior to weaning may attenuate any microbial changes that occur at weaning [14]. Faubladier *et al.* [13] evaluated the effect of feeding a fermented feed supplement to adult mares in late gestation to weaning and evaluated the outcomes of their foals. While foals from supplemented mares had higher body weights at several time-points in the first 2 months of age; foal fecal bacterial and VFA concentrations were influenced by time/age more so than mare supplementation [13].

Of the ~20 bacterial phyla [10,12] and ~150 bacterial species described in the feces/gut of adult horses, prominent members reside in the *Firmicutes*, *Bacteroidetes*, *Proteobacteria*, *Spirochaetes*, and *Actinobacteria*, and *Fibrobacteres* phyla[5,10-12,16]. Member of the *Firmicutes* phylum dominate (41-68% of total bacteria) the equine fecal/gut microbiota [375,10-12], which is similar to reports in other species [2,86-88]. Members of the *Clostridia* class appear to be prominent members of the *Firmicutes* phylum in equine fecal/gut microbiota [375,10,12]_ENREF_38. The *Firmicutes* phylum is a functionally diverse group in the horse as *Firmicutes* members range from saccharolytic to fibrolytic species. Willing *et al.*[17] reported a higher abundance of *Firmicutes* phyla members, particularly *Lactobacilli* spp. and *Streptococci bovis*, in the feces of horses fed a high starch plus forage vs. forage-only diet. Furthermore, there was a whole shift from *Bacteroidetes*-dominated (46% and 49% of T-RFLP clones represented by *Firmicutes* and *Bacteroidetes*, respectively) in the feces of horses fed forage-only to a *Firmicutes*-dominated (73% and 27%, respectively) in the feces of horses fed starch plus forage [17]. Interestingly, the dominance of *Bacteroidetes* in the feces of horses fed forage-only is unique from other equine reports as described elsewhere. Even a change in forage can result in changes within the *Firmicutes* phyla members in the colon of horses, particularly cultured counts of *Lactobacilli* spp. (increased) and *Streptococci* spp. (decreased) [17].

Fibrolytic bacteria, across different phyla, are prominent in the fecal/gut microbiota of horses_ENREF_53_ENREF_13. Fibrolytic bacteria established within the first week of life and appear to be relatively stable and developed by 2 months of age [13]. The most frequently studied fibrolytic bacteria in the adult horses are *Fibrobacter succinogenes*, a member of the *Fibrobacteres* phyla, and *Ruminococcus flavefaciens*, a member of the *Firmicutes* phyla. These fibrolytic species represent 12% and 4%, respectively, of total bacteria present in the equine hindgut [18-20]_ENREF_35. Drougal *et al.* [5] reported that the abundance of members within the *Ruminococcaceae* and *Fibrobacteraceae* families varied across the hindgut and fecal microbiota of horses [5]. de Fombelle *et al.* [21] reported that cellulolytic bacteria was higher in the hindgut than foregut, which is anticipated based on the hindgut being the primary site of fermentation. Other fibrolytic *Firmicutes* described in the equine hindgut include *Eubacterium* spp., and *Clostridiaceae* [20]. Several reports describe a reduction in hindgut cellulolytic bacteria when grains or high starch feeds are fed to horses vs. forage alone [21]_ENREF_103.

There are still microbes yet to be described in the equine fecal/gut microbiota. Despite the ever progressing technology of high throughput sequencing, recent studies continue to report a portion, albeit decreasing with improving methods and databases, of bacterial sequences as unclassified [5,10-12].

The archaeal and eukaryotic population of the equine fecal/gut microbiota has received much less attention than the bacterial population. Archaea are prokaryotes, like bacteria, but differ from bacteria in several characteristics including a more ridged lipid membrane and absence of peptidoglycan, the latter is shared with eukaryotes. Archaea include the methanogenic prokaryotes and represent < 1% of equine fecal/gut microbiota [10-12,22]. Lwin *et al.* [22] reported that the abundance of methanogens were higher in horse feces (0.04% total bacteria) than pony feces (0.02% total bacteria) and that *Methanomicrobiales* was the dominant methanogen in the feces of horses and ponies. Dougal *et al.* [4] reported that archaeal abundance and protozoal (eukaryotes) abundance was similar in the cecum and feces of horses and ponies, but the highest archaeal and protozoal abundance was reported in the right dorsal colon. The fungal abundance did not differ between the cecum, RDC, and feces of horses and ponies [4].

REFERENCES

[1]Bäckhed, F., et al. **Science** 2005;307:1915. [2]Xu, J. & Gordon, J. I. **Proc Natl Acad Sci** 2003;100:10452. [3]Suagee, J. K. et al. **Comp Biochem Physiol B Biochem Mol Biol** 2010;155: 322. [4]Dougal, K. et al. **FEMS Microbiol Ecol** 2012;82(3):642. [5]Dougal, K. et al. **PLoS One** 2013;8:_. [6]Milinovich, G. et al. **Environ Microbiol** 2007;9:2090. [7]Schoster, A., et al. BMC Res 2013;6:91. [8]Van Weyenberg, S., et al. **Livest Sci** 2006;99: 3. [9]Yuan, S., et al. **PLoS One** 2012;7(3):e33865. [10] Shepherd, M. L., et al. **FEMS Microbiol Lett** 2012;3(26):62. [11]Costa, M. C. et al. **PLoS One** 2012;7(7):e41484. [12]O' Donnell, M. et al. **Lett Appl Microbiol** 2013;57:492. [13]Faubladier, C., et al. **Br J Nutr** 2013;1(10):1040. [14]Faubladier, C., et al. **Anaerobe** 2014;25:61. [15]Earing, J. E., et al. **Adv Microbiol** 2012;2:79. [16]Steelman, S. M., et al. **BMC Vet Res** 2012;8:231. [17]Willing, B. et al. **Equine Vet J** 2009;41:908. [18]Hastie, P. M., et al. **Br J Nutr** 2008;100:561. [19]Lin, C. Z. & Stahl, D. A. **Appl Environ Microbiol** 1995;61:1348. [20] Daly, K., et al. **FEMS Microbiol Ecol** 2001;38:141. [21]de Fombelle, A. et al. **Anim Sci** 2003:77:293. [22]Lwin, K. & Matsui, H. **Archaea**. 2014;10:1.

EQUINE GUT MICROBIOME IN DISEASE

Megan Shepherd, DVM, PhD, DACVN
Blacksburg, VA

INTRODUCTION

The gastrointestinal tract (gut) of horses is inhabited by a complex population of microorganisms (gut microbes/microbiota), of which bacteria dominate. Members of the *Firmicutes* bacterial phylum dominate the equine fecal/gut microbiota. A hindgut (cecum & colon) dysbiosis has been described in the horse and appears to be associated with non-structural carbohydrates (NSC), gastrointestinal disease, and laminitis [1-3] _ENREF_4. In human medicine, gut dysbiosis has also been linked to several diseases and is the focus of some treatment methodologies [4]. Due to the reliance horses have on their gut microbes, addressing these dysbioses could be of valuable to the horse.

NON-STRUCTURAL CARBOHYDRATES

Feeding a high volume of sugar, starch, and/or fructan, collectively called non-structural carbohydrates (NSC), results in hindgut lactic acid accumulation in the horse [5] _ENREF_20. The production of lactic acid and resulting reduction in hindgut pH causes a shift in the hindgut microbiota and negative consequences to the gut mucosa. Changes in the hindgut microbiota are often associated with an increase in the abundance of members of the bacterial phylum, *Firmicutes*, in the equine hindgut [1]. Furthermore, members of the *Streptococcus* genera within the *Firmicutes* phylum have been most commonly associated with the diet-related change [1]. The increase in *Firmicutes* members associated with feeding a high volume of starch it not surprising as this phylum contains bacteria with diverse functions, from fibrolytic to saccharolytic. Recently, Daly *et al.* [5] reported changes in other bacterial phyla in horses fed concentrate plus forage vs. forage-only. In addition to changes within the *Firmicutes* phylum (increase in *Lachnospiraceae* and decrease in *Ruminococcaceae*), the authors reported that the relative abundance of *Bacteroidetes* was higher while the relative abundance of *Fibrobacteres* was lower in the colon of horses fed concentrate plus forage vs. just forage [5].

GASTROINTESTINAL DISEASE

Alterations in the equine gut microbiota has been linked to colic and colitis [6]. Colitis is associated with changes in the fecal microbiota of horses, characterized by reduced bacterial species diversity and a shift in the *Firmicutes:Bacteroidetes* phyla ratio of 4.9:1 to 0.8:1 [2]. Other phyla-level changes associated with colitis include reduced *Actinobacteria,* reduced *Spirochaetes,* increased *Proteobacteria*, and increased *Fusobacteria,* as compared to healthy horses [2]. Pathogenic bacteria, such as *Clostridium difficile, Clostridium perfringens*, and *Salmonella* spp. have been implicated in equine colitis [7]. However, the presence of pathogens does not always translate to disease as disease-free carriers of *Clostridium difficile* have been described [8] _ENREF_24. Disruption of the normal microbiome may allow pathogens to proliferate and cause disease [7].

LAMINITIS & COMORBIDITIES

Laminitis has been associated with and/or experimentally induced by consumption of large volumes of NSC (as starch or oligofructose)_ENREF_47_ENREF_28, hyperinsulinemia, and obesity. Changes in the equine hindgut microbiome have been linked to laminitis by the following proposed mechanisms: endotoxins, vasoactive amines_ENREF_41, lactic acidosis, and matrix metalloproteinases, and inflammation. Oligofructose-utilizing bacteria represent < 50% of the total anaerobic bacteria in the feces of healthy horses and this population has been shown to increase in abundance during the concurrent reduction in fecal pH associated with oligofructose administration and onset of laminitis [1]. The majority of bacteria implicated in laminitis are members of the *Firmicutes* phylum and include *Allisonella histaminiformis, Mitsuokella jalaludinii, Lactobacillus* spp., *Clostridium* spp., and *Streptococcus* spp. [1] _ENREF_2. *Streptococcus* spp., such as *Streptococcus bovis/equinus* and *Streptococcus infantarius/lutetiensis* ssp., have been most frequently associated with laminitis [1]. Recently, Steelman *et al.* [3] evaluated the fecal microbiota of horses with chronic laminitis and reported that chronically laminitic horses had higher bacterial diversity and higher abundance of members from the *Clostridiales* order. The higher bacterial diversity in laminitis is interesting as disease state is generally associated with reduced microbial diversity.

The fructan-overload laminitis model is considered to be the best model to date for the evaluation of carbohydrate-induced laminitis [1]. Inulin-type fructans from chicory root (*Cichorium intybus*) are rapidly fermented in the equine hindgut and can be used to induce laminitis [1] _ENREF_21. Fructans in forages are postulated to be the cause of pasture-associated laminitis in horses because cool-season grasses accumulate fructans and intake of pasture forage is not typically restricted. Grasses in the US contain 2-6% fructan on a dry matter basis (DMB) (calculated as WSC – ESC of grass hay dry matter from mean values from a commercial forage lab [9]). Therefore, a 500kg horse would consume 0.0.2-0.6kg fructan per day (estimated using 2% body weight in dry matter intake/day) or 0.4-1.1g fructan/kg BW/day. Therefore, total daily fructan consumption through US forages may provide the loading dose of fructan (1g oligofructose/kg BW), but would be unlikely to reach minimum induction concentrations (5g oligofructose/kg BW) as presented in laminitis-induction studies [10]. Furthermore, there is little published work evaluating the relationship between forage carbohydrates, the gut microbiome, and laminitis [11].

Furthermore, cool-season grasses produce mostly phlein or levan-type fructans vs. inulin-type fructans. Levan-type fructans are longer molecules (10-260 glucose and fructose polymers) with beta 2,6-linked bonds as compared to the shorter inulin-type fructan molecules ($\leq$10 glucose and fructose polymers) with beta 2,1 bonds . In rats, dietary levan was partially hydrolyzed by gastric acid and was fermented by the *Firmicutes* members, *Lactobacillus acidophilus, Clostridium sordellii,* and *Mitsuokella multiacidus,* along with several *Bifidobacterium* spp. (member of the *Actinobacteria* phylum) [12]. There are no studies to date evaluating the role of dietary levan-type fructans on the equine fecal/gut microbiota.

Laminitis is associated with obesity and insulin resistance, which may also have links to the gut microbiota. Obesity is a problem in the equine population, affecting nearly 50% of horses in some populations and has many negative downstream effects. An association between obesity has been established in humans and rodents [13], most commonly associated with an increase in members of the *Firmicutes* phylum_ENREF_23 [14]. Gut microbes promote weight gain in mice and appear to do so by enhancing energy extraction from the diet and augmenting host fat storage [13]. Humans obtain an estimated 10% of energy requirements from gut microbial products, yet horses obtain more than 50% of energy requirements from gut microbial products [15,16]_ENREF_80. A relationship between gut microbes or microbial products with equine obesity could influence future obesity prevention strategies especially since horses are becoming obese without consuming an energy dense western diet. An obese microbiome in the horse could explain, at least in part, why some horses maintain an obese phenotype on the same forage-based diet as lean counterparts.

Chronic consumption of highly digestible carbohydrates, in the form of grain/high starch concentrate, is positively associated with insulin resistance in horses. Fructan, however, is a fiber and thus should be minimally or not hydrolyzed in the small intestine. Baily *et al.*[11] reported that oral administration of levan-type fructan to non-obese (BCS 5/9) ponies with a history of laminitis resulted in reduced insulin sensitivity. In humans, a link between type-2 diabetes and gut microbes has been proposed [17]. While there is some evidence of gastric acid hydrolysis of fructans in rodents [12], the dynamic of equine gut microbes in fructan (fiber) fermentation and subsequent effects on insulin/glucose dynamics should be investigated.

DISEASE PREVENTION & MANAGEMENT

Identifying the complex microbial population in the equine gut is one piece of the puzzle. Inferring function through knowing what microbes are present is challenging because many bacterial groups contain individual bacteria with different functions and some bacteria can change function depending on the environment (i.e. shift from fibrolytic to saccharolytic). Functional changes in the microbiota as a whole may be of most importance to host health. What can we currently do to prevent/manage the negative changes described above? Certainly, there is a lot of evidence to support limiting high starch feeds in the equine diet. Furthermore, the gut microbiota can be negatively influenced by antibiotics [7]. The utility of probiotics in the treatment/management of colitis does not appear to support a substantial benefit to date [18]_ENREF_29. Of the probiotics studied in the horse to date, *Saccharomyces cerevisiae* appears to be the most beneficial and may be helpful in preventing negative diet-associated changes to the hindgut and associated microbiota and their products [19]. The use of fecal transplants or transfaunation has been used in both equine and human medicine to manage dysbiosis and restore the "normal" microbiome [4,20].

REFERENCES

[1]Milinovich GJ, et al. **Vet Clin North Am Equine Pract** 2010;26:79. [2]Costa MC, et al. **PLoS One** 2012;7:1. [3]Steelman SM, et al. **BMC Vet Res** 2012;8:231. [4]Petrof EO, et al. **Microbiome** 2013;1:3. [5]Daly K, et al. **Br J Nutr** 2012;107:989. [6]Shirazi-Beechey SP. **Equine Vet J** 2008;40:414. [7]Harlow BE, et al. **Vet Microbiol** 2013;166:225. [8]Schoster A, et al. **BMC Vet Res** 2012;8:94. [9]Equi-Analytical Inc. Ithaca, New York: 2004. [10]Kalck KA, et al. **Am J Vet Res** 2009;70:624. [11]Bailey S, et al. **J Am Vet Med Assoc** 2007;231:1365. [12]Yamamoto Y, et al. **J Nutr Biochem** 1999;10:13. [13]Ley R, et al. **Proc Natl Acad Sci** 2005:11070. [14]Verdam FJ, et al. **Obesity** 2013;21:E607. [15]Vermorel M, et al. **Livest Prod Sci** 1997;51:13. [16]McNeil N. **Am J Clin Nutr** 1984;39:338. [17]Larsen N, et al. **PLoS One** 2010;5:e9085. [18]Boyle AG, et al. **Vet Rec** 2013;172:128. [19]Faubladier C, et al. J Anim Sci 2013;91:1736. [20]Feary DJ & Hassel DM. **Vet Clin North Am Equine Pract** 2006;22:437.

ENERGY SUPPLY AND METABOLIC BACKGROUND OF TRAINING METHODS IN SPORTS – AN UPDATE BASED ON THE FINDINGS IN HUMAN PHYSIOLOGY

Prof. Dr. Ulrich Hartmann (PhD)
Leipzig, Germany

INTRODUCTION

About (metabolic) energy supply mechanisms in sport specific notions are existing in most of the textbooks. In the meantime it is well known that many of these perceptions need an overhauling revision. This implicates partially a revision of the expected (adaptation)effects on the active muscle tissue and consequential of the anticipated effects and benefits of the existing training methods and loads. Likewise it would be recommended to review a part of the momentary used terminology in training physiology and methodology, p.e. aerobic / anaerobic capacity, lactate tolerance training, lactate formation rate etc.. This includes to discuss and to redefine those actual perceptions and their influences to the sport practice.

Thus, the standard test methods in practice are step tests with relatively short step durations, the idea being to test substantially the full spectrum of an athlete's performance on the basis of several successive steps. The "anaerobic threshold" is determined under these submaximum loads, and is then often (tacitly) equated to the endurance or the "aerobic capacity".

METHODS/DISCUSSION

The ratio of the maximum performance to the maximum metabolic load, as measured by the maximum post-load lactate value, is used as a source of information about the anaerobic lactic acid component of the energy metabolism or the so-called "anaerobic capacity". In training practice itself, efforts are made to determine the maximum performance limit via the criterion of the maximum tolerable lactate acidosis. There is a general assumption that quasi-additive combination gives a sufficiently accurate indication of the metabolic capacity. It can be shown that results of this nature are relatively unsatisfactory and vague.

It is also regrettable that not enough is known about the relative contributions made by the different types of energy metabolism under competition loads specific to different sports. Consequently, recommendations regarding improvement targets for training are relatively inexact and vague as far as the energy aspect of performance is concerned. The literature gives different values for the contributions of the different types of energy metabolism in the various running events. Lack of knowledge and/or failure to make use of knowledge about the relationships between the metabolic parameters has led to considerable simplifications in performance diagnosis for endurance sports.

Therefore the direct experiment (ergometer testing) alone cannot provide an accurate analysis of the provision of energy during sport-specific (maximum) effort or the answer to the question whether and how utilization of the available metabolic capacities could be optimized or what increases in the proportions of the three metabolic resources would be best for a given competition.

The use of the lactate parameter in the regulation of training sometimes leads to the impression that effective training must always be associated with the attainment of certain lactate ranges in training or with the achievement of certain lactate values in relation to a corresponding load.

With a number of endurance sports as examples, it has been possible to show that the majority of training takes place in a range that is substantially free from lactate accumulation. Our own investigations have revealed that in long-distance loads a VO_2 corresponding to about 50 to 60% of the VO_2max of about 6 l/min can be expected without causing the lactate to rise above 1,5 to 2,0 mmol/min. Sporadic studies in connection with middle-distance and long-distance track-and-field events have led to a similar result.

On the assumption of slightly higher load intensities in accordance with the literature recommendations and simultaneous adherence to the indicated amounts of training, it would not be possible to reconcile the calorie demand for training, or the corresponding fraction of the glycogen consumption, with the available regeneration times. In connection with training for endurance sports of medium duration (duration of load > 3 minutes), it is sometimes argued that by shifting amount-oriented training to more intensity-oriented loads it is possible to keep the total effort unchanged or nearly so; however, this is questionable on energy grounds alone.

In general training methods in endurance training are classified in

1. distance training methods,
2. interval training methods and
3. competitive training methods.

Aim of all three methods is to stimulate the adaptation in a way to increase performance for short, middle and long term distances.

The problem is that existing perspectives are mainly focussing on the methodological implementation of training contents. They are orientated and based on the described and elapsed time of the competition profile and load. Additive, some biological aspects are hardly attended: 1. the oxidative based metabolic patterns in short and middle distances are underestimated, 2. there is a way of thinking, that each section of the competition distance can be trained isolated, and 3. the existing opinions about the energetic / metabolic backgrounds are insufficient or partly wrong.

Based on a more differentiated way of observation the effects of different training methods will be presented and discussed under expected aspects of adaptation. Examples from different disciplines will be given to clarify the problem for the practical implementation. Based on those explanations in conjunction with specific effects of adaptation possible mal-adaptations can be understood more easy. Consequences for aspects of the so called periodization in sport will be pointed out.

A more profound scientifically based interpretation of training load calls for a post-simulation of the training load in a suitably developed simulation model. However, this requires a rather more complex investigation methodology than the simple determination of a stepwise performance test.

Simulation calculations based on middle-distance and long-distance running events show that depending on the athlete's condition, it may be advisable to limit O2-uptake values in endurance training to about 50 to 60% of the VO_2max, or not more than 30 to 40% with a corresponding contribution from the burning of carbohydrates.

In horse racing mostly interval training methods are use to increase the physiological performance. For the complex understanding of the (positive and negative) effects for muscular adaptation the metabolic background of the different training methods will be shown and discussed. Based on a scientific background of the metabolic effects a better understanding of the correct application of the different training methods can be awaited. Examples based on the human physiology findings of top elite athletes will be given.

It is a popular fallacy to believe the so called "lactic capacity" could be trained in terms of an increase in a wide spectra. There is no measurable difference in the "lactic capacity" between trained and untrained athletes, therefore there is no trainable power benefit resultant, either. Thus, the secret of high power capability is not to try and increase the glycolytic power, but rather to decrease it while at the same time increasing oxydative performance can be awaited.

Another fallacy is to believe repeated (intensive) interval loading during training would retard fatigue resistance or enhance the so called "staying power". During the second half of medium term competition (4-8 min) it certainly is possible to achieve cellular pH values of less than 6.6 in humans. This is because of the relatively high activity of the glycolytic energy supply mechanism during the acceleration period and first half of the race. The over distance continuing, though lessening, glycolytic energy portion, shifts the already resultant acidosis further still into lower acid areas. With increasing acidosis, however, the phospho-fructo-kinase (a glycolysis enzyme) increasingly blocks out the anaerobic lactic energy supply mechanism to prevent a further onset of muscle lactate and thus prevent even lower pH values. If that would not be the case, severe interference within the physico-chemical milieu in a muscle cell up to destruction of structures of body protein and mitochondria would occur. Therefore, the only way to enhance the so called "staying power" is to enhance the oxidative (aerobic) power to keep the onset of lactate and its disadvantages within tolerable limits.

The possibilities offered by a computer-assisted post-simulation demonstrate how much the maximum aerobic and anaerobic work capacities affect one another when the duration of the event is taken into account and according to the condition of the individual. The resulting interpretations go far beyond the current range of routine performance diagnosis, which is dominated by a preoccupation with lactate. Though these interpretation models result from what appears to be a purely theoretical approach, they nevertheless lead, if used appropriately, to an explanation of the possible metabolic situations in the realization of the performance. Another effect of the computer-assisted interpretation method is that subjective influences due to the observer are pushed largely into the background, and this leaves room for a much more objective assessment of a given metabolic situation.

Another aspect which is necessary to take into account is the already existing level of muscular adaptation. This also has some important influences on the choice and variety of the training method and load structure during training.

Furthermore it is to conclude that many of the common laboratory test proceedings (mostly for endurance oriented sports) and the way how they are used for interpretation are partly or in total not fulfilling the basic quality criteria of a valid and reliable performance testing procedures. - For a differentiate interpretation the following testing procedures would be necessary:

1. A correct determination of the "maximal oxidative" (aerobic) capacity;

2. determination of the "maximal glycolytic" (anaerobic) capacity;

3. determination of the "endurance capacity";

4. (computer based) post exercise interpretation of test results.

SUMMARY

The general problem is that in physical orientated sports there is only a description of training contents of some parts of competition load which will be often defined and practised in this way. The insufficient result also can be seen how the terminology of "training science" is used in practice and particular practice aims and methods. This is also caused due to the dominant phenomenological and traditional description of performance demand without much respect to biology.

Many assumptions in sport training and test interpretation are done without regarding the complex dependency of each energy-producing/metabolic reaction due to work-load duration and intensity, without an interpretation of the complex process inside the muscle and very often without a sufficient explanation in adequate theoretical concepts.

REFERENCES

Di Prampero P E **Rev Physiol Biochem Pharmacol** 1981;89:144-222

Hartmann, U. et al in: Nolte V. (ed.): **Rowing Faster**. Human Kinetics, Champaign, 2004:9-23

Hoppeler H **Resp Physiol** 1990;80:137-146

Mader A et al **Int J Sports Med** 1986;(7):45-65

Margaria R et al **J Appl Physiol** 1963;18:367

ASPECTS OF PERFORMANCE DIAGNOSTICS IN RUNNING
– BENEFITS AND FAILURES FOR THE INTERPRETATION IN TRAINING

Prof. Dr. Ulrich Hartmann (PhD)
Leipzig, Germany

INTRODUCTION

By means of performance diagnostics methods the entire metabolic physiological capacity is defined due to the three energy supporting metabolic pathways and corresponding reactions. Normally each of the metabolic components is directly defined through the investigation by special laboratory performance tests and under specific conditions of the respective types of sport. – In accordance with these results, further conclusions are drawn concerning training monitoring, training methods and detailed proposals given for further training. Examining the literature there are only few findings of scientific based load profiles during competition and its corresponding specific adaptations in highly trained athletes. Therefore this study intends

- to question the existing imagination about the aims and objectives of current tests and performance methods,
- to examine current opinions in literature about the expected kinds of adaptation,
- to evaluate the resulting perspectives and conclusions
- to give recommendations for a possible validation and an advanced objectivity with the aim of a better and more specified interpretation of the given aspects.

METHODS/DISCUSSION

During endurance testing in a laboratory normally a stepwise increasing test is performed. Normally each step has the duration of 3 min with a 1 min rest between each step; the increase of each step is 50 watts on a cycle ergometer or 0,5 m/s on a treadmill, respectively. The decisive parameters for interpretation are the "endurance capability" / "anaerobic threshold" (power output at 4 mmol/l lactate (P4)), the maximum power output during the last (exhaustive) step (Pmax), the (maximum) lactate concentration after work-load, the cardiac-rate and, if at all, a parameter of the gas-exchange-metabolism. Traditionally from the determination of P4 the so called "aerobic capacity" is estimated, and from measuring Pmax the "anaerobic capacity" is indicated. - Both assumptions are incorrect:

An elite athlete's high absolute metabolic capacity and power results from the actively applicable muscle mass (act. MuM) which consists of 3 components: The alactic expenditure equals the amount of available creatin phosphate (CRPH), which has its resting state at ~20 mmol/kg MuM. This resting energy expenditure provides the energy for the high starting power to accelerate during the first 6 to 10 seconds. With that, 60-80% of the available energy supply is spent. The potential power generation during this period is 2 to 3 times higher than the power generated during the 2nd to 4th minute within a competition specific test. Clearly, such high power can best be achieved with muscles of mixed fibre structure.

The "maximum glycolytic power" is coupled with the ability, to generate large amounts of energy through the anaerobic lactic energy metabolism in best time. In this context, it may also be called a high or maximum lactate formation rate (VLA_{max} mmol/kg*s). In sprinting, for instance, it can only be mobilized between the 4th and 10th second. That so called "lactic (work-) capacity" complies with the accumulated lactate (CLa_{max} < 25 mmol/l) up to the physiologicly maximum possible acidosis (pH < 6.95 = high acidity) of the active MuM when the human body reaches complete exhaustion.

With a maximum mobilized lactate formation rate of approximately 1.2 mmol/s * kg MuM, a very high short-term power output would result. That power could theoretically be kept for several seconds (20-25 s). In practice, however, this is only realized and advisable for a few seconds, predominantly because of the remaining time and distance left in the race. More exact estimates can only be done via computer simulation of the energy metabolism dynamics for defined time and power (see **Figure 1**).

For longer load durations in the range of minutes (e.g. running; should be similar in horse races) the utilization of glycolytic power is restricted to the "lactic work capacity", that is the "accumulated lactate" under physiologic conditions. Measurement for that is the maximum of blood lactate (< ~25 mmol/l). A maximum post-exercise blood lactate concentration of ~22 mmol/l measured between the 7th and 12th minute after a competition specific load complies with a peak concentration of muscle lactate of ~35 mmol/kg just after termination of the (exhaustive) exercise. As a consequence of increased accumulated muscle lactate the pH value drops during loading.

The majority of a middle distance runner´s energy supply comes from the aerobic energy system; therefore, in order to estimate a runner´s performance, it is necessary to determine exactly how much energy this energy system provides. The more energy athletes can obtain aerobically, the better their basic conditions can be determined. By this reason it is necessary to determine the most significant criterion which is the maximal oxygen uptake (VO_2max) ability. Highest values in humans are above 6500 ml/min (or above 85 ml/min*s^{-1}) and must correlate with the associated mechanical power output and can be attained only under exceptional cases. Corresponding tests are performed in special laboratories or directly in the field by use of portable equipments. – Beside, lactate, heart rate and, if necessary, other parameters are measured and used for further interpretation.

In addition to the above mentioned possible achievements some more remarks to the general view of „endurance" and the terms "aerobic" and "anaerobic capacities" should be given: In sport practice the following terms are often used simultaneously:

- the *aerobic-anaerobic threshold* and the *oxidative / aerobic energy supply*,

as well as

- the *maximum tolerable lactate acidosis during maximum performance* and *the glycolytic / anaerobic energy supply (capacities).*

We should adhere to the statement, that endurance („aerobic capacity") is only a part of the oxidative energy supply mechanism; it is that part, which describes (mostly in %) the share of VO_2max, which is used in the range until the metabolic situation comes to that what is described and defined as the „aerobic-anaerobic threshold" (AT) or better the „maximum lactate steady state (MaxLass)".

That leads to the consideration and it also can be concluded that many of the common laboratory test proceedings and the way they are used for interpretation, do partly or in total not fulfill the basic quality criteria of valid and reliable performance testing procedures. - For a differentiated interpretation the following testing proceedings would be necessary:

1. A correct determination of the "maximal oxidative" (aerobic) capacity (VO_2max test) → short duration steps, rapid increase;
2. Determination of the "maximal glycolytic" (anaerobic) capacity → supramaximal load for short time;
3. Determination of the "endurance capacity" → long duration steps, small load increase, „steady state" conditions;
4. (Computer-based) Interpretation of test results. Examples based on the findings in human physiology of top elite athletes will be presented and discussed.

By the use of a simulation model (theoretical background see Mader (1996) the metabolic background of a given test result is demonstrated. By comparing the post exercise simulation with the traditional interpretation tools significant differences are obvious.

The problem is that in physically orientated sports there is mainly a description of training contents and of some parts of competition load. These will then be defined and practice in this way. The insufficient result can also be seen in how the terminology of "training science" is used in practice and in particular practice aims and methods. This is also caused by the dominant phenomenological and traditional description of the performance demand profiles without much respect to biology. - By means of the computer-based simulation programs of given loads all the inadequacy and limits of some reports and conclusions as well as their simple transferability into training practice can be shown and limited (**Figure 1**). – It will be discussed that a high performance in competition can be explained by different metabolic patterns. Based on those explanations in conjunction with the very specific effects of adaptation, possible maladaptations can be understood more easy.

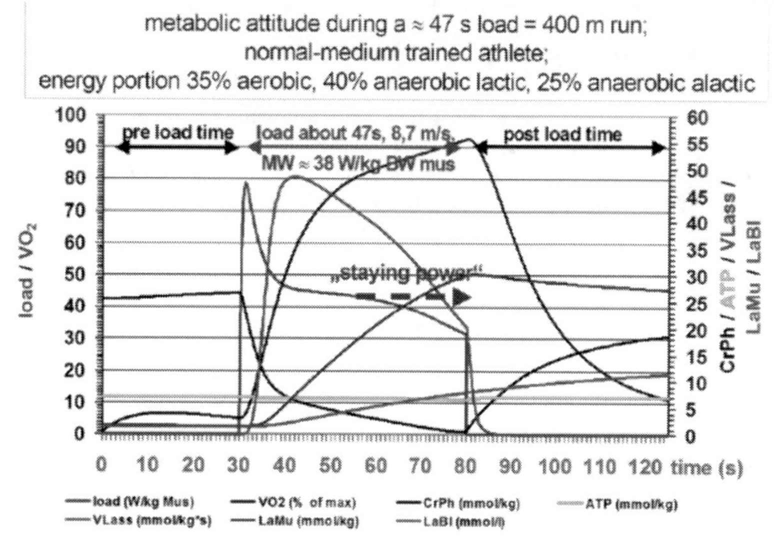

Figure 1. Metabolic proceedings during a 47 s load (≈400m run) of a medium trained athlete

CONCLUSION

Many assumptions in test interpretation and in sports and exercise training are made without regarding the complex dependence of each energy-producing/metabolic reaction due to work-load duration and intensity, without an interpretation of the complex process inside the muscle and very often without a sufficient explanation by adequate theoretical concepts. Beside that and almost for every sport and specific activity there are existing phenomenological descriptions for the expected adaptations. The problem results from the vague knowledge about physiological requirements in competitive loads and testing.

REFERENCES

Hartmann, U. et al in: Nolte V. (ed.): **Rowing Faster**. Human Kinetics, Champaign, 2004:9-23

Hartmann U et al in Rogozkin VA, Maughan RJ. (eds): **Current research in Sports Sciences: An international perspective**. 1996:179-185

Mader A. **J Sports Med Phys Fitness** 1996;31(1):1-19

Mader A et al in Mader A, Allmer H (eds). **Computersimulation – Möglichkeiten zur Theoriebildung und Ergebnisinterpretation**. 1996:124-162

PATHOGEN CONTAMINATION OF PET FOOD AND TREATS

J Scott Weese, DVM, DVSc, DACVIM
Guelph, Ontario, Canada

INTRODUCTION

Foodborne disease is poorly quantified in dogs and cats, yet it occurs.[1-3] Contaminated pet food and treats also pose a risk to humans that have contact with those items or animals that have ingested them.[2,4-7] As a result, pet food and treat manufacturers have a duty to minimize the risk of pathogen contamination of their products, and consumers have a duty to use basic practices to limit exposure to any contaminants that might be present. Pet food and treat manufacturing practices can play a major role in pathogen control, but are rarely considered by veterinarians and consumers.

PATHOGENS OF CONCERN

Most available information pertains to *Salmonella* because of well-documented human and animal disease from contaminated food and treats.[2,6,7] Yet, a variety of other pathogens are known or possible causes of foodborne disease, including a range of bacteria such as *Campylobacter, Escherichia coli, Listeria monocytogenes* and *Neorickettsia helminthoeca,* as well as parasites such as *Toxoplasma gondii.* The animal and public health relevance of these pathogens is discussed elsewhere.

PET FOOD AND TREAT MANUFACTURING

The pet food and treat manufacturing industry is highly variable. It includes large multinational corporations, small single-site manufacturers, small manufacturers that contract production to other companies and small scale (including in-home) commercial processing. Further, some pets receive food or treats prepared in the household. The foodborne pathogen challenges can vary greatly between these types, as do preventive measures.

Regulations vary between countries. In the US, pet food can fall under multiple legislative guidances and rules. The Food and Drug Association (FDA, via the FDA's Center for Veterinary Medicine) oversees with the Federal Food, Drug and Cosmetic Act, and under that Act, animal foods must contain no harmful substances, be truthfully labeled and manufactured under sanitary conditions. Recently, the US Food and Drug Association proposed a new Rule, *Current Good Manufacturing Practice and Hazard Analysis and Risk-Based Preventive Controls for Food for Animals.*[8] This establishes good manufacturing practices requirements for manufacturing, processing, packing and storage of animal food (including pet food). A written food safety plan, hazard analysis, preventive controls for identified hazards, monitoring, corrective actions, verification and associated record-keeping would be required as part of this Rule.

POTENTIAL SOURCES OF CONTAMINATION

Contamination can occur at many steps in the food preparation process. Key areas are outlined below:

Raw ingredients

Raw animal products can reasonably be assumed to be contaminated with pathogens. Contamination rates vary between species, sources and handling practices, but it is prudent to handle all raw animal-based ingredients as contaminated. Contamination rates of other food components are not well reported but contamination of vegetable-based ingredients or virtually any other type of food product can occur. While it is reasonable to assume that animal based products pose a greater risk than other products in pet food, numerous large fruit and vegetable foodborne disease outbreaks in humans clearly indicate that the risk is not only with meat products.

Palatability enhancers

Kibble (and some treats) is often coated with palatability enhancers. When this occurs after the cooking and extrusion steps (where heat would kill many pathogens), contamination of flavor enhancers could pose a risk. Some of these products would be excellent growth media so production and handling precautions must be used to reduce the risk of contamination from palatability enhancers.

Manufacturing equipment

Contaminants that are introduced from ingredients, personnel or other sources (e.g. vermin) can accumulate on processing equipment. Standard cleaning and disinfection practices are typically used in commercial facilities but deficiencies in practices can result in accumulation of pathogens. Some bacteria can multiply rapidly at room temperature, creating the potential for amplification of contamination. Poor quality or poorly maintained equipment can hamper disinfection steps, something that has been implicated in large outbreaks.

Post-manufacture storage

Contamination can occur after final processing but before packaging. Accumulation of contaminants in handling and packing equipment can result in widespread food contamination. This should be unlikely because of the low likelihood of contaminants entering the system from food; however, contamination from other sources (e.g. vermin) is possible. Properly maintained and inspected storage areas and facilities should reduce this risk.

Contamination can also occur at the retail level, in veterinary clinics or stores where food and treats are sold. Rodent infestation, with burrowing into (and defecating in) food bags is a possible concern.

The use of bulk bins for treats is a greater concern. If there is a low prevalence of contamination of single products or lots, but these are continually added to a bulk bin without regular emptying, cleaning and disinfection, contaminants (from treats or human hands) could accumulate.

Household

Contamination of dry food can occur in households, and rodent-contamination is probably of greater incidence in households versus point-of-sale. Continually refilling food or treat storage containers is also a common practice.

Preventive measures

Prevention of contamination and monitoring is a complex process that can involve multiple steps, each of which is designed to achieve a particular goal. The main approaches are reducing the risk of contaminated ingredients, reducing contamination during processing, killing pathogens during processing, inhibiting pathogen growth after processing and preventing contamination of finished product. Typically these are part of comprehensive manufacturing protocols or guidelines. Examples of specific areas are discussed briefly below.

Heat

Heat is one of the most effective and reliable pathogen elimination methods, and is a key component of foodborne pathogen elimination. Canned (wet) food undergoes high temperatures during the cooking and canning process, with temperatures that are adequate to render the food sterile. Pet foods are considered to be low-acid foods, which carry an increased risk of *Clostridium botulinum* growth if improperly cooked and canned, so canned food must follow low-acid canned food regulations that dictate time and temperature, along with the type of monitoring and monitoring equipment that is required. Dry foods undergo cooking, with heat and dessication eliminating many pathogens. The extrusion process also results in generation of heat and pressure that can inactivate many pathogens.

Pet treats are highly variable in the type of processing they receive (if any), with some being thoroughly cooked and others exposed to no heating process. Raw foods are obviously not cooked, with that being the main explanation for the high contamination rates that have been reported.

Irradiation

Irradiation is a highly effective means of eliminating pathogens, particularly from items that are not amenable to heat treatment. Irradiation is widely used for some food products (e.g. spices) but cost and consumer perceptions limit widespread use. Irradiation of pet foods is therefore rare, although it is possible (if not likely) that certain components of pet food have been irradiated. FDA regulations dictate that if ingredients accounting for <5% of the final product are irradiated, the product does not have to be labeled as irradiated. Irradiation of pet treats is more common. Irradiation of cat (but not dog) food was banned in Australia after an outbreak of neurological disease in cats fed a specific imported diet that had been irradiated,[9] potentially because of irradiation-induced reduction in vitamin A.[10] While this phenomenon has not been reported elsewhere, it is an important consideration if whole diets are being irradiated, as opposed to treats.

High-pressure pasteurization

This process, also referred to as high pressure processing, high hydrostatic pressure processing or ultra high pressure processing, involves pressurizing foods to 100-800 MPa, with only a slight (approximately 3^0C/100 MPa) increase in temperature.[11] Some raw food manufacturers use this process, largely to reduce *Salmonella* contamination. Bacteria vary in their susceptibility to pressure, with some being quite susceptible and some more tolerant. The bacterium, food matrix and pressurization conditions will impact efficacy of high-pressure pasteurization. Therefore, while potentially effective at eliminating pathogens, companies using this process should ensure that proper testing is done with their raw food, batch size and pressurization conditions, ideally with experimentally contaminated food. At the time of writing, objective data were lacking, although it is likely that this process does effectively eliminate *Salmonella* if done correctly. Bacterial spores (e.g. *Clostridium botulinum, Clostridium perfringens*) species are tolerant of high pressures. Parasites are assumed to be more susceptible to pressure than bacteria and this process should effectively eliminate parasites.

Preservatives

A variety of preservatives may be added to pet food. Regardless of whether they are 'natural' or 'artificial', something that is more of a marketing designation than a microbiologically relevant factor, preservatives are not typically intended to kill contaminants. Rather, they are mainly added to prevent oxidation of the food and associated discolouration or spoilage. Some preservatives may slow growth of any bacteria that might be present or become introduced, but bacterial inhibition by commonly used preservatives is not adequate to be relied on as a pathogen control method. The safety of preservatives comes under regular scrutiny and detailed discussion of that area is beyond the scope of these proceedings. That being said, evidence of a health risk of commonly used preservatives at commonly used doses is lacking.

Freezing

While not a method used in commercial food preparation, freezing will eliminate some pathogens and have little effect on others. Freezing is most often used to eliminate parasites. For example, freezing (e.g. 2 days at -20^{0}C) can effectively eliminate *Toxoplasma*.[12] *Campylobacter* spp are also relatively intolerant of freezing, although freezing is not a tool that can be used with confidence to eliminate the risk of campylobacteriosis.

Freeze-drying

Contrary to some lay perceptions (and some false advertising), freeze-drying does not eliminate most pathogens. In fact, freeze-drying is a common and effective tool for longterm preservation of bacteria. The freezing component of the process may eliminate pathogens such as parasites or *Campylobacter* that are susceptible to freezing, as discussed above.

Cleaning and disinfection

Cleaning and disinfection is a standard and important food safety method. It prevents accumulation of contaminants and growth of pathogenic microorganisms on environmental surfaces in the food production pathway. Detailed protocols and practices are used in well-run manufacturing facilities, but deficiencies can be present and can lead to contamination. Key aspects of cleaning and disinfection are the use of proper cleaners and disinfectants, the presence of surfaces amenable to cleaning and disinfection, provision of adequate time, and, perhaps most importantly, actually performing the task thoroughly.

Pathogen Testing

Testing finished product can allow for detection of contamination before food reaches the consumer, as well as provide an indication that something may have gone wrong in the food preparation process (with subsequent investigation). However, testing does not cover all potential pathogens. Sporadic testing may not be adequate to determine uncommon (but still relevant) contamination. The quality of results also depends strongly on the methods (i.e. if you don't really want to find *Salmonella,* you can use a protocol that is poorly sensitive). Specific pathogen testing practices and thresholds are typically proprietary. However, the FDA has a zero tolerance policy for *Salmonella* in pet food,[13] so any positive *Salmonella* test would result in a recall.

Hazard Analysis and Response

This is a standard food safety procedure that identifies and evaluates potential (foreseeable) hazards encountered during the manufacture, processing, packing and storage of food. The concept of hazard and critical control point (HACCP) analysis is well established and an expectation of any manufacturer. This approach is designed to identify key control points in the production system and to employ preventive and/or monitoring measures at those points to prevent and detect contamination. Thorough description of HACCP and individual control points is beyond the scope of these proceedings. HACCP is less relevant to raw foods that are not treated to eliminate pathogens, since there are few true control points that can be introduced or monitored. In situations where no control measures are implemented, the assumption must be that all product is contaminated and that all risk and risk mitigation is deferred to the consumer, a clearly undesirable approach.

CONCLUSIONS

Pet owners have myriad pet food and treat options and can, in general, have a high degree of confidence in the safety of these items. However, some risk will always remain and the endemic burden of foodborne disease is completely unknown. While consumers do not have a direct influence on food safety practices in products they purchase, increasing literacy in food preparation and food safety, and questioning manufacturers may help increase transparency and drive change if manufacturers realize that consumers will make purchase decisions based on perceptions of food safety. The high variability (and lack of transparency) in food safety practices and quality control certifications likely indicate differences in risk between different manufacturers. Consumers should expect that food and treat manufacturers maintain a food safety program and use recognized practices to minimize the risk of contamination, and provide clear information about the practices that are used. Consumers play a direct role in handling of food and treats that they have purchased, and there are some basic food storage and handling practices that can reduce the risk of disease.

REFERENCES

[1] Stiver SL, et al. ***J Am Anim Hosp Assoc*** 2003;39:538-542.
[2] Centers for Disease Control and Prevention. ***MMWR*** 2012;61:436.
[3] Sykes JE, et al. ***J Vet Int Med*** 2010;24:504-513.
[4] Clark C, et al. ***J Clin Microbiol*** 2001;39:3962-3968.
[5] Control CfD, et al. ***MMWR Morb Mortal Wkly Rep*** 2006;55:702-705.
[6] Centers for Disease C, et al. ***MMWR Morb Mortal Wkly Rep*** 2006;55:702-705.
[7] Centers for Disease C, et al. ***MMWR Morb Mortal Wkly Rep*** 2008;57:521-524.
[8]US Food and Drug Association. Current Good Manufacturing Practice and Hazard
Analysis and Risk-based Preventive Controls for
Food for Animals. Washington, DC, 2011.
[9] Child G, et al. ***Australian Veterinary Journal*** 2009;87:349-351.
[10] Caulfield CD, et al. ***Veterinary pathology*** 2009;46:1258-1269.
[11]US Food and Drug Association. Kinetics of Microbial Inactivation for Alternative Food Processing Technologies -- High Pressure Processing: US Food and Drug Association,, 2011.
[12] Kijlstra A, et al. ***Int J Parasitol*** 2008;38:1359-1370.
[13]US Food and Drug Association. Compliance Policy Guide Sec. 690.800, *Salmonella* in Food for Animals 2013.

ANIMAL AND PUBLIC HEALTH CONSEQUENCES OF PATHOGEN CONTAMINATION OF PET FOOD AND TREATS

J Scott Weese, DVM, DVSc, DACVIM
Guelph, Ontario, Canada

INTRODUCTION

Foodborne disease is common in humans, and it has been estimated that approximately 1/3 of people in developed countries acquire a foodborne illness every year. Basic principles of foodborne disease control are rather simple but infections continue to be common because of aspects such as the inherent risk of contamination of many foodstuffs and poor adherence to standard food handling and preparation recommendations. As with any type of food, pet food and pet treats can become contaminated with a range of pathogens. This poses a risk to both the pet that ingests contaminated items, as well as people that handle contaminated foods, the animal, its feces or other potentially contaminated items such as food bowls or food preparation surfaces. The incidence of pet food- and pet treat-associated infection in animals and humans is completely unknown. Most available data come from outbreaks, which, while enlightening, fail to provide any information about the endemic level of disease. These proceedings provide a general overview of health risks associated with bacterial and parasitic contamination of food. Toxins produced by fungi (e.g. aflatoxin) are also of concern but will not be considered further here.

ANIMAL HEALTH CONSEQUENCES

While there has been limited investigation of foodborne disease in pets, it undoubtedly occurs and can involve a wide range of infectious and non-infectious causes. Only microbial contamination will be considered further here, and that can be divided into three main categories.

Ingestion of a microorganism followed by growth in the intestinal tract and development of disease

This is the main concern with pet food and most often involves *Salmonella*. Other pathogens like *Campylobacter, E. coli* and *Staphylococcus aureus* could potentially cause disease in this manner, but this is much less likely and the role of these bacteria in disease in dogs and cats is unclear. Pet food-associated salmonellosis has been reported with raw meat and commercial dry diets,[1-3] and published reports are likely a profound underestimation of the incidence of disease.

For pathogens such as these, ingestion of viable bacteria does not necessarily result in disease. Various host (e.g. age, immune status, intestinal microbiota) and pathogen (e.g. species tropism, virulence factors…) factors are involved, and dose plays an important role. The infectious dose of enteropathogens is not known and likely varies between bacterial strains as well as between (and within) different host species. In people, the infectious dose for *Salmonella* has historically considered to be >100,000 cells, but some outbreaks have been caused by fewer than 100 cells per person, including reports of infection at estimated doses of <15-20 bacteria.[4,5] Canine and feline data are lacking but may be similar in that large doses are usually required to cause disease, with lower doses able to cause disease in the right circumstances but more often resulting in rapid elimination of the bacterium or transient shedding. A study of healthy adult dogs reported *Salmonella* shedding by 7/16 (44%) of dogs for 1-11 days following ingestion of a single meal of naturally contaminated raw meat, with no clinical abnormalities.[10] Host factors that influence the infectious dose probably include age, antimicrobial exposure and comorbidities.

Ingestion of food contaminated with pre-formed toxin that was produced in the food

Some bacteria can produce toxins in poorly handled food samples. This can occur with breakdowns in normal manufacturing processes or (more likely) with improper storage of home cooked food and raw food. *Staphylococcus aureus* and *Clostridium perfringens* are two bacteria that can produce potent enterotoxins in improperly handled food, and these are among the leading causes of 'food poisoning' in people. The incidence of this in animals is completely unknown. *Clostridium botulinum* can produce incredibly potent neurotoxins in an anaerobic environment. Botulism is very rare in pets but of concern because of the exceedingly low dose required for disease and the severity of disease. Botulism is predominantly a risk from improperly canned food, as well as ingestion of wildlife that have died of botulism.

Ingestion of a microorganism that translocates and causes extra-intestinal disease

This is an uncommon problem but can occur with pathogens such as *Listeria monocytogenes*. This is likely an exceedingly rare condition in dogs and of extremely low risk from commercial foods. The greatest risk is likely to compromised (elderly, immunocompromised) dogs and cats that are fed human foods that are at increased risk of *Listeria* contamination (e.g. raw or unpasteurized milk and dairy products, deli meats). A pathogen with a more complex and only partially understood pathophysiology but same ultimate pathway is *Neorickettsia helmintheoca*, the cause of salmon poisoning in dogs and a significant risk in dogs that are fed salmonids in certain.[6]

Outbreaks

Outbreaks of foodborne disease are rarely reported in animals. Whether this is because they are rare, rarely diagnosed or rarely reported is not known. Outbreaks of salmonellosis have been identified in racing greyhounds fed raw meat,[7] but most reports have involved commercial dry foods. A 2012 pet food-associated salmonellosis outbreak in people in the US was accompanied by numerous anecdotal reports of disease in dogs and cats that were fed recalled lots of food, as well as isolation of the outbreak strain from one sick and one healthy dog.[3] Clearly, dogs and cats are susceptible to different foodborne pathogens, but the relative risk compared to humans is unknown and the incidence of foodborne disease is completely unclear.

HUMAN HEALTH CONSEQUENCES

Many of the pathogens mentioned above for animals also pose a risk to humans. Human exposure can be from various routes, including handling food, cross-contaminating human food storage or preparation surfaces, handling food bowls, direct contact with the animal (e.g. contact with the animal's mouth after eating) or contact with feces. The relative risk of these is unknown and hard to elucidate.

Most information is available regarding salmonellosis, although various other known (*Campylobacter, Listeria monocytogenes, E. coli*) and potential (methicillin-resistant *Staphylococcus aureus, Clostridium difficile*) pathogens can be identified in food and treats.[8-12] Human disease has been reported associated with contaminated commercial dry diets[3] and pet treats (e.g. pig ears).[8,9] Contaminated commercial foods have the potential for widespread outbreaks because of the volume of product that can be contaminated and the widespread national (and sometimes international) dissemination of contaminated product. For example, a 2012 *Salmonella* Infantis outbreak associated with a single facility in South Carolina caused at least 20 human infections in 13 US states and 2 cases in Canada.[3] A 2006-2008 *S.* Schwarzengrund outbreak associated with one US plant resulted in at least 79 cases in 21 US states.[13] Feeding pets in the kitchen was identified as a risk factor in one investigation,[13] and factors such as poor hand hygiene probably also play important roles. The relative role of direct contact with contaminated food or treats versus indirect contact, cross-contamination or fecal shedding is not known. Dogs can shed *Salmonella* after ingestion of *Salmonella*-contaminated raw food,[14] so that route of transmission must also be considered.

Along with outbreaks of human disease, variable and sometimes high contamination rates of pet treats have been reported,[8-11,15] something that is not particularly surprising since they are often raw animal products that are dried (e.g. pig ears) and are not necessarily processed in a manner to kill pathogens. Human infection can occur from handling treats, packaging, cross contamination with other surfaces or contact with bulk bins that may become heavily contaminated over time. Moves by the industry in some regions to improve practices and reduce contamination may have improved the situation in recent years,[10] but pathogen contamination can still be found in commercial treats.[11]

Despite the high pathogen contamination rates associated with raw meat, there are no published outbreaks of human infections from contact with raw meat-based pet diets. Yet, this must be considered '*absence of evidence*' not '*evidence of absence*'. Indeed, the author is aware of multiple incidents of human salmonellosis that were most likely raw pet food-associated, yet inadequate data (typically a lack of the human isolate for side-by-side typing comparison with food and animal isolates) have been available for publication. Further, sporadic infections, as would be expected from contaminated raw meats, are less likely to be diagnosed or investigated, and a pet food link may not be queried. Commercial pet food outbreaks are more readily identifiable because of the potential for contamination of a large amount of food with the same strain, and subsequent identification of an issue after finding the same strain in multiple affected people. In contrast, raw meat contamination, while more common, would involve a large number of strains that likely vary between batches, with a lesser likelihood of a large clonal outbreak.

PREVENTION

A thorough description of preventive measures is beyond the scope of these proceedings. Numerous different areas or approaches need to be considered, including raw material sourcing and handling, commercial preparation practices (e.g. heat, preservative, irradiation), manufacturing quality control, post-manufacture practices (e.g. storage), consumer food sourcing (e.g. purchase of raw animal products), food storage in the household, food handling (e.g. where, how), hand hygiene practices and fecal handling practices. A single error at even one of these steps may be adequate to cause disease, although outbreaks are probably more often the result of a combination of errors. The pet food and pet treat industries obviously play a critical role and must implement a proper HACCP (hazard analysis and critical control point) approach to risk reduction rather than deferring preventive measures to consumers. Regardless, veterinarians can play a role in prevention of foodborne disease through client counseling, and pet owners have the duty to protect their animals, themselves, their families, through the use of practical infection control practices.

REFERENCES

[1] Chengappa MM, et al. ***J Vet Diagn Invest*** 1993;5:372-377.
[2] Stiver SL, et al. ***J Am Anim Hosp Assoc*** 2003;39:538-542.
[3] Centers for Disease Control and Prevention. ***MMWR*** 2012;61:436.
[4] Blaser MJ, et al. ***Rev Infect Dis*** 1982;4:1096-1106.
[5] Boring JR, et al. ***Am J Epidemiol*** 1971;93:49-54.
[6] Sykes JE, et al. ***J Vet Int Med*** 2010;24:504-513.
[7] Morley PS, et al. ***J Am Vet Med Assoc*** 2006;228:1524-1532.
[8] Centers for Disease C, et al. ***MMWR Morb Mortal Wkly Rep*** 2006;55:702-705.
[9] Clark C, et al. ***J Clin Microbiol*** 2001;39:3962-3968.
[10] Finley R, et al. ***Zoonoses and Public Health*** 2008:55:462-469.
[11] Freeman LM, et al. ***Can Vet J*** 2013;54:50-54.
[12] Pitout JD, et al. ***J Clin Microbiol*** 2003;41:4578-4582.
[13] Behravesh CB, et al. ***Pediatrics*** 2010:1-9.
[14] Finley R, et al. ***Can Vet J*** 2007;48:69-75.
[15] Kukanich KS. ***J Am Vet Med Assoc*** 2011;238:1430-1434.

FEEDING GERIATRIC PATIENTS: WHAT IS THE BEST NUTRITIONAL APPROACH?

Dr. Joe Bartges, DVM, PhD, DACVIM (Small Animal Internal Medicine), DACVN
Knoxville, TN

INTRODUCTION

Pets appear to be living longer and between 1/3 and 1/2 of pet dogs and cats are 7 years of age or older.[1] In the United States there has been an increase of pet cats older than 6 years of age from 24% to 47% over the past 10 years.[2] The Association of American Feed Control Officials, however, does not recognize a life stage called senior or geriatric and nutritional requirements are assumed to be similar to an adult dog or cat. This is understandable as pets, especially dog breeds, age at different rates and aging not only brings physiologic changes but also pathologic changes.

WHAT IS GERIATRIC?

What is meant by mature, senior, and geriatric? There have been several guidelines published that give definitions.[3-8] For the purposes of this presentation, we will use the following definitions: Growth = from birth to reproductive maturity; Junior = growth after reproductive maturity; Adult = reproductive maturity and growth completed; Mature = from approximately the middle of the expected life span to the last 25% of expected life span; Senior = the last 25% of the expected life span; Geriatric = life beyond expected life expectancy.[3] Thus, we will focus the remainder of the discussion on nutritional management of pet dogs and cats that are in the last 25% of their expected life span and beyond the expected life expectancy. There are several recent reviews on this topic as well.[2,9,10]

WHAT HAPPENS WITH AGEING?

There are many things that naturally occur or are assumed to occur with ageing of pet dogs and cats. Hair often whitens, there is a general decline in body and coat condition, and a decrease in acuity of senses such as sight and hearing. Other changes include alterations in digestive tract, immune system, renal, and other organ physiology and metabolism. This may result in loss of body and muscle condition as well as a decrease in activity and an increase in sleeping. One of the major effects of ageing is a decreased ability to meet the demands of the environment.[9] This results in a decreased ability to adapt to changes not only in environment but in nutrition and to a decreased reserve capacity to handle large nutritional excesses or deficiencies. Additionally, older patients may have subclinical diseases that affect nutritional status and health. However, there is a great deal of individual variation.

APPROACH TO THE SENIOR AND GERIATRIC PATIENT

As with all patients, nutritional assessment should be performed on all senior and geriatric patients including a more thorough evaluation.[4] The process involves assessment of the patient, diet, and husbandry and development and institution of nutrition. Patient assessment includes a thorough historical and physical examination. Body condition and muscle condition scoring is performed in association with body weight measurement.[4] A longitudinal study in cats indicated that weight loss is evident in ageing cats approximately 2 to 3 years before death from various causes often before clinical signs are apparent.[11] Body condition scoring provides an assessment of body fat content while muscle condition scoring provides an assessment of lean muscle mass. Changes in stool consistency and frequency of defecation, urination, breath odor, and overall activity are noted. Laboratory evaluation is an important part of health screening of older dogs and cats as older patients are prone to a variety of diseases. Dietary assessment includes collecting information on primary diet that is fed including the amount and frequency and any additional foods or treats. Additionally, information should be acquired concerning ability to eat, if food is left, and if there is interest or disinterest in eating. Assessment of husbandry includes information of the environment, exercise and activity, and interactions with other animals. Senior and geriatric patients fall under 4 categories: (1) those that are metabolically efficient, healthy, and maintain weight and condition, (2) those that are metabolically efficient, healthy, and gain weight, (3) those that metabolically inefficient, healthy, and lose weight and condition, and (4) those that are metabolically inefficient and have disease.[12] Information collected from the assessment phase is evaluated and a nutritional plan is formulated and implemented.

EFFECTS OF AGEING ON NUTRITIONAL REQUIREMENTS

Energy requirements

In most species, maintenance energy requirements (MER) decreases with age. In dogs, a decrease of approximately 20% in MER has been documented with the greatest decrease occurring in dogs greater than 7 years of age when compared to young adult dogs.[13] However, in another study evaluating Papillons, Labradors, and Great Danes, it was found that the smaller breed dogs had a higher MER than the other breeds associated with an increase in lean body mass but a decrease in fat mass.[14] Many cats appear to have a decrease in MER similar to other species;[15] however, after 11 years of age, MER per unit body weight increases.[16] Lean body mass is the primary driver of MER and decreases with age in many patients; however, as mentioned, smaller breed dogs may actually have an increase in lean body mass with age.[14] If MER decreases but energy intake does not, then obesity may result; however, the prevalence of obesity decreases as dogs and cats age and a greater proportion of dogs and cats over 12 years of age are underweight compared with other age groups.[11,17,18] Reduced digestibility of nutrients particularly fat and protein occurs with age, which impacts body and muscle condition and energy requirements.[11]

Protein requirements

Because MER is more closely related to lean muscle mass, dietary protein intake is an important. Protein restriction is unnecessary and may result in loss of lean muscle mass in older healthy pet dogs and cats.[19,20] When dietary protein is insufficient, protein is mobilized from lean body mass to support essential functional protein synthesis. Normal animals can adapt to the decreased protein intake and maintain nitrogen balance, yet be in a protein-depleted state associated with gradual loss of lean muscle mass. These animals appear healthy but have a decreased ability to respond to environmental insults including infections.[19,20] Ageing is also associated with loss of lean muscle mass and inadequate protein intake increases the rate of loss of lean muscle mass in aging dogs while adequate protein intake slows this loss.[21] Loss of lean muscle mass is a predictor of morbidity and mortality in ageing patients;[16] therefore, evaluation of muscle condition scoring is important in monitoring senior and geriatric pet dogs and cats. Another approach to improving lean muscle mass without increasing total dietary protein intake is to utilize certain amino acids. In a study involving cats > 12 years of age, increasing the dietary lysine-calorie ratio was correlated with variation in lean body mass. As the lysine-calorie ratio increased, loss of lean body mass was reduced. Cats fed a food with 36% protein (dry matter) and a lysine-calorie ratio of 6.30:1.0 maintained body weight and lean mass similar to that seen with consumption of a 45% protein (dry matter) diet that had a lysine-calorie ratio of 4.38:1.00.[22]

Fat requirements

As mentioned, fat digestibility may decrease with age; however, in patients that require increased energy intake, increased dietary fat may be required.

Fiber requirements

Dogs and cats do not have a dietary fiber requirement, per se; however, fiber affects not only nutrient availability but gastrointestinal function and the microbiota. Different types of fiber exert different effects and a blend of fibers may be more important than type of fiber. Older pet dogs and cats are prone to developing constipation and increased dietary fiber may help with this. However, diets with increased dietary fiber are often formulated to contain less fat and be less calorically dense, which may result in decreased energy intake and loss of body weight and condition.

Other nutrients

There is little evidence that requirements for vitamins and minerals are different in healthy older patients when compared with young adult patients. In patients with subclinical disease or in ageing patients that have decreased intestinal fat digestibility, vitamin, mineral, and electrolyte availability may also be decreased. Oxidative damage plays an important role in many diseases and in the ageing process, and deficiency of antioxidant nutrients can adversely affect antioxidant function, immune function, and markers of health.[23,24] Water intake is an important nutrient to monitor. Glomerular filtration rate declines with age as does concentrating ability; therefore, older pet dogs and cats are prone to dehydration. Omega-3 fatty acids may play an important role in senior and geriatric patients. Omega-3 fatty acids decrease inflammation, which may be beneficial with clinical or subclinical inflammatory diseases such as dermatologic issues and osteoarthritis and have been shown to be beneficial in maintaining glomerular filtration rate in dogs with induced renal disease.[25-28]

RECOMMENDATIONS

So, what is the best nutritional approach for feeding senior and geriatric pet dogs and cats? The answer is – it depends; however, a good nutritional assessment and following the iterative process promoted by the American College of Veterinary Nutrition is important. In patients that are metabolically efficient, healthy, and maintaining body weight and body and muscle condition, continuation of the diet they are currently eating is reasonable. In patients that are metabolically efficient, healthy, but gaining body weight and increasing in body condition, then caloric restriction and possibly feeding a higher fiber diet is indicated. Restriction of other nutrients, such as protein, at this time is not warranted. In patients that are metabolically inefficient but otherwise healthy and losing weight and body and/or muscle condition, feeding a more calorically dense and higher protein diet is indicated. In patients that are metabolically inefficient and not clinically healthy, dietary recommendations are directed towards modifying nutrients that impact upon the underlying disease. Suggested intake for key nutrients are presented in the table.

Recommended key nutritional factors for senior and geriatric dogs and cats.[9,10]

		Dogs		Cats	
Factor	Units	Normal body weight and condition	Inactive / obese prone	Normal body weight and condition	Inactive / obese prone
Water		Free access	Free access	Free access	Free access
Energy density	Kcal ME/gram	3.0-4.0	3.0-3.5	4.0-4.5	3.5-4.0
Crude fat	% DM	10-15	7-10	18-25	10-18
Crude fiber	% DM	≥ 2	≥10	≥5	5-15
Protein	% DM	15-23	15-23	30-45	30-45
Phosphorous	% DM	0.3-0.7	0.3-0.7	0.5-0.7	0.5-0.7
Sodium	% DM	0.15-0.4	0.15-0.4	0.2-0.4	0.2-0.4

ME = metabolizable energy; DM = dry matter

Once a diet is chosen based on the assessment phase, the amount to feed and route of administration is determined. In otherwise healthy senior and geriatric dogs and cats, oral consumption will be utilized. In patients with an inability to eat, assisted nutrition may be accomplished by using feeding tubes such as nasogastric, esophagostomy, or gastrostomy feeding tubes. The amount to feed is based on estimation of resting energy requirement (RER) and multiplying this by a factor to estimate MER. There are 2 formulae used to estimate RER:[29]

Linear: $30(BW_{kg}) + 70$

Exponential: $70(BW_{kg}^{0.75})$

The factor used to multiply the RER is based on the physiologic and pathologic status of the patient. If the patient is active and healthy, then a larger factor is used, but if the patient is inactive and prone to obesity, then a smaller factor is used. In many situations, a factor of 1.0-1.4 will be used, but as mentioned a greater or lesser factor may be chosen depending on the patient. Once a plan has been formulated and implemented, follow-up examinations are undertaken and the plan adjusted depending on response; this is the iterative process. For example, if the patient is gaining weight, then the amount of food provided may be decreased or perhaps a different less calorically dense diet may be chosen.

It is also important to realize the impact of nutrition during growth and adulthood on longevity. Caloric restriction has been shown to increase longevity in dogs and cats and to decrease risk of certain conditions such as osteoarthritis and glucose intolerance, and to improve immune function and lean muscle mass.[30-37] Therefore, incorporation of the nutritional guidelines into practice throughout the lifespan of a patient may increase life span and quality of life.

REFERENCES

1. Lund EM, Armstrong PJ, Kirk CA, et al. Health status and population characteristics of dogs and cats examined at private veterinary practices in the United States. *J Am Vet Med Assoc* 1999;214:1336-1341.
2. Laflamme DP. Nutritional care for aging cats and dogs. *Vet Clin North Am Small Anim Pract* 2012;42:769-791, vii.
3. Bartges J, Boynton B, Vogt AH, et al. AAHA canine life stage guidelines. *J Am Anim Hosp Assoc* 2012;48:1-11.
4. Baldwin K, Bartges J, Buffington T, et al. AAHA nutritional assessment guidelines for dogs and cats. *J Am Anim Hosp Assoc* 2010;46:285-296.
5. Hoyumpa Vogt A, Rodan I, Brown M, et al. AAFP-AAHA: feline life stage guidelines. *J Feline Med Surg* 2010;12:43-54.
6. Members WNAGTF, Freeman L, Becvarova I, et al. WSAVA Nutritional Assessment Guidelines. *J Small Anim Pract* 2011;52:385-396.
7. Epstein M, Kuehn NF, Landsberg G, et al. AAHA senior care guidelines for dogs and cats. *J Am Anim Hosp Assoc* 2005;41:81-91.
8. Pittari J, Rodan I, Beekman G, et al. American association of feline practitioners. Senior care guidelines. *J Feline Med Surg* 2009;11:763-778.
9. Debraekeleer J, Gross KL, Zicker SC. Feeding mature adult dogs: Middle aged and older In: Hand MS, Thatcher CD, Remillard RL, et al., eds. *Small Animal Clinical Nutrition*. 5th ed. Topeka, KS: Mark Morris Institute, 2010;273-280.
10. Gross KL, Becvarova I, Debraekeleer J. Feeding mature adult cats: Middle aged and older In: Hand MS, Thatcher CD, Remillard RL, et al., eds. *Small Animal Clinical Nutrition*. 5th ed. Topeka, KS: Mark Morris Institute, 2010;389-400.
11. Perez-Camargo G. Cat nutrition: What's new in the old? *Comp Contin Educ Small Anim Pract* 2004;26(Suppl2A:5-10.
12. Burkholder WJ. Age-related changes to nutritional requirements and digestive function in adult dogs and cats. *J Am Vet Med Assoc* 1999;215:625-629.
13. Harper EJ. Changing perspectives on aging and energy requirements: aging and energy intakes in humans, dogs and cats. *J Nutr* 1998;128:2623S-2626S.
14. Speakman JR, van Acker A, Harper EJ. Age-related changes in the metabolism and body composition of three dog breeds and their relationship to life expectancy. *Aging Cell* 2003;2:265-275.
15. Laflamme DP, Ballam JM. Effect of age on maintenance energy requirements of adult cats. *Comp Contin Educ Small Anim Pract* 2002;24(Suppl9A):82.
16. Cupp C, Perez-Camargo G, Patil A, et al. Long-term food consumption and body weight changes in a controlled population of geriatric cats. *Comp Contin Educ Small Anim Pract* 2004;26(Suppl2A):60.
17. Lund EM, Armstrong PJ, Kirk CA, et al. Prevalence and risk factors for obesity in adult cats from private veterinary practices. *Int J Appl Res Vet Med* 2005;3:88-96.
18. Lund EM, Armstrong PJ, Kirk CA, et al. Prevalence and risk factors for obesity in adult dogs from private veterinary practices. *Int J Appl Res Vet Med* 2006;4:177-186.
19. Laflamme DP. Pet food safety: dietary protein. *Top Companion Anim Med* 2008;23:154-157.
20. Wannemacher RW, Jr., McCoy JR. Determination of optimal dietary protein requirements of young and old dogs. *J Nutr* 1966;88:66-74.
21. Kealy RD. Factors influencing lean body mass in aging dogs. *Comp Contin Educ Small Anim Pract* 1999;21:34-37.
22. Frantz NZ, Yamka RM, Friesen KG. The effect of diet and lysine:calorie ratio on body compsotion and kidney health in geriatric cats. *Int J Appl Res Vet Med* 2007;5:25-36.
23. Freeman LM, Rush JE, Milbury PE, et al. Antioxidant status and biomarkers of oxidative stress in dogs with congestive heart failure. *J Vet Intern Med* 2005;19:537-541.
24. Yu S, Paetau-Robinson I. Dietary supplements of vitamins E and C and beta-carotene reduce oxidative stress in cats with renal insufficiency. *Vet Res Commun* 2006;30:403-413.
25. Brown SA, Brown CA, Crowell WA, et al. Effects of dietary polyunsaturated fatty acid supplementation in early renal insufficiency in dogs. *J Lab Clin Med* 2000;135:275-286.
26. Lascelles BD, Depuy V, Thomson A, et al. Evaluation of a Therapeutic Diet for Feline Degenerative Joint Disease. *J Vet Intern Med* 2010.
27. Bauer JE. Responses of dogs to dietary omega-3 fatty acids. *J Am Vet Med Assoc* 2007;231:1657-1661.
28. Budsberg SC, Bartges JW, Pazak HE, et al. Effects of different N6:N3 fatty acid diets on canine stifle osteoarthritis. Veterinary Orthopedic Society 28th Annual Meeting 2001.

29. Bartges JW, Kirk CA, Lauten S. Calculating a patient's nutritional requirements. *Vet Med* 2004;99:632.
30. Lawler DF, Larson BT, Ballam JM, et al. Diet restriction and ageing in the dog: major observations over two decades. *Br J Nutr* 2008;99:793-805.
31. Lawler DF, Ballam JM, Meadows R, et al. Influence of lifetime food restriction on physiological variables in Labrador retriever dogs. *Exp Gerontol* 2007;42:204-214.
32. Stowe HD, Lawler DF, Kealy RD. Antioxidant status of pair-fed labrador retrievers is affected by diet restriction and aging. *J Nutr* 2006;136:1844-1848.
33. Greeley EH, Spitznagel E, Lawler DF, et al. Modulation of canine immunosenescence by life-long caloric restriction. *Vet Immunol Immunopathol* 2006;111:287-299.
34. Lawler DF, Evans RH, Larson BT, et al. Influence of lifetime food restriction on causes, time, and predictors of death in dogs. *J Am Vet Med Assoc* 2005;226:225-231.
35. Larson BT, Lawler DF, Spitznagel EL, Jr., et al. Improved glucose tolerance with lifetime diet restriction favorably affects disease and survival in dogs. *J Nutr* 2003;133:2887-2892.
36. Kealy RD, Lawler DF, Ballam JM, et al. Effects of diet restriction on life span and age-related changes in dogs. *J Am Vet Med Assoc* 2002;220:1315-1320.
37. Greeley EH, Ballam JM, Harrison JM, et al. The influence of age and gender on the immune system: a longitudinal study in Labrador Retriever dogs. *Vet Immunol Immunopathol* 2001;82:57-71.

DIETARY PROTEIN AND CHRONIC KIDNEY DISEASE: HOW MUCH IS ENOUGH?

Dr. Joe Bartges, DVM, PhD, DACVIM (Small Animal Internal Medicine), DACVN
Knoxville, TN

INTRODUCTION

For decades, dietary protein restriction has been a cornerstone of nutritional management of pet dogs and cats with chronic kidney disease (CKD). Whether dietary protein restriction is beneficial or necessary; however, has been controversial for decades as well. Chronic kidney disease is defined kidney damage present for at least three months, with or without decreased glomerular filtration rate (GFR) or greater than 50% reduction in GFR persisting for at least three months.[1] Kidney damage is further defined as either 1) microscopic or macroscopic pathologic changes detected by histologic or direct visualization of the kidneys or 2) markers of damage detected by blood or urine tests or imaging studies. The International Renal Interest Society (IRIS; http://www.iris-kidney.com) recommends that the appropriate term is chronic kidney disease and proposes a dynamic staging system based on serum/plasma creatinine concentration, presence of proteinuria, and presence and degree of systemic arterial hypertension.[1] Before staging is undertaken, the patient must be proven to have kidney disease that is chronic in nature. Based on staging, therapeutic interventions may be undertaken including nutritional.

DIETARY PROTEIN AND PROTEIN METABOLISM[2]

Proteins are large, complex molecules composed of amino acids. Amino acids are composed of carbon, hydrogen, oxygen, nitrogen, and sometimes sulfur and phosphorous. Although hundreds of amino acids exist, only 20 are commonly found in protein of which 10 are essential for dogs and 11 are essential for cats. Essential amino acids must be supplied by diet as they cannot be synthesized endogenously in sufficient amounts. Proteins provide structure (e.g. collagen, muscle, keratin, hemoglobin), function (e.g. enzymes, hormones, antibodies), and are an energy source after the amino group is removed by deamination or transamination. Proteins also provide a source of nitrogen for synthesis of nitrogen containing compounds such as nucleic acids, creatinine, and some neurotransmitters.

Dietary proteins are digested and absorbed from the gastrointestinal tract with digestion beginning in the stomach by acid hydrolysis and pepsin. Pancreatic and brush border enzymes further metabolize protein so that amino acids, di-peptides, and tri-peptides are absorbed primarily. Amino acids are absorbed by a sodium-dependent energy-requiring process using one of 6 transporters for neutral, basic, acidic, tricarboxylic acid, imino, and beta- amino acids. Absorbed amino acids and di- and tri-peptides are reassembled into "new" proteins by the liver and other tissues. Amino acids are used for tissue protein synthesis; synthesis of enzymes, albumin, and hormones, and other nitrogen-containing compounds; and deamination and use of carbon skeletons for energy. Amino acids and proteins not absorbed in the small intestine may be fermented in the large intestine to produce fecal odor compounds and utilized by colonic bacteria. There is no storage pool of amino acids, per se, although structural and functional proteins can be catabolized and in essence represent a storage pool; however, their utilization results in depletion of such protein. In catabolic situations, the amine group is removed and the carbon skeleton of the amino acid used for glucose and/or ketone body production. The amine group can be converted to urea via the urea cycle or used for urinary ammonia excretion (glutamine) or purine synthesis.

When discussing dietary protein, it is not only the quantity provided, but the quality of the protein. Protein quality refers to the efficiency by which amino acids from food are converted into tissue. The efficiency depends on the protein source, concentration of essential amino acids in the food, and their availability.[3] High quality proteins provide all essential amino acids in high amounts and in proportion, whereas poor quality proteins have low amounts or lack one or more essential amino acids (limiting amino acid(s)). Poor protein quality can also result from an excess of certain amino acids that interfere with availability or usage of essential amino acids. Other factors affect protein quality as well.

Protein requirements for healthy pet dogs and cats depend on whether a high quality protein or commonly used protein sources are used. When using a high quality protein, growing dogs require approximately 18% protein (dry matter basis; DMB) and adult dogs require approximately 8% protein (DMB), while growing kittens require 18% protein (DMB) and adult cats require approximately 16% protein (DMB).[4] The Association of American Feed Controls Officials provides daily allowances for dietary protein where common dietary protein sources are used. Allowances for growing dogs are 22% protein (DMB), adult dogs are 22% protein (DMB), growing kittens are 30% (DMB), and adult cats are 26% (DMB).[5]

PROTEIN METABOLISM AND REQUIREMENTS WITH CHRONIC KIDNEY DISEASE

The kidney is a highly metabolic organ; therefore, CKD would be expected to have a variety of adverse systemic effects in addition to renal effects. In general terms, renal function includes filtration, reabsorption, and secretion of endogenous and exogenous substances; maintaining homeostasis of water, acid-base, minerals, electrolytes, and other substances; neurohumoral regulation of vitamins (such as vitamin D), red blood cell production, blood pressure, and others; and energy metabolism including gluconeogenesis and reabsorption of amino acids. There is little to no information on protein metabolism in dogs and cats with chronic kidney disease or the incidence of cachexia and sacropenia in these patients although it is assumed to occur in at least some.[6] It has been shown that dogs with CKD that are under-conditioned have a decreased survival compared to dogs who are optimally or over conditioned;[7] however, no information was provided in this study as to muscle condition scoring and lean body mass (LBM).

In uremic rodent models, 3-methylhistidine release, an index of myofibrillar protein degradation, is elevated to a greater degree than in rodents without CKD.[8] During *in vivo* infusion of ^{14}C leucine, protein synthesis following fasting was less in rats with CKD than in healthy rats.[9] Muscle protein synthesis was lower in rats with CKD following a period of stress than in rats without CKD.[10] There is also increased muscular alanine and glutamine release and a decrease in leucine incorporation in rats with CKD[11] and increased muscular tyrosine and phenylalanine release[12]. It has been shown in rodent studies that metabolic acidosis, excess angiotensin II, and inflammation increases muscular protein degradation and decreases muscular protein synthesis primarily through insulin resistance and impaired insulin/IGF-I signalling.[13-17] Data in human beings suggest that CKD, even when advanced, does not in itself engender net protein breakdown. Many nitrogen balance studies in human beings with CKD who are not undergoing chronic dialysis have demonstrated that they are able to maintain neutral or positive nitrogen balance studies with low protein intakes.[18-20] Thus, uremia, per se, does not stimulate net protein catabolism and human pre-dialytic patients fed low protein diets are able to conserve protein if metabolic academia is not present or if there is no concurrent illness. Protein turnover kinetic studies suggest that the mechanism of protein conservation include down-regulation of protein degradation, and amino acid oxidation and maintenance of protein synthesis at near normal levels.[13] If chronic inflammation, increased cytokine levels and activity, and/or metabolic acidosis exists, then protein requirements may be increased.

So, if protein restriction is not detrimental with CKD in many patients, the question is whether there is an advantage over not restricting dietary protein with CKD. In evaluating the veterinary literature, no clear consensus is derived. There are potentially several reasons for this. Studies evaluating only a dietary protein effect have used induced CKD models in dogs and cats, which may or may not adequately represent naturally occurring disease. Some studies demonstrate histologic changes while others have only evaluated blood biochemical parameters. Most induced CKD studies evaluated a small number of animals. Finally, CKD is a complex disease and optimal treatment of patients with CKD involves modifying multiple dietary factors. Nonetheless, dietary protein restriction if not detrimental has potential benefits. Decreased dietary protein intake inhibits secretion of TGF-β a cytokine involved in progression of CKD.[21] Decreased dietary protein intake reduces tubular hyperfunction by decreasing renal acid load and renal ammoniagenesis. Sulfur-containing amino acids in dietary protein contributes to renal acid load.[22] If proteinuria is present, dietary protein restriction is beneficial as proteinuria induces inflammatory and fibrogenic pathways and increases oxidative stress.[23]

Two studies evaluated effects of dietary protein on progression of induced CKD for one year in cats.[24,25] In one study, renal function did not progressively decrease, regardless of dietary protein amount and caloric intake; however, cats fed the higher protein diet had more severe glomerular and tubulointerstitial changes.[24] The cats in the higher protein diet group also consumed more calories. In the other study, no difference in renal function or glomerular lesions were found in cats consuming the high protein diet.[25] These studies were of short duration and progression of spontaneous CKD in cats may occur slowly.[26-28]

Clinical trials of naturally occurring CKD in pet dogs and cats have utilized dietary modification including but not limited to dietary protein restriction. The effectiveness of diet therapy in minimizing uremic episodes and mortality in dogs and cats with naturally occurring IRIS CKD stages 2 and 3 have been established in double-blinded randomized controlled clinical trials.[26,29,30] These studies compared a renal diet to a maintenance diet. The renal diets contained reduced quantities of dietary protein, but also contained reduced quantities of phosphorous and sodium and were supplemented with omega-3 fatty acids when compared with the maintenance diet. In the canine study, the risk of developing a uremic crisis was reduced by approximately 75% in dogs fed the renal diet when compared with dogs fed the adult maintenance diet.[29] The median symptom-free interval in dogs fed the renal diet was 615 days compared with 252 days in dogs consuming the maintenance diet. Furthermore, the risk of death irrespective of cause was reduced by 66% and the risk of death from renal causes was reduced by 69%. Median survival time for dogs consuming the renal diet was 594 days compared with 198 days for dogs consuming the maintenance diet. In the feline trial, cats in IRIS CKD stage 2 and 3 were evaluated.[26] The risks of uremic crises and renal deaths were significantly reduced. Among the cats fed the renal diet, there were no uremic crises or renal deaths and only three deaths from non-renal causes over the 2 year study period. In contrast, among the cats consuming the maintenance diet, 6 developed uremic crises, 5 died of renal causes, and 5 died of non-renal causes. In another study of cats with CKD, feeding a renal diet was associated with a significantly longer survival with a median survival time of 633 days for cats fed the renal diet versus 264 days for cats fed the maintenance diet.[28]

RECOMMENDATIONS

There are no clear cut recommendations concerning dietary protein with CKD; however, some recommendations can be made. Limiting dietary protein intake may help ameliorate clinical signs in patients. Although currently available evidence fails to support a recommendation for or against limiting dietary protein intake alone in non-uremic patients, with CKD, there are potential benefits assuming that patients maintain adequate caloric intake, body condition, and muscle condition. Patients may be more likely to accept a new renal diet if offered before uremia develops.[1] Although not discussed, dietary phosphorous restriction has been shown by itself to slow progression of CKD and it is difficult to achieve this degree of phosphorous restriction using typical ingredients without limiting dietary protein.[31] But it is not impossible. In determining how much protein to recommend for dogs and cats with CKD, patients should be monitored for signs of protein malnutrition and nutritional management adjusted to maintain optimal body condition and muscle condition minimally. For cats with CKD, the minimum dietary protein requirement was 20% of calories[32], which equates to 24% protein (on a dry matter basis; DMB). Similar studies have not been reported in dogs. The Association of American Feed Control Officials recommends a minimum of 18% (DMB) for dogs and 26% (DMB) for cats.[5] Recommendations for dogs with CKD is 14-20% (DMB) and for cats with CKD it is 28-35% (DMB).[33] It is emphasized that less total dietary protein can be fed if high biologic value proteins are fed.

Uremic gastrointestinal complications may be present in patients with CKD. These may be overt (e.g. vomiting, anorexia), but may be more covert with only hyporexia being present. Patients with CKD may benefit from anti-emetics, ant-acids, and/or gastroprotectant pharmacologic therapy.[34] Hypokalemia, which occurs more frequently in cats with CKD than in dogs, is not only associated with muscle weakness but hyporexia to anorexia.[35] If necessary, nutritional therapy can be facilitated through use of feeding tubes.[36]

In patients with CKD that experience sacropenia, a thorough physical examination and biochemical evaluation should be undertaken. Metabolic acidosis does not occur until late in progression of CKD in cats;[37] however, if present it induces protein metabolism and loss of lean body mass. In patients with sacropenia, an attempt should be made to increase dietary protein while monitoring response. If azotemia or acidosis worsens or if electrolyte and mineral imbalances occur, then the nutritional plan should be re-evaluated and adjusted.

There are several novel treatments for CKD. In pre-dialytic human patients with CKD, feeding a low protein diet supplemented with ketoanalogues of essential amino acids (keto-diet) proved effective in ameliorating metabolic disturbances of advanced CKD and delaying the initiation of dialysis without deleterious effects on nutritional status. Several recent studies report that the keto-diet could also slow down the rate of decline in renal function, with better outcomes after the initiation of dialysis.[38-40] Enteric dialysis using an oral sorbent may also permit feeding of higher protein diets.[41] A probiotic formula is currently available for use in dogs and cats with CKD and is marketed as enteric dialysis although no data substantiate this claim.[42] These treatments especially use of keto-acids represent unexplored territory in the treatment of CKD in pet dogs and cats.

REFERENCES

1. Polzin DJ. Chronic kidney disease In: Bartges J,Polzin DJ, eds. *Nephrology and Urology of Small Animals*. Ames: Wiley-Blackwell, 2011;433-471.
2. Gross KL, Yamka RM, Khoo CG, et al. Macronutrients In: Hand MS, Thatcher CD, Remillard RL, et al., eds. *Small Animal Clinical Nutrition*. 5th ed. Topeka, KS: Mark Morris Institute, 2010;49-105.
3. Brown RG. Protein in dog foods. *Can Vet J* 1989;30:528-531.
4. *Nutrient Requirements of Dogs and Cats*. Washington DC: National Academies Press, 2006.
5. *AAFCO Official Publication*. Champaign, IL: Association of American Feed Control Officials, 2013.
6. Freeman LM. Cachexia and sarcopenia: emerging syndromes of importance in dogs and cats. *J Vet Intern Med* 2012;26:3-17.
7. Parker VJ, Freeman LM. Association between Body Condition and Survival in Dogs with Acquired Chronic Kidney Disease. *J Vet Intern Med* 2011;25:1306-1311.
8. Li JB, Wassner SJ. Muscle degradation in uremia: 3-methylhistidine release in fed and fasted rats. *Kidney Int* 1981;20:321-325.
9. Holliday MA, Chantler C, MacDonnell R, et al. Effect of uremia on nutritionally-induced variations in protein metabolism. *Kidney Int* 1977;11:236-245.
10. Baliga R, George VT, Ray PE, et al. Effects of reduced renal function and dietary protein on muscle protein synthesis. *Kidney Int* 1991;39:831-835.
11. Garber AJ. Skeletal muscle protein and amino acid metabolism in experimental chronic uremia in the rat: accelerated alanine and glutamine formation and release. *J Clin Invest* 1978;62:623-632.
12. Harter HR, Karl IE, Klahr S, et al. Effects of reduced renal mass and dietary protein intake on amino acid release and glucose uptake by rat muscle in vitro. *J Clin Invest* 1979;64:513-523.
13. Lim VS, Kopple JD. Protein metabolism in patients with chronic renal failure: role of uremia and dialysis. *Kidney Int* 2000;58:1-10.
14. Bailey JL, Zheng B, Hu Z, et al. Chronic kidney disease causes defects in signaling through the insulin receptor substrate/phosphatidylinositol 3-kinase/Akt pathway: implications for muscle atrophy. *J Am Soc Nephrol* 2006;17:1388-1394.
15. Bailey JL, Price SR, England BK, et al. Signals regulating accelerated muscle protein catabolism in uremia. *Miner Electrolyte Metab* 1997;23:198-200.
16. Song YH, Li Y, Du J, et al. Muscle-specific expression of IGF-1 blocks angiotensin II-induced skeletal muscle wasting. *J Clin Invest* 2005;115:451-458.
17. Zhang L, Du J, Hu Z, et al. IL-6 and serum amyloid A synergy mediates angiotensin II-induced muscle wasting. *J Am Soc Nephrol* 2009;20:604-612.
18. Kopple JD, Swendseid ME. Nitrogen balance and plasma amino acid levels in uremic patients fed an essential amino acid diet. *Am J Clin Nutr* 1974;27:806-812.
19. Ford J, Phillips ME, et al. Nitrogen balance in patients with chronic renal failure on diets containing varying quantities of protein. *Br Med J* 1969;1:735-740.
20. Mitch WE. Beneficial responses to modified diets in treating patients with chronic kidney disease. *Kidney Int Suppl* 2005:S133-135.
21. Fukui M, Nakamura T, Ebihara I, et al. Low-protein diet attenuates increased gene expression of platelet-derived growth factor and transforming growth factor-beta in experimental glomerular sclerosis. *J Lab Clin Med* 1993;121:224-234.
22. Halperin ML. Metabolism and acid-base physiology. *Artif Organs* 1982;6:357-362.
23. Gorriz JL, Martinez-Castelao A. Proteinuria: detection and role in native renal disease progression. *Transplant Rev (Orlando)* 2012;26:3-13.
24. Adams LG, Polzin DJ, Osborne CA, et al. Influence of dietary protein/calorie intake on renal morphology and function in cats with 5/6 nephrectomy. *Lab Invest* 1994;70:347-357.
25. Finco DR, Brown SA, Brown CA, et al. Protein and calorie effects on progression of induced chronic renal failure in cats. *Am J Vet Res* 1998;59:575-582.
26. Ross SJ, Osborne CA, Kirk CA, et al. Clinical evaluation of dietary modification for treatment of spontaneous chronic kidney disease in cats. *J Am Vet Med Assoc* 2006;229:949-957.
27. Elliott J, Barber PJ, Rawlings JM, et al. Effect of phosphate and protein restriction on progression of chronic renal failure in cats. *J Vet Intern Med* 1998;12:221(abstract).
28. Elliott J, Rawlings JM, Markwell PJ, et al. Survival of cats with naturally occurring chronic renal failure: effect of dietary management. *J Small Anim Pract* 2000;41:235-242.
29. Jacob F, Polzin DJ, Osborne CA, et al. Clinical evaluation of dietary modification for treatment of spontaneous chronic renal failure in dogs. *J Am Vet Med Assoc* 2002;220:1163-1170.
30. Harte JG, Markwell PJ, Moraillon RM, et al. Dietary management of naturally occurring chronic renal failure in cats. *J Nutr* 1994;124:2660S-2662S.
31. Burkholder WJ. Dietary considerations for dogs and cats with renal disease. *J Am Vet Med Assoc* 2000;216:1730-1734.
32. Kirk CA, Hickman MA. Dietary protein requirements of cats with spontaneous renal disease. *J Vet Intern Med* 2000;14:351.
33. Forrester SD, Adams LG, Allen TA. Chronic kidney disease In: Hand MS, Thatcher CD, Remillard RL, et al., eds. *Small Animal Clinical Nutrition*. 5th ed. Topeka: Mark Morris Institute, 2010;865-810.
34. Bartges JW. Chronic kidney disease in dogs and cats. *Vet Clin North Am Small Anim Pract* 2012;42:669-692, vi.
35. Dow SW, Fettman MJ, LeConteur RA, et al. Potassium depletion in cats: Renal and dietary influences. *J Am Vet Med Assoc* 1987;191:1569-1575.
36. Elliott DA, Riel DL, Rogers QR. Complications and outcomes associated with use of gastrostomy tubes for nutritional management of dogs with renal failure: 56 cases (1994-1999). *J Am Vet Med Assoc* 2000;217:1337-1342.
37. Elliott J, Syme HM, et al. Acid-base balance of cats with chronic renal failure: effect of deterioration in renal function. *J Small Anim Pract* 2003;44:261-268.
38. Garneata L, Mircescu G. Effect of low-protein diet supplemented with keto acids on progression of chronic kidney disease. *J Ren Nutr* 2013;23:210-213.
39. Kovesdy CP, Kopple JD, Kalantar-Zadeh K. Management of protein-energy wasting in non-dialysis-dependent chronic kidney disease: reconciling low protein intake with nutritional therapy. *Am J Clin Nutr* 2013;97:1163-1177.
40. Mir S, Ozkayin N, Akgun A. The role of keto acids in the supportive treatment of children with chronic renal failure. *Pediatr Nephrol* 2005;20:950-955.
41. Friedman EA. Can the bowel substitute for the kidney in advanced renal failure? *Curr Med Res Opin* 2009;25:1913-1918.
42. Rishniw M, Wynn SG. Azodyl, a synbiotic, fails to alter azotemia in cats with chronic kidney disease when sprinkled onto food. *J Feline Med Surg* 2011;13:405-409.

FIVE NEW PUBLICATIONS YOU NEED TO KNOW ABOUT IN VETERINARY CLINICAL PHARMACOLOGY

Ted Whittem, BVSc, PhD, DACVCP, FACVSc and Sarah Robson BVSc (Hons)
Melbourne, Australia

INTRODUCTION

You probably need to know about papers that your colleagues think are landmark papers. So, what is a landmark paper in veterinary clinical pharmacology? At a basic level, landmark papers are a "must-read" for all in the discipline. But is a landmark paper one that breaks new therapeutic ground, introduces a new treatment modality, modifies forever the regulation of medicines, exemplifies ideal reporting of research, shows a new application of experimental methods or introduces a wholly new approach to an experimental design? Does an excellent review paper justify being called a landmark paper? Is it a landmark if it is a paper that challenges dogma, and debunks or confirms old myths? Should we include in the "must-read" category, papers that exemplify the awful, teach us how to befuddle peer reviewers or which attempt to justify the practice of eminence-based medicine?

How many landmark papers were there in 2013 and how did I pick just five? I asked three graduate students in clinical pharmacology to each identify five papers from 2013 publications that, to them, were landmarks for their learning. From these I selected the following…

1. A REVIEW ARTICLE

Court MH. Feline drug metabolism and disposition: pharmacokinetic evidence for species differences and molecular mechanisms. Vet Clin North Am Small Anim Pract 2013;43:1039-1054.

Every now and then a review article is published that becomes the go-to reference for decades: a paper that is lucid, ordered understandable, thorough and stands the test of time is a landmark of its time.

There are too few examples in veterinary clinical pharmacology. Erskine's paper on therapy for mastitis in dairy cows is an example, having been cited more than 100 times. [1]

Cited more than 240 times, Lin reviewed in 1995 the species differences in pharmacokinetics. Lin took a theoretical approach and cited examples, but very little attention was paid to the clinical veterinary point of view, as that was not the thrust of the paper. [2]

In 2013 Michael Court produced a review of drug metabolism in the cat [3] an area of interest in veterinary clinical pharmacology since the beginning of the discipline [4]. Court presents and article which categorizes the differences in metabolism between cats and dogs, lending order to our thinking about which drugs are (or might be) cleared more quickly, similarly or more slowly than would be expected based in inter-species comparisons. Reviewing data for more than 25 drugs in three major classes, Court illustrates a method for understanding the clinical ramifications of clearance pathways: drugs eliminated by conjugation pathways tend to be more slowly cleared, drugs which are oxidized tend to be more quickly cleared and drugs which are not metabolized tend to be similarly cleared to other species. This review paper also takes on board the task of myth-busting, by reviewing in an objective way the difference between text book dogma and peer-reviewed research into drug glucuronidation in cats, using aspirin's metabolism as an example. [3]The paper summarizes the genetic causes of changed drug metabolism, that have been recently elucidated, and it also refers to the phylogenetic origins of the cat's unique drug metabolism [5]. Whether this review paper becomes a landmark paper will depend on the extent to which it is accepted as a milestone: does it coordinate our knowledge of drug metabolism in the cat?

2. CHALLENGING DOGMA

Raidal SL, Edwards S, Pippia J, et al. Pharmacokinetics and safety of oral administration of meloxicam to foals. J Vet Intern Med 2013;27:300-307.

To publish findings that conflict with earlier work, or which incise between theory and practical outcome can lead to notoriety. Such papers can become particularly highly cited and landmark papers.

For example, the long perpetuated myth that competition for protein binding between concomitantly administered drugs will always result in changes to free drug concentration, and possibly therefore to efficacy or toxicity, exemplified by the misunderstanding over warfarin toxicity and NSAIDs, was well addressed by Toutain and Bousquet-Melou in a paper that is on the pathway to be one of our discipline's landmarks, already cited more than 35 times. [6]

The metabolism pathways of meloxicam in horses have been carefully described [7] and compare with the pathways known for adult humans. [8] The post natal expression of the enzymes to which meloxicam metabolism is attributed has been studied, and these enzymes have been shown to be less expressed in foals than adult horses. [9] Therefore, assuming no major changes in oral bioavailability or in drug distribution, one would predict that, like other NSAIDs, meloxicam would be eliminated more slowly in foals than in adult horses.

Raidal and colleagues conducted experiments to evaluate the pharmacokinetics of orally administered meloxicam in foals and they reported from their data that the clearance of meloxicam in foals is more rapid than is reported for adult horses. This was an unfortunate expression, since no intravenous dose was used and the clearance cannot be accurately calculated. However, the work did show that foals had a faster elimination rate for meloxicam than did adult horses. [10] This was a finding that conflicts with the earlier discussed theoretical conclusion based on metabolism of meloxicam.

Landmark papers can be those that provoke consternation, challenging theory: they oft require repetition.

3. INTRODUCING A NEW TREATMENT MODALITY OR EXEMPLIFYING IDEAL REPORTING OF RESEARCH?

Puigdemont A, Brazis P, Ordeix L, et al. Efficacy of a new topical cyclosporine A formulation in the treatment of atopic dermatitis in dogs. Vet J 2013;197:280-285.

A recent buzz has arisen surrounding the so-called new treatment modalities offered by "nanotechnology".[11] Nanotechnology focuses on developing active molecule presentations at less than 1 um. This is not a new field in veterinary medicine; for decades formulations have been marketed which are micellar solutions and colloidal solutions with particle sizes less than 100 nm.[12,13] The advent of nanotechnology as a field of galenic interest, has however, focused attention to its potential for solving drug delivery problems in a systematic way.

Puigdemont and his colleagues conducted a randomized clinical trial evaluation of a nano-capsule formulation of cyclosporine A for treatment of atopic dermatitis in dogs. This formulation is described as an oil-in-water emulsion incorporating the nanocapsules within one phase, the altered lipophillicity of which is said to aid penetration of the drug locally, after topical spray-on administration. Further, it was claimed that the nanoencapsulation prevented systemic absorption of the active moiety, thereby reducing systemic toxicity.[14]

Unfortunately, the Puigdemont manuscript does not present the non-absorption data. Further reporting problems with this manuscript prevent the landmark status it might otherwise have deserved. Indeed the report exemplifies some of the failings in reporting of clinical trials in veterinary medical literature that have been common in past years. Reporting of essential details enables subsequent re-evaluation and meta-analyses of clinical trials; in both human and veterinary medicine these details have been described.[15-17] A failure to follow these reporting guidelines diminishes the value of otherwise good work.

Landmark papers are expected to be flawless, but may overcome flaws if they ultimately result in adoption of their findings into practice. For this approach to formulating cyclosporine A, time will tell.

4. INTRODUCING NEW APPROACHES TO AN EXPERIMENTAL DESIGN

Liu Y, Imlay JA. Cell death from antibiotics without the involvement of reactive oxygen species. Science 2013;339:1210-1213.
Keren I, Wu Y, Inocencio J, et al. Killing by bactericidal antibiotics does not depend on reactive oxygen species. Science 2013;339:1213-1216.

Some recent work in the field of antimicrobial resistance has introduced to hypothesis that reactive oxygen species may be induced by antimicrobial drugs and cause stress related death or mutagenesis of bacteria.[18] This emerging hypothesis has been challenged by two consecutive papers published in Science in 2013.[19,20]

These workers used their knowledge of cell biochemistry to directly evaluate whether some antibiotics generated superoxide radicals, by replacing oxygen with nitrates as an alternative respiratory pathway in anaerobic conditions and showing that reactive oxygen species are not needed for bacterial killing by the aminoglycoside, kanamycin. Other approaches were also used to approach the 'proof' from a novel angle. Measurement of oxygen consumption showed that it declined in association with cell injury whereas an increase of oxygen consumption would have been expected if oxygen radical production was the mode of cell injury: the production of hydrogen peroxide was directly measured and found to not increase on exposure to the antibiotics. Bringing classical experimental design to the bench, these workers used positive controls to show that their assay systems were sensitive to the expected changes.[20]

Separate work published in the same issue of Science took alternative approaches to the same conclusion. Keren and colleagues especially compared several antibiotics' activity against the same strain of *Escherichia coli*, finding that there was no difference in activity under aerobic or anaerobic conditions, except for aminoglycosides at low concentrations.[19]

Papers that systematically and thoroughly address a disputed hypothesis can become landmarks.

5. REGULATORY SCIENCE IN CLINICAL PHARMACOLOGY

Lee SW, Markham PF, Coppo MJ, et al. Attenuated vaccines can recombine to form virulent field viruses. Science 2012;337:188.

Clinical Pharmacology is a broad discipline, the exponents of which populate regulatory roles for both small molecule therapeutics and biological products, such as vaccines. With the advent of protein therapeutics, gene therapy and DNA vaccines the historical gulf between drugs (FDA) and biologicals (USDA) is blurring. Clinical pharmacology needs to pay attention to these other newer disciplines that will impact our approach to diagnosis and treatment of animal disease.

This short paper was published in July 2012, but deserves inclusion as a landmark. The introduction of a new vaccine strain for infectious laryngotracheitis virus (ILTV) in Australia in 2006 coincided with multiple outbreaks of new and genotypically distinct virulent virus strains. This paper describes the genome sequencing of these strains and the vaccines, the evaluation of the strains in silico and by inoculation into specific pathogen free chickens. The authors concluded that the newly introduced vaccine strain and the long-standing vaccine strains had recombined to form two new, virulent viruses. This finding has implications for the regulation of live virus vaccines, particularly for ILTV but probably also for other live DNA virus containing vaccines.[21]

REFERENCES

1. Erskine RJ, Wagner S, DeGraves FJ. Mastitis therapy and pharmacology. Vet Clin North Am Food Anim Pract 2003;19:109-138, vi.
2. Lin JH. Species similarities and differences in pharmacokinetics. Drug Metab Dispos 1995;23:1008-1021.
3. Court MH. Feline drug metabolism and disposition: pharmacokinetic evidence for species differences and molecular mechanisms. Vet Clin North Am Small Anim Pract 2013;43:1039-1054.
4. Davis LE, Westfall BA. Species differences in biotransformation and excretion of salicylate. Am J Vet Res 1972;33:1253-1262.

5. Shrestha B, Reed JM, Starks PT, et al. Evolution of a major drug metabolizing enzyme defect in the domestic cat and other felidae: phylogenetic timing and the role of hypercarnivory. PLoS One 2011;6:e18046.
6. Toutain PL, Bousquet-Melou A. Free drug fraction vs free drug concentration: a matter of frequent confusion. J Vet Pharmacol Ther 2002;25:460-463.
7. Tevell Åberg A, Olsson C, Bondesson U, et al. A mass spectrometric study on meloxicam metabolism in horses and the fungus Cunninghamella elegans, and the relevance of this microbial system as a model of drug metabolism in the horse. Journal of Mass Spectrometry 2009;44:1026-1037.
8. Chesne C, Guyomard C, Guillouzo A, et al. Metabolism of Meloxicam in human liver involves cytochromes P4502C9 and 3A4. Xenobiotica 1998;28:1-13.
9. Nebbia C, Dacasto M, Carletti M. Postnatal development of hepatic oxidative, hydrolytic and conjugative drug-metabolizing enzymes in female horses. Life Sciences 2004;74:1605-1619.
10. Raidal SL, Edwards S, Pippia J, et al. Pharmacokinetics and safety of oral administration of meloxicam to foals. J Vet Intern Med 2013;27:300-307.
11. Mills P. Topical drug delivery and nanotechnology. Vet J 2013;197:519-520.
12. Brewster ME, Estes KS, Bodor N. Development of a non-surfactant formulation for alfaxalone through the use of chemically-modified cyclodextrins. J Parenter Sci Technol 1989;43:262-265.
13. Pak-Kan AL, Williams JB. Solubilization of ivermectin in water. USPTO US4389397 A Assignee; Merck & Co., Inc., Rahway, NJ.; 1983.
14. Puigdemont A, Brazis P, Ordeix L, et al. Efficacy of a new topical cyclosporine A formulation in the treatment of atopic dermatitis in dogs. Vet J 2013;197:280-285.
15. Moher D, Liberati A, Tetzlaff J, et al. Preferred reporting items for systematic reviews and meta-analyses: the PRISMA statement. Annals Of Internal Medicine 2009;151:264.
16. O'Connor AM, Sargeant JM, Gardner IA, et al. The REFLECT statement: methods and processes of creating reporting guidelines for randomized controlled trials for livestock and food safety. Prev Vet Med 93:11-18.
17. Begg C, Cho M, Eastwood S, et al. Improving the quality of reporting of randomized controlled trials. The CONSORT statement. JAMA 1996;276:637-639.
18. Kohanski MA, DePristo MA, Collins JJ. Sublethal antibiotic treatment leads to multidrug resistance via radical-induced mutagenesis. Mol Cell 2010;37:311-320.
19. Keren I, Wu Y, Inocencio J, et al. Killing by bactericidal antibiotics does not depend on reactive oxygen species. Science 2013;339:1213-1216.
20. Liu Y, Imlay JA. Cell death from antibiotics without the involvement of reactive oxygen species. Science 2013;339:1210-1213.
21. Lee SW, Markham PF, Coppo MJ, et al. Attenuated vaccines can recombine to form virulent field viruses. Science 2012;337:188.

EQUINE PHARMACEUTICALS: NEW PRODUCTS AND ADVANCES

Jennifer L. Davis, DVM, PhD, DACVIM (LA), DACVCP
Raleigh, NC

INTRODUCTION

During the last decade, several new products have been approved for use in the horse. In addition to label uses, multiple extralabel uses have been researched and used in practice. This lecture will detail the current information available on equine pharmaceuticals, including ceftiofur crystalline free acid, pergolide mesylate, firocoxib and ponazuril.

CEFTIOFUR CRYSTALLINE FREE ACID

The β-lactam antibiotics include the penicillins, cephalosporins and carbapenems. They have excellent activity against most gram-positive bacteria, and very few associated side effects. They are considered bactericidal and time-dependent. Post-antibiotic effects have been associated with some drugs in this class. The mechanism of action involves penetration of the outer cell wall and binding to penicillin binding proteins (PBP). This interferes with cell wall synthesis and opens channels through the cell wall to create pores that allow fluid into the cell, causing cell swelling and death. In general, the beta-lactam antibiotics have low plasma protein binding, distribute well to the extracellular fluid in most tissues, and are excreted renally. With a few exceptions, they have a very short half-life and require frequent dosing. Beta-lactams do not distribute well to protected sites, such as the CNS, the eye or the prostate.

The cephalosporin antibiotics have many advantages in animals, including a broad spectrum of activity and a good safety profile. Cephalosporins studied in horses for clinical use include the first-generation cephalosporins cephalexin, cefazolin, cephapirin, and cefadroxil; the second-generation cephalosporin cefoxitin; and the third-generation cephalosporins ceftiofur, cefpodoxime and ceftriaxone. Oral absorption of first-generation cephalosporins is somewhat limited, but this route can be used in some cases, as with cephalexin. First-generation cephalosporins have a spectrum that is limited to highly susceptible gram-negative bacilli (Enterobacteriaceae), streptococci, and Staphylococcus spp.

Perhaps the most frequently used cephalosporin in horses is ceftiofur. Ceftiofur is metabolized quickly to an active metabolite, desfuroylceftiofur (and other metabolites). Ceftiofur sodium (Naxcel®) was approved for use in horses for treatment of respiratory tract infections caused by Streptococcus equi subsp. zooepidemicus at a dose of 2.2 to 4.4 mg/kg q24h IM.

A new formulation of ceftiofur has been approved for use in horses (ceftiofur crystalline free acid (CCFA), Excede®). This formulation is designed to be a sustained release formulation that will provide 10 days worth of therapeutic concentrations against streptococci (MIC < 0.2 µg/mL) following a 2 dose regimen (6.6 mg/kg IM, repeated 96 hours later). The main adverse effect associated with this formulation has been injection site reactions, which can be minimized by splitting the dose into two different injection sites, using a large gauge needle (16g 1 ½ inch), and using one needle to draw up the medication and another for administration.[1] Subcutaneous dosing regimens have also been studied to see if this reduced the incidence of injection site reactions.[2] Although pharmacokinetics were favorable via this route, injection site reactions were more severe than those receiving IM injections. Extralabel dosing of the drug has been used in clinical practice, mainly through the use of prolonged dosing regimens. For treatment of streptococci, the recommended dosing regimen for CCFA is 6.6 mg/kg IM on day 1 and 4, then every 7 days after.[3] This regimen produced plasma and bronchoalveolar lavage concentrations above the MIC for streptococci for up to 32 days. For more resistant bacteria (MIC ≤ 1 µg/mL), the recommended dosing regimen is 6.6 mg/kg IM every 4 days, based on pharmacokinetic modeling.

All of the above regimens are for adult horses, as the pharmacokinetics differ in foals. Investigations in healthy neonatal foals showed that subcutaneous administration of CCFA had a lower AUC, shorter observed time to maximum concentration, and a higher observed Cmax compared to adult horses.[4] Based on that study, administration at a dose of 6.6 mg/kg body weight SC every 72 h would provide protection against bacteria isolated from neonatal foals, based upon an MIC of 0.5 µg/mL, and SC administration resulted in minimal injection site inflammation. Anecdotally, use of higher doses (13.2 mg/kg) more frequently (q48h) may be more effective. Weanling age foals (4-6 months old) also exhibit different pharmacokinetics, including higher peak plasma concentrations and area under the curve compared to adults, however these differences were not significant enough to necessitate a difference in dosing regimen.[5] Therefore, administration of 6.6 mg/kg IM in weanling foals provided plasma and PELF concentrations above the therapeutic target of 0.2 µg/mL for at least 4 days and would be expected to be an effective treatment for pneumonia caused by Streptococcus equi subsp. zooepidemicus at doses similar to the adult label.

CCFA has also been investigated for use in mares with placentitis. Concentrations in placenta and fetal tissues were below the MIC of relevant pathogens and drug was not detected in amniotic fluid or foal serum.[6] Additionally, treatment did not appear to improve foal survival. It did, however, reach appropriate endometrial tissue values to exceed the MIC of S. zooepidemicus, and should be effective in the treatment of equine bacterial endometritis caused by S. zooepidemicus and other susceptible bacterial pathogens in the mare.[7]

PERGOLIDE MESYLATE

Loss of dopaminergic innervation in the pituitary plays an important role in the development of ECD in horses, because dopamine inhibits ACTH secretion from the pars intermedia of the pituitary gland.[8] Pergolide mesylate is an orally administered ergot-derived dopamine agonist drug used to treat Parkinson's disease in humans and ECD. It is considered the treatment of choice for ECD and is effective at doses as low as 0.5-2 mg per horse per day (0.001-0.004 mg/kg/day).[9] If no response to low dose pergolide is seen, doses as high as 3-5 mg/day have been used safely. Adverse effects are rare, but may include diarrhea, colic and anorexia. Reports show that 80-85% of horses improve clinically on pergolide therapy. Pergolide tablets for human use have been pulled from the market since the discovery of an association with the development of valvular cardiac disease in humans being treated for Parkinson's disease.

Since no equine specific pergolide products were available in the US, horses requiring treatment were given compounded products. However, pergolide is not a stable drug, and stability can be affected by temperature and light. Stability studies indicate that pergolide suspensions should be stored in a light protected environment at refrigerated temperatures for no longer than 30 days when made in an aqueous solution.[10] Other studies have determined the stability of liquid, capsule and powder based products.[11] This confirmed that these formulations are also not stable, and also highlighted the inconsistencies of the compounded products in regards to initial potencies.

An equine specific pergolide formulation has recently received approval from the FDA and has been shown to be stable and efficacious and should be used in all cases of ECD. Results of clinical trials with 122 equids evaluated in a multi-site, field efficacy study are promising with this formulation.[12,13] Inclusion criteria for the studies were a hirsutism score ≥ 1 and either a plasma adrenocorticotropin (ACTH) concentration ≥ 50 pg/ml or a positive dexamethasone suppression test. Treatment with pergolide mesylate (2 µg/kg, PO, q 24 h) was initiated at study enrollment and animals were re-evaluated after 90 and 180 days of treatment. When endocrine test results remained abnormal at 90 d (n = 47/113), the dose was increased to 4 µg/kg, PO, q 24 h. Treatment success after 180 d was defined as either normalization of DST or a decrease in plasma ACTH concentration by 50% and improvement by a score of ≥ 1 in at least one clinical sign, without worsening of other clinical signs. Treatment was also considered successful when the sum of clinical scores improved by 3 or more, regardless of endocrine test results, as long as there was no worsening of any clinical signs. Seventy-six percent (86/113) of equids were treatment successes at day 180. Of the 122 equids enrolled, 8 animals died and 1 was euthanized due to worsening of pre-existing conditions or colic. These horses were excluded from evaluations. Both median scores of clinical signs and mean concentrations of ACTH and cortisol decreased during the study period. Transient inappetence was the most common adverse event observed in 40/122 (33%) equids, mostly during the initial 30 d of treatment. The conclusions of the clinical trials were that treatment with Prascend® resulted in improvement in clinical signs and endocrine test results and is efficacious in treatment of PPID.

Other current areas of interest for this drug involve the use of pergolide in pregnancy. There is concern that a dopamine agonist drug may prolong pregnancy and decrease milk production, similar to fescue toxicosis. Anecdotal evidence supports withdrawing pergolide treatment for 2-4 weeks prior to the expected foaling date in order to avoid these complications. However, further research is needed in this area. Prascend® is not labeled for use in pregnant mares.[13]

FIROCOXIB

Firocoxib is a non-steroidal anti-inflammatory drug introduced in the United States in 2005. It is labeled for use in the horse as a paste preparation for oral administration at a dose of 0.1 mg/kg for 14 days for the treatment of osteoarthritis.[14] In 2011, it was also approved as an injectible formulation for IV administration at 0.09 mg/kg for 5 days.[14] Firocoxib is a highly COX-2 selective, with COX 1:COX-2 IC_{50} ratios of 263-643.[15] Efficacy testing has shown a similar response to phenylbutazone in naturally occurring lameness.[16]

In safety studies, oral ulcers were detected at 3x and 5x the label dose, however no other adverse effects were noted.[14] However adverse effects may still occur, particularly in patients with pre-existing disease, or those that receive NSAID stacking. A recent study examining the effects of co-administration of firocoxib with phenylbutazone showed significantly higher serum creatinine, significantly lower total protein, and a significantly lower urine specific gravity following 10 days of treatment at commonly used doses.[17]

Pharmacokinetics in adult horses show good oral absorption with a long half-life and significant accumulation when the drug is administered once daily.[16] Based on these factors, it may take multiple days for steady concentrations and maximum efficacy to be reached. To overcome this, a loading dose of 0.3 mg/kg once has been recommended.[18] The administration of the single loading dose allowed achievement of average steady state drug concentrations faster, and after the LD, firocoxib at 0.1 mg/kg every 24 h was able to maintain a relatively constant average drug concentration which should produce less variability in onset of action and efficacy.

In foals, the pharmacokinetics differ from adult horses. Firocoxib is more rapidly absorbed, with a higher maximum concentration, but a shorter half-life, resulting in less drug accumulation.[19] Steady is achieved after approximately 3 doses. In healthy foals, no clinically apparent adverse effects were noted after 9 days of oral firocoxib at 0.1 mg/kg. Intravenous data in foals suggests the difference in pharmacokinetics is due to an increased clearance.

Another area of recent research with firocoxib is into the bioequivalence of the equine and canine formulations. Data suggests the canine formulation is absorbed in the horse and may also be used as an effective anti-inflammatory agent.[20,21] Multiple dose regimens produce similar plasma half-lives and comparable COX-2 inhibition. The use of the canine formulation is controversial, however. The canine formulation is often administered due to the decreased cost, but the AVMA has stated that they will not support the practice and has issued the following statement: "Since there is a firocoxib product, Equioxx®, labeled for use in horses, then the canine product cannot be used in an extra-label fashion. Selecting the canine product over the equine product for any reason, including economic, is not acceptable. It is only if or when the approved equine drug were judged clinically ineffective for that labeled use that a veterinarian could use another animal approved drug, in an extra-label manner."[22]

PONAZURIL

Ponazuril, Marquis®, was approved by the FDA for use in horses as a treatment of EPM in 2001. Label dosing is 5 mg/kg once a day for 28 days.[23] Extralabel use of the drug as a preventative for EPM has been investigated. Prophylactic administration of 2.5mg/kg and 5mg/kg PO once a week starting 7 days before challenge with *S. neurona* and continuing 28 days after challenge showed a decrease in the number of horses that developed clinical signs compared to control (untreated).[24] Seroconversion was significantly decreased in those horses that received 5.0 mg/kg (100 vs. 40%). Mild-to-moderate, multifocal signs of neuroinflammation were observed in those horses that were euthanized due to the development of neurologic signs, suggesting that treatment with ponazuril at 5.0 mg/kg minimizes, but does not eliminate, infection and clinical signs of EPM in horses.

More recent work has investigated the use of a loading dose of ponazuril to decrease the amount of time for drug to reach therapeutic and steady-state concentrations within the CSF. A loading dose of 15 mg/kg PO once is now recommended as part of the dosing regimen.[25]

REFERENCES

[1] Giguère S. et. al; J Vet Pharmacol Ther. 2011, 34, 193. [2] Fultz L. et. al; J Vet Pharmacol Ther. 2013, 36, 309. [3] Fultz L. et.al; Equine Vet J, 2013, Epub ahead of print. [4] Hall TL. et. al; J Vet Pharmacol Ther. 2011, 34, 403. [5] Credille BC. et. al; J Vet Pharmacol Ther. 2012, 35, 259. [6]Macpherson ML. et. al; J Vet Pharmacol Ther. 2013, 36, 59. [7]Scofield D, et.al; Equine Vet J. 2013, Epub ahead of print. [8]McFarlane D, et. al; J Neuroendocrinol. 2005, 17, 73. [9]Donaldson MT. et. al; J Vet Intern Med, 2002, 16, 742. [10] Davis JL. et. al; J Am Vet Med Assoc, 2009, 234, 385. [11] Stanley SD. et. al; Proceedings of the 56th AAEP, Baltimore, MD. [12]PRASCEND® Freedom of Information Summary. Boehringer Ingelheim Vetmedica, Inc.; 2011. [13]PRASCEND® package insert. Boehringer Ingelheim Vetmedica, Inc.; 2011. [14] Equioxx® package insert. Merial, LTD. 2005, 2011. [15]Letendre LT, et. al; Am J Vet Res, 2008, 69, 1399. [16]Doucet MY, et. al; J Am Vet Med Assoc, 2008, 232, 91. [17]Kivett L. et. al; J Vet Pharmacol Ther. 2013, Epub ahead of print. [18] Cox S. et. al; Vet J, 2013, 198, 382. [19]Hovannesian N, et. al; J Vet Pharmacol Ther. 2013, Epub ahead of print. [20] Knych HK. et. al; Equine Vet J, 2013, Epub ahead of print. [21]Barton MH, et. al; Equine Vet J, 2014, 46, 72. [22] https://www.avma.org/KB/Resources/Reference/Pages /Previcox-for-Equioxx.aspx. [23] Marquis® package insert. Bayer Animal Health. 2001. [24] Furr M. et. al; J Parasitol, 2006, 92, 637. [25] Reed SM, et. al; Proceedings AAEP, 2012, 58.

FOOD ANIMAL PHARMACEUTICALS: NEW PRODUCTS AND ADVANCES

Johann F. Coetzee, BVSc, Cert CHP, PhD, DACVCP
Ames, IA

INTRODUCTION

Sponsors of new pharmaceutical compounds intended for use in food animals can face significant regulatory challenges especially with respect to demonstrating consumer safety and drug efficacy. Furthermore, some approved antimicrobials have recently had limitations placed on their extra-label use in food animals due to concerns about resistance development.[1] In spite of this, recent chemotherapeutic drug approvals have been characterized by the development of formulations that exhibit long plasma elimination half-lives that reduce the frequency of drug administration and animal handling events in production systems. One concern with this trend is that the extended duration of drug exposure could potentially accelerate the development of antimicrobial and anthelmintic resistant organisms. The duration of activity of these drugs are typically extended by either identifying compounds with a high volume of distribution or high protein binding or by formulating existing compounds to be slowly released from the injection site. Specific applications of these technologies have been in the development of new therapeutics for the treatment of bovine respiratory disease (BRD) and parasitic infections. As a consequence of this trend, several pharmaceutical companies have released related compounds with similar pharmacokinetic characteristics especially for the treatment of BRD. For example, there are 3 new macrolide antimicrobials (tulathromycin, gamithromycin and tildiprosin) that maintain therapeutic concentrations in cattle in excess of 10 days. Veterinarians are thus faced with the challenging task of making treatment decisions without, in many cases, having the benefit of head-to head comparisons between these new drugs. We are also limited in our understanding of the relationship between tissue concentrations of these antimicrobials and the Minimum Inhibitory Concentrations (MIC) of common BRD pathogens. Determining the optimal post-treatment interval for determining treatment success has also proven to be challenging.

In addition to the development of these new compounds, there has also been a trend towards reformulating existing compounds into longer acting formulations. One example has been the reformulation of eprinomectin into a polymeric PLGA matrix for the extended treatment of parasitic infections. This formulation maintains effective plasma drug concentrations in excess of 100 days. In spite of the recent increase in the approval of anti-infective compounds, several unmet needs still exist, especially in the area of pain management where there are still no compounds specifically labeled for analgesia in food animals in the United States. In light of concerns about antimicrobial resistance, there is also a move towards the identification of alternative therapeutic and management strategies to minimize stress and therefore reduce disease in livestock production systems. In this presentation we will focus on the recent advances in the development of long acting anti-infectives in food animals in addition to discussing one way in which closely related formulations can be statistically compared in the absence of head-to-head comparisons to assist veterinarians in making treatment decisions. This paper will also focus on some of the unmet therapeutic needs in food animals and will highlight recent research aimed at reducing antimicrobial use in livestock production systems by reducing stress associated with long distance transportation and processing.

NEW COMPOUNDS WITH AN EXTENDED DURATION OF ACTIVITY

Treatment of internal and external parasites in cattle remains a challenge in cattle housed in extensive operations. In typical production systems with the potential for a high parasite load, cattle must be treated multiple times during the spring and summer to minimize the economic impact of intestinal parasitism. As a consequence, pharmaceutical companies have experimented with novel formulations that will result in either a sustained or pulse exposure of parasites to anthelmintic therapy after a single administration. One such formulation that has recently been approved involves the formulation of the avermectin eprinomectin, with a polymeric PLGA matrix that solidifies in vivo forming in situ gel that allows the gradual release of eprinomectin from the formulation (LongRange, Merial Ltd). The formulation is designed to immediately release of the drug after administration and then to produce a subsequent, smaller peak in plasma drug concentration at between 70 and 100 days after administration (**Figure 1; Package Insert**). Between pulses of drug release, the formulation maintains plasma drug concentrations between 0.5 and 1 ng/mL for over 100 days which reportedly represents the minimal drug level for optimal anti-nematodal activity. This pattern of drug release is purported to help minimize the emergence of avermectin resistance within a population but long-term studies are needed to substantiate this claim.

Figure 1. Plasma concentrations of Eprinomectin B over time[2]

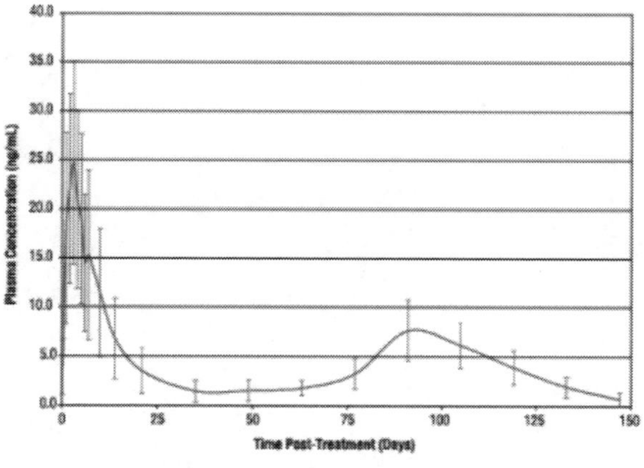

Bovine respiratory disease continues to be the primary cause of morbidity and mortality among cattle in the USA. Due to increased labor costs, concerns about the impact of handling stress on cattle wellbeing and to improve compliance, most of the new food animal pharmaceuticals have been formulated to have long plasma elimination half-lives in order to extend their duration of activity. Essentially this has come about through the identification of molecules that have a large volume of distribution such as the macrolides or molecules that are highly protein bound, such as the 3rd generation cephalosporins. In the case of the macrolides, the large volume of distribution of this class results in a large amount of the drug being located outside of the central compartment therefore, protecting this fraction from the process of elimination. Similarly in the case of the cephalosporins, drug bound to proteins is not exposed to elimination through the kidneys. Currently, ceftiofur crystalline free acid (CCFA) (Excede, Zoetis Animal Health) is the longest acting 3rd generation cephalosporin formulation approved for the treatment of BRD in the USA. However, the extra-label use of cephalosporins, with the exception of cephapirin, in food animals was recently prohibited due to concerns about antimicrobial resistance development. Recently, three related macrolides namely, tulathromycin (Draxxin, Zoetis Animal Health), gamithromycin (Zactran, Merial Animal Health) and tildipirosin (Zuprevo, Merck Animal Health), have been approved for the treatment of BRD. The recent increase in the number of new therapeutic options for BRD has created challenges for veterinarians in determining the most appropriate treatment in every case given the absence of head-to-head comparisons between all the possible treatment options that exist in each case.

The use of mixed treatment comparison meta-analysis to assess antibiotic treatments for BRD

In a recent paper, we used mixed treatment comparison meta-analysis to compare the efficacy of antibiotic treatments for bovine respiratory disease in beef cattle.[3] Studies were eligible for the meta-analysis if they were publically available and reported the assessment of antibiotic protocols registered for use in the United States (US) for bovine respiratory disease (BRD) in beef cattle and were conducted in North America. Three electronic databases, the proceedings of two bovine specific conferences, pharmaceutical company web sites and the US Food and Drug Administration website were searched to identify relevant trials. The network of evidence used in the analysis contained 194 trial arms from 93 trials. Of the 93 trials there were 8 with three arms. The network of evidence contained information for 12 antibiotics. The output from the analysis provided information about the risk ratio comparing all possible treatments for BRD including comparisons based only on indirect data. The output also included a relative ranking of the treatments and estimates of the probability that an antibiotic protocol was the worst treatment option (**Figure 2;** *Preventative Vet Med* (2013) 110 (2): 77- 87).

Figure 2. Ranking forest plot for treatment arms[3]

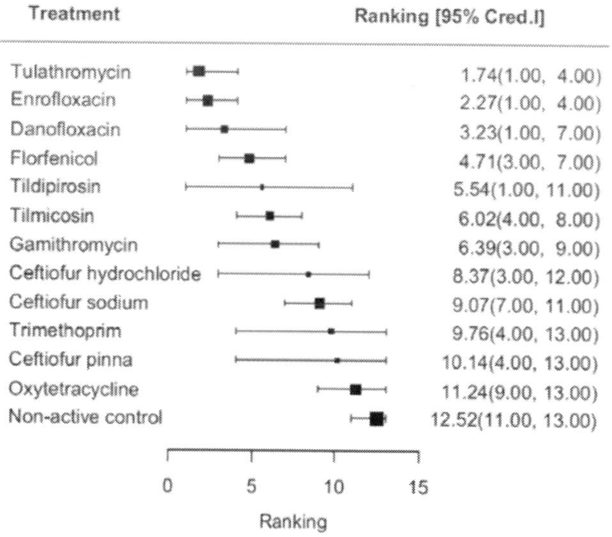

UNMET THERAPEUTIC NEEDS; IS THERE A CASE FOR ANALGESIC DRUG APPROVALS IN FOOD ANIMALS?

Societal concern about the moral and ethical treatment of animals is increasing.[4] In particular, negative public perception of pain associated with routine animal management practices such as dehorning and castration is mounting, with increasing call for the development of practices to relieve pain and suffering in livestock.[5] However, less that 20% of U.S. veterinarians currently report using analgesia routinely at the time of dehorning and castration.[6] There are currently several challenges associated with providing effective analgesia in food animals in the United States. The most significant of these are that there are currently no analgesic drugs specifically approved for the alleviation of pain in livestock.[7] Therefore, use of any drug for pain relief constitutes extra-label drug use (ELDU).[13] Under the Animal Medicinal Drug Use Clarification Act of 1994 (AMDUCA)[8], ELDU is permitted for relief of suffering in cattle provided specific conditions are met. These conditions include that (1) ELDU is allowed only by or under the supervision of a veterinarian, (2) ELDU is allowed only for FDA approved animal and human drugs; (3) ELDU is only permitted when the health of the animal is threatened and not for production purposes; (4) ELDU in feed is prohibited and (5) ELDU is not permitted if it results in a violative drug residue in food intended for human consumption. Therefore, use of an analgesic to alleviate pain associated with castration in calves in the United States would be required by law to comply with these regulations.

Another challenge to providing effective analgesia in cattle is that there is often a delay between the time of drug administration and the onset of analgesic activity. For example, local anesthetics require 2 - 5 minutes before a maximal effect is achieved.[9] This may slow animal processing as producers must wait for local anesthesia to take effect. This delay may serve as a disincentive for them to provide routine preemptive analgesia. Furthermore, the requirement for large numbers of animals to be processed quickly may result in procedures being initiated before optimal analgesia is achieved. A third challenge is that the route or method of analgesic drug administration may require specialized training and expertise or may be hazardous to the operator. For example, the NSAID flunixin meglumine is only approved for IV administration in the United States.

Taking these factors into consideration, there have been several recent studies that have investigated the potential for oral meloxicam to be administered to calves prior to undergoing routine dehorning and castration procedures. We published the pharmacokinetics of oral meloxicam in ruminant calves in 2009 demonstrating that plasma drug concentrations are maintained in an effective range for 3 – 5 days after a single administration of 1 mg/kg.[10] We have since demonstrated that this dose and route of meloxicam administration reduces stress associated with cautery dehorning irrespective of whether this is administered prior to or at the time of the procedure.[11] A single dose of meloxicam administered at 1 mg/kg PO prior to castration also reduced the incidence of BRD after processing in feedlot cattle.[12] We recently demonstrated that a single 1 mg/kg oral dose of meloxicam reduced cortisol concentrations in beef cattle when administered prior to long distance transportation to a feedlot facility.[13] In another study, we reported that meloxicam administered at 0.5 mg/kg PO alone or in combination with Gabapentin at 15 mg/kg PO once daily for 4 days reduced the severity of amphotericin B induced lameness in beef calves.[14] Taken together these studies demonstrate the benefits of analgesic drug administration in cattle and support efforts towards obtaining and analgesic drug approval for use in livestock.

REFERENCES

[1] 21 CFR Part 530 New Animal Drugs; Cephalosporin Drugs; Extralabel Animal Drug Use; Order of Prohibition. Available at http://www.gpo.gov/fdsys/pkg/FR-2012-01-06/pdf/2012-35.pdf. [2] LongRange Product label. Available at http://thelongrangelook.com/pdf/longrange-pi.pdf. [3] O'Connor et al. *Preventative Vet Med* 2013; 110 (2): 77- 87. [4] Rollin, B.E. *J. Anim. Sci*. 2004; 82: 955 – 964. [5] Weary, D.M., Fraser, D. In The well-being of farm animals: challenges and solutions. 1st edition. Edited by Benson GJ, Rollin BE. Ames IA: Blackwell Publishing, 325 – 338. [6] Coetzee, J.F. et al. BMC Veterinary Research 2010; 6 :12.[7] Compendium of Veterinary Products, 13th Edition. Bayley, A.J. (Publisher). Port Huron, MI: North American Compendiums INC, 2014. [8] Animal Medicinal Drug Use Clarification Act of 1994 (AMDUCA). [9] Spoormakers, T.J. et al.. Tijdschr Diergeneeskd. 2004. 129(17):548-551.[10] Coetzee, J.F. et al. Vet Ther. 2009; 10(4): E1 – E8. [11] Allen K.A.et al. J. Dairy Sci. 2013. 96 :1–12. [12] Coetzee JF et al. *Journal of Animal Science*. 2012. 90 (3):1026-39. [13] Van Engen, N.K., et al. Submitted to *Journal of Animal Science*. Accepted 24 November 2013. In Press. [14] Coetzee, J.F.,et al. Submitted to *Journal of Animal Science*, Accepted 7 December 2013. In Press.

ANTIMICROBIAL RESISTANCE FACTS VS. OPINION

Michael D. Apley, DVM, PhD, DACVCP
Manhattan, Kansas

INTRODUCTION

Resistance is resistance, right? The first thing to clear up about antimicrobial resistance is the definition of resistance. As clinicians, you are concerned about clinical resistance, based on clinically applicable breakpoints. Approved veterinary breakpoints are developed and approved by the Clinical and Laboratory Standards Institute Veterinary Antimicrobial Susceptibility Testing Subcommittee (CLSI VAST) based on the following.[1]

- Clinical outcomes coupled with pathogen susceptibility data
- MIC distributions of wild type isolate collections
- Pharmacokinetic/pharmacodynamic modeling

These breakpoints are intended to give guidance on the probability of an antimicrobial working within the context of a combination of pathogen, antimicrobial, disease, animal species, and specific treatment regimen. Once you deviate from any of these factors, the predictive value of the breakpoint is greatly diminished.

The second type of "resistance" is related to changes in population profiles of "wild type" susceptibility distributions. Instead of a clinical breakpoint, these are now referred to as an "epidemiological cutoff". These cutoffs are defined to indicate a change from the original population minimal inhibitory concentration (MIC) distribution, and may indicate the appearance of resistance genes. While the wild type cutoff is a component of a clinical breakpoint, without consideration of the other two components of the CLSI breakpoint process they are not necessarily correlated to clinical response. One of the outcomes of using both epidemiological cutoffs and clinical breakpoints in different monitoring systems may be declaring "resistance" at different MICs. This situation is complicated even further by the potential for resistance genes to be present that are not detected by phenotypic testing.

RESISTANCE CHALLENGES IN HUMAN MEDICINE

In veterinary medicine we are concerned with resistance in our patients and also how these resistance profiles may interact with resistance in human pathogens. Therefore, understanding the resistance challenges in human medicine informs our discussion of judicious antimicrobial use in veterinary medicine.

In 2013, the Centers for Disease Control and Prevention (CDC) released a report describing the major antibiotic resistance threats to human health.[2] In this report, the major threats were classified as threat levels of urgent, serious, and concerning. The urgent threats are high-consequence because of significant risks identified across several criteria. These threats may not be currently widespread but have the potential to become so and require urgent public health attention to identify infections and to limit transmission. The urgent threats are *Clostridum difficile*, carbapenem-resistant Enterobacteriaceae, and drug-resistant *Neisseria gonorrhoeae*. *Clostridium difficile* is included because of the increasing (in both prevalence and severity) incidence of pseudomembranous colitis in patients treated with antimicrobials.

The CDC serious classification includes threats that for various reasons are not considered urgent (e.g., low or declining domestic incidence or reasonable availability of therapeutic agents), but which will worsen and may become urgent without ongoing public health monitoring and prevention activities. These organisms are multidrug-resistant *Acinetobacter*, drug-resistant *Campylobacter*, fluconazole-resistant *Candida*, extended spectrum β-lactamase producing Enterobacteriaceae (ESBLs), vancomycin-resistant *Enterococcus* (VRE), multidrug-resistant *Pseudomonas aeruginosa*, drug-resistant non-typhoidal *Salmonella*, drug-resistant *Salmonella typhi*, drug-resistant *Shigella*, methicillin-resistant *Staphylococcus aureus* (MRSA), drug-resistant *Streptococcus pneumoniae*, and drug-resistant tuberculosis.

The CDC classification of "concerning" includes bacteria for which the threat of antibiotic resistance is low, and/or there are multiple therapeutic options for resistant infections. These bacterial pathogens cause severe illness. Threats in this category require monitoring and in some cases rapid incident or outbreak response. These organisms are vancomycin-resistant *Staphylococcus aureus* (VRSA), erythromycin-resistant Group A *Streptococcus*, and clindamycin-resistant Group B *Streptococcus*.

RESISTANCE CHALLENGES IN VETERINARY MEDICINE CLOSELY MIRROR THOSE IN HUMAN MEDICINE

Weese published a review of antimicrobial resistance issues in companion animals (2008), identifying these major organisms of concern.[3]

- *Staphylococcus aureus* and *Staphylococcus pseudintermedius*: both methicillin susceptible and resistant.
- Enterococci: *Enterocococcus faecium* and *Entercoccus faecalis*.
- *Streptococci*: *Strep. zooepidemicus* and *Strep. Equi* in horses, *Strep. canis*
- *Escherichia coli*
- *Salmonella*
- *Pseudomonas*

Methicillin-Resistant Staphylococcus aureus

While *Staphylococcus pseudintermedius* is a very real resistance concern in veterinary medicine, methicillin resistant *Staphylococcus aureus* (MRSA) highlights issues of zoonotic interactions in multiple veterinary species. A 2008 review article summarized MRSA occurrence in cattle, dogs, cats, sheep, chickens, horses, rabbits, seals, and psittacine birds.[4] Significant research has demonstrated the potential for exchange of isolates between people and their pets. Kottler, et al., evaluated the prevalence of MRSA in people and pets in the same household.[5] The sample consisted of one human nasal swab and one dog or cat nasal and fecal swab from 586 households. There was no difference in MRSA prevalence in households with human healthcare workers, veterinary healthcare workers, or without healthcare workers. The occurrence of MRSA in humans was 5.6%, with 3.4% in pets. In 4 of the 586 households (0.7%), the MRSA found in humans was the same strain as that found in the pet.

Faires, et al., evaluated the prevalence of concurrent infection in households where either a person or pet had a diagnosed MRSA colonization.[6] In part 1 of the study, 22 households were identified as having an MRSA infection in a pet (19 dogs and 3 cats). In these households, 10 of 56 humans (17.9%) were also colonized with MRSA. In part 2 of the study, 8 households were identified where humans had MRSA cultures from dermal abscesses. In only 1of these households was MRSA also isolated from a pet. In almost all cases of co-colonization or infection, the isolates were indistinguishable by PFGE.

O'Mahony, et al., evaluated MRSA isolates from dogs, horses, a cat, a rabbit, and a seal in Ireland along with isolates from 10 caregivers.[7] The PFGE results for the equine MRSA isolates were indistinguishable from the results for those isolates originating from the caregivers for the horses.

Several studies have evaluated risk factors for infection with MRSA in companion animals.
Faires, et al., evaluated risk factors for 40 MRSA infected dogs compared with 80 MSSA infected dogs.[8] The highest prevalence of both infections was in ears and skin. The statistically significant risk factors for MRSA infection as compared to MSSA infection included the use of any antimicrobial prior to diagnosis (odds ratio 2.84), use of fluoroquinolones (OR 3.58), use of β-lactams (OR 3.58), or intravenous catheterization (OR 3.72).

A retrospective study in horses in Canadian and American referral hospitals evaluated MRSA infections in 115 horses.[9] The infections originated both in the referral hospitals and in the community, with the frequency of both being approximately equal. Community acquired infections were significantly associated with previous hospitalization and previous gentamicin therapy. Hospital-acquired MRSA infections were significantly associated with infected incision sites.

Increasing attention in the literature has been paid to MRSA in swine and potential zoonotic concerns. While swine workers and veterinarians have been demonstrated to have nasal carriage of the MRSA type found in swine herds, epidemiological studies suggest that colonization is primarily limited to those working with the swine and further transmission is limited to familial communities of these exposed workers.[10] In the U.S., the human community-acquired outbreak strains are different from animal strains. In the Netherlands, a type of MRSA (ST 398) is epidemiologically associated with pig and cattle farmers and is said to be > 20% of carriage in humans.[11] MRSA has also been identified in bovine mastitis isolates.[12] The authors of a 2012 study using single nucleotide polymorphisms (SNPs) to evaluate 89 CC398 MRSA isolates proposed that this MRSA originated in humans as a methicillin-susceptible isolate and then acquired tetracycline and methicillin resistance in livestock, but also lost phage-carried human virulence genes.[13] MRSA CC398 has been documented to cause disease in humans, although it is not a major player in MRSA-associated disease in humans and appears to be a poor long-term colonizer.[14,15]

Carbapenemase-Producing Enterobacteriaceae

MRSA is an example of a resistant organism (which may also be multi-drug resistant) that brings the issue of treating our veterinary patients together with concerns about the effect of this pathogen's presence on our clients. A more recently emerging issue of shared resistance between human and animal pathogens is that of carbapenemase-producing enterobacteriaceae (CRE).

Human cases of CRE have been in the news lately where multidrug resistant pathogens have displayed resistance to this class of antimicrobials previously considered to be omnipotent. And, once again, the situation in veterinary medicine is mirroring the antimicrobial resistance challenges in human medicine. Carbapenemase-producing isolates of Klebsiella pneumoniae and E. coli have been identified in dogs from a single hospital in Germany, with the clonal nature of the isolates suggesting nosocomial spread.[16] Carbapenmase producing E. coli have also been confirmed in clinical isolates derived from dogs and cats in the United States.[17] An analysis of the literature related to potential sources of these organisms cites detection in dairy cows (France), horses (Belgium), a wild raptor (Germany), poultry and swine (China), dogs and cats (Germany and USA), and multiple instances in water and sewage throughout the world.[18] The pattern reasonably supports a hypothesis of spread to multiple veterinary species through environmental dissemination of human sources; this is further supported by no labeled carbapenems for food animals, a cost structure which makes extralabel use in food animals highly unlikely, and limited use in companion animals. While certainly not yet ubiquitous in occurrence, confirmation of these isolates in veterinary species strongly supports the need for continued evaluation of our use of carbapenems in veterinary species, and the potential for the occurrence of these organisms in clinical practice.

Bovine Respiratory Disease Pathogens

Lubbers and Hanzlicek published a retrospective analysis of Mannheimia haemolytica susceptibility results during 2009-2011 from the Kansas State Diagnostic Laboratory.[19] The percentage of isolates showing resistance to at least 3 of our main classes of antibiotics used for BRD were 42%, 46%, and 63% in 2009, 2010, and 2011 respectively. These data represent diagnostic laboratory isolates primarily from cattle of types associated with high risk for respiratory disease and do not represent a random sample across all cattle. However, this group is a dominant source for diagnostic bovine respiratory disease isolates, and the trend in multidrug resistance in this population of cattle is concerning.

Summary

Many of the resistance concerns in veterinary medicine mirror the challenges in human medicine. While it is possible for the discussion to devolve into one of assigning blame to antimicrobial use practices in one arena or the other, the solution is to adopt well-designed antimicrobial use strategies for all applications of antimicrobials.

REFERENCES

[1]CLSI document VET01-S2, Wayne, PA. 2013. [2]Centers for Disease Control and Prevention, 2013. [3]Weese JS. Animal Health Research Reviews. 9(2):169, 2008. [4]Morgan M. J Antimicrobi Chemother 62, 1181, 2008. [5]Kottler S, et al. J Vet Intern Med 24:132, 2010. [6]Faires MC, et al. JAVMA 235:540, 2009. [7]O'Mahony R, et al. Veterinary Microbiology 109:285, 2005. [8]Faires MC, et al. Emerging Infectious Diseases 16:69, 2010. [9]Anderson MEC, et al. Equine Veterinary Journal 41:401, 2009. [10]Cuny C, et al. PloS One, 4(8), 2009. [11] Van Loo I. Emerg Infect Dis 13, 1834, 2007. [12] Van Loo I. Emerg Infect Dis 13, 1834, 2007. [13] Vanderhaeghen W, et al. Veterinary Microbiology. E-pub article in press, accessed 3-31-2010. [14]Price LB, et al. MBio 3(1): 2012 [15] Graveland H, et al. In J Med Microbiol 301: 630, 2011. [16] Kock R, et al. J Hosp Infect 79:292, 2011. [16] Stolle I, et al. J Antimicrob Chemother 68:2802, 2013. [17]Shaheen BW, et al. Antimicrob Agents Chemother 57:2902, 2013. [18] Woodford N, et al. J Antimicrob Chemother 69:287, 2014. [19]Lubbers BV, Hanzlicek GA. J Vet Diagn Invest 25:413, 2013.

COMPOUINDING: WHERE DO WE STAND?

Mark G. Papich, DVM, MS, Diplomate ACVCP
Raleigh, North Carolina

INTRODUCTION

Compounding for veterinary patients has received a lot of attention in the past year. Because of problems with compounded preparations for human patients, there have been calls for more regulations on compounding. The recent problems that resulted from some human compounded formulations, including 64 human deaths, were highly publicized in the press. Subsequently, the U.S. Congress held hearings and proceeded with legislation to help prevent problems in the future. This act, named the "The Drug Quality and Security Act" was signed into law on November 27, 2013. It was intended to ensure better quality of compounded products. Unfortunately – or fortunately, depending on one's perspective – regulation of veterinary compounding was not included in the federal legislation that resulted from this action. The new federal legislation is somewhat limited in its scope and is intended to primarily improve quality of sterile compounded products. Some critics of the legislation suggest that it comes up short of ensuring safety and quality of compounded products. According to one perspective of the act, (Outterson, 2014) "Traditional compounders can now operate without fear of federal enforcement".

WHERE DO WE STAND WITH VETERINARY COMPOUNDING?

Because compounding for animals was not included in the new federal legislation, veterinary compounding still falls under the previous FDA Guidance issued in 2003. This guidance, called "Compounding of Drugs for Use in Animals, or Guidance 608.400" (http://www.fda.gov/iceci/compliancemanuals/compliancepolicyguidancemanual/ucm074656.htm), is currently active, but many parts of the guidance are not actively enforced. Regulation of compounding for veterinary patients is largely left up to the individual states.

Compounding, defined by the United States Pharmacopeia (USP) as "the preparation, mixing, and assembling, packaging, and labeling of a drug or device in accordance with a licensed practitioner's [the veterinarian] prescription of medication or under an initiative based on the practitioner/patient/pharmacist/compounder relationship in the course of professional practice." Drug compounding has always been a part of veterinary medicine. Historically, veterinarians have been known for preparing concoctions, mixtures, and remedies for their patients because there were few approved formulations on the market for animals. Now, there are more available drugs for animals, and a better understanding of the risks of drug instability and incompatible mixtures. Questions concerning this widespread practice have been raised, particularly with respect to the drug's stability, purity, and potency when the original dose form is altered, or when compounding is performed from the bulk drugs. "Bulk drugs" are defined as the active pharmaceutical ingredient (referred to as the API) used in the manufacture of finished dosage forms of the drug. Bulk API does not contain excipients, which are the inactive ingredients added as a preservatives, stabilizers, buffers, or to enhance solubility.

Compounding is performed for the purpose of ease of administration or because the original dosage form is unsuitable for the purpose intended. Compounding does not include the preparation of a drug by reconstitution or mixing that is according the manufacturer's instructions on an approved human or veterinary drug product.

Veterinary medicine has been debating the practice and regulation of compounding for over 20 years. In 1993, a symposium on Compounding in Veterinary Medicine was held by the American Academy of Veterinary Pharmacology and Therapeutics (AAVPT) (JAVMA, 1994). This symposium had representatives from AVMA, FDA/CVM, pharmacology and pharmacy groups, and USP. The symposium heard various views from practitioners, pharmacologists, regulatory officials, pharmacists, and lawyers. This symposium issued a Task Force Report that summarized the presentations and resulted in the Compliance Policy Guide published in 1996 (FDA 1996). The proceedings from this symposium are very informative and contained 115 pages of presentations, which cannot be adequately summarized here. The FDA followed several years later with a revised Compliance Policy Guide (CPG) of 2003 for compounding drugs that was cited above.

The FDA recognizes the importance of compounding in veterinary practice, but also must ensure that compounded drugs do not cause harm to the treated animals, produce ineffective potency, or residues in food animals. FDA regulations permit the compounding of formulations from approved animal or human drugs under the current federal code: 21 CFR 530.13. The FDA is concerned that allowance of some compounding on a patient-by-patient basis has gone over the limits intended in the original federal regulation and has been expanded to result in large-scale production and labeling of some products that may constitute manufacturing and distributing of new animal drugs.

The current CPG specifically was intended to clarify the regulations on compounding from unapproved or bulk drugs, or the bulk API. According to the FDA CPG, compounding from bulk drugs or unapproved drug substances is not allowed except for a few compounds provided on a list that are not subject to regulatory action. Despite this CPG, compounding from bulk drugs is commonly practiced by some pharmacies

THE NEED FOR COMPOUNDED DRUGS IN VETERINARY MEDICINE

The palatability, ease of administration, and dispensing factors are among the considerations when formulating drugs for animals. Drugs intended specifically for animals are designed with great care. Pastes and dosage syringes are available for some drugs used in horses. Flavored tablets are used commonly in dogs for tablets that are given by pet owners. Transdermal medications are available for dogs and cats to avoid the necessity of frequent administration to a pet that may be difficult to medicate. Sometimes compounding is a necessity. Despite the advances in new drugs available for animals, many needs are still not met. Therefore, many drugs are crossed over from one animal species to another, or are human drugs administered to animals. Some drugs require compounding simply because no approved form of the drug exists in the U.S. Drugs that are often compounded for veterinary medicine because approved forms are not available include: potassium bromide, metronidazole benzoate suspension, methimazole transdermal, diethylstilbestrol, cisapride, various antidotes, and other products that have either been discontinued in human medicine, or because of shortages in availability.

WHAT ARE THE CONCERNS FROM COMPOUINDED VETERINARY FORMULATIONS?

There are many compounded formulations that are compounded by reputable pharmacies that result in high quality medications with assurances of potency, stability, and beyond-use-dates (BUD). However, there are also concerns. Beyond-use-dates are provided in the USP general chapter on compounding (USP <795>). The BUD for aqueous (water-containing) oral formulations stored at controlled cold temperatures is 14 days to maintain the potency within 90-110% of the nominal formulation strength (+/- 10% is the USP standard for strength of formulations). But compounding pharmacies sometimes go beyond this date without studies to support their claim. There is evidence from published articles that some drugs are not suitable for compounding because they result in rapid inactivation or loss of potency, or lack of systemic absorption when administered to an animal. We now have evidence that administration of antibiotics from some compounded formulations are sub-therapeutic and can increase the risk of bacterial drug resistance, which is a risk both to the pet and animal owner. In some cases the compounded products may actually contain levels well above the labeled amount which presents a risk of toxicity for the pet, as well as the animal owner handling the medication.

Examples of potential problems

There are several relevant published examples in which drug stability and efficacy has been compromised through compounding. When protective coatings are disrupted, and the vehicles altered, the stability of the product may be compromised. When the formulation is prepared from the bulk API without excipients or preservatives added to maintain stability, strength, or to maintain a desired pH, the quality and strength of the drug may be compromised. In some instances, only a slight alteration of pH can affect the drug. According to the USP-NF, "improper pH ranks with exposure to elevated temperature as a factor most likely to cause a clinically significant loss of drug. A drug solution or suspension may be stable for days, weeks, or even years in its original formulation, but when mixed with another liquid that changes the pH, it degrades in minutes or days. It is possible that a pH change of only one unit could decrease drug stability by a factor of ten or greater." Addition of a water-based solution to a product to make a liquid solution or suspension can hydrolyze some drugs (beta-lactams, esters). Some drugs undergo epimerization (steric rearrangement) when exposed to a pH range higher than what is optimum for the drug (for example this occurs with tetracycline at a pH higher than 3). Other drugs are oxidized, catalyzed by high pH, which renders the drug inactive. Drugs most likely to be subject to oxidation are those with hydroxyl group bonded to a aromatic ring structure. Oxidation may occur from exposure to light and oxygen during reformulation and mixing. Oxidation is catalyzed by high pH and usually leads to drug inactivation.

For example, when omeprazole was compounded for oral use in horses, it was not as effective for treating gastric ulcers as the commercial formulation registered for horses (Gastroguard). Omeprazole is known for its instability unless administered in the original formulation intended for horses or people. Itraconazole is notorious for its instability and variable oral absorption. When the brand name itraconazole (Sporanox) and generic itraconazole were compared in research dogs, they produced similar plasma concentrations. But by comparison, the plasma concentrations from a compounded product were negligible. When doxycycline hyclate was examined in a compounded aqueous formulation made from crushed tablets, strength of the preparation depleted drastically after 7 days. Other compounded formulations known to produce formulations of poor strength and quality are pergolide, trilostane, pimobendan, clenbuterol, boldenone, amikacin, minocycline, and ketoprofen. On the other hand, some compounded formulations made from extemporaneous preparations have been shown to retain their strength for storage times of at least 28 days. These include enrofloxacin, carprofen, meloxicam, potassium bromide, sodium bromide, and metronidazole benzoate.

Veterinarians and pharmacists are obligated to be cognizant of the potential for interactions and interferences with stability. Oxidation is often visible through a color change (color change to pink or amber for example). Loss of solubility may be observed through precipitation. Some drugs are prone to hydrolysis from moisture. A rule-of-thumb for veterinarians is that if a drug is packaged in blister packs or moisture proof barrier, it is probably subject to loss of stability and potency if mixed with aqueous vehicles. If compounded formulations of solid dose forms show cracking or "caking", or swelling, the formulation has probably accumulated moisture and may have lost potency. Another rule-of-thumb is that if the original packaging of a drug is in a light-protected or amber container it is probably prone to inactivation by light. Vitamins, cardiovascular drugs, tetracyclines, and phenothiazines are labile to oxidation from light during compounding. Also, as a general rule, if an antibiotic is available in a powder that must be reconstituted in a vial or oral dispensing bottle prior to administration, it should not be mixed with other drugs.

Many drugs intended for one species (or humans) are frequently compounded for another veterinary species. In these instances, it is not only the compounding practice that may affect drug absorption, but also the species differences. Although one assumes that absorption may be similar, differences can exist that may result in poor efficacy. Grass & Sinko concluded that there is no apparent relationship when comparing bioavailability of orally-administered drugs, between humans, dogs, primates, and rodents (Grass & Sinko, 2002). Therefore, for drugs administered orally, it is very difficult to broadly extrapolate from studies performed in people to veterinary species. Specific studies are usually needed, unless it is known that the drug is highly stable, soluble, and well absorbed under a variety of conditions.

The problem with compounded transdermal gels

Several published studies, or studies presented at conferences in abstracts have demonstrated poor or unreliable systemic absorption from transdermal gels. To the author's knowledge, only two drugs have been shown to produce therapeutic effects when compounded into a transdermal gel and applied to the ear of cats: methimazole for feline hyperthyroidism, and amlodipine for feline hypertension. The skin is an efficient barrier and it is very difficult to facilitate the absorption of most drugs through the skin. Drugs have been combined with penetration enhancers to facilitate transdermal absorption. One popular example of a penetration enhancer, is pleuronic lecithin organogel (PLO), which is lecithin (derived from eggs or soybeans) mixed with isopropyl palmitate and a poloxamer (Pluronic). The ingredients in PLO are intended to act as surfactants, emulsifiers, and solubilizing agents. Although the use of PLO is popular among the veterinary compounding pharmacies, there are no successful commercial formulations that have combined PLO with systemic drugs.

The list of drugs that have been shown to be absorbed poorly or highly inconsistently in this form includes: cyclosporine, atenolol, glipizide, dexamethasone, buspirone, amitriptyline, fentanyl, morphine, fluoxetine, diltiazem, and enrofloxacin. Despite the lack of assurance of systemic absorption from transdermal gels, some veterinary compounding pharmacies advertise and promote transdermal gels on their web sites. One compounding pharmacy lists over 10 pages of compounded transdermal gels for cats. Development of transdermal technology in animals is very difficult and assurance of stability, potency, and systemic absorption for the vast majority of these products has not been demonstrated. Most compounded transdermals for animals are at best placebos. Because of the lack of effective absorption when these products are applied to the ears of cats, it presents a risk of exposure to the family of these pets, especially children. It is a common habit of cats to rub their ears against their owners and on furniture and bedding.

CAN COMPOUNDING PHARMACIES PROVIDE ASSURANCES?

The FDA-CVM uses regulatory discretion to allow veterinary drug compounding within the scope of clinical veterinary practice through the Compliance Policy Guideline. However, there are still some restrictions that apply, and each state may have more restrictive requirements than Federal Law. The source of the drug should be a USP or an NF grade substance. Drugs must be compounded from the original formulation, if an approved product exists. Compounding from bulk drugs is not allowed if a proprietary registered formulation is available. Therefore, if a veterinary pharmacy is in the practice of compounding drugs from a bulk source, it is engaging in an illegal practice according to the FDA CPG. If bulk drugs are used because there is no other available form, the pharmacist should use bulk substances registered with FDA, and accompanied by a valid certificate of analysis.

It is the responsibility of the veterinarian and pharmacist to ensure that regulations and guidelines are being followed for confidence in the compounded medication. The USP-NF lists specific guidelines in the General Chapter on Pharmaceutical Compounding <795> and <797>, (USP-NF, http://www.usp.org). Often overlooked in compounding practices is the guideline to ensure that the compounded formulation is not less than 90.0% and not more than 110% of the theoretically calculated and labeled quantity of active ingredient per unit weight or volume. There are also guidelines available for Quality Assurance in Pharmaceutical Compounding (USP <1163>).

Compounding pharmacies have a responsibility to provide veterinarians and their clients with assurances that their formulations meet compendial standards and are stable and potent for the specified beyond-use-date. Pharmacies can be accredited by the Pharmacy Compounding Accreditation Board (PCAB). This is a voluntary program for improving the quality of compounding operations and preparations. An accredited pharmacy that adheres to USP compounding standards provided in General Chapters <795> and <797> will have significant checks in place to ensure that compounded products meet a high standard of quality. On the other hand, some pharmacies may not follow these standards and instead, provide veterinarians and their clients with misleading information on the quality of their products. For example, some products may be promoted as meeting "stability standards" for 30, 60, 90, or 180 days after preparation. But, there is a difference between stability testing and potency testing (Kupiec, et al, 2008). True stability testing of products is extremely expensive and more than likely a stability-indicating method has not been used. According to the FDA, a stability indicating method is "a validated quantitative analytical procedure that can detect the changes with time in the pertinent properties of the drug substance and drug product. A stability-indicating method accurately measures the active ingredients, without interference from degradation products, process impurities, excipients, or other potential impurities." As described in the USP General Chapter <1191> there are five types of stability: chemical, physical, microbiological, therapeutic, and toxicological. Stability-indicating methods include forced degradation tests, high heat and humidity, UV radiation, acid exposure, base exposure, and peroxide exposure. This is much different from a simple HPLC assay that measures strength based on a peak response.

SUMMARY

Compounding for veterinary patients has not changed with the recent federal legislation. Compounding for veterinary patients is largely unregulated and without federal enforcement. Nevertheless, there may be changes enacted by specific states that could restrict some forms of compounding for animals. In the meantime, veterinarians are encouraged to be skeptical of broad claims of stability, quality, and systemic absorption from compounded products unless the pharmacy can provide some assurances based on studies conducted, or adherence to compendial standards.

REFERENCES

1. FDA-CVM. Compliance Policy Guide: Compliance Policy Guidance for FDA Staff and Insustry. Chapter 6, Subchapter 600, Sec. 608.400 – Compounding of Drugs for Use in Animals, July 2003. Food and Drug Administration, Department of Health and Human Services.
2. Grass GM, Sinko PJ. Physiologically-based pharmacokinetic simulation modeling. Advanced Drug Delivery Reviews 54: 433-451, 2002.
3. JAVMA. Symposium on Compounding in Veterinary Medicine. Journal of the American Veterinary Medical Association 205: 189-303, 1994. (This issue of the Journal contains several individual papers addressing compounding in veterinary medicine.)
4. USP. United States Pharmacopeia (2013). USP 36–NF 31. The United States Pharmacopeia, Thirty-Sixth Revision, and the National Formulary. Rockville, Maryland. General Chapters cited include <795> Pharmaceutical Compounding---Nonsterile Preparations, <797> Pharmaceutical Compounding---Sterile Preparations, and <1163> Quality Assurance in Pharmaceutical Compounding.
5. Kupiec TC, Skinner R, & Lanier L. Stability versus potency testing: The madness is in the method. International Journal of Pharmaceutical Compounding 12: 50-53, 2008.370: 97-99, 2014.
6. Outterson K. The drug quality and security act – mind the gaps. New Engl J Medicine

CANINE ALLERGIC DERMATITIS: NEW TARGETS, NEW THERAPIES

Douglas J. DeBoer, DVM, DAVCD
Madison, WI

INTRODUCTION

Allergic dermatitis is a complex of diseases with a varied and complex etiology and pathophysiology. Included within this complex are common diseases such as parasitic hypersensitivity (e.g., flea allergy dermatitis) and atopic dermatitis, as well as less common conditions such as food hypersensitivity and contact allergy There is frequently overlap between these syndromes, with, for example, patients frequently being allergic to fleas or food allergens along with environmental allergens. The complex pathophysiology of allergic skin diseases, in particular atopic dermatitis (AD),[1] dictates that clinicians must take a multimodal approach to therapy. Allergen avoidance, augmentation of epidermal barrier function, infection control, allergen-specific immunotherapy, and systemic anti-inflammatory drugs all are important therapies for AD which must be combined for each patient to provide an effective and safe management plan.

Glucocorticoids and ciclosporin historically have formed the foundation of systemic treatment for atopic dermatitis, with antihistamines and fatty acid supplements playing auxiliary roles despite more limited evidence for their effectiveness. One benefit of greater understanding of the pathophysiology of allergic skin disease is the uncovering of additional therapeutic targets. Recent mechanistic studies have uncovered a variety of new potential targets for anti-inflammatory and antipruritic therapies. Many of these approaches focus on modulation of cytokine release or action, through small-molecule inhibition of receptors or signaling pathways. The histamine H4 is another potential target,[2] as are neuropeptide receptors[3] and the sphingosine-1-phosphate pathway.[4] Clearly, we are only at the beginning of a new world of therapy approaches for our patients.

ANTIHISTAMINES RECONSIDERED

Medical management of the AD should begin with the safest drugs to administer longer-term. Antihistamines may benefit 5-20% of pets, and they are always worth trying, though actually from an evidence-based medicine standpoint there has been little or no proof that they are beneficial. Recent, better-designed and controlled studies have shown some success in dogs with the antihistamines dimetindine,[5] fexofenadine,[6] and a combination of hydroxyzine and chlorpheniramine.[5] However, the magnitude of relief that antihistamines appear to provide remains minimal.

CICLOSPORIN: TIPS FOR BETTER EFFICACY

The calcineurin inhibitors work by inhibiting production and action of cytokines, and through other mechanisms as well. Clinical trials of ciclosporin in dogs with AD demonstrate that this drug has efficacy equal to that of oral methylprednisolone.[7] The starting dose is 5 mg/kg/day, which can be given as a single dose or divided into multiple doses. After the first month of treatment, in some dogs the dose can be decreased. Perhaps 25% of patients will have some initial gastrointestinal discomfort from ciclosporin A modified (CsAM). Co-administration with maropitant for the first 5-10 days of treatment is often helpful if vomiting appears to be a concern. In most cases, GI effects will abate within a few weeks. Therapy with CsAM is remarkably free from long-term adverse effects. Gingival hyperplasia is a known adverse effect of longer-term, higher-dose CsAM therapy, but occurs only rarely (1-2%) at the doses typically necessary for AD. Protocols for combination of CsAM with ketoconazole for AD have not been developed. A major concern here would be potential for development of hepatotoxicity with long-term ketoconazole treatment. Long-term combination of CsAM and systemic corticosteroids has been associated with development of fatal opportunistic fungal infections. CsAM typically acts relatively slowly, often taking 2-4 weeks to reach maximum effectiveness. Therefore, many dermatologists begin a short (2-week) tapering course of oral prednisolone along with CsAM for faster patient relief; this appears safe to do. Therapeutic monitoring (serum chemistries, blood counts, or CsA serum concentrations) is neither recommended nor necessary when using CsAM for AD patients.

Ciclosporin is a large molecule, and is not readily absorbed through the epidermis. Other calcineurin inhibitors have been developed that are smaller, and active topically. Tacrolimus (Protopic™) is the most commonly used of these topicals. Topical tacrolimus cream is very effective at controlling inflammation in human AD and has been used to a limited extent in dogs[8]. Its use is limited in animals due to both the difficulties with using cream preparations on haired skin, and to its very high cost. A topical nanoemulsion formulation of CsA was recently shown to be effective for canine AD as well, though this is not yet a commercial product.[9]

KINASE INHIBITORS FOR ATOPIC DERMATITIS

One of the most exciting new classes of drugs in medicine is the kinase inhibitors (KIs). These drugs represent a new and completely different way to change cellular function. Development of KIs is only in very early stages of infancy. It is likely that many drugs in this group will be developed over the coming years, aimed at treating a variety of neoplastic, inflammatory, and immunologic diseases. Thus, it is important for veterinarians to understand the basis of these new and exciting molecules.

What are Kinases?

Very simply, kinases are enzymes that catalyze transfer of phosphate groups onto an amino acid within a protein molecule. Kinase enzymes function to grab an ATP molecule, and transfer a phosphate group from it onto an amino acid in their target protein.

When this amino acid becomes phosphorylated, the shape of the protein changes, in such as way that it goes from an "inactive" to "active" state so it can perform its function.

This phosphorylation is a basic way that cells regulate many different biochemical processes. There are many, perhaps hundreds, of different kinases that exist within cells; each is slightly different and controls a specific cell process. Some kinases are within the cytoplasm or nucleus of the cell. Others, termed "receptor kinases" are actually part of a cell surface receptor and form part of the "triggering" mechanism for that receptor. The kinase within the receptor is what transmits the signal from the outside of the cell to the inside. Thus, kinases are a diverse group of molecules that are at the very core of cellular signaling and regulation.

Why Are Kinases Interesting?

From a clinical standpoint, kinases are important for one simple reason: they are potentially great targets for drug development. Occasionally something goes wrong with a kinase, causing disease. Sometimes, a mutation will occur in a kinase that stops its normal functioning. Other times, a mutation will cause the kinase to become "overactive." Any of these situations can result in disease. In addition, in some disease processes it may be advantageous to inhibit normal functioning of the kinase, for example to downregulate an inflammatory process.

One common situation is for a cellular kinase to become mutated, and as a result become constantly activated – it loses the ability to regulate cell function because it is always "turned on." Many times, the result is uncontrolled cellular proliferation, perhaps with uncontrolled angiogenesis to the tissue as well – in short, CANCER. On the other hand, if a mutation occurs which stops all activation of the kinase– turns it "off" - then the cell may be unable to divide and function normally, leading to situations such as immunodeficiency diseases. You can imagine that if we could, pharmacologically, partially or selectively inhibit specific kinases, it is possible that we could create a rather dramatic effect in the patient!

Kinase Inhibitors

Small molecule drugs have been developed which are capable of inhibiting kinases. It is helpful to envision that the kinase enzyme has a "pocket" within its structure into which the ATP must fit. In most cases, these drugs fit into the ATP-binding "pocket" of the enzyme and thus prevent it from binding ATP and performing its normal action. Much research is centered on understanding exactly the shape of the pocket so that drugs can be developed that very specifically fit in the pocket, thus inhibiting only a small subset of kinases and leaving the rest unchanged. This seems to be the key: find a very specific kinase that is involved in the process you wish to inhibit, and find a drug that binds ONLY in the pocket of this very specific kinase enzyme, and not in others.

Some Specific Examples of Kinase Inhibitors

Many initially available kinase inhibitors were developed for treatment of cancer, and are secondarily being explored for their possible anti-inflammatory effects. Each has a unique profile of action and a unique set of potential toxicities. **Imatinib** (Gleevec, Novartis; human drug) blocks Abl and Kit. Many cases of human chronic myelogenous leukemia are caused by an "always-on" mutation in Abl, resulting in uncontrolled myeloid cell growth. Blocking this kinase with imatinib results in a 95% remission rate in such patients! Imatinib has been used to treat mast cell tumors, systemic mastocytosis, and sarcomas in dogs and cats. Other "human" drugs such as sunitinib (Sutent, Pfizer) and gefitinib (Iressa, AstraZeneca) are mostly unstudied in veterinary medicine. **Toceranib** (Palladia, Pfizer) is a veterinary drug that blocks VEGFR, PDGFR, and Kit – thus one would predict it has both anti-angiogenic and direct antitumor effects. It has demonstrated efficacy in mast cell tumors, sarcomas, carcinomas, melanoma, and myeloma and is currently in a wide range of clinical trials to further define its efficacy in animal cancers. There are no reports of its use in inflammatory disease. **Masitinib** (Kinavet or Masivet, AB Science) is a veterinary drug that targets Kit, PDGFR, and a few other kinases. Its primary indication is control of mast cell tumors in dogs, but it is also undergoing intensive study in other cancers such as melanoma and hemangiosarcoma, as a single agent or as combination therapy. Importantly for veterinary dermatology, it has shown preliminary effectiveness in cutaneous T-cell lymphoma. It is also being studied extensively for action in inflammatory disease, including arthritis, inflammatory bowel disease, and asthma. An initial study demonstrated some efficacy in canine atopic dermatitis,[10] though cost and adverse effects appear to be limiting practical use.

FOCUS ON JANUS KINASE, JAK/STAT SIGNALLING, AND OCLACITINIB

The most recent molecular target to receive attention is the family of kinases known as JAK. There are four Janus Kinases in the family – JAK1, JAK2, JAK3, and TYK2. The JAKs function to communicate a signal from a cell surface receptor to the nucleus, and many inflammatory pathways (especially cytokines and their receptors) use JAKs for signaling. JAKs function in pairs (e.g. JAK1/JAK2). When activated, JAKs next interacts with any of several STATs, another family of molecules, and the STAT proteins translocate to the nucleus to regulate transcription. There are many different JAK/STAT combinations functioning in many different tissues to signal a very large variety of end responses.

In considering function of JAK/STAT signaling, closer examination of the role of cytokines in inflammation and pruritus is useful. Cytokines (such as interleukins or interferons) are protein messengers used for intercellular communication. They are produced by cells and then travel to other cells where they bind to membrane receptors to convey their signals and induce biologic responses. It is well recognized that cytokines, such as IL-2, IL-4, IL-5, IL-12, IL-13, IFN- gamma, and many others, are important mediators of inflammation in allergic disease and are targets of therapies, such as the calcineurin inhibitors (ie, ciclosporin).

Recently, IL-31 has been identified as a key cytokine in pruritus. Several lines of evidence demonstrate its importance in a variety of pruritic skin diseases. IL-31 is secreted by the "proallergic" T helper 2 lymphocytes (among other cells) and works by binding to a receptor, which triggers a JAK/STAT pathway. When the pathway is activated, the result is a vicious cycle of neuroimmune interaction and pruritus. This raises the intriguing possibility that, if we could block this particular JAK/STAT pathway, we might be able to block the pruritic response.

Oclacitinib

Oclacitinib, a JAK inhibitor, has been studied in a variety of models of pruritus and has shown the ability to suppress pruritic responses rapidly and effectively, in some cases better and quicker than prednisolone.[11-13] At therapeutic doses, oclacitinib inhibits predominantly JAK1 and spares JAK2-dependent processes such as hematopoiesis. Initial controlled clinical trials of oclacitinib in the treatment of both canine allergic dermatitis (flea, food, contact) and canine AD have shown promising results. Results in head-to-head studies against either prednisolone or ciclosporin shows the drug to be equally effective in control of itch and inflammation, and to have a very rapid onset of action with relief sometimes apparent within hours of oral administration. The drug became available in the USA in early 2014 as Apoquel (Zoetis Animal Health). Apoquel is indicated for control of acute or chronic pruritus in dogs over 12 months of age. Recommended dosing consists of twice daily administration for 2 weeks, followed by once daily dosing for longer term use. A few dogs seem to worsen slightly when switched from twice to once daily, probably related to the short half-life of the drug (4 hr in the dog). Overall, it appears that at least 60-70% of allergic dogs receiving the drug have rapid, substantial, and prolonged relief of their clinical signs. Veterinary dermatologists involved in early clinical trials often comment that the drug is very useful as part of a multimodal treatment approach, and that some dogs with even very recalcitrant disease have shown remarkable response. Short-term adverse effects have been limited to GI disturbance in a very few of dogs, though the adverse effects occur nearly as frequently in placebo-treated dogs. There is no specific organ toxicity associated with Apoquel, therefore no specific laboratory monitoring is advocated, though good medical practice dictates that all dogs receiving this drug should be examined at least once annually. Apoquel has been administered for as long as 3 years in some dogs; in longer-term studies, occasional patients have developed benign or malignant neoplasms, but no more often than would be expected for dogs in the studied age range. The drug has been limited to use in dogs 12 months or older, mostly because in one high-dose safety study with young laboratory dogs, generalized demodicosis developed in some patients. Oclacitinib has not been evaluated in combination with other drugs such as systemic corticosteroids or ciclosporin; concurrent use should be avoided with these drugs at this point. However, it can be used safely along with antibiotics, antihistamines, antifungal drugs, NSAIDs, allergen-specific immunotherapy, and many other medications, and vaccination of treated dogs is effective. Apoquel does not appear to interfere with serologic or intradermal allergy tests. As with other immunosuppressive treatments, it should not be used in the face of a severe infection, demodicosis or with active malignancy.

REFERENCES

[1]Marsella R et al., **J Am Vet Med Assoc** 2012; 241(2):194. [2]Ohsawa Y et al., **Allergy** 2012; 67(8):1014. [3]Mikami N et al., **Crit Rev Immunol** 2012; 32(4):307. [4]Bäumer W et al., **J Invest Dermatol** 2011; 131(1):266. [5]Eichenseer M et al., **Vet Rec** 2013; 173(17):423. [6]Alja P et al., **Slov Vet Res** 2009; 46(1):5. [7]Steffan J et al., **Vet Dermatol** 2003; 14(1):11. [8]Marsella R et al., **Vet Dermatol** 2004; 15(5):294. [9]Puigdemont A et al., **Vet J** 2013; 197(2):280. [10]Cadot P et al., **Vet Dermatol** 2011; 22(6):554. [11]Gonzales AJ et al., **J Vet Pharmacol Ther** 2014; in press. [12]Collard WT et al., **J Vet Pharmacol Ther** 2013; in press. [13]Cosgrove SB et al. **Vet Dermatol** 2013; 24(5):479.

TRANSDERMAL AND TRANSMUCOSAL PRODUCTS FOR PAIN: NEW PRODUCTS AND ADVANCES

Butch KuKanich, DVM, PhD, DACVCP
Manhattan, KS

INTRODUCTION

The use of many analgesics in veterinary medicine is limited due to short durations of effect and limited routes of administration. For example, many opioids used in dogs and cats have a low oral bioavailability and as a consequence cannot be effectively administered per os (PO). The low PO bioavailability is not due to poor absorption, but due to extensive first pass metabolism. Most opioids are well absorbed from the gastrointestinal tract after oral administration. Orally administered drugs that are absorbed from the gastrointestinal tract must pass metabolizing enzymes within the gut walls. If the drugs pass intact through the gut walls, they enter the mesenteric vasculature, then the portal vein and enter the liver where the drugs must pass through the liver without being metabolized. The presystemic metabolism by the intestines and liver is termed first pass metabolism. Only drugs that make it through the first pass metabolism are available to circulate systemically and elicit a systemic effect. Additionally, many of the opioids have a short duration of effect (2-6 hours depending on the drug and dose) and as a result are inconvenient and difficult to administer at the frequency needed to maintain a desired effect, therefore alternative delivery mechanisms are often desired that will allow for increased dosing intervals. Another drawback of opioid use in veterinary outpatients is the risk of diversion, misuse and accidental exposure that may occur when they are dispensed.

Studies evaluating the efficacy of analgesics need to include control groups. A negative control demonstrates that the method of assessment and study conditions can differentiate a treated animal from an untreated animal (sensitive and specific for analgesia). A positive control demonstrates that the method of assessment and study conditions can truly detect a difference when it occurs with a treatment known to produce a positive effect. Therefore an ideal study will contain both negative and positive control groups. For example, a study design that includes a negative control group and treatment group results in no significant difference in the study outcomes. The interpretations of the study can be: 1) there was a difference between the two groups, but the method of assessment and study design could not differentiate the difference, 2) there was no true difference in the treatment groups. Inclusion of a positive control that demonstrated a significant difference would indicate that no difference in in the placebo and treatment group was detected, but the model was able to differentiate an analgesic effect. As another example, a study that includes a treatment group and positive control group results in no significant difference. The results could be interpreted as: 1) a true difference occurred, but the method of assessment or study design could not differentiate a true response from lack of a response, 2) neither drug was truly effective using the study methods, but the model could not detect a painful animal 3) both drugs were truly effective. Studies that include animals as their on control by comparing pre to post-treatment effects are not appropriate as factors other than the treatment may be affecting the assessment (see below).[1] Although often excluded from analgesia studies, negative control groups are needed to determine if the method of assessment is appropriate and can differentiate a true response from other variables such as concurrent drugs (see below), lack of appropriate assessment methods, or poor study design. Negative control groups are often excluded for humane reasons, but if a method of assessment is accurate and rescue analgesia is administered than there would be only short periods of uncontrolled pain in a small group of patients. Additionally, consistent administration of an analgesic premedication at appropriate doses could prevent unmitigated immediate postoperative pain. Conversely the lack of inclusion of a negative control group could result in the inappropriate conclusion that a treatment was effective when it was NOT truly effective resulting in potentially hundreds to thousands of animals administered an INEFFECTIVE drug or dose resulting in much greater welfare concerns. The ideal situation would be to develop an accurate and reliable method of analgesic effectiveness in clinical patients, but no method has been developed at the time of this writing.

Assessing the efficacy of perioperative analgesics in dogs and cats is difficult due to the lack of an assessment method that is specific for pain and not confounded by other variables.[2] Studies assessing the postoperative efficacy of analgesics are often confounded and difficult to interpret. Concurrent medications often administered in the perioperative period can interfere with most pain scoring systems (numerical rating scores, composite scoring systems, visual analogue scales, et al.). Sedatives often have confounding effects (sedation – which may have the animal appear to rest comfortably, result in lack of movement, lack of environmental response, hyperesthesia during palpation, and vocalization). Anticholinergic administration may also have confounding effects (tachycardia, pupil dilation, anxiety, urine and fecal retention which may decrease the animal's desire to stand and walk). Other confounding effects can include anxiety (resulting in aggression, decreased spontaneous movement, pacing, shivering, reluctance to walk, tachycardia, hyperesthesia during palpation, vocalization). Another factor such as an animal's attention towards a surgical site is also confounded as an animal may lick chew or bite the site for a variety of reasons independent of pain (pruritus, smell of surgical disinfectant, smell of blood or oozing of serum from the site, a bandage wrapped too tightly, and the mere presence of a bandage).

Assessing the efficacy of analgesics for outpatients, primarily osteoarthritis patients, is also very difficult. Placebo controlled studies have routinely identified a substantial placebo effect ranging from to 16 to 33% and up to 47% perceived response to the placebo.[3-5] Placebo effects in outpatient studies of osteoarthritis can be due to waxing and waning disease, client bias, veterinarian bias, variability in the scoring methods, animal excitement, agitation, and anxiety among other factors. Additionally, excitement during the trip to the veterinarian's office may release catecholamines and endorphins resulting in an analgesic effect independent of the treatment confounding the assessment.

TRANSDERMAL FENTANYL SOLUTION

A transdermal fentanyl solution is approved for use in dogs to provide postoperative analgesia for up to 4 days after administration, but should not be used in cats. The formulation is highly concentrated solution (50 mg/mL) and inadvertent human exposure should be avoided and requires special training and the RiskMAP – Risk Minimization Action Plan must be completed prior to use.

Fentanyl is a mu opioid agonist. The pharmacokinetics of transdermal fentanyl solution (TFS) has been determined in dogs. Application results in a rapid increase in plasma fentanyl concentrations with 0.6 ng/mL (estimated concentration eliciting an analgesic effect) reached in approximated 2 hours. [6] The mean average concentration of 1.9 ng/mL was achieved at 13.6 hours with an estimated terminal half-life of 74 h. Clinical efficacy of TFS (2.7 mg/kg as a single dose) was determined for postoperative pain in dogd using the positive control oxymorphone (0.22 mg/kg SC q 6 h for 90 hours), also a mu opioid agonist, and no negative control group. [7] No significant difference between the treatments were noted using a Glasgow composite pain score over the course of 96 hours. Lack of pain control was judged in 4/249 TFS and 8/253 oxymorphone treated dogs. Opioid associated adverse effects such as hypothermia, sedation, and postoperative weight loss occurred in both groups. The adverse effects of the TFS dogs were not considered severe enough to be classified as a treatment failure while 17 of the oxymorphone group dogs were considered treatment failures due to the adverse effects of treatments. None of the TFS were administered naloxone reversal, but 9 of the oxymorphone group were administered naloxone for opioid adverse effects. The study concluded both treatments were safe and effective for a 96 hour period, but opioid adverse effects were greater in the oxymorphone dogs.

The RiskMAP training is required prior to use of transdermal fentanyl solution. A brief summary of the RiskMAP training is included here, but complete information is available from the manufacturer and should be considered the source of information regarding safe use of the solution. Use of the approved solution requires protection including wearing gloves, safety glasses and a lab coat to decrease potential for human exposure. A minimum of two trained personnel are required to administer the drug. The owner must also sign a client information release prior to application of the dose. The dog must remain restrained for a minimum of 5 minutes after drug application and hospital staff handling the treated dogs should wear protective gloves when handling the dog in the peri- and postoperative periods. The site of application should not be touched for 72 hours after application. If a person inadvertently touches the application site, contact with the mouth should be avoided and soap and water should be used to wash immediately. The treated dog should not come in contact with children for 72 hours.

MELOXICAM TRANSMUCOSAL ORAL SPRAY

Meloxicam is a nonsteroidal anti-inflammatory drug approved for use as an injectable solution, oral suspension and transmucosal oral spray in dogs for the control of pain and inflammation associated with osteoarthritis. The pharmacokinetics of the transmucosal oral spray are similar to that of per os meloxicam suspension with a relative bioavailability of 98% and mean CMAXs of 619 ng/mL (transmucosal oral spray) and 608 ng/mL (oral suspension) and were not significantly different. [2] The efficacy of the transmucosal spray was demonstrated in a placebo controlled trial in dogs affected by osteoarthritis. [2] Dogs treated with meloxicam were significantly improved using client specific outcome measures with 73% of meloxicam dogs improved compared to 47% of placebo dogs.

Advantages of the transmucosal oral spray versus the PO solution could include ease of administration for clients, ease of dose calculation (number of pumps per dose) for clients, ease of administration to patients with difficulty with swallowing and for animals with an aversion to administration of per os administered drugs.

BUPRENORPHINE ORAL TRASMUCOSAL ADMINISTRATION

Buprenorphine is a mu opioid partial agonist most often administered by parenteral (subcutaneous, intramuscular and intravenous) injection. Administration of buprenorphine by the oral transmucosal (OTM) route has been investigated in cats and dogs. A recent pharmacokinetic study of buprenorphine administered IV and OTM in cats investigated the effects of sample site collection on pharmacokinetic parameters to provide more accurate parameter determinations. [8] The OTM bioavailability was 23-32% when determined using saphenous and carotid sampling sites compared to 47% using jugular samples. These data suggest the jugular samples overestimated systemic plasma concentrations by as much as 2 fold since the jugular vein directly drains from the site of administration (oral mucosa). These data suggest a lower amount of buprenorphine is absorbed from the OTM route in cats than initially thought. Buprenorphine OTM may be an effective dosing option for cats, but a higher dose of 0.06 mg/kg OTM would be needed to produce similar plasma concentrations as 0.02 mg/kg IV.

Buprenorphine OTM administration has also been assessed in dogs. However the study sampled dogs from the jugular vein which may have overestimated the actual systemic plasma concentrations. The relative bioavailability was 38 and 47% as 0.02 and 0.12 mg/kg. The study concluded a dose of 0.12 mg/kg would achieve similar plasma concentrations as 0.02 mg/kg IV. However, since this study sampled jugular blood, the jugular plasma concentrations may not be accurate assessments of systemic plasma concentrations due to the jugular vein directly draining the site of administration (oral mucosa). Therefore OTM buprenorphine is not a recommended route of administration at this time for dogs.

REFERENCES

[1] Shott S. 2011. **JAVMA**;238(1): 55.[2] KuKanich B. 2013. Vet Clin North Am Small Anim Pract;43(5):1109. [3] Anonymous. OroCAM transmucosal Oral Spray. NADA 141-346. [4] Anonymous. Rimadyl. NADA 141-053. [5] Anonymous. Adequan Canine. NADA 141-038. [6] Freise KJ et al. 2012. J Vet Pharmacol Ther;35(S2):65. [7] Martinez SA, et al. 2013 J Vet Pharmacol Ther; epub doi: 10.1111/jvp.12096. [8] Hedges AR, et al. 2013. J Vet Pharmacol Ther doi: 10.1111/jvp.12075. [9] Abbo LA, et al. 2008. Vet Ther;9(2):83.

CURRENT DIAGNOSTIC TESTING FOR OSTEOSARCOMA

Anne Barger, DVM, MS, Diplomate ACVP
Urbana, IL

Introduction

Neoplastic processes of bone can result from metastasis, local invasion or as a primary bone tumor. Primary bone tumors consist primarily of osteosarcomas, chondrosarcomas, synovial cell sarcomas, fibrosarcomas and less common tumors, such as multilobulated tumor of bone, hemangiosarcoma and liposarcoma. Additionally, multiple myeloma and lymphoma are classified as tumors of bone marrow which can result in boney lysis. Benign tumors of bone have been reported and include osteoma and ossifying fibroma.

Primary bone tumors

Cytologically, many of these tumors often have similar characteristics and can be difficult to distinguish with cytology alone. The cells of primary bone tumors consist of a population of round to spindle shaped cells with varying amounts of pale basophilic cytoplasm. Cells of osteosarcoma and chondrosarcoma often have eccentrically placed nuclei giving them a plasmacytoid appearance. Prominent and often multiple nucleoli are commonly observed. These samples are generally highly cellular compared to what would be expected if normal bone were aspirated. Often within the background is a pale, eosinophilic, proteinaceous material consistent with matrix. This is most commonly seen with osteosarcoma and chondrosarcoma but can also be observed with fibrosarcoma and synovial cell sarcoma. One of the most difficult challenges is differentiating neoplastic from reactive bone. A pure population of mesenchymal cells may be consistent with a neoplastic process or with reactive bone so close examination of individual cell features is essential. Primary bone tumors such as osteosarcoma show criteria of malignancy such as anisocytosis, anisokaryosis, variation in the nuclear to cytoplasmic ratio and binucleation. Prominent nucleoli can be observed reactive as well as neoplastic bone. A recent study compared specific cytologic differences between osteosarcoma and reactive bone and found that presence of mitoses, more than one nucleus/cell and increased N:C ratio were observed more commonly with osteosarcoma than reactive bone[1].

Cytochemistry and immunohistochemical staining

Cytologic differentiation of osteosarcoma from other mesenchymal tumors of bone is challenging however recent evaluation of a cytochemical stain shows promise. Staining of cells for alkaline phosphatase activity with a phosphate substrate allows osteosarcoma to be identified with 100% sensitivity and 89% specificity[2]. These results were supported by another study with a sensitivity of 100% and a specificity of 67%[3]. The ALP stain essentially consists of a phosphate salt (Nitroblue tetrazolium chloride/5-bromo-4-chloro-3-indolyl phosphate toluidine salt) which acts as a substrate for ALP and undergoes an oxidation/reduction reaction after dephosphorylation by the enzyme. This results in a color change of the insoluble product at the site of the reaction (the cell membrane). The primary limitation of this stain is that reactive osteoblasts will stain the same as neoplastic osteoblasts. Therefore, the ALP stain should only be used on samples where a cytologic diagnosis of sarcoma has been made. With such a high sensitivity, the stain is an excellent screening test, however other tumors and reactive bone will also stain positively with ALP, decreasing its specificity. In one study, 1 out of 4 chondrosarcomas, 1 of 2 amelanotic melanomas and one multilobulated tumor of bone stained positive for ALP activity. Generally the cytologic and clinical features of the tumors can assist in diagnosis. Often chondrosarcomas have excessive eosinophilic matrix, melanomas often will have pigment, (even small amounts in amelanotic melanoma), and multilobular tumors of bone have a distinct clinical appearance. The combination of cytology, clinical appearance, location of the tumor and radiographic changes are all important in the diagnosis of OSA. The additional information of ALP activity only improves the ability of the clinician and pathologist to appropriately diagnose primary tumors of bone.

Fine needle aspiration of lytic lesions of bone is becoming more common. Advantages of cytology over biopsy include rapid turnaround time (in some locations time from aspiration to evaluation can be less than 24 hours) lower risk of pathologic fracture, minimal patient discomfort, decreased procedure and recovery times, and decreased risk of wound infection. Cytology, however also has its limitations. These include difficulties in differentiating reactive from neoplastic tissues and poor cell exfoliation resulting in nondiagnostic specimens. A recent study tried to overcome these difficulties with a different method of sample acquisition for cytology. A core aspirate cytology was compared to fine needle aspiration and the accuracy of cytology compared to histopathology with ALP staining was evaluated[4]. The conclusion from this study was that there was no significant difference in the method but the diagnostic accuracy of cytology in the diagnosis of osteosarcoma with ALP staining was 85%-95%. The sensitivity of ALP staining in this study was 100%.

There are immunohistochemical stains available to assist with diagnosis of specific tumors of bone. In particular, hemangiosarcomas are positive for CD31[5]. CD31, therefore, can be useful in differentiating hemangiosarcoma from telangectatic osteosarcoma. Some synovial cell sarcomas will stain with both vimentin, a marker of mesenchymal tissue and cytokeratin, a marker of epithelial tissue. Positive staining with both will help support a diagnosis of synovial cell sarcoma but negative staining with cytokeratin does not rule out this diagnosis. Additionally, metastatic carcinomas will be positive with cytokeratin and negative for vimentin. Any of the primary bone tumors will be positive for vimentin so this staining pattern is useful in identifying metastatic epithelial tumors. No one stain or panel of stains has been found to differentiate osteosarcoma, chondrosarcoma or fibrosarcoma. The key to diagnosing these tumors is evaluating cellular morphology, architecture of the tumor and the production of matrix within the tumor. However, sometimes it is still impossible to distinguish these tumors even with biopsy. An antibody directed against tissue non-specific ALP has shown promise in the identification of osteosarcoma. ALP is comprised of two different isoenzymes that contribute the majority of the activity in the tissue and serum. The tissue nonspecific gene yields the liver ALP, kidney ALP and bone ALP isoforms[6]. Therefore, an antibody directed against the tissue nonspecific ALP (TNALP) would be expected to recognize bone origin ALP. Bone ALP is produced by osteoblasts and is anchored to outer membrane of the cell[7]. Because of this epicellular location, ALP activity cannot be detected by the NBT staining method. However, osteosarcoma cells have been shown to express ALP and the tissue nonspecific ALP antibody binds neoplastic osteoblasts[7]. Currently, we are evaluating different primary bone tumors to determine if TNALP will allow differentiation of osteosarcoma from other primary bone tumors as has been shown with the cytochemical stain.

REFERENCES

1. Reinhardt S, Stockhaus C, Teske E, et al. J Small An Pract 2005;46:65.
2. Barger A, Graca R, Bailey K, et al. Vet Pathol 2005;42:161.
3. Britt T, Clifford C, Barger A, et al. J Small An Pract 2007;48:145.
4. Neihaus S, Locke J, Barger A, et al. J An Hosp Assoc 2011;47:317.
5. Ferrer L, Fondevila D, Rabanal R, et al. J Comp Pathol 1995;112(4):319.
6. Wiedmeyer C, Solter P, Hoffmann WE. Am J Vet Res 2002;63:1089.
7. Sternberg R, Pondenis H, Yang M, et al. J Vet Intern Med 2013;27:955.

MESENCHYMAL STEM CELLS AND IMMUNOMODULATION IN VETERINARY MEDICINE

Dori L. Borjesson DVM, PhD, DACVP (Clinical Pathology)
Davis, CA

INTRODUCTION

Mesenchymal stem cells (MSC) are adult-derived multipotent stem cells that have been derived from almost every tissue.[1] For most current veterinary clinical applications, MSCs are isolated from adipose tissue, bone marrow, umbilical cord blood or umbilical cord tissue.[2] They are spindle-shaped, plastic-adherent cells capable of in vitro tri-lineage differentiation (differentiation into fat, cartilage and bone). There is no one MSC-specific antibody that defines an MSC. As such, panels of antibodies directed towards surface proteins are used to identify an MSC. Human and veterinary species MSCs are CD105, CD73, and CD90 positive and CD45, CD34, CD14, CD11b, CD79a, and CD19 negative.[3]

Both in human and veterinary medicine, MSCs promote tissue regeneration and healing via modulation of the immune response, including decreasing the cells and cytokines associated with inflammation and increasing blood flow to promote normal healing rather than scarring.[4-6] MSCs interact with cells of the immune system and change their function or phenotype. Although many of the details underlying the mechanisms by which MSC modulate the immune system have been defined for human and rodent MSCs, much less is known about MSCS from dogs, cats and horses. This is particularly important because the clinical use of MSCs in veterinary medicine is increasing and far exceeds MSC use in human medicine. A more thorough understanding of how MSCs regulate cells of the immune system will help 1) target cellular therapy to diseases and disorders that are most likely to respond to treatment and 2) define how local and systemic inflammation may alter MSC function and therapy outcome.

MESENCHYMAL STEM CELLS AND CELLS OF THE IMMUNE SYSTEM

T lymphocytes

Mesenchymal stem cells interact with both the innate and adaptive arms of the immune system. MSCs inhibit the proliferation of stimulated T lymphocytes in vitro. This is a well-described finding in companion animal species studied to date including horses and dogs as well as for nonhuman primates, sheep, and pigs, among others.[7,8] There are many proposed mechanisms by which MSCs inhibit lymphocyte proliferation including MSC-induced T cell cycle arrest or MSC-induced lymphocyte apoptosis.[9] Inhibition of lymphocyte proliferation is accompanied by decreased production of inflammatory cytokines including tumor necrosis factor alpha (TNFα) and interferon gamma (IFNγ). MSCs also decrease the expression of activation markers (CD25, CD38, and CD69) on T cells, preventing their activation and proliferation.[10] Stimulated MSCs also downregulate pro-inflammatory T cell subsets including cytotoxic CD8+ T cells, Th1 cells and Th17 cells. MSC induced suppression includes memory and naïve T cells. In the process of downregulating these T cells, MSCs promote the proliferation of immunomodulatory T cell subsets, notably T regulatory cells.[11,12] Although direct lymphocyte to MSC contact, via toll-like receptors, intracellular adhesion molecule 1, and vascular cell adhesion molecule 1 on the surface of MSCs, may play a part in MSC mediated immunosuppression, this contribution has not been examined in veterinary species.[13, 14] It is clear however that soluble factors secreted by the MSCs are critical to MSC-induced inhibition of lymphocyte proliferation. Cytokines implicated in the inhibition of T cell proliferation and the induction of regulatory T cells include prostaglandin E_2 (PGE_2), interleukin (IL)-10 and transforming growth factor beta (TGFβ) among others.[15]

B lymphocytes

The understanding of how MSCs interact with B cells is not nearly as far advanced as our understanding of MSC- T cell interactions. To date, there is no work with MSCs and B cells reported for any companion animal species. In vitro work with MSCs from humans and rodents demonstrate that MSCs increase B cell viability and block B cell proliferation via cell cycle arrest. MSCs block B cell differentiation towards plasma cells and decrease immunoglobulin production.[16] A deeper understanding of MSC- B cell interaction will be particularly important in Veterinary medicine as we approach the use of cell therapy for immune-mediated disorders.

Macrophages and Dendritic cells

As with MSCs and B lymphocytes, the delineation of MSC interaction with dendritic cells, antigen presenting cells and macrophages is still in its infancy in Veterinary species. Human and rodent MSCs modulate dendritic cell development from monocytes and impair dendritic cell function. MSC inhibit differentiation of monocytes into immature dendritic cells and their further maturation into mature DC. Impaired dendritic cell function includes modulation in major histocompatibility (MHC) II and T-cell costimulatory molecule expression, downregulation of cytokine production, and prevention of dendritic cell homing to lymph nodes.[17] Taken together, these changes ultimately limit the ability of dendritic cells to stimulate a T cell response and may be one indirect way in which MSCs modulate T cells. MSCs also induce regulatory antigen-presenting cells with T-cell suppressing properties. With mature antigen presenting cells, MSCs reduce the expression of co-stimulatory molecules and downregulate IL-12 secretion.

CONCLUSIONS

Mesenchymal stem cells are being increasingly investigated as a therapeutic option for a variety of inflammatory and immune-mediated disorders. The rationale for their use is based on a body of knowledge that has delineated profound effects of MSCs on the cells and mediators of inflammation. MSCs are activated in the context of inflammation and the inflammatory milieu in large part dictates MSC function and efficacy. In veterinary medicine, we are just beginning to understand the therapeutic potential of these cells. As veterinarians, with well-designed clinical trials, we have a unique opportunity to understand MSC function in the context of our patients with naturally occurring inflammatory and immune-mediated diseases.[5,15,18]

REFERENCES

[1]da Silva Meirelles L et al. 2006. J Cell Sci 119:2204. [2]Toupadakis CA et al. 2010. Am J Vet Res 71:1237. [3]Dominici M et al. 2006. The International Society for Cellular Therapy position statement. Cytotherapy 8:315. [4]Meirelles Lda S et al. 2009. Cytokine Growth Factor Rev 20:419.[5] Peroni JF et al. 2011. Vet Clin North Am Equine Pract 27:351.[6] Yi T et al. 2012. Arch Pharm Res 35:213.[7] Carrade DD et al. 2012. Cell Med 4:1.[8] Kang JW et al. 2008. Stem Cells Dev 17:681. [9]Glennie S et al. 2005. Blood 105:2821.[10]Le Blanc K et al. 2004. Scand J Immunol 60:307.[11] Lee WS et al. 2011. Biol Blood Marrow Transplant 17:465.[12] Akiyama K et al. 2012. Cell Stem Cell 10:544. [13]Ren G et al. 2010. J Immunol 184:2321.[14]Di Ianni M et al. 2008. Exp Hematol 36:309.[15]Carrade DD et al. 2013. Comparative Medicine 63:1.[16] Tabera S et al. 2008. Haematologica. 93.1301.[17] English K et al. 2011. J Cell Biochem 112:1963.[18] Borjesson DL et al. 2011. Clin Lab Med 31:109.

MESENCHYMAL STEM CELLS: TRANSLATION AND CLINICAL APPLICATIONS IN LARGE AND SMALL ANIMAL MEDICINE

Dori L. Borjesson DVM, PhD, DACVP (Clinical Pathology)
Davis, CA

INTRODUCTION

Mesenchymal stem cells MSCs) are in phase II and phase III clinical trials in humans for the treatment of disorders ranging from ischemic, metabolic, inflammatory and neurologic diseases to orthopedic disorders (http://clinicaltrials.gov/). In veterinary medicine, MSCs are considered safe and their use is widespread in equine practice but is more limited in small animal practice. To date, most studies have focused on evaluating MSC safety with only a few studies focused on efficacy. In addition, both small and large animals have been used to test stem cell therapy for pre-clinical models of human disease.

There is a paucity of prospective, well-controlled, blinded clinical trials using MSCs to treat well-defined lesions in veterinary medicine. Clinicians and researchers developing clinical trials in veterinary medicine should carefully consider where to obtain stem cells (company, university laboratory etc), what is in the stem cell product obtained and what steps are taken to assess stem cell quality. Additional considerations include stem cell tissue of origin, dose, route of administration and whether an autologous or an allogeneic product is preferred.

SMALL ANIMAL APPLICATIONS

In dogs, MSCs have primarily been used for tissue repair and regeneration and, to a lesser extent, for inflammatory lesions such as osteoarthritis. There is also some experimental work with dogs and spinal cord injury and repair. MSCs have been used as a sole therapy or in combination with scaffolding material for tissue repair. No significant adverse effects have been reported secondary to the use of MSCs in dogs with the exception of trials that attempted to target coronary venous or intracoronary arterial stem-cell delivery.[1,2] Both autologous (self) and allogeneic (non-self) MSCs have been used successfully in canine patients although the use of autologous bone marrow- or fat-derived MSCs is far more common.[3,4] Stem cells have been administered intravenously, intrathecally, and locally including renal injections, intraarticular injections or injections into the lacrimal gland.[5,6] MSC dose varies widely depending on the lesion and the route of administration. In cats, MSC therapy has been focused on renal disease[7] and feline chronic gingivostomatitis. This talk will be focused on recent therapeutic applications of MSC therapy in small animals highlighting areas with strong potential for clinical use.

LARGE ANIMAL APPLICATIONS

In horses, MSCs have been primarily used to treat a range of musculoskeletal disorders including tendinopathies/ tendonitis, stifle injuries (including cartilage damage), ligament injuries and bone disorders.[8] As with small animal applications, MSCs are isolated from a number of tissues including bone marrow, fat, cord blood and cord tissue and much of the work, to date, has been on the characterization and differentiation of MSCs from these tissue sources.[9,10] For tissue repair, MSCs are often applied with a scaffold matrix (such as fibrin or platelet rich plasma) to provide physical support and growth factors for the stem cells. Unlike human beings, in horses, very little work has focused on the use of MSCs for inflammatory or immune-mediated diseases however stem cells are being explored for use in laminitis and wound healing. This talk will overview some current uses of MSC therapy in horses with an emphasis on potential non-orthopedic uses for MSCs.[11]

REFERENCES

[1]Vulliet PR et al. 2004. Lancet 363:783. [2]Pogue B et al. 2013. J Small Anim Pract 54:361. [3]Aziz Aly LA et al. 2012. Int J Stem Cells. 5(2):104. [4]Kim JW et al. 2013. Vet Dermatol. 24:242. [5]Jung DI et al. 2009. J Neurol Sci 285:67. [6]Park SA et al. 2013. Cytotherapy 15:1498. [7]Quimby JM et al. 2013. Stem Cell Res Ther. 30:48.[8]Frisbie DD et al. 2010. Equine Vet J. 42:86. [9]Borjesson DL et al. 2011. Clin Lab Med.31:109. [10]Toupadakis CA et al. 2010. Am J Vet Res 71:1237 [11]De Schauwer C et al. 2013. Vet Q. 33:234

NONREGENERATIVE ANEMIA: RECENT ADVANCES IN UNDERSTANDING MECHANISMS OF DISEASE

Michael M. Fry, DVM, MS, DACVP
Knoxville, TN

INTRODUCTION

In most species (equids are an exception), uninhibited compensatory response to anemia is characterized by reticulocytosis, an increased concentration of reticulocytes in the blood, which is a peripheral manifestation of erythroid hyperplasia in the bone marrow. This type of anemia is conventionally referred to in veterinary medicine as *regenerative*, and is typical of anemia occurring due to hemorrhage or destruction of erythrocytes in circulation. A fundamental principle of clinical hematology is that reticulocytosis takes several days to develop after an acute hemorrhagic or hemolytic event. Patients in which a reticulocytosis is imminent are sometimes said to have "preregenerative" anemia.

Anemia without accompanying reticulocytosis is referred to as *nonregenerative*, and occurs as a result of decreased or ineffective erythropoiesis. Diminished erythropoiesis is associated with various specific mechanisms and is a recognized consequence of a wide array of primary diseases. *Ineffective erythropoiesis*, a relatively uncommon condition, is characterized by anemia despite normal to increased erythropoietic activity – a situation recognized to occur in veterinary species in certain types of neoplasia and immune-mediated disease.

This talk is not intended as a comprehensive review of nonregenerative anemia in veterinary species; rather, it will focus on mechanisms of nonregenerative anemia that have been elucidated recently or that have received minimal attention in companion animals.

ANEMIA OF INFLAMMATION – THE ROLE OF HEPCIDIN

The most common form of nonregenerative anemia is *anemia of inflammation* (aka anemia of chronic disease).
It has long been recognized that patients with infectious or inflammatory disease often develop anemia, and that this phenomenon is associated with functional iron deficiency. Of course, iron is an essential constituent of hemoglobin, and anemia of inflammation may be a side-effect of an evolutionary immunologic adaptation to limit microbial access to iron. In contrast to absolute iron deficiency anemia (see below), which in severe cases has a classic presentation of microcytosis and hypochromasia, anemia of inflammation is usually normocytic and normochromic.

Recent years have seen breakthroughs in the understanding of molecular mechanisms of anemia of inflammation, including new insights into the relationship between innate immunity and iron regulation. Many of the new insights involve the hormone, hepcidin. Some facts about hepcidin[1-3]:

- An acute phase protein, synthesized mainly in the liver.
- Bioactive form is a highly conserved 25-amino acid peptide.
- First recognized for its direct antimicrobial effects.
- Expression increases with inflammation or iron overload, and decreases with anemia or hypoxia.
- Exerts its effects by binding to the iron efflux protein, ferroportin, and inducing its internalization and degradation – an interaction that inhibits release of storage iron release of storage iron and absorption of dietary iron, thus resulting functional iron deficiency.
- Excreted mainly in the urine.

Dogs with given a therapeutic dose of iron dextran had evidence of increased serum hepcidin concentration, and dogs with experimentally-induced nutritional iron deficiency anemia had markedly decreased hepatic hepcidin gene expression – results consistent with known biology of hepcidin in humans and laboratory animals[4,5]. Translation of basic research findings related to hepcidin into diagnostic and therapeutic applications in people is underway, and a similar pattern seems likely to occur in veterinary medicine.

ANEMIA OF INFLAMMATION – OTHER MECHANISMS

Inflammation can also contribute to anemia via other mechanisms. For example, inflammation has been shown to cause decreased erythrocyte survival in some species[6,7]; some studies have implicated inflammatory cytokines as inhibitors of erythropoiesis via direct toxic effects on erythroid precursors, decreased expression of hematopoietic factors, or decreased expression of erythropoietin receptors; and some investigators have suggested that oxidants produced by activated neutrophils cause altered erythrocyte surface antigenicity, in turn lead to accelerated immunological destruction of erythrocytes[8-10].

CHRONIC RENAL FAILURE – MECHANISMS BESIDES DECREASED EPO PRODUCTION

Insufficient production of Epo is a widely recognized cause of anemia in patients with chronic renal failure. A partial list of other proposed mechanisms includes:

- Decreased urinary excretion of hepcidin[3].
- Dysregulation of the renin-angiotensin-aldosterone system[11].
- Neocytolysis (selective hemolysis of young erythrocytes)[12].
- Inflammation[13].

- Presence of inhibitory or myelotoxic factors, or lack of stimulatory factors, in uremic serum[14-16].
- Iron deficiency secondary to chronic blood loss associated with uremic toxin-induced platelet dysfunction or gastrointestinal ulceration[17,18].
- Parathyroid hormone-related effects (inhibition of Epo synthesis; myelofibrosis)[19,20].

NON-REGENERATIVE IMMUNE-MEDIATED HEMOLYTIC ANEMIA

Most cases of IMHA are associated with a strong regenerative response, but nonregenerative forms of IMHA also occur. The lack of an appropirate regenerative response may occur because of underlying disease that inhibits erythropoiesis or because of targeted destruction of developing erythroid cells. Nonregenerative IMHA is usually diagnosed based on exclusion of other causes of anemia and on response to immunosuppressive therapy.

Bone marrow findings in the erythroid lineage range from rare or absent erythroid precursors (as in pure red cell aplasia), to absence or decreased proportions of later-stage cells, to marked generalized erythroid hyperplasia (signifying ineffective erythropoiesis)[21]. In theory, bone marrow findings depend on the stage at which developing erythrocytes express the antigen(s) being targeted immunologically. Other types of bone marrow disease – including dysmyelopoiesis, myelonecrosis, myelofibrosis, inflammation, edema, hemorrhage, and hemophagocytic syndrome – have been found in dogs and cats with nonregenerative IMHA and marrow erythroid hyperplasia, suggesting additional mechanisms contributing to anemia[22].

In the last decade, bovine neonatal pancytopenia has become recognized as an important emerging disease in Europe; similar, perhaps related, cases have also been reported elsewhere[23,24]. The condition, sometimes called "bleeding calf syndrome", is characterized by pancytopenia and bone marrow aplasia in neonatal calves that frequently present with clinical signs related to bleeding tendency. There is strong evidence now that the disease is caused by Bovine Viral Diarrhea vaccine-induced maternal alloantibodies in colostrum that cross-react with blood and bone marrow cells in calves[25,26].

HEMOPHAGOCYTIC SYNDROME

Hemophagocytic syndrome is a condition characterized by hypercytokinemia, excessive stimulation and proliferation of nonneoplastic macrophages, and exuberant phagocytosis of blood cells[27]. It has been extensively described in humans and also reported in several veterinary species, including dogs, cats, cattle, and a macaque[28-32]. Nonregenerative anemia is a common feature in veterinary species and probably occurs due to bone marrow suppression secondary to hypercytokinemia and/or phagocytosis of erythroid precursors. An underlying cause (e.g., infection, systemic disease) is suspected to trigger the disease in both animals and people; a primary (inherited) form is also described in humans.

HEART FAILURE

Anemia is not a commonly recognized co-morbidity in veterinary patients with heart failure, but it is a common complication of HF in humans (cited prevalence of 15-61%) and is associated with increased morbidity and mortality and diminished quality of life[33]. Proposed mechanisms of anemia in patients with HF include those associated with inflammation, gastrointestinal dysfunction, neurohormonal activation, and cardiorenal syndrome[34-37].

PHYSIOLOGIC AND ENVIRONMENTAL FACTORS

Healthy humans develop increased Epo concentrations with age, and lack of a normal increase is suspected to play a role in unexplained anemia of elderly patients[38,39]. The physiologic erythrokinetic changes of geriatric animals are not well-characterized.

While exercise can lead to increases in erythrocyte mass, presumably as an adaptive measure to increase oxygen-carrying capacity, strenuous exercise may also be associated with anemia. Intense exercise can cause excessive iron losses, and human athletes are commonly diagnosed with iron deficiency. Established mechanisms of iron loss associated with strenuous exercise include sweating, hemolysis, hematuria, and GI hemorrhage[40]. Inflammation may also play a role in the development of sports anemia. Increased circulating levels of some inflammatory cytokines, especially IL-6, have been identified in humans following intensive exercise[41,42]. Strenuous exercise in rats also induced inflammation that was associated with increased circulating IL-6, increased hepatic hepcidin expression, and anemia[43].

Astronauts often experience a 10-15% decrease in red cell mass associated with space travel[44]. The microgravity environment causes central pooling of blood, known as acute central plethora. This localized hypervolemia initiates a sequence of events resulting in anemia: further redistribution of plasma to interstitial tissues (edema); increased central hematocrit (hemoconcentration); inhibition of Epo secretion; profound suppression of erythropoiesis; and promotion of neocytolysis. The end result is a decrease in red cell mass that is adaptive for space, but not suitable for terrestrial conditions. A similar red cell response has been identified in polycythemic, high-altitude dwelling individuals upon descent to sea level; the mechanisms that underlie anemia due to plethora have implications for changes in red cell mass associated with other conditions[45].

REFERENCES

1. Nemeth E, Valore EV, Territo M, Schiller G, Lichtenstein A, Ganz T. Hepcidin, a putative mediator of anemia of inflammation, is a type II acute-phase protein. *Blood.* 2003;101(7):2461-2463.
2. Nemeth E, Tuttle MS, Powelson J, et al. Hepcidin regulates cellular iron efflux by binding to ferroportin and inducing its internalization. *Science.* 2004;306(5704):2090-2093.

3. Zaritsky J, Young B, Wang HJ, et al. Hepcidin--a potential novel biomarker for iron status in chronic kidney disease. *Clin J Am Soc Nephrol.* 2009;4(6):1051-1056.
4. Fry MM, Kirk CA, Liggett JL, et al. Changes in hepatic gene expression in dogs with experimentally induced nutritional iron deficiency. *Vet Clin Pathol.* 2009;38(1):13-19.
5. Grimes CN, Giori L, Fry MM. Role of hepcidin in iron metabolism and potential clinical applications. *Vet Clin North Am Small Anim Pract.* 2012;42(1):85-96.
6. Rigby PG, Strasser H, Emerson CP, Betts A, Friedell GH. Studies in the anemia of inflammatory states. I. Erthrocyte survival in dogs with acute and chronic turpentine abscess. *J Lab Clin Med.* 1962;59:244-248.
7. Weiss DJ, Krehbiel JD. Studies of the pathogenesis of anemia of inflammation: erythrocyte survival. *Am J Vet Res.* 1983;44(10):1830-1831.
8. Roodman GD, Bird A, Hutzler D, Montgomery W. Tumor necrosis factor-alpha and hematopoietic progenitors: effects of tumor necrosis factor on the growth of erythroid progenitors CFU-E and BFU-E and the hematopoietic cell lines K562, HL60, and HEL cells. *Exp Hematol.* 1987;15(9):928-935.
9. Ganz T. Molecular pathogenesis of anemia of chronic disease. *Pediatric Blood & Cancer.* 2006;46(5):554-557.
10. Kruse A, Uehlinger DE, Gotch F, Kotanko P, Levin NW. Red blood cell lifespan, erythropoiesis and hemoglobin control. *Contrib Nephrol.* 2008;161:247-254.
11. Dunn A, Lo V, Donnelly S. The role of the kidney in blood volume regulation: the kidney as a regulator of the hematocrit. *Am J Med Sci.* 2007;334(1):65-71.
12. Rice L, Alfrey CP. The negative regulation of red cell mass by neocytolysis: physiologic and pathophysiologic manifestations. *Cell Physiol Biochem.* 2005;15(6):245-250.
13. Stenvinkel P, Alvestrand A. Inflammation in end-stage renal disease: sources, consequences, and therapy. *Semin Dial.* 2002;15(5):329-337.
14. Caro J, Erslev AJ. Uremic inhibitors of erythropoiesis. *Semin Nephrol.* 1985;5(2):128-132.
15. Macdougall IC. Role of uremic toxins in exacerbating anemia in renal failure. *Kidney Int Suppl.* 2001;78:S67-72.
16. Brunati C, Cappellini MD, De Feo T, et al. Uremic inhibitors of erythropoiesis: a study during treatment with recombinant human erythropoietin. *Am J Nephrol.* 1992;12(1-2):9-13.
17. Boccardo P, Remuzzi G, Galbusera M. Platelet dysfunction in renal failure. *Semin Thromb Hemost.* 2004;30(5):579-589.
18. Skorecki K, Green J, Brenner BM. Chronic renal failure. In: Brunwald E, ed. Harrison's Principles of Internal Medicine. New York: McGraw-Hill; 2001:1551-1561.
19. Washio M, Iseki K, Onoyama K, et al. Elevation of serum erythropoietin after subtotal parathyroidectomy in chronic haemodialysis patients. *Nephrol Dial Transplant.* 1992;7(2):121-124.
20. Rao DS, Shih MS, Mohini R. Effect of serum parathyroid hormone and bone marrow fibrosis on the response to erythropoietin in uremia. *N Engl J Med.* 1993;328(3):171-175.
21. Stokol T, Blue JT, French TW. Idiopathic pure red cell aplasia and nonregenerative immune-mediated anemia in dogs: 43 cases (1988-1999). *J Am Vet Med Assoc.* 2000;216(9):1429-1436.
22. Weiss DJ. Bone marrow pathology in dogs and cats with non-regenerative immune-mediated haemolytic anaemia and pure red cell aplasia. *J Comp Pathol.* 2008;138(1):46-53.
23. SAC C VS Disease Surveillance Report. More cases of bovine neonatal pancytopenia in Scotland. *Vet Rec.* 2012;171(1):11-14.
24. Pardon B, Steukers L, Dierick J, et al. Haemorrhagic diathesis in neonatal calves: an emerging syndrome in Europe. *Transbound Emerg Dis.* 2010;57(3):135-146.
25. Bastian M, Holsteg M, Hanke-Robinson H, Duchow K, Cussler K. Bovine Neonatal Pancytopenia: is this alloimmune syndrome caused by vaccine-induced alloreactive antibodies? *Vaccine.* 2011;29(32):5267-5275.
26. Kasonta R, Sauter-Louis C, Holsteg M, Duchow K, Cussler K, Bastian M. Effect of the vaccination scheme on PregSure® BVD induced alloreactivity and the incidence of Bovine Neonatal Pancytopenia. *Vaccine.* 2012;30(47):6649-6655.
27. Gupta S, Weitzman S. Primary and secondary hemophagocytic lymphohistiocytosis: clinical features, pathogenesis and therapy. *Expert Rev Clin Immunol.* 2010;6(1):137-154.
28. Weiss DJ. Hemophagocytic syndrome in dogs: 24 cases (1996-2005). *J Am Vet Med Assoc.* 2007;230(5):697-701.
29. Dunbar MD, Lyles S. Hemophagocytic syndrome in a cat with multiple myeloma. *Vet Clin Pathol.* 2012.
30. Walton RM, Modiano JF, Thrall MA, Wheeler SL. Bone marrow cytological findings in 4 dogs and a cat with hemophagocytic syndrome. *J Vet Intern Med.* 1996;10(1):7-14.
31. Naessens J. Bovine trypanotolerance: A natural ability to prevent severe anaemia and haemophagocytic syndrome? *Int J Parasitol.* 2006;36(5):521-528.
32. Cotroneo TM, Colby LA, Bergin IL. Hemophagocytic syndrome in a pancytopenic simian retrovirus-infected male rhesus macaque (Macaca mulatta). Vet Pathol. Vol. 48. United States; 2011:1138-1143.
33. Komajda M. Prevalence of anemia in patients with chronic heart failure and their clinical characteristics. *J Card Fail.* 2004;10(1 Suppl):S1-4.
34. Anand IS, Rector T, Deswal A, et al. Relationship between proinflammatory cytokines and anemia in heart failure. *European Heart Journal.* 2006;27(Supplement 1):485.
35. Naito Y, Tsujino T, Fujimori Y, et al. Impaired expression of duodenal iron transporters in Dahl salt-sensitive heart failure rats. *J Hypertens.* 2011;29(4):741-748.
36. Westenbrink BD, Visser FW, Voors AA, et al. Anaemia in chronic heart failure is not only related to impaired renal perfusion and blunted erythropoietin production, but to fluid retention as well. *Eur Heart J.* 2007;28(2):166-171.
37. Ronco C, Haapio M, House AA, Anavekar N, Bellomo R. Cardiorenal syndrome. *J Am Coll Cardiol.* 2008;52(19):1527-1539.
38. Ershler WB, Sheng S, McKelvey J, et al. Serum erythropoietin and aging: a longitudinal analysis. *J Am Geriatr Soc.* 2005;53(8):1360-1365.
39. Vanasse GJ, Berliner N. Anemia in elderly patients: an emerging problem for the 21st century. *Hematology Am Soc Hematol Educ Program.* 2010;2010:271-275.
40. Peeling P, Dawson B, Goodman C, Landers G, Trinder D. Athletic induced iron deficiency: new insights into the role of inflammation, cytokines and hormones. *Eur J Appl Physiol.* 2008;103(4):381-391.
41. Ostrowski K, Rohde T, Asp S, Schjerling P, Pedersen BK. Pro- and anti-inflammatory cytokine balance in strenuous exercise in humans. *J Physiol.* 1999;515 (Pt 1):287-291.
42. Pedersen BK, Bruunsgaard H, Ostrowski K, et al. Cytokines in aging and exercise. *Int J Sports Med.* 2000;21 Suppl 1:S4-9.
43. Liu YQ, Chang YZ, Zhao B, Wang HT, Duan XL. Does hepatic hepcidin play an important role in exercise-associated anemia in rats? *Int J Sport Nutr Exerc Metab.* 2011;21(1):19-26.
44. Udden MM, Driscoll TB, Pickett MH, Leach-Huntoon CS, Alfrey CP. Decreased production of red blood cells in human subjects exposed to microgravity. *J Lab Clin Med.* 1995;125(4):442-449.
45. Rice L, Ruiz W, Driscoll T, et al. Neocytolysis on descent from altitude: a newly recognized mechanism for the control of red cell mass. *Ann Intern Med.* 2001;134(8):652-656.

HELP ME HELP YOU! THE PATHOLOGIST'S APPROACH TO BONE MARROW

Bente Flatland, DVM, DACVIM (SA IM), DACVP (Clinical Pathology)
Knoxville, TN

INTRODUCTION

Optimal interpretation of bone marrow samples requires good communication between clinicians and pathologists, with inclusion of relevant hematologic information at the time of sample submission. Relevant hematologic and clinical data are particularly crucial when bone marrow is sampled for evaluation of unexplained cytopenia(s), but are also desirable when bone marrow is sampled for other reasons (e.g., tumor staging). The purpose of this presentation is to give an overview of the pathologist's approach to bone marrow evaluation, with a goal of facilitating optimal bone marrow sample submission and understanding of the pathologist's report. Specific lecture objectives include presenting considerations for bone marrow sampling and sample handling, conveying the importance of concurrent CBC data, outlining the pathologist's steps to bone marrow interpretation, and defining selected bone-marrow related terms.

HOW YOU CAN HELP THE PATHOLOGIST

1. Indicate the reason that bone marrow was sampled.
2. Provide concurrent CBC data.
3. Make good quality cytology smears (and/or take good quality core biopsy samples).
4. Keep cytology smears away from formalin. If relevant, ship cytology and biopsy samples in separate boxes (this applies not only to bone marrow, but other cytology samples too).

Knowledge of CBC data is required for optimal bone marrow interpretation, particularly when evaluating unexplained cytopenia(s). Bone marrow changes (e.g., hyper- or hypoplasia) can be absolute or relative, and ascertaining which is the case requires concurrent CBC data. At minimum, the pathologist should know PCV, total WBC count, and platelet count. Full CBC data (including blood smear review findings) are desirable. Ideally, the CBC is concurrent with the bone marrow sample (i.e., taken at the same time); however, in the author's opinion, a CBC within 24 hours of bone marrow sampling will suffice.

BONE MARROW SAMPLING AND SAMPLE HANDLING

Indications, sampling sites, and sampling procedure

Bone marrow is most commonly sampled for one of three reasons: (1) Hemic abnormalities detected on a CBC and blood smear review; (2) cancer staging; or (3) searching for occult infection or neoplasia. Hemic abnormalities include unexplained cytopenia(s) (e.g., persistent non-regenerative anemia, neutropenia, and/or thrombocytopenia), proliferative abnormalities (e.g., unexplained or inappropriate presence of immature cells in blood), and hemic cell morphologic abnormalities.[1] Commonly used bone marrow sampling sites in dogs and cats include iliac crest and proximal humerus. Iliac crest and sternum are often used in large animals. Sixteen- or 18-guage needles having a stylet (Rosenthal, Illinois sternal, or Jamshidi) are most commonly used to acquire the sample.[1] Samples may be obtained using only sedation and local anesthesia to restrain the patient if the clinician is experienced; general anesthesia may be desirable in small or fractious patients and in the teaching hospital setting.

Ideally (particularly when investigating unexplained cytopenia or when marrow fibrosis is suspected), both an aspirate and core biopsy are obtained and interpreted together. Advantages of an aspirate (cytology sample) include cytomorphologic detail and the ability to count a myeloid (granulocytic) to erythroid ratio (M:E). Advantages of a core biopsy (histology sample) include ability to assess tissue architecture, judge marrow cellularity, and identify fibrosis. In the author's opinion, if you are obtaining both an aspirate and a core biopsy, perform the aspirate first. Once satisfactory aspirate smears have been made (and removed from the immediate area to protect them from formalin fumes), proceed with core biopsy. Even if you did not originally plan to do a core biopsy, come prepared to do one in case you have a "dry tap". Sampling procedure is reviewed in detail in Harvey.[1]

Contraindications and potential complications

The main contraindication to bone marrow sampling is that it is not needed diagnostically. Although thrombocytopenic animals (and those with other bleeding disorders) may be predisposed to hemorrhage, excessive bleeding following bone marrow sampling for evaluation of unexplained thrombocytopenia is uncommon[1], particularly if appropriate pressure is placed on the biopsy site after the procedure. A skin suture at the incision site can also help minimize post-procedure hemorrhage. Other potential complications include infection at the biopsy site, damage to adjacent soft tissues (e.g., muscles or joints), needle breakage, and bone fracture. The first is unlikely if proper sterile procedure is followed, and the latter are unlikely if the individual performing the procedure is experienced. Risk of fracture is greatest in very small patients and those having bony lysis or decreased bone density.

Handling of marrow aspirate samples

The following instructions assume you have aspirated bone marrow into an EDTA-containing syringe and expelled the aspirated marrow sample into a clean petri dish.

1. Examine the petri dish and look for marrow "spicules", which are small, white flecks of hematopoietic and adipose tissue.
2. Using a thin, rubber-tipped glass pipette or plain microhematocrit tube, collect marrow spicules and transfer them to a clean glass microscope slide.

3. Angle the slide, such that excess blood runs off the short edge (away from the label end) and marrow spicules adhere to the glass. In the author's opinion, a good quality smear should contain 2 to 4 marrow spicules that have been smeared into a monolayer composed of intact cells.
4. Lay a second, clean glass slide over the first slide. Allow material sandwiched between the slides to spread (use no to minimal downward pressure), then draw the slides apart.
5. Repeat as needed to obtain three to five good quality smears.
6. **KEEP MARROW SAMPLE AND ALL SMEARS AWAY FROM FORMALIN FUMES!** Formalin (even fumes) causes fixation of the cells, which results in cell degeneration that significantly and adversely impacts cytomorphology.
7. Extra EDTA-anticoagulated marrow can be stored in the refrigerator for up to 24 hours (just like blood for a CBC).

PATHOLOGIST'S STEPS TO BONE MARROW EVALUATION

What is the pathologist looking at?

1. Marrow cellularity (is it decreased, normal, or increased?) and cell proportions (M:E ratio and relative proportions of other cell types). Increased cellularity suggests hyperplasia or infiltrative disease. Decreased cellularity suggests hypoplasia.
2. Hemic cell maturation (is it complete, orderly, and synchronous?)
3. Hemic cell morphology (is it normal?)

M:E (myeloid to erythroid) ratio is a.k.a. G:E (granulocytic to erythroid) ratio.[2] M:E is obtained by counting 500 nucleated granulocytic and erythroid cells (plasma cells, lymphocytes, and macrophages are excluded). M:E is simply the number of myeloid (granulocytic) cells divided by the number of erythroid cells. M:E reference values are given in other resources.[1,3]

Bone marrow evaluation steps:

1. Review CBC data. Are hemic cell abnormalities present on blood smear review?
2. From low power (4X and 10X objectives) – are marrow "unit particles" (the microscopic correlate of a spicule) seen? How cellular are these? Are megakaryocyte numbers adequate? Is storage iron (hemosiderin) observed?
3. From high power (50X and 100X objectives) – assess maturation of hemic cells, count M:E, assess hemic cell morphology, look for other cell types, look for things that don't belong (e.g., atypical cells of non-hemic lineage, microorganisms)
4. Interpret marrow cellularity and M:E ratio together; interpret bone marrow findings in light of CBC data.
5. Render a morphologic diagnosis for each cell line (erythroid, granulocytic, megakaryocytic) and write interpretive comments as needed.
6. Special stains can be used to further evaluate bone marrow, especially core biopsies. E.g., Perl's stain (if trying to demonstrate storage iron or ringed sideroblasts), masson-trichrome (for collagen, in cases of suspect fibrosis), silver stain (for reticulin, in cases of suspect fibrosis), toluidine blue (for mast cells), and CD markers (e.g., if immunophenotyping lymphoma).
7. Flow cytometric evaluation of bone marrow has been reported and may be available at select laboratories.[4]

Interpreting M:E and marrow cellularity together:

M:E	Hypocellular (Hypoplasia)	Normocellular	Hypercellular (hyperplasia)
Decreased	↓ granulopoiesis	↓ granulopoiesis ↑ erythropoiesis	↑ erythropoiesis
Normal	↓ granulopoiesis ↓ erythropoiesis	Normal	↑ granulopoiesis ↑ erythropoiesis
Increased	↓ erythropoiesis	↑granulopoiesis ↓erythropoiesis	↑granulopoiesis
	(one or both cell lines must be decreased)	**(both cell lines are normal OR altered in opposite directions)**	**(one or both cell lines must be increased)**

Definitions relevant to maturation:

- ***Maturation*** refers to a cell developing the morphologic changes indicative of full cell function. E.g., a neutrophil is mature, where as a myeloblast is immature.[5]
- ***Orderly maturation*** refers to the fact that normal marrow contains many more late-stage precursors than early-stage precursor (the "inverse pyramid" of hematopoiesis).[2]
- ***Synchronous maturation*** refers to the fact that the nucleus and the cytoplasm of a cell mature at the same rate. *Asynchronous maturation* is a manifestation of dysplasia. [An example of asynchronous maturation in a late-stage erythroid cell would be the presence of an immature nucleus in a cell with fully hemoglobinized (peachy-pink) cytoplasm.] Synchronous maturation is best assessed in cytology smears.[2]

- ***Complete maturation*** refers to the fact that maturation of the cell line progresses all the way to the most mature cell stage (e.g., RBC, segmented neutrophil).[2]
- ***Maturation arrest*** refers to cessation of cell maturation at a particular stage of development. Maturation does not progress normally beyond the arrested stage. E.g., "there is a maturation arrest at metamyelocytes" means that granulocyte (neutrophil) maturation is orderly and complete up to the point of metamyelocytes, but then no or only very few bands and segmented neutrophils are seen. Maturation arrest is usually accompanied by peripheral cytopenia of the affected cell line and may or may not be accompanied by dysplasia (depending upon the underlying cause).
- ***Ineffective hematopoiesis*** (ineffective erythropoiesis, ineffective granulopoiesis, ineffective megakaryocytopoiesis) refers to death or destruction of hematopoietic cells within the bone marrow such that mature cells are not released into circulation; it is ***defined by hypercellular marrow with concurrent cytopenia*** [i.e., increased cellularity of one or more cell lines, with concurrent cytopenia of the affected cell line(s)]. This is not a specific disease diagnosis and can be caused by many things, including immune-mediated disease, nutritional deficiencies, drugs/toxins, FeLV, leukemia/myelodysplastic syndromes/myeloproliferative disease, and congenital syndromes. Ineffective hematopoiesis may or may not be accompanied by dysplasia.

Terminology relevant to cytomorphology and dysplasia:

- ***Differentiation*** refers to a cell being committed to a particular cell lineage.[5]
- ***Dyshematopoiesis (dyserythropoiesis, dysmyelopoiesis, dysmegakaryocytopoiesis)*** refers to a defect in hemic cell production AND presence of cellular atypia (dysplasia). Dyshematopoiesis may or may not be accompanied by ineffective hematopoiesis; however, ineffective hematopoiesis is often present. This is not a specific disease diagnosis either and can be caused by many things, including nutritional deficiencies, drugs/toxins, FeLV, leukemia/myelodysplastic syndromes/myeloproliferative disease, and congenital syndromes.
- ***Myelodysplasia*** refers to presence of hematopoietic cells with atypical morphology (i.e., dysplastic hematopoietic cells). Strictly speaking, the term does not give information about whether cell production is appropriate (effective) or not; however, myelodysplasia is often accompanied by ineffective hematopoiesis. (By this definition, *myelodysplasia* and *dyshematopoiesis* are basically synonymous.) In human medicine, the term *myelodysplasia* is used more specifically as a synonym for *myelodysplastic syndrome (MDS)*. Usage of the term *myelodysplasia* in veterinary medicine is inconsistent and may or may not be a synonym for MDS. MDS is a group of neoplastic diseases occurring due to a clonal disorder of bone marrow stem cells and resulting in ineffective hematopoiesis, dyshematopoiesis, and susceptibility to acute myeloid leukemia. MDS is differentiated from leukemia by the percentage of immature (blast) cells present in the bone marrow. A detailed description of cellular morphologic features of dysplastic change in each of the three hemic cell lines can be found in other resources.[1,2,6]

Feature	MDS	Leukemia
CBC findings	Cytopenia of affected cell line	Cytosis of affected cell line
Marrow cellularity	Hypercellular, with < 20% blasts*; may see maturation arrest	Hypercellular, with > 20% blasts*
Hemic cell dysplasia	Present (dyshematopoiesis, asynchronous maturation, etc.)	Present, particularly in acute leukemias

*WHO criteria

REFERENCES

[1]Harvey JW. Veterinary Hematology A Diagnostic Guide and Color Atlas. Saunders Elsevier, 2012:234-259. [2]Stockham SL and Scott MA. Fundamendals of Veterinary Clinical Pathology, 2nd ed. Blackwell Publishing, 2008:323-368. [3]Jain NC. Essentials of Veterinary Hematology. Lea & Febiger, 1993:12. [4]Weiss DJ. Application of flow cytometric techniques to veterinary clinical hematology. Vet Clin Pathol, 2002: 31(3):72-82. [5]McMannus PM. Classificaiton of myeloid neoplasms: a comparative review. Vet Clin Pathol, 2005; 34(3):189-212. [6]Valli VE. Veterinary Comparative Hematopathology. Wiley-Blackwell, 2007:473.

INDICATIONS FOR CYTOCHEMICAL AND OTHER SPECIAL STAINS IN DIAGNOSTIC CYTOLOGY

Heather Priest, DVM, DACVP (Clinical Pathology)
Ithaca, NY

INTRODUCTION

Although routine cytology is often sufficient for a diagnosis in many cases, there are times where it is not enough. Additional non-immunologic stains can sometimes offer the needed clues to better characterize the lineage of a cell or to identify infectious agents that are either in very low numbers or possibly obscured by necrosis or cell debris on routine staining (e.g. Wright's, Diff-Quik® or other Romanowsky-type stains).

Cytochemical stains for neoplasia:

In contrast to immunocytochemical stains, which detect cellular antigens using immunologic techniques, cytochemical stains identify cellular constituents (enzymes, lipids, or carbohydrates) in order to better characterize the cell lineage[6]. These stains are used commonly in leukemia diagnosis along with flow cytometry, as well as in the diagnosis of other hematopoietic (e.g. histiocytic sarcoma) and non-hematopoietic (osteosarcoma) tumors. Historically these stains played a larger role in leukemia diagnosis in people as well as animals, but as more antibodies were and are being developed against cellular antigens, immunologic methods have become more popular and cytochemical staining has become less relied upon. However, given that immunologic reagents used in immunocytochemistry and flow cytometry are often species specific, we are often lacking in antibodies available to us. Cytochemical stains are not-species specific (though reactions do differ between the species) and are often the only tools we have left when a leukemia fails to react to most of the antibodies at our disposal. These stains become especially important in the diagnosis of acute myeloid leukemias, given the paucity of available markers for these tumors in veterinary species. A brief summary of some of the common cytochemical stains and the expected findings in certain diseases are listed below. A complete list of all available cytochemical stains is quite extensive. A good summary of the available stains and their applications are available in Schalm's Veterinary Hematology, 6th ed. Chapter 144: Cytochemical staining.

Myeloperoxidase (MPO): Enzyme present within primary granules of cells of neutrophil lineage, from promyelocyte to segmented neutrophils. Expected results (some species differences occur) in normal cells: positivity within neutrophils and in some monocytes and eosinophils. Other cells are generally negative. In myeloid leukemias: blasts positive for MPO in myeloblastic leukemia with or without differentiation (AML-M1 to M3), myelomonocytic leukemia (AML-M4) and erythroleukemia (AML-M6).

Specific esterases: These consist of a group of enzymes that are considered more specific for neutrophils than the other cytochemical stains (MPO and SBB). Chloroacetate esterase (CAE) is the most common stain used. Expected results (some species differences occur) in normal cells: positivity in neutrophils, basophils and mast cells, megakaryocytes may be mildly positive and lymphocytes can have some focal staining. Other cell lines are generally negative. Myeloid leukemias: blasts positive for CAE in myeloblastic leukemia with or without differentiation (AML-M1 to M3), myelomonocytic leukemia (AML-M4) and erythroleukemia (AML-M6).

Non-specific esterase (NSE): These are a group of esterases that have a wider spectrum of activity. The two most common are alpha-naphthyl butyrate esterase (ANBE) and alpha-naphthyl acetate esterase (ANAE). ANBE is the non-specific esterase offered in our clinical pathology laboratory. Expected results: Monocytes are diffusely positive, macrophages have more intense diffuse granular to globular positivity and platelets/megakaryocytes will be variably positive. In addition, T-lymphocytes display focal cytoplasmic staining. Myeloid leukemias: blasts positive for ANBE in myelomonocytic leukemias (AML-M4), monoblastic leukemias (AML-M5) and megakaryocytic leukemias (AML-M7). Given its positivity in cells of macrophage origin, it is often used as a marker of histiocytic lineage. ANBE staining can be a useful method to support a cytologic diagnosis of histiocytic sarcoma[1], though we have seen low numbers of neoplasms with a classic histiocytic morphology be negative or only weakly positive for this stain. Note: positive staining for ANBE only supports a histiocytic lineage and both neoplastic and non-neoplastic histiocytes will be positive, so diagnosis must be made with support from routine cytology or additional immunocytochemical stains.

Leukocyte Alkaline phosphatase (ALP/LAP): Enzyme present in the cytoplasm of certain leukocytes, osteoblasts and vascular endothelial cells. Expected result (several species differences found): Neutrophils of horses, cattle and humans stain positive (venous blood from horse is used as a positive control in our lab). Monoblasts stain positive in dogs and negative in other species. Osteoblasts and vascular endothelial cells also stain positive. Lymphocytes have been reported to stain with ALP (mantle zone B cells) and positive staining has been reported in dogs with lymphoma[7] but a recent abstract presented at ASVCP do not support this finding. This data suggests that ALP is a marker of monoblasts in the dog and positive staining in an acute leukemia supports myeloblastic, myelomonocytic or monoblastic leukemias in dogs and would argue against a lymphoid lineage[8]. This stain is also used to support a diagnosis of osteosarcoma in dogs. Barger et al. found that 33 out of 33 osteosarcomas were positive for ALP (100% sensitivity) and 3 other tumor types displayed positive staining (1 chondrosarcoma, 1 multilobular tumor of bone and 1 melanoma – specificity of 89%)[2]. Note: Since all osteoblasts will be positive for ALP, ALP positive results only tell us that the cells in question are osteoblasts and does not distinguish if the population is neoplastic or not. This is based on cytologic criteria of malignancy.

Stains for infectious agents:

Numerous stains are available to help detect and better characterize infectious agents in cytologic and histologic samples. A good summary of available stains is presented in Canine and Feline Cytology, 2nd edition, ed. Raskin and Meyer, Chapter 17: Advanced Diagnostic Techniques. Below is a list of the commonly used stains in our laboratory:

Gram stain: Likely the most well-known and well used stain apart from the routine Romanowsky-type stains we use. Indications: Characterization of bacterial populations, identification of mixed vs monomorphic bacterial populations within a sample. Differentiating debris or stain precipitates from bacteria – especially useful in urine samples. Identification of low numbers of Gram positive bacteria in a sample (note that Gram negative bacteria are best seen using routine stains). Tip: When looking for low numbers of bacteria search the feathered edge of a sample, bacteria and other infectious agents often get drawn out to the feathered edge.

Grocott/Gimori's methamine silver stain (GMS): Indications: Identification of fungal agents in a sample. Consider a GMS stain when history, clinical signs and/or the pattern of inflammation (e.g. mixed inflammation – neutrophils and macrophages or granulomatous inflammation). Tip: Fungal hyphae are very difficult to identify on routine stains especially when there is necrosis in the sample since the organisms may degrade and stain less intensely. Look in amongst the necrosis and request a GMS stain on cytologies if a fungal infection is on your differential diagnosis list. Periodic acid-Schiff (PAS) stain can also be used for fungal detection.

Acid fast stain (e.g. Ziehl-Neelsen stain): Indications: Identification of acid-fast bacteria (bacteria which resist decolorization during the staining procedure). Given the relatively small list of bacteria which are acid-fast, this test can be useful in narrowing down certain bacterial agents while awaiting culture results. Bacteria include *Mycobacteria* spp., as well as *Nocardia* spp. (can be variably acid-fast). Acid-fast staining can be used to help differentiate between *Nocardia* and *Actinomyces* spp. when branching filamentous rods are identified in a sample and to help identify *Mycobacteria* spp. within macrophages in granulomatous inflammation.

Miscellaneous stains:

Rhodanine: Indications: Copper accumulation within hepatocytes has been associated can have several causes including increased uptake, altered copper metabolism and decreased biliary excretion. Association has been made between copper accumulation in hepatocytes and chronic hepatitis[3]. Copper can be difficult to identify using routine cytology stains as it stains a light turquoise-blue and can resemble lipofuscin pigment (a common finding in liver aspirates of little clinical significance). Rhodanine stain applied to smears from fine needle aspirates of the liver stains copper a deep red color and increases the sensitivity of detection. The preferred methods for assessment of copper accumulation in the liver remains to be a combination of quantitative measurement of hepatic copper along with a qualitative assessment of the amount and location of copper accumulation on liver biopsy samples (rhodanine staining is also used on paraffin embedded formalin fixed samples)[4]. However, if a biopsy is either cost prohibitive or contraindicated for other reasons, the subjective assessment of copper within hepatocytes on cytology smears may aid in interpretation of clinicopathologic abnormalities involving the liver. Tip: Adequate numbers of cellular smears are needed for this additional stain. It is a good idea to indicate on your submission form that you suspect copper accumulation; attempts will be made to retain at least one unstained cellular smear for this test.

New methylene blue (NMB): This supravital stain is a rapid, inexpensive stain which can be useful in a few circumstances. Indications: It can be applied directly to unstained, un-fixed blood smears to assess for a regenerative response in an emergency situation while awaiting results from an automated CBC. NMB will stain immature RBCs (reticulocytes); it is often easier to detect reticulocytes with a NMB stain than to detect polychromasia on a routinely stained smear. Used in the same fashion by direct application to the blood smear NMB will stain Heinz bodies and can be useful to detect an oxidant induced anemia (e.g. zinc toxicosis in a dog or acetaminophen toxicosis in a cat).

REFERENCES

[1] Azakami, et. al **J Vet Med Sci** 2006;68:1343, [2]Barger, et. al. **Vet Path** 2005;42:161, [3] Hoffmann, **Vet Clinics of NA** 2009; 39:489, [4] Johnson et. al **JAVMA** 2013; 242:372, [5]Ramos-Vara et. al in **Canine and Feline Cytology. 2nd ed. Advanced Diagnostic Techniques** 2010: 395, [6] Raskin, in **Schalm's Veterinary Hematology. 6th ed., Cytochemistry** 2010:1141, [7] Raskin et. al **Vet Clin Path** 1992; 21:62, [8] Stokol et. al **Vet Clin Path** 2013; epub abstract.

IMMUNOCYTOCHEMISTRY: A VALUABLE TECHNIQUE IN DIAGNOSTIC CYTOLOGY

Heather Priest, DVM, DACVP (Clinical Pathology)
Ithaca, NY

INTRODUCTION

The detection of antigens within cells or tissues using specific antibodies is not a new process in human or veterinary diagnostics but the use of these techniques on cytology specimens or immunocytochemistry (ICC) is still somewhat underutilized. There are several reasons for this, stemming from a list of problems and limitations often encountered with ICC. Some of these problems are preventable and others are not so easy to avoid. These proceedings will provide a brief review of some of the technical aspects of immunocytochemical (ICC) staining, some of the advantages and disadvantages of the technique and will provide some scenarios where ICC has been a valuable resource in our diagnostic tool kit. Several comprehensive articles are also available for those who wish more detailed information on the technique[4, 6, 9, 10, 11].

Advantages and limitations of ICC

Many report very similar advantages when considering ICC, these include rapid results that can often be reported out the same day as sample receipt, ease and non-invasiveness of collection of cytologic specimens, the ability to re-sample if sample quality poor, often lacking the need for anesthesia which reduces overall diagnostic costs[2,3]. ICC is also able to identify certain antigens, such as T-cell subsets that are cannot be detected on formalin fixed sections[11]. Other diagnostic modalities which detect cellular antigens such as flow cytometry, lack the ability to assess the morphologic features of the cells in question, hindering diagnostic accuracy especially in samples of mixed cellularity[2].

However, the technique is not without limitations. One of the biggest disadvantages to ICC is the difficulty in obtaining a sufficient number of good quality cytology smears. Ideal smears will be of adequate cellularity, with minimal cell lysis and blood contamination. Although clinicians typically submit multiple smears from a lesion, often only a limited number of these will be diagnostic. A solution to this is to collect aspirates into a liquid media for shipment to the lab where cytospin smears can be made or to utilize cell blocks where multiple sections can be cut for analysis. This provides consistency between smears and reduces background staining. These techniques have been described by several authors[1,2,11]. This also allows for panels of antibodies to be applied to a given sample thereby increasing the diagnostic accuracy. ICC is an ancillary diagnostic test that should never be used in isolation and should always be interpreted along with a good quality routinely stained cytology smear (e.g. Wright's stain). The routine cytology should be used to develop a list of possible differential diagnoses and other ancillary diagnostics, such as ICC, flow cytometry, other special stains or molecular diagnostics, are then selected and run to confirm the diagnosis based on a positive result in one test/stain and a negative result in another. The use of a single antibody for diagnosis is strongly discouraged[4].

Some important points on ICC technique, standardization and validation

Most large diagnostic laboratories which include a histology laboratory have automated immunohistochemical stainers which are able to stain large volumes of submissions with little hands on time. Many of these stainers have been developed and optimized for use on human tissues, so additional steps are needed to validate the techniques in veterinary species. In addition, since staining of paraffin embedded formalin fixed tissue is the standard sample for these analyzers, staining protocols may need to be adjusted for cytology samples. Guidelines have been suggested for standardization of immunohistochemistry (IHC) testing in veterinary laboratories[7] but at this time similar guidelines are not available for ICC. Certain problems can arise when using an automated stainers for cytology samples, for example it has been our experience that cells on cytology smears, especially those when are thick, often detach during the staining procedure making the smear uninterpretable. Antibody dilutions and counterstaining often need adjusting and specific protocols put in place for cytology smears. For this reason, we do most of our ICC using a more time-consuming manual technique. Manual techniques also require standardization and validation and guidelines suggested for IHC need to be considered when developing an ICC test. This would include identification of optimum conditions for each step of the protocol including, but no limited to, antibody dilutions, incubation times, fixation, control tissues and confirmation of the specificity of the antibody under these conditions[7]. The latter should include confirmation of immunoreactivity in cells expected to be positive and negative staining in cells expected to not express the antigen. This requires a bank of control tissues.

Control tissues are critical when interpreting the results of the IHC and ICC. Controls need to be species specific and should be prepared and treated in the same manner as the diagnostic sample being tested. In the case of IHC, blocks of formalin fixed paraffin embedded tissues can be kept indefinitely to be used as positive controls, and many samples have enough normal tissue architecture as to have internal controls. Unfortunately, the same cannot be said in the case of ICC. We do keep a bank of tissues stored at -70 °C for commonly run antigens: impression and cytospin smears of canine and feline lymph nodes for positive controls for lymphoid markers, frozen sections of canine spleen for histiocytic markers and liver for a positive control for cytokeratin staining. Just as critical as the positive control is a negative control. We run a negative control on each case by applying either no antibody, an irrelevant antibody or non-immune serum from the same species as the primary antibody. We have seen non-specific staining of macrophages with certain test kits which can be problematic for interpretation since the negative control is critical to interpret the significance of immunoreactivity in these cells.

Indications

Despite the limitations listed above, ICC is frequently useful as an ancillary diagnostic test. There is a long list of cellular antigens that authors have reported detection of by ICC, but a complete list is beyond the scope of these proceedings. The most common include the following markers:

- Lymphoid: A variety of T cell (e.g. CD3, CD4, CD8, CD5) and B cell (e.g. CD79a, CD21, CD20, Pax5) markers are available for dogs and cats.
- Epithelial: Cytokeratins (AE1/AE3, Cam 5.2)
- Mesenchymal: Vimentin, von Willebrand Factor
- Histiocytic: CD1c, CD11c, CD11d, CD18, E-cadherin
- Endocrine/neuroendocrine: Synaptophysin, S-100, chromogranin
- Melanoma: Melan A, S-100

ICC has been a useful tool in our lab in the several situations

Immunophenotyping of lymphoma:

Historically biopsy and immunohistochemistry of formalin fixed tissue was the gold standard for lymphoma diagnosis and immunophenotyping following a cytologic diagnosis of lymphoma on fine needle aspirate (FNA). However, given the cost and time savings of performing the immunophenotyping on the cytology smears collected by FNA, more clinicians are requesting ICC. There have been several reports in both people and dogs that have shown excellent concordance between ICC results on FNA samples and IHC findings on tissue sections collected by biopsy[3,8]. Certain more mixed subtypes of lymphoma are more difficult to identify on FNA and other diagnostic tests will be needed in these cases. However, for the majority of canine lymphomas which are diffuse large cell forms, FNA cytology and ICC can offer clinicians and clients a rapid and cost effective diagnosis.

Sample to collect: At least 4 good quality unstained smears of at least moderate cellularity. One smear will be used to establish or confirm the diagnosis and will be routinely stained with Wright's stain. One smear will be stained for a B-cell marker (CD21 in our lab), one with a T-cell marker (CD3) and an irrelevant antibody will be applied to the last smear as a negative control. Additionally, a smear of normal canine (or feline) lymph node is also stained concurrently as a positive control for each antibody. Aspirated material can also be submitted in collection media. The collection media we recommend is: 0.9% NaCl + 0.1 ml of patient serum in a 3 ml EDTA tube, though several alternative exist[1,2,11]. ICC can then be performed on cytospin smears, though in most cases we would recommend flow cytometry when cells have been collected into a liquid media for the assessment of a larger panel of antibodies allowing better characterization of the neoplastic cells.

Identification of low numbers of neoplastic cells in a sample:

This situation arises when there are low numbers of cells in a very mixed cellular background, such as a metastatic carcinoma cells in a lymph node or a neoplastic infiltrate in bone marrow. ICC has been shown to be more sensitive to detect micrometastasis in lymph nodes than assessment with conventional staining[5]. In the case of bone marrow evaluation, flow cytometry is often the first diagnostic modality we reach for beyond routine cytology, but given the mix of cells in normal or even diseased bone marrow, it can be difficult to identify the neoplastic population on the background of other hematopoietic cells. ICC performed on bone marrow aspirates can be very helpful as an adjunct to flow cytometry to directly visualize the antigen immunoreactivity of the cells in question. ICC can also be performed on archived air-dried smears (up to 2 weeks or longer if frozen), whereas flow cytometry requires a fresh sample. We have used positive staining on ICC to identify infiltrates of neoplastic lymphocytes in bone marrow and we have used the lack of staining as for support acute myeloid leukemias.

Identification of the lineage of round or discrete cell neoplasia:

This is a situation when a panel of antibodies is needed. As stated above, the selection of antibodies should be based on the differential diagnosis for the lesion and making a definitive diagnosis involves observing positive immunoreactivity appropriate for this diagnosis and negative immunoreactivity for antigens of the alternate diagnosis/es. We often begin by staining the sample with T and B cell markers. If no staining is detected antibodies against histiocytic markers and/or cytokeratin or vimentin may be applied. Other markers such as endocrine or neuroendocrine antibodies can also be used. Once again, a disadvantage to ICC in these cases is the lack of adequate smears to stain for multiple markers and as such it is common that a biopsy and IHC are required for a definitive diagnosis.

REFERENCES

[1] Aulback, et. al **J Vet Diagn Invest** 2010;22:383, [2] Caniatti, et. al **Vet Path** 1996; 33:204, [3] Fisher, et. al **Vet Clin Path** 1995; 24:118, [4] Fowler, et. al **Arch Pathol Lab Med** 2008; 132:373, [5] Hoinghaus, et. al **J Comp Path** 2007; 137:1, [6] Ramos-Vara, et. al **Vet Path** 2014;51:42, [7] Ramos-Vara, et. al **J Vet Diagn Invest** 2008;20:393, [8] Sapierzynski, et. al **Pol J Vet Sci** 2012;15:247, [9] Sawa, et. al **Res Vet Sci** 2012; 93: 1341, [10] Skoog, et. al **Cytopathology** 2011; 22:215, [11] Valli, et. al **Vet Clin Path** 2009;38:261.

OCULAR CYTOLOGY, IT'S MORE THAN JUST CONJUNCTIVA

Anne Barger, DVM, DACVP
Urbana, Illinois

Introduction

Ocular cytology can be very useful to the clinician in establishing a diagnosis, determining if a patient is responding appropriately to therapy or determining the necessity for additional diagnostic testing. The clinician's familiarity with the appearance of normal tissue and resident cell types is essential for the accurate interpretation of a cytologic specimen. Before a pathologic process can be identified, an understanding of what is normal is critical. Each ocular structure has a distinct histology. Normal cytology will be reviewed in each section and followed by cytologic appearance of pathologic processes.

Conjunctiva

The conjunctiva is a mucous membrane that lines the inner surface of the eyelids and nictitating membrane and then reflects back onto the anterior portion of the globe. The palpebral conjunctiva consists of pseudostratified epithelium with goblet cells and lymphoid follicles[1]. The bulbar conjunctiva consists of nonkeratinized squamous epithelium[2]. As this tissue approaches the limbus it is common to see melanin granules coating the epithelial cells[3]. In horses, pigmented keratinzed squamous epithelial cells are normally located near the opening to the Meibomian gland. Low numbers of keratinized squamous epithelial cells can be considered normal when cytology is harvested from the palpebral conjunctiva in horses[4]. Epithelial cells from most tissues are cohesive and can be found in sheets and clusters however squamous epithelial cells can be more individualized. Columnar epithelial cells are elongate and on cytology can appear similar to spindle cells. The other types of epithelial cells are more likely to be round with a centrally placed nucleus and variable amounts of cytoplasm depending on the location within the conjunctiva[4].

Cytologic samples of conjunctiva can be obtained by multiple techniques, including, swabs, scrapes and imprints. The sampling technique will impact the types of epithelial cells observed. An imprint will likely sample the surface epithelium and result in squamous epithelial cells being the most prominent. A scrape will sample deeper areas of the tissue so goblet cells and small lymphocytes will also be observed in addition to epithelial cells. The epithelial cells from the deeper layers are more likely to have a centrally placed nucleus and moderate rim of pale basophilic cytoplasm[5]. Common findings from cytology from healthy conjunctiva include a predominance of epithelial cells with small amount of mucous and few small lymphocytes[6].

Cytologically, inflammation of the conjunctiva is characterized by the type of inflammatory cells seen. Non-specific changes that may accompany inflammation include lymphoid hyperplasia, epithelial hyperplasia, and squamous metaplasia with keratinization of squamous epithelial cells[7]. Excessive mucus is often produced resulting in thick eosinophilic strands of mucinous material. Small numbers of lymphocytes and plasma cells can be observed in normal conjunctiva but also can be associated with inflammation or immune stimulation. In true lymphoplasmacytic inflammation, the lymphoid inflammatory cell population predominates over the epithelial cells. Immune mediated diseases such as pannus, medial canthal blepharitis of long haired dachshunds and follicular conjunctivitis will have a very cellular infiltrate predominated by small lymphocytes with many plasma cells[2]. In other immune mediated diseases such as pemphigus, neutrophils and eosinophils are the primary inflammatory cells.

Cellular inclusions may be observed in conjunctival specimens. These inclusions can be identified in inflammatory cells and conjunctival epithelial cells and may be intracytoplasmic or intranuclear. Infectious organisms such as *Chlamydia sp.* or *Mycoplasma sp.* or distemper virus can be found in the cytoplasm of epithelial cells. However, more commonly the cytoplasmic inclusions are benign structures such as melanin granules or ophthalmic ointment which are easy to misinterpret as an infectious etiology. Viral inclusions have been reported occur with feline herpes virus and canine distemper but in reality are infrequently encountered in a cytologic specimens[8].

Conjunctival neoplasia may appear as a discrete mass or an irregular thickening of the conjunctiva. Many tumors have been described and include mast cell tumor, lymphoma, melanoma, hemangioma, hemangiosarcoma and squamous cell carcinoma, with squamous cell carcinoma as the most common. Cytologically, squamous cell carcinomas appear similar regardless of the location. The cells are round to polygonal with varying amounts of pale basophilic to keratinized cytoplasm. The nuclei vary greatly in size with a coarsely stippled to pyknotic chromatin. A ring of perinuclear vacuoles can be observed. Occasionally, neutrophils can be seen coating the cytoplasm. It is not uncommon to have a significant suppurative inflammatory response accompanying the tumor.

Nictitating Membrane

The nictitating membrane or third eyelid has a T-shaped cartilage with a large gland at its base and is covered in nonkeratinized stratified squamous epithelium[1,2]. Lymphoid tissue is also present on the bulbar conjunctiva of the third eyelid. The presence of this normal tissue should be considered when diagnosing inflammatory disease within the third eyelid.

Inflammatory disorders of the third eyelid are characterized cytologically by the cells that are identified. These can be an extension of conjunctivitis or follicular hyperplasia accompanying lymphoplasmacytic inflammation.

Neoplasms of the nictitating membrane are similar to those described in the conjunctiva but should also include adenocarcinoma due to the prominent glandular tissue. Adenocarcinoma is one of the most common tumors at this location[2]. These cells will exhibit prominent criteria of malignancy such as anisocytosis and anisokaryosis. The cells are glandular in origin so they will often contain multiple discrete vacuoles within the cytoplasm.

Sclera

The sclera consists of three layers; the lamina fusca which communicates with the outer choroid, the sclera proper which is composed of connective tissue and the episclera which consists of loose connective tissue and is highly vascular[1]. Sampling of normal sclera will be of low cellularity because healthy connective tissue does not exfoliate very well.

Scleral inflammation can be primary or secondary to other ocular inflammatory diseases. Primary canine episcleritis has been characterized immunohistochemically as a mixture of T and B lymphocytes and $CD18^{+}$ histiocytic cells[9]. Cytologically there will be a mixed inflammatory population predominated by small lymphocytes with high numbers of vacuolated macrophages. Low numbers of neutrophils can be observed.

Cornea

The cornea is the transparent anterior portion of the fibrous ocular tunic. It is composed of multiple layers. The most anterior layer consists of non-keratinized squamous epithelial cells. The stroma consists of mesenchymal cells, primarily fibrocytes also called keratocytes admixed with collagen bundles[3]. The posterior cornea consists of single-layered corneal endothelium and its basement membrane, Descemet's membrane[1]. Cytologically, the most anterior surface of the cornea is generally what is sampled, so non-keratinized squamous epithelial cells are the most frequent cells seen, occasionally admixed with few spindle shaped cells. With corneal injury some general changes that can occur include keratinization of the squamous epithelial cells and the presence of more basal appearing epithelial cells. Additionally, cells from the limbus and conjunctiva can be recruited to the cornea[7]. These cells are often coated in pigment granules.

Inflammation of the cornea can occur secondary to physical or chemical damage, infectious organisms, immune mediated corneal disease, uveitis and elevated ocular pressures[7]. Bacterial and fungal keratitis often causes a suppurative inflammatory response. Fungal keratitis is commonly seen in horses but has also been reported in dogs and cats[10]. Within the background are thick strands of mucinous material which should be differentiated from fungal hyphae.

Eosinophilic keratitis has been reported in cats. FHV-1 has been suggested as but not confirmed as a possible etiology[11]. Cytologically, eosinophils are seen in high numbers with a mixture of mature squamous epithelial cells, mucinous material and low numbers of neutrophils.

Primary neoplasia of the cornea is fairly uncommon however squamous cell carcinoma, hemangiosarcoma and melanoma have been reported in multiple species. Squamous cell carcinomas exfoliate well and consist of clusters of neoplastic epithelial cells, similar to those diagnosed in the conjunctiva.

Epithelial inclusion cysts have been reported and can be congenital or secondary to previous ocular injury[12]. Clinically they appear as white to yellow to tan smooth masses on the corneal surface associated with variable corneal vascularization. Aspiration of these structures reveals aggregates of mature keratinized epithelial cells, primarily anucleate with aggregates of keratinaceous debris. Cytologically, these greatly resemble follicular cysts of the dermis.

Intraocular Structures

Aspiration of intraocular structures may be deliberate or unintentional. Aspirates of aqueous and vitreous humors are useful diagnostic tools. Aqueous humor is a clear and colorless fluid. Anterior uveitis results in breakdown of the blood-aqueous barrier and transudation of inflammatory cells into the aqueous humor, increasing the cellularity and protein. The ciliary body and iris contain melanocytes so during an inflammatory process melanin granules and melanophages may exfoliate into the aqueous humor[2]. Neoplasms of the anterior uvea (iris and ciliary body) can be aspirated, however this type of diagnostic sampling is infrequently performed. Melanoma is a common anterior uveal tumor[13] and consists of individualized and aggregates of highly pigmented neoplastic melanocytes. Ciliary body tumors are most frequently ciliary body adenoma and adenocarcinoma[14]. These cells appear distinct from melanoma because they exfoliate in tightly cohesive clusters and are not pigmented. The adenocarcinomas will exhibit obvious cellular criteria of malignancy including large prominent nucleoli.

Vitreous humor is a gelatinous fluid of low cellularity which often contains melanin granules. Inflammation can result in a decreased viscosity of the fluid. Posterior uveitis caused by organisms such as *Blastomyces dermatitidis* cause severe inflammation, primarily neutrophils. The fluid is highly cellular and organisms are readily identified. Proteothecosis, histoplasmosis and cryptococcocus can also result in severe uveitis. When aspirating vitreous, other structures in the globe can be aspirated, especially in patients with intraocular diseases such as uveitis, rupture of the lens capsule or retinal detachment. The lens consists of an elastic capsul, tightly adherent squamous to cuboidal epithelial cells and lens fibers[1]. The lens fibers are elongate structures with scalloped edges that resemble lasagna noodles. Aspiration of the lens results in aspiration of many of these fibers. The retina, especially when detached can be easily aspirated. Aspiration of the retina often reveals a disorganized arrangement of cells. Distinguishing features of retinal cells include the unique shape of the photoreceptor nuclei and the rod shaped melanin granules of the retinal pigmented epithelium.

In conclusion, ocular cytology is a complex subject. A complete understanding of the cytology of the different structures of the healthy eye is essential so that appropriate recognition of ocular pathology including general changes associated with inflammation can be correctly identified.

REFERENCES

1. Samuelson DA. Eye and Ear. In: *Textbook of Veterinary Histology*. Saunders Elsevier, Philadelphia, PA, 2007;487-511.
2. Raskin RE. Eye and Adnexa. In: *Canine and Feline Cytology* (eds. Raskin RE and Meyer DJ) Saunders Elsevier, Philadelphia, PA, 2010;366-382.

3. Young KM and Prasse KW. The Eyes and Associated Structures. In: *Diagnostic Cytology and Hematology of the Dog and Cat 3rd ed* (eds Cowell RL, Tyler RD, Meinkoth JH and DeNicola DB) Mosby Elsevier, St. Louis, MO, 2007;149-171.
4. Bourges-Abella N, Raymond-Letron I, Diquelou A, et al. Vet Ophthal 2007;10:12.
5. Bolzan AA, Brunelli TJ, Castro MB, et al. Vet Ophthal 2005;8:401.
6. Lavach JD, Thrall MA, Benjamin MM, et al. J Am Vet Med Assoc 1977;170:722.
7. Wilcock BP. Eye and Ear. In: *Pathology of Domestic Animals Volume 1, 5th ed* (ed. Maxie MG) Saunders Elsevier, Philadelphia, PA, 2007;459-552.
8. Hillström A, Tvedten H, Källberg M, et al. Vet Clin Pathol 2012;41:283.
9. Breaux CB, Sandmeyer LS, Grahn BH. Vet Ophthal 2007;10:168.
10. Labelle AL, Hamor RE, Barger AM, et al. Vet Ophthal 2009;12:48.
11. Naisse MP, Glover TL, Moore CP, et al. Am J Vet Res 1998;59.
12. Simonazzi B, Castania M, Bosco V, et al. J Sm An Pract 2009;50:373.
13. Grahn BH, Peiffer RL, Cullen CL, et al. Vet Ophthal 2006;9:395.
14. Hendrix DVH, Donnell RL. Vet Pathol 2007;44:540-542.

FELINE IDIOPATHIC CYSTITIS: DIAGNOSTICS AND LATEST RESEARCH FINDINGS

C.A.T. Buffington, DVM, PhD, DACVN[1], Jodi L. Westropp, DVM, PhD, DACVIM[2], and Dennis J. Chew, DVM, DACVIM
Columbus, OH[1] and Davis, CA[2]

INTRODUCTION

Signs referable to the lower urinary tract (LUTS) are one of the most common reasons cats are presented for care to veterinarians. These signs can include variable combinations of stranguria, hematuria, periuria, and dysuria. The cause(s) for these clinical signs include some idiopathic disease, urolithiasis, urinary tract infection, neoplasia of the urinary tract, and anatomic abnormalities of the LUT. The name for this group of diseases has gone through several changes over the years. Osborne, et al.,[1,2] have repeatedly recommended replacement of the term "Feline Urologic Syndrome (FUS)"[3] with descriptive terms pertaining to the site, causes, and pathophysiologic mechanisms whenever possible to permit the same terminology and approach to diagnosis and treatment used for other species to be used for cats. Unfortunately, their 1984 chapter, subtitled "Feline Lower Urinary Tract Disease with Heterogeneous Causes", resulted in replacement of one acronym (FUS), with another (FLUTD). Such "urocentric" terms should likely be retired, as more recent research suggests that these signs sometimes reflect a disease affecting the LUT rather than a problem intrinsic to the LUT itself.

Researchers and clinicians must now look beyond the LUT to consider the whole individual when evaluating cats with LUTS. This change has come in part as the result of studies using cats with severe, recurrent LUTS (and variable comorbidities) as a model of a chronic pain syndrome in humans called interstitial cystitis (IC).[4] As in veterinary medicine, the names to describe this syndrome also are in flux in human medicine, with (also urocentric) suggestions including painful bladder syndrome/IC, bladder hypersensitivity syndrome, and bladder pain syndrome. Names such as FUS or FLUTD to describe recurrent LUTS may therefore be over simplified and focused on the end organ rather than reflecting current understanding of this disease. I have proposed use of a name like "Pandora Syndrome" for some of these cats, for at least two reasons; 1) this term does not identify any specific cause or organ, and 2) it seems to capture the dismay and dispute associated with the identification of so many problems outside the organ of interest of any particular subspecialty[5].

Clinical research on (F)IC has expanded to include genetic[6], and epigenetic influences[7], has documented that the time course of the appearance of comorbid disorders often occurs before the LUT symptoms[8], and documented the extent of systemic involvement that occurs in most patients.[9] Similar to humans, evidence has accumulated that additional problems outside the LUT commonly are present in cats with chronic LUTS.[5] This evidence has led to reconsideration of the cause(s) of the syndrome in these individuals, as well as to considerable debate about the most appropriate nomenclature, diagnostic approach, and treatment recommendations.

DIAGNOSTICS

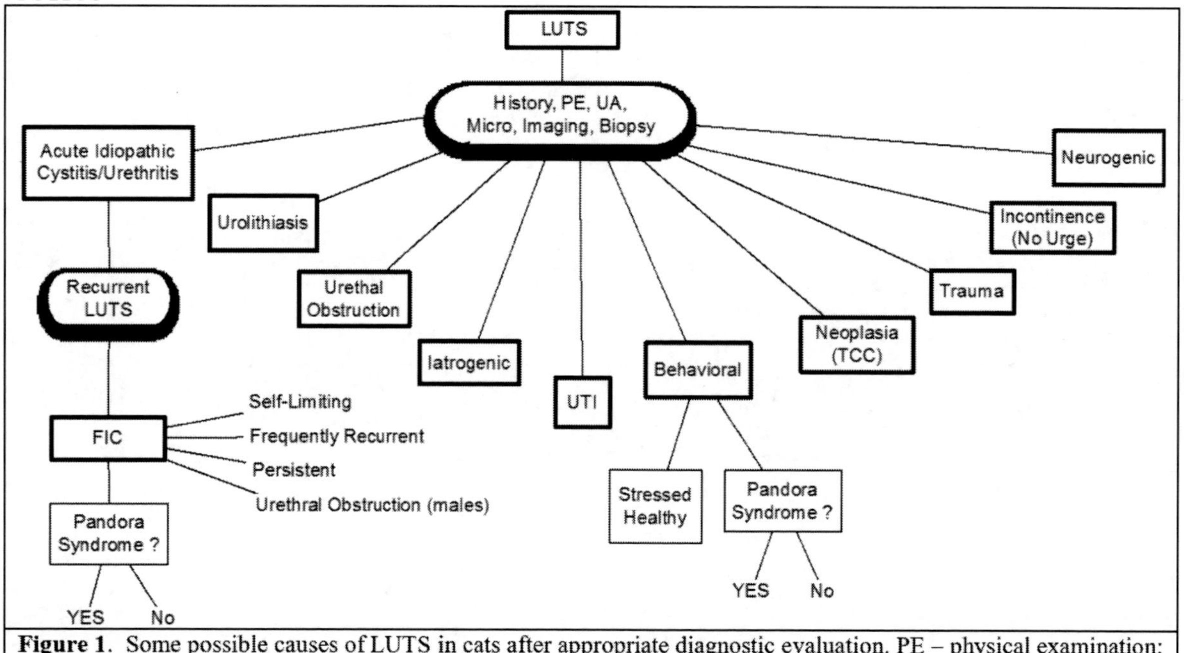

Figure 1. Some possible causes of LUTS in cats after appropriate diagnostic evaluation. PE – physical examination; UCS- quantitative urine culture (cfu/ml); Imaging – some combination of radiography, contrast urography, ultrasonography, and/or uroendoscopy. Not all tests are appropriate for every cat, so diagnostic evaluations tailored to each individual cat are most likely to arrive at the correct diagnosis.

Cats with FIC can be presented for an initial occurrence of clinical signs and may not return for care due to (an often presumed) improvement in the condition, or can return with recurrent signs of disease. Both genders appear to be affected equally. Although FIC can be obstructive or non-obstructive in its presentation, urethral obstruction is far more common in male cats, with no difference reported between intact and castrated males.[10] In addition to genetic and possible early adverse life events, other factors associated with an increased risk for chronic LUTS have been reported, and include excessive body weight, decreased activity, multiple cat households, indoor housing, and a variety of environmental stressors such as conflict with another cat in the household. [11,12]

Recommendations for diagnosis and treatment for the most common current causes of LUTS currently identified in unobstructed cats, (Figure. 1; although some 30 distinct causes of LUTS have been described[13]) are widely available. When presented with a cat with LUTS, clinicians also need to consider whether they are seeing the cat's initial episode, or whether the cat has chronic, recurrent disease, and what other health problems the cat may have.[14] This information permits one to judiciously utilize resources by choosing appropriate diagnostic tests to tailor treatment protocols to each individual cat.

Recent research further complicates the diagnostic challenge, however, because cats may have multiple reasons for their clinical signs as well as other medical and environmental conditions that need to be addressed. For example, some cats with severe, chronic LUTS seem to have a functional rather than a structural lower urinary tract disorder,[15] and have periuria has been found to occur in apparently healthy cats exposed to stressful circumstances.[16]

LATEST RESEARCH FINDINGS

Infectious etiologies

Although microorganisms in the LUT have not been identified as a common cause of (F)IC, they have been associated with chronic LUTS. A 2007 study from Norway of 134 cats with a variety of obstructive and non-obstructive causes of LUTS found a surprisingly high number of cats with bacteriuria exceeding 10^3CFU/mL in 44 (33%) cats, and exceeding 10^4 in 33 (25%), either alone or with variable combinations of crystals and uroliths.[17] These percentages are much greater than those reported from other studies. Interpretation of the general significance of Findings from this study is complicated by the proportions of the samples that were obtained from voided midstream (46%) or catheterized urine samples (21%) rather than from the gold standard of cystocentesis (21%; in 10% the method of urine collection was not recorded). 44 of 118 samples cultured on the same day isolated bacteria > 10^3 cfu/ml. In 33 of these 44 samples, growth was > 10^4 cfu/ml and in 20 growth was > 10^5 cfu/ml. The authors speculated that this higher rate for discovery of UTI in cats with LUTS might have resulted from differences between cases diagnosed at primary and tertiary care facilities, though geographical differences in occurrence in the development of UTI cannot be excluded. Quantitative growth from midstream voided samples from healthy cats also can be substantial; one study found > 10^3 cfu/ml in cultures from 55% of males and 40% of females.[18] In human beings, one recent study found evidence of UTI within the past 2 years in 38% of the patients with IC they studied, although they reported that,[9] " . . .the infection domain was not associated with any increased symptoms." It also has been speculated that intrinsic abnormalities of the LUT make it more vulnerable to microbial colonization,[19] which could be consistent with the observation of increased risk for bacterial UTI in patients with (F)IC.

The role of viruses also continues to be investigated.[20] The feline caliciviruses, FCV-U1 and FCV-U2, have been the most studied. Feline calicivirus (FCV) viruria has been detected in cats with chronic LUTS and in cats with upper respiratory infections; however, its etiologic significance has not been determined.[21] Serologic results suggested increased FCV exposure in cats with "FLUTD" compared with controls. A weak association between seropositivity for *Bartonella* spp. and FIC has also been reported.[22] What, if any, role these agents play in the etiopathogenesis of chronic LUTS, or in the systemic manifestations of the syndrome, remains unknown at this time. Lund, et al.,[23] recently reported the prevalence of various viral infections in client-owned cats, 102 with clinical signs of FLUTD and 73 healthy control cats, recruited from the Department of Companion Animal Clinical Sciences at the Norwegian School of Veterinary Science from 2006 to 2009. Urine samples were tested for the presence of FCV, feline coronavirus (FCoV) and feline herpesvirus-1 (FHV-1) by polymerase chain reaction. All urinary samples were negative for FCV and FCoV. Only one percent of urines from cats diagnosed with FLUTD cats were found to be positive for FHV-1, indicating that the viral infections examined were not associated with signs of FLUTD in the study sample.

Bladder and Systemic abnormalities in FIC

A variety of abnormalities have been reported in the urine, bladder and sensory apparatus of cats with chronic LUTS (reviewed in[15]). What is far less clear is what role(s) these abnormalities might play in the etiopathogenesis and progression of the syndrome, which are results of it, and which are unrelated "bystander" findings. In humans presenting with chronic LUTS, urodynamic evaluations often are performed to rule out other urinary tract diseases, such as overactive bladder, that could account for the clinical signs. Although a decrease in bladder compliance has been reported in cats with FIC, a recent study found no urodynamic evidence for spontaneous bladder contractions (overactive bladder) in female cats with FIC.[24] However, increased urethral closure pressures were noted in cats with FIC compared to healthy cats, despite a lack of clinical signs at the time the studies were performed.

Another recent study reported that the acoustic startle response, a brainstem reflex in response to unexpected loud stimuli, appears to be amplified in cats with FIC. The startle response was greatest and most different in cats with FIC from that of healthy cats during stressful situations, but was still greater in affected than in healthy cats even when adapted to enriched housing conditions.[25]

Epigenetic Studies

A range of studies support the concept that the central stress response of patients with (F)IC is unusually sensitive. Recent research suggests that one mechanism underlying the sensitization of the stress response system involves a process called epigenetic modulation of gene expression.[26] Epigenetic modulation of gene expression is a prominent candidate mechanism for the exaggerated stress responsiveness found in cats with FIC because it has been shown to occur in the offspring of pregnant females exposed to stressors, and to result in long term neuroendocrine abnormalities.[27] Importantly, research in both rodents[28] and cats[29-31] has demonstrated that effective environmental enrichment can mitigate much of the effects of early life adversity.

Comorbidities

A recent study of healthy cats and cats with FIC found that environmental stressors resulted in increased number of sickness behaviors (e.g., vomiting, lethargy, anorexia) in cats with FIC when the results were controlled for other factors. [32] Furthermore, cats with FIC had variable combinations of co-morbid disorders[11,31,33-35] such as behavioral, endocrine, cardiovascular, and gastrointestinal (GI) problems. Most human beings with IC also suffer from variable combinations of comorbid disorders that affect a variety of other body systems.[36] That patients with FIC and IC have variable combinations of other comorbid disorders, and the observation that no temporal relationship has been identified among them, raises the question of the extent to which a different etiology affects each organ versus the extent to which some common disorder affects all organs, which then respond in their own characteristic ways. Regardless of the eventual explanation(s), the breadth and consistency of these findings makes it that clinicians perform a complete physical examination rather than focusing entirely on the bladder, and obtain a detailed environmental history from owners of all cats with FIC.

Diagnostic Biomarkers

No well-accepted diagnostic test for (F)IC currently exists. In humans, various markers such as antiproliferative factor and heparin-binding epidermal growth factor[37] have been investigated, but are not clinically available. Diagnostic tests for FIC also have been investigated. For example, a decrease in Trefoil factor 2 was identified in both obstructed and non-obstructed cats with FIC, and the authors suggested that this might reflect an impaired ability for bladder repair and immune response, as well as a greater susceptibility to inflammation.[38] The same group of researchers also reported that fibronectin concentrations were increased in the urine of cats with FIC.[39] Fibronectin is involved in cell adhesion, migration, and wound healing and clot stabilization. These biomarkers may be part of the pathogenesis of FIC, be altered as a result of the secondary changes that occur in the bladder, particularly when urethral obstruction was present, and/or be "bystander" changes.

In humans and cats, a serum or blood spot-based test for (F)IC has been reported, with differences in tryptophan and its metabolites found in both patient groups.[40] The test also was able to differentiate cats with FIC from control cats regardless of whether they had clinical signs at the time of sampling, suggesting that this might represent a "trait" rather than a "state" marker of the syndrome. Interestingly, the kynurenine pathway for tryptophan metabolism appears to play a role in the pathobiology of a variety of central and peripheral syndromes.[41] However, the sensitivity and specificity of this test needs to be further evaluated, and it too is not available for clinical use at this time, so FIC currently remains a diagnosis of exclusion.

CONCLUSIONS

Clinical and basic science investigations over the past couple of decades have resulted in a more complex view of chronic LUTS, both in human beings and domestic cats, challenging the traditional view that the bladder is always the perpetrator of LUTS, in contrast suggesting that the bladder also can be one victim of a systemic process associated with a sensitized central stress response system. An overarching principle is that LUTS and a variety of other signs can occur as a consequence of placing a "sensitive cat" into a "provocative" environment. The relationship between the environment and health also is quadratic rather than linear, with both deficient and threatening environments leading to poor health outcomes.[42] With current technology, we have the opportunity to develop collaborative strategies to learn far more about "FUS", "FLUTD", "FIC", "Pandora Syndrome", or whatever one's preferred term is; potentially enough to come up with a truly evidence-based nosology, and treatment(s) based on this understanding.

Finally, one must keep in mind that even presumably healthy cats can develop sickness behaviors when exposed to sufficiently provocative environments,[16] and meeting the environmental needs of *all* pet cats to ensure their health and well-being is an animal husbandry responsibility we shoulder when we keep them as pets. While the cats studied to date appear to be more sensitive to their surroundings than healthy cats are, it may be that they are just further along on a continuum that likely includes all pet cats.

REFERENCES

1. Osborne CA, Johnston GR, Polzin DJ, et al. Redefinition of the feline urologic syndrome: feline lower urinary tract disease with heterogeneous causes. VCNA: Small animal practice 1984;134:409-438.
2. Osborne CA, Kruger JM, Lulich JPFlutdDotacTVcoNASap. Feline lower urinary tract disorders. Definition of terms and concepts. VCNA: Small animal practice 1996;20:169-179.
3. Osbaldiston GW, Taussig RA. Clinical report on 46 cases of feline urological syndrome. Vet Med/Small Anim Clin 1970;65:461-468.
4. Buffington CAT. From Animal Data to Human Practice. In: Baranowski A, Hughes J, eds. Abdominal and Pelvic Pain. Washington, DC: IASP Press; 2014:In Press.
5. Buffington CA. Idiopathic cystitis in domestic cats--beyond the lower urinary tract. Journal of veterinary internal medicine / American College of Veterinary Internal Medicine 2011;25:784-796.
6. Dimitrakov J, Guthrie D. Genetics and phenotyping of urological chronic pelvic pain syndrome. J Urol 2009;181:1550-1557.
7. Buffington CA. Developmental influences on medically unexplained symptoms. Psychother Psychosom 2009;78:139-144.
8. Warren JW, Howard FM, Cross RK, et al. Antecedent nonbladder syndromes in case-control study of interstitial cystitis/painful bladder syndrome. Urology 2009;73:52-57.
9. Nickel JC, Shoskes D, Irvine-Bird K. Clinical phenotyping of women with interstitial cystitis/painful bladder syndrome: a key to classification and potentially improved management. The Journal of urology 2009;182:155-160.
10. Hostutler RA, Chew DJ, DiBartola SP. Recent concepts in feline lower urinary tract disease. The Veterinary clinics of North America Small animal practice 2005;35:147-170, vii.
11. Buffington CA. External and internal influences on disease risk in cats. Journal of the American Veterinary Medical Association 2002;220:994-1002.

12. Buffington CA, Westropp JL, Chew DJ, et al. Risk factors associated with clinical signs of lower urinary tract disease in indoor-housed cats. Journal of the American Veterinary Medical Association 2006;228:722-725.
13. Osborne CA, Kruger JM, Lulich JP. Feline lower urinary tract disorders. Definition of terms and concepts. Veterinary Clinics of North America - Small Animal Practice 1996;26:169-179.
14. Westropp J, Buffington CAT. Lower Urinary Tract Disorders in Cats. In: Ettinger SJ, Feldman EC, eds. Textbook of Veterinary Internal Medicine, 7 ed. St. Louis: Elsevier-Saunders; 2010:2069-2086.
15. Buffington CA. Idiopathic cystitis in domestic cats-beyond the lower urinary tract. J Vet Intern Med 2011;25:784-796.
16. Stella JL, Lord LK, Buffington CAT. Sickness behaviors in response to unusual external events in healthy cats and cats with feline interstitial cystitis. Journal of the American Veterinary Medical Association 2011;238:67-73.
17. Eggertsdottir AV, Lund HS, Krontveit R, et al. Bacteriuria in cats with feline lower urinary tract disease: a clinical study of 134 cases in Norway. Journal of feline medicine and surgery 2007;9:458-465.
18. Lees GE, Simpson RB, Green RA. Results of analyses and bacterial cultures of urine specimens obtained from clinically normal cats by three methods. Journal of the American Veterinary Medical Association 1984;184:449-454.
19. Keay SK, Warren JW. Is interstitial cystitis an infectious disease? Int J Antimicrob Agents 2002;19:480-483.
20. Larson J, Kruger JM, Wise AG, et al. Nested Case-Control Study of Feline Calicivirus Viruria, Oral Carriage, and Serum Neutralizing Antibodies in Cats with Idiopathic Cystitis. J Vet Intern Med 2011;25:199-205.
21. Kruger JM, Osborne CA. The role of viruses in feline lower urinary tract disease. Journal of veterinary internal medicine / American College of Veterinary Internal Medicine 1990;4:71-78.
22. Sykes JE, Westropp JL, Kasten RW, et al. Association between Bartonella species infection and disease in pet cats as determined using serology and culture. Journal of feline medicine and surgery 2010;12:631-636.
23. Lund HS, Rimstad E, Eggertsdottir AV. Prevalence of viral infections in Norwegian cats with and without feline lower urinary tract disease. Journal of feline medicine and surgery 2012;14:895-899.
24. Wu CH, Buffington CA, Fraser MO, et al. Urodynamic evaluation of female cats with idiopathic cystitis. Am J Vet Res 2011;72:578-582.
25. Hague DW, Stella JL, Buffington CA. Effects of interstitial cystitis on the acoustic startle reflex in cats. Am J Vet Res 2013;74:144-147.
26. Jensen P. Transgenerational epigenetic effects on animal behaviour. Progress in biophysics and molecular biology 2013.
27. Reynolds RM, Labad J, Buss C, et al. Transmitting biological effects of stress in utero: implications for mother and offspring. Psychoneuroendocrinology 2013;38:1843-1849.
28. Russo SJ, Murrough JW, Han MH, et al. Neurobiology of resilience. Nature neuroscience 2012;15:1475-1484.
29. Westropp JL, Kass PH, Buffington CA. Evaluation of the effects of stress in cats with idiopathic cystitis. Am J Vet Res 2006;67:731-736.
30. Westropp JL, Kass PH, Buffington CA. In vivo evaluation of alpha(2)-adrenoceptors in cats with idiopathic cystitis. Am J Vet Res 2007;68:203-207.
31. Buffington CA, Westropp JL, Chew DJ, et al. Clinical evaluation of multimodal environmental modification (MEMO) in the management of cats with idiopathic cystitis. Journal of feline medicine and surgery 2006;8:261-268.
32. Stella JL, Lord LK, Buffington CA. Sickness behaviors in response to unusual external events in healthy cats and cats with feline interstitial cystitis. Journal of the American Veterinary Medical Association 2011;238:67-73.
33. Buffington CA. Comorbidity of interstitial cystitis with other unexplained clinical conditions. The Journal of urology 2004;172:1242-1248.
34. Buffington CA, Westropp JL, Chew DJ. A case-control study of indoor-housed cats with lower urinary tract signs. JAVMA 2006;228:722-725.
35. Freeman LM, Brown DJ, Smith FW, et al. Magnesium status and the effect of magnesium supplementation in feline hypertrophic cardiomyopathy. Canadian journal of veterinary research = Revue canadienne de recherche veterinaire 1997;61:227-231.
36. Warren JW, Wesselmann U, Morozov V, et al. Numbers and types of nonbladder syndromes as risk factors for interstitial cystitis/painful bladder syndrome. Urology 2011;77:313-319.
37. Erickson DR, Xie SX, Bhavanandan VP, et al. A comparison of multiple urine markers for interstitial cystitis. The Journal of urology 2002;167:2461-2469.
38. Lemberger SI, Dorsch R, Hauck SM, et al. Decrease of Trefoil factor 2 in cats with feline idiopathic cystitis. BJU Int 2010.
39. Lemberger SI, Deeg CA, Hauck SM, et al. Comparison of urine protein profiles in cats without urinary tract disease and cats with idiopathic cystitis, bacterial urinary tract infection, or urolithiasis. American journal of veterinary research 2011;72:1407-1415.
40. Rubio-Diaz DE, Pozza ME, Dimitrakov J, et al. A candidate serum biomarker for bladder pain syndrome/interstitial cystitis. Analyst 2009;134:1133-1137.
41. Walker AK, Kavelaars A, Heijnen CJ, et al. Neuroinflammation and Comorbidity of Pain and Depression. Pharmacological Reviews 2014;66:80-101.
42. Buffington CAT, Westropp JL, Chew DJ. From FUS to Pandora Syndrome; Where are we, How did we get here, and Where to Now? Journal of feline medicine and surgery 2014;in press.

TREATMENT STRATEGIES FOR GLOMERULAR DISEASES IN DOGS

David J. Polzin, DVM, PhD, DACVIM (Small Animal Internal Medicine)
St. Paul, MN

INTRODUCTION

Standard therapy for dogs with glomerular disease (GD) currently consists of feeding a renal diet and administration of angiotensin converting enzyme inhibitors and aspirin. In addition, management of edema and hypertension may be indicated in some dogs with GD. These therapies have been the primary approach for management of canine GD patients for over a decade.[1] While widely accepted, it is clear that this therapeutic approach rarely if ever fully reverses the course of canine glomerular diseases.

It is recognized that many of the mechanisms underlying kidney diseases are immune in nature.[4] Thus, interest has shifted toward considering immunomodulatory therapy (IM therapy) that may offer more promise in halting the typically progressive course of canine CD.[2,3] Recommending IM therapy requires a risk-benefit assessment comparing the legitimate risks of the proposed therapy versus the likelihood of achieving an important therapeutic benefit for the patient. In making this assessment, it is important to recognize the risk of not providing a potentially effective treatment because GD itself represents a significant risk to the patient.

A significant limitation to applying IM therapy to dogs with GD has been the lack of suitable diagnostic tests for establishing which patients should receive IM therapy and which patients are unlikely to benefit from such therapy. Until recently, it has been difficult to obtain a renal biopsy including light, electron and immunofluorescent microscopy in dogs with GD. Obtaining a renal biopsy is important because it may provide information that can be used to estimate the likelihood that IM therapy will benefit the patient. Specifically, a biopsy finding evidence supporting an immune mechanism contributing to GD may support intervention with IM therapy, whereas finding irreversible chronic damage would generally fail to support intervention with IM therapy.

MECHANISMS OF INJURY IN GLOMERULAR DISEASE

Glomerular disease in human beings is commonly described as presenting as either nephrotic or nephritic syndromes (or combinations of these two syndromes). Nephrotic syndrome has classically been defined as the combination of proteinuria, hypoalbuminemia, hypercholesterolemia, and edema. In contrast, the nephritic syndrome is characterized (at least in human patients) by the presence of active urine sediment, which may include red blood cells, white cells, and cellular and granular casts, and patients with nephritic syndrome typically have less severe proteinuria and greater impairment of glomerular filtration rate compared to patients with pure nephrotic syndrome. These clinical presentations often predict the lesions likely to be seen by the pathologist. Clinical presentations of dogs with GD can similarly be categorized as nephrotic, nephritic or a combination of these two forms.

The nephrotic syndrome (NS) results from non-inflammatory injury to the more distal elements within the capillary walls; the primary injury typically involves the parietal epithelial cells or podocytes (podocytopathy). Glomerular injuries at this site may be induced by one of three basic mechanisms: 1) Injury to podocytes as a consequence of cytokines released by mononuclear cells (occurs in minimal change disease or primary focal glomerulosclerosis (FGS) in human patients); 2) Immune complex formation and subsequent complement activation in the subepithelial space (occurs in membranous nephropathy in human patients); or 3) Deposition diseases (such as amyloidosis and light chain deposition disease).[4] Due to the lack of glomerular inflammation, severe acute tissue injury is not expected in the NS. As a consequence, the urine sediment is typically inactive and plasma creatinine concentration is usually normal or only mildly increased at the time of presentation.

In contrast to the non-inflammatory character of the NS, the nephritic syndrome (NES) is overtly inflammatory reflecting the influx of circulating inflammatory cells including neutrophils, macrophages, monocytes, and perhaps lymphocytes into the glomerulus. The primary site and target of injury in NES is the more proximal layer of the glomerular capillary wall including the endothelium and lamina rara interna (of the glomerular basement membrane). The reduction in GFR observed in NES patients with severe glomerular injury is due to a decrease in the surface area available for filtration subsequent to occlusion of glomerular capillary lumens by invading inflammatory cells or proliferating glomerular cells. With generalized involvement of most or all glomeruli, the onset of azotemia can be sudden (even overnight) and may be accompanied by sodium retention resulting in edema and/or systemic arterial hypertension (termed acute nephritic syndrome in humans). In other patients with NES, the development of the glomerular inflammatory process may be more protracted with a gradual onset of azotemia over days or weeks (termed rapidly progressive glomerulonephritis in humans) and lesser tendency toward sodium retention and hypertension. While these terms have not been generally applied to canine patients, a similar range of clinical presentations is observed in dogs with GD.

Structurally, lesions found in the NEC reflect injury to the endothelium resulting in active inflammation consequent to antibody-mediated binding and/or complement activation or formation and deposition of immune complexes in the subendothelial or mesangial areas. Four primary mechanisms have been recognized that may produce the glomerular inflammatory process and the NEC: 1) Immune complex formation and complement activation in the subendothelial space or in the mesangium (examples in humans include postinfectious GN and some lupus nephritis patients); 2) Circulating antibodies directed against the glomerular basement membrane (very uncommon in dogs); 3) Circulating antibodies directed against neutrophil cytoplasm antigens (ANCA); and 4) Activation of complement via the alternative pathway.[4] A role for cell-mediated immunity, if any, has not been confirmed in GD, although they do play a role in some renal interstitial inflammatory diseases.

GLUCOCORTICOIDS AND IMMUNOMODULATING AGENTS FOR MANAGEMENT OF GLOMERULAR DISEASE

Glucocorticoids and immunomodulatory agents are an important tool used to manage GD in human beings. However, the rationale underlying etiological treatment of primary glomerulonephritis remains lacking. Nonetheless, clinical studies have supported the effectiveness of glucocorticoids and immunomodulating agents in some subtypes of GD in human beings.[5] As stated earlier, it is highly desirable to choose patients with forms of GD likely to respond to such therapies to minimize the risk of exposing patients unlikely to respond to therapies that may have significant complications. Unfortunately, one of the principal justifications for considering using glucocorticoid and immunomodulating agents in dogs with GD is the success seen in some forms of GD in humans. The specific subtypes of canine GD likely to respond to immune therapy are not yet known. In the absence of such firm evidence, it is necessary to extrapolate what is known in humans to dogs, insofar as such extrapolations are reasonable.

Glucocorticoids (GCC) have been widely used with variable success in all subtypes of primary GD in humans.[5,8] They have been shown to have efficacy (with or without other immunomodulatory drugs) against minimal change disease, primary focal segmental glomerulosclerosis, idiopathic membranous nephropathy, and ANCA-associated vasculitis; however, they have failed to cure several progressive GD variants, although they extended patient and kidney survival.[8] They are the most effective anti-inflammatory agents and interfere with immune response, although they are more effective against cellular immunity than humoral immunity. Glucocorticoids suppress multiple inflammatory genes; interfere with the traffic of inflammatory cells; and at high doses may reduce generation of superoxide anion radicals, inhibit production of platelet-activating factor, modify the composition of the glomerular basement membrane, and inhibit complement-induced granulocyte aggregation. The principal problem with glucocorticoids is it their low therapeutic index. High dose intravenous "pulses" of methylprednisone (MPP) are used in human patients with rapidly progressive renal disease.[5] Theoretically, such therapy is thought to have more rapid and more effective interference with inflammatory and immune processes than other oral glucocorticoids. A short course of MPP is followed by moderate doses of oral prednisone. An alternate approach sometimes used in humans is administration of synthetic ACTH.

The primacy of glucocorticoids in managing GD may relate to its role as an anti-inflammatory agent. Many of the other immunosuppressive drugs seem to perform better when combined, at least for some period of time, with steroids. How important this is in managing dogs with GD is as yet unclear, but optimum therapy may require both anti-inflammatory effects and immunosuppression.

Glucocorticoids have generally been avoided in dogs with GD because of their tendency to promote proteinuria. However, dogs with spontaneous hyperadrenocorticism typically do not become azotemic or hypoalbuminemic despite their marked proteinuria. Empirically, some dogs with GD treated with glucocorticoids have appeared to go into remission for their GD. Nonetheless, when these drugs are used in dogs with GD, it is usually for limited courses in order to minimize the adverse effects and proteinuric enhancement. Further studies are needed to clarify the role of glucocorticoids in treatment of canine GD.

Mycophenolate and azathioprine are nucleotide synthesis inhibitors that have previously been used primarily in transplant and autoimmune diseases. They inhibit B/ and T-cell lymphocyte functions, B-cell proliferation and antibody synthesis and natural killer cell activity.[8] Principal toxicities of mycophenolate are bone marrow suppression, gastrointestinal signs and viral infections. Toxicities of azathioprine include bone marrow suppression, hepatic toxicity, infections and risk of neoplasia. Mycophenolate has been suggested for use for prolonged treatments due to its relative safety (primarily seen as diarrhea in dogs); however, the extent of its effectiveness has not yet been determined. In humans, mycophenolate may be most effective when combined with steroids.[5] An important benefit of mycophenolate may be its rapid onset of effect.

Cyclophosphamide and chlorambucil are alkylating agents that produce toxic effects on proliferating cells. They produce B- and T-lymphocytopenia and inhibit immunoglobulin synthesis. Because of their higher risk of toxicity (which is cumulative), it has been suggested that these drugs be used only in selected cases.[8]

Calcineurin inhibitors such as cyclosporine A and tacrolimus have been found to be effective in a selection of human GD including minimal change nephropathy, focal segmental glomerulosclerosis and membranous nephropathy. However, patients treated with these drugs tended to have a high rate of recurrence when the drugs were withdrawn. Cyclosporine is the only immunomodulatory drug that has been tested in dogs with GD by randomized, controlled clinical trial. Unfortunately, the study failed to find evidence of effectiveness with this drug, but it is possible that dosing may have been too low to be effective.

When Immunomodulatory Therapy Should Not Be Used

Immunomodulatory therapy should not be used when an immune mechanism is not thought to be a significant contributor to GD. Important examples include: canine hyperadrenocorticism, hemodynamically mediated glomerular injury associated with chronic kidney disease (CKD), deposition diseases such as amyloidosis or other fibrillary glomerular diseases, GD due to hereditary structural defects, and chronic glomerular disease where fibrosis is advanced and evidence of active immune processes cannot be established. These conditions are unlikely to respond to IM therapy and, in some instances, may be exacerbated by IM therapy (e.g. glucocorticoids for hyperadrenocorticism). Because hyperadrenocorticism can masquerade as proteinuric kidney disease (UPC values may exceed 10), it is important to exclude this cause for proteinuria. Glomerular hypertension accompanying CKD can generally be recognized as mild proteinuria (UPC typically < 2-3) in an azotemic patient with evidence of CKD (e.g. imaging studies). Amyloidosis may be confirmed by light microscopic examination when a complete renal biopsy cannot be performed. Hereditary glomerulopathies of a structural nature may be recognized from family history, but absent an adequate family history, a complete renal biopsy is required to confirm this diagnosis. Confirming chronic GD absent active immune lesions may require a renal biopsy.

REFERENCES

[1] Brown S et al: J Vet Int Med 27 (S1): S27, 2013; [2] Segev G et al: J J Vet Int Med 27 (S1): S44, 2013; [3] Pressler B et al: J J Vet Int Med 27 (S1): S55, 2013; [4] Rennke HG and Denker BM: Renal Pathophysiology, 4th Ed, Lippincott, 2014; [5] Ponticelli C and Glassock RJ: Treatment of Primary Glomerulonephritis, 2nd Ed, Oxford Univ Press, 2009; [6] Whitley NT, Day MJ: J Sm An Pract 52:70, 2011; [7] Smets PMY et al: J Vet Int Med 26:565, 2012; [8] Pani A: Autoimmunity Reviews 12:848, 2013.

ACKNOWLEDGEMENT OF CONTRIBUTIONS FROM MEMBERS OF THE IRIS GLOMERULAR DISEASE STUDY GROUP

Members of the IRIS Glomerular Disease Study Group: Claudio Brovida, Scott Brown, Larry Cowgill (co-chair), Sylvie Daminet, Josefa Fernandez del Palacio, Jonathan Elliot, Thierry Francey, Bernard Gerber, Richard Goldstein, Greg Grauer, Reidun Heine, Mary Labato, Cathy Langston, George Lees, Meryl Littman, David Polzin (co-chair), Barrak Pressler, Gilad Segev, Shelly Vaden, Astrid van Dongen, Andrea Zatelli.

EVIDENCE BASED MANAGEMENT OF STRUVITE RELATED DISEASE IN CATS

Jody P. Lulich, DVM, PhD, DACVIM (SAIM)
St Paul, Minnesota

Struvite was first described in the medieval sewer systems in Hamburg Germany in 1845 and named after the geologist, Heinrich von Struve, who characterized it. It is ironic that this mineral composed of magnesium, ammonium, and phosphate also precipitates in and clogs the urinary systems of cats. In the following sections, we describe three clinical syndromes in which struvite affects the lower urinary tract of cats.

STRUVITE CRYSTALLURIA

Increasing ionic concentrations of urinary minerals are the driving forces underlying crystal formation. As the mineral concentration increases, urine becomes thermodynamically unstable and struggles to keep its contents in solution. When a concentration is reached such that minerals can no longer remain dissolved that are forced out of solution into a solid phase (crystals) which decreases their concentrations in the solution phase. Now that the concentrations of minerals are lower, urine re-establishes a more favorable thermodynamic equilibrium necessary to retain its fluidic properties.

The urine concentrations of magnesium, ammonium, and phosphorus are the building blocks for struvite crystals. In addition, urine pH and its influence on the concentration of trivalent ionic phosphate (PO_4^{-3}) play a key role underlying struvite crystallization. As urine pH increases, H_3PO4, $H_2PO_4^{-1}$ and HPO_4^{-2} are rapidly deprotonated (i.e. removal of hydrogen ions), increasing the concentration of PO_4^{-3}, a principal component and driving force for struvite crystal formation.

Diagnostic Considerations

- Be skeptical of the clinical importance of crystalluria observed in samples improperly collected, or improperly stored (greater than 2 hours or refrigerated prior to analysis).[1] These in vitro phenomena may not represent what is occurring in the patient.
- Interpret the severity and persistence of crystalluria in light of urine pH and specific gravity. Increasing alkalinity and increasing urine specific gravity promote struvite crystal formation.
- Crystalluria is a marker of urine supersaturation and a risk factor for urolith and plug formation.
- Persistently alkaline urine and subsequent struvite crystalluria may be an indication to culture the urine for urease producing microbes including *Staphylococcus* sp., *Proteus* sp., and *Ureaplasma* sp, however, the urine of most cats is sterile.

Therapeutic Strategies

- Do not always assume that microscopic crystals are the cause for lower urinary tract signs; however, cat bladders appear to be more sensitive than most other species. Struvite crystals have been observed in urine of cats that are apparently healthy, have 1) sterile struvite uroliths, 2) infection-induced struvite uroliths, 3) non-struvite uroliths, 4) urethral plugs that contain a large quantity of non-crystalline matrix and struvite crystals, and 5) urinary tract disease without uroliths (e.g. feline idiopathic cystitis). Therefore, in some cats, crystalluria is theorized to contribute to clinical signs in some cats, but can be quickly eradicated nutritionally.
- To prevent life-threatening urinary obstruction caused by struvite crystalline-matrix plugs or uroliths in male cats with persistent struvite crystalluria, feed therapeutic foods that increase urine solubility of struvite. Between 2010 and 2012, the Minnesota Urolith Center analyzed urethral plugs from 1001 cats; 98% were struvite (53% were in cats ≤4 yr, 30% in 5 to 7yr, 13% 8 to 10yr, and 4% ≥10yr). On the basis of these epidemiologic data, we recommend feeding struvite preventive foods to male cats with persistent struvite crystalluria through at least their 7th year of life.

STRUVITE URETHRAL PLUGS

Urethral plugs, most often composed of struvite and non-crystalline matrix, are a common cause for urethral obstruction in male cats. These findings are significant because although the prevalence of calcium oxalate uroliths has increased dramatically in cats the prevalence of calcium oxalate urethral plugs has remained very low (1%). Between 1980 and 1999, the prevalence of urethral obstructions significantly decreased from approximately 20 cases per 1000 cats evaluated to 7.4 cases per 1000 cats.[2] We attribute this reduction in disease to the availability and use of specially formulated foods designed to minimize struvite crystalluria.

- Recognizing and localizing the underlying cause for urethral obstruction is essential for selecting appropriate therapy with a successful outcome. Medical imaging is essential to verify urinary bladder distension, localize disease, and recognize the underlying cause.
- To prevent recurrence of struvite plugs and urethral obstruction, recommend foods that dissolve and prevent struvite. Provide foods that prevent struvite crystalluria through at least the 7th year of the cat's life.

STRUVITE UROLITHS

The choice for urolith treatment method depends on clinician experience and expertise, patient factors, and client preferences. Several therapeutic foods marketed for dissolution of struvite uroliths are formulated to avoid excessive magnesium and phosphorus and to maintain acidic urine pH (i.e. maneuvers that increase struvite solubility). Despite effectiveness of nutritional management of struvite uroliths, some veterinarians prefer to remove uroliths surgically because of perceptions that surgical management is more effective, is less expensive, controls clinical signs quicker, and will not be associated with urethral obstruction as uroliths decrease in size. These perceptions are not supported by published evidence. We prefer nutritional dissolution because it is the compassionate choice that we would select for ourselves, has not been associated with urethral obstruction as stones become smaller, eliminates risks of anesthesia, eliminates the risk of a suture nidus contributing to stone recurrence, and is less expensive for the client. In our opinion, it should be the current standard of practice.

Diagnostic Considerations

- Urinary bladder palpation is not reliable to detect struvite uroliths. In two studies, 10% or less of cases were identified by palpation. Ultrasonography of the urinary bladder is a sensitive method for detecting uroliths; however, survey radiography affords the advantage of providing urolith characteristics for predicting their mineral composition. Moderately radiopaque, discoid uroliths are consistent with a mineral composition of struvite.

Therapeutic Strategies: which diet works best?

Table 1. Grade of evidence and description of several studies to dissolve struvite uroliths

Grade of Evidence[3]	Treatment Recommendation	Study and Outcome
I	Dry therapeutic dissolution foods (c/d mulitcare and s/d) are 100% effective in dissolving struvite stones.[4]	Randomization=Yes Masking=Yes Treatment=dry c/d multicare and s/d fed to client-owned cats with spontaneous struvite uroliths Control=yes N=32 client cats with sterile struvite bladder stones Duration=dissolution or 8 weeks Monitoring=Clinic visit and radiography weekly Outcome=100% dissolution. A few cats in both groups dissolved stones in 1 week. Mean dissolution time 13 (range=6-28) days (s/d), 27(range=7-52) days (c/d multicare) (p=0.0002)
III	Dry therapeutic dissolution foods (urinary SO) with additional sodium) are equally effective as canned dissolution foods (Urinary SO) in dissolving struvite stones.[5]	Randomization=No Masking = No Treatment=dry and canned dissolution food fed to client-owned cats with spontaneous struvite uroliths Control=yes N=17 client cats with sterile struvite bladder stones Duration=dissolution or 12 weeks Monitoring=Clinic visit and radiography weekly Outcome=100% dissolution. A few cats in both groups dissolved stones in 2 weeks. Median dissolution time was 13 (range=10-42) days (dry; n=9) and 16 (range=14-55) days (canned; n=8). No difference between diet forms

- As an alternative to surgery, consider nutritional dissolution for cats with suspected struvite uroliths (usually ≤7 years old, moderately radiopaque uroliths, neutral to alkaline urine pH, occasional struvite crystalluria) as a cat-friendly and cost-effective alternative. Dry prescription foods are very effective. Therefore select food formulation by owner and cat preference. After radiographic resolution of uroliths, continue the dissolution foods for an additional 2-4 weeks.
- Re-evaluate abdominal radiographs in 2 to 4 weeks; when feeding c/d multicare or s/d, struvite uroliths should be 33% to 100% (i.e. dissolved) smaller. With minimal change in size, the mineral composition of uroliths is unlikely struvite or nutritional recommendations are being ignored. If this occurs, consider surgical or other forms of stone removal.
- Several commercial foods dissolve and prevent struvite recurrence (e.g. Hills c/d multicare, Royal Canin feline SO, Purina Stox). There are several advantages of multipurpose foods: 1) can be used as a maintenance food (i.e. long-term), 2) can improve diet compliance as they can be fed to all other healthy cats in the household, 3) no need to transition cats to a different food for stone prevention.
- To dissolve infection-induced struvite uroliths, a dissolution food and antimicrobics need to be administered throughout the entire period of dissolution. This duration of therapy is necessary because bacteria remain viable within the matrix of uroliths. Antimicrobic therapy is unnecessary for dissolution of sterile struvite uroliths.

REFERENCES

1. Sturgess CP, et al. J Fel Med Surg 2001;3:81-85. 2. Lekcharoensuk C, et.al.. *J Am Vet Med Assoc,* 2002; 221: 502-505. 3. Forrester DS, et.al. Vet Clin Small Anim 2007;37:533-558. 4. Lulich JP, et. al. J Amer Vet Med Assoc. 2013;243:1147–1153. 5. Houston DM, et al. *Br J Nutr* 2011;106:S90–S92.

CLINICAL CORRELATES ASSOCIATED WITH GLOMERULAR PATHOLOGY IN DOGS

Larry D Cowgill, DVM, PhD, DACVIM and the WSAVA Renal Standardization Project Clinical Working Group*
Davis, CA

Glomerular diseases have become recognized increasingly as a major cause of acute and chronic kidney injury in dogs.[1,2,3] There are numerous causes and pathogenic mechanisms for glomerular diseases whose etiopathogenesis may be linked to genetic mutations, infectious agents, inflammatory diseases, neoplastic diseases, autoimmune diseases, and other causes. In many instances the underlying cause remains unknown, and the disease is classified as idiopathic. Glomerular diseases are detected easily on the basis of routine laboratory parameters and screening, as well as quantitative, urine tests.[4] The clinical consequences of glomerular diseases also are readily recognizable admixtures of weight loss, hypoalbuminemia, hyperlipidemia, progressive renal dysfunction, hypertension, and fluid disorders. What has remained poorly characterized is the spectrum and correlation of these laboratory and clinical features to the distinctive pathologic subclassifications of glomerular injury that have become recognized recently when glomerular pathology is evaluated by more comprehensive diagnostic imaging.[3] In human beings, specific glomerular diseases, identified by their light microscopic, immunopathologic, and electron microscopic features, often are correlated with clinical and clinicopathologic findings that further characterize the disease entity. In veterinary medicine, the clinical features of glomerular diseases are brushed broadly with general descriptors that preclude or prevent their potential for more specific classification and prognostic and therapeutic characterization.

In 2007 the WSAVA Renal Standardization Project (WSAVA RSP) was launched under the auspices of the World Small Animal Veterinary Association with generous support from Hill's Pet Nutrition and Bayer Animal Health. The overarching purpose of this ambitious project was to apply a broad array of pathologic and clinical assessments to promote a better clinicopathologic understanding of the behavior and outcomes of glomerular diseases in dogs.[5] The WSAVA RSP Clinical Working Group was tasked to correlate the clinical and clinicopathologic data with the comprehensive pathologic diagnoses to identify patterns of findings that may constitute clinical groupings characteristic of distinct glomerular disease entities. In combination with the pathologic features, these clinical patterns might be expected to better predict foreseeable outcomes or responses to treatment. Correlative pathologic and clinical features could be combined to render a formalized morphologic and clinicopathologic classification scheme that could be applied uniformly among veterinary nephropathologists and nephrologists once validated.

The Clinical Working Group established a scoring instrument founded on over 130 numerical and categorical clinical attributes of dogs with glomerular disease whose kidney biopsies were subjected to comprehensive diagnostic imaging and statistical characterization of the light, immunofluorescent, and electron microscopic morphologic features constituting distinct disease categories. Each clinical attribute was independently evaluated and scored by clinical team members from supplied medical records, consultation with the submitting clinician, or from data entered into an established medical database. Subsequently, a consensus score was established for each attribute by the team. This clinical attributes list is to be subjected to the statistical logistics (cluster analysis) applied to the morphologic features to test if specific groupings of clinical features associate with the morphologic characteristics or pathologic diagnoses. By this analytical approach, we anticipate canine glomerular diseases can be more precisely defined and classified on the basis of both the morphologic features (light, electron, and immunofluorescent microscopy) and the patterns or severity of the clinical and laboratory responses to the glomerular insult.

SELECTED FEATURES & PRELIMINARY ANALYSIS

An initial overview of nearly 90 dogs designated into 8 distinct pathologic categories based on comprehensive diagnostic imaging revealed some of the following clinical/laboratory features at the time of evaluation and kidney biopsy (% of all dogs):

Historical Findings: Decreased activity, 41%; decreased appetite, 37%, wt. loss, 40%; PU/PD, 41%; vaccinated, 75%;
Physical Findings: Subcutaneous edema, 21%; ascites 25%; periodontal disease, 15%; sarcopenia, 15%; abnormal demeanor, 20%
Hemogram: Anemia, 23%; leukocytosis, 8%; leukopenia, 6%; thrombocytosis, 18%; thrombocytopenia, 9%
Chemistries: Azotemia; 41%; hypoproteinemia, 55%; hypoalbuminemia; 78%; hypoglobulinemia, 17%; hypocalcemia, 31%; hypercholesterolemia, 31%
Urinalysis: Specific gravity (>1.030, 29%; 1.021-1.029, 22%; 1.013-1.020, 30%; 1.012-1.008, 15%; <1.007, 3%)
Dipstick protein (trace, 0%; +1, 5%; +2, 11%; +3, 43%; +4, 28%)
RBC/hpf (0-5, 75%; >6, 25%)
WBC/hpf (0-5, 91%; >6, 9%)
Up/c; mean=8.9 (range 0.7 to 37.7)
Blood Pressure: AP0, 30%; AP1, 10%; AP2, 30%, AP3, 30%
Associated Disorders: Infectious, 28%; allergic, 13%; endocrine, 13%; cardio/pulmonary, 23%; GI, 36%; hemo-lymphatic, 24%

Clearly, this broad analysis of clinical features associated collectively with morphologically disparate glomerular diseases of dogs depicts a wide range of variance and severity. It remains to be seen if more detailed analysis and statistical association of these clinical features to the discrete pathologic categories will reveal a more precise paring of the clinical and pathologic features that will better categorize this heterogeneous disease process in dogs. From these foundations, it is our hope future treatment strategies and outcomes assessments can be developed more rationally and tested more precisely than efforts of the past.

REFERENCES

[1] Littman MP *Vet Clin N Am (Small Animal)* 2011;41(1):31. [2]Schneider SM, et al *J Vet Intern Med* 2013;27:S67. [3]Cianciolo RE, et al *J Vet Intern Med* 2013;27:S10.. [4]IRIS Canine GN Study Group Diagnosis Subgroup, et al *J Vet Intern Med* 2013;27:S19. [5]Cowgill LD, et al J Vet Intern Med 2013;27:S5.

*Members of the WSAVA RSP Clinical Working Group: Claudio Brovida (Italy), Larry Cowgill (Co-chair WSAVA RSP, USA), Jonathan Elliott (UK), Reidun Heiene (Norway), George Lees (USA), David Polzin (Co-chair WSAVA RSP, USA), Hattie Syme (UK), Shelly Vaden (USA), Astrid van Dongen (Netherlands)

RECOMMENDATIONS FOR THE TREATMENT OF GLOMERULAR DISEASE BASED ON ESTABLISHED PATHOLOGY

Gilad Segev (Chair), Larry D. Cowgill, Reidun Heiene, Mary A. Labato, David J. Polzin
Rehovot, Israel

There are many causes and pathogenic mechanisms associated with glomerular diseases in dogs including genetic mutations, infectious agents, inflammatory, neoplastic, and immune mediated diseases. When kidney biopsy establishes evidence of an immune-mediated mechanism, use of immunosuppressive/anti-inflammatory therapy may be rational and indicated. Nevertheless, compelling evidence supporting this therapeutic practice in dogs is lacking and the recommendations presented herein and were developed with consideration of a comprehensive review of literature and on expert opinion. Official recommendations of the study group in this manuscript are italicized.

The empirical application of immunosuppressive/anti-inflammatory therapy is recommended for dogs with severe, persistent, or progressive glomerular disease in which there is renal biopsy supported evidence of an active immune pathogenesis and no identified contraindication to immunosuppressive therapy.

The immunosuppressive strategy selected to manage dogs with documented glomerular disease of an apparent immune mediated pathogenesis should be selected on the basis of the severity of the disease and its rate of progression. For diseases associated with high magnitude proteinuria, attendant hypoalbuminemia, nephrotic syndrome, and/or rapidly progressive azotemia, single drug or combination therapy consisting of rapidly acting immunosuppressive drugs is recommended.

The most definitive criteria implicating an immunopathogenesis in glomerular disease is identification of components of the immune system associated with active and ongoing pathologic injury in the glomerulus. By extension, a documented immunopathogenesis should be established prior to the use of immunomodulatory therapy. The evidence must come from the glomerulus itself based on light, immunofluorescent, and electronmicroscopic imaging of appropriately processed biopsies.

SELECTING AN IMMUNOSUPPRESSIVE PROTOCOL FOR DOGS WITH GLOMERULAR DISEASE

Clinical presentations of canine glomerular diseases can broadly be classified as: 1) peracute and/or rapidly progressive, or 2) stable and/or slowly progressive. Recommended protocols for peracute and/or rapidly progressive glomerular disease include:

1. *Mycophenolate alone or in combination with prednisolone*
2. *Cyclophosphamide (continuous or pulse therapy) alone or in combination with prednisolone*

Glucocorticoids have been largely avoided in the management of glomerular disease. The avoidance of glucocorticoid therapy likely stems from the characteristic adverse effects (relative to benefits). The use of glucocorticoids should be considered carefully in the presence of systemic or urinary tract infections to prevent worsening of these conditions (e.g., pyelonephritis). These clinical perceptions are weakly supported by an observational study demonstrating worse outcomes in dogs with glomerulonephritis when managed with prednisone.[1]

Short-term administration of glucocorticoids may be appropriated in fulminate cases where immediate immunosuppression is required if their use is adjusted to minimize their adverse effects. Potential benefits of alternative treatment regimens, i.e., pulse therapy vs. continuous therapy should be considered.[2] However, on the basis of current practice perceptions and anecdotal experience, the use of glucocorticoid therapy should be tapered to the minimally effective dose as quickly as possible due to predictable side effects. Glucocorticoids may also be appropriate in acute management of multisystemic immune-mediated diseases where there use has proven beneficial (e.g. concurrent polyarthritis, immune-mediated hemolytic anemia, etc.). The use of glucocorticoids as a sole therapy is not recommended, and when used concurrently with mycophenolate or cyclophosphamide, glucocorticoids should be tapered as quickly as possible to the minimally effective dose with the goal to discontinue the corticosteroid component within as soon as possible.

Mycophenolate is an immunosuppressive agent that inhibits inosine monophosphate dehydrogenase. Its immunosuppressive activity results from selective inhibition of both T-cell and B-cell proliferation. Gastrointestinal side effects including anorexia, vomiting and diarrhea are reported to be the main adverse effect of this drug.[3] Evidence for the effectiveness of mycophenolate in the management of glomerular diseases in human medicine is mounting;[4,5] however, the veterinary literature is limited to a single case report supporting its use in a dog with glomerulonephritis of uncharacterized pathology.[6]

Although randomized, controlled clinical trial evidence is lacking, based on preliminary, uncontrolled clinical experience with mycophenolate and its low rate of severe complications, mycophenolate is recommended as the first choice for therapy of dogs with peracute or rapidly progressive glomerular disease of an apparent immune mediated pathogenesis.

RECOMMENDED PROTOCOLS FOR STABLE AND/OR SLOWLY PROGRESSIVE GLOMERULAR DISEASE

Dogs with stable and/or slowly progressive glomerular disease may be managed with protocols including drugs with either rapid onset or more protracted onset.

1. *Mycophenolate*
2. *Chlorambucil alone or in combination with azathioprine on alternating days*
3. *Cyclophosphamide and glucocorticoids*
4. *Cyclosporine*

MONITORING IMMUNOSUPPRESSIVE THERAPY

Systematic clinical and laboratory monitoring is essential to optimize the therapeutic management. Initial assessments should be performed no later than 1-2 weeks after initiation of the therapy and every 2 weeks thereafter for the first 4 to 6 weeks of therapy. Thereafter, assessments are recommended at least every 4 weeks for the next 3 months and then at quarterly intervals until resolution of the disease. Therapeutic effectiveness can be assessed by the impact of therapy on: 1) proteinuria (as measured by UPC), 2) renal function (as measured by serum creatinine concentration), and 3) serum albumin concentration.

Responses to therapy may be characterized as: 1) complete response, 2) partial response or 3) therapeutic failure. Although performed uncommonly, the response to therapy may also be assessed by changes in renal biopsies obtained before and after the course of standard and/or immunosuppressive therapy.

Change in the magnitude of the proteinuria

Reduction in the magnitude of proteinuria is a marker of response to therapy and should be assessed to document improvement beyond standard therapy alone. It should be recognized that day-to-day variance in the UPC ratio may be substantial in dogs with glomerular disease and should be considered when determining whether a measured change represents a clinically significant response to the therapeutic intervention. One also has to consider the magnitude of proteinuria may decrease in animals in late stages of progressive kidney disease due to a reduction in the number of remaining nephrons. Response to therapy as measured by changes in UPC is defined as follows:

- A complete response is defined as a reduction of the UPC to less than 0.5.
- A partial response is defined as a reduction of the UPC by greater than 50% of the highest pretreatment UPC after standard therapy or with standard therapy if both were initiated simultaneously.
- Therapeutic failure is defined as a reduction in UPC of less than 50% of the highest pretreatment UPC after standard therapy or with standard therapy if both were initiated simultaneously.

Improvement or stabilization of renal function

In patients with acute presentation, renal function may return to reference ranges (complete response). In contrast, dogs with long-standing, persistent proteinuria and chronic kidney disease or dogs with acute presentations, the glomerular damage may not repair and may remain abnormal, progress, or resolve incompletely (partial response). The response to therapy is measured by changes in serum creatinine concentration and defined as follows:

- A complete response is defined as reduction in serum creatinine concentration to less than 1.4 mg/dL (124 μmol/L) (or the patient's last known serum creatinine concentration before onset of the glomerular disease).
- A partial response is a sustained reduction in serum creatinine concentration by 25% or greater than baseline serum creatinine concentration most proximate to starting therapy.
- Therapeutic failure is defined as a reduction of serum creatinine concentration less than 25% baseline serum creatinine concentration most proximate to starting therapy.

Improvement in serum albumin concentrations (improvement in serum globulins, total plasma protein, antithrombin, and fibrinogen may also represent improvement in glomerular permselectivity)

Although small increases in serum albumin concentration may reflect a response to therapy, the therapeutic goal is to achieve a serum albumin concentration greater than 2.0 gm/dL (20 g/L). Response to therapy as measured by changes in serum albumin concentration from baseline is defined as follows:

- A complete response is a sustained increase in serum albumin concentration to greater than 2.5 g/dL (25 g/L)
- A partial response is either: 1) a sustained increase in serum albumin concentration to 2.0-2.5 g/dL (20 to 25 g/L), or 2) a sustained increase of 50% or more in serum albumin concentration from the baseline serum albumin concentration.
- A therapeutic failure is defined as failure to increase serum albumin concentration to greater than 2.0 g/dL (20 g/L) or by less than 50% increase from baseline serum albumin concentration.

In the absence of overt adverse effects, at least 8 weeks of the rapidly acting non steroidal drug therapy (e.g., mycophenolate) and 8-12 weeks of slowly acting drug therapy (e.g., azathioprine) should be provided before altering or abandoning an immunosuppressive trial. If no response is evident or therapeutic goals are not achieved within these time intervals, consideration should be given to an alternative drug or dosing protocol. If no therapeutic response is noted after 3-4 months, consideration should be given to discontinue immunosuppressive therapy. Immunosuppressive therapy should be continued in dogs demonstrating a complete or partial response to initial therapy for a minimum of 12-16 weeks. Thereafter, consideration should be given to tapering the therapy to a dose/schedule that maintains the response without worsening of the proteinuria, azotemia or clinical signs.

REFERENCES

1. Center SA, Smith CA, Wilkinson E, et al. Clinicopathologic, renal immunofluorescent, and light microscopic features of glomerulonephritis in the dog: 41 cases (1975-1985). J Am Vet Med Assoc 1987;190:81-90. 2. Franchin G, Diamond B. Pulse steroids: how much is enough? Autoimmun Rev 2006;5:111-113. 3. Barnoon I, Shamir HM, Aroch I, et al. Evaluation of combined treatment with mycophenolate mofenil and prednisone for dogs with meningoencephalitis of unknown etiology: retrospective case series of 25 dogs. In: European society of veterinary neurology. Trier, Germany: 2011. 4. Appel AS, Appel GB. An update on the use of mycophenolate mofetil in lupus nephritis and other primary glomerular diseases. Nat Clin Pract Nephrol 2009;5:132-142. 5. Appel GB, Contreras G, Dooley MA, et al. Mycophenolate mofetil versus cyclophosphamide for induction treatment of lupus nephritis. J Am Soc Nephrol 2009;20:1103-1112. 6. Banyard MRC, Hassett RS. The use of mycophenolate mofetil in the treatment of a case of immune-mediated glomerulonephritis in a dog. Aust Vet Pract 2001;31:103-106.

RECOMMENDATIONS FOR THE TREATMENT OF GLOMERULAR DISEASE WITHOUT ESTABLISHED PATHOLOGY

Shelly Vaden (Co-chair), Barrak Pressler (Co-chair), Bernard Gerber, Cathy Langston, David Polzin
for the IRIS Glomerular Disease Study Group*
Raleigh, NC

The prevalence of chronic kidney disease (CKD) in dogs may be as high as 1.5% of dogs presenting to general practice.[1] Glomerular disease may be the cause of renal injury in 50% or more of dogs with CKD.[2] Because glomerular diseases are common in dogs and the outcomes are not always favorable, there has been a recent enhanced effort to improve early recognition, diagnosis and treatment of all renal diseases in dogs, with particular emphasis on proteinuria renal diseases. In support of this effort, the WSAVA Renal Standardization Project has been working to develop a classification system for glomerular pathologic findings in dogs. At the same time, the International Renal Interest Society (IRIS) appointed a working group charged with developing clinical guidelines for the management of dogs with glomerular disease. The latter initiative used a formal consensus method for developing several sets of recommendations, including one set about the use of immunosuppressive therapy absent an established renal pathologic diagnosis.

There is no substitute for a pathologic diagnosis in the formulation of therapeutic plans for dogs with glomerular disease. However, in there are times when it is not possible to collect a renal biopsy specimen and the decision must be made whether or not to administer immunosuppressive agents to the affected dog. Four recommendations were derived and accepted at a high level of consensus to address this situation. These recommendations can help guide the decision about performing a renal biopsy in patients with proteinuria as well as the use of immunosuppressive drugs in those patients where the decision was made not to perform renal biopsy. [3]

Proteinuric dogs suspected of having glomerular disease should initially be managed using standard therapy and regular monitoring. However, this therapy rarely leads to complete remission of disease and, in some cases, adverse effects of the drugs used for standard therapy limit their use. When the targeted reduction in proteinuria (i.e., urine protein creatinine ratio <0.5 or a 50% reduction from baseline) is not achieved, and immunosuppressive or anti-inflammatory therapy is being considered, it is reasonable to readdress renal biopsy. When a renal biopsy cannot be obtained, the risk-to-benefit assessment for the patient should consider the arguments for and against using immunosuppressive drugs in dogs with proteinuria renal disease.

WHEN NOT TO PERFORM RENAL BIOPSY

Recommendation 1: *"Renal biopsy should not be performed in dogs (1) with IRIS CKD Stage 4; (2) when other medical contraindications are present and cannot be mitigated (including coagulopathy, renal cystic disease, moderate to severe hydronephrosis, pyelonephritis, perirenal abscess, uncontrolled hypertension, severe anemia, and pregnancy); or when results of renal biopsy are deemed unlikely to alter treatment, outcome, or prognosis."*[3]

Ideally, a renal biopsy would be evaluated in all dogs that are being managed for proteinuric renal disease. However, many times renal biopsy cannot be performed due to medical, practical, or financial limitations. In addition to the contraindications specifically stated in recommendation 1, relative contraindications to renal biopsy include available experienced personal to perform renal biopsy and the lack of access to a qualified and experienced renal diagnostic pathology center. Furthermore, if results of renal biopsy are deemed unlikely to alter treatment, outcome, or prognosis, then renal biopsy should not be recommended (e.g., end stage renal disease). Other factors that may preclude performing a renal biopsy include financial constraints or ethical concerns of the owner.

EXCLUSION CRITERIA FOR USING IMMUNOSUPPRESSIVE THERAPY

Recommendation 2: *"Immunosuppressive/ anti-inflammatory therapy should not be administered to dogs with proteinuria prior to renal biopsy when (1) proteinuria is not definitively glomerular in origin; (2) immunosuppressive therapy is otherwise contraindicated; (3) the dog breed and age of disease onset suggest that a nonimmune-mediated familial nephropathy is likely; or (4) amyloidosis is the most likely histopathologic diagnosis."*[3]

The first step in the decision process for using immunosuppressive therapy absent a renal biopsy is to verify that the proteinuria is of glomerular origin. When the source of proteinuria has not been definitively localized, immunosuppressive drugs are not indicated. Collective anecdotal experience of this consensus panel suggests that dogs with chronic tubulointerstitial disease rarely have UPC values greater than 2.0-3.0, although occasionally, acute kidney injury may transiently be associated with higher UPC results (i.e. ≥ 5.0). It is believed that UPC ranges from dogs with glomerular disease versus acute or chronic tubulointerstitial disease overlap, and results therefore must be interpreted in conjunction with other clinicopathologic findings when predicting type of disease. Therefore, dogs with UPC values less than 2.0 in conjunction with increased serum creatinine concentration and persistent isosthenuria or absence of proteinuria at the time of initial diagnosis of kidney disease should not receive treatment with immunosuppressive/ anti-inflammatory therapy without biopsy-supported evidence of active immune-mediated glomerular injury.

Immunosuppressive therapy should not be administered to dogs with concurrent illnesses for which immunosuppression is contraindicated (e.g., diabetes mellitus, hyperadrenocorticism, infectious diseases). Additionally, specific immunosuppressive drugs may be contraindicated with particular conditions (e.g. glucocorticoids in dogs with pancreatitis or uncontrolled hypertension, azathioprine in dogs with bone marrow suppression, hepatic dysfunction, or pancreatitis, etc.).

Immunosuppressive therapy should not be given to dogs that are likely to have a nonimmune-mediated familial nephropathy. Familial disease should be suspected when multiple related dogs are diagnosed with similar proteinuric renal disease or when a dog is diagnosed with proteinuric renal disease that is characteristic of a familial disease reported to occur in that breed. The diagnosis of familial nephropathy should be considered presumptive until confirmed by renal biopsy.

Although indirect evidence suggests that reactive amyloidosis in dogs is associated with a dysregulated immune response, immunosuppressive therapy in people and dogs is either of no benefit or may contribute to more rapid progression of disease.[4,5] There is too much overlap in UPC values between dogs with amyloidosis and dogs with other glomerulopathies to reliably use the UPC value to predict histopathologic diagnosis. However, renal amyloidosis may be more likely in dog with glomerular disease when the affected dog is of a breed known to be predisposed to amyloidosis (e.g., Shar pei,), additional clinical signs associated with hereditary amyloidosis in Shar peis are present (i.e., cyclical fever, distal joint effusion), or when amyloid deposition has been confirmed in other organs, particularly the liver.

WHEN IMMUNOSUPPRESSIVE THERAPY SHOULD BE CONSIDERED

Recommendation 3: *"Immunosuppressive drugs should be considered in dogs with glomerular disease that are being given standard therapy and do not have a biopsy-confirmed renal pathologic diagnosis when (1) serum creatinine is >3.0 mg/dL, or azotemia is progressive; or (2) hypoalbuminemia is severe (i.e., <2.0 g/dL)."*[3]

Evidence of immune complex glomerular disease was found in only 241 of 501 (48.1%) renal biopsies obtained from dogs suspected of having clinical evidence of glomerular disease.[6] In other words, 1 out of every 2 dogs with clinical evidence of glomerular disease would likely be candidates for immunosuppressive/ anti-inflammatory therapy. Clinical trials of immunosuppressive agents in dogs with specific glomerular diseases have not been reported, but in people with select glomerular diseases these drugs have a clear role. Because survival can be short in dogs with glomerular disease characterized by either azotemia or nephrotic syndrome and nearly 50% of dogs with clinical evidence of glomerular disease have immune-complex glomerular disease possibly responsive to immunosuppressive drugs, a therapeutic trial might be warranted in some dogs with glomerular range proteinuria absent a pathologic diagnosis.[7] Immunosuppressive therapy might be indicated in dogs with rapidly progressive disease, in spite of standard therapy, that either cannot be biopsied or that have been biopsied but results are not yet available. Specifically, aggressive immunosuppression may be considered if: 1) azotemia is acutely severe and/or progressive (i.e., creatinine > 5 mg/dl, IRIS AKI grades 4 or 5) at the time of diagnosis and there is no evidence of chronic disease; or, 2) hypoalbuminemia is severe (serum albumin < 2.0 g/dl)). In these situations, the protocols for peracute and rapidly progressive diseases should be followed. Likewise, immunosuppressive therapy might be indicated in dogs with chronic glomerular proteinuria if biopsy is not possible, neither age nor breed are indicative of familial renal disease, and other contraindications to immunosuppressive therapy are not present. In this situation, the protocols for more protracted disease should be followed. In all cases listed herein, immunosuppressive therapy should be considered a therapeutic trial; if there is no response after 8 to 12 weeks, therapy should be discontinued and the previous decision to not perform a renal biopsy should be revisited.

PROS AND CONS OF IMMUNOSUPPRESSIVE THERAPY ABSENT A PATHOLOGIC DIAGNOSIS

Recommendation 4: *"Immunosuppressive drugs should be administered to dogs in the absence of a renal pathologic diagnosis only after thorough client communication regarding the arguments for and against the use of these drugs in this setting. These agents should be administered cautiously, with close and careful patient monitoring."*[3]

Above all do no harm. Will we harm the patient by recommending an unproven immunosuppressive or anti-inflammatory treatment for a dog with glomerular disease that may not benefit from the therapy? Alternatively, canine glomerular disease can lead to serious complications or death and failure to provide a potentially helpful therapy may result in more harm than the potential risks of the therapy. Thus it is important to consider both the potential risks and potential benefits of recommending immunosuppressive/ anti-inflammatory treatment for dogs with clinical evidence of glomerular disease absent the findings of a renal biopsy. As previously stated, there is approximately a 50:50 chance that we would appropriately recommend immunosuppressive therapy for a dog with clinical evidence of glomerular disease absent a renal biopsy. Likewise, there is a 50:50 chance that recommending such therapy could be inappropriate for the patient without a biopsy and the patient is being put at-risk for developing one of the many potential side effects of these agents. The decision to proceed with therapy requires a case-by-case consideration of the risks of therapy.

The view that immunosuppressive therapy may be effective in improving clinical outcomes in dogs with some forms of glomerular disease is based on observations in people with glomerular disease as well as recent anecdotal evidence. Cyclosporine, the only drug that has been studied prospectively in dogs with glomerular disease, was found to have no detectable benefit.[8] However, dogs were included in this study regardless of renal pathologic diagnosis and the cyclosporine dose may have been too low. A treatment effect might have been found if only dogs with documented immune complex mediate glomerular disease were studied. No other studies of the use of immunosuppressive agents in dogs with glomerular disease have been published. As a consequence, it is difficult to predict a positive treatment effect with any accuracy.

REFERENCES

[1]Brown SA. BSAVA Manual of Canine and Feline Nephrology and Urology 2007:223. [2]MacDougall DF, et al. Kidney Int 1986;29:1144. [3]Pressler B, et al. JVIM 2013;27:S55. [4]Yamada T, et al. Ann Rheum Dis 2001;60:124. [5]Martinez-Subiela S, et al. Vet Re 2004;154:814. [6]Schneider SM, et al. JVIM 2013;27:S67. [7]Klosterman ES, et al. JVIM 2011;25:206. [8]Vaden SL, et al. JVIM 2005;19:794.

*Members of the IRIS Glomerular Disease Study Group: Claudio Brovida, Scott Brown, Larry Cowgill (co-chair), Sylvie Daminet, Josefa Fernandez del Palacio, Jonathan Elliot, Thierry Francey, Bernard Gerber,** Richard Goldstein, Greg Grauer, Reidun Heine, Mary Labato, Cathy Langston,** George Lees, Meryl Littman, David Polzin (co-chair), Barrak Pressler,** Gilad Segev, Shelly Vaden,** Astrid van Dongen, Andrea Zatelli. **Members of the Immunosuppressive Therapy Absent a Pathologic Diagnosis subgroup.

COAGULATION FACTOR XII: FRIEND, FOE, OR DIAGNOSTIC DILEMMA

Marjory Brooks, DVM, DACVIM (Small Animal Internal Medicine)
Ithaca, NY

INTRODUCTION

The activated partial thromboplastin time (aPTT) and prothrombin time (PT) assays are often performed in tandem as a coagulation panel. Together with platelet count, these clotting time tests are included in the initial evaluation of patients suspected of having a bleeding disorder. Point-of-care coagulation instruments and routine pre-operative testing have extended the use of coagulation panels to broader patient populations. Prolongation of the aPTT may reflect a life-threatening coagulopathy or an incidental finding. By understanding test principles, recognizing patterns of abnormalities, and synthesizing test results with patient presentation, clinicians can appropriately interpret the coagulation panel to focus their work-up and improve patient outcomes.

COAGULATION PANEL TEST PRINCIPLE: COAGULATION PATHWAYS

The aPTT and PT measure enzymatic and cofactor activities of plasma coagulation factors based on the rate of fibrin formation. An activating agent is added to the patient's plasma sample to initiate coagulation and then the time (in seconds) is measured until a clot forms. The "trigger" reagent composition determines which distinct coagulation factors are initially activated until the cascade proceeds to the common, final reactions of coagulation. The common pathway ends in a burst of thrombin (Factor IIa) that cleaves soluble plasma fibrinogen to generate an insoluble fibrin clot (**Figure 1**).

The PT reagent contains phospholipid and high concentration tissue factor that directly activate Factor VII in the extrinsic pathway. While tissue factor is the primary initiator of in vivo fibrin formation, its large excess in the PT reagent is non-physiologic and renders the PT assay insensitive to deficiencies of Factor VIII and Factor IX, two critical factors for in vivo hemostasis.

The aPTT reagent contains phospholipids, but no tissue factor and initiates the intrinsic pathway. Coagulation is triggered in the aPTT by prolonged incubation of plasma with negatively charged particles that provide a surface for the reciprocal activation of Factor XII and other protein factors, referred to collectively as the contact pathway.

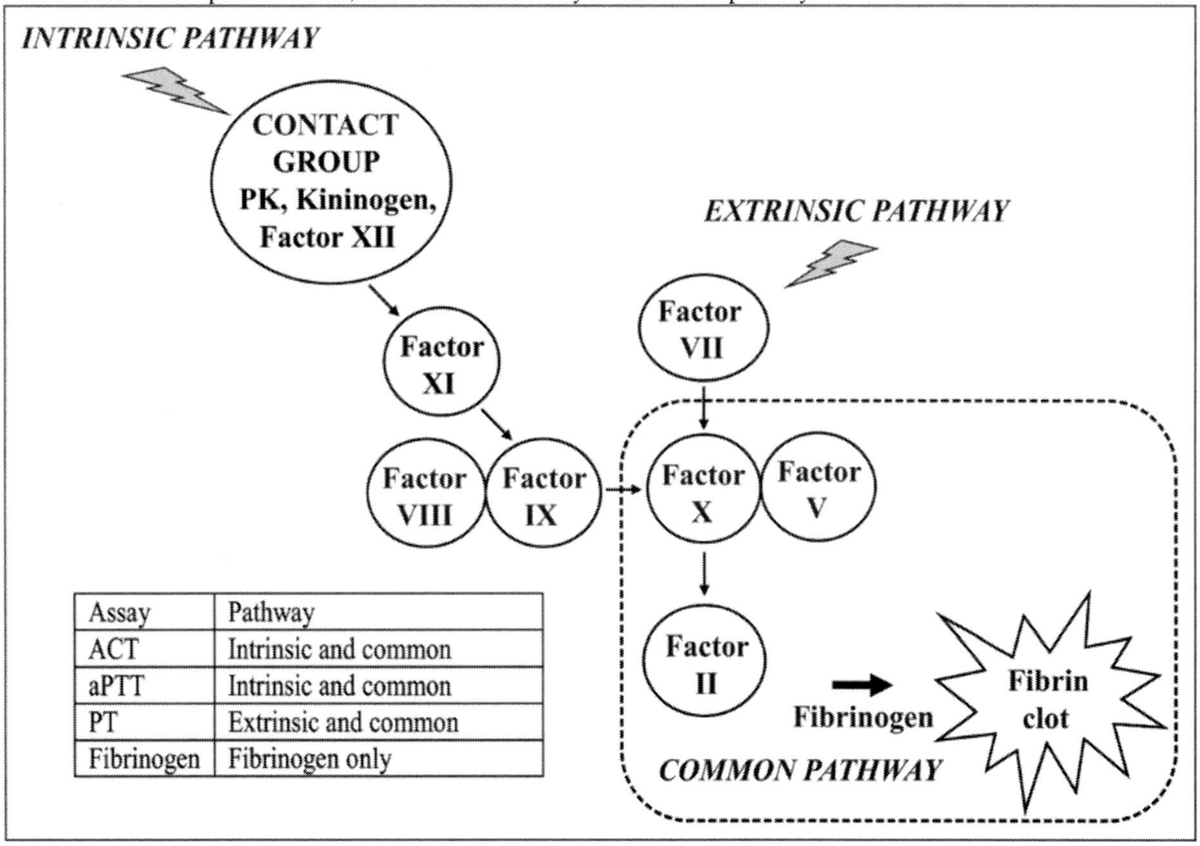

Assay	Pathway
ACT	Intrinsic and common
aPTT	Intrinsic and common
PT	Extrinsic and common
Fibrinogen	Fibrinogen only

Figure 1. Intrinsic, Extrinsic, and Common Coagulation Pathways

CONTACT ACTIVATION AND COAGULATION FACTOR XII

The contact pathway includes two serine protease proenzymes, Factor XII and Prekallikrein (PK), and a glycoprotein cofactor, High Molecular Weight Kininogen (HMWK) that interact to generate activated Factor XI. The aPTT reaction mixture provides charged surfaces and phospholipids for assembly of contact factor-cofactor complexes resulting in the formation of Factor

XIIa, kallikrein, and kininogen in a cycle of positive feedback amplification (**Figure 2**). Deficiency of Factor XII, or the other contact pathway factors, results in prolonged and variable delays in fibrin clot formation in the aPTT screening test.

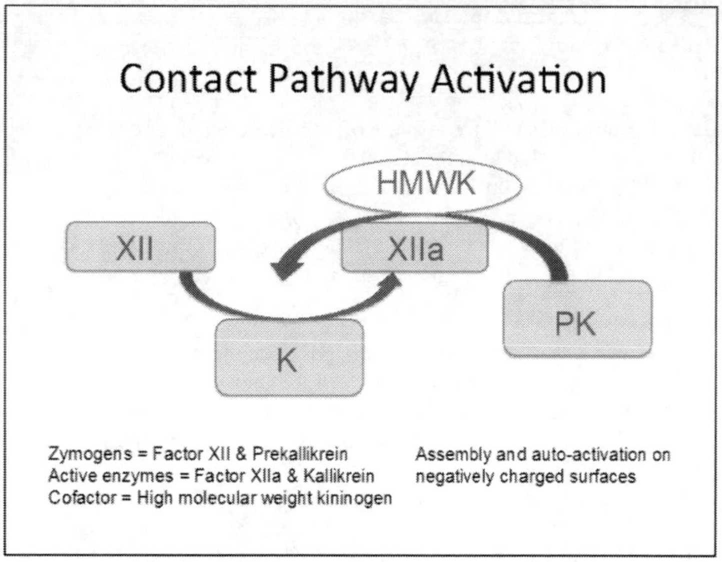

Figure 2. Contact Pathway Factors

The Contact Pathway and Inflammation

In spite of prolonged aPTT, the contact pathway's best defined in vivo role involves inflammation rather than coagulation.[1] The assembly of contact pathway factors on activated inflammatory cells and endothelium mediates the release of vasoactive and proinflammatory mediators such as kallikrein, bradykinin, and other kininogens, and results in complement activation. Dysregulation of the contact pathway is associated with tissue inflammation, increased vascular permeability, and hypotension but does not cause abnormal hemorrhage. While Factor XII participates in the activation of the contact pathway, to date, no clinical disease is attributed to a lack of Factor XII.

Factor XII: A Role in Physiologic Hemostasis?

Factor XII deficiency does not cause a bleeding disorder in any species. Factor XII deficiency has been identified in people, cats, and dogs and has been induced by knocking out the Factor XII gene in genetically engineered mice. Human Factor XII deficiency was first described in 1955 and named "Hageman Trait" for the index patient, John Hageman. Since that time, numerous patients have been diagnosed and their causative FXII mutations defined. None of these patients experience abnormal hemorrhage. Moreover, comparative genomics reveals a total lack, loss, or inactivation of the Factor XII gene in fish, birds, and marine mammals, respectively.[2,3] All of the current evidence rules out an essential role for Factor XII in supporting in vivo coagulation.

Factor XII: A Role in Pathologic Thrombosis?

Particulate substances such as kaolin, celite, silica, and ellagic acid support Factor XII activation in the aPTT, however its primary in vivo activator is unclear. Recent studies indicate that biologic substances at the site of thrombus formation may support Factor XII activation.[4,5] These compounds include extracellular nucleotides, long chain polyphosphates from degranulated platelets, aggregated proteins, and subendothelial collagen.[6] In murine models of arterial injury induced by chemical or photoactive compounds, Factor XII deficiency slows the rate of occlusive thrombus formation. This finding suggests that Factor XII may play a role in thrombus propagation or maturation. However, epidemiologic studies in people with Factor XII deficiency have not identified decreased rates of arterial thrombosis associated with atherosclerosis or decreased propensity to deep vein thrombosis.[4,5]

FELINE FACTOR XII DEFICIENCY

Hereditary Factor XII deficiency in cats is an autosomal recessive trait.[7] Male and female cats that are homozygous for the defect typically have Factor XII coagulant activity (FXII:C) less than 10% of normal, whereas heterozygotes have more moderate deficiencies up to 50% of normal. Prolongation of the aPTT is generally related to the severity of the Factor XII deficiency. Cats with severe Factor XII deficiency often have aPTT > 60 seconds in laboratory assays and > 180 seconds using point of care instruments.

Prevalence of Feline Factor XII Deficiency

Factor XII deficiency was first characterized in a related colony of domestic shorthair cats, with subsequent isolated descriptions of cases in unrelated domestic and Siamese cats.[7,8] In a retrospective review of feline coagulation testing at the Comparative Coagulation laboratory at Cornell University, Factor XII deficiency was by far the most common hereditary factor deficiency and represented 2% of all submissions.[9] The relative case ratio of Factor XII deficiency to hemophilia A (Factor VIII deficiency) was 2:1 and relative ratio to hemophilia B (Factor IX deficiency) was 4:1. Unlike hemophilia, an X-linked trait, Factor

XII deficiency was diagnosed in equal numbers of males and females. Factor XII deficient cats had a median age at diagnosis of 3 years, compared to hemophilic cats whose median age was < 1 year. Domestic shorthaired and longhaired cats represented approximately 80% of the Factor XII deficient cases and overall submissions, however Siamese cats were over-represented at 12% of Factor XII deficient cases and only 4% of total submissions. Additional cases have been identified in Himalayan and Maine Coon cats and a geographic skew toward the Midwest is also apparent.

Diagnosis and Management of Feline Factor XII Deficiency

Specific prolongation of the aPTT, with normal PT and fibrinogen are the hallmark features of intrinsic pathway factor deficiencies (**Figure 3**). Female gender, prolongation of aPTT beyond 60 seconds, and previous history of normal post-operative hemostasis are suggestive of a diagnosis of Factor XII deficiency. Definitive diagnosis is based on specific measurement of FXII:C in a citrate plasma sample.[10] Most young cats with Factor XII deficiency are diagnosed in the course of pre-operative coagulation screening for elective procedures and most adult cats are diagnosed in the work-up for an acquired disorder such as liver disease or lower urinary tract disease.

Regardless of the amount of residual Factor XII, cats with hereditary Factor XII deficiency do not require transfusion to replace Factor XII. Medical therapy, such as vitamin K, or hemostatic agents will not influence Factor XII levels and are not indicated for these cats. The PT assay, rather than aPTT, should be used as a screening test to detect acquired coagulopathies in this patient population. To date, Factor XII deficiency has not been associated with cardiac disease or shortened life-span in affected cats.

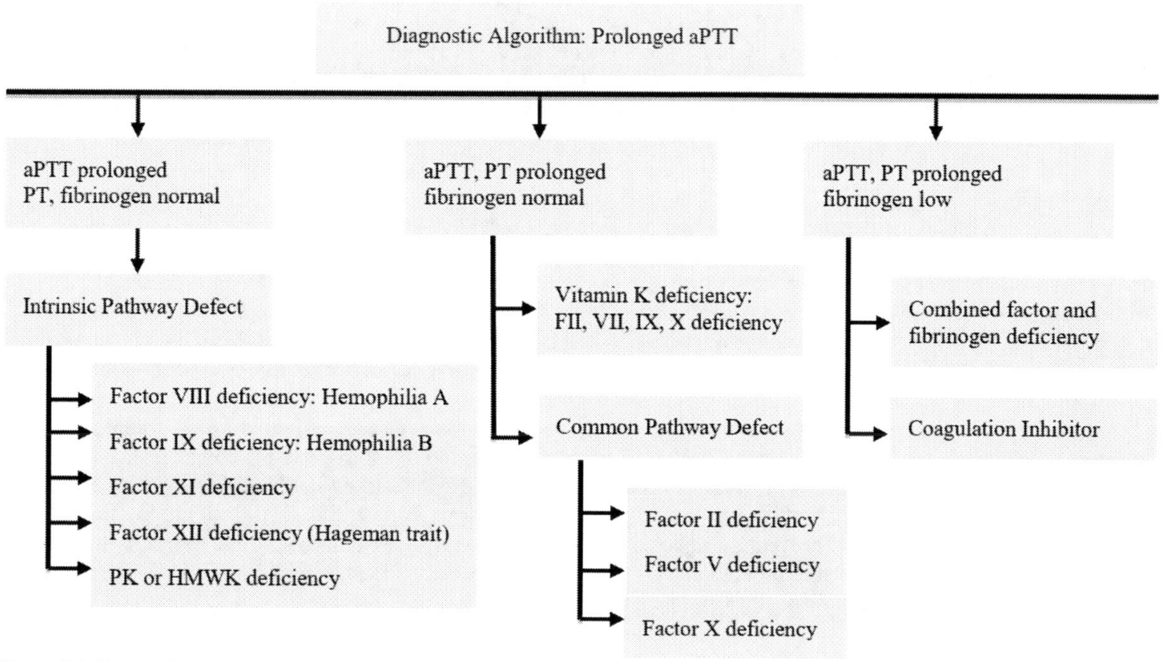

Figure 3. Differential diagnosis of prolonged aPTT

REFERENCES

[1]Maas C and Renne T. **Thromb Res** 2012;129:S73. [2]Ponczek MB, et al. **J Thromb Haemost** 2008;6:1876. [3]Davidson CJ, et.al. **J Thromb Haemost.** 2003;1:1487. [4]Stavrou E and Schmaier AH. **Thromb Res 2010**;125:210. [5]Renne T, et al. **Blood** 2012;120:4296.[6]Morrissey JH et al. PolyP **Blood** 2012;119:5972. [7]Kier AB, et al. Can J Comp Med 1980;44:309. [8]Green RA and White F. **A J Vet Res** 1977;38:893. [9]Brooks MB and DeWilde LA. **Compendium on Cont Ed Pract Vet.** 2006; 28:148. [10] Comparative Coagulation Laboratory https://ahdc.vet.cornell.edu/sects/Coag/

DISSEMINATED INTRAVASCULAR COAGULATION: PAST, PRESENT AND FUTURE

Tracy Stokol BVSc PhD DACVP
Ithaca, NY

INTRODUCTION

Disseminated intravascular coagulation (DIC) is a complex, dynamic state of hemostatic imbalance that results in thrombus formation throughout the small vessels. Many different primary disorders trigger DIC, particularly bacterial sepsis, severe inflammation (pancreatitis, immune-mediated hemolytic anemia) and cancer; it is never a primary disease. DIC should be thought of as a continuum of abnormal and excessive hemostasis activation. First there is the underlying disease that triggers activation of coagulation. This is likely mediated through cytokine-induced tissue factor expression on monocytes and endothelial cells, constitutive expression of tissue factor on cancer cells, tissue factor expression on dying cells, massive endothelial injury exposing subendothelial tissue factor on fibroblasts and endothelial matrix proteins, such as collagen (which support platelet adhesion and activation), and bacterial polyphosphates which activate prothrombin and factor IX (FIX) in the presence of FXII. These triggers activate secondary hemostasis, generating thrombin, and also activate platelets. Initially, the activated hemostatic system is controlled or restrained by inhibitors. This stage of DIC, which is characterized by an over-activated but restrained hemostatic system, is called **"non-overt" DIC**. During this stage, screening coagulation tests such as the prothrombin time (PT) and activated partial thromboplastin time (APTT) and platelet counts may be normal, but sequential testing may reveal worsening trends (e.g. decreasing platelet count). The animal is considered "hypercoagulable" (defined as excessive generation of thrombin) and could already be developing thrombi. With time or with severe disease, inhibitors become overwhelmed or are downregulated or inhibited by the underlying disease. Coagulation then becomes uncontrolled, disseminating throughout the vasculature, forming thrombi and, in dogs, ultimately leading to a hypocoagulable state, as platelets and coagulation factors are depleted. As hemostasis becomes uncontrolled, hemostasis tests become more and more abnormal and the syndrome becomes clinically apparent, either as thrombosis (more difficult to detect) or hemorrhage (far easier to detect). This stage of DIC is called **"overt" DIC**[1,2]. Thus, the clinical and laboratory manifestations of DIC vary among patients (depending on the initiating disease) and change over time, complicating its recognition and treatment.

DIAGNOSIS OF DIC: THE PAST

DIC has been traditionally diagnosed on a combination of clinical and laboratory criteria, including an underlying disease (must be present), clinical evidence of hemorrhage (usually from more than one site), and laboratory tests that reveal abnormalities in multiple hemostatic pathways, including thrombocytopenia (primary hemostasis), prolonged PT, prolonged APTT, hypofibrinogenemia (secondary hemostasis), low antithrombin (AT) activity (inhibitor) and high fibrin(ogen) degradation products (FDP) or D-dimer (fibrinolysis), and evidence of red blood cell fragmentation (acanthocytes, schistocytes). Alone, none of these tests are diagnostic for DIC (e.g. acanthocytes are frequently seen in cats with liver disease and do not indicate DIC in this setting) and their use, in combination, has sound rationale for the diagnosis of DIC.

However, this set of criteria restricts our ability to diagnose DIC to the end-stage of the process, where hemorrhage is manifest ("fulminant" or hemorrhagic phase of overt DIC). By this time, it may be too late to reverse the course of the syndrome, resulting in the unfortunate acronym of DIC, "Death is Coming". Ideally, DIC should be identified in the non-overt stage and certainly in the thrombotic phase of overt DIC. This would give us the greatest opportunity of preventing DIC from progressing (in those diseases where progression can occur) and to prevent thrombosis, which is the main cause of morbidity and mortality in affected animals, albeit difficult to recognize clinically. Furthermore, there is no uniformity between clinicians or investigators in the criteria used to diagnose DIC in published studies, making it difficult to compare across studies.

DIAGNOSIS OF DIC: THE PRESENT

In an attempt to overcome the above traditional criteria for diagnosing DIC, a DIC Scientific Subcommittee for the International Society of Thrombosis and Haemostasis (ISTH) developed a scoring system for diagnosis of non-overt and overt DIC[1,2]. There has also been an upsurge of interest in viscoelastic methods of clot detection, such as TEG® and ROTEM®. It was hoped that these methods would be able to detect hypercoagulability, particularly the non-overt stage and thrombotic phase of overt stage of DIC, since routine hemostatic tests are unable to detect these prothrombotic states.

The scoring system for non-overt DIC and overt DIC diagnosis in humans are based on readily available diagnostic assays, including the PT (more sensitive to DIC in humans than in dogs), fibrinogen concentration, platelet count, and fibrinolytic tests (D-dimer, FDP). These scoring systems, and those developed by others, such as the Japanese Association of Acute Medicine, all require the presence of an underlying disease that triggers DIC, although with some schemes (overt DIC from ISTH) no points are awarded for the disease. The various schemes differ in proposed tests, cut-offs used, and grading scheme applied to test cut-offs. In essence, the ISTH scoring scheme for overt DIC uses traditional laboratory criteria, but does not require bleeding symptoms (since they may not be present in the thrombotic phase of overt DIC). The non-overt scheme emphasizes the need for sequential testing (24-48 hour intervals) to detect worsening trends and gives most of the score (2 points, where $\geq$5 indicates non-overt DIC) to the initiating disease. The scoring systems have been tested in a few studies in human patients (usually those with gram negative bacterial sepsis) and do appear to detect those patients in overt DIC that have poor outcome (higher 28-day mortality) or those patients in non-overt DIC that progress to overt DIC with time.

There has been one preliminary study (abstract) on application of a modified ISTH scoring system for diagnosis of non-overt DIC in dogs[3]. In this study, the authors combined specific non-overt and overt DIC criteria (points were given for underlying disease, a prolonged PT, thrombocytopenia, high D-dimer and low AT activity) for diagnosing DIC and compared it to traditional assessment

(3 or more abnormal tests including thrombocytopenia and prolonged coagulation times) on daily blood samples, selecting the highest values (regardless of day of occurrence) for comparison. They then compared outcome (total hospitalization days and 28-day mortality rates) in 24 dogs suffering from diseases associated with DIC. They found that more dogs were diagnosed in DIC with traditional criteria but mortality rates were higher in the dogs classified in DIC by modified ISTH criteria. Total days of hospitalization were no different between groups. Subsequent studies in a larger cohort of dogs by the same authors showed that the modified ISTH scheme performed no better than traditional assessment. A scoring scheme may still have substantial benefit, the very least being the standardization of outcome measures and DIC definition across studies, permitting more valid across-study comparison. However, before any scheme is adopted, there needs to be consensus amongst investigators on which tests to include, the cut-offs that should be used, and the method of testing (since sensitivity and specificity of hemostasis testing is strongly influenced by reagents and method of clot detection).

The same authors that developed the modified ISTH scoring system also published a diagnostic scoring algorithm for DIC based on multiple logistic regression analysis of hemostatic assays[4]. The bottom line from this study was that they succeeded in developing an algorithm, but the developed algorithm performed no better (and in fact performed worse) than expert opinion which based the diagnosis of DIC (on the day of worst hemostatic test abnormalities) on traditional diagnostic assessment (i.e. the past). Although the model may have uses, it needs to be validated by other investigators, using different methods, since some of the methods used in the study are not universally available and the algorithm will only be applicable for the particular methods used in the published study.

Viscoelastographic testing was the hoped for best thing in DIC diagnosis, potentially being a test that was capable of detecting a hypercoagulable state (non-overt or thrombotic phase of overt DIC). Testing is done on whole blood and not plasma so the technique is considered more physiologic than screening coagulation assays, which lack the cells that support hemostasis (platelets and their membranes). Unfortunately, there have been limited studies evaluating the diagnostic utility of viscoelastic testing for DIC diagnosis[5] and it has become more apparent that the testing methodology is quite limited as to the information it provides and when it should be used. Recent studies have shown that results are affected by hematocrit, timing after collection, activators used and viscosity. In particularly, lower hematocrits (still within reference intervals) cause a trend towards hypercoagulability and higher hematocrits the reverse[6]. Since many dogs in DIC are concurrently anemic, the question is raised as to whether hypercoagulability in such dogs is a consequence of anemia (presumed an artifact of the system) versus an *in vivo* state. In addition, the test result most frequently used to diagnose hypercoagulability, i.e. clot strength or maximum amplitude, strongly correlates to fibrinogen concentrations and may actually yield no additional diagnostic information over fibrinogen measurement alone, i.e. in this system, high fibrinogen = hypercoagulability. A hypocoagulable tracing will be achieved in severely thrombocytopenic dogs, regardless of the status of coagulation factors, so there does not seem to be a valid reason for using this assay in severely thrombocytopenic dogs (i.e. a platelet count may provide the same information). Thus, the jury is still out as to whether this testing should be used for diagnosis of hypo- or hypercoagulable states, such as DIC (which can present as both). One arena where viscoelastic testing does show promise is detection of disorders of fibrinolysis[7], which may predispose dogs to both thrombosis and hemorrhage. We currently lack diagnostic tests that can measure inadequate or excessive fibrinolysis. Perhaps viscoelastic testing will be able to accomplish this.

DIAGNOSIS OF DIC: THE FUTURE

So where do we go from here? Well, there certainly is no distinct solution on the horizon and certainly no single diagnostic test that will detect thrombi or diagnose DIC. For the foreseeable future, it appears that we are destined to continue to diagnose DIC based on traditional clinical and laboratory criteria, as we have done in the past. However, by changing our thinking of DIC, i.e. that it is first and foremost a thrombotic and not a hemorrhagic syndrome, maybe we can change how we approach clinical cases. The most important thing is to recognize that an animal with a predisposing disease is at risk of DIC or already in DIC (we just cannot diagnose it yet), even if it is not showing signs of hemorrhage. Such a patient should be evaluated for thrombosis (to the best of our ability, e.g. use Doppler ultrasonography to detect large internal thrombi) and treated accordingly (although again, there is no consensus on treatment, with no evidence-based medicine on treatment efficacy in dogs). Serial monitoring of routine hemostasis test results (platelet count, PT, APTT, D-dimer, fibrinogen, AT activity) should be done to detect worsening trends in patients at risk of DIC, but is costly. Observing worsening trends would make a diagnosis of overt DIC highly likely and anti-thrombotic therapy (e.g. heparin) should be considered in such a patient. We should strive to achieve consensus and standardize, to the best of our ability, the definition of DIC, the best tests and test cut-offs used to diagnose DIC, and outcome assessment for clinical trials or research studies in dogs and cats. Then perhaps, we can tackle treatment.

REFERENCES

[1]Taylor et.al. **Thromb Haemost** 2001;86:1327, [2]Toh and Downey. **Blood Coagul Fibrinolysis** 2005;16:69, [3]Wiinberg et.al. **J Vet Intern Med** 2006;20:765 (abstract), [4]Wiinberg et.al. **Vet J** 2010;185:292, [5]Wiinberg et.al. **J Vet Intern Med** 2008; 22:357, [6]Smith et.al. **Am J Vet Res** 2012;73:789, [7]Spodsberg et.al. **Vet Clin Pathol** 2013; 42:281.

5 STEPS TO CONSTRUCTIVE CONFLICT RESOLUTION

Karyn Gavzer, MBA, CVPM
Springboro, OH

INTRODUCTION

Are you a conflict avoider? A silent sufferer? An exploder? There are more constructive ways to manage conflict! Properly handled, conflict can stimulate new thinking; foster healthy working relationships and build teamwork. Left unmanaged, however, conflict divides staff members and poisons goodwill.

It is a rare person that enjoys conflict. Most of us wish it would magically go away and that things would get better all by themselves. Conflict robs the day of joy and leads to gossip, bad feelings and poor morale. Confronting someone about a perceived wrongdoing, however, can be daunting. There is also the risk that you might win your point, only to loose the relationship and possibly make things worse for yourself.

It takes training, skill and strategy to constructively confront someone and have a successful conversation that achieves resolution and restores a healthy relationship.

FIVE STEPS FOR CONFLICT RESOLUTION

These are five practical steps to follow for constructive conflict resolution that should work even in a busy referral practice:

1. First, examine your thoughts – is it your interpretation of this person's motives that are making you upset, or is it what they actual did? Or is it both? It is important not to pre-judge the other person. Try to think of an acceptable reason why they would have done what they did.
2. Before you confront the other person, decide what you want to achieve. Do you want different behavior from this person? Do you want to restore the working relationship? Setting goals for the confrontation will keep you on track and limit the discussion to the one thing you want changed now (not every bad thing that they've done).
3. Choose to be curious, *not* confrontational in your approach. It is critical that you create a safe environment for the other person so that they can feel they can talk and be heard. Plan the conversation and pick a time when you won't be interrupted. Use triangulation to discuss the issue.
4. LISTEN! Do not accuse the other person. Once you have stated your observations and understanding of the incident, sincerely ask the other person what happened. Then be quiet and listen!
5. Plan a follow-up conversation after the initial confrontation to strengthen the relationship and ensure that the agreed outcome/behavior change endures. (Realize that you may have to change your behavior, too, to ensure a good outcome.) Change is hard work and check-in meetings will help the other person stay the course because they know they will be held accountable for making the change they promised.

Tip

If there is not time to talk when you see someone doing something they shouldn't be doing, call it when you see it. Think of it as throwing a "red flag" down to mark the behavior. Then quietly, without calling the attention of others, tell the other person you want to talk about it later. This aids clarity and makes the follow-up conversation easier because there is no doubt about the behavior you want to talk about when you meet. Make sure that you make "later" happen or there is no point in calling the behavior.

Mastering these steps can put you in charge of managing challenging confrontations to achieve behavior change, rebuild frayed relationships and bring back the joy of practice.

Constructive conflict resolution takes self-discipline and practice but it is well worth the work. If you want more information, many of the ideas in this talk were adapted from Crucial Confrontations, K. Patterson et al, 2005 McGraw Hill, a wonderful resource to have in your practice library, or even on your nightstand.

TEAM WORK: WHY IT WORKS AND WHAT TO DO WHEN IT DOESN'T

Karyn Gavzer, MBA, CVPM
Springboro, OH

INTRODUCTION

The goal of his program is for you to take home concrete ideas and tips to improve teamwork, profit and performance in your practice.

Today, it takes a winning team to run a successful practice. No one person can know it all or do it all alone. Today is the age of teamwork and getting it right has never been more important.

When it works, teamwork is magic! Work becomes more rewarding and fun. People enjoy and respect one another more. Morale, patient care, client service and practice finances all benefit. Yet, teamwork often breaks down or fails to develop in even the most well-intentioned practices. To get it right takes concrete skills, patience and practice.

In this program, we will cover the basic building blocks needed to support teamwork in a specialty practice as well as the four different stages of team development and the various roles and challenges that team leaders and team members face at each stage of a team's development. You will learn how teams go wrong, what to do to fix it, and how to set realistic team performance goals to achieve practice success.

INSTRUCTIONS

To get the most value from this talk, please add your notes to the space below to help you remember the four stages of teamwork and how the leader's role and the team members' roles have to change at each stage **of team** development for the team to achieve the next level of performance. When you are done, use these criteria to assess your own practice's team performance and target areas for improvement. The four stages of team development are:

1

Forming/Toddler

- **Team members still look to you to tell them what to do**
- **Team leader needs to provide direction and safe environment**

Add your notes here:

2

Storming/Adolescent

- **Team members start sharing different opinions and disagreeing about things**
- **Team leader needs to teach constructive conflict resolution**

Add your notes here:

3

Performing/Young Adult

- **Team members cooperation and ability to get things done improves**
- **Team leader needs to let go, but provide guidance and oversight**

Add your notes here:

4

Norming/Mature Adult

- **Reliable performance – success is normal and expected**
- **Team leader needs to trust team, delegate, act as a consultant to them**

Add your notes here:

SELF-ASSESSMENT QUESTIONS

What level of team work is your team? Where **are you as** a team leader? Where are you at as a team member? Whatever your position, what will you do differently to build team work in your practice?

SMART GOALS

"SMART" goals are very useful for team work. They help team members clarify their thinking and give them a structured way to organize their work. SMART goals are also useful to help determine metrics to measure success and to set progress checkpoints along the way to ensure that everyone stays on track. For our purposes, "SMART" stands for:

Specific – The specific thing that the team plans to accomplish. It cannot be a general goal, such as "improve r-dvm relationships." The goal could be: "improve case communication with r-dvms, before, during and after a patient is transferred to our specialty practice."

Measureable – The success of the goal could be measured by pre- and post- surveys of r-dvms to check their satisfaction on case communication before, during and after a patient is transferred to your specialty practice. (If it is too hard to figure out a metric to measure success, then you probably have a fuzzy "S" and need to re-work the first step to be more specific and clear.)

Accountable – Name the party(ies) who are responsible for the project. What is the name of the "go to person" that is heading up the project? One of the biggest reasons projects don't succeed is lack accountability. Naming the project leader, or listing the names of all team members, ensures accountability for action.

Related to Mission – If the goal is not related to the practice's Mission, then there is no point in doing it. For instance, if your Mission was to be the preferred specialty service provider in your community, the goal to improve the referral experience for the r-DVM would be a good fit.

Time-limited – Put a start date on the project. Do not say, "In six months," because six months is always six months away. Instead, put a hard date on it, e.g., "Beginning June 15, 2014." Also set dates for progress checks along the way to ensure that the project stays on track and gets done.

REFERENCES & RECOMMENDED READING ON TEAMWORK

Creating Effective Teams, A Guide for Members and Leaders, Susan Wheelan, SAGE Publications ISBN 0-7619-1817-5 www.sagepub.com; The Five Dysfunctions of a Team, Patrick Lencioni, Jossey-Bass, ISBN 0-7879-6075-6. High Five! The Magic of Working Together, Ken Blanchard and Sheldon Bowles, William Morrow Publishing, ISBN 0-688-17036-6; Crucial Confrontations: Tools for resolving broken promises, violated expectations and bad behavior, Kerry Patterson, Joseph Grenny, Ron McMillian, Al Switzler, McGraw-Hill, ISBN 0-07-144652-4.

WHY COMMUNICATION MATTERS IN THE DIGITAL AGE

Eric Garcia
Tampa, Florida

The latest enhancements in technology are forever changing the way we communicate. Although it has become easier and more efficient for us to communicate, there is still a mindset that is withholding the willingness to adapt to all of the advances in communication. There are several reasons for the hesitation to jump on board with new technology. For instance, there is the cost factor and return on my investment factor. At times we are unsure if using new methods of technology make client interaction less personal. It is very true to say there are many things to think about before getting on board with a new method of communication with clients. However, it is also very true that the longer you wait the potential of getting new clients, generating higher revenue and the satisfaction of your clients' needs will be slow to rise.

In our practices today we are seeing a new generation of clients who demand and crave a change in the way we communicate. Starting with Generation X who was exposed to the start of the .COM era or in basic terms "the start of the internet". The very start of online communication was initially slow getting started and adapting to. At the time having this "luxury" was not easily affordable at home or in the work place. Time elapsed and we started to see prices drop and even more enhancements come along with the rise of the Net Generation also known as Generation Y. The affordability of the computer and Internet at home made it easier for this generation to heavily rely on it as a resource and a convenient form of communication. With computers getting smaller and becoming part of our phones, it made it has made it easier for us to stay connected online wherever we are at.

It is easy to see that our current generation was exposed to online communication for a majority of their lives and prior generations are quickly catching on. It's time to focus on the specifics of how our clients are communicating on the web.

ONLINE PATIENT PORTAL

Let's start off by understanding what an online patient portal is. An online patient portal communicates information from your practice management software into an online secure patient portal. Information that is transmitted usually consists of the patient's name, birthdate, reminder dates, prescriptions recently filled and great educational material. Depending on whom you choose as your online patient portal provider will depend on the information transmitted online. Clients register for their online patient portals by registering online via the clinic website or by inputting an email address into the client's electronic file. Once your client registers and accesses the portal they will see their pet's up to date information.

The benefits of an online patient portal greatly exceed the total cost. Clients who have access to an online patient portal help increase the compliance percentages in your practice. They are able to see when their pet is due for services and read about the importance of the service due all in one place 24 hours a day 7 days a week. Reminders are automatically sent to the client's email address on file. Prescriptions are listed allowing clients to easily request a refill reminder by email and even request the refill online. An online patient portal is full of other fun features for clients. They can upload photos, blog with other pet owners, read about pet care and more. By having an abundance of information on the patient portal for clients to view you are increasing the likelihood of the number of visit. Your practice name and information is made prominent throughout the portal so clients can familiarize themselves with you and your practice.

EMAIL REMINDERS

It is difficult to think that 100% of all mail makes it to its desired destination. An incorrect address can be the key reason for mail being delivered to the wrong destination. A majority of practices still rely on the postal services to deliver reminder cards to their clients. There is a good reason why. Statistics show the return rates on post cards are still very high. It is not recommended to remove post card mailings in your practice. However, with our new generation of clients relying on e-mail as their primary form of communication, it is important to implement an e-mail reminder protocol for your practice. If the reminder post card never makes it to the client the probability of that client coming in to satisfy their reminders start to decrease. Even though pets are a priority in house holds across America they still rely on you to remind them of what medical attention their pet(s) needs.

To implement e-mail reminders in your practice start off by contacting your practice management software technical support team to see if there is a feature already offered to you at no additional cost. If not, consider moving toward a patient portal where clients will greatly benefit from not only e-mail reminders but a variety of other online services. Finally, if all else fails consider getting started with your own method of e-mail reminders. Start off by contacting an e-mail messaging provider. Talk to your practice management software technical support team about extracting reminders and client e-mail addresses in one file excel file. This file can easily be uploaded into the e-mail messaging provider's database and sent out as an e-mail reminder.

PROPERLY IMPLEMENTING EMAIL COMMUNICATION

Have you decided if another great way to communicate with clients is by email because of it being most convenient for your client? If so, you are taking a great step forward to adapting to our new generation of client's preferred way of communicating. The need to understanding how to properly communicate with pet owners using email is extremely important for formality and legal reasons. Like any important information you communicate verbally to clients you start by documenting it in the pet's medical record. Make sure you are using a branded or professional email address. You don't want to communicate with clients using an email that is intended for personal reasons or that is too creative (kewldoc23@vetsrule.com). Try setting up branded email with your web hosting company. If you do not have a website, setup an email address with a reputable web company such as Google, Hotmail or Yahoo and keep it simple (egarcia.tampa@gmail.com or eric@simplydonetechsolutions.com). When communicating do not use short phrases or terms only medical personnel tend to understand. Spell out exactly what you are trying to say so the client can understand it in an email. If you are not going to be sending the email yourself make sure whoever you rely on communicating this information for you is trust worthy and able to write a great amount of information with no errors. Make sure you keep each email that is sent and received in your inbox as this will serve as proof of communication. Any medical recommendations you make such as a new treatment route, change of treatment, medical dosage changes, should all be printed and put in the medical record. If you store your medical records electronically, print your email to a PDF printer and attach it to the patient file as a communication note. Make sure you check and reply to emails daily. If you do not check your email daily, see if you can enable an auto reply to all emails sent to you that explains you have a 5-6 hour reply turnaround time. Finally, as a major DO NOT in email communication... Do not allow your clients to use this to get out of a visit in your practice.

TEXTING

Texting is yet another great way to communicate with your clients. However, you should know there are times when text messages are best used and there are times a phone call should be made instead. Below is a list of ways you can utilize text messaging in your practice:

Dos

Quick update on how a client's pet is doing while boarding or dropped off for a surgical procedure

- "Elvis is doing great Mr. Garcia! We will keep you updated."
- "Elvis is recovering from his neuter nicely. We will call you soon!"

Quick reminder of a service becoming due or over due

- "This is a reminder Elvis is due for his Bordetella vaccine! - Simply Done Vet Staff"
- "This is a reminder blood work is required before Elvis' next medication refill"

Quick confirmation of a prescription filled

- "Elvis' meds are now ready to pick up."

Don'ts

Full doctor update on how a client's pet is doing while boarding or dropped off for a surgical procedure

- "Elvis is getting ready to under go anesthesia. We just put his catheter in and we have him hooked up on IV fluids. He is in his cage waiting to go next."
- "We gave Elvis his vaccines and exam. Everything is normal except his left eye has some discharge so I'm going to prescribe an ophthalmic ointment to clear that right up. His blood work results came back a bit elevated and I suggest medication ABC to go home and be given tonight. Call me if with any updates."

Abnormal Blood Work Results

- "Elvis' liver enzymes came back elevated. I suggest we take an ultrasound and x-rays to work up the case more and find out the cause of it. In the mean time I have some medications up front for you to pick up when you are able to."

SOCIAL MEDIA

Social media can easily help to communicate with a variety of generations of clients. Websites like Facebook and Twitter easily link up together to share some of the same information making it a one stop commucation shop. Information is easily transmitted to your client's pages and phones (if they have the feature enabled). Having this steady stream of communcation is key to keeping in touch with Generation X & Y clients. To properly implement a Facebook and Twitter page in your practice follow some of these basic rules. For starters, do not relay ANY medical information or information regarding the care and health of a patient on Facebook or Twitter. These websites are not intended for that purpose nor are recommended. Use these websites to <u>market your practice</u> to your current clients and potential new clients. Add a few photos of the practice and the bond with your staff and clients. In your information block add a mission statment or the values your practice was built on. A majority of this information and be copied and pasted right from your current website (if you do not have a website, do not use Facebook and Twitter as an alternative). Leave facts on your page that help educate clients and also lead to a question about a service you offer but don't make it obvious. For example: "Wow! We just recieved "Kobe's" DNA results back from the lab and it turns out he is mixed with a Yorkie and Maltese!", then show a photo of the patient and ask "What mix do you think your dog is?!". This will lead to dialogue on your page and possibly even have a client question if you offer DNA testing. I stress the importance of not making it seem obvious that you are making it known you offer a service. The last thing a client wants on their page is someone (especially their veterinarians office) trying to sell them into a service. Social media websites can help drive conversations with clients outside of their pets annual vaccination visits. You could even open the doors to clients', friends and family adding you to their page creating a web of free advertsing.

Website on the Go!

We discussed how clients rely on the web as a primary method of communication. We know clients access the web on their mobile devices. So now discover why a mobile website is beneficial for your practice and clients.

If you currently have a website it is most likely able to be accessed on a majority of mobile phone devices. What you may not know is the time it takes for a full blown website with high qualitiy images to load on a mobile phone. The speeds of broadband internet on a mobile phone and a computer (depending on provider, service and type of internet connection) cannot yet be matched or slightly compared for the most part. This makes your website useless on a mobile phone for clients. A website designed for mobile use can be opened up in almost any phone browser that has internet AND at a much faster speed with no page zooming required. The mobile version of your website will have basic information about your practice such as: hours of operation, mission statement, services offered and a few contact pages. The mobile contact web pages are great for clients being able to request refills on prescriptions and put in appointment request times on the go! It makes it easier for your clients to do this when they are out and about or sitting somewhere waiting and playing on their phone. If they visit your website and find it takes forever to load, and is too complicated to naviage on a phone, they may move on. If they visit your website and see a mobile they will continue and proceed with further communication. Mobile websites are yet another thing you can do to make life easier for your clients and allow them to get in touch with you quicker.

Don't Be The Odd Ball Out

Don't be the only practice in your area who does not offer these new methods of communication to your clients. Be one of the many practices in your area who offer these services. If you are the odd practice standing out and your competitors offer these services it won't be too hard for a new generation client move in the opposite direction.

SOCIAL MEDIA AND PUBLIC RELATIONS

Eric Garcia
Tampa, Florida

Public relations in a veterinary practice may sometimes seem overwhelming and difficult to manage. The importance of public relations for a practice is significant. This is how you get your name across in the community you serve, the clients you see and most importantly future pet owners. How you currently manage public relations in your practice will determine how clients view you. If you have poor public relations you may not have the best visibility amongst your competitors. If you have well managed public relations your clinic will stand out among the rest. Think about any recent public relations you have done recently. If the answer is little to none then this is a great place to start. However, if your answer consisted of a variety of public relations tasks then its time to bring out some new weapons from the web technology arsenal.

Cost Versus Value Factor

It is important to know that public relations does not have to cost your practice a fortune. In the past you may have implemented a variety of one-way communication methods of public relations and found it difficult to find a value for it. One-way methods of communication in public relations is where you communicate to the public but do not receive any communication back. For instance, you may have sent out newsletters by mail or have given out brochures at a variety of events. You can spend a few hundred dollars easily in postage and print alone. The end result was either a noticeably higher amount of new clients and/or increased revenue for a service area (such as vaccinations). Often this form of public relations has been seen as "risky" and at times even deemed useless. It is important to know that public relations is not ALL about generating high revenue, it's about cultivating relationships in a community and with your clients. Although, revenue does play a factor and is often expected when a PR expense is involved. Thanks to the development of social media and web technologies, public relations has been made easier for us to manage and made extremely affordable. When you reduce expenses by using social media (which we will discuss next) you have to understand in exchange you must give time. A majority of social media outlets require you to be... well... social. You have to communicate daily, upload photos, write stories and interact. The bond you create after time with clients and the community will greatly exceed any time or paid employee time spent. Yes, revenue will increase if you communicate correctly. Do you see your name in lights yet?

Social Media Types

For starters social media is media designed for social interaction. Social media does not have to be strictly online but for the purpose of this article we will focus on web based social media technologies.

The types of social media we will discuss are:

Content Based Social Media - Built around individual messages.

For example: Youtube is designed to host videos to share with others & FlickR was designed to host photos and share with others.

Personality Based Social Media - Allows each member to create a profile that includes a description and can be linked to other profiles of friends and colleagues eventually forming a network.

Interest-Based Social Media - Revolves around websites that are topic oriented such as Yahoo! Groups or Google Groups and even blogs on specific topics.

What's Happening?

Interest-Based Social Media revolves around websites that are topic oriented such as Yahoo! Groups or Google Groups and even blogs on specific topics.

If you never knew what Twitter actually meant, it simply asks you "What's Happening". Yes, it all makes sense now doesn't it? The catch to Twitter is it's not really asking you for a story or a blog entry of what's happening... it's asking you to reply in no more than 140 characters. Keep it quick and simple to follow. Twitter users can follow other users in numerous ways through a text messaging service, smart phone, laptop and PC to name a few. Sign up for a Twitter account, it's FREE. Use your full practice name (or abbreviate it if it is too long). For example: Simply Done Tech Solutions can be Simply Done Tech or ABC Animal Hospital can be ABC Animal. When customizing your page try to stick with the colors of your practice logo or practice. Simply state information about who you are and what is the purpose of your Twitter page. Your Twitter page can be used educate the public on pet health care or updates about your practice. Once you are all ready to go its time to Tweet (that means to tell everyone "what's happening"). Some practices have made their Twitter a place for their practice mascot. Basically, you can pretend your Tweets are coming from a dog or cat, which makes it cute and fun to follow. Like anything in veterinary public relations you NEVER want to come off as someone who is selling something. This doesn't mean you can't attempt to drum up some business, just don't be forward with it or you will turn people away from practice. Here are some sample Tweets:

"Getting ready to go to the dog park! Glad my shots are up to date, I am safe to play with all of my friends now ! Hope it doesn't rain..."

"I hope I stop itching!! Flea's are bad :o(I wish my mom didn't buy my flea stuff on-line, it doesn't work like the one from the vets!

Raining again! I hope I get to go for a walk today. I have to remember not to drink from the puddles of standing water, last time I got sick"

"FOOD RECALL... read more"

"we have plenty-o-kittens for adoption... call today to see one (813) 545-3044... tell 'em Bark referred you!"

"uh oh... hurricane season! I've better get my bordetella vaccine just in case I have to go to a hotel far away... make sure you are up2date too!"

Twitter is an effective and fun way of public relations and it cost you nothing. The idea is when you post, your clients will be able to follow your practice wherever they are located. If you post something extremely cute they will show friends and family. If you post something such as new services they weren't aware of they will call you to inquire about that service. Every Twitter is different... get creative with it.

Pro's

- The "Timeline" (or the Twitter feed) is public, which can help to give your business or organization greater exposure in web search results.
- The site is more open than Facebook, so it's easier to build a community of potential customers you don't personally know.
- Fans can "follow you" without you having to reciprocate.
- There is a quick way to share links to content.
- Users are very vocal, so if they are happy with your business or organization, they'll say so.

Con's

- The site is text only — pictures and video are shared through secondary links.
- It's challenging to say something of significance in only 140 characters.
- Lots of "noise." With so many identical-looking Tweets, it's hard to make a single one stand out in the crowd.
- Spammers are increasingly targeting the service.
- Users are very vocal, so if they have a problem with something, they'll say so.

Facebook

Personality Based Social Media allows each member to create a profile that includes a description and can be linked to other profiles of friends and colleagues eventually forming a network.

Facebook is a social networking website that allows members to create profiles about themselves and a business or organization they wish to represent. The amount of people on Facebook exceeds 400 million users, which answers your question: Yes, your clients are on Facebook. Facebook has many benefits for your practice's public relations plan. By creating a page for your clients to become a part of or "Like" your page, you are centralizing a place for clients to meet online. This makes it easier for you to relay any messages to clients you want to get across. Like Twitter you will want to never come off as selling something. You want to genuinely be there to care and communicate for your clients. The best way to get started with content on your Facebook is to copy and paste information from your practice website. Include a mission statement, a small list of services you offer and photos of the practice and practice employees with patients (after receiving consent from both the employees and pet owners). Starting dialogue with clients is simple by writing a small article on pet care they can relate to. Create a list of topics you can briefly write about and implement them. Some examples are: importance of vaccinations, heartworm prevention, yearly exams, dental care, microchipping, etc. This will allow clients to pipe in their own opinions and allow you to communicate back freely. On Facebook you can share websites you enjoy, videos and/or write short statements like Twitter. Facebook can also serve as a revolving door of reviews. When clients have a positive interaction with your practice they may post something on your page so other clients can see and even those who are not current clients but seeking can view as well. Any negative feedback can be removed immediately and you can handle the issue internally. Best of all you can allow Twitter and Facebook to link up. What does this means? Anything you type in Facebook automatically goes on Twitter. It's a 2 for 1 deal and it's free.

* It's important to know that before you can create a Facebook business page you must already have a Facebook account for yourself. If you don't really want to utilize it for yourself, just type in your information at the setup page and in the place of your name type the name of your practice and give your practice general email address.

Pro's

- The user base is huge, and that means many of your customers and constituents are already there.
- It's easy to use.
- Multimedia content can be integrated with your profile.
- You can separate your personal and professional use.

Con's

- You have limited ability to customize your Fan Page.
- Fan Pages do not have email alerts; you will have to check to see if there is any activity.
- It's a closed environment — only Facebook users can become fans or friends of your business or organization.

You Tube

Content Based Social Media - Built around individual messages. For example: You Tube is designed to host videos to share with others & FlickR was designed to host photos and share with others.

You Tube in your practice can be a great educational piece for your clients and those in the community. If you are not camera shy then this is definitely for you. Start off by picking a few topics to briefly speak about such as: Puppies & Kittens, Senior Pet Care, Behavior Training, etc. Then setup a camera or web cam and use an area of your practice (an area you don't mind showcasing online) and press record. Simply watch the video and confirm that it is to your liking (professional, not to long & point made). This will be a great place to direct your clients to learn more about animal health. The benefits of doing it yourself are that it allows you to place your practice information all over the page and video. Clients will watch the videos on their free time at home and then click the share button so their neighbor can get great information on pet care. Each time your video is shared the larger your network becomes and the higher number of views you get. A bigger network with your practice name all over it is definitely good public relations. You will make your clients proud, cultivate a smarter client & even gain a new client. This service is also free.

Why Social Media?

Now that you know the basics about which social media networks are which and what's notable about them. Here are some reasons you should try it out.

- They're free
- They're popular
- They're for all ages
- They allow you to be personal and professional
- They extend your brand and your relationship with customers
- They're everywhere!

E-Mail Marketing

Before we start discussing email marketing, lets take some time to understand the value of a client. First, you've already paid for them and it is 6-7 times more expensive to gain a client than it is to retain a client[1]. Repeat clients spend 67 percent more. Finally, they are your referral engine because after 10 visits, a client has already referred up to 7 people[2].

Now that we understand the value of a client let's discuss why we are choosing email as a method of sending newsletters.

- 91% of Internet users are between the ages of 18 & 64 send or read email[3]
- An even higher number of uses age 65 or older do the same [4]
- 147 million people across the country use email, most use it every day [5].

Email marketing is not only effective for client communication but it is also cost effective. For the same information to be communicated direct mail will cost you 20 times as much as email. The amount of time it takes to prepare an effective email newsletter is one hour or less (depending on content). Proper email marketing should deliver professional email communications to an interested client base that contains information they can find valuable.

Build your email database. Start off by placing clipboards with sheets that ask for the clients full name, email address and a place for the client to consent to signing up for receiving email from your practice as a form of communication. Place those emails with the client's file in your practice management software. Have your staff frequently asking clients for their email address. Collecting an email address should become apart of your checking for updated information routine. Add a "sign up for our newsletter" button on your website to collect email addresses 24/7. Most importantly, stay on top of collecting email addresses. The higher number of email addresses you have the larger your audience.

Send a welcome email for each group of new client email addresses you add to your email marketing database and be sure to include your practice logo, name and colors. Add a personalized message welcoming the client to the practices' email community. Discuss and reassure clients how important privacy is to you and that you will not sell, rent, trade or give your clients email address to anyone.

Now for the meat and potatoes of a routine email newsletter. Content has to have value to your clients. Three areas to try and focus on for each email is promotional, relational and informative. Promotional may include simple discounts or offers such as Frontline is buy 6 get 1 free or save 5% on a year supply of heartworm prevention. Relational are special privileges and acknowledgements for receiving the email for example discuss how email clients are the first to know about important pet care issues.

The last area to focus on is informative which gives clients advice on pet care, general information, updates on the latest advancements, facts, etc. Coming up with valuable email content is simple. Share your expertise, use facts, give guidance and direction, offer minimal discounts or coupons, and acknowledge your clients in each email. Determine the appropriate timing for your

[1] Source: Harvard Business Review
[2] Source: Bain and Company, 2002
[3] Source: eMarketer
[4] Source: eMarketer
[5] Source: eMarketer

newsletter. Will you send it out weekly? monthly? quarterly? etc. What area do you want to focus on? Will it be dental? allergies? spay and neuter? Whatever you choose make sure you site your article, check for legitimacy and don't choose anything controversial or political. The last thing you want to do is have your clients distrust you for not taking the time to put thought into your newsletter.

USING FACEBOOK FOR MAXIMIZED SUCCESS

Eric Garcia
Tampa, Florida

1. Why should a veterinary hospital use social media?

The real question should be, "Why wouldn't veterinary practices use social media?"

Social media is the form of communication that gives your message the largest reach, one that includes not only your clients, but your prospects, too.

What happens when you mail out postcards or give your business card to people who don't need your services right away?

We all like to think they save the information, but the reality is that when they do need you, that card is nowhere to be found. So they go to the Internet, and you can only hope they find you there.

If your practice has a social media (which you also listed on that business card), the prospective client can follow it or subscribe to its feed, and that puts your practice name in front of them several times a week.

This can help stimulate reminders about refilling heartworm and flea medications, coming in for annual vaccines and other things. Seeing your name is a reminder of the services you offer, and when they need those services, they can easily find you by logging into social media.

Used correctly, social media can actually increase compliance with pet owners.

2. What's the best way to use social media to stimulate business?

Most consultants will tell you not to use social media as a place to sell. I agree—the minute you start to sell to clients in "their" space, which is where they want to avoid advertising, they'll unsubscribe from your feed.

This was what businesses did when social media first took off, and they learned the repercussions quickly. But you can market services to clients through social media if you do it correctly.

I always tell people there are three parts to using social media successfully:

- A mention of a service you want to market
- Some sort of fun and engaging fact
- The most important thing: the social part

Here's an example: "Wow, we just received Penny's DNA results back from the Mars Veterinary Lab, and it turns out she's 40% Yorkie and 60% Maltese. What do you think your mixed breed is made of?"

You can easily see the service we're trying to market, but we made it cute. People look at the photo of Penny, think about her breed and socialize back with you. And then, they might call you to ask more about DNA testing, maybe schedule an appointment outside of a routine visit or share it on their neighbor's page or their own wall. And what's attached to that posting? The practice name and page, so now everyone can see the services you offer.

The average Facebook user has 140 friends. Say your practice has 100 followers, and three of those people share one of your posts. One hundred of your followers, plus 140 friends of each of those users equals a reach of 520 people! The possibilities are endless.

3. How do you ensure that social media messages are on-point and appropriate?

The best way to do that is make sure that a few people on your team go through training. Visit a conference and spend a little extra for the all-day social media lab, or hire a consultant to come in for one-on-one training. You may have a technician or receptionist who knows how to use social media, but knowing how to use it and how to communicate properly and effectively with clients are two totally different things.

I don't suggest outsourcing social media, by the way. When you outsource social media, it becomes an argument about return on investment. If you spend a few hundred dollars a month having another company communicate on your behalf, at some point you're going to start questioning the value of that communication.

It also makes no sense to outsource social media because the whole point is to honestly engage in conversation with your clients. How does it look if your clients find out you've hired an outside company to do that? It becomes fake social media—it's a lie.

4. How often should a veterinary practice post to social media? Which channels are best to focus on?

If you're just getting into social media, I suggest two postings per week. It's a realistic number. Start small and work your way up. If you've been doing this for a while, ramp up to three to five postings per week—not too few, but not overwhelming.

As far as the social channels:

Facebook:

Use it. Most users log on multiple times a day.

Twitter:

Use this, too. There are many users and you can easily link to Facebook if you don't have the number of followers to engage in true Twitter talk.

Pinterest:

This is still too new to be effective for veterinary practices. If you're an advanced social media user and want to dabble in this, go for it. If you don't have the time or are new to this, then hold off. Don't rush to be part of every new channel—you'll burn out.

Google + :

This is also still very new, plus there are many spammers. Here's how I gauge whether this channel is worth it or not: If Dr. Marty Becker has 881 people in his Google + circles and he's on top news channels and travels the nation to spread the good veterinary word. What would this look like for a veterinary practice? One distinct difference between Google+ and Facebook is Facebook appeals to every demographic with strong numbers. Google + can't say that yet. However, Google itself is an unstoppable force and you do not want to miss out on the opportunity to be part of the world's largest search engine's social network. I suggest mastering Facebook first then tap into Google+.

5. What's my 30 minute a week strategy for social media?

Make sure you preplan your marketing. This should include what your focus is (such as dental care, laser therapy, senior care). Pinpoint the dates you want to post and put them on a calendar. Finally, plan your resources for each topic: Visit a few websites in advance and copy/paste those links into a Word document. This way, you'll know when you will post, what you will post, and where to get a good reference to share when you do.

Most practices stop posting on social media because they wind up sitting there, trying to think of something productive to say. They waste time and lose sight of effective postings.

GETTING STARTED WITH INTERNET MARKETING

Eric Garcia
Tampa, Florida

Gone are the days when all you had to do was trust that word-of-mouth, a Yellow Pages ad and a few brochures or fliers would be enough marketing to bring the clients in your door.

Today's savvy business owners and managers are depending on technology specifically social media to keep their brands top-of-mind with their target audiences. If your practice has been trying to avoid jumping on the social media bandwagon, 2013 is the time to set your fears aside and go for it.

By now, you've no doubt learned the power of social media. It's something that companies of every size, in every part of the world, are embracing in order to remain competitive. For many of us, however, the thought of social media is quite intimidating.

How often should I post on Facebook? On Twitter? Do I need a blog? What if someone posts a negative comment on one of my social media platforms? How will I make the time for me and my staff to actively participate in social media? Does social media replace traditional forms of marketing?

The questions go on and on. Here are five social media tips to get you started on the right foot, with input from social media expert Eric Garcia.

1. Update your website.

Through social media, you'll often drive current and potential clients back to your website. Make sure your website is updated, easy to navigate and looking its best at all times. A bad website could turn off potential clients.

The expert (Eric Garcia) says:

"Prominently display all of your social media links on every page of your website so clients and prospects can easily link to the various social media channels you are part of."

2. Interact with followers.

It sounds crazy, but social media is about being social. You or someone from your staff should be engaging with your followers at least a few times per week. Comment on other Facebook pages regularly and respond to the comments your followers leave on your page.

Do not delete comments, even if they are negative. Unless comments are inflammatory in some way, simply respond to them and set the record straight for other followers who might have seen the initial comment.

Post things that initiate a conversation between your followers—link to interesting articles or photos and ask how your followers feel about them.

The expert (Eric Garcia) says:

"Engaging with your followers will constantly put your practice name in front of people. Clients and prospects don't follow you on social media to hurt you, but rather to support your practice. The likelihood of negative comments is very small. One way to engage followers is to post a photo of clients and their pets and ask your followers to caption the photo. Facebook, for any practice, should contain a fair mix of education, fun and personalized socialization.

3. Post daily.

Ideally, your business should be posting on at least one of the social media platforms every weekday. Content can, and should, vary.

Some posts could include new items related to pet health care in your area, interesting images from your practice (unique X-rays are always a big hit), fun pet-related videos, blog posts, etc. Sharing popular posts from other industry pages is a quick and easy way to come up with a post for the day.

The expert (Eric Garcia) says:

"Post daily by sharing the same content across all of your social media channels and remain consistent. If you post a blog on Facebook regarding dental awareness, make sure to post the same blog on Twitter and share a related video on YouTube to keep the message consistent."

4. Consider a blog.

Although a bit more time-consuming, a blog is a great way to connect with your clientele. It gives you the chance to show them who you are on a more personal level. Your blog can go in any direction you want it to, as long as it is a reflection of who you are, letting your clients feel like they are getting to know you, your staff and your practice.

Write posts in the first-person point of view, using many photos, videos and other forms of multimedia. Also, link to relevant sources and related articles as often as possible. Blog posts can feed your other social media efforts—after a blog is posted, link to it in a Facebook post and a tweet. Ideally, you should post to a blog weekly.

The expert (Eric Garcia) says:

"Blogs are a great way to enhance your online presence through organic search engine optimization. When posting blogs, make sure to tag and categorize each blog post under a related category. And don't stop after posting your blog – share it on various social media channels; be sure to tell your clients about the blog in person as well."

5. Supplement your marketing.

Social media should complement your traditional forms of marketing, not completely replace them. Make sure all of your marketing efforts reflect a consistent brand—colors, fonts and imagery should be similar throughout.

The expert (Eric Garcia) says:

"Don't try to become part of every social media channel at one time. Take the time to focus on one social media channel and do it really well, then start exploring additional channels to be part of." There's never been a better, or more important, time to make social media a part of your overall marketing strategy. Don't let it scare you. Start slowly and work your way up to meet your posting goals. Before you know it, it'll be second nature.

BLENDING OLD AND NEW MARKETING STRATEGIES

Amanda L. Donnelly, DVM, MBA
Tampa, FL

Veterinary practice leaders typically want to grow their practices and strive to implement marketing initiatives that will result in increased revenues. Unfortunately, efforts by many veterinarians and managers fall short due to a lack of proper planning. How do you decide where to spend money and your time? The answer is to engage in appropriate marketing planning. It's much better to have a strategic approach to marketing rather than just trying initiatives because they are popular or you hope they'll help you achieve the goal of increased revenues and enhanced profitability. Proper planning includes making assessments about the current status of the practice, evaluating the competitive environment and analyzing hospital data in order to develop marketing strategies that are right for your practice.

STRATEGIC PLANNING

Organization and planning is the foundation for successful marketing plans. Before writing a marketing plan, engage in strategic planning on an annual basis. Strategic planning starts with developing a clear vision and/or mission for the practice if you have not already developed these statements. The mission statement usually defines the core purpose of the organization while the vision is generally a loftier statement that creates a picture for where the practice owner wants to be- it is more long-term in focus. Branding is also part of this process-what do want to be known for? What is the practice's brand promise? Marketing initiatives should be aligned with the mission and vision of the business. Don't forget that the entire healthcare team needs to know and understand the vision of the practice so they can fulfill their critical roles in achieving practice goals.

Strategic planning also includes performing a situational analysis or SWOT analysis which refers to strengths, weaknesses, opportunities and threats of the business. Strengths and weaknesses involve an internal assessment of the business while opportunities and threats involve an external focus on the competitive environment. It's beneficial to involve staff in the SWOT analysis since team members often have knowledge, creative ideas for marketing and excellent feedback that can augment the assessments of the leadership team.

A thorough SWOT analysis helps determine appropriate and effective marketing objectives. Carefully consider the information gathered from the SWOT analysis. For example, what did you discover about your client service? Is it mediocre or exceptional? If the SWOT revealed that client service is a weakness due to some communication problems, then marketing efforts should center on improving service. Likewise, if a potential threat to the business is reduction in caseload for a particular service, then marketing efforts should center on how to minimize or overcome this threat. Perhaps the practice needs to consider enhanced education for referring veterinarians.

TRACKING KEY PERFORMANCE INDICATORS (KPIS)

Analyzing financial data and hospital data as well as trends in the data helps to guide marketing decisions. For example, are revenues increasing, decreasing or stagnant? Which profit centers are stagnant or decreasing? Is the number of transactions per month stagnant or declining? Answering these questions and many more will help you know which marketing goals make the most sense for your practice. Let's take a look at some of the key performance indicators or KPIs to routinely monitor.

Revenues

Track revenues on a monthly basis and compare to the same month from the prior year. Don't forget to look at trends. A slight dip in revenues for one month may not be a concern but a steady downward trend is a problem. The sooner you notice a decline in revenues or a downward trend, the sooner you can take action to improve revenues. Regardless of whether revenues are up, down or stagnant, further analysis needs to occur to see what is driving the numbers. For example, revenues might be up to a price increase that was instituted the prior month or quarter. But maybe the number of client transactions is down. Don't be complacent and miss the fact that the number of clients (referring veterinarians) or pets being seen is declining.

Doctor Production and Average Client Transaction

The amount of revenue generated by each doctor in the practice should be tracked on a monthly basis as well as their average client transaction (ACT). Changes in these figures may explain changes in the total revenue for the hospital. Sometimes practices that pay their doctors a salary without any bonus based on production are remiss in reviewing doctor production data. Don't make this mistake. Doctors must have adequate production of revenues for the hospital to be financially successful. The average client transaction can be an indicator of how effective doctors are at communicating the value of services to pet owners. If you notice a substantial decrease in doctor production or ACT, it is time to investigate the underlying cause for the change. If you notice considerable variations in the data for different veterinarians in the practice, see if you can determine why one doctor is able to produce more than his or her colleague.

Income For Profit Centers

When reviewing total revenues, analyze the revenue by service category or profit center to gain valuable information about which services are more or less utilized at your practice. Service categories include exams/consultations, imaging, laboratory testing, endoscopy, oncology, etc. By looking at the revenues for different services, you can begin to see areas of opportunity for establishing marketing goals.

Client Numbers

Be sure to track client data for the hospital each month. Since most of your caseload likely comes from referrals from area veterinarians, be sure to look at the number of cases sent each month from each practice and each veterinarian within the practice. Look for trends in the data-are referrals increasing or decreasing? Track the source of referral for all patients. For example, do you accept pet owners without a referral? If so, track how they decided to come to the practice. If you participate in community events, track whether this brings in pet owners who request your services.

ESTABLISHING MARKETING GOALS

Once you evaluate your practice management reports and hospital data, you can identify which marketing goals make the most sense for your hospital. For example, if your oncology caseload is decreasing, you should establish marketing goals specifically aimed at attracting new referrals from area practitioners. When establishing marketing goals, decide on a reasonable number of goals for each quarter or for the upcoming year. Practices that focus on two or three goals tend to do a better job of executing marketing tactics to reach these goals than practices that try to do too much. Team members become overwhelmed and frustrated trying to achieve an unreasonable number of marketing goals

BLEND OLD AND NEW MARKETING TACTICS

The next step after establishing written marketing goals is to formulate specific action steps or tactics to achieve each goal. The key to success is to stick with tried and true initiatives and combine these tactics with new strategies. Here are three rules to help blend old and new marketing tactics.

Stick With Marketing Strategies That Work and Get Rid Of What Doesn't

Carefully assess what's working and what isn't at your practice. List all your marketing actions looking for both successes and poor results. For example if you sent a referring veterinarian client mailing and had limited response then this initiative should be abandoned or tweaked to achieve better results. Don't forget that some initiatives are a must such as those related to improving service and client communications. Carefully evaluate whether your client communication and client service efforts are effective in gaining more referrals and enhancing pet owner compliance.

Modify or Update Existing Marketing Tactics

Many practices focus marketing efforts in the right place but don't achieve the desired results. In this instance, it's important to evaluate your tactics and modify or add tactics to your existing marketing efforts. For example, if the leadership team believes in the value of exceptional service but the practice is still getting client complaints (from rDVMs or pet owners), then it's time to determine the underlying problem so you can implement new protocols. You could survey clients to gain feedback and then implement new service standards. If you host continuing education meetings for community veterinarians and have limited attendance, then it's time to make a change. You may need to change the venue, time or content of your program. Another example is perhaps you realize the number of referrals has declined from 2 area practices. In this instance, your marketing tactics may include taking the owners to lunch as well as hosting a series of lunch and learns for the veterinarians and staff at these practices.

Add New Marketing Strategies

Look for opportunities to implement new marketing initiatives to attract more clients and increase service utilization. This may include offering a new service, purchasing new equipment, adding new programs or protocols, enhancing efforts to increase pet owner compliance with treatment plans or doing more community outreach. With all your marketing tactics explore ways to leverage the use of technology and take advantage of available resources outside the practice. Many companies have affordable and efficient tools to help you market your services and communicate better with clients.

MEASURE THE SUCCESS OF YOUR MARKETING INITIATIVES

How will you know if your marketing initiatives were successful if you don't measure your success? Why spend money or time on marketing efforts that yield minimal or no results for the practice? To determine the success of your marketing, calculate the return on investment for all major marketing expenditures. Total the amount of money spent as well as time involved and see if the amount of revenue generated or the number of new clients you gained was positive. This will help you decide if you want to continue similar marketing initiatives in the future.

TAKE THESE STEPS TO WOW CLIENTS AND GAIN MORE BUSINESS: SPECIALTY FRONT OFFICE STAFF

Amanda L. Donnelly, DVM, MBA
Tampa, FL

The delivery of exceptional client service doesn't just happen. Team members need to know and understand that creating a positive service experience for pet owners is part of their job. Employees have to be committed to serving clients. When team members provide excellent service clients are more likely to feel bonded to the practice and refer others. Pet owners with sick pets often need someone to explain complex medical conditions in a manner that they can understand, payment plan options, realistic information regarding patient outcomes, someone to validate their emotions and someone to convey empathy. Busy doctors and staff members sometimes forget to see the client's perspective when communicating with clients. Remember to focus on what clients may be feeling and what they need from the healthcare team as well as what their pet needs.

FOCUS ON CLIENT ENGAGEMENT

In their book, Human Sigma, authors John Fleming and Jim Asplund define customer engagement as addressing a person's functional requirements and emotional needs. Engaged customers they say have an emotional connection to the business. In the case of pet owners, their functional requirements are that they need pet healthcare services for their pets. Their emotional needs include a desire or need for empathy, peace of mind, understanding, trust and compassion. Anchor your team to meeting the emotional needs of clients, not just their functional requirements. Engaged clients will be more receptive to team members and more likely to hear client education messages. Train employees to first and foremost focus on building relationships and trust with clients. When pet owners trust the specialty team they are more likely to agree to treatment plans.

Ask Questions to Connect With Clients

Asking questions is done every day in veterinary practices. We ask clients questions about their pet's medical history, diet, and medications. We ask for contact information, best time to schedule recheck appointments, whether they want liquid or pills for take-home antibiotics and what form of payment they will use. We generally ask questions to gain medical information and to clarify expectations or processes. Where team members sometimes fail to ask questions is as a means to build rapport with pet owners and to develop stronger connections with clients.

If you think about social situations or even business meetings, are you more drawn to individuals who ask you about yourself or your opinion? Or do you prefer people who monopolize the conversation and appear to care less about you as a person? Most of us respect and appreciate family, friends, colleagues, co-workers, supervisors, sales personnel, vendors, service representatives and doctors who show an interest in us and ask questions to get to know us better.

It's appropriate and beneficial to ask questions during all client interactions. This is not to say that you should pepper clients with questions or make them feel they are "being grilled". But look for opportunities to gather more information, gain insight, and create stronger connections by asking questions. It often only takes a few minutes longer and this communication skill helps you develop more meaningful client relationships which leads to greater client loyalty.

One of the best ways to engage clients is to ask open-ended questions. Open-ended questions demonstrate interest which helps build relationships. Examples of open-ended questions include:

- Why did you name your cat "Peppermint"?
- What fun things have you been doing since we saw last saw you and Jake?
- You seem uneasy Mrs. Taylor. Can you tell me more about your concerns?
- I think Papillons are so adorable. How did you decide to get one?
- What toys does Sadie like to play with?

Give Clients Compliments

Everyone likes to receive compliments so look for opportunities to say something nice to clients. Always praise clients about how well they take care of their pet because this will reinforce your position on patient advocacy and show you care about keeping pets healthy and happy. For example, you may be able to comment on what a great job a client did committing to a weight loss program for their cat. Or you could compliment a client for being consistent with giving their elderly poodle 5 different medications for heart failure. Don't forget to compliment pets since owners love to hear team members show interest in their pets. You may be able to comment on how attractive the pet is, tell the owner you like their collar, or compliment the pet for being so cooperative and friendly. You don't have to limit your compliments to praise or comments about the pet. If you like something about the client let them know. You may give compliments about their clothes, jewelry, hairstyle, community involvement, quality of their work, or their sense of humor.

TALK WITH CONFIDENCE ABOUT MONEY

Often, the client relations staff receives the brunt of client complaints about fees since they collect deposits and receive payments. Don't think you need to apologize for the cost of services but instead remain committed to providing the best care possible to pets and focus on communicating the value of this care to pet owners. Inevitably pet owners express concern or complaints about fees when presented with treatment plan recommendations. When this occurs, convey empathy, identify possible other underlying causes for an owner's reluctance to agree to services and focus on how to overcome objections.

Since many pet owners express concern about the cost of veterinary services, it is important to differentiate an individual's ability to pay vs. willingness to pay. Often clients are open about their financial constraints and in these instances, discussion of treatment plans and fees is dictated by the client's financial limitations. Specialty practices routinely offer third party payment plans. Make sure you are familiar with the plans offered at your practice and comfortable presenting these plans.

COMMUNICATE THE VALUE OF SERVICES TO CLIENTS

When clients perceive a true need for veterinary services and understand the value of services, they are more likely to comply with recommendations. The front office team plays an integral role in reinforcing messages delivered by the technical team and in reassuring pet owners that they are making wise decisions. For example, if a pet owner leaves a pet for diagnostics, a client service representative could say "Mrs. Smith, I know you'll feel much better once you know what's going on with Scooter." And remember that much of the value clients find in visiting your practice is related to their service experience. Client service representatives play a pivotal role in creating a positive, memorable experience for pet owners.

PHONE SKILLS THAT ATTRACT CLIENTS

Anyone who has ever called a large association or company knows how frustrating it can be to listen to all the different options of the phone tree when all you want is to talk to a real person to answer your questions or help you. And then when you get a person on the phone, it is even more aggravating when they seem disinterested and unfriendly. Veterinary practices are small businesses and the phone can be an excellent tool to create WOW customer service especially since most people have come to expect mediocre phone interactions with businesses.

Developing outstanding phone skills is extremely valuable since it is often the first impression clients have regarding the practice. Clients begin to make assessments about the practice based on their interaction with staff on the phone. Communicating effectively with clients on the phone presents challenges since neither person on the phone can rely on visual cues to enhance communication efforts. It is helpful to create service standards for the phone to ensure consistency and to raise the level of client service. Here are examples of phone skills and phone standards that will improve client communications:

- Always answer the phone by the 3rd ring. When the phone rings more than 3 times, clients may become irritated and feel the practice does not value the importance of providing exceptional client service. To ensure the phone is answered by the fourth ring, consider dedicating specific employees to answer telephones.
- Do not quickly rattle off a greeting that the client cannot easily understand. Identify the name of the practice, your name and ask how you can help the client. This greeting should be articulated clearly and concisely.
- Use a friendly, relaxed and upbeat tone of voice when answering and speaking on the phone. The person on the other end of the line should immediately feel like they are talking to someone who is warm and interested in helping them.
- Ask if you may place someone on hold and don't put clients on hold until you have heard them verbalize "yes"
- Upon returning to calls that you have placed on hold, remember to apologize if you have to place someone on hold for any significant length of time. People are less agitated by being on hold if you say "I'm sorry for keeping you on hold" or "I'm so sorry you had to wait on hold. I didn't forget about you". In addition, thank the client for their patience.
- If you know that it may take some time to gather the information that a client wants, always ask if they would rather be on hold or have you call them back.
- Offer assurances to clients on the phone. For example, this may be an assurance that you will give a message to the doctor or an assurance that you will check on when their pet can be discharged and then call them back.

TAKE THESE STEPS TO WOW CLIENTS AND GAIN MORE BUSINESS: TECHNICAL STAFF

Amanda L. Donnelly, DVM, MBA
Tampa, FL

Technical staff members at a specialty practice play an integral role in creating an exceptional client service experience for pet owners. Technicians typically spend a significant amount of time with clients in exam rooms and on the phone. And pet owners know that technicians provide much of the nursing care for their pets so trust and peace of mind is vital to a positive relationship between clients and technical staff.

USE COMMUNICATION SKILLS TO BUILD TRUST AND RAPPORT

If you practice specific communication skills, you can build trust and rapport with clients which leads to greater client loyalty and increased client compliance. And an added benefit is that you can reduce the stress associated with some client interactions if you hone your communication skills.

Ask open-ended questions

Questions can be either close-ended or open-ended questions. Close-ended questions can be answered with a yes/no response or a simple phrase. There is nothing wrong with close ended questions. Just understand that they are not particularly helpful in creating dialogue or enhancing emotional connections with pet owners. On the other hand, open-ended questions demonstrate interest which helps build relationships. Examples of open-ended questions include:

- "Tell me how you are doing giving Sophie her medications?"
- "Tell me how Bella is doing since you were in last month?"
- "What concerns do you have about the treatment plan I've presented?"

Remember that when you ask open-ended questions, it's critical to allow clients time to respond and to listen attentively. This shows you care about what they have to say and weren't just asking the question to be polite.

Non-verbal Communication

Non-verbal communication can be just as important if not more powerful then verbal communication. Regardless of what we say, our body language also tells a story. It is imperative that team members be mindful of the message that is received by clients as a result of non-verbal communication. For example, folded arms, a frustrated sigh, or frowning may indicate impatience and frustration with the client or pet. On the other hand, smiling, nodding of the head, and leaning in toward the client demonstrate interest and patience when the client is talking.

Eye-contact is one of the most significant non-verbal communications. Communications can become impersonal and clients won't think you're very interested if you don't make eye-contact with them. Be sure to establish eye-contact with clients when you ask them to follow you to exam rooms, when you greet them in exam rooms, when you ask clients questions and when you say good-bye.

Try to sit down next to clients during important discussions. This helps make clients feel like they are a partner in the care of their pets and facilitates easier dialogue without distractions. When you stand while clients sit, clients may feel intimidated and be less likely to ask questions. In addition, it is easier for team members to focus on the client when sitting down to talk.

In addition to your own non-verbal communication, observe the non-verbal communication of clients which will give you clues about their feelings and what actions you may need to take to improve communication. For example, glancing at a cell-phone or watch, pacing, folded-arms, hands on the hip, standing instead of sitting in the exam room, standing near the door, and frowning can all indicate the client is in a hurry or unhappy about the wait time. Clients who are afraid or uncomfortable for some reason may display nervous behaviors such as clutching their pet, looking down, reluctance to talk, or fidgeting in their chair. Once you become more observant of client's non-verbal communication, you can take action to respond to their body language.

Practice Reflective Listening

Some people are naturally good listeners and some people have considerable difficulty with this communication skill. If you know you tend to always think about your response while some is talking or you tend to interrupt others when they talk, you may need to practice developing better listening skills. Listening to clients is important for two primary reasons. First, it is often critical to ensuring that you have gathered all the relevant medical history and information about the pet and client. In addition, attentive listening demonstrates interest and compassion towards the client.

Reflective listening involves using statements to let the client know that you heard what they said and ensure that you have accurately understood the client. Reflective listening helps to make sure that we have listened to clients, that we are processing the correct information and that we are cognizant of what pet owners are feeling and thinking. Examples of reflective listening statements include:

- "I'm hearing you say that you think Oliver's quality of life is not good, is that correct?"
- "I understand that you have some budget constraints for what you can spend today."
- "If I am hearing you correctly, it sounds like you aren't sure which treatment option is best for Jake?"

Reflective listening statements invite clients to affirm that you have understood them correctly. If you have made erroneous assumptions, clients have the opportunity to clarify their thoughts and feelings. In addition, reflective listening invites client to give you more information. This can be critical particularly when you are engaging in dialogue about serious medical conditions and need to ascertain the level of understanding by the pet owner and/or their willingness to care for the pet.

Use Empathy Statements

While veterinary teams generally do an excellent job of showing compassion to grieving clients and are routinely praised for being kind, many team members miss out on opportunities to convey empathy to pet owners. It isn't that doctors and staff don't want to let clients know they are empathetic; rather it is that teams often don't know what to say and when to use empathy statements. Let's begin by defining empathy statements. These statements convey to the client that you understand their perspective and feelings. Empathy statements are an acknowledgement of the client's emotions or their position. Here are some examples of empathy statements:

- "I can understand that this is a difficult time for you."
- "I appreciate that you were not expecting these expenses for Scooter's care today."
- "I bet it was frustrating to go through that experience."
- "That must have been very upsetting. I know I would be scared too."

Empathy statements don't have to be confined to discussions about the pet. You can express empathy about anything a client may tell you including family illness, job situations, personal problems, minor frustrations, etc. People appreciate affirmations and concern about what they are going through.

Communicating With Angry Clients

Pet owners visiting specialty practices are often stressed and emotional. Sometimes this can lead to anger which is directed towards team members. Here are some effective ways to respond to angry clients.

- Remain calm and resist the temptation to be defensive if clients make inflammatory statements.
- Listen with concern. Allow clients to vent their frustrations or to thoroughly explain what they are angry about. If you interrupt them too quickly, you may exacerbate their anger.
- Ask questions if necessary to clarify what the client is angry about. Sometimes when you ask questions, it will become apparent that the client is angry about something that has nothing to do with the client service and they don't need you to do anything other than be empathetic.
- Always attempt to validate clients' emotions. You can do this by saying, "I understand that you are angry" or "I know this is an upsetting situation" or "I realize this is a difficult time for you".
- Realize that some people use anger as a shield for fear or sadness. They may act angry when in reality they are really just afraid or sad about what is wrong with their pet. They may be afraid of how they will afford to pay for medical services, afraid of how family members will react or afraid that their pet is going to die. It is helpful to use phrases such as, "I know "Charlie" is a beloved member of your family and we will do everything we can to help him."
- The next step if at all possible is to take action to assist clients. This may be by getting additional information for them or you may need to tell them what you can do to help them. For example, if a client is upset that the doctor has not called them back, you may be able to offer an explanation and assure them that you will do everything you can to facilitate the return phone call.
- Many times it is helpful to ask the client what they want or what will make them happy. It is helpful to say phrases such as, "We value you as a client. What can I do to help you?"

COMMUNICATE THE VALUE OF SERVICES

Recommendations for care are made by the specialists but technicians routinely review treatment plans with clients and give medical updates to pet owners. Make sure everyone on the technical team is knowledge and well trained to talk to clients about your services and patient care. When team members are confident and knowledgeable, clients receive more education and compliance is likely to increase for treatment plans. And don't forget that the value of your services is closely tied to the quality of client service. For pet owners to become loyal clients and refer others, they need to be bonded to the practice. Clients will be bonded to the practice when the client service experience is on par with the high quality medical care.

TAKE THESE STEPS TO WOW CLIENTS AND GAIN MORE BUSINESS: SPECIALISTS

Amanda L. Donnelly, DVM, MBA
Tampa, FL

All service businesses can be differentiated based on the level and type of service that is provided. The service offered by McDonalds and Motel 6 is vastly different than the service offered by Morton's Steakhouse and the Ritz Carlton. No one would argue that the quality of the food and lodging, the physical environment, the amenities and the quality of service provided by employees are at different ends of the spectrum for these businesses. Just like other service industries, veterinary businesses offer different levels of service. Unfortunately many practice leaders don't consciously think about the level of service on a daily basis but instead make the assumption that the team is committed to providing quality service. Are you making a concerted effort to provide a specific level of service at your practice? If not, the service that clients (pet owners and referring veterinarians) receive is subject to inconsistencies and may not be what you intended.

DEFINE YOUR TEAM'S CLIENT SERVICE IQ

To define the client service experience you want to create and ensure consistency in the delivery of client service, it is helpful to begin by determining what I call your team's "client service IQ". How knowledgeable is everyone on the team about what pet owners want? Does everyone agree on what constitutes exceptional vs. average client service? Do team members know what exceptional client service looks like? Have they experienced outstanding customer service for themselves? Facilitate a team meeting and create a dialogue among team members by discussing the following questions:

- What are some examples of businesses that provide great customer service? What do they do to provide this level of service?
- Think of a time when you experienced above average or exceptional customer service. What was the business and why was the service exceptional?
- How do you want to be treated when you go to a service-oriented business? Be specific with your answers.
- How do you decide whether you will go back to a service-oriented business?
- Think of a time that you recommended a business to someone else. Why did you make the recommendation?
- What have businesses done to surprise and impress you? What could our practice do?
- What annoys you with respect to customer service?
- If you could change something about the service our hospital provides, what would it be?
- What is the practice vision? What client service experience do we want our clients to have each time they visit the practice?
- What are our strengths and weaknesses related to service?

The advantage of taking the time to discuss the team's "client service IQ" is to ascertain whether team members fully understand, appreciate and agree on what it means to deliver exceptional client service. This exercise affords the team the opportunity to offer feedback about service challenges.

FORMULATE A PLAN TO ENHANCE CLIENT SERVICE

You will need to formulate a specific plan if you want to take your client service to a higher level. This plan may include the following:

Client Surveys

To raise the level of client service, it makes sense to find out how clients perceive the current level of service. Are they unhappy, satisfied or wowed by your service? Surveys can help practices gain valuable information about what clients like or dislike and what additional services or changes they would like to see in service delivery.

Staff Training

The need for staff training in certain areas may be evident as a result of the team meeting on "What's your client service IQ". A plan needs to be established to address gaps in training. Client service training never stops for practices that want to provide exceptional service. Without a focus on client service training, it is easy to become complacent and deliver mediocre service. Brainstorm ideas to provide exceptional service. Team members are an excellent resource to brainstorm creative ideas to enhance client service. Routinely hold staff meetings to discuss service challenges and possible solutions.

Create Service Standards

Setting client service standards help to ensure consistency in the execution of service to your clients. Client service standards can differentiate your business especially if the level of service provided is superior to that offered by competitors. When implementing service standards, it is helpful to develop standards for specific client interactions as well as general standards for the entire team. I recommend veterinary teams develop client service standards by starting with 3-5 standards for various categories of client interactions such as the phone, exam room or patient discharge. It is helpful to have different teams in the practice brainstorm on the standards and reach consensus. For example, the client service representatives can work on standards involving phone interactions and the technical team can work on standards for exam room interactions. Once the standards are agreed upon, they need to be written and distributed to the entire team so you can train the staff to the standards.

PARTNER WITH CLIENTS TO PROVIDE THE BEST CARE

Pet owners may not accept treatment recommendations because they don't understand the benefits of care or may not fully trust the healthcare team. With more effective communication, veterinarians can help clients focus on patient advocacy to make decisions that are right for their pet and their family. Research in human medicine has shown that communication is a core clinical skill and better communication by physicians leads to better patient outcomes. Similar research is being done in veterinary medicine showing that communication skills of veterinarians play an integral role in the veterinarian-client relationship. Enhanced communication between pet owners and the entire veterinary healthcare team improves relationships while also leading to better patient outcomes for pets due to improved dialogue and decision-making about what is best for the pet.

Pet owners appreciate clinicians who include them in the decision making process. Rather than giving clients the lecture on heart failure or Cushings disease, begin by assessing their knowledge base and ask questions about their concerns. Asking open-ended questions to determine the client's perspective and to uncover all their concerns leads to more efficient appointments, happier clients and greater adherence to recommendations.

Make sure all aspects of the physical exam and procedures are thoroughly explained. While conversing with the client is desirable and necessary, don't make the mistake of having the pet owner completely miss out on the value of the exam because they didn't even realize the pet was being examined. The client should be informed of normal and abnormal findings and the reason for each part of the exam. Make sure clients know you palpated for lymph nodes and what you can feel when you palpate the pet's abdomen.

Use visual aids or models to reinforce verbal messages when possible. For example, use the body conditioning score chart to let clients see why you want to talk about weight loss for their pet. If you don't have client education brochures with pictures or anatomical drawings, look into obtaining these visual aids. In addition, use tablets or IPADs in exam rooms to show clients pictures, diagrams, videos and links to credible websites where they can learn more.

COMMUNICATING WITH REFERRING VETERINARIANS

The client service experience for referring veterinarians is critical to the success of most specialty practices. General practitioners tend to evaluate the service experience largely on the quality of their communications with the referral practice. Establishing and maintaining effective communication with referring veterinarians can be a challenge especially since much of the communication is done on the phone or with written reports. One of the best ways to improve communications is to be proactive and have an operational process in place to improve communications and enhance the specialist-rDVM relationship. This may include setting up routine hospital visits or lunches to meet with referring veterinarians, the use of periodic surveys, roundtable discussions, and routinely asking rDVMs questions on the phone whether they are happy with the service. One or two key employees should be assigned to assist in this process. Referring veterinarians should know who they can call besides the specialists to discuss client relations.

HEPATIC AND BILIARY PARASITES OF DOGS AND CATS

Dwight D. Bowman, MS, PhD, DACVM Parasitology (honorary)
Ithaca, NY

INTRODUCTION

There are a number of protist and helminth parasites that either utilize the liver specifically as part of their life cycle or which are found in the liver along with other areas of the body. The protistan parasites are typically those that occur within the liver parenchyma and which also parasitize other sites in the host as well. The trematodes tend to be confined to the gall bladder and bile ducts, and they can cause significant disease. The exception to this location of infection is the site of the blood flukes that live in abdominal veins; these parasites cause disease through their eggs being swept back into the liver. The cestodes stages found in the livers of dogs and cats are the larval forms, and their presence in this site, and actually within the bodies of dogs and cats in general, are very rare. There are a number of nematodes that migrate through the liver of dogs and cats to get to where they are going, but most cause little significant pathology unless they are there in massive numbers. The exception is the typically inconsequential presence of *Calodium hepaticum* that lives in the liver parenchyma.

PARASITES OF DOGS

PROTISTA

APICOMPLEXA

***Hepatozoon* species** Typically *Hepatozoon* in the United States is considered to be a very serious disease with muscle pain and periosteal thickening; this is the disease due to *Hepatozoon americanum*.[1] There is also disease caused by the more common form in Europe, *Hepatozoon canis*. With both of these parasites, the animals do often develop infections with in the liver, but typically, it is not associated with hepatitis, or at least, the manifestations is other tissues overshadow any disease cause by the stages in the liver.

Neospora caninum This parasite is usually considered to be a disease of the nervous system and musculature of dogs that occasionally presents with cutaneous dissemination. However, if dogs are immunocompromised or immunosuppressed, there is the chance that they can develop severe hepatic disease from this infection.[2]

Toxoplasma gondii This parasite is known to cause disseminate disease when the host is immunocompromised, and it is also known to disseminate in the presence of viral infections, such as canine distemper.[3] In a recent case, a severely ill puppy with canine distemper, infectious canine hepatitis, canine parvoviral enteritis, and canine infectious tracheobronchitis was humanely euthanized and found to have disseminated toxoplasmosis with tachyzoites multiplying in myocytes and hepatocytes.[4]

***Sarcocystis* species.** *Sarcocystis canis* is the single species of *Sarcocystis* that forms schizonts in the tissues of dogs.[5] It was originally described from 3 dogs, one of which had developed hepatitis with numerous schizonts in hepatocytes.[5] Disease with disseminated sarcocystosis, in some cases with liver schizonts, has also occurred in dogs with concomitant distemper infections or blastomycosis.[6]

EXCAVATA

Trypanosoma cruzi Although *T. cruzi* is primarily considered a disease of muscle and particularly myocytes, it can also be found in the fixed histocytes within the liver.[7] This is one of the reasons that in cases of leishmaniasis it is necessary to perform molecular or immunologic tests if organs are seen in liver in areas where the two diseases overlap. And, they do overlap in areas of the United States.

***Leishmania* species** Recently, canine visceral leishmaniasis has been noted to be a system disease that produces a characteristic intralobular liver fibrosis.[8] Collagen deposits have also been noted in the spleen, cervical lymph nodes, lungs and kidney in dogs with clinical signs. However, the liver is the organ in which the pathology is most pronounced.

HELMINTHS

PLATYHELMINTHES

Cestodes

Larval *Spirometra* Infection with a larval *Spirometra* can occur in two different ways.[9-17] It can occur via the ingestion of the first intermediate host, a freshwater copepod that contains a larval form that has developed after the copepod ate the stage that hatched from the egg of the tapeworm. Or, hosts can be infected with the larvae by paratenesis, i.e., the ingestion of other forms that have eaten the copepod. Thus, a dog might develop a larval *Spirometra* infection by drinking water containing copepods or by eating a frog or mouse that contains a sparganum. Proliferative sparganosis often involves organ involvement and displacement, and if in the peritoneal cavity is often treated by removing the sometimes massive numbers of small larval tapeworms, sterile lavage, and repeated treatment with anti-cestocidal drugs. The disease often recurs. Unfortunately, when intestinal walls, liver, or the lungs are involved with the massive larval tapeworms that do not respond to the various cestocidal therapies, the infections are often fatal.

Larval *Mesocestoides*

The larva of *Mesocestoides*, called a tetrathyridium, is typically found in the abdominal cavity of dogs. The stages can be present either a single forms, or like with sparganosis as proliferative forms. The disease with the proliferative form is virtually untreatable. Survivability of dogs with this disease has been examined as to survivability in 60 dogs undergoing treatment for this infection.[18] In this analysis, survival for 1 year was 60.5%, and dogs that were treated aggressively with surgery and peritoneal lavage along with high doses of fenbendazole were 5 times less likely to die than dogs not treated aggressively. These dogs presented with clinical signs that were ascites (60%), anorexia/weight loss (42%), vomiting (23%), diarrhea (9%) and tachypnea (9%); subclinical infections (22%) were incidentally detected, typically during ovariohysterectomy or neuter. In some cases of peritoneal mesocestodiasis, treatment with praziquantel has proven more efficacious than fenbendazole.[19]

Larval *Taenia pisiformis* There have been two cases where the cysticerci of *T. pisiformis* have been found in the liver, and the hooks on the cysticerci were measured and matched in number per scolex and size those of *T. pisiformis*.[20] In the third case, the cysticercus was found in the brain of the dog, and the hooks were not measured and the diagnosis presumptive at best.

Larval *Echinococcus multilocularis* Infections with the alveolar hydatid of *E. multilocularis* usually occur in rodents with the adult forms occurring in foxes. In Europe there were reports of larval alveolar hydatid disease occurring on occasion in dogs; most of the reports were from Switzerland, but there were also reports from Germany and Belgium. There has now been a report of alveolar hydatid disease in a dog from British Columbia.[21] This case was treated surgically by the removal of the majority of the hepatic mass and therapy with long-term albendazole. The dog was followed for over 150 days, and although there were some side effects from the long-term albendazole therapy, it recovered.

Trematodes

Heterobilharzia americana This is a schistosome-related trematode of raccoons in the United States that also regularly makes it way into dogs and other domestic and wild animals. The adults of this worm live in the mesenteric and intrahepatic portal veins, and the disease is caused mainly as a response to the eggs that become lodged in the intestinal wall and within the liver. These eggs can cause severe to deadly pathology and may become calcified and give the appearance of soft-tissue mineralization on ultrasound or radiography.[22]

Clonorchis sinensis* and *Opisthorchis species These trematodes acquired through the ingestion of raw freshwater fish are common in the bile ducts of dogs in Asia and parts of Europe. There are similar flukes that occur in North America, even in the very northern reaches of Canada, where it extends from Newfoundland to Saskatchewan and north as far as Hudson's Bay.[23-24] There has not been a significant amount of work on the pathology or disease that these infections cause in dogs.

Nematoda

Calodium hepaticum A good number of nematodes parasitic in dogs migrate as larvae though the liver of the canine host as they make their way to other tissues; however, these larval migrations do not for the most part cause any significant pathology. *Calodium hepaticum*, a capillarid nematode that lives within the parenchyma of rodent hosts, also has been found in dogs where it can cause lesions which are found incidentally during post-mortem examinations.[25-26]

PARASITES OF CATS

PROTISTA

Apicomplexa

Toxoplasma gondii This parasite is known to be a common parasite of the intestinal mucosa of cats when they serve as the oocysts-producing final host. Also, it is recognized that *T. gondii* also appears as an acute necrotizing hepatitis in cases with acute disease.

***Hepatozoon* species** The species *Hepatozoon felis* has recently been redescribed and PCR allowed the detection of the organism within the liver of cats.[27] The meronts of this species have been found mainly in the muscle tissues of cats. In a cat that originated in Hawaii, meronts were found in the liver,[28] and perhaps this species is not the same as the one redescribed because it had granulomatous hepatitis.

Excavata

***Trypanosoma cruzi* and *Leishmania* species.** Cats are being recognized as animals developing disease from both of these two genera of parasites. However, most of the work has concentrated on other organs, perhaps appropriately because only 1 of forty naturally infected and recently examined cats in Brazil had hepatomegaly.[29]

HELMINTHS

Platyhelminthes

Cestoda

Larval *Spirometra* A case of proliferative sparganosis has been reported from a cat with an abdominal mass.[30] In this case the spleen and liver were studded with numerous yellowish granules and there was a large node also on the serosal surface of the stomach.

Trematoda

Dicrocoelid flukes – Cats can be infected with *Platynosomum fastosum* which causes liver poisoning disease in the tropical portions of North America. These cases continue to occur and are being diagnosed more commonly.[31] These infections have also been associated with polycystic lesions in the liver and cholangiocarcinoma.[32-33] In Eurasia, cats can be infected with several species of *Euparadistomum*.

Opisthorchid flukes. Cats in the United States get infected with the bile-duct dwelling flukes *Amphimerus pseudofelineus*, *Metorchis conjunctus*, and *Parametorchis complexum*; these flukes cause disease very similar to that caused by *Clonorchis sinensis* and *Opisthorchis* species that causes disease in cats, dogs, and people in Asia.

Schistosoma japonicum is a *Heterobilharzia*-like schistosome that lives in cats in Asia, and is potentially capable of causing disease very much like the American heterobilharziasis in cats in Eurasia.

NEMATODA

Toxocara canis The migration of *Toxocara canis* through the liver of cats can caused marked pathologic changes.[34]
Calodium hepaticum has been reported from cats in the same fashion as it has for dogs.

CONCLUDING REMARKS

Cats and dogs have a number of parasites that live in their liver tissues and which can present as asymptomatic infections or as deadly infections that require surgical intervention and treatment with antiparasiticides to keep the pet alive. Praziquantel tends to be the drug of choice, but fenbendazole and febantel also have their place in treating these infections.

REFERENCES

[1]Baneth G **Vet Parasitol** 2011;181(1):3. [2]Hoon-Hanks LL, et al **J Vet Diagn Invest** 2013;25(6):807. [3]Dubey JP, et al **Vet Parasitol** 2006;137:36. [4]Headley SA, et al **J Vet Diagn Invest** 2013;25(1):129. [5]Dubey JP, et al **J Parasitol** 1991;77(4):522. [6]Trasti SL, et al **J Comp Path** 1999;121:179. [7]Almeida ABPF, et al **Rev Inst Med Trop Sao Paulo** 2013;55(4):287. [8]Silva LC, et al **Int J Exp Path** 2013;94:133. [9]Koch A, et al **Kleintierpraxis** 2012;57(5):245. [10]Simpson C, et al **Mol Cell Probes** 2012;26(1):60. [11]Ballweber LR **Vet Med** 2011;106(7):334. [12]Stief B, et al **Vet Pathol** 2011;48(4):911. [13]Drake DA, et al **J Am Vet Med Assoc** 2008;233(11):1756. [14]Naveen K, et al **Vet Practitioner** 2004;5(1)55. [15]Bannerjee PS, et al **J Vet Parasitol** 2001;15(2):159. [16]Little S, et al **Comp Cont Ed Pract Vet** 2000;22(4):299. [17]Beveridge I, et al **Austral Vet J** 1998;76(11):757. [18]Boyce WM, et al **Vet Parasitol** 2011;180(3/4):256. [19]Papini R, et al **Vet Parasitol** 2010;170(1/2):158. [20]Ivens V, et al **Am J Vet Res** 1969;30(11):2017. [21]Peregrine AS, et al **Can Vet J** 2012;53(8):870. [22]Corapi WV, et al **J Vet Diagn Invest** 2011;23(4):826. [23]Cameron TWM, et al **Can J Res** 1940;18:325. [24]Cameron TWM **Can J Res** 1944;22:6. [25]Lloyd S, et al **Vet Rec** 2002;151:419. [26]Palma HE,et al **Ciencia Rural Santa Maria** 2009;39(9):2642. [27]Baneth G, et al **Parasit Vectors** 2013;6:102. [28]Ewing, GO **Fel Pract** 1977;7(6):37. [29]Vicente Sobrinho LS, et al **Vet Parasitol** 2012;187:302. [30]Buergelt CD, et al **J Parasitol** 1984;70(1):121. [31]Headley SA, et al **Braz J Vet Pathol** 2011;4(3):227. [32]Andrade RLFS, et al **Vet Parasitol** 2012;190:277. [33]Daniel AGT, et al **Braz J Vet Pathol** 2012;5(3):137. [34]Parsons JC, et al **Int J Parasitol** 1989;19(5):479.

BLASTOCYSTIS HOMINIS

Dwight D. Bowman, MS, PhD, DACVM Parasitology (honorary)
Ithaca, NY

INTRODUCTION

Blastocystis hominis is an organism that has been place now within a group of Protista that goes by the name the Chromalveolates or the initials SAR (for Stramenopiles, Alveolates, and Rhizaria). Of these groups the only one really familiar to veterinary medicine is the alveolates, because this group contains the well-known coccidia, *Toxoplasma*, *Sarcocystis*, and *Babesia*. About the only other Stramenopiles that might be known to veterinarians are the two oomycete "fungal" disease agents *Saprolegnia* and *Pythium*. *Saprolegnia* species are water molds of aquaria and fresh water fish, and it looks like they are growing fine white hairs or clouds on their surface. *Pythium* species cause horrible dermatologic lesions in dogs and horses that get into water where the biflagellate swimming zoospores are present. One non-disease producing Stramenopile that veterinarians are definitely aware of even if they do not know that they are Stramenopiles are the giant kelp making up the huge forests in the oceans off the northwestern United States. So, these groups for the most part resemble each other in that they have flagella that bear fine hairs (these are from when the name Stramenopile is derived). *Blastocystis hominis* is now considers part of this group.

There is disagreement as to whether or not *Blastocystis* is pathogenic. The agent *B. hominis* is considered by some to be a commensal that overgrows during periods of diarrhea and by others to be the cause of diarrhea. This enigmatic parasite of the intestinal tract has cyst stages that are passed in the feces and infect the next host. *Blastocystis hominis* has previously been considered a protozoan, an amoeba, and a yeast; with the fungal association having been its most recent previous moniker. However, when molecular biology was performed on the enigmatic agent, *B hominis* with its long association with large bowel diarrhea, was found related to the Stramenopiles. Also, people got the agent to grow in culture so some of its vegetative stages could be examined in more detail. All this in combination with the very large increases of chronic large bowel disease has caused this organism to be examined in much greater detail in recent years.

Once people started looking, they realized that they could find it in hosts other than people. It was found in 70% of both cats and dogs in the Brisbane area of Australia.[1] It was found to be in the feces of chickens and pigs,[2] thus, not only did it exist as a "pathogen" in people, it was also a zoonotic agent. It was found to be very common in pigs in intensive production facilities.[3] It has been found in cattle.[4] It has also been suggested to be a potential zoonotic agent with a paper in Parasites & Vectors with the title "*Blastocystis* sp.: waterborne zoonotic organism, a possibility?"[5] Thus, it is found in animals, people, and in the water from rivers.

It turns out that like *Giardia*, there are different subtypes – some tend to be restricted to certain hosts, while some are spread between hosts. Molecular biologists have now divided the group into 9 to 14 different subtypes and are trying to classify where some are in one animal, some in another, and whether there is cross transmission between the two; especially if the two different animals are a pet and its human owner. Work in the Philippines looking at *Blastocystis* from animal and human hosts found that some subtypes were shared between different hosts, e.g., by monkeys, humans, and pigs.[6] In a survey of stools submitted to diagnostic laboratories within the state of Colorado, the most common agent detected in fecal examinations was *Blastocystis* at 4.3%, with *Giardia* being next at 1.4%.[7] And there were reports that concluded: "Conventional therapy is ineffective for symptomatic *Blastocystis* infection. The high prevalence of *Blastocystis* infection within households suggested transmission between humans and their pets. Subtyping analysis of SSU rDNA alone in *Blastocystis* does not appear to predict pathogenicity."[8] *Blastocystis hominis* was becoming a disease that was common and which would transmit between different hosts. This pathogen is of such interest at this point that the complete genome was sequenced showing that it has 15 chromosomes and was unambiguously placed within the Stramenopile group.[9]

More importantly, people began to look at the different pathologies caused by the different subtypes and whether or not they were shared between people and animals. It was shown that different subtypes caused different pathologies, and that people had the same subtypes as the animals that were their pets or for which they supplied care. Then it was shown that different subtypes would cause different pathologies. In nine people in Oregon who presented with chronic gastrointestinal illness with diarrhea, abdominal pain, fatigue, joint pain, skin rash and psychiatric co-morbidity, six were found to be infected with subtype 3 and one with subtype 1.[10] These people had all previously been treated with courses of antibiotics with no resolution of their symptoms. There was a characterization of the fecal samples of zoo animals in an Australian zoo, and the zookeepers within the zoo. Samples were collected and subtypes using molecular methods.[11] 42% and 63% of animals and zoo-keepers sampled were positive for *Blastocystis*, respectively. Most of the zookeepers at the zoo were infected, and for of the zookeeper isolates were identical to the isolates from the southern hairy nosed wombat and five primate species. And it was shown that in people with their pets that all pets were positive for *Blastocystis* with polymerase chain reaction testing, and seven out of eight had the same subtype as their symptomatic patient who cared for them.[8] So, the caretakers at the zoo were infected with the same subtype as found in the animals for which they supplied care, and people had the same subtype as their pet.

Might not be so bleak… A study has recently examined fecal samples on 240 stray dogs: 80 from Brisbane, Australia; 80 from Delhi and Mumbai, India, and 80 from Dong Village in Cambodia.[12] The infection rates were 2.5% for Australia, 1.3% in Cambodia, and 24% for India. This paper showed what they considered a geographic distribution of subtypes that they thought was similar to what has been seen in people. Secondly, and importantly, the paper states that the low prevalence of *Blastocystis* in dogs, with no predominant subtype, suggests that dogs are unlikely to be natural hosts of *B. hominis*. Rather, these authors believe that dogs are opportunistically infected via coprophagia or by drinking contaminated water.

CONCLUDING REMARKS

So, Zierdt has been a long term champion of *Blastocystis* being a pathogenic agent.[13] This issue has yet to be fully resolved; things are as they were in 1988. We now know there are different subtypes, we know how many chromosomes it contains, and we recognize that it is common in people. However, we are still not certain that it causes disease. It is found in the feces of dogs and cats, and at this point in time it seems that the risk of it being zoonotic is minimal. However, it seems highly likely that additional work will be generated over the next few years and people will continue to argue the two sides of the issue. The most important thing relative to this parasite right now with respect to dogs and cats is that there are almost no laboratories handling veterinary fecal sample examinations that can find this agent, yet alone subtype it. There is also a very good chance that clients will become aware of this agent through the giant cloud of media that surrounds us.

REFERENCES

[1]Duda A, et al **Vet Parasitol** 1998;76:9. [2]Tan KSW **Vet Parasitol** 2004;126(1-2):121. [3]Navarro C, et al **Vet Parasitol** 2008;153(3-4):347. [4]Fayer R, et al **Parasitol Res** 2012;111(3):1349. [5]Lee LI, et al **Parasit Vectors** 2012;5:130. [6]Rivera WL **Vet Parasitol** 2008;156(3-4):178. [7]Church C, et al **J Parasitol** 2010;96(1):194. [8]Nagel R, et al **Intern Med J** 2012;42(11):1187. [9]Denoeud F, et al **Genome Biol** 2011;12(3):R29. [10]Jones MS, et al **Parasitol Res** 2009;104(2):341. [11]Parkar U, et al **Vet Parasitol** 2010; 169(1-2):8. [12]Wang W, et al **Parasit Vectors** 2013;6:215. [13]Zierdt CH **Parasitol Today** 1988;4(1):15.

CARDIORENAL SYNDROME: CARDIOLOGIST VS NEPHROLOGIST

Joao S. Orvalho, DVM, DACVIM (Cardiology), Larry D. Cowgill, DVM, PhD, DACVIM (SAIM)
San Diego, CA

INTRODUCTION

It has become increasingly recognized by cardiologists and nephrologists that there are important bidirectional functional and pathological interactions between the heart and the kidney, wherein dysfunction of either organ promotes clinical worsening of the other.[1] Cardiovascular disease constitutes a significant threat for patients with renal disease, and renal dysfunction is also often present in patients with cardiac disease. The clinical consequences of these interactions have gained increasing focus and have prompted further definition, classification, and understanding. These interactions are the pathophysiological basis for the clinical entity termed Cardiorenal Syndrome (CRS) in human medicine. [1-4] Cardiorenal Syndrome *per se* has not been well characterized in veterinary medicine.

A DEFINITION

The definition of CRS includes a variety of acute or chronic conditions, in which the primary failing organ: the heart, the kidney, or both (due to a systemic condition), promotes dysfunction and/or failure of the other organ system.[4]

CLASSIFICATION

Five subtypes of CRS have been suggested in order to simplify the identification and the clinical approach. [3-5]

Type 1 CRS - Acute Cardiorenal Syndrome is characterized by a rapid impairment of cardiac function leading to acute kidney injury. There are multiple and complex mechanisms by which acute heart failure or an acute onset of chronic heart failure leads to acute kidney injury (AKI).[6] In humans, the onset of AKI is more pronounced in patients with decreased left systolic function, which may imply reduced renal perfusion.[7] The congestive state may also induce decreased diuretic responsiveness, which may lead to excessive use of diuretics and further kidney injury.[8,9] In CRS type 1, the early recognition of AKI remains a challenge due to the lack of early stage biomarkers. Serum creatinine (sCr) and blood urea nitrogen (BUN) have been the classic markers for AKI, but when the concentrations of these markers are detectably elevated the injury process is well established, and it is often too late to protect the kidney or prevent further damage. The discovery of novel urinary biomarkers such as neutrophil gelatinase-associated lipocalin (NGAL), cystatin c, symmetrical dimethylarginine (SDMA), may allow an earlier recognition of AKI and CRS.[10,11]

Type 2 CRS - Chronic Cardiorenal Syndrome consists of chronic cardiovascular disease causing progressive chronic kidney disease (CKD).[3-5] Chronic heart failure (CHF) is likely to cause persistently reduced renal perfusion, chronic renal congestion ("congestive kidney failure"), and neurohormonal changes associated with chronic sympathetic stimulation (production of epinephrine, angiotensin, endothelin, and release of natriuretic peptides and nitric oxide).[13,14] CHF therapy using diuretics and renin-angiotensin aldosterone system (RAAS) blockers can cause drug-induced hypovolemia, decreased renal perfusion, and hypotension.[6] In humans, the prevalence of renal dysfunction in CHF patients is high and constitutes an independent predictor of outcomes and mortality, therefore it is of the upmost importance to preserve the renal function in these patients.[15]

Type 3 CRS – Acute Renocardiac Syndrome is characterized by an acute primary worsening of kidney function that leads to acute cardiac dysfunction. AKI can affect the cardiac function through multiple mechanisms, such as fluid overload, electrolyte disturbances, neurohormonal activation, and myocardial depressant factors, potentially contributing to the development of arrhythmias, pericarditis and acute heart failure.[16,17] Diagnosis of AKI in patients concurrently treated for heart failure may force clinicians to reduce the dose or discontinue heart failure medications, further decompensating the cardiovascular system in detriment of preventing additional kidney injury.

Type 4 CRS– Chronic Renocardiac Syndrome consists of primary chronic kidney disease that contributes to cardiac dysfunction. Decreased systolic function, left ventricular hypertrophy, systemic hypertension, and a high output state (secondary to anemia) are some of the potential long-term cardiac sequelae of CKD.[5,12] The medical management of CKD and concurrent CHF is not as problematic as the acute forms of these conditions, but most CHF patients are likely undertreated, due to concerns of further worsening renal function and creating a vicious cycle of bidirectional damage secondary to the specific pathophysiology and pharmacotherapy of both conditions.

Type 5 CRS – Secondary Cardiorenal Syndrome is characterized by cardiac and renal dysfunction secondary to an acute or chronic systemic condition. Sepsis is the most common acute condition that affects both the heart and the kidney.[5,18,19] Diabetes mellitus and hyperadrenocorticism are typical chronic diseases in dogs that have a similar effect on the urinary and cardiovascular systems.[20-22]

Table 1. Cardiorenal Syndrome Classification

Type of Cardiorenal Syndrome	Brief definition	Conditions
Type 1 - Acute cardiorenal syndrome	Acute impairment of the cardiac function leading to acute kidney injury (AKI)	- Acute heart failure - Cardiogenic shock
Type 2 - Chronic cardiorenal syndrome	Chronic cardiovascular disease causing progressive chronic kidney disease (CKD)	- Chronic heart failure - "Congestive kidney failure"
Type 3 – Acute renocardiac syndrome	Acute primary worsening of kidney function that leads to cardiac dysfunction.	- Acute kidney injury - Hyperkalemia, uremia
Type 4 – Chronic renocardiac syndrome	Primary chronic kidney disease that contributes to cardiac dysfunction	- Chronic glomerular disease - Anemia, systemic hypertension
Type 5 – Secondary cardiorenal syndrome	Cardiac and renal dysfunction secondary to and acute or chronic systemic condition	- Diabetes mellitus - Sepsis

HYPOTHESIS

In dogs and cats like in people, intrinsic dysfunction of the cardiovascular system and/or its management promotes secondary injury and dysfunction of the kidney. In addition, intrinsic dysfunction of the kidney and/or its management promotes secondary injury and dysfunction of the cardiovascular system.

CARDIOVASCULAR DISEASE CONTRIBUTES TO KIDEY DYSFUNCTION

In patients with cardiovascular disease the renal hemodynamics are affected and considered the main contributor to secondary kidney disease.[13,23] A low output stage triggers neurohormonal activation through the sympathetic nervous system (SNS) and the RAAS. This process leads to initial preservation of the glomerular filtration rate (GFR), but decreased renal perfusion and secondary tubular hypoxic injury and loss of nephrons. These systems also promote water and sodium retention, elevating central venous pressure and causing organ congestion ('congestive kidney failure'). Renal congestion in turn also stimulates the SNS and RAAS contributing to progressive reduction of GFR.[13, 23-25]

Cardiac biomarkers such as N-terminal pro-B-type natriuretic peptide (NT-proBNP) and cardiac troponin-I (cTnI) are well established indicators of cardiac disease or injury.[26-30] Identification of tubular biomarkers of early kidney injury, such as NGAL, clusterin and cystatin c, are crucial to the recognition of CRS and management of both conditions.[5,10,11]

Classification of the heart disease, AKI and CKD is an important step to characterize the type of CRS. The recommended cardiac disease classification is the American College of Veterinary Internal Medicine (ACVIM) cardiac disease severity classification, which was adapted from the American College of Cardiology/American Heart Association classification system that uses an A through D categorization (Table 2).[31] The International Renal Interest Society (IRIS) proposed AKI and CKD classifications, which are the most widely accepted in veterinary medicine (Tables 3 and 4, respectively).[11,32,33]

Table 2. ACVIM cardiac disease classification.

ACVIM Classification	Class A	Class B_1	Class B_2	Class C_1	Class C_2	Class D_1	Class D_2
Brief Definition	Patients at risk	Asymptomatic No cardiomegaly	Asymptomatic Cardiomegaly	Heart failure Hospitalized	Heart failure At home	Refractory HF Hospitalized	Refractory HF At home

Table 3. IRIS acute kidney injury (AKI) grading criteria

AKI Grade	Serum creatinine	Clinical description
Grade I	< 1.6 (mg/dL) (<140 µmol/l)	**Non Azotemic AKI:** - Documented AKI: (Historical, clinical, laboratory, or imaging evidence of acute kidney injury, clinical oliguria/anuria, volume responsiveness) and/or - Progressive non azotemic increase in serum creatinine; ≥0.3 mg/dL (≥ 26.4 µmol/l) within 48 hours - Measured oliguria (<1 ml/kg/hr) or anuria over 6 hrs
Grade II	1.7 – 2.5 mg/dL (141-220 µmol/l)	**Mild AKI:** Documented AKI and static or progressive azotemia Progressive azotemic increase in serum creatinine; ≥0.3 mg/dL (≥ 26.4 µmol/l) within 48 hours), or volume responsiveness - Measured oliguria (<1 ml/kg/hr) or anuria over 6 hrs
Grade III	2.6 – 5.0 mg/dL (221-439µmol/l)	**Moderate to Severe AKI:**
Grade IV	5.1 – 10.0 mg/dL (440-880 µmol/l)	Documented AKI and increasing severities of azotemia and functional renal failure
Grade V	> 10.0 mg/dL (>880 µmol/l)	

Courtesy IRIS Board

KIDNEY DISEASE CONTRIBUTES TO CARDIOVASCULAR DYSFUNCTION

Abnormal kidney function may lead to metabolic and functional disturbances in multiple organs, which include the cardiovascular system. These processes may directly affect the heart, such as systemic hypertension causing decreased systolic function, left ventricular hypertrophy and diastolic dysfunction; or indirectly by reducing the production of erythropoietin leading to anemia.[34-36]

MANAGEMENT OF THE CARDIORENAL SYNDROME PATIENT

A significant number of patients with AKI and CKD are treated by cardiologists, but there is no established protocol for the treatment or monitoring of cardiovascular disorders in these patients. Patients with kidney dysfunction may receive a suboptimal treatment for concurrent cardiovascular conditions, even thought they may benefit from the standard therapies. This may account for part of the worse prognosis attributed to patients with renal disease. In human medicine, early referral to a nephrologist is associated with a delayed progression of CKD. CRS patients may benefit from co-management by cardiologists and nephrologists.[37]

Angiotensin converting enzyme-inhibitors and angiotensin receptor blockers (ARBs) are beneficial in cardiovascular and renal diseases, but patients with renal dysfunction are less likely to receive this type of drugs due to the concern of worsening renal function.[38-41] A better understanding of the relative risk of using these and other drugs may be very important in CRS patients.

Mineralocorticoid receptor antagonists (aldosterone blockers) have the potential for renal and cardiac protection, therefore the use of spironolactone in this subset of patients may be beneficial if the patients tolerate the drug.[31]

Loop diuretics may have conflicting effects on the renal function. By reducing renal congestion they may improve GFR and delay the progression of CKD, but on the other hand, excessive doses of diuretics may also decrease renal perfusion and consequently GFR. The combination of loop diuretics and thiazide diuretics has a synergistic effect that may cause excessive volume depletion and electrolyte disturbances, therefore they should be used with caution in the CRS patient.

Pimodendam improves the systolic function, which may increase GFR. Pimobendan does not enhance or suppress furosemide-induced RAAS activation.[39,42]

Digoxin and other drugs with predominant renal excretion may require closer monitoring and potential reduction of the dose.

Omega-3 fatty acids are a recommended oral supplement that has been used as an antioxidant and appetite stimulant in patients with heart and kidney disease.[22-35]

Table 4. IRIS chronic kidney disease (CKD) classification

CKD Stage	Serum creatinine (mg/dl)	Comments
Stage I	<1.4 Dogs <1.6 Cats	**Non-azotemic** Some other renal abnormality present, e.g: inadequate urinary concentrating ability without identifiable nonrenal cause; abnormal renal palpation and/or abnormal renal imaging findings; proteinuria of renal origin; abnormal renal biopsy results; increasing blood creatinine concentrations in samples collected serially
Stage II	1.4-2.0 Dogs 1.6-2.8 Cats	**Mild renal azotemia** (lower end of the range lies within reference ranges for many laboratories but the insensitivity of creatinine as a screening test means that animals with creatinine values close to the upper reference limit often have excretory failure) Clinical signs usually mild or absent
Stage III	2.1-5.0 Dogs 2.9-5.0 Cats	**Moderate renal azotemia** Many extrarenal clinical signs may be present
Stage IV	>5.0 Dogs and Cats	**Severe renal azotemia** Many extrarenal clinical signs are usually present

Courtesy, J. Elliott and A.D.J. Watson, and the IRIS Board

CONCLUSIONS AND FUTURE DIRECTIONS

An accurate appreciation of the kidney and the cardiovascular system and their interactions may have practical clinical implications. A multidisciplinary evaluation including the expertise of cardiologists and nephrologists may be the most appropriate approach for the patient at risk for CRS.[43] The outcome of CRS patients is likely to improve with the increasing awareness and ability to identify and understand the pathophysiological characteristics of cardiorenal syndrome.

REFERENCES

[1]Schrier RW. Nat Clin Pract Nephrol 2007;3:637. [2]Berl T, Henrich W. Clin J Am Soc Nephrol 2006;1:8 –18. [3]Ronco C. Int J Artif Organs 2008;31:1–2. [4]Berl T, Henrich W.. Clin J Am Soc Nephrol 2006;1:8 –18. [5]Ronco, C. et al.. J Am Coll Cardio. 2008; 52:1527–39 [6]Liang KV, Williams AW,et al. Crit Care Med 2008;36:S75– 88. [7]Jose P, Skali H, Anavekar N, et al. Am Soc Nephrol 2006;17:2886 –91. [8]Ellison DH. Semin Nephrol 1999;19:581–97. [9]Almeshari K, Ahlstrom NG, Capraro FE, Wilcox CS. J Am Soc Nephrol 1993;3:1878–83. [10]De Loor, J. et al. J Vet Intern Med 2013;27:998–1010. [11]Segev, G.,et al. J Vet Intern Med 2013;27:1362–1367. [12]Graziani, G. et al. Heart Fail Rev. 2013. [13]Mullens et al. J Am Coll Cardiol. 2009, 53:589-96. [14]Jessup, M. Costanzo,M. 2009;7:597–9. [15]Hillege HL, Nitsch D, Pfeffer MA, et al. Circulation 2006;113:671– 8. [16]Blake P, et al. ASAIO J.1996;42:M911–5. [17]Meyer TW, Hostetter TH. Uremia. N Engl J Med 2007;357:1316–25. [18]Langhorn, R. et al. J. Vet. Intern. Med. Am. Coll. Vet. Intern. Med. 2013; 27:895–903. [19]RJ et al. J Immunol 2002;168:5817–23. [20]Waldum, B. Os, I. Cardiology 2013;126:175–186 [21]Atkins, C. E. Vet. Clin. North Am. Small Anim. Pract. 1991. 21, 1035–1080. [22]Polzin, D. J. Vet. Clin. North Am. Small Anim. Pract. 2011; 41, 15–30 [23]Damman K,et al. Eur J Heart Fail 2007; 9: 872–878. [24]Ruiz-Ortega, M. et al. Kidney Int. Suppl. 2002; S12–22 [25]Remuzzi, G., Perico, N.et al. Kidney Int. Suppl. 2002; S57–65 [26]Sharkey, L. C. et al. J. Am. Vet. Med. Assoc. 2009; 234:767–770. [27]Häggström, J. et al. J. Vet. Cardiol. Off. J. Eur. Soc. Vet. Cardiol. 2000; 2, 7–16 [28]Serres, F. et al. J. Vet. Cardiol. Off. J. Eur. Soc. Vet. Cardiol. 2009;.11:103–121 [29]Schmidt, M. K. et al. J. Vet. Cardiol. Off. J. Eur. Soc. Vet. Cardiol. 2009.; 11 Suppl 1, S81–86 [30]Hezzell, M. J. et al. J. Vet. Intern. Med. Am. Coll. Vet. Intern. Med. 2012; 26:302–311 [31]Atkins C et al. J Vet Intern Med 2009.,23:1142-50. [32]Ross, S.. Vet. Clin. North Am. Small Anim. Pract. 2011; 41:1–14 [33]http://www.iris-kidney.com [34]Brown, S. et al. J. Vet. Intern. Med. Am. Coll. Vet. Intern. Med. 2007; 21:542–558. [35]Polzin, D. J. J. Vet. Emerg. Crit. Care San Antonio Tex 2001. 2013; 23:205–215. [36]Harison, E., Langston, C., Palma, D. et al.. Vet. Intern. Med. Am. Coll. Vet. Intern. Med. 2012; 26:1093–1098. [37]Campbell GA, Bolton WK:. Adv Chronic Kidney Dis 2011; 18: 420–427. [38]BENCH Study Group.J. Vet. Cardiol. Off. J. Eur. Soc. Vet. Cardiol.1999; 1:7–18. [39]Häggström, J. et al. the QUEST study. J. Vet. Intern. Med. Am. Coll. Vet. Intern. Med. 2008; 22, 1124–1135. [40]Ettinger, S. J. et al. (LIVE) Study Group. J. Am. Vet. Med. Assoc. 1998; 213:1573–1577. [41]The IMPROVE Study Group. J. Vet. Intern. Med. Am. Coll. Vet. Intern. Med. 1995; 9:234–242. [42]Lantis, A. C., Atkins, C. E. et al. Am. J. Vet. Res. 2011; 72:1646–1651. [43]Bock, J. S. Gottlieb, S. S. Circulation 2010; 121, 2592–2600

CARDIOVASCULAR-RENAL AXIS DISORDERS-TOWARDS A CONSENSUS

Mark A. Oyama, DVM, DACVIM (Cardiology), Shelly Vaden, DVM, PhD, DACVIM (Internal Medicine)
for the Cardiovascular Renal Axis Disorders Working Group*
Philadelphia, PA; Raleigh, NC

INTRODUCTION

A complex interaction between the renal and cardiovascular systems exists in both health and disease. In humans, the pathological interactions between these 2 organ systems are increasingly regarded as important and deserving of further definition, classification, and understanding. The term "cardiorenal syndrome" or CRS has been put forwards by physicians to describe this pathological interplay. Although kidney and cardiovascular disease are common in dogs and cats, the exact nature of this interaction is not well-defined or described within the veterinary profession. The members of the working group responsible for the current veterinary-focused presentation chose to call this cardiovascular-renal interaction "Cardiovascular-Renal Disorders" or CvRD to reflect the interplay of these organ systems in both health and disease, such that the entirety of the cardiovascular system (including both the heart and vasculature) is likely involved. In addition, the widely varied clinical manifestations of this interrelationship preclude legitimately terming the entity as a single syndrome.

METHODOLOGY

A working group comprised of 16 academic clinician scientists (8 cardiologists and 8 nephrologists)* from Europe and North America was created with the goal of achieving consensus definitions and recommendations for CvRD in dogs and cats. The specific purposes were to develop standard terminology, describe the current state of knowledge surrounding CvRD in dogs and cats, establish recommended minimum diagnostic databases for the evaluation of CvRD, provide general guidelines for the treatment of patients with co-existing cardiovascular and kidney disease with an emphasis on the potential interplay between the two organ systems, and finally, to stimulate research of CvRD as an important pathophysiologic mechanism.

A modified Delphi method was chosen for creation of a formal consensus around topics related to CvRD. The working group included a steering group, comprised of a chairman and 4 members, and an 11-member rating group. The steering group created a review of the scientific evidence surrounding human CRS through critical analysis and synthesis of the literature found using a PubMed search from 1995 to present. A draft set of statements regarding CvRD, accompanied by description of supporting evidence, was then reviewed and modified by the entire group. Group members deliberated on the appropriateness of the statements and iterative changes were made in order to coalesce input around points of agreement. Next, the modified Delphi method employed 2 rounds of voting during which each member rated each statement from 1 to 9 depending on the strength of their agreement with the statement, with 1 being no agreement and 9 being strong agreement. The 2 lowest ratings for the statement were discarded if the median score was > 5, and the two highest values if the median < 5. Thus, each statement was ultimately assessed on the basis of 14 individual rating scores. In order for a statement to be defined as achieving "consensus" statement none of the 14 individual ratings could be < 5 and the median rating of all member must have been 7 or greater. Between the first and second round of rating, the scores were reviewed and the statements and their supporting arguments were again amended to try to achieve the greatest consensus. The results of the second round of rating formed the basis of the final consensus document. For statements that achieved consensus, the strength of consensus was defined by the lowest value of the 14 individual rating scores. "Strong" consensus was achieve if the lowest individual score was ≥7); "good" consensus achieved if the lowest individual score was ≥5).

This presentation will present 13 different aspects of CvRD as evaluated by the working group, including a proposed classification system for CvRD in the dog and cat, etiology and pathophysiology of CvRD, diagnosis, staging, and clinical evaluation of patients with suspected CvRD, and management of CvRD.

DEFINING "CvRD"

CvRD was defined as "*disease, toxin, or drug-induced structural or functional damage to the cardiovascular system and kidney leading to disruption of the normal interactions between these systems, to the ongoing detriment of one or both*". The group proposed a classification system that described the directionality of CvRD with respect to the organ that was suspected to have primary disease vs. the secondary or target organ that was then being adversely affected. $CvRD_H$ was chosen to indicate instances of primary cardiovascular disease leading to kidney dysfunction; $CvRD_K$ was chosen to indicate instances of primary kidney disease leading to cardiovascular dysfunction; $CvRD_O$ was chosen to indicate concurrent impairment of both systems due to "other" independent disease processes external to the cardiovascular or renal systems.

EPIDEMILOGY

The pathophysiology of CRS in human patients is poorly understood, and in veterinary medicine, the existence and nature of CRS is largely speculative. Suspected mechanisms of CRS in humans include organ perfusion, activation of neurohormonal systems such as renin-angiotensin-aldosterone and autonomic nervous systems, anemia, inflammation, thromboembolism, hypertension, and congestion.

CLINICAL EVALUATION OF CvRD

Staging systems based on severity of cardiac (i.e., International Small Animal Cardiac Health Council [ISACHC], ACVIM-mitral valve disease consensus, modified New York Heart Association [NYHA]) and kidney (i.e., International Renal Interest Society [IRIS]) disease exist, but whether any of these systems describe dysfunction or injury related to CvRD is unknown. The working group achieved greater consensus around use of the IRIS system to describe both acute and chronic renal disease in $CvRD_H$ than did use of the ISACHC system to describe cardiac injury in $CvRD_K$. Biomarkers, such as creatinine, BUN, phosphorous, and urine specific gravity are commonly used to assess kidney function, but how these and other emerging biomarkers of kidney injury such as neutrophil gelatinase-associated lipocalin (NGAL) or symmetric dimethylarginine (SDMA) play a role in the evaluation of $CvRD_H$ requires further study. Similarly, the potential role of cardiac biomarkers such as cardiac troponin-I and N-terminal B-type natriuretic peptide in the assessment of $CvRD_K$ also requires additional study. Development of specific markers of CvRD or CRS remains a challenge in both veterinary and human medicine, respectively. Imaging modalities such as radiography and abdominal and cardiac ultrasound remain amongst the most useful diagnostics to evaluate organ structure. Considerable attention was paid to the subject of systemic hypertension as the heart, vessels and kidney are at risk for hypertensive damage, as well as the fact that systemic hypertension is a relatively common feature of chronic kidney disease. Definitions and guidelines surrounding hypertension were crafted with particular attention to the close relationship of the two organ systems. Discovering the best methods for detection and clinical evaluation of CvRD is a fertile subject area for future research.

MANAGEMENT OF CvRD

Management of concurrent cardiac and kidney disease can be highly challenging given the competing needs for fluid restriction in the case of congestive heart failure and the rehydration and volume expansion efforts that accompany renal failure. General management recommendations belie the complexities and nuances of treating any individual case, but despite this, the working group achieved consensus around several guiding principles. Standard care for heart failure that includes diuretics, ACE-inhibitors and vasodilators likely requires heightened monitoring and more frequent adjustment in cases of CvRD. Some members advocated for stepwise titration of certain agents when treating suspected $CvRD_H$ or concurrent primary disease of both organ systems. Unstable $CvRD_K$ or instances of concurrent primary disease, such as in cases of acute kidney injury and degenerative mitral valve disease, requires close attention to restoration of normal fluid balance and blood pressure while minimizing the risk for fluid and sodium overload (congestion). Use of stepwise titration with specific low sodium rehydration fluids, as well as recommendations for early detection of over-hydration and CHF, were presented. Nutritional aspects, such as sodium and protein intake, which are important to both heart and kidney disease were discussed. Finally, identification and treatment of systemic hypertension was again emphasized.

SUMMARY

In summary, the working group found that understanding if and how CvRD exists in veterinary medicine a challenging task. The proposed definition of CvRD includes instances of primary injury to either the cardiovascular system or kidney affecting the other organ, simultaneous primary disease of both organ systems, and instances of extra-renal/cardiovascular disease injuring both systems simultaneously. The mechanisms of CvRD in the dog and cat are largely speculative and discoveries involving the pathophysiology of CvRD are needed. Current diagnostic techniques might be inadequate to differentiate the subcategories of CvRD or to accurately guide therapy and predict outcome. Finally, treatment strategies specific to the various forms of CvRD are lacking and clinicians are often faced with difficult treatment decisions that risk further impairment of one or both of the organ systems. The working group acknowledged the closely intertwined relationship that exists between the heart and kidney, and hope that their deliberations encourage collaborative research between cardiac and renal.specialists.

REFERENCES

1. Braam B et al. Cardiorenal syndrome-current understanding and future perspectives. Nat Rev Nephrol 2014;10:48. 2. McCullough PA et al. Pathophysiology of the cardiorenal syndromes: executive summary from the eleventh consensus conference of the Acute Dialysis Quality Initiative (ASQI) Contrib Nephrol 2013;182:82. 3. Hatamizadeh P et al. Cardiorenal syndrome: pathophysiology and potential targets for clinical management. Nat Rev Nephrol 2013;9:99. 4. Ronco C et al. Cardiorenal syndrome type 1: pathophysiological crosstalk leading to combined heart and kidney dysfunction in the setting of acutely decompensated heart failure. J Amer Coll Cardiol 2012;18:1031. 5. Cruz DN et al. Role of biomarkers in the diagnosis and management of cardiorenal syndromes. Semin Nephrol 2012;32:79.

*Working group members:
Jean-Louis Pouchelon (Chair), Ecole Nationale Vétérinaire d'Alfort
Clarke Atkins (Steering Committee), North Carolina State University
Claudio Bussadori (Steering Committee), Clinica Veterinaria Gran Sasso
Mark Oyama (Steering Committee), University of Pennsylvania
Shelly Vaden (Steering Committee), North Carolina State University
John Bonagura, The Ohio State University
Valerie Chetboul, Ecole Nationale Vétérinaire d'Alfort
Larry Cowgill, University of California at Davis
Jonathan Elliot, The Royal Veterinary College
Thierry Francey, Universitat Bern
Virginia Luis-Fuentes, The Royal Veterinary College
Gregory Grauer, Kansas State University
Sydney Moise, Cornell University
David Polzin, University of Minnesota
Astrid Van Dongen, Utrecht University
Nicole Van Israel, Animal CardioPulmonary Consultancy

The Cardiovascular Renal Axis Disorders Working Group was sponsored by Vétoquinol.

CONTROVERSIES IN FELINE CARDIOMYOPATHY:
SAM OF THE MITRAL VALVE - WHAT CAUSES THAT FLIPPING THING?

Mark D. Kittleson, DVM, PhD, DACVIM (Cardiology)

Introduction

Dynamic left ventricular outflow tract (LVOT) obstruction (dynamic subaortic stenosis) due to the septal leaflet of the mitral valve (and/or the chordal structures associated with it) being dragged into the outflow tract is a common manifestation of hypertrophic cardiomyopathy (HCM) in cats. Although this phenomenon is called by several different names, systolic anterior motion (SAM) of the mitral valve is the generally accepted term.

The mechanism by which SAM occurs has been and remains controversial. The Venturi effect was an early theory. The Venturi effect is the reduction in fluid pressure (suction) that results when a fluid flows through a narrowed (stenotic) region, increasing its velocity. It was thought that the LVOT was narrowed due to HCM and that the increased velocity of blood flow in this region "sucked" the mitral valve into the LVOT. In the 1980s, the realization that SAM can start prior to the ejection of blood through the LVOT effectively disproved this theory.[1] However, one can occasionally still find this as a postulated mechanism in the literature.

While there may be other mechanisms, it appears to be currently well-accepted that 1) abnormal papillary muscles can play an important role and that 2) the anterior (septal) leaflet of the mitral valve is frequently elongated (in humans and in cats).[1,2,3,4] While it is easy to see how papillary muscles can be altered in HCM since they are made up of myocardium and can hypertrophy along with the rest of the LV, it is harder to fathom why a genetic disease that alters sarcomeres (HCM) would cause a mitral valve leaflet to elongate in humans. While this remains a mystery, it is the author's speculation that the leaflet elongates (grows longer) secondary to being stretched each time SAM occurs (i.e., with each beat). The more it elongates, the greater the propensity for SAM (vicious cycle). So the speculation is that this is a secondary finding that exacerbates the problem.

Mechanism of SAM

SAM occurs in both humans and cats with HCM. Although numerous variations in regional wall thickening occur in both species, asymmetric septal hypertrophy is the most common form seen in humans while cats more frequently have global hypertrophy.[3] Consequently, the mechanism of SAM production in the two species may be somewhat different. Papillary muscle hypertrophy is a frequent abnormality in cats with HCM and the papillary muscle hypertrophy is more severe in cats with HCM and SAM than in cats with HCM and no SAM.[3] Since the papillary muscles arise from the free wall, hypertrophy of these structures may not be as common in humans. Isolated papillary muscle abnormalities (e.g., displacement) have been shown to be capable of causing SAM in humans.[5] Surgically transplanting papillary muscles cranially (toward the LVOT) in dogs results in SAM.[2]

Based on observations of cats in a research colony with HCM and in clinical cases of feline HCM, the author believes that papillary muscles are the key component of SAM in cats with HCM. This is often not appreciated for two reasons - 1) Although SAM is a systolic phenomenon, frequently measurements that attempt to relate variables to its generation are made in diastole. 2) If standard views of the LV and LVOT are used (where the papillary muscles are purposefully not in view), the phenomenon is missed.

Normally, papillary muscles are positioned below the mitral valve where they tense and contract in systole in order to prevent the mitral valve from prolapsing into the left atrium. When they contract they thicken as well as move with the left ventricular free wall toward the interventricular septum. In a normal left ventricle they end up positioned toward the center of the chamber at end-systole. End-systolic cavity obliteration, especially in the apical and midventricular regions of the LV, is a common phenomenon identified in cats with HCM. A large part of the reason this happens is because of the enlarged (hypertrophied) papillary muscles that markedly thicken in systole. So in feline HCM the papillary muscles, instead of ending up in the middle of the LV chamber at the end of systole, end up touching the interventricular septum. That means the cranial papillary muscle end ups within the lower (apical) LVOT and below the remaining basilar LVOT. Based on this position alone, it is only logical that this papillary muscle must pull the tips of the septal (anterior) leaflet of the mitral valve toward and potentially into the LVOT. If one takes the time, while viewing the SAM, to move the 2D ultrasound beam forward and backward within the LV using a right parasternal long-axis view so that the papillary muscles come into view, one can often see the cranial papillary muscle, associated chords and a portion of the septal leaflet of the mitral valve in the LVOT. If one constrains oneself to the usual view where the papillary muscles are not in view, of course, this phenomenon is not appreciated. Instead all one sees are the mitral valve leaflet tip and/or it's associated chords in the LVOT. When the papillary muscles are included in the view one sees the papillary muscle below, the chords in the middle, and the malpositioned mitral valve leaflet (the SAM) above (figure 1). There are only two potential explanations for this - 1) the papillary muscles are dragging the chords and leaflet into the LVOT or 2) the mitral valve leaflet is dragging the papillary muscle into the LVOT. The force required to drag a mitral valve leaflet out of its normal position must be tremendous. The force needed to drag a papillary muscle into the LVOT must be even greater. Since the mitral valve and its associated chords cannot generate force, it is illogical to think that the mitral valve is the aggressor in this situation. And there is nothing else in the LV that can generate the sort of force needed to initially move the mitral valve so out of position - only the papillary muscles.

However, once the tip of the mitral valve leaflet is pulled into the LVOT, the LV itself is now ejecting blood with great force into the LVOT and out the aorta. Consequently, it is likely that once a tip of a leaflet is forced into the LVOT, blood flow gets underneath that leaflet tip and pushes it even further into the LVOT - up against the septum. This is termed the "dragging force" in the literature. This results in the tip of the leaflet extending into the LVOT beyond the position of the tip of the papillary muscle in some cases.

Then, of course, the thicker the interventricular septum is the closer it is to the papillary muscles and to the tip of the septal leaflet of the mitral valve and so the easier it is to displace the mitral valve into the LVOT.

SAM is not only caused by HCM. Dobutamine infusion in dogs and hypovolemia are examples of other methods of producing SAM.[6,7] Both can be explained again by the cranial papillary muscle encroaching on the LVOT in systole. However, primary mitral valve abnormalities can also result in SAM.[8,9]

Figure 1

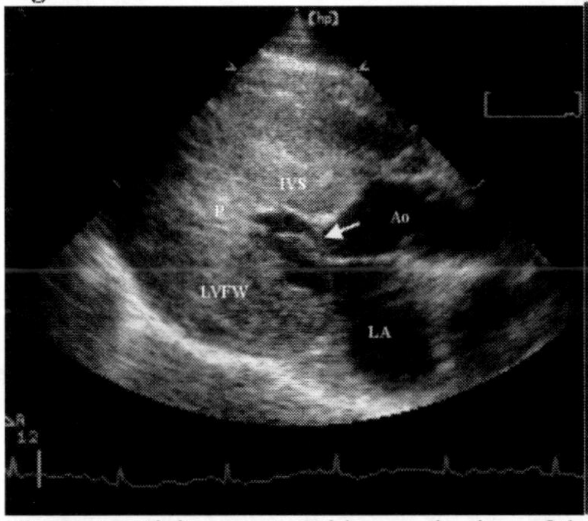

Figure 2

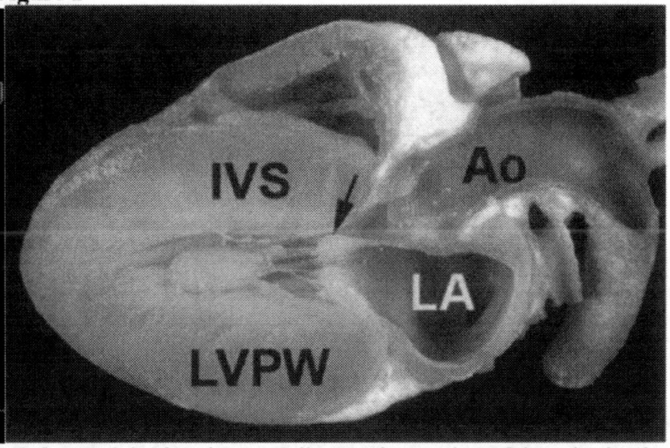

Figure 1 - Right parasternal long-axis view of the left ventricle showing systolic anterior motion of the mitral valve in a cat with severe HCM. Note the markedly enlarged papillary muscle pulling the mitral valve leaflet up against the septum (arrow). Ao - aorta; LA - left atrium; LVFW - left ventricular free wall; P - papillary muscle; IVS - interventricular septum

Figure 2 - Post mortem specimen of a heart from a cat with severe HCM. Myocardium commonly undergoes rigor after death and so simulates systole. The large papillary muscle looks as if it is pulling the mitral valve leaflet into the left ventricular outflow tract. Ao-aorta; LA - left atrium; LVFW - left ventricular free wall; P - papillary muscle; IVS interventricular septum; LVOT - left ventricular outflow tract

REFERENCES

1. Jiang L, et al. Am Heart J 1987;113:633–644.
2. Levine RA, et al. Circulation 1995;91:1189–1195.
3. Schober K, Todd A. J Vet Cardiol 2010;12:1–16.
4. Kim D-H, et al. Circulation 2010;122:1298–1307.
5. Izgi C, et al. Echocardiography 2010;27:E36–E38.
6. Sakurai S, et al. Circulation 1985;71:805–812.
7. Levisman JA. Chest 1976;70:687–688.
8. Kuijpers NW, et al. Tijdschr Diergeneeskd 2011;136:326–331.
9. Ohkado A, et al. J Heart Valve Dis 1997;6:60–62.

CONTROVERSIES IN FELINE CARDIOMYOPATHY: PIMOBENDAN -WHAT WE DO KNOW AND DON'T KNOW IN CATS

Mark D. Kittleson, DVM, PhD, DACVIM (Cardiology)

Introduction

Pimobendan (PB) is a pyridazinone benzimadazole derivative with positive inotropic and vasodilating properties.[1] The positive inotropic properties are thought to be due to a combination of sensitization of sarcomeric proteins to calcium and phosphodiesterase (PDE) III inhibition. It was approved for use in dogs in heart failure in the USA in 2007. Once a drug is approved for use in the USA there are few constraints on its use. Predictably veterinarians started to use the drug for non approved uses soon after it was marketed, including use in cats. The package insert states that the drug is contraindicated "in cases of hypertrophic cardiomyopathy (HCM)".

Retrospective Studies

There have been no prospective clinical trials looking at the use of PB in cats. There have been two retrospective analyses of cats in heart failure treated with PB. In the first, records from 170 cats with various forms of heart disease treated with PB were examined.[2] The investigators looked at records of cats that were in heart failure from a variety of cardiac diseases that were administered PB. The median dose used was 0.24 mg/kg q12 h (range 0.08 to 0.42mg/kg q12 h). Hypertrophic cardiomyopathy (HCM) was diagnosed in 40% of cases, unclassified cardiomyopathy (UCM) in 37%, dilated cardiomyopathy (DCM) in 16%, primary mitral regurgitation in 4%, and various other forms of cardiac disease in the remaining 7%. The echocardiographic variables were lumped together rather than split amongst the disease categories. The median LVIDs was 12 mm and 73% had a shortening fraction <40%, indicating that the majority had some degree of myocardial failure. The data on the group with HCM is sparse and so it is unknown how many had myocardial failure. Only 4 cats had dynamic left ventricular outflow tract obstruction (DLVOTO) so it would appear that clinicians in the various veterinary hospitals from which these records were obtained refrained from using PB in this population. The changes noted in echocardiographic variables from baseline to after PB administration in the 34% of cats that had a recheck echocardiogram were all well within measurement error and so appear to be clinically insignificant despite the fact that some attained statistical significance. A median survival time of 151 days was identified but there was no control group for comparison.

The second retrospective study looked at 27 cats in heart failure from one institution.[3] This population included only 1 cat with HCM that had regional LV wall hypokinesis while 41% had UCM, 30% had DCM, 15% had arrhythmogenic right ventricular cardiomyopathy, and 11% had congenital heart disease. All had evidence of left or right ventricular myocardial failure. The dosage administered was 0.25-0.3 mg/kg q12 h. One cat with DLVOTO developed systemic hypotension. There were no changes in echocardiographic variables after starting PB. Most clinical variables did not improve after PB but fewer cats were dyspneic and fewer were anorexic. However, there was no apparent attempt to maintain the other drugs and drug dosages constant so this improvement cannot be ascribed to PB. All 26 cats died. Median survival time was 167 days.

Based on these two studies, it would appear that one can say PB in cats in heart failure doesn't produce outward (apparent) harm in the short term.

Pharmacology

Ideally one would have pharmacokinetic and pharmacodynamic data on PB in normal cats as well as safety data followed by at least one randomized, placebo-controlled, and blinded clinical trial addressing efficacy in each of the cardiac disease seen in cats prior to its use. Recently (2012) pharmacokinetic data in cats were generated for the first time.[4] Background: In other species PB is metabolized (demethylated) in the liver by a cytochrome p450 isoenzyme (CYP1A2) to desmethylpimobendan (DMP).[5] DMP is an active metabolite that has approximately 8 and 100 times the PDE III inhibiting capability as PB in failing human and guinea pig myocardium respectively.[6,7] PB and DMP are both responsible for sensitizing proteins to calcium in dogs.[8] Consequently, it is important to know if cats produce DMP and, if so, to what degree and what its clearance is. DMP also inhibits PDE I and especially PDE II in human myocardium.[7] The positive inotropic effect of pimobendan is attenuated in human failing myocardium.[7] There are species differences when it comes to potency of both compounds and possibly mechanisms by which they increase contractility. Cats are less responsive to DMP than dogs.[9]

In the aforementioned study, 10 normal cats were initially used to study the pharmacokinetics of PB in cats. Two cats developed GI side effects (vomiting and ptyalism) and were removed from the study. Cats were administered PB chewable tablets at a dose of 0.25-0.3 mg/kg (1.25 mg/cat). Blood samples were obtained to measure PB and DMP concentrations over time. The drug concentration peaked (30-40 ng/ml) and subsequently declined (<1 ng/ml by 8 hours) in a log-linear fashion, characteristic of a simple one-compartment model (steady hepatic metabolism that is based on serum concentration presented to the liver). Half life was longer than reported in dogs (1.3 hours compared to 30 minutes) but there was no drug accumulation when the drug was given q12 h. An ion consistent with DMP was identified in some cats but not in others but a reference standard was not available so it is still unknown if cats produce DMP or not. The stunning result was that the peak serum concentration of PB was approximately 10 times that seen in dogs administered the same dose. Obviously this calls into question the safety of using this dose in cats. However, standard safety studies in dogs have been done by the company using doses of 2 mg/kg (7X dose) and 8 mg/kg (30X dose) IV q24 h for 4 weeks (http://www.fda.gov/downloads/animalveterinary/products/approvedanimaldrugproducts/foiadrugsummaries/ucm062328.pdf). Since this was an IV study the 2 mg/kg (7X) dose once a day study probably achieved a serum concentration similar to those seen in cats at least once a day. In these studies there was a low incidence of cardiac abnormalities on necropsy in the 2 mg/kg group and a high incidence in the 8 mg/kg group and consisted of mitral valve thickening and hemorrhage, and multifocal necrosis and scarring in papillary muscles, interventricular septum, and right ventricular free wall. The myocardial lesions are routinely seen in normal dogs administered a potent positive inotropic drug and so are expected. Overdose of a potent calcium sensitizing agent should cause severe diastolic dysfunction since calcium sensitizing drugs make it more difficult for myocardial proteins to dissociate from calcium. The fact that dogs can survive for 4 weeks on an approximately 30 times clinical dose calls into question the potency of PB when it comes to calcium sensitization. If PB is a relatively weak calcium sensitizer that would mean that its metabolite, DMP, may be the more potent of the two compounds. Hepatic CYP1A activity is approximately half in cats compared to dogs so it might be expected that DMP production in cats is lower than in dogs.[10] Maybe that means a serum concentration in cats that is 10 times higher than in dogs is needed to produce an effect. Maybe it means the drug has little to no effect in cats. Maybe it means there are subclinical toxic effects. It would appear that further studies are needed prior to recommending its routine use in cats to veterinary practitioners.

Even if the drug is safe and efficacious in cats, should it be used in cats with HCM and normal myocardial function? If nothing else it is counterintuitive that one could use it in HCM since there would appear to be no reason to increase contractility in a ventricle that is already hyperdynamic. Even though the hyperdynamic quality is due to decreased systolic wall stress (afterload) and the myocardium itself may be hypocontractile, it is still counterintuitive.[11] The primary problem with HCM is impaired diastolic function. PB is known to have positive lusitropic properties.[12] Could this be beneficial in a cat with HCM? Yes. But it is also counterintuitive that a drug that is a calcium sensitizer would have beneficial effects on diastolic function. What improves diastolic function in dogs - PB, the calcium sensitizer, or DMP, the PDE III inhibitor? Logic would suggest it's the PDE III inhibitor - the one that may not be produced in cats or produced at a lower level.

Survey

A survey of veterinary cardiologists/residents regarding their use of PB in cats with cardiomyopathy was performed by the author. Results show - 1. Few use PB in cats with HCM unless they have myocardial dysfunction or are refractory to conventional therapy. 2. In cats with HCM refractory to conventional therapy, around 50% use it only if SAM is not present and 50% use it regardless of whether it's present or not. 3. Almost all use PB in cats with HCM and myocardial failure, DCM, and UCM/RCM when the cat is in heart failure. 4. In cats with HCM, around 50-60% think that more than 50% of cats have a clinical response to pimobendan. 6. Approximately half had the impression that cats with HCM and DCM lived longer. Only around one-third thought cats with UCM/RCM lived longer. 7. Most use a dose of 0.25-0.3 mg/kg BID.

Conclusion

In conclusion, we know too little about PB in cats. Although the drug appears to be safe in the short term (weeks to months) it's not known what the long-term consequences of the drug are. And we know almost nothing with regard to its ability to increase contractility in failing feline myocardium or to dilate feline blood vessels. However, most veterinary cardiologists use PB in cats and many believe they see improvement in a significant number. Very few use it as first line therapy for heart failure due to HCM.

REFERENCES

1. Fitton A, et al. Drugs Aging 1994;4:417–441.
2. Macgregor JM, et al. J Vet Cardiol 2011;13:251–260.
3. Gordon SG, et al. J Am Vet Med Assoc 2012;241:89–94.
4. Hanzlicek AS, et al. J Vet Cardiol 2012.
5. Kuriya S, et al. Drug Metab Dispos 2000;28:73–78.
6. Westfall MV, et al. J Pharmacol Exp Ther 1992;260:58–63.
7. Bohm M, et al. CircRes 1991;68:689–701.
8. Takahashi R, Endoh M. J Cardiovasc Pharmacol 2001;37:209–218.
9. Smith PJ, et al. J Small Anim Pr 2005;46:121–130.
10. Van Beusekom CD, et al. J Vet Pharmacol Ther 2010;33:519–527.
11. Wess G, et al. J Vet Intern Med 2010;24:1375–1382.
12. Joho S, et al. J Card Fail 2002;8:423–430.

CONTROVERSIES IN FELINE CARDIOMYOPATHY:
FELINE LEFT VENTRICULAR NONCOMPACTION - REAL OR FANTASY?

Mark D. Kittleson, DVM, PhD, DACVIM (Cardiology)

Left ventricular noncompaction (LVNC) is a relatively new form of cardiomyopathy in humans. As with other cardiomyopathies during their emerging phase, it goes by several names including, but not limited to, isolated left ventricular noncompaction, noncompaction cardiomyopathy, noncompaction of the ventricular myocardium, and spongiform cardiomyopathy. The World Health Organization still classifies it as an unclassified cardiomyopathy while the American Heart Association recognized it as a distinct form of cardiomyopathy in 2006.[1] The diagnostic criteria vary somewhat from author to author. For example, one says "It is characterized morphologically by a severely thickened, 2-layered myocardium, numerous prominent trabeculations, and deep intertrabecular recesses (figure 1).[2] Another says it is "characterized by a distinctive ("spongy") morphological appearance of the LV myocardium. Noncompaction involves predominantly the distal (apical) portion of the LV chamber with deep intertrabecular recesses (sinusoids) in communication with the ventricular cavity."[1] It is speculated that it results from intrauterine developmental arrest of the normal compaction process of the myocardium during the first trimester leading to the formation of two layers of the myocardium: the compacted and the noncompacted layer. In brief, during embryonic development, the myocardium, when first formed, has deep recesses (trabeculations) that presumably allow for myocardial oxygenation. These recesses disappear (the myocardium compacts) during development as the myocardium becomes vascularized. In LVNC it is hypothesized that the inner myocardium fails to compact.

LVNC is identified as an isolated abnormality but is also seen in association with various congenital heart diseases and along with other forms of cardiomyopathy. In the isolated form, both familial and nonfamilial forms have been identified and mutations in sarcomeric genes (beta-myosin heavy chain (MYH7), alpha-cardiac actin (ACTC), and cardiac troponin T (TNNT2)), Z-line genes, mitochondrial genes, the alpha-dystrobrevin gene, the lamin A/C gene, and the tafazzin gene have been identified.[1,3] It has also been found that certain sarcomeric gene mutations can cause LVNC as well as hypertrophic cardiomyopathy (HCM) in different families and even LVNC, HCM and dilated cardiomyopathy (DCM) in the same family.[3,4]

The diagnosis is made most commonly using echocardiography although it can also be identified with MRI, CT, and angiography. Since there is no gold standard , the diagnostic criteria are somewhat variable from author to author.[5] This variability creates confusion. There are currently three different sets of criteria published. A recent study compared the three and found they all agreed with each other in only 30% of patients.[6] In all there is compacted (normal) myocardium toward the epicardium and trabeculated inner myocardium. The abnormality is most common apically. Variations in diagnostic criteria focus on echocardiographic views, number of trabeculations, and whether assessments should be made at end-diastole or end-systole.

Many patients with the isolated form of the disease are asymptomatic but LVNC can cause heart failure, arrhythmias, sudden death, and thrombo-embolic events. As expected, patients with a history of heart failure, ventricular tachycardia or an enlarged left atrium do worse than those without.[7]

To the author's knowledge, LVNC has only been reported in humans. This presentation will focus on echocardiograms from one 6-year-od cat from a research colony at UCDavis with suspected LVNC (figure 2). Echocardiograms from several other cats will also be presented for discussion.

Figure 1

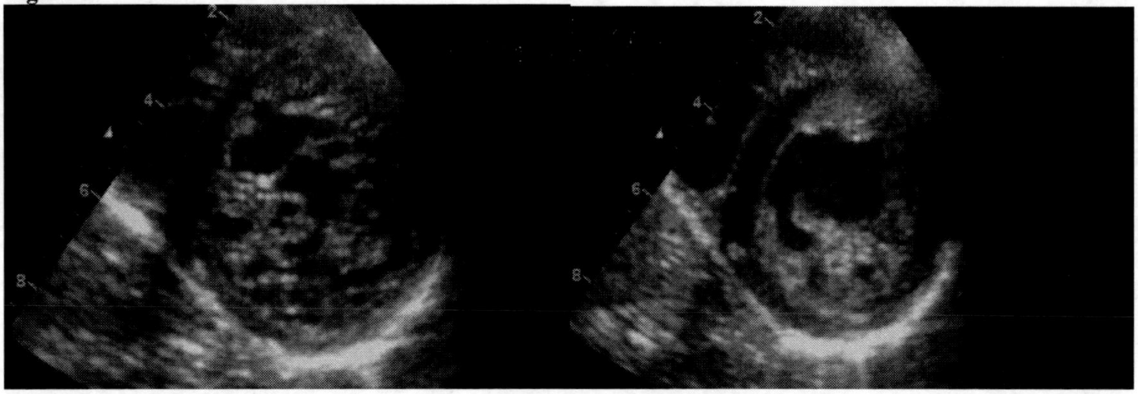

Echocardiographic views of the left ventricle from a human patient with LVNC showing outer compacted (normal) myocardium next to the pericardium and inner noncompacted (spongiform) myocardium. From Klaassen S, et al: Mutations in sarcomere protein genes in left ventricular noncompaction. Circulation 2008 Jun 3;117(22):2893-901.

Figure 2

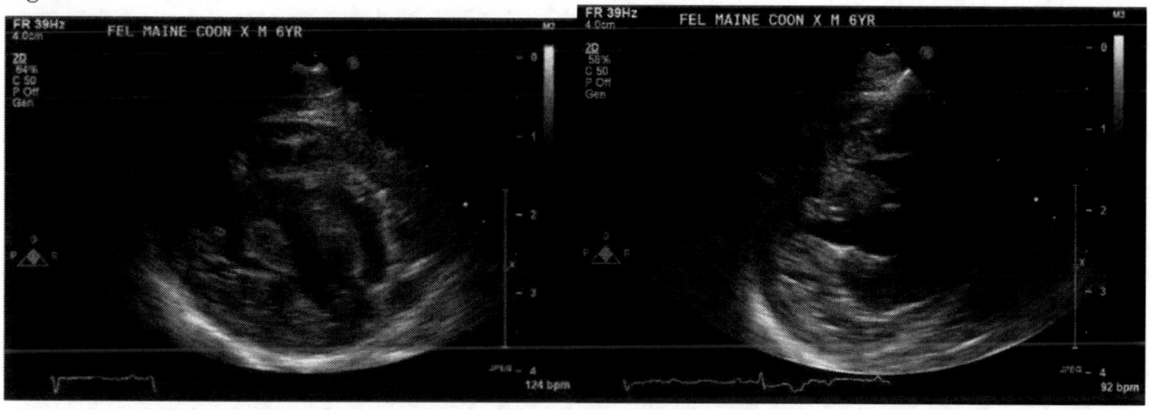

Echocardiographic views of the left ventricle from a Maine Coon cross cat that was heterozygous for the A31P mutation showing a thinner rim of normal myocardium next to the pericardium and a thicker spongiform ("moth-eaten") inner myocardium with deep recesses.

REFERENCES

1. Maron BJ, et al. Circulation 2006;113:1807–1816.
2. Udeoji DU, et al. Ther Adv Cardiovasc Dis 2013;7:260–273.
3. Klaassen S, et al. Circulation 2008;117:2893–2901.
4. Shieh JTC, et al. Am J Med Genet C Semin Med Genet 2013;163:141–143.
5. Paterick TE, et al. J Am Soc Echocardiogr 2012;25:363–375.
6. Kohli SK, et al. Eur Heart J 2008;29:89–95.
7. Stanton C, et al. Am J Cardiol 2009;104:1135–1138.

CRITICAL ASSESSMENT OF TREATMENT OF HYPERTROPHIC CARDIOMYOPATHY

Kristin A MacDonald, DVM, PhD, DACVIM (Cardiology)
Rohnert Park, CA

INTRODUCTION

There is a lack of studies evaluating the impact of treatment on the clinical course of disease in asymptomatic cats with HCM. The clinician is forced to form an opinion based often on personal experience, the anecdotal experience of an expert, or based on small studies looking at specific echocardiographic variables which may not translate to an impact in clinical progression of disease. Therefore, there is great controversy of whether asymptomatic cats should be treated, and with what specific medical therapy. Ideally, the fundamental goals of in treatment of asymptomatic cats would include: reduction in LV hypertrophy, improvement in diastolic function, reduction in risk of arterial thromboembolism, and prolongation to time of heart failure. However, there is little data that confirms or denies any particular treatment has a considerable effect on these goals. The purpose of this lecture will be to review the available literature pertaining to treatment of HCM and address clinical practices in HCM treatment.

TREATMENT DECISIONS

Given the scarcity of high level evidenced based data on treatment of feline HCM, how do we as clinicians make treatment decisions? This question was addressed in a survey that was sent to the ACVIM Cardiology list serve, which included 12 different clinical scenarios of cats with HCM ranging from mild subclinical to severe clinical disease.[1] Ninety-eight (29% of list serve) people responded, including cardiology diplomates in ACVIM and ECVIM (Cardiology), residents, and clinicians with a cardiology focus. There was a wide variation in treatment selections or abstention of treatment, which changed with the severity of the HCM in the clinical scenario, and treatment behaviors was not different based on diplomate status. Most people did not elect to treat very mild subclinical HCM (84%), but treatment was more commonly elected as the subclinical HCM worsened (i.e. severity of SAM, left atrial dilation). Once left atrial dilation was present and HCM was more severe, 50% of clinicians prescribed beta blockers, 25% chose ACE inhibitors, 10% chose diltiazem, and 20% chose an antiplatelet agent (aspirin or clopidogrel). Once systolic anterior of the motion became moderate to severe, the majority of respondents prescribed a medication (90-98%), with beta blockers favored (94%). Fewer clinicians prescribed beta blockers in the clinical scenarios with heart failure present, although diltiazem was more frequently used in the clinical heart failure scenarios in up to 30% of respondents. The reason why clinicians did not prescribe a medication was most frequently classified as due to the lack of evidence of benefit or that the treatment was not logical in that specific clinical scenario. However, the reasons why certain drugs were selected was more variable, including: theorhetical unproven benefit in cats or in other species (interspecies extrapolation), favorable personal experience, consensus among peers favoring use, and an opinion that the medication probably does not hurt and may be helpful. Probably the reasons for such variation and low class of evidenced base medicine clinicians utilize in making treatment decisions is due to the lack of prospective, controlled, high powered clinical trials in cats with HCM and lack of trials evaluating clinical outcome rather than physiologic or echocardiographic variables.

Veterinarians are not the only ones plagued by a lack of studies and evidenced base medicine for treatment of HCM. Although beta blockers are considered the "mainstay" of therapy in people with hypertrophic obstructive cardiomyopathy, the impact on clinical outcome is not defined, as stated in a 2012 article: "Whether long-term treatment with b-blockers ultimately impacts outcome in HCM patients remains still undefined."[2] Likewise, even though calcium channel blockers are considered standard therapy for people with non-obstructive HCM, there is also a lack of evidenced base medicine evaluating outcomes. "To date, there is no definite evidence that verapamil effectively improves functional capacity in HCM, although the drug has been used for decades to ameliorate quality of life in non-obstructive patients, and is considered standard treatment."[2]

SUBCLINICAL HCM

Beta blockers vs. Calcium channel blockers

Potential, theoretical goals of beta blocker therapy are to decrease severity of SAM of the mitral valve (if moderate or severe), decrease severity of left ventricular concentric hypertrophy which may improve ventricular diastolic compliance, treatment of concurrent supraventricular or ventricular tachyarrhythmias, and reduction in stress-induced tachycardia. These beneficial effects are mediated by sympathetic modulation of heart rate, ventricular contractility, prolonging diastolic filling time to improve myocardial perfusion, and reduced excitability. Atenolol is highly bioavailable (90%), has a half-life of 3.5 hours in the cat, and beta blocking effects last for 12 hours but less than 24 hours. Doses may range from 6.25-12.5 mg/cat q12-24 hr, often uptitrating the dose based on heart rate and murmur grade.

Diltiazem is a non-dihydropyridine (i.e. not vascular selective) calcium channel blocker used in HCM. It has mild negative inotropic effects, less than verapamil, and is used as an antihypertrophic agent in HCM, as well as for treatment of supraventricular arrhythmias. Calcium channel blockers improve early diastolic relaxation by reducing tau and isovolumic relaxation time.[3] Standard formulation of diltiazem requires every 8 hour dosing, which may cause challenges in long-term medication in cats who are notorious for resisting oral medications. Alternatives are two different sustained release formulas, which may have other pharmacologic short comings in cats. Cardizem CD is a sustained release formulation which is available as a 180 mg capsule that is filled with hundreds of microspheres, and can be divided by a compounding pharmacy to 45 mg doses placed in smaller gelatin capsules, to be administered as 45 mg PO q24 hr, or 10 mg/kg PO q24 hr.[4] Cardizem CD has half the bioavailability of the conventional preparation. The time to peak concentration is 340 minutes and terminal half-life is 410 minutes. Pharmacodynamic effects on heart rate and PR interval were studied in healthy cats, and neither conventional nor CD formulation reduced heart rate or prolonged PR interval, despite blood concentrations within the therapeutic range defined in people. Dilacor XR is another sustained-release form of diltiazem, and it is supplied as 120-, 180-, and 240-mg capsules, containing two, three, or four 60-mg sustained release tablets, respectively, which are removed from the gelatin capsule and then broken in half, and given at 30 mg PO q24 hr. A pharmacologic study of Dilacor XR in normal cats and cats with HCM showed that 30 or 60 mg PO q24 hr produced erratic blood levels of diltiazem, and many cats had higher than the recommended range of blood concentration for 12 hours, and less than recommended range at 18 and 24 hours post-pill.[5] Thirty six percent of cats with HCM had adverse reactions, consisting of decreased appetite, vomiting, lethargy, weight loss, and diarrhea.

A prospective, blinded, placebo controlled study evaluated the effects of atenolol versus diltiazem in asymptomatic cats with HCM, although the study remains unpublished and was presented as a presentation at an ACVIM forum. Results were relatively underwhelming. Atenolol had modest effects to improve one variable of diastolic function measured by echocardiography and slightly reduced interventricular septal thickness, and there were no significant effects of diltiazem on diastolic function, left atrial size, or LV hypertrophy. Atenolol also reduced the severity of SAM of the mitral valve, and diltiazem did not. In contrast, a small, unblinded, uncontrolled study found that diltiazem reduced LV free wall and interventricular septal thickness from 9 mm +/- 0.5 mm to 6 mm +/- 0.6 mm after 6 months of therapy, improved symptoms, and shortened LV relaxation time index after 3 and 6 months of treatment in 12 cats with HCM, yet these results have never been duplicated in other studies.

Effect of atenolol on biomarkers in subclinical HCM

Two small pilot studies evaluated the effect of atenolol on cardiac troponin I in subclinical cats with HCM, both finding opposing results. Cardiac troponin I is a cardiac isoenzyme that increases when there is cardiomyocyte damage or ischemia, and was shown to decrease over one to four weeks of atenolol therapy in five asymptomatic cats with HCM, [6] but another study reported that there was no change in cTnI or NT-proBNP in 6 cats with subclinical but severe (unobstructed) HCM when treated with atenolol 12.5 mg PO BID for 30 days. Recently, the AQUA study (unpublished) also showed no effect of atenolol on cTnI or NT-proBNP in cats with subclinical HCM.

Effect of Atenolol on clinical outcomes

The AQUA study evaluated the effects of atenolol on clinical variables including mobility and quality of life in cats with subclinical HCM, which included 19 cats treated with atenolol and 20 cats treated with placebo. Only 11 of the 19 atenolol treated cats and 10 of 20 placebo treated cats completed the study. The AQUA study reported atenolol improved owner's subjective overall assessment of their cat's quality of life compared to placebo but had no effect in mobility quantified by an accelerometer or in detailed questionnaire of quality of life. All cats were sedentary, with the average inactivity of 19 hrs a day, spending 78% of their day sleeping or resting regardless of treatment group. No studies have evaluated diltiazem on QOL measures.

The REVEAL study retrospectively evaluated a large cohort of asymptomatic cats with HCM (n=547) compared to normal cats (n=687) on time to heart failure, survival times, and impact of medications such as beta blockers, ACE inhibitors, diltiazem, and aspirin on survival. This is an unpublished study at the time of writing, and was presented at 2013 ACVIM Forum. It reported no effect of treatment with the aforementioned medications on survival time.

A non-randomized study evaluated all cause and cardiac mortality over 5 years in subclinical cats with HCM treated with either atenolol (n=42) versus no treatment (n=21). At baseline, cats placed on atenolol had more severe HCM based on larger left atrial size and more severe left ventricular concentric hypertrophy, were more likely to have SAM than did the cats in the no-treatment group, making it challenging to compare the effect of treatment group on survival.[7] The event rates were low (10/42 atenolol cats = cardiac death; 4/21 no treatment cats = cardiac death; 19/42 atenolol cats total deaths; 8/21 no treatment cats = total deaths), with an overall 5 year mortality rate of all HCM cats of 22%. There was no difference in time to death (all-cause) in HCM cats treated with atenolol vs. no treatment (P= 0.061), yet the study is underpowered and authors define it as pilot data rather than definitive evidence. Atenolol did not increase risk of heart failure, weakness, or syncope.

No studies have evaluated the effect of diltiazem on time to development of heart failure, mortality, or other clinical outcomes.

Renin angiotensin aldosterone system antagonists

ACE inhibitors

Angiotensin converting enzyme (ACE) inhibition has been shown to reduce LV hypertrophy in 2 small, uncontrolled or retrospective studies, where LV hypertrophy was assessed by either M-mode or 2-dimensional echo.[8,9] However, 2-dimensional echocardiographic assessment of LV hypertrophy is fraught with high inter-and intraobserver variability of 18-20% respectively. The speaker completed a double blinded, placebo controlled, randomized, prospective study evaluating the effect of ramipril on LV mass, diastolic function, neurohormones, and blood pressure in 26 Maine coon and Maine coon-cross cats with mild to severe familial HCM without CHF.[10] LV mass was quantified by cardiac magnetic resonance imaging, diastolic function was assessed by pulsed wave tissue Doppler imaging (TDI), and plasma concentration of aldosterone and brain natriuretic peptide were measured. There was no difference in LV mass, diastolic function, delayed contrast enhancement assessment of myocardial fibrosis, blood pressure, or neurohormones between the placebo and ramipril treatment groups.

Aldosterone antagonists

A small, double blinded, placebo controlled prospective study evaluating the effect of spironolactone on diastolic function assessed by TDI and on LV mass quantified by echocardiography was done on 26 Maine Coon cats with familial HCM without CHF.[8] All cats had diastolic dysfunction at baseline. Cats were given 2 mg/kg PO BID of placebo or spironolactone for 4-months, and echocardiography was performed every 2 months. There was no difference in diastolic function or systolic function assessed by TDI,

LV mass, or left atrial size. Serum aldosterone concentration was markedly elevated in cats treated with aldosterone. Approximately one-third of cats given spironolactone developed severe facial ulcerative dermatitis, which may have been caused by a drug reaction

Anticoagulant/ antiplatelet agents

Anticoagulant therapy is necessary in cats that have suffered from arterial thromboembolism or who have echocardiographic evidence of a thrombus or spontaneous contrast within the left atrium, and is typically recommended in cats with moderate or severe left atrial dilation. Aspirin (5 mg – 81 mg PO q 3 days) has relatively underwhelming and variable success in prevention of thromboembolism. In cats suffering from arterial thromboembolism, survival with or without thrombolytic therapy is only 30-40%. Recurrence rate with aspirin (standard or low dose) is variable and ranges from 28-90%.[15,16] Clopidogrel is a novel anti-platelet agent that irreversibly inhibits ADP receptors on platelet membranes. Clopidogrel has been shown to inhibit platelet aggregation, increase oral mucosal bleeding time, and reduce plasma serotonin concentration in normal cats at doses as low as 18.75 mg PO q 24 hours.[17] The FATCAT clinical trial evaluated clopidogrel (18.75 mg PO q24 hr) versus aspirin (81 mg PO q72 hours) in cats that recovered from arteriothromboembolism (ATE), with the primary end point of time to recurrent ATE within 1 year and secondary end point of cardiac mortality. Results were presented in ACVIM 2013 but have not yet been published at the time of this writing. Thirty six cats in each group were studied, and cats in the clopidogrel group had a median time to recurrent ATE of > 365 days compared to 192 days for aspirin- treated cats. Cardiac mortality was lower in clopidogrel- treated cats (median= 443 days) compared to cats treated with aspirin (median = 192 days), and combined ATE recurrence and cardiac death was longer in clopidogrel treated cats than aspirin treated cats (median 346 days vs. 128 days respectively).

Low molecular weight heparins (LMWH) such as dalteparin and enoxaparin are attractive alternatives to unfractionated heparin given their increased bioavailability and prolonged half- lives. Compared to unfractionated heparin, LMWHs work more specifically upstream in the coagulation cascade against Factor X with much lower activity against thrombin (Factor II). Since there is less anti-II activity, LMWHs do not alter PT and APTT times, and therapeutic efficacy must be assessed by measurement of anti-Xa activity. There are only a few pilot studies evaluating pharmacokinetics of dalteparin and enoxaparin in healthy cats. 1.5 mg/kg of enoxaparin SQ appears to adequately suppress factor Xa activity, but the optimal dosing interval is less clearly defined.[18] Dosing frequencies of BID to TID are likely necessary for sustained anti-Xa activity. In cats suffering from arterial thromboembolism, thrombolytic therapy with streptokinase, tissue plasminogen activator, or urokinase are generally not recommended as they are associated with approximately 50% mortality, the same mortality as conservative supportive therapy.

Ivabradine

Ivabradine is a novel negative chronotropic therapy that inhibits the pacemaker funny current channel (If) of the sinus node. Pharmacodynamic effects were evaluated in 6 normal cats and 6 cats with preclinical HCM, with invasive measures of systolic and diastolic hemodynamic function.[9] The physiologic rationale for using a negative chronotropic agent in HCM is that it could blunt the deleterious effects of tachycardia including: reduction in LV filling, decreased myocardial perfusion, and increased myocardial oxygen demand which may worsen preexisting diastolic dysfunction and myocardial ischemia. Treatment groups included esmolol and ivabradine, each with and without dobutamine. Ivabradine but not esmolol significantly reduced heart rate and reduced the heart rate – rate pressure product. Tachycardia induced by dobutamine was blunted by ivabradine not esmolol. However, ivabradine increased the left ventricular end diastolic pressure in both normal and HCM cats (4.9 mmHg, 4.2 mmHg respectively). This is a concerning finding given that elevated LV end diastolic pressure over a certain threshold (>25 mmHg) leads to development of left heart failure, and cats with subclinical HCM may be just beneath this threshold of heart failure.

CLINICAL HCM WITH DECOMPENSATED HEART FAILURE

The largest multicenter clinical study to date has evaluated the effect of atenolol, diltiazem, and enalapril on survival in symptomatic cats with CHF or arterial thromboembolism secondary to cardiomyopathy (mostly due to HCM).[12] Cats received a background therapy of furosemide if CHF was present. None of the drugs improved survival, and there was a trend in reduced survival in the atenolol group. For this reason, atenolol is often avoided or reduced once heart failure develops in cats.

Overall, what appears to be the standard treatment of heart failure secondary to HCM or other cardiomyopathies is furosemide and an angiotensin converting enzyme (ACE) inhibitor. Furosemide may be given parenterally (1-3 mg/kg every 1-4 hours) in cats with fulminant CHF, and the furosemide dose should be rapidly tapered once the respiratory rate and effort begin to improve. Although efficacy is unknown, the venodilator nitroglycerin may be applied transdermally every 6 hours for 2 days (until tolerance develops) for hospitalized patients. Oxygen therapy is necessary in severe cases, and the fractional inspired oxygen percentage should be decreased to 50% or less if possible within 12 hours to avoid further barotrauma. An ACE inhibitor can be slightly postponed in cats receiving aggressive diuresis, until the cat is eating, drinking, and not severely dehydrated since it may potentiate azotemia in dehydrated cats. Chronic oral dosing of furosemide may be started at 1 mg/kg PO q24 hr- BID and increased over time as the severity of CHF worsens to a maximal ceiling dose of 4 mg/kg PO TID. In cats receiving maximal furosemide doses that have refractory CHF, an additional diuretic may be added such as hydrochlorothiazide (1-2 mg/kg PO q24 hr – BID) or spironolactone (2 mg/kg PO q24 hr). Some degree of azotemia is expected in cats receiving aggressive diuretic therapy. In cats with refractory heart failure that continue to decompensate despite 4 mg/kg PO TID of furosemide, torsemide may be substituted for furosemide. The author utilizes torsemide at 1/10 the daily furosemide dose, split into BID dosing. For example, the torsemide dose to replace furosemide in a 4 kg cat receiving 16 mg furosemide PO TID would be torsemide 2.5 mg PO BID (tablets are 5 mg size). Likewise, cats may be given intermittent subcutaneous injections of furosemide by the owner several times a week.

REFERENCES

(1) Rishniw M, Pion PD. Is treatment of feline hypertrophic cardiomyopathy based in science or faith? A survey of cardiologists and a literature search. *J Feline Med Surg* 2011;13(7):487-497.

(2) Spoladore R, Maron MS, D'Amato R, Camici PG, Olivotto I. Pharmacological treatment options for hypertrophic cardiomyopathy: high time for evidence. *Eur Heart J* 2012;33(14):1724-1733.

(3) Lorell BH, Paulus WJ, Grossman W, Wynne J, Cohn PF. Modification of abnormal left ventricular diastolic properties by nifedipine in patients with hypertrophic cardiomyopathy. *Circulation* 1982;65(3):499-507.

(4) Johnson LM, Atkins CE, Keene BW, Bai SA. Pharmacokinetic and pharmacodynamic properties of conventional and CD-formulated diltiazem in cats. *J Vet Intern Med* 1996;10:316-320.

(5) Wall M, Calvert CA, Sanderson SL, Leonhardt A, Barker C, Fallaw TK. Evaluation of extended-release diltiazem once daily for cats with hypertrophic cardiomyopathy. *J Am Anim Hosp Assoc 2005 Mar -Apr;41 (2):98 -103* 2005;41:98-103.

(6) Cote E. Effect of atenolol treatment on serum cardiac troponin-I concentrations in cats with compensated hypertrophic cardiomyopathy: a pilot study. Proceedings, 17th Annual Congress of the European College of veterinary Internal Medicine (Budapest, Hungary) . 2007.

(7) Schober KE, Zientek J, Li X, Fuentes VL, Bonagura JD. Effect of treatment with atenolol on 5-year survival in cats with preclinical (asymptomatic) hypertrophic cardiomyopathy. *J Vet Cardiol* 2013;15(2):93-104.

(8) MacDonald KA, Kittleson MD, Kass PH. Effect of spironolactone on diastolic function and left ventricular mass in maine coon cats with familial hypertrophic cardiomyopathy. *J Vet Intern Med 2008 Mar -Apr;22 (2):335 -41 Epub 2008 Mar 10* 2008;22:335-341.

(9) Riesen SC, Schober KE, Smith DN, Otoni CC, Li X, Bonagura JD. Effects of ivabradine on heart rate and left ventricular function in healthy cats and cats with hypertrophic cardiomyopathy. *Am J Vet Res* 2012;73(2):202-212.

COMPARATIVE ASPECTS AND CLINICAL ASSESSMENT OF "OTHER" FELINE CARDIOMYOPATHIES

Kristin A MacDonald, DVM, PhD, DACVIM (Cardiology)
Rohnert Park, CA

INTRODUCTION

The term *cardiomyopathy* literally means "heart muscle disease" and designates a disorder of the heart which primarily affects the cardiac muscle (myocardium). "Primary" indicates the myocardial disease is not secondary to valvular disease, pericardial disease, coronary vascular disease, systemic or pulmonary hypertension, congenital abnormalities, or systemic disease. Primary cardiomyopathies (genetic, non-genetic, acquired) are confined to the heart muscle. Conversely, secondary cardiomyopathies show myocardial involvement as part of more widespread multi-organ systemic disorders, and were previously referred to as "specific" cardiomyopathies. Most primary cardiomyopathies are of unknown etiology (idiopathic). However, in human medicine, recent advances in genetic testing have identified causative mutations in many cardiomyopathies previously categorized as idiopathic. In 1995, the World Health Organization categorized the types of cardiomyopathies based on the dominant pathophysiology produced by the myocardial disease, which has since then been revised by the American Heart Association in 2006.[1, 2] Veterinary medicine has adopted much of the human categorization of cardiomyopathies, since they seem to clinically fit the diseases we see in feline cardiology.

Cardiomyopathies are classified into functional pathophysiologic categories based echocardiographic evaluation, including: hypertrophic cardiomyopathy (HCM; i.e. left ventricular concentric hypertrophy), dilated cardiomyopathy (i.e. systolic myocardial failure), restrictive cardiomyopathy (i.e. marked diastolic dysfunction in the face of relatively normal LV), arrhythmogenic right ventricular cardiomyopathy, and unclassified cardiomyopathy (i.e. atrial dilation without other functional myocardial abnormalities). A limitation in the functional classification of cardiomyopathies based on diastolic function, systolic function, and LV hypertrophy is that there may be considerable overlap encountered between categories, and the potential of shifting between categories based on cardiac remodeling.

In a retrospective study of 106 cats, HCM was the most common cardiomyopathy (57.5%), followed by RCM (20.7%), DCM (10.4%), unclassified cardiomyopathy (10.4%), and ARVC (<1%). Often times, RCM and UCM are combined together in the literature, since specific defining criteria are absent in veterinary medicine. This lecture will focus on cardiomyopathies other than HCM, as they are often overlooked and poorly understood.

RESTRICTIVE CARDIOMYOPATHY

RCM is a diverse group of conditions characterized by a common pathophysiologic abnormality defined as a restriction of diastolic filling caused by an increased ventricular stiffness. The clinical picture is of a normal ventricle with normal wall thickness, normal ventricular size and systolic function, but presence of left atrial or biatrial dilation with normal atrioventricular valves. Specific clinical and morphologic criteria for this diagnosis in the cat have not been as clearly defined as they have in humans.

Human Classification of RCM etiologies

Basic etiologic classifications include familial causes and non-familial causes. Within familial causes, idiopathic remains most common, yet new discoveries of causative sarcomeric mutations in troponin I, troponin T, α-cardiac actin, Β- myosin heavy chain, myosin light chain, myopalladin, α-tropomyosin genes have been found in approximately 1/3 of cases previously diagnosed as idiopathic.[1] Within families with these sarcomeric mutations, some people may have familial HCM, DCM, or RCM. Variations in the location of the mutations within specific functional domains of sarcomeric disease genes can result in different cardiac phenotypes of HCM, DCM, and RCM. Familial RCM usually has an autosomal dominant heritable pattern, although recessive patterns have been described. . In people with troponin I mutations causing RCM, there was greater calcium sensitivity than the troponin I mutations causing HCM, which leads to a more severe diastolic impairment, given cardiac troponin I's role of inhibiting interaction of myosin and actin to lead to muscle relaxation. In a group of 12 children diagnosed with RCM, four (1/3) had a family history of cardiomyopathy including RCM (n=2), DCM (n= 1), and left ventricular non-compaction (n=1).[1] Like in HCM, mutations causing familial RCM are in conserved and functionally important domains of the genes.

In people, primary RCM is a diagnosis of exclusion of other secondary causes of infiltrative cardiomyopathy typically with endomyocardial biopsy. Although primary RCM may affect people of any age, older women are more predisposed, and it has been associated with heart block and skeletal myopathy in people living into the fifth decade.[2] Treatment for primary RCM in people includes: diuretics, heart rate control to allow increased diastolic filling time, maintenance of sinus rhythm, and anticoagulation for people with atrial fibrillation. Often dual-chamber pacemakers are needed for patients with conduction disturbances such as third degree AV block. Cardiac transplantation is often the only intervention in patients with refractory heart failure. Like in cats, RCM in people is typically associated with a poor prognosis, with most suffering from symptoms referable to heart failure and having high incidence of premature mortality. Identification of restrictive filling pattern in people with heart failure confers a poorer prognosis, independent of the underlying heart disease present. People with heart failure and a restrictive filling pattern identified on pulsed wave Doppler interrogation of mitral inflow had a four-fold increase in mortality compared to heart failure patients without restrictive filling pattern.[3]

Secondary systemic diseases which cause RCM as well as other systemic effects include infiltrative diseases (amyloidosis, Gaucher disease, Hurler's disease, and Hunter's disease), storage diseases (hemochromatosis, Fabry's disease, glycogen storage disease, Niemann-Pick disease), scleroderma, carcinoid heart disease, metastatic cancers, radiation, and drugs (anthracycline).

Endomyocardial fibrosis is a subset of RCM, with etiologies in people including: hypereosinophilic syndrome, chromosomal cause, various drugs, or idiopathic.

Feline Restrictive Cardiomyopathy

The cause of RCM in cats is unknown, hence the term idiopathic. Like in humans, two forms have been described in cats: the endomyocardial fibrosis form and the myocardial fibrosis form.[4, 5]

Myocardial fibrosis form:

The myocardial fibrosis form of RCM has been described as an animal model of human RCM in 35 cats at the Animal Medical Center.[4] RCM represented 4.5% of all feline cardiac cases seen at the Animal Medical Center in that report. Seventy one percent of cats were male, with mean age of 10 +/- 4 years (range 1.5- 17.1 years). In contrast, a previous case series of RCM reported a female predisposition (71% of cats), and mean age of 7.1 years +/- 3 years. A majority of cats (32/35 in one report) present with clinical signs of heart failure.

Echocardiogram abnormalities:

Characteristic abnormalities seen on echocardiography are biatrial dilation, unremarkable left ventricular size and normal systolic function, and sometimes mild centrally arising atrioventricular valvular insufficiency. One of the defining criteria for diagnosis of RCM is evidence of a restrictive filling pattern seen on pulsed wave Doppler interrogation of the mitral inflow. A restrictive filling pattern occurs when there is severe diastolic impairment leading to markedly elevated left atrial pressure and a high left atrial to LV diastolic pressure gradient. It is characterized by an increased early diastolic filling velocity (> 1 m/s), a decreased atrial filling velocity (< 0.4 m/s), increased E:A ratio (>2), shortened deceleration time (normal is 59+/- 14 ms), and decreased isovolumic relaxation time (normal is 55 +/- 13 ms).[6] Not all cats with RCM have a restrictive filling pattern, and some may have delayed relaxation pattern or a pseudonormal pattern. In cats with tachycardia (HR >160-180 bpm), it is impossible to measure separate E and A waves for evaluating transmitral flow patterns. Tissue Doppler imaging may identify diastolic dysfunction as decreased early myocardial diastolic velocity. Left atrial thrombi or spontaneous echocardiographic contrast are common (reported in 23% of cases).

Thoracic radiographs

A majority (32/35 cats in one study) of cats with RCM have evidence of heart failure on radiographs.[4] Left heart failure (13/35 cats) or biventricular failure (13/35) is commonly seen on radiographs, followed by radiographic evidence of right heart failure including ascites, pleural effusion, hepatomegaly (6/35).

Histopathology

Myocyte disarray was seen in all 10 cats evaluated on necropsy, most cases had moderate myocyte disarray (n= 9/10). Interstitial fibrosis was also seen in all cases, and large scars of replacement fibrosis was also common (moderate in 3, mild in 6). Small vessel arteriopathy was common, and included thickened walls of intramural coronary arterioles with narrowed lumen (n= 8/10). No amyloid, eosinophilic infiltration, or hemochromatosis was identified in the 10 cats examined at necropsy.

Outcomes

Cats with RCM tend to have a grave prognosis, with short median survival times reported to range from 102 - 132 days .[4, 7] Heart failure is the most cause of death (18/35 cats in one study), followed by sudden death. Arterial thromboembolism is common in cats with RCM (reported in 28-45% of cats). [4, 7]

Endomyocardial fibrosis form

A separate entity of RCM is endomyocardial fibrosis (EMF), which causes the same pathophysiologic abnormality of increased ventricular stiffness. EMF is associated with massive replacement fibrosis involving the endocardium and endomyocardium. Unlike some people, cats do not appear to develop EMF secondary to eosinophilic infiltration.[5] The etiology of EMF in cats is unknown, with possible hypotheses including viral induced or immune mediated induced severe endomyocardial injury followed by reparative fibrosis.[8, 9]

Echocardiogram abnormalities

In cases of EMF, the endomyocardium appears thickened with a dense hyperechoic scar that is sharply delineated from the normal underlying myocardium. The pathognomonic feature of EMF is a large fibrotic bridging band that tethers the papillary muscles or LV free wall to the interventricular septum. There may be diastolic and systolic impedance to flow associated with the fibrotic bands causing turbulent blood flow and intraventricular pressure gradients on Doppler interrogation. Like people with EMF, cats may have apical obliteration from dense fibrotic scar, which may decrease LV chamber size. There may be mild to moderate atrioventricular valvular insufficiency as the papillary muscles may become distorted from the scar or chordae tendinae/ valve leaflet become tethered. There may be focal regions of mild concentric hypertrophy of the LV.

Pathology

The hallmark feature of EMF is severe and extensive endocardial scar of the left ventricle, rarely involving both ventricles. Often the scar bridges from the left ventricular free wall to the septum, which results in a fibrotic stenosis of the mid- to apical portion of the left ventricle.[5] An obliterative form of scar may occur and fuse papillary muscles, chordae tendinae, and distort the mitral valve to cause mitral regurgitation. Left atrial enlargement may be accompanied with left atrial endocardial thickening. Thrombi are commonly seen in the left atrium, left auricle, left ventricle, and arterial thromboemboli may also be present. Characteristic histopathologic abnormalities include: extreme endocardial thickening with hyaline, fibrous and granulation tissue. Myocardial interstitial fibrosis may also be present. The scar is predominantly fibrous tissue with hyaline and sparse amounts of elastic connective tissue. Endomyocarditis has been described, but without eosinophilia.

Unclassified Cardiomyopathy

This nebulous category consists of cats with left atrial enlargement +/- right atrial enlargement in the absence of significant concentric hypertrophy and normal to only mildly reduced systolic function. Essentially, it is a category of cardiomyopathy not fitting any of the other hemodynamic categories. Significant valvular insufficiency as a cause of atrial dilation must be ruled out, which may be difficult given the high heart rates of cats. This description is very similar to RCM, and many consider combining categories to UCM-RCM unless there is obvious endomyocardial fibrosis or a restrictive filling pattern on mitral inflow measured by PW Doppler. A restrictive filling pattern may be impossible to detect in cats with tachycardia and fusion of the early and late diastolic filling waves on PW Doppler interrogation of mitral inflow. Therefore, many clinicians use the term UCM to include cats with atrial or biatrial enlargement with normal LV size and function. There is a lack of specific criteria for diagnosis of UCM. The etiology of UCM in cats is unknown, but postulated causes of atrial dilation are: secondary to ventricular diastolic dysfunction or due to a form of atrial myopathy. Diastolic function assessed by tissue Doppler imaging echocardiography in feline unclassified cardiomyopathy may be normal or impaired.

Human UCM

Like feline UCM, human UCM is ill-defined. Several unusual forms of cardiomyopathy are placed in this category, including left ventricular non-compaction and Takotsubo cardiomyopathy. LV non-compaction is the most common cause of UCM in people, but is rare. LV non-compaction is characterized by prominent left ventricular trabeculae and deep inter-trabecular recesses, with a thickened myocardial wall and endocardium, and a thin compacted epicardium. Left ventricular systolic dysfunction and dilation may sometimes be present. LV non-compaction has not been reported in cats in the literature, but has been suspected in at least once cat in a colony of familial HCM. Takotsubo cardiomyopathy, also known as transient LV apical ballooning syndrome, has not been described in cats, and is characterized by transient regional systolic dysfunction of the LV apex and/or mid-ventricle that typically resolves within days to weeks.[10]

Treatment of Unclassified or Restrictive Cardiomyopathy

Restrictive cardiomyopathy is characterized by severe endomyocardial or myocardial fibrosis, which greatly reduces the compliance of the left ventricle and leads to diastolic heart failure. Beta blockers or calcium channel blockers are not useful to treat endomyocardial fibrosis, but may be used to treat significant tachyarrhythmias. Anticoagulant therapy (as discussed above) is often necessary in cats with moderate or severe atrial dilation as there is a high incidence of arterial thromboembolism of 45% in cats with restrictive cardiomyopathy. Standard heart failure treatment includes furosemide and an ACE inhibitor. Pimobendan may be considered in UCM or RCM cats since it is a preload and afterload reducer and a positive lusitrope (relaxes the heart), but is considered off-label use.

Dilated Cardiomyopathy

Dilated cardiomyopathy (DCM) consists of primary systolic myocardial failure. In response to reduced contractility, there is reduced cardiac output, which triggers volume expansion through the renin-angiotensin-aldosterone system, which leads to compensatory development of left ventricular eccentric hypertrophy. Less blood is ejected during systole and remains in diastole, which further increases left ventricular diastolic filling pressure. Pulmonary edema and pleural effusion may occur secondary to left sided CHF, and ascites and pleural effusion may occur secondary to right sided CHF. Other cardiac diseases that can lead to myocardial failure and may mimic DCM include severe volume overload of the left ventricle such as severe mitral regurgitation or left to right shunts. The myocardial failure and shortening fraction are not as severely reduced as in primary DCM.

Human DCM etiology

Thirty to fifty percent of people have evidence for familial DCM, which most often has an autosomal dominant mode of inheritance. Mutations in over 30 genes encoding a wide variety of proteins including: sarcomeric, Z disc, cytoskeletal, nuclear membrane, intercalated disc, and mitochondrial proteins.[10] X-linked diseases associated with DCM include muscular dystrophies (Becker and Duchenne) and X-linked DCM.

Canine DCM etiology

Although most cases of canine DCM are idiopathic, a causative mutation in pyruvate dehydrogenase kinase 4, a mitochondrial protein that regulates glucose metabolism, has been identified in Doberman pinschers with autosomal dominant heritable DCM.[11] Other modes of heritability have been identified including: X-linked in Great Danes, autosomal recessive with sex-specific alleles in Irish Wolfhounds (males are overexpressed), autosomal dominant with incomplete penetrance in Newfoundlands, and autosomal recessive in a juvenile form of DCM in Portuguese water dogs. Great Danes with heritable DCM likely have alterations in cellular calcium handling since they have been shown to have different RNA expression for calstabin 2 and triadin, (both are regulatory components of the cardiac ryanodine receptor). Secondary nutritional causes of canine DCM are rare, and include taurine deficiency in Cocker spaniels and other atypical breeds, and L-carnitine deficiency in a family of Boxer dogs.[12 13]

Feline DCM etiology

Prior to 1987, DCM was one of the most commonly diagnosed heart diseases in cats. Most cases at that time were likely a secondary cardiomyopathy associated with nutritional taurine deficiency. In current times, DCM is uncommon in cats, and is most often due to primary idiopathic DCM. Idiopathic DCM is a diagnosis of exclusion, ruling out secondary causes of myocardial failure such as taurine deficiency, tachycardiomyopathy, severe volume overload leading to secondary myocardial failure (i.e. severe mitral insufficiency or large left to right shunting congenital heart diseases), or toxicity such as doxorubicin cardiotoxicity.

Secondary nutritional DCM

Taurine deficiency induced DCM was discovered in 1987, and was secondary to inadequate dietary intake of this essential amino acid in cats. Only 30% of taurine deficient cats develop echocardiographic evidence of myocardial failure.[14] There is a prolonged pre-clinical phase where the cat is asymptomatic. Since modern feline diets are supplemented with taurine, these cases are rarely seen and the majority of cases are idiopathic. Cats fed exclusively one canned diet or vegetarian diets occasionally develop taurine deficiency induced DCM. Although myocardial failure secondary to taurine deficiency is now quite rare in cats, all cats with myocardial failure should be assumed to be taurine deficient until shown not to be taurine responsive. Whole blood +/- plasma taurine concentration must be measured in any cat with myocardial failure to evaluate for possible taurine deficiency, which is curative with taurine supplementation.

Echocardiogram abnormalities

Echocardiographic diagnosis of DCM is relatively straight-forward. Increased end systolic diameter (> 10 mm), reduced shortening fraction (<28 %, <20% in severe DCM), increased E point to septal separation (> 4 mm), and compensatory increased end diastolic diameter (> 18 mm) are abnormalities seen on echocardiography of the left ventricle. Left atrial enlargement is present in moderate to severe cases. Right ventricular and right atrial dilation may also be seen. TDI reveals reduced systolic myocardial velocity. Color flow Doppler often shows centrally arising mild mitral regurgitation secondary to annular dilation and lack of valve coaptation.

Treatment

Positive inotropic treatment with digoxin (1/4 of 0.125 mg PO q 24 -48 hours) and pimobendan (0.25 mg/kg PO BID, not labeled for use in cats) is recommended for severe DCM. Treatment of moderate DCM without CHF is debatable. Although taurine deficiency induced DCM is much less common now than idiopathic DCM, supplementation with taurine (250 mg PO BID) should be started while whole blood and plasma taurine concentrations are pending. Taurine supplementation in deficient cats leads to clinical improvement in several weeks, and measurable improvement in systolic function in 2-4 months (sometimes earlier). Myocardial function normalizes in these patients. Furosemide and an ACE inhibitor should be started if there is evidence of CHF. Thoracocentesis should be performed to remove pleural effusion. Antiarrhythmic drugs are not usually necessary, but can be used if there are severe tachyarrhythmias such as ventricular tachycardia, atrial fibrillation, or supraventricular tachycardia, with avoidance of strong negative inotropes such as beta blockers or high doses of calcium channel blockers. Severe ventricular arrhythmias may be treated with mexiletine 5 mg/kg PO TID +/- sotalol (2 mg/kg PO BID). Supraventricular arrhythmias may be treated with diltiazem or sotalol and digoxin.

ARRHYTHMOGENIC RIGHT VENTRICULAR CARDIOMYOPATHY (ARVC)

ARVC is a rare form of cardiomyopathy in cats, and is more common in Boxer dogs and people. It consists of fatty or fibrofatty infiltration of the right atrium and right ventricle, with subsequent severe dilation of those chambers.

Etiology in humans

ARVC is defined in people by the presence of right ventricular dysfunction +/- left ventricular disease, in the presence of histological evidence of adipose and fibrous tissue replacement of cardiomyocytes and/or characteristic electrocardiographic abnormalities. Autosomal recessive forms include Naxos and Carvajal syndromes, which are caused by mutations in the plakoglobin and desmoplakin genes respectively. Most cases, however, are caused by autosomal dominant heritable mutations in genes encoding plakophilin-2 and other proteins of the desmosome of cardiomyocytes, or mutations in TGF-B and ryanodine (calcium release) receptor.

Etiology in dogs

An autosomal dominant form of ARVC in Boxer dogs is caused by a deletion mutation in the striatin gene.[15] Recently, Boxer dogs with the echocardiographic phenotype of dilated cardiomyopathy were evaluated for presence of the striatin mutation, which was present in 30 of 33 Boxer dogs. The homozygous affected state and the striatin mutation were strongly associated with DCM, yet other mutations are likely present to cause DCM in Boxer dogs without the striatin mutation.

ARVC in cats

The etiology of ARVC in cats is unknown. In a case series of 12 cats, middle aged, male domestic shorthair cats were most commonly affected with ARVC.[16] Fatty replacement of the cardiomyocytes results in systolic myocardial failure +/- diastolic dysfunction.

Echocardiogram abnormalities

Echocardiography reveals right atrial and right ventricular dilation. The right ventricular wall is commonly thin and there may be right ventricular aneurysms. Centrally arising tricuspid regurgitation occurs secondary to tricuspid annular dilation and lack of valve coaptation. Changes mostly involve the right heart, but changes may also occur to a lesser extent in the left ventricle and left atrium. The main differential diagnosis of ARVC is tricuspid valve dysplasia, a congenital heart defect, with a better prognosis. Right sided CHF including pleural effusion and ascites was seen in 67% of cats with ARVC. Ventricular tachyarrhythmias are common (75%), as well as supraventricular tachyarrhythmias (42%). Prognosis of cats with ARVC is grave, with a median survival of 1 month (range 2 days to 4 months).

Treatment

Feline ARVC is typically treated similarly to idiopathic DCM, with use of positive inotropes (pimobendan +/- digoxin), furosemide, an ACE inhibitor, and usually anticoagulant therapy since atrial dilation is typically severe. Since ventricular tachyarrhythmias are common (75%), as well as supraventricular tachyarrhythmias (42%), antiarrhythmic therapy may be needed as described above.

REFERENCES

(1) Kaski JP, Syrris P, Burch M et al. Idiopathic Restrictive Cardiomyopathy in Children is Caused by Mutations in Cardiac Sarcomere Protein Genes. *Heart 2008 May 8 ;* 2008;.

(2) Ammash NM, Seward JB, Bailey KR, Edwards WD, Tajik AJ. Clinical profile and outcome of idiopathic restrictive cardiomyopathy. *Circulation* 2000;101(21):2490-2496.

(3) Whalley GA, Gamble GD, Doughty RN. The prognostic significance of restrictive diastolic filling associated with heart failure: a meta-analysis. *Int J Cardiol 2007 Mar 2;116 (1):70 -7 Epub 2006 Aug 8* 2007;116:70-77.

(4) Fox PR, Basso C, Thiene G, Maron BJ. Spontaneously occurring restrictive nonhypertrophied cardiomyopathy in domestic cats: a new animal model of human disease. *Cardiovasc Pathol* 2013.

(5) Fox PR. Endomyocardial fibrosis and restrictive cardiomyopathy: pathologic and clinical features. *J Vet Cardiol* 2004;6(1):25-31.

(6) Santilli RA, Bussadori C. Doppler echocardiographic study of left ventricular diastole in non-anaesthetized healthy cats. *Vet J* 1998;156:203-215.

(7) Ferasin L, Sturgess CP, Cannon MJ, Caney SM, Gruffydd-Jones TJ, Wotton PR. Feline idiopathic cardiomyopathy: a retrospective study of 106 cats (1994-2001). *J Feline Med Surg 2003 Jun ;5(3):151 -9* 2003;5:151-159.

(8) Meurs KM, Fox PR, Magnon AL, Liu S, Towbin JA. Molecular screening by polymerase chain reaction detects panleukopenia virus DNA in formalin-fixed hearts from cats with idiopathic cardiomyopathy and myocarditis. *Cardiovasc Pathol* 2000;9(2):119-126.

(9) Stalis IH, Bossbaly MJ, van Winkle TJ. Feline endomyocarditis and left ventricular endocardial fibrosis. *Vet Pathol* 1995;32:122-126.

(10) Elliott P, Andersson B, Arbustini E et al. Classification of the cardiomyopathies: a position statement from the European Society Of Cardiology Working Group on Myocardial and Pericardial Diseases. *Eur Heart J* 2008;29(2):270-276.

(11) Meurs KM, Lahmers S, Keene BW et al. A splice site mutation in a gene encoding for PDK4, a mitochondrial protein, is associated with the development of dilated cardiomyopathy in the Doberman pinscher. *Hum Genet* 2012;131(8):1319-1325.

(12) Kittleson MD, Pion PD, DeLellis LA, Tobias AH. Dilated cardiomyopathy in American cocker spaniels - taurine deficiency and preliminary results of response to supplementation. *Proceedings of the Ninth Annual Veterinary Medical Forum*.New Orleans: American College of Veterinary Internal Medicine; 1991. 879.

(13) Keene BW, Panciera DP, Atkins CE, Regitz V, Schmidt MJ, Shug AL. Myocardial L-carnitine deficiency in a family of dogs with dilated cardiomyopathy. *J Am Vet Med Assoc* 1991;198:647-650.

(14) Pion PD, Kittleson MD, Rogers QR, Morris JG. Myocardial failure in cats associated with low plasma taurine: a reversible cardiomyopathy. *Science* 1987;237:764-768.

(15) Meurs KM, Mauceli E, Lahmers S, Acland GM, White SN, Lindblad-Toh K. Genome-wide association identifies a deletion in the 3' untranslated region of striatin in a canine model of arrhythmogenic right ventricular cardiomyopathy. *Hum Genet* 2010;128(3):315-324.

(16) Fox PR, Maron BJ, Basso C, Liu SK, Thiene G. Spontaneously occurring arrhythmogenic right ventricular cardiomyopathy in the domestic cat: A new animal model similar to the human disease. *Circulation* 2000;102(15):1863-1870.

OPTIMAL NUTRITION IN CATS WITH HEART DISEASE

Lisa M. Freeman, DVM, PhD, DACVN
North Grafton, MA

Nutritional modifications are sometimes addressed only in cats with congestive heart failure (CHF) and few, if any, nutritional recommendations have been made for cats in earlier stages of heart disease. However, if one waits to address nutrition until this later stage of disease, the benefits of optimizing nutrition during the longer, and more common, earlier stages are lost. Optimizing diet may help to slow progression of disease and improve quality of life. Therefore, nutrition should be considered as an integral part of the overall care for cats in <u>all</u> stages of heart disease.

NUTRITIONAL GOALS

The nutritional goals for cats with heart disease include the following: 1)Maintaining optimal body condition; 2) Avoiding nutritional deficiencies; 3) Avoiding nutritional excesses; 4) Avoiding drug-nutrient interactions; and 5) Considering administration of certain nutrients in levels above those required for nutritional needs (i.e., nutritional pharmacology).

Optimal Body Composition

Achieving or maintaining optimal body composition is important for cats with heart disease. Goals for body condition in animals with CHF may be different (see below) but in asymptomatic heart disease, veterinarians should assess the animal's body weight, body condition score (BCS), and muscle condition score (MCS) at every visit as part of a nutritional screening. A BCS of 4-5 on a 9-point scale should be the goal. If the cat is above an optimal body condition (i.e., >5 on a 9-point scale), a gradual and comprehensive weight loss program should be instituted with careful monitoring.

When CHF develops, maintaining optimal body composition becomes even more important. Although obesity still can be present at this stage, cats with CHF commonly begin to demonstrate weight loss. This weight loss, or cardiac cachexia, is unlike that seen in a healthy animal, which loses primarily fat. In a cat with CHF, the primary tissue lost is lean body mass. The term *cachexia* does not necessarily equate with an emaciated, end-stage patient; and there is a spectrum of severity with cachexia. In the early stages it can be very subtle and may even occur in obese cats (i.e., a cat may have excess fat stores but still lose lean body mass). Loss of lean body mass usually is first noted in the epaxial, gluteal, scapular, or temporal muscles. Cardiac cachexia can occur with any underlying cause of CHF but typically does not occur until CHF has developed. Marked muscle loss in cats is most commonly observed when recurrent pleural effusion has developed. The loss of lean body mass in cardiac cachexia is caused by anorexia, increased energy requirements, and an increased production of inflammatory cytokines such as tumor necrosis factor (TNF) and interleukin-1 (IL-1). These cytokines cause anorexia, increase energy requirements, and increase the catabolism of lean body mass. Muscle loss is a common finding in cats with CHF and has deleterious effects on strength, immune function, and survival; thus it is important to recognize cachexia at an early stage for better opportunities to manage it effectively.

The nutritional management of cats with cardiac cachexia consists primarily of providing adequate calories and protein and modulating cytokine production. One of the most important issues for managing anorexia (complete loss of appetite), hyporexia (partial loss of appetite), or dysrexia (changes in food patterns and preferences) is to optimize medical therapy. An early sign of worsening CHF or the need for medication adjustment is a reduction in food intake in a cat that has previously been eating well. Medication side effects also can cause anorexia or hyporexia, so the current medications should be evaluated. Recurrent CHF is another cause for a reduction in food intake, thus thoracic radiographs should be considered.

Maintaining adequate calorie and protein intake as CHF progresses can be a challenge as appetite often becomes more variable (i.e., dysrexia) and it is helpful to warn owners of this issue. Communicating with the owner about appetite issues before they occur can help to reduce anxiety and to provide the owner with specific strategies to address the problem (and when to bring the cat in for reassessment). In addition to optimization of medical therapy, offering multiple choices of appropriate cat foods can be very useful. Palatability enhancers also can be very helpful for cats with severe CHF (e.g., cooked unsalted meat or fish for cats; not lunch meats, canned tuna fish, or rotisserie chicken). Encouraging the owner to try offering foods at different temperatures may increase food intake in some animals (e.g., warmed versus room temperature). Owners can also try feeding a balanced home-cooked diet (that has been formulated by a board-certified veterinary nutritionist; otherwise, it is highly likely to be nutritionally unbalanced). Supplementation of fish oil, which is high in omega-3 fatty acids sometimes can be helpful (see below). Other tips that may increase food intake include providing smaller, more frequent meals; feeding from a different dish or plate; or feeding the cat in a different area of the house.

In contrast to the healthy cat, obesity may actually be associated with a protective effect once CHF is present - this is known as the obesity paradox. Although there are a number of hypothesized reasons for the obesity paradox, the benefit of obesity in CHF is likely due more to a *lack* of cachexia, rather than to the obesity per se, given the adverse effects associated with cachexia. However, even in cats with CHF, BCS>7 were still associated with shorter survival times. Therefore, the author recommends maintaining a BCS of 5-7 on a 9-point scale for cats with CHF.

Avoiding Deficiencies, Excesses, and Interactions

<u>Protein and Amino Acids</u>

Adequate protein intake is recommended for cats with any stage of heart disease, unless severe concurrent kidney disease is present. The AAFCO minimum for protein for adult cats is 6.5 gm/100 kcal. Over-the-counter cat foods are often well over this amount, with some of the newer trend of "high protein, low carbohydrate foods" having protein contents well over 13.0 gm/100 kcal. In the healthy cat, there is no documented benefit to very high protein, low carbohydrate diets, and little is known about the long-term effects of these diets. Protein restriction is particularly detrimental to the cat with CHF because it can contribute to lean body mass loss and malnutrition. While protein-restricted renal diets are sometimes recommended for cats with heart disease because these diets often (but not always) are moderately sodium restricted, this is not a recommended practice. Unless concurrent advanced kidney disease dictates otherwise, diets fed to cats with cardiac disease should meet feline minimums; this is particularly important for cats with CHF and muscle loss.

Although dilated cardiomyopathy (DCM) is now uncommonly seen in cats, taurine deficiency and DCM can still occur, particularly in cats fed vegetarian, nutritionally unbalanced homemade diets, or commercial foods that are designed to be for "intermittent or supplemental use" or made by small companies that do not have stringent quality control practices. Collecting a complete diet history can identify cats eating diets that could predispose them to taurine deficiency (as well as a variety of other deficiencies). Plasma and whole blood taurine should be analyzed in cats with DCM, and treatment with taurine [125 to 250 mg orally q 12 h] should begin concurrent with medical therapy. If the cat is eating an unconventional diet, the owner also should be counseled to switch to a well-known, nutritionally balanced, meat-based commercial cat food.

Sodium

Severe sodium restriction is not recommended in early (asymptomatic) heart disease. In early heart disease, the goal should be to avoid excessive sodium intake and to educate the owner about treats and table foods high in sodium. The author counsels owner of cats with asymptomatic heart disease to avoid diets high in sodium (>100 mg/100 kcal) and to avoid high sodium treats and table food. However, this is an opportune time to begin talking to the owner about the animal's overall dietary patterns (i.e., the cat food, treats, and how medications are administered) since it is generally much easier to institute dietary modifications when the cat is asymptomatic. Also, owners sometimes do a great deal of Internet "research" when their cat is first diagnosed with heart disease and may make significant dietary changes or begin administering dietary supplements at this time. Providing owners with skills for searching the Internet about nutrition topics and useful websites can help them avoid the many common myths about pet foods and supplements.

More significant sodium restriction may have more benefits in cats with CHF. However, with the current availability of newer and more effective medications, the role of severe sodium restriction is no longer clear, although some degree of sodium moderation is appropriate for most cats with heart disease. The author currently recommends moderate sodium restriction (i.e., <80 mg/100 kcal) for cats with CHF, although as CHF worsens, additional sodium restriction may be useful.

There are currently no veterinary therapeutic diets marketed in the U.S. specifically for cats with heart disease. Diets designed for cats with kidney disease sometimes are recommended for cats with heart disease, but the author recommends against this practice because of the protein restriction (to varying degrees) inherent in these diets. Be careful not to recommend senior diets uncategorically for cats with heart disease. If a diet designed for senior cats is to be used, it is very important to look at the characteristics of the individual product. Since there is no legal definition for senior diets, the levels of calories, protein, sodium, and other nutrients can vary dramatically among products from different companies. The sodium content of senior cat foods can range from 64-346 mg/100 kcal (unpublished data; AAFCO minimum for cats = 50 mg/100 kcal).

Most owners do not know what foods are high or low in sodium. This information is not available on cat food or treat labels and, even when listed on human foods, it is difficult for owners to determine how much sodium is too much for a cat. Therefore, it is important to provide specific and clear information on appropriate cat foods, treats, table foods, and methods to administer medications. Owners also should be counseled on specific foods to avoid such as baby food, cheeses, lunch meats and cold cuts (e.g., ham, corned beef, salami, sausages, bacon, hot dogs), and many cat foods and treats.

Potassium and Magnesium

Potassium and magnesium are nutrients of concern in cats with heart disease because depletion of these electrolytes can cause cardiac arrhythmias, decreased myocardial contractility, and muscle weakness and can potentiate the adverse effects of cardiac medications. Cats with CHF can have low, normal, or high serum potassium concentrations, depending upon severity of disease and medications being administered. Some medications can contribute to hyperkalemia (e.g., ACE inhibitors and spironolactone), while others can lead to hypokalemia (e.g., loop diuretics). In addition, feline diets have a wide range of potassium content so using one appropriate for the individual patient is important (e.g., avoiding high-potassium diets would be recommended for cats with hyperkalemia). Consequently, serum potassium should be monitored, especially as more medications are administered to a cat with CHF. Recommended dietary potassium modifications will depend upon medications being administered and serum potassium concentrations. Serum magnesium concentrations also should be measured, although serum magnesium concentrations are a relatively poor indicator of total body magnesium stores. Nonetheless, serial evaluations in an individual patient may be useful, especially in cats with arrhythmias or in those taking high doses of diuretics. Diets high in magnesium or an oral magnesium supplement should be used in cats with hypomagnesemia.

Nutritional Pharmacology

At this time, there are no dietary supplements that have been shown to have a benefit for cats with heart disease. However, the supplement that may have the most potential is omega-3 fatty acids. The omega-3 fatty acids, eicosapentaenoic acid (EPA) and docosahexaenoic acid (DHA), are normally are found in very low concentrations in cell membranes compared to the omega-6 fatty acids, but they can be increased by a food or supplement enriched in omega-3 fatty acids. The benefit of having a higher concentration of omega-3 fatty acids in the membranes is that breakdown products of the omega-3 fatty acids (eicosanoids) are less potent inflammatory mediators than eicosanoids derived from n-6 fatty acids. Omega-3 fatty acids also reduce the production of the inflammatory cytokines, TNF and IL-1. The anti-inflammatory, antiarrhythmic, and antithrombotic effects of omega-3 fatty acids have many potential benefits in cats with cardiac disease. Omega-3 fatty acid supplementation can reduce muscle loss and improve appetite via its anti-inflammatory effects, and studies have shown benefits in dogs from omega-3 fatty acids. Although studies have been conducted in cats with omega-3 fatty acids to assess effects on coagulation, little work has been done to determine whether omega-3 fatty acid supplementation in cats would have similar benefits as in humans and dogs.

While dogs with CHF have been shown to have a relative deficiency of omega-3 fatty acids (which can be corrected by supplementation), a recent study in cats showed that cats with heart disease actually have significantly *higher* serum omega-3 fatty acid concentrations, particularly DHA.

Despite the lack of evidence in cats, many owners wish to supplement omega-3 fatty acids, particularly in cats with reduced appetite or muscle loss. The optimal dose of omega-3 fatty acids has not been determined; however, the author currently recommends a daily dosage of fish oil to provide 40 mg/kg EPA and 25 mg/kg DHA for cats with CHF. Most cat foods do not achieve this dose so supplementation is often necessary. Fish oil supplements vary in their concentration of EPA and DHA so the author recommends a one-gram capsule that contains 180 mg EPA and 120 mg DHA. At this concentration, fish oil can be administered at a dose of one capsule per ten pounds of body weight/day. The capsule is too large to be administered whole to most cats but the oil can be expressed from the capsule and given as a treat or in the food or a liquid fish oil can be used. While some cats appear to enjoy the taste, in cats that dislike the flavor, administration of omega-3 fatty acids should be discontinued due to adverse effects on food intake. Fish oil supplements should contain vitamin E as an antioxidant, but other nutrients should not be included to avoid toxicities. Cod liver oil and flax oil should not be used to provide omega-3 fatty acids.

IMPORTANT NUTRITIONAL CONSIDERATIONS FOR ALL CATS WITH HEART DISEASE

1. Perform a complete nutritional assessment on every cat at every visit: A nutritional assessment will help determine whether the current diet is optimized or what specific modifications might be beneficial for a cat with heart disease. A nutritional assessment includes an assessment of the cat and his or her current diet. Nutritional assessment consists of a screening evaluation and, if any "red flags" are identified, an extended evaluation. Standardized nutritional assessment forms make gathering complete information more efficient and ensure that the required information is collected for every patient. This screening evaluation can quickly identify nutritional issues that require further evaluation, such as unconventional diets, dietary supplements, inappropriate treats or table food, or high sodium foods used for medication administration. The screening can also quickly identify cats with weight loss, suboptimal body condition (either too thin or overweight), and muscle loss, or cats with concurrent medical conditions that also require dietary modifications.
2. Make individualized nutritional recommendations: Always make a specific recommendation for diet – this may be only to say that the diet the cat is currently eating is a good choice and that the owner is feeding an appropriate amount to keep the cat in ideal body condition. It is important to reinforce good nutritional choices that owners are making, since there are so many myths about cat nutrition. In addition to the importance of making a recommendation, it also is important to design it specifically for an individual cat based not only on the stage of disease, but also on the cat's nutritional assessment, physical examination, laboratory results, and individual food preferences. For example, cats with heart disease can be hyperkalemic, normokalemic, or hypokalemic, which influences the choice of diet. Therefore, the same diet will not meet the requirements of all cats with heart disease. Concurrent diseases also alter diet choice and were shown in one study to be present in 56% of cats with heart disease. The optimal diet will be very different for a cat with concurrent heart and chronic kidney disease compared to a cat with heart disease and struvite urolithiasis or inflammatory bowel disease. In complicated cases, consider referring the cat to a board-certified veterinary nutritionist.
3. Address all components of the diet: Be sure to address all components of the diet – not only the cat food, but also the treats, table food, and foods used for medication administration. In some cases, cats may be eating an ideal cat food but are getting large amounts of sodium from treats or table food. Therefore, a diet must be selected that has the desired nutritional properties and palatability, but it also is important to devise an overall dietary plan that includes appropriate treats, if desired by the owner. Many cat owners use food to administer medications, and the most commonly used foods are high in sodium (e.g., lunch meats). Including all sources of dietary sodium in the overall diet plan is important to achieve success with nutritional modifications. Cats with cardiac disease also may be receiving dietary supplements since "Internet surfing" for alternative treatments of cardiac disease is common among owners. Also, it is important to be aware that dietary supplements do not require proof of safety, efficacy, or quality control before being marketed. Therefore careful selection of type, dose, and brand is important to avoid toxicities or complete lack of efficacy.

RECOMMENDED READING

1. Finn E, Freeman LM, Rush JE, et al. The relationship between body weight, body condition, and survival in cats with heart failure. *J Vet Intern Med* 2010; 24: 1369-1374.
2. Freeman LM. Beneficial effects of omega-3 fatty acids in cardiovascular disease. *J Small Anim Pract* 2010; 51: 462-470.
3. Freeman LM. Cachexia and sarcopenia: Emerging syndromes of importance in dogs and cats. *J Vet Intern Med* 2012; 26: 3-17.
4. Freeman LM, Rush JE, Meurs KM, et al. Body size and metabolic differences in Maine Coon cats with and without hypertrophic cardiomyopathy. *J Fel Med Surg* 2013;15:74-80.
5. Torin DS, Freeman LM, Rush JE. Dietary patterns of cats with cardiac disease. *J Am Vet Med Assoc* 2007; 230: 862-867.
6. WSAVA Nutritional Assessment Guidelines Taskforce: Freeman L, Becvarova I, Cave N, et al. 2011 Nutritional Assessment Guidelines. *J Small Anim Pract* 2011; 52: 385-396.

RECOMMENDED WEBSITES

1. American College of Veterinary Nutrition. List of board-certified veterinary nutritionists. www.acvn.org. Accessed February 1, 2014.
2. Body condition score chart for cats. World Small Animal Veterinary Association Global Nutrition Committee Nutrition Toolkit. http://www.wsava.org/nutrition-toolkit. Accessed February 1, 2014.
3. Muscle condition score chart. World Small Animal Veterinary Association Global Nutrition Committee Nutrition Toolkit. http://www.wsava.org/nutrition-toolkit. Accessed February 1, 2014.
4. Savvy cat owner's guide to nutrition on the internet. World Small Animal Veterinary Association Global Nutrition Committee Nutrition Toolkit. http://www.wsava.org/nutrition-toolkit. Accessed February 1, 2014.
5. Selecting the best food for your pet. World Small Animal Veterinary Association Global Nutrition Committee Nutrition Toolkit. http://www.wsava.org/nutrition-toolkit. Accessed February 1, 2014.
6. Short diet history form. World Small Animal Veterinary Association Global Nutrition Committee Nutrition Toolkit. http://www.wsava.org/nutrition-

toolkit. Accessed February 1, 2014.
7. Tufts HeartSmart website (designed for owners of dogs and cats with heart disease): www.tufts.edu/vet/heartsmart. Accessed February 1, 2014.
8. Tufts Cummings School of Veterinary Medicine Nutrition Service FAQs on pet foods. http://www.tufts.edu/vet/nutrition/faq/ Accessed February 1, 2014.

MITRAL VALVE DISEASE: HUMAN VS. ANIMAL

Meg M Sleeper VMD, DACVIM (cardiology)
Philadelphia, PA

MITRAL VALVE DISEASE IN HUMANS

The management approach to mitral regurgitation (MR) in humans is highly dependent on the classification of the specific valvular lesion as either organic (i.e. the lesion affects the valve leaflets themselves) or functional (i.e. the valve is incompetent due to structural remodeling of the LV). Also important for management, is whether the cause is classified as acute or chronic. The most common causes of mitral regurgitation are:

Organic Mitral Regurgitation
- Degenerative mitral valve disease
- Congenital anomalies affecting the mitral valve
  - Cleft mitral valve leaflet; Mitral arcade anomaly
- Inherited metabolic diseases affecting the mitral valve
  - Lysosomal storage diseases

Functional Mitral Regurgitation
- Dilated cardiomyopathy (ischemic or nonischemic); Rheumatic mitral disease; Posterior wall myocardial infarction; Endocarditis; Hypertrophic cardiomyopathy with systolic anterior motion of the mitral valve leaflets

Organic mitral regurgitation

Degenerative mitral valve disease is the leading cause of organic MR in western countries. It is a degenerative connective tissue disease of the mitral valve leaflets characterized by leaflet thickening and redundancy, with a prevalence of 2% to 3% in the population. Redundancy of the leaflets most frequently leads to failure of coaptation with consequent mitral regurgitation. Dyspnea only develops when mitral regurgitation becomes severe. The primary complications associated with mitral valve prolapse are:
- Gradual worsening of regurgitation, typically over years
- Atrial fibrillation
- Spontaneous chordal rupture, with resultant acute severe regurgitation (up to 8% risk) with acute exacerbation of signs
- Congestive heart failure

The development of chronic severe mitral regurgitation typically is well tolerated initially, because the ventricle compensates for the volume overload state by dilation with preservation of cardiac output. There is a 7% to 10% per year incidence, however, of patients developing symptoms or cardiac decompensation requiring surgery.

Congenital deformation of the mitral valve occurs much less commonly than acquired mitral valve disease in humans. Various anomalies have been described including cleft mitral valve, parachute mitral valve and mitral valve arcade. Mitral arcade is an infrequent malformation characterized by an arc like configuration of 2 papillary muscles due to an interconnecting band of fibrous tissue that runs along the line of closure of the mitral valve

The most common inherited metabolic diseases that have been associated with mitral valve abnormalities are the lysosomal storage diseases (LSD). There are more than 40 different forms of inherited LSDs known to occur in humans with an aggregate incidence estimated at 1 in 7000 births. This family of diseases occurs due to disruption of normal lysosomal function and subsequent accumulation of incompletely degraded substrates. Most but not all of the LSDs are caused by loss of normal function of specific lysosomal acid hydrolases, which act to degrade large complex substrates that have been targeted for degradation after endocytosis or autophagy. Mitral valve thickening, regurgitation and stenosis has been recognized in several of the LSDs occurring in young humans.

Functional Mitral Regurgitation

The mitral valve is integrated anatomically into the LV, with a network of chordal attachments inserting into a pair of papillary muscles that then insert into the LV myocardium. If the LV chamber becomes dilated (eg, dilated cardiomyopathy) or the posterior wall of the ventricle (where the papillary muscles attach) is dysfunctional (eg, posterior wall myocardial infarction), the mitral valve leaflets are restricted from fully closing during systole and regurgitation occurs. Various cardiac diseases can lead to secondary or functional MR.

Assessment of mitral valve disease

The grading of mitral valve regurgitation is primarily determined with echocardiography. In routine clinical practice, the most commonly used technique relies on the ability of Doppler echocardiography to encode the regional velocity of blood flow using color flow Doppler imaging. The interpreter of the echocardiogram judges approximately what percentage of the left atrial volume is filled with regurgitant blood flow, and a grading scheme is used to determine the severity. Although this technique is practical and clinically useful, limitations include variable machine settings that have an impact on the apparent size of the regurgitant jet and the qualitative determination that depends on the experience and expertise of the interpreter. Other echocardiographic techniques are available for the quantitative determination of regurgitant volume but also suffer from geometric assumptions and substantial interobserver variability. Therefore, an integrative approach to interpretation should be used, including all available techniques when appropriate. Transesophageal echocardiography also may be helpful in more clearly defining the mitral valve anatomy and regurgitation severity.

Because of the inherent limitations and subjectivity of echocardiography, clinical correlation is always important in making management decisions. For example, a patient with mitral valve prolapse who is asymptomatic with normal heart size and function should display unequivocal severe mitral regurgitation before consideration of surgical intervention. In contrast, a symptomatic patient with reportedly moderate mitral regurgitation may warrant further testing with an exercise Doppler echocardiogram or a transesophageal echocardiogram to verify that the regurgitation is not underestimated. Exercise stress testing can also be useful in reportedly asymptomatic patients to confirm adequate functional capacity. In addition, Doppler echocardiography immediately post-exercise occasionally reveals more substantial regurgitation that is exercise induced and may explain the symptomatology.

Clinical Follow-up

Gradual progression of mitral regurgitation may eventually lead to severe symptomatic mitral regurgitation and cardiac decompensation. With routine clinical follow-up, however, progression to severe disease generally is readily detectable, either from symptom development (dyspnea or exercise intolerance) or from structural remodeling by echocardiography. Referral to a cardiologist is appropriate for moderate or severe regurgitation, although the primary care physician maintains a critical role in monitoring for symptoms.

Management of Organic Mitral Valve Regurgitation

Medical therapy with a vasodilator has not been shown beneficial in the treatment of mitral regurgitation, except for the treatment of concomitant hypertension or for stabilization in the setting of acute severe mitral regurgitation. The primary focus is to determine the timing and type of surgery to be performed. Severe mitral regurgitation resulting from mitral valve prolapse frequently can be treated with surgical repair rather than valve replacement. Because the long-term outcomes of valve repair are believed superior to replacement, the timing of surgery is influenced by whether repair is believed feasible. The major (class I) recommendations for mitral valve surgery are as follows:

-Symptomatic severe regurgitation (in the absence of severe LV dysfunction)
-Severe regurgitation with mild or moderate LV dysfunction
-In either of the above indications, mitral valve repair is recommended over replacement

MITRAL VALVE DISEASE IN ANIMALS

Organic mitral regurgitation

Degenerative valve disease is the most common form of acquired heart disease in the dog, occurring most frequently in small breed dogs. It occurs less commonly in the cat and the horse. The canine form of the disease is very similar to the human form with many authors suggesting its relevance as a model of the human disease.

Congenital abnormalities involving the mitral valve have been described in the dog, cat, and horse. However, these anomalies are rare in veterinary medicine compared to other congenital heart defects such as patent ductus arteriosus, pulmonic stenosis or subaortic stenosis.

Inherited metabolic diseases, in particularly LSDs, have been recognized as naturally occurring diseases in multiple species including the mouse, rat, dog, cat, guinea pig, emu, quail, goat, cattle, sheep and pig. In particular, several forms of mucopolysacchariosis have been identified in the dog and cat and result in mitral valve thickening and regurgitation.

Functional mitral regurgitation

As in humans, functional mitral regurgitation is commonly noted in animals due to various heart diseases other than degenerative mitral valve disease, including dilated cardiomyopathy (most often in dogs) and hypertrophic cardiomyopathy (most often in cats).

Assessment of mitral valve disease

The grading of mitral valve regurgitation is primarily determined with echocardiography in veterinary species and is very similar to the process for humans with mitral valve disease. As described earlier, an integrative approach to interpretation should be used, including all available techniques when appropriate. Transesophageal echocardiography and exercise stress testing can also be useful but is used less commonly in veterinary patients with heart disease than in human patients.

Clinical Follow-up

In many animals with degenerative valve disease, progression to severe symptomatic mitral regurgitation and cardiac decompensation is gradual, surgical intervention is not an option and there is not clear evidence that beginning cardiac medications before the onset of congestive heart failure is beneficial. Therefore, for many animals, regular monitoring of the resting respiratory rate and regular evaluation of heart size is recommended to recognize early signs of congestive heart failure.

Management of Organic Mitral Valve Regurgitation

In some canine patients with degenerative valve disease, cardiac surgery is an option and can be curative. Valvuloplasty procedures have also been successfully performed in dogs with congenital mitral valve disease. However, in the vast majority of dogs with acquired or congenital mitral valve disease and congestive heart failure, symptoms are controlled with diuretics (most often furosemide), an angiotensin converting enzyme inhibitor, and pimobendan due to financial constraints of the owners. Additional medications may be indicated in some patients. Therapeutic gene delivery has successfully reduced mitral valve thickening and leakage in purpose bred dogs and cats with LSDs and this approach may be a therapeutic option in dogs with other forms of valve disease in the future.

REFERENCES

Available upon request.

MITRAL VALVE DISEASE: NEW DEVELOPMENTS

Ingrid Ljungvall, DVM, PhD and Jens Häggström, DVM, PhD, DECVIM-CA (Cardiology)
Uppsala, Sweden

INTRODUCTION

Myxomatous mitral valve disease (MMVD) is by far the most common heart disease in dogs. Comparably extensive research has been conducted concerning different aspects of this disease in dogs and much has been learnt from these studies. However, many issues are still unresolved or incompletely studied. The presentation will review new aspects of myxomatous mitral valve disease (MMVD), such as recent molecular genetical findings associated with age of onset of the disease, new echocardiographic modalities for staging disease severity, and different techniques for improving the diagnosis of congestive heart failure, and results from recent clinical drug trials

GENETIC BACKGROUND AND THE SEROTONIN SIGNALING PATHWAY

It has long been known that mitral valve prolapse (MVP), which is an early sign of MMVD, often displays familial transmission in people. Studies in dogs have shown significant associations between cardiac status in parents and offspring both with regard to murmur intensity and MMVD severity. [1,2]

From the EU funded LUPA project, results have been published from a study where MMVD-affected and unaffected dogs were identified among more than thousand Cavalier King Charles Spaniels (CKCS).[3] MMVD affected dogs were defined as dogs with congestive heart failure (CHF) due to MMVD before 8 years of age or a significant systolic murmur over the mitral area before 4.5 years. Unaffected dogs were defined as dogs with no or mild heart murmur after 8 years of age. A genome-wide association study (GWAS) to identify loci associated with early onset of MMVD was performed. A 1.58 Mb region on CFA13 (pgenome= 4.0·10-5) and a 1.68 Mb region on CFA14 (pgenome= 7.9·10- 4) were found to be associated with early onset of MMVD.[3] A candidate gene has been identified within the locus, and this gene is currently studied further.

Another interesting development in the pathophysiology of MMVD is the suggested impact of increased activity in the serotonin (5HT) signaling pathway. Rats receiving 5-HT injections have been shown to develop valvular lesions.[4,5] Furthermore, 5-HT-producing carcinoid tumors,[6,7] or intake of serotonergic drugs have been shown associated with valvulopathy in people.[8-11] Two studies have demonstrated increased serum 5-HT, concentrations in dogs predisposed to MMVD in comparison to other breeds.[12,13] One of the studies reported, furthermore, that the serum concentrations of 5-HT decreased at more progressed stages.[13] The finding of higher serum 5-HT concentrations in dogs of breeds predisposed to early onset of MMVD (CKCS) and dogs with mild MMVD suggests that alterations in 5-HT signaling might play a role in progression of early stages of MMVD, and that 5-HT potentially could be genetically linked to the etiologic mechanisms of MMVD in CKCS.[14]

The results from the GWAS indicated an association between an early onset of MMVD and certain areas within the genome, whereas the 5HT hypothesis appears to contradict the GWAS findings. However, the results are not conflicting as the GWAS was performed *within* the CKCS breed, whereas the 5HT concentrations were shown to differ *between* breeds. The findings of increased 5HT in CKCS is likely to represent a fixed trait, which means that it is widespread in the breed, and this means that CKCS dogs must be compared to other breeds to find possible causative mutations. Other study designs are needed to identify such genetical differences between breeds, and such work is currently ongoing. The GWAS showed associations within the breed to certain areas in the genome for an early onset of MMVD. Likely, both these findings are likely to contribute to the situation of MMVD within the CKCS breed.

NEW ECHOCARDIOGRAPHIC MODALITIES

Standard echocardiographic means to estimate disease severity and stage MMVD severity include measurement of left atrial and ventricular diameters, evaluation of mitral valve morphology and motion, and color echocardiographic methods. Many other new methods have over the years been suggested to improve evaluation of the MMVD patient, but few of them survived the test of time. More recently, the introduction of tissue Doppler imaging (TDI), tissue tracking techniques, and real-time three-dimensional (RT3D) echocardiography have been introduced for evaluation of left ventricular function.[15-19] The TDI and tissue tracking techniques have been considered comparably independent of loading conditions, which are problematic in evaluating left ventricular (LV) function in mitral regurgitation, but recent studies suggest that this technique is also affected by loading condition and sympathetic tone activity to a greater extent than previously expected, which might limit the additional informative value obtained from this technique, compared to when using standard echocardiographic measurements. [17-19]

Real time three-dimensional echocardiography allows for superior outlining of the LV in real time and for accurate volume estimations of the left atrium and ventricle. Due to the ability to align the true short- and long-axes of the LV, the problems with chamber foreshortening and oblique imaging planes can be reduced; thus providing more anatomically correct views than conventional 2D echocardiography. [20] The potential of LV volume assessment by RT3D echocardiography has recently been demonstrated in dogs by our group.[16,21,22] The RT3D modality also allows evaluation of changes in LV shape, accompanying LV dilation, using a RT3D sphericity index.[23,24] A RT3D echocardiographically derived sphericity index, which can be obtained by dividing the end-diastolic volume by the volume of a sphere, has been demonstrated an earlier and more accurate predictor of remodeling compared to other echocardiographic variables following acute myocardial infarction in human patients.[25] Sphericity index may be calculated in the 2D image by dividing LV length with LV width, which means that this method is very sensitive to LV foreshortening. Studies indicate 3D sphericity index is approximately 0.8 in normal dogs, but may increase up to 1.4-1.5 in dogs with severe LV dilatation.[16] Measurement of 3D sphericity is, accordingly, a more accurate method, because it compared the LV volume with a volume of a sphere, and image planes may be optimized before volume estimations are performed. The sphericity index is usually not included in the routine echocardiographic protocol when assessing LV remodeling in dogs, but this index might have a potential when investigating progression of LV remodeling in dogs, as previously shown by our group in dogs affected by MMVD.[16]

DIAGNOSIS OF CHF

Because MMVD in small dogs is characterized by chronic progression, the owner and veterinarian often have no other alternative than to observe how the valvular lesions and MR progress slowly, often over years. At a certain point, the mitral regurgitation (MR) can no longer be compensated for; the forward cardiac output decreases, and the pulmonary capillary pressure exceeds the threshold for pulmonary edema. Thus, there is a need for clinical variables that have prognostic value. Several physical examination- echocardiographical- and hormonal variables have been evaluated in these aspects. At present, left atrial dimension (LA/AO ratio, LA/predicted AO etc.), body weight normalized LV end-diastolic dimensions ($LVIDd_{inc}$, $LVIDd_n$) , vertebral heart score index (VHS), and natriuretic peptides (Atrial and B-type natriuretic peptides) stands out as being most reliable in predicting the onset of decompensated CHF.as significant changes occur in all of them 0.5 to 1 year before the onset of clinical signs of decompensated heart failure. Recent studies show that these indicators change in a non-linear manner over time.[16,26-28] Small changes are seen over the comparably long period during which a dog progresses from very mild to moderate and severe MR, but the rate of change of the variables increase as the dog progresses into CHF.[16,27-29] It is our opinion that a number of these variables have the acceptable predictive value for groups, but less so for individuals. Furthermore, the performance of the test improves considerably if more than one observation point in time is available, and the closer in time to CHF the dog is.

All the above-mentioned tests require imaging and laboratory equipment and expertise. Studies in people have shown that the onset of dyspnea in people during exercise or sleep precedes the onset of overt signs of CHF. Furthermore, resting respiratory rate (RRR) in dogs in the clinical situation proved the most sensitive and specific single diagnostic test for identifying CHF as a cause of respiratory clinical signs in dogs with heart disease and was an independent variable in predicting CHF in multivariable regression analysis[30] A further study demonstrated that resolution of the CHF resulted in a reduction in respiratory rate into pre-CHF ranges.[31]Thus, respiratory rate has the potential of being a sensitive, albeit unspecific, indicator of developing or recurring CHF in dogs. Additionally, studies have recently shown that healthy dogs and dogs with subclinical left sided heart disease (many included MMVD dogs) had an average sleeping respiratory rates (SRR), measured by their owners at home, below 25 breaths/minute and rarely above 30 breaths/minute.[32,33] Furthermore, within-dog SRR did not vary much from day-to-day for most dogs, as evidenced by the low within-dog coefficient of variation. Thus, frequent monitoring of SRR could allow more timely therapeutic intervention or modulation in animals with severe subclinical heart disease (e.g. those with marked LA enlargement) or known prior history of CHF. Many veterinary cardiologists currently recommend their clients to monitor respiratory rates in subclinical canine and feline cardiac patients in their home environment to help determine either the onset of CHF in a previously subclinical patient, or to monitor effectiveness of CHF therapy in patients with prior history of CHF

CLINICAL TRIALS

The most recent clinical trials concerning MMVD therapy were aimed at investigating the effects of pimobendan. Previous veterinary clinical studies concerning pimobendan in dogs with CHF attributable to MMVD have shown improved quality of life (QoL) variables and a reduced risk for adverse events (death, recurrent CHF or additional veterinary visits etc.) when compared to a positive control (ramipril).[34,35] To study the effect of pimobendan on survival in dogs with CHF secondary to MMVD, the QUEST (Quality of life and Extension of Survival Time) study was untaken.[36] The aim of the study was to compare the time taken to reach the primary endpoint for dogs receiving either pimobendan or benazepril in conjunction with other therapy. The primary endpoint was a composite of spontaneous cardiac death, euthanasia for cardiac reasons or withdrawal from the study due to treatment failure. Two hundred and sixty dogs were recruited at 28 centers in Europe, Canada, and Australia, making it the largest prospective, blinded survival study so far undertaken in canine cardiology. Dogs were of small and medium breeds, and were only included in the study if they had, at some point, demonstrated convincing radiographic evidence of left-sided CHF. After enrolment to the study the dogs were randomized to receive either pimobendan, plus standard therapy or benazepril plus standard therapy. They were then re-evaluated at regular intervals and followed until they reached the study endpoint, were censored from the study for other reasons, or the study was concluded (whichever occurred first). Dogs receiving pimobendan plus standard therapy had a longer survival time compared to those receiving benazepril plus standard therapy (pimobendan 267 days versus. benazepril 140 days, P=0.0099). The benefit of pimobendan persisted after adjusting for all baseline variables (fig. 6). Several other baseline variables were shown to influence survival time. Most of these were direct or indirect indicators of disease severity.

Analysis of longitudinal data is far more complex than survival studies. There are several statistical methods that may be considered, but none of them are ideal. The QUEST trial data was analyzed using a method based on summary statistics, and the results showed that pimobendan results in a similar quality of life for the duration of the period of treatment.[37] However, results showed that pimobendan therapy reduced heart size (Vertebral heart scale, and left ventricular diameter in systole and diastole corrected for body weight) and resulted in a higher body temperature, sodium concentration, total protein and packed cell volume, by comparison with benazepril therapy. The differences between the treatment groups in the latter 3 variables indicate less retention of free water in the pimobendan treated dogs. Furthermore, the necessity for intensification of concurrent therapy in patients in the QUEST study was evaluated by looking at the time to intensification of therapy and the frequency of use of commonly used concurrent medications. The reason an investigator decided to increase the dose of one or more of the baseline therapies or add another agent was because he/she believed the dog would benefit from the change due to a deterioration in the condition of the dog or the owner being concerned that their pet was failing to achieve an adequate QoL on their current treatment regimen. Thus although the average values for QoL variables were similar in the two treatment groups over the duration of the study, dogs in the benazepril group required intensification of treatment earlier to maintain their QoL. The similarity of QoL in the two groups is therefore probably a consequence of investigators being at liberty to modify concurrent medication to maintain an acceptable QoL for their patients. Thus, the results showed that the benazepril group required alteration of their therapy sooner and more frequent administration of other medications (spironolactone and digoxin) to maintain the same quality of life as dogs receiving pimobendan.[37]

REFERENCES

[1]Swenson, et al. **J Am Vet Med Assoc** 1996;208:2009-2012, [2]Olsen, et al . **J Vet Intern Med** 1999;13:448-456, [3]Madsen, et al. **J Hered** 2011;102 Suppl 1:S62-67, [4] Droogmans, et al. **Cardiovasc Toxicol** 2009;9:134-141, [5]Gustafsson, et al. **Circulation** 2005;111:1517-1522, [6]Gustafsson, et al. **Int J Cardiol** 2008;129:318-324, [7]Lundin, et al. **Circulation** 1988;77:264-269, [8]Gardin, et al. **J Am Med Assoc** 2000;283:1703-1709, [9]Hendrikx, et al. J Heart Valve Dis 1996;5:235-237, [10]Rothman, et al. **Pharmacol The*r*** 2002;95:73-88, [11]Zanettini, et al. **N Engl J Med** 2007;356:39-46, [12]Arndt, et al. **J Vet Intern Med** 2009;23:1208-1213, [13]Ljungvall, et al. **J Vet Intern Med** 2013;27:1105-1112, [14]Oyama, et al. **J Vet Intern Med** 2010;24:27-36, [15]Carnabuci, et al. **Res Vet Sci** 2013;95:241-248, [16]Ljungvall, et al. **J Vet Intern Med** 2011;25:1036-1043, [17]Tidholm, et al **J Vet Intern Me***d* 2009;23:1197-1207, [18]Zois, et al. **J Vet Intern Me***d* 2013;27:875-883, [19]Zois, et al. **J Vet Intern Med** 2012;26:1309-1319, [20]Jacobs, et al. **Eur Heart J** 2006;27:460-468, [21]Tidholm, et al. **J Vet Intern Med** 2011;25:1320-1327, [22]Tidholm, et al. **J Vet Intern Med** 2010;24:1414-1420., [23]Di Donato, et al. **Eur J Cardiothorac Surg** 2006;29 Suppl 1:S225-230, [24]Monaghan **Heart** 2006;92:131-136, [25]Mannaerts, et al. **Eur Heart J** 2004;25:680-687, [26]Häggström, et al. **Am J Vet Res** 1997;58:77-82, [27]Lord, et al. **J Small Anim Pract** 2010;51:210-218, [28]Tarnow, et al. **Vet J** 2009;180:195-201, [29]Häggström, et al. Proc 15th ACVIM Forum 1997;202-204, [30]Schober, et al. **J Vet Intern Med** 2010;24:1358-1368, [31]Schober, et al. **J Am Vet Med Assoc** 2011;239:468-479, [32]Ohad, et al. **J Am Vet Med Assoc** 2013;243:839-843, [33]Rishniw, et al. **Res Vet Sci** 2012;93:965-969, [34]Lombard, et al. **J Am Anim Hosp Assoc** 2006;42:249-261, [35]Smith, et al. **J Small Anim Pract** 2005;46:121-130, [36]Häggström, et al. **J Vet Intern Med** 2008;22:1124-1135, [37]Häggström, et al. **J Vet Intern Med** 2013: 27: 1441-1451.

NEW DATA ON MITRAL VALVE TRANSCRIPTOMICS AND MICRO RNA EXPRESSION IN DOGS

Brendan Corcoran, MVB, DIP PHARM, PhD, MRCVS
Roslin, Midlothian, Scotland, UK

INTRODUCTION

There have been major advances in our understanding of myxomatous mitral valve disease (MMVD) of the dog in the last twenty years. Most efforts have been directed at examining the clinical aspects of MMVD with particular emphasis on diagnostics, therapeutics and predictors of disease progression. In parallel with increased clinical understanding of this disease has been a more recent advance in understanding disease pathogenesis. This has been possible through the acquisition and application of a dog research "molecular toolbox" that allows investigation of the structural, cellular and molecular events present in canine MMVD. A particularly interesting development has been the availability of credible canine annotated data bases that allow examination of protein and gene expression. The use of "omics" technologies has only been made possible through the major advances in experimental biological, computing and bioinformatics (mega data analysis), to such an extent that such technologies are routine, rapid, and relatively inexpensive and produce credible reproducible data. What the veterinary profession now needs is a trained cohort of research clinicians who can understand the application of such technologies to answer fundamental questions of disease biology and develop strategies for treatment and cure.

TRANSCRIPTOMICS

The transcriptome is effectively a measure of the gene expression (mRNA) at any one time in a particular cell, but will also include sequences for rRNA, tRNA, and non-coding sequences of RNA such as microRNAs. As a research tool it can give powerful information on the genes being expressed and biochemical pathways, but it is not a measure of the amount of protein being expressed (proteome), nor is protein expression proportional to mRNA expression. From that perspective alone the transcriptome would appear to have significant limitations when it comes to understanding the biological consequences of disease, but once those limitations are understood it can be a useful tool in identifying potential pathways worthy of further investigation.

Canine transcript studies use the Affymetrix Canine Gene 1.0 ST Array or the Agilent Canine Oligo Microarray. The Affymetrix chip is an improvement on the first microarray developed (Affymetrix Canine Gene 1.0 Array) with greater transcript annotation (27,681 genes) and transcript sensitivity (590,097 probes, 24 probes/gene). Coupled with improved array technology have been advances in bioinformatics with *in silico* network analysis, which allows functional analysis of pathways (e.g. Ingenuity Pathways Analysis database; IPA) identifying the biological functions that are most attributable to particular gene networks. Bioinformatics reduces the time needed to inspect and attribute gene function. However, it has added an additional problem in that gene ontology analysis platforms (IPA, DAVID, Biolayout Express) generate complex gene networks that can be over-whelming to interpret when trying to match gene expression with clinical phenotype.

In addition to examining gene expression, microRNA (miRNA) expression can also now be examined with microarrays. miRNAs are short (typically less than 22 nucleotides) RNA molecules that do not encode proteins, but have a very important role regulating gene expression. This important function of miRNAs has only been recognised recently, where they mainly have a negative control role of gene expression (miRNA ↑ and mRNA ↓) and so have a fundamental role in biology and disease. miRNAs are given a numeric suffix as they are identified e.g mir-23. This number also tends to suggest their ubiquity, such that a miRNA with a low number would have far more biological effects and so less tissue specificity than one with a large number (discovered later). This is important to appreciate when considering miRNAs as potential therapeutic targets.

MITRAL VALVE TRANSCRIPTOME

To date there has been only one transcriptome study of the dog mitral valve[1] (Oyama et al 2006) using the Affymetrix Canine Gene 1.0 array covering approximately 23,000 gene transcripts. This study identified 229 probe sets of interest (70% ↑and 30% ↓) of which 166 could be assigned to recognised genes, covering the functional classes (in descending order) of cell signalling, metabolism extracellular matrix, inflammation, cell defence, immunity, cell transport and cell structure. Of particular interest was a ~4 fold increase in 5-HT_2B receptor gene expression, which supported other hypothesis-driven studies looking at the role of 5-HT and the TGFβ signalling superfamily in MMVD pathogenesis. This is a clear example of how -omics technologies can inform hypothesis driven research, and while microarray data will never give categorical proof of the cause of any one disease, it will allow informed further research direction. Limited studies have also been undertaken in human MMVD with reported differential gene expression ranging between 400 and 2000, with most genes of interest showing a negative fold change, which contrasts to that reported for the dog[2,3]. With proper examination of such a large number of genes, the genes of interest to emerge from the human studies include members of the TGF-β superfamily (↑), including TGFβ-2 and BMP-4, and the adamalysins family of metalloproteinases (↓) (ADAMTS, proteoglycanases), the latter possibly contributing to an over-expression of versican. These are not surprising findings as they all have various roles to play in extra-cellular matrix synthesis and re-modelling, both during valve development and in adulthood. The human valve transcriptome also suggests the involvement of 5-HT with increased gene expression for the 5-HT_2B receptor and tryptophan hydroxylase 1, which is in agreement with canine studies. Changes in expression of genes associated with oxidative stress are also reported, in particular genes involved in protective pathways such as the metallothioneins and NADPH dehydrogenase (↓), but this contrasts with the dog where similar genes appear to increase[1]. Lastly, there is an interesting category of inflammatory pathway associate genes that are up-regulated in diseased valves. How these might contribute to disease pathogenesis is unknown, but there is no evidence of inflammatory cell infiltration in MMVD. From even the limited published data from the dog and human it can be appreciated that transcriptomics can give powerful insights into disease pathogenesis. However, the data are a snap-shot at a particular disease time point and do not inform on disease development, particularly if samples are from end-stage disease cases.

TRANSCRIPTOME UP-DATE

Recent work at our laboratory examining differential expression from cavalier King Charles Spaniels with moderate to severe disease has identified 5397 differentially expressed genes that fulfilled stringent cut-off setting (fold change, FDR and p-value). Enriching the data set through individual signal intensity value, 635 differentially expressed genes can be identified of which 593 are recognised as annotated canine genes (*Canis lupus familiaris*) of which 324 were up-regulated and 269 down-regulated. *In silico* analysis identified 10 biological functions as follows (Figure 1); cancer (177), cell movement (129), cell death and survival (120), cell-to-cell signalling (94), CV development (93), inflammation (88), connective tissue (79), immune cell trafficking (75), protein synthesis (57) and humoral immune responses (38). This is roughly comparable to previous reports[1,2,3]. For gene interaction and networking, 593 genes were analysed and the IPA software identified three gene networks of particular interest; (1) cellular assembly and organization, cellular function and maintenance, tissue development (2) connective tissue development and function, embryonic development, organ development (Figure 2) (3) cell death and survival, connective tissue disorders, inflammatory disease. 43 canonical pathways were also identified, two of which could be deemed relevant to MMVD, the endothelial-1 signalling pathway and caveolar mediated endocytosis, considering the potential role of endothelial damage in disease pathogenesis.

Many genes previously reported[1] as differentially expressed were confirmed as such, for example 5-HT$_2$B (↑5.7) and BMP6 (↑2.37), but with the exception of genes associated with cell movement, which is a novel finding. Taken together, and with identification of endothelium associated canonical pathways, differential expression of genes associated with basement membrane proteins and inflammation (Toll-like receptors and interleukins), a contribution of endothelial disarray and endothelial-to-mesenchymal transition in disease pathogenesis is worth considering.

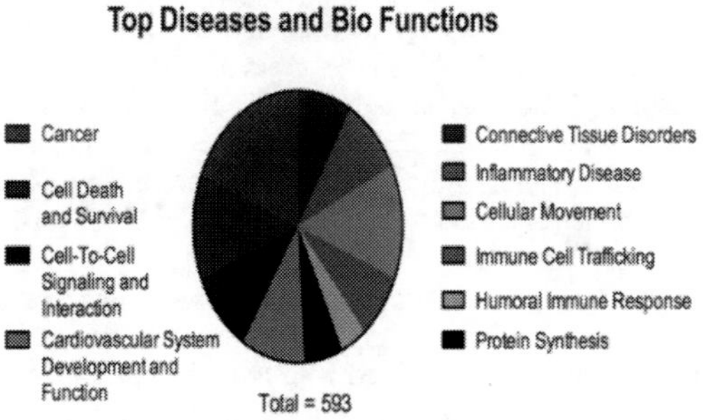

Figure 1: Top disease and biological functions associated with MMVD

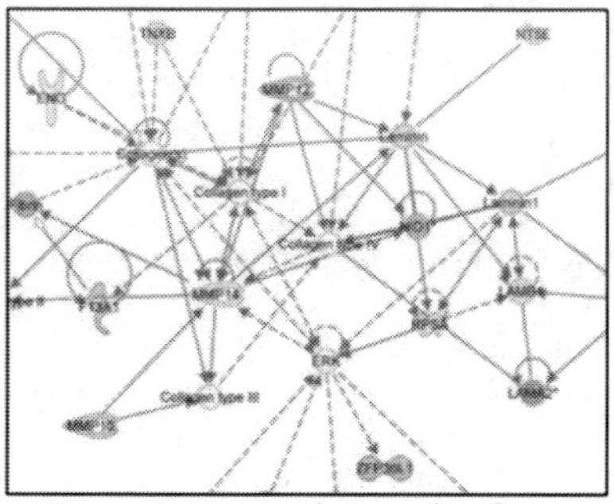

Figure 2: Cropped example of a gene network assigned to - Connective Tissue Development and Function, Embryonic Development, Organ Development (green↓red↑)

The coverage of the Affymetrix Canine Gene 1.0 ST Array also allows identification of miRNAs. In the CKCS study there were 301 detectable *Canis lupus familiaris* miRNAs (cfa-mir) of which 28 were differentially expressed (all down regulated). This would imply that the genes they interact are allowed to express. Not surprisingly, the miRNAs of interest have reported functions in heart development, extracellular matrix and cell phenotypic plasticity.

REFERENCES:

1. Oyama M, et al **AJVR** 2006; 67(8): 1307. Sainger R, et al **J Cell Physiol** 2012; 227: 2595. Hulin A, et al **Cardiovasc Res** 2012; 93: 480.

YEAR IN REVIEW: CONGESTIVE HEART FAILURE

John D. Bonagura, DVM, DACVIM
Columbus, OH

INTRODUCTION

Congestive heart failure (CHF) is a clinical syndrome triggered by impaired cardiac systolic or diastolic function and characterized by stereotypical changes in neurohormonal activation, renal sodium retention, and cardiac and vascular remodeling. Functional characteristics of heart failure include impaired exercise capacity, secondary pulmonary dysfunction, and metabolic disturbances related to impaired organ perfusion. Both quality and duration of life are limited by cardiac failure.

From the end of 2012 through early 2014 a number of manuscripts were published that expand our understanding of CHF in dogs and cats. These recent publications can be considered within the general categories of pathophysiology, epidemiology, diagnosis, prevention, therapy, staging, and prognosis. The purpose of this lecture is to highlight some of this interesting work and offer a personal perspective regarding the clinical impact of these reports within the context of the topics addressed. This presentation is directed to the generalist in clinical practice.

When assessing the clinical impact of published papers, a number of factors might be considered. Included are characteristics of study design that reduce bias such as a prospective design; independent subjects/uncorrelated data; appropriate analyses of repeated measurements; randomization to treatment groups; masking of investigators, clients, and those responsible for data measurement; use of state of the art techniques; *a priori* calculation of statistical power for the main effects; and adequate treatment controls. Busy journal readers often rely on just "reading the abstract". However many of abstracts offer readers an insufficient explanation of the study design; of the results and effects size; and of the clinical significance and impact of the study. Additionally there is an over-reliance on reporting the p-values of statistical significance with insufficient attention directed to the *magnitude* of these effects. Readers can gain a better understanding by focusing not just on the p-values attending significance testing, but also on the summary and raw data displayed in the graphs and figures. When correlation or regression analyses are involved, the r-squared (that relates to shared variance) should be emphasized, not the p-value (a nearly meaningless probability because two variables will nearly always correlate "significantly" if there are enough data points). When diagnostic tests are analyzed by receiver operator characteristic curve analysis, the reader should critically ask "what is the gold standard?" because this underpins whatever cut-offs are proposed to balance sensitivity with specificity of the test. Too often, the "gold standard" is the subjective interpretation of a small number of cardiologists (or other specialists) and rarely is that standard tested for reliability or repeatability across hospitals or departments. When diagnostic or treatment groups are compared using t-tests, ANOVAs, or linear mixed procedures, the confidence intervals for differences between means (or the interquartile ranges for nonparametric statistical tests) are important to understand. Readers should not be unduly influenced by relatively minor differences even when bolstered by "highly significant" p-values. In small studies – a problem common to most single-center trials – the confidence intervals for differences often approach zero, giving the reader less confidence about generalization of the study conclusions. Additionally, unless comparison groups are well-matched, results of small studies can be biased towards or against a diagnostic test or a treatment. Another issue is an over-reliance on retrospective study designs, especially for investigations of CHF therapy. While retrospective reports can provide some idea of *outcomes* in a selected group of patients, often these trials involve insufficient controls (including historical controls), unmasked participants or investigators, and lack of patient randomization. As a result, such trials are unlikely to be definitive. In this regard, the author would suggest that while retrospective studies of therapy can be interesting and identify areas for future investigation, only rarely (if ever) are these conclusive with respect to treatment. Another issue of concern relates to the diagnosis and classification of heart failure, which on face value should be straightforward, but in practice has confounded studies of both therapy and prognosis. For example, the survival data for all dogs with "heart failure" caused by degenerative valvular heart disease (DVD) in the pivotal (licensing) European trial of spironolactone was quite different than the median survival times reported in other European clinical trials (such as QUEST or BENCH). In the spironolactone study, even the dogs treated with placebo experienced very long median survival times. In cases like this, it is difficult to interpret the research results or to compare these to other studies. Some studies of diagnosis and prognosis are also hampered by the classification of patients, especially dogs with asymptomatic disease or borderline CHF or dogs with respiratory signs but without definite (and validated) evidence for CHF. Finally, the most pressing issue for any of us trying to perform clinical investigation is that near total lack of *independent* (foundation, university, or government) funding for veterinary clinical research. While industry funding of research is valued, the issues addressed are typically those of priority to the sponsor, not necessary to those practicing in the field. Although this is a relatively incomplete list of issues, it highlights some of the points readers might consider when evaluating journal articles that purport to change the way we practice cardiology.

PATHOPHYSIOLOGY

A number of recent studies have evaluated interesting aspects of pathophysiology in spontaneous heart disease of dogs and cats. Studies of pathophysiology provide a valuable back-story in terms of explaining potential mechanisms and targets for drug benefit or staging of disease. However, improvement in pathophysiologic markers of disease is neither a suitable regulatory target (for drug approval) nor an appropriate clinical endpoint for recommending therapy.

The role of neurohormonal and renal activation in heart disease is important to the development of progressive cardiac remodeling, clinical signs of heart failure, and development of heart rhythm disturbances. Three relatively recent studies of heart rate variability in dogs with DVD have shown enhanced sympathetic activity and parasympathetic withdrawal in dogs developing progressive disease or heart failure[1-3] while another has shown that enalapril therapy was associated with potentially beneficial withdrawal of sympathetic tone. Three reports of flow mediated vasodilatation[4-6] have demonstrated endothelial dysfunction in dogs with advanced DVD. These studies modified the duplex Doppler technique often used to evaluate brachial artery reactivity in humans (a technique to "uncover" endothelial dysfunction). This technique represents a noninvasive diagnostic approach to the study of cardiovascular diseases progression and responses to drugs. But first, more data and standardization of techniques are needed. In terms of biochemical and hormonal markers of heart failure, one study[7] showed a correlation between CHF and serum vitamin D (a relationship identified in humans). Any cause and effect relationship, as well as potential consequences of this relative deficiency, require further elucidation. In a study of dogs with DVD cardiac troponin I was measured prior to death and was correlated to arteriosclerotic and fibrotic changes at necropsy.[8] Previously published studies of dogs with DVD also have demonstrated altered circulating markers of myocardial collagen and indicators of ventricular remodeling.

EPIDEMIOLOGY & DIAGNOSIS

The diagnosis of CHF is initially entertained by finding historical and physical examination abnormalities compatible with cardiac disease and failure. CHF is further suspected by identifying objective signs of heart disease through auscultation and diagnostic imaging. While right-sided CHF is usually a physical diagnosis, the more common left-sided CHF is largely defined by evaluation of thoracic radiographs and identifying a response to appropriate initial therapies (oxygen, furosemide, pimobendan, ACE-inhibitors [ACEI], ±other treatments). Risk stratification can be further refined by combining measures of heart size (VHS or LV diastolic dimension on Echo) with measurements of natriuretic peptides (NT-proBNP in dogs and cats; BNP in dogs).[9] A recent retrospective study[10] even suggested that clinicians should expect a higher incidence of acute canine CHF presentations on Monday and Tuesday and during the autumn months.

One of the great challenges for clinicians is the identification of early CHF in dogs with advanced DVD and clinical signs of cough or tachypnea. Both past and recent reports[11,12] address the comorbidities of chronic airway and interstitial lung diseases (tracheal collapse, bronchial compression, bronchomalacia, chronic bronchitis, pulmonary fibrosis). Two approaches to identifying left-sided CHF involve the use of the resting or sleeping respiratory rates and the measurement of natriuretic peptides. A recent paper[13] involving 190 dogs with subclinical left-sided heart disease added more evidence to the value of this examination and the results are congruous with previously published, but smaller studies. Readers can review the specific data reported (that includes finding less variation in sleeping than in resting RR), but the general message was so long as the sleeping and resting RRs are <25 to 30/min, the likelihood of CHF is very low. A similar approach can be taken in cats based on experience. In both species, trends are especially valuable.

How might these findings be combined? In our own practice we often observed dogs with advanced DVD, cough, and a sleeping RR<25. These dogs often have large LV diastolic dimensions on Echo, often a very elevated NT-proBNP (>2000 pmol/L), and radiographically normal lungs (based on our interpretation and those of our radiologist colleagues). If untreated, most of these dogs continue to be comfortable for many months, suggesting a lack of cardiogenic pulmonary edema as the source of coughing. It is critical in our view to evaluate RR and radiographs (especially DV and VD projections) prior to initiating therapy in dogs with coughing and DVD. Alternative diagnoses also should be considered. A recent study indicated that veterinarians are more "confident" about the diagnosis of CHF when provided with BNP data[14]. No doubt the population of dogs with CHF have much higher NT-proBNP concentrations than normals or class B dogs with DVD; however, the experience in our hospital suggests that false positives for a "cutoff of CHF" are not uncommon (using cutoffs of >1,550 pmol/L), and caution is therefore suggested when evaluating individual cases. Furthermore, breed variation might be important when assessing natriuretic peptides[15].

A number of studies of echocardiography and radiography have been published related to risk or likelihood of CHF. One Echo study of dogs with advanced DVD used the advanced signal processing method of myocardial rotation and calculated "twist/untwist" rates in an attempt to predict onset of CHF. As typically found with studies of systolic function in smaller dogs with severe MR, even this sensitive method failed to identify global LV systolic dysfunction, again emphasizing the differences between DVD and DCM as causes of CHF.[16] In two feline studies the anticipated relationships between left atrial size, left atrial function, and transmitral filling velocities were demonstrated[17,18] Additional predictors in cats were the severity of LV wall thickening and impairment of LV systolic function. While 2D echo measurements of LA size are sensitive and useful for the recognition of advanced feline cardiomyopathies; in contrast, left atrial size is often underestimated by thoracic radiography and might be underwhelming even in the setting of CHF.[19] Another retrospective radiographic study of 67 cats used the vertebral heart scale for differentiation of cardiac and noncardiac causes of respiratory distress.[20] The investigators identified a VHS exceeding 8.0 vertebrae as the best cutoff for sensitivity (screening) and a VHS of >9.3 as relatively specific for the presence of heart disease. Another interesting feline study measured NT-proBNP in feline pleural effusion and identified higher values in cats with CHF.[21]

PREVENTION & THERAPY

From the author's perspective five things matter in terms of therapy: 1) Is the onset of CHF significantly delayed by a treatment; 2) Do the patients live longer; 3) Is quality of life better with treatment; 4) Are unplanned hospitalizations reduced; and 5) Are treatments cost-effective? Adverse drug effects relate to quality of life or mortality. Cost-effectiveness is often influenced by the locale of the client and local regulations regarding generic drug use in animals. Cost-effectiveness is especially important when drugs are used for *prevention* of CHF because a long duration of therapy and a potentially high cost of treatment might be involved (sometimes with marginal evidence for its use). When there is a potential for benefit based on favorable effects on markers of disease or ambiguous results of clinical trials, some clinicians will empirically recommend therapy; however, clients should understand the lack evidence and also appreciate the costs of such treatment as well as potential unintended adverse effects.

The prevention of CHF has been evaluated in a number of studies. In previous studies of ACEI, the two major prospective studies in dogs with DVD have been either negative (SVEP) or inconclusive (VETPROOF) with respect to delayed on set of CHF.

Although retrospective evaluations of ACEI have suggested a benefit of delaying clinical signs in dogs with DCM and DVD, these have not been substantiated with suitable prospective trials. This status is relatively unchanged over the past two years with the exception of the PROTECT trial (PROTECT) conducted in 76 Doberman pinschers.[22] This study compared pimobendan to placebo treatment of preclinical DCM and demonstrated a substantial improvement in both cardiac and all-cause mortality with pimobendan. The median time to the primary endpoint (CHF or sudden cardiac death) was 718 days (interquartile range of 441-1152 days) for dogs treated with pimobendan versus 441 days (IQR of 151-641 days) for dogs treated with the placebo. Although the study was small, it was well-designed and controlled. Importantly, as a group, ventricular ectopy did not seem to increase in the pimobendan treated dogs (based on 24h ECG) although some individual dogs did worsen in the raw data plots. This certainly recommends the use of pimobendan in this breed with clear evidence of pre-clinical DCM is evident based on echocardiographic findings. The major question is whether the study results can be *generalized* to other breeds with occult DCM. Furthermore, the benefits of potentially cardioprotective treatments (ACEI or beta-blockers or spironolactone) along with pimobendan were not investigated.

The second part of the QUEST trial data were also recently reported.[23] This study involved a well-constructed comparison of pimobendan versus benazepril in dogs with CHF due to DVD treated with a background of furosemide and spironolactone. The "survival" (endpoints) benefit of pimobendan was clearly evident when compared to the selected dosage of benazepril and this has previously been reported in the initial paper from this trial. The more recent report compared a number of quality of life indicators between the two groups. (In this study there was not a "combined" group receiving both pimobendan and an ACEI; such combined treatment would represent the typical standard of care for canine CHF). In this recent report, there was no demonstrable benefit in the quality of life indicators between groups, although the time to intensification of CHF therapy was prolonged in the pimobendan group and there was probably less free-water retention with inodilator therapy.

The short term effects of pimobendan were evaluated in another study[24] in dogs with DVD and treated with furosemide. Pimobendan demonstrated more benefit than benazepril on a number of hemodynamic and echo variables, although aldosterone concentrations were higher in the pimobendan group. The study was small and apparently not intended to assess early survival. However, the results provide further evidence for the immediate use of this inodilator in CHF of dogs. However, the continued "comparisons" of pimobendan and an ACEI might be better clarified for clinicians if a "combined" treatment group was included in these trials (considering that both drugs are typically used to treat CHF, at least in North America).

In terms of other treatment studies (for which more data are needed before any conclusions can be made): (1) A pilot study of imatinib for pulmonary hypertension in canine CHF was reported;[25] (2) An open-label pilot study of atorvastatin on clinical markers of inflammation and oxidative stress in dogs with CHF was performed[26]; the drug was well tolerated but more data will be needed to assess any potential value; (3) a retrospective report of very acceptable tolerability of pimobendan in cats with CHF was reported.[27]

STAGING & PROGNOSIS

A number of "biomarker" studies have been published in recent years and a full discussion would require a separate presentation. Briefly: (1) An interesting study of plasma ADH (vasopressin) concentrations in dogs with CHF was published[28]; this might provide a reference method for detecting neurohormonal activation and potentially point to dogs who could benefit from V2 receptor blockers; (2) A study of 26 dogs with CHF due to DVD showed a potential predictive value of NT-proBNP for determining 12-month survival when the value decreased (to <965 pmol/L); this small study[29] supports the qualitative data provided by previous studies of natriuretic peptides in advanced canine CHF.

A study of systolic blood pressure in dogs with DVD demonstrated the effectiveness of neurohormonal-renal compensations in advanced heart disease and CHF.[30] While SBP was statistically reduced relative to asymptomatic dogs the pressures were not dangerously low. Therapy for CHF did result in a further decline, pointing to the need to periodically evaluate BP (despite the technical limitations), especially in patients symptomatic for hypotension or those with azotemia.

Ultrasound-based studies are often used to stage the severity of heart disease, especially in cases of chronic DVD. In addition to papers previously mentioned, a study of serial Echo measurements (every 6 to 12 months) did show changes in dogs at higher risk for development of CHF. As might be expected, these included estimates of LV size as well as Doppler estimates of filling pressures and pulmonary hypertension. [31]

REFERENCES

[1]Rasmussen CE, et al. JVIM 2012, 26:76; [2]Oliveira MS, et al. Vet Rec 2012, 16:170; [3]Lopez-AlvareJ, et al. JVIM 2014, Epub ahead of publication Feb2014; [4]Cunningham SM, et al. JVIM 2013, 27:985; [5]Moesgaard SG, et al. JVIM 2012, 26:61; [6]Jones ID, et al. J Vet Cardiol 2012, 14:203; [7]Kraus MS, et al. JVIM 2014, 28:109; [8]Falk T, et al. JVIM 2013, 27:500; [9]Reynolds CA, et al. J Vet Cardiol 2012, 14:193; [10]Steinberg GR, et al. J Vet Emerg Crit Care 2012, 22:341; [11]Singh MK et al. JVIM 2012, 26:312; [12]Ferasin L, et al. JVIM 2013, 27:286; [13]Ohad DG, et al. JAVMA 2013, 243:839; [14]Singletary GE, et al. JVIM 2012, 26:542; [15]Sjostrank K, et al. JVIM Epub ahead of publication, Feb 2014; [16]Zois NE, et al. JVIM 2013, 27:875; [17]Payne JR, et al. JVIM 2013, 27:1427; [18]Linney CJ, et al. J Small Anim Pract 2014, Epub ahead of publication Feb; [19]Schober KE et al. Vet Radiol Ultrasound 2013, Epub ahead of publication; [20]Sleeper MM, et al. JAVMA 2013, 242:366; [21]Humm K, et al. J Small Anim Pract 2013, 54:656; [22]Summerfield NJ, et al. 2012 JVIM, 26:1337; [23]Häggström J, et al. JVIM 2013, 27:1441; [24]Häggström J, et al., JVIM 27:1452 ; [25]Aita S, et al., Can Vet J 2013, 54:255; [26]Cunningham SM, et al. JVIM 2013, 27:985; [27]Gordon SG, et al. JAVMA 2012, 241:89; [28]Scollan KF, et al. AJVR 2013, 74:1206; [29]Wolf J, et al. J Vet Cardiol 2012, 14:399; [30]Petit AM, et al. Vet J 2013, 197:830; [31]Hezzell MJ, et al. 2012, J Vet Cardiol 14:269.

YEAR IN REVIEW: CARDIAC CATHETERIZATION

Bruce W. Keene, DVM, MSc, DACVIM (Cardiology)
Raleigh, NC

Interventional cardiac catheterization is performed with increasing frequency in veterinary medicine in a variety of clinical settings, utilizing a variety of vascular approaches to access the heart, and a variety of imaging modalities for guidance. The purpose of this brief review is to summarize significant recent developments in veterinary interventional cardiac catheterization.

Patent Ductus Arteriosus (PDA)

Device closure of left to right shunting PDA is in many respects the "low hanging fruit" of veterinary cardiac interventional procedures; PDA is relatively easy to diagnose and treat, with successful ductal closure generally yielding an excellent prognosis, returning the patient to a normal life untroubled by clinically significant residual heart disease.

Interventional ductal closure is most commonly accomplished by deploying one of 3 intravascular devices – the Amplatz canine ductal occluder, an Amplatz vascular plug, or one or more vascular embolization coils.

The vascular approach (venous or arterial) to PDA varies by institution and operator preference. In retrospective reviews published to date, the arterial approach has generally proven to be more straightforward, faster, and associated with fewer complications. The size of the patient (and more importantly, the patient's blood vessels) impacts the choice of interventional device closure and vascular approach; patients generally need to weigh more than 2.5 – 3kg to accept the size sheath introducers or guiding catheters needed to deploy the Amplatz ductal occluder or plug. Embolization coils can be delivered through smaller vessels (e.g. through a 4 Fr. catheter).

Success and complication rates vary with training, experience, patient size, ductal conformation and device. Patients large enough to accept an appropriately sized Amplatz canine ductal occluder have the best of all possible worlds, although a recent cautionary tale exists in the literature of late device embolization, possibly associated with exercise.

It is clear from both the human and veterinary literature that relatively large PDAs in relatively small (e.g. <2.5kg) patients are still best closed surgically, or perhaps by a hybrid method (e.g. surgical exposure with device implantation), since device deployment in this population is associated with a higher complication rate than is present with larger patients.

There are (some recent) reports of unusual left to right shunts that may mimic PDA to varying degrees on both physical examination and transthoracic echocardiography, and awareness of these reports and this possibility should inform the diagnostic work up of all patients, but especially those that do not fit the signalment and clinical profile of the typical PDA case (e.g. male dogs, atypical breeds). Transesophageal echocardiography is especially useful in evaluating ductal morphology for choosing and sizing closure devices, and in guiding PDA closure by any interventional method.

Pulmonic Stenosis

Balloon valvuloplasty of valvar pulmonic stenosis (PS) is also among the most common interventional procedure in veterinary medicine (along with PDA closure and pacemaker implantation). The vascular approach to PS is venous, with operator preference (and to a lesser degree, patient size) determining whether the jugular or femoral approach is used. Recent retrospective reviews confirm the general consensus that dogs with moderate to severe stenosis (i.e. peak instantaneous pressure gradients across the pulmonic valve of ≥ 70 mmHg with at least moderate concentric right ventricular hypertrophy and echocardiographic evidence of chronic pressure loading) generally benefit from the procedures (i.e. they seem to lead longer, less symptomatic lives). In our cath lab, the following patient characteristics are associated with better results: age less than 6 months at the time of the procedure, smaller body size (within reason – as with PDA, complication rates rise in patients less than 2kg), and more purely valvar lesions. Dogs (or cats) with significant subvalvar lesions and marked valvular dysplasia are more difficult to dilate successfully, and those patients appear more likely to experience significant restenosis. Dogs with severe infundibular hypertrophy may occasionally experience "suicide" right ventricle – a condition in which relief of their valvar obstruction worsens the subvalvar infundibular obstruction to the point where the right ventricular outflow tract is compromised during systole, and the pressure gradient actually increases (or output severely falls). Such cases should be treated aggressively with beta blockade (at NCSU, patients undergoing pulmonic valvuloplasty are pretreated with a beta blocker, and in addition they receive a post-anesthetic induction fluid bolus (approximately 10ml/kg) along with intravenous lidocaine (2mg/kg)).

Coronary artery defects can complicate the interventional or surgical treatment of PS, and R2A and other variants have been described in English Bulldogs and other breeds (e.g. a Brittany spaniel). If questions regarding the coronary anatomy exist following transthoracic ultrasound (e.g. both coronary ostia are not clearly visible, a coronary artery appears to traverse the pulmonic valve), either coronary angiography or CT angiography can be used to determine the best therapeutic approach. Conservative valvuloplasty approaches have been described with varying clinical outcomes in these unusual cases. In animals (e.g. English bulldogs) in whom the venous anatomy is uncertain on physical examination, vascular ultrasound is useful to clarify the anatomy and aid in percutaneous catheter sheath introduction.

A recent paper confirms the general clinical impression that approximately 40% of dogs with severe pulmonic stenosis have at least some right to left atrial shunting through a patent oval foramen. These patients can occasionally experience severe (usually transient) cyanosis following balloon valvuloplasty if their tricuspid regurgitation worsens significantly, especially if the regurgitant jet is directed toward the patent oval foramen.

Pacemaker implantation: while pacemaker implantation is a common interventional cardiac procedure, it will not be covered in this review because of time / space constraints, and the different and specialized nature of the issues surrounding cardiac pacing.

Miscellaneous Interventional Procedures: there have been recent reports of several successful interventional or hybrid surgical-interventional procedures that have been used to address a variety of uncommon cardiac defects. These include a small series of cases in which cutting balloon valvuloplasty was used to address cor triatriatum dexter in dogs, and also cor triatriatum sinister in cats via a hybrid surgical – interventional approach.

Atrial septal defect closure has been described in patients (mostly standard poodles) with appropriate anatomy (i.e. ostium secundum ASDs with an adequate rim of tissue above and below the defect to anchor the device).

Muscular VSD closure has been described with both Amplatz devices as well as embolization coils, although the clinical impact of these procedures is less clear.

Mitral and tricuspid stenosis are also relatively uncommon defects that have occasionally been successfully addressed by balloon valvuloplasty.

The results of attempts to address another common congenital heart defect, severe subaortic stenosis, with a combination of cutting and high pressure balloon valvuloplasty, show more promise than previous attempts to maintain reduced pressure gradients using standard balloon valvuloplasty without the cutting balloon; however the long term clinical significance of these pressure gradient reductions remain to be evaluated.

Suggested Reading

Achen, S. E., Miller, M. W, Gordon, S. G., Saunders, A. B., Roland, R. M., & Drourr, L. T. (2008). Transarterial ductal occlusion with the Amplatzer vascular plug in 31 dogs. *Journal of Veterinary Internal Medicine / American College of Veterinary Internal Medicine, 22*(6), 1348–52

Arndt, J. W., & Oyama, M. A. (2013). Balloon valvuloplasty of congenital mitral stenosis. *Journal of Veterinary Cardiology : The Official Journal of the European Society of Veterinary Cardiology, 15*(2), 147–51. doi:10.1016/j.jvc.2013.03.002

Blossom, J. E., Bright, J. M., & Griffiths, L. G. (2010). Transvenous occlusion of patent ductus arteriosus in 56 consecutive dogs. *Journal of Veterinary Cardiology : The Official Journal of the European Society of Veterinary Cardiology, 12*(2), 75–84

Caivano, D., Birettoni, F., Fruganti, A., Rishniw, M., Knafelz, P., Moïse, N. S., & Porciello, F. (2012). Transthoracic echocardiographically-guided interventional cardiac procedures in the dog. *Journal of Veterinary Cardiology 14*(3), 431–44.

Carlson, J. A., Achen, S. A., Saunders, A. B., Gordon, S. G., & Miller, M. W. (2013). Delayed embolization of an Amplatz(®) canine duct occluder in a dog. *Journal of Veterinary Cardiology15*(4), 271–6

Chamberlin, S. C., Sullivan, L. A., Morley, P. S., & Boscan, P. (2001). Evaluation of ultrasound-guided vascular access in dogs. *Journal of Veterinary Emergency and Critical Care, 23*(5), 498–503

El-Said, H. G., Bratincsak, A., Foerster, S. R., et al. (2013). Safety of percutaneous patent ductus arteriosus closure: an unselected multicenter population experience. *Journal of the American Heart Association, 2*(6), e000424

Fonfara, S., Martinez Pereira, Y., Swift, S., et al: (2009). Balloon valvuloplasty for treatment of pulmonic stenosis in English Bulldogs with an aberrant coronary artery. *Journal of Veterinary Internal Medicine 24*(2), 354–9.

Francis, A. J., Johnson, M. J. S., Culshaw, GG et al (2011). Outcome in 55 dogs with pulmonic stenosis that did not undergo balloon valvuloplasty or surgery. *The Journal of Small Animal Practice, 52*(6), 282–8.

Fujii, Y., Aoki, T., Takano, H., Ishikawa, R., & Wakao, Y. (2009). Arteriovenous shunts resembling patent ductus arteriosus in dogs: 3 cases. *Journal of Veterinary Cardiology 11*(2), 147–51.

Fujii, Y., Nishimoto, Y., Sunahara, H. et al. (2011). Prevalence of patent foramen ovale with right-to-left shunting in dogs with pulmonic stenosis. *Journal of Veterinary Internal Medicine 26*(1), 183–5.

Goodrich, K. R., Kyles, A. E., Kass, P. H., et al. (2007). Retrospective comparison of surgical ligation and transarterial catheter occlusion for treatment of PDA in two hundred and four dogs (1993-2003). *Veterinary Surgery : VS, 36*(1), 43–9.

Gordon, S. G., Saunders, A. B., Achen, S. E., Roland, et al. (2010). Transarterial ductal occlusion using the Amplatz Canine Duct Occluder in 40 dogs. *Journal of Veterinary Cardiology 12*(2), 85–92.

Henrich, E., Hildebrandt, N., Schneider, C., Hassdenteufel, E., & Schneider, M. (2010). Transvenous coil embolization of patent ductus arteriosus in small (≤3.0 kg) dogs. *Journal of Veterinary Internal Medicine 25*(1), 65–70.

Johnson, M. S., Martin, M., Edwards, D., French, A., & Henley, W. (2012). Pulmonic stenosis in dogs: balloon dilation improves clinical outcome. *Journal of Veterinary Internal Medicine 18*(5), 656–62.

Kleman, M. E., Estrada, A. H., Maisenbacher, H. W., Prošek, R., Pogue, B., Shih, A., & Paolillo, J. A. (2012). How to perform combined cutting balloon and high pressure balloon valvuloplasty for dogs with subaortic stenosis. *Journal of Veterinary Cardiology 14*(2), 351–61.

Locatelli, C., Spalla, I., Domenech, O., Sala, E., Brambilla, P. G., & Bussadori, C. (2013). Pulmonic stenosis in dogs: survival and risk factors in a retrospective cohort of patients. *The Journal of Small Animal Practice, 54*(9), 445–52.

López-Alvarez, J., Dukes-McEwan, J., Martin, M. W. S. et al (2011). Balloon dilation of an imperforate cor triatriatum dexter in a Golden Retriever with concurrent double-chambered right ventricle and subsequent evaluation by cardiac magnetic resonance imaging. *Journal of Veterinary Cardiology 13*(3), 211–8.

Meijer, M., & Beijerink, N. J. (2012). [Patent ductus arteriosus in the dog: a retrospective study of clinical presentation, diagnostics and comparison of interventional techniques in 102 dogs (2003-2011)]. *Tijdschrift Voor Diergeneeskunde, 137*(6), 376–83.

Saunders, A. B., Gordon, S. G., Boggess, M. M., & Miller, M. W. (2013). Long-Term Outcome in Dogs with Patent Ductus Arteriosus: 520 Cases (1994-2009). *Journal of Veterinary Internal Medicine*

Saunders, A. B., Winter, R. L., Griffin, J. F., Thieman Mankin, K. M., & Miller, M. W. (2013). Surgical management of an aberrant left subclavian artery originating from a left patent ductus arteriosus in a dog with a right aortic arch and abnormal branching. *Journal of Veterinary Cardiology 15*(2), 153–9.

Scansen, B. A., Schober, K. E., Bonagura, J. D., & Smeak, D. D. (2008). Acquired pulmonary artery stenosis in four dogs. *Journal of the American Veterinary Medical Association, 232*(8), 1172–80.

Scollan, K., Salinardi, B., Bulmer, B. J., & Sisson, D. D. (2011). Anomalous left-to-right shunting communication between the ascending aorta and right pulmonary artery in a dog. *Journal of Veterinary Cardiology : The Official Journal of the European Society of Veterinary Cardiology, 13*(2), 147–52.

Silva, J., Domenech, O., Mavropoulou, A., Oliveira, P., Locatelli, C., & Bussadori, C. (2013). Transesophageal echocardiography guided patent ductus arteriosus occlusion with a duct occluder. *Journal of Veterinary Internal Medicine 27*(6), 1463–70.

Singh, M. K., Kittleson, M. D., Kass, P. H., & Griffiths, L. G. (2011). Occlusion devices and approaches in canine patent ductus arteriosus: comparison of outcomes. *Journal of Veterinary Internal Medicine, 26*(1), 85–92.

Tanaka, R., Soda, A., Saida, Y., Sugihara, K., Takashima, K., Shibazaki, A., & Yamane, Y. (2007). Evaluation of the efficacy and safety of coil occlusion for patent ductus arteriosus in dogs. *The Journal of Veterinary Medical Science, 69*(8), 857–9.

Visser, L. C., Scansen, B. A., & Schober, K. E. (2013). Single left coronary ostium and an anomalous prepulmonic right coronary artery in 2 dogs with congenital pulmonary valve stenosis. *Journal of Veterinary Cardiolog 15*(2), 161–9.

Kleman ME, Estrada AH, Tschosik ML et al (2013): An update on combined cutting balloon and high pressure balloon valvuloplasty for dogs with severe subaortic stenosis. ABSTRACT. ACVIM Forum, 2013, pg 632.

YEAR IN REVIEW: ARRHYTHMIAS

Roberto Santilli, DVM, PhD, DECVIM-CA (Cardiology)
Samarate, Varese, Italy

PERMANENT JUNCTIONAL RECIPROCATING TACHYCARDIA IN A DOG[1]

Atrioventricular accessory pathways (Aps) have been previously described in the dog with a typical location along the right posterior and right postero-septal tricuspid valve annulus. [2] In this species APs presented with an all-or-none pattern of conduction and in 75 % of cases were concealed.[2] Last year our group describe a particular type of AP located, similarly to the most common Aps, in the right postero-septal area but with unique conducting features: unidirectional and decremental retrograde conduction. [1] The AP mediated an uncommon form of atrioventricular reciprocating tachycardia described in human beings called permanent junctional reciprocating tachycardia (PJRT) or Coumel tachycardia. [3-6] Permanent JRT is a circus movement tachycardia in which the anterograde limb of the circuit is the AV node and the retrograde limb is a concealed slowly conducting fiber with decremental properties usually located in the postero-septal region. [3-6]

Permanent junctional reciprocating tachycardia is an incessant tachycardia with long RP' interval, narrow QRS complex and most frequently inverted P waves in the inferior leads of the surface electrocardiogram[7]. In the dog herein reported, surface electrocardiogram presented similar features: a narrow complex tachycardia with a ventricular cycle length of 260 msec, negative P' waves in the inferior leads (II, III and aVF) with an electrical axis on the frontal plane of -45°, indicating right-sided infero-to-superior atrial activation, a long RP' (140 ms) and RP'/P'R of 1.16. The endocardial mapping revealed a long RP' narrow complex tachycardia with inverted P waves in the inferior leads, which matched previous surface recordings. First beat, fixed VA interval post-overdrive atrial pacing ruled out focal atrial tachycardia arising from the triangle of Koch area and showed that the tachycardia was AV nodal dependent. The entity of prolongation of VA conduction time during programmed ventricular stimulation (from 140 and 185 ms) permitted to showed unidirectional retrograde decremental properties of the AP and to rule out the most common type of orthodromic atrioventricular reciprocating tachycardia (OAVRT).[1] To differentiate atypical atrioventricular nodal reciprocating tachycardia from PJRT, a tachycardia reset was performed releasing a ventricular paced beat during the refractory period of the His bundle. The paced beat anticipated the tachycardia and terminated it without atrial retro-activation. Other criteria that can be used to distinguish PJRT with this test is pre-excitation of the atria with the same sequence of activation and prolongation (> 50 ms) of the local VA interval.[5] This case report describes the electrocardiographic and electrophysiologic features of PJRT in a dog, an uncommon arrhythmia in humans, that can mimic a focal atrial tachycardia arising from the ostium of the coronary sinus. [8]

ANALYSIS OF THE ATRIAL DEPOLARIZATION WAVE IN DOGS WITH THIRD-DEGREE AV BLOCK[9]

Electrical potentials generated by atrial recovery process are called T_a waves. This phase is projected to the body surface during the PQ segment and QRS complex. During sinus rhythm, T_a wave is usually hard to recognize because potential amplitudes are small and overimposed on the P-QRS complex, therefore the evaluation of much of the recovery phase is usually difficult to analyse.[10-12] During third-degree atrioventricular (AV) block the situation changes, since the T_a wave and QRST complex are uncoupled. [12] T_a wave analysis is an important diagnostic tool to evaluate atrial infarction, injuries, chamber enlargement and atrial propensity to develop arrhythmias.[11-13] In this study we described the electrocardiographic characteristics of T_a wave in dogs with third-degree AV block.

T_a wave was visible in all dogs as a "dome-shaped" deflection following the P wave. In agreement with the previous study, T_a wave was found to be substantially longer in duration and smaller in amplitude than P wave , and T_a wave deflections were always negative in leads I, II, III, aVF and in all precordial leads (V1-V6) and with a polarity opposite to the one of P wave. [10] In our study, the mean amplitude and duration of the T_a wave, the T_a/P amplitude ratio and duration ratio showed some differences if compared with previous studies,[10] The discrepancy of these results can probably be attributable to the fact that in the present study atrial depolarization and repolarization phases were analyzed during sinus rhythm, while in previous studies the analysis of the T_a wave was conducted during atrial pacing with the electrical stimulus originating from the coronary sinus ostium[1] or both from the right and the left atrium.[10,15] Mean duration of the P-T_a interval was 187.99 (172.13-203.84) msec. This value is in contrast with previous studies, in which reported a mean P-T_a interval duration of 210 msec[1] and 248.0±25.3 msec in case of right atrial pacing and 256.0±38.4 msec in the case of left atrial pacing.[15] Shorter duration of the P-T_a interval reported in this study can be due to the different activation modalities of the atrial myocardium. In agreement with previous experimental study, we found a correlation between T_a wave and P wave amplitude but not a correlation between T_a wave and P wave duration. This study found a correlation between the amplitude of the T_a wave and the lead examined, with a mean electrical axis of the T_a wave on the frontal plane of -114,26°, and with higher values of amplitude of the T_a wave in lead II and lead aVF and the lower values reported in lead I and lead aVL. We also found a correlation between the P-T_a interval duration and the T_a wave duration since the duration of this interval is related to the repolarization phase duration. Similarly to previous analysis,[14] we also showed that the P-T_a interval duration is correlated with the atrial cycle length. It has been shown that this relationship is crucial to describe atrial vulnerability to arrhythmias when atrial refractory period fails to shorten with increasing rate.[16] This study presents some limitations since the presence of third degree AV block was a prerequisite and consequently, the study population was older than the population in general and the impact of the T_a wave on the ST segment was not assessed.

The study herein reported provides a detailed electrocardiographic description of the atrial repolarization wave in dogs with third degree AV block with reproducible measurements that can be used as reference values in the population of dogs with AV conduction disturbances and normal cardiac size.

RADIOFREQUENCY CATHETER ABLATION OF ATYPICAL ATRIAL FLUTTER IN DOGS[17]

The current classification of atrial flutters (AFLs) and regular atrial tachycardias is based on the mechanism of arrhythmia and isthmus dependence rather than surface electrocardiographic features.[18] According to this classification the term macroreentrant tachycardia (MRT) refers to all reentrant tachycardias with a large central obstacle, including typical AFL and other macroreentrant circuits in either the right or left atrium. Atrial flutters are classified into typical and atypical: typical AFL, when cavo-tricuspid isthmus (CTI) – dependent, regardless of the electrocardiographic pattern, and atypical when a non-CTI-dependent circuit is documented. Our group first described CTI-dependent atrial flutter in the dog, classifying AFL in typical when the wavefront travels counterclockwise around the tricuspid valve annulus in the left anterior 30 degree view, and in typical reverse when the wavefront travels clockwise.[19]

The term atypical AFL is used in human beings to describe non-CTI-dependent MRT, a more rapid and irregular arrhythmias, with variable flutter wave morphology, and often prefibrillatory.[18, 20] In humans these types of MRT can be secondary to previous atrial surgery or secondary to radiofrequency catheter ablation scars. [18, 20] Other forms of atypical AFL circuits that involve the left atrium (left atrial MRTs) have been described and include circulation of wavefronts around unexcitable areas over the posterior left atrial wall, around the pulmonary veins, the mitral annulus,[18,22] or over the left atrial septum (left septal AFL). [23] A last variant of atypical AFL is type II AFL of Wells's classification, which present shorter cycle length with no gap, it cannot be entrained, is transient and common under vagal stimulation, transitioning spontaneously to baseline flutter or fibrillation. [24]

In the study herein reported detailed endocardial mapping and entrainment showed two right atrial macro-reentrant circuits around electrically silent areas with two critical isthmus areas over the right atrial free wall (RAFW), and the right atrial septum. The right posterior free wall is a well recognized arrhythmogenic substrate in patients with AFL since it includes a line of transverse conduction block adjacent to the crista terminalis. [25] Activation mapping of the RAFW in the dogs described in our study showed the presence of an electrically silent area characterized by low amplitude and fractionated and/or delayed components as described in human patients without a history of atriotomy as well as in sterile pericarditis and models of atrial enlargement. [26-27]

The right atrial septum is another arrhythmogenic site in human patients with AFL since it displays a lower mean bipolar voltage and both longitudinal and transverse conduction velocity delays compared to normal patients. The atrial septum has a bilaminate muscle structure constituting the thin membranous primary atrial septum that may contribute to the protection and maintenance of septal flutters.[28] Atypical AFLs have been associated with different structural heart diseases such as coronary artery disease, idiopathic dilated cardiomyopathy and valvular heart disease, but have also been mapped in a small number of human patients with no underlying heart disease. Conversely, in case of left septal atrial flutter, 65 % of the patients showed no concurrent heart disease. [20-26] In the dogs herein reported, right septal AFLs were not associated with cardiac disease, while RAFW AFLs were associated with chamber dilation and ventricular dysfunction. In these cases, although tachycardia-induced cardiomyopathy must be considered, the etiology of the cardiac disease could not be completely elucidated.

Atypical MRT circuits are more difficult to map since there is not a fixed location for the circuit or its critical isthmus. In our case series the AFL circuits were stable and concealed entrainment could be achieved at the right atrial midseptal area and along the right atrial free wall.

The surface electrocardiogram may be used to predict the type of AFL in human patients. [29-30] Typical CTI-dependent AFL, are characterized by negative F waves in the inferior leads and positive in V1, while typical reverse AFL shows positive F waves in the inferior leads. Atypical AFL presents a highly variable flutter wave morphology and may appear as typical reverse AFL, resembling typical or focal AFL.[29-30] A broad-based upright V1 is highly predictive of a left-sided flutter, while when V1 has an initial isoelectric or inverted component followed by an upright component, this is consistent with right flutter. [29-30] Unfortunately, the variety of chest conformation in the dog did not permit to apply the precordial lead criteria for AFL diagnosis in the dog. Septal flutter circuits usually show large, positive F waves in V1 with almost flat F waves in other leads. [23] In dogs seen clinically and in experimental models CTI-dependent AFL circuits showed a saw-tooth pattern in inferior leads in case of typical or counterclockwise rotation, and positive deflection in inferior leads in case of typical reverse or clockwise rotation. [19] In the dogs herein reported both AFLs circuits showed large and positive F waves in the inferior leads, located in previous ST segment as previously described in dogs with typical reverse atrial flutter,[19] with a longer isoelectric interval, which might reflect wide areas of slow conduction.[27]

Reported success rate of radiofrequency catheter ablation of atypical AFLs in human is lower than that in case of CTI-dependent circuits, ranging from 67 to 73 %, and deterioration into atrial fibrillation was observed. [23,26] In the dogs reported in the present study complete resolution of the arrhythmias were achieved in three out of 5 cases, while the remaining two cases (1 with right septal AFL and 1 with RAFW) showed respectively recurrence of the arrhythmia or development of persistent atrial fibrillation. One of the possible causes for this failure is the anatomy of the critical isthmus which, unlike the CTI, are not completely protected by anatomical barriers and therefore linear lesions performed with RFCA may not completely interrupt the circuits.

In conclusion atypical right AFL may occasionally occur in dogs, endocardial mapping and entrainment techniques allowed to identify two right atrial macroreentrant circuits not yet reported in the dog. Atypical AFL circuit could be definitively eliminated with radiofrequency catheter ablation.

REFERENCES

[1] Santilli RA, et al. **J Vet Cardiol** 2013;15:225. [2] Santilli RA et al **J Am Vet Med Assoc** 2010;237:1142. [3]Coumel P. et al. **Arch Mal Coeur** 1967;60:1830. [4]Gallagher JJ et al. **Eur J Cardiol** 1978;8:413. [5]Critelli G et al. **J Am Coll Cardiol** 1984;4:601. [6]Farrè J et al. **Am J Cardiol** 1979;6:1099, [7]Gaita F, **J Am Coll Cardiol** 1995;25:655, [8] Santilli RA et al. **J Vet Intern Med** 2008;22 915, [9] Perego M et al. **Am J Vet Res** 2014;75:54, [10] Hayashi H et al. **Jpn Heart J** 1970;11:91. [11] Holqvist F et al. **Ann Nonivasive Electtrocardiol** 2009;14:13, [12] Hayashi H et al. **Am Heart J** 1976;91:689, [13] Hayashi H et al **Am Heart J** 1976;91:492. [14]Debbas NM et al. **J Am Coll Cardiol** 1999;33:358, [15] Mirvis DM et al. **J Electrocardiol** 1986;19:99, [16] Attuel P et al. **Int J Cardiol** 1982;2:179. [17]Santilli RA et al. **J Vet Cardiol** 2014;in press, [18] Saoudi N et al. **Eur Heart J** 2001;22:1162, [19] Santilli RA et al. **J Vet Cardiol** 2010;12:59, [20] Yang Y et al. **Circulation** 2001;103:3092, [21] Cosio FG et al, **J Cardiovasc Electrophysiol** 1996;7:60, [22] Ricard P et al. **Europace** 2002;4:229, [23] Marrouche NF et al, **Circulation** 2004;109:2440, [24] Wells JL et al, **Circulation** 1979;60:665, [25] Okumura et al **Circ J** 2007;71:636, [26] Kall JG et al, **Circulation** 2000:101:270, [27] Uno K et al. **Circulation** 1999;100:1354, [28] Yamaguchi N et al., **J Exp Med** 2008;215:13, [29]Medi C et al. **Europace** 2008;10:786; [30]Bochoeyer A, et al. **Circulation** 2003;108:60,

YEAR IN REVIEW: CARDIOMYOPATHY

Kathryn M. Meurs, DVM, PhD, DACVIM (Cardiology)
Raleigh, NC

INTRODUCTION

Cardiomyopathy, particularly dilated and arrhythmogenic, continue to be important diseases for the dog. An inability to develop a therapeutic regiman that leads to a significant increase in life span has led to research that focuses on trying to determine the underlying etiology or pathophysiologic mechanisms. These include evaluation of the role of serotonin and Wnt pathways.[1,2] Significant interest has also gone into early diagnosis and screening.[3] Updates in feline cardiomyopathy also evaluated similar topics in pathophysiology and screening.[4,5]

REFERENCES

[1]Fonfara S, et al. **Vet J** 2013:11, s1090, [2]Oxford EM et al. **J Vet Intern Med** 2014, 28, 92. [3]Holler P et al **J Vet Intern Med** 2014 28:123 [4]Hall DJ et al **J Feline Med Surg** 2013 (epub ahead of print).[5]Maclea HB et al. **J Vet Intern Med** 2013, 27:1416.

WHAT WE KNOW ABOUT THE COMPANION ANIMAL GI MICROBIOME IN HEALTH AND DISEASE

Jan S. Suchodolski, med. vet., Dr. med. vet, PhD, DACVM
College Station, TX

INTRODUCTION

The intestinal microbiota is defined as the collection of all living microorganisms (i.e., bacteria, fungi, protozoa, and viruses) that inhabit the gastrointestinal tract. With the development of novel molecular analysis tools (based commonly on sequencing of the bacterial 16S rRNA gene), it is now appreciated that the gastrointestinal microbiota of animals is highly complex, comprising several hundred to thousand bacterial phylotypes.[1,2] It is estimated that the mammalian intestine harbors a total of 10^{10}-10^{14} microbial cells, approximately 10 times more than the number of host cells. It is, therefore, obvious that this highly complex microbial ecosystem will play a crucial role in health and disease. Gut microbes are useful to the host by acting as a defending barrier against transient pathogens, they aid in digestion, and help to harvest energy from the diet. Furthermore, they provide nutrition for enterocytes and play an important role in the development and regulation of the host immune system. However, in the last few years convincing evidence has been gathered associating alterations in the composition of the intestinal microbiota with acute and chronic enteropathies of dogs and cats.[1,3]

ASSESSMENT OF INTESTINAL BACTERIA

Bacterial culture

Based on traditional bacterial culture, the small intestine harbors total bacterial counts ranging from 10^2 to 10^9 cfu/g feces. Cats appear to have higher counts of anaerobic bacteria compared to dogs. The total bacterial count in the colon ranges between approximately 10^9 and 10^{11} cfu/g feces. The most abundant cultivable groups are *Bacteroides, Clostridium, Lactobacillus, Bifidobacterium,* and *Enterobacteriaceae* spp.

Until a few years ago, traditional bacterial culture was the most commonly used method for describing the bacterial groups present in the GI tract of companion animals.[4] Culture is still a very useful technique for the detection of specific intestinal pathogens (e.g., *Salmonella, Campylobacter jejuni*). However, bacterial culture has limitations for assessing the complex bacterial diversity found in the gastrointestinal tract of mammals. Samples must be processed immediately to preserve aerobic and anaerobic species, which are prone to damage during sample handling. Selective culture media followed by phenotypic and biochemical identification systems often fail to accurately classify all bacteria. Therefore, it is now well accepted that culture is not a good method for in-depth characterization of the complex ecosystem present in the mammalian GI tract.[5] A culture-based method underestimates total bacterial numbers in the gut, and also does not allow identification of the majority of bacterial groups present in the GI tract. Some reasons for our inability to culture most intestinal bacteria include our lack of knowledge regarding their optimal growth requirements and that the canine and feline gastrointestinal tract harbors predominantly anaerobic bacteria, which are prone to sampling and handling damage. Furthermore, many selective culture media lack sufficient specificity and often other organisms than the target is enumerated.

Molecular tools

Molecular methods are now the standard in assessing gut microbiota. Based on molecular fingerprinting it has been shown that each individual animal harbors very unique microbial communities, with a pronounced spatial variation along the gastrointestinal tract.[7] Various studies have utilized high-throughput sequencing of the 16S rRNA gene, and have described the canine and feline microbiome harboring on average 10 different bacterial phyla. *Proteobacteria, Firmicutes, Actinobacteria, Bacteroidetes, Spirochaetes,* and *Fusobacteria* make up the vast majority of the gut microbiota. Minor abundant members are the bacterial phyla *Tenericutes, Verrucomicrobia, Cyanobacteria,* and *Chloroflexi.* The *Firmicutes* is a diverse bacterial phylum and consists of *Clostridiales* and *Lactobacillales.* Based on phylogenetic analysis, the order *Clostridiales* is comprised of many different phylotypes, which are organized into so called *Clostridium* clusters.[3,4] The various clusters differ in their locations within the intestine: sequences affiliated with *Clostridium* cluster XI and *Clostridium* cluster XIVa dominate in the proximal small intestine and colon, respectively.

In addition to bacteria, the GI tract harbors various fungi, archaea, protozoa, and viruses. Recent molecular studies have provided more in depth analysis of the diversity of these microorganisms in healthy animals, but their interactions, their influences on the host, and their role in disease remain unclear. A study reported the prevalence and identification of fungal organisms in the small intestine of healthy dogs and dogs with chronic enteropathies.[6] The results showed a high prevalence (up to 76.1% of dogs) and high degree of diversity of fungal organisms in the canine duodenum. Furthermore, dogs with gastrointestinal disease harbored opportunistic fungal pathogens. A total of 51 different phylotypes were identified. The most commonly observed phylotypes were *Pichia* spp., *Cryptococcus* spp., *Candida* spp., and *Trichosporon* spp.

A recent study has characterized the fungal microbiome (mycobiome) of 19 dogs (12 healthy dogs and 7 dogs with acute diarrhea) using fungal tag-encoded FLX-titanium amplicon pyrosequencing.[7] Five distinct fungal phyla were identified, with Ascomycota (median: 97.9% of obtained sequences) and Basidiomycota (median: 1.0%) being the most abundant. A total of 219 fungal genera were identified across all 19 dogs with a median (range) of 28 (4–69) genera per sample. Candida was the most abundant genus found in dogs. No significant differences were observed in the relative proportions of fungal communities between healthy and diseased dogs.

MICROBIOME IN HEALTH

Gut microbes benefit the host through many mechanisms. They offer a nutritional advantage to the host by producing short chain fatty acids (SCFA, such as butyrate, propionate, and acetate). These stimulate mucosal growth and epithelial cell proliferation.

Lactate, produced by microbial fermentation, serves as an energy source for the host. The microbial metabolites produced by the resident microbiome are thought to be one of the most important driving force behind the coevolution of the gastrointestinal microbiota with their host. More recent studies have attempted to study the functional properties of the intestinal microbiota using metagenomic (i.e., shot-gun DNA sequencing) and metabolomics approaches to better study the role of the intestinal microbiota in GI health. Metagenomic analysis of the canine fecal microbiome revealed the most represented functional categories of the microbiome as carbohydrate metabolism; protein metabolism; cell wall and capsule; cofactors, vitamins, prosthetic groups and pigments; DNA metabolism; RNA metabolism; amino acids and derivatives; and virulence.

In addition, the intestinal microbiota plays a major role in protection of the host from invasion by harmful bacteria through exclusion of potentially pathogenic organisms. The implicated defense mechanisms against these harmful microbes include competition for oxygen and nutrient substrates, competition for mucosal adhesion sites, creation of a physiologically restrictive environment for non-resident bacterial species (e.g., production of substances that are toxic to other bacteria, changes in pH and redox potential, or hydrogen sulfide production), and secretion of antimicrobial substances (e.g., bacteriocins).

MICROBIOME IN DISEASE

There is strong evidence that the residential gut microbiota plays an important role in the pathogenesis of inflammatory bowel disease (IBD) in several species including humans, dogs, and cats. Studies in humans have shown an association between IBD and microbial dysbiosis in the intestine. In these studies, a decrease in the bacterial phyla *Firmicutes* and *Bacteroidetes*, and an increase in *Proteobacteria* and *Actinobacteria* were associated with IBD. In most studies evaluating the microbiota between controls and IBD patients, a reduction in the diversity of *Clostridium* clusters XIVa and IV (i.e., *Lachnospiraceae* and *C. coccoides* subgroups) was associated with IBD, suggesting that this bacterial group may play (possibly due to production of SCFA) an important role in maintenance of gastrointestinal health.

In our own studies, we have recently demonstrated imbalances in bacterial communities in dogs with IBD. In the small intestine, the proportions of Fusobacteria, Bacteroidaceae, Prevotellaceae, and Clostridiales were significantly increased in healthy dogs. In contrast, specific bacterial genera within Proteobacteria, including Diaphorobacter and Acinetobacter, were either more abundant or more frequently identified in dogs with IBD.[8]

In another study, fecal samples from healthy dogs (n = 32), dogs with acute non-hemorrhagic diarrhea (n = 12), dogs with acute hemorrhagic diarrhea (n = 13), and dogs with active (n = 9) and therapeutically controlled idiopathic IBD (n = 10) were analyzed by sequencing of the 16S rRNA gene.[9] Dogs with acute diarrhea, especially those with acute hemorrhagic diarrhea, had the most profound changes in bacterial groups in their microbiome. Dogs with acute hemorrhagic diarrhea had significant decreases in Blautia, Ruminococcaceae including Faecalibacterium, and Turicibacter spp., and significant increases in genus *Sutterella* and *Clostridium perfringens* compared to healthy dogs.

In another recent study, we evaluated the fecal microbiome of healthy dogs (n=180), dogs with chronic enteropathies (CE; n=87), and dogs with acute hemorrhagic diarrhea (AHD; n=48).[10] Significant differences in the abundance of the evaluated bacterial groups were observed for the disease groups when compared to the healthy dogs. *Faecalibacterium* spp., *Turicibacter* spp., and Ruminococcaceae were significantly decreased in CE and AHD. Bacteroidetes were significantly decreased in CE, but not different in AHD. *E. coli* and *C. perfringens* were significantly increased in CE and AHD. *Bifidobacterium* spp. and γ-Proteobacteria were significantly increased in CE, but not different in dogs with AHD. *Lactobacillus* spp. and *Streptococcus* spp. were significantly increased in dogs with CE and decreased in dogs with AHD.

Future studies will need to evaluate how these microbial changes alter functional aspects within the GI tract. It has been hypothesized that such microbiome changes may lead to altered intestinal barrier function, damage to the intestinal brush border and enterocytes, an increased competition for nutrients and vitamins, and to an increased deconjugation of bile acids. Of interest is that commonly depleted groups in GI disease are *Lachnospiraceae, Ruminococcaceae,* and *Faecalibacterium*. These bacterial groups, which are important producers of short-chain fatty acids, may play an important role in maintenance of gastrointestinal health, as this leads to decreased production of SCFA (e.g., butyrate, acetate), which may impair the capability of the host to down-regulate an aberrant intestinal immune response. The importances of some of these bacterial groups that are depleted in IBD have recently been demonstrated in humans. For example, *Faecalibacterium prausnitzii* is consistently reduced in human IBD and this bacterium has been shown to secrete metabolites with anti-inflammatory properties, thereby down-regulating IL-12 and IFNγ and increasing IL-10 secretions.

To understand the impact of bacterial dysbiosis on host immune and metabolic function, we have recently evaluated the serum metabolic profile of dogs with idiopathic IBD before and after treatment. Serum and fecal samples were collected from healthy dogs (n=10) and from dogs with IBD (n=12) before and after 21 days of standard medical therapy (e.g., elimination diet and administration of anti-inflammatory drugs). Fecal DNA was evaluated for fecal microbial composition using 454-pyrosequencing of the 16S rRNA gene and quantitative PCR (qPCR). Serum metabolites were profiled using a GC-TOF/MS platform and analyzed using MetaboAnalyst 2.0. Before treatment, the abundance of *Blautia, Faecalibacterium*, and *Turicibacter* was significantly decreased in dogs with IBD relative to healthy dogs. Unifrac analysis revealed clustering between healthy and diseased dogs. While the CIBDAI was significantly decreased, no significant differences in the proportions of bacterial groups were found before and after treatment. Serum concentrations of several metabolites differed significantly between healthy dogs and dogs with IBD (decreased in IBD: tyrosine, aminomalonic acid, 4-hydroxyproline, trans-4-hydroxyproline, naphthalene, and proline; increased in IBD: xylitol and allantoin). Future studies are needed to evaluate more in-depth the functional changes that occur in dogs with chronic enteropathies.

REFERENCES

[1]Frank DN, et al. Molecular-phylogenetic characterization of microbial community imbalances in human inflammatory bowel diseases. *Proc Natl Acad Sci USA* 2007;104:13780-13785; [2]Suchodolski JS, et al. The effect of the macrolide antibiotic tylosin on microbial diversity in the canine small intestine as demonstrated by massive parallel 16S rRNA gene sequencing. *BMC Microbiol* 2009;9:210; [3]Suchodolski JS. Microbes and gastrointestinal health of dogs and cats. *J Anim Sci* 2010:1-34; [4] Mentula S, et al. Comparison between cultured small-intestinal and fecal microbiotas in Beagle dogs. *Appl Environ Microbiol* 2005;71:4169-4175; [5]Zoetendal EG, et al. Molecular microbial ecology of the gastrointestinal tract: from phylogeny to function *Curr Issues Intestinal Microbiol* 2004;5 31-48; [6]Suchodolski JS, et al. Prevalence and identification of fungal DNA in the small intestine of healthy dogs and dogs with chronic enteropathies. *Vet Microbiol* 2008;132:379-388; [7]Foster ML, et al. Characterization of the fungal microbiome (mycobiome) in fecal samples from dogs. *Vet Med Int* 2013;2013:658373; [8]Suchodolski JS, et al. 16S rRNA Gene pyrosequencing reveals bacterial dysbiosis in the duodenum of dogs with idiopathic inflammatory bowel disease. *Plos ONE* 2012;7:e39333; [9]Suchodolski JS, et al. The fecal microbiome in dogs with acute diarrhea and idiopathic inflammatory bowel disease. *Plos ONE* 2012;7:e51907; [10]Markel M, et al. Characterization of fecal dysbiosis in dogs with chronic enteropathies and acute hemorrhagic diarrhea. *J Vet Intern Med* 2012;26:765-766.

THE GI MICROBIOME AND INTESTINAL INFLAMMATORY DISEASE: HOST-MICROBE INTERACTIONS

Karen L. Madsen, PhD
Edmonton, Alberta, Canada

INTRODUCTION

Mammals harbour microbial communities in their gut that outnumber somatic and germ cells by about 10 to 1000 times.[1] Gut microbes have co-evolved with their eukaryote hosts to form a symbiotic relationship. Commensal microbes have fine-tuned their functions to not only aid in their colonization, but also to influence immunological, nutritional, and physiological processes of the host. Millions of years ago, the gut tube developed as a means of capturing, concentrating, and extracting soluble nutrients from the environment. This tube naturally became an attractive niche for 'free-living' bacteria who sought to benefit from the enriched ecosystem. This spawned a mutual coevolution of host and microbe; a result of the microbes' biochemical specialization to utilize the energy source eaten by the host and the host's ability to assimilate and utilize the novel metabolic processes conferred by the microbes.[2] Since then, both microbes and hosts have maintained a delicate mutualistic interdependence. In fact, any disruption in this mutualism or symbiosis between microbes and hosts appears to play a role in gastrointestinal disorders. Microbes interact with the gut epithelium through cell surface components, fermentation products, and extracellularly secreted compounds. Host cells that respond to these mediators include intestinal epithelial cells, goblet cells, paneth cells, macrophages, dendritic cells, and T and B cells.[3] Host-microbial interactions primarily involve TLRs (toll-like receptors) and NLRs (nucleotide-binding oligomerization domain (NOD) and leucine-rich repeat (LR) containing proteins). Gut microbes can also respond to signals from other microbes[4] and mediators from the host.[5]

Microbiota of the gut

The large intestine is the primary site of microbial colonisation with $10^{10} - 10^{11}$ bacteria per gram feces found. The gut is dominated almost entirely by only six phyla: *Bacteroidetes*, *Firmicutes*, *Actinobacteria*, *Verrucomicrobia*, *Fusobacteria*, and *Proteobacteria*.[6] Along with the diversity of microbes that exists in the colonic lumen, there are also different microbial micro-habitats along the horizontal length of the intestinal tract. There are substantial differences in gut microbes between individuals and changes in microbiota can also occur over time within an individual.[7, 8] Using 16S ribosomal RNA (rRNA) and metagenomic sequence-based methods, several studies have identified *Bacteroidetes* and *Firmicutes* to be the most dominant distal gut microbes in humans[9, 10]; however, in canines and felines, the most dominant groups also include *Fusobacteria* in canines[11] and *Proteobacteria* and *Actinobacteria* in felines.[8]

Establishment of the gut microbiota

During and following birth, exposure to microbes from the mother and surrounding environment leads to establishment of a complex microbiome.[12] The normal pattern of colonization in infants involves initial colonization by facultative anaerobes such as *E. coli* and other *Enterobacteriaceae*.[13] These organisms gradually deplete the initial oxygen supplies and allow for the colonization of strictly anaerobic bacteria, including *Bifidobacteria, Clostridium, and Bacteroides*.[13] From an initial low diversity and low complexity, the intestinal microbiota of the infant will slowly develop and mature, reaching an adult state around 3 years of age in humans.[14, 15] Evidence from animal and human studies supports the concept that gut colonization during infancy has a critical role in modulating immune development and maintaining mucosal homeostasis throughout life.[16] It has been postulated that a critical window of time exists during which immune development is initiated, and that disruption of this programmed development due to alterations in microbial colonization could predispose the host to chronic diseases associated with a dysregulated immune response.[17] In humans, establishment of the gut microbiome is influenced by early diet, antibiotic exposure, and method of birth, breast milk composition, and genetic background.[18] Antibiotic use during this period can significantly alter the composition of the gut microbiota and result in long term decreases in diversity and altered trajectories of immune development.[15, 19] Even temporary antibiotic-induced shifts in the composition of the gut microbiota in infants can disturb microbial metabolism and balance, thus allowing for outgrowth of potentially pathogenic organisms, which can remain as integral members of the community. Antibiotic exposure and infections during the first year of life have been identified as independent risk factors for inflammatory bowel disease in humans.[18, 20, 21]

Development of immune homeostasis

During neonatal development, the immune system learns to recognize non-pathogenic commensal microbes while maintaining the ability to recognize and respond to pathogens. This state of non-responsiveness to commensals is associated with a requirement for the host-specific microbiota for complete innate and adaptive immune maturation.[22] Innate sensing of microbes involves cross-talk between Toll-like receptors (TLRs) and NLRs (nucleotide-binding oligomerization domain (NOD)-like receptor - and leucine-rich repeat (LRR)-containing proteins) and is essential for maintenance of host-microbial homeostasis.[23] The TLR family is a group of pattern-recognition receptors that recognize conserved motifs referred to as microbe-associated molecular patterns (MAMPs). TLRs detect numerous MAMPs, including lipopolysaccharide (LPS) by TLR4; lipoproteins and lipoteichoic acids by TLR2; flagellin by TLR5; CpG DNA by TLR9; dsRNA by TLR3; and ssRNA by TLR7. Activation of TLRs results in the stimulation of cellular pathways to preserve cellular integrity as well as pathways aimed at defense against microbes. These responses can include the induction of tissue repair and cell survival pathways, cytokine and antimicrobial peptide production and secretion, and maturation of antigen-presenting cells.[24] In addition to ligand specificity, individual TLRs differ in their expression patterns and the signal transduction pathways they activate appear to be cell-type dependent. Intestinal homeostasis and repair is dependent upon the sensing of the intestinal microbiota by TLRs.[25]

Within the NOD/CARD family of proteins are NOD1 (Card4) and NOD2 (Card15), which areintracellular proteins that respond to distinct bacterial peptidoglycan substructures: NOD2 detects muramyl dipeptide, a molecular motif common to Gram-negative and Gram-positive bacteria, while NOD1 detects peptidoglycans containing *meso*-diaminopimelic acid (*meso*-DAP), which is more commonly found in Gram-negative bacteria.[26] Activation of NOD receptors triggers inflammation by the activation of NFκB and mitogen-activated protein kinase (MAPK) pathways.[26]

Alterations in expression, function, and localization of TLRs and NLRs have been linked with aberrant responses to both pathogens and commensals in humans and animals with inflammatory bowel disease.[27, 28] In healthy animals, a strong secretory IgA response in the gut together with the epithelial cell barrier limits the exposure of commensal microbes to the systemic immune system. Thus, the systemic immune system remains ignorant or naive to these organisms.[29] This IgA response develops during the neonatal period, and involves dendritic cells delivering microbes to the mesenteric lymph nodes where they interact with B and T cells to induce antigen-specific IgA responses.[30, 31] Homeostasis is maintained through interactions between epithelial cells, CD4+Foxp3+ Treg cells, tolerogenic dendritic cells (DCs), myeloid suppressor cells, innate lymphoid cells (ILC), and IL10-producing B cells.[32-35] Microbiota exert a high degree of control over naive T cell differentiation to T effector and regulatory cells.[22, 36] While some bacteria and their products induce tolerogenic anti-inflammatory Treg cells, other species lead to the development of more inflammatory Th17 cells. In mice, colonization with segmented filamentous bacteria, *Bacteroides fragilis*, and Clostridium has been shown to increase the number of intestinal Foxp3+ Treg cells and/or enhance the expression of IL-10 and CTLA-4 in Treg cells.[36-39] Thus, the overall composition of the microbiota can give it an inherently greater or lesser ability to activate pro-inflammatory responses. Studies have shown that during an intestinal infection, tolerance to commensal microbes is lost and bacteria-specific T cells against commensals become activated and differentiate to inflammatory effector cells.[40] Furthermore, these effector T cells form memory cells that are functionally identical to pathogen-specific T cells.

Gut microbes in inflammatory bowel diseases

Inflammatory bowel diseases (IBD) are characterized by chronic intestinal inflammation causing abdominal pain, diarrhea, intestinal bleeding, weight loss, malabsorption, and nutritional deficiencies. The current consensus is that IBD occurs in genetically-predisposed individuals who are exposed to as yet to be determined microbial, dietary, and environmental triggers. Humans and animals with IBD exhibit a "dysbiosis" of the gut microbiota.[27, 41] This has led to the hypothesis that CD occurs due to an imbalance between inflammatory-inducing and protective bacterial taxa that is accompanied by dysregulated innate and adaptive immune responses towards specific gut microbes. Patients with IBD have increased numbers and decreased diversity of microbes associated with the gut mucosa as compared with healthy individuals.[42] This loss of diversity is associated with decreased numbers of *Firmicutes*, in particular *Faecalibacterium prausnitzii* and *Roseburia*, and an increase in *Ruminococcus* and *Enterobacteriaceae*, especially adherent invasive *Escherichia coli* (AIEC).[43] Several studies have identified the increased prevalence of adherent-invasive *E.coli* (AIEC) in ileal biopsies of patients with Crohn's disease.[44-46] This adherent-invasive strain of *E.coli*, unlike other commensal strains, has the unique ability to evade macrophage killing, and survive and replicate within its vacuoles.[47] In a similar fashion, studies in dogs with IBD have also shown that members of the families *Enterobacteriaceae* and *Clostridiaceae* are increased in the diseased gut.[48, 49] It is not yet clear whether it is the overall low diversity or the increase or decrease in specific microbial taxa that are the more important factors in IBD, and whether these changes are causative or associative.

Role of Genes

Human genetic studies strongly implicate microbes in CD pathogenesis. Genome-wide association studies have linked Crohn's disease to several genes involved in innate immune responses to intestinal microbes (*e.g.,* Nucleotide Oligomerization Domain 2, *NOD2*) and clearance of intracellular microbes through autophagy (*e.g.,* ATG16 autophagy related 16-like protein 1, *ATG16L1*; and immunity-related GTPase M, *IRGM*).[50] These recent discoveries in IBD genetics crystallize the microbial basis of IBD; implicating defective containment, killing, and clearance of invasive or intracellular bacteria by the host being associated with the onset of IBD.

Conclusions

Current understanding of interactions between gut microbes and intestinal epithelial and immune cells suggests that these interactions are necessary for the development and maintenance of intestinal homeostasis. Immune and epithelial responses to microbes depend on the species under study, the concentration of microbes utilized, and the type of cell studied. A barrier between luminal microorganisms and the host immune system is maintained through intestinal epithelial cells, mucus and anti-microbial production, and IgA secretion. Disruptions in this symbiosis between microbes and hosts appears to play a critical role in gastrointestinal inflammatory disorders.

REFERENCES

[1]Turnbaugh PJ et al. **Nature** 2007;449:804-10; [2]Tremaroli V et al. **Nature** 2012;489:242-9; [3]Maynard CL, et al. **Nature** 2012;489:231-41; [4] Sonnenburg JL, et al. **PLoS Biology** 2006;4:e413; [5]Lyte M, et al **Cell Tissue Res** 2011;343:23-32; [6] Arumugam M, et al. **Nature** 2011;473:174-80; [7]Adlerberth I, et al. **Acta Paediatr** 2009;98:229-38; [8]Minamoto Y, et al. **Anim Health Res Rev** 2012;13:64-77; [9]Eckburg PB, et al. **Science** 2005;308:1635-8; [10]Qin J, et al. **Nature** 2010;464:59-65; [11] Hooda S, et al. **Anim Health Res Rev** 2012;13:78-88; [11]Matamoros S, et al. **Trends Microbiol** 2013;21:167-73;[13]Penders J, et al. **Pediatrics** 2006;118:511-21; [14]Palmer C, et al. **PLoS Biology** 2007;5; [15]Sharon I, et al. **Genome Res** 2013;23:111-20; [16]Rautava S, et al. **Nat Rev Gastroenterol Hepatol** 2012;9:565-76; [17]Collado MC, et al. **Gut Microbes** 2012;3:352-65; [18]Neuman MG, et al. **Transl Res** 2012;160:29-44; [19]Fouhy F, et al. **Antimicrob Agents Chemother** 2012;56:5811-20; [20]Shaw SY, et al. **Am J Gastroenterol** 2010;105:2687-92; [21]Hviid A, et al. **Gut** 2011;60:49-54; [22]Chung H, et al. **Cell** 2012;149:1578-93; [23]Uematsu S, et al. **Journal of Molecular Medicine** 2006;84:712-25; [24]Medzhitov R. **Nature** 2008;454:428-35; [25]Saleh M, et al. **Nat Rev Immunol** 2011;11:9-20;

[26]Le Bourhis L, et al. **Biochem Soc Trans** 2007;35:1479-84; [27]Allenspach K. **Vet Clin North Am Small Anim Pract** 2011;41:345-60; [28]Zaki MH, et al. **Trends Immunol** 2011;32:171-9; [29]Konrad A, et al. Gastroenterology 2006;130:2050-9; [30]Cebra JJ. **American Journal of Clinical Nutrition** 1999;69:1046S-1051S; [31]Groux H, et al. **Immunology Today**1999;20:442-5; [32]Hansen CH, et al. **PLoS One** 2012;7:e34043; [33]Hrncir T, et al. **BMC Immunol** 2008;9:65; [34]Cording S, et al. **Mucosal Immunol** 2013; [35]Cebula A, et al. **Nature** 2013;497:258-62; [36]Ivanov, I et al. **Cell Host & Microbe** 2008;4:337-49; [37]Atarashi K, et al. **Science** 2011;331:337-41; [38]Gaboriau-Routhiau V, et al. **Immunity** 2009;31:677-89; [39]Mazmanian SK. **Journal of Pediatric Gastroenterology & Nutrition** 2008;46 Suppl 1:E11-2; [40]Hand TW, et al. **Science**12;337:1553-6; [41]Morgan XC, et al. **Genome Biol** 2012;13:R79; [42]Gevers D et al. **Cell Host Microbe** 2014;15:382-392; [43]Manichanh C, et al. **Nat Rev Gastroenterol Hepatol** 2012;9:599-608; [44]Darfeuille-Michaud A, et al. **Gastroenterology** 2004;127:412-21; [45] Masseret E, et al. **Gut** 2001;48:320-5; [46]Darfeuille-Michaud A, et al. **Gastroenterology** 1998;115:1405-13; [47]Bringer MA, et al. **J Bacteriol** 2007;189:4860-71; [48]Suchodolski JS, et al. **Vet Microbiol** 2010;142:394-400; [49]Xenoulis PG, et al. FEMS Microbiol Ecol 2008;66:579-89; [50]Hisamatsu T, et al. **Pharmacol Ther** 2013;137:283-97.

Probiotics in Intestinal Inflammatory Disease: Mechanisms and Benefits

Karen L. Madsen PhD
Edmonton, Alberta, Canada

INTRODUCTION

The gut microbiome is increasingly being recognized as having a significant impact on human health. Changes in composition and decreased biodiversity of the gut microbiota have been associated with numerous diseases, including obesity, type 2 diabetes, irritable bowel syndrome, inflammatory bowel disease, metabolic syndrome, autism, and cancer.[1-6] Thus, therapies aimed at modulating the gut microbiota may help improve health status, and reduce the incidence of disease or attenuate disease symptoms. Probiotics are living microbial organisms that are administered in supplements or foods to enhance the well-being of the host. A growing body of work describing beneficial effects of various probiotic strains in ameliorating symptoms in a number of gastrointestinal and liver diseases, as well as in systemic disorders such as obesity and allergy now exists.[7, 8] Mechanisms involved in beneficial effects of probiotics involve both direct effects on the host as well as indirect effects through modulation of commensal gut microbes. A substantial amount of research has demonstrated that in a strain and dose-dependent manner, probiotics can enhance the barrier function of the gut, modulate systemic and mucosal immune function, antagonize pathogens through anti-microbial mechanisms, and exert metabolic effects on the host through production of metabolites.

Intestinal barrier function and disease

The ability of the gut epithelium to act as a barrier between the external environment and the closely regulated internal milieu is absolutely essential for human health.[9] When the gut barrier is disrupted, not only does a greater amount of microbial antigenic material transverse the gut, but the altered route of passage can also significantly alter the immunogenicity of the microbial antigen. A breakdown in barrier function of the gut has been associated with several different diseases, including inflammatory bowel disease (IBD), diabetes, celiac disease, and irritable bowel syndrome (IBS).[10] In patients with Crohn's disease, increased intestinal permeability has been shown to precede a clinical relapse[11, 12] and it is thought that this increase in permeability allows for inappropriate interactions between gut antigens and the mucosal immune system. Supporting a role for increased permeability and exposure of gut microbes to the host immune system are the findings that patients with inflammatory bowel disease have systemic immune responsiveness to specific microbial antigens, including OmpC from *E. coli*, OmpW from *Bacteroides*, a *Pseudomonas fluorescens*-associated sequence I2, a *Bacteroides caccae* TonB-linked outer membrane protein, and CBir1 flagellin.[13] In animal models of colitis it has been convincingly demonstrated that abnormal permeability is present prior to onset of histological disease,[14] and that therapies, including probiotic treatment, aimed at targeting small intestinal permeability can lead to improvement in colonic disease.[15, 16] Barrier function of the gut is maintained by intestinal epithelial cells, mucus secretion, anti-microbial peptide production, and secretory IgA. Intestinal epithelial cells (IECs) are the primary cell type that interacts with the luminal environment and these cells act as the first line of defense of the host against potential pathogens. IECs participate in innate immunity by recognizing pathogens through the expression of innate immune receptors, by releasing anti-microbial peptides and by the secretion of cytokines and chemokines that link innate with adaptive immune responses. Specific probiotic strains have been shown to enhance intestinal barrier function by stimulating mucus and antimicrobial peptide production; by increasing mucosal IgA responses; by enhancing tight junction protein expression and/or localization; by preventing epithelial cell apoptosis; and by the induction of cytoprotective molecules.[17]

Competitive exclusion

Several strains of *Lactobacillus* and *Bifidobacterium* are able to compete with pathogens for binding to intestinal epithelial cells and are able to displace pathogens even if the pathogens have already attached.[18-20] Some strains of lactobacilli express human mucus-binding pili, which enhance their ability for colonization.[21] Probiotic inhibition of pathogen adherence to epithelial cells is mediated partially by competition for lectin binding sites on glycoconjugate receptors on the brush border surface.[22] Probiotics express microorganism-associated molecular patterns (MAMPS) that bind to the same pattern recognition receptors as pathogens do, thus preventing access via competitive exclusion.[21] Probiotics are also able to disrupt biofilm formation through competing, excluding and displacing pathogens from the intestinal lumen.[23]

Effects on Apoptosis

Probiotics are able to prevent cytokine- and oxidant-induced epithelial damage by promoting cell survival. Lactobacillus GG and soluble factors (p75 and p40) released from LGG prevented epithelial cell apoptosis through activating anti-apoptotic Akt in a phosphatidylinositol-3'-kinase (PI3K)-dependent manner and inhibiting pro-apoptotic p38/MAPK activation.[24] This reduction in apoptosis would help to maintain epithelial barrier integrity and increase resistance to pathogens by reducing breaks in the mucosal barrier. In an *in vivo* study, *Lactobacillus plantarum* 299v increased expression of members of the inhibitor of apoptosis protein family (HIAP2/cIAP).[25] In another study, administration of *Bifidobacterium bifidum* increased expression of TLR-2, COX2, and prostaglandin E2 and significantly reduced apoptosis in the intestinal epithelium in a necrotizing enterocolitis (NEC) model.[26] Further, inhibition of COX2 signalling blocked the protective effect of *B. bifidum* suggesting that a probiotic-induced increased expression of COX2 and elevated production of PGE2 was responsible for thisprotection. This elevation of PGE2 production subsequently protected against epithelial cell apoptosis associated with NEC, thus preserving intestinal barrier function.[26]

Probiotics have direct effects on pathogens

Specific probiotic strains produce and release antibacterial peptides that are able to directly kill or inhibit the growth of several pathogens. *Lactobacillus acidophilus* was shown to disrupt lipopolysaccharide (LPS) within the bacterial cell wall of enterovirulent *Salmonella typhimurium*, thus permeabilizing the membrane.[27] Additionally, *L. acidophilus* also inhibited the swimming capability of *Salmonella* without affecting flagellum expression.[28] *Bifidobacterium* strains were found to not only decrease the viability of several pathogens but also to inhibit the entry and enhance the killing of intracellular *Salmonella typhimurium* by epithelial cells.[29] Several other studies have shown that probiotics have bactericidal properties and/or inhibit growth of pathogens through the production of short-chain fatty acids (SCFA) and subsequent lowering of the luminal pH. Similarly, probiotics can also influence the gene expression of pathogens by reducing the expression of virulence factors, especially those required by the pathogen for attaching and colonizing.[30] These studies illustrate how probiotics directly act on pathogens by disrupting cell walls, reducing mobility, and decreasing viability.

Probiotics induce release of antimicrobial peptides from intestinal epithelial cells

In the gut, bacteria are controlled by antimicrobial peptides (e.g., lysozyme, defensins, RegIII) produced by enterocytes and Paneth cells.[31] In addition to having direct antimicrobial effects, probiotics can also modulate the hosts' gene expression to increase antimicrobial peptide production. It is interesting to note that some patients with inflammatory bowel disease (IBD) have low levels of defensin production, and these patients are more susceptible to bacterial infections.[32] In a controlled clinical trial, subjects receiving the probiotic *E.coli* Nissle 1917 (EcN) twice daily for 3 weeks, had significantly increased levels of fecal β-defensin (hBD)-2 peptide than individuals receiving placebo. Interestingly, the levels of fecal hBD-2 peptide were still elevated at 9 weeks following cessation of treatment.[33]

Probiotic effects on immune function

Probiotics have the remarkable ability to not only attenuate inflammatory processes, but also to induce regulatory responses. Dendritic cells (DCs) are one of the key regulators of the mucosal immune system and are modulated through interactions with gut microbes. Various probiotic cell wall components and secreted factors are recognized by pattern recognition receptors found on DCs. As a result, probiotics can exert multiple immunoregulatory activities on DCs. These activities depend on the type of bacterial stimuli, other co-existing stimuli, and the local environment. Probiotics are able to induce DC maturation, increase DC survival, increase both anti- and pro-inflammatory cytokine production, and induce DC-regulated differentiation of naïve T cells into T helper (Th)-1, Th2, Th17, or T regulatory (Treg) cells.[34] These modulations of DCs are done through various surface and intracellular molecules of probiotics. In one study, *B.breve* was found to stimulate mouse intestinal $CD103^+$ DCs to induce an IL-10 producing subset of regulatory T cells (Tr1). [35] Conversely, soluble factors from a specific strain of *B.breve* and *L.paracasei* were found to prime human DCs, so that when the cells were challenged with *Salmonella*, the primed DCs had decreased pro-inflammatory cytokine and chemokine expression than unprimed DCs. [36] Interestingly, DCs react differently to probiotic components based on if they encounter probiotics directly or in the presence of other cell types. For example, in the presence of intestinal epithelial cells, luminal exposure to probiotics or probiotic DNA resulted in reduced inflammatory cytokines and increased anti-inflammatory responses. [37] Conversely, a recent study using human cultured biopsies found some strains of probiotics actually increased inflammation.[38] These studies suggest that intestinal epithelial cells also play a vital role in educating the mucosal immune system, and probiotics exposed to DCs alone may give rise to varying expression of cytokines and thus different effector functions. Therefore, it is vitally important to consider the state and integrity of the intestinal barrier prior to offering probiotics as therapeutics.

Metabolic effects of probiotics

Probiotics can exert beneficial effects through enzymatic activities and the release of specific metabolites. Complex carbohydrate fermentation by bacteria leads to the production of short-chain fatty acids (SCFA) including acetate, butyrate, and propionate.[39] Acetate maintains gut barrier function and can prevent pathogen translocation.[40] Butyrate is the primary energy source for colonocytes and also has numerous anti-inflammatory effects.[41] A lack of butyrate results in colonocyte cell death and autophagy.[42] Production of SCFA can lower the colonic pH, which can directly antagonize pathogens and reduce growth. Bile acids are synthesized from cholesterol in the liver and further metabolized by the gut microbiota into secondary bile acids. Probiotics can influence the host through modulating bile acid synthesis by altering host expression of genes involved in bile acid synthesis, conjugation, and reabsorption. Bile salts serve both a metabolic role and a bacterial signal to the host for microbiota status and microbial composition in the gut. Probiotics can also degrade lactose into glucose and galactose, which is beneficial to patients with lactose intolerance. Transitory probiotics can modulate activity of existing commensal strains directly through the production of anti-microbial compounds and/or indirectly through probiotic-induced changes in immune system or gut barrier function. The production of lactate during probiotic transit can affect specifically the commensal microbiota by promoting lactate users such as *Roseburia* or *Eubacterium*. These commensals will then produce different types of SCFA by secondary fermentation of lactate and other primarily metabolites primarily fermented.

Conclusion

A barrier between luminal microorganisms and the host immune system is maintained through intestinal epithelial cells, mucus and anti-microbial production, and IgA secretion. In a strain-dependent manner, probiotics have the capacity to enhance gut barrier function and modulate innate and adaptive immune function through numerous different mechanisms. It is clear that host-microbial interactions at the gut mucosal surface are critical for health and overall homeostasis and further, that probiotics may be harnessed in order to maintain health and protect against disease. In the future, larger clinical trials will need to be carried out to determine how best to use probiotics in the prevention or treatment of various diseases. The particular strains of probiotics best suited to the treatment of specific diseases will need to be determined, along with the amount of bacteria to use, the time of dosing, and whether combinations or single strains are preferable.

REFERENCES

[1] Kang DW, et al. **PLoS One** 2013;8:e68322; [2]Morgan XC, et al. **Genome Biol** 2012;13:R79; [3]Ng SC, et al. **J Gastroenterol Hepatol** 2013; [4]Qin J, et al. **Nature** 2012;490:55-60; [5]Turnbaugh PJ, et al. **Nature** 2006;444:1027-31; [6]Wu N, et al. **Microb Ecol** 2013;66:462-70; [7]Sekirov I, et al. **Physiol Rev** 2010;90:859-904; [8]Ceapa C, et al. **Best Pract Res Clin Gastroenterol** 2013;27:139-55; [9]Arrieta MC, et al. **Gut** 2006;55:1512-20; [10]Camilleri M, et al. **Neurogastroenterol Motil** 2012;24:503-512; [11]Hollander D, et al. **Annals of Internal Medicine** 1986;105:883-5; [12]Casellas F, et al. **American Journal of Gastroenterology** 1986;81:767-70; [13]Lodes MJ, et al. **Journal of Clinical Investigation** 2004;113:1296-306; [14]Wirtz S, et al. **Advanced Drug Delivery Reviews** 2007;59:1073-83; [15]Arrieta MC, et al. **Gut** 2009;58:41-8; [16]Ewaschuk J, et al. **Am J Physiol** 2008;295:G1025-34; [17]Rao RK, et al. **Curr Nutr Food** Sci 2013;9:99-107; [18]Collado MC, et al. **Curr Microbiol** 2007;55:260-5; [19]Sherman PM, et al. **Infect Immun** 2005;73:5183-8; [20]Lebeer S, et al. **Nat Rev Microbiol** 2010;8:171-84; [21]Kankainen M, et al.**Proc Natl Acad Sci U S A** 2009;106:17193-8; [22]Tallon R, et al. **Journal of Applied Microbiology** 2006;102:442-451; [23]Woo J, et al. **Lett Appl Microbiol** 2013;56:307-13; [24]Yan F, et al. **Gastroenterology** 2007;132:562-75; [25]Dykstra NS, et al. **Pediatr Res** 2011;69:206-11; [26]Khailova L, et al. **Am J Physiol Gastrointest Liver Physiol** 2010;299:G1118-27; [27]Coconnier-Polter MH, et al. **Appl Environ Microbiol** 2005;71:6115-20; [28]Lievin-Le Moal V, et al. **Antimicrob Agents Chemother** 2011;55:4810-20; [29]Lievin V, et al. **Gut** 2000;47:646-52; [30]Medellin-Pena MJ, et al. **Appl Environ Microbiol** 2007;73:4259-67; [31]Bevins CL, et al. **Nat Rev Microbiol** 2011;9:356-68; [32]Wehkamp J, et al **Mucosal Immunol** 2008;1 Suppl 1:S67-74; [33]Mondel M, et al. **Mucosal Immunol** 2009;2:166-72; [34] Bron PA, et al. **Microb Cell Fact** 2012;11:123; [35]Jeon SG, et al. **PLoS Pathog** 2012;8:e1002714; [36]Bermudez-Brito M, et al. **PLoS One** 2012;7:e43197; [37]Campeau JL, et al. **Infect Immun** 2012;80:2632-44; [38]Tsilingiri K, et al. **Gut** 2012;61:1007-15; [39]Flint HJ, et al. **Nat Rev Microbiol** 2008;6:121-31; [40]Fukuda S, et al. **Nature** 2011;469:543-7; [41]Zimmerman MA, et al. **Am J Physiol Gastrointest Liver Physiol** 2012;302:G1405-15; [42]Donohoe DR, et al. **Cell Metab** 2011;13:517-26.

THERAPEUTIC MANIPULATION OF THE SMALL ANIMAL GI MICROBIOME: WHAT EVIDENCE DO WE HAVE FOR EFFICACY?

Stanley L. Marks, BVSc, PhD, DACVIM (Small Animal Internal Medicine, Oncology), DACVN
Davis, CA

The mammalian intestinal tract contains a complex, dynamic, and diverse population of non-pathogenic bacteria. Researchers have estimated that the average human contains 100 trillion microbes in the gut, which is 10 times more than the cells of the human body.[1] The intestinal microbiota influences the health of the host by providing nutritional substrates, modulating the immune system, and providing a support in the defence against intestinal pathogens.[2] The term *microbiome* refers to the total number of microorganisms and their genetic material and is contrasted from the term *microbiota*, which is the microbial population present in different ecosystems in the body.

There has been a plethora of research focusing on the mechanisms by which pathogenic bacteria influence intestinal function and induce disease; however, recent attention has focused on the indigenous non-pathogenic microbiota and the ways in which it may benefit the host. Initial colonization of the sterile newborn intestine occurs with maternal vaginal and fecal bacterial flora. The first colonizers have a high reduction potential and include species such as enterobacter, streptococcus, and staphylococcus. These bacteria metabolize oxygen, favouring the growth of anaerobic bacteria, including lactobacilli and bifidobacteria. Colonization with these bacteria is significantly delayed in caesarean deliveries,[3] leading to delayed activation of the efferent limb of the mucosal immune response.[4] Additional beneficial effects of developing a normal bacterial flora is seen in germ free mice that have small intestines that weigh less than their healthy counterparts. This effect occurs partly due to underdevelopment of lymphoid constituents, with a lack of plasma cells in the lamina propria and Peyer's patches, and subsequent reduction in IgA production. Exposure to bacteria results in a reversal of this phenomenon within 28 days of exposure.[5]

The intestinal microbiota has been associated with Crohn's disease and ulcerative colitis, as well as irritable bowel syndrome in humans.[6-8] In addition, the intestinal microbiota has also been implicated in the pathogenesis of various canine GI disorders, either associated with the presence of specific enteropathogens such as Salmonella, *Clostridium perfringens*, and viruses in acute episodes of diarrhea or a non-specific dysbiosis precipitating inflammatory bowel disease.[9,10] Molecular –phylogenetic studies have revealed a bacterial and/or fungal dysbiosis in the duodenum of dogs with IBD. A decrease in the proportion of Clostridiales and an increase in Proteobacteria is most commonly observed.[10] Only a few studies have described the fecal microbiota of dogs with acute and chronic GI disorders. Dogs with acute diarrhea, particularly those with acute hemorrhagic diarrhea (AHD) have the most profound alterations in their microbiome characterized by decreases in Blautia, Ruminococcaceae including *Faecalibacterium* and *Turicibacter* spp, and significant increases in genus Sutterella and *Clostridium perfringens* compared to healthy dogs. Dogs with clinically active IBD had decreased *Faecalibacterium* spp. and Fusobacteria that increased during resolution of the IBD.[10] The bacterial species that are commonly decreased during diarrhea are thought to be important short-chain fatty acid producers and could promote intestinal health. A deeper understanding of the gut microbiome will provide reference values for healthy populations and assist in diagnosing and treating diseased animals. In addition, manipulation of the intestinal microbiome via dietary intervention, administration of probiotics, prebiotics, or synbiotics, and fecal transplantation is currently being performed to maintain gut health in people and companion animals.

Manipulation of canine GI microbiota to improve health via dietary intervention did not begin until the early 1990's. Dietary fiber, prebiotics and probiotics have been the major nutritional strategies studied to modulate the canine and feline GI microbiota. Unfortunately, most of the research studies published to date have evaluated the effects of dietary manipulation of the GI microbiome in clinically healthy dogs and cats, and many of these studies have used traditional plating techniques or qPCR to quantify a limited number of bacteria (e.g., *Lactobacillus*, *Bifidobacteria*, *C. perfringens*, and *E. coli*) to assess efficacy.

PROBIOTICS

Probiotics refer to live microorganisms which when administered in adequate amounts confer a health benefit on the host. The term probiotic was derived from the Greek, meaning "for life." The Food and Agricultural Organization of the United States (FAO) and the World Health Organization (WHO) have stated that there is adequate scientific evidence to indicate that there is potential for probiotic foods to provide health benefits and that specific strains are safe for human use.[11] There has been a literal explosion of interest and research on the subject in recent years. Despite this activity, much still remains to be done to determine the specific indications and applications of probiotics in dogs and cats. There has been tremendous interest among veterinary pet food companies and manufacturers of animal health and wellness products to market probiotic formulations that are safe, pure, stable, and confer a beneficial effect in dogs and cats. These products are generally preferred to the multitude of over-the-counter probiotics marketed for veterinary use, given the concerns pertaining to quality control of the over-the-counter products.[12] A number of criteria are essential for efficacy and safety of probiotics. These include resistance to gastric acid and bile, ability to colonize the gastrointestinal tract, efficacy against pathogenic microorganisms, and modulation of the immune system.[13] Several potential mechanisms have been proposed for how probiotics reduce the severity or duration of diarrhea: competition with pathogenic bacteria or viruses for nutrients, competition for receptor sites, modification of the metabolic activity of the intestinal microflora, and the direct antagonism through the action of antimicrobial metabolites.[14,15]

Evidence for the Benefits of Probiotics in People

There is currently level 1 evidence (i.e., data from either high-quality, randomized controlled trials with statistically significant results and few design limitations or from systematic reviews of trials) for effectiveness of probiotics in treating lactose intolerance/maldigestion, treating acute infectious or nosocomial diarrhea in children, preventing or treating antibiotic-associated diarrhea, preventing and maintaining remission of pouchitis in adults, and maintaining remission of ulcerative colitis in adults.[16] In addition, there is level 2 evidence (evidence obtained from randomized trials that have limitations in methodology or results that have wide confidence intervals) for using probiotics to treat traveler's diarrhea, prevent sepsis secondary to severe acute pancreatitis, and prevent infections in postoperative patients.[16] Unfortunately, a similar level of evidence critically evaluating the benefits of specific probiotic strains in dogs and cats is currently lacking in the veterinary profession.

Evidence for the Benefits of Probiotics in Dogs

To date, only a relatively small number of studies have been published evaluating the effects of probiotics in dogs, and many of these have focused on the intestinal microbiota in apparently healthy dogs. Probiotic strains of human or canine origin (*Lactobacilli, Bifidobacter,* and *Enterococcus*) were used in healthy adult dogs or dogs with food-responsive diarrhea to assess their effects on intestinal microbial populations, their ability to reduce specific pathogens in feces, and effectiveness as immunomodulators.[17-22] In many of these studies, probiotics added to the food in healthy dogs had an equivocal effect on fecal microflora and pathogens.[19,22] However, it is important to note that most of these studies were not randomized, controlled trials, and the strains of probiotic varied from study to study, making interpretation of findings more challenging. In addition, many studies focused on fecal isolation and quantitative cultures of putative pathogenic bacteria such as *C. perfringens,* rather than on the evaluation of more meaningful end points such as phylogenetic characterization of the microbiota, mucosal immunopathology, and alterations in intestinal integrity. Only two studies addressing the role of probiotics in management of dietary sensitivity and food-responsive diarrhea have been published to date, with overall positive results.[17,18] Only one of those studies was a randomized, placebo-controlled clinical trial,[17] and the results of that study, although clinically positive (all of the dogs in the study improved when they were placed on the elimination diet) showed no specific changes in the inflammatory cytokine patterns or a specific benefit of the probiotic. The immunomodulatory effects of *Enterococcus faecium* SF68 have been studied in dogs, and the probiotic was associated with increased fecal IgA concentrations and increased vaccine-specific circulating IgG and IgA concentrations.[23] Although increased immune globulins may suggest enhanced immune response, the clinical relevance of this finding is not known.

Additional studies are warranted in dogs to further assess the immunomodulatory effects of probiotics and to evaluate their safety. The latter issue is particularly important given the recent finding of increased intestinal adhesion of *Campylobacter jejuni* in an *in vitro* model of canine intestinal mucus following incubation with *Enterococcus faecium*.[24] It should be noted that this *E. faecium* strain is different from the *E. faecium* SF68 strain available commercially; moreover, to date there has been no clinical or anecdotal evidence of *Campylobacter*-associated diarrhea in dogs associated with *E. faecium* administration. Short-term treatment (6 weeks) with *E. faecium* SF68 to 20 dogs with chronic naturally acquired subclinical giardiasis failed to affect giardial cyst shedding or fecal giardial antigen and did not alter innate or adaptive immune responses at multiple time points.[25] These results are in contrast to those shown following the oral feeding of *E. faecium* strain SF68 to mice experimentally infected with *Giardia intestinalis* trophozoites.[26] Oral feeding of *E. faecium* strain SF68 starting 7 days before inoculation with *Giardia* trophozoites significantly increased the production of specific anti-*Giardia* intestinal IgA and blood IgG. This humoral response was mirrored at the cellular level by an increased percentage of CD4+ T-cells in the Peyer's patches and in the spleens of SF68-fed mice. The improvement of specific immune responses in probiotic-fed mice was associated with a diminution in the number of active trophozoites in the small intestine as well as decreased shedding of fecal *Giardia* antigens (GSA65 protein). The latter findings underscore the importance of carefully evaluating the animal model, the timing of probiotic administration (prior to infection or following infection), and the specific end-points assessed.

Evidence for the Benefits of Probiotics in Cats

Unfortunately, there is little published information pertaining to probiotic use in cats, and only one clinical study has reported a beneficial effect of probiotic therapy for any feline disease to date. In that study, administration of *Enterococcus faecium* SF68 to 217 cats housed in an animal shelter was associated with a significantly lower percentage of cats with diarrhea ≥ 2days (7.4%) compared with a placebo group (20.7%).[27] One study evaluating the effect of dietary supplementation with the probiotic strain of *Lactobacillus acidophilus* DSM 13241 (2×10^8 CFU/d for 4.5 weeks) administered to 15 healthy adult cats demonstrated that recovery of the probiotic from the feces of the cats was associated with a significant reduction in *Clostridium* spp. and *Enterococcus faecalis*.[28] However, the immunomodulatory effects were reported based on decreased lymphocyte and increased eosinophil populations and increased activities of peripheral blood phagocytes. The relevance of these findings is unclear, because this study was not a randomized trial and the changes reported in the populations of peripheral blood cells cannot be extrapolated into evidence of systemic health benefits. Evaluation of the effect of supplementation with *Enterococcus faecium* strain SF68 on immune function responses following administration of a multivalent vaccine was evaluated in specific pathogen-free kittens.[29] This prospective, randomized, placebo-controlled study resulted in the recovery of *E. faecium* SF68 from the feces of seven of nine cats treated with the probiotic, and a nonsignificant increase in feline herpesvirus 1–specific serum IgG levels. Concentrations of total IgG and IgA in serum were similar in the probiotic and placebo groups, and the percentage of CD4+ lymphocytes was increased significantly only in kittens at 27 weeks and not at any other time points. Probiotics also have been evaluated in juvenile captive cheetahs, a population with a relatively high incidence of bacteria-associated enteritis. Administration of a species-specific probiotic containing *Lactobacillus* Group 2 and *Enterococcus faecium* to 27 juvenile cheetahs was associated with a significantly increased body weight in the treatment group, with no increase in the control group.[30] In addition, administration of the probiotic was associated with improved fecal quality in the probiotic group.

PREBIOTICS

A prebiotic is defined as a "nondigestible food ingredient that beneficially affects the host by selectively stimulating the growth of and/or activates the metabolism of one or a limited number of health promoting bacteria in the intestinal tract."[31] The most common prebiotics studied are fructans, although other prebiotics such as mannans, lactosucrose, and lactulose are also being evaluated.

Evidence for the Benefits of Prebiotics in Dogs and Cats:

Unfortunately, there is a paucity of information evaluating the clinical benefits of prebiotics in dogs and cats. Most of the outcomes published in the literature are limited to nutrient digestibility, microbial concentrations in feces, and fecal protein catabolites that may not necessarily denote a health benefit to the patient. The effects of short chain fructooligosaccharides (scFOS) were evaluated in a group of German Shepherd dogs suspected to have IgA deficiency.[32] Although the scFOS supplemented dogs had decreased aerobic and anaerobic bacteria in intestinal biopsies, the findings of the study were clouded in light of the fact that anaerobic bacterial counts did not decrease in the intestinal fluid samples of the dogs supplemented with scFOS. The effects of various oligosaccharides have been tested in adult ileal cannulated dogs to evaluate the effects on ilial and total tract nutrient digestibilities, microbial populations, ileal pH, ammonia, blood glucose, fecal consistency, and SCFA concentrations. Oligosaccharides (oligofructose, mannanoligossacharides, and xylooligosaccharides) were each given at 0.5% of the diet in a Latin square design. The only significant finding was a decrease in fecal *Clostridium perfringens* populations in dogs fed MOS.[33] The effects of 3% inulin supplementation of elimination and hydrolyzed protein diets to healthy dogs was associated with a slight increase in fecal moisture content (not clinically relevant), decreased apparent nutrient digestibility coefficients of crude protein in dogs on the elimination diet, and no effect on fecal IgA concentrations.[34] Adult beagles fed diets containing cellulose or beet pulp plus oligofructose for 6 weeks were found to have similar fecal concentrations of total anaerobes and aerobes; however, the dogs fed oligofructose had fewer Enerobacteriaceae and clostridia and greater numbers of lactobacilli. In addition, dogs fed oligofructose had longer and heavier small intestines (35% heavier), and 37% more mucosal mass with consequent greater absorptive surface area.[35] Administration of 1% scFOS or 1% inulin to weanling puppies (12-weeks-old) during a pathogen (*Salmonella* Typhimurium) challenge was associated with a lower severity of enterocyte sloughing in puppies consuming the fructans versus the control diet. In addition, puppies fed inulin also had higher fecal acetate, total SCFA concentrations and lactobacilli, indicating that prebiotics appear to attenuate some of the adverse effects of Salmonella challenge, and may provide protection against infection in weanling puppies. Cats fed diets containing 0 or 0.75% oligofructose had significantly increased fecal concentrations of lactobacilli and decreased concentrations of *C. perfringens* and *E. coli* compared with controls.[36] A study evaluating the effects of short-chain fructooligosaccharides (0.5%) and galactooligosaccharides (0.5%) in healthy cats showed no effect on fecal protein catabolites, including ammonia, 4-methylphenol, indole, and biogenic amines, underscoring the fact that concentrations of oligosaccharides > 0.5% should be used to elicit a positive response.[37]

The first nutritional intervention study in dogs that used pyrosequencing to evaluate the effects of beet pulp fiber on fecal microbial composition was performed in 2010.[38] Dog fed a control diet were compared to dogs fed a diet containing 7.5% beet pulp in a crossover design with 14 day periods. *Eubacterium balii* and *Faecalibacterium prausnitzii*, both of which are butyrate producers, were overrepresented in the dogs fed the beet-pulp containing diet, suggesting a possible anti-inflammatory effect of the beet-pulp. In contrast, Fusobacteria was under-represented in dogs fed he beet-pulp-containing diet. The effects of a synbiotic formulation to privately owned dogs for 21 days was evaluated via fecal 454 pyrosequencing.[39] The most abundant phylum in feces of all dogs was Firmicutes, followed by Actinobacteria and Bacteroidetes, regardless of synbiotic treatment. Synbiotic administration was associated with increased abundance of family Eubacteriaceae and phylum Fusobacteria. Recent studies have attempted to characterize the fecal microbiota of diarrheic dogs, as well as dogs with IBD. Reduced bacterial diversity as well as significantly higher proportion of Enterobacteriaceae were observed in duodenal brush borders from dogs with IBD compared to healthy controls.[9] Suchodolski et al. confirmed a bacterial dysbiosis in fecal samples of dogs with chronic diarrhea (IBD) and acute hemorrhagic diarrhea, and observed changes in the microbiome between acute and chronic disease states. The bacterial groups that were commonly decreased are important producers of short-chain-fatty acids and may play an important role in canine intestinal health.[10]

Fecal Microbiotia Transplantation (FMT):

Fecal microbiotal transplantation or infusion of a fecal suspension from a healthy individual into the gastrointestinal tract of another person to cure a specific disease, is best known as a treatment for recurrent *Clostridium difficile* infection (RCDI) in people,[40] and experience with FMT for ulcerative colitis and Crohn's disease is somewhat limited. Re-establishment of the wide diversity of intestinal microbiotia via infusion of donor feces into the colon is the proposed mechanism in patients with RCDI and IBD. FMT has been performed in dogs with a variety of chronic enteropathies (Scott Weese, personal communication), and the author (SLM) is currently completing a clinical trial evaluating the efficacy of FMT in Macaques with chronic colitis. There are a variety of application methods to inoculate the donor feces into the patient, and most studies have relied upon colonoscopy over retention enemas or nasogatric tubes to facilitate inoculation of the feces in the ileum and colon.

Donor stool is most often used within 8 hours of passage; however, frozen stool samples from standardized donors have been thawed and colonoscopically administered 1-8 weeks after passage for treatment of RCDI with similar success rates to fresh stool.[41] Donor fecal samples must be carefully screened for bacterial, viral, and parasitic enteropathogens, and human donors are excluded if they have taken antibiotics within the preceding 3 months or are on immunosuppressive or chemotherapeutic agents. In addition, patients with IBD, atopy, GI malignancy, and chronic diarrhea are excluded from being donors. The amount of donor stool used has varied; however, in a recent review, relapse was four-fold greater when < 50 g of stool was used in people with RCDI.[42] Stool is most commonly suspended in nonbacteriostatic saline; however, other diluents (e.g., yoghurt and milk) have been successfully used. The donor stool is mixed with diluent to a consistency that can be injected via the biopsy channel of a colonoscope. The suspension should be filtered through gauze pads or strainer to remove large particulate matter before aspiration into the syringe. The volume of stool suspension that is deposited in the colon varies tremendously, although volumes of 300-500 mL are commonly used. A larger volume allows the clinician to deposit aliquots of 90-100 mL into multiple locations within the intestinal tract, including the jejunum, ileum, ascending colon, transverse colon, and upper descending colon.

Conclusions and Future Directions:

The potential benefits and specific indications for the administration of pro- and prebiotics to dogs and cats have yet to be fully defined, although our knowledge and understanding of the nature and diversity of the feline and canine intestinal microbiome during health and disease has expanded rapidly following the advent of high-throughput DNA-sequencing platforms. Defining a role for pro- and prebiotics as well as FMT in dogs and cats will require completion of prospective, randomized, placebo-controlled studies that rely on clinically relevant end points related to a particular physiological or pathological condition. Further studies are warranted to determine the need for probiotics to be live microorganisms following the provocative studies of Rachmilewitz et al., who documented that the beneficial effects of probiotics are mediated by their DNA, circumventing the need for live, viable bacteria.[43] Pro- and prebiotics do appear to have a potential role in the prevention and treatment of various gastrointestinal illnesses, but it is likely that benefits achieved are specific to the bacterial species used and to the underlying disease context.

REFERENCES

[1]Savage DC. **Annu Rev Microbiol** 1977;31:107-33; [2]Hooper LV, et al. **Science** 2001;291:881-884; [3]Gronlund MM, et al. **J Pediatr Gastroenterol Nutr** 1999;28:19-25; [4]Insoft RM, et al. **Pediatr Clin N Am** 1996;43:551-71; [5]McCraken VJ, et al. **Cell Microbiol** 2001;3:1-11; [6]Seksik P, et al. **Gut** 2003;52:237-242; [7]Sokol H, et al. **Inflamm bowel dis** 2006;12:106-111; [8]Nobaek S, et al. **Am J Gastroenterol** 2000;95:1231-1238; [9]Xenoulis PG, et al. **FEMS Microbiol Ecol** 2008;66:579-589; [10]Suchodolski JS, et al. **Plos ONE** 2012;7:e51907; [11]Food and Agricultural Organization of the United States and World Health Organization (Online), 2001; [12]Weese JS. **Can Vet J**, 2003;44:982-3; [13]Tuomola E, et al. **Am J Clin Nutr** 2001;73:393S-398S; [14]Coconnier MH, et al. **FEMS Microbiol Lett** 1993;110:299-305; [15]Rolfe RD, et al. **J Infect Dis** 1981;143:470-475; [16]Guarner F. In Versalovic J, Wilson M, editors: **Therapeutic Microbiology: Probiotics and Related Strategies**. ASM Press, Washington DC 2008; p.255; [17]Sauter SN, et al. **Anim Nutr** 2006;90:269; [18]Pascher M, et al. **Arch Anim Nutr** 2008;62:107; [19]Vahjen W, et al. **Arch Anim Nutr** 2003;57:229; [20]Swanson KS, et al. **J Nutr** 2002;132:3721; [21]Biagi G, et al. **Vet Microbiol** 124:160, 2007; [22]Baillon MLA, et al. **Am J Vet Res** 65:338, 2004. [23]Benyacoub J, et al. **J Nutr** 133:1158-1162, 2003; [24]Rinkinen M, et al: **Vet Microbiol** 2003;92:111; [25]Simpson KW, et al. **J Vet Int Med** 2009;23:476-481; [26]Benyacoub J, et al. **J Nutr** 2005;135:1171-6; [27]Bybee SN, et al. **J Vet Int Med** 2011;25:856-860; [28]Marshall-Jones ZV, et al. **Am J Vet Res** 2006;67:1005; [29]Veir JK, et al: **Vet Ther** 2007;8:229; [30]Koeppel KN, et al: **J So Afr Vet Assoc** 2006;77:127; [31]Gibson GR, et al. **J Nutr** 1995;125:1401-12; [32]Willard MD, et al. **Am J Vet Res** 1994;55:654-659; [33]Strickling JA, et al. **Anim Feed Sci Technol** 2000;86:205-219; [34]Verlinden A, et al. **Br J Nutr** 2006;96:936-944; [35]Buddington RK, et al. **Am J Vet Res** 1999;60:354-358; [36]Sparkes AH, et al. **Am J Vet Res** 1998;59:436-440; [37]Kanakupt A, et al. **J Anim Sci** 2011;89:1376-1384; [38]Middlebos IS, et al. **PloS ONE** 2010;5:e9768; [39]Garcia-Mazcorro JF, et al. **FEMS Microbiol Ecol** 2011;78:542-554; [40]Song Y, et al. **Plos ONE** 2013;8:e81330; [41]Hamilton MJ, et al. **Am J Gastroenterol** 2012;107:761-767; [42]Gough E, et al. **Clin Infect Dis** 2011;53:994-1002; [43]Rachmilewitz D, et al. **Gastroenterol** 2004;126:520-528.

RETHINKING EGUS: EQUINE SQUAMOUS GASTRIC ULCER SYNDROME

Ben Sykes, BSc, BVMS, MS, DACVIM, DECEIM, MBA
Upper Orara, NSW, Australia

TERMINOLOGY

The term Equine Gastric Ulcer Syndrome (EGUS) has been in use for some time but it is commonly misrepresented which leads to significant confusion. By definition the term refers to a syndrome, within which numerous disease entities exist and distinguishing between the separate diseases and/or disease syndromes is important. The proposed pathophysiology of disease of the squamous and glandular mucosa are different, different risk factors have been described for squamous[1–5] and glandular[6] disease, the presence of squamous and glandular ulceration within an individual are unrelated[7–9] and the response of glandular ulceration to treatment with omeprazole is inferior to that of squamous ulceration[10–12].

As such, the direct extrapolation of findings from the squamous mucosa, which is by far the better studied of the two, to the glandular mucosa is inappropriate. Instead, disease of the squamous mucosa should be considered as a distinct entity to disease of the glandular mucosa. Further, it is likely that within both the squamous and glandular mucosa that there is more than one primary etiological process that causes disease and as such the use of the terminology 'syndrome' is appropriate. In line with this over the next two lectures diseases of the squamous and glandular mucosa will be discussed separately. The term Equine Squamous Gastric Ulcer Syndrome (ESGUS) will be used to discuss diseases of the squamous mucosa, while the term Equine Glandular Gastric Ulcer Syndrome (EGGUS) will be used to discuss diseases of the glandular mucosa.

EQUINE SQUAMOUS GASTRIC ULCER SYNDROME

The predominant form of ESGUS is primary disease with ulceration attributable to the squamous mucosa being exposed to gastric acid. Underlying factors for this are largely management related including exercise and feeding. Secondary ESGUS, resultant from delayed gastric emptying does occur, but it is relatively uncommon. Management of secondary ESGUS is focused on treatment of the primary disease and acid suppression in line with primary disease. Detailed discussion of secondary ESGUS is beyond the scope of this first lecture. Instead the first lecture will focus on primary squamous disease with an emphasis on some controversial areas.

Prevalence

The prevalence of ESGUS mirrors exercise intensity with the risk of disease increasing as the intensity of work increases. This effect is consistent across a wide range of horse types. The highest prevalences have been reported in Thoroughbred racehorses with > 70% of animals affected across a wide range of studies[7,13–15]. Further the risk of ESGUS increases with time in work[6]. However, it is the author's opinion that these studies overstate the prevalence of clinically significant ulceration. Many of the "affected" animals are likely to have only had small focal lesions that are unlikely to be of clinical significance but by definition meet the criteria of ulceration. Although the limit for defining clinically significant vs. non-significant ulceration is poorly defined the author generally considers grade 2/4 ESGUS to be unlikely to be of clinically significance vs. grade 3-4/4 lesions which are more likely to be relevant and cause clinical disease.

As an example of the overstatement of prevalence, analysis of a data set of mixed breed pleasure horses presenting to a University clinic for evaluation for suspicion of EGUS revealed an overall prevalence of ESGUS of 50% consistent with previous reports. However, further evaluation of the data set reveals that only 26 % of horses had ulceration of grade 3 or 4 (/4) and the overall distribution of ESGUS in this population was; grade 0-1 50% (49/98), grade 2 24% (24/98), grade 3-4 26% (25/98). A similar effect is seen when enrolment data from a number of studies[10–12] in Thoroughbred racehorses is pooled and evaluated where the ESGUS distribution was; grade 0-1 21% (15/74), grade 2 33% (25/78) and grade 3-4 46% (34/78).

This over representation of the rate of the ESGUS is potentially important and likely has had a significant impact on our understanding of disease. In particular it has likely lead to a distortion of the impact of ESGUS when performance and/or welfare parameters have been examined and potentially explains why, to date, it has been difficult to consistently correlate clinical signs with the presence/absence of ESGUS. Similarly, it has led many clinicians to dismiss ESGUS as a "normal" finding in the performance horse along the lines of the logic 'if they all have it then and most seem clinically OK then it must be normal'. Further works into the effect of different grades, vs. just the presence or absence of lesions, are needed but importantly large enough data sets that can realistically be expected to detect an effect must be used.

Pathophysiology

The pathophysiology of ESGUS is well understood with a variety of management factors contributing to the development of disease. All of these factors share the common trait that they increase the exposure of the squamous mucosa, which has inherent poor defenses, to acid. The purpose of this review is not to review the well accepted causative factors although the direct effect of diet warrants comment. Although volatile fatty acids have been shown to be potential contributors to ESGUS[16] it, in the author's opinion, is unlikely that they are enough to incite disease in their own right. Instead it is likely that by far the dominant factor in the development of ESGUS is the horses' enormous capacity for the production of gastric acid. Indirect evidence for this exists in that the removal of gastric acid, through the use of omeprazole, consistently results in high rates of healing regardless of whether dietary management is changed or not. As such the management of diet plays less of a role during the treatment phase and as discussed later, it may in fact be counterproductive during the treatment phase due to the effect of feeding on omeprazole absorption.

Management of diet, and in particular the reduction of grain content and increasing roughage availability is likely to be beneficial in the prevention of squamous disease but to date this magnitude of this effect has not been documented. The author has observed countless individuals where significant ESGUS is present in the face of 'good' management practices, primarily in Thoroughbred racehorses. In contrast, the author's experience in riding and sport horses (show jumpers and dressage horses) suggests that 'good' management practices are much more effective in these populations. This suggests that the efficacy of management in the prevention of ESGUS is somewhat dependent on additive effect of other risk factors. Along these lines it is reasonable to expect that management reduces the risk of ESGUS, especially in lower risk horses, but it is important to recognize that at an individual level disease can still occur in well managed horses. Equally important is recognition that the expected effect of management changes on the development of ESGUS cannot be translated to EGGUS because, although some risk factors are shared, the effect of diet and exercise and the development of EGGUS is not a clear cut as it is in the development of ESGUS. This will be discussed further in the next lecture.

Diagnosis

The clinical signs of ESGUS are well described. Disease is common wherein ulcers are identified in 'normal' horses (i.e. those considered to be free of symptoms by the owner/trainer) on screening programs. However, many of these horses change in response to treatment with improved appetite and/or temperament suggesting that sub-clinical disease is common. As such, the presence of overt clinical signs increases the clinical index of suspicion, while their apparent absence does not discount EGUS as a differential.

Gastroscopy remains the main form of achieving a definitive diagnosis. Alternative means of diagnosis including a therapeutic trial, fecal blood testing and the use of a sucrose absorption test have been proposed and warrant discussion. If gastroscopy is not readily available then a therapeutic trial with omeprazole can be useful and a positive response to treatment supports the diagnosis of EGUS. However, it does not differentiate between ESGUS and EGGUS. Such as distinction is important as the treatment recommendations for ESGUS and EGGUS vary and as such, in the author's opinion, a positive response to treatment should be viewed as a clear indication for gastroscopy to obtain a definitive diagnosis. A negative response to trial therapy with omeprazole reduces the likelihood of EGUS, but does not completely discount it as some animals are slow to respond clinically and symptoms may not completely resolve until healing has occurred, especially with EGGUS. Further, the accuracy of therapeutic trials is predicated on the assumption that effective acid suppression is achieved in the individual. As discussed later there is significant individual variation in the absorption of omeprazole with some animals absorbing it very poorly. As such, in light of a negative response to omeprazole, other more likely differentials should be discounted first, but gastroscopy remains indicated if no specific diagnosis is made.

The importance of follow-up gastroscopy has been underplayed, largely predicated on the assumption that omeprazole is highly efficacious. However, as discussed below the treatment response of ESGUS is commonly less than 80% and, as discussed in the next lecture the treatment response of EGGUS is significantly worse than that. The author recommends follow up gastroscopy at 3-4 and 6-8 weeks for ESGUS and EGGUS, respectively. Further, horses on prophylaxis should be periodically examined (4 weeks after starting therapy). The reasons for this are two-fold; firstly, it allows the identification of horses that are poor absorbers of omeprazole (approximately 10% of the population). In this population with poor absorption the ongoing use of omeprazole is pointless, an unnecessary costs to the owner, in that they are paying for a drug that doesn't work, and clinically misleading, in that it is assumed that acid suppression is being achieved when indeed it is not. In the author's experience the presence of ESGUS in the face of appropriate omeprazole therapy is a clear indicator that that individual may be absorbing poorly. The use of test dosing to identify poor absorbers, as discussed later, it logical in this population. The second reason to routinely repeat gastroscopy and to monitor horses on prophylactic therapy is that the efficacy of omeprazole in the prevention of EGGUS is questionable as discussed in the next lecture.

Fecal occult blood testing is appealing in terms of ease of use and cost, with kits specifically designed for the equine market available. Preliminary results from a study in Thoroughbred racehorses performed by the author suggest that the diagnostic accuracy of a commercially available fecal blood test in the diagnoses of ESGUS is poor. Specifically, in that study, fecal haemoglobin was positive in 70% and 44% of horses with squamous ulceration $\leq$ grade 1/4 and $\geq$ grade 2/4, respectively (OR 0.34 (0.08-1.5); p=0.28).

The use of sucrose as a marker of gastrointestinal permeability has been reported and a blood test developed for the purpose[17,18], however the test has not found favor and gastroscopy remains the diagnostic of choice due to its accuracy and ability to distinguish ESGUS from EGGUS. Ongoing work into the sucrose test suggests that it may be a potentially useful screening test once further validated.

Treatment

Omeprazole has been the cornerstone of ESGUS treatment for 15 years. Contrary to popular belief the therapeutic response of ESGUS to omeprazole is not 100%. In fact reported healing rates are consistently only 70 – 80%[10,19–24]. Despite the less than 100% efficacy of the drug, quite remarkably, the pharmacokinetics of the drug and factors that affect treatment response are poorly reported. Similarly, the role of variable doses, feeding and timing relative to exercise have, until recently remained unexplored. Lastly the imminent expiration of the GastroGard® patent has brought another factor into consideration, namely formulation, that warrants consideration.

Gender has previously been described as a risk factor for failure of healing with females healing less well. However this effect has not consistently been shown and analysis of a combined data set from across a number of studies[10–12] does not identify an effect (healing – male 45/57 (79%) vs. female 27/35 (77%)). As such it is likely that the previous report was a type I error and should be discounted.

The effect of feeding and the timing of administration are potentially significant. The effect of feeding on buffered formulations is marked (fed absorption is 33% of fasted absorption)[25] whereas the effect with enteric coated formulations appears to be less pronounced (fed absorption is 67% of fasted absorption) (Sykes – Unpublished data). Direct comparison studies are ongoing to verify this effect, however if the difference between formulations is truly present then the use of enteric coated formulation may be preferred. This is especially true as the feeding of unrestricted roughage is a common recommendation for ESGUS management but, ironically, potentially counterproductive relative to pharmacological interventions.

It has also been suggested that the duration of acid suppression with omeprazole in the horse may be less than 24 hours[26]. As such the administration prior to exercise may be beneficial as this is considered the period in which squamous injury is most likely to occur. Although the results of a single study[10] failed to demonstrate an effect, analysis of a larger, combined data set of studies performed under similar conditions[10,12] suggests that administration prior to exercise is advantageous in the treatment of ESGUS (administration of 4 mg/kg PO SID: pre-exercise 12/13 (92%) vs. Post-exercise 14/21 (67%). At the least this small, no cost management change may be advantageous. Further, doses as low as 1 mg/kg have been shown to be highly efficacious in the treatment of ESGUS when administered after a brief fast and prior to exercise, although the role of the formulation used in that study[11] warrants consideration.

REFERENCES

1. Dionne, R., Vrins, A. & Doucet, M. Gastric ulcers in Standardbred racehorses: prevalence, lesion description, and risk factors. *J. Vet. Int. Med.* 17, 218–222 (2003). **2.** Luthersson, N., Nielsen, K. H., Harris, P. & Parkin, T. D. H. Risk factors associated with equine gastric ulceration syndrome (EGUS) in 201 horses in Denmark. *Equine Vet. J.* 41, 625–630 (2009). **3.** White, G., McClure, S. & Sifferman, R. Effects of short-term light to heavy exercise on gastric ulcer development in horses and efficacy of omeprazole paste in preventing gastric ulceration. *J. Am. Vet. Med. Ass.* 230, 1680 – 1682 (2007). **4.** Nieto, J. E. *et al.* Comparison of omeprazole and cimetidine in healing of gastric ulcers and prevention of recurrence in horses. *Equine Vet. Educ.* 13, 260–264 (2001). **5.** McClure, S., Carithers, D. & Gross, S. Gastric ulcer development in horses in a simulated show or training environment. *J. Am. Vet. Med. Ass.* 227, 775–777 (2005). **6.** Habershon-Butcher, J. L., Hallowell, G. D., Bowen, I. M. & Sykes, B. W. Prevalence and risk factors for ulceration of the gastric glandular mucosa in Thoroughbred racehorses in training in the UK and Australia. *J. Vet. Int. Med.* 26, 731 (2012). **7.** Begg, L. M. & O'Sullivan, C. B. The prevalence and distribution of gastric ulceration in 345 racehorses. *Aust. Vet. J.* 81, 199–201 (2003). **8.** Luthersson, N., Nielsen, K. H., Harris, P. & Parkin, T. D. H. The prevalence and anatomical distribution of equine gastric ulceration syndrome (EGUS) in 201 horses in Denmark. *Equine Vet. J.* 41, 619–624 (2009). **9.** Murray, M. J., Nout, Y. S. & Ward, D. L. Endoscopic findings of the gastric antrum and pylorus in horses: 162 cases (1996-2000). *J. Vet. Int. Med.* 15, 401–406 (2001). **10.** Sykes, B. W., Sykes, K. M. & Hallowell, G. D. A comparison between pre- and post-exercise administration of omeprazole in the treatment of EGUS: A randomised, blinded clinical trial. *Equine Vet. J.* Early View, 1–5 (2013). **11.** Sykes, B. W., Sykes, K. M. & Hallowell, G. D. A comparison of three doses of omeprazole in the treatment of gastric ulceration in the horse: A randomised, blinded clinical trial. *J. Vet. Int. Med.* 27, 652 (2013). **12.** Sykes, B. W., Sykes, K. M. & Hallowell, G. D. A comparison of two doses of omeprazole in the treatment of EGUS: A blinded, randomised, clinical trial. *Equine Vet. J.* Early View, 1–6 (2013). **13.** Bell, R. J. W., Kingston, J. K., Mogg, T. D. & Perkins, N. R. The prevalence of gastric ulceration in racehorses in New Zealand. *New Zealand Vet. J.* 55, 13–18 (2007). **14.** Murray, M. J., Schusser, G. R. F., Pipers, F. S. & Gross, S. J. Factors associated with gastric lesions in thoroughbred racehorses. *Equine Vet. J.* 28, 368–374 (1996). **15.** Vatistas, N. J. *et al.* Cross-sectional study of gastric ulcers of the squamous mucosa in Thoroughbred racehorses. *Equine Vet. J.* 31, 34–39 (1999). **16.** Nadeau, J., Andrews, F. & Mathew, A. Evaluation of diet as a cause of gastric ulcers in horses. *Am. J. Vet. Res.* 61, 784 – 790 (2000). **17.** Hewetson, M. *et al.* Sucrose Concentration in Blood: A New Method for Assessment of Gastric Permeability in Horses with Gastric Ulceration. *J. Vet. Int. Med.* 20, 388–394 (2006). **18.** O'Conner, M. S. *et al.* Evaluation of urine sucrose concentration for detection of gastric ulcers in horses. *Am. J. Vet. Res.* 65, 31–39 (2004). **19.** Andrews, F. M. *et al.* Efficacy of omeprazole paste in the treatment and prevention of gastric ulcers in horses. *Equine Vet. J., Suppl.* 31, 81–86 (1999). **20.** Doucet, M. Y., Vrins, A. A., Dionne, R., Alva, R. & Ericsson, G. Efficacy of a paste formulation of omeprazole for the treatment of naturally occurring gastric ulcers in training Standardbred racehorses in Canada. *Canadian Vet. J.* 44, 581–585 (2003). **21.** Lester, G. D., Smith, R. L. & Robertson, I. D. Effects of treatment with omeprazole or ranitidine on gastric squamous ulceration in racing Thoroughbreds. *J. Am. Vet. Med. Ass.* 227, 1636–1639 (2005). **22.** MacAllister, C. G. *et al.* Effects of omeprazole paste on healing of spontaneous gastric ulcers in horses and foals: a field trial. *Equine Vet. J., Suppl.* 31, 77–80 (1999). **23.** Murray, M. J. *et al.* Effects of omeprazole on healing of naturally-occurring gastric ulcers in Thoroughbred racehorses. *Equine Vet. J.* 29, 425–429 (1997). **24.** Sykes, B. W., Sykes, K. M. & Hallowell, G. D. A comparison of two doses of omeprazole in the prevention of squamous gastric ulceration: A randomised, blinded clinical trial. *J. Vet. Int. Med.* 27, 656 (2013). **25.** Daurio, C. P. *et al.* Effect of omeprazole paste on gastric acid secretion in horses. *Equine Vet. J.* 31, 59–62 (1999). **26.** Merritt, A. M., Sanchez, L. C., Burrow, J. A., Church, M. & Ludzia, S. Effect of GastroGard and three compounded oral omeprazole preparations on 24 h intragastric pH in gastrically cannulated mature horses. *Equine Vet. J.* 35, 691–695 (2003).

RETHINKING EGUS: EQUINE GLANDULAR GASTRIC ULCER SYNDROME

Ben Sykes, BSc, BVMS, MS, DACVIM, DECEIM, MBA
Upper Orara, NSW, Australia

Recently, interest in distinguishing EGGUS as a separate entity from ESGUS has increased[1] and a series of recent studies have clearly demonstrated that the risk factors for[2], and response to treatment of[3-5], EGGUS differ dramatically from that of ESGUS. This is not surprising given the differences in the anatomy of the glandular mucosa and pathophysiology of the conditions. The implications of this awareness of distinguishing ESGUS from EGGUS is significant as it should be recognized that the large body of literature that has been built up over the years for ESGUS cannot be directly applied to EGGUS. Instead EGGUS should be thought of a separate disease and one for which investigations into epidemiology, effect of performance, and welfare are needed.

The terminology for EGGUS is yet to be resolved but unpublished data[6] suggests that a number of different disease processes may be present. As such the use of the term "syndrome" is appropriate until the disease or diseases involved are further identified. It has been proposed that erosion may be a more appropriate term than ulceration as histologically the lesions are not full thickness mucosa in the majority of cases[6]. However in human medicine a similar situation is present in that many lesions are defined clinically as ulcers even though histologically they may not be full thickness. In human medicine the term ulcer is preferred for simplicity and to avoid confusion[7]. As such the term EGGUS will be used for the purposes of this presentation.

PREVALENCE

Prevalences of 47%[8] and 62.5%[2] have been reported in Thoroughbred racehorses in Australia. The prevalence in Thoroughbred racehorses in the UK is significantly lower (8%)[2], however, the low prevalence reported UK horses should be interpreted with caution as only a small number of horses (n=25) were examined. A population of mixed horses in Denmark had an overall EGGUS prevalence of 57%[9], endurance horses have reported prevalences of 27%[10] and 33%[11] whereas abattoir studies have reported rates of 57%[12] and 67%[13] in a variety of horse types. Importantly the relationship between increased exercise intensity and increased prevalence that is observed with ESGUS is not present in EGGUS. Further, the risk of EGGUS does not appear to increase with time in work[2]. Similarly, a Danish study found that horses in work were no more likely to have EGGUS than young horses, breeding mares or mature horses at pasture[14]. As such, while ESGUS is primarily, although not exclusively, a disease of high performance horses, the risk of EGGUS appears to be fairly consistent across a wide range of horse usage types. In previous lecture, the relevance, or otherwise, of grade 2/4 ESGUS lesions was discussed. Equally, the relative importance of grade 2/4 vs. more severe grade 3-4/4 lesions warrants consideration. Hard evidence is lacking at present but the author's opinion is that, in contrast to ESGUS where grade 2/4 lesions are often considered clinically insignificant, it is likely that grade 2/4 EGGUS are clinically relevant. Consistent with this the author has repeatedly observed symptomatic horse with mild grade 2/4 EGGUS lesions that improve in response to treatment.

PATHOPHYSIOLOGY

The pathophysiology of EGGUS is poorly understood. The glandular mucosa differs fundamentally from the squamous mucosa in that under normal physiological conditions it is exposed to highly acidic gastric contents with the pH in the ventral portion of the stomach relatively stable at between 1 and 3[15]. As such, whereas ESGUS results from exposure of mucosa unaccustomed to acidity, EGGUS is believed to result from a breakdown of the normal defense mechanisms that protect the mucosa from acidic gastric contents. Under normal conditions the glandular mucosa of the stomach is lined by a continuous protective layer that is a complex mixture of glycoproteins, water, electrolytes, lipids and antibodies[6]. The factors that contribute to breakdown of this protective layer are yet to be elucidated in the horse, but in humans bacterial agents and NSAIDs are the predominant causes of gastric ulceration with *Helicobacter pylori*-negative NSAID-negative gastric ulceration rare[7]. As such, current research in the horse is focused primarily on these two mechanisms.

To date there remains significant conflict in the literature as to the role of bacteria in EGGUS. Both gastric-adapted bacteria and opportunistic pathogens may play a role in squamous ulceration[16] but whether the situation is similar in the glandular mucosa is unknown. Further, although such bacteria are present in ESGUS their role appears to be secondary as the response to acid suppression alone is good (assuming adequate suppression is achieved). *Helicobacter*-like organisms have been identified in horses affected with EGGUS in some studies[17-19], whereas other studies have failed to indentify such organisms[12,13]. Based on the current knowledge it is the author's opinion that evidence for bacteria as a direct, causative agent is lacking.

The potential for NSAIDs to cause EGGUS under clinical conditions is controversial. An ulcerogenic capacity has been demonstrated for flunixin, phenylbutazone and ketoprofen at doses only 50% higher than typically recommended[20] while, at clinical doses phenylbutazone and suxibuzone did not induce gastric ulceration when administered for 15 days[21]. Further, the administration of NSAIDs was not identified as a risk factor in a recent study[2]. In a series of studies performed by the author EGGUS prevalences of 60 – 65% have consistently been observed, yet few of the horses examined regularly receive NSAIDs. As such, the author believes, that while NSAIDs may cause EGGUS in individual patients they do not explain the high prevalence observed in many populations.

DIAGNOSIS

The clinical signs of EGGUS appear to be different from those of ESGUS. Many horses maintain a good appetite and, in fact it has been suggested that EGGUS horses may be polyphagic[22]. Further, contrary to popular perception horses with EGGUS do not appear to be aggressive to either other horses or their handlers. On the contrary, there is some evidence that horses with EGGUS are less aggressive[2] suggesting that they may be withdrawing as a pain response. Lastly, it has been recognized that symptoms such as pain on girth tightening may be indicators of EGGUS[6].

As discussed in the previous lecture gastroscopy remains the main form of diagnosis. It is critical that a complete examination of the stomach is performed as the majority of glandular lesions occur in the pyloric antrum and the horse must be well fasted with a stomach completely empty of food (although the author does not routinely restrict water before gastroscopy). Adequate insufflation is also especially important. Repeat gastroscopy to monitor treatment is especially important due to the poor response rates observed with current dosing regiments. A commercial fecal blood test performed equally poorly for EGGUS as it did for ESGUS in a Thoroughbred racehorse population. Specifically, fecal hemoglobin was positive in 50% and 50% of horses with glandular ulceration ≤ grade 1 and ≥ grade 2, respectively (OR 1.0 (0.29–3.4); p=1.0).

TREATMENT

The response to treatment of EGGUS is significantly worse than that of ESGUS. In three recent studies only 25% of glandular ulcers healed with 28-35 days of omeprazole therapy at 4.0 mg/kg PO SID in direct contrast to an ESGUS healing rate of 78%[3–5]. Further, in the same studies, 23% of horses worsened despite omeprazole therapy at doses ranging from 1–4 mg/kg PO SID[3–5]. A number of factors might be important in the poor response observed for EGGUS but including the intra-day duration of acid suppression that is achieved following once daily administration of omeprazole in the horse. In other species the duration of acid suppression, and more specifically the abolition of acid production for at least 20 hours each day[23], is recognized as a prerequisite factor for ulcer healing to occur. In humans, similar durations of acid suppression are required for healing of the reflux esophagitis (analogous to ESGUS) and ulcerative gastritis (analogous to EGGUS)[23] and it has been proposed that 20 hours of acid suppression is required for healing of ESGUS in the horse to occur. However, this 20 hour timeframe is extrapolated directly from the human literature and has not been validated in the horse. Importantly, it ignores the influence of species-specific factors, such as the significant buffering effect of ingesta within the stomach of the horse and the influence of the horse being a constant grazer vs. a meal feeder. As such, the extrapolation is probably not appropriate.

One possibility for the disparity in healing between ESGUS and EGGUS is that the duration of intra-day acid suppression required for ESGUS healing may be less than the 20 hours cited from the human literature, and less than that required for healing of EGGUS. This possibility potentially explains why a good clinical response for ESGUS has been widely reported at this dose[24–28] even though a 2003 study suggested that the duration of acid suppression with administration of omeprazole at 4 mg/kg may be less than 12 hours[15]. As such, the possibility exists that the primary reason for failure of omeprazole to work for EGGUS might be that the duration of intra-day acid suppression with once daily administration, or at the doses currently used, is inadequate for healing to occur.

Another potential explanation is that adequate intra-day acid suppression is achieved with 4 mg/kg, and even potentially 1.0 mg/kg, but that the standard duration of therapy of 28 days is inadequate for EGGUS healing to occur. This is supported by early studies on omeprazole where doses as low as 0.7 mg/kg were demonstrated to be potent inhibitors of acid suppression for at least 27 hours[29]. In humans, the duration of therapy for glandular ulceration is dependent on primary cause of the ulceration[7]. For *Helicobacter pylori*-associated ulcers, triple therapy combining antimicrobials and acid suppression consistently yields first line eradication rates of greater than 80% with 7 – 14 days of therapy[7]. In contrast, NSAID-induced glandular ulceration requires longer treatment duration with 84% and 100% of patients healing at 8 and 12 weeks, respectively, in response to acid suppression therapy and discontinuation of NSAID therapy in one study[30]. No difference in NSAID-associated ulcer healing rates has been observed between ranitidine, conventional dose esomeprazole (the *S*-enantimer of omeprazole) and high dose esomeprazole, despite the fact that proton pump inhibitors are more effective at controlling intra-gastric acidity than ranitidine[31]. This suggests that the duration of therapy may be more significant than the magnitude of acid suppression within each 24 hour period.

Of these factors the potential of a dose effect has been the best investigated however, before discussing the effect of dose in different studies the effect of formulation warrants discussion. Throughout the majority of the world GastroGard® has been protected by a patent, however in some regions such as Oceania and the Middle East this protection has not been afforded. Further, the GastroGard patent is due to expire shortly stimulating interest in the use of alternative formulations. Preliminary results from an ongoing study suggest that significant differences between formulations exist with the bioavailability of enteric coated formulations approximately 30% higher than GastroGard (Sykes – Unpublished data). Further, the Cmax and Tmax of enteric coated formulations appear to be higher and earlier than buffered formulations which may be relevant under certain conditions. In humans area under the curve is the primary determinant of efficacy for omeprazole and as such the higher bioavailability of enteric coated formulations suggests that they may be more efficacious at equivalent doses. However, the magnitude of the increase in bioavailability observed does not justify the dramatic dose reduction (down to as low as 1 mg/kg) recommended by some manufacturers.

Two studies have investigated the use of lower doses in the treatment of EGUS, one using a buffered formulation[3] and one using a enteric coated formulation[4]. Taking into consideration the effect of formulation and the timing of administration the results of the studies suggest that the use of doses of 1 – 2 mg/kg (depending on the formulation) may be as efficacious as higher doses (4 mg/kg) under specific conditions. The role of individual variation in therapeutic response is also important. An ongoing study (Sykes – Unpublished data) has demonstrated that a wide degree of variation exists in the ability of different individuals to absorb omeprazole. Some horses are very high absorbers while some appear to absorb omeprazole very poorly. The latter group potentially explains the less than 100% therapeutic response consistently observed in ESGUS studies (i.e. the subset of horses in the population that are poor absorbers may not be effectively acid suppressing thus explaining why some horses don't respond). Taking this into consideration, individually tailoring of therapy warrants discussion. In particular, the identification of non-responders that may benefit from higher doses or alternative pharmacological approaches.

Omeprazole has been shown to be more effective than ranitidine at a population level, but care should be taken in applying this finding to individual animals as some (those that do not absorb omeprazole) may respond better to ranitidine. The use of test dosing and subsequent testing of gastric pH fluid can be useful in identifying such patients. As discussed in the previous lecture the author considers the persistence of ESGUS in the face of omeprazole therapy as particularly suspicious that the animal may be a poor observer. The author also uses test dosing as a way of documenting the efficacy of lower doses in the treatment of EGGUS and, given the long duration of treatment required, the expense of test dosing is readily offset but the ability to use lower doses. The protocol for test dosing is as follows: 22 hours prior to presentation the owner is instructed to administer the dose to be tested. Importantly this must be done under normal feeding conditions to accommodate the impact of feeding on bioavailability. The horse is then fasted as per normal and gastroscopy is performed the next day (22-24 hours after administration of the test dose). At the time of gastroscopy gastric fluid is sucked back through the biopsy channel of the endoscope and the pH tested with either a pH probe or litmus paper. If the omeprazole is being effective then the pH should be > 4 and a pH of < 4 suggests that either a higher dose, or an alternative drug (ranitidine) should be used.

Lastly, the role of adjunctive therapeutic agents warrants discussion. The use of antimicrobials in the treatment of EGGUS has been recommended[32] and is popular. To date, minimal evidence exists to support their use and the only study to date to investigate the efficacy of antimicrobials in the treatment of EGGUS found no clear beneficial effect[33]. As such the widespread use of antimicrobials for the treatment of EGGUS should be discouraged. That said, the raw data from the aforementioned study[33] does suggest that some antimicrobials may be beneficial in some animals and as such, although the widespread use of antimicrobials should be discouraged, the author believes that they may have a role in the treatment of refractory cases. The use of other adjunctive agents such as sucralfate or pectin-lecithin complexes is logical and although their benefit in treatment is, at this point, largely unproven the author believes that their routine use is justified. Recently the combination of a unique pectin-lecithin complex (Apolectol®), live yeast (*Saccharomyces cerevisiae*) and magnesium hydroxide has been shown to be an effective prophylactic for both ESGUS and EGGUS[34] and its use as an adjunctive in the treatment of EGGUS is logical. Sucralfate is also popular and commonly recommended[32]. Where finances allow, the author recommends the use of a 'triple-therapy' approach of omeprazole (at an appropriate dose confirmed by test dosing), Apolectol® (ideally formulated as described[34]) and sucralfate. The author does not use antimicrobials until therapy of at least 8 weeks of 'triple-therapy' has failed to resolve, or at least significantly improve, the EGGUS present.

REFERENCES

1. Merritt, A. Appeal for proper usage of the term "EGUS": Equine gastric ulcer syndrome. *Equine Vet. J.* 41, 616 (2009). **2.** Habershon-Butcher, J. L., Hallowell, G. D., Bowen, I. M. & Sykes, B. W. Prevalence and risk factors for ulceration of the gastric glandular mucosa in Thoroughbred racehorses in training in the UK and Australia. *J. Vet. Int. Med.* 26, 731 (2012). **3.** Sykes, B. W., Sykes, K. M. & Hallowell, G. D. A comparison of two doses of omeprazole in the treatment of EGUS: A blinded, randomised, clinical trial. *Equine Vet. J.* Early View, 1–6 (2013). **4.** Sykes, B. W., Sykes, K. M. & Hallowell, G. D. A comparison of three doses of omeprazole in the treatment of gastric ulceration in the horse: A randomised, blinded clinical trial. *J. Vet. Int. Med.* 27, 652 (2013). **5.** Sykes, B. W., Sykes, K. M. & Hallowell, G. D. A comparison between pre- and post-exercise administration of omeprazole in the treatment of EGUS: A randomised, blinded clinical trial. *Equine Vet. J.* Early View, 1–5 (2013). **6.** Hepburn, R. Equine glandular ulceration: Pathophysiology and epidemiology. in *American College of Veterinary Internal Medicine Annual Congress* (2012). **7.** Malfertheiner, P., Chan, F. F. K. L., McColl, K. E. L. & Leung, W. Peptic-ulcer disease. *Lancet* 374, 1449–1461 (2009). **8.** Begg, L. M. & O'Sullivan, C. B. The prevalence and distribution of gastric ulceration in 345 racehorses. *Aust. Vet. J.* 81, 199–201 (2003). **9.** Luthersson, N., Nielsen, K. H., Harris, P. & Parkin, T. D. H. The prevalence and anatomical distribution of equine gastric ulceration syndrome (EGUS) in 201 horses in Denmark. *Equine Vet. J.* 41, 619–624 (2009). **10.** Nieto, J. E. *et al.* Prevalence of gastric ulcers in endurance horses – a preliminary report. *Vet. J.* 167, 33–37 (2004). **11.** Tamzali, Y., Marguet, C., Priymenko, N. & Lyazrhi, F. Prevalence of gastric ulcer syndrome in high-level endurance horses. *Equine Vet. J.* 43, 141–144 (2011). **12.** Husted, L., Jensen, T. K., Olsen, S. N. & Mølbak, L. Examination of equine glandular stomach lesions for bacteria, including Helicobacter spp by fluorescence in situ hybridisation. *BMC Microbiology* 10, 1–8 (2010). **13.** Martineau, H., Thomson, H. & Taylor, D. Pathology of gastritis and gastric ulceration in the horse. Part 1: Range of lesions present in 21 mature individuals. *Equine Vet. J.* 41, 638–644 (2009). **14.** Luthersson, N., Nielsen, K. H., Harris, P. & Parkin, T. D. H. Risk factors associated with equine gastric ulceration syndrome (EGUS) in 201 horses in Denmark. *Equine Vet. J.* 41, 625–630 (2009). **15.** Merritt, A. M., Sanchez, L. C., Burrow, J. A., Church, M. & Ludzia, S. Effect of GastroGard and three compounded oral omeprazole preparations on 24 h intragastric pH in gastrically cannulated mature horses. *Equine Vet. J.* 35, 691–695 (2003). **16.** Al Jassim, R., McGowan, T., Andrews, F. & McGowan, C. Gastric ulceration in horses: The role of bacteria and lactic acid. *Australian Government: Rural Industries Research and Development Corporation* 1–26 (2008). **17.** Contreras, M. *et al.* Detection of Helicobacter-like DNA in the gastric mucosa of Thoroughbred horses. *Letters in applied microbiology* 45, 553–557 (2007). **18.** Morales, A., Garcia, F. & Bermudez, V. Detection of Helicobacter-like organisms in Thoroughbred horses from Venezuela. *Braz. J. Vet. Path.* 3, 52–55 (2010). **19.** Fox, J. The non-H pylori helicobacters: their expanding role in gastrointestinal and systemic diseases. *Gut* 50, 273–283 (2002). **20.** MacAllister, C. G., Morgan, S. J., Borne, A. T. & Pollet, R. A. Comparison of adverse effects of phenylbutazone, flunixin meglumine, and ketoprofen in horses. *J. Am. Vet. Med. Ass.* 202, 71–77 (1993). **21.** Andrews, F., Reinemeyer, C. R. & Longhofer, S. L. Effects of top-dress formulations of suxibuzone and phenylbutazone on development of gastric ulcers in horses. *Vet. Ther.* 10, 113–120 (2009). **22.** Malmkvist, J. *et al.* Behaviour and stress responses in horses with gastric ulceration. *Applied Animal Behaviour Science* 142, 160–167 (2012). **23.** Shin, J. M. & Sachs, G. Pharmacology of proton pump inhibitors. *Current Gastroenterology Reports* 10, 528–534 (2008). **24.** Andrews, F. M. *et al.* Efficacy of omeprazole paste in the treatment and prevention of gastric ulcers in horses. *Equine Vet. J., Suppl.* 31, 81–86 (1999). **25.** Doucet, M. Y., Vrins, A. A., Dionne, R., Alva, R. & Ericsson, G. Efficacy of a paste formulation of omeprazole for the treatment of naturally occurring gastric ulcers in training Standardbred racehorses in Canada. *Canadian Vet. J.* 44, 581–585 (2003). **26.** Lester, G. D., Smith, R. L. & Robertson, I. D. Effects of treatment with omeprazole or ranitidine on gastric squamous ulceration in racing Thoroughbreds. *J. Am. Vet. Med. Ass.* 227, 1636–1639 (2005). **27.** MacAllister, C. G. *et al.* Effects of omeprazole paste on healing of spontaneous gastric ulcers in horses and foals: a field trial. *Equine Vet. J., Suppl.* 31, 77–80 (1999). **28.** Murray, M. J. *et al.* Effects of omeprazole on healing of naturally-occurring gastric ulcers in Thoroughbred racehorses. *Equine Vet. J.* 29, 425–429 (1997). **29.** Andrews, F. & Jenkins, C. Effect of oral omeprazole on basal and pentagastrin-stimulated gastric secretion in young female horses. *Equine Vet. J.* 24, 80–83 (1992). **30.** Lancaster-Smith, M. J., Jaderberg, M. E. & Jackson, D. A. Ranitidine in the treatment of non-steroidal anti-inflammatory drug associated gastric and duodenal ulcers. *Gut* 32, 252–5 (1991). **31.** Goldstein, J. L., Johanson, J. F., Hawkey, C. J., Suchower, L. J. & Brown, K. A. Clinical trial: healing of NSAID-associated gastric ulcers in patients continuing NSAID therapy - a randomized study comparing ranitidine with esomeprazole. *Alimentary pharmacology & therapeutics* 26, 1101–11 (2007). **32.** Hepburn, R. Equine glandular ulceration: Comparative aspects and management. in *American College of Veterinary Internal Medicine Annual Congress* (2012). **33.** Sykes, B. W., Sykes, K. M. & Hallowell, G. D. Administration of once daily trimethoprim-sulphadimidine does not improve healing of glandular gastric ulceration in horses receiving omeprazole: A randomised, blinded clinical trial. in *Proceedings of the American Veterinary Internal Medicine Annual Congress* (2013). **34.** Sykes, B. W., Sykes, K. M. & Hallowell, G. D. Efficacy of a combination of Apolectol®, live yeast (CNCM I-1077) and magnesium hydroxide in the prevention of EGUS and fecal acidosis in thoroughbred racehorses: A randomised, blinded, placebo controlled clinical trial. *Equine Vet. J.* 45, 13 (2013).

RHODOCOCCUS EQUI IN ADULT HORSES: A REVIEW

Peter R. Morresey, BVSc, DACT, DACVIM
Lexington, KY

INTRODUCTION

Rhodococcus equi is a Gram-positive pleomorphic rod or coccobacillus that is a free-living soil saprophyte with widespread distribution being present on all continents except Antarctica.[1] The *R. equi* organism, then known as *Corynebacterium equi*, was first proposed by Magnusson in 1923 as a cause of purulent pulmonary disease in foals.[2] Disease of adult horses has been previously described; in these horses concurrent medical conditions were present, chiefly those that resulted in generalized immunosuppression.[3, 4]

BIOLOGY OF INFECTION

Horses come into contact with the *R. equi* organism by ingestion and inhalation, with dust aerosol considered the major route of infection.[5, 6] Of significance regarding pathogenicity, the bacteria becomes a facultative intracellular pathogen within macrophages.[7, 8] The phagolysosomes containing virulent *R. equi* are not as acidic as those containing avirulent bacteria, being suppressed by virulence factors enabling *R. equi* to survive and replicate in phagocytic cells.[9]

Due to the profound clinical and economic effect of *R. equi* infection, infectivity and pathogenesis of disease is best characterized in the juvenile foal. Pneumonic foals characteristically are infected with R. equi containing a plasmid encoding for a particular surface protein (vapA, virulence associated protein). Foals experimentally infected with the virulent vapA containing organism develop severe abscessing pneumonia, in contrast to those infected with R. equi in the absence of vapA which rapidly clear the organism without the onset of disease.[10]

The plasmid encoding vapA is therefore considered responsible for organism virulence and progression of *R. equi* infection to fulminant disease. Expression of vapA alone in the absence of the 85-kb plasmid was not sufficient to allow *in vitro* intracellular macrophage replication of *R. equi* or development of host disease, therefore for virulence isolates must both contain the 85-kb plasmid and express vapA.[10] Exposure of virulent, plasmid containing *R. equi* to sublethal acid pH during the exponential growth phase altered several membrane and cytoplasmic protein expressions that protected the organism from subsequent acid exposure in comparison to the avirulent type.[11] This would enable the organism to survive the acid environment of the phagolysosome.[10]

Clearance of *R. equi* relies on both humoral and cellular components of the immune system. Cytokine secretion, chiefly interferon gamma (IFN-γ) is also important. Experimental models show both CD4+ and CD8+ T lymphocytes can reduce bacterial counts in the lung. Type 1 immune responses (cell mediated) appear efficacious in removing infection, whereas a predominance of type 2 responses (antibody mediated) are associated with failure to clear infection.[12] The CD8+ cells have been shown to play the major role in clearance of infection.[13] Examination of bronchoalveolar lavage fluid (BALF) cells showed an increase in IFN-γ producing T-lymphocytes with no corresponding increase in peripheral blood IFN-γ positive cells, suggesting protection was related to a Type 1 recall response at the site of challenge (lung).[14] Additionally, deficiencies in complement components, phagocytic and NK cells have been demonstrated experimentally not to impair the pulmonary clearance of *R. equi* reinforcing that a competent cellular immune system and local pulmonary response is required to prevent pneumonia and death.[15] The presence of increased CD8+ cells is suggestive of inflammation secondary to infection causing general immune stimulation, candidate processes including viral, intracellular bacterial or tumor cell presence.

Compared to adults and healthy foals, those foals infected with *R. equi* have higher absolute and proportional circulating B cell counts, these alone being responsible for the increased lymphocyte fraction seen in their leukograms.[16] In antigen recall studies performed on adult horses, changes were seen in the BALF cells (a significant increase in the pulmonary T-lymphocyte count) without a noticeable alteration in peripheral blood mononuclear cell count making blood evaluation a poor indicator of pulmonary response to *R. equi* challenge in adults.[17] However, age related changes in the lymphocyte population have been demonstrated in the Andalusian horse, with adults having an age-related decline in absolute numbers of B and T lymphocytes, and CD2, CD4+ and CD8+ cell markers.[18] This is a common finding in many species. There is no evidence in this case of a decline in absolute or relative lymphocyte numbers compared to established values.[19]

DISEASE IN FOALS

Pneumonia resulting from *R. equi* infection is a common condition of foals and weanlings. A number of management factors have been found to be associated with infection including large farm size, high numbers of mares and foals, high stocking densities, dusty conditions resulting from reduced pasture cover, and transient horses.[20, 21] The particular susceptibility of the foal is not fully understood but can be explained in part by a combination of heavy challenge through the respiratory route coinciding with declining maternally derived antibody in the absence of fully competent foal cellular immune mechanisms.[22, 23] *Rhodococcus equi* is most commonly considered a cause of pneumonia in foals at or under weaning age, with most foals showing clinical signs before 4 months of age, however incubation period is not well defined, ranging from 6 to 18 days in experimentally infected foals.[24, 25] Theoretical calculations resulted in an incubation period of 49 days based on age of onset of clinical signs and age at death in another study assuming a point source of infection.[26] The difference between shorter incubation experimentally and observed occurrence can be explained by consideration of the ubiquitous nature of the organism exposing foals immediately after birth but countered by the protective effects of colostrum and maternally derived antibodies against virulent *R. equi*.

Clinical respiratory disease in foals is caused by chronic suppurative bronchopneumonia with abscessation and lymphadenitis. Extrapulmonary disease in foals has also been widely reported.[27] Up to half of affected foals have intestinal lesions (mesenteric lymphadenitis) found during necropsy evaluation.[8] In one retrospective study, one-third of affected foals displayed immune-mediated polysynovitis, most often affecting the tibiotarsal and femorotibial joints.[28] *Rhodococcus equi* septicemia, during pneumonic or intestinal disease, has also led to septic arthritis or osteomyelitis necessitating the investigation of any periarticular swellings affecting foals from endemic farms.[29]

Healthy foals on endemic farms have been shown to exhale virulent R. equi suggesting that these foals may harbor the organism in the absence of clinical problems.[30] This may have significance in the transmission and persistence of disease on endemic farms.

DISEASE IN HUMANS

Rhodococcus equi has also been reported as an opportunistic infection in chiefly immunosuppressed humans where it causes abscessation or cavitating pneumonia.[31] The first report of human disease was in 1967.[32] About one-third of human patients who have developed *R. equi* infections had contact in some way with herbivores or herbivore manure, with others likely having acquired infection from contact with soil or wild bird manure.[1, 33] Infection has been reported concurrently with neoplastic disease including leukemia.[34] and lymphoma.[35] In cases of human immunodeficiency virus/acquired immune deficiency syndrome (HIV/AIDS) pulmonary involvement was always the first manifestation of *R. equi* infection but extrapulmonary involvement was also present.[36] Iatrogenic immunosuppression secondary to renal transplantation has resulted in a number of reports of *R. equi* infection.[37-39] Treatment has involved multiple antimicrobial agents and surgery with variable success.[40] Infection in immune competent patients has also been reported with both localized and systemic disease noted, with pulmonary infections only accounting for 42% of cases as opposed to 84% in immune compromised patients.[41]

DISEASE OF ADULT HORSES

Rhodococcus equi infection is rarely reported in adult horses, even on properties where clinical disease in foals is commonplace.[17] When present in adult horses, pulmonary *R. equi* infection is almost invariably fatal in its own right or results in euthanasia. Reported cases have involved immune suppression.[3, 4] or concurrent pregnancy.[42]

Pulmonary *R. equi* infection is commonly not diagnosed until necropsy where it is an incidental finding.[8, 43-45] In one adult case, *R. equi* was an incidental finding in a horse that succumbed to a gastrointestinal accident.[8] Signs referable to respiratory disease (dyspnea, pyrexia) were present in some cases.[8, 44, 46] whereas other cases had non-specific clinical signs without overt respiratory disease, including chronic weight loss, anorexia, and lethargy.[3, 43] Osteomyelitis with tibial abscessation and a remote pectoral abscess was found in one horse.[47] An outbreak of multiple subcutaneous abscesses and pulmonary disease was reported in several adult ponies.[48] A multifactorial etiology was suspected, with a virulent organism, climatic conditions, and biting arthropods considered to be contributory to the clinical presentations.

Residual pulmonary pathology is not present in recovered horses that were diagnosed as infected while foals.[49] Subsequent racing performance is indistinguishable from that of the general horse population.[50] As previously published reports of adult infections consist of incidental necropsy findings or mortal disease, subsequent performance data for adult horses is lacking.

Abortion from seven months of gestation.[51, 52] to late term.[53] and fetal pneumonia induced by *R. equi* has been reported. Pulmonary pathology in aborted fetuses was similar to that seen in foals affected with *R. equi* pneumonia. The fetuses cultured *R. equi* from the lungs and stomach content suggesting infection through swallowing of amniotic fluid or inhalation during fetal breathing movements.[51-53] Placental evaluation of one case had histopathologic lesions consistent with hematogenous rather than ascending placental infection, with perivascular inflammatory cell and Gram-positive coccobacilli (consistent with *R. equi*) infiltration.[53]

Immunosuppression has been reported in one case of multi-system infection with *R. equi* in an adult horse.[3] Decreased immunoglobulin levels, lymphopenia and inadequate lymphocyte response to cell mitogen stimulus was present indicating failure of both the humoral and cell mediated components of the immune response. Pulmonary, hepatic and renal infection was demonstrated in an aged 5 month gestation pregnant mare that succumbed to abscessing pneumonia, with the suspicion that immune suppression was contributory.[45]

The potential for iatrogenic immune suppression should be considered in all cases where opportunistic infections occur. In one report of successful treatment of *R. equi* the horse was receiving dexamethasone for persistent inflammatory airway disease, however dose rates used in during management of the inflammatory airway disease were considered to be at the anti-inflammatory not immune suppressive level.[54] Both immunoglobulin levels and the lymphocyte population were found appropriate in this case report. However, a majority of the non-HIV affected human *R. equi* infected patients were receiving corticosteroids around the time of diagnosis in one retrospective study, though all had underlying immune suppression from concurrent disease including neoplasia.[55]

The prognosis for affected adults is generally considered poor due to the likelihood of facilitating immune suppression or serious disease. However, *R. equi* may be a treatable condition in horses free from concurrent serious disease in contrast to the cases previously reported in the literature, and should be considered as a causative agent of bacterial disease in horses that experience a change in clinical condition while undergoing corticosteroid based anti-inflammatory treatment for inflammatory airway disease. Evaluation of immune function is warranted in these cases.

REFERENCES

[1] Prescott JF. Clin Microbiol Rev 1991;4:20-34. [2] Magnusson H. Arch Wiss Prakt Tierhelkd 1923;50:22-38. [3] Freestone JF, et al. J Am Vet Med Assoc 1987;190:689-691. [4] Verdonck F, et al. Vet Rec 2004;154:149-150. [5] Giguere S, et al. Vet Microbiol 1997;56:313-334. [6] Muscatello G, et al. Appl Environ Microbiol 2006;72:6152-6160. [7] Woolcock JB, et al. Vet Microbiol 1987;14:215-224. [8] Zink MC, et al. Can Vet J 1986;27:213-217. [9] Toyooka K, et al. J Med Microbiol 2005;54:1007-1015. [10] Giguere S, et al. Infect Immun 1999;67:3548-3557. [11] Benoit S, et al. Res Microbiol 2001;152:439-449. [12] Hines SA, et al. Vet Microbiol 1997;56:177-185. [13] Nordmann P, et al. Infect Immun 1992;60:2748-2752. [14] Hines SA, et al. Clin Diagn Lab Immunol 2003;10:208-215. [15] Yager JA, et al. Vet Microbiol 1991;28:363-376. [16] Flaminio MJ, et al. J Vet Intern Med 1999;13:206-212. [17] Hines MT, et al. Vet Immunol Immunopathol 2001;79:101-114. [18] Satue K, et al. Vet Immunol Immunopathol 2009. [19] Stull CL, et al. Equine Vet J 2004;36:583-589. [20] Chaffin MK, et al. J Am Vet Med Assoc 2003;222:476-485. [21] Chaffin MK, et al. J Am Vet Med Assoc 2003;222:467-475. [22] Prescott JF, et al. Can Vet J 1991;32:421-425. [23] Chaffin MK, et al. J Am Vet Med Assoc 2003;223:1791-1799. [24] Barton MD, et al. Aust Vet J 1987;64:332-339. [25] Wada R, et al. Vet Microbiol 1997;56:301-312. [26] Horowitz ML, et al. J Vet Intern Med 2001;15:171-175. [27] Chaffin MK, et al. AAEP 1997 79-80. [28] Sweeney CR, et al. Vet Microbiol 1987;14:329-336. [29] Firth EC, et al. Aust Vet J 1993;70:304-307. [30] Muscatello G, et al. J Clin Microbiol 2009;47:734-737. [31] Drancourt M, et al. J Infect 1992;24:123-131. [32] Golub B, et al. Ann Intern Med 1967;66:1174-1177. [33] Sladek GG, et al. South Med J 1993;86:244-246. [34] Allen VD, et al. Pediatr Infect Dis J 1989;8:656-658. [35] Marsh JC, et al. Cancer 1973;32:147-149. [36] Arlotti M, et al. Scand J Infect Dis 1996;28:463-467. [37] Arya B, et al. Clin Transplant 2004;18:748-752. [38] Barsotti M, et al. Nephrol Dial Transplant 1997;12:2002-2004. [39] Hoque S, et al. Nephrol Dial Transplant 1996;11:2340-2341. [40] Shapiro JM, et al. Thorax 1992;47:62-63. [41] Kedlaya I, et al. Clin Infect Dis 2001;32:E39-E46. [42] Ozgur NY, et al. Vet Rec 2002;151:613. [43] Genetzky RM, et al. Mod Vet Pract 1982;43:706-708. [44] Roberts MC, et al. Aust Vet J 1980;56:96-97. [45] Waldridge BM, et al. Equine Vet Educ 2008;20:67-71. [46] Vengust M, et al. Can Vet J 2002;43:706-708. [47] Kilcoyne I, et al. Equine Vet Educ 2013; DOI: 10.1111/eve.12038. [48] Tapprest J, et al. Journal of Equine Veterinary Science 2012;32:S15-S16. [49] Ainsworth DM, et al. Am J Vet Res 1993;54:2115-2120. [50] Ainsworth DM, et al. J Am Vet Med Assoc 1998;213:510-515. [51] Fitzgerald SD, et al. J Vet Diagn Invest 1995;7:157-158. [52] Szeredi L, et al. Vet Pathol 2006;43:208-211. [53] Patterson-Kane JC, et al. J Vet Diagn Invest 2002;14:157-159. [54] Morresey PR, et al. J Vet Intern Med 2010;24:436-438. [55] Harvey RL, et al. Rev Infect Dis 1991;13:139-145.

NOVEL EQUINE HEPATITIS VIRUSES: ARE THEY CLINICALLY RELEVANT?

Thomas J. Divers, DVM, DACVIM, DAVECC and Bud C. Tennant, DVM, DACVIM
Ithaca, NY

INTRODUCTION

Acute hepatitis in horses (serum hepatitis, Theiler's disease) has frequently been associated with the antecedent administration of biologics of equine origin. The etiology is unknown but the occurrence of disease following parenteral administration of biologics produced in horses, the significant delay between exposure to biologics and the onset of acute hepatitis, and the histopathologic lesions in the liver are remarkably similar to those associated with acute human hepatitis A virus or hepatitis B virus infections, suggesting that acute hepatitis in horses may have a similar viral etiology. The human hepatitis viruses are now well characterized and tests are available for diagnosis and for preventing transmission via blood or plasma products. Recently, an outbreak of serum hepatitis in horses was associated with an infection of a highly divergent, previously unrecognized Flavivirus (Chandriani et al. 2013) with phylogenic closeness to human hepatitis C virus (HCV). This Theiler's disease associated virus (TDAV) is one of three, new, blood-borne Flaviviruses that have been described in horses since June 2012 (Burbelo PD et al 2012; Kapoor A et al 2013). All have been identified by direct metagenomic analysis using unbiased, high-throughput sequencing (next generation sequencing [NGS]). The following is a brief summary describing TDAV and the other two related horse viruses, the non-primate hepacivirus (NPHV) and the equine pegivirus (EPgV).

THEILER'S DISEASE REVIEW

In the first half of the 20th century, large epidemics of homologous serum hepatitis were observed in horses. The first of these was reported by Sir Arnold Theiler in South Africa, who described hundreds of cases of a highly fatal form of hepatitis that occurred in horses following vaccination against African horse sickness by the simultaneous administration of infectious virus and convalescent equine antiserum (Theiler, 1919). In the United States, the first cases were observed during the pandemic of western equine encephalomyelitis (WEE) that began in California in 1930 (Meyer, 1933) and then spread rapidly throughout the western states. Madsen (1934) and Marsh (1937a,b) described a second wave of disease that occurred two to three months after an initial outbreak of western equine encephalomyelitis during which affected horses had been immunized using anti-WEE antiserum alone or in combination with live WEE virus. The second disease was more severe than the initial encephalomyelitis outbreak and also was associated with signs of encephalopathy. Ultimately, the second wave of disease was shown to be associated with severe hepatitis similar to that reported by Theiler (Marsh, 1937b). Since these reports, hepatitis following the administration of equine serum or plasma has been described in horses from around the world (Hjerpe, 1965; Thomsett, 1971; Rose, et al., 1974; Robinson, et al., 1975; Messer and Johnson, 1994a,b; Guglick, et al., 1995; Zientara, et al., 1994 ,Aleman et al., 2005 and others). Epidemics also have been reported in which there was no association with prior administration of equine biologics (Stenius, 1941; Forenbacher and Topolnik, 1959). A recent review of 13 Theiler's cases at our hospital in the past 25 years found that in the records of only 4 of the 13 adult horses with Theiler's disease was there a history of receiving a serum or plasma product of equine-origin 30 to 70 days prior to the onset of clinical disease. All 9 of the horses that were diagnosed with Theiler's disease, but had no history of previous administration of an equine biologic product, occurred in June- November. Of the horses that died and had brain and liver tissue examined, severe centrilobular or massive liver necrosis and Alzheimer type II astrocytes in the brain were found in all cases. Evidence of vasculitis, which has been occasionally reported in horses with Theiler's disease (Hjerpe, 1975; Smith et al., 1991), was found in 3 of our cases.

THEILER'S DISEASE ASSOCIATED VIRUS (TDAV)

In August 2011, we were asked to consult on an outbreak of acute hepatitis in horses. Two and a half months earlier, horses from the affected farm had developed botulism. Twenty-two horses were treated with botulinum antitoxin from two separate sources. Eight of 17 horses treated with one lot of botulinum antitoxin developed clinical and/or biochemical evidence of hepatitis within 7 to 9 weeks after intravenous treatment with a standard dose of botulinum antitoxin obtained from hyperimmunized horses. Hepatitis was not observed in the five horses that were treated with a separate lot of botulinum antitoxin or in 53 other farm horses that had not received antitoxin.

Using high throughput, pyrosequencing, a new virus belonging to the family Flaviviridae (the family of human HCV) was identified in the serum of two clinical cases of acute hepatitis/Theiler's disease and in the botulinum antitoxin the two horses received 47 and 52 days prior to the onset of illness. (Chandriani,et al., 2013). The nucleic acid sequences of the viruses in the two hepatitis cases and in the botulinum antitoxin were virtually identical. The quantitative reverse transcriptase – polymerase chain reaction (qRT-PCR) then was used to test the serum of other horses for the newly identified equine hepatitis associated virus. All 17 horses that received the one lot of botulinum antitoxin were test positive for the new virus, including all 8 with acute hepatitis. There was not an association of level of viremia with disease as 15 of the 17 infected horses had 10^7 to 10^8 genomes per ml. One of the 3 horses used in production of the botulinum antitoxin was test positive for TDAV. Serum from 5 recipients of another lot of botulinum antitoxin, prepared from different donor horses, serum from 15 farm horses that had not been treated with botulinum antitoxin, and serum from 20 unrelated horses housed at a distant location were all RT-PCR negative for the virus. The single positive donor horse was TDAV negative six months following the outbreak and there have been no further reports of Theiler's disease associated with use of this product.

TDAV persisted chronically in a small percentage of infected horses. The original 17 TDAV positive horses were retested after one year by qRT-PCR and TDAV was undetectable in 13 of the previously infected horses. In 4 horses, persistent TDAV viremia was demonstrated (Chandriani,et al., 2013). Samples from two of the TDAV carriers were obtained two years after antitoxin treatment, and both remained TDAV positive by qRT-PCR. The other two horses that were TDAV positive after one year were lost to follow-up. The level of viremia has remained unchanged over time in chronically infected horses. Horses from the index farm have been re-tested since the outbreak and no additional horses have become PCR positive, suggesting direct horse to horse transmission or transmission by insect vectors may not occur or is not common.

To further confirm an association between TDAV and Theiler's disease, additional clinical cases are needed. We are continuing to recruit new cases of Theiler's disease and to test serum from such cases for the presence of TDAV. Since Theiler's disease occurs infrequently, time will be required to determine how often TDAV infection is associated with the disease. Data is not sufficient at this time to estimate the prevalence of TDAV infection in the horse population. To determine seroprevalance of the virus, an ELISA is being developed using the expressed TDAV NS5A protein.. We recently have acquired 500 equine serum samples from horses presumed to be healthy and residing in different areas of North America for TDAV antibody testing. Transfection studies with the cloned TDAV genome also are planned.

NONPRIMATE HEPACIVIRUS

NPHV was the first of the three new blood-borne flaviviruses including TDAV to be reported. Interestingly, NPHV first was detected in dogs (Kapoor, A et al., 2011; Bukh, J, 2011). NPHV later was found in pooled equine serum and in the serum of 8 of 103 (7.8%) normal horses, of which 35% were NPHV antibody positive (Burbelo, PD et al., 2012). Since description in the United States by Burbelo, et al., NPHV has been reported in 3 of 142 horses (2.1%) in Great Britain (Lyons S, et al., 2012) and in 7 of 210 horses (3.3%) from Germany (Drexler JF,et al 2013). NPHV is a member of the *Hepacivirus* genus and of the three new viruses, NPHV is most closely related genetically to HCV. Both are members of the Hepacivirus genus of which HCV is the prototype virus . Parallel qRT-PCR assays for NPHV RNA were performed on the serum of horses residing on the farm with TDAV associated hepatitis and on the control farm of 20 horses. The virus was identified in only three horses, two asymptomatic TDAV-positive horses from the farm with the hepatitis outbreak and one TDAV-negative horse with no history of antitoxin treatment residing on the geographically distant control farm. Currently, there is no documented association between NPHV infection and hepatic disease in horses, although in the initial report of the virus there was no testing for hepatic disease and the horses were presumed to be healthy (Kapoor A et al. 2011).

EQUINE PEGIVIRUS

The most recent of the new equine, blood-borne flaviviruses was reported by Kapoor, A. et al. (2013a) and has been named the equine pegivirus (EPgV). EPgV and TDAV are members of the Pegivirus genus but are genetically distinct. Pegaviruses, although a common cause of infection in humans and other species, are not known to cause clinical disease. Three of 12 EPgV infected horses (25%) were said to have had "elevated liver enzymes" while 4 of 62 horses (6.5%) with normal liver enzyme activities were EPgV positive. A closed herd of healthy horses were tested for EPgV at three separate time points over a period of 4 years. At the first test, 6 of 19 (32%) were EPgV positive by PCR. At the second test, 6 of 29 (21%) were positive and in the third test, 4 of 27 (15%) were EPgV positive. Two of the horses were demonstrated to be carriers of EPgV for at least 3.5 years. Nine liver samples obtained from random equine necropsies in California were tested for EPgV and one was positive (Kapoor A, 2013[a]). Recently, hepaciviruses and pegiviruses also have been described in wild rodents (Kapoor A., et al., 2013[b]).

FINAL COMMENTS

The association of TDAV with Theiler's disease remains inferential and testing of future cases will be necessary to strengthen any association between TDAV infection and Theiler's disease. The inability to associate level of viremia with disease in horses from the outbreak study makes it clear that if TDAV is a cause of Theiler's disease, host factors or additional, as-yet unidentified infectious agents may influence the presence of clinical disease. Additional metagenomic testing for additional hepatitis viruses should be conducted because equine hepatitis viruses may not be limited to TDAV, NPHV and EPgV viruses. Although the data to support a causal relationship between NPHV (or EPgV) infection and equine liver disease is limited at this time, the possibility cannot be ruled out and further epidemiological and experimental research is required. Our long-term research goals are to advance the understanding of the role these viruses may play in equine liver disease, in the etiology, pathogenesis and epidemiology of Theiler's disease, and, finally, to apply knowledge of the natural history of the disease to the control, prevention, and possibly eradication of serum hepatitis in horses.

REFERENCES

Nieto A, et. **J Vet Intern Med** 2005;19:120.
Burbelo PD, et al. **J Virol** 2012;86:6171.
Bukh J, et al. **J Med Virol** 2001;65(4):694.
Chandriani S, et al. **Proc Natl Acad Sci USA** 2013;110(15):E1407.
Cox H, et al. **J Am Med Vet Assoc** 1938;93:225.
Drexler J, et al. **PLoS Pathol** 2013;9(6):e1003438.
Forenbacher B, et al. **Veterinarski Archiv** 1959;29:322.
Guglick M, et al. **J Am Vet Med Assoc** 1995;206:1737.
Hjerpe C. **J Am Vet Med Assoc** 1964;144:734.
Kapoor A, et al. **Proc Natl Acad Sci USA** 2011;108(28):11608.
Kapoor A, et al. **J Virol** 2013(a);87(12):7185.
Kapoor A, et al. **mBio** 2013(b);4(2):e00216.
Lyons S, et al. **Emerg Infect Dis** 2012;18(12):1976.

Madsen D. **Utah Academy of Sciences, Arts, and Letters** 1934;11:95.
Marsh H. **J Am Vet Med Assoc** 1937;91:88.
Marsh H. **J Am Vet Med Assoc** 1937;91:330.
Messer N, et al. **J Am Vet Med Assoc** 1994;204:1790.
Messer N, et al. **J Am Vet Med Assoc** 1994;204:1934.
Meyer K. **N Amer Vet** 1933;14:30.
Records E, et al. **J Am Vet Med Assoc** 1933;32:608.
Robinson M, et al. J Comp Pathol 1975;85:111.
Rose J, et al. **Proc Annu Conven Am Assoc Equine Practit** 1974;175.
Shahan M, et al. **Vet Med** 1939;34:354.
Smith H, et al. Can Vet J 1991;32:362.
Stenius P. **Skand Vet Tidsskr** 1941;31:193.
Theiler A. **Report of Director of Veterinary Research** 1919;5&6:7.
Thomsett L. **Equine Vet J** 1971;3:15. Zientara S, et al. **Vet Rec** 1994;134:18.

DISSECTING GENE AND CELLULAR EXPRESSION IN EQUINE INFECTIOUS DISEASES

Maureen T. Long, DVM, PhD, DACVIM (Large Animal Internal Medicine)
Gainesville, FL

Systems Biology

Systems biology is an interdisciplinary study that utilizes global data from genomics, transcriptomics, proteomics, and, more recently, metabolomics that can connect the relationships between underlying genetics, cellular expression and metabolism, and clinical and anatomic pathology with infectious disease.[1] This is based on the underlying assumption that microorganisms, even without producing clinic disease cause changes in cellular expression which varies from basic cellular metabolism to overt injury. Challenge by a microorganism results in expression of cassettes of disease or organ specific markers that once known, can be utilized for creation of new diagnostics and therapeutics. Through the use of basic science and high through- and output technologies, high performance computing and computational biology, a global analysis of gene expression can be used to generate models or networks which explain and provide new hypotheses for many biomedical problems.

A holistic approach is utilized to understand the function and dysfunction leading to changes in the behavior of that system. For instance in sepsis (reviewed in 2), not only are inflammatory cytokines upregulated, but what are *all* of the genes that express cytokines doing?[2] How does this change in regulation, if present, affect the entire gamut of cell signaling pathways? Are there individual susceptibilities based on genetic markers that explain disease. Rather than looking at a single pathway such as tumor necrosis factor alpha (TNF-α), all of the pathways are examined in sepsis. These pathways are examined at all levels of expression including mRNA, isoforms of mRNAs, protein products and small products of metabolism.

Spontaneously occurring diseases in all of our domestic animals offer enormous potential to contribute new information regarding gene and protein expression and function during infection.[3;4] Adding a systems biology approach will create not only new information but also enhance potential funding opportunities in veterinary internal medicine. This is important for rapid and specific development of diagnostics and therapeutics for emerging pathogens, currently undiagnosable diseases, and unmanageable diseases that veterinary practitioners see every day. An explosive increase in gene sequence information is now available, but the actual characterization of genes, their expression, and protein function is largely biased towards a reductionist approach. In addition, most research focuses is on genes and their products that have a known function. This is typically due to one or more of the following: a failure to use appropriate tissues in the generation of initial libraries, a lack of case criteria or disease definition in selecting tissues for analysis, and limited availability of databases which focus on particular tissues and disease categories of interest. This review will focus on defining the "-Omics Pipeline" with review of the strengths, weakness of available methods within the pipeline. Published examples of global systems approaches in equine infectious diseases will be highlighted.

The Omics Pipeline

The pipeline starts with a target tissue, cell population, and/or body fluid (blood, plasma, serum, urine) and next generation analysis is performed.[5] The most efficient pipelines are built around fully sequenced organisms. At present 2,663 eukaryotic organisms have at least partial genome information; surprisingly complete genomes are listed for only 221 eukaryotes.[6] For whole genome sequencing of mammals, the genome of 25 animals are listed as complete. Of the domestic nonprimate species, *Bos taurus*, *Equus caballus*, *Canis lupus familiaris*, *Capri hircus* (Yunnan black goat) are sequenced. The bighorn sheep, *Ovis aries*, can provide a genomic scaffold for future sequencing of domestic sheep genome. In 2007, the full genome of the horse was released and the assembly EquCab2 was released in 2008.[7;8] Over the last five years, updates and additional analysis has identified 20,436 protein-coding genes and 4400 pseudogenes.[6] The availability of EquCab2 is starting to transform several areas of equine research in a mechanistic fashion. In 2014, EquCab3 will be released with many more genes that are fully predicted.

For tissues and cells, genomics, transcriptomics, proteomics and metabolomics can be performed. The raw data composed of either deep sequencing for gene sequences for genomics and transcriptomics or liquid chromatography/mass spectrometry (LC-MS) data for proteomic/metabolomics is obtained and initial cleaning of the raw data is performed.[5] Intermediate processing is performed so that initial gene sequences, transcripts, or protein data is annotated so that strings of genes and amino acids, respectively, are linked. Intermediate processing leads to annotation of gene strings for predicted identification of genes or proteins. The data is then analyzed to mine for biological meaning. If quantitative, data is normalized and statistically analyzed. The final step is biological network building so the data can be further mined for new hypotheses and new relationships. Downstream investigation consists of functional genomics where either cellular or animal responses are further validated.

Genomics: Definition, Current Techniques, Use in Equine Infectious Disease Research

The definition of genomics for this overview is the analysis with DNA for markers of inheritance and disease traits. The basis for high level investigation of equine genomics is the full sequencing of *Equus caballus*. From this point, one can use the sequence data to disease susceptibility traits via either basic low yield techniques such as **microsatellite mapping** to high yield technology techniques such as the use of **single nucleotide polymorphisms** (SNPs).[9-15] For infectious disease susceptibility studies, most susceptibility to infectious disease are complex trait analyses, and these are usually population based studies. This is further compounded by the fact that evolution since domestication of the horse has result in intense breed selection phenotypes such as muscle fiber type (race performance), gait, size, and coat color.[16-24] It is unknown what disease susceptibility characteristics have been selected with these breed traits. In plants and, other animal species such as the chicken and cattle, research into selection for disease susceptibility is a part of dissecting production performance.[25;26] There is limited work in the equine because of the high cost sampling and sampling analysis for population-wide studies. Nonetheless, most information can economically be derived for single targets or limited gene pathways or for breed specific investigations. In addition, there is extensive genetic analysis software and web-derived programs that help offset the cost of the analytical pipeline. In the horse, there are limited infectious disease SNP or microsatellite studies detectable in reference search engines such as PUBMED (NCBI) or ISI Web of Science (Rutgers).[9;12;13;27;28] In comparison, there are disease association-SNP studies with infectious disease in cattle (*e.g. Mycobacterium paratuberculosis*, Infectious bovine keratoconjunctivitis, infectious udder disease) and swine (e.g. Porcine reproductive and respiratory syndrome, *Mycoplasma pneumonia* but even these studies number under 50 in total for correlative experiments with particular infectious diseases (http://www.ncbi.nlm.nih.gov/pubmed).

Transcriptomics: Definition, Current Techniques, Use in Equine Infectious Disease Research

Transcriptomics is analysis of the total set of mRNA transcripts in a tissue or cell population.[29] Transcriptome generation is not the same as transcript profiling, which is analysis of a set of transcripts, usually through reverse transcription polymerase chain reaction (rt-PCR). The entire repertoire of transcripts is targeted and this global data is provides a complete characterization at the cellular level of gene response, allowing for direct and unbiased linking of genotype with phenotype. Given that a whole transcriptome contains around 5-8 X 10^{23} RNA bases, analysis of disease at this level would be overwhelming if not for first the development of the microarray and more recently, deep sequencing technology.[29;30] Through these two techniques, tissue gene expression can now be better categorized and prioritized to generate unique targets with the advent of tissue specific databases and the automation of high throughput sequencing strategies.

Even though the first microarray was described in 1995 and this technology has been developing for almost 30 years, this is still one of the most successful techniques for high performance analysis of the transcriptome by cell type, tissue, within organism, and for comparative genomics.[30] Thirty years of this technology have greatly improved this technique so that quality control (MicroArray Quality Control consortium). Today's microarrays are available for the equine through several commercial platforms and these are extremely useful and cost effective. This technique can now generate repeatable data that can even be normalized across experiments with historical controls and even across laboratories. Commercial array platforms are based on probes designed on either a known genome sequence or predicted open reading frame. mRNA from tissues is hybridized to array and usually scanned via laser that picks up each hybridized probe in a quantitative manner.

Presently, microarray and gene chip technology have been developed for bovine, equine, and porcine.[31;32;32;33] One tissue specific array is available for horses, and this is based on RNA expression of joint cells stimulated with endotoxin.[32] The second commercial chip is available based on the *Equus caballus* genome project in which probes were generated from deep sequencing data. Equine tissue specific microarrays are described in the literature.[34 35-37] One array is a 44,000 probe microarray biased toward brain sequences generated from a pool of normal and diseased horses infected with West Nile virus (WNV), and these probes were generated from deep sequencing data. Another is a 21,000 probe whole genome microarray. A third array is described for influenza. The use of these arrays has been limited for analysis of equine infectious diseases but thus far include age related changes in *Rhodococcus equi* peripheral blood mononuclear cell responses, in vitro responses to influenza, and the equine brain transcriptome in WNV infection.[35]

Direct sequencing of RNA in a massively parallel fashion is called deep sequencing.[29] Several techniques exist and center around fragmenting DNA or cDNA and then performing repetitive sequencing to generate qualitative and quantitative data in a very short time. Computer analysis is used to compare these sequences with the *Equus caballus* genome. Deep sequencing can provide qualitative direct data on exomes and thus can yield information on differences in alternative splicing. There are some major drawbacks that still exist even for this newer method that for small projects can outweigh the advantages as a way to quantitate the transcriptome while at the same time sequencing. The first is that the backbone data of the genome must be excellent, with EquCab only on its second version; the initial bioinformatics analysis can be flawed. Even though sequencing is *de novo* with the goal of discovery of new genes, etc., the initial large data set requires excellent bioinformatics. Second, with repeated sequencing of the small areas of the genome, there can be coverage biases introduced by limiting the rounds of sequencing to save expense. Thus, key but low expressed genes may be missed. Third, the file formats are extremely large and require skilled bioinformatics expertise, access to high performance computing, and other software to move sequence data into other genomics analysis programs. Most of the work thus far utilizing deep sequencing in the horse has been to map differences in gene expression to non-infectious disease susceptibility or physical traits as described for genomics.[38-45]

Proteomics: Definition, Current Techniques, Use in Equine Infectious Disease Research

The "proteome is the protein complement of the genome".[46] Proteomics is plasma, serum, cells, and tissue, digesting the sample for protein and analyzing the quantity, quality and function of the proteins in the cells. The end result of gene expression of proteins is not a one-to-one process. Once an RNA transcript is generated in a eukaryote, the message is alternatively spliced resulting in translation of different species of the same protein. Both background genomics (SNP) and infectious agents can alter the quality and quantity of alternative splice pathways resulting in differently translated proteins that differ often in functionality. Then once a protein is translated there are many forms of posttranslational modification. Proteomics utilizes high performance technology to subject complex mixtures to separation (2-D gel electrophoresis) and proteolysis.[47;48] Newer techniques label the protein mixture with isotopes (ICAT).[47] This is followed by mass spectrometry to generate a peptide mass fingerprint overlaid with peptide sequences to generate a putative protein sequence. Protein identification is performed by a computer algorithm using high performance computing. Like transcriptome technology, "chips" or glass slide formats are available. There are several versions of these chips, but limited development in the equine. The most commonly used technology is based on mass spectrometry.[4;49-56] Plasma protein proteomics has been used in equine laminitis, non-infectious joint diseases, exercise, and uveitis, and on body fluids such as milk.[49-56]

Metabolomics: Definition, Current Techniques, Use in Equine Infectious Disease Research

Metabolomics is a relatively new area of "omics". Metabolomics is primarily involved in generating global metabolite profiles. Metabolites are small molecules which are products of cellular metabolism. Frequently they are very small and can cross membrane barriers so that body fluids such as plasma, serum, urine, CSF, alveolar lavage, peritoneal, and joint fluid. Thus many of these can be used as biomarkers of disease. The process for identification is much like the previous proteomic pipeline only capillary electrophoresis can be used for the separation phase. There are thousands of metabolites so better spectrometry techniques have been added to the repertoire of metabolomics analysis which includes nanostructure-initiator MS, secondary ion MS, and nuclear magnetic resonance. Several equine investigators have utilized this new technology; however there are now publications that have investigation equine infectious diseases other than that derived from sepsis.[2;57] The fields of equine medicine and surgery have included laminitis, respiratory disease, and medication in sport horses. [54;55;58-60]

From Omics to Translational Results

The paucity of –Omics data in the equine and that used to characterize infectious diseases reflect 1) limited emphasis in residency programs on genetics, biochemistry, cell biology, and metabolism as a basis for understanding clinical disease; 2) limited support for extramural research that emphasized global and de novo strategies, and 3) limited veterinary college resources for animal genomics and cell biology. Those Universities and Veterinary Clinical departments, and veterinary clinical specialties that recognize the need for mechanistic research are our current leaders in the field; these include veterinary equine surgeons, equine sports medicine physiologists and clinicians. In the basic scientists the current literature is dominated by animal sciences and toxicologists. Nonetheless, the limited investigation thus far create an as opportunity for veterinary large animal internists because of the access to clinical case material. Emphasis on infectious diseases provides development of clear cut case definitions (presence of pathogen, presence of clinical signs, and presence of clinicopathologic abnormalities.

REFERENCES

1. Chadeau-Hyam M, Campanella G, Jombart T et al., Deciphering the complex: methodological overview of statistical models to derive OMICS-based biomarkers, *Environ.Mol.Mutagen.* 2013;54: 542-557.
2. Liu X, Ren H, Peng D , Sepsis biomarkers: an omics perspective, *Front Med.* 2014;8: 58-67.
3. Ceciliani F, Restelli L, Lecchi C , Proteomics in farm animals models of human diseases, *Proteomics.Clin.Appl.* 2014;
4. Bundgaard L, Jacobsen S, Sorensen MA et al., The Equine PeptideAtlas: A resource for developing proteomics-based veterinary research, *Proteomics.* 2014;14: 763-773.
5. Sung J, Wang Y, Chandrasekaran S et al., Molecular signatures from omics data: from chaos to consensus, *Biotechnol.J.* 2012;7: 946-957
6. U.S.National Library of Medicine. Eukaryotic Genome Annotation at NCBI, Available at: http://www.ncbi.nlm.nih.gov/genome/browse/. Accessed Jan 15, 2014.
7. Chowdhary BP, Bailey E , Equine genomics: galloping to new frontiers, *Cytogenet.Genome Res.* 2003;102: 184-188.
8. Chowdhary BP, Raudsepp T , The horse genome, *Genome Dyn.* 2006;2: 97-110.
9. Shiue Y-L, Millon LV, Skow LC et al., Synteny and regional marker order assignment of 26 type I and microsatellite markers to the horse X- and Y-chromosomes, *Chromosome.Res.* 2000;8: 45-55.
10. Swinburne JE, Hopkins A, Binns MM , Assignment of the horse grey coat colour gene to ECA25 using whole genome scanning, *Anim Genet.* 2002;33: 338-342.
11. Lampe V, Dierks C, Komm K et al., Identification of a new quantitative trait locus on equine chromosome 18 responsible for osteochondrosis in Hanoverian warmblood horses, *J.Anim Sci.* 2009;87: 3477-3481.
12. Mittmann EH, Lampe V, Momke S et al., Characterization of a minimal microsatellite set for whole genome scans informative in warmblood and coldblood horse breeds, *J.Hered.* 2010;101: 246-250.
13. Mittmann EH, Wrede J, Pook J et al., Identification of 21 781 equine microsatellites on the horse genome assembly 2.0, *Anim Genet.* 2010;41: 222
14. Zabek T, Golonka P, Fornal A et al., IHH gene polymorphism among three horse breeds and its application for association test in horses with osteochondrosis, *Hereditas* 2013;150: 38-43.
15. Orr N, Back W, Gu J et al., Genome-wide SNP association-based localization of a dwarfism gene in Friesian dwarf horses, *Anim Genet.* 2010;41 Suppl 2: 2-7.
16. te Pas MF, Wijnberg ID, Hoekman AJ et al., Skeletal muscle transcriptome profiles related to different training intensities and detraining in Standardbred horses: a search for overtraining biomarkers, *Vet.J.* 2013;197: 717-723.
17. McGivney BA, Browne JA, Fonseca RG et al., MSTN genotypes in Thoroughbred horses influence skeletal muscle gene expression and racetrack performance, *Anim Genet.* 2012;43: 810-812.
18. Promerova M, Andersson LS, Juras R et al., Worldwide frequency distribution of the 'Gait keeper' mutation in the DMRT3 gene, *Anim Genet.* 2014;45: 274-282
19. Metzger J, Schrimpf R, Philipp U et al., Expression levels of LCORL are associated with body size in horses, *PLoS.One.* 2013;8: e56497
20. Rendo F, Iriondo M, Manzano C et al., Identification of horse chestnut coat color genotype using SNaPshot, *BMC.Res.Notes* 2009;2: 255
21. Cook D, Brooks S, Bellone R et al., Missense mutation in exon 2 of SLC36A1 responsible for champagne dilution in horses, *PLoS.Genet.* 2008;4: e1000195.
22. Haase B, Jude R, Brooks SA et al., An equine chromosome 3 inversion is associated with the tobiano spotting pattern in German horse breeds, *Anim Genet.* 2008;39: 306-309.
23. Bellone R, Lawson S, Hunter N et al., Analysis of a SNP in exon 7 of equine OCA2 and its exclusion as a cause for appaloosa spotting, *Anim Genet.* 2006;37: 525.
24. Brooks SA, Bailey E , Exon skipping in the KIT gene causes a Sabino spotting pattern in horses, *Mamm.Genome* 2005;16: 893-902.
25. Sorensen LP, Madsen P, Mark T et al., Genetic parameters for pathogen-specific mastitis resistance in Danish Holstein Cattle, *Animal.* 2009;3: 647-656
26. Zekarias B, ter Huurne AA, Landman WJ et al., Immunological basis of differences in disease resistance in the chicken, *Vet.Res.* 2002;33: 109-125
27. Rios JJ, Perelygin AA, Long MT et al., Characterization of the equine 2'-5' oligoadenylate synthetase 1 (OAS1) and ribonuclease L (RNASEL) innate immunity genes, *BMC.Genomics* 2007;8: 313.
28. Kakoi H, Kijima-Suda I, Gawahara H et al., Individual identification of racehorses from urine samples using a 26-plex single-nucleotide polymorphism assay, *J.Forensic Sci.* 2013;58: 21-28.
29. Malone JH, Oliver B , Microarrays, deep sequencing and the true measure of the transcriptome, *BMC.Biol.* 2011;9: 34.
30. Schena M, Shalon D, Davis RW et al., Quantitative monitoring of gene expression patterns with a complementary DNA microarray, *Science* 1995;270: 467-470
31. Yao J, Burton JL, Saama P et al., Generation of EST and cDNA microarray resources for the study of bovine immunobiology, *Acta Vet.Scand.* 2001;42: 391-405
32. Gu W, Bertone AL , Generation and performance of an equine-specific large-scale gene expression microarray, *Am.J.Vet.Res.* 2004;65: 1664-1673
33. Medhora M, Bousamra M, Zhu D et al., Upregulation of collagens detected by gene array in a model of flow-induced pulmonary vascular remodeling, *Am.J.Physiol Heart Circ.Physiol* 2002;282: H414-H422
34. Bourgeois MA, Denslow ND, Seino KS et al., Gene expression analysis in the thalamus and cerebrum of horses experimentally infected with West Nile virus, *PLoS.One.* 2011;6: e24371
35. Kachroo P, Ivanov I, Seabury AG et al., Age-related changes following in vitro stimulation with Rhodococcus equi of peripheral blood leukocytes from neonatal foals, *PLoS.One.* 2013;8: e62879
36. Horin P, Sabakova K, Futas J et al., Immunity-related gene single nucleotide polymorphisms associated with Rhodococcus equi infection in foals, *Int.J.Immunogenet.* 2010;37: 67-71

37. Horin P, Osickova J, Necesankova M et al., Single nucleotide polymorphisms of interleukin-1 beta related genes and their associations with infection in the horse, *Dev.Biol.(Basel)* 2008;132: 347-351
38. Thomas KC, Hamilton NA, North KN et al., Sequence analysis of the equine ACTN3 gene in Australian horse breeds, *Gene* 2014;538: 88-93
39. Coleman SJ, Zeng Z, Wang K et al., Structural annotation of equine protein-coding genes determined by mRNA sequencing, *Anim Genet.* 2010;41 Suppl 2: 121-130
40. Wang W, Wang S, Hou C et al., Genome-wide detection of copy number variations among diverse horse breeds by array CGH, *PLoS.One.* 2014;9: e86860
41. Dupuis MC, Zhang Z, Durkin K et al., Detection of copy number variants in the horse genome and examination of their association with recurrent laryngeal neuropathy, *Anim Genet.* 2013;44: 206-208
42. Metzger J, Philipp U, Lopes MS et al., Analysis of copy number variants by three detection algorithms and their association with body size in horses, *BMC.Genomics* 2013;14: 487
43. Doan R, Cohen N, Harrington J et al., Identification of copy number variants in horses, *Genome Res.* 2012;22: 899-907
44. Doan R, Cohen ND, Sawyer J et al., Whole-genome sequencing and genetic variant analysis of a Quarter Horse mare, *BMC.Genomics* 2012;13: 78
45. Sundstrom E, Imsland F, Mikko S et al., Copy number expansion of the STX17 duplication in melanoma tissue from Grey horses, *BMC.Genomics* 2012;13: 365
46. Wasinger VC, Cordwell SJ, Cerpa-Poljak A et al., Progress with gene-product mapping of the Mollicutes: Mycoplasma genitalium, *Electrophoresis* 1995;16: 1090-1094
47. Di GF, Lante I, Muraca M et al., The Role of Mass Spectrometry in the "Omics" Era, *Curr.Org.Chem.* 2013;17: 2891-2905
48. Fliser D, Novak J, Thongboonkerd V et al., Advances in urinary proteome analysis and biomarker discovery, *J.Am.Soc.Nephrol.* 2007;18: 1057-1071
49. Steelman SM, Chowdhary BP , Plasma proteomics shows an elevation of the anti-inflammatory protein APOA-IV in chronic equine laminitis, *BMC.Vet.Res.* 2012;8: 179
50. Desjardin C, Balliau T, Valot B et al., A method for proteomic analysis of equine subchondral bone and epiphyseal cartilage, *Proteomics.* 2012;12: 1870-1874
51. Hinz K, O'Connor PM, Huppertz T et al., Comparison of the principal proteins in bovine, caprine, buffalo, equine and camel milk, *J.Dairy Res.* 2012;79: 185-191
52. Chiaradia E, Pepe M, Tartaglia M et al., Gambling on putative biomarkers of osteoarthritis and osteochondrosis by equine synovial fluid proteomics, *J.Proteomics.* 2012;75: 4478-4493
53. Wolf CA, Maslchitzky E, Gregory RM et al., Effect of corticotherapy on proteomics of endometrial fluid from mares susceptible to persistent postbreeding endometritis, *Theriogenology* 2012;77: 1351-1359
54. Racine J, Gerber V, Feutz MM et al., Comparison of genomic and proteomic data in recurrent airway obstruction affected horses using Ingenuity Pathway Analysis(R), *BMC.Vet.Res.* 2011;7: 48
55. Bright LA, Mujahid N, Nanduri B et al., Functional modelling of an equine bronchoalveolar lavage fluid proteome provides experimental confirmation and functional annotation of equine genome sequences, *Anim Genet.* 2011;42: 395-405
56. Fahiminiya S, Labas V, Dacheux JL et al., Improvement of 2D-PAGE resolution of human, porcine and equine follicular fluid by means of hexapeptide ligand library, *Reprod.Domest.Anim* 2011;46: 561-563
57. Steelman SM, Johnson P, Jackson A et al., Serum metabolomics identifies citrulline as a predictor of adverse outcomes in an equine model of gut-derived sepsis, *Physiol Genomics* 2014;
58. Boyard-Kieken F, Dervilly-Pinel G, Garcia P et al., Comparison of different liquid chromatography stationary phases in LC-HRMS metabolomics for the detection of recombinant growth hormone doping control, *J.Sep.Sci.* 2011;34: 3493-3501
59. Keller MD, Pollitt CC, Marx UC , Nuclear magnetic resonance-based metabonomic study of early time point laminitis in an oligofructose-overload model, *Equine Vet.J.* 2011;43: 737-743
60. Kieken F, Pinel G, Antignac JP et al., Development of a metabonomic approach based on LC-ESI-HRMS measurements for profiling of metabolic changes induced by recombinant equine growth hormone in horse urine, *Anal.Bioanal.Chem.* 2009;394: 2119-2128

FIVE YEAR RETROSPECTIVE OF NEONATAL INTENSIVE CARE UNIT FOAL ADMISSION, HOSPITALIZATION AND SURVIVABILITY DATA (2009-2013)

Peter R. Morresey, BVSc, DACT, DACVIM
Lexington, KY

INTRODUCTION

The financial viability of the equine industry has been severely challenged as a result of the global financial crisis beginning 2009. Where established as a mainstream economic activity, the equine industry can have a multi-billion dollar impact on state economic productivity.[1] Central to the viability of the equine breeding industry is the reproductive success of the brood mare as measured by live foal production, which is approximately 80% for mares bred in various locations.[2, 3] Economic modelling of the brood mare has determined that 6 foals must be produced over a 7 year period to maintain financial viability for the individual, this including the sale of viable progeny.[4] Survival of each foal delivered is therefore critical to the health of the overall equine industry.

A number have previous studies have investigated the prognostic value of clinical and clinical pathological findings in neonatal foals at time of admission to intensive care units, and mathematical models have been created to assist clinician decision making with respect to survival.[5-8] Repeatability between different locations or with different groups of foals in the same location where the model was created was however found to be poor. Prognostic equations were also found less useful for conditions that developed after the initial assessment.[8] Field use of such prognostic models is difficult as many researched parameters in academic settings are not measurable in the field, and for those that are results are typically not available in a reasonable timeframe. Furthermore any delay in hospital referral may alter both clinical and laboratory findings, and affect prognosis.[9] Once admitted, duration of hospitalization has a direct bearing on the cost of treatment.

REVIEW

Published studies vary widely in the factors deemed prognostic at initial neonatal foal evaluation, with seemingly conflicting findings in some cases. Prematurity or dysmaturity of the neonate is often a presenting sign, however in one study gestational age did not differ between survivors and non-survivors.[8] Maternal disease (both systemic and placental) was higher amongst non-surviving foals.[8] A history of dystocia has also been suggested as a negative indicator.[10, 11] Age at admission did not affect survival.[8, 12] However, duration of clinical signs prior to admission negatively affected outcome[8], and in one study a 5.8 fold increase in the risk of death for each 24 hour period of delay was noted.[12]

Heart rate $\geq$ 70 beats per minute, respiratory rate $\geq$ 60 breaths per minute, rectal temperature above 99°F, normal appearance of mucous membranes, and the ability to stand at admission also positively affected survival in one study.[8] However in another report of respiratory disease the majority of clinical examination variables assessed at initial evaluation were found to be unassociated with case outcome.[10] Fever was found to be an inconsistent finding in septic foals in one study.[13]

The presence of signs of neonatal sepsis has been a negative indicator in many studies. Sepsis was identified as the most common problem in both survivors and non-survivors in one study.[14] Diagnosis was assisted by sepsis scoring, this being considered more reliable that any individual parameter.[5] Bacteremia in one study was most consistently due to *Escherichia coli*, although mixed infections were also found.[15] Of 423 bacteremic foals in this study 254 survived, this being affected by a number of historical and clinical findings. A positive blood culture is associated with non-survival.[16]

Hematological and clinical chemistry values have been integral in many neonatal foal prognostic models. A low segmented neutrophil count[5, 7, 8, 12], low total erythrocyte count[7], and elevated fibrinogen concentration[7] were significant indicators of mortality in a number of studies. Immunoglobulin G concentration also factored in a number of studies involving survival and prevention of infection.[17, 18]

Lactate has received considerable attention as a prognostic indicator and has been widely reported. Lactate is considered a marker of circulatory competence, with the rate of clearance predicting survival.[19, 20] Serial blood lactate can predict development of multiple organ failure following human sepsis[21], and in sick neonates blood lactate elevations were associated with increased mortality, with increasing levels preceding clinical markers of deterioration.[22] Admission arterial blood lactate in neonatal foals has been shown associated with survival and occurrence of SIRS.[23] Venous blood lactate elevation, and persistent hyperlactatemia, has similarly been shown associated with survival and severity of illness.[24, 25] The highest admission lactate concentrations were present in hemorrhagic shock, sepsis and complicated perinatal asphyxia in one study[24], but in younger, premature or encephalopathic foals in another.[25] Regardless of initial diagnosis, lactate is significantly increased in non-survivors versus survivors.[16]

Blood glucose has been shown in other species to be indicative of survival. Hyperglycemia, both degree and duration, has been associated with non-survival in human medicine.[26-28] Hypoglycemia also is problematic, and although critical glucose levels have not been determined, prolonged human neonatal hypoglycemia is associated with negative outcomes, both neurological and survival.[29] In neonatal foals, hypoglycemia has been associated variably with survival, being associated with non-survival in septic foals[12] but not overall foal admissions.[7] In one study, hypoglycemia at admission was associated with sepsis, bacteremia and SIRS, and blood glucose outside the range of 50 -180mg/dL associated with decreased survival.[30] Blood glucose >120 mg/dL was associated with survival in another study.[12] It is uncertain from these studies whether perturbations in blood glucose cause, or are merely reflective of, pathophysiological processes decreasing survival.

Academic studies regarding the prognosis for neonatal survival have advanced to involve variables difficult to measure at point of care in routine clinical practice. Septic non-survivors have been shown to have elevated arginine vasopressin and adrenocorticotrophic hormone, with a subcategory of neonates having concurrent normal to decreased cortisol implying a relative adrenal insufficiency.[31, 32] Anion gap elevation and decreased venous pO_2 were also associated significantly with non-survival.[14]

RETROSPECTIVE

The goal of this study was to assess information that could be readily obtained in a field preadmission setting that could assist in the making of a decision to refer a neonatal foal for specialist care. Additionally, time to discharge or death for various diagnostic categories could be determined. Medical records of 514 foals up to 4 days of age admitted to the neonatal intensive care unit over the period of 2009-2013 were eligible for inclusion. Neonatal admissions were categorized on the basis of primary admitting disorder: dystocia, gastrointestinal, neurological, orthopedic, sepsis and multiple diagnoses. Information recorded included date of admission, gestational age, age at presentation, gender, rectal temperature, immunoglobulin G status, erythrocyte count, total and segmented neutrophil count, lactate, and glucose. Results of blood cultures obtained by sterile technique at admission were subsequently also available. Presenting mare treatment information was also collected.

For each of the diagnostic categories, time to discharge (survival) or non-survival for the 50th and 90th percentile were calculated. Diagnostic categories were selected based on retrospective determination of the primary cause of the foal's condition. Where concurrent conditions of similar severity were present, foals were placed in the multiple diagnosis category. Relative risk for non-survival for each of the recorded items was also calculated. The non-survival group consisted of foals that died or were euthanased due to clinician determination of poor prognosis due to unfavorable clinical progression or client financial reasons.

RESULTS

Foals presenting following dystocia without signs at admission consistent with concurrent disease had the shortest periods of hospitalization and highest survival rates. Multiple comorbidities at presentation led to longer periods of hospitalization for survivors and more rapid death for non-survivors in comparison to those foals with a solitary diagnosis. Sepsis and neurological disorders were common comorbidities with dystocia.

DISCUSSION

Although in non-surviving foals single factors may be identified as occurring at a higher frequency than in survivors, prognosis should not be based on single factors alone. Within each diagnostic category a range of survival occurs as a result of the severity of the condition, constitution of the foal, and extraneous factors (economic). Consistent with other studies evaluating more complex examination variables present at admission, multisystemic disease decreases likelihood of survival to discharge.[33]

REFERENCES

[1] University of Kentucky. http://uknowledge.uky.edu/equine_reports/1 2013. [2] Bosh KA, et al. Equine Vet J 2009;41:883-888. [3] Hanlon DW, et al. New Zealand Vet J 2012;60:329-334. [4] Bosh KA, et al. Equine Vet J 2009;41:889-894. [5] Brewer BD, et al. Equine Vet J 1988;20:18-22. [6] Brewer B, et al. Equine Vet J 1990;20:23-24. [7] Peek SF, et al. AAEP 2004:60-62. [8] Furr M, et al. J Vet Intern Med 1997;11:183-188. [9] Rohrbach BW, et al. J Amer Vet Med Assoc 2006;228:1748-1756. [10] Bedenice D, et al. J Vet Intern Med 2003;17:876-886. [11] Wohlfender FD, et al. Equine Vet J 2009;41:186-191. [12] Gayle JM, et al. J Vet Intern Med 1998;12:140-146. [13] Koterba AM, et al. Equine Vet J 1984;16:376-382. [14] Hoffman AM, et al. J Vet Intern Med 1992;6:89-95. [15] Sanchez LC, et al. J Amer Vet Med Assoc 2008;233:1446-1452. [16] Borchers A, et al. Equine Vet J 2012;44:57-63. [17] Tyler-McGowan CM, et al. Aust Vet J 1997;75:56-59. [18] Robinson JA, et al. Equine Vet J 1993;25:214-219. [19] Nichol A, et al. Critical Care 2011;15:R242. [20] van Beest P, et al. Annals of Intensive Care 2013;3:6. [21] Bakker J, et al. The American Journal of Surgery 1996;171:221-226. [22] Deshpande SA, et al. Archives of Disease in Childhood - Fetal and Neonatal Edition 1997;76:F15-F20. [23] Corley KT, et al. Equine Vet J 2005;37:53-59. [24] Castagnetti C, et al. Theriogenology 2010;73:343-357. [25] Henderson ISF, et al. J Vet Emerg Critic Care 2008;18:496-502. [26] Wintergerst KA, et al. Pediatrics 2006;118:173-179. [27] Bochicchio GV, et al. Amer surg 2005;71:171-174. [28] Faustino EV et al. J Pediatr 2005;146:30-34. [29] Rozance PJ, et al. Neonatol 2006;90:74-86. [30] Hollis AR, et al. J Vet Intern Med 2008;22:1223-1227. [31] Gold JR, et al. J Vet Intern Med 2007;21:791-796. [32] Hurcombe SD, et al. J Vet Intern Med 2008;22:639-647. [33] Smith LJ, et al. Equine Vet J 2004;36:452-456.

CORRECTING PROFOUND SODIUM IMBALANCES

Rose D. Nolen-Walston, DVM, DACVIM (LAIM)
Kennett Square, PA

INTRODUCTION

Sodium is the dominant electrolyte of the extracellular fluid compartment, and thus the principal determinant of plasma osmolality and total body water (TBW) content. Although the specific causes of sodium imbalances are legion, most dysnatremias can be attributed to an interruption in thirst or water access, antidiuretic hormone (ADH), also known as arginine vasopressin (AVP), aldosterone, or renal sodium transport.

Control of serum sodium levels

In the face of increasing osmolality, sodium levels are tightly controlled by stimulation of the hypothalamic osmoreceptors and subsequent pituitary secretion of AVP from the pituitary. Concurrently, there is an instigation of thirst and water-seeking behavior, and activation of the renin-angiotensin-aldosterone axis. AVP leads to the stimulation of the V2 receptor located the principal cells of the renal collection ducts, leading to insertion of aquaporin-2 into the luminal membrane which allows the re-absorption of free water at the terminus of the nephron. Meanwhile, the thirst response is unrelenting in the face of continued hyperosmolality, and it is rare for mentally-appropriate animals to develop hypernatremia in the face of access to free water. In contrast, hyponatremia can be caused by sodium depletion, dilution, or a combination of both, and thus assessment of overall TBW is essential to its correction. However, other than reducing thirst and the suppression of AVP release, there are no specific mechanisms that explicitly increase free water excretion.[1,2]

HYPONATREMIA

The normal serum sodium levels values differ significantly between mammalian species, and thus global definitions of an exact value signifying hyponatremia cannot be stated. However, below 135mEq/L in humans the risk for clinical hyponatremia increases, and at levels less than 110 mEq/L, becomes almost a certainty. Signs of hyponatremia are predominantly associated with cerebral edema. Changes in extracellular osmolality cause a water shift from the extracellular space to the intracellular space due to Starling's forces, and physiologic neuronal cell swelling inhibits both AVP secretion and the hypothalamic thirst center. Rapid, severe drops in tonicity, however, precipitate frank edema that is confined within the calvarium and can result in cerebral and brainstem signs that terminate with death due to tentorial herniation. A more chronic loss, as is often seen in diarrhea patients, leads to more moderate signs for a given level of hyponatremia. Presenting signs of hyponatremia include dullness, seizures, head-pressing, and other non-specific central signs.[1,3]

Hypovolemic hypotonic hyponatremia

The most common form of hyponatremia seen in large animal practice, hypovolemia hypotonic hyponatremia is usually secondary to massive GI losses associated with the infectious colidities[4,5,6] similar to that seen with cholera in humans, and results from losses of both free water and sodium salts. The pathophysiology of hypovolemic hypotonic hyponatremia involves the interplay of carotid baroreceptors, the sympathetic nervous system, the renin-angiotensin system, AVP, and renal tubular function. With decreased intravascular volume, the baroreceptors signal decreased arterial pressure which results in sympathetic activation and subsequent increased arterial tone. Reduced renal perfusion results in increased renin and angiotensin excretion that increases sodium absorption in the kidney and eventually impairs renal free water excretion which is exacerbated by increased levels of AVP. Concurrently hypothalamic receptors and angiotensin stimulate thirst, and free water ingestion aims to replete the circulation volume at the expense of sodium concentration, leading to hyponatremia.[1]

Hypertonic hyponatremia

Hypertonic hypernatremia is rare, and associated with normal total body sodium that is diluted through the presence of high concentrations of osmotically active solutes such as glucose, or mannitol such as diabetic ketoacidosis or hyperosmolar hyperglycemic syndrome.[1] In humans, glucose reduces serum sodium by 1.6 mEq/L for each 100 mg/dL of serum glucose over 100 mg/dL, increasing to 2.4 mEq/L for each 100 mg/dL increase when the glucose is greater than 400 mg/d L.[7]

Hypervolemic hypotonic hyponatremia

This manifestation is characterized associated with an increase in TBW characterized by extravasation of plasma volume resulting in edema or ascites and leading to a decrease in circulating volume and hence reduced tissue perfusion that leads to impaired renal water excretion. Commonly encountered examples include severe hypoproteinemia in large animal species an [1]congestive heart failure and liver cirrhosis in humans.[1]

Euvolemic hypotonic hyponatremia

In humans in developed countries, this is the most common form of hyponatremia in hospitalized patients, and is associated with a syndrome of inappropriate ADH secretion (SIADH) secondary to a variety of clinical conditions, including CNS disturbances, major surgery, trauma, pneumonia, infection, and some medications. In these cases, urine osmolality is generally above 400-500 mOsm/kg with normal renal function.[1]

Normotonic/pseudo-hyponatremia

Severe hyperlipidemia can lead to artifactually low measured serum sodium concentrations when flame photometry analyzers are used. It can be avoided by direct ion-selective electrode measurement.[1]

HYPERNATREMIA

Hypernatremia is almost always associated with a lack of access or ability to imbibe free water, rather than abnormalities in physiologic sodium handling, although some cases of diabetes insipidus may exceed their polydipsic drive to replace renal water losses. As plasma tonicity rises, the main clinical effect is an osmolar shift of water out of the cells causing a reduction in intracellular water and thus neuronal cell volume. The cell bodies respond within hours by electrolyte uptake which causes a partial restoration, but over the next one to two days, idiogenic osmoles including glycine, glutamine, sorbitol, and inositol build up within cell bodies, normalizing brain volume.[8] Patients with severe hypernatremia present with obtundation, lethargy, seizures, and abnormal behavior.[2,3]

CAUSES OF SODIUM IMBALANCES IN LARGE ANIMALS

Losses associated with diarrhea[4,5,6] are probably the most common causes of hyponatremia in large animal species, which may be exacerbated by renal insufficiency from the subsequent vasomotor nephropathy. In addition, long-distance endurance horses,[9] foals with uroperitoneum from bladder rupture,[10] and occasionally cases of adrenal insufficiency will present with significantly low serum sodium. Hypernatremia is more commonly seen in pigs and ruminants who have lost access to free water, been fed a diet with excessive sodium (such as mixing errors with milk replacers), or in hospitalized patients who are maintained for long periods on isotonic intravenous fluids without access to a free water source.[11] In camelids, a hyperosmotic is described that begins with hyperglycemia secondary to poor insulin response and is then compounded by glucose diuresis and additional free water loss, leading to profound hypernatremia,[12] and a case of intracranial malignancy has also been the subject of one case report in a horse with hypernatremia.[13]

PRINCIPLES OF SAFE CORRECTION

The main concern with correction of dysnatremias is causing sudden fluid shifts within the neuronal tissues of the brain. The treatment of hyponatremia is predicate on identification of the chronicity of the change and the underlying fluid status of the patient (*vide supra*). In cases where the derangement occurred within 24-48 hours replacement can be quicker, but in humans and some veterinary[14] and lab animal models, rapid correction of chronic hyponatremia is associated with serious neurologic sequelae from osmotic demyelination syndrome (central pontine myelinolysis) characterized by focal demyelination in the pons and extrapontine areas.[15] These signs do not occur immediately, as is seen with changes attributable to sudden shifts in brain volume from overzealous correction, but are manifested 1-3 days afterwards, with signs in man including irreversible, severe, and sometimes fatal central neurologic effects.[15] With hypernatremia, presence of intracellular electrolyte shifts and idiogenic osmoles result in profound brain edema and herniation if correction occurs too rapidly.[2]

The following principles[1,2,16] are generally applied in human patients and extrapolated to veterinary species; they should be followed when treating animals with presenting sodium concentrations of approximately <125 mEq/L or >155 mEq/L.

- In cases where the patient presents with neurologic signs (seizures, coma, obtundation) from dysnatremia of greater than 24-48 hours duration, the sodium level should be corrected by 4-6 mEq/L in the first 2 hours, with the rate reduced when seizure activity ceases.
- Otherwise, and subsequent to initial correction as described, sodium correction should be aimed at 0.5 mEq/L /hr (10-12 mEq/L in the first 24 hours) and not exceed 18 mEq/L in the first 48 hours.
- The Adrogue formula (Change in serum Na = [(infusate Na + infusate K) - serum Na] / [Total body water +1])[1] is used in humans with some efficacy at predicting appropriate replacement therapy. However, in veterinary species where hyponatremia is more often caused by overwhelming and ongoing GI losses, prediction of response to replacement should not rely too heavily on calculations. Sodium levels should be checked every 2 hours until the rate of correction has stabilized and every 2-4 hours there onwards.
- In this author's experience, relying heavily on calculations for correcting hyponatremia in the face of varying and unmeasured losses is often unrewarding.
  - An alternate approach is to first select an appropriate fluid rate that is commensurate with volume resuscitation and estimated fluids losses. The electrolyte concentration can then be modified using custom combinations of commercial electrolyte solutions, sterile water/D5W, hypertonic saline, or sodium bicarbonate to optimize desire tonicity and acid-base (strong ion difference) goals.
  - Early and repeated re-checking of serum sodium levels is the key to successful replacement.
- For hyponatremic patients, often a starting sodium concentration of 10-30 mEq/L above the patient's presenting serum sodium value is effective, but there is huge variation in response. Some patients will correct dramatically fast even with very judicious sodium replacement, and others will be refractory and require infusions of much higher sodium load to show any upward movement in plasma tonicity. Patients that present with hypernatremia have a tendency to over-correct with startling rapidity, especially if given access to drinking water.

- In lab animal models it has been shown that re-lowering of the serum sodium after rapid over-correction of chronic hyponatremia was beneficial if performed within 12-24 hours,[17] and that dexamethasone may be protective against neurologic sequelae when over-correction occurs accidentally.[18]

In summary, accurate and timely identification of clinically important sodium imbalances are an essential part of large animal critical care medicine, and they should be treated with meticulous attention to rate and composition of replacement fluids to avoid fatal complications. Although formulae exist that have been validated for human patients, their validity in veterinary species has not been corroborated. While the predicative potential of both the Adrogue formula and its precursors has been found to correlate overall with predicted changes in serum sodium, individual variation was extreme,[19] re-confirming the need to perform frequent and routine monitoring of the rate of correction, patient hydration status, and the interaction of other medications, infusions, and co-morbidities.

REFERENCES

[1]Adrogue HJ et al. **N Engl J Med.** 2000;342(21):1581; [2]Adrogue HJ et al. **N Engl J Med** 2000;342(20):1493; [3]Angelos SM et al. **Vet Clin North Am Food Anim Pract** 1999;15:587; [4]Johnson PJ. **Vet Clin North Am Equine Pract** 1995;11:491–514; [5]Stewart MC et al. **Aust Vet J.** 1995 Feb;72(2):41; [6]Lakritz J, **J Am Vet Med Assoc** 1992;200(8):1114; [7]Hillier T et al. **Am J Med** 1999;106(4):399; [8]Lien YH et al. **J Clin Invest** 1990;85:1427; [9]Scott HC et al. **Equine Vet J Suppl.** 2006;(36):37; [10]Dunkel B et al. **J Vet Intern Med.** 2005;19(6):889; [11]Ollivett TL et al. **J Vet Intern Med** 2013;27:592; [12]Cebra CK. **J Am Vet Med Assoc** 2000;217(11):1701; [13]Heath SE et al. **J Am Vet Med Assoc** 1995;207(6):738; [14]Churcher RK. **J Am Anim Hosp Assoc** 1999;35(6):493; [15]Murase T et al. **Am J Med** 2006;119(7 Suppl 1):S69; [16]Spasovski G et al. **Intensive Care Medicine**2014;40,(3):320; [17]Gankam Kengne F et al. **Kidney Int** 2009;76(6):614; [18]Sugimura Y et al. **Exp Neurol** 2005;192(1):178; [19]Lindner G et al. **Nephrol Dial Transplant** 2008;23(11):3501.

MULTIPLE ORGAN DYSFUNCTION SYNDROME: DOES IT EXIST IN THE HORSE?

Erin L. McConachie, DVM, DACVIM (Large Animal Internal Medicine)
Athens, GA

INTRODUCTION

Multiple organ dysfunction syndrome (MODS) is a phenomenon of modern times and reflects advancements in critical care medicine.[1] In the setting of the human intensive care unit, sepsis remains the most important predisposing factor that culminates in MODS. While sepsis is not commonly encountered in adult horses it remains one of the most common causes of mortality in neonatal foals.[2] In order to answer the question at hand, "Does MODS exist in adult horses?" we must first define MODS, review how and under what circumstances it is recognized in other species and ultimately determine what it means for our equine patients.

SYSTEMIC INLAMMATORY RESPONSE SYNDROME (SIRS) AND MODS

Terminology

SIRS is a *clinical diagnosis* based upon abnormalities in vital parameters and changes in the leukogram. In the adult horse, a consensus has not been reached as to what clinical criteria should be used, thus the criteria proposed here are based on extrapolation from human data and criteria used in equine clinical studies and are likely to change. SIRS criteria might be met when 2 or more of the following abnormalities are encountered: tachycardia > 60 beats/minute, tachypnea >30 breaths/minute, core body temperature abnormality (<98°F or >101.5°F) and/or a change in the leukogram (<4,500 WBC/uL or >12,500 WBC/uL or >10% immature band neutrophils).[3] *Sepsis* refers to clinical evidence of SIRS *and* a confirmed infection or strong index of suspicion of infection. *MODS* refers to potentially reversible dysfunction of two or more organs remote from the initial disease process (and not pre-exisiting) and typically occurs as a sequelae of a systemic disease process (SIRS or sepsis) (for example a horse with colitis that develops renal dysfunction and coagulopathy).

Etiology and Pathophysiology

SIRS is a non-specific, clinical phenomenon that when encountered simply alerts the attending clinician that there has been an insult to the body. It must be remembered that an underlying disease process is at hand when these signs are recognized and the clinician should immediately initiate an appropriate diagnostic plan. The underlying disease process may vary greatly, ranging from infection, to sterile inflammation, to trauma, to neoplasia, etc. Irrespective of the initial insult, the clinical manifestation of the immune system's response to that insult defines SIRS. The pathophysiology of SIRS is complex and dynamic. Taken in its simplest form, cells of the innate immune system (among others) interact with pathogen or host derived molecules via surface or cytosolic receptors deemed pattern recognition receptors. The interactions between these molecules (often via complex and redundant pathways) initiate a cascade of both pro-inflammatory and anti-inflammatory mediators. In healthy horses exposed to lipopolysaccharide (LPS) a classic pro-inflammatory cytokine profile (tumor necrosis factor- alpha TNF-α, IL-6 and IL-1β) is expected upon interaction with Toll like Receptor-4 (TLR-4) on innate immune cells. However, receptor polymorphisms, age, environmental and metabolic influences all may account for stark differences in physiologic responses to the same pathogen or disease process.[4-6]

Multiple organ dysfunction syndrome, as alluded to previously, is ironically the result of a job well done stabilizing the critically ill patient. Organ dysfunction is multifactorial but underlying microcirculatory dysfunction and derangements in cellular metabolic and energetic pathways are key contributors resulting in tissue hypoxia/dysoxia, cellular membrane damage and release of damage associated molecular patterns that intensify the inflammatory response and perpetuate organ dysfunction.[7] Alterations in autonomic nervous system tone and neurohormonal input can result in "organ isolation" and this disconnect between organ systems that normally rely on constant cross talk results is sequential organ dysfunction. [8] Histopathologic lesions in organ failure from MODS include varying degrees of edema, inflammatory cell infiltration, microthrombosis and fibrin deposition.[9]

Clinical recognition of SIRS, Sepsis and MODS in the horse

Acute gastrointestinal disease is likely the most common presenting complaint of critically ill adult equine patients and is the example used here to illustrate a clinical scenario. In general, horses with ischemic or inflammatory gastrointestinal lesions commonly have clinical or hematological findings that fulfill the proposed SIRS definition. In principle, a horse with a large colon volvulus provides a perfect clinical vignette. Severe cardiovascular compromise and sympathetic overdrive from pain and severe gastrointestinal distension set the stage for organ dysfunction. Local ischemic damage and edema from arterial occlusion and venous congestion respectively, cause severe disruption of the normal gastrointestinal barrier. As the lesion is surgically reduced an onslaught of inflammatory mediators, bacterial products and perhaps viable bacteria are inevitably released into the blood stream. Widespread immune system activation ensues. As of yet, the incidence of SIRS, the propensity to develop MODS, and the likelihood of mortality when this syndrome is encountered is unknown in horses. Proposed criteria to define individual organ dysfunction in adult horses are presented in Table 1 and have been adapted from recent publications.[37-39] These criteria are generally adapted from MODS scoring in human critical care in combination with relevant equine clinical case studies or research.[1,10-12,37] While laminitis can be considered a unique manifestation of organ dysfunction in the horse, this organ differs from other parenchymal organs in that there is irreversible structural damage rather than transient cellular dysfunction.

EVIDENCE FOR ORGAN DYSFUNCTION IN THE HORSE

MODS should be viewed as a physiologic loss of organ function that requires some intervention to maintain homeostasis. As there are no clear definitions proposed for equine patients it is important to consider whether a patient truly has MODS or has suffered complications associated with medical or surgical interventions, especially for research purposes. As an example, complications such as aspiration pneumonia, septic thrombophlebitis and renal dysfunction due to NSAID or aminoglycoside toxicity should not fall under the MODS umbrella. Organ dysfunction should only be considered as a manifestation of MODS if there is no other reasonable explanation for the organ dysfunction or other mechanisms of organ insult have been ruled out. Thus far in horses there are reports in the literature of individual organ dysfunction. The majority of the studies have been retrospective studies or small pilot studies most of which have a narrow focus on one individual organ system or reports of organ dysfunction as a post-operative complication. [3,11-16] A thorough discussion on all individual organ system dysfunction is beyond the scope of this session. For the remainder of this session dysfunction of the cardiovascular system will be discussed.

TABLE 1. Proposed Physical and Biochemical Criteria for MODS in the Horse

Organ system	Criteria for Dysfunction
Neurologic	Severe obtundation (stupor, semicoma, coma) or seizure
Renal	Creatinine concentration >2 mg/dL after 20 mL/kg IV crystalloid fluids, or increase $\geq$ 0.3 mg/dL since last measurement
Hemostatic	Thrombocytopenia, prolonged prothrombin or activated partial thromboplastin time, decreased fibrinogen concentration or prolonged thrombin time, increased fibrin degradation products or D dimer concentrations, decreased antithrombin III activity
Respiratory	Diffuse interstitial pattern radiographically, PaO_2/FIO_2 ratio $\leq$300 (if FIO_2 is unknown, assume a value of 0.21), hypercapnea, proteinacious fluid in the conducting airways, transtracheal aspirate or bronchoalveolar aspirate demonstrating neutrophilic inflammation[37]
Intestinal	Absent gut sounds, or absent motility on ultrasound examination
Hemodynamic	Hypotension,mean arterial pressure <65 mm Hg after 20 mL/kg IV crystalloid fluids or unexplained metabolic acidosis, lactic acidosis, reduced fractional shortening, reduced cardiac output, chamber dilation, increased serum cardiac troponin concentration
Hepatic	GGT >60 U/L; with no other explanation or bile acids concentration >25 μmol/L (BA should return to normal after surgery in colic cases[11])
Musculoskeletal	Increased serum creatine kinase activity if muscular involvement; For laminitis, hoof tester positive, pain eliminated by abaxial nerve block, radiographic evidence of laminar edema or separation of the hoof wall from the coffin bone (rotation or sinking)

CARDIOVASCULAR

In septic humans, myocardial depression is associated with an increased risk of death.[17] This seems logical as left ventricular function is central to maintaining adequate blood supply to the peripheral tissues and major parenchymal organs responsible for basic metabolic and fluid balance functions. In addition, the neurohormonal cross-talk that exists between the cardiovascular system and renal system is vital in maintaining adequate perfusion pressures (mean arterial pressure), blood flow and venous resistance. A failure of the plasticity of the cardiovascular system results in sequential failure of all major organ systems. Cardiovascular perturbations are often clinically apparent and generally manifest as tachycardia with an increased incidence of ectopic beats; but these findings are not well understood and are often interpreted as evidence of pain and endotoxemia. In order to understand the meaning of persistent tachycardia and ectopic foci in critically ill equine patients a more in depth and multifaceted approach is necessary to understand what the heart is doing in relation to the rest of the body.

Left ventricular function

With the introduction of echocardiography into the clinical setting of ICU patients, recent studies have demonstrated several consistent abnormalities in the critically ill patient including ventricular hypokinesia, reduction in left ventricular ejection fraction, and an increase in left ventricular end diastolic volume.[18] Echocardiographic evidence of left ventricular dysfunction has been correlated with increased serum troponin concentrations and death in people.[19,20] Non-invasive, stall side echocardiographic methods of cardiac output measurement are possible in both the adult horse and neonatal foal.[21,22] The noninvasive nature of echocardiography makes it an attractive diagnostic tool for assessment of cardiac dysfunction in critically ill patients and in the hands of trained personnel provides additional information regarding other structural and functional abnormalities.[23,24]

Heart rate variability (HRV)

HRV describes the short-term variations between consecutive heart beats and is a valuable tool to characterize the influences of the sympathetic and parasympathetic nervous systems on the sinoatrial node.[25] In general, decreases in HRV variables reflect a shift toward sympathetic dominance. There are numerous HRV parameters that can be easily determined using specifically designed software that integrates and interprets digitally stored telemetric electrocardiography data. Several HRV studies have been performed in the horse under various conditions including transport, laminitis, and pregnancy and have demonstrated correlation of HRV to other measures of stress and reduced vagal tone, such as serum cortisol concentration.[26-29] With the ease of obtaining HRV data and its excellent correlation to measures of cardiovascular health, MODS, and outcome in critically ill people in which acute coronary disease is not the primary disease process,[26,30-33] it appears to be an ideal parameter to monitor in critically ill patients. In horses with acute intestinal disease requiring celiotomy, a reduction of the standard deviation of normal to normal intervals (SDNN) was correlated with an increased serum cardiac troponin concentration as well as an increased risk of death when a cut off of < 27.6 msec was set (McConachie unpublished data).

Cardiac Biomarker

Cardiac troponin I (cTnI) is a regulatory protein of the thin actin filaments of cardiac muscle that is released into the circulation as a result of acute myocardial cell injury.[18] It is a highly sensitive and specific biomarker of myocardial health. Several studies have demonstrated increased serum troponin levels in critically ill patients.[18] Furthermore, increased troponin concentrations have been correlated with cardiac dysfunction and mortality in people.[19,20,34] Despite its excellent sensitivity and specificity to the heart, its utility in people is somewhat hindered by the high incidence of acute coronary syndrome which also causes a dramatic increase in serum troponin. Coronary heart disease is exceptionally rare in the horse, perhaps making it an ideal biomarker of cardiac damage in critical illness. Indeed, there have been multiple prospective studies in critically ill septic foals and adult horses with strangulating intestinal lesions, all demonstrating significant increases in serum cardiac troponin concentrations relative to healthy individuals.[13,14,35] Further study on the correlation of cTnI to measures of cardiovascular function and outcome is needed in the horse. Preliminary results of a large prospective clinical research project at the University of Georgia suggest that a troponin concentration of > 0.22 ng/mL within the first 48 hours of hospitalization in horses with acute intestinal disease that undergo exploratory laparotomy are 37 times more likely to not survive.

DIAGNOSIS AND CREATING A SCORING SYSTEM

The diagnosis of MODS is based on identifying an underlying disease process and documenting organ dysfunction based on physical or biochemical evidence of impairment requiring a specific intervention. Creation of a MODS scoring system might prove useful in prognostication and detection of dysfunction. However evolution of such a scoring system can't occur in the absence of large prospective clinical studies to help establish organ dysfunction criteria.

PROGNOSIS

When MODS develops for any reason in human patients the prognosis worsens significantly. In people, when four organs are involved, the mortality reaches 100%.[36] The same is likely true in equine patients, however, more work is needed to make it possible to report the incidence of secondary organ dysfunction and to determine whether the same increased risk of mortality exists.

REFERENCES

1. Mizock BA. **Dis Mon** 2009;55:476. 2. Corley KT et al. **Equine Vet J** 2005;37:53. 3. Epstein KLet al. **J Vet Intern Med** 2011;25:307. 4. Werners AH et al. **Vet Immunol Immunopathol** 2006;111:165. 5. Tadros EM et al. **Am J Vet Res** 2013;74:1010. 6. McFarlane D et al. **J Vet Intern Med** 2008;22:436. 7. De Backer D et al. **Virulence** 2013;5. 8. Werdan K et al. **Can J Physiol Pharmacol** 2009;87:266. 9. Cotovio M et al.. **J Vet Intern Med** 2007;21:308. 10. Khwannimit B.. **J Med Assoc Thai** 2008;91:1336. 11. Underwood C et al. **J Vet Emerg Crit Care** 2010;20:578. 12. Groover ES et al. **J Am Vet Med Assoc** 2006;228:572. 13. Radcliffe RM et al. **J Vet Emerg Crit Care** 2012;22:313. 14. Hallowell G et al. **J Vet Intern Med** 2007;21:604. 15. Davis Wet al. **Equine Vet J** 2013;45:224. 16. Mair TS et al. **Equine Vet J** 2005;37:310. 17. Annane D et al. **Lancet** 2005;365:63. 18. Hunter J et al. **Brit J Anaesthesia** 2010;104:3. 19. Ammann P et al. **J Am College of Card** 2003;41:2004. 20. Fenton Ket al. **Pediart Crit Care Med** 2004;5:533. 21. McConachie E et al. **J Vet Intern Med** 2013;27:324. 22. Valverde A et al. **Am J Vet Res** 2007;68:141. 23. Schwarzwald CC et al. **J Vet Intern Med** 2009;23:643. 24. Underwood C et al. **J Vet Intern Med** 2011;25:563. 25. Thayer JF et al. **Biomedical Sciences Instrumentation** 1997;33:482. 26. Pontet J et al. **J Crit Care 2003**;18:156. 27. Nagel C et al. **Theriogenology** 2012;78:759. 28. Schmidt A et al. **Domest Anim Endocrinol** 2010;39:205. 29. Rietmann TR et al. **J Vet Med A Physiol Pathol Clin Med** 2004;51:218. 30. Xu L et al. **Chinese Critical Care Medicine** 2007;19:160. 31. Kovatchev BP et al. **Pediatric Research** 2003;54:892. 32. Schmidt H et al. **Biomedizinische Technik Biomedical Engineering** 2006;51:264. 33. Schmidt H et al. **Crit Care Med** 2008;36:967. 34. Mehta N et al. **Int J Cardio** 2004;95:13. 35. Slack J et al. **J Vet Inter Med** 2005;19:577. 36. Deitch et al. **Surgical Clin North Amer** 1999;79:471-488. 37 Wilkins, PA. et al. **J Vet Emerg Criti Care** 2007;17:333. 38. Barton MH. et al. **Equine Emergency and Critical Care Medicine**. Wilkins Ed. In Press 2014. 39. Hart KA, et. al. Endotoxemia, **Large Animal Internal Medicine** Smith BP. ed, 2013

NOVEL BIOMARKERS IN EQUINE SEPSIS

Sandra D. Taylor, DVM, PhD, DACVIM (LAIM)
West Lafayette, IN

INTRODUCTION

Bacterial sepsis, defined as the systemic inflammatory response to bacterial infection, is a common cause of morbidity and mortality in neonatal foals and adult horses.[1-3] Early and accurate diagnosis of sepsis is critical to provide optimal care to critically ill patients, because delay in treatment is an independent risk factor for non-survival.[4] Diagnosis of sepsis relies on identification of the systemic inflammatory response syndrome (SIRS) with evidence of infection. Definitive diagnosis of bacterial sepsis is obtained through microbial culture; unfortunately, culture and antimicrobial susceptibility results are usually not available until more than 48 hours after admission and consequently, empirical treatment is often instituted prior to diagnosis. Early antimicrobial treatment, although warranted in suspect cases, may impede the sensitivity of culture as a diagnostic test for sepsis. In addition, false negative blood culture results may be due to low numbers of circulating bacteria or a relatively low volume of blood for culture, and varying blood culture systems may yield inconsistent results.[5] A weighted sepsis scoring system is currently used as a diagnostic tool for sepsis in neonatal foals, but demonstrates a relatively low sensitivity and specificity.[6,7] Although clinicopathological findings such as leukopenia, neutropenia with a degenerative left shift, and neutrophil toxicity have been associated with sepsis in horses, lack of test specificity limits their usefulness in contributing to a diagnosis of sepsis.[8] However, with regard to predicting survival, white blood cell (WBC) counts > 6,000 cells/µl and neutrophil counts > 4,000 cells/µl are significantly associated with increased odds of survival in septic foals.[4] Elevated plasma fibrinogen has been associated with *in utero* infection and is associated with non-survival, but is also a non-specific indicator of inflammation.[2] Increased blood L-lactate concentration and persistence of hyperlactatemia are associated with non-survival in horses with systemic illness, but a clear association with bacteremia is lacking.[9] Evaluation of blood L-lactate concentration at a single time point is a poor predictor of sepsis.[10] Cytokine profiles in septic horses have varied widely among studies, and are not currently utilized for diagnosis or prognostication of equine sepsis.[11-14] Thus, a simple, rapid and reliable test that provides an early diagnosis of sepsis and predicts survival has the potential to greatly reduce morbidity, mortality and treatment costs in affected horses.

In recent years, biological laboratory markers (biomarkers) have become important in diagnosing sepsis, guiding antimicrobial therapy, and assessing response to therapy in human medicine.[15] Over 170 different compounds have been tested as potential biomarkers in human sepsis, and biomarker development is an ongoing area of research within the global sepsis field.[16] Although cytokines, blood L-lactate concentration, plasma fibrinogen concentration and coagulation markers may be considered "biomarkers" in equine medicine, only acute phase proteins and pro-hormone/hormones will be discussed here. Proteins that have been recently investigated for their potential use as biomarkers for sepsis include include serum amyloid A (SAA), C-reactive protein (CRP), haptoglobin (Hp), soluble CD14 (sCD14), adrenomedullin (AM), procalcitonin (PCT) and activin A (AA).

BIOMARKERS

Serum amyloid A

Serum amyloid A is a major acute phase reactant in horses, with hepatic production rapidly increasing in response to inflammation.[17] Concentrations of SAA have been shown to be elevated in horses diagnosed with bacterial pneumonia, influenza, and septic arthritis.[18-20] There is evidence that infections are particularly effective in inducing a strong SAA response, suggesting that SAA concentration may be useful in distinguishing infectious from non-infectious disease.[21-23] In clinically abnormal horses, a 20.5-fold increase in median SAA was reported, compared to a relatively low (< 2-fold) increase in median WBC count and median plasma fibrinogen concentration.[24] In addition, the short half-life of SAA facilitates more accurate monitoring of response to treatment using serial measurements of SAA. Elevated concentrations of SAA have been documented in adult mares with experimentally-induced endometritis, with serum concentrations correlating well with endometrial SAA expression.[25,26] In equine grass sickness, which is often characterized by clinical and clinicopathological signs of systemic inflammation, SAA concentrations may be elevated; thus, SAA may be useful in differentiating inflammatory from non-inflammatory abdominal diseases.[27] Finally, SAA has been advocated as a useful biomarker in the diagnosis of equine neonatal sepsis, but peri-parturient events may influence SAA concentrations.[17,28-30] Initial studies in neonatal foals suggest that quantification of SAA can differentiate infectious from non-infectious illness.[17,28] This is consistent with findings in human pediatrics and supports its candidacy as a diagnostic tool for foal sepsis; however, studies with increased numbers of foals and serial measurements of SAA are needed.[31]

C-reactive protein

C-reactive protein is a well-established biomarker of infection and inflammation in human sepsis.[32,33] During the acute phase response to microbial infection, serum CRP concentration increases within hours and can reach 1,000 fold levels.[32] Levels remain elevated during persistent infection and then decrease rapidly, with a half-life of 19 hours.[34] Despite this relatively rapid increase in serum CRP concentration in response to sepsis, the sensitivity of serum CRP concentration is known to be the lowest during the early stages of infection. [35] In a study comparing serum CRP concentrations in infants at admission and 24 hours later, serum CRP concentration was a better predictor of sepsis at the later time point with a sensitivity of 78% and a specificity of 94%.[36] In humans, serum CRP concentration may increase with many non-infectious inflammatory conditions, including rheumatoid arthritis, acute myocardial infarction and pancreatitis.[37] However, serum CRP concentrations have not been shown to increase in adult horses with recurrent airway obstruction or exercise-induced inflammation.[38,39] A recent study evaluating plasma CRP concentrations in 40 septic, 40 sick non-septic, and 39 healthy control foals failed to demonstrate an increase in any group (Zabrecky et al, unpublished data). The accuracy of a single measurement for diagnosis of sepsis varies widely, with the ideal time point for measurement and peak levels in horses unknown.[35,36]

Haptoglobin

Haptoglobin is a protein that is synthesized in the liver, and functions to bind free hemoglobin following intravascular hemolysis. Serum Hp concentrations have been shown to decrease following hemolysis and increase following acute inflammation in several species, including horses.[40,41] In a human study, bacteremic newborns had higher serum Hp concentrations compared to newborns without infection, suggesting that serum Hp concentration may have clinical utility in the diagnosis of neonatal sepsis.[42] In horses, serum Hp concentrations have been shown to increase after elective and non-elective surgery, and in horses affected with recurrent airway obstruction.[38,43] A recent study investigating the acute phase response in critically ill neonatal foals found that 40 septic foals exhibited significantly lower plasma Hp concentrations than 37 healthy foals, but there was no evidence of increased hemolysis in the septic group (Zabrecky et al, unpublished data). Thus, more studies are warranted to investigate the role of Hp in sepsis.

Soluble CD14

Membrane-bound CD14 is found on the surface of monocytes and macrophages. Binding of CD14 to the LPS/LPS-binding protein complex initiates activation of TLR4, which triggers transcription of pro-inflammatory cytokines. A soluble form of CD14 (sCD14) can be found in serum, and is thought to be a product of membrane-bound CD14 that has been shed from cell surfaces.[44,45] Soluble CD14 can neutralize circulating LPS and transfer LPS to plasma lipoproteins, thereby serving as an anti-inflammatory molecule.[46] Elevated serum concentrations of sCD14 have been reported in humans with acute inflammation and sepsis, and are likely a reflection of LPS-induced cell activation.[47] In a recent study, sCD14 was found to be significantly elevated in serum of 15 septic foals compared to 15 healthy foals.[45] Results of a second study demonstrated a significant increase in sCD14 in 55 adult horses with diseases commonly associated with endotoxemia compared to 23 healthy controls; however, sCD14 concentrations did not correlate with outcome.[48] Further studies with larger sample sizes are needed to determine the usefulness of this marker as an indicator of sepsis in clinical practice.

Adrenomedullin

Adrenomedullin (AM) is a circulating hormone that is synthesized by endothelial cells, vascular smooth muscle cells and the adrenal medulla.[49,50] This polypeptide facilitates tissue perfusion via potent vasodilatory and inotropic properties,[51,52] and has been shown to increase rapidly following the onset of sepsis in humans and rodents.[53,54] AM is thought to contribute to hyperdynamic circulation which is characterized by decreased systemic vascular resistance, decreased mean arterial pressure and compensatory increases in cardiac output and stroke volume.[53,54] Infusion of AM to healthy human volunteers resulted in similar hemodynamic changes, and administration of anti-AM antibodies to rodents prevented hypotension in a sepsis model.[53,55] In humans, plasma AM levels increase with various clinical diseases, with up to 50-fold increases observed in septic patients.[51,56,57] Circulating AM also exhibits anti-inflammatory and bactericidal properties.[56,58] Although there is little known about AM in horses, a recent study investigated plasma AM concentration in three groups of neonatal foals: septic, sick non-septic and healthy controls (Toth et al, *J Vet Intern Med*, in press). Ninety critically ill foals (42 septic, 48 sick non-septic) and 61 healthy foals were included in the study. There was a 6-fold increase in the median plasma AM concentration in critically ill foals compared to healthy foals ($P < 0.0001$), but AM levels were not associated with sepsis ($P = 0.71$) or survival ($P = 0.051$). A plasma AM concentration of 0.041 ng/ml provided a test sensitivity of 91% and specificity of 54% to predict critical illness, suggesting that plasma AM concentration may serve as a marker of health in neonatal foals. Currently, the ELISA testing kit used to measure equine AM concentration is cost-prohibitive for a single patient. Further studies are needed to evaluate age-related changes in plasma AM concentration in healthy and sick foals, as well as the prognostic value of serial measurements in critically ill foals. Adrenomedullin has not been evaluated in adult horses.

Procalcitonin

In human medicine, serum PCT concentrations are consistently elevated in patients with sepsis, and are particularly useful in distinguishing bacterial sepsis from other forms of infection.[59] As with CRP, increased PCT concentrations can be associated with non-infectious inflammatory diseases, but human studies have suggested that PCT is a more reliable marker of sepsis that CRP.[60] Thus far, PCT concentrations in veterinary species have not been useful in the diagnosis of sepsis. A study investigating the expression of molecular markers in blood of neonatal foals with sepsis found no difference in PCT between 21 healthy foals and 49 sick foals.[14] Further studies are needed to determine if PCT may be a useful biomarker in equine medicine.

Activin A

Activin A is a member of the transforming growth factor (TGF)-β family of proteins and is produced by endothelial cells in response to inflammation. Elevated serum AA levels have been shown to be associated with human sepsis and inflammatory bowel disease, and may be predictive of outcome.[61,62] Few studies have investigated AA in horses. Serum AA was found to be increased in horses with inflammatory or strangulating intestinal lesions compared to healthy horses,[63] while a second study found serum AA to be elevated in horses with equine grass sickness.[27]

SUMMARY

Although several acute phase proteins and pro-hormones/hormones have been identified as potential sepsis biomarkers in horses, there are few studies that have measured protein concentrations in septic populations. One reason for this is that the inclusion criteria for sepsis is difficult to define, as currently available diagnostic tests for sepsis are insensitive and criteria for SIRS is non-specific. Thus, most available biomarker data comes from populations of horses with the less vigorous diagnosis of "inflammation." While useful, this limits our ability to draw conclusions regarding the use of biomarkers specifically for the diagnostic and prognostic value in cases of equine sepsis. Although SAA appears promising as a reliable and accurate marker for infectious disease in horses, further studies are warranted to investigate currently studied biomarkers and new, potential markers for equine sepsis.

REFERENCES

1. Cohen ND. **J Am Vet Med Assoc.** 1997;204:1644. 2. Koterba AM, et al., **Equine Vet J.** 1984;16:376. 3. Roy MF. **Vet Clin North Am Equine Pract.** 2004;20:41. 4. Gayle JM, et al., **J Vet Intern Med.** 1998;12:140. 5. Wilson WD, et al., **J Am Vet Med Assoc.** 1989;195:1759. 6. Brewer BD, et al., **Equine Vet J.** 1988;20:18. 7. Corley KTF. **J of Vet Emerg Crit Care.** 2003;13:149. 8. Hurcombe SD, et al., **J Vet Intern Med.** 2008;22:639. 9. Castagnetti C, et al., **Theriogenology.** 2010;73:343. 10. Borchers A, et al., **Equine Vet J Suppl.** 2012;44 Suppl 41:57. 11. Castagnetti C, et al., **Am J Vet Res.** 2012;73:1418. 12. Gold JR, et al., **J Vet Intern Med.** 2007;21:482. 13. Burton AB, et al., **Vet Immunol Immunopathol.** 2009;132:122. 14. Pusterla N, et al., **Am J Vet Res.** 2006;67:1045. 15. Mussap M, et al., **Semin Fetal Neonatal Med.** 2013;18:56. 16. Pierrakos C, et al., **Crit Care.** 2010;14:R15. 17. Chavatte PM, et al., **Proc Sixth Int Conf, Eq Inf Dis.** 1992;6:33. 18. Hulten C, et al., **Acta Vet Scand.** 1999;40:323. 19. Jacobsen S, et al., **Am J Vet Res.** 2006;67:1738. 20. Hulten C, et al., **Equine Vet J.** 2002;34:693. 21. Pepys MB, et al., **Equine Vet J.** 1989;21:106. 22. Kjelgaard-Hansen M, et al., **Clin Lab Med.** 2011;31:51. 23. Vandenplas ML, et al., **Am J Vet Res.** 2005;66:1509. 24. Belgrave RL, et al., **J Am Vet Med Assoc.** 2013;243:113. 25. Christoffersen M, et al., **Vet Immunol Immunopathol.** 2010;138:95. 26. Coutinho da Silva MA, et al., **Equine Vet J.** 2013;45:619. 27. Copas VE, et al., **Vet Rec.** 2013;172:395. 28. Stoneham SJ, et al., **Equine Vet J.** 2001;33:599. 29. Duggan V. **Vet J.** 2008;176:267. 30. Paltrinieri S, et al., **Vet J.** 2008;176:393. 31. Edgar JD, et al., **BMC Pediatr.** 2010;10:22. 32. Hofer N, et al., **Neonatology.** 2012;102:25. 33. Keshet R, et al., **Am J Med Sci.** 2009;337:248. 34. Vigushin DM, et al., **J Clin Invest.** 1993;91:1351. 35. Pourcyrous M, et al., **Pediatrics.** 1993;92:431. 36. Laborada G, et al., **Am J Perinatol.** 2003;20:491. 37. Vincent JL, et al., **Crit Care Clin.** 2011;27:241. 38. Lavoie-Lamoureux A, et al., **Vet Intern Med.** 2012;26:1419. 39. Cywinska A, et al., **Res Vet Sci.** 2012;93:1402. 40. Eurell TE, et al., **Can J Vet Res.** 1993;57:42. 41. Hulten C, et al., **Equine Vet J.** 2002;34:699. 42. Chavez-Bueno S, et al., **J Perinatol.** 2011;31:500. 43. Pollock PJ, et al., **Vet Rec.** 2005;156:538. 44. Bazil V, et al., **Eur J Immunol.** 1986;16:1583. 45. Wagner B, et al., **Vet Immunol Immunopathol.** 2013;155:124. 46. Kitchens RL, et al., **J Endotoxin Res.** 2005;11:225. 47. Kitchens RL, et al., ***J Clin Invest.*** 2001;108:485. 48. Silva A, et al., **Vet Immunol Immunopathol.** 2013;155:264. 49. Kitamura K, et al., **Biochem Biophys Res Commun.** 1993;192:553. 50. Hinson JP, et al., **Endocr Rev.** 2000;21:138. 51. Gibbons C, et al., **Mol Endocrinol.** 2007;21:783. 52. Nishikimi T, et al., **Hypertens Res.** 2003;26 Suppl:S121. 53. Wang P, et al., **Arch Surg.** 1998;133:1298. 54. Koo DJ, et al., **J Surg Res.** 2001;95:207. 55. Lainchbury JG, et al., **J Clin Endocrinol Metab.** 2000;85:1016. 56. Ueda S, et al., **Am J Respir Crit Care Med.** 1999;160:132. 57. Oncel MY, et al., **Pediatric Research.** 2012;72:507. 58. Isumi Y, et al., **FEBS Lett.** 1999;463:110. 59. Aikawa N, et al., **J Infect Chemother.** 2005;11:152. 60. Luzzani A, et al., **Crit Care Med.** 2003;31:1737. 61. Michel U, et al., **Eur J Endocrinol.** 2003;148:559. 62. Hubner G, et al., **Lab Invest.** 1997;77:311. 63. Forbes G, et al., **Vet J.** 2011;190:e154.

ARTICULAR PROCESS JOINT DISEASE: A CAUSE OF CERVICAL VERTEBRAL COMPRESSIVE MYELOPATHY?

Richard Hepburn, BVSc, MS, CertEM(IntMed), DACVIM(LAIM), MRCVS
Gloucestershire, UK

INTRODUCTION

Cervical spinal cord compression has mainly been thought to refer to intravertebral canal narrowing due to vertebral malformation, however this assumption underestimates the relevance of disease within the vertebral articular process joints (APJs) and the contiguous ligamentaum flavum (LF)[1], which together are in close proximity to the dorsolateral white matter of the spinal cord[2–5]. Osteoarthritis (OA) and osteochondrosis (OC) share many pathophysiologic similarities and commonly occur within the APJs. There is increasing pathological, diagnostic imaging and clinical evidence of the importance of dorsolateral compression not just in combination with vertebral canal malformation, but as a separate entity in both young and old horses.

ANATOMY OF THE ARTICULAR PROCESS JOINTS

The paired zygapophyseal synovial articular process joints, known more colloquially as facet joints[6], are located dorsolaterally on each vertebra, dorsal to the intervertebral foramen. Within each articulation the caudal articular process of the cranial vertebra sits axially to the cranial articular process of the caudal vertebra[7]. The intervertebral foramen represents the space between the articular process of the cranial vertebra and the vertebral body of the caudal vertebra. Multiple outpouches extend from the APJ margin: lateral outpouch (from the lateral gap between cranial and caudal articular processes, it extends laterally and dorsolaterally); medial outpouch (positioned dorsolaterally within the vertebral canal, it extends ventromedially from the medial gap between articular processes); cranial outpouch (wedge shaped, positioned along the lateral curve of the cranial vertebra); and caudal outpouch (extends caudally from the tip of the cranial articular process along the dorsal surface of the body of the caudal vertebra). From C2 to C7 the articular surfaces are located progressively more laterally; their angle relative to the ventral floor of the vertebral canal becomes gradually steeper; the lateral separation of the synovial surfaces progressively widens; a significant increase in cervical APJ volume occurs as does an increase in intervertebral foramina length and height[8,10-12]. The result is an increasingly laterally positioned, upright, medially oriented and wider joint space in the caudal articulations. Individual right and left APJs have variable joint margins, but are significantly related in shape, volume and medial outpouch into the vertebral canal[8]. The synovial surfaces of the cervical APJs are almost flat and covered with articular cartilage[11]. Variously sized intra-articular synovial folds, which contain synovial membrane, synoviocytes and adipose, are commonly found in the equine cervical spine[13], and in man have been described throughout the vertebral column[14]. They are though to function as passive space fillers, ensuring the articular cartilage is covered with synovial fluid during all ranges of motion. Their adipose tissue is innervated and may be important in proprioceptive feedback[15].

PATHOLOGICAL CHANGES WITHIN CERVICAL APJs

As with any other diathrodial joint, APJ disease can be broadly split into developmental (osteochondrosis, OC) and degenerative (osteoarthritis, OA) etiologies. Interestingly it has been observed that many of the general morphological characteristics of OA closely resemble the deranged endochondral differentiation processes seen in OC, leading to similar clinical outcomes[2]. Post mortem studies in horses of varying ages with spinal ataxia have shown dorsolateral spinal cord compression associated with APJ/LF pathology[3–5,17]. The contiguous nature of the APJ and LF is also seen pathologically, with degeneration occurring simultaneously in both structures as part of both OA and OC. Within OA moderate fraying and fibrillation of articular cartilage, villous proliferation of the synovial membrane, joint capsule hypertrophy, and production of new bone at the articular margins occurs[3,4]. In younger horses (<6 years old), OC lesions are more varied with profound malformation and instability of the APJs, OC dissecans lesions, subchondral bone cysts and extra-dural synovial cysts described. The LF becomes increasingly rigid due to hypertrophy, fibrovascular proliferation and thickening of the confluent dorsomedial aspect of the joint capsule. Hemorrhage within the LF, and necrosis of marginal fat, enlargement and osteosclerosis of the fibrocartilagenous attachment of the LF to vertebral dorsal lamina have all been reported. Extradural synovial cysts can cause spinal compression either alone, or as part of the APJ/LF process[3,4,19]. In man these are termed degenerative intraspinal cysts and arise from within degenerated APJs but also from the ligamentum flavum, leading to myelopathy or compression of spinal outflow[20]. The most likely etiology for the degenerative changes within the APJ/LF is cyclical articulation stress placed upon the vertebrae, particularly repeated loading strain (flexion-extension, and torsion)[21]. This is a logical part of OA in older horses; however repeated strain may also be a component of OC, causing production of excessive fibrocartilage that after endochondral ossification resembles osteophytosis[2]. Retention of cartilage matrix at the insertion site of the LF has been shown to occur in young horses with spinal ataxia, but not in controls[3]. Paracrine release of growth factors and pro-inflammatory cytokines from adipose within the synovial folds of the APJs may contribute to development of OA[14].

POSITION AND MOVEMENT OF THE CERVICAL APJs

Between C3-C7 the greatest lateral movement occurs at C6, the greatest range of flexion-extension occurs at C3-C4 and the least at C5-C6, which could make C5-C6 a "low motion, high pressure" joint, predisposing it to osteoarthritic change[10,15]. The general shape of the cervical APJs in vitro is not affected by neck flexion, however neck extension (raising) results in a significant reduction in intervertebral foramina dimensions from C4-T1[8,12]. Vertebral canal dimensions are also affected, with the size reduction occurring dorsolaterally and ventrolaterally, rather than dorsoventrally, reflecting the non-circular nature of the vertebral canal and APJs[16]. *Ex vivo* the typical head-neck position used in sport horses (neck raised, bridge of the nose around the vertical) leads to steepening of the intervertebral canal angle when compared to a neutral position: C5-C6 (66% increase), C6-C7 (36% increase)[30].

CLINICAL PRESENTATION OF CERVICAL APJ DISEASE

It is important to appreciate that whilst horses with a developmentally narrowed vertebral canal would be predisposed to compression from APJ/LF pathology; compression can occur *without* evidence of vertebral canal malformation[1,4], *i.e.* it is *possible* for spinal compression in younger horses to be solely due to APJ/LF pathology, leading to a clinical presentation similar to the compressive myelopathy seen in older horses with APJ/LF OA[5]. In this author's experience this appreciation is vital to interpretation of orthogonal cervical radiographs, as alteration in mean saggital ratios (MSR) will not occur with APJ/LF compression *alone*. How sub-clinical narrowing of the vertebral canal interacting with low grade APJ/LF enlargement to produce spinal cord compression affects the age of presentation, and also treatment choice warrants further investigation. The definition of two broad types of cervical vertebral compressive myelopathy (CVCM) can make appreciation of the relative significance of APJ/LF pathology difficult[22,23]. Type 1 CVCM describes developmental disease where caudal extension of the vertebral arch and caudal epiphysis produce a dorsoventrally narrowed vertebral canal, (together termed vertebral malformation) and are the primary compressing lesions. Secondary soft tissue compression from OC changes within the APJ/LF can be present. Type 2 describes a similar clinical presentation in older horses (8-25 years old at onset) due to APJ/LF OA, this has also be termed caudal cervical arthropathy[24,25]. Exactly where younger horses with compression solely due to APJ/LF OC fit into this classification, is unclear: previously young horses (<4 years old) with APJ/LF changes described as OA have been included in type 2[14]. Working in an area free of equine protozoal myeloencephalopathy (EPM) this author appreciates spinal cord compression due to APJ/LF disease in three scenarios, neurologically their presentation is very similar: young horses with APJ/LF disease and concurrent vertebral canal malformation; young horses with APJ/LF disease as the sole cause of spinal compression; and older horses with APJ/LF disease as the sole cause of spinal compression. Any age delineator for these scenarios is likely to be arbitrary given the individual variation in the timing of vertebral physeal closure, however it is this author's experience that young horses with APJ/LF compression alone are typically older than those with APJ/LF pathology in addition to vertebral canal malformation. In all cases neurological examination describes deficits that neuroanatomically localize to the cervical spine. A multicentre study of type 2 CVCM found a median thoracic limb ataxia grade of 2/5, and a pelvic limb grade of 3/5[26]. Over 40% cases were described as neurologically asymmetric, however this was not associated with any specific pathologic asymmetry. Cervical hyperesthesia was associated with the presence of APJ osteophytosis, but not with any other type of gross CVCM lesion. Cervical hyperesthesia is seen as an exaggerated skin prick response over the neck base and shoulder when compared to the trunk and rump; and is variably present in CVCM cases[26]. Hyperesthesia is significantly associated with APJ osteophytosis, but also is found more commonly in horses with acute onset neurological signs[24]. The median duration of clinical signs prior to diagnosis was 28 days (range 1-730). This author's experience of APJ/LF cases would reflect this, with a typical presentation being grade 2-3 thoracic and pelvic limb ataxia and dysmetria; and grade 2-3 pelvic limb upper motor neuron paresis. Clinically these deficits are seen as forelimb hypermetria and hindlimb spasticity that is exaggerated during backing up and hill walking; forelimb pivoting and hindlimb circumduction during tight circling; and weakness and uncontrolled distal limb pronation during tail pulling whilst walking. Gait deficits can be symmetric or asymmetric. Muscle atrophy is typically minimal and when present localised to the dorsal midline and rump. Horses may present with a history of being neurologically normal in the preceding 12 months or they may have a more prolonged, but non-progressive history of neurologic dysfunction including tripping and stumbling, particularly downhill, and altered performance (unbalanced, refusing to jump, weak behind, "clumsy")[24,27]. Some cases report a single bout of shoulder/neck trauma that may have occurred previously. Thorough orthopaedic examination is typically unremarkable, although concurrent mild lameness can occur. In this author's experience the differences between young horses with sole APJ/LF pathology from those with concurrent vertebral canal malformation are: sole APJ/LF disease tends to occur in older animals (3-6 years); it causes neurological gait deficits that are less severe and more similar in grade and presentation to those seen in older type 2 horses; muscle atrophy is less obvious; and the onset of disease is more insidious and less progressive.

DIAGNOSIS OF APJ/LF SPINAL CORD COMPRESSION

When considering the possibility of cervical spinal cord compression *solely* due to APJ/LF pathology it is important to apply consistent decision criteria and critically analyze all results. There should be: suitable neurological deficits that localise to the cervical spinal cord, with or without signs of spinal segment nerve compression; significant changes in the appearance of individual APJs on lateral-lateral plain radiographs; variable APJ symmetry on oblique lateral plain radiography; and an absence of vertebral canal malformation shown by normal mean saggital ratios and an absence of significant dorsal laminar extension/caudal epiphyseal body flair, or if these are abnormal then an absence of dorsoventral myelographic compression at that site. In regions where EPM is prevalent, an appropriate negative test result could be added to these criteria.

CERVICAL RADIOGRAPHY

Cervical radiography is performed with the horse standing and standard techniques for lateral-lateral and lateral-oblique projections have been described[7,27]. Careful positioning of the neck is important to minimize rotation, which will artefactually reduce intervertebral foramina size and increase APJ size. Lateral-lateral images showing rotation should be repeated, and a complete set of images from the occiput to T1 obtained. The majority of the data assessing the utility of plain radiography and its derived measurements in CVCM has focussed on dorsoventral compression[23], whereas APJ/LF compression occurs in a dorsolateral plane. In order to exclude the possibility of vertebral canal malformation as far as possible, this author calculates intravertbral mean saggital ratios at all articulations, using a normal cut off value of >0.52 for C3-C4 to C5-C6, and >0.56 for C6-C7, although the individual inaccuracy of this technique must be acknowledged[23]. This author recommends a greater focus on more critical and detailed description of the change in radiographic appearance of the cervical APJs by applying a suitable grading scheme[25]. Arthropathy and irregular radiographic enlargement of the margins of the C6-C7 APJs has been anecdotally described as a normal variation in mature horses[9], but an age effect on worsening APJ grade has only been shown at C5-C6[25]. IN this study no significant association between the grade of radiographic APJ appearance and presence/severity of ataxia has been reported, however the numbers of horses small. Preliminary analysis of a larger population of horses with spinal ataxia that fulfill the authors decision criteria for sole APJ/LF compression, in a region free of EPM, has shown an increased proportion of horses to have grades 3b and 4b at C5-C6 and C6-C7, and grade 4b at C6-C7 (Hepburn R, Butler J unpublished data), suggesting that ventral osteophytosis *could* be a more useful radiographic predictor of neurologically significant APJ/LF disease. In this population C6-C7 was the most commonly implicated articulation. Oblique lateral views of the caudal cervical articulations (C5-T1) should be routinely taken[7], and are particularly important in cases where there is asymmetry of the neurological deficits, or where possible spinal outflow compression is suspected. The primary beam is positioned tangential to the joint space of the APJ furthest from the x-ray machine, allowing assessment of its width and regularity; the axial margin of the closer APJ and the intervertebral foramen can also be identified[7]. Marked facet asymmetry can be found on oblique lateral projection that was not apparent on the lateral-lateral view. In this author's experience marked asymmetry has a negative effect on prognosis.

CERVICAL MYELOGRAPHY

Cervical myelography under general anesthesia can be used to better define the relative significance of APJ/LF pathology and exclude as best as possible vertebral canal malformation. For instance if an individual articulation has significant APJ/LF changes, and a separate site has an abnormal MSR or signs of malarticulation, myelography can assign significance and guide medical *as well as* surgical treatment. APJ/LF disease most commonly affects the caudal cervical articulations, where myelography is more sensitive and specific on neutral views rather than flexed[29]. Specific decision criteria should be used to identify compression during extension and flexion at individual articulations[29]. In addition the use of anatomically realistic extended views is recommended as reduced vertebral canal dimensions occur during *in vivo* neck extension, and extension of the caudal articulations occurs during exercise[12,30,31]. The degree of extension should mimic the head/neck positions that occur when ridden: where the neck is raised and the bridge of the nose is vertical; and where is neck is raised and extended, and the bridge of the nose is in front of the vertical[30]. Unfortunately specific values of compression are not available for these extended views; instead this author applies neutral view criteria, but only assigns significance when dye column alteration is >1.5x that found at adjacent articulations to account for positional artefacts. When concurrent APJ changes and a narrow MSR are found at a single articulation, it may be possible, though technically challenging, to assign relative significance to dorsoventral compression (from vertebral canal malformation) versus dorsolateral compression (from APJ pathology), by attempting oblique lateral or ventrodorsal projections. As no decision criteria exist for these views, critical comparison with adjacent articulations is recommended.

TREATMENT OF CERVICAL APJ/LF DISEASE

The response of ataxic horses with APJ OA to NSAIDs and reduced exercise or rest is poor[9]. Instead intra-articular administration of corticosteroid under ultrasound guidance is recommended, although there is little published evidence of its efficacy. In man corticosteroids can reduce synovial membrane size and joint effusion volume[30], which could abolish APJ/LF compression. A study of 33 sport horses with spinal ataxia attributable to CVCM, in addition to 26 with other cervical disease (neck pain, obscure lameness), reported that overall 32% returned to full function and 39% improved[33]. The overall effect was of variable duration (<1month to 5years). It is difficult to draw conclusions from this study, as the number with ataxia was small and not analysed separately, the diagnostic criteria were poorly defined, the response was owner assessed and the treatment was variable in dose and drug used. Preliminary analysis of a larger population of UK sport horses aged >8years with spinal ataxia, that fulfill this authors APJ/LF compression inclusion criteria has shown a >80% positive response rate (defined as all limbs neurological gait grade ≤1) (Hepburn unpublished data). All horses received either triamcinolone acetonide when treating 2 APJs (8mg/joint, 16 mg max dose) or methyl-prednisolone acetate when treating 4 APJs (40 mg/joint, 160mg max dose); with follow up neurological examination performed at 1, 4, and 12 months. Neurological resolution occurred over 1-4months, and lasted 1-5 years; with repetition of the treatment yielding a similar response. Application of the same approach to a smaller population of horses aged 4-8 years, but with the inclusion of myelography if a MSR is abnormal to exclude vertebral canal malformation, has shown a >60% positive response rate, with a duration of response lasting between 2 and 8 years (Hepburn unpublished data). The reduced response rate in younger horses with sole APJ/LF compression reflects the variety and increased severity of pathological changes that occurs in OC. Further study is required in horses of all ages to describe which cases are most appropriate for conservative therapy, in order to better understand the relevance of cervical APJ/LF pathology in the absence of the advanced imaging which has defined this disease in other species.

REFERENCES

[1]Hahn CN et al. **Vet Radiol Ultrasound**. 2008 Jan 10;49(1):1–6. [2]Dreier R. **Arthrit Res Ther**. 2010;12:216–27. [3]Powers BE et al. **Vet Pathol** 1986 Jul 1;23(4):392–9. [4]Trostle S et al. **J Vet Diag Invest** 1993 Jul 1;5(3):423–31. [5]Mitchell CW et al **Vet Radiol Ultrasound** 2012 May 25;53(6):613–20. [6]Dyson SJ *in* **Diagnosis and Management of Lameness in the Horse** 2003. p. 522–31. [7]Withers J et al. **Eq Vet J**. 2009;41(9):895–902. [8]Claridge H et al **EqVet J**. 2010;42(8):726–31. [9]Dyson SJ Vet **Clin N Am Eq Pract** 2011 Dec;27(3):417–37. [10]Zsoldos RR et al. **Eq Vet J Suppl.** 2010 Dec;42(38):516–22. [11]Clayton HM et al **Eq Vet J** 1989 May;21(3):189–92. [12]Sleutjens J et al **Eq Vet J Suppl** 2010 Nov;42(38):425–30. [13]Thomsen LN et al **Eq Vet J** 2013 Jul;45(4):448-53. [14]Webb AL et al **Man Ther** 2011;16(2):118–24.

[15]Benjamin M et al **Annals Rheum Dis** 2004 Dec;63(12):1549–55.[16]Clayton H et al **AJVR** 2012;73(8):1153–9. [17]Schmidburg I et al **Vet J** 2012 Jul;192(3):525–8. [18]Nixon AJ *in* **Equine Fracture Repair** 1996. p. 299–312. [19]Fisher LF et al **Vet Pathol** 1981;18:407–10. [20]Machino M et al **BMJ Case Reports**. 2012 Nov 29:1–5. [21]Laynon L et al **J Biomech**. 1984;17:897–905. [22]Mayhew IG. **Proc AAEP**. 1999. p. 67–84. [23]Van Biervliet J et al **Clin Tech Eq Prac** 2006 Mar;5(1):54–9. [24]Levine JM et al **JVIM** 2007;21(4):812–9. [25]Down SS et al **Eq Vet J** 2009 Jul 5;41(6):518–24. [26]Levine J et al **JAVMA** 2010;237(7):812–22. [27]Maher O et al Proc **XIV SIVE/FEEVA**. 2008. [28]Butler JB *in* **Clinical Radiology of the Horse** 2008:505–39. [29]Van Biervliet J et al **Eq Vet J** 2004 Jan;36(1):14–20. [30]Elgersma A et al **Eq Vet J Suppl** 2010 Nov;42(38):436–43. [31]Wijnberg ID et al **Eq Vet J Suppl** 2010 Nov;42(38):261–7. [32]Raynauld J-P (2003) **Arthritis Rheum** (48)2:370. [33]Birmingham SSW et al **Eq Vet Educ** 2010 Feb 12;22(2):77–82.

UPDATE ON DIAGNOSIS AND TREATMENT OF EQUINE PROTOZOAL MYELOENCEPHALITIS

Martin Furr, DVM, DACVIM (Large Animal Internal Medicine), PhD
Leesburg, VA

INTRODUCTION

The clinical diagnosis of EPM remains a challenging task for veterinarians. Variable and sometimes subtle presentation of clinical signs and the inherent complexities of the diagnostic tests make a clear cut and confirmed diagnosis of EPM in the clinical patient difficult. Given these constraints, careful interpretation of clinical and ancillary diagnostic laboratory testing are key to making a diagnosis. Diagnosis is dependent upon:

1. Confirming the presence of clinical signs which are consistent with EPM
2. Ruling out other potential causes of the observed clinical signs by appropriate means (spinal tap, radiography, serology, etc)
3. Confirming the presence of *S. neurona* specific antibodies by one of a variety of means (i.e., immunodiagnostic testing)

CLINICAL EXAMINATION

A thorough physical and neurological examination remain the keystone of the EPM diagnosis. A full clinical examination to document the presence or absence of other clinical abnormalities, and collection of historical information, are always warranted. A lameness examination is very important, and should be carefully done to eliminate disease of the musculoskeletal system, as it is much more commonly affected than the CNS. For the diagnosis of EPM to be supported conclusive evidence of CNS disease must be present. The classic clinical signs of EPM have been well described and include gait abnormalities, and/or signs of cranial nerve, brainstem or cerebral disease. Asymmetry of clinical signs is a hallmark of the disease, although symmetrical deficits can be seen. In addition, muscle atrophy is common. The finding of ataxia completes the diagnostic triad of the "three A's of EPM"- ataxia, asymmetry and atrophy. While these may not exist in every case, they are compelling findings in the EPM suspect when present.

Eliminate Other Potential Causes

A differential diagnosis list should be constructed to ensure that additional diagnoses are considered, and appropriate tests conducted to support, or eliminate, their consideration. Numerous diseases of the CNS can present similarly to EPM, including bacterial or verminous meningitis, neoplasia, spontaneous epidural hemorrhage, EHV- 1 myelitis, etc. Survey radiographs, serology, and cerebrospinal fluid (CSF) collection are all important components of a full evaluation. If all such testing is negative, the probability of EPM increases, particularly if immunodiagnostic testing supports the diagnosis.
Cervical radiographs are indicated for the evaluation of any horses which have neuroanatomic localization of the clinical signs to the cervical region. While EPM does not cause radiographic changes, it is important to rule out other causes of spinal ataxia, such as cervical compression, fracture or cervical facet arthritis, which may be present.

Collection and analysis of CSF is a common component of the neurologic evaluation and should not be ignored in evaluating the EPM suspect. The observation that CSF false positive tests can occur has led some to question the value of analyzing CSF in horses being evaluated for EPM. It is important to remember that the CSF is not collected for the sole purpose of doing an EPM test; additional information is obtained from the sample. In the authors experience, several neoplasia and meningitis cases were detected when fluid was collected "for an EPM test" in a neurologic horse. If the sample is not collected and evaluated, the correct diagnosis may be missed. Therefore, while it is true that some false positive tests occur using CSF, this should not preclude the use of this diagnostic test. Horses exposed to *S neurona* will not necessarily and invariably develop a false positive test on CSF. This appears to be associated with a high serum titer, and is often transient. In addition, a negative CSF test has great value in eliminating EPM in most cases. A weak or suspect positive reaction in CSF is not compelling evidence for a diagnosis of EPM, while a stronger test reaction on immunoblot, or a titer exceeding the reference laboratories high-cut-off is strong support for the diagnosis.

Cerebrospinal fluid collection and evaluation is also recommended for full evaluation of the horse with CNS disease. The CSF in horses with EPM is usually normal--although in fulminant cases abnormalities may be seen--this appears to be exceedingly rare, however. However, one should remember that much information about the CNS may be gained from evaluation of the CSF, and testing for EPM is not the only information obtained when a CSF sample is analyzed.

Immunodiagnostic Testing

The most important aspect of the diagnosis of EPM is the detection of *S neurona*-specific antibody.[1,2,7] The western blot (WB) was the first specific immunodiagnostic test for EPM developed. Immunoblot testing of CSF samples demonstrated greater than 90% specificity and sensitivity among approximately 100 horses with neurologic disease that received postmortem examination.[3] Only approximately one half of the cases were histologically confirmed as EPM, however, and there is some question about the accuracy of the reported test performance due to this fact. The non-quantitative nature of the WB test makes some clinicians uncomfortable, and some clinicians prefer tests with definable end-points.

Presence of antibodies in the serum only indicates exposure of the horse to the parasite, but not necessarily active disease. The presence of *S. neurona* specific antibodies in CSF is more suggestive of infection, but is not definitive. It was originally presumed that the presence of *S. neurona* antibodies in CSF could only be present if the organism was present within the CNS. Several years of experience have determined that this is not completely accurate. While *S. neurona* specific antibodies are present in the CSF in active disease, false positive results have been widely reported.[4,5] Blood contamination during collection of the sample is commonly incriminated as causing false positives. More importantly, however, antibodies to non-replicating protein can be found in the CSF of vaccinated horses, directly correlating to plasma concentrations.[6] This shows that in the normal, healthy horse, serum-derived antibody can pass the normal blood-brain barrier, giving rise to false positive tests. This observation suggests that examination of paired serum and CSF WB results may be beneficial. If the serum reaction is strong, the corresponding CSF response should be robust to support a diagnosis of EPM. This approach has been formally examined in several studies, and has been found to provide strong discrimination between exposed and actively infected horses.

Horses with clinical EPM (i.e. active infection of the CNS) will have local production of S neurona-specific antibody in the CNS, with antibody present within the cerebrospinal fluid. This is referred to as "intrathecal" antibody production. The challenge, therefore, is to determine if the antibody present within the CNS (and CSF) is there from local production (e.g. infection) or simply from diffusion from the serum (non-infected exposed). The BBB provides significant restriction to the movement of immuoglobulins into the CSF, and hence we can estimate how much antigen specific antibody should be present in the CSF based upon how much there is in the serum. These are simple tests of proportionality, and two major techniques have been described: the C-value (Goldman-Witmer coefficient) or the antibody index (AI).[7,8] The performance of these tests have been tested in several studies in the horse and applied to the specific diagnosis of EPM with strong results. These techniques require the determination of a specific antigen specific titer hence cannot be used with the WB or other non-quantitative measures. In addition, one must determine the concentration of some non-specific protein marker in the serum and CSF, the ratio of which serves as an indicator of BBB permeability. From this information, it is relatively easy to determine if the amount of antibody of interest (i.e. the *S neurona* titer) in the CSF is more than one would predict based upon the amount of the same antibody in the serum. If the CSF concentration is greater than that predicated by the serum concentration, then intrathecal antibody production is present and indicates infection and active disease. The mathematical construct for the C-value and AI results in a theoretical value of 1 or below for normal horses; values greater than 1 are consistent with intrathecal antibody production. These techniques differ only in the protein standard used; C –value used total IgG, while the AI uses albumen as the endogenous protein standard.

The C-value and AI have been tested in the horse with very positive results. An initial "test of concept" found that healthy normal horses vaccinated intramuscularly with ovalbumin became CSF antibody positive, but the index calculation was less than 1 using either technique, as expected. When normal horses which had been primed by peripheral vaccination with ovalbumin where challenged with intrathecal ovalbumin, however, the CSF ovalbumin titer became very high, and the index calculation was much greater than 1, confirming intrathecal production of antibody. The test was found to be highly specific as well, as the CSF EHV titers also went up (due to BBB inflammation), but the calculated C- value and AI remained below 1 – confirming that even though the CSF titer was increased, it did not arise from intrathecal production.[8]

Another manuscript reported on the use of the C-value and AI in the specific diagnosis of EPM caused by *S neurona*. In this paper, 14 well-characterized cases of EPM were tested using the C –value and AI. Further 15 horses with non-EPM neurologic disease (cervical compression) were also tested. Using a cut point of greater than 1 as indicating intrathecal production and active infection, the use of the C –value and AI demonstrated a sensitivity and specificity of 86 and 100% respectively for the C-value. In addition, it was found that this test remained accurate even in the face of substantial blood contamination.[9]

Hence, the C-value and AI appear to be robust tests of intrathecal infection. However, the calculations require that CSF and serum IgG or albumen concentrations be determined. This adds to the complexity, time, and expense of the test, hence a proxy measurement or approximation which does not require their determination is desirable. Fortunately, it has been determined that using a simple ratio of the serum to CSF titers is a good approximation. The serum contains much more total protein and immunoglobulin than the CSF, in a ratio that is approximately 100-150/1. Therefore, a ratio of the *S neurona* titers in serum and CSF should be the same. A ratio of serum to CSF *S neurona* titer less than 100 suggests that there is more antibody in the CSF than can be explained by the serum titer, and is consistent with active infection. This has been tested in a study using 128 horse -- 44 with EPM, 35 horses with cervical vertebral compression, 39 non- EPM and non-CVM horses with neurologic disease, and 10 normal horses. Using the Sag-2,4/3 ELISA the ratio method had a sensitivity and specificity of 93 and 81%, respectively for the diagnosis of EPM caused by *S neurona*.[10] If a more rigorous cut-off point of <50 was used, the sensitivity and specificity increased to 86 and 96%, respectively. At present, this testing approach is considered by the author to be the optimal approach to the diagnosis of EPM caused by *S neurona*. Other testing platforms which generate a specific antibody titer can also be used in this manner but to date have not been validated. Their overall performance when used as a ratio, however, will be influenced by the inherent sensitivity and specificity of the test method used to generate the titer. Hence, tests based upon the snSAG-1 antigen are likely to be less accurate.

TREATMENT

At the present time there are 3 FDA approved treatments for EPM; ponazuril, diclazuril , and sulfadiazine/pyrimethamine (S/P) combination. Ponazuril and diclazuril are chemically very similar, and hence have similar efficacy and toxicity profiles. The overall efficacy of both these drugs in a rigorous FDA approves study was roughly 60% success when treated for one month, with very little toxicity observed.

The time to reach steady state concentrations of ponazuril in the CSF of horses is approximately 1 week. For this reason, it has become commonplace to give an initial loading dose in an effort to achieve therapeutic concentrations more quickly. While the clinical value of this approach has not yet been demonstrated, maximum CSF concentrations of Ponazuril were achieved within 28 hours when horses were given a loading dose of 15 mg/kg body weight (3X the normal dosage).[11] Additionally, feeding 2 ounces of corn oil immediately before administering ponazuril results in blood concentrations which are 25% higher than if no corn oil is given (Furr, unpublished data). The author routinely recommends this approach.

Historically, a combination of sulfonamide and pyrimethamine (S/P) has been used to treat EPM. A field efficacy study reported a success rate of 57% after several months of treatment.[12] Weaknesses of S/P suspension in the treatment of EPM include the prolonged duration of treatment required to affect a positive response, and the toxicity of the compound. Toxicity includes anemia, fetal loss, and fetal abnormalities. A benefit of S/P combinations is the lower cost, yet this may be offset by the extended dosing interval required.

The broad-spectrum antimicrobial nitazoxanide (NTZ) was approved for the treatment of EPM but has been removed from the market and is no longer available. Decoquinate is an anti-protozoal drug that has been used to treat coccidiosis in poultry, cattle, sheep, and goats. Decoquinate in combination with the immunomodulator levamisole is currently being investigated for treatment of EPM, but results are not available.

Immunostimulants have been recommended by some on the presumption that immunosuppression is a component of the pathophysiology of EPM. Levamisole (1 mg/kg PO daily), EqStim (5 ml IM on day 1,3 and 7, then monthly), or Equimune IV, (1.5 ml IV weekly for 3 weeks) have all been advocated, however there is no specific information to suggest that these have any positive effect. One potential beneficial immune modulator, however, is Parapox Ovis Virus Immunomodulator (Zylexis, Pfizer Animal Health). While licensed as an aid in the treatment of horses with EHV-1 and EHV-4, parapox ovis has been demonstrated to increase IFN-□secretion in treated horses. As IFN-□is considered a key cytokine in the treatment of EPM, this compound might have use in select animals.

In conclusion diagnosis of EPM in a clinical patient is best supported at the present time by the finding of definitive CNS abnormalities on a neurologic exam, the elimination of other neurologic diseases by radiography and CSF evaluation, and the finding of a serum/CSF ratio of less than 100 using the SAG2,4/3 ELISA. Successful treatment is available with several approved medications with good safety and efficacy profiles.

REFERENCES

[1]Granstrom D. *Proc 21 Annual Meet ACVIM*, 1993:587; [2]Miller MM, *et al. J Am Vet Med Assoc* 1999;215:67; [3]Granstrom DE. *Proc 21 Annual Meet ACVIM* 1993:587; [4]Bentz BG, *et al. Compend Contin Educ Pract Vet* 1999;21:975; [5]Cook AG, *et al. J Am Vet Med Assoc* 2002;220:208; [6]Furr M. *J Vet Intern Med* 2002;16:588; [7]Furr M, *et al. J Vet Intern Med* 2002;16:618; [8]Furr M. *J Vet Intern Med* 2007;21:806; [9]Furr M., et al. *J Vet Intern Med* 2011;25:138; [10]Reed, SM, et al. *J Vet Intern Med* 2013;27:1193; [11]Reed SM, et al. *Clin Tech Equine Prac* 2006;5:9.

EQUINE LYME NEUROBORRELIOSIS: PERCEPTIONS AND REALITY

Amy L. Johnson, DVM, DACVIM (LAIM & Neurology) and Laura Johnstone, BVSc MVSc
Kennett Square, PA

INTRODUCTION

Infection with *B. burgdorferi*, the causative agent of Lyme disease, is commonly recognized in horses. However, diagnosis of clinical disease is very challenging. Unlike in people or dogs, there are more reports in horses describing neurologic disease than any other syndrome related to *Borrelia* infection, although "more" refers to a small number of cases! These proceedings describe some of the known and unknown aspects of this controversial disease.

EXPERIMENTAL INFECTION MODELS

Chang's group at Cornell performed 3 experimental infection trials with the goals of establishing an infection model and assessing the efficacy of vaccination and treatment.[1-3] The original model involved placing 30 adult *Ixodes scapularis* ticks with a 55-60% *B. burgdorferi* infection rate onto the clipped side of the pony and allowing the ticks to feed for 7 days. The ponies received high doses of dexamethasone (0.44 mg/kg q 24) for the first 5 days of tick exposure and again for 5 day starting at 5 months post-tick exposure. In the second study, the second course of dexamethasone was not administered since the ponies were euthanized 4 months after tick exposure. In the third study, the tick challenge was repeated a second time immediately after the first.

Infection was confirmed through isolation of *B. burgdorferi* via culture and PCR techniques. Combining results from all 3 studies, a total of 27 ponies were confirmed to be infected. (Eight additional ponies were exposed but not infected; seven of these were vaccinated ponies and one was not.) Of the 27 infected ponies, 26 of them seroconverted as assessed by kinetic ELISA and WB techniques; 1 of the ponies did not seroconvert despite isolation of *B. burgdorferi* from a lymph node. *B. burgdorferi* was isolated from skin biopsies in 26/27 infected ponies (not from the seronegative pony). *B. burgdorferi* was also identified in deeper tissues via culture, PCR, or both in many ponies: skeletal muscle (n=17), synovial membrane (n=16), lymph node (n=15), fascia (n=13), heart (n=7), bladder (n=7), kidney (n=5), thyroid gland (n=1), and meninges (n=1). Ponies that were euthanized at later time points (9 months after exposure) tended to have *B. burgdorferi* isolated from a greater number of tissues, and 7/12 antibiotic-treated ponies did not have *B. burgdorferi* isolated from any tissues. Histologic lesions, if present, were limited to the skin and lymph nodes in the majority of ponies and consisted of lymphohistiocytic dermatitis, occasionally with pseudolymphomatous reaction, and marked lymphoid hyperplasia. Occasional ponies had non-suppurative synovitis affecting various joints and skeletal muscles. A total of 3 ponies had histologic lesions in the nervous system; the pony with *B. burgdorferi* isolated from meninges did not have any lesions detected. One pony had mild lymphocytic neuritis and perineuritis affecting peripheral nerves and spinal nerve roots as well as several mild perivascular cuffs in the thalamus. Two ponies had mild lymphocytic perivascular cuffing within the meninges overlying the cerebellum and cerebrum.

Importantly, despite confirmation of infection and histologic lesions, none of the 27 infected ponies showed any clinical signs attributed to infection. This result supports the assertion that the majority of infections in horses are subclinical and also means that Koch's postulates have not been fulfilled for this disease.

NATURALLY OCCURRING CASES

Six cases of equine LNB have been reported in the literature. In the first, a 6-year-old Morgan gelding from Wisconsin showed signs consistent with encephalitis (constant aimless walking, head tilt, dysphagia, and flaccid tail). The horse was euthanized without treatment and diagnosis was made via a combination of positive serology and direct immunofluorescence of spirochetes isolated from the brain.[4] Another case report documents an 8-year-old Thoroughbred mare in the United Kingdom that showed lethargy, anorexia, pyrexia, ataxia, hyperesthesia, uveitis, and polysynovitis. Despite treatment with oxytetracycline for over 2 weeks, the mare's signs progressed to stupor, recumbency, and dangerous behavior necessitating euthanasia. This horse was positive for anti-*Borrelia* antibodies in serum and synovial fluid. Necropsy revealed lymphohistiocytic leptomeningitis and vasculitis, and PCR performed after culture enrichment confirmed the presence of *Borrelia* in the brain and other tissue samples.[5] A 12-year-old Thoroughbred gelding in the mid-Atlantic region was reported to show behavior changes, neck stiffness, and poor performance. Antemortem diagnosis of LNB was based on CSF analysis (neutrophilic pleocytosis with increased total protein), seroconversion, and positive *B. burgdorferi* PCR on CSF. The gelding showed an initial response to two months of treatment with doxycycline but had an acute recurrence of neurologic signs 2 months after treatment discontinuation; these signs included neck stiffness, ataxia, tremors, unilateral vestibular disease, menace response deficits, and abnormal behavior. Rapid progression of signs necessitated euthanasia, and necropsy revealed lymphohistiocytic leptomeningitis and vasculitis as well as lymphocytic cranial neuritis and radiculoneuritis.[6] In contrast, a 14-year-old Thoroughbred gelding from the northeast US with neck, back, and tail pain as well as paraparesis and ataxia appeared to respond well to doxycycline treatment. This gelding was diagnosed with LNB based on abnormal CSF (lymphocytic pleocytosis with increased total protein), evidence of intrathecal *Borrelia*-specific antibody production, and positive *B. burgdorferi* PCR on CSF.[7] An additional two horses were diagnosed with LNB post-mortem via direct observation of spirochetes, PCR (in 1 case), and immunohistochemistry. These horses showed similar clinical signs to the previously mentioned cases.[8]

Retrospective analysis of horses diagnosed with LNB post mortem at the University of Pennsylvania concurs that horses might present for a variety of concerns including weight loss, gait abnormalities, neck stiffness, behavioral changes, uveitis, seizures, dysphagia, atrial fibrillation, dyspnea and collapse. Various neurologic deficits were also recorded: mental status varied including dullness, increased anxiety and hyperesthesia; while gait abnormalities ranged from none to generalized stiffness to severe general proprioceptive deficits and paresis. Although not present in all horses, neck stiffness, muscle atrophy, muscle tremors or fasciculations and cranial nerve deficits were commonly recorded. The latter included reduced lip tone, bilateral arytenoid paralysis and dysphagia. Histopathologic lesions included neutrophilic, lymphocytic, lymphohistiocytic or lymphoplasmacytic inflammation of the meninges, spinal nerve roots, cranial nerves and ganglia, and the parenchyma of the brain and spinal cord. Inflammatory lesions of other organ systems were also observed, including lymphoplasmacytic tubulointerstitial nephritis, lymphoplasmacytic iridocyclitis, lymphohistiocytic hepatitis, interstitial pneumonia, lymphohistiocytic epicarditis and interstitial myocarditis.

DIAGNOSTIC TESTING

Human Medicine

Criteria for diagnosing LNB in people have been well-established. Published guidelines from the American Academy of Neurology (AAN; www.aan.com) state that diagnosis of definite LNB requires three criteria:

1. *possible tick exposure in an endemic area*
2. *one or more of the following:*
    a. *erythema migrans, Borrelia lymphocytoma, or acrodermatitis*
    b. *immunologic evidence of Borrelia exposure*
    c. *culture, histologic, or PCR identification of Borrelia*
3. *neurological disease, with exclusion of other potential etiologies (testing for intrathecal antibody production is indicated).*

Guidelines from the European Federation of Neurological Societies (EFNS; www.efns.org) are similar, with three criteria:

1. *neurological disease suggestive of LNB without other obvious reasons*
2. *cerebrospinal fluid pleocytosis*
3. *intrathecal anti-Borrelia antibody production.*

LNB is considered possible if 2 criteria are met and definite if 3 criteria are met.

Equine Medicine

Antemortem diagnosis of LNB is rarely accomplished, with post-mortem examination frequently required. Serology alone has limited value, since positive results only indicate previous or current infection (or perhaps vaccination) and cannot confirm that infection is responsible for clinical signs. Although specific criteria have not been established, criteria that assist in diagnosis include potential exposure to the organism (via time spent in an endemic area), presence of neurologic signs, and evidence of nervous system infection. In reported cases, cerebrospinal fluid analysis generally yields abnormal findings, with increased cell count and protein level and either a neutrophilic or lymphocytic pleocytosis. However, cases with minimal meningeal involvement might not show CSF abnormalities. Additionally, encephalopathic-type signs might be possible in the absence of CNS involvement; the mechanism in people is thought to involve soluble neuroimmunomodulator molecules that diffuse across the blood-brain barrier and alter brain physiology.[9] Specific *B. burgdorferi* testing should yield supportive results. Those most suggestive of LNB include confirmation of the organism in CSF via PCR or documentation of intrathecal antibody production. Seropositivity or isolation of *B. burgdorferi* from other sites (eyes, skin) might also be supportive.

To assess whether intrathecal antibody production is likely present, paired serum and cerebrospinal fluid titers are necessary. Although most accurate results would be obtained through calculation of an antibody index or C-value, the authors use an approach similar to that utilized in current EPM SnSAG 2,4/3 testing,[10] whereby the ratio between a horse's serum and CSF values for a specific antibody is compared to the "normal" or "expected" ratio for any antibody. Assuming a normal blood-brain barrier, serum:CSF antibody ratios generally range from 130:1 to 250:1. If the ratio drops below 130:1, intrathecal antibody production might be present (ie, more antibody than expected is present in the CSF). For EPM testing, using a cutoff of 100:1 provided optimal sensitivity and specificity. For Lyme testing, the optimal cutoff is not established. Additionally, interpretation of results can be complicated by dilution. For example, many practitioners rely on the Multiplex assay at Cornell for Lyme testing. Serum is run at a dilution of 1:400 whereas CSF is run undiluted. Therefore, to estimate the serum:CSF ratio, the serum result must be multiplied by 400 then divided by the CSF result. Looking at it another way, a CSF value up to 3 times higher than a serum value is NOT likely to indicate intrathecal production of antibody due to the dilution factor. The authors do not become suspicious of intrathecal antibody production unless the CSF result is approximately 4x higher than the serum result for any specific antibody level (OspA, OspC, or OspF).

TREATMENT OPTIONS

The ideal treatment regimen for equine LNB is unknown. Investigation is hampered by lack of a disease model as well as lack of a reliable antemortem diagnostic test. Although one of the experimental trials investigated antibiotic treatment,[3] it is difficult to draw firm conclusions since there were small numbers in each treatment group, none of the ponies displayed clinical signs, and the drug dosages utilized are somewhat different than what is used in current clinical practice. In this study, ponies were experimentally infected and then treated with antibiotics for 28 days starting approximately 3 months after tick exposure. Four ponies were administered tetracycline (5 mg/kg/day IV), 4 ponies were administered doxycycline hyclate (10 mg/kg/day PO), 4 ponies were administered ceftiofur sodium (2.2 mg/kg/day IM), and 4 ponies were untreated. Based on serology and postmortem tissue samples, infection was eliminated in 4/4 tetracycline-treated ponies, 2/4 ceftiofur-treated ponies, 1/4 doxycycline-treated pony, and none of the untreated ponies. These results led many practitioners to recommend a month-long course of IV tetracycline or oxytetracycline for suspect cases of equine Lyme disease.

In the authors' experience and based on reports in the literature, successful treatment of equine LNB is difficult. This assertion is certainly biased as definitive diagnosis is often based on post-mortem findings. However, the horses that have succumbed to LNB have often been treated for long periods of time (months) prior to death or euthanasia with drugs that in theory should be effective (oxytetracycline, doxycycline, minocycline). Although speculative, it seems reasonable to use similar approaches as in people. If financially feasible, using parenteral regimens of high-dose penicillin (44,000 U/kg IV q 4-6h), cefotaxime (25-50 mg/kg IV q 6-8h), ceftriaxone (25-50 mg/kg IV q 12h), or ceftazidime (20-40 mg/kg IV q 6-12h) might be most effective. If these options are cost-prohibitive, minocycline (4 mg/kg PO q 12h) is likely to be more effective than doxycycline or oxytetracycline due to more favorable bioavailability and blood-brain barrier penetration, respectively.

PROGNOSIS

Of the 6 reported cases, only 1 responded to treatment with long-term success. This poor long-term prognosis was reflected in the clinical progression of horses diagnosed at post-mortem with LNB at the University of Pennsylvania. Of the horses treated for LNB, clinical signs either continued to progress or, despite an initial improvement, showed recurrence and subsequent progression. However, the authors recognize that this might be a biased group and that there are anecdotal reports of horses that fulfill the diagnostic criteria used in human medicine for LNB that do respond to treatment.

PREVENTION

There are no Lyme vaccines labeled for horses. However, practitioners sometimes administer canine Lyme vaccines to horses with the intent of preventing infection. The authors are currently aware of four commercially available canine vaccines: three are two-strain bacterin vaccines (LymeVax, Zoetis; Duramune, Boehringer-Ingelheim; Nobivac-Lyme, Merck) and one is a recombinant pure non-adjuvanted protein (Recombitek Lyme, Merial). The three bacterin vaccines induce antibody production (in theory) against multiple antigens, including OspA and OspC (to varying degrees), whereas the recombinant vaccine only includes the OspA antigen.

There is very little evidence for vaccine efficacy in horses. One vaccine trial has been performed using a recombinant OspA product with adjuvant.[2] In this study, seven ponies received a series of three vaccinations (days 0, 20, and 82) and were compared to 4 control ponies that received only adjuvant. All ponies were experimentally infected via tick bite on days 112-119. All of the vaccinated ponies seroconverted after vaccination and before experimental infection, whereas 3 of the 4 control ponies seroconverted after infection. Live organisms were not isolated from vaccinated ponies but were isolated from control ponies. Although the vaccine protected ponies against infection, many questions are left unanswered, including duration of immunity, vaccine safety, and cross-protection against other *Borrelia* strains. Additionally, an OspA vaccine is unlikely to benefit horses that are infected prior to vaccination.

With such limited information on use of vaccination in horses, can we generalize from other species? No human vaccine is currently available, although a new multivalent OspA vaccine showed encouraging results in a phase 1/2 study.[11] A previously available OspA vaccine showed reasonable efficacy in a clinical trial, preventing Lyme disease from developing in 78% of vaccinates, but was withdrawn from the market after only a few years due to unfounded concerns about adverse events.[12] Although efficacy of a monovalent bacterin vaccine in dogs was reported as 78% for preventing illness and 90% for preventing seroconversion, the majority of SAIM diplomates do not recommend vaccination due to the subclinical nature of most infections, questionable efficacy of available vaccines for preventing disease, and concerns about safety.[13]

REFERENCES

[1]Chang Y-F, et al. Vet Pathol 2000; 37:68-76. [2]Chang Y-F, et al. Vaccine 2000;18:540-8. [3]Chang Y-F, et al. Vet Microbiol 2005;107:285-294. [4]Burgess EC, et al. J Am Vet Med Assoc 1987;191:1457-8. [5]Hahn CN, et al. Equine Vet J 1996;28:84-8. [6]James FM, et al. J Am Vet Med Assoc 2010;237:1180-5. [7]Wagner B, et al. AAEP Proceedings 2011;57:70-75. [8]Imai DM, et al. Vet Pathol 2011;48:1151-7. [9]Halperin JJ. Handbook of Clinical Neurology 2014;121:1473-83. [10]Reed SM, et al. J Vet Intern Med 2013;27:1193-2000. [11]Wressnigg N, et al. Lancet Infect Dis 2013;13:680-9. [12]Lathrop SL, et al. Vaccine 2002;20:1603-8. [13]Littman MP, et al. J Vet Intern Med 2006;20:422-34.

DIAGNOSIS AND MANAGEMENT OF INSULIN DISORDERS OF THE HORSE WITH ENDOCRINOPATHIC LAMINITIS

Nicholas Frank, DVM, PhD, DACVIM
North Grafton, MA and Sutton Bonington, Leicestershire, UK

INTRODUCTION

In this postgraduate course, we will discuss the diagnosis and management of insulin disorders in horses with endocrinopathic laminitis. Although the premise for this discussion is that laminitis has already developed, clinicians are advised to recognize risk factors ahead of time and proactively manage endocrine problems to prevent laminitis.

ASSESSMENT OF THE LAMINITIC HORSE

The most important question when presented with a horse suffering from laminitis is, "What caused the problem to develop?" Broadly speaking, laminitis cases fall into two categories: those arising from systemic inflammation and those associated with endocrine disorders (termed *endocrinopathic laminitis*). As risk factors for endocrinopathic laminitis have been identified, it has been useful to group together hyperinsulinemia/insulin resistance, obesity, regional adiposity, and dyslipidemia as equine metabolic syndrome (EMS).[1] Just as Metabolic Syndrome in humans is a cluster of risk factors used to predict the occurrence of coronary artery disease, the concept of an equine metabolic syndrome has helped us identify horses at risk for laminitis. The advantage of defining a syndrome is that it prompts more thorough investigation – if a horse with laminitis has regional adiposity or obesity, the clinician is more likely to think of EMS and perform insulin testing. Discussions of EMS have also allowed us to incorporate concepts such as the *easy keeper* metabolism that are hard to define, but fit well with our experiences managing obesity in horses. More recently, the term *weight loss resistance* has been used to describe the horse or pony that is slow to lose body mass when placed on a diet and exercise regimen.[2] A genetic predisposition can also be assumed from the higher prevalence of endocrinopathic laminitis in certain breeds and from observations that related horses are affected. There is much more to be learned about the genetics of endocrine disorders in horses and this is the subject of another presentation at this meeting.

If EMS is a cluster of risk factors first recognized in younger horses, then how does pituitary pars intermedia dysfunction (PPID) factor in? This endocrine disorder has also been associated with laminitis,[3] and develops as the horse ages. Horses with EMS can develop PPID, so both conditions occur concurrently and contribute to the development of laminitis in some animals. These disorders may have independent effects on laminitis development or be related through exacerbation of hyperinsulinemia by PPID and the effects of hyperadrenocorticism on tissue insulin sensitivity.

Insulin became the focus of intense research when laminitis was experimentally induced in healthy ponies and Standardbred horses by infusing insulin intravenously.[4,5] Fasting hyperinsulinemia and excessive insulin responses to oral sugars are detected in horses and ponies with naturally-occurring laminitis and *insulin resistance* (IR) can also be documented by measuring tissue insulin sensitivity using an insulin tolerance test, euglycemic hyperinsulinemic clamp, or frequently-sampled intravenous tolerance test with minimal model analysis. The presence of one or all of these abnormalities (fasting hyperinsulinemia, excessive insulin responses to oral sugars, or insulin resistance) is referred to as *insulin dysregulation* (ID). One reason for using this term is that positive results from any of the available insulin tests leads us to the same conclusion – that hyperinsulinemia is occurring as the horse is grazing on pasture and this raises the risk of laminitis. It is not clear whether tissue IR induces hyperinsulinemia to maintain euglycemia as conventionally thought or if hyperinsulinemia is the initiating problem and IR develops as a consequence. More research is required to resolve this question, but the take home point at present is that insulin testing is recommended for all horses with laminitis if ID is detected, it should be managed to reduce the risk of subsequent laminitis episodes. The emphasis has shifted so much towards insulin playing a role in laminitis that testing is recommended for all affected horses, even in the absence of obesity, regional adiposity, or other components of EMS.

Unfortunately, laminitis can also develop after administration of exogenous corticosteroids and the concept of endocrinopathic laminitis should be expanded to include these cases. Dexamethasone lowers insulin sensitivity in healthy horses and higher insulin responses were detected in insulin-resistant ponies during dexamethasone suppression tests.[6-8] It can therefore be assumed that injection of joints with long-acting corticosteroids or administration of immunosuppressive doses of corticosteroids increases the risk of laminitis in horses with endocrine disorders. Corticosteroid use cannot be avoided in many cases, but insulin testing should be performed before treatment so that owners can be apprised of the risks. The impact of endogenous stress associated with stress, intense training, or competition on glucose and insulin dynamics should also be considered before multiple joint injections are performed in horses of high-risk breeds such as ponies, Arabians, Morgan horses, Paso Finos, and Warmbloods. If ID is detected beforehand, treatments can be adjusted. A more conservative approach can be taken to joint injections by selecting a shorter-acting corticosteroid or injecting one joint at a time. Care should also be taken to limit sugars in the diet immediately preceding and following joint injections with corticosteroids. Consumption of high-sugar/high-starch feeds at the same time that multiple joint injections are performed may contribute to the development of hyperinsulinemia and laminitis. Performance-enhancing drugs are contributing factors in some cases.

We have discussed systemic inflammation and hyperinsulinemia separately, but it is conceivable that both mechanisms are involved in some cases. Anecdotal evidence suggests that the horse or pony with EMS is more likely to develop laminitis when systemic inflammation develops and a more proactive approach should be taken to laminitis prevention in these animals. Horses with endocrine disorders may be more susceptible to laminitis because hoof laminae have been damaged during previous laminitis episodes or because of the combined impact of systemic inflammation and stress- and inflammation-induced hyperinsulinemia on nutrient delivery to tissues. Systemic inflammation develops secondary to retained placenta, colitis, strangulated intestine, or pneumonia and laminitis is a concern in these situations. Cryotherapy should be initiated as soon as systemic inflammation is detected and a newer ice boots make this a practical option in the field. Once the horse has been placed in ice boots, it should remain there until 24 hours after clinical signs of systemic inflammation have subsided. Ice should be added to the boots regularly so that continuous cooling is achieved. Flunixin meglumine (1.1 mg/kg IV or PO q12h) should be administered when systemic inflammation is first detected and polymixin B (1,000-6,000 IU/kg IV q12h) is recommended to bind circulating endotoxin. It has recently been shown that inflammatory cytokine expression within the blood is prolonged in horses with EMS after intravenous injection of lipopolysaccharide, and this provides evidence that the endocrine/metabolic status of the animal impacts immune responses.[9] This association highlights the importance of routine insulin testing so that background information is on hand when major medical problems develop.

DIAGNOSTIC TESTING FOR INSULIN DISORDERS

Testing for insulin disorders is recommended for all horses with laminitis. It is also advisable to screen horses during routine wellness examinations in the same way that humans are screened for hypercholesterolemia. Insulin dysregulation is more likely to occur in ponies, Morgan horses, Paso Finos, Arabian, and Warmbloods, and any horse that exhibits generalized obesity, regional adiposity, or divergent hoof rings should be tested. Farriers will often identify at-risk horses.

An *oral sugar test (OST)* has been introduced to detect postprandial hyperinsulinemia in equids,[10] and has greater sensitivity than fasting insulin concentrations. The test is performed by fasting the horse and then administering corn syrup[1] at a dosage of 0.15 mL/kg PO (75 mL for a 500-kg horse). Fasting conditions are achieved by leaving only one flake of hay with the horse after 10 PM and performing the test the next morning, or feeding as normal in the morning and then testing after 4 PM. Corn syrup is given by mouth using 60-mL catheter-tip syringes and blood samples are collected for glucose and insulin measurements 60 and 90 minutes later. The owner can administer the corn syrup so that the veterinarian is only required for the 60- and 90-minute blood collections.

Interpretation:

- Normal if the insulin concentration is < 45 µU/mL (radioimmunoassay) at 60 and 90 min.
- Positive (postprandial hyperinsulinemia) if the insulin concentration is > 60 µU/mL at 60 or 90 min.
- Equivocal result if the insulin concentration is 45 to 60 µU/mL at 60 or 90 min. Repeat testing at a later time.
- Excessive glucose response if the glucose concentration is > 125 mg/dL at 60 or 90 min

Owners sometimes express concern about administering corn syrup to a horse with active laminitis. Problems have not been encountered in the author's experience, but a two-step approach can be adopted in these cases. The first step is to measure the fasting insulin concentration, and then the OST is performed if the resting insulin concentration is < 20 µU/mL or only mildly increased. It is still helpful to perform an OST in a horse with modest hyperinsulinemia in order to assess insulin responses to oral sugars and make diet recommendations, but not essential.

Fasting insulin concentrations were previously recommended for assessing insulin status,[1] but the sensitivity of this test should be considered before it is selected. It remains a useful test for confirming insulin problems in a horse with moderate of advanced ID, but is likely to return a normal result in the mildly affected animal. A general cut-off value of 20 µU/mL has been recommended in the past when measuring fasting insulin concentrations, but breed-specific reference ranges are being developed. The glucose concentration should also be measured in the same sample to detect hyperglycemia, and diabetes mellitus has been associated with PPID in aged horses.[11]

Other blood tests for assessing endocrine and metabolic status in horses include plasma triglyceride, leptin, and high molecular weight (HMW) adiponectin concentrations. Most clinical pathology laboratories measure triglyceride concentrations, but leptin measurements are only available from specialty laboratories such as the Animal Health Diagnostic Laboratory at Cornell University. Higher triglyceride concentrations are detected in horses with EMS and hyperleptinemia has been associated with obesity and IR.[12] Hypertriglyceridemia has been identified as a predictor of laminitis risk in ponies, with cut-off values of 57 and 94 mg/dL established from two studies of the same pony herd.[12,13] Hyperleptinemia (> 4 ng/mL) is detected in horses with increased adiposity and altered fat metabolism, and can be included as an additional component of EMS. Lower HMW adiponectin concentrations are detected in obese horses[14] and preliminary results from a comprehensive study performed by the University of Minnesota and Michigan State University indicate that this adipokine is a biomarker for assessing laminitis risk.

Tests of tissue insulin sensitivity can also be performed when more thorough investigation is required. An *insulin tolerance test (ITT)* can be used to measure insulin sensitivity and diagnose IR, but this test does not assess pancreatic insulin secretion, which could prevent mildly affected horses from being identified. An ITT can be performed by fasting the horse overnight and then injecting 60 mU/kg regular (soluble) insulin intravenously, with blood collections at 15, 30, 60, 90, 120, 150 minutes. Blood glucose concentrations are measured using a hand-held glucometer and the time taken for a 50% reduction in glucose concentration is recorded. This test has also been performed by injecting insulin at a dosage of 100 mU/kg and collecting blood at 0 and 30 minutes; horses with normal insulin sensitivity show a 50% decrease within 30 minutes.[15]

A *combined glucose-insulin test (CGIT)* can also be selected to assess insulin sensitivity. The horse is fasted overnight and an intravenous catheter placed to minimize stress associated with multiple blood collections. A pre-infusion (baseline) blood sample is collected and then 150 mg/kg body weight (bwt) 50% dextrose solution is infused (150 mL for a 500-kg horse), immediately followed by 0.10 units/kg bwt regular insulin (0.50 mL for a 500-kg horse). Blood samples are collected at 1, 5, 15, 25, 35, 45, 60, 75, 90, 105, 120, 135, and 150 minutes post-infusion, but these intervals can be adjusted to 0, 15, 30, 45, and 60 minutes in the field.

[1] Karo Light®; ACH Food Companies, Inc., Cordova, TN 38016

Insulin resistance is defined by the maintenance of blood glucose concentrations (measured with a hand-held glucometer) above baseline for 45 minutes or longer. The area under the insulin concentration curve should also be examined and reflects both the beta cell response and clearance rate of endogenous and exogenous insulin. Measuring the insulin concentration at 45 minutes provides information about the height of this insulin peak and concentrations > 100 μU/mL (by radioimmunoassay) are abnormal. There is a small risk of hypoglycemia with both the ITT and CGIT, so two 60-mL syringes containing 50% dextrose should be kept on hand and administered if sweating, muscle fasciculations, or weakness are observed; or if the blood glucose level drops below 40 mg/dL.

Tissue insulin resistance also affects oral sugar test results, with higher glucose and insulin concentrations detected at 60 and 90 minutes.

MANAGEMENT OF INSULIN DISORDERS

Before a management plan is formulated, it must first be determined whether exacerbating factors for hyperinsulinemia are a concern. The three major factors for consideration are obesity, high dietary sugar intake, and PPID.

Relationships between obesity and insulin sensitivity in equids are not as straightforward as it once seemed. Body fat mass is negatively correlated with insulin sensitivity in equids,[16] yet some obese horses suffer from profound hyperinsulinemia whereas others have normal insulin values. Genetics are therefore likely to play an important role in determining the magnitude of ID associated with obesity. Whether obesity is a cause or exacerbating factor for hyperinsulinemia, it should be managed to prevent other health problems, including formation of lipomas and increased biomechanical stress on laminae. Equine obesity is a major concern in the United States, with a study of horses in Virginia revealing prevalence rates of 32% for the overconditioned state and 19% for obesity.[17] Obese horses should be placed on a weight reduction diet and exercise increased when possible. Obese equids should be maintained on a diet of hay plus a ration balancer to provide essential nutrients. Amounts of hay can be calculated using digestible energy requirements or a simple approach can be followed in which hay is fed in amounts equivalent to 1.5% of current body weight. If body fat mass does not decrease after 4 weeks, the daily amount of hay should be lowered to 1.25% of body weight. A further reduction in hay ration to 1.0% of body weight is needed in some cases.

Dietary starch and sugar intake should be controlled in horses with ID by restricting pasture access and providing hay with lower (< 10% on a dry matter basis) non-structural carbohydrate (NSC) content. Hay with higher NSC values should soaked in cold water for 60 minutes before feeding to lower the sugar content. If additional energy is required for leaner horses, a low-NSC pelleted feed should be provided with added vegetable oil; and molasses-free beet pulp can be substituted for commercial low-sugar feeds if finances are limited. Placing a grazing muzzle on the horse and restricting the area for grazing are strategies for limiting access to sugars in grass. Pergolide should be administered to control pituitary pars intermedia dysfunction if this endocrinopathy is detected.

There are two indications for pharmacological intervention in the management of ID. The first is the acceleration of weight loss in obese animals and the second is control of postprandial hyperinsulinemia if other management strategies are unsuccessful.

Levothyroxine sodium – Levothyroxine accelerates weight loss in horses that are placed on a controlled diet and this is accompanied by increased insulin sensitivity.[18-20] Pretreatment with levothyroxine for 14 days also prevented healthy horses from developing IR following endotoxin infusion.[21] Levothyroxine is administered at an approximate dosage of 0.1 mg/kg, which is rounded to 48 mg/day for horses weighing 450 to 525 kg. A mild state of hyperthyroidism is induced and sustained for 3 to 6 months until body fat mass decreases.

Metformin hydrochloride – Since the oral bioavailability of the available metformin formulation is low (7.1 ± 1.5% in fasted horses),[22] it has been questioned whether the established effects of this drug on insulin sensitivity in humans occur in horses. Additional information was recently provided to resolve this question when Rendle et al.[23] demonstrated that metformin (30 mg/kg) given orally 30 minutes before an oral glucose tolerance test significantly lowered glucose and insulin concentrations. These findings suggest that metformin can limit postprandial hyperinsulinemia, even if its effects on insulin sensitivity are weak. The current recommendation is therefore to administer metformin at a dosage of 30 mg/kg, given 30-60 minutes before feeding. Metformin is available as 1 gram tablets and is well tolerated with the exception of oral irritation in some horses.

REFERENCES

[1]Frank N, et al. **J Vet Intern Med** 2010;24:467-475. [2]Argo CM, et al. **Vet J** 2012;194:179-188. [3]Donaldson MT, et al. **J Am Vet Med Assoc** 2004;224:1123-1127. [4]Asplin KE, et al. **Vet J** 2007;174:530-535. [5]de Laat MA, et al. **Equine Vet J** 2010;42:129-135. [6]Haffner JC, et al. **J Anim Sci** 2009;87:131-135. [7]Bailey SR, et al. **J Am Vet Med Assoc** 2007;231:1365-1373. [8]Tiley HA, et al. **Am J Vet Res** 2008;69:51-58. [9]Tadros EM, et al. **Am J Vet Res** 2013;74:1010-1019. [10]Schuver A, et al. **J Equine Vet Sci** 2013 in press. [11]Durham AE, et al. **Equine Vet J** 2009;41:924-929. [12]Carter RA, et al. **Equine Vet J** 2009;41:171-178. [13]Treiber KH, et al. **J Am Vet Med Assoc** 2006;228:1538-1545. [14]Wooldridge AA, et al. **Am J Vet Res** 2012;73:1230-1240. [15]Bertin FR, Sojka-Kritchevsky JE. **Domest Anim Endocrinol** 2012. [16]Vick MM, et al. **J Anim Sci** 2007;85:1144-1155. [17]Thatcher CD, et al. **J Vet Intern Med** 2012;26:1413-1418. [18]Frank N, et al. **Am J Vet Res** 2008;69:68-75. [19]Frank N, et al. **Am J Vet Res** 2008;69:76-81. [20]Sommardahl CS, et al. **Am J Vet Res** 2005;66:1025-1031. [21]Toth F, et al. **Am J Vet Res** 2010;71:60-68. [22]Tinworth KD, et al. **Am J Vet Res** 2010;71:1201-1206. [23]Rendle DI, et al. **Equine Vet J** 2013.

DIAGNOSIS AND MANAGEMENT OF PITUITARY DISORDERS OF THE HORSE WITH ENDOCRINOPATHIC LAMINITIS

Nicholas Frank, DVM, PhD, DACVIM
North Grafton, MA and Sutton Bonington, Leicestershire, UK

INTRODUCTION

In this postgraduate course, we will discuss diagnosis and management of pituitary pars intermedia dysfunction (PPID) in horses with endocrinopathic laminitis. This is a common condition in horses that is increasing in incidence as horses live longer and geriatric care improves. McGowan *et al.* [1] reported a prevalence rate of 21.2% for PPID in horses ≥ 15 years of age from a cross-sectional survey of horse owners in Queensland, Australia and there was a 22% prevalence rate for haircoat changes associated with PPID in aged horses in the United Kingdom.[2] Clinical signs of PPID develop when melanotropes within the pars intermedia of the pituitary gland secrete excessive amounts of hormones derived from the prohormone proopiomelanocortin (POMC). In the pars intermedia of a healthy horse, the end products of the POMC pathway are alpha melanocyte-stimulating hormone (αMSH), beta-endorphin, and corticotrophin-like intermediate peptide (CLIP). When PPID develops, secretion of αMSH and CLIP increases and other hormones derived from POMC are also released, including adrenocorticotropin hormone (ACTH), which is the primary product of the POMC conversion pathway in the pars distalis. Secretion of ACTH from the pars intermedia induces hyperadrenocorticism as hormone secretion becomes unregulated. The pars intermedia secretes only small amounts of ACTH under normal circumstances and is not under negative feedback control via the hypothalamic-pituitary-adrenal axis. Secretion of hormones by melanotropes of the pars intermedia is controlled instead by dopamine. This neurotransmitter is secreted by dopaminergic neurons that have their cell bodies within the paraventricular nuclei of the hypothalamus and extend down to the pars intermedia. These dopaminergic neurons degenerate over time as a result of oxidative damage,[3,4] and it is assumed that degeneration of dopaminergic neurons is enhanced in horses with PPID, which leads to pituitary hyperplasia, hypertrophy, and the formation of functional pituitary adenomas. Horses and ponies with equine metabolic syndrome (EMS) may be predisposed to PPID and anecdotal evidence suggests that they develop pituitary dysfunction at a younger age. This concern has led to the recommendation that all horses with EMS over 10 years of age be closely monitored for signs of PPID.

ENDOCRINOPATHIC LAMINITIS AND PPID

Mechanisms of endocrinopathic laminitis are still being elucidated, but current research focuses on hyperinsulinemia. Laminitis has been experimentally induced in ponies[5] and Standardbred horses[6] by infusing insulin intravenously and it is hypothesized that naturally-occurring laminitis develops when genetically predisposed horses experience high insulin concentrations when grazing on pasture. Insulin disorders are discussed in greater detail in the companion presentation and proceedings, but hyperinsulinemia might also be an important link between PPID and laminitis. High insulin concentrations are detected in some (but not all) horses and ponies with PPID;[7-9] and laminitis has been associated with this endocrinopathy.[8,10] Development of PPID in a horse or pony with EMS increases the risk of laminitis and high insulin concentrations are detected in animals concurrently affected by both disorders. Management of PPID with pergolide lowers insulin concentrations and anecdotal evidence suggests that the frequency and severity of laminitis episodes decreases. These observations, which require further investigation through controlled studies, suggest that PPID is a modifying factor that exacerbates hyperinsulinemia and increases the risk of laminitis. If this is the case, horses and ponies that are genetically predisposed to insulin dysregulation will have a greater risk of developing laminitis when PPID develops. Alternatively, PPID is an independent risk factor for laminitis because excess cortisol alters hoof structures and inhibits repair of laminar tissues.[11] Early recognition and management of PPID is recommended to prevent laminitis.

DIAGNOSIS OF PPID

It is important to consider the stage of disease when discussing PPID and the terms 'early PPID' and 'advanced PPID' are used here. Advanced PPID is easily diagnosed and testing is only required to confirm the diagnosis and assess severity. Clinical signs of advanced PPID include generalized hypertrichosis (previously referred to as hirsutism), lethargy, loss of skeletal muscle mass, rounding of the abdomen, polyuria and polydipsia, abnormal sweating, and laminitis in a subset of cases. Regional adiposity is detected in many, but not all affected horses, with expansion of adipose tissue depots in the neck and tailhead regions. Interestingly, horses with a history of EMS often exhibit pronounced regional adiposity as muscle loss occurs and these regions become more prominent. The easiest approach to testing horses with advanced PPID is to collect blood and measure plasma ACTH concentrations. Higher plasma ACTH concentrations are associated with more advanced/severe disease, so values obtained from this test can be used to assess severity and guide medical treatment. However, PPID differs from pituitary hyperadrenocorticism in other animals because numerous POMC peptides are released in excess and ACTH is only one of the hormones involved. Clinical signs should always be considered as PPID is diagnosed and treatment responses are assessed.

Plasma ACTH concentrations can fluctuate in horses and can vary by as much as 100 pg/mL in horses with very high (> 1,000 pg/mL) concentrations. It is surprising that these horses can tolerate such high ACTH concentrations and one explanation for this finding is that some of the ACTH measured by the assay has lower biological activity. This hypothesis was supported by results of a recent study performed in the laboratory of Dr. Dianne McFarlane in which cortisol production by adrenal explants treated with plasma from horses with advanced PPID was measured. Cortisol production increased, but not by the magnitude expected for plasma with such a high ACTH concentration.

Early PPID is more challenging to recognize and a thorough history and physical examination is required. Presenting complaints for early PPID are nonspecific and include reduced performance, loss of muscle tone and mass, and change in attitude.

Owners sometimes report that the horse is quieter than before, has a different personality, or appears lethargic. Loss of body condition occurs without changes in feed and it is common to see a reduction in epaxial muscle mass across the topline. Careful questioning of the owner reveals that the horse is taking a few weeks longer than normal to shed its winter haircoat or longer hairs are detected in specific locations. This regional hypertrichosis is common behind the elbow, beneath the mandible, and along the backs of the legs. Higher insulin concentrations may be detected as early PPID develops in horses with preexisting insulin dysregulation.

The thyrotropin-releasing hormone (TRH) stimulation test is recommended to diagnose early PPID because resting ACTH concentrations (and dexamethasone suppression test results) are often normal at this stage. When this test is performed, a baseline blood sample is collected and then 1.0 mg TRH (1 mL) is injected intravenously. A second blood sample is collected 10 minutes later and both samples are submitted for measurement of ACTH. In the past, blood was collected after 30 minutes, but a shorter time interval is preferred when this test is used in the field. Injected TRH stimulates ACTH secretion in healthy horses and those with PPID, and higher concentrations (> 110 pg/mL) are detected with PPID. Melanotropes possess TRH receptors and more ACTH is released as hyperplastic or neoplastic cells are stimulated. Use of this simulation test has facilitated earlier detection of PPID and this approach is strongly recommended for horses with only subtle clinical signs. The limited availability of TRH has restricted use of this test in the past, but protirelin is now available to practitioners from a compounding pharmacy at an approximate cost of $90 per mg. Transient yawning, coughing, mouth chomping, and display of the Flehmen response can follow administration of TRH, but signs subside within a few minutes. The effects of season on TRH stimulation test results have not been investigated, so this test should only be performed from January to June at present.

Another strategy for diagnosing early PPID is to measure ACTH concentrations in August, September, or October. Natural stimulation increases ACTH secretion at this time of the year and concentrations are higher than (seasonally adjusted) reference intervals if PPID is present. Horses with advanced PPID also remain responsive to changes in season and have markedly higher ACTH concentrations in the late summer and fall than healthy horses. Copas and Durham[12] established seasonal reference intervals for plasma ACTH concentrations in horses and reported upper limits of 29 pg/ml between November and July and 47 pg/ml between August and October for horses in the United Kingdom. Seasonally-adjusted reference intervals are also available from the Animal Health Diagnostic Center at Cornell University and a general recommendation for interpreting results in August, September, and October is to consider an ACTH concentration < 50 pg/mL to be negative for PPID, a value of 50-100 pg/mL to be equivocal, and a value > 100 pg/ml to be positive for PPID. Alpha melanocyte-stimulating hormone concentrations are also higher with PPID and increase in the late summer and fall,[13] but this assay is not offered by laboratories. When horses are being treated with pergolide, plasma ACTH concentrations can be measured biannually in the autumn and spring to determine whether the appropriate dosage has been selected.

MANAGEMENT OF PPID

Pergolide is recommended for the treatment of PPID in horses at an initial dosage of 0.002 mg/kg and is available as 1 mg tablets (Prascend®; Boehringer Ingelheim Vetmedica Inc., St. Joseph, MO). This ergot alkaloid dopamine receptor agonist is administered to restore dopaminergic inhibition of melanotropes. Interaction with D2 receptors inhibits hormone secretion and clinical signs improve. It has not been determined whether pergolide treatment also inhibits the development of pituitary hyperplasia or reduces the size of pituitary adenomas, but these beneficial effects are plausible considering the mechanism of action of the drug. Pergolide was previously marketed as Permax® (previously manufactured by Eli Lilly Co) and used for the treatment of Parkinson's disease. However, this product was voluntarily withdrawn from the market in March 2007 after the Federal Drug Administration issued a warning that pergolide was associated with increased incidence of valvular regurgitation in people.[14,15] Cardiac problems have not been encountered in horses and Prascend® was introduced in December 2011 as an FDA-approved drug for the treatment of PPID. Compounded pergolide is still available, but is not recommended because of concerns about quality control and stability. Davis et al.[16] demonstrated that pergolide was unstable over a 35-day period when prepared as an aqueous suspension, with higher temperatures and exposure to light enhancing degradation. Anorexia can be encountered when pergolide treatment is initiated, so it is advisable to start horses on half the dose every other day for 4 days and then half the dose daily for 3 days to avoid this complication. Cyproheptadine is also used for the management of PPID, but reserved for horses with advanced PPID on higher dosages of pergolide. This drug inhibits the action of serotonin, an excitatory transmitter that stimulates melanotropes. When administered with pergolide, cyproheptadine is prescribed at a dosage of 0.25 mg/kg orally every 12 hours.

The stage of disease must also be considered when monitoring treatment responses. In early PPID, clinical signs are monitored and it is reasonable to expect the horse to return to a normal appearance. Follow-up TRH stimulation tests are not required if an adequate clinical response is observed. When PPID is more advanced and diagnosed by detecting a high resting ACTH concentration, the value should be rechecked 30 days after treatment is initiated to make sure that the horse has responded adequately. If the value is still above reference interval, the dosage should be increased by 0.5 mg/day and a plasma ACTH concentration rechecked again after 30 days. It is a reasonable goal to return plasma ACTH concentrations to within reference range in most cases, but some horses with advanced PPID have exceptionally high ACTH concentrations. It is not possible to return plasma ACTH concentrations to normal in these cases, so other approaches must be considered. The first approach is to adjust the pergolide dosage until a significant downward shift in ACTH concentrations is detected. For example, a downward shift to 250 pg/mL represents a significant improvement in a horse with a plasma ACTH concentration of 1,000 pg/mL. If the owner can afford to administer pergolide at higher doses, the dosage should be incrementally increased to lower plasma ACTH concentrations. In the author's experience, pergolide can be administered to a maximum dosage of 0.01 mg/kg (5 mg for a 500-kg horse) without adverse health effects. Although this is the preferred approach, it is also practical to consider the finances of the owner and select a dosage that elicits a significant downward shift in plasma ACTH concentration, even if they remain above reference interval. Owners can be told that some of the ACTH produced in horses with advanced PPID has lower biological activity, so the clinical response is as important as the laboratory values. In the same example, lowering the ACTH concentration to 250 pg/mL could therefore be associated with a significant improvement in clinical signs. Clinical responses to treatment include increased alertness and activity, resolution of polyuria and polydipsia, lower incidence of laminitis, improved immune function and fewer bacterial infections, mitigation of hyperglycemia and hyperinsulinemia, increased muscle mass, and improvements in haircoat quality, hair length, and shedding. If a downward shift in ACTH concentrations is detected and clinical signs improve, owners should be encouraged to continue long-term treatment. Some horses show gradual improvement in clinical signs over time, even when receiving pergolide at a dosage is suboptimal from a laboratory standpoint.

Dietary recommendations for horses with PPID should be based upon body condition score and glucose and insulin results. An oral sugar test (OST) is recommended to assess the horse and determine responses to simple sugars in the diet. If there are concerns about inducing laminitis (or exacerbating laminitis in a horse that is having an episode) by performing the OST, resting glucose and insulin concentrations can be measured first. However, it is important to proceed with the OST if resting glucose and insulin concentrations are normal because the insulin response to ingested sugars must be assessed. If a high OST insulin response is detected or fasting hyperinsulinemia (> 20 mU/L) exists, care must be taken to select low-sugar feeds. Commercial low-sugar/low-starch pelleted feeds are recommended or molasses-free beet pulp can be fed with vegetable oil added as a more economical alternative. Hay with lower non-structural carbohydrate content should be provided. It is important to note, however, that many horses with PPID have normal OST insulin responses and can be fed senior feeds without concern. The stage of disease does not guide dietary recommendations in this regard; horses with advanced PPID have no greater likelihood of insulin problems. Breed susceptibility and a history of obesity are better guides for identifying high-risk horses.

OTHER CONSIDERATIONS

Hypertriglyceridemia is a concern when horses become stressed and enter negative energy balance, and this problem can develop secondary to colic, dental disease, or systemic illness.[17] Pain associated with laminitis induces stress and stimulates mobilization of fats, and a vicious cycle can develop if high lipid concentrations further suppress appetite. Type 2 diabetes mellitus contributes to hyperlipemia and is also a concern in some horses with PPID[18] if they encounter stress associated with illness or hospitalization. Blood glucose concentrations should be measured in all horses with PPID, and insulin therapy should be considered if hyperglycemia and hypertriglyceridemia are encountered.

Immunosuppression is an important component of PPID and sole abscesses contribute to lameness in some horses with endocrinopathic laminitis. White line disease can also develop and these conditions should be considered if lameness is more severe in one limb. Excessive urine production can contribute to the risk of sole abscesses and thrush by increasing moisture content in the bedding. Some horses with PPID are sounder than expected when diagnosed with laminitis develops, even with radiographic evidence of third phalanx rotation. This can be explained by the increased production of beta-endorphins in PPID that contribute to lethargy and provide pain relief.

REFERENCES

[1]McGowan TW, et al. **Equine Vet J** 2013;45:74-79. [2]Ireland JL, et al. **Equine Vet J** 2012;44:101-106. [3]McFarlane D, et al. **J Neuroendocrinol** 2005;17:73-80. [4]McFarlane D. **Ageing Res Rev** 2007;6:54-63. [5]Asplin KE, et al. **Vet J** 2007;174:530-535. [6]de Laat MA, et al. **Equine Vet J** 2010;42:129-135. [7]Reeves HJ, et al. **Vet Rec** 2001;149:449-452. [8]Walsh DM, et al. **J Equine Vet Science** 2009;29:87-94. [9]McGowan CM, et al. **Equine Vet J** 2004;36:295-298. [10]Karikoski NP, et al. **Domest Anim Endocrinol** 2011;41:111-117. [11]Johnson PJ, et al. **Vet Clin North Am Equine Pract** 2002;18:219-236. [12]Copas VE, Durham AE. **Equine Vet J** 2012;44:440-443. [13]Mc Gowan TW, et al. **Equine Vet J** 2013;45:66-73. [14]Zanettini R, et al. **N Engl J Med** 2007;356:39-46. [15]Schade R, et al. **N Engl J Med** 2007;356:29-38. [16]Davis JL, et al. **J Am Vet Med Assoc** 2009;234:385-389. [17]Dunkel B, et al. **Equine Vet J** 2013. [18]Durham AE, et al. **Equine Vet J** 2009;41:924-929.

RE-DEFINING EQUINE METABOLIC SYNDROME

Molly McCue[1], DVM, MS, PhD, DACVIM (LAIM), Raymond J. Geor[2], BVSc, PhD, DACVIM (LAIM), DACVSMR; Nichol Schultz[1], DVM
[1]St. Paul, MN, [2]East Lansing

Equine Metabolic Syndrome (EMS) refers to a cluster of clinical abnormalities associated with an increased risk of laminitis[1;2]. Johnson[1] recognized that primary features of a laminitis-prone phenotype (i.e. obesity, insulin resistance) were analogous to those described for the metabolic syndrome (MetS) in humans which is a constellation of abnormalities, including obesity, dyslipidemia, glucose intolerance and hypertension, associated with increased risk of cardiovascular disease and perhaps also type 2 diabetes mellitus[3]. The 2010 American College of Veterinary Internal Medicine (ACVIM) large animal consensus statement[2] listed several criteria for EMS based on research data available at the time, the three main criteria included:

1) Documented or suspected *insulin resistance* (IR), ie hyperinsulinemia and/or abnormal glycemic and insulinemic responses to oral or IV glucose or insulin challenges;
2) *Generalized obesity* and/or increased adiposity in specific locations (*regional adiposity*) including the nuchal ligament ("cresty neck"), the tail head, behind the shoulder, in the prepuce or mammary gland region;
3) A predisposition toward *laminitis* that develops in the absence of other recognized causes, such as grain overload, retained placenta, colitis, colic or pleuropneumonia.

Additional suggested components included:

4) Hypertriglyceridemia, dyslipidemia, and increased low density lipoprotein concentrations[4-6]
5) Hyperleptinemia[7]
6) Arterial hypertension[5;8]
7) Altered reproductive cycling in mares[9;10]
8) Increased systemic markers of inflammation in association with obesity[11]

Despite this earlier consensus statement, the features that define EMS are a subject of ongoing debate in the equine veterinary community. Descriptions of the metabolic phenotype of laminitis-prone horses and ponies have varied among published studies (Table 1)[4-6;8], making a unifying phenotypic definition difficult. The lack of consensus between reports may result from unmeasured explanatory variables, insufficient sample size to detect significant effects in the presence of confounding variables, and differences in experimental design including differences in test cohort, breed, time of sample collection (time of day and or season). Lack of consensus also likely reflects the complexity of the 'EMS' phenotype with multiple factors at both individual and environmental levels likely contributing to variation in metabolic traits.

Insulin resistance is the primary criteria used by clinicians and investigators to define EMS. However, true IR is rarely documented in EMS cases. Insulin resistance is defined as the reduced ability of a given concentration of insulin to lower blood glucose concentration. The diagnosis of IR in horses is complicated by the fact that "gold standard" methods for assessment of insulin sensitivity are not readily used in clinical practice. Further, the methods substituted for true measurement are either highly variable (e.g. measurement of fasting insulin and glucose concentrations) or have not been thoroughly evaluated as valid measures of insulin sensitivity (e.g. fasting insulin; combined glucose-insulin tolerance test [CGIT]). Thus, while there has been much discussion about IR in EMS, very few studies have reported quantitative data on insulin sensitivity and other aspects of glucose and insulin dynamics in affected animals[4;6;8]. Two studies in ponies that used minimal model analysis of a frequently-sampled glucose tolerance test (FSIGTT) have provided evidence of compensated IR, namely lower insulin sensitivity (SI), higher acute insulin response to glucose administration (AIRg; a measure of insulin secretion) and no difference in glucose tolerance when comparing laminitis-prone to non-laminitic ponies[12;13].

The majority of reports have used "fasting" or "resting" measures of insulin and glucose and/or indices derived from these measurements as surrogate indicators of IR, with the current consensus view that a fasting insulin concentration >20 mIU/L indicates IR. Fasting hyperinsulinemia is typically (but not always) accompanied by normoglycemia, which suggests compensated IR, i.e. an increase in pancreatic insulin secretion occurs in response to reduced tissue insulin sensitivity, resulting in maintenance of glucose homeostasis. The prevalence of hyperinsulinemia (insulin concentration > 20 mIU/L) was 18% in one study of 300 randomly selected mixed-breed horses in southwest Virginia[14], while another study in Australia reported a 28% prevalence of hyperinsulinemia in randomly selected ponies[15]. In the pony study, age, BCS, supplementary feeding and a history of laminitis were idnetified as risk factors for hyperinsulinemia[15]. Hyperinsulinemia has been shown to be a feature in the EMS phenotype in ponies[4;8] and Morgan horses[6], although there is wide variation in values among studies that may in part be explained by differences in sampling conditions (e.g. feed withholding vs. pasture grazing prior to collection of blood samples).

Evidence of obesity and/or the presence of one or more enlarged subcutaneous fat deposits (regional adiposity) has been regarded as one of the defining characteristics of EMS. Horses and ponies diagnosed with EMS may be obese, defined as a Henneke scales BCS of ≥7/9, and/or have regional adiposity. Common sites of regional fat accumulation include the nuchal ligament region (cresty neck), behind the shoulder (uni- or bilateral), around the tail-head, and in the preputial or mammary gland regions. Generalized obesity and/or regional adiposity are thought to play a role in the EMS phenotype, with the idea that expansion in adipose tissue mass invokes a local and systemic inflammatory response that, in turn, results in the development of IR and other metabolic abnormalities[16;17]. Studies in horses have demonstrated an inverse relationship between BCS and specific measures of insulin sensitivity[11] and positive correlations between adiposity, resting insulin concentrations and blood markers of inflammation (e.g. serum amyloid A, TNF-α, mRNA encoding for IL-1β and TNF-α)[11;18]. Induction of a ~20% weight gain in Arabian geldings has been shown to result in a reduction in insulin sensitivity,[44] while weight loss in obese ponies resulted in improvement in glucose tolerance and decreased fasting insulin concentrations (Geor et al unpublished data); while another study of horses and ponies reported no change in insulin sensitivity after an~20% increase in fat mass[19]. Interestingly, moderate weight gain (~15% increase in body weight) in Thoroughbred geldings failed to alter insulin sensitivity[20]. These observations reinforce the idea that factors other than BCS and adipose tissue mass contribute to differences in insulin sensitivity, but also raise the possibility that there is breed variation in regards to the impact of obesity on metabolic function.

Increased plasma triglyceride concentrations were a feature of the EMS phenotype in a closed herd of Welsh ponies[4] and was also documented in an out-bred population of ponies with a history of recurrent laminitis[8], although hypertriglyceridemia was variable with season. In a small, mixed-breed group of obese, insulin-resistant horses increased serum non-esterified fatty acid (NEFA) concentrations (but *not* triglycerides) were detected[6], whereas serum NEFAs were not useful in the differentiation of an EMS phenotype in ponies[4;5]. In one report, increased very low-density lipoprotein triglyceride concentrations were observed in horses with EMS[6].

Hypertension, altered serum adipokine concentrations, and systemic inflammation, all of which are features of metabolic syndrome in humans, may be additional components of the EMS phenotype. Bailey and colleagues[8] detected arterial hypertension in mixed-breed, recurrent laminitic ponies during summer but not in winter, while in our own studies mean blood pressure did not differ between groups of EMS vs. non-EMS ponies when measured in early spring, summer and fall[8]. Increased serum leptin has been reported in Welsh ponies with other features of the EMS phenotype[5], while other studies have reported a positive association between BCS and circulating inflammatory biomarkers and a significant inverse relationship between apparent adiposity and serum adiponectin[21;22].

In an effort to address some of the apparent discrepancies between previous reports we have recently completed phenotyping of a large across-breed cohort of horses and ponies representing a range of metabolic phenotypes. We have collected phenotypic, epidemiologic and environmental data from 610 horses/ponies from 166 farms. The predominant breeds sampled were Morgan horses, Tennessee Walking Horses (TWH), Arabs, Quarter Horses (QH) and Welsh ponies. Phenotypes include: measures of generalized and localized obesity (neck- and girth- to height ratio, body condition score), fasting blood glucose (GLU), insulin (INS), ACTH, triglyceride, non-esterified fatty acid, leptin and adiponectin (APN) concentrations, and post oral-sugar test insulin (INS_OST) and glucose. Additional data include medical history, laminitis status, age, breed, and sex, as well as environmental variables including month of sampling, latitude, housing, diet (hay, pasture grain and supplement samples all submitted for complete dietary analysis), exercise, and other management practices.

Multivariate, multilevel, multiple regression analysis was used to evaluate the 11 phenotypic responses and 16 individual (i.e., age, breed, etc) and environmental (i.e., diet, management, exercise) explanatory variables. Although complex, our statistical model allows relationships between explanatory and response variables to be evaluated at 3 levels. Level 1 is the multivariate responses within individual: for example normal physiologic control of glucose homeostasis is a complex system with many regulatory/counter-regulatory factors. Multivariate analysis allows for interpretation of the biochemical traits (i.e., insulin, glucose, triglycerides, adiponectin, etc) together, estimation of the correlation between traits, and identification of perturbations of the normal physiologic response. Level 2 is analogous to a multiple regression for differences between individuals: this level allows for elucidation of the effect of individual predictors (such as age, gender or breed) on metabolic traits. Level 3 accounts for similarities between individuals on the 166 farms: correlations between individuals at the farm level suggest that phenotypes are influenced by the shared environment. This statistical modeling also allows for partitioning of the phenotypic variability that is explained or not explained by the 16 measured predictors into unmeasured individual factors and environmental (farm) factors. From this analysis we have been able to asses several key aspects of EMS including:

1) How measured phenotypes and biochemical traits influence each other, and how these relationships differ in horses that are obese, have a history of laminitis or both;
2) How individual physiologic variables such as age, breed and gender affect trait measures;
3) The influence of environmental variables such as dietary adaptation, housing and exercise influence trait measures;
4) The relative importance of each of these confounding factors; and
5) How the measures differ in horses when pared into clinical groups (ie non-obese/non-laminitic (NO-NL), non-obese/ lamintic (NO-L), obese/non-lamnitic (O-NL), and obese/laminitic (O-L).

We believe that these data will allow for a more comprehensive understanding of all the factors that are in play in the measured metabolic phenotype in EMS 'suspect' horses; and further help define the key measures that are consistently different in horse with a previous history of laminitis in relation to those without. Our data indicate that elevated fasting insulin and triglycerides, elevated insulin 75 minutes post oral sugar challenge (OST insulin) and low serum adiponectin are the four measurements most consistently present in individuals with a history of laminitis, regardless of obesity status, age, gender or breed. Hypertriglyceridemia associated with laminitis history confirms previous findings in ponies[4;5], and demonstrates that hypertriglyceridemia, but not elevated NEFAs, are a key feature in horses, in contrast with previous reports[6]. Fasting insulin, OST insulin and triglycerides measurements are also higher in obese individuals without a history of laminitis, albeit not at the levels seen in previously laminitic horses. One interpretation of these findings is that obesity is not a requisite feature of the laminitis-prone (i.e. EMS) phenotype, but components of the phenotype such as IR may be exacerbated when affected animals become obese. Our preliminary observations in EMS mirror these findings in humans – IR and hypertriglyceridemia can occur in non-obese horses while, conversely, the metabolic profile of some obese horses does not differ from that in lean individuals. This interpretation is consistent with anecdotal clinical observations of the laminitis-prone phenotype – not all affected animals express obesity or regional adiposity while, conversely, not all obese animals appear to be at increased risk of laminitis.

The assertion that obesity is not a canonical feature of EMS also has implications for the use of adipokines as reliable diagnostic markers. In horses there is a positive correlation between circulating leptin concentrations and fat mass and/or body condition score (BCS)[5;21;23;24], and leptin concentrations increase with weight gain [25] and decrease with dietary restriction and weight loss[26]. Frank et al 2006[6] showed that resting leptin concentrations were significantly higher in EMS 'cases' when compared to healthy controls, and also observed that resting leptin concentrations were positively correlated to insulin area under the curve (AUC) values during a combined glucose-insulin test. Carter et al 2009[5] demonstrated that leptin concentrations, in combination with basal insulin and measures of obesity in the prediction of laminitis in ponies. However, other factors have been shown to affect leptin measurements in horses; large inter-individual variation in horses and ponies of similar body condition[7;9;27], and differences due to gender[7;9;23;27] and breed[24;28] have been reported. In our cohort, leptin levels were strongly correlated to measures of obesity, but not laminitis status, suggesting that leptin is primarily a marker for adiposity. In contrast, adiponectin levels were significantly lower in previously laminitic horses than in normal or obese non-laminitc horses; suggesting that adiponectin may be a more reliable diagnostic marker.

In addition to differences in obese and/or laminitic horses, we have also identified significant differences in the relevant biochemical measurements between breeds regardless of clinical phenotype. Several review articles have raised the possibility of breed predisposition to EMS, with Welsh and Dartmoor ponies, Morgans, TWH, Saddlebreds, Arabian and Paso Fino breeds thought to be more susceptible. Moreover, the observation that primary phenotypic characteristics of EMS (i.e. abnormal insulin dynamics, hypertriglyceridemia) are shared across breeds; and that some features are different between breeds raises the possibility of specific genetic risk factors; this aspect is the focus on ongoing studies at the University of Minnesota.

Analysis of this large and comprehensive dataset in a cross-sectional cohort of horses and ponies has highlighted the complexity of the relationships and interactions between measured phenotypes in EMS and the wide range of factors that affect these measures. Despite a comprehensive consensus statement defining EMS, interpretation of phenotypic measures and the assignment of the EMS diagnosis are confounded by multiple factors including obesity (BCS), gender, breed and age as well as environmental factors. We believe that the complexities of these relationships draw attention to several key ideas:

1) Insulin dysregulation (hyperinsulinemia and exaggerated response to oral challenge in this cohort), elevated serum triglycerides and low serum adiponectin are the most consistent measurable differences in horses with a history of laminitis.
2) It is possible that obesity is likely not a primary cause of EMS but instead it may be a marker of an underlying metabolic dysfunction that, depending on other environmental factors (e.g. diet, level of physical activity), drives adipose tissue accretion and development of obesity. If so, there are two very important implications in the identification and management of EMS. First, the presence or absence of obesity cannot be used as a diagnostic criterion. Second, whereas dietary restriction and weight loss may result in some improvement in insulin sensitivity etc. in affected animals, the underlying metabolic dysfunction is likely to persist.
3) Known physiologic relationships and the strong correlations between biochemical measures means that each has to be considered in light of the others when using experimental investigation into EMS pathophysiology or assessing the utility of proposed diagnostic criterion.
4) Many factors impacting metabolic phenotype have yet to be explored; these factors be they physiologic, genetic, or environmental variables, will hopefully shed light on EMS pathophysiology.

Table 1: A summary of findings related to obesity, regional adiposity and endocrine/metabolic variables in published studies of the equine metabolic syndrome (EMS) phenotype.

**Data obtained from the same population of Welsh and Dartmoor ponies; **serum insulin and triglyceride concentrations and RISQI differed between ponies with and without a history of laminitis in summer but not in winter*

BCS = body condition score; RISQI = reciprocal of the square root of the serum insulin concentration; CGIT = combined glucose-insulin tolerance test; NEFAs = nonesterified fatty acids

	Treiber et al. 2006*	Frank et al. 2006	Bailey et al. 2008	Carter et al. 2009*
Breed(s)	Welsh and Dartmoor ponies	6 breeds	Mixed-breed ponies	Welsh and Dartmoor ponies
Sample size	160	12	80	74
Obesity (BCS) in EMS	Yes	Yes	No	Yes
Regional adiposity in EMS	Yes	Yes	No	Yes
Hyperinsulinemia in EMS	Yes (21.6 vs. 10.7 mU/L)	Yes (50.5 vs. 9.1 mU/L)	Yes** (69.5 vs. 21.5 mU/L)	Yes (20.5 vs. 8.8 mU/L)
Insulin resistance in EMS	Yes (RISQI)	Yes (CGIT)	Yes** (RISQI)	Yes (RISQI)
Fasting glucose	Not different	Higher in EMS (83.9 vs. 66.9 mg/dl)	Not different	Not different
Triglycerides	Higher in EMS (97.2 vs. 52.3 mg/dl)	Not different	Higher in EMS** (0.55 vs. 0.38 mmol/L)	Higher in EMS (53.0 vs. 39.0 mg/dl)
NEFAs	Not different	Higher in EMS (366.5 vs. 197.1 μmol/L	Not evaluated	Not evaluated

REFERENCES

1. Johnson P (2002), The equine metabolic syndrome: peripheral Cushing's syndrome, *Vet.Clin.North Am Equine Pract.* 18: 271-293
2. Frank N, Geor RJ, Bailey SR, Durham AE, Johnson PJ (2010), Equine metabolic syndrome, *J.Vet.Intern.Med.* 24: 467-475
3. Reaven GM (2011), Insulin Resistance: the Link Between Obesity and Cardiovascular Disease, *Medical Clinics of North America* 95: 875-892
4. Treiber KH, Kronfeld DS, Hess TM, Byrd BM, Splan RK, Staniar WB (2006), Evaluation of genetic and metabolic predispositions and nutritional risk factors for pasture-associated laminitis in ponies, *J.Am.Vet.Med.Assoc.* 228: 1538-1545
5. Carter RA, Treiber KH, Geor RJ, Douglass L, Harris PA (2009), Prediction of incipient pasture-associated laminitis from hyperinsulinaemia, hyperleptinaemia and generalised and localised obesity in a cohort of ponies, *Equine Vet.J.* 41: 171-178
6. Frank N, Elliott SB, Brandt LE, Keisler DH (2006), Physical characteristics, blood hormone concentrations, and plasma lipid concentrations in obese horses with insulin resistance, *J.Am.Vet.Med.Assoc.* 228: 1383-1390
7. Cartmill JA, Thompson DL, Jr., Storer WA, Gentry LR, Huff NK (2003), Endocrine responses in mares and geldings with high body condition scores grouped by high vs. low resting leptin concentrations, *Journal of Animal Science* 81: 2311-2321
8. Bailey SR, Habershon-Butcher JL, Ransom KJ, Elliott J, Menzies-Gow NJ (2008), Hypertension and insulin resistance in a mixed-breed population of ponies predisposed to laminitis, *Am.J.Vet.Res.* 69: 122-129
9. Gentry LR, Thompson DL, Jr., Gentry GT, Jr., Davis KA, Godke RA, Cartmill JA (2002), The relationship between body condition, leptin, and reproductive and hormonal characteristics of mares during the seasonal anovulatory period, *Journal of Animal Science* 80: 2695-2703
10. Vick MM, Sessions DR, Murphy BA, Kennedy EL, Reedy SE, Fitzgerald BP (2006), Obesity is associated with altered metabolic and reproductive activity in the mare: effects of metformin on insulin sensitivity and reproductive cyclicity*, *Reproduction, Fertility and Development* 18: 609-617
11. Vick MM, Adams AA, Murphy BA, Sessions DR, Horohov DW, Cook RF, Shelton BJ, Fitzgerald BP (2007), Relationships among inflammatory cytokines, obesity, and insulin sensitivity in the horse, *Journal of Animal Science* 85: 1144-1155
12. Bailey SR, Menzies-Gow NJ, Harris PA, Habershon-Butcher JL, Crawford C, Berhane Y, Boston RC, Elliott J (2007), Effect of dietary fructans and dexamethasone administration on the insulin response of ponies predisposed to laminitis, *J.Am.Vet.Med.Assoc.* 231: 1365-1373
13. Treiber KH, Kronfeld DS, Hess TM, Byrd BM, Splan RK (2005), Pre-laminitic metabolic syndrome in genetically predisposed ponies involves compensated insulin resistance, *Journal of Animal Physiology and Animal Nutrition* 89: 430-431
14. Pleasant RS, Suagee JK, Thatcher CD, Elvinger F, Geor RJ (2013), Adiposity, Plasma Insulin, Leptin, Lipids, and Oxidative Stress in Mature Light Breed Horses, *Journal of Veterinary Internal Medicine* 27: 576-582
15. McGowan CM, Geor RJ, McGowan T. The prevelance and incidence of hyperinsulinemia in ponies. Journal of Veterinary Internal Medicine 22, 734. 2008. Ref Type: Abstract
16. Rasouli N, Kern PA (2008), Adipocytokines and the metabolic complications of obesity, *Journal of Clinical Endocrinology & Metabolism* 93: s64-s73
17. Despres JP, Lemieux I (2006), Abdominal obesity and metabolic syndrome, *Nature* 444: 881-887
18. Suagee JK, Corl BA, Crisman MV, Pleasant RS, Thatcher CD, Geor RJ (2013), Relationships between Body Condition Score and Plasma Inflammatory Cytokines, Insulin, and Lipids in a Mixed Population of LightGÇÉBreed Horses, *Journal of Veterinary Internal Medicine* 27: 157-163

19. Quinn RW, Burk AO, Hartsock TG, Petersen ED, Whitley NC, Treiber KH, Boston RC (2008), Insulin sensitivity in Thoroughbred geldings: effect of weight gain, diet, and exercise on insulin sensitivity in Thoroughbred geldings, *Journal of Equine Veterinary Science* 28: 728-738
20. Quinn RW, Burk AO, Hartstock TG, et al (2009), Insulin sensitivity in Thoroughbred geldings: effect of weight gain, diet and excercise, *Journal Equine Veterinary Science* 28: 728-738
21. Kearns CF, McKeever KH, Roegner V, Brady SM, Malinowski K (2006), Adiponectin and leptin are related to fat mass in horses, *Vet.J.* 172: 460-465
22. Wooldridge AA, Edwards HG, Plaisance EP, Applegate R, Taylor DR, Taintor J, Zhong Q, Judd RL (2012), Evaluation of highGÇômolecular weight adiponectin in horses, *American journal of veterinary research* 73: 1230-1240
23. Buff PR, Dodds AC, Morrison CD, Whitley NC, McFadin EL, Daniel JA, Djiane J, Keisler DH (2002), Leptin in horses: Tissue localization and relationship between peripheral concentrations of leptin and body condition, *Journal of Animal Science* 80: 2942-2948
24. Pratt-Phillips SE, Owens KM, Dowler LE, Cloninger MT. Assessment of Resting Insulin and Leptin Concentrations and Their Association With Managerial and Innate Factors in Horses. 30[3], 127-133. 2010.
Ref Type: Abstract
25. Carter RA, McCutcheon LJ, George LA, Smith TL, Frank N, Geor RJ (2009), Effects of diet-induced weight gain on insulin sensitivity and plasma hormone and lipid concentrations in horses, *Am.J.Vet.Res.* 70: 1250-1258
26. Van WS, Buyse J, Janssens GP (2009), Increased plasma leptin through l-carnitine supplementation is associated with an enhanced glucose tolerance in healthy ponies, *J.Anim Physiol Anim Nutr.(Berl)* 93: 203-208
27. Caltabilota TJ, Earl LR, Thompson DL, Jr., Clavier SE, Mitcham PB (2010), Hyperleptinemia in mares and geldings: assessment of insulin sensitivity from glucose responses to insulin injection, *Journal of Animal Science* 88: 2940-2949
28. Carter RA, Geor RJ, Burton SW, Cubitt TA, Harris PA (2009), Apparent adiposity assessed by standardised scoring systems and morphometric measurements in horses and ponies, *Vet.J.* 179: 204-210

GENETIC BASIS OF METABOLIC TRAITS IN HORSES

Molly McCue, DVM, MS, PhD, DACVIM (LAIM), and Nichol Schultz, DVM
St. Paul, MN

Equine metabolic syndrome (EMS) is a devastating disease characterized by three main features: obesity, insulin resistance, and laminitis. Certain breeds or individual horses are predisposed to EMS, and are often referred to as "easy keepers". These horses are very efficient at utilizing calories and often require a lower plane of nutrition to maintain body weight than other horses. This tendency to be an easy keeper likely begins with a genetic pre-disposition.

COMPLEX GENETIC DISEASE

Unlike the more familiar genetic diseases in horses that are inherited as simple (or Mendelian) traits, for example: Hyperkalemic Periodic Paralysis (HYPP), Severe Combined Immunodeficiency Syndrome (SCIDS), Overo Lethal White Syndrome (OWLS)[1], Junctional Epidermolysis Bulosa (JEB) in Belgians[2] and Saddlebreds[3], Glycogen Branching Enzyme Deficiency (GBED)[4], Malignant Hyperthermia (MH)[5], Hereditary Equine Regional Dermal Asthenia (HERDA)[6], Lavender Foal Syndrome (LFA) [7], Fell Pony Syndrome (Foal Immunodeficiency Syndrome, FIS) [8]; EMS is likely a complex genetic trait. A complex genetic disease is the result of one or more inherited genetic mutations (typically incompletely penetrant) and influenced by the environment. When an individual inherits a genetic mutation in a complex disease, it is not a guarantee they will develop disease, but rather, it just means they are at increased risk. A common example of a complex disease is breast cancer risk associated with the *BRCA* (1 or 2) mutation in humans. If a woman inherits a *BRCA* mutation, she has a five times greater risk of developing breast cancer. The *BRCA* mutation is not a guarantee that breast cancer will develop, as other factors including diet, exercise and overall health play a role in whether or not cancer develops. However, because of the increased risk, women with a *BRCA* mutation are screened for breast cancer more frequently starting at a younger age.

The variability in the phenotypic expression of type 1 Polysaccharide Storage Myopathy (PSSM1) in horses with the mutant *GYS1* allele provides a relatively 'simple' example of a complex genetic disease in the horse. Phenotypic expression of PSSM1 in horses with the *GYS1* mutation varies widely from subclinical disease, to severe clinical disease resulting in recumbency and in some cases euthanasia. Within Quarter Horses some individuals are completely incapacitated by this disease and cannot be exercised due to the frequent occurrence of muscle pain and cramping. In contrast, other individuals have very mild signs and are able to be functional competitive athletes[9]. Phenotypic variability in clinical signs and histopathologic abnormalities, cannot fully be explained by *GYS1* genotype (ie homozygous vs. heterozygous)[10]. Environmental factors and modifying genes may all contribute to variable phenotypic expression in Mendelian traits[11;12]; and this is clearly the case with PSSM1 in Quarter Horses. Providing dietary fat as an alternative energy substrate to carbohydrate, limiting the daily consumption of starch, and gradually increasing daily exercise decreases muscle stiffness and eliminates episodes of exertional rhabdomyolysis in 75% of Quarter Horses with PSSM[13;14]. When these environmental factors are controlled either in research settings or on large breeding operations where numerous individuals are housed and managed identically, phenotypic variation in the severity of PSSM1 is still evident[14;15]. Part of this variation in Quarter Horses can be explained by modification of the PSSM1 phenotype in the presence of the *RYR1* mutant allele responsible for malignant hyperthermia[5;10]. Clinical signs are also variable between breeds with exertional rhabdomyolysis and elevated serum creatine kinase activities (CK) frequently reported in Quarter Horses, whereas muscle atrophy, weakness and recumbency (sometimes without dramatic elevations in CK) reported more frequently in draft breeds[14;16-19]. All PSSM1 horses from different breeds with a *GYS1* defect share the same mutation; thus disease heterogeneity does not explain the variability in clinical presentation. It is likely that these breed differences are due to differences in 'genetic background' or the presence of other alleles that modify muscle and/or metabolic phenotype between breeds. Thus, while the *GYS1* allele has a dominant mode of inheritance with 95% penetrance when the phenotypic outcome is the presence of abnormal polysaccharide at greater than 2 years of age,[20] and an incompletely dominant model of inheritance when histopathologic changes in skeletal muscle are used as the phenotype;[21] clinical expression of the disease itself is a complex trait, where the horse's *GYS1* genotype, environment, breed and presence of modifying genetic loci must all be taken into account when estimating the likelihood of expressing the PSSM1 phenotype (Figure 1).

Figure 1. Risk model for showing clinical disease for PSSM1 based on known genotypic and environmental factors.

Risk of PSSM1 clinical disease	=	*GYS1 genotype (ie normal [N/N], heterozygous [N/P] or homozygous [P/P]*	+	*RYR1 genotype (ie normal [N/N], heterozygous*	+	*Environmental risk: high starch diet vs low starch diet*	+	*Environmental risk: consistent daily exercise vs limited*	+	*Genetic background (ie breed)*	+	*genotype x genotype or genotype x environment interaction*	+ ...	

PROPOSED RISK MODEL FOR EMS

Environmental factors including excessive nutrition, often from pasture, have been linked to EMS, obesity, insulin resistance and laminitis. Clinical experience with EMS animals suggests that episodes of laminitis often occur in pastured animals coincident with an increase in the pasture forage nonstructural carbohydrate (NSC, which includes sugars, fructans and starches) content, i.e.

rapidly growing pastures in the spring, after summer and fall rains, or after frost. Episodes of laminitis in pastured ponies have been associated with exacerbation of hyperinsulinemia[22;23], likely due to the increase in NSC intake although an effect of season per se on insulin sensitivity and dynamics is also possible. However, the high planes of nutrition and/or changes in the pasture do not result in EMS in all horses. Recent studies indicate this individual variability may be due to an underlying genetic predisposition[22;24;25]. We propose that the risk for EMS and laminitis is a complex disease where the risk of developing the disease is a result of an individual's unique combination of genetic alleles, in concert with environmental influences are responsible for disease risk (Figure 2). Precedent for this model also comes from work in other species. Although regulation of insulin, glucose and lipid metabolism in health and disease is complex and not completely understood, work in humans and rodent models[26;27] has identified multiple primary genetic risk alleles that result in alterations in insulin, glucose and/or lipid metabolism. These genes interact in large common pathways and a related group of metabolic phenotypes (obesity, IR, dyslipidemia) that increase the risk of cardiovascular disease and type 2 diabetes mellitus, with the relative importance of a given risk allele varying between populations and ethnic groups.

Figure 2. Proposed complex genetic risk model for EMS

Risk of metabolic disease	=	*Risk conferred by genetic risk allele 1 multiplied by copies of risk allele 1*	+	*Risk allele 2 x 0, 1, or 2*	+	*Risk allele 3 x 0, 1, or 2*	+	*Risk conferred by environment factor 1*	+	*Risk conferred by environment factor 2 (eg exercise)…*	+	*Risk conferred by genotype by environment interaction (eg allele 1 x diet)*	+	*Risk conferred by genotype by genotype interaction (eg allele 1 x allele 2)*	+	…..

While complex genetic disease risk can be thought of as a summation of all risk factors (and the effects of their interactions), it is important to remember that the relative importance of each risk factor can vary between individuals or groups of individuals. Figure 3 depicts this concept. For example two individual horses may have a similar (moderate) risk for developing EMS due to genetics and environment. However, in one individual, this risk may be the summation of a few genetic alleles and many environmental factors (Figure 3A), whereas another individual may have many risk genetic factors, but be housed in a more optimal environment (Figure 3B). The ideal scenario is in Figure 3C, where the individual has few genetic and environmental risks.

Figure 3

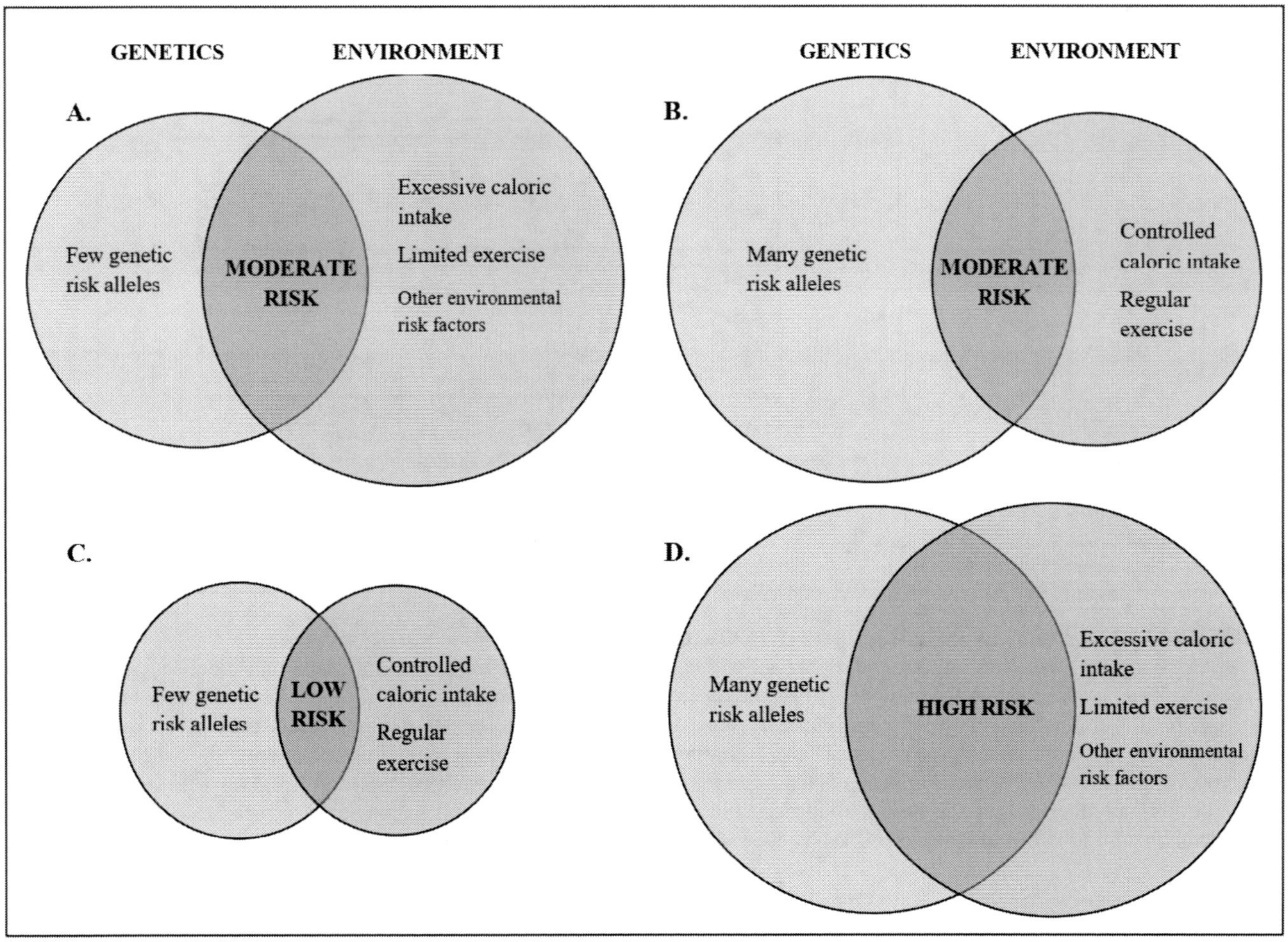

Manipulation of the environment would decrease the risk in scenarios A and D; moving the individual risk to low and moderate respectively. In scenario B environmental manipulation would have less impact.

EMS GENETIC RISK ACROSS BREEDS

Anecdotally, certain horse breeds such as Morgans, Arabians, Paso Finos, Tennessee Walking Horses, and most ponies, including the Welsh Pony, are more susceptible to EMS and subsequent laminitis. Other breeds such as the Quarter Horse and Thoroughbred appear to be less susceptible, though objective data in most of these breeds is lacking[28-30]. Affected horses and ponies appear to have high metabolic efficiency, meaning they require fewer calories for maintenance of bodyweight when compared to unaffected animals (i.e. they are "easy keepers"). Ponies in general have lower insulin sensitivity[25] and a higher prevalence of hyperinsulinemia when compared to horses[31,32]. Our preliminary data indicate that breeds of horses can share key metabolic features of EMS, such as hyperinsulinemia, exaggerated response to oral glucose, and elevation in serum triglycerides, but breeds can differ in magnitude of these responses or other features, such as fasting insulin, triglyceride, non-esterified fatty acid and adipokine levels. We propose that major genetic risk factors leading to EMS and laminitis susceptibility are shared across breeds, while differences in the severity and secondary features of the EMS phenotype between breeds, or between individuals within a breed, result from modifying genetic risk factors with variable frequencies. Figure 4 depicts a hypothetical model for EMS/laminitis risk allele sharing across breeds. In this model five alleles combine to confer EMS risk (circle sizes indicated relative frequency of each allele in each breed). All breeds share two major alleles resulting in hyperinsulinemia (**allele A**) and hypertriglyceridemia (**allele C**). Each breed also has a unique combination of modifying alleles (**B**, **D**, **E**, **F**) with different hypothetical effects explaining the breed differences in phenotypic means; e.g., **allele B→** altered insulin release and/or clearance and increased fasting/OST insulin, **allele D→** altered tissue utilization or lipolysis and increased triglycerides, **allele E→** enhanced insulin sensitivity/improved clearance and lower fasting insulin/OST insulin, **allele F→** elevated NEFAs. In this model: allele B frequency in Breed 3 is greater than Breed 2 which is greater than Breed 1 (ie Breed3 > Breed 2> Breed 1), which could explain higher mean insulin in Breeds 2 and 3 compared to Breed 1. In contrast, allele E which is unique to Breed 4 results in lower fasting insulin and lower insulin responses to glucose challenge. The high frequency of allele D and allele F (nearing 100%) in Breed 1 could explain differences in breed 1 means for other measures such as triglycerides and NEFAs.

Figure 4. Hypothetical polygenic genetic model explaining differences in breed means in measured metabolic traits

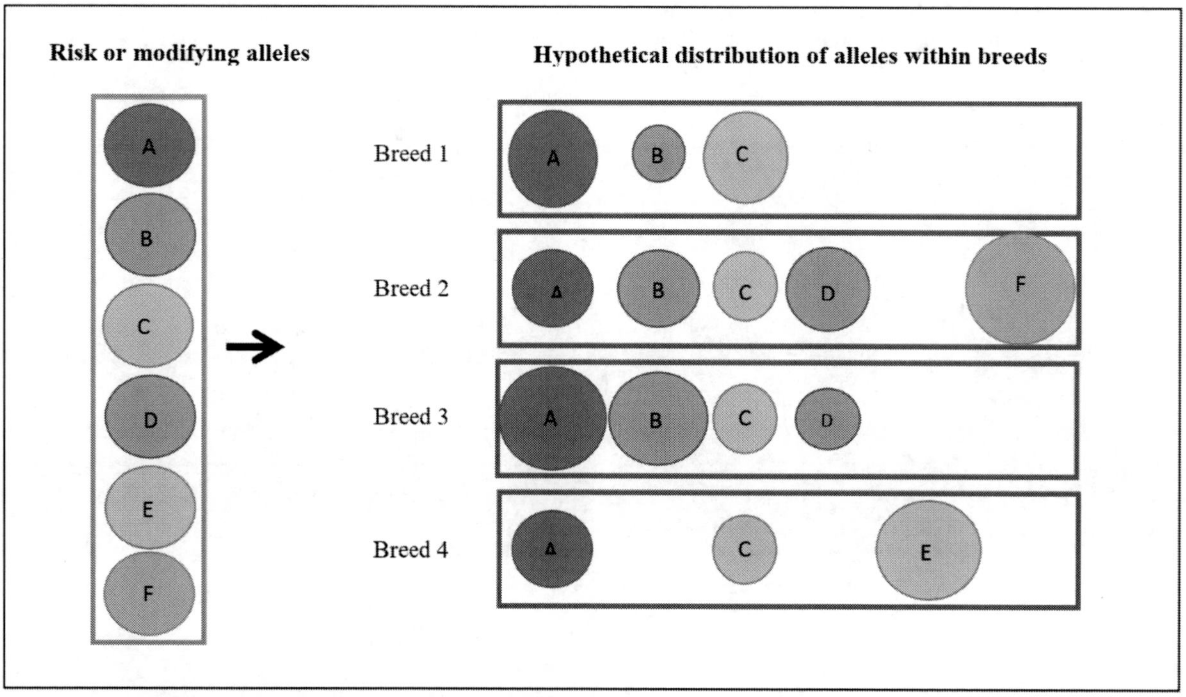

DISSECTING THE GENETICS OF EMS AND METBOLIC TRAITS

Identifying the genes and alleles underlying individual variability in the measured hormonal and biochemical measures of EMS susceptibility, and unraveling the complexity of gene by environmental interactions and variable phenotypic expression between breeds, requires a combination of genetic approaches. We have recently completed a genome wide association analysis in 286 Morgan horses for which 11 phenotypic responses and 16 individual (i.e., age, breed, etc) and environmental (i.e., diet, management, exercise) explanatory variables were collected. This genome wide association analysis has identified several genomic regions underlying traits including fasting and post-challenge insulin concentrations, and leptin, adiponectin and triglyceride concentrations. We are currently investigating these associations by identifying the putative functional mutations present in these genomic regions using whole-genome sequence data from 18 Morgans. To date we have identified 2 putative functional variants in a single gene associated with insulin responses. Follow-up with functional analysis and genetic epidemiology of these variants and the detection of gene-by-environment interaction is on-going. A similar genome wide association approach in Welsh Ponies is being used to detect breed/pony-specific alleles.

In addition, candidate genomic regions harboring putative EMS risk alleles have been identified through discovery of genomic regions unique to breeds that are anecdotally at high risk for EMS[33]. One of the initial genomic regions discovered harbored two genes integral in the incretin response. Sequencing of these genes found two putative functional variants. Quantification of incretin hormones and insulin and glucose responses in a cohort of Morgans and Welsh Ponies allowed for correlation of these genotypes to phenotype in the incretin pathway. This same approach has also identified putative EMS protective alleles that warrant further investigation.

SUMMARY

The genetics of metabolic trait variation and EMS/ laminitis risk in horses is complex, with variability in risk alleles, allele penetrance, environmental risk factors, gene-by environment and gene-by-gene interaction all playing a role in an individual's risk. Further, the variable phenotypes across breeds likely reflect the alleles segregating within each breed and their relative importance. While we have made progress in identifying the breed differences in metabolic phenotypes and have begun to identify alleles putatively associated with measured metabolic traits, dissecting out the all factors depicted in Figure 2 and the alleles depicted in Figure 4 will likely require several additional investigations.

REFERENCES

1. Santschi EM, Purdy AK, Valberg SJ, Vrotsos PD, Kaese H, Mickelson JR (1998), Endothelin receptor B polymorphism associated with lethal white foal syndrome in horses, *Mamm.Genome* 9: 306-309
2. Baird J.D., Millon LV, Dileanis S et al. Junctional Epidermolysis Bullosa in Belgian Draft Horses. 2003.
3. Graves KT, Henney PJ, Ennis RB (2009), Partial deletion of the LAMA3 gene is responsible for hereditary junctional epidermolysis bullosa in the American Saddlebred Horse, *Anim Genet.* 40: 35-41
4. Ward TL, Valberg SJ, Adelson DL, Abbey CA, Binns MM, Mickelson JR (2004), Glycogen branching enzyme (GBE1) mutation causing equine glycogen storage disease IV, *Mamm.Genome* 15: 570-577
5. Aleman M, Riehl J, Aldridge BM, LeCouteur RA, Stott JL, Pessah IN (2004), Association of a mutation in the ryanodine receptor 1 gene with equine malignant hyperthermia, *Muscle Nerve* 30: 356-365
6. Tryon RC, White SD, Bannasch DL (2007), Homozygosity mapping approach identifies a missense mutation in equine cyclophilin B (PPIB) associated with HERDA in the American Quarter Horse, *Genomics* 90: 93-102
7. Brooks SA, Gabreski N, Miller D, Brisbin A, Brown HE, Streeter C, Mezey J, Cook D, Antczak DF (2010), Whole-Genome SNP Association in the Horse: Identification of a Deletion in Myosin Va Responsible for Lavender Foal Syndrome, *PLoS Genet* 6: e1000909
8. Fox-Clipsham LY, Carter SD, Goodhead I, Hall N, Knottenbelt DC, May PD, Ollier WE, Swinburne JE (2011), Identification of a mutation associated with fatal foal immunodeficiency syndrome in the fell and dales pony, *PLoS.Genet.* 7: e1002133
9. McCue ME, Valberg SJ (2007), Estimated prevalence of polysaccharide storage myopathy among overtly healthy Quarter Horses in the United States, *J Am Vet.Med Assoc.* 231: 746-750
10. McCue ME, Valberg SJ, Jackson M, Borgia L, Lucio M, Mickelson JR (2009), Polysaccharide storage myopathy phenotype in quarter horse-related breeds is modified by the presence of an RYR1 mutation, *Neuromuscul.Disord.* 19: 37-43
11. Nadeau JH (2001), MODIFIER GENES IN MICE AND HUMANS, *Nat Rev Genet* 2: 165-174
12. Badano JL, Katsanis N (2002), Beyond Mendel: an evolving view of human genetic disease transmission, *Nature Reviews Genetics* 3: 779-789
13. Firshman AM, Valberg SJ, Bender JB, Finno CJ (2003), Epidemiologic characteristics and management of polysaccharide storage myopathy in Quarter Horses, *Am.J.Vet.Res.* 64: 1319-1327
14. Ribeiro WP, Valberg SJ, Pagan JD, Gustavsson BE (2004), The effect of varying dietary starch and fat content on serum creatine kinase activity and substrate availability in equine polysaccharide storage myopathy, *J.Vet.Intern.Med.* 18: 887-894
15. De La Corte FD, Valberg SJ, Mickelson JR, Hower-Moritz M (1999), Blood glucose clearance after feeding and exercise in polysaccharide storage myopathy, *Equine Vet.J.Suppl* 30: 324-328
16. McCue ME, Valberg SJ, Lucio L, Mickelson JR (2008), Glycogen Synthase 1 (*GYS1*) Mutation in Diverse Breeds with Polysaccharide Storage Myopathy, *J.Vet.Intern.Med.* in press:
17. McCue ME, Ribeiro WP, Valberg SJ (2006), Prevalence of polysaccharide storage myopathy in horses with neuromuscular disorders, *Equine Vet.J Suppl* 340-344
18. Valentine BA, de LA, Divers TJ, Ducharme NG, Orcutt RS (1999), Clinical and pathologic findings in two draft horses with progressive muscle atrophy, neuromuscular weakness, and abnormal gait characteristic of shivers syndrome, *J.Am.Vet.Med.Assoc.* 215: 1661-5, 1621
19. Naylor RJ, Livesey L, Schumacher J, Henke N, Massey C, Brock KV, Fernandez-Fuente M, Piercy RJ (2012), Allele copy number and underlying pathology are associated with subclinical severity in equine type 1 polysaccharide storage myopathy (PSSM1), *PLoS.One.* 7: e42317
20. McCue ME, Valberg SJ, Miller MB, Wade C, Akman O, DiMauro S, Mickelson JR (2008), Glycogen synthase (*GYS1*) mutation causes a novel skeletal muscle glycogenosis, *Genomics* doi:10.1016/j.ygeno.2008.01.011:

21. Naylor RJ, Livesey L, Schumacher J, Henke N, Massey C, Brock KV, Fernandez-Fuente M, Piercy RJ (2012), Allele copy number and underlying pathology are associated with subclinical severity in equine type 1 polysaccharide storage myopathy (PSSM1), *PLoS.One.* 7: e42317
22. Treiber KH, Kronfeld DS, Hess TM, Byrd BM, Splan RK, Staniar WB (2006), Evaluation of genetic and metabolic predispositions and nutritional risk factors for pasture-associated laminitis in ponies, *J.Am.Vet.Med.Assoc.* 228: 1538-1545
23. Carter RA, Treiber KH, Geor RJ, Douglass L, Harris PA (2009), Prediction of incipient pasture-associated laminitis from hyperinsulinaemia, hyperleptinaemia and generalised and localised obesity in a cohort of ponies, *Equine Vet.J.* 41: 171-178
24. Bailey SR, Habershon-Butcher JL, Ransom KJ, Elliott J, Menzies-Gow NJ (2008), Hypertension and insulin resistance in a mixed-breed population of ponies predisposed to laminitis, *Am.J.Vet.Res.* 69: 122-129
25. Treiber KH, Kronfeld DS, Geor RJ (2006), Insulin Resistance in Equids: Possible Role in Laminitis, *Journal of Nutrition* 136: 2094S-22098
26. Teran-Garcia M, Bouchard C (2007), Genetics of the metabolic syndrome, *Appl.Physiol Nutr.Metab* 32: 89-114
27. Benyamin B, Sorensen TI, Schousboe K, Fenger M, Visscher PM, Kyvik KO (2007), Are there common genetic and environmental factors behind the endophenotypes associated with the metabolic syndrome?, *Diabetologia* 50: 1880-1888
28. Frank N, Elliott SB, Brandt LE, Keisler DH (2006), Physical characteristics, blood hormone concentrations, and plasma lipid concentrations in obese horses with insulin resistance, *J.Am.Vet.Med.Assoc.* 228: 1383-1390
29. Johnson P (2002), The equine metabolic syndrome: peripheral Cushing's syndrome, *Vet.Clin.North Am Equine Pract.* 18: 271-293
30. Johnson P, Messer N, Slight S, Wiedmeyer C, Buff P, Ganjam V (2004), Endocrinopathic laminitis in the horse, *Clinical Techniques in Equine Practice* 3: 45-56
31. Pleasant RS, Suagee JK, Thatcher CD, Elvinger F, Geor RJ (2013), Adiposity, Plasma Insulin, Leptin, Lipids, and Oxidative Stress in Mature Light Breed Horses, *Journal of Veterinary Internal Medicine* 27: 576-582
32. McGowan CM, Geor RJ, McGowan T. The prevelance and incidence of hyperinsulinemia in ponies. Journal of Veterinary Internal Medicine 22, 734. 2008.
Ref Type: Abstract
33. Petersen JL, Mickelson JR, Rendahl AK, Valberg SJ, Andersson LS, Axelsson J, Bailey E, Bannasch D, Binns MM, Borges AS, Brama P, da C+ómara Machado A, Capomaccio S, Cappelli K, Cothran EG, Distl O, Fox-Clipsham L, Graves KT, Gu+¬rin G+, Haase B, Hasegawa T, Hemmann K, Hill EW, Leeb T, Lindgren G, Lohi H, Lopes MS, McGivney BA, Mikko S, Orr N, Penedo MC, Piercy RJ, Raekallio M, Rieder S, R++ed KH, Swinburne J, Tozaki T, Vaudin M, Wade CM, McCue ME (2013), Genome-Wide Analysis Reveals Selection for Important Traits in Domestic Horse Breeds, *PLoS Genet* 9: e1003211

RESEARCHING BRD – PART I: GETTING OUR HOUSE IN ORDER

Sébastien Buczinski[1], Dr Vét, DÉS, MSc, DACVIM (LAIM) and Joanne Hewson[2], DVM, PhD, DACVIM (LAIM)
[1]Quebec and [2]Ontario, Canada

INTRODUCTION

Despite the extensive body of literature available, representing massive prior research efforts, animal use, and expenditure of millions of research dollars, Bovine Respiratory Disease (BRD) remains a leading cause of wastage in the cattle industry. To strategically move research forward in understanding BRD of beef and dairy cattle, it is critical to first assess the evidence upon which current knowledge of BRD is founded. In Part I, we analyze the existing literature supporting the present understanding of BRD diagnosis, treatment and prevention.

BRD DIAGNOSIS

The diagnosis of BRD under field situations is mainly achieved through clinical inspection and examination of beef and dairy calves. Most of the time, the initial diagnosis is performed by non-veterinary personnel based (or not) on standard protocols that define cases of BRD previously established by the herd veterinarian. The lack of a clear and consistent definition of BRD is mainly due to the absence of a Gold Standard method to diagnose BRD ante mortem, and the fact that this syndrome results from different interactions between the host (feedlot calf, cow-calf, dairy calf), the pathogens (different viruses and bacteria) and the environment. The multiple accepted clinical definitions of BRD include (but are not limited to) various combinations of: respiratory signs (abnormal breathing pattern, nasal discharge, cough) and/or less specific signs (depression, decreased appetite, abnormal head carriage, drooping ear, rough hair coat) and/or increased rectal temperature, as well as the absence of other clinical signs relating to alternate body systems.[1,2] Undifferentiated fever (UF) has also been utilized as another proxy for BRD diagnosis.[3] In epidemiologic studies focusing on BRD, the diagnosis for case inclusion can therefore vary from clear (but variable from study to study) clinical definitions as above,[4] or BRD case inclusion may only be defined by the recording of BRD treatment being performed.[5] These case definitions have limitations because they highly depend on the care-taker in terms of clinical acumen or subjective treatment threshold, and this operator dependence may yield large inter-observer variation if the process is not standardized. As well, their accuracy has not been truly assessed given the absence of a Gold Standard method to diagnose BRD in the living animal. Beyond diagnosing BRD, the definition of "BRD relapse" after treatment similarly lacks consistency between studies. Different definitions exist, ranging from a systematic follow-up clinical assessment at varying periods of time after enrolment and treatment, to a passive clinical assessment during subsequent pen riding. More systematic approaches using clinical scoring systems have been used in both dairy[1] and beef[6] cattle. They have the advantage of being easily used by producers, and allow (even if imperfect) direct comparison between studies in trying to standardize outcome assessment. Despite imperfect test accuracy, focusing on BRD proxies that can be more objectively assessed can also be a solution to allow direct comparison from study to study. Examples include rectal temperature, or production outcomes such as death losses, average daily gain, carcass quality, etc. While alterations in these parameters have been associated with BRD, other causes of morbidity may also lead to reduced performance and may be difficult to discriminate from losses due to BRD.

Various diagnostic tests for BRD have been reported.[7,8] These tests can be related to isolation of pathogens, documenting seroconversion, detection of an associated immune response by the host, imaging methods (ultrasound, radiography), monitoring of body temperature (rectal temperature, rumen temperature, near infra-red thermography), detection of inflammatory biomarkers in blood, respiratory secretions, or exhaled breath, and monitoring of cattle behavior or activity. More recently, the proteomic approach also allows for the assessment of multiples proteins as biomarkers that could be potentially used for prediction of future BRD or as prognostic indicators. Diagnostic tests range from dichotomous tests to those with a continuous scale that are ultimately dichotomized using an optimized threshold value. The different tests available could be used at different moments including (1) during processing to help in the prediction of cattle with higher risk of developing BRD, or (2) at the time of suspected diagnosis (e.g. pulled during daily clinical inspection by the pen rider or by the calf team in a dairy farm). The test will therefore be a screening test to predict an outcome that has not happened in the first situation, or will be used as an add-on to an initial examination for confirmation of BRD in the second situation.

Development of diagnostic tests for BRD traditionally requires assessing their accuracy despite the absence of a Gold Standard test. This can be achieved by different methods including 2*2 or X*Y tables in dichotomous or polytomous tests or outcome levels. For continuous markers, various methods have been proposed including determination of the area under the Receiver Operating Characteristic curve (or concordance characteristic) or data-driven thresholds using the threshold that minimizes misclassification [Youden's index J=Max (Se+Sp-1)]. The optimal cutoff can therefore also be assessed in a 2*2 table. For predicting the outcome of tests applied to the stage of processing upon entry into the feedlot, the test can still be assessed as a diagnostic test, but can also be studied using survival analysis. It is important to remember that multiple biases can affect diagnostic accuracy.[9]

The development of a diagnostic test (after analytical validation) typically follows a 4-stage pattern that includes (1) determining the normal range of values of the diagnostic test in patients without the condition, (2) trying to assess accuracy in a case-control study which includes healthy patients and patients with the disease (assessed by a Gold Standard), (3) determining the clinical impact of introducing the test through randomized controlled trials, and (4) determining the effects of this test with large cohort studies. Most available diagnostic studies on BRD have focused on parts (1) and (2) of test assessment. Internal and external validation of these tests is not usually performed. Due to this, it is difficult to directly apply research findings to other settings. It is also important to assess tests in contexts other than in case-control studies, since this design frequently overestimates test accuracy. The most relevant question is how a test performs when used under the natural conditions for which the test was intended. In BRD research specifically, this is a determination of how well a diagnostic test can detect naturally occurring BRD in cattle encountering the typical husbandry, management practices and pathogen challenges of the feedlot or dairy barn setting. To date, no consensus has emerged concerning the optimal assessment scheme for diagnosing BRD that could be used for research as well as for being practical when used in commercial settings. This has led to the unfortunate circumstance that tremendous variability in case definition exists between studies, making comparison across studies difficult if not impossible to do in a meaningful manner. The following review of BRD treatment and prevention attempts to synthesize the available literature, but recognizes the limitations of variable case definition when doing comparisons across different studies.

BRD TREATMENT AND PREVENTION

Treatment and prevention methods used to reduce morbidity from BRD in feedlots were recently reviewed,[10] including the following interventions that are commonly performed in beef cattle: preconditioning, vaccination practices, processing (when processing is done, and what interventions are included), nutritional practices implemented on entry into the feedlot, and antimicrobial use for prophylaxis or metaphylaxis. Of these interventions, preconditioning and parenteral antimicrobial metaphylaxis were the only practices that appear to offer likely efficacy in reducing BRD morbidity based on consistency in study outcomes. In particular, weaning several weeks before sale as part of a preconditioning program was felt to be associated with a significant reduction in BRD in the early feedlot period, although potential confounding factors were identified that require further investigation. Vaccination prior to sale may also reduce BRD morbidity. Metaphylactic administration of various antimicrobials by injection at the time of entry into the feedlot was also effective in reducing BRD in the early feedlot period based on the studies reviewed. In contrast, no clear opportunity to reduce BRD morbidity could be ascribed to vaccination after arrival, various processing activities (growth promotant implants, anthelmintic or parenteral vitamin administration, aborting pregnant heifers), or prophylactic antimicrobial administration (in feed or water). In some instances, too few studies were available on the particular intervention, or study design was too variable, to permit comparisons to be made and thus conclusions could not be drawn. Despite this, many of these interventions are still accepted as standard practices across the beef industry. While the above review was extensive, how the search for publications to include in the review was performed, and how quality of the information within the studies was assessed were not defined, leaving the reader to determine how to ultimately interpret the conclusions of the review.

Evidence-based conclusions require a systematic approach to research methods, and readers need to be able to independently assess the validity of the study (review) design as it pertains to the conclusions reached. Evidence-based conclusions should inform the methods used to treat or prevent BRD in cattle. The remainder of this proceedings manuscript aims to provide bovine researchers with an overview of what is currently known in the areas of BRD treatment and prevention, based on the best evidence available in the published literature.

The Power of Systematic Reviews / Meta-analyses in Understanding BRD

The vastness of available literature precludes a singular veterinarian from reviewing all studies on respiratory disease in cattle before making decisions that will frame their clinical management of BRD. However, the use of systematic reviews of the literature allows individuals a more efficient way to understand the existing literature as they move forward. Systematic reviews allow for a meaningful synthesis of existing research, and provide evidence-based information upon which to act. "Systematic" refers to the process undertaken by the authors in amassing all the available literature pertaining to a particular research question, standardizing how the quality of that literature was assessed, and then including only studies of acceptable quality in their final review. By knowing these methods, the reader is able to evaluate whether author biases were prevented during collection and interpretation of the information. When information of suitable quality is systematically reviewed, comparison of results between multiple studies is possible in order to gain a stronger understanding of the results. Patterns among studies may help to confirm hypotheses, or such comparisons may allow the reader to understand where conflicts in results are originating from (e.g. confounding factors). When systematic reviews include a quantitative analysis of existing data from the various studies, the resulting meta-analysis is described using statistical methods. Meta-analysis requires that the study population is uniform across the various studies, the intervention or treatment is clearly defined, and a consistent outcome measure is available for comparison. It is essential that this information is therefore available in the original publication prior to including a study in a meta-analysis. Studies involving greater sample sizes yield greater power in detecting effects on the outcome measure, and as such, these studies receive greater weighting during amalgamation of results across studies in a meta-analysis. The final estimation of effect from the meta-analysis is typically more accurate and precise compared to the results obtained by each individual study.[11]

While the above features make systematic reviews and meta-analyses appealing to understand the available literature, their downfall is also precisely that – the available literature. In considering the conclusions reached during these reviews, one must understand the limitations of that literature in terms of publication biases, elimination of studies from the comparison due to variability in study design, study population, methodology, etc., and acknowledge that the selection of the studies used in the comparison will directly dictate the quality of the conclusions that can be made.

In the context of systematic reviews and meta-analyses, few exist in the study of BRD to date, although a recent surge of such studies in production medicine suggests that this concept is gaining traction and more people are becoming familiar with it as a standard of practice in understanding the literature. Systematic reviews were recently discussed in the context of BRD by Vandeweerd and colleagues.[11] Readers are directed to this paper to understand the steps required to conduct a systematic review, as well for interpretation of information put forth from a systematic review process.

Below are summaries of the existing systematic reviews that have been published on BRD treatment and prevention. In an effort to capture all reports of this type across the available literature, these papers were identified using several searchable databases (Appendix 1) and an extensive list of search terms (Appendix 2).

Treatment of BRD – Systematic Reviews of Antimicrobial Use

The cornerstone treatment of BRD is antimicrobial administration, based on the assumption that all clinical BRD may involve bacteria as either a primary or secondary pathogen in the disease process. While a multitude of studies report efficacy of individual products, few products have been directly compared to each other over multiple publications in the free-access literature. A systematic review directly comparing efficacy for individual products is currently limited to a comparative efficacy study between tulathromycin and either florfenicol or tilmicosin in treating clinical BRD in the feedlot setting.[12] In this review, inclusion criteria were: randomized clinical trials investigating the treatment efficacy of tulathromycin, conducted on naturally-occurring BRD in North American feedlot cattle. Of 782 potentially relevant manuscripts that were originally identified, only 4 manuscripts (2 peer reviewed publications and 2 technical bulletins from pharmaceutical companies, representing 12 high quality studies) were appropriate for eventual use in the meta-analysis. The outcome measure compared was "re-treatment for BRD" as an indicator of treatment failure, since this was the only outcome measure consistently reported across the 12 available studies. Comparing tulathromycin with tilmicosin, the summary Mantel-Haenszel relative risk was 0.56 (95% Confidence Interval: 0.48 – 0.64). Results of the meta-analysis therefore indicated approximately 50% reduction in re-treatment associated with tulathromycin when compared with tilmicosin. Due to insufficient studies involving florfenicol, it was not possible to estimate a relative risk comparing tulathromycin with florfenicol during this systematic review. The utility of a meta-analysis to understanding the existing literature was highlighted here by the fact that the impact of tulathromycin ranged from ineffective to highly effective across the 12 individual studies, leaving unresolved discrepancies as to its clinical efficacy prior to the meta-analysis.

It is also possible to indirectly compare drugs using a mixed treatment comparison meta-analysis. Mixed-treatment comparison (MTC) meta-analysis allows for quantification of differences in efficacy between treatments when no direct prior comparison has been made, reducing the need to perform multiple trials to rank efficacy when multiple treatment options exist (as is the case for BRD treatment in North America). Indeed, many such comparative studies would never be performed when several drugs are marketed by the same pharmaceutical company, and as a result such data would not be available to guide veterinary and producer treatment decisions. Mixed treatment comparison meta-analysis is appropriate provided that all products have, in some way, been indirectly compared within the overall group.

A recent meta-analysis using an MTC approach now provides information on the relative efficacy of the various antimicrobials labeled for the treatment of BRD in the United States of America.[13] Studies included in the MTC meta-analysis were limited to those involving treatment of naturally-occurring BRD in North American feedlot cattle, using only studies that were found through publically available sources. Ninety-three trials studying 12 common antimicrobials were used in the MTC meta-analysis; 60 trials represented drug efficacy compared with a non-active compound (placebo) whereas the remaining 33 trials were comparisons of one antimicrobial versus another. The analysis provided an estimated risk ratio for requiring re-treatment for BRD (as a measure of drug efficacy) between each pairing of antimicrobials for comparison, as well as a 95% confidence interval for each estimate. Using this study, it is now possible to make antimicrobial treatment decisions that are evidence-based, although final decisions will also be guided by other factors in cattle production such as cost-benefit analysis and consumer interests.

Treatment of BRD – Systematic Reviews of Ancillary Therapies

Similar to antimicrobials, there have been a multitude of studies investigating the impact of ancillary therapies such as anti-inflammatory agents on clinical outcome for BRD. Many of these studies have utilized small sample sizes, experimental infection models, or non-target animals (dairy calves) to determine whether control of inflammation reduces morbidity and mortality from BRD, making extrapolation of the findings to the feedlot setting tenuous. The array of drugs, dosages, routes of administration and timing of intervention also make both direct and indirect comparisons difficult between ancillary treatment approaches. However, a systematic review of anti-inflammatory drugs used in combination with an antimicrobial to manage naturally occurring BRD was recently reported to help understand the existing information.[14] Fourteen studies were identified that investigated either a non-steroidal anti-inflammatory drug (NSAID; 12 studies), a corticosteroid (1 study) or both (1 study). NSAIDs investigated included flunixin meglumine, ketoprofen, carprofen, tolfenamic acid, meloxicam, diclofenac, and aspirin. Corticosteroids included flumethasone, prednisolone, and methylprednisolone. Outcome measures included rectal temperature, clinical score, relapse rate, average daily gain, and mortality due to BRD. Not all measures were reported for each drug studied. Inconsistencies between the methodologies of the 14 studies made meta-analysis impossible, so the reports were compared qualitatively. Of the studies pertaining to NSAIDs, a faster return to normal rectal temperature was the most consistent clinical effect of the NSAID over control (antibiotic alone) treatment. However, often there was no difference between the groups by the end of the study period despite this early difference in rectal temperature by treatment group. While earlier resolution of fever can be claimed as a beneficial outcome from the perspective of animal welfare, the long-term effects on feedlot performance, carcass quality, and cost-benefit analysis remain undefined. There were too few studies on corticosteroid treatment in combination with an antimicrobial to draw firm conclusions, and therefore the clinical efficacy of corticosteroids to manage naturally-occurring BRD is still unknown. The authors of this systematic review noted many challenges in attempting to synthesize meaning from the existing literature, namely due to the extreme variability in details of the ancillary intervention, lack of a unified method to calculate clinical score as an outcome measure, as well as inadequate use (or reporting) of an appropriate study design.

Prevention of BRD – Systematic Reviews of Vaccination

A systematic review of the literature to understand the impact of vaccination on BRD was recently reported.[15] This meta-analysis compared morbidity and mortality rates between unvaccinated animals and cattle vaccinated against common bacterial causes of BRD (*Mannheimia haemolytica*, *Pasteurella multocida*, and/or *Histophilus somni*). The review was restricted to studies involving naturally-occurring disease in North American feedlots. Fifteen studies (representing 22 clinical trials) were used in the analysis. From this comparison, vaccination against *M. haemolytica* (alone or in combination with *P. multocida*) yielded a reduced risk of BRD morbidity (Mantel-Haenszel risk ratio 0.93, 95% Confidence Interval: 0.89 – 0.98) compared with unvaccinated cattle. No beneficial effect on mortality was observed for *M. haemolytica* vaccinates over controls. While the review did not identify evidence of efficacy for vaccinating against *Histophilus somni*, there were insufficient studies available to yield meaningful conclusions regarding *H. somni* vaccination. An additional challenge to evaluating the evidence for vaccination efficacy was that many studies housed vaccinates and controls together in the same pen, potentially biasing the results for both groups (i.e. less disease seen in the unvaccinated cattle due to "herd immunity" and more disease challenge presented to the vaccinated cattle). These confounding effects could not be teased out during the current systematic review.

Prevention of BRD – Systematic Reviews of Other Management Approaches

Prevention of BRD extends beyond vaccination to include a variety of management practices that have been adopted based on real or perceived factors predisposing to BRD. An example of this is the metaphylactic administration of an antimicrobial to "high-risk" groups of cattle on arrival at the feedlot. Meta-analysis was performed to assess whether the practice of metaphylaxis did indeed impact BRD morbidity or mortality.[16] Randomized clinical trials comparing metaphylaxis (with any antimicrobial) against untreated controls were included if they were conducted on beef cattle in a North American feedlot. Outcome measures included morbidity and mortality due to BRD, and feedlot performance indices (average daily gain, ADG; dry matter intake, DMI; gain-to-feed ratio, G:F). Nine manuscripts, describing 20 studies where an antimicrobial given metaphylactically was compared with a control group, were included. Antimicrobials included tilmicosin, oxytetracycline given intramuscularly, chlortetracycline or sulfamethoxazole given in the feed, or florfenicol. The majority of studies involved tilmicosin. Combining results of these studies, an improved ADG was seen in cattle receiving metaphylaxis (0.11 kg/day more than untreated animals; 95% Confidence Interval: 0.10 – 0.13, $P < 0.01$). Morbidity was also reduced in the metaphylaxis group (29%, 95% CI: 21.2 – 38.6%) compared with controls (55%; 95% CI: 44.5 – 65.1%, $P < 0.01$) as was mortality (metaphylaxis 1.8%; 95% CI: 1.05 – 3.12% versus controls 3.8%; 95% CI: 2.3 – 6.5%, $P < 0.01$).

CONCLUSIONS

In consideration of the significant economic consequences of BRD on beef production, and the extensive research over the past half-century to combat BRD, it seems that we have made little progress. Many of the existing publications, whether they study causative agents, predisposing factors, treatment efficacy, impact of preventative measures, or BRD implications on cattle production, present conflicting conclusions. This limits our ability to clearly understand any dimension of BRD. While systematic reviews enable readers to synthesize the multiple studies into meaningful conclusions, few currently exist in the field of BRD treatment and prevention. Those that have been performed have attempted to provide information relevant to the clinical situation, namely being focused on studies pertaining to feedlot cattle in North American feedlots that experience naturally occurring BRD. However, the criteria used to define BRD cases in such studies were highly variable and yet comparisons were still made. In order to move forward in a strategic manner, we must first improve and standardize the definition of BRD (case definition, appropriate diagnostic tests, criteria defining a BRD relapse, relevant production outcomes related to BRD…). We must also improve our global approach to BRD research, such that future studies better support systematic reviews and meta-analyses to enrich our evidence-based understanding of this challenging syndrome in the cattle industry. Specific to systematic reviews and meta-analyses on BRD, we need to reflect on the population studied, the outcome measures and how well they represent clinical and subclinical BRD, and publication biases in this field. These are discussed in greater detail in Part II of the proceedings.

REFERENCES

[1]McGuirk SM. **Vet Clin Food Anim** 2008;24:139. [2]Edwards TA. **Vet Clin Food Anim** 2010;26:273. [3]Hannon SJ, et al. **Vet Ther** 2009;10:E1. [4]Windeyer CM, et al. **Prev Vet Med** 2013; in press. [5]Bach A. **J Dairy Sci** 2011;95:1052. [6]Holland BP, et al. **J Anim Sci** 2010;88:2486. [7]Caswell JL, et al. **Vet Clin Food Anim** 2012;28:419. [8]Fulton RW, et al. Can Vet J 2012;53:754. [9]Leeflang MM, et al. **Clin Chem** 2008;54:729. [10]Taylor JD, et al. **Can Vet J** 2010;51:1351. [11]Vandeweerd JM, et al. **Vet Clin Food Anim** 2012;28:13. [12]Wellman NG, et al. **J Vet Pharmacol Therap** 2007;30:234. [13]O'Connor AM, et al. **Prev Vet Med** 2013;110:77. [14]Francoz D, et al. **Vet Clin Food Anim** 2012;28:23. [15]Larson RL, et al. **Vet Clin Food Anim** 2012;28:97. [16]Wileman BW, et. al. **J Anim Sci** 2009;87:3418.

Appendix 1: Databases searched (and years available).

Database	Years Searched
PubMed	1966 - present
Agricola	1970 - present
Commonwealth Agricultural Bureaux International (CABI)	1973 - present

Appendix 2: Keywords Used to Query BRD-related Publications (adapted from prior studies[12-14,16]). Words within columns were linked with "OR".

(Population)	AND	(Disease)	AND	(Study Design)	AND	(Treatment	OR	Prevention)
Bovine		BRD		Systematic		Antibiotic*		Prevention
Ruminant*		Respiratory		Meta-analysis		Antimicrobial*		Prophylaxis
Cattle		Pneumonia		Critical appraisal		Amoxicillin		Prophylactic
Calf		Undifferentiated fever				Ampicillin		Metaphylaxis
Calves		Bronchopneumonia				Ceftiofur		Metaphylactic
Cow*		Mannheimia haemolytica				Cloxacillin		Mass medication
Weanling*		Pasteurella haemolytica				Danofloxacin		Vaccin*
Yearling*		Pasteurellosis				Doxycycline		Preconditioning
Beef		Pasteurella multocida				Enrofloxacin		Conditioning
Dairy		Haemophilus somnus				Erythromycin		Value added
Heifer*		Histophilus somni				Florfenicol		Nutrition*
Steer*		Bovine herpesvirus 1				Gamithromycin		
Bull*		Infectious bovine rhinotracheitis				Gentamicin		
Feedlot*		IBR				Lincomycin		
Cow-calf		Parainfluenza virus 3				Marbofloxacin		
		PI-3				Moxifloxacin		
		Bovine respiratory syncytial virus				Penicillin		
		BRSV				Spectinomycin		
		Bovine Viral Diarrhea				Sulfadia*		
		BVD*				Sulfameth*		
		Bovine Corona virus				*tetracycline		
		Mycoplasma				Tildipirosin		
						Tilmicosin		
						Triamilide		
						Trimethoprim		
						Tulathromycin		
						Tylosin		
						Ancillary		
						Non-steroidal		
						NSAID*		
						Acetylsalicylic		
						Flunixin		
						Ketoprofen		
						Carprofen		
						Meloxicam		
						*Steroid		
						Dexamethasone		
						Predniso*		
						Isoflupredone		

RESEARCHING BRD – PART II: APPLES, ORANGES, OR PEARS?
DEFINING THE INGREDIENTS NEEDED FOR FUTURE RESEARCH IN BRD…

Joanne Hewson[1], DVM, PhD, DACVIM (LAIM) and Sébastien Buczinski[2], Dr Vét, DéS, MSc, DACVIM (LAIM)
[1]Ontario and [2]Quebec, Canada

INTRODUCTION

Despite the fact that numerous studies have been done investigating Bovine Respiratory Disease (BRD), several limitations to this prior research exist as seen in the first part of these proceedings. In particular, improvement in defining BRD cases during field studies, and improving overall reporting of BRD-related research, are areas that need immediate attention. In Part II, we propose some recommendations arising from veterinary and human research consensus or initiatives, in order to improve the quality of research data as well as the reporting of BRD research to be better able to appropriately compare different studies on the same topic. This will include discussion of potential directions and research methods that would strategically support future investigations in the field of BRD.

IMPROVING AND ATTEMPTING TO STANDARDIZE BRD DIAGNOSIS AND BRD-RELATED OUTCOMES

Reviewing the literature surrounding BRD over the past four decades, it is quickly evident to a reader that the definition of BRD has shifted over time, and that there are almost as many methods used to define (diagnose) cases as there are publications in this area. The absence of a Gold Standard definition of BRD diagnosis is one of the key components of research needing addressing in this area – how can we compare different studies on BRD (e.g. treatment, prevention, or disease impact studies) when no clear definition of case inclusion exists or when inclusion criteria are different between studies? In these cases, are we making appropriate conclusions when we compare or synthesize results from multiple studies, or are we actually "comparing apples with oranges"? It does not mean that one study is necessarily better than the other, but only that we cannot directly compare both studies. This lack of a Gold Standard diagnosis can be addressed by different strategies that have been recently reviewed by Rutjes and colleagues.[1] In these situations, the solutions consist of (1) inputting or adjusting for missing data on the reference test (which means that a Gold Standard does exist, but due to its cost or availability it is only performed in a subset of patients), (2) correcting for an imperfect reference standard, (3) constructing a reference standard using the association of different imperfect tests to improve classification, using a consensus process based on an expert panel, or using a statistical model to deal with the recorded data (latent class analysis), and (4) monitoring other types of outcomes that may be more easily measured (e.g. mortality, average daily gain, carcass characteristics…) since the true disease state is difficult to define. All of these methods have their pros and cons but can serve as future references for constructing studies that could be more easily compared together.

Current Tools That Can Be Used By Producers or Researchers for BRD Diagnosis

Any clinical scoring system should be statistically validated for the attribution of the different points' scores as well as for the weights of each parameter included in the score (internal validation). Its intra- and inter-observer validity should also be assessed and be sufficiently high to promote its future use by different research groups (external validation). One of the advantages of using standardized inclusion criteria is that, even if these tests are imperfect, they are repeatable. When using latent class analysis to assess test accuracy, it is possible to assess the test's accuracy despite inherent imperfections.[2,3] Therefore, such assessment and validation of the scoring systems would enable researchers to reach a consensus on a single scoring system, such that future information from research would be generated using a consistent assessment tool, thus allowing for more appropriate comparisons to be made across studies. In addition to clinical scoring systems, other diagnostic tests also need to be standardized (e.g. thoracic ultrasound) for the same purposes. Additionally, raw datasets using these measurement outcomes could be made available upon request to compare results of different studies.

How Can These Diagnostic Tools Be Used and Assessed?

The optimal assessment method for diagnosing BRD would ideally utilize a test having Sensitivity (Se) and Specificity (Sp) characteristics that are robust (i.e. not dependent on disease prevalence). In the search for such tests, the apparent Se/Sp of any new test (when compared to an imperfect resolver test in which Se/Sp is known) needs to be taken into account in order to calculate true Se/Sp. Recent research has shown the multiple potential biases in Se/Sp assessment.[4] Extensive research in human medicine has shown that different study characteristics have been associated with overestimation/underestimation of test accuracy (Se/Sp). These situations are (1) spectrum bias (for example, in a case-control study design when cases are those with advanced stages of the disease whereas controls are healthy animals), (2) verification bias (e.g. not all of the animals have the same verification or classifier test), (3) limited challenge (the test is not performed in some patients because of difficulties, such as the animal is too intractable to handle, too fat to allow a good imaging technique, etc.), (4) high prevalence of the disease, (5) small datasets and data-driven cut-offs (which frequently overestimate diagnostic accuracy), and (6) using imperfect tests as true Gold Standards against which to assess index test accuracy (in these cases the apparent accuracy is generally lower than the true accuracy). While there is currently much scientific interest in evaluating the various diagnostic tests available for BRD, these reports need to include discussion of the above study characteristics when test performance is being evaluated. As well, the community of BRD researchers needs to come together to define specifically what aspects of study design should be formalized across studies, and in what manner, so that BRD is being evaluated in a fashion deemed relevant to the cattle industry and at a suitably stringent level of Se and Sp for the conclusions to have meaning to the end users.

REPORTING GUIDELINES TO AVOID BIAS IN BRD STUDIES

The rationale for clear reporting is multifold. Clear reporting allows the study to be reproduced, verifying that results obtained were not due to chance. Reporting details of the study population, case definition, trial design, interventions performed, and methods used to measure outcomes allow the reader to independently assess both the internal validity and the external validity of the study to their clinical question. It also removes any ambiguity in interpretation of the study by the reader, such as whether potential sources of bias impacted the results. Finally, clear and complete reporting allows for data from the study to be extracted during synthesis of existing publications into larger analyses such as systematic reviews or meta-analyses. Through these benefits, the data generated by the study (and the underlying resources that were required to conduct the study) are optimally utilized, addressing the ethical and funding issues surrounding the use of animals in research.

Diagnostic and Prognostic Studies

As we have previously shown, several biases may occur throughout the various steps for accuracy assessment of a new index test for BRD. An evidence-based approach is therefore recommended when adopting new diagnostic methods, in order to take into account these potential pitfalls of test assessment. There are currently two different recommended procedures for reporting diagnostic studies that take into account the potential biases previously presented: the STARD (STAndard for the Reporting of Diagnostic accuracy studies)[5] and QUADAS (QUality Assessment of Diagnostic Accuracy included in Systematic reviews)[6] statements are both based on expert consensus and use a checklist in order to systematically address the potential biases which may occur at the different steps of diagnostic accuracy study design. The same initiative has also been proposed for reporting assessment of prognostic tests (PROGRESS: PROGnosis RESearch Strategy group) since these types of studies are generally more difficult to assess.[7]

Using these checklists (oriented on human medicine by default, but which can certainly be applied to animal research) would lead to improved methodology and reporting by researchers in the field of BRD diagnosis, and a better understanding of the presented results by end users (veterinarians, industry). These steps are also critical before trying to pool the results of different studies on the same test to generate a more realistic estimation of test accuracy. Having done this, researchers will be better able to define which methods to diagnose BRD are truly appropriate. Systematic reviews in the future can embed quality assessment of BRD diagnosis into their criteria for selecting studies to review, such that the systematic reviews utilize only data from those studies meeting accepted standards in their method to diagnose BRD.

Treatment and Prevention Studies

From the overview of existing BRD literature, and the challenges in identifying strong conclusions about BRD treatment or prevention based on systematic reviews/meta-analyses in Part I of these proceedings, it is clear that much of our prior research results are not being used to their fullest potential due to issues surrounding case definition or reporting of methodology and results to a degree that allows the reader to appreciate the strengths and weaknesses of the studies during interpretation. The same was historically true in human medicine, leading two groups in the early 1990s to simultaneously propose that reporting guidelines be established in human medicine (as reviewed by Altman[8]). The Journal of the American Medical Association therefore charged these groups to combine efforts in order to create a single, evidence-based series of recommendations regarding reporting of clinical trials. From this movement arose the CONSORT Statement (CONsolidated Standards Of Reporting Trials Statement), first published in 1996,[9] with later updates in 2001[10-12] and 2010.[13] This statement has been published across multiple journals, accompanied by an explanation and elaboration document to facilitate its understanding and use by authors, editors and peer-reviewers. To date, over 600 medical journals have endorsed use of the CONSORT Statement in reporting studies.[8] More recently, extensions to this document have been developed to account for the unique characteristics of reporting that arise from study design (e.g. observational studies - STROBE, systematic reviews of randomized trials - PRISMA, etc.) and for reporting studies initially in abstract form during continuing education events such as conference proceedings.[14] The CONSORT Statement and extensions for other study types are available at http://www.consort-statement.org.

A review of publications in veterinary medicine evaluated how well reporting of veterinary trials adheres to the CONSORT Statement recommendations.[15] A random selection of 100 publications involving clinical trials in livestock species showed that many reporting features from the CONSORT Statement were poorly represented in these veterinary publications. The association between characteristics of reporting and bias on treatment effects was also investigated. Results suggested that low adherence with reporting guidelines from the CONSORT Statement may be associated with biased treatment effects being identified. This publication also made additional recommendations for reporting veterinary trials beyond those used for human medicine, since veterinary reports often contain characteristics unique to animal studies such as experimental induction of disease, variation in animal housing and management, and interventions being allocated and assessed at varying levels such as the individual, pen, or herd.

Characteristics of reporting that are important for a reader in order to detect biases and to assess validity in clinical trials of BRD interventions include a description of the study enrollment criteria (e.g. was the study conducted under natural conditions and in the relevant population, was the diagnosis of BRD defined and appropriate), procedures used for randomization and blinding, numbers and reasons for any withdrawals, declaration of a primary outcome measure and statement of a null hypothesis for this outcome, use of clinically relevant outcome measures, justification of sample size as calculated using the primary outcome measure, adequate duration and intensity of follow-up, source of study funding, and declaration of any conflict of interest by authors.[16,17] A systemic review on reporting of trials involving antimicrobial treatment for BRD in North American feedlots was done, spanning all publications from 1970-2005.[16] This investigation was performed to characterize the reporting of key study design features, in order to determine whether there was an association between reporting and study bias towards identifying differences between treatments. This systematic review identified 29 manuscripts (representing 41 studies) that described at least one clinical trial, and were performed on cattle experiencing naturally occurring BRD in North American feedlots. While randomization was reported for 36 of 41 (88%) studies, only 9 of these 36 (25%) actually described the method of randomization to allow for independent scrutiny by the reader of whether randomization truly occurred. Less than half (20 of 41, 49%) mentioned that blinding was performed, but few necessarily provided the description of how blinding was done and at what stages it was applied (treatment allocation, treatment administration, animal caregivers, data collection and analysis). Reasons for inclusion and exclusion criteria were well reported (38/41 and 39/41 respectively). Few studies clearly identified a primary outcome measure, and only 3/41 (7%) reported calculation of a sample size based on this primary outcome to allow the reader to determine the power of the study to generate meaningful conclusions. No study defined a null hypothesis, making it impossible to determine whether the researchers were assessing that a treatment was not inferior to the comparison group (non-inferiority study) or that the treatment was assessed for being better than the comparison group (superiority study) – this distinction is an important one in terms of study design and data analysis, and can impact results obtained. Of the 41 BRD studies reviewed, 37 (90%) reported 3 or more outcomes being assessed, with the number of outcome measures in studies ranging from 2 to 14 outcomes per study. In the human literature, the recommendation is to limit outcome measures being evaluated in a study to less than three, since assessment of more outcomes increases the risk of analytical bias in favor of finding a significant difference. This suggests that the majority of BRD studies on treatment with an antimicrobial currently do have a reporting bias towards finding differences between treatments evaluated, and overall that many BRD studies currently do not provide sufficient information for the reader to independently assess the strength of the conclusions and the external validity of the study to their own situation.

Another important finding from the O'Connor systematic review on BRD treatment[16] was the source of available published research to date. While 12 of the 29 manuscripts (representing the 41 studies) did not state a funding source for the study, 21 of the 29 manuscripts had at least one author employed by a pharmaceutical company. The importance of pharmaceutical industry funding to continued research in this field is recognized as being key to moving our understanding of BRD forward. Historical concern of reporting bias by private companies in favor of their products has generated a skeptic bias in today's readers in favor of rejecting information generated from industry sources. A more structured reporting system would allow readers to independently evaluate which of these studies were performed using valid methods, so that the data generated by such studies has greater impact in the scientific community.

In the context of biased information, another key source is publication bias (studies with negative results not being published as frequently as those with positive results). In an effort to assess whether this occurs in veterinary medicine, abstracts presented at annual conferences (as an indicator of research studies being performed) were compared to publications (as an indicator of accepted information). Vaccine trials in swine and cattle were evaluated through review of the abstracts from forum proceedings of the American Association of Swine Veterinarians (AASV) and American Association of Bovine Practitioners (AABP) from 1988 – 2003.[18] These were compared against the final publications to determine publication rate, consistency of results between abstract and final publication, and the quality of reporting according to recommended guidelines.[14] Of 89 swine abstracts, 5 (5.6%) were ultimately published. Similarly, a low publication rate was seen for bovine studies, with 6 of 65 abstracts (9.2%) ultimately yielding a publication. This is in stark contrast to human medicine, where publication rate for conference abstracts was identified to be approximately 53%.[19] With such a low publication rate in these areas of veterinary research, there is an inherently high risk of duplicating studies, leading to the concern of inappropriate use of animals in research and wastage of research dollars. Negative results did not appear to bias a study against publication, and results were similar between abstracts and the final publication, however the low sample size (number of publications) reported here did not give sufficient power to draw strong conclusions on these questions. Reporting style in abstracts had poor adherence to recommended reporting guidelines for abstracts, with limited information on the primary outcome, randomization, and blinding such that a reader would have difficulty determining the study's internal validity by reading conference abstracts or research proceedings.

Recognizing these limitations during reporting in the veterinary literature, veterinary equivalents to the CONSORT Statement have been developed for various species. The REFLECT Statement represents guidelines directed at the reporting of randomized clinical trials in veterinary medicine involving livestock and food safety. While this document incorporates most recommendations from the CONSORT statement used in human medicine, it places further emphasis on aspects unique to animal studies such as animal housing, allocation of an intervention to the group level versus the individual animal, and design of experimental challenge studies. The REFLECT statement was crafted through extensive collaborative efforts of specialists across multiple veterinary fields. Expertise engaged during this collaboration included biostatisticians, epidemiologists, researchers across the industry sectors involved in food production (beef, dairy, poultry, swine, and aquaculture) as well as those involved in food safety, livestock production specialists, government agencies, and journal editors or assistant/associate editors. The REFLECT statement was published in 2010 in 5 separate journals,[20-24] to permit wide dissemination of the recommendations across the different specialties in the veterinary profession. As with the CONSORT statement, an explanatory document was also created for the REFLECT statement, providing users with more detailed insight for each recommendation as well as examples.[25,26] With resources such as the REFLECT Statement, the field of BRD research is well situated to advance our knowledge with greater momentum through improved study design and reporting in the future.

CONCLUSIONS

Our review of the prior literature in BRD shows that there continues to be big challenges in the assessment of diagnostic and prognostic tests for BRD, as well as our ability to evaluate any interventions (treatment/prevention/management changes) on reducing BRD. The definition of BRD between studies and current lack of a Gold Standard for diagnosis make it difficult to compare studies directly. It is also clear that previous methods in study design and reporting are hindering our ability to synthesize meaningful evidence-based conclusions based on this existing data. As such, many of our present day conclusions about BRD are tenuous, and some still stem from "dogma" in the industry rather than being evidence-based. New methods developed in biomedical research allow for a better understanding of the potential pitfalls of different study designs and reporting methods. The key steps in BRD research are to adhere to such guidelines in order to avoid potential bias. Standardization in the outcome assessment (disease status or production outcome) will also be a pivotal step in this process, since this is core to our ability to use and compare primary research studies of BRD in cattle.

An opinion statement several years ago summarized some of the needs of BRD research over the next two decades.[27] The needs included a call to establish an over-arching consortium of expertise to coordinate research efforts, in order to optimize research dollars, prioritize research needs, and to guide peer-review of research at the stages spanning from conception to publication. As well, McVey identified a need to define the "best practices" in conducting and reporting BRD research, and to establish methods of ongoing support for future BRD researchers through the training of bovine practitioners and specialists. Building upon this opinion statement, plus our review of the literature in these proceedings, we propose the following list of recommendations regarding future research in BRD:

Recommendations

General Actions

- Establish a community of BRD researchers

Improving Quality of Data

- Improve methods for diagnosing BRD under field situations
    - Do not use crude morbidity / crude mortality measures in BRD-specific research
    - Define the Se and Sp of all BRD tests
    - Standardize future methods to diagnose BRD that use tests of high Se, Sp
- Define standardized outcome measures
    - Consult key stakeholders
    - Define what would constitute a "significant difference" for the primary outcome(s)
    - Calculate appropriate sample sizes based on selected primary outcomes
- Address confounding factors during study design and data analysis
    - Pen allocation
    - Document pathogen(s) by study for naturally occurring disease
- Assess medications consistently
- Generate an open-access repository of raw data

Improving Quality of Reporting

- Advocate for increased adherence to existing reporting recommendations
- Include both a context expert and a methodology expert (for the study design) during peer review
- Standardize how outcome measures are reported
- Encourage more publication of negative results
- Repeat systematic reviews regularly
- Continue advancing recommendations in the area of study reporting

What Can the ACVIM Do To Improve BRD Research?

- Advocate for supporting infrastructure to facilitate a community of BRD researchers
- Generate a consensus statement on BRD – existing knowledge and directions for future research
    - Include a range of expertise
    - Train members on the processes of systematic reviews and meta-analyses
    - Utilize existing veterinary (and medical) reporting guidelines to further assess the quality of the information during review of the literature
    - Identify further strategic systematic reviews to reassess some conclusions
- Nurture the quality of future research:
    - Train residents to use appropriate study design and reporting methods
    - Improve statistical training of residents
    - Incorporate reporting standards into the general and/or specialty examinations
    - Host ACVIM forum interactive sessions on designing feedlot studies

REFERENCES

[1]Rutjes AW. **Health Tech Assess** 2007;11:1. [2]Branscum AJ. **Prev Vet Med** 2005;68:145. [3]White BJ, et al. **J Vet Diag Invest** 2009;21:446. [4]Leeflang MM, et al. **Clin Chem** 2008;54:729. [5]Smidt N, et al. **BMC Med Res Methodol** 2006;6:12. [6]Whiting P, et al. **BMC Med Res Methodol** 2003;3:25. [7]Hemingway H, et al. **BMJ** 2013;346:e5595. [8]Altman DG, et al. **Statist Med** 2012;31:2985. [9]Begg C, et al. **J Am Med Assoc** 1996;276:637. [10]Moher D, et al. **J Am Med Assoc** 2001;285:1987. [11]Moher D, et al. **Lancet** 2001;357:1191. [12]Moher D, et al. **Ann Int Med** 2001;134:657. [13]Schulz KF, et al. **Ann Int Med** 2010;152:726. [14]Hopewell S, et al. **Lancet** 2008;371:281. [15]Sargeant JM, et al. **Prev Vet Med** 2009;91:107. [16]O'Connor AM, et al. **J Am Vet Med Assoc** 2010;237:701. [17]Larson RL, et al. **Vet Clin Food Anim** 2012;28:97. [18]Brace S, et al. **Vaccine** 2010;28:5306. [19]Scherer RW, et al. **Cochrane Database Syst Rev** 2007;2:MR000005. [20]O'Connor AM, et al. **Prev Vet Med** 2010;93:11. [21]O'Connor AM, et al. **J Swine Health Production** 2010;18:18. [22]O'Connor AM, et al. **J Vet Intern Med** 2010;24:57. [23]O'Connor AM, et al. **Zoonoses Public Health** 2010;57:95. [24]O'Connor AM, et al. **J Food Prot** 2010;73:132. [25]Sargeant JM, et al. **Zoonoses Public Health** 2010;57:105. [26]Sargeant JM, et al. **J Food Prot** 2010;73:579. [27]McVey DS. **Animal Health Research Reviews** 2009;10:165.

INFLUENZA VIRUS IN CATTLE

Ben M. Hause, MS, PhD
Worthington, MN

INTRODUCTION

Influenza C virus (ICV) is a common pathogen of humans, with most people infected by adolescence. Unlike influenza A and B viruses (IAV and IBV, respectively), disease is usually mild or unapparent. Besides humans, ICV have been isolated from pigs in China in 1981. Genetic analysis of Chinese swine ICV showed high homology with contemporary human viruses, suggestive of interspecies transmission. More recently, an ICV-like virus was isolated from a pig in Oklahoma in 2011[1]. This virus, C/swine/Oklahoma/1334/2011 (C/OK), had only approximately 50% homology to human ICV. Additionally, C/OK failed to react with antibodies generated against human ICV in hemagglutination inhibition (HI) assays. C/OK was able to infect and transmit via direct contact in both pigs and ferrets, a surrogate for human influenza infection studies. Serological surveys found that 9.5% and 1.3% of pigs and humans, respectively, had a measurable HI titer to C/OK in the HI assay. Based on these results, C/OK was provisionally classified as a new sub-type of ICV. Despite the isolation of C/OK and finding of anti-C/OK antibodies in swine, we suspected an alternate viral reservoir in nature. Using reverse transcription PCR and HI assays, we found that C/OK-like viruses are widespread in bovines. Detailed genetic and antigenic analyses further suggest that C/OK-like viruses are sufficiently distinct from ICV to warrant classification as a new genus in the Orthomyxoviridae family, influenza D virus (IDV), with D/OK (formerly C/OK) the type species of IDV[2].

METHODS AND RESULTS

Identification of swine C/OK viruses in bovines

Nasal swabs from cattle displaying respiratory disease were submitted to Newport Laboratories for diagnostic testing. Using a previously described primer and probe assay, 45 samples were tested by relative reverse transcription PCR for C/OK virus. Samples were collected in early 2013 from 6 different states. Eight of the samples (18%) were positive. Positive samples originated from Minnesota and Oklahoma. Virus isolation was positive for 5 of the 8 PCR-positive samples on human rectal tumor cells (HRT-18G) as confirmed by hemagglutination assays and rRT-PCR. Full genome sequencing was performed for three of the bovine viruses. All genomic segments had >96% identity to C/OK. There was no evidence of reassortment with human ICV. This finding led us to hypothesize that the C/OK-like viruses are unable to reassort with human ICV.

Seroprevalence in bovines

HI assays were performed on bovine sera submitted to South Dakota State Laboratories for unrelated diagnostic testing using C/OK and C/660 as antigens. With the exception of one herd, all herds (n=7) had geometric mean titers greater than 40 to both viruses. These results suggest that bovine infection with C/OK-like viruses is common and that antibodies generated against C/OK and C/660 are cross reactive in the HI assay. The widespread presence of C/OK virus and anti-C/OK antibodies in bovines suggest that bovines likely represent the reservoir to this new group of viruses.

In vitro reassortment

A series of in vitro reassortment experiments were performed using two human ICV (C/Taylor/1947 (C/Tay) and C/Johannesburg/66 (C/JHB)), swine C/OK and a bovine isolate (C/660). Swine testicle cells were co-infected with all pairwise combinations of virus, will all four viruses or singly with each virus. Ten plaques from each infection condition were analyzed by full genome sequencing to determine segment parentage. Reassortant viruses were detected between the human ICVs C/Tay and C/JHB. Likewise, reassortant viruses were identified between C/OK and C/660. In contrast, analysis of plaques following co-infection of cells with human and animal ICV found only parental C/OK or C/660. Infection of cells with all four viruses led to C/OK and C/660 reassortants. The finding of reassortant viruses between the human viruses C/JHB and C/Tay as well as between the animal viruses C/OK and C/660 and the lack of reassortment between human and animal ICV suggests that human and animal ICV are unable to reassort and generate viable progeny.

Antibody cross-reactivity

Agar gel immunodiffusion (AGID) was used to investigate the antigenic relationship between human and animal ICV. Importantly, AGID has long been used to type influenza viruses, especially prior to modern molecular biology methods. A panel of high titer, detergent split IAV, IBV and ICV antigens, along with their respective polyclonal antisera, were studied by AGID. Antisera generated against representative IAV, IBV and ICV recognized intra-genus, not inter-genus virus antigens. Polyclonal antibodies against C/Tay did not recognize C/OK or C/660. Similarly, C/OK antiserum failed to react with C/Tay and C/JHB but did react with C/660. These results demonstrate the C/OK-like viruses are antigenically distinct from human ICV.

RNA sequencing

To study possible transcriptional differences between human ICV and C/OK-like viruses, Illumina-based deep RNA sequencing was performed on ST cells infected with C/OK. Read coverage for the polymerase genes (PB2, PB1 and PA) were biased to the 5' and 3' ends of the genomic segments, suggestive of the presence of internal deletions in the copies of these polymerase genes. HEF and NP had relatively evenly distributed read coverage. In contrast, a large reduction in reads over a well-defined region was evident for the M and NS segments, indicative of the presence of introns. Confirmatory RT-PCR and Sanger sequencing identified the splice junctions for the M and NS segments. While the splicing for the NS2 transcript for C/OK was similar to that of human ICV, the splicing of C/OK M1 was different that human ICV. The stop codon of human ICV M1 is derived from splicing. In contrast, C/OK derives a second exon of 4 amino acids in addition to a stop codon from splicing. The splicing strategy to derive M1 also illustrates differences between human ICV and C/OK.

CONCLUSIONS

The high seroprevalence (87.5% of herds with GMT HI>40) and rRT-PCR positive bovine respiratory disease samples (18%) suggest C/OK-like viruses are common in bovines. Phylogenetic analysis of C/OK and similar bovine viruses found no evidence of reassortment with human ICV. This finding was further supported by our inability to identify reassortant viruses when cell cultures were co-infected with human and animal ICV. Deep RNA sequencing showed an overall similar transcriptional strategy for C/OK and human ICV however we did identify differences in for M1 splicing. The previously demonstrated ability of C/OK to infect and transmit in both pigs and ferrets, along with our data on its prevalence in bovines, suggests that C/OK-like viruses can infect numerous mammalian species. The potential threat to human health is unknown but warrants further study.

REFERENCES

[1]Hause BM, et al. PLoS Pathog 2013;9(2):e1003176. [2]Hause BM, et al. mBio 2014; accepted.

PHARMACOLOGY AND CLINICAL RELEVANCE OF MACROLIDE ANTIMICROBIALS

Ronald K. Tessman, DVM, PhD, DACVIM, DACVPM
Hallsville, MO

Introduction

Bovine respiratory disease is the most common cause of morbidity and mortality in beef cattle. The bacterial pathogens most frequently recovered from cattle with respiratory disease include *Mannheimia haemolytica*, *Pasteurella multocida*, *Histophilus somni*, and *Mycoplasma bovis*. Administration of an effective antimicrobial agent is one of the most effective methods of preventing respiratory disease and the only effective method of treatment.

The plasma pharmacokinetics of most of these antimicrobial agents have been studied extensively in cattle. Optimal dosing of antimicrobial agents is dependent not only on the pharmacokinetics, but also on the pharmacodynamics of the drug. Pharmacodynamic properties of an antimicrobial drug address the relationship between drug concentration and antimicrobial activity. Pharmacokinetic-pharmacodynamic studies play an important role in drug development and drug evaluation. At the moment, the most widely used pharmacokinetic-pharmacodynamic approaches for antimicrobial agents rely on plasma concentrations as the pharmacokinetic input and minimum inhibitory concentration (MIC) as the pharmacodynamic input value. For many drugs, microorganisms, and disease processes, measurement of plasma concentration is an adequate surrogate marker of efficacy. However, for some drugs and disease states, concentration at the site of infection is more relevant and directly related to the desired therapeutic effect. This presentation reviews current information regarding pulmonary disposition of antimicrobial agents on bovine respiratory disease.

Minimum inhibitory concentration and interpretation of *in vitro* susceptibility data

In vitro bacterial susceptibility is determined by disk diffusion, concentration-gradient agar diffusion, or broth dilution methodologies. Disk diffusion provides qualitative susceptibility data, whereas broth-dilution methods and the concentration-gradient tests generate a minimum inhibitory concentration (MIC) expressed quantitatively in µg/mL. The MIC is the lowest concentration of an antimicrobial agent, which inhibits the growth of the target bacteria. Both the disk diffusion and methods generating MIC values assess inhibition of bacterial growth rather than killing of the pathogen as the endpoint. Susceptibility designations are determined by comparing the microorganism's MIC (or zone of inhibition if the disk diffusion method is used) to clinical breakpoints established by the Clinical Laboratory Standards Institute (CLSI). By itself, *in vitro* susceptibility of a specific pathogen does not guarantee clinical outcome. Other factors, such as the host animal's age, immune status, severity of disease at the time of initiation of therapy, distribution of the drug at the site of infection, microenvironment at the site of infection, and presence of mixed infections can all contribute to individual clinical response.

Pharmacokinetic-pharmacodynamic data to determine the optimal dosage regimen

Determination of the appropriate dose and dosing interval of an antimicrobial agent requires knowledge of and integration of its pharmacokinetics and pharmacodynamic properties. The pharmacokinetic properties of a drug describe its movement and disposition within the body and include drug absorption, distribution, metabolism, and excretion. Pharmacodynamic properties address the relationship between drug concentration and antimicrobial activity. Drug pharmacokinetic features, such as maximum plasma concentrations, the time concentrations remain above a target concentration and area under the plasma concentration-time curve (AUC), when integrated with MIC values, can predict the probability of bacterial eradication and clinical success. Pharmacodynamic studies can be done *in vitro*, *in vivo* in laboratory animal models of infection, or using experimental challenge models or natural disease in the species of interest.

The most significant parameter predicting the efficacy macrolides is the length of time that plasma concentrations exceed the MIC of the pathogen (T > MIC). Increasing the concentration of the drug several-fold above the MIC does not significantly increase the rate of microbial killing. Rather, it is the length of time that bacteria are exposed to concentrations of these drugs above the MIC that dictates killing effect. Therefore, optimal dosing of such antimicrobial agents typically involves frequent administration. However, how much above the MIC the plasma concentration should be maintained and for what percentage of the dosing interval therapeutic concentrations should be maintained, are sources of debate and likely vary depending on a given drug and specific microorganisms.

Some antimicrobials exert concentration-dependent killing or bactericidal characteristics. Their rate of killing increases as the plasma drug concentration increases above the MIC for the pathogen. As a result, it is not necessary or even beneficial to maintain drug levels above the MIC between doses. Thus, optimal dosing of concentration-dependent drugs involves administration of high doses with long dosing intervals. Some drugs exert characteristics of both time and concentration dependent activity. The best predictor of efficacy for these drugs is the 24-hour plasma AUC-to-MIC ratio where AUC is the AUC calculated over a 24 hour period following treatments. Some macrolides exhibit both characteristics. Very few studies have explored the pharmacokinetic/pharmacodynamics of antimicrobial agents as it relates specifically to bovine respiratory disease.

Concentrations of antimicrobial agents in the lungs

For pulmonary infections caused by extracellular bacteria such as *M. haemolytica* and *P. multocida*, concentrations of antimicrobials in the extracellular or interstitial space within the lungs would provide additional relevant information. Several methods have been used to estimate distribution of antimicrobial agents in the respiratory tract. These methods include measurement of drug concentration in tissue homogenates, microdialysis, nasal secretions, bronchial secretions, and pulmonary epithelial lining fluid (PELF). Measurement of drug concentration in pulmonary epithelial lining fluid (PELF) collected by bronchoalveolar lavage (BAL) is relatively easy, minimally invasive, and allows for repeated sampling of the same animals over time. The high ratios of PELF to plasma concentrations achieved with macrolides and azalides considerably exceed that which would be predicted solely on the basis of their high lipophilicity and good penetration across the alveolar epithelium. Macrolides and azalides are potent weak bases that become ion-trapped within acidic intracellular compartments, such as lysosomes and phagosomes. A beneficial consequence of macrolide accumulation within cells is increased activity against intracellular pathogens. In addition, phagocytes have been shown to act as a vehicle for the delivery of macrolides to the site of infection. Therefore, the higher than predicted PELF concentrations measured are likely the results of drug delivery to the respiratory tract by white blood cells *in vivo*.

In a recent study, groups of 8-10 calves were infected with *M. haemolytica* 1, 5 and 10 days after a single injection of gamithromycin (6 mg/kg, SC). A group of calves was used as untreated controls. The proportion of calves with complete bacterial eradication was significantly greater in calves infected 1 and 5 days after administration of gamithromycin (times at which drug concentration in PELF would be expected to be above the MIC of *M. haemolytica*) compared to calves infected 10 days after treatment (time at which only lung and BAL cells but not PELF concentrations would be expected above the MIC of *M. haemolytica*. These findings support the notion that drug concentrations in PELF is a better predictor of bacterial killing than either lung tissue homogenate or plasma concentrations in pulmonary infections caused by extracellular pathogens such as *M. haemolytica*. Clinical scores of all treated groups were statistically significantly improved when compared to the control group, even in the calves infected 10 days after treatment, but there was no difference seen in lung lesion scores.

Conclusions

The factors associated with identifying appropriate doses and dosing regimens for the treatment of bovine respiratory disease include the activity of a given drug against common respiratory bacterial pathogens and the ability of the drug to gain access to the site of infection. Plasma concentrations drugs such as macrolides are not useful surrogate markers of efficacy. For macrolide, recent findings support the notion that drug concentrations in PELF are better predictors of efficacy than either lung or plasma concentrations for the treatment of pulmonary infections caused by extracellular pathogens. Ultimately, the results of randomized prospective field trials are the best means of determining the relative efficacy of antimicrobial agents for the treatment of bovine respiratory disease in cattle.

REFERENCES

These proceedings were adapted from the previously published manuscript.

Giguère S, Tessman RK. Rational Dosing of Antimicrobials for Bovine Respiratory Disease: The Use of Plasma Versus Tissue Concentrations in Predicting Efficacy. *Intern J Appl Res Vet Med*. 2011;9(4):342-355.

MOLECULAR DIAGNOSTICS AND BOVINE MASTITIS

John R. Middleton, DVM, PhD, DACVIM (LAIM)
Columbia, MO

INTRODUCTION

During the last decade diagnostic tests based on molecular techniques have become increasingly more commonplace, but what are molecular diagnostics and what is the utility of the results? Are these methods more sensitive and specific, and what are the advantages and pitfalls associated with molecular methods versus conventional diagnostics?

By definition, molecular diagnostic tests detect DNA, RNA, proteins, or other metabolites associated with either the host or an inciting pathogen to help us differentiate health from disease. In the case of mastitis, these methods are currently employed to define whether or not a cow or mammary quarter has an intramammary infection (IMI) or to study the epidemiology of groups of pathogens. Note that, while IMI and mastitis are often used synonymously, each has a different definition. An IMI is defined as an infection occurring in the secretory tissue and/or the ducts and tubules of the mammary gland that is generally diagnosed based on the detection of a microbial pathogen, whereas mastitis is defined as inflammation of one or more quarters of the mammary gland that is almost always caused by an IMI, but is quantified by measures of inflammation such as somatic cell count or visible changes in milk quality. For many years, conventional microbiological methods based on morphology and phenotype expression on non-specific or specific growth media have been considered the "gold standard" for detection of an IMI. These methods are still in widespread use in veterinary practice and are recommended by the National Mastitis Council.[1] In contrast, molecular methods based on detection of microbial DNA, RNA, proteins, or other metabolites have historically been used in research laboratories to understand pathogen epidemiology. However, the landscape is changing and there are now commercially-available, DNA-based, detection systems marketed for the diagnosis of IMI. Understanding how these techniques work and their limitations is key to making a diagnosis and subsequent management decisions.

GOALS OF DIAGNOSTIC TESTING

Broadly speaking, diagnostic tests are performed to decide whether there is an infection present and if there is an infection present, what, specifically, is causing it.[2] In the case of mastitis, the goal of milk microbiology is to define broad groups of pathogens to facilitate the implementation of mastitis control and/or treatment measures. Mastitis pathogens can be classified based on their reservoir of infection as contagious or environmental or based on disease manifestations as clinical or subclinical. Clinical mastitis can be further divided into subacute, acute, and peracute. Subacute mastitis usually manifests as alterations in the milk such as flakes and clots, but rarely with swelling, heat, redness, and pain of the udder. Acute mastitis usually presents as a sudden onset of redness, heat, swelling and pain of the udder along with alterations in the milk. Peracute mastitis is similar to acute mastitis, but is accompanied by systemic clinical signs such as shock, electrolyte imbalances, sepsis, decreased rumen motility, and hypocalcemia. Subclinical mastitis is inflammation of the udder with no overt clinical signs. In clinical disease, broad categorization into Gram-positive and Gram-negative pathogens can be useful in defining treatment as Gram-positive organisms more often respond to therapy. With subclinical mastitis, defining pathogens as contagious or environmental can be useful in making decisions on where and how control measures need to be implemented. This has historically been done by identifying specific genera and species of bacteria and lumping them into contagious or environmental categories based primarily on phenotype and empirical knowledge (Table 1), e.g., if we identified *Staphylococcus aureus* on a blood agar plate from a cow with subclinical mastitis, it was considered contagious and control measures would have been focused on milking time hygiene, segregation and milking cows last, and culling of chronically infected cows, whereas if we identified an *Escherichia coli* from a cow with clinical mastitis, the cow would be treated with symptomatic and supportive care and prevention measures would be focused on the cow's environment.

Table 1. Common Gram-positive and Gram-negative bovine mastitis pathogens grouped according to the classical definitions of contagious and environmental pathogens.

Contagious	Environmental
Gram-positive	Gram-positive
Staphylococci	Streptococci
Coagulase negative *Staphylococcus spp.*	*Streptococcus dysgalactiae*
Coagulase positive *Staphylococcus spp.*	*Streptococcus uberis*
S. aureus	*Enterococcus spp.*
S. hyicus (agnetis)	
Streptococci	Gram-negative
Streptococcus agalactiae	Coliforms
Corynebacterium bovis	*Escherichia coli*
Arcanobacter pyogenes	*Klebsiella spp.*
	Enterobacter spp.
Other	*Pseudomonas spp.*
Mycoplasma spp.	*Pasteurella spp.*

While the same principles can be applied when using molecular diagnostic methods, molecular methods can also be employed to fine tune our understanding of mastitis pathogen epidemiology, e.g., strain-typing methods may be used to document whether a specific pathogen is found in more than one cow thus giving an indication of a contagious reservoir, or conversely if multiple strains are found in very few cows this might indicate an environmental reservoir of infection. Using molecular methods has demonstrated that some pathogens classically defined as contagious or environmental may exhibit one or both characteristics and hence mastitis pathogens may be better referred to as host-adapted or non-host-adapted.

Finally, the goals of the research scientist may differ from those of the clinician. In clinical medicine, as discussed above, our diagnostic goal is usually to facilitate the implementation of treatment or control measures. In the research laboratory, there may be an additional goals including definitive speciation of pathogens, understanding pathogen specific virulence, and an in-depth understanding of host-pathogen interactions.

SAMPLE COLLECTION

Milk is the usual medium being examined for the presence of microbes when diagnosing the cause of mastitis. To make a definitive diagnosis, milk samples are harvested aseptically at the cow or quarter-level. In contrast, if the goal is herd-level monitoring, bulk milk samples or milk collected using an in-line sampling device may be utilized. In the latter case, the goal is to get a picture of what might be causing of mastitis in a herd, but with such a strategy management practices cannot be implemented at the cow-level as the affected cows cannot be definitively identified. Further, interpretation of the results from bulk milk or in-line samplers needs to take into account possible sources for bacteria in a non-aseptically collected, comingled milk sample. For pathogens such as *S. aureus*, *Strep. agalactiae*, and *Mycoplasma spp.*, it can likely be assumed that there are infected cattle in the herd as the primary reservoir for these pathogens is the cow's mammary gland. However, the source of pathogens such as *E. coli* in non-aseptically, co-mingled milk samples is open to question as these pathogens can be on the udder or teat skin surface (dirty cows), in the mammary gland, or enter milk via post-harvest contamination.

As highlighted above, the key to a successful cow or quarter-level diagnosis, regardless of whether conventional or molecular diagnostic methods will be applied, is aseptic sample collection. Supplies needed include 70% isopropyl alcohol, cotton swabs, disposable gloves, and sterile sample vials. Milk samples are usually collected at the beginning of milking. Routine pre-milking hygiene is applied and teats are dried with a single-use towel. Each of the teats is then scrubbed with an alcohol soaked swab using a separate area of the swab for each teat. The far teats are scrubbed first and the near teats last with particular attention being paid to cleaning the teat end. Samples are collected from the near teats first and far teats last with the sample vial held away from the teat end to minimize sample contamination.

MOLECULAR DIAGNOSTIC METHODS

A variety of molecular methods have been applied to the study of mastitis epidemiology in research laboratories for a number of years. Two main approaches have been utilized 1) detection of microbial DNA using polymerase chain reaction (PCR) to diagnose IMI and 2) using molecular methods for genus and species identification or to strain-type bacteria within a specific genus and species. In the first case, PCR-based methods have been applied to directly to milk samples negating the need for time-consuming conventional culture. These techniques thus have particular appeal because, in many cases, they are more rapid than conventional bacterial culture methods. In addition, these methods may be used to detect DNA from organisms that are difficult to grow under routine culture conditions, e.g., *Mycoplasma spp.* or anaerobic bacterial genera. In the second case, bacterial culture is still needed to harvest an isolate for further manipulation. A variety of methods including PCR, DNA-sequencing, or proteomic methods (e.g., matrix-assisted laser desorption ionization time-of-flight analysis (MALDI-ToF)) have been employed for genus and species identification. Grouping of pathogens within a given genus and species according to strain has been performed using techniques such as pulsed-field gel electrophoresis, randomly amplified polymorphic DNA typing (RAPD), amplified fragment length polymorphism (AFLP), and multi-locus sequence typing (MLST). With the advent of high throughput inexpensive DNA sequence analysis, sequence-based methods are becoming quite commonplace especially for bacterial genus and species identification and even whole genomes are being studied to identify factors associated with pathogen virulence.

Today, molecular methods are no longer only in the realm of the research laboratory. There are now commercial test kits marketed for diagnosis of IMI that utilize real-time PCR methods to detect bacterial DNA in milk. One such test available through Thermo-Fisher (Pathoproof™) can be applied directly to milk without routine culture. This test detects DNA from up to 16 different commonly occurring mastitis pathogens, and boasts results in 4 hours with "unparalleled sensitivity and specificity" and applicability to various samples including mammary quarter milk samples, Dairy Herd Information Association (DHIA) samples, and bulk tank samples. While these claims can be substantiated, as with conventional culture, one must put the results in the context of where and how the milk sample was collected. As discussed above, non-aseptically collected samples from in-line samplers (DHIA samples) or bulk milk have their limitations. Further, as with conventional culture, detection of multiple genera and species ($\geq$ 3 types) in a mammary quarter or 4-quarter composite aseptically collected milk samples still indicates a contaminated sample. Also bear in mind that this test will only detect the 16 pathogens for which PCR primers are included. So, unlike conventional culture where, depending on the culture media used, there is the potential to detect an infinite number of genera and species, the PCR has its diagnostic limitations. Finally, the test only detects DNA and therefore there is no "proof of life" in the udder calling into question whether an active IMI is present.

In addition to PCR-based mastitis diagnostics, proteomics-based (MALDI-ToF), bacterial identification systems are finding their way into veterinary diagnostic labs for genus and species identification of bacteria and these inexpensive high throughput methods may ultimately replace classical phenotypic identification systems based on biochemical profiling (e.g., Sensititre AP90™).

ADVANTAGES AND DISADVANTAGES OF MOLECULAR DIAGNOSTICS

When evaluating a "new test", its sensitivity (Se), specificity (Sp), positive predictive value (PPV), negative predictive value (NPV), and accuracy are calculated by comparing the results of the new test to a "gold standard". One of the difficulties in diagnosing bacterial IMI is we do not have a perfect "gold standard" test. Milk culture is often considered the "gold standard" test, but manipulation of the milk inoculum size and growth media will impact Se and Sp. The National Mastitis Council recommendation for routine bacterial culture is to inoculate solid media (usually a blood-based agar) with 10 µl of milk.[1] Therefore, the limit of detection of this test is 100 colony forming units (CFU)/ml. Hence, if an individual mammary quarter is shedding < 100 CFU/ml of bacteria, it will be declared culture negative. Similarly, if the organism has specific growth conditions not used in routine screening, e.g., *Mycoplasma spp.* or anaerobes, the sample will likewise be declared negative. Inhibitors in milk such as antimicrobial treatments can also lead to a negative test result. These are examples of false negative test results that will impact Se. The opposite is true if, for example, the teat is not aseptically prepared prior to sample collection and bacteria from the outside of the teat are washed into the sample during collection. In this instance, we may get a false positive result impacting test Sp.

Similar issues arise when diagnosing and IMI using PCR-based methods. Because the PCR assay detects only those bacteria for which there are primers in the assay profile, organisms not included in the assay profile will be missed leading to false negative test results. An example of such a situation was recently documented in Europe using PCR detection of methicillin-resistant *Staphylococcus aureus* (MRSA). A series of isolates were found to have an oxacillin resistant phenotype based on conventional antimicrobial susceptibility testing, but were negative for the *mec*A gene, the gene encoding methicillin resistance, during PCR screening. Further investigation of the bacterial genome revealed that these isolates were a *mec*A variant, now termed *mec*C, that were missed because the original PCR primers did not detect this variant.[3] False positives will occur with contamination or when the assay detects bacterial DNA from organisms that the cow's immune system or antimicrobial therapy has already killed. As with conventional culture, contamination of the sample will also lead to false positives. The issues associated with bulk milk and in-line sampler collected milk samples are discussed in the section above and are also potential examples of false positive results. Recent work has shown that while carryover of bacterial organism (DNA) between cows when collecting milk from an in-line sampler (DHIA sample) is apparently low, carryover nonetheless occurs and may lead to a cow being falsely identified as having an IMI.[4] The false positive and negative rates can also be influenced with the cycle threshold (Ct-values) used to make a diagnosis. Similar to manipulating the inoculum size in conventional culture moving the Ct-value threshold up or down will impact the Se and Sp.[4]

When using molecular methods to make a definitive genus and species diagnosis, one must consider whether the techniques have been appropriately validated against bovine reference strains. By example, recent work has clearly demonstrated the superiority of DNA-sequence based methods over phenotypic methods for speciation of coagulase negative staphylococcal isolates.[5] That said, labs are now beginning to use bacterial proteomics (MALDI-ToF) to speciate bacteria, and while new technologies have appeal from a speed and expense standpoint, it is critical that the results be validated before being placed in widespread use.

With regard to strain-typing methods, it is important to understand the discriminatory power of the tests being used. Does the test lump isolates into one group or split them into many groups? How will the results be used to manage mastitis cases? If we are trying to determine contagiousness of *S. aureus* isolates in a group of cows, for example, a discriminatory test will be needed so that we do not needlessly cull cows falsely identified as possessing the contagious strain because the strain-typing test failed to adequately differentiate the strains.

FUTURE DEVELOPMENTS

The discussion so far has focused primarily on DNA-based (genomic) methods. There are other modalities which can be applied to the study of infectious disease including microbiomics, transcriptomics, proteomics, and metabolomics. These techniques are being used by researchers to understand bovine mastitis and some have promise as future diagnostic methods. For example, recent studies evaluating the milk microbiome have revealed some interesting results to further our understanding of mastitis epidemiology. The milk microbiome consists of all the microorganisms in the cow's mammary gland environment.[2] Using DNA extraction, PCR amplification, and subsequent DNA sequencing of a fragment of the 16S gene, multiple genera and species of bacteria are identified using bioinformatics. By studying the microbiome of healthy versus infected cattle, it can be determined where differences in the microbial population lie to better understand what may be causing inflammation in the udder.

CONCLUSIONS

As with any test, it is critical that the interpreter know the limitations of the test and view the results in the context of how and from where the sample was collected. While new technologies add to our toolbox for mastitis diagnosis, and in some cases improve the Se and Sp of diagnosis, it is important that the clinician or researcher have a clear understanding of the techniques, how they were validated, and the potential pitfalls.

REFERENCES

[1]Laboratory Handbook on Bovine Mastitis. **National Mastitis Council, Madison, WI** 1999. [2]Zadoks RN. **Proceedings of the National Mastitis Council 53rd Annual Meeting** 2014;111. [3]Garcia-Alverez L, et al. **Lancet Infect Dis** 2011;11:595. [4]Klaas IC, et al. **Proceedings of the National Mastitis Council 53rd Annual Meeting** 2014;167. [5]Zadoks RN & Watts JL. **Vet Microbiol** 2009;134:20.

EMERGING PATHOGENS IN BOVINE MASTITIS

Christopher Luby, VetMB, PhD, DACVIM (LAIM)
Saskatoon, SK

INTRODUCTION

Mastitis remains a major production-limiting disease on dairy operations worldwide. Treatment of mastitis results in more antimicrobial use in dairy cattle than for all other reasons combined[1]. Control measures for the major mastitis pathogens such as *Staphylococcus aureus*, *Streptococcus agalactiae*, *Escherichia coli* and *Streptococcus dysgalactiae* are well established. Implementation of these control measures is commonly successful in controlling these pathogens and therefore lowering bulk tank somatic cell count (SCC). Many milk processors offer premiums for bulk tank SCC that drops below a set level. As a result, attention has focused on emerging mastitis pathogens that may prevent dairies that implement effective control measures for major mastitis pathogens from taking advantage of these premiums. Furthermore, novel pathogens are emerging in all areas of infectious disease. It is important to have a better understanding of these pathogens to enable producers to take advantage of available premiums and to be able to respond to challenges with emerging pathogens that are potentially zoonotic. Emerging pathogens that will be covered in this presentation include the coagulase-negative staphylococci (CNS), *Prototheca* spp., *Klebsiella* spp., *Serratia* spp., *Streptococcus canis* and *Enterococcus faecalis*.

COAGULASE-NEGATIVE STAPHYLOCOCCI

The coagulase-negative staphylococci are a group of over 50 different bacterial species. They have traditionally been considered incidental findings on milk culture. In North America 15% of new intramammary infections (IMI) following calving are due to CNS[2]. Studies report differing effects of the CNS on udder health and milk quality. Evidence exists suggesting that the CNS caused a moderate increase in SCC which was lower than that of mammary quarters infected with a major mastitis pathogen[3]. In the same study, milk production was higher in CNS infected cows than in controls. The CNS contributed to 17.9% of the bulk tank SCC in herds with a bulk tank SCC less than 200,000 cells/ml. This evidence is supported by work showing that heifers with CNS IMI early in lactation have increased lactational milk production when compared with uninfected controls[4]. Other studies suggest that CNS do not significantly affect milk production in heifers or mature cows[5].

One major limitation of studies into CNS mastitis is the fact that these studies treat the CNS as a single pathogen rather than a collection of over 50 different bacterial species. Before the advent of readily available DNA sequencing, speciation was based on phenotypic characteristics of CNS. Studies into genotypic speciation have revealed that the majority of CNS isolates are misclassified if speciated phenotypically. Genotypic speciation of the CNS is commonly carried out through sequencing of the *rpoB* gene although other techniques such as amplified fragment length polymorphism (AFLP) or partial sequencing of the *cpn60* gene have also been used. There is little current work regarding the effects of different CNS species on milk quality. Some evidence suggests that different populations of CNS species are isolated from milk of infected cows and the environment, suggesting a contagious mode of transmission[6]. If control of CNS is attempted, it would be logical to implement control measures appropriate for control of contagious mastitis pathogens.

Infections with CNS appear to be susceptible *in vivo* to intramammary antimicrobial treatment with bacteriological cure rates ranging from 80%-90%. Cure rates decrease when isolates are resistant to β-lactams. In addition, the spontaneous cure rate of CNS without antimicrobial treatment is relatively high.

PROTOTHECA

Prototheca are unicellular algae that can cause granulomatous mastitis. They have been isolated from cases of mastitis in North America, South America, Europe and Asia. The two major causes of mastitis are *Prototheca zopfii* and *Prototheca blaschkeae*. These pathogens survive well in moist areas with high amounts organic material such as mud, manure storage and spoiled feed. They are rarely isolated from bovine milk but prevalence can reach over 30% during an outbreak. Human disease due to *Prototheca zopfii* and *Prototheca blaschkeae* has been reported but is rare. A link to bovine mastitis has not been reported although the low prevalence of this disease in both humans and dairy cattle would make a link difficult to prove.

A recent study identified risk factors for *Prototheca* mastitis on farms in Ontario[7]. These included intramammary administration of a drug not labeled for intramammary usage, the number of different injectable antimicrobials used, teat sealant usage and recent treatment of displaced abomasum. Hence, excessive antimicrobial use and inappropriate routes of antimicrobial administration are associated with outbreaks of *Prototheca* mastitis.

KLEBSIELLA

Klebsiella are Gram-negative mastitis pathogens. Clinical mastitis due to *Klebsiella* appears to be more prevalent in herds with a low bulk tank SCC[8]. This may reflect the fact that these herds have successfully implemented control programs for major mastitis pathogens.

The challenge with *Klebsiella* is that these pathogens cause severe clinical mastitis and are not susceptible to antimicrobial treatment. Severity of clinical mastitis is less severe in animals vaccinated with a J5 coliform mastitis vaccine. Infections tend to be considered to have an environmental source although evidence of clonal outbreaks is reported in the literature[9], suggesting a significant contagious mode of transmission.

The major risk factor for these pathogens was previously considered to be bedding animals on wood-based products. Recent work has shown that 70% of clinically normal cows shed *Klebsiella* in feces therefore control must focus on preventing fecal contamination of the udder through appropriate management of free stalls, manure alleys and pre-milking teat disinfection. Fecal contamination of the udder may be quantified and monitored by implementing a cleanliness scoring program. Higher cleanliness score (i.e. dirty) cows have been demonstrated to have greater *Klebsiella* contamination at the teat end.

SERRATIA

Serratia species are ubiquitous in the dairy farm environment with the major species isolated from milk being *Serratia marcescens* and *Serratia liquefaciens*. They are rare causes of mastitis which are less clinically severe than *Klebsiella*. The decreased severity of *Serratia* species is associated with the number of acyl groups within lipid A. *Serratia* has five acyl groups whereas *Klebsiella* has seven. A decreased number of acyl groups are associated with decreased biological activity of lipid A[9].

Despite its ubiquitous nature, *Serratia* mastitis is rare on dairy operations. A study into an outbreak of *Serratia* mastitis in the United States revealed an association with on-farm contamination of chlorhexidine-based teat disinfectants[9] therefore control should focus on hygienic teat dip management. The majority of *Serratia* infections appear to cure spontaneously.

STREPTOCOCCUS CANIS

Streptococcus canis is part of the commensal respiratory flora of cats and dogs. It has been reported as causing subclinical mastitis in dairy cattle[10]. *Streptococcus canis* is an emerging pathogen in humans and has been reported to cause soft tissue infections, osteomyelitis, pneumonia and sepsis[11]. Lateral gene transfer between *Streptococcus canis* and human streptococcal species has been reported[11] which may increase the zoonotic risk associated with this pathogen.

ENTEROCOCCUS FAECALIS

Enterococcus faecalis is a rare pathogen of the bovine udder but can cause infection in humans. Virulence traits appear to be shared between isolates from hospital patients with nosocomial infections and those isolated from food animals. These have been demonstrated to be acquired by cattle from their environment and are a heterogeneous, genetically diverse pathogen in dairy cattle[12]. Therefore, the human risk appears to be as a result of fecal exposure rather than from consumption of milk.

SUMMARY

Emerging pathogens represent one area in the challenge of mastitis to the dairy industry. These pathogens are of greatest concern to producers who have successfully controlled the major mastitis pathogens such as *Staphylococcus aureus*. As incentives to produce high quality milk become more widespread, these pathogens will be of interest to increasing numbers of dairy producers.

REFERENCES

[1]Pol, M, et al **J Dairy Sci** 2007;90:249. [2]Dingwell, RT, et al **Prev Vet Med** 2004;63:75. [3]Schukken, YH, et al **Vet Microbiol** 2009;134:9. [4]Piepers, S, et al **J Dairy Sci** 2013;96:5095. [5]Reksen, O, et al **J Dairy Sci** 2007;90:4670. [6]Piessens, V, et al **J Dairy Sci** 2011;94:2933. [7]Pieper, L, et al **J Dairy Sci** 2012;95:5635. [8]Barkema, HW, et al **J Dairy Sci** 1998;81:411. [9]Schukken, Y, et al **Vet Clin N Am-Food A** 2012;28:239. [10]Chaffer, M, et al **N Z Vet J** 2005;53:261. [11]Richards, VP, et al **BMC Microbiol** 2012;12:293. [12]Petersson-Wolfe, CS, et al **J Dairy Sci** 2008;91:615.

RADIOGRAPHIC AND ULTRASOUND IMAGING OF THE BOVINE UDDER

Lisa G. Britt, DVM, MS, DACVR; Isabelle Masseau, DVM, DACVR, PhD; John R. Middleton, DVM, PhD, DACVIM (LAIM)
Columbia, MO

INTRODUCTION

Radiography of the bovine udder was first studied intensely in 1968.[1] Historically, application of radiography in the field was not practical due to the need for developing films. However, with the advent of digital and computed radiography, images can be generated almost immediately cow-side on a laptop sized viewing system. Such modalities are becoming commonplace in large animal practice, and hence these techniques may be field applicable for the diagnosis of certain mammary gland conditions.

Ultrasound of the bovine mammary gland and teat was described as early as 1967 using A mode imaging and by 1986, modern B mode imaging was utilized.[2,3] Early ultrasound image quality, particularly of the teat, was hampered by lack of availability of affordable high frequency probes. However, ultrasound machines have grown in diagnostic quality with high frequency probes of 10 to 12 MHz now being readily available at a reasonable cost. As with radiography, modern ultrasound machines are highly portable and available in lap top configurations allowing cow-side, real-time detailed imaging of the teat and mammary gland tissue.

ANATOMY

Before discussing application of imaging modalities to the study of the bovine udder, an understanding of mammary gland anatomy is needed to understand the limitations mammary gland location, tissue density, and the presence of milk present to diagnostic imaging (Fig 1). The "gland" is referred to as the parenchyma, gland cistern, and the glandular part of the lactiferous sinus. It consists of the true parenchyma, spaces where milk pools, vessels, connective tissue/laminae. The "teat" includes the papillary part of the lactiferous sinus which is also known as the teat cistern. The teat canal is also known as the papillary duct or streak canal.

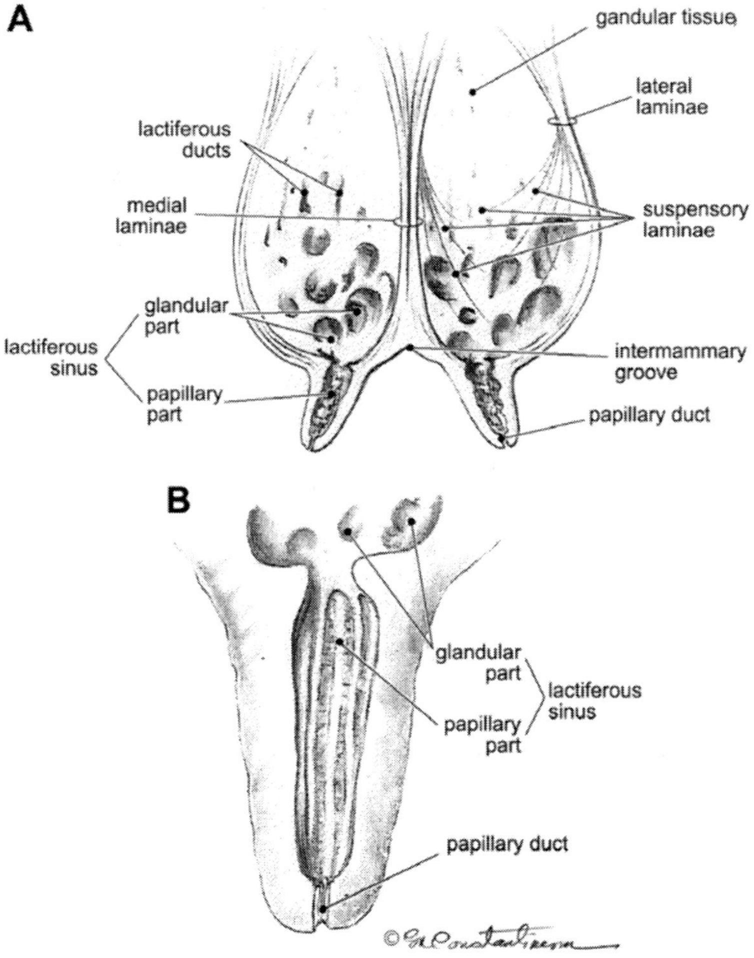

Figure 1 – Cross section of the bovine mammary gland (**A**) showing the gland cistern (glandular portion of the lactiferous sinus), teat cistern (papillary portion of the lactiferous sinus), and teat canal (papillary duct, also referred to as the streak canal). Close up of the bovine teat (**B**). *Image adapted from the Clinical Dissection Guide for Large Animals used with the permission of author Gheorge Constantinescu.*

INDICATIONS FOR DIAGNOSTIC IMAGING

Radiography and/or ultrasonography are well indicated when cows are presented for teat anomalies, milking difficulties, protuberances along the teat or palpation of a mass. The benefits of radiography over ultrasound are its ease of use with minimal technical handling skill to learn. Acquiring an image of each teat can be performed rapidly and repeatedly throughout a study.

Ultrasound is often preferred to radiography due to its portability. It is non-invasive with milk pooling in the teat cistern providing adequate separation and differentiation of internal tissues so infusion of a contrasting substance is not necessary. Ionizing radiation is also no longer necessary so images can be acquired in any environment. Although there is improved anatomic detail of both the soft tissues and lumen of the mammary gland and teat, the user must invest time into understanding how to operate the ultrasound equipment as well as interpret the images.

RADIOGRAPHY

The great variation between the thickness of the gland cistern and the teat make it impossible for one radiographic exposure to appropriately penetrate both of the areas. The high radiographic technique that must be used for the gland causes severe overexposure of the teat. The location of the hind mammary quarters between the pelvic limbs further limits the ability to image anything but the distal glandular tissue and the teat with conventional radiography. Since most of the surgically treatable abnormalities are located within the teat, radiographic techniques are generally aimed at optimizing visualization of this structure. The first radiographic studies utilized nonscreen dental film packs attached to individual teats with a rubber band.[1] The teats were stabilized with a stainless steel rod of adjustable height. Radiographs were made with a 0.6-second exposure and a focal spot-film distance of 66 cm at 85 kVp and 15 mA. With digital imaging, greater latitude in gray scale is possible and we have used our standard 8 x 10 DR receiver plate for each image with a technique of 0.1 mAs and 54 kVp (Fig 2A). Cows must be kept still using a head gate, squeeze chute, or other form of restraint and the DR plate must be held as close to the teat as possible often with upward pressure to help lift the glandular portion so that it does not interfere with exposure of the teat cistern. Survey radiography alone does not commonly provide diagnostic information as both mammary gland tissue and milk are of similar radiographic opacity resulting in the absence of contrast and delineation of structures. Positive contrast radiography can be safely performed using iodinated contrast medium such as iohexol or iopamidol, diluted 1 part contrast to 2 parts of saline. When infusing sterile contrast into the teat, strict aseptic technique must be used to prevent iatrogenic intramammary infection. The diluted contrast solution can be slowly injected through the opening of the papillary duct into the papillary part of the lactiferous sinus using a small teat cannula made of plastic. Dilution of contrast medium allows reduction of the viscosity, which makes the injection easier as well as it tends to decrease the irritation of the tissue by the contrast medium. Additionally, better quality radiographs are obtained when they are taken immediately after injection. Positive contrast radiography may be useful to demonstrate a connection between the papillary part of the lactiferous sinus (or teat cistern) and an accessory gland as well as stenosis within the sinus itself. Angiography can also be performed by injecting a sterile solution of iodinated contrast medium (dilution is not necessary but can be performed) into the papillary vein of the abnormal teat or within the papillary venous plexus at the junction between the glandular and papillary part of the lactiferous sinus. Angiography may be particularly useful for diagnosis of varicose veins which are often reported to create problems in milking of a particular quarter. Nodules or masses palpated along the teat as well as suspicion of inversion of the rosette of Fürstenberg may be better observed with double contrast studies. This study can be performed by injecting a small amount of iodinated contrast medium (5-10 mL) followed by 10-20 mL of room air. Radiographs should be taken immediately after injection. These studies can be repeated in a normal quarter for comparison. It is important to remember to obtain milk sample for culture if desired prior to injection of iodinated contrast medium since it is bacteriostatic and may yield negative culture results despite the presence of infection.

Radiography can also be helpful in a research setting in detecting if teat canal sealants are present.[3] In order for the sealant to be visible, a substance must be included that will appear as a mineral or metallic opacity on survey radiographs. OrbeSeal™ is a commercially available internal teat sealant (Zoetis Animal Health) that appears similar to positive contrast within the teat when imaged radiographically (Fig 2B).

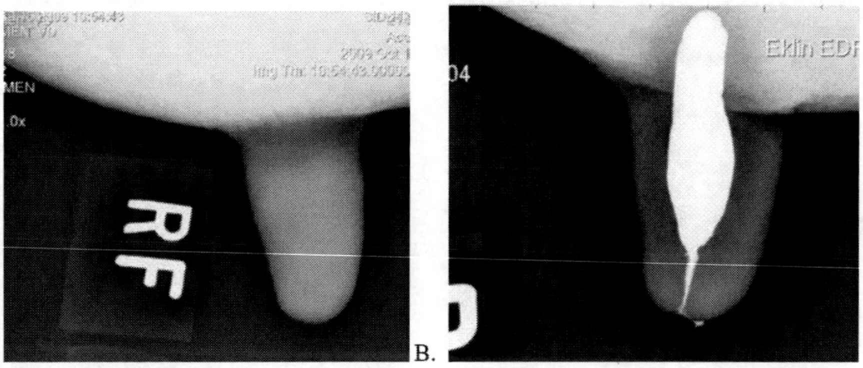

A. B.

Figure 2 - Digital image of a right front quarter teat prior to infusion A) and following the infusion of the internal teat sealant OrbeSeal™ B).

ULTRASONOGRAPHY

Similar to radiography, cows must be restrained in some sort of head gate or chute to facilitate ultrasound evaluation of the udder. The ultrasonographer must find a position alongside the caudal abdomen /flank that is safe and comfortable enough to perform the complete ultrasound examination. If possible, imaging of the mammary gland and teat should be performed at least 2 hours after milking. Hair covering the mammary gland must be shaved and any debris should be cleaned off the gland and teats. An acoustic coupling gel is applied to the skin surface of the gland and either the teat or the transducer depending on the technique used. Lower frequency transducers (3.5 to 8 MHz) are necessary to adequately penetrate the large mammary glands. These transducers can be linear, curvilinear/convex, or sector in configuration. If imaging the teat, a higher frequency transducer (7.5, 8, 10, or 12 MHz) is optimal. If a 5-8 MHz transducer is used, a water bath (water filled plastic cup with the teat surrounded by water) or gel stand off is necessary to obtain desired detail. Longitudinal images often give the best overview of the teat, and are comparable to what is seen on a contrast radiograph, but typically transverse images are also obtained over areas of interest.

The gland cistern appears as homogenous parenchyma interspersed with anechoic zones caused by blood vessels or lactiferous ducts.[4] Milk should appear anechoic with some echogenic foci. If harmonics are used, milk will appear more echogenic. Abnormalities such as edema, abscesses, hematomas, foreign bodies, connective tissue buildup, and tumors can be discovered within the gland cistern. Some pathogens cause mastitis with a characteristic appearance, such as *Staphlococcus aureus, Trueperella pyogenes* and Gram-negative bacteria of the Enterobacteriaceae family. Mass lesions can also be seen within the teat cistern as well as stenosis, obstructions, and fibrous changes of the teat canal and rosette of Fürstenberg. In addition, measurements of the teat canal length and diameter, teat cistern diameter and teat wall thickness can be taken and evaluated in light of breed, number of lactations, milking machine parameters etc.[4,5] Similar to radiography, ultrasound can also be used to confirm the normal anatomy of a teat prior to the infusion of an internal teat sealant (Fig 3A). The presence and retention of a teat sealant can be verified using ultrasound quickly and repeatedly without the concerns for safety that radiography would necessitate (Fig 3B).

A.

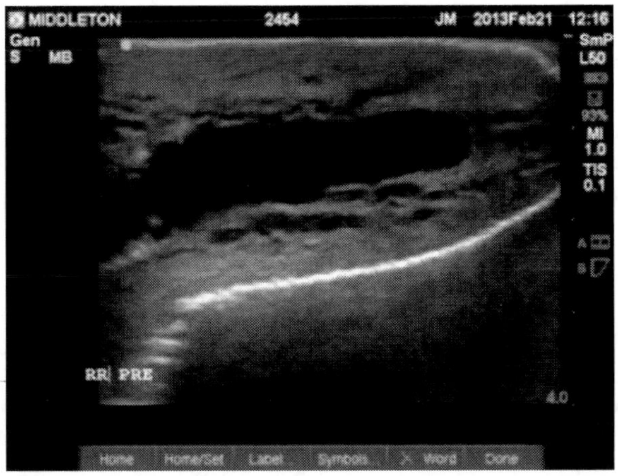

B.

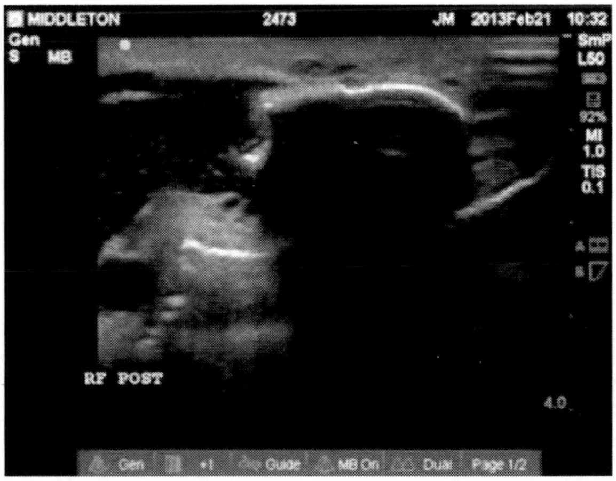

Figure 3 – Ultrasound image of a normal teat (A) and a normal teat containing an internal teat sealant (OrbeSeal™). Images recorded using a 6-13 MHz linear probe. Note the shadowing caused by the internal teat sealant in B.

CONCLUSIONS

Radiography and ultrasound both allow visualization of the teat cistern and teat canal lumens. Ultrasound has the additional advantages of facilitating examination of the gland cistern, which might lead to guided fine needle aspirates or biopsies if appropriate. In addition, there is no ionizing radiation with ultrasound, so images can be obtained in almost any location. Each modality has its place in evaluation of the udder, although more recent literature shows the preponderance of studies including ultrasound.

REFERENCES

[1]McDonald, JS **AJVR**. 1968; 29(6): 1315. [2]Franz, S, et al **Vet Clinics of NA** 2009; 25(3): 669. [3]Woolford MW, et al **New Zealand Vet J** 1998; 46: 12. [4]Fasulkov JR **Bulg J Vet Med** 2012; 15(1): 1. [5]Szencziov I, et al **Slovak J Anim Sci** 2012; 45(3): 96.

INFECTIONS AND ANTIBIOTICS IN NEW WORLD CAMELIDS

Christopher K. Cebra VMD, MA, MS, DACVIM
Corvallis, OR

INTRODUCTION

Due to frustration in trying to identify a primary disorder and over-interpretation of leukogram changes, veterinarians are often tempted to treat sick camelids with a course of antibiotics. As little information is available either on drug pharmacokinetics or the frequency of various infections, the treatment course is often as "shoot from the hip" as the diagnosis. This should summarize the current state of knowledge.

PHYSIOLOGIC CONSIDERATION

Camelids are anatomically and physiologically unique, making extrapolation of dosages dangerous. No medications are approved for use in New World camelids, and dosages found in clinical reports can differ from each other as much as 25-fold. To their advantage, because of their "pet" status, we are often able to use medications which would be prohibited or prohibitively expensive in food animals. However, as most drug use is empirical, and as the concept of camelids entering the human food change becomes more acceptable, the attending veterinarian must assume the responsibility for extra-label drug use, residues, and potential adverse effects on the animal.

As a general rule of thumb, antibiotics excreted in the urine have longer elimination half-lives in camelids than in domestic ruminants, possibly due to camelids' lower rate of urine production. This could lead to a longer therapeutic effect or also greater toxicity, the latter especially with aminoglycosides. The effects of concurrent fluid therapy have not been studied; presumably interventions which increase urine production will increase drug elimination. In some cases, particularly with subcutaneous dosing, the perceived long elimination half-lives are consequences of a slow, inconsistent absorption phase.

As another general rule, volume of distribution for aqueous antibiotics is larger in camelids than cattle or sheep, but varies tremendously among individual camelids. This presumably relates to relative body fat, muscle, and organ weights, as well as age and breed. Recent data suggest that adult alpacas have an extracellular fluid compartment that is approximately 37% larger than adult llamas, and that neonatal llamas have an ECF compartment that is 30% larger than adults. Lipophilic antibiotics (florfenicol) will be less affected than hydrophilic antibiotics (penicillins, cephalosporins, tetracyclines), which may have to be dosed higher in alpacas or crias than llamas. Whether the cria and alpaca effects are cumulative has not been tested scientifically. Aminoglycosides are a special case: they diffuse so slowly out of the vascular compartment that they should not be dosed at a higher rate.

Whereas the slow excretion suggests less frequent dosing may be acceptable, the higher volume of distribution suggests larger individual doses may be necessary. This remains theoretical however, because pharmacokinetic trials uniformly use bovine or ovine doses. The extreme individual animal variation in volume of distribution further complicates the issue. The best way around these problems is to use high-end doses of antibiotics with a high margin of safety.

ANTIBIOTIC AGENTS

Penicillins appear to be safe, with minimal toxic effects and reactions most common only after long treatment courses. Slow renal excretion suggests that lower dosages or less frequent dosing intervals may achieve sufficient therapeutic effect. Ceftiofur sodium is similar. Long-acting ceftiofur preparations tend not to maintain high concentrations for very many days. Aminoglycoside antibiotics generally have poor lipid solubility and move into the extracellular space slowly, thus decreasing dosage differences between species and age of camelid. They have long half-lives and are likely to be safer at once daily dosing rather than smaller dosages at smaller intervals. Nephrotoxicity has been reported. Except for penicillin, which is administered in different forms by different routes, these antibiotics appear to have similar pharmacokinetic properties whether administered subcutaneously, intramuscularly, or intravenously.

Sulfa antibiotics have been controversial. Gastrointestinal absorption is unreliable and reports on elimination after intravenous injection are conflicting. Usually, we recommend alternatives to oral sulfas. Macrolides have been used infrequently, mainly due to a high rate of adverse reactions to some agents.

Oxytetracycline, enrofloxacin, and florfenicol have also been studied in llamas and alpacas. All are currently recommended at bovine doses. Florfenicol has seen a fair amount of use, but recent trials suggest that it must be dosed more frequently than in cattle. Prolonged use may lead to reductions in red and white blood cells compatible with marrow suppression. Enrofloxacin is the only tested antibiotic that appears to have fair absorption (30% bioavailability) after oral administration. There is one report of retinopathy with enrofloxacin.

BACTERIAL ISOLATES

Very little information is available concerning the frequency and importance of bacterial isolates from New World camelids. Except for abscesses, specific, focal internal infections are rare. Most affected camelids have general signs such as fever, obtundation, and weakness; some have more localizing signs such as diarrhea or dyspnea. Collection and culture of pertinent body fluids (blood, peritoneal fluid, pleural fluid, cerebrospinal fluid, urine, feces, aspirates, et cetera) may yield information about specific pathogens and allow refinement of antibiotic selection.

Most of these septic syndromes have a relatively even division between gram-negative and gram-positive causes. Thus, broad-spectrum antibiotics coverage is recommended. Most commonly used are combinations of an aminoglycoside with a beta-lactam antibiotic or ceftiofur alone or in combination. Other choices include oxytetracycline, enrofloxacin, or florfenicol.

Some localized infections have a preponderance of certain isolates. Uterine infections frequently involve gram-negative enteric bacteria. Tissue and tooth abscesses more frequently involve gram-positive or anaerobic bacteria. These would dictate antibiotic choices. Oxytetracycline also remains the only antibiotic shown to have efficacy against *Mycoplasma haemolama*, and is presumed to work against *Ehrlichia phagocytophila* as well.

Table 1. Common Agents, Routes, and Dosages (For the most part, these are anecdotal, and the user assumes all responsibility for undesired effects.)

Ceftiofur	2.2-4.4 mg/kg	q12-24h IV, SC, or IM
	up to 8 mg/kg in neonates	
Ceftiofur CFA	6.6 mg/kg	q2-8d SC
Procaine Pen	22-44K U/kg	q 12-24h SC or IM
Crystal. Pen	22-44K U/kg	q 6 h IV
Ampicillin	10-120 mg/kg	q 12-24 (q 8h if IV)
Amikacin	18 mg/kg	q 24 h IV, SC, or IM
Gentamicin	5 mg/kg	q 24 h IV, SC, or IM
Enrofloxacin	5 mg/kg	q 24h IV or SC
	10 mg/kg	q 24h PO
Oxytet (200 mg/ml)	20 mg/kg	q 48-72h SC or IM
Oxytet (300 mg/ml)	20 mg/kg	??? SC or IM
Florfenicol	20 mg/kg	q 24h SC or IM?
Metronidazole	15-25 mg/kg	q 12h PO in neonates, PR in adults

REFERENCES

Ali BH, et al. Vet Res Commun 2003;27(6):475, Chakwenya J, et al. J Vet Pharmacol Ther 2002;25(5):321, Christensen JM, et al. J Vet Pharmacol Ther 1996;19(6):431, Dowling PM, et al. J Vet Pharmacol Ther 1996;19(2):161, Drew ML, et al. J Vet Pharmacol Ther 2004;27(1):13, Gandolf AR, et al. Am J Vet Res 2005;66(5):767, Hadi AA, et al. J Vet Pharmacol Ther 1994;17(1):48, Junkins K, et al. J Zoo Wildl Med 2003. 34(1):9, Lackey MN, et al. Am J Vet Res 1996;57(8):1193, Oukessou M, et al. Am J Vet Res 1992;53(9):1658, Snook CS, et al. J Vet Pharmacol Ther 2002;25(5):383, Dechant JE, et al. J Vet Pharmacol Ther. 2013;36(2):122, Holmes K, et al. J Vet Pharmacol Ther. 2012;35(4):382

IS THERE A ROLE FOR THE VETERINARIAN IN CLEANING AND DISINFECTION?

Brandy A. Burgess, DVM, MSc, DACVIM (LAIM)
Fort Collins, CO

IMPORTANCE

Historically, infectious disease prevention in veterinary medicine has focused on vaccination and antimicrobial use with little regard for more fundamental prevention methods – cleaning and disinfection. As leaders in the practice, veterinarians must have foundational knowledge that allows for effective implementation of cleaning and disinfection techniques within the practice of veterinary medicine. Veterinary hospitals are continually faced with the potential introduction of infectious disease into the hospital population. A survey of infection control programs (ICP) at American Veterinary Medical Association (AVMA) accredited Veterinary Teaching Hospitals (VTHs) found epidemics of healthcare-associated infections (HCAIs) to occur frequently and identified zoonotic infections as an important occupational hazard to personnel.[1] In order to provide excellent patient care, veterinary practices must implement an infection control program to protect the personnel, the patients, and the practice.

Not only do veterinarians have an ethical responsibility to consider infection control in daily practice, but there is a recognizable standard of care with respect to infection control.[2] Veterinarians must take appropriate precautions to mitigate the risk of infectious disease in patients and personnel. This is becoming ever more important as the world population ages and advancements in medicine improve survival for those with debilitating diseases and provides treatment regimens causing profound immune suppression. The veterinarian has a clear duty to recommend preventive measures to reduce the risk of zoonotic disease – this includes cleaning and disinfection – and to counsel clients to seek medical attention in the event of zoonotic disease exposure.[3]

CLEANING

The general goals of an ICP are to eliminate sources of infectious agents and to break the cycle of transmission. A key component in this endeavor is the proper use of cleaning and disinfection – a multistep process involving the removal of visible debris, scrubbing with a detergent, followed by rinsing and application of an appropriate disinfectant at the correct dilution and contact time. While rigorous cleaning alone can reduce the bacterial load on a concrete surface by 90%, applying a disinfectant after cleaning can reduce this by an additional 6%.[4] In the environment, many bacteria can form biofilms (organized bacterial communities which secrete a protective extracellular matrix) that enhance bacterial survival from a variety of environmental insults, including disinfectants, providing a continuous source of pathogens that can cause HCAIs.[5] Reducing the environmental load of potential pathogens is critical in an environment where animals from many different herds in varying states of immune compromise are comingled – creating an increased risk for HCAIs. A recent report of epidemic Salmonellosis in a VTH cited ineffective cleaning and disinfection as an important factor in the outbreak which resulted in patient fatalities, hospital closure, and an estimated financial cost of US$4.1 million. [6]

DISINFECTION

Disinfection is the reduction of viable microorganisms – and a disinfectant is a chemical applied to an inanimate object or surface for this purpose. This should be distinguished from an antiseptic – which is applied to living tissues. Ideally, a chemical disinfectant would have a broad spectrum of activity, be non-irritating and non-toxic, non-damaging to the environment, and efficacious on many different surfaces and in many different environments.[7] Realistically, available disinfectant classes have limitations in efficacy against diverse pathogens and environmental conditions. It is important for personnel responsible for selecting a disinfectant for use in veterinary practice to be familiar with major classes of disinfectants, their efficacy against different pathogens, and conditions that limit their effectiveness. In general, vegetative bacteria and enveloped viruses are sensitive to many disinfectants whereas bacterial and fungal spores, non-enveloped viruses, and protozoal oocysts are relatively resistant.

Disinfection is a chemical reaction, thus it is imperative that all surfaces are thoroughly cleaned – removing all visible organic debris such as feces, urine, uterine fluid, nasal exudate, and blood – prior to disinfectant application as presence of organic material may reduce product efficacy. Environmental conditions including UV light (i.e., sunlight) and temperature can greatly affect the efficacy of many products, thus consideration of these factors is important when disinfectants are applied in areas that are not climate controlled. In addition to traditional surface disinfection, aerosol application or directed-misting with a disinfectant can be considered for control of difficult-to-reach spaces such as ceilings or overhead areas. This approach does not negate the importance of appropriate cleaning, but it may be a useful adjunct to cleaning and more traditional methods of disinfection. After appropriate cleaning and rinsing, a disinfectant should be applied that is effective against the microorganism(s) of concern and for the appropriate concentration and contact time, in accordance with label directions. It is important to ensure that products being used are registered with the U.S. Environmental Protection Agency (EPA) and that recommended manufacturer safety precautions are followed.

APPLICATION IN CLINICAL PRACTICE

In clinical practice, proper cleaning and disinfection should be employed regularly in areas used to house and manage all patients, not just patients with suspected infectious diseases. Environmental contamination is common near where patients are managed and samples are processed, and commonly reported during times of epidemic disease.[6,8,9] Routine cleaning and disinfection helps to minimize the bacterial burden in the environment that may serve as a source for HCAIs. While all housing areas (i.e., stalls) should be cleaned and disinfected between patients, special attention should be paid to areas known to have housed patients with suspected infectious diseases and tended to immediately after use to minimize inadvertent trafficking of pathogens from contaminated to clean areas. Equipment in contact with animals having an infectious disease should be cleaned and disinfected before reuse and handled in a manner to prevent contamination of hospital surfaces. Keep in mind that for equipment contacting mucous membranes (e.g., endotracheal tube, nasogastric tube) disinfection should be attained via dry/moist heat sterilization or application of an antiseptic (a product approved for use on living tissue). It is critical that cleanable surfaces be maintained throughout veterinary hospitals. Dirt floors or porous surfaces (e.g., untreated wood) cannot be effectively disinfected and therefore should be avoided. If porous surfaces are present, they can be rendered less so by painting or sealing the surface – which may need to be done routinely throughout the lifetime of the practice.

CLEANING AND DISINFECTION EFFICACY

Cleaning and disinfection efficacy should be routinely monitored to ensure the ICP is comprehensive and effective in an ever changing landscape. There are many methods for accomplishing this including surveillance for syndromes commonly associated with HCAIs or routine environmental monitoring using electrostatic wipes or sterile sponges to culture specific organisms of concern or contact plates (i.e., Rodac plates) to enumerate non-specific bacterial growth.[8 10-13] Incorporating environmental surveillance into an ICP requires consideration of sample type, sample collection, detection method, laboratory selection, and available resources (both financial and personnel); also methods should be appropriately validated and optimized for their intended use. The type of sample collection device and testing method should be carefully considered as sensitivities of each will likely differ. For example, electrostatic wipes have been found to be a more sensitive collection method compared to sterile sponges for detection of *Salmonella enterica* in a large animal veterinary hospital environment.[11] These differences can be attributed to both the collection method (the device used and the size of the surface area sampled) as well as the culture method. In general, sampling a larger surface area will not only provide a more representative sample but will likely be a more sensitive method for organism detection. To gain meaningful information, environmental testing should be performed regularly to establish a baseline level of environmental bacterial contamination to which future findings can be compared. In this way potential environmental reservoirs of microorganisms can be detected and cleaning effectiveness can be continually monitored.

CONCLUSION

Current data on infection control practices in veterinary hospitals is limited however the veterinary profession can draw on the human healthcare experience to improve upon veterinary infection control efforts. The Study on the Efficacy of Nosocomial Infection Control (SENIC), conducted in U.S. human healthcare facilities from 1970-1976, found that implementation of an ICP reduced HCAIs by an estimated 32% – specifically identifying employment of trained infection control personnel, conducting surveillance, and reporting findings back to stakeholders, as important factors in the success of the program.[14] Veterinarians are key partners in this effort – having the knowledge base and practical experience to identify potential risks and employ both epidemiologically and scientifically sound prevention strategies.

REFERENCES

1. Benedict KM, et al. **J Am Vet Med Assoc** 2008; 233(5):767. 2. Morley PS, et al. **Equine Vet J** 2012; 45:131. 3. Babcock S, et al. **J Am Vet Med Assoc** 2008; 233(10):1556. 4. Dwyer RM. **Vet Clin North Am Equine Pract** 2004; 20(3):531. 5. Hall-Stoodley LJ, et al. **Nat Rev Microbiol** 2004; 2(2):95. 6. Dallap Schaer BL, et al. **J Vet Intern Med** 2010; 24(5):1138. 7. Quinn PJ, et al. **Disinfection, Sterilization, and Preservation**, S.S. Block, Editor. 2001:1069. 8. Burgess BA, et al. **J Am Vet Med Assoc** 2004; 225(9):1344. 9. Steneroden KK, et al. **J Vet Intern Med** 2010; 24(3):606. 10. Ruple-Czerniak AA, et al. **Equine Vet J** 2013. DOI: 10.1111/evj.12190. 11. Ruple-Czerniak AA, et al. Equine Vet J, 2013. DOI: 10.1111/evj.12193. 12. Stockton KA, et al. **J Am Vet Med Assoc** 2006; 228(7):1068. 13. Ruple-Czerniak AA, et al. **J Vet Intern Med** 2013; 27(6):1392. 14. Haley RW, et al. **Am J Epidemiol** 1985; 121(2):182.

GRAYANOTOXICITY IN SMALL RUMINANTS

Marie-Eve Fecteau, DVM, DACVIM (Large Animal Internal Medicine)
Kennett Square, PA

INTRODUCTION

Small ruminants commonly encounter and ingest materials not meant for consumption, including toxic plants. This may occur due to intentional (such as discarded lawn clippings) or unintentional feeding of these plants to the animals. Although many toxic plants are unpalatable, there have been reports of animals readily consuming them, even when other feed sources are available.[1-7] Some reports, however, have occurred during changes in weather patterns, when normal feed sources are compromised.[8-9]

Among the plants that are toxic to small ruminants are those belonging to the heath family (*Ericaceae*), including Rhododendron or Azalea (*Rhododendron* spp.), Mountain Laurel (*Kalmia* spp.), Japanese Pieris (*Pieris* spp.), Fetter-Bush (*Leucothoe* spp.), and Maleberry (*Lyonia* spp.). These plants contain andromedotoxins, also referred to as grayanotoxins, which act directly on voltage-gated sodium channels. The toxin is able to bind to intracellular receptor sites in such a way that it alters the configuration of the channel itself. This, in turn, prevents sodium channel inactivation, locking the cell in a depolarized, activated state.[10] The primary effects are on the heart, nervous system, and gastrointestinal tract. All parts of the plant including the nectar contain the toxin and animals eating approximately 0.2% of their body weight of leaves are likely to develop signs of poisoning. Greater than 25 isomers of these toxins have been isolated, but grayanotoxin I and III appear to be the primary isomers associated with clinical toxicities.[11]

CLINICAL SIGNS AND DIAGNOSIS

Although death may be sudden, clinical signs of toxicosis usually occur within 36 hours after ingestion. In small ruminants, they include depression, bruxism, hypersalivation, vomiting, bloat, colic, diarrhea, pyrexia, agalactia, abortion, weakness, fine muscle tremors, head pressing, hypotension, recumbency, seizures, coma and death. Profuse vomiting may lead to aspiration pneumonia, which is the primary cause of death in some animals. Affected animals may be bradycardic (or tachycardic as the syndrome progresses), and may develop cardiac arrhythmias such as heart block and ventricular tachycardia. Electrocardiogram alterations may include widening of the QRS complex, ST segment depression, and enlarged P waves.

Clinical pathology abnormalities may include hemoconcentration, acidemia, hyperkalemia, hyperchloremia, hyperglycemia, and elevation in serum creatinine. A presumptive diagnosis of grayanotoxicity is made based on the history of exposure, the presence of typical clinical signs, and by the identification of plant parts in vomitus or rumen contents at surgery. Where available, confirmation of grayanotoxin exposure can be made by liquid chromatography/mass spectrophotometry on urine, feces or gastrointestinal contents.

TREATMENT

Treatment of small ruminants with suspected grayanotoxicity varies greatly depending on the severity of the clinical signs. The main focus generally lies in supportive care, as there is no antidote for the toxin. Medical treatment should include intravenous fluids to correct acid-base, electrolyte disturbances, and to counter hypotension; judicious use of systemic anti-inflammatories, atropine if bradycardia or heart block is present, antimicrobials for secondary aspiration pneumonia, toxin adsorbents and nutritional support. Rumenotomy under general anesthesia may be indicated to remove plant material in severe cases. It is the author's experience that prompt surgical intervention shortens the course of the disease and improves the overall outcome. The decision for medical therapy only versus surgical and medical therapy should be based on the severity of clinical signs, the amount of plant ingested (if known), and finances available. It is also important to note that many of these animals may not be systemically stable enough to tolerate general anesthesia.

The surgical technique employed may vary by surgeon but should include suturing of the rumen wall to the skin to prevent abdominal contamination. The entire contents of the rumen should be evacuated and the rumen lavaged copiously. Healthy rumen contents from a donor cow should then be used to replace the evacuated contents. In addition to providing healthy rumen microflora, the fresh rumen contents will also provide nutritional support in the early post-operative phase. Prior to closing the rumenotomy incision, activated charcoal may also be directly administered into the rumen to help adsorb any remaining toxins within the gastrointestinal tract.

PROGNOSIS/OUTCOME

There is little information published on the short and long term prognosis of grayanotoxicity in small ruminants and most publications on the topic are limited to individual cases or small groups of animals.[1-9] Animals with mild clinical signs typically recover uneventfully with medical therapy alone. However, small ruminants with severe clinical signs leading to severe electrolyte and/or acid-base disturbances or cardiac signs may not survive despite aggressive medical and surgical intervention. Because it is often impossible to estimate the amount of grayanotoxin ingested, it may be very difficult to predict the course of the syndrome. In a retrospective study of 34 small ruminants (30 goats, 4 sheep) that had presented to the New Bolton Center Hospital for Large Animals of the University of Pennsylvania following ingestion of grayanotoxin-containing plants, 6 (18%) died or were euthanized following admission. Of those 6 animals, 3 had been managed with medical therapy only while the other 3 had undergone a rumenotomy and medical therapy. Of the 28 animals discharged alive, 18 were treated surgically and 10 were managed with medical therapy only. There was no significant difference in survival when modes of treatments (surgical versus medical) were compared. Complications noted in the small ruminants included in the study included aspiration pneumonia and incisional infection.

REFERENCES

[1]Humphreys DJ et al. *Vet Rec* 1983;113(21):503-504. [2]Hosie BD, et al. *Vet Rec* 1986;118(4):110. [3]Hollands RD et al. *Vet Rec* 1986;118(14):407-408. [4]Black DH. *Vet Rec* 1991;128(15):363-364. [5]Casteel S et al. *Vet Hum Toxicol* 1989;31(2):176-177. [6]Puschner B et al. *J Am Vet Med Assoc* 2001;218(4):573-575. [7]Plumlee KH et al. *J Vet Diagn Invest* 1992;4(3):363-364. [8]Higgins RJ et al. *Vet Rec* 1985;116:294-295. [9]Shannon D. *Vet Rec* 1985;116(16):451. [10]Power SB et al. *Vet Rec* 1991;128(25):599-600. [11]Thiemann AK. *Vet Rec* 1991;128(17):411.

GASTROINSTESTINAL MICROBIOME IN JOHNE'S DISEASE: A LINK TO CROHN'S DISEASE?

Marie-Eve Fecteau, DVM, DACVIM (Large Animal Medicine)
Kennett Square, PA

INTRODUCTION

The first case of Johne's disease in the United States was reported in 1906 by Dr. Leonard Pearson, then Dean of the University of Pennsylvania School of Veterinary Medicine. Unfortunately, over the next century *Mycobacterium avium* subsp. *paratuberculosis* (MAP) has spread throughout the United States, and most recent surveys suggest 90% of all dairy herds in this country have infected cattle, costing producers hundreds of millions of dollars.[1] Furthermore, MAP is shed in the milk of infected cows, is able to survive pasteurization, and has been identified in retail milk samples from grocery shelves in the United States.[2] Although MAP has not traditionally been considered a human pathogen, concerns about human exposure to MAP through milk and other products are now being raised in the context of its potential role in Crohn's disease.

CROHN'S DISEASE IN HUMANS AND A POTENTIAL ROLE FOR MAP

The definitive cause of Crohn's disease, a chronic, inflammatory gastrointestinal disease of humans remains elusive. It is thought to result from a complex interaction of host susceptibility factors and an abnormally intense immune response to bacteria or other antigens in the intestines. Because Crohn's disease has many clinical and pathologic similarities to Johne's disease, numerous studies have investigated the role of MAP in Crohn's disease.[3-5] These studies have shown conclusively that MAP can be isolated from intestinal tissue of Crohn's patients (significantly more than controls), but the medical community still debates whether the MAP organisms cause the intestinal inflammation, or merely are able to colonize already-compromised intestinal tissues of afflicted individuals.[3,5]

GUT MICROBIOME AND DYSBIOSIS

The gastrointestinal microbial ecosystem is fundamental to the overall health and production performance of most mammals. The mammalian gastrointestinal tract is colonized by hundreds of different species of bacteria, living in a delicate balance in a symbiotic relationship with the host. The feces of cattle contain as many as 10^{11} bacterial organisms per gram.[6] In the past, study of the bacterial "community" or "microbiome" of the gastrointestinal system required identification of the various bacterial species through culture. However, it is estimated that these traditional culture methods are only able to identify 20% of the bacterial species within the gastrointestinal tract.[7] With the recent advent of powerful molecular biological tools, including "next -generation" DNA sequencers, and computer software that can analyze DNA sequencing data, the microbiome can now be more completely characterized. Recent studies have shown that this phylogenomic approach more completely and accurately characterizes the gastrointestinal microbial community than the older methods that relied on culture.[6]

As mentioned above, this diverse community of gastrointestinal bacteria exists in a delicate balance with the host-- a balance that can be disrupted by changes in diet, antibiotic treatment, or infection with pathogenic bacteria. Disruption of this balance, termed "dysbiosis", can result in gastrointestinal dysfunction, including inflammation of the intestinal lining.[7] Studies have shown evidence of dysbiosis in patients with Crohn's disease, including a shift in the species of bacteria present and an overall reduced diversity of bacterial species.[8,9] These changes are speculated to contribute to the chronic inflammatory condition in the intestines of these patients.

CONCLUSIONS

In summary, Crohn's disease is a chronic intestinal inflammatory disease of humans, associated with an increased likelihood of infection of intestinal tissue by MAP, and also associated with a change in the gastrointestinal microbiome. The relative importance of dysbiosis, MAP infection, and other factors is unclear. Similarly, Johne's disease is a chronic intestinal inflammatory disease of cattle, caused by infection of intestinal tissue by MAP. The role of dysbiosis in cases of Johne's disease has not been studied, but one study suggested probiotic therapy was beneficial in MAP-infected animals.[10] Given the similarities between the two syndromes, dysbiosis in cattle with Johne's disease, similar to that reported in patients with Crohn's disease, is a reasonable expectation. The principal objective of the study described here was to investigate the intestinal microbiome of cattle with Johne's disease, to determine if dysbiosis similar to that of Crohn's disease can be found in those animals. This finding would a) strengthen the link between MAP and Crohn's by showing similar dysbiosis exists in a disease known to be caused by MAP (Johne's) and one suspected to be caused by MAP (Crohn's); b) provide potential targets for the treatment of MAP through probiotic agents designed to restore the microbiome to it's proper balance; c) provide fodder for future longitudinal studies on the development of dysbiosis following experimental MAP infection, correlating these changes with parameters of intestinal inflammation.

REFERENCES

[1]Lombard JE et al. Prev Vet Med 2013; 108:234-238. [2]Ellingson JL et al. J Food Prot 2005; 68: 966-972. [3]Abubakar I et al. Inflamm Bowel Dis 2008; 14:401-410. [4]Chiodini RJ, et al. Crit Reviews in Microbiology 2012; 38:52-93. [5]Feller M et al. Lancet Infect Dis 2007; 7:607-613. [6]Dowd SE et al. BMC Microbiogy 2008; 8:125. [7]Scaldaferri F et al. Gastroenterol Clin N Am 2012; 41: 581-602. [8]Kang S et al. Inflamm Bowel Dis 2010; 16:2034-2042. [9]Manichanh C et al. Gut 2006; 55: 205-211. [10]Click RE et al. Virulence 2010; 1(3):145-55.

CONTROLLING SALMONELLAE IN CALVES

Geof W Smith, DVM, PhD, DACVIM
Raleigh, NC

INTRODUCTION

Calf health should be a priority on both beef and dairy farms. Despite this importance, the USDA Dairy 2007 study shows a preweaned heifer calf mortality rate of 8.7% and reports that only 40% of farms can supply an adequate number of replacements from their own herd. The major diseases of both beef and dairy calves continue to be diarrhea, pneumonia, and septicemia. Colostrum management continues to be a significant area of concern on many modern dairies. There are many components to an effective control program for calf diarrhea including removing the calf from cow immediately after birth, ensuring adequate colostrum ingestion, proper housing, and sanitation of feeding equipment. There are also vaccines available that can help boost the calf's immunity to some of the more common causes of diarrhea.

Salmonella organisms are endemic on most large intensive farms and salmonellosis is a common cause of diarrhea and mortality in neonatal calves. Calves may be infected with a diverse array of *Salmonella* serotypes within hours of birth. The subsequent manifestations of disease are variable reflecting the balance between host immunity, pathogen dose, and virulence. Diarrhea outbreaks are frequently observed in calves between 4 and 28 days of age, however older calves may be affected (Anderson, 2001). Diarrheic feces range from watery, to voluminous, to profuse muco-fibrinous and hemorrhagic character. Differences are observed between infections with different serovars and potentially between different strains of the same serovar. This is a result of differences in virulence factors between infecting strains. *Salmonella enteritica* serovar Typhimurium is commonly incriminated in outbreaks of enteric disease in calves less than 2 months of age. In contrast, *Salmonella enteritica* serovar Dublin is associated with disease of similar frequencies in young and adult cattle.[9]

The primary route of infection is fecal oral transmission; however other reported routes include the mucosa of the upper respiratory tract and conjunctiva. Following ingestion, *Salmonellae* colonize the intestinal tract and invade through specialized cells in intestinal lymphoid tissue called M-cells, enterocytes and tonsilar lymphoid tissue.[9] In the lymphoid tissue, *Salmonellae* gain entry into mononuclear phagocytes and are rapidly disseminated throughout the body.

The basic virulence mechanisms of *Salmonella* species include: the ability to invade the intestinal mucosa, to multiply in lymphoid tissues, and to evade host defense systems resulting in systemic disease. The diarrhea associated with salmonellosis is largely believed to be mediated by the inflammatory response to infection. Release of endotoxin, prostaglandins, and pro-inflammatory cytokines (such as IL-1 and TNF-α) also promote increased vascular permeability and hypersecretion. Sloughing of the intestinal epithelial cells leads to acute hemorrhage, fibrin production, maldigestion and malabsorption. The resulting hyperosmotic state within the lumen of the intestine draws fluid into the intestinal tract contributing to a net loss of water, sodium, potassium, and bicarbonate. Mucosal damage also contributes to protein loss and hypoproteinemia. The pathophysiology of salmonellosis in calves is described in much greater detail in a recent review.[9]

CONTROLLING SALMONELLAE

The approach to infectious disease control described by Radostits over 3 decades ago includes: 1) removing the source of infection from the calf's environment; 2) removing the calf from the contaminated environment; 3) increasing the non-specific immunity of the calf; 4) increasing the specific immunity of the calf; and 5) reducing stress. These approaches are still applicable today and represent the core of Salmonella control programs. As partly summarized from a thorough review article on Salmonella infections in calves,[9] below is an overview of critical control points for controlling infections due to Salmonellae on dairy farms.

1) Sources of infection

Salmonellae may be introduced onto a farm by contaminated feedstuffs, water, fertilizer, livestock, wildlife, insects, people or equipment. In intensive farming systems where there is significant exposure to and recycling of effluent, livestock are frequently exposed to *Salmonellae.* Management is directed at reducing the risk of adverse health and production consequences through minimizing the challenge dose and promoting host immunity. Strategies to mitigate the potential adverse consequences of introduced and endemic *Salmonellae* come back to good feed management and implementation of good husbandry practices that avoid compromised herd immunity. Important areas include waste management, forage preparation and storage, commodity sourcing and storage, as well as nutritional, environmental, and reproductive management. In regards to feed management, the risk posed by forages irrigated with *Salmonellae* contaminated effluent can be reduced by ensuring effective ensilage. *Salmonellae* are eliminated when the pH drops below 5, and ensilage failure is associated with persistence of the organism. Rodent control in commodity barns is also recommended to prevent dissemination and amplification of contamination in the feed storage area.

2) Maternity pens and management

Calves have numerous opportunities to become infected with Salmonellae. On dairy farms experiencing problems with neonatal Salmonella infections, it is not uncommon to find evidence of disease and a high prevalence of Salmonella fecal shedding in postpartum cows.[1] Calves may be exposed to Salmonellae by exposure to manure from the dam following birth when the calf contacts it environment, or when calves have contact with the udder while attempting to nurse. Salmonellae may also proliferate in organic bedding material used in calving pens. The presence of moisture in the environment favors *Salmonella* proliferation. It is important to appreciate that significant pathogen exposure can occur within the first few hours of life. In a longitudinal study, 65% of calves were determined to be shedding *Salmonellae* in feces within 24 hours of birth.[7] Control points in maternity pens are directed at minimizing contamination (time cows spend in the pen, choice of bedding material, and frequency of bedding changes) and exposure risk (time calves spend in the maternity pen).

3) **Colostrum**

While cows chronically infected with Salmonellae can shed the organism in colostrum and milk, most cows do not. Despite this, several studies have identified contaminated colostrum as a significant source of infection.[4] Pooling colostrum increases the risk of *Salmonella* infection by disseminating the organism in a larger colostral volume and subsequently infecting a larger number of calves.[12] Pooling colostrum also reduces the efficiency of colostral transfer and subsequently compromises calf immunity. An increase in the incidence of Salmonella infections in calves and an earlier onset of clinical signs following increased volume of colostrum administration to newborn calves is consistent with Salmonella contamination of colostrum. Control points for reducing the risk of disease transmission associated with the feeding of colostrum include: 1) effective cleaning of equipment used in colostrum collection and storage; 2) avoid pooling of colostrum; 3) verify refrigeration units for storing colostrum are working correctly; 4) make sure colostrum is cooled rapidly if not fed immediately; 5) record date of collection on refrigerated colostrum and discard after 2 days; and 6) maintain dedicated equipment for administering colostrum. Fresh colostrum fed to calves should contain less than 100,000 cfu/mL total bacterial count and less than 10,000 coliforms/mL.[8]

4) **Bottlenecks**

To evaluate the risk of calf exposure to Salmonellae, it becomes necessary to trace the path from the place where calves are born to the location where they are housed. Bottlenecks that have the potential to concentrate pathogen exposure (like calf warmers, holding pens, transport trailer, etc) should be avoided when possible and regularly disinfected when they must be used. *Salmonellae* are susceptible susceptible to most disinfectants when they are used according to their directions. Removal of organic material (manure) reflects 99% of the cleaning process and is important to avoid neutralization of the disinfectant. Scheduling cleaning activities to a consistent routine helps to establish a consistent routine that ultimately gets done. Salmonellae are susceptible to ultraviolet radiation, so it's a good idea to leave hutches upside down between housing new calves to get sunshine and help reduce pathogen load.[9]

5) **Housing**

Ideally calves should be placed in a clean, dry environment in an area that has good drainage and is not exposed to manure or runoff from adult cattle. Individual calf hutches provide an effective means of minimizing calf to calf pathogen transfer. Hutches should be moved to clean ground before the next calf enters to avoid pathogen build up. Group housing provides efficient utilization of space when done properly and automatic feeders can reduce labor requirements. The age range of calves in group housing should be kept as tight as possible – and preferably managed as close as possible to "all-in all-out" flow. Mixing age groups increases the risk of pathogen exposure in young calves (for several pathogens including *Salmonellae*) and can compromise their access to milk at feeding time. In regards to calf comfort, if calves get wet and are exposed to cold and windy conditions, energy reserves are rapidly depleted and sick calves rapidly succumb to disease. Placing straw in the hutches of newborn calves helps to keep them warm and dry. In hot climates, calves can be subjected to relatively extreme temperatures, and it becomes critical to provide continuous access to fresh, clean water.

6) **Nutrition**

Adequate calf nutrition is critical for host immunity, and energy deprived calves are more likely to have increased morbidity and mortality due to diarrhea.[5,10] When assessing nutritional requirements it is important to factor in seasonal conditions that increase the calf's energy demands. Lack of refrigeration carries significant risk when residual milk from one feeding is kept for subsequent feedings. This is particularly important during hot weather when proliferation of *Salmonellae* is increased. Designing calf milk storage and delivery systems to conform to safe food standards reduces the risk of disease transmission. Pasteurization is also effective at reducing *Salmonella* contamination, and when implemented properly, provides significant risk management. Inclusion of quality control monitors is important to detect failures in the pasteurization process. Pasteurizers can also be difficult to clean so it is appropriate to implement routine bacterial counts for calf milk (before and after pasteurization) to verify that the system is functioning according to expectations. Feeding utensils and personnel often play a significant role in transmitting *Salmonellae* between calves. Salmonella infects the salivary glands and is shed in saliva and nasal secretions contaminating buckets, nipples, bottles and esophageal feeders. Adequate cleaning and disinfection of feeding utensils is necessary to remove Salmonella contamination.

7) **Sick calf management**

Sick calves (with diarrhea) can shed up to 109 Salmonella organisms per gram of manure increasing bacterial numbers in the environment as well as contaminating equipment and people. There are two different philosophies regarding the management of calves with salmonellosis.[9] One school of thought advocates the rapid removal of infected calves into a "sick" pen for treatment. The objective is to minimize contamination of the environment and subsequently the challenge to other calves in the area. A high prevalence of salmonella shedding has been observed in calves at 24 hours of age in calves with disease manifesting at 4 – 7 days of age, so salmonella exposure may have been significant by the time sick calves are removed.[7] A potential risk associated with using a sick calf pen is that it can be a source of mixed pathogen infection if strict quarantine and appropriate management procedures are not maintained.[3] An alternate philosophy says that by the time infected calves are identified, the other calves in the group have already been exposed. Therefore movement of calves may be contraindicated due to the potential for dissemination of salmonella contamination. The correct decision on whether or not to move calves will depend on the number of affected calves and space constraints. In the face of a disease outbreak, it is desirable to be able to break the infection cycle by avoiding the placement of newborn calves into a contaminated environment. Prior to moving sick calves, consideration should be given to clean space availability. If clean space or hutches are limited, it is better to preserve these for newborn calves.

Broad-spectrum antimicrobials are usually indicated in calves suspected of having Salmonella infection. Susceptibility testing has shown variable resistance patterns to ampicillin, amoxicillin, amoxi-clavulonic acid, ceftiofur, florfenicol, neomycin, sulfonamides, tetracycline, and trimethoprim-sulfa and general resistance to penicillin, erythromycin, and tylosin. Since *Salmonellae* are facultative intracellular pathogens, selecting an antimicrobial with good tissue penetration and the ability to attain intracellular therapeutic drug concentrations within macrophages is desirable. In an experimental challenge study, extralabel use of ceftiofur at 5 mg/kg was shown to attenuate the severity of clinical disease and reduce fecal shedding of *Salmonellae.*[2] However the use of ceftiofur at extralabel doses is no longer legal in the United States. With the current prohibition on extralabel use of cephalosporins and fluoroquinolones, finding an effective antibiotic that can be legally used for the treatment of Salmonella infections in calves can be difficult.

8) Environmental Control

The reality of environmental management is that it is feasible to reduce the number of Salmonella in the environment but unrealistic to expect elimination of the bacteria. Calf pens and hutches should be scraped and scrubbed to remove organic material prior to disinfection between groups. Bedding material should be maintained to keep calves clean and dry. Feed and water buckets/bottles are positioned to reduce the risk of fecal contamination and should be cleaned daily. Equipment used to handle milk or milk replacer products should be thoroughly clean to remove milk residue which will support the growth of bacteria.

9) Vaccination

There are three general types of Salmonella vaccines: killed whole cell (bacterin), bacterial fractions (subunit), and attenuated modified live (MLV). Most commercial salmonella vaccines are bacterins, and there are conflicting reports regarding the efficacy of these bacterins in cattle. Under field conditions, neonatal exposure often occurs during the first few days of life, limiting the opportunity to stimulate acquired immune mechanisms through vaccination of calves.[7] Passive immunity acquired from maternal *Salmonella* bacterin vaccination and colostral transfer is limited, however partial protection has been reported in some experimental challenge trials. The use of modern commercially available vaccines (ie. SRP®) in the late gestation or dry cow and their ability to concentrate antibodies in colostrum to provide some early immunity against Salmonella infections in calves is currently being studied. Preliminary results of this research will be discussed during the presentation. Adverse reactions in the form of anaphylactic reactions are occasionally reported in cattle vaccinated with *Salmonella* bacterins. Subunit vaccines are composed of bacterial fractions or surface antigens. The Salmonella Newport bacterial extract vaccine (SRP,® Zoetis) is a relatively new subunit cattle vaccine composed of purified extracts of siderophore receptors and porins. These proteins are common to all strains of Salmonella and offer the potential to induce immunity to multiple serovars. At this time, there are no controlled clinical trials documenting the efficacy of this product at preventing salmonellosis in calves. Entervene®-d is a modified live S. dublin vaccine sold in the United States by Boehringer Ingelheim that is approved for use in healthy calves over 2 weeks of age. This vaccine also has some cross protection against S. typhimurium as the B and D serogroups have antigenic similarities. This vaccine has also been used in an extra-label manner in newborn calves to protect against Salmonella mortality occurring around 5-10 days of age. Although it can be effective in this manner, some producers have reported adverse reactions to vaccination in newborn calves. Oral use of this vaccines has been reported (in order to limit anaphylaxis) but was not successful in one study.[6] Experimental studies with this vaccine in calves demonstrated protective immunity to Salmonella when calves were challenged within 3 weeks of vaccination.[11] Protection typically disappears about 1 month after vaccination following clearance of the organisms from the reticuloendothelial system.

CONCLUSION

Calf diarrhea and Salmonellosis are both largely management diseases that can be prevented by having a good colostrum program in place and maintaining good cleanliness on the farm. Since fecal-oral is the main route of contamination for most pathogens and infected calves are the reservoir for enteric pathogens, limiting contact between young calves is critical. Risk factors for calf diarrhea include dirty calving pens, inadequate colostrum ingestion, nursing dirty teats, unsanitary feeding utensils (nipples, bottles), group housing, particularly with continuous rearing (not all in, all out), overcrowding, and failure to isolate sick calves.

REFERENCES

1. Anderson RJ, et al. J Am Vet Med Assoc 2001; 219:310.
2. Fecteau M, et al. Am J Vet Res 2003; 64:918.
3. Fossler C, et al. Prev Vet Med 2005; 70:279.
4. Giles N, et al. Epidemiol Infect 1998; 103:235.
5. Godden SM, et al. J Am Vet Med Assoc 2005; 226:1547.
6. Habing GG, et al. J Am Vet Med Assoc 2011; 238:1184.
7. House JK, et al. Am J Vet Res 2001; 62:1897.
8. McGuirk SM, et al. Vet Clin Food Anim 2004; 20:593.
9. Mohler VL, et al. Salmonella in calves. *Vet Clin Food Anim* 2009; 25:37.
10. Ollivett T, et al. J Am Vet Med Assoc 2012; 241:1514.
11. Smith BP, et al. Am J Vet Res 1984;45:59.
12. Veling J, et al. Prev Vet Med 2002; 54:157.

RUMINANT AND CAMELID CYTOLOGY: A CASE-BASED REVIEW

Tamara Gull, DVM, PhD, DACVIM (LAIM), DACVPM, DACVM, Stillwater, OK; Allison J. Stewart, BVSc(hons), MS, DACVIM (LAIM), DACVECC; Elizabeth Spangler, DVM, PhD, DACVIM (SAIM), DACVP; Auburn, AL; Robin W. Allison, DVM, PhD, DACVP, Stillwater, OK; Jenna Bayne, DVM. Auburn, AL

INTRODUCTION

Both primary care and referral food animal and camelid practitioners should take advantage of readily-available cytologic samples in their practice. Many cytology samples are easy to obtain using a needle and syringe following a simple scrub, and acquisition of most cytologic samples is fairly simple, fast, and associated with minimal morbidity. Cytologic examination by a clinical pathologist is always recommended, but with appropriate sampling, careful slide staining technique and practice, a provisional diagnosis can often be made in clinic allowing for earlier initiation of appropriate therapy. The presenters will offer a selection of food animal and camelid cases illustrating the usefulness of cytology from various sources in the diagnosis and prognosis of both routine and unusual cases. The cases may include, but are not limited to the following examples.

CASES

Case 1. Acute ataxia in a 2-year-old female Huacaya alpaca.

The alpaca presented for acute onset of left hindlimb lameness per the owner. Upon examination, the alpaca showed severe ataxia and weakness in both hind limbs, worse on the left. She knuckled over on the hind fetlocks and her patellar tendon reflexes were mildly increased. She had abnormal foot placement and delayed reactions when forced to hop on her hind limbs. TPR were within normal limits. Ticks were evident in both ears. A lumbosacral cerebrospinal fluid sample was obtained and showed an eosinophilic pleocytosis. *Paraelaphostrongylus tenuis* infection was suspected and the alpaca was treated with fenbendazole, ivermectin, dexamethasone and vitamin E. She made a full recovery over several weeks.

Case 2. Right displacement of the abomasum in a 3-year-old Holstein cow.

The cow presented with marked bilateral abdominal distension. Low-pitched pings were ausculted bilaterally (from the 11th rib caudally on the right and over the majority of the left side). A high pitched ping was heard caudal to the 13th rib on the right. An exploratory laparotomy through the right paralumbar fossa confirmed a right displaced abomasum. Chemistry profile revealed metabolic alkalosis (bicarbonate = 47.1 mmol/L), hypochloremia (61 mmol/L), hypocalcemia (7.5 mg/dL), hyperphosphatemia (10.7 mg/dL), hypermagnesemia (3.2 mg/dL), hyponatremia (126 mmol/L), hyperglycemia (663 mg/dL), and azotemia (creatinine = 4.0 mg/dL). CK and GGT were slightly elevated, and serum iron was decreased (48 µg/dL). Abdominal fluid had a nucleated cell count of 285,000 cells/µL, with 65% non-degenerate neutrophils and 35% macrophages, a few small lymphocytes and some reactive mesothelial cells. No infectious organisms or foreign material were observed. The cytologic interpretation was purulent inflammation, suggestive of bacterial infection secondary to bacterial translocation and culture was recommended. The abomasum was decompressed, and a right paralumbar fossa omentopexy was performed. She was treated with flunixin, oxytetracycline, hypertonic saline and calcium IV, and oral fluids. Due to her continued decline despite surgical and medical treatments, euthanasia was elected. Necropsy showed ulcerative abomasitis with multifocal peracute perforation and peritonitis.

Case 3. 15-month-old heifer with mass on neck.

The heifer had recently been introduced to grain and hay which resulted in bloat. The owners had removed gas from her rumen 5 times prior to referral. The heifer presented with a "papple" shaped abdomen and brisket edema. A mass on her ventral neck and shoulders had increased in size over 2 weeks. Ultrasound showed a cavitary mass of heterogeneous echogenicity extending from the left submandibular area to right shoulder. Cytologic examination of a fine needle aspirate showed a mixed cellular population with a predominance of large lymphocytes. The large lymphocytes had nuclei 1-2 times the size of a neutrophil, stippled chromatin and variably distinct nucleoli. A few mitotic cells were seen. The cytologic interpretation was probable lymphoma. Necropsy confirmed disseminated lymphoma of thymic origin. The mass along the ventral neck extended from the thoracic inlet to the 4th cervical vertebra and measured 75 x 26 cm. Masses were also present in the cervix, liver and submandibular lymph nodes. Vagal indigestion had resulted from compression of the esophagus by the mass, making eructation impossible.

Case 4. 11-month-old Holstein heifer with acute onset of lameness.

The heifer had been lame in the right hindlimb for 10 days prior to presentation. At presentation she was grade 4/5 lame and the right stifle joint was swollen and firm. Radiographs did not show any osseous involvement. Synovial fluid contained 75,500 nucleated cells/uL with 7.1 g/dl protein and specific gravity of 1.040. There were ~90% neutrophils, 10% macrophages, and rare lymphocytes and plasma cells. The protein/specific gravity was falsely increased owing to a small volume of joint fluid in the EDTA. Cytologic interpretation: although no infectious agents were observed, septic arthritis was considered likely. A heavy growth of *Trueperella pyogenes* was cultured. Prognosis was poor, and although arthroscopic debridement was offered, euthanasia was recommended. The heifer was treated with meloxicam (300 mg PO) until the owners approved euthanasia at which time the heifer was 5/5 lame and the joint was 2.5 times the size of the opposite stifle. Necropsy confirmed diffuse, chronic fibrosis of the joint capsule and synovial hyperplasia secondary to septic arthritis.

Case 5. 3-year-old Brangus cow with a sublingual abscess.

The cow had been unable to eat for 2 days and was drooling. There was a sublingual mass. Radiographs showed osteolysis of the rostral aspect of the mandible. A CBC showed mild neutrophilia (4,900 cells/µL). Fibrinogen concentration was 800 mg/dL and plasma protein was 9.0 g/dL. The abscess was lanced and surgical curettage of the mandible performed. The wound was packed with amikacin beads and dental packing. Cytologic examination of an impression smear showed moderate numbers of neutrophils, small lymphocytes and macrophages with a background of blood. Moderate numbers of long, slender rod-shaped bacteria with a beaded appearance suggestive of *Actinomyces* were present both extracellularly and intracellularly within neutrophils. Large spindle cells suggestive of fibrocytes were also present. Cytologic opinion: septic, mixed-cell inflammation; bacteria have an appearance suggestive of *Actinomyces* and culture with susceptibility is recommended. Histopathology of the bony lesion revealed osteomyelitis with a mixed population of bacteria, mostly Gram-positive cocci. No sulfur granules or organisms consistent with the filamentous to branching *Actinomyces bovis* were observed histologically. Culture of the bony lesion revealed overgrowth of *Trueperella pyogenes*. The cow was treated with sodium iodide IV (once) and florfenicol SQ- 3 doses given at 1 week intervals.

Case 6. 1.5-year-old barrow.

The pig had been attacked by dogs 1 year ago which resulted in a rectal stricture. Over the last 8 weeks, he had been anorexic, and vomiting commenced 3 days ago. The rDVM had placed a stent in the rectum (10ml syringe barrel) the previous week. The owner had administered trimethoprim- sulfonamide, flunixin and detergent enemas (approximately 1 gallon) daily, for two weeks. Upon presentation, the pig was depressed with cold extremities and tachycardia. He was severely dehydrated (~10%). The rectal stent was removed. Rectal palpation revealed a stricture 5 cm orad to the rectum, with a diameter of ~6mm. Melena was present. IV fluids were commenced. The pig was hypoproteinemic (3.4 g/dL), hypoalbuminemic (0.7 g/dL), mildly azotemic (creatinine = 2.7 mg/dL), and had marked electrolyte abnormalities (low calcium, bicarbonate [16.3 mmol/L], sodium [123 mmol/L]) and hypoferremia (27 µg/dL). Ultrasonographic imaging of the caudal ventral abdomen demonstrated a moderate amount of free fluid. Abdominocentesis yielded malodorous, straw colored fluid with WBC = 2,240 cells/µL, protein = 3.0 g/dL and specific gravity = 1.022. Direct smears had low cellularity, and a cytospin was performed. There were 60% degenerate neutrophils, 32% lymphocytes and 8% macrophages. Neutrophils were markedly karyolytic and frequently contained intracellular cocci with fewer rods. Cytologic opinion: septic suppurative inflammation. *Escherichia coli* was cultured. The pig was treated with misoprostol, omeprazole, oxytetracycline, sulphadimethoxine and dextrose supplemented plasmalyte. The pig died that evening and necropsy confirmed septic peritonitis due to a ruptured (and markedly dilated [~40 cm in diameter]) colon orad to the rectal stricture.

Case 7. 6-day-old Brangus bull calf presenting with acute lameness.

The calf had been mildly lame in his right rear limb for a few days, and then became severely lame and was toe dragging. The stifle was extremely swollen, with palpable fluid pockets. There was atrophy of the gluteal and quadriceps muscles. Radiographs of the stifle showed osteomyelitis of the lateral femoral condyle. The urachus was firm and a purulent, foul-smelling exudate drained from a fistula. Aspirated stifle synovial fluid was cloudy and orange in color, with a nucleated cell count of 10,400 cells/µL, protein = 5.7 g/dL and specific gravity of 1.034. Approximately 95% of the cells were neutrophils, with rare macrophages and small lymphocytes. The neutrophils were markedly degenerate and several had engulfed bacteria which were mostly short rods that appeared bipolar. There were a few free bacteria. Cytologic opinion: marked purulent septic inflammation. *Pasturella multocida* was cultured. Arthrotomy, joint lavage and curettage of the osteomyelitic lesion were performed. A collagen-based gel containing amikacin (100 mg) and clindamycin (50 mg) was infused and intrarticular morphine administered. The infected umbilicus was also excised. The joint was lavaged daily and intrarticular amikacin administered. Florfenicol was administered until discharge. The arthrotomy site was closed after 25 days and the calf was discharged.

Case 8. 6-year-old Anglo-Nubian doe goat presented for wheezing of 1 week duration.

A mass with a pink cobblestone appearance was observed within the right naris that was ~4cm in diameter. Skull radiographs showed a soft tissue opacity in the right naris, with no osseous involvement. The cytology smear was very cellular and had a large accumulation of macrophages, occasional neutrophils and a few eosinophils. The macrophages were often epithelioid type and many were multinucleated. Abundant fungal hyphae were scattered throughout. The hyphae were variably septated and branched at varied angles. They had non-parallel sides with bulbous ends. Round/oval fungal forms were often observed within macrophages. Cytologic opinion: Fungal infection, granulomatous inflammation. Histopathology results provided a provisional diagnosis of a fungal granuloma (*Conidiobolus coronatus*), which was confirmed by a heavy growth on culture. Hyphae were also observed on KOH prep. Surgical debulking of the mass was performed and clotrimazole was instilled in the right nostril. Clotrimazole infusion was repeated 1 week later. The owner was instructed to apply clotrimazole cream to the nasal cavity daily for 10-14 days. After 2 months the mass had reoccurred and the goat was open mouth breathing. Rhinoscopy confirmed the presence of the mass, and cytology and histopathology reconfirmed *C. coronatus*. Intravenous fluconazole was administered for 13 days, but the goat developed renal azotemia. The respiratory stridor improved. Weekly IV sodium iodide was administered for 3 treatments, and then oral fluconazole was commenced. Repeat rhinoscopy was performed after 2 months and no masses were observed. Chemistry profile was normal. Treatment was continued for a further 3 months. All clinical signs have resolved.

Case 9, 5-year-old Angus cow with acute weight loss.

The cow was 5 months pregnant. She had been losing weight over the past 3-4 weeks despite a good appetite, with onset of dull mentation during the past week. At presentation she had no physical examination abnormalities except that she was slightly thin with a dull coat and had mild enlargement of submandibular, prescapular and prefemoral lymph nodes. A fine needle aspirate of a prescapular lymph node revealed a population of primarily very large lymphocytes with prominent nucleoli and few normal small lymphocytes. A hematocrit done at admission showed a very thick buffy coat, and examination of a blood film showed circulating neoplastic lymphocytes (leukemia). Cytologic opinion: stage V lymphoma with leukemia. The cow was euthanized and necropsy revealed multicentric lymphoma in the lymph nodes, hemal nodes, heart, reticulum, rumen, abomasum and uterus.

Case 10, 2-week-old Longhorn bull calf with recumbency.

The calf was born normal but began to stumble and seemed weak at 1 week of age. The calf presented for failure to thrive and being unable to rise. At presentation, the calf was recumbent and minimally responsive to stimuli with mild extensor rigidity. Physical exam revealed a thin, dehydrated calf with a large umbilical abscess. Aspirate of the abscess yielded pus containing a large number of degenerate cells and a mixed population of bacteria seen on a Diff-Quik-stained smear. A Gram-stained smear revealed numerous Gram-positive cocci as well as a large number of Gram-positive rods with bulging terminal endospores consistent with *Clostridium tetani*. Cytologic opinion: septic suppurative inflammation. A mixed bacterial population is evident. The calf was treated with penicillin, wound debridement and flushing, tetanus antitoxin, IV fluids and nutritional support. Over 24 hours of hospitalization the calf developed opisthotonus and classic tetanus signs and methocarbamol, diazepam and acepromazine were added to the treatment regimen. The calf was euthanized at the owner's request after 24 hours of treatment. Culture of the abscess aspirate grew *Propionibacterium granulosum, Streptococcus uberis, Escherichia coli, Morganella morganii, Enterococcus durans, Enterococcus faecium, Clostridium difficile* and *Clostridium tetani*.

Case 11, a 5-year-old Angus bull with lameness.

The bull presented for acute non-weight-bearing lameness of one week's duration. The referring veterinarian had diagnosed a sand crack and had treated the bull with hoof wall resection, antibiotics and NSAIDs without improvement. At presentation, the bull was 5/5 lame on the RF and the limb was grossly swollen to mid-metacarpus. Examination revealed fetlock joint effusion and pain at the toe of the RF lateral claw, which was pared out and a block applied to the medial claw. Radiographs showed osteomyelitis of the lateral P3 and no bony changes in the fetlock joint. Joint fluid cytology of the RF fetlock revealed a highly cellular smear with predominantly degenerate neutrophils. Both intra- and extracellular bacteria were seen. Cytologic opinion: septic suppurative inflammation. The bull was treated with multiple arthrotomies and joint flushes in addition to intraarticular and systemic antimicrobials and NSAIDs. Culture of joint fluid grew *Trueperella pyogenes* and an unidentified anaerobe. He failed to improve despite multiple joint flushes and developed azotemia, so antibiotics were changed and NSAIDs were discontinued in favor of fluids and a pain management protocol that included morphine, detomidine and ketamine. The bull developed melena and became anemic (PCV 19%) and febrile on Day 7 of hospitalization. Ultrasound of his abdomen displayed no visible abomasal ulcers but fluid in the abomasum appeared to be blood. The bull received omeprazole and aminocaproic acid. Worsening osteomyelitis of the lateral P3 resulted in additional curettage of the bone, but fetlock joint total protein was decreasing. Anemia progressed despite multiple blood transfusions and both lung consolidation and distended small intestine were seen on ultrasound on day 11. Two days later the bull collapsed and died. Necropsy demonstrated hemorrhagic enteritis consistent with hemorrhagic bowel syndrome, some abomasal ulcers and bronchopneumonia. The right front fetlock evaluation indicated that synovitis was present but resolving and the right lateral claw osteomyelitis was still present. The fetlock and claw lesions did not appear to communicate with each other. Gut content was culture-negative for both *Clostridium perfringens* and *Salmonella*, and negative for clostridial toxins.

Case 12, 7-year-old crossbred bucking bull with weight loss.

The bull had been transported to many events over the past few months and had been noted to be losing weight and not bucking as well despite a good appetite. The bull also had soft and occasionally fluid feces, but the handler reported that this was normal for the bulls during the performance season. On presentation, the bull was thin and had a dull haircoat with evidence of dried feces on the perineum and hocks. No additional abnormalities were noted on physical examination except for a dull rumen ping. A PCV and total protein showed mild anemia and hypoproteinemia. Feces were collected for Johne's culture, and a rectal scraping was obtained. Acid-fast staining of the rectal scraping demonstrated acid-fast bacilli within the rectal mucosa, predominantly within macrophages. Cytologic opinion: acid-fast bacteria concentrated within macrophages, supportive of a diagnosis of Johne's disease. The bull was transported to a reproduction facility for semen collection and freezing. The fecal culture was positive for *Mycobacterium avium paratuberculosis* after 4 weeks of incubation.

DECISION MAKING IN CAMELID COLIC: A CASE-BASED APPROACH

Christopher K. Cebra VMD, MA, MS, DACVIM (Large Animal Internal Medicine)
Corvallis, OR

INTRODUCTION

The purpose of these cases is to outline thought processes involved in diagnosing and treating alpacas with abdominal pain. Important factors are defining that the animal actually has pain, trying to identify the likely cause, and determining the best treatment for that condition. Historically, determining these factors has been problematic. The subjectivity of interpretation of pain signs has affected both sensitivity and specificity of certain physical and behavioral findings. Biases from early experiences regarding disease frequency and treatment modalities have persisted and continue to affect modern case management. The considerable experiences in working with alpacas gained over the last 15 to 20 years should enable us to refine and improve in all these areas.

CASE 1

A 2-year-old male castrated alpaca is examined for a 3-day history of lethargy and inappetance. The owners noted that the alpaca is more approachable than usual and now remains recumbent most of the time. Physical examination reveals a body condition score (BCS) of 2/9, a rectal temperature of 38.5 C, a pulse rate of 78 beats/minute, and a respiratory rate of 24 breaths/minute. Thoracic and abdominal auscultation reveal no cardiac murmurs, no abnormal lung sounds, and near-absent forestomach contractions or other gastrointestinal sounds. Gastric fill appears to be low, but the abdomen has a mild pot-bellied appearance. Mucous membranes are pink and somewhat moist with a capillary refill time of < 2 seconds. A brief locomotor and neurologic assessment reveals no abnormalities. The alpaca spends most of the time in the kush position. It can stand, but lies down again within 10 minutes. No spontaneous urination or defecation is noted over a 1 hour period.

This case represents the archtype of the inappetant camelid. It has no specific organ signs, including signs of colic. Although it remains recumbent most of the time, this may be more a sign of weakness than something specific such as discomfort. A more thorough diagnostic evaluation is indicated, but there is no particular reason to think abdominal pain is part of the picture.

MODIFYING CASE 1

A very similar presentation to above, except that the alpaca lies down immediately and vocalizes whenever pressure is placed on the right cranioventral abdomen. It also vocalizes and shifts its position when pressure is put on this spot while recumbent, and occasionally during thoracic auscultation and procedures under the tail.

Interpreting these findings is always subjective. Some feel that pressure on this particular spot on the abdomen exacerbates pain associated with ulcers of the third compartment. However, many camelid exhibit a protective response to this region, because it is one of the places that they attack each other when they fight. Hence, it may be difficult to specify the cause of the reaction.

Teeth-grinding is another sign that may be difficult to interpret. Some think it arises from abdominal discomfort. Others feel it is a non-specific sign of nervousness, and reflects the animal's mental discomfort at its current situation. Observing or hearing this when the animal is allowed to relax, by being left alone for a period of time, is more convincing evidence of its importance, but even this remains subjective and may not specifically be linked to abdominal pain.

Lateral recumbency is another such finding. Colicky camelids can certainly lie in lateral recumbency and are more likely to do so than non-colicky herdmates, but it is important to remember that abdominal pain usually elicits an active response: the painful alpaca will shift weight every few minutes as it seeks to find a comfortable position. A sick alpaca that remains still in lateral recumbency is most likely too weak to maintain another position, rather than that it is doing so preferentially. This may be the progression of colic signs as the animal goes into shock, and hence may have some overlap, or may be related to some entirely different problem.

CASE 2

An 8-month-old intact male alpaca is presented for evaluation. For 2 days, it was noted to intermittently lie on one side. It took bites of hay, but chewed them very slowly. For the following 2 days prior to admission, it appeared dull and ate nothing. Now, it tends to stay in sternal recumbency, occasionally sliding the hind legs out to the side for 1 to 2 minutes at a time. Physical examination reveals a BCS of 5/9, a rectal temperature of 38.2 C, a pulse rate of 84 beats/minute, and a respiratory rate of 20 breaths/minute. Thoracic and abdominal auscultation reveal no cardiac murmurs, no abnormal lung sounds, and near-absent forestomach contractions or other gastrointestinal sounds. The abdomen is rounded with a moderate pot-bellied appearance. Mucous membranes are pink and tacky with a capillary refill time of 2.5 seconds. A brief locomotor and neurologic assessment reveals no abnormalities. The alpaca continues spending most of the time in the kush position, with frequent shifting of the hind end and outward projection of the hind limbs. It can stand, but lies down again within 2 minutes. No spontaneous urination or defecation is noted over a 1 hour period.

In this case, there is more specific information suggesting abdominal pain, namely the restlessness and the abnormal recumbent posture. The low temperature, lack of urination, and high heart rate suggest shock in addition to pain, so this patient is in critical condition. The abdominal distention and lack of fecal passage may suggest an obstruction is present, but this may be functional (such as a ruptured or severely inflamed viscus) or may more benignly suggest lack of feed intake combined with hypoproteinemia. In any case, further evaluation is warranted, but unlike Case 1, there is strong evidence to support abdominal diagnostics. Camelids, especially the smaller specimens such as this one, are difficult to palpate per rectum, and often only the most caudal abdominal viscera may be evaluated. In a young male, this would include assessment of the urinary tract, but without stranguria, examination beyond digital palpation is probably not worth the risk to the patient. Imaging studies, including radiography, ultrasonography, and computed tomography allow far more comprehensive and safer abdominal assessments. In this case, ultrasonography was chosen and revealed little ingesta within the small intestine, little free fluid and a subjectively enlarged third compartment. Blood work revealed moderate hypochloremia, metabolic alkalosis, azotemia and mild hyperalbuminemia. Blood count revealed changes most compatible with stress. With the history of colic turning into dullness, physical evidence of dehydration, lack of fecal production, and abdominal fluid distention, and lab work supporting dehydration and pyloric outflow issues (also supported by ultrasonography), the presumptive diagnosis in this case would be intraluminal obstruction at or near the pylorus, most commonly related to a phytobezoar.

A variety of further tests could be used to support this diagnosis or evaluate the health of the patient. These include further imaging tests to try to identify the obstruction, assessments of forestomach fluid to confirm internal vomition, or abdominocentesis to assess bowel wall health and abdominal inflammation. Exploratory surgery would also allow direct examination of the involved tissues and potentially a means to correct the problem. Given the extended history (up to 4 days), there should be legitimate concerns about bowel health. Coupled with the relative inaccessibility of the presumptive obstruction site, compromised bowel wall complicates the treatment of cases such as this and worsens the prognosis. If the owners are unwilling to perform surgery with a poor prognosis, abdominocentesis may provide useful information. If they are determined to proceed regardless, exploration through a right paracostal incision offers best access to the presumptive site of obstruction.

CASE 3

A 2-year-old unbred female alpaca is examined for a 2-day history of inappetance and odd behavior. The owners noted that the alpaca appears restless, frequently lying down and getting up, kicking its legs out to the side and kicking at its belly when up. It eats mouthfuls of hay, but allows most of it to drop back out of its mouth. Physical examination reveals a body condition score (BCS) of 4/9, a rectal temperature of 39.3 C, a pulse rate of 66 beats/minute, and a respiratory rate of 24 breaths/minute. Thoracic and abdominal auscultation reveal no cardiac murmurs and no abnormal lung sounds. Forestomach contractions are heard twice/minute and other gastrointestinal gurgling is heard on the right side. Gastric fill appears to be low, and the abdomen has a mild pot-bellied appearance. Mucous membranes are pink and somewhat moist with a capillary refill time of < 2 seconds. A brief locomotor and neurologic assessment reveals no abnormalities. The alpaca continues to exhibit the behavior observed by the owners. Spontaneous urination and defecation are noted within a 1 hour period. The feces are soft, clumping pellets.

This animal is undoubtedly showing signs of abdominal pain, but also has evidence of continued GI function and patency. It is bright, afebrile, and vigorous. Of the main causes of colic (inflammation, ischemia, stretch receptor activation, muscle spasms), inflammation appears the most likely, so localizing this and identifying the cause are the major roles of the following diagnostic approach. The major inflammatory causes of colic in camelids include parasitic disorders, some ulcerative conditions, bacterial peritonitis, and fat necrosis.

Basic blood work reveals hypoalbuminemia, hyponatremia, and other minor changes. Imaging studies reveal fluid distended but motile loops of intestine. A moderate amount of free abdominal fluid is seen. Unlike the previous case, understanding the nature of the inflammation is of paramount importance, so abdominocentesis becomes part of the core work-up. In this case, it reveals a modified transudate, which makes bacterial peritonitis of a gut perforation less likely. A fecal examination for parasites is also indicated, with *Eimeria macusaniensis* and *Trichuris* spp. being the most important parasitic causes of colic. Flotation for either of these parasites requires a fairly dense flotation solution (SG = 1.27+). Either can also cause prepatent disease or disease associated with minor shedding. In this case, the fecal reveals 50 *Trichuris* eggs/gram. With evidence of patency, and a likely culprit identified, surgery can be avoided, and treatment efforts directed toward pain control, elimination of the parasites, and correction of hypoproteinemia. A number of options are available for pain control; constant infusions, such as lidocaine, appear to offer more reliable analgesia in camelids than NSAIDs or sedatives.

MODIFICATIONS TO CASE 3

If CBC and peritoneal tap had revealed inflammatory changes, fecal production was absent, or the imaging study had revealed fluid-distended, atonic bowel, establishing the lead differentials and best avenues of treatment becomes considerably more difficult. These inflammatory changes may indicate that the disease started as bacterial peritonitis (*Streptococcus zooepidemicus),* or that an obstructive, inflammatory, or ulcerative GI disease has compromised the integrity of the bowel wall. If fecal production is absent, or the feces passed appear older, the possibility of a lesion requiring surgical correction becomes greater. Alternatively, if this GI slow-down reflects progression of bowel compromise, it worsens the prognosis and increases the likelihood of non-therapeutic surgical procedures. Historically, many camelids have had this combination of problems, often leading to a disappointing outcome.

CASE 4

A 4-year-old, mid-gestation female llama is examined for a 1-day history of inappetance and restlessness. She frequently lays down and gets up, she rolls intermittently into lateral recumbency, and she sticks all 4 legs out to the side. She refuses all feed. Physical examination reveals a body condition score (BCS) of 6/9, a rectal temperature of 39.2 C, a pulse rate of 72 beats/minute, and a respiratory rate of 18 breaths/minute. Thoracic and abdominal auscultation reveal no cardiac murmurs and no abnormal lung sounds. Forestomach contractions are absent, but some gastrointestinal gurgling is heard on the right side. Gastric fill appears to be normal, and the abdomen has a mild pot-bellied appearance. Mucous membranes are pink and somewhat moist with a capillary refill time of < 2 seconds. A brief locomotor and neurologic assessment reveals no abnormalities. Small amounts of urine and formed feces are passed within 1 hour.

This animal is undoubtedly showing signs of abdominal pain. There may be evidence of continued GI function and patency, but the history is short enough that that is questionable. It is bright, afebrile, and vigorous. Of the main causes of colic (inflammation, ischemia, stretch receptor activation, muscle spasms), all are possibilities. If inflammation is present, it has not lead to shock, diarrhea, or GI shutdown. Ischemia likewise often leads to shock and GI shutdown well within this time frame. Blood work, imaging studies, and potentially abdominocentesis will help define the problem. In this case, blood work is unenlightening, making inflammation and ischemia appear less likely, and placing an obstructive lesion either in the forestomach (rare) or further downstream from the pylorus. Imaging studies support this. Ultrasonography can reveal bowel distention and lack of progressive motility. Fluid-distended small intestinal and spiral colon appear fairly similar in size and wall thickness. Spiral colon loops tend to appear more organized, with multiple loops coursing in a parallel direction, whereas jejunum tends to have a fairly chaotic appearance. It is rare to be able to identify a specific intraluminal obstruction ultrasonographically, unless it is a large linear impaction, but other causes of obstruction that affect wall thickness may be identified. Tomography yields a more complete abdominal picture, and can be useful in identifying the actual size, nature, and location of the lesion. In this case, a colonic phytobezoar was identified.

The most common intervention at this point is usually exploratory/corrective surgery. Owners wishing to avoid the expense may wish the veterinarian to attempt to resolve the obstruction using other means, including fluids, surface-acting lubricants, and analgesics. Although those attempts are usually unsuccessful, these patients, with their shorter clinical course, lack of inflammatory changes, and lack of severe metabolic derangements offer better candidates for these attempts than colic patients with many other lesions. However, excessive delay may lead to development of bowel or systemic compromise and hence adversely affect outcome. Thus, conservative treatment efforts may be attempted, but should not be continued too long with evidence of success.

REFERENCES

Cebra CK, et al. Vet Surg 1998;27(2):112. Sullivan EK, et al. Vet Surg 2005;34(5):524. Newman KD et al. Vet Clin North Am Food Anim Pract 2009;25(2):495.

THE TEN BEST FOOD ANIMAL MEDICINE PAPERS OF 2013

Ingrid Lorenz, Dr. med. vet., Dr. med. vet. habil., DECBHM[1], J T. McClure DVM, MSc, DACVIM[2]
[1] Dublin, Ireland, [2]Prince Edward Island, Canada

SELECTION PROCESS

Peer-reviewed articles in ruminant medicine published in 2013 were examined. A search of the PubMed database was performed in January 2014 using the search term: (medicine [tiab] OR surger* [tiab] OR treatment [tiab] OR prognosis [tiab] OR diagnos* [tiab] OR diseas* [tiab] OR clinic* [tiab]) AND (cattle [tiab] OR calf [tiab] OR calves [tiab] OR heifer [tiab] OR bov* [tiab] OR cow* [tiab] OR sheep [tiab] OR goat [tiab] OR camelid* [tiab] OR llama [tiab] OR alpaca [tiab]). [tiab] restricts the search to title and abstract.

This initial search found 4279 potential manuscripts. One author (IL) screened these initially by title and selected for manuscripts that clearly fell into the scope of ruminant medicine and surgery and reduced the list to 304 papers. From these 304 papers, both authors independently created a shortlist of ~ 40 papers based on title and abstracts out of the remaining pool of 304 papers. Papers that were in the top 20 of one author and in the top 40 of the other were then chosen for detailed review of the entire paper. The final selection was done cooperatively with a deliberate effort on diversity of topics and regional distribution. Review articles, retrospective studies and case reports were given lower emphasis while randomized field or clinical trials and controlled randomized, blind studies were given higher emphasis. Of course our own bias in what we found most interesting was a factor in selection.

SELECTED PAPERS BY ALPHABETICAL ORDER OF FIRST AUTHOR

1. Bhattarai B, Fosgate GT, Osterstock JB, Fossler CP, Park SC, Roussel AJ. Comparison of calf weaning weight and associated economic variables between beef cows with and without serum antibodies against or isolation from feces of *Mycobacterium avium* subsp paratuberculosis. *J Am Vet Med Assoc.* 2013 Dec 1;243(11):1609-15. doi: 10.2460/javma.243.11.1609.

2. Boulay G, Francoz D, Doré E, Dufour S, Veillette M, Badillo M, Bélanger AM, Buczinski S. Preoperative cow-side lactatemia measurement predicts negative outcome in Holstein dairy cattle with right abomasal disorders. *J Dairy Sci.* 2014 Jan;97(1):212-21. doi: 10.3168/jds.2013-6898. Epub 2013 Nov 1.

3. Cameron M, McKenna SL, Macdonald KA, Dohoo IR, Roy JP, Keefe GP. Evaluation of selective dry cow treatment following on-farm culture: Risk of postcalving intramammary infection and clinical mastitis in the subsequent lactation. *J Dairy Sci.* 2014 Jan;97(1):270-84. doi: 10.3168/jds.2013-7060. Epub 2013 Nov 1.

4. Constable PD, Hiew MW, Tinkler S, Townsend J. Efficacy of oral potassium chloride administration in treating lactating dairy cows with experimentally induced hypokalemia, hypochloremia, and alkalemia. *J Dairy Sci.* 2013 Dec 27. pii: S0022-0302(13)00870-9. doi: 10.3168/jds.2013-6982. [Epub ahead of print]

5. Eicher SD, Lay DC Jr, Arthington JD, Schutz MM. Effects of rubber flooring during the first 2 lactations on production, locomotion, hoof health, immune functions, and stress. *J Dairy Sci.* 2013 Jun;96(6):3639-51. doi: 10.3168/jds.2012-6049. Epub 2013 Apr 12.

6. Silper BF, Lana AM, Carvalho AU, Ferreira CS, Franzoni AP, Lima JA, Saturnino HM, Reis RB, Coelho SG. Effects of milk replacer feeding strategies on performance, ruminal development, and metabolism of dairy calves. *J Dairy Sci.* 2013 Dec 13. pii: S0022-0302(13)00838-2. doi: 10.3168/jds.2013-7201.[Epub ahead of print]

7. Taurel AF, Guatteo R, Lehebel A, Joly A, Beaudeau F. Vaccination using phase I vaccine is effective to control *Coxiella burnetii* shedding in infected dairy cattle herds. *Comp Immunol Microbiol Infect Dis.* 2014 Jan;37(1):1-9. doi: 10.1016/j.cimid.2013.10.002. Epub 2013 Oct 14.

8. Thomson DU, Wileman BW, Rezac DJ, Miesner MD, Johnson-Neitman JL, Biller DS. Computed tomographic evaluation to determine efficacy of euthanasia of yearling feedlot cattle by use of various firearm-ammunition combinations. *Am J Vet Res.* 2013 Nov;74(11):1385-91. doi: 10.2460/ajvr.74.11.1385.

9. Werne S, Isensee A, Maurer V, Perler E, Drewek A, Heckendorn F. Integrated control of gastrointestinal nematodes in lambs using a bioactive feed×breed approach. *Vet Parasitol.* 2013 Dec 6;198(3-4):298-304. doi: 10.1016/j.vetpar.2013.09.021.Epub 2013 Sep 27.

10. Windeyer MC, Leslie KE, Godden SM, Hodgins DC, Lissemore KD, Leblanc SJ. Factors associated with morbidity, mortality, and growth of dairy heifer calves up to 3 months of age. *Prev Vet Med.* 2014 Feb 1;113(2):231-40. doi:10.1016/j.prevetmed.2013.10.019. Epub 2013 Nov 1.

Table 1: Characteristics of selected best papers of 2013

First Author	Animal studied	Journal	Country	Study design	Organ system/ Topic
Bhattarai	Beef cattle	Journal American Veterinary Medical Association	USA	Retrospective study	Gastroenterology
Boulay	Dairy Cattle	Journal of Dairy Science	Canada	Prospective field study	Gastroenterology
Cameron	Dairy Cattle	Journal of Dairy Science	Canada	Prospective field study	Mammary
Constable	Dairy Cattle	Journal of Dairy Science	USA	Randomized controlled experiment	Metabolic
Eicher	Dairy Cattle	Journal of Dairy Science	USA	Prospective randomized field trial	Musculoskeletal
Silper	Dairy Calves	Journal of Dairy Science	Brazil	Controlled clinical trial	Calf nutrition
Taurel	Dairy Cattle	Comparative Immunology Microbiology of Infectious Diseases	France	Prospective randomized field trial	Infectious disease
Thomson	Beef Cattle	American Journal of Veterinary Research	USA	Randomized controlled experiment	Euthanasia/Welfare
Werne	Sheep	Veterinary Parasitology	Switzerland	Prospective randomized field trial	Gastroenterology
Windeyer	Dairy Calves	Preventative Veterinary Medicine	Canada	Observational study	General calf health

MANAGEMENT OF TRANSITION DAIRY COWS TO REDUCE KETOSIS RISK

Charles T. Estill, VMD, PhD, Dip. ACT
Corvallis, OR

INTRODUCTION

Fresh cow performance is critical to the success of a dairy enterprise. Good performance includes not only delivery of a healthy calf and a rapid increase in production of high quality milk, but freedom from infectious and metabolic disease, the ability to breed back quickly after freshening, and herd longevity. A majority of health condition in dairy cattle occur in the early postpartum period[1]. Transition cow management will directly impact subsequent production and reproductive performance and an integrated approach is required to ensure success. When transition cow management is deficient the result is reduced milk production and deranged components, metabolic and infectious disease, premature culling or death, reduced calf viability, and reduced reproductive success. Dramatic increases in energy requirements during late gestation and early lactation, superimposed on an animal with a profound drop in dry matter intake (DMI) just before calving, makes the dairy cow highly susceptible to the metabolic diseases ketosis and hepatic lipidosis[2]. It appears the reduction in DMI is the most important risk factor for ketosis[3]. Cows do respond to the increased energy demand by increasing DMI but the response is typically delayed, resulting in negative energy balance and mobilization of adipose reserves. Ketosis and hepatic lipidosis are associated with an increased incidence of infectious diseases such as metritis and mastitis through immune dysfunction and as well as displaced abomasum, reduced milk production, and poor reproductive performance. This paper will discuss each area of transition cow management and feeding that can be controlled to reduce ketosis mitigation and ultimately contribute to cow health and well-being.

The transition period is loosely defined as the period from approximately 3 weeks prior to calving until 3 weeks post calving. The dairy cow is extremely fragile from a health and metabolic standpoint and at the onset of lactation she must make tremendous homeostatic adjustments to maintain her well-being. The incidence of clinical and subclinical ketosis serves as a biomarker of transition success or failure. When husbandry, feeding, and general management are optimized to reduce ketosis incidence, successful transition is likely.

This report will examine the physiological changes the transition cow experiences and the management options available to "help" these cows get off to a good start in early lactation. This paper will address the following areas of transition cow management: physiology of transition, ketosis risk factors, transition diets, housing and environment for transition cows, and herd monitoring for ketosis incidence.

Definition of ketosis

Ketosis is characterized by abnormally elevated concentrations of ketone bodies: acetoacetic acid (AcAC), acetone (Ac), and β-hydroxybutyric acid (BHB) in tissues and body fluids[4]. It is a metabolic condition that occurs when intake of nutrients, especially energy, is inadequate to meet production demands. The incidence of ketosis is difficult to estimate from the literature due to varying parameters used in case definitions. Ketosis may be classified as clinical or subclinical based on concentrations of ketone bodies in blood, urine, and milk and whether or not clinical signs are evident. Incidence rates (cases/year) reportedly range from 1.87% to 13% for clinical ketosis[5] and 7.3% to 12.1% for subclinical ketosis[6]A Canadian study found that 12.1% of postpartum cows had subclinical ketosis[7]. An estimated 6-8% of dairy cows in the USA are clinically affected each lactation. In a large North American study, ketonuria was reported in 30.4% of fresh cows[8]. It usually occurs a few days to six weeks after calving in high-producing cows with peak incidence about three weeks after calving[5,9]. Most high-producing cows go through a subclinical type of ketosis (BHB>1.40 mM/L) in early lactation when they are unable to consume enough energy to meet the energy output in milk[10]. The mostly commonly used cut-point for ketosis is $\geq$ 14.4 mg/dl (>1400 μmol/L) of blood BHB[11].

Ketonemia becomes a disease condition when production of ketones exceeds utilization and is accompanied by elevated concentrations of free fatty acids, nonesterified fatty acids (NEFAs) and reduced blood glucose. Ketosis may be primary or secondary. Primary ketosis results from negative energy balance during early lactation which is associated with a reduction of glucose in the blood and liver resulting in increased fat mobilization. When hepatic production and absorption of ketone bodies exceeds their use as an energy source, the result is elevated blood ketones and nonesterified fatty acids (NEFAs) and reduced blood glucose. The bovine liver has limited capacity to completely oxidize fatty acids which leads to fatty liver so oxidation to ketones or storage a triglycerides results in liver impairment and ketonemia[12]. Primary ketosis may be subclassified as Type I or Type II[13]. Type I primary ketosis occurs during peak lactation (3-6 weeks) as a result of negative energy balance. Type II primary ketosis occurs early in lactation (1-10 days) and is usually associated with hepatic lipidosis. Secondary ketosis results from reduced appetite associated with other problems that have occurred during the periparturient period such as retained placenta, displaced abomasum, hardware, mastitis, or metritis. The clinical signs of ketosis tend to be vague and nonspecific[14].

PHYSIOLOGY OF THE TRANSITION COW

There are various metabolic adaptations which must occur during the transition period. These are due to the metabolic demands for maintenance, and in the case of heifers, growth as well as the rapidly growing fetus[15,16]. At the onset of lactation nutrient demand increases dramatically[1]. With the onset of lactation there is a fourfold increase in energy demands of the dam. Within four days of calving glucose production doubles. In muscle tissue, glucose use is reduced and instead, use of fatty acids and ketone bodies increases. This is not sufficient to offset the increased energy demand so gluconeogenesis in liver and kidney tissue is increased and blood glucose concentration is reduced[17]. This result is about a 500 g/d deficit which must be "paid" for by drawing on body reserves. Protein is mobilized from skeletal muscle to be utilized as a substrate for gluconeogenesis and milk protein synthesis[1] which results in a 25% reduction in muscle fiber diameter post-calving. Non-esterified fatty acids (NEFA) are bound to albumin and are oxidized in the liver and peripheral tissues to CO_2 or ketones. DMI depression this culminates in a negative energy balance of about 12 Mcal/day and a negative protein balance of about 560 g/day. Successful adaptation to this increased nutrient demand requires a coordination of appropriate endocrine, neural, and immune homeostatic and homeorhectic mechanisms[18,19] in different tissues resulting in metabolic changes that attempt to ensure that the cow's genetic potential for milk yield is exploited. For example, estradiol increases through late pregnancy and peaks 1-2 weeks prior to calving. Estradiol may be partially responsible for the decreased appetite[20]. Somatotropin, which plays a role in nutrient partitioning, increases in late pregnancy and the early postpartum period. It has a powerful galactopoietic influence and enhances liver gluconeogenesis. Somatotropin also alters tissue responsiveness to insulin and catecholamines.

Due to increased energy requirements in the face of declining dry matter intake in the last 7-10 days of gestation[21], there is decreased lipogenesis and increased lipolysis accompanied by a reduction in insulin receptor responsiveness and reduced circulating insulin concentrations[22]. In the early postpartum period the ratio of growth hormone to insulin is high and stimulates mobilization of long-chain fatty acids from adipose tissue in support of lactation[1]. Fatty acids released from adipose tissue circulate as non-esterified fatty acids (NEFA), which are a major source of energy to the cow during this period. The concentration of NEFA in blood reflects the degree of adipose tissue mobilization[23]. NEFA are bound to albumin and are oxidized in the liver and peripheral tissues to CO_2 or ketones. They can also be esterified in the liver and released as VLDL or, when production is very high, stored in the liver leading to fatty liver. In turn, increased lipid deposition leads to reduced protein output, reduced insulin clearance, reduced gluconeogenesis, and reduced ureagenesis. Taken together, reduced circulating insulin and insulin resistance promotes fat and amino acid mobilization. As an additional factor contributing to lipolysis, catecholamines, via β-adrenergic receptors, increase lipolytic responsiveness in the early postpartum period; sensitivity to lipolytic signals (epinephrine and norepinephrine) is greatly enhanced[24,25]. Consequently, stressors and poor nutritional management causing decreases in voluntary DMI will result in large increases in NEFA immediately after calving [26-28]. The reduction of DMI prepartum may be due to increased concentrations of sex hormones and reduced rumen capacity due to the space occupied by the developing fetoplacental unit[28]. It has been noted that cows with elevated NEFA concentrations are at higher risk of fatty liver, metritis, DA, and reduced fertility. Fatty liver is characterized by a high content of lipids and triglycerides in the liver. When severe, clinical fatty liver disease, which is often refractory to treatment, results.

Risk factors for primary ketosis (or elevated NEFA/BHB) amenable to management

- Parity – primiparous heifers had lower risk of ketosis[29]
- Body condition score greater than 3.5/5[3,29,30]
- Season of calving – lower risk for summer calvings at a northern latitude[29] and cool season in warmer climates[31]
- Prepartum DMI – severe drop in DMI during late gestation is associated with ketosis[32]
- DMI immediately prepartum – DMI at day -1 is highly correlated with DMI at day 21 postpartum[33]
- NEFA prepartum – elevation are associated with higher risk of ketosis[34]
- Animal grouping – heifers housed with mature cows (fed same diet) in the close-up dry period experience more drop in DMI[3]
- Days in maternity pen – cows which spend more than3 days in the maternity pen have greater ketosis incidence[35]

There is an excellent tool available for calculating transition group sizes from the University of Wisconsin at:
http://www.vetmed.wisc.edu/dms/fapm/fapmtools/5house/TransitionCowpenSizeCalculator.xls

- Ketogenic silage – silage high in butyric acid and/or putrescine[36,37]

It has been noted that cows which are relatively thin (BCS < 3.0) have better DMI in the last three weeks of gestation[38] and subsequently produce more milk in early lactation than their "fatter" herdmates. Cows with high body condition scores at calving experience greater weight loss in early lactation and have reduced DMI during weeks 6-12 postpartum. Approximately three weeks prior to calving dry matter intake of cows and first-calf heifers will be about 1.9 and 1.7 percent of their body weight respectively. By the time calving occurs DMI is typically down about 25%. The most precipitous drop occurs in the last week prior to calving. In cows carrying twins, the reduction in dry matter intake occurs sooner and is more severe that in cows carrying a single fetus. NRC energy models underestimate the requirement for cows carrying twins. These cows are often in severe negative energy and protein balance prepartum which results in poor performance postpartum. During transition, the cow goes from an anabolic state to a catabolic state.

Recent studies indicate that maximizing DMI of close up dry cows may not be as important as minimizing the prepartum reduction in DMI in the week prior to calving. The amount of fat in the cow's liver is correlated to the amount DMI drops prepartum and not the average intake during the three weeks prepartum. This suggests that high intakes are not required for good fresh cow performance but preventing the prepartum drop in DMI is critical. This is not to imply that high DMI during the dry period is detrimental but if metabolic problems are to be avoided then the intake must be maintained as high as possible right up until calving.

NUTRITIONAL MANAGEMENT OF THE TRANSITION COW TO REDUCE KETOSIS RISK

Late gestation nutrition has a large impact on multiple aspects of the cow's productivity after freshening. The health and viability of the neonate are directly affected. In addition, colostrum quality and quantity are influenced by prepartum nutrition. Early lactation milk yield and milk components are largely a reflection of prepartum nutritional management. The incidence of metabolic and infectious disorders can be minimized through proper nutritional management of the transition cow. Fertility of the dam is strongly correlated with late gestation and early postpartum nutrition.

The start of the transition period begins three weeks before anticipated calving date and ends three or four weeks after freshening. Protein intake must be adequate to support fetal growth and maintenance of the dam. As a monitor, BUN can be examined and should be 9-12 mg/dL. In general, absorbed protein should be +200-400 g/d (Spartan Ration Balancer). Excessive protein intake prepartum (>12%) is associated with reduced milk yield postpartum[39] and increased BHB[40]. The NSC concentration should be at 30-35% to stimulate microbial protein synthesis and rumen papillae development. Papillae development is stimulated by propionate (from fermentable starch) and takes about 4 weeks to occur. It is also very important to avoid excessive NSC or fatty liver will result. There are several serious consequences of low protein/energy intake during the close-up dry period. Insulin concentrations will decline leading to increased protein catabolism and reduced protein synthesis. Somatotropin concentration will be elevated which increases protein synthesis but redirects protein away from muscles towards other tissues such as the conceptus and mammary gland which results in a depletion of body protein reserves. It is important to avoid extremely high fiber diets during the close up dry period because DMI will be reduced due to physical fill and reduced transit time. Very high energy diets resulted in a substantial decline in DMI prior to calving and should also be avoided. General guidelines for a close up dry cow ration are:

- ✓ 12-15% CP (14% min if heifers in with close up cows) – increase RUP to 35-38%
- ✓ NE_L,Mcal/lb 0.68-0.72
- ✓ 33-35% NSC
- ✓ 40-50% NDF
- ✓ Forage >70% of diet
- ✓ Ca <.6%, Mg 0.35-0.45%
- ✓ K>1.5%, P<0.4% (<50 g/day)
- ✓ Increase trace minerals 20-50% above NRC and force feed

When evaluating rations compare the diet to NRC recommendations and make adjustments based on the environment, observed intake, and cow response. It is essential to not only look at the ration "on paper" but to also determine the nutrient density of the TMR and compare to the "paper" value. Also evaluate mixing consistency, evaluate particle size, and measure the diet actually consumed (fed - refusals). It is advisable to use a blend of CHO and protein sources to ensure adequate rumen degradable and rumen undegradable fractions. It is not recommended to feed added fats unless they will be included in the fresh-cow diet. In general, fats fed prepartum tend to reduce DMI and may increase the incidence of metabolic disease. Dry cow feed should be fed to 10% refusal and should be pushed up 2-4 times daily. The refusals should be removed daily and lactating cow refusals should not be given to close up dry cows. If anionic salts are fed urine pH should be monitored. If the incidence of periparturient diseases is unacceptable, a program to evaluate the close up program should be initiated and include metabolic profiling of selected groups of cows.

A review of the literature on feeding the dry cow to reduce ketosis risk produces conflicting recommendations. Several studies suggest a link between prepartum DMI while others do not[3]. Data indicates that, for Holstein cows, if DMI falls below 11 kg/d the week prior to parturition, NEFA will become elevated[41]. The preponderance of studies indicate the DMI should be maximized prepartum but other studies do not support this concept. One line of evidence is based on the observation that prepartum feed-restricted cows may be less prone to fatty liver. Restriction can be accomplished either by limit feeding or by designing a high fiber, low energy ration. There is the suggestion that cows fed prepartum diets that result in elevated NEFA afford the liver of those animals an adaption period such that rates of liver fatty acid esterification are reduced postpartum[42] but results from Grummer did not support this idea[3]. The use of high fiber bulky diets prepartum improved metabolic status postpartum and reduced ketosis risk[43]. The consistent factor in prepartum diets that are associated with reduction in postpartum ketosis risk is they prevent the decline in DMI at parturition. Therefore, a low energy diet or low DMI could be associated with reduced ketosis risk if those cows do not experience a decline in DMI at and just after calving. On the other hand, cows on a high energy diet prepartum or those with high DMI prepartum could be prone to ketosis if DMI drops severely just prior to parturition and in the early postpartum period. The precise magnitude of change in DMI around parturition that is associated with increased ketosis risk has not been determined[3].

Feed additives associated with reduced risk of ketosis

Methionine[44]
Propylene glycol[45], sodium propionate [46]
Monensin[47]
Niacin[48,49]
Choline[50-52]

HOUSING/ENVIRONMENT CONSIDERATION FOR TRANSITION COWS

Transition cow facilities need to be clean, dry, well ventilated, and comfortable to minimize stress and pathogen exposure. Pens should not be crowded and stocking rate should only be around 90%. It is a good idea to avoid excessive movement between groups since regrouping causes social stress. This is particularly important as calving nears. If possible, separate facilities should be available for heifers and cows. Nutrient requirements are different for cows and heifers and cows may tend to bully heifers at the feed bunk. Transition diets must be made very palatable and an attempt should be made to minimize introduction of unpalatable ration components such as fat and bypass protein sources. Feeds should be fresh and free from molds. A TMR is recommended and should be fed in area with adequate bunk space. An option to consider is restricting intake during the last three weeks to avoid the severe drop in DMI in the week before calving. This can be achieved either by limiting the feed available to the cows or by increasing fiber concentration in the close up ration. Adequate fresh water must always be accessible. In the summer months, heat abatement measures should be instituted. The barns should be well ventilated and provide good footing. In warm times of the year, fans/sprinklers should be used over the feeding area. It is a good exercise to calculate the Cow Comfort Quotient (CCQ) where CCQ=cows lying properly install/cow "in" stall * 100. A reasonable goal is 80%. There should be a separate calving area which is clean, dry, and stress-free. The calving pens should be at least 9 X 9 m and have lock up capability. Sand is becoming increasingly popular as a bedding material for maternity pens. The bedding should be comfortable, dry, and provide good footing. The maternity pens should be cleaned between each use.

HERD MONITORING

Serum samples (individual or pooled 10/pool) can be assayed for NEFA and BHB in cows after calving. Testing for circulating concentrations of these metabolites reflect the success of adaptation to a state of negative energy balance and values increase as part of a normal adaptive response to increased energy demands. Poor adaptation can be reflected in excessive concentrations of NEFAs and BHAs. Circulating BHA concentrations > 1000 to 1400 μmol/L or NEFA >600 to 1,000 μmol/L in the first week after calving are at risk for clinical disease and reduced milk production. Generally, threshold values for individual animal should be the higher of the stated range while pooled threshold values should be set closer to the lower end of the range. Tests are available to check the ketone concentrations of blood, milk, and urine. The ketone level of milk is about half the blood level, whereas the ketone level of urine exceeds the level in blood by about four times. This makes the urine test for ketone levels overly sensitive for diagnoses, with most high-producing cows showing positive urine tests in early lactation, often without need for treatment. The milk test is considered more accurate for indicating when there may be a problem. Total blood ketones are > 30 mg/dl and urine ketones are > 84 mg/dl while milk ketones are > 10 mg/dl. In clinical cases blood glucose concentrations are 20-40 mg/dl. In cows with subclinical ketosis blood glucose is low normal with total blood ketones of 10-30 mg/dl and milk ketones of 2 mg/dl. Secondary ketosis cases may have values intermediate between clinical and subclinical primary cases depending on the duration and severity of the concurrent disease. Commercial tests that detect Ac, AcAC, or BOB are widely used as field screening tests. Urine ketone levels are often high in normal cows since ketones are concentrated 2-20X in the urine. A specific and sensitive electronic test device is available for use with blood which makes the diagnosis more reliable[53-55] When greater than 10% of the at-risk postpartum cows have blood BHA ≥1400 μmol/L or prepartum cows have NEFA ≥0.400 μmol/L management practices and adjustments are warranted.

SUMMARY

Even though the transition period is the most stressful time of a dairy cow's life we now have the knowledge and skills to reduce that stress to a level that will be compatible with good health, high production, and acceptable reproductive efficiency. When designing a transition cow program, the needs of the cow and the convenience of labor force who will meet those needs must be considered. The entire dairy management team must be involved with this critical group of cows to ensure sustainability and profitability of the dairy enterprise. Attention to detail is what makes the difference between success and failure in transition cow management. All of the essential factors including grouping, nutrition, housing, treatment, and monitoring must operate smoothly and consistently to promote success. Producers can avoid ketosis outbreaks by preventing obesity and reduced dry matter intake in far-off dry cows. The most important factor in prevention is to prevent a drop in DMI in late gestation and early lactation. Over-conditioning causes depressed appetites at freshening and increases fatty liver problems. Changes in the ration should be made slowly and poor quality feeds should be avoided.

REFERENCES

[1]Ingvartsen KL. An Feed Sci Tech 2006;126:175.[2]Gerloff BJ. Vet Clin North Am Food Anim Pract 2000;16:283.[3]Grummer R, et al. The Vet Clin North Am. Food Animal Pract 2004;20:447.[4]Taylor E. Dorland's Illustrated Medical Dictionary WB Saunders, Philadelphia, 1988.[5]Kauppinen K. Acta Vet Scand 1983;24:349.[6]Dohoo IR, et al. Canadian J Comp Med. Revue canadienne de medecine comparee 1984;48:1.[7]Dohoo I, et al. Canadian Journal of Comparative Medicine 1984;48:1.[8]Markusfeld O. J Dairy Sci 1987;70:158.[9]Dohoo I. Prev Vet Med 1984;2:671.[10]Baird GD, et al. The British Vet Journal 1974;130:318.[11]McArt JAA, et al. The Vet J 2013;198:560.[12]Kleppe BB, et al. J Dairy Sci 1988;71:1813.[13]Herdt TH. Vet Clin North Am, Food Anim Pract 2000;16:215.[14]Foster LA. Vet Clin North Am Food Anim Pract 1988;4:253.[15]Bell AW, et al. J Dairy Sci 1995;78:1954.[16]Forbes J. An Prod 1977;24:91.[17]Butler WR. A Repro Sci 2000;60–61:449.[18]Herdt TH. Vet Clin North Am. Food An Pract 2000;16:215.[19]Bauman DE, et al. J Dairy Sci 1980;63:1514.[20]Grummer RR, et al. J Dairy Sci 1990;73:1537.[21]Marquardt J, et al. J Dairy Sci 1977;60:929.[22]Hayirli A. Vet Res Comm 2006;30:749.[23]Pullen DL, et al. J Dairy Sci 1989;72:49.[24]Theilgaard P, et al. An Sci 2002;75:209.[25]Underwood J, et al. J Dairy Sci 2003;86:106.[26]Bertoni G, et al. Zootec. Nutr. Anim;24:17.[27]Drackley JK. J Dairy Sci 1999;82:2259.[28]Ingvartsen KL, et al. J Dairy Sci 2000;83:1573.[29]Gillund P, et al. J Dairy Sci 2001;84:1390.[30]Markusfeld O. Vet Record 1985;116:489.[31]Skaar TC, et al. J Dairy Sci 1989;72:2028.[32]Zamet C N, et al. Therio 1979;11:229.[33]Grummer RR. J Anim Sci 1995;73:2820.[34]Cameron RE, et al. J Dairy Sci 1998;81:132.[35]Cook NB, et al. Vet Clin North Am Food Anim Pract 2004;20:495.[36]Andersson L. Vet Clin North Am Food Anim Pract 1988;4:233.[37]Lingaas F, et al. J Dairy Sci 1992;75:2433.[38]Hayirli A, et al. J Dairy Sci 2002;85:3430.[39]Greenfield RB, et al. J Dairy Sci 2000;83:703.[40]Doepel L, et al. J Dairy Sci 2002;85:2315.[41]French PD. J Dairy Sci 2006;89:1057.[42]Grum D, et al. J Dairy Sci 1996;79:1836.[43]Janovick NA, et al. J Dairy Sci 2011;94:1385.[44]Osorio JS, et al. J Dairy Sci 2013;96:6248.[45]Fisher LJ, et al. Can J An Sci 1973;53:289.[46]Schultz LH. J Dairy Sci 1958;41:160.[47]Green BL, et al. J Dairy Sci 1999;82:333.[48]Dufva GS, et al. J Dairy Sci 1983;66:2329.[49]Jaster EH, et al. J Dairy Sci 1983;66:1039.[50]Zom RLG, et al. J Dairy Sci 2011;94:4016.[51]Goselink R M A, et al. J Dairy Sci 2013;96:1102.[52]Lima FS, et al. The Vet J 2012;193:140.[53]Panousis N, et al. Res Vet Sci 2012;93:435.[54]Iwersen M, et al. J Dairy Sci 2009;92:2618.[55]Wilson D, et al. J Vet Sci Tech 2013;4.

CYSTIC OVARIAN DISEASE

Charles T. Estill, VMD, PhD, Dip. ACT
Corvallis, OR

INTRODUCTION

Cystic ovarian disease (COD), also described as cystic ovarian follicles, is an important ovarian dysfunction in dairy cattle, especially during the early postpartum period and results in reduced fertility and associated economic losses. COD may negatively impact dairy profitability mainly by delaying conception. COD arises as the result of ovulatory failure. Folliculogenesis and ovulation are orchestrated by the hypothalamic-pituitary-gonadal axis. The primary factor is thought to be a deficient luteinizing hormone surge resulting in anovulation. However, the endocrine disruption that results in cyst formation is likely of multifactorial etiology and remains poorly understood. A dysfunction can occur at different levels of this neuroendocrine system, causing COD. Clinical signs of COD vary due to the different forms of ovarian cysts and the dynamic nature of the disorder. Recent progress in understanding the pathogenesis has facilitated development of more predictable treatment options. Because several management-related risk factors increase the incidence of COD, proactive intervention can reduce the incidence.

DEFINITION

There is currently no clear and unambiguous definition of COD. Cystic ovarian follicles develop when one or more follicles fail to ovulate and subsequently do not regress but maintain growth and steroidogenesis[1]. Traditionally, cysts have been defined as anovulatory follicular structures (diameter >25 mm) that persist for 10 or more days in the absence of a functional corpus luteum and are accompanied by abnormal estrous behavior (irregular estrus intervals, nymphomania or anestrus)[2]. However, more recent data using ultrasonography indicate that ovulatory follicles typically ovulate by the time they reach 17-20 mm in diameter, so follicles that persist and are 17mm or greater may be considered to be pathologically "cystic." Therefore, a revised definition of COD is; the presence of follicles larger than 20 mm in diameter, in the absence of an active corpus luteum that interferes with normal cyclicity[1].

CLINICAL SIGNS AND DIAGNOSIS

Outward clinical signs associated with COD are variable and generally unremarkable. Most (60-85%) cows with COD do not show obvious clinical signs and are likely to be anestrous[1]. Nymphomania, irregular expressions of estrus, and development of a masculine appearance are less commonly associated with COD[3].

Diagnosis of COD is typically during the first 60 days of lactation but cases can also occur much later[4]. COD has even been identified inbreeding age heifers[5] and during pregnancy[5,6]. It should be noted that spontaneous recovery from COD in the first 60 days postpartum is relatively high[7] but the economic impact is still substantial[8]. This is likely due to delayed resumption of cyclicity associated with COD and the impact of not having one or more ovulations prior to first postpartum breeding on subsequent conception rate[9].

Diagnosis is usually made on the basis of transrectal palpation and/or ultrasonographic imaging of the ovaries and reproductive tract. Finding follicular structures greater than 20 mm in the absence of estrual signs (uterine tone, clear viscous mucous vaginal discharge) leads to a presumptive diagnosis. Diagnosis is often supported by a history of anestrus irregular cycle lengths. Based on morphological and endocrine differences, ovarian cysts are subdivided into either follicular or luteal types although a continuum of intermediate types is found in practice. Ultrasound evaluation of the ovaries is helpful in determining if a cyst is follicular or luteal or if it is a non-pathological cystic corpus luteum[10]. Luteal cysts are typically thought of as being in a later stage as evidenced by the development of luteal tissue in the cyst wall which is > 3 mm and visible as being more echogenic that the surrounding ovarian tissue. Luteal cysts may also have web-like structures within the lumen[10]. Finding progesterone concentration above baseline in blood or milk support a diagnosis of luteal cyst[11]. Follicular cysts have a thin wall ($\geq$ 3 mm) and the lumen contains uniformly anechogenic fluid[10,12]. Progesterone concentration in blood or milk is at baseline[12]. Intermediate forms occur with patches of luteinized tissue in the follicular wall and variable progesterone concentrations in blood and milk, which defy classification[13,14].

ECONOMIC IMPLICATIONS

Cystic ovarian disease in dairy cattle occurs most frequently during the postpartum period, 30 to 60 days after calving, when normal ovarian activity usually resumes. The reported incidence of COD varies between 6 and 30 % [7,15,16;4] but is generally reported as 15% . Each incidence of ovarian follicular cysts increases the calving interval by 22-64 days [17] and experience lower first-service conception rates and require more services per conception compared with cows that do not develop COD[4]. Because of variability in dairy operational economics it is not possible to accurately predict the cost of days open but estimated cost of days open past 120 days is generally between $3.19 and $5.41/day with higher costs associated with longer days open [4,18]. There are also costs associated with diagnosis, treatment, and culling.

RISK FACTORS FOR DEVELOPMENT OF COD

There are a wide variety of risk factors associated with COD. The major predisposing factors leading to the occurrence of COD are high milk yield, stage of lactation, season, stress[19], negative energy balance, and genetic predisposition. It has been observed that high milk production is associated with COD[20-23]. A one kg increase in milk yield raised the risk of cysts by a factor of 1.05. However, other studies have failed to support the notion that milk production and COD are directly linked[24]. It seems more plausible that failure of the cow to adapt to the physiologic demands of high milk yield is more likely involved in the pathogenesis of COD than is high milk production *per se*. It appears that some cows are more able to increase dry matter intake to support high milk production and avoid severe or prolonged negative energy balance[22]. Energy balance is defined as the difference between net energy intake and net energy requirement. In the case of a high producing postpartum dairy cow, there is always some degree of negative energy balance during early lactation. Negative energy balance results in a myriad of endocrine and metabolic adaptation, all with potential to influence ovarian function[25]. In general, negative energy balance is associated with reduced reproductive efficiency postpartum [26]. More specifically, it has been shown that elevated average peak milk yield and milk fat and protein yield in early production are risk factors for COD but milk fat:protein ratio (a proxy for energy balance) is not [27]. Collectively, several studies indicate that when negative energy balance is severe or when the nadir occurs later in the postpartum period, the incidence of COD is increased[28,29]. In contrast, cows with COD tended to mobilize less body reserve than non-cystic cows[25]. In spite of inconsistent findings, it seems clear that a link exists between negative energy balance and COD but the mechanisms involved are not clear. It seems likely that metabolic hormones and metabolites including IGF-1, insulin, glucose, NEFAs, BHB, and leptin alter hypothalamic-pituitary-ovarian axis function. Is established that IGF-1, which is depressed during negative energy balance, together with reduced insulin, leads to down regulation of follicular granulosa LH receptors which may result in unresponsiveness to the preovulatory LH surge and ovulation failure resulting in COD[30,31]. Leptin appears to be an important link between metabolic status and endocrine function in cows and a threshold level may exert a permissive effect on postpartum ovarian activity[32]. Elevation in plasma NEFAs, associated with negative energy balance, has the potential to impact ovarian function through multiple mechanisms. Elevated NEFAs, in rats, were shown to cause reduced insulin secretion in response to glucose[33] and NEFAs are cytotoxic to bovine granulosa and theca cells[34,35]. Despite the potential for NEFAs to disrupt ovarian function and lead to COD, a correlation between NEFA concentrations in cystic and normal cows has not been identified [31,36]. This finding tends to argue against negative energy balance as a significant risk factor for COD. Ketosis is associated with fat mobilization and its' occurrence is correlated with COD [21]. Thus, the direct effects of negative energy balance in development of COD are unclear but metabolic derangements which are time-sensitive may ultimately be involved. Additional risk factors for development of COD in the early postpartum period have been identified. Cows calving in summer were 2.6 times more likely to develop ovarian cysts than those giving birth in winter[23]. The risk of having a cyst was 1.9 times higher in cows with an abnormal puerperium. A one unit increase in body condition score (scale 1 to 5) from prepartum day 60 to parturition increased the risk of cyst development 8.4 times [37]. It has been known for decades that COD is heritable [13]. Heritability estimates for COD, using on-farm records in Holstein herds is 0.05 [38]. In a genetic study on Dutch Black and White cattle the heritability of COD was estimated at 0.087. Additionally, in the same study, the genetic correlations between COD and 305-d milk, fat, and protein yields were 0.345, 0.379, and 0.441, respectively. The investigators concluded that a genetic predisposition for COD exists in Dutch Black and White dairy cattle. Because a genetic correlation between high milk production and development exists, continued selection for milk yield will inadvertently also select for cows that tend to develop COD[39]. However, genetic selection to reduce the incidence of COD based on on-farm recording systems has the potential to be effective[38].

ETIOLOGY AND PATHOGENESIS

COD results from failure of ovulation of an ovulatory or larger sized follicle [40]. When the dominant follicle persists and does not ovulate, consecutive follicular waves also do not reach ovulation so that cows with COD have ovarian follicular waves similar to those of normally ovulating cows that ovulate but these follicles fail to ovulate. The foundational cause of follicular persistence and COD is lack of an appropriately timed LH surge or failure of the follicle to respond to the LH surge[41]. It is plausible, that in some cases, both mechanisms are at work. In the transition period between the prepartum condition and resumption of ovarian cyclicity postpartum the aberrant LH release is not caused by reduced hypothalamic GnRH content or inability of the pituitary to respond to GnRH with a LH release[42]. Failure to release LH is thought to be due to estrogen insensitivity of the hypothalamic surge center to positive feedback of follicular estradiol such that surge amounts of GnRH are not released and, in turn, inadequate LH is released. Progesterone priming of the hypothalamic surge center is required to impart responsiveness to estrogen[43,44]. It appears that progesterone up-regulates the estrogen receptor in the GnRH surge center[18]. Experimentally, a large anovulatory follicle, similar to COD, can be induced by estradiol induction of a GnRH/LH surge in the absence of an ovulatory follicle, and this condition prevents a subsequent GnRH/LH surge in response to high doses of estradiol[45]. Development of COD in this model can be prevented by administration of progesterone after the initial estradiol/ GnRH/LH surge [46]. Therefore, the endocrine dysfunction that disrupts the LH surge is abnormally low progesterone concentration during the preceding luteal phase [40,47]. Restoration of hypothalamic responsiveness to estradiol positive feedback requires 7 days of progesterone exposure[44]. If progesterone is iatrogenically provided at suprabasal concentration at the time of the preovulatory rise in estradiol the LH surge will be blocked and ovulation inhibited resulting in formation of a follicular-type cyst[48]. This appears to be the case in many cases of naturally occurring COD[49]. Once established cysts tend to persist due to the continued secretion of LH which results in cyst growth and maintenance[50]. A role for matrix metalloproteinases in COD has been proposed[51,52]. Higher proMMP-2 and -9 levels were present in the follicular fluid of cysts than in normal follicular fluid of dominant follicles. The conversion of proMMP to the active form, which plays a role in follicular wall breakdown at ovulation, is triggered by LH[52]. Therefore, if the LH surge is lacking, insufficient, or mistimed, ovulation may not occur due to failure of follicular wall breakdown.

TREATMENT OPTIONS

Treatment involves both management and pharmacological intervention. Correction of nutritional and metabolic disorders must be encouraged to prevent recurrence of existing cases and development of new ones. GnRH has been the most widely accepted treatment for over 35 years. In 1975, it was reported that 74.4% of cows with COD (follicular or luteal cyst) responded favorably to a single im injection of 100 µg GnRH [53]. The response to 10,000 i.u hCG iv was similar [53]. Subsequently, PGF was reported effective in restoring cyclicity in cows with luteal cysts [54]. However, clinically differentiating follicular from luteal cysts is imprecise without ultrasonography [12] and limits usefulness of this treatment. GnRH has been used in conjunction with PGF in numerous protocols. One such protocol consists of initially injecting GnRH followed by PGF 14 days later. This protocol was improved by simultaneous treatment with GnRH and PGF on day 0 followed by PGF on day 14 [55]. The same study also compared their most successful protocol to the traditional Ovsynch protocol and found their treatment superior to Ovsynch. Another study compared the Ovsynch protocol (with TAI) to treatment with GnRH followed by PGF 7 days later and breeding on observed estrus during the 7 day period following PGF. There was no significant difference in pregnancy rates [56]. More recently, several studies have investigated the effectiveness of incorporating progesterone treatment in COD treatment protocols. The rational is that exposure to progesterone, via negative feedback on the hypothalamus and pituitary, will reduce pulsatile LH secretion and hence deprive the cyst of luteotropic support. Cystic follicles then undergo atresia and decreased production of estradiol and inhibin which, in turn, permits the emergence of a new follicular wave [50]. Progesterone exposure will also restore responsiveness of the surge center to estradiol-stimulated secretion of GnRH. A Florida study compared pregnancy rates following treatment with either the Ovsynch protocol or placement of a CIDR for 7 days with injection of PGF at time of removal and insemination based on observed estrus within 7 days after PGF [57]. There was no difference in pregnancy rates due to treatment protocol. While progesterone can be used as the sole treatment with moderate success most of the recently developed protocols use progesterone in combination with other hormonal treatments. In a comparative study, lactating dairy cows diagnosed with follicular cysts were randomly assigned to two treatments: (1) a single injection of GnRH at diagnosis (Day 0) and AI at estrus (AIE) within 21 days (GnRH group, $n = 70$), or (2) insertion of a CIDR device containing progesterone and an injection of GnRH on Day 0, $PGF_{2\alpha}$ injection at the time of CIDR removal on Day 7, GnRH injection on Day 9, and TAI 16 h after the GnRH injection (CIDR-based TAI group, $n = 65$). Conception rate after the CIDR-based TAI protocol (52.3%) was greater ($P < 0.05$) than that after AI following a single GnRH injection (26.9%)[58]. Experimental evidence also suggests a higher positive response in cow diagnosed with follicular cysts compared with those having luteal cysts [59].

REFERENCES

[1]Vanholder T, et al. Repro Nutr Devel 2006;46:105.[2]Day N. Vet Med 1991;86.[3]Fourichon C, et al. Therio 2000;53:1729.[4]Bartlett P C, et al. Prev Vet Med 1986;4:15.[5]Buoen LC, et al. J Am Vet Medl Asso 1981;179:808.[6]Carroll D, et al. Therio 1990;34:349.[7]Kesler DJ, et al. J Anim Sci 1982;55:1147.[8]López-Gatius F, et al. Therio 2002;58:1623.[9]Shrestha HK, et al. Therio 2004;61:1559.[10]Edmondson A, et al. An Repro Sci 1986;12:157.[11]Nakao T, et al. Am J Vet Res 983;44:888.[12]Jeffcoate I, et al. Vet Rec. 1995;136:406.[13]Garverick HA. J Dairy Sci 1997;80:995. [14]Kesler DJ, et al. Therio 1981;16:207.[15]Lopez-Diaz MC, et al. Therio 1992;37:1163.[16]Hooijer GA, et al. Vet Rec. 2001;149:383. [17]Hatler TB, et al. Bio Repro 2003;69:218.[18]Blache D, et al. J Neuroendo 1994;6:329.[19]Dobson H, et al. J Repro Fert 2000;120:405.[20]Laporte H, et al. Livestock Prod Sci 1994;38:191.[21]Heuer C, et al. J Dairy Sci 1999;82:295.[22]Lucy M. J Dairy Sci 2001;84:1277.[23]Gröhn Y, et al. An ReproSci 2000;60:605.[24]Nanda AS, et al. Br Vet Jo 1989;145:39.[25]Beam S, et al. J Repro Fert Suppl1998;54:411.[26]Butler WR, et al. J Dairy Sci 1989;72:767.[27]Hooijer GA, et al. Livestock Prod Sci 2003;81:25.[28]Zulu VC, et al. J Vet Med Sci/Japanese Soc Vet Sci 2002;64:879.[29]Sovani S, et al. In: Soc Vet Epid Prevent Med. 2000;137.[30]Davoren JB, et al. Endocrinology 1986;119:2155.[31]Vanholder T, et al. Repro Dom An 005;40:460. [32]Huszenicza G, et al. BSAS OCCASIONAL PUBLICATION 2001;2:405.[33]Mason TM, et al. Diabetes 1999;48:524.[34]Vanholder T, et al. An Repro Sci 2005;87:33.[35]Vanholder T, et al. An Repro Sci 2006;92:51.[36]Cairoli F, et al. Vet Res Com 2008;32:123.[37]Lopez FJ, et al. ndocrinology 1992;130:1986.[38]Zwald NR, et al. J Dairy Sci 2004;87:4287.[39]Hooijer GA, et al. J Dairy Sci 2001;84:286.[40]Wiltbank MC, et al. Therio 002;57:21. 41 Yoshioka K, et al. Res Vet Sci 1996;61:240.42 Cook D L, et al. J Repro Fert 1991;91:19.43 Ozturk M, et al. J Repro Fert 1998;114:1. 44 Gümen A, et al. Therio 2005;63:202.45 Gümen A, et al. Repro 2005;129:737.46 Gümen A, et al. J Dairy Sci 2002;85:43.47 Silvia W J, et al. Domestic Animal Endocrinology 2002;23:167.48 Stock A E, et al. Endocrinology 1993;132:1108.49 Hamilton S, et al. Biology of Reproduction 1995;53:890.50 Todoroki J, et al. The Journal of Reproduction and Development 2006;52:1.51 Imai K, et al. Domestic Animal Endocrinology 2003;24:171.52 Robker R L, et al. Steroids 2000;65:559.53 Bierschwal C J, et al. J. Anim Sci. 1975;41:1660.54 Kesler D J, et al. J. Anim Sci. 1978;46:719.55 Lopez-Gatius F, et al. Therio 2002;58:1337.56 Bartolome J A, et al. Therio 2000;53:815.57 Crane M B, et al. Therio 2006;65:1563.58 Kim I-H, et al. Animal Reproduction Science 2006;95:206.59 Amer H A M A B. Livestock Research for Rural Development 2008;20.

COLOSTRUM REPLACEMENT PRODUCTS FOR CALVES

Keith P Poulsen, DVM PhD, DACVIM
Corvallis, Oregon

Due to the cotyledonary placenta, the bovine fetus does not receive immunglobulins in utero. Calves are born immunocompetent, yet immuno-naïve, and require ingestion of immunoglobulin in colostrum for passive transfer of maternally derived antibodies. Failure of passive transfer (FPT) has long been known to be a risk for increased morbidity and mortality for large animal species. For calves, FPT has also been linked to decreased production, which can diminish their lifespan and economic output. The solution to minimize risk of FPT seems simple by feeding adequate amounts of clean, good quality colostrum to a calf. However, in reality, there are many critical control points in the standard operating procedures on a farm or ranch that can lead to a calf not receiving adequate amounts of IgG for effective immunity against environmental and infectious microbial pathogens. The estimated prevalence of FPT on US dairies has been reported to be 19.2% in 2009, which is a drastic improvement from 40% in 1993.[1,2] In 2009, from the 2007 USDA National Animal Health Monitoring System (NAHMS) survey data, the greatest increase in odds of FPT were from: lack of monitoring of serum proteins (x13.8), >4 hours of life when receiving colostrum (x2.7), dystocia without assistance of a veterinarian (x2.6), allowing the calf (dairy) to nurse (x2.4), feeding pooled colostrum (x2.2), and lack of additional heat for calves experiencing dystocia in cold weather (x1.6).[1] Lack of supply of clean, good quality colostrum can be due to multiple reasons as commonly seen on farms with Johne's disease control programs or decreased availability of labor during the 2nd and 3rd shifts. To fill the gap on these farms to minimize FPT, colostrum replacement products have been commercially available for approximately the past 20 years. This talk will focus on the types of colostrum replacement products available and the published data reviewing their appropriate use on farms.

Several different attempts have been made to supplement calves with IgG for passive transfer. Dr. Jim Quigley is, in the author's opinion, the "father" of colostrum supplements and has been quite active in development of interventions to bolster the calf's immune system to improve passive transfer and pre-weaning health. The first studies of adding supplements to colostrum date back to 1993, but supplementation of maternal colostrum has not been shown to increase calf [IgG].[3] Importantly, adding colostrum supplements directly to maternal colostrum could decrease efficiency of absorption due to proteins other than IgG.[4] Since the original work, the source of IgG in available colostrum replacement products available are based on spray dried bovine serum or freeze dried bovine colostrum. Multiple products are available on the market and can be sold under several different brand names depending on the licensing in any particular region of North America. Two key points are critical for veterinarians and calf raisers when choosing a particular product: 1. IgG mass (look for 150g or more) and 2. Published, independent study of absorption efficiency.[5] Initial reports concluded that 100g of IgG was sufficient to reach the accepted threshold of PT of 10.0 mg/mL, but later reports suggest this threshold should be higher (13.4 mg/ml), requiring larger doses of IgG (150-200 g). For reference, 1 gallon of good quality maternal colostrum should have a minimum of 50 g/L or 190 g of IgG (reviewed in 6). The method of delivery of colostrum replacement products, as with maternal colostrum, via bottle or esophageal feeder does not significantly alter passive transfer.[7,8] Also important to note is that serum total protein measurements may be lower for colostrum replacement calves compared to calves fed maternal colostrum and this is dependent on the colostrum replacement product used. Gold standard testing for IgG in calves remains to be radial immunodiffusion (RID).

The preponderance of published work has shown that colostrum replacement products are effective to provide passive transfer, but effectiveness is dependent on the product as a whole, not the ingredients on the label. However, little work has been done to look at calf health and production later in life. In the pre-weaning period, Priestly et al. concluded that calves fed maternal colostrum had lower morbidity and mortality with higher weaning and daily gain compared to calves fed plasma or colostrum derived colostrum replacement products.[9] In the long-term, Pithua et al. focused on one of the most important indications for use of colostrum replacement products, Johne's Disease control, and showed use of a colostrum replacement product to decrease disease incidence, compared to calves fed maternal colostrum.[10] In general, more data is needed to see if the initial findings by Priestly et al. and Pithua et al. hold up to farm and product variability. This may prove to be very difficult as multiple variables are hard to control, similar to how it has been difficult to link FPT with later production in the milking herd.[11]

Overall, opting for colostrum replacement products compared to maternal colostrum is dependent on 1. The available product and 2. The specific needs of the farm. Remember that no product will overcome poor management of a farm or ranch. Regular veterinary consultation and presence on a farm is key to implement strategic colostrum replacement use on farm to minimize FPT and optimize calf health.

REFERENCES

[1]Beam AL, et al. J Dairy Sci 2009;92:3973. [2]USDA. #N118.0293. USDA-APHIS-VS, CEAH Fort Collins, CO 1993. [3]Abel Francisco SF et al. Am J Vet Res 1993;54:1051. [4]Quigley JD et al. J Dairy Sci 1998;81:1936. [5]Smith GW et al. J Dairy Sci 2007;90:2905.[6]Godden SM et al. J Dairy Sci 2009;92:1750.[7]Godden SM et al. J Dairy Sci 2009;92:1758.[8]Kaske M et al. J Anim Physiol Anim Nutr (Berl) 2005;89:151.[9]Priestley D et al. J Dairy Sci 2013;96:3247.[10]Pithua P et al. J Am Vet Med Assoc 2009;234:1167.[11]DeNise SK et al. J Dairy Sci 1989;72:552.

TREATMENT METHODS OF BOVINE COLOSTRUM TO IMPROVE QUALITY

Keith P Poulsen, DVM PhD, DACVIM

Oregon Veterinary Medical Association, March 2014
Corvallis, Oregon

The solution to minimize risk of failure of passive transfer (FPT) seems simple by feeding adequate amounts of clean, good quality colostrum to a calf. However, in reality, there are many critical control points in the standard operating procedures on a farm or ranch that can lead to a calf receiving poor quality or low quantities of colostrum. This creates problems for the calf because these breakdowns in colostrum feeding programs frequently lead to transfer of disease and the calf receiving inadequate amounts of IgG for effective immunity against environmental and infectious microbial pathogens. Those calves with [IgG] > 10 g/L have lower morbidity, mortality, and reach breeding weights faster compared to calves with [IgG] < 10 g/L.[1]

When troubleshooting problems with passive transfer or infectious disease, colostrum feeding programs can be broken down into five parts: 1. The cow, 2. Collection equipment, 3. Storage systems, 4. Delivery methods, and 5. The calf. Colostrum can be contaminated from the source with infectious diseases shed by the dam in the colostrum including Paratuberculosis, *Salmonella* Dublin, and *Mycoplasma bovis*. Colostrum can also be grossly contaminated with bacteria in cases of mastitis or improperly cleaned collection, storage, and delivery equipment. Storage of colostrum, or significant delay from collection to storage, frequently results in log-phase growth of bacteria, which has been suggested to be a risk for high rates of FPT.[2] Finally, management of newborn calves can affect FPT rates because dystocia, cold weather, and prolonged time to colostrum feeding can decrease intestinal IgG absorbtion.[3] Logistical limitations on farms may make microbial contamination of colostrum, for whatever reason, unavoidable. The focus of this talk will discuss multiple methods, specifically heat pasteurization, to improve microbial quality of colostrum and the published research behind their use.

The Pasteurized Milk Ordinance dates back to 1924, resulting in foodborne disease from milk to drop from 25% of all food and waterborne illness to less than 1%. Chicago was the first city to require all milk for human consumption to be pasteurized in 1908 and Michigan was the first state to require pasteurization, starting in 1947. Pasteurizing waste milk for feeding calves started gaining attention in the 1980's by manufacturer of on-farm processing systems and Jamaluddin et al. estimated to be $8.13 gain per calf by pasteurizing milk *and* colostrum in 1996. Jamaluudin et al. also estimated feasibility to be at a minimum of 1,260 cows with 315 calves on milk.[4] This study used High Temperature Short Time (HTST -75° for 2 min) and resulted in lower [IgG] concentration, but similar FPT rates, and a major negative side effect of HTST, congealing of colostrum as it cools to form a thick pudding that is virtually impossible to feed to calves. Because of this, the industry has turned to batch pasteurization (60°C for 60 minutes), which has minimal effects on viscosity while significantly decreasing microbial load. [IgG] concentrations to decline from pasteurization, but with a single 4 L feeding, calves have the same rate of FPT compared to calves fed unprocessed colostrum.[5-7] The decrease in [IgG] is dependent on how much globulin is in the pre-pasteurization sample.[8] However, the 4 L meal should make up for the loss, similar to why 1st and 2nd lactation colostrum can be fed to calves despite lower [IgG] compared to 3rd lactation colostrum. For this same treatment protocol (60°C for 60 minutes), calves fed heat-treated colostrum had greater serum [IgG] compared to control, but no significant differences in other globulin types or nutritional markers. Elizondo-Salazar and Heinrichs showed that calves fed batch pasteurized colostrum (60°C for 30 minutes) to have slightly higher serum [IgG] and apparent efficacy of absorption, with no differences in growth, calf starter intake, health scores, and blood lymphocyte counts until 6 months of age.[9] In another study, Elizondo-Salazar and Heinrichs found heat-treated colostrum-fed calves to have higher serum [total protein] and [IgG] compared to control and calves fed high-bacteria colostrum.[10] In terms of infectious pathogens, initial findings are promising but more work is needed. Risk of Paratuberculosis transmission has been shown to be decreased, but not eliminated in calves in a study (with limited numbers) by Stabel and Godden et al. showed variability in recovery of the fastidious organism.[11,12] However, in the same study, Godden et al. also showed that other pathogens of interest, including *Mycoplasma bovis*, to be eliminated with batch heat treatment (60°C for 30 minutes).

Heat treatment of colostrum is the most widely studied method to improve quality, adapting technology from the human food industry. Other technologies exist and could prove to be effective and more efficient than batch heat treatment, but are not currently commercially available. Microwaving and gamma-irradiation of milk has some interest, but has slowed by logistical issues and considerable political opposition. Ultraviolet light and high pressure have also shown promise as potential alternatives, but are at this time, not as effective as heat treatment. Phenothyiazine dyes (methylene blue) have also been used to inactivate bovine viral diarrhea virus in goat colostrum, but have not been tested in bovine colostrum.

Overall, batch pasteurization has high potential to improve colostrum quality on the farm with continual problems with high microbial load. There is some training and additional overhead required, but gains in calf health can outweigh costs. Future work should focus on specific pathogen elimination while maintaining globulin and non-specific immune factor concentrations intact.

REFERENCES

[1]Furman-Fratczak K et al. J Dairy Sci 2011;94:5536.[2]Poulsen KP et al. J Vet Intern Med 2002;16:339.[3]Beam AL et al. J Dairy Sci 2009;92:3973.[4]Jamaluddin AA et al. J Am Vet Med Assoc 1996;209:751.[5]Godden SM et al. J Dairy Sci 2003;86:1503.[6]Elizondo-Salazar JA et al. J Dairy Sci 2010;93:961.[7]McMartin S et al. J Dairy Sci 2006;89:2110.[8]Donahue M et al. J Dairy Sci 2012;95:2697.[9]Elizondo-Salazar JA et al. J Dairy Sci 2009;92:3265.[10]Elizondo-Salazar JA et al. J Dairy Sci 2009;92:4565.[11]Stabel JR. J Dairy Sci 2008;91:3600.[12]Godden SM et al. J Dairy Sci 2006;89:3476.

PROCOAGULANT MICROPARTICLES IN DOGS WITH IMMUNE-MEDIATED HEMOLYTIC ANEMIA

Linda Kidd, DVM, PhD, DACVIM; Michael Sueda, DVM, DACVIM,
Tim Concannon, DVM, DVECCS; Nigel Mackman, PhD, FAHA
Pomona, CA

Immune-mediated hemolytic anemia (IMHA) is an important cause of morbidity and mortality in dogs.[1] Hospitalization stays are lengthy, and long term treatment is required for dogs who survive. Thromboembolic disease is a major factor that affects short and long-term survival.[1] Despite aggressive therapy that often includes thromboprophylaxis, the mortality rate for idiopathic IMHA in dogs is high.[1] Establishing more effective thromboprophylactic protocols is necessary if survival rates in dogs with IMHA are to be improved.

In addition to determining which treatments are most effective, identifying patients at risk for thromboembolism is also important because thromboprophylactic therapy may be expensive, require extensive monitoring, and be associated with side effects. Guidelines for appropriate thromboprophylaxis have not been determined for dogs with IMHA. Heparin, ultra-low dose aspirin and clopidogrel have all been used.[1,2] Due to a lack of controlled clinical trials and a high likelihood of inadequate dosing of heparin and possibly aspirin in individual dogs, it is not known if any of these agents is effective or if one is superior to another.[1] Recent studies using individually adjusted doses (IAD) of unfractionated heparin based on patient anti-factor Xa activity are encouraging.[2] However, IAD heparin requires extensive monitoring which can be expensive and time consuming. No reliable clinical biomarkers have been identified that predict thrombosis and help clinicians decide which dogs should receive anticoagulant therapy. Understanding mechanisms of thrombosis may help identify clinical markers that allow veterinarians to identify which dogs would benefit from thromboprophylaxis, provide a pathophysiologic basis to help choose promising thromboprophylactic agents for clinical trials, and lead to new therapeutic options.

Recent studies using thromboelastography have shown that the majority of dogs with IMHA are in a generalized hypercoagulable state.[1] Dogs with IMHA exhibit excessive platelet activation and activation of coagulation.[1,3] Tissue factor (TF) initiates coagulation during normal hemostasis. TF is not normally expressed on quiescent vascular endothelial cells and circulating blood cells, but rather, is expressed on perivascular cells such that damage to vessels initiates coagulation. Subsequent thrombin production results in platelet activation and further amplification and propagation of coagulation.

TF expression is induced on the surface of activated monocytes and endothelial cells. Studies strongly suggest that TF expression is a major instigator of pathologic thrombus formation in inflammatory states such as disseminated intravascular coagulation (DIC) and sepsis. Proinflammatory cytokines, heme, CD40 ligand and ischemia/reperfusion induce TF expression.[3,5] Because IMHA in dogs is also associated with an inflammatory state, platelet activation and hypoxia, TF may be an important instigator of coagulation in these patients. Indeed a recent study found that dogs with IMHA had increased levels of TF messenger RNA (mRNA) expression in peripheral blood.[6] However, TF mRNA expression levels do not always correlate with protein expression and activity. Whether TF activity is increased in the blood of dogs with IMHA has not been determined.

Circulating microparticles (MPs) are derived from cell membranes of RBCs, platelets, neutrophils, endothelial cells, and monocytes.[7,8] They express cell surface and other molecules derived from their cell of origin, and are able to interact with, and induce cell signaling and gene expression in other cell types. Phostidylserine (PS) exposing MPs are procoagulant. Normal membranes have an asymmetric distribution of phospholipids, with anionic phospholipids, such as PS located in the inner leaflet. Loss of asymmetry and exposure of PS on the outer surface occurs after activation of platelets.[9] Anionic phospholipids facilitate assembly of the prothrombinase and tenase complexes of the coagulation cascade.[9] Activated platelets release PS exposing MPs.[10] Importantly, defects in platelet exposure of PS and MP formation result in bleeding tendencies in people and dogs.[9,11] A recent study showed that platelet MPs were found to be elevated in association with platelet activation in dogs with IMHA.[4] Whether these MPs express PS was not investigated.

As is the case for platelet activation, senescence of red blood cells results in loss of asymmetry and exposure of PS on the outer RBC membrane surface.[9] PS-positive RBCs bind to macrophage PS receptors. PS receptor mediated binding induces signaling that results in phagocytosis. Activated platelets, damaged RBC and their associated MPs are pro-coagulant because they have anionic phospholipids, such as PS, on their surfaces.[8,12] Recent in vitro studies have shown that PS receptors are constitutively expressed on the surface of endothelial cells.[13] Erythrocyte PS also binds to thrombospondin, a protein that is present in sub-endothelial matrix.[14] Endothelial cell PS receptor expression is up-regulated in response to inflammatory cytokines, hypoxia and free hemoglobin. Furthermore, RBC MPs have been shown to scavenge NO and impair endothelial dependent vasodilation.[15] Thus, PS-positive MPs may contribute to coagulation and thrombosis in IMHA by providing surfaces for the tenase and prothrombinase coagulation complexes to bind, by localizing their pro-coagulant membranes to inflamed endothelium, and by inhibiting vasodilation. Increased PS expression on the surface of RBC occurs in human hemolytic diseases associated with thrombosis, including sickle cell anemia, beta thalassemia and hereditary spherocytosis.[9] The risk of thrombosis correlates with the magnitude of PS expression on the surface of RBC in humans with sickle cell anemia.[16] Plasma levels of MPs are increased and associated with thrombosis in human patients in sickle cell crisis, and RBC MP number correlates inversely with hemoglobin levels in these patients.[8,17] Triggers for increased RBC microvesiculation include complement attack, sheer stress and oxidative injury, conditions that are also present in dogs with IMHA.[1,8]

In addition to the procoagulant role of PS expression, co-expression of TF on the surface of MPs can further activate endothelium and the extrinsic coagulation pathway.[7,8] MPs derived from monocytes and endothelial cells can express TF. Many recent studies suggest that TF expression on the surface of MPs may play an important role in pathologic thrombus formation.[7,8] For example, TF bearing MPs (TFMPs) are increased and associated with thrombosis in human patients with sickle cell crisis.[10,17,18] TFMPs have been associated with markers of activation of coagulation in some, but not other studies of human patients with sickle cell disease.[19,20,17] Thus the role of TFMPs in the pathogenesis of thrombosis in sickle cell disease in people requires further investigation. Elevated levels of TFMPs occur in other prothrombotic hemolytic diseases in people, including hemolytic uremic syndrome, and paroxysmal nocturnal hemoglobinuria.[21,22] Whether elevated levels of TF exposing MP occurs in human auto-immune hemolytic anemia has not been investigated.

In addition to increased production of MPs due to inflammation and hemolysis, therapeutic interventions could increase circulating MP levels in dogs with IMHA. It is thought that MPs are cleared by phagocytic cells in the spleen that contain receptors for PS or opsonins on the MP surface, such as complement.[8] It is therefore possible that immunosuppressive and immunomodulatory therapy directed at inhibition of phagocytosis such as hIVIG and splenectomy might increase levels of circulating MPs and contribute indirectly to the prothrombotic state in dogs with IMHA. Some therapies may directly increase circulating MP as well. For example, cyclosporine was recently shown to induce complement activating endothelial cell MPs in vitro, and endothelial cell MP's increased in human renal transplant patients treated with tacrolimus.[23] Other therapies such as transfusion of PRBC, may directly increase circulating microparticles and contribute to thrombotic risk. A recent study revealed marked formation of PS-expressing MPs during storage beyond seven days in canine red blood cell concentrates.[24]

Because evidence strongly suggests that MPs play an important role in pathologic thrombus formation, therapies that target MPs directly or decrease MP formation are being investigated as thromboprophylactic agents in people.[7,8,25,26]

The evidence reviewed here suggests that measurement of MPs may have clinical utility as a diagnostic tool and therapeutic targets in preventing thrombosis and death in dogs with IMHA. The purpose of this pilot study was to determine if circulating levels of PS+ and TF+MPs are elevated in dogs with IMHA compared to healthy dogs. Results will be presented. Additional studies investigating the cellular origin and composition of circulating MPs, and whether MP subsets are associated with markers of activation of coagulation or increased risk of thrombosis in dogs with IMHA are warranted.

REFERENCES

1. Kidd L, Mackman N. J Vet Emerg Crit Care. 2013;23(1):3.
2. Helmond SE et al. J Vet Intern Med. 2010; 24(3): 597.
3. Weiss DJ, Brazzell JL. J Vet Intern Med. 2006; 20(3): 682.
4. Ridyard et al J Small Anim Pract. 2010; 51(6):296.
5. Kjelgaard-Hansen M, et al. J Vet Intern Med. 2011; 25(1): 76.
6. Piek CJ, et al. Vet Immunol Immunopathol. 2011; 144: 346.
7. Mackman N. Circ Res. 2009; 104(8): 925.
8. Burnier L, et al. Thromb Haemost. 2009; 101(3): 439.
9. Zwaal RF, et al Cell Mol Life Sci. 2005; 62(9): 971.
10. Owens AP, Mackman N. Circ Res. 2011; 108(10):1284.
11. Brooks MB, et al. Blood. 2002; 99(7): 2434.
12. Ataga KI, Haematologica. 2009; 94(11): 1481.
13. Setty BN, Betal SG. Blood. 2008; 111(2): 905.
14. Gayen Betal S, Setty BN. Transl Res. 2008; 152(4): 165.
15. Liu et al. Free Radic Biol Med 2013; 65 1164.
16. Setty BN, et al Blood. 2002; 99(5): 1564.
17. Shet AS, et al Blood. 2003; 102(7): 2678.
18. Kasar et al J Thromb Thrombolysis 2013epub.
19. Ataga et al. PLOS One 2012; 7(1) e29786.
20. Setty et al Br J Haematol. 2012; 157(3) 370.
21. Ståhl AL, PLoS One. 2009; Sep 11;4(9):e6990.
22. Liebman et al Thromb Res. 2003; 111 235.
23. Renner et al J Am Soc Nephrol. 2013; Nov;24(11):1849.
24. Herring et al Vet Clin Pathol. 2013; Jun;42(2):163.
25. Adiguzel C, et al. Clin Appl Thromb Hemost. 2009; 15(5): 501.
26. Rosenblade et al. J Pharm Pharmaceut Sci 2013; 16(2) 238.

CRITICAL ILLNESS RELATED CORTISOL INSUFFICIENCY IN NEONATAL FOALS

Michelle Henry Barton DVM, PhD, DACVIM (LAIM) and Kelsey A. Hart DVM, PhD, DACVIM (LAIM)
Athens, GA

INTRODUCTION

The body's primary adaptive response to the homeostatic and metabolic disturbances produced in critical illness includes activation of the hypothalamic-pituitary-adrenal (HPA) axis. This results in secretion of corticotropin-releasing hormone by the hypothalamus, and the subsequent release of adrenocorticotropic hormone (ACTH) by the anterior pituitary gland. ACTH, in turn, enters the systemic circulation and stimulates the adrenal cortices to synthesize and secrete cortisol. Cortisol exerts a variety of systemic effects that are vital for adaptation to the physiologic stresses associated with severe illness, including maintenance of adequate blood pressure, provision of nutrients to tissues, and control of an appropriate inflammatory response.[1,2] These effects ultimately serve to modulate existing physiologic stressors and restore homeostatic balance.

Despite its vital role in the physiologic response to the stress of disease, dysfunction of the HPA axis is well documented in critically ill humans; this is particularly true in patients with sepsis, septic shock, and systemic inflammatory response syndrome (SIRS).[3-7] This may be manifested by absolute adrenal insufficiency due to complete and permanent destruction of the adrenal glands by the primary illness, though this is relatively uncommon in both human and veterinary medicine.[7,8] Less severe and transient relative adrenal insufficiency, however, is much more common during critical illness; this is typically defined as an *inadequate* cortisol response for the existing degree of illness.[8] As a result, the patient is rendered incapable of mounting an appropriate and sufficient cortisol response to cope with the increased physiologic demands of acute severe disease and clinically was coined critical illness related cortisol insufficiency or CIRCI.[3,4,7,8] Because of the important immunomodulatory and anti-inflammatory effects of cortisol, HPA axis dysfunction during critical illness can ultimately result in immunoendocrine dysregulation, allowing an initially appropriate inflammatory response to become dangerously exaggerated.[9] Disruption or dysregulation of the immune/inflammatory response due to HPA axis failure may result in imbalance between pro- and anti-inflammatory cytokines, with two important consequences: impairment of the body's ability to combat infection, and/or overwhelming inflammation that damages host tissues and organs.[10] In addition, because cortisol plays a vital role in the maintenance of blood pressure, HPA axis dysfunction can result in significant hypotension. This overwhelming inflammatory response and systemic hypotension is most often clinically manifested as septic shock, SIRS or multiple organ dysfunction syndrome (MODS). In addition, an overzealous and inappropriate inflammatory response can result in continued and additional HPA axis suppression, as described above. This further perpetuates inadequate tissue perfusion and impairment of the host's ability to clear the infectious organism, often culminating in the death of the patient.[3,10]

SIGNIFICANCE OF HPA AXIS DYSFUNCTION IN CRITICAL ILLNESS

An extensive body of evidence in the human literature clearly links HPA axis dysfunction with an increased incidence of shock, MODS, SIRS, and death in critically ill patients.[7,11] The potential importance of adrenal insufficiency in critically ill patients is further underscored by the results of two comprehensive meta-analyses evaluating the effect of low (physiologic or replacement) dose corticosteroid supplementation in patients with septic shock.[10,12] These reviews, as well the results of several smaller, single-center trials, showed a significant beneficial effect of low-dose corticosteroid supplementation in critically ill patients with CIRCI.[3,10,12] In these studies, patients meeting criteria for CIRCI that were supplemented with low (physiologic) doses of corticosteroids showed more rapid shock reversal, earlier cessation of vasopressor therapy, shorter ICU stays, more rapid ventilator weaning, and improved survival than patients with CIRCI that did not receive steroid supplementation.[3,10,12,13] Thus, the detection of HPA axis dysfunction is clearly an important diagnostic consideration in human critical care, and provides a means by which to identify patients that may benefit from corticosteroid replacement therapy.

Despite the importance of HPA axis dysfunction in critical illness, definitive diagnosis of CIRCI can be problematic. Accurate assessment of an adequate cortisol response to severe disease in each individual patient is inherently difficult, due to wide individual variations in systemic cortisol levels during the course of disease and the many host, pathogen, and environmental factors that play a role in the physiologic response to infection. Current suggestions for a diagnosis of HPA axis dysfunction in septic human patients is based on one or more of the following findings: 1) basal serum cortisol concentrations less than those commonly induced by stress in health [3,4] 2) a blunted cortisol response (delta cortisol) to administration of a high (supraphysiologic) dose of synthetic ACTH (125-250 µg) in the classic high dose ACTH stimulation test;[4,14] and/or 3) a blunted response to administration of a more "physiologic" low dose (1 µg) of synthetic ACTH.[3,15] Other methods for evaluating HPA axis function include measurement of serum free cortisol levels (the biologically active fraction of cortisol), the insulin tolerance test, and the metapyrone test.[14] The later tests require specialized and tedious methodologies that are not used commonly in clinical settings.

EVIDENCE FOR CIRCI IN NEONATAL FOALS

Documentation of CIRCI in the equine literature is limited, but may occur with significant frequency in some populations of critically ill horses. Neonatal foals are likely at the greatest risk for HPA axis dysfunction and CIRCI during illness, for several reasons. First, compelling evidence over the last 30 years has shown that maturation of the HPA axis in the neonatal foal occurs very late in the foal's development, beginning during the last 4 to 5 days prior to parturition and continuing into the first two weeks of life.[15-17] This HPA axis immaturity is certain to impact the foal's ability to respond to physiologic stress and disease in the neonatal period, and may make the HPA axis more vulnerable to dysfunction during periods of critical illness. Furthermore, the leading cause of morbidity and mortality in neonatal foals, both under field conditions and in tertiary referral centers, is bacterial sepsis; as described above, sepsis is the illness most commonly associated with concurrent HPA axis dysfunction in human critical care.[2,6]

The potential incidence and significance of HPA axis dysfunction in critically ill neonatal foals was further investigated in our laboratory through a 2008 grant from the ACVIM Foundation. HPA axis function was assessed in 71 neonatal foals < 7 days of age at admission to two tertiary referral centers, and correlated with indicators of disease severity and outcome.[18] Basal plasma endogenous ACTH and total serum cortisol concentrations were measured using chemiluminescent immunoassays in 71 foals, and a paired low-dose (1 μg) / high-dose (100 μg) ACTH stimulation test was performed in 59 of these foals. The paired low-dose/high ACTH stimulation protocol was based on previous synthetic ACTH dose-response assessment in healthy neonatal foals,[19] and evaluated in a healthy, age-matched foals to determine appropriate cortisol responses to exogenous ACTH in foals during the neonatal period.[20] HPA axis dysfunction was defined using similar criteria as described in the human literature, as: 1) an inappropriately low basal serum cortisol concentration, and/or 2) an inadequate increase in serum cortisol concentration (delta cortisol) after administration of synthetic ACTH. Specifically, an inappropriately low basal cortisol concentration was defined as a basal cortisol concentration less than the lowest cortisol concentration (mean – 1 standard deviation) achieved after administration of a physiologic dose (10 μg) of synthetic ACTH to healthy age-matched foals. An inadequate delta cortisol was defined as a delta cortisol less than the mean delta cortisol achieved in healthy age-matched foals using the same synthetic ACTH stimulation protocol.

Using these two criteria independently, HPA axis dysfunction was diagnosed in 49% and 51% of foals, respectively. Seventy-six percent of foals met criteria for sepsis (sepsis score $\geq$ 11 and/or positive blood culture); HPA axis dysfunction was diagnosed in 37% and 41% of septic foals, respectively, with the above criteria. As is reported in critically ill human beings, dogs, and adult horses, increased baseline cortisol concentrations were significantly correlated with disease severity and non-survival (death or euthanasia for prognostic reasons during hospitalization) when all foals were considered. In the septic foals, low delta cortisol was also significantly correlated with non-survival. In addition, 90% of the septic foals with an increased basal serum cortisol concentration and a low delta cortisol did not survive, as compared to a non-survival rate of only 9% in foals with a high basal cortisol concentration and an appropriate delta cortisol. This finding suggests that exhaustion of cortisol synthetic capacity ("loss of adrenal reserve") may contribute to HPA dysfunction in septic foals. A significantly higher incidence of MODS and shock was also observed in septic foals with a low delta cortisol than in septic foals with an appropriate delta cortisol. A low-dose (10 μg ACTH) delta cortisol cutoff of at least 3.2 μg/dl and a high-dose (100 μg ACTH) delta cortisol cutoff of at least 5.1 μg/dl were determined to be the most accurate predictors of survival to discharge, while a baseline cortisol concentration cutoff of at least 6.7 μg/dl was most accurate for predicting the occurrence of shock and MODS.

SUBSEQUENT WORK IN FOALS FOLLOWING THE ACVIM FOUNDATION GRANT

Collectively, the above evidence strongly suggests that HPA axis dysfunction occurs in septic foals with significant prevalence, similar to the approximate 50% prevalence reported in septic adult humans and infants.[4,11] In addition, as is reported in people,[4,11] septic foals with evidence of HPA axis dysfunction are more likely to have shock, multiple organ failure, and are less likely to survive to discharge than septic foals with intact HPA axis function. Although septic neonatal foals with CIRCI may potentially benefit from cortisol replacement therapy in the form of low-dose hydrocortisone, such treatment has not been critically evaluated in foals. Given the marked differences in HPA-axis function and cortisol dynamics in foals, compared with humans, hydrocortisone replacement regimens used in people may not be directly applicable to foals.

In order to determine a physiologically appropriate low-dose hydrocortisone regimen for potential use in foals, we measured daily endogenous cortisol production and hydrocortisone pharmacokinetics in healthy horses and foals.[21] Both adult horses and foals exhibited pulsatile cortisol secretion over a 24 hour period, but a circadian rhythm was only evident in horses. Mean 24-hour cortisol concentration was slightly but significantly lower in foals (20 ± 1 ng/mL) than adult horses (26 ± ng/mL), despite a significantly higher daily endogenous cortisol production rate in foals. This is likely due to substantially increased hydrocortisone/cortisol clearance rates in foals, presumably associated with much higher percentages of free cortisol in foals as compared to adult horses. These data suggest that approximately 1.3 mg/kg/day, or two times the daily endogenous cortisol production rate in healthy foals, might be an appropriate hydrocortisone dose in septic foals with CIRCI. Division of this daily dose into q 4 hour IV boluses would best mimic cortisol secretion patterns in healthy foals.

Corticosteroid-related immunosuppression is an important concern when considering the use of steroids in septic patients with CIRCI. In healthy 2-to-6-day-old foals, a 3.5 day tapering course of hydrocortisone at this initial dose of 1.3 mg/kg/day IV divided q 4 hours has potentially beneficial anti-inflammatory effects without significantly impairing innate immune function.[22] Specifically, isolated leukocytes from foals treated with this low-dose hydrocortisone regimen had significantly decreased expression of the pro-inflammatory cytokines TNF-α, IL-6, and IL-1β after *ex vivo* exposure to LPS than leukocytes from untreated foals. Oxidative burst activity and phagocytic function were not significantly different between treated and untreated foals. Further study is needed, however, to confirm these effects *in vivo* and in septic foals in a clinical setting.

REFERENCES

1. Gibson S, et al. **Emer Med Clin N Am** 2005;23:909. 2. Marik P, Zaloga G. **Chest** 2002; 122:1784. 3. Aneja R, Carcillo J, et al. **Arch Dis Child** 2007;92:165. 4. Annane D, et al. **Am J Respir Crit Care Med** 2006; 174:1319. 5. Annane D, et al. **J Amer Med Assoc** 2000; 283:1038. 6. Cooper M, Steward P. **New Eng J Med** 2003; 328:727. 7. Loisa P, et al . **Acta Anaesthesologica Scandanivia** 2002; 46:145. 8. Marik P, Zaloga G. **Crit Care Med** 2003; 31:141. 9. Marik P. **Curr Opin Crit Care** 2007; 13:363. 10. Keh D, Sprung C. **Crit Care Med** 2005; 32:S527. 11. Pizarro C, et al. **Crit Care Med** 2005; 33:855. 12. Minneci P, et al. **Ann Intern Med** 2004; 141:47. 13. Huang C., Hong-Chyuan L. **Am J Respir Crit Care** 2006; 173:276. 14. Asare K. **Pharmacotherapy** 2007; 27:1512. 15. Nathanielsz P, et al. **J Reprod Fertil Suppl** 1975; 23:625. 16. Rossdale P, Ousey J. **Equine Vet J** 1984; 16:300. 17. Silver M, et al. **Equine Vet J** 1984; 16:278 18. Hart K, et al. **J Vet Int Med** 2009; 23:901. 19. Hart K, et al. **J Vet Int Med** 2007; 21:314. 20. Hart K et al. **J Vet Int Med 2009;** 23:344. 21. Hart K, et al. **Am J Vet Res** 2012; 73:408. 22. Hart KA, et al. **Pediatr Res** 2011; 70:72.

MATRIX METALLOPROTEINASE EXPRESSION AND THE IMMUNE RESPONSE IN CANINE INTRACRANIAL TUMORS

Christopher L. Mariani, DVM, PhD, DACVIM (Neurology)
Raleigh, NC

INTRODUCTION

Brain tumors are common in the dog, occurring at a frequency greater than other domestic species and humans.[1] Although a variety of primary tumors have been described, the most common brain tumors in the dog are meningiomas, which arise from the arachnoid layer of the meninges, the soft tissue layers overlying the central nervous system. Gliomas, which arise from oligodendroglia (cells responsible for myelination in the central nervous system) or astrocytes (cells with a multitude of functions including maintenance of extracellular ion balance, augmentation of neurotransmission, composition of the blood-brain barrier, and tissue repair after injury) are also common. Options for definitive therapy of intracranial neoplasms in canine patients include surgical removal of the tumor, radiation therapy and chemotherapy.

Chemotherapy remains an unproven modality for brain tumors in veterinary patients. Despite often having a relatively benign cellular appearance on histopathological examination, many meningiomas in the dog are biologically aggressive, displaying invasion into adjacent brain parenchyma.[2,3] At surgery, an obvious demarcation between the tumor and normal tissue is frequently lacking, and recurrence often occurs after surgical therapy alone.[4] Because of this, most veterinary neurologists and oncologists recommend adjunctive radiation therapy, which appears to improve survival time, but is still not curative, and dogs almost inevitably succumb to their tumors.[4] The reason for this aggressive behavior in canine meningiomas remains unclear. Several groups have examined the proliferative index of these tumors, and although some tumors have shown high proliferation rates, the majority have been compatible with benign, slow-growing tumors.[5] As the aggressive behavior of these tumors is associated with invasion of adjacent brain parenchyma, one likely contributing mechanism is the secretion of enzymes that break down extracellular matrix, allowing subsequent tumor cell invasion. Several enzymes have been associated with this activity in both human and canine cancers, including matrix metalloproteinases (MMPs) and cathepsins.[6] Gliomas in dogs display similar invasion into surrounding brain tissue and are also incurable, even with aggressive surgical, chemotherapeutic or radiotherapeutic intervention.

Major improvements in brain tumor outcomes in dogs are likely to require novel therapies and different interventional strategies. One such strategy that may be worthy of consideration is interference with the activity of enzymes that may be promoting the invasion of tumor cells into adjacent brain tissue. A second strategy, considered by some to be the "holy grail" of tumor therapy, is immunotherapy, or manipulation of the immune system in order to effect tumor cell eradication. The huge promise of immunotherapy is tied to the immune system's ability to seek out and eliminate cells in remote areas that are inaccessible to conventional therapy; in this case, those cells invading the normal brain parenchyma.

INVASION PROTEINS WITHIN TUMOR TISSUE

Matrix Metalloproteinases

The MMPs are a family of zinc-dependent endopeptidases that degrade various components of the extracellular matrix. Both membrane-associated and secreted forms of these enzymes have been described, and they are typically released as inactive pro-enzymes, requiring proteolytic activation after interaction with other proteinases. There are over 20 known MMPs, with a wide variety of different substrate specificities.[7] The gelatinases MMP-2 (gelatinase A) and MMP-9 (gelatinase B) degrade denatured collagen and type IV collagen, as well as other extracellular matrix components such as laminin, vitronectin and fibronectin.[8] These two enzymes are expressed by a number of different human and canine tumors. Expression of MMP-2 and MMP-9 has been reported in malignant gliomas and meningiomas from human patients, and has been associated with aggressive behavior.[6,9] Previous studies have correlated expression of MMP-2 and MMP-9 with recurrence[10,11] or invasion[12,13] of benign meningiomas or with higher meningioma grade.[10,13]

Latent MMP-2 is constitutively expressed in nervous system tissue, but active forms of this enzyme and both forms of MMP-9 are not detectable in healthy CNS tissue. The Comparative Neuroimmunology and Neurooncology Laboratory (CNNL) at North Carolina State University has evaluated brain tumor tissue from a number of canine patients using gelatin zymography and immunohistochemistry. As expected, we found expression of latent MMP-2 in all samples. However, active MMP-2 was noted in a small proportion of tumors and MMP-9 was expressed in most tumors evaluated.

Cathepsins

Cathepsin B is one of a family of cysteine peptidases found within lysosomes of a variety of cell types.[6] These enzymes can degrade collagen, laminin and other extracellular matrix molecules, and have been implicated in promoting invasion and progression of a variety of human tumors, including meningiomas.[6,14,15] The role of cathepsin B in facilitating neoplastic invasion in canine tumors has been poorly studied to date. However preliminary studies from the CNNL suggest that cathepsin B is expressed by canine meningiomas, although its role in promoting the invasive properties of these tumors remains speculative at this time.

IMMUNE RESPONSES TO INTRACRANIAL NEOPLASMS

The information available in the literature on the immune response to canine intracranial neoplasms is very limited. To further investigate potential mechanisms of tumor invasion and as a prelude to considering potential immunotherapies for these tumors, we were interested in characterizing the immune cell populations within and surrounding brain tumors in dogs. We focused first on meningiomas, although we have evaluated a number of glial cell neoplasms as well. Canine meningiomas elicit a robust microglial response, with large numbers of these cells surrounding and infiltrating into the neoplasms.[16] CD18+ macrophages were found within the tumors and at the tumor-brain interfaces; some of the macrophages in this latter location expressed toll-like receptors (TLR) 4 and 9. CD3+ T cells were found in 22/23 meningiomas (95.7%), often in robust numbers, while B cells were much less common and noted in only 4/23 cases (17.4%). FoxP3+ lymphocytes, suspected to be regulatory T cells, were found in small numbers in 9/23 cases (39.1%).

The CNNL has also analyzed cytokine and chemokine concentrations within the serum and cerebrospinal fluid (CSF) of dogs with a variety of intracranial tumors.[17] Considering all tumors together, chemokines promoting immune cell infiltration (CXCL1-like, CCL2, IL-8) were increased in the CSF of dogs with tumors, while cytokines with anti-inflammatory effects (IL-4, IL-10) were decreased relative to controls. Concentrations of TGF-β2, a cytokine generally considered to be immunosuppressive in nature, were elevated in the CSF of tumor-bearing dogs.

TUMOR BIOMARKERS

Apart from demonstrating the presence of these proteolytic enzymes within tumor tissue, identification of these proteins in blood or CSF samples from tumor-bearing patients may provide clinically relevant diagnostic and prognostic information. Both MMPs and cathepsin B have been detected in the peripheral blood of patients with a variety of cancers, and have correlated with survival in some studies.[18,19] The presence of activated MMP-2 and MMP-9 in the CSF distinguished patients with intracranial tumors from those without in one human study,[20] and MMP-9 has been detected in the CSF of patients with malignant gliomas.[21] Cathepsin B concentrations were elevated in the CSF of human patients with leptomeningeal metastasis in one study.[22]

We have evaluated MMP activity in the CSF of dogs with various intracranial tumors using gelatin zymography.[23] Similar to humans, the CSF of normal dogs contains the latent form of MMP-2, but not active MMP-2, or either latent or active MMP-9.[24] Our control dogs similarly expressed only latent MMP-2, and active MMP-2 was not detected in any sample (tumor-bearing or control). However, whereas MMP-9 was not found in any control CSF, it was significantly more likely to be detected in the CSF of dogs with tumors. In addition, there seemed to be a distinct difference between tumor types, as dogs with lymphoma and choroid plexus tumors had detectable MMP-9 in a higher proportion of cases than those with meningiomas or gliomas.[23]

The CNNL has also evaluated the utility of CSF vascular endothelial growth factor (VEGF) as a diagnostic biomarker to distinguish intracranial neoplasms from other disease processes. We found that VEGF concentrations were higher in the CSF of dogs with tumors than in those with most other disease processes, although in the analysis completed to date this analyte could not consistently distinguish tumor-bearing dogs from those with inflammatory CNS disease.[25]

CONCLUSIONS AND THERAPEUTIC IMPLICATIONS

Although evidence of a direct role in tumor invasion remains to be demonstrated, the expression of MMPs and cathepsin B by canine intracranial tumors are supportive of this possibility, and may provide a rationale for inhibition of these enzymes as a novel therapeutic intervention. Canine meningiomas and gliomas have extensive infiltration of microglia, macrophages and lymphocytes, which may influence tumor growth and suggest a rational basis for consideration of immunotherapy for these tumors. The analysis of MMP-9 and VEGF within the CSF of dogs with intracranial tumors may prove useful, providing potential biomarkers for both diagnostic and therapeutic purposes.

REFERENCES

1. Moore MP, et al. **Vet Clin North Am Small Anim Pract** 1996;26:759. 2. Patnaik AK, et al. **Vet Pathol** 1986;23:369. 3. Patnaik AK, et al. **Vet Pathol** 1986;23:362. 4. Axlund TW, et al. **J Am Vet Med Assoc** 2002;221:1597. 5. Long S, et al. **Neuropathol Appl Neurobiol** 2006;32:662. 6. Levicar N, et al. **Acta neurochirurgica** 2003;145:825. 7. Egeblad M, et al. **Nat Rev Cancer** 2002;2:161. 8. McCawley LJ, et al. **Curr Opin Cell Biol** 2001;13:534. 9. Siddique K, et al. **Int J Oncol** 2003;22:289. 10. Das A, et al. **Neuropathology** 2003;23:275. 11. Okada M, et al. **J Neurooncol** 2004;66:29. 12. Fritz J, et al. **Neuropathol Appl Neurobiol** 2005;31:339. 13. Nordqvist AC, et al. **J Neurosurg** 2001;95:839. 14. Levicar N, et al. **J Neurooncol** 2002;58:21. 15. Strojnik T, et al. **Neurosurgery** 2001;48:598. 16. Boozer LB, et al. **Vet Pathol** 2012;49:784. 17. Boozer LB, et al. **J Vet Intern Med** 2010;24:741. 18. Kos J, et al. **Int J Biol Markers** 2000;15:84. 19. Zucker S, et al. **Ann N Y Acad Sci** 1999;878:212. 20. Friedberg MH, et al. **Cancer** 1998;82:923. 21. Wong ET, et al. **Cerebrospinal Fluid Res** 2008;5:1. 22. Nagai A, et al. **Clin Chim Acta** 2003;329:53. 23. Mariani CL, et al. **Am J Vet Res** 2013;74:122. 24. Bergman RL, et al. **Am J Vet Res** 2002;63:1359. 25. Niman ZE, et al. **J Vet Intern Med** 2010;24:697.

GENETIC RISK FOR LYMPHOMA AND CHEMOTHERAPY RESPONSE: THE GST STORY

Lauren A. Trepanier, DVM, PhD, DACVIM, DACVCP
Madison, WI

ENVIRONMENTAL RISK FOR LYMPHOMA

Lymphoma is one of the most common cancers diagnosed in dogs, and resembles non-Hodgkin lymphoma in humans. The risk for developing lymphoma has been linked in humans to industrial waste and environmental pollutants, to include chlorinated hydrocarbons, [1] benzene, [2] various pesticides, [3] urea herbicides, and triazole fungicides. [4]

Comparable environmental risks have been documented in dogs with lymphoma, although there are relatively few studies. Demonstrated risk factors include exposure to pesticides, [5] household chemical use (especially paints and solvents), [6] and living in industrial areas or in proximity to polluted sites. [6,7] One group also reported an association between the commonly used broad leaf herbicide 2,4-dichlorophenoxy acetic acid (2,4-D) and canine lymphoma, [8,9] but these results were quite controversial and were refuted by other researchers. [10,11]

Overall, these findings suggest a possible causal relationship between exposure to certain environmental chemicals and lymphoma in dogs. However, it is unclear which chemicals are important and what other factors may modulate risk with exposure.

GSTs AND CANCER RISK IN HUMANS

Glutathione-S-transferase (GST) enzymes are a family of Phase II biotransformation enzymes that detoxify many xenobiotics, including pesticides and environmental pollutants. A number of GST substrates are carcinogens that have been associated with the risk of various human cancers, including lymphoma (Table 1).

Table 1: Carcinogens that are substrates for glutathione-S-transferases (GSTs)

Carcinogen	**Source**	**Cancer risk in humans**
Aflatoxin [12]	Contaminated peanuts, corn, meat, and milk	Hepatocellular carcinoma [13]
Benzene [14]	Automobile exhaust, cigarette smoke, gasoline	Leukemia, non Hodgkin lymphoma, myelodysplastic syndromes [15] Multiple myeloma [16]
Butadiene [17]	Gasoline, industrial waste, cigarette smoke	Leukemia [18]
Chlorpyrifos [19]	Organophosphate pesticide	Hodgkin lymphoma [20] Gliomas [21]
2,4-Dichlorophenoxy acetic acid (inhibitor of GSTs?)	2,4-D, found in lawn herbicides for broad leaf weeds Agent Orange (defoliant used during the Vietnam War)	Non Hodgkin lymphoma [22-25]
Dichloromethane (methylene chloride) [26]	Found in paint strippers, manufacturing solvents, and fumigants on fruits and grains	Multiple myeloma [27] Lymphoma [28]
Trihalomethanes [29] (note: *bioactivation* by *GSTT1*)	Chlorinated drinking water	Bladder cancer [21]

Multiple classes of human GSTs have been identified, to include GST-alpha (GSTA), GST-mu (GSTM), GST-pi (GSTP), GST-theta (GSTT), and others. Various polymorphisms in the genes encoding these enzymes lead to impaired GST function. [30] In fact, homozygous deletions (null genotypes) in two human genes, *GSTM1* and *GSTT1*, lead to complete loss of enzyme expression. [31,32] These null variants are found in 30-50% and 15-25% of Americans, respectively. [33]

There have been dozens of studies on the relationship between GST variants and the risk of human cancers. Defective GST genotypes are over-represented in patients with various solid tumors, [33-35] and hematopoietic malignancies. [36-39] In particular, the *GSTT1* null genotype was found in multiple human studies to be a risk factor for Hodgkin and non-Hodgkin lymphoma, [39-43] as well as myelodysplastic syndromes. [44] This increased risk has been attributed to impaired detoxification of environmental pollutants. For example, exposure to chlorinated hydrocarbons is a significant risk factor for lymphoma in humans, [1] and the chlorinated hydrocarbon dichloromethane (a.k.a. methylene chloride) is a GSTT1 substrate. [26] This compound is found in paint strippers, manufacturing solvents, and in fumigants on fruits and grains.

GSTs AND DRUG RESPONSE IN HUMANS

In addition to the biotransformation of environmental contaminants, GSTs enhance the clearance of various therapeutic drugs by rendering them more water soluble for excretion. GSTs neutralize reactive moieties on chemotherapeutic drugs, to include chlorambucil, melphalan, nitrogen mustard, CCNU, and cyclophosphamide metabolites (acrolein and phosphamide mustard). [45-47] GSTs can also mediate chemotherapeutic drug resistance through a mechanism that is independent of drug clearance. Specifically, GSTP and GSTM inhibit the MAP kinase apoptotic pathway in tumor cells, through direct protein-protein interactions between GSTs and JNK1. [48] These GSTs, when over-expressed, can lead to tumor resistance to doxorubicin, cisplatin, and carboplatin. [45]

Because of the role of GSTs in drug clearance, low activity GST polymorphisms have been associated greater drug toxicity in cancer patients. For example, lymphoma patients with the *GSTT1* null genotype show earlier and greater drug toxicity than patients with functional GSTT1, [49] and pediatric *GSTT1* null leukemia patients have higher mortality during intensive chemotherapy. [50] In patients with SLE, a low activity GSTP1 variant is associated with both myelosuppression and GI toxicity from cyclophosphamide. [47]

Some studies have also shown a better long-term response to chemotherapy in cancer patients carrying low activity GST variants. [41,51-53] For example, both *GSTT1* [41] and *GSTM1* deletions [49] have been linked to improved disease-free survival in human patients undergoing treatment for Hodgkin lymphoma. The *GSTT1* null has also been associated with improved response to prednisone and decreased risk of relapse in pediatric patients with acute lympoblastic leukemia (ALL). [54] However, low activity GSTs may also predispose to secondary cancers after chemotherapy. [55]

GSTs IN DOGS: RISK OF LYMPHOMA

Compared to the hundreds of studies in humans, relatively little is known about GSTs in dogs. In one small study, plasma GSTA was higher in dogs with lymphoma compared to controls, and increased further at the time of relapse. [56] The mechanism for this increase was not explored.

GSTT1 expression has been evaluated in dog livers, with 45-fold variability in activity among individual dogs in research colonies. [57] Notably, 12% of beagle dogs had very low hepatic GSTT1 activity and virtually no protein expression. [57,58] This demonstrated that GSTT1 was polymorphic in dogs as in humans. Given what is known in humans, and the observation that GSTT1 is quite variable in dogs, we hypothesized that variants in this carcinogen detoxification pathway might also be associated with lymphoma risk in dogs. We re-sequenced the canine *GSTT1* gene, including part of the promoter, all exons and splice junctions, and the 3' untranslated region (UTR), in 93 dogs with LSA and 86 age-matched controls of comparable breed distribution. [59] We found a significant association between a variant in the second intron of canine *GSTT1*, I2+28 G>A, and lymphoma risk, with an odds ratio of 6.3 (95% CI, 1.8-22.2). The AA genotype was found in 18.3% of dogs with lymphoma but only in 3.5% of controls (P=0.002). [59] This intronic variant was predicted to alter splicing of the *GSTT1* gene, which could lead to a truncated, dysfunctional protein.

In this survey, we also found an insertion variant in the 3' untranslated region of canine *GSTT1*, *101_*102 ins T, which was observed with a minor allele frequency (MAF) of 0.129 in dogs with LSA, compared to 0.048 in healthy dogs (P = 0.009). However, this difference did not reach significance when adjusted for multiple comparisons. This polymorphism was predicted to delete miRNA binding sites, which could alter gene expression. Functional characterization of both I2+28 G>A and *101_*102 ins T variants is underway in our laboratory. We also observed that the *101_*102 ins T variant was found in 5 of 17 Golden retrievers with lymphoma but in none of 14 healthy Golden retrievers. This preliminary observation suggests that this variant could be associated with lymphoma in this high-risk breed; however, this needs to be evaluated in a larger population.

GSTs IN DOGS: RESPONSE TO CHEMOTHERAPY

In addition to lymphoma risk, we are also evaluating the relationship between *GSTT1* genotype and response to chemotherapy in the same population of dogs. Fifty-five of the 93 genotyped dogs began induction chemotherapy with the CHOP protocol. In this population, the I2+28 AA genotype was significantly associated with a longer disease free interval (median DFI 359 days, range 196 to 472; compared to the GG genotype, DFI 207 days, range 0 to 641; P = 0.04). However, this association was no longer significant (P = 0.33) when examined in a multivariate analysis that included disease stage, immunophenotype, and the presence of hypercalcemia. There was no apparent association with either variant and bone marrow or GI toxicity from cyclosphophamide; however, the study was underpowered to evaluate this.

ONGOING WORK

We are working on sequencing the genes encoding canine GSTM1, GSTP1, and GSTA in order to determine the degree of variability in these additional GSTs. This will allow us to include variants in all four major GST isoforms in risk assessment, since there are overlapping substrate specificities among GSTs, and some human studies have found additive effects for low activity GST genotypes and cancer risk. Our GST work may also provide a panel of genetic polymorphisms for evaluation in dogs with other cancers that have been associated with environmental carcinogens, [21] such as transitional cell carcinomas and gliomas. These results together will allow more refined molecular epidemiologic studies of the relationship between specific environmental chemicals and cancers in dogs, which may lead to better screening, prevention, and avoidance strategies.

REFERENCES

1. Seidler A, et al. *J Occup Med Toxicol* 2007;2:2. 2. O'Connor SR, et al. *J Pathol* 1999;189:448-453. 3. Fritschi L, et al. *Am J Epidemiol* 2005;162:849-857. 4. Orsi L, et al. *Occupational and environmental medicine* 2009;66:291-298. 5. Takashima-Uebelhoer BB, et al. *Environ Res* 2012;112:171-176. 6. Gavazza A, et al. *J Vet Intern Med* 2001;15:190-195. 7. Pastor M, et al. *J Vet Intern Med* 2009;23:301-310. 8. Hayes HM, et al. *J Natl Cancer Inst* 1991;83:1226-1231. 9. Hayes HM, et al. *Environ Res* 1995;70:119-125. 10. Carlo GL, et al. *Regul Toxicol Pharmacol* 1992;16:245-252. 11. Kaneene JB, et al. *Vet Hum Toxicol* 1999;41:164-170. 12. Chen SY, et al. *Nutr Cancer* 2000;38:179-185. 13. Magnussen A, et al. *World J Gastroenterol* 2013;19:1508-1512. 14. Carrieri M, et al. *Toxicol Lett* 2012;213:63-68. 15. Hayes RB, et al. *J Natl Cancer Inst* 1997;89:1065-1071. 16. Infante PF. *Annals of the New York Academy of Sciences* 2006;1076:90-109. 17. Landi S. *Mutat Res* 2000;463:247-283. 18. Soeteman-Hernandez LG, et al. *Toxicol Sci*;136:566-580. 19. Fujioka K, et al. *Chem Res Toxicol* 2007;20:1211-1217. 20. Karunanayake CP, et al. *J Agromedicine* 2012;17:30-39. 21. Clapp RW, et al. *Rev Environ Health* 2008;23:1-37. 22. Miligi L, et al. *Ann NY Acad Sci* 2006;1076:366-377. 23. Hoar SK, et al. *JAMA* 1986;256:1141-1147. 24. Hardell L, et al. *Br J Cancer* 1981;43:169-176. 25. Fontana A, et al. *Arch Environ Health* 1998;53:384-387. 26. Thier R, et al. *Arch Toxicol* 1998;72:622-629. 27. Liu T, et al. *Cancer Causes Control*;24:2037-2049. 28. Barry KH, et al. *Am J Epidemiol* 2011;173:404-413. 29. Ross MK, et al. *Chem Res Toxicol* 2003;16:216-226. 30. Hayes JD, et al. *Annu Rev Pharmacol Toxicol* 2005;45:51-88. 31. Seidegard J, et al. *PNAS* 1988;85:7293-7297. 32. Pemble S, et al. *Biochem J* 1994;300 (Pt 1):271-276. 33. Rebbeck TR. *Cancer Epidemiol Biomarkers Prev* 1997;6:733-743. 34. Strange RC, et al. *Arch Toxicol* 1998;20:419-428. 35. Sorensen M, et al. *Int J Cancer* 2004;110:219-224. 36. Soucek P, et al. *Int Arch Occup Environ Health* 2002;75 Suppl:S86-92. 37. Krajinovic M, et al. *Pharmacogenetics* 2002;12:655-658. 38. Ye Z, et al. *Eur J Cancer* 2005;41:980-989. 39. Al-Dayel F, et al. *Leuk Lymphoma* 2008;49:122-129. 40. Kerridge I, et al. *Br J Haematol* 2002;118:477-481. 41.Hohaus S, et al. *Clin Cancer Res* 2003;9:3435-3440. 42. Ruiz-Cosano J, et al. *Mol Carcinog* 2012;51:508-513. 43. Abdel Rahman HA, et al. *J Cancer Res Clin Oncol* 2012;138:1363-1368. 44. Chen H, et al. *Lancet* 1996;347:295-297. 45. Michael M, et al. *J Clin Oncol* 2005;23:205-229. 46. Tew KD. *Cancer research* 1994;54:4313-4320. 47. Zhong S, et al. *Br J Clin Pharmacol* 2006;62:457-472. 48. Townsend DM, et al. *Oncogene* 2003;22:7369-7375. 49. Lourenco GJ, et al. *Leukemia & lymphoma* 2010;51:2215-2221. 50. Davies SM, et al. *J Clin Oncol* 2001;19:1279-1287. 51. Hohaus S, et al. *Clin Cancer Res*

2005;11:2175-2179. 52. Stanulla M, et al. *Blood* 2000;95:1222-1228. 53. Ekhart C, et al. *Cancer Treat Rev* 2009;35:18-31. 54. Anderer G, et al. *Pharmacogenetics* 2000;10:715-726. 55. Allan JM, et al. *Proc Natl Acad Sci U S A* 2001;98:11592-11597. 56. Hahn KA, et al. *In Vivo* 1999;13:173-175. 57. Watanabe T, et al. *Arch Toxicol* 2004;78:218-225. 58. Watanabe T, et al. *Arch Toxicol* 2006;80:250-257. 59. Ginn J, et al. *Vet Comp Oncol* 2012;Epub Oct 2012.

FOSFOMYCIN FOR TREATMENT OF URINARY TRACT INFECTIONS IN DOGS

Dawn Merton Boothe, DVM, PhD, DACVIM, DACVCP, Patrick Hubka BS
Auburn, AL

Multidrug-resistant pathogenic E. coli s an emerging national disease in veterinary companion animal medicine. *E coli* is an organism which responds to antimicrobial selection pressure by developing resistance (Miller 2002). Multidrug resistant *E coli* has been documented as a cause of canine urinary tract infections (Cooke 2002), nosocomial infections in a veterinary teaching hospital (Sidjabat 2006); and in fecal isolates in ICU patients (Ogeer-Gyles 2006).We have previously reported in abstract form (Boothe, 2005a) a high incidence of mdr in *E. coli* uropathogens isolated from canine patients admitted to Auburn University's Veterinary Teaching Hospital (n=175). Overall *E.coli* resistance to 17 antimicrobials included in our standard antibiogram was 31%. More problematic, resistance approximated 40% to 50% toward first (amoxicillin, amoxicillin-clavulanic acid, cephalexin and sulfonamide-trimethoprim) and second choice (enrofloxacin) empirical antimicrobals. A similar pattern was found in other southern Veterinary Teaching Hospitals as well (Boothe 2005b). As a follow-up to these findings, our report in progress (initial presentation in abstract form, Smaha 2006), *Escherichia coli antimicrobial Resistance in Small Animals: The scope of the Problem,* we prospectively determined the susceptibility of canine E. *coli* pathogens (n=376) causing spontaneous disease from different geographical areas of the United States. Of these isolates, approximately 50% were characterized by MDR, from the urinary tract (Shaheen 2010, Boothe 2010). Based on this data, we have implemented a prevalence study, funded by the Morris Animal Foundation, that focuses on the phenotyping of *E. coli* associated with spontaneous disease in dogs and cats. Isolates are subjected to C&S using custom made plates (Trek Diagonistic®). To date, we have studied close to 1000 isolates. Approximately 75% of the isolates are from the urinary tract. **Figure 1** demonstrates the distribution of those isolates collected thus far expressing NDR (no drug resistance), SDR (single drug or drug class) , MDR (multiple; 2 to 4 drug or drug classes), XDR, (extreme; 5 to 6) and ZDR (all 7 drug classes tested).

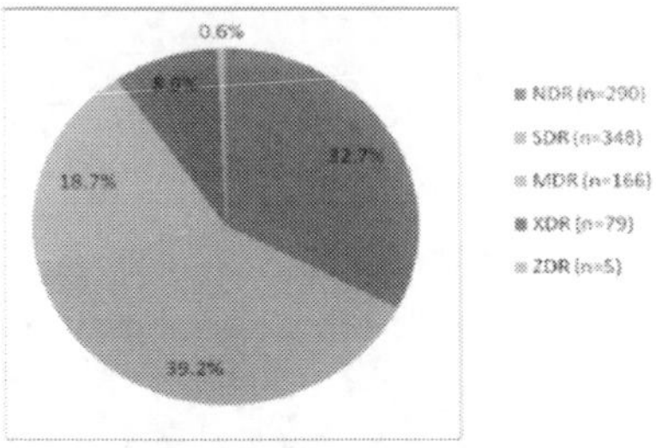

Figure 1. Number of feline or canine pathogenic *E. coli* expressing no (NDR), single (SDR), multiple (2-4 drug classes; MDR), extreme (5-6 drug classes, XDR) or total (all drug classes tested; ZDR) resistance.

Although MDR and XDR isolates largely retain susceptibility to aminoglycosides, treatment of these with aminoglycosides is costly, requiring hospitalization. Further, aminoglycoside therapy is associated with the risk of nephrotoxicity. Although mdr-Eco continues to express susceptibility to nitrofurantoin, this drug also is associated with undesirable side effects as well as potential adverse reactions (peripheral neuropathy).

The importance of integration of pharmacokinetic-pharmacodynamic parameters (from the species of interest) as the basis of antimicrobial therapy in the clinical patient is a widely accepted necessity among medical communities and is increasingly being applied to clinical medicine (Vinks 2009). Relevant pharmacokinetic parameters for concentration dependent drugs include maximum drug concentration (C_{max}) and for time dependent drugs, C_{max} and plasma and elimination half-life. Relevant population pharmacodynamic statistics include the MIC_{90}, which represents the 90th percentile within the range of MIC which characterize the isolate's susceptibility to the drug of interest.

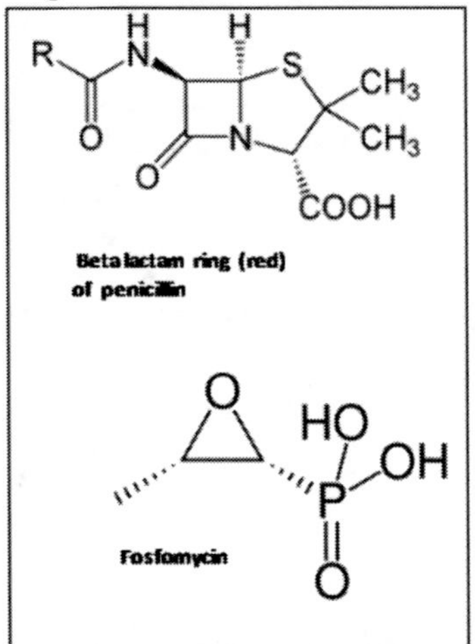

Figure 2. Structures of the cell wall inhibitors penicillin (top) and fosfomycin (bottom)

There is a need for a safe, convenient, effective alternative drug for treatment of canine multidrug resistant E coli uropathogens. The need is exemplified by three clinical cases presented to the PI 1. Shadow K, is a 6 yr old Scottish Terrier with a chief complaint of chronic pyoderma, who has been treated with antimicrobials (cephalexin, amoxicillin-clavulanic acid, or enrofloxacin at low to mid doses) and immunosuppressive therapy for 2 years. Her serum creatinine and blood urea nitrogen are 3.5 and 75 mg/dl, respectively. Three weeks ago, she developed clinical signs indicative of UTI and urinalysis reveals clumps of white blood cells; C&S reveals > 10^5 CFU/μl of MDR-*E coli* susceptible only to amikacin. 2. Nanouk H. is a 12 year old female Malamute with a chief complaint of recurrent UTI, associated with poor vulvular confirmation. Her *E coli* infection progressively developed resistance to amoxicillin, amoxicillin clavulanic acid, and most recently, enrofloxacin. Currently, the organism is susceptible only to nitrofurantoin and amikacin. Nanouk has undergone vulvuloplasty and is now being re-presented for treatment. 3. Wyn Wyn L. is a 10 year old female mixed breed dog presented in her third year of treatment for pemphigus foliaceus. Immunosuppressive therapy including glucocorticoids and cyclophosphamide was given to her. Urinalysis reveals 10,000 CFU/μl of MDR-*E coli* susceptible only to amikacin. Although clinical signs are absent, a previous episode of septicemia has led both client and clinician to pursue treatment of the infection.

Fosfomycin tromethamine (Monurol®) offers a potentially safe, effective and convenient alternate treatment for these patients with MDR-E coli UTI. Fosfomycin, a phosphonic acid which contains a carbon-phosphorous bond, is a natural antibiotic produced by *Streptomyces fradiae* (**Fig 2**). Its *in vitro* spectrum is broad, with potential efficacy toward isolates expressing MDR, including *E coli* and Gram positive isolates. Approved for human use in the USA, its indication is as a one time (or up to 3 days) treatment of *E coli* UTI in humans (Schito 2003).

Fosfomycin is a phosophonol pyruvate analogue that irreversibly inhibits phosphoenol pyruvate transferase, an enzyme that catalyzes the first step of cell wall peptidoglycan synthesis of microbial cell walls (Fillgrove 2007). Although its mechanism of action is similar to the beta-lactams, unlike the beta-lactams, fosfomycin is not susceptible to destruction by any class of beta-lactamases. Rather, resistance to fosfomycin, which is unusual, reflects the FosX or FosA enzyme which hydrolyzes the drug. The gene for this protein is chromosomally mediated. Thus, when resistance does occur, it is usually SDR, and generally not associated with MDR (Fillgrove 2007). Further, compared to susceptible strains, fosfomycin-resistant mutants are impaired, exhibiting poorer growth rates, and reduced adherence to uroepithelial cells (particularly in the presence of -acetylcystein (Marchese 2003a) and
urinary catheters (Marchese 2003b). As a cell wall inhibitor, fosfomycin is bactericidal when present at the site of infection at therapeutic concentrations. Cell wall inhibition indicates efficacy to be time dependent and facilitated if drug concentrations exceed the MIC of the infecting microbe for at least 50% of the dosing interval (T>MIC=50%). However, fosfomycin also exhibits a concentration-dependent effect (Kawaguchi 2002), which facilitates efficacy in the face of a short elimination half-life as long as concentrations are substantially above (eg, 10 X) the MIC of the infecting microbe. Studies in humans have demonstrated that fosfomycin distributes well to soft tissues, reaching therapeutic breakpoints (Sauermann 2007). Other attributes of fosfomycin that support its use for treatment of *E coli* UTI include renal excretion, synergistic interaction with several other classes of antimicrobials (Olay 1974, 1989), and preparation as a 3 g sachet (granules) which is mixed with water to orally deliver approximately 40 mg/kg (in humans). Kill studies in our laboratory suggest that the efficacy of fosfomycin is not concentration dependent, suggesting, as might be expected of a cell-wall inhibitor, time-dependent effects. We anticipate that a week of treatment with Monurol® at 40 mg/kg bid will cost approximately $12 /kg for 1 week of therapy. Alternative therapies are based generally on injectable drugs (aminoglycosides, carbepenems) and thus hospitalization, and for aminoglycosides, intensive monitoring. Side effects of fosfomycin appear to be limited to diarrhea.

Fosfomycin tromethamine *Pharmacodyanmics.* Fecal *E. coli* isolates exhibiting NDR (n=50), SDR or MDR (n=25 each; isolates resistant to 14/16 drugs on the antibiogram) or non-resistant (n=50) fecal *E coli* (see Background) were subjected to fosfomycin *E test.* Isolates were obtained either from an experimental study in which multidrug resistant fecal *E. coli* were induced following administration of enrofloxacin (5 mg/kg/day PO for 7 days or from an ongoing surveillance study in partnership with IDEXX laboratories (1000 isolate *E. coli* isolates/year from dogs or cats with spontaneous disease). The E-test® is a solid agar test to which is applied a strip containing the drug of interest at increasing concentrations; for fosfomycin, glucose-6-phosphate is included. Compared to standard antibiograms, the E-test® has a distinct advantage when studying microbial pharmacodyanmics: the range of MIC studied is very broad (compared to 2 to 6 fold for commercial antibiograms), and susceptibility and resistance measurements are more precise (ie, not limited to serial two fold increments). Thus low or high level resistance can be detected because concentrations exceed the resistant breakpoint MIC (MIC_{BP}). The fosfomycin breakpoints are ≤ 6e mcg/ml (susceptible) and ≥ 128 mcg/ml (resistant).

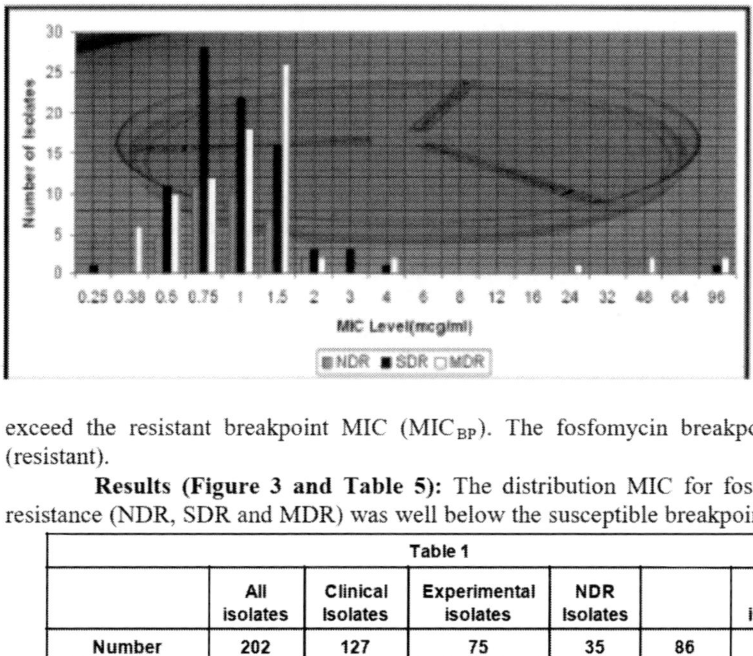

Results (Figure 3 and Table 5): The distribution MIC for fosfomycin for all of the isolates, regardless of the type of resistance (NDR, SDR and MDR) was well below the susceptible breakpoint (≤ 64 μg/m) for fosfomycin. The MIC range was 0.25 to 4 μg/ml; the MIC 50 and 90, were, respectively, 1 and 1.5μg/ml, respectively, except for MDR, for which MIC90 was 2 μg/ml. Although it is not known if these canine fecal *E coli* are uropathogenic, the data none-the-less suggests that MDR *E coli* should remain susceptible to fosfomycin. This should be confirmed with MIC testing of *E coli* isolates associated with spontaneous disease whose serotypes are known. Kill studies in our laboratory suggest that the efficacy of fosfomycin is not concentration dependent, with *in vitro* killing of mdr-Eco being similar after exposure to 4, 8 and 10 X the MIC. ***Kill studies.*** Preliminary studies in our lab demonstrate that the rate and extent of mdr-Eco killing *in vitro* are greater at 2 compared to 1 X the MIC of the microbe, but killing is similar at 2, 4, and 8 X the MIC. This suggests that fosfomycin efficacy is not concentration dependent. Dosing regimens will thus target at least 4 times the MIC for 75% of the dosing interval.

Table 1

	All isolates	Clinical Isolates	Experimental isolates	NDR Isolates		MDR isolates
Number	202	127	75	35	86	81
% Susc	98.5	97.6	100	100	98.8	97.5
MIC_{90}	1.5	2	1.5	1.5	1.5	2
MIC_{50}	1	1	1	1	1	1
Geo mean	1.09	1.12	1.03	0.93	1	1.25

Pharmacokinetics of fosfomycin tromethamine in normal dogs. Pharmacokinetics have been previously determined in dogs using the diphosphate salt of fosfomycin. This study established safety of fosfomycin when administered daily for 3 days at 80 mg/kg orally and IV; however, the drug was a preparation compounded from bulk substance. The diphosphate salt of fosfomycin is only 40% bioavailable in dogs. We have determined the pharmacokinetics of fosfomycin tromethamine, the commercially available drug, when administered as a single oral dose of 80 mg/kg (Figure 3). After oral administration, C_{max}, elimination half-life and mean residence time were 66±21 (µg/ml), 2.5±1.09 h and 5.1 ±1.7, h respectively. Drug was detected at concentrations exceeding the MIC 90 of fosfomycin for multidrug resistant (MDR) *E. coli* (1.5 mcg/ml) until 7 (2.5 mcg/ml) and 12 hr (9 mcg/ml) following IV and oral administration, respectively. Oral bioavailability was 88± 32%. Gastrointestinal upset manifested as mild to moderate diarrhea was observed in four of the twelve dogs. Food increased oral bioavailability: with food, 109±31% (95% CI: 84-135) and without food, 66±16% (95%CI: 52-79%). Gender had no impact on oral bioavailability. From this data, we conclude that food facilitates absorption of fosfomycin, and that 40 mg/kg should result in plasma drug concentrations maintained well above the MIC 90 of mdr-Eco established in our laboratory for over 12 hrs. Urine concentrations should be maintained at more than 10 times the inhibitory MIC for at least 12 hrs. Based on these studies, fosfomycin tromethamine appears to be a drug which may reasonably be effective for treatment of canine UTI, including that caused by MDR *E. coli.* However, clinical trials are now indicated to determine the most appropriate dosing regimen (bid versus once daily) and the duration of therapy.

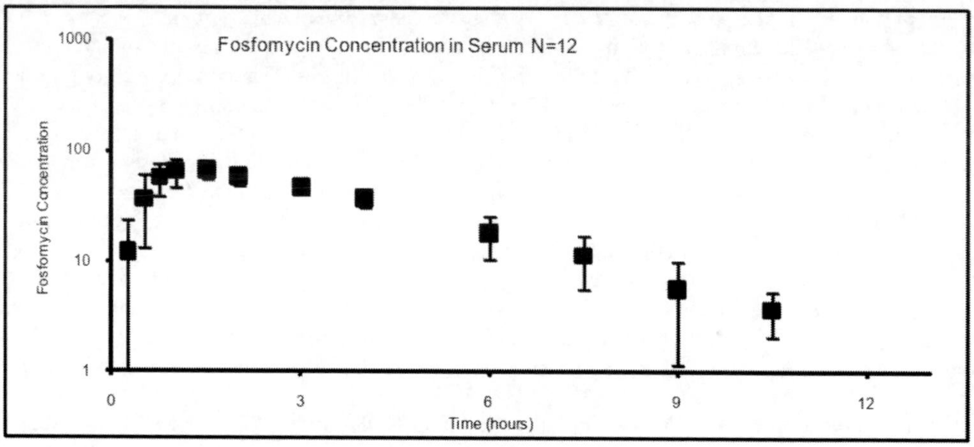

Figure 4. Plasma drug concentration vs time curves for fosfomycin after administration of fosfomycin tromethamine (80 mg/kg) po in 12 apparently healthy dogs

SELECTED REFERENCES (OTHERS AVAILABLE UPON REQUEST)

Shaheen BW, Boothe DM, Oyarzabal O, et al. The contribution of *gyrA* mutation and efflux pumps to fluoroquinolone and multi-drug resistance in canine and feline pathogenic *Escherichia coli* Isolates from the US. Accepted Am J Vet Res 2010; in press. Shaheen BW, Boothe DM, Oryzabal O, Smaha T. Antimicrobial resistance profiles and clonal relatedness of canine and feline escherichia coli pathogens expressing multidrug resistance in the United States. Accepted J Vet In Med 2009; in press.Fillgrove KL, Pakhomova S, Schaab MR, Newcomer ME, Armstrong RN. Structure and mechanism of the genomically encoded fosfomycin resistance protein,FosX, from Listeria monocytogenes.Biochemistry. 2007 Jul 10;46(27):8110-20. Kawaguchi K, Hasunuma R, Kikuchi S, Ryll R, Morikawa K, Kumazawa Y. Time- and dose-dependent effect of fosfomycin on suppression of infection-induced endotoxin shock in mice. Biol Pharm Bull. 2002 Dec;25(12):1658-61.

BACTERIA: INTERPRETATION OF MICs

Mark G. Papich, DVM, MS, Diplomate ACVCP
Raleigh, North Carolina

INTRODUCTION

Empirical treatment of common infections in animals can be accomplished by selecting the antimicrobial agent with a high likelihood of success for the suspected clinical infection. Many guidelines have appeared in published proceedings, review papers, consensus documents, and textbooks. But when empirical treatment fails, or when resistance is suspected, a culture and susceptibility test is needed to guide therapy. The empirical choice for initial treatment is based on the assumption that the infection is not complicated and the infection is caused by wild-type bacteria. *Wild-type* strains of bacteria are those that have an absence of acquired and mutational resistance mechanisms. Whereas *Non-wild-type* strains of bacteria are those that have the presence of an acquired or mutational resistance mechanism to the drug in question. Wild-type strains may include bacteria that have inherent resistance to antimicrobials. For example, wild-type anaerobic bacteria are inherently resistant to aminoglycosides by virture of a lack of an oxygen-dependent drug entry to the bacteria. Gram-negative bacteria of the Enterobacteriaceae and *Pseudomonas aeruginosa* are inherently resistant to macrolide antibiotics.

Culture and antimicrobial susceptibility testing (AST) will provide the best guidance for drug selection. Nevertheless, it is also understood that even when an agent is selected from the "susceptible" category, treatment may not always be successful. The prediction of whether the bacteria will, or will not respond to treatment is commonly referred to as the "90/60 rule" (Doern & Brecher, 2011). The 90/60 rule was derived from the observation that, in general, bacteria treated with antimicrobials to which the strain is sensitive will have a favorable therapeutic response in approximately 90% of the patients. On the other hand, when the bacteria is resistant to the antimicrobial administered, despite the susceptibility result, approximately 60% of patients will respond to therapy. In veterinary medicine, we have no data to confirm or challenge the 90/60 rule.

SUSCEPTIBILITY TESTING

The most important information for the clinician is simply which drugs have an "S" and which ones have an "R". These results then guide their treatment. What really goes into this interpretation? The standards for interpretation are available from the Clinical and Laboratory Standards Institute – CLSI, (CLSI, 2013, also found at http://www.clsi.org/). Not all laboratories in the United States use CLSI standards. It is a voluntary program. However, if a laboratory does not adhere to a public standard such as CLSI, breakpoints may vary and interpretation may be inconsistent from laboratory to laboratory, or among different regions of the country.

Microdilution Test for Determination of MIC

It is becoming more common for laboratories to directly measure the minimum inhibitory concentration (MIC) of an organism with an antimicrobial dilution test. Zone inhibition (also known as the Kirby-Bauer test) also is performed but provides only qualitative information. The MIC dilution test is performed by inoculating the wells of a plate with the bacterial culture and dilutions of antibiotics are arranged across the rows. The test is usually performed in modern laboratories using high-throughoutput plates, but individual tubes or plates can be used for dilution tests also. Antibiotic drug concentrations are arranged in serial dilutions, with each concentration doubled from lowest to highest in a range. The MIC is not a measure of efficacy, but instead it is simply an in vitro measurement of drug activity and bacterial susceptibility. The lower the MIC value, the more susceptible the isolate is to that drug. The MICs are determined using serial two-fold dilutions of drug to which is added a standardized inoculum that is incubated for a prescribed time. Concentrations are always listed in µg/mL. For example, if one were to start at a concentration of 256 µg/ml, the MIC dilution series would be as follows: 128, 64, 32, 16, 8, 4, 2, 1, 0.5, 0.25, 0.12, 0.06 µg/ml, etc. If, for example, bacterial growth occurs at a dilution of 0.12 µg/ml for a specific drug, but not at 0.25 µg/ml and above, the MIC is determined to be 0.25 µg/ml. The MIC dilution test is only semi-quantitative because there are gaps between each dilution. Realistically, the true MIC lies somewhere between these values, but the MIC is recorded as the lowest value. In some laboratories other methods to measure the MIC are being used such as the E-test (epsilometer test) by AB Biodisk. The E-test is a quantitative technique which measures the MIC by direct measurement of bacterial growth along a concentration gradient of the antibiotic contained in a test strip.

WHY REPORT ONLY THE MIC?

The MIC is the Minimum Inhibitory Concentration, which is the lowest concentration that inhibits visible bacterial growth. In surveys, or surveillance data, it is frequently this is expressed as MIC_{50} or MIC_{90}, which is the MIC that inhibits 50% or 90% of the bacteria, respectively. It is sometimes cited in error that the MIC_{50} and MIC_{90} are the average concentrations for 50% and 90% efficacy. These values should not be confused with clinical efficacy (more on that later).

The MBC is the Minimum Bactericidal Concentration, which is the lowest concentration that kills 99.9% of the bacteria. Standards are available to measure the MBC, the test is more complicated and difficult to perform than the MIC determination. Therefore, the MBC is rarely measured or reported in clinical laboratories.

The MPC is the Mutant Prevention Concentration. This is lowest antibiotic concentration that prevents growth of the least susceptible first-step resistant mutant among a large bacterial population (eg, 10^7 or 10^{10} cfu) (Epstein et al, 2004). It also may be defined as the MIC of the most resistant first-step cell present in a bacterial population. The mutant selection window (MSW) is the concentration between the MIC of susceptible organisms, and the MPC. The MPC test is not standardized and is more difficult to perform in a clinical laboratory. Large inoculums are required. The interpretation of the MPC value for clinical dose determinations is difficult and has not been established for veterinary antimicrobials.

INTERPRETATION OF SUSCEPTIBILITY TESTS

Resistance and susceptibility are determined by comparing the organism's MIC to the drug's breakpoint as established by the Clinical and Laboratory Standards Institute (CLSI) – formerly known as the National Committee for Clinical Laboratory Standards (NCCLS) (CLSI 2013). After a laboratory determines an MIC, it may use the CLSI "SIR" classification for breakpoints (S, susceptible; I, intermediate, or R, resistant). In practice, if the MIC for the bacterial isolate falls in the susceptible category, there is a greater likelihood of successful treatment (cure) than if the isolate were classified as resistant. It does not assure success; drug failure is still possible owing to other drug or patient factors (for example, immune status, immaturity, or severe illness that compromises the action of antibacterial drugs), and interactions. If the MIC is in the resistant category, bacteriologic failure is more likely because of specific resistance mechanisms or inadequate drug concentrations in the patient. However, a patient with a competent immune system may sometimes eradicate an infection even when the isolate is resistant to the drug in the MIC test.

The intermediate category is intended as a buffer zone between susceptible and resistant strains. This category reflects the possibility of error when an isolate has an MIC that borders between susceptible and resistant. If the MIC value is in the intermediate category, therapy with this drug at the usual standard dosage is discouraged because there is a good likelihood that drug concentrations may be inadequate for a cure. However, successful therapy is possible when drug concentrates at certain sites – in urine, or as the result of topical therapy, for example – or at doses higher than the minimum effective dose listed on the label. For example, prescribing guidelines for some antimicrobials allow for an increase in dose when susceptibility testing identifies an organism in the intermediate range of susceptibility. For example, fluoroquinolone antimicrobials have been approved with a dose range that allows increases in doses when susceptibility testing identifies an organism in the Intermediate range of susceptibility. In these cases higher drug concentrations make a cure possible if the clinician is able to safely increase the dose above the minimum labeled dose. (For example, in the case of enrofloxacin in dogs, this would be equivalent to a dose of 10 to 20 mg/kg/day, rather than the minimum dose of 5 mg/kg/day.)

MIC data should not be used in isolation, but by coupling the MIC from a laboratory report with CLSI breakpoints and other important information such as the virulence of the bacteria and the pharmacology of the antibiotics being considered, the clinician can make a more informed selection of an antibacterial drug.

Does the susceptibility test provide tissue-specific interpretation?

The susceptibility interpretation is based on plasma/serum concentrations. No tissue-specific interpretation can be provided that accounts for differences in drug distribution among tissues. For example, even though it is anticipated that many antibiotics concentrate in the urine, which may be beneficial for treating a urinary tract infection, the susceptibility interpretation is based on achieving adequate concentrations in the blood. (There are two exceptions to this because amoxicillin and amoxicillin-clavulanate interpretations allow for high concentrations in urine.) One shouldn't assume that concentrations in urine – even when they are high due to concentration by the nephrons – are sufficient to eradicate infections of the urinary tract. Infections may involve the deeper layers of the mucosa, the renal tissue, or the prostate tissue. In these instances, it is the tissue concentration – which is correlated to the plasma concentration – that will be predictive of a bacteriologic cure (Frimodt-Møller, 2002).

A frequent mistake in MIC interpretation is to compare the MIC with published tissue concentrations that are derived from whole-tissue homogenized samples. Tissue concentration data is often published by pharmaceutical companies in their product information. These concentrations may be misleading because they may either underestimate or overestimate (depending on the drug's affinity for intracellular sites) the true drug concentration at the site of infection.

In most instances the clinician should not be concerned with the question of whether or not there are tissue-specific susceptibility interpretations. For most tissues, antibiotic drug concentrations in the serum or plasma approximate the drug concentration in the extracellular space (interstitial fluid). This is because there is no barrier that impedes drug diffusion from the vascular compartment to extracellular tissue fluid (Cars, 1991; Nix et al, 1991). There is really no such thing as "good penetration" and "poor penetration" when referring to most drugs in most tissues. Pores (fenestrations) or microchannels in the endothelium of capillaries are large enough to allow drug molecules to pass through unless the drug is restricted by protein binding in the blood. Tissues lacking pores or channels may inhibit penetration of some drugs (discussed below).

If adequate drug concentrations can be achieved in plasma, it is unlikely that a barrier in the tissue will prevent drug diffusion to the site of infection as long as the tissue has an adequate blood supply. Clinicians should be concerned when treating tissues that have poor or impaired blood supply. ¬Drug diffusion into an abscess or granulation tissue is sometimes a problem because in these conditions, drug penetration relies on simple diffusion and the site of infection lacks adequate blood supply. In an abscess, there may not be a physical barrier to diffusion – that is, there is no impenetrable membrane – but low drug concentrations are attained in the abscess or drug concentrations are slow to accumulate.

In some tissues a lipid membrane (such as tight junctions on capil¬laries) presents a barrier to drug diffusion. In these instances, a drug must be sufficiently lipid-soluble, or be actively carried across the membrane in order to reach effective concentrations in tissues. These tissues include: the central nervous system, eye, and prostate. A functional membrane pump (p-glycoprotein) also contributes to the barrier. There also is a barrier between plasma and bronchial epithelium (blood:bronchus barrier). This limits drug concentrations of some drugs in the bronchial secretions and epithelial fluid of the airways. Lipophilic drugs may be more likely to diffuse through the blood-bronchus barrier and reach effective drug concentrations in bronchial secretions.

Is susceptibility interpretation by CLSI specific for veterinary species?

In past years, the veterinary diagnostic laboratories had to rely heavily on the CLSI interpretation from the human standards. There were not enough veterinary-specific interpretive criteria available to establish breakpoints for veterinary drugs and veterinary species. This is now changing. The current edition of the CLSI standard document (VET 01-S2) for veterinary drugs has the tables clearly separated into those drugs with veterinary interpretive criteria, and drugs that still rely on human standards for interpretation. In the last several years, CLSI has tremendously expanded the list of drugs for which there are veterinary-specific breakpoints. For companion animals, veterinary-specific MIC breakpoints have now been established for the four licensed fluoroquinolones (enrofloxacin, difloxacin, marbofloxacin, and orbifloxacin), gentamicin, cefpodoxime proxetil, ampicillin/amoxicillin, amoxicillin-clavulanic acid, first-generation cephalosporins (cephalexin and cefazolin), amikacin, doxycycline, and clindamycin. Important changes were also made for the interpretation of Staphylococcus resistance. Until other veterinary-specific breakpoints are established for other antibiotics used in companion animals, we will continue to rely on the human breakpoints for drugs such as chloramphenicol, erythromycin, carbapenems (imipenem), penicillins, sulfonamides, potentiated sulfonamides, and tetracyclines other than doxycycline. The CLSI committee is working on filling in these gaps. But in the meantime, similarities in pharmacokinetics and pathogen susceptibilities between humans and animals allow for an acceptable approximation to extrapolate human breakpoints to animal situations for many drugs until veterinary-specific standards are available. The human equivalent of the veterinary CLSI standard (M 100) also has made substantial changes in recent years. Because of concerns for mis-identifying extended-spectrum β-lactamase producing Enterobacteriaceae, the cephalosporin breakpoints have been lowered compared to previous criteria (Dudley, et al, 2013). Carbapenem breakpoints also have been recently lowered.

HOW ARE BREAKPOINTS DERIVED?

The paper by Turnidge and Paterson (2007) describe in detail the process of setting breakpoints. The CLSI subcommittee for Veterinary Antimicrobial Susceptibility Testing (VAST), uses strict criteria to establish and evaluate breakpoints. Sponsors are required to follow guidelines provided by CLSI and must submit data to support a proposed breakpoint. The data includes pharmacokinetic data in the target species, MIC distributions for the pathogens targeted, clinical data from the drug used under field conditions at the approved dose, and pharmacokinetic-pharmacodynamic (PK-PD) analysis, using Monte Carlo Simulations (Ambrose, 2006) to show that at the approved dose the drug attains PK-PD targets for the labeled pathogen (Drusano 2007).

Are these Standards, or Guidelines?

The CLSI is a consensus-driven process and after approval by the subcommittee the standards become public documents. The consensus process involves the development and public open review of documents, revision of documents in response to discussion, and, finally, the acceptance of a document as a consensus standard or guideline. The CLSI documents used for culture and susceptibility testing should be regarded as a public standard, not a guideline.

A Standard is a document developed through the consensus process that clearly identifies specific, essential requirements for materials, methods, or practices for use in an unmodified form. A standard may, in addition, contain discretionary elements, which are clearly identified. A Guideline is a document developed through the consensus process describing criteria for a general operating practice, procedure, or material for voluntary use. A guideline may be used as written or modified by the user to fit specific needs.

REFERENCES

Ambrose PG. Monte Carlo simulation in the evaluation of susceptibility breakpoints: predicting the future: insights from the society of infectious diseases pharmacists. Pharmacotherapy. 2006;26(1):129-34.

Cars, C. (1991) Pharmacokinetics of Antibiotics in tissue and tissue fluids:A review. Scandinavian Journal of Infectious Diseases Supplement, 74, 23-33.

Clinical Laboratory Standards Institute. CLSI document VET 01-S2. Wayne, PA: Clinical and Laboratory Standards Institute; 2013 *(available from www.clsi.org)*. Doern GV & Brecher SM. J Clin Microbiol 49(9 suppl): S11-S14, 2011.

Drusano GL, et al. Clin Infect Dis 2007; 45, 753-760.

Dudley MN, Ambrose PG, Bhavnani SM, Craig WA, Ferraro MJ, Jones RN (for the Antimicrobial Susceptibility Testing Sub-Committee of the Clinical Laboratory Standards Institute). Background and Rationale for Revised Clinical Laboratory Standards Institute (CLSI) Interpretive Criteria (Breakpoints) for Enterobacteriaceae and Pseudomonas aeruginosa: I. Cephalosporins and Aztreonam. Clinical Infectious Disease (in press, 2013).

Dudley MN, et al. (2013). Background and rationale for revised Clinical and Laboratory Standards Institute interpretive criteria (breakpoints) for Enterobacteriaceae and Pseudomonas aeruginosa: I. Cephalosporins and aztreonam. Clinical infectious diseases, 56(9), 1301-1309.

Epstein BJ, Gums JG, & Drlica K. (2004). The changing face of antibiotic prescribing: the mutant selection window. The Annals of pharmacotherapy,38(10), 1675-1682.

Frimodt-Møller N. Correlation between pharmacokinetic/pharmacodynamic parameters and efficacy for antibiotics in the treatment of urinary tract infections. Int J Antimicrobial Agents 19: 546-553, 2002.

Nix DE, Goodwin SD, Peloquin CA, et al. Antibiotic tissue penetration and its relevance: Impact of tissue penetration on infection response. Antimicrob Agents Chemo¬ther 35: 1953-1959, 1991.

Turnidge, J., & Paterson, D. L. (2007). Setting and revising antibacterial susceptibility breakpoints. Clinical Microbiology Reviews, 20(3), 391-408.

ANTIFUNGAL SUSCEPTIBILITY TESTING: WHAT DOES THE MIC MEAN AND HOW DO I APPLY THIS INFORMATION?

Nathan P. Wiederhold, Pharm.D.
San Antonio, TX

Fungi are important causes of infections in both humans and animals. The clinical spectrum of infections caused by these organisms can range from dermatologic infections, such as those caused by dermatophytes, bloodstream infections caused by yeasts such as *Candida* species, pulmonary infections caused by filamentous fungi such as *Aspergillus* and member of the order *Mucorales*, to life-threatening disseminated infections involving multiple organ systems, which can occur in immunocompromised hosts and may be caused by numerous types of fungi. In addition, systemic fungal infections can also occur in immunocompetent individuals, such as those caused by endemic fungi (e.g., *Blastomyces dermatitidis, Coccidioides* spp, and *Histoplasma capsulatum*). The rapid diagnosis and prompt initiation of appropriate therapy have been shown to improve outcomes in the treatment of these infections (1, 2). However, antifungal resistance can develop upon exposure to these drugs, and in some important species increased antifungal resistance has been reported (3, 4). Thus, clinicians may request antifungal susceptibility testing in order help guide appropriate therapy or determine potential reasons for failure when certain drugs are used.

Antifungal susceptibility testing is a useful tool to help guide clinicians in the selection of appropriate therapy against specific fungal pathogens. As against bacteria, antifungal susceptibility test results describe the in vitro potency of an agent against a particular fungal isolate, and this measure is referred to as the minimum inhibitory concentration, or MIC. The MIC is defined as the lowest concentration of a drug that inhibits the growth of an isolate. This can be measured visually, by optical density, or via colorimetric means. Against bacteria, the endpoint that is used is 100% inhibition of growth, or optically clear wells. In antifungal susceptibility testing, the endpoints may vary depending on the drug and the pathogen against which it is being tested (5-7). In addition, different methods can be used to determine the MIC of an antifungal against a particular pathogen. These can include broth microdilution (the reference method), disk diffusion (e.g., Kirby-Bauer), colorimetric assays (e.g., YeastOne Sensititre), diffusion gradient testing (e.g., Etest), and automated systems (e.g., Vitek 2). However, the values that are reported are called MICs regardless of the method and endpoint that are used.

In the treatment of fungal infections, identification of the infecting species can help to guide therapy before susceptibility results are available. This knowledge can be paired with the either local or regional susceptibility patterns to determine if certain drugs are more likely to have activity against the infecting pathogen. Several diagnostic assays that can rapidly identify certain *Candida* species directly in biological fluids have recently become available. These include peptide nucleic acid fluorescence in situ hybridization (PNA FISH) and magnetic resonance technology (8, 9). In addition, other technologies, such as matrix assisted laser desorption ionization time of flight mass spectrometry (MALDI-TOF), are reducing the time needed to identify cultured isolates (10). In susceptibility testing for both bacteria and fungi, the "90-60" rule may apply (11, 12). In general, infections due to susceptible isolates respond to therapy ~90% of the time, while infections due to resistant isolates respond to therapy ~60% of the time. There are several key things to remember regarding antifungal susceptibility testing. First, the MIC results may be dependent on the testing variables. Thus, the method that is used may influence the in vitro results. Second, a low MIC, indicating potent in vitro activity, does not necessarily predict clinical success (11-13). Other factors, including the immune status of the host and certain drug-specific factors, are also important predictors of outcome. Conversely, a high MIC does not predict a poor outcome. However, in vitro resistance does often predict clinical failure.

Susceptibility testing can provide two pieces of information: 1) evidence of in vitro activity of a compound against a microbe, and 2) a measure of the degree of potency of a drug against a particular isolate (11, 12). This potency is as the MIC. For the first line of evidence, an isolate may be classified as either susceptible or resistant. Whether an organism is susceptible or resistant to a particular drug depends on the MIC that is measured and if this value is below or above a specific threshold, or breakpoint. This breakpoint serves to classify microbes as resistant or susceptible to specific agents. In the United States, the breakpoints used to classify drugs as susceptible or resistant are set by the Clinical and Laboratory Standards Institute (CLSI; clsi.org). The CLSI also has established the reference methods for susceptibility testing against both bacteria and fungi. In Europe, this is done by the European Union Committee on Antimicrobial Susceptibility Testing (EUCAST; www.eucast.org). The reference methods for susceptibility testing can be found in the CLSI documents M27-A3 and M38-A2 for yeast and filamentous fungi, respectively (5-7). The current CLSI antifungal breakpoints are shown in Table 1, and are both species and drug specific. However, there are several limitations to breakpoints. The current breakpoints only apply to specific *Candida* species and not to moulds or endemic fungi. In addition, the CLSI has not set breakpoints for some commonly used agents, including amphotericin B, itraconazole, posaconazole, or ketoconazole.

There are several lines of evidence used to establish these clinical breakpoints. These include evaluations of the MIC distributions of antimicrobial agents against specific species, pharmacokinetic characteristics and pharmacokinetic/pharmacodynamics (PK/PD) parameters of the individual drugs, safety and tolerability data, and evidence from clinical trails. Breakpoints should optimize the detection of non-wild type or resistant strains of individual species. One key piece of evidence in setting clinical breakpoints is the epidemiologic cut-off value (ECV or ECoff) (14, 15). To determine this value for a specific species against a specific drug, the MIC distribution against a large number of wild type, or non-resistant, isolates must be determined. There are various ways to determine the epidemiologic cut-off value, but in general this value encompasses the MIC value of a specific agent against a specific species for 95 – 97.5% of the wild-type isolates. This value serves to separate wild type isolates with no mechanisms of resistance from non-wild type isolates, or those with acquired mechanisms of resistance and identify isolates less likely to respond to therapy. The epidemiologic cut-off value also helps to ensure that the clinical breakpoint does not divide the wild-type population, thereby preventing the classification of truly susceptible isolates as resistant. Thus, the overall purpose of the epidemiologic cut-off value is to serve as a sensitive measure of the emergence of resistance, and these values are often used to set the clinical breakpoints for both antibacterial and antifungal agents. For many species, the epidemiologic cut-off value may be the strongest piece of evidence to help guide therapy. However, it is important to remember that they are not the same as clinical breakpoints.

The pharmacokinetics of an agent, or concentrations of the drug within various bodily fluids or tissue in relation to time, is also an important feature that can help guide therapy. This is information is especially useful in the treatment of systemic infections or those occurring in anatomically privileged sites, as the use of drugs with low concentrations at the sites of infection can lead to clinical failure despite potent in vitro activity. Pharmacokinetic/pharmacodynamic (PK/PD) relationships are studied in vivo in order to determine the relationship between the drug concentration and the pharmacodynamic characteristics of the drug, for which the MIC is often used. There are three PK/PD parameters that are used to characterize antimicrobial agents. These include the Cmax/MIC and the AUC/MIC ratios, and the time the concentration of the agent is above the MIC (16, 17). The Cmax is the peak concentration of an agent following administration, where as the AUC, or area under the curve, is a measure of total drug exposure. For many antifungals, the AUC/MIC is the PK/PD parameter most closely associated with in vivo efficacy. Studies have shown that outcomes in in vivo models of invasive fungal infections are improved when the PK/PD parameter associated with efficacy is maximized. While these PK/PD parameters are often used as evidence in the establishment of clinical breakpoints, they are often difficult to measure in individual patients. For the azoles, the trough concentration of the azoles itraconazole, voriconazole, and posaconazole are often measured (18-22). These levels are often convenient to obtain, and there have been several studies that have reported associations between trough concentrations and clinical outcomes in the treatment of or prophylaxis against invasive mycoses. Azole concentrations associated with efficacy and potential toxicity are shown in Table 2. For other antifungal agents with wide therapeutic indices, such as fluconazole and the echinocandins caspofungin, anidulafungin, and micafungin, therapeutic drug monitoring may not be warranted. In order for therapeutic drug monitoring of antifungals to be effective, the concentration results must be available in a timely fashion (23). In addition, well-established pharmacokinetic characteristics correlated with improved outcomes are needed. Unfortunately, in many animal species, the pharmacokinetics of many drugs, including antifungals, are not well characterized.

One of the most important lines of evidence used to establish breakpoints comes from clinical reports, including individual case studies, case series, and if available data from clinical trials. Although randomized controlled clinical trials are the gold standard that helps to guide changes in practice, such types of studies are often not possible for many invasive fungal infections other than invasive candidiasis and invasive aspergillosis due to issues with patient accrual. In addition, in many studies of invasive mould infections, the patients that are enrolled often are classified as having probably invasive fungal infections based on risk factors and associated signs and symptoms. However, in these patients a definitive diagnosis cannot be made, as no organism is able to be cultured. Thus, in many clinical trials susceptibility testing and a correlation with efficacy cannot be made. Although not ideal due to inherent biases, individual case reports and case series are used in the determination of breakpoints, and are often used to provide guidance in making treatment decisions when MIC results are available but clinical breakpoints or results from clinical trials are not available.

Antifungal susceptibility testing can be a valuable tool to help clinicians choose appropriate antifungal therapy. Breakpoints that classify fungal pathogens as susceptible or resistant to specific drugs have been established, but these are not available for all antifungals and against all clinically relevant pathogens. When breakpoints are not available, then the interpretation of the MIC value obtained from susceptibility testing must be interpreted in the proper clinical context. Other pieces of information that can aid in these situations include the MIC distributions of potential therapeutic agents against the species causing the infection, pharmacokinetic data to help ensure that appropriate concentrations are being achieved, and previous clinical experience as reported in the literature.

Table 1. Current CLSI antifungal breakpoints against *Candida* species. Breakpoints have not been established for amphotericin B, posaconazole, itraconazole or other agents, nor have they been set against other fungal species. S-DD - susceptible dose-dependent.

Agent	Species	MIC Breakpoint (μg/ml)		
		Susceptible	Intermediate or S-DD	Resistant
Fluconazole	*C. albicans*	≤ 2	4	≥ 8
	C. glabrata	---	≤ 32	≥ 64
	C. tropicalis	≤ 2	4	≥ 8
	C. parapsilosis	≤ 2	4	≥ 8
Voriconazole	*C. albicans*	≤ 0.12	0.25 – 0.5	≥ 1
	C. tropicalis	≤ 0.12	0.25 – 0.5	≥ 1
	C. parapsilosis	≤ 0.12	0.25 – 0.5	≥ 1
	C. krusei	≤ 0.5	1	≥ 2
Anidulafungin	*C. albicans*	≤ 0.25	0.5	≥ 1
	C. glabrata	≤ 0.12	0.25	≥ 0.5
	C. tropicalis	≤ 0.25	0.5	≥ 1
	C. parapsilosis	≤ 2	4	≥ 8
	C. krusei	≤ 0.25	0.5	≥ 1
Caspofungin	*C. albicans*	≤ 0.25	0.5	≥ 1
	C. glabrata	≤ 0.12	0.25	≥ 0.5
	C. tropicalis	≤ 0.25	0.5	≥ 1

	C. parapsilosis	≤ 2	4	≥ 8
	C. krusei	≤ 0.25	0.5	≥ 1
Micafungin	*C. albicans*	≤ 0.25	0.5	≥ 1
	C. glabrata	≤ 0.06	0.12	≥ 0.25
	C. tropicalis	≤ 0.25	0.5	≥ 1
	C. parapsilosis	≤ 2	4	≥ 8
	C. krusei	≤ 0.25	0.5	≥ 1

Table 2. Azole concentrations associated with in vivo efficacy and toxicity.

Outcome	Itraconazole	Voriconazole	Posaconazole
Efficacy	Prophylaxis – trough ≥ 0.5 μg/mL	Treatment - trough ≥ 1 μg/mL	Prophylaxis – trough 0.5 – 0.7 μg/mL Treatment – trough >1.25 μg/mL
Toxicity	Not well established	Trough >5.5 μg/mL	Not well established

REFERENCES

1. **Garey KW, Rege M, Pai MP, Mingo DE, Suda KJ, Turpin RS, Bearden DT.** 2006. Time to initiation of fluconazole therapy impacts mortality in patients with candidemia: a multi-institutional study. Clin Infect Dis **43**:25-31.
2. **Morrell M, Fraser VJ, Kollef MH.** 2005. Delaying the empiric treatment of candida bloodstream infection until positive blood culture results are obtained: a potential risk factor for hospital mortality. Antimicrob Agents Chemother **49**:3640-3645.
3. **Pfaller MA, Messer SA, Woosley LN, Jones RN, Castanheira M.** 2013. Echinocandin and triazole antifungal susceptibility profiles for clinical opportunistic yeast and mold isolates collected from 2010 to 2011: application of new CLSI clinical breakpoints and epidemiological cutoff values for characterization of geographic and temporal trends of antifungal resistance. J Clin Microbiol **51**:2571-2581.
4. **Verweij PE, Howard SJ, Melchers WJ, Denning DW.** 2009. Azole-resistance in Aspergillus: proposed nomenclature and breakpoints. Drug Resist Updat **12**:141-147.
5. **CLSI.** 2012. Reference Method for Broth Dilution Antifungal Susceptibility Testing of Yeasts; Fourth Informational Supplement. CLSI document M27-S4. Clinical and Laboratory Standards Institute, Wayne, PA.
6. **CLSI.** 2008. Reference method for broth dilution antifungal susceptibility testing of yeasts; Approved standard - Third Edition. CLSI Document M27-A3. Clinical and Laboratory Standards Institute, Wayne, PA.
7. **CLSI.** 2008. Reference method for broth dilution antifungal susceptibility testing of filamentous fungi: Approved standard - Second edition. CLSI document M38-A2. Clinical and Laboratory Standards Institute, Wayne, Pennsylvania.
8. **Hall L, Le Febre KM, Deml SM, Wohlfiel SL, Wengenack NL.** 2012. Evaluation of the Yeast Traffic Light PNA FISH probes for identification of Candida species from positive blood cultures. J Clin Microbiol **50**:1446-1448.
9. **Beyda ND, Alam MJ, Garey KW.** 2013. Comparison of the T2Dx instrument with T2Candida assay and automated blood culture in the detection of Candida species using seeded blood samples. Diagn Microbiol Infect Dis **77**:324-326.
10. **Yang S, Jin Y, Zhao G, Liu J, Zhou X, Yang J, Wang J, Cui Y, Hu X, Li Y, Zhao H.** 2014. Improvement of matrix-assisted laser desorption/ionization time-of-flight mass spectrometry for identification of clinically important Candida species. Clin Lab **60**:37-46.
11. **Pfaller MA, Rex JH, Rinaldi MG.** 1997. Antifungal susceptibility testing: technical advances and potential clinical applications. Clin Infect Dis **24**:776-784.
12. **Rex JH, Pfaller MA, Galgiani JN, Bartlett MS, Espinel-Ingroff A, Ghannoum MA, Lancaster M, Odds FC, Rinaldi MG, Walsh TJ, Barry AL.** 1997. Development of interpretive breakpoints for antifungal susceptibility testing: conceptual framework and analysis of in vitro-in vivo correlation data for fluconazole, itraconazole, and candida infections. Subcommittee on Antifungal Susceptibility Testing of the National Committee for Clinical Laboratory Standards. Clin Infect Dis **24**:235-247.
13. **Pfaller MA.** 2012. Antifungal drug resistance: mechanisms, epidemiology, and consequences for treatment. Am J Med **125**:S3-13.
14. **Pfaller MA, Diekema DJ, Andes D, Arendrup MC, Brown SD, Lockhart SR, Motyl M, Perlin DS, Testing CSfA.** 2011. Clinical breakpoints for the echinocandins and Candida revisited: integration of molecular, clinical, and microbiological data to arrive at species-specific interpretive criteria. Drug Resist Updat **14**:164-176.
15. **Turnidge J, Paterson DL.** 2007. Setting and revising antibacterial susceptibility breakpoints. Clin Microbiol Rev **20**:391-408.
16. **Andes D.** 2004. Antifungal pharmacokinetics and pharmacodynamics: understanding the implications for antifungal drug resistance. Drug Resist Updat **7**:185-194.
17. **Smith J, Andes D.** 2008. Therapeutic drug monitoring of antifungals: pharmacokinetic and pharmacodynamic considerations. Ther Drug Monit **30**:167-172.
18. **Glasmacher A, Hahn C, Leutner C, Molitor E, Wardelmann E, Losem C, Sauerbruch T, Marklein G, Schmidt-Wolf IG.** 1999. Breakthrough invasive fungal infections in neutropenic patients after prophylaxis with itraconazole. Mycoses **42**:443-451.
19. **Glasmacher A, Hahn C, Molitor E, Marklein G, Sauerbruch T, Schmidt-Wolf IG.** 1999. Itraconazole trough concentrations in antifungal prophylaxis with six different dosing regimens using hydroxypropyl-beta-cyclodextrin oral solution or coated-pellet capsules. Mycoses **42**:591-600.
20. **Pascual A, Calandra T, Bolay S, Buclin T, Bille J, Marchetti O.** 2008. Voriconazole therapeutic drug monitoring in patients with invasive mycoses improves efficacy and safety outcomes. Clin Infect Dis **46**:201-211.
21. **Walsh TJ, Raad I, Patterson TF, Chandrasekar P, Donowitz GR, Graybill R, Greene RE, Hachem R, Hadley S, Herbrecht R, Langston A, Louie A, Ribaud P, Segal BH, Stevens DA, van Burik JA, White CS, Corcoran G, Gogate J, Krishna G, Pedicone L, Hardalo C, Perfect JR.** 2007. Treatment of invasive aspergillosis with posaconazole in patients who are refractory to or intolerant of conventional therapy: an externally controlled trial. Clin Infect Dis **44**:2-12.
22. **Dolton MJ, Ray JE, Chen SC, Ng K, Pont L, McLachlan AJ.** 2012. Multicenter study of posaconazole therapeutic drug monitoring: exposure-response relationship and factors affecting concentration. Antimicrob Agents Chemother **56**:5503-5510.
23. **Summers KK, Hardin TC, Gore SJ, Graybill JR.** 1997. Therapeutic drug monitoring of systemic antifungal therapy. J Antimicrob Chemother **40**:753-764.

BARTONELLA UPDATE

Edward B. Breitschwerdt, DVM, DACVIM and Michael R. Lappin, DVM, PhD, DACVIM
Raleigh, North Carolina and Fort Collins, Colorado

INTRODUCTION

Bartonella species are fastidious gram-negative bacteria that are highly adapted to a mammalian reservoir host and within which the bacteria usually cause a long-lasting intraerythrocytic bacteremia. These facts are of particular importance to veterinarians and physicians, as an increasing number of animal reservoir hosts have been identified for various *Bartonella* species. Among numerous other examples, *Bartonella henselae* has co-evolved with cats, *Bartonella vinsonii* subsp. *berkhoffii* has co-evolved with dogs and wild canines, and *Bartonella bovis* has co-evolved with cattle. Importantly, the list of reservoir-adapted *Bartonella* species, including a large number of rodent species that might serve as "pocket pets", continues to grow exponentially, as new *Bartonella* spp. are discovered. Prior to 1990, there were only two named *Bartonella* species, whereas there are now at least 22 named and numerous unnamed or candidatus species, based upon deposited Gen Bank sequences or preliminary reports, respectively.

In the natural reservoir host, chronic bacteremia with a *Bartonella* species can frequently be detected by blood culture or PCR in outwardly healthy individuals. In contrast, the diagnostic detection of a *Bartonella* spp. in a non-reservoir adapted host can be extremely difficult. Most, although not all diseases caused by *Bartonella* spp. occur in accidental hosts and these organisms are being increasingly implicated as a cause of zoonotic infections. Until recently, mechanisms that facilitate persistent *Bartonella* bacteremia in mammals were not well understood. Recent reports have identified an intra-endothelial and intra-erythrocytic localization for these bacteria, which represents a unique strategy for bacterial persistence. Non-hemolytic intracellular colonization of erythrocytes and endothelial cells would preserve the organisms for efficient vector transmission, protect *Bartonella* from the host immune response, and potentially contribute to decreased antimicrobial efficacy. Other *in vitro* studies indicate that *Bartonella* spp. can infect dendritic cells, microglial cells, pericytes, monocytes and CD34+ bone marrow progenitor cells.

EPIDEMIOLOGY

Bartonella vinsonii (berkhoffii) was isolated from a dog with endocarditis in our North Carolina laboratory in 1993. Retrospectively, long-term administration of immunosuppressive doses of corticosteroids for a presumptive diagnosis of systemic lupus erythematosus may have facilitated the isolation of the original type strain of *B. vinsonii (berkhoffii)* from this dog that subsequently developed endocarditis. Due to the relatively recent recognition that dogs can be infected with *B. vinsonii* (*berkhoffii*), *B. henselae* and potentially other *Bartonella* spp., seroprevalence data is limited. Seroprevalence was determined in 1,920 sick dogs from North Carolina or surrounding states that were evaluated at a veterinary teaching hospital. Using a reciprocal titer of >32, only 3.6% of sick dogs had antibodies to *B. vinsonii (berkhoffii)*. Risk factors that could be associated with seroreactivity included: heavy tick exposure (Odds ratio 14.2), cattle exposure (OR 9.3), rural vs. urban environment (OR 7.1) and heavy flea exposure (OR 5.6). These data were interpreted to support the possibility that exposure to *B. vinsonii (berkhoffii)* was more likely in dogs in rural environments that were allowed to roam. In addition, these dogs were likely to have a history of heavy tick and flea infestations. Using sera from dogs experimentally infected with *R. rickettsii* or *Ehrlichia canis*, cross reactivity to *Bartonella* antigens was not detected. However, 36% of serum samples derived from dogs naturally infected with *E. canis* were reactive to *B. vinsonii* antigens. As *E. canis* is transmitted by *Rhipicephalus sanguineous*, this tick may be involved in the transmission of *B. vinsonii*. The possibility of tick transmission was further supported by two additional studies involving dogs infected with one or more *Ehrlichia* spp. from the same geographic region, in which seroreactivity to *B. vinsonii (berkhoffii)* antigens was 30% and 89%, respectively. Seroprevalence, using *B. vinsonii (berkhoffii)* antigens, was 10% (4/40 dogs) in dogs with suspected tick-borne illness from Israel and 36% in dogs with fever and thrombocytopenia from Thailand. Using an ELISA assay, 35% of 869 samples, derived from coyotes in California, contained antibodies to *B. vinsonii (berkhoffii)* antigens. Current data indicates that exposure to *B. vinsonii (berkhoffii)* can be found throughout much of the United States and most tropical and subtropical regions of the world.

Studies from Hawaii, the United Kingdom and Japan identified *B. henselae* seroprevalences of 6.5% (2/31dogs), 3.0% (3/100 dogs) and 7.7% (4/52) In our laboratory, *B. henselae* is the most common *Bartonella* species found in sick dogs using the BAPGM enrichment blood culture platform. The pathogenicity of all *Bartonella* spp. in dogs is poorly understood; however, like humans, dogs can be infected with numerous *Bartonella* spp. *B. henselae* was amplified and sequenced from the liver of a dog with peliosis hepatis, a unique pathological lesion also reported in *B. henselae* infected people. *B. henselae* DNA was also amplified from a dog with granulomatous hepatitis, a histopathological lesion reported with some frequency *B. henselae*-infected children. Similarly, *B. clarridgeae* DNA has been amplified and sequenced from the liver of a Doberman pincher with copper storage disease and from the aortic valve of a dog with vegetative valvular endocarditis. *Bartonella elizabethae*, a species that infects rodents, was found in a dog that had experienced chronic weight loss culminating in sudden unexplained death.

In cats, *B. henselae* and *B. clarridgeiae* have been amplified or grown most frequently from cats and the fleas collected from cats. Flea-associated transmission has been well documented amongst cats. In flea endemic areas, *Bartonella* spp. seroprevalence rates can be greater than 90% and bacteremia rates can be greater than 50%. Granulomatous myocarditis has been reported in cats, naturally or experimentally-infected with *B. henselae*. *B. vinsonii* (*berkhoffii*) caused recurrent osteomyelitis in a cat.

PATHOGENESIS

Although as yet unproven, *B. vinsonii (berkhoffii)* is presumably transmitted to dogs by the bite of an infected tick. Based upon antidotal evidence, dogs may be infected with *B. henselae* by a cat bite or scratch, analogous to cat scratch disease in people. *B. vinsonii* appears to cause chronic intraerythrocytic and endothelial cell infections dogs for extended periods of time. Similar to other highly adapted intracellular vector-transmitted pathogens, the factors that ultimately result in these bacteria causing disease manifestations are yet to be determined. If similar to babesiosis, another intraerythrocytic pathogen, stress, hard work, parturition, or concurrent infection with other organisms may contribute to the development of pathology. Following experimental inoculation of SPF dogs with culture grown *B. vinsonii (berkhoffii)*, there was sustained suppression of peripheral blood CD8+ lymphocytes, accompanied by an altered cell surface phenotype and an increase in CD4+ lymphocytes in the peripheral lymph nodes. Therefore, infection with *B. vinsonii (berkhoffii)* might induce a degree of chronic immunosuppression that could predispose dogs to other infectious agents, resulting in a wide array of clinical manifestations in naturally-infected dogs. Less is known about pathogenesis of disease in cats but some mechanisms are likely similar. Cats infected with *Bartonella* spp. are commonly co-infected with hemoplasmas and more than one *Bartonella*. However, whether co-infections magnify disease manifestations of either genera is unclear and in most studies co-infections did not appear to potentiate illness. From an evolutionary perspective, it is obvious that vectors, vector-borne organisms, and animal and human hosts have developed a highly adapted form of interaction. In general, vectors need blood for nutrition; bacterial, rickettsial and protozoal organisms need an intracellular environment to survive, and immunologically, most hosts appear to be able to support chronic infection with many vector-borne organisms for months to years without obvious deleterious effects. These factors serve to illustrate the potential difficulty in establishing causation in cats, dogs or people co-infected with multiple tick-transmitted pathogens.

CLINICAL FINDINGS

The spectrum of disease associated with *Bartonella* infection in dogs and most other animal species is currently unknown. Endocarditis has been reported in cats, dogs and humans infected with a spectrum of *Bartonella* spp. In some dogs, intermittent lameness, bone pain, epistasis or fever of unknown origin can precede the diagnosis of endocarditis for several months, whereas other dogs will present with an acute history of cardiopulmonary decompensation. Cardiac arrhythmias secondary to myocarditis can be detected in dogs without echocardiographic evidence of endocarditis. Granulomatous lymphadenitis has been associated with *B. vinsonii (berkhoffii)* and *B. henselae* in dogs. *B. vinsonii (berkhoffii)* and other *Bartonella* species appear to contribute to the development of dermatologic lesions indicative of a cutaneous vasculitis, panniculitis, anterior uveitis, polyarthritis, meningoencephalitis and immune-mediated hemolytic anemia. Additional research efforts, using carefully designed case controlled studies will be necessary to establish the frequency and extent to which *Bartonella* spp. contribute to ocular, orthopedic, neurological or hematological abnormalities in dogs.

Many disease manifestations have also been attributed to *Bartonella* spp. infections in cats. However, it is very difficult to prove disease associations in cats in the field because of the high prevalence rates in non-clinical carriers. In research cats that are infected by exposure to *C. felis*, fever, endocarditis, and myocarditis are the most common disease manifestations. As discussed for dogs, additional case controlled studies are needed in cats.

DIAGNOSIS

Thrombocytopenia, anemia, which frequently can be immune-mediated, and neutropenia or neutrophilic leukocytosis are the hematological abnormalities in dogs that are seroreactive or BAPGM enrichment blood culture/PCR positive. Thrombocytopenia is found in approximately half, eosinophilia approximately one third of infected dogs and monocytosis frequently occurs in Bartonella endocarditis. Hematological abnormalities have been rarely reported in cats. Serum biochemical abnormalities are usually very mild or nonexistent in both cats and dogs. In cats, *Bartonella* spp. antibodies were correlated to polyclonal hyperglobulinemia and hypoglycemia.

As *B. vinsonii (berkhoffii)* antibodies are infrequently detected (<4%) in sick or healthy (<1%) dog populations in endemic regions, detection of *B. vinsonii (berkhoffii)* antibodies in a sick dog provides strong clinical evidence for prior exposure and potentially active infection. For this reason, treatment of seroreactive dogs or dogs from which any *Bartonella* spp. DNA is detected in blood or tissue samples would be recommended.

ISOLATION AND MOLECULAR DETECTION OF *BARTONELLA* SPECIES

Because conventional microbiological isolation techniques lack sensitivity, bartonellosis is usually diagnosed by PCR amplification of organism specific DNA sequences and/or through serological testing. Recently, the development of a more sensitive isolation approach, using BAPGM (*Bartonella* alpha *Proteobacteria* growth medium) followed by real time PCR has greatly facilitated the molecular detection or isolation of *Bartonella* species from the blood of sick or healthy animals, including dogs and human beings. Obviously, the relative sensitivity of the diagnostic methods used to detect *Bartonella* species infection greatly influences an investigator's ability to establish disease causation or a clinician's ability to initiate appropriate treatment. The use of this optimized microbiological approach has facilitated the recognition of blood-borne *Bartonella* spp. infections in dogs, horses, human beings and porpoises. Diagnostic testing for *Bartonella* species (serology, PCR and BAPGM Enrichment Blood Culture/PCR) is available through Galaxy Diagnostics, Inc. (contact@galaxydx.com). In cats, PCR or culture combined with serology is recommended and can be procured at Galaxy Diagnostics Inc. and Colorado State University (www.dlab.colostate.edu)

PATHOLOGIC FINDINGS

In dogs, pathologic findings associated with *Bartonella* spp. infection include endocarditis, myocarditis, granulomatous lymphadenitis, granulomatous hepatitis, osteomyelitis, bacillary angiomatosis and peliosis hepatitis. Multifocal areas of severe myocardial inflammation can be found in dogs with *B. vinsonii* (*berkhoffii*) endocarditis. Although not specific for *Bartonella* spp. infections, organisms can be detected in diseased tissues using silver stains, particularly in acute bartonella infections, such as acute regional lymphadenitis (cat scratch disease). During chronic infections, organisms are too few in number to be detected in tissues by silver staining, unless the fulminate infection is localized to heart valves. The cardiac abnormalities noted in cats to date are similar to those described for dogs. It seems likely that the spleen plays an important immunomodulatory role in controlling persistent *Bartonella* spp. bacteremia in animals and people. The extent to which *Bartonella* spp. induce splenic pathology deserves additional research consideration.

THERAPY

To date, an optimal protocol has not been established for the treatment of *Bartonella* spp. infections in cats, dogs, or people. Regardless of the antibiotic that is used for treatment, a long duration of antibiotic administration (4-6 weeks) may be necessary to eliminate the infection. Due to the rapid development of resistance, macrolides (azithromycin) are no longer recommended for treating *Bartonella* infections. Fluoroquinolones in combination with doxycycline are currently being used by the authors to treat clinical cases of bartonellosis, while exploring efficacy through experimental infection studies. Doxycycline alone does not appear to eliminate *B. vinsonii (berkhoffii), B. henselae* or *B. clarridgeae* in cats, dogs or other animal species. Serum antibody titers decrease rapidly (3-6 months) and are generally no longer detectable in dogs that recover following antimicrobial therapy. Therefore, post-treatment serology may be a useful adjunct to BAPGM/PCR to determine if therapeutic elimination of *Bartonella* infections has been achieved. Whether there is clinical benefit to follow serologic or molecular assay results in cats has not been widely studied, but most treated cats do not become seronegative in the short term. However, bacteremia can resolve after treatment or spontaneously in some.

PREVENTION

Although somewhat circumstantial, there is increasing evidence that *Bartonella* species can be transmitted by fleas and ticks to cats, dogs or human beings. Based upon scientific evidence generated during the past several decades, vector-transmitted pathogens can induce clinical manifestations ranging from acute fatal illness (i.e. Rocky Mountain spotted fever, ehrlichiosis, babesiosis and bartonellosis) to chronic debilitating disease states (ehrlichiosis, babesiosis, borreliosis, and bartonellosis). Therefore, minimizing or eliminating flea and tick exposure is perhaps of greater veterinary and public health importance today, than during any previous time in history. When rigorous flea and tick control measures are instituted, it is highly probable that transmission of *Bartonella* species will be greatly reduced or eliminated.

PUBLIC AND OCCUPATIONAL HEALTH CONSIDERATIONS

Due to extensive contact with a spectrum of animal species, veterinary professionals appear to have an occupational risk of infection because of frequent exposure to *Bartonella* spp., therefore these individuals should exercise increased precautions to avoid arthropod bites, arthropod feces (i.e. fleas and lice), animal bites or scratches and direct contact with bodily fluids from sick animals. As *Bartonella* spp. have been isolated from cat, dog or human blood, cerebrospinal fluid, joint fluid, aqueous fluid, seroma fluid and from pleural, pericardial and abdominal effusions, a substantial number of diagnostic biological samples collected on a daily basis in veterinary practices could contain viable bacteria. The increasing number of defined *Bartonella* spp., in conjunction with the high level of bacteremia found in reservoir-adapted hosts, which represent the veterinary patient population, ensures that all veterinary professionals will experience frequent and repeated exposure to animals harboring these bacteria. Therefore, personal protective equipment, frequent hand washing and avoiding cuts and needle sticks have become more important as our knowledge of this genus has improved and various modes of transmission have been defined. Physicians should be educated as to the large number of *Bartonella* spp. in nature, the extensive spectrum of animal reservoir hosts, the diversity of confirmed and potential arthropod vectors, current limitations associated with diagnosis and treatment efficacy, and the ecological and evolving medical complexity of these highly evolved intravascular, endotheliotropic bacteria.

Readers are referred to the following reviews for additional detail and specific references:

Chomel BB, et al. **Vet Res** 2009;40:29, Breitschwerdt EB, et al. **J Vet Emerg Crit Care** 2010:20:8, Chomel BB, et al. **Ann N Y Acad Sci** 2009;1166:120, Harms A and Dehio C. **Clin Microbiol Rev** 2012;25:42, Deng H, et al. **Vet Res** 2012;43:1,Stutzer B and Hartmann K, **J Fel Med Surg** 2012:14:612, Breitschwerdt EB, et al. **J Comp Pathol** 2013;148:115.

INTERACTION OF THE MICROBIOME AND THE IMMUNE SYSTEM

J Scott Weese DVM DVSc DACVIM
Guelph, Ontario, Canada

INTRODUCTION

The body hosts (or is hosted by) a diverse, complex and abundant microbial population, referred to as the microbiota. This complex polymicrobial system outnumbers its mammalian host by a factor of 10-100:1. When the totality of the microbial populations' genetic makeup, the microbiome, is considered, the microbial dominance of host-microbiome complex becomes even more profound. The microbiota and microbiome have tremendous interactions with the body and are absolutely required for health. They are also key determinants of a wide range of diseases, including diverse infectious, inflammatory, neoplastic and metabolic conditions. The complexity of the microbiota, the involvement of multiple groups (bacteria, viruses, Archaea, fungi, parasites), limited information about the function of many microbial genes, functional redundancy between microorganisms of vastly different phylogeny and the complexity of the immune make this a difficult area to study. But, it is clear that an understanding of this polymicrobial population and its role in health and disease is critical.

Evolution and tolerance

One must consider that some components of the microbiota presumably co-evolved with humans and were probably part of the microbiota of pre-humans. The same applies to different animal species. This has allowed abundant time for complex relationships to develop, largely in a synergistic fashion whereby the mammalian host provided nutrients and protection and the microbial partners participated in a range of activities such as vitamin synthesis, digestion, exclusion of opportunistic pathogens and beneficial interaction with the immune system. A key component of this evolutionary relationship is the concept of *tolerance*. Hosts must be able to tolerate their microbial inhabitants. An inability to tolerate the massive antigenic exposure from the microbiota would result in disease (if not rapid death). Tolerance is a well-developed and critical component of immunological development and deficiencies in tolerance are associated with a variety of inflammatory and allergic diseases.[1] Indeed, this area has lead to the 'Hygiene Hypothesis" in humans, whereby there are concerns that decreased antigen exposure early in life has lead to faulty immunoregulation and an immune system that is more apt to respond abnormally over time.[2] While controversial, it focuses needed attention on the complex interaction between the body and its microbial partners, and the positive and negative impacts of this important (and unavoidable) interaction. This has caused a reevaluation of dogma (e.g. parasites – bad), with a particular focus on helminths, since they have essentially been eliminated from individuals in developed countries and development of parasitic infection has been associated with some positive impacts such as decreased multiple sclerosis exacerbations.[3] Improvement in ulcerative colitis and Crohn's disease has been achieved with ingestion of *Trichuris suis*.[4-6] Yet, the original basis of the Hygiene Hypothesis (decreased exposure to microbes early in life leads to increased disease risk) has evolved, with thoughts that ongoing impacts on the microbiota (e.g. antimicrobial exposure, diet) throughout life may also be critical. While large and better-controlled studies are needed, these studies provide further support to the notion that alteration of the local intestinal environment can have significant effects on inflammation. While not specifically studied, it is reasonable to consider the same issues in companion animals, particularly urban dwellers that may have limited exposure to the outdoor environment and other animals.

Gastrointestinal Tract

Most attention has been paid to the gastrointestinal (GI) microbiota, something that is logical because of the abundance of this microbial population, the complexity of the microbiota and clear evidence of a strong influence of the GI microbiota on the immune system. There is overwhelming evidence that the gut microbiome has a complex and continuous interaction with the immune system, as would be expected in a body site with trillions of bacteria from thousands of different microbial species. Germ free animal studies have clearly shown that the immune system cannot develop properly in the absence of the microbiota and that the microbiota can have a profound influence (positive or negative) on the pathogenesis of various inflammatory bowel diseases.[7]

Mechanisms by which the body interacts with the microbiome are complex, only partially understood and beyond the scope of this review, but a few general aspects should be considered. Gut-associated lymphoid tissue (GALT) plays a key role in immunoregulation in the intestinal tract, and these tissues obviously have a close physical relationship with the intestinal microbiota.[8] Dendritic cells in the lamina propria project dendrites through tight junctions of intestinal epithelial cells, allowing direct contact with (and sampling of) the intestinal microflora.[9] These dendritic cells express toll-like receptors that can have marked and variable effects on the immune system. Overall, exposure to the commensal microbiome drives T_{reg} vs Th1 or Th2 effector cells.[8]

Wide-ranging studies have evaluated the impact of the microbiota on the immune system and inflammatory diseases, but the intestinal microbiota can also influence extra-intestinal diseases. For example, one landmark study demonstrated that the susceptibility of rats to adjuvant-associated arthritis depended on their gut microflora.[10] Subsequent studies in conventional and laboratory animals expanded on this initial association and demonstrated a complex (and sometimes contradictory) relationship between the gut microbiota and extra-intestinal inflammation. The wealth of laboratory animal studies that have evaluated microbial/immune system interactions are beyond the scope of this review; however it is important to consider insights that have been derived from study of germ free animals. As mentioned above, it has been well-established that germ free animals develop a relatively 'immature' immune system with decreased lymphoid cell number and function, decreased circulating immunoglobulins and limited mucosal immunity.[8] Further, a bacterial microbiota is required for oral tolerance induction,[11] and deficiencies in tolerance may be associated in risk of various inflammatory and allergic diseases. As an example, germ free mice maintained Th2-mediated responses such as IgE, IgG and IL-4 production in response to ovalbumin administration while the Th2 responses of SPF mice (who had a commensal microflora) allowed them to develop tolerance, but treatment of germ free mice with *Bifidobacterium infantis* (a major component of the mouse microbiota), restored the ability of these mice to develop tolerance. [11] These short proceedings cannot even begin to capture the complex relationship between the microbiota and the immune response, and the role of the microbiota in the development (and prevention or treatment) of disease.

Other microbiotas

While most of the attention is (not unreasonably) focused on the gastrointestinal tract, every body site that communicates with the external world has a microbiota, and these have some degree of interaction with the body. The relative importance of different microbiotas (e.g. skin, upper respiratory tract, oral cavity, urogenital) is impossible to determine, but each of these areas will interact with the body's immune system. Two examples are discussed below, but other microbiotas (e.g. vaginal) are likely also of significance.

Respiratory Tract

The respiratory tract is comprised of a number of diverse ecological niches that can support abundant and diverse microbial populations. This population has been shown in animal models to modulate immune responses (increased expression of pro-IL-1B and pro-IL-18) that are important for protection against influenza virus infection, and that neomycin (but not vancomycin or metronidazole) administration diminished the protective immune response.[12] This suggests that not only does the microbiota influence local protective immune responses, but that this is driven by a specific (in this case, neomycin-susceptible) subset. Similarly, another mouse study reported administration of a microbial supplement and neomycin reduced lung damage after influenza infection compared to neomycin alone.[13]

There has been limited study in animals, yet an equine respiratory microbiota has recently been described (unpublished data). The oral microbiota has also been described in dogs[14] and cats (unpublished data), and differences in the oral microbiota in cats with FIV infection compared to controls have been identified (unpublished data). How these impact on (or are impacted by) the immune system remain to be seen. The nasal microbiota has also been investigated and differences have been identified in pigs carrying methicillin-resistant *Staphylococcus aureus* (MRSA) vs MRSA-negative pigs (unpublished data), but whether this is cause or effect, and whether there is an immunological basis to any differences are unclear.

Skin

The skin is sometimes referred to as the largest organ and it has both a complex microbiota and significant relationship with the immune system. Inflammatory and allergic diseases are also common. The human skin microbiota is highly diverse and rich, containing >1000 bacterial species.[15] The canine skin microbiota is also rich and diverse, and differences have been identified in dogs with allergic skin disease compared to healthy controls.[16] Specific microbiota-immune (either local or systemic) relationships have not been well investigated, but there has been study of the immunomodulatory role of the common skin commensal *Propionibacterium acnes* in various species.

Influence of the immune system on the microbiota

Most study has focused on the impact of the microbiota on the immune system. However, the influence is not necessarily unidirectional. There is limited information about how the immune system itself modulates the body's microbiotas, yet this must certainly occur. This could be through subtle modifications through minor changes in the local immune response or more profound changes such as overgrowth of opportunistic pathogens in response to immunosuppression. For example, a recent study indicates that cats with FIV infection have significant differences in their oral and ocular microbiotas compared to healthy controls (unpublished data), presumably because of alterations in the immune system. But, in many situations, one major difficulty studying the microbiota in different disease states is differentiating cause and effect. Namely, are microbiota alterations the cause of a disease or have they occurred as a result of the disease or its effects (e.g. treatment, diet change…). This is a major challenge in the study of microbiotas and one that is hard to overcome at times.

Using the microbiota to modify the immune response

Questions of cause and effect aside, it is clear that the microbiota impacts the immune response and that alteration of the microbiota (through disease, normal development or treatment) can have a clinical impact. Harnessing the immunomodulatory effects of the microbiota to prevent or treat disease is an appealing area of study and one that is fraught with challenges. Different approaches have been tried. For many years, probiotics have been used to treat inflammatory and allergic diseases in various species, with a lack of a clear understanding of mechanisms of action. While some probiotic studies have shown promise (for certain conditions in some species), results have been at best mixed. Many factors could account for this, including an inability of a relatively small dose of a single (or small number) of different bacteria that are minor components of the normal microbiota to significantly modify an 'abnormal' microbiota. This has lead to resurgence in interest in polymicrobial approaches that more accurately mimic the normal, desired microbiota. Specifically, fecal transplantation has been used with remarkable success for the treatment of recurrent *Clostridium difficile* infection in humans.[17] Promising results have been achieved in preliminary studies involving inflammatory diseases such as ulcerative colitis.[18] A small trial in dogs and cats reported high cure rates in animals with inflammatory bowel disease or idiopathic chronic diarrhea. Though the mechanism of these effects is unclear, it is likely that modification of local inflammatory response in response to modification of the intestinal microbiota plays at least a partial role in clinical response. More study is required to characterize the mechanisms behind response to treatment and refine treatment, but this is an encouraging field of study for intestinal (and perhaps extra-intestinal) disease.

REFERENCES

[1] Hörmannsperger G, et al. *J Allergy Clin Immunol* 2012;129: 1452-1459.
[2] Rook GAW. *Clin Rev Allergy Immunol* 2012;42:5-15.
[3] Correale J, et al. *Ann Neurol* 2007;61:97-108.
[4] Summers RW, et al. *Gut* 2005;54:87-90.
[5] Summers RW, et al. *Gastroenterology* 2005;128:825-832.
[6] Summers RW, et al. *Arch Pathol Lab Med* 2006;130:1753; author reply 1753-1754.
[7] Yi P, et al. *Vet Microbiol* 2012;157:1-7.
[8]Rook GAW, et al. *Therapeutic Microbiology*. 2008;231-247.
[9] Rescigno M, et al. *Nat Immunol* 2001;2:361-367.
[10] Kohashi O, et al. *Microbiol Immunol* 1985;29:487-497.
.[11] Sudo N, et al. *J Immunol* 1997;159:1739-1745.
[12] Ichinohe T, et al. *Proceedings of the National Academy of Sciences of the United States of America* 2011;108:5354-5359.
[13] Wu S, et al. *Current microbiology* 2013;67:414-422.
[14] Sturgeon A, et al. *Veterinary Microbiology* 2012.
[15] Grice EA, et al. *Nat Rev Microbiol* 2011;9:244-253.
[16] Hoffmann A, et al. *PLoS ONE* 2014.
[17] Shahinas D, et al. *MBio* 2012;23:e00338-00312.
[18] de Leon LM, et al. *Clin Gastroenterol Hepatol* 2013; 11: 1036-1038.

THE IMPACT OF GLOBAL WARMING ON INFECTIOUS DISEASES MORBIDITY AND MORTALITY

Michael J. Day, BSc, BVMS(Hons), PhD, DSc, DECVP, FASM, FRCPath, FRCVS
Langford, United Kingdom

INTRODUCTION

Climate change presents one of the major global challenges for now and for the future. This presentation will review the impact of climate change on infectious diseases in wildlife, human and domestic animal populations. The effective management of issues related to climate change requires a One Health approach.

There are two slightly different definitions of climate change. The Intergovernmental Panel on Climate Change (IPPC) defines climate change as a 'change in the state of the climate that can be identified (e.g. using statistical tests) by changes in the mean and/or the variability of its properties, and that persists for an extended period, typically decades or longer'. It refers to any change in climate over time, whether due to natural variability or as a result of human activity. In contrast, the United Nations Framework Convention on Climate Change (UNFCCC) defines climate change as 'a change of climate that is attributed directly or indirectly to human activity that alters the composition of the global atmosphere and that is in addition to natural climate variability observed over comparable time periods.'

There can no longer be doubt that over the past century there has been appreciable overall global warming with increased temperatures of both land and oceans. In fact, global temperature has been relatively stable for the past millennium, but there are modelled predictions for significant rises to 2100. The IPCC provides a range of modelling scenarios for key points over the next century (i.e. 2020, 2050 and 2080) and these modelling tools are often used in studies that aim to predict the effects of climate change on infectious disease prevalence and distribution. Of course, there are still sceptics that contest the science of climate change and this remains an active political and scientific debate.

It is of note that the effects of climate on disease were recorded as early as 2,500 years ago by Hippocrates. He noted that people living in cities with different climates were affected by different diseases and that sudden climatic alterations appeared to affect the severity and incidence of a range of infectious diseases[1]. Throughout the course of history there have been extremes of climate that have impacted on human civilization and led to famine, disease and political unrest[2].

CONSEQUENCES OF CLIMATE CHANGE

The consequences of climate change are wide and include: floods, rain, winds, hurricanes, bushfires, drought, heat waves and rising sea level. These natural phenomena, in parallel with climate change *per se*, impact on the prevalence and distribution of infectious disease. However, it is not simply a climatic event that alters the epidemiology of an infectious disease. In almost all cases, climate change in conjunction with other factors underlies these changes. The range of climate change-associated factors includes: travel and migration, socioeconomic factors, changes in agricultural practice, deforestation and habitat change, the incursion of industry, use of fossil fuels, atmospheric pollution, increased solar radiation and the effects of all of these on the host immune system. The most accurate predictive models of the effects of climate change take into account the relevant confounding factors and do not simply attempt to associate changes in the pattern of disease with climatic factors.

The direct consequences of such climatic shifts for the human (and animal) population will include lack of food and drinking water, reduced sanitation, lack of shelter, enforced population migration, exposure to polluted air and excessive heat, and increased exposure to infectious disease[3].

CLIMATE CHANGE AND NON-INFECTIOUS DISEASE

Although the focus of this presentation is on infectious disease, climate change is also predicted to impact on a spectrum of non-infectious disorders. In particular, there is expected to be a greater prevalence of cardiovascular and respiratory disease, skin cancer, traumatic events (during natural disasters) and to be more widespread malnutrition and mental illness. Allergic diseases are predicted to increase in prevalence as there are longer pollen seasons and with incursion of new pollen-bearing species into novel geographical areas[4]. Those individuals that are most at risk are predicted to be the young, the elderly and women (particularly pregnant women)[5].

CLIMATE CHANGE AND INFECTIOUS DISEASE

In 2008, the Wildlife Conservation Society published a list of 12 pathogens that were predicted to spread into new regions as a result of climate change, with potential impacts on human and wildlife health and on global economies. The list comprises: (1) avian influenza, (2) babesiosis, (3) cholera, (4) Ebola virus, (5) internal and external parasites, (6) Lyme disease, (7) plague (Yersiniosis), (8) 'red tides' (algal blooms), (9) Rift Valley fever, (10) trypanosomiasis, (11) tuberculosis and (12) yellow fever. A number of these are discussed in more detail below.

Wildlife Disease

There are several excellent examples of infectious disease affecting endangered wildlife species that is driven by climate change. One of the most important is chytridiomycosis caused by the fungus *Batrachochytrium dendrobatidis*. This organism infects the outermost skin of amphibians and is a key emerging pathogen of amphibian species, which are the most threatened of vertebrate species worldwide due to habitat loss and infection. The infection has caused numerous mass mortality events in amphibian species. The skin of amphibians is permeable and important in maintaining homeostasis. The fungal colonization of the skin inhibits hydration, which coupled with increased environmental temperature, leads to desiccation and death[6, 7].

A further example of climate change affecting wildlife is the northwards incursion of avian *Plasmodium* infection. Studies have shown an increase in mean annual temperature in Alaska of 1.6°C over the past 60 years and recent surveys have shown *Plasmodium* in new-hatched wild birds up to 64° north[8]. The Arctic and Antarctic are particularly susceptible to the effects of global warming.

Climate change also has particular effects on the world's coral reefs. Rising sea temperatures is associated with the phenomenon of 'bleaching' of coral leading to increased susceptibility to infection and to death of the reef[9].

The Asian elephant is at risk of extinction and a recent study has shown that these animals are particularly susceptible to heat stroke and infectious disease occurring in warmer temperatures. Predictions for global warming are likely to exacerbate these susceptibilities[10].

Infectious Disease of Man and Animals

One of the most significant disease risks for human and animal populations posed by climate change is an increase in water-borne infections. These are predicted to occur with a rise in sea temperature leading to widespread brackish water in coastal regions. These conditions are suitable for increased presence of a range of *Vibrio* species (e.g. cholera) that might be taken in directly or via infected seafood by the human population[11]. This change in coastal water is also predicted to increase the numbers of mosquitoes with saline-tolerant larvae.

There has been increasing prevalence of human infection by the fungal pathogen *Cryptococcus gattii* as this organism has extended its range from tropical to temperate climates. The spores of the fungus are inhaled and establish pulmonary infection which can then disseminate haematogenously – often to the central nervous system[12].

One of the major effects of climate change has been increased incidences of major flooding, particularly in Europe and Asia. Following on from recent floods, there has been an increased incidence of human infection with a range of water-associated pathogens, in particular leptospirosis.

The spread of monkey pox infection has also been related to climate change in combination with deforestation which has driven the spread of this infection from wildlife to man. Predictive modelling in the Congo region of Africa shows a range shift of the infection at the key IPCC points of 2050 and 2080[13]. Monkey pox cycles naturally in rodents and squirrel rats and only occasionally extends to involve non-human primates. Human exposure in Africa comes through the bush meat trade and this may then lead to human – human transmission. In 2003, an infected African rat was imported into the USA leading to the spread of monkey pox to prairie dogs captured for the pet trade and then to people.

In the case of schistosomiasis in Africa, global warming is predicted to reduce the geographical range of the intermediate host snail (*Biomphalaria* spp.), presumably due to loss of the wet habitats of the snail. However, the prevalence of the human disease is not predicted to decline as it is related to other cofactors in addition to the climate[14].

Vector-Borne Disease

Climate change is predicted to have a significant impact on the vector-borne diseases, related to (1) changes in the geographical distribution of vectors, (2) changes in the rate of vector development and seasonal persistence, (3) increased contact with man and animals, (4) altered transmission dynamics and (5) increased pathogen load per vector. There are already numerous examples of the incursion of vectors and infectious diseases carried by them into traditionally non-endemic areas.

Bluetongue virus is now well established in Europe and recent outbreaks have involved more northern European countries and a greater range of strains of the virus. This effect is related to persistence of the virus during the winter months, the northern expansion of the *Culicoides* vector and the involvement of new *Culicoides* vector species[15].

Predictive modelling of trypanosomiasis in Africa suggests that the infection and the Tsetse fly vector will acquire a 60% greater geographical range with climate change, leading to exposure of 46 – 77 million more people (and their livestock)[16].

A major global threat is posed by the spread of Dengue virus, which is transmitted by *Aedes* mosquitoes, primarily *Aedes aegypti* and *Aedes albopictus*. The disease is estimated to affect >100 million people with the greatest focus in SE Asia and the western Pacific. The range of the vector mosquitoes is increasing due to climate change, travel and urbanization[17]. *Aedes albopictus* is now present in Europe after being imported in used car tyres and in 2012 there was an outbreak of Dengue affecting over 1300 people on the island of Madeira. Autochthonous cases of malaria have also been reported in Europe, with outbreaks in Italy and Greece.

Aedes albopictus is also the vector for Chikungunya virus, which has spread from Africa into Asia and now into Europe with an outbreak affecting 161 people in NE Italy in 2007[18]. The same mosquito is also a competent vector for *Dirofilaria immitis*, which is extending its range within Europe.

Leishmaniosis (*Leishmania infantum*) is also extending its range with spread of the *Phlebotomous* sandfly vector related to climate change. The vector is now present in European countries that lie north of the historical Mediterranean endemic region and modelling in the USA suggests similar spread of competent vectors with a prediction for doubling of human exposure by 2080[19].

Babesiosis in dogs is also extending northwards in Europe with spread of one of the vector ticks (*Dermacentor reticulatus*) and autochthonous cases being reported in traditionally non-endemic countries such as the UK.

Climate change is also one part of the underlying reasons for the spread of West Nile virus infection, with modelling focusing on the northward spread of the disease through Canada. These predictions show extension of the range of *Culex tarsalis* with longer seasonal availability, leading to an estimated 18-times greater prevalence of infection by 2050[20]. Similarly, the vector of Lyme disease (*Ixodes scapularis*) is extending its range in Canada due to climate change and dispersal by birds and resident hosts. Predictions are for the range to extend by 46 km/year, increasing the susceptible human population from 18% in 2010 to >80% in 2020[21].

Outbreaks of Rift Valley fever are directly associated with increased rainfall and flooding leading to increased numbers of the primary *Aedes* mosquito vectors that transmit the disease to livestock. Human exposure may come via secondary vectors (e.g. *Culex*) or by direct exposure to infected animals. The disease has spread within Africa and to the Arabian Peninsula and there is potential for spread to Europe related to climate change[22].

THE ONE HEALTH APPROACH TO MANAGING CLIMATE-ASSOCIATED INFECTIOUS DISEASE RISKS

The complex nature of the climate change-associated infectious diseases described above are excellent examples of the necessity for a One Health approach to their management and control. Individuals from a wide range of scientific disciplines (e.g. epidemiology, parasitology, microbiology, ecology) should work together with meteorologists to investigate and model predicted scenarios. In the field, a joined-up approach with veterinarians working together with human physicians and public health officers will be required to identify outbreaks and instigate treatment and control in affected human and animal populations. It is encouraging that within the EU strategy for the management of climate change-associated infectious disease there is already a highlighted need for strengthening the cooperation between the 'three branches of health' (human, animal and plant)[23].

REFERENCES

[1]Fagagas ME et al. **Enferm Infec Microbiol Clin** 2010; 28:716. [2]McMichael AJ et al. **Proc Natl Acad Sci USA** 2012; 109:4730. [3]Kjellstrom T, McMichael AJ **Global Health Action** 2013; 6. [4]Ring J et al. **New Trends Allergy Atopic Eczema** 2010; 96:7. [5]Rylander O et al. **Global Health Action** 2013; 6. [6]Rohr JR, Raffel TR **PNAS** 2010; 107:8269. [7]Voyles et al. **Ecol Evolution** 2012; 2:2241. [8]Loisseau C et al. **PLOS One** 2012; 7. [9]Ruiz-Morenol et al. **Dis Aquatic Organ** 2012; 100:249. [10]Mumby HS et al. **Ecology** 2013; 94:1131. [11]Vezzulli L et al. **Microbial Ecol** 2013; 65:817. [12]Raffa RB et al. **J Clin Pharm Ther** 2012; 37:502. [13]Thomassen HA et al. **PLOS One** 2013; 7. [14]Stensgaard AS et al. **Acta Tropica** 2013; 128:378. [15]Purse BV et al. **Nat Revs Microbiol** 2005; 3:171. [16]Moore S et al. **J Royal Soc Interface** 2012; 9:817. [17]Thomas SM et al. **Erdkunde** 2011; 65:137. [18]Poletti P et al. **PLOS One** 2011; 6. [19]Moo-Llanes D et al. **PLOS Neg Trop Dis** 2013; 7. [20]Chen CC et al. **Int J Environ Res Pub Health** 2013; 10:3053. [21]Leighton PA et al. **J Appl Ecol** 2012; 49:457. [22]El Vilaly et al. **Prog Phys Geog** 2013; 37:259. [23]Bezirtzoglou C et al. **Anaerobe** 2011; 17:337.

HEMOSTASIS AND COAGULATION IN HUMAN LIVER DISEASE: BLEEDING AND CLOTTING - BALANCE AND IMBALANCE

Stephen Caldwell, MD
Nashville, TN

INTRODUCTION

Bleeding and pathological clotting are common problems in chronic liver disease and poorly predicted by conventional testing and often challenging to predict even with refined testing. Work has been ongoing in this field for many years but laboratory advances especially over the past ten years have resulted in a paradigm shift and discrediting of the concept of 'auto-anticoagulation' in cirrhosis. The complexity of the balance and imbalance of hemostatic mechanisms in liver disease patients is increasingly apparent and clinically relevant.

CELL BASED MODEL OF HEMOSTASIS

There have been numerous outstanding contributions to this field over the years but several stand out in my opinion. One is the emergence of the Cell Based Model of hemostasis much of which emerged through the work of Utta Hedner in the development of recombinant factor VIIa. In this model, activation of the clotting cascade can be divided into three phases; primary hemostasis, coagulation or clot formation and fibrinolysis. Activation of the cascade involves initiation (or priming), propagation and amplification. Hemostasis is initiated at the site of vascular injury when Tissue Factor (TF) forms a complex with Factor VIIa and results in a priming amount of thrombin. Thrombin (Factor II) cleaves fibrinogen (Factor I) to fibrin and participates in activation of platelets. Activated platelets, serving as a phospholipid scaffold on which the cascade occurs, bind to a complex of multimeric endothelial-derived von-Willibrand factor and Factor VIII to form a platelet plug. Counter-regulatory elements (protein C, protein S, anti-thrombin) are also activated to counter balance the coagulation cascade. Plasminogen is activated to plasmin which degrades fibrin and breaks down the clot. This entire process is highly regulated. For example, the plasminogen to plasmin ratio is balanced by tissue plasminogen activator (t-PA) and inactivating agents such as plasminogen activator inhibitor (PAI), plasmin inhibitor, and thrombin-activatable fibrinolysis inhibitor (TAFI). Immunothrombosis, an emerging area of investigation highlighting the role of the innate immune system in pathological thrombosis, along with the field of microparticle metabolism can be superimposed on the cell based model in addition to the more conventional intrinsic and extrinsic pathways of hemostasis.[1]

HEMOSTATIC MECHANISMS IN CIRRHOSIS

Bleeding is a common problem in cirrhosis patients especially those undergoing procedures.[2] The classic study by Boks et al demonstrated the extensive derangements of clotting indices in hospitalized cirrhosis patients and the relationship between mucosal bleeding and an index of fibrinolysis.[3] However, stable cirrhosis is characterized by relatively preserved hemostatic mechanisms and even relative hypercoagulability despite prolongation of the prothrombin time (PT) and INR (international normalized ratio).[4] This has been established especially by work from Armando Tripodi and colleagues using modifications of the thrombin generation assay (endogenous thrombin production or ETP) which measures the capacity of blood to produce thrombin or activated factor II. Prolongation of the PT/INR indicates decreased liver synthesis of *pro*-coagulant factors, but these changes are offset by impaired hepatic synthesis of *anti*-coagulant factors such as protein C, S and antithrombin (AT) and other changes in relevant factors synthesized by the vascular endothelial cells. Impaired hepatic synthesis of protein C with normal levels of endothelial-derived thrombomodulin (a protein C co-factor) and increased levels of endothelial-derived factor VIII and von Willebrand factor, lead to enhanced thrombin production and enhanced aspects of platelet function such as adhesion.[1,5,6] Thrombomodulin (TM) is a transmembrane glycoprotein which exists mainly in the endothelium (less so in a soluble form) which promotes protein C activation. Its use in thrombin generation assays permits more physiological study of the pro- and anti-coagulant systems in cirrhosis. Normally, activated protein C (APC) together with TM inactivates factors V and VIII thereby limiting thrombin production (thus promoting ***anti***coagulation). In laboratory testing, the addition of a physiological amount of TM reveals that, thrombin production is significantly limited by the presence of APC when a normal amount of protein C is present. Low levels of liver-derived protein C and high levels of endothelial-derived factor VIII augment thrombin production. Notably, past series have shown fibrinogen levels are often preserved or only mildly decreased in cirrhosis (see Boks et al). These compensatory changes in stable cirrhosis preserve and even accentuate the hemostatic pathway of clot formation.[7]

Hemostatic mechanisms in stable cirrhosis are best seen as rebalanced but in a fragile state that can be easily disturbed by multiple exogenous factors.[8] Bleeding problems are dominant, especially among hospitalized patients, however thrombotic processes such as portal vein thrombosis can result in liver atrophy and failure.[9,10] Bleeding is favored by superimposed conditions such as infection (endogenous 'heparinoids') and renal failure (uremic platelets).[11,12] Hyperfibrinolysis resulting in premature clot dissolution may occur. Portal hypertension obviously drives bleeding in many situations when the collateral system is engorged – this warrants careful consideration of the adverse effects of volume expansion on portal pressure.[13,14] Thrombosis may dominate as a result of the balance of pro- and anti- coagulant factors and especially by stasis of blood flow.[15] Microparticles (phospholipid vesicles) produced by the sick liver may provide another stimulant of coagulation and warrant further study.[16,17]

THE PRECARIOUS BALANCE

The term 'DIC' is sometimes used to describe the coagulopathy of liver disease but the disorder usually lacks features of classical DIC such as rapidly declining platelets and fibrinogen and especially decreased FVIII. The term accelerated intravascular coagulation and fibrinolysis or 'AICF' has been proposed as more accurately describing cirrhotic coagulopathy which is characterized by low but usually stable platelet levels, mildly decreased or normal fibrinogen, decreased pro- and anti-coagulant factors, **increased** Factor VIII and von Willebrand factor, increased clot degradation products and increased fibrinolytic capacity.[18,19]

HYPERFIBRINOLYSIS

Although the topic remains controversial even in 2014, increased clot dissolution (hyperfibrinolysis) in liver disease was described 100 years ago in Dr EW Goodpasture's classical 1914 study. Sixty years later, Boks et al demonstrated the difficulty in predicting bleeding in both acute and decompensated chronic liver disease based on conventional measures of hemostasis among hospitalized patients.[2] In that study, mucosal bleeding in hospitalized cirrhosis patients could only be predicted by factoring in a measure of fibrinolysis or clot dissolution. In my experience, hyperfibrinolysis is most evident in the decompensated cirrhotic where it is characterized by mucosal and puncture wound bleeding and sometimes by delayed post-procedure bleeding. In prior studies, it is estimated to be apparent in around 5-10% of decompensated cirrhotics.[20] The mechanism remains poorly defined but abnormal metabolism of TAFI (Thrombin Activatable Fibrinolysis Inhibitor) has been implicated[21] as has the presence of ascites – a provocative association. Expressed as the 'Global Fibrinolytic Capacity', a step-wise increase in clot dissolution is evident in progressively more severe liver disease.[22] Importantly, one should consider the potential role of anti-fibrinolytics, such as Amicar (aminocaproic acid), to slow clot dissolution when this condition is proven or highly suspected. Most body cavities have fibrinolytic capacity including the oral cavity, bladder and peritoneal cavity so therapy of bleeding from these sources should especially involve consideration of anti-fibrinolytics. Thromboelastography (TEG) is probably the most convincing laboratory test to confirm the diagnosis of hyperfibrinolysis but this approach really needs more clinical translational study because of variation in laboratory techniques.

APPROACH TO BLEEDING AND BLEEDING RISK MANAGEMENT

Clinically bleeding remains the most significant problem facing clinicians whether as a result of procedure complications or otherwise.[2,9] Recent advances discussed above have revealed the limitations of conventional INR as a risk measure in cirrhosis. However, of the ETP test (endogenous thrombin production) have mostly been performed in stable patients and lack clinical/translational studies in decompensated in-patients. Thromboelastography (TEG) may provide a more global measure of risk and has precedent in defining hypercoagulability based on the rate of clot formation and maximal clot strength.[23] However, there is inter-center variation in nuances of the test and variability in performance to detect hyperfibrinolysis between devices (unpublished experience). At our center, in collaboration with Hematology and Radiology, we have developed the following approach which is increasingly gaining acceptance:

1. Identify the risk of procedure and clinical situation: Assess patient's bleeding history and carefully examine for physical clues (i.e. ecchymosis, epistaxis or other mucosal bleeding). Determine if prophylaxis necessary - realizing most routine procedures (EGD with band ligation, colonoscopy with polypectomy, large volume paracentesis) do not require routine prophylaxis. In cases of active bleeding, once initial resuscitation occurs, avoid immediate volume expansion with unnecessary blood products or fluid.
2. Optimize the patient prior to procedure:
   Ensure appropriate medical therapy is administered (i.e. anti-portal hypertensive therapy in some conditions)
   Correct platelets > 50,000 k/μL if evidence of ongoing bleeding or high-risk procedure.
   Correct fibrinogen > 120 mg/dL (with cryoprecipitate) if evidence of ongoing bleeding or high-risk procedure.
   Avoid over-transfusion of blood due to effects on portal pressure and optimize renal function to minimize uremia.
   Avoid using INR as a barometer of bleeding risk and do not use FFP to attempt to correct it.
   Vitamin-K replacement can be occasionally helpful in malnourished patients.
   Consider pre-procedure prophylactic use of DDAVP (intranasal preparation) as an adjunct
3. Pre-plan **rescue** therapies in case of complication in severe bleeding:
   Arrange backup with interventional radiology or surgery if indicated (i.e. need for TIPS, hepatic angiogram)
   Consider use of alternative agents like aminocaproic acid, recombinant FVII, prothrombin complex concentrates
4. Recognize the risk of hypercoagulability in cirrhosis patients and unintended thrombotic complications these therapies may have even if patient demonstrates bleeding.

While much of the foregoing discussion applies mainly to chronic liver diseases, recent work from RT Stravitz and T Lisman has shown that a similar process is evident in fulminant liver failure. Prolongation of the INR as a measure of increased bleeding risk in these cases is not supported by thromboelastography which often reveals the patients to be remarkably hypercoagulable. These findings have substantial implications for what until recently has been conventional therapy including the common use of procoagulant therapy which is now called into question.

HYPERCOAGULABILITY, DVT AND THERAPEUTIC APPROACH IN LIVER DISEASE

Hypercoagulability is not infrequently evident in chronic liver disease patients. Hospitalized cirrhosis patients have an incidence of peripheral deep venous thrombosis of around 1-2%.[24,25] Stasis plays an important role in thrombotic risk both peripherally and with portal vein thrombosis (see below) due to the relative relaxed state of vascular tone in liver disease and within the liver due to capillarization of the sinusoids. Recently, an important role has been identified for the innate immune system and especially for neutrophils which accumulate early along venous walls in stasis and elaborate 'NETs' (neutrophil extracellular traps) which are composed of negatively charged fibrils of DNA and provide a scaffold on which to activate platelets and clotting factors.[26] Such findings raise questions about thrombotic risks in general and some issues regarding over correction of neutropenia in patients undergoing therapy for hepatitis C for example. DVT prophylaxis is generally safe in the absence of bleeding.[27] However, how best to measure drug effect remains uncertain due to the overall changes in hemostasis and clotting co-factors. In vitro studies comparing cirrhosis to control subjects and using thrombin production as an end-point indicate significant differences in activity in liver disease compared to healthy subjects. For example, dibigatran (thrombin inhibitor) appears to have more potent effect while rivaroxaban (a factor Xa inhibitor) effect is somewhat diminished and heparin and low molecular weight heparin showed only a slightly increased anti-coagulant effect in liver disease.[28]

PORTAL VEIN THROMBOSIS (PVT)

PVT develops in 5% to 15% of cirrhosis patients annually with a prevalence of 25% or more among patients without HCC (which increases the risk).[29] The clinical spectrum is broad ranging from partial to occlusive PVT, overtly symptomatic to subclinical, localized or extensive thrombus, and acute versus chronic with cavernous transformation (cavernoma). Additional thrombophilic risk factors like Factor V Leiden and Prothrombin mutation are present in about 10-15% of cases.[30] Slow portal vein blood flow (<15 cm/s) is among the most significant risks for PVT.[31] This has implications for use of one of the most common therapies of portal hypertension - beta blockers - which are known to decrease portal flow. Although there are conflicting reports, the association of PVT in cirrhosis with smaller liver size, ascites and hydrothorax, recurrent variceal bleeding, increased Child-Pugh/MELD scores and especially the effect on survival convincingly argue for its clinical significance.[32,33,34] While also remaining controversial, the occurrence of PVT has an impact on transplant surgery: in one of the larger studies, PVT pre-transplant shifted the survival benefit such that the net benefit of transplant on survival was evident at higher pre-transplant MELD.[35] This effect is complicated but likely real and favors therapeutic intervention in PVT.

PVT: CLINICAL MANAGEMENT

When PVT is discovered at screening ultrasound cross-sectional venous phase imaging with CT or MRI is important to confirm (or refute) the finding since slow flow can masquerade as occlusion in the PV. Such imaging commonly reveals a spontaneous splenorenal or gastrorenal shunt which in most cases appears to have antedated the PVT and probably contributes to PVT risk through steal of portal blood promoting stasis in the PV (unpublished opinion). A hypercoagulable workup is usually undertaken including testing for FV Leiden (resistant for protein C activity), prothrombin 20210, and MTHFR (methylenetetrahydrofolate reductase) mutations to help guide possible anti-coagulant.[36] Measurement of protein C, S and AT is less helpful as their interpretation in cirrhosis is challenging. Early endoscopy should be undertaken with band ligation if significant varices are evident. Efforts to reverse the thrombus with anti-coagulant therapy are increasingly accepted. Anti-coagulation with coumadin or low molecular weight heparin results in recanalization in 40-60% of patients compared to infrequent spontaneous recanalization.[37,38] The decision to anti-coagulate requires consideration of bleeding and fall risk, symptoms, acuity, extent, thrombophilic risks, and possible transplant candidacy. It is generally less effective after 6 months or if cavernous transformation has occurred. The 'on treatment' bleeding risk is approximately 5 - 10%. The role of newer anti-coagulants such as factor Xa or II inhibitors warrants consideration (see above), but clinical data is yet lacking. Anecdotally, Xa inhibitors appear less effective than lovenox in producing recanalization. Their role may ultimately be in prophylaxis although clinical trials are needed to clarify this. TIPS (usually along with post-TIPS anti-coagulation) is an option in these patients even with occlusive PVT provided intrahepatic portal branches can be visualized.[39] This approach has the advantage of offering thrombectomy and varices decompression but carries technical challenges in some patients and increased risk of hepatic decompensation with higher MELD scores.

WHAT IS THE ROLE OF PROPHYLACTIC ANTI-COAGULATION IN CHRONIC LIVER DISEASE WITHOUT PVT?

Because of the general impression that PVT is detrimental, studies have been undertaken to examine the role of prophylaxis. In an initial study of enoxaparin in patients with PVT presenting with esophageal variceal bleeding (anti-coagulation started after varices eradication), Amitrano et al noted no benefit in variceal re-bleeding but they did observe improvement in survival in the treated group.[40] These results have been supported by a subsequent prospective, controlled, and partially blinded trial by Villa et al.[41] In that study, 70 cirrhosis patients without PVT were randomized to prophylactic dosing of enoxaparin (34) versus no treatment (36) for 48 weeks with an additional year of follow-up consisting of q3mos US and q6mos CT. Not unexpectedly, the authors showed improvement in the rate PVT in the treated group but they also demonstrated remarkably diminished episodes of decompensation and improved actuarial survival with therapy. The results were attributed to improved gut perfusion and resulting decreased gut microbial translocation. Alternatively, one could speculate on another possible mechanism which needs further investigation. In particular, the proposed role of intrahepatic microvascular thrombosis in 'parenchymal extinction' put forward by Ian Wanless et al as a mechanism of liver atrophy in cirrhosis may provide a rational and target for prophylactic anti-coagulation. The concepts involved with parenchymal extinction are further supported in work by Anstee and Thursz et al showing that thrombin activates fibrosis pathways in stellate cells via the PAR-1 receptor (PAR: protease activated receptor) and in unpublished work by Y Ikura et al showing evidence of intra-hepatic activated platelets in advanced cirrhosis. Further studies are needed to examine these potentially important questions.

CONCLUSION

The fundamental concepts underlying the coagulopathy of liver disease have undergone extensive refinement such that the notion of 'auto-anticoagulation' has been discredited. Although bleeding problems dominate in the clinical setting, liver disease patients may be relatively hypercoagulable despite prolongation of the INR. This has been demonstrated in laboratory studies of human blood samples comparing liver disease patients to healthy controls comparing the ability of the samples to generate thrombin (ETP: endogenous thrombin production) and in more global assays of blood clot formation such as thromboelastography. The ETP assay has utilized reactions with and without thrombomodulin, an endothelial derived protein C activator, to demonstrate the key importance of liver-derived protein C deficit in producing a relative hypercoagulable state. Indeed, it is striking that the risk of developing portal vein thrombosis (PVT) appears to be inversely proportional to protein C levels such that lower levels of this important anticoagulant system increase the risk of PVT. It is very plausible that this situation combined with stasis (slow flow) in the portal vein and secondary activity of fibrilar neutrophil-derived 'NETS' (neutrophil extracellular traps) underlie a substantial proportion of PVT in liver disease patients. Among the challenges now facing clinicians is to translate recent laboratory progress into clinically viable means of risk stratification in order to guide prophylactic and interventional therapy. The exciting possibility that long term anti-coagulant therapy in stable cirrhosis staves of disease progression opens a new and potentially significant therapeutic avenue to prevent liver atrophy and decompensation. However, recognition of conditions promoting premature clot dissolution such as hyperfibrinolysis or the formation of weak clots possibly prone to abnormal regulation such as in dysfibrinogenemia needs much more investigation to allow appropriate management.

REFERENCES

[1] Engelmann B. Massberg S. Thrombosis as an intravascular effector of innate immunity. Nature Reviews 2013;13:34-45

[2] Giannini EG, Greco A, Marenco S, Andorno E, Valente U, Savarino V. Incidence of bleeding following invasive procedures in patients with thrombocytopenia and advanced liver disease. Clin Gastroenterol Hepatol 2010;8:899-902

[3] Boks AL et al., Hemostasis and fibrinolysis in severe liver failure and their relation to hemorrhage. Hepatol, 1986;6:79-86.

[4] Tripodi A, Mannucci PM. The coagulopathy of chronic liver disease. N Engl J Med. 2011 Jul 14;365(2):147-56.

[5] Lisman T et al. Elevated levels of von Willebrand Factor in cirrhosis support platelet adhesion despite reduced functional capacity. Hepatology 2006;44:53-61

[6] Biguzzi E et al. Endothelial protein C receptor plasma levels increase in chronic liver disease, while thrombomodulin plasma levels increase only in hepatocellular carcinoma. Thromb Res. 2007;120:289-93.

[7] Tripodi A et al. Evidence that low protein C contributes to the procoagulant imbalance in cirrhosis. J Hepatol. 2013 Apr 11. pii: S0168-8278

[8] Lisman T, Porte RJ. Rebalanced hemostasis in patients with liver disease: evidence and clinical consequences. Blood. 2010;116:878-85

[9] Shah NL et al. A clinical survey of bleeding, thrombosis, and blood product use in decompensated cirrhosis patients. Ann Hepatol. 2012;11:686-90

[10] Anstee QM et al. Parenchymal extinction: coagulation and hepatic fibrogenesis.Clin Liver Dis. 2009;13:117-26.

[11] Montalto P et al. Bacterial infection in cirrhosis impairs coagulation by a heparin effect: a prospective study. J Hepatol. 2002;37:463-70

[12] Caldwell SH et al. Coagulation disorders and hemostasis in liver disease: pathophysiology and critical assessment of current management. Hepatology. 2006;44:1039-46.

[13] Massicotte L et al. Effects of phlebotomy and phenylephrine infusion on portal venous pressure and systemic hemodynamics during liver transplantation. Transplantation. 2010;89:920-7.

[14] Zimmon DS, Kessler RE. The portal pressure-blood volume relationship in cirrhosis. Gut 1974; 15: 99-101

[15] Northup P, Caldwell S. Coagulation in Liver Disease: A Guide for the Clinician. Clin Gastroenterol Hepatol. 2013 Mar 16. doi:pii: S1542-3565(13)00317-0. 10.1016/j.cgh.2013.02.026. [Epub ahead of print]

[16] Kornek M et al. Circulating microparticles as disease-specific biomarkers of severity of inflammation in patients with hepatitis C or nonalcoholic steatohepatitis. Gastro 2012;143:448-58.

[17] Lacroix R et al. Leukocyte- and endothelial-derived microparticles: a circulating source for fibrinolysis. Haematologica. 2012;97:1864-72.

[18] Joist JH. AICF and DIC in liver cirrhosis: expressions of a hypercoagulable state. Am J Gastroenterol. 1999;94:2801-3.

[19] Ben-Ari Z et al. Disseminated intravascular coagulation in liver cirrhosis: fact or fiction? Am J Gastroenterol. 1999;94:2977-82.

[20] Hu KQ et al Hyperfibrinolytic activity in hospitalized cirrhotic patients in a referral liver unit. Am J Gastroenterol. 2001;96:1581-6.

[21] Colucci M et al. Deficiency of thrombin activatable fibrinolysis inhibitor in cirrhosis is associated with increased plasma fibrinolysis. Hepatology. 2003;38:230-7.

[22] Rijken DC et al. Evidence for an enhanced fibrinolytic capacity in cirrhosis as measured with two different global fibrinolysis tests. J Thromb Haemost. 2012;10:2116-22.

[23] Ben-Ari Z et al Hypercoagulability in patients with primary biliary cirrhosis and primary sclerosing cholangitis evaluated by thrombelastography. J Hepatol. 1997;26:554-9.

[24] Northup PG et al. Coagulopathy does not fully protect hospitalized cirrhosis patients from peripheral venous thromboembolism. Am J Gastroenterol. 2006;101:1524-8

[25] Gulley D et al. Deep vein thrombosis and pulmonary embolism in cirrhosis patients. Dig Dis Sci. 2008;53:3012-7.

[26] von Bruhl M-L, Stark K, Steinhart A et al. Monocytes, neutrophils, and platelets cooperate to initiate and propagate venous thrombosis in mice in vivo. J Exp Med 2012;209:819-35

[27] Intagliata NM et al. Prophylactic anticoagulation for venous thromboembolism in hospitalized cirrhosis patients is not associated with high rates of gastrointestinal bleeding. Liver Int. 2013 May 13. doi: 10.1111/liv.12211

[28] Potze W, Arshad F, Adelmeijer J et al. Differential in vitro inhibition of thrombin generation by anti-coagulant drugs ib plasma from patients with cirrhosis. PLOS One 2014;9:1-9

[29] Tsochatzis EA et al. Systematic review: portal vein thrombosis in cirrhosis. Aliment Pharm Ther. 2010;31:366-74

[30] Garcia-Pagan JC, Valla DC. Portal vein thrombosis: a predictable milestone in cirrhosis? J Hepatol. 2009;51:632-4.

[31] Zocco MA et al. Thrombotic risk factors in patients with liver cirrhosis: correlation with MELD scoring system and portal vein thrombosis development. J Hepatol. 2009;51:682-9

[32] John BV, Konjeti R, Aggarwal A, et al. Impact of untreated portal vein thrombosis on pre- and post-liver transplant outcomes in cirrhosis, Ann Hepatol 2013;12:952

[33] Amitrano L et al. Risk factors and clinical presentation of portal vein thrombosis in patients with liver cirrhosis. J Hepatol. 2004;40:736-41.

[34] Englesbe MJ et al. Portal vein thrombosis and survival in patients with cirrhosis. Liver Transpl. 2010;16:83-90.

[35] Englesbe MJ et al. Portal vein thrombosis and liver transplant survival benefit. Liver Transpl. 2010;16:999-1005

[36] Deleve LD et al. Vascular disorders of the Liver. Hepatol 2009;49:1729-64

[37] Francoz C et al. Splanchnic vein thrombosis in candidates for liver transplantation: usefulness of screening and anticoagulation. Gut. 2005;54:691-7.

[38] Delgado MG et al Efficacy and safety of anticoagulation on patients with cirrhosis and portal vein thrombosis.
Clin Gastroenterol Hepatol. 2012;10:776-83

[39] Senzolo M et al. Prospective evaluation of anticoagulation and transjugular intrahepatic portosystemic shunt for the management of portal vein thrombosis in cirrhosis. Liver International 2012;32:919-27

[40] Amitrano L et al. Splanchnic vein thrombosis and variceal rebleeding in patients with cirrhosis. Eur J Gastroenterol Hepatol. 2012;24:1381-5.

[41] Villa E et al. Enoxaparin prevents portal vein thrombosis and liver decompensation in patients with advanced cirrhosis. Gastroenterology. 2012;143:1253-60

HEMOSTASIS IN SMALL ANIMAL LIVER DISEASE
A REVIEW OF THE LITERATURE

Armelle deLaforcade DVM, DACVECC
North Grafton, MA

INTRODUCTION

A complex balance between pro- and anti-coagulation systems exists in hepatobiliary disease. Classically, one thinks of patients with liver disease as being hypocoagulable, but hypercoaguability may also occur. Hemostasis consists of primary hemostasis, secondary hemostasis, and fibrinolysis. Primary hemostasis, the initial interaction between platelets and the endothelial surface to form the platelet plug, is activated by vascular injury. Secondary hemostasis is the series of steps through which coagulation factors become activated and leads to the generation of thrombin. Thrombin in turn catalyzes the formation of a cross-linked fibrin network that stabilizes the initial platelet plug. Fibrinolysis, the degradation of fibrin, is mediated by multiple factors. Plasminogen is normally incorporated into the clot. Thrombin generated during hemostasis stimulates endothelial cells to produce tissue plasminogen activator (tPA). When t-PA binds to fibrin in the clot, it develops a strong affinity for plasminogen and cleaves it into plasmin. Plasmin is responsible for the breakdown of fibrin into fibrin degradation products (FDP's).

Several regulatory mechanisms regulate normal hemostasis. First, since platelets cannot adhere to normal vascular endothelium, maintenance of vascular integrity itself prevents inappropriate activation of primary hemostasis. Endothelial cells produce nitric oxide (NO) and prostacyclin (PGI2), which reduces platelet aggregation and causes vasodilatation. Endothelial cells also produce Von Willebrand Factor (vWF), which is initially secreted as biologically active large multimers and cleaved into smaller multimers by ADAMTS13. These smaller multimers are less efficient in mediating platelet aggregation and activation. ADAMTS13 can also cleave vWF at the edges of the thrombus preventing clot expansion. Endogenous anticoagulant systems downregulate hemostasis through regulation of thrombin. Antithrombin (AT) limits thrombin formation by inactivating any FXa that dissociates from the vessel surface. Thrombin also binds to thrombomodulin, a plasma membrane protein on endothelial cells, which activates protein C. Activated protein C forms a complex with protein S that then cleaves and inactivates FVa and FVIIIa, effectively stopping the generation of thrombin. Excess fibrinolysis is prevented by alpha-2-antiplasmin and tissue plasminogen activator inhibitor (PAI-1). Alpha-2-antiplasmin inactivates free plasmin and to a lesser extent fibrin bound plasmin while PAI-1 inactivates tPA. Conversely, activated protein C can promote fibrinolysis by inactivating PAI-1.

The liver plays a crucial role in all aspects of these coagulation pathways. The majority of factors that regulate pro-coagulation, anti-coagulation and fibrinolysis are produced and/or activated in the liver. Fibrinogen, prothrombin, FV, FVII, FIX, FX, FXI, FXII,and FXIII are all synthesized at least to some extent in the liver. Hepatocytes are also the site of vitamin-K-dependent carboxylation of prothrombin, FVII, FIX, FX, protein C, and protein S. Many anticoagulant proteins including AT, protein C and protein S, are synthesized in the liver. In addition to the liver's role in the production of pro and anticoagulant factors, it also synthesizes factors that control fibrinolysis including plasminogen, PAI-1, and alpha-2-antiplasmin. Lastly, the liver is responsible for the clearance of many activated pro and anticoagulant factors and their by-products (e.g FDPs and D-dimers). With the liver playing such a critical role in both the formation and clearance of both pro-coagulants and anti-coagulants as well as many components of fibrinolysis, it is not surprising that a wide range of coagulation disturbances can accompany abnormal hepatobiliary function.

Primary hemostasis in hepatic disesase

Platelet abnormalities occur in veterinary patients with hepatobiliary disease. In a large retrospective study evaluating over 800 dogs with thrombocytopenia, 4% had hepatobiliary disease. Several retrospective studies have documented mild to moderate thrombocytopenia in approximately 25% of dogs with chronic hepatitis (CH). Dogs with cirrhosis are more likely to be thrombocytopenic than dogs with CH without cirrhosis. The presence of thrombocytopenia also has been was correlated with decreased survival in Labrador retrievers with CH. Mild thrombocytopenia is seen in a small proportion of dogs with congenital portosystemic shunts (CPSS). Conversely, massive hepatocellular carcinoma has been associated with thrombocytosis (> 500,000/ul). The mechanisms responsible for changes in platelet number in dogs with hepatobiliary disease have not been elucidated. Potential causes include gastrointestinal hemorrhage, occult thrombosis, autoimmune destruction, increased consumption from low grade DIC, an hepatic synthetic failure of thrombopoeitin. Thrombocytopenia is rare in cats with hepatobiliary disease and usually reflects the presence of another underlying disorder such as DIC or sepsis.

The contribution of thrombocytopathia to coagulation disorders in veterinary patients with hepatobiliary disease remains poorly characterized due to inherent difficulties in monitoring these changes. In one study dogs with various liver diseases including CPSS, CH, hepatocellular degeneration or neoplasia, there was evidence for thrombocytopathia on the basis of decreased platelet aggregation in response to arachidonic acid and collagen despite normal platelet numbers. Three out of 5 of the dogs with abnormal platelet function had concurrent bleeding tendencies suggesting that alternations in function may be important clinically. Additional studies are necessary on platelet function in veterinary patients with hepatobiliary disease.

Emerging evidence supports a role for vascular endothelial dysfunction in coagulation changes related to hepatobiliary disease. Portal hypertension and resulting sheer stress leads to production of vasoactive substances (NO, Prostacyclin) which impact platelet aggregation.

Secondary Hemostasis and Hepatic Disease

Defects in secondary hemostasis related to hepatobiliary disease can be due to reduced production of coagulation factors, failure of vitamin-K mediated carboxylation of coagulation factors, and altered production of endogenous anticoagulants. Multiple veterinary studies have documented changes in coagulation factor levels in veterinary patients with liver disease. Altered PT and aPTT are documented in roughly 1/3 of dogs with chronic hepatitis. Decreased FXI has been documented in dogs with hepatic degeneration and decreased FIX, FX, and FXI have been seen with cirrhosis. In a study of 47 dogs with liver disease, abnormalities in at least one coagulation test were documented in 57% of dogs. Abnormal prothrombin time (PT) and activated partial thromboplastin time (aPTT), which primarily reflect coagulation factor function, were found in 50% and 75% of dogs respectively. In one study measuring FII, FV, FVII, FVIII, FIX, FX, FXI, and FXII levels (as compared to 15 control dogs), all factors were decreased in patients with cirrhosis, but not in those with CH. Overall, 53% of dogs had a coagulation factor abnormality. None of these studies however have reported that these coagulation abnormalities were associated with bleeding tendencies.

Several studies have suggested that many cats with acute and chronic hepatobiliary disease have some component of vitamin K deficiency. In two studies looking at a mixed group of feline patients with HL, cholangitis, or hepatic neoplasia, the most frequently reported coagulation changes were PT prolongation and reduced FVII levels or increased PIVKA (protein invoked by vitamin K absence), both of which are suggestive of vitamin K deficiency. It is important to note that in these studies that some cats with abnormal coagulation tests did have increased bleeding tendencies, all of which were associated with provocative procedures such as venipuncture or IV catheter placement.

Fibrinolytic System in Liver Disease

There is some evidence supporting the presence of hyperfibrinolytic states in veterinary patients with liver disease. In one study, increased serum fibrin degradation products were found in 5/8 of the animals with liver disease. This increase could be reflective of reduced hepatic clearance of FDPs. In a study using thromboelastography in 50 dogs with DIC, 3 dogs with acute hepatitis were classified as having DIC, with 2 being hypercoagulable and 1 hypocoagulable. In our preliminary work we have seen several dogs with acute liver failure that have a profound hyperfibrinolyis present on TEG analysis. Decreased fibrinogen levels have been reported in feline patients with HL (12/19 patients in one study) and in dogs with CPSS, acute liver disease and CH/cirrhosis. Whether these changes in fibrinogen correlate with altered coagulation status has not been investigated.

Hypocoagulability in liver disease

Spontaneous bleeding in patients, even those with reported coagulation test abnormalities, is rare. It most often occurs in veterinary patients with acute liver failure. Acute liver failure is a severe variant of acute liver disease defined by the onset coagulopathy (prolongation of PT and aPTT) and encephalopathy. Dog presented with ALF due to xylitol and aflatoxin have been reported to have spontaneous bleeding disturbances (hemoabdomen, gastrointestinal bleeding) which may be a manifestation of their disease, but also could reflect the onset of DIC. Bleeding following provocative procedures is possible with hepatobiliary disease although the incidence of bleeding complications after invasive procedures such as catheter placement, jugular venipuncture or percutaneous biopsy has not been well document. In one study, a significant correlation between thrombocytopenia (less than 80,000/uL) and bleeding after ultrasound-guided biopsy of the kidney and/or liver has been demonstrated in both dogs and cats. This study evaluated patients for minor (drop of hematocrit by greater than 10% but no intervention necessary) and major complications (intervention required or death) after biopsy. Of the 434 cases, 254 had liver biopsies. Dogs with a prolonged PT (above the normal reference range) and cats with a prolonged PTT (1.5 times the normal reference range) were more likely to develop major or minor complications post-procedure. The time to the lowest post-procedure hematocrit in this study was 5.2 hr and all major complications occurred within 10 hr. Another retrospective study in cats found that liver disease was a common cause for blood transfusion (7/93) and FFP transfusion (7/19), although the study did not specifically indicate whether the transfusions were therapeutic or prophylactic. Another study evaluated one dog with cirrhosis and one with CH which had prolonged BMBT and coagulation times. When these dogs were treated with DDAVP, the values improved but did not normalize. None of the dogs in this study developed complications after liver biopsy.

Hypercoagulability in liver disease

A hypercoagulable state can accompany hepatobiliary disease in people. In humans, chronic liver disease is complicated by the occurrence of portal vein thrombosis (PVT) and the presence of intrahepatic microvascular thrombi. In addition, portopulmonary hypertension, a syndrome seen in cirrhotics is associated with intrapulmonary microthrombi. Hypercoagulability in liver disease arises from increased procoagulant activity (vWF), reduced activity of endogenous anticoagulants (AT, protein C), altered vascular flow (due to portal hypertension), or disordered fibrinolysis. Increases in endothelial cell production of vWF occur due to sheer stress and reduced clearance by the liver. Sheer stress secondary to portal hypertension also increases endothelial cell production of NO resulting in vasodilation and slowing of blood flow creating an environment for clot formation. In humans, decreased portal vascular flow secondary to portal hypertension is an important risk factor for thrombotic complications in cirrhotics. Hypercoagulability may also result from decreased synthesis of antithrombotic molecules such as AT, protein C, and protein S. Loss of AT, protein C or protein S leaves the pro-coagulation potential of thrombin unchecked resulting in hypercoagulability.

Microvascular thrombosis occurs in chronic liver disease in humans and contributes to ongoing pathology in a phenomenon called parenchymal extinction. Parenchymal extinction is the concept that microthrombi in the hepatic vasculature lead to tissue ischemia and activation of hepatic stellate cells. Stellate cell activation is closely associated with platelet activation and aggregation around sites of hepatic injury. Microvacular thrombi formation itself leads to thrombin generation and thrombin in turn binds to and activates hepatic stellate cells. In murine models of hepatotoxicity or cholestasis, inhibition of thrombin, thrombocytopenia, or administration of low molecular weight heparin attenuates the development of hepatic fibrosis, but not inflammation/necrosis. These studies suggest that anticoagulant therapy might be beneficial in slowing the progression of hepatic fibrosis.

Hemostatic changes potentially contributing to a hypercoagulable state have been documented in veterinary patients. Decreased protein C and AT activities have been documented in dogs with ALD, cirrhosis and CPSS, although none of the dogs with these alterations were reported to have thrombotic complications. A combination of decreased AT, low protein C, and hyperbilirubinemia did, however, did carry a grave prognosis in ALD probably reflected hepatic synthetic failure. In CPSS the decreased PC and AT may be due to decreased hepatoportal perfusion itself and not synthetic failure as factor levels and fibrinogen levels were normal. In addition, restoration of portal flow post shunt ligation normalizes protein C and AT activity in these dogs. Although this study did not report on thrombotic complications, PVT has been reported after ligation of CPSS in dogs. A recent study using TEG in dogs with CPSS identified hypercoagulability in 9/21 dogs, with hypercoagulable dogs shown to be more likely to have hepatic encephalopathy.

Portal vein thrombosis (PVT) has been reported as a complication in cats and dogs with hepatic disease A retrospective evaluation of 6 cats with PVT revealed underlying hepatic disease in each cat (3 with CPSS, 2 with hepatic neoplasia, and 1 with acute cholangitis). All cats had clinicopathologic evidence of hepatobiliary disease. Prothrombin (PT) and partial thromboplastin (aPTT) times were obtained in 5 cats. Two cats had a mild increase in PT (<20% above upper limit of normal), and 1 cat with acute liver failure had a marked increase in both PT and aPTT. All cats had abnormal findings on hepatic histopathology. Small to absent terminal portal venules and increased numbers of hepatic arterioles were seen in 3 cats, consistent with a CPSS. One of these cats had a pancreatic biopsy which revealed evidence of chronic lymphocytic and eosinophilic pancreatitis with fibrosis. Two cats had neoplasia in the liver. One cat had a tubulopapillary adenocarcinoma and the other a poorly differentiated carcinoma (along with severe multifocal chronic neutrophilic cholangitis with fibrosis). One cat had acute hepatic necrosis, pancreatitis, and cholangitis. Two cats were suspected to have acute thrombi and 4 cats had chronic thrombi.

In a series of 33 dogs with PVT, 14/33 (42%) had underlying hepatobiliary disease. Of these 14 dogs, 4 had chronic PVT and 10 acute PVT. Eight of 14 dogs with hepatic disease had been diagnosed with CH. Other diagnoses included hepatic neoplasia (cholangiocarcinoma, undifferentiated sarcoma, neuroendocrine tumor, unclassified cystic liver mass), congenital portosystemic shunt (CPSS), and congential hepatoarteriovenous fistula. Many dogs had other pre-existing potentially thrombotic conditions. Five out of 8 dogs with CH were on glucocorticoids, 2 of which had at cirrhosis on necropsy. Two additional dogs with CH had a history of an invasive procedure. Similarly 2/3 dogs with vascular disease and 2/4 dogs with neoplasia had a history of recent surgical intervention. Overall 12/14 dogs with PVT and hepatobiliary disease had 1 or more pre-existing potentially prothrombotic condition. Additional thrombi were identified on ultrasound in 4/8 dogs with CH (splenic vein (2), mesenteric vein (2), and aorta (1)). Interestingly, 7/14 dogs with hepatobiliary disease and PVT survived to discharge and this was not different from the overall proportion of dogs that survived (19/33). Three of 4 dogs with chronic PVT survived but only 4/9 with acute PVT lived and this difference was significantly different. A diagnosis of PVT should be considered in dogs with hepatobiliary disease that present with abdominal effusion and/or abdominal pain particularly dogs that are on glucocorticoids or who have recently undergone an invasive procedure.

REFERENCES

1. Botsch V, Kuchenhoff H, Hartmann K. Retrospective study of 871 dogs with thrombocytopenia. Vet Rec 2009; 164:647-651
2. Willis S, Jackson M, Meric S. Whole blood platelet aggregation in dogs with liver disease. Am J Vet Res 1989; 50(11): 1893-1989
3. Badylak S, Van Vleet J. Alterations in prothrombin time and activated partial thromboplastin time in dogs with hepatic disease. Am J Vet Res 1981;42(12):2053-2056
4. Prins M, Schellens C, van Leeuwen et al. Coagulation disorders in dogs with hepatic disease. Vet J 2010;185:163-168
5. Kavanagh C, Shaw S, Webster CL. Coagulation in hepatobiliary disease. J Vet Emerg Crit Care 2011;21(6): 589-604
6. Kelley D, Lester C, deLaforcade A, et al. Thromboelastographic evaluation of dogs with congenital portosystemic shunts. J Vet Intern Med 2013;27(5): 1262-1267
7. Respess M, O'Toole TE, Taeymans O, et al. Portal vein thrombosis in 33 dogs:1998-2011. J Vet Intern Med 2012;26(2):230-7
8. Rogers CL, O'Toole TE, Keating JH, et al. Portal vein thrombosis in cats: 6 cases (2001-2006). J Vet Intern Med 2008;22(2):282-7

TO BLEED OR TO CLOT: MANAGEMENT OF HEMOSTATIC DISORDERS IN SMALL ANIMALS WITH LIVER DISEASE

Cynthia RL Webster, DVM, DACVIM (Small Animal Internal Medicine)
Grafton, MA

CLINCIAL AND DIAGNOSTIC EVALUATION OF PATIENTS WITH LIVER DISEASE

Establishing the coagulation status of a patient with hepatobiliary disease can be quite challenging due to the wide range of coagulation alterations that can occur with hepatobiliary dysfunction. The liver plays a crucial role in all aspects of these coagulation pathways.[1] The majority of factors that regulate pro-coagulation, anti-coagulation and fibrinolysis are produced and/or activated in the liver. Fibrinogen, prothrombin, FV, FVII, FIX, FX, FXI, FXII, and FXIII are all synthesized at least to some extent in the liver. Hepatocytes are also the site of vitamin-K-dependent carboxylation of prothrombin, FVII, FIX, FX, protein C, and protein S. Many anti-coagulant proteins including AT, protein C and protein S, are synthesized in the liver. In addition to the liver's role in the production of pro and anticoagulant factors, it also synthesizes factors that control fibrinolysis including plasminogen, PAI-1, and alpha-2-antiplasmin. Lastly, the liver is responsible for the clearance of many activated pro and anticoagulant factors and their by-products (e.g. FDPs and D-dimers). With the liver playing such a critical role in both the formation and clearance of both pro-coagulants and anti-coagulants as well as many components of fibrinolysis, it is not surprising that a wide range of coagulation disturbances can accompany abnormal hepatobiliary function

All patients presenting for liver disease should have baseline coagulation testing including platelet count, PT, and aPTT as these tests can help guide diagnostic and treatment decisions, and perhaps even more importantly carry prognostic significance.[2-4] The PT represents the time it takes plasma to clot after the addition of tissue factor, and can be prolonged secondary to a lack of vitamin K, decreased factor synthesis (especially FVII), or increased consumption such as with DIC. The aPPT may become prolonged secondary to factor VIII, IX, XI, or XII deficiency, vitamin K deficiency (later in the disease course than PT), the use of heparin, or secondary to VWF deficiency if it is affecting FVIII.

The clinical relevance of prolonged PT and aPTT in human patients with liver disease has been evaluated at length.[5,6] The biggest limitation of PT and aPTT is the poor correlation with actual clinical bleeding, which is likely due to the failure of these tests to measure abnormalities in the vascular endothelium or fibrinolytic system or to detect alterations in the level of anti-coagulants. PT and aPTT are more suitable for evaluation of congenital factor deficiencies of pro-coagulants than the acquired deficiencies of both pro and anticoagulants that occur with liver disease. None-the less, evaluation of PT and aPTT is of value in assessing the coagulation status of patients. Severe increases (>1.5 x upper limit of normal) do represent increased risk for post-procedure bleeding. Since bleeding is a serious complication of invasive procedures such as surgery or biopsy in liver disease patients, the identification of patients with high risk becomes important even if the test is somewhat imprecise.

In both human and veterinary patients, prolonged clotting times have prognostic significance. A study with 151 dogs with CH showed that both PT and PTT elevations predicted death within the first week.[2] Additionally in a smaller study of Labrador retrievers with CH, a prolonged PT was associated with a higher probability of dying in the first 2 months, while a prolonged PTT was associated with an overall shorter survival time.[3] A similar association between PT/PTT and shortened survival time was noted in a study evaluating dogs with both acute and chronic hepatitis.[5] Increases in clotting times also predict non-survival in ALF.[4] In humans PT and PTT prolongation, along with the occurrence of encephalopathy, are used as part of the diagnostic criteria to differentiate acute liver disease from acute liver failure. The coagulation parameters are part of the scoring systems (King's College Criteria) used to determine the severity of ALF and to decide on the need for possible liver transplantation.

Feline studies have not been able to document an association between prolongation of PT and/or aPTT and survival. Two separate studies, however, have demonstrated an association between abnormal coagulation tests and the degree of elevation of serum liver enzymes (aspartate aminotransferase and alkaline phosphatase) and bilirubin in cats,[1] suggesting that coagulation abnormalities are markers of more severe disease. Consistent with this observation, all cats reported with acute liver failure due to diazepam toxicity had coagulation abnormalities and all died. Further studies are needed evaluating the prognostic significance of coagulation abnormalities in feline patients with liver disease.

Measurements of protein C and AT, each of which can be decreased in hepatic disease, are selectively evaluated in patients with hepatobiliary disease particularly those perceived to be at an increased risk for thrombosis.[4,7,8] These include cirrhotics and patients with liver disease who need therapy with drugs (corticosteroids) or have concurrent diseases (immune-mediated hemolytic anemia, hyperadrenocorticism, or protein losing enteropathy) that might shift coagulation status toward hypercoagulability.

The buccal mucosal bleeding time (BMBT) has also been used to assess primary hemostasis. Prolongation of the BMBT indicates qualitative or quantitative platelet defects or defective vasoconstriction at the site of vascular injury. In humans with hepatic disease increases in BMBT have been associated with a 5 times increased risk of a drop in hemoglobin after biopsy.[1] However, correction of the deficit with platelet transfusion in people, is not always associated with less risk of bleeding.

Additional coagulation testing such as determination of FDPs and D-dimers, which are markers of hyperfibinolysis, have not proved useful in classifying coagulation status in human patients with liver disease. In patients with decreased liver function, elevations in these values most often represent decreased hepatic clearance and not increased production from fibrin breakdown. It can be challenging to differentiate the coagulation abnormalities accompanying liver failure from those seen with DIC. In humans, the parameter used to distinguish these states is the presence of moderate to severe thrombocytopenia with DIC.

The application of more global coagulation tests, such as TEG, may ultimately provide a better understanding of coagulopathy and liver disease. TEG is a coagulation test performed on whole blood that allows for a global assessment of hemostatic function as it yields a graphic representation of clot formation through evaluation of primary hemostasis, secondary hemostasis and clot lysis. Clot formation is evaluated as a dynamic process, measuring clot time and strength, as well as the kinetics of clot formation. For many years, TEG has proven useful to help predict coagulation status and plan for intervention in liver transplant and cardiac surgery patients. Although very few studies using TEG to evaluate the coagulation status of human patients with liver disease exist, isolated case series have shown utility in directing coagulation therapies during transplant. TEG has also been used to document hypercoagulability in patients with biliary disease and PVT, hypocoagulability in patients with infection and early rebleeding from esophageal varicies. Although TEG may provide valuable information about coagulation status, it is important to realize that this test is still in the early stages of evaluation even in human patients with liver disease. The other limitations of TEG include availability, time constraints (test must be run shortly after blood is obtained), and its inability to measure EC dysfunction.

A recent study in veterinary literature documents hypercoaguability in dogs with exrahepatic bile duct obstruction.[9] We have conducted preliminary analsyis of TEG profiles in dogs and cat with hepatic disease. In a recently published study, TEG profiles in dogs with congenital portosystemic shunts showed a tendency to be hypercoagulable.[10] Hypercoaguability was correlated with hepatic encephalopathy perhaps reflecting the pro-inflammatory state of the latter. In preliminary results, dog with acute liver injury, G value defined 11 dogs as hypocoagulable, 3 as hypercoagulable, and 8 as normocoagulable. Eight dogs were hyperfibrinolytic, 7 of which had normal platelet counts. All dogs with ALF were hypocoagulable on TEG and 6/13 were hyperfibrinolytic. In contrast, 3/9 dogs with ALI were hypercoaguable and 1 was hypocoaguable. A study in dogs with chronic hepatitis and cats with cholestatic disorders are currently in progress.

TREATMENT OPTIONS FOR COAGULOPATHIES

Once a coagulation status is established for a patient with hepatobiliary disease, treatment may be initiated to help prevent the complications associated with alterations in hemostasis. Ideally treatment should be aimed at reversal of clinically relevant problems (bleeding, thrombosis) and not at reversing abnormal laboratory parameters which may or may not provide an accurate assessment of bleeding/thrombotic risk.

Bleeding risk assessment is an important part of clinical evaluation of patients with hepatic disease. The decision as to whether it is safe to perform invasive procedures (fine needle aspirate, biopsy, surgery) in a patient with liver disease can be complicated. Although there is no clear consensus in human as to what coagulation parameters represent a risk too severe to perform liver biopsy, or warrant the prophylactic use of blood products, some guidelines exist. These suggest that the number of platelets required for near normal thrombin generation in cirrhotics is around 56 X 10^9/L, and discourage biopsy with a PT >1.5 ULN.[11] A single study in the veterinary literature suggests that platelet counts less than 86,000 ul/L in the dog and cat, any prolongation of PT in the dog and aPPT prolongation greater than 1.4 X ULN in cats may be risk factors for bleeding from percutaneous US guided biopsy procedure, but risks associated with surgery or other invasive procedures have not been assessed.[12]

In the human literature, multiple studies have demonstrated that attempted correction of prolonged PT and PTT values or thrombocytopenia with FFP administration or platelet transfusions, respectively, does not always have an effect on the incidence of post-procedural bleeding. These observations suggests that the reason for the underlying bleeding tendency in patients with hepatobiliary disorder may not be entirely due to defects in coagulation factors or platelet function , but may instead be secondary to some component of the coagulation cascade that is not impacted by replenishment of soluble factors or platelets, such as vascular endothelial dysfunction.

Despite the limitations in current coagulation testing and often inadequate response to pro-coagulant therapy, due to ethical and legal considerations, the practice of prophylactic therapy remains common among human hepatologists.[11] Other factors that figure into the decision to give prophylactic blood product therapy are history of bleeding disorder in the patient, the perceived bleeding risk associated with procedure (aspirate vs biopsy vs surgery), likelihood of rescue therapy being successful after the proposed procedure, the cost of the intervention, and ease of managing potential complications or side effects of prophylactic treatment. Any benefit of prophylactic therapy needs to be weighed against potential side effects of blood product support which include transmission of infectious agents, occurrence of transfusion related lung injury(TRALI) or other immune-related transfusion reaction and volume overload. Indeed the latter may be particularly deleterious as there is literature in the human field to support the observation that volume overload potentiates portal hypertension which actually promotes greater bleeding. While administration of blood products has not decreased transfusion requirements in human cirrhotics undergoing hepatic resection (usually for HCC or liver transplantation), the maintenance of low normal central venous pressure by fluid restriction, forced diuresis, venous dilatation or even phlebotomy has been associated with a striking reduction in blood loss. Presumably the benefit is related to a reduction in portal hypertension and fits with the emerging theory that bleeding tendencies in human cirrhotics might best be explained by the presence of portal hypertension and vascular endothelial dysfunction. Whether the same relationship is true in veterinary patients is unknown. Clearly, this treatment strategy would be associated with other risks such as renal failure.

MANAGEMENT OF NON-BLEEDING PATIENTS

Despite the inherent limitations of bleeding risk assessment and the evidence for lack of efficacy in interventional management, there are some guidelines that can be applied to the management of coagulation status in patients with hepatobiliary disease. All patients with prolonged coagulation times should be treated with subcutaneous vitamin K1 to ensure that this vitamin deficiency is not contributing to the coagulopathy. Considering the poor sensitivity the PT and aPTT in detecting bleeding abnormalities however, it may be prudent to prophylactically treat all cholestatic (hyperbilirubinemic) patients with VK1. This may be particularly true in feline patients. Since there is a high likelihood that these patients may need provocative procedures during their hospitalization, prompt initiation of VK1 therapy is warranted. Parenteral administration of a 0.5 to 1 mg/kg dose of VK1 subcutaneously should correct the deficiency within 24 to 36 hours. Excess VK1 administration should be avoided as it is pro-oxidant. Typically therapy is given for 3-5 days and then weekly if the primary cholestatic disorder is not corrected.

In patients with prolonged clotting times, but no evidence of spontaneous bleeding in which no provocative procedures are planned, there is no need to administer blood products solely for the sake of improving bleeding times. Even if provocative procedures are necessary there is growing feeling in human hepatologists that in the face of mild alterations in clotting times, these patients should not receive prophylactic therapy, but instead be provided with rescue therapy, if necessary. Prior to performing invasive procedures in these patients, the clinician should be prepared to address any post procedure hemorrhage by ensuring that the patient is blood typed and that blood products are available.

Cats with obstructive jaundice present a unique challenge. Multiple studies have demonstrated that these patients have significant bleeding associated with surgery and often require blood product support. In two retrospective studies, 40 and 57% of undergoing surgery for extra-hepatic bile duct obstruction required transfusion therapy peri-operatively, despite a low incidence of prolonged pre-operative PT/aPTT. [13,14] Possible explanations for the increased need for transfusions in these patients may be the relatively low sensitivity of PT/aPTT for vitamin K deficiency (the most common coagulopathy with obstructive biliary disease), the high vascularity of the feline hepatobiliary system, the high incidence of hypotension intra-operatively which may contribute to the decision to transfuse, and the dilutional coagulopathy that is likely to occur after administration of high volumes of fluids to maintain blood pressure in these patients. Additionally preliminary analysis of TEG profiles in cholestatic cats have identified some cats with hyperfibrinolysis. Studies which closely examine global coagulation status and/or the benefit of prophylactic blood products in the peri-operative management of in cats with obstructive jaundice are urgently needed.

MANAGEMENT OF PATIENTS WITH ACTIVE HEMORRHAGE

Fresh frozen plasma (FFP), it is the first line treatment modality for spontaneous hemorrhage or iatrogenic bleeding as it supplies the widest range of coagulation factors and is readily available. The side effects of FFP in veterinary patients are poorly documented, but volume overload (with exacerbation of portal hypertension), infections, transfusion reactions, and acute lung injury are all potential problems associated with its use. The recommended dose range in humans is 5-15 ml/kg, and this recommendation has been suggested in veterinary medicine as a starting point, using discontinuation of bleeding or correction of coagulation times as potential endpoints. Given the previously discussed poor correlation of clotting times to bleeding tendency, the authors recommend using discontinuation of bleeding as an endpoint for FFP transfusions and not correction of PT and PTT. In one study looking at the transfusion of blood product to cat, liver disease was the 3rd most common reason for transfusion with fresh frozen plasma.[15] In 2 studies in dogs, 46% and 80% of the FFP transfusions given to dogs were for coagulation tests abnormalities.[16,17] One study documented significant improvement of PT and aPTT with transfusion. Cryoprecipitate is indicated for hypofibrinogenemia (fibrinogen less than 100 mg/dl in humans), and can also be used to supply FVIII, vWF, and FXIII. The advantages of cryoprecipitate include a decreased volume of administration, and the provision of targeted correction. Platelet transfusions are not typically used due to the expense and very short duration of effect.

A newer treatment option for coagulopathic humans with hepatobiliary disease is administration of recombinant factor VIIa (rFVIIa).[18] Increasing circulating FVIIa works to enhance thrombin generation at the site of vascular injury either through increased TF-FVIIa complex (thus increased FXa production) or by the binding of rFVIIa to the surface of activated platelets (which may also increase the activation of FX). Human studies have shown rFVIIa can improve PT, but the current consensus is that this should be reserved for cases nonresponsive to or too volume overloaded for FFP or cryoprecipate due to expense and relatively limited clinical evaluation of the product. Ideally recombinant FVIIa should be given in combination with FFP to replenish deficient coagulation factors. The obvious risk of rFVIIa use is thrombosis, but this risk is limited by the products short half live (2-3 hrs).

Antifibrinolytic therapy would be indicated if hyperfibrinolysis is documented with TEG or suspected based on delayed bleeding from mucosal surfaces after invasive procedures. The most frequently used therapies in human patients are aminocaproic acid and tranexamic acid, which inhibit plasmin. In humans undergoing liver transplantation, these therapies have been shown to reverse the evidence of hyperfibrinolysis seen on TEG, and case series suggest they can be effective in controlling hemorrhage in patients with cirrhosis. Veterinary studies evaluating these anti-fibrinolytic agents, and their possible use in patients with liver disease are needed.

MANAGING HYPERCOAGUABLE STATES

At this time, the role of treatment for hypercoagulability in liver disease is poorly defined. In a recent consensus statement from the American Association for the Study of Liver disease it was recommended that patients with acute thrombosis should be treated with anticoagulation (warfarin or heparin) for at least 6 months. The treatment of chronic thrombosis with anticoagulants is more controversial, as studies show little benefit in thrombus resolution with treatment. Other treatment options for hypercoagulable states include platelet inhibitors such as clopidogrel or NSAIDS. Since veterinary patients with hepatic disease are already predisposed to gastrointestinal ulceration, NSAIDs are relatively contraindicated in these patients due to their ulcerogenic potential. Clopidogrel, which works by inhibiting ADP receptors on the platelet membrane, may be a safer choice in veterinary patients with liver disease. Treatment is recommended in patients with current evidence of thrombosis, or high risk of thrombosis based on TEG or underlying disease process. Anti-platelet drugs should be discontinued at least several (ideally ten) days prior to liver biopsy. A recent study in dogs with portal vein thrombosis suggested that treatment with anti-coagulants improved survival.[19]

Emerging evidence suggests that anti-coagulation with LMWH may be beneficial in patients with cirrhosis.[20-24] Enoxapariin was effective in rat models of toxic or cholestatic disease in decreasing fibrosis. In a small randomized controlled trial, a 12-month course of enoxaparin was safe and effective in preventing PVT in patients with cirrhosis and appeared to delay the occurrence of hepatic decompensation and improve survival. Administration of LMWH also had an anti-fibrotic effect in a small study of patients with chronic hepatitis B.

REFERENCES

1.Kavanagh C, et.al. J Vet Emerg Crit Care 2011;21(6):589-604. **2.** Shih JL, et. al. J Vet Intern Med 2007;21:33-39. **3**.Strombeck DR et. al. J Am Vet Med Assoc 1988;193(9):1109-1113. **4**.Dereszynski DM, et al J Am Vet Med Assoc 2008;232(9):1329-1337. **5**.Tripodi A, et. al. Aliment Pharmacol Ther. 2007;26(2):141-8 **6.** Lisman T, et. al J Hepatol. 2010 ;53(2):362-71. **7**. Poldervaart JH, et. al. J Vet Intern Med. 2009;23(1):72-80. **8**. Prins M et. al. Vet J. 2010;185(2):163-8. **9.** Mayhew PD,et. al. . J Am Vet Med Assoc. 2013;242(6):778-85 **10.** Kelley D,et. al. J Vet Intern Med. 2013;27(5):1262-7. **11.**Rockey DC, et.al. Hepatology 2009;49:1017-1044. **12.** Bigge LA, et. al. J Am Anim Hosp Assoc 2001;37(3):228-233. **13.** Boute NJ, et al. J Am Vet Med Assoc 2006;9:1376-1382. **14** Mayhew PD,et. al. J Small Anim Pract 2002;43(6):247-253. **15.** Castellanos I,et.al. J Vet Intern Med 2004;18:529-532. **16**. Logan JC, et al. J Am Vet Med Assoc 2001;218(9):1449-1455. **17**. Snow SJ et. al. J Vet Emerg Crit Care 2010;20(4):441-5. **18.** Sajjad S, et. al. Dig Dis Sci 2009;54(5):1115-1119. **19.** Respess M, et. al. J Vet Intern Med. 2012;26(2):230-7. **20**. Lee JH, et. al. Biomaterials. 2011;32(5):1438-45. **21.** Abe W, et. al. J Hepatol. 2007;46(2):286-94. **22.**Abdel-Salam OM et. al. Pharmacol Res. 2005;51(1):59-67. **23.**Villa E, et. al..Gastroenterology. 2012 ;143(5):1253-60. **24**.Shi J, et. al. World J Gastroenterol. 2003;9(7):1611-4

NEW AND OLD DRUGS FOR CANINE STATUS EPILEPTICUS

Edward (Ned) E. Patterson D.V.M., Ph.D., DACVIM (SAIM)
St. Paul, Minnesota

INTRODUCTION

Most seizures in dogs are self-limiting and last a couple of minutes or less. Seizures that last more than a few minutes or occur back-to-back without recovery in-between are emergency situations and should be treated promptly, aggressively, and with a systematic plan.[1] When 30-60 minutes of continuous seizure activity occurs, irreversible neuronal damage begins, mainly due to excito-toxic cell injury related to excessive glutamate release.[2] This can lead to loss of normal brain homeostasis, kindling, functional and structural hippocampal changes, neuro-degeneration and altered distribution of ion channels and neurotransmitter receptors on a cellular level.[3] Systemically, status epilepticus (SE) can result in hyperglycemia, hypertension,[4] neuronal necrosis, hyperthermia, cardiac arrhythmias, kidney damage, metabolic acidosis, disseminated intravascular coagulation, cardiorespiratory failure, and a predisposition to further seizure episodes.[5] Consequently, aggressive and safe early intervention is important. This review will cover old drugs and also drugs ***new*** or **"potentially new*"** for canine SE.

DEFINITIONS

Status epilepticus in the current veterinary and human literature is most often defined as a single seizure lasting 5 minutes or longer, or 2 or more seizures without recovery in between.[6] Previous definitions had often been seizure activity lasting for at least 30 minutes due the clear cut potential for permanent damage at that point. The rational for a shorter duration for the definition has been the need for urgent and aggressive treatment early in the process well before the 30-minute mark.[7]

Convulsive status epilepticus (CSE) is SE with convulsions that are associated with rhythmic jerking of the extremities such as generalized tonic–clonic movements and/or mental status impairment and/or neurological deficits.[8]

Non-convulsive status epilepticus (NCSE) is defined as prolonged seizure activity seen on electroencephalogram (EEG) without the clinical findings associated with CSE with either a wandering and confused patient or an with severely impaired mental status.[8] Since EEGs are not common practice in veterinary medicine at this time, NCSE is not well recognized and this review will be limited to CSE.

Refractory status epilepticus (RSE) in people is defined as patients who continue to experience either clinical or EEG seizures after receiving adequate doses an initial benzodiazepam followed by a second acceptable AED.[8] Patients with RSE usually require anesthetic agents at anesthetic doses as third line treatment. Experimental studies have shown that there are changes in GABA receptors[3] during RSE and that is likely why traditional benzodiazepams likely are no longer effective at this stage.

Super refractory status epilepticus (SRSE) is SE that goes on 24 hours or more after the onset of anesthesia.[9] Patients with SRSE usually require days to weeks of anesthetic agents at anesthetic doses as fourth line treatment. In the author's experience, due to financial constraints and other considerations, companion animals are almost always invariably euthanized before SRSE occurs.

The initial management of SE should utilize basic principles of life support and drug administration to stop the seizures. Sedating anti-seizure drugs can lead to loss of pharyngeal tone and risk of aspiration. Oral or nasal administration of oxygen is therefore needed in some patients, and when anesthetic drugs are used, intubation and ventilator support is required. Intravenous access should be obtained as soon as possible and blood collected for glucose measurement and therapeutic monitoring in patients already on AED therapy. Temperature, pulse oximetry, EKG, and blood pressure should be monitored and any abnormalities, such as hyperthermia, hypoglycemia or hypocalcemia should be quickly treated.[4] As the SE is being treated with drugs, efforts should also be made to diagnose the underlying cause as quickly as possible and start treatment for any causes with specific therapies.

HISTORICAL DRUG THERAPY FOR STATUS EPILEPTICUS IN PEOPLE

In the 19th century, starting in the 1860's bromides were given hypodermally, rectally, or by direct injection into the mouth. Intravenous barbiturates, including phenobarbital, were introduced in the 1920's. Phenytoin was first used by intravenous injection in 1958. Benzodiazepines were introduced for SE in 1965 and the therapy of SE, surprisingly remained mainly unchanged for more than 40 years until the recent introduction of IV levetiracetam, IV valproate, and IV lacosamide.[10]

CURRENT THERAPY OF STATUS EPILEPTICUS

Despite the high levels of morbidity and mortality associated with SE in people, there is a relative paucity of prospective, controlled research to determine the optimum treatment regimens for people. To date, 4 well-planned, prospective, randomized, double-masked studies have been performed. These have been limited to first line therapy for SE in people. The most notable findings from these studies are: IV lorazepam is superior to IV diazepam for first line treatment[11]; IV lorazepam is superior to IV phenytoin alone, and is comparable to IV phenobarbital treatment and combination IV diazepam and phenytoin treatment[12]; IM midazolam is equivalent to IV lorazepam in the pre hospital treatment by paramedics.[13]

To the author's knowledge there has not yet been an expert panel consensus statement for the treatment of companion animal SE. There does seem, however, to be fairly similar recommendations from a number of sources,[4,15,16] that can be generally summarized by this author as: <u>First line therapy</u> should be with a benzodiazepam which is most often is IV **diazepam,** but can be by other routes[14] if IV access is not available, and/or midazolam, or lorazepam. (There have not been any published studies comparing the benzodiazepams to each other in dogs or cats as there has been for people). Shortly after the benzodiazepam there should be IV loading or mini loading doses of IV **phenobarbital** or IV ***levetiracetam*** to start chronic therapy, for when the short acting benzodiazepams wear off. In <u>second line therapy</u> for continuing seizure activity, IV phenobarbital or IV levetiracetam or a CRI of diazepam are most often recommended In this author's experience, two or more of these second line therapies can potentially be given to the same patient. Third line therapy for refractory SE (RSE) to induce general anesthesia can be with IV **propofol** or IV **pentobarbital**. In some instances IV ketamine or inhalant anesthesia has been administered. The pharmacokinetics of intranasal midazolam and intranasal lorazepam have also been studied in dogs for possible use in the hospital and/or at home by owners for SE or cluster seizures. It should be noted that midazolam and lorazepam have poor rectal absorption. [15]

Recent Studies in Dogs

Levetiracetam (LEV) was approved in the US in 1998 for the oral treatment of partial onset seizures in adult people. An IV formulation was approved in 2006 for use as bridge therapy in people.[7] Since that time, there have been increasing reports of the off-label use of LEV in people for the treatment of seizure emergencies with variable response rates. In adults, success rates for the treatment of SE have ranged from 44-71%, with LEV as second line therapy with few reported side effect. [17-19] A wide safety margin and minimal drug-drug interactions, has led to LEV use in children, the elderly and critically ill patients with SE. Our group completed a randomized clinical trial of LEV for SE and cluster seizures for dogs in 2012 indicating that 30-60mg/kg LEV IV for second line therapy is safe and might be effective.[20]

Fosphenytoin (FOS) is the most often used second line agent in people.[8] Our group has published that IV FOS at 15mg/kg IV phenytoin equivalent (PE) at a rate of 50mg/min PE infusion is generally well tolerated in normal dogs, with some minor vomiting in a few instances.[21] We have just completed a randomized clinical trial of diazepam and placebo vs. diazepam and 15mg/kg PE of FOS for canine SE in 31 cases and we found that of the 22 who received FOS, 59.1% had no additional seizures versus 22.0% of the placebo group, which was statistically significant difference.[22] The diazepam/FOS response was similar to the diazepam/PHT results of 55.8% in the 1998 Trieman study in people.[10] The results of this study are in preparation for full manuscript submission, and we expect they should be published in the near future. Based on the full results of our study and possibly other future studies, FOS might be a potential additional non-anesthetizing choice for 2nd line therapy of SE for dogs.

Combination first line therapy

It has been suggested by some that first line therapy be with two non-anesthetizing drugs immediately rather than benzodiazepines first and second line therapy later, and should be studied.

Synergy between LEV and diazepam has been shown in both rodent models of SE, and in a clinical population of people with acute repetitive or prolonged seizures. In rats with electrical-stimulation induced SE, the combination of LEV and diazepam, as first line therapy was superior to either drug alone, even when plasma concentrations were well below the therapeutic range of either drug alone.[30] Two reasons for possibly utilizing IV LEV combined with benzodiazepines as first line therapy are the very high safety profile of LEV, and favorable pharmacokinetics. Compared to all other AEDs, LEV has the widest safety margin of an AED.[31] Doses as high as 1200mg/kg/day PO have been shown to cause only mild side effects in long-term oral dosing studies in dogs, (data on file UCB Pharma). Future controlled prospective trials are needed to determine whether early combination therapy might be more effective and safe compared to benzodiazepam therapy only.

FUTURE POSSIBILITIES

With the IV formulations for people of **valproate*** and **lacosamide*** being used 2nd line for HSE, there could be the potential of use of these drugs in companion animals. These drug have too short of a half life to be useful as chronic therapy in dogs, but either might still be proven useful and safe as non-anesthetizing drugs for canine SE, if tested in future veterinary studies.

In the last ten years there are other evolving potential therapies for SE beyond the benzodiazepams, which open the hope for new treatments, and if successful there could also be paradigm shifts; these include but are not limited to the next generation of piracetams such as **brevetiracetam***, **neuro-steroids***[32] , gene therapy, optogenetics, use of antagomirs[10], and new biochemical targets such as adenosine.[33]

FINAL CONCLUSIONS

Status epilepticus in companion animals is an emergency situation and should be quickly treated by recommended first line (emergent) therapy with benzodiazepines followed by loading doses of chronic therapy drugs, and then second, and third line (refractory) therapy when needed.[1]

REFERENCES

1 Patterson EE. Status Epilepticus and Cluster Seizures. *Vet Clin Small Anim.* 2014:In press.
2 Huff JS and Fountain NB. *Emerg Med Clin North Am* 2011; 29(1):1
3 Raith K et al. Journ. *Vet. Emerg.* Crit. Care. 2010;20(4):446
4 Thomas, W. Vet Clinics of NA. *Sm An Pract.* 2010;40(1): 161
5 Bateman SW & Parent JM. *J Am Vet Med Assn.* 1998; 215(10):1463
6 Mariani CL. *Topics in Companion An Med.* 2013;28:34
7 Hardy, Brian Thomas. M.S. Thesis - University of Minnesota Graduate School. 2012 http://conservancy.umn.edu/bitstream/122959/1/Hardy_Brian_March2012.pdf
8 Brophy GM, Bell R, *Neurocrit Care.* 2012;17:3
9 Shorvon S. & Ferlisi M. *Brain.* 2012;135:2314
10 Shorvon S. *Epilepsia.* 2013;54(Suppl. 6): 64–7
11 Leppik, IE et al. *JAMA.* 1983; 249(11):1452

[12] Treimam, DM et al. *N Engl J Med.* 1998; 339(12):792
[13] Silbergleit R et al. *N Engl J Med.* 2012;366(7):591 Platt S
[14] Platt S et al. *Am J Vet Res* 2000; 61(6): 651
[15] Podell, M. *Clin Tech Small Anim Pract.*1998;13(3):185
[16] Platt S. and Garosis L Small Animal Neurological Emergencies. Manson Publishing. 2012:425-426 Google eBook.
[17] Gamez-Leyva et al. *CNS Drugs*. 2009;23(11): 983.
[18] Moddel GS et al. *J Neurol Neurosurg Psychiatry*. 2009;80(6): 689-92.
[19] Alvarez, V et al. Epilepsia. 2011; 52(7):1292
[20] Hardy BT et al. *J Vet Intern Med.* 2012;26:334
[21] Leppik IE et al. *Epilepsia* 2011;52 Suppl 8:31
[22] Patterson EE et al. The canine translational platform; proof of concept study of fosphenytoin for status epilepticus in canine clinical patients. American Epilepsy Society Annual Meeting. Abstract, Washington DC, December 6-10, 2013.
[23] Mariani CL et al. *Vet Intern Med* 2003;17:402.
[24] Platt SR. Status Epilepticus: Life after Diazepam. Proceedings 26th Annual ACVIM Forum, San Antonio Texas June 4-7, 2008.
[25] Eagleson JS et al. *Am J Vet Res* 2012;73:539-45.
[26] Podell M et al. *Vet Pharmacol Ther* 1998;21:158
[27] Mariani CL. Treatment of Cluster Seizures and Status Epilepticus. Proceedings 28th Annual ACVIM Forum, San Antonio Texas June 4-7, 2008.
[28] Dewey CW. *Vet Clin Small Anim.* 2006;36:1107
[29] Serrano SD et al. *J Vet Intern Med.* 2006; 20(1):194
[30] Mazarati, AM et al. *Epilepsy Res.* 2004; 58(2-3): 167-74.
[31] Patterson EE et al. *J Vet Pharm. Ther.* 2008 31(3):253
[32] Rogowksi M et al. *Epilepsia,.* 2013; 54(Suppl. 6):93-8.
[33] Bolson D. *Epilepsia.* 2013; 54(Suppl. 6):20

Emergency Treatment of Canine Seizures – General Practice Guidelines

Status Epilepticus

- seizure > 5mins, or 2 or more without recovery in-between

1. Check: pulse and respiration, Place IV line, start IV fluids

FIRST LINE: diazepam 0.5 – 1.0 mg/kg IV - can be repeated twice[24] or 0.5mg/kg IN[23] or 1-2mg/kg PR[24]
Alternatives midazolam 0.06-0.3 mg/kg IV or IM[24], 0.2mg/kg IN[25], not PR
lorazepam 0.2mg/kg IV, or IN[23], not PR

Abbreviations:
PB: phenobarbital
KBr: potassium bromide
LEV: levetiracetam
IN: Intranasal
PR: per rectum

2. Question the owner.
-prior antiepileptic therapy?
-treatment for diabetes mellitus?

3. Check: if decreased give:
a) glucose......... 1 - 2ml/kg 50% dextrose iv
calcium......... 0.5 – 1.5ml/kg 10% Ca gluconate 20-30min, monitor EKG
b) body temp...... Cool down if > 104F

*Even if seizures stop - temporarily or permanently
Give increased doses of ongoing maintenance doses of meds and/or partial or full loading dose of new chronic drug:
PB – PO or IM, **KBr** - PO or rectal, **LEV**- PO or IV or IM

If seizures do not stop or continue
Did diazepam stop the initial event and has been given < 3 times?

***Yes** (still potentially benzodiazepam responsive?)

diazepam CRI:
0.5 – 1.0mg/kg bolus followed by
0.1 – 0.5mg/kg/hr in D5W or 0.9% NaCL
0.1-2.0 mg/kg/hr[24]
(wrapped iv line – light sensitive)
NOT in LRS – Ca can cause precipitation)

Seizures continue despite high diazepam rate

***No** (benzodiazepam refractory)

SECOND LINE (monitor vitals carefully but intended to be non- anesthetizing)
phenobarbital 4-5mg/kg IV q30-120 mins to effect[27] - max dose 24mg/kg/24hrs[24]
and/or
levetiracetam 30-60mg/kg IV[20]
or alternative
? fosphenytoin 15mg/kg phenytoin equivalent (PE) IV[22]

If seizures do not stop or continue

THIRD LINE – Refractory Status Epilepticus – General Anesthesia
propofol 1-4mg/kg IV bolus slowly to effect[24] 0.1-0.6 mg/kg/hr CRI[28]
or
pentobarbital 3-15mg/kg IV to effect[24], 0.5-4mg/kg/hr CRI[28]
(may need diazepam CRI when weaning off)

Other anesthetic/alternative therapies if still refractory
isoflurane 1-2% MAC[16]
or
ketamine[29] 2-10mg/kg IV?

If on CRI - No seizures for 4-6 hours – slowly decrease CRI over 6-24 hours to prevent withdrawl seizures.

PHARMACORESISTANT EPILEPSY: STRUCTURAL OR FUNCTIONAL? THE CLINICAL PERSPECTIVE

Holger A. Volk, PhD, DECVN, MRCVS
London, UK

Introduction

The most common chronic neurological disorder in people and dogs is epilepsy with an estimated prevalence of 0.4-1% (Cowan, 2002; Sander and Shorvon, 1996) and 0.6-2% respectively (Schwartz-Porsche, 1986; Kearsley-Fleet et al., 2013). A bit more than twenty per cent of human patients with epilepsy will continue to have seizures despite chronic medical treatment with two adequate antiepileptic drugs (AEDs) (Picot et al., 2008). Around 20-80% of dogs with suspected idiopathic epilepsy (depending on the examined hospital population) will not get into remission and continue to have recurrent seizures (Heynold et al., 1997; Berendt et al., 2002; Arrol et al., 2012; Boothe et al., 2012) and around 20–30% will remain inadequately controlled (less than 50% reduction of seizure frequency) despite adequate treatment with the standard AED phenobarbitone (PB) and/or potassium bromide (KBr) (Trepanier et al., 1998; Schwartz-Porsche et al., 1985; Podell and Fenner, 1993). Dogs suffering from epilepsy, especially drug-refractory epilepsy frequently characterized by a high seizure frequency and cluster seizures, have an increased risk of developing behavior changes, premature death, and a reduced owner-perceived quality of life (Berendt et al., 2007; Shihab et al., 2011; Chang et al., 2006; Wessmann et al., 2012). Seizures can appear to the owner uncontrollable and unpredictable. This does not only have an impact on the quality of life of the affected pets, but also on the quality of life of the 'care-taker', the owner (Wessmann et al., 2012; Chang et al., 2006). It is therefore pivotal to improve our understanding of which patient will respond appropriately to pharmacological interventions and which patient might not, potentially requiring a more aggressive medical treatment approach early on in the disease process.

Epilepsy is not one disease with one unique and simple pathophysiology. The term embraces a heterogeneous group of chronic conditions with seizures being its clinical manifestation. Complex simplicity - the brain is composed by a complex structure, however, it can only function and dysfunction in certain limited ways. This explains why a wide variety of disturbances of brain structure and function can result in seizures and epilepsy. It is thought that similar pathophysiological pathways cannot only lead to epilepsy, but also to other neurodevelopmental and neurobehavioral disorders (Johnson and Shorvon, 2011; Shihab et al., 2011). These disorders share pathophysiological commonalities at various levels. Understanding the common and discrete pathophysiological pathways can help improve our understanding of these diseases and how to diagnose, monitor and treat them appropriately. Furthermore, pharmacologically characterizing of these disorders and understanding why treatment does fail in some cases cannot only improve our knowledge of the mechanisms behind pharmacoresistance and how to overcome them, but potentially even more importantly deepen our understanding of their pathophysiology and natural clinical course.

DEFINITION

The definition of pharmacoresistant epilepsy remains a matter of debate. Formerly, epilepsy was considered refractory to medical management if seizures were not reduced by ≥50% on one or two adequate AED(s) achieving therapeutic serum concentration(s) and steady state (Regesta and Tanganelli, 1999). In recent years, this definition was used in veterinary medicine in many epilepsy drug trials to establish AED response (e.g.: Volk et al., 2008; Platt et al., 2006; Dewey et al., 2004, 2009; von Klopmann et al., 2007; Muñana et al. 2010, 2012b). The limitation of establishing response by defining it as more than 50% seizure frequency reduction is that even such a reduction in seizures can still mean that the animal experience a too high seizure frequency and therefore has a reduced quality of life (Wessmann et al., 2012; Chang et al., 2006). In human medicine, it has been established that freedom of seizure activity is the best outcome measure, as this has been associated with the best improvement in quality of life (Elsharkawy et al., 2009).

A task force of the International League against Epilepsy (ILAE) has established a consensus statement on granted outcome measures for therapeutic interventions (Level 1) and definition of pharmacoresistant epilepsy (Level 2) considering at the same time the incidence of drug-induced adverse effects and seizure control (Kwan et al., 2010).

Level 1 categorizes the outcome of each therapeutic intervention as either *free of seizures* or *treatment failure* (not seizure free). The task force of the ILAE recommended for therapeutic trials a follow-up time which is at least three-times longer than the inter-ictal period prior of the therapeutic intervention; e.g. if the inter-ictal period was a week then the follow-up time should be at least 3 weeks. Nonetheless, as this is not a real practical approach and this would require the calculation of different follow-up times for each patient, the term 'seizure freedom' can usually be considered if the patient had no seizures for a 12-month period. As drug-induced adverse effect can also impact on the quality of life of a patient, Level 1 does not only consider efficacy, but also the occurrence of drug-induced adverse effects.

Level 2 defines pharmacoresistant epilepsy as a failure to achieve seizure freedom despite adequate trials of two (or more) well-tolerated, correctly chosen and used AED regimens (whether administered as mono-therapies or in combination) (Kwan et al., 2010). The working group suggested to reduce the number of AED trialed to "two" based on the assumption that adequate seizure control is not as likely once the patient has not responded to two correctly chosen AEDs (Kwan and Brodie, 2000; Arts et al., 2004).

Reflecting on the revised ILAE classification of pharmacoresistant epilepsy, many of the dogs with idiopathic epilepsy could be classified as resistant to medical AED treatment. Around 63% of human epilepsy patients will either spontaneously or following AED treatment cease having recurrent seizures (Kwan and Brodie, 2000). The spontaneous and drug-induced remission rate is far less in dogs with idiopathic epilepsy, which usually ranges between 15 and 24% (Heynold et al., 1997; Berendt et al., 2007; Arrol et al., 2012). However, in a more recent drug trial comparing the efficacy and tolerability of phenobarbitone with potassium bromide as a first line treatment, 85% and 52% respectively became free of seizures during the 6-months study period (Boothe et al., 2012). It is possible, however, that some of the treated dogs could have had further seizures with a year or more follow-up time.

Drug-resistance should not be thought of as a static state but rather as a dynamic process, which is dependent on the underlying pathophysiology (Kwan et al., 2010). Seizure severity and frequency can fluctuate over time (Berg et al., 2009; Muñana et al., 2010) and the 'state' of drug-resistance is determined on the time point of evaluation (Kwan et al., 2010). Like in humans, some epilepsy syndromes might change even to the better. In the Lagotto Romagnolo dog a benign "childhood" or juvenile epilepsy form has been described with the first seizures occurring when they are around 6 weeks of age (Jokinen et al., 2007). Nearly all of the affected dogs come into remission at the age of 10 weeks generally without receiving medical treatment. Future, large-scale and long-term epidemiological studies are needed to establish the prevalence of drug response more accurately in the general (first-opinion) epileptic dog population, in addition to investigating potential risk factors such as breed, age, sex, neuter status and duration and type of seizure history.

Factors associated with remission and drug response

Genetic factors

Many dog breeds have been suggested to have a higher than normal prevalence of epilepsy (Jaggy et al., 1998; Kathmann et al., 1999; Berendt et al., 2002; Casal et al., 2006; Gullov et al., 2011; Berendt et al., 2008). This does not only suggest the likelihood of genetic risk factors playing a role for the development of epilepsy, but potentially also for the likelihood of drug-response. The aforementioned Lagotto Romagnolo breed can be affected by a benign childhood or juvenile epilepsy. Ninety-six percent of dogs stop having recurrent seizures at around 10 weeks of age. The *Lgi2* gene is altered in this breed, which is involved in the immediate post-natal phase of neuronal pruning. The function of the *Lgi2* gene in cerebral synaptic remodeling becomes of less significant with age. The later part of pruning is mainly regulated by the *Lgi1* gene and therefore the condition improves (Jokinen et al., 2007; Seppala et al., 2011).

Improved seizure control has also been associated with a gene mutation located on the *ABCB1* gene (multidrug resistance [*MDR*] 1) encoding the transmembrane ATP-dependent multidrug transporter, P-glycopprotein (P-gp). P-gp is expressed at the blood brain barrier (BBB) and is thought to be one of the protection mechanisms of the brain for potential CNS toxins (Schrickx and Fink-Gremmels, 2008; Schinkel et al., 1994, 1996). P-gp is non-specific and can transport a big variety of substances back into the blood stream, including AEDs such as phenobarbitone (Basic et al., 2008; Potschka et al., 2002). Failure of this CNS-protective ATP-dependent efflux pump can lead to CNS intoxication (Schinkel et al., 1994, 1996). Collies and other herding breeds have been reported to have P-gp altered which is caused by a four base pair deletion (c.296_299del) in exon 4 of the *ABCB1* gene (Mealey et al., 2001). These dogs have been reported to be more sensitive to drugs such as ivermectin causing CNS-intoxication, but this defect could also lead to improved seizure control in affected herding dogs as certain AEDs might penetrate the BBB easier (Muñana et al., 2012a). Although, Border Collies are generally perceived to have a difficult to control epilepsy phenotype characterized by cluster seizures, a high seizure frequency and severity. Around three-quarters of epileptic Border Collies were classified as drug-resistant in a current study (Hulsmeyer et al., 2010). The same group found that Phenobarbitone-resistance in Border Collies with epilepsy was associated with a single nucleotide polymorphisms (SNP or snips) at intron 1 near the 5-end of the *ABCB1* gene, which was suggested to alter the function of the promoter elements of this gene and lead to an overexpression of P-gp at the BBB (Alves et al., 2011). This SNP was, however, not associated with pharmacoresistance in epileptic Australian Shepherd dogs which are closely related to Border Collies (Weissl et al., 2012).

Five potential genes were associated with lack of response to phenobarbitone in a recent study (*KCNQ3*, *SCN2A*, *GABRA2*, *EPOX HYD*, and *ABCB4*), unfortunately, after correction for multiple comparisons they did not reach statistical significance (Kennerly et al., 2009). Three of these genes encode ion channels and are reported targets for AEDs ('*KCNQ3* - voltage gated potassium ion channel important for post excitatory membrane re-polarization, *SCN2A* – sodium ion channel, *GABRA2* – GABA receptor')(Armijo et al., 2005). The other two reported genes are involved in phenobarbitone metabolism (*EPOX HYD*) or transportation (*ABCB4*) (Kennerly et al., 2009).

Clinical factors

In recent years, epidemiological and basic science studies have investigated which clinical factors of the disease phenotype might predict poor or good response to AED(s). There continuous to be a debate in human medicine, if seizures can be controlled or stopped when treated early, best after the first seizure. Epilepsy is thought to be a chronic condition and, as aforementioned, drug-responsiveness can be dependent on factors independent of the disease process itself. Epidemiological studies in third-world countries, in which AEDs are not readily available or affordable, revealed remission rates being similar to those in industrial countries (Placencia et al., 1993). AEDs have been developed to suppress seizure activity and therefore were often referred to 'anticonvulsant'. However, not every seizure will be associated with motor activity and therefore the term 'antiepileptic' is commonly used. Nevertheless, the current AEDs have little or no effect on the disease process itself (epilepsy). Promising AEDs such as levetiracetam which have been thought to have disease modifying and/or antiepileptogenic effect have failed in chronic epilepsy models and have not prevented the development of epilepsy (Brandt et al., 2007).

In an epidemiological study from 1997, Labrador Retrievers, which stopped seizuring, received medical treatment for longer after their first seizure than those epileptic patients, which continued to have epileptic activity (Heynold et al., 1997). This could mean that timing of treatment in relation to the first seizure might not be the most important factor for drug-responsiveness. We have, however, good data from studies in dogs, rodents and humans showing that a severe disease phenotype, such as prolonged seizure activity (status epilepticus), high density of seizures (cluster seizures) and a general high seizure frequency, are risk factors for the development of AED pharmacoresistance (Kwan and Brodie, 2000; Heynold et al., 1997; Hulsmeyer et al., 2010; Weissl et al., 2012;,

Löscher and Brandt, 2010). The severity of epilepsy will determine how fast and aggressive a clinician will treat the disease, e.g. a dog presented with a high seizure frequency will be treated earlier and potentially more aggressively than a dog which has a very low seizure frequency. This needs to be considered when evaluating epidemiological data sets.

Young dogs with epilepsy have been thought to have a worse outcome, e.g. in a study Labradors Retrievers had a better response to medication when they developed epilepsy later in life (Heynold et al., 1997). A recent study did not echo these findings and showed that the onset of seizures before the age of one year had no influence on outcome (Arrol et al., 2012). Intriguingly, the researchers did not find that the presence of focal seizures was a risk factor for poorer outcome, which is the opposite in humans with epilepsy (Kwan et al., 2011).

CONCLUSION

Pharmacoresistant epilepsy continuous to be a major challenge for veterinary medicine and further studies are needed to unravel its pathophysiology and determine clinical risk factors, so that new treatment strategies can be developed which are individually tailored for the patient to maximize responds to AEDs. Future studies should consider that seizure freedom and quality of life might be important outcome measures for drug trials.

REFERENCES

Alves L, et al., **JVIM** 2011;25:484, Armijo J, et al., **Cur Pharm Des** 2005;11:1975, Aronica E, et al., **Neurosci** 2003;118:417; Arrol L. et al., **Vet Rec** 2012, 170:335, Arts WF, et al., **Brain** 2004;127:1774, Basic S, et al., **Seizure** 2008;17:524, Berendt M, et al., **JVIM** 2002; 16:262, Berendt M, et al., **JVIM** 2007;21:754, Berendt M, et al., **Acta Vet Scand** 2008;50:51, BergAT, et al., **An Neurol** 2009;65:510, Boothe DM, et al., **JAVMA** 2012;240:1073, Brandte C, et al., **Neuropharm** 2007;53:207, Chang Y, et al., **JSAP** 2006;47:574, Cowan LD, **Mental Retard Dev Dis Res Rev** 2002;8:171, Deqey CW, et al., **JAVMA** 2009;235:1442, Dewey CW, et al., **JAAHA** 2004;40:285, Heynold V, et al., **JSAP** 1997;38:7, Hulsmeyer V, et al., **JVIM** 2010;24:171, Jaggy A, et al., **JSAP** 2011;39,275, Johnson MR and Shorvon SD, **Epilep Behav** 2011;22:421, Jokinen TS, et al., **JVIM** 2007;21:464, Kathmann I, et al., **JSAP** 1999;40:319, Kersley-Fleet L, et al., **Vet Rec** 2013; 171(13):338, Kennerly EM, et al., **Pharm Gen** 2009;19:911, Kwan P, et al., **Epilepsia** 2010;51:1069, Kwan P and Brodie MJ, **New Eng J Med** 2000;342,314, Kwan P, et al., **Epilepsia** 2002;43:1318, Kwan P, et al., **Pharm Gen** 2008;18:989, Kwan P, et al., **New Eng J Med** 2011;365:919, Löscher W and Brandt C **Epilepsia** 2010;51:89, Mealey KL, et al., **Pharmacogen** 2001;11:727, Muñana KR, et al., **JVIM** 2010;24:166, Muñana KR, et al., **JVIM** 2012a;26:1358, Muñana KR, et al., **JVIM** 2012b;26:341, Picot MC, et al., **Epilepsia** 2008;49:1230, Placencia M, et al., **Epilep Res** 1993;14:237, Platt SR, et al., **Vet Rec** 2006;159:881, Podell M and Fenner WR, **JVIM** 1993;7:318, Potschka H, et al., **Neurosci Let** 2002;327:173, Regesta G and Tanganelli P, **Epilep Res** 1999;34:109, Sander JW and Shorvon, **J Neurol Neurosurg Psych** 1996;61:433, Schinkel AH, et al., **Cell** 1994;77:491, Schinkel AH, et al., **J Clin Invest** 1996;97:2517, Schwartz-Porsche D, et al., **J Vet Pharm Therap** 1985;8:113, Schwartz-Porsche D, **Proc ACVIM** 1986;1161, Seppala N, et al., **PLos Gen** 2011;7, Shihab N, et al., **Epilep Behav** 2011;21:160, Trepanier LA, et al., **JAVMA** 1998;213:1449, Volk HA, et al., **Vet J** 2008;176:310, von Klopmann T, et al., **JSAP** 2007;48:134, Weissl J, et al., **JVIM** 2012; 26:116, Wessmann A, et al., **JVIM** 2012;26:1.

PHARMACORESISTANT EPILEPSY: STRUCTURAL OR FUNCTIONAL? THE NEUROPATHOLOGY PERSPECTIVE

Kaspar Matiasek, DVM, DrMedVet, Assoc Memb ECVN, MRCVS
Munich, Germany

Introduction

The failure of antiepileptic drugs (AED) to achieve seizure control, broadly, can be traced back to disease related, drug related or patient related factors that either interfere with the bioavailability or the efficacy of the drugs. With exception of very rare monospecific tolerance development, pharmacoresistance impedes with the action of multiple compounds. Hence, resistance to one drug cannot be easily overcome by switching to the next. The consequences for the patient´s health and the quality of life for both the pet and owner are severe. Multidisciplinary exploration of the various mechanisms of pharmacoresistance is essential to lower the percentage of AED non-responders and to overcome 50% seizure frequency reduction thresholds, mimicking clinical success (Volk et al., 2008; Platt et al., 2006).

PATHOMECHANISTIC CONSIDERATIONS

Persistent epileptogenic triggers

Most of seizures result from hypersynchronized excitation of neuronal clusters within the cerebral grey matter. The zones involved during the event, however, may not resemble the ictal onset area or the epileptogenic trigger. For example in structural epilepsies, epileptogenic brain lesions may be remote from the ictal cortex, identified on MRI or EEG. Via distant mechanisms such as mass shifting, vascular compromise or inhibitory deafferentiation, these brain lesions may render the cortical areas hyperexcitable and non-responsive to AED.

If the primary lesion is identified, neurosurgical resection of the trigger can be attempted, which may result in subsequent AED efficacy and seizure control. Neurosurgical removal of a lesion, primary and postoperative brain injury, however, may leave glial scars that can serve as epileptogenic trigger themselves (Steinmetz et al., 2013).

To date, magnetic resonance imaging (MRI) remains the mainstay of intravital detection of epileptogenic lesions in animals. Lack of respective MRI features is one of the hallmarks for Epilepsy of Unknown Cause (EUC) (Wahle et al. 2013). Large studies on human patients have explored the possibility of missing microscopic lesions, triggering epilepsy. Amongst these, focal cortical dysplasia, gliosis, harmatia, hippocampal sclerosis and low grade glioma are most common (Wang et al. 2013; Erikkson et al. 2005). The detection rate can be significantly increased using positron electron microscopy (Kuba et al. 2011). No such data are available yet in epileptic animals.

Neuronal network modifications

Regardless of their nosology, recurrent seizures result in significant rearrangement of neuronal networks due to excitotoxic loss of both inhibitory and excitatory neurons, axonal sprouting, adult neurogenesis and excitatory synaptogenesis. These changes may occur focally, depending on topography and impact of a primary brain lesion. On the other hand, generalising neuronal discharges in the vulnerable hippocampus may converge to hippocampal sclerosis, a loss of pyramidal neurons accompanied by astrogliosis, and mossy fibre sprouting that in humans have been associated with progression and perpetuation of epilepsy as well as pharmacoresistance (Bluemcke et al., 2013). Similar observations have been made anecdotally in dogs and in a large group of epileptic cats.

Neuronal channel modifications & functional tolerance

Ongoing excitatory discharges, in particular during status epilepticus, have been proven to not only disrupt the inhibitory GABAergic neurotransmission but also to reverse the function of the GABA-A receptors, contributing to hyperexcitation and ectopic discharges in the hippocampus (Barmashenko et al., 2011). As we know from rodent models of pharmacoresistance, GABA-A receptors of phenobarbitone non-responders also can lose diazepam-sensitivity as an adaptation to treatment (Volk et al., 2006). In contrast to seizure-related GABA receptor downregulation, these studies also extrapolate the sophisticated algorithm necessary to draw sound conclusions from receptor expression analyses.

Multidrug resistance

In order to protect the brain tissue from toxicity by xenobiotics, the blood brain barrier and choroid plexuses are equipped with membranous efflux molecules of the ATP-binding cassette superfamily that push lipophilic and cationic compounds back into the blood stream and plexus stroma (Fromm, 2000). Moreover, astrocytes are equipped with inducible multidrug resistance proteins that fetch xenobiotics from the perineuronal, perisynaptic and perinodal space in order to guide them back to an exporter blood vessel. Amongst those drugs, subjected to early repulsion, are some AED such as valproate, gabapentin, topiramate, phenytoin, carbamazepine, phenobarbitone, felbamate and lamotrigine (Löscher and Potschka, 2002). Hence, upregulation of multidrug resistance proteins, as p-glycoprotein, may limit the intracerebral bioavailability of these drugs. Indeed, multidrug transporters p-glykoprotein and MRP1 were upregulated in epileptogenic brains of human pharmacoresistant patients and in animal models of pharmacoresistant epilepsy (Hoffmann et al., 2006; Volk and Löscher, 2005; Sisodiya et al., 2002). Over-expression of p-glycoprotein also was demonstrated in dogs after increased seizure activity, status epilepticus and cluster seizures and, hence, may limit the drug availability in clinical patients (Pekcec et al., 2009).

Metabolic tolerance

Metabolic or pharmacokinetic tolerance results from induction of AED-metabolizing enzymes as a response to chronic/repetitive drug exposure. Although possible, the risk of developing metabolic tolerance to AED is considered low, with exception to the benzodiazepine family. For other AED with known enzyme induction, such as phenytoin, phenobarbitone and carbamazepine, the tolerance can be overcome easily by higher dosages (Löscher et al. 2006). Metabolic tolerance resembles an extracranial cause of drug failure and does not display any surrogate markers in the brain.

CONCLUSION

Pharmacoresistance in epilepsy patients may result from persistent epileptogenic brain lesions, from reorganization of neuronal circuits, from neurotransmitter and receptor modifications, from drug resistance and rarely from metabolic tolerance. Morphological evaluation can help to specify the set of involved factors retrospectively or after lobectomy. The search should implement algorithms and markers for both neurobiological changes and multidrug resistance.

REFERENCES

Barmashenko G, et al. **Epilepsia**. 2011;52(9):1570-8. Bluemcke I, et al. **Lancet Neurol**. 2013;12(9):844-6. Eriksson SH, et al. **Acta Neurol Scand** 2005: 112: 151–156. Fromm MF. **Pharmacol Ther**. 2000 ;38(2):69-74. Hoffmann K, et al. **Epilepsy Res**. 2006;69(1):1-14. Kuba R, et al. **Epilepsy Behav**. 2011;22(3):537-41. Löscher W, et al. **Epilepsia**. 2006;47(8):1253-84. Löscher W et al. **J Pharmacol Exp Ther**. 2002;301(1):7-14. Pekcec A, et al. **Epilepsy Res**. 2009;83(2-3):144-51. Platt SR, **Vet Rec**. 2006; 159(26):881-4. Sisodiya SM, et al. **Brain**. 2002;125(Pt 1):22-31. Steinmetz S, et al. **Epilepsia**. 2013;54(4):580-8. Wahle AM, et al. **J Vet Intern Med**. 2014;28(1):182-8. Wang ZI, et al. **Mod Pathol**. 2013;26(8):1051-8. Volk HA, et al. **Vet J**. 2008;176(3):310-9. Volk HA, et al. **Neurobiol Dis**. 2006;21(3):633-46. Volk HA, et al. **Brain**. 2005;128(Pt 6):1358-68.

CANINE MENINGOENCEPHALITIS: WHAT DO WE KNOW?

Christopher L. Mariani, DVM, PhD, DACVIM (Neurology)
Raleigh, NC

INTRODUCTION

Meningoencephalitis is a common condition in dogs. Although there have been a multitude of infectious organisms documented to cause meningoencephalitis in this species, in the vast majority of cases an infectious organism cannot be documented. There are a number of specific syndromes within this group of so-called "sterile" meningoencephalitides, including steroid-responsive meningitis-arteritis (SRMA), granulomatous meningoencephalitis (GME), necrotizing meningoencephalitis (NME) and necrotizing leukoencephalitis (NLE, also known as necrotizing encephalitis). A less common variant is a primarily eosinophilic meningoencephalitis (EME). Although there are characteristics that differentiate these disease conditions from one another, there are also cases that blur the lines between these diagnoses, and clinicians and researchers are divided on whether to "lump" or "split" these conditions.[1,2] In addition, despite having some characteristic distributions (for example, GME preferentially affects the medulla, cerebellum and cervical spinal cord, while NME affects the forebrain), without histopathology of affected tissue (or even with tissue in some cases), arriving at a firm diagnosis can be very difficult. For this reason, many clinicians use the term "meningoencephalitis of unknown etiology" (MUE) or "meningoencephalitis of unknown origin" (MUO) to refer to clinical cases diagnosed with meningoencephalitis with imaging and/or cerebrospinal fluid (CSF) evaluation.

THE HUNT FOR INFECTIOUS ETIOLOGIES

There are a number of well-recognized infectious organisms that can cause meningoencephalitis or meningomyelitis in dogs, including viral, bacterial, rickettsial, fungal and protozoal organisms. However, in most canine cases of inflammatory central nervous system (CNS) disease evaluated clinically with advanced imaging (magnetic resonance imaging [MRI] or computed tomography [CT]) and cerebrospinal fluid (CSF) analysis, an infectious etiology cannot be documented. Several groups have performed broad-based screening on tissue and/or CSF in meningoencephalitis cases with sensitive molecular techniques (i.e., polymerase chain reaction [PCR]) to detect infectious organisms. In one study an intriguing association with *Mycoplasma canis* in pugs with NME was noted that requires further validation.[3] However, in the majority of patients, no infectious etiology could be documented.[3-5] Thus, the weight of the evidence suggests that infectious organisms that directly infect the central nervous system are a relatively uncommon cause of meningoencephalitis in dogs.

GENETICS OF CANINE MENINGOENCEPHALITIS

It has long been recognized that certain breeds are predisposed to developing meningoencephalitis. The prevalence of SRMA is higher in a number of breeds including the beagle, Bernese mountain dog, boxer and Nova Scotia duck tolling retriever.[6-8] Eosinophilic meningoencephalitis has been disproportionately recognized in Golden retrievers and Rottweilers.[9,10] A unique meningoencephalitis of greyhounds has also been described.[11] Although it was likely initially recognized as part of the reticulosis complex, the term GME was first used in 1978 by Braund et al.[12] A larger series of dogs published shortly afterwards suggested a predilection in poodles.[13] However, unlike some of the other meningoencephalitides, GME is frequently described in many different dog breeds and mixed breed dogs of varying sizes. Pug encephalitis was first described in 1983,[14] with a more detailed account following in 1989.[15] It was later reported that the Maltese and then Chihuahua breeds developed a similar condition, and the term NME was proposed to encompass the disease process in these breeds, as they appeared similarly histologically.[16,17] Select other breeds were subsequently reported to develop NME.[1,18] A similar disorder was documented in Yorkshire terriers in 1993 but involved the white matter of the forebrain and brainstem while sparing the meninges, and became known as NLE (or just necrotizing encephalitis).[19] French bulldogs were later also documented to develop NLE.[20] Interestingly, all dogs reported with either NME or NLE have been small or toy breeds.

Pugs with NME have been the most intensely studied group from a genetic perspective, with several investigations published to date. Although long suspected, Greer et al. first demonstrated the heritability of NME in this breed.[21] Subsequent studies showed a strong association of affected dogs with single nucleotide polymorphisms (SNPs) within the MHC II region (DLA), located on chromosome 12.[22,23] Different haplotypes were identified in these pugs that conferred varying risk for development of the disease.[23,24] Such an association with the MHC II region is common in immune-mediated or suspected autoimmune diseases in humans, including multiple sclerosis (MS).[25] However, more recent genetic studies of Chihuahuas and Maltese dogs with NME have failed to show this same association with the MHC II region,[26,27] suggesting that different genetic pathways may result in a phenotypically similar disease.

Fewer genetic studies are available for other meningoencephalitides in dogs. However, genome-wide association and candidate gene mapping studies of Nova Scotia duck tolling retrievers with SRMA have identified candidate loci in these dogs on chromosome 32 and within the MHC class II locus respectively.[28,29]

IMMUNOPATHOLOGY

Histology and Immunohistochemistry

Steroid responsive meningitis arteritis is probably the best-studied canine inflammatory CNS disorder. The inflammatory response predominates in the region of the cervical spinal cord and consists of neutrophilic meningitis and necrotizing arteritis of the leptomeningeal vasculature.[2] Smaller numbers of macrophages, lymphocytes and plasma cells may be present. Although the pathology and resulting clinical signs indicate primarily involvement of the meninges overlying the cervical spinal cord, other areas of the CNS can be involved (e.g., brain) and vessels within other organ systems such as the heart or thyroid gland may also have similar lesions, reflecting the systemic distribution of this disease process.[8] Tipold and colleagues have suggested that SRMA is driven by a Th2-dominated response, based on a high CD4:CD8 lymphocyte ratio and increased B cells within the blood and CSF, upregulation of Th2 cytokines (IL-4) and robust immunoglobulin (Ig)-A production in both the serum and CSF.[8,30]

Granulomatous meningoencephalitis is characterized by a mixed cell inflammation that predominates in the brainstem, cerebellum and cervical spinal cord. There is a characteristic perivascular accumulation of macrophages admixed with lymphocytes, plasma cells and occasionally neutrophils; these perivascular cuffs can be quite substantial in size and occasionally coalesce to form large granulomatous lesions.[1,12,13] Segregation of GME into disseminated and focal forms has been proposed, with some investigators suggesting that the focal form may represent a neoplastic population of dendritic cell origin.[1,2] The most widely referenced immunohistochemical study of GME suggests that lymphocyte populations consist predominantly of T cells,[31] although other studies including our own have suggested a substantial B cell presence within the lesions.[32]

Necrotizing meningoencephalitis can occur as an acute or chronic process; in the latter, significant necrosis and malacia can be seen, sometimes with collapse of brain parenchyma. Lesions are more severe in the forebrain and typically become less severe or disappear altogether in the brainstem, cerebellum and cervical spinal cord.[1,2] A mixed cell infiltration of histiocytic cells, lymphocytes and plasma cells is found within the parenchyma of the gray and white matter accompanied by perivascular cuffing and meningeal inflammation. In NLE, necrotizing malacic lesions involve the white matter of both the forebrain and brainstem, with sparing of the cerebral cortex and meninges. Histologically, a histiocytic infiltrate of macrophages and microglia is present along with lymphocytic perivascular cuffing.[1,2]

Autoantibodies

The presence of autoantibodies against astrocytic proteins has been documented within the CSF and serum of dogs with NME and GME.[33,34] Further analysis suggested that these antibodies target glial fibrillary acidic protein (GFAP), an intermediate filament protein important to astrocyte function.[35] Elevated concentrations of the GFAP protein itself have also been documented in dogs with NME and interestingly in some clinically normal pugs.[36] The role of such autoantibodies in the pathogenesis of these diseases is still being investigated, and the extent to which they actively contribute to lesion generation is unclear. Autoantibodies have also been documented in dogs with SRMA, but appear to play a minor role and are considered to be an epiphenomenon rather that a predominant driver of pathology.[37]

Cellular Proliferation, Migration and Cytokine Production

In order to become activated and sensitized against specific antigens, lymphocytes require presentation of these antigens by antigen presenting cells (dendritic cells, macrophages, B cells) in the context of the appropriate MHC molecules. This typically occurs in lymphatic organs (usually lymph nodes) draining the organ of interest. Activated lymphocytes then travel to the organ of interest to exert their effect (e.g., killing of microbes). Restimulation at the site of interest can also occur, with further lymphocyte proliferation in order to expand the effector cell population. Activated T cells are known to routinely enter and surveil the CNS. A question concerning inflammatory canine CNS disease with potentially important therapeutic implications is whether some of the proliferation of activated lymphocytes occurs directly within the CNS compartment or whether this is mainly a peripheral phenomenon. At the Comparative Neuroimmunology and Neurooncology Laboratory (CNNL) at North Carolina State University, our data suggest that there is active cellular proliferation occurring within inflammatory CNS lesions (including NME and GME).

Matrix metalloproteinases (MMPs) are zinc-dependent endopeptidases that have important roles in cellular migration, development, wound healing and other related processes. Certain MMPs are important in the migration of cells across vascular and tissue borders and into the CNS or CSF. Specifically, MMP-9 appears important for this activity. Dogs consistently have detectable concentrations of MMP-2 within the CSF, but MMP-9 is not found in clinically normal animals.[38] We have analyzed CSF from dogs with a variety of inflammatory CNS disorders and found detectable MMP-9 concentrations in a subset of these dogs, suggesting that this enzyme may be involved in the migration of inflammatory cells into the CNS in these patients.

We have analyzed the concentrations of a number of cytokines and chemokines within the CSF and serum of dogs with meningoencephalitis or meningomyelitis. One must be cautious in overinterpreting data taken from a set time point when analyzing proteins that are in constant dynamic change. However, patterns of overt inflammatory change or Th1 skewed responses were not noted. Rather, many dogs showed elevations in cytokines or chemokines associated with lymphocyte proliferation and monocyte or lymphocyte trafficking into tissues (e.g., IL-2, IL-7, IL-15 and CCL2). There was considerable variation between patients, supporting the notion that meningoencephalitis in dogs is a heterogeneous entity, with multiple mechanisms and disease subtypes.

DIAGNOSTIC BIOMARKERS

The use of biomarkers to aid in the diagnosis of a condition or to monitor therapeutic efficacy is just beginning to be investigated for inflammatory CNS diseases in dogs. Several groups have shown that dogs with SRMA have elevated serum concentrations of acute phase proteins including C-reactive protein and serum amyloid-A, reinforcing the contention that SRMA is a condition associated with systemic inflammation.[39,40] In addition, these markers decrease with effective glucocorticoid therapy and may be useful in identifying those dogs predisposed to a relapse in clinical signs.[39] Serum and CSF IgA concentrations did not perform as well in predicting response or relapse in these patients, as concentrations remained elevated throughout the course of the disease.[39] The use of GFAP and anti-GFAP autoantibodies as biomarkers within CSF has been suggested to aid in a diagnosis of NME.[36] There is little information available on the serial monitoring of these analytes, although one preliminary report suggests some promise.[41]

The CNNL has investigated the utility of several biomarkers within CSF, serum and urine with several potential goals in mind: 1) to identify analytes that would augment standard CNS imaging and CSF evaluation in the diagnosis of inflammatory CNS disease, particularly in differentiating such disease from neoplastic or vascular disorders; 2) to monitor the change in these analytes after therapy to determine if effective therapy is accompanied by a change in biomarker concentrations, and 3) to determine if there is a correlation between analyte concentrations and prognosis. Biomarkers investigated have included C-reactive protein, vascular endothelial growth factor (VEGF), lactate, neopterin and MMP-9.

IMPLICATIONS FOR THERAPY

Although glucocorticoids have been used for decades to control clinical signs associated with canine meningoencephalitis, more recently a number of additional immunosuppressive agents have been utilized in these patients, including cytosine arabinoside, cyclosporine A, azathioprine, leflunomide, lomustine, and procarbazine.[42-46] Although rigorous evaluation of the efficacy of most of these agents in this context is lacking, anecdotally they appear to provide a benefit in the management of dogs with inflammatory CNS disorders, either by providing additional immunosuppression or helping to lower glucocorticoid doses and ameliorate adverse effects associated with these latter drugs. Although these agents act through a variety of specific mechanisms, one shared mechanistic endpoint is an interference with cellular proliferation. The observations concerning cellular proliferation described above indicate one potential method by which these immunosuppressive medications may help to control inflammation and suggest that medications that penetrate the blood-brain or blood-spinal cord barriers might be preferred in the management of these conditions. Additional therapeutic interventions that might be considered in the future include interference with inflammatory cell migration into the CNS compartment through MMP inhibition or blocking of specific cytokines or chemokines.

CONCLUSIONS

From a mechanistic standpoint, canine meningoencephalitis appears to comprise a diverse group of disease processes, even within conditions that appear phenotypically distinct using conventional tests. Proliferation of inflammatory cell populations occurs within the CNS compartment in many patients, suggesting that medications with adequate penetration into this compartment may be preferred for therapy. The use of biomarkers in clinical samples is just starting to be investigated, but may yield useful tests for diagnostic, therapeutic and prognostic purposes. Continued investigation into the underlying mechanisms of this group of diseases may yield important insights informing the development of future medications or other therapeutic interventions.

REFERENCES

1. Higgins RJ and LeCouteur RA. **Proc of 20th Annual Symposium of the European Society and College of Veterinary Neurology** 2007;35. 2. Vandevelde M, et al. **Veterinary Neuropathology. Essentials of Theory and Practice**. Wiley-Blackwell; 2012. 3. Barber RM, et al. **J Vet Intern Med** 2012;26:962. 4. Jaderlund KH, et al. **J Vet Intern Med** 2009;23:669. 5. Schatzberg SJ, et al. **J Vet Intern Med** 2005;19:553. 6. Anfinsen KP, et al. **Can J Vet Res** 2008;72:350. 7. Behr S, et al. **J Am Anim Hosp Assoc** 2006;42:277. 8. Tipold A, et al. **J Small Anim Pract** 2010;51:150. 9. Bennett PF, et al. **Aust Vet J** 1997;75:786. 10. Smith-Maxie LL, et al. **J Vet Intern Med** 1989;3:167. 11. Shiel RE, et al. **Vet Rec** 2010;167:333. 12. Braund KG, et al. **J Am Vet Med Assoc** 1978;172:1195. 13. Cordy DR. **Vet Pathol** 1979;16:325. 14. de Lahunta A. **Veterinary Neuroanatomy and Clinical Neurology**. W.B. Saunders; 1983. 15. Cordy DR, et al. **Vet Pathol** 1989;26:191. 16. Higgins RJ, et al. **Vet Pathol** 2008;45:336. 17. Stalis IH, et al. **Vet Pathol** 1995;32:230. 18. Cooper JJ, et al. **J Vet Intern Med** 2014;28:198. 19. Tipold A, et al. **J Small Anim Pract** 1993;34:623. 20. Timmann D, et al. **J Small Anim Pract** 2007;48:339. 21. Greer KA, et al. **Res Vet Sci** 2009;86:438. 22. Barber RM, et al. **J Hered** 2011;102 Suppl 1:S40. 23. Greer KA, et al. **Tissue Antigens** 2010;76:110. 24. Pedersen N, et al. **J Vet Diagn Invest** 2011;23:68. 25. Baranzini SE. **Curr Opin Immunol** 2009;21:596. 26. Vernau KM, et al. **J Vet Intern Med** 2010;24:736. 27. Barber RM, et al. **Proceedings of the Forum of the American College of Veterinary Internal Medicine**, 2014. 28. Wilbe M, et al. **Immunogenetics** 2009;61:557. 29. Wilbe M, et al. **Nat Genet** 2010;42:250. 30. Maiolini A, et al. **Vet J** 2012;191:219. 31. Kipar A, et al. **Vet Pathol** 1998;35:43. 32. Mariani CL. **Proc of 21st Annual Symposium of the European Society and College of Veterinary Neurology** 2008. 33. Matsuki N, et al. **J Vet Med Sci** 2004;66:295. 34. Uchida K, et al. **Vet Pathol** 1999;36:301. 35. Shibuya M, et al. **J Vet Med Sci** 2007;69:241. 36. Toda Y, et al. **Vet Rec** 2007;161:261. 37. Schulte K, et al. **Berl Munch Tierarztl Wochenschr** 2006;119:55. 38. Bergman RL, et al. **Am J Vet Res** 2002;63:1359. 39. Lowrie M, et al. **J Vet Intern Med** 2009;23:862. 40. Bathen-Noethen A, et al. **J Vet Intern Med** 2008;22:1149. 41. Matsuki N, et al. **J Vet Med Sci** 2009;71:99. 42. Adamo PF, et al. **J Small Anim Pract** 2007;48:486. 43. Coates JR, et al. **J Vet Intern Med** 2007;21:100. 44. Zarfoss M, et al. **J Small Anim Pract** 2006;47:588. 45. Wong MA, et al. **J Am Vet Med Assoc** 2010;237:929. 46. Smith PM, et al. **J Vet Intern Med** 2009;23:520.

ACUTE SPINAL CORD INJURY: FACT, FICTION, AND FUTURE DIRECTIONS

Sarah Moore, DVM, DACVIM (Neurology)
Columbus, OH

INTRODUCTION

There is a high incidence of naturally occurring spinal cord injury (SCI) secondary to intervertebral disc extrusion (IVDE) in the thoracolumbar spine of dogs, resulting in a combination of compressive and concussive injury to the spinal cord[12]. Histopathologically, spinal cord lesions in these dogs consist of central gray matter hemorrhage, necrosis and cavitation with sparing of subpial axons, axonal degeneration, variable degrees of demyelination, prominent microglial activation and a pronounced innate immune response[23, 25]. Traumatic spinal cord injury (SCI) also affects 12,000 new human patients in the United States each year. Significant recovery of function in patients with severe injury is rare, with the cost of medical care for patients living with chronic SCI ranging from $1.5 to 4.5M per person over the lifetime of the individual[15]. The gross, histopathologic and ultrastructural similarities between the injured human and canine spinal cord underscores the importance of dogs as a spontaneous model of SCI[10, 23]. Additionally, information gained from human clinical trials related to this disease likely has implications for canine patients.

PRIMARY AND SECONDARY CAUSES OF INJURY

In dogs, the most common cause of acute SCI is IVDE, although fibrocartilagenous embolism (FCE) and spinal fracture are also seen in the clinical setting. In cats, spinal fractures and thromboembolic myelopathy are most common, but IVDE can occur. *Primary damage* to the spinal cord in SCI is caused by contusion, shearing, and in some cases laceration[13]. Compression of the spinal cord by extruded disc material or by a displaced spinal fracture can affect spinal cord blood flow by interfering with arterial supply or blocking venous drainage. These primary mechanical injuries to the spinal cord initiate a cascade of *secondary events* including hemorrhage, free radical damage, inflammation, and apoptosis. While it is all but impossible to intervene to prevent the primary damage associated with SCI, secondary causes of tissue injury occur in the hours to days (even weeks) after the primary insult, making them an attractive target for therapeutic intervention. Routine interventions to limit secondary injury include prompt decompression of the spinal cord when IVDE or fracture is the cause of injury and maintaining spinal cord perfusion via hemodynamic support. Emerging, experimental, or controversial treatments to limit secondary injury are discussed below.

METHYLPREDNISOLONE SODIUM SUCCINATE

Methylprednisolone sodium succinate (MPSS) was previously considered standard of care for human patients with acute SCI. This was based on results in multiple experimental models of SCI where MPSS showed a treatment benefit. The beneficially effects of MPSS were hypothesized to be related to free-radical scavenging effects, which are unique to the drug and are unrelated to its glucocorticoid steroid receptor activity. These experimental results paved the way for human clinical trials which concluded that MPSS provided some modest benefit[1-3]; however, the design and interpretation of data from these trials has been challenged making treatment with MPSS controversial. Recent studies have shown that MPSS causes a significant increase in morbidity among human SCI patients, and is no longer considered standard of care for treating humans with SCI[4, 8]. A blinded, placebo-controlled veterinary clinical trial is currently evaluating MPSS in dogs with IVDE.

POLYETHYLENE GLYCOL

Polyethylene glycol (PEG) is a hydrophilic polymer that acts as a “fusogen” to target damaged cell membranes in the injured spinal cord, preventing intracellular ionic disturbances. Treatment with PEG limited secondary axonal damage and enhanced neurologic function in an experimental guinea pig model of SCI[22]. A phase I clinical trial of PEG in deep pain negative dogs with IVDE was published in 2004, and showed a significant increase in neurologic recovery for dogs treated with PEG when compared to a group of historical controls[9]. However, the group of historical controls had a very low rate of recovery compared to what is reported in most other veterinary studies. When comparing PEG treatment to surgery alone in other studies evaluating deep pain negative dogs with IVDE, the recovery rate is approximately 60% for both groups[5, 12, 21]. A blinded, placebo-controlled veterinary clinical trial is currently evaluating PEG in dogs with IVDE.

MINOCYCLINE

Minocycline, an antibiotic belonging to the tetracycline group, has garnered recent interest from neuroscientists because of its potential role to modulate post-traumatic neuroinflammation. Minocycline also appears to have some activity as an inhibitor of apoptosis[7]. Numerous rodent studies have evaluated the effects of minocycline on neurologic recovery, and several have reported reduced secondary damage at the injury site along with improved functional outcome[18, 20]. However, a recent initiative from NIH to replicate encouraging results in laboratory models of SCI attempted to reproduce positive results with minocycline and was not successful. The efficacy of minocycline has not been evaluated in dogs.

RILUZOLE

Riluzole (Rilutek, Sanofi Aventis) is an FDA-approved sodium channel blocker that can be orally administered. It is currently approved for the treatment of amyotrophic lateral sclerosis (ALS) in humans. Because secondary axonal damage after SCI is mediated in part by abnormal sodium channel permeability, investigators have evaluated the efficacy of riluzole to attenuate white matter damage after SCI[11, 24, 26]. Results have been encouraging in rodent models and a human clinical trial is currently underway. The efficacy of riluzole has not been evaluated in dogs.

CELL-BASED THERAPIES

Stem cells: As it is routinely used in scientific literature, the term "stem cell" encompasses several cell types. These include true totipotent or pluripotent embryonic stem cells (ESC) with the capability to differentiate into any cell type in the body, as well as somatic stem cells (SSC) present in differentiated tissues which have the capability to dedifferentiate in to a restricted number of cell types. Stem cells of each type have been explored for cell-based therapies to treat SCI, including several recently published veterinary studies. All canine studies to date have included only a small number of dogs, but safety and feasibility have been promising. Variable amounts of improvement in neurologic function have been observed in dogs receiving autologous bone marrow stromal cells injected in to CSF[16] or mesenchymal stem cells derived from fat[17, 19].

Autologous olfactory ensheathing cells

Olfactory ensheathing cells (OEC) are progenitor cells associated with axons of olfactory receptor neurons in the nasal cavity[14]. They function as a hybrid of central nervous system glia and peripheral Schwann cells to facilitate and support regrowth of damaged olfactory axons. Intraspinal implantation of OEC has been shown to improve recovery in rodent models of SCI. A phase I trial recently evaluated the effects of OEC in chronically deep pain negative dogs with SCI caused by IVDE[6]. Results were mixed in that dogs in the treatment group gained some improvements in forelimb-hindlimb coordination; however, no improvement in long-track signs was noted when comparing the treatment group to a contemporaneous control group. The safety and feasibility of treatment in the aforementioned trial indicate that the utility of OEC to improve functional recovery after SCI in dogs warrants further study in a larger population.

REFERENCES

[1]Bracken MB, et al. **N Engl J Med** 1990; 322 (20): 1405. [2]Bracken MB, et al. **JAMA** 1997; 277 (20): 1597. [3]Bracken MB, et al. **J Neurosurg** 1998; 89 (5): 699. [4]Bydon M, et al. **World Neurosurg** 2013; doi: 10.1016/j.wneu.2013.02.062. [5]Duval J, et al. **Vet Surg** 1996; 25: 6. [6]Granger N, et al. **Brain** 2012; 135 (Pt 11): 3227. [7]Heo K, et al. **Neurosci Lett** 2006; 398: 195. [8]Hurlbert et al. **Can J Neurol Sci** 2008; 35: 41. [9]Laverty PH, et al. **J Neurotrauma** 2004; 21 (12): 1767. [10]Levine J, et al. **J Neurotrauma** 2011; 28: 675. [11]Mu X et al. **J Neurotrauma** 2000; 17: 773. [12]Olby NJ, et al. **J Am Vet Med Assoc** 2003; 222: 762. [13]Olby, NJ. **Vet Clin North Am Small Anim** 2010; 40: 791. [14]Nandoe Tewarie RS, et al. **J Spinal Cord Med** 2009; 32 (2): 105. [15]National Spinal Cord Injury Statistical Center. 2012. Spinal cord injury facts and figures at a glance. **www.nscisc.uab.edu.** University of Alabama, Birmingham. Birmingham, Alabama. [16]Nishida H, et al. **Am J Vet Res** 2011; 72 (8): 1118. [17]Park SS, et al. **Cytotherapy** 2012; 14 (5): 584. [18]Pinzon A, et al. **Brain Res** 2008; 1243: 146. [19]Ryu HH, et al. **J Vet Med Sci** 2010; 74 (12): 1617. [20]Saganova K, et al. **Neursosci Lett** 2008; 433: 246. [21]Scott HW, McKee WM. **J Small Anim Pract** 1999; 40: 417. [22]Shi R, Borgens RB. **J Neurophysiol** 1999; 81: 2406. [23]Smith PM, Jeffery ND. **Brain Pathol** 2006; 16: 99. [24]Springer JE, et al. **J Neurochem** 1997; 69: 1592. [25]Spitzbarth I, et al. **J Neuropathol Exp Neurol** 2011; 70: 703. [26]Stutzmann JM, et al. **Neuroreport** 1996; 69: 1592.

CHIARI-LIKE MALFORMATION AND SYRINGOMYELIA IN CAVALIER KING CHARLES SPANIELS

Holger A. Volk, PhD, DECVN, MRCVS
London, UK

INTRODUCTION

Many studies about Chiari-like malformation (CM) and syringomyelia (SM) have been published in recent years (Driver et al., 2013a). The CM/SM complex is a challenging disease to understand, most likely multifactorial and might vary between species. This report will focus on CM and SM in Cavalier King Charles Spaniels (CKCS).

Dogs do not have a cerebellar tonsil, nevertheless, CM shares similarities with Chiari type I malformation (CMI) in people, which is characterized by a reduced cranial posterior fossa volume. This results in a caudal displacement of the cerebellar tonsils (Milhorat et al., 2010) changing the flow of cerebrospinal fluid (CSF) at the level of the foramen magnum and the cranial cervical spinal cord (Levine, 2004; Greitz, 2006). The term '*Chiari-like*' malformation (CM) was established by an international working group (Cappello and Rusbridge, 2007) and has been used since then to describe a mismatch between caudal cranial fossa (CCF) volume and the CNS structures within leading to a caudal cerebellar and brainstem displacement into or through the foramen magnum. CM has a high prevalence in CKCS. One study reported that 92% of CKCS had at least one morphologic abnormality consistent with CM (Cerda-Gonzalez et al., 2009a).

The development of SM is thought to cause neurologic dysfunctions and neuropathic pain (Cerda-Gonzalez et al., 2009a; Rusbridge et al., 2007; Rutherford et al., 2012). Larger and more asymmetric syringes have a stronger likelihood to cause neuropathic pain (Rusbridge et al., 2007). However, in contrast to human medicine, there is still a debate if CM alone can cause clinical signs (Lu et al., 2003; Cerda-Gonzalez et al., 2009a; Driver et al., 2013a). In CKCS, SM secondary to CM is caused by a complex oligogenic trait with a moderately high estimate of hereditability (Lewis et al., 2010). SM appears to be a late-onset disease process. Twenty-five per cent of asymptomatic CKCS aged 12 months have imaging findings compatible with SM, ranging to 70% in CKCS aged >71 months (Parker et al., 2011). SM can affect multiple spinal cord segments (Loderstedt et al., 2011) and its width increases over time (Driver et al., 2012). It is not always easy to predict which dog will develop or will have clinical signs when looking at SM's radiological appearance. This is not only a challenge for clinical decision-making, but also complicates study designs. Even more frustratingly is that in dogs, which develop clinical signs of neuropathic pain, the clinical signs can be progressive despite medical or surgical intervention (Vermeersch et al., 2004; Dewey et al., 2005, 2007; Rusbridge, 2007; Plessas et al., 2012). CM appears to be the risk factor for the development of SM, but other morphologic abnormalities such as atlantooccipital overlapping could also alter the subarachnoid space and CSF flow and may contribute to the development of SM. Atlantooccipital overlapping could be considered to be part of the CM phenotype (Marino et al., 2012).

MORPHOMETRIC ANALYSIS OF CM/SM

CKCS is a brachycephalic breed with an extremely wide braincase (Schmidt et al., 2011). Breeding selection pressures towards a paedomorphic head shape could have resulted in a miniaturized and brachycephalic skull shape potentially causing the craniocerebral mismatch seen in CM. As in people with CMI, occipital bone hypoplasia has been suggested as the basis of CM in dogs (Rusbridge et al., 2000; Rusbridge and Knowler, 2004).

The occipital bone hypoplasia causes a reduction of the volume of the posterior fossa in CMI (Trigylidas et al., 2008), which has a significantly reduced size in children with CMI and SM (Sgorous et al., 2006). Similar changes were expected in CKCS, however, initial studies did not find an association of CCF area or volume with SM (Carruthers et al., 2009; Cerda-Gonzalez et al., 2009a; Driver et al., 2010a; Schmidt et al., 2009), which might be explained by a limitation of these studies not considering age of the dogs. SM can be more commonly seen in older dogs (Parker et al., 2011) and dogs could have been included which might develop SM later in life. In a study of our group, we compared the two 'extreme phenotypes', the volume of CCF in skeletally mature CKCS with CM and SM (<2years) versus older CKCS with CM only (>5years), which showed a small but significant difference with the SM affected dogs having a reduced CCF (Driver et al., 2010b). Using the same 3-D volumetric technique we found that CKCS appeared to have a similar sized CCF to comparable small breed dogs, but had a parenchymal volume within the CCF that was similar to Labradors (Cross et al., 2009). In a subsequent study, we showed that a relative increase of brain parenchyma within the CCF was associated with the presence of SM on imaging (Driver et al., 2010a) and more specifically it appears to be related to an increase in size of the cerebellum (Shaw et al., 2011). The abnormal cerebellar structure could cause an alteration of cerebellar function. And indeed CKCS have a higher variability in certain gait parameters compared to other dogs with a similar body frame, which is suggestive for ataxia (Suiter et al., 2012).

VASCULAR MORPHOLOGY

CKCS with SM also appear to have enlarged ventricles (Driver et al., 2013a). Arachnoid villi act as a one-way valve. The degree of CSF drainage is dependent on the pressure gradient between CSF and venous blood pressure (Di Chiro et al., 1976). Humans with achondroplasia, have a shortened braincase due to earlier closure of cranial base synchondroses (Di Mario et el., 1995), which also can affect the cranial foramina leading to venous congestion and increased venous pressure (Driver et al., 2013a). Children with craniosynostosis and ventriculomegaly had significantly smaller jugular foramina (Rich et al., 2003). Similar findings were reported in CKCS with SM, which had significantly smaller jugular foramina than those without SM (Schmidt et al., 2012). Using a magnetic resonance venography technique we showed that CKCS with CM and SM had a reduced venous blood volume in the CCF compared to CKCS with CM only (Fenn et al., 2013). A reduced venous drainage has also been suggested as part of the pathogenesis of SM in Griffon Bruxellois without CM (Rusbridge et al., 2009).

SPINAL MORPHOLOGY

SM is thought to be caused by regular re-occurring distending centrifugal forces exerted on the spinal cord (Iskander et al., 2004; Greitz, 2006). The narrowing of the subarachnoid space in CM causes a Venturi effect changing the CSF flow and pulse pressure gradients between the spinal cord and subarachnoid space driving the formation and propagation of syringes (Rusbridge et al., 2006). It was thought that vertebral canal narrowing in the region of C1-3 might be a risk factor for the development of SM (Rusbridge et al., 2006). A recent study, however, did not find an association between vertebral canal narrowing or angulation at C2/3 and SM (Carruthers et al., 2009). Other stenoses of the subarachnoid space at the C1 region have been described associated with SM, including dorsal impingement by soft tissue structures, atlantooccipital overlapping and atlantoaxial instability (Cerda-Gonzalez et al., 2009a; Marino et al., 2012). Although atlantoaxial morphology in CKCS appears to be different from other breeds, it does not correlate with the appearance of syringes (Stalin et al., 2008).

CSF HYDRODYNAMICS

CSF flow changes can be assessed using phase contrast cardiac gated cine MRI in human patients with CMI and/or SM, which appears to be a good technique to study the pathophysiology. The severity of symptoms is correlated with the severity of CSF flow changes (Lee et al., 2002). In Veterinary Medicine, CSF flow velocity in dogs is less than in humans and therefore can be more challenging to measure. One group managed to assess CSF flow in CKCS with CM and SM, and demonstrated turbulent CSF flow at the foramen magnum being correlated with the severity of syrinx formation and CSF velocity dorsal to C2/3 was inversely correlated to the existence of SM (Cerda-Gonzalez et al., 2009b).

CEREBELLAR PULSATION

With every pulse wave of the cerebral arteries the cerebellum moves caudally (piston action) into the foramen magnum (Greitz et al., 1992). This piston action could result in larger CSF pulse waves in patients with CMI, which can result in pressure gradients between the subarachnoid space and spinal cord, creating 'pulsating' centrifugal forces on the spinal cord distal to the blockage and causing the development of syringes (Greitz, 2006). In a former study the degree of caudal cerebellar herniation was not associated with SM (Lu et al., 2003), however, a recent study showed that the degree of caudal cerebellar herniation is dependent on head positioning (Upchurch et al., 2011) and cardiac cycle. Upchurch et al. (2011) did find a difference in the degree of caudal cerebellar herniation between CKCS with CM and CM/SM in a flexed head position, but not in a 'neutral' or extended head position. Human patients with CMI and SM have a greater caudal displacement during the cardiac cycle (Bunck et al., 2012). As in people, CKCS with CM and SM also show more cerebellar pulsation than CKCS with CM only (Driver et al., 2013b).

NEWER THEORIES OF CM/SM PATHOGENESIS

Morphometric studies have shown differences in CKCS with CM/SM and CM only, but looking at it critically, it is not clear if this is an epiphenomena or the changes seen are causative. There is, however, sufficient data showing that the CCF and associated soft tissue structures are abnormally formed in CKCS with CM (Driver et al., 2013a). The pathogenesis of SM between humans with CMI and dogs with CM share similarities, such as changes in cerebellar pulsation and abnormal CSF hydrodynamics.

CHANGES IN CRANIOSPINAL COMPLIANCE

Mechanical compliance determines the ability of a system to accommodate a change in volume in the face of pressure changes (Driver et al., 2013a). Craniospinal compliance is the sum of the ability of both the cranial and spinal compartments to accommodate changes in parenchymal, blood or CSF volumes which all exist in a state of dynamic equilibrium (Marmarou et al., 1975). Craniospinal compliance is primarily determined by the morphometric dimensions of the surrounding bone structures (skull and vertebral column), but is influenced by several other factors including CSF and blood volume, CSF and blood flow, the integrity of vasculature and potential space occupying lesions.

The unifying theory for hydrocephalus, CMI and SM has been proposed by Williams (2008) and is based on the assumption of CM causing narrower foramina and a reduced venous drainage resulting in an altered craniospinal compliance (Williams, 2008). In this case the CSF absorption into the venous sinuses is reduced, and hydrocephalus and communicating SM develop. The morphologic abnormalities previously discussed in CKCS with CM have the combined effects of reducing the amount of blood in the venous sinuses and its drainage (Schmidt et al., 2012; Fenn et al., 2013), and reducing the subarachnoid space and amount of CSF surrounding the brain parenchyma in the CCF (Driver et al., 2010a, 2010b) and at the foramen magnum (Upchurch et al., 2011). These changes could not only result in CSF outflow obstruction at the foramen magnum, but also in CSF flow changes resulting in the development of SM and ventriculomegaly. Reduced craniospinal compliance may affect the transition of pressures between the cranial and spinal systems' blood and CSF compartments (Driver et al., 2013a). The cerebellar pulsation changes, which we recently reported in CKCS, can also be explained by this, however, it is unclear if this is part of the pathogenesis or an effect of the aforementioned changes in craniospinal compliance.

SM FORMATION

CSF pressure pulses are propagated along the subarachnoid space during the cardiac cycle; the timing (Oldfield et al., 1994) and shape of this pressure wave appears to change subsequently to decompressive surgery in human CMI patients (Heiss et al., 1999). A new theory proposed that syrinx formation is the result of delay in the peak of CSF pulse pressure such that it arrives out of phase with the peak of the arterial pulse (Bilston et al., 2010). This delay could cause the CSF peak pressure wave to arrive between arterial pulses. At this time point the arteries are at their smallest and the perivascular spaces at their largest and the resistance to CSF flow along perivascular spaces at its lowest (Driver et al., 2013a). This could explain the one-way flow of fluid and its accumulation first in the spinal cord parenchyma (pre-syrinx edema) and then forming syringes (Fishbein et al., 1999). In CM cases, the delay in formation and propagation of the CSF waveform may occur due to reduced craniospinal compliance (Driver et al., 2013a).

CONCLUSION

CM/SM can be a debilitating disease process in CKCS with a complex, multifactorial pathogenesis, sharing some similarities to the human condition. Clinically, CM/SM remains frustrating to treat taking into consideration its variable presentation and natural progression (Driver et al., 2012, Plessas et al., 2012). In theory, surgical treatment of CM in dogs may be appropriate to restore craniospinal compliance, which is however a major surgery and the majority of cases appear to remain to have a good quality of life with medical therapy (Plessas et al., 2012).

REFERENCES

Bilston LE, et al., **J Neurosurg** 2010;112: 808, Bunck AC, et al. **E Radiology** 2012;22:1860, Capello R and Rusbridge C, **Vet Surg** 2007;36:509, Carruthers H, et al., **JSAP** 2009;50:394, Cerda-Gonzalez S, et al., **Vet Radiol Ultrasoun** 2009a;50:37, Cerda-Gonzalez S, et al., **Vet Radiol Ultrasoun** 2009b;50:467, Cross HR, et al., **JSAP** 2009;50:399, Dewey C, et al., JAVMA 2005; 227:1270, Dewey CW, et al., **Vet Surg** 2007;36:406, Di Chiro G, et al., **Neurology** 1976;26:1, Di Mario FJ, et al. **Neurology** 1995;45:519, Driver CJ, et al., **JSAP** 2010a;51:382, Driver CJ, et al., **Vet Rec** 2010b;167:978, Driver CJ, et al., **BMC Vet Res** 2012:8, Driver CJ, et al., **Vet J** 2013a;198(3):551, Driver CJ, et al., **Vet J** 2013b;198(1):88, Fenn J, et al., **Vet J** 2013;197(3):896, Greitz D, et al., **Neurorad** 1992;34:370, Greitz D, **Neurosurg Rev** 2006;29:251, Heiss JD, et al., **J Neurosurg** 1999;91:553, Lee JH, et al., **Spin Cord** 2002;40:501, Lewis T, et al., **Vet J** 2010;183:345, Levine DN, **J Neurol Sci** 2004;220:3, Loderstedt S, et al., **Vet J** 2011;190:359, Lu D, et al., **Vet Rec** 2003;153:260, Marino DJ, et al., **JAVMA** 2012;73:105, Marmarou A, et al., **J Neurosurg** 2010;43:523, Milhorat TH, et al., **Acta Neurochir** 2010;152:1117, Odfield EH, et al., **J Neurosurg** 1994;80:2, Parker JE, et al., **Vet Rec** 2011;168:667, Plessas IN, et al., **Vet Rec** 2012;171:501, Rusbridge C, et al., **JAAHA** 2000;36:34, Rusbridge C and Knowler SP **JVIM** 2004;18:673, Rusbridge C and Knowler SP, **JSAP** 2006;47:603, Rusbridge C, et al., **JVIM** 2006;20:469, Rusbridge C, et al., **JSAP** 2007;48:432, Rusbridge C, **Vet Surg** 2007;36:396, Rusbridge C, et al., **JSAP** 2009;50:386, Schmidt MJ, et al., **Am J Vet Res** 2009;70:508, Schmidt MJ, et al., **Vet Radiol Ultrasoun** 2011;52:436, Schmidt MJ, et al., **BMC Vet Res** 2012;8:158, Shaw TA, et al., **PLoS One** 2011:7, Stalin CE, et al., **Am J Vet Res** 2008;69:89, Suiter EJ, et al., **JVIM** 2013;27:390, Trigylidas T, et al., **Child Nerv Sys** 2008;24,329-336, Upchurch JJ, et al., **Vet Rec** 2011;169:277, Vermeersch K, et al., **Vet Surg** 2004;33:355, Williams H, **Cerebrospin Fluid Res** 2008;5:7.

FELINE HYPERTENSIVE ENCEPHALOPATHY (FHE): WHEN MECHANICS MEETS CHEMISTRY

Kaspar Matiasek, DVM, DrMedVet, Assoc Memb ECVN, MRCVS
Munich, Germany

INTRODUCTION

The term hypertensive encephalopathy was introduced in 1928 to summarize the effects of systemic arterial hypertension on brain function and structure (Oppenheimer & Fishberg, 1928). In humans, systemic blood pressure values > 140/90 mm Hg are associated with an increased risk of so-called target organ damage involving the CNS (Price & Kasner, 2014). One of the most frequent manifestations of hypertensive encephalopathy in people is the posterior reversible encephalopathy (PRE), a syndrome presenting with a typical spatial pattern of vasogenic edema. PRE, however, also is seen in patients with bone marrow and solid organ transplantation, preeclampsia/eclampsia, autoimmune disorders and chemotherapy. Independent of the primary condition, the common understanding is that the breakthrough edema is caused by loss of autoregulation of the brain vasculature with subsequent hyperperfusion (Bartynski 2008).

Apart from these pervasive effects, hypertension appears to be the most important risk factor for transient ischemic attacks, cerebral infarcts and focal intracerebral hemorrhage. Predisposition to these cerebrovascular accidents is due to aggravation of atherosclerosis in large arteries, promotion of heart disease and widespread small vessel damage.

HYPERTENSION IN CATS

Hypertension is a common clinical problem in feline medicine. In contrast to humans, where primary hypertension accounts for nearly 90% of affected cases, often involving young and middle aged individuals, hypertension in felids mostly is secondary to systemic disease and culminates with age. The most common diseases associated with feline hypertension are chronic renal disease and hyperthyroidism, followed by hyperaldosteronism and diabetes mellitus. Hence, feline hypertension rarely is a hemodynamic problem only. Instead it is accompanied by factors altering blood chemistry. Hence metabolic, endocrine and toxic factors come into effect and may aggravate brain damage and dysfunction (Jepson, 2011).

FELINE HYPERTENSIVE ANGIOPATHY VERSUS ENCEPHALOPATHY

In most long-standing feline cases of systemic hypertension > 180 mmHg, intracranial vascular damage is evident histologically. Three types of vasculopathies can be distinguished with variations regarding the mode of damage, size/type of affected blood vessel and extraaxial versus intraaxial localization. The effect of systolic pressure rise is most obvious in large leptomeningeal arteries of the ventral aspects of the brain and of the dorsal longitudinal fissure. Exaggerated systolic loading results in reactive media hyperplasia followed by necrosis with multifocal transmural infiltrates. Even though the lumen is partially narrowed, significant hypoperfusion is unlikely due to multiple collaterals. Slightly more widespread in adjacent areas, a multifocal arteriolosclerosis is evident with marked to severe obliteration of the vascular lumens. Also these occlusive changes do not appear to translate directly into acute focal or regional hypoperfusion of the brain tissue. Parenchymal lesions, instead, perfectly colocalize with and are adjacent to small vessel changes and plasma leakage resembling human hypertensive encephalopathy. These affect predominantly the caudate nucleus, splenial, suprasplenial and marginal gyri and the associated subcortical white matter, as well as the cerebellar hemispheres. In spite of the hyperperfusion theory, these areas are hot spots for ischemic and hemorrhagic infarcts during hypertension as well (O´Neill, 2013). Infarcts can be seen in about 40% of cats with hypertension and neurological signs. Since more than 75% of all cats with confirmed hypertensive intracranial angiopathy also show vasogenic parenchymatous lesions it would be more appropriate to refer to a Feline Hypertensive Angiopathy-Encephalopathy Complex.

CHEMICAL AND METABOLIC FACTORS

Before doppler sonography and vascular resistance measurement identified systemic arterial hypertension being the most important risk factor for neurological complications of kidney dysfunction, electrolyte dysturbances and retention of neurotoxic compounds were considered major players in neuronal dyshomeostasis (Mahoney et al., 1982). Indeed, uremic toxins, hyper- and hyponatremia can precipitate encephalopathy in chronic kidney disease, as can hypercalcemia and hypermagnesemia (Davenport, 2008). In particular the contribution of uremic toxins cannot be assessed easily via blood work. In this regard, a study ideally would compare non-toxic hypertensive conditions such as hyperthyroidism and pheochromocytoma to renal encephalopathy. Hyperthyroid cats however often present with multimorbidity and are affected by kidney dysfunction as well. Pheochromocytoma on the other hand is extremely rare in cats.

One possible approach to discriminating pervasive effects, by hyperperfusion, from metabolic effects, by ischemia, is by comparing the expression profiles of the inducible water channel molecule aquaporin-4 in astrocytes to scavenger receptors on brain vasculature, astrocytes and microglia. By this method, we documented that water influx is evident in all neurological cats affected by FHA whereas scavenger receptors are upregulated rarely and in restricted areas only.

CONCLUSION

Systemic hypertension is an important risk factor of vasogenic encephalopathy in cats. Hyperperfusion and widespread breakdown of the blood brain barrier appear to be key factors throughout all stages of the disease. Hypoperfusion instead is restricted to advanced stage and vulnerable areas. On the other hand, it may result in cerebral infarcts that then dominate the clinical presentation. Early control of systemic hypertension in cats suffering from kidney disease will minimize the risk of neurological complications.

REFERENCES

Bartynski WS. **Am J Neuroradiol**. 2008;29(6):1043-9. Davenport, A. **Blood Purif;**26:526-536. Jepson RE. J Feline Med Surg. 201;13(1):25-34. Mahoney CA, et al. **Am J Kidney Dis**. 1982;2(3):324-36. O'Neill J, et al. **J Am Anim Hosp Assoc**. 201;49(6):412-20. Oppenheimer BS et al. **Arch Intern Med.** 1928;41:264-78. Price RS, et al. **Handb Clin Neurol**. 2014;119:161-7.

CANINE CENTRONUCLEAR MYOPATHIES: AN UPDATE AND COMPARATIVE REVIEW

Gemma Walmsley, MA, VetMB, PhD, DECVN, MRCVS
Liverpool, UK

CONGENITAL MYOPATHIES

An increasing number of breed-associated myopathies are described in veterinary species. These include inherited structural disorders – muscular dystrophies (such as X-linked or dystrophin-deficient muscular dystrophy) and congenital myopathies – as well as channelopathies, primary metabolic and mitochondrial myopathies and breed-predispositions for immune-mediated muscle disease (such as Vizsla polymyositis and dermatomyositis in Shelties).

Congenital myopathies are a nebulous grouping of inherited structural disorders that often have an onset in early life. Common characteristics are histopathological or ultrastructural abnormalities within myofibres. Unlike muscular dystrophies, these occur without a pronounced degeneration-regeneration picture. In the human medical literature they are divided into the following groups based on histopathological phenotype[1]:

- Congenital myopathies with central/internal nuclei (centronuclear myopathies (CNM))
- Congenital myopathies with protein accumulations (in particular nemaline myopathy)
- Congenital myopathies with cores (e.g. central core disease)
- Congenital myopathies with fibre-size variation (i.e. congenital fibre type disproportion)
- Congenital myopathies with vacuoles (e.g. sarcotubular myopathy)

However, mutational analysis is now considered the gold standard for diagnosing and classifying these conditions in human patients. Conditions from the first two pathological groups have been reported in dogs, and genetically characterized CNM are now recognized affecting a number of canine breeds.

CENTRONUCLEAR MYOPATHIES IN HUMANS

Centronuclear myopathies were first described in the human medical literature in the 1960's. They are characterized clinically by paresis and skeletal muscle atrophy and are defined by common histopathological features of which abnormal, central or internally-displaced nuclei (within $\geq$ 25% of myofibres) are the most prominent. Classically, the conditions have been organized based on inheritance and disease severity into X-linked myotubular myopathy and autosomal dominant, recessive and sporadic CNMs. More recent advances have identified genetic defects associated with each of these forms (see Table 1); however, for around 20% of cases, the molecular basis remains unknown[2,3].

CENTRONUCLEAR MYOPATHIES IN DOGS

As in humans, centronuclear myopathies also cause debilitating muscle weakness and atrophy in canine patients. Whilst these disorders are an uncommon cause of the myopathic syndrome in general, in certain breeds they are an important differential for a juvenile onset neuromuscular disorder. CNM associated mutations in dogs have been found in some of the same genes implicated in CNM in humans (MTM1 and BIN) and also an additional locus (PTPLA).

AUTOSOMAL RECESSIVE CNM IN LABRADOR RETRIEVERS

Canine CNM (previously known as hereditary Labrador retriever myopathy (HLRM), type II fibre deficiency of Labrador retrievers, autosomal recessive muscular dystrophy of Labrador retrievers and floppy Labrador syndrome) was first described in 1976 and has been recognized worldwide. Labradors of both sexes and all three colors can be affected and typically are presented to veterinarians at around 3-6 months of age displaying a marked paresis, hypotonia, exercise intolerance and generalized muscle atrophy[4]. Myotactic reflexes are reduced or absent. Clinical signs are often progressive and at later stages may include megaesophagus, secondary regurgitation and aspiration pneumonia. Muscle from CNM-affected dogs exhibits several characteristic and progressive features including loss of type II fibres, fat infiltration, fibrosis, and centralization of nuclei. There is currently no treatment, and whilst milder cases can be carefully managed, many owners opt for euthanasia[4,5].

A colony of affected dogs was established at Alfort Veterinary School in 1992, which allowed further characterization and the condition was redesignated canine CNM based on numerous shared clinical and pathological features with human CNM[5,6] and the underlying genetic defect was ultimately determined as a mutation in the PTPLA gene, not previously associated with CNM[6]. A genetic test is available and a recent article confirmed that descriptions of canine CNM from European groups and HLRM in the USA cover the same disease entity with a spreading founder mutation that is disseminated throughout the breed globally[7]. Interestingly, the PTPLA gene has recently been associated with a similar congenital myopathy in a human family but further work is necessary to characterize this fully[8].

MYOTUBULAR MYOPATHY IN DOGS

Myotubular myopathy is an X-linked disorder that is the most common and most severe CNM in humans – the majority of affected boys display severe muscle weakness that results in a high mortality before 1 year of age[3]. The disease has also been described in 2 families of Labrador Retrievers in Canada[9,10]. Dogs presented at around 3-4 months of age with severe tetraparesis and generalized muscle atrophy associated with absent myotactic reflexes. In all patients, clinical signs progressed rapidly over a 3-4 week period requiring euthanasia by 4-6 months of age. Pathological features included the presence of numerous hypotrophic myofibres with centralized nuclei. A conserved mutation was found in the myotubularin gene in all male affected dogs and obligate female carriers, and genetic analysis suggested a recent founder in the local Labrador population[10]. A colony of these dogs has been established and is currently being used for important research into gene replacement therapies for this disorder[11].

Myotubular myopathy has also been reported in a litter of Manchester Terriers[12]. In a published abstract, the authors described 3 dogs that were presented at 2 months of age with generalized paresis and failure to thrive. Histopathological and ultrastructural findings were consistent with myotubular myopathy and immunoblotting of extracted proteins from affected dogs revealed absence of myotubularin protein.

INHERITED MYOPATHY OF GREAT DANE DOGS

An autosomal recessive inherited myopathy in Great Dane dogs (IMGD) was first reported in the UK in 1987 and has since been reported worldwide. It was originally described as a Central Core Myopathy but subsequently reclassified. Male and female affected dogs have fawn and brindle coat colors and usually present around 1 year of age (range 6-19mo). Clinical signs include progressive paresis, exercise intolerance with fatigue and atrophy. Muscle biopsy reveals variable myofibre diameters, type 1 fibre predominance, centralized nuclei and characteristic central areas of cytoplasmic disorganization. The condition appears to be progressive and around 50% of affected dogs are euthanized by 1 year of age[13].

Recent evidence indicates that this condition is a centronuclear myopathy with a BIN1 mutation similar to progressive AR CNM in humans. The mutation, found in 5 dogs with inherited myopathy of Great Danes from 2 continents, reduces mRNA and protein levels in muscle[14]. A number of dogs diagnosed with IMGD did not possess this mutation or any other mutation in the BIN1 coding sequence therefore there may be several disorders with a similar phenotype in Great Danes[14].

CNM PATHOGENESIS

Centronuclear myopathies are a grouping of phenotypically and genetically distinct inherited muscle diseases. Their pathogenesis is incompletely understood however shared skeletal muscle pathological features between the different diseases in the group suggest a common underlying mechanism. Advances have been made in recent years following the identification of several genetic defects associated with these conditions: many of these genes/proteins are linked to membrane trafficking processes and structural abnormalities in the t-tubular and/or sarcoplasmic reticulum systems appear to be consistent features[15]. Further work on this area in humans and dogs, and in cell culture and animal models of these conditions, will hopefully be able to benefit both species.

Table 1: Genes implicated in classical CNM in humans and canine CNM. (Modified from Romero and Laporte[2]).

Gene	Protein	Function	Inheritance	Epidemiology in humans	Epidemiology in dogs
MTM1	Myotubularin	Phosphoinositide phosphatase	X-linked recessive	1:50,000 live male births ~300 affected families	Mutation reported in 2 families of Labradors (suspected common founder). Suspected in a family of Manchester Terriers
BIN1	Amphiphysin 2	Membrane remodelling	Autosomal recessive	5 families affected	Mutation confirmed in 5 Great Danes and evidence of disseminated mutation in breed.
DNM2	Dynamin 2	GTPase in membrane trafficking	Autosomal dominant	40 families affected	Not reported
RYR1	Ryanodine receptor 1	Sarcoplasmic reticulum calcium channel	Sporadic AR/AD?	14 families affected	Not reported
MTMR14/JUMPY/MIP	Myotubularin related 14	Phosphoinositide phosphatase	Sporadic unclear - AR/digenic	2 families affected	Not reported
PTPLA/HACD1	Protein Tyrosine Phosphatase-like A	Very long chain fatty acid hydroxyacyl-co A dehydratase	Autosomal recessive	Not reported. 1 family with congenital myopathy with fibre type disproportion.	Disseminated mutation with founder effect in Labradors (15-20% of dogs carry the mutation)

REFERENCES

[1]North K. Neuromuscular Disorders 2008; 18: 433-442 [2]Romero N and Laporte J. Muscle disease: pathology and genetics 2nd Edition 2013 [3]Dubowitz V, et al. Muscle biopsy:a practical approach 4th Edition 2013 [4]Shelton GD. Veterinary Comparative Orthopaedics and Traumatology 2007; 20:249-255 [5]Tiret L, et al. Human Genetics 2003; 113: 297-306 [6]Pele M, et al. Human Molecular Genetics 2005; 14: 1417-1427 [7]Maurer M, et al. PLoS One 2012; 7:e46408 [8]Muhammad E, et al. Human Molecular Genetics 2013; 22:5299-5236 [9]Cosford KL, et al. Canadian Veterinary Journal 2008; 49:393-397 [10]Beggs AH, et al. Proceedings of the National Academy of Sciences USA; 107:14697-14702 [11]Childers MK, et al. Science Translational Medicine 2014: 220ra10 [12]Robinson F, et al. Journal of Veterinary Medicine 2008: 687-824 [13]Feliu-Pascual AL, et al. Journal of Small Animal Practice 2006; 47:249-254 [14]Bohm J, et al. PLoS Genetics 2013; 9:e1003430 [15]Al Qusiari L and Laporte J. Skeletal Muscle 2011; 1:26

WHAT IS CANINE EPILEPSY?

Mette Berendt, DVM, PhD
Copenhagen, Denmark

INTRODUCTION

Due to its heterogenic nature canine epilepsy is complex to approach clinically. Most of the time no outward signs of disease are present. The tendency for epilepsy to become inactive for a period of time, or even disappear, makes it difficult to study (1,2).

Epilepsy is defined by recurrent seizures originating from the brain. Seizure activity arises as a sequel to abnormal hypersynchronous electrical activity which stems from a group of local cerebral neurons or from the whole cortex in collaboration with subcortical structures (2). Factors involved in epileptogenesis may be neuroplasticity/synaptic plasticity, decreased GABA (gamma amino butyric acid) activity, increased glutamate activity, new formation of excitatory circuits, receptor alterations, ion channelopathies, dysfunction of glia cells, cortical dysgenesis and primary brain disease among others (3,4).

Epilepsy is characterized by a sudden loss of control and paroxysmal manifestations with seizures which suddenly occur and suddenly cease. Seizures are transient and seldom last longer than 2-3 minutes. The clinical seizure signs in the individual patient typically follow a repetitive pattern where signs and sequence are identical from episode to episode. Signs may include motor, sensory, and autonomic elements, behavioural disturbances and altered or loss of consciousness (2,5). In some individuals with epilepsy, seizures can be provoked by external factors such as excitement, agitation, flashes of light, and sleep deprivation.

PREVALENCE

Canine epilepsy is not an uncommon condition. Prevalence has been estimated to be 1-2% in a hospital referral population (6). It is however recognized, that in breeds where a genetic background is associated with epilepsy (idiopathic epilepsy), there is an accumulation of epileptic individuals and consequently a higher breed prevalence, which have been estimated for Labrador retriever (3.1%), Belgian shepherd (9.4%), and Petit basset griffon vendeen (8.9%) (7-9). A possible or identified genetic background for epilepsy has been reported for a number of breeds, among others Beagle, Keeshound, Belgian shepherd, Labrador Retriever, Vizla, Shetland Sheepdog, Irish Wolfhound, English Springer Spaniel, Australian shepherd, Border collie, Petit basset griffon vendeen and Lagotto Romagnolo (7-20).

CLASSIFICATION OF EPILEPSY AND EPILEPTIC SEIZURES

There are many similarities between human and canine epilepsy. Veterinary epilepsy terminology has traditionally been based on the human ILAE (International League Against Epilepsy) classification system (21-22). Human, as well as veterinary classification and terminology are continuously being debated (21-31). However, the increasing identification of etiologies and mechanisms leading to epilepsy in people, has brought the ILAE terminology and classification system to a sophisticated level that is yet not directly applicable in dogs (30-31).

At present, a standardized veterinary classification system for epilepsy universally accepted by all does not exist, and therefore authors may use different terminologies to describe epilepsy types and epileptic seizures (2,23-26). The present author will, in the following, use the classification system based on previous ILAE guidelines from 1981 and 1989, which is still widely used in veterinary medicine.

Classification of epilepsy

The 1981 and 1989 ILAE guidelines operate with three categories: idiopathic epilepsy, symptomatic epilepsy, and probable symptomatic epilepsy - previously known as cryptogenic epilepsy (21-22). Idiopathic epilepsy refers to epilepsy of unknown cause, where no structural cerebral pathology is identified, and with a suspected or known genetic component involved. Symptomatic epilepsy refers to epilepsy caused by a known/identified disorder of the CNS (cerebral pathology e.g. identified with MRI or CSF examination). The term probable symptomatic epilepsy may be used as a designation for epilepsy with a suspected symptomatic cause, which however remains obscure.

Classification of Epileptic Seizures

The clinical manifestation of epileptic seizures is directly related to the amount and distribution of abnormal electrical activity in the brain. Epileptic seizures may be classified as primary generalized seizures, focal seizures (previously termed partial seizures), or focal seizures with secondary generalization.

With primary generalized seizures, there is a sudden outburst of abnormal electrical activity throughout both cerebral hemispheres. Without warning, the patient collapses with tonic, clonic, or tonic-clonic seizures. With focal seizures , the abnormal electrical activity arises in a localized group of neurons (the epileptic focus). The clinical signs reflect the functions of the area or areas involved. In dogs focal motor, sensory, autonomic or behavioural signs can manifest alone or in combination (2,5,23-25). Focal motor signs may include rhythmic jerks of an extremity, repeated jerking head movements, abnormal rhythmic blinking and twitching of facial musculature. Focal seizure activity believed to represent psychic and/or sensory seizure phenomena in the brain may in dogs result in a short lasting episodic change in behaviour such as anxiousness, restlessness and abnormal clinging to the owner. Focal seizures with parasympathetic/epigastric components may present as hyper salivation, vomiting and dilated pupils (5,23). More than one sign may be present during a seizure event. In the 1981/1989 human ILAE classification, focal/partial seizures were classified into simple focal/partial seizures (where the state of consciousness is maintained) and complex focal/partial seizures (where the state of consciousness is impaired). This sub-classification has now been abandoned by the ILAE. For dogs it is difficult to appraise different states of consciousness as this is limited to subjective interpretation made by the person observing the dog. In the authors opinion it is therefore not meaningful to use this sub-classification for dogs with epilepsy.

If the electrical activity in the epileptic focus becomes sufficiently massive, focal seizures can spread through subcortical structures to involve more areas and eventually the entire brain, a focal seizure with secondary generalization. In that case the initial localized seizure signs will subsequently be followed by tonic, clonic, or tonic-clonic convulsions. This is the most common seizure type observed in the dog (1,2,23). The onset of the focal seizure is often very short (few seconds to minutes) and the secondary generalization with convulsions follows rapidly. The focal seizure onset may be difficult to detect due to its brief nature.

What is aura? Commonly owners report that they can foresee a convulsive seizure when specific and to the owner well known signs repeatedly appear within seconds or minutes prior to convulsions. The term aura has in the past been used to describe such a forewarning of convulsions. This term originated from human epileptology where aura in early ILAE classifications was used to denote symptomatology that encompasses subjective sensory phenomena as well as vegetative signs (for example, the epigastric sensations accompanying mesial temporal epilepsy,) and thus did not include motor phenomena. This author recommends that the term aura is not used in the veterinary clinic. The signs occurring as the first indication of seizure activity (marking the beginning of ictus) and interpreted by the dog owner as a warning sign is indeed a focal seizure onset and should be referred to as such.

SEIZURE PHASES

The epileptic seizure is organized as ictus (seizure activity) and postictal phase (where the normal brain function is restored). Ictus may consist of a primary generalized seizure, a focal seizure, or a focal seizure with secondary generalization. In the postictal phase, the brain restores its normal function. The postictal phase may be very short or last for several hours to days. Typically, the animal is disoriented, ataxic, hungry or thirsty, express a need to go in the garden or appear exhausted. Postictal blindness or aggression may be present.

In a few dogs, ictus may be preceded by a so-called prodrome, a long-term change in disposition and indicator of forthcoming seizures (23). Humans may experience days of irritability, withdrawal or other emotional aberrations (32). In dogs the most common prodromal signs described are restlessness and attention-seeking which is known to the owner as a long-term marker of a forthcoming seizure episode.

PROGNOSIS ASSOCIATED WITH EPILEPSY

Canine epilepsy has been associated with premature death (1). Prognosis with epilepsy is influenced by the type of epilepsy experienced (idiopathic versus symptomatic), the tendency to develop cluster seizures and status epilepticus, seizure control and the balance between seizure control and side-effects of medication. By nature idiopathic epilepsy holds a better prognosis than symptomatic epilepsy originating from intracranial pathology. Prognosis may be influenced negatively in some breeds experiencing genetic epilepsy, and where pharmaco-resistance, cluster seizures and status epilepticus dominate , as reported for the Irish wolfhound, Border Collie and Australian shepherd (17,33,34). Other breeds may display a milder form of epilepsy and therefore have a better chance of a full lifespan as in the Belgian shepherd (35). Several studies have reported that status epilepticus and/or cluster seizures influence survival negatively (33,36-38).

Owners often fear that epilepsy may cause their dog to die spontaneously. In humans the phenomenon "sudden expected death" (SUDEP) refers to death directly related to epilepsy in patients with chronic epilepsy. SUDEP is defined as "*Sudden, unexpected, witnessed or unwitnessed, nontraumatic and nondrowning death, occurring in benign circumstances, in an individual with epilepsy, with or without evidence for a seizure and excluding documented status epilepticus (seizure duration >30 min or seizures without recovery in between), in which postmortem examination does not reveal a cause of death*" (39). A similar phenomenon has been suspected to occur in dogs, although sparsely reported (1,35).

Importantly, epilepsy may go into remission. Both spontaneous remission and remission with treatment has been reported in dogs with remission rates around 15% (1,7,19,35).

REFERENCES

1. Berendt M et al. **J Vet Int Med** 2007;21:754.
2. Chandler K. **Vet J** 2006;172:207.
3. Diechter MA. **Arch Neurol** 2009;66:443.
4. Raymond AA et al. **Brain** 1995;118: 629.
5. 30. Berendt et al. **Epilepsy Res** 2004;61:167.
6. Schwartz-Porsche D. **Proc. ACVIM** 1986;4:1161.
7. Berendt M et al. **J Vet Int Med** 2002;16:262.
8. Berendt M et al. **Acta Vet Scand** 2008;50:51.
9. Gulløv et al. **J Vet Int Med** 2011;25:1372.
10. Bielfelt SW et al. **Am J Vet Res** 1971;32:2039.
11. Wallace ME. **J Small Anim Pract** 1975;16:1.
12. Jaggy A et al. **J Small Anim Pract** 1998;39:275.
13. Morita T et al. **Can J Vet Res** 2002;66:35.
14. Oberbauer AM et al. **J Hered** 2003;94:57.
15. Patterson EE et al. **J Vet Int Med** 2003;17:319.
16. Patterson EE et al. **J Am Vet Med Assoc** 2005;226:54.
17. Casal ML et al. **J Vet Int Med** 2006;20:131.
18. Berendt M et al. **J Small Anim Pract** 2009;50:655.
19. Hülsmeyer V et al. **J Vet Int Med** 2010;24:171.
20. Seppälä EH et al. **PLoS One** 2012;7(3):e33549.
21. Commission on Classification and Terminology of the International League Against Epilepsy. **Epilepsia** 1981; 22:489.
22. Commission on Classification and Terminology of the International League Against Epilepsy. **Epilepsia** 1989;30:389.
23. Berendt M et al. **J Vet Int Med** 1999;13:14.
24. Licht BG et al. **Epilepsy and Behavior** 2003;3:460.
25. Podell M. Seizures. **BSAVA manual of canine and feline neurology**. Fourth ed. 2013:117.

26. Platt S. Seizures. **Small Animal Neurological Emergencies** 2012:155.
27. Lüders H et al. **Epilepsia** 1998;39:1006.
28. Blume WT et al. **Epilepsia** 2001;42:1212.
29. Engel J. **Epilepsia** 2001;42:796.
30. Fisher R.S et al. **Epilepsia** 2005;46:470.
31. Berg AT et al. Epilepsia 2010;51:676.
32. Aicardi et al. History and Physical examination. **Epilepsy a comprehensive textbook**, second ed. 2008:785.
33. Saito M et al. **J Am Vet Med Assoc** 2001;219:618.
34. Weissl J et al. **J Vet Int Med** 2012;26:116.
35. Gulløv CH et al. **J Vet Int Med** 2012;26:1115.
36. Bateman et al. J Am Vet Med Assoc 1999;215:1463.
37. Arrol L et al. **Vet Rec** 2012;170:335.
38. Monteiro et al. **J Small Anim Pract** 2012;53:526.
39. Nashef L et al. **Epilepsia** 2012;53:227.

THE DIAGNOSTIC CHALLENGE OF SEIZURING DOGS

Mette Berendt, DVM, PhD
Copenhagen, Denmark

INTRODUCTION

Various medical conditions may cause unexplained paroxysmal events and, as no gold standard test exists to confirm the diagnosis of canine epilepsy, the diagnostic approach to such patients is often challenging. The clinician must build her/his case from a platform where historical information of seizures and seizure phenomenology may identify certain characteristics typically associated with epilepsy, or point in a different direction (1,2). This information is combined with a variety of clinical and paraclinical investigations which can help the clinician to discriminate between epilepsy and other potential seizure provoking conditions.

EPILEPSY OR SOMETHING MIMICKING EPILEPSY

The clinical approach to dogs suspected of epilepsy always takes its outset in the owners' report of the history and observations of the clinical signs occurring before, during and after seizure events. The term seizures may be used by the owner to cover a variety of episodic phenomenon including e.g. convulsions, fainting, flaccid collapse and behavioral changes. Video recordings of seizures are therefore very helpful in order to confirm the owners' observations.

The basic clinical work-up is targeted against determining if the signs described are of an epileptic nature (originating from excessive discharge of cerebral neurons) or of a nonepileptic origin pointing towards differential diagnosis causing paroxysmal events such as sleeping disorders, cardiovascular disturbances, metabolic disease or intoxication (3). The clinical work-up includes historical information, a thorough description of the observed events, clinical and neurological examination, routine hematology and blood biochemistry and urine examination. If any of these investigations suggest a nonepileptic condition, this should be pursued further (e.g. in the case of cardiovascular disease with blood pressure measurements, ECG, ultrasound of the heart and maybe Holter monitoring). Useful tables of nonepileptic differential diagnosis which may cause paroxysmal events with or without convulsions can be found in textbooks (3).

HISTORY

A thorough history includes detailed information of age at seizure debut, previous history of head trauma or intracranial disease, known relatives with epilepsy, seizure proximity, the circumstances under which seizures occur, duration of seizures, clinical signs during seizures and the sequence and succession of such signs, and finally the clinical status following closely after seizures and between seizure events.

Certain characteristics define epilepsy, among others that seizures are transient, seldom last longer than 2-3 minutes and seldom occur while physically active like on walks and playing. In some epileptic dogs, high performance situations associated with stress may trigger seizures (1,2). Most commonly epileptic dogs experience seizures while at home, at rest or upon awakening. Epileptic episodes are unique in the individual dog and will follow a repetitive pattern of clinical signs and sequence. Clinical signs may include localized motor activity, autonomic signs and/or behavioural disturbances alone (focal seizures) (1,2). Such signs may be followed by convulsions (focal seizures with secondary generalization). Finally seizures may simply manifest as convulsions coming on instantly (primary generalized seizures). Consciousness may be altered or lost (1,2).

Seizure episodes associated with physical activity, and where the owner reports that the animal is fainting or losing tonus, becomes cyanotic, or episodically exhausted and reluctant to walk, should raise a suspicion of a diagnosis other than epilepsy (e.g. cardiovascular, upper respiratory or neuromuscular disease). So-called anoxic convulsions may be associated with syncope.

IDIOPATHIC OR SYMPTOMATIC EPILEPSY

Intracranial pathology as the cause for seizures is routinely being investigated with computerized tomography (CT), magnetic resonance imaging (MRI) and cerebrospinal fluid (CSF) examination. In any epilepsy patient, seizures may be "the top of the iceberg" and thus the only indication of intracranial abnormalities. A normal neurological examination does not exclude that brain pathology is present. It has been shown in epileptic dogs with a normal interictal neurological examination, that the likelihood of an abnormal MRI is low (2,2%) in individuals <6 years of age, whereas the risk of an abnormal MRI is relatively high (26,7%) in dogs >6 years of age (4). Diagnostic imaging and CSF should therefore always be considered.

A diagnosis of idiopathic epilepsy implies that no structural intracranial pathology can be identified and that a suspected or known genetic component may be involved in the pathogenesis. Many dog breeds have been suspected or confirmed to suffer from epilepsy with a genetic background involving mutations or complex interactions among multiple genes, among others the Irish wolfhound, Vizla, English springer spaniel, Lagotto Romagnolo and Belgian shepherd (5-13). It has been reported that seizure onset occurs early in life (between 1-5 years) in dogs with idiopathic epilepsy (7,13-16).

In the case of symptomatic epilepsy, intracranial pathology must be confirmed with advanced diagnostic tests. In dogs with seizures, interictal abnormal findings on the neurological examination localized to the brain are a strong indicator of intracranial pathology. Other diagnostic clues could be a general change of personality expressed as changes in behaviour, daily routines and/or interaction with people and other animals. A higher probability for symptomatic epilepsy has been reported if onset of seizures occurs before the age of one or after the age of seven (13).

RECOGNITION OF EPILEPTIC SEIZURE PHENOMENOLOGY AS A DIAGNOSTIC INDICATOR

The attempt to classify seizures is not only of academic interest, but is truly helpful when trying to identify if the clinical signs reported by the owner are in agreement with the seizure patterns associated with focal or generalized epileptic seizures (2,3,17-20).

Electroencephalography (EEG) is beneficial and routinely used in people to support a diagnosis of epilepsy and to localize and identify focal and generalized seizure activity in the brain. However certain limitations are also associated with EEG as some individuals with epilepsy will have a normal EEG. EEG has never been introduced as a routine procedure for dogs and, in veterinary medicine, has mostly been used for clinical epilepsy research (21,22). Although EEG is desirable for diagnostic purposes, the observations reported by the dog owner, and the clinician's interpretation of this information and recognition of the typical clinical signatures of epilepsy, remains the most central diagnostic tool in the clinical approach to patients with seizures.

REFERENCES

1. Chandler K. **Vet J** 2006;172:207.
2. Berendt M et al. **Epilepsy Res** 2004;61:167.
3. Platt S. Seizures. **Small Animal Neurological Emergencies** 2012:155.
4. Smith et al. **Vet J** 2008;176(3):320.
5. Casal et al. **J Vet Int Med** 2006;20:131.
6. Patterson EE et al. **J Vet Int Med** 2003;17:319.
7. Patterson EE et al. **J Am Vet Med Assoc** 2005;226:54.
8. Jokinen TS et al. **J Vet Int Med** 2007;21:464.
9. Berendt M et al. **J Small Anim Pract** 2009;50:655.
10. Seppälä et al. **PLoS One** 2012;7(3):e33549.
11. Ekenstedt KJ et al. **BMC genetics** 2011;12:38.
12. Ekenstedt KJ et al. **Mamm Genome** 2012;23:28.
13. Podell M et al. J Am Vet Med Assoc1995;206:1721.
14. Licht BG et al. J Am Vet Med Assoc 2007;231:1520.
15. Hülsmeyer V et al. **J Vet Int Med** 2010;24:171.
16. Gulløv et al. **J Vet Int Med** 2011;25:1372.
17. Berendt M et al. **J Vet Int Med** 1999;13:14.
18. Licht BG et al. **Epilepsy and Behavior** 2003;3:460.
19. Commission on Classification and Terminology of the International League Against Epilepsy. **Epilepsia** 1981; 22:489.
20. Commission on Classification and Terminology of the International League Against Epilepsy. **Epilepsia** 1989;30:389.
21. Berendt et al. **Acta Scand Neurol** 1999:99:276.
22. Brauer C et al. J Vet Int Med 2012;193:185.

ANTIEPILEPTIC DRUGS IN DOGS: USE AND MISUSE

Joane Parent DMV, MVetSc, DACVIM (Neurology)
St-Hyacinthe, Quebec

INTRODUCTION

The diagnosis and treatment of the dog with epilepsy remain a challenge for veterinarians. Idiopathic/Genetic epilepsy is often erroneously diagnosed. Treatment failure can be associated not only to wrong diagnosis, but also to wrong choice of antiepileptic drug (AED), wrong use of the chosen AED, or simply, false expectations of seizure control. The goal of this conference is to explain how seizure control is optimized with an accurate diagnosis, how the diagnostic work up is intimately linked to seizure pattern, and how seizure control is improved if antiepileptic treatment takes into account the seizure pattern.

TERMINOLOGY

The commission of the International League Against Epilepsy (ILAE) has proposed in 2010 some changes to the previous human classification of seizures and epilepsies.[1] Prior to 2010, human epilepsies were classified as idiopathic, symptomatic or probably symptomatic.[2] The proposed 2010 terminology classifies epilepsy as Genetic, Structural/Metabolic or of Unknown Cause. Mariani, in a review of the subject this past year, proposed that similar classification be used in veterinary medicine.[3] Ideally, the term 'idiopathic' should be eliminated, being a source of great confusion among veterinarians as to its meaning. In the present text, the epilepsy is **Genetic (GE)** when the direct result of a known or presumed genetic defect(s) in which seizures are the core symptom of the disorder[1] with no underlying structural brain lesion or other neurological signs or symptoms[2]. It is usually age-dependant[1,2]. In this text, the term Genetic has replaced the word Idiopathic. The epilepsy is **Structural** when the seizures are the result of identifiable lesion(s) of the brain.[2] Structural lesions in dogs include disorders such as inflammation, tumour, trauma, congenital abnormality, etc. Some of these may be genetic in origin (e.g., pug encephalitis); however, as currently understood, there is a separate disorder interposed between the genetic defect and the epilepsy.[1] The epilepsy is of **Unknown Cause** when seemingly structural, but no aetiology has been identified.[1] In this instance, a thorough diagnostic work up has been unremarkable.

A seizure is epileptic when originating primarily from the brain. Seizures are generalized, focal or focal with secondary generalization. The seizure is generalized when bilaterally engaging cortical structures.[1] The animal does not need to be unconscious or recumbent for the seizure to be generalized. It only needs to involve both cortices. The seizure is focal when discreetly localized or widely distributed but, limited to one hemisphere.[1] The focal seizure may secondarily spread to the other hemisphere. The seizure is then classified as focal with secondary generalization. The rapidity of generalization varies in time.

PROPER DIAGNOSIS

Ascertain Seizure Activity

The most important step toward seizure control is an accurate diagnosis. To do so, one must first ascertain presence of seizure activity. In most cases of generalized tonic-clonic seizures (convulsion), it is rather safe to diagnose that convulsion has occurred from the family's recollection. However, a detailed description must always be requested. Even health professionals should be asked to describe what they have witnessed. Acute cervical spasms may be mistaken for seizure activity. It can also be difficult to differentiate epileptic seizures from other episodic events such as movements or cardiac disorders. As a rule, cardiac seizures, or seizures associated with hypoglycaemia (ex: insulinoma), occur at exercise, whereas most epileptic seizures are observed when the animal is resting. Occasionally, the first epileptic seizure occurs at exercise but, as the disease develops, most seizures occur at rest. The presence of salivation, most common autonomic sign observed, helps at times ascertaining seizure activity. Sometimes, it is the post-ictal phase that brings clues as to the presence of seizure activity; dogs may sleep following a seizure but they often are unresponsive, restless, pacing aimlessly, appearing blind, collide with objects, are uncoordinated, thirsty or hungry. Previous to the first witnessed seizure, urine may have been found in the house in inappropriate places. Many of the seizures are focal, the animal only showing lids and ear twitching, mainly unilaterally. The aura is a short period (seconds to minutes) prior to generalization of a seizure where the dog seeks helps, looks lost or has a glaze look. The aura is the beginning of the seizure. If it precedes the tonic-clonic generalized seizure, the seizure has a focal onset. The generalized seizure may be mild, the animal remaining sternal or even standing during the event. At times, these episodes are quite long, lasting 20 minutes or more. Excessive licking of surfaces and fly biting are not seizure activity.[4,5]

Evaluate Breed, Age and Seizure Pattern

A diagnosis of GE is often given when the dog with recurrent seizures is normal on physical and neurological examinations and has unremarkable blood results. In fact, it is the breed of dog, its age at onset of seizures and, the type and frequency of seizures that are the most useful toward diagnosis. When presented with a dog with recurring seizures, the breed and seizure pattern of the dog should be compared to the 'typical' dog with GE. Doberman dogs are rarely presented with seizures. Regardless of their age, GE should be ruled out on breed alone. Typically, the dog with GE is 10 months to 3 years of age, with a range of 6 months to 5 years at time of the first seizure. Although GE is frequent in the golden retriever, it should not be diagnosed if the dog is >6 years of age at onset of seizures. Indeed, seizure control can be erratic if a diagnosis of GE is given to an animal with cerebral tumour. Treatment must address the primary cause if optimal seizure control is to be achieved. In the case of a brain tumour, radiation, surgical removal or steroids may be better at controlling seizures than AED alone.

During history taking, detailed information about seizure description and frequency of seizures since their onset is obtained. The seizure symptomatology remains the same throughout the course of the disorder, although its severity may vary. An aura (focal onset) is often present at the beginning of the seizure with rapid secondary generalization.[6] The generalized seizure is usually convulsive (tonic-clonic) with the animal falling on its side, paddling with all limbs or holding the limbs in extension, the head and neck in extension. Some of the GE dogs have milder forms of generalized seizures, remaining standing or sternal. Convulsive seizures are not always the norm in GE. Some breeds of dogs with GE experience mainly focal seizures.[7,8,9,10] The frequency of seizures may not change throughout the animal's life, such as observed in GE poodles (all sizes), but often the seizures increase in frequency, with variable rapidity depending on the breed. However, as a rule, in GE, there should not be more than two seizures within the first week of seizure onset. Typically, the first two seizures are separated in time by weeks to months. In Border collies and Australian shepherds where GE is severe, cluster seizures and status epilepticus are common occurrence despite treatment.[11,12] These dogs may have cluster-seizures (>1 seizure/24h) at onset. If there are more than two seizures during the first week of appearance of seizures, GE should be ruled out, and a diagnostic work up recommended. In this speaker's opinion, it is very helpful to adhere closely to the description of the characteristic GE dog. With this approach, more patients fall into the category of Epilepsy of Unknown Cause, but this is beneficial to the patient, as every single time the animal is presented to its veterinary clinic for follow up, it forces the veterinarian to challenge his/her diagnosis with the latest published data.

An epileptic seizure is a diagnostic entity with etiologic, therapeutic and prognostic implications.[1,2] Genetic epilepsy is ruled out when there are acute cluster-seizures or status epilepticus at onset of seizures, regardless if the seizures are focal or generalized. Most seizures in dogs are intracranial in origin. Metabolic and endocrine disorders are rare causes of seizure activity at the exception of hypoglycaemia. Bile acids should not be routinely done in dogs presented with seizures. Miniature breeds are exception as they often have congenital liver dysfunction concomitantly to brain disease. Bile acid measurements in these breeds are part of database testing to better orient the choice of AED. Hypothyroidism is difficult to diagnose because many conditions and drugs affect T4 concentration. Nevertheless, this author has never been able to associate hypothyroidism with epilepsy. Dogs presented with concomitant acute status epilepticus and inflammatory central nervous system disease have poor survival rate. The outcome of young dogs with GE (< 2 years old) is guarded given the breed.[11,12] Structural epilepsy is to be expected in dogs presented with focal seizures that progress, over a few days, into generalized seizures or convulsive status epilepticus. Healthy dogs with acute generalized status epilepticus as first manifestation, intoxication should always be considered.

The symptomatology of focal seizures varies in relation to the location of the epileptic focus. Facial and jaw motions are far more frequent than limbs. When limb motions are present, it is important to observe the head/face attentively before diagnosing seizure activity. Frequently, focal seizures involve facial musculature. The cortical motor parenchyma is more concerned with facial and jaw than limb function. A given epileptic focus always leads to the same symptomatology. If there are two types of focal seizures, there are at least two foci of seizure activity. Regardless of the dog's age, it is wise to assume that presence of focal seizures, or seizures with focal onset, harbours focal cerebral lesions.

Diagnostic work up

A physical examination should always be carried, with attention to pulse deficits and arrhythmias, if there is suspected syncope/seizure. A complete neurological examination is then performed with attention to mental status, menace responses, responses to nasal septum stimulation and paw placement (proprioceptive positioning). These tests evaluate the thalamocortex and as such, the environment and the patient's anxiety level may greatly affect their evaluation. The examination should be performed in a calm environment where the animal is given time to quiet down. The mental status is evaluated during history taking, the veterinarian observing carefully the animal left on the floor of the examination room while speaking with the owner over the details of the history. Of the neurological tests targeting the cerebral hemispheres, the mental status is the most difficult to evaluate. Clinical experience has shown that owners of dogs, with significant MRI cerebral lesions, involving frontal or temporal lobes, are poor at evaluating their pet's behaviour. As a consequence, the neurological examination is assessed as unremarkable.

Once the seizure pattern is established and the examinations done, CBC, biochemical profile, urinalysis and endocrine panel including thyroid profile (T4 and TSH) and baseline cortisol are recommended, to rule out concomitant presence of metabolic/endocrine disorders and as baseline prior to initiating antiepileptic treatment if the latter is introduced. In inflammatory diseases, titers or PCR of infectious diseases encountered in the geographical area are pursued as well. If the database is suggestive of metabolic, endocrine or neoplastic disease, thoracic radiographs (3 views) and abdominal ultrasound are added prior to anaesthesia for neuroimaging.

In dogs suspected of Structural Epilepsy, a thorough diagnostic work up is recommended that includes MRI of the brain +/- cerebrospinal fluid analysis. If inflammatory infectious or metastatic disease is suspected, abdominal ultrasound and thoracic radiographs are in order prior to neuroimaging. In dogs > 5 years of age, these tests are also recommended to rule out metastatic disease and evaluate concomitant presence of other significant illnesses related or not to primary presentation.

In dogs with suspected Epilepsy of Unknown cause, MRI of the brain should be done early in the course of the disease if antiepileptic treatment fails. Such an example would be an 8 year-old Labrador dog, normal physically, neurologically and on blood work, with generalized seizures for the last six months that progress despite treatment. Unfortunately, a significant proportion of epileptic dogs remain undiagnosed despite all effort.

ANTIEPILEPTIC TREATMENT

Treatment should be customized to patient and its family. The AED choice is based on seizure types, seizure frequency, animal's general health and age, family's lifestyle and cost of the drug. Not all areas of the brain have the same propensity to seize and cause kindling effect. The more frequent and severe are the seizures, the earlier treatment introduction to avoid snowball effect. By order of frequency, the AEDs used presently in maintenance therapy of canine epilepsy are phenobarbital (PB), potassium bromide (KBr), zonisamide (ZNS), levetiracetam (LEV), gabapentin (GB) and clonazepam or clorazepate. In a recent survey, 82% of the respondents used a combination of phenobarbital and potassium bromide to manage refractory seizure disorders in dogs.[13] Although more costly, there are now newer options available with lesser side effects. The recommended treatment approach herein is based on the speaker's experience and drugs available in her country.

When to treat

If the epilepsy is structural, with an obvious demonstrated lesion, the epileptic focus rarely disappears spontaneously. In these instances, AEDs are continued indefinitely. In GE and epilepsies of unknown cause, antiepileptic treatment is introduced at the 2nd convulsive epileptic seizure in a dog < 2 years of age, in convulsive seizures with decreasing interictal periods when seizures are > 2 seizures/6 months, and in acute onset cluster-seizures/status epilepticus of focal or convulsive generalized seizures. To avoid the sedation often present at onset of treatment, treatment can be initiated gradually, treating once daily or giving half of the dose for a few days, if the seizure pattern allows it.

When not to treat

Quality of life is equilibrium between seizure control and side effects. When it becomes clear that the small seizures are not developing into major seizures, seizures can be tolerated if they do not affect the dog's quality of life. Not all brief non-convulsive generalized epileptic seizures need to be abolished, especially if the AED side effects are significant.

Discontinuation of treatment

An attempt at tapering off AED(s) can be done when the animal has been seizure-free > 2 years, depending on the seizure disorder. If it was difficult to achieve control, the waiting can be longer. The tapering of medication(s) should be done slowly over several months. In patients with refractory generalized convulsive seizures, a number of seizures may be preferable to high AED serum levels causing significant sluggishness and difficulty ambulating. Once pharmacological attempts at controlling seizures have failed over many months to years, quality of life should prevail over seizure control. The same is true when intolerable side effects occur, even at drug levels below those recommended. In some dogs, discontinuing medication can sometimes improve epileptic conditions, while in others withdrawal of medication may not cause deterioration of the condition, yet the animal is freed from side effects.

ANTIEPILEPTIC DRUGS

Phenobarbital is the most potent AED for convulsive and non-convulsive generalized seizures in veterinary medicine. It is available in oral and parenteral forms making its use possible in all situations. It is the **1st line** AED when the seizure frequency does not allow utilization of other less toxic agents. The frequent secondary hepatotoxicity necessitates close monitoring of liver parameters. Beside the common clinical side effects (sedation, polyuria, polydipsia, polyphagia), its activation of the hepatic microsomal enzymes (p450 system) alters the metabolism of other AEDs metabolized by the liver such as ZNS and LEV. Its pitfalls are the activation of its own metabolism necessitating multiple serum level measurements in the first 6 months of use and metabolic tolerance, forcing gradual increments of the dosage to maintain control, until higher end of optimal therapeutic levels or hepatotoxicity are reached. The drug is administered orally q12h. The first serum level measurement is done at 4 weeks, then two weeks after every increment, until optimal therapeutic levels are reached (100-120µmol/L; 23-28µg/mL). Hepatotoxicity is likely with chronic use <u>and</u> levels >140µmol/L (>33µg/mL). In patients that continue seizing, another AED is added. Although not always possible, the use of LEV and ZNS are optimized when given prior to PB in reason of induction of activation of p450 system.

Potassium bromide should be restricted to intractable convulsive generalized epilepsy when everything else has failed. Its long half-life, the significant variations of the serum levels with dietary salt concentrations (1% increment reduces half-life from 69 to 24 days[14]), the ataxia and sluggishness, the polyuria, polydipsia, and polyphagia with weight gain render its use problematic.[15]

Zonisamide is unavailable in Canada but is easily obtained from Canadian veterinary compounding pharmacies. It is efficacious against most types of seizures. It is used as **1st line** whenever antiepileptic treatment must be initiated and seizures are < 1 seizure/week. Efficacious against generalized seizures. Mild sedation, decreased appetite and gastrointestinal signs are reported. One dog died of acute hepatotoxicity following the first few days of treatment. It is available as oral form, is given at 5mg/kg q12h and at 10mg/kg when added to PB. It is important to measure serum levels when given concomitantly with PB. The dosage is often underestimated especially in small breeds of dogs. Serum level should be measured at trough 7 days post-treatment (15 - 45µg/mL.)

Levetiracetam[16] is the safest of the AEDs in dogs with only mild transient sedation observed. Use as **1st line** in focal seizures, mild generalized seizures and in convulsive generalized seizures when there is < 1 seizure/week. Its efficacy is mitigated as add-on AED in canine refractory epilepsy. It is available in oral form in Canada and in oral, oral extended-release and parenteral forms in US and Europe. It is administered orally at 20mg/kg q8h, and it must be given q8h to optimize treatment. The extended-release form can be used twice daily but the clinical efficacy is to be shown. Serum levels are measured before midday treatment, 7 days after treatment introduction. The therapeutic range advocated is as in human 10 – 40µg/mL. When added to PB, levels must be measured and dosages increased accordingly.

Gabapentin does not seem efficacious against canine epileptic seizures according to this author. **Clonazepam/clorazepate** are used for occasional treatment of in-home cluster-seizures. Their use as maintenance therapy may diminish diazepam efficacy when the latter is used as home-emergency measure. Epileptic dogs frequently become refractory to the benzodiazepine AED group when used long term.

CONCLUSION

Accurate diagnosis and AED that target seizure pattern are essential to optimize treatment response. Despite significant advances in the field of epilepsy, 30-40% epileptic dogs are refractory to antiepileptic treatment.[16]

REFERENCES

1. Berg AT et.al, **Epilepsia** 2010;51:676. 2. Engel J Jr, et.al, **Epilepsia** 2001;42:796. 3. Mariani Cl, **Topics in Compan An Med** 2013;28:34. 4. Bécuwe-Bonnet V, et.al, **J Vet Behavior** 2012;7:194. 5. Frank D et.al, **Can Vet J** 2012;53:1270. 6. Berendt M et.al, **J Vet Intern Med** 1999;13:14. 7. Berendt M, et.al, **J Small Animal Prac** 2009;50:655. 8. Gulløv CH, et al, **J Vet Intern Med** 2011;25:1372. 9. Licht BG, et.al, **J Am Vet Med Assoc** 2007;231:1520. 10. Patterson EE, et.al, **J Vet Intern Med** 2003;17:319. 11. Hülsmeyer V, et.al, **J Vet Intern Med** 2010;24:171. 12. Weissl J, et.al, **J Vet Intern Med** 2012 ;26 :116. 13. Kluger EK, et.al, **Aust Vet J** 2009;87:44. 14. Trepanier LA, et.al, **Res Vet Sci** 1995;58:252. 15. Baird-Heinz HE et.al, **J Am Vet Med Assoc** 2012;240:705. 16. Muñana KR, **Topics in Compan An Med** 2013;28:67.

FELINE EPILEPSY : RECOGNITION, EVALUATION AND TREATMENT

Joane Parent DMV, MVetSc, DACVIM (Neurology)
St-Hyacinthe, QC

INTRODUCTION

The symptoms and causes of seizures in the cat have little in common with those of the dog. This conference presents a practical approach to the diagnosis and treatment of feline seizures in light of the latest developments regarding epilepsy in this species. In the context of this presentation, and to follow the now widely accepted International League Against Epilepsy (ILAE) terminology, the epilepsy is considered **genetic** when there "is only epilepsy, with no underlying structural brain lesion or other neurological signs or symptoms; it is presumed genetic and usually age dependant".[1,2] According to the ILAE, the genetic origin must have been proven through family studies, gene isolation or other specific forms of evidence.[2] The epilepsy is **structural** when "the epileptic seizures are the result of one or more identifiable structural lesions of the brain".[1,2] The epilepsy is **of unknown cause** when "the epilepsy is believed to likely be symptomatic but no etiology has been identified".[1,2] Most cats with seizures have structural epilepsy (SE) or epilepsy of unknown cause (EUC).[3] Genetic epilepsy (GE) has never been recognized in client-owned cats.[3] The only report of GE is from an experimental breeding colony.[4]

SEIZURE SEMIOLOGY

According to the ILAE terminology, the seizure is focal when discreetly localized or widely distributed but limited to one hemisphere.[1] The seizure is generalized when bilaterally engaging cortical structures.[1] The animal does not need to be unconscious or recumbent for the seizure to be generalized. If the focal seizure secondarily spreads to the other hemisphere, it is then classified as focal with secondary generalization. Without electroencephalographic data, it is unknown if some of the frequent seizure manifestations in cats involve one or both hemispheres. The use of the terms focal, partial complex, and generalized seizure for the same seizure semiology by authors renders analysis of the feline literature difficult.

Subtle repetitive ear, eyelid or whisker twitching is common in cats, the owners often not recognizing that the subtle twitches they frequently observe are seizure activity. The twitches may occur countless time over the day. In some cats, these have been present for months to years. Since eyelid closure is a bilateral reflex, it is difficult to tell if this type of seizure is focal (contralateral hemisphere) or generalized (both cortices). Occasionally, one can see more specific involvement of one side because the palpebral fissure on this side is smaller, the ear may be slightly held back, or the eyelid twitches occur more frequently. Other times, ear and eyelid twitches are accompanied with salivation, which can be mild, or profuse. This clinical picture can also be accompanied by a low growl. Behavioural abnormalities can be the main, or the only manifestation of the seizure. It can be subtle where the cat is immobile as if staring, but clearly mentally altered. Other times, the animal is mentally altered, on his sternum or sitting on his rear end, and has aggressive posture, screeching and growling at the least stimulation. Other times, the behaviour is much more dramatic, the animal running frantically and aimlessly, propelling itself in the air, running and colliding with objects, biting its tongue, avulsing its claws. Convulsive generalized seizures are usually not difficult to recognize from owners' description. There is loss of consciousness with tonic-clonic movements of the limbs and autonomic signs, with salivation being by far the most frequently observed. However, in this author's opinion, non-convulsive seizures are more frequent in the cat. Occasionally, cats admitted to the intensive care unit with on-going seizures continue seizing even though they seem deeply sedated by the antiepileptic treatment. When touched, their muscle mass is vibrating with activity. The muscular necrosis that ensues can be marked (creatine kinase > 50,000U/L).

The seizure pattern includes (1) the age of the patient at onset of seizures, (2) the seizure type (focal, generalized, or focal with secondary generalization) and (3) the seizure frequency. In the epileptic cat, the seizure pattern is not as useful in establishing the differential diagnosis as it is in dogs. However, it is helpful in orienting the antiepileptic treatment and occasionally in localizing the lesion. The seizure with excessive salivation and aggressive behaviour may be indicative of hippocampal involvement. In this author's opinion, the clinical manifestation of the seizure does not help in establishing differential diagnosis. Indeed, any disease affecting the thalamus or the cortex can cause a seizure, and the seizure can be of any type.

CAUSES OF SEIZURES

Although most seizures in the cat are intracranial in origin, metabolic causes are well represented. In 3 retrospective studies of seizures in cats, metabolic causes were diagnosed in 16/91 (18%)[5], 3/17 (18%)[6] and 25/81 (31%)[3] of the cats. The different causes included hepatic encephalopathy, uraemia, insulin overdose, severe hyperthyroidism, polycythemia vera, hypocalcaemia and severe hypertriglyceridemia.[5,6] In the only prospective study of epileptic cats, only 2/30 (7%) cats had metabolic disease, both cats suffering from polycythemia vera.[7] The differences are likely due to different inclusion criteria. Nevertheless, the important point is that in most cases of seizures secondary to metabolic causes, the diagnosis can be made on history, physical examination and blood test results.

Intracranial causes are the most frequent causes of seizures in the cat. They include infectious and non-infectious inflammatory diseases, neoplasms, vascular events, congenital defects, degenerative diseases and EUC. No specific disease seems to predominate as causative of seizures. The location of the disease to thalamus and cortex is more important to seizure generation. In one study, there were no age differences between SE and EUC groups, the age of the cats varying between 4 months and 19 and 14 years of age respectively.[3] In SE, the history and the physical, ophthalmological and neurological examinations are the most helpful tools in establishing a list of possible causes. Moreover, in most cats with SE, neurological deficits are present at time of presentation.

Inflammatory Infectious Causes. In a study of cerebrospinal fluid (CSF) analysis of cats with inflammatory diseases, 24% (15/62) of the cats had seizures but their occurrence was not associated with a specific inflammatory disease.[9] The most frequent inflammatory diseases in this study were feline infectious peritonitis (FIP) (18%; 11/62) and cryptococcosis (6%; 4/62). In a study with confirmed central nervous system (CNS) FIP, 26% (14/55) of the cats had seizures. It was found that cats that had seizures and FIP were more likely to have extensive forebrain inflammation, suggesting that the presence of seizures in FIP may indicate an unfavorable prognosis.[10] Central nervous system FIP is likely if the cat is < 3 years of age, has protracted history, systemic signs, and presence of an elevated serum anti-coronavirus antibody titre (1:1600).[9,11] Cryptoccosis, the most common systemic mycosis of domestic cats, occurs worldwide, but mostly in Australia, western Canada (C. *gattii*) and western California (C. *gattii*).[12] Central nervous system involvement in cats results from cryptococcal meningitis or multifocal granulomatous encephalomyelitis.[12] There is no study documenting the frequency of seizures in Cryptococcus-infected cats. However, it was found that if seizures occur, long-term prognosis is better if there is no mental status change.[13] Toxoplasmosis is an infrequent cause of seizures in cats. There was a total of 6% (5/91) of cases in one study[5] and none in the two others[6,7]. In none of the studies is there mention of feline immunodeficiency virus (FIV) as cause of seizures.

In a recent study of cats with seizures, 18/81 (22%) cats were diagnosed with EUC.[3] In another study, 7/17 (41%) cats also had unremarkable diagnostic work up.[6] However, in the only prospective study of epileptic cats, all cats had a diagnosis of SE, the most frequent cause being non-suppurative, possibly viral, encephalitis (12/30; 40%).[7] The cats with viral non-FIP encephalitis were young to middle-aged.[7] The seizures were, at times, preceded by mild unspecific transient systemic signs (fever, anorexia, cough, vomiting, diarrhea) in the three weeks preceding seizure onset, with the animal clinically normal at time of the first seizure.[14] The seizure onset was acute and frequently severe, progressing over a few days. The disease was often self-limiting but lead occasionally to frequent focal and/or generalized tonic-clonic seizures, occurring occasionally as clusters and even *status epilepticus*.[14] A few cats were euthanatized due to uncontrolled non-convulsive status. These cats' seizure pattern is similar to the pattern observed in the EUC cats of the two other studies. An explanation of the difference between these studies may reside in the reference range used for the CSF WBC values which, for the prospective study, was based on the published reference parameters of feline CSF, established at the same institution.[15] In the prospective study, the CSF WBC was considered abnormal if the WBC was >2 cells/ µl, whereas >5 cells/µl were used in the two other studies. The strength of the magnetic field for the prospective study was 4T versus 1.5[3] and 0.23T[5] in the other studies. Brain MRI may be a more appropriate gold standard than post-mortem examination to detect brain lesions because only a few sections of the cerebrum are studied in routine necropsy. In the prospective study, the great majority of cats had abnormalities suggestive of viral infection, including mononuclear pleocytosis on CSF analysis, multifocal MRI abnormalities or mononuclear perivascular cuffs in the brain at necropsy. The acuteness of the seizures and the recovery of many of the cats, in all three studies, suggest some insult to the brain.

Brain tumors. Two studies obtained similar results regarding the occurrence of seizures with brain tumors: 23% (14/61)[8] and 22.5% (36/160)[16]. The age of cats at onset of seizures varied between 2 and 13 years of age with a mean of 7.5 years.[8] 50% were ≤ 8 years of age. The inclusion of CNS lymphoma was responsible for this lower age bracket. Among the cats with seizures, 3 cats had meningioma, 5 had lymphoma and 3 had astrocytoma, this latter group having the highest incidence of seizures. Generalized convulsive seizures occurred in 79% of the cats. Another study looking at CNS lymphoma reported seizures in 71% (10/14) of the cats.[17] Looking at all cats in this study, mean age for meningioma was 10.3 years (range 5 – 13 years) and 5.6 years (range 2 – 11yrs) for lymphoma. The first clinical sign was seizures in 9/14 cats and the reason for presentation in 7/14 cats. The interval between the first seizure and presentation was <14 days (range 1 – 14 days) for lymphoma and >14 days (14 -365 days) for meningioma. Of importance is that all cats (13/14) examined in the context of the study (1 cat died at home after a cluster of seizures) had neurological deficits. Brain MRI is the best diagnostic tool for meningioma because of its imaging characteristics, whereas for lymphoma, the combination of MRI and CSF analysis is best because atypical lymphoid cells are frequently found on CSF cytology. Similar tumors and location do not necessarily all lead to seizures. The interface between tumor and brain parenchyma seems to have a greater relationship to generation, maintenance and propagation of seizures.[8]

Hippocampal necrosis is reported as a clinical entity with a specific seizure pattern.[18] The seizures are described as orofacial (facial twitches, salivation) with altered mentation and aggressive behavior. However, the seizures described with the syndrome are not unique to the disease. The diagnosis of hippocampal necrosis necessitates MRI. The cause is unknown but could be secondary to severe seizures, or may be a primary epileptic syndrome. It has been recognized in Europe where it seems frequent, whereas this is not the case in North America. The only two cases this author has encountered involved two cats, already known epileptic patients. Both developed acute and severe clusters of orofacial seizures with marked aggressive behavior. For both cats, these seizures were different than their previous seizure semiology. Both cats were diagnosed with bilateral symmetrical hippocampal necrosis on MRI. Both were euthanatized because of uncontrolled seizures and aggression within days of onset.

Feline ischemic encephalopathy (FIE). In a retrospective study of 11 cats with suspected cuterebra larvae myiasis of the CNS, 3 cats developed seizures.[19] In the prospective study of cats with seizures[7], 6/30 cats had diagnostic data suggestive of FIE.[7] In FIE, the seizures are often infrequent and may not necessitate antiepileptic treatment. The outcome is usually good if the animal survives the initial cerebral ischemia. Strangely, a few cats never displayed the acute signs of cerebral ischemia yet were presented later on with seizures. In these cats, the presumptive diagnosis was made on MRI demonstrating bilateral but asymmetrical atrophy of the cerebrum.[7,14]

Thiamine deficiency. Thiamine concentration was found deficient in over 15% of 90 feline commercial diets evaluated.[20] Thiamine deficiency should be part of the differential diagnosis in any cats presented with acute neurological signs, including seizures. Neurological deficits usually accompany seizures.

Head trauma. Contrary to what is seen with dogs, there does not seem to be a link between head trauma and epilepsy in cats.[21]

Renal failure, diabetes mellitus, and hyperthyroidism are frequent diseases causing hypertension in the cat. However, for the frequency of these diseases in the aged cat, cerebrovascular events are relatively rare findings.

DIAGNOSTIC WORK-UP

A thorough diagnostic work up including physical, neurological and ophthalmological examinations, CBC, biochemical profile (with CK and T4 measurements), FeLV, FIV and urinalysis, thoracic radiographs (3 views), abdominal ultrasound, MRI (1.5 tesla) of the brain, and CSF analysis should be advocated in all cats presented with seizures. The mental status, responses to menacing gesture, responses to nasal septum stimulation and proprioceptive positioning are tests of the neurological examination that target cerebral hemispheres. They must be done with attention to details, especially as it relates to the cat's mental status. In cats presented with acute and severe seizures, uneventful history and unremarkable physical, neurological and ophthalmological examination and blood work, MRI is indicated to help determine prognosis.

TREATMENT

Response to antiepileptic drug (AED) is erratic when the primary cause of seizures is not addressed. Focal seizures can be very difficult to abate. It is important to adequately measure serum AED levels whenever available. The recommended treatment approach herein is based on the speaker's experience and AEDs available in her country.

There are many AEDs available to treat seizures in cats. By order of use, they include phenobarbital, gabapentin, levetiracetam, zonisamide and diazepam. As a rule, if the seizure frequency allows it, it is preferable to introduce AED gradually to avoid overt sedation. **Phenobarbital**[22] remains the drug of choice. It is well tolerated, serum level measurement is readily available in most commercial laboratories and the hepatotoxicity reported in dogs does not occur in cats. It is effective against focal and generalized seizures. The optimal serum therapeutic concentration is 100-130 μmol/L. The dosage is not calculated by weight but by cat. Most cats necessitate 22.5mg/day, given in two treatments (15mg and 7.5 mg) to reach optimal serum concentration. Usually, there is no polyuria, polydipsia or polyphagia, although these are possible.

Gabapentin[23] is used as second line AED in the treatment of focal seizures and generalized seizures with focal onset. Its use is safe and well tolerated in cats. Based on a pharmacokinetic study, a dosage of 3-8mg/kg q6h is advocated to reach serum levels reported efficacious in human with seizures. The author uses 10 to 40 mg q8h per cat, introducing the drug gradually to avoid sedation, if the seizure pattern allows it.

Levetiracetam[24] is efficacious for focal seizures and generalized seizures with focal onset. It seems well tolerated but experience in cats is limited. A dosage of 20 mg/kg orally q8h is recommended to reach plasma concentrations within therapeutic range established for humans (10 – 40μg/ml).

Zonisamide is reported as efficacious against focal and generalized seizures. It was reported as efficacious in electrophysiological and seizure-suppressing effects in kindled cats. The experience is still limited with this drug in cats. The dosage advocated is 5 to 10mg/kg once daily per os (15-45μg/ml).[25] A toxicity study in cats using repeated dosing of 20mg/kg/day reported anorexia, diarrhoea, vomiting and ataxia.

Diazepam, a long time first or second AED of choice in cats has been relegated to treatment of refractory epilepsy because of the 17 cases of acute hepatic necrosis and liver failure reported in the 90s.

Potassium bromide should not be used in cats in reason of the life-threatening lung disease it frequently causes.

PROGNOSIS

Contrary to what is seen in dogs, the severity of seizures at onset of disease does not have a determining effect on outcome. Indeed, it is not rare that a cat, severely epileptic secondary to encephalitis or EUC, stops seizing once treatment is instituted. In cats, with unremarkable diagnostic work up, many of the acute causes of seizures are self-limiting. With structural epilepsy or EUC, the antiepileptic treatment is continued for 6 months seizure-free; then the patient is weaned off treatment very gradually.

CONCLUSION

Feline epilepsy is usually intracranial in origin. The epilepsy in most cases is structural or of unknown cause. A thorough diagnostic work up including MRI of the brain and CSF analysis is advocated in all cats with seizures. It is only once a final diagnosis has been made that the appropriate treatment can be applied. If the epilepsy is structural, it is crucial to treat the primary cause to optimize seizure control. Among the AEDs available, phenobarbital and gabapentin are the treatments of choice. The prognosis for arrest or adequate seizure control in cats is superior to what is observed in dogs. Cats with seizures and no other neurological abnormalities can have a good prognosis for long-term survival.

REFERENCES:

1. Berg AT et al. **Epilepsia** 2010 ;51:676–685. 2. Berg AT et al. **Epilepsia** 2011;52:1058–1062. 3. Wahle AM et al. **J Vet Intern Med** 2014;28:182. 4. Kuwabara T et al. **Epilepsy Res** 2010;92:85–88. 5. Schriefl S et al. **J Am Vet Med Assoc** 2008;233(10):1591-7. 6. Barnes HL et al. **J Am Vet Med Assoc** 2004;225:1723-1726. 7. Quesnel AD et al. **J Am Vet Med Assoc** 1997;210:65–71. 8. Tomek A et al. **J Feline Med Surg** 2006;8:243-253. 9. Singh M et al. **J Feline Med Surg** 2005;7:77-93. 10. Timmann D et al. **J Feline Med Surg** 2008;10(1):9-15. 11. Hartmann K et al. **J Vet Intern Med** 2003;17:781-790. 12. Trivedi SR et al. J Feline Med Surg 2011;13:163-172. 13. Sykes JE et al. **J Vet Intern Med** 2010;6:1427-1438. 14. Quesnel AD et al. **J Am Vet Med Assoc** 1997;210:72-77. 15. Rand J et al. **Am J Vet Res** 1990 ;51:1044-1048. 16. Troxel MT et al. **J Vet Intern Med** 2003;17;850-859. 17. Noonan M et al. **Comp Cont Ed Pract Vet** 1997;19:497-504. 18. Pakozdy A et al. **J Feline Med Surg** 2011;13:687–693. 19. Glass EN et al. **J Vet Intern Med** 1998;12:365. 20. Markovich JE et al. **J Am Vet Med Assoc** 2014 ;244 :175-179. 21. Grohmann KS et al. **J Am Vet Med Assoc** 2012;241:1467–1470. 22. Finnerty KE et al. **J Am Vet Med Assoc** 2014;244:195–199. 23. Siao KT et al. **Am J Vet Res** 2010;71:817–821. 24. Carnes MB et al. **Am J Vet Res** 2011;72:1247–52. 25. Hasegawa D et al. **J Feline Med Surg** 2008;10:418-21.

PROCEED WITH CAUTION: CONTRAINDICATIONS IN ONCOLOGY

Jackie Wypij, DVM, MS, DACVIM (Oncology)
Urbana, IL

INTRODUCTION

Both cancer itself and its successful treatment can result in adverse consequences. To minimize impairment of quality of life (QOL) and maximize effectiveness, it is imperative to consider accepted standard medical protocols in conjunction with assessment of the individual patient's concurrent medical issues as well as the implications of multimodal therapy and polypharmacy. A textbook "recipe" for cancer therapy can turn into a recipe for disaster in an individual patient.

CO-MORBIDITY: RADIATION THERAPY, SURGERY, AND CHEMOTHERAPY

Many cancer patients undergo surgical resection as a component of their treatment protocol. In human cancer patients, complications after radiation therapy occur in up to 60 percent of surgical patients.(1) Clinical sequela include skin atrophy, tissue fibrosis, desquamation, epithelial ulceration, and fistula formation. Impaired peri- and postoperative wound healing is of concern. Acute radiation effects typically appear midway to at the end of radiation therapy but are usually self-limiting and resolve within 2- 6 weeks. Early side effects affect tissues with rapid turnover (e.g. epithelial) and include erythema, moist and dry desquamation, and hair loss. Late effects affect tissues with slower turnover (e.g. connective tissues, stroma, vasculature, nerves) and include skin atrophy, dryness, telangiectasia, dyspigmentation, leukotrichia, fibrosis. Late effects are more likely with palliative radiation protocols because of the higher the dose per fraction; however these patients typically do not live long enough to develop clinically significant side effects.

At a cellular level, response to radiation is determined by inherent radiosensitivity of tissue stem cells, adaptability, and duration of the cell cycle. Radiation therapy may disrupt cytokines (primarily growth factors and proinflammatory interleukins) in the inflammatory and proliferative phases of wound healing. Radiation-induced overexpression of these cytokines may lead to increased matrix accumulation and fibrosis Irradiated wounds have impaired strength, thought to be a response to altered proliferative phase of wound healing, characterized by granulation tissue formation, re-epithelialization and neovascularization. This phase is regulated by various growth factors; in addition nitric oxide levels are thought to be reduced, which may reduce collagen deposition and lead to impaired strength of irradiated wounds. Matrix metalloproteinases (MMPs) and their tissue inhibitors are central to the remodeling process; MMP-1 is decreased after radiation therapy which may contribute to altered soft tissue mechanics. (1)

Advantages of preoperative irradiation include delivery of a smaller total radiation dose and smaller volume of normal tissue irradiation. In addition, microscopic seeding during surgical resection may be reduced. Pre-operative radiation may also result in macroscopic tumor shrinkage, improving surgical odds of complete resection. The main disadvantage of preoperative radiotherapy is increased risk of wound healing complications. Advantages of postoperative radiotherapy are accurate histopathologic assessment of tumor grade and size and adequacy of surgical margins without the confounding influence of prior radiotherapy. Adjunctive radiotherapy also decreases the likelihood of delayed wound healing. Yet postoperative radiotherapy requires higher total doses and irradiation of a larger volume of normal tissue, and the tissue hypoxia from the surgical bed may theoretically reduce efficacy of postoperative radiotherapy. In a human study, survival times with both pre- and postoperative radiotherapy were comparable however there were less late effects of with preoperative radiotherapy. (2) Given adequate knowledge of tumor histology and size, preoperative radiotherapy may be beneficial for veterinary patients to reduce normal tissue irradiation and minimize late side effects.

Post-operative chemotherapy may have adverse effects on wound healing and infection risk and must be considered carefully. This is of particular concern with situations such as intestinal resection/anastomosis, body wall healing, severe infections (e.g. MRSA), and altered albumin levels. Most veterinary chemotherapy drugs are not highly protein bound, but polypharmacy with drugs such as warfarin, NSAIDs, etc. can be complicated. Chemotherapy itself has minimal impact on T-cell numbers and ability to mount antibody responses in dogs with lymphoma and OSA, although these dogs may be comparatively T-cell deficient before initiation of chemotherapy. (4) Particularly in the setting of severe dental disease, dental prophylaxis is recommended prior to radiation of the oral cavity to minimize adverse effects of mucositis and risk of deep tissue infection. On the other hand, although aseptic technique is paramount, surgical infection may improve outcomes, as evidenced in several studies with osteosarcoma, likely via mediated of tumor recognition infection acting as an immunostimulant.(3) In human cancer patients, conventional radiotherapy causes statistically significant reductions in peripheral blood cell counts and in dogs treated with half body radiation, there is Mild to moderate hematologic toxicity.(5) For canine cancer patients treated with definitive external bean radiotherapy for localized disease, there are statistically significant reductions in hematocrit, total white blood cell count, neutrophils, platelets, and other cells however they remained within laboratory reference intervals and were clinically insignificant. Head/neck and pelvic sites, were more likely to experience significant decreases in WBC compared to limb sites.(6) Limited prospective studies have evaluated the effects of combination radiation therapy and chemotherapy but increased toxicity is a potential concern.(7)

CARDIOTOXICITY

Doxorubicin is a standard of care chemotherapy agent for many cancers (e.g. lymphoma, osteosarcoma, various carcinomas, histiocytic sarcoma) and is a potentially cardiotoxic chemotherapy agent. No optimal predictive screening has been identified, and ECG changes are often non-specific. Consideration should be taken for breed predispositions to dilated cardiomyopathy (DCM). (8) Many of these breeds are also commonly affected by cancer (e.g. golden retriever, boxer). Additional risk factors for doxorubicin cardiotoxicity include a history of pre-existent cardiac disease and cumulative doxorubicin dose exceeding 240 mg/m^2. Treatment alternatives include liposome-encapsulated doxorubicin, doxorubicin with concurrent dexrazoxane, epirubicin, actionmycin-D, mitoxantrone, and others. Infusion time is key to minimizing arrhythmias with longer infusions of 30-60 minutes suggested for at-risk patients.

Top ten breeds to develop DCM(8)	
1.	Doberman
2.	Boxer
3.	Great Dane
4.	Cocker spaniel
5.	German shepherd dog
6.	St Bernard
7.	Labrador retriever
8.	Irish wolfhound
9.	Golden retriever
10.	Newfoundland

HEPATOBILIARY DISEASE

Direct chemotherapy-related hepatotoxicity as well as hepatobiliary dysfunction related to primary cancer or pre-existing condition must be considered. In the setting of altered liver function with elevated bilirubin, the liver is not able to metabolize drugs such as vinca alkaloids and doxorubicin, resulting in increase systemic drug exposure and potential for severe and life-threatening illness. Liver antioxidant capacity, including that provided by glutathione production, may protect against free radical injury from doxorubicin but theoretically may also alter efficacy. In a standard CHOP-style protocol, vincristine is routinely replaced with cyclophosphamide in the setting of hepatic bilirubinemia. The liver cytochrome P450 system converts cyclophosphamide to 4-hydroxycyclophosphamide and aldophosphamide. Intracellular cleavage of aldophosphamide yields active cytotoxic metabolites phosphoramide mustard and acrolein. Despite its hepatic metabolism, cyclophosphamide can be given in the face of elevated liver enzymes and/or bilirubin. In humans, the few reports of elevated hepatic enzymes attributed to the drug are thought to be idiosyncratic rather than direct toxicity. Escherichia coli L-asparaginase hydrolyzes L-asparagine in serum and is reported to induce hepatotoxicity in people. Impaired protein synthesis from asparagine depletion may reduce serum levels of albumin, γ-globulins, and some coagulation factors and is of concern in pets with low albumin levels or clotting deficiencies. Hyperammonemia may occur as asparagine is broken down but is rarely clinically significant. CCNU is hepatotoxic in dogs, and contraindicated with evidence of hepatic disease and/or elevated liver enzymes (ALT). Elevated ALT is reported in ~50% of patients receiving CCNU (9) with >5x elevation in 29% of dogs, usually after 1-3 doses. (10) In approximately half of cases, moderate ALT elevations are abrupt. Clinical hepatopathy/hepatic failure is reported in 1.2-2.8% of cases. (10) Risk factors include ≤ 5 years of age.(10) Denamarin reduces the prevalence of increased liver values (ALT, AST, alkaline phosphatase, and bilirubin), and improves reduces ALT-related treatment delays. (11) Clinically significant hepatic injury is seemingly uncommon in cats treated with CCNU. (12)

POLYPHARMACY AND DRUG METABOLISM

It is reported that 20–30% of all adverse reactions in people are caused by drug interactions.(14)Drug interactions may result from pharmacokinetic, pharmacodynamic or a combination of mechanisms. Pharmacokinetic interactions occur when one drug alters the absorption, distribution, metabolism or elimination of another drug, such as when two drugs compete for the same metabolic pathway. When the pathway is "saturated" neither drug can be fully metabolized resulting in higher serum concentrations of both drugs than expected. Pharmacodynamic interactions occur when two drugs have similar molecular targets while not affecting the pharmacokinetic parameters of each other. In this situation, co-administration of the drugs may result in additive effects with increased toxicity, or conversely drugs with opposing effects can reduce the response to one or both drugs. (e.g. cyproheptadine/ mirtazapine).

The primary site of drug metabolism is the liver. Drug interactions that involve a metabolizing enzyme can be either induction or inhibition reactions. Some drugs inhibit p450 metabolism (e.g. cimetidine, metronidazole, omeprazole, erythromycin, ketoconazole), resulting in INCREASED levels of drugs metabolized by this enzyme system. Some drugs (e.g. phenobarbital) potentiate p450 metabolism, resulting in DECREASED levels of drugs (increased clearance). (14,16) Cytochrome p450 activities are also variable between species and in the cat, and potentially different between male and female cats. (15) Cats also have low levels of glucuronyl transferase and thus glucuronide conjugation is very slow, leading to high toxicity of acetaminophen, most NSAIDs, and other drugs. In dogs, phase II acetylation reactions are absent, however this has an advantage as acetylated compounds can induce renal damage in other species. (16)

As an example, serotinin syndrome is a rare but potentially fatal adverse side effect of drug overdose or polypharmacy. Clinical signs include agitation, tachycardia, hyperthermia, gastrointestinal distress, and mydriasis. Some drugs should not be used in combination or should be used with caution, e.g. selegiline, tramadol, mirtazapine. Selegiline is used for canine cognitive dysfunction in geriatric animals, who may also be receiving pain medications for arthritis. Tramadol is routinely used for canine pain and mirtazapine is an appetite stimulant; all of these effect serotonin levels. If serotonin syndrome suspected, discontinue drugs and treat with cyproheptadine.

NEPHROTOXICITY

Many geriatric dogs currently receive NSAIDs for arthritis or are prescribed piroxicam as a component of their cancer therapy. NSAIDs are historically nephrotoxic and cause gastrointestinal irritation and ulcers, and should be monitored closely. This is especially important in situations in which there are multiple insults to the kidneys or intestine. When assessing the potential for nephrotoxicity, pre-existing chronic renal insufficiency, acute tumor lysis syndrome, TCC of the bladder, concurrent nephrotoxic drugs such as chemotherapy (platinum agents) or sedation (radiation, surgery, etc.) may all exacerbate the effects of NSAIDs. When diagnosed with cancers such as mast cell disease, lymphoma, or other tumors, glucocorticosteroids such as prednisone are routinely recommended. These drugs should never be given concurrently, and a washout period should also be instituted. Clients should be closely questioned about home care, as aspirin is often administered at home to dogs thought to be painful by their owners. In situations in which steroids are needed immediately (e.g. IMHA, ITP associated with acute lymphoma), misoprostol, a PGE2 analog, can be used to minimize gastrointestinal side effects. Platinum agents, particularly cisplatin, are nephrotoxic. In dogs, cisplatin-related renal changes are characterized as proximal tubule dilatation, vacuolization, degeneration, and interstitial inflammation. Inflammatory cytokine elevations (IL-2, IL-8, others) precede biochemical evidence of azotemia. (13) Carboplatin is commonly substituted for cisplatin in practice.

ABCB1/MDR1 GENE MUTATION

The traditional "white feet, don't treat" adage no longer applies for MDR (multi-drug resistance) defects, or "ivermectin-related" sensitivity. The ABCB1 gene codes for p-glycoprotein, a drug transport pump that plays a significant role in the multidrug-resistance (MDR) phenotype in cancer cells as well as hepatic and renal drug metabolism. MDR1 is known to play a role in conferring resistance to a number of anticancer agents, including the vinca alkaloids, taxanes and anthracyclines. Dogs with p-glycoprotein deficiency are increased risk of life-threatening severe adverse effects with certain classes of chemotherapy drugs, even at reduced doses. Dysfunctional P-glycoprotein in dogs with homozygous ABCB1-1Δ mutation impairs drug clearance resulting in prolonged exposure. If a dog is homozygous for the mutant allele, it will be affected and at risk; if heterozygous, it is a carrier and, although risk may be modestly increased, clinical impact is similar to wild type dogs. P-glycoprotein function is deficient in several breeds of dogs, classically herding breeds. Common North American breeds with the highest percentage of affected dogs include Collies, Australian/Mini Australian Shepherds, Shetland Sheepdog, German shepherd dog, and herding breed crosses. (17-19) Drug cautions include: acepromazine, butorphanol, erythromycin, loperamide, ivermectin, vincristine, vinblastine, doxorubicin. A genotyping assay for mutation status is recommended for at-risk breeds prior to administration of P-gp substrate drugs. In practice, many oncologists substitute cyclophosphamide for vincristine in a standard CHOP-style protocol for lymphoma while awaiting results of genotyping.

CATS AND MISCELLANEOUS

Large/giant breed dogs tend to be "undertreated" based on the traditional veterinary BSA chart, which also does not account for variation in body type and bone mass. Similarly lower percent body fat seen in the Greyhound can lead to a lower than expected volume of distribution for lipophilic compounds and subsequently higher serum drug concentrations as compared to that seen in breeds with a higher percent body fat. Accordingly, Greyhounds are associated with a greater risk of toxicity for lipophilic compounds. Other breed variations also exist. (20) In addition to the previously noted contraindications in cats, cisplatin and 5-FU are contraindicated in cats, although intralesional use of cisplatin appears safe. This species difference is suspected to be related to high cortical mitochondrial AST activity and high cysteine-S-conjugate β-lyase in cats compared with dogs, which may lead to greater production of reactive metabolites.(21) In addition chronic use of CCNU is implicated in pulmonary fibrosis in cats, although this risk appears rare. (22)

REFERENCES

(1) Haubner F, Ohmann E, Pohl F, Strutz J, Gassner H. Wound healing after radiation therapy: review of the literature. Radiation oncology 2012;7:162-162.

(2) Wolfson A. Preoperative vs postoperative radiation therapy for extremity soft tissue sarcoma: controversy and present management. Curr Opin Oncol 2005;17(4):357-360.

(3) Lascelles BD, Dernell W, Correa M, Lafferty M, Devitt C, Kuntz C, et al. Improved survival associated with postoperative wound infection in dogs treated with limb-salvage surgery for osteosarcoma. Annals of Surgical Oncology 2005;12(12):1073-1083.

(4) Walter, Claudia Biller, Barbara Lana, Susan Bachand,Annette Dow, Steven. Effects of chemotherapy on immune responses in dogs with cancer. Journal of veterinary internal medicine 2006;20(2):342-347.

(5) Lurie DM, Kent MS, Fry MM, The [illegible] and chemotherapy in dogs with lymphoma. Veterinary and comparative oncology 2008;6(4):257-267.

(6) Clermont, T Leblanc, A K Adams, W H Leblanc, C J Bartges,J W. Radiotherapy-induced myelosuppression in dogs: 103 cases (2002-2006). Veterinary and comparative oncology 2012;10(1):24-32.

(7) Hume, K R Johnson, J L Williams,L E. Adverse effects of concurrent carboplatin chemotherapy and radiation therapy in dogs. Journal of veterinary internal medicine 2009;23(1):24-30.

(8) Martin, M W S Stafford Johnson,M J Celona, B. Canine dilated cardiomyopathy: a retrospective study of signalment, presentation and clinical findings in 369 cases. J Small Anim Pract 2009;50(1):23-29.

(9) Heading, K L Brockley, L K Bennett,P F. CCNU (lomustine) toxicity in dogs: a retrospective study (2002-07). Aust Vet J 2011;89(4):109-116.

(10) Hosoya, K Lord, L K Lara Garcia, A Kisseberth, W C London, C A Couto,C G. Prevalence of elevated alanine transaminase activity in dogs treated with CCNU (Lomustine)*. Veterinary and comparative oncology 2009;7(4):244-255.

(11) Skorupski, K A Hammond, G M Irish, A M Kent, M S Guerrero, T A Rodriguez, C O Griffin,D W. Prospective randomized clinical trial assessing the efficacy of Denamarin for prevention of CCNU-induced hepatopathy in tumor-bearing dogs. Journal of veterinary internal medicine 2011;25(4):838-845.

(12) Musser, Margaret Quinn,Hilary Chretin, John. Low apparent risk of CCNU (lomustine)-associated clinical hepatotoxicity in cats. Journal of feline medicine and surgery 2012;14(12):871-875.

(13) McDuffie, J E Sablad, Marciano Ma,JingYing Snook, Sandra. Urinary parameters predictive of cisplatin-induced acute renal injury in dogs. Cytokine 2010;52(3):156-162.
(14) Scripture CF, William. Drug interactions in cancer therapy. Nature Reviews.Cancer 2006;6(7):546-558.
(15) Shah, S S Sanda, S Regmi, N L Sasaki,K Shimoda, M. Characterization of cytochrome P450-mediated drug metabolism in cats. J Vet Pharmacol Ther 2007;30(5):422-428.
(16) Jerzsele A. CYP1A2 in dogs. In: Noreddin A, editor. Comparative Veterinary Pharmacokinetics, Readings in Advanced Pharmacokinetics - Theory, Methods and Applications. 978-953-51-0533-6 ed.: 2012.
(17) Mealey KL, Meurs KM. Breed distribution of the ABCB1-1Delta (multidrug sensitivity) polymorphism among dogs undergoing ABCB1 genotyping. Journal of the American Veterinary Medical Association 2008 Sep 15;233(6):921-4.
(18) Mealey KL, Fidel J, Gay JM, Impellizeri JA, Clifford CA, Bergman PJ. ABCB1-1Delta polymorphism can predict hematologic toxicity in dogs treated with vincristine. Journal of veterinary internal medicine / American College of Veterinary Internal Medicine 2008 Jul-Aug;22(4):996-1000.
(19) Gramer, Irina Leidolf, Regina DÃ [illegible]. Breed distribution of the nt230(del4) MDR1 mutation in dogs. The veterinary journal 2011;189(1):67-71.
(20) Fleischer, Steven Sharkey, Michele Mealey, Katrina Ostrander,Elaine Martinez, Marilyn. Pharmacogenetic and metabolic differences between dog breeds: their impact on canine medicine and the use of the dog as a preclinical animal model. The AAPS journal 2008;10(1):110-119.
(21) Katayama, Rieko Nagata, Saori Iida, Hiroko Yamagishi, Norio Yamashita,Tetsuro Furuhama, Kazuhisa. Possible role of cysteine-S-conjugate À [illegible]lyase in species differences in cisplatin nephrotoxicity. Food and chemical toxicology 2011;49(9):2053-2059.
(22) Skorupski, K A Durham, A C Duda, L Sorenmo,K U. Pulmonary fibrosis after high cumulative dose nitrosourea chemotherapy in a cat. Veterinary and comparative oncology 2008;6(2):120-125.

DRUG REPOSITIONING: EVIDENCE FOR ANTI-CANCER EFFECT OF NON-CANCER DRUGS

Jackie M. Wypij, DVM, MS, DACVIM (Oncology)
Urbana, IL

INTRODUCTION

Drug repositioning (also called drug repurposing, re-profiling, and therapeutic switching) is the mechanism of identifying new purpose for old drugs. This approach is one method of minimizing risk and capitalizing on prior investment. The field is primarily led by the pharmaceutical industry, although academics are also researching candidate drugs. The idea is to take an approved drug, with optimized safety and efficacy for a particular indication or disease, and obtain regulatory approval for novel therapeutic applications.

This is a near-complete reversal to the main approach to drug discovery that has dominated progress in the past 20 years which has been to identify a single selective drug for a single mechanism while minimizing "off-target" effects.

Two main approaches to drug repositioning:

- Known compound-new target: one drug can act on multiple targets resulting in 'off target' effects, which have traditionally been undesirable. However, these off-target effects enable companies to pursue new indications when the secondary (or tertiary) target is medically and/or commercially relevant.
- Known mechanism-new indication/disease: one biological process/pathway/target may be relevant to more than one disease and thus enable a company to establish the relevance of a known drug to a new disease via molecular similarity.

CURRENT STATE OF BIG PHARMA

The profitability and growth prospects of pharmaceutical companies are declining, in part due to:

- Healthcare budgets are become increasingly strained, and competition in the healthcare insurance field is putting pressure on drug costs.(1)
- Generic drugs, while helping to keep drug prices in check for consumers, are currently approaching 70% of all prescriptions written in the United States.(1)
- Key patent expirations between 2010–2014 have been estimated to put more than $209B in annual drug sales at risk, with nearly $113B sales lost to generic substitution.(1)
- For every dollar lost in declining revenues from patent expirations, estimates forecast that large pharmaceutical companies will only be able to replace on average 26 cents with new product revenues.(1)
- In the last decade, only about 20-30 new molecular entities are approved by the FDA annually. (1)
- Of new FDA-approvals, only ~30% are considered "first in class" or mechanistically innovative. (1)
- Although the number of approved NMEs has been stable, the potential revenue-generating drugs as a percentage of R&D expenditures has fallen drastically.(1)
- It is estimated to cost $1.8B and 13 years to bring a new drug to market.(1)
- Preclinical success rate is estimated at 20-40%, and subsequent drug development clinical success rate is <10%.(2)

CHALLENGES TO DRUG DISCOVERY

- Although the traditional approach to drug discovery has been single drug/single target, we know that a clinically used drug is virtually never truly "selective" for a single target. Biological signaling mechanisms are inherently redundant. For example, yeast studies demonstrate that 85–90% of pathway knockouts are phenotypically silent.(2)
- Drugs, particularly novel anti-cancer drugs, fail in pre-clinical and/or clinical trials primarily due to inadequate safety and/or efficacy.(3)
- Safety issues are related to the toxicological and pharmacokinetic profiles addressed primarily in preclinical studies and phase I and IIa clinical trials. These evaluate therapeutic regimen (i.e., dose and frequency) and safety.(3)
- Efficacy issues are related to the putative protein target/biological pathway being targeted, and indirectly linked to the clinical efficacy of the drug; intermediary endpoints are often used as a clinical readout of drug targeting.(3)

WHY DRUG REPOSITIONING?

- It's all about the money. (see industry challenges above)
- Drug repositioning has been growing over the last decade. Of the 50 top selling brand drugs, 90% have additional indications approved or in development since their initial launch.
- Enhance or renew a product lifecycle or bring a discarded/cousin drug back to life.
- A common theme among the approaches is that they represent the reverse of conventional, targeted approaches to drug development.(4)
- This provides an opportunity to focus on orphan, neglected, and rare diseases that are otherwise rarely addressed directly due to limited commercial gain, and fulfills unmet societal needs. (4)

SCREENING METHODS

- It is estimated that there are ~10K commercially available drugs. (5)
- Although initially identified serendipitously, new methods of computerized and automated analysis are driving the industry.
- For older, generic drugs, information is garnered from peer review scientific literature.
- Patent information is available for many drugs from the last few decades.
- Non-hypothesis driven phenotypic screening such as non-mechanistically biased in vivo rodent screens. This is similar to drug discovery in the 1960s-1970s before the advent of target-based mechanistic screening. (2)
- Data mining including public resources.
- Cheminformatic/bioinformatic analysis of large corporate databases; often performed by small, niche companies.
- High throughput and high content cell-based screening.
- Novel computational methods which estimate small molecules target profiles with increasing levels of precision. (4)

ADVANTAGES

- Meet unmet clinical and social need and provide broader service to community by expanding into fields such as rare diseases.
- Cheaper, quicker, efficient utilization of limited resources.
- Safety/toxicity data and potentially proof-of-concept of biologic target already performed.
- Enabling shorter times to market and lower investment costs compared with traditional drug discovery methods.
- Translational focus – bring together clinicians and scientists for collaboration.

DISADVANTAGES

- Pre-clinical studies may need to be repeated for proof-of-concept or if the dose is higher than the maximum dose already established and approved by regulatory agencies. (3)
- Patient compliance may be an issue, as compliance subsequent to side effect tolerance varies widely with therapeutic area (e.g. side effects may be more acceptable in cancer vs. less morbidity/cosmetic conditions).
- Current regulatory requirements may be more stringent particularly for older drugs and thus new phase I trials may be required.
- Other concerns include: intellectual property strategies, legal considerations, regulatory strategies, identifying useful archived information to manage resources, and patent issues. (4)
- New indication may require a different delivery capability (e.g. need localized delivery system) that may require repeating preclinical/phase I trials.

EXAMPLES

- The most widely known repurposed drug is sildenafil (Viagra, Pfizer, New York, NY, USA) which was initially tested for hypertension but switched to treating erectile dysfunction following observations in phase I trials.

Examples of drug repositioning in human/veterinary oncology-related field(3-5)		
Drug name	**Indication #1**	**Re-purposed**
Bisphosphonates	Hypercalcemia of malignancy/osteoporosis	Skeletal neoplasia
Cytarabine	Anti-cancer	GME
Metoclopramide	Anti-emetic	Prolactin release (breast-feeding)
Misoprostol	Abortifacient	Gastric protectant/PGE2 analog
Nitrogen mustards	Bioweapon	Anti-cancer
Octreotide	Secretory diarrhea	Anti-cancer (neuroendocrine)
Phenothiazines	Anti-malarial, anti-histamine, sedative, others	Anti-adhesion via VLA-4 for inflammation and cancer
Piroxicam	NSAID	Anti-cancer
Raltegravir	HIV-1 integrase inhibitor	Anti-cancer
Sildafenil	Pulmonary hypertension	Erectile dysfunction, jet lag
Thalidomide	Sedative, anti-nausea	Leprosy, anti-cancer
Tretinoin	Acne	Differentiation agent /anti-cancer
Valproic acid	Seizures	Anti-cancer
Vincristine	Anti-cancer	Immunosuppressant

REFERENCES

(1) Paul S, Mytelka D, Dunwiddie C, Persinger C, Munos B, Lindborg S, et al. How to improve R&D productivity: the pharmaceutical industry's grand challenge. Nature reviews.Drug discovery 2010;9(3):203-214.

(2) Lipinski CA. Drug repurposing. Drug Discovery Today: Therapeutic Strategies 2012;8(3):57-59.
(3) Oprea TIM, J. Drug repurposing: far beyond new targets for old drugs. The AAPS journal 2012;14(4):759-763.
(4) Sleigh SH, Barton CL. Repurposing strategies for therapeutics. Pharm Med 2010;24(3):151-159.
(5) Blatt JC, Seth. Drug repurposing in pediatrics and pediatric hematology oncology. Drug Discov Today 2013;18(1-2):4-10.

EPITHELIOTROPIC T-CELL LYMPHOMA IN DOGS

Angela E. Frimberger, VMD, MANZCVS, DACVIM (Oncology)
Wauchope NSW, Australia

INTRODUCTION

Lymphoma is usually considered a systemic disease; however in humans as well as dogs, primary cutaneous variants can arise. The nomenclature and classification systems of the cutaneous lymphomas in people are complex but are summarized below. In dogs, cutaneous lymphomas may be either epitheliotropic on non-epitheliotropic, and may be either T-cell or B-cell. The most common form of cutaneous lymphoma in dogs is epitheliotropic T-cell lymphoma, which is somewhat analogous to mycosis fungoides (MF) in humans and has been proposed as spontaneous animal model for that condition. [1-3] However, there are phenotypic and clinical differences (see below, "Comparative Aspects") between canine epitheliotropic T-cell lymphoma and human mycosis fungoides. [3, 4]

PRESENTATION AND NATURAL BEHAVIOR

Dogs with epitheliotropic T-cell lymphoma are older dogs with no sex predilection.

The presentation of cutaneous lymphoma in dogs may be a skin condition that has appeared to be a longstanding refractory dermatitis, starting as erythroderma or patches and plaques; or can initially appear as single or multiple obvious cutaneous nodules or masses which may be alopecic or crusted. Pruritus is a prominent feature in many dogs, as it is in many people. Secondary infection of lesions is extremely common. Epitheliotropic and non-epitheliotropic lymphomas are distinguished by histologic criteria, so histopathology (below) is needed for classification as well as diagnosis. [6] Epitheliotropic lymphoma is considered to be typically T-cell disease comprised of CD8+ cells.

Classically epitheliotropic T-cell lymphoma is described as progressing through three apparent clinical stages, however not all patients progress through all stages:

1. *Premycotic / Patch Stage (Erythroderma):* Lesions typically appear flat, erythematous and telangiectatic, hypo- or hyperpigmented, may be scaly, pruritic and can be self-traumatized
2. *Mycotic Stage (Plaque Stage):* Lesions typically appear raised and thickened but not nodular; more erythematous and pruritic, and can be exudative, ulcerated and crusted.
3. *Tumor Stage:* Frank nodules and masses become obvious.

Extracutaneous involvement of epitheliotropic T-cell lymphoma can be present initially or develop over the course of the disease, eventually occurring in the majority of dogs. Sites that may become involved include the lymph nodes, liver, spleen, bone marrow and blood, intestinal tract, lungs, and brain. [5] However, in early stages mild lymph node enlargement is more likely to be reactive than neoplastic. [5] Nevertheless enlarged lymph nodes or lesions elsewhere should be aspirated for cytology for staging.

Epitheliotropic T-cell lymphoma can affect primarily the oral mucosa or rarely other mucosal sites. Again, this may a solitary lesion, or multiple or almost diffuse involvement. The skin may or may not also be affected. Other relatively commonly affected mucocutaneous sites include eyelids including conjunctiva and footpads.

Sézary syndrome has been described in dogs as circulating neoplastic cells accompanying epitheliotropic T-cell lymphoma. [8] Erythroderma is prominent and waxes and wanes with the circulating neoplastic cell count, with flare-ups of erythroderma following increases in the circulating neoplastic cells, as the neoplastic cells migrate and home from the blood to the skin. The prognostic importance of Sézary syndrome in dogs is unclear. In humans, Sézary syndrome is now classified as a separate disease entity that presents with diffuse erythroderma and leukemia, and carries a worse prognosis than mycosis fungoides does. [7, 9]

DIAGNOSIS AND STAGING

As discussed above, because this disease is defined by histologic and immunophenotypic characteristics, biopsy must be performed, and multiple biopsies may sometimes be needed to confirm the diagnosis. Both histopathology and immunohistochemistry are needed. The hallmark is the focus of neoplastic infiltration in the epidermis and periadnexally (epitheliotropism). The presence of Pautrier's micro-abscesses is classical but not uniform.

In dogs, the most common immunophenotype of epitheliotropic lymphoma is CD3+ CD8+, however CD3+CD4-CD8- and CD3-(and B-cell marker-) (null or NK cell) phenotypes have also been identified. [3, 10] It is not known whether there is any prognostic or behavioral difference between these phenotypes.

Because early cutaneous lymphoma may appear inflammatory and may sometimes be difficult to differentiate from a true dermatitis, one might wonder whether PCR for T-cell receptor gene rearrangement (PARR) would be a useful addition to the diagnostic process in some dogs. In humans, this may not be helpful because several conditions classes as non-neoplastic may demonstrate T-cell clonality. However in dogs cutaneous lymphocytosis has recently been suggested to be better considered as an indolent (nonepitheliotropic) T-cell lymphoma on the basis of clonal T-cell receptor gene rearrangements present in lesional cells in 8 dogs, a very long clinical course, and eventual progression of skin lesions or development of nodal high grade lymphoma. [11]

Staging requires careful physical examination and palpation of all peripheral lymph nodes, liver and spleen, thoracic radiographs and possibly abdominal ultrasonography. As for all dogs with cancer, a general health panel consisting of CBC, serum biochemistry panel and urinalysis (plus culture if any active sediment is seen) should be done. Bone marrow cytology can also be considered. Enlarged lymph nodes, liver or spleen or lesions elsewhere should be aspirated for cytology for staging. These are to evaluate the extent of disease and the overall health of the patient and fitness for therapy; and to determine the presence of any comorbidities.

COMPARATIVE ASPECTS

In humans, cutaneous lymphomas are mostly T-cell, [7, 9] as in dogs. Cutaneous T-cell lymphoma (CTCL) is a complex of several diseases classified by the World Health Organization (WHO)/European Organization for Research and Treatment of Cancer (EORTC) classification (WHO-EORTC classification) system [12] of which mycosis fungoides is the most common. Other members of the group include lymphomatoid papulosis, CD30+ anaplastic large cell lymphoma, subcutaneous T-cell lymphoma, and natural killer cell lymphoma. The cells of mycosis fungoides are most commonly of a CD3+, CD4+, CD45RO+, CD8- memory T-cell phenotype, but it is not uncommon for the neoplastic cells to lose antigen expression. .

Mycosis fungoides is usually a disease of middle age, with a median age at diagnosis of 55-60 years, but pediatric cases also occur. No causative agent has been identified, although chronic antigenic stimulation may play a role in the accumulation of mutations.

The utility of clonality testing for T-cell receptor gene rearrangement in diagnosis of mycosis fungoides in people can be helpful but is not entirely definitive because several non-neoplastic conditions including lichen planus, pityriasis lichenoides, and pseudolymphoma, have may express clonality, whereas early mycosis fungoides may be negative for clonality. [7]

Mycosis fungoides in people is classically divided into 3 stages: patch (atrophic or nonatrophic), plaque, and tumor. However acceleration can occur at any time in the disease process to a suddenly much more aggressive form, which is typically accompanied by the appearance of large atypical CD30+ cells ("large cell transformation"). [7]

The patch stage may be preceded by another dermatopathy such as atopic dermatitis or parapsoriasis which may be present for as long as decades before a diagnosis of mycosis fungoides is made. It is unclear whether large-plaque parapsoriasis should be considered as a separate entity that may precede mycosis fungoides, or as a stage of mycosis fungoides itself.

As in dogs, the patch stage may be very longstanding, and appears as a nonspecific patchy dermatitis often on sun-protected areas. Pruritus may be mild or severe. Lesions may sometimes regress spontaneously. Histopathology in this stage shows lymphocytes in the lower epidermis and along the dermo-epidermal junction.

In the plaque stage people demonstrate sharply demarcated scaly plaques. Pathology in this stage demonstrates epidermal hyperplasia, dermal fibrosis, and infiltration of larger, solitary or clustered atypical lymphocytes in the epidermis.

In the tumor stage, nodules and masses which may be ulcerated predominate, and secondary infection is common. It is thought that this stage may reflect a shift to more aggressive neoplastic cell behaviour and patients in the tumor stage usually have a worse outcome. Histopathology in this stage shows diffuse infiltration of pleomorphic neoplastic lymphocytes through the dermis and into the subcutis. The sudden development of tumor stage mycosis fungoides without apparently passing through the patch or plaque stage is referred to as "d'emblee" presentation (translation - "immediately").

As mentioned above, some authors [9] classify Sézary syndrome as a separate entity from mycosis fungoides, although still a member of CTCL. Sézary syndrome is characterized erythroderma, lymphadenopathy, and >1000 neoplastic cells per mm^3 of blood. Hepatosplenomegaly may also be present. Sézary syndrome commonly transforms to a more aggressive form. Histopathology of the skin shows changes similar to mycosis fungoides, including epidermotropism and Pautrier's microabscesses. This is an aggressive entity with reported a disease-specific 5-year survival rate of 24%.

Non-classical types of mycosis fungoides include hypopigmented type, Pagetoid reticulosis (Woringer-Kolopp disease), granulomatous slack skin syndrome (GSSS), and follicular mucinosis type. [4] Solitary lesions may also occur that are otherwise similar to the lesions of classical mycosis fungoides; and these tend to respond very well to conservative management.

Unlike in dogs, mycosis fungoides in people rarely involves the mucosa and when it does occur is usually in the context of large cell transformation, where it is considered to carry a particularly poor prognosis. [7]

In addition to the "patch, plaque, tumor" stages, a TNM type staging system has been proposed for human mycosis fungoides, as follows:

Table 1. TNM staging system for human mycosis fungoides (adapted from [13])

Stage		T	N	M	Prognosis / 10-year survival [7]
I	A	< 10% of body area affected by patches or plaques (T1)			Similar to controls / 97%
	B	>10% BSA patch or plaque (T2)			83%
II	A	Patch or plaque (T1-2)	Clinical lymphadenopathy (biopsy negative)		
	B	Cutaneous tumors (T3)			42%
III		Erythroderma (T4)			
IV	A	T1-4	Nodal involvement (biopsy positive)		20%
	B	T1-4		Visceral involvement	Poor

Treatment of mycosis fungoides in humans is initially skin-targeted, including topical corticosteroids, topical nitrogen mustard and topical carmustine, topical bexarotene or tazarotene, imiquimod, radiotherapy including total skin electron beam irradiation, and psoralen UVA (PUVA) therapy. For more advanced diseased, etoposide, interferon alfa, systemic retinoids, pentostatin, histone deacetylase inhibitors, immunomodulators, and targeted toxin fusion proteins are being developed and various combinations of therapies. For nodal or systemic involvement (stage 4 above), combination chemotherapy is used. Other strategies sometimes used include extracorporeal photopheresis and allogeneic hematopoietic transplantation. Surgical excision may be useful for early solitary lesions, although recurrence is common. Radiotherapy may also be beneficial for localized disease. [7, 9]

Thus, while there are similarities between human mycosis fungoides and canine epitheliotropic T-cell lymphoma, there are also specific differences. These are summarized here:

Table 2. Summary of major differences between human mycosis fungoides and canine epitheliotropic T-cell lymphoma

Characteristic	**Canine**	**Human**
Immunophenotype	Mostly CD8+	Mostly CD4+
Oral mucosal form	Yes and possibly favorable	Rare and unfavorable
Clinical behavior	Typically considered aggressive	Typically indolent
Sézary syndrome	Occasionally	Separate entity with worse prognosis

TREATMENT AND OUTCOMES IN DOGS

As in humans, solitary and early lesions may respond well to local or skin-targeted therapies. Cures of such lesions by surgery and radiotherapy have been reported however other dogs with solitary localized lesions treated in this way have been reported to experience recurrence and disease progression. [5, 14] Topical nitrogen mustards pose practical problems in veterinary medicine. Psoralen UVA (PUVA) therapy has not yet been reported in veterinary medicine.

For more aggressive disease, corticosteroids are usually palliative initially but resistance generally develops quickly when used as a single agent.

The most widely reported and used chemotherapeutic in this condition is CCNU (lomustine), with reported response rates of (78%) for a median duration of 106 days, [15] and 83% for median duration of response of 94 days. [16]

Other drugs that have had responses reported include L-asparaginase and DTIC. [5] Combination chemotherapy has been suggested [5, 6] but has not been evaluated in a clinical trial or retrospective case report. It is likely that more emphasis on alkylating agents may be more rewarding, as in other forms of T-cell lymphoma in dogs.

Less aggressive therapies that can be considered [5] include retinoids and linoleic acid [17] and these may play a role in adjuvant therapy of aggressive epitheliotropic T-cell lymphoma and primary therapy of less aggressive variants.

As noted above, radiotherapy may provide excellent responses for solitary or localized lesions, or for palliation of individual lesions causing a quality of life impact in more generalized disease. However, the area to be treated is a limiting factor because of toxicity. Whole-body electron beam radiotherapy is extremely technically challenging but greatly reduces toxicity so that wide areas can be treated; a protocol has recently been described in dogs. [18]

SUMMARY

Epitheliotropic T-cell lymphoma in dogs is considered a model for mycosis fungoides, a member of the cutaneous T-cell lymphomas in humans. It does have many aspects in common with this disease but there are also some clinical points of difference. The disease in dogs has a reputation for being aggressive and difficult to manage therapeutically. In some cases this is true but the prognosis may not be as grave as has generally been thought. Presently therapy of aggressive variants is centered on CCNU, but masitinib may be a promising drug for this disease. The best treatment option for this condition has yet to be determined.

REFERENCES

1. Fivenson DP et al. Cancer 1992;70(8):2091. 2. Fivenson DP et al. J Invest Dermatol 1994;102(2):227. 3. Moore PF et al. Am J Pathol 1994;144(2):421. 4. Goldenberg G. Medscape Reviews 2006. http://www.medscape.org/viewarticle/529351 5. Moore AS. In Ogilvie & Moore. Manag Can Cancer Pt. 2006. Vet Learning Systems. 6. Vail DM. In Withrow et al. Small An Clin Oncol. 2013. Elsevier. 7. Schwartz RA et al. Medscape Reviews 2011. http://www.medscape.org/viewarticle/727380_2. 8. Thrall MA et al. Vet Pathol 1984;21(2):182. 9. Foss et al. In DeVita et al. Cancer Prin Prac Oncol. 2008. Lippincott. 10. Day MJ. J Comp Pathol 1995;112(1):79. 11. Affolter VK et al. Vet Dermatol 2009;20(5-6):577. 12. Willemze R et al. *Blood* 2005;105(10):3768. 13. Herald PW et al. Program Abs 64 Ann Am Acad Derm 2006. 14. Fontaine J et al. Vet Comp Oncol 2009;7(1):1. 15. Williams LE et al. J Vet Intern Med. 2006;20(1):136. 16. Risbon RE et al. J Vet Intern Med. 2006;20(6):1389. 17. Iwamoto KS et al. Cancer Lett 1992 30;64(1):17. 18. Rechner KN. Vet Radiol Ultrasound 2011;52(3):345.

TREATMENT OF T-CELL LYMPHOMA IN DOGS

Antony S Moore, BVSc, MVSc, DACVIM (Oncology)
Wauchope NSW, Australia

INTRODUCTION

General statistics for survival of canine lymphoma patients can be derived from reports of the efficacy of different chemotherapy protocols. Studies have evaluated multiple factors in an effort to provide a prognosis for the individual patient; currently lymphoma is one of the best characterized of canine tumors with regards to prognostic factors. It is clear that good clinical staging and ancillary laboratory testing provide important prognostic information. In addition to the extent of disease (stage) and physical "wellness" of the patient (substage), the most consistently important prognostic factor has been identification of the T-cell phenotype.

CLINICAL PRESENTATION AND SIGNS

The majority of canine lymphomas in all studies have been B-cell. The percentage of T-cell lymphomas in clinical studies ranges from 17% to 28%, (1-7) with a small number of studies reporting a high frequency of T-cell lymphoma; 38% in the Nederlands(8) and 48% in Mexico.(9)

Breeds that had higher than expected prevalence of T-cell lymphoma (not graded, so includes indolent) were huskies (89%) and Shar Peis (67%) as well as Airedale terriers (80%),Yorkshire terriers (80%), Australian shepherds (64%), boxers (55%), cavalier King Charles spaniels (80%), golden retrievers (54%), Irish wolfhounds (100%), and Shih Tzus (81%). That same study also found a higher prevalence of T-cell lymphoma in dogs less than 3 years of age. (10) The higher prevalence of T-cell lymphoma in the boxer breed has also been described separately.(11)

Dogs with T-cell lymphoma appear more likely to present with hypercalcemia. Hypercalcemia in dogs with T-cell lymphoma appears to be mediated primarily by upregulation of parathyroid hormone related protein (PTHrP), but other cytokines such as TNF-alpha appear to be involved to some extent in increased osteoclast activity. There is probably no common mechanism, but rather a combination of different cytokines that differs in different patients.(12) Only dogs that expressed CD-4 on lymphoma cells had hypercalcemia in one study.(5)

Another correlation has been made between the finding of a mediastinal mass /mediastinal lymphadenopathy and T-cell immunophenotype, and this often accompanies hypercalcemia. In one study there was correlation between radiographic finding of mediastinal lymphadenomegaly and immunophenotype. Dogs with B-cell lymphoma and no mediastinal lymphadenopathy had a median survival of 10.8 months compared to 6.4 months for B-cell and enlarged mediastinal lymph nodes. Dogs with T-cell lymphoma and no mediastinal lymphadenopathy had a median survival of 6.4 months compared to 3.4 months for T-cell and enlarged mediastinal lymph nodes.(13)

PATHOLOGY AND DIAGNOSIS

Cell lineage has most commonly been determined by inmmunoperoxidase staining of paraffin embedded biopsy specimens for CD-3 (T-cell) and CD-79a (B-cell); other methods to determine cell lineage include immunocytochemistry, as well as PARR/ clonality and flow cytometry. Histopathology and immunohistochemistry has the advantage of allowing histologic evaluation of the specimen that is not always possible with cytology or other methods, and allow the pathologist to grade the lymphoma and measure of other potential prognostic factors such as mitotic rate; one study showed worse outcome for dogs with B-cell or T-cell lymphoma that had >21 mitotic figures per hpf (14).

T-cell lymphoma in human patients has been classified in different systems; the World Health Organization system from 2008 recognizes more than 20 subtypes of peripheral T-cell lymphoma that includes the cutaneous forms of T-cell lymphoma. The term peripheral refers to "mature" or "extra-thymic" T-cell disease, and does not reflect the location in the body. Many of these entities are human specific (ATL; adult T-cell leukemia and lymphoma, associated with HTLV-1), or unrecorded in dogs (lethal midline granuloma/nasal NK lymphoma). The largest subgroup is those that do not fit an anatomic or etiologic subset, and are termed peripheral T-cell lymphoma not otherwise specified (NOS). This subset of systemic peripheral T-cell lymphoma (PTCL) tends to be aggressive disease, and is probably the most akin to the non-cutaneous, T-cell lymphomas that we see in dogs.

It also seems that histologic classification of canine lymphoma in combination with immunophenotype may allow identification of subtypes with differing prognoses.(15) This has been formalized in a pathology survey that used the WHO classification to categorize canine lymphoma, there were 3 broad categories of non-cutaneous and non-enteric, T-cell lymphoma identified.(16) The largest subgroup is (as it is for people) peripheral (extra-thymic) T-cell lymphoma-not otherwise specified: PTCL-NOS). In dogs this subtype is histologically distinguished by a variable, but often high mitotic rate; cell size is usually large (but occasionally small). PTCL-NOS most commonly affects peripheral lymph nodes. The other commonly encountered subtype was T-cell lymphoblastic lymphoma (T-LBL); histologically this was characterized by dispersed chromatin that often obscured a high mitotic rate. Clinically, dogs affected with T-LBL were likely to be unwell, and show rapid progression to involve the marrow.

The third category described was T-zone lymphoma; an indolent nodal lymphoma that is histologically characterized by an expanding paracortical population of T-cells that compress the germinal centres causing atrophy throughout the medulla. Morphology of the neoplastic cells is small to intermediate with absent to rare mitoses. (16) In a follow-up study of 992 patients; high-grade, peripheral T-cell lymphoma accounted for 16.6% of the cases, while indolent (T-zone) lymphomas accounted for 11.2%. (14) One study showed lymphocytosis occurred in a high proportion of TZL patients. (17)

A further study found they could combine the group of non-cutaneous, PTCL-NOS and lymphoblastic TCL based on gene expression profiling that identified them as belonging to an identical molecular group, and called these high-grade T-cell lymphoma (HGTCL). Samples from dogs with T-zone lymphoma (TZL) had a different molecular profile. That study showed the combined (HGTCL) group as being an aggressive disease with worse outcome with treatment than for dogs with B-cell lymphoma, or those with indolent T-cell lymphoma, a poor prognosis when treated. That same study showed those with TZL had a markedly better prognosis mirroring other studies.(14;17;18) They also showed that expression of CD21 (or complement receptor-2) was high in indolent, T-zone lymphoma, and co-expression with CD3 may act as a marker for this disease when histology is not available. (18)

Finally, a recent study identified a subset of CD4 high-grade T-cell lymphomas that were CD4+, but CD45-, with high MCH-II expression that appeared to have an improved outcome compared to other dogs with CD4 positive lymphoma.(26)

TREATMENT

In outcome analyses for human patients with lymphoma prior to the mid-1990's the effect of T-cell lineage was controversial. Only when non-indolent lymphomas were studied separately did the adverse effect of T-cell phenotype become apparent statistically. In human patients those with aggressive/systemic PTCL have reduced disease –free survival and decreased overall survival rates.

In a recent clinical study of 75 dogs with indolent lymphoma; the majority had T-zone lymphoma (61%), and of those dogs, more than half had lymphocytosis. The magnitude of that lymphocytosis was prognostic, with dogs that had lymphocyte counts >9.2 having a median survival of 15.4 months while dogs with lower counts did not reach a median (>50% were still alive). (17) In that study, stage of disease, substage, presence of lymphocytosis (not magnitude, above) and type of chemotherapy did not alter survival times. Dogs with T-zone lymphoma had a median survival time of 33.5 months, and those with B-cell (marginal zone) lymphoma had a median survival time of 21.2 months; both considerably better than for dogs with high-grade lymphoma.

The inclusion of indolent T-cell lymphoma patients in series of treated dogs may also have altered the outcome analyses of early studies, but most recent studies have confirmed that, as with human patients, dogs with high-grade B-cell lymphoma are more likely to achieve a complete remission (81-84% complete remission rate) than dogs with high-grade T-cell lymphoma (50-67% complete remission rate) when treated with the same chemotherapy protocol.(5;6;19) Some studies have also shown dogs with B-cell lymphoma have a much longer remission duration and survival than dogs with T-cell derived lymphoma.(1;5;13;15) Often these studies found that immunophenotype was a stronger predictor of remission duration than any other potential factor studied.

In human treatment trials for PTCL, the most widely used chemotherapy protocol has been CHOP (cyclophosphamide, hydroxydaunorubicin/doxorubicin, oncovin/vincristine, prednisone). In various studies the overall response rates are around 80% (79-84%), with complete response (CR) rates more variable (40-65%). The responses do not appear very durable with 30 – 35% of patients still progression free at 5 years. The addition of autologous or allogeneic stem cell transplantation improves long term survival rates to around 40%.(20;21) The benefit in canine T-cell lymphoma is less clear; a small study showed that autologous transplant added a median of 8 months to survival in treated dogs, although toxicity was high, and the death rate was 13%. (27)

Otherwise in dogs the situation appears to be similar to that in humans. In a pathology survey which included data for 70 patients with PTCL, the median survival 5.3 months, which was worse than for TZL and B-cell lymphoma. (14) Historically, in the clinical literature, remission durations in dogs with PTCL are also short (median 1.7-5 months).(22) A retrospective study of 24 dogs with T-cell multicentric lymphoma treated with a CHOP-based protocol produced a very high CR rate (88%) although the PFS period was short at 3.4 months. (23)

Further highlighting the poor response of multicentric T-cell lymphoma to chemotherapy was a study that same year, showing that 50% (6/12) of dogs with multicentric T-cell lymphoma did not respond to a single dose of doxorubicin (compared to 100% overall response rate of dogs with B-cell lymphoma).(24)

A further study analysed the response of 50 dogs with T-cell multicentric lymphoma treated with L-asparaginase and MOPP (Mustargen, Oncovin/vincristine, procarbazine, prednisone) chemotherapy, and found it gave a superior survival and response rate compared to historical data for those treated with CHOP based protocols. The CR rate was 78% and the median PFS (Progression Free Survival) was 6.2 months with 25% progression free at 30 months. (22) These data were consistent with an observation we had made previously regarding the use of alkylating agents to treat lymphoma (to follow).

In the last twenty years we have made a move towards the addition of a consolidation phase where animals are continued on therapy (either the same as induction or different) whilst in clinical remission. A number of papers have assessed the value of different consolidation and maintenance phase therapies but none specifically assesses the effectiveness of these therapies with respect to high-grade T-cell lymphoma. Our previously published protocol (VELCAP-SC; vincristine, L-asparaginase, cyclophosphamide, lomustine (CCNU), doxorubicin, prednisolone, procarbazine and mechlorethamine) utilizes a combination of alkylating agents for a consolidation phase. In one study using this protocol, an overall CR of 70% and median survival time of 10 months was reported. B-cell and T-cell populations were not independently reported, but dogs with T-cell lymphoma did not fare statistically worse than those with B-cell lymphoma, implying that there may be some advantage to this approach. (25)

We recently re-analyzed those data and found that there were 36 dogs with T-cell lymphoma that had a median survival of 7.4 months, with 36% alive 1 year and 6.3% alive 2 years after treatment was started. The complete remission rate was low at 53%, but those 19 dogs had a median survival of 13.4 months, with 52.6% alive 1 years and 9.5% alive 2 years after starting chemotherapy. By way of comparison, in the same study, 48 dogs with confirmed B-cell lymphoma had a median survival of 11.4 months, with 46% alive 1 year and 14% alive 2 years after treatment was started. The complete remission rate was higher at 81%, but interestingly those 39 dogs had a median survival of 12.4 months, with 50.3% alive 1 year, but 16.9% alive 2 years after starting chemotherapy. These figures led us in 2003 to start using a similar alkylating consolidation phase after a VELCAP induction using the drugs above (actinomycin-D substituted for mechlorethamine as the latter is not available in Australia) for treating T-cell lymphoma, while we continued to use a VELCAP protocol only for dogs with B-cell lymphoma. The results of treatment in 75 dogs with T-cell lymphoma are the subject of a research abstract (Goodman, 2014) at this meeting.

REFERENCES

1 Dobson JM, et al. *J Small Anim Pract* 2001;42:377. 2 Fournel-Fleury C, et al. *J Comp Pathol* 1997;117:35. 3 Appelbaum FR, et al. *Hematol Oncol* 1984;2:151. 4 Caniatti M, et al. *Vet Pathol* 1996;33:204. 5 Ruslander DA, et al. *In vivo* 1997;11:169. 6 Williams LE, et al. *J Vet Intern Med* 2004;18:703. 7 Elliott JW, et al. Vet & Comp Oncology 2013; 11: 185. 8 Teske E, et al. *Exp Hematol* 1994;22:1179. 9 Alvarez-Berger FJ, et al. 2003 Proc 23rd Annual Conf VCS: p. 28. 10 Modiano JF, et al. *Cancer Res* 2005;65:5654. 11 Lurie DM, et al. *Veterinary and Comparative Oncology* 2004;2:171. 12 Nadella MV, et al. *Vet Comp Oncol* 2008;6:39. 13 Starrak GS, et al. *Vet Radiol Ultrasound* 1997;38:411. 14 Valli VE, et al. *Vet Pathol* 2013;50:738. 15 Ponce F, et al. *Vet J* 2004;167:158. 16 Valli VE, et al. *Vet Pathol* 2011;48:198. 17 Flood-Knapik KE, et al. *Vet Comp Oncol* 2013;11,272. 18 Frantz AM, et al. *Vet Pathol* 2013;50,693. 19 Kiupel M, et al. *Vet Pathol* 1999;36:292. 20 Reimer P, et al. *J Clin Oncol* 2009;27:106. 21 Savage KJ,et al. *Crit Rev Oncol Hematol* 2011;79:321. 22 Brodsky EM, et al. *J Vet Intern Med* 2009;23:578. 23 Rebhun RB, et al. Vet & Comp Oncology 2011; 9: 38. 24 Beaver L, et al. J Am Vet Med Assoc 2010;237: 1052. 25 Morrison-Collister KE, et al. *Vet & Comp Oncology* 2003;1:180.26 Avery et al. *J Vet Intern Med* 2014. 27 Suter et al. *J Vet Intern Med* 2014

BIOMARKERS FOR SCREENING AND MONITORING DISEASES

Kim A. Selting, DVM, MS, DACVIM (Oncology), DACVR (Radiation Oncology)
Columbia, MO

Introduction

Biomarkers are products of biological processes that can be measured in a minimally invasive manner, reflecting the state of health or disease. A wide variety of products can be used as biomarkers including DNA, proteins (including enzymes and hormones), lipids, volatile molecules, and anything that could be produced in a normal or pathologic process. Biomarkers are usually detected and measured in bodily fluids and excreta including blood, urine, saliva, and breath. Diagnosis of a suspected disease process can be confirmed with characteristic biomarker derangements, and response to treatment as well as relapse can be monitored by serial measurement of the biomarker. Tests should be both sensitive and specific.

Classic examples of biomarkers in veterinary medicine include monitoring of urea nitrogen and creatinine as markers of renal function, liver enzyme measurements for liver disease, and leukogram changes to reflect inflammation and infection, among other processes. Beyond routine bloodwork, biomarkers specific to certain disease states are also well accepted. Cortisol (with ACTH stimulation) is used to indicate adequacy of control of hyperadrenocorticism or to diagnose this disease, or hypoadrenocorticism. When evaluating pet animals for cancer, hypercalcemia caused by lymphoma or anal sac adenocarcinoma, or increased alkaline phosphatase in dogs with osteosarcoma are familiar examples.

In addition to measurement of endogenous biomarkers, exogenous biomarkers can be administered to detect cancer or other disease processes. Again, ACTH stimulation to diagnose adrenal disorders is an easy example. Less familiar examples include the use of F-18 labeled fluorodeoxyglucose (FDG) followed by positron emission tomography (PET) scanning to identify tissues in the body with an abnormally high utilization of glucose, a characteristic of neoplastic tissue.

This talk will focus on biomarkers of cancer (both endogenous and exogenous), with reference to "false positives," ie. biomarker derangements that are supportive of a diagnosis of cancer but may be influenced by a non-neoplastic disease. Biomarkers, both endogenous and exogenous, and in any bodily process or fluid, can be used to confirm a diagnosis and monitor response to therapy. They should never be used in the absence of appropriate clinical data.

Endogenous biomarkers of cancer on routine bloodwork

Hypercalcemia- Considered one of the most common paraneoplastic syndromes, hypercalcemia of malignancy is classically associated with certain cancers, but could occur with any kind of cancer. Calcium homeostasis is affected by intestinal absorption, renal function (conversion of storage forms of vitamin D to active form), parathyroid gland (parathyroid hormone) and thyroid gland (calcitonin), and bone as a storage site of calcium. Hypercalcemia occurs when there is whole body derangement of calcium homeostasis. As a brief review, cancers associated with hypercalcemia include but are not limited to lymphoma, apocrine gland anal sac adenocarcinoma, thyroid and other carcinomas (lung, mammary, nasal, pancreas, thymus/thymoma, vaginal, and testicular), multiple myeloma, and hyperparathyroidism. In cats, lymphoma and squamous cell carcinoma are common causes. Interestingly, a recent investigation of ionized calcium compared to total calcium in dogs with B and T cell lymphoma showed a surprising number of dogs with high ionized calcium but normal total calcium, and ionized calcium was prognostic in both B and T cell lymphoma.(Boudreaux, VCS abstract 2012)

Alkaline phosphatase- In cancer medicine, ALP is best known for its prognostic value in canine osteosarcoma. In a study of 61 dogs with osteosarcoma, median survival with high vs. normal total ALP was 5.5 vs. 12.5 months, respectively.[1] Similar results were seen with bone isoenzyme of ALP. This was confirmed in another contemporaneous study. Though some studies did not find a correlation, many veterinary reports are notoriously underpowered to detect a difference. Additionally, in a recent meta-analysis confirmed ALP as a negative prognostic indicator when increased.[2]

Hyperglobulinemia- The globulin fraction of blood proteins can be increased for many reasons including antigenic stimulation due to infection disease, or neoplastic disease. When globulins are markedly increased, multiple myeloma must be the most likely diagnosis, with lymphoma as a differential. Moderate increases in globulins can result from ehrlichia or other infectious/inflammatory diseases. With multiple myeloma, this can be used to assess response to therapy, with normalization correlating with a clinical remission.

Neutrophilia- Neutrophils can be increased as a paraneoplastic process. While infectious causes of neutrophilic leukocytosis (example Babesia) are of concern with a sick or febrile animal, paraneoplastic neutrophilia can be present in an asymptomatic animal. This has been described in many cancers including hemangiosarcoma and anecdotally observed with nasal tumors. Other reported tumors include pulmonary papillary carcinoma, rectal adenomatous polyp, renal transitional cell carcinoma, renal carcinoma, metastatic fibrosarcoma, lymphoma, mast cell tumor, lymphoid leukemia, ovarian carcinoma, mesothelioma, thymoma, pulmonary adenocarcinoma, and Leydig cell tumor. This likely relates to cytokine signaling mediated by the primary cancer. Additionally, myelogenous leukemia should also be considered and is distinguished from leukemoid response by atypia and lack of proper maturation, as well as circulating atypical blasts.

Endogenous biomarkers of cancer on advanced testing

Thymidine kinase- Thymidine kinase type 1 (TK) is an enzyme that is activated as part of the salvage pathway for DNA synthesis. In populations of rapidly dividing cells, exceeding normal capacity for DNA synthesis, TK provides an alternate pathway for the generation of deoxythymidine. Rapid or uncontrolled DNA synthesis beyond normal cellular capacity is a hallmark of cancer and thus, TK can correlate with the presence of a neoplastic population in the body. TK is soluble and can be detected in serum, which distinguishes it as a diagnostic tool from other markers of proliferation such as AgNORs and Ki67 which must be detected in tissue via biopsy and immunohistochemistry. TK in dogs with hemoabdomen was evaluated because of earlier findings of marked increases in dogs with hemangiosarcoma. Comparing 61 dogs with hemoabdomen of various causes to controls, the test showed high specificity and positive predictive value when distinguishing hemangiosarcoma from normal dogs.[3]

Because TK can be measured in a minimally invasive manner, it is an attractive biomarker for early detection of cancer. In a recent study in which apparently healthy dogs (n=360) in high risk breeds (Golden retrievers and German shepherd dogs over the age of 5 years) were screened for increased TK and canine C-reactive protein (c-CRP, see below). All dogs were followed by health questionnaire for a minimum of 6 months, and dogs for which biomarkers were above normal were evaluated with serial measurements when possible, and thoracic and abdominal imaging. A small number of dogs were diagnosed with cancer within 3-5 months of initial testing and increases in biomarkers correlated well with an impending diagnosis of cancer.[4] TK and cCRP were also recently evaluated in a large population of dogs with various cancers, and dogs with most types of cancer had significant increases in biomarkers.(Selting VCS abstract, 2013) This suggests that these biomarkers in combination may be used to monitor response to therapy in a minimally invasive manner for a wide variety of cancers.

TK has been well studied in human and canine lymphoid malignancies. It correlates with stage of disease with greater increases with advancing stage of lymphoma in both species. TK is prognostic and higher TK correlates to shorter remission times.[5] Current efforts are investigating the utility of TK to monitor response to chemotherapy for lymphoma with the ultimate goal of using TK to tailor first and second line protocols to improve response and overall survival. Anecdotally, TK can become increased prior to loss of clinical remission. It is not yet known whether altering or restarting chemotherapy based on biomarker assessment offers improved survival over lymph node palpation alone. Initial investigations in cats have shown similar findings with increased TK in cats with lymphoma.

<u>Canine C-reactive Protein (cCRP) and others-</u> CRP is a well accepted biomarker of inflammation. Produced as an acute phase protein from the liver, increases in CRP have been seen in a wide variety of diseases and post-surgical states. Inflammation is a hallmark of cancer, and inflammation is an enabling characteristic that can promote tumor angiogenesis, tissue invasion, and promote proliferation (via cytokine signaling such as IL-1α). As a logical partner to TK, cCRP was combined with TK to provide improved specificity for detecting occult neoplasia in a commercially available panel called the Neoplasia Index, which is based on a proprietary algorithm. Combinations and panels of biomarkers are usually more powerful than single biomarkers in detecting pathologic processes. <u>Haptoglobin</u> is another biomarker of inflammation that has been investigated in various inflammatory and post-surgical settings. Haptoglobin is currently being evaluated in horses and cats for its utility in combination with TK in detecting and monitoring occult neoplasia and other diseases.

<u>Lymphoma blood test (LBT)-</u> This commercially available blood test is a proteomics-based assay that initially investigated nearly 20 proteins that were often increased in dogs with lymphoma. After refining data and applying proprietary algorithms, 2 proteins proved to be most consistently associated with the diagnosis.[6] Past studies demonstrated good correlation, and this test can be used to help confirm an elusive diagnosis. Current investigations have shown that the LBT results can increase before loss of clinical remission, and that higher LBT results correlate with shorter remission time. Again, it is unknown whether restarting chemotherapy prior to recurrent lymphadenopathy will improve survival time. One of the proteins in the LBT algorithm is <u>alpha-1 acid glycoprotein</u> (AAG), which was previously shown to correlate with lymphoma remission, and to increase prior to the return of lymphadenopathy when remission is lost.[7] AAG has also been investigated in cats with cancer and was increased compared to healthy controls.[8] AAG did not predict loss of remission in cats with lymphoma.

Other endogenous biomarkers of importance in cancer therapy

<u>Cardiac troponin I (cTnI) and T (cTnT)-</u> Troponin is part of the thin filament of the contractile apparatus and has three subunits and two isoforms, cardiac and skeletal. Troponin C binds calcium to initiate muscle contraction by causing a conformational change in cTnI to release its inhibition, and the coding sequence is highly similar between cardiac and skeletal myocytes (the major difference is one additional calcium binding site in skeletal compared to cardiac muscle). Troponin T keeps the troponin molecule attached to tropomyosin, and the coding sequence is highly unique between cardiac and skeletal myocytes, but with some homology. Troponin I mediates the interaction between actin and myosin, inhibiting muscle contraction, and the coding sequence for the cardiac isoform is 100% specific.[9] Canine and feline cTnI are highly (95-96%) homologous to human cTnI which allows use of commercially available assays.[10] cTnI has been evaluated for its utility in detecting cardiac hemangiosarcoma, and was able to distinguish dogs with cardiac hemangiosarcoma from dogs with non-cardiac hemangiosarcoma and non-neoplastic causes of pericardial effusion.[11] cTnI has also been investigated for its potential ability to quantify cardiac damage from doxorubicin chemotherapy.[12]

Collection of bodily fluids for biomarker assessment

Blood collection to harvest plasma or serum is the most common method of biomarker assessment. However, urine has been used to screen for bladder tumors (Veterinary Bladder Tumor Antigen test) as well as measure byproducts of collagen degradation to monitor response of bone tumors to bisphosphonates using N-telopeptide. Additionally, in humans sputum is used to assess the airway. Collection of exhaled breath condensate has been investigated in humans and dogs as a way to detect primary and metastatic pulmonary neoplasia.

Exogenous biomarkers

Functional scanning can be combined with anatomic scanning to provide superior information about cancer activity and location, respectively. Some tumors can deform normal tissue architecture (ex. nasal and bone tumors) which does not revert to normal even with successful cancer treatment. Functional scans can demonstrate tumor quiescence despite persistent anatomic distortion.

Scintigraphy- The most common example of gamma scintigraphy is technetium scans for thyroid, bone, and renal perfusion. Technetium-99m-MDP scintigraphy is often used in staging of canine osteosarcoma, and determining candidacy for samarium treatment, as a measure of osteoblastic activity.

Positron emission tomography (PET)- PET scanning has become commonplace in human medicine, and is used for cancer staging and to detect occult cancer. Positron emitters are radioisotopes that decay via positrons which, once released, travel a short distance before encountering an electron in a cloud around an adjacent atom, resulting in an annihilation reaction. Photons that can be imaged are released 180 degrees to one another, and detected in the PET scanner. The choice of positron emitter relates to decay half life, and the messenger into which it can be incorporated. The choice of messenger depends on the metabolic process of interest. The most common radioisotope used is fluorine-18 labeled FDG, a glucose analog. Cancer cells use deranged glucose metabolism so this molecule is preferentially taken up in higher than normal amounts. Inflammation can cause false positive results. The decay half life is 110 minutes which is long enough to produce the radioisotope and transport it to a regional recipient, and short enough to result in limited holding times. In lymphoma in people, PERCIST criteria incorporate PET results into response assessment, and PET results are used to determine whether chemotherapy treatment is truncated or extended. In veterinary medicine, PET scanning is in its infancy but data are accumulating rapidly. Preliminary investigations confirm the use of PET to correlate staging and clinical remission in canine lymphoma.[13] A small number of academic institutions, including the University of Missouri, have acquired PET scanners and use this to aid diagnostic and treatment decisions.

Other radioisotopes include copper-64, carbon-11, nitrogen-13, oxygen-15, and iodine-121. Additional messengers include FLT (3'-fluoro-3'-deoxy-L-thymidine) which images proliferation (see TK discussion above), and ATSM (diacetyl-bis(N(4)-methylthiosemicarbazone) to image hypoxia.

REFERENCES

[1]Garzotto CK, et al. **J Vet Intern Med** 2000;14(6):587-92. [2]Boerman I, et al. **Vet Res** 2012;8:56. [3]Thamm DH, et al. **Vet Comp Oncol** 2011;0(0). [4]Selting KA, et al. **Vet Comp Oncol** 2013;0(0). [5]von Euler H, et al. **J Vet Int Med** 2004;18: 696-702. [6]Ratcliffe L, et al. **Vet Comp Oncol** 2009;7(2):92-105. [7]Hahn KA, et al. **J Am Vet Med Assoc** 1999;;214(7):1023-5. [8]Selting KA, et al. **J Vet Intern Med** 2000;14: 503-506. [9]Apple F. **Clin Chim Acta** 1999;284:151-159. [10]Rishniw M, et al. **Am J Vet Res** 2004;65:53-58. [11]Chun R. **J Am Vet Med Assoc** 2010;237(7):806-11. [12]Selting KA, et al. **Vet Comp Oncol** 2004;2:142-156. [13]Blanc AK, et al. **Vet Radiol Ultrasound** 2009;50(2):215-23.

REVIEW AND UPDATES ON THE ROLE OF VITAMIN D IN CANCER

Kim A. Selting, DVM, MS, DACVIM (Oncology), DACVR (Radiation Oncology)
Columbia, MO

Introduction

Traditionally associated with bone health, vitamin D functions as a (pro)hormone with a wide variety of effects on gene transcription and regulation (more than 2000 genes), immune function, and cardiac health, among other effects. Vitamin D insufficiency has been associated with cardiovascular disease, cancer, immune-mediated disease, and more. In veterinary medicine, low vitamin D (most often measured by the storage form, 25-hydroxyvitamin D3) has been associated with mast cell tumor development in Labrador retrievers,[1] congestive heart failure,[2] inflammatory bowel disease,[3] and kidney disease.[4] Vitamin D has an anti-inflammatory effect, and abrogating chronic inflammation can potentially limit the consequences of such inflammation such as fibrosis or cancer. Vitamin D receptors are present on many different cell types in the body.

While humans can derive vitamin D from both sun exposure and diet, dogs and cats rely almost exclusively on dietary intake. Since companion animals usually eat a regular diet with little variation in content and amount, this presents a unique opportunity to more accurately assess the impact of diet on vitamin D status, and the effect of manipulating dietary or supplementary vitamin D content on disease pathogenesis, progression, and response to therapy. Because of the effects on blood calcium, and the effect of hypercalcemia on physiologic functions, supplementation must be approached with caution. Monitoring of serum vitamin D before and after supplementation is critical. Additionally, attempts at augmenting vitamin D in the diet have proven dangerous in some instances, leading to toxicity and product recall. It is likely that vitamin D status may vary among individuals, and that monitoring is important in addition to evaluation of vitamin D dietary content. While administration of the active form, 1,25-hydroxyvitamin D can lead to hypercalcemia, evidence exists to support local conversion by cells from storage to active forms which may allow supplementation of less toxic formulations.

Definition of vitamin D concentrations is evolving. The concentration required for skeletal health may be different than that required for good immune health and decreased inflammation. In animals, gastrointestinal health will impact vitamin D absorption, and there may be breed or sex related differences in food intake or micronutrient absorption.

Vitamin D effects on cancer cell metabolism

Vitamin D has been shown to exert transcriptional regulation of p21 (CDKI), a tumor suppressor gene. Vitamin D can also cause upregulation of PTEN, another tumor suppressor gene, and can promote apoptosis. It can also interfere with insulin-like growth factor, and has effects on TGFβ and the Wnt/β-catenin pathway. Additional indirect effects on cancer by vitamin D include inhibition of angiogenesis via inhibition of VEGF and HIF-1, and modulation of immune effector cells. Cancer can have an effect on vitamin D as well, to limit the restrictions that vitamin D brings by downregulating expression of the vitamin D receptor and of CYP27b1 (an enzyme expressed by many epithelial cells that can locally convert vitamin D to its active form, 1,25hydroxyvitaminD as an autocrine loop). This is further evidence of a direct target effect by vitamin D on cancer cells. Additionally, in vitro work has shown increased sensitivity to anticancer drugs with the addition of vitamin D. As an anti-inflammatory agent, vitamin D can suppress prostaglandin, inhibit p38 stress kinase signaling, which inhibits production of inflammatory cytokines and NF-kB signaling.

Vitamin D sufficiency

Optimal vitamin D concentration for cellular health has not been adequately defined. In people, biomarkers that are influenced by vitamin D status such as intact parathyroid hormone (iPTH), calcium, and phosphorous, have been used to determine sufficiency. When feedback from these biomarkers plateaus, suggesting no further stimulation to increase vitamin D, then this is felt to reflect sufficient vitamin D stores.[5] This approach was repeated in healthy (n=282) and tumor-bearing dogs (n=63). iPTH variation plateaued when 25hydroxyvitaminD concentrations were 100-120 ng/mL, suggesting this as a lower limit of sufficiency. The upper and toxic limits remain to be determined. Additionally, at these concentrations, there was a significant decrease in canine C-reactive protein, a classic inflammatory biomarker when compared to dogs with lower concentrations of 25hydroxyvitaminD. Further, when vitamin D concentrations were compared among healthy dogs and dogs with both neoplastic and benign hemoabdomen, the relative risk of neoplasia increased as vitamin D concentrations decreased.(Selting VCS 2012) Similar investigations in cats with good health, cancer, and other diseases are ongoing.

Risk of cancer

The table below depicts findings in the vitamin D study in which dogs with hemoabdomen were compared to controls.

25(OH)D	Relative Risk	P value
< 40 ng/ml	3.5	<0.0001
< 60 ng/ml	1.9	<0.0001
< 85 ng/ml	1.3	<0.0001
< 100 ng/ml	1.1	0.0005
>100 ng/ml	0.3 (benefit)	0.127

Vitamin D and cancer risk

In addition to the possible impact on hemangiosarcoma development, vitamin D insufficiency has been investigated in dogs with other cancers. When 25hydroxyvitaminD3 concentrations in 33 Labrador retrievers were compared to 53 control cases of healthy Labradors, vitamin D was significantly lower, even when controlling for dietary intake and body condition.[1] In vitro and in vivo investigations support an anticancer effect of vitamin D in dogs affected by mast cell tumors, which is further supported by the presence of vitamin D receptors on mast cells.[6] In human medicine, low vitamin D is considered a risk factor as well as a poor prognostic factor in people with non-Hodgkin's lymphoma and chronic lymphocytic leukemia, and has been linked to the development of a variety of solid tumors. A wealth of research is ongoing to define the role of vitamin D in breast cancer.

Vitamin D in a wide variety of cancers

Our group has recently completed an evaluation of vitamin D status (as measured by serum 25hydroxyvitaminD3) in dogs with a wide variety of cancer. Dogs at high risk for common cancers, Golden retrievers and German shepherd dogs over 5 years of age (total 157 dogs), were recruited through breed groups and dog shows. These were used as controls for comparison to 346 dogs with cancer. Differences in vitamin D status were noted with neuter status but it is not clear whether this related to the choice of control group or other factors. Dogs with malignant cancer had lower vitamin D concentrations than control or dogs with benign tumors, and this was true for both hematologic and solid cancers.(Husbands VCS 2013) Of interest, dogs with histiocytic sarcoma were all below the established level for sufficiency, and in a concurrent investigation, these dogs had high thymidine kinase and C-reactive protein. Thus, various biomarkers could improve detection, monitoring, and therapy for some dogs with solid tumors such as histiocytic sarcoma.

It is important to note that despite a large and growing body of supportive and circumstantial evidence, an absolute cause and effect has not been shown for vitamin D as a factor in the pathophysiology of cancer. It is possible that cancer develops for unrelated reasons in all cases, and low vitamin D is an effect, rather than a cause, of these observations. This seems less likely.

Vitamin D supplementation

Recent studies using canine mastocytoma C2 cells in vitro found calcitriol to exhibit antiproliferative and synergistic activity when combined with commonly used anticancer drugs for canine mast cell tumor (CCNU, vinblastine, and the receptor tyrosine kinase inhibitors imatinib and toceranib). Calcitriol, however is the active form of vitamin D and attempts at supplementation with DN101, a concentrated formulation, lead to frequent hypercalcemia and other toxicity.[6] Therefore other formulations should be investigated for supplementation. Since vitamin D receptors exist on many cells, and 25hydroxyvitaminD may be further hydroxylated to an active form, supplementation with an inactive, pro- form may be adequate, but this requires further investigation.

When 320 pet owners were questioned about diet and supplements at the time of vitamin D measurement in their apparently healthy dogs, most were fed commercial diets (n=292). There was a statistically significant difference in serum vitamin D concentrations among dogs eating food from different manufacturers, suggesting that varying dietary content can contribute to varying serum vitamin D concentrations. The greatest variability in serum vitamin D was found among dogs eating homecooked diets, which is also intuitively rational, assuming more variability in ingredients and preparation. When evaluating regular supplements, dogs receiving salmon oil had significantly higher vitamin D than those that did not. (Sharp VCS 2012)

REFERENCES

[1]Wakshlag JJ, et al. **Br J Nutr** 2011;106(Suppl 1):S60-3. [2]Kraus MS, et al. **J Vet Intern Med** 2014; 28:109-115. [3]Gow AG, et al. **J Sm Anim Pract** 2011;52(8):411-8. [4]Gerber B, et al. **Am J Vet Res** 2003;64(9):1161-6. [5]Hollis BW. **J. Nutr.** 2005;135:317–322. [6]Malone E. **Vet Comp Oncol** 2010;8(3):209-20

OVERVIEW OF MODERN TECHNIQUES IN RADIATION THERAPY

Michael W. Nolan, DVM, PhD, DACVR (Radiation Oncology)
Raleigh, NC, USA

INTRODUCTION

The last decade has witnessed a tremendous advancement in the techniques used to deliver external beam radiotherapy (EBRT) to small animal veterinary cancer patients. The development of these techniques has been driven by the availability of new technologies. This lecture will focus on techniques such as intensity-modulated radiation therapy (IMRT), image-guided radiation therapy (IGRT) and stereotactic radiation therapy (SRT), review the technologies used to deliver such therapies, and conclude with a summary of clinical application for these modalities.

CONVENTIONAL EXTERNAL BEAM RADIATION THERAPY

Conventional megavoltage (MV) EBRT involves delivery of ionizing radiation using cobalt teletherapy (which emits gamma rays) or linear accelerators (X-rays and electrons). Much less commonly, superficial X-rays are delivered using orthovoltage irradiators. In most cases, megavoltage beams are preferred because they are deeper penetrating, result in a more predictable pattern of dose deposition, and are inherently more sparing of tissues such as skin and bone than are lower energy (orthovoltage) beams.

The most common means for planning radiation therapy is to use either manual setup techniques or computerized planning. In either case, radiation beams are collimated around a defined target. Generally, three tumor-related volumes are generated per patient. The gross tumor volume (GTV) includes any identifiable macroscopic tumor; the clinical target volume (CTV) includes a field of tissue surrounding the gross tumor, which is at risk for harboring microscopic tumor; the planning target volume (PTV) is an expansion around the CTV which accounts for errors and uncertainties in positioning of the target. The ultimate goal is to conform the high dose region as closely as possible to the shape of the PTV. As such, the number and position of beams is determined by the shape and location of the tumor. Additional beam modification can be employed (blocks, wedges, bolus, etc.) to further improve the quality of the RT plan, with the goal being delivery of a homogenous radiation dose to all targets, while minimizing the volume of tissue outside these targets which is exposed to doses of radiation at or above the prescription.

Pre-treatment portal radiography can be used to visualize regional anatomy, and ensure the radiation field is appropriately centered on the target volume. Though possible to image more frequently, weekly portal imaging is often performed during the course of RT for purposes of quality assurance.

Definitive-intent radiotherapy is commonly delivered using fine fractionation. High total doses (often in the range of 48-54 Gray [Gy]) are given to the target over the course of 3-4 weeks. Daily treatments allow for low fractional doses (2-4 Gy per fraction). The combination of a high total dose with low dose per fraction limits the risk for developing severe late radiation-induced side effects in normal tissues lying within the radiation field. Risks drastically increase when fractional doses are increased relative to the total dose. These so-called "hypofractionated" treatment protocols are typically reserved for palliative-intent therapy, where the goal is improving quality of life, but risk for inducing a late effect can be cautiously disregarded since application of such therapy has been limited to patients with a generally poor prognosis for survival.

INTENSITY-MODULATED RADIATION THERAPY

IMRT was developed to improve the conformity of radiation fields to the PTV. Whereas conventional radiation therapy is planned such that a uniform dose is delivered from each field, IMRT uses inverse treatment planning to develop non-uniform fluences. In other words, IMRT relies on complex computer algorithms to create dose distributions in which the high dose regions tightly conform to the shape of target volume – this is "dose-painting". The radiation beam can be shaped and delivered using helical tomotherapy, dynamic multileaf collimation and/or volumetric arc therapy. IMRT has been reported for use in dogs with intranasal tumors (both in a prospective clinical trial[1] and a retrospective case series[2]) and carcinomas of the lower urinary tract[3]. It is useful for limiting radiation-induced toxicity and escalating the dose of radiation which can safely be delivered to a variety of tumors.

IMAGE-GUIDED RADIATION THERAPY

Portal imaging has conventionally been performed using a film-based system, which is exposed using the MV treatment beam. Because Compton interactions predominate in MV irradiation, target localization via conventional portal imaging is plagued by poor image contrast; it is often difficult to see confidently see more than bony landmarks.

Several advances in imaging have been made, each of which has incrementally improved image quality and confidence in patient positioning. Digital electronic portal imaging allows for rapid acquisition and evaluation of MV port films, without having to process film. The ability to window/level images acquired on an electronic portal imaging device (EPID) also improves readability of such films. Some newer linear accelerators are built with on-board imaging (OBI), which employs a kilovoltage X-ray source and electronic detector to facilitate acquisition of kilovoltage planar radiographs. These images provide contrast and quality closer to that of diagnostic radiography studies. Computed tomography (CT) has been built into some RT systems. Tomotherapy® employs MV CT, which allows for online viewing of 3D images, which can be compared directly with the simulation CT used in RT planning. Other systems use kilovoltage cone beam CT (CBCT), which has the benefit of not evaluating patient/target positioning using 3D images, but also the soft tissue contrast afforded by kilovoltage imaging.

Several other means for target localization have been developed which use mechanisms other than anatomic imaging to facilitate precise patient and/or target localization. An example of such a system is one that uses radiofrequency technology to track small electromagnetic sensors that have been implanted into tumors and surrounding tissues. Other examples will be discussed.

STEREOTACTIC RADIATION THERAPY

SRT is a technique which can be used to deliver definitive-intent radiotherapy using a convenient treatment schedule. The requirements for SRT are that there is a well-defined bulky target, a system for creating an RT plan which provides a steep dose gradient, capable of administering high doses of ionizing radiation to the tumor, while minimizing dose delivered to the surrounding normal tissues, and a means for localizing the target prior to treatment. Typical courses involve 1-5 fractions given daily, or spread out with a day or two between fractions. The radiobiologic hindrances to hypofractionated RT are overcome by targeting the lesion and sculpting the beam to such a high degree that there is only a very limited volume of normal tissue exposed to doses nearing their tolerance. SRT has also been referred to as stereotactic radiosurgery (SRS), stereotactic body radiation therapy (SBRT), stereotactic ablative radiotherapy (SABR); the latter terms are not rigorously defined, though SRS generally refers to single fraction irradiation of the central nervous system (CNS), whereas SBRT refers to single fraction or modestly fractionated SRT given outside of the CNS.

SRT can be delivered with using a specialized cobalt irradiator (as is done with Elekta's GammaKnife® system), linear accelerator-based systems (e.g., Varian's Trilogy®, Novalis® and TruBeam™), and accelerators that have been mounted on robotic arms (i.e, Accuray's Cyberknife®). SRT has been reported for use in canine appendicular osteosarcoma[4], intracranial tumors of dogs[5], as well as acromegaly[6], oral squamous cell carcinoma[7] and injection site sarcomas in cats[8].

WORDS OF CAUTION

As radiation technologies advance, and the ability to conform the high dose regions of radiation to the target volumes improve, the importance of accurate definition of those treatment volumes also increases. Additionally, and perhaps more importantly, with an expanding range of RT options, it is also important to be thoughtful and precise in selecting cases and stating therapeutic goals. We will discuss treatment protocol design for a variety of cases to better understand the pros and cons of using IMRT, IGRT and/or SRT versus conventional EBRT.

REFERENCES

1. Lawrence, J. A.; Forrest, L. J.; Turek, M. M.; Miller, P. E.; Mackie, T. R.; Jaradat, H. A.; Vail, D. M.; Dubielzig, R. R.; Chappell, R.; Mehta, M. P., Proof of principle of ocular sparing in dogs with sinonasal tumors treated with intensity-modulated radiation therapy. *Veterinary Radiology & Ultrasound* **2010,** *51* (5), 561-570.
2. Hunley, D. W.; Mauldin, G. N.; Shiomitsu, K.; Mauldin, G. E., Clinical outcome in dogs with nasal tumors treated with intensity-modulated radiation therapy. *Canadian Veterinary Journal-Revue Veterinaire Canadienne* **2010,** *51* (3), 293-300.
3. Nolan, M. W.; Kogan, L.; Griffin, L. R.; Custis, J. T.; Harmon, J. F.; Biller, B. J.; LaRue, S. M., Intensity-modulated and image-guided radiation therapy for treatment of genitourinary carcinomas in dogs. *Journal of Veterinary Internal Medicine* **2012,** *26* (4), 987-995.
4. Farese, J. P.; Milner, R.; Thompson, M. S.; Lester, N.; Cooke, K.; Fox, L.; Hester, J.; Bova, F. J., Stereotactic radiosurgery for treatment of osteosarcomas involving the distal portions of the limbs in dogs. *Javma-Journal of the American Veterinary Medical Association* **2004,** *225* (10), 1567-+.
5. Mariani, C. L.; Schubert, T. A.; House, R. A.; Wong, M. A.; Hopkins, A. L.; Barnes Heller, H. L.; Milner, R. J.; Lester, N. V.; Lurie, D. M.; Rajon, D. A.; Friedman, W. A.; Bova, F. J., Frameless stereotactic radiosurgery for the treatment of primary intracranial tumours in dogs. *Vet Comp Oncol* **2013**.
6. Sellon, R. K.; Fidel, J.; Houston, R.; Gavin, P. R., Linear-accelerator-based modified radiosurgical treatment of pituitary tumors in cats: 11 cases (1997-2008). *J Vet Intern Med* **2009,** *23* (5), 1038-44.
7. Yoshikawa, H.; Ehrhart, E. J.; Charles, J. B.; Custis, J. T.; Larue, S. M., Assessment of predictive molecular variables in feline oral squamous cell carcinoma treated with stereotactic radiation therapy. *Vet Comp Oncol* **2013**.
8. Nolan, M. W.; Griffin, L. R.; Custis, J. T.; LaRue, S. M., Stereotactic body radiation therapy for treatment of injection-site sarcomas in cats: 11 cases (2008-2012). *J Am Vet Med Assoc* **2013,** *243* (4), 526-31.

INNOVATIVE AND INTERESTING APPLICATION OF RT IN THE THORAX

Michael W. Nolan, DVM, PhD, DACVR (Radiation Oncology)
Raleigh, NC, USA

INTRODUCTION

Thoracic irradiation is plagued by a number of physical and biological challenges. But armed with a good understanding of the radiobiological constraints of normal tissues in the canine thorax, and an ever-expanding range of therapeutic options, radiation oncologists are actively redefining the role of radiation therapy (RT) in around the chest. From rib to right atrium, we will discuss how a variety of irradiation techniques have been used to treat malignant diseases in the thorax of small animal veterinary patients.

NORMAL TISSUE TOLERANCE TO RADIOTHERAPY

While acute radiation toxicity can be unpleasant, such side effects are rarely life-threatening, and without intervention, typically resolve within weeks of finishing RT. Late radiation effects often result in fibrosis and tissue necrosis. They are rarely self-limiting, can be difficult to treat, can be associated with significant morbidity, and may even result in death. It is therefore the risk of late radiation injury which defines the RT prescription.

The risk of a late radiation injury is proportional to the total dose of radiation, the dose per fraction and the volume of tissue irradiated. The esophagus, trachea and great vessels are all very tolerant of conventionally fractionated, definitive-intent RT.[1] Lung is far less tolerant, but has a large functional reserve volume.[2] It is therefore tolerance of the heart and spinal cord which often drives the radiation prescription when using full-course RT in and around the chest.[1] Much larger doses per fraction are used in stereotactic body RT (SBRT) than in conventional RT. This results in a dramatic shift in normal tissue tolerance. Trachea and esophagus often become the dose-limiting structures when considering SBRT for heart-base and lung tumors.[3]

MANAGEMENT OF RESPIRATORY MOTION DURING RADIOTHERAPY

Conventionally fractionated RT is usually well-tolerated by most thoracic tissues. It is therefore safe and appropriate to apply a planning target volume expansion around the tumor that is large enough to account for respiratory motion during normal breathing. SBRT usually involves prescribing a dose to the tumor which exceeds the tolerance of surrounding normal tissues. In this case, it is important to consider how the phase of the respiratory cycle may alter the position of both tumor and normal tissues. Respiratory motion can be managed using either respiratory gating or breath-holding techniques. Both techniques will be discussed as means for minimizing the volume of normal tissue irradiated during thoracic SBRT.

MEDIASTINAL TUMORS

The most common cranial mediastinal tumors in small animal patients are lymphoma and thymoma. Most dogs and cats with mediastinal lymphoma are treated with cytotoxic chemotherapy. Some cats or dogs with mediastinal lymphoma-associated pleural effusions benefit from palliative-intent radiotherapy. Furthermore, although the prognosis for young FeLV positive cats is typically quite poor even with aggressive therapy, older FeLV negative cats can have localized mediastinal lymphoma which can be associated with reasonably long survival times if responsive to therapy. Many of these localized feline lymphomas are amenable to irradiation.[4]

Thymomas are epithelial in origin and are typically infiltrated with lymphocytes. They are commonly associated with the occurrence of myasthenia gravis (and secondary megaesophagus and aspiration pneumonia) as a paraneoplastic syndrome in dogs (and occasionally in cats). With appropriate therapy, outcomes in patients without megaesophagus can be good. Thymomas are often amenable to surgical resection, but larger, more adherent and/or invasive thymomas can be difficult to completely excise. Radiation therapy can be a useful adjuvant to surgery; it can also result in measurable responses when used as monotherapy in dogs and cats.[5] Radiation therapy has also been successfully used in rabbits with thymoma.[6]

CARDIAC AND HEART-BASE TUMORS

Canine aortic body tumors are often slow to metastasize, but as they grow, they can cause significant functional changes in cardiovascular function. Because canine aortic body tumors are difficult and often unsafe to resect, management of the primary tumor remains challenging. Therapy is therefore most often palliative in nature. Because dogs with these heart-base tumors are predisposed to developing pericardial effusions, the most effective palliative therapy is considered pericardectomy. Radiation therapy has occasionally been used. A single case report describes a dog doing well 42 months after full-course 3-D conformal RT (3DCRT) for a heart base chemodectoma.[7] Many other dogs have been successfully treated with either 3DCRT or stereotactic radiation therapy.

Cardiac hemangiosarcoma is difficult to manage. The distant metastatic rate is high, and although surgical resection of some right atrial/auricular tumors is possible, not all are amenable to surgery. RT has been used to treat canine cardiac hemangiosarcoma; currently, a prospective clinical trial of high dose single fraction RT for this disease is being conducted at North Carolina State University.

LUNG TUMORS

Radiation therapy was historically reserved for management of unrectable lung tumors in people. Advanced radiation techniques have made RT more feasible, and SBRT is now commonly used in management of non-small-cell lung cancer (NSCLC).[8] While NSCLC is rare in dogs and cats, SBRT is becoming more readily available, and there is no reason to think that it would not be useful in at least a subset of small animal veterinary patients presenting with primary lung carcinomas. Proper case selection is important, but SBRT can also be considered in cases of oligometastatic pulmonary neoplasia.

THORACIC WALL TUMORS

Although the list of differential diagnoses for rib tumors is quite long, the most common primary rib tumors are osteosarcoma, chondrosarcoma and fibrosarcoma.[9] These types of tumors are best treated with surgery, but adjuvant RT can be considered if incomplete surgical margins are achieved; either electron therapy or intensity-modulated and image-guided RT can be considered in such cases. Bulky rib tumors can often be treated with SBRT, which may provide significant analgesia, and result in a durable measurable response.

REFERENCES

1. Gillette, S. M.; Gillette, E. L.; Shida, T.; Boon, J.; Miller, C. W.; Powers, B. E., Late radiation response of canine mediastinal tissues. *Radiotherapy and Oncology* **1992,** *23* (1), 41-52.
2. Poulson, J. M.; Vujaskovic, Z.; Gillettet, S. M.; Chaney, E. L.; Gillette, E. L., Volume and dose-response effects for severe symptomatic pneumonitis after fractionated irradiation of canine lung. *International Journal of Radiation Biology* **2000,** *76* (4), 463-468.
3. Benedict, S. H.; Yenice, K. M.; Followill, D.; Galvin, J. M.; Hinson, W.; Kavanagh, B.; Keall, P.; Lovelock, M.; Meeks, S.; Papiez, L.; Purdie, T.; Sadagopan, R.; Schell, M. C.; Salter, B.; Schlesinger, D. J.; Shiu, A. S.; Solberg, T.; Song, D. Y.; Stieber, V.; Timmerman, R.; Tome, W. A.; Verellen, D.; Wang, L.; Yin, F. F., Stereotactic body radiation therapy: The report of AAPM Task Group 101. *Medical Physics* **2010,** *37* (8), 4078-4101.
4. Elmslie, R. E.; Ogilvie, G. K.; Gillette, E. L.; McChesneygillette, S., Radiotherapy with and without chemotherapy for localized lymphoma in 10 cats. *Veterinary Radiology* **1991,** *32* (6), 277-280.
5. Smith, A. N.; Wright, J. C.; Brawner, W. R.; LaRue, S. M.; Fineman, L.; Hogge, G. S.; Kitchell, B. E.; Hohenhaus, A. E.; Burk, R. L.; Dhaliwal, R. S.; Duda, L. E., Radiation therapy in the treatment of canine and feline thymomas: A retrospective study (1985-1999). *Journal of the American Animal Hospital Association* **2001,** *37* (5), 489-496.
6. Andres, K. M.; Kent, M.; Siedlecki, C. T.; Mayer, J.; Brandao, J.; Hawkins, M. G.; Morrisey, J. K.; Quesenberry, K.; Valli, V. E.; Bennett, R. A., The use of megavoltage radiation therapy in the treatment of thymomas in rabbits: 19 cases. *Veterinary and Comparative Oncology* **2012,** *10* (2), 82-94.
7. Rancilio, N. J.; Higuchi, T.; Gagnon, J.; McNiel, E. A., Use of three-dimensional conformal radiation therapy for treatment of a heart base chemodectoma in a dog. *Javma-Journal of the American Veterinary Medical Association* **2012,** *241* (4), 472-476.
8. Fakiris, A. J.; McGarry, R. C.; Yiannoutsos, C. T.; Papiez, L.; Williams, M.; Henderson, M. A.; Timmerman, R., Stereotactic body radiation therapy for early-stage non-small-cell lung carcinoma: four-year results of a prospective phase II study. *International Journal of Radiation Oncology Biology Physics* **2009,** *75* (3), 677-682.
9. Montgomery, R. D.; Henderson, R. A.; Powers, R. D.; Withrow, S. J.; Straw, R. C.; Freund, J. D.; Ogilvie, G. K.; Klausner, J. S.; Caywood, D. D.; Norris, A. M.; McCaw, D.; Tomlinson, J. L.; Fowler, J. D., Retrospective study of 26 primary tumors of the osseous thoracic wall in dogs. *Journal of the American Animal Hospital Association* **1993,** *29* (1), 68-72.

CLINICAL TRIALS: PLANNING FOR SUCCESS

Jeffrey N. Bryan, DVM, MS, PhD, DACVIM (Oncology)
Columbia, Missouri

INTRODUCTION

Prospective, blinded, randomized trials are necessary to confirm the therapeutic efficacy of the treatments we recommend to clients on a daily basis. Practical and ethical considerations demand that health professionals have a clear idea of the true efficacy of the potentially morbid or lethal treatments that they recommend clients pursue, often at significant financial and emotional cost. However, the practicalities of funding, accruing, interpreting, and publishing trials of sufficient size and quality are a major roadblock to improving evidence-based knowledge in veterinary medicine. In the author's opinion, oncology as a specialty has a longer and more effective history of evaluating outcomes-based treatment approaches than most specialties. Much of the literature is hampered by retrospective designs and lack of randomized allocation to the treatment groups, but focus on outcome pre-dates much other literature generated in the management of chronic diseases. The cost of randomized, blinded, prospective trials is very high with few funding agencies or industry partners to share or bear the cost with veterinarians seeking to generate such knowledge. A deluge of anecdotal information is available to the pet-owning public through internet searches which may contradict what evidence-based knowledge has been generated. To combat anecdotal reports and better guide therapy, it is imperative that the clinical research that is done in the effort to improve outcomes for companion animals is of sufficient quality to assist in clinical decision-making. Designing, funding, and executing quality clinical trials in veterinary medicine are huge challenges that require a great deal of prior planning to overcome. Based on the author's experience, oncology trials will be the primary focus of this discussion, but the principles can be generalized to any trial effort.

Prior to planning for or engaging in a clinical trial, consider developing a Clinical Trials SOP for your service. This SOP will define the protocols for preparation, data collection, data maintenance, data review, service communication, and participant expectations for all future clinical trials. This should be a living document that is updated as challenges or errors arise to attempt to systematically eliminate likelihood of such problems. During the preparation of the trial, have both clinicians and nurses review the proposed procedures and trial budget to be sure that they are adequate to answer the question proposed. Be sure that your institutional Animal Care and Use Committee will approve the proposed trial as written. If your practice does not have an ACUC, designate a director of clinical research and consider legal and ethical evaluation from independent sources before agreeing to the trial. Generate clinical trial forms with group input to ensure that data is collected efficiently, uniformly, and accurately. Once preparation is complete, meet to go through all trial procedures and ensure that all team members understand procedures, data forms, expected data entry format, and deadlines associated with the proposed trial. For the execution of the trial, define data collection procedures, photo-documentation procedures, timeliness expectations, and personnel responsibilities for that project. Establish intervals for data evaluation throughout the trial to be sure that data has been collected and recorded appropriately. Establish protocols for data security and accuracy of transcription into electronic formats from written data collection sheets. Agree on the frequency of regular meetings to discuss ongoing trial activities to address concerns quickly and ensure success of the effort. Information on good clinical practices is available through the FDA, covering everything from trial performance to data storage and security (http://www.fda.gov/ScienceResearch/SpecialTopics/RunningClinicalTrials/).

PLANNING CLINICAL TRIALS

As you begin to plan a clinical trial, spend significant time framing the clinical question. Ask simple questions for which there are concrete answers. Consider carefully both inclusion and exclusion criteria. Consider whether there are known prognostic factors for the disease in question that could significantly confound outcome interpretation. If so, consider stratifying based on the prognostic factor. However, in so doing, consider that no evaluation of the factor's impact on outcome will be possible, as that factor has been essentially eliminated from affecting the outcome by stratification. The danger of so doing is that the factor could be associated with another stronger factor that affects outcome and not an independent influence itself. Association may bias the selection of the treatment groups. Meet with a statistician in the planning phase. Be sure that the proposed questions can be answered by the study design, and that the necessary sample size can be accrued in a reasonable amount of time. Sample size for an efficacy trial will be predicated on the expected efficacy of the treatment, the variability of measurement of the end-points of the trial, and some rate of spontaneous response from which a treatment response must be differentiated. Prepare a literature search for each of these factors to present to the statistician so that an accurate estimation of needed sample size can be made. In Phase I clinical trials, consider carefully whether it is appropriate to consider response in the treated patients. Consider pure Phase I or Phase II trial designs. Phase I/II combined clinical trials have the limitation of undertreating the dogs in the early determination of dose by Phase I and underestimating efficacy for Phase II. Additionally, if the trial is enrolling heavily pre-treated patients, efficacy may be underestimated from a Phase II perspective due to prior induced resistance. Work with a statistician to determine the most efficient and effective means to answer a very focused question.

When selecting the study population, a balance must be struck between defining response for a disease in general and having variability that makes the results challenging to interpret. Worse, the results may be confounded by disease factors that prevent the finding of statistically significant results. As an example, many trials evaluating response of canine lymphoma to chemotherapy have been published. Most have been essentially all-comer trials with any immunophenotype or histotype included. Further, among these types have been a range of stages with lower stage dogs highly under-represented and potential higher stage dogs undetected from lack of evaluation. As a result, many oncologists have drawn the conclusion that stage has a fairly weak impact on prognosis of lymphoma in dogs.[1-3] Perhaps the truth is that no trial of significant size limited to a single immunophenotype or histotype has been performed to evaluate stage as a factor. Such a trial would be a significant enrollment challenge and would ideally be multi-center. The interaction of immunophenotype, histotype, and stage in humans has great impact on prognosis and outcome. Perhaps the same is true for dogs. As future trials are planned, considering these variables *a priori* may allow a greater understanding of their real impact on response and outcome. As a practical matter, the more variables to be considered in analysis, the larger the sample size necessary to adequately evaluate the impact of those variables.

SAMPLE SIZE CALCULATION FOR PHASE I TRIALS

Phase I trials are designed to identify the optimal, often maximal, dosing scheme to carry into Phase II trials. The classic Phase I design is a 3 + 3 design beginning at the dose of 10% of the mouse LD10 dose. Three patients are enrolled sequentially per cohort and the dose is escalated according to a Fibonacci sequence (e.g. 100%, 65%, 50%, 40%, 30% and to 35% thereafter). Once a single dose-limiting toxicity (DLT) is encountered, the cohort is expanded to six. If no further DLTs are seen in the cohort of six, the dosage is escalated further until 2/3 or 2/6 patients experience DLT at a particular dose level. This approach has both ethical and efficiency concerns.[4] Beginning at such a low dose and including 3 patients per cohort exposes a relatively large number of patients to doses that are unlikely to be therapeutic. Also, this design often utilizes a relatively large number of patients to answer the question of maximally tolerated dose (MTD). These patients could be utilized in other Phase I or II clinical trials if the question of MTD could be answered with fewer enrolled patients.

Many alternatives have been proposed to improve the efficiency of Phase I trials. In a large analysis of Phase I trials, increasing the starting dosage to 20% of the mouse LD10 would have reduced the number of dose levels from an average of 7 to an average of 5. A further increase to 30% of the mouse LD10 would have reduced the number of dose levels to an average of 3. The tradeoff to this increase in patient efficiency was a parallel increase in trials deemed unsafe as defined by three or fewer dose levels to reach the MTD. The authors concluded that 20% of the mouse LD10 is probably a reasonable starting dosage.[4]

Internal, within-patient dose escalation can decrease the number of patients who receive significant under-dosage, but is somewhat confounded by individual or breed pharmacogenomics. There are described protocols for these accelerated titration designs, many of which depend on pharmacokinetic modeling to predict safe dosing. Such data may not exist to make these schemes as safe for dogs as other schemes.[4] If pharmacokinetic analysis is planned, consider minimum body size for blood collection volume.

Other novel dose-escalation schemes include the Escalation With Overdose Control (EWOC) approach and the modified Continual Reassessment Method (mCRM). In the EWOC design, dose level is selected based on a curve approximating the most likely probability of dose-limiting toxicity. The curve is adjusted with each patient prior to selection of the next dose. An advantage of this statistical approach is that a confidence interval for MTD is defined by the end of the trial.[5] In the mCRM design, a dose-response curve is estimated prior to the trial, based on preclinical toxicity data and prior experience with the class of drug under study in the planned species. As patients are added, the dose-level is increased to better define this curve.[6] Because most dose levels have only a single patient, the patient characteristics must be carefully selected at enrollment to be as representative as possible. In both cases, these approaches require statistical support. While these approaches utilize fewer patients in their execution, dose escalations are not permitted until it is determined whether a DLT is seen at any dose level. As a result, the total time for completion of the trials may not be significantly different than for standard 3 + 3 designs. The decision of which approach to consider may be in part dependent on the relative prevalence of the disease under study. A plethora of patients may support a traditional 3 + 3 design. Infrequent patients may match an EWOC or mCRM design well.

SAMPLE SIZE CALCULATION FOR PHASE II TRIALS

Appropriate sample size should be determined prior to writing the proposal for the clinical trial. Sample size is determined by expected difference between treatment groups; between treatment group and historical efficacy; tolerance of Type I error; tolerance of type II error; and variability of measurement/outcome. Need for sample size is compounded by unmeasured variability and inhomogeneity in the study population. Addition of variables significantly complicates sample size calculation and needed sample size.

Several variables must be determined *a priori* for most effective trial design. First, the clinical question to be answered must be framed. Which disease and which forms of the disease are of interest to the investigators? For each additional variable of biological behavior added to the trial, a larger sample size will be necessary. Ignoring those variables could result in a negative trial outcome that is confounded by unevaluated causes.

Second, trial end-points must be defined. Molecular end-points can be some of the earliest and most definitive outcome assessments. However, to have direct clinical applicability, molecular end-points must have documented clinical relevance to response or outcome. Response to therapy can also yield rapid outcome assessment. Without duration of response, however, the clinical significance of this measure can be questionable. For veterinary trials, duration of response to therapy may be the most robust outcome. Unlike overall survival, it is not biased by euthanasia decisions unless the treatment or disease is excessively morbid. In both cases of response evaluation, the measure of determination is critical. Following standardized criteria is recommended.[7,8] Three-dimensional imaging to determine response can be optimal in some cases.

Finally, to determine effectiveness in a Phase II setting, a drug must be compared either to a presumed rate of spontaneous result or to a comparator of prior accepted therapy or placebo. It should be noted that the objective response rate to placebo control in one recent study of mast cell tumors in dogs was 7.9%.[9] On the other hand, the response to a well-accepted active therapy control (lomustine for mast cell tumors) was only 1%.[10] This is markedly different from a response rate of mast cell tumors to lomustine of

42% published previously.[11] For this reason, many statisticians recommend two-arm Phase II trials to more robustly identify the rate of response of patients to new treatments.[12]

The practical result of current statistical designs of one- and two-arm Phase II trials are tables that define expected parameters and projected outcomes of trial results. In a one-arm Phase II trial design, rates of expected spontaneous results can be predicted and then differentiated from likely response rates following testing in appropriate sample sizes. For example, assuming a 5% rate of spontaneous response to any intervention, a minimum 20% response rate could be predicted in a one-arm Phase II trial observing 4 objective responses in 23 patients with an alpha of 0.026 and a beta of 0.7.[13] Other investigators argue that two-arm trials are necessary to have confidence in the results. For example, in cancer types with a poor response to standard of care, given a 5% rate of response to current therapy, and assuming a 10% rate of response to the novel agent, patients could be enrolled in the study at a 1:2 control:investigational drug ratio with 9 in the control arm and 18 in the investigational arm, giving 95% confidence that the standard error of the arms yields a true difference.[12] Designs in which fewer patients receive control can be better accepted by clients, improving enrollment and exposing fewer patients to placebo. Lookup tables are practical for sample size selection and minmize the required numbers of patients for the study. Using Chi-square or Fisher's exact tests, the required sample sizes would be far larger.[12]

PITFALLS OF CLINICAL TRIALS

The challenges of clinical trials are both practical and theoretical. On a practical level, experience, planning, and attention to detail in execution contribute to success and quality of results. Enrolling a homogenous population of patients can be particularly challenging if inclusion and exclusion criteria have not been well defined prior to trial initiation. Patient heterogeneity escalates rapidly in loosely defined conditions, making interpretation of results challenging. It is a commonly laughed at phenomenon that the start of a clinical trial often means the end of a steady stream of appropriate cases of the disease of interest. Consider trial publicity carefully in advance. The shorter the enrollment window, the larger the effort necessary to publicize the trial and optimize accrual. Trial complexity may also dictate that enrollment be carefully paced, so as not to overwhelm the staff responsible for the trial and interfere with sample or data collection. Sometimes, publicity can result in accrual too rapid to comfortably manage. In budget planning, consider that patients who travel for evaluation may not meet trial criteria. Budgeting for enrollment failures can improve accrual rate by reducing financial risk to the client. Establish clear subsidy protocols to manage the costs associated with the trial. Inadvertently sending a bill to a client for trial-associated costs can damage good will and relationship with the hospital. This is a bigger concern for institutional practices, but could be a problem in smaller practices as well.

On a theoretical level, arriving at an interpretable outcome is the primary goal of any clinical trial. For trials testing straightforward hypotheses, consider the variables to be analyzed in advance. Limiting those variables and selecting them from prior published literature allows clear analysis of their impact on the outcome. Discuss with your statistician whether accounting for multiple comparisons would be necessary in factors selected *a priori*. If the trial is more exploratory in nature, and the variables of interest are harder to define, be sure to account for multiple comparisons in the statistical analysis. Choosing to analyze a large number of variables after the fact will almost certainly result in P values less than 0.05. However, the multiple comparisons may cause false discovery rather than identify true differences.[14] As an example, evaluating 100 variables at a significance level of 0.05, one would expect 5 significant tests to occur with all true null hypotheses. For 20 variables, we would expect one to be falsely positive.

Not all clinical trials will be subsidized. Reasonable clinical research can be conducted using drugs with proven efficacy in new combinations or patterns, but the project may not be attractive to most funding sources. When considering this approach, all of the above guidelines apply. Give careful consideration in planning to which data will be important to collect and at what time points. Pre-plan dose reductions and escalations to maintain uniformity throughout the study period. Carefully consider which patients will be enrolled for later evaluation to have a consistent patient population. The result, although not blinded or randomized, will likely yield a better answer if careful planning and data collection occurs.

REFERNCES

1. Garrett LD, Thamm DH, Chun R, et al. J Vet Intern Med 2002;16:704-709. 2. Moore AS, Cotter SM, Rand WM, et al. J Vet Intern Med 2001;15:348-354. 3. Burton JH, Garrett-Mayer E, Thamm DH. Vet Comp Oncol 2013;11:306-315. 4. Eisenhauer EA, O'Dwyer PJ, Christian M, et al. J Clin Oncol 2000;18:684-692. 5. Babb JS, Rogatko A. Stat Med 2001;20:2079-2090. 6. Goodman SN, Zahurak ML, Piantadosi S. Stat Med 1995;14:1149-1161. 7. Vail DM, Michels GM, Khanna C, et al. Vet Comp Oncol 2010;8:28-37. 8. Nguyen SM, Thamm DH, Vail DM, et al. Vet Comp Oncol 2013;10. 9. London CA, Malpas PB, Wood-Follis SL, et al. Clin Cancer Res 2009;15:3856-3865. 10. Vail DM, von EH, Rusk AW, et al. J Vet Intern Med 2012;26:598-607. 11. Rassnick KM, Moore AS, Williams LE, et al. J Vet Intern Med 1999;13:601-605. 12. Mayo MS, Mahnken JD, Soong SJ. OJ Biopharm Stat 2010;20:106-124. 13. A'Hern RP. Stat Med 2001;20:859-866. 14. Curran-Everett D. Am J Physiol Regul Integr Comp Physiol 2000;279:R1-R8.

STEM CELLS AND THE OSTEOSARCOMA CELL OF ORIGIN: BIOLOGIC AND THERAPEUTIC IMPLICATIONS

Anthony J. Mutsaers, DVM, PhD, DACVIM (Oncology)
Guelph, ON, Canada

INTRODUCTION

There are a variety of primary bone sarcomas encountered in veterinary medicine: osteosarcoma, chondrosarcoma, fibrosarcoma, hemangiosarcoma, among others. Osteosarcoma is by far the most common in both humans and dogs. Osteosarcoma is a cancer with complex genetic abnormalities but very few well defined genetic drivers of its carcinogenesis. In addition to the various primary bone sarcoma types, osteosarcoma itself consists of a number of histologic subtypes, such as osteoblastic, chondroblastic, fibroblastic, epithelioid, giant cell-rich, and telangiectatic. Although outcomes after treatment with conventional local control options (surgery and/or radiation therapy) followed by adjuvant chemotherapy do not substantively differ among the subtypes, (cohort size could be a limiting factor in the analyses) a better understanding of the cellular origin of each type and subtype of bone tumor may permit more individualized therapy and improve survival for bone cancer patients. It is possible that ongoing sequencing efforts in human oncology will reveal unique genetic entities between osteosarcoma subtypes. A precedent for distinct genetic drivers of osteosarcoma can be seen from human familial syndromes that are predisposed to osteosarcoma and the mutations found in sporadic osteosarcoma. For example, Li-Fraumeni and Hereditary Retinoblastoma patients are highly predisposed to osteosarcoma and mutations in the p53 and Rb pathway are common events in sporadic osteosarcoma.

Tumor Heterogeneity

Heterogeneity is a common feature of many cancers, including osteosarcoma. The cellular and molecular determinants of tumor heterogeneity are key research areas in the larger field of cancer biology. There are two general models of tumor heterogeneity. In the "genetic mutation" model, the tumor phenotype is entirely determined by the number and type of mutations sustained by a susceptible cell type, resulting in a variety of tumor subtypes based on individual genetic makeup. In the "cell of origin" model, different tumor subtypes result from oncogenic events sustained by different cell populations within the lineage hierarchy of the organ or tissue[1] (for example, mesenchymal stem cell, pre-osteoblast, or osteoblast in the case of osteosarcoma). Tumor-associated matrix (osteoid production) is one example of heterogeneity in osteosarcoma, which can vary widely in different portions of the tumor, and determines diagnostic subtype (high presence equals osteoblastic). This heterogeneity in osteosarcoma undoubtedly extends beyond histological characteristics to the molecular properties of the malignant cells themselves. What then may be gleaned from these observations of heterogeneity? Does osteosarcoma begin in a particular cell of origin or can multiple different cells along the differentiation pathway from mesenchymal stem cell to committed osteoblast generate the disease? In addition, does osteosarcoma conform to a stem cell model of tumorigenesis, where only a subset of the malignant cell population can self-renew and initiate/propagate the tumor? Is this tumor initiating cell population different from the cell of origin? Finally, is it possible that osteosarcoma represents a heterogeneous group of diseases with differing cells of origin that culminate in a family of cancers that produce osteoid?

THE CANCER CELL OF ORIGIN

There has been much interest in defining the specific cell populations that <u>initiate</u> cancers – the cell of origin, and how these cells might relate to the population that <u>sustains</u> tumors – the cancer stem cell[1]. These two cell populations may not be synonymous, but an understanding of their role in disease development will be crucial to making meaningful clinical progress with treatment. It is plausible that therapeutic advances will be realized by focusing on the characteristic features of the cell of origin, rather than those that may be reflected in the phenotype of the bulk of the tumor mass. Indeed, the histopathology of tumors does not necessarily reflect the cell of origin, as has been demonstrated for other cancers, such as prostate, breast, basal cell, and pancreatic carcinomas[1]. In bone, the osteoblast develops from the mesenchymal stem cell through a series of reasonably well-defined intermediate precursor cell populations (pre-osteoblasts) that attain osteoblast-specific lineage commitment as they differentiate. Characteristic genes are highly expressed at defined stages of mesenchymal commitment to the osteoblast lineage and subsequent differentiation along the lineage. Such genes include Runx2, Osterix-1, Osteocalcin and ultimately osteocyte specific genes such as Sclerostin and Phex. This genetic information has been useful for generating lineage-restricted inducible mouse models of osteosarcoma by engineering genetic insults to cells at varying stages of osteoblastic differentiation. Comparison of the tumor types that develop in these mice may shed light on the potential cell of origin for this disease[2-5]. One challenge that exists when attempting to decipher the cancer cell of origin through gene expression experiments is selection of the appropriate normal cell for comparison. In the case of osteosarcoma, should this normal cell population be the mesenchymal stem cell, pre-osteoblast, or mature osteoblast? Studies have been reported that utilize each of these options, and it is not clear which is the best approach to take for genetic comparison experiments.

Evidence From Mouse Models

Can genetically engineered mouse models inform the osteosarcoma cell of origin? Advances in genetic manipulation in the mouse have allowed the relatively recent development of murine models of osteosarcoma that show much improved fidelity to the human condition[2-5]. The majority of these have utilized loss of the tumor suppressor gene p53, with or without disruption of the Rb suppressor gene pathway, to generate highly penetrant osteosarcoma models. The most recent iterations of these models have used conditional gene deletion approaches restricted to the mesenchymal or osteoblast populations or osteoblast specific transgenic overexpression. These models have demonstrated the unique role for p53 mutation in the initiation and maintenance of osteosarcoma. Additional relevant observations regarding likely cells of origin in osteosarcoma can be drawn from these models as they have made use of different cell types to delete p53, namely multipotent mesenchymal progenitors, early committed osteoblasts (pre-osteoblasts) and in the more mature osteoblast population.

Mesenchymal Stem Cell of Origin

Mesenchymal stem cells give rise to osteoblasts so it is reasonable to propose that these cells may constitute the origin of osteosarcoma. If this were the case would the likelihood of forming osteosarcoma be balanced by the alternate possible fates of a multipotent mesenchymal precursor population into adipogenic or chondrogenic fates? Mesenchymal stem cells from both mouse and human have been shown to undergo induced transformation[6]. The resulting sarcomas include osteosarcoma but other sarcoma types as well, such as poorly differentiated soft tissue sarcoma. This lack of specificity for osteosarcoma has been replicated in transgenic mouse studies that target mesenchymal progenitor cells of the limb bud, with a 60% incidence of osteosarcoma with the remainder of tumors being mainly poorly differentiated soft tissue sarcomas. The definition of human mesenchymal stem cells and multipotent mesenchymal progenitors has become more standardized but the exact nature of the populations defined as mesenchymal stem cells is still relatively heterogeneous and contains cells with differing potentials for the various lineages. The same problems currently exist in veterinary mesenchymal stem cell studies.

(Pre)Osteoblastic Cell of Origin

Mouse studies have provided the most direct evidence to date for osteosarcoma arising due to mutations occurring within the osteoblast-committed cell population at either the pre-osteoblast or osteoblast stage of differentiation. Using promoters that are relatively restricted to the osteoblast lineage, such as *Osterix1, Col1a1 and Osteocalcin*, investigators have generated mouse models of osteosarcoma with significantly higher incidence – up to near 100%. Interestingly, when transgenic shRNA was used to suppress p53, rather than knock it out, the subtype of the resulting osteosarcomas switched from predominantly fibroblastic to osteoblastic[5]. These findings point to the potential for differing insults to the same driver gene(s) influencing cancer subtype development, increasing our understanding of how individual osteosarcomas may originate.

CANCER STEM CELLS IN OSTEOSARCOMA

The field of cancer stem/initiating cells has been an intense area of research focus over the last decade. There is evidence from a variety of studies, including veterinary reports[6], that suggest osteosarcoma conforms to a hierarchical cancer stem cell model, rather than a stochastic model. Tumor initiating cells have the unique and exclusive capability to regenerate the tumor and may be a rare component of the overall tumor mass, however, at the moment it is not certain exactly how rare tumor initiating cells are within the osteosarcoma cell population. The hierarchical tumor stem cell model implies that the majority of cells in the tumor represent a differentiation state that is incapable of self-renewal. Several groups have tried to address this question, however, they have all to date relied on long established cell lines rather than primary tumor material. The lessons from other cancer types suggest that the robust and systematic *in vivo* analysis of limiting cell transplants coupled with primary tumor material is the only way to identify tumor initiating cells from tumors. As mentioned earlier, the existence of a cancer stem/initiating cell does not necessarily reflect the cell of origin for osteosarcoma[1]. If osteosarcoma conforms to a cancer stem cell model, the cell of origin could still be distinct, such as a differentiated osteoblast, or its precursor cell that has acquired cancer stem cell properties during its path from cancer initiation to malignancy. One would surmise that if/when the cell of origin for osteosarcoma is the mesenchymal stem cell, this cell would already have the required characteristics, namely self-renewal and differentiation abilities. In this case the cell of origin and cancer-initiating cell would be synonymous, however this cell may still only represent a minority of the tumor cell population because of a retained ability to differentiate along the osteoblastic lineage and produce osteoid. In addition, one would assume that tumors derived from a mesenchymal stem cell of origin would also include tumors that differentiated along alternate cell lineages, and with mesenchymal stem cells this list of possibilities is large, including but not limited to fibrosarcoma, chondrosarcoma, liposarcoma, hibernoma, and any other malignant counterpart to a cell originally derived from a mesenchymal stem cell. If the mesenchymal stem cell is indeed the cell of origin for osteosarcoma, important research questions relate to what drives these malignant cells to differentiate specifically along the osteoblastic lineage pathway, rather than remaining undifferentiated or differentiating along other lineages during malignant progression.

Veterinary Studies

Studies in canine osteosarcoma have revealed that this disease may follow a stem cell model. Wilson and colleagues used canine cell lines cultured in non-adherent conditions to successfully form primitive "sarcospheres"[7]. These colonies were characterized using expression of the embryonic stem cell associated genes Nanog, Oct4 and STAT3, confirming a primitive phenotype. These sarcospheres could be reproduced consistently when passaged multiple times and produced adherent cultures when returned to normal conditions. Isolation of primary cells also resulted in tumorspheres in another study, with demonstrated resistance to chemotherapy in these cells[8]. Using Affymetrix canine microarray, canine osteosarcoma stem cells and daughter adherent cells had many significant differences in gene expression, and COX-2 was shown to have a role in sphere formation and therefore possibly tumor initiation[8].

The Future

Where do we go from here to improve our understanding of the cell of origin of osteosarcoma? Further advances in tumor sequencing capacity will be critical to understanding the molecular genetics of this tumor. However, equally important will be increasing our knowledge of the cellular biology and cell(s) of origin of osteosarcoma. This information will significantly refine our attempts to understand osteosarcoma tumor initiation and progression in the individual. The use of more sophisticated murine models of osteosarcoma has enabled an initial glimpse into the susceptibility of cells at different stages of differentiation within the osteoblastic lineage to give rise to osteosarcoma. Increased use of primary tumor material to investigate the cell of origin should allow for improved outcomes through the identification of new stratified therapeutic targets.

REFERENCES

[1]Visvader JE. Nature 2011;469(7330):314. [2]Mutsaers AJ, et al. Bone 2014;62C:56-63. [3]Ng AJ, et al. Clin Sarcoma Res 2012;2(1):19. [4]Walkley CR, et al. Genes Dev 2008;22(12):1662-1676. [5]Mutsaers AJ, et al. Bone 2013;55(1):166. [6]Xiao W, et al. Clin Sarcoma Res 2013;3(1)10. [7]Wilson H, et al. Vet J 2008;175(1):69-75. [8]Pang LY, et al. PLoS One 2014;9(1).

IDENTIFYING THE DRIVERS OF BONE CANCER PAIN IN CANINE OSTEOSARCOMA

Timothy M. Fan, DVM, PhD, DACVIM (Oncology, Small Animal Internal Medicine)
Urbana, IL

INTRODUCTION

The true prevalence of cancer pain in dogs and cats is unknown; however given the conserved biology of cancer between companion animals and people[1-3], it is plausible that the incidence of cancer pain is comparable for these two species. Pain is a common ailment in human cancer patients. Based upon epidemiologic studies, the incidence of cancer pain at initial diagnosis approaches 30%, and upon disease progression up to 65-85% of human cancer patients will experience pain at some point.[4-8] Correlating with its high reported incidence, pain is the most common physical symptom in people diagnosed with terminal cancer.[9-10]

Cancer pain negatively affects quality of life as well as many important physiological functions, and its alleviation in patients should be an utmost clinical and humane priority. No cures exist for many patients suffering from advanced cancer; however effective analgesic strategies can diminish the discomfort and suffering associated with terminal disease progression. Estimates indicate that more than 70% of human cancer patients suffering from pain can be relieved with opioid-based regimens[4, 11-13], and it is justifiable to believe that equally effective cancer pain management is achievable for companion animals too. However, for cancer pain to be adequately managed, it must be recognized early and frequently reassessed by veterinary caregivers and pet owners. Despite the treatability of cancer pain, many barriers obstruct its optimal management in animals, including poor recognition associated with many cancers, difficulty in response assessment, limited knowledge regarding the usage of analgesics, and suboptimal communication between veterinary caregivers and pet owners.[14, 15]

RECOGNITION OF PAIN

One of the biggest problems in controlling bone cancer pain with veterinary patients is that of its recognition by the clinician. Unlike human beings, companion animals tend to be more stoic when it comes to the outward expression of painful sensations. As such, pet owner and veterinarians are often oblivious to painful conditions affecting companion animals until skeletal diseases become very advanced with corresponding extreme pain levels. Although subjective pain recognition is useful in a clinical setting, certain physiological parameters, such as heart and respiratory rates, temperature, pupil size, and blood pressure, have been used as surrogate measures of pain in veterinary patients, should also be integrated into determining the extent and intensity of pain. Importantly, changes in clinical behavior provides a better estimate of painful conditions, and pet owner or veterinarian observations regarding changes in movement, posture, grooming, appetite and thirst, licking of the painful area, vocalization, growling when approached, among others overt changes in behavior may provide a semi-objective information on the degree of pain experienced by a given dog or cat suffering from malignant osteolytic diseases.

BONE CANCER PAIN MECHANISMS

Bone is a living organ, rich in blood supply and nerves. Like any other organ, sensations of pain affecting the skeleton can dramatically decrease quality-of-life scores in companion animals. Given its principal anatomic function for bearing weight, any significant compromise in bone mineral density and quality, poses a risk for severe pain and pathologic fracture. Neoplasms which involve the skeleton can arise primarily from the bone, or secondarily invade or metastasize to the skeleton. In dogs, osteosarcoma is the most common tumor histology associated with focal skeletal pain; however, several additional tumor types have a predilection to involve bone including metastatic carcinoma and hematopoietic neoplasms such as multiple myeloma. In cats, primary bone tumors are much less frequent in comparison to dogs; however, involvement of bone from secondary invasion is common for oral squamous cell carcinoma. Despite the diverse tumor histologies which may pathologically affect bone, the mechanisms for how tumor cells invade and cause skeletal pain are universally shared between dogs and cats. Understanding how bone cancer pain is generated and sustained remains paramount for developing newer and more effective treatment strategies to improve quality-of-life scores in companion animals suffering from skeletal tumors.

Bone cancer pain is attributed to specific host responses occurring within the bone microenvironment. First, the presence of cancer cells results in the release of chemical mediators by neoplastic and non-neoplastic stromal cells, which in turn stimulate sensory afferent nociceptors and cause painful sensations. Specific ligands secreted by both cancer and stromal cells capable of nociceptor activation include endothelin-1, nerve growth factor, and prostaglandin E_2. Additionally, trafficking immune cells within the tumor microenvironment secrete proinflammatory cytokines including IL-1β, TNF-α, and bradykinin that also stimulate nociceptors. Second, derived from preclinical studies, the generation and maintenance of bone cancer pain is directly attributed to pathologic osteoclastic bone resorption. Mechanistically, osteoclastic bone resorption is mediated by the coordinated secretion of protons and cathepsin K, a cysteine protease. Evidence suggests that the localized acidic environment created by osteoclasts stimulates afferent nociceptors through transient receptor potential vanilloid receptor-1 (TRPV-1) and acid sensing ion channels 2/3 (ASIC 2/3). Third, bone cancer pain can be generated as a consequence of bone erosion and subsequent mechanical instability, which allows for distortion of putative mechanotransducers belonging to the transient receptor potential vanilloid receptor family which innervate bone. Based upon these unifying pathologic mechanisms, effective management of bone cancer pain requires a multi-pronged therapeutic approach which addresses the fundamental drivers that contribute to pain generation.

CANINE OSTEOSARCOMA

The most common and striking presenting sign of dogs with osteosarcoma (OS) is lameness associated with severe pain of the affected bone. The mechanisms of bone cancer pain have been characterized through the use of murine preclinical models. Given the evolutionary importance of adaptive pain, it is plausible that the same mechanisms responsible for pain generation in mouse models are also operative in spontaneously-arising bone tumors in companion animals, such as canine OS. Since tumor cells and osteoclasts appear to play preponderant roles in the genesis of bone pain associated with malignant osteolysis, therapies that reduce the viability of resident cancer cells and inhibit pathologic osteoclastic activities are expected to be effective in decreasing bone pain in canine OS patients. In addition, several key cytokines are likely drivers of bone cancer pain, including endothelin-1, nerve growth factor, and prostaglandin E2.

POTENTIAL TARGETABLE DRIVERS OF BONE CANCER PAIN

Recent advancements have been made in clarifying the fundamental underpinnings which drive different forms of pain, especially bone cancer pain through the use of murine preclinical models. With a greater understanding for what tumor microenvironmental ligands are responsible for peripheral afferent nociceptor stimulation, it has recently become possible to rationally identify and design molecularly targeted therapeutics expected to provide analgesia in companion animals.

Endothelin-1 (ET-1) is gaining interest over the past decade as a regulator of bone pain, specifically with respect to infiltration of cancer to bones. Further investigation in animal models and in humans suggested that release of ET-1 has a major role in stimulation of bone cancer pain due to its direct excitation of C-nociceptors in sensory pain fibers. Among carcinomas, prostatic tumors are especially known to cause bone pain with progression to skeletal metastasis and have been used as a model for research of ET-1 and bone pain. A few compounds were evaluated in pre-clinical mouse models and have shown promise as ETAR antagonists. For example, in a model of prostate carcinoma-induced bone pain the use of atrasentan (ABT-627) and YM598, both ETAR specific antagonists, significantly attenuated bone pain and specifically blocked ET-1 induced nociception. Also, in a mouse model of bone sarcoma it was found that ET-1 induces pain specifically through ETAR since pain alleviation occurred only with ABT-627 and not with the endothelin-B receptor (ETBR) specific antagonist. Pre-clinical evaluation of atrasentan was also done in people with prostate cancer and found it to be well tolerated and effective for prolonging survival. These findings lend for ETAR antagonism as a promising avenue for bone-specific pain blockade and further research of this pathway in animals with cancer could be promising and may lead to clinical use of these drugs.

Nerve growth factor (NGF) is another molecule implicated in the pathophysiology of bone cancer pain. NGF has specific importance since it appears to be involved in multiple layers of inflammatory and neuropathic pain stimulation at health including trophic stimulation for growth of sympathetic and sensory neurons of the dorsal root ganglion, release of substance P, calcitonin gene-related protein (CGRP) and brain-derived neurotrophic factor (BDNF) for central processing of pain stimulation, and modulation of activity of multiple ligand- and voltage-gated ion channels involved in nociception. Mechanistically, sensory stimulation by NGF is signaled preferentially by binding to its cognate receptor, TrkA, which is expressed by the majority of bone sensory nerve fibers [27]. NGF is also involved in the pathophysiology of chronic bone pain since it stimulates abnormal sprouting and reorganization of sensory nerve fibers in the area of tumor infiltration; this function appears to be regulated by release of NGF from local stromal cells and not by tumor cells and causes abnormal organization and activation of TrkA-expressing sensory nerve fibers. Indeed, it was found in a mouse model of osteolytic sarcoma that pain is generated by tumor-induced injury to local primary afferent nerve fibers. Further investigation found that an NGF blocking antibody was able to significantly reduce ongoing and movement-evoked bone sarcoma pain, greater than that achieved with acute morphine treatment. Importantly, TrkA was already found to be expressed in samples of primary and metastatic canine OS, making it a possible target for further investigation for pain control improvement and overall, the above findings make NGF and TrkA good targets for pain suppression in animals.

Non-steroidal anti-inflammatory drugs have been used for many years as analgesics in the treatment of bone cancer pain. Cyclooxygenase-2 (COX-2) inhibition is particularly useful for this purpose since the products of this inducible enzyme, especially prostaglandin E_2 (PGE_2), are known to be involved in local inflammation and stimulation of sensory nerve fibers in cancer mediated bone damage. It was found that the release of prostaglandins after injury and inflammation leads to sensitization of afferent fibers and hyperexcitability of spinal cord neurons which than causes states of hyperalgesia and allodynia. COX-2 inhibition was also shown to be very helpful for pain control in models of bone cancer pain. For example, chronic COX-2 inhibition reduced hyperalgesia in a rat model of breast carcinoma bone metastasis and in a mouse model of osteolytic sarcoma both acute and chronic treatment with a COX-2 inhibitor limited ongoing and movement-evoked pain responses. The enzyme directly responsible for PGE_2 production, prostaglandin E synthase-1 (PGES-1), is another major target for pain alleviation after inflammation and cancer invasion and it was shown that this enzyme is directly involved in bone cancer growth and increased pain responses in a mouse model of metastatic lung carcinoma. The concept of COX-2 blockade is specifically appealing due to the known pro-carcinogenic effects of this enzyme as well as its expression in a wide variety of tumors in companion animals and further understanding of the of involvement of PGES-1 and PGE_2 could be valuable for improved control of cancer-induced pain.

REFERENCES

[1]Breen, M. and J.F. Modiano, Evolutionarily conserved cytogenetic changes in hematological malignancies of dogs and humans--man and his best friend share more than companionship. Chromosome Res, 2008. **16**(1): p. 145-54. [2]Paoloni, M., et al., Canine tumor cross-species genomics uncovers targets linked to osteosarcoma progression. BMC Genomics, 2009. **10**: p. 625. [3]Paoloni, M. and C. Khanna, Translation of new cancer treatments from pet dogs to humans. Nat Rev Cancer, 2008. **8**(2): p. 147-56. [4]Cleary, J. and P.P. Carbone, Pharmacologic management of cancer pain. Hosp Pract (1995), 1995. **30**(11): p. 41-9. [5]Cleeland, C.S., et al., Pain and its treatment in outpatients with metastatic cancer. N Engl J Med, 1994. **330**(9): p. 592-6. [6]Foley, K.M., Improving palliative care for cancer: a national and

international perspective. Gynecol Oncol, 2005. **99**(3 Suppl 1): p. S213-4. [7]Jacox, A., D.B. Carr, and R. Payne, New clinical-practice guidelines for the management of pain in patients with cancer. N Engl J Med, 1994. **330**(9): p. 651-5. [8]Sykes, N.P., Pain control in terminal cancer. Int Disabil Stud, 1987. **9**(1): p. 33-7. [9]Coyle, N., et al., Character of terminal illness in the advanced cancer patient: pain and other symptoms during the last four weeks of life. J Pain Symptom Manage, 1990. **5**(2): p. 83-93. [10]Grond, S., et al., Prevalence and pattern of symptoms in patients with cancer pain: a prospective evaluation of 1635 cancer patients referred to a pain clinic. J Pain Symptom Manage, 1994. **9**(6): p. 372-82. [11]Lussier, D., A.G. Huskey, and R.K. Portenoy, Adjuvant analgesics in cancer pain management. Oncologist, 2004. **9**(5): p. 571-91. [12]Slavin, K.V., E.P. Tesoro, and J.J. Mucksavage, The treatment of cancer pain. Drugs Today (Barc), 2004. **40**(3): p. 235-45. [13]Yennurajalingam, S., V. Peuckmann, and E. Bruera, Recent developments in cancer pain assessment and management. Support Cancer Ther, 2004. **1**(2): p. 97-110. [14]Kyles, A.E. and D. Ruslander, Chronic pain: osteoarthritis and cancer. Semin Vet Med Surg (Small Anim), 1997. **12**(2): p. 122-32. [15]Lester, P. and J.S. Gaynor, Management of cancer pain. Vet Clin North Am Small Anim Pract, 2000. **30**(4): p. 951-66, ix.

CLINICAL EVALUATION OF A HER2/NEU TARGETING VACCINE TO PREVENT METASTATIC OSTEOSARCOMA

Nicola J. Mason B.Vet.Med., PhD (Immunology), DACVIM (Small Animal Internal Medicine)
Philadelphia, PA

Osteosarcoma (OSA) is a highly aggressive mesenchymal bone tumor that affects approximately 10,000 dogs per year in the USA and accounts for approximately 85% of all canine bone tumors. It most frequently affects the long bones of adult large and giant breed dogs including Rottweilers, Labradors, Greyhounds, Newfoundlands, Irish Wolfhounds, Great Danes and Scottish Deerhounds. It is estimated that 90-95% of dogs have microscopic metastases at the time of diagnosis. Standard of care includes removal of the primary tumor either by amputation, limb sparing surgery or stereotactic radiation followed by chemotherapy using platinum based drugs, either alone or in combination with adriamycin. Systemic chemotherapy is given after amputation to delay or prevent development of pulmonary and/or skeletal metastatic disease. However, despite treatment, most dogs succumb to metastatic disease within one year of diagnosis.

There have been no significant advances in the treatment of this disease over the past 2 decades. Given that 90-95% of patients are considered to have microscopic metastatic disease at the time of diagnosis, the first steps to improve outcome should focus on the prevention of metastatic disease. "Cancer immunotherapy" describes employment of the patient's immune system to specifically recognize and eliminate tumor cells. Over 100 years ago William Coley, an orthopedic surgeon at Memorial Sloane Kettering, recognized that human sarcoma patients presenting with natural bacteria infections of their tumors had improved survival compared to those without infection. This led him to develop a therapeutic concoction of live bacteria (Coley's toxins) that he injected into his patients with sarcomas of the bone[1]. He documented improved survival and in some cases complete remission in people that were known to have aggressive disease using this early form of "immune therapy." In a remarkable parallel with Coley's observations, researchers have recognized that dogs developing bacterial infections at the surgical site after limb sparing surgery tend to have longer survival times than dogs that do not develop infections[2]. These findings together suggest that activation of the immune system (by bacterial infection in these examples) may help to delay or prevent metastatic osteosarcoma[2]. More recently other investigators have used different strategies to activate the immune system in dogs with osteosarcoma to try to achieve the same immune stimulatory effect[3-5]. In these studies, the use of "immune therapy" had a positive effect on outcome and prolonged overall survival in a number of dogs following amputation. Taken together, these findings indicate that osteosarcoma is a cancer that is amenable to immune therapy.

Her2/neu as a molecular target

Central to the success of cancer immunotherapy is the identification of biologically relevant molecules expressed by the tumor that can be targeted either by antibodies or $CD8^+$ cytotoxic T cells. Her2/neu (also known as EGFR2 and erbB-2) is one such target. Her2/neu is one of four membrane bound receptor tyrosine kinases belonging to the erbB family. Her2/neu preferentially forms heterodimers with the other erbB receptors when they encounter their ligands. Dimerization leads to autophosphorylation of receptor tyrosine residues and initiates signaling through the MAPK, PI3K/Akt, PKC and STAT signaling pathways. Thus engagement of erbB receptors promotes cellular proliferation and prevents apoptosis.

Amplification and over-expression of the erbB2 gene that encodes Her2/neu occurs commonly in human breast, ovarian, stomach and uterine cancer and is thought to contribute to tumorigenesis. Over-expression of the Her2/neu protein has also been demonstrated in ~50% of human OSA samples[7-9]. In pediatric OSA the expression of Her2/neu correlates with a more aggressive phenotype, increased risk of metastasis and a significantly worse histological response to pre-operative chemotherapy, which correlates with decreased event-free survival.[8] Consistent with these observations, expression of Her2/neu is higher in human OSA cell lines with greater metastatic potential than in those with lower metastatic potential and metastatic OSA lesions are more frequently Her2/neu positive compared to primary lesions in humans.

Approximately 40% of canine primary OSA express Her2/neu as determined by IHC and qRT-PCR[6] and Western blot (unpublished data) and, as in human OSA, over-expression of Her2/neu appears to correlate with poor prognosis in the dog[6]. Recently other groups have suggested that EGFRs are over expressed in canine primary OSA and may drive tumorigenesis leading to a more aggressive metastatic phenotype with shorter progression free survival and overall survival [10,11]. Taken together, these findings in both pediatric and canine OSA suggest that Her2/neu expression confers a beneficial metastatic advantage to malignant osteoblasts and therefore may be a relevant immune target for prevention of metastatic disease.

A recent clinical trial evaluating the use of trastuzumab (a monoclonal antibody that targets Her2/neu) in the adjuvant setting for women with hormone receptor-positive breast cancer supports this hypothesis. The surprising results of this clinical trial showed that patients with Her2/neu negative breast cancer benefited as much from trastuzumab as patients with Her2/neu positive cancers[12]. Several groups investigated this unexpected finding and found that Her2/neu is selectively expressed and regulates self-renewal in cancer stem cells (CSC) of patients with breast cancer[13]. The same group demonstrated that engagement of RANKL on the surface of CSC by osteoblast derived RANK, leads to NF-kappaB activation and up regulation of Her2/neu on the CSC enabling them to be targeted with Her2/neu immune therapy. Thus, elimination of Her2/neu positive CSC may be responsible for the clinical benefit seen in human breast cancer patients with Her2/neu negative primary tumors. In concordance with these results and with respect to OSA, recent data have revealed that OSA tumor initiating cells can be eliminated by Her2/neu specific T cells *in vitro* and that T cells specific for Her2/neu can produce significant therapeutic effects in mouse OSA xenograft models.[15] Taken together these results suggest that as in breast cancer, immune therapeutic approaches that generate Her2/neu specific cytotoxic T cells may be able to eliminate OSA cells at their source.[16]

ADXS31-164.

Listeria monocytogenes is a gram-positive intracellular bacteria that is capable of inducing potent innate and adaptive immune responses[14]. Through its expression of listeriolysin (LLO), a pore forming lysin that enables it to escape from the phagosome prior to lysosome fusion, it rapidly gains access to the cytoplasm of the cell and hence the MHC I processing pathway (Fig. 1). This is essential for the bacteria to induce $CD8^+$ cytotoxic T cell responses. A large body of literature supports the use of a highly attenuated form of *L. monocytogenes* to introduce target antigens into antigen-presenting cells such as monocytes, macrophages and dendritic cells[15-19]. By genetically fusing the target antigen of interest to LLO, potent $CD8^+$ T cell responses can be generated against the target antigen[19]. This is particularly useful in the case of tumor immunotherapy, where the goal is to induce potent CD8+ T cell responses against tumor antigens. ADXS31-164 is a highly attenuated strain of *L. monocytogenes* that has a chimeric human Her2/neu antigen (that comprises portions of the protein that contain the majority of HLA-A2 immunodominant epitopes of Her2/neu) fused to its LLO. ADXS31-164 breaks peripheral tolerance and induces potent $CD4^+$ and $CD8^+$ T cell responses against Her2/neu that eliminate implanted Her2/neu$^+$ positive tumors in mice and prevent their spontaneous growth in Her2/neu transgenic mice[18].

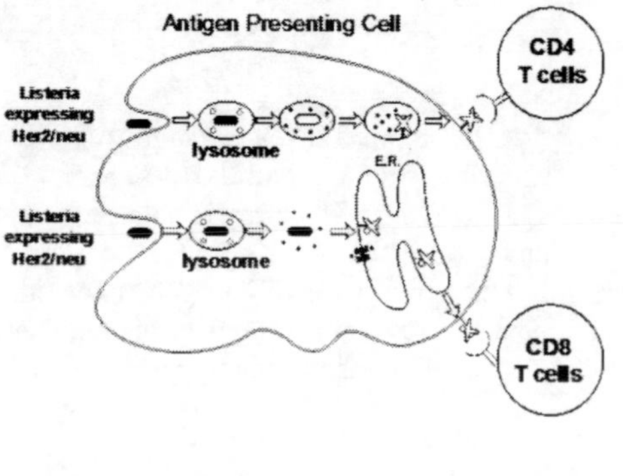

Fig. 1. Genetically modified listeria enables access of targeted antigen to the MHCI and MHCII antigen-presenting pathway

We have taken advantage of this technology and performed a phase I dose escalation clinical trial to determine whether ADXS31-164 can safely elicit Her-2/neu specific T cell immune responses and prevent recurrence of OSA when administered in the adjuvant setting, to dogs with Her2/neu$^+$ appendicular OSA. Dogs that had undergone standard of care amputation and carboplatin chemotherapy were enrolled. Dogs received intravenous infusions of ADXS31-164 and were monitored every 2 months thereafter. Results have been highly encouraging and will be discussed in the presentation. We hypothesize that the clinical efficacy observed with this approach is associated with ADXS31-164 mediated induction of Her2/neu specific cytotoxic T cells that eliminate metastatic Her2/neu$^+$ OSA cells, preventing disease recurrence and prolonging overall survival.

Combination immune therapy and radiation therapy in the treatment of primary osteosarcoma

Preliminary results from our phase I clinical trial evaluating the ability of ADXS31-164 to prevent metastatic disease when administered to dogs with Her2/neu$^+$ appendicular OSA in the setting of minimal residual disease are encouraging. However, some dogs are unable to tolerate limb amputation due to their large size or concurrent orthopedic or neurological problems. Treatment options for these dogs are limited and usually consist of palliative radiation and pain management using NSAIDs, opioids and GABA analogs such as gabapentin. Overall survival of dogs that undergo palliative radiation alone (with concurrent pain medications) is about 4-6 months. We are now interested in determining whether ADXS31-164 could synergize with radiation therapy to induce potent anti-tumor immunity and eliminate primary tumor cells.

In pre-clinical murine models the listeria platform has shown marked immunomodulatory effects on the tumor microenvironment, reducing regulatory T cells and myeloid derived suppressor cells[18]. These effects serve to augment vaccine-induced antigen-specific cytotoxic T cell function within the tumor. Ionizing radiation induces immunogenic cell death via up-regulation of MHC class I, co-stimulatory molecules and death receptors on tumor cells. It also enhances antigen presentation by APCs and releases chemokines that promote T cell trafficking to the tumor. As such, combination RT and immune therapy using recombinant listeria technology leads to synergistic anti-tumor effects as shown recently in a murine model of prostatic carcinoma[20]. We are now evaluating the synergistic effects of RT with ADXS31-164 in the treatment of spontaneous primary Her2/neu$^+$ appendicular OSA in dogs that have not undergone amputation. In this trial, all patients receive palliative radiation (2 doses on 2 consecutive days)[21] and then receive ADXS31-164 every three weeks for a total of 8 vaccinations. Dogs are being evaluated every three weeks and assessed for lameness and pain, overall well-being and any evidence of tumor spread. This is a unique trial that combines two cancer treatment modalities (radiation and immune therapy) in an attempt to improve the outcome of large and giant breed dogs with OSA. Preliminary results of this approach will be discussed in the presentation.

Summary

Previous work has shown that OSA may be an "immune responsive" tumor and that the concept of delaying or preventing metastatic disease using the immune system holds much promise. Our preliminary results from a phase I clinical trial support this approach. We are now extending this work to address the hypothesis that a combination of palliative radiation and immune therapy will synergize to generate potent anti-tumor immune responses that will eliminate primary as well as metastatic disease. Our studies will continue and as more patients are treated in both the adjuvant and neo-adjuvant setting, survival trends will become apparent and statistically significance for outcome will be determined. The future for this approach is exciting and it looks set to offer an additional potent weapon in our arsenal against OSA and other Her2/neu$^+$ tumors.

REFERENCES

1. Coley WB. Sarcoma of the Long Bones: The Diagnosis, Treatment and Prognosis, with a Report of Sixty-Nine Cases. Ann Surg 1907;45:321-368. **2.** Lascelles BD, Dernell WS, Correa MT, et al. Improved survival associated with postoperative wound infection in dogs treated with limb-salvage surgery for osteosarcoma. Ann Surg Oncol 2005;12:1073-1083 **3.** Dow S, Elmslie R, Kurzman I, et al. Phase I study of liposome-DNA complexes encoding the interleukin-2 gene in dogs with osteosarcoma lung metastases. Hum Gene Ther 2005;16:937-946. **4.** Kurzman ID, MacEwen EG, Rosenthal RC, et al. Adjuvant therapy for osteosarcoma in dogs: results of randomized clinical trials using combined liposome-encapsulated muramyl tripeptide and cisplatin. Clin Cancer Res 1995;1:1595-1601. **5.** Modiano JF, Bellgrau D, Cutter GR, et al. Inflammation, apoptosis, and necrosis induced by neoadjuvant fas ligand gene therapy improves survival of dogs with spontaneous bone cancer. Molecular therapy : the journal of the American Society of Gene Therapy 2012;20:2234-2243. **6.** Flint AF, U'Ren L, Legare ME, et al. Overexpression of the erbB-2 proto-oncogene in canine osteosarcoma cell lines and tumors. Veterinary pathology 2004;41:291-296. **7.** Gorlick R, Huvos AG, Heller G, et al. Expression of HER2/erbB-2 correlates with survival in osteosarcoma. J Clin Oncol 1999;17:2781-2788. **8.** Rainusso N, Brawley VS, Ghazi A, et al. Immunotherapy targeting HER2 with genetically modified T cells eliminates tumor-initiating cells in osteosarcoma. Cancer gene therapy 2012;19:212-217. **9.** Zhou. Her-2/neu expression in osteosarcoma increases risk of lung metastasis and can be associated with gene amplification. J Pediatr Hematol Oncol 2003;25:27-32. **10.** McCleese JK, Bear MD, Kulp SK, Mazcko C, Khanna C, London CA., Met interacts with EGFR and Ron in canine osteosarcoma. Vet Comp Oncol. 2013 Jun;11(2):124-39. **11.** Expression of epidermal growth factor receptor in canine osteosarcoma: association with clinicopathological parameters and prognosis. Selvarajah GT, Verheije MH, Kik M, Slob A, Rottier PJ, Mol JA, Kirpensteijn J. Vet J. 2012 Aug;193(2):412-9. **12.** HER2 status and benefit from adjuvant trastuzumab in breast cancer. Paik S, Kim C, Wolmark N. N Engl J Med. 2008 Mar 27;358(13):1409-11. **13.** HER2 drives luminal breast cancer stem cells in the absence of HER2 amplification:implications for the efficacy of adjuvant trastuzumab. Ithimakin S, Day KC, Malik F, Zen Q, Dawsey SJ, Bersano-Begey TF, Quraishi AA, Ignatoski KW, Daignault S, Davis A, Hall CL, Palanisamy N, Heath AN, Tawakkol N, Luther TK, Clouthier SG, Chadwick WA, Day ML, Kleer CG, Thomas DG, Hayes DF, Korkaya H, Wicha MS. Cancer Res. 2013 Mar 1;73(5):1635-46. **14.** Pamer EG. Immune responses to Listeria monocytogenes. Nat Rev Immunol. 2004;4:812-23. **15.** Le DT, Dubenksy TW, Jr., Brockstedt DG. Clinical development of Listeria monocytogenes-based immunotherapies. Semin Oncol. 2012;39:311-22. **16.** Maciag PC, Radulovic S, Rothman J. The first clinical use of a live-attenuated Listeria monocytogenes vaccine: a Phase I safety study of Lm-LLO-E7 in patients with advanced carcinoma of the cervix. Vaccine. 2009;27:3975-83. **17.** Seavey MM, Pan ZK, Maciag PC, Wallecha A, Rivera S, Paterson Y, et al. A novel human Her-2/neu chimeric molecule expressed by Listeria monocytogenes can elicit potent HLA-A2 restricted CD8-positive T cell responses and impact the growth and spread of Her-2/neu-positive breast tumors. Clin Cancer Res. 2009;15:924-32. **18.** Shahabi V, Seavey MM, Maciag PC, Rivera S, Wallecha A. Development of a live and highly attenuated Listeria monocytogenes-based vaccine for the treatment of Her2/neu-overexpressing cancers in human. Cancer Gene Ther. 2011;18:53-62. **19.** Singh R, Dominiecki ME, Jaffee EM, Paterson Y. Fusion to Listeriolysin O and delivery by Listeria monocytogenes enhances the immunogenicity of HER-2/neu and reveals subdominant epitopes in the FVB/N mouse. J Immunol. 2005;175:3663-73. **20.** Hannan R, Zhang H, Wallecha A, Singh R, Liu L, Cohen P, et al. Combined immunotherapy with Listeria monocytogenes-based PSA vaccine and radiation therapy leads to a therapeutic response in a murine model of prostate cancer. Cancer Immunol Immunother. 2012;61:2227-38. **21.** Knapp-Hoch HM, Fidel JL, Sellon RK, Gavin PR. An expedited palliative radiation protocol for lytic or proliferative lesions of appendicular bone in dogs. J Am Anim Hosp Assoc. 2009;45:24-32.

USING TYROSINE KINASE INHIBITORS AND METRONOMIC CHEMOTHERAPY TO IMPROVE ANTITUMOR IMMUNITY IN DOGS

Barb Biller, DVM, PhD, DACVIM
Fort Collins, CO

INTRODUCTION

Capturing the ability of the immune system to fight cancer holds significant promise for the treatment of highly aggressive malignancies, particularly for prevention or control of metastatic disease. This is the rationale for considerable research over the past decade focused on the development of immunotherapeutics for cancer patients. But despite many encouraging findings in mouse tumor models the results of immunotherapy clinical trials have generally been disappointing. Unfortunately it is becoming increasingly apparent that the efficacy of tumor vaccines, adoptive T cell transfer and other forms of immunotherapy are limited by potent mechanisms of tumor induced immune suppression. Combination with treatment strategies to overcome tumor tolerance is clearly needed to improve the outcome for humans and animal with cancer.

WHY MOST TUMOR VACCINES FAIL TO CONTROL CANCER PROGRESSION

Tumor vaccines are an especially attractive, and potentially one of the most effective, strategies for preventing metastatic or relapsing disease because of their ability to stimulate both tumor specific cytotoxic T lymphocytes (CTLs) and long lasting memory T cells. Although tumor vaccines have been shown to have activity against certain types of cancer such as melanoma, their overall efficacy as a stand-alone therapy appears minimal.

Two of the biggest hurdles to their success are tumor induced recruitment of inhibitory immune cells and production of immunosuppressive cytokines. The key players are regulatory T lymphocytes (Treg) and myeloid derived suppressor cells (MDSC), immune cells that normally function to prevent autoimmunity and limit chronic inflammation. A significant amount of data shows that Treg and MDSC accumulate in most malignant tumors of mice and humans.[1,2] Increased numbers of Treg and MDSCs are also frequently found in the blood or draining lymph nodes of human cancer patients; a finding that correlates with a poorer prognosis or advanced stage disease.[3-5] While our understanding of the role of Treg and MDSC in tumor bearing dogs is in its infancy, both cell types have been identified in higher numbers in the blood of dogs with cancer compared to healthy ones.[6-9] Our own studies have contributed to this body of knowledge through the initial identification and characterization of Treg in dogs and demonstration of their association with outcome in osteosarcoma (OSA).[10] Others have confirmed and extended this work by demonstrating the immunosuppressive functions of Treg and their correlation with outcome for other canine tumors including mammary carcinoma and oral melanoma.[11-13] Less information is currently available regarding the role of canine MDSC; however they have been shown to be present in greater numbers in dogs with advanced or metastatic cancers compared to those with earlier stage disease. [8-9]

HOW CAN WE GET RID OF THE BAD GUYS?

Because of their central role in tumor tolerance, Treg and MDSC are attractive therapeutic targets. One strategy is the use of sunitinib, an oral tyrosine kinase inhibitor with immune stimulatory effects in addition to its better known tumor cell cytotoxic and antiangiogenic properties. For example sunitinib was found to decrease MDSC accumulation, prevent Treg development and improve the efficacy of a tumor immunotherapy protocol in a mouse tumor model. [14] In humans with renal cell carcinoma administration of sunitinib was associated with stimulation of type-1 (Th1) T-cell mediated immune responses and reversal of immune dysfunction mediated through reduction in MDSC and Treg number and function. [15-16] In agreement with this data, we recently demonstrated the ability of toceranib (Palladia), a structural analog of sunitinib, to selectively decrease circulating Treg when administered to dogs with cancer. [17] (See Figure 1).

Another approach to target Treg is the use of the alkylating chemotherapy agent cyclophosphamide (CYC). Although both high and low doses of CYC can inhibit Treg function in mice and humans, a lower, continuous dosing schedule (such as metronomic dosing) appears to allow for more effective and prolonged Treg inhibition. This was recently demonstrated in women with metastatic breast cancer in which metronomic CYC resulted in a transient (4 to 6 weeks) decrease in Treg numbers but a durable increase in tumor reactive T cells; a finding that also correlated with stable disease and increased overall survival. [18] In a clinical trial for dogs with soft tissue sarcoma, daily administration of low-dose CYC was associated with a selective decrease in circulating Treg, suggesting that CYC may have similar immunomodulatory effects on canine Treg.[19] In a separate study these investigators also found that addition of metronomic CYC to toceranib led to significantly increased serum concentrations of the pro-inflammatory cytokine IFN-γ as Treg numbers declined, consistent with a shift from immune suppression towards a more favorable stimulatory immune environment.[17] (Figure 2).

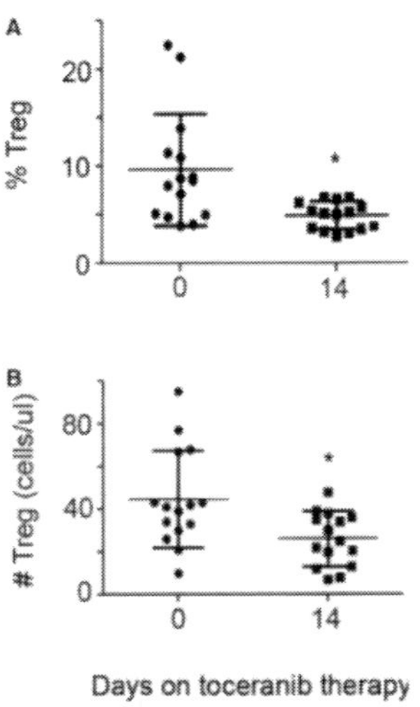

Figure 1. Relative and absolute numbers of Treg in the peripheral blood of dogs with cancer treated with toceranib. The mean percentages **(A)** and absolute numbers **(B)** of Treg are compared pretreatment and after 14 days of toceranib therapy.

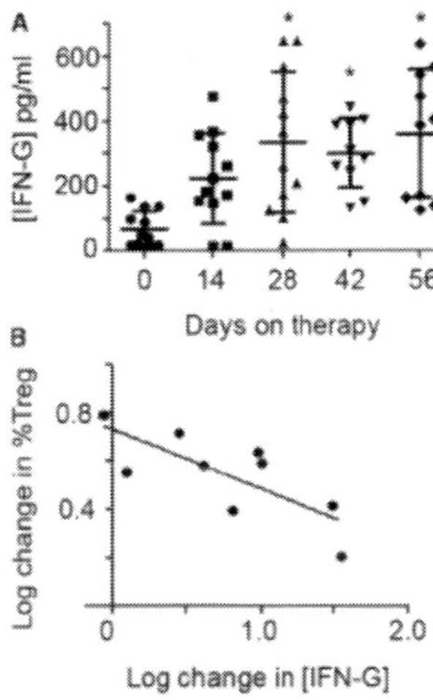

Figure 2. IFN-γ production and correlation with changes in Treg numbers in dogs receiving toceranib and CYC. All dogs received toceranib alone for the first 14 days then metronomic CYC was added. The mean serum concentration of IFN-γ is plotted in (A). In (B) the correlation between the log change of IFN-γ and Treg number at day 0 and day 56 is plotted.

CONCLUSION AND FUTURE DIRECTIONS

It is exciting to witness the recent development of immunotherapies for veterinary cancer patients including the DNA vaccine now available for dogs with malignant melanoma. We anticipate that other treatments such as monoclonal antibody therapy for dogs with lymphoma may soon be available as well. As our experience with canine cancer immunotherapy grows it will be critical to avoid underestimating the potency of tumor induced immune suppression that has led to failure of many similar products for humans. Towards this end we are currently investigating whether metronomic CYC, toceranib or the combination increases the immunogenicity and effectiveness of a tumor vaccine in dogs with osteosarcoma. To do this we are asking the question whether each therapy or the combination stimulates a more robust adaptive immune responses compared to that elicited with the vaccine alone. We are also exploring the ability of this approach to abrogate immune suppression based on changes in Treg and MDSC numbers and serum concentrations of immune suppressive cytokines. As in human clinical trials, the success of immunotherapy for companion animals will likely depend on the identification of treatment strategies to overcome immune suppression associated with cancer.

REFERENCES

1. Byme, W. et al. Cancer Res 2011;71:6915. 2. Toh, B. et al. Oncoimmunology 2012;1:1360. 3. Liyanage V, et al. J Immunol 2002;169:2756. 4. Ormandy L, et al. Cancer Res 2005;65:2457. 5. Umansky V, et al. Cancer Microenvironment 2012;5:1821. 6. Abrams, V. et al. Vet Immunol Immunopath 2010;135:257. 7. Biller, B. et al. Vet Immunol Immunopath 2007;116:69. 8. Goulart, M. et ala. PLoSOne 2012;7:3327. 9. Sherger, M. BMC Vet Res 2012;8:209. 10. Biller, B. et al. JVIM 2010;24:1118. 11. Pinheiro, D. et al. Immunology 2011;132:111. 12. Kim, J. et al. Vet J 2012;193:222. 13. Horiuchi, Y. Microbiol Immunol 2010; 54: 152. 14. Ozao-Choy, J. Cancer Res 2009;69:2514. 15. Finke, J. Clin Cancer Res 2008;14:6674. 16. Ko, J. Cancer Res 2010;70:3526. 17. Mitchell, L. et al. JVIM 2012;26:355. 18. Ge, Y. et al. Cancer Immun Immunother 2012;61:353. 19. Burton, J. et al. JVIM 2011;25:920.

MULTIDISCIPLINARY APPROACH TO ONCOLOGIC DISEASES. PERSPECTIVES FROM AN ONCOLOGIST, A SURGEON AND A PATHOLOGIST. ORAL TUMOR: MALIGNANT MELANOMA

CA Clifford[1]; CJ LeBlanc[2], Brian Bretz[3]
[1]Malvern, PA; [2] Knoxville, TN; [3]Langhorne, PA

Introduction

The goal of this lecture is to provide a multidisciplinary approach to the diagnosis, staging, treatment and prognosis associated with oral melanoma in a case based manner. This overview will specifically highlight the input of a medical oncologist, clinical pathologist and surgeon.

Melanoma is the most common oral tumor in dogs and commonly occurs in Poodles, Dachshunds, Scottish terriers, Golden retrievers, Schnauzers and Rottweilers. The tumor is most commonly located on the gingiva (~50%), buccal/labial (~30%) surface, tongue (~5%) and hard palate (~15%). Biologically, these are locally aggressive with a high metastatic rate, although recent evidence suggests that some locations may carry a better prognosis. There is also reported a "low grade" subtype that may carry a more favorable prognosis. The history for most dogs involves signs such as halitosis, abnormal chewing, weight loss, hypersalivation, bleeding from the mouth, or the visualization of a mass.

Staging

Dogs with oral melanomas are generally staged via:

- Tumor cytology: Oral melanomas typically exfoliate easily resulting in highly cellular slides. The slides are typically predominated by many irregularly round to polygonal to spindle-shaped cells with a variable amount of cytoplasm that is typically lightly basophilic and will contain either no or many greenish-black granules or globules. Many high-grade malignant melanomas will contain no visible melanin histologically (hence the term amelanotic melanoma), but even in these cases a small number of cells will contain a slight dusting of melanin that can be seen cytologically. Malignant melanomas typically exhibit numerous criteria of malignancy, such as large round to irregular nuclei with very coarsely stippled nuclear chromatin, multiple prominent nucleoli of variable size and shape, binucleation, moderate to marked anisokaryosis, and abnormal mitotic figures. Macrophages (melanophages) are commonly present and may consist of many large vacuoles filled with melanin pigment.
- Routine labwork (CBC, chemistry panel, urinalysis)
- Three-view thoracic radiographs
- Regional lymph node cytology: A metastatic lymph node can only be confirmed by the presence of melanocytes displaying malignant criteria. In contrast, melanophages are commonly found in benign lymph nodes and their presence does not warrant a diagnosis of metastatic melanoma.
- Advanced imaging (Computed Tomography)

Therapy

Treatment for oral malignant melanoma generally consists of a two-pronged approach aimed at providing both local and systemic disease control. Local control is typically achieved through surgical excision and tumor location usually determines the type of surgery that is indicated.

Oral tumors are treated with the same principles of surgical oncology as apply to masses in other locations of the body. As mentioned previously, oral tumors should have, at a minimum, a fine needle aspirate performed to obtain a diagnosis prior to surgery. If the results are non-diagnostic, incisional biopsy is necessary to allow for appropriate surgical planning. There are multiple locations from which these tumors can arise in the mouth, including the mandible, maxilla, soft palate and even the surface of the tongue. The primary goal of surgical resection is to obtain complete surgical margins, however in some cases (surface of the tongue, soft palate) complete resection may not be possible and surgery is performed with palliative intent. The goal of surgery and corresponding plan of action must be decided upon with the owners prior to the procedure.

For definitive treatment, aggressive surgical resection is recommended and generally requires either a mandibulectomy or maxillectomy. Resecting only the grossly visible portion of the mass may provide short-term palliative benefit but is not adequate for long-term success. In our practice, a CT scan is performed in many of these patients immediately prior to surgery to help determine surgical margins. Computed tomography is much more sensitive than radiographs for assessing bone lysis and many of these tumors will have contrast enhancement in the soft tissue which extends well beyond the visible mass. Ideal margins are 2-3cm, as in any other oncologic surgery, however this may not always be achievable in the confined space of the oral cavity.

Maxillectomy and mandibulectomy are each approached in the same fashion, with a complete circumferential, full thickness gingival incision. After the gingiva is incised, an elevator can be used to elevate the surrounding healthy gingiva and provide visualization of the underlying bone. Following gingival elevation, a power saw or sharp osteotome and mallet are used to resect the appropriate margins of bone. The mass should always be removed with gingiva and bone as an en bloc sample.

Examples of maxillectomy include hemimaxillectomy, rostral maxillectomy, central hemimaxillectomy, and caudal hemimaxillectomy.

Maxillectomy incisions tend to have moderate hemorrhage due to the number of vessels lining the nasal cavity. Larger vessels should be ligated with suture or sealed with electrocautery or vascular clips.

Mandibulectomy is started in the same fashion as maxillectomy with a full thickness gingival incision. Again, in this case, a power saw or osteotome and mallet should be used to transect the mandible at the appropriate location based on preoperative planning. In a complete hemimandibulectomy, the mandibular artery and vein are ligated caudal to the vertical ramus of the mandible. In the partial mandibulectomies, the artery can be ligated within the mandible.

Examples of mandibulectomy include hemimandibulectomy, rostral mandibulectomy, rostral hemimandibulectomy, and caudal hemimandibulectomy.

As with any incision closure, oral oncologic surgery requires appropriate instruments, gentle tissue handling and alleviation of tension to allow the gingiva to heal appropriately. Gloves and instruments should be changed prior to beginning closure of the site. For suture material, many surgeons prefer vicryl for the closure of oral incisions because it is absorbable, soft and has very little suture memory that is helpful when working in confined spaces in the mouth. Simple interrupted sutures are preferred to allow for accurate apposition of tissue.

After completion of the primary resection, the mandibular lymph node should ALWAYS be resected and submitted for histopathology. In several cases, a nodal metastasis has been noted in patients despite a negative fine-needle aspirate and unremarkable lymph node palpation.

All samples should be submitted in formalin for histopathology. A rule of thumb is that "if it is important enough to remove, it is important enough to submit for histopathology." A complete history should be provided and the gross margins of the sample should be marked. This can be accomplished with special tissue dye or suture tags so that the completeness of excision can be assessed by the pathologist.

Although surgery can provide adequate local control, when used alone, surgery does not appear to demonstrably improve survival (reported median survival of 7-9 months with surgery alone). In cases where the only a marginal resection has been achieved, additional local therapy, with either a second more aggressive surgery or radiation therapy (6 weekly treatments to the site and draining lymph node), is recommended.

Although the majority of oral melanomas are straightforward in their diagnosis, a challenge arises for the clinician in cases in where the tumor is considered amelanotic or poorly differentiated. Histologically, the distinction between melanoma and sarcoma can be difficult. Since the ultimate systemic therapy is distinct for each histology, further delineation through the use of a panel of immunohistochemical markers is warranted (PNL, Melan-A, TRP-1, TRP-2).

In cases where the tumor is deemed nonresectable or the owner is unwilling to pursue surgery, radiation therapy can be utilized, as it has been associated with a high response rate (70-90%) for malignant melanoma. Most protocols are classified as "hypofractionated", in that they require only a few high dose-per-fraction treatments. Generally, the protocol utilized is clinician-dependent but consists either of three to four weekly fractions of 8 Gray (Gy) or six weekly fractions of 6 Gy. Side effects (oral mucositis) are generally mild and self-limiting with hypofractionated protocols and the majority of radiation oncologists will include not only the primary tumor site or residual scar but also the draining lymph node bed in the radiation fields. Similar to outcomes with surgery alone, patients treated with radiation therapy may achieve local control but generally succumb to metastatic disease within 7-9 months of diagnosis.

Due to the high metastatic rate associated with melanoma, systemic therapy is warranted. Responses to chemotherapy (Carboplatin, Cisplatin, Dacarbazine, Melphalan and Doxorubicin) have been noted in the measureable disease setting, but generally these are partial responses and are not durable. Carboplatin was evaluated for dogs that had adequate locoregional control for malignant melanoma of various anatomic locations, and no significant survival benefit was documented. Uniquely for malignant melanoma, a highly immunogenic tumor type, the patient's own immune system can be trained to target the tumor. Immunotherapy is not a new therapeutic modality in cancer treatment, but because overcoming immune tolerance has historically been a challenge, anti-cancer vaccine therapy has failed to flourish. A novel therapeutic DNA vaccine (OnceptTM) has been developed for dogs and this xenogenic model (using DNA from one species to elicit a response against another species) significantly improved survival compared to historical controls not treated with immunotherapy. This is the first therapeutic vaccine approved for the treatment of cancer in either animals or humans. Importantly, the vaccine is safe, with rare side effects noted. The vaccine is administered via a "needleless" transdermal device. The vaccine protocol comprises four treatments on an every other week basis for four doses and booster vaccines every 6 months.

Prognostic Factors

The use of established "prognostic factors" aids in assessing each individual patient, thereby enabling a clinician to more adequately predict survival during discussions with owners. Prognostic factors for oral melanoma include but are not limited to:

The size of the tumor (<2cm poor vs. <2cm favorable)
Spread to lymph nodes or lungs (poor)
Site of the tumor (tumors along gum line or lip or tongue may be more favorable)
Grade (low grade more favorable)
Ability of first treatment to provide local control (favorable)
Proliferation indices of the tumor (High indices is associated with poorer prognosis).
Cyclo-oxygenase expression (Low is favorable)
Addition of the melanoma vaccine (OnceptTM) in patients with adequate local-regional control (favorable)

Novel strategies

The use of tyrosine kinase inhibitors, metronomic (low dose chemotherapy) and alternative medicines are all interesting approaches for this cancer that warrant further large-scale evaluation.

REFERENCES

1. Bergman PJ, et al. In: Withrow and MacEwen's Small Animal Clinical Oncology. 5th ed. 2013:321.
2. Spangler WL, Kass PH: Vet Pathol 43:136, 2006.
3. Williams LE, Packer RA: J Am Vet Med Assoc 222:1234, 2003.
4. Smedley RC, et al. Vet Pathol. 2011;48(1):54.
5. DeNicola DB. Round cells. In Cowell and Tyler's Diagnostic Cytology and Hematology of the Dog and Cat, 4th Ed. Valenciano AC and Cowell RL, eds. Elsevier Mosby, St. Louis, MO, 2014.
6. Busam KJ, Jungbluth AA. Adv Anat Pathol 6: 12-18, 1999.
7. Rassnick KM, et al. J Am Vet Med Assoc. 2001;218:1444.
8. Brockley LK, et al. New Zeal Vet J. 2013;61(1):25.

MULTIDISCIPLINARY APPROACH TO ONCOLOGIC DISEASES. PERSPECTIVES FROM AN ONCOLOGIST, A SURGEON AND A PATHOLOGIST. INTRATHORACIC TUMOR: FELINE THYMOMA

CA Clifford[1]; CJ LeBlanc[2], Brian Bretz[3]
[1] Malvern, PA; [2] Knoxville, TN; [3] Langhorne, PA

Introduction

The goal of this lecture is to provide a multidisciplinary approach to the diagnosis, staging, treatment and prognosis associated with feline thymoma in a case based manner. This overview will specifically highlight the input of a medical oncologist, clinical pathologist and surgeon.

Cranial mediastinal tumors are uncommon in cats, with lymphoma and thymoma as the most common differentials. Others including ectopic thyroid carcinoma, aortic body tumors, rib and sternal sarcomas extending into mediastinal space, and non-neoplastic diseases, such as mediastinal and thymic branchial cysts, are less common to rare. Thymomas arise from thymic epithelial tissue, although mature lymphocyte involvement is variable and can be dominant. For this reason, and because the lymphoid component of thymomas exfoliates more readily than the epithelial component, differentiation from cranial mediastinal lymphoma can be difficult. However, the thymic epithelium is the malignant component, not the lymphocytes. Thymomas can also be cystic and mimic non-neoplastic cranial mediastinal cysts.

Thymomas are classified as invasive or non-invasive, but this is usually based on surgical findings and not preoperative imaging. Thymomas are non-invasive in 50-100% of cats. Non-invasive thymomas are well-encapsulated, however, in contrast, invasive thymomas invade into adjacent structures such as the cranial vena cava, thoracic wall, lung, and pericardium, and are more common in the dog. Metastatic thymoma has been rarely reported in cats.

Clinical presentation

The clinical signs for cats with thymoma often depend on whether the tumor is invasive or not. For non-invasive thymomas, the thymoma may be clinically silent and an incidental finding or associated with non-specific clinical signs. These include exercise intolerance, coughing, dyspnea, dysphagia, and weight loss, and are mostly caused by the space-occupying thoracic mass. Coughing and dyspnea are caused by either pleural effusion or extraluminal compression of the trachea and/or mainstem bronchi by the tumor. Laryngeal paralysis can be seen with peripheral nerve entrapment. The clinical signs in cats with invasive thymomas are often the same as for those with non-invasive thymoma. Cranial vena cava syndrome is common with invasion or thrombosis of the cranial vena cava. Cranial vena cava syndrome is characterized by edema of the submandibular area, neck, thoracic inlet, and thoracic limbs, and is associated with pleural effusion, particularly chylothorax.

Autoimmune diseases, paraneoplastic syndromes, and second non-thymic malignancies occur in cats with thymoma. Myasthenia gravis, characterized by dysphagia, weakness and the presence of megaesophagus, is uncommon in cats compared to dogs. Polydipsia and polyuria, although frequently observed in dogs with paraneoplastic hypercalcemia, is also uncommon in cats. An interesting paraneoplastic disease noted in cats is called Feline Thymoma-Associated Exfoliative Dermatitis (FTED). This is a T-cell–mediated process (CD3+): abnormal antigen presentation of neoplastic thymic epithelial cells cause immature auto-reactive T-cells to attack keratinocytes). Lesions are often noted on the head, cervical region, ears, chest and trunk. Initially considered a non-pruritic, erythematous dermatitis with exfoliative scaling this can later develop into a keratosebaceous debris along the digits, claw beds and ears. Surgical removal of the thymoma often yields complete resolution of all clinical signs.

Clinical Staging

Clinical signs, physical examination findings, and minimum database abnormalities are usually non-specific and do not necessarily assist in the diagnosis of thymoma. Since the majority of cranial mediastinal tumors are large and displace the heart and lungs, thoracic auscultation is often abnormal and characterized by muffled heart and bronchovesicular sounds. The presence of clinical features consistent with cranial vena cava syndrome is highly suggestive of an invasive cranial mediastinal tumor.

Imaging is paramount in the diagnosis of thymoma. Thoracic radiograph findings include: Identification of a space-occupying mass with dorsal elevation of trachea and esophagus, caudal displacement of the cardiac silhouette, megaesophagus with secondary aspiration pneumonia seen with paraneoplastic myasthenia gravis, and pleural effusion which may obscure the cranial mediastinal mass. Ultrasonography can be useful in helping to differentiate thymoma from lymphoma. Specifically, the majority of thymomas have a mixed echogenic pattern with occasional cavitation, while lymphomas are often homogenous and hypoechoic. Computed tomography scans are useful in determining respectability, invasion into adjacent structures such as the pericardium, lungs, and thoracic wall, and screening for pulmonary metastasis. Ultrasound can be used to obtain guided aspirates for diagnostic purposes.

Cytology is helpful in diagnosing thymomas and differentiating thymoma from lymphoma. Highly cellular aspirates of a thymoma are typically predominated by many small, mature lymphocytes; thymic epithelial cells are most commonly absent or present in very low number. When present, epithelial cells are found individually or in small cohesive clusters; they are typically uniform, have a small to moderate amount of lightly basophilic cytoplasm, and lack criteria of malignancy (such as anisokaryosis and prominent nucleoli). A low number of well-granulated mast cells are a common finding; intermediate-sized to large lymphocytes can also be found in low concentrations. In contrast, thymic lymphoma will look like lymphoma found in other lymphoid organs. The predominant cell type will be intermediate-sized to large immature lymphocytes, with fewer small lymphocytes, and no mast cells or thymic epithelium. In equivocal cases, flow cytometric immunophenotyping can be used to differentiate thymomas from other lymphocyte-rich mediastinal lesions. Specifically, ten percent or greater of the small lymphocytes in thymomas should characteristically co-express CD4 and CD8. Classification of thymomas as benign or malignant is best accomplished histologically.

Treatment

Surgery and radiation therapy are the two main treatment options for the management of thymomas in cats (and dogs). Median sternotomy is the standard surgical approach because of the large size of most cranial mediastinal tumors, the limited space in the cranial thorax, and the numerous surrounding nerves and vascular structures. Although CT scans can be helpful in assessing resectability, they are not always reliable for thymomas because it can be difficult to differentiate between invasive and non-invasive masses.

Non-invasive thymomas do not adhere to adjacent structures and are resected using a combination of blunt and sharp dissection. Electrocautery can be especially helpful in minimizing hemorrhage and providing precise dissection. The brachiocephalic trunk, cranial vena cava, and phrenic nerves are often located along the craniodorsal aspect of the cranial mediastinal mass and should be preserved. The internal thoracic artery and/or costocervical trunk may occasionally need to be ligated to complete tumor removal, but this does not have any significant impact on postoperative recovery.

Surgical resection of invasive thymomas is more difficult and may not be possible, depending on which structures are invaded by the tumor. In addition to resection of the tumor, pericardectomy, lung lobectomy, and chest wall resection may be required. Involvement of the cranial vena cava can be managed with temporary vascular occlusion, partial or segmental caval resection, and either primary caval reconstruction or partial or *en bloc* vascular grafts. For thymomas invading the cranial vena cava, venous grafts are commonly used in humans and have been reported in a dog..

Radiation therapy is indicated when the resection of an invasive thymoma is either not possible or would result in unacceptable postoperative morbidity. Because radiation therapy is an effective treatment for thymomas, the benefits and risks of resection of an invasive thymoma should be carefully discussed with owners prior to surgery.

Radiation Therapy

Thymomas are radiation-sensitive tumors. Radiation therapy is recommended for the treatment of animals that are either poor surgical candidates, have invasive thymomas, or as a method of downstaging disease prior to attempted curative-intent resection. Round cells, such as lymphocytes, are very sensitive to the effects of radiation therapy and the lymphoid component of the thymoma may determine the completeness of response. A response rate of up to 75% is observed in cats, with 20% having a complete response following radiation therapy. Pneumonitis and pericarditis are reported acute side effects.

Chemotherapy

The role of chemotherapy in this disease is undetermined, CHOP-based protocols have been used anecdotally with mixed results. Platinum agents are commonly used for treatment of human thymoma, however, data is lacking in cats treated with carboplatin. This author has noted mixed results for thymomas treated with chemotherapy.

Prognosis

The prognosis is good to excellent in cats following either surgical resection or radiation therapy. In 10 cats surviving surgery, the median survival time was greater than 21 months with no cat having either local tumor recurrence or metastasis. Myasthenia gravis has been reported postoperatively in two cats. The median survival time following a coarse fractionated radiation protocol, which is usually considered a palliative radiation protocol, was 720 days.

REFERENCES

1. Souza CH, In: Withrow and MacEwen's Small Animal Clinical Oncology. 5th ed. 2013:688.
2. Singh A. Can Vet J 2010:757-60.
3. Grimes CN, Fry MM, LeBlanc CJ, Hecht S. The lung and intrathoracic structures. In Cowell and Tyler's Diagnostic Cytology and Hematology of the Dog and Cat, 4th Ed. Valenciano AC and Cowell RL, eds. Elsevier Mosby, St. Louis, MO, 2014.
4. Lana S, Plaza S, Hampe K, et al. J Vet Intern Med 20: 1161-1165, 2006.
5. Zitz JC, et al. JAVMA 2008; 1186-92.
6. Smith, A, et al. J Am Anim Hosp Assoc. 2001;37:489-96.

MULTIDISCIPLINARY APPROACH TO ONCOLOGIC DISEASES. PERSPECTIVES FROM AN ONCOLOGIST, A SURGEON AND A PATHOLOGIST. INTRAABDOMINAL TUMOR: FELINE GASTROINTESTINAL LYMPHOMA

CA Clifford[1]; CJ LeBlanc[2], Brian Bretz[3]
[1]Malvern, PA; [2]Knoxville, TN; [3]Langhorne, PA

Introduction

The goal of this lecture is to provide a multidisciplinary approach to the diagnosis, staging, treatment and prognosis associated with feline gastrointestinal neoplasia, specifically lymphoma, in a case based manner. This overview will specifically highlight the input of a medical oncologist, clinical pathologist and surgeon.

Lymphoma (LSA) is the not only the most common neoplasia in cats (~25%) but also the most common hematopoietic malignancy in this species, with the gastrointestinal form representing nearly 80% of cases. Since lymphoma is often considered a "liquid cancer", combination chemotherapy is standard of care and can lead to complete remission rates of approximately 30-65%, with a median duration of remission of 3-7 months.

The paradigm of LSA has changed, as this is no longer a virally induced cancer in the majority of patients. A percentage of FeLV positive cats will develop lymphoma during their lifetime, typically at a younger age, the anatomic location of the mediastinum and often associated with a poorer prognosis. With the advent of more routine vaccinations, FeLV induced LSA has diminished remarkably. More recent factors that may play a role in the development of LSA include breed, environmental exposures (tobacco smoke), altered immune system function including that associated with organ transplantation and potentially inflammatory bowel disease.

Clinical Presentation and Staging

Lymphoma in cats is most often described according to its anatomic location. The gastrointestinal/abdominal form is the most common in cats and makes up 80% of cases. This form includes alimentary, hepatosplenic, renal, and/or abdominal lymph nodes. Generalized peripheral lymphadenopathy, which is the most common form in dogs, is rare in cats. The mediastinal and extranodal forms (nasal, cutaneous, CNS), make up the remaining 20% of cases.

The clinical history is dictated by the form of LSA affecting the patient. Common clinical signs of the abdominal/gastrointestinal form include: weight loss, anorexia, lethargy, vomiting, and/or diarrhea. Multicentric cases typically present clinically with peripheral lymphadenopathy.

Recommended staging consists of a physical examination, complete blood count, ancillary bloodwork (cobalamin if gastrointestinal) serum biochemical profile, urinalysis, three view thoracic radiographs, and abdominal ultrasound.

Cytology can often provide an expedited diagnosis. Feline GI lymphoma has three main cytologic presentations: large cell lymphoblastic lymphoma, large granular lymphoma, and small cell lymphocytic lymphoma. Large cell lymphoblastic lymphoma is typical of most lymphomas in dogs and cats; the predominant cell type is a large immature lymphocyte (larger than a neutrophil) with a scant amount of lightly to moderately basophilic cytoplasm and a large round to indented nucleus with coarsely clumped chromatin and prominent to indistinct nucleoli. The unique feature about these immature lymphocytes in cats is the presence of several small, clear, punctate vacuoles within the cytoplasm (this feature is not common in canine lymphoma).

Large granular lymphoma cells can appear similar to those described above, with one exception - instead of vacuoles, these intermediate to large lymphocytes contain few to many small fine to large chunky magenta granules. These cells are of cytotoxic T cell (CD8+) or NK cell origin. Because they are predominated by large immature lymphocytes, these first two types of feline lymphoma can typically be definitively diagnosed cytologically. In contrast, small cell lymphocytic lymphoma typically poses a diagnostic dilemma. The cellularity will be predominated by a uniform population of small lymphocytes (smaller than a neutrophil) with clumped chromatin and indistinct nucleoli. A very homogeneous mature lymphocyte population lacking plasma cells and intermediate to large lymphocytes may result in a 'probable lymphoma' diagnosis via cytology. However, to definitively differentiate small cell lymphoma from a lymphocytic inflammation (IBD) or aspiration of a Peyer's patch, histopathology (ideally full thickness biopsies) +/- immunohistochemistry and/or PCR for antigen receptor rearrangement (PARR) are usually necessary. Typical immunohistopathologic characteristics of neoplastic lymphocytes include epitheliotropism and CD3 expression. Although histopathology is the gold standard for diagnosing small cell lymphocytic lymphoma, aspiration of enlarged intra-abnormal organs that typically lack lymphocytic cell populations (ie. kidneys, liver) may help support a cytologic diagnosis of lymphoma in cases where surgery is an unfavorable option.

Treatment

Surgical treatment for intestinal LSA is mostly commonly performed to obtain full-thickness biopsy samples or to resect obstructive masses. In some cases, endoscopic biopsies may provide a diagnosis of lymphoma but in many cases, full-thickness biopsies are required for accurate diagnosis.

If surgery is performed for the purpose of obtaining biopsies, multiple GI tract locations should be sampled. At a minimum, samples should be obtained from the stomach, duodenum, jejunum, and ileum. Liver and mesenteric lymph node samples should also be obtained. These locations should be sampled regardless of whether or not there is gross evidence of neoplasia. The stomach biopsy is obtained via a small gastrotomy incision. Intestinal biopsies are performed in the same fashion as any other full-thickness biopsies. The segment of jejunum should be isolated with non-traumatic forceps and a full thickness sample taken with a no. 15 scalpel blade. Alternatively, a small punch biopsy (~4mm) can be used which results in a more symmetrical biopsy site that is easier to close. The site is closed with absorbable simple interrupted sutures.

Liver biopsy has a low risk of morbidity and can be performed with either vascular clips or a "guillotine" suture around the end of a liver lobe. The mesenteric lymph nodes are sampled via an incisional biopsy over the largest part of the node. Prior to the incision, care should be taken to identify any mesenteric arcade vessels in the region as the mesenteric lymph nodes lie on top of the vessels at the root of the mesentery. Damage to these vessels will result in intestinal necrosis and perforation postoperatively. The lymph node biopsy site is closed with small (usually 4-0) absorbable interrupted sutures.

Surgery for obstructive masses consists of resection of the primary site with wide margins of intestinal tissue. Closure of the site is performed with simple interrupted sutures while making sure to achieve good mucosal and serosal apposition. The site should be leak-checked with a 25 gauge needle and several milliliters of saline prior to abdominal closure. Omental or serosal patching can be performed if there is a concern about tissue at the anastomosis site. While this will not overcome poor tissue handling or a devitalized anastomosis site, it can increase the blood supply to region and support the healing process.

Risk of intestinal dehiscence is elevated in the face of intestinal lymphoma because of the generalized nature of the disease throughout the intestinal tract and because many of these patients are hypoproteinemic. However, it is difficult to quantify this risk and is generally not a reason to avoid resection and anastomosis. Recent advances in laparoscopic-assisted surgery and wound retraction now allow for removal of full-thickness intestinal biopsies and resection of focal intestinal masses without the need for a full abdominal incision. This is especially important in debilitated patients and can prevent chemotherapy delays due to poor healing.

Chemotherapy is considered standard therapy for the treatment of feline lymphoma. Single agent chemotherapy with agents such as prednisone, doxorubicin, and mitoxantrone reports response rates ranging from 9-40% and associated remission durations of 3-12 months. Multiple studies have documented improved remission rates, duration of remission, and overall survival with combination chemotherapy. The most commonly used protocols are COP or CHOP. This author always uses a rotating protocol consisting of prednisone, L-asparaginase, vincristine, cyclophosphamide, and doxorubicin, administered over a 25 week time frame. Although the literature suggests complete response rates of 30-65% and associated remission rates of 5-9 months, it is the author's impression that 50% of patients will respond to treatment and the average survival for responder is approximately 1 year. Furthermore, there seems to be a small subset of patients who enjoy long term survival (> 2 years). Non-responders, however, typically succumb to their disease in 4-8 weeks.

Radiation therapy (RT) may be considered for intestinal LSA either during initial chemotherapy protocol or possibly in the rescue setting. The standard protocol, however, has yet to be determined and prospective studies are lacking.

Small cell lymphoma (lymphocytic/indolent LSA) is considered a low grade or less aggressive variant of lymphoma. Although unproven, there is certainly evidence that suggests a possible link to inflammatory bowel disease and the potential that the two syndromes represent a continuum of the same disease process. In general, clinical signs occur over a prolonged time frame (several months to even years) and typically include weight, diarrhea, intermittent vomiting and loss of appetite. Unlike the aggressive high-grade form which is composed of large "blastic" lymphocytes, this form is composed of small lymphocytes that invade the intestines, lymph nodes, stomach and liver. Often, bloodwork is normal and ultrasound findings may yield thickened intestines or a mass effect. As discussed, aspirates can be unrewarding in providing a definitive diagnosis as lymph nodes normally contain mostly small lymphocytes, thus histopathology is needed to definitively diagnose the disease. In general, this requires either an endoscopy, or more preferred by this author, an exploratory surgery, whereby, multiple organs can be biopsied (full thickness) in a effort to determine the true extent of the disease.

Treatment for small cell lymphoma, unlike lymphoblastic lymphoma, does not involve multidrug protocols, rather, daily prednisolone and every other day chlorambucil (Leukeran) is the gold standard. Clinical signs (resolution of gastrointestinal signs and weight gain), and abdominal ultrasound are used to monitor efficacy. If a relapse occurs, rescue therapy with cyclophosphamide can be considered, as this protocol has been associated with a high response rate. In general, the median survival time is nearly 2 years for low grade LSA.

Large granular lymphoma represents an uncommon but very aggressive form of feline lymphoma, with a high percentage of patients presenting clinically ill with very advanced disease at diagnosis. This form appears to have low overall response rate and is considered uniformly fatal.

REFERENCES

1. Vail DM, Pinkerton ME, and Young KE. Withrow and MacEwen's Small Animal Clinical Oncology. 5th ed. 2013:608.
2. Haddad JL, Marks Stowe DA, Nee JA. The Gastrointestinal Tract. In Cowell and Tyler's Diagnostic Cytology and Hematology of the Dog and Cat, 4th Ed. Valenciano AC and Cowell RL, eds. Elsevier Mosby, St. Louis, MO, 2014.
3. Andreasen CB, Jergens AE, Meyer DJ. Oral cavity, gastrointestinal tract, and associated structures. In Canine and Feline Cytology, A Color Atlas and Interpretation Guide, 2nd Ed. Raskin RE and Meyer DJ, eds. Saunders Elsevier, St. Louis, MO, 2010.
4. Kiupel M, et al. Vet Pathol 48;2011:212-22
5. Kiselow MA, et al. J Am Vet Med Assoc 232; 2008: 405-10
6. Krick EL, et al. Vet Comp Oncol 6;2008:102-10

MULTIDISCIPLINARY APPROACH TO ONCOLOGIC DISEASES. PERSPECTIVES FROM AN ONCOLOGIST, A SURGEON AND A PATHOLOGIST. PERIRECTAL TUMOR: ANAL SAC, ANAL GLAND ADENOCARCINOMA (AGASACA)

CA Clifford[1]; CJ LeBlanc[2], Brian Bretz[3]
[1] Malvern, PA; [2] Knoxville, TN; [3] Langhorne, PA

Introduction

The goal of this lecture is to provide a multidisciplinary approach to the diagnosis, staging, treatment and prognosis associated with AGASACA in a case based manner. This overview will specifically highlight the input of a medical oncologist, clinical pathologist and surgeon.

Tumors of the anal sac are uncommon a represent a small percentage of all skin tumors in dogs (<2%). The most common malignant tumor of the perianal region is the anal sac (gland) carcinoma accounting for 15-20% of all perianal tumors. These tumors are locally invasive and metastasize early in the course of disease. The average age of dogs diagnosed with this disease is 10-11 years and there does not appear to be a breed or sex predilection. To date, no consistent etiology has been noted.

In many cases these tumors are noted as an incidental finding on a routine rectal examination in approximately 50% of dogs diagnosed with AGASACA; as such a rectal examination should be part of all routine examinations. In dogs presenting with clinical signs: Perianal swelling, straining to defecate, licking at the perianal region, and bleeding as a result of a large anal sac tumor or severely enlarged regional lymph nodes are commonly noted. It is important to measure the size of the tumor (to the best of one's ability) as this will play a role in prognosis. In another subset of patients, polyuria and polydipsia may be the only presenting complaint as a result of hypercalcemia. Hypercalcemia is the most common paraneoplastic syndrome seen in dogs with AGASACA and occurs in nearly 27% of cases.

Diagnostics

A diagnostic workup for this cancer includes:

- Labwork: CBC, serum chemistry, and urinalysis
- Chest radiographs: 3 view thorax: pulmonary metastasis is less common but should be performed all cases
- Abdominal ultrasound: regional lymph nodes are most commonly affected and ultrasound is clearly superior to plain radiographs.
- Computed tomography (CT) of the thoracic and abdominal cavities: CT is a superior imaging modality and thusoffers an advantage over radiographs and ultrasound; cost and availability are a factor.
- Aspirates/Biopsy of the tumor or regional lymph nodes: Aspirates of anal sac adenocarcinomas typically result in slides that consist of a moderate to large amount of blood, many free, round nuclei, and many small to large loosely cohesive clusters that commonly exhibit acinar structures. The epithelial cells typically have a small to moderate amount of lightly basophilic cytoplasm with indistinct borders, resembling endocrine or neuroendocrine tumors. Despite their malignant behavior, these cells exhibit minimal criteria of malignancy. The nuclei are round to oval in shape with finely-stippled chromatin and small indistinct nucleoli. Anisocytosis and anisokaryosis are typically mild, but can be moderate. Sublumbar/intrapelvic lymph node metastasis is commonly associated with this tumor. An important differential diagnosis for a mass in this region of the body is a perianal gland or hepatoid tumor. Aspiration of these tumors typically results in numerous well defined clusters of uniform polygonal cells with abundant moderately basophilic to pink, granular cytoplasm. The nuclei of these 'hepatoid' cells are round, centrally to eccentrically located, and contain one to two distinct nucleoli.

Treatment

The first stage of treatment for anal gland adenocarcinoma is surgical resection followed by chemotherapy and, in some cases, radiation therapy. The main goal of surgery is to remove the perianal mass because if it is not resected, the primary mass will affect the quality of life long before lymphadenopathy or distant metastasis. As a result, surgery is recommended in all cases where the mass is potentially resectable.

Surgery to the gland is approached with an incision directly over the mass and careful dissection is used to maintain the integrity of the tumor capsule. Maintenance of the capsule is extremely important because this is one of the locations in the body where wide resection is never possible due to the proximity of surrounding structures including the rectal nerve, anus, rectum, sciatic nerve and local blood vessels. For larger masses, a combination of blunt, sharp and digital dissection is often required and electrocautery is useful because these masses are highly vascular. All of these patients should be placed on broad spectrum antibiotics

Lymph node biopsy is useful in these cases to further stage the tumor but for most practitioners it is difficult to access the affected nodes without a large abdominal incision. The increasing availability of laparoscopy does allow some practices to obtain these samples in a minimally invasive manner with low morbidity but the need for biopsy should be assessed on a case by case basis with the consulting oncologist.

In the case of gross lymphadenopathy on ultrasound exam, extirpation of the nodes is strongly recommended. The medial iliac and hypogastric lymph nodes are located in the region of the distal vena cava and bifurcation of the aorta. With mild amounts of lymphadenopathy, removal is straightforward but as the degree of lymphadenopathy increases, surrounding structures including the distal aorta, ureters, and iliac vessels can all become involved. The keys to increasing the chances for successful resection in this location are adequate exposure and a thorough knowledge of the surrounding structures. The standard celiotomy incision will need to be extended caudally to the pelvic brim to allow for appropriate exposure to the region. In addition, before beginning the extirpation, the surgeon should clearly identify important surrounding vital structures. In some cases, surgical exploratoration may determine that the nodes are not resectable without causing undue risk to the patient and in these cases, the nodes should always be biopsied prior to abdominal closure.

In cases where surgery is not complete, then radiation therapy to the primary tumor site and regional lymph nodes will improve local control. Radiation therapy protocols range from 16-20 fractions on a Monday through Friday schedule. Acute side effects of radiation therapy can be moderate to severe and may result in colitis and rectal mucositis. During this time period and e-collar must be worn to prevent licking of the area. The benefits of chemotherapy for this cancer is not well characterized, however, chemotherapy is often incorporated into treatment protocols. The most commonly used chemotherapy agents include; doxorubicin, carboplatin, mitoxantrone, melphalan and gemcitabine. NSAIDs are often utilizedfor putative anticancer effects. In some cases all three modalities are utilized (surgery, chemotherapy, radiation therapy) in the treatment of dogs with anal sac carcinoma. One study found dogs treated with surgery, radiation therapy (15 treatments) and chemotherapy (mitoxantrone) resulted in an overall survival of >900 days. In nonresectable tumors, hypofractionated radiation therapy has anecdotally been useful in palliation with a relatively high rate of response noted. Side effects such as rectal stricture may occur with this type of protocol.

Novel therapy

Toceranib phosphate (Palladia) is an orally bioavailable multi-targeted receptor tyrosine kinase (RTK) inhibitor active against several members of the split-kinase RTK family (VEGFR, PDGFR and Kit). Although approved for use in canine mast cell tumors, due its mechanism of action in targeting new blood vessel growth, activity in a broad range of tumors might be expected. Following approval in 2009, toceranib was used off label to treat a variety of canine cancers, most often in patients having failed multiple treatment modalities. A recent retrospective study of 32 dogs with ASAGACA treated with toceranib, 8 dogs (25%) experienced partial response (PR) and 20 (62.5%) experienced stable disease (SD) for a clinical benefit (CB) rate of 87.5%. The median duration of treatment for all 32 dogs treated with toceranib was 25 weeks (range, 0–>47 weeks).

Prognosis

Prognostic factors established in dogs with AGASACA include:

Hypercalcemia (Poor)
Pulmonary metastasis (Poor)
Surgery as part of therapy (improved prognosis) vs. no surgery
Tumor >10cm^2 (Poor)

REFERENCES

1. Turek, M and Withrow SJ. In: Withrow and MacEwen's Small Animal Clinical Oncology. 5th ed. 2013;503.
2. Fisher DJ. Cutaneous and subcutaneous lesions. In Cowell and Tyler's Diagnostic Cytology and Hematology of the Dog and Cat, 4th Ed. Valenciano AC and Cowell RL, eds. Elsevier Mosby, St. Louis, MO, 2014.
3. Williams LE, JM Gliatto, RK Dodge, et al. Canine carcinoma of the apocrine glands of the anal sac: 113 cases (1985-1995). JAVMA 2003;223:825-831.
4. London C; Mathie T; Stingle N, et al. Preliminary evidence for biologic activity of toceranib phosphate (Palladia(®)) in solid tumours. Vet Comp Oncol. 2012;10:194-205.

DIAGNOSIS AND MANAGEMENT OF FELINE URETERAL OBSTRUCTIONS: PAST, PRESENT AND FUTURE

Dr. Allyson Berent, DVM, DACVIM (SAIM)
New York, NY

Introduction

Ureteral disease can create a significant dilemma in our feline patients. The relatively common incidence of ureteral disease, combined with the invasiveness and morbidity associated with traditional surgical techniques, makes the investigation of minimally invasive alternatives appealing. Interventional radiologic (IR) and interventional endoscopic (IE) techniques, like ureteropyelography, percutaneous nephrostomy tube placement, ureteral stenting and the use of a subcutaneous ureteral bypass (SUB) device, have aided tremendously in the ability to simultaneously diagnosis and treat ureteral obstructions in a minimally invasive manner with long-term ureteral protection from future obstruction, while concurrently lowering the morbidity and mortality when compared to traditional surgerical techiques (ureterotomy, neoureterocystostomy, ureteronephrectomy or renal transplantation). Endourologic procedures are considered standard of care in human adult and pediatric surgery and are currently being applied in veterinary patients regularly. This talk will focus on the diagnosis, management and treatment of cats with ureteral obstructions from various causes and will review data from what has been reported in the past, the more recent internventional modalities, and the future options on the horizon. Data from over 250 consecutive cases treated interventionally will be reported and discussed.

Human Ureteral Intervention

In human urology, the development and improvements in ureteroscopy, ureteral stenting, extracorporeal shockwave lithotripsy (ESWL), laser lithotripsy, laparoscopy, and percutaneous nephroureterolithotomy (PCNUL) has nearly eradicated the need for open ureteral surgery for stone disease, strictures, trauma, and neoplasia. In humans, ureteroliths < 5 mm have a 98% chance of spontaneous passage with medical management alone. For larger stones, or those that do not pass spontaneously, ESWL is effective in 50-81% of cases and ureteroscopy has a near 100% success rate when Holmium:YAG laser lithotripsy is used. PCNUL has also been successful for large proximal impacted ureteral stones. Ureteral stenting was first introduced in 1967 for evaluation of human patients with malignant ureteral obstructions. They are still widely used to treat both benign and malignant obstructive disease, but are more commonly considered a temporary fixation, rather than a permanent alternative for benign disease. Ureteral stenting for stone disease is typically done after ureteroscopy for post-scoping spasm and edema, and in children a stent is routinely placed prior to ureteroscopy to allow for passive ureteral dilation in anticipation of immediate ureteral decompression, spontaneous stone passage, or the ease of extracorporeal shockwave lithotripsy. When stents are used in humans they typically are either removed within 2 weeks after a permanent fixation procedure has been employed (i.e ureteroscopy and laser lithotripsy or ESWL) or left in place long-term with scheduled stent exchanges every 3-6 months to prevent stent encrustation and ureteral reocclusion.

The use of a subcutaneous nephrovesicular ureteral bypass device has been reported in approximately 50 human patients. The first reports were focused on the use of this device for malignant neoplasia where reconstructive surgery or ureteral stenting had failed or was not possible. More recently, this device is reported for use after renal transplantation induced ureteral strictures and in cases where stenting or surgery is not a feasible option due to patient stability, abnormal anatomy, or irreparable damage. So far, the literature would suggest that the use of this device has been met with few complications and long-term patency rates are considered excellent in the human literature.

Feline Ureteral Obstructions

Feline ureteral obstructions occur most commonly secondary to ureterolithiasis (~80%), ureteral stenosis (~20%), and trigonal neoplasia (<3%). Additional causes can be from accidental ureteral ligation during surgery, ureteral trauma, ureteral edema following ureteral surgery, and secondary to purulent material associated with pyelonephritis/pyonephrosis. Obstruction can result in life-threatening azotemia, particularly when presenting bilaterally (~15%) or in cats with concurrent pre-existing renal insufficiency/failure (>75-90%). Greater than 92% of feline ureteroliths have been documented to be composed of calcium oxalate, which is a trend from decades prior. In a recent study the presence of dried solidified blood stones was seen in a small handful of cats as well. This means that these stones will not dissolve medically, and either need to pass spontaneously, remain in place, or be removed.

Once medical management fails to encourage stone passage (traditionally: intravenous fluid therapy, mannitol CRI and alpha-adrenergic blockade +/- amitrypilline or glucagon therapy for 24-48 hours) the immediate resolution of the obstruction should be considered. The physiologic response the body has to a ureteral obstruction is very complex. If there is a complete ureteral obstruction, decompression of the renal pelvis becomes imperative to preserve renal function to the ipsilateral kidney. Studies in dogs have shown that after a complete ureteral obstruction the renal blood flow diminishes to 40% of normal over 24 hours, and to 20% of normal by 2 weeks. This increase in pressure generated by the ureteral obstruction is transmitted to the entire nephron resulting in a precipitous decrease in the glomerular filtration rate (GFR). After 7, 14, and 40 days the GFR was found to permanently decline by 35%, 54% and 100%, respectively. These numbers were in a dog model with no evidence of pre-existing azotemia, chronic interstitial nephritis, or chronic obstruction/nephroureterolithiasis, so extrapolation of a worse outcome might be expected in our feline patients. Interestingly, in contrast to the irreversibility of a complete obstruction, partial obstructions result in less severe/dramatic destruction and more return to function after the obstruction is relieved. In one dog model it was found that the GFR returned to normal after a partial obstruction was present for 4 weeks. Knowing that many of our feline patients are partially obstructed for weeks to months with concurrent renal compromise (CKD), aggressive management and obstruction relief is recommended. Discouraging fixation due to chronicity is not recommended, as improvement in renal function is routinely documented after fixation, regardless of chronicity.

In cases of ureteral obstruction secondary to a ureterolith, some patients can be managed medically with supportive care until a ureterolith passes. It is important to realize that over 20% of feline ureteral obstructions are reported to be associated with strictures (+/-stones), where medical management and traditional surgery are not expected to be highly effective. Medical management has been than emergency intervention should be considered to avoid permanent damage. For cases that are oliguric, anuric, hyperkalemic, or overhydrated, than immediate intervention or intermittent hemodialysis therapy should be considered for immediate stabilization. The main interventional options considered in the author's practice include: nephrostomy tube placement, ureteral stenting, and the placement of a subcutaneous ureteral bypass (SUB) device. Decompression of the renal pelvis will decrease the hydrostatic pressure, and stop the permanent damage that is occurring to the renal parenchyma, allowing for immediate renal recovery.

Traditional Surgery: The PAST

In the **past**, traditional intervention for ***ureterolithiasis*** has been accomplished surgically via ureterotomy, neoureterocystostomy, ureteronephrectomy, and renal transplantation. Kyles et al. in 2005 reported two retrospective studies in a large number of cats. There were high procedure associated complications (over 30%) and perioperative mortality rates (21%) documented. This study was from two universities that have extensive expertise with ureteral surgery. The morbidity and mortality may be higher in environments where operating microscopes and microsurgical experience is not as readily available. Many of the associated complications with surgery are due to site edema, recurrence of obstruction from stones that pass from the renal pelvis to the surgery site, stricture formation, and ureterotomy-associated or nephrostomy tube-associated urine leakage. In one study, over 10% of cats that survived their complications required a second surgical procedure during the same visit and 30% of those cats were then subsequently euthanized or died from serial complications associated with their ureteral stones/surgeries. Of the cats that had long term imaging follow-up, 40% had evidence of ureteral stone recurrence, 85% percent of which had evidence of ipsilateral nephrolithiasis at the time of the first ureteral surgery. The number of animals that did not have stone recurrence with prior nephrolithiasis was not evident in that study. Chronic kidney disease was found to be common at the time of diagnosis (>75% were azotemic with a unilateral obstruction), and persistent azotemia was commonly seen after a successful surgery (over 50-80% of cats). Though, with all of the surgical concerns, the survival rates were dramatically higher for cats that had surgical intervention performed when compared to those treated with medical management alone (33% mortality prior to discharge; 87% failure to renal function improvement). Medical management has been shown to be effective in a minority of cases (7.7-13%), but should be considered prior to any intervention, particularly for small stones (1-2 mm) in the distal ureter. This large series was in cats in which surgery was performed or attempted, hence excluding patients that were not considered good surgical candidates. Over the past 8 years the number of stones found in the ureter and kidney have been much greater (median of 4 stones per ureter in our recent study of 79 ureters, Berent et al, JAMVA 2014), with over 60% of patients in that study not being considered good traditional surgical candidates due to the number of stones, location of stones, presence of a proximal stricture, etc. In this same study over 86% of cats had evidence of ipsilateral nephrolithiasis, making the risk of ureteral re-obstruction higher than that of previous studies.

Feline ***ureteral strictures*** were reported in 10 cats in 2011 by Zaid et al, and have since been reported in a larger series of 79 feline ureteral obstruction (Berent, JAVMA 2014) and in a series of 22 cats with circumcaval ureters (Steinhaus, abstract, JVIM 2013). Ureteral strictures can occur for various reasons, and are most commonly seen: secondary to a previous surgery of the ureter for stone disease, from a stone getting embedded in the ureteral mucosa, as a congenital abnormality, associated with idiopathic renal hematuria, as an idiopathic process, and possibly associated with a circumcaval (retrocaval) ureter. Most ureteral strictures were found to occur in the very proximal ureter (<2.5 cm from the renal pelvis), in which traditional surgical approaches were difficult requiring re-implantation with renal desensus and psoas cystopexy or a ureteral resection and anastomosis. With a cat ureter being 0.4 mm in luminal diameter, this can be a very challenging situation, particularly in a potentially unstable patient.

Finding an alternative that results in immediate decompression and stabilization of the kidney, while concurrently allowing patency to be established quickly and effectively, for any cause of a ureteral obstruction (stone, numerous stones, stricture, debris, blood clots, purulent debris, blood stones, etc), which can decrease post-operative complications, mortality and long-term re-obstruction is ideal. The use of interventional radiologic techniques, either minimally invasively, or with surgical assistance, has aided in circumventing the complications of ureteral surgery (peri-operative leakage, stricture, re-obstruction, long anesthesia times, worsening renal function, etc), particularly in cases that are not considered good surgical candidates. IR techniques have allowed successful and efficient stabilization of the patient regardless of the cause of ureteral obstruction, while decreasing renal pelvic pressure and stopping the cycle of pressure induced nephron damage and loss.

Interventional Options: The PRESENT

Percutaneous Nephrostomy Tube Placement

Ureteral obstructions secondary to ureteroliths or malignancy can result in severe hydronephrosis and/or life-threatening azotemia when presenting bilaterally, or in animals with concurrent renal insufficiency/failure. Fixation procedures can be relatively prolonged and complicated in these debilitated patients, especially in facilities where expertise in treating feline ureteral obstruction is not as vast. One possibility is to place a nephrostomy tube to quickly relieve the obstruction and determine whether adequate renal function remains before prolonged anesthesia for ureteral surgery, ureteral stent or SUB device placement can be performed. A locking loop pigtail catheter is recommended (5 or 6 french) to decrease the risk of inadvertent tube removal or urine leakage. Historically, nephrostomy tubes have been met with much resistance due to the high risk of post-placement complications (over 50%). These complications were usually due to premature removal or dislodgement, urine leakage, or poor drainage. With the advent of sturdy, multi-fenestrated tubes, that form a loop that will lock the catheter in the renal pelvis, these complications seem to have declined dramatically. We recently reported on the use of the locking-loop pigtail catheter in 20 patients. One patient developed leakage into the subcutaneous space and the remaining catheters worked without complication. (Berent et al, JAVMA, 2012). These catheters were deemed to be safe, effective and well tolerated in both dogs and cats. This is typically done percutaneously in dogs using ultrasound and fluoroscopic guidance, and surgically assisted with fluoroscopy in cats.

The Subcutaneous Ureteral Bypass Device (SUB)

The SUB device (Norfolk Vet Products) was initially created for feline patients as an alternative to ureteral stents when either a stricture was present (higher rate of stent occlusion [>50%]), or a stent was unable to be successfully placed due to excessive stones, very narrow lumen, or patient stability. Over the course of the following 3 years, the long-term outcomes documented with the SUB device, as far as long-term complications and patient comfort (primarily stent occlusion and recurrent dysuria), were found to be superior to that of stents. Due to these findings, the trend has moved toward using the SUB device as a primary treatment option of feline ureteral obstructions in the author's practice. It is very important to understand the risks and benefits of traditional surgery, ureteral stenting and SUB device placement device prior to deciding the best treatment option for your patient.

The placement of nephrostomy catheters in veterinary medicine was described above, and has been met with excellent success when the appropriate device is used. The biggest limitation is the externalized drainage, requiring careful management and hospitalization to prevent infection and dislodgement. The development of an indwelling SUB device using a combination locking-loop nephrostomy catheter and cystostomy catheter connected to a subcutaneous shunting port has been highly successful for the treatment of all cause feline ureteral obstructions (stones, stricture, tumors and obstructive pyelonephritis). In humans, a similar device has been shown to reduce complications associated with externalized nephrostomy tubes and ureteral stents and improve quality of life.

The post-operative management of the SUB device is similar to that described below for all cases of ureteral obstruction. Care should be taken to avoid fluid overload as a post-obstructive diuresis is common. Abdominal palpation is discouraged for the first 2 weeks to prevent manipulation of the device. Following discharge, patients are monitored similarly to all feline ureteral obstructions including serial ultrasounds and urine cultures. Urine is generally sampled from the SUB device access port using a special non-coring Huber needle, during which time the device can be flushed using sterile saline under ultrasound guidance. This is recommended every 3-6 months to help ensure and maintain patency. A cystocentesis should be discouraged in these patients.

The use of the feline SUB device has recently been described in various abstract and peer-reviewed formats. To date, in the authors practice, over 125 of these devices have been placed over a 5 year period with good short and long-term success. The median pre-operative creatinine was 6.8 mg/dL and 2.4 mg/dL post-SUB placement. The peri-operative mortality after surgery was 5.6%, with 1 death associated with a procedure related complication early in the course of the study. This compares well to traditional surgery (21% peri-operative mortality) and ureteral stenting (7.5%). To date the median follow-up time is over 19 months. Peri-operative complications are typically rare and include leakage (at the port or nephrostomy tube insertion site) (<5%), kinking of the catheters (or failure of the renal function to recover (<3%), and blockage of the system with either blood clots, debris, purulent material, or future stones (~8%; 0% for strictured ureters and ~10% for stones). Dysuria is rarely seen in cats with a SUB device (<2%) compared to the ureteral stents (38%). The SUB device is considered a functional option and is the author's treatment of choice in cats with ureteral obstructions.

Ureteral Stenting

Ureteral stenting has been performed for a variety of disorders in both dogs and cats (over 400 cases to date). The goal of using a feline double pigtail ureteral stent (Infiniti Medical, LLC) is to divert urine from the renal pelvis into the urinary bladder during a ureteral obstruction (ureterolithiasis, obstructive neoplasia, ureteral stricture/stenosis, dried solidified blood clots or severe obstructive pyelonephritis or pyonephrosis). Stents also encourage passive ureteral dilation (for ureteral stenosis/strictures or future ureteroscopy, extracorporeal shockwave lithotripsy [ESWL], etc). The double pigtail stent, which is the stent of choice in cats, is completely intracorporeal, and can remain in place long-term maintaining ureteral patency and passive ureteral dilation (a 0.4 mm ureter was documented to dilate to 1.5-2.0 mm within 3-7 days in over 90% of cats). Each loop of the pigtail is curled (one in the bladder and one in the renal pelvis), allowing for direct urinary diversion from the kidney to the urinary bladder, around the stones, or through the stricture.

In most cats (male and female) the procedure is done using fluoroscopy and surgical assistance using nephrostomy needle access in an antegrade manner, though it can be attempted endoscopically in female cats with a reported 20% success rate (compared to over 95% success with antegrade surgical-assistance). This procedure requires special training and experience using wires, catheters and stents and is not recommended in all patients with a ureteral obstruction. The learning curve is very steep and the procedure should not be performed until the operator is comfortable with the risks that are associated. If the operator is not comfortable placing a locking-loop pigtail catheter for a nephrostomy tube then this procedure should not be attempted because if stent placement is not successful the ureter will experience severe edema and spasm and will be acutely obstructed, requiring another mechanism of drainage.

A recent study evaluating 69 feline patients (79 ureters) with ureteral obstructions treated with a ureteral stent had a 96% placement success, 14% were bilaterally obstructed, 28% were associated with strictures (+/- concurrent ureteroliths), and 85% of cases had concurrent nephroliths. (Berent et al. JAVMA, 2014) The median number of stones in the obstructed ureter was 4. Approximately 1/3 of cats had documentation of a urinary tract infection prior to stent placement, and 30% of cats had at least 1 positive culture in their lifetime post-stent placement. The median pre-operative creatinine was 5.3 mg/dL and 2.1 mg/dL post-stent placement. The peri-operative mortality after surgery was 7.5%, with no death associated with a ureteral obstruction or the procedure. This compares to 21% mortality rate with traditional surgery. The most common long-term (>1 month) complications after ureteral stent placement was dysuria (38%, but only persistent in 2%), chronic hematuria (18%), stent occlusion (19%), stent migration (6%), and ureteritis/mucosal proliferation (4%). The dysuria typically resolved with short-term steroid therapy. A ureteral stent exchange with either a bigger stent or a SUB device was needed in 27% of cats for either occlusion or migration. This complication was most commonly associated with ureteral strictures (over 50% of strictured ureters expereicned reocclusion), or an occlusion at a previous ureterotomy site. Ureters remained patent long-term in most cats with the longest stent in place and functional for over 5 years. The median survival time was 498 days. The median survival time when renal cause of death was over 1250 days, with only 21% of cats dying of suspected CKD. Another study evaluating pre-operative prognostic factors for short- and long-term survivals and renal recovery did not identify any clinical, biochemical or imaging findings to suggest overall survival in patients.

Extracorporeal Shockwave Lithotripsy (ESWL) for ureterolithiasis

ESWL is another minimally invasive alternative for the removal of ureteral calculi. ESWL delivers external shockwaves through a water medium directed under fluoroscopic guidance in 2 planes. The stone is shocked at different energy levels to allow for implosion and powdering. This procedure has been effective in only a small number of feline cases and is ideal for very distal ureteroliths, or small stones <3 mm in diameter. It is important to remember that the feline ureter is normally only 0.4 mm in diameter and the ESWL unit will break stones to approximately 1 mm in diameter. Without ureteral dilation (concurrent ureteral stent) most of these stones will not pass. In addition, feline stones are typically embedded into the ureteral mucosa, so even if the stones fragment well with ESWL, they may not pass as effectively as seen in dogs.

Percutaneous Nephroureteroscopy

Ureteroscopy can be performed in an antegrade manner via renal access for proximal ureteroliths or nephroliths. This technique is more commonly performed in dogs, but can be attempted in cats if the ureter and pelvis is significantly dilated to accommodate a flexible ureteroscope (2.7 mm). With the use and success of ureteral stenting and a SUB device this procedure has rarely been necessary in the author's experience.

Post-operative management

Feline patients can have a substantial post-obstructive diuresis after relieving a ureteral obstruction necessitating a large quantity of fluids to maintain hydration. This should be carefully monitored for and managed. The leading complications post-operatively, regardless of treatment elected, was the development of congestive heart failure associated with fluid overload. Most of these patients had a normal echocardiogram prior to their procedure, but the volume of fluids required to maintain their hydration was overloading to their heart. Because of this finding the author has advocated using a combination of a maintenance rate of enteral hydration (unflavored electrolyte solution-Pedialyte) through the use of an esophagostomy tube (placed at the time of ureteral decompression); a maintenance rate of a low sodium maintenance fluid (0.45% NaCl +2.5% dextrose); and the remainder of fluids necessary as a replacement fluid. This has been found to lower the sodium load and decrease the risk of fluid overload in the author's experience. Once the creatinine level reaches a plateau an the patient can be discharged they are monitored at 1 and 4 weeks post-operatively (with blood work and a urinary tract ultrasound) and then every 3 months thereafter for blood work, focal urinary tract ultrasound, urinalysis, urine culture and a SUB flush (when applicable).

FUTURE

The SUB device has nearly replaced the use of ureteral stents in cats in the author's practice due to the ease of placement, fast anesthesia times (45-60 minutes), lower mortality rates (5.6%), and lower short and long-term complication rates (dysuria <2% versus 38%; re-occlusion <10% versus 27%). The issues still experienced with these devices are the fact that occlusion can still occur and mechanisms to prevent this, like serial flushing, re-designing the catheter size and shape, and using different materials is underway. As more of these devices are being used future changes to perfect the design will occur.

Conclusion

The feline ureter is a frustrating area to gain access to for both diagnosis and treatment of disease. With the recent advances in veterinary interventional endourologic techniques, diagnosis and treatment has become less invasive, more effective, and safer. Proper training and the availability of specialized equipment are needed to help these procedures become more available in the future. With these new devices in veterinary medicine, we are hoping to find better alternatives for these problematic conditions, as we have seen in our human counterparts. These techniques offer an alternative option to traditional surgery, and as the data is published the veterinary community can decide if, and when, using interventional techniques will replaced traditional surgery, as is the case in human medicine.

Conflict of Interest Disclosure

The author is a consultant for Infiniti Medical, LLC and Norfolk Vet Products.

REFERENCES

More available upon request

1. Kyles A, Hardie E, Wooden B, et al. Clinical, clinicopathologic, radiographic, and ultrasonographic abnormalties in cats with ureteral calculi: 163 cases (1984-2002). JAVMA, 226(6): 932-936.
2. Kyles A, Hardie E, Wooden B, et al. Management and outcome of cats with ureteral calculi: 153 cases (1984-2002). JAVMA, 226(6): 937-944.
3. Roberts S, Aronson L, Brown D. Postoperative mortality in cats after ureterolithotomy. Vet Surg; 2011; 40: 438-443.
4. Berent A. Weisse C, Bagley D. et al. Ureteral stenting for benign feline ureteral obstructions in 79 ureters. J Am Vet Med Assoc. 2013, in press.
5. Horowitz C. Berent A, Weisse C. et al. Predictors of outcome for cats with rueteral obstructions after interventional management using ureteral stents or a subcutaneous ureteral bypass device. J Feline Med Surg. 2013. May Epub.

6. Steinhaus J, Berent A, Weisse C, et al. Circumcaval ureters in cats with and without ureteral obstructions. A comparative study. Abstract, American College of Veterinary Internal Medicine, Seattle WA ACVIM 2013.
7. Berent A, Weisse C, Bagley D, et al. The use of a subcutaneous ureteral bypass device for the treatment of feline ureteral obstructions. ECVIM, Seville Spain, 2011
8. Berent A. Ureteral obstructions in dogs and cats : a review of traditional and nrew interventinal diagnostic and therapeutic options. J Vet Emerg Crit Care. 2011; 21(2): 86-103.

CASES AND CONTROVERSIES: MANAGEMENT OF URETERAL OBSTRUCTIONS: MEDICAL, SURGICAL, OR INTERVENTIONAL....YOU DECIDE

Dr. Allyson Berent, DVM, DACVIM (SAIM)
New York, NY

INTRODUCTION

An increasing incidence of problematic nephroliths and ureteroliths has been seen in veterinary practice over the past decade. The invasiveness and morbidity associated with traditional surgical techniques like nephrotomy, ureterotomy, ureteral re-implantation and ureteral resection and anastomosis, makes these conditions a management dilemma. Interventional radiologic (IR) and interventional endoscopic (IE) techniques have enabled clinicians to simultaneously diagnose and treat upper tract stone disease in an effective, and minimally invasive manner. The technology requires the use of fluoroscopy, often with flexible and/or rigid endoscopy, lithotripsy, and various wires, catheters and stents. The development and improvements in endourology in human medicine (like ureteroscopy, ureteral stenting, ESWL, laser lithotripsy, laparoscopy, and percutaneous nephroureterolithotomy [PCNUL]) have almost eradicated the need for open surgery for upper urinary tract stone disease, as well as other etiologies (strictures, tumors, congenital anomalies, etc). The same trend is starting to occur in veterinary medicine and this talk will focus on the controversies that exist in veterinary medicine in making these decisions, both diagnostically and therapeutically.

It is important to understand that most of the data pertaining to IR/IE treatment of upper tract stones is solely the experience of the author, and some of this has only been published and/or presented in abstract form. Traditional therapeutic options will be briefly reviewed, and was expanded upon in more detail in the prior lecture. In addition, some of the new alternatives in dealing with these challenging cases will be discussed.

ETIOLOGY

Greater than 90% of feline, and 50% of canine, upper tract stones are reported to have a composition of calcium oxalate, meaning that they will not dissolve medically. They either need to pass spontaneously, be removed, or the urine needs to be diverted when an obstruction is present. Additionally, obstructions can occur for other reasons beyond stone disease, like dried solidified blood clots (cats), strictures (20% or more of cats; rare but possible in dogs), and tumors (<5% of dogs or cats with ureteral obstructions). For each of these conditions different options exist and should be considered.

Regardless of the cause of the obstruction treatment needs to be considered in an expedited manner and should not be delayed. The physiologic response to a ureteral obstruction is documented to be very complex. In dog models it is shown that after a complete ureteral obstruction an immediate increase in renal pelvic pressure occurs and the renal blood flow diminishes by 60% over the first 24 hours, and 80% within 2 weeks. This excessive pressure decreases the glomerular filtration rate (GFR). The contralateral kidney will increase its GFR in response, as long as that kidney is normal and has the potential for hypertrophic compensation. In these normal dog models it was shown that the longer the ureter remains obstructed, the more progressive the damage (after 7 days the GFR permanently diminished by 35%, and after 2 weeks by 54%). Since this research model was performed in normal dogs with a complete ureteral obstruction, without pre-existing azotemia, chronic interstitial nephritis, fibrosis, or chronic obstruction, an extrapolation of a worse outcome might be expected in a clinical patient once the hypertrophy mechanisms are exhausted. Additionally, since over 30% of cats will develop renal azotemia as an adult, leaving less than 25% of renal function, any preventable loss in GFR should be avoided. This encourages aggressive and timely intervention whenever possible. Interestingly, in contrast to the irreversibility of a complete obstruction, partial obstructions have been shown to result in less severe and slower nephron destruction, giving more time for intervention when needed. In a dog model it was found that after decompression the GFR returned to normal after 4 weeks of a partial obstruction. Knowing that many patients are partially obstructed, with concurrent renal compromise, aggressive diagnosis, management, and obstruction relief is recommended whenever possible.

Traditionally, partial obstructions were not treated, but it is clear that partial obstructions result in permanent renal damage and treatment should still be considered. Knowing the cause of the obstruction is very important when deciding on traditional surgical fixation methods (like ureterotomy or ureteral reimplantation), so that no stone is left within the ureter, if a stricture is present than a resection and anastomosis may need to be prepared for, and if it is for a tumor that there is a reasonable amount of normal bladder to allow for a reimplantation of each ureter. For the newer interventional treatment options (double pigtail ureteral stent, a subcutaneous ureteral bypass device, or a nephrostomy tube), these details are less important, as all can safely be used to treat most causes of a ureteral obstruction (stone, stricture, or tumor).

<u>CONTROVERSIES in the DIAGNOSIS of a ureteral obstruction</u>

1) *Presenting signs:* Feline patients with a ureteral obstruction(s) typically present with vague signs associated with vomiting, lethargy, weight loss and inappetance.[4] Unless there are concurrent bladder or urethral stones, signs of dysuria are not commonly observed. It is important to realize that cats with a unilateral ureteral obstruction may not show any signs at all and will continue to urinate normally, as they have one non-obstructed ureter and a normal urethra, is monitoring progress based on clinical signs is difficult. Concurrent urinary tract infections are documented in approximately 33% of cats and 60-80% of dogs. Pain on palpation of the affected kidney is more commonly seen in acute obstructions and in dogs with severe pyelonephritis. If this is not present one should not rule out a ureteral obstruction. It is important to remember that some pets have a normal physical examination, are urinating normally, and the only thing reported by the owner is weight loss, decreased energy and/or vague signs.
2) *Biochemical data:* Cats are often anemic (48%) at diagnosis and this is either due to concurrent CKD, chronic disease, or excessive blood sampling during previous hospitalizations. Dogs often have a moderate to severe neutrophilia associated with concurrent pyelonephritis and 44% of dogs with ureterolithiasis-induced obstructions were reported to have some degree of

thrombocytopenia, which can be severe (<40,000 platelets). The thrombocytopenia may be due to sepsis or immune-mediate thrombocytopenia. Azotemia is common at the time of diagnosis, even with a unilateral obstruction (83-95% of cats and 50% of dogs). The degree of azotemia at presentation does not appear to be associated with outcome if successful decompression occurs and should not be a reason to discourage treatment. Hyperphosphatemia (54%), hyperkalemia (35%), hypercalcemia (14%), and hypocalcemia (22%) were reported in a large series of cats with ureteral obstructions. Crystals were observed in the urine of 29% of cats (amorphous crystals and calcium oxalate being most common) with calcium oxalate stones.

3) *Radiographs:* Bilateral ureteral obstructions occur in about 15-25% of cats and 12% of dogs. Radiopaque calculi are typically visible in patients (>90%) with calculi induced obstructions, making radiographic evaluation mandatory during a workup. This helps to document stone size, number, location and the presence of concurrent nephrolithiasis, but does not replace the need for an ultrasound. Over 20% of cats have a stricture causing their ureteral obstruction, in which case a radiograph would not provide much information on the cause of the obstruction. When an obstruction is suspected, and no stones are seen they are either too small to be visualized (<1mm), radiolucent stones (dried solidified blood stones, urate or cystine stones), or the cause is a soft tissue obstruction due to either a ureteral stricture or tumor.
4) *Ultrasound* is ideal for the documentation of hydroureter, hydronephrosis, and the exact location of the most proximal obstructive lesion. If hydroureter is present and very proximal, with no evidence of a shadowing stone at the junction of normal and abnormal ureter, then a ureteral stricture may be the cause of the obstruction. If there is severe hydroureter to the level of the ureterovesicular junction than the cause of the obstruction is likely either a UVJ stenosis (classic in dogs with ectopic ureters), a trigonal tumor (should see a tumor in the bladder on ultrasound), or a distal stone that is difficult to see due to air in the colon on ultrasound. In a recent study 60% of cats with a ureteral stricture had evidence of peri-ureteral hyperechoic tissue at the stricture site on ultrasound, and these are most common to be seen in the proximal right ureter and associated with a circumcaval ureter. Identifying concurrent nephroliths and other ureteroliths is vital in therapeutic decision-making when considering traditional surgery or ESWL. This is less important for interventional treatment like a ureteral stent or SUB device because the entire ureter will be bypassed, regardless of stone number, location or size. The sensitivity for abdominal radiographs for feline ureteral calculi was 81% (vs 88% in canine) and for abdominal ultrasound was 77% (vs 100% in canine). In combination the sensitivity of both was 90% in cats. Studies have documented 60-86% of cats had concurrent nephroliths, and 9-20% cystic calculi, whereas 50% of canine patients had concurrent nephroliths. Additionally 15-25% of dogs will have a ureteral obstruction due to non-stone related disease, like a tumor, stricture, or ectopic ureter-induced stenosis, in which the ureteral dilation is seen all the way down to the UVJ. Another key piece of information is knowing the exact diameter of the dilated renal pelvis based on the ultrasound. It is important to know this in deciding which interventional option is best for the patient. When hydroureter and hydronephrosis are seen to a focal region in the ureter, a ureteral obstruction is present, and it is either partial or complete. Knowing the degree of obstruction (partial or complete) should not guide whether or not a ureter should be treated for their obstruction.
5) *Intravenous pyelography (IVP)* has been traditionally used to diagnose the degree of the nephrogram phase, as a prediction of renal function. Additionally it can aid in detecting the degree of renal pelvic and ureteral dilation, the location of a ureteral obstruction, and potentially, the location of ureteral leakage. IVP's have been shown to inaccurately predict the lack of renal function based on poor filling and should not be used to decide whether treatment is indicated. With severe hydrostatic pressure from a ureteral obstruction the kidney may not fill with contrast, even in cases where renal function can be preserved. If a complete obstruction is present for more than 12 hours, which is the case in many patients, the diagnostic rate of a ureteral obstruction can be as low as 36-75% due to the inadequate excretion of contrast material. Additionally, since intravenous contrast material can be nephrotoxic, performing this procedure in renal compromised patients should be carefully considered prior to using this diagnostic tool. The value of antegrade pyelography, if a ureterogram is needed, should be considered over an IVP.
6) *Antegrade Pyelography:* Percutaneous antegrade pyelography provides good visualization of the renal pelvis and ureter, and allows localization of the ureteral obstruction, aiding in determining whether a complete or partial obstruction is present. Ideally, this procedure should be performed under fluoroscopic guidance, which allows visualization of the filling of the ureter and the point of obstruction in real time and documents the degress of contrast passage into the urinary bladder. In a study of 11 cats the sensitivity and specificity was excellent in the cases where the study could be interpreted. Forty-four percent of the studies had evidence of leakage and 30% were not diagnostic, making its use somewhat more limited. With the improvement in ultrasound imaging, the confidence of obstruction based on ultrasound and radiographs, the safety of treatment of partial obstructions with newer options, as well as the need for pyelography during treatment (ureteral stenting, nephrostomy tube placement or SUB placement), the use of purely diagnostic antegrade pyelography has declined tremendously in the author's practice, and is instead performed intra-operatively during obstruction relief, aiding in guide wire passage during device placement.
7) *CT-IVP* can be performed prior to surgery if a ureterotomy is going to be performed and the stone location and number is not clear based on radiographs and ultrasound. Intravenous contrast administered during the CT scan can aid in concurrent differentiation of partial or complete obstructions. IVP is often not useful in animals with ureteral obstructions due to the poor filling of an obstructed kidney and the nephrotoxicity risk of the contrast material that is being filtered by the kidneys during the nephrogram phase, particularly considering a significant number of both canine and feline patients are azotemic at the time of diagnosis. This risk is considered theoretical at this time but should be considered. In certain instances an IVP might be indicated. Since the advent of interventional management, the location of the stones are slightly less important if the entire ureter will be bypassed by a stent or SUB device, and most of the stones will not ultimately be removed. This is not the case when traditional surgical therapy is being performed.
8) *Retrograde Ureteropyelography* can be used to detect the cause and location of a ureteral obstruction. This is most useful in dogs with ureteral stricture/stenosis where stones are not clearly visualized. This is typically done using endoscopic and fluoroscopic guidance, using a guide wire and catheter. Most commonly this technique is done prior to the placement of a ureteral stent or

cystoscopic-guided laser ablation endoscopically.

9) *GFR studies/Scintigraphy:* The GFR of individual kidneys can be measured by technetium Tc 99 diethylenetriamine pentacetic acid (DTPA) scintigraphy. The GFR of an obstructed kidney is most often reduced, and the predictability of return to function based on scintigraphy is often unclear, potentially significantly underestimating post-relief renal function. The measurement of the GFR of the contralateral kidney may assist in the decision whether to treat unilateral obstructions, or potentially perform a nephrectomy if absolutely necessary (not recommended unless resection of renal neoplasia).

CONTROVERSIES in the TREATMENT of a ureteral obstruction

1) **Medical:** Dissolution of obstructive ureteroliths is contraindicated, regardless of the stone composition. The time required will result in excessive kidney damage. The traditional approach to a partial ureteral obstruction has been medical management (intravenous fluid therapy, mannitol diuresis via constant-rate infusion [CRI] and α-adrenergic blockade with or without amitriptyline therapy) when stone disease is the cause. This will obviously not be helpful for strictures, stenosis or tumors. If this approach is unsuccessful many clinicians would have historically elected conservative monitoring without fixation due to the presumed low risk:benefit ratio of attempted surgical removal, which is associated with a high morbidity and morality rate. Based on the literature in cats, it is clear that surgical intervention provides a higher success rate (~75-80%) than medical management alone (~13%). With medical management being effective in some feline cases (7.7-13%), medical therapy should be considered before any intervention for 24-48 hours. The author finds beyond that time the chance of stone passage is low, and the risk of permanent renal damage is high. Medical management is contraindicated in cases with persistent hyperkalemia, worsening renal pelvic dilation, the presence of oliguria/anuria, and worsening azotemia despite medical management.

2) **Traditional surgical intervention** is reported to include ureterotomy, neoureterocystostomy, ureteronephrectomy, or renal transplantation. In a small study of dogs, results after a ureterotomy for stone disease were associated with a mortality rate of 6.25%, and 15% required an additional surgery for re-obstruction within 4 months. 21% failed to show improvement in renal function after surgery and 15% had worsening renal function. In cats, the procedure-associated complication and mortality rates are reported to be over 30% and 20%, respectively. The morbidity and mortality rates may be higher in environments where operating microscopes and microsurgical experience and expertise are not available. Many of the complications associated with surgery are due to site edema, recurrence of stones that pass from the renal pelvis to the surgery site, stricture formation, persistent obstruction, missed ureteroliths, and surgery-associated or nephrostomy tube–associated urine leakage.

   It is important to realize that a ureteronephrectomy is not ideal in any stone forming patient, especially in those with concurrent renal azotemia (50% to 97% cats and 25-50% dogs). In a more recent study 97% of cats were azotemic at the time of ureteral obstruction diagnosis, even when unilateral. Persistent azotemia is a common problem, even after a successful intervention (40% to 70% of cats), but it is often minor and remains stable for years. In one study after traditional surgery, 40% of feline patients developed a second ureteral obstruction within 2 years of surgery. Eighty-five percent of these cats had evidence of nephrolithiasis documented at the time of the first surgery. Nephroliths have the potential to pass into and obstruct the ureter during a post-obstructive diuresis, and can occur immediately after surgery, but a concurrent nephrotomy increases the risks for postoperative complications and will likely worsen renal function. Traditionally, in cases where stones are not the cause of the obstructin, than surgery to reimplant the ureter, or perform a ureteral resection and anastomosis, is typically performed (strictures, tumors, distal stenosis).

3) **Interventional Treatment**

   a. *Nephrostomy tubes:* Since fixation procedures can be relatively prolonged and complicated, especially in facilities where expertise in treating a ureteral obstruction is not as vast, the placement of a nephrostomy tube can quickly relieve the obstruction and determine whether adequate renal function remains before prolonged anesthesia for ureteral surgery, ureteral stent or SUB device placement can be performed. A locking loop pigtail catheter has been shown to decrease the risk of inadvertent tube removal or urine leakage, which was the main complications seen historically. With the advent of sturdy, multi-fenestrated tubes, that form a loop that will lock the catheter in the renal pelvis, these complications seem to have declined dramatically, which was recently reported in 20 dogs and cats. This is typically done percutaneously in dogs using ultrasound and fluoroscopic guidance, and surgically assisted with fluoroscopy in cats.

   b. *Extracorporeal shock-wave lithotripsy* can be considered for the removal of upper tract calculi in the renal pelvis or ureters. This utilizes external shockwaves that pass through a water medium and it is directed under fluoroscopic guidance. The stone is shocked between 1000 and 3500 times at different energy levels to fragment the stone into small pieces (typically around 1mm). The debris is then left to pass down the ureter into the urinary bladder over a 1-2 week period. It has been seen to take as long as 3 months. This procedure can be performed safely for nephroliths smaller than 10 mm, and ureteroliths smaller than 5 mm. Because the fragments are rarely less than 1 mm, and the feline ureter is 0.3 mm in diameter, this is not considered an effective treatment for cats. For large stone burdens an indwelling double pigtail ureteral stent is typically placed prior to ESWL. This will aid in preventing a ureteral obstruction during stone passage, and allows the ureter to passively dilate around the stent, accommodating more stone fragments to pass. This may also hinder ureteral peristalsis, making the stone fragments take longer to pass. ESWL is believed to be safe and well tolerated by canine kidneys. The renal changes observed in dogs after ESWL typically result in minimal decreases in GFR, returning to baseline within 1 week. ESWL treatment was reported to be successful in approximately 85% of dogs with calcium-containing nephroureterolithiasis. Successful fragmentation of renal stones was achieved in 90% of dogs, but some required two treatments, and 30-50% requiring more than 1 treatment. The newer dry units are more powerful, with a more focused beam, resulting in lower stone retreatment rates (15-20%, in the author's experience) when used appropriately. Success with ESWL is highly stone and patient dependent. Stone size and composition are clearly associated with success in human urology, and the same seems to be true in veterinary medicine. Struvite, urate and calcium oxalate stones are considered more amenable to ESWL than are cystine stones.

c. *Double pigtail ureteral stents* have many benefits: (1) immediate decompression of the renal pelvis, (2) passive ureteral dilation for urine and stone passage, (3) prevention of strictures and leakage associated with surgical or non-surgical procedures, and (4) prevention of a ureteral obstruction after extracorporeal shock-wave lithotripsy (ESWL). All of these benefits have been clearly documented in human medicine and, more recently, in veterinary medicine, with the placement of ureteral stents in more than 125 feline and 150 canine ureters for various indications in the author's practice. The use of double pigtail ureteral stents in dogs and cats has been investigated as a potentially less invasive alternative to traditional surgery, resulting in immediate ureteral decompression, stabilization of the associated azotemia, a decreased risk of ureteral stricture or ureteral leakage/uroabdomen, and a decreased rate of ureteral obstruction recurrence. Placement of a double-pigtail ureteral stent, via either minimally invasive techniques (endoscopy and interventional fluoroscopy in most dogs) or with surgical assistance using fluoroscopic guidance (nephrostomy access in most cats), seem to have minimized the complications of surgery alone (e.g., leakage, stricture, re-obstruction), and has resulted in expedited stabilization. There are currently veterinary ureteral stents available, which were designed for use in dogs and cats. Prior to considering this option, the operator must be appropriately trained in stent placement, the risks of stent placement needs to be understood, and owners must be educated on these risks. The operator must understanding the nuances of stent placement, which can be very difficult, and these outcomes reported are all from the same operators, where over 250 stents have been placed to date. Once experience is obtained, than stent placement is highly successful. The main concerns with stents are typically seen months to years after placement. They are not usually life threatening, and are relatively easy to address on an outpatient basis. These include: dysuria, stent migration, stent occlusion (seen most commonly with ureteral strictures), and stent encrustation. These problems are all more common in cats than in dogs, which has pushed the move toward using the subcutaneous ureteral bypass (SUB) device preferentially in cats and ureteral stents in dogs.

In dogs with ureterolithiasis or strictures, ureteral stenting is almost always performed with endoscopic and fluoroscopic guidance in a retrograde manner, and is considered an outpatient procedure. This has been successful in 98% of 150 dogs to date in the author's practice. The complications observed in dogs are far fewer than that seen in cats in all time periods (peri-operative, short and long-term). These included recurrent urinary tract infections (<20%, with over 75% having infections pre-stent), proliferative tissue growth around the distal loop of the stent at the ureterovesicular junction (~15%), stent migration (<5%), stent occlusion (<5%), and encrustation (<5%). Dysuria is much less common in dogs than in cats after stent placement (<1%), and both species are typically glucocorticoid responsive if it does not resolve spontaneously.

A recent study evaluating 69 feline patients (79 ureters) with ureteral obstructions treated with a ureteral stent had a 96% placement success, 14% were bilaterally obstructed, 28% were associated with strictures (+/- concurrent ureteroliths), and 85% of cases had concurrent nephroliths. The median number of stones in the obstructed ureter was 4. Approximately 1/3 of cats had documentation of a urinary tract infection prior to stent placement, and 30% of cats had at least 1 positive culture in their lifetime post-stent placement. The median creatinine was 5.3 mg/dL pre- and 2.1 mg/dL post-stent placement. The peri-operative mortality after surgery was 7.5%, with no death associated with a ureteral obstruction or the procedure. This compares to 21% mortality rate with traditional surgery. The most common long-term (>1 month) complications after ureteral stent placement was dysuria (37.7%, but only persistent in 1.7%), chronic hematuria (18%), stent occlusion (19%), stent migration (6%), and ureteritis/mucosal proliferation (4%). The dysuria typically resolved with short-term steroid therapy. A ureteral stent exchange with either a bigger stent or a SUB device was needed in 27% of cats for either occlusion or migration. This complication was most commonly associated with ureteral strictures (over 50% of strictured ureters experienced reocclusion), or an occlusion at a previous ureterotomy site. Ureters remained patent long-term in most cats with the longest stent in place and functional for over 5 years. The median survival time was 498 days. The median survival time when renal cause of death was over 1250 days, with only 21% of cats ultimately dying of suspected CKD over the course of the entire followup period. Another study evaluating pre-operative prognostic factors for short- and long-term survival and renal recovery did not identify any clinical, biochemical or imaging findings to suggest or predict overall survival times in feline patients.

This preliminary data suggests that ureteral stenting, in both dogs and cats, is safe and effective, resulting in immediate decompression of the renal collection system. Few major procedural or peri-operative complications occurred, particularly in dogs, but the learning curve was steep. In cats long-term stent exchange or manipulation may be necessary if dysuria, a ureteral reaction, stricture-induced stent occlusion, or stent migration occurs. Because of the management issues associated with feline ureteral stents, which are tolerable but can be expensive and frustrating, the author currently recommends the subcutaneous ureteral bypass device (SUB) for cats with ureteral obstruction and ureteral stents for dogs.

d. *Subcutaneous ureteral bypass (SUB) device.* The biggest limitation to nephrostomy tubes is the externalized drainage, requiring careful management and hospitalization to prevent infection and dislodgement. The development of an indwelling SUB device using a combination locking-loop nephrostomy catheter and cystostomy catheter can remain indwelling long-term. In humans, a similar device has been shown to reduce complications associated with externalized nephrostomy tubes and improve quality of life as a long-term treatment. The SUB has been placed in over 125 cats to date for the treatment of all causes of ureteral obstructions and met with great success.

Initially this was reserved for proximal ureteral strictures or ureteral stent failure, but more recently is the first choice for all causes of feline ureteral obstructions. This procedure requires fluoroscopic guidance and is done with surgical assistance. Patency of the device has been evident in ~94% of cats and 90% of dogs over a median of 2.5 years (range, 1 month to 5 years). The use of the feline SUB device has recently been described in various abstract and peer-reviewed formats. The median pre-operative creatinine was 6.8 mg/dL and 2.4 mg/dL post-SUB placement. The peri-operative mortality after surgery was 5.6%, with 1 death associated with a procedure related complication early in the course of the study. This compares well to traditional surgery (21% peri-operative mortality) and ureteral stenting (7.5%). To date the median follow-up time is over 19 months. Peri-operative complications are typically rare and include leakage (at the port or nephrostomy tube insertion site) (<5%), kinking of the catheters (or failure of the renal function to recover (<3%), and blockage of the system with either blood clots, debris, purulent material, or future stones (~8%; 0% for strictured ureters and ~10% for stones). Dysuria is rarely seen in cats with a SUB device (<2%) compared to the ureteral stent (38%). The SUB device is considered a functional option and is the author's treatment of choice in cats with ureteral obstructions.

e. <u>*Miscellaneous*</u>. *Percutaneous nephroureterolithotomy (PCNUL)* can be used for nephroliths and proximal ureteroliths causing a ureteral outflow obstruction at the ureteropelvic junction. This procedure is done using ultrasound and fluoroscopic guidance and is considered more renal sparing for stone removal than a nephrotomy or ESWL. Cases of distal ureteral stenosis associated with intramural ectopic ureters has been seen in a small group of male and female dogs. This is most commonly seen in male Labrador retrievers that may, or may not, be incontinent. They have a thick intramural ureteral tunnel that enters the proximal prostatic urethra ectopically. Due to the long urethra they are often not incontinent but will have severe hydroureter and hydronephrosis on ultrasound. This can be treated with *cystoscopic guided laser ablation* by opening the distal ureteral tunnel to allow for improved urine flow. Finally, the author has been recently treated bladder/trigonal neoplasia that is localized to the proximal urethra/bladder more aggressively with bilateral SUBs and total bladder cystectomy. The owner must be prepared for an incontinent dog, as the distal catheter will be placed into the urethral lumen for bilateral kidney drainage, but the gross disease of tumor can be removed enbloc, potentially providing a longer survival. This is still under investigation in our practice.

4) <u>**Recurrence/ Prevention.**</u> Since the most common cause of ureteral obstruction is stone disease it is important that regardless of the cause prevention is discussed with the owner. This is typically addressed with using medications like potassium citrate for calcium oxalate stone disease, appropriate dietary therapy for stone formers if non-azotemic, and serial SUB device flushing if this is available. All patients with a history of obstruction have the risk of re-obstruction on the contralateral side, especially for stones and tumors. This, and the risk of re-occlusion of a device or from traditional surgical stricture/stone formation or migration, necessitates serial ultrasounds to be performed. It is recommended to have a renal ultrasound every 3 months after treatment of a ureteral obstruction to address any issues that arise prior to a crisis developing.

Conflict of Interest Disclosure

The author is a consultant for Infiniti Medical, LLC and Norfolk Vet Products.

REFERENCES

More available upon request

1. Kyles A, Hardie E, Wooden B, et al. Clinical, clinicopathologic, radiographic, and ultrasonographic abnormalties in cats with ureteral calculi: 163 cases (1984-2002). JAVMA, 226(6): 932-936.
2. Kyles A, Hardie E, Wooden B, et al. Management and outcome of cats with ureteral calculi: 153 cases (1984-2002). JAVMA, 226(6): 937-944.
3. Roberts S, Aronson L, Brown D. Postoperative mortality in cats after ureterolithotomy. Vet Surg; 2011; 40: 438-443.
4. Berent A. Weisse C, Bagley D. et al. Ureteral stenting for benign feline ureteral obstructions in 79 ureters. J Am Vet Med Assoc. 2013, in press.
5. Horowitz C. Berent A, Weisse C. et al. Predictors of outcome for cats with rueteral obstructions after interventional management using ureteral stents or a subcutaneous ureteral bypass device. J Feline Med Surg. 2013. May Epub.
6. Steinhaus J, Berent A, Weisse C, et al. Circumcaval ureters in cats with and without ureteral obstructions. A comparative study. Abstract, American College of Veterinary Internal Medicine, Seattle WA ACVIM 2013.
7. Berent A, Weisse C, Bagley D, et al. The use of a subcutaneous ureteral bypass device for the treatment of feline ureteral obstructions. ECVIM, Seville Spain, 2011
8. Berent A. Ureteral obstructions in dogs and cats ; a review of traditional and nrew interventinal diagnostic and therapeutic options. J Vet Emerg Crit Care. 2011; 21(2): 86-103.

UPDATES ON DOG ERYTHROCYTE ANTIGEN 1, BLOOD TYPING, AND CROSSMATCHING

Urs Giger, DACVIM-SA, DECVCIM-CA, DECVCP, Michelle M Acierno, Klaudia Polak, Karthik Raj,
Philadelphia, PA

Introduction

Veterinary clinicians and clinical pathologists play a key role in providing safe and effective transfusion therapy. Blood typing is clinically important to ensure blood compatibility and therefore is recommended for any dog in need of a transfusion or considered to become a blood donor. Moreover, previously transfused dogs also should be crossmatched. Unless blood typing is performed regularly in practice, blood should be sent to a clinical pathology laboratory for typing. Different viewpoints exist regarding the extent and methods used for compatibility testing, and various techniques for laboratory or point-of-care use have been applied or are being developed. This presentation will present the current knowledge of the canine blood types and their clinical importance and review the different typing and crossmatching techniques for the in-practice and clinical pathology laboratory.

Canine Blood Types

Blood types are genetic markers on erythrocyte surfaces that are antigenic and species specific. A set of blood types of two or more alleles makes up a blood group system. Dogs have likely more than a dozen blood group systems known as dog erythrocyte antigens (DEA); however, there is no DEA 2 blood group. Some may be rather labeled high frequency or common red blood cell (RBC) antigens (e.g. DEA 4) and some have not yet received a DEA designation (e.g. Dal). Canine erythrocytes are either positive or negative for a blood type (e.g., DEA 4+ or DEA 4-), and these blood types are thought to be inherited codominantly. The DEA 1 system was thought to be an exception with DEA 1.1 (A1), DEA 1.2 (A2) and potentially DEA 1.3 (A3) being allelic. Thus, a dog can apparently be DEA 1.1+ or DEA 1.1- and DEA 1.1- dogs can be DEA 1.2+ or DEA 1.2-. However, these studies were based upon weak polyclonal antibodies (DEA 1.1 and 1.X) requiring Coombs' reagents. The authors' most recent studies with a monoclonal antibody showed that the DEA 1 blood group is a continuum from DEA 1- to weakly to strongly DEA 1+ rather than having 2-3 blood types. The degree of DEA 1 expression is constant when blood is stored at 4°C and in each dog and appears to be codominantly inherited. The biochemical structure of the DEA 1 remains still unknown. Thus we refer subsequently only to DEA 1+ and DEA 1-.

The clinically most important canine blood type is DEA 1, which elicits a strong alloantibody response after sensitization of a DEA 1- dog by a transfusion and thus can be responsible for a transfusion reaction in a DEA 1- dog previously transfused with DEA 1+ blood. It is currently unknown if DEA 1- dogs are equally sensitized by weakly to strongly DEA 1+ blood. Furthermore, transfusion reactions against other blood types or common antigens have rarely been described. They include reactions against the DEA 4, Dal, and another common RBC antigens, and other clinically important blood types may be found in the future. No reagents currently are available against many antigens, and additional blood types continue to be recognized. Only limited surveys on the frequency of these blood types have been reported, which suggest possible geographic and breed-associated differences.

Strongly antigenic blood types are of great clinical importance because they can elicit a potent alloantibody response. These alloantibodies may be of the immunoglobulin G (IgG) or IgM class and may be hemagglutinins or hemolysins. Based upon experimental and clinical data, dogs can become sensitized after receiving a mismatched transfusion (i.e., a blood unit positive for one or more blood types not found on the recipient's RBCs).

There are no clinically important, naturally occurring alloantibodies (also known as isoantibodies) present before sensitization of a dog with a transfusion. Sensitizing dogs in experimental studies in the 1950s led to the documentation of some transfusion reactions caused by blood group incompatibilities and to the characterization of new blood types. Clinically the most antigenic blood type in dogs is the DEA 1.1. Transfusion of DEA 1+ RBCs to a DEA 1- dog invariably elicits a strong alloantibody response. Following a first transfusion, anti-DEA 1 antibodies develop after more than 4 days and may cause a delayed transfusion reaction. However, a previously sensitized DEA 1- dog develops an acute hemolytic reaction after transfusion of DEA 1.1+ blood. Transfusion reactions also may occur after a sensitized dog receives blood that is mismatched for a RBC antigen other than DEA 1 (e.g. DEA 4 and Dal). However, in most cases the incompatible blood type has not been determined. Because administration of a small (<1 ml) amount of incompatible blood can result in life-threatening reactions, the practice of giving small "test volumes" of donor blood to assess blood-type compatibilities is unacceptable. In contrast, pregnancy does not cause sensitization because of a complete placenta in dogs and does not induce alloantibody production; thus dogs with prior pregnancies can be used safely as donors.

Canine Blood-Typing Procedures

Because of the strong antigenicity of DEA 1, typing of donors for DEA 1 is recommended. Whenever possible, the recipient also should be typed to allow the use of DEA 1+ blood for DEA 1+ recipients. Canine blood typing was based generally on serologic identification by agglutination reactions but chromatographic strip methods are also offered. Originally serum from sensitized dogs has been used for typing, but such polyvalent alloantibodies vary from batch to batch, may require Coombs' reagent to show agglutination, and may not be always available and are therefore not optimal. Monoclonal antibodies against DEA 1 have been developed at Kansas State University and at the University of Lyon. The gel column technology, widely used in human blood banking, was found to be an excellent standardized laboratory method (DiaMed), but is unfortunately no longer commercially available. A blood typing card has been available since the mid-1990s (and recently modified) as a simple in-practice kit to classify dogs as DEA 1- or DEA 1+ (degree of reaction can vary). A cartridge became available in the mid-2000s from Alvedia that relies on a standardized simple chromatographic technique. This method is now also available as a strip technique for larger laboratories. Another cartridge with a similar strip technique was introduced by DMS/AgroLabo, but has not been evaluated. Moreover, a third cartridge method in which blood flows through the cartridge is also available (DMS/Abaxis) but was producing inconsistent results.

Caution should be exercised whenever the patient's blood is autoagglutinating or has a low hematocrit (<10%). If autoagglutination is not too severe, it does not appear to affect the Alvedia cartridge technique because only free RBCs are moving up the strip. Clinicians should check for autoagglutination of blood with buffer/saline on a slide or the card. Autoagglutinating blood may be first washed three times with saline to overcome the apparent autoagglutination similar to what is done for the Coombs' and crossmatch tests. However, if autoagglutination after three washes persists at more than 1+, it is considered to reflect true autoagglutination, which may preclude typing (as well as Coombs' testing and crossmatching), because it always looks like DEA 1+ blood. In such circumstances, DEA 1- blood should be used, until the patient does not agglutinate anymore and can be retyped. DEA 1+ blood from anemic animals may not agglutinate when exposed to the anti-DEA 1 or other reagents because of a prozone effect. In these cases some of the patient's plasma may be discarded before applying a drop of blood onto the card. Finally, recently transfused dogs may display a mixed field reaction, with only the transfused or recipient cells agglutinating if they were DEA 1 mismatched.

Blood Crossmatching Test

Whereas blood typing tests reveal the blood group antigens on the red blood cell surface, blood crossmatching tests assess the serologic compatibility or incompatibility between donor and recipient. Thus the crossmatch test checks for the presence or absence of naturally occurring and induced alloantibodies in serum (or plasma) without determining the blood type and thus does not replace blood typing. These antibodies may be hemolysins and/or hemagglutinins and can be directed against known blood groups or other RBC surface antigens. A standardized tube crossmatching procedure has been proposed and is used by many laboratories. The crossmatching test requires some technical expertise, may be accomplished through a veterinary laboratory along with blood typing, and is done with washed EDTA-anticoagulated blood from recipient and potential donor(s). The DiaMed gel column technique and more recently the in-clinic DMS gel tube assay have been evaluated and were found to be simple, sensitive, and standardized methods to crossmatch dogs and cats. In addition, Alvedia introduced this year a simple strip crossmatch test with a Coombs' phase which is being evaluated.

The major crossmatch tests search for alloantibodies in the recipient's plasma against donor cells, whereas the minor crossmatch test looks for alloantibodies in the donor's plasma against the recipient's RBCs. Generally tube segments from collection bags are used for this purpose in dogs. Blood in a small EDTA tube is saved at the time of the blood collection by a small open or specially prepared closed system in cats. The presence of autoagglutination or severe hemolysis may preclude the crossmatch testing. A major crossmatch incompatibility is of greatest importance because it predicts that the transfused donor cells will be attacked by the patient's plasma, thereby causing a potentially life-threatening acute hemolytic transfusion reaction. Because fatal reactions may occur with less than 1 ml of incompatible blood compatibility testing by administering a small amount of blood is not appropriate; this has been shown in experimental studies to result in fatal reactions. A minor crossmatch incompatibility should not occur in dogs if canine donors have not been transfused previously and is of lesser concern because donor's plasma volume is small, particularly in packed red cell products, and is diluted markedly in the patient.

The initial blood crossmatch between two dogs that have never before received a transfusion should be compatible, because dogs do not have naturally occurring alloantibodies. Therefore a crossmatch may be omitted before the first transfusion in clinical practice for dogs. Because the crossmatch does not determine the blood type of the patient and donor, a compatible crossmatch does not prevent sensitization of the patient against donor cells within 1 to 2 weeks. Therefore, previously transfused dogs should always be crossmatched, even when receiving again blood from the same donor. The time span between the initial transfusion and incompatibility reactions may be as short as 4 days and the alloantibody lasts for many months to years (i.e., years after the last transfusion alloantibodies may be present). Obviously, a blood donor never should have received a blood transfusion to avoid sensitization. The practice of transfusing patients with the least compatible unit does not have any scientific basis. Nevertheless, some minor agglutination results in crossmatching a patient may be unrelated to alloantibodies and unspecific (e.g., patient's RBC damage by uremia and other illnesses, donor cells after extended storage of unit in the refrigerator). Of course any patient with true/persistent autoagglutination may not be matched to any donor.

Although transfusion of blood and its components is usually a safe and temporarily effective form of therapy, there is always a risk for potential hazards. Adverse reactions usually occur during or shortly after the transfusion and can be due to any component of whole blood. Most transfusion reactions can be avoided by carefully selecting only healthy donors; using appropriate collection, storage, and administration techniques; performing blood typing and crossmatching; and administering only the needed blood components.

REFERENCES

Acierno MA, Raj K, Giger U (2014). DEA 1 expression on canine erythrocytes analyzed by immunochromatographic and flow cytometric techniques. J Vet Intern Med in press.

Blais MC, Berman L, Oakley DA, Giger U (2007). Canine Dal blood type: a red cell antigen lacking in some Dalmatians, J Vet Intern Med 21: 281-286.

Blais MC, Rozanski EA, Hale AS, Shaw SP, Cotter SM (2009). Lack of evidence of pregnancy-induced alloantibodies in dogs. J Vet Intern Med. 23:462-465.

Blois SL, Richardson DM, Abrams-Ogg AC (2013). Comparison of a gel column blood typing method and a point-of-care cartridge for dog erythrocyte antigen 1.1.J Vet Emerg Crit Care. 23:340-343

Ergul Ekiz E, Arslan M, Ozcan M, Inal Gultekin G, T Kirmizibayrak, Giger U (2011): Prevalence of dog erythrocyte antigen DEA 1.1 in 4 breeds native to different areas in Turkey. Vet Clin Pathol, 40:518–523.

Ferreira RR, Gopegui RR, Matos AJ (2011). Frequency of dog erythrocyte antigen 1.1 expression in dogs from Portugal. Vet Clin Pathol. 40:198-201.

Giger U, Stieger K, Palos H (2005): Comparison of various canine bloodtyping methods, Am J Vet Res 66:1386-90.

Giger U (2014). Blood-typing and crossmatching to ensure blood compatibility. Chapter 26 in Kirk's Current Veteriarny Therapy. In press.

Hohenhaus AE (2004): Importance of blood groups and blood group antibodies in companion animals , Transfus Med Rev 18 : 117-22.

Iazbik MC, O'Donnell M, Marin L, Zaldivar S, Hudson D, Couto CG (2010). Prevalence of dog erythrocyte antigens in retired racing Greyhounds. Vet Clin Pathol.39:433-435.
Kessler RJ, Reese J, Chang D, Seth M, Hale AS, Giger U (2010): Dog erythrocyte antigens 1.1, 1.2, 3, 4, 7, and Dal blood typing and cross-matching by gel column technique. Vet Clin Pathol, 39:306-16.
Kohn B, Classe G, Weingart C (2012). Clinical evaluation of the QuickVet/ RapidVet canine dog erythrocyte antigen 1.1 blood-typing test. J Vet Diagn Invest. 24:539-45.
Lucidi C, Takahira RK, Gerlach JA, Davis JM, Schwartz KA, Scott MA (2011). Flow cytometric assessment of canine erythrocytes and platelets for dog erythrocyte antigen 1.1. Vet Clin Pathol. 40:435-43.
Patterson J, Rousseau A, Kessler RJ, Giger U (2011): In vitro lysis and acute transfusion reactions with hemolysis caused by inappropriate storage of canine red blood cell products. J Vet Intern Med. 25:927-933.
Riond B, Schuler E, Rogg E, Hofmann-Lehmann R, Lutz H (2011). Prevalence of dog erythrocyte antigen 1.1 in dogs in Switzerland evaluated with the gel column technique. Schweiz Arch Tierheilkd. 153:369-74.
Seth M, Jackson K.V, Winzelberg S, Giger U (2012): Comparison of gel column, card and cartridge techniques for Dog Erythrocyte Antigen. Am J Vet Res, 73:213–219.

Supported in part by a grant from the NIH (OD 010939).
Alvedia and DMS Laboratories kindly provided reagents and kits for these studies.

HEREDITARY ERYTHROCYTE DEFECTS CAUSING ANEMIA

Urs Giger, DACVIM-SA, DECVCIM-CA, DECVCP, Inal Gultekin, Sarah Lehman, Karthik Raj,
Philadelphia, PA

Anemia is one of the most common clinical signs and abnormal laboratory test results in companion animals. Although acquired conditions such as infections, immune disorders, intoxication, blood loss, and chronic organ failure represent the main causes of anemia, hereditary erythrocyte defects leading to anemia are becoming increasingly important in clinical practice. Several hereditary erythrocyte defects have been clinically reported in companion animals, and much new information, including the molecular basis of some disorders, has emerged. Frequently they are only considered after treatments for immune and infectious diseases have failed or the anemia is recurring, despite the fact that in some breeds these conditions are common and well recognized.

Canine and Feline Erythrocytes

Although the major structural features and functions of erythrocytes are similar among mammals, some noticeable differences exist between canine and feline erythrocytes. Feline erythrocytes are smaller than canine erythrocytes. Normal variations observed in many Akitas and Miniature Poodles are erythrocyte microcytosis and macrocytosis, respectively. The normal life spans of erythrocytes in dogs and humans are similar (100-120 days), but erythrocyte life span is only 70-75 days in cats.

Devoid of a nucleus and mitochondria, erythrocytes have a limited and specialized metabolism that enables them to survive in circulation and to adequately transport oxygen. Energy is generated almost exclusively through anaerobic glycolysis, also known as the Embden-Meyerhof pathway. The hexose monophosphate shunt reduces pyridine nucleotides and glutathione, which are necessary for degradation of oxidants, thereby preventing membrane damage and hemoglobin denaturation. In addition, the methemoglobin or cytochrome-b5 reductase system reduces heme iron from the ferric (Fe^{3+}) to the ferrous (Fe^{2+}) form. The Rapoport-Luebering pathway is responsible for the synthesis of 2,3-diphosphoglycerate (DPG), which influences the oxygen affinity of canine - but not feline - hemoglobin. The erythrocytic DPG concentration is similar in dogs and humans but very low in cats.

Dogs and cats apparently have embryonic, but no fetal hemoglobin. With the exception of some Japanese breeds, which have two hemoglobins (HbA and HbB), only one adult hemoglobin has been found in dogs. However, further studies are needed to characterize canine hemoglobin. In cats, one α-globin and six different β-globins have been found. With each cat having one to four different β-globins, many hemoglobin patterns are recognized.

The erythrocyte membrane consists of a lipid bilayer, affixed to a membrane skeleton, which determines cell shape and deformability. Various transmembrane glycoproteins function as receptors or transporters. Canine and feline erythrocytes lose their Na^+,K^+-ATPase during late maturation in the bone marrow, owing to proteolysis (with the exception of erythrocytes from some Japanese dogs). Therefore, erythrocytes' high sodium and low potassium concentrations are similar to those of serum electrolytes. Consequently, hyperkalemia generally does not occur after intravascular hemolysis or storage of unseparated clotted blood unless stress reticulocytes or high-potassium erythrocytes from Japanese dogs are lysed. In fact, Akitas' erythrocytes are "leaky" in vitro, and pseudohyperkalemia has been observed in serum that has not been separated immediately from its clot. Finally, canine erythrocytes have been noted to be uniquely fragile under alkaline conditions, compared with erythrocytes from cats and other species, presumably because of a facilitated calcium entry under these conditions. This pH sensitivity may explain the tendency of canine erythrocytes to lyse in uncapped tubes in the laboratory.

Classification of Erythrocyte Defects

Hereditary erythrocyte defects form a large and clinically heterogeneous group of diseases. Each erythrocyte disorder is observed only rarely, although some enzyme deficiencies occur frequently within certain breeds. The mode of inheritance is autosomal recessive, with the exception of feline porphyria, which can be inherited as a dominant trait. While the degree of characterization is varied, many seem to be homologous to hereditary diseases in humans. These disorders have been classified into four groups: (1) heme defects and hemoglobinopathies, (2) cytosolic enzyme deficiencies, (3) membrane abnormalities, and (4) production and maturation defects; a few specific disorders are discussed in more detail.

Although the signalment, type, and severity of the anemia and the pleiotropic effects observed may provide clues to an inherited erythrocyte defect, a full laboratory evaluation is essential in validating the diagnosis or discovering new inherited disorders. In a few breeds, more than one erythrocyte disorder has been recognized. Routine hematological laboratory tests are used to detect hematological abnormalities and to rule out acquired anemia. An inherited erythrocyte defect should be considered in animals with Coombs-negative hemolytic anemia, a lack of toxin exposure or infection, and adequate kidney as well as liver function. A careful examination of a peripheral blood smear is pivotal in recognizing any poikilocytosis, such as elliptocytosis, spherocytosis and stomatocytosis, although most erythrocyte defects cause no change in cell shape and are historically known as nonspherocytic hemolytic anemias. The degree of reticulocytosis is often marked, even in the case of PK deficiency with osteosclerosis, and is generally proportional to the shortened survival of defective erythrocytes. Thus, a bone marrow examination rarely provides new information. The signs of hemolysis may be mild owing to the chronicity and low-grade hemolysis. Hyperbilirubinuria and -emia are generally noted, but may be mild due to adaptation. Low serum haptoglobin concentrations, hemoglobinemia, or hemoglobinuria, which all indicate intravascular hemolysis, have been reported in PFK deficiency but are unlikely to be associated with other defects. Some defective erythrocytes appear extremely fragile in vitro, resulting in artificial in vitro lysis in blood tubes. Furthermore, cats with porphyria exhibit erythrodontia and pigmenturia.

Special laboratory tests for defining the nature of an intrinsic erythrocyte defect can be divided into general screening tests, used to characterize unknown erythrocyte disorders, and specific screening tests for known defects in certain breeds. Both are performed only in specialized laboratories such as the Metabolic Genetics and Josephine Deubler Genetic Disease Testing Laboratories at the School of Veterinary Medicine, University of Pennsylvania, Philadelphia (http://www.vet.upenn.edu/penngen) and several veterinary DNA testing laboratories.

Hemolysis resulting from erythrocyte defects may be well compensated for by marked erythropoiesis, thereby causing no or only minimal clinical signs and allowing the animal to have a normal life expectancy. Furthermore, affected animals may have adapted well to the chronic anemia. In contrast, some disorders are associated with severe hemolytic crises for which animals may need to receive supportive therapy, including blood-typed and, if previously transfused, crossmatch-compatible transfusions. Cats with increased erythrocytic osmotic fragility or PK deficiency develop marked splenomegaly and may be helped with splenectomy by removing a major site of erythrocyte destruction; however, it appears that PK- and PFK-deficient dogs have not benefited from this procedure. Finally, affected and carrier animals should not be used for breeding in order to prevent the further spread of these disorders. However, in case of a narrow gene pool or due to many desirable traits, a carrier may be bred to a clear animal, as long as all offspring considered to be used for breeding are also screened.

Hemoglobin Defects including Porphyrias

In contrast to the common occurrence of thalassemia and sickle cell anemia in people, no hemoglobinopathies have been documented in dogs and cats. Isolated cases of methemoglobinemia associated with cytochrome-b5 reductase deficiency were found among dogs of various breeds and domestic shorthair cats. Note, hereditary methemoglobinopathies are associated with erythrocytosis rather than anemia.

Porphyrias are a group of inborn errors resulting from accumulation of porphyrins due to deficient activities of specific enzymes in heme biosynthesis seen thus far only in cats but not dogs. In humans, they are clinically classified as either erythroid with cutaneous involvement or hepatic with acute neurovisceral attacks. All porphyric cats had erythrodontia (brown discolored teeth which fluoresced pink), porphyrinuria, and a mild compensated hemolytic disorder, but no evidence of acute life-threatening neurovisceral attacks or cutaneous lesions (with near normal life-expectancies). Urinary uroporphyrin I and coproporphyrin I concentrations were increased in the porphyric cats. However, cats with **acute intermittent porphyria (AIP)** had high urinary 5-aminolevulinic acid and porphobilinogen concentrations, while the cat with **congenital erythroid porphyria (CEP)** had normal or low levels of these porphyrin precursors. Cats with AIP had <50% of normal hydroxymethylbilane synthase (HMBS) activity in erythrocytes and tissues. In contrast, the erythrocyte and hepatic uroporphyrinogen III-synthase (UROS) activity was markedly decreased in the cat with CEP. Sequencing the feline *HMBS* gene revealed different mutations in each of the 8 lines of cats with AIP including duplication, deletions, and missense mutations, making it thus far the most mutated gene in companion animals. Interestingly, 7 cats were heterozygous for the *HMBS* mutations, while one was homozygous for a mild missense mutation. The cat with CEP was homozygous for a missense mutation in *UROS* resulting in drastically reduced enzyme activity and stability. Thus, cats with discolored teeth and normal or mild hemolysis may have either deficient HMBS or UROS activity and represent homologues of human AIP or CEP, respectively. The biochemical and molecular characterization facilitates screening of affected cats. DNA testing is offered through the PennGen laboratory.

Membrane Defects including Stomatocytosis and Increased Osmotic Fragility

Elliptocytosis and **spherocytosis** resulting from a deficiency of the protein band 4.1 and spectrin of the cytoskeleton have been characterized at the molecular level in mongrel dogs. Other presumed membrane abnormalities include stomatocytosis in Alaskan malamutes and in Miniature and Middle Schnauzers, spherocytosis in Golden retrievers with gastritis, nonspherocytic anemia in Beagles, and erythrocytes with increased osmotic fragility in Abyssinian, Somali domestic shorthair cats.

Increased osmotic fragility of erythrocytes suggests a membrane or ion transport defect. Chondrodysplastic Alaskan malamute dwarfs with **stomatocytosis** were the first dogs described with fragile erythrocytes, but the exact mechanism is still unknown. Miniature and Middle Schnauzers with stomatocytosis had no skeletal abnormalities, and they had only mild anemia based on hemoglobin measures. Interestingly while the macrocytosis is severe, the anemia is only mild.

A marked osmotic fragility of erythrocytes associated with intermittent anemia, severe splenomegaly, and hyperglobulinemia has been observed in Abyssinian, Somali and domestic shorthair cats. These erythrocytes are macrocytic, extremely fragile in vitro, and lyse when stored overnight in the refrigerator. Although the cause has not been identified, a membrane defect is suspected. Affected cats with marked splenomegaly may benefit from anti-inflammatory doses of prednisolone. If anemia is severe and frequently recurring, and if there is massive splenomagly, splenectomy may help; splenectomized animals are particularly prone to develop sepsis for the first month post-operatively. Osmotic fragility testing is offered through the PennGen laboratory.

Erythroenzymopathies including PK and PFK Deficiency

Deficiencies of the two key regulatory glycolytic enzymes result in distinctly different forms of hemolytic anemia in several breeds of dogs, while only pyruvate kinase (PK) deficiency has been seen in many breeds of cats. Although PK deficiency was first, and best, characterized in the Basenji breed >4 decades ago, the typical clinical features and biochemical abnormalities appear identical in other canine breeds. In contrast PK-deficient cats seem to have signs more similar to phosphofructokinase (PFK) deficiency in dogs.

Phosphofructokinase (PFK) Deficiency - Despite its discovery 3 decades ago and the availability of enzymatic and DNA screening tests, this glycolytic enzyme deficiency is still common in field-trial English Springer spaniels in the United States, Great Britain, and Denmark but has also been reported in some bench/show dogs as well as Cocker spaniels, Wachtelhunds, and mixed-breed dogs. PFK deficiency is caused by a missense mutation of the muscle-type PFK that results in a truncation and instability of the PFK enzyme protein in all breeds reported except for the Wachtelhund which has another PFK mutation.

The disorder is characterized by hemolytic crises and exertional myopathy. Sporadic dark pigmenturia resulting from severe hemoglobinuria and bilirubinuria is a key feature and commonly develops after episodes of excessive panting and barking, intensive exercise, and fever or high environmental temperature. Consequently, hyperventilation-induced alkalemia results in intravascular lysis of PFK-deficient erythrocytes. During these crises, affected dogs may become severely anemic and icteric, and show fever, lethargy, and anorexia, which usually resolve within days. Furthermore, affected dogs totally lack PFK activity in muscle; therefore, they have a metabolic myopathy characterized by exercise intolerance, occasional muscle cramps, and mildly increased serum creatine kinase activity. As they cannot run hard and fast for any length, they will perform poorly as field-trial dogs.

Simple mutation-specific tests are offered by PennGen and other laboratories to accurately diagnose PFK-deficient and carrier dogs. For those breeds without a known PFK mutation an enzyme activity or hemoglobin-oxygen dissociation curve test can be helpful. Situations triggering hemolytic crises should be avoided such as excessive panting, barking, exercise, and heat. When experiencing a crisis, their lethargy will help in their recovery as they will be resting, but they may require symptomatic and transfusion therapy. PFK-deficient dogs may have a normal life span, but have persistent hyperbilirubinuria and reticulocytosis despite a near normal hematocrit because of the high hemoglobin-oxygen affinity of PFK-deficient erythrocytes. Many are presumptively treated for immune-mediated hemolytic anemia for months to years and thereby are subjected to unnecessary diagnostics and potentially harmful treatments.

Pyruvate Kinase (PK) Deficiency - Despite the severity and persistency of the anemia in PK-deficient dogs, the clinical signs, except for pallor, are mild until a triggered hemolytic crises at any age, or onset of the terminal phase at a few to several years of age. The anemia is highly regenerative, with numerous circulating metarubricytes and reticulocyte counts as high as 90%. An unexplained progressive myelofibrosis and osteosclerosis of the bone marrow and generalized hemosiderosis/-chromatosis with associated hepatic failure develop, causing death usually before 6 years of age, though some PK-deficient West Highland white and Cairn terriers and Beagles have reached 10 years of age. Erythrocytes completely lack the adult erythrocyte isozyme form of PK known as R-PK. Instead, they express a fetal M-PK form that is also present in the spleen and white blood cells. However, M-PK appears to be completely malfunctioning in erythrocytes in vivo, as shown by the shortened erythrocyte survival and metabolic block. The molecular genetic basis of PK deficiency has been identified in Basenjis, Beagles, Labrador Retrievers, Pugs, and West Highland White terriers, and mutation-specific DNA tests are available for these breeds, but not for others. PK deficiency has also been reported in Miniature Poodles, Eskimo toys, Miniature Dachshunds, and Chihuahuas. It appears probable that the previously described nonspherocytic hemolytic anemia and osteosclerosis in Poodles was caused by a PK deficiency. In these breeds, a cumbersome PK-enzyme test with isozyme characterization is required to define PK deficiency. Carriers do not express the M-PK form and have half the normal PK activity; however, differentiation between carriers and homozygous normal dogs based on enzyme activity can be difficult. Clinical signs are mild, likely due to chronic adaptation to severe anemia and high erythrocytic DPG concentrations facilitating oxygen delivery. Hepatosplenomegaly may result from severe extravascular hemolysis, extramedullary hematopoiesis and hemosiderosis/-chromatosis. Splenectomy has not been shown to be effective in dogs. Bone marrow transplantation has been experimentally proven effective for both enzymopathies in dogs, but because of the likely lack of a major histocompatibility complex compatible donor and the need for severe bone marrow suppression, this treatment is not clinically offered.

In cats, PK deficiency causes intermittent anemia with a mild to moderate regenerative response, but cats do not develop osteosclerosis. Affected cats may develop bilirubin gall bladder calculi and can have mild splenomegaly. Anti-inflammatory doses of prednisolone and in severe cases splenectomy appears to ameliorate the clinical signs of intermittent anemia, with the oldest such cat reaching 11 years of age. Erythrocyte PK activity is severely reduced, and there is no M-type PK expression, thereby simplifying the biochemical diagnosis. In the late 1990s, we documented that PK deficiency is caused by a single splicing defect resulting in a 13 base deletion and a DNA screening test for PK deficiency in Abyssinian, Somali, and several other breeds of cats as well as domestic shorthair cats and testing has been available since then from PennGen and more recently many other laboratories. Based upon the screening data this specific PK-mutation is widespread on different continents. Any anemic Abyssinian and Somali should be screened for PK deficiency, as this is a much more likely cause of anemia than infectious and immune-mediated disorders.

Reduced Erythropoiesis including Hereditary Cobalamin Malabsorption

Whereas the previous defects resulted in shortened erythrocyte survival and regenerative anemias, the production and maturation disorders are reflected not only in a nonregenerative anemia but also in changes in other bone marrow-derived cells. Selective cobalamin (vitamin B_{12}) malabsorption, also known as Imerslund-Gräsbeck syndrome, resulting from the ileal intrinsic cobalamin receptor defect has been reported in several breeds of dogs. Giant Schnauzers and Australian shepherds have mutations in the *AMN* gene and Beagles and Border collies have mutations in *CUBN*. Affected animals fail to thrive, and have varied degrees of cachexia, neurologic signs, leukopenia, thrombocytopenia, anemia, low serum cobalamin levels and a methylmalonic aciduria. They respond completely to parenteral cobalamin administration every 2-4 weeks. A detailed description is given in another presentation.

REFERENCES

Available from authors.

Supported in part by a grant from the NIH OD 010939.

A MULTIMODAL IMAGING APPROACH FOR CHRONIC SMALL ANIMAL RESPIRATORY DISORDERS

Isabelle Masseau, DVM, DACVR, PhD; Carol Reinero, DVM, DACVIM, PhD
Columbia, MO

OVERVIEW

Chronic respiratory disease is commonplace in dogs and cats, and can be broadly broken into upper airway disease, lower airway disease, pulmonary parenchymal disease (including pulmonary vascular disease), and pleural cavity disease. There are numerous diagnostic tests available to help determine the specific type of respiratory disease. Survey radiography is an important initial step in the work up of chronic respiratory disorders, but is complemented by other imaging modalities including fluoroscopy, ultrasonography, computed tomography, magnetic resonance imaging, nuclear scintigraphy and positron emission tomography. This lecture will focus on clinical cases of chronic respiratory disorders in dogs and cats in a case based format.

INITIAL DIAGNOSTIC APPROACH

A thorough history and physical examination are key initial steps in the diagnostic work up of chronic respiratory disease. While many of these disorders will have longstanding clinical signs which are obvious to owners, chronic respiratory disorders may also manifest more acutely, especially in cats. Waxing and waning clinical signs may lead to owner perception that a particular episode of respiratory clinical signs is acute, making it important that clinicians get a complete history including prior occurrence of respiratory signs. Type of clinical signs, duration and severity should be assessed as well as other systemic signs (anorexia, weight loss, lethargy, etc.). Owners should be questioned about the pet's environment and if there are any triggers to the clinical signs. If there are other pets in the household, are they similarly affected? Is there a history of travel to areas of the country endemic for certain infectious diseases? Physical examination can reveal clues to help localize the site of disease and should include a visual and auditory examination of the respiratory tract. Visual examination includes abnormal discharge or airflow of the nasal passages; facial asymmetry; easily inducible cough on tracheal palpation; labored inspiration (indicative of impedance to airflow due to disease in the upper airways or pleural cavity, or an enlarged abdomen impairing the ability of the diaphragm to contract normally), and labored exhalation (indicative of impedance to airflow from lower airway/intrathoracic airway disease) or both. An auditory exam includes sound heard without or with the use of a stethoscope. The presence of stertorous or stridorous breathing can localize the problem to the upper airways. Wheezing can localize disease to the lower airways. Attempts to elicit a cough by firm but gentle tracheal compression may lead to paroxysmal coughing in dogs and cats most commonly with primary airway disease; however, any cause of prolonged coughing can lead to secondary tracheal irritation and an easily elicited cough. Using a stethoscope, adventitial sounds (wheezing or crackles) as well as cardiac abnormalities (murmurs, arrhythmias) should be noted. The history and physical examination should be adequate to narrow down the site of respiratory disease to the upper airways, lower airways, pulmonary parenchyma or pleural cavity, and subsequent diagnostic testing can proceed from there.

SURVEY RADIOGRAPHY

Survey radiography may be helpful to rule out obvious cardiac disease or pleural abnormalities as causes of respiratory disorders. Superimposition of structures, lack of patient cooperation leading to difficulty in positioning and the need for significant bone loss [reported to be at least 30 % to 50% loss of calcium before radiographic detection] often limits accurate evaluation of the nasal cavities. Pharyngeal and laryngeal masses (nasopharyngeal polyp, tonsilar neoplasia, nasal neoplasms extending caudally) may sometimes be discerned radiographically when lesions are severe enough and cause mass effects although advanced imaging (CT or MRI) is often needed to better determine extent of lesions and involvement of surrounding structures. As discussed in cases of acute respiratory disorders, evaluation of tracheal collapse may also be limited on survey radiography if the collapse is intermittent. Compared to fluoroscopy, radiography tends to underestimate the degree of collapse throughout the length of the trachea.[1] Additional factors such as redundancy of the trachealis membrane may be confounding in the interpretation of the collapse; fluoroscopy or computed tomography are more ideal modalities in these scenarios. Moreover, evaluation of the lower airways with radiography often lacks sensitivity and specificity in both dogs and cats.[2,3] Pulmonary parenchymal patterns are often non-specific and other diagnostics are often necessary.

FLUOROSCOPY

Assessment of patients with chronic respiratory disorders related to the trachea, the bifurcation and larger bronchi is greatly improved from fluoroscopic exam compared to radiography alone. Fluoroscopic provides "real-time" evaluation of these structures and this can be done in different conditions (at rest, right after running, while coughing) depending on what the owner reports being the trigger for the clinical signs. Fluoroscopy also allows documentation of intermittent collapse and cranial lung lobe herniation, this latter not usually being appreciated on radiography. Fluoroscopy allows guiding of tracheal stent placement and monitoring of their placement and function. In dogs with muscular dystrophy or other neuromuscular disorders affecting the diaphragm, fluoroscopy may also provide monitoring of diaphragmatic motion during spontaneous breathing and document progression of disease.

ULTRASONOGRAPHY

Ultrasound may be particularly beneficial to allow evaluation of pulmonary nodules/masses and differentiation between abscesses and solid masses particularly when lesions are in proximity to the thoracic wall. Abscesses will tend to demonstrate a thick hyperechoic capsule while the center of the lesion will be filled with echogenic fluid, sometimes swirling. Solid masses may be homogenous or heterogenous in echogenicity, and sometimes may be cavitated. Observation of blood supply to the mass usually strongly supports the presence of a neoplastic process in contrast to abscessation. Pleural diseases may also be easily documented with ultrasound as the costal parietal pleura is intimately connected to the endothoracic fascia and intercostal muscles. The presence and amount of fluid as well as location (unilateral vs. bilateral; cranial vs. caudal) may be easily assess as fluid constitutes the ideal window for transmission of ultrasound waves. Pleural nodules may also be differentiated sonographically from fluid in contrast to the similar opacity seen on radiography between soft tissue and fluid. Ultrasound may also be useful to guide sample collection for cytologic assessment or culture with or without the need for sedation and thus avoiding general anesthesia. Interestingly, non cardiac ultrasound-guided fine needle aspirates can provide a 91% yield for cytologic assessment.[4]

COMPUTED TOMOGRAPHY

With advances in computed tomography (CT) technology, modern CT scanners can now provide faster and more accurate assessment of patients related to presence and location of pathology, as well as the extent of disease and involvement of surrounding structures than radiography.[5] Additionally, it provides cross-sectional anatomic information. Traditionally, the requirement of general anesthesia for CT limited this modality in patients with chronic respiratory disease or presenting in acute respiratory distress. Nowadays, faster acquisition of images with 32-row or 64-row detector CT scanners permits rapid evaluation of critical patients without the need of general anesthesia or with mild sedation. Information gained by CT imaging in comparison to radiography has been shown to lead to a change of diagnosis in several cases.[5] Furthermore, CT is more sensitive than radiography for the detection of osteolysis involving the nasal cavities and cribriform plate and infiltration of nasal neoplasms into the calvarium or orbit. The information gained by CT in patients affected with nasal disease helps facilitate discrimination between neoplastic and non-neoplastic disorders, allowing for more appropriate treatment options and a more accurate prognosis.

MAGNETIC RESONANCE IMAGING

Magnetic resonance imaging (MRI) may be useful to determine extent of disease in terms of distribution and severity. It is particularly valuable in patients with nasal and nasopharyngeal disease. MRI may also allow better detection of foreign objects localized to the nasal cavities or that have migrated into deeper soft tissues than CT or radiography due to the localized inflammatory reaction of the body triggered by the presence of foreign material. Artifacts related to respiratory and cardiac motion as well as low proton density of the lungs has limited the use of MRI for evaluation of intra-thoracic respiratory disorders. Some of these limitations may be overcome by cardiac and respiratory gating. Elaboration of protocols suited for evaluation of lower airways and pulmonary parenchymal disease, particularly those resulting in increased lung density, will probably provide valuable information in terms of more accurate diagnosis and elaboration of appropriate therapeutic plan in the upcoming years. Nonetheless, the necessity for general anesthesia, longer acquisition time and more expensive cost may remain important limiting factors for the use of MR imaging in veterinary patients with chronic respiratory disorders in comparison to other imaging modalities.

SCINTIGRAPHY

Nuclear scintigraphy is particularly useful to document ciliary dyskinesia in veterinary patients, a condition in which the respiratory cilia lining the luminal surface of the airways have an impaired function leading to absence of clearance of mucus containing fluid, bacteria and particles from the airways. Veterinary patients with ciliary dyskinesia are often presented with continuous mucoid nasal discharge, sneezing and coughing and signs may be observed early in life. Chronic rhinitis, sinusitis, bronchiectasis and bronchopneumonia that reoccur after discontinuous use of antibiotics are common in patients with this condition. In brief, absence of proper mucociliary clearance can be demonstrated by depositing a small volume of radioactive particle into the distal portion of the trachea near the bifurcation and measuring the time it takes for the radioactive material to travel along the trachea towards the larynx using sequential imaging with a gamma camera. In patients with ciliary dyskinesia, the radioactive material often remains immobile or is observed to have significantly delayed travel time in comparison with normal dogs.

POSITRON EMISSION TOMOGRAPHY

Positron emission tomography (PET) is a functional imaging technique in which a positron-emitting radionuclide (radiotracer) is injected intravenously and paired gamma rays emitted by the tracer are detected and represented in three-dimensional images. One of the most common radiotracer used is ^{18}F-fluorodeoxyglucose (FDG), which is an analogue of glucose that concentrates into tissues with increased metabolic activity such as those affected with infections or neoplasms. ^{18}F-FDG PET is often combined with CT or MRI to provide precise anatomic location of abnormal tissue. This technique can be particularly useful to differentiate quiescent processes (granulomas) from tumors or infections affecting the respiratory tract. It can also be used to monitor response to treatment as the activity of the cells will likely decrease with appropriate treatment sometimes prior to measurable changes in size on CT or MRI.

REFERENCES

1. Macready DM et al. *JAVMA* 2007;230:1870-1876.

2. Mantis P et al. *JSAP* 1998;39:518-520.
3. Gadbois J et al. *JAVMA* 2009;234:367-375.
4. Reichle JK and Wisner ER. *Vet Radiol Ultrasound* 2000;41:154-162.
5. Prather AB et al. *Vet Radiol Ultrasound* 2005;46:114-121.

A MULTIMODAL IMAGING APPROACH FOR ACUTE SMALL ANIMAL RESPIRATORY DISORDERS

Isabelle Masseau, DVM, DACVR, PhD; Carol Reinero, DVM, DACVIM, PhD
Columbia, MO

OVERVIEW

Dogs and cats commonly seek veterinary care for acute onset of respiratory clinical signs including but not limited to sneezing, nasal discharge, stertor, stridor, cough, increased rate or effort of breathing, overt respiratory distress and cyanosis. These respiratory disorders range from uncomfortable to life-threatening. A wide variety of diagnostic tests are available to veterinarians to more specifically identify the underlying abnormality. While nasal, cervical and thoracic radiographs have traditionally played a central role in respiratory disease diagnosis, there are many other types of imaging that can be employed in diagnosis and management of small animal respiratory diseases. These imaging modalities will be highlighted in this lecture based on their contributions to a series of clinical cases.

CLINICAL SIGNS HELP DICTATE ANATOMIC LOCATION OF LESIONS

Imaging techniques allow for visualization of anatomic structures to determine if the tissues are grossly normal or abnormal. If abnormal, imaging can characterize the type of lesion, precisely pinpoint its location, determine if it is dynamic or static and allow for targeted sampling. Dogs or cats presenting for sneezing or nasal discharge have disease present in the nasal cavity. Sterterous breathing can reflect a lesion in the nasal passages, nasopharynx or pharynx. Animals that have reverse sneezing generally have nasopharyngeal disease. Dogs or cats with stridor have laryngeal obstruction or extrathoracic tracheal obstruction. Wheezing reflects intrathoracic airway obstruction. Coughing may originate from the respiratory tract, including the upper or lower airways or the pulmonary parenchyma. Coughing may also originate from cardiogenic causes (i.e., left-sided or biventricular congestive heart failure leading to cardiogenic pulmonary edema or mitral insufficiency and its sequela of left atrial enlargement compressing the left mainstem bronchus). Respiratory distress may be due to upper or lower airway obstruction, flail chest, severe abdominal enlargement, pulmonary parenchymal disease, pleural cavity disease, pulmonary thromboembolism or "look-a-like" disorders which increase the rate or effort of breathing without being associated with respiratory dysfunction (e.g., drugs, severe anemia, acidosis, pain, etc.).

WHAT IMAGING CONTRIBUTES TO THE DIAGNOSIS OF RESPIRATORY DISORDERS

In general, imaging techniques can be two or three dimensional and allow capture of static or dynamic images. Thus it is a means of global assessment of anatomic structures of the respiratory tract either at a moment in time or with inhalation and exhalation. Some lesions may only be present during a particular phase of respiration—e.g., intratracheal collapse is best appreciated on exhalation. In acute small animal respiratory disorders it is important to remember that early lesions may not be effectively captured with some imaging techniques. Serial monitoring may be necessary if initial imaging has negative results; alternatively, complimentary diagnostics may need to be performed. However, especially in cases of respiratory distress, survey radiography represents a relatively quick and global means to rule in or out important disorders which is critical to streamline the diagnostic plan. A wide variety of other diagnostic imaging techniques including fluoroscopy, ultrasonography, computed tomography, magnetic resonance imaging, nuclear scintigraphy and positron emission tomography potentially have a role in acute respiratory disorders of dogs and cats.

SURVEY RADIOGRAPHY

Radiography may provide a quick assessment of the thoracic cavity without the need for chemical restraint and often allows ruling out cardiogenic and pleural causes associated with acute respiratory signs. In cases of upper respiratory signs associated with the nasal cavities, nasopharynx, pharynx or larynx, radiography is often limited due to superimposition of structures and generally requires heavy sedation or general anesthesia to obtained radiographic images of diagnostic quality. Tracheal collapse can often be missed on radiographic images due to the common dynamic nature of the collapse and the lack of significant change in tidal volume during respiratory phases. Moreover, sensitivity and specificity of radiography for the diagnosis of lower respiratory disease is low.[1]

FLUOROSCOPY

Fluoroscopic imaging has proven to be particularly useful for documentation of dynamic collapse of the trachea, tracheal bifurcation and principal (mainstem) bronchi as well as for the presence of cranial lung lobe herniation. In contrast to radiography, fluoroscopy provides a "real-time" assessment of anatomical structures and allows clinicians to recapitulate conditions in which patients have been reported to demonstrate clinical signs. For example, some dogs will demonstrate tracheal collapse only when coughing is elicited which can be induced by irritation of the trachea using a wooden spoon to apply gentle external tracheal pressure while simultaneously observing the airways with fluoroscopy. In addition to providing assessment of the major airways, fluoroscopy may allow evaluation of the diaphragmatic motion during spontaneous breathing and currently remains the best tool to document unilateral or bilateral diaphragmatic paralysis in small animals. Limitations to fluoroscopy include superimposition of structures due to the two-dimensional display of anatomical structures, poor assessment of pulmonary parenchymal lesions and radiation exposure to personal restraining the patient.

ULTRASONOGRAPHY

Ultrasonography has gained tremendous popularity among veterinary practitioners in the recent years due to several advantages over other imaging modalities including real-time imaging, absence of radiation exposure, accessibility and relatively inexpensive cost for ultrasound units as well as lack of need for chemical restraint in most patients. Ultrasonography is limited for assessment of upper airway disorders at the exception of the larynx for which ultrasonography has been shown to be beneficial in cases of laryngeal paralysis. Ultrasound examination of the thorax is limited to superficial structures (thoracic wall, parietal and visceral pleura, cranial mediastinum) as the normally gas-filled lungs create a barrier for penetration of ultrasound waves. However, sonographic imaging of the lungs is usually possible if peripheral lesions are present and air is replaced with fluid or tissue (e.g. abscesses, primary or metastatic neoplasia, hemorrhage, inflammatory cellular infiltration and atelectasis). In cases of acute respiratory signs associated with pleural effusion or pneumothorax, ultrasonography may provide guidance for needle placement during fluid or air collection. Ultrasound may also help in monitoring recrudescence of pleural abnormalities. Ultrasonography may also be useful in cases of acute lung lobe torsion for which lack of blood flow in the affected lung lobe may be evaluated with color Doppler. In critical patients with respiratory distress for which general anesthesia is contraindicated and therefore diagnosis cannot be reached using bronchoscopy and bronchoalveolar lavage, ultrasonography may allow cytologic evaluation and culture/sensitivity of the peripherally affected lungs through fine-needle aspirates.

COMPUTED TOMOGRAPHY

With the advance of computed tomography (CT) technology, CT has proven to be particularly useful for the assessment of acute respiratory disorders. Advantages of modern CT scanner over radiography include fast acquisition of images (e.g. 1 mm thickness images of an entire feline thorax can be acquired in less than 5 seconds using a 64-row detector CT scanner), avoidance of chemical restraints critical in patients with respiratory distress, lack of superimposition of anatomical structures, acquisition of 3 plane (e.g. horizontal, sagittal, transverse) images allowing for 3D volume rendering and virtual endoscopic imaging. The 3D volume rendering may provide surgeons with more information in preparation for surgery in cases of acute upper airway obstruction related to traumatic events (displacement of bone fragments, hemorrhage, etc.). The high resolution algorithms also provide unprecedented diagnostic quality images allowing more precise localization and extent of lesions compared to radiography. CT has been shown to be more accurate in documenting acute lung lobe torsion[2] and pulmonary thromboembolic disease when combined with angiography compared to radiography.[3]

MAGNETIC RESONANCE IMAGING

Although not commonly used for the diagnosis of acute respiratory disorders in small animals mostly due to the longer time of image acquisition and necessity for general anesthesia, magnetic resonance imaging has been shown to better determine the extent of nasal neoplasia[4] and nasopharyngeal disease. Gated MRI has also the potential to be useful for the diagnosis of specific acute respiratory disorders related to the lower airways such as pulmonary thromboembolic disease. In human medicine, development of functional lung imaging using hyperpolarized gas or ^{19}F atoms is promising and perhaps could be applied to veterinary patients in the near future and alleviate the use of radioactive agents or ionizing radiation.

SCINTIGRAPHY

Traditionally pulmonary scintigraphy has been the gold standard for evaluation of pulmonary thromboembolic disease in small animals. Lung scintigraphy consists of inhalation of radioactive aerosols for assessment of ventilation and intravenous administration of radiopharmaceuticals for evaluation of lung perfusion ("ventilation-perfusion scans"). Disadvantages related to this method include low temporal and spatial resolution and the need for patients to be temporarily kept in radiation isolation which can be a limiting factor in critical patients needing immediate medical attention. In recent years, scintigraphy has been somewhat replaced by CT angiography which offers faster assessment, improved spatial resolution and no need for radiation isolation.

POSITRON EMISSION TOMOGRAPHY

Positron emission tomography can be useful for the detection of masses involving the airways. The most common positron emitter used in veterinary medicine is ^{18}F fluorodeoxyglucose which is uptaken by metabolically active cells often found in tumors or site of infections.

REFERENCES

1. Gadbois J et al. *JAVMA* 2009;234:367-375.
2. Schultz RM et al. *JSAP* 2009;50:360-363.
3. Jung J et al. *Vet Radiol Ultrasound* 2010;51:288-293
4. Petite AF and Dennis R. *JSAP* 2006;47:529-536

SYMMETRIC DIMETHYLARGININE (SDMA) AS A KIDNEY FUNCTION BIOMARKER

Jean A. Hall, DVM, PhD, DACVIM (Small Animal Internal Medicine)
Corvallis, OR

BACKGROUND

Early indicators of kidney damage include a reduction in glomerular filtration rate (GFR). Later indicators include impaired urine concentrating ability and azotemia [elevated serum creatinine and blood urea nitrogen (BUN) concentrations]. Measurement of GFR is the gold standard method for estimating renal function and staging kidney disease.[1] Because measuring GFR is technically cumbersome and expensive, serum creatinine concentration remains the standard surrogate for GFR.[2] However, serum creatinine has limitations as a marker of kidney function, most notably insensitivity because it remains in the normal reference interval (with a flat slope over much of the GFR range) until GFR is reduced approximately 75%.[3] In addition, other non-renal factors influence serum creatinine, including endogenous production by muscle such that muscle mass, breed, and sex may influence serum creatinine concentration. Although serial evaluations of serum creatinine within the same animal (if available) increase its sensitivity for detecting progressive changes in GFR,[2] there is need for a better biomarker for earlier detection of CKD.

Serum symmetric dimethylarginine (SDMA) is a byproduct of protein methylation. Although phosphorylation is the best understood post translational modification of proteins, arginine methylation is gaining prominence.[4] Arginine methylation occurs when the nitrogens of arginine within polypeptides are post translationally modified to contain methyl groups. There are three main species of methylated arginine: monomethylarginine (MMA), asymmetric dimethylarginine (ADMA), and SDMA.[4] The methyl group is transferred from S-adenosylmethionine to a nitrogen of arginine by one of several protein arginine methyltranferases (PRMT).[4] Subsequent protein degradation of methylated proteins by hydrolysis yields individual methylated arginine amino acids, which are biologically active. ADMA functions as an endogenous nitric oxide synthase (NOS) inhibitor,[5] preventing production of nitric oxide from L-arginine. Although SDMA is not a direct inhibitor of NOS, it can compete with arginine for transport across membranes,[5] and thus, indirectly reduces nitric oxide synthesis by limiting L-arginine supply.[5,6] ADMA is primarily degraded to dimethylamine and citrulline (≥ 80%) after metabolism by dimethylarginine dimethylaminohydrolase (DDAH), whereas only a small amount of intact ADMA (≤ 20%) is eliminated by the kidneys.[7,8] SDMA is excreted primarily (≥ 90%) by renal clearance.[8,9] Because SDMA is excreted by the kidneys, plasma concentrations are affected by changes in GFR. ADMA accumulates in patients with renal dysfunction, although it is less abundant than SDMA, and its accumulation may be related to renal parenchymal damage, resulting in reduced DDAH expression and activity, rather than to reduced glomerular filtration of ADMA.[8]

Symmetric dimethylarginine (SDMA) has been shown to be an accurate and precise biomarker for calculating estimated GFR in humans,[5] as well as a more sensitive biomarker than serum creatinine for assessing renal dysfunction.[10] In healthy human subjects from the Framingham offspring cohort, dietary components were not associated with plasma SDMA concentration.[11,12] SDMA concentrations correlated positively to age and serum creatinine concentration, and negatively to body mass index, effective GFR, and diastolic blood pressure.[13] A meta-analysis of 18 studies involving human patients showed that SDMA concentrations correlated highly with inulin clearance (r = -0.85) and serum creatinine (r = 0.75).[14] In another study with human subjects, a close correlation among GFR (measured by an iodothalamate clearance technique) and serum creatinine with SDMA was reported (r = -0.84 and 0.89, respectively).[15] Plasma SDMA concentrations have also been shown to be increased in cats with chronic kidney disease and to correlate with plasma Cr concentration (r = 0.74).[7]

SDMA, UNLIKE CREATININE, INCREASES IN GERIATRIC CATS AS GFR DECLINES WITH AGE

A recent study in older cats (n=32), mean age 14.0 years (range 8.3 to 19.6 years), that were fed complete and balanced foods that met the nutrient requirements for adult cats is summarized. Serum concentrations of renal function markers (serum creatinine and SDMA) and GFR were measured at baseline, and at 1.5, 3, and 6 months. Body composition was determined by dual-energy x-ray absorptiometry. There was no significant change in renal function or total-lean-body weight in individual cats over time of study (6 months). There was, however, a benefit of using serum SDMA vs. serum creatinine to monitor renal function in older cats with less total-lean-body weight. Compared with cats <12 years of age, cats >15 years had a lower total-lean-body weight (P<0.01), lower GFR (P<0.05), and lower serum creatinine concentration (P<0.05). However, serum SDMA concentrations were higher in the older cats (P<0.01). This study shows that serum SDMA concentration is a sensitive indicator of change in renal function in cats, and unlike serum creatinine, which declines with age in cats with muscle wasting, SDMA increases as GFR declines with age.

SDMA ALLOWS EARLIER DETECTION OF CHRONIC KIDNEY DISESE IN CATS COMPARED WITH SERUM CREATININE

In another study in cats with chronic kidney disease (n=21), SDMA and serum creatinine concentrations were determined retrospectively from historical data or banked serum samples. Both SDMA and serum creatinine concentrations were significantly correlated to GFR. SDMA increased above the upper reference limit of 14 µg/dL before (mean 14.6 months; range 1.5 to 48 months) creatinine increased above the upper limit of the reference interval (2.1 mg/dL). Earlier detection of chronic kidney disease is desirable for initiating renoprotective interventions that may slow progression of disease.

COMPARISON OF SERUM SDMA AND CREATININE AS KIDNEY FUNCTION BIOMARKERS IN HEALTHY GERIATRIC DOGS

A recent study in older dogs (n=81), mean age 10.4 years (range 8 to 14 years), that were fed complete and balanced foods that met the nutrient requirements for adult dogs is summarized. Serum concentrations of renal function markers (serum creatinine and SDMA) and GFR (determined by iohexol clearance) were measured at baseline, and at 1.5, 3, and 6 months. Body composition was determined by dual-energy x-ray absorptiometry. There was no significant change in total-lean-body weight in individual dogs over time of study (6 months). All dogs showed an increase in GFR (P<0.001), which was correlated to a decrease in SDMA (P<0.001) and a decrease in serum creatinine (P<0.001) concentrations over the 6-month feeding period. Of note, older dogs (10.4 to 14.2 years) had significantly lower total-lean-body weight (P<0.001), lower GFR (P<0.001), and higher SDMA concentrations (P<0.001) compared with younger dogs (7.9 to 10.3 years). Serum creatinine concentrations were not influenced by age (P=0.3), likely because serum creatinine was positively correlated to total-lean-body weight (P<0.01), such that the decline in total-lean-body weight with age offset any effect of decreasing GFR on serum creatinine. This study shows that serum SDMA concentration is a sensitive indicator of change in renal function in dogs, because it is not affected by change in muscle mass with aging.

SDMA ELEVATES EARLIER THAN SERUM CREATININE IN DOGS WITH CHRONIC KIDNEY DISEASE

In a retrospective study, 8 dogs with CKD (diagnosis based on >30% reduction in GFR) and with normal serum creatinine and SDMA at the time of diagnosis, were followed for 3 years. The aim of the study was to determine the point at which these dogs developed elevated serum SDMA and creatinine concentrations, in order to determine if SDMA increases earlier than creatinine. Two of the 8 dogs currently have normal SDMA and creatinine concentrations. In the other 6 dogs, SDMA increased above the upper reference limit of 14 µg/dL before (mean 17 months; range 11 to 26 months) creatinine increased above the upper limit of the reference interval (1.5 mg/dL). These results suggest that serum SDMA is an earlier biomarker than serum creatinine for diagnosing and monitoring CKD in dogs also.

REFERENCES

1. Von Hendy-Willson VE, Pressler BM. An overview of glomerular filtration rate testing in dogs and cats. **Vet J** 2011;188:156-165.
2. Polzin DJ. Evidence-based step-wise approach to managing chronic kidney disease in dogs and cats. **J Vet Emerg Crit Care** (San Antonio) 2013;23:205-215.
3. Finco DR, Brown SA, Vaden SL, et al. Relationship between plasma creatinine concentration and glomerular filtration rate in dogs. **J Vet Pharmacol Ther** 1995;18:418-421.
4. Bedford MT, Richard S. Arginine methylation an emerging regulator of protein function. **Mol Cell** 2005;18:263-272.
5. Bode-Boger SM, Scalera F, Kielstein JT, et al. Symmetrical dimethylarginine: a new combined parameter for renal function and extent of coronary artery disease. **J Am Soc Nephrol** 2006;17:1128-1134.
6. Closs EI, Basha FZ, Habermeier A, et al. Interference of L-arginine analogues with L-arginine transport mediated by the y+ carrier hCAT-2B. **Nitric Oxide** 1997;1:65-73.
7. Jepson RE, Syme HM, Vallance C, et al. Plasma asymmetric dimethylarginine, symmetric dimethylarginine, l-arginine, and nitrite/nitrate concentrations in cats with chronic kidney disease and hypertension. **J Vet Intern Med** 2008;22:317-324.
8. Schwedhelm E, Boger RH. The role of asymmetric and symmetric dimethylarginines in renal disease. **Nat Rev Nephrol** 2011;7:275-285.
9. Kielstein JT, Boger RH, Bode-Boger SM, et al. Marked increase of asymmetric dimethylarginine in patients with incipient primary chronic renal disease. **J Am Soc Nephrol** 2002;13:170-176.
10. Dixon JJ, Lane K, Dalton RN, et al. Symmetrical dimethylarginine is a more sensitive biomarker of renal dysfunction than creatinine. **Critical Care** 2013;17:P423.
11. Kakimoto Y, Akazawa S. Isolation and Identification of Ng,Ng-Dimethylarginine,Ng,N'g-Dimethylarginine, Nepsilon-Monomethyllysine, Dimethyllysine, Glucosylgalactosyl-Hydroxylysine and Galactosyl-Delta-Hydroxylysine from Human Urine. **J Biol Chem** 1970;245:5751-&.
12. Paiva H, Lehtimaki T, Laakso J, et al. Dietary composition as a determinant of plasma asymmetric dimethylarginine in subjects with mild hypercholesterolemia. **Metabolism** 2004;53:1072-1075.
13. Schwedhelm E, Xanthakis V, Maas R, et al. Plasma symmetric dimethylarginine reference limits from the Framingham offspring cohort. **Clin Chem Lab Med** 2011;49:1907-1910.
14. Kielstein JT, Salpeter SR, Bode-Boeger SM, et al. Symmetric dimethylarginine (SDMA) as endogenous marker of renal function--a meta-analysis. **Nephrol Dial Transplant** 2006;21:2446-2451.
15. Fliser D, Kronenberg F, Kielstein JT, et al. Asymmetric dimethylarginine and progression of chronic kidney disease: The mild to moderate kidney disease study. **J Am Soc Nephrol** 2005;16:2456-2461.

CAVEATS OF CONVENTIONAL TESTS FOR ESTIMATING GFR AND THE ROLE OF SDMA

Mary B. Nabity, DVM, PhD, DACVP
College Station, TX

INTRODUCTION

Glomerular filtration rate (GFR) can be estimated using a variety of techniques, including plasma and urinary clearance of an injected substance (*e.g.*, iohexol, creatinine) and "spot" or single time point tests measuring endogenous indicators of GFR (e.g., serum creatinine (sCr), urea nitrogen (UN)). For both methods of GFR estimation, various biological, technical, and analytical factors must be considered when using these tests to diagnose and monitor chronic kidney disease (CKD) in dogs and cats. This session will focus on the limitations of conventional tests for estimating GFR and summarize several recent studies of symmetric dimethylarginine (SDMA) in dogs.

CONVENTIONAL ESTIMATES OF GFR

Plasma clearance

Renal or plasma clearance of an injected substance is thought to be the most accurate estimate of GFR and therefore a more sensitive means for detecting early CKD than spot methods of GFR estimation. However, this technique suffers from several considerable limitations, with the major one being the need for serial, precisely timed blood draws. The most accurate clearance calculation requires as many as 8 post-injection blood samples over 6 hours or longer, although reasonable estimates can be obtained with limited sampling (*i.e.*, 2 or 3 post-injection samples). Timing of these limited sample collections varies depending on the substance used.[1] Some studies have found that calculation of plasma clearance based on a single post-injection sample is strongly correlated with 3-sample techniques, as long as an estimated volume of distribution can be determined.[2] This would be particularly helpful in cats, where multiple collections can prove difficult.

Another limitation with plasma clearance is the large amount of variability in what is considered to be "normal" in dogs and cats. For example, in one study of 118 healthy dogs, mean iohexol plasma clearance was approximately 2.5 mL/min/kg. However, the calculated reference interval included values as low as 0.95 mL/min/kg and as high as 4.25 mL/min/kg, depending on the size of the dog.[3] When using a weighted mean analysis to evaluate previously published studies of various clearance techniques in healthy dogs and cats, mean clearance demonstrated a narrower range: 3.05-3.91 ml/min/kg (dogs) and 2.52-3.05 ml/min/kg (cats).[4] However, 95% confidence intervals were much wider. Therefore, it is difficult to define a normal GFR in a particular animal without a baseline for that patient, and it limits the ability of plasma clearance to detect early reductions in GFR.

In addition to biological considerations, analytical considerations in plasma clearance calculations are important. For instance, when using a limited sampling technique, a correction formula must be applied to correct for the initial distribution phase in order to avoid overestimation of the GFR. Correction formulas for both dogs and cats are available when using iohexol.[5,6] Normalization to body weight, surface area, or extracellular volume has been recommended, but it is not clear which normalization technique should be used in dogs and cats. Also, one must consider the analytical variability of each post-injection measurement, which contributes to overall variability of the GFR estimation.

Determination of plasma clearance can often be time and cost-prohibitive, particularly when monitoring renal function of a patient over time. In addition, because of the numerous technique-specific, biological, and analytical factors that must be considered when performing plasma clearance techniques, it is best performed by practitioners with adequate experience with and knowledge of these factors. Given these significant limitations in determining plasma clearance, it is desirable to have an endogenous marker that can accurately reflect GFR.

Serum creatinine (sCr)

The routine endogenous marker currently used for estimating GFR in both humans and animals is sCr. Its major limitations in veterinary medicine include its dependence on muscle mass and breed variations in dogs. Therefore, sCr may overestimate renal function in cachectic, geriatric, and very young patients, and, as with plasma clearances, a reference interval is often not appropriate. With ongoing muscle wasting, monitoring of sCr over time is a less reliable indicator of declining GFR. However, with early renal disease, muscle mass is usually stable, and careful monitoring and trending of fasted sCr concentrations can allow for early detection of renal disease manifested by a declining GFR. Trending is the serial determination of sCr in an individual patient to look for significant increases that are likely to reflect worsening renal function. Serum creatinine concentration nicely lends itself to trending because it demonstrates very little intra-individual variation in healthy, adult animals, even over several years.[7] Therefore, small increases within the reference interval can reflect significant decreases in GFR in an individual patient, and a reference interval will not be helpful when using sCr to identify the earliest declines in GFR due to kidney disease in most patients. Serial measurements of sCr correlate strongly with renal clearance of an exogenous substance,[8,9] and subtle increases in sCr can identify a decrease in GFR at a similar point in disease progression (unpublished observations). While trending of sCr can be useful in the early detection of renal disease, many patients present to a clinic with no prior bloodwork. In this case, the various factors that can influence sCr need to be considered (e.g., breed, age, muscle mass, diet/fasting status, hydration status) when evaluating a single value. Any concerning values, even within the reference interval, should prompt further evaluation.

When trending sCr, one must be aware of both its biological and analytical variability. Biological variability in an individual patient is best determined by bloodwork performed during routine health checks. As long as at least 3 measurements have been obtained, one can calculate the reference change value (RCV) for a patient.[10] This value represents that at which one can be 95% confident that an increase or decrease in the analyte has occurred. Some reference laboratories are incorporating similar statistical analyses into their reports to allow easy trending of various blood analytes. If enough previous bloodwork results are not available for a particular patient, another guide that can be used is the RCV determined in a population of dogs with mild (sCr < 2 mg/dl) but stable renal disease. Total (biological + analytical) variability of sCr was determined over a 3-week period in these dogs, and the RCV was 0.2 mg/dl, meaning that an increase in sCr of 0.2 mg/dl would indicate a statistically significant increase in this marker when sCr < 2 mg/dl.[11]

In addition to biological variability, analytical variability is another important factor in the assessment of sCr. Most reference laboratory instruments have excellent precision in their sCr measurement, having a coefficient of variation < 5%. However, the difference between sCr measurements in the same sample can be as much as 0.2 mg/dl (18 µmol/L) in mildly azotemic samples using the same instrument[12] and much higher in moderately to markedly azotemic samples or if different instruments are used.[12,13] These studies illustrate the high imprecision and bias possible with some instruments. Further, many in-clinic instruments have minimal, if any, quality assurance programs in place to ensure optimal performance. Based on both the bias and imprecision found in instruments commonly used in veterinary laboratories, the ASVCP Quality Assurance and Laboratory Standards Committee has set their total allowable error (TE_a) guideline for sCr at 20%.[14] This means that analytical variability alone could potentially account for an increase or decrease in sCr of ≥ 0.2 mg/dl. For these reasons, it is particularly important that serial determinations of sCr are measured on the same instrument, ideally one that is subjected to a strict quality assurance program.

SYMMETRIC DIMETHYLARGININE (SDMA)

SDMA has shown great promise as an endogenous marker of GFR in humans with CKD as it appears to be exclusively eliminated by glomerular filtration, and significant extra-renal influences on its production and elimination have not yet been identified.[15,16] An accurate and precise assay for SDMA has been developed for measurement in dogs and cats.[17] A recent study found no differences in SDMA concentration in whole blood (EDTA), serum (red top and serum separator tubes), and plasma (lithium heparin, EDTA, and sodium citrate) obtained from dogs and cats.[18] In addition, SDMA was stable in these samples for at least 7 days at both 4°C and room temperature, and it was not altered with freezing in serum or plasma.[18]

Another recent study determined the reference limit for SDMA in a population of 120 healthy dogs.[19] These dogs were all > 1 year old and represented a variety of breeds. The reference interval for SDMA was calculated to be 6-13 µg/dl, with an upper reference limit set at 14 µg/dl,[19] which is similar to that determined in healthy human subjects (6.4-13.1 µg/dl)[20] and cats (upper reference limit: 14 µg/dl (unpublished data)). In this same study, the reference interval for sCr in this population of dogs was 0.6-1.5 mg/dl.[19]

Week-to-week variability of SDMA, sCr, and iohexol clearance was determined in 22 female dogs that were carriers for X-linked hereditary nephropathy (XLHN), a population of dogs with mild but stable CKD.[11] Measurements were obtained each week for 3 weeks, and the RCV was calculated. No significant differences in the week-to-week variability of SDMA, sCr, and iohexol clearance were found.[11]

In another study, SDMA was measured serially in male dogs with rapidly progressing CKD due to XLHN and compared with iohexol clearance and sCr.[9] Measurements commenced when the dogs were clinically healthy, with normal renal function, and continued until dogs reached end-stage renal disease. Both SDMA and sCr correlated strongly with GFR estimated using iohexol clearance.[9] Notably, when using reference intervals, SDMA identified a decrease in GFR earlier than sCr. However, when both were trended over time, no major differences in initial identification of declining GFR were observed, noting that their body condition score was normal at this relatively early stage (unpublished observations).

The similar reference limit in multiple species and ability to detect declining GFR earlier than sCr when based on a reference interval supports that SDMA might be more suitable to evaluate renal function within the context of a reference limit than plasma clearance or sCr. In addition, a single measurement of SDMA might be sufficient to detect mild decreases in GFR when trending is not possible. SDMA could prove especially useful in the initial diagnosis of CKD in those patients with poor muscle mass, for which sCr might not provide a reliable estimate of GFR. Additional studies must be performed to verify whether or not there are any significant extra-renal influences in both dogs and cats.

REFERENCES

[1]Paepe et al. **J Vet Intern Med** 2010;24(3):679. [2]Katayama, et al. **J Vet Intern Med** 2013;27(1):17. [3]Bexfield, et al. **J Vet Intern Med** 2008;22(1):66. [4]Von Hendy-Willson & Pressler. **Vet J** 2011;188(2):156. [5]Heiene et al. **J Vet Intern Med** 1999;13:587. [6]Finch et al. **Am J Vet Res** 2011;72(12):1652. [7]Braun et al. **Vet Clin Pathol** 2003;32(4):162. [8]Nabity, et al. **J Vet Intern Med** 2012;26(2):282. [9]Nabity, et al. **J Vet Intern Med** 2013;27(3):733. [10]Walton. **Vet Clin Pathol** 2012;41(2):175. [11]Nabity, et al. **J Vet Intern Med** 2013;27(3):734. [12]Ulleberg et al. **Acta Vet Scand** 2011;53(1):25. [13]Braun et al. **Vet Rec** 2008;162(7):215. [14]Harr et al. **Vet Clin Pathol** 2013;42(4):424. [15]Schwedhelm et al. **Nat Rev Nephrol** 2011;7(5):275. [16]Pedersen et al. **Res Vet Sci** 2006;80(3):336. [17]Padmanabhan, et al. **J Vet Intern Med** 2013;27(3):736. [18]Yerramilli, et al. **J Vet Intern Med** 2013;27(3):737. [19]Rentko, et al. **J Vet Intern Med** 2013;27(3):750. [20]El-Khoury, et al. **Anal Bioanal Chem** 2012;402:771.

CANINE PANCREATITIS: NO SUCH THING AS A TYPICAL CASE

Michael Willard, DVM, MS, DACVIM-SA
College Station, TX

DIAGNOSIS

History and physical examination are critically important, but are not sufficient to diagnose canine pancreatitis. History and physical examination are most useful for finding other problems that may be mimicking acute pancreatitis. Minature Schnauzers and Yorkshire terriers are well known for being prone to develope acute pancreatitis. However, these breeds get a lot of other diseases besides pancreatitis that cause vomiting. In addition, acute pancreatitis can develope in any dog of any breed. Canine pancreatitis is classically described as presenting with acute vomiting and anorexia. Abdominal pain is frequently present, but it is easy to miss during physical examination. Fever is occasionally seen. Some dogs (especially those with pancreatic abscesses) may have relatively mild, intermittent, unimpressive vomiting and continue to eat a reasonable amount. Severely ill patients may present in classic systemic inflammatory response syndrome (SIRS – which used to be called septic shock). Many dogs with very severe acute pancreatitis present as though they had an acute, septic abdomen. Some dogs with pancreatitis have no abdominal fluid while others have small amounts that can be seen on ultrasound. A few have copious amounts of abdominal fluid. Many dogs with acute pancreatitis seem to develope the disease in association with ingestion of fat or lipemia or with developement of diabetic ketoacidosis. Trauma and drugs can also cause canine pancreatitis, but are uncommon causes.

The immunoreactive canine pancreatic lipase assay (i.e., cPLI or Spec cPL) appears to be the most sensitive test for canine pancreatitis (approximately 85-90% sensitive). The SNAP PL is relatively comparable in sensitivity.[1] There are a few false negative results with both tests; however, the cPLI is more sensitive than serum cTLI or serum amylase or serum lipase.[1] The hard question is how specific the cPL is for clinically important disease (i.e., lesions of the pancreas that are causing clinical disease as opposed to microscopic lesions that are clinically unimportant).[2,3] Many of us believe that the cPLI test can be positive in animals with clinically unimportant histologic changes in the pancreas. The biggest advantage of the cPLI test is that if it is negative, pancreatitis is probably not the important clinical problem. In such a case, you need to look very hard for extra-pancreatic disease in the dog.

The serum pancreatic specific elastase-1 test has been suggested to have roughly similar sensitivity to the cPL test, at least for the more severe forms of pancreatitis.[4] However, this test is not commonly performed in the US and we tend to have minimal experience with it.

Imaging is very valuable in the diagnosis of acute pancreatitis. Probably the most greatest value of abdominal radiographs is that they help eliminate other diseases that could be causing signs similar to those caused by pancreatitis. Occasionally, one finds radiographic signs which specifically suggest acute pancreatitis. A sentinel loop (i.e., a dilated, air-filled segment) in the descending duodenum, and/or lack of serosal detail in the upper right abdominal quadrant, and/or lateral displacement of the descending duodenum on the ventro-dorsal projection, and/or a mass medial to the descending duodenum (on the ventro-dorsal projection) and/or a mass just behind the liver and just below the pylorus (on the lateral projection) are somewhat suggestive of acute pancreatitis. These findings are only meaningful if present; many dogs and cats with acute pancreatitis do not have any of these radiographic findings.

Ultrasonography is the most useful imaging modality for looking for canine pancreatitis. It may either find changes that are suggestive of acute pancreatitis and/or it may help eliminate other potential causes of vomiting and abdominal pain. Ultrasonography has been suggested to be about 40-70% sensitive in finding canine pancreatitis. One may sometimes detect hypoechogenicity surrounded by hyperechoic fat in the region of the pancreas that is due to pancreatitis. At other times, a markedly thickened pancreas may be found. Both findings are very specific evidence of pancreatitis. Evidence of extrahepatic biliary tract obstruction is very suggestive of acute pancreatitis in the dog. Any dog with extra-hepatic biliary tract obstruction that is not due to a mucocoele should be assumed to be affected with acute pancreatitis until proven otherwise. Please note that the ultrasonographic appearance of the pancreas can change dramatically within a few hours, so repeating abdominal ultrasound on the same day is not necessarily a bad idea if you strongly suspect pancreatitis but are finding nothing. Combining ultrasound with blood tests is probably the best combination for diagnosis.[4]

Chronic pancreatitis can be much more challenging to diagnose than acute pancreatitis. Diagnosing exocrine pancreatic insufficiency (EPI) with a low cTLI value in a breed of dog that is not commonly affected with pancreatic acinar atrophy (e.g., German shepherd, rough-coated Collie) makes chronic pancreatitis the most likely diagnosis. This is probably the easiest way to diagnose chronic pancreatitis. Otherwise, dogs with episodic vomiting due to recurrent bouts of pancreatitis may not have any other signs of disease, and they invariably are admitted to your clinic for a work up after the last bout has run its course or is on the mend. Episodes of vomiting and anorexia due to recurrent pancreatitis can be random and unpredictable. In such patients, the previously mentioned diagnostics may be attempted, especially when acute exacerbations occur. Very rarely, upper gastrointestinal barium contrast radiographs may rarely reveal duodenal abnormalities (e.g., dilatation, stricture) which suggest that recurrent bouts of acute pancreatitis have caused scarring of the pancreas which in turn have compromised the maximum size of the duodenal lumen. Ultrasonographic changes are nice if they are present, but they can be minor making it difficult to accurately interpret them. Feeding an ultra-low fat diet for 3-4 times longer than what was previously the longest interval between episodes may sometimes be helpful in making a presumptive diagnosis.

Pancreatic abscesses in dogs are invariably sterile. Affected dogs can have a much more chronic, smoldering course (e.g., vomiting for a month or more, mild loss of appetite) than most dogs with more typical acute pancreatitis although some have acute, severe course of disease. We have even found a few dogs which had pancreatic abscesses that were completely asymptomatic. Abdominal pain may be present or absent. CBC and serum biochemistry findings are unpredictable. Diagnosis requires ultrasound. Treatment may be surgical marsupalization, percutaneous ultrasonographic drainage or just observation.

THERAPY

As of this writing, there is not a well-designed, robust, prospective, stratified study on the treatment of canine acute pancreatitis. Therefore, all any of us has is opinions, period. Preliminary work from Australia suggests that it is safe and perhaps beneficial to feed dogs with acute pancreatitis per os or with an esophagoscopy tube as soon as they can tolerate it. By "tolerate", I specifically mean that they do not get worse. They are fed even if they are still vomiting.[5] I recommend that a) you feed as low a fat content as possible, and b) if the feeding is associated with worsening of the vomiting, that you stop it and either try again in a day or two or go to jejunostomy feeding. Do not try to get full caloric intake into the patient initially; rather, start with small amounts to see if the patient will hold down the food. Obviously, if feeding is associated with worsening of the vomiting or general condition, stop the feeding.

Fluid therapy is critical, and subcutaneous administration of fluids is clearly inferior to IV fluids for all but mildly affected animals. IV fluid administration is often sufficient, even in dogs in which a pancreatic granuloma has temporarily blocked the main bile duct. Adequate pancreatic circulation is probably crucial for healing damaged pancreatic tissue; therefore, it is probably better to provide a little too much fluid rather than a little too little fluid unless the patient has congestive heart failure or oliguric renal failure. Obese and fat dogs (which describes a lot of dogs with pancreatitis) do not necessarily have skin tenting when they are dehydrated. Classically, you would expect dry, tacky oral mucus membranes in a dehydrated dog; however, a nauseated animal may be salivating enough to make the mucus membranes moist even though it is dehydrated. If a vomiting dog is not eating or drinking, then it is dehydrated regardless of how well hydrated it appears on physical examination. However, if you give substantially too much crystalloid and dilute the serum protein concentrations, this could be detrimental.

One should monitor the serum albumin concentration (NOT the serum total protein concentration or the plasma total solids concentration) during fluid therapy in these patients. If the serum albumin concentration decreases significantly (i.e., to < 2.0 gm/dl), then the plasma oncotic pressure likewise decreases. Since perfusion of the pancreatitis is critical to treating dogs with pancreatitis, one should probably become concerned whenever the serum albumin concentration falls below 2.0 gm/dl. It is very hard to administer enough plasma to significantly raise the plasma albumin concentration. Hetastarch is probably a better choice because it will stay in the circulation and raise the plasma oncotic pressure for much longer than plasma. Human albumin can be used effectively, but it occasionally causes anaphylactoid reactions that can kill the patient; therefore, it is not recommended. Canine albumin is safer, but it does not last as long as hetastarch. One retrospective study has shown that plasma did not help treat dogs with pancreatitis; however, this study suffers from the problems inherent in all retrospective studies.

If the patient cannot tolerate early refeeding (i.e., the vomiting becomes worse with any oral intake), then jejunostomy feeding is another option. It is safer, less expensive, and less dangerous than parenteral nutrition, and has been associated with a better prognosis. In particular, it should be considered if an exploratory laparotomy was performed when the pancreatitis was diagnosed because a J-tube can be placed at that time. Alternatively, one can place a jejunostomy tube via laparoscopy, through a G-tube, and via the nose (naso-jejunostomy).

I prefer to only use antiemetics for short periods of time because I want to see if the patient is improving enough so that it no longer needs the antiemetic to stop vomiting. If the patient is vomiting multiple times per day or obvious feels terrible due to severe nausea, then maropitant (1 mg/kg SQ) is probably the first choice of antiemetic. Maropitant might also have the advantage of providing some analgesia because it blocks Substance P binding. Dolasetron (0.3-1.0 mg/kg qd) and ondansetron (0.25 mg/kg qd) can also be effective, but are usually not as effective as maropitant.

H-2 receptor antagonists have been used to prevent gastric ulceration and erosion. It is doubtful that ulceration/erosion is a common problem in all but the sickest acute pancreatitis patients. If you desire to offer protection to the gastric mucosa, then the proton pump inhibitors are superior to the H-2 receptor antagonists. Pantoprazole (1 mg/kg IV qd) or omeprazole (1.5-2 mg/kg PO bid) are the most commonly used drugs. I suspect that they provide more benefit by being antidyspepsic in nature (thereby making the patient feel better) than they do by preventing ulcers, but that is a guess.

Antibiotics have been used to "prevent" secondary bacterial infection of the inflamed pancreas, which is supposed to be "fertile ground" for infection. However, there is minimal evidence that bacterial infection is of any clinical importance in routine canine pancreatitis. Antibiotics do not hurt these patients, but it is very questionable how helpful they are. However, dogs in SIRS due to pancreatitis may be a different story. Any dog in SIRS from any reason is potentially at increased risk of infection due to severely compromised mesenteric circulation.

Corticosteroids are very controversial in the treatment of pancreatitis. While they increase serum amylase and lipase activities, they do not cause pancreatitis. It is possible (and this is very controversial) that they may be useful in treating patients that are in Systemic Inflammatory Response Syndrome (i.e., SIRS, which used to be called "septic shock").[6] At this time, there are good data showing that it is probably reasonable to give physiologic doses to dogs in SIRS that are not responding well to typical fluid therapy. These dogs may have what has been termed "Critical illness related corticosteroid insufficiency" (CIRCI), that is to say that they are relatively hypoadrenal. This is a controversial statement. If steroid therapy for pancreatitis is contemplated, it should probably be reserved for the severely ill dog which is not responding to fluid resuscitation. If steroid therapy will be administered, you need to warn the owners of the unknown nature of this therapy. A larger question is whether steroids may be useful in treating the inflammation found in severe pancreatitis. At this time we might have minimal evidence that anti-inflammatory doses of steroids might (?) be helpful in some patients (at least the patients did not clearly become a lot worse). However this is very preliminary and anecdotal.

Although heparin therapy used to be the treatment for DIC (which can probably make acute pancreatitis worse), it has not been shown to be useful. If DIC appears to be a major problem, aggressive administration of fresh frozen plasma to replace clotting factors and anti-thrombin III concentrations is probably more effective than administering heparin.

Dogs are clearly much "tougher" than people, and they often hide their pain well. Unless there is some good reason to the contrary, it is best to routinely assume that dogs with acute pancreatitis are in pain and will benefit from analgesic. In very mild cases, butorphanol is usually sufficient. In moderate cases, methodone is reasonable. As the pain becomes more severe, we progress to constant rate infusions of fentanyl. Is the most severe cases, we administer a constant rate infusion of fentanyl, lidocaine and ketamine.

The best to determine if the patient is improving is to look at the patient. CBC, cPLI, and ultrasound may all look worse in a patient that is clinically better. Trust the patient, not the imaging or the lab work, to know if it is improving.

Do not be in a hurry to take dogs with pancreatitis to surgery. That is why it is important to provide the best possible medical management when first confronted with the patient with suspected acute pancreatitis. I will usually try aggressive medical therapy for 5-8 days, depending upon how quickly the patient is decompensating. If I have no indication that optimal medical therapy is helping after 7-9 days, then I might consider surgery. If the patient has an extrahepatic bililary tract obstruction that appears to be due to acute pancreatitis, then I will wait longer, even a week or more, assuming that the patient is improving, before I consider surgery to relieve the obstruction. Almost all of these patients will eventually have the biliary obstruction resolve if aggressively treated medically. If it appears important to relieve biliary tract obstruction (i.e., the problem has been going on for well over two weeks or the biliary tract might be infected), then percutaneous drainage using ultrasound guidance might be appropriate. Rarely a stent can be placed in the bile duct, but stents can have problems. If a pseudocyst or abscess is not found, lavaging the area and resecting obviously necrotic tissue might help the patient, but this is uncertain.

Prognosis is difficult to predict. Hyperbilirubinemia is not necessarily a poor prognostic sign; pancreatic granulomas causing icterus due to obstruction of the bile duct typically resolve if the patient receives appropriate supportive therapy. Hypocalcemia, while a poor prognostic sign in people, is infrequently found in dogs and cannot be used to predict the outcome. Pancreatic abscesses are usually sterile. We cannot correlate the ultrasonographic appearance of the pancreas with prognosis. Finding a degenerative left shift and/or a marked thrombocytopenia (probably due to DIC) are not known to be negative prognosticators, but one intuitively fears that a poor outcome is more likely with such findings. Evidence of SIRS is probably the most important reason to give a poor prognosis.

When to send the patient home is the last question. The patient is sent home when it is doing well. Monitoring the ultrasonographic appearance or the cPLI is much less valuable than observing the patient. If the patient feels good and is eating and is not vomiting, then you can probably send it home regardless of what the ultrasound looks like or how high the cPLI is. When the dog is sent home, it should be on a very strict diet that is free of fat . If the dog ever gets sick again, they should come in and see you sooner rather than later, in case it is pancreatitis that is recurring.

REFERENCES

1) McCord K, et al., **J Vet Int Med** 2012; 26: 888, 2) Trivedi S, et al. **J Vet Int Med** 2011; 25:1241, 3) Mansfield CS, et al. **J Vet Diag Invest** 2012; 24:312, 4) Mansfield CS, et al., **J Vet Diag Invest** 2011; 23:691, 5) Mansfield CS, et al, **J Vet Int Med** 2011; 25: 419, 6) Yarema TC, et al, Crit Care Nurse 2011;31: 16.

DRUG RESIDUES AND ABSORPTION OF TRANSDERMAL METHIMAZOLE IN CATS

Kate Hill, BVSc, MANZCVSc, DACVIM
Palmerston North, New Zealand

INTRODUCTION

Drug penetration and absorption through the skin can be assessed by *in vitro* or *in vivo* methods. In vitro studies are important as experimental conditions can be carefully controlled. Skin layers can be separated and analysed for drug residue. Limitations of in vitro studies that the effects of skin blood flow cannot be evaluated and also the effects of skin metabolism on drug permeation. This talk will discuss the use of diffusion cells to study the transdermal permeation of methimazole in cats.

Diffusion cells

There are several types of diffusion cells for in vitro transdermal penetration kinetics. All of the cells consist of a donor chamber and a receptor chamber between which the skin is positioned, with the stratum corneum uppermost.

The simplest two-chambered diffusion cell incorporating excised skin, is the Franz – type diffusion cell (Figure 1), descrived by Franz in 1975.[1] Diffusion cells must be made of inert material such as glass, Teflon or stainless steel, so that the test material will not be adsorbed during the experiment.

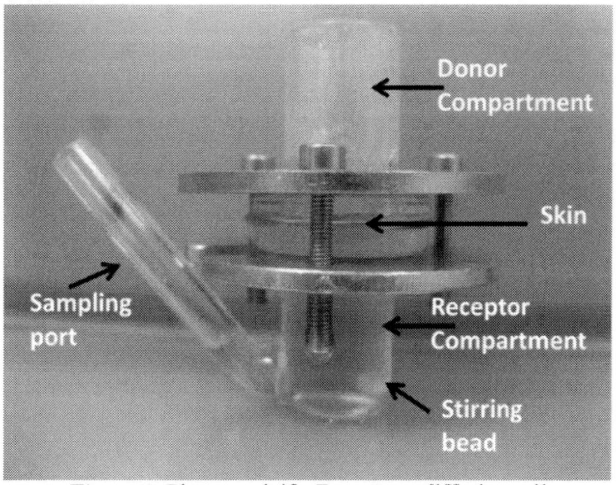

Figure 1. Photograph if a Franz-type diffusion cell

The Franz-type cell was developed to study percutaneous absorption under conditions that simulated real life.[1] The skin is clamped between the upper donor compartment, which is open to the ambient laboratory conditions and lower receptor compartment. The receptor compartment simulates the physical conditions surrounding the subcutaneous tissues. Inside the receptor compartment, a donor solution in which the drug will dissolve in, such as buffered isotonic saline, comes in contact with the dermal surface of the excised skin. To mimic *in vivo* skin temperatures of 32°C (for dogs), the cell is submersed in a water bath, thermostatically controlled to 35°C.[2] To maintain homogenous temperatures in the receptor compartment, and prevent a layer of higher drug concentration forming under the skin, a Teflon coated magnetic stirring bar is inserted. The stirring bar is driven by a magnetic field plate within the water bath. (Figure 2)

An advantage of this type of cell, is that a finite (therapeutic) or infinite (excess) dose of drug can be applied to the stratum corneum within the donor compartment. A finite amount of drug is used when the test ingredient is replicating a clinical application or clinically relevant scenario, ie a marketed formulation of drug at a set dosage. A finite dose may result in a marked depletion in the donor concentration over the duration of the experiment. A finite dose experiment will result in a reduction of the rate of permeation and an eventual plateau effect in the cumulative permeation profile.

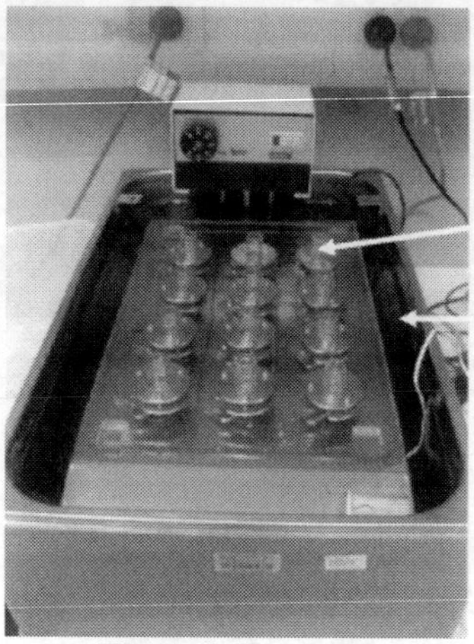

Figure 2. Franz-type cells in a water bath

Receptor solutions

The receptor solution must have adequate solubility for the compound being studied. Adequate sink conditions must be maintained throughout the experiment, which will allow the absorption rate to continue as it would with a functioning circulatory system *in vivo*. Isotonic or buffered saline solutions (pH 7.4) are suitable for water soluble compounds. Preservatives such as sodium azide can be added to prevent microbial build up if experiements are of long duration or if microbes could contribute to skin metabolism. The Organisation for Economic Co-operation and Development (OECED) guidance notes on dermal absorption, state that saline alone may underestimate in vitro absorption for lipophilic compounds. For lipophilic compounds, the receptor fluid should contain solvent mixtures such as ethanol and water (50% aqueous ethanol), <6% polyoxyethelene (20) oleyl ether in water, or 5% bovine serum albumin.[3]

Storage conditions

Storage of skin samples is inevitably required when conducting an *in vitro* experiment. Freezing of skin in dogs has shown there may be a decrease in barrier properties after thawing and a subsequent increase of permeation of hydrocortisone compared to non-frozen skin.[4] However as long as skin is not overly hydrated when frozen, it has been stated that it is unlikely that the permeation characteristics will be significantly different from non-frozen skin.[5]

The permeation experiment

The number of skin replicates recommended in permeation studies is 12. This is due to the high intra and intersubject variability. If groups are being compared, matched skin samples should be used.[5] The Organisation for Economic Co-operation and Development Guideline 428 states:

> *When finite dose conditions of exposure are used, the quantity washed from the skin, the quantity associated with the skin (and in the different skin layers if analysed) and the amount present in the receptor fluid (rate, and amount or percentage of applied dose) should be calculated. Skin absorption may sometimes be expressed using receptor fluid data alone. However, when the test substance remains in the skin at the end of the study, it may need to be included in the total amount absorbed (see Guidance Document, paragraph 66).*

Samples of the receptor solution aree taken at regular intervals, most commonly every 2 hours for 24 up to 48 hours.

Early samples (between 1-4 hours) can be important to determine damaged skin permeability.[5] A permeation profile (ie cumulative amount of drug penetrating per defined skin area versus time) is then plotted.

REFERENCES

1. Franz TJ. Percutaneous absorption - relevance of invitro data. J Invest Dermatol 1975;64:190-195.
2. Mills PC, Magnusson BM, Cross SE. Effects of vehicle and region of application on absorption of hydrocortisone through canine skin. American Journal of Veterinary Research 2005;66:43-47.
3. Organisation for Economic Co-operation and Development : Guidance notes on dermal absorption, . In. http://www.oecd.org/chemicalsafety/testingofchemicals/48532204.pdf: 2011.
4. Ahlstrom LA, Cross SE, Mills PC. The effects of freezing skin on transdermal drug penetration kinetics. J Vet Pharmacol Ther 2007;30:456-463.
5. Finnin B, Walters K, Franz T. *In Vitro* Skin Permeation Methodology. In: Benson H, AC W, eds. Topical and Transdermal Drug Delivery. New Jersey: John Wiley and Sons Inc; 2012:85-130.

TRILOSTANE: BEYOND 3β-HSD

Audrey K. Cook, BVM&S, MRCVS, DACVIM-SAIM, DECVIM-CA, DABVP (Feline Practice)
College Station, TX

INTRODUCTION

Trilostane is a synthetic steroid analogue. Its pharmacological uses reflect its ability to competitively and reversibly inhibit 3β-hydroxysteroid dehydrogenase (3β-HSD), an enzyme required for the production of all steroid hormones.

Trilostane (as Vetoryl®) is licensed for the treatment of hyperadrenocorticism (HAC) in dogs, and is widely used for this purpose. It is now clear however that this agent has complex metabolic effects that extend beyond its recognized impact on 3β-HSD, including alterations in other enzyme activities, effects on cortisol metabolism, and changes in expression of intracellular glucocorticoid receptors (GRs). In addition, routine use in dogs with HAC has been associated with adrenal gland necrosis and permanent hypoadrenocorticism (HOC). This lecture will review the current understanding of the effects of trilostane on endocrine processes beyond its recognized impact on 3β-HSD.

BACKGROUND

The potential pharmacological role of trilostane was first described in 1978 and its ability to mitigate the clinical signs of HAC in humans with Cushing's disease was reported that same year.[1,2] However subsequent studies reported an inconsistent response with no increase in the expected precursors, and trilostane fell out of favor for this use.[3] In contrast, trilostane is predictably effective for the management of spontaneous HAC in dogs, although the dose required to control clinical signs is variable. Is also seems as though the response to treatment cannot be reliably predicted by the adrenocorticotrophic hormone (ACTH) stimulation test, with some dogs doing well despite robust circulating cortisol levels and others showing poor control despite tight control of adrenal synthetic function. It seems likely that some of this variation may be due to the effect of trilostane on other intracellular events, and reflects the difficulties in assessing the metabolic effects of a hormone by simply measuring circulating concentrations. In addition to its use in dogs with HAC, trilostane has been shown to improve hair growth in dogs with hair cycle arrest (HCA; aka alopecia X), although the mechanism for this is unclear.

PHYSIOLOGY OF CORTISOL

Like many hormones, cortisol circulates as both a free (metabolically available) and protein-bound (unavailable) molecule. In healthy animals, > 95% of circulating cortisol is bound to a high-affinity, limited capacity protein carrier (namely corticosteroid binding globulin [CBG]) or to albumin.[4] CBG is made by the liver and is a negative acute phase protein, meaning that levels drop during times of illness, thereby making cortisol more available to the tissues. Routine serum cortisol assays measure total cortisol and do not differentiate between the two forms.

Free cortisol passes easily through cellular membranes; specific transport mechanisms have not been identified. Free cortisol may be converted within the cell to cortisone, which is essentially the inactive form. This process is catalyzed by 11β-hydroxysteroid dehydrogenase (11β-HSD). This enzyme exists in two forms; type 1 which converts cortisone to cortisol (and is essentially a glucocorticoid promoter), and type 2 which converts cortisol to cortisone (an inactivator).[4] Tissue activities of these isoforms are known to vary between species. A recent canine study (using mRNA expression as a marker for enzyme activity) showed high levels of type 1 in the liver, lung and renal medulla, with high levels of type 2 in the renal cortex and adrenal gland.[5] Although both types were found in all tissues examined, higher type 1 expression was reported in all sites except for the adrenal gland, colon and rectum. It is thought that the variable expression of these two forms is required for appropriate physiological responses. For example, high levels of type 2 11β-HSD within the distal tubular cells of the nephron are thought to prevent attachment of cortisol to aldosterone receptors. In essence, the 11β-HSDs are the intracellular gate-keepers of glucocorticoid action.

In order to produce its numerous metabolic effects, cortisol must bind to intracellular GRs. This interaction is complex, and the overall magnitude of the cell's response to glucocorticoids depends both on intracellular cortisol concentrations and the sensitivity of the GR itself, i.e., the efficiency of the GR-mediated signal transduction.[6]

EFFECTS OF TRILOSTANE

Steroidogenic enzymes

For several years, it has been suggested that trilostane impacts more than just 3β-HSD activity. This hypothesis is supported by work looking at the levels of steroid precursors both before and after the proposed site of action. Based on the assumption that trilostane competitively inhibits 3β-HSD, accumulation of pregnenolone, 17α-OH-pregnenolone, dehydroepiandrosterone (DHEA) and androstenediol is expected (see Figure 1). This has been proven in various studies in dogs with both HAC and HCA.[7,8] However, increased or unchanged levels of both 17α-OH-progesterone and androstenedione have also been reported in association with trilostane administration. As these molecules sit below the site of action of 3β-HSD, these results are somewhat unexpected but may be explained by up-regulation of substrate synthesis. Trilostane (a competitive inhibitor of 3β-HSD) has more affinity for the enzyme than the usual substrate (e.g., 17α-OH-pregnenolone), and will decrease production of the product (e.g., 17α-OH-progesterone) unless substrate is markedly increased, in which case the concentration of the product may remain constant. It seems likely that this happens in dogs on trilostane, as endogenous ACTH concentrations predictably increase and will drive the formation of abundant substrate from cholesterol.[7,9] Based on these findings, trilostane must impact one or more steps below the 3β-HSD step, as cortisol production is predictably limited by drug administration. Looking at effects of trilostane on 11- and 21-dexoxycortisol concentrations, it seems likely that trilostane concurrently inhibits 11β-hydroxylase.[7]

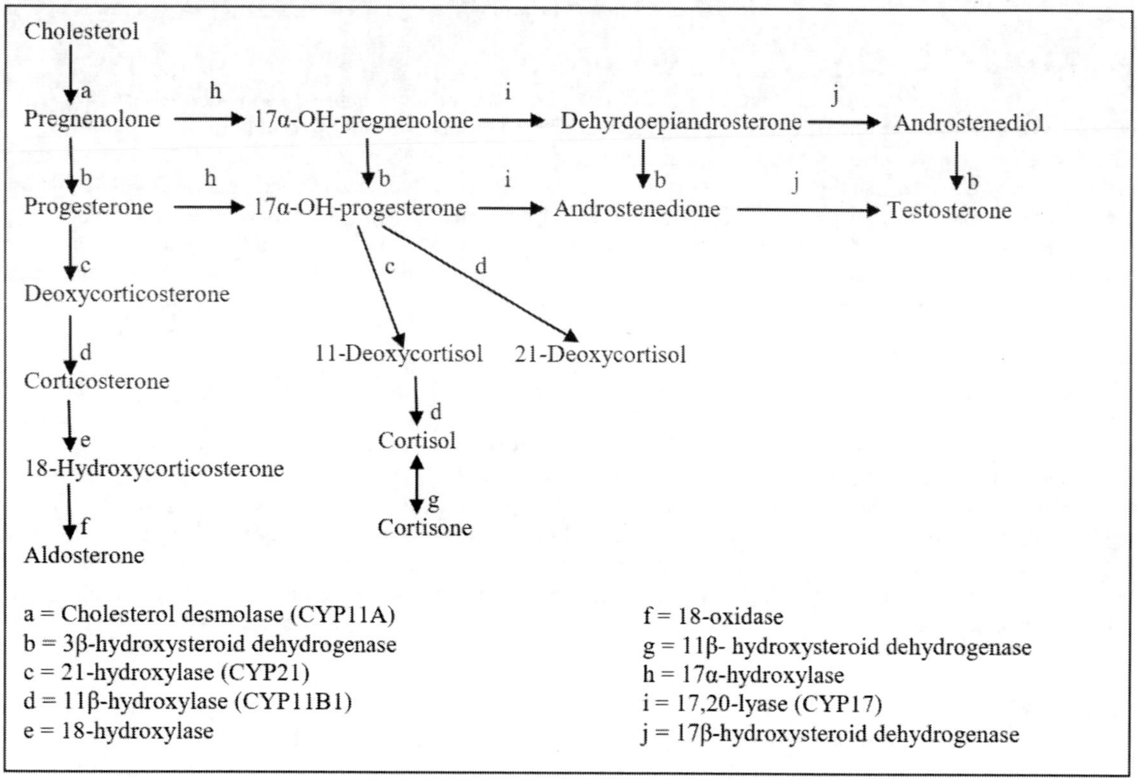

Figure 1. Biochemical pathway for the production of aldosterone and cortisol in the adrenal cortex

Cortisol metabolism

Work performed many years ago suggested that trilostane accelerates the oxidation of cortisol to cortisone by 11β-HSD.[10] This raises the possibility that trilostane may also impact the local tissue endocrine environment in dogs with HAC, independent of its effects on circulating cortisol concentrations. In rodent studies, trilostane administration has been shown to increase type 2 11β-HSD mRNA levels, whilst type 1 levels were unchanged.[11] This shift would reduce intracellular concentrations of cortisol, shifting it to the inactive cortisone. Little is known about the physiology of 11-HSD type 2 gene regulation, although it has been shown to be up-regulated by dexamethasone.

Serum cortisol and cortisone levels in healthy dogs and dogs with pituitary-dependent HAC receiving trilostane have been described.[12] Although both cortisol and cortisone levels were higher in the dogs with HAC before treatment, trilostane administration altered the ratio between the two, with more impact on cortisol concentrations. This finding supports the idea that trilostane may drive the conversion of active cortisol to inactive cortisone. This supports the hypothesis that trilostane potentiates 11β-HSD type 2 activity in dogs. Alternatively, ACTH may impact the activity of the 11 HSD isoforms, with a subsequent relative increase in cortisone production. This would contradict findings in humans, but is still a possibility. It is knows that endogenous ACTH levels rise substantially during trilostane therapy, and it is certainly possible that this impacts intracellular cortisol metabolism.

Glucocorticoid receptors

It has been known for some time that trilostane directly inhibits the function of intracellular estrogen receptors, leading to its use in patients with tamoxifen-resistant breast cancer.[13] If similar effects occur on the GR, tissues may be less sensitive to cortisol, thereby mitigating the signs of HAC even if serum hormone levels are still increased. Studies performed in rats given trilostane showed that GR mRNA levels in rat liver tissue were significantly decreased by trilostane administration, whilst mineralocorticoid receptor mRNA levels increased.[11] It is interesting to note that this change occurred in conjunction with increased 11β-HSD type 2 transcription and translation, further limiting the impact of available cortisol on cellular events.

At the present time, the reason for hair regrowth in dogs with HCA receiving trilostane is unclear. Although increased concentrations of the cortisol precursor 17α-OH-progesterone have been reported in dogs with this disorder, concentrations actually increase with trilostane administration, despite a clear clinical improvement.[8] Potentially, GR expression within hair follicles may be different in affected dogs, resulting in an aberrant local tissue endocrine environment. Administration of trilostane may mitigate this abnormality, with subsequent improvement in hair growth.

Adrenal necrosis

As trilostane is a competitive enzyme inhibitor, initial reports describing acute, severe and persistent hypoadrenocorticism in dogs receiving the drug were unexpected and suggested a direct toxic effect upon the adrenal cortex. Examination of adrenal tissue in one affected dog revealed cortical necrosis and inflammation.[14] Triggers for this event have not been identified, and permanent hypoadrenocorticism may occur after days or months of therapy. In a report describing the histological appearance of the adrenal glands in dogs with HAC receiving trilostane, variable degrees of necrosis were reported, mixed in with the expected cortical hyperplasia.[15] Detailed examination of affected areas suggested an apoptotic component to the necrosis.

Rodent studies now suggest that adrenal necrosis is a consequence of sustained exposure to high levels of endogenous ACTH.[16] Chronic administration of exogenous ACTH to rats resulted in dose-dependent hemorrhage, vacuolization and apoptosis in the adrenal glands. These findings were in contrast to those in control rats and rats receiving trilostane. Although dogs with pituitary-dependent HAC may have endogenous ACTH concentrations above the reference limits at the time of diagnosis, the synthesis of ACTH usually increases substantially with trilostane administration, presumably due to a lack of feedback on the pituitary tumor +/- some degree of hypocortisolemia.[7,17] It seems likely that sustained stimulation of the adrenal cortex by high levels of ACTH may be responsible for the acute adrenal necrosis reported in dogs on trilostane, although direct comparisons between rodent studies and canine patients should be made with caution.

SUMMARY

At this point, our understanding of the effects of trilostane in still incomplete, although it is clear that it has more extensive effects on the production, metabolism and cellular effects of cortisol than first thought. It is also clear that assessing trilostane therapy by simply measuring circulating cortisol concentrations may not provide a complete picture of the effects in a particular patient.

REFERENCES

[1]Potts GO, et al. **Steroids** 1978, 32(2):257. [2]Komanicky P, et al. **J Cin Endocrinol Metab** 1978;47(5):1042. [3]Dewis P, et al. **Clin Endocrinol** 1983;18(6):533. [4]Chapman K, et al. **Physiol Rev** 2013;93(3)1139. 5Seiber-Rucksthul NS, et al. **Horm Metabl Res** 2007;39(8):548. [6]Bamberger CM, et al. **Endocr Rev** 1996;17:245. [7]Sieber-Ruckstuhl NS, et al. **Dom Anim Endocrinol** 2006;31(1):63. [8]Cerundolo R, et al. **Vet Derm** 2004;15(5):285. [9]Witt AL, et al. **Vet Rec** 2004;154:399. [10]Touitou Y, et al. **J Steroid Biochem** 1984;20(3);763. [11]Maluuitre SDM, et al. **J Steroid Biochem Mol Biol** 2006;101(2):97. [12]Seiber-Ruckstuhl NS, et al. **Vet Rec** 2008;163:477. [13]Puddefoot JR, et al. **Cancer** 2002;101:17. [14]Chapman PS, et al. **J Small Anim Pract** 2004;45:307. [15]Reusch CE, et al. **Vet Rec** 2007;160:219-224. [16]Burkhardt WA., et al. **Dom Anim Endocrinol** 2011;40:155. [17]Galac S, et al. **Vet J** 2010;183:75.

ORAL CYCLOSPORINE USE IN DOGS IN INFLAMMATORY AND IMMUNE-MEDIATED DISEASES

Todd M Archer, DVM, MS, DACVIM
Mississippi State, MS

INTRODUCTION

Cyclosporine is a potent immunosuppressive agent with treatment applications in both human and veterinary medicine. It is a cyclic polypeptide derived from the soil fungus *Tolypocladium inflatum*. Cyclosporine was discovered in the Sandoz laboratory by Borel in Switzerland in the 1970s, who demonstrated *in vitro* inhibition of human immune reactions as well as prolonged survival of skin grafts.[1] This data was presented in 1976 at the British Society of Immunology, paving the pathway towards utilization of its immunosuppressive abilities with organ transplantation in human medicine.[1,2] Cyclosporine was first described in human medicine by Calne in 1978 with published data documenting prevention of organ rejection in those receiving renal allografts.[2] Once these results were shown to be repeatable, cyclosporine became and still remains a cornerstone of immunosuppression in human transplant medicine. Cyclosporine gained U.S. Food and Drug Administration approval in 1983 for the treatment and prevention of transplant rejection in human medicine. Novartis Animal Health received FDA approval in 2003 for oral cyclosporine capsules marketed as Atopica® for the treatment of atopic dermatitis in dogs. Oral cyclosporine is currently being applied to a spectrum of inflammatory and immune-mediated diseases in dogs, including but not limited to atopic dermatitis, autoimmune skin disorders, perianal fistula, inflammatory bowel disease, granulomatous meningoencephalitis, and immune-mediated blood disorders such as immune-mediate hemolytic anemia and immune-mediated thrombocytopenia. Cyclosporine has also been used in canine organ transplantation.

ORAL FORMULATIONS

Two oral formulations of cyclosporine exist. The initial formulation of cyclosporine was a vegetable-oil based preparation called Sandimmune®. Sandimmune® demonstrates significant variation in absorption from patient to patient, causing wide intra-individual and inter-individual differences in pharmacokinetics. Because of this, a newer formulation was introduced into the human market by Novartis in 1996. The newer formulation is an ultramicronised formulation which forms a microemulsion upon contact with aqueous fluids and has a much more consistent and predictable absorption.[3] Novartis manufactures the approved human ultramicronised formulation (Neoral®) as well as the veterinary approved ultramicronised formulation product (Atopica®). There are now also generic formulations available, both as the ultramicronised formulation and the non-ultramicronised formulation. The generic ultramicronised formulations demonstrate equivalent product bioavailability to the FDA approved product, but only the FDA approved veterinary product (Atopica®) has undergone extensive pharmacokinetic testing in dogs to demonstrate its clinical utility, and Atopica® is therefore the recommended product. Because of the poor absorption and extreme variability of the non-ultramicronised Sandimmune®-type preparations, they are not recommended in dogs.

MECHANISM OF ACTION

Cyclosporine is a calcineurin inhibitor, thus modulating the adaptive immune system and specifically inhibiting T-cell function, potently inhibiting cell-mediated immunity with a relatively lesser effect on humoral immunity.[4] Calcineurin inhibitors such as cyclosporine begin their biological effects by binding to intracellular proteins called cyclophilins. The most abundant cyclophilin found in T cells is cyclophilin A. The binding of cyclosporine to cyclophilin A facilitates the complex's affinity for calcineurin. When this complex binds to calcineurin, inhibition of calcineurin occurs by inhibition of its phosphatase activity, thus preventing its binding to and subsequent activation of nuclear factor of activated T-cells (NFAT). This facilitates decreased cytokine gene expression, with the most notable cytokine being IL-2. Other cytokines shown to have decreased expression in response to exposure to cyclosporine include IL-4, INF-γ, and TNF-α. The role of IL-2 in the inflammatory process is complex and multifactorial, involving both pro-inflammatory as well as regulatory functions.[5] In T-cells, IL-2 recognition by the IL-2 receptor initiates a signaling cascade which ultimately promotes cytokine transcription, cell survival, cell-cycle entry, and proliferation. IL-2 works within T-cells to further induce the expression of IL-2 receptors. IL-2 remains the most effective cytokine for T-cell clonal expansion. By decreasing IL-2 expression in CD4+ Th1 cells, proliferation and activation of both T-helper and T-cytotoxic lymphocytes is inhibited and the immune response is blunted.

THERAPEUTIC DRUG MONITORING

The assay used to measure cyclosporine blood concentrations as well as the sample type must be considered when interpreting results. Most labs measuring cyclosporine blood concentrations recommend for the measurement of whole blood concentrations. Current methods for measuring cyclosporine blood concentrations include high-pressure liquid chromatography (HPLC) and specific monoclonal antibody radioimmunoassay (RIA) methods. HPLC measures only the parent drug without measuring metabolite levels. RIA methodology measure parent drug as well as metabolites, and the blood concentrations will therefore be higher compared to the same sample being analyzed using HPLC. Although HPLC is labor intensive and not routinely offered for patient monitoring, it is considered the gold standard for measuring cyclosporine blood concentrations. RIA is most often clinically employed, with the laboratory providing the assay typically providing recommendations regarding ideal target blood drug levels.

Veterinary pharmacology laboratory recommendations vary regarding the type of assay used to determine cyclosporine concentrations, the ideal target cyclosporine concentration range for controlling disease, and whether the target should be based on trough blood concentrations, peak blood concentrations or both. Measurement of trough cyclosporine concentrations prevailed for many years in veterinary medicine based on initial work done in renal transplant studies. However, in human medicine, peak cyclosporine concentrations are now well-recognized to provide information regarding therapeutic efficacy that is superior to trough concentrations.[6] Recent recommendations for therapeutic drug monitoring in dogs receiving cyclosporine therapy are based on the evaluation of both peak and trough levels, although specific recommended target concentrations still depend on the laboratory reference ranges as well as the assay used to measure cyclosporine concentrations.

The disposition of cyclosporine in clinical canine patients, coupled with variables such as differing diseases and differences in drug preparation, all can contribute to markedly variable cyclosporine blood concentrations among patients receiving similar dosages and even within the same patient. And even though cyclosporine is gaining favor for treating a wide number of inflammatory and immune-mediated diseases in veterinary medicine, optimal oral dosing protocols in dogs are still unclear. Therapeutic drug monitoring through the measurement of blood cyclosporine concentrations can not only help avoid toxic concentrations but also allow the establishment of the therapeutic range for the individual patient. Then subsequent monitoring allows dosage adjustments to maintain a patient's blood cyclosporine concentration within that range.

PHARMACODYNAMICS

Therapeutic drug monitoring provides evidence that cyclosporine blood concentrations in patients may fall within the estimated therapeutic range, which is established to capture successful therapy in the majority of patients for a given disease or condition. For some patients, this estimation may be valid while, for others, there may not be clinical efficacy and response to therapy despite having blood cyclosporine concentrations within the 'therapeutic range'. This variation in clinical responses to comparable blood concentrations can be attributed to individual variability in biological responses to cyclosporine. Studies in human transplant medicine are now emerging that demonstrate the extra benefits of pharmacodynamic monitoring above those provided by monitoring of blood cyclosporine concentrations alone for adjusting immunosuppressive therapy in individual patients.[7,8]

Pharmacodynamic monitoring allows direct quantitation of the effects of cyclosporine on target T-cells, allowing adjustment of therapy based on measurable immunosuppressive effects even when individual patient responses to comparable blood levels are variable. Several pharmacodynamic biomarkers have been studied in human medicine, including lymphocyte proliferation, enzyme (calcineurin) activity, lymphocyte surface antigens, and intracellular cytokine quantification. In human medicine, pharmacodynamic monitoring has shown individually distinct degrees of calcineurin inhibitor sensitivity in patients. These methods show great promise for optimizing cyclosporine therapy as well as delivering individualized therapy.

Studies performed by our research group have documented measurable suppression of select biomarkers of immunosuppression in T-cells of normal dogs, particularly IL-2 and INF-γ, when T-cells are exposed to cyclosporine. An initial *in vitro* study documented suppression of a panel of biomarkers when T-cells were exposed to cyclosporine.[9] During a subsequent *in vivo* study, measurable suppression of T-cell cytokine expression was noted to occur not only at high dosages of cyclosporine (Atopica®) but also in some dogs at the low dosage which is used for the treatment of canine atopic dermatitis.[10] Some dogs receiving the low dosage had suppression of T-cell function despite very low trough blood cyclosporine concentrations that were in fact well below the target immunosuppressive ranges recommended by most reference laboratories. Our findings call into question the ability of pharmacokinetic monitoring of blood drug concentrations to be the sole technique for assessing the efficacy of cyclosporine therapy in individual clinical patients. Our initial studies utilized flow cytometry to evaluate T-cell expression of cytokines. Flow cytometry, which requires rapid and specialized sample processing, may not however be the ideal tool for the development of pharmacodynamic monitoring in clinical patients, and a quantitative real-time PCR assay may in the long term hold more promise as a method for clinical pharmacodynamic monitoring of patients. We have now validated a qRT-PCR assay for use in clinical patients in which the blood must be processed within 48 hours of collection.[11] We are just in the beginning phases of clinical testing, and over time, the clinical utility of this assay will be determined.

OVERVIEW OF CLINICAL USE

Our pharmacodynamic research findings in dogs are similar to those seen in human medicine, with individual responses being quite variable from dog to dog receiving the same oral dosage as well as in dogs having oral dosages adjusted to maintain comparable blood cyclosporine concentrations. Because of the inherent variable individual responses to cyclosporine, protocols should differ based on the disease severity in the individual patient. Those patients with chronic, nonlife-threatening diseases should have dosing protocols different from more life-threatening immune-mediated diseases.

In dogs being treated with cyclosporine for chronic inflammatory nonlife-threatening diseases such as immune-mediated skin diseases, mild inflammatory bowel disease, and anal furunculosis, patients are often started at lower once daily or twice daily oral dosages. In canine atopy for example, a low once daily dosage (5 mg/kg) has been shown in many studies to be an effective treatment. Therapy is typically delivered long term until remission of disease is achieved.[12,13] Cyclosporine therapy is then slowly tapered down until the lowest effective dosage to maintain disease remission is found. Often patient response to therapy is the main criteria used to decide if therapy is adequate, with blood cyclosporine concentrations usually not measured. Measurement of cyclosporine blood concentrations may be considered in patients who fail to respond to standard cyclosporine dosages.

In more critical and life-threatening diseases such as immune-mediated hemolytic anemia and immune-mediated thrombocytopenia, the best way to utilize cyclosporine therapy is not known. Cyclosporine is often started at a much higher dosages than that used to treat canine atopic dermatitis, with therapy aimed at delivering effective immunosuppression as rapidly as possible. Multiple possible side effects can sometimes affect our patients at these higher dosages, and the drug can be cost-prohibitive for many clients. In contrast to dogs with skin diseases such as atopy, in dogs with more serious illnesses therapeutic drug monitoring is recommended. Cyclosporine blood concentrations are measured during the initial treatment period to ensure that drug concentrations are within a recommended therapeutic window, being high enough for disease control but still no higher than needed in order to minimize adverse drug side effects and unnecessary expense. Monitoring drug blood concentrations is recommended in more critical patients because, in the individual patient, cyclosporine pharmacokinetics are inherently variable and unpredictable even with the ultramicronised formulations, largely due to issues with oral bioavailability. A number of studies demonstrate a wide variability in cyclosporine pharmacokinetics, with differing blood concentrations in individual dogs administered the same dosage. Therapeutic drug monitoring attempts to optimize the dosage for individual patients with varying oral bioavailability. However, there are only a limited number of clinical studies in veterinary medicine which address ideal target therapeutic cyclosporine blood concentrations. Unfortunately, target blood concentrations do not necessarily correlate with clinical outcome or response to therapy. Some of our canine patients respond to therapy although their blood concentrations are less than the recommended cyclosporine target range, and some have not responded even when their cyclosporine blood concentrations were within the recommended target range. With further investigation into the pharmacokinetics and pharmacodynamics of cyclosporine in veterinary medicine, a better assessment of the immune status may be gained when combining these two techniques and allow for optimization of therapy and delivery of individualized treatment.

REFERENCES

[1]Calne R. **Transplant Proc** 2004; 36:13S. [2]Graeb C, et.al. **Transplant Proc** 2004; 36:125S. [3]Robson D. **Vet Rec** 2003;152:739. [4]Robson D. **Vet Rec** 2003; 152:768. [5]Lan RY, et.al. **Journal of Autoimmunity** 2008; 31:7. [6]Cantarovich M, et.al. **Transplant Proc** 1998; 30:1462. [7]Barten MJ, et.al. **Cell Prolif** 2007, 40, 50. [8]Kuzuya T, et.al. **Biol. Pharm. Bull**. 2009, 32, 604. [9]Fellman CL, et.al. **Vet Immunol Immunopathol** 2011; 140:175. [10]Archer TM, et.al. **J Vet Intern Med** 2011, 25:1391. [11]Riggs C, et.al. **Vet Immunol Immunopathol** 2013; 156:229. [12]Steffan J, et.al. **J Am Vet Med Assoc** 2005; 226:1855. [13]Olivry T, et.al. **J Am Vet Med Assoc** 2002; 221:370.

CURRENT UNDERSTANDING OF STOMATITIS IN CATS AND DOGS

Jamie G. Anderson, DVM, MS, AVDC, ACVIM
Concord, CA

INTRODUCTION

"Stoma" is a medical term referring to a "mouth-like" opening. Stomatitis is an inflammation inside of the mouth, usually a small sore or ulcer. It is not a gastric or intestinal problem. Internal medicine specialists are likely to see the refractory cases of stomatitis, when patients have failed routine therapy. Whether it is a cat or a dog these pets are painful. Stomatitis is a chronic condition, with co-morbidities including poor appetite, weight loss, dehydration and behavioral changes. The development of side effects, including diabetes, from chronic steroid therapy is also debilitating. Stomatitis continues to be a frustrating condition of unclear etiology with generally only fair results to invasive treatment options

The discussion will focus on the current understanding of stomatitis in cats and dogs. Current standard of care for diagnostics, dental radiographs, medical and surgical therapies will be presented. Theories on etiology and a description of the histopathology of both diseases will be shared. Thoughts on prognosis, new therapies and current research investigations will round out the presentation.

FELINE CHRONIC GINGIVOSTOMATITIS (FCGS)

Understanding the name

In 1983, one of the first reports of Feline plasma cell gingivitis-pharyngitis was published. It has been variously called Plasmacytic stomatitis-pharyngitis complex, chronic lymphocytic plasmacytic stomatitis complex, Feline lymphocytic plasmacytic gingivitis stomatitis (LPGS) and most recently, Feline chronic gingivostomatitis or FCGS. Inflammation of the glossopalatine arch area is now referred to as caudal stomatitis. Now we recognize two different forms based on location of inflammation: Type 1: cases with alveolar and labial/buccal mucositis/stomatitis only, and Type 2: cases with caudal mucositis/stomatitis (with or without alveolar and labial/buccal mucositis/stomatitis)

Description of the clinical picture

The cat with gingivitis stomatitis complex will usually present to the veterinarian with weight loss, difficulty eating, hemorrhagic ptyalism, unkempt hair coat, halitosis, and behavioral changes which reflect pain. Physical examination abnormalities may include poor body condition, mandibular lymphadenopathy, and signs of dehydration. Non-anesthetized oral examination may be difficult due to pain, especially when attempting to open the mouth wide when there are caudal stomatitis lesions. The examination may reveal a spectrum of involvement from severe ulceration to ulcerative proliferation of the attached gingiva extending onto the adjacent buccal and sublingual mucosa. The tissues may bleed spontaneously. Often there is involvement of the glossopalatine arches and pharynx. Occasionally there will be ulceration and cracking at the commissures of the mouth. The dentition may be either normal or show evidence of significant gingival recession and tooth mobility Tooth resorption may be clinically evident as a cavitation in the coronal portion of the tooth filling in with granulation tissue or as a pink discoloration of the crown. Partial hypodontia is fairly typical and most likely indicates the presence of subgingival root fragments.

Consensus on etiology

The specific etiology of this condition has remained elusive. Currently there is no direct cause and effect for FCGS. Numerous theories have been put forth to explain the initiation or maintenance of this chronic oral inflammatory disease. These include oral bacterial pathogens and periodontal disease, *Bartonella spp.*, Feline Calicivirus (FCV), Herpes virus, tooth resorption, retroviral diseases, and food or other allergens. An immune system aberration is thought to play a vital, as yet poorly defined, role. Recent studies have evaluated the bacteria associated with FCGS. Oral swabs obtained from the palatoglossal folds of eight FCGS cats, revealed *Pasteurella pneumotropica* and *Pasteurella multocida* subsp. *multocida* were the most prevalent species cultured in both normal and FCGS patients. Bacteria identified using bacterial 16S rRNA gene sequencing revealed that *Capnocytophaga canimorsus* was the most prominent species in normal cats. Researchers wondered about the influence of bacteria on the immune response in FCGS and if change in the expression of Toll-Like Receptors (TLRs) might inform about the possible pathogenic agents involved in this disease. They found that FCGS was associated with an increase in the expression of mRNAs encoding several cytokines and TLRs. Putative pathogens *Tannerella forsythia* and FCV appear particularly capable of increasing TLR and cytokine mRNA expression.

Viruses implicated in the initiation or maintenance of FCGS include; Calicivirus, the retroviruses and Herpes virus. The prevalence of Calicivirus in cats affected with stomatitis is reported between 50 and 92%. Eight-five percent of cats with chronic stomatitis are long-term carriers of the virus Years ago, inoculation of healthy specific pathogen free cats with isolates of feline Calicivirus from cats with stomatitis failed to reproduce the disease. The conclusion remains that chronic stomatitis is not induced by Calicivirus alone.

Consensus on diagnostic evaluation

Relative consensus exists on the appropriate diagnostics to perform in a cat with FCGS. Though the clinical presentation typically allows a diagnosis, other differentials should be considered and ruled out. A CBC, Chemistry profile, Urinalysis, FeLV and FIV, and Calicivirus isolation is suggested. There is no consensus on testing for *Bartonella spp.* Microbiologic testing is best performed at a tertiary level institution with individuals experienced in oral microbiology. As reported, the most common clinical pathology abnormality is hyperglobulinemia (polyclonal). Most cases are Calicivirus positive.

Under general anesthesia, a complete periodontal evaluation should be performed and the following assessed: gingival index, plaque index, calculus index, furcation index, periodontal pocket depth, mobility score, attachment loss derived from the summation of pocket depth (mm) and gingival recession (mm), Stomatitis Disease Activity Index Score (SDAI). A full mouth set of dental radiographs is the standard of care. Radiographically, the following parameters should be assessed: alveolar bone loss (mm), furcation involvement, root tips, resorptive lesions, periapical abscess or unexplained bone loss.

Routine histopathology should be submitted if an area looks particularly aggressive compared to the rest of the mouth. Findings from a series of 75 stomatitis cats revealed a strong correlation between FCGS and SCCa in general, and in particular for geriatric felines. Routine histopathologic characterization reveals a heavy infiltrate of plasma cells, ulceration and loss of/or a very thin remaining epithelium with lesser lymphocyte populations and neutrophils. Virus induced cytopathic changes in epithelial cells were not seen. The underlying connective tissue was very vascular with an intense surrounding plasma cell infiltrate.

A recent immunohistochemical characterization of FCGS caudal stomatitis lesions identified increased expression of CD3, CD4, CD8, CD79a, IgG, IgM, IgA, leucocyte antigen 1 (L1) and class II molecules of the major histocompatibility complex (MHC). CD79a+ cells, L1+ cells and CD3+ T cells, and the level of MHC class II expression, correlated with severity of inflammation. CD8+ T cells were more numerous than CD4+ T cells which may indicate induction of an underlying cytotoxic cell-mediated immune response.

Consensus on management

Consensus agreement includes oral hygiene measures including topical formulations, extraction of the teeth caudal to the canines (with removal of all the root fragments), local and systemic analgesia, perioperative antibiotics, and nutritional support through the treatment phase. In essence, treat the mouth first and extract diseased teeth. Steroids are no longer considered a first line course of treatment. Other therapeutics with variable success: NSAIDs, laser fulguration, ESFA supplements, Cyclosporine, Doxycycline, Feline Interferon Omega and most recently pentoxifylline.

Prognosis

Currently there is no consensus on how to assign a prognosis in cats with chronic gingivostomatitis. It is believed that a large percentage of cats (more than 50%) will improve or require no further treatment with extraction of the teeth caudal to the canines. For those cases with Caudal Stomatitis, or those with no dental disease, this rule does not seem to apply.

Current research studies

The Dentistry and Oral Surgery Service at UC Davis School of Veterinary Medicine is accepting FCGS cases with refractory disease after all teeth have been removed. The objective of the study is to determine if injection of stem cells will improve the clinical disease.

CANINE CHRONIC ULCERATIVE STOMATITIS (CCUS)

Understanding the name

Canine chronic ulcerative stomatitis (CCUS), also known as chronic ulcerative paradental stomatitis (CUPS), is a painful, chronic disease condition of the oral cavity.

Description of the clinical picture

CCUS is a condition clinically characterized by chronicity, and pain associated with diffuse oral ulceration and often necrosis. The ulcers typically occur on the labial/buccal mucosal surfaces opposite to plaque on the dentition. Ulcers may also occur in the tongue, glossopalatine folds and mucocutaneous border of the lips. Considerable morbidity occurs with chronic ulceration, as pets refuse to eat, loose weight, drool and are chronically painful. Home care is often difficult due to extreme patient discomfort.

Breed predilection

Breed related ulcerative stomatitis has been reported in Cavalier King Charles spaniels and Maltese dogs. Other breeds commonly affected include greyhounds, Labrador retrievers, and cocker spaniels. In Maltese dogs clinical oral ulceration is common. Unpublished histopathology studies from the 1980's revealed lymphocyte and plasma cell infiltration and complement activation in the lamina propria and epithelium.

Consensus on etiology

The pathogenesis for CCUS remains poorly defined, though an abnormal immune system response to dental plaque is suspected. Uremic stomatitis, Erythema multiforme, Mycosis fungoides, desquamative gingivitis, and autoimmune vesiculobullous diseases are differential diagnoses for CCUS.

Consensus on diagnostic evaluation

There are very few reports of CCUS in the literature. My personal recommendations include complete blood count, chemistry screen, urinalysis, and thyroid testing. A detailed medication history should be taken. Canine Ulcerative Stomatitis Disease Activity Index (CUSDAI) is an objective scoring system that is helpful in determining response to therapy. A set of full-mouth dental radiographs and periodontal indices as per FCGS should be recorded. Other diagnostic considerations are ANA, B12/folate levels, bacterial cultures, and histopathology

Consensus on management

There is no consensus amongst diplomats of the American Veterinary Dental College; though many believe the standard of care is to extract teeth opposite buccal or labial mucosal chronic ulceration. Resolution of some lesions is seen with extraction of the teeth; however, in other cases ulceration does not improve and a vicious cycle of ulceration and scarring ensues. Other important medical and therapeutic management entails oral hygiene at home, antibiotics, pain control, frequent general anesthesia periodontal treatment, and topical treatments at the time of the GA exam. Alternate therapies with marginal success: corticosteroids, sucralfate, metronidazole, and doxycycline. A therapeutic trial of pentoxifylline, doxycycline and niacinamide (PDN) is currently underway and appears to have promise.

Prognosis

Different from FCGS; extraction of all the teeth in a canine can be mutilating, is not always curative and is costly. Prognosis in the future is linked to better knowledge of the disease. I use the following in determining a prognosis for patients: response to periodontal treatment and follow-up home care, then response to PDN triple therapy and finally response to extractions.

Current research study

2013 Foundation for Veterinary Dentistry Grant was awarded to evaluate CCUS.
The goal is to evaluate clinical and radiographic findings, microscopic observations and immunohistochemistry.

REFERENCES

1. Johnessee JS, Hurvitz AI. Feline plasma cell gingivitis-pharyngitis. *J Am Anim Hosp Assoc* 1983; 19:179–181.
2. Harley R, Gruffydd-Jones TJ, Day MJ. Salivary and serum im- munoglobulin levels in cats with chronic gingivostomatitis. *Vet Rec* 2003 152:125–129.
3. Harley R, Gruffydd-Jones TJ, Day MJ. Immunohistochemical characterization of oral mucosal lesions in cats with chronic gingivostomatitis. *J Comp Pathol* 2011; 144:239–250.
4. Dolieslager SMJ, Riggio MP, Lennon A, et al. Identification of bacteria associated with feline chronic gingivostomatitis using culture-dependent and culture-independent methods. *Vet Mi- crobiol* 2011; 148:93–98.
5. Sims TJ, Moncla BJ, Page RC. Serum antibody response to an- tigens of oral gram-negative bacteria by cats with plasma cell gingivitis-pharyngitis. *J Dent Res* 1990; 69:877–882.
6. Anderson JG. Diagnosis and management of gingivitis stomati- tis com plex in cats. *Waltham Focus* 200313(3): 4–10
7. Hennet P. Chronic gingivo-stomatitis in cats: long-term follow-up of 30 cases treated by dental extractions. *J Vet Dent* 1997; 14:15–21.
8. Lommer MJ, Verstraete FJM. Concurrent oral shedding of felinecalicivirus and feline herpesvirus 1 in cats with chronic gingivo-stomatitis. *Oral Microbiol Immunol* 2003; 18:131–134.
9. Dolieslager SMJ, Bennett D, Johnston NW, Riggio MP. Novel bacterial phylotypes associated with the healthy feline oral cavity and feline chronic gingivostomatitis. *Res Vet Sci.* 2013 Jun; 94(3): 428-32.
10. Dolieslager SM, Lappin DF, Bennett D, Graham L, Johnston N, Riggio MP. The influence of oral bacteria on tissue levels of Toll-like receptor and cytokine mRNAs in feline chronic gingivostomatitis and oral health. *Vet Immunol Immunopathol.* 2013 Feb 15; 151(3-4): 263-74.

COBALAMIN MALABSORPTION IN DOGS AND CATS

John C. Fyfe, DVM, PhD
East Lansing, MI

INTRODUCTION

Cobalamin (a.k.a. vitamin B_{12}) is an essential cofactor of two enzymes of mammalian intermediate metabolism. In nature, cobalamin is synthesized only by certain microorganisms, and mammals must obtain the vitamin from their diet. Cobalamin is a micronutrient classified among the water-soluble vitamins, but it is always protein-bound during absorption, plasma transport, and intracellular activity. Cobalamin concentrations in natural or adequately formulated commercial diets for carnivores are never limiting, so cobalamin deficiency in dogs and cats is caused by GI malabsorption. Cobalamin malabsorption can result from a variety of acquired or inherited conditions and can be cobalamin-specific or part of a more generalized malabsorption disorder.

GASTROINTESTINAL COBALAMIN ABSORPTION

Gastrointestinal absorption of dietary cobalamin is a complex process comprising a sequence of protein-binding events, the success of which each depend on the longitudinal secretory and absorptive organization of the GI tract.[1,2] Dietary cobalamin is released from foodstuffs in the stomach by digestive enzymes. In this low pH milieu cobalamin binds to haptocorrin, a heavily glycosylated, non-specific, corrinoid-binding protein found in salivary, stomach, and pancreatic secretions, bile, neutrophils and monocytes, and milk. Pancreatic proteases degrade haptocorrin in the proximal small intestine and as pH of the intestinal chyme rises, cobalamin is released from haptocorrin and binds to intrinsic factor. Intrinsic factor is a highly specific cobalamin-binding glycoprotein secreted by gastric epithelial cells in humans, but it is also secreted by pancreatic ducts cells in dogs, and exclusively by pancreas in cats. In all 3 species, there is a specific endocytic receptor called cubam on the epithelial brush border of distal small intestine. Cubam is so named because it is a heteromeric protein composed of two gene products, cubilin and amnionless.[3] Cubam expression is restricted to the ileum in humans and cats, but extends to mid-jejunum in dogs. Cubam binds the IF-cobalamin complex at near-neutral pH, but will not bind IF or cobalamin alone. Cubam-mediated endocytosis delivers the complex to lysosomes, from which cobalamin is released by a specific exporter. Cobalamin released on the basolateral side of enterocytes is bound to transcobalamin, a cobalamin-specific plasma transport protein, for distribution throughout the body. Approximately 15 μg/day of recirculating cobalamin is taken up by hepatocytes and secreted in bile bound to haptocorrin for enterohepatic recirculation.

The general outline of GI cobalamin absorption and responses to cobalamin deficiency are common among dogs, cats, and humans, but there are species-specific differences that are diagnostically and therapeutically noteworthy. For instance, early investigations mistakenly assumed 1) gastric mucosa was the sole source of intrinsic factor (IF) in dogs, and 2) cobalamin deficiency in dogs would cause macrocytic anemia. Such errors led to misinterpreted experimental data and published conclusions that an intrinsic factor- and ileal receptor-mediated process of cobalamin absorption was not important in dogs and cats. Better experimental design in the 1980s and discovery of inherited cobalamin malabsorption attributable to dysfunction of specific proteins have corrected this historical misstep. Though complicated, GI cobalamin absorption is efficient and specific; daily dietary requirements are 2-3 μg.

Primary (inherited) GI cobalamin malabsorption occurs when gene mutation causes absence or malfunction of intrinsic factor, cubilin, amnionless, the lysosomal cobalamin exporter, or transcobalamin. Of these, cubilin or amnionless mutations causing cobalamin malabsorption have been reported in dogs (see below) and none so far in cats. These cobalamin-specific malabsorption disorders are uncommon but breed-specific and autosomal recessive, and DNA-based carrier tests are available for each. Secondary cobalamin malabsorption may occur in congenital or acquired abnormalities of gastric, pancreatic, or intestinal function that disrupt operation of the intrinsic factor/cubam-mediated system of cobalamin uptake. Such conditions include gastrinoma, pancreatic insufficiency, surgical short-bowel, intestinal dysbiosis, inflammatory bowel disease, and inflammatory or infiltrating distal intestinal lesions such as lymphoma. An underappreciated complication of any condition causing cobalamin malabsorption is that untreated cobalamin deficiency will lead to megaloblastic and megalocytic changes of the intestinal epithelium that in turn cause generalized malabsorption. Thus, chronic intestinal disease of whatever cause may not be otherwise treated successfully until cobalamin stores are replenished by parenteral cobalamin administration. Because biliary cobalamin secretion continues in the face of failing GI absorption, total body stores of the vitamin can be exhausted rapidly. The estimated biological half-life of cobalamin in dogs is 50-100 days, whereas in humans it is 330-400 days.

Historically, direct assessment of GI cobalamin absorption in patients has been by oral administration of radiolabeled cobalamin and time course quantification of radioactivity in plasma, urine, and/or feces.[4,5] Unfortunately, radiolabeled cobalamin is no longer available and such clinical research is greatly hampered.[6] Indirect measurements in humans currently rely on quantification of the cobalamin-transcobalamin complex (holo-transcobalamin) in plasma that reflects newly absorbed cobalamin. In dogs and cats, indirect assessment relies on determining the half-life of plasma cobalamin for several weeks after a parenteral megadose (1-2 mg).[7] Better methods are needed.

COBALAMIN DEFICIENCY

Biomarkers

During GI cobalamin absorption, the newly absorbed vitamin binds to transcobalamin for distribution to cells and is rapidly cleared from plasma by an endocytic receptor (CD320) that is specific for holo-TC. Therefore, the serum holo-transcobalamin concentration is reflective of GI cobalamin absorption. In humans, ~80 % of serum cobalamin is a non-labile pool bound to haptocorrin, so the clinician must measure holo-transcobalamin to indirectly assess cobalamin absorption. In dogs and cats, however, transcobalamin is the only plasma cobalamin-binding protein, so the total serum cobalamin concentration is equal to holo-transcobalamin, and the determination does not require transcobalamin fractionation. A caveat is that serum samples can be contaminated by haptocorrin released from neutrophil secondary granules, but this is not diagnostically important unless there is neutrophilic leukemia causing a falsely high serum cobalamin concentration. In sum, low serum cobalamin concentrations in dogs and cats indicate current or insipient cobalamin deficiency due to cobalamin malabsorption. As cobalamin deficiency develops in dogs, serum cobalamin concentrations fall below the normal range several days or weeks before metabolic biomarkers of deficiency (see below) begin to rise (unpublished observations, N. Berghoff and J. Fyfe).

Cobalamin in the 5'deoxyadenosyl form is an essential cofactor for the mitochondrial enzyme, methylmalonyl-CoA mutase, whereas in the methyl form it is a cofactor for the cytoplasmic enzyme, 5-methyltetrahydrofolate-homocysteine methyltransferase (a.k.a. methionine synthase). Cobalamin malabsorption reduces activity of both enzymes as intracellular cofactor concentrations drop. The substrate concentrations of each enzyme increase, and cells release methylmalonic acid (MMA; conserving the CoA moiety) and homocysteine to plasma. Increased serum concentrations of these molecules are biomarkers of cobalamin deficiency, although MMA is more specific than homocysteine. MMA also rises more rapidly than homocysteine as cobalamin deficiency develops. Cats typically maintain normal serum homocysteine concentrations during cobalamin deficiency through the activity of betaine-homocysteine methyltransferase, a cobalamin independent enzyme, and high serum concentrations of betaine. Being an abnormal metabolite, plasma MMA is filtered at the glomerulus and not recovered by tubules and is readily detected in urine.

Low serum cobalamin concentrations do not correlate perfectly with elevated MMA in serum or urine when each is measured at a single point in time. Elevated MMA reflects reduced mutase activity due to cobalamin deficiency within mitochondria that may not develop until plasma cobalamin has been low for some time. In some individuals or breeds, there may be benign polymorphisms of mutase activity leading to moderately higher MMA concentrations when serum cobalamin is normal. Additionally, imprecision in laboratory measurements of serum cobalamin and MMA, and normal ranges that by definition do not reflect the full range of normal in a population may contribute. Regardless, low serum cobalamin reflects failing GI cobalamin absorption, which if unrelieved will lead to cellular cobalamin deficiency indicated by elevated MMA and other metabolic disruptions.

Clinical signs

Signs of cobalamin deficiency are nonspecific and vary depending on the age when it occurs, the severity of deficiency, and the duration of the condition. Secondary metabolic disruptions may mimic other conditions, and signs of cobalamin deficiency are often mixed with others caused by the primary condition. Cobalamin deficiency in young and growing animals, such as occurs in primary cobalamin malabsorption or may occur in congenital pancreatic insufficiency or distal intestinal resection, causes chronic, relapsing inappetence, lethargy, and failure to thrive. Linear growth may be near-normal, but weight gain is greatly retarded (low body scores). There may be chronic diarrhea and/or intermittent vomiting, and seizures may occur if diagnosis and treatment are delayed.

Signs of simple cobalamin deficiency in adult animals are often occult. Dogs with documented complete blockade of cobalamin absorption due to *AMN* mutation were, as adults, held off routine parenteral cobalamin replenishment for 6 months. There were no clinical signs other than slightly reduced food intake, despite serum cobalamin concentrations that fell below the limit of detection and highly elevated serum MMA and homocysteine. These findings suggest that clinical signs observed in adult animals exhibiting secondary cobalamin malabsorption are largely due to the primary disorder until, as mentioned above, cobalamin malabsorption is itself sufficiently long standing to alter mucosal function. Demyelinating lesions of the spinal cord as often observed in cobalamin deficient humans have not been observed in cobalamin deficient dogs but were reported in one adult cat exhibiting hypocobalaminemia and concurrent serum folate deficiency

Clinical laboratory findings

Peripheral hematologic changes caused by cobalamin deficiency in dogs and cats include mild nonregenerative anemia with red blood cell morphologic abnormalities, metarubricytosis, neutropenia, and hypersegmented neutrophils. The red cell distribution width increases, but macrocytosis as occurs in human patients is not observed. Thrombocytopenia and rare giant platelets may be observed. Peripheral changes are due to megaloblastosis of the bone marrow wherein nuclear maturation is delayed relative to cytoplasmic maturation (nuclear cytoplasmic asynchrony). Bone marrow aspirates or biopsies often demonstrate a hyperplastic but ineffective erythroid lineage and concomitantly decreased myeloid:erythroid ratio. In the granulocytic series there are giant metamyelocytes and bands. Nuclear maturation of bone marrow cells is slowed by decreased activity of methionine synthase that traps tetrahydrofolate intermediates in the 5-methyl form and, thereby, slows synthesis of nucleotides. Cell replication and DNA repair are both impaired, so the effects are most evident in rapidly dividing cell populations, including bone marrow lineages, intestinal epithelium, and spermatogonia.

Reduced methylmalonyl-CoA mutase activity secondarily inhibits a number of metabolic enzymes and transporters that operate in the urea cycle, glycine cleavage, and gluconeogenesis.[8] Thus, severe and long standing cobalamin deficiency may result in signs and serum biochemical indications of hyperammonemia, hypoglycemia, and/or ketoacidosis and a life-threatening metabolic crisis. Hypoproteinemia is often observed in primary cobalamin malabsorption disorders for unknown reasons. Elevated AST likely reflects ongoing red cell precursor destruction. Unique to the primary cobalamin malabsorption disorders caused by mutations affecting cubam expression is a benign, mild mid- to low-molecular weight proteinuria that persists despite cobalamin treatment.

Treatment

There are two aspects to treatment of cobalamin deficiency, reestablishing normal intermediate metabolism and replenishing cobalamin stores. Experimental treatments have demonstrated that parenteral doses (2.5-5 μg/day) near the established daily cobalamin requirement will also cause metabolic and hematologic remission in cobalamin-deficient, 8-week old puppies, but daily administration is clinically inconvenient. A single 1 mg dose of parenterally administered cyanocobalamin will rapidly restore the MMA-CoA mutase and methionine synthase activities to normal, with clearance of elevated plasma MMA and reactivation of the urea cycle within a few days. Appetite often returns to normal within 12-24 hours, and weight gain ensues over the following weeks. A reticulocyte response will typically occur 3-4 days post-administration and last for 10-14 days, and there is often a similar burst of neutrophil production. Red blood cell and neutrophil numbers normalize in 2-3 weeks. The single 1 mg parenteral megadose will maintain clinical, metabolic and hematologic remission for up to 8 weeks, even in a rapidly growing dog.

Strictly speaking, there is no organ storage of the vitamin other than as cofactors bound to the enzymes of cobalamin-dependent metabolism, but these and their transport proteins may be cobalamin saturated or unsaturated. The apparent half-life of cobalamin cofactor binding and normal activity of the cognate enzyme is much longer than the residence time of the vitamin in plasma circulation. Therefore, cobalamin-dependent metabolism will remain normal for several weeks after parenteral administration of cobalamin, even when serum cobalamin concentrations rapidly fall below normal due to ongoing malabsorption. The unsaturated cobalamin binding capacity of plasma may be exceeded by a dose of 1 mg in a small dog or cat, with unbound cobalamin being lost immediately by glomerular filtration. Therefore, a reasonable parenteral cobalamin dosage scheme for animals in the deficiency state is 40-50 μg/kg administered once a week for 4 weeks. This will reestablish normal metabolism and saturate cobalamin-binding proteins throughout the body. However, measuring serum cobalamin concentrations after treatment to determine dosage frequency is misleading because there is no medical exigency to maintaining a "normal" serum concentration. If a clinician feels the need to do so, it is more rational to monitor metabolic normalcy by measuring serum or urine MMA concentrations. Dosage frequency after the initial replenishment period need be determined by whether the cause of cobalamin malabsorption has been alleviated. If not, a larger dose administered less frequently (1 mg monthly) is often more convenient, and owner compliance is more likely.

One should not use injectable multivitamin preparations to treat cobalamin deficiency because the cobalamin concentration is too low. Cyanocobalamin is typically supplied as a single ingredient at 1 mg/mL, making the dosage volume appropriate for small animals. Cyanocobalamin has an immeasurably high margin of safety and is inexpensive and readily available, so there is no cogent reason to use hydroxocobalamin or adenosylcobalamin preparations.

PRIMARY COBALAMIN MALABSORPTION IN DOGS

It is more than 30 years since the author first observed inappetence and failure to thrive in a 5 month old giant schnauzer suffering from cobalamin malabsorption.[8] Collaborative research on that disorder led to fundamental understanding of ileal cubam expression and function and better recognition of cobalamin deficiency morbidities in dogs and cats. As indicated above, discrete cubilin and amnionless mutations are demonstrated to cause autosomal recessive cobalamin-specific malabsorption in dogs.[5,10,11] In each breed, the disorder typically presents in young dogs as selective cobalamin malabsorption, life-threatening hematological and metabolic derangements, and benign proteinuria. The combination of intestinal cobalamin malabsorption and selective proteinuria consigns these disorders to a subset of congenital cobalamin malabsorption called Imerslund-Gräsbeck syndrome (I-GS). Each mutation is breed-specific, and while low serum cobalamin concentration and response to parenteral cobalamin administration are suggestive, the breed-specific mutation tests allow for precise diagnosis. The genetic tests also allow diagnosis prior to the development of clinical signs and detection of asymptomatic carriers to guide breeding decisions.

Amnionless and cubilin, the two components of cubam, are expressed in renal proximal tubule cells in addition to ileal enterocytes. The benign proteinuria of I-GS is a reflection of failed cubam function in the nephron to mediate reabsorption of a number of proteins filtered at the glomerulus, including albumin, vitamin D-binding protein, transferrin, and haptoglobin. As in the ileum, both components must be expressed and fold together in order for the receptor to be transported from the rough endoplasmic reticulum and expressed on the apical plasma membrane. Thus, mutations of either gene encoding these two functionally dependent proteins cause the same dysfunction of cubam in both ileum and kidney, but the signs of each organ dysfunction are defined by the cubam ligands present in the luminal fluids that are not absorbed or reabsorbed.

I-GS in giant schnauzers and Australian shepherds is caused by distinct amnionless mutations, whereas in beagles and border collies I-GS is caused by distinct cubilin mutations. Mutations in the first 3 breeds abrogate apical membrane cubam expression entirely, but the border collie cubilin mutation allows for a slight amount of residual cubilin expression. While there is no documentation of residual intrinsic factor-cobalamin absorption, the border collie disorder is unique. In the first 3 breeds, the signs of cobalamin deficiency are readily recognized in the early growth period between 8 and 16 weeks of age. In I-GS border collies, owners and veterinarian are often unaware of problems until several months or a year of age. I-GS border collies may be chronic poor-doers with intermittent vague signs confused by intercurrent disorders. In at least two cases to date, the specific diagnosis was not made until ~4 years of age, and in one, chronic neutropenia and episodic pyrexia were thought to indicate trapped neutrophil syndrome (TNS) of border collies. Because the signs of cobalamin deficiency are often non-specific, it is crucial that clinicians recognize the various breed predispositions for primary cobalamin malabsorption so that specific and potentially lifesaving therapy is instituted in a timely manner and maintained for the dog's life. In border collies, I-GS and TNS may mimic each other[12], but there are mutation-specific DNA tests for each. As recessively inherited disorders, diagnosis of an affected dog implicates both parents as carriers. Appropriate genetic counseling and DNA testing can rescue the breeding program in a kennel where I-GS appears.

REFERENCES

[1]Fyfe JC. Schalm's Veterinary Hematology, 5th ed, 2000; p 965; [2]Nielson MJ *et al*. Nat Rev Gastroenterol & Hepatol 2012;9:345; [3]Fyfe JC *et al*. Blood 2004;103:1573; [4]Fyfe JC *et al*. Pediatr Res 1990;29:24; [5]Fyfe JC *et al*. Mol Genet Metab 2013;109:390; [6]Carmel R. J Nutr 2007;137:2481; [7]Simpson KW *et al*. J Vet Intern Med 2001;15:26; [8]Fenton WA *et al*. The Metabolic & Molecular Basis of Inherited Disease, 8th ed. 2001; p 2165; [9]Fyfe JC *et al*. JAAHA 1989;25:533; [10]He Q *et al*. Blood 2005;106:1447; [11]Fyfe JC *et al*. J Vet Intern Med 2014; doi: 10.1111/jvim.12284[epub ahead of print]; [12]Mason SL *et al*. J Small Anim Prac 2014;55:57.

THE INTESTINAL VIROME OF DOGS

Dr Caroline Mansfield
Melbourne, Australia

INTRODUCTION

Viruses are genetic material surrounded by a protein shell called a capsid. The nucleic acid genomes may be single or double stranded DNA or RNA, have different capsid structures and can infect both eukaryotic and prokaryotic cells. Phage (bacteriophage) is the term used to describe virus-like particles that infect bacteria. In most microbial communities, viruses and phages outnumber bacteria 10 to 1.[1] Human feces contain at least 10^9 virus-like particles per gram.[2] Early studies of the human gut virome suggest that most identifiable DNA viruses are phages and that many of the RNA viruses are plant viruses.[3] Although viruses outnumber microbes 10:1 in most environments, viral DNA often only represents 2-5% of the total DNA in a microbial community. [3] In people, it also appears likely that viruses play an important role in maintaining a healthy gut environment. The vast majority of viral data isolated from human feces is unmatched in databases, with the majority of these unreadable sequences being phages. [2,3] The predicted hosts of the phage populations coincided with the bacterial groups detected by 16S ribosomal RNA gene analysis in the same samples.[4,5]

The viral genome appears to be unique to each individual, with little to no similarity between individuals.[4] Dietary intervention has associated with a change in the virome- leading to convergence of individuals receiving the same diet, although the virome remained specific to the individual.[4] In a study of feces from one healthy individual at various time-points over 2.5 years by the same research group, 80% of the viral contigs persisted throughout, indicating long-term stability.[6] The viruses themselves showed diversity within each time point, and accumulated variation over time. Microviridae, which package ssDNA genomes and are lytic bacteriophages, showed high levels of substitution. Lowest substitution rates were seen for temperate bacteriophages (which can integrate into host bacterial genome). Similarly, in a study of twins the viral gut population remained stable and specific to the individual, with no clustering observed between twins or their mother.[7] This is in contrast to bacterial populations, which tend to cluster within families, and is potentially one explanation for why there is variable disease expression of Crohn's disease in monozygotic twins despite identical genetic and environmental factors being present.[8]

BIOLOGICAL FUNCTIONS OF BACTERIOPHAGES (PHAGES) IN THE INTESTINE

Phages have really diverse roles, and one example from nature that demonstrates this are the cyanophages that infect the cyanobacterial clades of Prochlorococcus and Synechococcus in the oceans. These phages are essential for driving diversity of their host and to adapt according to the surrounding level of nutrients, and so provide a beneficial effect.[3] Lysogenic or temperate phages are able to integrate their chromosomes into the bacterial genome- thereby altering the host bacterial phenotype by lysogenic conversion. As above, this promotes adaptation to the host environment. Temperate phages may protect the bacterial host from super-infection with the same or similar phage; and genes encoded by the phage may expand the niche of the bacterial host by altering the function of the bacteria. Methods by which this expansion can occur include enabling use of new nutrient sources, development of antibiotic resistance or production of virulence factors.[3]

Using different methodologies such as looking at CRIPSR (clustered regularly interspaced short palindromic repeats), which encode for resistance to phages, or sequencing of integrated prophages in bacterial genomes, it has been shown that phages are highly selective for different bacterial lineages.[6] This is further demonstrated by a study of *Enterococcus faecalis*, a composite phage with two distinct prophages was shown to confer an advantage to the *E facealis* V583 strains during competition with other *E faecalis* strains.[9] This study demonstrated that phages are able to cooperate to perform tasks, and that nutrient availability influences prophage induction and lytic replication.

'Kill the winner' theory

This theory postulates that phages 'kill' or destroy bacteria when they are in over-abundance. However, in a longitudinal study, the bacteriophage and bacterial abundances did not oscillate widely, which seems to refute this simple predator-prey theory.[6] However, changes in the virome and microbiome may still have occurred but not been observed because these changes happen too rapidly for detection or the sample set was not dense enough to detect changes.

Bacteriophage Adherence to Mucus (BAM) Model

Mucus is present on all mucosal surfaces and is heavily colonized by bacteria and viruses. This mucus community provides nutrients and other services to the host as well as being a vital component of innate immunity that is not host-derived.[10, 11] The mucus contains large mucin glycoproteins (200-900 kDa), and the thickness and physical characteristics of mucus varies with anatomical location.[12] Mucus provides structure and nutrients to enable microbial growth; and so additional mechanisms are necessary to protect the host from bacterial invasion. These protective mechanisms include the biophysical properties of the mucus, and trefoil peptides that are secreted by the goblet cell and mucus neck glands.[12]

There has recently been proposed a theory that the presence of phage within the mucus may also confer a protective effect. One in vitro study showed that phage abundance is dependent on mucus, and increased phage numbers protected the underlying epithelium from bacterial infection, independent of other factors.[11] The enrichment of phage in mucus is via interaction between mucus glycoproteins and Ig-like protein domains that are exposed on phage capsids. This is a great example of symbiosis whereby the phage protects the eukaryotic host from bacteria, and the phage is then able to interact with bacterial hosts within the mucus.

ROLES IN DISEASE

Because the GI tract is a primary site of exposure of host organism to microbes, there is development of extensive crosstalk between the host and the luminal environment of the intestine. Toll-like receptors (TLRs) in the gut detect and initiate inflammation against viruses. In particular, TLR3 recognizes dsRNA and signal via adaptor proteins the production of type 1 interferons and pro-inflammatory cytokines.[13] There are also retinoic acid inducible gene I (RIG-I)-like receptors (RLRs) that have been shown to play an important role in the innate immune response against rotavirus (which is a dsRNA virus).[13] Nucleotide binding domain and leucine-rich repeat containing receptors (NLRs) including cytosolic NOD1 and NOD2 have been well characterized in the gut, and again detection of these ligands leads to activation of pro-inflammatory cytokines and antimicrobial peptides. NOD2 has been shown to be involved in the detection of viral products as well, especially ssRNA viruses.

It has been shown experimentally that bacterial factors (derived from *Bacteroides thetaiotaomicron* and *Lactobacillus casei*) can block infection of rotavirus in intestinal epithelial cells in vitro.[14] Additionally, an experimental model of influenza in mice suggest that commensal microbiota are necessary to prime cytokines for second signal release following viral infection.[15] Conversely, the intestinal microbiota may play role in enteric virus transmission and replication. Although observed clinically, it is not known by what mechanism viral infection predisposes to secondary bacterial infections.

Inflammatory bowel disease: cause or cure?

In human gastroenterology, the role of viral infections in IBD, and particularly in human Crohn's disease (CD), was proposed more than 40 years ago and has been the focus of many investigations. Early microscopic studies detected "virus-like particles" in cell culture passages of CD ileal filtrates and biopsy specimens.[16-18] Since then, numerous viruses have been investigated in CD by polymerase chain reaction (PCR), enzyme linked immunoabsorbent assay (ELISA), histochemistry, and in situ hybridization techniques, but have failed to show a definitive link. A recent meta-analysis of viruses associated with CD reported a significant association ($p<0.05$) between CD and 3 of 9 (33%) Epstein-Barr virus studies, 3 of 11 (27%), cytomegalovirus studies, 7 of 16 (44%) measles virus studies, and in 1 of 3 (33%) rotavirus studies.[19] In 1of 4 (25%) human herpes virus studies, 1 of 3 (33%) rubella virus studies and 1 of 2 (50%) norovirus studies, the individual virus was detected more often in controls than affected samples.[19]

One potential mechanism by which viruses could play a role in IBD is to dramatically augment NOD2 signaling and increase production of pro-inflammatory cytokines.[20] However, recent work would suggest that the role viruses play in the pathogenesis of CD of disease is more complex than that. In one recent study, a lineage of mice with a mutation in the CD susceptibility gene ATG16L1 was concomitantly infected with murine norovirus (MNV).[21] The ATG16L1 gene is an autophagy gene, and the disease variant is frequent in people, conferring < 2-fold increase in susceptibility to CD. In this mouse study, intestinal pathology was present only in the face of concomitant MNV infection. The intestinal pathology was mediated by TNF-α and IFNγ, and appeared to be strain specific, associated with a strain of MNV that persisted. Although it is plausible that the mutation in the autophagy gene alters innate immunity and so leads to enhanced viral replication and pathogenicity in Paneth cells, there was no viral RNA detected in Paneth cells negating that hypothesis. Additionally, the mice did not have increased MNV replication, and the ATG16L1 gene played no role in MNV replication in vitro. The intestinal pathology was prevented by administration of a broad spectrum antibiotic, leading to the conclusion that commensal bacteria plus environmental factors plus genetic susceptibility plus specific viral strains may all contribute to the development of CD. This is a very important concept when looking at studies that assess viral association with CD, as a particular genotype may be required to develop the disease. There are other animal disease models whereby viruses affect or cause disease only in specific genetic backgrounds.[22]

On the flip side to this, phages may have a beneficial effect on the development of particular strains of bacteria, and be protective against pathogenic strains of bacteria. As a result, there may be the potential for phage therapy in the future of inflammatory bowel disease (IBD). The absence of phage in probiotics, as well as genetic variation, may be a possible reason for different responses in different studies to probiotics in CD.[23] Similarly, the success of fecal transplantation for treatment of chronic *Clostridium difficile* infection and IBD (particularly ulcerative colitis) may be in part due to the presence of phage.[24,25]

APPROACHES TO IDENTIFYING THE VIROME

Part of the difficulty in identifying many viruses and phages is that they do not have a conserved protein or gene, and are genetically diverse due to horizontal transfer of genes and recombination events. The latest technology including viral microarray and high throughput sequencing (Next Generation Sequencing) provides a thorough, fast and economical approach to identify viral candidates associated with IBD etiology. Microarray technology still relies on known viral sequence information but can be very efficient in screening a large number of viruses from clinical sample[1]. One of the leading cutting edge viral detection methodologies is based on virus discovery by sequence-independent genome amplification - SISPA (sequence-independent single primer amplification[2]. The incorporation of a DNase and or RNase step to remove non-viral nucleic acid before viral capsid lyses enhances the viral nucleic acid purity, which is crucial to enhance the recovery of viral sequence during sequencing.

WHAT DO WE KNOW ABOUT THE CANINE VIROME?

We have in our group optimized the protocols for removing non-viral DNA/RNA in canine feces, and have sequenced a number of samples from a variety of healthy dogs and dogs with diarrhea. The predominant group of viruses in all dogs was bacteriophages, similar to what is found in people. However, with disease there is a difference in the percentage of phages that were found. Overall (Table 1), dogs with diarrhea had a lower percentage of viral contigs compared to dogs with acute diarrhea and chronic diarrhea. Further work is required to determine the functions of the viruses present, and their relationship with the bacterial population in the same individual dog. There were a number of eukaryotic viruses present in all sample sets, particularly Herpesviridae and Retroviridae (Figure 1). Some viral families such as Astrovirdae and Caliciviridae were not detected in samples from healthy dogs, but were in dogs with diarrhea.

Table 1: Percentage of contigs obtained from dogs with chronic diarrhea (n=10), acute diarrhea (n= 10) and healthy dogs (n=8)

Classification	Chronic diarrhea group	Acute diarrhea group	Healthy dogs
Cellular organisms	94.31	80.74	56.76
No Hits (nonDOG/nonBacterial sequences)	**5.36**	**18.38**	**41.81**
Low complexity	0.26	0.77	0.98
not assigned	0.08	0.11	0.46
Bacteria reads	94.09	80.57	56.66
Archaea	0.21	0.17	0.09
Eukaryota	0.00	0.00	0.00

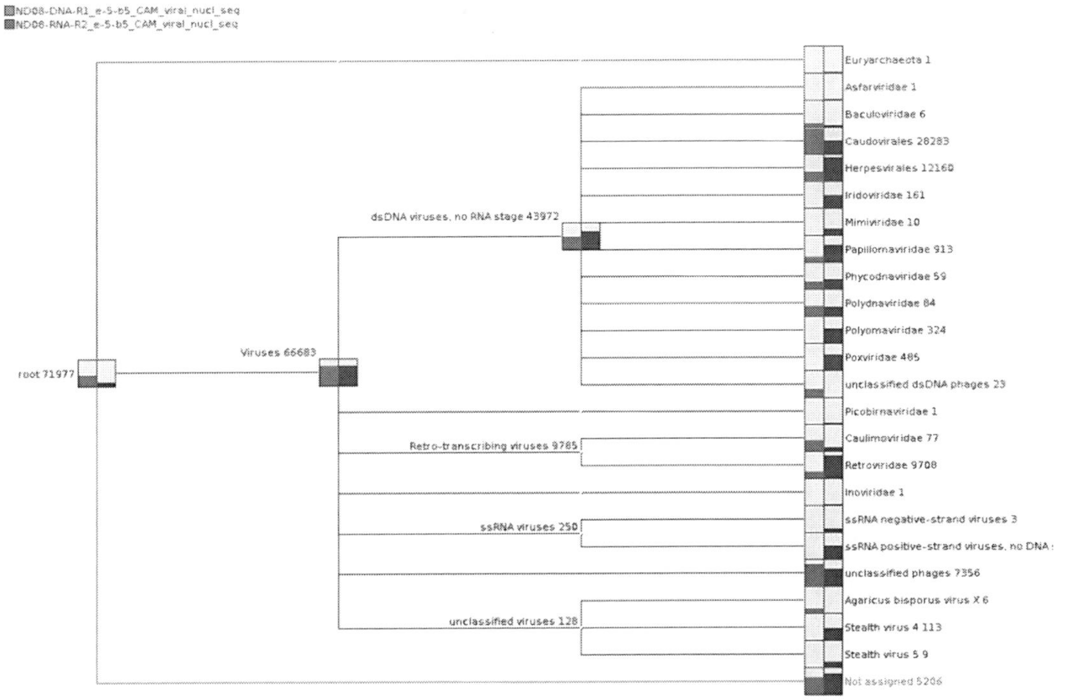

Figure 1: An example of the viral sequencing from the feces of a healthy dog.

CONCLUSION

The intestinal virome of dogs is predominated by prokaryotic viruses similar to the human gut. The interplay between viruses (especially phages), bacteria and host immunity is still being determined. There is a strong argument for phages to play a role in innate immunity of the gut; and equally that viruses may play a role in the etiology of inflammatory disease. To date there is no definitive link between a viral etiology and CD. Failure to prove a definitive link may be due to a number of factors, namely a low viral titer may be present in patients at the time of sample acquisition; the viral load may be below detection levels for many technologies; and most viral studies have relied on known viral information such as virus specific antibody, PCR or probes targets. Next generation sequencing is uncovering a large number of previously unknown or unidentified viruses. Other compounding factors include the possibility that a genetic predisposition is necessary for there to be a viral cause of inflammation, and that there may be a difference in pathogenicity between viral strains. Additionally, many studies are unable to conclude whether viral infections were present before the onset of CD or whether they became established as a secondary infection during disease onset, and during tissue inflammation and ongoing reduced immunological status. Further evaluation of the canine virome will be greatly beneficial to improve our global understanding of the development of chronic inflammatory disease in the gut.

REFERENCES

[1]Samson JE et al. **Nature** 2013; 11:675 [2]Schoenfeld T, et al. **Trends Microbiol** 2010; 18:20 [3]Reyes A et al. **Nature** 2012; 10: 607- 617 [4]Minot, S et al. **Genome Research** 2011; 21:1616 [5]Stern A et al. **Genome Research** 2012; 22:1985 [6]Minot S et al. **PNAS** 2013; 110:12450 [7]Reyes A et al. **Nature** 2010; 466:334 [8]Scharl M & Rogler G 2012 **Curr Opin Gastroenterol** 28:301 [9]Duerkop A et al. **PNAS** 2012; 109:17621 [10]Chow J et al **Adv Immunol** 2010; 107: 243 [11]Barr JJ et al, **PNAS** 2013; 110:10771 [12]McGuckin MA et al **Nature Reviews Micro** 2011; 9: 265 [13]Moon C & Stappenbeck TS **Current Opinion in Immunology** 2012; 24: 405 [14]Varyukhina S et al **Microbes Infect** 2012; 14: 273 [15]Ichinohe T et al **Proc Natl Acad Sci** 2011; 108: 5354 [16]Gitnick GL et al **Lancet** 1976;2:215 [17]Gitnick GL et al

Lancet 1976;2:217 [18] Riemann JF. **Acta Hepatogastroenterol (Stuttg)** 1977;24:116 [19] Wagner, J et al **Reviews in Medical Virology** 2013; 23:145 [20] Kim YG et al **Cell Host Microbe** 2011; 9:496 [21] Cadwell K et al **Cell** 2010; 141:1135 [22] Dyrberg et al **J Clin Invest** 1988; 81:928 [23] Prantera C et al 2002 **Gut** 51:405 [24] Gough E et al 2011, **Clinical Infectious Diseases** 53: 994 [25] Borody T et al **J Clinical Gastro** 2003; 37:42

UPDATES ON CYSTINURIA AND FANCONI SYNDROME: AMINO ACIDURIAS IN DOGS

Urs Giger, DACVIM-SA, DECVIM CA, DECVCP, Ann-Kathrin Brons, Caitlin A Fitzgerald, Jeffrey Slutsky, Karthik Raj, Victor Stora, Adrian C Sewell and Paula S Henthorn
Philadelphia, PA

Introduction

Disorders of the renal proximal tubules can cause selective or generalized aminoaciduria and may be associated with urinary losses of other solutes such as glucose, lactate, electrolytes and bicarbonate. Two renal tubular defects involving amino acids have long been recognized in dogs, namely cystinuria, leading to cystine calculi and urinary obstruction, and Fanconi syndrome, progressing to renal failure if untreated. Both hereditary disorders have been investigated at the molecular level and are more complex than originally anticipated. Furthermore, the ingestion of Chinese jerky treats has recently been found to be associated with Fanconi syndrome in many dogs and rarely cats. The current understanding of pathophysiology, clinicopathological findings, diagnosis, and therapeutic options will be presented.

Fanconi Syndrome

Fanconi syndrome, named after the Swiss pediatrician Guido Fanconi and also known as Fanconi's syndrome or Fanconi disease, should not be confused with Fanconi anemia, a bone marrow disorder in humans. Fanconi syndrome represents a majorproximal renal tubular defect, which hampers the adequate reabsorption of glucose, amino acids, bicarbonate, sodium, calcium, phosphate, lactate, ketones, and carnitine. This rather general loss of multiple functions of the proximal renal tubules can be associated with renal tubular acidosis and lead to progressive renal failure if left untreated. In the renal tubules there are multiple co-transporters for sodium and glucose, amino acids, calcium, and inorganic phosphorus and a sodium/hydrogen ion antiporter, which, depending upon the concentration gradient established by the sodium-potassium pump, move hydrogen ions into the urine. Defects in all tubular transporters, peroxisomes, the membrane sodium/potassium ATPase and thus tubular gradient or structural membrane abnormalities may be responsible. For most types of Fanconi syndrome, the genetic defects remain unknown. In humans, Fanconi syndrome is most often associated with hereditary cystinosis (not to be confused with cystinuria), fructose intolerance, galactosemia, glycogenosis, as well as Lowe and Wilson disease. Most recently, aberrant localization of the peroxisomal *EHHADH* gene product with disruption of mitochondrial metabolism, has been found to lead to renal Fanconi syndrome in one kindred, thus indicating a central role of mitochondria in proximal tubular function. The mechanisms involved in Fanconi syndrome in dogs may include multiple genetic predispositions, acquired causes, or combinations of both. The molecular defect associated with Fanconi syndrome in Basenjis has recently been elucidated. Basenjis are genetically predisposed to Fanconi syndrome, which is inherited as an autosomal recessive trait in the breed. As many as 10% of Basenjis are affected, and they typically develop signs in middle-age (4-7 years). A mutation in the *Fan1* gene on CFA3 has recently been discovered. Norwegian Elkhounds also often develop glucosuria early in life, which may progress to Fanconi syndrome, but no molecular studies have been reported. Other breeds that may also be genetically predisposed to develop Fanconi syndrome include Labrador Retrievers, Cocker spaniels, West Highland White Terriers, Shetland Sheepdogs, and Schnauzers in which Fanconi syndrome has been secondary and related to a copper-associated hepatopathy in these breeds.

There are several acquired causes of Fanconi syndrome such as exposure to heavy metals (e.g. lead, mercury, and cadmium), drugs (gentomycin, tetracycline, and azathioprine) and infections (leptospirosis). Fanconi syndrome has also been reported in cancer and hepatic copper disease in several dogs. Particularly, recent studies indicate that Labrador Retrievers with copper-associated hepatopathy develop Fanconi syndrome. More recently Fanconi syndrome has been associated with the ingestion of chicken and duck and even veggie jerky treats of different brands that contain Chinese products. Currently no specific ingredient or contaminant responsible for the intoxication has been identified but the FDA and state governments are actively investigating the jerky treats. Six antibiotics (trimethoprim, tilmicosin, enrofloxacin, sulfaclozine, sulfamethoxazole, and sulfaquinoxaline) have been found in excessive amounts in some chicken jerky treats, however, there is no evidence to associate their presence with the observed illnesses. Nearly all the dogs affected are from North America and Australia and include toy to small breed dogs, such as Chihuahuas, Maltese, and Yorkshire, Jack Russell and West Highland White Terriers, but not brachycephalic breeds. Their predisposition may be related to the proportionally larger amount of jerky treats ingested by small breeds compared to large breed dogs, or may reflect a broader hypersensitivity of the smaller breeds.

Clinical signs of intoxication may occur within days to months of ingesting jerky treats, are unspecific and may include lethargy, inappetence, vomiting, and diarrhea. Polydipsia and polyuria are the most common clinical signs of Fanconi syndrome, while other signs of Fanconi syndrome are highly variable and may be related to the ensuing renal failure. In the past, a diagnosis of Fanconi syndrome resulted in a guarded to poor prognosis. However, in the present day, treated Basenijs may reach a near normal life expectancy. Moreover, if the trigger is removed, such as withdrawal of Chinese jerky treats, the renal damage may be more minimal and reversible. Routine serum chemistry and urinalysis as well as venous blood gas analysis are indicated to define the extent of the tubular defect and acidosis. Liver enzymes may be elevated suggesting a hepatopathy. Venous blood gases may show a metabolic acidosis (with blood pH normal to near 7), low bicarbonate concentration and a reduced base excess. A total CO_2 level on a chemistry screen is not sensitive enough to detect the metabolic acidosis due to compensatory hyperventilation. Serum chemistry also will reveal deficiencies in sodium, potassium, calcium and phosphorous. Osteopenia is apparently not a feature of Fanconi syndrome in dogs as it is in humans. Some dogs develop azotemia, which worsens over time, with renal failure being the most common cause of death.

Urinalysis typically reveals glucosuria. In fact, the marked glucosuria in light of a normal blood glucose level is generally the reason to diagnostically pursue a Fanconi syndrome. Urine metabolic screening from dogs with Fanconi syndrome shows often massive generalized aminoaciduria and lactic aciduria (and the glucosuria is also confirmed). Metabolic screening at the University of Pennsylvania (http://research.vet.upenn.edu/penngen/PennGenHome/tabid/91/Default.aspx) can document the aminoaciduria, the hallmark finding of Fanconi syndrome. Dogs with Fanconi syndrome have much higher urinary concentrations of cystine and dibasic amino acids in comparison to cystinuria, as well as exhibit massive excretion of all other amino acids. Paradoxically, despite a severe degree of cystinuria in any dog with Fanconi syndrome, they rarely if ever develop cystine calculi and urinary obstruction; the dilute and alkaline urine may prevent the crystal formation. Moreover, while there is severe aminoaciduria, there is either no or only mild proteinuria. The University of Missouri offers DNA testing for Basenijs (http://www.caninegeneticdiseases.net). However, the molecular basis in other breeds is unknown .

The therapeutic protocol (https://www.basenji.org/ClubDocs/fanconiprotocol2003.pdf) developed and modified over the years by Steve Gonto, a human anesthesiologist as well as a critical care/life support clinician and owner of Basenjis, has been used widely. While there are no formal comparative trials of different therapeutic modalities, the life expectancy and quality of life have apparently greatly improved by using this protocol. It focuses on aggressive correction of the various losses with particular attention to bicarbonate, normalizing electrolytes and minerals, and providing a high quality diet. Bicarbonate is dosed to increase pCO_2 and HCO_3 levels. Unless the animal is in renal failure, a high protein diet is recommended to replenish the amino acid (and protein) losses and to rebuild muscle mass. Renal dialysis may be considered in the acute management of severe renal failure.

In a survey of medical records, Basenjis with Fanconi syndrome received between 2.6-15.6 grams of sodium bicarbonate in tablet form divided per day. Multivitamin and calcium/phosphorus may be supplemented as needed. Treatment of Fanconi syndrome can be rewarding if the disease is recognized early. The median survival time for Basenjis was 5 years after the diagnosis was made. This medical protocol can be used to manage dogs with the acquired form, as long as the trigger is removed. The treatment must be adjusted for the breed size and severity of the metabolic derangements. Dogs that ingested Chinese jerky treats are known to have improved and even resolved their metabolic derangements within a few months after withdrawing the treats, unless they had already developed severe renal failure by the time of diagnosis. Labrador Retrievers and other dogs with copper-associated hepatopathy and Fanconi syndrome may also require copper chelation and low copper diets.

Cystinuria

In contrast to the rather broad transmembrane transport deficiency causing hereditary and acquired forms of Fanconi syndrome, cystinuria is a limited hereditary renal transport disorder involving cystine and the dibasic amino acids ornithine, lysine, and arginine, collectively known as COLA. Instead of the normal >99% reabsorption in the proximal renal tubules, these four amino acids are lost in the urine, but only cystine causes a problem. The low solubility of cystine in acidic urine predisposes the formation of cystine crystals and uroliths in the urinary tract, resulting in the typical clinical signs of cystinuria. While the transport of these amino acids also occurs in the gut and thus cystinuria also affects intestinal COLA absorption, the impaired intestinal absorption of the COLA amino acids is not associated with any clinical deficiencies in animals, with the exception of hyperammonemia due to arginine deficiency in cats.

Cystinuria is one of the first reported "inborn errors of metabolism" described in humans more than 200 years ago, and has been described in several other species including dogs, cats, ferrets, maned wolves, and servals. Two genes, *SLC3A1* and *SLC7A9* encode the polypeptide subunits of the $b^{o,+}$ basic amino acid transporter system and its dysfunction results in cystinuria. *SLC3A1* encodes a protein referred to as rBAT and *SLC7A9* encodes a protein called $b^{o,+}$AT. $b^{o,+}$AT heterodimerizes with rBAT exclusively to form the COLA amino acid transporter. Cystinuria in humans has been classified phenotypically into two types differing by mode of inheritance (autosomal recessive or incomplete autosomal dominant) are caused by mutations in the *SLC3A1* and *SLC7A9* genes which encode the $b^{o,+}$AT basic amino acid transporter system; mutations have not been identified for approximately 10% of cystinuric human patients.

Since 1823, when cystinuria was first described in dogs, dogs of many breeds have been diagnosed, but only recently have studies revealed the genetic heterogeneity. We have proposed a new classification system for canine cystinuria based upon that in humans. A severe cystinuria showing autosomal recessive inheritance was first characterized in Newfoundlands and Landseers, and has more recently been identified in additional breeds. Although both males and females are affected, males more frequently show clinical signs of urinary obstruction. Different mutations in *SLC3A1* resulting in an early stop codon and loss of $b^{o,+}$AT function have been identified in Newfoundlands and Labrador Retrievers (Type I-A). A 6 bp deletion removing 2 threonine residues in *SLC3A1* was found in autosomal-dominant (AD) cystinuria with a more severe phenotype in homozygous than in heterozygous Australian Cattle Dogs and a mixed breed dog (Type II-A). A missense mutation in *SLC7A9* was identified in autosomal dominant cystinuria in Miniature Pinschers from Europe, with only heterozygous affected dogs observed to date (Type II-B).

Cystinuria has also been described in an additional 70 canine breeds. In several of these breeds from which more data is available, such as the Mastiff, French Bulldog, Basset hound and Irish terrier, the disease predominantly occurs later in life, is less severe, more variable, and only involves males. We now refer to this as canine androgen-dependent or type III cystinuria but still do not know the specific genetic defects. Interestingly, Irish Terriers in Europe and Australia seem to be more commonly affected than in the US, where cystinuria in Irish Terriers is rarely diagnosed. Type III cystinuric dogs (with or without cystine calculi) also seem to have variable degrees of cystine and COLA in their urine. These variations could be due to cystine precipitation, diet, age-related variability, the castration effects, and genetic heterogeneity. Molecular studies in Irish Terriers, Mastiffs and several other breeds have failed to detect disease-causing mutations in the exons of either *SLC3A1* or *SLC7A9* genes, however an *SLC3A1* missense mutation appears to be associated with cystinuria in Mastiffs and related breeds but not Irish Terriers or Scottish Deerhounds. Only mature males are cystinuric and surgical (and even medical) castration resolves excessive cystine and COLA excretion, but the mode of inheritance is still unclear. On average, dogs with type I and II cystinuria have a several fold higher urinary COLA excretion than type III cystinuric male dogs.

Cystine uroliths can form in the kidney (rarely), ureters, bladder or urethra and can lead to urinary blockage, renal failure and associated life-threatening clinical manifestations. A clinical diagnosis of a cystine uroliths can be suspected based upon the breed and yellow-brown color of the uroliths, and can be confirmed by crystallographic or chemical analysis of calculi in specialized stone laboratories such as those at the Universities of Minnesota and California, as well as commercial laboratories worldwide. While in the US approximately 1% of analyzed urinary calculi are cystine stones, in Europe the proportion appears to be as high as 3% . Cystine calculi are usually radiolucent and thus may be missed by radiography, but are readily visualized by ultrasound. Furthermore, urinalysis often reveals the presence of characteristic hexagonal cystine crystals.

A simple urinary screening test, to determine if a dog is cystinuric, is available through the Metabolic Genetic Laboratory at the University of Pennsylvania (http://research.vet.upenn.edu/penngen/PennGenHome/tabid/91/Default.aspx). This test can detect any type I and II cystinuric animal, but not necessarily all dogs with type III cystinuria. Amino acid analyzers can be used to determine the amount of cystine and the other amino acids in urine, however this quantitative amino acid analysis is currently restricted to a few laboratories and is relatively expensive. Inclusion of the dibasic amino acids (ornithine, lysine, and arginine) in the quantitation of cystine is preferred due to the possibility that urinary cystine may have precipitated, causing lower concentrations than were originally present in the urine, and therefore leading to false negative interpretations. Based upon our studies, dogs with either cystine levels of >200 µmol/g creatinine or COLA values of >700 µmol/g creatinine are considered cystinuric. Moreover, for several breeds with type I and II cystinuria, a breed-specific mutation test is available that not only detects cystinuric dogs, but also the asymptomatic carriers (for recessive traits). While in the 1990s as many as 5% and 30% of Newfoundlands were cystinuric (homozygous affected) and carriers, respectively, very few Newfoundlands are cystinuric or carriers in 2013, thanks to the implementation of this DNA screening test, which is available through our and many commercial laboratories. In contrast, the mutant alleles in Labrador Retrievers, Australian Cattle Dogs, and Miniature Pinscher dogs appear to be limited to a small subset of the breed population. Finally, in Mastiffs and closely related breeds, a DNA marker test to identify dogs that are at risk for early stone formation is being offered on a limited basis at the University of Pennsylvania (http://research.vet.upenn.edu/WSAVA-LabSearch).

Urinary tract obstructions due to cystine calculi can present as life-threatening emergency situations. If the calculi are in the urethra, catheterization and endoscopy may allow their passage, permit them to be moved back into the bladder (retrograde hydropulsion). Laser therapy as well as lithotripsy may break them up, and surgery can be used to remove them entirely. In contrast, attempts to dissolve cystine stones medically by drugs and diet have generally failed. Dogs with type I and II cystinuria are persistently strongly cystinuric from puppyhood and remain at high risk of developing cystine calculi throughout life, independent of gender. Alpha-mercaptopropionylglycine (Thiola), a chelating agent that has fewer side-effects and is more potent than D-penicillamin, may lower the risk of calculi formation in cystinuria but is very expensive. Forced diuresis, urine alkalinization (pH>7.5), and reduced protein diets (no chicken) may lower the recurrence risk, while high protein diets and protein supplements tend to increase the risk for calculi formation. Male dogs with type III cystinuria have more variable and usually lower cystine excretion. Thus, they may or may not develop calculi and their recurrence rate is lower. Moreover, castration of Mastiffs and related breeds and Irish Terriers with type III cystinuria diminishes the excretion of cystine and dibasic amino acids to the normal female range, indicating a testosterone-mediated modulating effect on the renal tubular amino acid reabsorption. In these breeds, we recommend castration to reduce the formation and risk of blockage by cystine calculi, which will also contribute to reducing the incidence of this genetic predisposition.

REFERENCES

Appleman EH, Cianciolo R, Mosenco AS, Bounds ME, Al-Ghazlat S (2008). Transient acquired Fanconi syndrome associated with copper storage hepatopathy in 3 dogs. J Vet Intern Med. 22:1038-1042.

Bovee KC. Fanconi syndrome in the dog (1980). In: Kirk RW, ed. Current Veterinary Therapy VII. Philadelphia: WB Saunders Co, 1075-1078.

Bovee KC, Joyce T, Reynolds R, et al (1978). The Fanconi syndrome in Basenji dogs: a new model for renal transport defects. Science 201:1129-1131.

Brons AK, Henthorn PS, Raj K, Fitzgerald CA, Liu J, Sewell AC, Giger U (2013). SLC3A1 and SLC7A9 mutations in autosomal recessive or dominant canine cystinuria: A new classification system. J Vet Intern Med, 27:1400–1408

Casal ML, Giger U, Bovee KC, Patterson DF (1995): Inheritance of cystinuria and renal defect in Newfoundlands. Journal of the Am Vet Med Assoc 207:1585-89.

Escolar E, Perez-Alenza D, Diaz M, et al. (1993) Canine Fanconi syndrome. J Small Anim Pract. 34:567-570.

Farias F, Mhlanga-Mutangadura T, Taylor JF, O'Brien DP, Schnabel RD, Johnson GS (2012): Whole genome sequencing shows a deletion of the last exon of *Fan1* in Basenji Fanconi syndrome. Proceedings of the Advances in Canine and Feline Genomics and Inherited Diseases Conference, Visby 51.

FDA Update (Oct 22, 2013) http://www.fda.gov/AnimalVeterinary/SafetyHealth/ProductSafetyInformation/ucm360951.htm.

Fieten H, Hooijer-Nouwens BD, Biourge VC, Leegwater PA, Watson AL, van den Ingh TS, Rothuizen J (2012). Association of dietary copper and zinc levels with hepatic copper and zinc concentration in Labrador Retrievers. J Vet Intern Med. 26:1274-80

Gonto S. Fanconi disease management protocol for veterinarians (2013). Available at: https://www.basenji.org/ClubDocs/fanconiprotocol2003.pdf

Giger U, Fitzgerald C, Liu J, Erat-Huff A, Sewell AC, Henthorn PS (2011): Characterization of non-type I cystinuria in Irish Terriers. J Vet Intern Med. 25.

Harvenik L, Hoppe A, Söderkvist P (2006). SLC7A9 cDNA cloning and mutational analysis of SLC3A1 and SLC7A9 in canine cystinuria. Mamm. Genome 17:769-776.

Henthorn PS, Liu J, Gidalevich T, Fang J, Casal ML, Patterson DF, and Giger U (2000): Canine cystinuria: Polymorphism in the canine SLC3A1 gene and identification of a nonsense mutation in cystinuric Newfoundland dogs. Human Genetics 200;107:295-303.

Henthorn PS, Giger, U (2006): Cystinuria, in The Dog and Its Genome, eds. Ostrander, Giger, and Lindblad-Toh (Cold Spring Harbor Laboratory Press), 349-64.

Henthorn PS, Raducha M, Brons AK, Raj K, Fitzgerald C, Liu J, SewellA, Giger U (2013): Unraveling the phenotypic and genetic complexity of canine cystinuria. Proceedings of the Advances in Canine and Feline Genomics and Inherited Diseases Conference, Boston, MA.

Hill TL, Breitschwerdt EB, Cecere T, Vaden S (2008). Concurrent hepatic copper toxicosis and Fanconi's syndrome in a dog. J Vet Intern Med. 22:219-222.

Hoffmann G, Jones PG, Biourge V, van den Ingh TS, Mesu SJ, Bode P, Rothuizen J (2009). Dietary management of hepatic copper accumulation in Labrador Retrievers. J Vet Intern Med. 23:957-63.

Hooper AN, Roberts BK (2011). Fanconi syndrome in four non-basenji dogs exposed to chicken jerky treats. J Am Anim Hosp Assoc. 47(6):e178-87.
Hostutler RA, DiBartola SP, Eaton KA (2004). Transient proximal renal tubular acidosis and Fanconi syndrome in a dog. J Am Vet Med Assoc. 224:1611-1614.
Klootwijk ED et al (2014) Mistargeting of peroxisomal EHHADH and inherited renal Fanconi's syndrome. N Engl J Med. 370:129-38.
Langlois DK, Smedley RC, Schall WD, Kruger JM (2013). Acquired proximal renal tubular dysfunction in 9 Labrador Retrievers with copper-associated hepatitis (2006–2012). J Vet Intern Med. 27:491-9.
Noonan CH, Kay JM (1990). Prevalence and geographic distribution of Fanconi sydrome in Basenjis in the United States. J Am Vet Med Assoc. 197:345-349.
Palacin M, Nunes V, Font-Llitjos M, Jimenez-Vidal M, Fort J, Gasol E, Pineda M, Feliubadalo L, Chillaron J, Zorzano A (2005) The genetics of heteromeric amino acid transporters. Physiol. 20:112-124.
Sheridan R, Mirabile, J, Hafler K (2014) Determination of six illegal antibiotics in chicken jerky dog treats. J Agri Food Chem, Jan 29 Epub ahead of print.
Slutsky J, Raj K, Yunke S, Bell J, Fretwell N, Hedhammar A, Wade C, Giger U (2013) A web resource on DNA tests for canine and feline hereditary diseases. Vet J. 197:182-187.
Slutsky J, Fitzgerald CA, Store V, Sewell A, Giger U (2014): Results of urinary metabolic screening in dogs with acquired and hereditary Fanconi syndrome. In preparation.
Yearly JH, Hancock DD, Mealey KL (2004). Survival time, lifespan and quality of life in dogs with Fanconi syndrome. J Am Vet Med Assoc. 225:377-383.

Supported in part by grants from the National Institutes of Health (RR 002512 and OD 010939) and the AKC Canine Health Foundation. LaboKlin assisted in sample collection.

LEPTOSPIROSIS: OVERPLAYED OR OVERLOOKED?

George E. Moore DVM, MS, PhD, DACVPM, DACVIM
West Lafayette, IN

INTRODUCTION

Leptospirosis is a well-documented infectious disease in dogs. This disease in dogs however can, and often does, have wide variation in reservoir sources, geographical prevalence, annual prevalence, seasonal prevalence, serovar prevalence, severity of disease, organ systems affected, clinical signs, clinicopathological abnormalities, accuracy of diagnostic tests, and prognosis. Knowledge of these variations can help the astute clinician appropriately diagnose and treat the disease.

RESERVOIRS AROUND US

Leptospirosis has been found in more than 150 mammalian species, and therefore the risk of this disease to dogs or cats must be dependent on the prevalence of leptospirosis in different species in a geographical area and subsequent exposure risk to dogs. The importance of several wildlife reservoirs has been well documented. In 1999, researchers in Illinois reported in the *Journal of Wildlife Diseases* that 48% (222/459) of live-trapped raccoons were seropositive for leptospirosis; and almost all (220/222) serological reactions were due to serovar Grippotyphosa. A study in Connecticut published in 2003 (*Vectorborne and Zoonotic Diseases*) found evidence of exposure to leptospirosis in 36% of raccoons, although Icterohaemorrhagiae was the most common reactive serovars in these animals. In skunks, exposure was less common (13%) but Grippotyphosa was the most common serovar. Seropositivity to leptospirosis has also been noted in coyotes although studies are limited.

Most wildlife studies (actually most leptospirosis studies, regardless of species) are based on serological testing via the microscopic agglutination test. Recent unpublished research at Purdue University, evaluating raccoons trapped in Indiana, found that very large proportions of the population (>80%) were positive for leptospirosis by PCR or dark-field microscopy yet were seronegative by antibody testing. Thus, serological testing of wildlife that are natural reservoir hosts may (greatly?) underestimate the prevalence of disease in these same species. With this information, the pertinent patient history question is not "Does your dog have potential exposure to rats or livestock?" but rather "Does your dog have potential exposure to raccoons, deer, or skunks (or specifically their urine)?"

CHANGING RISK?

Approximately 10 years ago research at Purdue University found that leptospirosis cases were more likely to be dogs from suburban, or recently urbanized, areas than from rural settings.[1] Wildlife studies at other universities indicate that population densities of peri-urban wildlife may be 8-12 times greater than in their rural counterparts due to increased availability of food and lack of predators, thus increasing disease exposure risk. A recent review of leptospirosis cases from the VMDB (Veterinary Medical Database) from university teaching hospitals over the last 40 years documents a change in the signalment in diagnosed cases, with dogs less than 15 pounds proportionately more likely to be diagnosed with the disease than dogs in other weight groups.[2] These small dogs in our experience have not been vaccinated against leptospirosis, leaving them susceptible to infection.

Other recent studies in the US and in Canada have documented increasing disease in small breed dogs. In fact, one study showed Labrador retrievers at reduced risk. These findings may merely be a reflection of vaccination protocols, i.e. sporting breed dogs are being vaccinated but small breed (so called 'backyard') dogs are not. Tying this finding with previous studies suggests: 1) urban wildlife such as raccoons are known carriers of leptospirosis, 2) wildlife populations are often denser in suburban/urban than in rural areas, and 3) small dogs can be exposed to this disease in their backyard due to wildlife activity. This hypothesis seems reasonable, in part due to the increase in clinical cases in the fall and early winter (Sep-Dec) when raccoon activity increases due to dispersion of the year's weaned offspring and securing territory/habitat for winter.

CHANGING SEROVARS?

Which serovars are important? The veterinary literature from the 1950s and 60s documented serosurveys of stray unvaccinated dogs in the US, and antibodies against *Leptospira* serovars Canicola and Icterohaemorrhagiae were most common in these dogs. Maintenance hosts for these two serovars are dogs and rats/rodents, respectively. Vaccines were therefore developed in the 1960s to protect dogs against these two serovars. From the mid-1970s to early 1990s, there were few published reports, in peer-reviewed literature, of canine leptospirosis in the US. This may have been due to reduced infection in reservoir hosts or due to the effectiveness of the bivalent (Canicola and Icterohaemorrhagiae) bacterin. The bacterin was typically marketed as the liquid component of a multivalent canine vaccine. Vaccine-associated adverse events were sometimes attributed to this bacterin, and post-vaccinal antibody concentrations were reported to decline after 3-6 months, thus leading to questions of the efficacy of this bacterin. Nevertheless, no apparent resurgence of these 2 serovars appeared.

Through the 1990s, case series reports of canine leptospirosis began to document (usually based on serology) canine infections caused by nonvaccinal serovars. In 1991, Nielsen in Indiana reported 2 lepto cases caused by serovar Bratislava. In 1992, clinicians in Massachusetts reported the canine leptospirosis cases from 1985-1989 in their hospital were predominantly due to serovars Pomona and Grippotyphosa. In 1996, serovar Grippotyphosa was also reported from 11 dogs with acute renal failure in Georgia. The same year, serovars Pomona and Grippotyphosa were reported to cause canine disease in New Jersey and serovars Pomona, Autumnalis, and Grippotyphosa were incriminated in Michigan. Two years later, researchers in New York would report the same 2 serovars as New Jersey, and in 2000 researchers in Davis, CA, would report a predominance of serovars Pomona and

Bratislava in their canine leptospirosis cases from 1990-98. They noted that the majority of cases were identified in the last 3 years of their study.

Thus, in the face of possible protection against serovars Canicola and Icterohaemorrhagiae, clinical cases were increasingly attributed (serologically) to serovars Pomona, Grippotyphosa, Bratislava, and Autumnalis. Serovars Pomona and Grippotyphosa were recognized pathogens in livestock (cattle and pigs) although not previously pathogens of concern to dogs. The other two serovars (Bratislava and Autumnalis) were documented pathogens on other continents but had not been recognized as serovars of concern to livestock or companion animals in the US. Continuing supportive evidence has led to a change in serovars in vaccines (see section below). The current role of serovar Bratislava in canine infections in the US is poorly defined, as is its wildlife reservoir. Serovar Autumnalis has still not been adequately documented as a cause of disease, and titers to this serovar (often seen in initial stages of disease due to IgM) should be considered as a cross reaction. Convalescent titers typically reveal a rising titer to another serovar.

CHANGING CLINICAL PRESENTATION?

Clinical signs in dogs are usually attributable to localization of infection to the renal, hepatic, and/or vascular system. Clinical findings can be quite varied in severity, ranging from acute oliguric renal failure, polyuric renal failure, renal and hepatic disease, fever of unknown origin (FUO), or even no clinical signs (asymptomatic). Less recognized by practitioners is presentation with hepatic disease alone or radiographic abnormalities of the lungs due to vasculitis.[3] A general classification of the frequency of organ involvement in canine leptospirosis, based on clinical and diagnostic findings, is: renal only: 30-50%, renal and hepatic: 25-35%, hepatic only: 10-20%, and other (uveitis, myalgia, FUO): 5-10%. Some serovars may be more likely to cause disease in certain organs (as noted in some studies), but the evidence to confirm this is not consistent across the clinical literature.

IMPROVED DIAGNOSTICS??

Determination of the infective serovar and even clinical diagnosis is hindered by lack of a sensitive, specific, low-cost, rapid and widely available diagnostic test for leptospirosis. Most cases of leptospirosis are diagnosed by serology, and the reference method is the microscopic agglutination test (MAT). The MAT is difficult to standardize and requires live organisms for antigens. Cross-agglutination is also common. Despite these drawbacks, the MAT is still the diagnostic norm for most laboratories. Clinicians must presume that the serovar with the greatest antibody titer is the infective serovar, although paradoxical reactions to un-infective serovars have been noted. This is believed to occur most commonly early in infection, due to a non-specific IgM response; the MAT primarily measures IgM rather than IgG. Likewise, clinicians are in a quandary when 2 serovars have equal titers as dual infection is probably unlikely. The use of paired sera (2-4 weeks apart) is often required to confirm the diagnosis and clarify the infective serovar. Problematic however is the capability of leptospiral serovars to alter their outer membrane proteins. This is done in the natural host environment in order to reduce the host immune response to the invader. This transformational ability in laboratory-maintained serovars also could reduce the MAT correlation between laboratories and compared to the infective serovar.

PCR of urine and/or blood is also used to diagnose leptospirosis, but its use and impact have raised new questions. Although PCR is increasingly available through many laboratories, controlled studies have not defined the correlation between PCR and MAT, via a true "gold standard" in a large number of cases. One limitation of PCR-based diagnosis is the inability of most PCR assays to identify the infecting serovar. While this may not be important for individual patient management, serovar identity has important epidemiological and public health value. Not all PCR tests are performed with the same methodology, and sensitivity and specificity may vary. Generally PCR tests are highly sensitive. False negatives are considered uncommon, but can occur with low/zero levels of leptospiruria or leptospiremia. Certain methodologies however may be more prone to false positive test results, or reduced test specificity. A comparison study of two PCR methods yielded 0% false-positives in one methodology but the same samples had 13% false-positives via an alternative methodology (manuscript in preparation). PCR true positive results however do not necessarily mean there is viable organism or transmission risk.

TREATMENT – ANY CHANGES?

Fortunately the *Leptospira* organism, a spirochete, does not seem to acquire antibiotic resistance. Doxycycline (5 mg/kg q12hr) is still the recommended antibiotic, but recent problems in market availability/cost of doxycycline have raised interest in alternative antimicrobials. Minocycline is in the same drug family as, and is similar to, doxycycline; it may be more available but not necessarily cheaper than doxycycline. Dosages for minocycline (7-10 mg/kg q12hr) are reported to be slightly higher or similar to doxycycline. Parentally administered ampicillin (22 mg/kg q6hr) is often the first choice for vomiting, or critically ill, dogs but it does not clear the renal carrier state. Early use of oral doxycycline is encouraged and should be continued for at least 2 weeks after gastrointestinal signs abate. Quinolones, chloramphenicol, and sulfonamides are not effective against leptospiral infection. Selected cephalosporins, i.e. ceftriaxone and cefotaxime, have been effective in other species, as has azithromycin, but studies in dogs are lacking.

CANINE VACCINATION

Serological evidence in the US clearly supports the use of 4-serovar (Icterohaemorrhagiae, Canicola, Grippotyphosa, and Pomona) vaccines, rather than 2-serovar vaccines; and the 2010 ACVIM Small Animal Consensus Statement on Leptospirosis and the 2011 AAHA Canine Vaccination Guidelines do not recommend the use of 2-serovar products. Leptospiral vaccines are generally considered serovar-specific, and cross-protection between serovars is not believed to occur. This concept is being challenged in some research, however, and some (albeit not total) cross-protection may occur between selected serovars. Recent research suggests that MAT seropositivity for serovars Autumnalis, Grippotyphosa, Bratislava, and Pomona are strongly correlated. Thus, there appears to be some molecular mimicry between these serovars. At Purdue, we have not documented a case of leptospirosis attributable to serovar Bratislava or Autumnalis in a dog properly vaccinated with a 4-serovar product, again suggesting some cross-protection may occur between some serovars.

All leptospiral vaccines are similar in that they are bacterins (recombinant products do not exist). Different vaccines can vary in the quantity of whole inactivated bacteria or cell wall antigens present, or in quantity of vaccine excipients (such as bovine serum albumin) remaining from vaccine production. This variation in exogenous protein/antigen most likely explains the occurrence – or lack of occurrence – of allergic reactions following leptospirosis vaccination and observed differences in the rate of these reactions among vaccines by different manufacturers. Nevertheless, current vaccines are improved in safety compared to the biologicals produced more than a decade ago.

Manufacturers typically recommend 2 initial vaccines, and so these can be administered as part of final (rather than all or initial) puppy vaccinations. Annual revaccination is recommended, and supported by epidemiological studies. Because some dogs have low or undetectable leptospiral antibodies at 12 (or even 6) months post-vaccination, the duration of immunity from leptospiral bacterins has been considered by some to be less than 1 year – even though manufacturers' recommendations are for annual revaccination intervals. In 2003, a study in Holland assessed the duration of immunity in dogs attained with a commercial inactivated bivalent leptospirosis vaccine, using challenge with serovars Canicola and Icterohaemorrhagiae. Two vaccinations induced a high rate of protection from generalized infection from these serovars at 5, 27, and 56 weeks after the second vaccination. Several dogs protected from clinical challenge had no detectable agglutinating antibodies. The study used a small number of dogs and perhaps could have used a stronger infective challenge, but it demonstrates the potential fallacy of relying on antibody concentrations to measure true immune status. Evidence of 12-15 months of disease protection following leptospiral vaccination has been reported by some investigators recently. Although infection can occur in a properly vaccinated dog, this is quite uncommon and may reflect a high but less than 100% efficacy in large populations.

Concern has occasionally been raised regarding data from vaccine trials indicating that vaccinated dogs may develop renal carriage if challenged/exposed to infection. If this is true, there should be documentation of transmission of infection from vaccinated dogs; such evidence from more than 40 years of vaccine use is lacking. Closer examination of experimental data however indicates that the challenge dose of leptospires in vaccine trials probably far exceeds the natural exposure dose. Also, documented renal carriage or leptospire shedding has been noted exclusively or predominantly with serovar Canicola, the serovar for which the dog is a natural reservoir. Serovar Canicola is very rarely encountered in dogs in the US, but is more common in other countries and continents. Although a chronic-shedding state for dogs remains to be proven by culture, viable organism can be shed in acutely ill dogs - presenting a public health risk to owners, families, and veterinary staff. After 48 hours of effective antibiotic therapy, any shed organisms are non-viable – yet would result in a positive PCR test of the urine.

LEPTOSPIROSIS IN CATS? OVERLOOKED?

What about cats? Does leptospirosis affect cats, causing clinical disease? Although cats can have exposure to leptospires via rodents or urine-contaminated water (and cats are well known to get kidney disease), published diagnoses of feline leptospirosis are rarely found in veterinary medical literature.[4] Cats can, and do, produce titers against leptospirosis; but seropositivity rates in cats are generally much lower than found in dogs. Leptospires are also rarely documented in histologic renal sections from cats with chronic interstitial nephritis. Although earlier studies have noted there was no significant difference in seropositivity for cats with kidney disease compared to cats without a history of kidney disease – again suggesting no significant disease impact in this species, a recently published study found slightly greater rates in cats with kidney disease than in healthy cats.[5] Potential explanations for lower seropositivity rates include greater innate immune protection against initial infection and subsequent reduced viability of the organism in the feline renal/urinary system. The hypothesis that leptospiral infection may be only temporary in most infected cats but may initiate a sterile chronic nephritis is interesting, yet is without parallel in other species.

The testing of cats for leptospires by PCR has recently renewed some professional interest in, and concern for, feline infection.[6] The risk of false-positive PCR results has been recently documented however.[7] A critical review of the literature does note that ascites was a physical examination finding on some of the reputed feline cases of leptospirosis.[4,8] Whether this is a manifestation of a leptospiral-induced vasculitis remains to be determined. Nevertheless, clinical disease in cats due to leptospiral infection is uncommon. If diagnostic testing is believed warranted, the use of paired acute and convalescent sera to document a 4-fold rise in titer (and thus recent infection) is recommended. Vaccination of cats against leptospirosis is not warranted at this time.

REFERENCES

[1]Ward MP, et al. **J Am Vet Med Assoc** 2004;225:72. [2]Lee, et al. **J Vet Intern Med** 2013 Dec 26. doi: 10.1111/jvim.12273. [3]Tangeman, et al. **J Am Vet Med Assoc** 2013;243:1316. [4]Bryson, et al. **J Small Anim Pract** 1976;17:459. [5]Rodriguez, et al. **J Vet Intern Med** 2014 Jan 13. doi: 10.1111/jvim.12287. [6]Hartmann, et al. **J Feline Med Surg** 2013;15:576. [7]Fenimore, et al. **J Vet Intern Med** 2013;27:726. [8]Arbour, et al. **J Am Anim Hosp Assoc** 2012;48:256.

TRANSDERMAL MEDICATIONS IN CATS: A REVIEW OF THE LITERATURE AND EVIDENCE

Kate Hill, BVSc, MANZCVSc, DACVIM
Palmerston North, New Zealand

INTRODUCTION

As cats are notoriously hard for owners to medicate with oral drugs many compounding pharmacies have started formulating drugs into gels. These gels are applied to the inner side of the ear of the cat and are thought to be absorbed through the skin to allow there systemic action. Most of these pharmacies combine the drug of interest with the vehicle Pluronic Lecithin Organogel (PLO) to aid transdermal drug delivery. Transdermal drug delivery is a rapidly developing field in both veterinary and human medicine. However, much of the evidence for transdermal drug delivery is anecdotal. The aim of transdermal drug delivery, is for the drug to enter into the systemic circulation. The aim of topical drug therapy is for the drug to remain on local organs. To understand which drugs are suitable for transdermal delivery and which vehicle is required to aid their delivery, we must first review skin physiology and the pharmacology of transdermal drugs.

ANATOMY OF THE SKIN IN CATS

An overview of the structure and function of the skin is required to understand transdermal drug penetration and the factors that can affect absorption. The skin is comprised of the epidermis, dermis and subcutaneous tissue. Each layer of the skin is anatomically and physiologically different Figure 1. The skin is the largest and most visible organ in the body. The skin of cats is thinner than dogs, with an average thickness of 0.4-2.0mm. It is thickest in the dorsal regions and proximal limbs, and thinnest on the ventrum and distal limbs.[1] Cats have different hair follicles than dogs, both have a compounded hair follicle, the cat has a single primary follicle surrounded by up to five compound follicles, each with three primary and up to 12 secondary hairs. The epidermis is an avascular layer that has distinct layers; stratum corneum, stratum lucidum, stratum granulosum, stratum spinosum and the inner most stratum basale which creates the epidermal-dermal junction. (See figure 1). Cells of the epidermis are constantly being renewed, with formation of new keratinocytes at the stratum basale. Cells migrating to the outer stratum corneum layer, gradually lose their nucleus and form dessicated proteinaceous corneocytes.

The stratum corneum regulates water loss and is an efficient physical barrier to chemicals and microbes penetrating the skin. This layer also regulates water loss and disruption of the stratum corneum can result in easy penetration of chemicals. The stratum corneum is the skins most important barrier for the prevention of absorption of substances such as drugs.

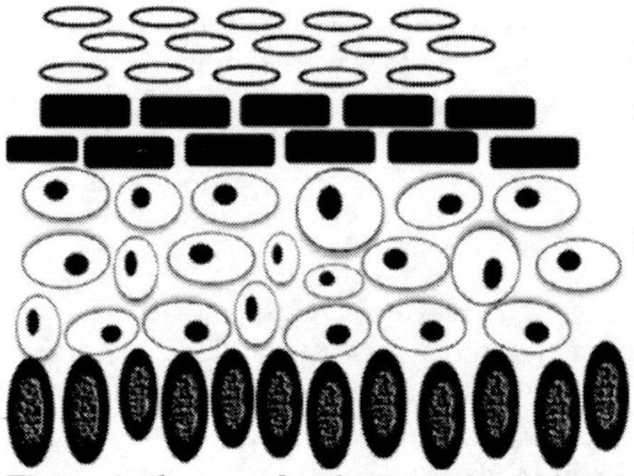

Figure 1: There are four layers to the epidermis. The stratum corneum is comprised of dead cells and a lipid bilayer. Disruption to this lipid bilayer facilitates transdermal drug delivery.

PERMEATION OF TRANSDERMAL DRUG THROUGH THE SKIN

The skin is a good barrier to drug permeation. The stratum corneum (epidermis) is the most important barrier to drug permeation due to the lipid bilayer. Partition of this lipophilic subcutaneous layer, will allow drugs to move into the more hydrophilic viable epidermis before they reach the systemic circulation. There are several factors that affect the transdermal delivery of drugs.[2]

Firstly, the drug molecule itself needs to be small (< 500Da), have few atoms available for hydrogen bonding, be lipophilic and have a low melting point. An ideal drug is one that is potent and has a total daily dose of only a few milligrams. The vehicle must be sufficiently soluble for the drug to dissolve, and the drug itself needs to be able to diffuse through the subcutaneous lipids. For transdermal absorption, the drug must be highly lipophilic and formulated in the correct vehicle.[3] Skin factors will also affect drug penetration, with age, body site and skin pathology all playing a role.

Vehicles

The major role of the vehicle, or penetration enhancer, is to help the drug permeate through the lipid bilayer of the subcutaneous tissue. The ideal properties of a chemical penetration enhancer are listed in Table 1.[4]

Table 1: Properties of an Ideal Chemical Penetration Enhancer
Pharmacologically inert
Non irritating, nonallergenic, nontoxic
Non-damaging to viable cells
Rapid onset of effect
Effects are completely and rapidly reversible on removal

Effects do not cause loss of endogenous materials from the body
Physically and chemically compatible with drugs and excipients in the dosage form
Cosmetically acceptable when applied to skin
Odourless, inexpensive, tasteless and colourless

The vehicle most commonly studied in cats is Pluronic lecithin organogel (PLO). PLO was developed in the 1990's by a compounding pharmacist Marty Jones and his collegue Lawson Kloesel.[5] PLO is an emulsion of lipid and aqueous substances, that acts like and feels like a gel. The aqueous part of PLO is composed of 20-30% of Pluronic F-127. The lipid phase of the formula is equal parts lecithin and either isoporophyl palmitate or isopropyl myristate (20% by weight or 22% by volume). To prepare the aqueous phase, Pluronic F-127 is mixed with ice cold water and agitated overnight in a refrigerator.[6] The aqueous phase must remain at low temperatures to remain in a liquid state, as at higher temperatures it forms a gel.[5] The lipid phase is also allowed to stand overnight to ensure complete dissolution. If a preservative is required, then sorbic acid at 0.2% can be added to both phases. PLO is then combined using shear force such as the syringe to syringe method or an electronic mortar and pestle.[5] Shear force will result in uniform micelle production (figure 2).

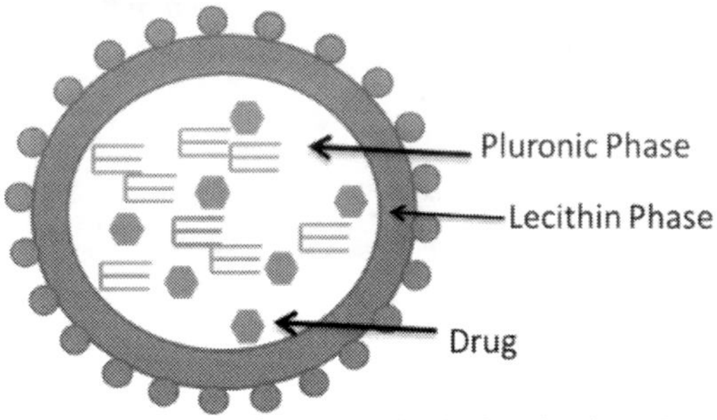

Figure 2 Micelle of a water soluble drug in Pluronic Lecithin Organogel.

To incorporate PLO with the active drug, the drug can be added to the prepared PLO or depending on the solubility of the drug, can be added to either the lipid or aqueous phase, before the two phases are mixed together. A hydrophilic drug can be mixed with a small quantity of water before mixing with the aqueous phase, and a hydrophobic drug can be mixed with propylene glycol prior to mixing with the lipid phase.[6]

Lipoderm® is an alternative to PLO as a vehicle. Lipoderm is manufactured by Professional Compounding Association of America (PCAA) and according to the manufacturer is superior to PLO. Lipoderm has a smooth and creamy texture and has the advantage that the transdermal formulation is stable and does not separate on refrigeration. Lipoderm contains a proprietary liposomal component that may increase the permeation of a variety of actives.[7]

EVIDENCE FOR TRANSDERMAL DRUG DELIVERY IN CATS

PLO as a vehicle

Despite PLO being widely used in veterinary and human medicine as a penetration enhancer, scientific evidence is limited about the physicochemical structure of PLO, such as the stability, rheology, structure, drug incorporation[6]. A preliminary study into the formulation of methimazole in PLO and Pluronic gel has recently been published.[8] This study concluded that the vehicles tested were adequate as a transdermal release formulation. A second conclusion from the study was that the vehicles had a large bioadhesive and absorptive capacity and did not leave any residue on the skin.

After the development of PLO gel, Jones and Kloesel collaborated with local physicians to treat patients with a variety of drugs incorporated within PLO. This study resulted in anecdotal evidence of the efficacy of the use of this vehicle in transdermal drug delivery.[5] There are still only a handful of systematic evaluations of the efficacy of PLO as transdermal drug delivery vehicle in the veterinary and medical literature.

Not all drugs are suitable for transdermal penetration. A variety of research methods with varying degrees of evidence have been used to assess the efficacy of transdermal drugs in cats. A few studies in cats have determined whether therapeutic plasma concentrations are reached after transdermal absorption. [9-17] Other studies used a single application to assess efficacy [9,10,13,15,17] and others multiple applications[11,12,16,18-21]

The European Medicine Agency recommends that pharmacokinetics of transdermal medications should be assessed after single and multiple doses, therefore many of these studies in cats may be underestimating the bioequivalence compared to oral dosing. (Anonymous 2010) All future transdermal pharmacokinetic studies in cats should include full pharmacokinetic profiles after single and multiple doses.

Clinical efficacy or biological end-points, are a good way of assessing transdermal drug efficacy. For example, the endpoint of successful treatment of anti-thyroid drugs can be determined by their biological effect, i.e. the decrease in TT4 concentration and resolution of clinical signs rather than the plasma concentration of methimazole alone. Some studies in cats have used clinical parameters or biological endpoints to determine efficacy of transdermal absorption.[9,11,12,18-21]

Evidence for other transdermal vehicles in cats

Lipoderm has been used as the vehicle for amlodipine to successfully treat hypertension in 8 cats.[11,22] An *in vitro* study performed by the manufacturer demonstrated good penetration through feline skin of tramadol in Lipoderm, however a clinical study has yet to be performed. Diltiazem was stable in Lipoderm for up to 60 days at a concentration of 99.6mg/ml. [23] However when diltiazem in Lipoderm was assessed in a single dose study in healthy cats relative bioavailability of transdermal compared to intravenous diltiazem was only 10%.[24]

A novel lipophilic for methimazole has also been studied in cats with hyperthyroidism.[25,26] All previous studies of transdermal methimazole in cats have used PLO gel as the vehicle, [17,18,20,21] however methimazole is a lipophilic drug, and PLO gel might not be the most suitable vehicle for a lipid soluble drug.[2] A suitable vehicle should be sufficently soluble to contain the active drug without the drug precipitating out of solution.[2] As a vehicle for methimazole, PLO gel has problems with drug precipitation and also development of a a non-homogenous texture to the gel..[20] Pharmacokinetic studies of methimazole in PLO gel showed that it was poorly absorbed when applied transdermally but did show clinical efficacy in some cats after repeated applications.[17,18,20,21] This novel lipophilic vehicle showed clinical efficacy and reasonable pharmacokinetic parameters when compared to oral carbimazole.[25,26]

SUMMARY

There is still a lot to be learned about transdermal drug therapy in cats. Few studies are available, and yet many drugs are used to treat cats by transdermal application. In cases where drugs have not been studied, the clinician should ensure there is a biological endpoint to assess efficacy (ie thyroxine concentrations, blood pressure values etc). Antimicrobial drugs should not be used transdermally unless there is efficacy to support their application by this method. Drugs that have a dose larger than 50mg should not be compounded for transdermal use. Communication is essential with the compounding pharmacist, as some drugs are unsuitable for transdermal penetration. It is recommended that future studies in veterinary medicine follow the guidelines set by the European Medicine Agency for *in vivo* and *in vitro* studies on transdermal medications.

REFERENCES

1. Strickland JH, Calhoun ML. INTEGUMENTARY SYSTEM OF CAT. American Journal of Veterinary Research 1963;24:1018-&.
2. Mills PC, Cross SE. Transdermal drug delivery: Basic principles for the veterinarian. Veterinary Journal 2006;172:218-233.
3. Riviere JE, Papich MG. Potential and problems of developing transdermal patches for veterinary applications. Advanced Drug Delivery Reviews 2001;50:175-203.
4. Benson HA, Watkinson AC. Topical and Transdermal Drug Delivery. Hobeken: John Wiley & Sons; 2012.
5. The history of pluronic lecithin organogel: an interview with Marty Jones. International Journal of Pharmaceutical Compounding 2003;7:180-183.
6. Murdan S. A review of pluronic lecithin organogel as a topical and transdermal drug delivery system. Hospital Pharmacist 2005;12:267-271.
7. PCAA. http://www.pccarx.com/pcca-products/pcca-exclusives/bases/lipoderm.
8. Morales ME, Clares B, Lopez-Viota M, et al. Preparation, characterization, and in vitro release of new transdermal methimazole as alternative to oral delivery. Drug Deliv 2009;16:1-10.
9. Bennett N, Papich MG, Hoenig M, et al. Evaluation of transdermal application of glipizide in a pluronic lecithin gel to healthy cats. American Journal of Veterinary Research 2005;66:581-588.
10. Ciribassi J, Luescher A, Pasloske KS, et al. Comparative bioavailability of fluoxetine after transdermal and oral administration to healthy cats. American Journal of Veterinary Research 2003;64:994-998.
11. Helms SR. Treatment of feline hypertension with transdermal amlodipine: A pilot study. Journal of the American Animal Hospital Association 2007;43:149-156.
12. MacGregor JM, Rush JE, Rozanski EA, et al. Comparison of pharmacodynamic variables following oral versus transdermal administration of atenolol to healthy cats. American Journal of Veterinary Research 2008;69:39-44.
13. Mealey KL, Peck KE, Bennett BS, et al. Systemic absorption of amitriptyline and buspirone after oral and transdermal administration to healthy cats. Journal of Veterinary Internal Medicine 2004;18:43-46.
14. ScherkNixon M. A study of the use of a transdermal fentanyl patch in cats. Journal of the American Animal Hospital Association 1996;32:19-24.
15. Willis-Goulet HS, Schmidt BA, Nicklin CF, et al. Comparison of serum dexamethasone concentrations in cats after oral or transdermal administration using Pluronic Lecithin Organogel (PLO): A pilot study. Veterinary Dermatology 2003;14:83-89.
16. Miller R, Schick AE, Boothe DM, et al. Absorption of Transdermal and Oral Cyclosporine in Six Healthy Cats. Journal of the American Animal Hospital Association 2014;50:36-41.
17. Hoffman SB, Yoder AR, Trepanier LA. Bioavailability of transdermal methimazole in a pluronic lecithin organogel (PLO) in healthy cats. Journal of Veterinary Pharmacology and Therapeutics 2002;25:189-193.
18. Hoffmann G, Marks SL, Taboada J, et al. Transdermal methimazole treatment in cats with hyperthyroidism. Journal of Feline Medicine and Surgery 2003;5:77-82.
19. Buijtels J, Kurvers I, Galac S, et al. Transdermal carbimazole for the treatment of feline hyperthyroidism. Tijdschrift Voor Diergeneeskunde 2006;131:478-482.
20. Lecuyer M, Prini S, Dunn ME, et al. Clinical efficacy and safety of transdermal methimazole in the treatment of feline hyperthyroidism. Canadian Veterinary Journal-Revue Veterinaire Canadienne 2006;47:131-135.
21. Sartor LL, Trepanier LA, Kroll MM, et al. Efficacy and safety of transdermal methimazole in the treatment of cats with hyperthyroidism. Journal of Veterinary Internal Medicine 2004;18:651-655.
22. Mixon WRMF, Helms SRDDA. Transdermal Amlodipine Besylate in Lipoderm for the treatment of Feline Hypertension: A Report of Two Cases. Int J Pharm Compd 2008;12:392-397.
23. Buur JL, Baynes RE, Yeatts JL, et al. Analysis of diltiazem in Lipoderm transdermal gel using reversed-phase high-performance liquid chromatography applied to homogenization and stability studies. Journal of pharmaceutical and biomedical analysis 2005;38:60-65.
24. DeFrancesco T. Transdermal Cardiac Therapy in Cats: the NCSU experience. In: 21st Annual ACVIM Forum, Charlotte, NC 2003.
25. Hill K, Gieseg MA, Bridges J, et al. The pharmacokinetics of methimazole in a novel lipophilic formulation administered transdermally to healthy cats. New Zealand veterinary journal 2013.
26. Hill KE, Gieseg MA, Kingsbury D, et al. The Efficacy and Safety of a Novel Lipophilic Formulation of Methimazole for the Once Daily Transdermal Treatment of Cats with Hyperthyroidism. Journal of Veterinary Internal Medicine 2011;25:1357-1365.

THE HISTORY AND PHYSIOLOGY OF HYPERBARIC OXYGEN THERAPY

Diane M. Levitan, VMD, DACVIM (SAIM)
Commack, New York

The value of hyperbaric oxygen has become increasingly evident in human medicine and the knowledge gained has been valuable to establish effective and rational uses for hyperbaric oxygen therapy (HBOT) in small animal internal medicine. Hyperbaric oxygen therapy in humans (HBOT) fell out of favor in the mid-1900s due to misuse and false claims as well as a lack of controlled studies to support its effectiveness in certain conditions. In the past 40 years, thousands of evidence based medicine papers have help to elucidate not only what conditions it is helpful in, but also the cellular and physiologic mechanisms for its actions.

Hyperbaric Oxygen Therapy (HBOT) is a medical treatment in which the patient is entirely enclosed in a pressure chamber breathing 100% pure oxygen (O2) at greater than one atmosphere of pressure. It is the use of oxygen under pressure as a drug. HBOT is delivered in a hyperbaric chamber. Multiplace chambers are capable of treating multiple individuals simultaneously. Monoplace chambers are the most common type used in veterinary medicine. These chambers are best suited to one person, however, up to 3 pets could potentially be treated together in a 'monoplace' chamber. The risks and benefits of this must be determined on a case by case basis. There are no attendants and no monitoring can occur inside the monoplace chamber.

Pressure terminology

At sea level, or normobaric conditions, the pressure is 1 atmosphere. The pressure on an individual is equal to the weight of the atmosphere around them. 1 atmosphere of pressure is equal to 14.7 pounds per square inch (PSI) or 760 mmHg or 760 Torr or 0 ft of sea water. The sum of pressures that something is exposed to is called atmospheres absolute (ATA). Therefore, if an animal is in a chamber at sea level (14.7 PSI or 0 feet of sea water), and the chamber pressure is brought up to 2 ATA, then the pressure is raised in the chamber an additional atmosphere to 14.7 PSI. The effect of 2 ATA of pressure equals the pressure of being surrounded by 33 feet of sea water. Most hyperbaric treatments in small animal medicine involve pressures of 1.5 to 3.0 ATA—one atmosphere from the earth's atmosphere plus an additional one half to two atmospheres exerted by pressurizing the hyperbaric chamber.

Physiologic effects of hyperbaric oxygen

These are discussed here in detail as these effects will be sited when discussing rationale of HBOT for any internal medicine applications. There are two basic effects of HBOT on tissues—(1) effects related to the mechanical forces of the pressure itself and (2) the effects related to the increased oxygen in blood and tissues.

Mechanical effects of pressure

Pressure gets gas into solution, shrinks bubbles and decreases diffusion distances. At normal atmospheric pressure arterial oxygen tension is 100 mmHg and tissue oxygen 55 mmHg. In 100% Oxygen at 3 ATA, arterial oxygen tension is 2000 mmHg and tissue oxygenation is 500 mmHg. Instead of 3 ml of oxygen per liter, there is 60 ml oxygen per liter. (1) This results in oxygen transport and tissue survival without the need for hemoglobin, as was demonstrated by Boerema in 1955. (2) (3) (4) High oxygen gradients result in diffusion of oxygen into areas of low oxygen. This provides immediate help to ischemic and compromised tissue even with marginal or no blood flow. Oxygen dissolved in plasma can be delivered from capillaries to tissues at least three times further than delivered when carried only by hemoglobin. (3) (4) HBOT decreases the volume of any gas in proportion to the increase in atmospheric pressure delivered by the chamber (Boyle's law). At elevated pressures, oxygen and nitrogen diffuse into solution.

HBOT helps scuba divers with decompression sickness by dissolving large toxic nitrogen bubbles back into solution. The same concept applies in the case of a venous or arterial gas embolism in any animal. At 2.8 atmospheres, any bubble volume reduces by almost two thirds. Hyperbaric oxygen speeds the dissolution of the inert-gas bubble by replacing the inert gas in the bubble with oxygen, which is then rapidly metabolized by the tissues. The use of hyperbaric oxygen also prevents the formation of new bubbles. (1) Compression in a chamber will decrease gas from ileus, bloat or severe intraluminal gas accumulation. HBOT is particularly useful in perioperative gastrointestinal obstruction, ileus, gas gangrene, emphysematous biliary or urinary bladder tissues, etc. due to a reduction in gas volume, nitrogen diffusion into tissues and blood and a high gradient for nitrogen removal via respiration. The decrease in pressure from gas in tissues or cavities results in less injury from vascular compromise, decreased bacterial translocation, necrosis, pain and swelling. (5) (6) (7) (8) (9)

Effects related to the increased oxygen in blood and tissues

Hyperbaric Oxygen therapy decreases inflammation by many mechanisms

We will have a detailed discussion of the mechanisms involving cytokines, prostaglandins and nitric oxide. (10)

Vasoconstriction

High levels of oxygen cause vasoconstriction, limiting blood flow and oxygen transport via red blood cells. There are organ specific vasoconstrictive responses to hyperbaric oxygen- but the most profound vasoconstriction is in the brain, retina and skeletal muscle. For the most part, oxygen induces smooth muscle contraction in all muscularized vessels on both the arterial and venous sides of circulation. (11) This seems counterintuitive, but overall oxygen delivery to tissues is increased regardless of the vessel sizes. (12) (13) (14)

Reduction of swelling and edema

Blood and lymphatic vessel disruption, dilation and widening following damage to tissue results in decreased blood and lymphatic flow. Increase in blood flow from damaged tissue (vasogenic) coupled with increased vascular permeability from cellular injury (cytogenic) leads to tissue swelling/edema. (15)This creates increased local pressure further compromising blood flow. If this pressure approaches or exceeds that in the blood vessels, blood flow will slow or cease altogether. Swelling contributes to tissue hypoxia by increasing capillary distances within tissues. With HBOT, there is decreased filtration of fluid from the capillary to the extracellular space as a consequence of vasoconstriction while reabsorption of fluid at the capillary level is maintained. Reducing edema decreases intercapillary distances, therefore increasing oxygen perfusion. Fluid reabsorption helps inflammatory cells remove cell debris and micro-organisms that are in the vicinity. This decreases edema and is important in the treatment of burns, crush injuries and injured tissues in general. Edema reduction has been mostly attributed to oxygen induction of vasoconstriction, but is also likely due to reduction in inflammation. (12) (13) (14) (16) (17)

HBOT Enhances efficacy of leukocytes

White blood cells use oxygen-independent and oxygen-dependent mechanisms to kill microorganisms. HBOT's effects on neutrophil activity result in decreased inflammation and increased bactericidal capabilities. In tissues where oxygen tensions fall below 30 to 40 mmHg, leukocytes lose their effectiveness against gram-positive and gram-negative organisms, allowing bacteria to flourish. HBOT improves phagocytosis impaired by hypoxia. HBOT enhances granulocyte production of antimicrobial agents such as free radicals and reactive/toxic oxygen species. (18) (19) (20) Phagocytosis of pathogens results in an "oxidative burst" comprised of oxygen radicals such as hydroxyl radicals, peroxides, and superoxides to kill microorganisms. The oxidative burst increases the oxygen consumption of the leukocyte 30 times; therefore production of these oxygen radicals requires an oxygen rich environment which is restored with HBOT. Anaerobic bacteria are very susceptible because they have poor defenses against free radicals and other toxic oxygen species created by neutrophils in that environment. (18) (19) (20)

HBOT will decrease inflammation by decreasing inflammatory cytokines that result in neutrophil rolling, activation, transmigration into tissues and local adherence by impairing cGMP synthesis in activated leukocytes. (21) High oxygen concentration results in reduced expression of endothelial adhesion molecules and increases nitric oxide production which also inhibits neutrophil adhesion. (22) (23) (24) HBOT reduces pro-inflammatory cytokine induction which also dampens the inflammatory response. (25)

Macrophages, like neutrophils, are affected by tissue oxygen tension. In hypoxic conditions macrophages are unable to scavenge effectively or produce peroxides (26) Hypoxia induces macrophages to produce the inflammatory cytokines TNF-alpha, IL-1, IL-8, and intracellular adhesion molecule-1, which decrease response to infection. (27) HBOT increases oxygen to normal levels so that macrophages can function normally. (20)

Direct Antibacterial effects of HBOT

HBOT has direct bacteriostatic and bactericidal effects against gram positive, gram negative aerobic and anaerobic microorganisms equal to some antibiotics. HBOT increases the concentration of superoxides both intracelluarly and extracellularly. Increased superoxide levels lead to increased production of hydrogen peroxide and other toxic oxygen radicals all of which impair microbial metabolic reactions. Anaerobic microorganisms cannot survive the amount of reactive oxygen species/oxygen based free radicals created in hyperoxic environments because they lack the superoxide dismutase, the superoxide degrading enzyme and catalse, the hydrogen peroxide degrading enzyme. Free radicals oxidize proteins and membrane lipids, damage DNA and inhibit metabolic functions (protein and nucleic acid synthesis) needed for bacterial growth and so are lethal for microorganisms that lack protection from them (particularly anaerobes). Clostridium perfringens is an anaerobe easily killed in the hyperbaric environment. (28) (29) HBOT will suppress secretion of alpha toxin in *C. perfringens* infections, which is the major source of tissue injury and death. When HBOT is utilized in the early course of infection with appropriate antibiotics and surgical attention, many of the deleterious effects of bacterial endotoxin release are minimized. HBOT treatment of anaerobic infections caused by Clostridial perfringens has increased patient survival, reduced number of surgeries, shortened hospital stays, reduced morbidity, extension of necrosis, and systemic toxicity. (30) (29) (31) (32) Actinomycosis is also very susceptible to HBOT. Some *Actinomyces* spp. are obligate anaerobes while others are facultative anaerobes. HBOT combined with proper surgical debridement and antibiotics greatly enhances healing and speeds clearing of infection. (33)

Leukocyte mediated killing of several aerobic bacteria

Proteus vulgaris, Salmonella typhimurium, Klebsiella pneumoniae, Serratia marcescens, Pseudomonas aeruginosa and Staphylococcus species - is significantly reduced in hypoxia. Increasing the concentration of oxygen over ambient levels has been shown to reduce infection (34) (35) (36)HBOT enhances phagocytic clearance of S. aureus, including Methicillin-resistant S. aureus and helps potentiate the bactericidal effects when used with appropriate antibiotics. (37) (38) (39) Additionally HBOT aids in the treatment of mycobacteria such as Mycobacterium ulcerans (40), Mycobacterium tuberculosis (41), and Mycobacterium fortuitum (42) Several studies have suggested effective treatment of Borrelia Burgdorferi (Lyme disease) with HBOT. The studies have not proven eradication of the disease but show excellent evidence that HBOT improves quality of life for patients who have failed all other treatment regimes. (43) (44)

Increases efficacy of antibiotics

When oxygen tension falls below 30 mmHg, bacteria quickly destroy tissue. Evidence suggests that hyperoxygenation and antibiotics combine to yield a synergistic effect. (45) (46) HBO raises oxygen tension in hypoxic tissue to level necessary for killing of bacteria by PMNs and optimal activity of certain antibiotics. HBOT aids in the penetration, efficacy and action of aminoglycosides and the antimetabolites/sulfonamides trimethoprim, sulfamethoxazole, and sulfasoxazole. Efficacy of cephalosporins, nitrofurantoin and amphotericin B are also improved with HBOT. Other antibiotics, like vancomycin and fluoroquinolones, become less efficient under hypoxic conditions but recover their usual effectiveness when normal oxygen tension is restored. Presumably their efficacy may be increased with increased oxygen tension. (47) (48) (29)

HBOT as an antifungal

HBOT has also been shown to inhibit the growth of some fungi and to potentiate the antifungal effect of amphthericin B. Antifungal effects are attributed to membrane damage, enzyme inactivation, and reduced protein synthesis caused by increased production of oxygen-based free radicals during exposure to HBOT (49) HBOT use as proven efficacious as adjunctive therapy for treating invasive rhinocerebral mucormycosis, (50) (51) aspirgillosis (52),*Candida albicans*, (53) (54) Pneumocystis carinii has been considered to be a fungus. In vitro studies show oxygen to be lethal to P.Carinii due to low levels of antioxidant enzymes. (29)

Hyperoxia facilitates healing by interfering with the inflammatory cascade

Most importantly, oxygen relieves hypoxia to damaged tissues. Studies have proven that HBOT decreases neutrophil rolling, activation, transmigration into tissues and local adherence by imparing cGMP synthesis in activated leukocytes. (21) Hyperoxia results in reduced expression of endothelial adhesion molecules and increases nitric oxide production which also inhibits neutrophil adhesion (22) (23) (24) HBOT significantly reduces proinflammatory cytokine induction which also contributes to minimizing inflammatory response (25)

Hyperoxia reduces free radical damage of reperfusion injury

Ischemia/reperfusion injury occurs in many disease states when the restoration of blood flow to tissues causes dangerous vascular and tissue damage, further impairing blood flow and healing. Ischemia/reperfusion injury occurs in many instances such as stroke, crush injury, embolic disease, myocardial infarctions, CO poisoning, and any other disease state suffering an ischemic insult followed by the restoration of blood flow. During reperfusion, the restoration of blood flow actually leads to progressive vascular damage and an expanding area of poor blood flow. (55) (56) HBOT interferes with the sequestration of neutrophils on injured endothelium and lessens the initiation of the reperfusion injury. It affects neutrophil function, endothelial cellular adhesion molecule expression, nitric oxide production, nitric oxide synthetase expression, cellular energetics, lipid peroxidation and microvascular blood flow. (57) (58) HBOT protects tissues from the harmful effects of toxic oxygen radicals or reactive oxygen species in many ways by providing enough oxygen for the reperfused tissues to generate scavengers (by increasing local endothelial surface superoxide dismutase and peroxidase activity) to detoxify the reactive oxygen species. (59) (56) HBOT antagonizes lipid peroxidation of cell membranes from toxic oxygen radicals by preventing the conversion of endothelial xanthine dehydrogenase to xanthine oxidase, the essential step required for lipid peroxidation. (59) (57)

Longer term effects (days to weeks)

Effects of HBOT are prolonged after HBOT sessions end and tissue oxygen levels return to pre-treatment values. Gas and tissue levels of O_2 decrease rapidly after HBOT but are still elevated for at least 4 hours. (60) (20) Marx and Johnson demonstrated

that for irradiated wounds, HBOT induces neovascularization, which becomes significant after about 14 treatments and continues for years after the HBOT treatments have ended. (61)Other important beneficial effects of HBOT are attributed to improved tissue repair by improving necrotic tissue proteolysis, fibroblast proliferation, collagen deposition, angiogenesis, migration of epithelial cells, and bone remodeling by osteoblastic/osteoclastic activity, which are all significantly diminished in hypoxic tissues (62) A number of days following injury there is migration of fibroblasts (connective tissue cells responsible for collagen production) into the areas of damage. These cells then divide and replicate producing large amounts of collagen-(connective tissue used to repair damage to tissue) that acts as the building block for the healing of tissue and wounds. (63) When the environment of fibroblast cells has an oxygen tension of less than 10mmHg, the cells can divide, but can no longer make collagen. It also cannot migrate to where it is needed for healing. When the oxygen tension is increased, the fibroblast can again carry out these wound healing functions. The collagen produced by these cells forms a fibrous matrix, and new capillaries grow into this. HBOT increases fibroblast growth, production of collagen and cellular matrix via cytokine down regulation and growth factor up-regulation. HBOT up-regulates collagen synthesis through pro-al(I) mRNA expression. (64) In ischemic flaps HBOT up-regulates fibroblast growth factor (FGF) causing an increased effect over that seen with FGF alone. (65) In situations where FGF is ineffective, HBOT can make it more effective (66) (20) Research has shown that increasing oxygen concentration in tissues above normal will increase collagen production rate above normal. (67) (62) (68)
HBOT increases osteoclast and osteoblast production, proliferation, differentiation and activity. This aids in bone regeneration and healing. Bone healing as well as resorption of necrotic bone by osteoclasts is enhanced with HBOT. (69) (70) (71)A recent study suggests that the proliferative effects of HBOT on osteoblasts may contribute to the recruitment of osteoblasts at fracture sites. The same study demonstrated that HBOT stimulated significantly increased mRNA expression of fibroblast growth factor (FGF)-2 as well as many other bone promoting proteins promoting osteoblast proliferation and cell cycle progression after 3 days of treatment. (72)

Angiogenesis

Destruction or injury of tissues results in damage of the supporting blood vessels. The healing process relies on supply of the blood containing the cells and nutrients that carry out and enable healing as well as remove damaged cells, debris and foreign microorganisms. Angiogenesis occurs in response to high oxygen concentrations (73) HBOT helps create collateral blood flow by increased angiogenesis (new vessel growth budding from existing intact vessels) and vasculogenesis (when bone marrow derived endothelial precursor cells populate the tissue and differentiate and grow into new vessels). Recent evidence shows that hematopoietic stem cell derived vessels occur much more than previously believed. (74) (75) (76)
The increased oxygen supply and increased pressure employed by hyperbaric oxygen supply are both responsible for the stimulation of new capillary growth and the repair of existing capillaries. Collateral blood vessels are a result of increased fibroblast production of collagen and are oxygen dependent and collagen is the foundation matrix for angiogenesis. Growth of new blood vessels in the micro-circulation and repair of damaged capillaries are stimulated within the damaged tissues by hyperbaric oxygen therapy, providing increased oxygen availability to these areas. (77) (78) (73) (79) (76)

Hyperbaric Oxygen Mobilizes Stem Cells

One of the most exciting effects of HBOT is the increased mobilization of stem cells in circulation of patients. This finding could explain much of what has been seen in the physiologic activity of HBOT. American Journal Physiology - Heart and Circulatory Physiology reports a single 2-hour exposure to HBOT at 2 ATA doubles circulating CD34+ progenitor stem cells (primordial cells targeted to salvage and restore damaged structures); and at approx. 40-hours of HBOT; circulating CD34+ cells increases eight fold (800%). (80)

There are pharmaceutical options for increasing stem cell levels in the body such as erythropoiesis stimulating agents (erythropoietin, darbepoetin) and colony stimulating factors such as granulocyte colony stimulating factors (G-CSF). According to Thom, et al publication in the American Journal of Physiology-Heart and Circulation Physiology, exposure to HBOT mobilizes stem/progenitor cells from the bone marrow by a nitric oxide (·NO) -dependent mechanism. The population of $CD34^+$ cells in the peripheral circulation of humans doubled in response to a single exposure to 2.0 atmospheres absolute (ATA) O_2 for 2 hours. Over a course of 20 treatments, circulating $CD34^+$ cells increased eightfold, although the overall circulating white cell count is not significantly increased without any pharmacologic intervention or underlying disease processes. This ground breaking study offers a new, safe, therapeutic option for stem cell proliferation with little risk of side effects. (80) This was demonstrated to be particularly useful in diabetic patients where increases in stem cells were directly correlated to improved refractory lower extremity neuropathic ulcers. Nitric oxide synthase activity was found to be acutely increased in patient's platelets following HBOT and remains elevated for at least 20 hours. HBOT aided healing may be due to increased vasculogenic stem cell mobilization to skin wounds. (81)

REFERENCES

1. **Patrick M.Tibbles, John S. Edelsberg.** Hyperbaric-Oxygen Therapy. *NEJM.* June 1996, Vol. 334, 25, pp. 1642-1648.

2. **Boerema I, Meijne NG, Brummelkamp WH, et al.** *Life without blood. J Cardiovasc Surg 182: 133, 1960.*
3. **Weibel, ER.** Delivering oxygen to the cells. [book auth.] Weibel ER. *The Pathway for Oxygen.* Boston : Harvard University Press, 1984, pp. 175-210.
4. **AL Gill, CNA Bell.** Hyperbaric oxygen: its uses, mechanisms of actions and outcomes. *Q J Med.* 2004, 97, pp. 385-395.
5. *The striking effect of hyperbaric oxygenation therapy in the management of chronic idiopathic intestinal pseudo-obstruction.* **Yokota T, Suda T, Tsukioka S, Takahashi T, Honma T, Seki K, Matsuzawa J, Miura M, Aoyagi Y, Asakura H.** 1, Jan 2000, Vol. 95, pp. 285-8.
6. *Hyperbaric oxygen therapy for intestinal obstruction in children: an exceptional experience in a compromised child.* **Ohno Y, Kanematsu T.** 33, Oct 1998, J Pediatr Surg., Vol. 10, pp. 1543-5.
7. *Hyperbaric oxygen ameliorates bacterial translocation in rats with mechanical intestinal obstruction.* **Akin ML, Uluutku H, Erenoglu C, Ilicak EN, Elbuken E, Erdemoglu A, Celenk T.** 45, Jul . 2002 Jul;45(7):967-72., Dis Colon Rectum, Vol. 7, pp. 967-72.
8. **LM, Ludwig.** The role of hyperbaric oxygen in current emergency medical care. *J Emerg Nurs. 1989 May-Jun;15(3):229-37.*
9. **Ambiru S, Furuyama N, Aono M, Kimura F, Shimizu H, Yoshidome H, Miyazaki M, Shimada H, Ochiai T.** Hyperbaric oxygen therapy for the treatment of postoperative paralytic ileus and adhesive intestinal obstruction associated with abdominal surgery: experience with 626 patients. *Hepatogastroenterology.* Oct-Nov 2007, Vol. 54, 79, pp. 1925-9.
10. **Butler, Al-Waili and.** Effects of Hyperbaric Oxygen on Inflammatory Response to Wound and Trauma: Possible Mechanism of Action. *TheScientificWorldJOURNAL* . 6, 2006, pp. 425-441.
11. **Piantadosi, Claude A.** Pulmonary Gas Exchange, Oxygen Transport, and Tissue Oxygenation. [book auth.] TS Neuman and SR Thom. *Physiology and Medicine of Hyperbaric Oxygen Therapy.* Philadelphia : Saunders, 2008, pp. 133-185.
12. . **Nylander G, Nordstrom H, Eriksson E.** Effects of hyperbaric oxygen on oedema formation after a scald burn. *Burns Incl Therm Inj.* 1984, Vol. 10, 3, pp. 193-196.
13. **Shupak A, Gozal D, Ariel A, Melamed Y, Katz A.** Hyperbaric oxygenation in acute peripheral posttraumatic ischemia. *J Hyperbaric Med.* 1987 , Vol. 2, pp. 7-14.
14. **JL., Phillips.** Air as medicine. [book auth.] Phillips JL. *The Bends.* New Haven : Yale University Press, 1998, pp. 197-203.
15. **Strauss, Michael.** Crush Injury Compartment Syndrome and Other Acute Traumatic Peripheral Ischemias. [book auth.] Harry T Whelan Eric P Kindwall. *Hyperbaric Medicine Practice.* 2nd. Flagstaff : Best Publishing Company, 2004, 30, pp. 753-778.
16. **Skyhar MJ, Hargens AR, Strauss MB, Gershuni DH, Hart GB, Akeson WH.** Hyperbaric oxygen reduces edema and necrosis of skeletal muscle in compartment syndromes associated with hemorrhagic hypotension. *J Bone Joint Surg Am.* 1986, Vol. 68, 8, pp. 1218-1224.
17. **McFarlane RM, Wermuth RE.** The use of hyperbaric oxygen to prevent necrosis in experimental pedicle flaps and composite skin grafts. *Plast Reconstr Surg.* 1966, Vol. 37, 5, pp. 422-430.
18. **Allen DB, Maguire JJ, Mahdavian M, Wicke C, Marcocci L, Scheuenstuhl H, Chang M, Le AX, Hopf HW, Hunt TK.** Wound hypoxia and acidosis limit neutrophil bacterial killing mechanisms. *Arch Surg.* 1997, Vol. 132, 9, pp. 991-996.
19. **Knighton DR, Halliday B, Hunt TK.** Oxygen as an antibiotic. The effect of inspired oxygen on infection. *Arch Surg.* 1984, Vol. 119, 2, pp. 199-204.
20. **Wright, James.** http://www.worldwidewounds.com/2001/april/Wright/HyperbaricOxygen.html#ref14. *Hyperbaric oxygen therapy for wound healing* . [Online] May 2001.
21. **Chen Q, Banick PD, Thom SR.** Functional inhibition of rat polymorphonuclear leukocyte B2 integrins by hyperbaric oxygen is assoicated with impared cGMP synthesis. *J Pharmacol Exp Ther.* 1996, Vol. 276, pp. 929-933.
22. **Buras JA, Reenstra WR.** Hyperbaric oxygen decreases endothelial cell E-selectin protein expression in an in-vitro model of ischemia/reperfusion. 32:S17. *Ann Emerg Med* . 1998, 32, p. S17.
23. **Banick PD, Chen Q, Xu YA, Thom SR:** Nitric oxide inhibits neutrophil β β2 integrin function by inhibiting membrane-associ- ated cyclic GMP synthesis.1997, 172:12-24. *J Cell Physiol.* 1997, Vol. 172, pp. 12-24.
24. Bitterman, Haim. Bench to Bedside review: Oxygen as a drug. *Critical Care.* 2009, Vol. 13, 1.
25. Masatomo Yamashita, Mamuro Yamashita. Hyperbaric oxygen treatment attenuates cytokine induction after massive hemorrhage. *Am J Physiol Endocrinol Metab.* 2000, 278, pp. E811-E816.
26. Mrotek JJ, Henderson R, Kiel J, et al. Normobaric oxygen concentrations affect cultured mouse macrophage responses. . *Fed Proc.* 1986, Vol. 45, p. A958.
27. G., Scannell. Leukocyte responses to hypoxic/ischemic conditions. . *New Horiz.* 1996, Vol. 4, 2, pp. 179-183.
28. IA, Silver. Cellular microenvironment in healing and non-healing wounds. [ed.] Heppenstall RB, Pines M. Edited by Hunt TK. *Soft and Hard Tissue Repair.* New York : Praeger, 1984, pp. 50-66.
29. Park, Matthew. Effects of Hyperbaric Oxygen in Infectious Diseases: Basic Mechanisms. [book auth.] Harry T. Whelan Eric P. Kindwall. *Hyperbaric Medicine Practice.* Second. Flagstaff : Best Publishing Company, 2004, 10, pp. 205-243.
30. Korhonen K, Hirn M, Niinikoski J. Hyperbaric oxygen in the treatment of Fournier's gangrene. *Eur J Surg.* 1998, Vol. 164, 4, pp. 251-255.
31. Korhonen K, Klossner J, Hirn M, Niinikoski J. Management of clostridial gas gangrene and the role of hyperbaric oxygen. *Ann Chir Gynaecol.* 1999, Vol. 88, 2, pp. 139-142.
32. Demello FJ, Haglin JJ, Hitchcock CR. Comparative study of experimental Clostridium perfringens infection in dogs treated with antibiotics, surgery, and hyperbaric oxygen. 1973, 73, p. 936.
33. Manheim S.D., Voleti C., Ludwig, A., and Jacobson J.H. JAMA . *Hyperbaric oxygen in the treatment of actinomycosis.* 1969, Vol. 210, 3, pp. 552-553.
34. DC., Hohn. Host resistance of infection: Established and emerging concepts. [ed.] Hunt TK. *Wound Healing and Wound Infection.* New York : Appleton-Century-Crofts, 1980, pp. 264-280.
35. Knighton DR, Halliday B, Hunt TK. Oxygen as an antibiotic. A comparison of the effects of inspired oxygen concentration and antibiotic administration on in vivo bacterial clearance. *Arch Surg.* 1986, Vol. 121, 2, pp. 191-195.
36. Greif R, Akca O, Horn EP, Kurz A, Sessler DI. Supplemental perioperative oxygen to reduce the incidence of surgical-wound infection. Outcomes Research Group. . *N Engl J Med.* 2000, Vol. 342, 3, pp. 161-167.
37. Tsuneyoshi I, Boyle WA 3rd, Kanmura Y, Fujimoto T. Hyperbaric hyperoxia suppresses growth of Staphylococcus aureus, including methicillin-resistant strains. *J Anesth.* 2001, Vol. 115, 1, pp. 29-32.
38. Knighton DR, Fiegel VD, Halverson T, Schneider S, Brown T, Wells CL. Oxygen as an antibiotic. The effect of inspired oxygen on bacterial clearance. *Arch Surg.* Jan;125(1):97-100. 1990, Vol. 125, 1, pp. 97-100. .
39. Jain, K.K. HBO Therapy in Infections. [ed.] K.K. Jain:. *Textbook of hyperbaric medicine.* 5. Cambridge : Hogrefe& Huber Publishers, Inc., 2009, 13, pp. 135-148.
40. Krieg RE, Wolcott JH, Confer A. Treatment of Mycobacterium ulcerans infection by hyperbaric oxygenation. *Aviat Space Environ Med. 1975 Oct;46(10):1241-5.* Oct 1975, Vol. 46, 10, pp. 1241-1245.
41. Warren J, Sacksteder MR, Thuning CA. Therapeutic effect of prolonged hyperbaric oxygen in adjuvant arthritis of the rat. . *Arthritis Rheum.* Apr 334-339, 1979, Vol. 22, 4.
42. GL., Goodhart. J Orthop Trauma. Mycobacterium fortuitum osteomyelitis following trauma. 1993, Vol. 7, 2, pp. 142-145.
43. Fife, W. P., and D. M. Freeman. "[abstract] TREATMENT OF LYME DISEASE WITH HYPERBARIC OXYGEN THERAPY.". Undersea and Hyperbaric Medical Society, Inc. (http://www.uhms.org) http://archive.rubicon-foundation.org/6541998. [Online] 1998.
44. Taylor RS, Simpson. In Review of treatment options for lyme borreliosis. *J Chemother.* 17, Sep 2005 , Vol. Suppl 2, pp. 3-16.
45. Jonsson K, Hunt TK. Ann Surg. *Oxygen as an isolated variable influences resistance to infection.* 1988, Vol. 208, 6, pp. 783–787.
46. Knighton D, Halliday B, Hunt TH. Oxygen as an antibiotic: a comparison of inspired oxygen concentration and antibiotic administration on in vivo bacterial clearance. *Arch Surg. 1986 Feb;121(2):.* Feb 1986, Vol. 121, 2, pp. 191-5.

47. Gottlieb SF, Solosky JA, Aubrey R, et al. Synergistic action of increased oxygen tensions and PABA-folic acid antagonists on bacterial growth. *Aerosp Med.* Aug 1974, Vol. 45, 8, pp. 829-33.
48. M K Park, K H Muhvich, R A Myers, and L Marzella. Hyperoxia prolongs the aminoglycoside-induced postantibiotic effect in Pseudomonas aeruginosa. *Antimicrob Agents Chemother.* April 1991, Vol. 35, 4, pp. 691–695.
49. Bitterman, Haim. Hyperbaric Oxygen for Invasive Fungal Infections. *Isr Med Assoc J.* May 2007, Vol. 9, 5, pp. 387-388.
50. Yohai RA, Bullock JD, Aziz AA, Markert RJ. Survival factors in rhino-orbital-cerebral mucormycosis. *Surv Ophthalmol. 1994 ;39(1):3-22.* Jul-Aug 1994, Vol. 39, 1, pp. 3-22.
51. Ferguson BJ, Mitchell TG, Moon R, Camporesi EM, Farmer J. Adjunctive hyperbaric oxygen for treatment of rhinocerebral mucormycosis. *Rev Infect Dis.* May-Jun 1988, Vol. 10, 3, pp. 551-559.
52. García-Covarrubias L, Barratt DM, Bartlett R, Metzinger S, Van Meter K. Invasive aspergillosis treated with adjunctive hyperbaric oxygenation: a retrospective clinical series at a single institution. *South Med J.* Apr 2002, Vol. 95, 4, pp. 450-6.
53. Gudewicz TM, Mader JT, Davis CP. Combined effects of hyperbaric oxygen and antifungal agents on the growth of Candida albicans. *Aviat Space Environ Med. Jul;58(7):673-8.* Jul 1987, Vol. 58, 7, pp. 673-678.
54. WJ., Cairney. Effect of hyperbaric oxygen on certain growth features of Candida albicans. *Aviat Space Environ Med.* Aug 1978 ;49(8):956-8., Vol. 49, 8, pp. 956-958.
55. J., Buras. Basic mechanisms of hyperbaric oxygen in the treatment of ischemia-reperfusion injury. *Int Anesthesiol Clin.* Winter, 2000, Vol. 38, 1, pp. 91-109.
56. Buras JA, Reenstra WR. Endothelial-neutrophil interactions during ischemia and reperfusion injury: basic mechanisms of hyperbaric oxygen. *Neurol Res.* Mar 2007, Vol. 29, 2, pp. 127-131.
57. SR, Thom. Antagonism of CO-mediated brain lipid peroxidation by hyperbaric oxygen. . *Toxicol Appl Pharmacol.* 1990, Vol. 105, p. 340.
58. Yogaratnam JZ, Laden G, Guvendik L, Cowen M, Cale A, Griffin S. Hyperbaric oxygen: a novel technology for modulating myocardial ischemia-reperfusion via a single drug. *Adv Ther.* Jul-Aug 528-33, 2006, Vol. 23, 4.
59. Kaide, Sorabh Khandelwal. Colin G. Hyperbaric Oxygen Therapy. s.l. : McGraw-Hill Companies, Inc, 2010.
60. CH, Wells. Tissue Gas measurements during hyperbaric oxygen exposure. *6th International Congress of Hyperbaric Medicine 1977: 118-124.* . 1977, pp. 118-124.
61. Marx RE, Johnson RP. Problem wounds in oral and maxillofacial surgery: The role of hyperbaric oxygen. [book auth.] Hunt TK Davis JC. *Problem Wounds - the role of oxygen.* New York : Elsevier, 1988, pp. 65-123.
62. Gimbel ML, Hunt TK. Wound healing and hyperbaric oxygenation. [ed.] Whelan HT. Kindwall EP. *In Hyperbaric Medicine Practice.* 2nd. Flagstaff : Best Publishing Company, 1999, pp. 169-204.
63. S.C., Hunt TK. In: H. Columbia: Oxygen and wound healing. [book auth.] 8th Annual Advanced Symposium yperbaric Medicine 2000. s.l. : Palmetto Richland Memorial Hospital and the University of South Carolina School of Medicine, 2000.
64. Ishii Y, Myanaga Y, Shimojo H, Ushida T, Tateishi T. Effects of hyperbaric oxygen on procollagen messenger RNA levels and collagen synthesis in the healing of rat tendon laceration. *Tissue Eng.* 1999, Vol. 5, pp. 279-86.
65. Bayati S, Russell RC, Roth AC. Stimulation of angiogenesis to improve the viability of prefabricated flaps. *Plast Reconstr Surg.* 1998, Vol. 101, 5, pp. 1290-1295.
66. Wu L, Pierce GF, Ladin DA, Zhao LL, Rogers D, Mustoe TA. Effects of oxygen on wound responses to growth factors: Kaposi's FGF, but not basic FGF stimulates repair in ischemic wounds. *Growth Factors.* 1995, Vol. 12, 1, pp. 29-35.
67. Asano T, Kaneko E, Shinozaki S, Imai Y, Shibayama M, Chiba T, Ai M, Kawakami A, Asaoka H, Nakayama T, Mano Y, Shimokado K. Hyperbaric oxygen induces basic fibroblast growth factor and hepatocyte growth factor expression, and enhances blood perfusion and muscle regeneration in mouse ischemic hind limbs. *Circ J.* 71, Mar 2007, Vol. 3, pp. 405-411.
68. Kang TS, Gorti GK, Quan SY, Ho M, Koch RJ. Effect of hyperbaric oxygen on the growth factor profile of fibroblasts. *Arch Facial Plast Surg.* Jan-Feb 2004, Vol. 6, 1, pp. 31-35.
69. Strauss MB, Malluche HH, Faugere MC. Effect of hyperbaric oxygen on bone resorption in rabbits. . *Seventh Annual Conference on Clinical Application of HBO.* June 1982 , pp. 8-18.
70. Penttinen R, Niinikoski J, Kulonen E. Hyperbaric oxygen and fracture healing: a biomechanical study with rats. . *Acta Chir.* 1972, 138, pp. 39–44.
71. Ueng SW, Lee SS, Lin SS, Wang CR, Liu SJ, Yang HF, Tai CL, Shih CH. Bone healing of tibial lengthening is enhanced by hyperbaric oxygen therapy therapy: a study of bone mineral density and torsional strength on rabbits. *J Trauma.* 1998, 44, pp. 676–681.
72. Hsieh CP, Chiou YL, Lin CY. Hyperbaric oxygen-stimulated proliferation and growth of osteoblasts may be mediated through the FGF-2/MEK/ERK 1/2/NF-κB and PKC/JNK pathways. *Connect Tissue Res.* . Dec 2010, Vol. 51, 6, pp. 497-509.
73. Knighton DR, Silver IA, Hunt TK. Regulation of wound-healing angiogenesis-effect of oxygen gradients and inspired oxygen concentration. *Surgery.* Aug 1981, Vol. 90, 2, pp. 262-270.
74. Milovanova TN, Bhopale VM, Sorokina EM, Moore JS, Hunt TK, Hauer-Jensen M, Velazquez OC, Thom SR. Hyperbaric oxygen stimulates vasculogenic stem cell growth and differentiation in vivo. *J Appl Physiol.* Feb 2009, Vol. 106, 2, pp. 711-728.
75. OC., Velazquez. Angiogenesis and vasculogenesis: inducing the growth of new blood vessels and wound healing by stimulation of bone marrow-derived progenitor cell mobilization and homing. *J Vasc Surg.* Jun 2007, Vol. 45, Suppl A, pp. A39-47.
76. Gallagher KA, Goldstein LJ, Thom SR, Velazquez OC. Hyperbaric oxygen and bone marrow-derived endothelial progenitor cells in diabetic wound healing. *Vascular.* Nov-Dec 2006, Vol. 14, 6, pp. 328-37.
77. Sheikh AY, Gibson JJ, Rollins MD, et al. Effect of hyperoxia on vascular endothelial growth factor levels in a wound model. *Arch Surg.* . Nov 2000, Vol. 135, 11, pp. 1293-7.
78. Hunt TK, Pai MP. The effect of varying ambient oxygen tensions on wound metabolism and collagen synthesis. *Surg Gynecol Obstet.* Oct 1972 , Vol. 135, 4, pp. 561-567.
79. Sander AL, Henrich D, Muth CM, Marzi I, Barker JH, Frank JM. In vivo effect of hyperbaric oxygen on wound angiogenesis and epithelialization. *Wound Repair Regen.* Mar-Apr 2009, Vol. 17, 2, pp. 179-84.
80. Stephen R. Thom, Veena M. Bhopale, Omaida C. Velazquez, Lee J. Goldstein, Lynne H. Thom, and Donald G. Buerk. Stem cell mobilization by hyperbaric oxygen. *Am J Physiol Heart Circ Physiol.* 2006, 290, pp. H1378-H1386.
81. Thom SR, Milovanova TN, Yang M, Bhopale VM, Sorokina EM, Uzun G, Malay DS, Troiano MA, Hardy KR, Lambert DS, Logue CJ, Margolis DJ. Vasculogenic stem cell mobilization and wound recruitment in diabetic patients: Increased cell number and intracellular regulatory protein content associated with hyperbaric oxygen therapy. *Wound Repair Regen.* 2011, Vol. 19, 2, pp. 149–161.
82. Hardy, Kevin. The physics of Hyperbaric Oxygen Therapy. [book auth.] Stephen R. Thom Tom S. Newman. *Physiology and Medicine of Hyperbaric Oxygen Therapy.* Philadelphia : Saunders, 2008, pp. 57-64.
83. Kindwall, Eric. The Physics of Diving and hyperbaric Pressures. [book auth.] Harry T. Whelan Eric P. Kindwall. *Hyperbaric Medicine Practice.* Flagstaff : Best Publishing Company, 2004, pp. 21-36.
84. Hyperbaric oxygen therapy reduces severity and improves survival in severe acute pancreatitis. *J Gastrointest Surg.* Aug 2007, Vol. 11, 8, pp. 1008-15.
85. Christophi C, Millar I, Nikfarjam M, Muralidharan V, Malcontenti-Wilson C. Hyperbaric oxygen therapy for severe acute pancreatitis. *J Gastroenterol Hepatol.* Nov 2007 , Vol. 2211, pp. 2042-2046.
86. Cuthbertson CM, Su KH, Muralidharan V, Millar I, Malcontenti-Wilson C, Christophi C. Hyperbaric oxygen improves capillary morphology in severe acute pancreatitis. *Pancreas.* . Jan 2008, Vol. 36, 1, pp. 70-75.
87. Yu X, Li YG, He XW, Li XR, Din BN, Gan Y, Xu M. Hyperbaric oxygen reduces inflammatory response in acute pancreatitis by inhibiting NF-kappaB activation. *Eur Surg Res.* . 2009, Vol. 42, 2, pp. 130-135.

HYPERBARIC OXYGEN: CASE EXAMPLES AND PRESENTATION OF RELEVANT NEW SCIENTIFIC DATA

Diane M. Levitan, VMD, DACVIM (Small Animal Internal Medicine)
Commack, New York

The rationale for HBOT in many different internal medicine and emergency scenarios will be discussed in this lecture. The following are some of the disease processes that are responsive to HBOT.

Central Nervous System Brain trauma Spinal cord trauma Stroke Cerebral ischemia Compressive cord diseases GME Fibro-cartilaginous emboli Peripheral nerve injury Head trauma Acute cerebral edema Peripheral nerve damage Polyradiculononeuritis Intracranial abscess	**Inflammatory, infectious and immune mediates diseases:** Allergic reactions/anaphylaxis GME RMSF induced vasculitis/ meningitis/arthritis Immune mediated polyarthritis IMHA ITP Inflammatory bowel disease Pancreatitis Cholangitis: Osteomyelitis Septic arthritis Septicemia Endotoxemia Mold exposure Aspergillosis Mucormycosis, Candidiasis pneumocystis Lyme disease Anaerobic infections: Actinomyces, clostridium, etc Necrotizing fasciitis Intracranial and abdominal abscess	**Gastrointestinal/hepatic** Post-op enterotomy or GI surgery Ileus Pancreatitis Inflammatory bowel disease Peritonitis Gastric ulceration Reperfusion Hepatic inflammatory diseases Biliary tract inflammation Biliary tract infection Hepatic necrosis Oral ulceration or wounds Esophageal inflammation	**Emergency Medicine** Carbon monoxide Cyanide toxicity Smoke inhalation Lymphangitis Burns Thromboembolic disease Gas embolism Acute traumatic ischemia Bloat Necrotizing faciitis Blood loss anemia Tetanus Septicemia Head trauma Acute cerebral edema Skin burns Fractures and acute trauma Envenomation- spider, snake	**Musculoskeletal** Athletic injuries Tendonitis Periostitis Fracture Myositis Crush injuries All wounds Refractory osteomyelitis Severe degloving Wounds Burns Delayed wound healing Acute cruciate injuries Delayed fracture healing Skin grafts
Dermatology: Infection control: deep pyoderma and MRSA Immune mediated skin dz Improved skin healing Hair regrowth Decrease otitis inflammation and infection +/-myringotomy	**Oncology:** Post-operative healing Pre and post radiation therapy Palliative care for neoplasia, pain and swelling Radiation burns/necrosis Osteoradionecrosis Soft tissue radionecrosis Chemotherapy extravasation	**Cardiovascular** Hypotension Shock Cardiac infarction Acute anemia Thromboembolic disease	**Renal** nephrotic syndrome, ischemia reperfusion, inflammation, hypoxia Lyme disease	

A review of recent literature and ongoing research will be discussed. There is a great deal of room for research in the field of veterinary hyperbaric oxygen therapy. Be the CHAMPION of this therapy.

COMPLETE GLOBAL ISCHEMIA AND HYPERBARIC OXYGEN THERAPY IN SMALL ANIMALS

Ronald Lyman DVM, DACVIM
Fort Pierce, Florida

Introduction

Hyperbaric oxygen therapy [HBOT] has transcended from its traditional place in treatment of decompression illness ["the bends"] to a variety of clinical applications globally. Basic science studies have demonstrated HBOT effects and mechanisms of action. Evidence based reviews and meta-analyses have been published on HBOT's beneficial effect on humans with extreme anemia and in wound healing [22, 24]. Clinical studies on practical applications are abundant in the human medical literature. Book chapters describing clinical applications of HBOT have been included in the most recent editions of Ettinger's "Textbook of Veterinary Internal Medicine" and Bonagura's "Current Veterinary Therapy XV". The June 2010 Veterinary Emergency and Critical Care Journal featured HBOT as the subject in it's 'State of the Art' articles [17]. Recently, Dr. Colin Kaide [from The Ohio State University Medical Center] discussed HBOT and its role in the therapy of human infectious diseases in The Emergency Medicine Clinics of North America [18]. He concluded with the statement "in our opinion withholding HBOT because of a lack of randomized trials in the face of a huge body of clinical experience and in vitro studies, especially in these devastating disease processes, would be bordering on unethical". The Veterinary Hyperbaric Medicine Society [VHMS] has developed didactic material and test questions for hyperbaric oxygen therapy technical certification. HBOT research has reached the point that studies are demonstrating the specific treatment pressures which are most effective at altering inflammatory gene expression (32). This presentation will focus on HBOT in complete global ischemia of small animals. Specific citations of the literature and examples of patients treated with HBOT will be presented.

HBOT Chambers and Ancillary Hardware

Modestly priced therapeutic chambers are now available from medical equipment suppliers. Small animal sized veterinary models have been manufactured. Experienced professionals are available to assist in designing systems, obtaining and installing chambers, training hospital personnel regarding operation and safety issues, and servicing HBOT equipment. Clinical chambers range in size from monoplace [single patient] design, to chambers accommodating multiple patients / caregivers, and operating suites. Most chambers require an oxygen supply system capable of large gas volumes delivered at a rapid flow rate. Liquid oxygen tanks and a medical grade oxygen pipeline of large diameter satisfy these requirements. A passive discharge pipeline directs the exhaust gases outside the hospital building. Some chambers have ports allowing [specific] IV lines and/ or tubing for assisted ventilation to be connected to the patient during an HBOT session. Current oxygen costs to operate a small animal monoplace chamber may range from approximately $4 to $20 per treatment session depending on chamber design, length of session, pressure employed, and oxygen supply system. The clinician should recognize that a distinct difference exists between Hyperbaric Oxygen Therapy and "Minimal" Hyperbaric Oxygen Therapy. The portable, [usually vinyl, plastic, and fabric] balloon type chambers marketed for "Minimal" HBOT do not operate at pressures or oxygen concentrations in the typical therapeutic ranges of the medical grade hyperbaric oxygen chambers.

Treatment Protocol ("The Dive")

The patient is prepared by examination and body temperature determination. HBOT is generally not given when the small animal's patient's body temperature is greater than 103.5 degrees Fahrenheit. Theoretically, excessive hyperthermia may increase oxygen uptake by tissues, and enhance the chance of oxygen toxicity. The vast majority of patients are quiet and relaxed during HBOT. Occasional patients are given sedation. Acepromazine is the usual drug of choice for small animals. Occasionally, benzodiazepines or barbiturates are employed in canines. Cotton bandages cover IV catheters. Metal collars are removed. A static inhibitor ground plate is placed under the patient. The session begins with a gradual pressure increase to the designated treatment pressure over approximately 10-15 minutes. The desired treatment pressure may range from 1.5 to 3 atmospheres absolute [ATA] in a 100% oxygen environment. The treatment period "at pressure" may range from 45 minutes to 2 hours. Then, a gradual decompression over approximately 10-15 minutes is made, and the patient is removed from the chamber. Currently, most of our small animal patients are treated at 1.5 ATA for 45 minutes "at pressure". Patients with Lyme disease are treated with progressively increasing pressures, and then followed up with treatments at 2.4 ATA, which is bacteriocidal to the spirochete. In patients perceived to be at reduced threshold for seizures, one should employ a pressure of 1.5 ATA for the "at pressure" treatment period. The patients are rested at room atmospheric pressure for at least 4 hours between HBOT sessions to minimize the chances of CNS or pulmonary oxygen toxicity. We treat most patients with BID sessions, and continue HBOT acutely until marked and progressive improvement occurs [for example, the tetraparetic, paraparetic, or monoparetic patient is using the affected limb[s] with improving strength and function over several days, or the wound is well on its way to resolution]. We treat selected patients with follow up sessions after initial discharge from the hospital. In many cases dogs or cats may be treated as outpatients. Patients may receive as little as 1 session [for example, after orthopedic surgery or venomous insect bite when administered to reduce inflammation and swelling] to as many as 40 or more sessions in extremely critical or chronic disease entities. One canine patient with multifocal CNS inflammatory disease received over 700 treatment sessions in 5 ½ years as her sole method of therapy for this clinical issue.

Physiology of HBOT

At normal atmospheric pressures [1 ATA] and while breathing room air, insignificant amounts of oxygen are dissolved and delivered to the tissues via the plasma. Mild increases are possible when oxygen is administered via mask, oxygen cage, intranasally, or intubation. The effects of HBOT are based upon the fact that oxygen delivered to the alveoli under increased atmospheric pressure results in large increases in the amount dissolved in plasma. This can increase the plasma arterial oxygen content from 0.32 vol % [ml 02/ dl whole blood] breathing room air at 1 ATA , to 4.44 vol % at 2 ATA breathing 100% oxygen, and up to 6.8 vol % at 3 ATA breathing 100% oxygen. This dramatic increase in dissolved oxygen [for example, breathing 100% oxygen at 3 ATA] allows diffusion and delivery via the plasma up to 4 times further through the tissues than is possible while breathing room air at 1 ATA. This delivery is independent of the hemoglobin system. Tissue PO2 measurement devices have demonstrated that oxygen tension remains elevated at > 110% of normal for up to 3 hours following a 1 hour HBOT session.[1] Thus, cells and mitochondria in areas where nearby capillary blood flow has been compromised by traumatic, ischemic, or inflammatory processes may receive oxygen subsequent to HBOT which would otherwise be denied by the underlying pathological disease process. Hyperbaric oxygen at 2 ATA has been shown to reduce cerebral blood flow up to 25% [by effecting vasoconstriction], yet concurrently increase [X 10] cerebral oxygen content. Intracranial pressure and cerebral edema are reduced following HBOT. In addition to the direct action of supplying oxygen necessary for mitochondrial metabolic processes, HBOT creates short pulses of increased reactive oxygen species [ROS] and reactive nitrogen species [RNS]. These molecules act in conjunction with redox systems involving glutathione, thioredoxin, and pyridine nucleotides to coordinate cell signaling and anti-oxidant, protective pathways [23]. Some of the results of these reactions are listed below:

- An increase in neovascularization [new blood vessel growth] . HBOT induces the production of vascular endothelial growth factor [VEGF]. [3, 19, 20]
- Enhancement of white blood cell oxidative killing and antibacterial effects [29]
- Potentiation of the cidal or static antibacterial effects of antibiotics such as the flouroquinolones, aminogycosides, vancomycin, and sulfas [29]
- Stimulation of nitric oxide synthetase [20]
- Modification of cytokine effects by regulating production and affecting receptor sites [anti-inflammatory]. HBOT Modulates MMP balance [anti-inflammatory][5]
- Inhibition of neutrophil adherence to microvascular endothelia, resulting in SUPPRESSION of the deleterious cascade of events that follows in ischemia- reperfusion injury. HBOT has been found to inhibit free radical formation. [4]
- Multiplication and mobilization of vasculogenic stem/progenitor cells [SPC's] which can be demonstrated in blood, tissue and in certain matrix materials implanted throughout the body[23].
- An increase of a key subset of SPC's, endothelial progenitor cells [EPC's]. EPC's have the ability to "home" to the site of tissue injury or dysfunction, differentiate into mature cells, and participate in vascular repair. This augmentation of SPC recruitment and differentiation is ascribed to HBOT's engagement of a physiological autocrine activation loop involving a complex set of responses initiated by the intracellular proteins thioredoxine, hypoxia inducible factor – 1, and hypoxia inducible factor – 1 – dependent growth factors. [19, 20]
- Modification of cellular responses at the level of genes [23,32]
- Increased ERK 1/2 levels and up-regulation of PDGFR-R expression leading to accelerated wound healing [25]
- Down regulation of ICAM-1 expression [27]
- Reduction in the size of intravascular gas bubbles [air emboli] by the direct effect of Boyle's Law
- Reduction in intracranial pressure [33]

Three recent textbooks and other publications offer detailed discussions of the physiology of oxygenation, its relation to HBOT, and effects on the CNS, PNS, and pathological disease states including the wound healing process [2,3,4,19, 20, 29].

Therapeutic Effects and Clinical Uses

The most common case types treated with HBOT at our small animal facility include intervertebral disc disease, pre and post op spinal and orthopedic surgical cases, acute severe anemia, ischemic syndromes, post cardiopulmonary arrest, immune mediated diseases, fibrocartilagenous emboli, pancreatitis, head trauma, snake bite, sepsis, and wound care. We describe hyperbaric oxygen therapy in complete global ischemia in this forum session.

Complete Global Cerebral Ischemia

A 1992 paper[2] described a controlled study from Japan where 20 dogs with complete global cerebral ischemia of 15 minutes duration were randomized and divided into two groups. One group received 1 hour sessions of HBOT at 3 ATA in 100% oxygen at 3, 24, and 29 hours after recirculation was established. The control group received room air at 1 ATA. Both groups received nursing support. The HBOT treated group's mean neurologic recovery score was > 65/100 after 5 days, with neurologic recovery scores of > 95/100 reached in two dogs from the HBOT group [range: 0=brain dead to 100= normal]. In contrast, in the group not receiving HBOT, the mean neurologic score remained < 60/100 throughout the 14 day follow up, and only 1 of 10 dogs from the non HBOT group recovered to a score of > 65/100. 7 of the HBOT group survived to the 14 day follow up, whereas 3 of the non HBOT group survived to the end of the 14 day follow up.

Case example: A 1 year old M/I Shih-Tzu canine fell off a grooming table while the groomer was distracted, and was accidentally hung by the lead around the neck. CPR was successful. Acute cortical blindness followed. The patient was given IV methylprednisolone immediately following the resuscitation, and referred for hyperbaric therapy the following day. Low dose oral prednisolone was continued. BID HBOT [45 minute sessions at 2 ATA] was given for 7 days. After the second session vision began to return as evidenced by obstacle course performance. Two week follow up confirmed improving vision based on owner interview and obstacle course examination. Consultation by an ophthalmologist confirmed resolving cortical blindness. In human cases [near hanging, near drowning, neonatal asphyxia, stroke, cardiopulmonary arrest, etc.] SPECT brain blood flow imaging has been used to demonstrate abnormalities in cerebral perfusion and subsequent improvement following immediate and long term HBOT. HBOT is often employed in cases of birth asphyxia in foals.

Other case examples in this presentation include patients who experienced cardiopulmonary arrest and near drowning. Referral for HBOT therapy after cardiopulmonary arrest during anesthesia or after injury has become more frequent as primary care veterinarians recognize the option exists.

Other Clinical Applications of HBOT in Small Animal Medicine

Exceptional Anemia Secondary to Blood Loss or Hemolytic Etiologies

HBOT should be considered in patients with severe anemia while cross-matching before transfusion, and after the transfusion of stored blood cell preparations. HBOT can provide necessary oxygen to ischemic systemic tissues through plasma delivery of oxygen. HBOT effectively reduces the critical "Oxygen Debt" which develops in severe anemia prior to presentation and treatment. Studies have shown that in hemolytic anemia endogenous carbon monoxide [CO] is produced as heme is catabolized. This blocks oxygen binding sites on viable remaining RBC's. CO also binds to myoglobin and to mitochondrial cytochrome, and reduces vital red blood cell deformabilty. Poorly deformable RBC's cannot pass freely into and through the deep tissue microcirculation. CO causes a shift of the oxygen dissociation curve to the left. HBOT markedly shortens CO half-life, and reverses these adverse physiological processes. Stored blood products used for transfusion have poor RBC deformability, and their oxygen dissociation curve is shifted to the left as well. These transfused "stored" RBC's can carry oxygen, making the mucous membranes appear "pink", but cannot effectively release the oxygen to the tissues until the curve shifts to the right. It can be many hours after transfusion before the viable RBC's in stored blood become deformable and shift their oxygen dissociation curve toward the right. HBOT can hasten these processes while it supplies oxygen to the systemic tissues in need. [21] Hyperbaric oxygen therapy has saved the lives of many severely anemic Jehovah's Witnesses, who do not allow transfusion in the practice of their religion. There is an extensive evidence-based review demonstrating the benefit of HBOT in the treatment of severe anemia in human medicine. [22] HBOT can be employed in situations where compatible blood is difficult to obtain or not available, as in felines with "Type B" blood.

Sepsis

Humans and mice with sepsis exhibit reduced red blood cell deformability as a part of the pathophysiology of this disease process [30]. A course of HBOT can supply the deep tissue beds with oxygen, which cannot be delivered into the microcirculation by the non-malleable RBC's. Adjuvant HBOT also potentiates certain antibiotic classes [for example, quinolones and aminoglycosides], and inhibits neutrophil adherence to endothelial membranes.

Snakebite

Snakebite pit viper envenomation is the classic example of the clinical situation where the anti-inflammatory, tissue sparing and tissue healing effects of HBOT come together to create dramatic changes in many patients. If a canine snakebite patient is rapidly placed into an HBOT chamber within a few hours of the bite event, grossly visible swelling is often markedly reduced during and after the treatment session [with or without anti-venin administration]. HBOT may also be effective in many snakebite cases in the face of delayed treatment.

Head Trauma

Many studies have shown that HBOT reduces elevations in intracranial pressure resulting from head trauma in animals and in human head trauma patients.[4] Clinical trials have demonstrated improved survival rates and neurologic recovery scores in human head trauma patients[3]. In addition, improved glucose utilization in rat brain tissue has been demonstrated to persist for up to 24 hours following the last HBOT.[4] There is much contemporary focus and research on the role of matrix metalloproteinases [MMPs] in traumatic brain injury and inflammatory CNS and PNS diseases, including Multiple Sclerosis [MS] in humans. A change in the balance of certain types of these proteases is associated with cellular infiltrates, apoptosis, and gliosis in traumatic, inflammatory, and neoplastic CNS disease. A 2006 rat model study of traumatic brain injury demonstrated that subjects treated with HBOT had reduced expression of certain MMPs, milder neutrophilic inflammatory infiltrates, and decreased secondary cell death when compared to controls.[5]

Case example: 3 month old F/I canine Yorkshire Terrier hit by a recliner chair. The patient presented obtunded and able to posture only when placed sternally. IV mannitol and ICU oxygen unit therapy were given, and a neck brace placed overnight by the E/CC clinicians. BID HBOT was begun the following morning [45 minute sessions at 2 ATA] and continued for 5 days. Marked neurologic improvement was obvious as the patient exited the chamber after the first session, and progressed rapidly until discharge.

Fibrocartilagenous Emboli

Some HBOT studies performed on dogs date back several decades. A 1965 study demonstrated that spinal cord pO2 could be raised by HBOT as long as 72 hours after tissue hypoxia was created in a model of spinal cord injury in canines.[6] Canine cases of fibrocartilagenous emboli often show rapid improvement when promptly treated by HBOT. It is especially important for nursing care and physical therapy purposes to get these primarily large breed patients ambulatory as soon as possible.

Case examples: 1] 8 year old M/N Labrador mix . Acutely monoparetic, then paraparetic. 2] 8 year old F/S Boxer. Acutely monoparetic. Clinical diagnoses were based on neurologic examinations, CSF analysis, and myelograms within a few hours of emergency presentation. Both patients received one dose of parenteral corticosteroids, then low dose short term oral prednisolone. Both patients were given BID HBOT [45 minute sessions at 2 ATA] . The Lab received 14 HBOT treatments, the Boxer 10. It is interesting to note in this acute ischemic myelopathy that patients will often improve markedly as they exit the chamber session, then regress, presumably as p02 in the affected cord region drops off. Patients who exhibit this pattern typically improve for longer periods after each session, and eventually maintain their neurologic improvement.

Spinal Cord Trauma, Intervertebral Disc Disease, and other Compressive Spinal Cord Lesions

In a 1972 canine model of spinal cord compression, spinal cord tissue p02 in the normal cord rose when the subjects were breathing 100% oxygen at 1 ATA (Normobaric Oxygen). After compressive trauma, the cord p02 dropped to near zero on room air, and did not rise when the subjects were breathing 100% oxygen at 1 ATA. When HBOT at 2 ATA was given with 100% oxygen, tissue p02 rose sharply even during the compressive period. Neurologic recovery of the HBOT treated group was improved compared to the control group.[7]

Case example: 11 year old Bichon M/N. Acute upper motor neuron tetraparesis with severe neck pain. Normal CSF protein and cytology. Spinal stenosis at a C2-3 fusion site [no history of previous surgery or trauma] and a C3-4 disc extrusion were found on myelographic examnination. Initial treatment was with HBOT [2 ATA, 45 minutes], prior to the diagnostics. At this point the patient became strongly ambulatory. Diagnostics were performed, followed immediately by a limited continuous dorsal laminectomy. Low dose oral prednisolone and HBOT were given post operatively for 6 days. Progressive clinical improvement was rapid. A desirable effect in neurosurgical cases treated with post op HBOT is rapid surgical wound healing, with reduced swelling, inflammation, seroma, and scar formation. HBOT has become standard pre and post neurosurgical adjunctive therapy in our practice, for both spinal and brain surgical cases. We have used HBOT in conservative treatment of documented or presumed spinal cord compressive diseases.

Inflammatory CNS Diseases

We have discussed some of the mechanisms of HBOT anti-inflammatory effects above. The reader should note that in the UK, there are 64 centers [distributed to cover the population density] which are dedicated solely to the treatment of MS in human patients. Over 1.7 million sessions have been completed in the last 24 years without serious incident.[8] Patients generally manage their own long term treatment schedule, and many find that their symptoms, especially frequency of micturition, are controlled only by regular HBOT sessions. Some schedule their holidays to be near a "Center". The reader is referred to a detailed chapter on rationale for HBOT in inflammatory disease, including MS in a recent HBOT textbook.[2] A 2006 symposium presentation from a Russian HBOT research clinician described the use of combined HBOT and cyclosporine in the treatment of inflammatory CNS disease in humans[9]

Case example: 9 ½ year old M/N mixed canine presented for persistent fever, sudden loss of coordination and slight personality change. Neurologic exam supported multifocal CNS lesions. CSF cisternal and lumbar puncture protein and cytology were normal. MRI revealed multiple small mixed size hyperintense T2-weighted lesions throughout the thalamus and rostral brainstem. Most remained hyperintense on T2-weighted FLAIR sequencing. Multiple serologies and PCR's, PARR [LSA] were submitted. A 1/164 positive serology for E equi was the only notable laboratory finding. BID treatments of HBOT [45 minute sessions at 2 ATA over 7 days] were initiated along with cyclosporine, prednisolone, imidocarb, and doxycycline. Neurologic signs improved over a 7 day period, with the owner commenting on a marked improvement at home at the two week recheck exam. The immunosuppressive therapy was tapered very slowly over several weeks.

Major Organ Ischemia

HBOT may be employed in cases with major organ ischemia, such as post-surgical correction GDV in the canine, where the ischemia- reperfusion injury complex often results in serious clinical complications.

Pancreatitis

This inflammatory process is often severe, effusive, and fatal to canine and feline patients. Recent publications describing experimental rat pancreatitis models demonstrate reduced histopathologic severity and improved survival rate in the groups exposed to HBOT when compared to controls. [14,15,16] Current pathophysiology extrapolated from experimental models and human medicine incriminates ischemia, along with neutrophil migration and adhesion to the endothelial walls following the initial activation of pancreatic enzymes locally. During the disease process upregulation of intercellular adhesion molecule -1 [ICAM-1] occurs, promoting neutrophil adhesion and the subsequent inflammatory cascade [26]. In addition to increasing oxygen to the swollen and inflamed pancreas, HBOT has been demonstrated to downregulate ICAM-1 expression [27], thus reducing the tendency for neutrophils to adhere to the vascular endothelium. Our clinical experience is that now even cases with severe, effusive pancreatitis can be given a prognosis which includes a good chance for survival when HBOT is employed along with intensive supportive care. Case example… 12 year old Fe spayed Lab who ate a "rotted carcass" in a nearby yard. History, clinical signs, physical exam, radiographs, ultrasound exam, fluid analysis, and PLI all supported a diagnosis of pancreatitis. Severe pleural effusion was also present. BID HBOT for 45 minutes at 2 ATA was given. Traditional supportive care was continued. By hospitalization day 3 the pleural and peritoneal effusions were nearly resolved. A total of nine sessions were given, and rapid clinical improvement resulted in discharge by day 5. Physical exam one week after discharge was normal. Follow up PLI and ultrasound exam of the chest and abdomen 3 months later were normal.

Smoke Inhalation/ CO Toxicity/ Burn Wounds from Exposure to a Fire

This combination of pathology is a unique opportunity to treat three different clinical issues with one form of adjuvant therapy, HBOT.

Osteomyelitis/ Infectious Agents

The anti-inflammatory and angiogenic effects of HBOT are ideal here. HBOT potentiates the effect of certain antibiotics [for example, the quinolones and aminoglycosides], and is directly bacteriocidal to anaerobic bacteria. At 2.4 ATA HBOT is bacteriocidal to Borrelia burgdorfi [the Lyme disease spirochete][31]. Patients with Lyme disease are managed by gradually and progressively increasing treatment pressure from 1.5 ATA to 2.4 ATA, and continued with a course of 2.4 ATA sessions. This helps to reduce the severity of the Jarish-Herxheimer reaction as the organisms are destroyed.

Delayed or Non-Healing Wounds, Accelerated Healing of Exceptional or Large Wounds, Crushing Wounds

HBOT is widely accepted in human medicine for its adjuvant effect in wound healing, especially in diabetes mellitus or vascular occlusive disorders. It is a coded and reimbursed treatment modality for radiation induced tissue damage and non-healing wounds. As previously discussed, meta-analyses have demonstrated the benefit of HBOT in wound healing [24]. In addition to its enhancement of local wound tissue oxygen tension and vasculogenic effects, HBOT directly modulates intracellular signal pathways [comparable to peptide growth factors], and up-regulates PDGF-R expression, activating fibroblasts using a similar transduction cascade as traditional peptide growth factors, resulting in accelerated wound healing [25].

Osteoarthritis

The effects of HBOT on multiplication of circulating stem cell numbers and up-regulation of PDGF-R expression can be used to the patients' advantage when employing "Platelet Enhancement Therapy "in the management of osteoarthritis[20,25].

Miscellaneous Clinical Applications of HBOT

Other cases treated with HBOT therapy include acute tetraparesis secondary to Atlanto-axial [A/A] luxation, acute peripheral nerve injury, and brain lesions pre and post operatively. Decubital ulcers, thermal burns, shearing wounds, and near drowning are a few of the other clinical applications for which we have employed HBOT. Human medical practice is often limited by "approved indication lists" and diagnosis "codes". For example, in the U.S., there are currently 14 indications approved by Medicare, and thus typically funded by third party payers. HBOT for Non- healing wounds of certain types is funded by some third party payers. The U.K. has 13 "approved" indications, Japan 20, China 53, and Russia 85. In the latter three countries the "approved" lists contain disease indications which translate as "severe spinal cord disorders", "acute and severe hypoxic brain disorders", "severe spinal cord injury", "Bell's palsy", "Multiple Sclerosis", "cranial traumas", "viral encephalitis", "illeus", "sepsis", "non - healing wounds, " "osteomyelitis", "burns", "mountain sickness", etc.

Bubbles

HBOT is the treatment of choice for arterial, venous, pulmonary, or cerebral air embolism. In veterinary medicine air embolism may occur as the result of complications resulting from central venous lines, interventional procedures, surgical procedures [such as C-section], and the ingestion of hydrogen peroxide or the use of hydrogen peroxide on well vascularized wounds.

Safety Issues

HBOT should not be administered to patients with pneumothorax, certain older type pacemakers, within 7-10 days after doxorubicin, or in a patient currently receiving cis-platinum or bleomycin. Acetazolamide, Zonisamide, or other carbonic anhydrase inhibitors should be tapered prior to HBOT, as they inhibit oxygen induced vasoconstriction and increase blood flow to the brain, increasing the chance of oxygen toxicity. Transdermal patches should be temporarily removed. Conservative dosages of narcotic analgesics should be employed prior to HBOT, as increased CO2 tension secondary to respiratory depression can cause cerebral vasodilatation and predispose to oxygen induced seizures. Insulin dosages should be tapered prior to HBOT therapy. Patients with hyperthyroidism also may have an increased chance for oxygen induced seizures. Vestibular signs may be aggravated due to barostimulation of the ear structures. This said, at the AERC we have witnessed less than 40 short motor seizures in over twenty-seven thousand individual HBOT treatment sessions at our hospital over the past decade. The seizures all ended rapidly, and a gradual reduction in treatment pressure after the seizure activity finished was initiated in these individuals. No sequellae were recognized as a result of those seizures. The previously cited textbooks are a good source of detailed information on potential contraindications and specific interactions of HBOT with other therapeutic measures [28,29].

The chamber contains 100 % oxygen under pressure. Measures should be taken to reduce the chance of fire. The chamber should be grounded. A static grounding plate should be used in small animals. Hair coats should be gently wiped with a wet cotton towel. Cotton towels and bandages generate less static than polyester fibers. No flammable materials or electronic devices should be placed in the chamber. Metal cages should not be used in the chamber. No metal collars, polyester collars, or air-filled collars should be used. Skin staples should not be used, or should be wrapped/ covered in 100% cotton bandaging or stockinette. Metal external fixators must be covered in the same manner. No alcohol, Vaseline, or flammable based dressings should be used. An emergency master oxygen shut off valve should be installed and known by all personnel. A water based fire extinguisher should be close and functional. A textbook on hyperbaric safety is recommended.[10]

HBOT in Equine Racing Athletes

Under the care and direction of veterinarians some horses receive post-race HBOT as a method to reduce the inflammatory sequellae of a supreme racing effort, or in the treatment of orthopedic or neurologic injuries.

In the spring of 1999 equine treatment chambers in trailers began appearing near racetracks in New York and California, as some trainers of equine athletes around the world began to use HBOT in their horses. No governing body was consulted initially, which resulted in New York racing officials issuing an immediate 7 day ban on HBOT for racing horses. After consultation with veterinarians familiar with HBOT, Maryland and New York have settled on a 48 hour pre- race moratorium on HBOT. Kentucky requires a 24 hour HBOT- free period prior to any race.

Conclusion

Over the past several years it has been very rewarding for the doctors and staff to observe the clinical improvement of many cases treated by HBOT at our hospital. The addition of this treatment modality has resulted in an obvious enhancement of our ability to successfully treat many critically ill patients.

REFERENCES

1 Wells, CH,et al. Tissue gas measurements during hyperbaric oxygen exposure, in Smith G [ed]. Proceedings of the Sixth International Congress on Hyperbaric Medicine. Aberdeen, Scotland, Aberdeen Press, 1977, pp 118-124

2 Takahashi, M., et al, Hyperbaric oxygen therapy accelerates neurologic recovery after 15 – minute complete global cerebral ischemia in dogs. Critical Care Medicine, 1992;Vol.20,No.11:1588-1593

3 Holbach KH,et al. "Improved reversibility of traumatic mid-brain syndrome with application of hyperbaric oxygen pressure"Acta Neurochir. 1974;30:247-256

4 Contreras Fl, et al. "The effect of hyperbaric oxygen on glucose utilization in freeze-traumatized rat brain" J Neurosurg. 1988;65:615-624

5 Vlodavsky, E et al. "Hyperbaric therapy reduces neuroinflammation and expression of matrix metalloproteinase-9 in the rat model of traumatic brain injury" Neuropathol Appl Neurobiol. 2006; Vol32, No. 1: 40-50

6 Maeda, N. "Experimental studies on the effect of decompression procedures and hyperbaric oxygenation for the treatment of spinal cord injury" J Natl Med Assoc. 1965 Vol 16: 429-447

7 Kelly, DL. Et al. "Effects of hyperbaric oxygenation and tissue oxygen studies in experimental paraplegia" Neurosurg. 1972; Vol 36: 425-429

8 Neubauer, RA, 1st International Symposium for the Use of Hyperbaric Oxygenation in the Neurosciences, 2006, pp 78-9

9 Kazantseva, NV. 1st International Symposium for the Use of Hyperbaric Oxygenation in the Neurosciences, 2006, pp32-3

10 Workman, W. Hyperbaric Facility Safety: A practical guide, Flagstaff, Az. Best publishing, 1999

14 Mas et al, "Hyperbaric oxygen induced changes in bacterial translocation and acinar ultrastructure in rat acute necrotizing pancreatitis" J Gastroenterol, 2005, Vol 40, No. 10, 980-98

15 Cuthberson, CN et al, "Hyperbaric oxygen reduces severity and improves survival in acute pancreatitis ", abstract from the 36th annual meeting of the American pancreatic association, Nov 3-4, 2005

16 Chen et al, "Hyperbaric oxygen therapy attenuates pancreatic microcirculatory derangement and lung edema in acute experimental pancreatitis in rats", Pancreas, 1998 Jul: 17[1]: 44-9

17 Edwards, ML.."Hyperbaric oxygen Therapy, History, Principles, Applications in Disease", J Vet Emerg Crit Care, Volume 20, Nimber 3, June 2010, 289-298

18 Kaide CG. , "Hyperbaric Oxygen therapy: Applications in Infectious disease" , Emergency Medicine Clinics of North America, 26[2008], 571-595

19 Milovanova, TN, et al. "Hyperbaric oxygen stimulates vasculogenic stem cell growth and differentiation in vivo", J Appl Physiol 106: 711-728, 2009

20 Thom, SR, et al. "Vasculogenic stem cell mobilization and wound recruitment in diabetic patients: Increased cell number and intracellular regulatory protein content associated with hyperbaric oxygen therapy" Wond Rep Reg (2011) 19 149-161

21 Myking O. Schreinen A. Hyberbaric oxygen in hemolytic crisis. JAMA 1974;227:1161-1162

22 Van Meter KW. A systemic review of the literature reporting the application of hyperbaric oxygen in the treatment of exceptional blood loss anemia: an evidence based approach. Undersea hyperbaric medicine, Vol 32, No1, 2005, 61-79

23 Thom SR. "Hyperbaric Oxygen-Its Mechanisms and Efficacy, Plast Reconstuct Surg 2011 January, 127 [Suppl 1]: 131S-141S

24 Goldman RJ, "Hyperbaric Oxygen Therapy For Wound Healing and Limb Salvage: A Systemic Review", Physical Medicine And Rehabilitation. 2009:1:471-489

25 Saulis AS et al, Hyperbaric Oxygen Modulates PDGF Receptor Beta Expression and ERK 1/2 Activation in Human Dermal Fibroblasts In Vitro, Proceedings, Undersea and Hyperbaric Medical Society, 2000

26 Mansfield, C, "Pathophysiology of Acute Pancreatitis: Potential Application from Experimental Models and Human Medicine to Dogs", J Vet Intern Med 2012;26:875-887

27 Buras JA, et al, "Hyperbaric oxygen downregulates ICAM-1 expression induced by hypoxia and hypoglycemia: the role of NOS", Am J Physiol Cell Physiol. 2000 Feb;278[2]:C292-302

28 Jain KK, Textbook of Hyperbaric medicine, 5th revised edition, Hogrefe, 2009

29 Kindwall and Whelan, Hyperbaric Medicine Practice, second edition revised, Best, 2004

30 Baskurt OK et al, Red Blood Cell Deformability in Sepsis, Am J Respir Crit Care Med Vol 157. pp 421-427, 1998

31 Fife WP, Freeman DM, Treatment of Lyme Disease with Hyperbaric Oxygen Therapy, Abstract, UHMS proceedings 1998

32 Kendall, AC et al, "Different oxygen treatment pressures after inflammatory gene expression in human endothelial cells", VHM 2013, Vol 40, No. 2, Pates 115-123.

ADVANCES IN DIABETES THERAPY: GLP-1 BASED DRUGS IN VETERINARY MEDICINE

Chen Gilor, DVM, PhD, DACVIM
Columbus, OH, USA

INTRODUCTION

Incretin-based therapies are revolutionizing the field of diabetes therapy by replacing insulin therapy with a safer and more convenient, long-acting drugs. Incretin hormones (GLP-1 and GIP) are secreted from the intestinal tract in response to the presence of food in the intestinal lumen. GLP-1 augments insulin secretion and suppresses glucagon secretion during hyperglycemia in a glucose-dependent manner. Clinical data have revealed that incretin-based drugs are as effective as insulin in improving glycemic control while reducing body weight (GLP-1 analogs, specifically) in patients with type 2 diabetes. Furthermore, incidence of hypoglycemia is relatively low with these drugs because of their glucose-dependent mechanism of action. Another significant advantage of these drugs is their duration of action. While insulin injections are administered at least once daily, long-acting GLP-1 analogs have been developed as once-a-week injections and could potentially be administered even less frequently than that in diabetic cats. This review considers the physiology of incretin hormones, and the pharmacology and use of GLP-1 analogs in people. We will also review up-to-date research data on GLP-1 analogs in veterinary medicine.

INCRETIN PHYSIOLOGY

Incretin-secreting cells (K and L cells) are interspersed in the epithelium of the gut. Their physiological role is to sense the type and quantity of digested nutrients in the gut. They then secrete incretin hormones as preparatory signals to other remote organs (e.g. brain, pancreas etc.). In the pancreas, the incretin signal is translated to increased sensitivity to the stimulatory effect of glucose. This effect is responsible for the observed difference in insulin secretion between oral and intravenous glucose and is defined as the "incretin effect"; oral glucose leads to much greater insulin secretion compared to intravenous glucose even when blood glucose concentrations are equal.[1,2] The incretin effect is thought to be exclusively mediated by 2 peptide hormones: Glucagon-like peptide-1 (GLP-1) which is secreted from L cells, and glucose-dependent insulinotropic polypeptide (GIP) which is secreted from K cells. Incretin hormones also promote expansion of pancreatic beta-cell mass: They promote differentiation of pancreatic ductal cells into beta cells and increase beta cell proliferation. Importantly, incretin hormones protect beta cells from apoptosis induced by various cytotoxic agents or dexamethasone. They also prevent apoptosis secondary to glucotoxicity and lipotoxicity.[3]

Incretin hormones affect glucagon secretion (GLP-1 inhibits, GIP stimulates), slow the rate of gastric emptying (GLP-1), increase satiety (GLP-1), and increase the sensitivity of adipose tissue to insulin (GIP). Thus, they play a major role in glucose homeostasis.[3,4] Active GIP and active GLP-1 are degraded by the enzymes dipeptidyl peptidase-4 (DPP-4, also known as CD26) and neutral endopeptidase 24.11 (NEP-24.11) into inactive forms, thereby modifying or inhibiting their activity.[5] DPP-4 and NEP are ubiquitous in tissues, and DPP-4 is also present in a soluble form in the blood. Consequently, when injected intravenously, active GLP-1and GIP are quickly degraded into inactive forms. When injected intravenously, the half-life of GLP-1 is only about 1-2 minutes and the half-life of GIP is only about 5 minutes. The inactive GLP-1 and GIP are quickly cleared by the kidneys.

INCRETIN HORMONES IN DIABETES AND OBESITY

The incretin effect is severely reduced or absent in people with type 2 DM, contributing to glucose intolerance and post-prandial hyperglycemia. In diabetics, the secretion of GIP is normal or slightly reduced but its effect on the pancreas is markedly decreased. In contrast, GLP-1 retains its insulinotropic effects in type 2 DM (at least in supraphysiologic concentrations), but its secretion is decreased.[4] Whether a blunted incretin response is a primary process in type 2 DM or a secondary process caused by diabetes, is debatable. There is some evidence that obesity contributes to attenuation of the incretin effect independently of diabetes. In a recent study, obesity was associated with decreased incretin effect including decreased beta-cell response to GLP-1 but also decreased GLP-1 secretion.[6] In some studies, GLP-1 secretion was normal in obese people but GIP secretion was increased during fasting and early after a meal.[7] Insulin resistance in itself, regardless of obesity or diabetes, has been associated with abnormal secretion of GIP and GLP-1.[8] Increased insulin concentrations in the insulin-resistant subjects might down-regulate incretin secretion.

Traditional treatments for diabetes are either just partially effective (diet change, oral drugs) or associated with significant side effects (insulin therapy). Owner compliance is a major problem because twice-a-day injections are required with most commonly used insulin preparations. Weight gain may indicate a good response to therapy initially but can eventually become a problem. Hypoglycemia is a common complication of insulin therapy and can be life-threatening. These side effects have been significantly reduced in human medicine with the introduction of incretin-based treatments. GLP-1 stimulates glucose-dependent insulin secretion. Its effect is blunted in euglycemia and therefore they are less likely to cause hypoglycemia. GLP-1 also promotes weight loss. GLP-1 analogs are not only effective in controlling blood glucose in diabetics but they are also associated with weight loss in diabetic and non-diabetic obese people. One major advantage of incretin-based therapies is their potential to reverse the course of the disease. Incretin-based treatments in diabetic people correct not only hyperglycemia but also all the major markers of beta-cell dysfunction.

The very short half-life of GLP-1 limits its clinical use. There are numerous treatment strategies that overcome this problem and take advantage of incretin-hormone physiology. Novel GLP-1 analogs have a prolonged half-life and are used as SQ injections at intervals of up to one week.

INCRETIN-BASED THERAPY STRATEGIES

There are several treatment modalities that overcome the problems associated with direct administration of incretin hormones. Long-acting, DPP-4-resistant, synthetic GLP-1 mimetic peptides are most commonly used. Available formulations are all for SQ injections but other delivery systems are being developed (including enteral, pulmonary, or sublingual). Also under development are non-peptidic GLP-1 receptor agonists.[9] Oral DPP-4 inhibitors that prolong the half-life of incretin hormones are also commercially available.[3] Non-peptidic drugs that activate free fatty acid receptors in L cells and stimulate GLP-1 secretion are being developed (GRP120 and GRP119 agonists).[10,11,12] Diet manipulation should be investigated.

Drugs that are commercially available or that are currently in advanced clinical trials are described below.

GLP-1 MIMETICS AND ANALOGS

Exenatide

The peptide exendin-4 was first isolated from the poisonous venom of the Gila Monster (*Heloderma suspectum*). Exendin-4 is a 39-amino acid peptide that shares only a 53% sequence homology with GLP-1 but its affinity for the GLP-1 receptor is 1000 times greater than the affinity of GLP-1. Unlike GLP-1, exendin-4 is not a substrate for DPP-4 and NEP.[4] Exenatide is a synthetic exendin-4. Resistant to degradation, exenatide is eliminated by the kidneys and has a half-life of 3-4 hours in people. Its biological effect lasts about 8 hours after subcutaneous injection and it can be detected in the plasma for up to 15 hours.[13] Multiple studies, both *in vitro* and *in vivo*, have shown that, in general, exendin-4 has the same physiologic effects as GLP-1 in the pancreas, GI tract, and brain.[3] Exenatide is associated with improvement in some of the earliest and most fundamental abnormalities of type 2 diabetes: diminished "first-phase insulin response" and proinsulin/insulin ratio. Acute administration of exenatide in type 2 diabetic patients corrects the abnormal insulin secretion pattern after an IV glucose bolus (first phase and second phase insulin responses) and restores the ability of beta cells to respond to rapid changes in blood glucose concentrations.[3] Exenatide also improves proinsulin/insulin ratio after 30 weeks of treatment.[14]

Exenatide has been shown to be as effective as insulin glargine in the treatment of DM but with less side effects (e.g. hypoglycemia and weight gain).[15] In a 2-year follow-up of patients receiving exenatide, patients achieved sustained and significant reductions in glycosylated hemoglobin, accompanied by significant weight loss (instead of weight gain commonly seen in diabetics receiving insulin) and improvement in serum liver enzyme activity and blood pressure. Most importantly, treatment with exenatide improved beta cell function as measured by homeostasis model assessment of beta cell function (HOMA-B).[16]

Exenatide has minimal side effects in people. It is mostly associated with nausea and less frequently with vomiting. Infrequently, it might cause hypoglycemia. Severe hypoglycemia (requiring assistance) was reported rarely (only 5 of 2781 patients) and only in patients who also received sulfonylurea drugs. Antibodies to exenatide developed in 67% of patients but this did not affect outcome and was not associated with side effects.[17]

When first discovered, exendin-4 was shown to potentiate amylase release from rat acinar pancreatic cells in response to other hormones such as cholecystokinin. This was shown *ex vivo* and in high doses. A possible association between GLP-1 analogs and pancreatitis as well as medullary thyroid cancer has been suggested and lead to issuing of a warning by FDA. However, multiple studies have refuted these concerns and it is widely accepted that the proven enormous benefits of GLP-1 analogs outweigh its hypothetical risks.[18,19]

Exenatide is commercially available in the USA under the trade name Byetta®.

Exenatide in cats

In healthy cats, exenatide was quickly absorbed after a SQ injection and caused glucose-dependent insulin secretion. Increased glucose tolerance, however, was not observed after a single SQ injection.[20,21] At a dose of 1.0 mcg/kg SQ (about 10 times the dose that is used in diabetic people), exenatide injection did not cause any side effects in healthy cats, except for hypoglycemia in 1 out of 9 cats.[20] Exenatide has led to significant weight loss in healthy cats of 7.0 ± 4.9% (from 4.78 ± 1.5 kg to 4.48 ± 1.5 kg) with a dose of 1.0 mcg/kg SQ BID for 28 days.[21]

Exenatide Extended-Release

A long-acting sustained-release formulation of exenatide (Bydureon®) has recently been approved by the FDA as the first once-weekly subcutaneous injection for treatment of type 2 diabetes people. It consists of injectable microspheres of exenatide and poly(D,L-lactic-co-glycolic acid), a common biodegradable medical polymer with established use in absorbable sutures and extended-release pharmaceuticals, that allows gradual drug delivery at controlled rates. In people, Exenatide plasma concentrations are sustained at an effective concentration (50 pg/mL) for longer than 60 days after a single injection at doses of 5mg, 7mg or 10mg.[22]

It has been shown in a recent clinical study to be more effective than once-a-day insulin glargine in achieving glycemic control with decreased risk of hypoglycemia and with reduction (instead of gain) in body weight.[23] This extended release formulation was also more effective than regular exenatide (Byetta) in achieving glycemic control with no increased risk of hypoglycemia, decreased side effects like nausea, and with similar reductions in body weight.[24]

Exenatide Extended-Release in cats

We have studied Exenatide Extended-Release in healthy cats at a dose of 0.13 mg/kg. Three weeks after a single subcutaneous injection fasting BG was decreased, glucose tolerance improved and median glucagon concentrations decreased. Insulin concentrations did not differ significantly and no side effects were observed throughout the study. Further evaluation is needed to determine its efficacy and duration of action in diabetic cats.

Liraglutide

Liraglutide is a GLP-1 synthetic analog with 2 amino acid substitutions and a fatty acid acyl group that enables noncovalent binding to albumin, thereby extending the pharmacokinetic profile of the GLP-1 molecule. Liraglutide exhibits a prolonged pharmacokinetic profile after a single injection, and exhibits all of the actions of native GLP-1.[4] Liraglutide (once-a-day) was recently compared to exenatide (twice-a-day). Liraglutide provided significantly greater improvements in glycemic control than did exenatide and was generally better tolerated.[25] Liraglutide once-daily was also more effective than Exenatide-ER once-weekly in achieving glycemic controlling but nausea, vomiting and diarrhea occurred less frequently with Exenatide-ER. [26] Liraglutide has been used successfully to treat obesity in non-diabetic patients.[27]

Liraglutide is commercially available in the USA and Europe under the trade name Victoza®.

Liraglutide in cats

We have studied liraglutide in healthy cats at a dose of 0.6 mg/cat once daily for 7 days. Liraglutide caused significant weight loss in all cats at day 8. Appetite was subjectively decreased in all cats and one cat was withdrawn on day 4 because of 48 hours of anorexia. Liraglutide was associated with a trend towards improved glucose tolerance and higher insulin concentrations. Glucagon concentrations and baseline glucose concentrations were not affected.

Liraglutide in dogs

Liraglutide have been used in healthy and diabetic dogs in one study. Liraglutide was associated with improved glycemic control post-prandially although overall it was not associated with increased insulin concentrations.[28] It is likely that the effect of Liraglutide in diabetic dogs is not related to stimulation of insulin secretion but to rather to suppressed glucagon secretion or delayed gastric emptying.

Albiglutide

Albiglutide is a recombinant GLP-1-albumin fusion protein that exhibits a reduced affinity for the GLP-1R, but displays a broad spectrum of GLP-1R-dependent actions in preclinical studies, including inhibition of food intake and gastric emptying and reduction of glycemia excursion after meal ingestion. Importantly, it has the potential to be used as a very long-acting drug: It has been investigated in people for use as a weekly, biweekly and monthly injection.[29] Albiglutide awaits FDA approval (as of 2014).

DPP-4 INHIBITORS

DPP-4 inhibitors (e.g. sitagliptin, vildagliptin) are administered orally. In people, vildagliptin and sitagliptin are well-tolerated and not associated with hypoglycemia when used alone. Both agents increase plasma concentrations of GLP-1 and GIP after meal ingestion, enhance glucose-stimulated insulin secretion and reduce ratios of proinsulin:insulin, consistent with an improvement in beta cell function. They are, however, less potent than other oral hypoglycemic drugs.[3] In contrast to GLP-1 analogs, DPP-4 inhibitors are not associated with nausea or vomiting in people, but they are associated with weight gain. DPP-4 inhibitors are also associated with increased risk of nasopharyngitis, urinary tract infections and headaches.[17] Increased risk of infections might be related to the action of DPP-4 in T-cells as a co-stimulatory molecule (CD26).

DPP-4 inhibitors are commercially available alone or in combination with other oral hypoglycemic drugs (e.g. sitagliptin is marketed alone as Januvia® and in combination with metformin as Janumet®).

A DPP-4 inhibitor has been used experimentally in cats and was effective in enhancing insulin secretion and inhibiting glucagon secretion after an intravenous glucose challenge. Glucagon inhibition was also observed after a meal challenge.[30] In that study, the drug was administered subcutaneously and not orally.

REFERENCES

1. McIntyre N, et al. J Clin Endocrinol Metab. 1965;25:1317-24.
2. Nauck MA, et al. J Clin Endocrinol Metab 1986;63:492-498
3. Kim W, Egan JM. Pharmacol Rev 2008;60:470-512
4. Holst JJ, et al. Mol Cell Endocrinol. 2009;297:127-36.
5. Baggio LL, Drucker DJ. 2007;132:2131-57.
6. Muscelli E, et al. Diabetes 2008;57:1340-8.
7. Vilsboll T, et al. J Clin Endocrinol Metab. 2005;88:2706-13.
8. Rask E, et al. Diabetes Care. 2001;24:1640-5.
9. Sloop KW, et al. Diabetes, 2010 Dec;59(12):3099-107
10. Adachi T, et al. Biol Pharm Bull. 2008;31:1021-3.
11. Morishita M, et al. J Control Release. 2008;132:99-104.
12. Overton HA, et al. Br J Pharmacol. 2008;153:S76-81.
13. Kolterman OG, et al. Am J Health Syst Pharm. 2005;62:173-81.
14. DeFronzo RA, et al. Diabetes Care. 2005;28:1092-100.
15. Glass LC, et al. Curr Med Res Opin. 2008;24:639-44.
16. Buse JB, et al. Clin Ther. 2007;29:139-53.
17. Amori RE, et al. J Am Med Assoc. 2007;298:194-206.
18. Alves C, et al. Diabetes Res Clin Pract. 2012 Nov;98(2):271-84.

19. Chiu WY, et al. Exp Diabetes Res. 2012;2012:924168.
20. Gilor C, et al. Domest Anim Endocrinol 2011 Jul;41(1):42-9
21. Seyfert TM, et al. Intern J Appl Res Vet Med 2012 ;10(2) :147-156
22. Fineman MS, et al. Clin Pharmacokinet. 2011 Jan;50(1):65-74.
23. Diamant M, et al. Diabetes Care. 2012 Apr;35(4):683-9.
24. Buse JB, et al. Diabetes Care. 2010 Jun;33(6):1255-61
25. Buse JB, et al. Lancet. 2009 Jul 4;374(9683):39-47.
26. Buse JB, et al. Lancet. 2013 Jan 12;381(9861):117-24.
27. Astrup A, et al. Lancet. 2009;374:1606-16.
28. Oda H, et al. Res Vet Sci. 2013 Oct;95(2):381-8
29. Rosenstock J, et al. Diabetes Care. 2009 Oct;32(10):1880-6.
30. Furrer D, et al. Vet J. 2010;183:355-7.

ATYPICAL CUSHING'S DISEASE: IS IT REALLY ATYPICAL?

Rhett Nichols, DVM, ACVIM (Small Animal Internal Medicine)
Smithtown, NY

WHAT IS ATYPICAL CUSHING'S DISEASE?

In 2002, clinical researchers from the UK described the first cases of what is often referred to as atypical Cushing's disease (1). Dogs with atypical Cushing's disease (ACD) have a history, physical examination findings, CBC, biochemical chemistry profile and urinalysis results, and adrenal imaging consistent with hyperadrenocorticism (HAC), however, routine screening tests (i.e., the low-dose dexamethasone suppression test (LDDST) and ACTH response test are both negative. In addition, one or more sex steroids are elevated (e.g., androstenedione, progesterone, or 17-hydroxyprogesteone [17-OHP]) following administration of ACTH. The UK study evaluated 24 dogs with HAC; the cause of the HAC was not determined in 2 dogs. In the remaining 22 dogs, pituitary-dependent Cushing's disease (PDH) was diagnosed in 16 dogs and 6 dogs had an adrenal tumor (AT). Of the 16 dogs with PDH, 3 dogs had ACD; of the 6 dogs with an AT, 4 dogs had ACD. All the dogs with ACD had elevated post-ACTH 17-OHP concentrations. In addition, all dogs in this study treated with standard therapeutic regimes of trilostane or mitotane showed resolution of clinical signs.

ROUTINE DIAGNOSTIC TESTING FOR CLASSIC HAC: WHAT ARE THE EXPECTATIONS?

The 2012 ACVIM Consensus Statement regarding the diagnosis of the classical form of spontaneous canine hyperadrenocorticism (HAC) considers the LDDST the screening test of choice (2). The reported sensitivity and specificity of the LDDST ranges from 85 to 100% and from 44 to 73% respectively (2). The ACTH response test, because of its poor sensitivity, is considered diagnostically inferior to the LDDST as a screening test for spontaneous HAC. For dogs with AT and PDH the sensitivity is 57-63% and 80-83% respectively and the specificity ranges between 50 and 93% (2). While neither screening test is 100% accurate, most dogs with spontaneous HAC have one positive screening test. For example, in a series of 64 dogs with HAC, no dog was negative on both screening tests, but such cases clearly exist (3). ***Summary impact point:*** *The majority of dogs with classical Cushing's disease screened with an ACTH response or LDDST and will have at least one positive test.*

ARE NONCORTISOL STEROID HORMONES ONLY ELEVATED IN DOGS WITH ATYPICAL CUSHING'S DISEASE?

In addition to the UK study, there are four other relatively large published studies that have evaluated multiple noncortisol adrenal steroid hormones (sex steroids and other cortisol precursors) and cortisol concentrations post-ACTH in dogs diagnosed with HAC (4, 5, 6, 7). The diagnosis of HAC was based on results of history, clinical signs, routine biochemistry and urinalysis findings, assessment of adrenal ultrasound imaging, a positive ACTH response test and/or LDDST, and in some cases an endogenous ACTH level. In two of these studies dogs with adrenal tumors and dogs suspected of HAC but diagnosed with another condition were also included (4, 5). Overall, when the information from all these studies is combined, 76 dogs were diagnosed with HAC and 59 of these dogs had PDH. The remaining 17 dogs had either an adrenal cortical adenoma or carcinoma or the HAC was unclassified. In addition, and most importantly, one or more noncortisol adrenal steroid hormones were elevated in all but one dog diagnosed with HAC and 42-50% of dogs suspected of having HAC but ultimately diagnosed with a nonadrenal illness. The question often arises why noncortisol adrenal steroids are commonly elevated in dogs with HAC and nonadrenal illness. The rate-limiting step in the synthesis of cortisol and the major site of action of ACTH is the conversion of cholesterol to pregnenolone. It is therefore not surprising that *any* disease causing increased concentration of ACTH results in increased concentration of cortisol precursors as well as cortisol itself, and other noncortisol steroid hormones. ***Summary impact point:*** *Noncortisol steroid hormones are commonly elevated in dogs with classic HAC and nonadrenal illness and therefore, elevated levels are not specific for ACD.*

WHY IS THE ACTH RESPONSE AND LDDST NEGATIVE IN DOGS WITH ATYPICAL CUSHING'S DISEASE?

Since dogs with classical HAC usually have one positive screening test, several questions regarding ACD are often asked. 1) Is ACD the result of dogs with classical HAC that just happen to have a normal LDDST and ACTH *response test at the testing time? 2) is ACD really classical Cushing's in disguise because the current screening tests lack sensitivity and reference ranges and cut-offs need to be re-evaluated? Or 3) is ACD a distinctly different syndrome that shares similarities to other disorders such as congenital adrenal hyperplasia in people or food-induced Cushing's disease which is caused by an aberrant expression of functional receptors in adrenocortical tissue other than ACTH?*

Current reference ranges and cut-off levels may be misleading and need to be re-evaluated

Because the screening tests for Cushing's disease were introduced in the 1970's and 1980's, the 2012 ACVIM consensus statement for diagnosing Cushing's disease suggests that the current reference ranges and cut-offs should be re-evaluated to avoid a misdiagnosis (2). In addition, the original clinical trials to evaluate screening tests were performed at referral centers with high disease prevalence, but today the tests are often performed in first opinion clinics with low disease prevalence. Moreover, the incidence of mild cases of HAC appears to have increased over time, possibly because of increased awareness and earlier patient presentation. Milder cases would be expected to have lower cortisol hypersecretion and an increased sensitivity of the HPA axis to dexamethasone.

Cut-offs and the LDDS

The cut-off for the 8 hour cortisol following the administration of a low dose of dexamethasone can be quite variable from one veterinary laboratory to another. For example, the cut-off for the 8 hour cortisol at Antech Diagnostics, Michigan State University, and UC Davis ranges from 1.4 ug/dl, 1 ug/dl, and 0.8 ug/dl respectively. To further add to the confusion, an "inverse pattern" in which the cortisol concentration 8 hours after dexamethasone is below the cut-off value, but the cortisol concentration 4 hours post-dexamethasone is above the cut-off has been described in 5 dogs with PDH (8). Because this pattern is highly suspicious for HAC, further testing should be pursued.

Cut-offs and the ACTH response test

At Antech Diagnostics and Michigan State University diagnostic laboratories a post-ACTH cortisol concentration above 20 ug/dl is consistent with HAC, while at the University of Tennessee Endocrine diagnostic laboratory the reference range for cortisol in healthy dogs is different depending on the whether the dog is a female or an intact or neutered male. For example, a post-ACTH cortisol > 10.85 ug/dl in an intact male is consistent with HAC, while a post-ACTH cortisol > 10.85 but < 17.5 ug/dl would be considered normal for an intact or spayed female and therefore not consistent with HAC. ***Summary impact point:*** *The cut-off values previously established for screening tests may be misleading, especially with milder cases and cases worked up at clinics with low disease prevalence. In other words, because the cut-offs are too high, ACD may be in actuality misdiagnosed cases of classic Cushing's disease.*

CONGENTIAL ADRENAL HYPERPLASIA AND FOOD-INDUCED CUSHING'S DISEASE: ARE THEY LINKED TO ACD?

What is congenital adrenal hyperplasia?

Congenital adrenal hyperplasia (CAH) is a group of autosomal recessive disorders in people characterized by impaired synthesis of cortisol, aldosterone, or both (9). The most common form of CAH is caused by mutations in CYP21A2, the gene encoding the adrenal steroid 21-hydroxylase enzyme. This enzyme converts 17-OHP to 11-deoxycortisol and progesterone to deoxycorticosterone, which are the respective precursors of cortisol and aldosterone. When the pituitary gland senses relative cortisol insufficiency it releases ACTH which stimulates the adrenal cortex to produce more cortisol. However, patients with CAH have insufficient amounts of the enzyme 21-hydroxylase, needed to convert a precursor molecule (i.e., 17-OHP) into cortisol. As a result, the pituitary gland continues to sense the need for cortisol and secretes more ACTH. This leads to an overabundance of 17-OHP, which is converted in the adrenal cortex into excess androgens (e.g., androstenedione). ***Summary impact point***: *Because ACD in dogs is often associated with adrenal hyperplasia, elevated 17-OHP and androstenedione concentrations, and occasionally low to low-normal cortisol concentrations post-ACTH administration, it has been suggested that ACD may be a congenital adrenal hyperplasia-like syndrome.*

What is food-induced Cushing's disease?

This disorder is an ACTH-independent form of HAC that was first reported in the dog in 2008 (10). Interestingly, once considered a rare disorder, food-induced Cushing's disease is seen on a regular basis by the endocrinology service at the Veterinary School in Utrecht, Netherlands (11). In humans, various adrenocortical membrane-bound receptors functionally coupled to steroidogenesis have been reported, including gastric inhibitory peptide (GIP), catecholamine, and luteinizing hormone (LH) receptors. In the dog, both normal and neoplastic adrenal tissue has also been shown to contain cellular receptors other than ACTH such a LH and GIP (12). Due to the aberrant expression of functional adrenocortical GIP receptors, cortisol secretion is stimulated by food intake, leading to the development of HAC. The dogs with food-induced Cushing's disease have signs and symptoms consistent with HAC, ultrasound evidence of bilaterally enlarged adrenal glands, normal ACTH response and LDDS tests, *suppressed* endogenous ACTH levels, and an increase in the urine cortisol:creatinine ratio > 50% following a meal (10). ***Summary impact point:*** *Food-induced Cushing's disease, like some cases of adrenocortical neoplasia, is an ACTH-independent variant of ACD.*

TREATMENT CONSIDERATIONS FOR ATYPICAL CUSHING'S DISEASE

To date there are several medical treatment options for ACD which include melatonin and lignans, trilostane, ketoconazole, mitotane, and adrenalectomy. It is not known whether high concentrations of cortisol precursors or other noncortisol steroid hormones contribute to the clinical signs of HAC. In previous studies it was demonstrated that during treatment with trilostane concentrations of 17-OHP, a cortisol precursor, did not change (13). Treatment control was achieved by suppressing post-ACTH cortisol levels into a well defined range where cortisol concentrations were highly correlated with the results of treatment (1, 14, 15). This high correlation and the fact that other cortisol precursors that were measured did not decrease, suggests that noncortisol steroid hormones are unlikely to be associated with clinical signs of the disease. Unfortunately, the treatment for ACD, especially with trilostane, is fraught with confusion. It is a common belief that because trilostane increases 17-OHP concentrations that it is not the preferred treatment, and may in fact have zero efficacy with cases of ACD. Despite these beliefs, most dogs treated with trilostane have resolution of clinical signs. However, trilostane may not be the treatment of choice in dogs, especially in Scottish Terriers, that have extremely high concentrations of both androstenedione and 17-OHP. It is theorized that these dogs as adults have a congental adrenal hyperplasia-like syndrome whereby elevated 17-OHP acts as an androgen precursor and fuels the rise of androstenedione which may be a risk factor for hepatocellular carcinoma. In these cases, mitotane may be the preferred treatment because it will consistently lower both androstenedione and 17-OHP concentrations.

SUMMARY

The major question is the following - is atypical Cushing's disease a distinctly different disorder than classic Cushing's disease? Interestingly, only 14 cases of atypical Cushing's have been reported in the veterinary literature and the majority of cases are ATs (1,16,17,18). Published and unpublished data would indicate that it is not uncommon for ATs to be associated with excessive sex steroid production and normal screening tests. In other words, many ATs often fit the description of ACD, but this description can be considered "normal" for an adrenal tumor, especially an adrenal carcinoma. Of even more interest is the expression of 21-hydroxylase is normal in ATs, and, therefore, a deficiency of 21-hydroxylase is not the cause of excessive steroid production (11). ***Summary impact point:*** *Some ATs fit the description of ACD, but this presentation is not uncommon for ATs.*

Primary bilateral adrenal hyperplasia caused by aberrant expression of receptors in the adrenal cortex is a relatively new form of Cushing's disease reported in the dog which fits the definition of ACD. The primary example is food-induced Cushing's disease. These cases present with obvious signs of HAC, negative screening tests, suppressed endogenous ACTH concentrations, and symmetrically enlarged adrenal glands.

A theory for non-suppressed ACTH cases (ACTH-dependent ACD) is the cut-off levels for the current endocrine screening tests are simply too high, especially in mild cases. In other words, some or possibly many cases of classic Cushing's disease are misdiagnosed as ACD. And lastly, the possibility that ACD is a congenital adrenal hyperplasia-like syndrome is unclear at this time. To prove or disprove this theory, adrenal glands from dogs with ACD, preferably not treated with mitotane or trilostane, would have to be tested for deficiencies of the various enzymes involved in the cortisol synthetic pathways.

And finally, and maybe most important, no matter what the underlying cause of ACD, medical therapy with trilostane or mitotane is efficacious in most cases, even in dogs with an AT that are not good surgical candidates.

REFERENCES

1.Ristic JME, et al. J Vet Intern Med 2002;10:433. 2 .Behrend EN, et al. J Vet Intern Med 2013;27:1292. 3. Feldman EC and Nelson RW. In Canine and Feline Endocrinology and Reproduction. 3rd ed. St Louis, MO: Saunders; 2004 .4. Sieber-Ruckstuhl NS, et al. Vet Rec 2008;162:673. 5. Monroe WE,Panciera DL, Zimmerman KI. J Vet Intern Med 2012;26:945. 6. Frank LA, Schmeitzel LD, Oliver JW. J Am Vet Med Assoc 2001;218:214. 7. Hill KE, et al. J Am Vet Med Assoc 2005;226:556. 8. Mueller C, et al. Vet Rec 2006;159:489. 9. White PC, Speiser PW. Endocrine Rev 2000;21:245. 10. Galac S, et al. Vet J 2008;177:141. 11. Galac S. Personal communication. 12. Galac S, et al. Domest Anim Endocrinol 2010;39:63. 13. Sieber-Ruckstuhl NS, et al. Domest Anim Endocrinol 2006;31:63 14. Ruckstuhl NE, et al. Am J Vet Res 2002;63:506. 15. Wenger M, et al. Am J Vet Res 2004;65;1245 16. Benitah N, et al. Am J Vet Med Assoc 2005;227:1095. 17. Syme HM, et al. J Am Vet Med Assoc 2001;219:1725. 18. Norman EJ, Thompson H, Mooney CT. Vet Rec 1999;144:551

ADVANCES IN THE TREATMENT OF FELINE HYPERTHYROIDISM: A STRATEGY TO SLOW THE PROGRESSION OF CKD

Mark E. Peterson, DVM, DACVIM
New York, NY

INTRODUCTION

Hyperthyroidism and chronic kidney disease (CKD) are both common disorders in older cats. Therefore, it should not be surprising that both disorders frequently develop together in the same cat. The prevalence of concurrent renal disease in cats with hyperthyroidism is estimated to be approximately 30-40%.[1-3]

Hyperthyroidism tends to "artificially" increase the renal blood flow (RBF), as well as the glomerular filtration rate (GFR).[1-7] When the GFR is increased in a hyperthyroid cat with underlying CKD, it can mask renal insufficiency; serum concentrations of urea nitrogen and creatinine may be normal despite mild to moderate kidney disease. Decreased muscle mass, a common feature of hyperthyroidism, also contributes to the lowered serum creatinine level in these cats (since creatinine is derived from muscle tissue).[1,3]

Successful treatment of hyperthyroidism restores the high serum T_4 concentration to normal and, in cats without CKD, also returns the high GFR back to normal values. In cats with CKD, however, the GFR will fall to the low-normal or subnormal levels expected with moderate renal dysfunction.[1-7] Therefore, this decrease in GFR can result in the apparent worsening of the serum kidney function tests or the development of renal disease.[1-7] It is important to remember, however, that treating the hyperthyroidism itself does not cause the CKD in these cats. The renal disease was already present before treatment but was masked by the hyperdynamic state of the hyperthyroidism.[1-3]

PHYSIOLOGICAL INTERACTIONS BETWEEN THYROID HORMONE & THE KIDNEY

Hyperthyroidism results in increased renal blood flow (RBF) and GFR due to both pre-renal and renal effects.[1-3,8] Among the pre-renal factors, thyroid hormones increase the cardiac output by positive chronotropic and inotropic effects, as well as a reduction in systemic vascular resistance by dilating the arterioles of the peripheral circulation.[1-3] This indirectly contributes to an increase in RBF. There is an increased endothelial production of nitric oxide in the renal cortex and medulla by induction of nitric oxide synthase, directly by the thyroid hormones and indirectly by high arterial pressure related endothelial shear stress.[9] This is accompanied by a reduction in renal vasoconstrictor endothelin.[10] Thus, an increased intrarenal vasodilatation and decreased vasoconstriction ensues, contributing to a net increase in RBF.

The GFR increases by about 25% among hyperthyroid cats.[1-7] This increase in GFR results in part from the increase in RBF, but activation of the renin-angiotensin-aldosterone system and increased β-adrenergic activity play a role.[1-3,8,11] Overall, this results in afferent arteriolar vasodilatation and efferent arteriolar vasoconstriction, leading to increased filtration pressure and glomerular hyperfiltration. This adds to the magnitude of increase in GFR over-and-above that contributed by the increase in RBF alone.

These systemic hemodynamic factors (i.e., increased cardiac output, decreased systemic vascular resistance) combined with intrarenal vasodilation lead to increases in RBF, GFR, and glomerular hydrostatic pressure. Thyroid hormones also influence renal tubular function, including electrolyte handling.[1-3,8,11] With time, hyperthyroidism can lead to renal changes including glomerular hypertension, glomerulosclerosis, proteinuria, and hyperplasia and hypertrophy of the renal tubules.[1-3]

HYPERTHYROIDISM ITSELF CONTRIBUTES TO CHRONIC KIDNEY DISEASE (CKD) IN CATS

The prevalence of CKD in general feline population has been reported to be 7.7% in cats >10 years of age and 15.3% in cats >15 years of age.[2] In contrast, the prevalence of CKD in hyperthyroid cats has been estimated to be as high as 30-40%.[1,2] This much higher-than-expected prevalence of CKD supports the supposition that untreated thyrotoxic state is initiating or worsening the CKD in these hyperthyroid cats. If that is true, one may want to always try to correct the cat's hyperthyroidism, even in cats with azotemia and known pre-existing CKD.

Recent research provides further evidence that hyperthyroidism may contribute to the development or progression of CKD in cats. First of all, a number of recent reports indicate that many untreated hyperthyroid cats develop proteinuria, which resolves within 4 weeks of successful treatment.[12-14] This proteinuria, which reverses after treatment, could be a reflection of glomerular hypertension and hyperfiltration, changes in tubular protein handling, or a change in the structure of the glomerular barrier. Whatever the cause of the proteinuria, no treatment is generally needed other than treatment of the hyperthyroid state itself. The proteinuria will resolve after correction of the hyperthyroid state.

Secondly, cats with untreated hyperthyroidism have high levels of retinol binding protein (RBP), a urinary marker for tubular dysfunction or damage.[12,15] This high urinary RBP excretion may reflect tubular damage or dysfunction resulting from the thyroid-induced hypertrophy and hyperplasia of the tubular cells. After treatment, these high urinary RBP levels fall in cats without azotemia but may remain slightly high in cats with pre-existing CKD. This too suggests that hyperthyroidism can cause reversible renal dysfunction; however, the renal tubular changes may become irreversible with time as CKD progresses.

Thirdly, many cats with untreated hyperthyroidism have high values for urinary N-acetyl-b-D-glucosaminidase (NAG), a lysosomal glycosidase found primarily in epithelial cells of the proximal convoluted tubule.[16] Like RBP, NAG is a specific marker of active proximal tubular damage. After treatment, these high urinary NAG levels decrease, again suggesting that these renal changes can be reversed, at least in cats without pre-existing CKD. Overall, these studies all suggest that that leaving a hyperthyroid cat untreated (or poorly regulated with methimazole) may be detrimental to long-term kidney function. Treating and curing hyperthyroidism may help to both reverse renal damage and preserve remaining kidney function.

PREDICTING WHICH HYPERTHYROID CATS HAVE CONCURRENT (MASKED) CKD

Predicting which untreated hyperthyroid cats have clinically significant underlying CKD is difficult.[1,2,17] In the absence of methods for accurately measuring GFR in veterinary practice, serum urea and creatinine concentrations, urine specific gravity, and UPC should be carefully evaluated in individual cases. If any parameter is abnormal, then renal failure may become apparent upon resolution of the hyperthyroid state.

Because of these difficulties, trial therapy using methimazole has been recommended as a test of renal function.[18,19] If deterioration in renal function does not occur after medical normalization of the T_4 values, then a more definitive therapeutic option for hyperthyroidism may be selected. Except for advanced CKD, the necessity of this approach is questionable, given that treatment for the hyperthyroidism would still be recommended in most cats.[20] In support of that reasoning, the survival of cats that develop azotemia is no different to those that remain non-azotemic after treatment of hyperthyroidism.[3,13,21]

In most hyperthyroid cats that manifest overt renal disease after antithyroid drug treatment, permanent treatments with surgery or radioiodine can be given once the concurrent renal disease is stabilized with proper treatment (subcutaneous fluids, dietary therapy). It is important to realize that the decline in GFR following initial treatment of hyperthyroidism is not progressive with long-term euthyroidism; once serum T_4 concentrations have fallen into the normal range, the GFR tends to stabilize at that new rate.[7]

IATROGENIC HYPOTHYROIDISM CONTRIBUTES TO CKD IN CATS

Iatrogenic hypothyroidism can develop during therapy with antithyroid drugs,[22-25] after surgical thyroidectomy,[22,25] or after treatment with radioiodine (^{131}I).[27-30] Although early reports suggested that clinical signs associated with iatrogenic hypothyroidism in cats were uncommon and that most cats did not require treatment, it is now realized that milder degrees of iatrogenic hypothyroidism are relatively common and that these cats may benefit from thyroid replacement therapy (especially if concurrent CKD is present).[3,20,25,31] Cats with severe or overt iatrogenic hypothyroidism can easily be confirmed by finding low serum T_4 and T_3 concentrations in conjunction with high serum TSH values, as measured by the cTSH assay.[22,24,25,31]

However, based on recent studies, it is not uncommon for high serum TSH to also develop in cats treated for hyperthyroidism despite maintenance of completely normal concentrations of T_4, free T_4, or both.[24,31] In a recent study, high serum TSH values developed in over a third of hyperthyroid cats treated with methimazole.[24] However, 17% of these cats had a high TSH value despite low-normal T_4 and completely normal free T_4 concentrations. We have seen similar findings in cats treated with radioiodine, i.e., a subgroup of cats with low-normal T_4 values (<1.5 µg/dl) will develop high cTSH levels, suggestive of mild or partial hypothyroidism.[31] These cats remain totally asymptomatic and show no clinical signs suggestive of hypothyroidism (e.g., lethargy, hair loss), although some may develop azotemia.

Do these cats with low-normal T_4 values but high TSH concentrations really have mild degrees of hypothyroidism? Is this a form of "biochemical" subclinical hypothyroidism, as reported in man?[32,33] Should the methimazole dosage (or normal ^{131}I dose) be reduced or should they be treated with levothyroxine despite their normal circulating thyroid hormone values? Or are these serum TSH values falsely high, simply representing a TSH laboratory artifact? At this time, the answers to these questions remain unknown. However, our current studies do suggest that these cats may indeed have a form of subclinical hypothyroidism.

If These Treated Cats Do Not Show Any Clinical Signs of Hypothyroidism, Why is Any of This Important?

In recent studies by Williams *et al*,[3,13,25,34] cats with "overt" iatrogenic hypothyroidism were more likely to develop azotemia in the 6 months after treatment than cats that remained euthyroid. Hypothyroid cats with azotemia also had shorter survival times than nonazotemic cats, whereas no difference in survival between euthyroid cats with or without azotemia could be detected.

Overall, these findings suggest that the development of iatrogenic hypothyroidism contributes to the development of newly detected azotemia, at least in cats with mild underlying CKD that has been masked by the hyperthyroid state. More importantly, the hypothyroidism may shorten survival after treatment of hyperthyroidism.[3,13,25]

Recommendations for Monitoring and Treating Hypothyroidism

Whatever treatment option for hyperthyroidism is considered, we now believe it is extremely important to try to avoid iatrogenic hypothyroidism, as it may have its own detrimental effects on GFR. Hypothyroidism in both humans and dogs has been showed to reduce GFR,[8,11,35] and it is likely that similar changes occur in cats that develop iatrogenic hypothyroidism. In cats with normal renal function, such a fall in GFR may not be clinically important; however, hypothyroid cats with underlying CKD may develop moderate to severe azotemia, which is generally progressive.

If a cat with post-treatment azotemia develops a low to low-normal T_4 concentration (T_4<1.5 µg/dl), a serum TSH level should be measured to help exclude overt or subclinical hypothyroidism. If iatrogenic hypothyroidism is diagnosed based on the finding of a high TSH concentration, the methimazole dosage should be reduced. If the cat has been treated with surgery or radioiodine, then treatment with L-T_4 should be considered.[2,19,36] Although once daily administration (0.1 mg) may be adequate in some cats, most cats respond better to twice daily treatment (0.075 mg, BID) given on an empty stomach. The dosage should be adjusted based on post-pill serum T_4 and cTSH determinations.[31,36] Most cats treated will show improvement in their azotemia as the hypothyroidism resolves and euthyroidism is restored.[34,36]

PROGRESSION OF CKD IN TREATED CATS THAT MAINTAIN EUTHYROIDISM

In most cats that develop post-treatment azotemia, the CKD is not that severe or life threatening, especially if euthyroidism is maintained (i.e., serum T_4 kept at 1.5-3.0 μg/dl). In euthyroid cats, it is unusual to see a jump of more than one IRIS stage after treatment.[14] In other words, hyperthyroid cats with IRIS Stage I-II CKD may develop overt azotemia after treatment, but one would not expect those cats' CKD to advance to more than IRIS Stage II-III after treatment. In addition, the decline in GFR after successful treatment of a cat's hyperthyroidism is not very progressive. Rather, the fall in GFR is detectable within 1 month but then remains stable at this level for months thereafter.[1,2,7] The rise in serum urea nitrogen and creatinine values in cats with CKD follow the decrease in GFR, so that azotemia, when it does occur, would be expected to develop within 1 month of treatment but remain relatively stable over many months. However, if overt or subclinical hypothyroidism develops in these cats, treatment with L-T_4 can help raise the cat's GFR and may result in lowered serum renal values.

USING L-T4 SUPPLEMENTATION IN CATS WITH PRE-EXISTING CKD AFTER I-131 OR SURGERY

Managing cats that present with concurrent hyperthyroidism and overt azotemia are more problematic. Again, treatment of the hyperthyroid state will lead to a rise in serum creatinine values, but this worsening in azotemia will be related to the magnitude of the suppression of circulating T_4 and T3 concentrations.

Occasionally cats with pre-existing chronic kidney disease will demonstrate overt deterioration in kidney function following the initiation of therapy for hyperthyroidism leading to a potentially life-limiting uremia. It is my firm belief that the majority of these cases are caused by periods of iatrogenic hypothyroidism that contribute to further decline in renal function. Therefore, it is of critical importance to avoid even temporary hypothyroidism in this subgroup of cats with overt CKD (serum creatinine ≥ 2.0 mg/dl), in order to maintain RBF and GFR as much as possible. Iatrogenic hypothyroidism has been shown to both worsen azotemia and shorten the life expectancy in cats with pre-existing CKD.[2,25]

Following radioiodine therapy or surgical thyroidectomy, a transient period of iatrogenic hypothyroidism is very common.[37] This period of hypothyroidism develops in the period following the radiation-induced death (or surgical removal) of the adenomatous thyroid tissue responsible for the hyperthyroidism but before regeneration of any suppressed normal thyroid tissue is possible. The duration of this iatrogenic hypothyroidism is usually a few weeks and may be completely clinically silent. However, depending on the degree of suppression or damage to normal thyroid tissues, this transient period may be clinically detected.

In most treated cats that have relatively normal renal function, this period of temporary hypothyroidism is not clinically significant and resolves without any intervention. Therefore, unless cats with normal renal function demonstrate other overt symptoms of this transient hypothyroidism (e.g., anemia, lethargy), thyroid hormones (e.g., L-T_4) supplementation is generally avoided. However, cats with preexisting renal dysfunction may show a clinically significant worsening azotemia during this period of iatrogenic hypothyroidism. As a result, thyroid hormone supplementation in these cats with iatrogenic hypothyroidism and overt CKD may be indicated.

To avoid iatrogenic hypothyroidism when treating hyperthyroid cats with concurrent CKD, we have adapted a therapeutic protocol that ensures resolution of the cat's thyrotoxicosis while avoiding even transient hypothyroidism. This therapeutic approach involves the use of customized radioiodine therapy doses (as low as possible), followed by supplementation with oral L-T_4 that is initiated at discharge from the hospitalization required for radioiodine therapy.[38,39]

Preliminary evaluation of cats with stage 2-3 IRIS chronic kidney disease and pretreatment serum creatinine levels of 2.0-3.5 mg/dl treated with customized doses of radioiodine ranging between 1-5 mCi and discharged on oral L-T_4 (0.075-0.1 mg q12-24h) confirmed the avoidance of iatrogenic hypothyroidism and minimal associated worsening in azotemia.[37,38] In these cats the supplementation with physiologic levels of thyroxine on discharge from the hospitalization needed for radioiodine therapy ensures that the incremental decrease in GFR that accompanies resolution of thyrotoxicosis will not be accompanied by additional renal injury caused by iatrogenic hypothyroidism.

Most cats with concurrent CKD and hyperthyroidism will also benefit from the initiation of other standard therapies including subcutaneous fluids, dietary phosphorus restriction, phosphate binders, potassium gluconate, antihypertensive therapy (as indicated by confirmation of hypertension), benazepril (as indicated by confirmation of proteinuria), and supplemental B- vitamin administration.

REFERENCES

[1]Syme HM. *Vet Clin North Am Small Anim Pract* 2007;37:723. [2]Langston CE et al. *Clin Tech Small Anim Pract* 2006;21:17. [3]Williams TL. PhD thesis, Royal Veterinary College, University of London, 2013. [4]Graves TK, et al. *Am J Vet Res* 1994;55:1745. [5]Adams WH, et al. *Can J Vet Res* 1997;61:53. [6]Adams WH, et al. *Vet Radiol Ultrasound* 1997;38:231. [7]Boag AK, et al. *Vet Rec* 2007;161:711. [8]Basu G, et al. *Indian J Endocrinol Metab* 2012;16:204. [9]Quesada A, et al. *Eur J Endocrinol* 2002;147:117. [10]Singh G, et al. *Life Sci* 1994;54:1901. [11]Mariani LH, et al. *J Am Soc Nephrol* 2012;23:22. [12]van Hoek I, et al. *Domest Anim Endocrinol* 2009;36:45. [13]Williams TL, et al. *J Vet Intern Med* 2010;24:863. [14]Harley LS, et al. *J Vet Intern Med* 2011;25:678. [15]van Hoek I, et al. *J Vet Intern Med* 2009 23:1031. [16]Lapointe C, et al. *J Vet Intern Med* 2008;22:1103. [17]Riensche MR, et al. *J Feline Med Surg* 2008;10:160. [18]Peterson ME. *Proceedings 29th Annual ACVIM Veterinary Medical Forum* 2011;104. [19]Mooney CT, et al. *Manual of Canine and Feline Endocrinology*. 2012; 92. [20]Baral R, et al. *The Cat: Clinical Medicine and Management*. 2012;571.
[21]Wakeling J, et al. *J Vet Intern Med* 2006; 20:1523. [22]Graham P. *J Vet Intern Med* 2000;14:342. [23]Fischetti AJ, et al. *Vet Radiol Ultrasound* 2005;46:267. [24]Chciuk K, et al. *J Vet Intern Med* 2013;27:691. [25]Williams TL, et al. *J Vet Intern Med* 2010;24:1086. [26]Welches CD, et al. *Vet Surg* 1989;18:392. [27]Meric SM, et al. *J Am Vet Med Assoc* 1990;197:621. [28]Jones BR, et al. *N Z Vet J* 1991;39:71. [29]Peterson ME, et al. *J Am Vet Med Assoc* 1995;207:1422. [30]Nykamp SG, et al. *J Am Vet Med Assoc* 2005;226:1671. [31]Peterson ME. *Compend* 2013;35:E1. [32]Biondi B, et al. *Endocr Rev* 2008;29:76. [33]Khandelwal D, et al. *Drugs* 2012;72:17. [34]Williams TL, et al. *J Vet Intern Med* 2010;24:684. [35]Gommeren K, et al. *J Vet Intern Med* 2009;23:844. [36]Peterson ME. *August's Consultations in Feline Internal Medicine*. 2014. [37]Peterson ME, et al. *Kirk's Current Veterinary Therapy*, Volume XV , 2014.[38]Broome MR, et al. *J Vet Intern Med* 2013;27:685.

BILE ACIDS: BEYOND THE LIVER

Cynthia RL Webster, DVM, DACVIM (Small Animal Internal Medicine)
Grafton, MA

WHY ARE BILE ACIDS IMPORTANT IN CLINICAL MEDICINE?

For many years in veterinary medicine we have used the measurement of total serum bile acids (TSBA) in the diagnosis of hepatobiliary disease. TSBA are a sensitive screening test for congenital portosystemic shunts and a specific diagnostic test to confirm hepatobiliary functional disease. Likewise we routinely use the hydrophilic bile acid, ursodeoxycholic acid (UDCA) as a cytoprotective agent to treat dogs and cats with hepatobiliary disease. With the recent discovery that bile acids can bind to nuclear and plasma membrane receptors and act as hormones to control glucose and lipid metabolism as well as energy expenditure, the therapeutic potential of bile acids has expanded.

WHAT ARE BILE ACIDS?

Bile acids are organic anions synthesized exclusively from cholesterol in the liver. The rate-limiting step is the addition of a hydroxyl (OH) group in the 7th position of the cholesterol steroid nucleus, a reaction catalyzed by 7-alpha-hydroxylase (CYP7A1). Since cholesterol already contains a 3-alpha-OH group, the simplest bile acids are the 3-alpha-7-alpha di-OH bile acids such as chenodeoxycholate (CDCA). Additional hydroxylation at the 12-alpha position creates the tri-OH bile acids in the cholate group. Bile acids can also be synthesized by an alternative pathway in which cholesterol is oxidized at C27 by CYP27A1 prior to hydroxylation by CYP8B1 to form CDCA. In general, the greater the degree of hydroxylation, the less hydrophobic is the bile acid. Primary bile acids are conjugated in the liver to glycine or taurine. In humans, the major circulating bile acid is the di-OH bile acid, glycochenodeoxycholate (GCDC), while in dog and cats it is the tri-OH bile acid, taurocholate (TCA). Conjugation of bile acids further increases their water solubility and thus decreases their passive absorption in the bile ducts and small intestine. In the intestine, primary bile acids can be metabolized by bacteria to secondary bile acids. They can be deconjugated or dehydroxylated at the 7 alpha position. The most common modifications yield the secondary bile acids, lithocholate (LCA) and deoxycholate (DCA). A large family of gram positive bacteria (primarily Clostridium and Firmicutes) deconjugate bile acids while a smaller population of Clostridium sp have the enzymes necessary to carry out the dehydroxylation.

Bile acids undergo efficient enterohepatic circulation. After secretion into the bile, bile acids are stored in the gallbladder. Post-prandially, fat in the duodenum stimulates the release of cholecystokinin which causes gallbladder contraction and the release of bile acids into the intestinal lumen. The amphipathic bile acids self-associate to form micelles which solubilize lipids facilitating the absorption of dietary fat and fat soluble vitamins. Conjugated bile acids undergo minimal passive intestinal re-absorption. Instead, an efficient ileal sodium coupled bile salt transporter (ABST) recaptures them and deposits them into the portal circulation. The bile acids circulate back to the liver and are actively transported back into hepatocytes by a distinct sodium coupled bile salt transporter, Ntcp. Bile acids move through the hepatocyte cytosol to the canalicular membrane where an ATP dependent bile salt export pump, BSEP, transports them against a steep concentration gradient into the bile. Transport of bile acids across the canalicular membrane is the rate-limiting step in the enterohepatic circulation of bile acids. The transhepatic movement of bile acids provides the osmotic driving force for bile formation.

BILE ACIDS AS TESTS OF HEPATIC FUNCTION

This enterohepatic circulation of bile acids operates at 98% efficacy so that circulating serum bile acids values are typically very low after fasting. Following a 12 hour fast, only small amounts of bile acids spill over into the systemic circulation (serum values < 15 uM). Endogenous challenge to serum bile acid handling occurs in the post-prandial period and serum bile acids increase transiently in this setting (up to 25-30 uM). Disturbances in hepatobiliary extraction, conjugation or excretion of bile acids or interruption in the portal delivery of bile acids to the liver result in elevations in TSBA.

BILE ACIDS CONTROL INTRACELLULAR SIGNALING

Bile acids control intracellular signaling events several ways.[1,2] In 1999, it was discovered that bile acids bind to their own nuclear receptor, the farnesoid X receptor (FXR), through which they regulate their own biosynthesis, as well as glucose and lipid metabolism, liver regeneration and the intestinal bacterial population. Bile acids can also activate other nuclear receptors including the vitamin D receptor and the pregnane X receptor. In 2006 it was reported that bile acids bind to a cell surface G protein coupled receptor, TGR5, through which they control energy and glucose metabolism.[3] The various members of the bile acid family have different affinity for the bile acid receptors. For FXR activation, the potency is CDCA > LCA = DCA > CA, and for TGR5 the potency is LCA > DCA > CDCA > CA. Conjugated forms of the bile acids are less potent agonists at both receptors. LCA is the predominate bile acid ligand for vitamin D and the pregnane X receptors.

BILE ACID SIGNALING THROUGH FXR

FXR activation controls bile acid synthesis as well as glucose and lipid metabolism.[4] FXR is present in high levels in the liver, intestine, kidney and adrenal with smaller amounts in adipose tissue, pancreas and heart. When bile acids bind to FXR, the receptor translocates to the nucleus where it binds directly to specific response elements in the promoter of genes. FXR activation also signals by increasing the expression of the atypical nuclear hormone receptor, SHP. SHP has no DNA binding activity itself but can bind to and inhibit other nuclear receptors.

FXR activation controls many facets of bile acid metabolism including synthesis, conjugation, secretion and uptake.[5] When bile acid levels are high as in cholestatic liver disease, bile acid binding to FXR decreases the transhepatic flux of bile acids. This is accomplished by increasing hepatic SHP expression which in turn decreases the transcription of basolateral hepatocyte transporters, such as Ntcp and OATp (organic anion transporter) and increases the expression of canalicular transporters, BSEP, phospholipid transporter MDR3, and multiple drug related protein (the bilirubin transporter), MRP2. In addition, SHP inhibits the expression CYP7A1 and thus turns off bile acid synthesis. FXR activation also increases the detoxification of bile acids by up-regulating the enzymes necessary for bile acid oxidation, sulfation and glucuronidation. In the intestine, FXR activation inhibits expression of the ileal bile salt transporter, ABST. All of these changes have the net effect of decreasing intracellular bile acid concentrations. Thus it is not surprising that treatment with FXR agonists protects rodents in experimental models of cholestasis while knockout of FXR function (FXR-/- mice) result in an increased susceptibility to hepatobiliary injury.

There is considerable interest in the pharmaceutical industry to develop non-toxic FXR agonists for therapeutic use in treating cholestatic disorders. A 6-ethyl derivative CDCA (obeticholic (OCA) manufactured by Intercept Pharmaceuticals) is currently undergoing clinical development for the treatment of cholestatic disease.[6] Two Phase II, multicenter, double-blind randomized controlled study of OCA have been conducted in humans with primary biliary cirrhosis (PBC), an immune mediated cholangiopathy. In both studies patients on OCA had substantial reductions in serum liver enzymes. Pruritus was the most common adverse event. A Phase III, multicenter study of OCA in PBC is currently underway (clinicaltrials.gov, NCT01473524). A recent study also suggests that OCA may be beneficial in reducing portal pressure as it was effecting in doing so 2 animal models of portal hypertension.[7]

Experiments in mice lacking FXR expression, FXR-/- mice, have revealed important roles for the nuclear receptor in controlling glucose and lipid metabolism. FXR-/- mice have high circulating free fatty acid and triglyceride levels and develop hepatic lipidosis. They also have an increased tendency to develop atherosclerosis. This phenotype is due to the fact that FXR activation results in: 1.) increased expression of apolipoprotein CII (a co-activator of lipoprotein lipase) and the peroxisomal proliferator-activated receptor (which enhances mitochondrial fatty acid oxidation), 2.) decreased expression of apo CIII (an inhibitor of lipoprotein lipase) and 3.) through SHP leads to inhibition of the sterol regulatory element binding protein-1C (SREBP) (which controls the expression of many genes involved in lipogenesis). The net result of FXR activation is increased clearance and decreased production of triglycerides. Apoe -/- atherogenic mice treated with OCA have a 95% reduction in lesions presumably due to the cholesterol and triglyceride lowering effects of the drug.[4]

$FXR^{-/-}$ mice also have impaired glucose tolerance, decreased insulin sensitivity, and significantly blunted insulin responsiveness in both skeletal muscle and liver. In wild type mice, activation of FXR by dietary CA suppresses the expression of multiple genes involved in gluconeogenesis and lowers glucose levels. Both of these responses to CA feeding are lost in $FXR^{-/-}$ and $SHP^{-/-}$ mice. These results suggest that FXR acting through SHP is required for the maintenance of normal glucose metabolism. In diabetic animals activation of FXR lowers serum glucose and increases insulin secretion. Activation of FXR represses hepatic gluconeogenic genes phosphoenolpyruvate (PEPCK) and glucose -6-phosphate (G6Pase). A range of preclinical studies have shown that OCA increases insulin sensitivity and regulates glucose homeostasis in animal models. A recent human clinical trial using OCA as monotherapy for the treatment of nonalcoholic fatty liver disease/nonalcoholic steatoheaptitis (NAFLD/NASH) and associated type 2 diabetes showed reduced markers of liver inflammation and fibrosis as well as an increase in insulin sensitivity.[8] NAFLD/NASH is a hepatic manifestation of the metabolic syndrome in humans and currently is the most common hepatic disorder in people. The syndrome starts with hepatic lipidosis and insulin resistance and in some people progresses to an inflammatory hepatitis that can lead to liver failure from cirrhosis.

Some of the anti-diabetic effects of bile acids may be mediated by binding of bile acids to FXR receptors on the pancreas. Treatment of primary cultures of pancreatic islets with bile acids or FXR agonists increases glucose-stimulated insulin secretion, an effect not seen in islets from FXR(-/-) mice.[9,10] In these cells FXR acts as a cytosolic and non-genomic effector of K_{ATP} currents. Interestingly, the cellular location of FXR in β cells seems to be dependent on the body mass with lean mice having the capacity for cytoplasmic signaling while obese mice only have nuclear localization of FXR.

BILE ACID SIGNALING THROUGH TGR5

Bile acids activate a G protein coupled plasma membrane receptor, TGR5.[3,11] Bile acid binding to TGR5 results in increased production of the second messenger, cAMP, which in turn up-regulates both post-translation and translational events.

In the liver TGR5 is expressed in gallbladder and biliary epithelial cells, Kupffer cells and sinusoidal endothelial cells. Bile acid activation of the TGR5 receptor is involved in chloride secretion in biliary epithelial cells and promotes gallbladder filling. In Kupffer cells activation of the receptor suppresses the release of inflammatory cytokines. In sinusoidal endothelial cells, TGR5 activation is associated with release of eNOS and production of the potent vasodilatory compound, NO. The net effect of TGR5 receptor signaling in hepatocytes is choleretic, anti-inflammatory, vasodilatory and GB emptying. Interestingly TGR5 is not present on hepatocytes.

The TGR5 receptor is also expressed on adipocytes, vascular endothelial cells, intestinal epithelial cells, cardiac myocytes, enteric neurons, and in the brain.[12] In non-hepatic tissue activation of TGR5 has been associated with regulation of energy expenditure, and glucose and lipid metabolism. Binding of bile acids to TGR5 in adipocytes results in cAMP mediated increases the activity of the Type 2 deiodinase enzyme (D2).[3] This is the enzyme responsible for the conversion of T4 to the active form of thyroid hormone, T3, in peripheral tissues. The net result of bile acid binding to the TGR5 receptor is an increase in metabolic rate. Consequently, mice fed CA have increased energy expenditure associated with decreased weight gain.

TGR5 also affects hepatic lipid metabolism.[12] TGR5 activation decreases hepatic steatosis and plasma levels of NEFA and TG in rodent obesity models. TGR5 receptor binding is also anti-atherogenic in mice. Administration of TGR5 agonists to mice genetically predisposed to develop atherosclerosis attenuates lesion size due to a decrease in plaque macrophage content and intraplaque inflammation. These effects are likely mediated by TGR5 dependent inhibition of NF kappa beta in macrophages and the resultant decrease in production of inflammatory cytokines

Activation of the TGR5 receptor on enteroendocrine L-cells stimulates the release of glucagon like peptide-1 (GLP-1). GLP-1 is an incretin that is released post-prandially. It is a potent anti-hyperglycemic hormone, inducing glucose-dependent stimulation of insulin secretion while suppressing glucagon secretion. Once in the circulation, GLP-1 has a half-life of less than 2 minutes, due to rapid degradation by the enzyme, dipeptidyl peptidase-IV. Drugs that increase activity of the GLP-1 axis such as inhibitors of dipeptidyl peptidase-IV (ie. the gliptins) are already successful therapies for type 2 diabetes in humans and studies have shown that these drugs can suppress glucagon secretion in cats. In obese mouse models stimulation of TGR5 activity enhances glucose tolerance. Recent studies suggest that bile acid mediated release of GLP-1 may explain remission of type 2 diabetes following bariatric surgery.[13] A recent study in rats which had bile diverted to the distal gut in the absence of bariatric surgery demonstrated increased serum bile acids, enhanced post-prandial GLP1 response, improved glucose tolerance and decreased hepatic steatosis. A similar study in healthy humans who had their small intestine infused with bile acids showed a reduced glycemic response to small intestinal glucose which was associated with increased GLP1 secretion.[14]

There is considerable interest in the development of TGR5 agonists to treat diabetes.[15] INT-777, a semisynthetic, steroidal derivative of CDCA is a TGR5 agonist. INT-777 is well absorbed in vivo and undergoes enterohepatic cycling, largely without conjugation. In rodents, the compound exerts a choleretic effect, increases GLP-1 release from intestinal L-cells and improves glucose tolerance. INT-777 also increases energy expenditure in mice fed a high-fat diet. The compound is still in preclinical phase of development. Another steroidal derivative of CDCA INT-767 has dual agonist properties for TGR5 and FXR. INT-767 decreases plasma TG and cholesterol in diabetic mouse models. Interestingly INT-767, but not INT-747 or INT-777, has hepatoprotective effects in a genetic mouse cholangiopathy. Whether this combined beneficial effect holds in metabolic disease has not been tested.

In neurons and astrocytes, the TGR5 receptor is responsive to the binding of bile acids and neurosteroids. The receptor is down-regulated in human and animal models of hepatic encephalopathy and thus may play a role in the pathophysiology of this disorder.[15] TGR5 activation in the enteric neurons inhibits GI mobility.[15] In the cardiovascular system, bile acids bind to TGR5, as well as muscarinic (M2) and large conductance calcium activated potassium channel receptors to regulate heart rate and vascular tone.[16]

BILE ACIDS AND THE GUT MICROBIOME

Intestinal microbial organisms play an important role in bile acid metabolism.[17] Intestinal microbes deconjugate and 7 alpha dehydroxylate primary bile acids that are not re-sorbed in the distal ileum and convert them to secondary bile acids (DCA, LCA and UDCA). The bacterial bile salt hydrolases which catalyzes bile acid deconjugation are necessary to maintain normal bile metabolism, normal BA signaling through FXR and TGR5, and ultimately maintenance of glucose homeostasis. Alteration of the gut microbiome by diet, antibiotics and disease is accompanied by changes in bile acid composition. For example, antibiotic treatment of humans with vancomycin leads to a marked decrease in secondary bile acids likely related to a decrease gram positive bacterial of the Firmicules phylum which is associated with a decrease in peripheral insulin sensitivity.[18] In healthy humans the degree of insulin resistance has been correlated with higher ratios of 12-alpha-hydroxylated bile acids (CA and DCA) to non-12 hydroxylated bile acids (CDCA). This data suggests that higher activity of CYP8B1 (the enzyme necessary for 12-alpha-hydroxylation of bile acids) is important in the development of insulin resistant.[19,20]

Just as intestinal microbes can modulate bile acids, bile acids can also modulate the gut microbiome. In rat feeding of CA triggers a significant increase in the Firmicutes (particularly Clostridium) and a decrease in Bacteroidetes. It is interesting to note that the observed shift triggered by CA feeding resembles the phylum-level population shifts reported in high-fat diet–induced changes in mice. Because dietary fat increases bile secretion, it is tempting to speculate fat induced alterations in the bacterial population were promoted by increased DCA levels.[21]

Bile acid can suppress intestinal bacterial growth both by direct physiochemical bacteriostatic effects as well as through activation of FXR. When wild type and FXR -/- mice undergo bile duct ligation, both develop bacterial overgrowth. Treatment with an FXR agonist can restore normal bacterial flora in the wild type bile duct ligated mice but not in FXR -/- mice. Microarray analysis in wild type mice treated with an FXR agonist shows up-regulation of several genes whose products promote innate intestinal defense against bacteria.

CLINICAL MODULATION OF BILE ACID SIGNALING

Bile acid sequestrants (BAS) such as the first generation drug cholestyramine have mild to moderate glucose lowering ability. The mechanism for this effect is through both TGR5 and nuclear receptors.[22] The major effect of BASs is to increase circulating GLP-1 levels by binding to TGR5 receptors on enteroendocrine cells. BAS actually decrease FXR activity in liver. This results in a decrease in SHP production which in turn activates of another nuclear receptor liver X receptor (LXR). LXR activation inhibits PEPCK and therefore inhibits gluconeogenesis. A second generation sequestrant, colestimide, is been approved for use as adjunct therapy for glycemic control in type 2 diabetics in humans. The main detriment to the use of first generation BAS in humans was gastrointestinal side effects including dyspepsia, nausea, and constipation. Colesevelam and colestimide are second-generation sequestrants with reduced gastrointestinal side effects in humans. There are no reports on the use of these drug in veterinary patients with diabetes.

The cytoprotective action of UDCA in hepatobiliary cells has been well documented.[23] Emerging evidence suggests that this cytoprotective effect is also seen in neurons (including the retina), cardiac myocytes, renal tubular cells, and pancreatic beta cells. In non-hepatic tissue UDCA's cytoprotective effects have been linked to stabilization of endoplasmic reticulum and mitochondrial function. UDCA is a poor FXR agonist, but can activate TGR5 and thus some of its protective effects in non-hepatocytes may be associated with activation of this receptor.

UCDA can preserve mitochondrial integrity. This effect can be due to direct interaction of UDCA with mitochondrial membranes as well as an ability to decrease the production of reactive oxygen species. UCDA's anti-apoptotic actions may also involve stimulation of cellular survival signaling. In neurons and cardiac myocytes, UDCA protection against cell death requires activation of the lipid kinase, phosphoinositide-3-kinase, and downstream activation of the serine/threonine kinase, Akt. UDCA mediated activation of ERK's and p38 MAPK have also been implicated in its protective effect. Clinical trials for the use of UDCA in the treatment of chronic heart failure and neurodegenerative disease such as ALS and Huntington's disease are underway. [24,25]

Endoplasmic reticulum (ER) stress is associated with obesity, insulin resistance, and type 2 diabetes.[26] Treatment of obese diabetic mice with UDCA reverses ER stress and results in normalization of hyperglycemia, restoration of systemic insulin sensitivity, resolution of fatty liver disease, and enhancement of insulin action in liver, muscle, and adipose tissues. Enhanced insulin sensitivity has also been demonstrated in obese humans after oral administration of UDCA.[27] In animal models, UDCA can attenuate retinal glucose toxicity and diabetic nephropathy suggesting a multimodal beneficial effect of treating diabetes with UDCA.[28,29] A recent study in humans demonstrated that an oral meal with UDCA stimulates GLP-1 release.[30]

REFERENCES

1. de Aguiar Vallim TQ, et. al. Cell Metab. 2013;17:657-69. **2.** Proez G, Et.. al. J Lipid Res 2012;53:1723-37. **3.** Watanabe M, et. al. Nature. 2006 26;439:484-9. **4.** Zhang Y et. al. Drug Future 2010;35:635-642 **5.** Mason A, et al. J Hepatol 2010;52(Suppl 1):745-58. **6.** Silveira MG, et. al. Expert Opin Pharmacother. 2014;15(3):365-72. **7.** Verbeke L, et al. Hepatology. 2013. doi: 10.1002/hep.26939. **8.** Mudaliar S et.al. Gastroenterology 2013:145: 574-582. **9.** Seyer P et. al. J Clin Invest. 2013;123(4):1662-76. **10.** Düfer M et al, Islets. 2012;4(5):333-8. **11.** Thijs WH et. al. J Hepatol 2011;54:1263-1272. **12.** Stepanov V et. al. J Rec Sig Trans 2013;33;213-223. **13.** Kohli R, et. al. Endocrinology 2013;154:2341-2351. **14.** Wu T et. al. J Clin Endocrinol Metab 2013;98:E718-E722. **15** Porez G, et. al. J Lipid Res. 2012;53(9):1723-37. **16.** Khurana S, et. al. Clin Transl Sci. 2011;4(3):210-8 . **17.** Jones ML,et. al. Expert Opin Biol Ther. 2014 Jan 30 **18.** Vrieze A et. al J Hepatol 2013 **19.** Haeusler RA,et al. Diabetes. 2013;62:4184-91 **20.** Haeusler RA, et. al. Cell Metab. 2013;15:65-74. **21.** Islam KB et al. Gastroenterology 2011; 141:1773-81. **22.** Out C et. al. Curr Opin Lipidol. 2012;23:43-55. **23.**Roma M et al. Ckin Sci 2011;121:523-544. **24.** von Haehling S, et. al. J Am Coll Cardiol 2012;59:585-92. **25.**Gaspar JM, et. al. Neuroscience. 20133;253:380-8. **26.** Ozcan U, et al Science. 2006;313:1137-40. **27.** Kars M, et. al. Diabetes. 2010;59:1899-905. **28.** Kars M, et. al. Diabetes. 2010 ;59:1899-905 **29.** Gupta S, et al. PLoS One. 2012;7:e48950. **30.** Murakami M et. al Springerplus 2013;2:20.

MANAGING FELINE PAIN: AN INTERISTS APPROACH

Lisa Moses, VMD, DACVIM (SAIM)
Boston, MA

INTRODUCTION

As an internist who practices exclusively pain medicine, I find that my internal medicine training both shapes and informs my pain practice. Internists are well suited to treating pain, as the complexities of diagnosis and therapy favor a systematic and creative approach. I surveyed a healthy number of internists to ask what questions they still have about feline pain. The content of these proceedings and the presentation reflect my answers to those questions. My goal is to persuade the listener that the assessment and treatment of feline pain is usually possible applying tools that we already have. Additionally, I hope to convince the listener that separate, structured pain management can be the key to improving inadequate response to treatment of some feline diseases.

These proceedings will introduce key concepts in feline pain assessment and therapy. The associated lecture is part of a pain management track at the forum which will consist of interactive, case based presentation with panel discussions of the case material.

PAIN PATHOPHYSIOLOGY

Pain physiology and pathophysiology will be discussed in more detail at the beginning of this track. For purposes of this lecture, there are key concepts to know.

Pain is a multidimensional experience, defined by the International Association for the Study of Pain (IASP) as an unpleasant sensory and emotional experience (perception) associated with actual or potential tissue damage or is described in terms of such damage[1]. Pain is the sum total perception of the sensory, affective/emotional ("suffering") and functional components. Being a subjective perception distinguishes pain from other clinical signs that can usually be externally and concretely measured. Nociceptive pain (transient pain without tissue damage) and pathological pain (pain with tissue damage) are well understood physiological processes[2]. These types of acute pain are physically linked to a noxious stimuli, generally serve a protective function, and resolve when tissue damage heals.

Treating pain can improve response in the many patients whose clinical signs are only slowly or incompletely resolved with treatment of their primary disease. In particular, visceral cavity pain is implicated in poor appetite, ileus, nausea, and lethargy.

Chronic pain is a maladaptive pathological process in which pain may persist after tissue healing. There are several important implications to this: pain can be present when no structural lesion exists and persistent or severe, inadequately treated acute pain are precursors to chronic pain states. This is because the nervous system actually changes in response to pain stimulation (neuroplasticity).[3] For patients with chronic illness, realize that chronic illness often means chronic pain. Chronic pain will not always resolve when the primary disease process resolves. This distinguishes chronic pain from most other pathological processes we treat and explains why it requires separate treatment for full resolution of clinical signs.

Chronic pain is characterized by hyperalgesia (enhanced perception of pain in a local, often expanding area) and allodynia (pain from a stimulus that does not normally provoke pain). [1, 2] Chronic pain is often episodic, can be spontaneous, and subject to waxing and waning severity. [1,2,4] These features challenge our typical approach to assessment since this type of pain may not be elicited on physical examination. The pathological changes that occur in pain signaling and processing appear to be closely conserved in many different mammalian species studied including dogs, cats, rodents and humans, so we reasonably expect that pain states are similar in different species.

Our best defense against debilitating chronic pain states are prompt and aggressive treatment of acute pain, specifically aimed at reducing peripheral and central sensitization.[5] Chronic pain is a disease of the central nervous system that depends upon three main mechanisms: development of hypersensitivity of the spinal cord, activation of nociceptive pathways projecting to the medulla and midbrain, and activation of descending pain facilitatory pathways.[2] To reverse or change the reorganization in the nervous system requires specific therapeutic interventions separate from treatment of the original disease process. Animals that develop chronic pain states will not have spontaneous resolution of their pain when their primary disease process is improved or resolved. We are lacking methods to determine what percentage of feline patients who have chronic disease will develop persistent pain states. The numbers vary in human patients and are heavily influenced by genetic factors and past history of pain, besides the original disease process.

FELINE PAIN ASSESSMENT

When queried, most of the internists I asked cited difficulties in pain assessment as a major obstacle to treatment. As internists, we are very dependent upon a diagnostic process consisting of history taking and objective information gathering to confirm our clinical suspicions. Obviously, pain does not easily fit in this paradigm. Although we usually assess pain based on cursory palpation of the affected area, this gives us a very incomplete picture of pain. Most of us are better at pain assessment than we know, however. We observe feline behavior and posture in the outpatient setting and hospital every day. We usually intuitively know whether a cat's resting posture is abnormal, whether their facial expression is changed and whether they greet us or not when approached. When we combine those observations with objective information about appetite, eliminations, etc., they form a solid basis for pain assessment. Although it seems anthropomorphic, our intuitive and often almost subconscious assumption of how a cat is "feeling" is often based on these automatic observations. We tend to discount this information as lacking objectivity, but indeed, subjective pain assessment is all we have. Since the keys to pain assessment are consistency of observer and frequent observations, we can increase our assessment accuracy simply by observing our patients more frequently (both from a distance and while interacting with them), recording changes and asking technical staff or owners to do the same.

I am frequently asked about the use of pain scales as a way to improve accuracy of pain assessment. Several research groups have been involved in an effort to produce validated pain scales for use in dogs, but those do not yet exist for cats. Multidimensional validated scales exist to assess canine post-operative pain, chronic osteoarthritis pain and health related quality of life in cancer patients. Many hospitals have adapted simple acute pain scaling tools (like numerical rating systems and simple descriptive scales) for use in their hospitals. As with all pain scaling tools, these are not sensitive enough to be used to withhold pain management. They may, however, encourage hospital staff to more regularly observe and record trends is pain behavior. This information may be useful in setting minimum levels of pain treatment.

Dr. Sheilah Robertson of Florida and Dr. Jacqueline Reid of Glasgow are currently collaborating on the development of a validated pain scale for cats. Dr. Robertson, in particular, is involved in identifying feline specific pain behaviors. Work done on cats with osteoarthritis demonstrated that owners are able to distinguish changes in behavior associated with reduced pain in cats given analgesics compared to those given placebo.[6, 7] Until this work is completed, the use of client/patient specific outcome measures is a useful approach. This method of assessment depends upon careful history taking in order to identify behavior changes that have occurred since pain began and the return of more "normal" behaviors with therapy. For example, cats with osteoarthritis may stop jumping on the bed, but may resume with treatment of their pain. The author uses an intake form that asks owners to comment on functional mobility and social behavior (i.e. grooming, jumping, sleeping position and location, stretching, playing, etc.). The client is asked to repeat filling out this form at regular intervals during therapy.

Another useful approach is having owners keep a pain diary. This can be in the form of a standardized questionnaire or a few questions specifically tailored to the patient. It is important to limit it to a few, mostly y/n questions, and that the owners record entries about the same time every day to capture trends. A pain diary is particularly useful for clients who have trouble assessing their cat's overall progress (or lack thereof) or whose medical problems are sufficiently complex to confuse pain assessment.

Given that pain is, by definition, subjective, the most useful method of pain assessment is still clinical suspicion and response to treatment. We are often not comfortable with this approach since it counters our usual diagnostic process. Since the consequences of untreated pain are so significant for our patient's quality of life and the treatments are generally low risk, our discomfort does not dismiss our responsibility to address the potential of pain in our patients.

FELINE MUSCULOSKELETAL PAIN AND DEGENERATIVE JOINT DISEASE

For many reasons, veterinarians tend to underdiagnose joint disease in cats. Some of the problem in recognition is due to differences in both the disease process and signs in cats compared to dogs. Additionally, a problem may simply be that internists are usually focused on visceral disease rather than musculoskeletal disease. Despite this, joint disease appears to be a common problem in cats that significantly affects their quality of life.

Degenerative joint disease (DJD) including osteoarthritis (OA) of synovial appendicular joints deserves particular attention when considering chronic pain in cats. Radiographic evidence of DJD is exceedingly common in cats. The reported prevalence ranges from 90% to 22% in various studies.[8] In recent years multiple researchers and authors have written about the mismatch between radiographic prevalence of feline OA and lameness.[6-9] Investigation into this central question has led to multiple conclusions: Lameness does not appear to be the most common or reliable sign of DJD in cats, radiographic evidence of DJD is not necessarily associated with pain in cats, but maybe most importantly, radiographically normal joints may actually be painful. Joint thickening, effusion, and crepitus correlate well with radiographic DJD in contrast to pain on manipulation of joints.[9] Lascelles reminds us of an important point; radiographs should be used to confirm our suspicions of DJD, but not to guide treatment.

DJD appears to have other features in cats that distinguish it from the disease in dogs. Cats form less new bone and have less radiographic pathology with DJD than dogs do. Clinical features are also different. Changes in ability to jump up or down, changes in the height of jumps, decreased overall movement, avoiding contact with others and irritability when handled are better indicators of DJD associated pain in cats than lameness. Improvements in all these signs and return to using the litter box, grooming, stair climbing, claw sharpening and playing are indicators of treatment success.

Treatment of DJD in cats is very rewarding in spite of the complexities of NSAID therapy in this species. In 2010, The International Society of Feline Medicine and the American Association of Feline Practitioners published excellent, comprehensive guidelines on the use of long term NSAID therapy in cats.[10] The guidelines summarize recent work on the pharmacology of NSAIDs in cats and make specific recommendations on use with concurrent disease and medications. Down load these guidelines for free at www.isfm.net/toolbox/info_sheets/NSAIDs_guidelines.pdf .

The author uses combinations of Adequan® subcutaneously, acupuncture, anti-convulsants (i.e. gabapentin) and environmental modification for treatment of feline DJD. NSAIDs are incorporated into therapeutic plans as required, even in cases with other diseases typical of elderly cats. Case examples will be presented during the lecture.

COMMON FELINE DISEASES ASSOCIATED WITH PAIN REQUIRING TREATMENT

Almost all of the serious visceral and somatic diseases diagnosed in cats may be associated with enough discomfort to cause observable changes in behavior. Remember that pain behaviors are frequently what trigger veterinary care. Recognition of associated pain depends upon asking careful questions of a caretaker about the daily routine and being open to the idea that some clinical signs of disease may be direct pain behaviors. An easy place to start is in cases where appropriate therapy has not resulted in expected resolution of clinical signs. For cats, this often means persistent poor appetite, sleeping more than usual/lethargy and decreased grooming and social behavior.

While many chronic diseases are easily recognized as manifesting as pain states (i.e. osteoarthritis, stomatitis), others may be overlooked. Diseases of visceral abdominal organs cause episodic or constant discomfort, but clinicians typically assess abdominal pain only with palpation. Palpation has a low sensitivity for detecting chronic visceral pain, in part because of poor localization and the variable intensity of this type of pain. Cancer pain is multifactorial in nature; there are usually multiple, changing mechanisms of pain generation in an individual as cancer progresses. Inflammatory mammary carcinoma is a typical example of this. Besides the obvious inflammatory pain in the tumor, neuropathic pain develops as nerves are impacted and hypersensitivity and/or allodynia develops in surrounding, seemingly normal tissue. Many clinicians rarely consider treating pain as a separate problem in cases of visceral abdominal disease or even cancer until the evidence for pain is unequivocal. Since cats are very capable of hiding signs of pain, under treatment of chronic pain is likely more prevalent than it is in dogs. The challenge given here is to consider treating pain as a specific clinical problem, worthy of being addressed early in the progression of disease, before you have glaring evidence of pain. Below is a list of diseases that although obviously painful, often prompt only minimal direct treatment of pain.

Consider trying more aggressive and longer pain management for:

- Lower urinary tract disease
- Chronic pancreatitis
- IBD/chronic hepatitis
- Chronic constipation/megacolon
- Degenerative neurological disorders (e.g. diabetic polyneuropathy)
- Chronic ocular disease
- Neoplasia (e.g. fibrosarcoma, GI LSA, oral and nasal SCC)
- Periodontal disease, stomatitis

During the lecture, we will discuss case examples of treatment options for visceral abdominal pain.

FELINE POST-SURGICAL AND POST-TRAUMA PAIN

Acute pain in hospitalized cats can be managed similarly to that in dogs, with some differences in choice of drugs and doses. Clinical preferences and experiences with routine acute pain management will be discussed during the case presentation in the lecture.

Persistent post-surgical, therapeutic, or traumatic pain are clinical problems that have traditionally been difficult to manage. Examples of these cases are post-onchyectomy neuropathy and persistent pain after fibrosarcoma excision or radiation therapy. Persistent post-surgical pain syndromes can happen after any type of surgery, but are most often associated with nerve trauma (either due to the disease process or surgery or both). These cases often do not have demonstrable structural lesions and pain may not be elicited on palpation. Owners will sometimes report odd behavior, like chewing fur off a localized area unrelated to the original problem or spontaneous vocalization and running. This is usually due to neuropathic pain syndromes where the pain may be spontaneous and episodic. These cases may show signs of opioid hyperalgesia and usually respond better to treatment with anti-convulsants, anti-depressants, acupuncture, and anti-inflammatory medications. They generally require layered therapy and response to treatment may take several months, but can be very successful.

GENERAL THERAPEUTIC STRATEGIES FOR TREATING FELINE PAIN

Although the problem of pain may be more often brought to surgeons, internists are uniquely qualified to manage it. These cases often require manipulation of multiple therapies and monitoring of the impact of therapy on organ function. This is the case even in the straightforward case of feline DJD without co-morbidities. The ultimate goal of this session is to convince our colleagues that treatment of chronic pain is possible with current knowledge and will improve outcome for our patients with chronic diseases.

Key Points:

- Response to treatment may be the only "test" we have for pain; treat for *expected* pain, not confirmed pain
- Tailor acute pain management plans and drug doses to both response to test doses and time since trauma or surgery
- Co-morbidities and concomitant drug therapy does not justify withholding pain management
- Ensure that all potential discomfort causing clinical problems are addressed by the pain management plan
- Multimodal therapy is usually required to successfully treat chronic pain
- Non-pharmaceutical therapy is particularly useful for chronic pain and reduces adverse drug events
- Use nutraceuticals and disease-modifying OA agents (DMOAD) rationally and carefully
- Use adjunctive pain medications to reduce reliance on medications more likely to cause adverse drug events
- Design client and/or patient specific outcome measures and give clients strategies to assess progress
- Explicitly discuss therapeutic goals and educate clients about reasonable expectations of chronic pain management

- If therapy becomes ineffective, look for new problems
- Collect baseline CBC, chemistry profile and urinalysis before beginning drug therapy and at regular intervals after.
- Any patient receiving drug therapy that becomes sick should be examined and have lab work evaluated

REFERENCES

[1] Mersky H, et al.Classifications Of Chronic Pain, 1994, IASP Press
[2] Vanderah TW. **Med Clin N Am** 2007;91:1.
[3]Grubb T. **Top Comp Anim Med** 2010;25(1):45.
[4]Roberston S, et al **J Feline Med Surg** 2010;12:188
[5]Pozzi A, et al. **JAVMA** 2006;228(1):53.
[6]Lascelles et al. **JVIM** 2007;21:410.
[7]Bennett D et al. **J Feline Med Surg** 2009;11:997.
[8]Kerwin SC. **Top Comp Anim Med** 2010;25(1):218.
[9]Lascelles D, et al. **J Feline Med Surg** 2010;12:200.
[10]Sparks A, et al. **J Feline Med Surg** 2010;12:521.

GAINING ON PAIN: AN ONCOLOGIST'S APPROACH

Philip J. Bergman DVM, PhD, DACVIM (Oncology)
Bedford Hills, NY

INTRODUCTION

Pain is unfortunately a common sequelae in human cancers (~25-50%) and is extremely common in advanced and/or terminal human cancers (~75-95%). The exact percentages of veterinary cancer patients experiencing pain is poorly understood due to the dearth of investigations in this area to date. A recent PubMed search of "human cancer pain" returned over 67,000 citations, whereas a similar search on companion animals resulted in only ~ 300 citations, with the vast majority of these not having pain as a primary specific endpoint of investigation (e.g. "CT findings in dogs and cats with TMJ disorders").

Thankfully, investigations into the mechanisms of companion animal cancer pain have begun within the last 10 years and are now beginning to appear in the veterinary literature, but far more investigations are desperately needed. Until then, we are forced to extrapolate information from the human side, which may or may not fully translate into our companion animal patients. Even if these human studies currently fully extrapolated, it is highly likely a fair statement that we currently under-appreciate and/or poorly detect pain in companion animal patients with cancer. This may be because we may: 1) Focus more on the treatment than the pain (and forget that some cancer treatments temporarily cause pain in of themselves), 2) Poorly assess pain and dogs/cats can have an innate ability to hide pain and disease, 3) Poorly teach our clients and our staff on how to assess pain in their pets, 4) Have a variable understanding of the possible drugs and other pain relief techniques and 5) Have a limited appreciation that many different tumors are significantly associated with pain. Therefore, we also likely sub-optimally treat the pain being experienced in our companion animal patients with cancer. As this pain significantly affects quality of life and furthermore has important secondary physiologic sequelae, it is important that we have as a priority to aggressively and appropriately detect and treat cancer-associated pain in our patients.

GENERAL APPROACH

While every veterinary patient with cancer is different, the general approach one can take specifically for their pain management is quite similar. It is generally centered around the following large aspects: 1) Listen to the owner and ask open-ended questions around pain-associated behaviors at home, 2) Incorporate a pain evaluation into your standard physical examination, 3) Determine and deliver the best treatment(s), often recognizing this requires a multimodal approach and 4) Repeat by going back to #1 as the management of a veterinary cancer patient's pain will be fluid and continually need to evolve over time based on the success or failure of that cancer treatment and concurrent pain treatments.

While any tumor type may potentially cause pain, the tumor types which are most commonly associated with pain in order of descending severity are primary and metastatic bone tumors, inflammatory mammary carcinoma, oropharyngeal or nasal tumors especially with bony destruction, ulcerative and/or invasive skin tumors, CNS tumors (especially extradural), brachial plexus tumors, prostate and genitourinary tumors, intra-thoracic or intra-abdominal tumors, invasive and/or large soft tissue sarcomas and disseminated intra-cavitary tumors. It is also worthy to note that in addition to the inherent pain associated with the above tumors, iatrogenic pain can be associated with the various treatments of cancer. This is most commonly and severely associated with surgical extirpation and external beam radiation, but can also temporarily be noted with chemotherapy and some of the newer rationally targeted therapeutics such as TKI's (tyrosine kinase inhibitors). Thankfully, many of the newer radiation modalities (e.g. CyberKnife, GammaKnife and other stereotactic methods) more precisely target cancerous tissue and minimizes normal tissue dose, which reduces, but does not eliminate, the pain secondary to these radiation treatments.

The behaviors of veterinary cancer patients that can be associated with pain include decreased appetite and/or activity, changes in attitude, increased hiding, increased respirations or overt panting, poor self-grooming, adverse/aversion responses to touching and/or palpation of the affected area, self-trauma, changes in facial expressions and changes in urination and/or defecation. It is important to note that vocalizations are seen rarely in response to chronic pain in dogs and cats and we further have a duty to inform our clients of this since this is not intuitive for the lay public. That said, cats may have changes in their meows or may hiss or purr when experiencing cancer-associated pain, whereas dogs may display grunting and/or whining as their methodology of vocalization secondary to significant cancer pain. The physiologic parameters noted in cancer-associated pain may include tachypnea, tachycardia, pupil dilation and increased blood pressure and/or body temperature.

TREATMENT

Tumor-directed Therapy

The most rational strategy for managing pain in companion animal patients with cancer is to whenever possible treat the underlying tumor causing the pain. When indicated, this most quickly and least expensively is through surgical removal. When local treatments are indicated and surgery is not an option due to logistical, financial and/or emotional constraints, radiation is often a next best local treatment option. For those tumors which are systemic and/or likely to metastasize, then chemotherapy and/or rationally targeted therapies (e.g. TKI's, melanoma vaccine, etc.) are typically indicated. In those situations whereby the tumor is both locally aggressive and has a high propensity to metastasize, then combinations of local (ie Sx or RT) and systemic (ie chemo, TKI's, melanoma vaccine) treatment would be indicated to best treat the tumor and therefore most effectively reduce the patients cancer-associated pain.

Pharmacologic Therapies

The pharmacologic therapy of pain is also often utilized for the treatment of cancer-associated pain as well as cancer treatment-associated pain. A general approach to the management of cancer pain through various analgesics which has worked generally well over time is the 1986 version of the WHO (World Health Organization) "pain ladder." Step 1 of the ladder is for mild pain and utilizes a non-opioid (e.g. an NSAID or similar non-opioid) with an adjuvant non-opioid when necessary (see possible adjuvant analgesics outlined below). Step 2 of the ladder is for patients with moderate pain and combines those treatments used in step 1 alongside a low dose opioid. Step 3 of the ladder is for patients with severe pain and combines step 1 treatments with a regular dose opioid. Importantly, one should not routinely utilize two products belonging to the same category and/or mechanism of action simultaneously. For the last 25 years, this 3-tiered ladder system has worked well, but with the advent of newer analgesic treatments, a 4-tier ladder is being used more and more frequently in the treatment of human cancer-associated pain. Step 4 in this new system incorporates all the treatments used in steps 1, 2 and 3 and then also utilizes one or more treatments such as spinal stimulators, epidurals, nerve blocks, patient-assisted pain pumps, etc. The other unique aspect to the 4-tier system is that instead of unidirectional moving up the ladder steps, one may navigate the ladder bi-directionally depending on the type, level and chronicity of the pain being experienced by the patient (ie step UP or step DOWN as needed). These modifications to the original 3-step WHO ladder are very much not intended to advise against or negate the use of the original ladder, but more importantly are meant to be an upgrade or an evolution of the incredibly helpful original ladder. Graphical representations of the 3 and 4-tier pain ladders will be presented in the lecture.

The WHO ladder also makes 5 extremely simple recommendations for the practitioner to ensure the therapeutics chosen have the greatest chance for success. These simple recommendations are as follows: 1) Use oral analgesics whenever possible (unfortunately some recent data in dogs suggests oral opioids at standard doses may not equate to effective plasma concentrations), 2) Regular intervals of analgesics are imperative, 3) Analgesic dosing should be individualized to the patient, 4) the scale and breadth of analgesic use should match the patient's level of pain and 5) analgesics should be prescribed and described in detail for the medical staff and owner (in hospital and at home use, respectively).

Pharmacologic Therapy Categories

- **Non-Opioids (NSAID's & Acetominophen)** – NSAID's are excellent first line analgesics for cancer pain. All NSAID's are most likely equally efficacious analgesics, but only piroxicam and deracoxib have been described to date to have potentially anti-cancer benefits as well. Cox-2 selective NSAID's appear to have less GI side effects, but equal hepatic and/or renal side effects to more Cox-1 selective NSAID's. Use of acetaminophen in cats continues to be contraindicated and concurrent use of NSAID's and corticosteroids is contraindicated across species. Robenacoxib and meloxicam use in cats is currently FDA-approved for postoperative short term analgesia, whereas mildly and moderately longer term use has been described in the EU, respectively.
- **Opioids** – Opioids are typically indicated for moderate to severe pain and are commonly used as part of a multimodal approach outlined in the aforementioned cancer pain ladder. Side effects can be significant and include vomiting, diarrhea, sedation, dysphoria (esp northern breeds and cats) and with chronic use, constipation and hyporexia. Even though very few studies have been performed investigating the efficacy and/or toxicity of opioids in dogs or cats for chronic cancer pain, the most commonly utilized agents are oral butorphanol/codeine/tramadol/morphine, sublingual buprenorphine in cats or transdermal fentanyl. Oral butorphanol is a mixed mu antagonist and kappa agonist with a relatively short duration of analgesia (but long duration of sedation) and is therefore NOT routinely recommended for chronic cancer pain. Weak opioids such as codeine and/or tramadol may be useful for mild cancer pain. Tramadol has the advantage of being better tolerated in people compared to other opioids with similar analgesic potency. Tramadol also has norepi and serotonin reuptake inhibition, so it must be carefully utilized in patients already on serotonin reuptake inhibitors (e.g. fluoxetine, paroxetine, sertraline, trazadone, etc.) to thereby prevent serotonin syndrome. Buprenorphine is a partial mu agonist & antagonist, which when given by an oral transmucosal route is well tolerated in cats and is mildly to moderately effective for treating breakthrough pain. Mu agonists such as morphine, hydromorphone, oxymorphone and fentanyl are the best agents for moderate to severe chronic cancer pain. Unfortunately, a large percentage of dogs achieve sub-therapeutic plasma concentrations when fentanyl patches were investigated in a recent paper and cats do not convert codeine to its active form so it is not recommended for use in cats. A new liquid transdermal topical fentanyl product from Elanco (Recuvyra™) appears to offer more consistent dosing and therapeutic concentrations for postoperative orthopedic analgesia in dogs, but no studies have been reported to date using it off label for cancer pain. Methadone is being used more commonly for human chronic cancer pain due to its mu agonist and NMDA antagonist properties alongside reduced side effects and costs.
- **Adjuvant analgesics** – these are a diverse group of agents with a primary use other than analgesia but are useful in painful situations. These include corticosteroids, NMDA antagonists such as amantadine or ketamine, NK1 antagonists such as Cerenia, tricyclic antidepressants such as fluoxetine, anticonvulsants such as gabapentin, local anesthetics such as lidocaine or mexiletine, calcitonin, radiopharmaceuticals like strontium^{89} or samarium^{153}, bisphosphonates like pamidronate, α2 agonists such as medetomidine, muscle relaxants, topical agents such as lidocaine and capsaicin, benzodiazepines, duralactin, Adequan, etc. The use of adjuvant analgesics is strongly recommended with neuropathic pain due to this type of pain being comparatively less sensitive to opioids. Furthermore, consistent use of adjuvant analgesics in patients with chronic cancer pain may enable reduced dosing of other analgesics and/or allow the clinician to less quickly need to take "steps up" the aforementioned pain ladder.

Non-Pharmacologic Therapies

A variety of adjunctive non-pharmacologic therapies for pain is available and appears to be potentially increasingly more important in the multi-modal treatment of cancer-associated pain. These modalities include acupuncture, rehabilitation therapies (such as hydrotherapy, massage, stretching, superficial cold or heat therapy, etc.), electrical nerve stimulation, pulsed magnetic field therapy, cold laser therapy, ultrasound, etc. These modalities should not be used in the place of analgesics, but may complement the benefits seen with standard-of-care analgesics.

New Analgesics on the horizon?

New molecular-based analgesics are currently an extremely hot research topic. Items which may have clinical promise in the future include anti-NGF (nerve growth factor) monoclonal antibodies, intrathecal substance P, cannabinoid receptor agonists, T-type calcium channel blockers, endomorphins (natural mu ligands) and opioid peptide esters.

Can Analgesics Promote Cancer?

While this issue is not an intuitive primary concern, it is worth reviewing that some analgesics may be cancer promoting. Fentanyl has been found to promote growth of a number of cancer cell lines in vitro. Similarly, numerous opioids have been found to stimulate angiogenesis through human endothelial cell proliferation. Furthermore, cancers of the larynx, esophagus, lung, stomach and bladder have as a risk factor the chronic use of opium. Some of the very newest ion channel-based analgesics target the same ion channels involved in tumorigenesis. For now the use of analgesics when indicated easily trump any worries in current cancer patients, but more importantly this author felt it was worth highlighting this issue to promote a dialogue about the possible molecular and cellular effects of chronic analgesic use in non-cancer patients.

Cases with various types of cancer pain and the strategies for their treatment will also be presented at the lecture.

REFERENCES

1. Macfarlane PD, Tute AS, Alderson B. Therapeutic options for the treatment of chronic pain in dogs. *J Small Anim Pract* 2014 January 28.
2. Martinez SA, Wilson MG, Linton DD, Newbound GC, Freise KJ, Lin TL et al. The safety and effectiveness of a long-acting transdermal fentanyl solution compared with oxymorphone for the control of postoperative pain in dogs: a randomized, multicentered clinical study. *J Vet Pharmacol Ther* 2013 December 18.
3. Gearing DP, Virtue ER, Gearing RP, Drew AC. A fully caninised anti-NGF monoclonal antibody for pain relief in dogs. *BMC Vet Res* 2013;**9**:226.
4. Brown DC, Agnello K. Intrathecal substance P-saporin in the dog: efficacy in bone cancer pain. *Anesthesiology* 2013 November;**119**(5):1178-1185.
5. Teixeira RC, Monteiro ER, Campagnol D, Coelho K, Bressan TF, Monteiro BS. Effects of tramadol alone, in combination with meloxicam or dipyrone, on postoperative pain and the analgesic requirement in dogs undergoing unilateral mastectomy with or without ovariohysterectomy. *Vet Anaesth Analg* 2013 August 5.
6. Beckman B. Anesthesia and pain management for small animals. *Vet Clin North Am Small Anim Pract* 2013 May;**43**(3):669-688.
7. Morgaz J, Navarrete R, Munoz-Rascon P, Dominguez JM, Fernandez-Sarmiento JA, Gomez-Villamandos RJ et al. Postoperative analgesic effects of dexketoprofen, buprenorphine and tramadol in dogs undergoing ovariohysterectomy. *Res Vet Sci* 2013 August;**95**(1):278-282.
8. Lee SK, Dawson J, Lee JA, Osman G, Levitin MO, Guzel RM et al. Management of cancer pain: 1. Wider implications of orthodox analgesics. *Int J Gen Med* 2014;**7**:49-58.
9. Mearis M, Shega JW, Knoebel RW. Does Adherence to National Comprehensive Cancer Network Guidelines Improve Pain-Related Outcomes? An Evaluation of Inpatient Cancer Pain Management at an Academic Medical Center. *J Pain Symptom Manage* 2014 January 16.
10. Towler P, Molassiotis A, Brearley SG. What is the evidence for the use of acupuncture as an intervention for symptom management in cancer supportive and palliative care: an integrative overview of reviews. *Support Care Cancer* 2013 October;**21**(10):2913-2923.
11. Mercadante S. Cancer pain. *Curr Opin Support Palliat Care* 2013 June;**7**(2):139-143.
12. Santini D, Lanzetta G, Dell'Aquila E, Vincenzi B, Venditti O, Russano M et al. 'Old' and 'new' drugs for the treatment of cancer pain. *Expert Opin Pharmacother* 2013 March;**14**(4):425-433.
13. Looney A. Oncology pain in veterinary patients. *Top Companion Anim Med* 2010 February;**25**(1):32-44.
14. Gaynor JS. Control of cancer pain in veterinary patients. *Vet Clin North Am Small Anim Pract* 2008 November;**38**(6):1429-48, viii.
15. Lester P, Gaynor JS. Management of cancer pain. *Vet Clin North Am Small Anim Pract* 2000 July;**30**(4):951-66, ix.

MANAGING PAIN: A NEUROLOGIST'S APPROACH

Stephanie Kube, DVM, CCRT, CVPP
DACVIM (Neurology)
Middleboro, MA

Pain management is a major part of what we deal with every day as neurologists. Dealing with the central nervous system directly we have the ability to select our analgesics most efficiently. Therefore, knowing the pain pathways, the drugs and their mechanisms of action can be an integral component in treating our patients.

Proper pain management comes from an in depth understanding of the physiology of pain. Nociception is the transduction, conduction, and central nervous system processing of signals generated by a noxious insult. The conscious, cognitive processing of nociception results in pain.

Neuropathic pain is defined as pain that originates from injury to or involvement of the central nervous system (CNS) or peripheral nervous system (PNS). It can be associated with motor, sensory or autonomic deficits. With neuropathic pain, our patients can also present with allodynia (pain caused by a stimulus that doesn't typically cause pain) and dysesthesias (unpleasant, abnormal sensations) which increase the challenge involved with managing their pain.

When dealing with pain in neurologic patients we are also dealing with both inflammatory and incisional pain, in addition to the neuropathic pain. In addition, there can be an acute and/or chronic component.

Some of the syndromes/diseases we deal with are specific "pain" diseases, i.e. Feline orofacial pain (trigeminal neuralgia) and syringohydromyelia, we also deal with many problems directly impacting the nervous system; meningitis, brain tumors, disk disease, LS pain. In addition to those problems, there are secondary issues caused by the consequences of a CNS lesion – esophagitis, corneal ulcers, urine scalding, and abrasions. All of these issues should be taken into account when devising a protocol to treat pain in our patients.

A multifaceted approach to managing pain in the neurologic patient is recommended. Cases will be discussed.

REFERENCES

Available upon request.

PRACTICAL GASTROINTESTINAL FUNCTION TESTS AND MOLECULAR TOOLS

Albert E. Jergens, DVM, PhD, DACVIM
Ames, IA

INTRODUCTION

Chronic enteropathies are common in dogs and cats and include adverse reactions to food, idiopathic inflammatory bowel diseases (IBD), and antibiotic responsive diarrheas. While the definitive diagnosis often requires collection of mucosal biopsies for histopathologic evaluation, non-invasive laboratory tests are useful in the diagnosis of these disorders and in monitoring the response to treatment.

LABORATORY TESTS FOR SMALL INTESTINAL FUNCTION AND DISEASE

Cobalamin (Vitamin B_{12}) is an essential water-soluble vitamin and an important co-factor for many biochemical reactions. Canine and feline diets typically contain abundant cobalamin, making dietary insufficiency unlikely. The physiologic mechanism of cobalamin is complex and requires a functioning digestive system. Major disorders that interfere with cobalamin update include exocrine pancreatic insufficiency (EPI), distal or diffuse small intestinal disease, and excessive bacterial utilization of cobalamin associated with bacterial dysbiosis. Hypocobalaminemia is most often seen with long-standing and severe intestinal disease (i.e. IBD, lymphoma, or fungal enteritis) affecting the small intestine. The measurement of serum cobalamin concentrations in animals with chronic enteropathies is important in that failure to recognize this deficiency will result in delay of clinical recovery. Importantly, most serum cobalamin concentrations have been shown to be a risk factor for negative outcome in dogs with chronic enteropathies, including IBD and PLE. Appropriate dosages of cyanocobalamin are typically administered subcutaneously once weekly with serum concentrations reassessed at 4-6 week intervals.

Folate is a water soluble vitamin which is abundant in canine and feline diets making nutritional deficiencies unlikely. Folate concentrations are typically measured in concert with cobalamin and may provide information indicative of small intestinal dysbiosis (SIBO/ARD) or proximal small intestinal mucosal disease. Note that serum folate may be influenced by bacterial production of this vitamin and is therefore a less sensitive indicator of small intestinal disease vs cobalamin.

Citrulline is an amino acid synthesized predominately by intestinal enterocytes. Citrulline has been shown to be a marker of functional enterocyte metabolic mass and preliminary studies have shown that plasma citrulline concentrations are significantly decreased in dogs with parvovirus enteritis. Thus, citrulline might serve as a potential marker of spontaneous and acute intestinal dysfunction; however, additional clinical trials evaluating this marker are warranted.

EVALUATION OF INTESTINAL PROTEIN LOSS

Protein losing enteropathies (PLE) define a heterogeneous group of disorders characterized by enteric loss of plasma proteins. Common causes for PLE include inflammatory enteropathies, infectious enteritis, lymphangiectasia, intestinal neoplasia, and significant microbial imbalances. Most animals' exhibit classic clinical signs of emaciation, weight loss, and will often have panhypoproteinemia on laboratory evaluation. It is important to first rule out other causes for hypoalbuminemia due to glomerular injury and hepatic failure or insufficiency. Definitive diagnosis of PLE requires collection and histopathology assessment of mucosal biopsy specimens. An ELISA assay for the measurement of ***fecal alpha (α)1–PI*** has been validated for dogs and may be used in assessing protein loss in the GI tract. Due to the cumbersome criteria for collection of multiple fecal specimens, the laboratory assessment of α1-PI as a diagnostic test for intestinal protein loss is performed infrequently. This laboratory assay may be most useful in assessing response to therapy in dogs with protein losing enteropathies.

Decreased serum albumin has been associated with negative outcome in separate canine IBD studies. Severe IBD (and other chronic enteropathies) may predispose to hypoalbuminemia due to enteric plasma protein loss as well as nutritional deficiencies and nutrient malabsorption. Craven et al retrospectively reported low serum albumen to be associated with negative outcome (ie, poor clinical response to dietary/drug therapies or euthanasia) in 80 dogs diagnosed with IBD. Similarly, Allenspach et al showed that hypoalbuminemia was a significant risk factor for negative outcome along with severity of clinical disease, duodenal endoscopy score, and hypocobalaminemia. In our experience, dogs having severe hypoalbuminemia (ie, value <1.5 g/dL) with cavity effusion and/or salient gastrointestinal signs are at significant risk for euthanasia.

LABORATORY MARKERS FOR ASSESSMENT OF INTESTINAL INFLAMMATION AND DAMAGE

Several ***serologic markers for inflammation*** have been designed to predict the disease course and response to therapy of canine chronic enteropathies. Titers for ***perinuclear anti-neutrophil cytoplasmic antibodies (pANCA)*** have been evaluated as diagnostic markers in canine IBD in separate studies. In one study of 31 dogs, results indicated that pANCA was a highly specific marker for IBD, although the sensitivity assay was too low to be of value to be a screening test. This immunofluorescence assay was shown to be most useful in the detection of dogs with food-responsive enteropathy at the time of diagnosis. More recently, pANCA was shown to be a highly specific serologic marker versus ANA for differentiating dogs with IBD from other gastrointestinal disorders. Lastly, others have shown that pANCA was of value as a diagnostic marker for familial PLE in Soft-Coated Wheaten Terriers. Unfortunately, this assay is not readily available in North America and is presently limited for use in research investigations.
C-reactive protein (CRP) is a sensitive, but nonspecific inflammatory marker (ie, acute phase protein produced by the liver in response to inflammation) shown to correlate with moderate-to-severe clinical disease activity (CIBDAI) in canine IBD. This marker is the best characterized inflammatory marker where no fewer than four different studies have shown that dogs with IBD have

heterogeneous serum elevations of this marker. New studies have evaluated the index of variability for CRP, suggesting that the use of a population-based reference range is not appropriate for evaluating changes in CRP in individual patients. Therefore, serial CRP measurements should be performed and modest changes in serum CRP may indicate the severity of GI disease.

Fecal calprotectin and other S100 gene proteins have been reported to be increased in the serum of human patients and dogs with IBD. Calprotectin and S100 A8/A9 and A12 are calcium-binding proteins found predominantly within neutrophils and other immune cells in inflamed mucosa. Fecal concentrations of these proteins have been recently shown to be increased in humans and dogs with IBD, when compared to healthy controls. Importantly, the S100 A12 fecal marker has greater sensitivity for the detection of intestinal inflammation and has been positively associated with endoscopy and clinical scores. ***The expression of P-glycoprotein*** (P-GP) in mucosal lymphocytes has also been investigated and shown to be up-regulated in IBD dogs treated with prednisolone. This earlier study showed that in dogs with steroid-responsive enteropathy, low P-GP expression was associated with a good therapeutic response.

MOLECULAR TOOLS OF PRACTICAL VALUE FOR CLINICIANS

Polymerase chain reaction for antigen receptor rearrangements (PARR) amplifies variable T or B cell antigen genes and is used to detect the presence of a clonally expanded population of lymphocytes. This molecular technique is useful in differentiating severe mucosal inflammation (severe IBD) from alimentary lymphoma. This test is often performed with histopathology (routine H&E stains) and immunophenotyping return ambiguous results. Data derive from feline studies shows that this molecular technique is highly sensitive for differentiating LSA vs IBD; however, further studies are needed to assess the value and sensitivity of this technique in dogs.

Fluorescence in situ hybridization (FISH) is a molecular technique used to identify microbial populations in tissues. FISH analysis has been applied to a variety of gastrointestinal tissues, including the pancreas, liver, and alimentary tract in dogs and cats. This technique utilizes culture independent analysis of bacterial 16S or 23S rDNA genes targeted by DNA probes tagged with a fluorophore. Bacterial populations are then imaged via fluorescence microscopy. Microbial imbalances in diseased gastrointestinal tissues (eg, Helicobacter gastritis, idiopathic IBD, feline cholangitis) of dogs and cats have been demonstrated using FISH techniques. Specifically, FISH has identified an association between gastrointestinal inflammation and a shift in the microbiome by means of bacterial translocation (eg, Helicobacter gastritis) or dysbiosis of mucosa-associated microbial populations IBD, granulomatous colitis of Boxers).

REFERENCES

Allenspach K, Bergman PJ, Sauter S, et al. P-glycoprotein expression in lamina propria lymphocytes of duodenal biopsy samples in dogs with chronic idiopathic enteropathies. J Comp Pathol. 2006 Jan;134(1):1-7.

Suchodolski JS, Dowd SE, Wilke V, et al. 16S rRNA gene pyrosequencing reveals bacterial dysbiosis in the duodenum of dogs with idiopathic inflammatory bowel disease. PLoS One. 2012;7(6):e39333. doi: 10.1371/journal.pone.0039333. Epub 2012 Jun 15.

Allenspach K, Wieland B, Gröne A, Gaschen F. Chronic enteropathies in dogs: evaluation of risk factors for negative outcome. J Vet Intern Med. 2007 Jul-Aug;21(4):700-8.

Mancho C, Sainz Á, García-Sancho M, Villaescusa A, Rodríguez-Franco F. Evaluation of perinuclear antineutrophilic cytoplasmic antibodies in sera from dogs with inflammatory bowel disease or intestinal lymphoma. Am J Vet Res. 2011 Oct;72(10):1333-7.

Dossin O, Rupassara SI, Weng HY, et al. Effect of parvoviral enteritis on plasma citrulline concentration in dogs. J Vet Intern Med. 2011; 25: 215-21

PROBIOTICS IN CANINE IBD- THE SCIENCE AND THE NITTY GRITTY PRACTICALS

Karin Allenspach DECVIM-CA, Royal Veterinary College, University of London, UK

INTRODUCTION

Probiotics are defined by the World Health Organisation (WHO) to be "live microbes which when administered in adequate amounts confer a health benefit to the host". They are frequently, though not necessarily, commensal bacteria. The probiotic genera most often used in small animal practice are *Enterococcus, Lactobacillus* and *Bifidobacterium*, which are all normal inhabitants of the colonic flora.

The human gastrointestinal tract contains > 100 trillion organisms comprising approximately 500 species, many of which have not been identified. These bacteria are parasites, commensals, or mutualists. Within hours after birth, the gastrointestinal tract is populated with numerous bacterial species, but bifidobacteria predominate in mammals. The change to a more complex diet after weaning allows the microbial population to diversify. Critical differences exist between resident bacteria of humans and other animals. Each organism strain appears to be associated with specific effects on other organisms and on the host, and specific cytokine profiles and biological effects result from these interactions. Gastrointestinal bacteria have coevolved with their host and have become highly specialized over time. Composition of fecal microbiota in individual human beings and animals is unique but stable over time. Recent studies show evidence that a healthy composition of the microbiome in the intestine is important for the normal development of the gastro-intestinal immune system, and especially for adequate production of mucosal IgA.

The composition of the microbiome has been shown to be significantly altered in specific diseases in human beings as well as in dogs and cats. Specifically, changes in the numbers of *Enterobacteriaceae* species have been associated with inflammatory bowel disease in people as well as in dogs. It has therefore been hypothesized that the microbiome composition can be influenced by feeding probiotics, by shifting the microbiome to a more healthy or "normal" composition.

PROPOSED MECHANISM OF ACTION OF PROBIOTICS

Probiotics are able to modulate the host's defences. They may also have direct effects on microbial products like toxins, host products e.g. bile salts and food ingredients[1]. Probiotic products are recognized by host cells expressing pattern recognition receptors such as Toll like receptors (TLRs) on their surface. Binding of bacteria to the receptors triggers an intracellular signalling cascade leading to immune modulation. [13] TLR binding has been shown to be species- and activity-specific, with the most important receptors in this family including TLR2, TLR9 and NOD2. However, some in vitro and in vivo effects of probiotics were found not to require TLRs so far. Deficiency in MyD88 abolished the investigated activities, thus it remains to be established whether other TLRs or a MyD88-dependent, but TLR-independent, pathways are involved[2]. It has also been proposed that probiotics suppress the NF-κB activation pathway[3]. They have an effect on the levels of IL-12, the IL-10:IL-12 ratio, IL-6, TNF-α or IL-8[3].

Several *in vitro* studies have shown that probiotics can modulate the balance of Th1, -Th2- and Th17 –derived cytokines in the intestine. To this end, much attention has been paid to the abilities of probiotics to induce the development of T regulatory cells (T_{regs}). For instance, two *Lactobacillus strains* have been shown to stimulate human monocyte-derived immature dendritic cells (Dcs) and to trigger their differentiation into regulatory DCs; these cells were then able to induce IL-10 producing Tregs[4].

PROBIOTICS IN DOGS WITH IBD

Probiotics have to date mainly been investigated in cases of acute diarrhea in dogs and cats. Several studies have shown a mild effect in such cases, and it is probably safe to say that in many mild cases their use is to date preferred over the non-justified use of antibiotics. However, their efficacy in dogs with chronic diarrhea is much less well documented.

One study performed in 2005 used a cocktail of 5 different lactobacillus strains in dogs with food-responsive chronic enteropathy. This probiotic formulation never reached the market and so is not available commercially. 10 dogs received placebo in addition to dietary treatment for 4 weeks, and 11 dogs received probiotics in addition to diet. Duodenal IL-10 mRNA levels decreased and colonic IFN-gamma mRNA levels increased after probiotic treatment, but cytokine patterns were not associated with the clinical features irrespective of the dietary supplementation. No clinical effect of the probotic treatment could be discerned, although all dogs improved significantly on treatment[5].

In another more recent study which is only available as an abstract so far, the human probiotic cocktail VSL#3 (VSL Pharmaceuticals, Fort Lauderdale, FL; consisting of 4 strains of lactobacillus: *L. casei, L. Plantarum, L. acidophilus, L. delbrueckii subsp. bulgaricus,* 3 strains of Bifidobacterium: *B. longum, B. breve, B. infantis,* and 1 strain of *Streptococcus salivarius subsp. thermo- philus*) was udes in a study on IBD in dogs. This probiotic formulation has previously been shown to induce IL-10 production in dendritic cells in the peripheral blood and in *ex vivo* cultured intestinal biopsies from people with ulcerative colitis (UC)[6]. The extrapolation of these *in vitro* findings to the patient is supported by the clinical improvement in people with UC treated with VSL#3 versus placebo[6]. Interestingly, VSL#3 has species-spanning effects on reducing intestinal inflammation, as noted in several mouse models of IBD[7-9]. This study therefore evaluated the efficacy of VSL#3 in 10 dogs with chronic relapsing IBD (n=5 dogs on placebo alone and n=5 dogs on VSL#3), and was shown to have a moderate effect on clinical signs as measured by (CIBDAI)[10]. This clinical amelioration of signs was accompanied by an increase in TGF-beta and FoxP3+ cells in the mucosa of dogs 2 months after treatment, implying that T regulatory cells and an anti-inflammatory cytokine milieu could be responsible for the observed clinical effect.

We have recently performed a clinical trial on the efficacy of *Enterococcus Faecium (EF)* (Symbiotics) in dogs with FRD. After studies *in-vitro*[11], in canine whole blood and *ex-vivo* cultured canine duodenal biopsies, the aim of the clinical trial was to investigate the *in-vivo* effects of EF on gene expression in duodenal and colonic biopsies of dogs diagnosed with FRD[12]. Dogs were recruited to receive EF in a double-blinded, placebo-controlled manner in addition to an exclusion diet based on a hydrolysed protein. Duodenal and colonic biopsies were taken before and 6 weeks after start of treatment, and RT-PCR assays performed for the following genes: TLR2/4/5/9, IL-17A, IL-22, IL-23p19, RORC, IL-2, IL-12p35, TNFα, IL-4, IFNγ, IL-10, TGFβ, IL-1β, IL-18,

NLRP3, casp-1 and PPARγ. Relative quantification of gene expression was performed using four reference genes for the duodenum and 5 for the colon. Twelve dogs finished the clinical trial, 7 received EF and elimination diet and 5 dogs elimination diet and placebo. Six weeks after the beginning of treatment, significantly reduced gene expression was found in duodenal tissues of the placebo group for TLR4, TNFα, RORC, IL-1β, IL-18 and PPARγ. In the colon, significant increases in the expression of all tested genes apart from Th17- and Th1 cell cytokines and TGFβ, RORC and casp-1 were seen in the placebo group. There was therefore a site-specific effect of treatment in canine FRD: down-regulation of several genes of the innate and adaptive immune system were found in the duodenum, whereas gene expression for nearly all examined genes was increased in colonic tissues of the placebo group, despite clinical improvement seen in all dogs. There was no effect of the EF on the clinical activity index as measured by canine chronic enteropathy clinical activity index (CCECAI).

It is therefore possible that similar to findings in people with IBD, clinical effects beyond the treatment with diet are difficult to elicit using probiotics, possibly because the effect would be too small to demonstrate with numbers typical for a clinical trial. However, further treatment trials are warranted to evaluate the efficacy of other currently licensed probiotics in chronic diarrhea in dogs, and in dogs with more severe clinical signs.

REFERENCES

1. Oelschlaeger TA. Mechanisms of probiotic actions - A review. Int J Med Microbiol 300:57-62. 2. Borchers AT, Selmi C, Meyers FJ, et al. Probiotics and immunity. J Gastroenterol 2009;44:26-46. 3. Shida K, Kiyoshima-Shibata J, Kaji R, et al. Peptidoglycan from lactobacilli inhibits interleukin-12 production by macrophages induced by Lactobacillus casei through Toll-like receptor 2-dependent and independent mechanisms. Immunology 2009;128:e858-869. 4. Tsuji M, Komatsu N, Kawamoto S, et al. Preferential generation of follicular B helper T cells from Foxp3+ T cells in gut Peyer's patches. Science 2009;323:1488-1492. 5. Sauter SN, Benyacoub J, Allenspach K, et al. Effects of probiotic bacteria in dogs with food responsive diarrhoea treated with an elimination diet. J Anim Physiol Anim Nutr (Berl) 2006;90:269-277. 6. Ng SC, Plamondon S, Kamm MA, et al. Immunosuppressive effects via human intestinal dendritic cells of probiotic bacteria and steroids in the treatment of acute ulcerative colitis. Inflamm Bowel Dis 2010;16:1286-1298. 7. Bassaganya-Riera J, Viladomiu M, Pedragosa M, et al. Probiotic bacteria produce conjugated linoleic acid locally in the gut that targets macrophage PPAR gamma to suppress colitis. PLoS One 2012;7:e31238. 8. Bassaganya-Riera J, Viladomiu M, Pedragosa M, et al. Immunoregulatory mechanisms underlying prevention of colitis-associated colorectal cancer by probiotic bacteria. PLoS One 2012;7:e34676. 9. Reiff C, Delday M, Rucklidge G, et al. Balancing inflammatory, lipid, and xenobiotic signaling pathways by VSL#3, a biotherapeutic agent, in the treatment of inflammatory bowel disease. Inflamm Bowel Dis 2009;15:1721-1736. 10. Rossi G, Pengo, G., Caldin, M., Steiner, J.M., Suchodolski, J.S. Effects of the Probiotic VSL#3 on Microbiological, Histological, and Immunomodulatory Parameters in Dogs with Inflammatory Bowel Disease. Journal of Veterinary Internal Medicine 2012;Abstract. 11. Schmitz S, Henrich M, Neiger R, et al. Comparison of TNFalpha responses induced by Toll-like receptor ligands and probiotic Enterococcus faecium in whole blood and peripheral blood mononuclear cells of healthy dogs. Vet Immunol Immunopathol 2013. 12. S. Schmitz DW, K. Allenspach. A pilot study on the effect of probiotic treatment in gene expression in intestinal biopsies form dogs with food-responsive chronic enteropathy. Abstract, ECVIM Congress 2013, Liverpool, UK.

A NEW CLINICAL PROTOCOL FOR CONTRAST VIDEOFLUOROSCOPIC ASSESSMENT OF DYSPHAGIA IN FREELY BEHAVING DOGS

Teresa E. Lever, PhD, Joan R. Coates, DVM, MS, Mitchell Allen, BS, Vanessa Gaiser, MHS
Columbia, MO

The current protocol for contrast videofluoroscopic assessment of dysphagia in dogs requires physical restraint in sternal or lateral recumbency to minimize head/body movement and retain positioning within the fluoroscopy field of view (Pollard et al. 2000 & 2007; Bonadio et al. 2009, Pollard 2012). Restraint is accomplished by manually holding the dog's head/body against the examination table, which results in substantial radiation exposure to individuals performing the procedure. Barium sulfate is the oral contrast agent commonly used for this test, which is administered in two forms: thin liquid barium (60% w/v) and barium-soaked kibble. Self-feeding from a bowl is encouraged but often refused by dogs, likely due to a combination of the aversive taste/smell of the contrast agent and anxiety/struggling from being restrained in an unfamiliar environment. As a result, syringe and hand feeding methods are routinely necessary, which further contributes to anxiety and struggling behaviors and causes profound head movements that significantly degrade image quality. Moreover, the majority of dogs let the liquid dribble from the mouth and/or spit the kibble out. The end result is an examination of unnatural feeding activity rather than the dog's typical feeding behaviors and swallow function.

A restraint device was recently designed to position dogs in sternal recumbency during this test (Bonadio et al. 2009). The device has a rectangular, polycarbonate frame that is open on both ends. One side panel is adjustable to accommodate dogs of different widths. After a dog assumes sternal recumbency within the device, the side panel is positioned so that it squeezes against the dog to prevent escape; the head remains unrestrained for feeding purposes. Struggling behaviors are still common in this device; therefore, head and body motion remains problematic. Furthermore, radiation exposure to individuals cannot be minimized because syringe and hand feeding methods are still required. To overcome these limitations, we have developed an innovative clinical protocol for contrast videofluoroscopic assessment of dysphagia in freely behaving dogs.

Our protocol consists of three components: 1) a custom-designed kennel that permits self-feeding from a bowl while standing unrestrained in a confined space, 2) special formulated recipes that mask the aversive taste/smell of oral contrast agents, thereby eliminating the need for force-feeding techniques in food-motivated dogs, and 3) a 12-step test protocol to quickly assess swallowing of several consistencies included in a typical canine diet (e.g., thin liquid, ground/pureed food, chunky wet food, and dry kibble), without the need for behavioral conditioning. The combined effect produces a comfortable, low stress, no struggle, self-feeding examination environment that more closely resembles a dog's daily mealtime routine. Head and body movements are sufficiently self-stabilized while the dog's muzzle is engaged in the food bowl, thereby permitting high quality videofluoroscopic and spot film imaging. Additionally, individuals performing the procedure can administer the examination hands-free, which dramatically reduces the risk of radiation exposure. To date, we have used this protocol with approximately 30 healthy and 15 dysphagic dogs of various breeds and ages from the local community and an existing research colony. Testing was performed without sedation after overnight fasting of food to further motivate participation; water was not restricted. Test items were standardized for contrast density, volume, and order of presentation. The self-feeding nature of this test prevented standardization of bolus volume within and between dogs. We expect bolus volumes during spontaneous eating and drinking may be considerably larger than bolus volumes produced by syringe and hand-feeding methods. As a result, the clinical signs of dysphagia identified in the home environment may be more readily evoked when using our self-feeding test protocol. Moreover, frame by frame analysis of the digital videos revealed several swallow measures that were not quantifiable using the current restraint protocol. We are in the process of developing a multi-center study to further refine and standardize our self-feeding protocol for ease of use in research and clinical practice. While we realize this protocol may not be suitable for all dogs, it does provide a beneficial alternative to the currently used restraint protocol, especially when testing food-motivated dogs. Future directions include adapting this protocol for research and clinical use with other companion animals, such as cats, small ruminants, and foals.

REFERENCES

Pollard et al. **Veterinary Radiology & Ultrasound** 2000;409-412. Pollard et al. **Veterinary Radiology & Ultrasound** 2007;48(3):21-226. Bonadio et al. **J Vet Intern Med** 2009;23:801-805. Pollard **ISRN Veterinary Science** 2012;15 pages.

RATIONAL APPROACH TO DISORDERS OF THE CANINE ESOPHAGUS: 15 YEARS OF EXPERIENCE

Stanley L. Marks, BVSc, PhD, DACVIM (Small Animal Internal Medicine, Oncology), DACVN
Davis, CA

INTRODUCTION

The normal swallowing reflex is a four-stage process, characterized by the oral preparatory phase, oral phase, pharyngeal phase, and esophageal phase.[1] Disorders affecting the oropharynx and esophagus are relatively common in dogs, and are typically associated with dysphagia. Abnormalities of the pharyngeal phase of swallowing are associated with pharyngeal weakness secondary to neuropathies or myopathies, pharyngeal tumors or foreign bodies, and obstruction of the proximal esophageal sphincter (PES) secondary to hypertrophy of the cricopharyngeus muscle. Synchrony between constriction of the pharyngeal muscles and relaxation of the cricopharyngeus muscle is essential to allow passage of the bolus into the esophagus. The esophageal phase is involuntary and begins with the relaxation of the PES and movement of the bolus into the esophagus.

Dogs with an abnormal oral phase of swallowing typically have difficulty with prehension and/or abnormal transport of a bolus to the tongue base. These disorders can often be diagnosed on physical or neurological examination, or by watching the animal eat. Oropharyngeal dysphagias affecting the pharyngeal phase of swallowing can be more challenging to diagnose, and often present with non-specific signs such as gagging, retching and the necessity for multiple swallowing attempts prior to the successful movement of a bolus into the proximal esophagus.[2,3] These patients have abnormal transport of bolus from the oropharynx to the hypopharynx or hypopharynx to the esophagus. Cricopharyngeal dysphagia is associated with the abnormal transport of a bolus through the PES. Signs are similar to those seen with pharyngeal disorders. Cricopharyngeal dysphagia may result from incomplete or lack of opening of the PES (true cricopharyngeal achalasia) or from abnormal timing of PES opening (cricopharyngeal dyssynchrony) such that PES opening lags behind bolus presentation from pharyngeal contraction.[2,3] In addition, a subset of dogs have combined disorders affecting both the pharyngeal and cricopharyngeal phases of swallowing, necessitating the implementation of videofluoroscopic swallow studies to accurately diagnose this disorder.

ASSESSMENT PROCESS

Assessment of dogs with signs of feeding and/or swallowing disorders encompass multiple dimensions that include, but may not be limited to: a) review of the signalment; b) review of drug history and history of recent anesthesia; c) physical examination (prefeeding assessment); d) clinical feeding and swallowing evaluation; and e) instrumental evaluation of swallowing.

Signalment

The importance of the animal's signalment cannot be overemphasized. Puppies can be diagnosed with a variety of congenital functional disorders affecting swallowing, including pharyngeal weakness, cricopharyngeal dysphagia, esophageal dysmotility, and sliding hiatal hernias. Breeds that have a hereditary predisposition or a high incidence of oropharyngeal or esophageal dysphagia include the Golden Retriever, Cocker spaniel, Springer spaniel, Maltese, and miniature Dachshund (cricopharyngeal dysphagia), English bulldog, French bulldog, Chinese Char Pei, Boston terrier, and Pug (sliding hiatal hernia), Bouvier des Flandres, Golden Retriever, and Cavalier King Charles Spaniel (muscular dystrophy), Boxer and Newfoundland (inflammatory myopathy), and German Shepherd (vascular ring anomaly). In addition, large breed dogs are predisposed to megaesophagus and masticatory muscle disorders.

Physical Examination

Physical examination of the animal must include careful examination of the pharynx using sedation or anesthesia if necessary. The pharynx and neck should be carefully palpated for masses, asymmetry, or pain. The chest should be carefully auscultated for evidence of aspiration pneumonia. Evaluation of cranial nerves should be performed including assessment of tongue and jaw tone, and abduction of the arytenoid cartilages with inspiration. Complete physical and neurological examination may identify clinical signs supportive of a generalized neuromuscular disorder, including muscle atrophy, stiffness, or decreased/absent spinal reflexes. The gag reflex should be evaluated by placing a finger in the pharynx; however, the presence or absence of a gag reflex does not correlate with the efficacy of the pharyngeal swallow nor adequacy of deglutitive airway protection.[4]

Observation of Animal Eating and Drinking

The importance of the clinician carefully observing the dysphagic animal while it is eating (kibble and canned food) and drinking in the hospital cannot be overemphasized, and is instrumental in helping to localize the problem to the oral cavity, pharynx, or esophagus. Dogs with an abnormal oral phase of swallowing typically have difficulty with prehension and/or aboral transport of a bolus to the tongue base, and these disorders can often be diagnosed by watching the animal eat. Oropharyngeal dysphagias affecting the pharyngeal phase of swallowing can be more challenging to diagnose, and often present with non-specific signs such as gagging, retching and the necessity for multiple swallowing attempts prior to the successful movement of a bolus into the proximal esophagus. These patients have abnormal transport of bolus from the oropharynx to the hypopharynx or hypopharynx to the proximal esophagus. Cricopharyngeal dysphagia is associated with the abnormal transport of a bolus through the PES, and signs are similar to those seen with pharyngeal disorders.

Cervical and Thoracic Radiographs

The pharynx of healthy animals is evident on radiographs because it is air filled. The size of the air-filled space can be decreased by local inflammation or neoplasia, laryngeal edema, or elongation of the soft palate. Pharyngeal size can also appear increased with dysfunction of the pharynx or upper esophageal sphincter, chronic respiratory (inspiratory) disease, and chronic severe megaesophagus. The normal esophagus is not visible on survey radiographs. An exception occurs following aerophagia due to excitement, nausea, dyspnea, or anesthesia.

Neuromuscular Minimum Database

A comprehensive database is warranted in animals with OPD, and should consist of a complete blood count, serum chemistry including creatine kinase (CK) and electrolyte concentrations, urinalysis, evaluation of thyroid function, and the acetylcholine receptor (AChR) antibody titer for acquired myasthenia gravis. A persistently elevated CK could be an indication of an inflammatory myopathy, whereas markedly elevated CK concentrations may suggest a necrotizing or dystrophic myopathy. A normal CK does not rule out a myopathy, particularly when the myopathy is focal (masticatory muscle myositis) or in the chronic stage of disease. Acquired myasthenia gravis is an important neuromuscular cause of OPD, and can be associated with focal signs including pharyngeal, esophageal, and laryngeal weakness without clinically detectable limb muscle weakness. Pharyngeal weakness, as the only clinical sign of myasthenia gravis, was described in 1% of the myasthenic dogs.[5]

Videofluoroscopic Swallow Study

Contrast videofluoroscopy involves real time image capture of the animal as it is swallowing liquid barium or barium-soaked kibble, and is one of the most important procedures for assessing the functional integrity of the swallow reflex and esophageal motility. Videofluoroscopy is used to determine the normal sequence of events that make up a swallow and to measure the timing of these events in relation to one another. Additionally, the movement of certain anatomic structures is measured in relation to a fixed point to further assess function. Swallowing events that occur out of sequence, at inappropriate times or with reduced vigor can cause significant morbidity. One problem with videofluoroscopy is that animal positioning is not standardized in veterinary medicine. Alterations in body position (sternal versus lateral recumbency) do not appear to affect measurements of pharyngeal constriction ratio or the timing of swallowing in healthy dogs, however; cervical esophageal transit is significantly delayed when dogs are imaged in lateral recumbency. Swallow studies performed with the dog in sternal recumbency are significantly more likely to result in generation of a primary peristaltic wave for both liquid and kibble boluses. Thus, it is important to recognize that the retention of liquid or kibble boluses in the cervical esophagus may not be considered abnormal when clinically ill dogs are imaged in lateral recumbency because this may be related to body position. The fluoroscopic swallow study typically involves assessment of 5 swallows each of 5-10mL liquid barium (60% w/v) followed by 5 swallows of kibble soaked in barium.

The timing of the swallow can be easily determined when viewing the swallow study frame by frame in the video, where each frame represents 1/30th of a second in the NTSC system, the analog television system used in the US. The frame in which the epiglottis is observed to close over the larynx is considered as the starting point for all time measurements, and frames are counted until the observation of maximal contraction of the pharynx, opening of the PES and closing of the PES. The swallow is considered completed when the epiglottis is observed to re-open, and usually takes 5-6 frames in healthy dogs.[2] More recently, a contrast videofluoroscopy method for quantifying pharyngeal contractility has been described in the dog.[6] The pharyngeal constriction ratio (PCR) is calculated by dividing the pharyngeal area at maximum contraction by the pharyngeal area at rest. As pharyngeal contractility diminishes, the ratio approaches 1.0.[6] This simple procedure provides important information regarding the strength of pharyngeal contraction in dysphagic dogs, and facilitates the improved selection of dogs diagnosed with cricopharyngeal dysphagia for surgical intervention.

Laryngoscopy/Pharyngoscopy and Esophagoscopy

Thorough laryngeal examination is important in all animals with OPD to rule out laryngeal paralysis associated with a polyneuropathy. Geriatric, large-breed dogs can experience a progressive generalized neuropathy, with associated pharyngeal weakness, OPD, and esophageal dysmotility.[7] Pharyngoscopy and esophagoscopy provide anatomic information about the structures involved in the oropharynx and esophagus, but both procedures are of limited diagnostic utility for evaluating functional disorders in anesthetized animals. Unsedated transnasal videoendoscopy is an easily accomplished and useful procedure in people that is often performed as an outpatient procedure. The author has assessed the feasibility of fiberoptic endoscopic evaluation of swallowing (FEES) via transnasal intubation in fully awake dogs; however, the procedure is limited to larger dogs that can be readily restrained for the procedure. Although esophagoscopy and survey radiography provide anatomic information about the structures involved with the swallowing reflex and esophagus, neither provides information about esophageal function. This is an important limitation of these diagnostic procedures, particularly in animals that are dysphagic secondary to dynamic disorders such as cricopharyngeal disease or esophageal dysmotility. Esophagoscopy is helpful for diagnosing esophagitis, esophageal strictures (that can be missed on barium swallow studies), and hiatal hernias.

Esophageal Manometry

Esophageal manometry measures pressure within the esophageal lumen and sphincters, and provides an assessment of the neuromuscular activity that dictates function in health and disease. Manometric techniques have improved in a step-wise fashion from a single pressure channel to the development of high-resolution manometry (HRM) with up to 36 pressure sensors.[8] Advances in computer processing allow pressure data to be presented in real time as a compact, visually intuitive "spatiotemporal plot" of esophageal pressure activity. This spatiotemporal plot provides objective measurements of the forces that drive food and fluid from the pharynx to the stomach. This diagnostic modality is limited to academic institutions conducting clinical research studies on esophageal motility and reflux. HRM has been successfully implemented in awake dogs following intranasal insertion[9]

Esophageal pH/Impedance Testing

Esophageal pH/impedance testing is a useful diagnostic tool that is used to diagnose acid and non-acid reflux in animals with suspected gastroesophageal reflux (GER), unexplained esophagitis, or hiatal hernias. The technology of esophageal pH testing has advanced tremendously in recent years, and clinicians have several choices when selecting esophageal pH probes. The catheter-free Bravo pH Monitoring System from Medtronic is the first catheter-free system used to measure esophageal pH levels in human patients who are suspected of having GER, and is revolutionizing the way esophageal pH testing is done, because it is allows people to maintain their regular diet and activities during pH testing. The Bravo system is an alternative to the traditional pH trans-nasal pH catheter that can cause patient discomfort, and is easily dislodged by dogs and cats if the animal is not closely monitored. The Bravo System consists of two primary components: a) A small pH capsule the size of a gelcap that is attached to the wall of the esophagus and transmits data to a receiver; and b) a pager-sized receiver worn by the patient or animal that receives pH data from the Bravo capsule. After the test is completed, data from the receiver is downloaded to pH analysis software (Polygram™ Net pH Analysis Software) using infrared technology. The main disadvantage of the Bravo system is that one can only record esophageal pH, and the system does not utilize impedance technology that allows one to measure both acid and non-acid reflux. Esophageal pH testing has been extensively utilized in anesthetized dogs in an effort to identify risk factors for GER, and the effects of body position and prokinetic agents on GER.[10]

Electrodiagnostic Testing

Electrodiagnostic evaluation, including electromyography and measurement of motor and sensory nerve conduction velocities, does not provide a specific diagnosis in most cases, but can provide important information as to the severity, distribution, and character of a myopathic or neuropathic disease process, and assist with guiding the optimal anatomic site for biopsy. Electrodiagnostic testing should also include evaluation of the pharyngeal muscles and tongue. The health status of the animal must be taken under consideration as the procedure is performed under general anesthesia.

Muscle and Nerve Biopsies

Muscle and nerve biopsies are usually integral to reaching a specific diagnosis. This procedure should be performed after the serum AChR antibody titer has been determined to be negative. If the onset of clinical signs is recent and the antibody test negative, retesting in 4-6 weeks is suggested, as a significant number of dogs with early clinical signs can have antibody titers below the detection limits of the test at initial testing, but test positive 4-6 weeks later. Muscle biopsies are usually obtained from a large proximal pelvic limb such as the vastus lateralis or a thoracic limb such as the triceps muscle; however, biopsies of the pharynx and cricopharyngus muscle should also be obtained in dogs with oropharyngeal dysphagia. Muscle biopsies should ideally be performed relatively early in the disease process when warranted before irreversible muscle fibrosis and myofiber loss is extensive.

The veterinary profession has made tremendous strides in our ability to evaluate and diagnose disorders of the upper gastrointestinal tract in dogs. The diagnosis of cricopharyngeal and esophageal functional disorders is challenging, but can be facilitated with a comprehensive and structured approach, including observation of the animal eating and drinking, a thorough physical and neurological examination, evaluation of survey radiographs of the chest and neck, and selection of optimal advanced diagnostic modalities based on initial diagnostic findings, clinical suspicion, and availability of these tools.

TREATMENT OF ESOPHAGEAL DISORDERS

Cricopharyngeal Dysphagia

Definitive treatment for cricopharyngeal dysphagia has been reported with myotomy or myectomy of the cricopharyngeal muscles; however, a recent study documented a high failure rate in 14 dogs with cricopharyngeal dysphagia following cricopharyngeal myotomy and myectomy.[11] Clinically, six dogs showed no improvement after surgery, three of which worsened postoperatively. Eight of the 14 dogs were euthanized due to problems related to cricopharyngeal dysphagia, including persistent dysphagia (8 dogs) and aspiration pneumonia (5 dogs). This study underscored the importance of critically assessing dogs for surgical intervention, and has revised our therapeutic approach to dogs with cricopharyngeal dysphagia. Likely reasons for the high failure rate observed in the dogs of the study following myotomy or myectomy of the cricopharyngeus muscle included incomplete work-up of dogs prior to surgical intervention, resulting in failure to accurately diagnose dogs with focal myasthenia gravis, laryngeal paralysis, and other underlying polyneuropathies/myopathies. In addition, a few dogs with severe aspiration pneumonia had exacerbation of their pneumonia following the surgical procedure, and these dogs should have been better stabilized prior to surgery. Finally, dogs with concurrent pharyngeal weakness and cricopharyngeal dysphagia likely had exacerbation of their dsyphagia following the myotomy or myectomy procedure. For these reasons, all dogs with cricopharyngeal dysphagia are currently worked up comprehensively to assess pharyngeal function via fluoroscopy, rule out concurrent or systemic polyneuropathies/myopathies. Dogs that are diagnosed with underlying neuropathies are managed conservatively with alterations of feeding practice or low-profile gastrostomy devices if the underlying neuropathy or myopathy cannot be specifically managed.

In veterinary medicine, the standard surgical approach for myotomy or myectomy has remained constant over the years. The cricopharyngeal and thyropharyngeal muscles can be approached by either a standard ventral midline approach, with 180° rotation of the larynx on its longitudinal axis or via a lateral approach with 90° rotation of the larynx.[12] Cricopharyngeal myotomy involves transecting the cricopharyngeal muscle to the level of the pharyngeal mucosa. Endoscopic CO_2 laser cricopharyngeal myotomy is being increasingly utilized in people, and this procedure has been shown to be safe and effective, with decreased anesthesia time and morbidity compared with the more traditional transcervical cricopharyngeal myotomy.[13] This procedure has been performed in a handful of dogs at UC Davis and is less invasive than the traditional open myectomy procedure; however, the potential for recurrence of dysphagia as a consequence of incomplete myectomy of the cricopharyngeus muscle appears greater compared to the open procedure.

An alternative and less invasive procedure for the management of cricopharyngeal dysphagia involves the injection of botulinum toxin injection into the cricopharyngeus muscle.[14] Botulinum toxin A (BTA) is a neurotoxin synthesized from the bacillus *Clostridium botulinum*. It acts at the presynaptic cholinergic nerve terminals to block the release of acetylcholine at the myoneuronal junction. In a dose-related manner, it weakens contraction when injected into the target muscle. The toxin has been used successfully in people for the treatment of esophageal achalasia, a condition characterized by hypertonicity of the lower esophageal sphincter. It has also been reported to be of benefit for the management of people with cricopharyngeal achalasia. For clinical use, botulinum toxin activity is defined in units such that one unit represents the median lethal dose in mice. Because of its short half-life, the toxin is reconstituted shortly before injection with 0.9% sterile saline to a concentration of 25 units/cc. The average injection dose in people ranges between 25-50 units.[14] In people with cricophayryngeal achalasia, physicians frequently balloon dilate the proximal esophageal sphincter during the botulinum injection procedure. We have successfully managed dogs with severe cricopharyngeal dysphagia with injection of botulinum toxin, although the benefits of the toxin wore off after approximately 4 months. This is an expected and well-documented phenomenon in people, and can be easily reapplied if warranted. Alternatively, a more permanent myotomy or myectomy of the cricopharyngeus muscle can be performed at this time. The limited duration of botulinum toxin's effect is a benefit, as animals that respond favorably to the toxin should do well following surgical myotomy or myectomy. In contrast, animals that do poorly following botulinum injection can be supported with an enteral feeding device until the effects of the toxin have worn off. Dogs diagnosed with cricopharyngeal dyssynchrony are challenging to manage, because there is no direct evidence of an obstructing "bar" (hypertrophied cricopharyngeal muscle) on fluoroscopy or surgery. A comprehensive search for an underlying neuropathy/myopathy should be performed prior to intervention, and if unable to be documented, these dogs should be managed with injection of Botox prior to consideration of cricopharyngeal myectomy.

Gastroesophageal Reflux Disease

Gastroesophageal reflux disease (GERD) is a common chronic disorder in the Western world. The basic cause of GERD has been well characterized—the fundamental defect is a loss of integrity of the gastroesophageal barrier. What is less clear is the most appropriate means of addressing this reflux. GERD has a variety of symptoms in people, ranging from typical presentations of heartburn and regurgitation (without esophagitis) to atypical presentations, such as severe erosive esophagitis and its associated complications. Because of its symptomatic diversity, physicians may select from a variety of therapeutic approaches.

In dogs, most cases of GERD occur secondary to hiatal hernia; however, the author has documented cases of severe flaccidity of the lower esophageal sphincter in the absence of a hiatal hernia. Medical therapy aims at decreasing acidity by suppressing proton secretion and has been well established. Available medications include antacids and alginates, H2-receptor antagonists, motility agents, and proton pump inhibitors (PPIs). All of these medications do little to prevent the reflux of gastric contents, and antireflux surgery is usually warranted. In people, antireflux surgery is commonly performed laparoscopically, and is aimed at reinforcing and repairing the defective barrier through plication of the gastric fundus.[15] The earliest performed successful procedures were the Nissen and Toupet fundoplications, to which several modifications have since been made. It has been demonstrated in preliminary studies and long-term outcomes of such open surgery and preliminary studies of such laparoscopic surgery that antireflux surgery is an effective approach, with overall outcomes superior to those achieved with medications.[16] Endoscopic antireflux therapies are currently being utilized in people, although clinical trials in dogs are lacking. Three main methods are currently employed: endoscopic intraluminal valvuloplasty,[17] endoscopic radiofrequency therapy,[18] and endoscopic injection or implantation of foreign material.[19] The endoluminal suturing method is highly demanding technically, and its short-term results are encouraging, although largely dependent on the experience of the endoscopist. Several prospective cohort studies have shown that the radiofrequency procedure (Stretta®) significantly improves GERD symptoms and quality of life while reducing esophageal acid exposure and eliminating the need for antisecretory medications in the majority of patients within 6–12 months. Most recently, some researchers have studied the endoluminal implantation of polymers, such as Plexiglas™ (polymethyl-methylacrylate), Gatekeeper® hydrogel, and Enteryx® (ethylene vinyl alcohol copolymer).[20] The preliminary results of these studies showed that the implantation method was feasible and safe; however, the only multicenter trial related to outcome that has been published has included just 1 year of follow-up.

It is clear that there are a plethora of research studies using animal models to evaluate less-invasive approaches to prevent reflux in people. These models provide an ideal template to further our investigations in an effort to provide additional options for the surgical and endoscopic management of congenital or acquired hiatal hernias and reflux disease in the dog.

Megaesophagus

Idiopathic megaesophagus is the most common type of megaesophagus in the dog and cat. The syndrome may be manifested either in puppies at the time of weaning or in adulthood. The *etiology* of idiopathic megaesophagus is unknown. The congenital form of the disease may be due to a delay in maturation of the esophageal neuromuscular system; a theory that explains why young dogs may improve with careful feeding management. Idiopathic megaesophagus has been shown to be inherited in the wire-haired fox terrier and the miniature schnauzer. A breed predisposition also exists for the German Shepherd, Great Dane and Irish Setter. The site and pathogenesis of the lesion in idiopathic megaesophagus is unknown. Suggested hypotheses include abnormalities of the afferent limb of the reflex arc (receptors, neurons) or of the swallowing center in the CNS. Idiopathic megaesophagus may also occur rarely in the cat. Secondary megaesophagus may result from a large number of systemic diseases including, myasthenia gravis, SLE, polymyositis, polymyopathies, dermatomyositis, polyneuropathies, dysautonomia, botulism, distemper, neoplasia, brain stem disease, lead and thallium toxicity, Addison's disease, hypothyroidism, pituitary dwarfism, and thymoma. Many obstructive esophageal diseases (neoplasia, granuloma, vascular ring anomaly, stricture, periesophageal masses and foreign bodies) can also lead to megaesophagus if they are of sufficiently chronic duration.

Repeat acetylcholine receptor antibody testing is important in dogs, as many dogs with focal myasthenia gravis have been misdiagnosed with idiopathic megaesophagus on the basis of a single normal acetylcholine receptor antibody test. Repeat testing should be done approximately 6-8 weeks following the first test. Additional diagnostic procedures that can be performed based on the animal's signalment, history, and neurological examination include an EMG, nerve conduction velocities, and muscle biopsies. Videofluoroscopy is essential for the diagnosis of functional esophageal disorders (esophageal dysmotility) not associated with esophageal dilation and has some prognostic value in megaesophagus via assessment of the severity of peristaltic dysfunction. Esophagoscopy is less reliable than radiography and fluoroscopy, although it can be used to rule out underlying causes of megaesophagus such as esophagitis, neoplasia, and radiolucent foreign bodies.

Medical management of generalized megaesophagus involves modification of feeding practices. Treatment of the underlying cause (secondary megaesophagus) is of paramount importance. Dogs with megaesophagus generally tolerate a liquid or semi-liquid gruel better than solid food. Feeding from an elevated position allows gravity to help move the liquid into the stomach. If possible the animal should be held in a vertical position for 5 - 10 minutes after eating. This can often be accomplished with the advent of a Bailey chair or similar device. Multiple feedings rather than one large single meal may also help minimize food accumulation in the esophagus. We have successfully placed low-profile-gastrostomy tubes for feeding in many dogs with idiopathic megaesophagus in an effort to minimize aspiration pneumonia. The silicon tubes used are extremely durable and are usually replaced on a yearly basis. The frequency of aspiration pneumonia has been markedly reduced in comparison to oral feeding and this therapeutic modality should be considered when a client is willing to dedicate the time to tube maintenance and feeding. The prognosis for dogs with megaesophagus is very variable depending upon the underlying etiology, the degree of dysfunction and the systemic status of the dog. The long-term *prognosis* is poor in most cases, although some cases can be managed successfully for years. The prognosis is improved if treatment of an underlying disease is possible.

Esophageal Strictures

Esophageal strictures are a relatively common problem in dogs and less commonly cats, and can be caused by benign and malignant causes, although the latter are relatively uncommon in dogs and cats. *The most common cause of esophageal stricture formation is gastroesophageal reflux in association with general anesthesia.* This phenomenon has been reported to occur in up to 65% of cases of esophageal stricture, with a median onset of clinical signs occurring 7.5 days post-anesthesia. The incidence of gastroesophageal reflux (GER) in dogs during anesthesia varies from 16-55%, and occurs secondary to a decrease in lower esophageal sphincter (LES) pressure. General anesthesia is associated with transient lower esophageal sphincter relaxation (TLESR) that is mediated by a vago-vagal neural pathway. In addition, reduction of LES pressure has also been shown to occur secondary to a variety of anesthetic agents, including atropine, morphine, acepromazine, thiopentol, xylazine, and isoflurane.

The second most common cause of esophageal stricture formation is from esophagitis induced by administration of doxycycline or clindamycin. The proposed mechanism of tablet-induced esophagitis and stricture formation is from tablet retention in the esophagus due to poor esophageal clearance with a "dry" swallow. Other important causes of esophagitis and secondary stricture formation include chronic vomiting of acid contents from the stomach, foreign body ingestion, and swallowing of caustic substances. Esophagitis is associated with a weakening of the LES pressure that can result in further reflux of gastric contents and increased damage to the esophageal mucosa. Damage to the muscularis layer of the esophagus is often associated with fibroblastic proliferation and contraction leading to stricture formation. The clinical signs associated with severe esophagitis and/or esophageal stricture formation include odynophagia (painful swallowing), dysphagia, increased salivation, regurgitation, anorexia, coughing (secondary to aspiration pneumonia), and weight loss. These signs are often insidious at the onset and are often missed by owners, but are progressive as the esophageal lumen gets progressively narrowed.

Mechanical dilation of the stricture is best accomplished using balloon dilation or bougienage. The theoretical advantage of balloon dilation is that the forces applied to the stricture are a radial stretch, in contrast to the longitudinal forces applied with the rigid bougienage instrument. However, a retrospective case series in 20 dogs and 8 cats with benign esophageal strictures that underwent bougienage treatment suggested that this procedure was safe and effective for most dogs and cats with benign esophageal strictures, with outcomes similar to balloon dilation.[21] The balloons used for balloon dilation are available in various diameters (up to 20-mm) and are made of a rigid plastic material that can withstand a relatively high pressure up to 45 psi. The balloons are manufactured to either pass through the biopsy channel of the endoscope, or alongside the endoscope with the use of a guide-wire. If a guide-wire is used, it is passed through the channel of the endoscope (or imaged with fluoroscopy) and advanced beyond the stricture into the stomach or caudal esophagus. The scope is removed as the guide-wire is advanced through the channel, thus leaving the guide-wire near its original position. The balloon catheter is then passed over the guide-wire (with the balloon deflated) until it is positioned within the stricture. The position of the balloon is visualized through the endoscope or via fluoroscopy. An inflating device that has a pressure reading is attached to the balloon and the pressure is slowly increased to the pressure specified by the manufacturer for that particular balloon. The balloon is kept inflated for 1-2 minutes before it is deflated. Sequentially larger balloons are used or increasing pressure is applied to the balloon until the desired amount of mucosal tearing and stricture dilation has been attained (generally $\leq$ 50% of the original esophageal stricture diameter). The procedure is repeated 3-5 days later in an effort to maximally dilate the stricture.

The administration of triamcinolone into the stricture site using a four-quadrant approach before the balloon dilation procedure has been associated with a reduced rate of restructure formation.[22] We generally inject approximately 2.5 mg triamcinolone into each of the quadrants using a Wang needle (or similar transbronchial needle) that can be threaded down the biopsy channel of the endoscope. The steroid is generally used for the first 2-3 dilation procedures. Topical mitomycin C has also been shown to be beneficial for preventing restricture formation.[23] Clinicians generally apply 5 mg of Mitomycin using a soaked gauze sponge that is placed endoscopically at the stricture site for approximately 5 minutes. The site is then rinsed with 60 mL of water following the removal of the sponge.

Intraluminal stents are being used with increasing frequency in veterinary medicine for patients that have failed balloon dilation or for patients with recurrent stricture formation. Stents are available both covered (polypropylene) and uncovered. The covering helps prevent the ingrowth of tissue within the stent. Available stent materials include Nitinol (nickel plus titanium), Elgiloy (cobalt, nickel, plus chromium), stainless steel, polyester plastic/silicone, or a biodegradable material such as PDS. The selection of a particular stent is based upon the characteristics of the stricture such as its location and length, and the need for removal of the stent. Once the stent is deployed, it must be anchored in place or it will rapidly migrate into the stomach. The stent can be secured in place using a suturing device (GI Stitch, Pare Surgical) that can be used through a double channel endoscope.

Esophagitis

Medical management of esophagitis involves a combination of proton pump inhibitors such as omeprazole, sucralfate suspension, and a potent prokinetic agent such as cisapride to minimize further gastric reflux and facilitate gastric emptying. Proton pump inhibitors can initially be administered q 12 hrs for 3-4 days to decrease acid production more rapidly, before continuing with once daily administration. H2-receptor antagonists can also be administered at night to help minimize nocturnal acid breakthrough. Proton pump inhibitors should be given approximately 30 minutes before a meal.

REFERENCES

[1]Dodds WJ et al. **Am J Roentgenology** 1990;154:953-963; [2]Pollard RE et al. **Vet Radiol Ultrasound** 2000;41(5):409-412; [3]Davidson AP et al. **Am J Vet Res** 2004;65(3):344-349; [4]Cook IJ. **Dig Dis** 1998;16:125-133; [5]Shelton GD et al. **J Am Med Vet Assoc** 1997;211:1428-1431; [6]Pollard RE et al. **Vet Radiol Ultrasound** 2007;48(3):221-6; [7]Stanley B et al. **Vet Surg** 2010;39:139-149; [8]Fox MR et al. **Gut** 2008;57:405-423; [9]Kempf J et al. **Am J Vet Res** 2013;74:895-900; [10]Zacuto AC et al. **J Vet Int Med** 2012;26:518-525; [11]Warnock JJ et al. **J Am Vet Med Assoc**. 2003;223(10):1462-8; [12]Niles JD et al. **J Small Anim Pract** 2001; 42: 32-35; [13]Pitman M, et al. **Laryngoscope** 2009;119:45-53; [14]Haapaniemi JJ et al. **Dysphagia** 2001;16(3):171-5; [15]Peters JH et al. **Ann Surg** 1998;228:40-50; [16]O'Connor KW et al. **Gastrointest Endosc** 1988;34:106-112; [17]Mason RJ et al. **Gastrointest Endosc** 1997;45(3):283-90; [18]Comay D et al. **Can J Gastroenterol** 2008;22(6):552-8; [19]Domagk D et al. **Am J Gastroenterol** 2006;101(3):422-30; [20]Zhi XT et al. **J Long Term Eff Med Implants** 2005;15(4):375-88; [21]Bissett SA et al. **J Am Vet Med Assoc** 2009;235(7):844-850; [22]Orive-Calzada A et al. **Surg Laparosc Endosc Percutan Tech** 2012;22(6):518-522; [23]El-Asmar et al. **J Pediatr Surg** 2013;48(7):1621-1627.

THE ROLE OF MICROPARTICLES IN HEMOSTASIS AND THROMBOSIS

Linda Kidd DVM PhD DACVIM
Pomona, CA

INTRODUCTION

Microparticles (MPs) are small (0.1-1 uM) vesicles that bud from cell membranes in response to various forms of stimulation. Circulating microparticles are derived from cell membranes of RBCs, platelets, neutrophils, endothelial cells, and monocytes.[1,2] MPs express cell surface and cytoplasmic molecules derived from their cell of origin. MPs are able to interact with, and modulate cell signaling and gene expression in other cell types. Recent studies have illustrated their importance as cellular messengers that modify the biologic properties of neighboring and distant cells through delivery of membrane-associated receptors, proteins, cytokines, biologically active lipids, transcription factors, and nucleic acids including mRNA. They are involved in normal hemostasis, vascular homeostasis, apoptosis, innate and adaptive immunity and inflammation, and have prominent roles in pathologic thrombus formation for a number of proinflammatory disease states. Because of their role in pathologic thrombosis, MPs are being investigated as biological markers of thrombotic risk, and as therapeutic targets for thromboprophylaxis for a variety of diseases. This review will summarize recent information regarding the role of microparticles in hemostasis and thrombosis.

THE ROLE OF MICROPARTICLES IN HEMOSTASIS

The most well described role that MP play in normal hemostasis is to provide anionic phospholipid surfaces that localize coagulation factors to the site of clot formation. Normal membranes have an asymmetric distribution of phospholipids, with anionic phospholipids, such as phosphatidyl (PS) primarily located in the inner leaflet, and phostatidylcholine and sphingomyelin enriched in the outer membrane. This distribution if maintained by specific transporters (flippase, floppase and scramblase). Upon activation outward transport of PS occurs. The rate of this transport varies between cell types. Upon activation, platelets rapidly expose PS on their outer membrane and release PS exposing MPs.[3,4] The MPs released from platelets provide phospholipid surfaces with 50-100 fold more procoagulant activity than activated platelets.[4] Anionic phospholipids, now exposed, facilitate assembly of the prothrombinase and tenase complexes of the coagulation cascade.[4,5] Recent in vitro studies have shown that PS receptors are constitutively expressed on the surface of endothelial cells.[6] Thus, PS exposure may also facilitate localization of a forming thrombus to the vessel wall.

The importance of platelet PS exposure and MP formation in normal hemostasis is illustrated by a bleeding disorder called Scotts syndrome. In people, Scott syndrome is characterized by a defect in platelet, red blood cell and B lymphoblast ability to externalize PS and shed MP.[3] A similar disorder has recently been described in German shepherd dogs.[7] Other bleeding disorders associated with decreased platelet MP formation in people have been described.[2]

Because platelet and other cell types such as neutrophils, endothelial cells and monocytes form microparticles that expose PS, adhesion molecules, and proteins involved in coagulation and fibrinolysis, it is thought that MP contribute to normal hemostasis by additional means. However, most research has focused on the role of microparticles in pathologic thrombus formation.

THE ROLE OF MICROPARTICLES IN PATHOLOGIC THROMBOSIS FORMATION

Diseases that are commonly associated with thrombosis in small animals include immune-mediated hemolytic anemia, neoplasia, sepsis, protein losing nephropathies and cardiac disease.[8] There are few studies investigating the role of microparticle formation in thrombosis in dogs and cats. The following highlights the role of microparticles in thrombosis for select diseases in people and laboratory animal models. It should be noted that microparticles play a very important role in the pathogenesis of atherosclerosis in people and the reader is referred to several comprehensive reviews for more information on that topic.[9,10]

Hemolytic Anemias

Both PS expressing and tissue factor (TF) expressing MPs play a role in pathologic thrombus formation in people with hemolytic anemias. As is the case for platelet activation, senescence of red blood cells results in loss of asymmetry and exposure of PS on the outer RBC membrane surface.[5] PS-positive RBCs bind to macrophage PS receptors. PS receptor mediated binding induces signaling that results in phagocytosis. Activated platelets, damaged RBC and their associated MPs are pro-coagulant because they have anionic phospholipids, such as PS, on their surfaces.[2,11] Recent in vitro studies have shown that PS receptors are constitutively expressed on the surface of endothelial cells.[12] Erythrocyte PS also binds to thrombospondin, a protein that is present in sub-endothelial matrix.[13] Endothelial cell PS receptor expression is up-regulated in response to inflammatory cytokines, hypoxia and free hemoglobin. Furthermore, RBC MPs have been shown to scavenge NO and impair endothelial dependent vasodilation.[14] Thus, PS-positive MPs may contribute to coagulation and thrombosis in hemolytic anemias by providing surfaces for the tenase and prothrombinase coagulation complexes to bind, by localizing their pro-coagulant membranes to inflamed endothelium, and by inhibiting vasodilation. Increased PS expression on the surface of RBC occurs in human hemolytic diseases associated with thrombosis, including sickle cell anemia, beta thalassemia and hereditary spherocytosis.[5] The risk of thrombosis correlates with the magnitude of PS expression on the surface of RBC in humans with sickle cell anemia.[15] Plasma levels of MPs are increased and associated with thrombosis in human patients in sickle cell crisis, and RBC MP number correlates inversely with hemoglobin levels in these patients.[2,16]

In addition to the procoagulant role of PS expression, co-expression of TF on the surface of MPs can further activate endothelium and the extrinsic coagulation pathway.[1,2] Tissue factor (TF) initiates coagulation during normal hemostasis. TF is not normally expressed on quiescent vascular endothelial cells and circulating blood cells, but rather, is expressed on perivascular cells such that damage to vessels initiates coagulation. Subsequent thrombin production results in platelet activation and further amplification and propagation of coagulation.

MPs derived from monocytes and endothelial cells can express TF. Many recent studies suggest that TF expression on the surface of MPs may play an important role in pathologic thrombus formation.[1,2] For example, TF bearing MPs (TFMPs) are increased and associated with thrombosis in human patients with sickle cell crisis.[3,16,17] TFMPs have been associated with markers of activation of coagulation in some, but not other studies of human patients with sickle cell disease.[16,18,19] Thus the role of TFMPs in the pathogenesis of thrombosis in sickle cell disease in people requires further investigation. Elevated levels of TFMPs occur in other prothrombotic hemolytic diseases in people, including hemolytic uremic syndrome, and paroxysmal nocturnal hemoglobinuria.[20,21]

Thromboembolic disease is a major factor that affects short and long-term survival in dogs with IMHA.[22] Dogs with IMHA exhibit excessive platelet activation and activation of coagulation.[22,23] Whether elevated levels of TF exposing MP occurs in human auto-immune hemolytic anemia in people has not been investigated. However, a recent study showed that platelet MPs were found to be elevated in association with platelet activation in dogs with IMHA.[24] In addition, triggers for increased RBC microvesiculation include complement attack, sheer stress and oxidative injury, conditions that are also present in dogs with IMHA.[2,22] Therefore, RBC PS exposure could potentially contribute to the hypercoagulable state and vascular occlusion in dogs with IMHA. Proinflammatory cytokines, heme, CD40 ligand and ischemia/reperfusion induce TF expression.[23,25] Because IMHA in dogs is also associated with an inflammatory state, platelet activation and hypoxia, TF may be an important instigator of coagulation in these patients. Indeed a recent study found that dogs with IMHA had increased levels of TF mRNA expression in peripheral blood.[26] However, TF mRNA expression levels do not always correlate with protein expression and activity. Whether TF activity is increased in the blood of dogs with IMHA has not been determined. We are currently investigating the role of PS and TF exposing MPs in dogs with IMHA.

Neoplasia

Hypercoagulable states and thrombosis, particularly venous thromboembolic events (VTE) (including pulmonary thromboembolism) are commonly associated with neoplasia in people and in dogs.[27,8] One of the most important mechanisms of activation of coagulation is upregulation and increased expression of TF. TF surface expression and associated procoagulant activity has been associated with pancreatic, breast, colorectal and squamous cell carcinomas in people, and high rates of TF expression have been associated with the occurrence of VTE. Indeed, high levels of TF exposing MPs are commonly found in cancer patients as compared to healthy controls. In some, but not all studies, circulating MPTF levels and associated MPTF procoagulant activity have been found to be associated with and predictive of thrombosis.[3] Recently, MP formation and MPTF activity was found to be increased in canine cancer cell lines in vitro.[28]

Sepsis and septic shock

Microparticles appear to play both procoagulant or anticoagulant/anti-inflammatory roles in sepsis. LPS generates TF bearing monocytes. Evidence in mouse models of endotoxemia and in people with meningococcal septic shock suggest that LPS induced MPTF formation contribute to activation of coagulation and thrombosis during sepsis.[3,29] However, microparticles may also convey protective and anti-inflammatory effects in sepsis. Activated protein C binds to endothelial protein C receptor expressed on endothelial cells and monocytes and limits inflammation in sepsis. Endothelial cells and monocytes have receptors for endothelial protein C (EPCR). Endothelial cells and monocytes release EPCR microparticles, and these microparticles appear to have protective effects on the vascular endothelium.[2,30] Recent studies have shown that microparticles derived from leukocytes expressing alpha-2 macroglobulin (A2MG) are elevated in septic patients, but interestingly, were significantly more elevated in patients that survived sepsis as compared to non-survivers. In mouse models of sepsis it was shown that incorporation of A2MG into microparticles upregulated anti-inflammatory lipid mediators and immunomodulatory cytokines.[31]

Nephrotic syndrome

Thrombosis is associated with nephrotic syndrome in people and has been associated with protein losing nephropathies and nephrotic syndrome in dogs.[8,32] A recent study showed that PS expressing MP from RBC, platelet and endothelial cells were increased in nephritic syndrome patients as compared to controls. MP shedding and PS exposure of RBC and platelets were highly procoagulant, and most of the procoagulant activity was attributed to PS exposure.[32] The role of MP in thrombosis in dogs with protein losing nephropathy or nephrotic syndrome has not been investigated.

Cardiac disease

Most research involving microparticles and thrombus formation in the context of cardiac disease in people is focused on their role in proinflammatory atherosclerotic lesions. Given that venous stasis, increased shear stress, and hypoxia can induce microparticle formation, common companion animal cardiopulmonary disease that result in atrial enlargement, regurgitation of blood flow across valves, and endothelial dysfunction might be associated with elevated microparticle levels that would contribute to a prothrombotic state. Whether microparticles play a role in these diseases in dogs and cats has not been investigated to date. However, in a human study, circulating platelet microparticles were found to be increased in patients with rheumatic mitral stenosis and atrial fibrillation when compared to healthy controls.[33] In another study of patients with precapillary pulmonary hypertension, levels of circulating endothelial and leukocyte MPs were increased compared with control subjects. Interestingly, levels of PECAM(+) and VE-cadherin(+) endothelial MPs predicted hemodynamic severity of the disease.[34]

Targeting microparticles for thromboprophylaxis

Because evidence strongly suggests that MPs play an important role in pathologic thrombus formation, therapies that target microparticles directly and/or decrease MP formation are being investigated as thromboprophylactic agents in people.[1,2,35,36]

SUMMARY

Microparticles arising from platelets, red blood cells, endothelial cells and leukocytes likely play an important role in hemostasis and pathologic thrombosis. Their utility as biological markers of risk of thrombosis and as therapeutic targets is currently an area of intensive research in human medicine. The role of MP in proinflammatory prothrombotic disease in companion animals deserves further investigation.

REFERENCES

1. Mackman N, Circ Res. 2009; 104(8): 925.
2. Burnier L, et al. Thromb Haemost. 2009; 101(3): 439
3. Owens AP, Mackman N, Circ Res. 2011; 108(10):1284.
4. Morel O, et al Semin Immunopathol 2011; 33:469.
5. Zwaal RF, et al Cell Mol Life Sci. 2005; 62(9): 971.
6. Setty BN, Betal SG, Blood. 2008; 111(2): 905.
7. Brooks MB, et al. Blood. 2002; 99(7): 2434.
8. DeLaforde A, Topics in Compan An Med 27 (2012) 59.
9. Shantsila et al, Thromb Haemost. 2010 8(11):2358.
10. Angelillo et al. Circ Res. 2012;110:356
11. Ataga KI, et al, Haematologica. 2009; 94(11): 1481.
12. Setty BN, Betal SG, Blood. 2008; 111(2): 905
13. Gayen Betal S, Setty BN, Transl Res. 2008; 152(4): 165.
14. Liu et al, Free Radic Biol Med 2013 65 1164.
15. Setty BN, et al Blood. 2002; 99(5): 1564.
16. Shet AS, et al Blood. 2003; 102(7): 2678.
17. Kasar et al J Thromb Thrombolysis 2013epub.
18. Ataga et al. PLOS One 2012 7(1) e29786.
19. Setty et al, Br J Haematol. 2012 157(3) 370.
20. Ståhl AL, PLoS One. 2009 Sep 11;4(9):e6990.
21. Liebman et al, Thromb Res. 2003 111 235.
22. Kidd L, Mackman N. J Vet Emerg Crit Care. 2013;23(1):3.
23. Weiss DJ, Brazzell JL. J Vet Intern Med. 2006; 20(3): 682.
24. Ridyard A, et al, J Small Anim Pract. 2010 51(6):296.
25. Kjelgaard-Hansen M, et al. J Vet Intern Med. 2011; 25(1): 76.
26. Piek CJ, et al. Vet Immunol Immunopathol. 2011.
27. Date K, et al, Thrombosis Research 131 2013; 109–115
28. Stokol T, et al, Am J Vet Res. 2011 72(8):1097.
29. Hellum M, et al, Thromb Res 2013 epub.
30. Perez-Casal M, et al, Haematologica. 2009 94(3):387.
31. Dalli J, et al, EMBO Mol Med. 2014;6(1):27
32. Gao C, et al, Thromb. Haemost. 2012; 107: 681–689
33. Azzam H, et al, Hematology. 2009 14(6):357-60
34. Amabile N, et al, Am J Respir Crit Care Med. 2008 1;177(11):1268
35. Adiguzel C, et al. Clin Appl Thromb Hemost. 2009; 15(5): 501.
36. Rosenblade A, et al, J Pharm Pharmaceut Sci 2013 16(2) 238.

FELINE ARTERIAL EMBOLISM: CLINICAL FEATURES AND MANAGEMENT CHOICES

Janice M. Bright, BSN, MS, DVM, DACVIM (Cardiology, Small Animal Internal Medicine)
Fort Collins, CO

INTRODUCTION

Arterial thromboembolism (ATE) occurs when thrombus formed in one part of the circulation embolizes to a peripheral artery. Mechanical obstruction of the artery and constriction of collateral vessels from the embolic event impair blood flow to distal tissues causing ischemic neuromyopathy as well as potential reperfusion injury and gangrenous necrosis. Systemic ATE is a common complication of any type of primary or secondary cardiac disease in cats (with exception of arrhythmogenic right ventricular cardiomyopathy), and is most frequent in those with severe myocardial disease. Occasionally, ATE also occurs in cats without heart disease.

RISK FACTORS

Identifying cats at increased risk for ATE relies greatly on echocardiographic findings. Nevertheless, cats with a history of previous ATE are undeniably at risk for recurrent embolism even if the previous clinical signs were mild or transient. Cats with severe enlargement of the left atrium are also at risk for ATE, and clearly those with visible thrombus or spontaneous echo contrast are at even greater risk.[1] Impaired left atrial systolic function is also associated with increased ATE risk as is poor left ventricular systolic function.[2, 3] Low velocity flow in the left auricle (maximal flow velocity < 0.20 m/s) has been shown to be an independent predictor for the occurrence of spontaneous echo contrast and, therefore, is predictive of a prothrombotic state in cats.[2] Asymptomatic cats with hypertrophic cardiomyopathy and a left atrium of normal size do not appear to be at high risk for ATE.

PATHOPHYSIOLOGY

Intra-atrial thrombus formation in cats with ATE is believed to involve all three components of Virchow's triad: stasis of blood, hypercoagulability, and endothelial injury. The thrombus typically originates within the left auricle (left atrial appendage). The unique anatomy of the auricle as a blind pouch with prominent trabeculations predisposes this site to both blood flow stasis and endothelial contact and, hence, to thrombogenesis. Low velocity flow and low shear promote erythrocyte and platelet aggregation and accumulation of clotting factors. The role of endothelial dysfunction or damage in thrombus formation within the left atrium is somewhat more speculative. However, atrial stretching and fibrosis is believed to result in exposure of subendothelial collagen and tissue factor leading to platelet adhesion and activation as well as activation of the clotting cascade.[4] Interestingly, one study in rats has shown a reduction in endocardial thrombomodulin expression with subsequent increase in local thrombin production as a result of high atrial pressure.[5] Cats with myocardial disease are likely to exist in a hypercoagulable state based on studies documenting increased platelet reactivity and abnormally high circulating concentrations of thrombin-antithrombin complexes in cardiomyopathic cats.[6, 7]

The importance of impaired collateral circulation in the clinical presentation of ATE in cats is has been documented by surgical studies of distal aortic ligation that failed to produce the clinical syndrome of ischemic neuromyopathy. The clinical syndrome is only reproduced when thrombus is created within the terminal aorta after ligation.[8, 9]

CLINICAL FEATURES

The clinical signs produced by a thromboembolic event will vary depending on the site of occlusion, whether the occlusion is total or partial, and the amount of collateral circulation. The distal aorta (aortic trifurcation) is the most frequent site of embolism producing the "saddle thrombus" syndrome. However, other vascular beds may be affected including the renal or mesenteric circulation, the cerebral vasculature, or the forelimb vasculature. Saddle thrombus in cats is characterized by sudden onset of pain and paresis (or paralysis) of the pelvic limbs. Affected cats show signs of acute pain and anguish including vocalization, agitation, and tachypnea. It is not unusual for one limb to be more severely affected than the other. The clinical signs of saddle thrombus are often described as "the five P's" – pain, paresis (or paralysis), pulselessness, poikilothermy (hypothermia of affected extremities), and either pallor or purple discoloration (cyanosis) of the skin and nail beds. Motor function is usually absent or diminished distal to the stifles with absent skin sensation distal to the tarsus. Either forelimb may be affected with these signs if embolization is to a brachial artery. If the embolus is small or rapidly lysed, or if collateral circulation quickly regained, motor function may be present by the time physical examination is performed. Signs of embolization to the abdominal organs or brain are more variable but may include acute abdominal pain, vomiting, or central nervous system signs.

Although fewer than 12% of cats with ATE have a history of heart disease prior to the embolic event[10], cats may present with significant respiratory distress as a result of left-sided congestive heart failure (CHF). If crackles or pleural effusion line are apparent during auscultation, it is likely that heart failure is present. Many cats (up to 72%) will be severely hypothermic due to systemic hypoperfusion.[11] Most cats with ATE have no known history of cardiac disease, and signs due to the embolism may be the initial indication of heart disease. Cardiac auscultation may be normal in as many as 40% of affected cats.[12]

DIAGNOSTIC TESTING

Like the clinical signs, abnormalities found on ancillary diagnostics are highly variable depending greatly on the site and severity of vascular occlusion. Most cats have some degree of azotemia and hyperphosphatemia from reduced renal perfusion; however, these findings may also be due to embolism of a renal artery. Although hyperkalemia is a potentially lethal complication of ATE that may arise from reperfusion injury or acute renal failure, some cats are actually hypokalemic on presentation. AST and ALT are often increased subsequent to poor hepatic perfusion. Typically, serum creatine kinase is profoundly increased. Coagulation testing is usually unremarkable with exception of the presence of D-dimers.

While radiographic studies are best avoided in cats with severe respiratory distress, thoracic radiographs are important for confirming or refuting CHF in tachypneic cats with normal respiratory sounds. Echocardiographic examination may be too stressful to safely perform in cats with respiratory compromise, and, fortunately is rarely important for initial management of cats with ATE, even those with possible pulmonary edema. Echocardiographic confirmation of left atrial enlargement is present in most cats with ATE, and this finding does not confirm CHF in tachypneic animals.

Ultrasound may be used to identify thrombus in the terminal aorta; however, this diagnostic is seldom necessary since clinical signs are pathognomonic.[1] Furthermore, absence of visible thrombus in the distal aorta does not rule out ATE, especially if signs have been present for more than 24 hours.

ATE MANAGEMENT STRATEGIES: THROMBOLYTIC THERAPY

The use of pharmacologic agents such as streptokinase or tissue plasminogen activator (TPA) to lyse the thrombus is a tempting management strategy. Several trials of intravenous TPA administration in cats with saddle thrombus have shown rapid restoration of distal perfusion in 64% of affected limbs within 24 hours.[13, 14] However, only 50% of the cats in these trials survived to hospital discharge, a mortality that is no better than other therapies.[12] The high mortality rate of thrombolytic therapy is attributed to reperfusion injury, and this complication may, perhaps, be less severe in cats treated within several hours of the embolic event. High mortality rate and the need for intensive monitoring have decreased enthusiasm for thrombolytic therapy in cats with ATE. However, the potential of achieving a rapid recovery of function and of preventing gangrenous necrosis with thrombolytic therapy may be attractive to some owners.

ATE MANAGEMENT STRATEGIES: ANTICOAGULANT THERAPY

An alternative to the thrombolytic management is the use of anticoagulants. The goal of anticoagulant therapy is to prevent extension of the thrombus while allowing the endogenous fibrinolytic system to remove it. Antithrombin (ATIII) is the major plasma inhibitor of thrombin and the other activated coagulation factors. ATIII forms an inactive complex with these factors, and the rate of formation of these inactivating complexes is increased by a factor of several thousand in the presence of heparin.[4] Unfractionated heparin requires ATIII for its anticoagulant effect, and this drug has been traditionally used for initial treatment of acute ATE. However, the efficacy of heparin in this setting has not been established. If unfractionated heparin is used, it should be dosed to effect based on coagulation monitoring.

Low molecular weight heparin (LMWH) has a longer half-life and a more predictable dose response than unfractionated heparin and can be dosed according to body weight in people without coagulation monitoring. However, the optimal dose schedule for the low molecular weight heparins in cats has not been established, and these agents are currently very expensive.

At this time, the primary anticoagulant used for prevention of stroke in people is warfarin. Warfarin acts by inhibiting production of the vitamin K dependent coagulation factors (II, VII, IX and X) as well as several pro-coagulant substances (protein C and protein S). Warfarin has unpredictable pharmacokinetics and pharmacodynamics and is quite difficult to use safely and effectively, even with extensive coagulation monitoring. Overall, the use of warfarin for management of ATE is both impractical and expensive. The intensive monitoring required and the recommendation that warfarin-treated cats be kept exclusively indoors, compromises quality of life for both cat and owner.

Several newer oral anticoagulants are FDA approved for use in human patients with increased thrombotic risk. These agents include dabigatran, a direct inhibitor of thrombin, and apixaban, a factor Xa antagonist. Apixaban has been shown to have superior efficacy compared to warfarin for deterring thrombus formation in clinical trials of human patients with atrial fibrillation, and is dosed based on body weight without coagulation monitoring.[15] Studies determining the pharmacokinetic and pharmacodynamics properties of apixaban in normal cats are currently in progress.

ATE MANAGEMENT STRATEGIES: ANTIPLATELET THERAPY

As with anticoagulants, the goal of antiplatelet therapy in cats with ATE is to prevent extension of the thrombus. Aspirin irreversibly inhibits platelet COX-1 activity thereby inhibiting the conversion of arachidonic acid to thromboxane A_2 and preventing platelet activation by this mechanism. This inhibition of platelet activity by aspirin lasts for the lifetime of the platelet and is achieved at low doses. Unfortunately, platelet turnover is very rapid in cats with heart disease, and activation of platelets by thromboxane A_2 is only one of numerous pathways by which platelet activation occurs. Low dose aspirin (5 mg/cat q48-72h) is no more effective than high dose aspirin (20 mg/cat q72h)[12], but fewer side effects may result with the lower dose. Aspirin does not prevent ATE in cats at risk, and it is not known whether aspirin reduces the incidence of ATE in cats.

Clopidogrel exerts its antiplatelet effects by irreversibly inhibiting the platelet ADP receptor P2Y12. Like aspirin, clopidogrel inhibits only a single pathway of platelet activation; however, this drug further reduces the risk of thrombosis when combined with aspirin in people. A randomized, multi-center clinical trial comparing aspirin to clopidogrel for preventing re-thrombosis in cats with documented ATE has been undertaken (FAT CAT trial) using a primary end point of ATE recurrence and secondary end points of cardiac mortality and all-cause mortality. Unpublished results suggest superiority of clopidogrel.

Finally, it has been suggested that early, combined use of aspirin and clopidogrel in cats with ATE may be preferable to heparin therapy for early management.[1]

INITIAL MANAGEMENT OF ATE

Cats with ATE are typically in extreme pain, and, unless adequate analgesia using opiates (such as methadone, fentanyl, or oxymorphone) is available, euthanasia should be suggested. Cats with two limbs affected and cats with rectal temperature <98.9°F (<37.2°C) have a poor prognosis (40% and 50% survival to hospital discharge respectively), and this information may be helpful.[12, 16] Since the initial 24 hours following the embolic event are the most painful, if euthanasia is likely, it is most humane not to delay.

The first priority is pain management which may be needed even before the decision to treat has been made. Intravenous methadone or constant rate infusion of fentanyl are suggested.[1] Although buprenorphine is inappropriate for cats in severe pain from bilateral rear limb occlusion, this agent may provide adequate analgesia for cats with partial ATE. Cats with respiratory distress should receive supplemental oxygen, and thoracic radiographs are indicated for determining whether congestive heart failure is present in tachypneic cats without crackles or pleural fluid line on auscultation. If auscultation or radiographs indicate pulmonary edema, intravenous furosemide should be administered (1-2 mg/kg IV) and repeated as needed until the respiratory status stabilized. Antithrombotic therapy should also be administered using heparin and/or antiplatelet agents. Once these treatments have been initiated, the patient should be monitored (vital signs, demeanor, renal function, mobility, electrolytes), and echocardiography may be more safely performed. Long-term antithrombotic therapy is indicated for cats surviving the ATE.

REFERENCES

[1]Luis Fuentes V. **J Feline Med Surg** 2012;14:459. [2]Schober KE et al. **J Vet Inter Med** 2006;20:120. [3]Rush JE et al. **J Am Vet Med Assoc** 2002;220:202. [4]Konkle BA *In* **Harrison's Hematology & Oncology** (Longo DL Ed.) 2010;22. [5]Kapur NK et al. **Circ** 2007;115:67. [6]Helinski CA et al. **J Vet Intern Med** 1987;1:24. [7]Bedard C et al. **Vet Clin Pathol** 2007;36:167. [8]Imhoff RK. Nature 1961;192:141. [9]Schaub RG et al. **Circ Res** 1976;39:736. [10]Laste NJ et al. J Am Anim Hosp 1995;31:492. [11]Moore KE et al. J Vet Emerg Crit Care 2000;10:245. [12]Smith SA et al. **J Vet Intern Med** 2003;17:73. [13]Pion PD et al. **Proc ACVIM** 1987;5:295 [14]Welch KM et al. **J Feline Med Surg** 2010;12:122. [15]Granger CB et al. N Engl J Med 2011;365:981. [16]Schoeman JP. **J Feline Med Surg** 1999;1:221.

THE IMPACT OF EICOSANOIDS ON HEMOSTASIS

John M. Thomason, DVM, MS, DACVIM
Starkville, MS

INTRODUCTION

Normal blood hemostasis is a complicated process that involves platelets, vascular endothelium, numerous coagulation factors, and several interacting cells. Platelets are a critical component to the normal hemostatic process; from recognizing the site of injury, to recruiting additional platelets and interacting with coagulation factors, platelets ensure the proper formation of a platelet plug and blood clot. When exposed to injured or damaged endothelium, either due to normal cell turnover or vessel trauma, platelets undergo several functional responses, including platelet adhesion, shape change, release of granules and pro-coagulant substances, surface expression of procoagulant receptors, that culminates in platelet aggregation and the formation of a platelet plug.[1] The formation of a platelet plug during primary hemostasis is the first step in coagulation, and not only provides local hemostatic control, but also provides the foundation for coagulation proteins to generate fibrin and a blood clot during secondary hemostasis. Without coagulation factors and clot formation, the platelet plug will weaken, a blood clot may not form, and excessive hemorrhage would develop. Alterations in any phase of hemostasis can result in ineffective hemostasis, leading to hemorrhage or clot formation.

PLATELET PHYSIOLOGY

When exposed to injured or damaged vascular endothelium, either due to normal cell turnover or vessel trauma, platelets will adhere via multiple adhesive ligands and receptors.[1] Under high shear conditions, the adherence of platelets to the exposed vascular subendothelium is primarily regulated by the receptors that bind collagen and von Willebrand factor (vWF), glycoprotein (GP) Ib-IX-V and GP VI. The GP Ib-IX-V complex is the major platelet receptor mediating the interaction of collagen with vWF, while the GP VI receptor is a low affinity collagen receptor, but has a high potency in terms of initiating platelet activation.

Following platelet adhesion, rapid signal transduction will cause platelet activation, which typically consists of a change in platelet shape, expression of surface receptors, synthesis of thromboxane A_2, and release of intra-platelet granules. When activated, platelet morphology changes from a simple disk to a sphere with many long pseudopodia that provide additional surface area for coagulation factors to initiate the generation of thrombin, deposition of fibrin, and clot formation.[1-3] Activated platelets will also express several surface integrins necessary for adhesion and aggregation. Additionally, platelets will synthesis thromboxane A_2 and release granules to aid in the recruitment of additional platelets needed to enhance aggregation. The combined effects of platelet activation are to cause vasoconstriction and decrease blood flow, ultimately increasing contact time between activated platelets, coagulation factors and damaged vascular subendothelium.[4] Intra-platelet granules contain multiple proteins that enhance platelet adhesion, promote cell-to-cell interactions, and stimulate vascular repair.[5,6] Alpha-granules are the main secretory component of platelets, but dense (delta) granules and lysosomes are also present.[6] Substances located in α-granules include adhesion molecules (vWf, P-selectin, and fibronectin), angiogenic factors, and coagulation factors (fibrinogen, factors V and VIII), while dense granules contain adenosine diphosphate (ADP), epinephrine, and serotonin.[6]

The binding of agonists to surface receptors will initiate platelet activation, even if there is no adhesion with the negatively charged phospholipids of the vasculature. Regardless of whether the initiating pathway is platelet adhesion or receptor binding, the final step in platelet activation, and mediator of platelet aggregation, is a conformational change of the surface GPIIb/IIIa complex (also known as glycoprotein integrin αIIbβ3), enabling the binding of fibrinogen to stabilize the platelet plug.[7] Expression of the GPIIb/IIIa complex is dependent on agonist/receptor binding, known as inside-out signaling, to initiate platelet activation. Platelet receptors and pathways of activation have become targets for medications that inhibit platelet function.

CYCLOOXYGENASE

Cells, including platelets, do not store eicosanoids, but these prostaglandins and thromboxanes are synthesized from arachidonic acid that has been released from phospholipid membranes through the actions of several phospholipases that include phospholipase A_2, prostaglandin H synthase, and thromboxane synthase.[8,9] Once mobilized, arachidonic acid is converted into prostaglandin H_2 by the cyclooxygenase (COX) enzyme. At least three COX isoforms have been described in dogs. COX-1 is constitutively expressed in most body systems, and functions to maintain normal cellular activities including platelet thromboxane A_2 production, maintenance of gastric mucosal protection, and preservation of renal blood flow.[9,10] Relative to COX-1, COX-2 has a far more limited tissue distribution, and expression is often at much lower levels. COX-2 expression occurs in the brain, kidney, thymus, circulating monocytes, tissue macrophages, and the vascular endothelium.[10] COX-2 expression can be induced by various endogenous and exogenous mediators including inflammatory cytokines, growth factors, and endotoxins.[9] The COX-3 isoform is transcribed from the COX-1 gene as a splice variant and has similar catalytic features, but only about 20% of the activity of COX-1.[11,12]

Both COX enzymes play key roles in normal hemostasis, and lead to the production of several biologically active prostaglandins, mainly thromboxane A_2 and prostacyclin.[13] Unlike thromboxane A_2, which is mainly produced by platelets to enhance aggregation, prostacyclin (prostaglandin I_2) originates from the vascular endothelium and has the opposite physiologic effect on primary hemostasis (vasodilation and inhibiting platelet aggregation). These two prostaglandins work in concert to maintain a balanced, normal hemostatic system.

Since mature platelets do not contain a nucleus and are therefore incapable of mRNA synthesis, it was long believed that the constitutively expressed COX-1 was the only COX isoform expressed in circulating platelets, and that COX induction could happen only at the level of the bone marrow. As a result, it was originally believed that COX-1 was the primary mediator of platelet thromboxane A_2 production, while COX-2 was the main isoform that controlled endothelial prostacyclin production.[8,14] Recently, using flow cytometry, COX-2 expression was identified in both human and canine platelets.[14,15] In humans, platelet COX-2 expression is greater during times of increased platelet turnover and higher levels of reticulated platelets.[13,16] These recent discoveries suggest that platelet COX-2 could in fact play a role in platelet function.[10,14]

THROMBOXANE METABOLITES

Once synthesized and released by activated platelets, thromboxane A_2 (TXA_2) triggers vasoconstriction, enhances platelet aggregation, and induces further platelet activation. Thromboxane concentration is a reliable indicator of COX function, both in hypercoagulable states and during aspirin therapy. Urinary concentrations of 11-dehydro-thromboxane B_2 (the most abundant urinary metabolite of thromboxane) and 2,3-dinor thromboxane B_2 are increased in human diseases that are associated with a hypercoagulable state.[17] Elevated urinary concentrations of 11-dehydro-thromboxane B_2 have also been identified in dogs with gastric dilatation-volvulus.[17] Measurement of urinary 11-dehydro-thromboxane B_2 has been utilized in previous studies as an indicator of aspirin-induced COX inhibition.[17-19] Urinary assessment of 11-dehydro-thromboxane B_2 and/or 2,3-dinor thromboxane B_2 are useful indicators of *in vivo* thromboxane production, however, the 2,3-dinor thromboxane metabolite is a more sensitive indicator of aspirin induced platelet dysfunction.[17,19,20] In humans, greater than 95% reduction in thromboxane synthesis is required for aspirin to have a sufficient clinical effect on platelet function.[14]

Levels of thromboxane metabolites in plasma or serum can be artifactually increased due to platelet activation during sample collection and processing. The concentration of 11-dehydro-thromboxane B_2 in plasma is too low to measure reliably. The use of urine to measure thromboxane metabolites avoids the problem of *in vitro* platelet activation associated with blood collection, and is therefore considered to be a more reliable indicator of platelet thromboxane synthesis.[18,21] Unfortunately, assessment of thromboxane metabolites currently requires access to specialized equipment that is not available in practice, limiting the effectiveness of these assays in a clinical setting. Urinary 11-dehydro-thromboxane B_2 and 2,3-dinor thromboxane B_2 levels can be measured via gas chromatograph/tandem mass spectrometry, ELISA, and radioimmunoassay. Even though there is significant physiological effect on platelet function and hemostasis, thromboxane is also synthesized by neutrophils, activated monocytes and endothelial cells, thus, thromboxane A_2 metabolites are not a specific marker of platelet activation. Additionally, the kidney also is capable of synthesizing thromboxane, but this thromboxane usually is excreted into the urine as thromboxane B_2 and not 11-dehydro-thromboxane B_2.[17,20]

ASPIRIN

Aspirin inhibits platelet function by irreversibly acetylating a hydroxyl group on the COX enzyme, preventing arachidonic acid from reaching the active site of the enzyme.[9,13] By acetylating the COX enzyme, aspirin prevents platelet thromboxane A_2 synthesis, and inhibits platelet activation and function.[9,22] Since platelets do not contain a nucleus, and are unable to synthesize new and functioning COX enzymes, aspirin-induced platelet dysfunction will persist for the entire lifespan of the platelet, and function will not return until the drug is discontinued and new platelets, not exposed to aspirin, are released from the bone marrow. The actions of COX-2 at the vascular endothelium, which is less sensitive than COX-1 to inhibition by aspirin at anti-platelet doses, allows for continued COX-2 dependent PGI_2 synthesis of anti-platelet doses, in spite of the early suppression of COX-1.[23] Prostacyclin produced by the vasculature can be derived through the actions of both COX-1 and COX-2,[8,22] and the actions of COX-2 at the level of the vascular endothelium are less sensitive to inhibitory actions of aspirin at anti-platelet doses.[24] This allows continued COX-2 dependent prostacyclin synthesis in the face of anti-platelet doses of aspirin, in spite of the early suppression of COX-1 production of prostacyclin.[22] A high dose of aspirin permanently inhibits COX function in all cells that express this enzyme, including vascular endothelial cells. However, unlike the higher, anti-inflammatory dose, a 'low' dose of aspirin will inhibit platelet thromboxane synthesis without permanently inhibiting COX function of the vascular endothelium, allowing prostacyclin synthesis to continue.[25]

ADDITIONAL PROSTAGLANDIN EFFECTS ON COAGULATION

Thromboxane and prostacyclin are not the only prostaglandins that influence platelet function and coagulation. Prostaglandin D_2 is synthesized from arachidonic acid by a cyclooxygenase enzyme, and acts on a platelet receptor that stimulates adenylate cyclase, increases concentrations of cyclic adenosine monophosphate (cAMP) and suppresses platelet function.[26] Inhibition of the intra-platelet second-messaging proteins involved in signal transduction, in contrast, will block platelet activation regardless of the inciting agonist or stimulus. The cyclic nucleotide cAMP is an essential inhibitory intra-platelet second messengers that is capable of interfering with several facets of platelet activation, including degranulation, cytoskeletal rearrangement, and fibrinogen receptor activation.[27] An increase in platelet cAMP levels inhibits signal transduction and platelet activation.[27,28] The effect of prostaglandin E_2 represents a balance between enhancement and inhibition of platelet function via different surface receptors and intra-platelet second messaging proteins. In fact, the low concentration of PGE has a proaggregatory effect on platelets, while a high concentration of this prostaglandin is capable of inhibiting platelet function.[29]

REFERENCES

1. McMichael M. *J Vet Emerg Crit Care* 2005;15(1):1. 2. Lee D, et al. *Biophys J* 2012;102:472. 3. Andrews R, et al. *Thromb Res* 2004;114:447. 4. Sobieszczyk P, et al. *Circulation* 2002;(106):1748. 5. Italiano J, et al. *Blood* 2008;111(3):1227. 6. Sehgal S, et al. *J Thromb Haemost* 2007;5:2009. 7. Rivera J. *Haematologica* 2009;94:700. 8. McAdam B, et al. *Pro Natl Acad Sci USA* 1999;96:272. 9. Bjorkman D. *Am J Med* 1998;105(1B):8S. 10. Rocca B, et al. *Proc Natl Acad Sci USA* 2002;99:7634. 11. Munoz J, et al. *Inflammopharmacol* 2010;18:65. 12. Chandrasekharan N, et al. *Proc Natl Acad Sci USA* 2002;99(21):13926. 13. Guthikonda S, et al. *J Thromb Haemost* 2007;5:490. 14. Weber A, et al. *Br J Haematol* 2002;117:424. 15. Thomason J, et al. *J Vet Intern Med* 2011;25:1106. 16. Tanaka N, et al. *Arterioscler Thromb Vasc Biol* 2004;24:607. 17. Baltzer W, et al. *Am J Vet Res* 2006;67(1):78. 18. Perneby C, et al. *Thromb Res* 1999;96:427. 19.McConnell J, et al. *Neurosci Lett* 2001;313:88. 20. Hoh C, et al. *Am J Vet Res* 2011;72:1038. 21. Catella F, et al. *Proc Natl Acad Sci USA* 1986;83:5861. 22. Patrono C, et al. *Chest* 2001;119:39S. 23. Clarke R, et al. *N Engl J Med* 1991;325:1137. 24. Topper J, et al. *Proc Natl Acad Sci USA* 1996;93:10417. 25. Rackear D, et al. *J Am Anim Hosp Assoc* 1988;24(1):23. 26. Vezza R, et al. *Blood* 1993;82:2704. 27. Gresele P, et al. *Br J Clin Pharmacol* 2011;72(4):634. 28. Yamamoto H, et al. *Circ J* 2008;72:1844. 29. Iyu D, et al. *Platelets* 2010;21:329.

THROMBOLYSIS AND THROMBECTOMY: INITIAL INTERVENTIONAL EXPERIENCE

Chick Weisse, VMD, DACVS
New York, NY

Advanced imaging techniques, the management of extremely debilitated patients with a host of hypercoagulable risk factors, and the growing use of indwelling catheters and medical devices will likely increase the number veterinary patients identified with clinically detectable thrombosis. While many of these patients will not require intervention for these problems, a growing proportion will as the underlying disease processes are now being better managed. The most common underlying diseases to investigate in a complete evaluation include cardiac, endocrine, inflammatory, hepatic, renal, and neoplastic processes. In addition, idiopathic thrombosis can also occur when an underlying condition is not identified. The complete diagnostic work-up for a patient with thrombosis is beyond the scope of this lecture, which will focus primarily on interventional management options when the clinician feels they are indicated. Standards-of-care have not yet been determined or evaluated in veterinary patients with symptomatic thrombosis so the procedures discussed are based solely on the author's experience.

Differentiating Thrombus Type and Guiding Therapy

This lecture will be divided into three separate parts based upon anatomical location and thrombus type, including arterial, venous and pulmonary thrombi. Arterial thrombi form under high shear forces are therefore comprised mostly of platelets with fibrin strands binding them together. Alternatively, venous thrombi form under much lower shear forces and therefore are composed mostly of fibrin and red blood cells. Pulmonary thrombi are generally an intermediate form or mixed thrombi. The development of pulmonary thrombi are likely very different from those encountered in humans and therefore the treatment options will likely vary.

Understanding the nature of the thrombi, as well as the age or chronicity, will help the clinician determine the appropriate therapy and likelihood of success with treatment. For instance, it is generally recognized that anti-platelet drugs are generally not necessary for low-platelet venous thrombi, although veterinarians often prescribe these drugs. In addition, the location of the thrombus, and completeness of vascular obstruction, will also play an important role in determining the role of interventions. For instance, an arterial thrombus leading to impaired bloodflow to an organ or anatomical location is an emergency whereas a venous obstruction of the caudal vena cava or jugular vein may be either subclinical or only result in swelling and edema. On the other hand a jugular thrombus may grow down to the cranial vena cava and result in pleural effusion, chylothorax, and severe head swelling with respiratory compromise or a portal vein thrombus may result in severe portal hypertension, ascites, and a compromised gastrointestinal tract requiring urgent intervention.

Lastly, the patient's physical status, underlying comorbidities, prognosis, and the client's financial concerns will likely all be important considerations in the care provided to the individual. Often, even when the thrombus can be safely and effectively addressed, if the underlying condition is not treatable (ie cannot be permanently resolved), then the patient will remain at risk for future thrombosis. In addition, these patients may even be at higher risk for recurrence if medical devices such as intravascular stents are placed.

Interventional Radiology Treatment Options

The failure of medical therapies to be effective alone, combined with the overall high morbidity associated with surgical thrombectomy in hypercoagulable, sometimes post-surgical patients soon to be receiving anti-coagulation, makes minimally-invasive treatment options a reasonable consideration in many of these patients. The excitement associated with performing these procedures must not overshadow the most important part of the therapy; The underlying condition must be addressed and the hypercoagulable state must be resolved or any intervention is likely to fail. Any intervention meant to treat these problems must be associated with a comprehensive medical plan to manage the underlying disease process and hypercoagulable state prior (if possible) to the intervention. This means aggressive anticoagulation and antiplatelet therapy if indicated. Recently available human oral anticoagulation (Rivaroxaban) may facilitate this process in animals as well.

Arterial Thromboembolism

Arterial obstructions often present with more severe clinical signs depending upon the amount of tissue risking devitalization. Peripheral arterial thrombi are less commonly diagnosed in veterinary patients, likely due to the vast and abundant collateral perfusion throughout much of the body. The classic arterial thromboembolism scenario in veterinary medicine is the saddle thrombus occluding unilaterally or bilaterally the pelvic limbs. In severely affected animals the attending clinician often does not have the luxury of time to see if systemic anticoagulation and/or lytic therapy will be effective. Historically, surgical thrombectomy has been performed and rheolytic thrombectomy has been described. Both techniques are effective but unfortunately the underlying cardiac disease and fragile systemic condition of the patient often results in poor longer term outcomes in approximately 50% of the patients. In addition, acute reperfusion of both pelvic limbs would be associated with malpractice litigation if performed in humans due to the anticipated risk of death from reperfusion injury. Before performing such a procedure the attending clinician should have a long conversation with the pet owner about such a sequela. It has been the author's experience that dogs (and likely cats) can likely tolerate occlusion of both the internal iliac arteries without clinical signs. Once the thrombus grows and extends beyond this bifurcation, clot begins travelling down the external iliacs arteries and into the femoral arteries. This is the point of clinical presentation. As long as both external iliac arteries can remain patent the animal can recover. The use of infusion thrombolysis with vascular stenting has been successful in a small number of these patients treated by the author. This technique appears to provide a more delayed reperfusion than the hyperacute reperfusion encountered during complete clot removal.

Interestingly, the author has recognized more chronic distal aorta thrombi in dogs between the renal and external iliac arteries; the chronicity is identified by collateral circulation and the clinical signs associated with claudication rather than acute hindlimb ischemia. These patients should not be "lumped in" with saddle thrombi patients. These patients can respond to vascular therapy very well as the risk of acute reperfusion injury is not expected. Aggressive and early physical therapy should be performed as the author has seen muscle contracture following arterial ischemia the pelvic limb in a dog.

Venous Thromboembolism (VTE)[1]

Treatment of venous thrombi must be based upon consideration of the overall clot burden and extent, the resulting clinical signs associated with the venous obstruction, and the ability to control the underlying cause of the clot if it is known. In addition, one of the most problematic complications in humans with venous obstruction is "Post Thrombotic Syndrome (PTS)", a group of conditions including swelling, pain, skin ulceration and discoloration resulting from chronic untreated DVTs. The cause is unclear but likely associated with chronic inflammation and damaged venous valves. We are unclear if this occurs in animals. The major goals of VTE treatment include resolution of clinical signs, avoidance of PTS, and reduction in risk of PTE. While anticoagulation and systemic thrombolysis continues to play a predominate role in VTE, catheter-based techniques may help improve outcomes when considering the goals stated above. These techniques are generally divided into three categories including (1) Passive infusion of thrombolytics into the clot through infusion catheters called "catheter directed thrombolysis (CDT), (2) "Percutaneous mechanical thrombectomy (PMT)" involving mechansims of clot aspiration and maceration, and (3) Lytic-assisted devices providing pharmacomechanical and sonically enhanced thrombolysis techniques.

CDT is technically easy and relatively inexpensive to perform however improved outcomes have not yet been documented and prolonged infusion times and increased risk of bleeding are the problems. PMT provides an elegant approach and have been demonstrated to work well with hyperacute thrombi, however they tend to fall short in more chronic thrombi that are 2-3 weeks old. The lytic-assisted devices are the new frontier and work well in smaller vessels, however larger vessels may also require adjunctive CDT or vascular stenting.

Venous stenting is often employed in processes in which the large veins may be externally compressed lading to recurrent thrombosis however stent are often avoided at the confluence of veins or in the limbs where stent fracture and other complications can occur. Stenting has been employed more often at the author's institution due to the relative ease and rapidity of placement, immediate resolution of clinical signs, and general chronicity of most of the clots encountered in our practice.

Pulmonary Thromboembolism (PTE)[2]

As mentioned above, the cause of PTE in veterinary patients is likely different from those in humans who often suffer from migration of deep vein thrombosis (DVTs); As such the treatments are likely to be different. In people, PTE carries an approximate 30% mortality rate if left untreated compared to ~8% when treated. Not all PTEs are the same, however, so careful risk stratification must be pursued in order to determine which patients will benefit from such an intervention and which patients may be harmed. As such, the terms "massive PE" and "submassive PE" have been used. In massive PE, hemodynamic instability (systemic hypotension and circulatory collapse) is likely an absolute indication for intervention (thrombolysis for instance). In submassive PE, right ventricular dysfunction (RV:LV ratio greater than or equal to 1 in humans), high clot burden, or concomitant DVTs may raise the consideration for intervention.

The presence of DVTs (highly uncommon in veterinary patients) makes the use of vena cava filters a rare consideration in our patients while it is a very common, rapid, minimally-invasive and safe procedure in humans. These implantable basket-like devices allow bloodflow to persist through the vena cava but catch any blood clots launching up the vena cava, preventing further PTE. In addition, the clot is trapped in an area of high bloodflow, facilitating lysis.

Systemically administered thrombolysis is contraindicated in patients with recent surgery, known or recent hemorrhage, intracranial disease (bleeding, tumor, trauma, or stroke). While this treatment is appealing, in humans there has been statistically significant clinical evidence that systemic thrombolysis is associated with improved mortality over anticoagulation therapy alone. There is growing evidence that shorter infusion times (<2hrs) may be preferred over the historically longer infusion times (~12hrs for instance) in terms of more rapid clot lysis and fewer bleeding sequelae. For PTEs, there is evidence that systemic thrombolysis does not provide sufficient clot exposure to the compound and more localized therapy is therefore warranted (consider an open bottle of wine "breathing" without a decanter).

Surgical embolectomy is associated with high morbidity and mortality in human hospitals without substantial experience and multi-disciplinary approaches. As such, the risks associated with the procedures performed in veterinary patients is like to be at least as high but may be considered in those patients with massive PE and contraindications to thrombolysis. Catheter-based techniques delivering localized tPA through infusion catheters, pigtail fragmentation catheters, thrombectomy devices, etc. offer a more elegant and theoretically improved clot removal approach but little evidence exists supporting an advantage at this time.

REFERENCES

[1] Razavi MK. Catheter-based therapies for DVT. *Endovascular Today* 10:2;45-48, 2011. [2] Arellano, MP, Tapson, VF. Pulmonary embolism treatment strategies. *Endovascular Today* 11:11;74-78, 2012.

THE IMPORTANCE OF PROTEINURIA IN CHRONIC KIDNEY DISEASE

Gregory F. Grauer, DVM, MS, DACVIM (Small Animal Internal Medicine)
Manhattan, KS

INTRODUCTION

The healthy glomerular capillary wall restricts the filtration of most plasma proteins based on their size and, to a lesser extent, their charge. Type IV collagen within the lamina densa of the glomerular basement membrane and slit-diaphragm formed by the foot processes of podocytes are the major structures responsible for size and charge semi-permeable nature of the glomerular capillary wall. Plasma proteins that do enter the glomerular filtrate are largely reabsorbed by the proximal tubular epithelial cells via endocytotic uptake driven by the multi-ligand receptors megalin and cubulin. Proteinuria in dogs and cats with chronic kidney disease (CKD) can occur due to glomerular and/or tubular lesions. Glomerular proteinuria can be caused by loss of integrity of/damage to the capillary wall (e.g., immune complex disease and x-linked hereditary nephropathy). It is also likely that increases in glomerular capillary pressure increases the amount of filtered plasma protein. Intraglomerular hypertension may result from loss of nephrons (loss of autoregulation) and from systemic hypertension being transmitted into glomerular capillaries. Structural glomerular disease and CKD are often accompanied by systemic hypertension which can exacerbate intraglomerular hypertension and glomerular proteinuria. Tubular proteinuria occurs when tubular reabsorption of protein from the glomerular filtrate is compromised. Tubular proteinuria is typically of lesser magnitude compared with glomerular proteinuria. Reduced tubular reabsorption of protein in dogs and cats with CKD can occur with tubulointerstial injury and decreased numbers of functioning tubules. Whether caused by capillary wall lesions, tubular lesions, or intraglomerular hypertension, excessive quantities of protein in the glomerular filtrate may contribute to additional glomerular and tubulointerstitial lesions leading to loss of more nephrons.

Renal proteinuria is a diagnostic marker of the severity of renal disease and potentially a mediator of glomerular and tubular injury. Recent findings have demonstrated that proteinuria is associated with increased risk of developing azotemia in aged cats and progression of renal disease in both dogs and cats with azotemic CKD. Proteinuric renal disease and systemic hypertension are often co-existent and therefore, it is difficult at times to separate the effects of high systemic and intraglomerular pressures and proteinuria.

DIAGNOSIS OF RENAL PROTEINURIA

Diagnosis of proteinuria in cats and dogs with CKD should be accomplished in a step-wise fashion. The specificity of the dipstick screening test for albuminuria is poor (especially in cats) and therefore confirmation of traditional dipstick positive proteinuria should be accomplished with a more specific follow-up test such as the sulfosalicylic acid (SSA) turbidimetric test, UP/C, or species specific albuminuria assays. The second step is assessment of proteinuria is to determine its origin. Renal proteinuria can adversely affect the prognosis of dogs and cats with CKD and therefore, physiologic or benign proteinuria and pre- and post-renal proteinuria should be ruled out. Subsequently, via serial monitoring, it should be determined if the proteinuria is persistent or transient and if persistent – is the magnitude stable, increasing, or decreasing over time? Renal proteinuria is persistent and associated with a benign or inactive urine sediment (hyaline casts may be observed in the urine sediment in cases of renal proteinuria). Persistent proteinuria is defined as at least two positive tests at two week intervals. Relatively mild proteinuria in dogs and cats with spontaneous/naturally-occurring CKD appears to be a negative predictor of survival. In dogs and cats with the remnant kidney model of CKD, proteinuria is associated with nephron hypertrophy and increased intraglomerular pressures. Persistent proteinuria of renal origin of a magnitude $\geq$ UP/C of 0.4 in cats and $\geq$ 0.5 in dogs with azotemic CKD should be treated with a renal diet (high quality, reduced quantity dietary protein) and an angiotensin converting-enzyme inhibitor (ACEI). Borderline proteinuria is defined as a UP/C between 0.2 and 0.5 in dogs and 0.2 and 0.4 in cats and warrants increased monitoring.

PROTEINURIA: POTENTIAL MEDIATOR OF RENAL DISEASE PROGRESSION?

Proteinuria in dogs and cats with CKD is associated with disease progression and shorter survival times. It is unknown however if proteinuria is a mediator of renal disease progression or if proteinuria is simply a marker of more severe disease that is more likely to be progressive. If the former is true, treatments aimed at reducing proteinuria should be associated decreased disease progression and improved survival. Excessive filtration of plasma proteins could lead to progression of CKD via enhanced glomerular and tubulointerstitial damage. Within Bowman's capsule plasma proteins may cause mesangial cell proliferation and expansion, adhesions of glomerular capillaries to Bowman's capsule, and ultimately, glomerulosclerosis. Proteinuria that overloads tubular reabsorption may also cause inflammatory and fibrotic damage to the tubulointerstital compartment. Excessive reabsorption of plasma proteins may cause activation of transcription factors like NF-κβ which can in turn up-regulate production and release of cytokines, vasoactive peptides, and growth factors. Epithelial cell trans-differentiation into myofibroblasts is mediated by TGF-β and may be an important step in the genesis of tubulointerstital fibrosis. TGF-β also promotes synthesis of extracellular matrix. Urine TGF-β/creatinine ratios are increased in cats with CKD (compared with healthy cats) and correlate with serum creatinine concentrations.

WHAT EVIDENCE EXISTS THAT PROTEINURIA IS DETRIMENTAL IN CANINE AND FELINE CKD?

1. Glomerular capillary hypertension, glomerular enlargement, and proteinuria have been documented in dogs with the remnant kidney model of renal failure. Brown SA, et al. Am J Physiol 1990;258:F495.
2. In cats with the remnant kidney model of CKD, increases in single nephron GFR occur in association with glomerular hypertrophy, increasing intraglomerular pressures, hyperfiltration, mesangial matrix expansion, and proteinuria. Brown SA, et al. Am J Physiol 1995;269:R1002.

3. In dogs with naturally occurring CKD, the relative risk of uremic crises and mortality was approximately three time greater in dogs with UP/C's > 1.0 (n=25) compared with dogs with UP/C's < 1.0 (n=20). In this study the risk of an adverse outcome was approximately 1.5 times greater for every 1 unit increase in UP/C and the decline in renal function was greater in dogs with higher UP/C's. Jacob F, et al. Am J Vet Res 2005;226:393.
4. In cats with naturally occurring CKD, relatively mild proteinuria (UP/C's > 0.4) appear to be negative predictors of survival. Increasing proteinuria was associated with increasing serum creatinine concentrations and increasing systolic blood pressure (presumably related to glomerular hyperfiltration). UP/C, age, and serum creatinine concentration (but not blood pressure) were independently associated with mortality. Syme HM, et al. J Vet Intern Med 2006;20:528.
5. Proteinuria has been associated with increased risk of mortality due to all causes in cats that have normal renal function when their proteinuria is first detected. Walker D, et al. J Vet Intern Med 2004;18:417.
6. In 141 client-owned cats with naturally-occurring systemic hypertension, amlodipine treatment decreased both blood pressure and proteinuria. Proteinuria (UP/C) (before and after treatment as well as the change in UP/C) was the only variable related to survival in these cats. Jepson RE, et al. J Vet intern Med 2007;21:402.
7. In cats with naturally-occurring CKD, shorter survival time was associated with an increased UP/C as an independent risk factor. King JN, et al. J Vet Intern Med 2007;21:906.
8. In a prospective, longitudinal cohort study of non-azotemic cats ≥ 9 years, 95 cats (median age = 13) were followed for 12 months or until death or azotemia developed. 29/95 (30.5%) developed azotemia (Sr Cr > 2.0 mg/dl). Proteinuria at presentation (median UP/C of 0.19 vs. 0.14) was significantly associated with development of azotemia. Jepson RE, et al. JVIM 2009; 23:806-813.
9. In 59 cats with CKD (serum creatinine concentrations > 2.0 mg/dl, USG < 1.035, and history and clinical signs compatible with CKD vs. AKI) with postmortem data and biochemistry and urine protein data (obtained during the last 2 months of life), the UP/C was positively correlated with both renal interstitial fibrosis score and maximal glomerular volume. 34 of the 59 cats (58%) were classified as proteinuric. Chakrabarti, et al. Diagnostic Pathol 2013;50:147-155.
10. When cats with CKD surviving for more than one month (surviving group, n = 34) were compared with cats with CKD surviving less than one month (non-surviving group, n = 16), UP/C was significantly higher in the non-surviving group. In the surviving group, UP/C was the only clinicopathologic variable that exhibited a consistent alteration (increase) in relation to first visit data and was most likely to be associated with mortality. Kuwahara, et al. J Small Anim Pract 2006;47:446-450.
11. When client-owned cats with stable CKD (n=112) were compared with client-owned cats with progressive CKD (n=101), median UP/Cs in the progressive group were higher when compared with the stable group (0.27 vs. 0.14). A 0.1 increase in UP/C was associated with a 24% increase in risk of progression of CKD. Chakrabarti, et al. J Vet Intern Med 2012; 26:275-281.

ACE INHIBITORS AND RENAL DISEASE: THE GOOD, BAD, AND UGLY

Gregory F. Grauer, DVM, MS, DACVIM (Small Animal Internal Medicine)
Manhattan, KS

INTRODUCTION

By altering pre-glomerular resistance, healthy kidneys can maintain relatively stable glomerular capillary pressures despite variations in systemic blood pressure. This pressure regulatory process is termed "renal autoregulation". Autoregulation can be reduced when renal disease results in loss of nephrons. Compromised autoregulation allows high systemic blood pressure to be transmitted to glomerular capillaries. This glomerular hypertension has been documented by micropuncture studies in dogs and cats with surgically reduced renal mass. In these models, glomerular hypertension was associated with glomerular hypertrophy, sclerosis and proteinuria. Systemic hypertension is relatively common in dogs and cats with renal disease. In a recent study of dogs with spontaneous chronic kidney disease (CKD), 29/45 (64%) had systolic blood pressure $\geq$ 144 mm Hg and 14/45 (31%) had systolic blood pressure $\geq$ 161 mm Hg. In cats with naturally-occurring CKD, systemic hypertension has been observed in 19-65% of cases depending on the definition of hypertension.

Proteinuria in dogs and cats with chronic kidney disease (CKD) can occur due to glomerular and/or tubular lesions. Glomerular proteinuria can be caused by loss of integrity of/damage to the capillary wall (e.g., immune complex disease and x-linked hereditary nephropathy). It is also likely that increases in glomerular capillary pressure increases the amount of filtered plasma protein. Intraglomerular hypertension may result from loss of nephrons (loss of autoregulation) and from systemic hypertension being transmitted into glomerular capillaries. Structural glomerular disease and CKD are often accompanied by systemic hypertension which can exacerbate intraglomerular hypertension and glomerular proteinuria. Tubular proteinuria occurs when tubular reabsorption of protein from the glomerular filtrate is compromised. Tubular proteinuria is typically of lesser magnitude compared with glomerular proteinuria. Ruduced tubular reabsorption of protein in dogs and cats with CKD can occur with tubulointerstial injury and decreased numbers of functioning tubules. Whether caused by capillary wall lesions, tubular lesions, or intraglomerular hypertension, excessive quantities of protein in the glomerular filtrate may contribute to additional glomerular and tubulointerstitial lesions leading to loss of more nephrons. Renal proteinuria is a diagnostic marker of the severity of renal disease and potentially a mediator of glomerular and tubular injury. Recent findings have demonstrated that proteinuria is associated with increased risk of developing azotemia in aged cats and progression of renal disease in both dogs and cats with azotemic CKD.

HYPERTENSION

Described and potential etiologies of secondary hypertension include acute and chronic renal disease, hyperthyroidism, hypothyroidism, hyperadrenocorticism, hyperaldosteronism, pheochromocytoma, diabetes mellitus, and obesity. Chronic kidney disease has the greatest association with hypertension and may often be causal. A recent report suggested approximately 29% of elderly cats with CKD were hypertensive, with the range reported in 4 studies being 19-65%. In dogs with CKD, approximately one-third will be normotensive, one-third will have borderline hypertension, and one-third will be hypertensive.

Systemic hypertension may contribute to progressive nephron loss by causing irreversible glomerular damage via increased intraglomerular pressures and glomerulosclerosis. Inappropriate dilation of the afferent glomerular arteriole occurs in dogs and cats with CKD and diminishes the ability of the afferent arteriole to protect the glomerulus from variations in systemic blood pressure. Although the exact mechanism of the CKD-associated hypertension is not known, a combination of glomerular capillary and arteriolar scarring, decreased production of renal vasodilatory prostaglandins, increased responsiveness to normal pressor mechanisms, and activation of the renin-angiotensin-aldosterone system (RAAS) may be involved. The increased renin secretion leads to increased production of angiotensin II and aldosterone. In addition to its direct pressor effects, angiotensin II also has a stimulatory effect on the sympathetic nervous system, increasing vascular tone, and, in CKD vasoconstriction of the efferent arteriole which further contributes to the intraglomerular hypertension. Angiotensin and aldosterone may also stimulate renal tissue remodeling via increased matrix production and fibrosis.

The consequences of systemic hypertension are usually dependent on the magnitude and duration of the blood pressure elevations. Acute ocular and central nervous system abnormalities can occur associated with hemorrhage or edema formation. Renal damage associated with hypertension tends to be more chronic and characterized by glomerular lesions (e.g., glomerulosclerosis) and proteinuria. Finally, functional/adaptive changes like ventricular hypertrophy can occur due to increased after-load in patients with hypertension. Diagnosis and treatment of hypertension in dogs and cats with CKD may prevent development of retinal and CNS lesions or may limit or slow progression of renal and cardiac lesions.

PROTEINURIA

Renal proteinuria is a diagnostic marker of the severity of renal disease and potentially a mediator of glomerular and tubular injury. Recent findings have demonstrated that proteinuria is associated with increased risk of developing azotemia in aged cats and progression of renal disease in both dogs and cats with azotemic CKD. In addition, studies have shown that therapies that reduce the magnitude of proteinuria may be renoprotective. Proteinuric renal disease is often associated with systemic hypertension, which can conversely exacerbate renal proteinuria and therefore, it is difficult at times to separate the effects of high systemic and intraglomerular pressures and proteinuria. Prior to treatment of proteinuria, renal proteinuria should be confirmed (persistent proteinuria with a normal urine sediment evaluation) and the average of 2-3 UP/Cs should be used to quantitate baseline proteinuria.

WHAT EVIDENCE EXISTS THAT HYPERTENSION IS DETRIMENTAL IN CKD? (See proceedings "The Importance of Proteinuria in Chronic Kidney Disease" for evidence that proteinuria is detrimental to dogs and cats with CKD)

1. In dogs with the page/remnant kidney model of chronic renal failure (CRF), systemic hypertension had adverse effects on renal function and morphology. Finco DR. *J Vet Intern Med* 2004; 18:289-294.
2. In dogs with remnant kidneys, 40-45 mmHg increases in systolic blood pressure over baseline resulted in enhanced renal damage. Brown SA, et al. *J Vet Intern Med* 2000; 14:351.
3. In dogs with spontaneous chronic renal failure, initial high systolic blood pressure was associated with increased risk of developing a uremic crisis and of dying. Jacob F, et al. *J Am Vet Med Assoc* 2003; 222:322-329.
4. In cats with the remnant kidney model of CKD, hypertension was associated with more severe glomerular histologic lesions. Mathur, et al, *Am J Vet Res* 2004; 65:1006-1013.
5. In 69 cats with CKD (serum creatinine concentrations > 2.0 mg/dl, USG < 1.035, and history and clinical signs compatible with CKD vs. AKI) with postmortem data and time-averaged systolic blood pressure (SBPOT) (obtained over a mean of 284 days) the SBPOT was positively correlated with maximal glomerular volume, hyperplastic arteriolosclerosis, and glomeruloscerosis. 34 of the 69 cats (49%) were classified as hypertensive (SBPOT of 159 mm Hg vs 136 mm Hg in normotensive cats). Chakrabarti, et al. *Diagnostic Pathol* 2013;50:147-155.

MANAGEMENTOF HYPERTENSION AND PROTEINURIA WITH ACEI

Several studies document the antiproteinuric/antihypertensive effects of enalapril and benazepril in dogs and cats. Most of the canine studies have been accomplished with enalapril whereas most of the feline studies have used benazepril. Because of their antiproteinuric effects, largely associated with normalization of intraglomerular pressures, the use of ACEI is considered part of the standard of care in dogs and cats with proteinuric CKD. ACEI treatment is recommended when the UP/C is > 0.4 and > 0.5 in azotmic cats and dogs, respectively or there is mild to moderate systolic hypertension (BP > 160 but < 180 mmHg). The starting dose for both enalapril and benazepril is 0.5 mg/kg PO q 24 hours. The goal of treatment is a UP/C < 0.4/0.5 in cats and dogs, respectively or at least a 50% reduction in the baseline UP/C. Maintenance of systolic blood pressure < 160 mmHg is the antihypertensive treatment goal. Enalapril and benazepril can be increased by 0.5 mg/kg/d if necessary to obtain treatment goals. Doses greater than 2.0 mg/kg/d should be used with caution and increased monitoring (doses greater than 1.0 mg/kg/d may result in further reductions in proteinuria; however there are no controlled studies proving these higher doses are beneficial). Enalapril and benazepril appear to be equally effective in their antiproteinuric and antihypertensive effects but there are important differences in excretion between the two compounds. The active metabolite of benazepril, benazeprilat, is largely excreted via the biliary system and therefore excretion is not affected by decreases in GFR. Enalapril and its active metabolite, enalaprilat, are largely excreted by the kidney and may accumulate when GFR is reduced. This difference in excretion is likely not clinically relevant until serum creatinine concentration is > 3.0-3.5 mg/dl. Use of ACEI is likely associated with reduced efficacy and increased risk of adverse drug events as azotemia increases. Increased monitoring should be employed in dogs and cats with serum creatinine concentrations greater than 3.0 mg/dl. Use of ACEI in dogs and cats with serum creatinine concentrations greater than 5.0 mg/dl is usually not recommended. Dehydration should be avoided/quickly corrected in any dog or cat being treated with an ACEI. Further recommendations regarding ACEI treatment in dogs can be found in: Brown SA, et al. Consensus recommendations for standard therapy of glomerular disease in dogs. *J Vet Intern Med* 2013; 27:S27-S43.

In addition, transitioning patients onto renal diets with reduced quantity/high quality protein may have an additive effect on proteinuria when used with ACEI (Burkholder WJ, et al, *J Vet Intern Med* 2004; 18:165-175 and Cortadellas O, et al, ACVIM Forum Proceedings, 2012). These diets are also typically supplemented with n-3 fatty acids which have also been shown to reduce/normalize intraglomerular pressure and proteinuria in dogs.

Gradual reduction of dietary salt is often recommended as the first line of treatment for hypertension; however there are no studies that document the efficacy of this treatment. The calcium channel antagonist (CCA), amlodipine is often recommended as the first choice anti-hypertensive treatment for cats. Recent studies have shown that treatment with a CCA alone reduces proteinuria in cats with CKD suggesting that intraglomerular pressures were also reduced by this treatment (Jepson RE, et al. *J Vet Intern Med* 2007; 21:402-409). In those cases where systemic hypertension persists after initiation of ACEI treatment, or is initially greater than 180 mm Hg, a CCA can be added or used as sole therapy. The overall risk of target organ damage to the eyes, brain, heart, and kidneys is thought to be minimal if systolic blood pressure is maintained at ≤ 150 mm Hg.

WHAT EVIDENCE EXISTS THAT ACEI ARE BENEFICAL IN DOGS AND CATS WITH CKD? (THE GOOD)

1. ACEI treatment dogs with induced diabetes mellitus and unilateral nephrectomies resulted in reduced proteinuria, reduced glomerular hypertrophy, and reduced intraglomerular pressures. Brown SA, et al. *Kidney International* 1993; 43:1210-1218.
2. ACEI treatment in Samoyed dogs with X-linked hereditary nephritis resulted in improved survival, renal function, and renal histology compared with no treatment. Grodecki KM, et al. *J Comp Pathol* 1997; 117:209.
3. ACEI treatment that lowered systolic blood pressure and decreased proteinuria was associated with improved outcome in dogs with spontaneous glomerulonephritis. There was a 91.2% chance that treatment with enalapril prevented an increase in serum creatinine concentrations > 0.2 mg/dl (attributable risk reduction). Categorical response to treatment was significantly better in the enalapril treatment group. Grauer GF, et al. *J Vet Intern Med* 2000; 14:526-533.
4. ACEI treatment in dogs with remnant kidney model was effective in modulating progressive renal injury, which was associated with a reduction in glomerular and systemic hypertension. Brown SA, et al. *Am J Vet Res* 2003; 64:321-327.

5. In 61 cats with naturally-occurring CKD, treatment with benazepril decreased proteinuria (UP/C) and decreased progression of the CKD to advanced stages (although the latter may not be clinically significant). Mizutani H, et al. *J Vet Intern Med* 2006; 20:1074-1079.
6. In 192 cats with naturally-occurring CKD, treatment with benazepril significantly reduced proteinuria in cats with all levels of proteinuria. Survival time was increased in the benazepril group compared with the control group but the effect was not significant. King JM, et al. *J Vet Intern Med* 2006; 20:1054-1064.
7. In 26 client-owned dogs with CKD, treatment with benazepril (vs. placebo) decreased UP/C and increased GFR (plasma clearance of iohexol) and health status score over a 6 month study. Tenhundfeld J, et al. *J Am Vet Med Assoc* 2009; 234:1031-1037.

ACEI ARE NOT ALWAYS EFFECTIVE IN MODULATING PROTEINURIA (THE BAD)

Although there are multiple studies documenting the antiproteinuric/antihypertensive effects of ACEI in dogs and cats, clear evidence that ACEI treatment slows the progression of CKD or improves survival is minimal. In addition, ACEI treatment is not always effective in reducing hypertension and/or proteinuria. It's well known that CCA (vs. ACEI) are usually more effective in reducing systemic hypertension in cats compared with dogs. In some cases renal proteinuria may initially respond to treatment with an ACEI and then "escape". This escape may be associated with up-regulation of alternative pathways for production of angiotensin II. For example, elastase and cathepsin G may directly convert angiotensinogen to angiotensin II. In other cases, renal proteinuria may fail to respond, even initially, to ACEI treatment. Potential reasons for this lack of efficacy may involve dosing (i.e., suboptimal doses) or advanced renal disease. Very few of our patients with proteinuric CKD undergo a thorough histologic evaluation. Dogs and cats with advanced histologic disease may be less likely to be responsive to any pharmacologic influence. Similarly, there may be certain structural pathologies that are more likely to be responsive than others. Histologic correlation with response to ACEI treatment has not been evaluated in dogs or cats..

In those proteinuric cases where there is a poor response to treatment (i.e., less than 50% reduction in baseline UP/C), the use of angiotensin receptor blockers (ARB) (either alone or in combination with ACEI) has been recommended. Losartan appears to be the ARB most commonly used in dogs even though dogs do not appear to be able to produce one of the active losartan metabolites. The pharmacodynamics of losartan have been described in normal dogs (Christ DD, et al. *J Pharm Exp* Therapeutics 1994; 268:1199-1205) and there is anecdotal evidence that losartan has antiproteinuric effects in dogs (SL Vaden) and that telmisartan has antihypertensive effects in normal dogs (SA Brown). A recent meta-analysis on the effects of monotherapy and combination therapy with inhibitors of the renin–angiotensin system compared the efficacy of ARB and ACEI in people with proteinuric renal disease. Forty-nine studies involving 6181 participants reviewed randomized trials of ARB vs. placebo, ACEI vs. placebo, or the combination of ARB and ACEI in patients with microalbuminuria or proteinuria for whom data were available on urinary protein excretion at baseline and at 1 to 12 months. Reduction in proteinuria from ARB and ACEI was similar, but their combination was more effective than either drug alone. However, the authors stated that the uncertainty about adverse effects and outcomes that are important to patients limits applicability of these findings to clinical practice (Kunz R, et al, *Ann Intern Med.* 2008; 148:16-29). There are no studies in veterinary medicine evaluating the safety and efficacy of ARB (alone or in combination with ACEI) for the treatment of hypertensive/proteinuric renal disease.

Aldosterone also appears to be associated with progressive fibrotic renal injury, in part by increasing production and release of TGF-β. Aldosterone may also mediate renal damage via hemodynamics effects resulting in intraglomerular hypertension. Treatment with ACEI will often decrease aldosterone concentrations initially but the effect may not be sustained (aldosterone breakthrough). It has been suggested that aldosterone antagonists used in conjunction with ACEI may have enhanced renoprotective effects, but this has not been evaluated in dogs or cats.

ACEI TREATMENT MAY BE ASSOCIATED WITH WORSED AZOTEMIA IN CKD (THE UGLY)

In some cases treatment of proteinuric renal disease with ACEI is associated with worsening azotemia. Fortunately, this adverse event is relatively rare. Most adverse effects occur in dogs with congestive heart failure (CHF) and the relative contributions of the ACEI vs. decreased cardiac output and Lasix therapy on renal function are unknown. It should be noted however that ACEI treatment of CHF had relatively few adverse effects on kidney function (The COVE study, *J Vet Intern Med* 1995; 9:243-252 and Atkins CE, et al, *J Am Vet Med Assoc* 2002; 221:654-658). Adverse effects on renal excretory function would be expected to be more commonplace as azotemia increases. Increased monitoring for worsening of azotemia is recommended in patients with baseline serum creatinine concentrations greater than 3.0 mg/dl; use of ACEI in dogs and cats with serum creatinine concentrations greater than 5.0 mg/dl is usually not recommended.

ALDOSTERONE BREAKTHROUGH (ABT) IN DOGS: PREVALANCE IN CLINICAL CARDIAC AND RENAL PATIENTS RECEIVING ACE INHIBITORS

M. Ames DVM, DACVIM, C. Atkins, DVM, DACVIM, A. Lantis, DVM, DACVIM, G. Grauer, DVM, MS, DACVIM, S. Vaden, DVM, PhD, DACVIM
Fort Collins, CO

INTRODUCTION

Chronic exposure to high concentrations of aldosterone results in excessive sodium retention with expansion of extracellular volume, favors potassium and magnesium wasting, inhibits myocardial norepinephrine uptake, diminishes heart rate variability (a sign of cardiac health), produces cardiac arrhythmias, decreases baroreceptor sensitivity, contributes to endothelial dysfunction and vascular inflammation, and is independently associated with renal, vascular and cardiac remodelling and heart failure.[1-5] The interruption of RAAS is a key therapeutic strategy in the therapy of chronic heart and kidney disease and is achieved via the administration of angiotensin converting enzyme-inhibitors (ACE-I), angiotensin II receptor blockers (ARB), and mineralocorticoid receptor blockers (MRB), alone or in combination. Incomplete pharmacologic blockade of the RAAS system by ACE-I (±ARB) has been referred to as both aldosterone escape and aldosterone breakthrough (ABT). The authors prefer the latter term, as it more accurately reflects the phenomenon in which circulating aldosterone levels rise despite ACE-I or ARB therapy. Aldosterone escape refers to the physiologic phenomenon that allows the kidneys to 'escape' the sodium and water retaining effects of excessive aldosterone levels, preventing severe volume expansion and edema.

CLINICAL RELEVANCE

Increasing evidence links aldosterone excess and/or activation of mineralocorticoid receptors to the development and progression of various cardiovascular disease processes in humans.[6,7] Plasma aldosterone levels at presentation are known to be significantly predictive of mortality after myocardial infarction in humans.[6] The benefit of sequential RAAS blockade with a mineralocorticoid receptor blocker added to either an ACE-I or ARB was shown in three large clinical trials evaluating patients with diminished left ventricular function, and severe heart failure (RALES, EPHESUS) and mild to moderate heart failure (EMPHASIS-HF). These three studies demonstrated that the addition of mineralocorticoid receptor blockade to standard heart failure therapy, including an ACE-I or ARB, led to a significant reduction in mortality and hospitalization for cardiac reasons.[7-9]

In humans with proteinuria due to idiopathic glomerulonephritis and diabetic nephropathy, the addition of spironolactone to either an ACE-I or ARB leads to a greater reduction in proteinuria when compared to cohorts receiving either an ACE-I or ARB alone.[10-12] Also, in patients with chronic kidney disease due to either diabetic nephropathy or IgA nephropathy, rates of urine albumin/protein excretion were found to be higher in patients that experienced ABT.[13] The results of another study of patients with diabetic nephropathy indicated no correlation between these parameters, yet patients with ABT had a greater rate of decline of glomerular function when compared to patients that did not experience ABT.[14] Overall, there is no over-arching consensus regarding the definition, mechanism, and clinical implications of ABT in patients with either chronic renal or heart disease, though existence and importance of ABT are uniformly accepted.

Evidence of the harmful effects of RAAS activation and the benefits of RAAS suppression are evident in veterinary patients as well. In a prospective veterinary study by Hezzell et al.[a], urinary aldosterone concentrations were found to be negatively associated with survival (P=0.005) in 54 dogs with mitral valve disease. A recent double-blind field study in 212 dogs demonstrated a 69% reduction in risk of cardiac morbidity and mortality in dogs with chronic degenerative mitral valve disease that were treated with spironolactone, in addition to an ACE-I with furosemide, ± digoxin, when compared to furosemide, an ACE-I, ± digoxin, without spironolactone.[15] Previous studies in cats[16], experimental models of heart failure in dogs[17,18], and dogs with naturally occurring chronic degenerative valve disease[19] have revealed persistent aldosterone secretion despite ACE-inhibition. Studies in the authors' laboratory have revealed that furosemide-induced RAAS activation of 10 days duration was not always attenuated with concomitant benazepril administration.[b] Similar results were found with enalapril.[c] In both of these studies, failure to suppress aldosterone excretion was associated with effective inactivation of angiotensin converting enzyme. Overall, our experimental data suggests that ABT occurs in over 70% of dogs with amlodipine- or furosemide-activated RAAS, treated with benazepril or enalapril. While we have demonstrated ABT in this experimental model of RAAS activation, the prevalence of ABT in canine patients is unknown. We are currently investigating the prevalence of aldosterone breakthrough in canine patients receiving ACE-I and/or angiotensin receptor blockers for heart failure and for proteinuric renal disease.

DETECTION, DEFINITION, AND INCIDENCE OF ABT IN VETERINARY PATEINTS

An assay for urine aldosterone in the dog has been validated and the urine aldosterone to creatinine ratio (UA:C), using "spot tests" and RIA methodology, has been correlated with 24 hour urine aldosterone excretion[20] and aldosterone secretion (plasma aldosterone concentration).[d] This assay measures free aldosterone. Further validation of the UA:C as a measure of RAAS activation comes in the expected response of the UA:C to perturbations, such as volume reduction, blood pressure reduction, changes in sodium intake, and stress. The pharmacologic interruption of RAAS is studied using a model of RAAS activation with furosemide or amlodipine.[20-24] We have demonstrated ABT in this experimental model of RAAS activation in our laboratory.[23,b,c]

There is no consensus regarding how to define ABT. In human studies of ABT, definitions usually involve 1) a blood aldosterone concentration that increases relative to baseline despite therapy with ACE-I or ARB or 2) a blood aldosterone concentration that exceeds an absolute cutoff value. Most cut-off values are determined from samples of healthy control subjects and therefore vary between studies.

The authors have used two strategies in an effort to define ABT in clinical patients. The first is a baseline definition, when a UA:C is available prior to the initiation of an ACE-I. In these patients, ABT is defined as a UA:C, after initiation of ACE-I, that exceeds the baseline value for that patient. A cut-off definition, used when a baseline UA:C measurement is not available, was determined using the mean UA:C in a population of young (1-2 years of age) normal dogs from NCSU-VTH (0.44 μg/g, ± 0.26). These data were then used to determine two 'cut-off' values, the first being a UA:C greater than the upper 99% confidence limit of the reference interval of our normal population (>0.54 μg/g). The second cut-off is a UA:C greater than the mean plus two standard deviations (>0.96 μg/g). ABT is defined as a UA:C during ACE-I therapy that exceeds this pre-determined cut-off value.

The incidence of aldosterone breakthrough in humans with heart and renal disease varies depending on the definition of aldosterone breakthrough used by investigators. The incidence of ABT is generally lower in studies that use an absolute cutoff value of serum aldosterone.[1] The time period over which patients are evaluated also vary between studies, further complicating assessment of the true incidence of ABT in people. Overall, the prevalence is somewhere between 10% and 53% of people receiving an ACE-I and/or ARB, as treatment for heart failure, chronic kidney disease, or systemic hypertension.

The results from our on-going study, in which dogs with cardiac disease and with protein-losing renal disease are being prospectively enrolled and evaluated for ABT, will be presented. Preliminary results from the author's laboratory suggest that ABT occurs in dogs both with heart disease and proteinuric renal disease. Preliminary data include 35 dogs treated at NCSU for heart disease. All of these dogs were receiving an ACE-I, the majority had been receiving long-term ACE-I therapy, and 83% were in congestive heart failure and were receiving furosemide. The majority of these dogs had degenerative mitral valve disease, yet dogs with dilated cardiomyopathy, congenital heart disease and pulmonary hypertension were also included. Twenty nine percent of these dogs fit the more stringent (mean + 2SD) cut-off value definition of ABT and 51% of dogs fit the less stringent (upper 99% CI) definition of ABT. In seven dogs with proteinuria and serial measurements of UA:C including a baseline measurement, our data indicate a prevalence of ABT in 3/7 (43%) of these dogs, using a baseline definition. In an additional 9 dogs with proteinuria, for which only a single UA:C measurement was available, 3/9 (33%) fit our cut-off definition of ABT (UA:C > 0.96 μg/g). While these numbers are small, we are currently analyzing data on an additional number of dogs (>100) and will present these data at the 2014 ACVIM FORUM. Our laboratory studies and early clinical data suggest that ABT occurs in dogs with experimental RAAS activation and in clinical cases with both cardiac and proteinuric renal disease.

REFERENCES

[1]Bomback AS, et al. **Nat Clin Pract Nephrol** 2007;3:486-492. [2]Weber KT. **N Eng J of Med** 2001;345(23):1689-1697. [3]Wang W, et al. **Hypertension** 1992;19:270-277. [4]Brilla CG, et al. **Eur Heart J** 1995;16 (Supp O):107-109. [5]Martinez, FA. **Cardiovasc Drugs Ther** 2010; 24:345-350. [6]Palmer BR, et al. **Europ Heart J** 2008; 29:2489-2496. [7]Pitt B, et al. **N Eng J Med** 1999;341:709- 717. [8]Pitt B, et al. **N Engl J Med.** 2003;348(14):1309-21. [9]Zannad F, et al. **N Eng J Med** 2011;364(1):11-21. [10]Esteghamati A, et al. **Nephrol Dial Transplant** 2013;28:2823-2833. [11]Bianchi S, et al. **Kidney Int** 2006;70:2116–2123. [12]Nielsen SE, et al. **Diabetic Medicine** 2012;29:e184-e190. [13]Horita Y, et al. **Nephrology** 2006;11:462–466. [14]Schjoedt KJ, et al. **Diabetologia** 2004;47:1936–1939. [15]Bernay F, et al. **J Vet Intern Med** 2010;24:331-341. [16]MacDonald KA, et al. **J Vet Intern Med** 2006;20:1093-1105. [17]Suzuki, G. et al. **Circulation** 2002;106(23):2967-2972. [18]Cataliotti A, et al. **Circulation** 2002;105:639-644. [19]Häggström J, et al. **Am J Vet Res** 1996;57(11):1645-1651. [20]Gardner SY, et al. **J Vet Cardiol** 2007;9:1-7. [21]Atkins CE, et al. **J Vet Pharm Therap** 2007;30:394-400. [22]Sayer MA, et al. **J Vet Intern Med** 2009;23:1003-1006. [23]Lantis, AC, et al. **Amer J Vet Res** 2011;72:1646-1651. [24]Ames MK, et al. **Am J Vet Res** 2013;74(8):1084-1090. [a] Hezzell MJ, et al. (abst) **J Vet Intern Med** 2010;24(3):672. [b] Lantis A, et al. (abst) **J Vet Intern Med** 2010;24(3):672. [c] Lantis, et al. (abst) **J Vet Intern Med** 2012; 26:715. [d] Ames MK, et al. 2014; unpublished, study in progress.

UNUSUAL FELINE ENDOCRINE DISEASES

Cynthia Ward, VMD, PhD, DACVIM
Athens, GA

Thyroid Storm and Acute Thyroid Toxicosis – see previous notes

Hyperadrenocorticism

Hyperadrenocorticism is a rare disorder in cats. It is usually caused by a pituitary adenoma, but like the dog can result from an adrenal adenoma. Functional adrenal carcinomas have also been reported in the cat. Cats usually present in mid-age with hyperadrenocorticism. They are 5-16 years of age with no breed or sex predilection. Most (~90%) have concurrent insulin-resistant diabetes mellitus. An outstanding clinical feature of a cat with hyperadrenocorticism is extremely thin and friable skin. It will often rip when attempting to get a blood sample. Often these cats will present with non-healing wounds. Other clinical signs include polyuria/polydipsia; polyphagia, recurrent infections of the upper respiratory tract, cutaneous systems, and urinary tract. Cats can also be lethargic with significant muscle atrophy and weakness. Their abdomens may be enlarged and they may have alopecia or lack the ability to regrow hair after being clipped.

Clinicopathologic findings from baseline bloodwork include: elevated ALP and ALT, hypercholesterolemia, hyperglycemia, and stress leukogram (mature neutrophilia, lymphopenia, eosinopenia). A urinalysis may reveal glucosuria, lack of urine concentrating ability, proteinuria, and bacteriuria with or without pyuria.

The diagnosis of hyperadrenocorticism in the cat requires a clinical suspicion and the appropriate historical and clinical signs. Baseline bloodwork should be consistent with hyperadrenocorticism. Specific adrenal function testing is similar to that used in screening dogs suspected of having hyperadrenocorticism, but with several nuances. Both the low-dose dexamethasone suppression test and ACTH-stimulation tests are used to assess adrenal function in the cat. The low dose dexamethasone suppression is done similarly as for dogs with a serum taken for baseline cortisol and at 4 and 8 hours following dexamethasone given IV. The difference for cats is that the dose of dexamethasone is 10x that for dogs at a dose of 0.1 mg/kg IV. Hyperadrenocorticism in the cat is supported by lack of suppression of cortisol below the normal range at 4 and/or 8 hour after injection. Unlike in the dog, the Low dose dexamethasone suppression test cannot be used as a differentiating test to indicate pituitary versus adrenal disease. The low dose dexamethasone suppression test is the more sensitive in the cat. The ACTH-stimulation test can also be used to screen cats for hyperadrenocorticism; however, 40-50% of affected cats will have a normal response to ACTH. Synthetic ACTH (cortrosyn) is given IV and serum samples are collected at 0, 60 and 90 minutes. An exaggerated response to ACTH of >19 mcg/dl is consistent with a diagnosis of hyperadrenocorticism.

Differentiating tests are performed to distinguish pituitary versus adrenal disease. In cats endogenous ACTH can be measured using the assay for dogs; however, there is considerable overlap between levels in normal cats and those with pituitary-dependent hyperadrenocorticism. However, if the cat has adrenal disease the endogenous ACTH levels should be low. Adrenal imaging in the cat usually involves ultrasound. It pituitary dependent hyperadrenocorticism is present, both adrenal glands will be enlarged. If there is an adrenal tumor, then one adrenal should be enlarged with a smaller adrenal present on the other side. Mineralization of adrenal glands can be seen in normal cats. MRI can be used to detect pituitary tumors.

Trilostane (Vetoryl^{TM}) is used to medically treat hyperadrenocorticism in cats. Mitotane is not safe for use and should not be given to cats. Trilostane should be given at 5 mg/kg PO q 24 hours and increased or decreased as necessary. As with dogs, trilostane therapy is monitored with an ACTH-stimulation test with the goal of the pre- and post- ACTH cortisol levels being within the normal range. It is important to monitor serum Na/K levels in cats. The ratio may become low signaling a decrease in mineralocorticoid activity. Trilostane dosage should be reduced if this occurs even if the cortisol levels are acceptable. Most cats with hyperadrencorticism have concurrent diabetes mellitus. In these cats, the insulin dosage should be reduced by at least 25% before starting trilostane. The reduction in cortisol levels by trilostane will decrease insulin resistance and make the insulin more potent.

Unilateral adrenalectomy should be considered in cats with an adrenal tumor. Clinical signs of hypercortisolism should be controlled with trilostane for at least 2 weeks before surgery in order to reduce skin fragility, heal skin lesions, and control concurrent diabetes mellitus if present. Glucocorticoid and/or mineralocorticoid therapy may be necessary for several weeks while the healthy adrenal gland regenerates.

Hyperaldosteronism (Conn's Syndrome)

Another uncommon endocrinopathy in cats is hyperaldosteronism or Conn's syndrome. It is caused by overactivity of mineralocorticoid producing cells in the adrenal gland. This is usually caused by a unilateral adrenocortical tumor, although bilateral tumors that both oversecrete aldosterone have been described. This may be a benign or malignant disease.

This is a disease of middle-older aged cats (range 6-20 years) with no breed or sex predilection. The clinical presentation includes a history of episodic lameness or muscle weakness, polyuria/polydipsia, vision loss (due to hypertension-induced retinal detachment), polyphagia, and weight loss. Clinical signs include hypokalemic polymyopathy with ventroelxion of the neck, hindlimb weakness, and paresis. Retinal abnormalities including detachment, tortuous vessels, henorrhages, bullae, and edema may be present. Hypertension is also a common clinical finding.

Clinicopathologic changes include hypokalemia (common), hypernatremia (uncommon), azotemia, elevated CK, and elevated plasma aldosterone. Ultrasound will reveal unilateral or bilateral adrenal masses. Extension of the tumors into the vena cava is common and should be investigated before surgery in considered.

Diagnosis of hyperaldosteronism relies on the presence of appropriate history and clinical signs. Supportive biochemical data should be present; often only hypokalemia is present. Definitive diagnosis is made on the demonstration of elevated plasma aldosterone levels and the presence of adrenal masses on imagine.

Treatment centers around correction of hypokalemia, blockage of aldosterone, and treatment of hypertension. Hypokalemia may be treated with oral potassium gluconate (0.5 mEq/kg PO) or potassium chloride IV not to exceed 0.5 mEq/kg/hr. Aldosterone activity is blocked by spironolactone at 2-3 mg/kg PO q12. Hypertension may be treated with ace inhibitors enalapril or benazepril at 0.5-2 mg/kg PO q 12 hr or amlodipine at 0.625-1.25 mg PO or rectally q 12-24 hours. For severe hypertension, nitroprusside at 0.5-5 mcg/kg/min as a constant rate infusion may be used. Definitive treatment involves surgical removal of the tumor.

Prognosis is good with successful surgical removal of a non-malignant tumor at up to 60 years. If medical therapy is pursued, clinical signs will improve, but blood values may not normalize. Quality of life can be improved for 1-3 years, although chronic renal failure usually prompts euthanasia.

Acromegaly

Feline acromegaly is caused by an oversecretion of growth hormone from a pituitary tumor. Feline acromegaly is being diagnosed more frequently leading to the thought that it may not be as rare in cats as once supposed. Clinical signs are less pronounced than in people or dogs, but cats present with massive heads and the owner notices a widening of the head between the ears. These cats also have big features elsewhere including increased interdental widening. Often these cats will present with heart failure due to hypertrophic cardiomyopathy. As growth hormone causes insulin resistance in the cat, most of these cats will have diabetes mellitus that is resistant to insulin. They may also have neurological signs from a pituitary macroadenoma.

Diagnosis is definitively made by measurement of elevated serum growth hormone. Unfortunately, this assay is not always available since no commercial laboratory runs it routinely. Feline growth hormone can be measured by the canine assay and in acromegalic cats the serum basal levels are often greater than 6 times the normal serum level. The practitioner should recognize that growth hormone can be normal in the early stage of the disease; however, this is not usually a problem since clinical suspicion for acromegaly does not usually occur until the disease has progressed to cause significant clinical signs. If growth hormone assays are not available, insulin-like growth factor, a hormone produced by the liver and secreted in response to growth hormone, can be measured. The human assay can be used and is more reliably available than growth hormone. There is a report of IGF-1 being elevated in diabetic cats unrelated to acromegaly, so levels should be interpreted with caution. CT or MRI imaging of the pituitary to localize a tumor is also helpful.

Treatment of acromegaly is important to control insulin-resistant diabetes mellitus and hypertrophic cardiomyopathy. Medical treatment has centered around long acting somatostatin analogues like octreotide. The results have not been exciting. Recently the use of hypofractionated radiation has been advocated for the treatment of feline acromegaly. This treatment avenue appears to be more successful in controlling the insulin resistance and clinical signs.

Hypothyroidism

Unlike in dogs, naturally-occurring hypothyroidism is uncommon in cats. The most common reason for feline hypothyroidism is over-treatment with methimazole or ^{131}I therapies. Naturally occurring hypothyroidism may present in young to middle-aged cats. If congenital, the cat may have skeletal abnormalities including short, stunted legs. Additionally congenital hypothyroid cats may present as very lethargic and mentally dull. Often a hallmark clinical feature is chronic constipation/obstipation.

Diagnosis is usually achieved by measurement of a T_4 level. Because the condition is so unusual in cats, the clinical signs are usually not recognized until the disease has progressed so that the T_4 will be very low. That finding plus the clinical signs usually leads to a straightforward diagnosis.

Treatment for feline hypothyroidism is supplementation with levothyroxine. Half of the smallest dose (0.1 mg) twice per day is a place to start with T4 measurement in 2 weeks to gauge effect. It may take several weeks/months for all the clinical signs to resolve, so it is important to continue constipation therapy and wean slowly.

Hypoadrenocorticism

Primary hypoadrenocorticism is an uncommon disease in cats, with approximately 40 reported cases in the veterinary literature. This disease is commonly described in humans and dogs, and the etiology is most commonly due to an immune-mediated destruction of the adrenal cortex. Additional causes in cats are described in isolated case reports and include trauma, adrenal infiltration by lymphoma, or iatrogenic due to rapid withdrawal of exogenous corticosteroids or megestrol acetate. The cause of primary hypoadrenocorticism in cats is unknown, although a previous case series did show lymphocytic inflammation of the adrenal glands, suggesting a similar immune-mediated process to that suspected in humans and dogs.

Diagnosis of hypoadrenocorticism is made by use of an ACTH-stimulation test revealing pre- and post-cortisol levels less than 2 mgc/dl. In one case report of a cat with hypoadrencorticism, aldosterone levels were measured pre- and post-ACTH and were found to be undetectable.

Treatment of hypoadrencorticism is by glucocorticoid and mineralocorticoid supplementation. Glucocorticoids are used similarly to dogs in that a physiological dose of 0.2-0.4 mg/kg of prednisone is usually adequate. Prednisolone is recommended for use in the cat as it does not require liver metabolism. Fludrocortisone or DOCP may be used for mineralocorticoid supplementation. DOCP dosing intervals may be longer than those required for the dog.

REFERENCES

Valentin SY, Cortright CC, Nelson RW, Pressler BM, Rosenberg D, Moore GE, Scott-Moncrieff JC. J Vet Intern Med.,2014; Mellett Keith AM, Bruyette D, Stanley S., J Vet Intern Med. 2013 Nov;27(6):1471-7;Lo AJ, Holt DE, Brown DC, Schlicksup MD, Orsher RJ, Agnello KA., J Vet Intern Med. 2014 Jan;28(1):137-43;Combes A, Pey P, Paepe D, Rosenberg D, Daminet S, Putcuyps I, Bedu AS, Duchateau L, de Fornel-Thibaud P, Benchekroun G, Saunders JH., J Feline Med Surg. 2013 Jun;15(6):445-57; Djajadiningrat-Laanen S, Galac S, Kooistra H. J Feline Med Surg. 2011 Sep;13(9):641-50;Lamb CR, Ciasca TC, Mantis P, Forcada Y, Potter M, Church DB, Niessen SJ., J Feline Med Surg. 2014 Feb;16(2):99-108;Niessen SJ, Church DB, Forcada Y.; Vet Clin North Am Small Anim Pract. 2013 Mar;43(2):319-50; Gunn-Moore D. Feline endocrinopathies. 2005; 35: 171-210;Parnell NK, Powell LL, Hohenhaus AE, Patnaik AK and Peterson ME. JAVMA 1999; 214: 1208-11;Smith SA, Freeman LC and Bagladi-Swanson M. J Am Anim Hosp Assoc. 2002; 38: 41-4; Peterson ME, Greco DS and Orth DN. JVIM 1989; 3: 55-8; Woolcock AD, Ward CR, submitted for publication, 2014.

TROUBLESHOOTING DIFFICULT DIABETIC CASES IN DOGS AND CATS

Linda M. Fleeman BVSc PhD MANZCVS
Melbourne, VIC, Australia

Difficulty monitoring glycemic response – results do not match the animal's clinical signs

Glucose homeostasis is a dynamic process that can change very rapidly. Cats and dogs with diabetes have dysfunctional glucose homeostasis and so there is even more pronounced variability of blood glucose concentrations, especially in response to stress or illness. A common error when managing diabetes is to assume that the condition can be 'stabilized' and that measurements taken on one day are representative of diabetic control on other days. A more appropriate approach is to establish an ongoing monitoring regimen that will detect the changing trends in the animal's response to treatment. This is especially important for cats because improvement in glycemic control and resolution of glucose toxicity can result in substantially decreased insulin requirement and diabetic remission. Improved activity, resolution of polydipsia and polyuria, and weight gain are all indicators of improved glycemic control. In cats, it is important to realize that negative glucosuria can occur at any time once clinical signs have resolved, and may occur within days, weeks, or even months of the last insulin dose adjustment. Once the blood glucose concentration is consistently below the renal threshold, insulin requirement can decrease substantially in both cats and dogs. However, the rate of diabetic remission in cats appears to be optimized if the insulin treatment is withdrawn gradually, with a dose decrease every 1-2 weeks. In contrast, sudden cessation of insulin can result in recurrence of hyperglycemia, return of glucose toxicity, and increased requirement for exogenous insulin.

Micro-management with frequent insulin dose adjustment should be avoided. In general, insulin dosage adjustments should not be made any more frequently than once every 1-2 weeks. The exception is following an episode of hypoglycemia, which should always prompt reduction of the insulin dose (unless there was a dosing error). When presented with a patient that has shown a very variable response to insulin and inconsistent glycemic results, it is often helpful to carefully review the history with the owner to determine if there were any periods when there was resolution of clinical signs. It is important to also note the insulin dose administered before any occurrences of hypoglycemia. A reasonable approach is to resume the treatment regimen that the animal was receiving when there was good glycemic control and to carefully monitor the response for a minimum of 2 weeks. This information can then provide baseline data for an ongoing regimen of monitoring and treatment adjustment.

Unexpected hypoglycemia

Recording urine glucose at home provides a useful 'early warning system' for the risk of clinical hypoglycemia. Decreasing the insulin dose within 2 weeks of the onset of negative glucosuria will ensure that hypoglycemia is avoided in the majority.

The owner's insulin administration protocol should be carefully evaluated whenever hypoglycemia occurs to identify dosing errors. It is strongly recommended that insulin injections are administered at strict 12-hour intervals. Irregular timing of insulin injections can lead to overlap of insulin action with that of the previous dose. If it is not possible to administer an insulin injection on time, then the best approach is to miss that injection and resume insulin administration at the next injection time. Missing a single injection will usually have negligible consequences. In contrast, late administration of insulin can lead to increased insulin action (and therefore over-dose) if the following insulin injection is administered on time. The usual meals can be fed whenever an insulin injection is missed. For cats that are prone to hypoglycemia, longer-acting insulin preparations such as glargine insulin are recommended as these will likely have a smoother action and so minimize periods of both hyper- and hypo-glycemia.

It is important to note that feeding in cats will not typically cause a postprandial increase in blood glucose concentration and so is not a reliable treatment for hypoglycemia. In particular, low carbohydrate foods such as those typically recommended for diabetic cats will not be helpful in this situation. However, liberal application of glucose syrup or honey to the oral mucosa will usually effectively control clinical signs although repeated dosing might be required. The glucose syrup or honey can also be administered per rectum, which is usually the safest option in an animal with severe seizures. Once recovered, dogs should be fed a meal of their usual food. Once a cat has recovered, glucose syrup or honey can be added to cat food. Cats cannot taste sweet foods and usually appear not to notice honey if it is mixed thoroughly with food. If oral glucose is not successful, the hypoglycemia should be managed initially with an intravenous bolus of 1 mL/kg (0.5 g/kg) 50% dextrose administered slowly. This can be followed with a constant rate infusion of 2.5% dextrose solution at 6 mL/kg/hr with the flow rate adjusted based on subsequent blood glucose measurements.

Finicky appetite at meal times

Because cats do not have a clinically relevant postprandial blood glucose response, feeding does not need to be matched with insulin injections. In contrast, it is crucial in diabetic dogs that insulin dosing is matched to the feeding of consistent meals, with each meal containing exactly the same quantity and type of ingredients. Most diabetic dogs will readily consume meals twice-daily if the meals are highly palatable and contain half their daily caloric requirement. Usually the best approach is to feed standardized meals based on the dog's usual diet. In the majority of cases, this is one brand and flavor of commercial dog food formulated for adult maintenance with <10% of the calories from treats or other foods. Note that increased dietary fiber content in excess of that routinely recommended for healthy dogs might be associated with unwanted effects such as weight loss and inappetence. When polyphagia resolves, both caloric requirement and appetite typically decrease and so the meal size must also be decreased to avoid over-feeding.

For finicky eaters, the meal should be fed at the time of insulin administration and remain available until the expected end of the period of maximal insulin activity (approximately 4 hours for lente and NPH insulin). Appetite can be tested with a small, highly palatable treat prior to insulin injection. If eaten readily, the dog should be encouraged to sit and stay, receive the injection, and then be immediately rewarded with another small treat. The dog should then be given the meal in a private place and allowed up to 4 hours to consume the food. It is important to note the quantity of food consumed at each meal so that the meal size can be adjusted to suit the individual's requirements. It might take up to 3 weeks to establish a practical feeding routine, but once established, diabetes management will become simpler and more successful.

Continued weight loss despite good glycemic control

Continued weight loss despite good glycemic control is commonly a problem in diabetic dogs. Approximately one-third of diabetes in dogs is caused by chronic pancreatitis which progressively destroys both endocrine and exocrine pancreatic tissue. Loss of body weight and condition despite polyphagia are presenting signs for both diabetes mellitus and exocrine pancreatic insufficiency (EPI). Treatment of diabetes usually results in weight loss being arrested before optimal glycemic control is achieved. The possibility of concurrent EPI should be considered if weight loss continues despite adequate glycemic control, and serum TLI concentration assayed and/or treatment with pancreatic enzyme supplementation trialed. Note that active pancreatic inflammation in dogs with EPI can increase TLI concentration to within the reference range. Affected dogs typically lose weight despite consumption of 2 or more times their expected caloric requirement and pass feces 3 or more times daily. Feces may have normal consistency. The goal is to identify EPI early in the clinical course so that pancreatic enzyme supplementation can be started before there is excessive loss of body condition. A positive response to pancreatic enzyme supplementation comprising substantial weight gain and decreased gastrointestinal signs is typically seen within days in affected dogs. For example, a thin 7-8 kg dog might gain 0.5 kg in the first week and have a marked reduction of gastrointestinal signs. They will typically still require a high calorie intake and defecate 3 or more times a day, but they gain weight and are much more comfortable. Therefore it is not usually necessary to monitor anything other than body weight/condition and the owner's observations. Over time, it might be necessary to increase the dose of enzyme supplementation.

Insulin resistance

The majority of uncomplicated diabetic dogs require an insulin dose of approximately 0.5-1.0 U/kg q12h and few require doses greater than 1.5 U/kg q12h. Insulin resistance in dogs can thus be defined as poor diabetic control at insulin doses exceeding 1.5 U/kg. Insulin doses should always be based on estimated ideal body weight rather than actual body weight in underweight or overweight animals. Most diabetic cats require doses of 5 U or less of glargine insulin administered q12h. Occasionally an individual cat will need 6 or 7 U before good glycemic control is achieved. Once persistent negative glucosuria is achieved and insulin requirement decreases, many cats will go into remission and the remainder will typically require 3 U or less of insulin q12h. An unexpected increase in insulin requirement is recognized when a cat that has been in long-term diabetic remission needs insulin therapy again, or when a cat that has needed only a low insulin dose for months unexpectedly requires higher doses.

The major differential diagnoses for diabetic dogs requiring an insulin dose >1.5U/kg q12h or for a diabetic cat to require either >7U q12h or an unexpected increase in insulin requirement are:

1. Error in insulin handling or administration
2. Obesity
3. Concurrent disease or drug therapy
4. Compensatory hyperglycemia secondary to insulin overdose (Somogyi phenomenon).

1. Error in insulin handling or administration

Insulin can become inactivated if exposed to temperatures >30°C or light for prolonged periods. An expedient method of ruling out the possibility of inactivated insulin when investigating insulin resistance is to change to a new vial of insulin. Insulin suspensions must be thoroughly mixed prior to administration or doses might vary greatly.

Although experienced owners rarely report difficulty with administration of insulin to their pet, it is important to review their injection technique for errors whenever insulin resistance is investigated. A wide range of insulin syringes are available and inadvertently changing to a different type of syringe can lead to dosing errors. Dosing errors are less frequent with insulin dosing pens than with needles and syringes. However, insulin dosing pens must be primed prior to administration of each dose to ensure there is no air in the system. It is also important to check that the dosing dial has returned to the 'zero' position after each dose.

2. Obesity

Obesity is a common reason for a diabetic cat to have increased insulin requirement. Reversible insulin resistance is induced and if weight loss occurs insulin requirement can rapidly return to normal and care must be taken to avoid hypoglycemia.

Weight gain is initially a desirable outcome when treating diabetic animals as it is an indicator of good glycemic control. However, once this is achieved, overweight and obese-prone diabetic cats should be carefully managed to attain and maintain an ideal body condition. Both body weight and body condition should be monitored long-term, and calorie restriction will sometimes be necessary.

3. Concurrent disease or drug therapy

Concurrent disease or drug therapy causing insulin resistance may be suspected based on history and physical examination findings and can be further investigated by performing hematology, serum biochemistry, urinalysis, and microbial culture and sensitivity of the urine. Almost any concurrent disease including dental disease might cause insulin resistance that affects diabetic control. Once the concurrent disease state has resolved, insulin sensitivity can be expected to improve and there will be increased risk of hypoglycemia unless the insulin dose is decreased.

Hyperadrenocorticism and hypothyroidism are two important causes of insulin resistance in diabetic dogs. Both can present a diagnostic challenge in a diabetic dog. Systemic and topical corticosteroids are the drugs most commonly associated with insulin resistance. In cats, hyperthyroidism and chronic renal failure are commonly recognized causes of insulin resistance. Acromegaly should be considered if all other diseases are ruled out. The classic textbook signs of acromegaly are not always present in affected diabetic cats, with some showing no sign other than high insulin requirement. Diagnosis is typically based on increased serum IGF-1 concentration. To optimize the chance of detecting increased IGF-1 concentration, cats must first be treated with insulin for at least 4-6 weeks. This is because portal insulin induces IGF-1 production and acromegalic cats can give normal IGF-1 results when they are insulin deficient. It has been suggested that the incidence of acromegaly in diabetic cats might be underestimated.

4. Compensatory hyperglycemia secondary to insulin overdose (Somogyi phenomenon)

Compensatory hyperglycemia secondary to insulin overdose (Somogyi phenomenon) is one of the most common causes of insulin resistance in diabetic dogs and cats. There is typically a period of good glycemic control that is followed by deteriorating glycemic control despite increasing insulin doses. The period of good glycemic control may be very brief and is sometimes missed, especially if dose adjustment is based only on results of serial blood glucose concentration curves without careful consideration of the clinical signs.

In insulin-treated diabetic people, hypoglycemic events occur more frequently during the night than during the day and the same is likely true for dogs and cats. Therefore insulin-induced hypoglycemia is likely to be missed with day-time monitoring. Continuous subcutaneous glucose monitors are a useful diagnostic aid for detecting night time hypoglycemia. Compensatory hyperglycemia often appears to persist for several days following insulin-induced hypoglycemia.

For diabetic dogs with poor glycemic control that are receiving insulin doses greater than 1 U/kg and diabetic cats with poor glycemic control that are receiving insulin doses greater than 1.5 U/kg, where administration/dosing errors and concurrent disease/drugs have been ruled out, it is recommended that the insulin dose be decreased to 0.5 U/kg and the response to this change monitored. If insulin resistance was due to compensatory hyperglycemia secondary to insulin overdose, there is typically marked clinical improvement within 1-2 weeks. If there is another cause of insulin resistance, clinical signs typically become much worse within a few days and the previous insulin dose can be resumed and the investigation for another cause continued.

Management of diabetes in animals with concurrent disease or drug therapy

Diabetes that is complicated by concurrent disease and/or medications that are causing insulin resistance can sometimes present with very unpredictable blood glucose results that vary anywhere from 30-550 mg/dL (1.7-30 mmol/L) despite consistent insulin dosing. In cats, such cases will often roughly follow a 3-day cycle fluctuating between hyper- and hypo-glycemia and might present with intermittent hypoglycemic seizures. Diabetic dogs with chronic pancreatitis might have 'brittle' glycemic control as a result of variation in the degree of pancreatic inflammation and increased risk of complications such as diabetic ketoacidosis, insulin resistance, and hypoglycemia. Measurement of blood glucose concentration by owners immediately prior to each insulin injection can greatly help to guide insulin dosing in these cases. An example of dosing recommendations that were developed for an individual diabetic cat that has chronic pancreatitis/triaditis and requires prednisolone and chlorambucil therapy is as follows. Note that these recommendations are based on blood glucose measurements using a veterinary glucose meter and treatment with glargine insulin.

Measure blood glucose prior to each insulin injection:
If blood glucose is 270 mg/dL (15.0 mmol/L) or greater, give 5 U
If blood glucose 180-269 mg/dL (10.0-14.9 mmol/L), give 4 U
If blood glucose is 108-179 mg/dL (6.0-9.9 mmol/L), give 3 U
If blood glucose is 68-107 mg/dL (3.8-5.9 mmol/L), give 2 U
If blood glucose is 67 mg/dL (3.7 mmol/L) or less, give no insulin.

Such an approach in cases where diabetes has proved very difficult to manage provides owners with a practical tool to avoid clinical hypoglycemia and achieves reasonably good control of the clinical signs of hyperglycemia. If owners of such complicated diabetic animals are unwilling to perform blood glucose testing at home, then the safest approach is to recommend consistent treatment with an insulin dose lower than that which has previously caused clinical hypoglycemia and to closely monitor clinical signs and urine glucose.

Short- or long-duration of insulin action

Short- or long-duration of insulin action may be suspected based on results of serial blood glucose testing. It is important to recognize that there is great day-to-day variability of serial blood glucose concentration curves and apparent duration of action varies considerably from day to day.

A diagnosis of short-duration of insulin action must be supported by compatible clinical signs such as polydipsia that repeatedly occurs for only a few hours before an insulin injection is due. Short-duration of insulin action must also be differentiated from compensatory hyperglycemia secondary to insulin-induced hypoglycemia. Short-duration of insulin action can be associated with insulin resistance and investigation for concurrent disease is indicated. If short-duration of insulin action is confirmed and insulin overdose and concurrent disease have been ruled out, then the same insulin preparation may be administered every 8-hours or a different insulin preparation with an expected longer duration may be trialed.

Long-duration of insulin action usually responds to a decrease in the 12-hourly insulin dose. It is not usually necessary to decrease the frequency of insulin administration.

TREATING SEPSIS IN DOGS AND CATS: WHAT ARE THE GOALS?

Amy E. DeClue DVM, MS, DACVIM (SAIM)
Columbia, MO

INTRODUCTION

The severity of bacterial sepsis varies from patient to patient. Some patients will require advanced life support measures to survive while others will require only basic supportive care. The key aspects of sepsis treatment include 1.) initial resuscitation, 2.) alleviating the inciting cause and 3.) aggressive supportive care. The treatment of sepsis can be roughly organized into initial resuscitation (i.e., first 1-6 hours) and long-term management (i.e., > 6 hours to discharge from hospital). Initial resuscitation includes not only restoration of hemodynamic homeostasis, but also alleviating the inciting cause through administration of antimicrobial drugs and/or debridement. Recently, the concept of early goal directed therapy has become more prevalent in human medicine. Goal directed therapy is based on the idea of tailoring treatment to specific physiologic parameters thought to be associated with restoration of homeostasis and improved outcome.

INITIAL RESUSCITATION

Restoration of hemodynamic stability

Initial hemodynamic resuscitation of animals with sepsis is important for restoration of homeostasis. While there are no generally accepted, standardized initial resuscitation protocols for animals, guidelines have been established for people. These guidelines are outlined in the Surviving Sepsis Campaign: International Guidelines for Management of Severe Sepsis and Septic Shock. These guidelines were initially established in 2004 and then updated in 2008 and 2013. The Surviving Sepsis Campaign guidelines were developed by a group of physicians based on an extensive literature review. Recommendations were graded based on the quality of scientific evidence available. The authors recommended protocolized, quantitative resuscitation of patients with sepsis-induced tissue hypoperfusion. This approach is also termed early goal-directed therapy. Early goal directed therapy resulted in a 17.7% absolute reduction in 28-day mortality in a multicenter trial of people with sepsis. The goals of early goal-directed therapy in people are to achieve each of the following during the first 6 hours of intervention:

- CVP 8-12 mmHg
- MAP ≥65 mmHg
- Urine output ≥ 0.5 ml/kg/h
- Superior vena cava oxygenation saturation of 70% or mixed venous oxygen saturation of 65%
- Normalization of blood lactate concentrations in patients with increased blood lactate concentrations.

While these goals have not been specifically tested in randomized, multicenter trials in animals, the general concept behind the goals could have value in veterinary patients. Resuscitation goals should be tailored to the unique physiologic parameters of the species. Parameters that could have value in developing a standardized treatment protocol in your practice include:

- Central venous pressure: Central venous pressure is an indicator of volume status. The most ideal CVP to achieve in dogs with sepsis has not been determined, but many veterinarians aim for a CVP of 7-10 cm. It is important to note that there are many non-volume related factors that can influence CVP and thus monitoring trends in CVP and developing patient specific goals might be more helpful than using a generalized guideline.
- Blood pressure: A mean arterial blood pressure of less than 60-65 mmHg or a systolic blood pressure < 90 mmHg is considered hypotension in the dog and cat. While the exact target blood pressure goals have not been determined, proposed target Doppler systolic blood pressure post resuscitation is > 90 mmHg. Achievement of a Doppler systolic blood pressure > 90 mmHg within the first hour of fluid resuscitation was associated with increased survival in one study of dogs presenting to an emergency room with a variety of conditions. Induction of hypertension (systolic blood pressure >150 mmHg) should be avoided.
- Urine output: In a dog or cat with normal hydration status, oliguria is defined as urine output less than 0.5-1.0 mL/kg/h and anuria is defined as urine output less than 0.3-0.5 mL/kg/h. Avoidance of oliguria or anuria is a reasonable goal for dogs and cats with sepsis.
- Additional parameters: Superior vena cava oxygen saturation ($ScvO_2$), lactate and base deficit have been evaluated as prognostic markers for dogs with sepsis in a prospective clinical trial evaluating goal-directed therapy for sepsis. The sensitivity and specificity of $ScvO_2$, lactate and base deficit for predicting in-hospital mortality in dogs with severe sepsis or septic shock are listed in table 3. A higher $ScvO_2$ (odds ratio – 0.76) and a smaller base deficit (odds ratio = 0.55) was associated with a lower probability of death (odds Ratio= 0.76). These data suggest that these parameters may have value in developing goal-directed therapeutic protocols in the dog although specific recommendations for target values have not been determined.

Table 3: Sensitivity and specificity of ScvO2, lactate and base deficit concentrations at ICU admission for predicting in-hospital mortality in dogs with severe sepsis or septic shock (adapted from Conti-Pantara, 2012).			
Parameter	Cut-off points	Sensitivity	Specificity
ScvO2	52%	64.5%	100%
Lactate	3.5 mmol/L	54.5%	89.5%
Base deficit	-9.5 mmol/L	100%	68.4
	-16.5	54.5%	94.7%

There is a continuum between early goal directed therapy and later management. While it is important to consider early resuscitation goals, maintenance of these goals during continued care outside of the initial few hours of management is important. After initial resuscitation, it is not uncommon for animals to have recurrence of hemodynamic derangements.

Achieving hemodynamic support goals can be accomplished using boluses of isotonic crystalloids (Dogs 15-20 ml/kg, IV to effect; Cats 10-15 ml/kg, IV to effect) or colloids (Dogs, 5 ml/kg to effect; Cats 2 ml/kg to effect). Then, ongoing fluid therapy should be tailored to the needs of the patient to provide for maintenance requirements, correct interstitial hydration deficits, replace ongoing losses and continue to correct any hemodynamic derangements. Patients with sepsis will have a propensity to develop interstitial edema, including subcutaneous and pulmonary edema, because of increased vascular permeability and decreased blood colloid osmotic pressure. Administration of colloid fluids (e.g., hetastarch) may help prevent interstitial edema and should be considered for volume resuscitation and during maintenance of sepsis patients, in addition to crystalloid fluid support. However, there is evidence in people that hetastarch may increase the risk for acute kidney injury. No evidence supporting or refuting this phenomenon is available in dogs and cats with sepsis at this time. Additionally, the use of 0.9% NaCl has fallen out of favor in people with sepsis because of the amount of chloride in the solution. Administration of a chloride rich solution induces shock in animal models, promotes metabolic acidemia, induces a pro-inflammatory shift, reduces renal perfusion, increases the risk of acute kidney injury in people with sepsis and increases mortality in adult and pediatric people with sepsis. For these reasons, the use of a lower chloride solution like plasmalyte is recommended.

Despite aggressive volume resuscitation, some animals will require additional support to maintain normal blood pressure and tissue perfusion. The safest and most efficacious treatment for septic shock has not been determined for dogs or cats. Prior to administration of any sympathomimetic drug, care should be taken to ensure that the patient is not hypovolemic. Since the goal of treating septic shock is to maintain tissue perfusion, medications that cause vasoconstriction should be used only if absolutely necessary. Positive inotropic drugs (e.g., dobutamine) may be a good initial management choice for septic shock since they help combat decreased cardiac output caused by myocardial dysfunction without inducing peripheral vasoconstriction. However, care should be taken to avoid positive inotropic drugs in cats with evidence of HCM. If volume resuscitation and positive inotropic support has failed to restore adequate blood pressure, a vasopressor agent (e.g., dopamine, norepinephrine, epinephrine or vasopressin) could be added. Vasopressors should always be administered as a CRI and ideally, direct blood pressure evaluation should be used for monitoring. Although there are no clinical trials evaluating these drugs, epinephrine is generally considered the least ideal vasopressor. Critical illness related adrenal cortical insufficiency has been documented in dogs and cats with sepsis and refractory hypotension and should be considered in any animal requiring vasopressor therapy during sepsis. Judicious use of corticosteroids may address therapeutic failure of vasopressor therapy in these patients.

Alleviate the cause

The most important aspect of treating bacterial sepsis centers on the identification and eradication of the inciting cause. Although stringent effort should be made to identify the cause of sepsis, early antimicrobial treatment is critical for survival and should not be withheld pending culture results. In humans, early antibiotic administration has been shown to decrease mortality. Broad spectrum, bactericidal antimicrobial agents (e.g., fluoroquinolone + penicillin) administered IV should be instituted as quickly as possible in patients with suspected sepsis, ideally within the first 30-60 minutes. Antibiotic therapy should be selected based on the most likely type of organism given the site of infection. Depending upon the source of infection, surgical debridement may be necessary for infection control. Once culture specific antibiotic sensitivity is determined, the antibiotic with the narrowest spectrum of activity should be chosen and administered until there is complete clinical resolution.

LONG TERM MANAGEMENT

Supportive care

GI/Nutrition- Bacterial translocation from the GI tract can contribute to systemic inflammation during sepsis. Early placement of a feeding tube and initiation of enteral or, in patients with vomiting, parenteral nutrition will help maintain GI barrier function. Additionally, medication aimed at maintaining normal GI protective mechanisms (e.g., omeprazole, famotidine, sucralfate) could be considered. Hyperglycemia can be a complication of nutritional therapy, especially parenteral nutrition. Hyperglycemia has been associated with increased inflammation and a poorer prognosis in people with sepsis and SIRS. Although the importance of glucose homeostasis in cats with sepsis is unknown, iatrogenic hyperglycemia should be avoided.

Cellular oxygenation- Oxygenation, acid-base status, PCV or hemoglobin concentration and organ function should be closely monitored. Along with maintaining good tissue perfusion, maximizing cellular oxygenation will help maintain tissue viability and avoid multiple organ dysfunction. This may be accomplished by ensuring good oxygen carrying capacity, providing supplemental oxygen when indicated and preventing microthrombi formation. For the majority of patients, acid-base abnormalities are related to lactic acidosis secondary to poor tissue perfusion. Typically these abnormalities will resolve with resolution of hypotension and normalization of perfusion. Therefore, bicarbonate administration is rarely needed and in fact may be contraindicated.

Nursing care/Pain management- Care should be taken to ensure adequate patient comfort including management of pain, careful catheter maintenance, bladder/bowel care and frequent patient turning/movement to prevent decubital ulcers. Almost all patients with sepsis will require analgesic administration for pain management. Although many analgesics like ketamine and buprenorphine may offer specific anti-inflammatory advantages during endotoxemia, some may be detrimental. Morphine, for instance, augments the inflammatory response to endotoxin, has a detrimental effect on mean arterial pressure and increases mortality in endotoxemic rats. While it is not known if morphine is detrimental during feline sepsis its routine use in septic patients should be avoided where possible.

REFERENCES

Conti-Patara et. al **J Vet Emerg Crit Care** 2012;22(4):409-18. Dellinger et. al **Crit Care Med** 2012; 41(2):580-637. Ertmer et. al **Crit Care Med** 2011; 39(4):783-92. Kellum et. al **Chest** 2006 Oct;130(4):962-7. Kellum et. al **Chest** 2004 Jan;125(1):243-8. Mutter et. al **Cochrane Collaboration** 2013; 1-27. Silverstein et. al J **Vet Emerg Crit Care** 2012; 22(6);666-73. Young et. al **Ann Surg** 2014;259(2):255-62.

CURRENT CONCEPTS OF SEPSIS IN DOGS AND CATS

Amy E. DeClue DVM, MS, DACVIM (SAIM)
Columbia, MO

INTRODUCTION

Sepsis is defined as the systemic inflammatory response to infection. Sepsis is a common cause of morbidity and mortality in dogs and cats. Understanding the complex pathophysiology of sepsis and developing novel diagnostic tests and treatments for sepsis are active areas of research currently. This lecture will review some of the novel concepts emerging in the field of sepsis research and the implication of these findings for dogs and cats with sepsis.

NEW CONCEPTS IN PATHOPHYSIOLOGY

Classically sepsis has been considered an imbalance between hyper-inflammatory and hypo-inflammatory responses. The hyper-inflammatory response has largely been blamed for the morbidity and mortality associated with sepsis. The hyper-inflammatory response is observed in the beginning of sepsis and is characterized by activation of the innate immune system and production of a cytokine storm. This leads to cellular recruitment and the classical manifestations of sepsis. Millions of dollars and years of time have been spent trying to identify treatments to dampen the hypo-inflammatory response during sepsis. Little attention was given to the hypo-inflammatory response, also known as the compensatory anti-inflammatory response syndrome (CARS). Recently the importance of the anti-inflammatory aspects of the immune response to sepsis has been realized.

The compensatory anti-inflammatory response is characterized by immunoparalysis or lack of immune response to infection. Immunoparalysis is the result of a loss of lymphocytes, dendritic cells, endothelial cells, GI epithelial cells and thymocytes. The predominant mechanism for this loss is apoptosis. During the hyper-inflammatory response, activation of inflammatory cells leads to nuclear translocation of NF-kappaB and IL-1β activation both of which induce apoptosis of adaptive immune cells. People with sepsis have depletion of CD4 and CD8 T cells, B cells and dendritic cells. The anti-inflammatory cytokine IL-10 is produced in abundance leading to suppressed CD8 T cell function and reduced monocytic production of pro-inflammatory cytokines. A reduction in cytokine production by and expression of inhibitory receptors on T cells is also observed during immunoparalysis. There is an increase in T regulatory cells and myeloid-derived suppressor cells which contributes to immune cell quiescence as well. Clinically, immunoparalysis leads to an inability to fight an eliminate infection and the development of new infections from opportunistic or otherwise normally weakly virulent organisms. Reactivation of latent viruses is another possible sequela.

The classical pathophysiologic model of sepsis is a systematic progression through a hyper-inflammatory phase followed by resolution during the hypo-inflammatory phase. However, this paradigm has changed. It is now recognized that sepsis is a continuum, where the animal may vacillate between a hyper- and hypo-inflammatory state and that the hypo-inflammatory stat itself plays a role in morbidity and mortality. Many people with sepsis ultimately succumb to secondary infections permitted by immunoparalysis.

Clinical importance in the dog and cat: Currently, there are no definitive clinical markers that can be used to identify if a dog or cat with sepsis is in the hyper- or hypo-inflammatory phase of sepsis. Dynamic cycling through hyper- and hypo-inflammatory phases of sepsis is an important consideration if the use of immunomodulating drugs is being considered for a dog or cat with sepsis. The nature of the immunologic response should also be factored in to the design and interpretation of clinical trials evaluating immunomodulating therapies for sepsis in the dog and cat. Immunoparalysis should be considered in any dog or cat with sepsis that appears to have a worsening clinical course or that has initial improvement followed by decline. Diagnostic evaluation aimed at identifying secondary infection is important in these patients.

NEW CONCEPTS IN THE TREATMENT OF SEPSIS

Appropriate resuscitation fluid selection

For many years electrolyte solutions used for intravascular volume expansion have been generally considered equivalent in the context of restoring hemodynamic stability. Recent evidence in people and animal models of sepsis suggest that the composition of the electrolyte solution used for resuscitation impacts physiology and clinical outcome. Rapid administration of large volumes of 0.9% NaCl results in hyperchloremia, decreased blood pH and reduced base excess compared to administration of a similar volume of plasmalyte. The use of plasmalyte, compared to 0.9% NaCl, has the advantages of improving acid base status and reducing the incidence of hyperchloremia in people. In people with sepsis, hyperchloridemia is associated with higher mortality rates. This is because chloride increases production of cytokines, induces hypotension, promotes shock and increases the risk for acute kidney injury. Hyperchloremic acidosis causes activation of the immune system through induction of NF-kappa B and production of TNF, IL-6 and IL-10. Increased risk of acute kidney injury is thought to be secondary to alterations in renal blood flow. In a study of healthy people, administration of 0.9% NaCl reduced renal blood flow by 40%, decreased urine production and reduced Na excretion. In comparison, plasmalyte does not promote cytokine production, induce hypotension, promote shock or increase the risk of kidney injury. Plasmalyte also produces significantly greater diuresis than 0.9% saline. For these reasons, chloride rich solutions like 0.9% saline should be avoided for volume resuscitation during sepsis.

Clinical importance in the dog and cat: The composition of electrolyte solutions used for intravascular volume expansion should be considered in any dog or cat with sepsis. Selection of electrolyte solutions with chloride concentrations in a more physiologic range (i.e., 95-110 mEq/L) that contain a buffer should be used when possible. Relative sodium and chloride concentrations of electrolyte solutions commonly used in veterinary patients are listed in table 1.

Table 1: Comparison of the composition of commonly used electrolyte solutions.			
Fluid	**Na (mEq/L)**	**Cl (mEq/L)**	**Buffer**
0.9% NaCl	**154**	**154**	**None**
LRS	**130**	**109**	**Lactate**
Normosol R	**140**	**98**	**Acetate/gluconate**
Normosol M	**40**	**40**	**Acetate**
Plasmalyte	**140**	**98**	**Acetate/gluconate**
Plasma	**140**	**100**	**Bicarbonate**

Administration of colloid fluids (e.g., hetastarch) may help prevent interstitial edema and should be considered for volume resuscitation and during maintenance of sepsis patients, in addition to crystalloid fluid support. In several clinical trials in people with sepsis, hetastarch has increased the risk of acute kidney injury and mortality. However in goats with experimental endotoxemia, hetastarch did not alter creatinine clearance or ultrastructural tubular integrity indicating that there may be species variation in response to colloids.

Clinical importance in the dog and cat: No evidence supporting or refuting this phenomenon is available in dogs and cats with sepsis at this time. Additionally, there are readily available, affordable and safe alternative colloids (e.g., human albumin) that are used in people while no such species specific products are readily available and affordable for dogs and cats. Thus, while the judicious use of hetastarch should be considered, the possible risk of kidney injury should not impede its use in animals requiring colloid oncotic pressure support.

Optimizing pathogen clearance

Shock is a key contributor to morbidity and mortality during sepsis. The concept of irreversible shock was first introduced by Wigger et al. in 1950. Wiggers suggested that shock can only be tolerated for a defined period of time and that if not reversed, and by extension the cause of shock is not reversed, the animal will die. This led to many of the current philosophies in emergency medicine such as early goal directed therapy and the concept of the “golden hour”. While there is great emphasis placed on reversing shock and establishing hemodyanmic homeostasis, classically there has been less focus placed on alleviating the inciting cause in a rapid fashion. However, since the pathogen is the driving force for immunostimulation which leads to septic shock, eradication of the pathogen should reduce further immune system stimulation. In recent years, the concept of pathogen clearance optimization has come into favor. Optimizing pathogen clearance involves early administration of appropriate antimicrobials and source control when indicated. Indeed, rapid initiation of appropriate antimicrobial therapy reduces mortality in people with sepsis.

Clinical importance in the dog and cat: Veterinarians should consider how to improve the rapidity of pathogen clearance in dogs and cats with sepsis. Since identification of infection may be challenging in some animals, early administration of antimicrobials should be considered any time sepsis is suspected. Factors to consider when developing protocols for optimizing pathogen clearance in the dog and cat are noted in table 2.

Table 2: Factors to consider when optimizing pathogen clearance in dogs and cats (adapted from Kumar, 2014).	
Goals	Considerations
Early administration	Administration within 30 min-1 hour of presentation or identification of sepsis
	Maximize the bioavailability (IV)
	Maximize the speed at which the antimicrobial becomes bioavailable
	Consider loading doses for some antimicrobials
Maximizing antimicrobial potency	Use -cidal antimicrobials when possible
	Utilize pharmacokinetics to maximize potency; time vs concentration dependency
	Consider combining multiple antimicrobials with different mechanisms of action
	Provide broad spectrum coverage
	Consider antibiotic penetration into the infection microenvironment and anatomic location; community vs hospital acquired infection
Pathogen removal/reduction in pathogen exposure	Source control via surgical debridement

NEW CONCEPTS IN DIAGNOSING SEPSIS

There are several innovative new techniques for identifying infection. One of these involves measuring volatile metabolites of pathogens. These volatile compounds can be identified in various biologic samples from patients or in exhaled air. Volatile compound detection has a high sensitivity and negative predictive value so it could be useful in ruling out bacterial infection. Other goals of volatile compound detection include identifying the specific strain of bacteria and separation among bacterial antimicrobial resistance pattern phenotypes. More research is needed before these techniques are routinely used in the clinic, but the general concept holds promise as a diagnostic test for sepsis.

Surface acoustic wave (SAW) biosensors are another emerging technology that might have value in the early detection of sepsis. Surface acoustic wave biosensors consist of an electrode on a piezoelectric crystal surface. Molecules interact at the sensor surface which initiates an acoustic wave which is then detected by a sensor, analyzed and reported. This type of technology can be used to detect bacterial products, inflammatory markers, and products of hypoxemia and allows for detection of multiple molecules concurrently. SAW biosensors can be designed for use as bedside, point of care instruments.

Clinical importance in the dog and cat: Currently, these techniques are used predominately in a research setting. Hopefully, in the future, these types of technologies will be available for bed-side evaluation of dogs and cats with suspected sepsis.

REFERENCES

Boniatti et. al. **J Crit Care** 2011; 26(2):175-9. Boomer et. al **Virulence** 2014; 5(1);45-56. Bos et. al **PLOS Path** 2013;9:1-8. Csete et. al J **Clin Monit Comput** 2013; 27:427-31. Dellinger et. al **Crit Care Med** 2012; 41(2):580-637. Ertmer et. al **Crit Care Med** 2011; 39(4):783-92. Hutchins et. al **Trends Mol Med** 2014: Epub ahead of print. Kellum et. al **Chest** 2006 Oct;130(4):962-7. Kellum et. al **Chest** 2004 Jan;125(1):243-8. Mutter et. al **Cochrane Collaboration** 2013; 1-27. Young et. al **Ann Surg** 2014;259(2):255-62. Zhou et. al **Crit Care Med** 2013: Epub ahead of print.

ACROMEGALY IN HUMANS

Andrea L. Utz, MD, PhD
Nashville, TN

INTRODUCTION

Acromegaly is a rare and insidious disorder. Considering the diagnosis is often the major hurdle, as patients frequently accumulate long-term, disease-associated morbidities for an extended time prior to initial diagnosis. Acromegaly disease related co-morbidities are common in the general population and include hypertension, insulin resistance/diabetes mellitus, sleep apnea, carpal tunnel syndrome, arthralgias, and headaches. It is the astute clinician who recognizes the concurrence of these findings, often in addition to distinctive facial and acral growth, and begins the evaluation for acromegaly. Due to the long delay between symptom onset and diagnosis, the medical implications can be significant, including destructive arthropathy, cardiovascular disease, neoplasms, and irreversible bony changes.

The biochemical diagnosis of acromegaly can be challenging in some situations due to mildly abnormal levels and IGF-1 and GH assay variability and inadequately defined reference ranges. Discordance between GH and IGF1 levels may occur and correlation with clinical and radiologic findings is necessary, and in some cases serial measurements, to determine trends, are required.

After establishing the acromegaly diagnosis biochemically, pituitary imaging, generally with MRI, is indicated. Figure 1 shows a typical appearance of a macroadenoma with chiasm compression and cavernous sinus invasion. Once a pituitary tumor has been identified, surgical debulking to relieve chiasmal compression and to minimize excess growth hormone is often the first step in management. As many of these lesions are locally invasive, complete resection is often not possible and a multi-modality approach including surgical adenoma resection, medical treatment, and radiation therapy may be necessary.

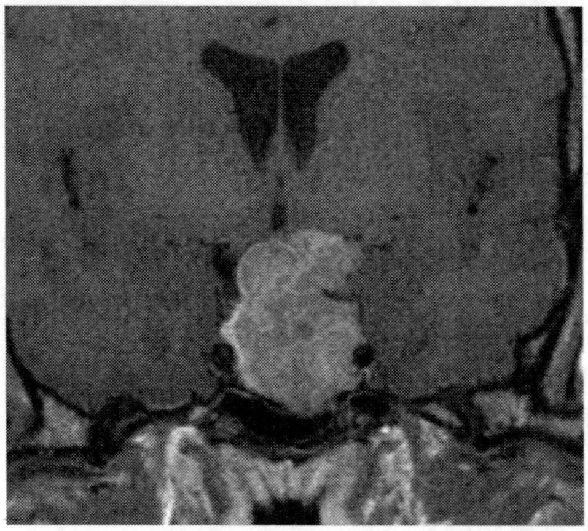

Figure 1. Typical coronal T1 post-gadolinium MRI appearance of a somatotroph adenoma

ESTABLISHING THE DIAGNOSIS OF ACROMEGALY

In an individual with suggestive clinical findings, the following assessments are indicated for screening and confirming the diagnosis of acromegaly:

- Elevated serum insulin-like growth factor type 1 (IGF-1)
- Elevated oral glucose-suppressed nadir serum growth hormone (GH measurements every 30 minutes for 120 minutes following 75 grams oral glucose)
- Pituitary MRI
- Rarely: growth-hormone releasing hormone (GHRH) level if a pituitary adenoma is not apparent

Variability of IGF-1 and GH assays and reference ranges has complicated the diagnosis and post-treatment monitoring of acromegaly. IGF-1 reference ranges are dependent on age with peak values around puberty. BMI also predicts IGF-1 with lower values at the extremes of BMI (< 21 and > 38 kg/m^2). Malnutrition, anorexia nervosa, severe liver disease, exogenous estrogen, and poorly controlled diabetes mellitus lower IGF-1 production.

As GH assay sensitivity has increased, the cut-off for normal GH nadir following oral glucose suppression has decreased. Most studies have defined a normal glucose-suppressed GH nadir as less than 1 ng/ml; however, using a sensitive assay, normal values are likely less than 0.4 ng/ml. Oral glucose GH suppression testing is inaccurate in those on somatostatin analogs and thus the goal in this population is a random GH < 1 ng/ml. In patients with discordant IGF-1 and GH levels, clinical assessment of symptoms and tumor burden are necessary to determine the need for initiation or intensification of therapy.

ACROMEGALY THERAPY

The goals of acromegaly therapy include the following:

- Lower GH and IGF-1 to clinically safe levels to improve symptoms of GH excess
- Minimize tumor compressive symptoms
- Replace pituitary hormone deficiencies

Table 1 details the available therapies for acromegaly. Transsphenoidal pituitary surgery is often the initial treatment as this may provide cure, particularly with small lesions. Surgical debulking can relieve compression symptoms and may minimize the doses of medication needed for biochemical control and may limit the extent of the field if radiation is required.

Table 1. Acromegaly Treatment Options

Pituitary Surgery		
Medical Therapies		
	Somatostatin Analogs	
		subcutaneous octreotide
		octreotide long-acting release (Sandostatin LAR)
		lantreotide depot (Somatuline depot, lanreotide Autogel)
		pasireotide – not currently FDA approved
	Growth hormone receptor antagonist	pegvisomant (Somavert)
	Dopamine agonists	
		cabergoline
		bromocriptine – limited efficacy
Radiation Therapy		
	Fractionated radiation	
	Stereotactic radiosurgery	LINAC, Gamma Knife, proton beam

When surgery does not provide biochemical normalization, medical management is often the next step in management. Somatostatin analogs have a dual effect to potentially both lower GH and IGF1 as well as to shrink/minimize growth of responsive somatotroph adenomas. However, when octreotide LAR was used as primary therapy, only 34% achieved IGF-1 normalization. This percentage was increased to 67% when used as secondary therapy after surgery or radiation.

While pegvisomant is often very effective in normalizing IGF-1 level (approximately 90% with a 20 mg/day dose), there is no beneficial effect on somatotroph tumor bulk and lowering IGF-1 may promote growth of the somatotroph adenoma.

Cabergoline may be effective as single agent therapy in a small population of patients – the ease of use and relatively lower cost make this a reasonable consideration. Cabergoline may be particularly useful in tumors with high prolactin co-secretion. The addition of cabergoline to somatostatin analog or pegvisomant therapy may reduce the amount of amount drug required for biochemical control. In cases of very large tumors, combination therapy with a somatostatin analog and pegvisomant may be necessary to control size as well as hormone hypersecretion. Combination therapy may limit side effects from individual drugs as well as decrease total medication cost.

The medical management of acromegaly is often very expensive and the medications may have side effects. Table 2 details the common side effects related to the medications.

Table 2. Acromegaly Medication Side Effects

Somatostatin Analogs	
	gastrointestinal symptoms
	gallbladder dysfunction / gall stone formation
	suppressed insulin release → hyperglycemia
	bradycardia
	hair loss
	vitamin B12 deficiency
	injection site discomfort
Pegviosmant	
	hepatoxiticy
	injection site lipodystrophy
	possible somatotroph tumor enlargement
Cabergoline	
	nausea
	headache, lightheadedness
	cardiac valvulopathy

Radiation therapy may provide a definitive treatment for acromegaly. In most cases, radiation therapy successfully prevents growth of the somatotroph adenoma. Biochemical remission is approximately 50% by 10 years with more rapid remission obtained with radiosurgery relative to fractionated radiation. Radiation does not work immediately to lower GH/IGF-1 levels and oftentimes medical management is necessary in the interim between radiation dosing and subsequent decline in hormonal hypersecretion. Radiation therapy frequently leads to anterior pituitary hormone deficiency and in rare cases can cause secondary tumor formation or neurologic dysfunction.

LONG TERM MANAGEMENT ISSUES

In addition to normalizing the biochemical indicators of growth hormone excess, it is important to consider other aspects in the management of these patients. Pituitary surgery and radiation may lead to hypopituitarism and appropriate testing and replacement is indicated. Lifelong monitoring for hypopituitarism after radiation is required. Periodic pituitary imaging for somatotroph tumor growth or development of secondary tumors following radiation therapy is recommended.

As acromegaly can lead to a number of co-morbidities, it is important to evaluate for and address each of these issues. In many cases, tumor debulking and biochemical control improve the co-morbidities; however, some persist even after GH/IGF1 levels are normalized. The following lists the most common acromegaly co-morbidities:

- Impaired glucose / diabetes mellitus
- Hypertension
- Sleep apnea
- Joint / musculoskeletal pain and dysfunction, carpal tunnel
- Cardiovascular disease – cardiac valve dysfunction, cardiac hypertrophy, cardiomyopathy, coronary atherosclerosis
- Colon polyps
- Thyroid nodules
- Fatigue
- Depression
- Sweating
- Headache
- Dental abnormalities
- Altered bone density / fractures
- Cosmetic issues

CONCLUSIONS

Acromegaly is rare disorder that often requires several treatment strategies to provide both biochemical and tumor control. It is important to know the testing characteristics and reference ranges of local IGF-1 and GH assays to best interpret the data for diagnosis and to monitor long-term disease control. In most cases surgical resection is the initial treatment; however in select individuals primary medical therapy can be considered when the majority of the tumor is not surgically accessible or if the patient is not a good surgical candidate. Surgical debulking may decrease amount of medicine required, minimize tumor compressive effects, and may decrease tumor to an extent to allow stereotactic radiosurgery. Combination medical therapy may reduce cost and side effects. It is important to consider and address the co-morbidities and side effects related to growth hormone excess and its treatment.

REFERENCES

Carmichael JD, et al. J Clin Endocrinol Metab 2009;94:523.
Caron P, et al. J Clin Endocrinol Metab 2010;95:4680.
Freda P. J Clin Endocrinol Metab 2005;90:4465.
Freda P. Clin Endo 2009;71:166.
Frystyk J. Growth Hormone & IGF Research 2010;20:8.
Giustina A, et al. J Clin Endocrinol Metab2010;95:3141.
Higham CE, et al. J Clin Endocrinol Metab 2012;97:1187.
Loeffler JS, et al. J Clin Endocrinol Metab 2011;96:1992.
Melmed S, et al. J Clin Endocrinol Metab 2009;94:1509.
Mercado M et al. Clin Endo2007; 66:859.
Sandret L, et al. J Clin Endocrinol Metab2011; 96:1327.
Swearingen B, et al. J Clin Endocrinol Metab 1998;83:3419.
van der Lely AJ. J Clin Endocrinol Metab 2012;97:1589

FELINE ACROMEGALY

Dr. Stijn J.M. Niessen, DVM, PhD, DECVIM (Medicine), PGCVetEd, FHEA, MRCVS, London, UK

INTRODUCTION

When it comes to knowledge about feline hypersomatotropism or acromegaly, the contrast between the last decade and previous decades could not be greater. The journey of feline acromegaly research could be summarised as "From a rarely diagnosed freak of nature disease with an extreme phenotype to a relatively common explanation for diabetes mellitus seen in general practice". This session aims to provide a cutting-edge update on feline acromegaly. Specific points of discussion will include recent revelations suggesting a role for environmental and genetic factors in its etiology. Additionally, novel diagnostic tools, including serum ghrelin and procollagen-propeptide, as well as the effective use of novel somatostatin analogues and hypophysectomy, will be discussed.

UPDATE ON PREVALENCE

After initial studies published in 2007 proposed feline acromegaly to be a relatively common underlying cause for feline diabetes mellitus, reporting a 26-32% prevalence rate[1,2], two further prevalence studies have been conducted recently.

One study was conducted among first opinion veterinarians in Switzerland and the Netherlands who were asked to provide blood samples from diabetic cats treated with insulin for at least 4 weeks[3]. Serum IGF-1 was analyzed by IGFBP blocked radioimmunoassay (RIA) and like the Author's 2007 study[1], who used a different RIA, an IGF-1 > 1000 ng/dl was considered suggestive of acromegaly. Blood samples were collected from 225 cats and IGF-1 concentrations ranged from 15–2471 ng/ml (median 584); interestingly, 17.8% of the cats had a serum IGF-1 > 1000 ng/ml. These data therefore once again confirm that it is likely that feline acromegaly represents a frequently encountered underlying etiology for the diabetes mellitus in cats seen in primary practice. The data also suggest there to be potential for geographical differences and/or differences generated through differing recruitment methods. The same investigator group are currently conducting CT studies to further qualify the meaning of those elevated IGF-1 values, i.e. to confirm hypersomatotropism. However, even if a proportion of these diabetic cats with elevated IGF-1 prove to have an unremarkable pituitary gland during CT-imaging, it would be prudent to bear in mind that some structural pituitary abnormalities can be missed by CT and MRI may be needed. In addition, CT and MRI negative cases have been described which emphasises that feline acromegaly should, above all, be seen as a hormonal dysfunction. As such, it should be diagnosed by documenting endocrine dysfunction, rather than be defined by the presence or absence of a macroscopically identifiable pituitary structural abnormality. This will be especially relevant when screening newly diagnosed diabetic cats, given that the pituitary somatotropinomas tend to be slow growing and therefore will be very small at such early stage of the disease process.

The second prevalence study represented a considerable extension of the Author's 2007 study and is by far the largest screening study conducted at this moment in time[4]. Free fructosamine determination for diabetic cats was offered to first opinion veterinary practices across the United Kingdom. Once again, spare serum was used to measure IGF-1 (RIA) and if found to be > 1000 ng/ml, free contrast-enhanced pituitray computed tomography (CT) was offered. In total 1222 diabetic cat samples were submitted. IGF-1 measurement suggested possible acromegaly (>1000 ng/ml) in 323 cats (26.4%). CT confirmed acromegaly in approximately 90% of patients with IGF-1 > 1000ng/ml, once again making the point that serum IGF-1 elevations at this level are reasonably specific for establishing the presence of hypersomatotropism.

The scientific debate about the exact prevalence of feline hypersomatotropism/ acromegaly is currently ongoing and remains interesting. Differences might relate to the use of differing IGF-1 assays, recruitment methods, as well as geographical influences. Nevertheless, from a clinical point of view, the outcome of such debate will not change the overall message coming from all studies conducted in the last decade: feline hypersomatotropism/ acromegaly should be considered as a possible cause of diabetes mellitus in the cat. Since it is generally accepted to screen for urinary tract infections (UTI) in feline diabetics (UTI prevalence estimate 12%) and it is also recommended to differentiate adrenal-dependent (ADH) from pituitary dependent hyperadrenocorticism in dogs (ADH prevalence estimate 15%), it therefore seems only logical to also recommend screening newly diagnosed diabetic cats for the presence of hypersomatotropism/ acromegaly (prevalence estimate 18-32%), especially given the tremendous implications on optimal treatment method, potential for diabetic remission and prognosis.

UPDATE ON THE POSSIBLE CAUSES OF FELINE ACROMEGALY

As with many endocrine cancers, the exact cause of human and feline acromegaly remains to be elucidated. Nevertheless, work is currently ongoing to help understand the triggers for pituitary oncogenesis in both species. In fact, this might represent an area where human medicine can learn from the progress made in veterinary medicine.

Organohalogenated contaminants (OHCs) have been implicated in pituitary oncogenesis in rodent models. OHCs are persistent and bioaccumulative chemicals and include organochlorine pesticides, industrial chemicals, such as polychlorinated biphenyls (PCBs) and brominated flame retardants (BFRs) added to various materials to reduce flammability, such as polybrominated diphenyl ethers (PBDEs). Both cats and humans are exposed to such chemicals via a multitude of routes including food and indoor contamination, mainly through dust ingestion which is especially relevant in light of cats' grooming behaviour. It has therefore been hypothesized that companion animals may serve as sentinels to assess the relationship between human exposure to chemicals via indoor environments and adverse health effects[5,6]. Comparison can therefore be drawn with feline hyperthyroidism where a similar link has been previously suggested between its emergence and the presence of OHCs in domestic materials and/ or food[5,7,8].

In light of this, the author and colleagues compared OHCs levels in the plasma of a group of elderly cats without identifiable endocrinopathy, with those with primary diabetes mellitus and those with secondary diabetes mellitus as a consequence of hypersomatotropism. Significantly higher levels for all contaminants were observed in cats suffering from hypersomatotropism. Particularly, total PBDE levels, brominated flame retardants, were significantly higher in the plasma of acromegalic cats. When additionally looking at the metabolites of these OHCs, data suggested that cats with hypersomatotropism have a lower capacity to metabolize persistent chemicals like PCBs than other cats.

It remains prudent, however, to point out that cause and effect have not yet been established. Indeed, it could be hypothesized that differences in OHC-metabolites are in fact related to the effects of excess growth hormone on hepatic metabolism. The activity of CYP enzymes, in particular of CYP3A4 and CYP2B6, is stimulated by excess growth hormone, which can lead to higher amounts of metabolites (HO-PCBs) formed in these cats. However, this would not account for the documented higher concentration of parent compound (not metabolized) OHCs encountered in these cats. Therefore, a causative relationship could exist between an excess in contaminants and feline hypersomatotropism, and thus pituitary oncogenesis. The latter would not be surprising, since xenoestrogens, such as PCBs and DDT, have previously been linked with a range of neoplastic diseases in man and specifically growth hormone producing pituitary adenomas[9,10].

Apart from environmental factors, the role of genetic factors in the etiology of feline acromegaly is currently also being investigated. A particular gene of interest here is the aryl hydrocarbon receptor interacting protein (AIP) gene. Previously, human pituitary adenomas were widely considered to be monoclonal in origin[11], supporting a hypothesis of an acquired single cell, somatic, mutation and subsequent formation of a neoplasm. However, this has since been questioned[12], when it was demonstrated that an additional mechanism of somatotropinoma formation involved germline mutations causing inactivation of tumour suppressor genes[13-16]. Particularly, mutations of the human aryl hydrocarbon receptor interacting protein (AIP) gene have been strongly associated with up to 40% of isolated familial and spontaneous human somatotropinomas. Forty-nine sequence variants of the AIP gene have been identified in humans; 70% of mutations are located at the C-terminal end of the protein which encodes three tetratricopeptide double helix motifs and a terminal 7 amino acid helix, considered highly important for the function of AIP. AIP forms a complex within the cytosol with a single aryl hydrocarbon receptor and two heat shock protein molecules. The resultant complex subsequently translocates into the nucleus, binds to aryl hydrocarbon receptor nuclear translocator, leading to activation of sensitive genes including xenobiotic metabolising enzymes[13-16]. This makes the link with exposure to environmental OHCs even more interesting since it could be hypothesised that AIP-mutations, and therefore AIP functional deficits could lead to accumulation of OHCs, which in turn can then exert their oncogenic potential.

Interestingly, human AIP mutation positive acromegalics are known to develop somatotropinomas at a younger age, and have a larger somatotropinoma size than other acromegalics[13-16]. Additionally, AIP mutated somatotrophinomas are known to cause more extreme GH secretion compared to non-AIP mutated somatotrophinomas, and respond less well to medical management[13-16]. The latter point might be particularly of interest, since thus far it has proven difficult to treat feline hypersomatotropism medically. The Author's research group has now sequenced the feline AIP-gene and is currently screening for AIP-mutations in cats coming through the RVC Acromegalic Cat Clinic. This work is anticipated to shine light on the role of AIP-gene aberrations in feline hypersomatotropism in late 2014.

UPDATE ON THE EXPECTED PHENOTYPE

The largest case assessment to date included 323 suspect acromegalic diabetic cats[4]; of these cats, most were male neutered (70%); mean age+/-SD was 11.3+/-2.7 years (range: 4-19); most were Domestic Short Hair cats (87%), though a variety of breeds were affected including Maine Coon cats (2%). This is unfortunately largely comparable to cats with primary diabetes mellitus and can therefore not be used to help form an index of suspicion. Nevertheless, Maine Coons seem overrepresented in the acromegalic population which poses interesting questions.

Another interesting aspect of this study was that clinicians were asked to indicate whether they suspected acromegaly at time of submission of the blood sample for fructosamine quantification. Of all submitting clinicians only 24% strongly suspected acromegaly in these cases which subsequently were shown to have an IGF-1 greater than 1000 ng/ml[4]. Body weight and fructosamine were significantly higher in cats with IGF-1> 1000 ng/ml (compared to those with IGF-1< 1000 ng/ml), despite significantly higher daily insulin dosages (mean 15 iu/day versus 6 iu/day). Nevertheless, the acromegalic cats displayed a range of body weights and a range of insulin demands (from modest to high). These data therefore confirm the 2007 study data of the author's group and suggest that historical and phenotypical clues for the presence of acromegaly are rarely (convincingly) present, which further validates the practice of screening newly diagnosed diabetic cats for the presence of hypersomatotropism. It also suggests it is more appropriate to use the term hypersomatotropism and not acromegaly, the latter indicating a syndrome which includes phenotypical changes, which however do not appear consistently present.

When looking at the cat with hypersomatotropism in more detail and with the benefit of more advanced imaging methods, more subtle clues can be spotted giving away the possible presence of this endocrinopathy. Apart from the presence of a pituitary tumor on CT imaging of the cranium in the majority (though not all) of cats, cats with advanced hypersomatotropism have been shown to have significantly thicker skull bones, and thicker skin and subcutis than unaffected cats (assessed through of three-dimensional surface-rendered CT images)[17]. Explaining the relatively high incidence (53%) of upper respiratory stridor in the history of acromegalic cats, the nasopharynx was found to be narrower in cats with hypersomatotropism than in control cats. Nevertheless, all of these changes would be extremely difficult to spot with the naked eye, in parallel with the experience in human acromegaly. In cats, additional difficulties may occur because of the presence of facial hair and the tendency for owners to examine their pet's face less frequently or thoroughly than their own. Furthermore, variations in skull conformation with breed and the lack of changes in early cases complicate assessment of signs of acromegaly in cats. Of the various differences identified in this latest CT-study, increased width of the head may be the most readily identifiable on physical examination in advanced cases. In summary, presence of these physical changes might help the clinician, though their absence should not decrease the suspicion of hypersomatotropism in the diabetic cat.

UPDATE ON NEW DIAGNOSTIC TOOLS

Serum GH and IGF-1

For most veterinarians, serum total insulin-like growth factor-1 (IGF-1) assessment represents the most feasible and accessible means of performing screening for hypersomatotropism in the (diabetic) cat[18]. However, hepatic IGF-1 production is dependent on presence of sufficient portal insulin, which can be deficient in newly diagnosed diabetic cats, resulting in false-negative results. Additionally, elevation of IGF-1 has been reported in non-acromegalic diabetic cats. Finally, differences in the performance of IGF-1 assays exist[19]. Growth hormone assays remain difficult to access commercially, despite their potential value[20]. Alternative or additional diagnostic tests for hypersomatotropism, as well as to evaluate the success of treatment of hypersomatotropism, are therefore desirable.

Serum procollagen propeptide

Since feline HS is associated with tissue growth, serum type III procollagen propeptide (PIIIP), a peripheral indicator of collagen turnover, was recently shown to be a useful indicator of active disease or growth hormone bioactivity in the cat[21]. Median serum PIIIP was five times higher in hypersomatotropism-induced diabetic cats compared to cats with primary diabetes mellitus. There was also a significant correlation between serum IGF-1 and PIIIP. Given that PIIIP is not dependent on portal insulin availability, this parameter could prove useful, perhaps as part of a panel, to increase the possibility of reliably diagnosing feline hypersomatotropism without the explicit need for intra-cranial imaging.

Serum Ghrelin

The orexigenic peptide ghrelin is a growth hormone secretagogue and has been shown to be down-regulated in humans with hypersomatotropism, whilst it increases following successful therapy. Serum ghrelin was shown to be significantly higher in control cats compared to cats with hypersomatotropism-induced diabetes mellitus[22]. However, the latter group had similar concentrations as those with regular diabetes mellitus, rendering this test not useful in the initial diagnostic phase. However, it has also been shown that post-radiotherapy serum ghrelin increases significantly in cats with hypersomatotropism-induced diabetes mellitus, whilst IGF-1 changes prove not significant[22]. This suggests a possible role for serum ghrelin as a marker of treatment effect.

Computed tomography beyond the pituitary gland

Although intra-cranial imaging is often merely used to prove or disprove the presence of a pituitary enlargement, CT or MRI likely has a role beyond mere assessment of pituitary morphology. As mentioned above, useful hints can also be found when looking at presence or absence of prognathia inferior (prevalence 24%), presence of TMJ malformation, increased thickness of calvarium bone, subcutis and skin[17]. However, once again, relying on these phenotypical changes is fraud with problems, since it will be confounded by various factors, including duration of the condition prior to diagnosis, variations in magnitude of endocrine dysfunction by pituitary neoplasms, the disease's gradually progressive nature and individual / breed variations in head conformation.

UPDATE ON MOST EFFECTIVE TREATMENT MODALITIES

Hypophysectomy

With increasing experience of using hypophysectomy to definitively treat feline hypersomatotropism / acromegaly, it also becomes increasingly clear that this constitutes the gold-standard treatment option, provided there is no contra-indication[23]. Contra-indications can include excessive tumour size, significant co-morbidities that prevent a safe anaesthetic and post-operative recovery (particularly significant renal and cardiovascular disease) or an owner who is completely risk-averse. The initial financial commitment forms another obstacle although it is important to realise that a year of ineffective insulin treatment or radiotherapy could have similar cost implications. The hypophysectomy experience in acromegalic cats is therefore in line with the human acromegaly experience. In the author's Hypophysectomy Clinic, feline acromegalics have been treated since April 2012, and approximately 80% of cases will enter diabetic remission within 1 month after surgery[18]. This demonstrates the enormous resilience of the cats' pancreatic beta-cells which seem unaffected by, often, more than a year of severe insulin resistance. The peri- and post-operative mortality rate in our clinic is currently 16%, which is acceptable to most cat owners given the paucity of equally effective alternative treatment modalities. Post-operatively, cats are treated with a low dose of hydrocortisone and levo-thyroxine (both lifelong) and DDAVP, which can be discontinued in most cats.

Advent of direct medical management

Medical treatment options remain desirable although somatostatin (sst) analogues and dopamine agonists have thus far proven largely ineffective[18]. Pasireotide (SOM230), a novel multi-receptor ligand sst analogue with high binding affinity for sst receptor subtypes 1, 2, 3 and 5 has been shown to suppress growth hormone (GH) and insulin-like growth factor-1 (IGF-1) in diabetic cats with hypersomatotropism. All eight cats treated with BID injections of a short-acting form showed a significant decrease in serum IGF-1, average 12-hour blood glucose and insulin requirements[24]. Unfortunately, the cost of the short-acting form of this drug (trade name Signifor®, Novartis, Basel, Switzerland) is considerable. A long-acting release formulation (LAR) has also been recently developed and is expected to be marketed (for human use) soon. Our research group tested this once monthly formulation for 6 months in 12 diabetic cats with hypersomatotropism and recorded a significant decrease in IGF-1, insulin dose and fructosamine. A quarter of cats went into diabetic remission, though 5 retracted from the study in light of lack of sufficient effect or mild adverse events (particularly loose stools).

Update on radiotherapy

Although radiotherapy has initially been suggested to be the best available therapy for this endocrinopathy in the cat, even the advent of stereotactical or gamma-knife radiotherapy does not rival the results of hypophysectomy as a treatment modality[18]. The effect of radiotherapy remains unpredictable in terms of a. whether it will occur at all; b. when it occurs; c. to what extent it will make a difference in insulin requirements and d. how long it will last for. Lack of response in terms of IGF-1 normalisation puts further doubt over its efficacy even in those cases with reduced insulin requirements. Radiotherapy still serves a valuable purpose though in those cases where there is a large pituitary tumour, increasing the risk of surgery, or when there are contra-indications for hypophysectomy (see above).

REFERENCES

1 Niessen S, et al. JVIM 2007;21(5):899; 2 Berg R, et al. JVIM 2007;21(5):892; 3 Schaefer S, et al. Proc Annual ECVIM Congress 2013: 220; 4 Niessen S, et al. JVIM 2011; 25: 1489; 5 Dye J, et al., Environ Sci Technol 2007;41:6350; 6 Mensching D, et al. J Toxicol Environ Health A 2012;75:201; 7 Olczak J, et al. N Z Vet J 2005;523:53; 8 Scarlett J, et al. Prev Vet Med 1988;6: 295; 9 Garcia M, et al. J Endocrinol Invest 1995;18:450; 10 Gooren L, J Clin Endocrinol Metab 1988;66:444; 11 Herman V, et al. J Clin Endocrinol Metab 1990;71:1427; 12 Clayton R, et al. Brain pathology 2001;11(3):313; 13 Ozfirat Z, et al. Mol and cell endocrinol 2010;326:71-9; 14 Lin B, et al. J Biol Chem 2007;282:35924; 15 Cazabat L, et al. Eur J Endocrinol. 2007;157(1):1; 16 Daly A, et al. J Clin Endocrinol Metab 2010;95(11):E373; 17 Lamb C, et al. J Feline Med Surg 2014;16(2):99; 18 Niessen S, et al. Vet Clin North Am Small Anim Pract 2013;43(2):319; 19 Tschuor F, et al. Am J Vet Res 2012;73(12):1925; 20 Niessen S, et al. Vet Rec 2007;160(26):902; 21 Niessen S, et al. Proc Annual ECVIM Congress 2012: 220; 22 Jensen K, et al. Proc Annual ECVIM Congress 2012: 219; 23 Meij B, et al. J Feline Med Surg. 2010 May;12(5):406; 24 Niessen S, et al. JVIM 2013;27(3):685

DISORDERS OF WATER BALANCE IN HUMANS

Andrea L. Utz, MD, PhD
Nashville, TN

INTRODUCTION

Hyponatremia is a commonly encountered clinical problem. In most cases hyponatremia is due to excessive production of anti-diuretic hormone (ADH) or ADH release in response to absolute or effective volume depletion. In normal situations, ADH is released in response to increasing plasma osmolality. Additionally, ADH is released by a non-osmotic stimulated mechanism in response to significant decline in blood pressure or blood volume. In situations of hypovolemia or effective volume depletion (such as congestive heart failure, cirrhosis, or nephrotic syndrome), ADH is released to retain water in an attempt to preserve perfusion volume. This excess water preservation relative to sodium may lead to hyponatremia. Euvolemic hyponatremia is most commonly caused by syndrome of inappropriate antidiuretic hormone (SIADH). The rate and degree of sodium decline determines subsequent symptoms. Raising the serum sodium should be rapid in the setting of severe symptoms; however overly rapid correction in the setting of chronic hyponatremia with mild to moderate symptoms should be avoided to prevent osmotic demyelination syndrome.

On the other end of the water balance spectrum are disorders of polyuria and polydipsia. After eliminating hyperglycemia and renal dysfunction, it is necessary to distinguish between the possibilities of diabetes insipidus (DI) and primary polydipsia. Diabetes insipidus may be of central origin due to ADH deficiency or renal origin due to ineffective ADH action. Primary polydipsia is often associated with psychiatric disease but may be a side effect of various medications or more rarely a symptom of a central nervous system lesion.

HYPNONATREMIA PRESENTATION

The symptoms of hyponatremia depend upon the rapidity of the decline in sodium, the degree of sodium decline, and underlying medical comorbidities. Acute hyponatremia, which occurs within a 48 hour timeframe, can lead to severe life-threatening symptoms due to cerebral edema. The symptoms include seizures, impaired consciousness, and respiratory arrest. These symptoms can progress quickly even with moderate degrees of low sodium if the rate of decline is rapid. Conversely in individuals who develop a gradual decline in sodium, resulting in chronic hyponatremia, much greater declines in sodium can be tolerated, sometimes with few apparent symptoms. This discrepancy is due to the brain's ability to compensate when sodium levels decline gradually. The brain responds to serum hypo-osmolality by extruding intracellular effective osmoles, which prevents brain edema - this process occurs within 48 hours of a decline in serum sodium. Subsequent rapid increases in serum sodium lead to intracellular water efflux, which may cause the osmotic demyelination syndrome.

CLASSIFICATION OF ETIOLOGY OF HYPONATREMIA

Figure 1 shows the algorithm for determining the etiology of hyponatremia.

Figure 1. Etiologies of Hyponatremia

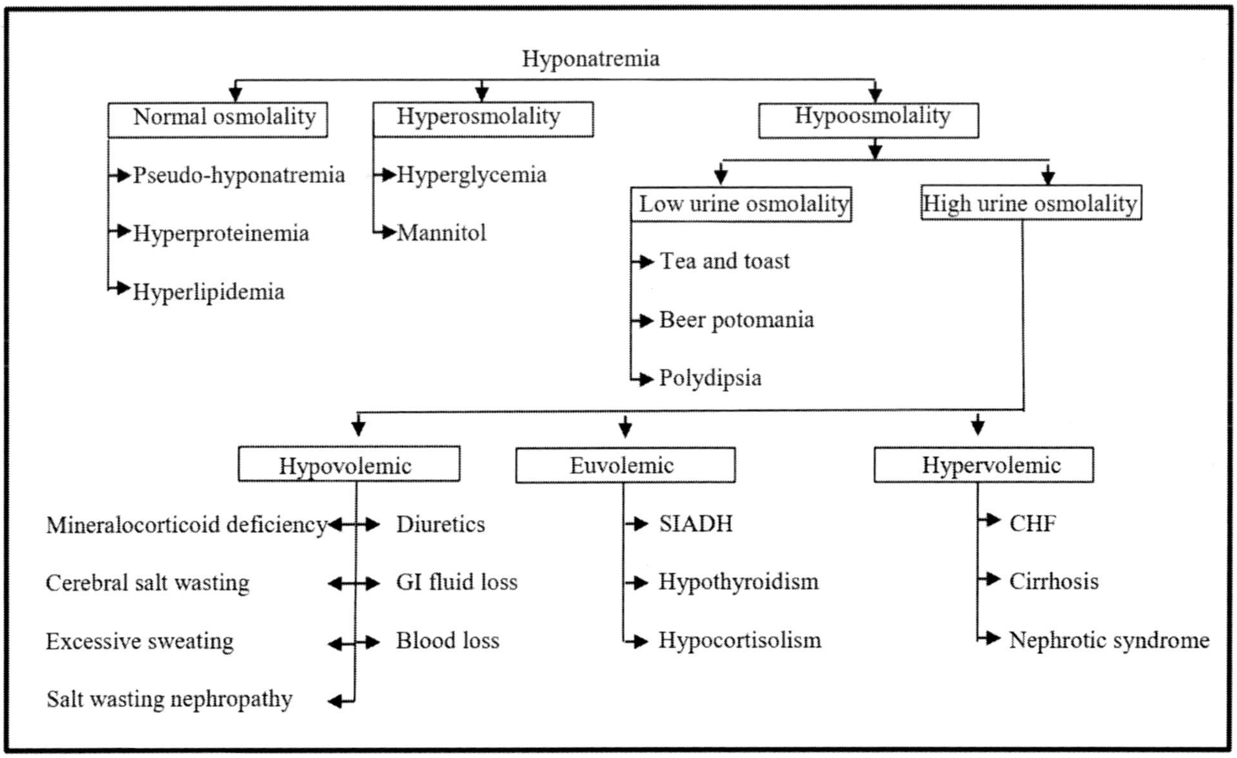

DIAGNOSIS AND MANAGEMENT OF SIADH

SIADH is diagnosed in individuals who are euvolemic, have non-maximally dilute urine osmolality, and a relatively high urine sodium concentration. It is important to rule out or treat hypothyroidism, adrenal insufficiency, and diuretic use prior to diagnosing SIADH as these conditions can cause similar biochemical findings. ADH release may be due to ectopic production by tumors or inappropriate stimulation of pituitary ADH release by a non-osmotic stimulus, such as nausea, pain, certain medications (Table 1), pulmonary disorders, and a variety of central nervous system disorders.

Table 1. Medications that May Cause SIADH

carbamazepine	nicotine	phenothiazines	NSAIDs	desmopressin
chlorpropamide	opiates	SSRIs / SnRIs	prostaglandin synthesis inhibitors	oxytocin
vincristine	barbiturates	MAOI inhibitors		vasopressin
cyclophosphamide	MDMA (ecstasy)	tricyclic antidepressants		
ifosfamide				

Treatment of hyponatremia depends on several factors including chronicity of hyponatremia, degree of low sodium, and symptom severity. Chronicity is divided into acute (0-48 hours) and chronic. The degree of low sodium can generally be divided into mild (130–135 mmol/L), moderate (120–130 mmol/L), and severe (less than 120 mmol/L). Symptoms are divided into minimally symptomatic, mildly symptomatic, moderately symptomatic, and severe symptoms. Mild symptoms include mood changes, irritability, fatigue, impaired concentration, and gait instability. Moderate symptoms include anorexia, headache, weakness, nausea, and confusion. Severe symptoms include vomiting, seizure, obtundation, respiratory depression, and coma. Treatments for hyponatremia are chosen based on the above criteria. In an individual with severe symptoms or acute onset moderate symptoms, rapid bolus hypertonic saline is recommended to minimize the risk of cerebral edema. For moderate symptoms, hypertonic saline with a slow to moderate infusion rate is given to raise the serum sodium 8 to 12 mmol/L in the first 24 hours and a total of 18 mmol/L in the first 48 hours. Alternatively, treatment with a vaptan (ADH receptor antagonist) medication, such as conivaptan or tolvaptan can be used to produce a water diuresis and subsequent rise in serum sodium. Frequent sodium assessment with hypertonic saline infusion rate or vaptan dose adjustment is necessary. Vaptan medications must be initiated in a hospital setting to minimize the risk of rapid serum sodium elevation. Conivaptan is an infusion and tolvaptan an oral medication, both of which are metabolized by CYP3A4 and thus there is a significant potential for drug-drug interactions. For mild or minimally symptomatic hyponatremia, fluid restriction and increased sodium intake may be effective. Fluid restriction often provides a relatively slow increase in serum sodium of approximately 1-2 mmol/L/day and may be intolerable. Demeclocycline may be added but it can cause renal dysfunction. In some situations, long-term hyponatremia can lead to a reset osmostat which is often minimally symptomatic and may not respond well to fluid restriction due to increased thirst.

MANGEMENT OF HYPOVOLEMIC AND HYPERVOLEMIC HYPONATREMIA

Hypovolemic hyponatremia is frequently due to excessive gastrointestinal fluid loss or diuretic use, particularly thiazides. Hypovolemia is a clinical diagnosis that is supported by finding a low urine sodium (except in the setting of thiazide use), indicative of aldosterone receptor activation. Hypovolemic hyponatremia is standardly treated with isotonic saline infusion. It is important to note that overly rapid correction of chronic hyponatremia, no matter the cause, increases the risk of osmotic demyelination syndrome and the rate of serum sodium rise should be carefully monitored and titrated. If the rate of sodium increase over-shoots the goal, oral water or intravenous dextrose in water can be provide to return the serum sodium to the target level.

Increasing sodium in the setting of hypervolemic disorders, such as heart failure and cirrhosis, is very challenging. Fluid restriction is a central component of management. In heart failure ACE inhibitors, ARBs, and loop diuretics may be beneficial. The oral ADH receptor antagonist, tolvaptan, may have a role in the future but safety and efficacy have not been clearly established. Hyponatremia in the setting of cirrhosis is a prognosticator of poor outcome. If severe symptoms related to hyponatremia are present, the sodium level should be increased slowly due to an increased risk of osmotic demyelination.

DIAGNOSIS AND ETIOLOGY OF DIABETES INSIPIDUS

Distinguishing between diabetes insipidus and primary polydipsia is done with fluid deprivation testing. This process involves restricting all fluid intake and measuring the ability to concentrate urine. Diabetes insipidus is diagnosed when urine is inadequately concentrated in the setting of elevated serum osmolality as shown in Figure 2. Once the diagnosis of diabetes insipidus is established, treatment with ADH is used to distinguish between central and nephrogenic DI, as patients with central DI will begin to concentrate urine following treatment with ADH, while those with nephrogenic DI will have minimal response. Central DI is caused by the following parasellar disorders: inflammatory or infiltrative (hypophysitis, sarcoidosis, Wegener's granulomatosis), neoplastic (craniopharyngioma, germinoma, metastases, lymphoma, Langerhans cell histiocytosis), traumatic (pituitary region surgery, head injury), vascular (subarachnoid hemorrhage, cerebrovascular accident), congenital, infectious (TB, fungal), or idiopathic. Nephrogenic DI is seen in the following settings: certain chronic renal disorders (polycystic kidney disease, renal infarcts, infiltrative renal disease, bilateral ureteral obstruction), hypercalcemia, hypokalemia, medications (lithium, demeclocycline), and congenital (vasopressin receptor defect, aquaporin defect).

Figure 2. Urine Concentration in Response to Increasing Serum Osmolality in Normal and Pathologic Situations

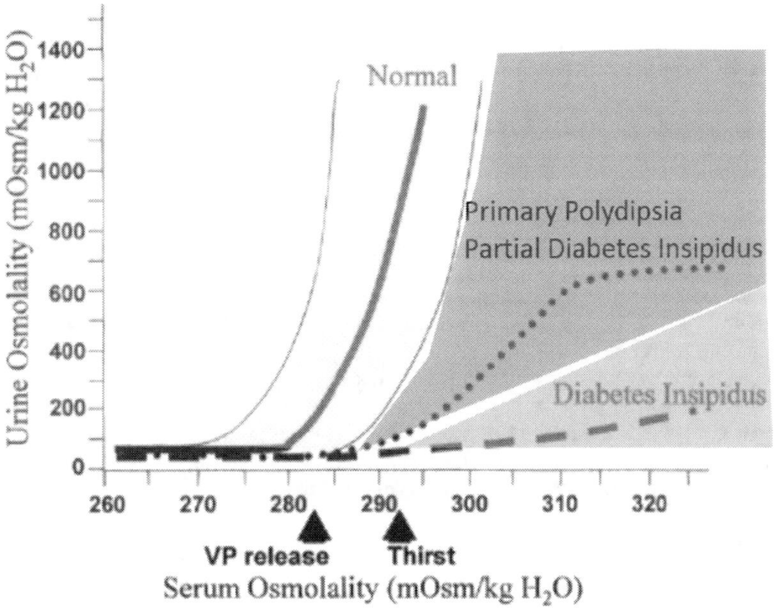

MANAGEMENT OF DIABETES INSIPIDUS

The mainstay of treatment for central DI is desmopressin (dDAVP), a synthetic analog of ADH. Titration of desmopressin and fluid intake are very effective to minimize polyuria and to maintain normal serum sodium levels. Care must be taken to prevent excessive fluid intake following desmopressin treatment as this can lead to hyponatremia. Treatment of nephrogenic DI is often much less effective and trials of high dose desmopressin, hydrochlorothiazide, NSAIDs, and decreased solute intake may be beneficial.

SODIUM FLUX FOLLOWING PITUITARY SURGERY

A unique situation exists following pituitary surgery, during which water balance can fluctuate from SIADH to DI or vice versa. This is felt to be due to disruption of normal ADH release caused by pituitary stalk or posterior pituitary injury. In the majority of cases, normal water balance is restored within two weeks, but a small proportion of patients develop permanent DI.

CONCLUSIONS

Disorders of water balance are frequent and due to a myriad of causes. It is imperative to understand the biochemical patterns associated with each disorder in order to approach treatment in an effective and expedient manner. Treatment of hyponatremia requires a rapid increase in serum sodium in the setting of severe symptoms but avoidance of overly rapid correction when symptoms are mild or moderate and hyponatremia is chronic. Diabetes insipidus is rare and the more common central form can be readily treated with the ADH analog, desmopressin.

REFERENCES

Berl T, et al. J Amer Soc Nephrology 2010;21:705
Ellison DH,et al. NEJM 2007;356:2064
Hensen J, et al. Clin Endoc 1999;50:431
Laville M,etal. Clin Kidney J 2013;6:i1
Loh JA, et al. Nat Clin Pract End Met 2007;3:489
Loh JA, Verbalis JG. Endo and Meta Clinics of NA 2008;37:213
Oiso Y, et al. JCEM 2013;98:3958
Schrier RWet al. NEJM 2006;355:2099
Thurman J,et al. Core Concepts in the Disorders of Fluid, Electrolytes and Acid-Base Balance: Springer US; 2013:29
Verbalis JG. Best Prac & Res Clin Endo & Metab 2003;17:471
Verbalis JG. J of Hosp Med 2010;5:S18
Verbalis JG, et al. The Amer J Med 2013;126:S1

CANINE AND FELINE HYPOPHYSECTOMY

Björn Meij, DVM, PhD, DipECVS
Utrecht, The Netherlands

Introduction

Transsphenoidal selective adenomectomy is the primary therapy for Cushing's disease in humans. The most common approach in humans is by the standard microsurgical submucosal transseptal transsphenoidal procedure using the neurosurgical operating microscope. There are many virtues of the midline transsphenoidal approach. Most importantly, it is the least traumatic route of surgical access to the sella. The lack of visual scars, lower morbidity and mortality as compared with transcranial procedures, the necessity of only a brief hospital stay, the relatively brief recuperative period add to the procedure's appeal. More and more human pituitary surgeons employ the pure endoscopic endonasal transsphenoidal surgical approach for pituitary tumor removal using rigid endoscopes. The pure endoscopic approach is facilitated by the air-filled sphenoid sinus in humans that is only separated by a thin bony floor from the pituitary fossa. Although there are profound differences between humans and canines/felines in the pituitary morphology, localization and surrounding bony sellar structures, there are important lessons to be learned from the vast experiences from pituitary surgeons written down in numerous publications on pituitary adenomas and related sellar lesions in humans.

In dogs, the most common method of treatment for Cushing's disease or pituitary-dependent hypercortisolism (PDH) remains medical treatment with mitotane (o,p'-DDD) or trilostane. However, medical therapy leaves the pituitary adenoma untreated. Also, it may be hypothesised that the removal of the chronic negative feedback exerted by the glucocorticoid excess at the pituitary level, may actually stimulate pituitary tumor proliferation and expansion.

At the Utrecht University transsphenoidal hypophysectomy was re-started in 1993 and has become an important addition in the management of Cushing's disease in the Netherlands (and occasionally for patients coming from other European countries). At present more than 300 dogs and 20 cats have undergone pituitary surgery. Surprisingly, pituitary surgery in dogs and cats is, besides the Netherlands, only advocated in a few other institutions (Japan and only very recently the USA, UK, and Italy). In dogs the indications for pituitary surgery include pituitary corticotroph adenomas (causing Cushing's disease), debulking of clinically non-functioning pituitary macroadenomas (causing diabetes insipidus or central neurological signs by the tumor mass effect) and occasionally sellar meningiomas. In cats the indications are pituitary corticotroph adenomas (Cushing's disease), pituitary somatotroph adenomas (acromegaly) and clinically non-functioning pituitary macroadenomas.

Pituitary Imaging

Computed tomography (CT) and magnetic resonance imaging (MRI) visualize the pituitary size and the surgical landmarks that are a prerequisite for pituitary surgery. Since the surgical landmarks are bone structures, they are more difficult to discern on MR images than on CT images. CT has proven to be practical, fast and accurate for detection of pituitary abnormalities. Surgical localisation of the pituitary gland in the various canine and feline skull types is dependent on the continuous visual assessment of typical bone features during the transsphenoidal approach and relating those, real-time, to the CT images in the operating room. The enhancement pattern of the neurohypophysis during dynamic contrast enhanced CT has been called the 'pituitary flush'. The displacement, distortion, or disappearance of the pituitary 'flush sign' in dynamic CT examinations can be used to confirm left or right-sided lateralization of (micro)adenomas.

Low field MRI (0.2 Tesla) has been proven to give insufficient information on the bone surgical landmarks and therefore contrast-enhanced CT is preferred. However, with the introduction of high field MRI (1.5 to 3 Tesla) machines in veterinary settings, MRI will certainly replace CT as the preferred method of pituitary imaging, as it has done in humans. High field MRI allows accurate assessment of the bone surgical landmarks together with improved visualization of pituitary structures. The T1 signal hyperintensity is a common finding at magnetic resonance imaging of the sellar region. This signal intensity pattern has different sources, and its significance depends on the clinical context. Normal variations in sellar T1 signal hyperintensity are related to vasopressin storage in the neurohypophysis, hyperactive hormone secretion in the anterior pituitary lobe (e.g. during pregnancy). Pathologic variations in T1 signal hyperintensity may be related to clotting of blood (in hemorrhagic pituitary adenoma and infarction (pituitary apoplexy) or the presence of a high concentration of protein (Rathke cleft cyst, craniopharyngioma).

Surgical Technique

Pituitary surgical techniques include selective removal of the pituitary adenoma (adenomectomy), removal of the adenohypophysis (adenohypophysectomy), removal of a significant part of pituitary tumor mass in the case of a macroadenoma (pituitary debulking), or complete removal of the pituitary gland including the tumor (hypophysectomy). Hypophysectomy in the dog and cat is performed by the midline transoral, transnasopharyngeal, transsphenoidal, microsurgical approach with the animal in sternal recumbency. Access to the pituitary fossa is obtained with a burr. An operating loupe or videoscope is used to provide magnification. Bone punches (Kerrison) are used to enlarge the opening created in the inner cortical lamina of the sphenoid bone. Following incision of the dura mater, fine ball-tipped probes are used to separate the pituitary tissue from the surroundings and the pituitary adenoma is then extracted through the dural opening using fine neurosurgical grasping forceps with controlled suction. In most cases the complete adenohypophysis is usually affected by the tumor and there is no sharp definition between adenoma and normal pituitary tissue. In all cases of hypophysectomy there will be loss of cerebrospinal fluid during surgery due to the incomplete diaphragm sellae. Unlike humans with Cushing's disease, well-defined pituitary (micro)adenomas are rare in dogs and cats. The hypophysectomy is considered complete when 1) there is an unobstructed view of the ventral hypothalamic surface and the opening to the third ventricle, and 2) there are no pituitary remnants upon exploration of the extensions of the hypophyseal fossa. The pituitary fossa can also be inspected for pituitary tumor remnants using rigid endoscopes, e.g., and endoscope with a diameter of 2.7 mm and a 30 degree viewing angle.

Postoperative intensive care includes close monitoring of vital functions, plasma electrolytes (sodium and potassium), plasma osmolality, and central venous pressure. Oral water intake is encouraged as soon as possible. Postoperative medication includes antibiotics and analgesics. Hormone replacement consists of hydrocortisone (1 mg/kg IV every 6 hours) and desmopressin, a vasopressin analogue (4 µg administered as a drop into the conjuntival sac every 8 hours for 2 weeks). When the dog has resumed eating and drinking, oral replacement therapy is started: cortisone acetate (1 mg/kg every 12 hours) and thyroxine (15 µg/kg every 12 hours). Over a period of 4 weeks the dose of cortisone acetate is gradually tapered to 0.25 mg/kg every 12 hours. Desmopressin (0.01%) is administered for 2 weeks, 1 drop into the conjunctival sac every 8 hours.

In dogs with giant (>2 cm) pituitary adenomas, the aim is to remove as much of the tumor tissue as possible to reduce the mass effect (debulking). Pituitary (macro)adenomas may undergo sudden infarction and hemorrhaging with rapid increase of the tumor volume, a condition which is known as pituitary apoplexy. Dogs with pituitary apoplexy show sudden onset of neurological worsening, with visual deficits, circling, depression, and eventually may enter a comatose state due to rapid increase in intracranial pressure and mass effect. When diagnosed in time, these dogs may benefit from emergency transsphenoidal debulking.

Results and Complications

The efficacy of transsphenoidal hypophysectomy in the treatment of dogs with PDH has been investigated in a prospective study in 181 dogs with a median age of 9 years. The 1-, 2-, 3-, and 4- year estimated survival rates were 86%, 83%, 80%, and 79%, respectively. Treatment failures included postoperative mortalities (= death within 4 weeks after surgery irrespective of the cause of death, 14 dogs), and incomplete hypophysectomies (12 dogs). The 1-, 2-, 3-, 4-year estimated relapse-free fractions were 90%, 77%, 72%, and 62%, respectively. Survival and disease-free fractions after hypophysectomy were markedly higher in dogs with nonenlarged pituitaries than in dogs with enlarged pituitaries. The main postoperative complications after hypophysectomy are reduction in tear production (31%) and prolonged diabetes insipidus (53%). Tear production restored to normal values in 79% of the affected dogs over a median period of 9 weeks. Diabetes insipidus occurred more frequently in dogs with enlarged pituitaries than in dogs with nonenlarged pituitaries and was permanent in 22% of the dogs. The results compare favourably with those of 129 dogs treated with o,p'-DDD in the same institution in another time frame. With longer follow-up time, hypophysectomy leads to better results than o,p'-DDD treatment.

Adrenocortical Function after Hypophysectomy

Adrenocortical function after hypophysectomy can easily be measured using the basal urinary cortisol/creatinine ratio (UCCR) in samples collected at home. It is important note that measurement of UCCRs are only reliable in laboratories that have established their own reference ranges, in the Utrecht Veterinary Diagnostic Laboratory (UVDL) the reference ranges for heathy dogs are $< 10 \times 10^{-6}$. When the patient leaves the hospital, usually 3 days after surgery, owners receive tubes for urine collection at 2 weeks, 8 weeks, 6 months, and 1 year after surgery. Thereafter yearly assessment of adrenocortical function is advised. Urine samples are collected at home when the dog is 24 hours free of cortisone medication. The early (<8 weeks) UCCR has prognostic value when considering long term survival and disease free fractions. In dogs with early postoperative UCCR $< 5 \times 10^{-6}$, the survival and disease free fractions are greater than in dogs with early postoperative values between 5 and 10×10^{-6}.

Pituitary Surgery in Cats

The indications for transsphenoidal hypophysectomy in cats are: Cushing's disease caused by an ACTH-cell pituitary adenoma or acromegaly caused by a GH-cell pituitary adenoma. Both conditions in cats are usually accompanied by diabetes mellitus requiring insulin administration but chronic cases develop typically severe insulin resistance which is the trigger that leads the clinician to the diagnosis of an underlying pituitary adenoma. The surgical technique of hypophysectomy in cats is similar to that in dogs. However, cats have an air-filled sphenoid sinus just rostral to the pituitary fossa that facilitates localization and enables somewhat more rostral extension of the approach than in dogs.

In cats with Cushing's disease, transsphenoidal hypophysectomy has disappointing results compared with dogs with Cushing's disease; higher postoperative mortality due to frequent co-morbidity in cushingoid cats make hypophysectomy not the first choice of treatment. However, this is strikingly different for cats with acromegaly and concurrent diabetes mellitus due to a pituitary somatotroph adenoma. Eight cats with acromegaly have been treated by hypophysectomy with excellent outcome. Diabetes mellitus disappeared and insulin administration was discontinued within 1 to 4 weeks after surgery, and in the long term GH and IGF-1 levels normalized.

REFERENCES

1. Meij BP; Voorhout G; Rijnberk A. Progress in transsphenoidal hypophysectomy for treatment of pituitary-dependent hyperadrenocorticism in dogs and cats. Molecular Cellular Endocrinology 2002; 197: 89-96.
2. Hanson JM; van 't Hoofd MM; *et al.* Efficacy of transsphenoidal hypophysectomy in treatment of dogs with pituitary-dependent hyperadrenocorticism. Journal of Veterinary Internal Medicine 2005; 19: 687-694.
3. Meij BP; Voorhout G; et al. Transsphenoidal hypophysectomy for treatment of pituitary-dependent hyperadrenocorticism in 7 cats. Veterinary Surgery 2001;30:72-86.
4. Hanson JM; Teske E; *et al.* Prognostic factors for outcome after transsphenoidal hypophysectomy in dogs with pituitary-dependent hyperadrenocorticism. Journal of Neurosurgery 2007;107: 830-840.
5. Hara Y; Teshima T; et al. Efficacy of transsphenoidal surgery on endocrinological status and serum chemistry parameters in dogs with Cushing's disease. The Journal of Veterinary Medical Science. 2010; 72: 397-404.
6. Meij BP; Auriemma E; *et al.* Successful treatment of acromegaly in a diabetic cat with transsphenoidal hypophysectomy. Journal of Feline Medicine and Surgery. 2010; 12: 406-410.
7. Bertolini G; Rossetti E; Caldin M. Pituitary apoplexy-like disease in 4_dogs. Journal of Veterinary Internal Medicine 2007;21:1251-1257.

Figure 1. Transverse contrast-enhanced computed tomography of the skull of a dog with Cushing's disease due to a pituitary adenoma, before (A) and 8 weeks after transsphenoidal hypophysectomy (B).

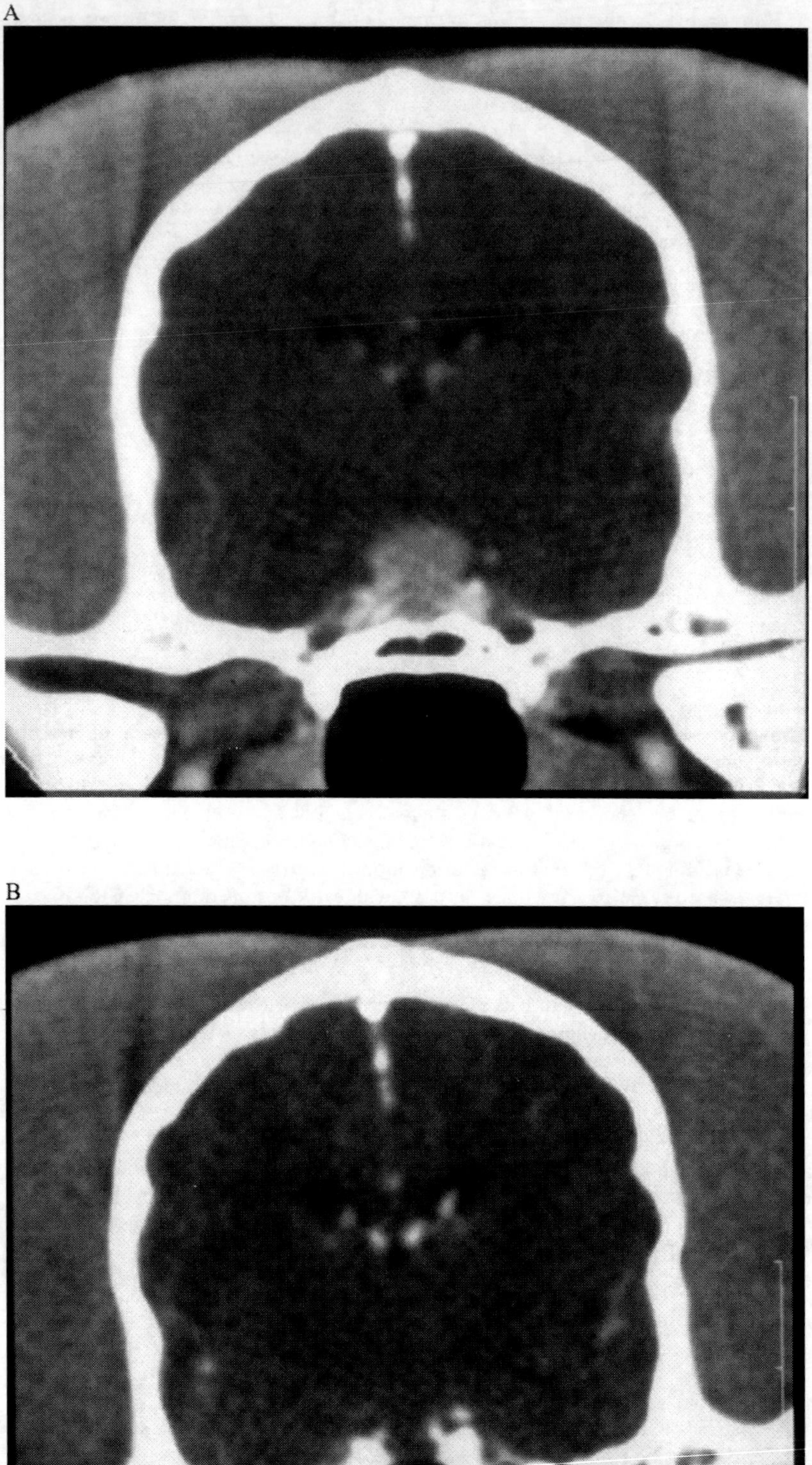

LUNG PATHOLOGY AND CLINICAL CORRELATES: SMALL ANIMALS

Kurt Williams, DVM, DACVP, PhD East Lansing, MI; Carol Reinero, DVM, DACVIM, PhD Columbia, MO

OVERVIEW

Dogs and cats are susceptible to a variety of infectious, non-infectious inflammatory/fibrotic, vascular and neoplastic causes of pulmonary parenchymal disease. Classically the diagnostic approach begins with a thorough history and physical examination and progresses to include thoracic radiography. Radiography is critical to help discern the location, severity, distribution and type of lesion but provides no diagnosis on a cellular or tissue level. Some ancillary tests may provide diagnostically valuable information without sampling the lung directly (e.g., serology for certain types of infectious disease), while collection of cells via trans- or endo-tracheal wash, fine-needle aspiration, bronchoalveolar lavage or bronchial brushing can be effective means for microscopic identification of neoplastic or inflammatory cells and some infectious agents. However, a cytologic diagnosis is only successful if infectious organisms or neoplastic cells are seen in the sample.[1,2] In certain cases, histologic evaluation of lung tissue is the sole means to making a diagnosis. In this lecture, illustrative examples of canine and feline patients with spontaneous respiratory diseases in which histopathology was critical to diagnosis will be presented.

WHEN AND HOW SHOULD A LUNG BIOPSY BE PERFORMED?

Because antemortem lung biopsy is invasive, expensive and poses some risk, it is often considered a "last resort". There are several situations in which lung biopsy should be considered. These include: when less invasive diagnostic tests have failed to provide a definitive answer, when empirical therapeutic trials have failed, when therapies based on other diagnostics resolve or greatly improve clinicopathologic abnormalities but disease recurs with discontinuation of therapy, and when characterization of the underlying pathology (e.g., fibrosis, emphysema, other remodeling changes) would provide important prognostic information.

Samples of respiratory tissue may be obtained in different ways. Bronchoscopy can be used to obtain guided mucosal pinch biopsies of the airways. These samples are small and superficial and are most diagnostic when specific lesions are visually noted (e.g., nodules or masses extending in the airways). Transbronchial biopsies are rarely performed in dogs and cats, although they are fairly routinely performed using endoscopic or fluoroscopic guidance in humans. They are used in humans in place of open lung biopsy for interstitial lung diseases. Other methods to procure tissue include thoracoscopy, and thoracotomy using a "key-hole" biopsy technique[3], a larger intercostal approach or a median sternotomy.

ADVANTAGES AND LIMITATIONS OF LUNG BIOPSY

A lung biopsy allows for histologic examination of tissue to provide information on the type, distribution, chronicity and anatomic location of pulmonary lesions. Situations in which obtaining lung tissue for histopathology is critical include infectious disease in which there are few organisms, the organisms are "walled off", or the organisms require special stains to be readily visible; cancer which is poorly exfoliative, is associated with inflammation making cytologic evaluation non-definitive (e.g., dysplasia vs neoplasia), or in which special stains are required to identify the exact type of neoplastic lesion; all of the non-infectious inflammatory/fibrotic interstitial lung diseases; as well as vascular diseases.

Animals with severe respiratory disease that often are most in need of the information that a lung biopsy would provide are paradoxically poor candidates for this procedure. Anesthesia is associated with respiratory depression and may be relatively contraindicated. Anesthesia may also dry the airways and impair mucociliary function. Pain associated with surgery may further compromise the work of breathing in dogs or cats that are unwilling/unable to fully expand their thoracic cage. Narcotics used for pain management may also compromise effective breathing. Techniques to obtain smaller sections of lung tissue (e.g., mucosal biopsies or key-hole biopsies) are generally associated with decreased morbidity. However, there is a trade-off in that the size of biopsy may preclude definitive diagnosis. Other risks associated with surgery include hemorrhage, pneumothorax, infection and surgical dehiscence.

WHY OBTAIN A LUNG BIOPSY: A PATHOLOGIST'S PERSPECTIVE

In human medicine, the use of open lung biopsies began in the 1960s; it is now a routine diagnostic procedure widely used in a variety of lung diseases. Lung biopsy has been particularly important in facilitating the classification of human interstitial lung diseases and expanding the understanding of the pathogenesis of these diseases. In spite of this, in 2005 the pathologist Dr. Talmadge King wrote in the American Journal of Respiratory and Critical Care Medicine that, "Surgical lung biopsy remains the investigation with the greatest overall diagnostic sensitivity for interstitial lung diseases. Many clinicians remain reluctant to subject their patients to surgical lung biopsy, likely reflecting their pessimism that findings on lung biopsy will alter the proposed treatment plan. Consequently, surgical lung biopsy continues to be underused". If, as Dr. King asserts, lung biopsy is underutilized in human medicine, then from the perspective of the pathologist in veterinary medicine, lung biopsy is woefully underutilized. Lung biopsy promises to fundamentally alter our understanding of the pathogenesis, and even identification, of respiratory disease in our profession. The indications and value of lung biopsy in veterinary medicine are similar to human medicine.

HOW TO OBTAIN AND HANDLE LUNG BIOPSIES: A PATHOLOGIST'S PERSPECTIVE

Surgical lung biopsies in veterinary medicine will likely only be performed once in the lifetime of the patient so it is incumbent on the clinicians and pathologists to maximize the amount of information that can be obtained from each biopsy. Site selection for the biopsy is of paramount importance, thus communication between the clinician, surgeon, radiologist, and pathologist is important (though rarely done). If a lesion is determined clinically to be solitary then biopsy of the single site is appropriate, though it is important to include the border between normal lung and lesion in the biopsy sample. In cases with more widespread disease selection of the site(s) to be biopsied may determine the value of the information obtained from the biopsy. In general the following recommendations should be considered:

1) Biopsy more than one lobe if possible.
2) If radiographic or CT information suggests heterogeneity of the process, then multiple sites reflecting the various stages of progression of disease should be biopsied.
3) DO NOT only biopsy lesions considered to be the oldest / most well developed: also sample lesions deemed intermediate in progression as well as lung considered to be normal.
4) Avoid obtaining the apex of lung lobes as these may have incidental lesions, i.e. sub-pleural fibrosis, that may be unrelated to the disease process of interest.
5) Keep a section of fresh unfixed lung for potential microbiologic examination.
6) If an entire lung lobe is removed, it is advisable to inflate the lobe with fixative (usually 10% neutral buffered formalin) via the lobar bronchus using a 60cc syringe until the pleura is uniformly expanded and smooth.
7) Thoracoscopically-obtained biopsies can be gently inflated with fixative using a needle and syringe inserted through the pleura.
8) It may be valuable to consult with the pathologist prior sectioning the biopsy material to discuss proper regions to be processed for histologic evaluation.

HOW TO UTILIZE THE INFORMATION OBTAINED FROM THE LUNG BIOPSY

Communication between the clinicians and pathologist should continue following the histologic evaluation of the biopsy. In settings where it is possible, i.e. an academic teaching hospital, the histopathology should be reviewed with the clinician and radiologist. Beyond the obvious benefit to all in the management of the case at hand, reviewing the histopathology provides an important means to understand and correlate structural changes in the tissues with decrements in health and lung function documented clinically. The histologic findings provided by the pathologist must be consistent with the clinical findings; if they don't it may reflect biopsy site selection which doesn't reflect the underlying disease process, inappropriate interpretation of the clinical findings or the pathologist's interpretation of the tissue changes is not correct. Clinicians should not be shy about expressing concern if the reported histopathology findings do not meet their clinical impressions. Meeting with the pathologist to review the case is the most effective way to address such concerns.

SUMMARY

Pathology provides a powerful tool to better understand respiratory disease. In veterinary medicine the lung has historically been one of the least biopsied organs: in part this is because of the clinical fragility of many patients that might be candidates for biopsy, but also because of a perception that the information gleaned from a biopsy won't change the management of the case, making the perceived risk of biopsy not worth the potential benefit. Because of this, there is dearth of clinical - pathology correlate information available on most veterinary respiratory diseases. Clinical – pathology correlate information is important to better understand the pathogenesis of the disease being managed at the time, and also provides a predictive context for future cases that present with similar clinical signs that may direct management of these cases before biopsy is necessary.

REFERENCES

1. Norris CR et al. J Am Anim Hosp Assoc 2002; 38:337-45.
2. Norris CR et al. . J Am Vet Med Assoc. 2001;218:1456-61.
3. Norris CR et al. J Am Vet Med Assoc. 2002;221:1453-9.
4. King Jr. TE. Am J Respir Crit Care Med 2005; 172:268-279

ANGIOSTRONGYLUS VASORUM: WHY YOU SHOULD BE WORRIED AND WHAT TO DO

Jakob L. Willesen DVM, PhD & Manuela Schnyder DVM, DEVPC
Copenhagen, Denmark and Zürich, Switzerland

INTRODUCTION AND BIOLOGY

The French heartworm (*Angiostrongylus vasorum*) is an emerging snail borne disease belonging to the superfamily of metastrongyloidea. Dogs, foxes, wolves (definitive hosts) and other related species can all be infected with this parasite through oral ingestion of infective third stage larvae (L3). The life cycle of *A. vasorum* is indirect, requiring snails or slugs as intermediate hosts, in which the first stage larvae (L1) excreted by definitive hosts develop into L3. A large number of gastropod species have the capability of being an intermediate host, which allows the parasite to establish itself into different geographical areas. Recently developed ELISAs (Enzyme-linked immunosorbent assays)[1,2] were applied for epidemiological surveys and a new serological diagnostic test has been commercialized, which, in addition to copromicroscopic methods, has increased the diagnostic options to identify infections with this important and potentially fatal canine parasite.

EPIDEMIOLOGY

Since its first description in the south of France in the 19th century[3], *A. vasorum* has been reported as a common parasite of foxes and less frequently in dogs in several countries in Europe, Africa, Atlantic Canada and South America. Traditionally, *A. vasorum* has been considered endemic in distinct isolated foci in France[4], Denmark[5], Ireland[6] and in south east England and Wales.[7,8] However, in the last two decades the number of reports from previously considered non-endemic areas in Europe sharply increased, with further detection of *A. vasorum* in dogs from Italy, Switzerland, Greece, The Netherlands and Hungary and Poland, and in foxes from some more distant countries, such as on the Iberian Peninsula as well as in Croatia.[9] In North America, the spread from the Canadian province of Newfoundland-Labrador was anticipated with high probability[10], and so, very recently, the first case in a fox in USA has been reported.[11] Prevalences in dogs vary between <0.5 and 10%, depending on the tested area (endemic vs. non-endemic), dog population (i.e. with or without clinical signs indicative of canine angiostrongylosis) and the adopted diagnostic methods. In addition to better diagnostic tools and increased disease awareness, different reasons for the apparent spread of *A. vasorum* have been discussed. Regional climate changes in vector epidemiology and movements in animal populations were taken into account.[12,13] The evidence of higher prevalence in foxes compared to dogs[14] accounts for the fact that wildlife, in the absence of obvious geographic barriers, plays an important role in expansion and establishment of *A. vasorum*. An increasing fox population close to human settlements[15] generally also increases the infectious threat for dogs. Therefore, migration of infected wildlife, most importantly the foxes, and movement of dogs, may cause local establishment and long distance spread of disease.

CLINICAL PRESENTATION

The presence of adult worms in the pulmonary artery and the migration of L1 through the pulmonary parenchyma trigger clinical signs related to the respiratory tract such as cough, dyspnea and exercise intolerance. However, a wide range of clinical signs may be seen, ranging from acute life threatening bleeding problems to very subtle and non-specific clinical signs such as anorexia or vomiting.[14]

Infected dogs

The mean age of infected dogs is reported in several studies to be about 18 month of age (range 4 months to 15 years) with no sex predisposition. A few studies have suggested that some breeds have an increased risk, but so far insufficient data exists to confirm this conclusion. However, miniature breeds very seldom are diagnosed with this disease, which may be explained by the fact that these dog breeds seldom exercise without a leash and thereby avoid coming into contact with infective material.[14]

Clinical signs

The most common reported clinical signs are cough, dyspnea and exercise intolerance (70%), caused by the migration of L1 through the lung parenchyma and subsequent verminous pneumonia. Pulmonary hypertension and right sided heart failure may be an associated complication. Life-threatening clinical signs such as coagulopathy are reported in 15-35% of cases in referral practices.[14,16] Bleeding may occur anywhere, with the oral cavity, eyes, skin and lungs being commonly reported sites. The underlying pathological mechanism is not fully understood but disseminated coagulopathy, coagulation factor deficiencies or immune-mediated thrombocytopenia have all been suggested, potentially relating to the release of worm products. Analyses of traditional coagulation parameters, such as APTT or PT, have been without consistent findings. Recently however, thromboelastography (TEG) has been shown to be a useful diagnostic tool in dogs and can be used diagnostically and to assess response to therapy in bleeding patients.

Bleeding in the CNS and spinal cord are also well described and commonly lead to neurological signs related to the anatomical location of these bleedings.[17,18] Seizures are most often reported but varying degrees of lameness and paralysis are also frequent. Neurological signs are reported in about 4% of all cases of canine angiostrongylosis. Non-specific clinical signs, such as depression, anorexia or gastro-intestinal signs, may be the only presenting clinical sign. Of particular importance are cases with no clinical signs (which in one study have been reported to occur in 7% of the cases)[19] and, obviously, the fatal cases. This high variability of the clinical picture constitutes a diagnostic challenge to the clinician.

Paraclinical parameters

Thoracic radiographs may be the most informative diagnostic tool to assess severity of infection and should be performed in all diagnosed cases. Changes are most frequently seen in the periphery of the lung field with focus in the dorso-caudal lung field. Broncho-alveolar patterns predominate, whereas a vascular pattern is less frequently reported.[20] Thoracic CT allows very detailed and accurate characterization of pulmonary parenchymal lesions, bronchi, and lymph nodes and closely reflected histopathological changes.[21] Hematological changes show inconsistent findings, with blood eosinophilia, neutrophilia and mild thrombocytopenia being most commonly reported. [22] Absence of these findings cannot be used to exclude a diagnosis of canine angiostrongylosis. Serum biochemistry is often unrewarding with hyperglobulinemia and reduced serum fructosamine as the most frequent finding.[16,23,24]

DIAGNOSIS

Currently, diagnosis of *A. vasorum* infections is mainly based on the isolation of L1 in faecal samples by larval migration methods such as the Baermann-Wetzel technique, followed by morphological identification of the larvae. Direct fecal smears[16,25] are also in use, having lower sensitivity. Larval excretion (patency) is described to start after 38-57 days post infection[16,22,26], when damage to the lung parenchyma is already present. Furthermore, it has to be considered that the sensitivity of copromicroscopic methods is limited in cases of intermittent excretion of larvae[27], low parasite load, and during prepatency. Therefore, repeated testing of faecal samples is suggested, in order to increase diagnostic sensitivity. In order to avoid misdiagnosis, experienced staff are required for accurate morphological differentiation of larvae, in particular; from other lungworm larvae, such as *Crenosoma vulpis* and *Filaroides* spp.[28] or larvae that may be present in delayed samples (i.e. of *Strongyloides* and/or hookworms) or in samples collected from the ground (containing free-living or plant parasitic nematodes). This latter consideration is valid also for FLOTAC, a technique based on the counting of larvae in chambers after spinning faecal samples onto a surface and is suggested to improve diagnosis of *A. vasorum* especially when larval migration techniques cannot be used.[29] Bronchoalveolar lavage is another technique for the detection of lung worm larvae but is not frequently used because of its low sensitivity and invasive nature. Additional non-pathognomonic laboratory parameters and diagnostic imaging techniques for the diagnosis of *A. vasorum* infections have already been mentioned. Alternative diagnostic tools, such as molecular methods[30,31] or highly specific serological tests, have been developed to overcome these problems. In particular, ELISAs for the detection of circulating antigen[1] and specific antibodies against *A. vasorum*[2] represent a valid and affordable tool for diagnosis in both individual and population studies, and have been very recently validated for *A. vasorum* infection in field studies.[9,32,33] The combined use of antigen and antibody detection ELISA offers great potential for early and sensitive diagnosis of these infections and for further epidemiological studies. As a consequence, a rapid, in-clinic assay (Angio Detect™ Test, IDEXX Laboratories, Westbrook, Maine, USA) for the serological detection of circulating antigen and intended for routine in-clinic diagnosis was developed. Its specificity and sensitivity have been evaluated with sera of naturally and experimentally infected dogs, providing evidence of good sensitivity and very high specificity of the rapid device for detecting circulating *A. vasorum* antigen in dogs with suspected canine angiostrongylosis.[34] It represents a very easy and useful tool to be broadly applied in veterinary practices. The rapid detection of infection in dogs is a key point for initiating an indispensable and urgent therapy, improving prognosis after adequate anthelmintic chemotherapy.

TREATMENT

Until recently, no drugs were available with a license to treat *A. vasorum* infections. A range of treatment protocols had been published based on single cases or smaller case series, but no larger scale prospective studies had been conducted. Today there are two treatment protocols licensed to treat *A. vasorum* infections. Milbemycin oxime, at a dosage of 0.5 mg/kg BW per os, weekly for four weeks, has been proven efficacious and equally so has a spot-on solution of imidacloprid 10%/moxidectin 2.5%.[35,36] The use of fenbendazole (25 mg/kg BW per os for 3 weeks) has also been well documented. Prophylactic treatment has also been documented using imidacloprid 10%/moxidectin 2.5% as spot-on solution, and spinosad/milbemycin per os according to labeled dosages.[37,38] Depending on clinical signs, supportive treatment is often needed. Most important is the use of transfusion therapy with fresh frozen plasma or whole blood products. When TEG is used to guide transfusions, this diagnostic tool can be used to determine the amount of plasma needed to stop bleeding and to assess its efficacy after treatment. Further treatment, such as supplemental oxygen, antibiotics, corticosteroids and fluids, is indicated on a case-by-case basis. Restricted exercise for a period of 4-6 weeks after treatment is often recommended. [14,16]

PROGNOSIS

When reviewing literature, a fatal outcome is reported in about 25% of the cases. This outcome is an overestimation, since fatal cases in new geographical areas are more often reported. Therefore, the authors suggest that the prognosis should be based on the present clinical signs. In this respect, cases with severe clinical signs, such as coagulopathy or neurological signs, may have a fatal outcome in as many as 25-30% of the cases. Conversely, dogs with an absence of these signs have a positive outcome in about 95% of the cases[(16,36)]. Long-term outcome is hard to predict as it depends on duration and magnitude of infection, which are unknown factors in most naturally infected cases.

CONCLUSION

Canine angiostrongylosis is an important parasitic disease. Failure to make a timely diagnosis and commence treatment may result in severe complications, such as chronic granulomatous pneumonia and coagulopathy, and may potentially lead to death. In recent years there has been an increase in the number of diagnosed cases and from previously unknown endemic areas in Europe and North America. Increased knowledge concerning the epidemiological situation contributes to increased disease awareness, sustained by the use of newly-available diagnostic tools such as ELISAs (for large-scale studies and individual follow-ups), and such as a commercial blood based rapid device for in-clinic use. Accordingly, adequate anthelmintic and new evidence-based treatment protocols can be implemented. In known endemic areas prophylactic anthelmintic treatment or routine screening of dogs for *A. vasorum* infection are recommended.

REFERENCES

[1]Schnyder, M et. al., **Vet Par** 2011;179:152. [2]Schucan, A et. al., **Vet Par** 2012;185:216. [3]Serres, F, **J Vet du Midi**1854;7:70. [4]Guilhon, J et. al., **Comp Ren Hebd des Seances de l Acad des Sci D** 1969;269:2377. [5]Rosenlund, P et. al., **Danish Vet Jour** 1991;74:451. [6]Dodd, K, **Vet Rec** 1973;92:195. [7]Jacobs, DE et. al. **Vet Par** 1976;1:377. [8]Simpson, VR et. al., **Vet Rec** 1982;111:303 [9]Schnyder, M et. al. **Parasitol Res** 2013;112 Suppl 1:109. [10]Conboy, GA, **Vet Par** 2011;176:382. [11]Kistler, WM et. al. **Vet Par** 2014; http://dx.doi.org/10.1016/j.vetpar.2013.12.007 [12]Morgan, ER et. al. **Parasitol Int** 2009;58:406 [13]Traversa, D et. al. **Parasites & Vectors** 2010;3:62 [14]Koch, J et. al. **Vet Jour** 2009;179:348 [15]Deplazes, P et. al. **Trends Parasitol** 2004;20:77 [16]Chapman, PS et. al. **J Small Anim Pract** 2004;45:435 [17]Garosi, LS et. al. **J Small Anim Pract** 2005;46:93 [18]Gredal, H et. al. **Acta Vet Scand** 2011;53:43 [19]Koch, J et. al. **Proc Int Conf Emerging Snail Borne Diaseas** 205:10 [20]Boag, AK et. al. **Vet Rec** 2004;154:426 [21]Dennler, M et. al. **Vet Radiol Ultrasound** 2011;52:289 [22]Schnyder, M et. al. **Parasitol Res** 2010;107:1471 [23]Willesen, JL et. al. **Jour Vet Med Series A** 2006;53:266 [24]Willesen, JL et. al. **Vet Jour** 2009;180:106 [25]Humm, K et. al. **J Small Anim Pract** 2010;51:200 [26]Bolt, G et. al. **Vet Rec** 1994;135:447 [27]Oliveira-Junior, SD et. al. **Vet Par** 2006;141:101 [28]McGarry, JW et. al. **Vet Rec** 2009;165:258 [29]Schnyder, M et. al. **Parasitol Res** 2011;109:63 [30]Al-Sabi, MN et. al. **Vet Par** 2013;196:366 [31]Jefferies et al.,**Vet Par** 2009;166:112 [32]Guardone, L et. al. **Vet Par** 2012;192:192 [33]Schnyder, M et. al. **Parasitology** 2013;140:1442 [34]Schnyder, M et. al. **Parasites & Vectors** 2014:In press:corrected proof [35]Conboy, G, **Vet Rec** 2004;155:16 [36]Willesen, JL et. al. **Vet Par** 2007;147:258 [37]Bohm, C et. al. **Vet Par** 2013:199:272 [38]Conboy, G et. al. **Proc AAVP** 2004:122

CHRONIC CANINE NASAL DISEASE: DIAGNOSTIC METHODS AND DISEASE DIAGNOSIS

Leah A. Cohn, DVM, PhD, DACVIM (SAIM)
Columbia, MO

INTRODUCTION

Clinical signs of nasal disease may be similar regardless of underlying disease cause. Indeed, a number of systemic diseases can also result in "nasal" sings. Efforts should be made to rule out systemic disease before undertaking a specific investigation into possible differential diagnosis for nasal disease, especially when the signs of nasal disease are either epistaxis, or bilateral mucopurulent nasal discharge. Primary and secondary coagulopathy, severe hypertension, hyperviscosity syndromes, viral infection, dysautonomia, and vomiting or regurgitation are all important non-nasal considerations for nasal signs. Once non-nasal causes are ruled out, it can be helpful to keep in mind which nasal diseases are most common. There are limited retrospective studies available, and all have come from secondary and tertiary care facilities, likely biasing results. For instance, oral/dental disease is very likely more common a reason for nasal signs than reflected in the literature simply because veterinarians typically do not refer these cases for evaluation. In most studies, nasal neoplasia is the most common cause of chronic nasal signs, with inflammatory rhinitis close behind. Fungal rhinitis and structural disease (including foreign bodies) occur less often. Depending on the study, dogs without a confirmed diagnosis range from fewer than 5% of those evaluated to more than 35%. Importantly, primary bacterial rhinitis is believed to be an extremely rare cause of chronic nasal signs.

HISTORY AND PHYSICAL

Nasal disease is more common in monocephalic and doliocephalic dog breeds. Nasal neoplasia is more likely in large breed dogs, and more than 2.5 times more likely in doliocephalic dogs. Nasal neoplasia is also more common in older dogs, but periodontal disease is also more likely in old dogs, and inflammatory rhinitis is found in dogs of every age.

History can be extremely useful in narrowing the differential diagnosis list. Key questions for dog owners include: Are there signs of illness that are not directly related to the nose, such as anorexia, weight loss, lethargy, bleeding, or bruising?; What is the duration of nasal signs, and have any therapies been attempted? If so, what was the response?; Did the dog's environment change before the clinical signs began (including trips to a boarding facility or a move to a new location)?; If nasal discharge is present, does it come from 1 nostril (and if so, which one) or both?; Has discharge changed from unilateral to bilateral?; What is the predominant character of nasal discharge, and has it changed over time?; If nasal signs are long-standing, is there a seasonal change? Systemic illness is particularly likely with acute nasal signs, but can occasionally be chronic (eg, vomiting or regurgitation can cause chronic mucopurulent nasal discharge). With chronic signs, seasonal variation can suggest inflammatory disease. Although signs of neoplasia can be present for days, weeks, or month, many months of static signs is more suggestive of structural or inflammatory disease than neoplasia or even fungal rhinitis. Discharge due to structural defects or foreign bodies is usually unilateral, while inflammatory rhinitis can be unilateral or bilateral. It is common for discharge associated with both neoplasia and fungal rhinitis to progress from unilateral to bilateral over time. Response to prior antibiotic use does not confirm a primary bacterial infection as secondary infections are common, and many antibiotics (eg, doxycycline) have anti-inflammatory properties.

Physical examination can be very helpful in prioritizing differential diagnoses. Examination of the entire dog can help identify systemic disease. More specific examination should include assessment of symmetry of the face and muzzle; character of nasal discharge, with notation of side(s); facial deformity or ulceration; patency of airflow through each nostril; condition of the teeth and gums; examination of the roof of the mouth to the pharynx (to degree possible); ability to retropulse the eyes; pain on opening the mouth or manipulating the muzzle; epiphora; percussion of the frontal sinuses and muzzle; pigmentation of the nasal planum; size and texture of submandibular lymph nodes. The oral cavity must be examined thoroughly, which often requires sedation or anesthesia. Tooth root abscess, oronasal fistulae, fractured teeth, osteomyelitis, and intranasal migration of displaced teeth are all potential causes of nasal discharge.

Common physical exam clues to disease causation may direct the most useful next steps. Deformity of the face, epiphora, or inability to retropulse the eyes all suggest a space occupying lesion such as a tumor. Absence of airflow also suggests occlusion of the nasal passages, but this might be either by tissue (tumor) or mucus accumulation. Loss of pigment on the nasal planum suggests aspergillosis, and airflow is often preserved in dogs with this diagnosis. Ulcers on the nasal planum might be due to either cancer or aspergillosis, but ulcers on the muzzle are more often seen with neoplastic disease. Character of the discharge can also provide clues, but is never pathognomonic. Serous discharge is often due to early viral infection, allergy, nasal mites, or stress (stress results in acute serous discharge). Mucoid and mucopurulent discharge are often due to systemic disease, oronasal/periodontal disease, nasal neoplasia, inflammatory nasal disease (reactive or primary), fungal rhinitis, foreign body, or secondary bacterial infection. Epistaxis is often due to systemic disease, trauma, nasal neoplasia, fungal rhinitis, or inflammatory nasal disease.

LABORATORY STUDIES

Routine laboratory studies such as CBC, chemistry profile, and urinalysis are especially useful when nasal signs are due to systemic disease rather than nasal disease. Evaluation of platelet count is the very first step for dogs with epistaxis! Even if platelet count is normal, thrombocytopathia or secondary hemostatic defects should be considered if there is epistaxis without any other signs of nasal disease. These routine parts of the "minimum data base" seldom identify a nasal cause of nasal signs, but can be important to assess the animal's over-all health prior to invasive diagnostic tests that require general anesthesia.

Cytology is very rarely useful in finding nasal disease, but is simple, minimally invasive, and inexpensive. If regional lymph nodes are enlarged, FNA can reveal metastasis of nasal neoplasia thus achieving a simple and cheap diagnosis. Cytologic assessment of nasal discharge may very, very rarely identify fungal elements or abnormal cells. The same is true for nasal lavage fluid, although the actual act of lavage may dislodge foreign material and be therapeutically useful.

Other laboratory studies are useful in certain cases. Serologic agar-gel immunodiffusion or IgG enzyme-linked immunosorbent assay tests for aspergillosis have good sensitivity and better specificity, but are still best used in cases where fungal rhinitis is suspected rather than in every dog with nasal signs. Blood and urine tests for galactomannan are not helpful for sinonasal fungal disease. Other serologic tests, like tests for Ehrlichiosis in dogs with epistaxis, are of occasional value. The utility of bacterial and fungal cultures of the nose are controversial. Since the nasal passages are not sterile, it is possible to grow bacteria completely unrelated to disease, and primary bacterial rhinitis is a very rare condition. For this reason, the author almost never performs bacterial cultures. Fungal cultures are time consuming and somewhat costly; if obtained from an area with visible fungal plaques, culture has as high as 80% sensitivity, but random swabs are much less likely to provide meaningful information.

IMAGING STUDIES

Radiographic imaging

Radiographic imaging techniques useful in dogs with nasal disease include dental radiographs, skull radiographs, thoracic radiographs, computerized tomography, and magnetic resonance imaging. Dental radiographs may be the only imaging study required if dental disease is suspected based on exam, and then confirmed on dental films Thoracic radiographs can be used to look for metastatic disease, but nasal neoplasia is generally locally invasive and metastasizes to regional lymph nodes rather than to the lungs. More useful are imaging studies of the skull/muzzle itself. Although more expensive and only available at some veterinary hospitals, advanced imaging with CT or MRI offers numerous advantages over skull radiographs and should be strongly considered as the initial imaging modality if the dog's owners have any interest in pursuing therapy. A table comparing modalities is included here.

	Skull radiographs	**Computerized tomography**	**Magnetic resonance imaging**
Availability	Readily available	Moderate availability	Least available
General anesthesia	Required	Anesthesia or sedation	Required
Cost	Least costly	Moderate cost	Most costly
Show cribriform plate integrity	Poor	Excellent	Excellent
Ability to discriminate between tissue and mucus	Poor	Excellent (with contrast)	Excellent
Sensitivity to detect soft tissue changes	Poor to moderate	Good	Excellent
Sensitivity to detect bony changes (lysis or hyperostosis)	Moderate	Excellent	Good
Ability to evaluate sinuses	Moderate	Excellent	Good to excellent

Rhinoscopic imaging

Rhinoscopy, usually performed with a rigid endoscope allows visualization of the nasal passages. A flexible scope allows visualization of the caudal choanae as well. Rhinoscopy is best performed immediately after imaging studies as it will cause bleeding that can alter radiographic appearance of the nose. Not every "nook and cranny" of the nose can be examined, and it is possible for rhinoscopy to miss small foreign bodies. It is most useful in visualization of fungal plaques and in helping to collect guided biopsies, but it can also show abnormalities such as hyperemia, excessive mucus, nasal mites, and sources of bleeding.

HISTOPATHOLOGIC STUDIES

Histopathologic examination of tissue recovered from the nose is often necessary for the diagnosis of nasal disease, and is always necessary to confirm tumor type in dogs with nasal neoplasia and for the diagnosis of inflammatory rhinitis. Adenocarcinoma is by far the most common tumor type, but undifferentiated carcinomas, chondrosarcoma, squamous cell carcinoma, fibrosarcoma, lymphoma, osteosarcoma, and others are possible as well. Treatment and prognosis may differ for these tumors. Inflammatory rhinitis may be lymphoplasmacytic, eosinophilic, granulomatous, or suppurative. Each of these different types of inflammation would suggest different treatments. The clinician must be cognizant that inflammation ca represent the reaction to a true underlying disease rather than disease itself; an inflammatory biopsy might simply be that the disease site was "missed".

Appropriate sample collection is crucial. While "blind" biopsies collected via a small cup type biopsy instrument or a larger pinching forceps (eg, mare endometrial biopsy instrument) can give a correct diagnosis, they often miss the mark. For large tumors, hydropulsion may dislodge tumor volume that can not only provide a specific diagnosis, but may result in improved clinical signs in the short term. The site of biopsy is best guided by imaging studies, rhinoscopy, or both. On occasion, more aggressive sampling via rhinotomy may be necessary to find an underlying cause of disease.

REFERENCES

Cohn LA, Vet Clin N Am, 2014:44:75; Johnson LR,. J Am Vet Med Assoc 2006;228(5):738; Miles MS, (2001-2004). J Am Vet Med Assoc 2008;232(12):1844; Meler E, J Small Anim Pract 2012;53(8):434; Bissett SA, J Am Vet Med Assoc 2007;231(12):1843; Pomrantz JS, J Am Vet Med Assoc 2007;230(9):1319.

MANAGEMENT OF TRACHEAL COLLAPSE: SURGERY, STENTS, AND MEDICAL

Stephan A. Carey, DVM, PhD, DACVIM
East Lansing, MI

INTRODUCTION

Tracheal collapse is a clinical syndrome associated with progressive flattening of the tracheal rings, leading to cough, airflow obstruction, and exercise intolerance. Affected dogs may be minimally symptomatic or even asymptomatic for years [1]. Typically, once clinical signs become significant, the symptoms progress as a result of a cycle of cough- and collapse-induced airway damage, airway epithelial repair and metaplasia, progressive loss of mucociliary clearance function, hyperplasia of the mucus secretory apparatus, progressive airway obstruction, and obstruction-induced clinical signs, including cough. The syndrome is due in part to a progressive softening of the tracheal ring cartilage, a condition known as tracheomalacia. When the cartilage supporting the principal bronchi, lobar bronchi, and segmental bronchi is also involved, the condition is termed tracheobronchomalacia. Bronchomalacia and bronchial collapse can also occur in the absence of tracheal involvement [2, 3].

Surgical intervention (external ring prostheses) and intraluminal tracheal stenting techniques are both excellent palliative options for patients with tracheal collapse. Each procedure has the potential to provide immediate improvement of clinical signs. However, neither provides a cure for tracheal collapse, and both have the potential for post-procedural complications [4, 5]. As a result, most experts agree that early and aggressive medical management of cough should be the first line of therapy for tracheal collapse, as well as adjunctive therapy for persistent cough following intraluminal stenting or surgical intervention [1, 6, 7]. Exceptions to this recommendation would include dogs exhibiting airflow obstruction as their principal presenting complaint, or dogs presenting in respiratory distress. This session will discuss the roles of stenting, surgical interventions, and medical management in the treatment of tracheal collapse, and will focus on the tools available for medical management and environmental modifications for the tracheal collapse patient.

SURGICAL INTERVENTION FOR TRACHEAL COLLAPSE

Surgical intervention for tracheal collapse involves placement of extraluminal prosthetic rings around the cervical trachea to the level of the thoracic inlet [8]. This is a technically challenging surgical procedure, and its use and timing of its use are a source of controversy. When performed by adept surgeons, the procedure is associated with significant improvement of clinical signs and quality of life, and has excellent reported outcomes [8]. The use of extraluminal prosthetic rings has also been associated with potentially serious perioperative complications, including post-operative laryngeal paralysis [4] and disruption of tracheal blood supply leading to tracheal necrosis [1, 8, 9]. Perioperative mortality rates of up to 18% have been reported [8, 9].

A recent retrospective study reported survival times of dogs following extraluminal prosthetic ring surgery, and found an overall median survival time of 4.6 years. Furthermore, this study reported that dogs exhibiting solely cervical tracheal collapse had a reported median survival time of >6.8 years, while dogs with cervical and intrathoracic (tracheal, bronchial, or both) airway collapse had a median survival time of 4.1 years [4]. Another study found that younger dogs had better perioperative outcomes, despite having more severe collapse at the time of diagnosis [8]. While the risk and severity of perioperative complications are both considerable, it is this author's opinion that external ring prosthetic surgery for tracheal collapse is a good option for young patients with cervical collapse and cough that cannot be adequately managed medically, and for young patients with tracheal collapse exhibiting airflow obstruction or respiratory distress.

INTRALUMINAL STENTING FOR TRACHEAL COLLAPSE

The introduction of intraluminal stents for management of tracheal collapse provides an important option that was previously unavailable for patients with thoracic inlet or intrathoracic tracheal collapse. Tracheal stenting is a simple, relatively brief, and rapidly effective procedure when performed by experienced clinicians. While the actual deployment of the tracheal stent is a relatively simple operation, substantial pre-procedural planning and operator skill are necessary to determine the appropriate stent size and location for placement. In contrast to extraluminal ring prosthetic surgery, the perioperative morbidity and mortality associated with tracheal stenting are minimal [10]. Long term complications associated with the first generations of tracheal stents included granuloma formation, stent migration, and stent fracture [5, 11]. These initial stents were biliary, gastrointestinal and vascular stents designed for use in people [12]. Modifications in intraluminal stent options specific for veterinary applications have evolved during the past 10 years, including variable diameter stents, flared end stents, and the use of shape memory alloys (e.g., nickel-titanium alloys). Although long term follow-up studies with the newer stent designs are lacking, these modifications may reduce long-term complications and prolong the life and utility of a properly sized and properly placed tracheal stent.

While intraluminal stenting provides an option for addressing intrathoracic collapse and airflow obstruction, some coughing may persist, either due to collapse of airways beyond the trachea or co-morbidities. Because ongoing coughing may increase the risk of stent fracture and may contribute to airway inflammation leading to granuloma formation, lifelong medical management is indicated for patients following tracheal stent placement.

MEDICAL MANAGEMENT OF TRACHEAL COLLAPSE

My approach to the newly diagnosed patient with tracheal collapse includes identification and management of co-morbid cardiopulmonary conditions, control of cough and airflow obstruction, control of airway inflammation, and environmental modification. These same strategies should be employed for continued cough and airway management for patients following tracheal stenting or extraluminal ring prosthesis surgery.

CO-MORBID CARDIOPULMONARY CONDITIONS

Tracheal collapse commonly affects middle-aged toy and miniature breed dogs, including Toy Poodles, Yorkshire Terriers, Pomeranians, and Chihuahuas [13]. This distribution also includes many breeds that are also commonly afflicted with chronic bronchitis and has significant overlap with breeds commonly affected by primary cardiac disease, including mitral valve endocardiosis. These conditions should be considered important differential diagnoses in the evaluation of the coughing dog, but should also be considered as possible co-morbidities for the tracheal collapse patient with persistent cough [3, 14]. Cough of lower airway origin or as a result of cardiomegaly can exacerbate tracheal or mainstem bronchial collapse, while congestive heart failure can not only trigger cough, but may also alter the tenacity of airway secretions.

COUGH SUPPRESSANTS

Because of the importance of coughing in the cycle of cough-induced airway inflammation and the tendency of cough to exacerbate intrathoracic tracheal collapse, antitussive therapy is often used in the medical management of tracheal collapse. I commonly employ narcotic cough suppressants, as some forms combine both antitussive and sedative properties. For dogs requiring only mild cough suppression, I use hydrocodone (Hycodan) at 0.22 mg/kg PO up to every 6 hours as needed. For dogs requiring sedation as well as cough suppression, such as dogs exhibiting excitement-induced coughing, I prefer to use butorphanol (Torbutrol) because of its wide dose range and sedative properties. I typically start low at 0.25 mg/kg PO every 6-12 hours, and gradually increase the dose if necessary, up to 0.5-0.75 mg/kg. In dogs who are exhibiting cough-induced cycling, I like to start treatment at higher doses for 2-3 days on a Friday, so that owners can observe their pets during the weekend on the first couple of days of sedation, and reduce the dose as control of coughing is gained. Smaller dogs may experience constipation at higher doses of butorphanol. For small dogs starting at 0.5 mg/kg or higher, I also recommend adding canned pumpkin or cantaloupe as treats to minimize this effect.

BRONCHODILATORS

The use of bronchodilators in the management of tracheal collapse remains controversial. For some authors, the evidence that beta-agonists and methylxanthines exhibit relaxant action on trachealis muscle *in vitro* suggests that bronchodilators might exacerbate tracheal flattening and worsen airflow obstruction [15, 16]. For others, the rationale behind their use is that dilation of small airways may reduce respiratory effort, decrease expiratory intrathoracic pressure, and minimize dynamic collapse of the central airways. Methylxanthine bronchodilators may have the added benefit of enhancing mucokinesis by stimulating ciliary beat frequency. I do not routinely use bronchodilators in the management of tracheal collapse, except in patients exhibiting evidence of small airway obstruction (exercise-induced wheeze, radiographic evidence of small airway disease, etc). In these cases, long-acting preparations of theophylline are most commonly used. Because of the narrow therapeutic window of these drugs, I tend to start at low doses (5-10 mg/kg PO BID), and gradually raise the dose to effect.

CONTROL OF AIRWAY INFLAMMATION

While few would argue that the control of airway inflammation is critical to the management of cough in tracheal collapse, there is no current consensus on the best way to achieve and monitor this control. The short-term use of oral glucocorticoids appears to be associated with reduction of clinical signs in tracheal collapse [1]. Chronic use may put patients at risk of bacterial infection (bacterial tracheobronchitis, bronchopneumonia), and the potential side effects of corticosteroid therapy (hepatomegaly, weight gain, abdominal fat redistribution) may also serve to worsen or exacerbate cough and airflow obstruction in tracheal collapse.

The ideal patient for corticosteroid therapy is one in whom non-septic tracheal inflammation can be documented via airway cytological sampling. Since it is impractical to routinely perform airway sampling in tracheal collapse patients, most patients are treated empirically with anti-inflammatory courses of corticosteroids following exacerbations of previously controlled conditions. I typically start with prednisone at 1 mg/kg/day divided BID for 5-7 days, followed by a gradual dose reduction with the goal of stopping therapy within 2-3 weeks. In dogs experiencing relapse during drug tapering, the goal is to reduce the dose to the lowest effective dose that maintains control. In dogs requiring long-term anti-inflammatory therapy for tracheal collapse, for patients that are intolerant of the side effects of short courses of oral steroids, or for dogs with concurrent chronic bronchitis, inhaled corticosteroids may be used as adjunctive or alternative therapy. In my experience, inhaled fluticasone propionate (110 mcg every 12 hours) therapy via facemask and spacer device is well tolerated by patients with tracheal collapse requiring prolonged anti-inflammatory therapy.

ENVIRONMENTAL MODIFICATIONS

Many dogs with tracheal collapse are overweight, and the distribution of fat in the abdomen, around the heart, and within the mediastinum can interfere with ventilatory function. Weight loss goals should be conservative and achievable, as even small changes in airway caliber can have significant effects on airflow and pressures. Modifications to the environment designed to minimize stress are often behavioral changes, including altering "hello" and "goodbye" routines, changing the timing and length of daily walks, and keeping other pets and/or kids in separate environments during exacerbations. The use of chest and body harnesses instead of collars should be encouraged to minimize irritation of the cervical trachea. Methods to improve indoor air quality, including eliminating exposure to cigarette smoke, increasing ventilation (especially outside ventilation), and minimizing exposure to irritants (aerosols, aromatic cleaning products) are also important adjunctive aids in the management of tracheal collapse.

REFERENCES

1. White, R.A.S. and J.M. Williams, Journal of Small Animal Practice, 1994. 35(4): p. 191-196.

2. Adamama-Moraitou, K.K., et al., Vet J, 2010. 191(2): p. 261-6.
3. Johnson, L.R. and R.E. Pollard, J Vet Intern Med, 2010. 24(2): p. 298-305.
4. Becker, W.M., et al., Vet Surg, 2010. 41(4): p. 501-6.
5. Moritz, A., M. Schneider, and N. Bauer, J Vet Intern Med, 2004. 18(1): p. 31-42.
6. Mittleman, E., et al., J Am Vet Med Assoc, 2004. 225(8): p. 1217-21, 1196.
7. Sun, F., et al., The Veterinary Journal, 2008. 175(2): p. 186-193.
8. Buback, J.L., H.W. Boothe, and H.P. Hobson,. J Am Vet Med Assoc, 1996. 208(3): p. 380-4.
9. White, R.N., J Small Anim Pract, 1995. 36(4): p. 151-8.
10. Sura, P.A. and D.J. Krahwinkel, J Am Vet Med Assoc, 2008. 232(2): p. 228-36.
11. Radlinsky, M.G., et al., Vet Surg, 1997. 26(2): p. 99-107.
12. Durant, A.M., et al., Vet Surg, 2012. 41(7): p. 807-17.
13. Johnson, L., Vet Clin North Am Small Anim Pract, 2000. 30(6): p. 1253-66, vi.
14. Marolf, A., M. Blaik, and A. Specht, Vet Radiol Ultrasound, 2007. 48(3): p. 199-203.
15. Brichetto, L., et al., J Appl Physiol, 2003. 95(2): p. 735-41.
16. Miyamoto, K., et al., Biochem Pharmacol, 1994. 48(6): p. 1219-23.

ENVIRONMENTAL EXPOSURE AND AIRWAY DISEASE IN HORSES WITH INFLAMMATORY AIRWAY DISEASE

Laurent Couetil, DVM, PhD, DACVIM (LAIM)
West Lafayette, IN

INTRODUCTION

Stabling results in exposure of horses to airway irritants such as dust, endotoxin, ammonia and β-glucan. Indoor housing and feeding of hay in particular, has been associated with chronic respiratory diseases including inflammatory airway disease (IAD) and recurrent airway obstruction (RAO). The role played by inhaled dust and endotoxin in RAO pathogenesis is well established however, similar evidence regarding IAD is currently lacking.

The purpose of this presentation is to review the pathophysiology of IAD and RAO, discuss methods available to assess environmental exposures and present recent findings that suggest a role for respirable dust in a subset of cases of IAD.

CHRONIC AIRWAY DISEASES OF THE HORSE

Inflammatory airway disease is defined as non-septic inflammation of the lower airway, characterized by excess tracheal mucus, chronic cough, and poor performance.[1] Presumptive diagnosis of IAD is often based upon history and airway endoscopic findings. Diagnosis is confirmed by demonstration of increased inflammatory cells upon bronchoalveolar lavage (BAL) fluid cytology or mild alteration in lung function. Horses with RAO usually have similar historical and physical examination findings however, contrary to IAD, horses with RAO exhibit increased respiratory efforts and the degree of BAL fluid inflammation is often moderate to severe. Diagnosis confirmation may be obtained by documenting reversibility of lung dysfunction following administration of bronchodilators, glucocorticoids, or decreasing exposure to dust.

The etiology of IAD remains poorly understood. In mature to older horses, mild increase in the proportion of BAL fluid neutrophils is the most common cytologic finding. However, in young horses, BALF mast cells and eosinophils are often increased, suggesting hypersensitivity. [1-4] It is possible that the different cytologic manifestations of IAD reflect differing disease mechanisms. This hypothesis is supported by divergent cytokine messenger RNA (mRNA) profiles for each of these cytologic phenotypes. Neutrophilic airway inflammation has been associated with increases in interleukin-1 beta expression, while BALF mastocytosis has been associated with increased expression of both interferon-gamma and interleukin-4 mRNA expression, indicating a mixed T helper lymphocyte type 1/ type 2 profile. [3-4] Clinically, neutrophilic IAD has been associated with cough in the absence of derangement in lung function[2] while eosinophilic and mastocytic airway inflammation are accompanied by increased airway hyper-responsiveness. [1,5]

Exposure to airborne dust and other irritants appears to play an important role in triggering IAD. Upon barn confinement, horses are exposed to higher concentrations of dust[6], endotoxin[6-7] and ammonia[8]. Housing of healthy horses in stalls is sufficient to induce airway inflammation[9-11] and higher dust environments are associate with increased tracheal mucus accumulations.[12] Despite convincing circumstantial evidence, research directly relating BALF cytology to exposure is sparse. Furthermore, investigation of the relationship between airway inflammation and environmental exposures requires measurement in the horse's breathing zone to capture individual variations in exposure.[13]

Similarly, the barn environment plays an integral role in the development and exacerbation of RAO. Clinical signs of RAO are reliably elicited upon introduction of susceptible horses to stall confinement with straw and dusty hay and removal from such environment results in resolution of clinical signs. [14] As such, the disease is believed to be a manifestation of hypersensitivity to molds and other components of organic dust. Analysis of BALF recovered from horses suffering from RAO has revealed higher mold-specific IgE and IgG levels when compared to BALF antibody levels in control horses. [15-16]

The role of the various horse barn aeroallergens and other irritants in RAO and IAD pathophysiology remains poorly explored. An understanding of the methods available to assess exposure to respiratory irritants is necessary to properly detect and eventually mitigate chronic respiratory diseases in horses.

ENVIRONMENTAL EXPOSURE ASSESSMENT

Dust size fraction and respiratory health

Inhaled particles deposit on the airways due to impaction, settling, interception, diffusion, and electrostatic interactions. The anatomical location where a particle is deposited depends on characteristics of the airflow, geometry of the airway, the rate and pattern of breathing, and particle characteristics. For particles larger than approximately 0.3 μm deposition is essentially determined by aerodynamic diameter (AD). The latter describes the diameter of a spherical particle of unit density that settles with the same terminal velocity. Three main size fractions are recognized in human environmental health: the inhalable fraction, the thoracic fraction, and the respirable fraction. The likelihood of each fraction to reach different level of the human respiratory tract is based upon the AD of the particles. By definition, particle having 50% penetration that are less than 100, 10 and 4 μm AD represent the inhalable, thoracic and respirable fractions, respectively. It is likely that the relationship between size fractions and particle deposition in the equine respiratory tract differs from humans. However, dust sampling equipment is designed to capture those size fractions implicated in human environment health studies.

Dust sampling

Various size-selective samplers have been designed to collect either the inhalable, thoracic, or respirable fractions. The Institute of Occupational Medicine (IOM) inhalable sampler provides the best approximation of the inhalable fraction. The IOM sampler is connected to a pump sampling air at a given flow rate and inhalable dust is trapped on a filter secured in the sampling cassette. Gravimetric analysis is performed by weighing the filter and assembled cassette before and after sampling and calculating the difference. Respirable particles are collected with cyclones such as the 10 mm Dorr-Oliver designed to remove larger particles from the airstream by centrifugal motion and trap particles with a 50% cut-point at AD = 4 μm. The advantages of filter sampling are that exposure to the various size fractions can be quantified and dust collected is available for further testing (e.g. microscopic examination, culture, endotoxin and β-glucan measurement, etc.). The disadvantages are that only cumulative exposure is assessed and sampling in low dust environments often yields measurement below the detection limit of the instrument.

Real-time particulate exposure can be measured by specialized equipment. Some sensors for example draw air in a chamber where light scatter detection is used to quantify dust. The advantage of such instrument is that particle exposure is determined continuously allowing evaluation of various activities (e.g. stall cleaning, hay shaking) on dust exposure and the limit of detection is lower than with gravimetric sampling. The disadvantage is that most of those instruments do not actually collect dust thereby preventing further testing.

EFFECT OF AIRBORNE IRRITANTS ON HORSE'S AIRWAY

Organic dust

When subjected to organic dust challenge by exposure to straw and moldy hay, healthy mature horses respond with airway neutrophilia. [11,17] However, lung function remains normal and there is a trend for increased airway responsiveness and airway smooth muscle turnover. [17] Exposure of RAO horses to dust from moldy hay and straw results in the classic clinical signs and lung dysfunction. [14] The magnitude of the airway neutrophilia in RAO horses is far greater than that seen in organic dust-challenged healthy horses. Experimental exposure studies using hay-dust suspension and its various fractions showed that airway neutrophilia is dose-dependent. [18]

Endotoxin

Experimental inhalation of increasing doses of endotoxin do not result in any appreciable clinical signs or airway mucus accumulation in healthy controls or horses with RAO however, they both exhibit a dose-dependent BALF neutrophilia. [19] Attenuation of endotoxin activity in hay dust suspension severely blunts the neutrophilic airway responses in both RAO and healthy horses and abolishes the lung dysfunction observed in RAO horses.

Study in young Thoroughbred racehorses

The purpose of the study was to evaluate changes in BALF cytology and exposure to particulates, endotoxin, and ammonia in of horses during their first month in training. Forty-nine client-owned, 12-36-month-old, Thoroughbred horses entering race training were recruited.

In this prospective cohort study, horses were assigned to be fed hay from a net (n=16), while the remaining horses were fed hay from the ground (n=33). BALF was collected at enrollment and after 14 days and 28 days in training. Respirable particulate, inhalable particulate, respirable endotoxin and ammonia concentrations were measured at the breathing zone of each horse weekly.

Median respirable particulate were significantly higher when horses were fed from hay nets than when fed hay from the ground. Likewise, inhalable particulate and respirable endotoxin exposures were significantly higher when horses were fed from hay nets. Feeding hay from a net resulted in significantly higher BALF eosinophil proportions over time. BALF eosinophils were significantly related to respirable particulate exposure.

We concluded that pulmonary eosinophilic inflammation develops in response to respirable particulate exposure in young Thoroughbreds, indicating a potential hypersensitivity to inhaled particulate allergens.

REFERENCES

[1]Couetil L, et al. **J Vet Int Med** 2007;21:356; [2]Benedice D, et al. **J Vet Int Med** 2008;22:1022; [3]Lavoie JP, et al. . **J Vet Int Med** 2011;25:322; [4]Beekman L, et al. . **J Vet Int Med** 2012;26:153; [5]Hare JE, et al. . **J Vet Int Med** 1998;163; [6]McGorum BC, et al. **Equine Vet J** 1998; 30:430; [7]Berndt A, et al. **Vet J** 2010;54; [8]Malikides N, et al. **RIRDC** 2003;03/089; [9]Holcombe S, et al. **Equine Vet J** 2001;33:244; [10]Gerber V, et al. **J Vet Int Med** 2004;18:92; [11]Robinson NE, et al. **Am J Vet Res** 2003;64:550; [12]Millerick-May ML, et al. **Equine Vet J** 2011;43:599; [13]Ivester KM, et al. **Equine Vet J** 2012;44:51; [14]Tesarowski DB, et al. **Am J Vet Res** 1996;57:1214; [15]Haliwell REW, et al. **Vet Immunol Immunopath**1993;38:201; [16]Schmallenbach KH, et al. **Vet Immunol Immunopath** 1998;66:245; [17]Leclere M, et al. **Am J Respir Cell Mol Biol** 2011;45:181; [18]Pirie RS, et al. **Clin Exp Allergy** 2003;33:676; [19]Pirie RS, et al. **Equine Vet J** 2001;33:311;

LUNG PATHOLOGY AND CLINICAL CORRELATES: LARGE ANIMALS

Rose D. Nolen-Walston, DVM, DACVIM (LAIM)
Kennett Square, PA
Kurt Williams, DVM, PhD, DACVP
East Lansing, MI

INTRODUCTION

Humans and domestic large animal species, especially equids, are afflicted with a variety of pulmonary diseases that share a number of similarities in their clinical expression and pathology. For example, the steroid-responsive airway disease Recurrent Airway Obstruction in horses is somewhat similar to asthma in humans while ARDS in horses is clinically and pathologically similar to the same disease in people.

Interstitial lung disease (ILD), especially chronic ILD in equids and humans are often difficult to diagnose and manage clinically, and very often are of unknown cause.[1] In humans viral infections, especially lung gammaherpesvirus infections, have been implicated in playing a role in the pathogenesis of the most clinically important of these diseases, idiopathic pulmonary fibrosis (IPF).[2] Chronic idiopathic ILD characterized by progressive fibrosis of the lung have recently gained recognition in veterinary medicine as an important cause of respiratory disease in horses, donkeys, cats, and dogs, and also share similarities with some forms of idiopathic ILD in people[1]. Equine Multinodular Pulmonary Fibrosis in adult horses and Donkey Pulmonary fibrosis in adult donkeys are two such diseases. Evidence from both species suggests that gammaherpesviruses may play a role in the pathogenesis of their respective diseases; this information may have important implications in understanding chronic ILD in humans.

EQUINE MULTINODULAR PULMONARY FIBROSIS AND HUMAN IDIOPATHIC PULMONARY FIBROSIS

Pathophysiology: horses versus humans

Equine multinodular pulmonary fibrosis (EMPF) is a relatively recently recognized progressive fibrosing interstitial lung disease of adult horses. First described in 2007, it was at that time tightly associated with lung infection with EHV 5, and was the first report of a disease being associated with this common gammaherpesviral infection.[3] The pathology of EMPF is quite characteristic and explains the clinical signs and physiologic perturbations associated with the disease. Grossly, as the name suggests, the disease is typified by the development of nodules of fibrosis; these foci develop within the alveolar parenchyma leading to significant decrements in gas exchange. The nodules vary from large (multiple centimeter) discreet tumor-like masses to numerous coalescent nodular foci of fibrosis that may encompass nearly the entirety of the lung. Histologically there is retention of an alveoli-like architecture. The remodeled airspaces are characterized by marked expansion of the interstitium with collagen. The lumen of the spaces is lined by cuboidal epithelial cells and there is a moderate inflammatory infiltrate comprised mostly of neutrophils and macrophages. Evidence of the viral infection can be found in the form of large eosinophilic viral inclusion bodies within small numbers of enlarged macrophages, which can be seen to consist of herpesviral particles using transmission electron microscopy.[3]

EHV 5 infection in horses is extremely common in horses, with infection occurring early in life.[4] Horses likely remain infected throughout their life, harboring latent virus in peripheral blood mononuclear cells as well as possibly epithelial cells of the upper respiratory tract. The apparent ubiquity of infection in horses with EHV 5 is incongruous with the relative rarity of EMPF in the horse population.

As noted above, the initial description of EMPF associated lung infection with EHV 5 with the development of the disease; association does not mean causation without additional evidence in support of the pathogenicity of the virus in the equine lung. Recently a study was performed to evaluate the ability of EHV 5 from cases of EMPF to induce disease in clinically normal horses.[5] Inoculation of such virus into the lungs of clinically normal horses induced nodular fibrosis similar to naturally occurring EMPF. Unlike spontaneous EMPF, virus was not detected in the lung using PCR or virus isolation, though viral antigen could be detected using immunohistochemistry. The results of this study provides strong evidence that EHV 5 is the cause of EMPF and that the disease and may begin while the virus is in a latent state in the lung.[5]

The discovery of the relationship between lung infection with the gammaherpesvirus EHV 5 and the development of EMPF may be important to understanding the pathogenesis of some idiopathic ILD in humans, especially IPF. While the cause(s) of IPF in people isn't known, there is wide acceptance that the disease develops as a result of abnormal healing in the lung following injury from a variety of insults, including viruses.[6] Surveys of human IPF lung have identified a greater rate of infection in the lung with herpesviruses, especially the human gammaherpesvirus Epstein Barr virus (EBV).[2] Like EHV 5, EBV is a common subclinical infection in people. What role, if any, that EBV or other viral infections may play in the pathogenesis of IPF remains to be determined. The data from horses infected with EHV 5 and its relationship to the development of EMPF provides 'proof of concept' regarding the ability of a gammaherpesvirus to induce progressive fibrosing interstital lung disease, which may prove to be important in the pathogenesis of some forms of idiopathic ILD in people.

Clinical signs: horses versus humans

There are a host of chronic fibrosing interstitial pneumonias in man, the most well known being idiopathic pulmonary fibrosis (IPF). Patients with IPF are generally older (>50 years of age) and present with a dry, non-productive cough brought on by exercise that progresses to dypnea and exercise intolerance as the disease worsens. Pulmonary function tests show evidence of restrictive disease, with decreases in vital capacity, but gas exchange (as measured by the diffusion capacity of carbon monoxide) is uniformly affected.[7] The median survival for IPF of 2.8–4.2 years, making it more deadly than many cancers.[8] Treatment of the disease is frustrating; traditionally, acetylcysteine, azathioprine and prednisone have been used as a triple therapy, but recent large clinical trials do not support their use. Pirfenidone, an anti-inflammatory, anti-oxidant, and anti-fibrotic medication demonstrated a significantly reduced risk of disease progression in patients with moderate-to-severe IPF by 30%.[9] Definitive treatment is available only through lung transplant.

Equine Multinodular Pulmonary Fibrosis (EMPF) is an interstitial lung disease seen in mature horses that is similar in many ways to IPF in man. The disease is generally characterized by severe tachypnea and increased respiratory effort with profound hypoxemia, tachycardia, and fevers of over 104°F. These signs are usually accompanied by significant loss of body condition and lethargy, and profound hyperfibrinogenemia and a mature neutrophilia characterize the hematalogic abnormalities in many cases. Thoracic radiographs and sonographic examination of the pleural lung parenchyma show a severe nodular interstitial pattern, of which the main differential diagnoses are fungal pneumonia or neoplasia.[10] Progression is variable, and no controlled studies have been performed to evaluate the various treatment options in common use. However, a combination of corticosteroids and anti-viral agents is now commonly used. Systemic prednisolone or dexamethasone at anti-inflammatory or immune-suppressive doses is administered in a tapering course over weeks to months. The addition of acyclovir, which has poor oral bioavailability and short half-life in the horse, is now used less in favor of its prodrug, valacyclovir, although valacyclovir has not been shown to be effective in reducing the effects or shedding of EHV 1 infection in horses.[11] Prognosis is fair-to-guarded, with some horses showing a durable clinical and radiographic cure, while others succumb to a failure of gas exchange and respiratory failure.

DONKEY PULMONARY FIBROSIS AND HUMAN PLEUROPARENCHYMAL FIBROELASTOSIS

Pathophysiology: donkeys versus humans

A distinctive form of interstitial pneumonia in donkeys was described in 2002 associated with two previously unknown gammaherpesviruses – Asinine herpesviruses 4 and 5 (AHV 4, AHV 5).[12] The gross pathology described for this disease is mostly localized to the cranioventral lung field. The affected lung is consolidated and discolored and fails to collapse. Histologically there is lymphohistiocytic inflammation within the alveolar interstium, inflammation surrounding bronchioles, as well as inflammation within the lumen of affected bronchioles. Large syncytial cells may be present within the bronchioles and alveoli of the animals and there is variable interstitial fibrosis present.

More recently a syndrome of pulmonary fibrosis distinct from the aforementioned interstitial pneumonia has been documented in aged donkeys. This disease, known as Donkey Pulmonary Fibrosis (DPF), has been reported in approximately 35% of aged donkeys that underwent necropsy at the time of death in a large survey in the UK. The pathology of DPF is grossly and histologically similar to a rare idiopathic ILD of humans, pleuroparenchymal fibroelastosis (PPFE).[13,14] The gross pathology of DPF is evident along the visceral surface of the dorsal lung. The lesion consists of pleural fibrosis that is arranged in coalescing linear to vermiform bands with contraction of the underlying lung in these regions. On cut section, the fibrosis is present within the pleura and extends along interlobular septa; fibrosis is also less commonly found within the parenchyma. The histologic findings of DPF confirm the fibrosis evident grossly. In addition to the fibrosis there is abundant elastin fiber accumulation (i.e. elastosis).

PPFE in humans is a rare form of idiopathic ILD first described in 2004.[15] As in DPF the distribution of the lesions is primarily within the upper (dorsal in donkeys) zone of the lung. The histologic findings of PPFE are very similar to DPF, with a mix of fibrosis and elastosis affecting the pleura, interlobular septa, and adjacent parenchyma. The cause(s) and pathogenesis of the human and donkey forms of this disease are not known at this time. AHV 4 and 5 were identified in cases of DPF as well as control donkey lungs, so the role of gammaherpesviruses in DPF is not clear. The identification of this disease in donkeys may provide important information into the etiopathogenesis of PPFE in humans.

Clinical signs: donkeys versus humans

The main defining characteristic of PPFE is "upper-lobe-dominant progressive pulmonary fibrosis", although its symptoms and clinical progression closely mimic other diseases within the same spectrum. The initial presenting complaints are usually dyspnea and dry cough, with fewer than half of the patients having crackles audible on auscultation. Patients presenting with PPFE are often slender in build with a flattened thoracic cage, which may be congenital or secondary to shrinkage of the fibrotic upper lung lobe. It can be either truly idiopathic, or secondary to diseases such asbestosis, lung or stem-cell transplant, or mechanical restriction (ankylosing apondylitis), but unlike IPF, it presents in patients from the third decade of life all the way into old age. Diagnosis is based on imaging studies of the pulmonary parenchyma, but can be made only definitively with biopsy or autopsy. Early radiographs may show hilar opacities, but a lateral projection can demonstrate the dorso-ventral thoracic flattening early in the disease.[16]

KL-6 is a mucin-like glycoprotein released by regenerative ATII cells, and is a reliable biomarker for interstitial lung diseases in humans, but is often only at the high end of the reference range in PPFE until very late in the course of the disease. Likewise, surfactant protein D (SP-D) may be elevated in some cases. Lung function studies show restrictive ventilatory impairment, with forced vital capacity and total lung capacity decreased. An increase of the ratio of residual volume/total lung capacity is observed distinguishes this syndrome from IPF. Treatment options are limited in this disease, with minimal improvement seen with immunosuppressive of corticosteroid use. The progression of disease is slower than that of IPF, usually over 10-20 years with death due to hypoxemia and respiratory failure as the predominant outcome.[16]

Donkeys demonstrate a very similar syndrome known as Donkey Pulmonary Fibrosis, which has been shown to affect 35% of aged donkeys in one post-mortem study,[17] although it is frequently an incidental finding. This is most likely because of the low athletic demands put of donkeys in first world countries. Early in the disease, crackles and mild tachypnea without abdominal effort is noted. Often donkeys are only presented for evaluation for an episode of severe respiratory distress possibly characterized by inspiratory effort and marked hypoxemia. Fever and cough are variable signs, with the main differential diagnoses being Recurrent

Airway Obstruction ("heaves"), and bacterial or verminous pneumonia. Radiographs show an interstitial pattern and evidence of cor pulmonae may be evident. Lung biopsy is suggested as a diagnostic aid but is rarely performed.[18]

REFERENCES

[1]Roman J et al. **Ann Am Thorac Soc** 2013 10:S224; [2]Tang YW et al. J **Clin Microbiol** 2003 41:2633; [3]Williams KJ et al. **Vet Pathol** 2007 44:849; [4]Wang L et al. **Vet Microbiol** 2007; 121:18; [5]Williams KJ et al. **PLoS One** 2013 8(10):e77754; [6]Selman M et al. **Ann Intern Med** 2001; 134:136; [7]King et al. **Lancet**. 2011 378:1949; [8]Maher TM et al. **Eur Respir J** 2007 30:835;[9]Richeldi L. **Respir Res** 2013;14 Suppl 1:S4;[10]Garré B et al. **Vet Microbiol** 2009 135(3-4):214;[11]Wong DM et al. **J Am Vet Med Assoc** 2008 232(6):898;[12]Kleioboeker SB et al. **J Vet Diagn Invest** 2002 14:273;[13]Miele A et al. **Chest** 2014 Mar 6 (epub);[14]Thusen JH. **Curr Resp Med Rev** 2013 9:238;[15]Frankel SK et al. **Chest** 2004 126:2007;[16]Watanabe K. **Curr Respir Med Rev** 2013 9:299;[17]Morrow L et al. **J Comp Pathol** 2010 144:145;[18]Thiemann A et al. **IVIS.org** www.ivis.org/special_books/Lekeux/bell/ivis.pdf.

STANDARDIZATION OF BRONCHOALVEOLAR LAVAGE IN HORSES CONSIDERATIONS

Julie Fjeldborg, DVM, Ph.D.
Sanni Hansen, DVM, Ph.D. and Keith E. Baptiste, BVMS, Ph.D., DACVIM, DECEIM
Copenhagen, Denmark

INTRODUCTION

Bronchoalveolar lavage (BAL) has been a standard diagnostic procedure in horses since the 1980´s and entails the sampling of pulmonary epithelial lining fluid (PELF) from the terminal bronchioles and alveoli.

The BAL procedure is typically well tolerated by adult horses and a valuable diagnostic supplement for investigating horses presented with lower airway symptoms and, exercise intolerance. However, still more than thirty years after its introduction, there is no international standardized method for performing the equine BAL, including agreed reference intervals (RIs) based on scientific studies. In comparison recommendations for standardization of all steps of the BAL procedure have been evaluated in human medicine and published by the European Respiratory Society in 1999 (Haslam and Baughman, 1999) and the British Thoracic Society in 2001(British Thoracic Society Bronchoscopy Guidelines Committee, 2001). These recommendations include considerations of possible sources of variation, including instilled volume, number of aliquots instilled, minimization of dwelling time, aspiration suction pressure, lung surface area lavaged, handling and storage of samples and finally the reporting of measurements. Furthermore, increasing expectations are placed on the diagnostic utility of equine BAL for certain lower airway diseases (e.g. inflammatory airway disease [IAD]), where recommendations of cuf-off values are given for BAL total cell counts and percentages of cell types. However, it is possible that such reference ranges require further evaluation for different populations of horses (e.g. stabled versus non-stabled, performance horses, age etc.) The presentation aims to outline considerations for a standardized BAL procedure in the horse for cytology. Separate considerations may need to be evaluated for the equine BAL procedure for microbiological purposes and for cellular measurements (e.g. drug concentrations)

PROPOSAL FOR A STANDARDIZED BAL PROCEDURE

Endoscopic evaluation and BAL sampling

Prior to BAL sampling, a history and clinical examination should always be performed with a focus on the upper and lower airways. The horse is then sedated typically with an α_2-agonist + opioid. However the effect of sedation on BAL results should be further evaluated, since sedative products also cause muscle relaxation and/or anti-tussive effects that could influence the total lung surface area sampled as well as aspiration volume (e.g. loss of elastic recoil). Following sedation, the horse is placed in a stock, and the left inner nostril is cleaned with saline soaked gauze. A 3 meter endoscope is passed through the nasal passages, larynx and into the trachea. For large horses, an endoscope of a minimum of 2.5 meter is necessary to reach the lower bronchioles.

The amount of tracheal mucous can be evaluated and a tracheal mucous scoring system is described by Gerber et al. (2004) where: 0 = none, 1 = little, 2 = moderate, 3 = marked, 4 = large and 5 = extreme. A mucous score ≤ 2 is considered normal.

The endoscope is then guided typically to the right caudo-dorsal lung-area and gentle wedged into a 4-6 generation bronchiole. A standardized BAL volume with sterile saline has not been determined, whereby volumes ranging from 100-500ml have been reported in the literature. BAL instilled volumes have the potential to create the significant variation on BAL results. Also, the influences of other additions to sterile saline (e.g. lidocaine) have not been evaluated as to their influence on aspiration volumes and BAL cytology. A standardized BAL volume has yet to be investigated that reliably samples enough alveolar surface area, results in adequate aspiration volumes, minimal cytology variation and provides enough PELF material that is not too diluted. Typically, this volume is lavaged into the lungs as one bolus infused manually through the biopsy channel of the endoscope, as quickly as possible, and immediately manually aspirated. However, infusing BAL volumes as a single bolus or multiple lavage boluses has yet to be investigated. The endoscope is gentle removed, and washed in an endoscope disinfectant cleaner before the next horse.

The aspirated volume is subjectively evaluated for transparency and amount of mucus and cellular debris. The aspirated volume is recorded and approximately 5ml of the pooled BAL aspirate is placed in a 10ml EDTA (fixative) tube for cytological evaluation. The whole procedure can be recorded for later evaluation of e.g. further mucus scoring. During the procedure the length from the nares to respectively *birfurcatio trachealis* and the flushing site in the caudo-dorsal lung part can evaluated and noted.

Endoscope versus BAL Bivona® tube

The endoscope has some obvious advantages over the blindly placed Bivona® tube, being able to visualize the amount and character of tracheal mucus as well as septum thickness (e.g. *birfucatio trachealis*). Another advantage is to guide and fixate the endoscope to a specific lung site and almost always the same lung location, which in most cases will be the caudo-dorsal lung part. The blindly placed Bivona® tube is easy to handle, cheap and studies have shown that the tube will often (e.g. >90%) wedge in the right caudo-dorsal part of the equine lung field, which is the predilection site for certain airway diseases (e.g. recurrent airway obstruction [RAO], IAD, exercise induced pulmonary haemorrhage [EIPH]) (McGorum et al., 2007)

Volume instilled

Based on experience, larger BAL volumes instilled (e.g. 300-500ml) will give a lower amount of neutrophils (Sweeney et al., 1992) and mast cells in horse BAL differential cell counts, from cytospin compared to small volumes (Hoffman, 2008; Jean et al., 2011), however this is not the case in human BAL differential cell counts (Heron et al., 2012). Since neutrophils and mast cells are important markers of equine lower airway diseases, then an agreed standardized BAL volume is needed.

By using the same volume every time, the potential problems with variation in differential cell counts will be reduced.

Volume aspirated

I human medicine, the percentage of aspirated BAL fluid has been linked to aging and loss of elastic recoil, which predisposes to obstructive lung diseases (Olsen et al., 2012). In horses, no loss of elastic recoil has been found with increasing age (Martin et al., 1977). However, the authors as well as others have found lower BAL volume aspirates in RAO horses compared to control horses (Jean et al., 2011). Aspirated BAL volumes could have significant influences on BAL cytology results since it needs to represent an adequate amount of alveolar surface area sampled as well as PELF.

Internal or external marker

A less appreciated fact about the BAL technique is that the actual quantity of PELF aspirated as a part of the BAL sample in often unknown and quite variable (Dargaville et al., 1999) this means that total cell counts and differential cell counts are, at best semi-quantitative values with variation between individual horses. Dilution markers can be used to better quantify the cytology between horses, and either endogenous or exogenous biomarkers have been described in the literature.

For example, a common endogenous biomarker is urea, and it is assumed that the PELF concentration of urea is equal to urea concentration in plasma and also that urea is distributed uniformly throughout the bronchiolar tree (McGorum et al., 1993). However, a challenge of using urea as a biomarker is possible blood contamination or increased permeability from the alveolar epithelium caused by inflammation (Kiem and Schentag, 2008).

Inulin is an inert polysaccharide of high molecular weight, that is not metabolised nor taken up by cells and does not penetrate the bronchoalveolar epithelium. An equine study, comparing PELF recovery between healthy and RAO horses, using inulin was published by Kirschvink et al. (2001), the results demonstrated that the aspirated BAL fluid is heavily diluted product. For example, in both human and horses PELF recovery consistently accounts for less than 10% of the aspirated BAL sample.

The use of biomarkers in the BAL procedure is not routinely utilised and will increase the cost of the samples. Biomarkers have the potential as a useful research tool to investigate a standardized equine technique. Further evaluation is needed to address this point, that can also assist with determining the sensitivity and specificity of BAL for diagnostic purposes.

BAL CYTOLOGY

The BAL EDTA sample is mixed typically in the laboratory for 10 minutes and a 200µl aliquot of the BAL EDTA sample is applied to the funnel of a disposable cytospin and centrifuged at 93 x g. for eight minutes in a StatSpin® Cytofuge. Preparation of cytospin should take place within 30 minutes of sampling, and then fixated in methanol and stained with May-Gründwald-Giemsa for microscopic evaluation. However, different staining techniques likely influence the accuracy of BAL cytology results.

Five hundred cells are classified by use of light microscopy under high magnification (400x) as neutrophils, lymphocytes, alveolar macrophages, mast cells or eosinophils. Each cell type is expressed as a percentage of the 500 cells. Epithelial cells, erythrocytes, and fungal spores can be semi-quantified subjectively using a scoring system, e.g. 0 = none, 1 = few, 2 = moderate or 3 = many.

Total cell count can be performed manually using a bürkerTürk counting chamber. An aliquot of 50µl of the sample is diluted in 950 µl of methyl-violet vinegar-acid and mixed for five minutes. Approximately, 50 µl of the mixed dilution is transferred to both sides (A and B) of the counting chamber. After enumerating both sides then the total cell count was calculated as: $(((A+B)/2) \times 50) / 1000 = 10^9$ cells/L

REFERENCES

[1]British Thoracic Society Bronchoscopy Guidelines Committee, **Thorax** 2001;56,Suppl 1, i1-21. [2]Dargaville P.A, et al **Am J Respir Crit Care Med** 1999;160,778-784. [3]Gerber V, et al **Eq Vet J** 2004;36,576-582. [4]Haslam P.L, Baughman R.P **Eur Respir J** 1999;14, 245-248. [5]Heron M, et al **Clin Exp Immunolo** 2012;167,523-531. [6]Hoffman A.M, **Vet Clin North Am Eq pract** 2008;24, 423-435, vii-viii. [7]Jean D, et al **Am J Vet Res** 2011;72,838-842. [8]Kiem S, Schentag J.J **Antimicrob Agents Chemother.** 2008;52,24-36. [9]Kirschvink N, et al **Vet Res** 2001;32,145-154. [10]Martin C.J, et al **Am J Respir Crit Care Med** 1977;115,981-988. [11]McGorum, B.C, et al **Res Vet Sci** 1993;55,52-56. [12]McGorum B.C, et al **Eq Res Med and Sur** 2007. [13]Olsen H.H, et al **PloS one** 2012;7,e43644. [14]Sweeney C.R, et al **Am J Vet Res** 1992;53,1376-1379.

STANDARDIZATION OF BRONCHOALVEOLAR LAVAGE IN HORSES

BAL REFERENCE INTERVALS

Sanni Hansen, DVM, Ph.D.
Julie Fjeldborg, DVM, Ph.D. and Keith E. Baptiste, BVMS, Ph.D., DACVIM, DECEIM
Copenhagen, Denmark

INTRODUCTION

Bronchoalveolar lavage (BAL) has been a standard procedure in horses since the 1980´s and entails the sampling of pulmonary epithelial lining fluid (PELF) from the terminal bronchioles and alveoli. The BAL procedure is well tolerated by adult horses and used widely as a diagnostic supplement in investigating horses presented with coughing, exercise intolerance or lower airway problems. A BAL cytology sample aspirated from a clinically healthy horse contains only few epithelial cells and erythrocytes and consist thus primarily of the resident population of immune cells, predominantly alveolar macrophages and lymphocytes.

However, still more than thirty years after its introduction, there is no international standardized method for performing BAL, including agreed reference intervals (RIs) based on scientific studies. BAL is an invaluable tool for equine lower airway diagnostics, whereby better scientific justified RIs will improve the diagnostic value of the test results.

The objectives of this study were to 1) establish RIs for equine BAL cytology samples valid for adult stabled horses, and 2) investigate intra- and inter-observer agreement of the BAL cytology evaluations.

MATERIAL AND METHODS

Seventy-eight stabled clinically healthy adult horses, aged 5-27 years were included. This was a part of a yearly health control. The horses were Danish Warm blood and two Friesian horses. The herd were a homogeneous group with regard to stabling environment, management procedures and daily exercise. They were all stabled on wood shavings and feed 'hay-lage' and grain.

A full clinical examination was performed on each horse in the morning before the BAL procedure. A blood sample for haematology was taken and the horses were sedated before the BAL procedure. A 3 meter long endoscope was used and the amount of mucus in trachea was evaluated using the mucus scoring system described by Gerber et al 2004. When the endoscope wedged into a bronchi of the right dorso-caudal lung field 180 ml (3x60 ml) sterile sodium chloride was infused and immediately aspirated. 5 ml of the pooled BAL samples were transferred to an EDTA tube (fixative).

BAL cytospin samples were prepared within 30 minutes of sampling, and stained in May-Gründwald Giemsa for microscopic evaluation. Five hundred cells were classified by use of light microscopy under high magnification (400x) as neutrophils, lymphocytes, alveolar macrophages, mast cells or eosinophils.

Two experienced and three inexperienced observers were evaluating the BAL cytology slides (n=50) counting 200 cells, for the observer agreement evaluation. The intra class correlation method was used to calculate the agreement.

The RIs were calculated using the non-parametric 95% confidence interval method described by the Clinical and Laboratory Standards Institute (C28-A3, 2008), and using dedicated software (Geffre et al., 2011). Initial investigating for potential partitioning with regard to age was done by investigating the ratio between sub-groups, but further studies are needed to determine whether partitioning of the established RIs is justified statistically (Harris and Boyd, 1990) as well as clinically (Kjelgaard-Hansen, 2010).

RESULTS

BAL cytology RIs includes: alveolar macrophages [47.0; 88.0%], lymphocytes [6.5; 44.5%], neutrophils [<15.0%], eosinophils [<1.5%] and mast cells [<6%] (Table 1).

High intra-observer agreement for the two experienced observers and adequate to poor intra-observer agreement for the three inexperienced observers were found (Table 2). The observer agreement showed high inter-observer agreement between experienced observers, except for mast cells, and adequate inter-observer agreement between inexperienced observers. The overall agreement between all observers was found to be high (Table 3).

DISCUSSION

The RIs for BAL fluid cytology estimated in this study are the first prospectively constructed RIs for stabled clinically healthy horses and valid for stabled adult horses between 5-27 years. Both the BAL procedure and subsequent laboratory methods for cytological evaluation used in this study are easy to adopt and thus relevant for equine field practice conditions. When having said that, it is important to only use these RIs along with the standardized procedure and using an instilled volume of 180 ml sterile saline.

The volume instilled varies a lot between studies, typically between 300-500ml, and previous studies have found different amount of neutrophils and mast cells using varying volume infused (Hoffman, 2008; Jean et al., 2011; Sweeney et al., 1992). Likewise, the volume aspirated, which is a function of the dwelling time and subsequently suction pressure, can give an indication of lower airway problems. Previous studies have found a lower BAL volume aspirated in horses suffering from recurrent airway obstruction (RAO) compared to control horses (Jean et al., 2011).

Stabled horses are exposed to a large amount of dust components (aero-allergens) which will create a variable degree of subclinical or clinical airway irritation and cytological this will result is a larger amount of inflammatory cells in the alveolar lumen of especially neutrophils. Previous accepted RIs established from several small groups of healthy control horses, encompassing eighteen different studies conducted during 1990-2008 (Hoffman, 2008). These RIs are most likely based on BAL samples obtained from horses at pasture, however they are not followed by a standardized BAL procedure with regard to volume infused, patient history (e.g. environment, age or gender), storage time, laboratory procedure nor statistical prerequisites.

The BAL fluid cytology from horses suffering from RAO show moderate to severe neutrophilia; >20% (Couetil et al., 2007) to >25% (Leclere et al., 2011) in BAL differential cell counts. The present RIs for neutrophils established in this study is >15% which makes the RIs sufficient for diagnosing stabled horses suffering from RAO based on the before mentioned cut-off values for RAO horses.

Horses with IAD will present with a BAL cytology profile characterized by an increase in total cell counts, mild neutrophilia (5-20%), lymphocytosis and monocytosis (Fogarty and Buckley, 1991; Rush Moore et al., 1995), or a BAL cytology profile, which is more specific for young horses with IAD and characterized by eosinophilia (>0.1%) and or an increase in mast cells (>2%) (Hare and Viel, 1998; Hoffman, 2008). The RIs presented in this thesis are incompatible with the cut-off values for diagnosing horses with IAD. The RIS are valid for stabled healthy horses and further studies are needed to differentiate healthy stabled horses from diseased IAD stabled horses.

Based on the observer agreement study, laboratory variation will be reduced by using experienced observers to perform BAL cytology. A recent study by Fernandez et al (2013) also explored observer agreements, and the results concurred with the results for the experienced observers in this study; they found high agreements for neutrophils, lymphocytes and macrophages, but only adequate agreement for eosinophils and mast cells. Focus on observer agreement also exists in human, where a study by De Brauwer et al (2002) found poor agreement for mast cells. The inexperienced counters showed adequate agreement, with bigger variations in both intra- and inter-observer agreements. Laboratory variation should be avoided whenever possible and thus experience amongst observers is important.

Table 1: Bronchoalveolar lavage (BAL) cytology reference intervals for macrophages, lymphocytes, neutrophils, eosinophils and mast cells together with total cells counts and percentage of BAL aspirated.

	n	Median	Range	95% RI	90% CI lower ref. limit	90% CI upper ref. limit
Macrophages(%)	78	66.0	40.5-89	47.0-88.0	40.5-52.4	83.6-89.0
Lymphocytes(%)	78	22.8	6.5-46.2	6.5-44.5	6.5-8.0	37.7-46.2
Neutrophils(%)	78	4.5	0-19.5	0.5-15.0	0.0-1.0	12.5-19.5
Eosinophils(%)	77	0.0	0.0-2.0	0.0-1.5	0.0-0.0	0.5-2.0
Mast cells(%)	78	2.0	0.0-6.0	0.0-6.0	0.0-0.0	5.5-6.0

Table 2: Intra-observer for two experienced and three inexperienced observers of bronchoalveolar lavage (BAL) cytology slides. The same slide was examined three times at the same microscope coordinates by all observers. Results are expressed as intraclass coefficient correlations (95% confidence intervals).

		Experienced		**Inexperienced**		
Cell Type	Agreement	Observer 2	Observer 5	Observer 1	Observer 3	Observer 4
Neutrophils	Intra-observer	0.91 (0.87-0.94)	0.94 (0.91-0.96)	0.82 (0.74-0.88)	0.83 (0.75-0.88)	0.83 (0.76-0.88)
Lymphocytes	Intra-observer	0.90 (0.85-0.93)	0.92 (0.88-0.95)	0.84 (0.77-0.89)	0.85 (0.78-0.90)	0.74 (0.64-0.82)
Macrophages	Intra-observer	0.91 (0.87-0.94)	0.90 (0.85-0.93)	0.85 (0.78-0.90)	0.87 (0.81-0.91)	0.82 (0.74-0.88)
Mast cells	Intra-observer	0.71 (0.60-0.80)	0.68 (0.56-0.77)	0.63 (0.49-0.74)	0.76 (0.66-0.84)	0.57 (0.42-0.69)
Eosinophils	Intra-observer	0.96 (0.94-0.97)	0.92 (0.88-0.95)	0.89 (0.84-0.93)	0.82 (0.74-0.88)	0.52 (0.36-0.65)

Table 3: Inter-observer and total agreement between two experienced and three inexperienced observers of bronchoalveolar lavage (BAL) cytology slides. The same slide was examined three times at the same microscope coordinates by all observers. Results are expressed as intraclass coefficient correlations (95% confidence intervals).

		Experienced		**In-experienced**		
Cell Type	Agreement	Observer 2	Observer 5	Observer 1	Observer 3	Observer 4
Neutrophils	Overall		0.91 (0.90-0.92)			
	Inter-observer	0.93 (0.91-0.95)			0.89 (0.87-0.91)	
Lymphocytes	Overall		0.93 (0.92-0.94)			
	Inter-observer	0.89 (0.86-0.91)			0.83 (0.80-0.86)	
Macrophages	Overall		0.91 (0.90-0.92)			
	Inter-observer	0.90 (0.87-0.92)			0.84 (0.81-0.87)	
Mast cells	Overall		0.90 (0.89-0.91)			
	Inter-observer	0.67 (0.60-0.74)			0.60 (0.53-0.66)	
Eosinophils	Overall		0.34 (0.27-0.40)			
	Inter-observer	0.97 (0.96-0.98)			0.82 (0.78-0.85)	

REFERENCES

[1]Couetil L.L. et al **J Vet Int Med** 2007;21,356-361. [2]De Brauwer E.I, et al **AQCH** 2002;24,337-341. [3]Fernandez N.J, **Vet Clinical Pathol** 2013;42,92-98.[4]Fogarty U, Buckley T, **Eq Vet J** 1991;23,434-437. [5]Geffre A, et al **Vet Clinical Pathol** 2011;40,107-112. [6]Hare J.E, Viel L, **J Vet Int Med** 1998;12,163-170. [7]Harris E.K, Boyd J.C, **Clin Chem** 1990;36,265-270. [8]Hoffman A.M, **Vet Clin North Am Eq prac** 2008;24,423-435,vii-viii. [9]Jean D, et al **Am J Vet Res** 2011;72,838-842. [10]Kjelgaard-Hansen M, **Vet J** 2010;185,245-246. [11]Leclere M, et al **Respirology** 2011;16,1027-1046. [12]Rush Moore B, et al **Am J Vet Res** 1995;56:562-567. [13]Sweeney C.R, et al **Am J Vet Res** 1992;53, 1376-1379.

INFLAMMATION, COAGULATION, AND HCM: WHAT'S THE LINK?

Karl E. Jandrey, DVM, MAS, DACVECC
Davis, CA

INTRODUCTION

Measurements of biomarkers [e.g. B-type natriuretic peptide (BNP)] are performed in cases of human cardiac disease to improve diagnostic accuracy, assess disease severity and disease progression. BNP has been shown to be correlated to severity of idiopathic hypertrophic cardiomyopathy (HCM) in humans[1]. Studies currently examine the role of inflammatory cytokines as biomarkers and mediators of cardiac disease progression. More specifically, elevated levels of TNF-α[1,2], IL-1[3], and IL-6[1] have been demonstrated in many human cardiac diseases and provide increased diagnostic accuracy compared to BNP alone.

Feline HCM is the predominant cardiac disease of cats[4] and is frequently present in overtly healthy cats[5,6]. Several studies have identified elevated BNP levels in feline HCM patients[7]. One study in cats determined that, similarly to humans, a degree of correlation exists between BNP and severity of disease[8]. The only study to date which has investigated inflammatory cytokine levels identified elevated plasma levels of TNF-α in cats with HCM compared to normal patients[9].

CARDIAC DISEASE AND INFLAMMATION IN HUMANS

In human medicine, biomarkers have been investigated for their usefulness in clinical assessment of cardiac disease. B-type natriuretic peptide (BNP), a polypeptide secreted by the ventricular myocardium, is currently used clinically to assess the severity of heart failure in humans[10]. The role of inflammation in the progression of human cardiac disease has been a topic widely investigated over the past two decades, focusing on inflammatory cytokines and their role in the development of congestive heart failure[11,12]. The inflammatory mediator TNF-α is of interest, not only because it has been found at elevated concentrations in patients with cardiac disease,[2] but also because it is involved in pathways that contribute to the clinical manifestations of the disease[13,14]. Interleukins, specifically IL-1 and IL-6, are other inflammatory mediators that interact with TNF-α and can affect cardiac performance when elevated[15].

As in humans, cardiac disease is common in cats, and similarly, biomarkers are an active area of research in veterinary cardiology. Hypertrophic cardiomyopathy (HCM) is also prevalent in apparently healthy cats[5,6], making a biomarker test useful for the early detection and screening of HCM in some patients. Studies have been done in cats to determine if BNP has a clinical role in detecting cardiomyopathy[7]. While BNP can be correlated with the degree of heart disease, it has some limitations as a screening diagnostic tool. For example, concurrent diseases commonly seen in cats such as hyperthyroidism and renal dysfunction can also elevate BNP concentrations in the blood[7]. Furthermore, while BNP levels are significantly elevated on a population level, there is significant overlap between groups that makes the test of limited use for accurate and specific disease screening purposes. This concern has led to development of multimarker approaches in humans for disease diagnosis, whereby multiple unique biomarkers are assessed and an algorithm employed to increase sensitivity and specificity of diagnosis superior to that achieved with any single biomarker alone.

CARDIAC DISEASE AND INFLAMMATION IN CATS

Comparatively, little research has been undertaken evaluating the role of inflammatory cytokines in feline hypertrophic cardiomyopathy. One study of cats with symptoms of congestive heart failure showed a clinically significant proportion (38.5%) of cats with HCM had increased TNF-α concentrations as compared to those without HCM[9]. There are, however, no studies taking the global approach to quantify and compare multiple inflammatory markers in cats with and without feline HCM or to compare their significance as biomarkers to BNP. Multimarker studies in human cardiac disease have proven beneficial to predict risk, disease progression and mortality[16–19], thus we plan to investigate this approach in feline cardiac disease.

CURRENT RESEARCH INTO ABNORMAL COAGULATION IN CATS WITH HEART DISEASE

Virchow's triad suggests that hypercoagulability is caused by the interplay of 3 parameters: endothelial damage, alterations in blood flow, and coagulation factor and/or platelet hyper-aggregability. Biomarkers (i.e. endothelin-1, atrial natriuretic peptide) may be indicators of endothelial damage. Alterations in blood flow, especially flow stasis, have been the long held tenet of thrombus formation in the cat. Schober *et al*[20] reported that low flow velocities in the left atrial appendage of cats identified a subgroup at increased risk of spontaneous echocardiographic contrast and possible thromboembolism. This study examined 89 cats with various types of heart disease as well as hyperthyroid heart disease. This may be strong evidence for the likelihood of thromboembolism, but this study did not follow patients forward to that outcome.

There is recent interest into the measurement of hemostatic markers and measurements of coagulation in cats with HCM. The majority of studies have not investigated inflammatory biomarkers while measuring hemostatic abnormalities.

Bedárd *et al*[21] concluded that thrombin-antithrombin complexes (TAT), D-dimers, prothrombin time (PT), and activated partial thromboplastin time (PTT) were not different between the group of 18 normal cats and the group of 20 cats with HCM. However, despite a lack of statistical difference between the entire groups, they did find that there were 45% of individual cats with HCM in their study that had increased markers of hypercoagulability. They found that the left atrial/aortic ratio and the left atrial size did not correlate to the coagulation results.

Brazzell *et al*[22] studied antithrombin activity and D-dimer concentrations in 3 groups of cats (normal, n=30; ill, n=30; and cats with cardiomyopathy, n=13). They scored several parameters and judged some cats to be in disseminated intravascular coagulation when ≥3 results of the following were abnormal: thrombocytopenia, prolonged PT, prolonged PTT, and abnormal erythrocyte morphology. They found that antithrombin was statistically increased in ill cats compared to the normal group. They also found that the D-dimer assay used was not sensitive or accurate in any group. From these 2 studies, we may conclude that D-dimers, TAT, antithrombin activity may be elevated in cats with critical illness or heart disease. However, the correlation to HCM is weak and any individual cat may have normal measures despite the presence of severe HCM.

Stokol *et al*[23] demonstrated hypercoagulability in 43 cats with cardiomyopathy using other markers of the coagulation system. They defined hypercoagulability as having 2 or more of the following: increased fibrinogen, Factor VIII:C activity, low antithrombin activity, TAT, or D-dimer elevations. They also measured von Willebrand factor (vWF) activity; cats with arterial thromboembolism had higher concentrations than any other group. They concluded that vWF elevation is a marker of endothelial injury in this group. They found hypercoagulability criteria in cats with spontaneous echocardiographic contrast or arterial thromboembolism, not left atrial enlargement alone. This is the second study to show that left atrial enlargement by itself does not correlate with hypercoagulability. Another interesting finding despite significant overlap of the median measurements of each group, was that the fibrinogen concentrations were significantly elevated between the normal group and all of the groups with cardiomyopathy.

Most recently, the author's laboratory[24] used the Platelet Function Analyzer-100® (PFA) to assess platelet function in both normal cats and cats with HCM. Even though a reference interval was established for this analysis in cats, the closure time measured by the PFA did not distinguish between normal cats and cats with HCM. The PFA has been used more commonly for the diagnosis of reduced platelet function and currently is not be the best test to study hypercoagulability.

There is currently no simple analytical tool available for the evaluation of the hypercoagulable state in cats. While the PFA is very likely a useful clinical tool, it is essential that the data it generates correlate well with other assessments of feline platelet physiology.

Thromboelastography (TEG) is another tool to assess all phases of coagulation from instigation of coagulation through fibrinolysis. Many emergency and critical care departments in veterinary teaching hospitals currently have the essential equipment. TEG has been used to detect hypercoagulability in many diseases in which dogs and cats are prone to thrombus formation. Kaolin-activated TEG was used to assess for hypercoagulability in cats with HCM[25]. Data indicate that although individual cats may be hypercoagulable there is significant overlap in data between healthy cats and those with HCM.

CONCLUSION

A broader study to investigate multiple cytokines as biomarkers of HCM has not been completed, nor correlated with hemostatic measurements. The identification of multiple plasma biomarkers that can identify cats with HCM and correlate with disease severity could allow for earlier cost-effective screening and monitoring of HCM severity in feline patients.

REFERENCES

1. Zen K, Irie H, Doue T, Takamiya M. *Int Hear. J.* 2005;46:231. 2. Levine B, Kalman J, Mayer L. *N. Engl. J. Med.* 1990;323:236. 3. Matsumori a, Yamada T, Suzuki H, Matoba Y, Sasayama S. *Br. Heart J.* 1994;72(6):561. 4. Ferasin L. *J. Feline Med. Surg.* 2003;5(3):151. 5. Paige CF, Abbott J a, Elvinger F, Pyle RL. *J. Am. Vet. Med. Assoc.* 2009;234(11):1398. 6. Côté E, Manning AM, Emerson D, Laste NJ, Malakoff RL, Harpster NK. *J. Am. Vet. Med. Assoc.* 2004;225(3):384. 7. Oyama M, Boswood A. *J. Am. Vet. Med. Assoc.* 2013;243(1):71. 8. Wess G, Daisenberger P, Mahling M, Hirschberger J, Hartmann K. *Vet. Clin. Pathol.* 2011;40(2):237. 9. Meurs KM, Fox PR, Miller MW, Kapadia S, Mann DL. *Am. J. Vet. Res.* 2002;63(5):640. 10. McCullough PA, Nowak RM, McCord J, Al. E. *Circulation.* 2002;106(4):416. 11. Anker S, Haehling S von. Inflammatory mediators in chronic heart failure: an overview. *Heart.* 2004;90(4):464–470. 12. Mann DL. *Circ. Res.* 2002;91(11):988. 13. Sivasubramanian N, Coker ML, Kurrelmeyer KM, et al. *Circulation.* 2001;104(7):826. 14. Pagani F, Baker L, Hsi C. *J. Clin. Invest.* 1992;90(August):389. 15. Francis G, Tang W, Walsh R. 26. Pathophysiology of Heart Failure. In: Fuster V, Walsh R, Harrington R, eds. *Hurst's The Heart.* 13th ed. McGraw-Hill; 2011. 16. Nozaki T, Sugiyama S, Koga H, et al. *J. Am. Coll. Cardiol.* 2009;54(7):601. 17. Sabatine M, Morrow D, Lemos J de. *Circulation.* 2002;105(15):1760. 18. Zethelius B, Berglund L, Sundström J, et al. *N. Engl. J. Med.* 2008;358(20):2107. 19. Wang TJ, Gona P, Larson MG, et al. *N. Engl. J. Med.* 2006;355(25):2631. 20. Schober KE, Maerz I. J Vet Intern Med 2006;20:120. 21. Bédard C, Lanevschi-Pietersma A, Dunn M. Vet Clin Pathol 2007;36:167. 22. Brazzell JL, Borjesson DL. Vet Clin Pathol 2007;36:79. 23. Stokol T, Brooks MB, Rush JE, *et al.* J Vet Intern Med 2008;22:546. 24. Jandrey KE, Norris JW, MacDonald KA, *et al.* Vet Clin Pathol 2008;37:385. 25. Jandrey KE, Norris JW, Kittleson MD, *et al.* In *Proceedings.* 8th European Emergency and Critical Care Society Congress 2009;99.

REVIEW OF THE TOP 5 COAGULATION ARTICLES

Karl E. Jandrey, DVM, MAS, DACVECC
Davis, CA

INTRODUCTION

The wealth of information published in veterinary medical journals on the topic of coagulation and hemostasis has risen considerably over the past few years. The author has provided 5 of his favorite articles from the past year that have significant impact and application to daily practice for general practitioners and specialists alike. No attempt was made to search and score the medical literature as a whole; the information explosion in human medicine and the basic and translational research areas for coagulation and hemostasis is huge. Many of those studies are referenced in these articles. The interested reader is encouraged to independently investigate some of the best journals for this: *Blood; Journal of Thrombosis and Haemostasis; Blood, Coagulation and Fibrinolysis; Current Opinions in Hematology.*

ARTICLE #1 TITLE: Anticoagulant rodenticide screening in dogs: 123 cases (1996-2003)
AUTHORS: Waddell LS, Poppenga RH, Drobatz KJ
REFERENCE: J Am Vet Med Assoc. 2013, 242(4):516-521.

HYPOTHESIS/OBJECTIVES: The purpose of this study was to evaluate the use of anticoagulant rodenticide screening in dogs, determine the frequency of positive results, identify any correlation between prolongation in the prothrombin (PT) and activated partial thromboplastin time (aPTT) and concentration and type of anticoagulant rodenticide, and evaluate the prognostic value of both positive and negative results for anticoagulant screening.

STUDY DESIGN: Retrospective case series

RESULTS: One hundred twenty-three dogs had antemortem anticoagulant rodenticide (AR) screens submitted. Seventy-five were positive (61%) and 48 were negative (39%). A known or possible exposure was noted in 48.8%. In the negative group, 72.9% (35/48) had no known exposure and 25% (12/48) had a possible exposure. In the positive group, 28% (21/75) had no known exposure whereas 65.3% (49/75) had possible or known exposure. No exposure history was recorded in 1/48 in the negative group and 5/75 (6.7%) in the positive group. There were several AR found in positive samples (range= trace to 1120 ppb): brodifacoum, diphacinone, and chlorphacinone. There was no correlation found between the prolongation in PT or aPTT and concentration of AR. Dogs in the positive group received more fresh frozen plasma than dogs in the negative group; however, both groups received similar amounts of packed red blood cells. There was no difference in median hospital length of stay. Nearly 98% of the positive group survived; all of the dogs in the negative group that did not survive were euthanized (most common diagnoses: neoplasia, immune-mediated disease, and severe gastrointestinal bleeding).

DISCUSSSION/CLINICAL RELEVANCE: AR concentrations were not correlated to severity of clinical signs or degree of prolongation of the PT or aPTT. Patients with severe coagulopathies but negative AR results had a poor prognosis. Initial clinical signs were similar between both groups of dogs.

ARTICLE #2 TITLE: Reference values and repeatability of buccal mucosal bleeding times in healthy sedated cats
AUTHORS: Alatzas DG, Mylonakis ME, Kazakos GM, Kostoulas P, Kritsepi-Konstantinou M, Polizopoulou ZS
REFERENCE: J Fel Med Surg 2014, 16(2):144-148.

HYPOTHESIS/OBJECTIVES: The purpose of this study was to establish the reference interval of the buccal mucosal bleeding time (BMBT) in a population of healthy sedated cats and to investigate the intra-observer repeatability of the test by means of a commercially-available, disposable, single-blade device.

STUDY DESIGN: Prospective clinical study

RESULTS: Fifty-six healthy cats were studied. Ketamine (5mg/kg IM), dexmedetomidine (30ug/kg IM) and morphine (0.2mg/kg IM) were administered prior to reflection of the right maxillary labium following a standardized approach for the BMBT in each cat. The BMBT was completed and then repeated on the opposite side. Four of the initial incisions created in the buccal mucosa lead to no bleeding; an area adjacent to this non-bleeding incision was immediately selected and used to create a new incision that did bleed. One hundred twelve measurements were thus available for statistical analysis. Bootstrap estimation was carried out using 56 random values as well as Bland Altman plots to assess intra-observer repeatability. The reference interval determined for the BMBT from this study found the range of 34-105s with a mean to be 58.6s. Intra-observer repeatability ranged from -34.59 to +52.05, meaning that the BMBT from one side may be 35s less or 52s more the BMBT from the opposite side. Any two sequential readings of the BMBT from the same cat can vary by as much as 87s.

DISCUSSSION/CLINICAL RELEVANCE: All completed BMBT measurements were less than 2 minutes in this study comparing favorably with other studies (using various sedative protocols) that measured BMBTs in cats. The authors suggest that any BMBT greater than 3 minutes in a cat should be considered abnormal. They also suggest that this sedative combination is unique and appropriate for BMBT measurement, as it does not affect platelet function.

ARTICLE #3 TITLE: Coagulation factor and hemostatic protein content of canine plasma after storage of whole blood at ambient temperature
AUTHORS: Walton JE, Hale AS, Brooks MB, Boag AK, Barnett W, Dean R
REFERENCE: J Vet Int Med 2014, epub accessed 27Jan14.

HYPOTHESIS/OBJECTIVES: The aim of this study was to compare the coagulation factor and hemostatic protein content (CF&HPC) of plasma produced from blood stored at ambient temperature for 8, 12, and 24 hours. Another aim was to compare the CF&HPC between Greyhound types and other breeds.

STUDY DESIGN: *in vitro* (convenience sampling) study

RESULTS: Regardless of time of processing, the units contained therapeutic CF&HPC. Frozen plasma prepared after 24 hours had significantly higher factor VIII ($P = .014$) and factor X ($P = .03$) when compared with the frozen plasma prepared at 8 hours. Factor X ($P < .01$), fibrinogen ($P < .01$), and von Willebrand Factor (vWF, $P = .04$) were significantly lower in plasma collected from Greyhound types than in plasma collected from other breeds.

DISCUSSSION/CLINICAL RELEVANCE: This study indicated that storing whole blood for up to 24 hours is a suitable method for production of fresh frozen plasma (FFP) (compared to the conventional current standard recommendation to produce FFP within 8 hours of collection). Lower values for some coagulation factors and hemostatic proteins in plasma produced from Greyhound types (vWF, factor X, fibrinogen) would not preclude these dogs as FFP donors.

ARTICE #4 TITLE: Prospective multicenter evaluation of coagulation abnormalities in dogs following severe acute trauma
AUTHORS: Holowaychuk MK, Hanel RM, Wood RD, Rogers L, O'Keefe K, Monteith G
REFERENCE: J Vet Emerg Crit Care 2014, 24(1):93-104.

HYPOTHESIS/OBJECTIVES: The objective of this study was to describe coagulation abnormalities in dogs following severe acute trauma and to evaluate the association of coagulation and thromboelastography (TEG) variables with clinical and laboratory variables, including disease and injury severity, as well as the ability of coagulation variables to predict the presence of body cavity hemorrhage (BCH), necessity of blood product administration, and outcome. The primary hypothesis was that severely traumatized dogs would demonstrate hypocoagulability that would be predictive of the presence of BCH, the need for blood product administration, and non-survival.

STUDY DESIGN: Prospective, multicenter, observational study

RESULTS: Forty client-owned dogs that sustained severe blunt or penetrating trauma were studied at 2 university teaching hospitals. Blood samples were collected within 12 hours of the traumatic incident for measurement of blood gases, lactate concentration, platelet count, activated clotting time, prothrombin time, activated partial thromboplastin time (aPTT), fibrinogen concentration, antithrombin activity, D-dimer concentration, protein C activity, plasmin inhibition, plasminogen activity, and kaolin-activated thromboelastography. Decreased platelet count was a risk factor for the presence of BCH ($P = 0.006$). Decreased platelet count ($P < 0.001$), protein C activity ($P = 0.001$), angle (α) ($P = 0.001$), maximum amplitude (MA) ($P < 0.001$), and clot strength (G) ($P = 0.002$) were risk factors for blood product administration. Nonsurviving dogs were hypocoagulable with prolonged aPTT ($P = 0.008$), decreased plasmin inhibition ($P = 0.033$), decreased α ($P = 0.021$), and decreased MA ($P = 0.038$) compared to surviving dogs. Multivariate analysis accounting for disease severity showed that prolonged aPTT ($P = 0.004$, $OR = 1.74$) was the strongest predictor of nonsurvival. Prolonged aPTT was positively correlated with APPLE-fast score ($P < 0.001$, $r(2) = 0.35$), lactate concentration ($P < 0.001$, $r(2) = 0.35$), and negative base excess ($P = 0.001$, $r(2) = 0.27$). Acute traumatic coagulopathy, as defined by 2 or more abnormal coagulation tests, was diagnosed in 15% of dogs at hospital admission and was more common in dogs with increased disease severity ($P = 0.002$), decreased systolic blood pressure ($P = 0.002$), and increased lactate concentration ($P = 0.011$).

DISCUSSSION/CLINICAL RELEVANCE: Hypocoagulability and decreased clot strength were risk factors for the presence of BCH or the need for blood product administration after severe trauma as studied in these 40 dogs from 2 academic centers. Prolonged aPTT was strongly predictive of nonsurvival and was also correlated with increased disease severity and decreased perfusion as assessed by increased lactate and negative base excess. Acute traumatic coagulopathy, as defined by ONE or more abnormal coagulation or TEG variables, predicted the requirement of blood product administration. Acute traumatic coagulopathy, as defined by TWO or more abnormal coagulation or TEG variables, in a small subset of dogs, found a mortality rate nearly three times those traumatized dogs without ATC. In dogs with severe traumatic injuries and hypoperfusion, measurement of thromboelastography and aPTT should be considered to support clinical assessments in predicting the need for blood product administration and nonsurvival.

ARTICLE #5 TITLE: Evaluation and modification of the overall hemostasis potential assay for use with canine plasma

AUTHORS: Dengate AL, Morel-Kopp M-C, Beatty JA, Barrs V, Braddock JA, Churcher RK, Wilson BJ, Ward CM

REFERENCE: Am J Vet Res 2013, 74(12):1493-1498.

HYPOTHESIS/OBJECTIVES: The objectives of this study were to optimize the overall hemostasis potential (OHP) assay for use with canine platelet-poor plasma and determine reference intervals in healthy dogs.

STUDY DESIGN: in vitro laboratory investigation

RESULTS: Modifications to the published methodology for the OHP assay were required for use with canine plasma, with less coagulation activator (thrombin) and more fibrinolysis activator (tissue plasminogen activator) than used with human plasma. Male dogs had a higher OHP than did females. High fibrinogen concentrations were associated with increases in maximum optical density, OHP, and overall coagulation potential, and reduced prothrombin time was associated with increases in maximum optical density, overall coagulation potential, OHP, and maximum slope.

DISCUSSSION/CLINICAL RELEVANCE: The OHP evaluates fibrin generation and lysis over time and can be used to detect hypercoagulability in human plasma by calculating the area under a fibrin generation curve as well as assessing fibrinolysis. The OHP is used in humans to detect hypercoagulability in diseases with high potential for thrombosis. The results of this study indicate that the OHP may be used as a cost-effective global coagulation assay for canine plasma. Clinical application as an alternative to thromboelastography or thrombin generation assays remains to be determined.

RESPIRATORY PATTERN RECOGNITION IN DOGS AND CATS WITH ACUTE RESPIRATORY DISTRESS

Gretchen Lee Schoeffler, DVM, DACVECC
Ithaca, New York

INTRODUCTION

Animals in severe respiratory distress are critically ill and may be at risk of respiratory arrest even with minimal handling. These patients frequently exhibit orthopneic postures, postures that ease the work of breathing. More often than not, they will stand, sit, or lie in sternal recumbency rather than in lateral recumbency. They will lift and extend their head and neck in order to maximally open the airway. They breathe through an open mouth in order to reduce nasal and pharyngeal airway resistance. They may also abduct the elbows to allow for maximal movement of the chest wall with each breath. Timely assessment and intervention are crucial in the management of these patients. An abbreviated, targeted physical examination based upon observation of the character of respiration and thoracic auscultation combined with an approach based upon the eight basic sources of respiratory distress can afford the clinician the ability to rapidly identify a small number of possible disorders and provide appropriate, directed, life-saving care.

CLINICAL APPROACH

When presented with a patient in acute respiratory distress it is important to follow a standard and consistent protocol to assess and stabilize the patient's respiratory and cardiovascular systems (i.e., airway, breathing, circulation). Dogs and cats presented in overt respiratory distress should be immediately triaged to a quiet workspace and supplemented with oxygen. Initially, brief assessment of general demeanor, mucus membranes, and auscultation of the thorax is prioritized. Afterwards the clinician should take time to study the patient's respiratory pattern. This must be done in the least stressful environment possible. For small dogs and cats an oxygen cage is ideal, larger dogs may benefit from flow-by oxygen supplementation while on an exam table. When observing the pet's breathing pattern, the clinician should note the following:

1. Is the patient's effort on inspiration, expiration or both?
2. Does the chest wall have a large excursion, or does this appear to be a restrictive breathing pattern?
3. Is the pet calming down and is its respiratory status improving or is an emergent intervention necessary?

If the clinician remains unsure of the etiology of the patient's respiratory distress, a dose of furosemide (2 mg/kg IM) can be administered and a water bowl placed within easy reach of the patient. Pets whose respiratory distress is severe and who do not stabilize, or those that roll into lateral recumbency (frequently a sign of impending arrest) may need more aggressive intervention including emergent anesthetic induction, intubation, and positive pressure ventilation.

Using the information gained in this brief physical exam, the observations made while characterizing the breathing pattern, and the patient's signalment and history will oftentimes allow the clinician to make an accurate assessment. Additional data including a rectal temperature, blood pressure, and evaluation of jugular vein distention may also prove useful, but ONLY if this information can be collected without undue stress to the patient. If the cause of the respiratory distress remains elusive the clinician may need thoracic imaging and other diagnostic tests to determine the cause of the problem. Generally speaking there are eight universal sources of respiratory distress including:

1. Upper airway
2. Lower airway
3. Pulmonary parenchyma
4. Pleural space
5. Chest wall
6. Pulmonary thromboembolism
7. Abdominal distention
8. Look-likes

LOCALIZING THE SOURCE OF DISTRESS

Upper Airway

The classic signs associated with upper airway disease are stridor and stertor. Some animals may also cough, and depending upon the location of the disease, distress may be inspiratory (with extrathoracic disease) or expiratory (with intrathoracic disease). There are many differential diagnoses for upper airway diseases and a systematic approach (such as the DAMNIT-V scheme) is useful.

Degenerative - laryngeal paralysis, collapsing trachea, left atrial enlargement
Anomalous - brachycephalic airway syndrome
Metabolic - ?
Neoplasia - primary vs. metastatic disease
Infectious and Inflammatory - abscess, granuloma, foreign body, sterile laryngitis
Trauma - hematoma, swelling, fracture
Vascular - coagulopathy

Treatment of upper airway obstruction consists of oxygen supplementation, and relief of the obstruction! In some cases of dynamic obstruction sedation is adequate. If unsuccessful, patients should be intubated or a tracheostomy tube placed.

Lower Airway

Wheezes and or whistles on auscultation are the hallmark finding in patients with lower airway disease. Patients may also cough, and their respiratory distress characteristically has a prominent expiratory component. Supplemental oxygen should be administered to all patients. Dogs rarely present acutely with lower airway distress. Cats with lower airway respiratory distress most likely have feline asthma.

Feline asthma is the most common lower airway disorder in cats. It can be diagnosed in cats of any age but is usually seen in young to middle-aged adults (mean 4 years, range 1-15 years). There is no sex predilection, but the Siamese breed appears to be overrepresented. Cats suffering an asthma attack may be presented in respiratory distress resulting from bronchoconstriction. Cats experiencing an asthma attack predictably have a pronounced expiratory push. They are typically normothermic and normotensive with normal jugular vein distention. Thoracic auscultation of an asthmatic patient frequently yields expiratory wheezes, and coughing is a common owner complaint. The emergency treatment of asthma in cats includes supplemental oxygen and administration of a bronchodilator [β2- agonists such as terbutaline (0.01 mg/kg IM) or an albuterol inhaler (90 mcg/cat), or a phosphodiesterase inhibitor such as aminophylline (5-10 mg/kg IM)] and corticosteroids (dexamethasone SP - 0.1 mg/kg IM).

Pulmonary Parenchymal

Fluid and inflammatory debris in the alveoli, leading to alveolar collapse during expiration and alveolar re-expansion during inspiration, cause the characteristic crackles heard on auscultation in animals with pulmonary parenchymal disease. It is important to note however, that not all animals with significant pulmonary parenchymal disease will have crackles and that areas of parenchyma that are completely consolidated may actually sound dull. The distress can be mixed, but inspiratory difficulty predominates. Alveoli may contain water (cardiogenic or non-cardiogenic edema), inflammatory exudates (pneumonia, ARDS), blood (contusion), or non-inflammatory exudates (neoplasia). Interstitial disease (e.g., pulmonary fibrosis) should also be considered, but is rare in small animals. Supplemental oxygen should be administered to all patients. Additional treatment varies and should be guided by the underlying cause.

Pets whose respiratory distress is due to the presence of cardiogenic pulmonary edema (heart failure), are frequently hypothermic and may be hypotensive secondary to low cardiac output and poor tissue perfusion. They may also exhibit prominent or distend jugular veins. The emergency treatment of pulmonary edema secondary to heart failure includes supplemental oxygen and intravenous or intramuscular administration of furosemide (2-4 mg/kg). Cutaneous application of nitroglycerin may enhance venodilation, however there is no evidence supporting its use and dosing is empirical.

Pleural Space Disease

Dull lung sounds or the absence of lungs sounds should raise the clinician's index of suspicion for pleural space disease. Classically, pleural space disease has been associated with a rapid, shallow breathing pattern oftentimes described as a restrictive breathing pattern. A recent study found that asynchronous breathing more accurately describes the pattern demonstrated by the majority of these patients. Regardless, pneumothorax (air secondary to trauma and ruptured bullae), hydrothorax (serous fluids secondary to congestive heart failure and hypoproteinemia), pyothorax (lung abscess rupture and penetrating wounds), hemothorax (trauma and coagulopathy), chylothorax (heart disease, neoplasia, heart-worm, and idiopathic), other exudative processes (FIP and neoplasia), and diaphragmatic hernia should be considered in these patients. Supplemental oxygen should be administered to all patients.

Thoracocentesis is indicated in most patients with pleural space disease. A brief cage side thoracic ultrasound can be used to confirm the presence of pleural effusion or air as the cause of the patient's respiratory distress. When performing this procedure, the pet should not be restrained, but rather is allowed to stand or sit in a sphinx position and simply have the probe placed on an area of chest wall that has been clipped free of hair. Pleural fluid usually appears as an anechoic (black) area surrounded by typical anatomic boundaries (chest wall, diaphragm, etc....). If ultrasound is not readily available, blind thoracocentesis should be performed. Thoracic radiographs are very stressful and should not be performed in unstable patients. In cases of pyothorax, tension pneumothorax, or rapidly recurring air or fluid accumulations, thoracostomy tube placement should be considered.

Chest Wall

Palpation of a chest wall defect, pain on palpation, or visualization of a flail segment is the most common finding. Penetrating wounds cause pleural space disease, and rib fractures are look-likes, leaving flail segment as the primary chest wall cause of respiratory distress. Supplemental oxygen should be administered to all patients. Providing sufficient analgesia and placing the animal with the affected side down will oftentimes partially stabilize a flail segment and allow maximal inflation of the opposite lung. Patients that cannot be stabilized should be anesthetized, intubated, and ventilated. Once they are more stable, the flail segment may be stabilized with external or internal coaptation.

Pulmonary Thromboembolism (PTE)

Clinically significant PTE usually results in respiratory distress that is often minimally responsive to oxygen supplementation. Identification of underlying, predisposing diseases such as Cushing's disease, protein-losing enteropathy/nephropathy, immune mediated diseases, cardiac disease, and other causes of endothelial damage such as sepsis and SIRS are crucial. Diagnosis is typically by exclusion, with V/Q scanning and CT angiography as well as other advanced imaging modalities required for definitive diagnosis. Regardless of response, supplemental oxygen should be administered to all patients. Thrombolytic (tPA, streptokinase), anticoagulant (heparin), and anti-platelet (aspirin, clopidogrel) therapies can be considered.

Abdominal Distention

This can usually be visualized or is detected via abdominal palpation on physical exam. Ascites (transudates, exudates, hemorrhage), organomegaly, neoplasia, and fat are common causes of abdominal distention. Acute respiratory distress due to abdominal distention is not common, but when it does occur it is typically due to acute onset of ascites. Supplemental oxygen should be administered to all patients. Abdominocentesis is indicated to allow restoration of appropriate tidal volumes until the underlying disease process leading to the effusion is identified and treated.

Look-likes

Look-likes are disease processes that make patients appear to be in distress, but no underlying respiratory dysfunction exists. Clinical signs vary depending on the underlying disease. There are 3 categories of look-likes including behavioral causes such as stress, fear, pain, or anxiety, metabolic causes such as respiratory acidosis or anemia, and environmental causes such as hyperthermia. Treatment is aimed at the underlying cause. Behavioral causes may be treated with anxiolytics or analgesics. Resolution of metabolic acidosis typically requires treatment of the underlying disease. Anemic patients should be treated with packed red blood cells.

CONCLUSION

Few emergencies are as stressful for the emergency clinician as respiratory distress. However, with a brief physical examination, the 8 potential sources of respiratory distress can be quickly reduced to 2 or 3, and appropriate, life-saving therapy can be administered. Patients exhibiting severe respiratory distress or cyanosis may require the clinician to rapidly gain control of the airway and provide intermittent positive pressure ventilation until the underlying etiology and definitive treatment can be provided.

REFERENCES

[1]Sigrist E, et al. JVECCS 2011;21(1):13. [2]Rozanski E, et al. Vet Clin Small Anim Pract 2005;35:307. [3]Tseng L, et al. Clin Tech Small Anim Pract 2000;15(2):53. Trzil JE, et al. Vet Clin Small Anim Pract 2014;44:91. [4]King L, et al. Veterinary Focus 2010;20(2):36. [5]Reinero CR. The Veterinary Journal 2011;190(1):28. [6]Beatty J, et al. J of Fel Med and Surg 2010;12(9):693.

DIFFERENTIATING CARDIOGENIC FROM OTHER CAUSES OF ACUTE RESPIRATORY DISTRESS IN SMALL ANIMALS

Gretchen Lee Schoeffler, DVM, DACVECC
Ithaca, New York

INTRODUCTION

Timely assessment and intervention are crucial in the management of dogs and cats presented to the emergency clinician in respiratory distress. These patients are critically ill and may be at risk of respiratory arrest, even with minimal handling. An abbreviated, targeted physical examination based upon observation of the character of respiration and thoracic auscultation can help differentiate the small subset of differentials allowing the clinician to optimize their intervention and therapy. Generally speaking there eight universal sources of respiratory distress including (1) upper airway, (2) lower airway, (3) pulmonary parenchymal, (4) pleural space, (5) chest wall, (6) pulmonary thromboembolism, (7) abdominal distention, and (8) look-likes. This session will focus on primary respiratory diseases that originate from the lower airway, pulmonary parenchyma, and pleural space, as these are the most challenging to differentiate from congestive heart failure.

CLINICAL ASSESSMENT

When presented with a small animal patient in acute respiratory distress, it is important to follow a standard and consistent protocol to assess and stabilize the patient's respiratory and cardiovascular systems (i.e., airway, breathing, circulation). They should be immediately triaged to a quiet workspace and supplemented with oxygen. Methods of oxygen supplementation include flow-by, face-mask, nasal cannula, oxygen cage, and intubation. Oxygen cages are nice for smaller patients as many will not tolerate handling for flow-by, mask, or placement of nasal cannula. Generally, a brief physical exam prioritizing assessment of general demeanor, mucus membranes, palpation of pulses, and auscultation of the thorax can be performed shortly after presentation. If the clinician is unsure of the etiology of the patient's respiratory distress, a dose of furosemide (2 mg/kg IM) can be administered and a water bowl placed within easy reach of the patient. Before walking away, the clinician should take time to study the patient's respiratory pattern and feel confident that the patient's respiratory status is improving. While observing the breathing pattern, the clinician should note the following:

1. Is the effort on inspiration, expiration or both?
2. Is there an abdominal component and if so, how do you characterize it? Does the chest wall have a large excursion or does it appear to be restricted?
3. Is the pet calming down and is its respiratory status improving?

Patients whose respiratory distress is severe and who do not stabilize, or that roll into lateral recumbency (frequently a sign of impending arrest) may need more aggressive intervention including emergent anesthetic induction, intubation, and positive pressure ventilation.

Using the information gained in this brief physical exam, the observations made while characterizing the breathing pattern, and the patient's signalment and history will oftentimes allow the clinician to make an accurate assessment. Additional data including a rectal temperature, blood pressure, and assessment of jugular vein distention may also prove useful, but ONLY if this information can be collected without undue stress to the patient. If the cause of the respiratory distress remains elusive the clinician may need thoracic imaging and other diagnostic tests to determine the cause of the problem.

Lower Airway

Lower airway disease resulting in overt respiratory distress is uncommon in dogs. While feline asthma is the most common lower airway disorder of cats, it must be distinguished from chronic bronchitis and a variety of parasitic infections. Feline asthma can be diagnosed in cats of any age but is usually seen in young to middle-aged adults (mean 4 years, range 1-15 years). There is no sex predilection, but the Siamese breed appears to be overrepresented. Cats experiencing an asthma attack typically have an expiratory respiratory distress, with a pronounced expiratory push. They are oftentimes normothermic and normotensive with normal jugular vein distention. Thoracic auscultation of an asthmatic patient frequently yields expiratory wheezes, and coughing is a common owner complaint. The emergency treatment of asthma in cats includes supplemental oxygen and administration of a bronchodilator [β2-agonists such as terbutaline (0.01 mg/kg IM) or an albuterol inhaler (90 mcg/cat), or a phosphodiesterase inhibitor such as aminophylline (5-10 mg/kg IM)] and corticosteroids (dexamethasone SP - 0.1 mg/kg IM).

Parenchymal

Most of the common parenchymal diseases that result in respiratory distress in dogs and cats affect the alveoli. Interstitial disease (e.g., pulmonary fibrosis) should also be considered, but as a cause of respiratory distress is comparatively rare in small animals. The alveoli may contain water (cardiogenic or non-cardiogenic edema), inflammatory exudates (pneumonia, ARDS), blood (contusion), or non-inflammatory exudates (neoplasia). Fluid and inflammatory debris accumulate in alveoli, leading to alveolar collapse during expiration and alveolar re-expansion during inspiration, cause the characteristic crackles heard on auscultation in animals with pulmonary parenchymal disease. Crackles on auscultation are specific for pulmonary parenchymal disease however it is not unusual for these patients to have normal or harsh bronchovesicular sounds and NO crackles. Additionally, areas of parenchyma that are completely consolidated may sound dull. The distress can be mixed, but inspiratory difficulty predominates. Supplemental oxygen should be administered to all patients. Additional treatment varies and should be guided by the underlying cause.

Differentiating cardiogenic pulmonary edema from other parenchymal causes of respiratory distress based on clinical signs alone can be challenging. While congestive heart failure (CHF) can occur in pets of any age, most are middle aged to older. Those with cardiogenic pulmonary edema commonly present with acute or progressive respiratory distress without an otherwise helpful history. Dogs in congestive heart failure frequently have a cough, heart murmur, or arrhythmia. In contrast, cats with heart disease typically do not have a history of cough and may not have a murmur. With that being said, auscultation of a gallop heart rhythm is nearly always indicative of significant heart disease. Patients in congestive heart failure are frequently hypothermic and hypotensive secondary to low cardiac output and poor tissue perfusion. They may also exhibit prominent or distended jugular veins. The emergency treatment of pulmonary edema secondary to heart failure includes supplemental oxygen and intravenous or intramuscular administration of furosemide (2 -4 mg/kg). Cutaneous application of nitroglycerin may enhance venodilation, however there is no evidence supporting its use and dosing is empirical.

Pleural

Pet's with significant volumes of pleural effusion typically present with an acute or progressive history of respiratory distress. Dull lung sounds or the absence of lungs sounds should raise the clinician's index of suspicion for pleural space disease. Classically, pleural space disease has been associated with a rapid, shallow breathing pattern oftentimes described as a restrictive breathing pattern. A recent study found that asynchronous breathing more accurately describes the pattern demonstrated by the majority of these patients. Regardless, pneumothorax (air secondary to trauma and ruptured bullae), hydrothorax (serous fluids secondary to congestive heart failure and hypoproteinemia), pyothorax (lung abscess rupture and penetrating wounds), hemothorax (trauma and coagulopathy), chylothorax (heart disease, neoplasia, heart-worm, and idiopathic), other exudative processes (FIP and neoplasia), and diaphragmatic hernia should be considered in these patients. Supplemental oxygen should be administered if it can be done without stress.

Thoracocentesis is indicated for most patients that have accumulated fluid or air in the pleural space. In patients with respiratory distress secondary to pleural space disease, it is imperative that the clinician evacuates the pleural space as quickly as possible. Thoracocentesis is not indicated for patients with diaphragmatic hernia and hemothorax secondary to coagulopathy unless significant compromise is present. A brief cage side thoracic ultrasound can be used to confirm the presence of pneumothorax or pleural effusion as the cause of the patient's respiratory distress. When performing this procedure, the patient should not be actively restrained, but rather is allowed to stand or sit in a sphinx position and simply have the probe placed on an area of chest wall that has been clipped free of hair. If ultrasound is not readily available or too stressful for the patient, blind thoracocentesis should be performed. Thoracic radiographs are very stressful and should not be performed in unstable patients.

Thoracocentesis

When performing a thoracocentesis in a small dog or cat, a butterfly catheter (21-23 g) attached to a three-way stopcock and a 6-20 ml syringe is usually used. In larger pets a 1-inch (20-22 g) needle or over-the-needle catheter attached to an extension set, three-way stopcock and 6-20 ml syringe may be used. Rarely, are 1.5-inch, larger gauge needle or catheter is indicated. The animal is allowed to sit in sternal recumbency (sphinx) or standing position. Most pets tolerate the procedure without sedation or local analgesia. The chest wall is clipped and surgically scrubbed in the area where the tap is performed. If possible, sterile gloves should be worn and strict aseptic technique maintained. Two to three people may be necessary to perform the tap with one handling the needle or catheter, the other the syringe and stopcock and the third holding and comforting the patient. The thoracocentesis is performed at the 7th, or 8th intercostal space, approximately 1-2 centimeters above the costochondral junction. Introduce the needle or catheter into the middle of the selected intercostal space (clipped and aseptically prepared). Use care to avoid the large vessels associated with the caudal aspect of the rib margins. Advance the needle or catheter into the pleural space. Aspirate fluid while the needle or catheter is being advanced to allow prompt recognition of the appropriate depth of needle placement. When using a needle, one can orient it flat against the rig cage with the bevel facing inward to minimize damage to the lung surface. Gently aspirate. If only a small amount of fluid is present or it is difficult to obtain a sample, ultrasound guidance may be required. The needle or catheter should be removed if the tap is unsuccessful or if the patient moves excessively. Thoracocentesis with on-site characterization of the fluid is indicated in all cases in which respiratory distress is secondary to pleural effusion. In cases of pyothorax or rapidly recurring effusions, thoracostomy tube placement should be considered. Additional treatment is variable and depends on the underlying disease process.

In addition to air, various types of fluids might accumulate in the pleural space and common differentials for each include hydrothorax (serous fluids secondary to congestive heart failure are most common), chylothorax (congestive heart failure, neoplasia, heart-worm disease, and idiopathic), pyothorax (lung abscess rupture and penetrating wound), hemothorax (trauma and coagulopathy), and other exudative processes including feline infectious peritonitis and neoplasia.

CONCLUSION

Acute respiratory distress is one of the most challenging and common complaints in the veterinary emergency room. Using the information gained in a brief physical exam, the observations made while characterizing the breathing pattern, and the patient's signalment and history will oftentimes allow the clinician to make an accurate assessment. Additional data including a rectal temperature, blood pressure, and assessment of jugular vein distention may also prove useful, but ONLY if this information can be collected without undue stress to the patient. The emergency clinician's ability to distinguish cardiogenic from non-cardiogenic causes and provide treatment on the basis of history, physical examination, and by simply watching the patient breathe, can mean the difference between life and death.

REFERENCES

[1]Sigrist E, et al. JVECCS 2011;21(1):13. [2]Rozanski E, et al. Vet Clin Small Anim Pract 2005;35:307. [3]Tseng L, et al. Clin Tech Small Anim Pract 2000;15(2):53. Trzil JE, et al. Vet Clin Small Anim Pract 2014;44:91. [4]King L, et al. Veterinary Focus 2010;20(2):36. [5]Reinero CR. The Veterinary Journal 2011;190(1):28. [6]Beatty J, et al. J of Fel Med and Surg 2010;12(9):693.

BACTERIAL TRANSLOCATION AND NUTRITION IN CRITICALLY ILL DOGS & CATS

Christopher G. Byers, DVM, DACVECC, DACVIM (SAIM)
Omaha, Nebraska USA

NORMAL INTESTINAL BARRIER

The intestinal barrier is composed of a monolayer of epithelial cells and displays a number of specialized protective adaptions, including tight junctions, secretions that coat the apical epithelial surface, commensal gastrointestinal bacteria, and a local immune system (gut-associated lymphoid tissue / GALT). The intestinal epithelium is considered the first line of defense for the intestinal barrier. Tight junctions seal the paracellular spaces. Secretions prevent direct access by microorganisms to the intestinal mucosa; specifically goblet cells secrete mucus at the epithelial surface, and Paneth cells secrete microbicidal alpha-defensins, lysozyme and cathelicidins upon exposure to bacterial antigens. Normal gastrointestinal microbiota helps the patient digest cellulose, salvage energy, and synthesize vitamin K. Anaerobes, the most abundant bacteria in the gastrointestinal tract (GIT), compete with potential pathogens for nutrients and mucosal attachment sites. The GALT is the largest lymphoid system in the body, composed of three compartments: intra-epithelial lymphocytes, lamina propria (containing solitary lymphoid follicles and Peyer's patches) and mesenteric lymph nodes. The majority of lymphocytes within Peyer's patches are B-cells that can form immunoglobin A (IgA) that is apically secreted to decreased adherence of pathogens.

SEQUELAE OF MALNUTRITION

Malnutrition is defined as any nutritional disorder resulting in inadequate or unbalanced nutrients. Undernutrition and overnutrition are the two forms of malnutrition, although malnutrition and undernutrition are often used interchangeably. One must recognize the difference between simple (uncomplicated) starvation and stressed (hypermetabolic) starvation. Simple starvation occurs in healthy animals that have experienced an acute disruption of caloric intake. In affected patients glycogen stores serve as the primary source of energy; these stores are rapidly depleted, and a metabolic shift to the preferential use of stored adipose occurs. During stressed starvation, a pro-inflammatory milieu induces a catabolic state, and energy predominantly comes from accelerated proteolysis; adipose is preserved in the face of lean muscle tissue loss. Cardiac arrest, pulmonary failure and immunosuppression occur when more than 25% of amino acids come from skeletal muscle and organ tissue.

Table #1: Major Differences Between Simple & Stressed Starvation

	Simple Starvation	Stressed Starvation
Energy Expenditure	Decreased	Increased
Fuel Source	Glucose / Fat	Protein
Gluconeogenesis	1+	3+
Protein synthesis	Decreased	Markedly Decreased
Catabolism	Not occurring	3+
Amino acid oxidation	+/-	3+
Ureagenesis	+/-	2+
Ketosis	+/-	+
Responsiveness	3+	1+
Rate of malnutrition	1+	3+

Adapted from Zoran DL, ACVIM Forum 2003.

Inadequate enteral nutrition induces enterocyte atrophy and cholestasis with subsequent down regulation of gastrointestinal and pancreatic enzymes, as well as reduced small intestinal absorption in as little as 48 hours. Major complications of malnutrition may be categorized as decreased tissue synthesis and repair, decreased immunocompetence, and altered intermediary drug metabolism.

Multi-Organ Dysfunction Syndrome

A patient's primary condition (i.e. trauma, shock, SIRS) can readily induce splanchnic hypoperfusion; this hypoperfusion is often called the "first hit" of the "two-hit" theory of multi-organ dysfunction syndrome (MODS). Tissue oxygen delivery (DO_2) is reduced by hypoperfusion, resulting in decreased production of adenosine triphosphate (ATP). Tissue ischemia induces the formation of reactive oxygen species commonly referred to as oxygen free radicals and pro-inflammatory cytokines that contribute to hemodynamic derangements and tissue injury. Bacterial endotoxin, infections, enzymes and environmental factors readily cause GIT mucosal immune dysregulation. Specifically, neutrophils and macrophages are activated as part of the "first hit." The pro-inflammatory cytokines tumor necrosis factor (TNF-alpha) and interleukin-6 (IL-6) induce the production of IL-1-beta that subsequently binds to IL-1 receptors located near the tight junction complexes within the GIT. This binding activates a potent inflammatory cascade propagated by nuclear factor kappa-beta (NF-KB) that facilitates the synthesis of myosin L-chain (MLC) kinase to degrade the cytoskeletal network and tight junction proteins. These changes are exacerbated by inadequate nutritional intake that leads to enterocyte atrophy caused by a lack of luminal nutrients, GALT dysfunction and altered permeability of the GIT. Subsequently bacteria and/or bacterial toxins translocate to induce systemic inflammation and organ dysfunction, thus serving as the "second hit" of the "two hit" theory of MODS.

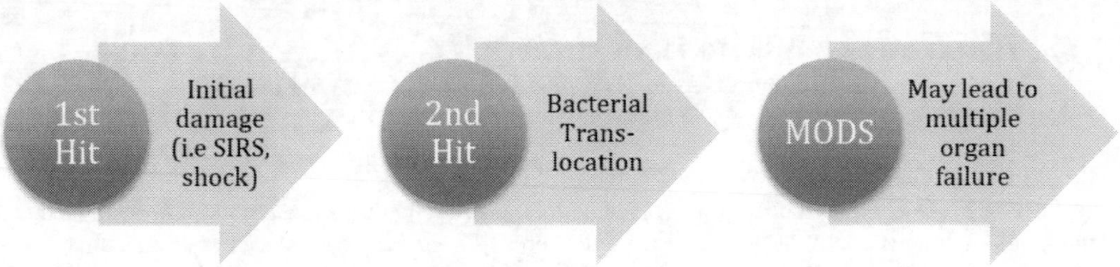

Common manifestations of MODS in dogs and cats include:

- Gastrointestinal – paralytic ileus, ulceration
- Cardiopulmonary – dysrhythmias, hypoxemia, hypoventilation
- Coagulation – prolonged partial thromboblastin time, prolonged prothrombin time, thrombocytopenia, thrombocytopathia
- Hepatic, Renal – hyperbilirubinemia, elevated aminotransferases, azotemia, proteinuria
- Vascular – vasodilation, hypotension

IMPLEMENTING NUTRITIONAL SUPPORT

Unfortunately, as a profession, we do not consistently provide adequate nutrition for hospitalized patients, particularly the critically ill; thus this population commonly becomes deficient in both protein and calories, further increasing their risk for developing systemic inflammatory response syndrome (SIRS) and MODS. Remillard *et al* showed only 27% of dogs studied had a positive energy balance during hospitalization. This is an unfortunate finding, as patient outcome and caloric intake have been directly related. In contrast Michel *et al* showed patients with temporary supplemental feeding tubes placed during hospitalization tended to receive their feeding prescription. A patient should not be allowed to enter a negative energy balance unless the owner imposes restrictions that prevent the recommended nutritional support.

Patients should be thoroughly evaluated prior to initiating nutritional therapy. Several risk factors that may indicate a predisposition to malnutrition are more than 5 days of hyporexia/anorexia, serious underlying disease, and marked protein losses. Body condition scores (BCSs) are clinically useful for nutritionally evaluating patients, as they are not influenced by fluid shifts that effect body weight. Commonly accepted indicators of malnutrition include poor hair quality, muscle wasting, unintentional weight loss (>10%), diminished wound healing, hypoalbuminemia, lymphopenia, and secondary coagulopathy.

Feeding should be provided as soon as possible to every patient for whom feeding is not contraindicated. There are few reasons nutritional intervention should be withheld, including dehydration, hypotension, hypothermia, electrolyte imbalances, glucose abnormalities, acid-base derangements, inability to protect airway, and a need for sedation/anesthesia. Any derangement should be corrected expeditiously, and then nutritional support should be initiated immediately. Patients unable to protect their airway should not be fed orally, but should be supported via a temporary supplemental feeding tube and/or parenteral nutrition. Recognizing most critically ill animals have been hyporexic or anorexic prior to hospitalization and small intestinal villous atrophy may occur within 24 hours, this patient population should be fed as soon as they are hemodynamically stable to help counter the catabolic stress of illness/disease. If a patient requires sedation or anesthesia for a diagnostic test and/or therapeutic procedure, one should consider concurrently placing a temporary supplemental feeding tube. There is mounting evidence early enteral nutrition (EN) in patients with major illnesses is well tolerated and positively affects patient outcome. Mansfield *et al* demonstrated early EN was well tolerated in canine patients with severe pancreatitis. Klaus *et al* documented similar results in cats with suspected acute pancreatitis. Mohr *et al* showed early enteral nutrition in canine parvovirus patients reduced the time for normalization of demeanor and appetite, more rapid resolution of vomiting and diarrhea, and significant weight gain compared to fasting until resolution of vomiting.

The site of feeding within the GIT is dependent on a variety of patient factors, and whenever possible, one should adhere to the adage, "if the gut works, use it!" However patients with a completely dysfunctional GIT may benefit from parenteral (partial vs. total) nutrition (sole therapy vs. concurrently with EN). Of course *per os* support is ideal, but many critically ill patients are not able or are unwilling to ingest adequate calories voluntarily. Thus temporary supplemental feeding tubes should be strongly advocated for patients with a functional or partially functional GIT. For those with a functional upper GIT who will likely require nutritional support for less than 3-5 days, one should consider placing a nasogastric (NG) or esophagostomy (E) tube. If enteral support will be needed for more than 3-5 days, one should consider placing a gastrostomy (G) tube or percutaneous endoscopic gastrostomy (PEG) tube. For patients with a dysfunctional upper GIT, a jejunostomy (J) tube is most appropriate. Feeding tubes are typically very well tolerated, and may truly be life-saving interventions. Patients with temporary feeding tubes require frequent monitoring for a few but potentially serious complications, including:

- Mechanical – obstruction, inadvertent removal
- Gastrointestinal – vomiting, diarrhea, regurgitation, abdominal discomfort secondary to cramping, aspiration pneumonitis
- Metabolic – refeeding syndrome, CHF, hyperglycemia

A patient's daily resting energy requirement (RER) may be calculated using the following equation: kcal/day = 70 x (body weight in kilograms)$^{0.75}$. Frequent patient reassessment is required, as energy requirements may be as high as 2RER in the critically ill and some may require 6-7g protein per 100 kcal to meet the goal of maximizing protein synthesis / preservation of lean body mass.

Microenteral nutrition (MEN) is defined as the delivery of small volumes of water, electrolytes and easily absorbed nutrients directly to the GIT to maintain mucosal integrity. It is commonly used in critically ill patients who are not able to tolerate full enteral nutrition, and is initially recommended for all patients with GIT dysfunction. Goals of MEN include maintenance of gastrointestinal mucosal blood flow to reduce incidence of GIT ulceration, prevent down regulation of gastrointestinal and pancreatic enzymes, reduce cholestasis, and to help preserve GALT. Microenteral feeding is commonly achieved via constant rate infusion (CRI) through temporary supplemental feedings. Isotonic crystalloid solutions are readily available and can easily be formulated to contain 5-20% dextrose. Pre-made electrolyte solutions (Rebound OES, Virbac Animal Health, Inc., Fort Worth, TX; Day 1 Critical Care, JAM Animal Health, Dungannon, N. Ireland) are also commercially available alternatives. The initial infusion rate is 0.25-0.5 mL/kg/hr, and should be gradually increased to 1-2 mL/kg/hr over a 24-hour period for tolerating therapy.

Once MEN infusion rates reach 1-2 mL/kg/hr, one should attempt to transition a patient to a gruel or liquid diet. Gruel and liquid diets may also be administered via CRI, but bolus feedings may also be viable. Campbell JA *et al* demonstrated no difference in tolerability or ability to reach nutritional targets between the two administration methods. Elemental (monomeric) diets are mixtures of amino acids, peptides (di- and tri-) and monosaccharide sugars, but to the author's knowledge there are no currently available monomeric diets specifically developed for use in dogs and cats. Polymeric diets (Clinicare, Abbott Laboratories, Chicago, IL) contain polypeptides, proteins and polysaccharides that are digested in the GIT prior to absorption. Infusion rates should be increased 25-30% every 8-12 hours until 100% RER is being delivered, ideally within 48-72 hours of initiating nutritional support. The canine stomach may expand to hold 80-90 mL/kg; the feline stomach may hold 60 mL/kg, but diseased cats often have lower gastric distensibility (20 mL/kg). Enteral fluid volume must be accounted for when determining intravenous fluid requirements for a patient.

For patients with a completely dysfunctional GIT, parenteral nutrition (PN), a mixture of amino acids, glucose and lipids, should be used either alone or ideally in conjunction with enteral nutrition to provide energy requirements. Parenteral nutrition is not without complications, and has a reputation of causing more harm than good. Devising a PN formulation that optimizes recovery and minimizes complications is currently challenging. Chan *et al* showed patients receiving enteral nutrition and partial PN survived more than those receiving only partial PN. In contrast Queau *et al* documented most complications accompanying PN did not affect survival. Some believe many of the misconceptions and complications may be due to over-estimation of caloric needs rather than the form of nutritional support. When calculating the number of calories required for a patient, some aim to meet all caloric requirements with lipids and dextrose; they then provide additional amino acids as required by the patient but do not take into account the calories this component provides (~4 kilocalories per gram once metabolized). The belief is an animal not provided with their entire energy requirements in the form of dextrose and lipids will use the provided amino acids for energy rather than protein synthesis. However there is little evidence to support this hypothesis. Overall there is evidence to show provision of more calories than the estimated RER is associated with the development of hyperglycemia and possibly decreased survival. The time at which PN should be implemented is currently controversial. Casaer *et al* demonstrated enteral support followed by PN within 48 hours of ICU admission was associated with high morbidity and mortality compared to enteral support followed by delayed PN. We do not yet know if such data should be extrapolated to veterinary patients.

REFERENCES

1. Baldwin K et al. **J Am Anim Hosp Assoc** 2010;46(4):285.
2. Bibbo C, et al. **J Invest Surg** 1996;9(4):293.
3. Brunetto MA, et al. **J Vet Emerg Crit Care** 2010;20(2):224.
4. Campbell JA, et al. **J Vet Emerg Crit Care** 2010;20(2):232.
5. Chan DL, et al. **J Vet Intern Med** 2002;16(4):440.
6. Chan DL, et al. **Vet Clin North Am Small Anim Pract** 2006;36(6):1225.
7. Dahlinger J, et al. **J Vet Intern Med** 1997;11(6):319.
8. Klaus JA, et al. **J Vet Emerg Crit Care** 2009;19(4):337.
9. Liu Q, et al. **Int J Pancreatol.** 2000;27(2):157.
10. Michel KE, et al. **J Vet Emerg Crit Care** 2006;16(S1):S2.
11. Qin HQ, et al. **Hepatobiliary Pancreat Dis Int** 2002;1(1):150.
12. Queau Y, et al. **J Vet Intern Med** 2011;25(3):446.
13. Mansfield CS, et al. **J Vet Intern Med** 2011;25(3):419.
14. Mohr AJ, et al. **J Vet Intern Med** 2003;17(6):791.
15. Remillard RL, et al. **Vet Ther** 2001;2(4):301.
16. Winkler KP, et al. **J Am Anim Hosp Assoc** 2003;39(4):361.

PRINCIPLES OF FLUID THERAPY IN SEPSIS & SIRS IN DOGS & CATS

Christopher G. Byers, DVM, DACVECC, DACVIM (SAIM)
Omaha, Nebraska USA

DEFINITIONS

Systemic inflammatory response syndrome (SIRS) is the terminology used to describe the body's inability to localize inflammation, and SIRS may occur with or without infection. Extrapolating from criteria used in human medicine, diagnostic criteria for SIRS in dogs and cats were proposed (Table 1). One must document a patient to have two or more of the proposed criteria to diagnose SIRS. Some questioned the validity of these criteria as they lack specificity; subsequently a newer approach, termed PIRO, has been proposed. The acronym *PIRO* incorporated four factors (Predisposition, Infection, Response of the host and Organ dysfunction) and effectively expanded the criteria for SIRS by including physical parameters and biomarkers.

Table 1: Diagnostic Criteria for SIRS (a patient must exhibit two of the criteria)

Variable	*Dog*	*Cat*
Heart Rate	>120 bpm	<140 bpm or >225 bpm
Respiratory Rate	>20 rpm	>40 rpm
Body Temperature	<100.6°F or >102.6°F	<100.0°F or >103.5°F
White Blood Cell Count	<6000/uL or >16000/uL	<5000/uL or >19500/uL

Sepsis is simply the body's systemic response to infection. Sepsis and SIRS are both associated with marked inflammatory, immunological and hormonal responses, but only sepsis is characterized by an identifiable source of infection. *Severe sepsis* is defined as sepsis associated with detectable organ dysfunction or hypoperfusion. *Septic shock* is severe sepsis with *sepsis-induced hypotension* (systolic blood pressure <90 mmHg, mean arterial pressure <70 mmHg, OR >40 mmHg drop from baseline) and hypoperfusion despite adequate fluid resuscitation. *Multi-organ dysfunction syndrome (MODS)* denotes the presence of altered organ function in the acutely ill to the extent that homeostasis cannot be maintained without intervention.

EARLY GOAL DIRECTED THERAPY

The profound inflammation associated with SIRS and sepsis induces vascular leakage secondary to endothelial dysfunction. Subsequently fluid extravasation results in relative and absolute hypovolemia, hypoperfusion and possibly septic shock. Fluid therapy is the mainstay of resuscitation of patients with SIRS, sepsis, severe sepsis and septic shock. In 2001 Rivers *et al* introduced the concept of early goal directed therapy (EGDT), a specific algorithm for managing patients presenting with septic shock. Aggressive fluid resuscitation to achieve specific endpoints was recommended. In-hospital mortality was 30.5% in the EGDT group and 46.5% in standard therapy group, a finding that was statistically significant. Patients in the EGDT group also had significantly higher mean $S_{cv}O_2$, lower lactate, lower base deficit, and higher pH than the patients assigned to standard therapy. Subsequently clinical guidelines, the Surviving Sepsis Campaign Guidelines (SSCG), for the management of human patients with severe sepsis and septic shock were initially published in 2004 and most recently updated in 2012.

Aggressive initial fluid resuscitation is considered the cornerstone of initial therapy for human and veterinary patients with severe sepsis and septic shock. The SSCG recommend immediately instituting volume resuscitation in affected people according to a specific protocol; specifically crystalloids are the recommended initial fluid of choice and the initial fluid challenge should be at least 30 mL/kg. Within the first six hours of therapy, one should target:

- Mean arterial pressure (MAP) ≥65 mmHg
- Central venous pressure (CVP) 8-12 mmHg
- Urine output (UOP) ≥0.5 mL/kg/hr
- Central venous oxygen saturation ($S_{cv}O_2$) ≥70%

Interestingly despite the SSCG there is marked variation in the type of resuscitation fluid used in human patients worldwide. In the United States, crystalloids are the initial recommended resuscitation fluid. More than 50% of fluid resuscitations in Australia are performed with mainly albumin and in European countries almost 50% of resuscitations are done with mainly hydroxyethyl starches. Vasopressors should be used within six hours for those patients with hypotension despite aggressive initial fluid resuscitation. Transfusion with packed red blood cells to maintain a packed cell volume (PCV) >30% and vasopressor therapy may also be indicated. Serial monitoring of perfusion parameters, most notably lactate, should be pursued, and one should seek to normalize lactate levels as rapidly as possible.

Despite the SSCG, to date the optimal composition and volume of resuscitation fluid is not known, and remains a highly controversial topic in human emergency and critical care medicine. Furthermore prospective randomized controlled trials (RCTs) have not been conducted in veterinary medicine; thus the ideal guidelines for resuscitation of veterinary patients with SIRS, sepsis, severe sepsis and septic shock are similarly unknown at this time.

CONTROVERSIES WITH STARCHES

Hydroxyethyl starches (HES) have been recommended for resuscitation in patients with sepsis given their reported ability to induce a more rapid and lasting circulatory stabilization than crystalloids. Dubin *et al* documented fluid resuscitation with 6% HES 130/0.4 may improve sublingual microcirculation better than normal saline solution in humans with septic shock. Zhu *et al* showed resuscitation with a combination of 7.5% hypertonic saline and 6% HES 130/0.4 was associated with a lower incidence of MODS compared to resuscitation with Ringer's lactate in patients with severe sepsis. Guidet *et al* showed human patients with severe sepsis resuscitated with 6% HES 130/0.4 required less volume than those receiving normal saline solution; this study was not powered sufficiently to assess differences in mortality. Lv *et al* compared lactate Ringer solution to 6% HES 130/0.4 in human patients with septic shock; resuscitation fluid volume and vasoactive drug dosage required to achieve endpoints of resuscitation were lower in those receiving HES, but patient outcome was not statistically different between the two groups.

The use of HES has been called into question in human patients given various adverse reactions reported in RCTs, most notably coagulation alterations, allergic reactions, increased incidence of acute kidney injury (AKI) and need for renal replacement therapy (RRT). Initially complications were thought to be isolated to the use of HES with high molecular weights and higher substitution ratios, but HES with lower molecular weights and substitution ratios have also been implicated in adverse reactions.

Acute Kidney Injury

Hydroxyethyl starches have received a tremendous amount of attention in human critical care due numerous studies documenting AKI in patients with severe sepsis and septic shock. Burkhorst *et al* documented an increased incidence of acute renal failure and need for RRT in human patients with severe sepsis or septic shock receiving 10% HES 200/0.5 compared to Ringer's lactate. Haase *et al* showed administration of 6% HES 130/0.38-0.45 to human patients with sepsis was associated with increased need for RRT and more serious adverse events that crystalloid or albumin. Perner *et al* identified 6% HES 130/0.42 to human patients with severe sepsis was associated with increased risk of death at 90 days, increased need for RRT, decreased days alive with RRT and decreased days alive out of the hospital; patients also had a trend toward increased breeding tendencies and a requirement for blood products. Bayer *et al* showed fluid resuscitation with 6% HES 130/0.4 was an independent risk factor for acute kidney injury, and patients receiving synthetic colloids received significantly more allogenic blood products. Myburgh *et al* identified patients with severe sepsis or septic shock that were resuscitated with 6% HES 130/0.4 required RRT more those receiving 0.9% sodium chloride; however there was no difference in mortality.

Coagulation Abnormalities

Hydroxyethyl starches have been implicated in inducing several coagulation abnormalities in various patient populations, not just those living with sepsis. Certainly the potential to induce a coagulopathy must be considered when contemplating the use of HES in septic patients. Smart *et al* evaluated 6% HES 670/0.75 in healthy dogs, and infusion was associated with prolonged closure time for up to 24 hours possibly due to both hemodilution and platelet dysfunction. Wierenga *et al* showed both 6% HES 600/0.7 and 6% HES 670/0.75 prolonged closure time to a greater extent than 0.9% sodium chloride. Chohan *et al* evaluated various hemostatic various in healthy canine patients receiving either 6% HES 600/0.75 or lactated Ringer's solution. Both fluids altered packed cell volume (PCV), total protein (TP), prothrombin time (PT), buccal mucosal bleeding time (BMBT), and colloid osmotic pressure (COP), but neither fluid was associated with increased clinical bleeding. In addition to an increased need for RRT, Haase *et al* also showed administration of 6% HES 130/0.38-0.45 to human patients with sepsis was associated with increased need for red blood cell transfusion. In their study Bayer *et al* concurrently showed patients resuscitated with 6% HES 130/0.4 received significantly more allogenic blood products than those who received normal saline. Lv *et al* documented those who received HES showed a gradual decrease in plasma plasminogen activator inhibitor (PAI) that was significantly different than before resuscitation and the Ringer's lactate patients 24 hours after resuscitation.

BOLUS FLUID RESUSCITATION

Patients with sepsis-induced hypotension, severe sepsis and septic shock commonly have hyperlactatemia, reduced UOP and other signs of hypoperfusion. Certainly there is concern for inadequate vital organ perfusion, reduced cardiac output (CO) due to vasodilatation and subsequently reduced preload due to diminished venous return. Understandably one administers a rapid fluid bolus to improve CO, vital organ perfusion and normalize arterial blood pressure, an intervention embraced and recommended by the SSCG.

Recently the validity of this logic has been probed, asking the following questions:

- Does one need to be concerned about increasing delivery of cytokine-rich and possibly toxic blood to organs?
- Could the hypotensive state, to a degree, be a protective mechanism?
- Does fluid bolus resuscitation in patients actually increased vital organ perfusion? If yes, how long does the effect last?

Zanotti-Cavazzoni *et al* and Smith *et al* documented increased survival in mice and rats, respectively, with fluid bolus resuscitation. Conversely Brandt *et al* showed reduced survival in pigs receiving higher volume fluid resuscitation compared to those receiving a fixed moderate volume. In the United States, emergency departments generally adhere to the principles of EGDT; however in other parts of the world these guidelines are not routinely followed. Despite this, there exists some mounting evidence mortality rates compare favorably and may even be considerably lower than those observed with EGDT. Reasonably these data call into question the validity of the recommendation for initial aggressive fluid bolus resuscitation.

At this time, no experts have yet recommended a change in the treatment of humans with severe sepsis and septic shock in the United States. Priorities remain the maximization of perfusion and oxygen delivery (DO_2) to tissues. Veterinary emergency and critical care clinicians are familiar with the concept of crystalloid "shock" volumes for dogs and cats (90 mL/kg/hr and 50-60 mL/kg/hr, respectively), but this entire volume is not automatically bolused to hypotensive patients. More appropriately, crystalloids are infused at 1/4 or 1/3 "shock" rates; after the delivery each bolus, endpoints of resuscitation (EORs) are evaluated to determine the need for additional boluses. Common EORs in veterinary medicine are listed in Table 2. Perhaps in veterinary medicine we are already embracing the concept of moderated fluid boluses through the serial evaluation of EORs in patients.

Table #2: Veterinary Endpoints of Resuscitation
Restoration of normal vital signs
Normalization of abnormal mentation
Restoration of normal blood pressure (systolic >80-90 mmHg)
Normal serum lactate (<2.0 mmol/L)
$S_{cv}O_2$ >70%
PCV >25%
UOP >1 mL/kg/hr
S_pO_2 >93% @ F_iO_2 = 21%
CVP = 5-10 cmH_2O

One should note all fluid resuscitation studies have inconsistencies, most notably the trigger for fluid infusion, scheme for fluid infusion, and endpoint(s) of resuscitation. These are important issues that should be addressed in future prospective RCTs that are sincerely needed in veterinary medicine, particularly those with severe sepsis and septic shock!

REFERENCES

1. American College of Chest Physiciains / Society of Critical Care Medicine Consensus Conference. **Crit Care Med** 1992;20(6):864.
2. Bayer O, et al. **Crit Care Med** 2012;40(9):2543.
3. Brady CA, et al. **Vet Clin North Am Small Anim Pract** 2001;31(6):1147.
4. Brady CA, et al. **J Am Vet Med Assoc** 2000;217(4):531.
5. Brandt S, et al. **Crit Care** 2009;13(6):R186.
6. Brunkhorst FM, et al. **N Engl J Med** 2008;358(2):125.
7. Dellinger RP, et al. **Crit Care Med** 2013;41(2):580.
8. Dubin A, et al. **J Crit Care** 2010;25(4):659.
9. Guidet B, et al. **Crit Care** 2012;16(3):R94.
10. Haase N, et al. **BMJ** 2013;15;346:f839.
11. Hauptman JG, et al. **Vet Surg** 1997;26(5):393.
12. Lv J. **Zhongguo Wei Zhing Bing Ji Jiu Yi Xue** 2012;24(1):38.
13. Myburgh JA, et al. **N Engl J Med** 2012;367(20):1901.
14. Peake SL, et al. **Resuscitation** 2009;80(7):811.
15. Perel P, et al. **Cochrane Database Syst Rev** 2011;16(3):CD000567.
16. Perner A, et al. **N Eng J Med** 2012;367(2):124.
17. Rivers E, et al. **N Engl J Med** 2001;345(19):1368
18. Smart L, et al. **J Vet Emerg Crit Care** 2009;19(5):444.
19. Smith EF 3rd, et al. **Pharmacology** 1993;46(5):254.
20. The Australian Resuscitation in Sepsis Evaluation (ARISE) Investigators. **Crit Care Resusc** 2007;9(1):8.
21. Westphal M, et al. **Anesthesiology** 2009;111(1):187.
22. Wierenga JR, et al. **Am J Vet Res** 2007;68(6):605.
23. Zanotti-Cavazzoni SL, et al. **Intensive Care Med** 2009;35(4);748.
24. Zhu GC, et al. **Zhongguo Wei Zhing Bing Ji Jiu Yi Xue** 2011;23(3):150.

SHOCK PRINCIPLES IN THE CAT

Allison Sande, DVM, DACVIM (SAIM), DACVECC
Tinton Falls, NJ

INTRODUCTION

Circulatory shock is a common clinical condition. It is a syndrome in which tissue perfusion is reduced and oxygen delivery to tissues is inadequate to meet cellular metabolic requirements. Low or unevenly distributed blood flow leads to a critical decrease in oxygen delivery (DO_2) in relation to oxygen consumption (VO_2) and inadequate cellular energy production.[1-4] The mortality rate for veterinary patients suffering from circulatory shock remains high, especially for patients in septic shock, among whom the overall mortality rates range from 33-50%.[5-7] Early recognition of hemodynamic instability by physical examination findings and point of care testing in conjunction with rapid, aggressive therapy and appropriate monitoring are critical to optimize survival.

CLASSIFICATION

Shock may be classified into four subsets: hypovolemic, cardiogenic, distributive, and obstructive shock.[8] While these categories function to help explain why oxygen delivery is not matching oxygen demand, it is imperative to remember that clinical cases of shock usually do not fit neatly into one category, but often overlap. Shock classification can also be simplified into two broad categories with typical hemodynamic profiles. The first category is hypodynamic shock, which includes hypovolemic, cardiogenic, and obstructive shock subsets. The second category is hyperdynamic shock, and includes distributive shock. The common features of hypodynamic shock are a low cardiac index and a high-resistance vasoconstricted state. Hypovolemic shock is characterized by a decrease in circulating blood volume; the development of organ dysfunction is directly related to inadequate global blood flow. Common causes of hypovolemic shock include hemorrhage, dehydration from gastrointestinal losses, and third spacing. Decreased cardiac filling pressures are the hallmark of hypovolemic shock, in contrast to cardiogenic shock, in which state they are elevated. Cardiomyopathies, arrhythmias, and drug overdoses (anesthetics, β-blockers, calcium channel blockers) may lead to decreased cardiac contractility and subsequent cardiogenic shock. Obstructive shock is related to a variety of causes that causes a physical impediment to cardiac venous return such as gastric dilatation volvulus (GDV) and pericardial effusion causing cardiac tamponade. Hyperdynamic circulatory shock is characterized by a high cardiac index and a low-resistance vasodilated state. Common causes of hyperdynamic shock include sepsis, anaphylaxis, and adrenal insufficiency. The underlying hemodynamic defect in each of these syndromes is maldistribution of blood flow and/or blood volume leading to compromised nutrient delivery. In contrast to hypodynamic shock, oxygen extraction is normal or decreased despite evidence of hypoperfusion. Direct mediator-related effects combine with tissue hypoperfusion to produce the cellular injury and organ dysfunction seen in patients with septic shock.[1-3,8,9]

PATHOPHYSIOLOGY OF SHOCK

The primary metabolic defect in circulatory shock is impaired oxidative metabolism. This impairment is most commonly due to decreases in oxygen supply caused by global decreases in blood flow or by maldistribution of blood flow. Oxygen delivery is determined by cardiac output (CO), hemoglobin concentration, and the arterial oxygen saturation. Under normal circumstances, oxygen consumption is independent of oxygen delivery and CO. However, during shock, as metabolic demand increases (VO_2) or DO_2 diminishes, oxygen extraction rises to maintain aerobic metabolism. When extraction is maximized, a critical level of oxygen delivery (DO_2crit) is reached, oxygen consumption falls, and anaerobic metabolism ensues.[1,2] Aerobic ATP generation is dependent on cytoplasmic glycolysis and oxidative phosphorylation in the mitochondria. Under anaerobic conditions, ATP generation is limited and results in lactic acid accumulation. As such, the presence of lactic acidosis serves as an indicator of critical deficits in high-energy phosphate metabolism. Oxidative metabolism may also be impaired by mechanisms independent of tissue hypoperfusion. Inflammatory mediators such as nitric oxide, endotoxin, reactive oxygen species, calcium and tumor necrosis factor impair mitochondrial function. Mitochondrial abnormalities and decreased mitochondrial activity have been observed in patients with septic shock.[1,10-12]

Compensatory Mechanisms

Regardless of the underlying cause for shock, tissue hypoxia elicits a complex set of neural and hormonal compensatory reflexes aimed at maintaining CO and vascular tone. Regional autoregulatory mechanisms and changes in capillary blood flow maintain perfusion to vital organs such as the heart, brain and kidney, at the expense of others such as the liver, muscle, and gastrointestinal tract.[1,13] Acute compensatory effects are mediated by the sympathetic nervous system; catecholamine release occurs within minutes and affects heart rate and venous return. As CO decreases, baroreceptor impulse generation at the carotid sinus and aortic arch in the heart decreases, leading to decreased inhibition of the vasomotor center within the brain, increased sympathetic stimulation, and ultimately, norepinephrine release from the adrenal gland. Norepinephrine binds to α-adrenergic receptors within blood vessels leading to vasoconstriction and β_1-adrenergic receptors in the myocardium to leading to increased heart rate and contractility.[9,14,15] Hypoxemia is also an important stimulus for catecholamine secretion. Chemoreceptors located in the carotid artery and aorta sense oxygen content in the blood. Decreased oxygen delivery sensed by these chemoreceptors also stimulates the vasomotor center to increase sympathetic stimulation independent of peripheral blood pressure.[9,15] Cortisol is rapidly mobilized in the acute shock phase. Corticotropin-releasing hormone and adrenocorticotropic hormone concentrations are increased during pain or physical stress leading to cortisol release from the adrenal gland. Increased concentrations of cortisol in turn lead to stimulation of glycogenolysis and fat and protein mobilization for gluconeogenesis. The resultant increased blood glucose concentrations supply energy to brain cells, vascular endothelial cells for contraction, and myocardial cells for contraction.[9,14] Finally, in extreme cases of hypovolemia, the transient absorption of fluid to the plasma volume results in 'autotransfusion,' however this effect is limited and only lasts for a few minutes. According to the original model of Starling's forces, it was believed that there was a significant transcapillary shifting of fluid from the interstitial space into the vasculature leading to an expanding blood volume. However, recent research has resulted in a revised Starling equation, which includes the contributions of the endothelial glycocalyx layer, the endothelial basement membrane, and the extracellular matrix. In this glycocalyx model, there is a no absorption rule. So, even in acute cases, absorption quickly reverts to filtration as proteins diffuse into the subglycocalyx from the interstitial space, diminishing the COP difference that opposes filtration.[16]

Other compensatory mechanisms occur within the first hour following the development of shock. The primary sympathetic response to shock is augmented by activation of the renin-angiotensin-aldosterone system (RAAS). Decreased renal blood flow secondary to decreased CO is sensed by baroreceptors within the juxtaglomerular cells, which release renin in response. Renin causes conversion of angiotensinogen to angiotensin I, which is then converted to angiotensin II by angiotensin-converting enzyme in the lungs. Angiotensin II binds to receptors on the blood vessels leading to vasoconstriction. Vasoconstriction improves vessel tone, but also forces blood from less critical peripheral tissues, such as the splanchnic circulation, to improve venous return to the heart, in turn leading to improved CO. Angiotensin II likewise causes renal arterial vasoconstriction, reducing blood filtration, and leading to water and sodium retention.[9,13-15] Increased osmolarity or decreased effective circulating blood volume is sensed by atrial baroreceptors and stretch receptors. A 1% change in osmolarity and a 10% decrease in blood volume results in vasopressin release from the posterior pituitary gland. Vasopressin binds to V1 receptors on arterioles, leading to vasoconstriction and improved vascular tone to help maintain tissue perfusion.[9]

As shock progresses, additional compensatory mechanisms function to replace blood volume. However, these mechanism may take up to 48 hours to reach full effect. In addition to the above effects, angiotensin II also stimulates the adrenal glands to secrete aldosterone. Aldosterone stimulates sodium reabsorption, along with water, from the distal convoluted renal tubule. Vasopressin also has an added effect as antidiuretic hormone (ADH). The antidiuretic effects are mediated via V2 receptors in the renal collecting ducts. Receptor activation leads to insertion of aquaporin channels into the collecting ducts and allows resorption of water from the ducts.[9,14]

CLINICAL ASPECTS

Clinical signs of shock vary depending on the type and clinical phase of shock, as well as by species. In compensated shock, due to the various physiologic mechanisms discussed above, the patient is generally able to maintain tissue perfusion to preserve normal cellular metabolism. Dogs in compensatory shock may have subtle clinical signs characterized by mild to moderate mental depression, pale mucous membranes with a normal or shortened capillary refill time (CRT), tachycardia, tachypnea, peripheral vasoconstriction with cool extremities, and a normal blood pressure. With ongoing compromise of systemic perfusion, compensatory mechanisms become inadequate and begin to fail. Patients in decompensatory shock develop severe mental depression, prolonged CRT, increasing tachycardia and tachypnea, poor pulse quality, and decreased arterial blood pressure.

Dogs with distributive shock can show clinical signs of hyperdynamic or hypodynamic shock. The initial hyperdynamic phase of sepsis is characterized by hyperemic mucous membranes, tachycardia, and bounding peripheral pulses. Hyperemic mucous membranes occur when peripheral vasodilation overwhelms the expected vasoconstriction due to mediators such as nitric oxide and cytokines. If SIRS and sepsis progresses, a decrease in CO and hypodynamic shock may due to the effects of cytokines on the myocardium or myocardial ischemia. Clinical findings at that time include dull mentation, hypothermia, pale mucous membranes with a prolonged CRT, tachycardia, poor pulse quality, and hypotension. In this state of decompensatory shock, the pallor seen in SIRS and sepsis patients is caused by a lack of blood delivery to the mucous membranes, not vasoconstrictive mechanisms. Without intervention decompensatory shock will result in organ damage and death. Terminal phases of shock are marked by vasomotor dysfunction characterized by loss of arteriolar tone with paradoxical increased venular resistance. The resultant increase in capillary hydrostatic pressure leads to intravascular volume loss and worsening of the shock state, eventually resulting in a state of irreversible shock. The gastrointestinal tract is considered the shock organ in dogs and often leads to ileus, diarrhea, hematochezia and/or melena.[1-3,9]

In cats, the hyperdynamic phase of shock is rarely recognized. Also, in contrast to dogs, cats with shock have unpredictable changes in heart rate; they may exhibit tachycardia or bradycardia. Typically, cats present with mental depression, hypothermia, pale mucous membranes, bradycardia or relative bradycardia, weak pulses, cool extremities, and generalized weakness. In cats, the lungs are the organ most susceptible to damage during shock, and so signs of respiratory dysfunction such as tachypnea or dyspnea are common.[2,3,17,18]

DIAGNOSTICS AND MONITORING

It is of vital importance that frequent assessment through physical examination and monitoring be performed for every patient suffering from shock. Basic diagnostic tests should be performed to identify the etiology of the shock state and to assess the degree of organ injury. A arterial or venous blood gas, lactate measurement, complete blood count, serum chemistry, coagulation panel, and urinalysis should be performed. Additional diagnostics such as thoracic radiographs, abdominal radiographs, abdominal ultrasound, and echocardiogram may be indicated once patient stabilization has occurred.[2,3] The magnitude of oxygen deficit is a key predictor in outcome in patients with shock. Optimizing perfusion and DO_2 are the primary objectives of therapy and comprehensive monitoring is imperative to achieve these goals. The adequately perfused patient maintains certain characteristics. These include: normal body temperature, heart rate, heart rhythm, and respiratory rate; pink mucous membranes with a CRT of less than two seconds; mean arterial pressure (MAP) of 70-120 mmHg and diastolic pressure above 40 mmHg; urine production of greater than or equal to 1 ml/kg/hr; and central venous pressure between 5 and 10 cm H_2O in dogs and 2 to 5 cm H_2O in cats. These parameters create a baseline for patient assessment.[2,3] Controversy exists over the best manner in which to monitor tissue perfusion in patients with circulatory shock. The commonly used parameters such as heart rate, blood pressure, and even CO, correlate poorly with survival in human critically ill patients. This is especially true in patients suffering from septic shock.[19,20] These findings have led to the use of more specific indices of tissue oxygen metabolism to guide resuscitation efforts.

Shock Index

The shock index (SI) is defined as the ratio of heart rate (HR) to systolic arterial blood pressure (SBP); that is, SI = HR/SBP. This was originally developed as a simple means of quantifying the severity of shock in humans on presentation to the ER, and has also been used to monitor treatment response in people. This calculation is thought to provide a more objective measure of shock than HR and SBP alone allowing for the identification of derangements in perfusion status in the face of outwardly normal cardiovascular parameters, specifically in cases of occult hypoperfusion. The ease of use and low cost of the SI make it an easily implemented monitoring tool for the identification of shock.[21,22] Two recent veterinary studies have evaluated the use of the SI in canine patients. The first study[21] evaluated dogs with confirmed hemorrhagic shock and concluded that the SI was significantly higher in patients with hemorrhage compared to healthy dogs. Using a cutoff of an SI value of 0.9 there was a high sensitivity but a low specificity as a triage tool. As such, an SI of > 0.9 should increase a clinician's concern for shock and further evaluation is indicated. The second study[22] evaluated dogs presenting to the emergency room with moderate to severe shock and concluded that dogs in shock had a median SI of 1.37, which was significantly higher than healthy dogs and dogs presenting to the ER not in shock. An SI of > 1.0 was found to be a highly sensitive and specific indicator to identify dogs in shock. While these veterinary studies did not evaluate the SI in cases of shock with occult hypoperfusion, it may prove to be a valuable and inexpensive tool for initial patient evaluation and screening. Further studies evaluating dogs with early, developing shock as well as studies evaluating the SI in cats are warranted.

Lactate

Patients with inadequate tissue perfusion, DO_2, or VO_2 often develop hyperlactatemia due to anaerobic metabolism. The normal lactate concentration in adult dogs and cats is less than 2.5 mmol/L.[23] Critically ill human patients with hyperlactatemia are at increased risk of developing multiple organ failure and have an increased mortality rate.[24] Elevated lactate levels in dogs may also help predict mortality.[23,25-27] Serial lactate measurements should be taken during shock resuscitation to gauge response to treatment and evaluate the resuscitation end points. Trends in lactate concentrations have been found to be a better predictor of outcome than single measurements. The ability to correct an elevated lactate is directly correlated with survival.[28-30]

Mixed Venous Oxygen Saturation and Central Venous Oxygen Saturation

Mixed venous oxygen saturation (SvO_2) can be utilized as an index of global tissue oxygenation. If VO_2 is held constant, then the SvO_2 is determined by CO, hemoglobin concentration, and systemic arterial oxygen tension. The SvO_2 decreases if DO_2 is reduced by low CO, hypoxia, or severe anemia, or if VO_2 is increased as seen in fever. A reduction in SvO_2 indicates the presence of tissue hypoxia and is an early sign of patient deterioration. However, this does not hold true in septic shock and other situations when the relationship between venous blood and tissue oxygenation is altered by maldistribution of blood flow. In these circumstances, the ability of the tissues to extract oxygen is limited by decreases in effective nutrient flow; SvO_2 may be increased or normal despite the presence of tissue hypoxia and anaerobic metabolism. Accordingly, normal SvO_2 levels do not preclude tissue hypoperfusion. Ideally the SvO_2 is measured in a blood sample derived from a pulmonary arterial catheter (PAC) as this represents a true mixed venous oxygen sample. However, PAC placement is not practical in most veterinary clinics. Instead, oxygen saturation can be determined within the central circulation using a central venous catheter within the cranial vena cava; this is termed the central venous oxygen saturation ($ScvO_2$). $ScvO_2$ values are generally higher than SvO_2, but the measurements closely parallel one another and the difference between the two is typically about 5%. This difference is explained by blood flow redistribution and differences in VO_2 across the hepatosplanchnic, coronary, and cerebral circulations during shock.[1-3] While $ScvO_2$ measurement in small animals is not widely used, clinical studies suggest that it is useful as a diagnostic, prognostic, and monitoring tool.[30,31]

THERAPY

Initial Approach

The approach to patients with circulatory shock involves rapid assessment of the underlying disease process and restoration of cardiopulmonary stability. Patients should be assessed by history and physical examination for clues as to the etiology of the shock syndrome and for evidence of organ hypoperfusion. Efforts to achieve cardiopulmonary stability should occur simultaneously with the initial diagnostics discussed above to maximize DO_2 and prevent irreversible shock. The VIP approach can be used to prioritize resuscitation efforts by focusing on *v*entilation, *i*nfusion, and *p*ump activity.[1-3]

Oxygen

Oxygenation and adequate ventilation must be ensured. If the patient is breathing then flow-by oxygen methods (50-150 ml/kg/min) or nasal oxygen cannulas (50-100 ml/kg/min) can be utilized for supplementation. However, evidence of respiratory muscle fatigue or refractory hypoxia should prompt intubation and the initiation of mechanical ventilation.[1,2]

Fluids

Critical hypovolemia is present in the majority of patients presenting with circulatory shock. The first and most important therapeutic goal for noncardiogenic shock is to restore the effective circulating volume. Appropriate vascular access is essential for the rapid administration of large volumes of fluids. Poiseuille's law of flow states that resistance to flow through a catheter is directly proportional to the length of the catheter and inversely related to the radius. Therefore, short, large-bore intravenous catheters should be placed in a peripheral vein for resuscitation; 18- or 20-gauge catheters for cats or small dogs and for larger dogs multiple venous catheters (14- to 18-gauge) may be placed.

Fluids should be infused in boluses and titrated to specific endpoints of heart rate, blood pressure, urine output, and clearance of arterial lactate.[1] Isotonic crystalloids remain the cornerstone of treatment of noncardiogenic circulatory shock. The shock dose of isotonic crystalloid solution is approximately one blood volume (90 ml/kg in the dog and 50-60 ml/kg in the cat). The fluid administered rapidly distributes into the extracellular space so that only 25% of the delivered volume remains in the intravascular space by 30 minutes after infusion.[32] In patients with active hemorrhage it may be advantageous to perform limited volume resuscitation until the hemorrhage is controlled. In this situation a MAP of no more than 60 mmHg is targeted.[33]

Hypertonic saline (HTS) administration reportedly resulted in a transient osmotic shift of water from the extracellular compartment into the vasculature via an osmotic gradient. However, transvascular fluid exchange is much less than predicted as illustrated by the contribution of the endothelial glycocalyx layer of the vasculature and the revised Starling equation.[16] Still, using HTS may be advantageous; HTS may function to reduce endothelial swelling, increase cardiac contractility, decrease peripheral vascular resistance, and decrease leukocyte adhesion and migration blunting the inflammatory response. However, due to the resulting osmotic diuresis and redistribution of sodium ions, the intravascular volume expansion is transient and so additional fluid therapy must be administered. The recommended dose is 3-5 ml/kg of 7.5% saline given intravenously over 5 mintes.[2,32-34]

Synthetic colloids are large molecules that do not readily cross the intact vascular membrane. These solutions are hyperoncotic and oppose filtration, helping to retain fluid within the intravascular space when normal vascular permeability exists.[16] The recommended dose of synthetic colloids is up to 20 ml/kg in the dog and 10-15 ml/kg in the cat.[2,34] However, the newer generation synthetic colloid, VetStarch, has been shown to be safe at doses of up to 50 ml/kg/day in dogs.[35] Synthetic colloids are appropriate for shock therapy in patients with hypoproteinemia. In addition, they may be used along with isotonic or hypertonic crystalloids to maintain adequate plasma volume expansion with lower interstitial volume expansion. Despite numerous clinical studies, there is no definitive evidence that colloids are superior to crystalloids for resuscitation.[2,3,36]

Vasopressors

Shock patients that remain hypotensive despite adequate intravascular volume resuscitation require vasopressor therapy to modify vascular tone. The choice of vasopressor agent is dependent upon the goal of therapy. Norepinephrine has mixed α- and β-adrenergic receptor agonist effects with preferential α-receptor activity, so that the effects on heart rate and contractility are less potent than the vasoconstrictive effects.[2] Norepinephrine is considered the first-choice vasopressor in human medicine and is administered to target a MAP of ≥ 65 mmHg. The recommended dose in veterinary medicine is 0.1-2.0 μg/kg/min intravenously. Vasopressin can be added to norepinephrine to either raise the MAP to the target value, or to decrease the norepinephrine dose, but it is not recommended as the initial vasopressor.[36] Vasopressin has direct and indirect effects on the vascular smooth muscle and induces vasoconstriction in most vascular beds. At low doses, however, this drug causes vasodilation in renal, pulmonary, mesenteric, and cerebral vasculature and may help maintain perfusion to these organs. Vasopressin also enhances sensitivity to catecholamines and so may allow the dose of concurrent catecholamine therapy to be lowered. The recommended dose in veterinary patients is 0.1-1 μU/kg/min intravenously.[2,3,37] When hypotension is accompanied by evidence of impaired cardiac contractility, a positive inotrope such as Dobutamine is indicated. Dobutamine is primarily a β-agonist and increases CO, DO_2, and VO_2 without causing vasoconstriction at lower doses. Caution should be exercised when using this drug in cats, however, as it may precipitate seizure activity. The intravenous dose range is 1-5 μg/kg/min in cats and 5-20 μg/kg/min in dogs.[2,3] Vasopressors should be infused through a dedicated intravenous catheter as continuous infusions at low fluid rates using syringe pumps or fluid pumps; strict attention should be paid to drug compatibilities.

Antimicrobials

Early antimicrobial therapy is crucial in shock patients with suspected or proven sepsis. Time from triage to administration of appropriate antimicrobial therapy and duration of hypotension before initiation of antimicrobial therapy are critical determinants of survival in human patients with severe sepsis and septic shock.[38-41] If possible, cultures of indicated body fluids should be collected before antimicrobial therapy is begun. However, antimicrobial therapy should never be delayed for diagnostic testing. Broad-spectrum antibiotic therapy should be initiated pending culture and sensitivity results. Empiric choices should be effective against gram-positive and gram-negative organisms and anaerobes.[2,3]

Gastrointestinal Protectants

Stress-related mucosal disease (SRMD) is a common complication in critically ill humans. The frequency of SRMD in veterinary medicine has not been described. However, identified risk factors for SRMD in critically ill people including mechanical ventilation, coagulopathy, shock, trauma, and multiple organ failure frequently occur in hospitalized critically ill veterinary patients. Splanchnic ischemia leading to hypoperfusion of the gastrointestinal mucosa during shock, excessive gastric acid secretion, and impaired mucosal defense mechanisms contribute to the development of SRMD. Although no standardized guidelines exist for the management of SRMD in veterinary patients, treatment in critically ill patients, such as those suffering from circulatory shock, is warranted. To reduce the consequences of SRMD, shock reversal via optimization of tissue perfusion and source control for septic patients are fundamental initial therapeutic steps. High-risk patients should also receive pharmacological prophylaxis for SRMD. Current evidence in human medicine indicates that proton pump inhibitors (PPIs) are superior to histamine-2 receptor agonists

(H_2RAs), which are both superior to sucralfate in the prevention of SRMD.[2,42] Evidence in veterinary medicine supports the use of PPIs over H_2RAs.[2,42-44] In one study[43] comparison of PPIs to H2RAs revealed that only omeprazole improved gastric acid suppression compared to H_2RA therapy. Another study[44] performed in racing sled dogs compared the efficacy of omeprazole to high-dose famotidine. There was a significant decrease in the prevalence of clinically significant gastric lesions in those dogs treated with standard dose omeprazole compared to high-dose famotidine. Available PPIs include oral omeprazole 0.7-1 mg/kg given orally and pantoprazole 1.0 mg/kg given intravenously.[2]

Nutrition

Following initial hemodynamic stabilization of the shock patient, nutritional status should be addressed. Adequate nutrition is essential for recovery in critically ill patients. A recent veterinary study[45] showed improved survival to hospital discharge with appropriate nutritional support. Enteral nutrition is preferable as long as the patient is normotensive, not vomiting, and mentally appropriate. Adequate nutrition is easily provided via nasogastric tube in the patient unwilling to eat.[2,3] Enteral nutrition has the benefit of providing direct nutrition to the mucosal cells of the gastrointestinal tract and may modulate the systemic inflammatory response. Patients receiving enteral nutrition have also been found to have a higher survival rate than those receiving parenteral nutrition.[45] However, if the enteral route is not feasible or if it is contraindicated then parenteral nutrition should be considered. This may be provided through peripheral/partial parenteral nutrition or via total parenteral nutrition. Hyperglycemia should be avoided as it has been associated with an increased incidence of infection and a worse prognosis.[2,3]

Critical Illness-Related Corticosteroid Insufficiency

The syndrome of critical illness-related corticosteroid insufficiency (CIRCI), previously referred to as relative adrenal insufficiency, is characterized by an inadequate production of cortisol in relation to increased demand during periods of severe physiologic stress. CIRCI is seen chiefly in critical illnesses such as sepsis or septic shock. These patients have increased morbidity and mortality in clinical studies.[46-49] CIRCI may be due to a decrease in glucocorticoid synthesis or peripheral resistance to glucocorticoids. It should be considered as a differential diagnosis in all critically ill patients requiring vasopressor support, especially those with refractory hypotension. Patients suffering from CIRCI have been traditionally recognized by having a blunted response to ACTH stimulation testing.[2,47,48] However, there is no current consensus in human or veterinary medicine regarding diagnostic testing and the identification of CIRCI in patients. In addition, the optimal dosage, type of corticosteroid, and duration of treatment with corticosteroids in patients with CIRCI has not been determined. Nonetheless, it is reasonable to start volume-resuscitated vasopressor-dependent patients on corticosteroid treatment. Both physiologic and supraphysiologic doses may be employed; 0.25-1.0 mg/kg/day IV of prednisone equivalent.[2,48]

REFERENCES

1. Astiz ME. Pathophysiology and classification of shock states. In: Fink MP, Abraham E, et al., eds. Textbook of Critical Care. 5th edi. Philadelphia, PA: Saunders Elsevier; 2005:897. 2. Santoro-Beer K, Silverstein DC. Shock. In: Bonagura JD, Twedt, DC. Kirk's Current Veterinary Therapy XV. 15th edi. Philadelphia, PA: Saunders Elsevier; 2013:18. 3. deLaforcade AM, Silverstein DC. Shock. In Silverstein DC, Hopper K, eds. Small Animal Critical Care Medicine. St. Louis, MO: Saunders Elsevier; 2009:41. 4. Mellema M. Vet Clin North Am Small Anim Prac 2001; 31(6):1175. 5. deLaforcade AM, et al. J Vet Intern Med 2003; 17(5):674. 6. Greenfield CL, et al. J Am Vet Med Assoc 1987; 191(1):100. 7. King LG.. J Am Vet Med Assoc 1994; 204(3):407. 8. Hinshaw LB, et al. The Fundamental Mechanisms of Shock. New York: Plenum Press; 1972. 9. Thomovsky E, et al. Compendium 2013; E1. 10. Crowell JW, et al. Am J Physiol 1964; 206:313. 11. Samsel R, et al. J Appl Phsiol 1988; 65:1377. 12. Weil MH, et al. Circulation 1970; 41:989. 13. Prittie JJ Vet Emerg Crit Care 2006; 16(4):329. 14. Hall JE. Circulatory shock. In: Guyton and Hall, eds. Textbook of Medical Physiology. 12th ed. Philadelphia, PA: Saunders Elsevier; 2011:273. 15. Bonanno FG. J Emerg Trauma Shock 2011; 4(2):222. 16. Woodcock, et al. Br J Anaesth 2012;108:384-94. 17. Brady CA, et al. J Am Vet Med Assoc 2000; 217:531.
18. Costello MF, et al. J Am Vet Med Assoc 2004; 225:897. 19. Shoemaker W, et al. Crit Care Med 1979; 7:424. 20. Cady L, et al. Crit Care Med 1973; 1:75. 21. Peterson KL, et al. J Vet Emerg Crit Care 2013; 23(5):545. 22. Porter AE, et al. J Vet Emerg Crit Care 2013; 23(5):538. 23. Boag A, et al. Vet Clin North Am Small Anim Practice 2005; 35(2);319. 24. Nguyen HB et al. Crit Care Med 2004; 32(8):1637. 25. de Papp E, et al. J Am Vet Med Assoc 1999; 215(1):49. 26. Nel M, et al. J Vet Intern Med 2004; 18(4):471. 27. Lagutchik MS, et al. J Vet Emerg Crit Care 1998; 8(2):117. 28. Zacher LA, et al. J Am Vet Med Assoc 2010; 236(8):892. 29. Green TI, et al. J Vet Emerg Crit Care 2011; 21(1):36. 30. Conti-Patara A, et al. J Vet Emerg Crit Care 2012; 22(4):409.
31. Hayes GM, et al. J Small Anim Prac 2011; 52(8):433. 32. Silverstein DC, et al. J Vet Emerg Crit Care 2005; 15(3):185. 33. Hammond TN, et al. Compendium 2009; 309. 34. Sande A, et al. J Vet Emerg Crit Care 2010; 20(2):177. 35. Muir W. Abbot Animal Health 2012. 36. Dellinger RP, et al. Crit Care Med 2013; 41(2):580. 37. Silverstein DC, et al. J Vet Emerg Crit Care 2007; 17(4):399. 38. Gaieski DF, et al. Crit Care Med 2010; 38:1045. 39. Kumar A, et al. Crit Care Med 2006; 34:1589. 40. Bochud P, et al. Crit Care Med 2004; 32[Suppl.]:S495. 41. Puskarich MA, et al. Crit Care Med 2011; 39:2066. 42. Monnig AA, et al. J Vet Emerg Crit Care 2011; 21(5): 484. 43. Bersenas AM, et al. Am J Vet Res 2005; 66:425. 44. Williamson KK, et al. J Vet Intern Med 2010; 24:285. 45. Brunetto MA, et al. J Vet Emerg Crit Care 2010; 20(2): 224. 46. Cooper MS, et al. N Engl J Med 2003; 348. 47. Burkitt HM, et al. J Vet Intern Med 2007; 21(2):226. 48. Martin LG. Vet Clin Small Anim 2011; 41:767. 49. Annane D, et al. Cochrane Database Syst Rev 2004. 1 CD002243.

ENDPOINTS OF RESUSCITATION AND EARLY GOAL DIRECTED THERAPY IN DOGS AND CATS

Allison Sande, DVM, DACVIM (SAIM), DACVECC
Tinton Falls, NJ

INTRODUCTION

Proper evaluation and treatment of critically ill patients can be a confusing and intimidating task. Tissue hypoxia is a causative factor in the development of multiple organ failure and contributes to morbidity and mortality.[1] As such, optimization of early resuscitative efforts is vital to successful treatment of hypoperfused patients. Indicators of illness severity, success of resuscitative efforts, and outcome are invaluable. Goal-directed therapy involves intensive monitoring and aggressive management of hemodynamics in high-risk patients. The aim of goal-directed therapy is to ensure adequate tissue oxygenation and patient survival.[2,3]

TRADITIONAL ENDPOINTS OF RESUSCITATION

Conventional vital signs such as heart rate (HR), blood pressure (BP), and urine output are commonly employed to assess hemodynamic stability and recognize shock in critically ill patients. [2,4] However, several studies have revealed that relying on the normalization of these measurements as endpoints of resuscitative efforts can conceal deficiencies in systemic oxygenation, leading to the accumulation of O_2 debt and increased morbidity and mortality.[2,5-7] Up to 85% of human patients have evidence of tissue hypoxia despite normal vital signs following significant trauma, suggesting the presence of oxygen debt and compensated shock.[5,8] Tachycardia, hypotension, and oliguria do not occur until early, decompensatory shock due to compensatory neurohumoral mechanisms in the earlier stages of injury. [2,5] Macrovascular monitoring parameters are related to systemic measures of cardiopulmonary status and include BP, central venous pressure (CVP), and urine output. As these parameters are measured before blood flows through the tissue beds, they are also referred to as upstream parameters. These parameters are the most commonly utilized by veterinarians and are considered standard of care in human medicine.[2-4,9]

Blood Pressure

Arterial BP is the pressure exerted on the vascular walls and it is derived from the ejection of blood from the left ventricle. The mean arterial pressure (MAP) is not the average of systolic and diastolic pressures, but instead represents the pressure in proportion to time spent in each phase of the cardiac cycle.[3] Mathematically, it is defined as diastolic pressure plus one-third of the pulse pressure (systolic pressure - diastolic pressure). MAP is superior to the systolic pressure for arterial pressure monitoring; it is the true driving pressure for peripheral blood flow.[10] The primary site of BP control is the terminal and small arterioles.[3,11] Constriction of these arterioles leads to an increase in systemic BP at the expense of reduced flow to the capillary beds. This is most pronounced in the hepatosplanchnic and skeletal muscle circulation as local autoregulatory mechanisms maintain flow in the cerebral, coronary, and renal circulations. Because vasoconstriction and vasodilation lead to changes in BP, a normal BP does not always correlate to normal tissue perfusion.[3]

BP can be monitored by invasive or noninvasive means. Direct arterial BP measurement via arterial catheterization is considered the gold standard.[3] However, several limiting factors in veterinary medicine exist including small patient size, lack of appropriate equipment, and client cost. Accordingly, the most common method of BP measurement in small animal medicine is through non-invasive blood pressure (NIBP) monitoring. Various devices exist for NIBP monitoring and all use the same method; a pressure cuff occludes arterial blood flow, then return of flow is measured. Numerous studies have been performed comparing the Dinamap, Cardell, Doppler, and high-definition oscillometric systems in dogs and cats. However, to date, there has not been one best NIBP monitor identified for use in critically ill veterinary patients as all tend to underestimate BP in hypotensive patients.[3,12-15]

BP is commonly used to gauge the adequacy of resuscitation and to define criteria for shock severity. Hypotension is a common complication in critically ill patients and may be related to hypovolemia, inadequate cardiac output (CO), or inappropriate vasodilation.[2,3] Hypotension is defined as a MAP of less than 60-65 mmHg. This is the range when renal and cerebral autoregulation is lost and blood flow to these organs becomes reliant on systemic pressure. However, human studies have revealed a poor correlation between MAP and CO in high-risk human trauma and post-operative ICU patients. The development of hypotension was delayed for up to 12 hours after a decrease in blood flow, suggesting that compensatory mechanisms transiently maintained BP despite decreasing flow.[2,16]

Central Venous Pressure

CVP is used to assess cardiac preload and volume status in critically ill patients and guide fluid resuscitation.[3,17,18] The CVP is equivalent to the right atrial pressure and thus right ventricular end-diastolic pressure, which indicates the amount of preload available to the heart. It is a measure of the hydrostatic pressure within the intrathoracic vena cava. The CVP can be measured using an electronic pressure transducer or a water column manometer. To convert a measurement in mmHg to cmH_2O, multiply the value by 1.36. Water manometers may overestimate the mean CVP. So, in critically ill patients, continuous evaluation of CVP using an electronic pressure transducer connected to the central venous catheter may be more accurate in assessing rapidly changing trends and response to therapy.[3]

While CVP is most commonly used as an estimate of volume status, multiple factors affect its measurement. Any anatomic, physiologic, or pathophysiologic factor affecting the pressure between the left ventricle and the right atrium can alter the relationship between the measured CVP and the left ventricular end-diastolic volume. False elevations in CVP occur from diastolic dysfunction, mitral and tricuspid regurgitation, decreased ventricular compliance, pulmonary hypertension, elevated intra-abdominal pressure, elevated intrathoracic pressure during positive pressure ventilation, pleural effusion, and pericardial effusion. Trends in CVP values are of more clinical significance than single measurements.[3,17,18] CVP values in a well-perfused patient are reported to be between 5 and 10 cm H_2O in dogs and 2 to 5 cm H_2O in cats.[19] If a previously normal CVP trends low, then the patient may be hypovolemic. Whereas a newly elevated CVP may indicate impending volume overload.

Urine Output

Acute kidney injury (AKI) is a common complication in critically ill patients. In human medicine the RIFLE criteria (Risk of renal dysfunction, Injury to the kidney, Failure of kidney function, Loss of kidney function, and End-stage kidney disease) were developed to define and classify renal injury. This system utilizes changes in creatinine, glomerular filtration rate, and urine output.[3,20] However, no such classification system exists in veterinary medicine at this time. Several scoring systems have recently been proposed to predict outcome for dogs and cats requiring hemodialysis for AKI, but additional studies are warranted for validation.[21,22]

Azotemia may be due to prerenal, renal, or postrenal causes. Oliguria is defined as urine output less than 0.5-1.0 ml/kg/hr in the normally hydrated patient, while anuria is defined as urine output less than 0.3-0.5 ml/kg/hr. Measuring urine output is most accurately performed by placement of an indwelling urinary catheter. In situations when catheter placement is not possible or practical, though, cage pads can be weighed and urine output calculated from this measurement. Urine output may be decreased in situations of hypovolemia, hypotension, and decreased CO.[3] However, it is not a sensitive measure of tissue hypoperfusion or hypoxia as compensatory mechanisms and renal autoregulation in early shock maintain renal blood flow.[2]

MICROVASCULAR MONITORING

Due to the unreliability of traditional vital signs in documentation of early shock states and assessment of the success of resuscitation efforts, recent research has focused on the identification of more sensitive markers of ongoing O_2 debt.[2] While the upstream, or macrovascular parameters are most commonly used, the microcirculation is likely the best determinant of tissue and organ survival in critical illness. Microvascular parameters are related to tissue oxygenation and include lactate and lactate clearance, central venous oxygen saturation ($ScvO_2$), and base deficit. These are also referred to as downstream parameters since they are measured after blood flows through the tissue beds. Additional more experimental measures of microvascular flow include tissue specific tonometry, direct measurements of tissue oxygen pressure, tissue oximetry, laser Doppler, and sidestream dark-field microscopy.[2,3,23]

Lactate and Lactate Clearance

Lactate production occurs in the cytosol of all cells by anaerobic glycolysis during periods of O_2 deficiency. The majority of lactate production is by skeletal muscle and the intestinal tract, but the brain, skin, erythrocytes, leukocytes, and thrombocytes may contribute to production. Blood lactate levels increase when lactate production by hypoxic tissue overwhelms its elimination by the liver and kidneys. Therefore, tissue hypoperfusion and resultant hypoxia is a significant cause of hyperlactatemia.[2,3,24] Normal lactate levels in dogs are less than 2.5 mmol/L.[3,25] Levels in cats are similar depending on the extent of struggling during sample collection.[3,26,27]

Arterial blood lactate levels have been shown to approximate the severity of shock.[2] Critically ill human patients with hyperlactatemia are at increased risk of developing multiple organ failure and have an increased mortality rate.[28] Elevated lactate levels in dogs also help predict mortality.[29-32] In addition to the prognostic value of admission lactate levels, the rate or effectiveness of lactate clearance has been shown to have predictive value. Lactate clearance is a more important prognostic marker for survival than a single initial measurement. This may be because the rate of lactate reduction reflects the rapidity of correction of tissue dysoxia.[28,33,34]

There are limitations to the reliance on blood lactate concentration as a downstream marker of tissue perfusion and oxygenation. Although the most common cause of hyperlactatemia is type A lactic acidosis resulting from systemic tissue hypoxia, the presence of lactic acidosis is not specific for global tissue hypoxia. Type B lactic acidosis occurs without tissue hypoperfusion and hypoxia. Type B hyperlactatemia occurs from underlying diseases, drugs, and toxins and is due to decreased clearance, dysfunctional carbohydrate metabolism, or mitochondrial dysfunction.[2,24] Sepsis is associated with both type A and B hyperlactatemia. Tissue hypoperfusion causing type A hyperlactatemia results from hypovolemia, cardiovascular derangements and microvascular thrombosis, while type B hyperlactatemia develops as a result of altered cellular metabolism and mitochondrial dysfunction.[2,3] Furthermore, restoration of blood flow to regional tissues may result in release of sequestered lactate where it had accumulated during the period of hypoperfusion. This is known as 'wash-out' phenomena and results in delayed lactate clearance.[2,35] Regardless of these limitations, lactate clearance has been shown to be a strong predictor of mortality.[2,3,33,34]

Base Deficit

Base deficit (BD) is defined as the amount of base in millimoles required to titrate 1 L of arterial blood to a pH of 7.40 at 37°C and a PCO_2 of 40 mmHg. BD serves as another method to determine the degree of O_2 debt and anaerobic metabolism present during shock states.[2,3] Recent studies in human patients have found that BD is a sensitive indicator of compensated shock, and the magnitude of BD during the early stages of illness correlates with development of organ failure and mortality.[36-38] Veterinary studies have shown that BD is predictive of systemic inflammatory response syndrome (SIRS) in dogs with pyometra[39] and mortality in dogs with severe sepsis and septic shock.[40]

Mixed Venous Oxygen Saturation and Central Venous Oxygen Saturation

Mixed venous oxygen saturation (SvO_2) and central venous oxygen saturation ($ScvO_2$) serve as markers of tissue oxygenation. The SvO_2 is measured in a blood sample derived from a pulmonary arterial catheter (PAC) as this represents a true mixed venous oxygen sample. However, PAC placement is not practical in most veterinary clinics. Instead, oxygen saturation can be determined within the central circulation using a central venous catheter within the cranial vena cava; this is the $ScvO_2$.[2,3] The SvO_2 and $ScvO_2$ are considered equivalent in human medicine.[41] The $ScvO_2$ reflects the balance of oxygen delivery (DO_2) minus oxygen consumption (VO_2). In health, DO_2 far exceeds tissue requirements. However, as DO_2 decreases through anemia, hypoxemia, or hypoperfusion, the O_2 extraction ratio (percent of oxygen consumed) increases to maintain tissue oxygenation.[3,19,42] In normal patients, the $ScvO_2$ is typically greater than 65%.[7] During times of tissue hypoxia, systemic O_2 extraction from blood increases, leading to a concurrent decline in $ScvO_2$. Venous desaturation is one of the chief compensatory mechanisms to maintain peripheral VO_2 in low flow states. Still, the presence of a low $ScvO_2$ indicates tissue dysoxia and oxygen debt.[2,3]

Hypotension and hypovolemia causing hypoperfusion are the most common causes of a low $ScvO_2$. A $ScvO_2$ that remains low despite fluid resuscitation and normalization of the macrovascular parameters (HR, BP, and CVP) indicates ongoing tissue oxygen debt. Additional causes include decreased oxygen content seen in anemia and hypoxia, decreased CO, and inappropriate vasodilation.[3] $ScvO_2$ has been shown to be a strong predictor of mortality in human patients.[43,44] Likewise, a recent veterinary study[40] demonstrated a significant difference in $ScvO_2$ between survivors and nonsurvivors; dogs with a lower $ScvO_2$ at ICU admission had an increased mortality. Limitations to the use of $ScvO_2$ for evaluation of critically ill patients exist. The $ScvO_2$ is influenced by systemic arterial oxygen tension and hemoglobin concentration. In addition, defects in oxygen extraction exist in certain clinical situations such as sepsis. This can lead to a normal or increased $ScvO_2$ despite the presence of significant O_2 debt.[2,3,16]

EARLY GOAL-DIRECTED THERAPY

The objective of early goal-directed therapy (EGDT) is to match DO_2 with tissue oxygen demand to prevent tissue hypoxia. This is accomplished by utilizing a series of predefined resuscitation end points to help clinicians effectively resuscitate shock patients as soon as the syndrome is recognized.[3,45] EGDT in human patients with severe sepsis and septic shock was evaluated in a landmark study by Rivers, et al. in 2001.[18] The investigators demonstrated a 16% reduction in in-hospital mortality for the patients receiving EGDT versus those receiving standard therapy. Due to the significance of these findings, EGDT has been incorporated in the Surviving Sepsis Campaign as a strong recommendation.[41]

The initial six-hour time period following diagnosis of severe sepsis is known as the "golden hours" as recognition and resuscitation during these hours is associated with reduced organ dysfunction and improved survival. In the Rivers, et al. study,[18] one group of patients were treated with six hours of EGDT for sepsis-induced hypotension. Patients were treated, in sequential fashion, with crystalloid fluid resuscitation to achieve a CVP of 8-12 mmHg. If MAP was less than 65 mmHg, vasopressors were given to maintain a MAP of at least 65 mmHg. If MAP was greater than 90 mmHg, vasodilators were given until it was 90 mmHg or below. If $ScvO_2$ was less than 70%, red blood cell transfusions were administered to achieve a hematocrit of at least 30%. After the CVP, MAP, and hematocrit was thus optimized, if the $ScvO_2$ was still less than 70%, dobutamine administration was started and the dose titrated until $ScvO_2$ was 70% or higher or until a maximum dose of 20 µg/kg/min was given. Dobutamine was decreased or discontinued if MAP was less than 65 mmHg or if the HR was greater than 120 beats/minute. Patients in whom hemodynamic optimization could not be achieved received mechanical ventilation and sedatives in order to decrease O_2 consumption. The other patient group was treated with standard therapy at their clinician's discretion.

Additional goals should be incorporated into EGDT. During volume resuscitation, urine output should increase to greater than 0.5 ml/kg/hr as CVP goals are reached. If the urine output remains inadequate, then the development of AKI should be considered.[3] Patients with sepsis should be evaluated for an infection focus. Antibiotic therapy should be started as soon as sepsis is recognized, and the administration of broad-spectrum antimicrobial therapy should be initiated within 1 hour of the recognition of septic shock and severe sepsis.[3,41] In humans with septic shock, mortality rates have been found to increase by 7.6% for every hour that antibiotic therapy is delayed.[46,47] Initiation of inappropriate antimicrobial therapy has also been associated with a 5-fold increase in mortality in human septic shock patients.[48] As such, the source and location of infection should be considered when initiating antimicrobial therapy. Broad-spectrum, combination antimicrobial therapy should be administered intravascularly. Antibiotics should be bacteriocidal, not bacteriostatic.[3,19] Reassessment of antimicrobial therapy with de-escalation based on culture and sensitivity results should also occur when possible with the aim of using the safest, and most narrow-spectrum antibiotic.[3,41]

A recent veterinary study by Conti-Patara, et al.[40] evaluated the response of dogs with severe sepsis and septic shock to EGDT at ICU admission. Resuscitation was guided by clinical parameters, $ScvO_2$, lactate, and BD. Patients were treated sequentially with crystalloid therapy, colloid therapy, and finally vasopressor therapy with dopamine to target a Doppler blood pressure greater than 90 mmHg, CVP between 8-12 mmHg, urine output above 2 ml/kg/hr, $ScvO_2$ above 70%, and to reduce lactate and BD from baseline concentrations. Red blood cell transfusions were administered as needed when the hemoglobin decreased below 7.0 g/dl. Results revealed that survivors had higher $ScvO_2$ values and lower lactate levels than nonsurvivors, and the nonsurvivors had higher BD than survivors. The $ScvO_2$ and BD were found to be the best discriminators between survivors and nonsurvivors.

SUMMARY

Optimization of early resuscitative efforts has proven to have a survival benefit in human shock patients, and significant advances have been made in determining which resuscitation endpoints to target in this patient population. The use of both macrovascular and microvascular parameters and EGDT allows the most complete patient assessment and improves patient survival. Clinical trials evaluating indices of global and regional tissue hypoxia as well as resuscitation endpoints in veterinary patients are scarce, and further studies are warranted.

REFERENCES

1. Shoemaker WC. Diagnosis and treatment of shock and circulatory failure, In: Shoemaker WC, Ayres SM, Grenjvik A, Holbrook PR. eds. Textbook of Critical Care, 4th ed. Philadelphia, PA: WB Saunders Co; 2000:92. 2. Prittie J. *J Vet Emerg Crit Care* 2006; 16:329. 3. Butler AL. *Vet Clin Small Anim* 2011; 41:817. 4. American College of Surgeons Committee on Trauma. Advanced Trauma Life Support Course of Doctors. Chicago, IL: American College of Surgeons; 1997. 5. Abou-Khalil B, et al. *Crit Care Med* 1994; 22:633. 6. Siegel JH, et al. *J Trauma* 2003; 54:862. 7. Rady MY, et al. *Am J Emerg Med* 1996; 14:218. 8. Marino PL. Tissue Oxygenation, In: Marino PL, ed. The ICU Book, 3rd ed. Philadelphia, PA: Lippincott Williams & Wilkins; 2007, pp193. 9. Trzeciak S, et al. *Crit Care Med* 2005; 9(S4):S20. 10. Marino PL. Arterial Blood Pressure, In: Marino PL, ed. The ICU Book, 3rd ed. Philadelphia, PA: Lippincott Williams & Wilkins; 2007, pp151. 11. Segal S. *Microcirculation* 2005; 12:33. 12. Brown S, et al. *J Vet Intern Med* 2007; 21:542. 13. Bosiak AP, et al. *J Vet Emerg Crit Care* 2010; 20:207. 14. Shih A, et al. *J Vet Emerg Crit Care* 2010; 20:313. 15. Acierno M, et al. *J Am Vet Med Assoc* 2010; 327:402. 16. Wo CCJ, et al. *Crit Care Med* 1993; 21:218. 17. Reems MM, et al. *Compendium* 2012:E1. 18. Rivers E, et al. *N Engl J Med* 2001; 345:1368. 19. deLaforcade AM, Silverstein DC. Shock. In Silverstein DC, Hopper K, eds. Small Animal Critical Care Medicine. St. Louis, MO: Saunders Elsevier; 2009:41. 20. Bellomo R, et al. *Crit Care* 2004; 8:R204. 21. Segev G, et al. *J Vet Intern Med* 2013; 27:830. 22. Lee YJ, et al. *J Vet Intern Med* 2012; 26:500. 23. DeBacker D, et al. *Intensive Care Med* 2010; 36:1813. 24. Sharkey LC, et al. *Vet Clin Small Anim* 2013; 43:1287. 25. Hughes D, et al. *Am J Vet Res* 1999; 60:521. 26. Rand JS, et al. *J Vet Intern Med* 2002:16:123. 27. Redavid LA, et al. *J Vet Emerg Crit Care* 2012; 22:580. 28. Nguyen HB, et al. *Crit Care Med* 2004; 32:1637. 29.Boag A, et al. *Vet Clin North Am Small Anim Practice* 2005; 35;319. 30. de Papp E, et al. *J Am Vet Med Assoc* 1999; 215:49. 31. Nel M, et al. *J Vet Intern Med* 2004; 18:471. 32. Lagutchik MS, et al. *J Vet Emerg Crit Care* 1998; 8:117. 33. Arnold RS, et al. *Shock* 2009; 32:35. 34. Stevenson CK, et al. *Vet Clin Pathol* 2007; 36:234. 35. Lagutchik MS, et al. *J Vet Emerg Crit Care* 1996; 6:81. 36. Davis JW, et al. *Surgery* 1991;173:473. 37. Davis JW, et al. *J Trauma* 1996; 41:769. 38. Rutherford EJ, et al. *J Trauma* 1992; 33:417. 39. Hagman R, et al. *Acta Vet Scand* 2009; 51:2. 40. Conti-Patara A, et al. *J Vet Emerg Crit Care* 2012; 22:409. 41. Dellinger RP, et al. *Crit Care Med* 2013; 41:580. 42. Astiz ME. Pathophysiology and classification of shock states. In: Fink MP, Abraham E, et al., eds. Textbook of Critical Care. 5th edi. Philadelphia, PA: Saunders Elsevier; 2005:897. 43. Pope J, et al. *Ann Emerg Med* 2010; 55:40. 44. Lima A, et al. *Crit Care* 2009; 13(S5):S13. 45. Santoro-Beer K, Silverstein DC. Shock. In: Bonagura JD, Twedt, DC. Kirk's Current Veterinary Therapy XV. 15th edi. Philadelphia, PA: Saunders Elsevier; 2013:18. 46. Kumar A, et al: *Crit Care Med* 2006; 34:158. 47. Gaieski DF, et al. *Crit Care Med* 2010; 38:1045. 48. Kumar A, et al. *Chest* 2009; 136:1237.

DIRECT BLOOD PRESSURE MEASUREMENT: A TARNISHED GOLD STANDARD

Anthony P. Carr, Dr. med. vet., DACVIM
Saskatoon, Saskatchewan

Blood pressure measurement is vital in assessing the health of a patient, especially in emergency situations or under anesthesia. In most cases this is accomplished with indirect methods such as oscillometry or the Doppler method. In very critical situations or when indirect methods are being tested for accuracy direct blood pressure measurement will often be used. Direct blood pressure measurement is considered the gold standard of blood pressure measurement. Direct blood pressure measurement can be carried out with the pressure transducer on the tip of the catheter (so called direct tip or Millar catheters) or the pressure transducer can be located outside of the patient and connected to the artery via a fluid filled catheter. The heart and the arteries are continuous leading to the assumption that direct blood pressure measurement in any of these arteries is accurate and varies little between site selected.[1] This is however incorrect and there can be significant deviations from central blood pressure.[1, 2, 3, 4, 5]

Water filled catheters and a external pressure transducer are most commonly used to measure direct blood pressure in veterinary medicine since they are inexpensive and easy to use. There are however technical limitations to this method. It is important the tubing be short and non-compliant and that all air bubbles are purged from the system. There are also inherent issues with the ability of these systems to record blood pressure accurately based on the fluid dynamics of the entire measurement system which becomes more important the higher the heart rate of the patient.[6] The ability of the system to accurately determine blood pressure can be partially assessed by determining the natural frequency and the damping coefficient of the entire system. Since the "entire" system includes the arterial tree, this calculation can only be done with the catheter in the artery of the patient being measured. A fast flush test is carried out and calculations are made on the oscillations after the initial flush. The natural frequency is how rapidly the system oscillates[2] Damping coefficient refers to how quickly the oscillations return to baseline.[2,3] In an overdamped system oscillations are small which results in underestimation of systolic and overestimation of diastolic blood pressure, in systems that are underdamped the opposite is true. These oscillations are around mean blood pressure, meaning that mean pressures will still be accurate, however systolic and diastolic will not be. The lower the natural frequency the less likely the system can accurately record blood pressures with heart rates that are common in out patients, especially cats.

Damping is a feature of resonant systems. In a resonant system, the variation of the systolic and diastolic readings yields a mean arterial pressure (MAP) that is consistent with the central blood pressure readings.[1] A central artery, such as the aorta or the carotid will have a more muscular wall that is less compliant and more elastic. The brachial artery will be slightly less muscular and more compliant, allowing further expansion of the artery during systole.[5] In some instances the arterial pressure pulse can be 50% greater in the femoral artery than in the aorta.[3] In patients undergoing coronary bypass surgery there was a significant difference and inaccuracy when comparing the radial artery to the aorta. There was an overshoot up to 35 mmHg and an average of 10 mmHg overshoot in the radial artery compared to the aorta. However, the DAP and MAP were consistent (1-3 mmHg difference).

In a resonant system, even if systolic or diastolic pressures are inaccurate, the mean should be comparable. However, the vascular tree cannot however be described as classically resonant because it has pulsatile flow, friction within the system, and dissipation of energy due to this friction.[1] The pressure measured at a single point or location in the vascular tree is dependent on systemic vascular resistance (SVR) and the degree of impedance in the cannulated artery.[1,4,5] The greater the impedance in the artery the greater the decrease in the measurement of the blood pressure at that point.[1] This artificially decreases systolic, diastolic and mean blood pressure and is referred to as attenuation of the signal. Attenuation is seen with thrombi, vasospasm, tortuous artery, or sharp turns proximal to the catheter as well as with small diameter catheters.[1] In many studies the dorsal pedal artery has been used. Based on the anatomy of this vessel, significant attenuation should be expected. Unfortunately it is not possible to detect or quantify attenuation.

The accuracy of various direct blood pressure measurement methods has rarely been compared in veterinary medicine. In one study there was comparison between the lingual arteries and dorsal pedal in 16 beagles using a catheter/external transducer system. Systolic blood pressure was higher using the dorsal pedal artery than the lingual artery, during hypertension this discrepancy was over 14 mmHg, whereas DAP and MAP were very closely correlated over all pressure ranges.[7]

REFERENCES

1. Ercole A. Attenuation in invasive blood pressure measurement systems. BJA 2006; 96: 560-562.
2. Todorovic M, Jensen EW, Thogersen C. Evaluation of dynamic performance in liquid-filled catheter systems for measuring invasive blood pressure. Inter J of Clin Mon and Comp. 1996; 13: 173-178.
3. Rothe CF, Kim KC. Measuring Systolic Arterial Pressure: Possible Error from extension tubes or disposable transducer domes. Crit Care Med. 1980. Vol. 8. 11.
4. Promonet C, Anglade D, Menaour A, Bayat S, Durand M, Eberhard A, Grimbert F. Time-dependent Pressure Distortion in a Catheter Transducer System. Anesthesiology; 2000; 92: 208-218.
5. Pauca AL, Wallenhaupt SL, Kon ND, Tucker WY. Does Radial Artery Pressure Accurately Reflect Aortic Pressure? Chest. 1992, 192: 1193-1198.
6. Gardner RM. Direct blood pressure measurement - dynamic response requirements. Anesthesiology. 1981:54; 227-236.
7. McMurphy RM, Stoll MR, McCubrey R. Accuracy of an oscillometric blood pressure monitor during phenylephrine-induced hypertension in dogs. Am J Vet Res. 2006; 67:1541-5.

SUPRAVENTRICULAR TACHYARRYTHMIAS IN THE DOG

Stacey B. Leach, DVM, DACVIM (Cardiology)
Columbia, MO, USA

INTRODUCTION

Supraventricular tachyarrhythmias (SVTs) are rapid heart rhythms that originate and/or utilize the atrial or atrioventricular (AV) junctional structures tissues as part of their electrical circuit. Understanding the normal cardiac electrical conduction system is vital for accurate diagnosis and management of arrhythmias. Electrical impulse formation normally starts within the sinoatrial (SA) node, located near the junction of the cranial vena cava and right atrium. From here, the electrical impulse depolarizes and passes through the atrial myocardium, indicated by the P wave on the surface ECG. The electrical impulse eventually travels to the AV node, where the conduction velocity markedly slows down to allow time for the atria to contract and eject blood into the ventricles. Once the impulse has slowly traversed the AV node, it then rapidly spreads through the His bundle, bundle branches, and Purkinje cells to electrically depolarize the ventricular myocardium, resulting in generation of the QRS complex on the surface ECG. Following ventricular depolarization, the ventricles slowly repolarize, indicated by the T wave on the surface ECG.

Cells of the SA node have automaticity, meaning that they spontaneously depolarize due to leakage of ions across their cells membranes at rest. The SA node is the normal pacemaker of the heart with an intrinsic firing rate of 60-180 beats/min in the dog, depending on underlying sympathetic and parasympathetic tone. Other cells that demonstrate automaticity include the AV node and the Purkinje cells; however, their intrinsic firing rates are much slower at approximately 40-60 beats/min and 20-40 beats/min, respectively. The AV node and Purkinje cells are considered subsidiary pacemakers as they normally do not control the heart rate unless the SA node fails to function appropriately. Under normal conditions, the myocytes of the atria and ventricles do not display automaticity. Instead, these cells must be stimulated to depolarize from a propagated wave of electrical activity.

The broad mechanisms for arrhythmogenesis include disorders of impulse formation and/or impulse conduction. Disorders of impulse formation include altered automaticity of the normal pacemaker cells of the heart, development of automaticity in cells that normally should not display automaticity, or triggered activity due to oscillations in the resting membrane potential. Abnormalities of impulse conduction can occur due to altered autonomic tone or due to structural anatomic abnormalities such as accessory bypass tracts, ischemic myocardium, and areas of fibrosis.

SVTs most commonly occur due to structural heart disease[1], but can also occur secondary to non-cardiac disorders including metabolic/electrolyte disturbances, endocrinopathies, autonomic dysfunction, as well as various drugs and toxins. Clinical signs of SVT may be vague and non-specific and include lethargy, generalized weakness, exercise intolerance, and syncope. These signs are usually attributed to a reduction in cardiac output from excessively fast heart rates. Left untreated, persistent or recurrent SVTs can lead to myocardial failure with resultant left and/or right-sided congestive heart failure referred to as tachycardia-induced cardiomyopathy. Tachycardia-induced cardiomyopathy is indistinguishable dilated cardiomyopathy (DCM); however, it is potentially reversible with adequate arrhythmia control.

CLASSIFICATIONS

Broadly, SVTs can be classified as either atrial or junctional in nature. Atrial tachyarrhythmias are initiated and maintained by the atrial myocardium, whereas junctional tachyarrhythmias require the AV node for initiation or maintenance. Specific types of atrial tachyarrhythmias include sinus tachycardia, atrial tachycardia, atrial flutter, and atrial fibrillation. Whereas junctional tachyarrhythmias include AV nodal tachycardia (AVNT) and AV reciprocating tachycardia (AVRT).

DIAGNOSIS

The signalment, history, and physical examination findings may provide helpful clues as the presence, cause, and effects of the arrhythmia. Hemodynamically important SVTs can lead to altered mentation status, mucous membrane color, capillary refill time, and pulse quality. Blood pressure measurement can be helpful to further evaluation the hemodynamic consequence of the arrhythmia. Echocardiography is invaluable in determining if underlying structural heart disease such as DCM or valvular heart disease is present. Electrocardiography (ECG) is essential for the diagnosis of an SVT and may be helpful in determining the type of SVT present[2].

Electrocardiographic Features

Classically, SVTs have narrow QRS complexes. Rarely, SVTs can have a wide QRS complex and resemble ventricular tachycardia if they are conducted with aberrancy (ie. bundle branch blocks or conduction across an accessory pathway). The ECG should be systematically evaluated and includes determination of the heart rate, regularity of the rhythm, presence and morphology of P waves, relationship of the P waves to the preceding QRS complex (ie. short or long RP interval), initiation and/or termination of the SVT, and the response to physiologic maneuvers or pharmacologic agents. Physiologic maneuvers include vagal maneuvers (carotid sinus massage, digital ocular pressure) or precordial thumps that may restore a normal sinus rhythm or temporarily break the SVT. Pharmacologic agents such as calcium channel blockers or beta blockers can also be useful slowing the ventricular response rate or converting the SVT.

THERAPY

Pharmacologic Management

Emergency intervention may be required in patients that are hemodynamically compromised. Intravenous anti-arrhythmic therapy with calcium channel blockers (ie. diltiazem, verapamil) or rapid acting beta blockers (ie. esmolol) can be used to slow conduction across the AV node and thus slow the ventricular response rate (ie. "rate control"). Both calcium channel blockers and beta blockers exhibit negative inotropic properties and may be deleterious if significant myocardial dysfunction is present. Lidocaine has been traditionally reserved for the management of ventricular tachycardia but may be useful in acutely converting some SVTs[3]. Chronic anti-arrhythmic therapy can be attempted with calcium channel blockers, beta blockers, and/or digoxin.

Non-pharmacologic Management

Alternatively, direct current synchronized cardioversion can be used to restore a sinus rhythm (ie. rhythm control). This procedure is painful and requires brief general anesthesia. Maintenance of sinus rhythm following cardioversion may be dependent on the chronicity/duration of the SVT and whether or not structural heart disease is present[4]. Transvenous catheter radiofrequency ablation is available in some veterinary referral hospital and may be particularly useful in curing bypass mediated AVRT and ectopic atrial tachycardias[5].

REFERENCES

[1]Finster ST, et al. J Vet Emerg Crit Care 2008; 18(5):503 [2]Santilli RA, et al. J Vet Intern Med 2008; 22(4):915 [3]Johnson MS, et al. J Vet Intern Med 2006; 20:272 [4]Bright JM, et al. J Vet Intern med 2008; 22(1):114 [5]Wright KN, Vet Clin Small Anim 2004; 34:1171

CLINICAL GENETICS FOR THE TECHNICIAN

Kathryn M. Meurs, DVM, PhD, DACVIM (Cardiology)
Raleigh, NC

INTRODUCTION

There is increasing evidence that many forms of congenital and acquired disease in veterinary medicine are of a familial origin. As of 2005, 430 canine diseases and 180 feline diseases have been identified as familial. It has been estimated that 5-10 new familial diseases are identified every year.[1]The significant number of familial veterinary diseases has been thought to be associated with the improved control of diseases caused by environmental factors as well as the desire to breed animals to maintain an appearance and selecting animals from a small group of popular founders (founder effect). It is reasonable to consider a familial etiology for common feline and canine heart diseases, particularly those with strong breed predispositions.

Many forms of congenital and acquired disease have a strong breed predisposition suggesting a familial etiology. The small animal practioner may be asked to provide consultation on genetic issues in order to provide information to owners of breeding animals to aid them in breeding decisions and to provide pet owners about etiologies of disease. Breed specific lists of known and presumed inherited diseases in the dog and cat can be found at the following web sites:

Cats: http://www.fabcats.org/breeders/inherited_disorders/index.php

Dogs: http://www.vet.cam.ac.uk/idid/ and http://www.upei.ca/~cidd/intro.htm

Even though the number of genetic diseases in small animals is significant, the actual genetic cause of it has not commonly known. Even in the absence of knowledge of a molecular cause a great deal of information can be provided just from an understanding of pattern of inheritance.

Genetic diseases are most commonly observed in pure breed animals. Breed organizations have intentionally limited their gene pool by preventing breeding to other breeds. Therefore, the gene pool is of a limited, closed size and aggressive removal of breeding animals because of the identification of a defect will make the gene pool even smaller. One must CAREFULLY recommend removal of breeding animals based on severity of defect, mode of inheritance and importance of a particular animal to the breed.

Genetic mutations rarely have an all or nothing. For some diseases, a genetic cause has now been identified and a genetic test can be done. The presence of the mutation does NOT mean that all animals will show the trait, or the same severity of the trait, this depends on disease "penetrance" , a poorly understood phenomenon that likely involves genetic modifiers.

MODES OF INHERITANCE

Knowledge of the mode of inheritance of a specific trait can be used to provide guidance about reducing the prevalence of a trait within a particular line of animals. The determination of the pattern of inheritance will be useful for determining the best recommendations for breeders and will allow the development of plans to exclude or include affected animals in breeding programs. The most common include autosomal recessive, autosomal dominant, X-linked recessive and X-linked dominant.

X-linked

X-linked traits are almost always recessive and should have the following criteria: more affected males than females, an affected male crossed with a normal (non carrier) female should produces silent carriers, silent carrier females should have a 50:50 chance of passing the trait on to male offspring and affected females are the result of a cross between a silent carrier female and an affected male (not a frequent occurrence).

Example- Progressive retinal atrophy in the Akita.

Recommendation: If animals with an X-linked trait are used for breeding, they should always be bred to an unrelated line, this will decrease the presence of the trait in the line.

Autosomal recessive

Autosomal recessive traits should have the following criteria: the disease should appear to "skip" a generation (parents do not usually show the trait), males and females should equally show the trait, the mating of two silent carriers (heterozygotes, carry one copy of the abnormal gene) should produce offspring that show the trait in a 3:1 ratio and if both parents are show the trait, all offspring should show the trait.

Example- Diabetes mellitus in Keeshond

Recommendation: If animals with an autosomal recessive trait are used for breeding, they should always be bred to an unrelated line, this will decrease the presence of the trait in the line.

Autosomal dominant

Autosomal dominant traits should have the following criteria: males and females are should equally show the trait, every affected individual should have at least one affected parent, and all heterozygotes (one copy of the abnormal gene) are affected and transmit a mutant gene to half of their offspring. Affected animals may carry the genetic mutation on one or both copies of the gene. If it is on both copies of the gene (one inherited from each parent) they are considered homozygous for the mutation and will pass one copy of the mutation to all of their offspring.

Example- Dilated cardiomyopathy in Doberman pinscher

Recommendation: If animals with an autosomal dominant trait are used for breeding, they will have a 50-100% chance of passing on the trait (depending on if they are heterozygous or homozygous for the trait). Therefore, only those with the most mild form of the disease and the most positive attributes should be selected to be used.

Utilization of molecular information for screening and therapeutic issues

In some cases, a molecular genetic cause for a specific trait has been identified and a genetic test is available. Genetic tests are generally a PCR test that identifies either a marker for the disease or the identifies the actual genetic mutation. PCR is a method that takes a small amount of DNA provided by the clinician or owner and amplifies a region of interest so it can be carefully inspected. DNA can be usually provided in a blood sample in an EDTA tube, a buccal swab or even a semen sample. The DNA will be inspected for the abnormality by the lab and the presence or absence of the marker or mutation identified. However, breeders and owners should be cautioned and advised how to best use the information. The results should be carefully considered and should be weighed against the severity of the trait, the size of the breed's gene pool, the mode of inheritance of the trait and the positive traits that this individual animal brings to a breed. In some cases, strict screening and removal programs may be very detrimental to small gene pools in specific breeds; breeding recommendations should be carefully designed.

REFERENCES

[1]Giger U, Clinical Genetics in Textbook of Veterinary Internal Medicine, 6th edition, 2005,264

GENETIC TESTING FOR CARDIAC DISEASES IN THE SMALL ANIMAL PATIENT

Kathryn M. Meurs, DVM, PhD, DACVIM (Cardiology)
Raleigh, NC

INTRODUCTION

Many forms of cardiac disease have a strong breed predisposition suggesting a familial etiology.

FELINE HYPERTROPHIC CARDIOMYOPATHY

Hypertrophic cardiomyopathy is the most common form of heart disease in the cat. It is an adult onset myocardial disease known to be inherited in the Maine Coon and Ragdoll breeds and thought to be inherited in Norwegian Forest, Siberian, Sphynx and Bengal cats among others. Causative genetic mutations have now been identified in the Maine Coon and Ragdoll.

Hypertrophic cardiomyopathy is inherited as autosomal dominant trait in the Maine Coon.[1] Autosomal dominant traits should have the following criteria: males and females are generally equally affected, every affected individual should have at least one affected parent, and the trait is generally observed in every generation. In the Maine Coon, a genetic mutation has been identified in the myosin binding protein C (MYBPC3) gene.[2] Myosin binding protein C is the second most commonly mutated gene responsible for the human form of the disease. In the Maine Coon cat, the mutation is a single base pair change from a guanine to a cytosine in the 31st codon of the gene. This mutation changes the structure of myosin binding protein C and alters the ability to interaction with other contractile proteins. The Maine Coon mutation appears to be quite breed specific. It is unlikely to be associated with hypertrophic cardiomyopathy in other breeds of cats unless they are closely related to the Maine Coon breed.

A substitution mutation has also been identified in the myosin binding protein C gene in the Ragdoll cat.[3] However, the Ragdoll mutation is different from the Maine Coon mutation and is a cytosine to thymine substitution in the 820th codon. It is extremely unlikely that the Maine Coon and Ragdoll mutations were inherited from a common ancestor since the mutations are different and are located in such different regions of the gene. Although the mode of inheritance of this mutation in the Ragdoll has not been identified through breeding studies, it is most likely an autosomal dominant trait as well. In the Ragdoll, homozygous cats appear to be very severely affected with development of heart failure and thromboembolic episodes often before 2 years of age. Heterozygous cats appear to have a much more mild form of the disease that may include only mild papillary muscle hypertrophy. The Ragdoll mutation also appears to be breed specific to the Ragdoll.

Genetic testing is now available to test a cat for either mutation by submitting a DNA sample to a reputable screening laboratory (http://www.ncstatevets.org/genetics/http://www.ncstatevets.org/genetics/). Good quality DNA samples can be obtained either from a blood sample in an EDTA tube or by brushing the oral gums of the cat with a special buccal swab, although many labs will even accept samples submitted on a cotton swab.

The test results should verify that the cat is negative, heterozygous or homozygous for the mutation. Cats that test negative do not have the mutation. This does not mean that they cannot ever develop hypertrophic cardiomyopathy, it simply means that they will not develop the form of the disease caused by the specific genetic mutation. Although these mutations have been shown to be the cause of hypertrophic cardiomyopathy in many Maine Coons and Ragdoll cats, there are some cats that are echo positive for hypertrophic cardiomyopathy that do not have the specific mutations.

Due to an apparently fairly high prevalence of the mutation in both breeds, it would seem to be unwise to recommend that all cats with the mutation be removed from the breeding programs since this could result in dramatically altering the genetic makeup of these breeds. Additionally, it should be emphasized that not all cats that have the mutation, particularly if they are heterozygous, will develop a clinical form of the disease. Our current recommendations for both breeds are to not use cats that are homozygous for the mutation for breeding purposes since they will certainly pass on the mutation and they have the highest risk of developing the disease. Heterozygous cats should be carefully evaluated. Cats that have many strong positive breed attributes and are disease negative at time of breeding could be bred to a mutation negative cat. Their lack of clinical disease may suggest that they have a less penetrant form of the disease or that they just do not show evidence of this adult onset clinical disease yet. Therefore these cats should only be used if they are exceptional for the breed and they should be clinically evaluated for the disease every year. If they develop the clinical disease, they should no longer be kept in the breeding program. The offspring of the mating of a positive heterozygous and a negative should be screened for the mutation, and if possible, a mutation negative kitten with desirable traits should be selected to replace the mutation positive parent in the breeding colony. Over a few generations this will decrease the prevalence of the disease mutation in the population, hopefully without greatly altering the genetic makeup of the breed too significantly. Finally, disease negative but mutation positive cats should be evaluated annually for presence of disease.

CANINE CARDIOMYOPATHIES

Dilated cardiomyopathy is characterized by cardiac enlargement and impaired systolic function of one or both ventricles. In human beings, the disease has been shown to be familial in at least 20-40% of the cases and causative mutations have been identified in twenty-four genes. Dilated cardiomyopathy appears to be familial in many breeds of dogs as well although significant variation in the presenting complaint, clinical evaluation and rate of progression has been observed depending on the breed of dog and there is likely to be genetic variation with regards to the mutation as well.

Doberman pinschers

Dilated cardiomyopathy is a familial disease that often causes sudden death or heart failure. We have shown Doberman DCM to be inherited as an autosomal dominant trait. In humans, at least 24 causative genes for familial dilated cardiomyopathy have been identified, including cystoskeletal, sarcomeric and mitochondrial genes. A deletion mutation has been identified on canine chromosome 14 in the splice donor site of exon 14 in the PDK4 gene in affected dogs.[4] The PDk4 gene is known to be important in cardiac mitochondrial metabolism. However, a European study demonstrated that some german Doberman pinschers with this disease did not have this mutation, thus suggesting that there may be multiple causes of the disease in the Doberman pinscher.[5]

Arrhythmogenic right ventricular cardiomyopathy in the boxer

Since the early 1980's, the term boxer cardiomyopathy has been used to describe adult boxer dogs that present with ventricular arrhythmias, and sometimes, syncope. Recent studies have demonstrated that the disease has many similarities to a human disease called Arrhythmogenic Right Ventricular Cardiomyopathy (ARVC). The similarities between the diseases include clinical presentation, etiology and a fairly unique histopathology that includes a fibrous fatty infiltrate of the right ventricular free wall. The disease is most commonly characterized by ventricular arrhythmias, syncope and sudden death. However, systolic dysfunction and ventricular dilation are seen in a small percentage of cases.

Arrhythmogenic right ventricular cardiomyopathy is a familial disease in the boxer and appears to be inherited as an autosomal dominant trait. Unfortunately, the disease also appears to be a disease of variable genetic penetrance and affected dogs can have many different presentations including asymptomatic, syncope, sudden death and systolic dysfunction with CHF. A genetic mutation has now been identified for boxer arrhythmogenic right ventricular cardiomyopathy.[6] (http://www.ncstatevets.org/genetics/).

In human beings with ARVC, there are multiple genetic mutations that can lead to the development of the disease. At this time we do not know if this mutation is the only cause in the Boxer, therefore, clinical screening is still recommended. Since ARVC presents as an electrical abnormality more often than one of myocardial dysfunction, screening efforts should be based on annual Holter monitoring as well as annual echocardiography. Unfortunately, clear criteria for the diagnosis of occult ARVC do not exist. However, dogs that are symptomatic (syncope, heart failure) or have evidence of ventricular tachycardia on a Holter should not be used for breeding. Additionally, dogs that have over 100 left bundle branch block morphology VPCs/24 hours are probably highly suspicious of being affected. However, not all affected dogs will ever develop clinical signs and many may live a normal lifespan. It is likely that there are multiple factors that may influence which dogs become symptomatic for the disease. To help decrease the risk of making an error when adding or removing a dog from a breeding program, owners should be encouraged to screen annually rather than putting significant emphasis on a single Holter monitor reading. Until a greater understanding of disease inheritance and disease progression exists, caution should be used when advising breeders to remove dogs from breeding programs. Overzealous removal of animals based on the results of a single Holter monitor may have a significant negative impact on the breed.

REFERENCES

[1]Kittleson et al **Circ**,1999,99,3172; [2]Meurs et al **Hum Mol Genet** 2005,14,3587;[3]Meurs et al **Genomics** 2007, 90:261,[4] Meurs et al **Human Genet** 2012,131,1319; [5] Mausberg et al **PLOS One**. 2011, 6, e20042 [6]Meurs et al, **Human Genet,** 2010 128, 315

Stabilization of the Dog and Cat in Respiratory Distress

Laura A. Nafe, DVM, MS, DACVIM (SAIM)
Madison, WI, USA

Introduction

Respiratory distress is a common reason for dogs and cats to present on an emergency basis to a veterinary clinic. The general approach to the veterinary patient in respiratory distress is dependent on signalment of the patient, localization of the respiratory pattern, evaluation of vital parameters, and cardiothoracic auscultation.

The goals of this presentation include: 1) Understanding of respiratory terminology 2) Localization of the respiratory pattern 3) Immediate stabilization of the respiratory distress patient 4) Recognition of physical exam abnormalities that may help guide specific treatments.

Respiratory terminology

When a dog or cat presents with increased respiratory effort, it is important to use proper terminology when describing the patient to other veterinary professionals. Common terms used to describe respiratory distress or difficulty in dogs and cats include:

Tachypnea – Increase in rate of respiration.
Hyperpnea – Increase in depth and rate of respiration.
Orthopnea – Dyspnea that can be improved with alterations in position.
Dyspnea – The sensation of difficult or distressed breathing; "air hunger."

Dyspnea is a term that is used often over-used clinically, as it should be reserved for patients that are truly in respiratory distress. These are dogs and cats that are truly "oxygen hungry" and therefore unable to interact with their environment normally. This is in contrast to a patient that has an increase in respiratory effort, but is not truly distressed. It is also important to note that not all patients in respiratory distress are tachypneic. A classic example is a dog with severe laryngeal paralysis may have a normal respiratory rate with severe inspiratory dyspnea.

Localization of the respiratory pattern

Approach to managing the dog and cat in respiratory distress is dependent on which area of the respiratory tract is affected. The following scheme is helpful when localizing the respiratory pattern:

Respiratory localization	Respiratory pattern
1. Upper airway	Increased effort on inspiration; stridor or stertor
2. Lower airway	Increased effort on expiration; wheezes
3. Abdominal distension	Severe abdominal enlargement
4. Pulmonary parenchyma	Mixed inspiratory and expiratory effort; pulmonary crackles
5. Pleural space disease	Shallow breathing with mixed respiratory effort; decreased lung sounds
6. "Look-a-like" diseases	Mixed inspiratory and expiratory pattern with normal oxygenation

The first 3 should be evident simply by watching the patient breath. Upper airway disease results in a prolonged inspiratory phase of respiration and generally results in loud upper airway noise (stridor and/or stertor) during inspiration. Stridor localizes the problem to the larynx or pharyngeal region, whereas stertor localizes to the nasopharyngeal region. Lower airway disease results in a prolonged expiratory phase of respiratory and often an "abdominal push" can be observed. Wheezes (musical airway sounds) may be heard with or without a stethoscope. Although uncommon, patients can present for respiratory distress secondary to severe abdominal enlargement (eg; ascites, GDV, abdominal mass) and this should be obvious when initially evaluating the patient.

In contrast, further investigation is needed to help differentiate between pulmonary parenchymal, pleural space, and "look-a-like" diseases. Pulmonary parenchymal diseases include pulmonary edema (cardiogenic or non-cardiogenic), pneumonia, pulmonary thromboembolism, and pulmonary fibrosis. These patients often have a mixed inspiratory and expiratory respiratory pattern and may have pulmonary crackles on auscultation of the lungs. Patients with pleural space disease also may have a mixed respiratory pattern with most cases having slightly more effort on inspiration. One consistent feature of pleural space disease is a shallow breathing pattern. Dogs and cats with pleural effusion will have decreased ventral lung sounds and may have muffled heart sounds (remember that pericardial effusion or a thoracic/mediastinal mass could also cause muffled heart sounds!) on cardiothoracic auscultation. Patients with pneumothorax will have decreased lung sound dorsally. Most patients that present for a "look-a-like" disease may have tachypnea without an appreciable increase in respiratory effort. These patients should NOT have decreased oxygenation based on SpO_2 or arterial blood gas evaluation.

Immediate stabilization

One of the first and most important steps in stabilization is providing *oxygen* therapy. Cats presenting in respiratory distress are best provided oxygen via an oxygen cage, because flow-by oxygen via mask often adds additional stress. Many dogs can be provided oxygen flow-by via mask while the patient is being assessed. Initial assessment of the patient is going to be important when determining the best stabilization management strategy.

Efforts should be made to place an intravenous (IV) catheter as soon as possible if the patient will tolerate flow-by oxygen and it does not place additional stress on the patient. Placing an IV catheter allows intravenous medications, including additional sedatives and potentially propofol if the patient requires intubation. The best way to manage most cats initially is to administer intramuscular sedation (butorphanol +/- acepromazine) and minimize stress by allowing them to calm down in an oxygen cage without the need for restraint. After the cat has been sedated, attempts can be made to then place a catheter quickly with flow-by oxygen and minimal stress and restraint.

Utilizing the localization scheme may help to determine if this is an upper airway or lower airway respiratory condition. Patients with upper airway obstruction are going to benefit the most from sedation (butorphanol +/- acepromazine) and possibly intubation in severe cases. For example, the laryngeal paralysis Labrador Retriever or brachycephalic airway syndrome English Bulldog may need to be intubated on initial presentation to capture and bypass their upper airway if sedatives and oxygen does not improve their respiratory status. Both dogs and cats with severe upper airway disease resulting in respiratory distress benefit from glucocorticoid therapy to help reduce inflammation and associated edema.

Alternatively, lower airway disease may be responsive to bronchodilator agents. The best example being the cat in status asthmaticus that presents with increased expiratory effort with an abdominal push and expiratory wheezes. This patient may benefit greatly from inhaled albuterol or injectable terbutaline (IM or SQ). Dogs have not been documented to have true bronchoconstriction like cats and humans do, and therefore, do not benefit from bronchodilator therapy in most cases. Anti-inflammatory therapy (eg; glucocorticoids) is an important aspect of chronic management of canine and feline airway disease, and should be included in the acute management of airway disease as well. Unless the abdominal distension is due to severe ascites or gastric dilatation volvulus, there is little that can be performed initially to stabilize the patient with abdominal distension. Sedatives and proper positioning may be helpful in these patients.

Other than oxygen, patients with pulmonary parenchymal disease can be difficult to stabilize initially without further diagnostics. If congestive heart failure is suspected, a bolus dose of furosemide (2-4 mg/kg IV or IM in dogs; 1-2 mg/kg IV or IM in cats) can be administered to help reduce pulmonary edema. Ideally, thoracic radiographs should be performed prior to furosemide administration to avoid situations where the diagnosis is clouded by previous furosemide administration. Remember that often a single dorsoventral thoracic radiograph may be enough to diagnose congestive heart failure and minimizes stress to the patient.

If pleural space disease is strongly suspected, a thoracocentesis should be performed to help stabilize the patient prior to additional diagnostics (eg; thoracic radiographs). With the addition of portable FAST scan ultrasound available in many veterinary clinics, this technique may be useful to quickly evaluate for pleural effusion. Alternatively, a single lateral or dorsoventral radiograph is diagnostic for effusion or pneumothorax if the patient is stable to perform this test.

Treatment	Dose	Indication
Butorphanol	0.2-0.4 mg/kg IV, IM, SQ	Sedation
Acepromazine	0.01-0.04 mg/kg IV, IM, SQ	Sedation
Furosemide	2-4 mg/kg IV or IM bolus (Dogs) 1-2 mg/kg IV or IM bolus (Cats) Consider CRI following bolus	Congestive heart failure
Terbutaline	0.01-0.04 mg/kg IM or SQ	Bronchoconstriction secondary to lower airway disease (cats)
Albuterol	Metered dose inhaler: 1-2 puffs - Spacer and facemask required Nebulization: 1.25 mg in 2 mL of 0.9% NaCl over 10-20 minutes	Bronchoconstriction secondary to lower airway disease (cats)
Glucocorticoids	Dexamethasone: 0.1-0.3 mg/kg IV, IM, SQ Prednisolone: 1-2 mg/kg IV, IM, SQ	Upper airway inflammation Tracheal collapse Lower airway disease

Recognition of physical exam abnormalities

Dogs and cats presenting in respiratory distress are challenging cases, because we are often forced to make therapeutic decisions with minimal diagnostic information (due to the need to reduce stress to the patient). That being said, there are specific aspects of a case that can help localize the cause of the respiratory distress without invasive diagnostic testing.

Signalment of the patient: Although any species, breed, or age of animal is capable of acquiring any respiratory disease, this information can be helpful when thinking about the "likely" differential diagnoses. For example, the Yorkshire Terrier is more likely have tracheal collapse and the Labrador Retriever is more likely to have laryngeal paralysis. Additionally, the Cavalier King Charles Spaniel and Doberman Pincher are more likely to have congestive heart failure. In general, cats are less likely to acquire aspiration pneumonia and more likely to have a true lower airway obstruction secondary to bronchoconstriction from asthma. Cats are also more likely to develop pleural effusion associated with congestive heart failure. Age and environment may be helpful in patients in which a viral or bacterial infectious disease is suspected, as these tend to be young animals with a history of exposure to other animals (eg; shelter, pet shop, boarding).

Vital parameters: With the exception of upper airway obstruction, most patients with respiratory distress have an increase in respiratory rate. In addition, patients that are hypoxic will often have concurrent tachycardia in an effort to improve oxygenation and/or the stress and anxiety associated with respiratory distress. In my opinion, temperature is perhaps the most important vital parameter, as hypothermia is often observed in patients with congestive heart failure (especially cats). In contrast, an increase in temperature associated with fever may be observed in patients with pneumonia, pyothorax or neoplastic disease. Hyperthermia is often documented in patients with airway obstruction (upper airway more often than lower airway).

Cardiothoracic auscultation: In addition to observing the respiratory pattern, pulmonary and tracheal auscultation can be useful when differentiating upper airway, lower airway, pulmonary parenchymal, and pleural space diseases. Most dogs that are congestive heart failure have a heart murmur heard on auscultation (exception might be a dog with dilated cardiomyopathy). However, many cats (up to 50%) do not have a heart murmur or gallop auscultated at the time of congestive heart failure diagnosis. If pulmonary crackles are heard, determining if they are diffuse or focal may help narrow the differential diagnoses list, as aspiration pneumonia may be focal whereas diffuse crackles could indicate pulmonary edema or other causes of pneumonia (viral, fungal). Decreased lung sounds ventrally would support pleural effusion, whereas decreased lung sounds dorsally indicates pneumothorax.

REFERENCES

Ettinger SJ, Feldman EC. *Textbook of Veterinary Internal Medicine*. 7th ed. St. Louis: Saunders, 2010. Rozanski EA, Rondeau MP. Respiratory pharmacotherapy in emergency and critical care medicine. *Vet Clin North Am Small Anim Pract* 2002 ;32: 1073-1086. Sumner C, Rozanski E. Management of respiratory emergencies in small animals. *Vet Clin North Am Small Anim Pract* 2013; 43: 799-815.

THE "GROWING" USE IN STEM CELLS

Kenichiro Yagi, BS, RVT, VTS (ECC, SAIM)
Los Altos, CA

INTRODUCTION

The current mode of therapy in traditional medicine involves the usage of chemical molecules targeting single receptors of pathways to create desired beneficial effects, while monitoring undesirable negative effects. These molecules are metabolized and eliminated through various pathways, resulting in a short term half-life necessitating the need for repeating doses on a regular basis, and sometimes indefinitely, in order to maintain desirable effects. In contrast, cellular therapies, while not the cure-all (a common misconception), are being investigated for their use in a multitude of pathologies and applications in human medicine and seem to have many favorable characteristics overcoming some of the limitations of traditional chemical therapies. Some of these therapies, such as adipose-derived mesenchymal stem cells (ADMSC) have been utilized in the treatment of osteoarthritis in dogs and tendinitis in horses with favorable evidence steadily accumulating. Promising evidence and reports regarding mesenchymal stem cell (MSC) use in treating chronic kidney disease, cutaneous wounds, dermatologic conditions, and muscle tears are emerging. Stem cell therapy remains an area of interest and holds the potential to provide future options for ailments currently difficult to treat.

STEM CELL BASICS

Stem cells are undifferentiated cells capable of multiplying, and developing into many different types of cells required in the body. These cells are important during the early life and growth stages of an animal, differentiating into specialized cells such as myocytes, erythrocytes, neurons, and many other necessary cells. The cells that can be harvested from a blastocyst during embryonic development, known as embryonic stem cells, contain three primary germ layers (ectoderm, endoderm, mesoderm) allowing differentiation into hundreds of different types of cells, and there is much ethical debate surrounding its use. Stem cells are found in various tissues even during adulthood, and repair tissue by replacing damaged cells. These stem cells are called somatic (or adult) stem cells, and are found in the bone marrow, muscles, adipose tissue, the brain, and a few other organs.

Stem cells can be described based on their ability to differentiate into various tissues. Totipotent stem cells are able to differentiate into embryonic and extraembryonic cells, which are capable of developing into a whole animal. Totipotent cells are found in very early stage embryos prior to the blastocyst stage (morula). The embryonic cells that will differentiate either into the ectoderm (gastrointestinal tract, lungs, etc), endoderm (blood, bone, muscle, etc.), or mesoderm (skin, nervous system, etc) layers are pluripotent, with the ability to differentiate into all types of cells aside from extraembryonic tissues (amnion, chorion, and parts of the placenta). Somatic stem cells are multipotent; they are able to differentiate into cells of the same line. An example would be hematopoietic cells capable of differentiating into megakaryocytes, erythrocytes, lymphocytes, and leukocytes. However, they are not likely capable of differentiating into neurons of glial cells, for example, and can be described as progenitor cells. Unipotent cells are what may be considered precursor cells, which will have the ability to differentiate into a single tissue type. In addition to its potency, stem cell therapy can be classified by the relationship between the donor and the recipient of cells. Autologous cell therapy refers to cells harvested from one animal and used for therapy in the same animal. Allogeneic cell therapy refers to cells harvested from one member of a species and used in another member of the same species. Heterologous cell therapy refers to cells harvested from one member of a species and used in a member of another species.

In veterinary medicine, the use of stem cells from bone marrow and adipose tissue is the most common. One form of bone marrow stem cell therapy is bone marrow hematopoietic stem cell transplantation (HSCT), performed to attempt treatment of lysosomal storage disease like mucopolysaccharidosis, leukemia, lymphoma, canine leukocyte adhesion deficiency, and X-linked severe combined immunodeficiency. HSCT for these conditions is performed experimentally to evaluate its effectiveness translatable to the human population, and is not performed therapeutically in veterinary practice. The use of mesenchymal stem cells derived from adipose tissue, however, has seen an increase in veterinary medicine, providing a form of multipotent, autologous cell therapy.

CLINICAL APPLICATIONS

Clinical application of the use of stem cells can be performed as what is known as cell therapy, involving the introduction of undifferentiated stem cells to an animal to treat a specific disease or condition. Cell therapy with MSC has received attention as a potential treatment for tendinitis in horses and osteoarthritis in dogs and has published evidence for its positive effects in wound healing, some dermatologic conditions, and muscle tears in dogs. MSC is commonly derived from bone marrow or adipose tissue.

ADMSC therapy is performed autologously, involving a process of harvesting adipose tissue from the patient, isolating the stem and regenerative cells within the tissue and administering them back into the patient. Adipose tissue is a good source of stem cells, having a 500x higher concentration when compared to bone marrow, and cryopreservation is possible. Stem cells harvested from adipose tissue have demonstrated the ability to differentiate into many types of tissue, including osteoblasts, osteochondrogenic progenitor cells, and myoblasts. The ability to harvest the tissue in a much less invasive manner at a higher concentration, eliminating the need for further processing beyond isolation such as culturing makes ADMSC a very reasonable source for stem cells. Tissue harvesting is performed under general anesthesia. A 5-10cm incision is made in the inguinal, abdominal, or thoracic wall region, and approximately 20-40g of adipose tissue is resected and placed in the sample collection tube. A volume of blood may also be collected and submitted (refer to each company protocol for amount). The samples are then processed for MSC isolation.

Isolation of stem cells from the adipose tissue is currently available in two forms in clinical practice. Stem cell processing services process stem cell concentrate from harvested tissue, and return it to the practice with a short turnaround time. Processing by these companies provides technicians experienced in processing samples utilizing validated procedures to remove contaminants at a laboratory dedicated to this purpose. Some disadvantages may include the turnaround time, which require that the patient make two visits for harvest and then for administration of the cells, and the possibility of samples being lost or damaged in shipping.

More recently, the option to purchase the equipment necessary to isolate the stem cells has become available. The option to purchase may carry a significant financial investment up front, but allows the patient to receive stem cell therapy on the same day it is harvested, as it only takes a couple of hours for the stem cells to be isolated. The isolation being performed in-house also eliminates any potential for loss or damage during shipment, but will require the space for the equipment, staff training, and establishment of internal quality control to ensure the process is being performed to adequate standards. The practice's osteoarthritis case load will likely play a significant role in the decision making between the two options.

The isolation process involves washing of the tissue with phosphate buffered saline (PBS) which is minced and further washed to remove debris and blood. The adipose tissue is then digested through incubation with collagenase and hyaluronidase. This mixture is then centrifuged at 400 x *g* for 15 minutes to yield the stromal vascular fraction (SVF), which is then resuspended for use. The SVF contains the ADMSC as well as fibroblasts, pericytes, endothelial cells, and circulating blood cells. The content of MSC in the process sampled can be determined through flow cytometry for the presence of cell surface antigens appropriate to MSC (CD9 and CD44).

Once the MSCs are isolated, the suspension is used in different manners depending on the therapy intended. In the case of treatment for osteoarthritis, the patient is placed under sedation and receives an intraarticular injection in the affected joints, which may be combined with an intravenous injection. Studies evaluating the effectiveness of intraarticular injection of ADMSC for treatment of osteoarthritis have observed for positive effects on lameness examination, pain on manipulation, range of motion, and positive responses to a standardized questionnaire on the Cincinnati Orthopedic Disability index (indicating ability for the dog to walk, run, jump, turn suddenly, get up, use the stairs, urinate/defecate, play, etc). Another study evaluated some objectively measurable parameters of peak vertical force and peak vertical impulse on the joint (indicating an animal's willingness to use the limb) to measure changes. Both studies concluded and commented on improvement after intraarticular injections of ADMSC noting a significant reduction in pain and discomfort, as well as increased functional ability.

Intradermal injections of MSC and its effect on healing of induced wounds were studied. The study involved the use of bone marrow derived MSC, evaluating direct injection into the wound bed as well as injection in a radial pattern around the wound edges. The two different methods and control subjects were evaluated for cellular proliferation, angiogenesis, and expression of wound healing-related factors. The study indicated that while wounds from treated and non-treated animals closed without infection and showed formation of a scar, the MSC treated wounds showed more rapid wound closure. Observations in the study included a faster re-epithelialization, increased collagen deposition, higher blood vessel density, and suppressed inflammation in the wounded skin when treated with MSC.

Further uses of MSC in dermatology include use in chronic non-healing wounds, inducing cell migration, cellular proliferation and differentiation, assembly of scaffolding, angiogenesis, and suppression of inflammation. Immune-mediated skin diseases may benefit from immunomodulation through suppression of T-cell reactivity, though definitive evidence is currently lacking. MSC therapy to limit scar tissue formation and treatment of alopecia is being investigated in animal models, and is driven by demand for human application.

MECHANISMS OF ACTION

Questions abound regarding the potential benefits of stem cell therapy in many of the non-curative ailments our patients suffer. Many of these questions of therapeutic application are driven by our current knowledge of the mechanism of action of stem cells and its theoretical benefit to each condition. MSC is directly delivered to the area of pathology and trauma such as in osteoarthritis, tendinitis, and wounds, or delivered systemically through intravenous injections to reach organs such as the liver or kidneys, and to have effects on immune-mediated diseases. The intent of the therapy with MSCs is to replenish damaged tissue by differentiating into necessary cells, and to have a paracrine effect, affecting the surrounding tissue through secretion of hormones or other cellular signals to promote regeneration in the area. The exact mechanisms of action are still under investigation, and current thought is that the benefit of MSC therapy is based on its ability to provide trophic support, provide anti-inflammatory and immunomodulatory effects, promote revascularization, reduce apoptosis, differentiate into various cell lineages, and promote homing capabilities. Stem cells have the ability to communicate with tissues in the environment by influencing local and remote cell populations through cell signals, and they react to signals provided from injured cells in the microenvironment.

Trophic Support: MSCs have been observed to secrete bioactive levels of cytokines and growth factors supporting angiogenesis, tissue remodeling, and anti-apoptosis. Secretion of factors such as vascular endothelial growth factor, hepatocyte growth factor, and basic fibroblast growth factors provide trophic support, stimulating the regeneration of tissue through recruitment of local and systemic cells, promoting more blood flow, and reducing destruction of tissue.

Anti-inflammatory and Immunomodulatory Effects: When MSCs are presented with an inflammatory environment, they are observed to play a role in altering the types and levels of cytokines in the environment from a pro-inflammatory mix to a less inflammatory or anti-inflammatory profile. ADMSCs do not express surface antigens which could induce an immune-mediated response, and also suppress T-cell proliferation. Adult stem cells also play a role in recruiting cells inhibiting lymphocyte reactions and T-cell proliferation. Both types of stem cells have been observed to aid in controlling immune-mediated rejection of transplanted tissue in mice.

Revascularization: ADMSC and endothelial progenitor cells are contained within the SVF derived from adipose tissue. These cells play a role in promoting angiogenesis and neovascularization through secretion of signaling factors such as vascular endothelial growth factor, transforming growth factor, fibroblast growth factor, angiopoietin, hepatocyte growth factor, and various cytokines.

Many of the current studies have shown increased vascularization in the area of MSC injection. An *in vivo* model of hind limb ischemia was observed to have increased blood flow and vascular density, and to have incorporated these stem cells into the renewed vasculature.

Anti-apoptosis: Apoptosis is often referred to as "cell suicide", and is a genetically controlled event ending in programmed cell death. Apoptosis plays an important role in removing aged cells to preserve the function of tissues in normal conditions, but may be induced when cells sustain certain insults through pathology. ADMSC is observed to reduce endothelial cell apoptosis during ischemic events preventing further ischemic damage and tissue injury. This effect is thought to be produced through factors suppressing apoptotic mechanisms, supporting survival of the cell through insult.

Differentiating Capacity: The ability of stem cells to differentiate into different types of cells was the initial basis for the therapeutic theory involving stem cell therapy. While other beneficial effects of paracrine function of stem cells have since been recognized, the ability of MSCs to differentiate and replenish or replace injured cells remains a significant benefit. MSCs have the plasticity to differentiate into adipogenic, myogenic, cardiomyogenic, chondrogenic, osteogenic, endothelial, epithelial, hematopoietic, hepatogenic, and neurogenic cell lines, thereby playing a significant role in creating beneficial effects. The regenerative capacity of autologous ADMSC and its use in injury of the superficial digital flexor tendon in horses was determined by a statistically significant level of improvement in repair on histologic samples. The regenerative capacity of stem cells remains a strong potential in future applications.

Homing: The ability of cells to direct movement and migration based on chemical signals, or chemotaxis, is an important property possessed by stem cells. Through this function, stem cells administered through intravenous injections are able to find the site of injury and intended therapeutic target. The exact mechanism of stem cell homing to sites of tissue injury is currently unclear, but thought to be related to chemokines stimulating receptors on injured cells to express ligands which facilitate routing of the stem cells. Adhesion molecule expression at sites of inflammation which allows leukocytes to be routed also seems to function similarly for stem cells. Chemokines and adhesion molecules are shown to be expressed on bone marrow MSCs, and the ability of MSCs to migrate to sites of cerebral and lung injury has been demonstrated.

CONSIDERATIONS

While MSC therapy shows promise and more evidence is becoming available pointing towards effectiveness of the therapy, there are some points to consider upon employment. Stem cell therapy should not be thought of as a replacement for conventional therapy proven to produce a positive outcome. In the case of osteoarthritis, for example, stem cell therapy seems effective in providing repair and overall pain relief, and should be considered as part of a multi-modal approach along with conventional approaches such as NSAID, pain management, weight control, and physical therapy. While thought to be beneficial in some spinal injuries or ligament ruptures, stem cell therapy should not be used to replace surgical solutions when applicable. Its effectiveness for use in neoplastic disease is currently unknown, and should be carefully considered as the results are currently unpredictable in terms of patient safety.

The harvesting of MSC containing tissues typically requires general anesthesia, and while usually a short procedure, will carry risks associated with anesthesia. Successfully performing intraarticular injections consistently is required to make certain the effort gone into MSC isolation is not wasted. Success rates of the physical injection may vary depending on clinician experience and may affect outcomes. Even in the cases of successful injections, injections may cause short term articular pain post-injection.

Stem cell therapy involves significant financial investment from the client end, seemingly increasing the cost of providing therapy along with other conventional means of treatment. Considerations should be made in discussing how much of the need for surgical, rehabilitative, and medical interventions may be alleviated with therapy and overall quality of life improvement is expected with implementation for each particular patient.

FUTURE OUTLOOK

Stem cell therapy is a relatively new form of therapy in veterinary medicine, and research in both human and veterinary fields will provide translational findings leading to better understanding and development of further applications. Tissue engineering is another form of stem cell therapy with a different approach from mentioned MSC therapies involving *ex vivo* growing of specific, functional tissues, organs and cells to deliver functional live material for replacement of injured or deficient functional units. Engineered skin for skin grafting in burn victims, and engineered RBC suspensions utilized in transfusions for anemic patients, are only a couple of the various investigations that are ongoing, with some in clinical trial stages. The US clinical trials registry for human studies has approximately 2000 open clinical trials on the topic of stem cells, and as positive results are seen, we should expect to see stem cell therapy become a valuable and established form of therapy to better serve our patients.

REFERENCES

[1]Harman RJ **Vet Dermatol** 2013;24:90. [2]Kim J, et al **Vet Dermatol** 2013;24:242. [3]Vilar JM, et al **BMC Vet Res** 2013;9:131 [4]Brown SG, et al **Stem Cell Disc** 2012;2(2):41. [5]Nixon AJ, et al **Am J Vet Res** 2008;69(7):928. [6]Black LL, et al **Vet Ther** 2007;8(4):272.

VALUE AND PERFORMANCE OF GI FUNCTION TESTS FOR GASTROINTESTINAL DISORDERS IN DOGS/CATS

Albert E. Jergens, DVM, PhD, DACVIM
Ames, IA

BACKGROUND

Chronic enteropathies are common in dogs and cats and include adverse reactions to food, idiopathic inflammatory bowel diseases (IBD), and antibiotic responsive diarrheas. While the definitive diagnosis often requires collection of mucosal biopsies for histopathologic evaluation, non-invasive laboratory tests are useful in the diagnosis of these disorders and in monitoring the response to treatment.

LABORATORY TESTS FOR SMALL INTESTINAL FUNCTION AND DISEASE

CASE 1: "Cisco", 6 year old, SF, Flame Persian, 3.7 kg

History: chronic watery diarrhea with weight loss; sequential therapies have not improved GI signs; referred for diagnostics

Cobalamin (Vitamin B_{12}) is an essential water-soluble vitamin and an important co-factor for many biochemical reactions. Canine and feline diets typically contain abundant cobalamin, making dietary insufficiency unlikely. The physiologic mechanism of cobalamin is complex and requires a functioning digestive system. Major disorders that interfere with cobalamin update include exocrine pancreatic insufficiency (EPI), distal or diffuse small intestinal disease, and excessive bacterial utilization of cobalamin associated with bacterial dysbiosis. Hypocobalaminemia is most often seen with long-standing and severe intestinal disease (i.e. IBD, lymphoma, or fungal enteritis) affecting the small intestine. The measurement of serum cobalamin concentrations in animals with chronic enteropathies is important in that failure to recognize this deficiency will result in delay of clinical recovery. Importantly, most serum cobalamin concentrations have been shown to be a risk factor for negative outcome in dogs with chronic enteropathies, including IBD and PLE. Appropriate dosages of cyanocobalamin are typically administered subcutaneously once weekly with serum concentrations reassessed at 4-6 week intervals.

Folate is a water soluble vitamin which is abundant in canine and feline diets making nutritional deficiencies unlikely. Folate concentrations are typically measured in concert with cobalamin and may provide information indicative of small intestinal dysbiosis (SIBO/ARD) or proximal small intestinal mucosal disease. Note that serum folate may be influenced by bacterial production of this vitamin and is therefore a less sensitive indicator of small intestinal disease vs cobalamin.

Citrulline is an amino acid synthesized predominately by intestinal enterocytes. Citrulline has been shown to be a marker of functional enterocyte metabolic mass and preliminary studies have shown that plasma citrulline concentrations are significantly decreased in dogs with parvovirus enteritis. Thus, citrulline might serve as a potential marker of spontaneous and acute intestinal dysfunction; however, additional clinical trials evaluating this marker are warranted.

EVALUATION OF INTESTINAL PROTEIN LOSS

CASE 2: "Storm", 3 year old, CM, Shar Pei, 22 kg

History: Polyphagia, severe weight loss, watery fetid diarrhea; physical exam reveals diffuse fluid-filled bowel loops

Protein losing enteropathies (PLE) define a heterogeneous group of disorders characterized by enteric loss of plasma proteins. Common causes for PLE include inflammatory enteropathies, infectious enteritis, lymphangiectasia, intestinal neoplasia, and significant microbial imbalances. Most animals' exhibit classic clinical signs of emaciation, weight loss, and will often have panhypoproteinemia on laboratory evaluation. It is important to first rule out other causes for hypoalbuminemia due to glomerular injury and hepatic failure or insufficiency. Definitive diagnosis of PLE requires collection and histopathology assessment of mucosal biopsy specimens. An ELISA assay for the measurement of ***fecal alpha (α)1–PI*** has been validated for dogs and may be used in assessing protein loss in the GI tract. Due to the cumbersome criteria for collection of multiple fecal specimens, the laboratory assessment of α1-PI as a diagnostic test for intestinal protein loss is performed infrequently. This laboratory assay may be most useful in assessing response to therapy in dogs with protein losing enteropathies.

Decreased serum albumin has been associated with negative outcome in separate canine IBD studies. Severe IBD (and other chronic enteropathies) may predispose to hypoalbuminemia due to enteric plasma protein loss as well as nutritional deficiencies and nutrient malabsorption. Craven et al retrospectively reported low serum albumen to be associated with negative outcome (ie, poor clinical response to dietary/drug therapies or euthanasia) in 80 dogs diagnosed with IBD. Similarly, Allenspach et al showed that hypoalbuminemia was a significant risk factor for negative outcome along with severity of clinical disease, duodenal endoscopy score, and hypocobalaminemia. In our experience, dogs having severe hypoalbuminemia (ie, value <1.5 g/dL) with cavity effusion and/or salient gastrointestinal signs are at significant risk for euthanasia.

LABORATORY MARKERS FOR ASSESSMENT OF INTESTINAL INFLAMMATION AND DAMAGE

CASE 3: "Justin" 7 year old, CM, mixed breed, 9 kg

History: Chronic intermittent vomiting and diarrhea for 3 weeks; dietary change shows no improvement

Several ***serologic markers for inflammation*** have been designed to predict the disease course and response to therapy of canine chronic enteropathies. Titers for ***perinuclear anti-neutrophil cytoplasmic antibodies (pANCA)*** have been evaluated as diagnostic markers in canine IBD in separate studies. In one study of 31 dogs, results indicated that pANCA was a highly specific marker for IBD, although the sensitivity assay was too low to be of value to be a screening test. This immunofluorescence assay was shown to be most useful in the detection of dogs with food-responsive enteropathy at the time of diagnosis. More recently, pANCA was shown to be a highly specific serologic marker versus ANA for differentiating dogs with IBD from other gastrointestinal disorders. Lastly, others have shown that pANCA was of value as a diagnostic marker for familial PLE in Soft-Coated Wheaten Terriers. Unfortunately, this assay is not readily available in North America and is presently limited for use in research investigations. ***C-reactive protein (CRP)*** is a sensitive, but nonspecific inflammatory marker (ie, acute phase protein produced by the liver in response to inflammation) shown to correlate with moderate-to-severe clinical disease activity (CIBDAI) in canine IBD. This marker is the best characterized inflammatory marker where no fewer than four different studies have shown that dogs with IBD have heterogeneous serum elevations of this marker. New studies have evaluated the index of variability for CRP, suggesting that the use of a population-based reference range is not appropriate for evaluating changes in CRP in individual patients. Therefore, serial CRP measurements should be performed and modest changes in serum CRP may indicate the severity of GI disease.

Fecal calprotectin and other S100 gene proteins have been reported to be increased in the serum of human patients and dogs with IBD. Calprotectin and S100 A8/A9 and A12 are calcium-binding proteins found predominantly within neutrophils and other immune cells in inflamed mucosa. Fecal concentrations of these proteins have been recently shown to be increased in humans and dogs with IBD, when compared to healthy controls. Importantly, the S100 A12 fecal marker has greater sensitivity for the detection of intestinal inflammation and has been positively associated with endoscopy and clinical scores. ***The expression of P-glycoprotein*** (P-GP) in mucosal lymphocytes has also been investigated and shown to be up-regulated in IBD dogs treated with prednisolone. This earlier study showed that in dogs with steroid-responsive enteropathy, low P-GP expression was associated with a good therapeutic response.

MOLECULAR TOOLS OF PRACTICAL VALUE FOR CLINICIANS

Case 4: "Sydney" 12 year old, DSH,

History: Diagnosed with IBD; treated with diet change and steroids but no improvement after 7 days and is still sick

Polymerase chain reaction for antigen receptor rearrangements (PARR) amplifies variable T or B cell antigen genes and is used to detect the presence of a clonally expanded population of lymphocytes. This molecular technique is useful in differentiating severe mucosal inflammation (severe IBD) from alimentary lymphoma. This test is often performed with histopathology (routine H&E stains) and immunophenotyping return ambiguous results. Data derive from feline studies shows that this molecular technique is highly sensitive for differentiating LSA vs IBD; however, further studies are needed to assess the value and sensitivity of this technique in dogs.

Fluorescence in situ hybridization (FISH) is a molecular technique used to identify microbial populations in tissues. FISH analysis has been applied to a variety of gastrointestinal tissues, including the pancreas, liver, and alimentary tract in dogs and cats. This technique utilizes culture independent analysis of bacterial 16S or 23S rDNA genes targeted by DNA probes tagged with a fluorophore. Bacterial populations are then imaged via fluorescence microscopy. Microbial imbalances in diseased gastrointestinal tissues (eg, Helicobacter gastritis, idiopathic IBD, feline cholangitis) of dogs and cats have been demonstrated using FISH techniques. Specifically, FISH has identified an association between gastrointestinal inflammation and a shift in the microbiome by means of bacterial translocation (eg, Helicobacter gastritis) or dysbiosis of mucosa-associated microbial populations IBD, granulomatous colitis of Boxers).

REFERENCES

Allenspach K, Bergman PJ, Sauter S, et al. P-glycoprotein expression in lamina propria lymphocytes of duodenal biopsy samples in dogs with chronic idiopathic enteropathies. J Comp Pathol. 2006 Jan;134(1):1-7.

Suchodolski JS, Dowd SE, Wilke V, et al. 16S rRNA gene pyrosequencing reveals bacterial dysbiosis in the duodenum of dogs with idiopathic inflammatory bowel disease. PLoS One. 2012;7(6):e39333. doi: 10.1371/journal.pone.0039333. Epub 2012 Jun 15.

Allenspach K, Wieland B, Gröne A, Gaschen F. Chronic enteropathies in dogs: evaluation of risk factors for negative outcome. J Vet Intern Med. 2007 Jul-Aug;21(4):700-8.

Mancho C, Sainz Á, García-Sancho M, Villaescusa A, Rodríguez-Franco F. Evaluation of perinuclear antineutrophilic cytoplasmic antibodies in sera from dogs with inflammatory bowel disease or intestinal lymphoma. Am J Vet Res. 2011 Oct;72(10):1333-7.

CANINE PANCREATITIS: NOT AS EASY AS THE TEXTBOOKS WOULD SUGGEST

Michael Willard, DVM, MS, DACVIM-SA
College Station, TX

DIAGNOSIS

History and physical examination are, as always, critically important. But, they are not that helpful for diagnosing pancreatitis. They are most useful for finding other problems that may be mimicking pancreatitis. Schnauzers and Yorkies are famous for acute pancreatitis, but these breeds get a lot of other diseases that cause vomiting. Furthermore, acute pancreatitis can be found in any breed of dog. Canine pancreatitis is classically considered to present with acute vomiting and anorexia. Abdominal pain is frequently present, but it is easy to miss during physical examination, and fever is occasionally seen. Some dogs (especially those with pancreatic abscesses) may have relatively mild, intermittent, unimpressive vomiting and continue to eat a reasonable amount. Severely ill patients may present in classic systemic inflammatory response syndrome (SIRS). Many dogs with very severe acute pancreatitis present as though they had an acute, septic abdomen. Some have no abdominal fluid while others have substantial amounts of abdominal fluid and there is everything in between. Most cases of canine pancreatitis seem to be temporally related to either ingestion of fat or lipemia associated with diabetic ketoacidosis. Trauma and drugs can also cause canine pancreatitis, but are uncommon causes.

The immunoreactive canine pancreatic lipase assay (i.e., cPLI or Spec cPL) appears to be the most sensitive test for pancreatitis available (approximately 85-90%). The SNAP PL is relatively comparable in sensitivity.[1] There are a few false negative results with this test, and it is clearly much more sensitive than serum amylase or lipase.[1] The real question is how specific the cPL is for clinically important disease (i.e., lesions of the pancreas that are causing clinical disease as opposed to microscopic lesions that are clinically unimportant).[2,3] Nobody knows the answer to this question for sure, but many of us believe that the cPLI test can be positive in animals with clinically unimportant changes in the pancreas. The biggest advantage of the cPLI test is that if it is negative, pancreatitis is probably not the important clinical problem. In such a case, you need to look very hard for extra-pancreatic disease in the dog.

The serum pancreatic specific elastase-1 test has been suggested to have roughly similar sensitivity to the cPL test, at least for the more severe forms of pancreatitis.[4] However, this test is not commonly performed in the US and we tend to have minimal experience with it.

Imaging is critical in the diagnosis of pancreatitis. Probably the most greatest value of abdominal radiographs is that they help eliminate other diseases that could be causing signs similar to those caused by pancreatitis. Occasionally, one finds radiographic signs which specifically suggest acute pancreatitis. A sentinel loop (i.e., a dilated, air-filled segment) in the descending duodenum, and/or lack of serosal detail in the upper right abdominal quadrant, and/or lateral displacement of the descending duodenum on the ventro-dorsal projection, and/or a mass medial to the descending duodenum (on the ventro-dorsal projection) and/or a mass just behind the liver and just below the pylorus (on the lateral projection) are somewhat suggestive of acute pancreatitis. These findings are only meaningful if present; many dogs and cats with acute pancreatitis do not have any of these radiographic findings.

Ultrasonography is the most useful imaging modality. It may find abnormalities that suggest pancreatitis as well as eliminate other potential causes of vomiting and abdominal pain. Ultrasonography has been suggested to be about 40-70% sensitive in finding canine pancreatitis. One may sometimes detect hypoechogenicity surrounded by hyperechoic fat in the region of the pancreas that is due to pancreatitis. At other times, a markedly thickened pancreas may be found. Both findings are very specific evidence of pancreatitis. Evidence of EHBO is very suggestive of pancreatitis. Any dog with extra-hepatic biliary tract obstruction (that does not have a mucocoele) and that is vomiting/anorexic should be assumed to have pancreatitis until proven otherwise. It is very important to note that the ultrasonographic appearance of the pancreas can change dramatically within a few hours, so repeating abdominal ultrasound on the same day is not necessarily a bad idea if you strongly suspect pancreatitis but are finding nothing. Combining ultrasound with blood tests is probably the best combination for diagnosis.[4]

Chronic pancreatitis can be challenging to diagnose. Anytime you find exocrine pancreatic insufficiency (diagnosed with TLI) in a breed that is not commonly affected with pancreatic acinar atrophy (e.g., German shepherd, rough-coated Collie), then chronic pancreatitis becomes a major concern. This is probably the easiest way to diagnose chronic pancreatitis. Otherwise, dogs with episodic vomiting due to recurrent bouts of pancreatitis may not have any other signs of disease, and they invariably are admitted to your clinic for a work up after the last bout has run its course or is on the mend. Episodes of vomiting and anorexia due to recurrent pancreatitis can be random and unpredictable. In such patients, the previously mentioned diagnostics may be attempted, especially when acute exacerbations occur. Very rarely, upper gastrointestinal barium contrast radiographs may rarely reveal duodenal abnormalities (e.g., dilatation, stricture) which suggest that recurrent bouts of acute pancreatitis have caused scarring of the pancreas which in turn have compromised the maximum size of the duodenal lumen. Ultrasonographic changes are nice if they are present, but they can be minor making it difficult to accurately interpret them. Feeding an ultra-low fat diet for 3-4 times longer than what was previously the longest interval between episodes may sometimes be helpful in making a presumptive diagnosis.

Pancreatic abscesses in dogs are invariably sterile. Affected dogs can have a much more chronic, smoldering course (e.g., vomiting for a month or more, mild loss of appetite) than most dogs with more typical acute pancreatitis although some have acute, severe course of disease. We have even found a few dogs which had pancreatic abscesses that were completely asymptomatic. Abdominal pain may be present or absent. CBC and serum biochemistry findings are unpredictable. Diagnosis requires ultrasound. Treatment may be surgical marsupalization, percutaneous ultrasonographic drainage or just observation.

THERAPY

As of this writing, there is not a well-designed, robust, prospective, stratified study on the treatment of canine acute pancreatitis. Therefore, all any of us has is opinions, period.

Preliminary work from in Australia suggests that it is safe and perhaps beneficial to feed dogs with acute pancreatitis per os or with an esophagoscopy tube as soon as they can tolerate it (i.e., they do not get worse, even if they are still vomiting).[5] I recommend that a) you feed as low a fat content as possible, and b) if the feeding is associated with worsening of the vomiting, that you stop it and either try again in a day or two or go to jejunostomy feeding. Do not try to get full caloric intake into the patient initially; rather, start with small amounts to see if the patient will hold down the food. Obviously, if feeding is associated with worsening of the vomiting or general condition, stop the feeding.

Fluid therapy is critical, and subcutaneous administration of fluids is clearly inferior to IV fluids for all but mildly affected animals. IV fluid administration is often sufficient, even in dogs in which a pancreatic granuloma has temporarily blocked the main bile duct. Adequate pancreatic circulation is probably crucial for healing damaged pancreatic tissue; therefore, it is probably better to provide a little too much fluid rather than a little too little fluid unless the patient has congestive heart failure or oliguric renal failure. The abdominal viscera is not "first in line" to receive circulation when the patient is dehydrated (which most dogs with pancreatitis are when they come to your office). Obese and fat dogs (which describes a lot of dogs with pancreatitis) do not necessarily have skin tenting when they are dehydrated. Likewise, although you might expect dry, tacky oral mucus membranes, a nauseated animal may be salivating enough to make the mucus membranes moist even though it is dehydrated. If a vomiting dog is not eating or drinking, then it is dehydrated regardless of how well hydrated it appears on physical examination. However, if you give substantially too much crystalloid and dilute the serum protein concentrations, this could be detrimental.

One should monitor the serum albumin concentration during fluid therapy in these patients. If the serum albumin concentration decreases significantly (i.e., to < 2.0 gm/dl), then the plasma oncotic pressure likewise decreases which diminishes the effective perfusion at the capillary level. Since perfusion is so critical to treating dogs with pancreatitis, one should probably become concerned whenever the serum albumin concentration falls below 2.0 gm/dl. It is very hard to administer enough plasma to significantly raise the plasma albumin concentration. Hetastarch is probably a better choice because it will stay in the circulation and raise the plasma oncotic pressure for much longer than plasma. Human albumin can be used effectively, but it occasionally causes anaphylactoid reactions that can kill the patient; therefore, it is not recommended. Canine albumin is safer, but it does not last as long as hetastarch. One retrospective study has shown that plasma did not help treat dogs with pancreatitis; however, this study suffers from the problems inherent in all retrospective studies.

If the patient cannot tolerate early refeeding (i.e., the vomiting becomes worse with any oral intake), then jejunostomy feeding is another option. It is safer, less expensive, and less dangerous than parenteral nutrition, and has been associated with a better prognosis. In particular, it should be considered if an exploratory laparotomy was performed when the pancreatitis was diagnosed because a J-tube can be placed at that time. Alternatively, one can place a jejunostomy tube via laparoscopy, through a G-tube, and via the nose (naso-jejunostomy).

Antiemetics are useful in patients that are vomiting repeatedly or that feel so nauseated that they feel terrible. I prefer to only use antiemetics for short periods of time because I want to see if the patient is improving enough so that it no longer needs the antiemetic to stop vomiting. However, if the patient is vomiting multiple times per day or obvious feels terrible due to the nausea, then maropitant (1 mg/kg SQ) appears to be very useful. Maropitant might also have the advantage of providing some analgesia because it blocks Substance P binding. Dolasetron (0.3-1.0 mg/kg qd) and ondansetron (0.25 mg/kg qd) can also be effective, but are usually not as effective as maropitant.

H-2 receptor antagonists have been used to prevent gastric ulceration and erosion. It is doubtful that ulceration/erosion is a common problem in all but the sickest patients. Furthermore, if it is desired to provide some level of protection to the gastric mucosa, the proton pump inhibitors are far superior to the H-2 receptor antagonists. Pantoprazole (1 mg/kg IV qd) or omeprazole (1.5-2 mg/kg PO bid) are the most commonly used drugs. I suspect that they provide more benefit by being antidyspepsic in nature (thereby making the patient feel better) than they do by preventing ulcers, but that is a guess.

Antibiotics have been used to "prevent" secondary bacterial infection of the inflamed pancreas, which is supposed to be "fertile ground" for infection. However, there is minimal evidence that bacterial infection is of any clinical importance in routine canine pancreatitis. Antibiotics do not hurt these patients, but it is very questionable how helpful they are. However, dogs in SIRS due to pancreatitis may be a different story. Any dog in SIRS from any reason is potentially at increased risk of infection due to severely compromised mesenteric circulation.

Corticosteroids are very controversial in the treatment of pancreatitis. While they increase serum amylase and lipase activities, they do not cause pancreatitis. It is possible (and this is very controversial) that they may be useful in treating patients that are in Systemic Inflammatory Response Syndrome (i.e., SIRS, which used to be called "septic shock").[6] At this time, there are good data showing that it is probably reasonable to give physiologic doses to dogs in SIRS that are not responding well to typical fluid therapy. These dogs may have what has been termed "Critical illness related corticosteroid insufficiency" (CIRCI), that is to say that they are relatively hypoadrenal. This is a controversial statement. If steroid therapy for pancreatitis is contemplated, it should probably be reserved for the severely ill dog which is not responding to fluid resuscitation. If steroid therapy will be administered, you need to warn the owners of the unknown nature of this therapy. A larger question is whether steroids may be useful in treating the inflammation found in severe pancreatitis. At this time we might have minimal evidence that anti-inflammatory doses of steroids might (?) be helpful in some patients (at least the patients did not clearly become a lot worse). However this is very preliminary and anecdotal.

Although heparin therapy used to be the treatment for DIC (which can probably make acute pancreatitis worse), it has not been shown to be useful. If DIC appears to be a major problem, aggressive administration of fresh frozen plasma to replace clotting factors and anti-thrombin III concentrations is probably more effective.

Analgesics are useful. One must remember that dogs are clearly much "tougher" than people, and they often hide their pain well. Unless there is some good reason to the contrary, it is best to routinely assume that dogs with acute pancreatitis are in pain and will benefit from analgesic. In very mild cases, butorphanol is sufficient. In moderate cases, methodone is reasonable. As the pain becomes more severe, we progress to constant rate infusions of fentanyl. Is the most severe cases, we administer a constant rate infusion of fentanyl, lidocaine and ketamine.

Taking a dog with pancreatitis to surgery is probably one of the hardest decisions in small animal internal medicine. Strict guidelines cannot be given, but some basic principals may be suggested. In general, one must realize that unless an abscess, pseudocyst, and/or an obstructed gall bladder with a bacterial infection (extremely rare in veterinary medicine) are found, surgery will probably not benefit the patient. In fact, it may even be detrimental because anesthesia is usually associated with some decrease in visceral perfusion unless great care is taken to maintain circulation. Try to find reasons why you should NOT do surgery, only doing surgery if you cannot find any good reason to avoid it. Ultrasonography usually finds the type of lesions that would be benefitted by surgery (i.e., abscess or pseudocyst). If one is uncertain whether or not a cyst or abscess might be present and might be responsible for appropriate medical therapy being ineffective, one may reasonably decide to explore the abdomen. However, this decision should be based on the finding that 5-7 days of excellent, aggressive supportive medical management has not been of any appreciable benefit.

The best to determine if the patient is improving is to look at the patient. CBC, cPLI, and ultrasound may all look worse in a patient that is clinically better. Trust the patient, not the imaging or the lab work, to know if it is improving.

Do not be in a hurry to take dogs with pancreatitis to surgery. That is why it is important to provide the best possible medical management when first confronted with the patient with suspected acute pancreatitis. I will usually try aggressive medical therapy for 5-8 days, depending upon how quickly the patient is decompensating. If I have no indication that optimal medical therapy is helping after 7-9 days, then I might consider surgery. If the patient has an EHBO, I generally will wait a longer time (i.e., a week or more, assuming that the patient is improving) before I consider surgery to relieve the obstruction. Almost all of these patients will eventually have the EHBO resolve if aggressively treated medically. If it appears important to relieve EHBO (the problem has been going on for 2+ weeks or the biliary tract might be infected), then percutaneous drainage using ultrasound guidance might be appropriate. Rarely a stent can be placed in the bile duct, but stents can have problems. If a pseudocyst or abscess is not found, lavaging the area and resecting obviously necrotic tissue might help the patient, but this is uncertain.

Prognosis is difficult to predict. Hyperbilirubinemia is not necessarily a poor prognostic sign; pancreatic granulomas causing icterus due to obstruction of the bile duct typically resolve if the patient receives appropriate supportive therapy. Hypocalcemia, while a poor prognostic sign in people, is infrequently found in dogs and cannot be used to predict the outcome. Pancreatic abscesses are usually sterile. We cannot correlate the ultrasonographic appearance of the pancreas with prognosis. Finding a degenerative left shift and/or a marked thrombocytopenia (probably due to DIC) are not known to be negative prognosticators, but one intuitively fears that a poor outcome is more likely with such findings. Evidence of SIRS is probably the most important reason to give a poor prognosis.

When to send the patient home is the last question. The patient is sent home when it is doing well. Monitoring the ultrasonographic appearance or the cPLI is much less valuable than observing the patient. If the patient feels good and is eating and is not vomiting, then you can probably send it home regardless of what the ultrasound looks like or how high the cPLI is. When the dog is sent home, it should be on a very strict diet that is free of fat . If the dog ever gets sick again, they should come in and see you sooner rather than later, in case it is pancreatitis that is recurring.

REFERENCES

1) McCord K, et al., **J Vet Int Med** 2012; 26: 888, 2) Trivedi S, et al. **J Vet Int Med** 2011; 25:1241, 3) Mansfield CS, et al. **J Vet Diag Invest** 2012; 24:312, 4) Mansfield CS, et al., **J Vet Diag Invest** 2011; 23:691, 5) Mansfield CS, et al, **J Vet Int Med** 2011; 25: 419, 6) Yarema TC, et al, Crit Care Nurse 2011;31: 16.

NEUROIMMUNOLOGY: FROM DEFENSE RESPONSE TO MALADAPTIVE MECHANISM

Sarah Moore, DVM, DACVIM (Neurology)
The Ohio State University, Columbus, OH

INTRODUCTION

The immune system plays an integral role in all systemic disease processes; however, that role is only recently been recognized for many neurologic diseases not classically labeled "immune-mediated". For most neurologic diseases, the immune system has the potential to play both a beneficial role in limiting disease, or a detrimental role by damaging nervous tissue. Because tissue in the central nervous system (CNS) has limited regenerative capacity, this can have severe implications. The goal of this review is to provide a broad overview of the normally functioning immune system as it related to the brain and spinal cord, as well as to describe some of the cutting edge work related to neuroimmunology and neuroinflammation in veterinary medicine.

AN OVERVIEW OF IMMUNITY

The immune system is a network of biological responses that protect against disease. Immune responses can be classified into two main subsystems: the innate immune system (basic immunity) and the adaptive immune system (acquired immunity). An overactive immune system is classically known to result in autoimmune diseases. Conversely, patients who are immunodeficient can have recurring infection.

The *innate immune system* is the body's first line of defense against pathogens and its responses are broad and non-discriminate. It is the predominant system of immunity in plants, insects, and primitive multicellular organisms but is also present and active in animals. Innate immunity is comprised of a number of non-specific responses to pathogens that do not confer long-lasting or protective immunity. A common example of an innate immune response is the body's reaction to lipopolysaccharide (LPS), an endotoxin found on the surface of bacterial cell membranes. LPS is an example of a pathogen-associated molecular pattern (PAMP), a molecule recognized by the innate immune system as "not self". The innate immune system requires no previous exposure to LPS to be able to mount a response to it. This is because cells of the innate immune system such as macrophages are inherently equipped with receptors that recognize LPS. These receptors are called pattern recognition receptors (PRRs), and for LPS, the specific receptor is toll-like receptor 4 (TLR4). There are many different PRRs and many different PAMPs, all of which allow the body to mount an initial immune response to a pathogen without having ever "seen" that pathogen before, because that pathogen expresses a molecule that the body is able to recognize as "not myself".

The *adaptive immune system* is found only in vertebrates, and is the system that allows an animal or person to create an immunologic "memory" after an initial exposure to a pathogen. When this occurs, subsequent encounters with the same pathogen will lead to enhances and more effective immune responses. The mechanism behind this response involves the development of pathogen-specific receptors on B and T lymphocytes, which produces a long-lived and specific immunity to that pathogen. These T and B lymphocyte responses are mediated primarily through antigen presenting cells (APCs) such as macrophages and dendritic cells that reside in many tissues throughout the body. Specialized APCs in the CNS include glia and choroid plexus epithelial cells[6,7]. The role of APCs is to engulf a pathogen and then migrate via the lymphatic system to regional lymph nodes. Here the APC "presents" the pathogen to T lymphocytes, which mount an adaptive immune response to the pathogen. A classic example of an adaptive immune response is vaccination.

In reality, this is an extremely oversimplified description of the roles of the innate and adaptive immune systems. Both systems work together to produce an immune response, and many cells throughout the body participate in one or both systems. Additionally, both the innate and adaptive systems can stimulate a response to "self", thus creating what is termed autoimmunity. Lastly, non-infectious diseases may initiate the production or release of damage-associated molecular patterns (DAMPs) which function similar to PAMPS, and can initiate innate immune responses in the absence of pathogens. When this happens, immune responses have the potential to become maladaptive processes that potentiate disease rather than eliminate it.

THE CNS AS AN IMMUNE-PREVILEGED SITE

The brain and spinal cord have long been thought of as immune-privileged sites. This is because under normal conditions the blood-brain or blood-spinal cord barrier is structured to limit entrance of many immune cells into the CNS. Evolutionarily, the concept of immune-privileged was thought to develop as a protective measure for vital structures such as the brain and eyes to limit damage caused by systemic inflammation. However, the concept of CNS immune-privilege has been overwhelmingly challenged in recent years. Both native macrophages (microglia) and immune cells from systemic circulation do traffic through the CNS and contribute to both health and disease in the brain and spinal cord. While the CNS is no longer thought to be immune-privileged, it is "immune-limited" in that certain APCs such as dendritic cells are not present in tissues of the normal brain and spinal cord parenchyma. Additionally, the CNS does not have a lymphatic drainage system as is classically noted in other organs, but drainage of antigens from CNS tissues to the cervical lymph nodes has been demonstrated.

DEFENSIVE RESPONSE OR MALADAPTIVE MECHANISM: THE SPINAL CORD INJURY EXAMPLE

Immune responses that are protective in one scenario can be maladaptive in another. An example of this is observed in the context of acute spinal cord injury (SCI). Acute SCI in dogs can be caused by spinal fracture, FCE, and most commonly intervertebral disc extrusion (IVDE). Primary tissue damage after SCI results from contusion, shearing, and in some cases laceration[12]. Compression of the spinal cord by extruded disc material or by a displaced spinal fracture can affect spinal cord blood supply by interfering with arterial blood flow or blocking venous drainage. These primary mechanical injuries to the spinal cord initiate a series of secondary events, one of the most significant being an immune-mediated inflammatory response. This post-traumatic inflammation after SCI is complex. Initially, inflammation is necessary to promote clearance of damaged neural tissues such that axonal regeneration can occur. However, continued waves of inflammation that occur over time coincide with secondary damage to axons and neurons spared by the initial mechanical insult. This is because infiltrating immune cells cause collateral damage to the healthy tissues that remain after injury. Modulation of the immune responses that occur after SCI is an active area of investigation. Recent promising results include studies that show improved outcome by inhibiting pro-inflammatory mediators such as IL-1β, or inhibiting compounds that enhance leukocyte infiltration into the spinal cord such as MMP-9.

OTHER EXAMPLES IN VETERINARY CLINICAL NEUROLOGY

Granulomatous meningoencephalitis (GME), necrotizing meningoencephalitis (NME) and necrotizing leukoencephalitis (NLE) are common idiopathic inflammatory conditions of the canine CNS[18] . Various studies have attempted to isolate an underlying infectious cause of disease in infected dogs, but to date none have been successful[3, 4, 15]. Some recent literature supports the hypothesis that these diseases represent a single autoimmune disorder with a spectrum of breed specific presentations [13], while other studies highlight key immunologic differences between the granulomatous and necrotizing inflammatory syndromes[14]. The underlying immunologic trigger for disease in dogs with these disorders remains an active area of investigation, although abnormalities in both IL-17 and IFN-γ expression have been noted in a small number of cases[14]. Additionally, autoantibodies against transglutaminase 2 (TG2)[19] and glial fibrillary acidic protein (GFAP)[11, 17] have been identified in the CSF of dogs diagnoses with NME, indicating the potential involvement of aberrant immune responses against those proteins in this syndrome.

Canine Degenerative Myelopathy (DM) is a multisystem degenerative axonopathy involving the central and peripheral nervous system of adult dogs. It is associated with mutation in the gene encoding for the protein superoxide dismutase-1 (SOD1), a feature shared with some familial forms of amyotrophic lateral sclerosis (fALS) in humans[2,8, 16]. Investigation of neuroinflammatory responses associated with canine DM has been previously limited, perhaps because DM has been classically described as a "non-inflammatory axonopathy"[1,9,5,10]. A moderate increase in macrophages has been previously noted in DM, but was presumed to be a secondary response to the presence of degenerating axons and myelin[10]. However, in light of recent studies highlighting the importance of inflammation in the progression of neurodegenerative diseases such as ALS, further investigation into the presence of neuroinflammatory responses associated with DM has been undertaken by our laboratory. Of note, dogs with DM overexpress a DAMP called heat shock protein 70 (hsp70) (Moore, unpublished data). This enhanced expression of hsp70 is concurrent with a significant increase in macrophage numbers, particularly in the gray matter of DM-affected dogs and may be an indicator of a maladaptive innate immune response in DM.

REFERENCES

[1]Averill, DR. **J Amer Vet Med Assoc** 1973; 162 (12): 1045. [2]Awano, T, et al. **PNAS** 2009; 106 (8): 2794. [3]Barber, RM, et al. **J Vet Int Med** 2010; 24 (2): 372. [4]Barber, RM, et al. **J Vet Int Med** 2012; 26 (4): 962. [5]Braund, KG et al. **Am J Vet Res** 1978; 39 (8): 1309.
[6]Canova C, Neal JW, Gasque P. 2006. **J Neuroinflam**. 3: 22-31. [7]Chakravarty S, Herkenham M. 2005. **J Neurosci**. 25 (7): 1788. [8]Coates, JR, et al. **Vet Clin North Am Small Anim** 2010; 40 (5): 929. [9]Griffiths, IR, et al. **J Small Anim Prac** 1975; 16 (8): 461. [10]March, PA, et al. **Vet Path** 2009; 46 (2): 241. [11]Matsuki, N, et al. **J Vet Med Sci** 2009; 71(1): 99. [12]Olby, NJ. **Vet Clin North Am Small Anim** 2010; 40: 791.
[13]Park, ES, et al. **Vet Path** 2012; 49(4): 682. [14]Park ES, et al. **Vet Path** 2013; 50 (6): 1127. [15]Schatzberg, SJ, et al. **J Vet Diag Invest** 2009; 21 (6): 844. [16]Shelton, GD, et al. **J Neurolog Sci** 2012; 318 (1-2): 55. [17]Shibuya, M, et al. *J Vet Med Sci* 2007; 69 (3): 241.
[18]Talarico, LR, et al. **J Small Anim Prac** 2010; 51(3):138. [19]Tanaka, M, et al. **J Vet Med Sci** 2012; 74(6): 733.

Rehabilitation: Treatment for the Neurologic patient

Stephanie Kube, DVM, CCRT, CVPP, DACVIM (Neurology)
Middleboro, MA

Many of our patients have clinical signs secondary to a problem affecting their nervous system. The inciting cause can be trauma, infections, degenerative diseases, structural anomalies, vascular problems and neoplasias. Although addressing the underlying cause is often the mainstay of treatment, many of the symptoms can be helped by the addition of rehabilitation. The goal of rehabilitation is to restore, maintain and promote optimal function as well as quality of life.

Spinal cord injuries are a common indication for beginning a rehabilitation program. Neuromuscular facilitation, proprioceptive training, and balance/coordination exercises can all be helpful to patients with spinal cord injuries. Working with patients with intrinsic (stable) injuries can often be more fun and rewarding to the practitioner because you don't need to worry about harming the patient. However, with fractures (post-repair) and IVDD (post-surgery or without), care must be taken not to put these patients in compromising positions.

Brain injuries, both traumatic and vascular, can be problems where the patients are ideal candidates for physical rehabilitation. Working with patients that are recumbent is beneficial to the patient and rewarding to the practitioner. Rehabilitation has been shown to help with vestibular patients as well as reconditioning patients post craniotomies or traumatic injuries. There are also techniques/exercises that may be done to help retrain and/or rebuild muscle of patients with cranial nerve deficits.

Rehabilitation is recommended to help with pain management and often times pain is associated with neurological injuries. Often times, manual therapy and therapeutic exercises are helpful alone. Hydrotherapy, LASER therapy, cryotherapy and thermotherapy all have their place in the rehabilitation practice.

REFERENCES

Available upon request.

Pain Management for the Neurologic Patient

Stephanie Kube, DVM, CVPP, CCRT, DACVIM (Neurology)
Middleboro, MA

Animals experience pain from many different ailments, of which, neurologic etiologies are many and are often are the most challenging to deal with.

Neuropathic pain is defined as pain that originates from injury to or involvement of the central nervous system (CNS) or peripheral nervous system (PNS). It can can be associated with motor, sensory or autonomic deficits. With neuropathic pain, our patients can also present with allodynia (pain caused by a stimulus that doesn't typically cause pain) and dysesthesias (unpleasant, abnormal sensations).

Pain from meningitis, brain tumors, and disk disease are just a few of the diseases we treat on a daily basis. These diseases and the neuropathic pain that we can see associated with them add to the challenge of managing the pain.

The first aspect to dealing with pain is to localize the origin of the pain. A complete physical exam should be done, including an orthopedic and neurologic exam. If the source of the pain is identified, dealing with the underlying cause directly is the best way to begin to resolve the pain. Sometimes, this is not possible due to the underlying cause or the limitations of the patients/owners. Then we often need to become more aggressive with our pain management.

A multifaceted approach is recommended. This could involve a combination of drugs with different mechanisms, manual therapies/rehabilitation, lifestyle changes, neutraceuticals, and acupuncture.

REFERENCES

Available upon request.

NEURO-NUTRITION: THE GOOD, THE BAD, THE FORGOTTEN

Korinn E. Saker, MS, DVM, PhD, DACVN
Raleigh, NC

Introduction

Insult to the central nervous system (CNS), from any number of causes (trauma, neoplasm, ischemia, metabolic disorders, autoimmune disease, or neuromuscular dysfunction), has a profound impact on the nutritional status of the patient.() The nutritionally mis-managed (mal-nourished) neurologic patient is put at a higher risk for delayed response to treatment(s) and subject to poorer outcomes. Nutritional support can be extremely beneficial to these patients if provided in a timely, appropriate manner. Inappropriate nutrition support can be as detrimental as no nutrition support to the neurologic patient. Therefore, supporting all components of the CNS with the correct nutrients and nutrient levels; and delivering these nutrients at the appropriate time and rate can optimize management of the neurologically-impaired patient. This overview will focus on nutritional support aspects of managing the common categories of neurological impairment in companion pets.

Nutrition Support* Goals** (can/should) be both *general* and *specific* to address the individual neurologic condition(s) of the patient. ***General Nutrition Support goals of all neurologically-impaired patients should include:

- provision of calorie-supplying nutrients to meet their specified daily energy (calorie) needs to prevent mal-nutrition;
- nutrient delivery via a method (voluntary, assisted) that ensures adequate delivery and maximizes nutrient (diet) utilization to the patient; and
- diet choice and feeding plan that "does no harm" to the patient or does not exacerbate the adverse current clinical picture.

Neurologic condition-*specific* nutrition support goals will be highlighted as the conditions are summarized below. The first step to adequately address the nutritional needs of any patient, including the neurologically-impaired patient, would be to asses that patients' current **nutritional status.** This can be accomplished through a **screening**, or in some cases, may require a more **comprehensive** assessment. The findings from your nutritional assessment would ideally be interpreted into a **Nutrition Support Plan** (NSP) for the patient. A NSP would provide specifics regarding when to initiate feeding; what diet to feed, what diet dose, feeding route and frequency; and finally, monitoring guidelines, including what to reassess and how often. The monitoring aspect is as important as choosing a diet and feeding method, in that it will provide feedback regarding the effectiveness of your NSP for that patient and allow for revisions to optimize the plan in a timely manner.

Nutritional Assessment of the Neurologic Patient

There are several suggested approaches to nutritional assessment of a patient. More recently AAHA, in conjunction with the ACVN, have published a basic screening evaluation, "Nutritional Screening: Risk Factors" **(Figure 1)**. It provides a check list of factors associated with the patients' clinical history and physical examination that would influence overall patient nutritional status. Many of the factors listed are components of a thorough patient assessment. They are included as part of the *Nutritional Screening* because, although nutrition may not be the direct cause of the factor listed, it is very likely that alterations in diet / feeding method can help mitigate or minimize adverse clinical signs associated with the identified factor.

If no 'red flags' are raised for the patient from the **nutritional screening**, one can assume they are not at high risk for mal-nutrition and that the current diet and feeding plan for that patient will be adequate. Once <u>ANY</u> changes occur in the patients' neurologic, physiologic or environmental status, the nutritional screening should be repeated (reassessed). Depending on the patient, this re-assessment could be within the same day, week, month or longer. If there is any indication from the screening that the patient is "at risk", a **comprehensive nutritional assessment** is indicated. A comprehensive nutritional assessment involves a detailed evaluation of diet and feeding specifics for the patient as concerns for mal-nutrition have been raised.

Enlisting the assistance of a veterinary clinical nutritionist, when possible, is recommended to facilitate and interpret a patients' comprehensive nutritional assessment, especially for the very high risk neurologic patient. This may not be possible in many clinic settings, but should not deter you from the instituting this process.

What are the steps of a comprehensive nutritional assessment?

- Determine current body weight (CBW), body condition score (BCS) and ideal body weight (IBW) of the patient. IBW can be based on:
  - Current BCS as compared to ideal BCS
- Determine actual daily calorie intake of the patient. This will be based on:
  - The caloric density of the diet being offered.
  - The actual amount (grams, cans, cups) of diet offered vs consumed.
- Determine desired daily calorie intake for the patient based on CBW and IBW.
- Identify *Key Nutrients of Concern* and *Diet Characteristics* for the patient.
  - Interpretation of available clinical history, diagnostics and diagnoses are integral to identifying key nutrients and diet characteristics that can be beneficial to the patient.
  - These nutrients or characteristics, if 'tweaked' in some manner, will help improve the clinical picture and/or better ensure adequate daily calorie intake.

Collectively, the information from your comprehensive nutritional assessment will assist you in devising a specific NSP to correct any mal-nutrition state of your patient. It outlines when to initiate feeding; what diet to feed, what diet dose, feeding route and frequency; and appropriate monitoring guidelines.

Common Neurologic Conditions and Associated Nutritional Concerns

Nutritional screening and comprehensive nutritional assessment may seem a bit daunting depending on your background in clinical nutrition concepts. A "user friendly context" is outlined below that summarizes the common neurologic conditions and highlights key nutritional concerns, with regards to what would be commonly noted on a comprehensive nutritional assessment of the neurologic patient with one of these common neurologic conditions.

Neurological disorders can affect any breed / breed mix of dog or cat. They affect the pets' central nervous system which includes the brain, the spinal cord, and the nerves. Some disorders are congenital and some develop with age or a traumatic event. Other disorders are idiopathic and the cause is not known. Disorders can be categorized in various ways such as: genetic, degenerative, infectious, inflammatory, trauma-induced, and other; and there may be overlap across categories for a specific disorder.

Table 1 below summarizes what this author considers to be the most commonly encountered neurologic-related disorders of dogs and cats in veterinary practice, the nutrition-related concerns and key nutrient and diet characteristics associated with these disorders.

Table 1.

Disorder	Nutrition-related concerns	Key nutrient & diet characteristics
Degenerative - Myelopathy Aging Trauma	Managing body weight Managing muscle mass/tone Ability to chew, swallow Oxidative damage Managing associated inflammation Stool volume, form, incontinence	Total calorie intake Adequate protein intake Diet form & delivery route (assisted?) Antioxidant sources Anti-inflammatory nutrients (omega(n3)-3 FA) Adjusting dietary fiber (soluble vs insoluble)
Trauma - Head Spinal cord Limbs Other	Managing associated inflammation Minimizing metabolic sequella Managing BW & muscle mass/tone Oxidative damage Incontinence	Anti-inflammatory nutrients (omega-3 FA) Total calorie intake (hypo- vs hyper-caloric) Adequate protein intake Diet form & delivery route (assisted?) Antioxidant sources Adjusting dietary fiber (soluble vs insoluble)
Cognitive Dysfunction – Aging (vestibular) Hepatic encephalopathy Seizure-related Inflammatory/Infectious Other (tumor, trauma, etc)	Managing BW & muscle mass Nutrient intolerances (protein?) Ability to chew, swallow Oxidative damage Incontinence	Total calorie intake Appropriate protein intake/sources Anti-cancer type diet profile (fat, CHO, n3-FA,antioxidant sources/timing) Antioxidants, flavonoids Diet form & delivery route
Seizure – Epilepsy	Ability to chew, swallow Oxidative damage Managing body weight Managing associated inflammation	Diet form & delivery route & timing (assisted?) Antioxidant sources Total calorie intake Anti-inflammatory nutrients
Inflammatory/Infectious Meningioencephalomyelitis	See cognitive dysfunction	See cognitive dysfunction Anti-inflammatory nutrients (omega-3 FA)

Considerations for appropriate nutrition support to the neurologically-impaired patient

I. An initial step in ensuring appropriate nutrition support for the neurologically-impaired patient is determination of desired daily calorie intake. This is referred to as daily energy requirement, (DER). Providing adequate DER is a fundamental aspect of preventing or correcting mal-nutrition in the neurologic patient. **How is DER determined?**

- DER is a calculation to determine the calorie intake, reported as kcal/day, required to maintain a desired body weight/condition. The desired weight/condition needs to be determined based on the patients' clinical picture. A critically ill patient should be maintained at current BW (CBW), no matter what current body condition, until deemed stable and into the recovery phase. Therefore, DER is calculated using CBW. DER for the non-critical patient should be calculated based on ideal BW and condition. The DER calculation is as follows:
  - DER kcal/day $= [(BW_{kg})^{0.75} \times 70] \times$ **DER factor** (1-3)
  - The chosen **DER factor** should reflect the patients' calorie needs to support such things as age (life-stage), gender, activity level, disease state, target BW and condition. A truncated DER factor guide for the neurologic patient is summarized in **Table 2.**

- Post-op patients should NOT be overfed during the initial recovery period. Overfeeding can result in the onset of undesired metabolic complications and simple gastrointestinal discomfort presenting as nausea, vomiting, diarrhea, or anorexia. Use a DER factor for critical care or the low end of adult maintenance for initial re-feeding of these patients

Table 2. DER factors for the neurologic patient

Condition	Canine DER factor	Feline DER factor
CRITICAL ILLNESS	**1.0**	**1.0**
Growth life-stage[a]	1.8 - 3.0	2.0 - 3.0
Adult maintenance life-stage[a]	1.2 - 1.6	1.1 - 1.2
Weight gain (at ideal BW)	1.8 - 2.5	1.4 – 1.8
Weight loss	1.0	0.8

[a] The younger the pet, the higher the DER factor.

II. Equally important, especially in the post-op neurologic patient and patients with degenerative neurologic conditions, is **adequate protein intake**. Protein has a multitude of functions in the neurologically-impaired patient. It supports lean muscle mass; immune system function; daily cell turnover; the acute phase response (acute phase proteins, hormones, inflammatory mediators); catabolism associated with mal-nutrition, and much more.

- Daily protein needs will be dictated by the patients age (life-stage), current nutritional status, disease state, surgical status, and degree of neurologic impairment. Protein needs can be estimated and an appropriate diet identified to supply those daily needs.
- Commercial prescription and over-the counter diets provide a wide range of protein levels (and sources) which allows for specific patient needs to be met. The protein content in diets can be evaluated based on an energy density (grams protein/100 kcal) or percent dry matter (%DM) basis. **Table 3** summarizes suggested target protein levels for the common neurologic conditions highlighted above.

Utilize both calculated DER and target daily protein intake to choose the appropriate diet or diet options for each neurologic patient. Protein levels for all prescription and many over-the-counter pet diets can be found in product reference guides. Guides are available, free of charge, from pet food companies.

Table 3.

Condition category	**Suggested protein target level - Canine -**		**Suggested protein target level - Feline -**	
	Grams/100 kcal	% Dry Matter	Grams/100 kcal	% Dry Matter
Degenerative	7 -11	30-44	9.5-13	38-55
Trauma (post-op) and Inflammatory/Infectious	8-14	32-50	10-16	44-60
Cognitive Dysfunction if encephalopathic	6-8 2.5-4	28-32 8-17	9-11 5-7	34-45 26-31
Seizure-related	6-8	28-32	9-11	34 -45

III. When to start feeding and how much to feed neurologic patients are two key considerations to minimize metabolic and physiologic complications that may arise from inappropriate nutritional support.

- By approximately 4 hours of nutrient deprivation, the tight junctions of gut endothelial cells start to loosen their connections, potentially compromising the gut mucosal barrier. By day 3-4 of nutrient deprivation, the intestinal villi are significantly blunted, crypt depth is decreased, and tight junctions are loosened, setting the stage for bacterial translocation, nutrient maldigestion-malabsorption, and ultimately a mal-nourished state for the patient. Delays in provision of nutrition support will exacerbate ongoing clinical concerns.
- Ideally, nutritional support should not be delayed longer than day 3 of anorexia-food deprivation (NPO).
- If hospitalized and critical, start re-feeding conservatively in terms of total calories (kcal/day), but address daily protein needs as effectively as possible.
    - Refer to **Table 4** for details
- If not hospitalized, provide calculated DER calories at the onset, and ensure protein intake is adequate.

Table 4.

Status	**Time to NS[a]**	**Start rate of NS**
Critical hospitalized	No later than day 3,	23-33% of $DER_{critical\ care}$

	within 48 hr preferred	
Hospitalized, not critical	No later than day 3, within 48 hr preferred	66% of $DER_{critical\ care}$ to 50% $DER_{maintenance}$ [b]
Not hospitalized, not critical	Immediately	50-100 % $DER_{maintenance/growth/weight\ loss/gain}$

[a] NS = nutrition support

[b] The start rate will depend on total days of reported food deprivation/NPO. The longer the period of food deprivation, the more conservative when starting NS.

Key nutrients and diet-feeding options focused on specific neurologic conditions

Degenerative (myelopathy, trauma, aging):

Key Nutrients - increased protein, antioxidants, omega-3 fatty acids (eicosapentanoic, EPA and docosahexanoic, DHA), fat adjusted to target optimal BW, fiber adjusted to target weight loss, improve stool consistency, B-vitamins (especially B_3) for neuroprotection, zinc for protein accretion and neuroplasticity.

Diet feeding options - over-the-counter or prescription growth, weight loss or higher protein senior life-stage diets.

eeding method - voluntary intake and/or assisted tube feeding when indicated to ensure adequate DER intake.

Trauma, post-op (brain, spinal cord, other):

Key Nutrients - increased protein, antioxidants, EPA& DHA, fat adjusted for hypo-caloric feeding immediately post-op, post-trauma, zinc for protein accretion and neuroplasticity, but take care not to cause neurotoxicity, B-vitamins (especially B_3) for neuroprotection.

Diet feeding options - high protein, calorically dense (higher fat) prescription diets, moist diets that can be easily blended.

Feeding method - offer voluntary intake, but assume some form of assisted feeding may be needed to ensure appropriate nutrition support. If assisted feeding required consider syringe feeding, placement of nasogesophageal, esophagostomy, gastrostomy tubes or parenteral feeding for hospitalized patients.

Cognitive dysfunction (aging, HE, tumor, trauma, other):

Key Nutrients – protein adjusted based on mental status and liver function, branched chain amino acids if HE, antioxidants (lipoic acid and carnitine), flavonoids, EPA & DHA, fat adjusted to offset lowered carbohydrate calories if cancer, total calories adjusted to target ideal BW, B-vitamins (especially B_3) for neuroprotection.

Diet feeding options - prescription diets formulated for cognitive dysfunction, liver disease, or senior life-stage (over-the counter).

Feeding method - voluntary intake and/or assisted tube feeding when indicated to ensure adequate DER intake.

Seizure (epilepsy, HE):

Key Nutrients - protein adjusted based on mental status and liver function, branched chain amino acids if HE, lowered chloride (sodium or potassium chloride), antioxidants, flavonoids, EPA & DHA, fat adjusted to offset lowered carbohydrate calories if cancer and/or to make patient ketogenic, total calories adjusted to target leaner BW.

Diet feeding options - prescription diets formulated for cognitive dysfunction, liver disease, or senior life-stage (over-the-counter).

Feeding method - voluntary intake and/or assisted tube feeding when indicated to ensure adequate DER intake.

Inflammatory/Infectious (encephalomyelitis, etc):

Key Nutrients - high protein based on mental status, EPA & DHA, fat adjusted to spare protein and limit carbohydrate calories, total calories adjusted to target ideal BW, vitamin A (beta carotene), B-vitamins (especially B_3) for neuroprotection, zinc for protein accretion, neuroplasticity and immune function.

Diet feeding options - prescription diets formulated for increased protein and palatability, fish protein sources or n3-FA supplemented .

Feeding method - voluntary intake and/or assisted tube feeding when indicated to ensure adequate DER intake.

Concluding Comments

Nutrition support for the neurologic patient (neuro-nutrition) can easily be overlooked or forgotten in a clinical setting. Consuming a bite or two of diet is **not** adequate or appropriate nutrition support for these patients. Performing a nutrition screening, followed by a more comprehensive nutritional assessment when necessary, is a basic standard of care process for any veterinary practice. In so doing you will identify the specific nutritional needs of your neurologic patient. Addressing these needs in a timely and appropriate manner will minimize adverse clinical complications, improve patient outcomes, and enhance the overall quality of life for these patients.

REFERENCES

Barkhoudarian G, Hovda DA, Giza CC. 2011. The molecular pathophysiology of concussive brain injury. Clin Sports Med. 30:33-48.

Chi Y, Sauve AA. 2013. Nicotinamide riboside, a trace nutrient in foods, is a Vitamin B3 with effects on energy metabolism and neuroprotection. Curr Opin Clin Nutr Metab Care. 16:657-661.

Chinopoulus C, Adam-Vizi V. 2006. Calcium, mitochondria and oxidative stress in neuronal pathology. Novel aspects of an enduring theme. FEBS J. 273: 433-450.

Gomez-Pinilla F, Tyagi E. 2013. Diet and cognition: interplay between cell metabolism and neuronal plasticity. Curr Opin Clin Nutr Metab Care. 16:726-733.

Gomez-Pinilla F. 2008. Brain foods: the effects of nutrients on brain function. Nat Rev Neurosci. 9:568-578.

Gomez-Pinilla F, Nyugen TT. 2012. Natural mood foods: the actions of polyphenols against psychiatric and cognitive disorders. Nutr Neurosci. 15:127-133.

Hall ED, Vaishnav RA, Mustafa AG. 2010. Antioxidant therapies for traumatic brain injury. Neurotherapeutics. 7:51-61.

Lukaski HC, Hardy G. 2013. Micronutrients in brain function, traumatic brain injury. Curr Opin Clin Nutr Metab Care. 16:700-702.

Mayurasakorn K, Williams JJ, Ten VS, Deckelbaum RJ. 2011. Docosahexaenoic acid: brain accretion and roles in neuroprotection after brain hypoxia and ischemia. Curr Opin Clin Nutr Metab Care. 14(2):158-67.

McConeghy KW, Hatton J, Huges L, Cook AM. 2012. A review of neuroprotection pharmacology and therapies in patients with acute traumatic brian injury. CNS drugs. 26:613-636.

Morris DR, Levenson CW. 2013. Zinc in traumatic brain injury: from neuroprotection to neurotocixity. Curr Opin Clin Nutr Metab Care. 16:708-711.

Figure 1Nutritional Screening: Risk Factors

Nutritional Screening Risk Factor	
History Check	**(√) if present**
Altered gastrointestinal function (e.g., vomiting, diarrhea, nausea, flatulence, constipation)	____
Previous or ongoing medical conditions/disease	____
Currently receiving medications and/or dietary supplements	____
Unconventional diet (e.g., raw, homemade, vegetarian, unfamiliar)	____
Snacks, treats, table food >10% of total calories	____
Inadequate information about or inappropriate feeding management	____
Inadequate or inappropriate housing	____
Physical Examination	
Body condition score	____
5-pt scale: any score other than a 3	____
9-pt scale: any score less than 4 or greater than 6	____
Muscle condition score: Mild, moderate or marked muscle wasting	____
Unintended weight loss of >10%	____
Dental abnormalities or disease	____
Poor skin or hair coat	____
New medical conditions/disease	____

THE HORSE DURING EXERCISE: CONTRIBUTIONS AND PITFALLS FROM THE HEART AND LUNGS

Anna M Firshman, BVSc, PhD, DACVIM, DACVSMR
Saint Paul, MN

The cardiovascular and respiratory systems work in a coordinated way to transport oxygen and energy products (e.g. glucose, fatty acids) to the muscle fibers, where they are used for aerobic energy production; and to remove waste products (e.g. lactate, carbon dioxide, water) from them. During exercise increased amounts of oxygen is required by working tissues. Oxygen delivery is improved by increases in the volume of air breathed, the amount of blood pumped by the heart, and the oxygen carrying capacity of the blood, and the redistribution of the blood flow to the working muscles.

CARDIAC RESPONSE TO EXERCISE

VO_2 Max

VO_2 max is the maximal volume of oxygen the horse's body can use per minute. The athletic capacity of horses is attributable to a large number of physiologic and anatomic adaptations that allow an extremely high VO_2 max (200 ml/kg/min vs. 85 ml/kg/min in an athletic man). The adaptations allow include a very large lung size, high cardiac output and stroke volume, high hemoglobin concentration and the capacity for splenic contraction. In addition to various anatomical adaptations that allow efficiency of gait, horses also have well developed mechanisms that allow the efficient release of the heat generated during exercise, such as the production of large volumes of sweat and evaporation from respiratory tract. Training in horses typically increases VO_2 max by 10-25%.

Heart Rate Monitoring Methods

Heart rate measuring technology is rapidly progressing in horses and is being used increasingly by trainers in their training programs. Several methods have been available for many years.

1. Feel the digital or mandibular pulse. A simple, low cost method, however impractical during exercise.
2. Stethoscope – impractical to use during exercise.
3. Electronic monitoring - Most reliable method during and while recovering from exercise.

Heart Rate = number of heat beats per minute

The normal resting heart rate of a mature horse is between 30 to 40 beats per minute. This rate is difficult to obtain in some situations, as certain horses become excited by external stimuli, which elevates the resting heart rate. Although resting heart rate in humans can decrease dramatically as a result of physical conditioning, the resting heart rate of horses does not appear to change appreciably with fitness. Athletic horses however often have low resting HRs and this is an indication of their large stroke volume capacity. During exercise the rise in HR is the major contributor to the increase in cardiac output. At exercise onset, HR increases rapidly from approximately 30 beats/min to approximately 110 beats/min, however when running at low speeds the heart rate may elicit an early over shoot. At faster speeds further elevations are achieved less rapidly and are driven by the sympathetic nervous system and circulating catecholamines.

The anaerobic threshold in horses is around 150 to 170 beats per minute. Heart rates below this threshold indicate that a large percentage of exercise is being performed aerobically. When exercise intensity or duration increase, the requirements of the cardiovascular system increase, which in turn results in an elevated heart rate. Heart rates above the anaerobic threshold characterize rates of metabolism that exceed the abilities of the oxygen dependant pathways supplying energy. Heart rates of 170 beats per minute or greater indicate that a large percentage of metabolism is occurring anaerobically, or without oxygen. Anaerobic metabolism is supported primarily with glucose and glycogen as the fuel source. The threshold heart rate, like resting heart rate, does not change dramatically with physical conditioning in horses and it is likely that the threshold HR for anaerobic exercise may be influenced by genetics.

The maximal HR (HR_{max}) of a horse is in the range of 210-280 beats/min, which represents a 7 fold increase over resting values. Each horse has its own HR_{max}, which is reached at a particular exercise intensity. Once a horse reaches its HR_{max} a further increase in speed is still possible, but it does not elevate the HR any more. A horse with a high HR_{max} is favored as an athlete because it can perform more work at a specific heart rate. Conditioning does not alter a horse's resting HR or the HR_{max}. After conditioning the horse reaches its HR max at a higher workload, and travels faster/works harder at a given HR. In humans HR_{max} decreases with age and an age related decline in HR_{max} has recently been described in horses. Maximal heart rates should not be used as a major part of conditioning programs; rather, they should be considered a danger zone where fatigue may occur quickly. The speed or velocity a horse can achieve or sustain at a submaximal heart rate of 140, 170, 200 beats/min (i.e. V_{140}, V_{170}, V_{200}) provides information about stroke volume and cardiovascular capacity and can be used as an indicator of fitness and racing potential i.e. as a horse gets fitter the velocity at which it travels at a heart rate of 170bpm should increase.

Stroke Volume (SV) = amount of blood pumped during each beat

At rest 450kg horse SV ~ 900ml (2-2.5ml/kg)
During exercise 450kg horse SV ~ 1200 ml (increases by 33%)

Typically SV increases sharply at exercise onset up to around 40% VO_{2max}. Maximal stroke volume may not coincide with maximal HR as during very high heart rates; diastole is insufficient, resulting in inadequate ventricular filling. SV will increase a certain amount with training.

Cardiac Output (CO) = HR X SV
At rest 450kg horse CO ~ 30-45 l/min
During exercise 450kg horse CO increases to a maximum of about 240 l/min.

The cardiac output is the amount of blood pumped by the heart each minute and is the most important means of increasing delivery of oxygen to the muscle and so is the principle determinant of $VO_{2\ max}$. Increases in CO are mostly due to an increase in HR and to a smaller degree by an increase in SV.

Cardiovascular Response to Exercise

HR increases in anticipation of exercise. The more excitable the horse is the more it will show an anticipatory rise in HR. With the onset of exercise there is a rapid elevation of the horse's HR. When steady submaximal work is performed the HR rate shows an initial overshoot before falling to a plateau after 2-3 minutes. The size of the overshoot and the steady HR depends on the horse's fitness and work intensity. In horses galloping at a steady speed on flat ground, there is a linear relationship between speed and HR at speeds in the range of 350-700 m/min (13-26mph), which are roughly equivalent to HR of 140-200 beats/min. As the horse approaches HR_{max}, the HR response curve tends to flatten. The graph of HR vs. speed shifts to the left when the horse is sick and to the right with increasing fitness. When high intensity exercise ceases, there is a rapid deceleration of the HR, with the greatest reduction occurring in the first minute post-exercise. The rate at which the HR declines depends on the intensity and duration of the work, the horse's fitness and the environmental conditions (heat, humidity). Generally the fitter the horse, the faster the HR returns to normal after a standard amount of exercise.

Effects of Conditioning

The cardiovascular system shows considerable adaptations in response to conditioning and the changes occur relatively rapidly in comparison to the slow rate of adaptation of the musculoskeletal system. As the horse gets fitter there are reductions in the HR and CO at a given level of exercise, although HR_{max} does not change. The horse's ability to consume oxygen improves due to improvements in the efficiency of the CV system in delivering oxygen to the skeletal muscles and in the ability of the muscles to extract oxygen from the blood. There is an increase in capillarization of the muscles which slows capillary transit time and enhances gas exchange by allowing a longer period of contact between the RBCs and the muscle fibers. Some of the exercise induced changes in the CV system vary with the nature of the regular work. Endurance conditioning raises the plasma volume by about 20% and the hemoglobin concentration by about 34%. High intensity sprint conditioning stimulates greater increases in the PCV, RBCs and Hb than endurance exercise.

RESPIRATORY RESPONSE TO EXERCISE

Horses are unusual in that they are obliged to breathe through the nose, unlike other animal species that have the option of breathing either through the nose or the mouth. During exercise the primary function of the respiratory system is gas exchange which involves supplying oxygen and removing carbon dioxide from the blood in the pulmonary capillaries. The respiratory system also plays an important role in thermoregulation and acid-base balance.

Respiratory Rate = number of breaths per minute

Horses have a normal resting respiratory rate of 12-20 breaths per minute. During exercise the respiratory rate rises as high as 180 breaths per minute. At a walk and to a certain extent at a trot and pace, the horse selects an appropriate respiratory rate for the intensity of exercise. In the canter and gallop, however, the respiratory rate is usually coupled to the stride rate with a 1:1 ratio. This is called respiratory:locomotor coupling.

Tidal Volume = Amount of air inhaled and exhaled at each breath

In a 450kg horse at rest the tidal volume is about 4-7 liters, rising to a maximum of about 10 liters during exercise.

Minute Volume = Amount of air passing in and out of the lungs per minute
MV = RR X TV

In a 450kg horse at rest the minute volume is about 100 liters/min, rising to a maximum of about 1500 liters/min during maximal exercise. This increase is mainly due to the increase in respiratory rate and to a lesser extent to the available increase in tidal volume.

Upper Airway

The resistance to airflow in the airways affects the energy expended to drive respiration. The greatest resistance occurs at the narrowest parts of the respiratory tract, which are the nostrils and the larynx. During exercise the resistance to airflow is reduced by flaring the nostrils and dilating the larynx. Other methods of reducing upper airway resistance during exercise include vasoconstriction of the nasal mucosa and straightening of the horse's head and neck, which decreases air turbulence.

Respiratory Response to Exercise

Rate and depth of breathing are controlled in part by chemoreceptors in the blood vessels which respond to changes in arterial oxygen tension, carbon dioxide tension and pH. A rise in respiratory rate is stimulated by a reduced oxygen tension, an increased CO_2

tension or a reduced pH. At the onset of exercise, the respiratory rate and TV rise rapidly in accordance with the body's need for oxygen. In the canter and gallop and to a lesser extent the trot and pace the respiratory rate is coupled to stride rate and so the mechanics of locomotion override the chemical control of breathing. When exercise ceases, the respiratory rate decreases due to the cessation of the locomotor forces that drive respiration. Typically the horse takes a few deep breaths, and then the respiratory rate settles in the range of 60-100 bpm with the horse breathing deeply until the oxygen debt is repaid. Following repayment of the oxygen debt the respiratory response depends largely on the horse's body temperature. Panting is rapid shallow breathing, where air passes through the nasal passage but the tidal volume is small. It is therefore important to observe both the rate and depth of respiration, together with HR and rectal temperature to assess whether a horse is overheated after exercise.

Locomotor-Respiratory Coupling

Horses use locomotor-respiratory coupling (LRC) in canter and gallop to minimize the muscular effort of respiration. In LRC respiration is driven by the locomotor forces associated with weight bearing on the front limbs, the pressure of the abdominal organs against the diaphragm and changes in orientation of the body axis. Thus during canter and gallop there is a strict 1:1 ratio between the respiratory and locomotor cycles. In the trot and pace there is greater flexibility and horses will select a 1:2, 1:3 or 2:3 ratio between the respiratory rate and stride rate. In the canter and gallop one of the major contributors to LRC is the compression of the thorax between the two scapulae as the front legs bear weight. The diaphragm is the principle muscle of inspiration and LRC augments this action through the inertial movements of the abdominal organs (known as the visceral piston). When the front legs impact the ground there is a moment of deceleration, and when the leading front leg pushes off there is a moment of acceleration and this moves the visceral piston. The overall effect is that expiration occurs as the forelimbs hit the ground and inspiration occurs as the forelimbs push off the ground into the suspension phase. The effect is also helped by the rocking action of the horse's head, neck and trunk during the canter gait.

Implications of LRC for Sport Horses

At a constant stride the visceral piston settles into the stride rhythm and the energy used for breathing is minimized. However, each time the stride rate changes it can take several strides before the visceral piston catches up with the new rhythm which creates extra work for the horse. Therefore in order to avoid this extra work, it is easier for a horse to change speed by adjusting its stride length, while maintaining the same rhythm. This is particularly relevant in which long periods of gallop are involved such as in eventing. In addition a horse with a longer stride length is able to travel at the same speed at a lower stride rate allowing it to breathe more deeply at each stride. Anything that shortens the stride length, such as a low-grade lameness will limit the time available for inspiration, and thus may reduce performance.

Effects of Conditioning

Conditioning improves the function of muscles that hold the upper airways open during exercise, particularly the muscles of the nostrils, pharynx and larynx. Some horses that have a slight respiratory noise at the beginning of a conditioning program often will lose it as they get fitter. In the lower respiratory tract, conditioning has relatively little effect; the alveoli, pulmonary capillaries, bronchioles, bronchi and chest wall show minimal adaptation in response to regular exercise. In addition there is little change in the MV and TV with training. This is in contrast to the marked adaptive responses of the CV and muscular systems and suggests that the respiratory system may be the limiting factor to athletic ability in a fit horse. In fact studies have shown that as horses get fitter they become more hypoxic and hypercapnic with exercise. Consequently the respiratory system is regarded as the weak link in the oxygen supply pathway in horses.

REFERENCES

Equine Sports Medicine and Surgery, Hinchcliff et al., Elsevier 2004. Conditioning Sport Horses, Hilary M. Clayton, Sport Horse Publications 1991.

PRACTICAL EQUINE FLUID THERAPY

Allison J. Stewart, BVSc(hons), MS, DACVIM(LAIM), DACVECC
Auburn, AL

INTRODUCTION

Intravenous fluid therapy can be life saving, but is very expensive for a full sized adult horse. Oral fluid therapy is often a much cheaper option when the horse is not refluxing. It is advantageous for veterinary technicians to have an understanding of the volumes, costs and equipment needed to safely and economically administer fluids to horses. An understanding of basic mathematics and metric conversion can make calculations very easy. There are 2.2 pounds in every kilogram. A 1100 pound horse weights 500 kg. There are approximately 4 liters in a gallon. I liter weighs 1 kilogram- which makes it very easy to convert from body weight to volumes of fluid required. The metric system makes calculation very easy, so do not be scared of it!

Fluids are given to restore and maintain hydration; correct electrolyte or acid-base imbalances and on occasion, to administer a limited amount of calories with dextrose supplementation. The total body water of an animal is ~60% of the total body weight, which is ~300 L in a 500 kg horse. The blood volume of a horse is ~8% of body weight (40 liters for 500kg horse) with ~1/3 red blood cells (ie the PCV) and 2/3 being plasma. Therefore the plasma volume is about 27 liters. If an animal is dehydrated, it is actually the cells (intracellular fluid) and fluid between the cells (interstitial fluid) that has been deprived in order to try and maintain the plasma volume, circulation and the delivery of oxygen to the tissues. Therefore when rehydrating an animal we must replace the fluid in the circulation and within and between the cells. For fluid therapy to be successful, we also need sufficient proteins in the blood to create sufficient drawing power to prevent all the fluid leaking out of the vasculature. This attractive force is call the oncotic pressure. If the total protein is <4.0 g/dL, then a plasma transfusion or administration of synthetic colloids is usually required. For example 6% (60mg/ml) hetastarch can be dosed up to 10 mLs/kg q 36-48 hr.

When fluids are replaced intravenously they should have a similar electrolyte composition to that of the blood (isotonic). We cannot give sterile water or diluted dextrose without severely diluting the body's electrolytes. The same is true for oral fluids unless that animal is eating or receiving food through the nasogastric route- in which case the animal simply needs the water. Hypertonic solutions are those with a higher osmolality than plasma. These solutions are given rapidly, restoring circulating volume by creating high concentrations of sodium and chloride within the blood vessel, which creating a higher osmotic force in the intravascular space thereby drawing water into the intravascular space from the extravascular space. This is only successful if the animal is not whole body water depleted- ie dehydrated. You cannot draw water out of a sun dried tomato! Hypertonic saline works very well after acute hemorrhage as the circulating volume is low from the hemorrhage, but the rest of the animal is normally hydrated. Sometimes hypertonic saline is administered to dehydrated animals, but it must be immediately followed by shock volumes of isotonic fluids. The dose of 7 % hypertonic is 2-5 mLs/ kg (for example- a dose of 4mLs/kg = 2 liters for a 500 kg horse).

Volume of fluid required

To calculate the approximate volume of fluid required for 24 hours, one must estimate the volume required to restore hydration, the volume to provide for the horse's daily maintenance and any additional ongoing losses.

Hydration status is assessed by physical examination and assisted by laboratory results. Dehydration of <5% is not clinically detectable. Dehydration >15% is not compatible with life. This would mean weight loss of 75 kg in a 500 kg horse!

With mild dehydration (5-7%) a horse may appear lethargic, have dry mucous membranes, slightly prolonged CRT and decreased urine production. With moderate dehydration (8-10%) there is also elevation in heart rate, decreased skin turgor, a weak pulse and prolonged jugular filling. With severe dehydration (>10%), there is also cold extremities, marked depression, sunken eyes and possibly recumbency. Laboratory data that is supportive of physical examination findings includes: increased PCV and total protein; elevated creatinine concentration and increased urine specific gravity. Concentrated urine tells us that the kidneys are working adequately, but markedly concentrated urine, (>1.035 for an adult horse and >1.025 for a foal) indicates the kidneys are in overdrive and the horse is probably dehydrated. If the horse's stall is dry or only 1-2 west spots are present in the bedding over a 12-24 hr period there is something wrong! The horse is too dehydrated to spare making much urine. This is an important warning sign that the horse is dehydrated. A normally hydrated foal that is consuming adequate milk should have a urine specific gravity of < 1.007.

% Dehydration	CRT (sec)	Heart Rate (BPM)	PCV/TP (%/g/L)	Creatinine (mg/dl)
6%	2	40-60	40/7	1.5-2
8%	3	61-80	45/7.5	2-3
10%	4	81-100	50/8	3-4
12%	>4	>100	>50/>8	>4

Example- a 500 kg horse is 8% dehydrated. Calculate the volume required for restoration. 500 kg x 8 % = 40 L or 500 kg x 8 / 100 = 40L. (Notice that with the metric system kilograms can be used interchangeably with liters, because 1 liter of water weighs 1 kilogram.)

Maintenance fluid account for normal urinary, fecal and metabolic water loss. The maintenance fluid requirement for adult horses is ~60 mL/kg/day (some suggest 50 mL/kg/day). Lower requirements are needed if the horse is off food, as fluid is not needed for digestion. Maintenance rates are increased for foals because they have a higher percentage of body water and less fat and muscle than an adult animal (80mLs/kg/day for a neonate). The maintenance rate for a 500 kg horse = 60 mL x 500 kg = 30,000 mLs = 30 L per day. (There are 1000 mLs in 1 liter).

Ongoing losses include pathologic losses such as reflux and diarrhea. It is very difficult to estimate the volume of diarrhea- you often have to guess- 20, 30, 40, 50 liters depending on the volume and frequency of each defecation over a day. Luckily the volume of reflux can be measured in 24 hours or the volume measured in say 4 hours and the total amount that might be lost in the next 24 hours estimated.

For example- if a horse refluxes 7 liters over a 4 hour period, then estimate that they will lose 42 liters in 24 hours. The volumes calculated for rehydration, maintenance and ongoing losses are added together and should be administered during the initial 24 hour treatment period. In our example of a 500 kg horse that is 8% dehydrated and is losing approximately 42 L a day in reflux would need 40L +30L + 42 L = 112 L in the next 24 hours.

Rate of fluid administration

If an animal is in shock, then the maintenance fluids- ie the first 40L can be given theoretically over the first hour. Realistically it is impossible to give this much fluid to an adult horse without two very large (10 gauge) jugular catheters and a pump. By placement of 1 (or 2) 14 gauge intravenous catheters, elevating the fluid bags and using a large bore 30 Fr (large animal) fluid administration set wide open, then 20 liters can be administered by gravity in a few hours. Practically, we can go "wide open" until this first 40 liters is administered if the animal is in shock (say over 10% dehydrated). If the horse is mild-to moderately dehydrated then approximately 20 liters can be given "wide open", or the rate can be constant over the initial 24 hours. To calculate a drip rate after the initial bolus is given, subtract the volume administered as a bolus from the calculated 24 hr requirement. For example 112L minus 40L bolus which was given over 6 hours. We must then administer 72 L in the remaining 18 hrs of the day = 4 L per hour. Two new 5 liter bags must be used up every 2.5 hours. You can give this as a bolus every 2.5 hours or calculate a constant rate infusion of 4L/hr or 4,000mLs/hr. Our administration sets generally make a drip that is 1/10 of a mL (ie 10 drips/mL). Therefore we need to have 40,000 drips per hour or 667 drips per minute (by dividing by 60 minutes in the hour) or 11 drips per second (by dividing by 60 seconds in the minute). This is too fast to count, so set the drip rate at "dripping, but too fast to count". The main thing is that the horse gets its 10L every 2-3 hours. Speed up or slow down the rate as needed. If the desired fluid volume has not been administered then catch up by delivering a bolus of fluids.

For a second example- say we have a horse with an impaction colic that the doctor decides to treat only with intravenous fluids. The horse may not appear dehydrated. There is no ongoing loss, so maintenance = 30L. However, the doctor may decide to add an additional 30L to help "rehydrate the impacted intestine", so the horse is to receive 60L in 24 hours. You need to add two 5L bags (10L) 6 times in the day, or 10L every 4 hours. The horse needs to receive 60L in 24 hr = 2.5 L/hr = 2,500 mLs/hr or 25,000 drops per hour = 417 drops per minute or 7 drops per second. Set the drip rate to a fast , but still countable drip rate and count how many drips in 10 seconds= the correct rate would be about 70 drips in 10 seconds (6-8 drips per second is OK for barn work!)

Isotonic crystalloid fluids are ideal for resuscitation, and these contain electrolytes in sterile water at a strength such that the number of electrolyte molecules is similar to that of plasma. Crystalloid fluids will rapidly exude from the vasculature and move into the fluid between and then finally within cells. This is fine as we are trying to rehydrate the entire body and not just fill up the depleted plasma volume. One hour after crystalloid fluid administration only about 25% of the fluids will remain within the vasculature. Examples that have electrolyte compositions similar to plasma are: Lactated Ringer's solution, Normosol-R® or Plasma-Lyte-A® which contain a combination of sodium, chloride, a buffer (lactate, acetate or gluconate) and a small amounts of calcium and/or magnesium. Any of these fluid can be used for initial rehydration. Over the long term, additional electrolytes may be required if the horse is not eating. Although 0.9% sodium chloride is still isotonic, it has more sodium and chloride ions than plasma. It is useful for foals that have a ruptured bladder or animals with renal failure that have hyponatremia, hypochloremia and hyperkalemia.

If horses are not eating and especially if they are refluxing, then additional potassium, calcium and magnesium should be added to the fluids. These 3 electrolytes are essential for normal intestinal motility, and especially in treating or preventing ileus (reflux) in horses with anterior enteritis or after colic surgery. These can be supplemented in response to monitoring of electrolytes, on blood work, except that potassium and magnesium are mostly intracellular cations and therefore their plasma concentration is not an accurate reflection of the concentrations within the cells. For calcium and magnesium, ionized concentrations should be used in preference to total concentrations, as the ionized cations are biologically active and not dependent on plasma protein concentrations. If blood work is not available, then the low level of supplementation is recommended. If the kidney is functioning normally, excessive electrolytes can be eliminated from the body. Higher levels of supplementation should only be given if deficits have been confirmed by blood work. Potassium, calcium and magnesium can be fatal if large amounts are given rapidly. Do not supplement any fluids given at shock rates.

Potassium: Sick horses rapidly become depleted of potassium due to anorexia or fasting, refluxing or diarrhea, and dilution from administration of IV fluids. If the serum [K+] is < 3 mEq/L then potassium should be supplemented. Add 20-40 mEq/L of KCL to the IV fluids. KCl concentrate comes as 1-4 mEq KCl/mL depending on the manufacturer. For example- the serum potassium is 2.5 mEq/L so we decide to supplement a 5 liter bag of Normosol-R® at 30 mEq/L. How much of the 2mEq KCL/mL concentrate do we need to add? 30 x 5 = 150 mEq of KCL. Therefore we will need to add 75 mLs of the KCL concentrate to each 5L bag (or 15mLs to each 1L bag). Potassium can be supplemented orally if the horse is not refluxing. Give 10-20 grams PO every 6 hours to a 500 kg adult horse.

Calcium: Calcium depletion is common in sick horses, especially those that are refluxing, have diarrhea, Blister beetle poisoning, thumps and in lactating animals held off feed. Calcium is commonly supplemented using 23% calcium gluconate: give 1 mL or 230 mg of 23% calcium gluconate per kg body weight spread over the day. An easy approximate method is to add 50 mLs of the 23% calcium gluconate to each 5L bag. Example- we only have 1 L bags of fluids- how much calcium should we add? 10

mLs per 1L bag. For severe hypocalcemia, with clinical signs administer 100 mLs of the 23% calcium gluconate to each 5L bag (or 10 mLs per 1 L bag) for the first few bags. Care- calcium will precipitate with bicarbonate and several drugs, therefore carefully flush the IV fluid line when fluids have been supplemented with calcium, and never supplement calcium and bicarbonate in the same bag of fluids- you will get calcium carbonate, which is chalk!

Magnesium: Subclinical hypomagnesemia is common in critically ill horses. Magnesium sulfate ($MgSO_4$) can be safely added to the maintenance IV fluids at 150 mg/kg BW/day which is 0.3 ml/kg BW/day of the 50% $MgSO_4$ solution to provide the daily requirement. For example- a 500 kg horse requires 150 of the $MgSO_4$ solution per day. How much do we have to add to each bag of fluids? This depends on the volume of IV fluids you are going to give that horse per day- if he is receiving only maintenance fluids at 30L per day, then add 5mLs of the $MgSO_4$ solution to each IL bag or 25 mLs to each 5L bag. If he is receiving twice maintenance fluids at 60L per day then add 2.5mLs of the $MgSO_4$ solution to each IL bag or 12.5 mLs to each 5L bag. Oral supplementation can be cheaply given if the horse is not refluxing. Give 15.5 g/day of MgO, 46.5 g/day of $MgSO_4$ or 32 g/day of $MgCO_3$, for maintenance, slightly more for lactation.

IV catheters and complications:

Polytetrafluoroethylene (Teflon®) catheters should be replaced every 3 days and are associated with an increased rate of thrombophlebitis, especially in critically ill horses. Polyurethane (Mila®) or silicone rubber catheters can last over 2 weeks in many cases. Use of an extension that screws into the hub of the catheter is recommended to minimize manipulation of the actual catheter and reduce risk of dislodgement. A 3 meter (120 inch) long equine administration set (not coiled) has a cost price of about $10.50. This is suitable for a hose tethered in stocks or cross ties. A regular 72 inch small animal administration set is too narrow and only about 2-3 liters can be administered every hour. This is only useful for minis or foals. A 30 Fr coiled administration set is required for horses moving freely within a stall. An overhead pulley system with a rotating hook prevents fluids from getting tangled. Catheters are usually not covered with a bandage, so that any problem is quickly identified. Bandages may need to be applied in foals or adult horses that are frequently recumbent. A triple antibiotic ointment may be applied at the insertion site on the skin, to decrease infection. Catheters should be flushed with heparinized saline (10 IU/ml) four times a day if they are not being used for fluid administration. When administering a medication, the injection cap should be wiped with alcohol prior to insertion of the needle. The injection cap should be changed daily.

Thrombophlebitis is the most common complication of IV catheter use. Catheterized veins should be carefully inspected twice daily for heat, swelling, presence of discharge, or a "corded feel" of the vein, and promptly removed before extensive thrombosis occurs. The tip of the catheter can be transected in a sterile manner and placed into transport media for culture. If fever develops, then systemic antibiotic therapy is indicated. Otherwise the vein can be treated topically with nitroglycerin ointment, hot packs, DMSO/antibiotic ointments. Any scabs should be removed and abscesses carefully squeezed to the outside. If one jugular vein is thrombosed, do not catheterize the other jugular vein because bilateral jugular vein thrombosis will severely limit removal of blood flow from the head. The head may develop severe swelling, which can be fatal, and will likely limit athletic performance of survivors. If one jugular vein becomes occluded then a catheter can be placed in a cephalic, saphenous or lateral thoracic vein. When placed in any other location than the jugular vein, the catheter should be flushed more frequently, as these catheters tend to clot easier. Leg catheters are usually bandaged, because they are more prone to dislodgment than jugular catheters.

Oral fluids

This is a very inexpensive method to administer fluid therapy, but the horse must have a functional gastrointestinal tract. Oral fluids are contraindicated in horses that are refluxing, and fluid may not be well absorbed in horses that have diarrhea. The stomach volume of a 500 kg horse is ~12-16 liters. The volume that can safely be administered at one time is 8-10 L in a 500 kg horse. This can be given through a nasogastric tube, either using gravity flow, and a funnel or by use of a pump. This can be repeated every 30 minutes, after checking the stomach for reflux. If there is a distal intestinal obstruction- ie a large or small colon impaction, then if the obstruction is complete, the small intestine and stomach may eventually fill and lead to reflux. 60L of oral fluids could be administered to a horse with a large colon impaction by stomach tubing the horse 6-7 times per day. The stomach tube can be left in place or removed and replaced each time. When a regular stomach tube is left in place the horse should be muzzled. If a horse is going to receive oral fluids for more than 1 day, it is advisable to place a nasogastric feeding tube- which is slightly narrower than a regular stomach tube. Horses can even continue to eat with such a tube in place. This tube can be connected to a coiled fluid administration set and fluids administered as a constant drip from an elevated plastic container with a tap at the base (carboy). Loosen the top lid, else a vacuum will be created and the fluids will not flow. If the horse is off feed, then electrolytes should be mixed with the water to achieve an approximately isotonic solution. The table below can be used to make oral fluids.

Electrolyte	Source	mEq	Grams per 1 L water	Grams per 10 L water
Sodium	Table salt	135/L	5.3	53
Potassium	Lite salt	5/L	0.4	4
Bicarbonate	Baking soda	45/L	3.8	38
Magnesium	Epsom salts	8/L	1	10
Chloride	Table salt and Lite salt	95/L	Included above	

ZOONOTIC ABORTION IN SMALL RUMINANTS FOR VETERINARY TECHNICIANS

Shirley Sandoval, BAS, LVT VTS-LAIM
Palouse, WA

INTRODUCTION

Abortion is a major cause of economic and production loss in small ruminant flocks and herds. Although random, non-infectious causes of abortion do occur, it is not uncommon to have abortion outbreaks (or storms) due to infectious causes. The main focus of this seminar will be the more common infectious causes of abortion in small ruminants that are also zoonotic. We will discuss how veterinary technicians can provide key steps to protect the producers, farm workers, laboratory staff and veterinary personnel that may come into contact with these infectious agents.

CAUSES OF ABORTION IN SMALL RUMINANTS

Abortions in small ruminants occur from non-infectious and infectious reasons. Some non-infectious causes of abortion in small ruminants include, but are not limited to: trauma, stress, iatrogenic, prostaglandin use in does, ingestion of toxins, nutrition, and exposure to harsh environments, among others.

Infectious causes of abortion in small ruminants may be caused by bacterial, fungal, parasitic or viral disease. Some of these agents are considered to be zoonotic. Of the more common agents, bacterial causes include: Leptospirosis, Listeriosis, Q-fever, Brucellosis, and Campylobacterosis. Toxoplasmosis is the most common zoonotic abortive parasite of small ruminants.

BACTERIAL ZOONOTIC ABORTIVE DISEASES

Leptospirosis

Leptospirosis is a bacterial spirochete with many serovars. They may be either infectious or non-infectious. Common infectious agents in the sheep and goat include: *Leptospira icterohaemorrhagiae, L. Pomona, L. harjo, L. Bratislava and L. grippotyphosa.*

These agents are contracted through contact with contaminated food and water. When infected, the animal spreads the bacteria mainly in urine, however it is also found in abortive tissues and the subsequent vaginal discharge. Once in the environment, *L.*spp agents can survive for long period of time as long as they do not dry out.

Clinical signs in small ruminants vary, but may include pyrexia, anorexia, hemoglobinuria, anemia, jaundice, abortion or death. Producers may notice a decrease in milk production in ewes.

In humans, the clinical signs include flu like symptoms: fever, headache, chills, muscle aches or they may be asymptomatic. Chronic forms of the disease can manifest as kidney or liver failure, or meningitis.

Diagnosis can be made from serology of acute and convalescent titers; dark field microscopy immunofluorescence of tissue from the aborted fetus and placenta.

There are many biological products available to prevent leptospirosis. They should be administered 2-4 times a year dependent upon the region and the persistence of the agent in the environment. Control of the agent may be aided by limiting the intermingling of species, ensuring rodent control and providing clean water supply to animals.

Listeriosis

Listeria in small ruminants is mainly caused by *Listeria monocytogenes.* It may be spread to other species, including humans. For sheep, *L. ivanovii* can also cause abortions. The organism is a common finding in the environment (soil, water, plants, in facility drains, walls, or floors), as well as the GI tract of man and ruminants. It is a hearty agent that persists in certain environments for years. A primary source of contamination for *Listeria* appears to be improperly fermented silage that is fed during the winter and spring months. However, *L. monocytogenes* does not multiply well in properly fermented silage that has an anaerobic environment with a pH less than 4.5.

The transmission of *Listera* is through abrasions in the oral cavity when eating contaminated feed, nursing from soiled udders or ingesting infected milk from the dam (ingestion of contaminated raw milk in people).

Common clinical signs in small ruminants can manifest as the neurological disease of encephalitis; the reproductive disease of abortion, stillbirths or weak offspring followed by metritis; and finally septicemia resulting in overall depression, ill thrift, anorexia, diarrhea, and emaciation.

In humans, Listeriosis primarily affects the young, old, pregnant and immunocompromised individuals. Clinical signs may include GI disease, meningoencephalitis, septicemia, abortions and multiorgan infection.

Diagnostic sampling of abortive tissues associated with Listeriosis can be disrupted by autolysis; therefore it is important for an expedient necropsy and submission of samples for an accurate diagnosis.

Prevention and control of *Listeria* is difficult as it is ubiquitous in some environments. The judicious cleaning of feed and water containers is important. Limiting the availability of contaminated silage and disinfection of those areas contaminated by aborted materials are all ways to decrease the possibility of continued infection.

Q-Fever (Query Fever, Queensland Fever)

Q-Fever is a rickettsial disease caused by the organism *Coxiella burnetii.* It affects many species, including sheep and goats. Its zoonotic potential in humans is highest amongst people who work around livestock (farmers/ranchers, veterinary personnel, slaughter house workers etc.). The Center of Disease Control (CDC) reports that in documented cases of Q-fever between 2000-2010, the incidence of disease was highest in persons 60-64 years of age (http://www.cdc.gov/qfever/stats/index.html).

Spread through contaminated dust and bedding, *Coxiella burnetii* remains viable in the environment for long periods of time; it is easily disseminated by the wind. This agent is found in large numbers in the placenta and abortus of infected dams; but is also found in the milk, urine and feces of these females. The primary mode of transmission is inhalant, although humans have been infected through ingestion of contaminated raw milk. Ticks are also known to spread Q-fever, but they are not the major mode of transmission.

In small ruminants, open females infected with Q-fever are usually asymptomatic of clinical signs. Pregnant females can have a variety of symptoms including placentitis, late term abortion, or stillbirths. Often abortion storms (abortions in multiple animals over a short time) occur. Sometimes infected females appear normal, or may show depression and anorexia.

In humans, clinical signs may take 2-3 weeks to develop. Q-fever is mostly acute, with up to 5% of those affected becoming chronic. Acutely, there may be no clinical signs or they may have flu like symptoms. In immunosuppressed individuals, the disease can become quite serious and result in pneumonia, hepatitis, or myocarditis. Premature delivery or miscarriages are a risk factor when pregnant women become infected. Chronic disease can manifest weeks to years after the initial infection resulting primarily in endocarditis.

Identification of Q-fever in animals requires finding the agent. It can be found in high numbers in aborted tissues. Care should be taken when obtaining samples when Q-fever is suspected, the use of a respirator can assist in decreasing contamination of the person obtaining samples.

Brucellosis, *B. abortus* and *B. melitensis*;

Brucella abortus is a disease of cattle that can also infect co-housed sheep or goats. It causes Undulant Fever in humans. It can be transmitted by milk & abortus of infected animals. The agent enters sheep and goats via ingestion of pasture or water contaminated by secretions of infected cattle. The *B. abortus* organism can remain viable in the right environment for extended periods of time.

Brucella melitensis is a disease of goats and sheep. It is called Malta Fever in humans. *B. melitensis* is not commonly found in the United States or Canada, but it is known to be in Mexico. It is transmitted through carrier animals that excrete the agent in milk and reproductive secretions. Large quantities of the agent are also found in the placentas from both aborted and live birthed offspring. Naive animals (young and old) become infected by ingestion of the milk, contact with the agent via the respiratory tract or through skin lesions. Additionally, in infected does and ewes that do not abort, the resulting offspring may have a latent infection that can cause abortion during their first gestation. Conversely, there is evidence of preventing disease transmission to these same offspring if they are weaned early and removed from the contaminated environment.

Clinical signs in sheep and goats include late term abortion, although not all pregnant animals abort. Goats are more susceptible than sheep and clinical illness is usually noticed through abortion storms. In goats, systemic illness may present with pyrexia, depression, diarrhea, decreased weight, lameness, mastitis and orchitis in bucks. No apparent signs of disease are present upon gross examination of the placenta.

In humans, clinical signs include reoccurring fever, especially in those who eat unpasteurized goat dairy products. It also is found in those people who work in slaughterhouses, and herdsmen or veterinarians that assist in the delivery of offspring or deal with the abortive tissues.

Diagnosis is made by isolating the agent in aborted tissues. Unfortunately, there are no vaccinations currently available in the U.S. for use in sheep and goats; however, other countries where the disease is more prevalent do have access to biological agents. In the U.S., it may be prudent to use a flock/herd test and slaughter program, followed by judicious cleaning of the environment prior to reintroduction of livestock.

Campylobacteriosis (Vibriosis)

The bacteria *Campylobacter fetus* and *Campylobacter jejuni* cause Campylobacteriosis in small ruminants. These are common flora found in the intestinal tract of some apparently healthy ruminants, birds and dogs. *Campylobacter* is spread via the fecal-oral route in all species. In humans, the agent is often spread by cross contamination of kitchen counters and utensils when improper handling of contaminated animal carcasses, especially fowl, occurs. The clinical signs are more prevalent in sheep and rare in does. These small ruminants can be asymptomatic or demonstrate diarrhea, late term abortions and still births or weak neonates. Necrotic cotyledons and placentitis is a common sequela of camplyobacteriosis. After aborting, the affected females may become immune; however some can become persistently infected and will continue to shed the organism. Additionally, mortality due to complications of the abortions can result in metritis, retained fetuses or peritonitis.

Clinical signs in humans include gastrointestinal disease, but *Campylobacter* can also cause premature birth or abortion in women.

Diagnosis is made by isolating the agent in aborted tissues. There are biologicals available in the U.S. for this disease and they are recommended for the prevention and control of *Campylobacter*. These vaccines can be used in the face of an outbreak. For those farms with active disease, good sanitation of the lambing or kidding environment is important.

PARASITIC ZOONOTIC ABORTIVE DISEASE

Toxoplasmosis

Toxoplasmosis is a parasitic disease caused by the protozoan parasite, *Toxoplasma gondii*. This parasite is passed in the feces of kittens whose dams have consumed infected birds and rodents during gestation; this results in a transplacental infection.

Transmission to small ruminants is through the consumption of contaminated feed that occurs when the kittens defecate in the hay or grain, shedding oocytes from the intestinal parasite into the environment. Once consumed, the oocytes infect the blood stream of the sheep and goats, finally settling in the tissue of these animals. Raw milk can also be a source of infection. The clinical signs in sheep and goats are primarily associated with abortions; however systemic disease including fevers may occur. Neurological disease may develop in immunosuppressed animals that become infected. In pregnant females, the oocytes multiply in the placental tissue and invade the fetuses resulting in fetal death, still births or weak neonates depending upon the time and severity of infection. However, females infected later in gestation may give birth to normal appearing neonates that are infected but have established an immune response. Immunity to the infection occurs in post abortion or post-partum does & ewes. Those animals infected prior to breeding do not abort.

In humans, antibodies to toxoplasmosis have been found in 50% of the population. Most people are asymptomatic. However *T. gondii* in people can manifest with flu like symptoms in the normal healthy adult. In immunocompromised individuals, such as those with HIV or chemotherapy patients, the symptoms can be severe resulting in confusion, headache, coordination issues, seizures, and inflammation of the retina or signs consistent with pneumonia. Infection in pregnant women can result in miscarriage or stillbirth. Infected babies may show signs of disease or the disease may manifest later in life.

Diagnosis may be made via paired serum samples, acute and convalescent titers. Chalky brain lesions in aborted feti may be demonstrated upon necropsy. Prevention includes the elimination of cats and kittens from feed storage areas and pregnant does and ewes. Prevention in humans includes discouraging pregnant women from cleaning litter boxes or handling contaminated livestock feed. People should be weary of consuming unpasteurized milk or undercooked meat.

BIOSECURITY

Biosecurity should be of the utmost importance to veterinary personnel when handling any animal abortion cases because of potential zoonosis. It is of the utmost importance for pregnant or immunocompromised persons to refrain from handling pregnant, aborting or the aborted tissues and fluids of small ruminants. Biosecurity in the form of isolation or quarantine must be implemented to prevent the spread of infection throughout the remaining herd or flock. By isolating the affected group of animals, the producer can prevent further contamination of the environment and spread of disease. It is also important for personnel to wear protective clothing, shoe covers, gloves, and face coverings. In some cases the use of a mask will help prevent infection via inhalation. These animals and their housing area should be handled by specific personnel that only work with the infected herd/flock to prevent cross contamination, or the housing area should be cleaned and the animals treated last.

LABORATORY SUBMISSIONS

When dealing with small ruminant abortions, samples from the abortus (fetus, fluids, placenta) should be submitted to a diagnostic laboratory for evaluation. Most diagnostic laboratories have abortion panels specific to the species being evaluated. Check with the laboratory of choice for the samples required and method of preference for preserving them for evaluation. Fresh, frozen, in formalin or on ice are among the variety of ways different laboratories prefer the samples. It is also preferred to provide acute and convalescent whole blood samples for evaluation. Precautions should be well labeled both inside and out of the package prior to shipment to alert both the package handler and the laboratory staff of the contents and potential for zoonotic disease. Be aware of the rules for both the diagnostic laboratory and the shipping company when sending biological specimens as they differ between companies. Disposal of subsequent infected tissues (placenta & feti), contaminated bedding etc. should be buried deep or burned to prevent continued infection of the premises. All veterinary personnel should be educated and become familiar with standard operating procedures regarding prevention and control of zoonotic diseases. Large animal personnel must be vigilant during lambing & kidding as this is when complacency can occur due to the number and repetitiveness of calls dealing with pregnancy related issues. As most large animal calls are ambulatory in nature, veterinary technicians must take precautions working around small ruminants during parturition or abortions. It is important to have procedures in place to limit the contamination of equipment and clothing. Clean-up of equipment post procedure should occur on farm, removing all contaminated clothing and placing it in a plastic bag for decontamination later will protect the ambulatory vehicle, its occupants, and the cross contamination with other farms. Contaminated clothing should not enter the veterinary hospital upon return from the call, rather it should be placed in a laundry container reserved primarily for unknown contaminants and disinfected accordingly. The personnel responsible for disinfecting this laundry should be well educated on the precautions they need to take and the PPE they need to wear when working at this capacity. Credentialed veterinary personnel are responsible for educating the public to help prevent the spread of contagious diseases.

REFERENCES

Menzies, P, **Vet Clinics FA** 2011;27: 81. Pugh, DG, et. al **Sheep & Goat Med** 2012;224. Roest, H-J et. al **PLOS ONE** 2012;7(11) e48949. Roest, H-J et. al **Vet Quart** 2013; 33(3)148.

EQUINE OUTBREAK INVESTIGATION AND MANAGEMENT

J Scott Weese, DVM, DVSc, DACVIM
Guelph, Ontario, Canada

INTRODUCTION

Infectious diseases and disease outbreaks are an ever-present risk in the equine industry, which has much movement of horses, cross-contact of personnel, sporadic (and often ineffective) use of infection control measures, a limited culture of infection control and a general acceptance of the inevitability of infectious diseases. Outbreaks can be highly variable, ranging from small, unpublicized clusters of disease on single farms to international outbreaks causing tremendous morbidity, mortality, economic impacts and stress. A high profile and dramatic example was the 2007 outbreak of equine influenza in Australia that ultimately affected over 4500 premises,[1] cost billions of dollars and resulted in mass disruption. A more recent high profile event was a 2011 equine herpesvirus type I (EHV-1) outbreak from an American cutting horse competition ultimately involved at least 242 premises across 19 US states,[2] with further movement into two Canadian provinces. Numerous outbreaks in equine hospitals have been identified, such as a 2004 salmonellosis outbreak that infected 54 horses and cost the facility over US$4 million.[3]

GENERAL CONCEPTS OF OUTBREAK INVESTIGATION AND RESPONSE

A complete review of outbreak response is beyond the scope of these proceedings, and is otherwise impractical because of the variable nature of outbreaks, diseases and facilities. Yet, some key concepts remain.

Act early…it's better to be proactive than reactive

Often, outbreak response is initiated later than would be desired. The earlier a problem is identified, the sooner a response can be initiated. This can help contain the disease, prevent spread on or between farms, obtain earlier diagnostic testing results and foster effective communication. A "let's wait a couple days a see if things get better" approach rarely works and can allow a small outbreak to develop into a large outbreak

You don't need to confirm it's an outbreak before starting

Getting information about the scope of the problem is critical, as is discussed further below. Knowing the incidence of disease and related factors can guide the outbreak response, but solid data confirming that an outbreak is actually present and its scope are not always required to initiate a response.

You don't need a diagnosis to develop an intervention…..

Waiting for a definitive diagnosis almost invariably delays the onset of an outbreak response. While a definitive diagnosis is useful, the outbreak response can be initiated in the absence of a diagnosis. Usually, there are only a few likely causes of disease, and the response can be tailored to them. In reality, the core components of outbreak response are the same for most infectious diseases so an effective initial response can easily be developed while diagnostic testing is being pursued. The initial response should be based on the most likely pathogens.

…but get a diagnosis as soon as you can

Making a definitive diagnosis can allow for refinement of the outbreak response to improve its effectiveness and potentially relax some measures that are no longer necessary. From a psychological standpoint, making a diagnosis is also often important to people that are affected since dealing with 'unknowns' can accentuate fear and stress. Proper diagnostic testing of as many specimens as possible should be a priority.

Make a team

Outbreak investigation and response can be time consuming, difficult and draining. A team approach can help distribute the effort. More importantly, a team can be assembled with members with different strengths, backgrounds and experiences to make the investigation and response more effective. Ideally, an investigation team contains individuals with expertise in the infectious diseases or outbreak investigation, clinicians that are performing the ground-level response and interacting with owners, and relevant stakeholders. While the latter group (e.g. farm managers, racing commission, track managers, horse owners) may have little scientific background, they will likely have more knowledge of current practices on the facility and challenges (e.g. physical, personnel, logistical, compliance) that might be encountered. Having stakeholders is also important to facilitate good communication and to demonstrate to the people being affected that they are represented. A strict initiative (e.g. quarantine) that is developed by a team that includes stakeholders that are directly affected may be better accepted by others that are affected. Care must be taken when assembling the team, particularly lay members, to ensure that a productive environment can be created.

Get help

Sometimes, the team that is established has all of the required expertise, but external assistance is typically required. (Even in an academic hospital where widespread expertise may be present, getting an outside opinion is useful.) Consulting with equine infectious disease specialists, either remotely or integrating them as part of the outbreak team, is important and should be done early in the process. Recruitment of external experts is not an indication of an uncontrolled or uncontrollable outbreak, or lack of competent local personnel, it is an understanding that there are people with additional expertise and experience in these rare situations. Ego should never get in the way of getting proper advice.

Consultation with local regulatory personnel is also useful. At a minimum, it is prudent to keep them informed. However, in some situations, regulatory bodies may be able to help by providing expertise, funding (e.g. subsidized testing) or assistance implementing required measures. In some jurisdictions, local (e.g. state/provincial) authorities may have quarantine power for non-reportable diseases. While voluntary measures are preferred to directives from regulatory bodies, the unspoken 'threat' of enforced quarantine can be a motivating factor to get facilities to take a required response.

It's better to be overly aggressive at the start

It is much better to implement a comprehensive (and sometimes quite restrictive) set of measures at the start, with removal of some over time than to try a stepwise approach starting with a few measures and increasing them if (when) they do not work. It is more effective to be aggressive, particularly early when the scope of the problem may not be well known. Further, from a morale standpoint, it is better to gradually decrease the severity of the response (thereby making people's lives easier and showing that things are getting better) than to increase measures (leading to concern that things may not be under control or the team does not know what they are doing)

Determine the who, what, where, when, why and how

Each of these is critical. Identification of *who* allows for determination of the incidence of disease, its distribution and potential risk factors. It also identifies individuals for diagnostic testing and measures such as isolation, and identification of potentially exposed individuals for cohorting of risk groups. *What* is important, as described above, to identify specific measures that can be implemented and for effective communication. *Where* can involve multiple levels of investigation. One aspect is where of the facility cases have occurred, to identify potential causes and exposed individuals. It also involves determination of whether the outbreak is confined to the farm or potentially part of a broader local, regional, national or even international event. Careful investigation of infected horses can identify *when* the pathogen may have entered the facility, to modify the response and determine potentially exposed animals. *Why* an outbreak occurred is an important step, but often more relevant to post-outbreak synopsis. Understanding factors that led to the outbreak is important to help prevent future problems. In some situations though, the *why* may indicate an ongoing gap or risk that needs to be addressed during the active outbreak response. *How* can be taken in many ways, but perhaps most important are how the pathogen has been transmitted, how it can be further transmitted and how this can be contained, all factors that are beyond the scope of these proceedings.

Infection control isn't rocket science

Key aspects of outbreak response are simply and practical. Basic tools such as cohorting, use of barrier precautions, isolation, restriction of animal and personnel movement, and personal hygiene are almost always key components.

Get someone on the farm

While some members of the team can participate remotely if required, a ground level presence of some team members is needed to understand the layout, confirm reported information, observe standard practices, monitor implementation of recommendations, identify infrastructure or management challenges and to be able to talk directly with stakeholders.

Communicate, communicate, communicate

Effective communication is a critical and often overlooked aspect of outbreak investigation and response. Some important adages should be considered.

The truth is almost always better than rumours (also known as *control the news cycle*). It is inevitable that information will be disseminated, sometimes widely through social media. This can lead to excessive concern and false rumours. Rarely does the Internet rumour mill depict an excessively rosy situation. Rather, inaccurate information can incite fears amongst involved individuals as well as others in the area. Prompt and transparent communication is important, as it is much better for people realize they will get timely and accurate information from the investigation team, rather than to rely on rumours.

Never announce a problem without announcing a solution. Simply announcing a problem does little to soothe fears and foster an effective response. Announcing the problem is important, but it should always be accompanied with an explanation of what is being done. This can calm affected individuals as well as motivate them to help with any required actions.

Update all stakeholders regularly. Information must flow regularly to keep people informed, calm, motivated and compliant. If communication slows down, fears and rumours can rise, or compliance can decrease if the need for ongoing practices is not emphasized.

Talk to local colleagues. Updating colleagues can be useful to obtain information about circumstances that might be associated with the outbreak (e.g. a case of strangles that occurred at a show prior to a strangles outbreak at a facility) and to help them be on the lookout for disease outside of the affected facility.

Empathize…

Restrictions associated with outbreak response can cause considerable hardship. Whether it is disappointment at missing a show or potentially dire economic impacts of quarantine of a racing stable that is the sole income for an individual, it is important to recognize the hardship that people may face. If people feel that their concerns are at least being acknowledged, the hardship may be more palatable. With good communication, those concerns can be used for motivation to facilitate the outbreak response, by emphasizing the importance of a good response to allow for lifting of any restrictions as early as possible.

...but accept short term pain

While empathy is important, it should not interfere with a proper response. The adage '*short term pain for longterm gain*' is highly applicable to outbreak response.

Perform an outbreak post-mortem

While resolution of an outbreak is often accompanied by investigation-fatigue, reviewing the factors that led to the outbreak is important, for various reasons. One is that it is important to try to prevent further outbreaks. Another is that it is likely that improvements in infection control practices would be useful for the facility (whether they were a cause of the outbreak or not) and implementation of needed improvements is more effective when people are sensitive to the topic of infection control and infectious diseases. An opportunistic approach such as this can often be a way to improve general practices.

CONCLUSIONS

There is no 'recipe book' approach to outbreak investigation and management because of the variations between diseases, facilities and different outbreaks. However, many general principles apply to all situations and implementation of a logical and comprehensive approach can limit the impact of outbreaks. Early intervention, obtaining the input of relevant experts, applying basic infection control and epidemiological tools and effectively communicating are critical components that need to be applied to any investigation and response.

REFERENCES

[1]Callinan I. Report of the equine influenza inquiry: the August 2007 outbreak in Australia. Canberra, Australia: Commonwealth of Australia, 2008.
[2] Traub-Dargatz JL, et al. ***J Vet Int*** 2013;27:339-346.
[3] Dallap Schaer BL, et al. ***J Vet Int Med*** 2010;24:1138-1146.

CHEMOIMMUNOTHERAPY FOR THE TREATMENT OF CANINE LYMPHOMA

Gregory K. Ogilvie, DVM, DACVIM (SAIM & Oncology), DECVIM-CA (Oncology)
Carlsbad, CA

INTRODUCTION

Lymphoma is the most common hematopoietic tumor in the dog and cat that generally responds well to chemotherapy resulting in an excellent quality of life and happy clients. There are many types of lymphoma, each with a unique biological behavior, response to therapy and thus, prognosis. There are many differences between lymphoma in dogs and cats. The disease is usually not painful in either species, and it responds very well to systemic chemotherapy. Quality of life is generally excellent, especially if the veterinary health care team does everything possible to minimize adverse effects. The purpose of this brief manuscript is to review some key points about canine lymphoma and its treatment, including the amazing new world of chemoimmunotherapy.

Overview

Lymphoma accounts for approximately 90% of canine hematopoietic tumors in the dog. Genetics and the environment have both been cited as factors that can influence the prevalence of this disease in the dog.[1-5] Some believe this disease is increasing in prevalence, with boxers, bulldogs, bull mastiff, golden retrievers and rottweiler being over represented in several studies.[1-6] Boxers may have more T cell lymphoma than the general population.[7] Golden retrievers have been reported to have three times and rottweilers are 2.5 and 6 times more likely to develop lymphoma than other breeds. [8-13] C type virus particles have been observed in cases of canine lymphoma and leukemia although it is unknown if these viruses are innocent passengers or involved in causality.[14,15] Similarly, factors in the immediate environment have been implicated such as environmental factors, industrial paints, solvents, herbicides such as 2, 4-D, exposure to electromagnetic fields, and diseases of the immune system such as immune thrombocytopenia.[14-24]

Lymphoma is uncommon in young dogs; affected dogs are typically middle-aged. One controversial study suggested gender and neutering is a predisposing factor for developing lymphoma, at least in Golden Retrievers. In that study, the authors stated that nearly 10% of male dogs neutered early in their lives were diagnosed with lymphoma, 3 times more than intact males.

Diagnostics[1-24]

When evaluating an animal for treatment of lymphoma, it is important to not only obtain a definitive diagnosis but also assess the general health of the patient by clinical examination and ancillary diagnostics. The clinical stage of a dog with lymphoma is determined in part by performing a physical examination. All lymph nodes should be palpated; accurate measurements should be recorded to help the clinician assess response to therapy. Careful attention should be directed to examine the eyes for any evidence of iritis, uveitis or hyphema. A delayed identification of this finding can result in blindness. Other, extranodal sites should be examined as well. Each dog is clinically staged based on the results of physical examination, clinical laboratory testing (i.e., complete blood count [CBC], biochemical profile, urinalysis, and bone marrow cytology), and imaging procedures (i.e., radiography and ultrasonography). In each case, a biopsy or cytology and flow cytometry of the affected lymphoid tissue is necessary to confirm the diagnosis and determine the cell type (histologic grading and immunophenotype). The purpose of doing these tests is to confirm the diagnosis and extent of lymphoma, the health of the patient, and to determine which therapy is likely to be most effective which should result in the best quality of life. By knowing where the lymphoma begins, it is possible to subsequently reassess the patient to determine the response to therapy. The most important diagnostics include:

- **Cytology:** Samples obtained from lymph node, liver, spleen, skin or other organs that might be altered by lymphoma for cytological review is practical, cost effective, accurate and expeditious. These samples are easily obtained by repeatedly directing an 18-22 gage needle into the site in question without the needle attached to the syringe. The goal is to fill the needle with cells from the tissue in question. The sample is then expressed onto a slide and gently spread across the glass using a number of techniques. The slide is then stained and examined under the microscope. Cytology will always be the most common way of diagnosing this lymphoma despite the fact that histopathology and flow cytometry are generally more accurate and precise. This diagnostic test can be performed in house although confirmation by a board certified clinical pathologist is important. Occasionally, it may be difficult to differentiate a low-grade lymphoma from a reactive lymphoid disease. In this situation, the lymphocytes in the malignant disease would be expected to have a clonal population, but those in the reactive disease would not. Flow cytometry can help delineate the benign from the malignant disorder.[26]
- **Biopsy:** A histopathologic biopsy is generally considered more accurate than results from cytology. Needle core, wedge, or excisional biopsy samples should be placed into formalin (1 part tissue to 10 parts formalin) and then sent to the pathology laboratory. Special stains can be applied to determine the origin of the cells, however flow cytometry is becoming a preferred diagnostic test.
- **Flow cytometry:** This diagnostic test is performed in the laboratory from cells obtained via needle and syringe and placed in special media to keep the cells alive. The test not only confirms the diagnosis of lymphoma, it determines if the cells are B or T cell in origin, if they express class II major histocompatibility complex (MHC), and the cell size of the lymphoma. The clinician should check with the laboratory performing the test to determine the optimal sample handling technique.
- **Blood and urine tests:** Dogs with lymphoma can have bone marrow infiltration and its associated peripheral blood changes including thrombocytopenia, anemia and either neutropenia or neutrophilia and/or leukemic cells. Changes in biochemical profile parameters are also commonly seen in dogs with lymphoma including elevated liver enzymes and hypercalcemia and

subsequent elevations in ionized calcium, magnesium, blood urea nitrogen (BUN) and creatinine with an isosthenuric urine specific gravity. The presence of hypercalcemia, anemia, leukemic cells in the peripheral blood, presence of pan or bicytopenia are all negative prognostic indicators. As reported below, flow cytometry is a key part of directing effective chemoimmunotherapy.

- **Radiographs or computerized tomography of the chest:** The presence of hilar or sternal lymphadenopathy is seen fairly commonly, especially among dogs with T cell lymphoma. Some dogs with lymphoma will present with subtle respiratory signs that may be associated with a diffuse interstitial pattern of the lungs.
- **Ultrasound or computerized tomography of the abdomen:** Abdominal lymphadenopathy, hepatomegaly or splenomegaly may not be appreciated without these diagnostics. Ultrasonographic features of splenic and sometimes hepatic involvement includes a honeycombed or "chicken wire" like appearance to that organ.
- **Bone marrow aspirate:** The common misperception is that a bone marrow is not needed if the complete blood count is within normal limits. This is not true, as infiltration of the bone marrow, and thus a poor prognosis can occur when blood cells are normal or abnormal. Thus, the presence of lymphoma in the bone marrow suggests not only an unfavorable prognosis, but also concern about the exaggerated myelosuppression associated with the use of chemotherapeutic agents.

TREATMENT[1-18]

Treatment for malignant lymphoma must be systemic in nature since it is a multi-system disease. Chemotherapy is most frequently utilized; however, some immunotherapy has been effective. Within the immunotherapy field, monoclonal antibodies have shown the most promise. Untreated, dogs affected with malignant lymphoma live an average of only six weeks once a diagnosis has been made. With chemotherapy, dogs can survive for 6-10 months with an excellent quality of life. Dogs may receive Doxorubicin every 21 days at a dosage of 30 mg /m2.

Chemotherapy[15-32]

Malignant lymphoma is one of the most responsive forms of cancer presented to the practitioner. With appropriate chemotherapy or chemoimmunotherapy protocols, nearly 90% of animals placed on therapy should reach complete remission. Of those that reach remission, approximately 80-90% will maintain a reasonable (>6 mos) time frame of excellent quality life. Cost is not inexpensive; however, it is within reasonable financial reach of many clients, thus allowing therapy to be possible.

Although marginally effective, prednisone is inexpensive and often used in combination with other drugs to treat lymphoma. With prednisone therapy, the average pet lives 2 months. One-third of the dogs and cats treated with prednisone will go into complete remission, one-third will go into partial remission, and one-third will not respond at all.

Doxorubicin is one of the most effective single agent treatments for lymphoma in dogs.[29] Of the dogs treated with Doxorubicin, 81% developed a complete and partial remission. The duration of remission is approximately 9 months. Dogs treated with Doxorubicin and then switched to COP (cyclophosphamide, oncovin, prednisone) had a higher second remission rate compared to those started on COP and then switched to Doxorubicin.

The COP protocol is effective for inducing a remission in 75% of dogs with lymphoma.[30] A median duration of remission of 6 months is commonly seen. Approximately twenty percent of the dogs treated with the COP regimen are in remission at 1 year. The addition of Doxorubicin to the COP regiment resulted in longer remission time (7 months versus 6 months). A complete remission was attained in eighty-four percent of dogs treated with COPA. An additional 7% achieved a partial remission. Twenty-two percent of dogs treated with the COPA protocol were in remission at one year.

Garrett et al[28] compared a maintenance-free chemotherapy protocol based on CHOP to a similar protocol with a maintenance phase for the treatment of canine lymphoma. Fifty-three dogs with multicentric lymphoma were treated with a 6-month modified version of the University of Wisconsin (UW)-Madison chemotherapy protocol. Disease-free interval (DFI) and survival were compared to a historical control group of 55 dogs treated with a similar protocol with a prolonged maintenance phase. Remission rate for the study dogs was 94.2%. The remission and survival between the 2 groups did not differ significantly. Thus, we, and others, believe the 6-month chemotherapy protocol based on CHOP with no maintenance phase provided is equal to a similar protocol with a prolonged maintenance phase.

Chemoimmunotherapy[32-36]

One monoclonal antibody developed for the treatment of cancer in humans or animals is canine lymphoma/monoclonal antibody 231 (CL/MAb 231; Synbiotics Corporation, San Diego, California), which recognizes canine lymphoma cells.[38, 39] It mediates antibody-dependent cellular cytotoxicity against a canine lymphoma cell line and has been reported to prolong remission duration when used in combination with chemotherapy in dogs with lymphoma.[39] The median survival time of dogs treated with the monoclonal antibody and chemotherapy was 591 days, as compared to historical controls, which had a median survival time of 189 days. Additional studies are needed to further understand the clinical utility of this treatment modality.

Rituximab, a chimeric murine/human IgG monoclonal antibody that targets the B-cell antigen CD20 is now considered the standard of care for the management of human B-cell lymphoproliferative disorders including diffuse large B-cell lymphoma (DLBCL).[33,34] This aggressive form of DLBCL has many similarities to the most common form of lymphoma in dogs.[33-35] In both species, DLBCL generally responds well to combination chemotherapy, however the duration of remission is often limited. The addition of the species-specific antibody Rituximab in combination with chemotherapy to treat DLBCL in people is associated with ttle increase in toxicity, yet enhanced efficacy. The increased frequency in the use of Rituximab in the management of lymphoid malignancies has grown to the point that it presented a huge logistical challenge for the human medical system because the antibody was originally given by slow infusion, a time intensive procedure. Therefore, more cost and time effective procedures have been developed to safely and effectively infuse the antibody.[36] Thus, the availability of a similar antibody that targets the canine CD20 antibody combined with the lessons learned on how to safely and effectively incorporate this type of therapy could have a powerful impact for the management of DLBCL.

The discovery, development and commercial production of a Rituximab-like chimeric murine/canine monoclonal antibody that specifically and precisely targets the B-cell antigen CD20 in dogs is notable.[1] In fact, if this construct is confirmed to be safe and effective, then the potential to benefit dogs with a long list of benign or malignant diseases that are caused by B-cell lymphocytes that express CD20 is high. That is exactly what occurred with the development and subsequent use of Rituximab. We therefore embarked on this double blind, randomized, placebo controlled study to be the very first to test the hypothesis that the Rituximab-like chimeric murine/canine monoclonal antibody is safe and effective for the treatment of CD20 positive DLBCL in the dog. In that study, dogs were enrolled in a prospective, randomized, double blind, placebo controlled clinical trial. All dogs were treated with one 4-week cycle of an L-CHOP chemotherapy protocol followed by either the canine lymphoma antibody or a placebo. Once remission was lost, a single dosage of doxorubicin was given followed by the antibody. Assessment of antitumor efficacy was based on modified Response Evaluation Criteria in Solid Tumors (RECIST) guidelines and toxicity based on criteria delineated by Veterinary Cooperative Oncology Group's common terminology criteria for adverse events (VCOG-CTCAE). The median proliferation free interval and overall survival time respectively of dogs with DLBCL treated with one cycle of the L-CHOP protocol followed by the antibody or the placebo was much longer in the antibody arm compared to the placebo arm. Toxicities that were observed in this protocol were mild and restricted to the L-CHOP cycle with no significant adverse events in the antibody or placebo arms. On the contrary, there was a significant improvement in client perceived quality of life parameters in the antibody, but not the placebo treated dogs.

REFERENCES

1. Dorn CR et al: *Am J Vet Res* 28:993–1001, 1967. 2. Schneider R: *J Natl Cancer Inst* 70:971–977, 1983. 3. Teske E: *Vet Q* 16:209–219, 1994.
4. Dobson JMet al: *J Small Anim Pract* 43:240–246, 2002. 5. Clancy E et al: *Proc 19th Ann Conf Vet Cancer Soc* 71. 1999. 6. Priester WA: *J Natl Cancer Inst* 39:833–845, 1967. 7. Lurie DM et al: *Vet Comp Oncol* 2:171–175, 2004. 8. Onions DE: *J Natl Cancer Inst* 72:909–912, 1984. 9. Edwards DS et al: *Vet Comp Oncol* 1:200–206, 2003. 10. Moore AS et al: *J Vet Intern Med* 15:348–354, 2001. 11. Zemann BI et al: *J Vet Intern Med* 12:465–470, 1998. 12. Jagielski D et al: *J Vet Med A Physiol Pathol Clin Med* 49:419–424, 2002. 13.Teske E et al: *Vet Q* 16:134–136, 1994.
14. Onions D: *Eur J Cancer* 16:345–350, 1980. 15. Tomley FM et al: *Brit J Cancer* 45:644, 1982. 16. O'Brien DJ et al: *Prev Vet Med* 42:1–15, 1999. 17. Gavazza A et al: *J Vet Intern Med* 15:190–195, 2001. 18. Hayes HM et al: *J Natl Cancer Inst* 83:1226–1231, 1991. 19. Reynolds PM et al: *Cancer Epidemiol Biomarkers Prev* 3:233–237, 1994. 20. Carlo GL et al: *Regul Toxicol Pharmacol* 16:245–252, 1992.21. Kaneene JB et al: *Vet Human Toxicol* 41:164–170, 1999. 22. Hayes HM et al: *Environ Res* 70:119–125, 1995. 23. Reif JS et al: *Am J Epidemiol* 141:352–359, 1995. 24. Keller ET: Immune-mediated diseases as a risk factor for canine lymphoma. *Cancer* 70:2334–2337,1992. 25 Torres de la Riva G et al: PLoS One. 2013;8(2):e55937. 26. Burnett RC et al: *Vet Pathol* 40:32–41, 2003. 27. Morrison-Collister KE et al: *Vet Comp Oncol* 1:180–190, 2003.
28. Garrett LD et al: *J Vet Intern Med* 16:704–709, 2002. 29. Valerius KD et al: *JAVMA* 210:512–516, 1997. 30. Cotter SM et al: *J Am Anim Hosp Assoc* 19:159–165, 1983. 31. Rosales C et al: *Cell Immunol* 115:420–428, 1988. 32. Jeglum KA: Monoclonal antibody treatment of canine lymphoma. *Proc East States Vet Conf*: 222–223, 1992. 33. Vose JM et al: *J Clin Oncol.* 2001 Jan 15;19(2):389-97. 34. Vose JM et al: *Leuk Lymphoma.* 2005 Nov;46(11):1569-73. 35. Marconato L et al: *Hematol Oncol.* 2013 Mar;31(1):1-9. 36. Sehn LH et al: *Blood.* 2007 May 15;109(10):417.

WHAT'S NEW IN CANINE MELANOMA: AN UPDATE ON DIAGNOSTIC AND THERAPEUTIC OPTIONS

Heather Wilson-Robles, DVM, DACVIM (Oncology)
College Station, TX

INTRODUCTION

Malignant melanoma (MM) is the most common oral tumor of the dog. Melanomas occurring in the mouth and other mucosal locations have been associated with an aggressive behavior in the dog. This disease frequently affects older dogs and Scottish Terriers, golden retrievers, poodles and dachshunds are predisposed. These tumors frequently affect older dogs greater than 10 years of age and, unlike humans, UV exposure does not seem to play a prominent role in the development of malignant melanoma in the dog. Melanomas represent a diverse group of tumors with a variety of appearances, behaviors and treatment challenges[1].

Common Melanoma Presentations

Location can be one of the most predictable factors in determining the biologic behavior of MM in the dog. Frequently, those tumors seen on haired skin have a benign behavior that can be cured with surgery alone. However, those tumors that occur at muco-cutaneous junctions or in the mouth often have a more aggressive behavior and require more aggressive therapy. The biologic behavior of these tumors is unpredictable at best; however a few factors have been shown to predict behavior. These factors include location, size of tumor, stage of tumor and histologic parameters. Even with a thorough understanding of these factors, the behavior of a specific melanoma can be difficult to predict[1].

Oral melanomas are the most common presentation of malignant melanoma in the dog. These tumors are very rare in the cat. These tumors often appear in the mouths of dogs with pigmented oral mucosa. They may occur anywhere in the mouth and range in presentation from a small well circumscribed dark pigmented mass to a large ulcerated non-pigmented and friable mass. These tumors often invade into bone when associated with the mandible or maxilla. They are aggressive with metastatic rates of 60-80%. These tumors often spread to the regional lymph nodes first and then on to the lungs and other sites. A study by Williams and Packer noted that approximately 70% of enlarged regional lymph nodes near an oral melanoma were positive for metastasis. The other 30% were likely enlarged due to inflammation associated with tumor necrosis and secondary bacterial infection. However, the more worrisome finding from this study is that up to 40% of normal sized lymph nodes were metastatic[2].

Melanomas of the digit (nail bed) are also often malignant and have an aggressive behavior. These tumors often present initially as an infected or traumatized toe nail that does not heal or respond to therapy. Often there is boney lysis of the digit associated with this mass causing pain. These tumors may be pigmented or amelanotic and are often ulcerated. Up to 40% of these tumors will metastasize to the regional lymph nodes first and on to the lungs or other sites[3]. Some anecdotal information also suggests that melanomas of the foot pads may have a similar behavior to nail bed or digital melanomas rather than their more benign cutaneous counterparts and should be treated aggressively. A similar presentation in humans (melanomas of the foot pads or palms) has also been shown to have an increased risk of metastasis compared to other sites.

Melanomas of haired skin often have a more benign behavior and presentation. They are usually pigmented and well differentiated. A few tumors, however, may have an aggressive behavior. Those with rapid growth rates, ulceration or infiltrative presentations should be evaluated more thoroughly.

Other less common locations for melanomas in the dog include gastrointestinal tract, other mucocutaneous junctions, nasal cavity and the eye.

In cats, ocular melanomas are the most common presentation for this disease. Intraoccular melanomas represent up to 50% of the ocular tumors seen in cats[4] (withrow and vail, Miller ch 31 page 596). These tumors have a metastatic rate reported between 55-66% but this occurs rather late in the disease process. Often the pigmentary changes in the eye can take months to years before enucleation is necessary and metastasis often occurs 1-3 years after enucleation.

DIAGNOSIS

If accessible, fine needle aspirate may be all that is required for a diagnosis of this disease. This is especially true for pigmented tumors. Amelanotic melanomas may be easily determined as malignant neoplasia on cytology, but a diagnosis of melanoma cannot be made without special stains necessitating a biopsy. Melanomas have may variants histologically and cytologically. They may look like epithelial, mesenchymal or even round cell tumors initially. These tumors often stain positive for melanin A and s100.

Michigan State University offers diagnostic and prognostic panels for melanomas. Their diagnostic panel uses immunohistochemistry with a cocktail of antibodies against Melan-A, PNL-2, TRP-1, and TRP-2. This panel of antibodies is highly specific for diagnosing melanoma when standard histopathology is unable to do so. They also offer a prognostic melanocytic panel which includes immunolabeling with Ki67, and assessment of nuclear atypia, mitotic index, and degree of pigmentation for prognostic information[5]. This information may be very useful for those tumors in bad locations with fairly benign histopathologic descriptions or dermal melanomas with more atypia than expected. All of these tests can be performed on standard paraffin embedded tissue samples.

STAGING

Malignant melanomas have a fairly high metastatic rate. Because of this, all patients with malignant tumors should be staged. Staging tests for melanoma include lymph node aspirates of the draining lymph nodes even if they are normal in size, three view chest radiographs, baseline blood work (CBC, chemistry profile and urinalysis) as well as advanced of the head for oral lesions and radiographs of the digit for nail bed lesions. The stages of melanoma are listed below in Table 1.

Table 1: WHO TNM based staging scheme for dogs with oral melanoma[1]

T: Primary Tumor	N: Nodal Metastasis	M: Distant Metastasis
T1= < 2 cm in diameter T2= 2-4 cm in diameter T3= >4 cm in diameter	N0= no evidence of lnn involvement N1= Histologic or cytologic evidence of nodal metastasis N2=Fixed metastatic lymph nodes	M0= No evidence of distant metastasis M1= Evidence of distant metastasis
Stage I: T1, N0, M0 Stage II: T2, N0, M0 Stage III:T2, N1, M0 or T3, N0, M0 Stage IV:Any T, Any N and M1		

he clinical stage of disease does correlate with prognosis. The lower the stage of disease the better the long term prognosis is for the patient.

TREATMENT

Surgery

Surgical removal of the primary mass is the mainstay of therapy for many melanomas. In the case of dermal melanomas, removal is typically curative for these benign lesions. Wide surgical margins of 2-3 cm should be attempted for any malignant melanoma. For a digit melanoma this often means amputation of the affected digit (p1-P3). In some cases, multiple digits must be removed if the tumor is large and invading into the metacarpal/tarsal region. For oral melanomas this often means removal of a section of oral mucosa and underlying bone (maxilla, mandible, palate, etc). Rostral tumors of the mandible are the most accessible to surgical removal in the dog, however, hemimandibulectomies and bilateral mandibulectomies back to the molars have been performed with good success in dogs. Maxillary masses are more difficult to remove due to the close association of the maxilla to the nasal structures in the dog. Still, rostral and partial maxillectomies can be performed with good outcomes in many patients.

Radiation

Radiation is another form of local therapy for melanomas that are difficult or impossible to remove surgically. This form of local therapy is often delivered once to twice weekly for a total of 4-6 doses. Unlike many other tumors melanoma requires the application of a large dose per fraction to induce the highest amount of cell killing. This means that more protracted definitive protocols (15-20 doses over 3-4 weeks), are less likely to be effective in shrinking the tumor. With this type of protocol a patient is more likely to have late term side effects of the radiation such as fibrosis of the area, radio-osteonecrosis of the bone, cataracts and dry eye, oral nasal fistula, and tooth loss. However, up to 70% of cases will respond to radiation therapy objectively with reduction of the tumor volume and improved clinical symptoms.

Chemotherapy

Chemotherapy typically plays a small role in the management of this disease due to the inherent chemoresistance of most melanomas. One of the most successful chemotherapy agents reported is carboplatin. Overall response rates (ORR) in the gross disease setting are often reported to be between 25-30%. In most studies, there is no survival advantage conferred in patients treated with carboplatin in a microscopic disease setting. Other chemotherapy agents that have shown some benefit include melphalan, piroxicam and cisplatin.

Metronomic chemotherapy has also been used to treat several tumors in dogs. Though this method has not been shown to directly affect malignant melanoma growth, it may be tried against this tumor. The goal of this therapy is to decrease blood vessel formation in the tumor. Low dose chemotherapy is not cytotoxic to the tumor cells, but is cytotoxic to new forming endothelial cells. Commonly used drugs for this type of therapy include cyclophosphamide, piroxicam and chlorambucil. A study evaluating this technique in a mouse model with canine melanoma xenografts showed a significant decrease in blood vessel formation and microvascular density suggesting that this therapy may have some merit in the treatment of this disease[6]

Immunotherapy

Oncept is a DNA vaccine marketed for dogs with completely excised malignant melanomas with no gross evidence of disease, however it has been used in many species with many stages of melanoma. This vaccine aims to promote immune competence in canines against canine tyrosinase by exposing them to human tyrosinase. Two recent studies regarding the efficacy of the vaccine reported opposing results. Ottnod et al evaluated 45 dogs retrospectively that presented with malignant melanoma (most stage II or III disease) and determined that there was no survival advantage in dogs receiving the vaccine when compared to those dogs who did not receive the vaccine[7]. Grosenbaugh et al evaluated the Oncept in 111 dogs (58 treated and 53 historical controls) and determined that those dogs with Stage II and Stage III disease did have a significant improvement in median survival time, however no overall response rate was reported and the median survival time for treated dogs was not determined in the study[8]. Currently, the efficacy and utility of this therapy is still undetermined.

Therapeutic Investigations:

Several new and promising therapies are on the horizon and may soon become available for the treatment of malignant melanoma in dogs. One study recently evaluated the utility of adipose tissue derived mesenchymal stem cells loaded with cisplatin in a murine xenograft model of canine melanoma and found that these cell track to the tumor sites and deliver high doses of drug there making them more effective than traditional systemic delivery of chemotherapy[9]. This is an attractive therapy because it delivers higher doses of chemotherapy to the tumor than can be done when cisplatin is given IV while avoiding the traditional toxicities seen with classical administration of the drug.

Another promising study used the drug bortezomib, a drug commonly used in human medicine to treat multiple myeloma, against canine melanoma cells in vitro and in a murine xenograft model. This drug works by inhibiting degredation of I κB leading to inhibition of the NF κB pathway and proliferation. In this study, several cell lines and were evaluated and found to be sensitive to bortezomib. Additionally, the growth of xenograft tumors was significantly inhibited by this drug[10].

Another study ongoing at Texas A&M is evaluating the use of S100b inhibitors for the treatment of malignant melanoma in canines. S100b is a calcium binding protein that can bind and hold p53 (tumor suicide gene) in the cytoplasm of a cell making it ineffective. Certain drugs can inhibit S100b and prevent it from stopping cellular death. In this study, 4 of 9 dogs treated thus far have demonstrated objective tumor responses for varying amounts of time indicating that this class of drugs may have promise in treating this tumor (unplublished data).

CONCLUSIONS

Malignant melanoma is an aggressive tumor in the dog. Early intervention is the best way to improve survival times and complete surgical removal of the mass is the best way to achieve the longest tumor responses (when possible). Radiation can also be successful at achieving local control of the tumor. There are not many good options for the prevention or treatment of metastasis in this disease thus underscoring the importance of early detection and treatment.

REFERENCES

[1] Berman P, et. al **Small Animal Clinical Oncology** 2013; ch 19, 5th ed, p 321. [2] Williams LE, et al, **JAVMA**, 222(9):1234-1236 2003. [3] Henry CJ, et al, **JAVMA**, 19(5):720-724 2005. [4] Miller et al, **Small Animal Clinical Oncology**, 2013, Chap 31, 5th ed, pg 596.
[5] http://www.dcpah.msu.edu/bin/Catalog.exe?Action=Test&Id =2885 (Michigan state website for prognostic panel). [6] Choisunirachon N, et al. **VCO**, 2013, Jul 26. doi: 10.1111/vco.12059, [7] Ottnod JM, et al, **VCO**, 2013, 11(3):219-29. [8] Grosenbaugh DA, et al, **Am J Vet Res**. 2011 Dec;72(12):1631-8. [9] Ahn J, et al, **PLoS One**. 2013 Sep 9;8(9):e74897. [10] Ito K, et al, **Vet J**. 2013 Dec;198(3):577-82.

MANAGING CANINE MAST CELL TUMORS: AN UPDATE ON DIAGNOSIS AND TREATMENT

Jenna H. Burton, DVM, MS, DACVIM (Oncology)
Davis, CA

INTRODUCTION

Mast cell tumors are the most commonly occurring skin tumor in dogs, comprising 16 to 21% of all skin tumors.[1] Breeds that are at increased risk for mast cell tumor (MCT) development include Boxers, pugs, Boston terriers, English bulldogs, Golden retriever, Labrador retrievers, Staffordshire terriers and beagles; male and female dogs have a similar risk of developing MCT.[1] Mast cells are a white blood cell that normally function as important mediators of immediate (type I) hypersensitivity. These cells arise from the bone marrow and normally reside in many tissues of the body, including the skin, liver, spleen, and gastrointestinal tract. Mast cells have numerous granules that contain bioactive substances such as heparin, histamine, proteases and tumor necrosis factor-α (TNF-α) which are important for development of an immune/allergic response after detection of an antigen.[2] While important in health, these bioactive substances can cause complications for dogs with MCT; these complications are reviewed at the end of the material presented here.

MCT most commonly occur in the skin and subcutaneous tissues but have been reported to occur in other sites such as the oral cavity, gastrointestinal tract, and nasopharynx. While solitary MCT are most common, some dogs may present with multiple MCT. It is important to remember that MCT can have a wide range of clinical presentations and can truly look like *anything*. Mast cell tumors may be raised, alopecic, erythematous, well circumscribed, diffuse, and/or ulcerated. Subcutaneous MCT can feel fleshy on palpation and can easily be mistaken for lipomas. Owners may report that the mass has been present for months without changing or that it waxes and wanes in size or that it is growing rapidly. Sometimes the area around the MCT will appear bruised. Since mast cell tumors may be mistaken for more benign processes based on their clinical appearance and palpation alone, it is important to aspirate all skin and subcutaneous masses to ensure that they are not MCT.

APPROACHES TO DIAGNOSIS AND STAGING

Most MCT are easily diagnosed with a fine needle aspirate (FNA) and cytology. Cytology of a MCT aspirate reveals round small to medium round cells with numerous dark purplish granules within the cytoplasm.[3] There may be granules that appear in the background of the slide as well if cells have been ruptured during slide preparation; these some times may be mistaken for cocci bacteria. Generally, these granules readily uptake the Diff-Quik (or Romanowsky) stains resulting in the dark purple color but some a small number of MCT may have granules that don't readily uptake the stain. However, if rounds cells with numerous clear vacuoles in the cytoplasm are present and the neoplastic cells have a "fried egg" appearance, the slide should be stained with Wright-Giemsa or toluidine blue stain to reveal the granules and aid in diagnosis. Eosinophils are often seen in increased numbers on cytology of MCT and the presence of numerous eosinophils may also give a clue that the mass may be a poorly staining MCT. Occasionally, a biopsy will be required to make a definitive diagnosis.

Staging tests, such as local lymph node aspirates, thoracic radiographs, abdominal ultrasound, or bone marrow aspirate are routinely performed for many tumors we encounter in veterinary oncology, however, not all MCT will require extensive staging prior to surgical removal. If possible, an aspirate of the regional (nearest draining) lymph node should be performed for all mast cell tumors to ensure there is no evidence of lymph node metastasis is present prior to going to surgery. If the tumor is amenable to resection with wide surgical margins and no negative prognostic indicator are present, no additional staging prior to surgery may be needed. Negative prognostic indicators that indicate that additional staging may be required include tumors that have been reported to have recent and rapid growth, tumor ulceration, lymph node metastasis, large tumor size or clinical illness such as lethargy, vomiting, inappetence or melena. Any and all of these signs may all indicate that this could be a more aggressive MCT and additional staging with abdominal ultrasound and thoracic radiographs should be pursued prior to surgery. Bone marrow aspirates have not been demonstrated to have clinical utility in staging most canine mast cell tumors and are no longer routinely performed.[4]

PROGNOSTIC INFORMATION

MCT can have a wide range of biologic behavior; some dogs may be essentially cured of their MCT with surgery alone while other MCT may require additional treatment with chemotherapy and/or radiation therapy to prevent/delay the recurrence or spread (metastasis) of the MCT. There are many factors that may influence prognosis for a dog with MCT. Location of the MCT may provide some clue to prognosis as MCT that occur in the oral cavity, gastrointestinal tract, scrotum, preputial area or at mucocutaneous junctions have been reported to have a worse prognosis than MCT that occur at other sites.[5, 6] Dogs with an advance stage of disease that has spread to visceral organs such as the liver and spleen have also been shown to have a very poor outcome.[7, 8]

One of the more reliable ways to predict how aggressively a MCT is likely to behave is through determination of the histopathologic grade when the tumor (or biopsy sample) is removed and submitted to the pathology lab. MCT tumors have previously been graded using a three-tiered grading scheme classifying MCT as low grade, intermediate grade, or high grade, but more recently a grading scheme has been proposed that separates MCT into two tiers, low or high grade.[9, 10] Many pathologists will include results from both grading schemes in their biopsy report, as there is no consensus at this time as to which maybe more predictive of the biologic behavior of the MCT. Both the two-tier and the three-tier grading schemes include assessment of the mitotic index (MI), which is the number of dividing cells (or mitoses) that are seen in 10 high power (400x) fields on the biopsy specimen. MI is a measure of cellular proliferation. MI is strong predictor of outcome and it has shown that MCT with a high MI ($\geq$5) are more likely to have a shorter overall survival that those dogs with MCT with a low MI.[11]

There also exists a *mast cell tumor panel* that evaluates other markers of cellular proliferation (Ki67, AgNOR, and PCNA); this test may have value to better determine the prognosis for mast cell tumors than MI alone.[12, 13] The author generally reserves use of the mast cell tumor panels for MCT that have a MI around the cut-off for the grading schemes (5 to 7 MI) or in situations when the pathologists are having difficulty determining whether this tumor is a low or high grade MCT. Mast cell tumor panels also can evaluate tumor samples for aberrant expression of KIT, which is a protein known as a *receptor tyrosine kinase* (RTK) found on the surface of mast cells. KIT is responsible for transmitting signals for growth and differentiation of normal mast cells. It has been found that a percentage of MCT have a mutation in the gene that codes for KIT and presence of this mutation is associated with more aggressive MCT behavior. The gene that encodes for the protein KIT is known as *c-kit*. Mast cell tumor panels can look for mutations the in the gene *c-kit* as well as aberrant KIT expression within the mast cells which may provide additional prognostic information beyond what can be determine with MI alone.[14]

TREATMENT

Surgery

Surgery remains the mainstay of treatment for the majority of MCT. If wide excision can be achieved, surgery can be curative for many low to intermediate grade MCT. Generally, wide excision is considered to be 3 cm laterally around the tumor and a fascial plane deep to the tumor, although some have advocated that smaller margins (1 to 2 cm) may be appropriate for small, low-grade tumors.[15, 16] Occasionally, the if complete margins are not achieved with the first surgery, a second surgery may be recommended to obtain wider surgical margins and completely remove any residual tumor. Alternatively, radiation therapy can be used to prevent the recurrence of incompletely removed mast cell tumors for which a second surgery is not feasible. Not all MCT that are removed with incomplete surgical margins will ultimately recur, but it is impossible to predict which will recur and which will not.[17] Since incomplete resection of a MCT is associated with decreased survival, it is prudent to offer clients additional therapy for incompletely removed MCT. For MCT that have that spread to a local lymph node, the metastatic lymph node should be removed at the time of primary tumor excision as well to decrease the MCT burden as much as possible.

Chemotherapy/TKIs

For high grade MCT or MCT that have spread to a local lymph node at the time of diagnosis, chemotherapy should be considered. There are a number of protocols that have been evaluated for treating MCT in the microscopic disease setting and opinions between oncologists as to which protocol is preferred varies widely. In general, cytotoxic chemotherapy agents that are known to be effective against MCT include the vinca alkaloids (vinblastine, vincristine, vinorelbine) and alkylating agents such as CCNU (lomustine). Steroids are often included in chemotherapy protocols for MCT as these may have some efficacy against MCT as well. Dogs with high grade MCT +/- lymph node metastasis that undergo surgery to remove al sites of disease and then receive follow-up treatment with chemotherapy can experience prolonged survivals.[18, 19] Chemotherapy administered in the face of gross disease is unlikely to result in durable tumor remissions and is generally expected only to slow the growth and spread of the MCT.

There are two FDA approved drugs for the treatment of canine mast cell tumors, Palladia® (toceranib phosphate; Zoetis) and Kinavet-CA1® (masitinib mesylate; AB Science). Both of these drugs are receptor tyrosine kinase inhibitors (TKIs) and function by inhibiting KIT, thereby preventing the signaling for MCT growth and replication. These drugs are labeled for use in recurrent or non-resectable grade 2 or 3 mast cell tumors. Both Palladia® and Kinavet-CA1® are oral drugs which owners may assume are associated with fewer and less side effects that traditional cytotoxic chemotherapy drugs. However, both drugs have been associated with a number of side effects with GI signs such as diarrhea, vomiting, and/or inappetence, being most common.[20, 21] GI side effects associated with TKI administration tend to be mild and self-limiting with “drug holidays” or dose reductions. However, more severe side effects can occur and routine monitoring of blood work, blood pressure, urinalyses and assessment of urine protein creatinine ratios (UPC) should be performed at regular intervals while dogs are receiving therapy.

Radiation therapy

Radiation therapy is commonly used to prevent recurrence of MCT when clean surgical margins cannot be obtained. When treating MCT where all visible tumor has been removed and mast cells remain at the edge of the surgical margins, full course radiation can provide excellent long term control of low to intermediate grade MCT with 77-86% of cases remaining disease free at 2 years.[22, 23] Radiation therapy can also be performed in the gross disease setting when tumors cannot be resected. Clients should be aware than the response rates with full course radiation in this situation are much less, with only about a 50% control rate at one year.[1] Radiation therapy can also be performed in a palliative setting where larger fractions of radiation are administered on a less frequently basis (i.e. weekly for 4 fractions). The goals with palliative radiation therapy are generally to slow tumor growth, decrease inflammation and decrease pain. Administration of larger fractions of radiation to a tumor may increase the risk of MCT degranulation (see below).

COMPLICATIONS OF MAST CELL TUMOR TREATMENT

The bioactive compounds contained within mast cells can occasionally be released from the cells in large quantities into the surrounding tissues or systemic circulation or both, resulting in complications such as bleeding, hypotension, and/or anaphylaxis. These complications are commonly referred to as “degranulation episodes” and it is important to recognize events or treatment that may trigger MCT degranulation so they can hopefully be prevented. Manipulation of the tumor alone may cause release of these substances and result in erythema and swelling (wheal) at the site of the tumor; this is known as *Darier's sign*.

Heparin is one of the substances contained within MCT granules and prevents of clotting of blood. Because of the heparin that can be released from the granules, it is not uncommon to find that a MCT will ooze blood for a few minutes after a fine needle aspiration is performed. There is little we can do to prevent this from occurring other than to apply gentle pressure to the aspiration site for several minutes after procedure. If a MCT is to be biopsied for information regarding tumor grade prior to definitive removal, a punch or Tru-cut biopsy may be preferable to a larger incisional wedge biopsy as potential bleeding may be more easily controlled with a smaller biopsy site.

Histamine is also contained within granules of MCT and is an important mediator of allergic response in health. Histamine release from mast cell tumors may cause pruritus at the tumor site as well as contribute to gastrointestinal ulceration that can occur secondary to MCT. Histamine is released from the MCT into circulation and then binds to the parietal cells in the stomach; this causes the parietal cells to increase production of hydrochloric acid (HCl) results in gastrointestinal irritation and ulceration. Dogs with signs of inappetence, vomiting coffee-ground-like material or with melena may have GI ulceration secondary to their MCT. To mitigate the effects of histamine release, anti-histamines, such as diphenhydramine, are often prescribed to block the activity of histamine. Additionally, H_2 blockers such as famotidine and proton pump inhibitors like omeprazole may be prescribed to patients with MCT to decrease HCl production by the parietal cells thereby decreasing GI irritation.

One of the most critical and frightening complications of mast cell tumor degranulation is development of hypotension, resulting in sudden collapse of the patient or significant anesthetic complications. It is thought that both histamine and prostaglandin release from the MCT granules cause systemic vasodilation, resulting in hypotension. For this reason, dogs that are undergoing surgery for MCT removal are routinely pre-medicated with diphenhydramine (1 to 2 mg/kg IM) to minimize the effort of any degranulation that may occur with surgical manipulation. Some clinicians routinely administer diphenhydramine prior to fine needle aspirate of MCT; in this clinician's opinion, degranulation is unlikely to occur with the majority of MCT that are aspirated but cautioned may be warranted with very large tumors or dogs experiencing clinical signs secondary to their MCT. Opioids such as morphine have been reported to contribute to histamine release and are generally avoided in anesthetic protocols for MCT removal. Other opioids such as hydromorphone and butorphanol have not been associated with increased histamine release but are often still used with caution.

MCT degranulation and secondary hypotension can also occur when treatment with either chemotherapy or radiation therapy is initiated. This is a relatively uncommon occurrence and anecdotally, the risk appears to be higher for dogs with large tumor burden or those already experiencing clinical signs. To prevent hypotension from occurring, premedication with diphenhydramine and injectable steroids such as dexamethasone SP may be administered prior to treatment. Alternatively, dogs can be started on oral diphenhydramine and prednisone for several days before starting definitive therapy to potentially decrease the likelihood of a degranulation episode. Patients should be monitored closely for signs of hypotension following initiation of treatment; these may include weakness, disorientation, tachycardia or collapse. If any of these signs are noted, the patient should be assessed and temperature, pulse, respiration, and blood pressure checked. An IV catheter should be placed for intravenous access as IV crystalloid fluids may be indicated to support the patient. Injectable diphenhydramine and corticosteroids (dexamethasone SP) should be available for administration, if needed.

SUMMARY

Dogs with MCT will present in a wide variety of ways and the optimal treatment will be likely different for each patient that you see. Understanding the biologic behavior and associated prognostic factors will help you be aware of the treatment decisions available for each patient. An awareness of situations where MCT degranulation may occur and recognizing signs of MCT degranulation will help you provide the best care possible and avoid potentially life threatening situations for your patients.

REFERENCES

1. London, C.A. and D.H. Thamm, *Mast Cell Tumors*, in *Withrow & MacEwen's Small Animal Clinical Oncology*, S.J. Withrow, D.M. Vail, and R.L. Page, Editors. 2013, Elsevier: St. Louis. p. 335-355.
2. Abbas, A.K., A.H. Litchman, and S. Pillai, *Chapter 19: Immediate Hypersensitivity*, in *Cellular and Molecular Immunology*. 2010, Saunders, Elsevier: Philadelphia. p. 441-461.
3. Raskin, R.E., *Mast Cell Tumor*, in *Atlas of Canine and Feline Cytology*, R.E. Raskin and D.J. Meyer, Editors. 2001, Saunders: Philadelphia. p. 81-84.
4. Endicott, M.M., et al., *Clinicopathological findings and results of bone marrow aspiration in dogs with cutaneous mast cell tumours: 157 cases (1999–2002)**. Veterinary and Comparative Oncology, 2007. **5**(1): p. 31-37.
5. Sfiligoi, G., et al., *Outcome of dogs with mast cell tumors in the inguinal or perineal region versus other cutaneous locations: 124 cases (1990–2001)*. Journal of the American Veterinary Medical Association, 2005. **226**(8): p. 1368-1374.
6. Hillman, L.A., et al., *Biological behavior of oral and perioral mast cell tumors in dogs: 44 cases (1996-2006)*. Journal of the American Veterinary Medical Association, 2010. **237**(8): p. 936-942.
7. O'Keefe, D.A., et al., *Systemic Mastocytosis in 16 Dogs*. Journal of Veterinary Internal Medicine, 1987. **1**(2): p. 75-80.
8. Takahashi, T., et al., *Visceral mast cell tumors in dogs: 10 cases (1982–1997)*. Journal of the American Veterinary Medical Association, 2000. **216**(2): p. 222-226.
9. Patnaik, A.K., W.J. Ehler, and E.G. MacEwen, *Canine Cutaneous Mast Cell Tumor: Morphologic Grading and Survival Time in 83 Dogs*. Veterinary Pathology Online, 1984. **21**(5): p. 469-474.
10. Kiupel, M., et al., *Proposal of a 2-Tier Histologic Grading System for Canine Cutaneous Mast Cell Tumors to More Accurately Predict Biological Behavior*. Veterinary Pathology Online, 2011. **48**(1): p. 147-155.
11. Romansik, E.M., et al., *Mitotic Index Is Predictive for Survival for Canine Cutaneous Mast Cell Tumors*. Veterinary Pathology Online, 2007. **44**(3): p. 335-341.
12. Scase, T.J., et al., *Canine Mast Cell Tumors: Correlation of Apoptosis and Proliferation Markers with Prognosis*. Journal of Veterinary Internal Medicine, 2006. **20**(1): p. 151-158.

13. Simoes, J.P.C., P. Schoning, and M. Butine, *Prognosis of Canine Mast Cell Tumors: A Comparison of Three Methods.* Veterinary Pathology Online, 1994. **31**(6): p. 637-647.
14. Kiupel, M., et al., *The Use of KIT and Tryptase Expression Patterns as Prognostic Tools for Canine Cutaneous Mast Cell Tumors.* Veterinary Pathology Online, 2004. **41**(4): p. 371-377.
15. Simpson, A.M., et al., *Evaluation of surgical margins required for complete excision of cutaneous mast cell tumors in dogs.* Journal of the American Veterinary Medical Association, 2004. **224**(2): p. 236-240.
16. Fulcher, R.P., et al., *Evaluation of a two-centimeter lateral surgical margin for excision of grade I and grade II cutaneous mast cell tumors in dogs.* Journal of the American Veterinary Medical Association, 2006. **228**(2): p. 210-215.
17. Séguin, B., et al., *Clinical outcome of dogs with grade-II mast cell tumors treated with surgery alone: 55 cases (1996–1999).* Journal of the American Veterinary Medical Association, 2001. **218**(7): p. 1120-1123.
18. Lejeune, A., et al., *Aggressive local therapy combined with systemic chemotherapy provides long-term control in grade II stage 2 canine mast cell tumour: 21 cases (1999–2012)*.* Veterinary and Comparative Oncology, 2013: p. n/a-n/a.
19. Thamm, D.H., M.M. Turek, and D.M. Vail, *Outcome and Prognostic Factors Following Adjuvant Prednisone/Vinblastine Chemotherapy for High-Risk Canine Mast Cell Tumour: 61 Cases.* Journal of Veterinary Medical Science, 2006. **68**(6): p. 581-587.
20. London, C.A., et al., *Multi-center, Placebo-controlled, Double-blind, Randomized Study of Oral Toceranib Phosphate (SU11654), a Receptor Tyrosine Kinase Inhibitor, for the Treatment of Dogs with Recurrent (Either Local or Distant) Mast Cell Tumor Following Surgical Excision.* Clinical Cancer Research, 2009. **15**(11): p. 3856-3865.
21. Smrkovski, O.A., et al., *Masitinib mesylate for metastatic and non-resectable canine cutaneous mast cell tumours.* Veterinary and Comparative Oncology, 2013: p. n/a-n/a.
22. Turrel, J.M., et al., *Prognostic factors for radiation treatment of mast cell tumor in 85 dogs.* J Am Vet Med Assoc, 1988. **193**(8): p. 936-40.
23. Al-Sarraf, R., et al., *A Prospective Study of Radiation Therapy for the Treatment of Grade 2 Mast Cell Tumors in 32 Dogs.* Journal of Veterinary Internal Medicine, 1996. **10**(6): p. 376-378.

JUST WASH YOUR HANDS: IMPROVING HAND HYGIENE IN VETERINARY CLINICS

Maureen E.C. Anderson, DVM, DVSc, PhD, DACVIM
Guelph, ON, Canada

INTRODUCTION

Veterinary clinics are important community hubs where both animals and people with a wide variety of health conditions meet, with the potential for transmission and dissemination of many different infectious pathogens. Hand hygiene is said to be the single most important factor in infection control in hospitals and other healthcare facilities.[1] It is a similarly important measure for controlling the spread of pathogens between animal patients within veterinary hospitals, as well for preventing the spread of zoonotic pathogens (such as methicillin-resistant *S. aureus* (MRSA) and enteropathogens) from animals to humans and vice versa.[2-4] However, until recently hand hygiene has received very little attention in a veterinary context. Recommendations regarding hand hygiene, including when, how and how often it should be performed, appear in guidelines from multiple healthcare-associated organizations,[1,5,6] as well as some veterinary infectious disease control guidelines.[2-4]

HAND HYGIENE BASICS

The goal of hand hygiene in the context of modern healthcare is to reduce or eliminate the transient microflora of the hands as atraumatically as possible, and ideally to prevent rebound growth, in order to decrease the likelihood of transfer of pathogenic or opportunistic microbes from the hands to other surfaces or individuals/tissues. This may be accomplished by physical removal of the microbes from the skin, or use of antimicrobial compounds that kill or neutralize the microbes. The resident microflora of the skin is more difficult to remove but is also less likely to include pathogens. Maintaining the integrity of the skin even after repeated hand hygiene attempts is crucial, as skin damage provides niches for opportunistic pathogens to cause infection, resulting in pain and discomfort as well as increased risk of transmission due to the larger number of pathogenic bacteria present on the skin.

Soaps or detergents are primarily emulsifying agents that break up oils to help mechanically remove contamination from the hands via the flushing action of the water.[1,6] Plain soap has no residual activity against microorganisms, but is suitable for use in lower-risk situations (e.g. homes, general public restrooms), whereas use of antibacterial soap is recommended in healthcare settings, and this could be reasonably extended to include veterinary clinics.

Alcohol-based hand sanitizers (AHSs) rely on the action of the alcohol component to rapidly kill microorganisms, by denaturing proteins and disrupting the cellular membrane, which is most effective at a concentration of 65-90%. The alcohol then evaporates relatively quickly, but many AHSs also contain other active ingredients (e.g. chlorhexidine, quaternary ammonium compounds, octenidine, triclosan) that remain on the skin and provide some residual disinfectant action.[1] Alcohol-based hand sanitizers are very convenient because they do not require water (and therefore sinks/plumbing and a drying mechanism), so dispensers are easily placed and used in any point-of-care area, or even carried in a pocket. These products have been shown to be equally effective as hand washing with soap and water in the presence of minimal gross contamination,[7,8] and they are actually less damaging to the skin over time than soap and water. The main disadvantage of AHSs is that they do not physically remove microorganisms from the skin. There are some microorganisms that are inherently resistant to even the non-specific killing action of alcohol (e.g. clostridial spores, *Cryptosporidium* oocysts, certain non-enveloped viruses such as canine parvovirus), therefore AHSs are (theoretically) ineffective in terms of controlling contamination with these types of pathogens. Significant (i.e. visible) amounts of dirt/debris/discharge are not removed by using AHS, and may protect microorganisms from the action of the active ingredients in these products. If hands are visibly soiled, or if there is suspected contamination with an alcohol-resistant pathogen, it is recommended that soap and water be used instead of a AHS.[1,2,5] Another disadvantage of AHS is that the liquid is flammable due to its high alcohol content, which has been an issue in terms of placing AHS dispensers in some public places, and requires precautions for storage of large volumes and clean up of large-volume spills.

HAND HYGIENE TECHNIQUE AND TIMING

How well hand hygiene is performed when it is attempted is equally important to timing and frequency for hand hygiene to be effective. In order to effectively reduce or kill the transient microflora of the hands, it is generally recommended that soap be applied for a minimum of 10-20 s before rinsing, or for AHS that enough product be applied to cover all surfaces of the hands and then rubbed until dry (which should take approximately 10-20s as well).[1,2,6] Areas of the hand most often missed during washing are under the nails, back of fingers, back of the hands and parts of the thumbs.[6]

Individuals have a relatively easy time remembering to wash their hands when they are visibly soiled, with the possible exception of veterinary personnel working in a situation when their hands are expected (or accepted) to be dirty, such as in a barn environment. Unfortunately, in the majority of healthcare situations, as well as much of everyday life, microbial contaminants on the hands are not accompanied by visible dirt. Hand hygiene should be used after activities that may result in contamination of the hands, as well as before activities that may result in transmission of the transient flora of the hands to a common surface or another person or animal. Specifically, hand hygiene in healthcare (human and veterinary) should ideally be performed:[1,2,5,6]

- before patient contact
- before invasive procedures
- after potential exposure to bodily fluids, discharge, etc. (e.g. dirty procedures, specimen handling)
- after glove removal
- after patient contact

As much as possible, hand hygiene should be performed *immediately* before and after patient contact, in order to minimize the opportunity for contamination of the hands or by the hands of the person in question. This requires hand hygiene stations to be readily available in all patient contact areas, but this has not been a consideration in the design of many veterinary clinics (i.e. clinics that do not have a sink in all exam rooms and wards).

HAND HYGIENE COMPLIANCE

Compliance with hand hygiene protocols is the most challenging component of ensuring efficacy. Numerous studies have investigated hand hygiene compliance in hospitals, and in general, compliance is poor (<50%) among healthcare workers.[1] Major barriers to compliance in many situations include skin irritation (irritant contact dermatitis), lack of accessibility of hand hygiene stations, time constraints (i.e. too busy), and lack of perceived importance (i.e. that it is generally unimportant, or more likely that it is less important than other tasks/procedures which therefore take precedence over hand hygiene).[1,9] Another barrier to hand hygiene compliance is that in some cases people may simply forget to perform hand hygiene;[1] to prevent this, hand hygiene needs to become an automatic habit for healthcare and veterinary personnel, like covering a sneeze or a cough.

Gloves

The impression that use of gloves negates the need for hand hygiene is problematic, and has been reported as a barrier to hand hygiene compliance in human healthcare studies. Although gloves do provide an additional barrier between the skin of the wearer and the patient and/or objects the wearer may touch, pre-existing defects, damage/punctures that may occur during use and contamination of the hands during glove removal make gloves an imperfect barrier. Since gloves are most often used in situations when there is a need to take additional precautions to prevent contamination of a site with bacteria from the hands, or when heavy contamination of the hands or contamination with a serious infectious pathogen is expected to occur, hand hygiene before and after glove use remains necessary for infection control.

HAND HYGIENE STUDIES IN VETERINARY MEDICINE

Literature regarding hand hygiene in veterinary healthcare specifically is limited. Traub-Dargatz et al.[7] examined the effectiveness of traditional hand washing versus AHS following basic physical examination of a horse, and found the AHS to be more effective at reducing bacterial counts on the hands. However, it was pointed out that the hands of veterinary personnel working in equine practice may become more heavily soiled on a regular basis than personnel working in a small animal or human hospital, which may require hand washing to remove gross contamination, rather than use of AHS alone for hand hygiene. Anderson et al.[10] found that hand washing following contact with infectious cases and between farms had a protective effect against colonization with MRSA in veterinary personnel who work with horses. With regard to hand hygiene compliance, in two studies of companion animal veterinary personnel, self-reported compliance (i.e. always perform hand hygiene between patients) was 42-48%.[11,12] In the study by Wright et al.,[12] self-reported compliance among large animal and equine practitioners was even lower (<20%). Only slightly more than half of small animal practitioners in the same survey (590/1069) reported always washing their hands before eating, drinking or smoking while at work, and the same was true for less than a third of large animal and equine practitioners. Observed compliance in one companion animal veterinary teaching hospital was 21-42% before and after a hand hygiene promotional campaign, respectively.[13] Overall compliance was 27-29% in similar study in a small animal intensive care unit at another hospital, with compliance among technicians and assistants specifically ranging from 42-48%.[14] Nakamura et al.[11] reported the most common reason for not performing hand hygiene among veterinary support staff was being too busy (72%).

IMPROVING HAND HYGIENE COMPLIANCE

Providing better access to hand hygiene stations and supplies is often the first step to improving hand hygiene compliance, as it aids in minimizing the time required to comply with protocols, and acts as a visual reminder to perform hand hygiene. Because renovating facilities to improve the location of sinks for hand washing is often not feasible, improving access typically involves introducing or changing the location or number of AHS dispensers.[1] From an infection control standpoint, sink placement is a critical consideration in the design of new clinics, in order to facilitate hand hygiene as much as possible, particularly in the most common patient contact areas (i.e. exam rooms, treatment rooms, wards).

Monitoring and feedback systems can also have a significant impact on compliance. Feedback may be from fellow staff, patient owners (in the case of veterinary clinics), researchers or infection control personnel. Making people aware of how often they neglect to perform hand hygiene, and/or giving them positive feedback when their compliance improves (the "stick and carrot" approach), seems to provide additional incentive to further improve compliance. It has also been shown in human healthcare that hand hygiene compliance among physicians is often lower than among nurses.[15,16] The same may be true of veterinarians and technicians in some cases.[14] Empowering technicians to remind veterinarians to perform hand hygiene and observe other infection control protocols may help to improve overall compliance.

Convincing staff of the importance and utility of hand hygiene in curbing the spread of infectious agents is crucial to achieving compliance.[1] The perception that hand hygiene is relatively unimportant is likely more problematic in veterinary medicine compared to human medicine. Because there are fewer diseases that can be transmitted from animal-to-person compared to person-to-person, some veterinary personnel have a relatively cavalier attitude toward contamination of hands/clothing/equipment with animal blood, body fluids and excreta. However, hand hygiene is also a critical means of preventing potential indirect transmission of non-zoonotic pathogens from animal to animal, thus protecting veterinary patients. The potential of a zoonotic pathogen, brought into a clinic by a single person or animal, to spread first through the clinic and then into the community is significant. Examples of this kind of event have occurred,[17,18] and many more likely go unreported or simply unrecognized.

Ideally hand hygiene should be something that everyone wants to do, rather than something everyone must be asked to do, but education alone is frequently insufficient to change behaviour. Development of a clinic infection control culture, whereby infection control practices and principles become integrated into the group mentality through ongoing communication, conversation, training and strong support from team leaders/management, is the ultimate means of improving compliance with hand hygiene and other infection control measures.[19] A study by Murphy et al.,[20] that looked primarily at environmental disinfection practices, management of infectious patients and antimicrobial use in clean surgical procedures in veterinary clinics, identified that none of the 101 participating clinics had a formal infection control program. Wright et al.[12] found that only 247/797 (31%) small animal practices, and even fewer large animal and equine practices, had written infection control policies. Infection control practices, including hand hygiene, are more likely to be underutilized under these circumstances, and their effectiveness cannot be evaluated without established policies and a formal infection control program, even if the program is very basic. Murphy et al.[20] also showed that veterinarians and technicians generally consider certain aspects of infection control, such as environmental cleaning and disinfection, important measures, yet 40-60% did not know what products were used in their own clinics for disinfecting areas contaminated with infectious body fluids, nor how to properly prepare the products for use. It is likely that the same shortfalls hold true for hand hygiene practices – even among veterinary personnel who acknowledge the importance of hand hygiene, few likely perform it how and when they should. These findings illustrate the lack of an infection control culture in veterinary clinics, and in veterinary medicine in general; this issue must be addressed if sustained improvement in hand hygiene and other infection control practices is to be achieved.

CONCLUSION

A chain is only as strong as its weakest link, and the same can be said for any infection control program. In order to be effective, infection control measures such as hand hygiene need to be practiced by every member of the clinic team, from veterinarians and technicians to kennel staff and volunteers. By taking the time to perform hand hygiene and use other infection control measures, technicians serve as examples to other staff (including veterinarians). Several studies have suggested that involvement and support of upper-level management and administration are necessary for effective implementation of hand hygiene protocols in human health care facilities,[21-24] and it is likely that the same is true in veterinary clinics, for both hand hygiene and other routine infection control measures. Nonetheless, do not underestimate the power of one's own good (or bad) example when it comes to promoting infection control and other behaviours.

Without appropriate attention to infection control efforts, veterinary clinics can become amplifiers of both animal diseases and zoonotic diseases in the community, due to the high flow-through of out-patient animals and their caretakers, the mixing of both sick and susceptible animals (i.e. very young, very old or immunocompromised), the wide variety of patients seen at a clinic (e.g. medical, surgical, oncological, regular health check-ups), and the close interaction of animals, clinic personnel and members of the general public that occurs there. Improving hand hygiene practices in veterinary clinics has the potential to curb such spread of recognized pathogens within and outside these facilities, and could help to prevent widespread transmission of other emerging pathogens in the future before they are even recognized.

REFERENCES

[1]Boyce JM, et al **Am J Infect Control** 2002;30:S1–46. [2]Scheftel JM, et al **J Am Vet Med Assoc** 2010;237:1403–1422. [3]Australian Veterinary Association **Guidelines for Veterinary Personal Biosecurity** AVA, 2013 (http://www.ava.com.au/biosecurity-guidelines). [4]Canadian Committee on Antibiotic Resistance **Infection Prevention and Control Best Practices for Small Animal Veterinary Clinics** CCAR, 2008. [5]Pittet D, et al **Infect Control Hosp Epidemiol** 2009;30:611–622. [6]Larson EL **Am J Infect Control** 1995;23:251–269. [7]Traub-Dargatz JL, et al **Can Vet J** 2006;47:671–676. [8]Larson EL et al. **J Emerg Med** 1992;10:7–11. [9]Pittet D. **Infect Control Hosp Epidemiol** 2000;21:381–386. [10]Anderson M, et al **Vet Microbiol** 2008;129:410–417. [11]Nakamura RK, et al **J Small Anim Pract** 2012;53:155–160. [12]Wright JG, et al **J Am Vet Med Assoc** 2008;232:1863–1872. [13]Shea A, et al **J Am Vet Med Assoc** 2012;240:61–64. [14]Smith JR et al. **J Am Vet Med Assoc** 2013;243:1042–1048. [15]Pittet D **Emerg Infect Dis** 2001;7:234–240. [16]Pittet D et al. **Ann Intern Med** 1999;130:126–130. [17]Weese JS, et al **Vet Microbiol** 2006;114:160–164. [18]Wright JG, et al **Emerg Infect Dis** 2005;11:1235–1241. [19]Attard K et al. **Aust Vet J** 2012;90:438–441. [20]Murphy CP, et al **Zoonoses Public Health** 2010;57(6):429–438. [21]Whitby M et al. **J Hosp Infect** 2007;65:1–8. [22]Kretzer EK et al. **Am J Infect Control** 1998;26:245–253. [23]Larson EL et al. **J Hosp Infect** 1995;30 Suppl:88–106. [24]Pittet D et al. **Lancet** 2000;356:1307–1312.

TO CLOT OR NOT TO CLOT? HYPERCOAGULABILITY: LARGE & SMALL ANIMAL

Kenichiro Yagi, BS, RVT, VTS (ECC, SAIM)
Los Altos, CA

INTRODUCTION

To clot, or not to clot? That is the question. The hemostatic system is a complex protective mechanism sealing off avenues of blood loss through the vascular system upon damage. While cessation of blood loss is vital in preventing subsequent anemia and eventual death, coagulation of the blood is normally controlled, limiting it within the intravascular space to maintain normal blood flow and minimizing chances of thrombosis. This delicate balance and interaction of procoagulant and anticoagulant mechanisms are essential in the proper function of the hemostatic system. In emergency and critical care, dysfunctions of the hemostatic system lead to life-threatening situations through a variety of mechanisms. One facet of these dysfunctions lies in excessive coagulation induced by a hypercoagulable state.

HEMOSTASIS

The coagulation mechanism has been described traditionally by dividing the process into two phases of coagulation; primary hemostasis (platelet plug formation) and secondary hemostasis (coagulation cascade), followed by the fibrinolytic phase. Whether coagulation occurs is controlled through a balance of thrombosis and fibrinolysis, modulated through antithrombosis and antifibrinolysis.

Newer knowledge of thrombosis incorporates the roles of tissue factor (TF) bearing cells and platelets in less distinct, overlapping phases called the "cell based model" of coagulation. The cell based model consists of the initiation, amplification, and propagation phases. (Note: For the rest of the document, coagulation factors will be referred to with an "F" for "factor", followed by its number, and with or without an "a" to indicate inactive and active form, respectively).

Initiation: TF expressed by cells located outside of the vasculature serves as the initiator for coagulation. The localization of TF in the extravascular compartment prevents the activation of coagulation in normal circumstances where the endothelium is intact. TF is also expressed by some cells in circulation but is kept in an inactivated form. Injury of the endothelium exposes the TF, leading to the binding of FVIIa, which in turn activates more TF-FVII complexes. The complex activation leads to generation of FIXa, FXa, and FVa. FXa and FVa form the prothrombinase complex, leading to small amounts of thrombin generation through cleaving of prothrombin.

Amplification: Platelets are activated by the generated thrombin as they leave the intravascular space through the injured endothelium. Through changes in membrane surfaces incited by thrombin binding to receptors, platelets degranulate to release fibronectin, platelet factor, von Willebrand's factor (vWF), calcium, and other procoagulants. These factors further active platelets and promote binding of coagulation factors to the membrane surface. Thrombin then activates FXIa and FVa on the platelet surface, and cleaves vWF off of FVIII, releasing vWF and activating FVIII to FVIIIa. This phase is considered to have amplified the small signaling for coagulation through thrombin generation in the initiation phase.

Propagation: The effect of degranulation includes the expression of platelet surface ligands, which results in platelet aggregation. FIXa (generated during initiation and amplification) and FVIIIa (generated during amplification) form the tenase complex on the platelet surface. Tenase then rapidly generates FXa on the platelet surface, binds to FVa (generated in the amplification phase)- and cleaves prothrombin into thrombin. Thrombin is the protease responsible for fibrin formation. When there is a significant amount of thrombin generated, they serve to convert fibrinogen to fibrin, allowing for clotting. Thrombin further promotes coagulation by activating FVII, FXI, FVIII, and FV, activating platelets, and activation of FXIII (a transglutaminase responsible for crosslinking fibrin fibrils).

In a healthy animal, coagulation is inhibited to maintain good blood flow and prevent thromboembolism. Decreased thrombin levels or activity, entailing the body's normal antithrombotic mechanism, involve several events. These events include tissue factor pathway inhibitor (TFPI) inhibiting TF-FVIIa and FXa, C1-inhibitor inhibiting FXIIa, and FXIa, antithrombin inhibiting thrombin, protein C inactivating FVa and FVIIIa, thrombomodulin modifying thrombin to decrease its ability to convert fibrinogen to fibrin, and protein S aiding in protein C's ability to inactivate factors as well as TFPI's inhibition of FXa.

Fibrin clots seal the damage in vasculature to prevent intravascular fluid loss. Once the damage is repaired through healing mechanisms, removal of the clot is required to re-establish unimpeded blood flow. Fibrinolysis is the process of breaking down of the fibrin clot. Fibrin degradation is performed by an enzyme called plasmin. Plasmin is formed by the cleaving of plasminogen by plasminogen activators such as tissue-type plasminogen activator (tPA, active when fibrin is present) and urokinase plasminogen activator (uPA, active extravascularly and independent of fibrin presence). Fibrinolysis is normally inhibited by inhibitors of plasmin activity (thrombin activatable fibrinolysis inhibitor), plasminogen activation (α_2-antiplasmin), or plasminogen activator inhibitors (plasminogen activator inhibitor 1).

The hemostatic system is a complex system involving elements controlling thrombosis, antithrombosis, fibrinolysis, and antifibrinolysis. Antithrombosis, involving tissue factor inhibitors, platelet inhibitors, and antithrombin, act to provide local control of thrombosis to ensure normal blood rheology and prevention of thrombi formation when coagulation is unnecessary. When fibrin meshworks (clots) are no longer needed, fibrinolytic factors, including plasmin (activated from plasminogen), work to dissolve clots to provide recanalizing of blood vessels and dissolve unwanted thrombi to prevent thromboembolism. Local control of fibrinolysis is provided by antifibrinolytic factors including plasminogen activator inhibitor, antiplasmin, and thrombin activatable fibrinolysis inhibitor which inhibit plasmin activity and plasminogen activation. These processes interact with each other through complicated feedback systems to form clots where appropriate to maintain the vascular system.

DISORDERS OF HEMOSTASIS

Consequences of hemostatic dysfunction can be divided into two categories: hypercoagulation (thrombosis), or hypocoagulation (hemorrhagic coagulopathy). Thrombosis occurs from the development of a hypercoagulable state, and can be further divided into macrovascular thrombosis (leading to clinical manifestations of aortic thromboembolism and pulmonary thromboembolism or PTE), and microvascular thrombosis (leading to microvascular ischemia and coagulation factor depletion and resultant DIC). Hemorrhagic coagulopathies can arise from dysfunctions in the coagulation, occurring due to platelet depletion or dysfunction, as well as the coagulation factor inhibition or depletion.

Thrombosis

Thrombosis, or formation of matrices of platelets, fibrin, and cellular debris, occurs within the intravascular space potentially leading to thromboembolism due to three factors (described by Virchow's triad): hypercoagulability, blood stasis, and endothelial injury.

Hypercoagulability can result from platelet hyper-reactivity, excessive coagulation factor activation, natural anticoagulant deficiency, hypofibrinolysis, or a combination of any of these causes. Platelet hyper-reactivity can result from increases in platelet activators. This includes direct stimulation by inflammatory cytokines, release of other platelet agonists, and hypoalbuminemia-mediated increase in thromboxane A2 (a prothrombotic). Endothelial damage leading to decreased availability of platelet inhibitors (prostacyclin, nitrous oxide, and ADPase) or administration of antiplatelet drugs can lead to a reduced inhibition. Excessive coagulation factor activation occurs when increased expression of TF by endothelial cells, macrophages, and cell-derived microparticles is induced by endotoxins or inflammatory mediators. An increased exposure of already existing TF can also occur due to endothelial damage. Deficiencies in natural anticoagulants may be genetic or acquired in origin (we will focus on acquired origins as they are more prevalent in animals). Antithrombin (AT) deficiencies can arise from hepatic failure, protein-losing nephropathies, consumption due to a pathologic increase in thrombin (DIC or massive thromboembolism), or suppression from drugs. Protein C deficiencies can develop due to sepsis, malignancy, pancreatitis, DIC, and hepatic or cardiac failure. Tissue factor pathway inhibitor (TFPI) which inhibits coagulation can be deficient in animals with hypercholesterolemia, thus making hypercholesterolemia a risk factor for thromboembolism. Hypofibrinolysis has not been very well defined in veterinary medicine, but is thought to occur with an increased level of plasminogen activator inhibitor (PAI-1) and/or α2-antiplasmin. Hypoplasminogenemia has been reported in traumatized dogs and horses with strangulating obstructions. Areas of impaired blood flow can lead to varying degrees of blood stasis. Changes in cardiovascular anatomy and cardiovascular function can cause abnormal or reduced flow, being risk factors of thrombosis. Left atrial dilation causing an aberrant blood flow is a well-known cause of FATE. Other conditions associated with altered blood flow include hypovolemia, hyperviscosity disorders, neoplasia, vascular and cardiac abnormalities.

Protein-losing nephropathy: Glomerular diseases leading to protein loss through the urine is known to cause a hypercoagulable state, associated with venous and arterial thromboembolism in dogs. AT loss in the urine is the primary factor in thrombosis for these patients. Because AT and albumin have similar molecular sizes, hypoalbuminemia (< 2.0g/dL) can be thought of as a decreased AT level and increased risk of thromboembolism as well. As mentioned earlier, hypoalbuminemia in itself leads to increased risk of thrombosis due to increased thromboxane A2. Glomerular nephropathies are also associated with hyperfibrinogenemia, hypercholesterolemia, increased FVIII, increased PAI-1 (causing hypofibrinolysis), all contributing to hypercoagulability. Thrombin activation is thought to increase even with only a very modest amount of protein loss in the urine.

Neoplasia: An association between neoplasia and thrombosis is now well accepted, with some types to have a higher degree of association than others. Thrombosis is observed to be associated with lymphoma, acute leukemia, and other solid tumors. Occurrences of hypercoagulable states have been determined by thromboelastography (TEG) in significant percentages of patients with malignant neoplasia (45-66%) studies, supporting this connection. Patients with malignancies with metastasis show higher fibrinogen and D-dimer concentration compared to those with local neoplasia. The mechanism behind neoplasia associated hypercoagulability includes a multitude of factors. Malignant cells release procoagulant material and inflammatory cytokines leading to platelet activation, and can directly interact with endothelial cells, platelets, and monocytes. Malignant cells may also express TF, and release cancer procoagulants. Malignancy may also lead to activated protein C resistance and deficiencies in anticoagulants. Neoplasia can indirectly contribute to hypercoagulability by causing vascular invasion and anomalies, immobility, or prompt venous catheterization, corticosteroid therapy, surgery, and chemotherapy.

Immune-mediated Hemolytic Anemia: Thromboembolism is commonly associated in patients with IMHA, leading to significant mortality. Evidence of thromboembolism is seen on necropsy findings in 30-80% of IMHA patients. Other laboratory findings such as hyperfibrinogenemia, high D-dimer levels, thrombocytopenia, hypoalbuminemia, hyperbilirubinemia, and low AT level are also seen, which supports a hypercoagulable state. The mechanism behind the state is not fully understood, but is thought to involve an increased circulating level of activated platelets, a hemolysis mediated hyper-reactivity of the platelets to platelet agonists, and antiphospholipid antibodies.

Acute Pancreatic Necrosis: Pancreatitis is associated with venous and arterial thromboembolism in dogs, and venous thromboembolism in cats. Damaged pancreatic tissues release TF and inflammatory cytokines. There is also an increase in phospholipase A2, leading to increased platelet activating factor and cytokine synthesis. Both of these events trigger platelet activation and increase platelet hyper-reactivity. Platelet reactivity is further enhanced through inhibition of prostacyclin production and increased thromboxane A2 release, resulting from high plasma free fatty acid and hypoalbuminemia, respectively. In addition, an increased level of PAI-1 and decreased activity of tPA lead to hypofibrinolysis.

Hypercortisolism: Pulmonary thromboembolism, portal vein, and splenic vein thrombosis are known complications encountered by dogs with hyperadrenocorticism. A high cortisol level is thought to be involved in the development of a hypercoagulable state, indicated by similar effects seen with corticosteroid therapy, though a relationship between dosage and duration of therapy is not well defined. An increase in several coagulation factors (II, V, VII, IX, X, XII, fibrinogen), reduction in AT levels, and TEG changes consistent with a hypercoagulable state, support the connection. Dogs with pituitary-dependent hyperadrenocorticism were found to have increased clot strength, rate of clot formation, and possibly platelet reactivity, with trilostane therapy significantly decreasing clot strength.

Sepsis and DIC: Microvascular thrombosis and DIC are known to be triggered by sepsis and systemic inflammatory response syndrome (SIRS). An increase in proinflammatory cytokines induce TF expression on the monocytes and endothelial cells, leading to initiation of coagulation. In addition, several events contribute to a loss of anticoagulation. First, TFPI is broken down by granulocytic elastases. Second, TFPI expression is suppressed by cytokines. Lastly, TFPI is overwhelmed by the number of TF-FVIIa. As thrombin level increases through amplification, anticoagulation through upregulation of antithrombin and protein C becomes insufficient and leads to a hypercoagulable state. While factors normally leading to an increase in fibrinolysis (increased release of tissue plasminogen activator and inflammation related activation of contact pathway) are present, fibrinolysis is inadequate due to increase in plasminogen activator inhibitors and activation of thrombin-activatable fibrinolysis inhibitor. As anticoagulant factors are overwhelmed and thrombin activates more platelets, diffuse microvascular fibrin thrombi are generated. The perpetuation of coagulation in DIC contributes to consumption of coagulation factors. Eventual hemodynamic changes which decrease tissue perfusion in addition to secondary acidosis, which inhibit platelet function and coagulation factor activity, lead to a hemorrhagic coagulopathy.

CLINICAL MANIFESTATION OF THROMBOSIS

When balance of hemostatic forces is tilted towards the side of thrombosis, a hypercoagulable state exists, manifesting clinically as consequences of thromboembolism. In canine patients, thromboembolism is seen most commonly with protein losing nephropathy, neoplasia, immune-mediated hemolytic anemia, necrotizing pancreatitis, hyperadrenocorticism, and corticosteroid therapy. Having multiple predisposing conditions increase the likelihood of thromboembolism. Common manifestations involve arterial and venous thromboembolism with involvement of multiple organs. Single organ manifestations include arterial thromboembolism is in dogs with atherosclerosis, and PTE in dogs with IMHA.

Occurrence of thromboembolism in felines seem to be more rare, but manifest similarly to those in dogs. While venous thromboembolism may occur in cats with neoplasia, cardiac disease, and multisystemic diseases; aortic thromboembolism associated with cardiomyopathy is the most common thromboembolic event in cats. Typically due to hypertrophic cardiomyopathy, a thrombi form in the left ventricle and embolize at the aortic trifurcation causing what is known as a "saddle thrombus", or less commonly embolize proximal to the renal artery or brachial vessel. The occlusion leads to vasoconstriction induced by the embolus, leading to further reduced collateral flow and ischemia.

Equine patients commonly experience thromboembolism from vascular injury. Thrombosis resulting from jugular catheterization and thrombophlebitis is common with various risk factors including endotoxemia, hypoproteinemia, salmonellosis, anti-diarrheal therapy, and anti-ulcer therapy. Gastrointestinal vasculopathy, which can be induced by parasites, can cause secondary thromboembolism leading to acute and chronic colic. Endovascular migration of parasites is presumed to lead to distal limb thrombosis, aortic thromboembolism, and iliac thromboembolism. Other causes of distal limb thrombosis may be sepsis and gastrointestinal inflammation. Hypercoagulability syndromes in horses leading to multiple organ thromboembolisms can result from severe gastrointestinal diseases associated with endotoxemia and subsequent renal microvascular thromboembolism. Carbohydrate-induced laminitis is thought to be linked to vascular thrombosis in the dermal lamina. Toxin or endotoxin induced hypercoagulability may lead to PTE, which is thought to be an under-recognized type of thromboembolism in equine patients.

Treatment of thrombosis is directed at preventing thrombi formation and dissolution of existing thrombi with thrombolytics. Antiplatelet drugs and anticoagulants are used to provide inhibition of thrombogenesis. Cyclooxygenase inhibitors such as aspirin are commonly used as an antiplatelet therapy. Clopidogrel is a thienopyridine which inhibits ADP binding to platelet ADP receptors, reducing the platelet's ability to aggregate. Heparin is an anticoagulant commonly used, though anticoagulant effect is unpredictable. Low-molecular weight heparin produced by depolymerization of heparin, bind to antithrombin which enhances its antithrombotic activity. These drugs are used prophylactically in patients at risk for the thromboembolic event or have a history of thromboembolic disease. Thrombolytic drugs such as tissue plasminogen activator, streptokinase, and urokinase act to enhance fibrinolysis aiming to lead to dissolution of the clot. Use of thrombolytics has demonstrated variable outcomes in veterinary patients, and their efficacy is still under investigation.

Hemostasis is a complex system involving proper coagulation, anticoagulation, fibrinolysis, and antifibrinolysis occurring in a finely balanced manner. Many causes of disruption in these forces result in the formation of thrombi leading to thromboembolism or a coagulopathy leading to hemorrhagic disorders. As a veterinary technician, having thorough knowledge of risk factors and progression of hemostatic disorders will allow us to aid in our ability to minimize potentials for a pathological consequence. With potential clinical manifestations in mind, the improved ability to recognize consequences of hemostatic dysfunction will allow for a swift and appropriate action.

REFERENCES

[1]Boswood A, et al. **J Small Anim Prac** 2000;41:109. [2]Hogan DF, et al.**J Am Vet Med Assoc** 2004;225(9):1406.[3]Ettinger SJ, et al. **Text of Vet Int Med** 2010. [4]Weiss DJ, et al. **Schalm's Vet Hem** 2010.

CLINICAL PATHOLOGY: A PRACTICAL APPROACH TO FOOD ANIMAL CASES

Melissa G. Ackerman, DVM, MS
Pullman, WA

INTRODUCTION

Clinical cases in agricultural animal practice vary widely in presentation and nature. Sometimes an individual member of a herd or flock is involved, other times the presentation of an individual may represent a herd problem. The clinical presentation may be straightforward and a thorough history and physical examination will readily provide the diagnosis and guide the veterinarian in the proper direction of therapeutics and support. However, there are many occasions where the diagnosis might not be so forthcoming and a more thorough approach must be taken. A sound understanding of appropriate sampling and laboratory evaluation is essential to define the clinical scenario. This session will review some of the common laboratory measurements that veterinarians in agricultural animal practice request from the clinical pathology laboratory.

Diagnostic tests have several purposes; to confirm a suspected diagnosis, to give information that adds to the prognosis, to assess or look for subclinical disease, to guide and/or monitor response to therapy and supportive care, and to monitor the clinical course of disease in a patient.

LABORATORY DATABASE

A basic database often includes a complete blood count (CBC), a chemistry profile, and a urinalysis. In reality, not every animal receiving diagnostic testing will have a "complete" database. Here, we will review the varying components of ruminant hematology and clinical chemistry as they are commonly performed on individual animal cases. The CBC and chemistry profile will be divided for discussion and case examples will be presented to augment the laboratory materials.

COMPLETE BLOOD COUNT (CBC)

The CBC can be divided into the following components: hemogram, leukogram, and protein components.

Hemogram

The most common conditions assessed from the CBC regarding the hemogram are the potential for anemia and relative or absolute polycythemia. Polycythemia is an increase in packed cell volume (PCV) or erythrocyte count and hemoglobin concentration is above normal. This may be absolute (real) or relative (caused by hemoconcentration with dehydration or shock). Real polycythemia is rarely caused by myeloproliferative disorders in large animals. It is more commonly seen secondarily (relative polycythemia) due to hypoxia and increased erythropoietin stimulation, for example with cardiovascular shunts and pulmonary disease.

Anemia is defined as a decreased oxygen carrying capacity of the blood. The PCV is the most accurate laboratory measurement of anemia. Hydration and level of excitement may affect the PCV. Ruminant red blood cells (RBC) vary in size (very small in goats) and shape (elliptical in camelids), so a spun or microhematocrit centrifugation is the most accurate measure of the PCV. There are three mechanisms of anemia development: blood loss, increased erythrocyte destruction (hemolysis), and inadequate erythrocyte production. Blood loss and hemolysis are regenerative anemias, whereas inadequate erythrocyte production is non-regenerative. In ruminants, regenerative anemias are characterized by the appearance of immature erythrocytes (nucleated RBCs and reticulocytes) in circulation.

Erythrocyte indices are used to characterize anemia, RBC size, and hemoglobin concentration. These include mean corpuscular volume (MCV), mean corpuscular hemoglobin (MCH), and mean corpuscular hemoglobin concentration (MCHC). MCV is a measure of erythrocyte size. MCV will be increased in regenerative anemia as younger cells tend to be larger. This can also be seen in cobalt and vitamin B12 and folate deficiency. Decreased MCV occurs in iron deficiency anemia. Cells undergo an extra division as there is a lack of hemoglobin formation. MCH is a measure of the amount of hemoglobin in each erythrocyte. It is increased when immature RBCs (reticulocytes) are present or there is hemolysis, and is decreased in iron deficiency anemia. MCHC is the concentration of hemoglobin in percentage or grams of hemoglobin per deciliter. It is increased in hemolytic states and decreased in iron deficiency or when increased reticulocytes are present.

Further hemogram descriptors include anisocytosis, which means varying sizes of RBCs are present. Slight anisocytosis is normal in ruminants, but marked amounts are a sign of regeneration. Polychromasia means a varying color of stained RBC is present. This may be a bluish color when caused by reticulocytes; a pale color is seen in iron deficiency. Poikilocytosis is the presence of abnormal shapes of RBCs. This occurs with increased fragility or fragmentation and may be seen in iron deficiency or disseminated intravascular coagulation (DIC). Basophilic stippling is a normal feature of regenerative anemia in ruminants but is also seen in some cases of lead poisoning. Howell-Jolly bodies are nuclear remnants in some immature RBCs as a response to anemia. Nucleated RBCs are developing RBCs that are released prematurely from the bone marrow into the circulation as a response to severe anemia. Heinz bodies are denatured hemoglobin that precipitate in the RBC and is related to cellular oxidative stress. RBCs with Heinz bodies are prone to hemolysis while in circulation.

Leukogram

The leukogram is broken down into categories by cell type. There are two types of leukocytes in circulation: polymorphonuclear cells (neutrophils, eosinophils, basophils) and mononuclear cells (lymphocytes, monocytes).

Neutrophils originate in bone marrow and take 4-9 days to develop and mature. In large animals the marginated pool of neutrophils is approximately the same as the circulating pool. The half-life of a neutrophil once in circulation is approximately 9 hours. The entire circulating population is replaced 2.5 times a day. In the absence of loss to lesions or disease, neutrophils are removed or destroyed by macrophages in the bone marrow, spleen, liver or are lost to body secretions (mucosal surfaces) within 96 hours of leaving the bone marrow. Neutrophils function to phagocytize and destroy foreign material (like bacteria). Phagocytosis leads to fusion with the cellular lysosomes, degranulation and release of enzymes, which kill the organism and potentially caused damage to surrounding tissues. Many gram-negative bacteria are able to resist this process.

Eosinophils also originate in the bone marrow and exist in circulation from 0.5 to 10 hours. They exist in tissues for up to 12 days. They aid in control of parasitic infections and regulating inflammatory and allergic reactions such as type 1 (immediate–type) hypersensitivity reactions.

Basophils are relatively uncommon in large animal circulation. They are produced in the bone marrow and have a circulating half-life of 6 hours with a tissue half-life of 10-12 days. Their stored intracellular granular substances are associated with immediate hypersensitivity reactions and inflammatory processes.

Lymphocytes are produced in bone marrow, lymph nodes, thymus, spleen and Peyer's patches. In humans, lymphocytes can circulate for up to 4 years. Lymphocytes are divided into T and B cell types, both of which are in the large animal circulatory system (T cells are in greater quantity). B lymphocytes, under the right conditions, transform into plasma cells and make antibodies. T cells help regulate this process and are responsible for fighting intracellular pathogens and delayed-type hypersensitivity.

Monocytes are produced in bone marrow, and then circulate for 1-3 days before entering tissues. Once there, they transform into macrophages and survive for weeks to years. These "free" macrophages protect body cavities, joints, lung, spleen, bone and bone marrow, and lymph nodes. They are involved in removing and repairing dead and damaged tissue and act against bacteria, viruses, protozoa, and fungi. They are strong combaters of intracellular organisms that cause granulomatous inflammation like tuberculosis and Johne's. They are also involved in tumor defense, immune functions, and hematopoiesis.

THE RUMINANT LEUKOGRAM

There are unique aspects of the ruminant leukogram that are important for interpretation. In adult sheep and cattle, it is normal for lymphocytes counts to exceed neutrophils. The leukogram in the goat will usually reflect an equal or slightly larger number of neutrophils than lymphocytes. Therefore, some conditions characterized by leukocytosis in monogastric species might not be characterized as such in ruminants. Additionally, fibrinogen is used as a marker for active inflammation because ruminants (especially cattle) have great capacity for fibrinogen production.

CHEMISTRY PROFILE

It is useful to interpret the profile in light of the signalment, history and physical examination. When interpreting a chemistry profile, group together the parameters of a body system or physiologic process so that those test results will provide a pattern of dysfunction. Patterns of change within and among these groups can provide useful diagnostic information about disease manifestations of various organ systems. Examples of chemistry test groupings as would be seen or available on many chemistry profiles or by request: electrolytes (sodium, potassium, chloride), acid/base parameters (TCO2, bicarbonate, anion gap), minerals (calcium, phosphorus and magnesium), protein (total protein, albumin, globulin, A/G ratio, fibrinogen), kidney parameters (BUN, creatinine), liver parameters, carbohydrate parameters (glucose), lipid parameters (cholesterol, triglycerides, non-esterified fatty acids), muscle parameters (CK, AST, LDH).

Electrolytes and Minerals

Always interpret in light of the acid/base status. Diarrhea, blood loss, ruptured bladder, and renal disease can all result in hyponatremia, while water deprivation, hypertonic dehydration, salt toxicity, and hypertonic fluid administration can all contribute to hypernatremia. Herbivores take in 60-100 grams of potassium daily to maintain adequate body potassium. Hypokalemia can result quickly from anorexia, abomasal reflux, metabolic alkalosis, and excessive glucose or sodium bicarbonate administration. Kidney disease, metabolic acidosis, and hypovolemia can result in hyperkalemia. Chloride levels often follow sodium levels, but with certain acid/base disturbances it may not.

Ruminants with muscular weakness and tetany likely have decreased calcium, phosphorous, and/or magnesium levels. Total serum calcium is a measure of bound calcium, whereas ionized calcium is a measure of the free ion (the bioavailable form). Hypocalcemia is often a finding with milk fever, grass tetany, hypoalbuminemia, renal failure, anorexia, endotoxemia, and oxalate toxicity. Hypercalcemia may exist with vitamin D excess or excessive calcium administration. Serum phosphorus may be elevated with renal failure or normally in young, growing animals. Hypophosphatemia may be seen with milk fever, postparturient hemoglobinuria, starvation, vitamin D deficiency, and inadequate dietary intake. Hypomagnesemia occurs with grass or winter tetany, transit tetany, and in calves on an all milk diet. Hypermagnesemia can occur with excess magnesium administration.

Acid/Base Parameters

Disorders of acid/base physiology occur relatively common in ruminants in all aspects of production. It is important to interpret acid/base disorders within the context of electrolyte findings because they are associated. For example, metabolic acidosis is typically associated with hyperkalemia and often also with hypernatremia and hyperchloremia. Also, metabolic alkalosis is typically associated with hypokalemia and hypochloremia. TCO2 is a measure of all the blood CO2 that in reality is mostly HCO3 (20:1 ratio of HCO3 to PCO2). Therefore a measure of TCO2 can give insight into the metabolic side of an acid/base state but have little value as to the respiratory component. Increase in TCO2 would suggest metabolic alkalosis, where a decrease would suggest metabolic acidosis.

Proteins

Proteins of all types are involved in many physiologic processes within the body ranging from basic structure to hemostatis, fluid flux, immunology, disease resistance, and carrying or binding electrolytes and drugs. Measurement of total plasma or serum protein, albumin, globulin, and fibrinogen provide a baseline for interpretation of the state of the proteins in the body. Serum electrophoresis and measures of specific immunoglobulins may further define the state.

Hyperproteinemia may be a result of an increase in all classes of proteins or in a single class such as globulins. With all classes elevation, causes may include dehydration, acute or chronic inflammation, and even liver or kidney failure. Elevated globulins in ruminants often are polyclonal and a result of chronic inflammatory processes such as abdominal and pulmonary abscesses, pleuritis, pneumonia, and hepatic disease. Hypoproteinemia in ruminants is often due to hypoalbuminemia. Albumin is made in the liver and accounts for 75% of oncotic pressure. Hypoalbuminemia may be seen in diseases states with increased losses to the gut or kidneys, like pyelonephritis, intestinal and hepatic parasitism, lymphosarcoma, Johne's, and even with starvation. When all classes of proteins are decreased, consider blood loss by blood-sucking parasites such as *Haemonchus* species or by hemorrhage. Other causes include protein-losing enteropathy, gastrointestinal (GI) ulcers, and renal disease.

Kidney Parameters

BUN (blood urea nitrogen) and creatinine are the two common measures of glomerular filtration and therefore renal function. However, these parameters must be interpreted in light of the urinalysis findings and specifically urine specific gravity (USG). Azotemia is an increase in BUN and/or creatinine. USG < 1.025 is likely renal azotemia and >1.025 is likely pre-renal azotemia. Pre-renal azotemia is caused by decreased renal perfusion, as with dehydration. Renal azotemia is caused by renal failure and primary renal disease. Post-renal azotemia is caused by disease states such as urolithiasis and ruptured bladder. In ruminants, creatinine is a more specific indicator of renal function than BUN because ruminant intestinal microflora can metabolize urea nitrogen. Additionally, BUN is influenced by diet.

Liver Parameters

Disorders of the liver may be recognized as primary hepatocellular damage, disorders of the excretory tracts and loss of metabolic function. A liver biopsy may be necessary to fully assess the liver. Chemistry parameters concerning the liver are categorized as hepatocellular enzymes, cholestatic enzymes, and parameters that reflect liver function.

Aspartate aminotransferase (AST), sorbitol dehydrogenase (SDH), and lactate dehydrogenase (LDH) are enzymes that, when elevated, can indicate hepatocellular injury. SDH is more specific for the liver than AST, which can also rise with muscle damage. Increases in alkaline phosphatase (ALP) and gamma-glutamyl transpeptidase (GGT) are consistent with stasis in the biliary system (cholestasis). However, GGT is present in colostrum, and will be elevated in ruminant neonates that have ingested colostrum recently. Liver function work is partially assessed with measurements of the liver's involvement with hemoglobin metabolism (bilirubin), the enterohepatic circulation of bile acids and breakdown of ammonia. Total and indirect bilirubin are elevated in hemolytic states and in the presence of liver failure whereas direct bilirubin is more directly associated with the liver's ability to conjugate and excrete bilirubin. Total serum bile acids are a good test of liver function.

Energy Parameters: Carbohydrates, Lipids, Muscle

On a chemistry profile glucose is most representative of basic energy metabolism in the body. Hypoglycemia is one of the most common causes of hypothermia, which can be especially important in newborn ruminants. Hypoglycemia is a common finding with neonatal anorexia, negative energy balance states such as ketosis and pregnancy toxemia, hepatic failure, and shock states. Hyperglycemia may be a result of stress, excitement, diabetes mellitus, excessive dextrose administration, and corticosteroids.

Blood cholesterol and triglycerides are normal measures of lipid status in monogastric animals where fat is an important component of the diet. In herbivores cholesterol is less informative; however, triglycerides and non-esterified free fatty acids (NEFAs) can provide valuable information as to energy balance.

Creatine kinase (CK) is the most specific muscle enzyme that is present on routine profiles. AST and LDH relate to muscle but also have origins in other tissues (liver) and must be interpreted in light of other liver and muscle enzymes. Elevations in CK can occur in any situation where there is release of the enzyme from striated muscle, for example myositis (bacterial, viral, parasitic), recumbency (downer syndromes), muscle trauma including injections, and capture myopathy.

REFERENCES

Smith, B. P., 2009. *Large Animal Internal Medicine*, 4th ed. Mosby Elsevier, St. Louis, MO. Radostits, O.M. et al. 2007. *Veterinary Medicine: A Textbook of the Diseases of Cattle, Sheep, Goats, Pigs, and Horses, 10th ed.* Saunders, Elsevier. Presentation notes from Dr. S. Parish. *Diagnostic Testing using Clinical Pathology in Agricultural Animal Practice.* Washington State University, Professor Ag Animal Medicine.

WHAT YOU SHOULD KNOW ABOUT PULMONARY HYPERTENSION

Romain Pariaut, DVM, DACVIM (Cardiology), DECVIM-CA (Cardiology)
Baton Rouge, LA

INTRODUCTION

The pulmonary circulation is a low-pressure and low resistance system with systolic pressures of 20-25 mmHg on the arterial side. Pulmonary hypertension is present when systolic pulmonary artery pressure determined by Doppler echocardiography is greater than 30-35 mmHg. It usually occurs when more than 80% of the pulmonary vascular luminal area is lost.

Pulmonary hypertension is commonly secondary to left-sided heart failure and chronic lung disease. It is also a frequent complication of heartworm disease. Other causes of pulmonary hypertension include left-to-right cardiovascular shunts, pulmonary thromboembolism and neoplasia.

Three factors are necessary for the development of pulmonary hypertension: Vasoconstriction, thrombosis, and smooth muscle cell and endothelial cell proliferation. Endothelial cells are the major regulators of vascular function. In response to various stimuli, they will trigger the release of vasoconstrictors, prothrombotic factors, pro-inflammatory signals, and will activate smooth muscle growth.

PULMONARY HYPERTENSION SECONDARY TO LEFT HEART DISEASE

It is a common complication in dogs with severe left-sided heart failure, usually secondary to degenerative mitral valve disease. Simple backward transmission of left atrial pressure results in mild pulmonary hypertension. However, left heart failure may be associated with "out-of-proportion" pulmonary hypertension, when pulmonary pressures are much higher than what would be expected based on left atrial pressure alone.

PULMONARY HYPERTENSION SECONDARY TO HEARTWORM DISEASE

The presence of heartworms in the pulmonary vasculature results in inflammation, hypertrophy and stiffening of the pulmonary artery walls. Pulmonary hypertension can occur in association with heartworm infestation, but it is also observed in dogs free of heartworms but that were treated for heartworm disease earlier in their life.

CLINICAL SIGNS OF PULMONARY HYPERTENSION IN DOGS

Common signs include lethargy, weakness, dyspnea, cough, syncope and ascites.

Physical examination in dogs with pulmonary hypertension may reveal an accentuated pulmonary component of S2 (split), a precordial pulse on the right side, a pansystolic murmur of tricuspid regurgitation, a diastolic murmur of pulmonic insufficiency, jugular distension, jugular pulse, hepatomegaly and abdominal distension with ascites.

Two dimensional echocardiography reveals increased right ventricular free wall thickness, increased papillary muscle size, enlarged right ventricular chamber diameter, impaired right ventricular contractility, interventricular septum flattening in diastole and systole and pulmonary artery dilation. Signs of right atrial enlargement, and the degree of tricuspid valve regurgitation are important parameters that are correlated with the risk of developing right-sided congestive heart failure, and therefore influence long-term survival. In dogs with pre-existing left-sided heart failure, left cardiac chamber size is usually increased. In dogs with lung disease or heartworm disease, the left ventricle is usually small because of decreased venous return, which results in low cardiac output. It is important to assess left ventricular size, as it changes treatment strategies.

TREATMENT OF DOGS WITH PULMONARY HYPERTENSION

The goal of treatment is to decrease pulmonary vascular resistance, while sparing systemic vascular resistance, increase right ventricular contraction and control signs of right-sided congestive heart failure. If a thromboembolic event is suspected to be the cause for pulmonary hypertension, antithrombotic therapy may be indicated (Clopidogrel, aspirin, heparin).

Sildenafil, a phosphodiesterase 5 inhibitor, is a potent vasodilator of the pulmonary vasculature. It is used at a dose of 1 to 2 mg/kg q8-12 hours. Pimobendan, an inodilator and phosphodiesterase III inhibitor is used at a dose of 0.25 mg/kg q12 hours to improve right ventricular contractility.

In the presence of ascites, abdominocentesis is required to remove fluid and decrease intra-abdominal pressure. When pulmonary hypertension is secondary to left-sided heart disease ("overfilled" left ventricle), furosemide (1 to 3 mg/kg q8-12 hours) is used to prevent recurrence of ascites. However, when the left ventricle is underfilled, the use of diuretics may worsen clinical signs as cardiac output decreases, leading to pre-renal azotemia and frequently episodes of weakness or syncope.

REFERENCES

Kellihan HB, et al. J Vet Cardiol 2012;14(1):149. Stepien RL. J Small Anim Pract 2009;50 Suppl 1:34.

MANAGEMENT OF ARRHYTHMIAS : ADVANCED CONCEPTS
A PRIMER FOR VETERINARY TECHNICIANS

William Rausch DVM, DACVIM (Cardiology)
Portland, Oregon

AIMS OF THIS TALK

This lecture is intended for Veterinary Technicians who play an active role (or desire to play a more active role) in monitoring the ECG. Technicians are often tasked with being the primary guardian of the cardiac rhythm - both in surgery and in unstable hospitalized patients. As such, it is critical not only to recognize variances from a normal sinus rhythm, but also to understand that not all arrhythmias are created equal. It takes quick and accurate judgement to recognize which arrhythmias likely need to be acted upon and why. This lecture is designed to bring you closer to that point.

First, we will briefly review ECG basics. Second, we will discuss a wide selection of common arrhythmias with a focus on when and why they occur and the hemodynamic consequences that often determine how and if we intervene. Lastly, we will discuss our therapeutic options, why some treatment choices are better in particular situations and how we should measure success/failure of therapy.

ECG BASICS

This section will be covered briefly in the lecture and is provided here for review purposes. The ECG is a graphic representation of the depolarization and repolarization of cardiac myocytes as recorded from the body surface. Time is expressed in millimeters/second and is plotted on the horizontal axis versus voltage, which is expressed in millivolts on the vertical axis. The ECG is a superb diagnostic tool for determination of heart rate and rhythm. Many electrocardiographic recording devices will automatically count the heart rate, but these automatic counts can lead to inaccuracies - especially in veterinary medicine. The relatively wide breadth of morphologic normality for T waves and QRS complexes in our patients can lead machines to undercount or overcount the true heart rate. Likewise, ECG monitors that purport to automatically yield a rhythm diagnosis can often prove misleading. These machines have a particularly difficult time interpreting the variations in sinus rate that are often a normal variant in our patients. Thus, there is still no substitute for careful assessment of cardiac rhythm and an accurate manual heart rate. Acquisition of an accurate heart rate is especially important when managing bradycardic and tachycardic patients. Accurate manual heart rate assessment simply involves counting the number of QRS complexes in three or more seconds and multiplying appropriately in order to yield a per minute rate. If the cardiac rhythm is highly irregular or slow, then using at least six seconds of an ECG strip to extrapolate a heart rate will yield a more accurate count.

INDICATIONS AND LIMITATIONS

While the ECG is uniquely poised to give information regarding cardiac rate and rhythm, it is important to realize its limitations. A major limitation is that the ECG is a relatively insensitive test for detecting cardiac disease and is generally non-specific as to the underlying disease in the presence of abnormal ECG findings. There are other tests that are better suited to determine if primary heart disease is present, make a cardiac diagnosis, evaluate hemodynamic status and characterize cardiac function. The ECG is but one indispensable part of the cardiac evaluation. Another important limitation is that the transient nature of some rhythm disorders can lead to false negative ECG findings if the arrhythmia does not occur during the recording period. This is a limitation that can often be overcome with ambulatory ECG (Holter) monitoring and typical case examples utilizing these technologies will be discussed.

Accurate ECG assessment can be limited by the quality of the recording. Proper recording technique can help to avoid many of the pitfalls associated with artifact. The standard position for recording a small animal ECG is right lateral recumbency. If the patient is too dyspneic for lateral recumbency a standing or sternal ECG can be recorded. Gentle limb restraint will often help to minimize tremble artifact. It is important to avoid contact between the leads and to limit lead and wire movement with respiration. Such movement can be avoided by proper placement of the limb leads away from the chest (just above the olecranon and in the area of the patellar ligaments. Case examples will be presented that focus on methods to determine if abnormal "complexes" represent arrhythmia or artifact. A helpful guideline to keep in mind is that an actual arrhythmia has to make physiologic sense in the context in which it is seen. Thus, the complexes surrounding the suspect deflection can often yield important information that alters the likelihood that the deflection represents an arrhythmia. For example, a deflection that may represent a VPC is likely to be so if it is immediately followed by a non-conducted P wave, but more likely to be artifact if it appears to fall in the QT interval of a sinus beat without altering the T wave of the sinus complex. An ECG device that is capable of recording multiple leads simultaneously can also be helpful when trying to distinguish artifact from arrhythmia, as certain motion artifact will be detected in some leads but not others. Such a finding would suggest artifact, rather than arrhythmia, as at least some portion of an arrhythmic complex would be expected to be seen in each lead tracing.

PHYSIOLOGIC CORRELATES AND ECG ASSESSMENT

A regular sinus rhythm emanating from the sinus node is the normal rhythm of the heart. The sinus node is a highly organized cluster of specialized cells located in the right atrium near the junction with the cranial vena cava. The cells within the sinus node have the ability to spontaneously depolarize during phase 4 of the cardiac cycle. While this is not a trait exclusive to cells of the sinus node, the sinus node assumes the role of the dominant cardiac pacemaker under normal physiologic conditions because of two basic physiologic properties of sinus node pacemaker cells. These cells have both a low level of resting diastolic membrane potential and a relatively rapid rate of rise of phase 4 diastolic depolarization, as compared to other cells in the heart that have the ability to act as pacemakers. Thus, the sinus node normally controls the heart rate and rhythm, and by doing so suppresses the activation of subsidiary pacemakers. The importance of these phenomena in understanding bradyarrhythmia and tachyarrhythmia will be discussed in this lecture.

The initial depolarization of the cells of the sinus node is a relatively small electrical event and is not significant enough to be detected on a surface ECG. After emerging from the sinus node the cardiac impulse traverses both atria in its path to the atrioventricular (AV) node. The normal P wave represents the depolarization of the atria that results from sinus node firing. The sum of the wavelets depolarizing the atria is an electrical wave that predominantly moves caudally and from right to left, and thus the normal P wave is thus a positive deflection when viewed from lead II. Disease processes that structurally alter the atria can affect the height or width of the P wave. A tall P wave (P pulmonale) is classically associated with an enlarged right atrium and wide P waves (P mitrale) are often associated with an enlarged left atrium. These are not entirely specific findings however, and these ECG changes are relatively insensitive markers of chamber enlargement. The AV node is the only electrical pathway between the atria and the ventricles under normal circumstances. Electrical activation of the AV node begins somewhere during the period of the P wave on the ECG. Conduction proceeds relatively slowly through the AV node, which has implications in the medical management of arrhythmias as will be discussed in this lecture. Because it cannot be known precisely when AV nodal conduction begins, the PR interval (the time from the beginning of the P wave to the beginning of the QRS complex) is used as a surrogate for measuring AV conduction velocity. Normal ventricular activation occurs via the rapidly conducting specialized Purkinje system. Cardiac or pulmonary diseases that alter the relative distribution of myocardial mass can affect QRS morphology and the mean electrical axis of the ECG. Cardiac diseases that lead to eccentric or concentric myocardial hypertrophy can lead to changes complex size and sometimes cause conduction disturbances.

With this information as background, we will consider how it is that we decide if a particular arrhythmia warrants therapy. Many arrhythmias that we see in practice have severe manifestations which require intervention and less severe forms that may not. We will use case examples to explore in detail the following questions - the answers to which will determine if and how we choose therapy.

REFERENCES

1. Is the arrhythmia hemodynamically significant?
2. Does it appear to be causing clinical signs?
3. Can a correctable or self-limiting underlying cause for the arrhythmia be identified?
4. Does the arrhythmia appear likely to degenerate into a less stable rhythm?
5. Is there a consideration for interactions between anti-arrhythmic therapy and current therapy?
6. Is the patient under or soon to be under general anesthesia?
7. What are the risks of beginning therapy?
8. What measures will I use to determine if my therapy is effective?

ACUTE MANAGEMENT OF CONGESTIVE HEART FAILURE DUE TO DEGENERATIVE MITRAL VALVE DISEASE: A PRIMER FOR VETERINARY TECHNICIANS

William Rausch DVM, DACVIM (Cardiology)
Portland, Oregon

Acquired degenerative mitral valve disease (DMVD) is the most common cause of cardiac morbidity and mortality in our canine patients. Although this is a chronic disease that generally has a multi-year asymptomatic period, many patients will present for care because of acute symptoms associated with left-sided congestive heart failure. However, if we are able to successfully get them through these acute crises, they can return to a high-quality life for a significant amount of time.

The diagnosis and successful treatment of left-sided congestive heart failure resulting from DMVD depends upon timely and correct diagnostics and therapy. The role of the Veterinary Technician in proper identification, triage, observation and care of these unstable patients can not be understated. Although a certain percentage of patients with left-sided congestive heart failure resulting from DMVD will respond positively to a formulaic treatment approach, many will not. This is partially due to variation in individual patient parameters such as the degree of disease, the severity of pulmonary edema on presentation, the level of pre-existing cardiac therapy, the presence or absence of arrhythmias or pulmonary hypertension and the health of other organ systems among other factors. Additionally, the unique combination of this disease and it's therapies can lead to complications including systemic hypotension, azotemia, tachyarrhythmias or simply treatment failure. It is in these patients in particular that a more nimble individualized approach can lead to better outcomes. Such a nuanced approach requires careful observation of the patient and their response to our initial therapeutic choices. The Veterinary Technician is in a unique position to help shape and positively influence these choices. This live lecture will first review the natural history, pathophysiology and common clinical presentations seen with this disease. We will then discuss how skilled history taking and careful clinical observations can positively influence our choices at critical decision points. Your observations and judgements at these inflection points are a large determinant of better or worse patient outcomes.

The first challenge is correctly identifying these patients upon arrival at the hospital. There are clues in the presenting complaint, patient signalment, physical exam and clinical history that can raise or lower your index of suspicion for this diagnosis. Patients suffering the symptoms of left-sided congestive heart failure commonly present because their owners notice some combination of tachypnea, dyspnea, lethargy, syncope, cough or inappetance. Many dogs in this condition will show some, but not all, of these symptoms. These symptoms do not make the diagnosis though as they are not exclusive to dogs in congestive heart failure. Many dogs presenting with some combination of these complaints will be doing so for other medical reasons.

DMVD does occur commonly in both male and female dogs, although some reports suggest that males often acquire the condition at a slightly younger age. Although most dogs with DMVD will never develop congestive heart failure, those who do tend to be middle-aged to older. Importantly, they usually have had a multi-year period of exhibiting DMVD without serious symptoms prior to developing left-sided congestive heart failure. Therefore if they have been patients of your hospital during this time with regular visits there is a strong chance that a left-sided systolic heart murmur will have been noted during these past visits. Dogs with clinically significant DMVD should have an auscultable systolic heart murmur heard best at the left cardiac apex. In an emergency presentation it can occasionally be more difficult to hear because of sounds associated with increased respiratory effort or because of the presence of arrhythmias. Keep in mind that the presence of this murmur with the clinical symptoms described above is *consistent with* congestive heart failure but not diagnostic for it. However, the clear absence of a left-sided cardiac murmur is strongly suggestive that a dog does not have serious DMVD. There are other clues in the heart rate and rhythm that can suggest whether congestive heart failure from DMVD is more or less likely. Dogs in this condition most commonly will have either sinus tachycardia or a relatively fast sinus rhythm. The presence of a relatively slow sinus rhythm should raise serious suspicion that the clinical signs seen may not be the result of congestive heart failure. Ventricular and supraventricular (including atrial fibrillation) arrhythmias are not uncommon. Serious tachyarrhythmias may require anti-arrhythmic treatment at the same time the congestive heart failure is being treated in order to optimize patient outcomes. Thoracic radiographs are of course an indispensable part of diagnosing left-sided congestive heart failure in dogs with DMVD, but your initial observation can greatly improve the accuracy of your clinical suspicions and more appropriately guide initial diagnostics and therapy. With this as background we will in this live lecture discuss the risks and possible benefits of the many therapeutic options we can choose from. We will discuss common complications in treated patients (hypotension, tachyarrhythmias, azotemia, anxiety, worsening failure) and how we might react to these complications by re-calibrating our approach for a particular patients needs.

ISSUES SURROUNDING END OF LIFE DECISIONS AND EUTHANASIA IN PATIENTS WITH CHRONIC HEART DISEASE

Jo Anne Winget, CVT and Mark A. Oyama, DVM, DACVIM
Philadelphia, PA

INTRODUCTION

Euthanasia is a decision frequently faced by owners of pets with chronic heart disease. Having a better understanding of what factors go into an owner's decision to euthanize will help us better manage both our patient and client. A previous study reported that poor prognosis, recurrent signs of heart failure, and perceived poor quality of life were common reasons why pet owners elected to euthanize pet dogs with heart disease. Other factors included poor appetite, inappropriate urination or defecation habits, cost of therapy, and medication toxicity. Despite these data, relatively little is known about how owner perceived quality of life directly impacts decisions regarding euthanasia.

METHODS

A list of factors believed to be important to quality of life in dogs and cats with heart disease was compiled by the investigators and formed the basis of a questionnaire instrument that was offered to owners of dogs and cats presenting at one of 6 different cardiology specialty practices in the US and Canada. Data concerning the underlying diagnosis and severity of disease was recorded by the attending cardiologist. Anonymous pet owner data regarding age, gender, and number of pets in the household was recorded. Pet owners then completed a 3 page questionnaire that included questions about stress, perception of disease severity, and pet behavior, appetite and activity. The objective of this questionnaire was not only to determine what factors lead an owner to euthanize their pet, but to obtain more specific information about important factors. This session will be co-presented by a veterinary technician/nurse and cardiologist. We will discuss the specific results and how they can be used to better understand the decision process an owner goes through when contemplating euthanasia. Better understanding of these end-of-life decisions could potentially improve owner-veterinarian communication, focus healthcare on those parameters deemed most critical to good quality of life, and prolong longevity in dogs and cats with heart disease.

REFERENCES

1. Mallery KF et al. Factors contributing to the decision for euthanasia of dogs with congestive heart failure. JAVMA;1999:1201. 2. Oyama MA et al. Perceptions and priorities of owners of dogs with heart disease regarding quality of life vs quantity of life. JAVMA;2008;233:104. 3. Reynolds CA et al. Perceptions of quality of life and priorities of owners of cats with heart disease. J Vet Intern Med;2010:1421. 4. Nogueira Borden LJ et al. Use of the measure of patient-centered communication to analyze euthanasia discussions in companion animal practice. JAVMA 2010;237:1275.

FLUID THERAPY: ARE WE DROWNING OUR PATIENTS IN GOOD INTENTIONS?

Adam Eatroff, DVM, DACVIM
New York, NY

The role of fluid therapy in the perioperative setting and in various disease states is undergoing re-examination in both human and veterinary medicine. As is frequently the case, the veterinary profession must look to the human literature for cues on adjusting our own practices. Recent human literature has indicated that liberal administration of intravenous fluids can have deleterious effects in a variety of clinical scenarios, ultimately affecting patient mortality.[1-3]

In veterinary medicine, intravenous fluid administration is considered the cornerstone for the treatment of diseases such as acute kidney injury, chronic kidney disease, and acute bouts of pancreatitis. Additionally, high rates of fluid administration are prescribed in the perioperative period and for decontamination in cases of specific intoxications. Rarely is a clear distinction made for a resuscitative phase of fluid therapy versus a maintenance phase and high rates of fluids can be administered for the duration of a patient's hospital stay. Despite a general perception that high-volume fluid administration is relatively innocuous, a small, but growing body of data from the veterinary literature has shown that fluid accumulation occurs in cases of liberal fluid administration[4], and that this accumulation is associated with a poor outcome. [5]

Examination of certain clinical scenarios may shed light on the misperceptions surrounding the benefit and harm of liberal fluid therapy. In cases of acute kidney injury and chronic kidney disease, a belief exists that administering high rates of parenteral fluids "flushes" retained uremic toxins out of the kidney and into the urine. In reality, when high volumes of fluids are administered, the beneficial effect is generally limited to restoration and maintenance of normal fluid balance in cases of existing or potential dehydration, respectively. Fluid administered beyond that which is necessary to restore normal fluid balance and blood volume does not travel directly and exclusively to the kidney (within a vascular conduit) to promote excretion of wastes. If only a fluid with those properties existed!! Rather it redistributes from the vascular space into the interstitial and (depending on the fluid type) intracellular space with little regard for tissue type or physiologic needs. Therefore, high volume fluid administration typically leads to tissue edema and dysfunction, a problem that can have devastating effects, especially when the kidneys are not properly functioning to excrete this fluid load.

The type of fluid administered may also affect outcome. Synthetic colloids (e.g. Hetastartch) were once thought to be an attractive alternative to high volume crystalloid administration for restoration and maintenance of blood volume, as they were thought to be more effectively retained in the intravascular space. However, recent evidence has shown that not only is the reduction in total volume of fluid required to meet certain cardiovascular endpoints marginal compared to crystalloids,[6] but colloids can have significant adverse effects on renal function, coagulation, and probability of survival.[7] To complicate the role of this fluid type even further, a large amount of the human literature supporting the use of colloids has recently been retracted due to scientific misconduct.

While fluid therapy has a role in the day to day treatment of our veterinary patients, we must take cues from our colleagues in the human field, who are accumulating evidence that forces consideration of the adverse effects of a prescription that was once thought to be relatively innocuous.

REFERENCES

1. Wiedemann HP, Wheeler AP, Bernard GR, et al. Comparison of two fluid-management strategies in acute lung injury. *N Engl J Med* 2006;354:2564-2575. 2. Brandstrup B, Tonnesen H, Beier-Holgersen R, et al. Effects of intravenous fluid restriction on postoperative complications: comparison of two perioperative fluid regimens: a randomized assessor-blinded multicenter trial. *Ann Surg* 2003;238:641-648. 3. Payen D, de Pont AC, Sakr Y, et al. A positive fluid balance is associated with a worse outcome in patients with acute renal failure. *Crit Care* 2008;12:R74. 4. Boscan P, Pypendop BH, Siao KT, et al. Fluid balance, glomerular filtration rate, and urine output in dogs anesthetized for an orthopedic surgical procedure. *Am J Vet Res* 2010;71:501-507. 5. Mobley A, Sullivan L. Retrospective determinatin of fluid overload in critically ill dogs. *Journal of the Veterinary Emergency and Critical Care Society* 2012;22. 6. Hartog CS, Bauer M, Reinhart K. The efficacy and safety of colloid resuscitation in the critically ill. *Anesth Analg* 2011;112:156-164. 7. Zarychanski R, Abou-Setta AM, Turgeon AF, et al. Association of hydroxyethyl starch administration with mortality and acute kidney injury in critically ill patients requiring volume resuscitation: a systematic review and meta-analysis. *JAMA* 2013;309:678-688.

NEW APPROACHES TO DIABETES MONITORING

Cynthia R.Ward, VMD, PhD, DACVIM
Athens, GA

Diabetes mellitus is a common endocrinopathy in dogs and cats, affecting as many as 1 in 100 to 1 in 500 animals.[1] Monitoring of the diabetic patient can be a challenge for both owners and veterinarians. The 12-hour glucose curve has been the gold standard method for monitoring insulin therapy in dogs[2] and cats.[3] This technique involves hospitalizing patients to have blood glucose measurements typically every 2-4 hours, but up to every 1 hour if necessary. This traditional glucose curve has many limitations, including disruption of the patient's normal activity and eating routine, the introduction of stress-related hyperglycemia,[4,5] and labor intensiveness of the procedure. Several studies have emphasized that traditional blood glucose curves are inaccurate due to significant variations in day-to-day readings, even when done by owners at home. [7,8] Furthermore, intermittent blood glucose measurements may miss rapidly occurring glucose nadirs that may cause symogii responses.

Continuous glucose monitoring systems (CGMS) have been approved by the United States Food and Drug Administration for use in human diabetics and provide a minimally invasive method for continuously evaluating glycemic control for up to 72 hours. Interstitial glucose has been shown to correlate well with plasma glucose levels,[9] and the relationship between plasma glucose and interstitial glucose is not affected by insulin.[10]

Continuous glucose monitoring systems have been utilized in human medicine since 2000 and are considered the standard of care for monitoring glycemic control in people.[11] The systems are comprised of an external sensor with a flexible electrode that reacts with glucose when it is inserted into the subcutaneous tissue. Each sensor contains a glucose oxidase-coated platinum working electrode, which is kept in constant potential to a reference electrode. The sensor is encased in flexible polymer tubing with a small window of polyurethane that covers the active area of the electrode. This polyurethane window acts as a glucose diffusion limiting membrane so a linear relationship between interstitial glucose and sensor current is maintained.[10,13] Hydrogen peroxide, which is generated from the oxidation of glucose by glucose oxidase, is then oxidized on the electrode to produce a current with intensity that is proportional to the surrounding glucose concentration.[10,12] The sensor is then calibrated with blood glucose levels to produce the interstitial glucose readings in accordance with the correlating reaction intensity. As such, it is important to interpret absolute values generated by the CGMS with caution.[11] The manufacturer has set a working range of 40-400 mg/dl for the device. One limitation of the device is that it cannot be calibrated when the animal's blood glucose is out of this range. The sensor used in the CGMS Gold and Guardian systems was designed and developed by Minimed for three consecutive days of subcutaneous glucose monitoring. [12]

This sensor then communicates glucose measurements wirelessly to a pager-sized monitor. Several companies have manufactured these devices including Medtronic, Abbot, and Dexcom. The Medtronic systems have been available for a longer time period and are the ones used at the author's institution. They are also used most commonly in the veterinary community. Medtronic devices used by the author include the Guardian-Real Time and the Medtronic iPro2. The Guardian-Real Time model is used for animals while in the hospital. The monitor can be attached to the cage or collar of the animal. The sensor and the monitor can be separated for 45 minutes without loss of data, but must be within 6 feet of each other for most of the day. The Guardian-Real Time provides real-time glucose values that can be read on the monitor without disruption of the recording. It can display 3, 6, 12 or 24 hours of glucose measurements as a curve on the monitor. The Guardian-Real Time system requires calibration of the system at least every 12 hours. If this is not achieved, it will cease to record data.

The iPro2 CIGM monitor was developed by Medtronic for blinded human use. This system consists of the glucose sensor previously described; however, the measurements are stored in the transmitter and are unavailable for viewing until downloaded by the healthcare professional. The simplicity and ease of use of the iPro2 system has allowed for reliable at home use in veterinary patients.

Interstitial glucose measurements are obtained every 10 seconds with an average value recorded every 5 minutes. The recorded values are then downloaded via ComLink™ connector to a computer and a curve is generated for analysis. Current versions of analysis software are web-based, and each patient's information can be saved under an individual username and password. This allows easy information sharing with colleagues and owners.

CGMS placement may vary slightly. At the authors' institution, a 3 cm by 3 cm patch of hair over the interscapular region is clipped and aseptically prepared. Alternatively, a similar sized region can be prepared over the lateral thoracic region.[12-17] The sensor with an attached 22-gauge stylet is loaded into an insertion device (Sen-serter) provided by the manufacturer. The insertion device is spring-loaded and allows quick implantation of the sensor tip into the subcutaneous tissue. Quick implantation and the small stylet size preclude the need for sedation or anesthesia (general or local) for monitor placement. The stylet is then removed and discarded. The external portion of the sensor is attached to the patient by means of the sensor's own self-adhesive. Surgical glue is also often used to ensure good sensor fixation.

The sensor must undergo a 5-15 minute "wetting" period as the electrode becomes immersed in the interstitial fluid. After this period, the monitor is turned on and then the transmitter or cable is connected to the sensor. If the iPro2 is used, the light will turn green after sufficient "wetting" has occurred. A 2 hour (wireless models) initializing "warm-up" phase occurs before the first calibration can be entered. The glucose curve begins with the first calibration and ends automatically 72 hours later. The iPro2 will continue to collect glucose data after 72 hours; however, the readings may be less accurate. The sensor and attached transmitter are then covered with a protective dressing. The monitor is then attached to the collar of medium and large dogs, or in the case of cats and small dogs, the monitor is placed on or within the cage. The monitor may also be secured to the animal's dorsum by means of an elastic bandage and tape harness, or by a recording vest.[13] Dogs and cats that are sent home with the iPro2 system are fitted with a ThunderShirt™ to be kept on continuously while the sensor is in use.

The Guardian-Real Time CGMS requires calibration after the 1 or 2 hour initialization period. Calibrations are performed by drawing a blood sample, obtaining a glucose reading, and entering the glucose level obtained into the monitor. Glucose measurements must be entered within 5 minutes of acquisition to ensure proper calibration. A reliable portable glucometer [18-20] may be used for the purposes of obtaining quick glucose measurements. The patient's glucose must fall within the working range of 40-400 mg/dl to be used for CGMS calibration or to be recorded. Animals residing in the hospital during their recordings may have measurements taken via a central catheter or by peripheral venipuncture. The first calibration must occur as soon as possible after this initialization period or improper interstitial glucose measurements may occur.[13] Three separate calibrations are required per each 24 hour period the monitor is recording. It is not necessary to evenly space the timing of calibrations, but the interval between calibrations may not exceed 12 hours. It is recommended by the manufacturer to avoid performance of calibrations in the 2 hours following insulin administration or consumption of a meal.

Animals fitted with the iPro2 to be sent home do not need to have the device calibrated. However, to ensure accuracy of the resulting glucose curve, owners must record blood glucose concentrations approximately every 12 hours. These are recorded on a data sheet along with times the animal was fed and exercised. The amount, type, and timing of insulin delivery are also recorded. Owners may be educated on techniques for collection of a capillary blood sample from the ear pinna.[21] Good correlation has been documented between glucose concentrations in capillary blood and venous blood,[22] and this collection method has been shown to be well tolerated by pets and owners.[21] Owners are discharged with disposable lancets and a glucose-monitoring device if they do not already have one at home. After completion of home glucose curve, the owners detach the iPro2 device and sensor. The ThunderShirt™, glucose monitor, and iPro2 device, and monitoring log are mailed back in a postage-paid box. The glucose data is then downloaded from the iPro2 and insulin recommendations are made to the owner and referring veterinarian.

The use of continuous glucose monitoring systems has been evaluated in veterinary medicine.[12-17,23] Results of a study of 10 diabetic dogs showed good correlation when a CGMS was compared to a standard venipuncture generated glucose curve.[12] In general, glucose concentrations recorded by the CGMS were similar to, but lower than, glucometer readings. The authors did note in 6 of 10 dogs that a postprandial rise in blood glucose concentration was detected with measurement by a glucometer, but the CGMS failed to capture such an event. This phenomenon has not been demonstrated by results from other studies involving diabetic dogs,[13,14,16] but their results do confirm good correlation between blood glucose curves and CGMS recordings.

Researchers in another study evaluated 16 CGMS recordings in 14 diabetic cats.[15] Good correlation was noted between CMGS values and blood glucose concentrations. In 7 of 16 tracings, the 40-400 mg/dl working range was a limiting factor in the use of the monitor. Four cats experienced hyperglycemia greater than 400 mg/dl, 3 were hypoglycemic less than 40 mg/dl, and another was both hyper and hypoglycemic during the attempted curves. All cats tolerated the device well, although removal and kinking of the sensor required sensor replacement in one and two cats, respectively.

The Guardian-Real Time wireless device has been shown to be accurate in generating glucose curves in 10 diabetic dogs. [23] All dogs were hospitalized and had reliable curves for 72 hours. There was excellent correlation between the Guardian-Real Time and blood glucose curves. All dogs tolerated the device and no sensors were dislodged.

A CGMS has been used successfully in a diabetic horse.[14] Furthermore, good correlation between glucose measurements obtained from a CGMS and standard venipuncture collected samples has been documented in horses and a cow undergoing intravenous glucose tolerance testing.[14,16] A CGMS has also been used at the authors' institution to evaluate stress-induced hyperglycemia in a white-faced capuchin monkey and sepsis-induced hypoglycemia in a cria. A study to evaluation of blood glucose in rabbits by CIGM is now underway.

REFERENCES

[1]Feldman EC, Nelson RW. *Canine and feline endocrinology and reproduction*. 3rd ed. St. Louis: W.B. Saunders, 2004:486-579. [2]Fleeman LM, Rand JS.*Vet Clin North Am Small Anim Pract* 2001;31:855-880 [3]Rand JS, Martin GJ. *Vet Clin North Am Small Anim Pract* 2001;31:881-913.[4]Rand JS, Kinnaird E, Baglioni A, et al. *J Vet Intern Med* 2002;16:123-132.[5]Nelson R.*J Vet Intern Med* 2002;16:121-122.[7]Fleeman LM, Rand JS. *J Am Vet Med Assoc* 2003;222:317-321.[8].Alt N, Kley S, Haessig M, et al. *J Am Vet Med Assoc* 2007;230:1011-1017.[9]Rebrin K, Steil GM. *Diabetes Technol Ther* 2000;2:461-472.
[10]Rebrin K, Steil GM, van Antwerp WP, et al. *Am J Of Physiol* 1999;277:E561-571.[11]Melki V, Ayon F, Fernandez M, et al. *Diabetes Metab* 2006;32:123-129.[12]Davison LJ, Slater LA, Herrtage ME, et al. *J Small Anim Pract* 2003;44:435-442.[13]Wiedmeyer CE, Declue AE. *J Vet Intern Med* 2008;22:2-8.[14]Wiedmeyer CE, Johnson PJ, Cohn LA, et al. *J Am Vet Med Assoc* 2003;223:987-992.[15]Ristic JM, Herrtage ME, Walti-Lauger SM, et al. *J Feline Med Surg* 2005;7:153-162.[16]Wiedmeyer CE, Johnson PJ, Cohn LA, et al. *Diabetes Technol Ther* 2005;7:885-895.[17]DeClue AE, Cohn LA, Kerl ME, et al. *J Am Anim Hosp Assoc* 2004;40:171-173.[18]Wess G, Reusch C. *Am J Vet Res* 2000;61:1587-1592.[19]Cohn LA, McCaw DL, Tate DJ, et al. *J Am Vet Med Assoc* 2000;216:198-202.[20]Wess G, Reusch C. *J Am Vet Med Assoc* 2000;216:203-209.[21]Casella M, Wess G, Reusch CE. *J Am Anim Hosp Assoc* 2002;38:239-245.[22]Wess G, Reusch C. *J Small Anim Pract* 2000;41:60-66.

CANINE AND FELINE DIABETES:
EDUCATING AND PREPARING YOUR CLIENT TO ENHANCE CARE AND QUALITY OF LIFE

Rachel V. Poulin, RVT, VTS (SAIM)
Coral Springs, Florida

INTRODUCTION

Diabetes Mellitus is estimated to affect one in 500 dogs and one in 250 cats.[1,2] Diabetes mellitus (DM) is a chronic endocrine disorder that is characterized by hyperglycemia, glucosuria, polyphagia, polydypsia, polyuria and weight loss. Usually these signs occur because the pancreas is unable to produce enough insulin to meet the animal's requirements or the animal's body is resistant to insulin.

Diabetes Mellitus is an overwhelming diagnosis for most clients, several times clients have admitted to not even knowing that animals get diabetes. Not only will they need to understand the financial commitment involved in caring for a diabetic pet, they will also need learn about diabetes, insulin administration, home glucometry, and diabetic warning signs of potential emergencies. In addition, there is the daunting responsibility of blood sugar curves, proper diet and monitoring clinical signs. Time and patience are crucial in order to give the client the proper information and education to ensure they will be able to offer proper care at home.

DIAGNOSIS

A diagnosis is usually obtained because clinical signs of polyuria/polydypsia and weight loss are being noticed at home. Sometimes these signs will go unnoticed and the patient will present on emergency in a diabetic crisis, such as a diabetic ketoacidosis. On occasion the signs will not be noticed and the veterinarian will discover the diagnosis during a routine yearly exam. The diagnosis itself is made up of hyperglycemia, glucosuria, and elevated fructosamine. Other abnormalities can be seen on a biochemical profile such as elevated liver values, cholesterol and in cases of severe complications, acid/base imbalances and dehydration.

POTENTIAL COMPLICATIONS FROM DM

Approximately 80% of dogs diagnosed with DM will develop cataracts within 16 months of diagnosis.[4] Surgical correction is possible for some of these patients depending on good diabetic control and anesthetic risk. 10% of cats with chronic hyperglycemia will develop diabetic neuropathy[5], resolution of this neuropathy is possible in time with good diabetic control but resolution is not guaranteed. In addition, DM suppresses the immune system and makes fighting infections challenging for the patient. Skin infections and urinary tract infections are not uncommon developments for patients with DM. To make matters worse, an infection can cause inflammation and stress on the body that further deregulates diabetic control. Moreover, DM reduces the body's ability to heal. Therefore, if the diabetic patient were to sustain an injury or undergo surgery the recovery time would be prolonged and potentially complicated. For these reasons, and more, it is necessary to offer a very good diabetic education to the client, they should be able to know what warning signs are for a potential crisis and they should be aware of what risks this disease carries with it and why excellent control is so important on many different levels.

THE NURSES ROLE

The veterinary nurse can take an active role in education and preparation. The veterinary nurse can outline the definition of diabetes mellitus, can highlight what the clinical signs to be aware of to avoid emergencies, can teach how to properly handle insulin and administer it subcutaneously. In addition, the veterinary nurse can teach a client how to perform home glucometry and check daily urine readings for ketonurina and glucosuria. They can also discuss nutrition goals for diabetic animals and review the recommendations of the veterinarian. The veterinary nurse can play a vital role in the successful outcome of diabetic treatment and management by forming a trusting relationship with the client and teaching such vital information. By giving owners the time and tools and patience needed for them to understand the information they will feel more confident with their new responsibilities to their pet.

DIABETIC RED FLAGS

There are several "red flags" that clients should be aware of when taking on the responsibility of caring for a diabetic pet.

Lack of Appetite	This is a red flag because normally a diabetic patient is hungry and will not skip a meal. There are several reasons why a pet may not have a good appetite. A decrease in appetite among other things can lead to a necessary adjustment in the amount of insulin required. Clients should be made aware to not administer insulin if a patient does not eat well. They should instead call and report a change in appetite to their veterinarian. If the behavior continues for more than 24 hours an appointment should be scheduled to investigate the cause of the appetite change. It is important to remember tha infections, stress, inflammation, and man other factors can disregulate a stable diabetic. Therefore, appetite changes should not be taken lightly.
Polydypsia/ Polyphagia PU/PD	This is usually a sign that the blood glucose levels are high and there is a lack of diabetic control.
Polyphagia	This is usually a sign that the blood glucose levels are high and there is lack of diabetic control.

Ataxia	This is usually a sign of low blood sugar. An ataxic pet should be brought in right away. Karo syrup® or Maple syrup can be placed on a patient's gums to deliver sugar quickly as a short term solution. This will not last long and the patient should still be brought into the clinic right away.
Seizures	This is usually also because of low blood sugar. Seizuring pets should be brought in right away.
Inappropriate Urination	This can also be a sign of hyperglycemia secondary to diabetes but it can also be a sign of a urinary tract infection, which are common among diabetic patients. Infection can be ruled out with a urianlysis and urine culture.
Vomiting	A vomiting diabetic can be as simple as a one time upset stomach or the beginning of a crisis. If a diabetic patient is vomiting a call to the clinic is essential to help triage the situation.
Diarrhea	Diarrhea in a diabetic patient can be a simple one time upset stomach or, as with vomiting it can be the beginning of a crisis. The client should make the veterinary clinic aware so that the situation can be triaged and a potential crisis avoided.
Elevated Blood Glucose Readings	If daily blood glucose readings are trending higher the client should be aware that changing the insulin levels without the direction of a veterinarian is very dangerous. Instead, a 12 hour blood glucose curve may be appropriate, as well as a fructosamine level. Therefore, discuss the importance of a re-evaluation with the client so that the veterinary team is staying on top of diabetic control.
Ketones	Anytime ketones are detected it is a potential emergency. Clients should inform the clinic right away if ketones are detected in urine. The action taken will be up to the veterinarian but it will depend on clinical signs and blood glucose readings.

The more awareness and education the client has, the better they will able to control the disease and avoid life threatening and costly complications.

HOME GLUCOMETRY

One of the best tools we have to regulate this disease and achieve diabetic control is home glucometry. The AlphaTrak® is widely used among veterinary professionals as it is species- specific, cost effective, easy to teach and easy to use. If the importance of home monitoring is stressed by the doctor, then the act of teaching it to the client is much easier for the nurse. Blood glucose monitoring at home is one of the most effective ways for the client to understand the relationship between insulin and blood glucose and it a valuable tool in engaging the client in the management process and help them feel part of the treatment team.[6] It also provides valuable information to the veterinary team about the control of the disease state in a non stressful home setting. Twice daily readings prior to administering insulin give veterinarians an idea of the blood glucose trends, and allow the client to know the blood glucose levels are at a high enough level to safely administer insulin. Home glucometry also allows us to obtain more accurate readings as stress hyperglycemia is not a concern for animals being tested in their home environments. Physiologic hyperglycemia occurs post-prandially and in response to stress in all species. This can be mediated by epinephrine (and is transient, lasting 4-6 hours) or corticosteroids (results in a more sustained increase in glucose). Cats tend to produce marked stress hyperglycemias.[7]

Blood glucose readings should be taken at least twice daily prior to insulin administration for well regulated pets and up to 4-6 times a day for pets that are not well controlled. A blood glucose curve should be done every 4-6 weeks on a well-regulated diabetic and more frequently on those patients that are difficult to manage.[8] A blood glucose curve will allow the veterinarian to see the patient's response to insulin and adjust accordingly. Blood glucose curves should also be done 10-14 days after changing insulin dosage. It is important to let the client know that the blood glucose value will fluctuate throughout the day. The veterinarian will need to assess the full blood glucose curve before being able to offer recommendations on insulin therapy. One reading does not offer substantial information, therefore a client should again be reminded that these values will change throughout the day and how important this data is for diabetic control. Since we know that blood glucose levels will rise in response to stress it is important that the day a blood glucose curve is performed is a quiet, non-stressful day at home. During or following a party or out- of- town guests visiting is not the proper time to conduct a blood glucose curve.

Teaching a client how to obtain a blood glucose reading can start out as simply as asking them to watch a how-to DVD or directing them to websites that offer instructional videos. Asking them to review the task prior to hands on teaching tends to lesson the anxiety associated. Blood glucometry can be taught at a re-evaluation exam, it tends to be overwhelming when taught at the first diagnostic appointment. It is important to go over the various areas where a capillary sample is possible and then- demonstrating how to obtain it. The client should then be asked to go through the steps a few times and be successful in blood glucose acquisition with the nurse in the room. Repetition will increase confidence and the client will be more likely to be successful at home.

NUTRITION

Current concepts in dietary therapy for diabetic dogs and cats are described in order to enhance insulin therapy and improve glycemic control.7 The diet fed should primarily be palatable and nutritionally balanced. Improved glycemic control may be achieved in some dogs if the diet contains increased insoluble fiber.[8] The feline diabetic presents a challenge because of their need for protein and intolerance for carbohydrates. Previously fiber-fortified foods have been recommended for cats with DM, the advantages include weight control and slowed glucose uptake from the intestines. Recent studies have showed us that low-carbohydrate foods are considered to be superior for feline diabetic management.[9] More recently, another nutritional goal, achieving diabetic remission, has come to the forefront of the nutritional plan. A diabetic plan that includes an appropriate food and feeding plan that provides glycemic control as well as weight management will give the diabetic feline the best possible chance for diabetic remission.[10,11]

COMMUNICATION/EMPOWERMENT AND TOOLS

As stated before, DM is an overwhelming, complicated and expensive disease. The euthanasia rate for diabetic patients is 40%[12]. However, when a client is given the tools and education necessary to take care of their sick pet, when someone is able to offer them time, patience and knowledge of the disease, then we greatly increase that patient's chances of survival. A diabetic patient can live a long and happy life when three factors all come together: 1) client compliance, understanding and willingness to play a role in treatment, 2) the veterinary team's knowledge and ability, 3) patient response. It will not be possible to prevent every diabetic death; there are far too many complications and potential crises associated with this disease. However, with a strong client relationship based on respect, trust and care for the pet and a well educated client willing to be proactive at home, that diabetic animal has a very good chance of living a long happy life. In the end that is the ultimate goal for us all.

REFERENCES

1) Veterinary Team Brief: Diabetes and It's Management by Audrey K. Cook, BVM&S, DACVIM, DECVIM
2) Veterinary Team Brief: Diabetes and It's Management by Audrey K. Cook, BVM&S, DACVIM, DECVIM
3) VetLearn: Care and Management of Diabetes Mellitus by Ann Wortinger, BIS, LVT, VTS (ECC, SAIM, Nutrition) August 2009 (Vol 30, No 8)
4) PubMed: A retrospective-cohort study on the development of cataracts in dogs with diabetes mellitus by Michael Davidson, DVM, DACVO, Maria T. Correa, PhD: Vet Ophthalmol 1999, 2(3): 169-172
5) DVM 360: Managing complications in diabetic cats by Robin Sereno, BS, CVT, VTS (SAIM), May 2011
6) Veterinary Team Brief/Clinical Brief: A Protocol For Diabetic Management by Audrey K. Cook, BVM&S, DACVIM, DECVIM
7) https://ahdc.vet.cornell.edu/clinpath/modules/chem/hypergly.htm
8) Seminars from Veterinary Medicine and Surgery (Small Animal) Nutritional Management of Diabetes Mellitus by Nelson, R.W; Lewis, L.D: 1990 Vol. 5. 3pp 178-186
9) Purina Veterinary Clinics of North America (Small Animal Practice) Dietary Management by Clauda A. Kirk, DVM, PhD Joseph W. Bartges, DVM, PhD, DACVIM, DACVN: November 2006 Vol 36 Number 6
10) Purina Veterinary Clinics of North America (Small Animal Practice) Dietary Management by Clauda A. Kirk, DVM, PhD Joseph W. Bartges, DVM, PhD, DACVIM, DACVN: November 2006 Vol 36 Number 6
11) Veterinary Team Brief/Clinical Brief: A Protocol For Diabetic Management by Audrey K. Cook, BVM&S, DACVIM, DECVIM
12) PubMed: Diabetes Mellitus in a population of 180,000 Insured Dogs: Incidence, Survival, and Breed Distribution by Tove Fall, Helene Hansson Hamlin, Åke Hedhammar, Olle Kämpe, Agneta Egenvall: J Vet Intern Med 2007; 21:1209-1216

THE IMPORTANCE OF PROTEINURIA AND HYPERTENSION IN CKD: ACEI AND BEYOND

Gregory F. Grauer, DVM, MS, DACVIM (Small Animal Internal Medicine)
Manhattan, KS

INTRODUCTION

Chronic kidney disease (CKD) is a common problem that adversely affects both quality of life and survival time. Although the prevalence of CKD in the general small animal population is ill defined, CKD may affect up to 10% of dogs and 35% of cats in referral hospital populations (Polzin and Osborne 1986; Krawiec and Gelberg, 1989). Nephron damage associated with CKD is usually irreversible and can be progressive. Renal failure results when three-quarters or more of the nephrons of both kidneys are not functioning. Whether the underlying CKD primarily affects glomeruli, tubules, interstitial tissue, or renal vasculature, irreversible damage to any portion of the nephron renders the entire nephron nonfunctional. Healing of irreversibly damaged nephrons occurs by replacement fibrosis and therefore a specific etiology is often not determined. Chronic kidney disease occurs over a period of weeks, months, or years and since it is often not possible to improve renal function in CKD, treatment is aimed at stabilizing renal function. In addition to dietary therapy, there is increasing evidence that treatment with ACE inhibitors can decrease the progressive nature of CKD by attenuating systemic hypertension, intraglomerular hypertension, and proteinuria. There are also some new/evolving anti-fibrotic treatments that may also slow the progression of CKD.

By altering pre-glomerular resistance, healthy kidneys can maintain relatively stable glomerular capillary pressures despite variations in systemic blood pressure. This pressure regulatory process is termed "renal autoregulation". Autoregulation can be reduced when renal disease results in loss of nephrons. Compromised autoregulation allows high systemic blood pressure to be transmitted to glomerular capillaries. This glomerular hypertension has been documented by micropuncture studies in dogs and cats with surgically reduced renal mass. In these models, glomerular hypertension was associated with glomerular hypertrophy, sclerosis, and proteinuria. Systemic hypertension is relatively common in dogs with renal disease. In a recent study of dogs with spontaneous chronic kidney disease (CKD), 29/45 (64%) had systolic blood pressure ≥ 144 mm Hg and 14/45 (31%) had systolic blood pressure ≥ 161 mm Hg. In cats with naturally-occurring CKD, systemic hypertension has been observed in 19-65% of cases depending on the definition of hypertension.

Renal proteinuria can result from glomerular and/or tubular abnormalities in dogs and cats with CKD. Glomerular proteinuria may arise from immune complex disease or structural abnormalities involving the glomerular capillary wall (e.g., amyloidosis and x-linked hereditary nephropathy). Protein-losing nephropathy caused by glomerular capillary wall lesions is often accompanied by systemic hypertension and glomerular proteinuria can be exacerbated by intraglomerular glomerular hypertension that can result from systemic hypertension. Tubular proteinuria occurs when tubular reabsorption of protein from the glomerular filtrate is compromised. Whether caused by capillary wall lesions, tubular lesions, or intraglomerular hypertension, excessive quantities of protein in the glomerular filtrate may contribute to additional glomerular and tubulointerstitial lesions leading to loss of more nephrons.

DIAGNOSIS OF HYPERTENSION AND PROTEINURIA

Current recommendations are that blood pressure be measured in a quiet area prior to examining the patient, typically in the presence of the owner and after a 5-10 minute period of acclimation. The ACVIM Panel on Hypertension suggests discarding the first measurement, then obtaining a minimum of 3, preferably 5-7, consecutive measurements with less than 10-20% variability in systolic blood pressure. The animal's disposition, body position, and heart rate, the cuff size and measurement site as well as all measured values should be recorded in the medical record. Many clinicians suggest that hypertension be documented on more than one occasion before accepting the diagnosis (unless ocular lesions compatible with systemic hypertension already exist).

Diagnosis and management of proteinuria in cats and dogs with CKD should be accomplished in a step-wise fashion. The specificity of the dipstick screening test for proteinuria is poor and therefore confirmation of traditional dipstick positive proteinuria should be accomplished with a more specific follow-up test such as the sulfosalicylic acid (SSA) turbidimetric test, UP/C, or species specific albuminuria assay. The second step is assessment of proteinuria is to determine its origin. Renal proteinuria can adversely affect the prognosis of dogs and cats with CKD and therefore, physiologic or benign proteinuria and pre- and post-renal proteinuria should be ruled out. Subsequently, via serial monitoring, it should be determined if the proteinuria is persistent or transient and if persistent – is the magnitude stable or increasing or decreasing over time? Persistent proteinuria is defined as at least two positive tests at two week intervals. Relatively mild proteinuria in dogs and cats with spontaneous/naturally-occurring CKD appears to be a negative predictor of survival. In dogs and cats with the remnant kidney model of CKD, proteinuria is associated with nephron hypertrophy and increased intraglomerular pressures. Persistent proteinuria of renal origin of a magnitude ≥ UP/C of 0.4 in cats and ≥ 0.5 in dogs with azotemic CKD should be treated with an ACEI and/or dietary protein reduction. Borderline proteinuria is defined as a UP/C between 0.2 and 0.5 in dogs and 0.2 and 0.4 in cats.

MANAGEMENT

Gradual reduction of dietary salt is often recommended as the first line of treatment for hypertension; however there are no studies that document the efficacy of this treatment. In many cases, vasodilators (angiotensin-converting enzyme inhibitors [ACEI] and calcium channel antagonists [CCA] may be necessary to control systemic hypertension. The CCA amlodipine is often recommended as the first choice anti-hypertensive treatment for cats. Recent studies have shown that treatment with a CCA alone reduces proteinuria in cats with CKD suggesting that intraglomerular pressures were also reduced by this treatment. Initial treatment with an ACEI is also appropriate in dogs and cats with hypertensive/proteinuric CKD because of the beneficial effects on intraglomerular pressure and proteinuria. In those cases where systemic hypertension persists after initiation of ACEI treatment, or is initially ≥ 180 mm Hg, a CCA can be added or used as sole treatment. The overall risk of target organ damage to the eyes, brain, heart, and kidneys is thought to be minimal if systolic blood pressure is maintained at ≤ 150 mm Hg.

ACE inhibitors are not effective in all patients in attenuating proteinuria and hypertension and other patients experience ACEI "escape" (the ACE inhibitor loses its effectiveness with long-term treatment). This ACE inhibitor escape is thought to occur secondary to development of alternative pathways for angiotensin production. For example, elastase and cathepsin G may directly convert angiotensinogen to angiotensin II. For these reasons, alternative RAAS inhibition strategies may soon be employed in veterinary medicine. These strategies may include angiotensin receptor blockers (ARBs), aldosterone receptor antagonists, renin inhibitors, and combination treatment with ACE inhibitors and ARBs. Unfortunately at this time, the use of any of these alternative RAAS inhibitors has not been evaluated in dogs and cats with CKD.

Renal fibrosis is the non-specific end result of many renal diseases and results extracellular matrix (ECM) protein accumulation in the renal interstitium (tubulointerstitial fibrosis) and glomerulus (glomerulosclerosis). Independent of the underlying cause, progressive renal fibrosis results in the continuous decline of renal function. Transforming growth factor-β (TGF-β) is a profibrotic cytokine with anti-inflammatory and immunoregulatory properties that are instrumental in wound healing and organ development and function. TGF-β is over expressed in renal disease and contributes to renal fibrosis by increasing ECM synthesis, decreasing ECM degradation, and facilitating cell matrix interactions. There is increasing evidence that, at least some of the matrix producing cells come from within the kidney itself. The differentiation of tubular epithelial cells to fibroblasts or myofibroblasts is known as epithelial to mesenchymal transdifferentiation (EMT) and is recognized as an early and essential step in renal fibrosis. TGF-β has been shown to not only induce this cellular transdifferentiation but to be vital to the whole process.

WHAT EVIDENCE EXISTS THAT SYSTEMIC HYPERTENSION AND/OR PROTEINURIA ARE DETRIMENTAL TO CANINE AND FELINE KIDNEYS?

1. In dogs with the page/remnant kidney model of chronic renal failure (CRF), systemic hypertension had adverse effects on renal function and morphology. Finco DR. J Vet Intern Med 2004;18:289.
2. In dogs with remnant kidneys, 40-45 mmHg increases in systolic blood pressure over baseline resulted in enhanced renal damage. Brown SA, et al. J Vet Intern Med 2000;14:351.
3. In dogs with spontaneous chronic renal failure, initial high systolic blood pressure was associated with increased risk of developing a uremic crisis and of dying. Jacob F, et al. J Am Vet Med Assoc 2003;222:322.
4. Angiotensin-converting enzyme inhibition (ACEI) treatment that lowered systolic blood pressure and decreased proteinuria was associated with improve outcome in dogs with spontaneous glomerulonephritis. Grauer GF, et al. J Vet Intern Med 2000;14:526.
5. ACEI treatment in dogs with remnant kidney CRF was effective in modulating progressive renal injury, which was associated with a reduction in glomerular and systemic hypertension. Brown SA, et al. Am J Vet Res 2003;64:321.
6. Glomerular capillary hypertension, glomerular enlargement, and proteinuria have been documented in dogs with the remnant kidney model of renal failure. Brown SA, et al. Am J Physiol 1990;258:F495.
7. In cats with the remnant kidney model of CKD, increases in single nephron GFR occur in association with glomerular hypertrophy, increasing intraglomerular pressures, hyperfiltration, mesangial matrix expansion, and proteinuria. Brown SA, et al. Am J Physiol 1995;269:R1002.
8. In dogs with naturally occurring CKD, the relative risk of uremic crises and mortality was approximately three time greater in dogs with UP/C's > 1.0 (n=25) compared with dogs with UP/C's < 1.0 (n=20). In this study the risk of an adverse outcome was approximately 1.5 times greater for every 1 unit increase in UP/C and the decline in renal function was greater in dogs with higher UP/C's. Jacob F, et al. Am J Vet Res 2005;226:393.
9. In cats with naturally occurring CKD, relatively mild proteinuria (UP/C's > 0.4) appear to be negative predictors of survival. Increasing proteinuria was associated with increasing serum creatinine concentrations and increasing systolic blood pressure (presumably related to glomerular hyperfiltration). UP/C, age, and serum creatinine concentration (but not blood pressure) were independently associated with mortality. Syme HM, et al. J Vet Intern Med 2006;20:528.
10. Proteinuria has been associated with increased risk of mortality due to all causes in cats that have normal renal function when their proteinuria is first detected. Walker D, et al. J Vet Intern Med 2004;18:417.
11. In 141 client-owned cats with naturally-occurring systemic hypertension, amlodipine treatment decreased both blood pressure and proteinuria. Proteinuria (UP/C) (before and after treatment as well as the change in UP/C were the only variables related to survival in these cats. Jepson RE, et al. J Vet intern Med 2007;21:402.
12. In cats with naturally-occurring CKD, shorter survival time was associated with an increased UP/C as an independent risk factor. King JN, et al. J Vet Intern Med 2007;21:906.
13. In 61 cats with naturally-occurring CKD, treatment with benazepril decreased proteinuria (UP/C) and decreased progression of the CKD to advanced stages. Mizutani H, et al. J Vet Intern Med 2006;20:1074.
14. In a prospective, longitudinal cohort study of non-azotemic cats ≥ 9 years, 95 cats (median age = 13) were followed for 12 months or until death or azotemia developed. 29/95 (30.5%) developed azotemia (Sr Cr > 2.0 mg/dl). Proteinuria at presentation (median UP/C of 0.19 vs. 0.14) was significantly associated with development of azotemia. Jepson RE, et al. JVIM 2009; 23:806-813.
15. In cats with the remnant kidney model of CKD, hypertension was associated with more severe glomerular histologic lesions. Mathur, et al, AJVR 2004;65:1006-1013.

16. In 59 cats with CKD (serum creatinine concentrations > 2.0 mg/dl, USG < 1.035, and history and clinical signs compatible with CKD vs. AKI) with postmortem data and biochemistry and urine protein data (obtained during the last 2 months of life) the UP/C was positively correlated with both renal interstitial fibrosis score and maximal glomerular volume. 34 of the 59 cats (58%) were classified as proteinuric. Chakrabarti, et al. Diagnostic Pathol 2013;50:147-155.
17. When cats with CKD surviving for more than one month (surviving group, n = 34) were compared with cats with CKD surviving less than one month (non-surviving group, n = 16), UP/C was significantly higher in the non-surviving group. In the surviving group, UP/C was the only clinicopathologic variable that exhibited a consistent alteration (increase) in relation to first visit data and was most likely to be associated with mortality. Kuwahara, et al. J Small Anim Pract 2006;47:446-450.
18. When client-owned cats with stable CKD (n=112) were compared with client-owned cats with progressive CKD (n=101), median UP/Cs in the progressive group (0.27 vs. 0.14). A 0.1 increase in UP/C was associated with a 24% increase in risk of progression of CKD. Chakrabarti, et al. J Vet Intern Med 2012; 26:275-281.
19. In 69 cats with CKD (serum creatinine concentrations > 2.0 mg/dl, USG < 1.035, and history and clinical signs compatible with CKD vs. AKI) with postmortem data and time-averaged systolic blood pressure (SBPOT) (obtained over a mean of 284 days) the SBPOT was positively correlated with maximal glomerular volume, hyperplastic arteriolosclerosis, and glomeruloscerosis. 34 of the 69 cats (49%) were classified as hypertensive (SBPOT of 159 mm Hg vs 136 mm Hg in normotensive cats). Chakrabarti, et al. *Diagnostic Pathol* 2013;50:147-155.
20. In 26 client-owned dogs with CKD, treatment with benazepril (vs. placebo) decreased UP/C and increased GFR (plasma clearance of iohexol) and health status score over a 6 month study. Tenhundfeld J, et al. *J Am Vet Med Assoc* 2009; 234:1031-1037.

VIDEO ANALYSIS OF GAIT DISORDERS: TECHNICIAN PROGRAM

Karen Dyer Inzana, DVM, PhD DACVIM (Neurology)
Blacksburg, VA

Organization of the Nervous System

The ability to walk depends on the coordinated function of the axial skeleton, joints, tendons, skeletal muscles and the nervous system. From a neurological perspective, a complex system of motor neurons as well as a variety of sensory input from skin, joints and muscles must work together to produce a normal gait. The motor system is most easily divided into 3 components; lower motor neuron systems, which are controlled by upper motor neuron centers which are, in turn, controlled by higher centers including the cerebellum and forebrain.

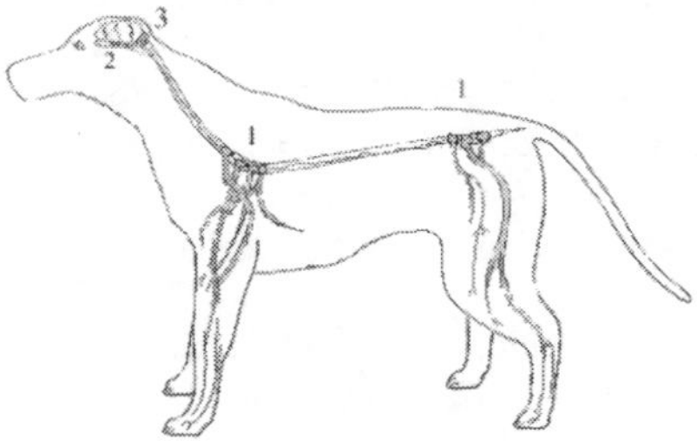

(1) Lower motor neurons have their origin in gray matter of the spinal cord (or brainstem for Cranial Nerves) and extend out of the nervous system to supply peripheral muscles. (2) Upper motor neurons originate primarily in the brainstem (a few in the cerebral cortex) and extend down white matter tracts in the spinal cord to regulate the firing of lower motor neurons. (3) Upper motor neurons are regulated by a series of feedback circuits through forebrain and cerebellum that make movement voluntary and coordinated.

The sensory system originates in skin, joints, and muscles and provides constant input to all components of the motor systems to provide information about where the muscles, joints, and limbs are in space, a physiological function termed proprioception.

Normal Gaits

When considering normal dog gaits, it is best to divide them into symmetrical or asymmetrical gaits. With symmetric gaits movement on one side of the body mirrors movement on the opposite side with near even placement of the feet with each movement. Examples of symmetric gaits are the walk, trot, and pace. In Asymmetric gaits such as a gallop, both sides move independently and stride length varies among limbs. A good example of normal gaits may be viewed at this website: http://vanat.cvm.umn.edu/gaits/info.html The walk is a four-beat gait, in which each limb steps (lift/swing/support/thrust) sequentially. Either two or three limbs are on the ground and the alternate limbs on the front and rear advance in order. The head and neck come up slightly during the support phases of the front limbs and down during the swing phases. Clearly, the characteristics of the gait vary with different breeds and limb length, but the important observation is that movement is coordinated and stride length is equal in each limb. The trot is a two beat gait where the right and left diagonal limbs move together to support weight. The head and neck should remain level. The canter and the gallop are three or 4 beat gaits depending on the speed with the front inside limb leading. The normal sequence is the forelimb followed by the opposite hind limb and then the diagonal. As the speed increases to a gallop there is more suspension between the final diagonal phase making it a 4 beat gait. The normal sequence is leading forelimb, opposite hind limb, diagonal with the rear landing before the forelimb with a period of suspension between.

Clinical Evaluation

Neurological examinations should always begin with observation of at a distance, preferably with the animal moving freely off leash in a space such as an examination room taking note of any abnormal behaviors, body posture, or movements. It is best to evaluate an animals gait on a non-slick surface such as carpet, mowed grass, or other textured surface. The symmetric gaits of walk and trot are easiest to evaluate. Circling, moving over objects such as blocks, curbs, steps can also help detect subtle deficits. Three distinctive abnormal gaits maybe observed; Ataxia, Paresis, and Lameness. Each may be seen in animals with neurological diseases.

1) Ataxia is difficulty performing normal coordinated movements. There are three types of ataxia.
   a) Cerebellar ataxia is caused by injury to portions of the cerebellum. The rate of movement as well as the range of movement is often affected creating a hypermetric or high-stepping gait. There is generally marked trunkal sway and goal directed tremor of the head and neck (intention tremor).
   b) Vestibular ataxia is created by an imbalance of input from normal and diseased vestibular systems. There generally is a head tilt and tendency to fall or roll to the side of the injury. Abnormal eye movements (nystagmus and strabismus) are frequently seen with vestibular ataxia.
   c) Somatosensory or Proprioceptive ataxia is caused by injury to sensory nerves, ascending sensory pathways in the spinal cord, brain stem and forebrain. It is characterized by a loss of sensory recognition of where the limb and trunk are in relations to other parts. This gait appears clumsy and uncoordinated with a base-wide stance and the limbs placed away from the center of gravity.

In the spinal cord, motor systems run in close proximity to sensory fibers, so it is difficult to have a pure sensory ataxia without some loss of motor control as well. Animals with upper motor neuron weakness often have a delayed onset of protraction (swing phase) increased stride length, and the limb is placed in an abnormal position when it lands. Muscle tone is preserved in UMN paresis so the gait often appears spastic due to poorly regulated or exaggerated reflex activity in the limbs.

2) Paresis or weakness is caused by injury to motor units (peripheral motor nerves, skeletal muscle, or the neuromuscular junction). Animals with these conditions appear weak, have a hard time supporting weight for any length of time and tend to sit or lay down frequently. Often the head and neck are ventroflexed due to weakness of the neck and back muscles. Paretic gait is manifested by

shortened strides and a tendency to move the rear limbs closer to the center of gravity. Hock weakness characterized by either excessive movement (over flexion) or plantagrade stance may be seen. When asked to stand paretic animals may have a fine generalized tremor simply due to weakness.

3) Lameness generally appears as if the limb was painful. Animals may hold the limb up, be reluctant to bear weight on it and therefore have a shortened stride length in the affected limb. The contralateral limb will be used to bear most of the weight so it will have a longer stance phase and longer stride length. If an animal has a bilateral orthopedic disease (for example bilateral cruciate disease) the gait of both affected limbs are shortened and the animal will tend to shift weight to the font limbs. As a rule, limbs tend to be carried with orthopedic diseases while neurological diseases causes the limb to drag or scuff the nails as the animal walks. The primary exception to this rule is if the animal has nerve root irritation causing shooting limb pain in which case the animal may carry that limb. Nerve root irritation is one example where neurological diseases may be difficult to differentiate from orthopedic lameness.

Abnormal Movements

Abnormal movements may occur constantly, or only be intermittent. They may be generalized, occurring throughout the head, neck and body or may be localized. These are best classified into the following categories.

1) Myotonia: results from a stimulus induced persistent, repetitive muscle contraction without relaxation. Abnormal movements appear during sudden movement especially is the animal is stressed or startled. There is limb stiffness that causes a short stiff appearing gait and occasionally collapse (or fainting). Affected animals may appear more heavily muscled
2) Myokymia is similar to myotonia. It results in repetitive contractions of small bands of muscle fibers (continuous muscle fiber activity) that causes episodic stiffness of the limbs often leading to collapse with the limbs remaining in a tetanic stiffness. Undulating involuntary movements may also be observed in the skin due to contraction of underlying bands of muscles. Myokymia may be exacerbated during excitement and episodes may last a few minutes to hours. This syndrome is most likely caused by abnormalities in the nerves supplying the muscle fibers so the term neuromyokymia may be more correct.
3) Tetanus or Tetany is caused by persistent muscle (tetanus) or intermittent (tetany) contraction most prominent in extensor muscles. It is most commonly seen in animals that have lost the ability to control extensor muscle tone due to loss of inhibitory neurotransmitters. The most common form is caused by Clostridium tetani infection but occasionally congenital abnormalities or immune mediated damage to receptors (stiff person syndrome) may be seen.
4) Myoclonus is a movement characterized by the sudden contraction of a group of muscles followed by relaxation. A sporatic form of myoclonus would look like a twitch of a group of muscles that is intermittent in origin. This can be caused by structural diseases in the forebrain and represent a form of epilepsy. It can also be idiopathic in origin. Repetitive myoclonus results in more consistant twitching of a group of muscles. This may be constant (constant repetitive myoclonus) and occurs during rest as well as during movement, or may be only action related (action-related repetitive myoclonus) affecting groups of skeletal muscles that are being activated and disappearing during rest. This type of myoclonus is often classified as tremors because of their appearance.
    a) Intention tremors as seen with cerebellar disorders is a tremor that is generally confined to the head and neck and occurs during periods of behavioral focus such as with eating or sniffing something.
    b) Whole body action tremors occur with myelin disorders, intoxications, electrolyte abnormalities, or abnormalities in neurotransmitter function.
    c) Postural repetitive myoclonus is a type of tremor that occurs only in muscles required to support weight. It may be episodic and confined to the head and neck in some breeds of dogs such as Doberman pinchers, Bulldogs, Boxer dogs were it is often referred to as head bobbing. The other form of postural repetitive myoclonus referred to as orthostatic tremors occurs in limb muscles, especially those needed to support weight. A benign form occurs in the pelvic limbs of aged dogs and disappears when the animal starts to move. A more severe form has been reported in Great Dane and Scottish deerhounds. This syndrome is most severe when the dog attempts to lie down or postures to eat, drink or eliminate. It disappears with rest and during normal ambulation. Phenobarbital and gabapentin have been beneficial in ameliorating clinical signs.
    d) Dyskinesia is a general term used to describe movement disorders characterized by involuntary contractions of groups of muscles in a conscious animal. The movements may take the form of fast contractions of groups of muscles creating a flailing appearance (Ballism and Chorea) or more prolonged contraction of muscles (dystonia or athetosis). Most reported cases of these type of movements tend to be familial and poorly characterized. Examples are Scottie cramps, Chinook seizures, Spike disease in Border terriers, Dyskinesia in Bichon frise, and Collapsing Cavalier King Charles spaniels.

During this session, video's of each of these abnormalities will be presented. This will be followed by a series of unknowns where the audience will be asked to provide their opinion. The overall goal is to recognize different forms of movement abnormalities in dogs.

REFERENCES

Datt VL and Fletcher TF. Gait Foot-Fall Patterns 2012 Supported by University of Minnesota College of Veterinary Medicine available online at http://vanat.cvm.umn.edu/gaits/info.html; Caprette H and Dyce J. 2002. Canine Lameness Learning Module available online at http://accad.osu.edu/~hcaprett/COTA_741_sp04/CanineLO_090504.html; de Lahunta A, Glass EN, Kent M. Classifying involuntary muscle contractions. Comp Cont Educ Pract 2006;28:516–530.; de Lahunta A, Glass E. Veterinary neuroanatomy and clinical neurology. 3rd ed. St. Louis: W.B. Saunders; 2009; Deuschl G, Elble R. Essential tremor—neurodegenerative or nondegenerative disease towards a working definition of ET. Mov Disord 2009;24:2033–2041; Donaldson I, CD Marsden,

Schneider S, and Bhatia K. Marsden's Book of Movement Disorders. Oxford University Press, Oxford 2012; Thomas WB. Evaluation of veterinary patients with brain disease. Vet Clin Small Anim 2010; 40: 1–19; Vite CH. Myotonia And Disorders Of Altered Muscle Cell Membrane Excitability. Vet Clin Small Anim. 2002; 32: 169-187; Gillette RL, Angle TC. Recent developments in canine locomotor analysis: A review. The Veterinary Journal 2008,178:165–176' Nunamaker DM, Balauner PD. Normal and Abnormal Gait in Newton CD and Nunamaker DM eds. Textbook of Small Animal Orthopaedics Philadelphia J.B. Lippincott Company 1985 available online at http://cal.vet.upenn.edu/projects/saortho/chapter_91/91mast.htm

NEURO-INFECTIOUS DISEASE: IT'S NOT ALL IN YOUR HEAD

Adam J. Birkenheuer DVM, PhD, DACVIM
Raleigh, NC

INTRODUCTION

Neurologists are commonly faced with inflammatory central nervous system diseases or CNS "look-a-likes" that have infectious diseases as potential underlying causes.

Table 1. Infectious Agents Causing Meningoencephalomyelitis in Small Animals

Category of Agent	Specific Organisms
Viral	Canine distemper virus, FIP, FeLV, Rabies, Pseudorabies, West Nile virus
Fungal	*Cryptococcus neoformans, Coccidiodes immitis, Blastomyces dermatidis, Aspergillus spp., Cladosporium spp.*
Rickettsial	*Ehrlichia canis, Rickettsia rickettsia, Anaplasma phagocytophilum*
Bacterial	*Streptococcus spp., Bartonella spp., Borrelia burgdorferi., Leptospira spp., Mycoplasma spp.*
Protozoal	*Neospora caninum, Toxoplasma gondii, Entamoeba histolytica, Babesia spp. Cytauxzoon felis*
Parasitic	*Cuterebra spp., Dirofiliaria immitis, Baylisascaris procyonis, Angiostrongylus spp., Prototheca spp.*

Unfortunately, sometimes despite aggressive diagnostic workups we are "stuck" with a diagnosis of idiopathic or presumptive auto-immune disease. [1-4]

BEWARE THE "BLACK SWAN"!

In 16th century the term "Black Swan" was used to denote an impossible event. This was corroborated for people on a daily basis because all of the swans in the old world were white. If every swan you saw throughout your entire life was white, what else would you think? In the late 16th century a Dutch explorer discovered black swans in Australia…melanistic forms of nearly every other species exist so why was it so hard to believe the same would be true of swans? In the 21st century a "Black Swan" event has been described as:

1. The event is a surprise (to the observer).
2. The event has a major effect.
3. After the first recorded instance of the event, it is rationalized by hindsight, as if it *could* have been expected; that is, the relevant data were available but unaccounted for in risk mitigation programs. The same is true for the personal perception by individuals.

Since infectious diseases are only rarely diagnosed in dogs with inflammatory CNS disease it is easy for us as clinicians to start believing that infectious diseases are "black swans" or… even worse can represent a "Black Swan" event.

Let's assume that 1% of inflammatory CNS diseases are caused by infectious disease and our infectious disease test has a 90% sensitivity (it will be positive for every infected animal-which is incredibly optimistic!). If ten neurology practices each saw one dog with inflammatory CNS disease every other week (i.e. 24 cases per practice per year) at the end of the year 8/10 practices would have never diagnosed a single dog with an infectious disease (i.e. only saw white swans). In these situations we can easily fall into the trap of "we don't have that here…we only have white swans here…" I actually think that it takes far fewer than 24 cases for most clinicians to start to have these thoughts and want to stop looking for or even treating black swans. How can infectious diseases turn into "Black Swan" events?

1. The event is a surprise (to the observer)-"…we don't have that infection here…"
2. The event has a major effect-"…immune-suppression can have fatal consequences for some infections…"
3. After the first recorded instance of the event, it is rationalized by hindsight, as if it *could* have been expected; that is, the relevant data were available but unaccounted for in risk mitigation programs. The same is true for the personal perception by individuals.-"…duh! of course Black Swans exist…"

I want my diagnostic and patient care team to always consider the existence of black swans.

ALWAYS COLLECT AND SAVE SAMPLES

Infectious disease diagnostics typically fall into one of three categories: microscopic identification, detection of antibodies or detection of pathogen "parts" (sugars, proteins or nucleic acids). Even though the signs may be most evident in the CNS, the diagnosis may be made more easily using another sample. All systemically ill animals should have whole blood, serum and urine collected in addition to cerebrospinal fluid. Due to the complexity of infections and host immune responses, very few diagnostics tests perform at the mythical 99% sensitivity and specificity levels. This means multiple diagnostics assays (microscopic, antibody and molecular) should often be used in parallel.

CLUES FOR INFECTIOUS DISEASES

Physical examination

Lymphadenopathy, splenomegaly, uveitis, polyarthritis, vasculitis, edema, cough, pale mm, petechial, cutaneous or subcutaneous masses (esp. draining)

Hematology

Anemia: Ehrlichia spp., Rocky Mountain Spotted Fever, Babesia spp., Bartonella spp., FeLV, Cytauxzoon felis
Thrombocytopenia: Ehrlichia spp., Rocky Mountain Spotted Fever, Babesia spp., Bartonella spp., CDV, FeLV, C. felis
Leukocytosis: RMSF, Ehrlichia spp. (lymphocytosis)
Leukopenia: Ehrlichia spp., FeLV, Cytauxzoon felis
Eosinophilia: Dirofilaria immitis, Systemic fungal

Biochemistry

Hypoalbuminemia: Ehrlichia spp., Rocky Mountain Spotted Fever, Bartonella spp., FeLV, Cytauxzoon felis
Hyperglobulinemia: Ehrlichia spp., Babesia spp.
Hypoglobulinemia: Bartonella spp.
Azotemia: Leptospira spp., Borrelia spp., Ehrlichia spp., Rocky Mountain Spotted Fever, Babesia spp.,
Creatinine kinase: Neospora spp., Toxoplasma gondii

Urinalysis

Proteinuria: Ehrlichia spp., Bartonella spp., Borrelia spp.,

CONCLUSIONS

While infectious diseases are rarely diagnosed in dogs or cats with inflammatory CNS disease, it is our duty as specialists to consider them diagnostically and therapeutically in all of our patients. We never want to look back and think "duh!..of course there are black swans!"

REFERENCES

1. Barber RM et. al. Vet Intern Med 2012;26:962–968
2. Barber RM et. al. J Vet Intern Med. 2010 Mar-Apr;24(2):372-8.
3. Nghiem PP et. al. J Vet Emerg Crit Care (San Antonio). 2010 Feb;20(1):46-61.
4. Schatzberg SJ et. al. J Vet Diagn Invest. 2009 Nov;21(6):844-9.

APHERESIS IN SMALL ANIMAL VETERINARY MEDICINE

Carrie A. Palm, DVM, DACVIM
Davis, CA

INTRODUCTION

Extracorporeal therapies, such as hemodialysis and apheresis, are procedures where blood is removed from patient circulation and processed before return to the patient. Apheresis is an extracorporeal therapy where blood is removed from patient circulation, separated into components, and one or more of the components is removed or processed before return of blood to the patient. Therapeutic plasma exchange (TPE) is an apheresis treatment where the plasma component of blood, (which is suspected to contain pathologic substances, such as excessive immunoglobulins) is removed and exchanged with replacement fluids. This is the most commonly applied apheresis therapy in veterinary medicine.

In human medicine, applications for apheresis are vast and are divided into four categories based on the defined or proven benefit that apheresis can provide for each particular disease process/condition. A category I condition, such as myasthenia gravis (MG), is defined as a disease for which apheresis is considered to be a standard of care and is therefore prescribed as a first line therapy. A category II condition, such as Amanita mushroom poisoning, is one in which apheresis is considered to be a supportive therapy, in combination with other therapies. A category III condition, such as acute hepatic failure, is a condition in which controlled trials are limited or inconclusive and a category IV condition is one in which therapy appeared to be ineffective in controlled clinical trials. As the application of apheresis in veterinary medicine has only recently been prescribed more commonly, similar categorization is not yet possible, however, the already established human studies can be used as a reference.

INSTRUMENTATION

Apheresis can be carried out with either centrifugal or filtration methodologies. Both techniques require venous access, which typically involves placement of a large bore double lumen jugular catheter. With centrifugation, once blood is removed from the patient, it is separated by density (similar to centrifugation for evaluation of packed cell volume). Blood components are listed here in order of increasing density: plasma, platelets, lymphocytes and monocytes, granulocytes and red blood cells. Once blood components are separated, outlet tubes are positioned within the apheresis machine so that a specific component(s) can be selectively removed from the patient based on this density variation. The other components can then be returned to the patient, and are mixed with replacement fluids, such as synthetic colloids, plasma and crystalloids during return. To prevent clotting in the extracorporeal circuit, anti-coagulation is necessary. Typically, for centrifugal apheresis, regional citrate is used, where only the extracorporeal circuit is anti-coagulated, and systemic anticoagulation does not occur. In the United States, centrifugation techniques are most commonly used.

Filtration apheresis techniques are not commonly used in the United States. Similar to hemodialysis, filtration apheresis utilizes membranes that can separate plasma out from the other blood components. Once separated, the plasma can be run over an adsorption column so that binding and removal of pathologic substances can be achieved. TPE can remove specific pathologic plasma components, such as low-density lipoproteins. Filtration techniques cannot be used for removal of specific cell components, whereas centrifugal apheresis can. With filtration apheresis, systemic anti-coagulation with heparin is most commonly used, while regional citrate anti-coagulation is utilized with centrifugal techniques.

To successfully carryout apheresis, it is critical to have a trained and experienced team. This includes clinicians and technicians. At our institution, our highly trained technical staff is integral to the success of our extracorporeal unit. They play a critical role in machine maintenance and troubleshooting, and also in the care and monitoring of patients, during treatments and in between treatments.

PRINCIPLES OF APHERESIS AND THERAPEUTIC PLASMA EXCHANGE

Successful therapeutic plasma exchange (TPE) treatments are based on the principle that the disease being treated is in large part due to pathologic substances (commonly immunoglobulins) within the plasma compartment. The ability for successful substance removal during TPE is dependent on the distribution of the substance within the body {the volume of distribution (Vd)}, the rapidity with which the substance equilibrates between body compartments, and the ability to remove adequate amounts of the pathologic substance. As with other extracorporeal therapies, the most efficient removal is achieved with substances that have a small Vd, and are primarily located within the intravascular space. As an example, immune-mediated diseases are commonly managed with TPE, whereby alleviation of, or improvement in clinical signs is achieved by removal of immunoglobulins present in the intravascular space. Immunoglobulin removal has a primary positive effect by decreasing secondary complementation fixation and other adverse systemic consequences, and also allows for adjuvant medical therapies to be more effective.

With each TPE treatment, approximately one plasma volume is removed and replaced with fluids, such as albumin preparations, colloids and crystalloids. Plasma is not typically used as a replacement fluid in human medicine, but rather, a combination of albumin and saline is most commonly used. Due to the potential for complications associated with repeated human albumin administration in dogs, and given the cost and lack of experience with canine albumin, albumin has not been used as a replacement fluid by the author. Instead, combinations of saline, hetastarch and plasma are used. Removal of one plasma volume per treatment is based on predicted clearance curves for various substances. For most larger molecular weight substances, such as immunoglobulins, there is no increased efficiency of pathologic substance removal when more than 1 or 1.5 plasma volumes are exchanged in a single treatment. If clinical improvement is not seen within 3-5 apheresis sessions, it is likely that no significant benefit will be gained. Likewise, if clinical remission is achieved after 3 treatments, there is likely no benefit in performing more treatments at that time. Total plasma exchange sessions are typically performed every other day. This allows for redistribution of the pathologic substance into the intravascular space, increasing the overall treatment efficiency.

As case numbers in veterinary patients are limited, information regarding complication rate and outcome is limited. Potential complications include adverse reactions to plasma, hypocalcemia or alkalosis secondary to citrate, blood loss via clotting of the extracorporeal circuit, as well as hypotension secondary to large extracorporeal blood volumes, and hypoproteinemia secondary to iatrogenic removal.

INDICATIONS FOR TPE IN VETERINARY MEDICINE

In the author's institution, apheresis has been used to treat patients with immune-mediated diseases that are severely affected or are refractory to, or minimally responsive to more traditional therapies. When utilizing TPE, traditional medical management is typically still instituted, and the TPE is considered as an adjuvant therapy to improve patient outcome. It is important to consider risk-benefit ratios when prescribing apheresis. In the critical care setting, apheresis techniques have been used for the treatment of immune-mediated diseases, such as myasthenia gravis (MG) and immune-mediated hemolytic anemia (IMHA). While case numbers are low, the anecdotal success of these cases is promising. Leukapheresis, or the removal of white blood cells, has also been used to carry out bone marrow transplantation in veterinary patients.

Myasthenia Gravis

Myasthenia gravis (MG) is an immune-mediated disease characterized by antibody-mediated inactivation of the neuromuscular endplate, manifesting as muscle weakness. Myasthenia gravis can occur as a focal or generalized disease and can lead to life-threatening complications, such as respiratory depression and aspiration pneumonia secondary to megaesophagus. Traditional medical management involves supportive care and the use of anti-cholinesterase medications. In cases refractory to medical management or in patients with rapidly progressive disease, apheresis, and specifically TPE, may improve outcome. As discussed above, MG is a category I condition in human medicine, and given the similarities between human and canine MG, it has been hypothesized that TPE may benefit canine patients. Five dogs with severe generalized and/or rapidly progressive MG have been treated with TPE at the author's institution. Four of the five dogs had resolution of clinical signs after 3 apheresis treatments. Of those 4 dogs, 2 dogs had complete remission, 1 dog redeveloped mild and intermittent regurgitation only (without radiographic evidence of megaesophagus) approximately one month after the final apheresis treatment and the fourth dog redeveloped significant regurgitation approximately one month after discontinuation of therapy. The fifth dog did not have any significant improvement with TPE. Antibody removal was confirmed for the first 3 dogs treated at our institution, via serial measurement of acetylcholine receptor antibody titers during and after apheresis sessions. It is important to note, that all patients were treated with concurrent medical therapy, in conjunction with TPE. No significant adverse events occurred in any of the treated dogs. Filtration apheresis has been reported for treatment of MG in a canine patient.

Immune Mediated Hemolytic Anemia

Therapeutic plasma exchange has also been used to treat refractory immune-mediated hemolytic anemia (IMHA) at our institution. All patients that were treated with TPE also received at least 4 days of appropriate immunosuppression prior to initiation of apheresis therapy. Use of TPE allowed patients that had previously been crossmatch incompatible due to severe agglutination, to receive blood transfusions as a reduction in antibody load occurred. In addition, use of TPE allowed for successful control of disease with the same immunosuppressive protocols that had been previously unsuccessful.

FUTURE APPLICATIONS

Large-scale studies regarding the benefits of apheresis in veterinary medicine are lacking, however, the limited data available suggests that there are likely benefits. By understanding the principles of apheresis, and by using the human medical experience as a guide, these advanced procedures have the potential to improve outcomes for veterinary patients. Apheresis techniques may be useful in treating a variety of fulminant immune-mediated diseases, such as immune-mediated thrombocytopenia, polyradiculoneuritis, hyperviscosity syndrome, lyme nephritis and amanita mushroom poisoning, among others. As more experience is gained in veterinary medicine, apheresis techniques will likely provide therapies for veterinary patients that may lack other treatment options.

REFERENCES

1. Burgstaler EA. Current apheresis instrumentation. In: McLeod BC, Price TH, Weinstein R, eds. Apheresis: principles and practice. 2nd ed. Bethesda, MD: AABB Press, 2003:95-130.
2. Gurland HJ Lysaght MJ, Samtleben W, et al. A comparison of centrifugal and membrane-based apheresis formats. Int J Artif Organs 1984;7:35-38.
3. Orlin JB, Berkman EM. Partial plasma exchange using albumin replacement: removal and recovery of normal plasma constituents. Blood 1980;56:1055-1059.
4. Mantegazza R, Bonanno S, Camera G, et al. Current and emerging therapies for the treatment of myasthenia gravis. Neuropsychiatric Dis Treatment 2011;7:151-160.
5. Palm C, Dickinson P, Sturges B, Cowgill LD. Plasmapheresis for the treatment of myasthenia gravis in 3 dogs. (abstr), ACVIM 2011, Denver Co.
6. Bartges JW, Klausner JS, Bostwick EF, et al. Clinical remission following plasmapheresis and corticosteroid treatment in a dog with acquired myasthenia gravis. J Am Vet Med Assoc 1990;196:1276-1278.
7. Nenov DS, Nenov KS. Therapeutic apheresis in exogenous poisoning and in myeloma. Nephrol Dial Transplant 2001;16:101-102.
8. Bektas M, Idilman R, Soykan I, et al. Adjuvant therapeutic plasma exchange in liver failure: assessments of clinical and laboratory parameters. J Clin Gastroenterol 2008;42:517-521.
9. Chhibber V, Weinstein R. Evidence-based review of therapeutic plasma exchange in neurological disorders. Seminars Dialysis 2012;25:132-139.
10. Stone MJ, Bogen SA. Evidence-based focused review of management of hyperviscosity syndrome. Blood 2012;119:2205-2208.

PROBIOTICS AND DIETARY SUPPLEMENTS IN HORSES: USEFUL OR NOT?

Angelika Schoster Dr.med.vet, DVSc, DACVIM, DECEIM
Vetsuisse Faculty University of Zurich, Switzerland

INTRODUCTION

Feed supplements are all substances fed to a horse in addition to a diet of forage, which technically would also include grain. However, the term is now most often used to describe additional nutrients (vitamins, minerals, extra protein, energy, herbal supplements and other neutraceuticals as well as probiotics). Neutraceuticals are defined as a food or a part of food that provides medical or health benefits, including prevention or treatment of disease. These feed supplements have become popular for the prevention or treatment of various diseases, such as gastrointestinal disease, osteoarthritis, equine gastric ulcer disease and others in recent years. Feed supplements are also promoted as performance enhancers and digestibility improvers. Many probiotic and nutraceutical formulations are available for horse owners over the counter in North America, claiming various health benefits for horses. Many contain a selection of nutrients while some only contains specific vitamins. Many horse owners administer, but also some veterinarians recommend these products based on anecdotal evidence or based on the claim on the label. Scientific, peer reviewed evidence behind these formulations is limited with only few products having been researched, with disappointing results. Additionally there is also the issue of quality control, which is not regulated by the authorities. Not only can this be a waste of money, but also dangerous, as not all products might be safe for use in horses.

In North America feed supplements are classified as neutraceuticals and are not regulated by the food and drug agency, which regulates the laws regarding food and feed safety. Probiotics for animal use are also called 'direct fed microbials' and can either be classified as a drug, in which case they have to be approved by the FDA (none currently approved for horses), or they can be classified as a dietary supplement and therefore do not need to go through the general process of drug approval. The federal government only requires that feed supplement products be labeled in a truthful and not misleading manner (www.fda.gov.com). Consequently in North America, there are numerous probiotic and feed supplements for equine use on the market, that can be obtained over the counter. Horse owners cannot assume that a product is safe and effective solely on the fact that it is produced and marketed.

SUPPLEMENTS

Because of the myriad of products all claiming different effects it is important to understand what a specific product is designed to do. Supplements cannot replace a good feeding program, good general health management or veterinary advice and should only complement these management factors.

Feed supplements administered to horses include vitamins and minerals, substances for joint and bone maintenance, performance enhancement, hoof and coat health, immune enhancement, treatment for behavioral issues, metabolic disorders, increase in weight gain, gastrointestinal health and others. The substances used are yeasts, live-bacteria (probiotics), specific vitamins (e-g- Vitamin E), macrominerals (e.g. calcium, phosphorus, sodium, chloride, potassium), microminerals (e.g. iron, zinc, copper, manganese, selenium), extra protein, energy and others. In general these are substances owners perceive as 'missing' or inadequate in a horses diet. Some of the products have been researched however others have little or no scientific evidence.

Probiotics and yeast

Metchnikoff first defined probiotics as 'live microorganisms which exhibit a health promoting effect' in 1908. This definition has since been modified by several authors until in 2008, the FAO and WHO defined probiotics as *'live microorganisms, that when administered orally at adequate concentrations, provide a beneficial effect beyond that of their nutritional value'*. Many reported mechanisms of action of probiotics are based on in-vitro studies only, and extrapolation of these results to in vivo conditions is controversial; however, some evidence has also been generated by in vivo studies. There are four main mechanisms of action by which probiotics exert their beneficial effect on the host:

- immune modulation of the host innate and acquired immune system
- antimicrobial production
- competitive exclusion of pathogenic bacteria
- inhibition or inactivation of bacterial toxins.

Probiotic mechanisms and effect are strain specific; therefore results from a study using a specific probiotic product cannot be extrapolated to another product.

Adverse effects of probiotic administration are rare. In humans the few reports available usually describe extra-intestinal infections, not enteric disease. There are no published reports of enteric disease in adult horses following probiotic administration[1-3]. Even when using high doses, adverse effects are not commonly seen[4]. Several published studies are also available that demonstrate safety of commercially available probiotics in foals[1,5], however one study showed that neonatal foals fed a probiotic had a higher incidence of diarrhea requiring veterinary intervention that foals fed a placebo, indicating that not all probiotics should be considered 'safe'[6]

As described earlier, manufacturers of over the counter products in North America have no obligations to perform quality control of their products. This results in most commercial veterinary probiotic preparations not being accurately represented by label claims. For example in a study by Weese et al, only 2/13 (15%) of veterinary and human probiotics contained the specified organism at the concentration propagated on the leaflet[7]. Probiotics have been evaluated as a means to treat or prevent a variety of gastrointestinal diseases in horses[4,6,8-10]. While some studies have shown some beneficial effects of probiotics, other studies could not corroborate these results. For example two studies have evaluated the efficacy of the yeast *S. boulardii* for treatment of acute enterocolitis. In one study, horses receiving *S. boulardii* had a shorter duration of diarrhea and watery diarrhea [10]. In the second study a beneficial effect could not be shown[11].

Probiotics have also been evaluated as a means to improve digestibility in several studies. Yeasts are used to improve feeding of the microbes of the hindgut to improve digestion of roughage and keep a healthier balance of microbes. While some improvement was noted for some parameters such as an increase in dry matter intake, there was limited influence on extraction of minerals from the diet in probiotic fed horses [12,13]. In each study only some of the factors were positively influenced while others remained unaffected.

Most probiotics consist of species of Lactobacilli. Recent investigations into the equine microbiome (overall bacterial population of the gastrointestinal tract) has shown that Lactobacilli only constitute a small amount of the overall population, and phyla such as Firmicutes predominate[14]. The dosages used in horses are extrapolated from human medicine, however horses are inherently different from humans and other animals as they are hind gut fermenters with an immense microbial population of their hindgut where diseases occur. It is therefore questionable whether a positive health effect can be achieved using Lactobacilli species and using the dosages currently proposed.

Overall the evidence for a positive health effect in horse for currently available probiotics is scarce. Larger scale controlled studies of different strains and products are necessary before conclusions can be drawn on the clinical efficacy of probiotics.

Vitamins and minerals

While supplementing vitamins can be beneficial in certain cases but over supplementation with vitamins has to be avoided as toxicities can occur. For example overfeeding of Vitamin D causes hypervitamoinosis D, a disease associated with calcification of soft tissues. Vitamin C is produced in a healthy horses liver and rarely needs to be supplemented. Vitamin B is produced by the microbes in the hindgut and supplementation should only be considered in extremely horses under extreme strenure. Vitamin E is effective as an anti-oxidant and can aid in healing of oxidative injury.

Macrominerals are most often supplemented in electrolyte supplements or due aid in bone formation (calcium and phosphorus). Salt should be made freely available to healthy horses. In cases of horses with kidney disease the recommendations of a veterinarian have to be followed.

Microminerals are mainly needed for biochemical reactions in the body but only trace amounts are needed. None of these are safe to feed free choice as toxicities can occur. Copper toxicity for example, results in fulminant renal and liver failure once toxic levels are reached. Specific regions of the US and Europa have various soil deficiencies. Knowledge of ones area is required for adequate supplementation. For example selenium deficient soil can be found in many areas over North America and Europe. Selenium is essential for muscle health. Selenium deficiency is associated with diseases such as white Muscle disease, Equine Motor Neuron Disease and Equine degenerative Myeloencephalopathy. Selenium toxicity, however, can also occur, therefore supplementation in the face of adequate contents in the diet are not recommended. Iodine deficiencies are associated with foals being born with contracted tendons.

Joint and Bone Supplements

Glucosamine (GS), chondroitin sulfate (CS), methyl sulfonyl methane (MSM), hyaluronic acid, silicon, soybean and avocado extract are examples of joint and bone supplements widely advertised and used in humans and horses to promote joint and bone health. Advocates propose that these supplements help the body repair cartilage. While there is in-vitro evidence that these substances have an effect on cartilage cell metabolism, there is few clinical studies evaluating an effect in the animal. A recent study evaluated the effect of a 3-month supplementation with GS, CS and MSM on improvement of stiff gait in geriatric horses in a randomized placebo controlled trial. No beneficial effect of the supplement over a placebo was seen[15]. In a recent systematic review to determine the effect of neutraceuticals on clinical signs of pain or abnormal locomotion 5 studies were reviewed. The systematic review concluded that was no effect of soybean and avocado and very weak evidence for all of the other substances evaluated. This low evidence suggest that these studies had major limitations, including used semi-subjective outcome measures[16].

Gastrointestinal disease

Equine gastric ulcer syndrome (EGUS) is a highly prevalent disease among performance horses. Various factors such as stress, feeding management and genetics predispose a horse to EGUS. Current treatment options are limited to medical treatment with omeprazole or histamine H2 blockers such as ranitidine and cimetidine. Treatment is expensive and required for prolonged periods of time during which the stress levels for the horse should be reduced (including high level exercise). Therefore, focus has turned to supplements as a means of preventing disease or as an adjunct treatment. Several supplements are marketed with such a health claim. Scientific evidence is limited and several recent studies have shown disappointing results. Sea Buckthorn berries and pulp were evaluated but showed only a limited effect for treatment of prevention of EGUS. Only certain types of ulcers were prevented, a treatment effect was not seen on these ulcers[17].

Equine Metabolic syndrome (EMS)

EMS is a recently emerging clinical syndrome of horses associated with management problems. The syndrome results from over feeding, inadequate physical activity and obesity. These factors culminate in insulin resistance and a predisposition to laminitis. Medical treatment options are limited and dietary restriction and weight loss is the mainstay of treatment. Forage should be restricted in these horses and hay is often soaked to release excessive amounts of sugar. As this practice also results in loss of minerals and vitamins, the use of forage stabilizer have been proposed. Short chain fructo-oligo saccharides (sc-FO) have been investigated, however a beneficial effect on insulin sensitivity in these horses could not be shown consistently in all studies[18]

CONCLUSION

In conclusion there is currently weak evidence for the use of probiotics to prevent or treat gastrointestinal disease in horses, and limited evidence for probiotic products to enhance digestibility. Given the lack of quality control and questionable content of commercial products it is doubtful whether any effects can be achieved. Care has to be taken with self-manufactures products, particularly in foals as negative effects have been reported. Supplements have to be divided into vitamin and mineral supplements for horses with a true lack of such in the diet and such where these would be fed in addition. If there is a lack of a particular vitamin or mineral due to deficiency of the soil, these should be added to the diet. Unfortunately companies have little incentive to perform research on their products, as studies are not required for marketing of the product and products sell well without results from such studies.

Clients of healthy horses are best advised to feed a balanced diet of forage supplemented with grain if the horse has high energy expenditures. As some vitamin and mineral deficiencies can be subclinical (not obvious) owners of high performance horses can also err on the side of caution and add a vitamin/mineral supplement to the diet. Clients of horses with specific diseases that are likely influenced by nutrition, such as PSSM, HYPP, Equine Metabolic syndrome, gastric ulcer disease, inflammatory bowel disease and others are best advised to seek advice from their veterinarian and follow their recommendations. There is no conclusive evidence in the current literature to support feeding commercial supplements as a prevention or treatment of diseases.

REFERENCES

1. Weese JS, Anderson ME, Lowe A, et al. Screening of the equine intestinal microflora for potential probiotic organisms. Equine Vet J 2004;36:351-355.
2. Botha M, Botes M, Loos B, et al. Lactobacillus equigenerosi strain Le1 invades equine epithelial cells. Appl Environ Microbiol 2012;78:4248-4255.
3. Weese JS, Anderson ME, Lowe A, et al. Preliminary investigation of the probiotic potential of Lactobacillus rhamnosus strain GG in horses: fecal recovery following oral administration and safety. Can Vet J 2003;44:299-302.
4. Parraga ME, Spier SJ, Thurmond M, et al. A clinical trial of probiotic administration for prevention of Salmonella shedding in the postoperative period in horses with colic. J Vet Intern Med 1997;11:36-41.
5. Yuyama T. Evaluation of a host-specific Lactobacillus probiotic in neonatal foals. In: Yusa S, ed. Journal of Applied Research in Veterinary Medicine: 2004:26-32.
6. Weese JS, Rousseau J. Evaluation of Lactobacillus pentosus WE7 for prevention of diarrhea in neonatal foals. J Am Vet Med Assoc 2005;226:2031-2034.
7. Weese JS. Microbiologic evaluation of commercial probiotics. J Am Vet Med Assoc 2002;220:794-797.
8. Kim LM, Morley PS, Traub-Dargatz JL, et al. Factors associated with Salmonella shedding among equine colic patients at a veterinary teaching hospital. J Am Vet Med Assoc 2001;218:740-748.
9. Landes AD, Hassel DM, Funk JD, et al. Fecal sand clearance is enhanced with a product combining probiotics, prebiotics, and psyllium in clinically normal horses. J Equine Vet Sci 2008;28:79-84.
10. Desrochers AM, Dolente BA, Roy MF, et al. Efficacy of Saccharomyces boulardii for treatment of horses with acute enterocolitis. J Am Vet Med Assoc 2005;227:954-959.
11. Boyle AG, Magdesian KG, Durando MM, et al. Saccharomyces boulardii viability and efficacy in horses with antimicrobial-induced diarrhoea. Vet Rec 2013;172:128.
12. Jouany JP, Gobert J, Medina B, et al. Effect of live yeast culture supplementation on apparent digestibility and rate of passage in horses fed a high-fiber or high-starch diet. J Anim Sci 2008;86:339-347.
13. Swyers KL, Burk AO, Hartsock TG, et al. Effects of direct-fed microbial supplementation on digestibility and fermentation end-products in horses fed low- and high-starch concentrates. J Anim Sci 2008;86:2596-2608.
14. Costa MC, Weese JS. The equine intestinal microbiome. Anim Health Res Rev 2012:1-8.
15. Higler MH, Brommer H, L'Ami J J, et al. The effects of three-month oral supplementation with a nutraceutical and exercise on the locomotor pattern of aged horses. Equine Vet J 2013.
16. Vandeweerd JM, Coisnon C, Clegg P, et al. Systematic review of efficacy of nutraceuticals to alleviate clinical signs of osteoarthritis. J Vet Intern Med 2012;26:448-456.
17. Huff NK, Auer AD, Garza F, Jr., et al. Effect of sea buckthorn berries and pulp in a liquid emulsion on gastric ulcer scores and gastric juice pH in horses. J Vet Intern Med 2012;26:1186-1191.
18. McGowan CM, Dugdale AH, Pinchbeck GL, et al. Dietary restriction in combination with a nutraceutical supplement for the management of equine metabolic syndrome in horses. Vet J 2013;196:153-159.

MANAGING CARDIAC DISEASE NUTRITIONALLY

Kara M. Burns, MS, MEd, LVT, VTS (Nutrition)
Wamego, KS

Cardiovascular disease is a common disorder in dogs and cats, with 11% of canines and up to 20% of feline populations affected by cardiac disease.[1,2] Chronic valvular disease has been found to be the most common acquired heart disease in dogs with an overall incidence greater than 40%.[3] Chronic mitral valvular disease is the most common acquired cardiac abnormality in dogs, affecting more than one-third of patients over 10 years of age.[4] In approximately 30% of cases the tricuspid valve is involved. However, disease of the tricuspid valve is usually less severe. Valvular disease is more prevalent in small breed dogs. Acquired valvular disease in cats is rare. Following the discovery that taurine deficiency was the principal cause of dilated cardiomyopathy in cats in 1987, the prevalence of this disease has decreased significantly.[3,4] Now hypertrophic and restrictive cardiomyopathies are more prevalent causes of myocardial failure in cats.

Several types of myocardial disease not recognized 40 years ago now appear commonly in dogs. Large-breed dogs, especially males, are predisposed to dilated cardiomyopathy, especially the Doberman pinscher breed. Hypertrophic cardiomyopathy occurs rarely in dogs. Arrhythmogenic right ventricular cardiomyopathy is common among boxer dogs.

Pulmonary vascular disease with secondary cor pulmonale is typically associated with *Dirofilaria immitis* infection (heartworm disease). This disease is more widespread in areas with higher mosquito and heartworm populations and is worsened when the dog does not receive appropriate preventive medication. Pulmonary hypertension appears to be more common than earlier believed with diagnosis increasing due to heightened awareness. Pulmonary thromboembolism is most commonly associated with renal disease, hyperadrenocorticism, corticosteroid therapy, neoplasia, nephrotic syndrome, pancreatitis, and immune-mediated hemolytic anemia. Primary systemic vascular disease is uncommon. Secondary aortic thromboembolism in cats may occur with any of the forms of cardiomyopathy and is the most frequently acquired feline vascular abnormality. Systemic hypertension in dogs and cats now appears to be more common than believed 40 years ago.

Dogs and cats with hypertension may present at any age, but typically hypertension is seen in middle-aged to geriatric patients. The mean age of dogs is nine years and the mean age of cats is 15 years. The strength of the peripheral pulses does not help detect systemic hypertension. Retinal hemorrhages and detachments are common end-organ changes in patients with moderate to severe hypertension. These ocular signs are frequently the first signs seen associated with hypertensive disease. A fundic examination should be part of the routine evaluation of all dogs and cats. Other clinical signs of hypertension are most often related to the underlying disease that causes systemic hypertension. The effects of long-term or severe systemic hypertension may cause significant heart disease (e.g. left ventricular concentric hypertrophy), and hypertension may complicate the treatment of chronic mitral valvular disease in dogs by worsening valvular regurgitation. The healthcare team should screen dogs and cats with significant heart disease for the presence of systemic hypertension. Also the healthcare team should search for underlying heart disease in patients with known hypertension, especially those exhibiting clinical signs that may be indicative of heart disease.

The most frequently encountered problems associated with cardiovascular disease that require nutritional modification are fluid retention states associated with chronic CHF, primary or secondary hypertension, obesity, cachexia and myocardial diseases related to a specific nutrient deficiency (taurine- and carnitine) and electrolyte disorders that may predispose to cardiac dysrhythmias.

Effective treatment requires a multifaceted approach, of which therapeutic nutrition is an important component. Foods designed for patients with cardiovascular disease should supply age-appropriate nutrition and specific nutrients that may help manage hypertension, decrease fluid retention and control the signs associated ascites, maintain heart muscle function, help slow the progression of concurrent kidney disease, and help counter the loss of nutrients in the urine of pets prescribed diuretics.

Heart failure is characterized by inadequate cardiac output and insufficient delivery of nutrients relative to tissue metabolic needs. Heart failure is a clinical syndrome which results from a variety of structural and functional disorders of the heart or great vessels. Clinical manifestations of heart failure are due to reduced cardiac output (weakness, exercise intolerance, syncope), pulmonary congestion (dyspnea, orthopnea, cough, abnormal breath sounds with crackles and wheezes), systemic fluid retention (jugular venous distention, hepatomegaly, ascites, pleural effusion) or a combination of these conditions. Obesity and chronic bronchitis often occur in dogs and cats with heart disease and cause clinical manifestations similar to those of heart failure, which can further complicate the diagnosis.

KEY NUTRITIONAL FACTORS

Sodium and Chloride

Retention of sodium, chloride, and water is associated with CHF. Consequently, the healthcare team should focus on these nutrients in patients with cardiovascular disease. A few hours after the ingestion of high levels of sodium, healthy dogs and cats easily excrete any excess in their urine. Patients in early cardiac disease may lose this ability to excrete excess sodium. As heart disease worsens and CHF arises, the ability to excrete excess sodium is worsened. Historically, sodium retention was primarily implicated in the pathogenesis of CHF and some forms of hypertension. A number of studies have examined the interaction of sodium with other ions, including chloride. Chloride may also act as a direct renal vasoconstrictor. According the NRC the minimum recommended allowance for sodium and chloride in foods for adult dogs is 0.08 and 0.12% dry matter (DM), respectively. In foods for adult cats it is 0.068 for sodium and 0.096% for chloride (DM).When dealing with patients with cardiovascular disease the sodium levels in foods designed to manage these patients should be limited to 0.08 to 0.25% DM for dogs and 0.07 to 0.3% DM for cats. Recommended chloride levels are typically 1.5 times sodium levels.

Avoiding excess sodium chloride in cat foods is more difficult than in dog foods as ingredients used to meet the higher protein requirement of cats also contain sodium and chloride and thus increase the sodium chloride content of cat food.

Taurine

Taurine is an important amino acid in dogs and cats with myocardial failure. The mechanism of heart failure in taurine-deficient cats and dogs is not well understood. Taurine may function in inactivation of free radicals, osmoregulation and calcium modulation. Taurine is also known for its direct effects on contractile proteins. Additionally, there may be other factors responsible for contributing to the development of myocardial failure in patients with taurine deficiency. Dilated cardiomyopathy and heart failure may result from an inciting or contributing factor combined with taurine deficiency. As we know, taurine is an essential amino acid in cats, therefore a minimum recommended allowance for taurine is necessary in cat foods. Taurine should be 0.04% DM. Taurine content of foods for cats with cardiovascular disease should contain at least 0.3% DM. Levels of taurine typically recommended for supplementation of feline cardiovascular patients (250 to 500 mg taurine/day) provide approximately twice that much.

Taurine is not an essential amino acid for dogs. Nevertheless, an association between dilated cardiomyopathy and plasma taurine deficiency and low myocardial taurine concentrations, has been observed. The association between taurine deficiency and dilated cardiomyopathy is strongest in American cocker spaniels and golden retrievers. An association between taurine deficiency and dilated cardiomyopathy has also been shown in Newfoundlands, Labrador retrievers, Dalmatians, English bulldogs, Portuguese water dogs and Irish wolfhounds. Even in canine dilated cardiomyopathy dogs with normal plasma and whole blood taurine levels, additional taurine may be warranted. Therefore foods for management of cardiovascular disease contain in dogs should contain added taurine. The level of taurine in foods for canine patients can be lower than for cats due to the fact that dogs can synthesize taurine. The recommendation for taurine in foods for canine cardiovascular disease patients is at least 0.1% DM. This is somewhat lower than would be supplied by the typical recommendation for taurine supplementation of foods for dogs with dilated cardiomyopathy (500 to 1,000 mg taurine/day). Studies support the fact that in dogs and cats, taurine is considered safe.

L-Carnitine

Deficiency of L-carnitine in dogs has been linked to dilated cardiomyopathy in dogs. Cardiac muscle function benefits from carnitine because carnitine is a critical component of the mitochondrial membrane enzymes responsible for transporting activated fatty acids in the form of acyl-carnitine esters. These are transported across the mitochondrial membranes to the matrix. From here b-oxidation and high energy phosphate generation occur. Free L-carnitine serves as a mitochondrial detoxifying agent.

Currently, the recommendation for carnitine supplementation in dogs with dilated cardiomyopathy is 50- to 100-mg L-carnitine/kg body weight three times daily. It is widely accepted that even if the cause of cardiomyopathy in a heart disease patient is not due to carnitine deficiency, supplementing dogs with carnitine does not appear to do any harm and may in fact be beneficial. Foods for heart disease patients should provide at least 0.02% DM of carnitine.

Phosphorus

Patients with cardiac disease are often suffering from concurrent disease conditions. It is understood that phosphorus is a nutrient of concern in patients with concurrent chronic kidney disease and that kidney disease is one of the more prevalent diseases seen concurrently with cardiac disease. Therefore nutritional management should avoid excess phosphorus in patients with concurrent chronic kidney disease. The recommended amount of phosphorous in nutritional management of cardiac disease is 0.2 to 0.7% DM in dogs and 0.3 to 0.7% DM in cats.

Potassium and Magnesium

Another concern in cardiac disease patients is the metabolism of potassium and magnesium. Hypokalemia, hyperkalemia, and hypomagnesemia, all have the potential for complications when medication therapy is introduced in patients with cardiovascular disease. Veterinary technicians should be aware that potassium or magnesium homeostasis abnormalities can:

1) cause cardiac dysrhythmias
2) decrease myocardial contractility
3) produce profound muscle weakness
4) potentiate adverse effects from cardiac glycosides and other cardiac drugs.

The amounts of potassium and magnesium recommended for adult maintenance in dogs and cats (0.4 and 0.52% DM potassium, respectively, and 0.06 and 0.04% DM magnesium) should be the minimum amounts included in nutritional management of CHF. If abnormalities in these electrolytes occur, the healthcare team should consider supplementation or switching to a different food.

Protein

Cardiac cachexia is a major concern in patients with cardiac disease. The protein requirements of patients with cardiac cachexia have not been investigated extensively to date. The metabolic changes associated with cachexia and their effect on overall nutrient requirements is only recently being investigated. Many patients with cachexia present with concomitant disease (i.e., chronic kidney disease), which also significantly affects nutrient requirements. Nutritionists do know that profound anorexia enhances protein-energy malnutrition in patients with cachexia. Subsequently, patients with cachexia or exhibiting signs potentially leading to cachexia should be encouraged to eat a complete and balanced food that contains adequate calories and adequate high-quality, highly digestible protein.

Omega-3 Fatty Acids

In cardiac cachexia, TNF and IL-1cytokines have been implicated as pathogenic mediators. Fish oil (known to be high in omega-3 (n-3) fatty acids) has been shown to alter cytokine production. Early investigations involving fish oil suggest that fish-oil-mediated alterations in cytokine production may help dogs with CHF. Consequently, it is believed that heart failure patients with cachexia may benefit from the alterations of cytokine production through omega-3 fatty acid supplementation.

It is believed that omega-3 fatty acids electrically stabilize heart cells through modulation of the fast voltage-dependent Na(+) currents and the L-type Ca(2+) channels which results in the heart cells becoming resistant to dysrhythmias. Clinical studies of fish oil as a source of long-chain omega-3 fatty acids have confirmed the reduction in frequency of ventricular arrhythmia in boxer dogs.

Omega- 3 fatty acids have been shown to have a significant effect on survival times when used in dogs diagnosed with DCM or chronic valvular disease. The effect of the omega-3 fatty acids may be attributed to: anti-inflammatory effects, cachexia prevention, improved appetite, or anti-arrhythmic effects. The veterinary healthcare team should also be aware of further effects of omega-3 fatty acids on the patient. The healthcare team must be mindful of the fact that omega-3 fatty acids have the potential to alter immune function. This alteration in immune function may contribute to the cardiovascular effects of omega-3 fatty acids. Also, omega-3 fatty acids reduce platelet aggregation resulting from the production of thromboxane B5. The reduction in platelet aggregation might be of benefit in cats with cardiac disease and at risk for thrombus formation. However, this effect is also important to be mindful of when using omega-3 fatty acids in animals with coagulopathies.

Additional studies and discussion is needed in the long term, but it is believed that dogs and cats with cardiac disease may benefit from omega-3 fatty acid supplementation. However, the healthcare team must consider a number of factors: 1) dose, 2) timing, and 3) omega-3 fatty acid form.

At this point in time, no optimal dose of omega-3 fatty acids has been established for humans, cats, or dogs. The current recommendation from nutritionists studying fatty acids and cardiac disease is a dose of 40 mg/kg EPA and 25 mg/kg DHA for both dogs and cats.

Timing also needs to be taken into consideration when supplementing omega-3 fatty acids. The healthcare team should remember and educate owners that the majority of omega-3 fatty acids benefits occur after peak plasma and tissue concentrations have been achieved. Although plasma concentrations may increase significantly in the first week of omega-3 fatty acid supplementation, typically 4 to 6 weeks are required to reach peak plasma concentrations.

EPA and DHA can be provided via the diet or as a dietary supplement. There are a few therapeutic pet foods with high levels of EPA and DHA, but the majority of foods manufactured today do not achieve the recommended level of EPA and DHA. The current recommended dose is 40 mg/kg EPA + 25 mg/kg DHA. Therefore a manufactured food would need to contain between 80 and 150 mg/100 kcal EPA + DHA. Other factors that would need to be taken into consideration would be the size of the pet and the amount of food consumed. If the pet is not prescribed one of the high fatty acid foods, a recommendation of fish oil supplementation would be necessitated. However, caution must be given when making a supplement recommendation, as fish oil supplements vary widely in the amount of EPA and DHA they contain. The healthcare team should be familiar with various brands of fish oil supplements and plan to make a recommendation based on a specific brand with which the concentration of EPA and DHA have been researched and confirmed.

Water

With all patients, veterinary technicians must remember to talk with clients about the importance of water for pets. Veterinary technicians need to remind clients that pets should be offered water free choice and it should be clean and fresh. Healthcare teams must also keep in mind that water quality varies considerably, even within the same community. We must be cognizant of the fact that water can be a significant source of sodium, chloride, and other minerals. Veterinary healthcare teams should be familiar with the mineral levels in their local water supply. Water samples can be submitted to state or other government laboratories for analysis. Also municipal water companies can be contacted to ask about mineral levels in local water supplies. Distilled water or water with less than 150 ppm sodium is recommended for patients with advanced heart disease and failure.

REFERENCES

1. Buchanan JW. Prevalence of cardiovascular disorders. In: *Textbook of Canine and Feline Cardiology. 2nd ed.* Eds P. R. Fox, D. Sisson and N. S. Moise. 1999. Saunders, Philadelphia, PA, USA. pp 457-470.
2. Paige C F, Abbott JA, Elvinger F, Pyle R L. Prevalence of cardiomyopathy in apparently healthy cats. *Journal of American Veterinary Medical Association*, 2009, 234, pp.1398-1403.
3. Rush JE. Chronic Valvular Disease in Dogs. In *Kirk's Current Therapy XIV, 14th ed.* Bongura JD, Twedt DC, eds. 2009. Suanders Elsevier, St. Louis, MO
4. Roudebush P, Keene BW. Cardiovascular Disease. In *Small Animal Clinical Nutrition, 5th ed.* Hand M, Thatcher C, Remillard R, Roudebush P, Novotny B., eds. 2010. Mark Morris Institute, Topeka, KS.
5. Freeman LM, Beneficial effects of omega-3 fatty acids in cardiovascular disease. *Journal of Small Animal Practice*. Vol 51. September 2010.pp. 462 – 470.
6. Lunn J, and Theobald HE. The health effects of dietary unsaturated fatty acids. 2006 *British Nutrition Foundation Nutrition Bulletin*, 31 , pp. 178–224.
7. Gross KL, Yamka RM, Khoo C., et al. Macronutrients. In *Small Animal Clinical Nutrition, 5th ed.* Hand M, Thatcher C, Remillard R, Roudebush P, Novotny B., eds. 2010. Mark Morris Institute, Topeka, KS.
8. Freeman L M, Rush J E, Kehayias J J, Ross, J N JR, et al. Nutritional alterations and the effect of fish oil supplementation in dogs with heart failure. *Journal of Veterinary Internal Medicine*, 1998, 12, pp.440-448.
9. Slupe J L, Freeman LM, Rush JE. Association of body weight and body condition with survival in dogs with heart failure. 2008. *Journal of Veterinary Internal Medicine*, 22.pp. 561-565

NUTRITION AND THE CANINE AND EQUINE ATHLETE

Ann Wortinger, BIS, LVT, VTS (ECC, SAIM, Nutrition) and Kara Burns, MS, MEd, LVT, VTS (Nutrition)
Allen Park, MI and Wamego, KS

Just like people who are athletes, training and nutrition can play a major role in the canine athletes' success. But nutrition cannot overcome deficits in genetics and training. Matching nutrition to exercise type allows a canine athlete to perform to its genetic potential and level of training. (1) In general, all working dogs have increased energy requirements over those of an adult dog during time of normal activity. (2) The type of work being done and the intensity of work may require modifications in the nutrient composition of the food and the feeding schedule. (2)

Exercise requires transfer of chemical energy into physical work. ATP (adenosine triphosphate) is the sole source of energy for muscle contraction. (1) ATP is formed from metabolic fuels stored in muscle (endogenous) and from other body stores (exogenous). The energy is converted to ATP using either aerobic pathways using oxygen, or anaerobic pathways that can work without oxygen. (1) The proportion of each pathway used is determined by duration and intensity of exercise, conditioning and nutritional status of the animal. (1)

Fats

The two primary fuels used by the body for working muscles are muscle glycogen and free fatty acids. An intermediate athlete would receive ~70-90% of their energy from fat metabolism, and only a small amount from carbohydrate metabolism. (2) Dogs rely more heavily on free fatty acids for energy generation at all exercise levels than do people. (1) Feeding a higher fat diet to endurance and intermediate trained athletes prepares the muscles to efficiently mobilize and use free fatty acids for energy. It also exerts a glycogen sparing effect that can help prolong glycogen use during work. (2)

Carbohydrate

Provided sufficient gluconeogenic precursors are available in the diet, no dietary requirements for carbohydrates exist except during gestation and neonatal development. (1) Adipose tissue supplies glycerol for glucose production (breaking down triglycerides) and fatty acids for oxidation to supply energy, whereas muscle catabolism releases glucogenic amino acids, lactic acid and pyruvate for glucose production by the liver. (1) Carbohydrates fed to athletes should be highly digestible to decrease fecal bulk in the colon.

Protein

Amino acids provide ~5-15% of the energy used during exercise, most of this energy comes from the branched-chain amino acids (leucine, isoleucine and valine). All of these are essential amino acids and cannot be synthesized from other amino acids; they must be included in the diet. Muscle and organ meat based proteins have the highest level of essential amino acids, and are also the most digestible and most bioavailable.

Amino acids are not stored as proteins in the body but are deaminated (broken down) to ketoacids. These ketoacids are either oxidized for energy or converted to fatty acids and/or glucose and stored as adipose tissue (fat) or glycogen. (1)

Water

Water is used as a solvent for biological solutes; it acts as a transport medium for nutrients, wastes and heat, absorbs physical shock and lubricates various internal and external surfaces. (1) Heat is the primary byproduct of muscle contraction and the respiratory tract through panting is responsible for dissipation of this heat. (1) Because evaporative heat loss is the primary way dogs dissipate heat, ensuring adequate hydration is crucial for the maintenance of normal body temperature. (1) Depending on the type of work done and environmental conditions, water losses can increase by 10-20 times normal during exercise. (2)

Supplements

Many breeders, exhibitors and trainers believe stressed dogs must also receive supplements of certain vitamins and minerals. There is no evidence to suggest that working dogs have increased requirements of these nutrients. (2) If a diet is nutritionally balanced and the dog is consuming enough to meet its energy requirements during work, then additional supplements should not be necessary. (1)

Diet Requirements

A diet needs to be highly digestible to limit the total volume of food consumed at each meal. An ideal diet would provide increased levels of high quality protein to meet anabolic requirements and enough non-protein energy nutrients (fats and carbohydrates) to meet energy requirements. The food needs to be calorically dense and palatable, highly digestible and practical, so that the dog can physically consume enough to meet their caloric requirements. (1)

Feeding Plan

Look at where the dog is housed (inside/outdoors), medications they may be taking, dietary history; amount fed, type of food fed and timing of meals in relation to exercise/training and the nutrient profile of the diet, exercise and training history (amount of exercise done, frequency and performance of exercise). (1)

Compare the current diets key nutritional factors, determine the amount to be fed and the timing of the meals, estimate energy expenditure using body condition scoring and exercise level.

Timing of meals is important to allow the most availability of nutrients to the athlete. Ideally 1 meal at least 4 hours before exercise, 1 meal within 2 hours after exercise and if necessary due to the duration of exercise small amounts during. The largest meal should be given post exercise. It is also very important to allow access to plenty of fresh clean water to prevent dehydration. (1)

Recommended Caloric distribution for Canine Athletes

Calories from protein: 30-35% ME
Calories from fat: 50-65% ME
Calories from carbohydrate: 10-15% ME (2)

NUTRITION OF THE EQUINE ATHLETE

A successful equine athlete needs to perform with high intensity and sustain this intensity longer than its competitors. The conditioning of the horse in conjunction with its individual soundness, are two factors that play a role in the success of the equine athlete. Additionally, nutrition and an effective nutritional program will also help the horse reach its peak athletic potential.

Energy

The goal of the equine athlete is to maintain ideal body condition while not sacrificing energy stores or the availability of the energy stores. Energy requirements will differ dependent upon age, body condition, amount of exercise, etc. If the energy intake is not adequate, even for a short time, the potential for rapid fatigue and reduced performance is highly likely. When determining proper energy amounts, the rider and equipment (if applicable to the sport) needs to be taken into account; because anything that increases the weight also increases the energy cost of locomotion. Energy requirement = maintenance requirements plus an allowance for the work being undertaken. Gaining or losing weight alters lean body mass in addition to body fat percentage. Speed is linked to fat free mass subsequently altering performance. Consequently, a balance is needed to maximize lean body mass while concurrently minimizing excessive body fat. Restricting caloric intake to produce a lean equine is not recommended. We must look at the exercise and nutritional regimen overall. (3)

A body condition score should be performed on every equine athlete, as a less than ideal BCS can impair performance. A score of 4- 6 on a 1-9 scale would be ideal, except for the racehorse where the BCS should be closer to 4-5/9.

According to the NRC the maintenance daily energy (DE) requirements for a 450kg horse that is working as an endurance athlete are approximately 13.6 – 16.3 Mcal/day. The NRC also uses a multiplier of 1.9 over maintenance to estimate requirements for elite endurance equine athletes. Each equine is an individual and actual DE requirements will vary. Weekly monitoring, along with BCS and BW will enable the equine athlete feeding regiment to be tailored appropriately. (4,5)

Protein

Protein requirements of athletic horses are higher versus sedentary horses. Physical conditioning results in anabolic changes in muscle, a need to repair muscle damage from exercise, and the enabling of muscle hypertrophy. However, as with humans it is hard to state the exact amount of protein as it is dependent upon the type and amount of conditioning and the sport in which the horse is involved. It is important to determine the amount of protein a horse is receiving to determine if the protein requirement is being met. High quality hay – especially legume hay (CP >20%) – provides significant amounts of protein; often equal to or greater than the grain portion of the diet. Too much protein on the other hand may not be desirable due to the possible adverse effects on heat production and acid base balance, especially when the horse is at maximal exercise. It has also been shown that high protein intake results in increased urine production and possible evaporative fluid losses. This may be an excessive challenge for horses during extended exercise.(3) A grass hay has a CP of 8 – 14% and is recommended as long as it is of good quality. If grass hay is mixed with alfalfa, it is recommended that the alfalfa not be more than 30% of the mix.

Evidence has also pointed out that nitrogen, in some instances substantial amounts, are lost through sweat – in the range of 1 – 1.5g of N/kg. This is equivalent to a loss of over 200g of crude protein. Thus protein requirements can also be influenced by the environment in addition to duration and intensity of the exercise.(4) Quality of the dietary protein plays a role in the performance of the horse. If the athletic horse's core diet has a strong protein/amino acid make-up, supplementation is not warranted. However, research is ongoing regarding certain amino acids to evaluate whether long-term effects of amino acid supplementation benefits muscle structure, repair and recovery from exercise. Arginine and ornithine have been shown to stimulate growth hormone release and promote increased lean tissue.

Forage

It is recommended that forage be the foundation of the diet for all horses. This should include feeding 1.5 – 2.0 kg (DMB) per 100kg of bodyweight of chopped fiber or long stem roughage. Overly mature forages reduce digestibility and have the potential to reduce water reservoir capacity and therefore should be avoided.

Calcium rich forages (e.g., alfalfa) should be avoided as discussed above and as calcium rich forages are perceived to increase the risk of "thumps' or synchronous diaphragmatic flutter, especially during endurance rides.

Antioxidants and Electrolytes

Adequate intake of antioxidants is recommended. Additionally increased intake of Vitamin E may be advantageous and studies are ongoing to see the correlation between increased Vitamin E and improved equine performance. Electrolytes must be monitored closely in equine athletes. As mentioned earlier increased sweat will result in electrolyte losses. If a horse loses body water the osmolarity of its body fluids increase and are hypertonic. The major electrolytes contributing to extracellular fluid in the equine are sodium and chlorine. The major electrolyte in intracellular fluid is potassium. Typically it is extracellular fluid loss for which the healthcare team needs to be concerned. However if the horse is dehydrated, we must be cognizant of the potential for intracellular loss. Replacement of electrolytes is best achieved through feeding – typically by placing the amount of electrolytes lost in the feed. During competition the best way to replace electrolyte loss is through electrolyte solutions. The goal of offering these solutions prior to competition is to insure the horse is hydrated before the competition and to insure the horse will drink the electrolyte solution the day of competition.

The equine athlete is evolving faster and faster in today's society and along with it the optimal nutritional management for equine athletes. Veterinary technicians must look at each athlete on an individual basis, as each equine will react to exercise differently. We must monitor the key nutritional factors to insure proper performance and equine health, just as we do in canine and human athletes.

Conclusion

As with any other species, excellent nutrition cannot overcome poor genetics or inadequate training for either dogs or horses. Improving nutrition can help to provide that little extra boost for an excellent animal that has been trained and conditioned to perform to its full potential. Improving nutrition can also allow these athletes to perform for a longer period of time and sustain fewer injuries, which is in everyone's best interest.

REFERENCES

1. Toll PW, Reynolds AJ. The Canine Athlete. In Hand MS, Thatcher CD, Remillard RL, Roudebush P (eds) *Small Animal Clinical Nutrition. Fourth Edition*. Marceline, Missouri: Walsworth Publishing, 2000: 261-283, 359
2. Case LC, Carey DP, Hirakawa DA, Daistotle L. Performance and Stress. In *Canine and Feline Nutrition, A Resource for Companion Animal Professionals*, Second Edition. St Louis, Missouri: Mosby Publishing, 2000: 259-273
3. Nielsen, BD. Practical considerations for feeding racehorses. In *Equine Applied and Clinical Nutrition*, Geor RJ, Harris PA, Coenen M, eds. Elsevier Saunders, 2013,pp. 261- 271. St. Louis.
4. Harris PA, Schott HC. Nutritional management of elite endurance horses. In Equine *Applied and Clinical Nutrition*, Geor RJ, Harris PA, Coenen M, eds. Elsevier Saunders, 2013,pp. 272- 288. St. Louis
5. National Research Council (NRC): Nutrient requirements of horses. 6th revised edition, 2007, National Academy Press, Washington DC.
6. Marlin D, Nankervis K. Equine Exercise Physiology. 2007. Blackwell, Ames, IA.

FELINE LOWER URINARY TRACT DISEASE: NEW NUTRITIONAL APPROACHES

Kara M. Burns, MS, MEd, LVT,VTS (Nutrition)
Wamego, KS

Feline lower urinary tract disease (FLUTD) is a term used to describe any condition affecting the urinary bladder or urethra of cats and is a common reason for hospital visits and veterinary evaluation of our feline patients. Regardless of underlying cause, FLUTD is characterized by the following signs: dysuria, pollakiuria, stranguria, hematuria, and/or periuria (urination in inappropriate places). It is important that veterinary technicians are aware of signs and symptoms of FLUTD when talking with clients.

Over the past 10 years knowledge of specific causes of FLUTD has increased in the veterinary profession, allowing diagnostic and therapeutic efforts to be directed toward identification and elimination of specific underlying disorders. The most common cause of FLUTD in cats < 10 years of age is feline idiopathic cystitis (FIC), followed by uroliths, and urethral plugs. A diagnosis of FIC is made by excluding all other causes of FLUTD. In older cats (> 10 years), urinary tract infection and/or uroliths are the most common cause of FLUTD.

In 1981, 78% of feline uroliths were composed of struvite and only 2% were calcium oxalate. In the mid- to late-1980s, the occurrence of calcium oxalate uroliths began to increase. Between 1994 and 2002, approximately 55% of uroliths were calcium oxalate and only 33% were struvite. Since 2001, however, the number of struvite uroliths has continued to increase while occurrence of calcium oxalate uroliths has decreased. Based on 10,093 feline uroliths analyzed at the Minnesota Urolith Center in 2006, the most common mineral types were struvite (50%) and calcium oxalate (39%), followed by purine (5%). These trends have continued into 2011. In 2006, 88% of urethral plugs evaluated at the Minnesota Urolith Center were composed of struvite, 9% were matrix, < 1% were calcium oxalate, and 2% were of other mineral compositions.

Diagnostic Evaluation

Diagnostic evaluation of cats with recurrent or persistent lower urinary tract signs should include a urinalysis and diagnostic imaging. If there is a history of urinary tract manipulation (e.g., urethral catheterization), evidence of urinary tract infection (e.g., pyuria, bacteriuria, malodorous urine), or the cat is older (usually > 10 years), a urine culture should be done. More advanced procedures (e.g., contrast radiography) are appropriate in some cases.

Urinalysis is an important part of evaluating patients with signs of lower urinary tract disease. Ideally, the veterinary technician should perform the urinalysis in-house since fresh urine samples analyzed within 30 minutes of collection are preferred. Urine specimens evaluated after this time may form crystals that are not in fact present in the patient. Samples may be refrigerated for up to 8 hours and then evaluated (after the sample has returned to room temperature). This method is not best for evaluating crystalluria and should be avoided as much as possible by the veterinary technician.

Although it may be tempting to only perform dipstick analysis, measure urine specific gravity, and omit urine sediment examination, it is very important to perform complete urinalysis. The only way to accurately detect pyuria, hematuria, bacteriuria, and crystalluria is by sediment examination. You cannot rely solely on urine dipstick analysis because results for detection of pyuria are often false positive in cats and the occult blood reagent pad on the dipstick is not specific for hematuria (in addition to red blood cells, it also becomes positive with hemoglobin and myoglobin). Pyuria (> 5 WBCs/hpf) indicates inflammation and it may be caused by several disorders (urolithiasis, bacterial infection); it is less commonly observed in cats with FIC. If you see increased numbers of white cells, you should look carefully for bacteria. Take care not to misinterpret presence of cellular debris and Brownian motion as bacteriuria. It is the veterinary technician that really has an impact on this portion of the FLUTD evaluation!

Several different types of crystals may be identified on urine sediment examination, but typically struvite (triple phosphate) and calcium oxalate are the most common. The presence of crystals indicates that the urine is supersaturated with that substance and the patient is at risk for forming uroliths. Remember that cats also may have crystals and never develop uroliths. In the absence of other findings such as uroliths or urethral plugs, the presence of crystals alone is not diagnostic of urolithiasis or struvite disease. Struvite crystals may be present in normal cats and cats with struvite uroliths (sterile or infection-induced), non-struvite urolithis (including some cats with calcium oxalate uroliths), urethral plugs, or other urinary disorders such as FIC.

Survey radiographs are helpful for identifying radiopaque uroliths and crystalline-matrix urethral plugs. Remember during positioning to include the caudal abdomen (urethra) in the radiograph, or you risk missing potentially important information. Normal survey radiographs do not exclude FIC, radiolucent uroliths (urate/purine), small uroliths (< 2 mm), neoplasia, blood clots, or anatomic defects. Abdominal ultrasonography and/or contrast urethrocystography is helpful in these cases. If no cause is identified after thorough diagnostic evaluation, a diagnosis of FIC is very likely.

Treatment of Cats with Feline Idiopathic Cystitis

The goals of managing cats with FIC are to decrease severity of clinical signs and increase the interval between episodes of lower urinary tract disease. Over the past 40 years, many different treatments have been recommended to control signs in cats with FIC, yet only a few have been evaluated in clinical trials of cats with FIC.

Nutritional Management

It has been found that feeding moist food (>60% moisture) has been associated with a decreased recurrence of clinical signs in cats with FIC. During a 1-year study, clinical signs recurred less often in cats with FIC when fed a moist food compared with cats fed the dry formulation of the same food. Beneficial effects have been observed in cats with FIC when urine specific gravity values decrease from 1.050 to values between 1.032 and 1.041. Veterinary technicians should be aware of and discuss with clients additional methods for increasing water intake (eg, adding broth to foods, placing ice cubes in the cat's water, and providing water fountains) as these may be useful for some cats.

A recent study shows that consistently feeding a therapeutic urinary food was associated with a reduction in recurrent episodes of FIC signs. This is the first study to definitively show that foods of different nutritional profiles impact the expression of acute episodes of FIC signs in cats. Additionally, the addition of L-tryptophan, a precursor of serotonin that inhibits neurotransmitters in the brain to balance mood, as well as hydrolyzed casein, a bioactive peptide that helps relieve anxiety in cats has been presented as nutrients that will aid in managing the stress component of FIC.

Increasing salt content of food is an effective method of causing urine dilution in cats, but the potential for adverse effects should be considered. At this time, there are differing opinions regarding role of sodium in cats with kidney disease. In a recent study, the effects of high-salt [1.2% sodium, dry matter basis (DMB)] intake for 3 months were evaluated in 6 cats with mild azotemia due to naturally occurring chronic kidney disease. These cats had progressive increases in BUN, serum creatinine, and serum phosphorus compared with consumption of food with 0.4% sodium (DMB). Based on all findings to date, further study is needed to better determine the role of sodium in healthy cats fed long-term as well as cats with hypertension, chronic kidney disease, and calcium oxalate uroliths. Pending further studies, it is sensible to avoid high-salt foods in cats with chronic kidney disease and monitor kidney function when high-salt foods are fed to cats at risk for kidney disease.

Inflammation plays a role in many causes of FLUTD, especially FIC and urolithiasis. Therefore, a key nutritional factor for managing cats with FLUTD includes omega-3 fatty acids, which are known to have potent anti-inflammatory effects. In addition, vitamin E and beta-carotene are helpful for counteracting oxidative stress and reducing free radical damage, conditions that often accompany inflammation.

Environmental Enrichment

In addition to nutritional management, the currently recommended treatment for cats with FIC also includes environmental enrichment and stress reduction. This is crucial in an FIC treatment plan and the veterinary technician should be readily able to discuss environmental enrichment with the client.

A recent prospective study evaluating effects of multimodal environmental modification was reported in 46 client-owned cats with FIC. The findings showed significant reductions in lower urinary tract signs, fearfulness, and nervousness after treatment for 10 months. With cats that are suffering with FIC, stressful situations (e.g., conflict with other cats in the home) should be avoided or minimized. Owners should provide opportunities for play/resting (horizontal and vertical surfaces for scratching, hiding places, and climbing platforms). Any changes (eg, switching to a new food) should be made gradually so the cat has adequate time to adapt and avoid becoming stressed.

Another critical component of managing cats with FLUTD, especially FIC, involves appropriate use and maintenance of litter boxes in the home. The majority of cats prefer clumping, unscented litter; however, it may be necessary to give cats several choices and let them select their preference. It may be possible to have cats within the home that prefer different types of litter or litter boxes. In general, uncovered litter boxes are recommended because they are less likely to trap odors inside. For older cats with mobility issues, the owner should select a litter box with low sides to facilitate the cat getting in and out of the box. Litter boxes should be scooped daily and washed every few weeks with warm, soapy water. Because plastic can absorb odors over time (months to years), owners should consider replacing litter boxes with new ones periodically. Finally, there should be an adequate number of litter boxes (the 1 + 1 rule = 1 more than the number of cats) in the home and they should be located on multiple floors where cats can enter and exit readily. More detailed information about environmental enrichment and litter box management is available in the suggested reading. It may be helpful to encourage owners to read this additional information as well because their involvement is critical for a successful outcome. Finally, health care team members, especially technicians, play a crucial role in educating cat owners about the importance of environmental enrichment and litter box management.

Managing Cats with Struvite Uroliths or Urethral Plugs

Treatment options for cats with struvite uroliths include physical removal of uroliths or dissolution via nutritional management. Only two foods have been evaluated in cats with struvite uroliths (Prescription Diet® s/d® Feline, Hill's Pet Nutrition, Topeka, KS and Medi-Cal® Dissolution Formula, Veterinary Medical Diets, Guelph, Ontario, CANADA); mean time required for dissolution of sterile struvite uroliths using these foods is approximately 1 month. For cats with suspected struvite uroliths, it is appropriate to transition to feeding a canned calculolytic food over a 7-day period. Cats should be re-evaluated every 2-4 weeks (urinalysis and abdominal radiographs). Urine pH should remain < 6.1 and specific gravity should be < 1.040 if canned food is being fed exclusively. Nutritional management (dissolution) should be continued 1 month beyond radiographic resolution of the urolith.

After dissolution or removal of struvite uroliths or urethral plugs, nutritional management should continue to prevent recurrence. There are several commercially available foods for struvite prevention; however, only one (Prescription Diet® s/d® Feline, Hill's Pet Nutrition) has been evaluated in cats with struvite disease. Several other foods formulated for struvite prevention have been evaluated in healthy cats by measuring urine saturation values of struvite. A dissolution (calculolytic) food is appropriate for initial management (1-3 months) after relieving urethral obstruction; this should be followed by feeding a struvite preventive food indefinitely, with the cat being evaluated routinely by the veterinary healthcare team.

Managing Cats with Calcium Oxalate Uroliths

The treatment of choice for calcium oxalate urolithiasis is urolith removal, followed by methods to prevent recurrence. At present, the standard of care for preventing calcium oxalate urolith recurrence is to feed moist therapeutic food and encourage water intake. There are several commercially available therapeutic foods for prevention of calcium oxalate uroliths in cats.

All cats should be monitored for recurrence including urinalysis every 3 months to detect calcium oxalate crystalluria and diagnostic imaging every 6 months to detect uroliths. If uroliths recur, less-invasive procedures such as voiding urohydropropulsion are more likely to be effective when uroliths are smaller.

Summary

Increased understanding of specific causes of FLUTD has allowed diagnostic and therapeutic efforts to be directed toward identification and elimination of specific underlying disorders. The most common cause of FLUTD in cats < 10 years of age is feline idiopathic cystitis (FIC), followed by uroliths, and urethral plugs. A diagnosis of FIC is made by excluding all other causes of FLUTD. In older cats (> 10 years), urinary tract infection and/or uroliths are the most common cause of FLUTD. It is imperative that veterinary technicians have a thourough understanding of FLUTD and the how the various treatments affect the different types of FLUTD. Veterinary technicians play a very important role in the treatment of FLUTD. The history obtained from discussions with the pet owner aids in the diagnosis of FLUTD. The technicians' discussion of the treatment plan with the client is key to the client's understanding and compliance with the veterinarian's recommendation and ultimately the health of the pet.

REFERENCES

1. Lulich JP. FLUTD: Are you missing the correct diagnosis? Proc 2007 Hill's FLUTD Symposium 2007:12-19 (www.hillsvet.com/conferenceproceedings).
2. Forrester, SD. FLUTD: How Important is it? Proc 2007 Hill's FLUTD Symposium 2007:5-11 (www.hillsvet.com/conferenceproceedings).
3. Burns KM, Forrester SD. Feline Lower Urinary Tract Disease. NAVTA Journal. July/August 2007.
4. Markwell PJ, Buffington CA, Chew DJ, et al. Clinical evaluation of commercially available urinary acidification diets in the management of idiopathic cystitis in cats. *J Am Vet Med Assoc* 1999;214:361.
5. Kirk CA, Jewell DE, Lowry SR. Effects of sodium chloride on selected parameters in cats. *Vet Ther* 2006;7:333-346.
6. Kruger JM, Lulich JP, Merrills J, et al. A Year-Long prospective, randomized, Double-masked study of nutrition on Feline Idiopathic Cystitis.Proceedings. American College of Veterinary Internal Medicine Forum 2013.
7. Buffington CA, Westropp JL, Chew DJ, et al. Clinical evaluation of multimodal environmental modification (MEMO) in the management of cats with idiopathic cystitis. *J Feline Med Surg* 2006;8:261-268.
8. Westropp JL, Buffington CAT, Chew D. Feline lower urinary tract diseases In: Ettinger SJ, Feldman EC, eds. Textbook of Veterinary Internal Medicine, 6th ed. Philadelphia: Elsevier Saunders, 2005;1828-1850.
9. Neilson JC. FLUTD: When should you call the behaviorist? *Proc 2007 Hill's FLUTD Symposium* 2007:20-28 (www.hillsvet.com/conferenceproceedings).
10. The Indoor Cat Initiative (www.vet.osu.edu/indoorcat) 2007.
11. Bohnenkamp G. *From the Cat's Point of View - The Complete Book on Cat Behavior*, 1991.
12. Lulich JP, Osborne CA, Lekcharoensuk C, et al. Effects of diet on urine composition of cats with calcium oxalate urolithiasis. *J Am Anim Hosp Assoc* 2004;40:185-191.
13. Grauer, G. Current Thoughts on Pathophysiology & Treatment of Feline Idiopathic Cystitis. Today's Veterinary Journal. Nov/Dec 2013, pp 38 – 41.
14. Osborne C, Lulich J, et al. Feline urolith epidemiology update: 1981 to 2011. Tracking the trends of mineral composition in cats with urolithiasis. DVM360. June, 2013.
15. Pereira GG, Fragoso S, Pires E. Effect of dietary intake of l-tryptophan supplementation on multi-housed cats presenting stress related behaviors. Proeedings, BSAVA 2010.
16. Beata C, Beaumont-Graff E, Coll V, et al. Effect of alpha-casozepine (Zylkene) on anxiety in cats. Journal of Veterinary Behavior, 2007; 2(2):40-46.

MODERN VETERINARY RADIATION ONCOLOGY

Michael W. Nolan, DVM, PhD, DACVR (Radiation Oncology)
Raleigh, NC, USA

INTRODUCTION

This lecture will introduce the veterinary technician to various types of radiotherapy (external beam radiotherapy, injectable radiation, surface applicators and brachytherapy) commonly used in clinical practice for veterinary patients. It will also provide an overview of the role veterinary technician's play as radiation therapists.

BASIC RADIATION BIOLOGY AND PHYSICS

Radiation therapy (RT) involves the transfer of energy from a radiation source to the patient. That energy (most often in the form of X-rays and/or electrons) targets the nucleus of cells, where it can damage DNA and prevent cellular division. Since cancer is a process of dysregulated cellular division, RT is an effective anti-cancer treatment.

All cells are at risk for radiation-induced damage. Fortunately though, the cells of most normal tissues are better able to repair radiation damage than the cells of tumors. Furthermore, in almost all cases, RT is delivered in a targeted manner, such that the tumor receives a high dose of radiation, while most normal tissues in the body remain unexposed. These two factors are what make RT not only effective, but also safe.

GOALS OF RADIATION THERAPY

RT can be delivered with either definitive- or palliative-intent.[1] **Definitive-intent RT** typically involves giving a high total dose of radiation, using daily treatments for 3-4 weeks. While the risk for early (acute) toxicity is high, and these side effects can be unpleasant for the patient, they tend to resolve quickly with supportive care. The dose per fraction (treatment session) is low, which keeps the risk for serious long term side effects low. The goal of such treatment is to provide local tumor control for as long as possible. **Palliative-intent RT** delivers a lower total dose, and fewer treatment sessions. Acute toxicity is usually mild, if at all present. This type of RT is intended to improve quality of life, without focusing on longevity. The goals of palliative-intent RT are often to decrease tumor-associated pain and inflammation, dry up effusions and bleeding, and shrink tumors to improve function.

TYPES OF RADIOTHERAPY

External Beam Radiotherapy

The most common way to deliver RT is through an external beam. High energy (megavoltage) radiation is given using specialized machines, such as a cobalt machine, or more commonly, a linear accelerator. External beam RT (EBRT) is versatile, and can be used to treat a variety of tumor types, in almost any location in the body. EBRT encompasses a wide range of therapies, including conventional EBRT, electron beam RT, 3-dimensional conformal RT, intensity-modulated RT, image-guided RT and stereotactic RT (also known as stereotactic radiosurgery).

Injectable Radiation

Radioactive iodine (I-131) can be used to destroy thyroid tissue, without harming other tissues in the body. It is usually given as a subcutaneous injection. Patients having received this treatment will be radioactive for period, but the iodine passes out of the body within a few days. It is commonly used to treat feline hyperthyroidism. It can also be used for treatment of canine thyroid carcinoma. Though many facilities are licensed to treat cats with I-131, few are capable of treating dogs.[2]

Surface Applicators

Plesiotherapy utilizes a probe with a small-diameter (typically <2 cm) radioactive tip. The probe is made of strontium-90, which emits low energy electrons that can only penetrate a few millimeters into tissue. These surface applicators are particularly useful for treatment of small, superficial tumors.[3]

Brachytherapy

Brachytherapy involves placement of a radioactive source on or in a tumor. The most commonly used material is iridium-192.[4] Like the beta rays given off by strontium, radiation given off by these sources cannot penetrate very far in tissue, meaning RT delivered by brachytherapy is essentially limited to the tumor. Although brachytherapy is commonly used in human oncology, and can be useful for a variety of veterinary cancers, its use is limited to only a few sites worldwide, because brachytherapy implants pose some safety hazards, and are tightly controlled by regulatory agencies. **Electronic brachytherapy** shares many of the benefits of traditional brachytherapy, but because it uses a low-energy, electronically-generated radiation, it is safer and easier to use. Several veterinary centers have acquired electronic brachytherapy systems, and a lot of work is currently being done to investigate how it can best be utilized to treat cancers in small animal patients, as well as horses.

ROLE OF THE RADIATION THERAPIST

RT is planned by dosimetrists, radiation oncologists and medical physicists. It is delivered by radiation therapists. Many veterinary radiation therapy centers provide veterinary technicians with training so they can act as a radiation therapist. The role of the therapist is to assist in patient setups, acquisition and analysis of port films (special radiographs used to verify appropriate patient positioning), and delivery of the prescribed radiation dose. Therapists are also responsible for periodic quality assurance testing, and record keeping. Clinically, they assist radiation oncologists in monitoring and managing acute radiation side effects.

REFERENCES

1. Farrelly, J.; McEntee, M. C., Principles and applications of radiation therapy. *Clinical Techniques in Small Animal Practice* **2003,** *18* (2), 82-87.
2. Turrel, J. M.; McEntee, M. C.; Burke, B. P.; Page, R. L., Sodium iodide I 131 treatment of dogs with nonresectable thyroid tumors: 39 cases (1990-2003). *Javma-Journal of the American Veterinary Medical Association* **2006,** *229* (4), 542-548.
3. Ware, K.; Gieger, T., Use of strontium-90 plesiotherapy for the treatment of a lingual plasmacytoma in a dog. *Journal of Small Animal Practice* **2011,** *52* (4), 220-223.
4. Walker, M. A., Interstitial implant brachytherapy in small animals. *Veterinary Clinics of North America-Small Animal Practice* **1997,** *27* (1), 59-&.

THE VETERINARY TECHNICIAN'S ROLE IN RADIATION ONCOLOGY

Michael W. Nolan, DVM, PhD, DACVR (Radiation Oncology)
Raleigh, NC, USA

INTRODUCTION

The field of veterinary radiation oncology and therapy (RO) is highly specialized, and involves diagnosis, staging and treatment of a variety of cancers. We will discuss the unique role of the veterinary technician in the RO clinic, including anesthesia specific to radiotherapy (RT) patients, nursing care, and management of side effects.

RADIATION SAFETY

The most common form of RT used in veterinary patients is external beam RT (EBRT). This involves use of an external source to deliver high energy (megavoltage) ionizing radiation to tumors in anesthetized patients. The two most commonly used machines to deliver such radiation are cobalt-60 teletherapy units, and linear accelerators. Cobalt units emit gamma rays that are produced during decay of the cobalt source, which is housed within the machine. Linear accelerators involve no radioactivity, and instead electronically produce X-rays and electrons, much like a diagnostic radiography unit.

Although beams of radiation pass through patients treated with EBRT, it is important to understand that those patients are never themselves radioactive, and as such, pose no radiologic danger to staff involved in their treatment. Additionally, while cobalt teletherapy units house a radioactive source, these machines are well-shielded; linear accelerators contain no radioactive materials, and only emit radiation when the beam is "turned on". Only the patient is allowed in the vault (a heavily shielded room housing the EBRT unit) while the beam is on; anesthetic monitoring must occur from outside the room, typically via closed-circuit cameras. All RO workers (radiation oncologists, therapists, technicians, physicists, etc.) must wear a dosimetry badge to monitor for potential exposure to radiation, and ensure that exposure does not exceed limits set by federal regulations. In reality, radiation exposures are typically extremely low in RO workers, far below those of personnel staffing diagnostic radiology and interventional cardiology departments.

It is important for RO staff to have a good understanding of radiation safety. Teaching visitors and colleagues about radiation therapy (and its risks) is an important component of the technician's job. Above all else, it is the shared responsibility of RO team members to maintain a culture of safety for both patients and people.

ANESTHESIA

Treatments are designed to ensure the tumor is adequately covered with a high dose of radiation; an equally important quality of a good radiation plan is that normal tissues surrounding the tumor are shielded as well as possible from those high doses. But a radiation plan is only as good as the ability of the therapist to deliver the radiation to the intended target. Because inaccurate patient positioning can result from small movements, and can lead to misadministration of the radiation dose, veterinary patients are anesthetized for treatment. There are several components of the typical anesthetic plan which are unique to RO patients.

It is ideal to have venous access throughout the anesthetic procedure. This can be achieved using a conventional peripheral venous catheter. However, because many definitive-intent RT protocols span several weeks, some clinics find it advantageous to use longer lasting techniques.[1] Some clinics achieve this using long centrally-placed intravenous catheters placed in the jugular vein or medial saphenous vein (PICC lines). Other institutions prefer subcutaneous venous access ports.[2,3] Clinics opting against these long-lasting catheters but still desiring venous access throughout the RT procedure may opt to leave shorter peripheral vein catheters in for 3-5 days at a time, teaching pet owners how to flush the catheters at home (most RT procedures are performed on an outpatient basis). Other clinics opt against maintenance of venous access during RT due to the short nature of most radiation treatments (5-10 minutes). In this case, they may use a temporarily placed butterfly needle or standard hypodermic needle for injection of induction agents.

Because most radiation treatments are short, it is ideal to design anesthetic protocols which allow for quick induction and recovery. Exceptions clearly exist, but propofol is the most commonly used induction agent, and either isoflurane or sevoflurane are typically used for maintenance. Pre-medications are used sparingly, and when possible, the preference is to use short-acting and/or reversible drugs (e.g., opiods and benzodiazepines).

Most small animal patients tolerate repeated daily anesthesia without a problem. One commonly discussed consideration is repeated use of propofol in cats, which has been shown to cause oxidative injury in red blood cells, which could lead to Heinz body anemia. Fortunately, most cats are amenable to gas induction, with injectable drugs used as needed. If there is a cat for which daily propofol induction seems otherwise ideal, the risk of clinically meaningful anemia is low.[4] The most common complication of daily anesthesia in RO patients is tracheitis, which results in a mild nonproductive cough and typically resolves within a few days of finishing treatments.

NURSING CARE AND SIDE EFFECTS

Radiation patients often develop diarrhea secondary to stress colitis. Though metronidazole is commonly used in general practice for stress colitis, it is often avoided in patients undergoing RT due to its actions as a radiosensitizer, which could theoretically worsen acute radiation side effects.[5] A commonly used alternative is tylosin.

Perhaps the most common side effects seen in dogs and cats during and immediately after RT are dermatitis and mucositis. Most acute radiation side effects will heal within a few weeks with no, or minimal, intervention. Supportive care is encouraged, and the mainstay of such care is appropriate analgesia, as many acute side effects can be painful. Patients are often treated with anti-inflammatories. Both corticosteroids and non-steroidal anti-inflammatory drugs can be successfully be used (though not at the same time!) in dogs and cats. Other commonly used adjunctive analgesics include tramadol, gabapentin, amantadine, acetaminophen (in dogs only) and fentanyl patches. In cases of severe acute effects, short-term hospitalization for local blocks and/or injectable pain medications may be needed at the end of a definitive course of RT. Sedatives, such as acepromazine, can be helpful for some patients.

Dermatitis rarely progresses beyond dry desquamation in cats. However, dog skin is more radiosensitive, and it is common to see moist desquamation.[6] Thought the skin will look raw and may even have serous exudates, bacterial pyoderma is rare, making it uncommon to need antibiotics. Bandaging and debriding are also discouraged, as they may worsen the effects and delay healing. Mucositis is difficult to manage. Some mouth washes may provide temporary relief of pain, and may help prevent colonization of the mouth with unwanted bacteria and fungi. Mucositis/pharyngitis is rarely bad enough to preclude patients from eating, though in the rare case of an anorexic patient, temporary feeding tubes can be useful.

RADIATION THERAPY

Many veterinary technicians working in the RO clinic will learn to function as radiation therapists. They will become intimately familiar with some interesting terminology. For example, linear accelerators emit radiation in amounts known as monitor units. But radiation is not prescribed as numbers of monitor units (MU); instead, it is prescribed in units of Gray (Gy), which equates to a Joule per kilogram of tissue. Dosimetrists, physicists and oncologists will convert Gy to MU, but the therapist must have a basic understanding of what these terms mean, and how they relate to each other. Therapists keep a written log of how many MU and Gy have been administered for each day of an RT protocol.

Technicians will also work with materials like bolus (a tissue-like material that is applied to the skin and can help distribute radiation dose more ideally), blocks, multileaf collimators (which shape the radiation beam), as well as wedges (which alter the intensity of the beam as it exits the head of the irradiator).

Finally, RO is undergoing a great evolution in technologies and techniques through which RT can be delivered to patients, so RO technicians need to become familiar with terms such as 3D-CRT, IMRT, IGRT and SRT.

REFERENCES

1. Evans KL, Smeak DD, Couto CG, et al. Comparison of 2 indwelling central venous access catheters in dogs undergoing fractionated radiotherapy. *Veterinary Surgery* 1994;23:135-142.
2. Culp WTN, Mayhew PD, Reese MS, et al. Complications associated with use of subcutaneous vascular access ports in cats and dogs undergoing fractionated radiotherapy: 172 cases (1996-2007). *Javma-Journal of the American Veterinary Medical Association* 2010;236:1322-1327.
3. Mayer MN, Grier CK, Yoshikawa H, et al. Complications associated with the use of vascular access ports in dogs receiving external beam radiation therapy. *Javma-Journal of the American Veterinary Medical Association* 2008;233:96-103.
4. Bley CR, Roos M, Price J, et al. Clinical assessment of repeated propofol-associated anesthesia in cats. *Javma-Journal of the American Veterinary Medical Association* 2007;231:1347-1353.
5. Stone HB, Withers HR. Metronidazole - Effect on radiosensitivity of tumor and normal tissues in mice. *Journal of the National Cancer Institute* 1975;55:1189-1194.
6. Flynn AK, Lurie DM. Canine acute radiation dermatitis, a survey of current management practices in North America. *Veterinary and Comparative Oncology* 2007;5:197-207.

DEALING WITH A "LUMP": WHEN IT'S OK TO WATCH IT VS. WHEN IT'S NOT!

Cecilia Robat, DVM, DACVIM (Oncology)
Madison, WI

INTRODUCTION

« All old dogs have fatty lumps, so this probably is one too! », « The dog is at the clinic for another problem so let's ignore the lump on his back /leg/head », « He doesn't seem bothered by it, so why should we do something ? », « The owner doesn't have much money so she probably won't want to do anything about the lump on her dog »

How many of these have you heard/said? These are a few of the common mistakes/misconceptions in the veterinary world. As a rule of thumb, it's always better to know what you're dealing with than to make assumptions, it's always better to do something early on if you want a higher likelihood of curing the patient, and finally, it's your/the DVM's role to educate clients, and let them make the final decision to do something or not.

There is a national move to raise awareness of the importance of checking out every lump on a dog/cat. This is headed by Dr Sue Ettinger, a veterinary oncologist in Yonkers, NY. Her motto is « See something, do something » early detection is key ! Her basic rules are: if a mass has been there for a month OR if it is larger than 1cm (« see something »), then « do something » (fine needle aspirate, or biopsy).

WHAT TO DO WHEN YOU FIND A LUMP

Inform the owner

If you notice a mass on a dog or cat, the first thing to do is point it out to the owner and ask them about it. Have they noticed it before? How long has it been there? Has it changed at all? Does the dog/cat seem bothered by it?

Body map

The next step is to « map » the mass. We use pre-drawn outlines of a dog to precisely locate the mass, and then measure it with calipers (see examples). It is very important to write the date and measurements down, so that size comparison is possible at future visits. Each time the dog has a physical examination, the masses should be re-measured, and those values noted on the body map. If a mass is growing, further diagnostics need to be performed.

Fine Needle Aspirate (FNA)

The first step in working up a mass is performing a FNA. As always, owner consent needs to be obtained prior to pursuing this procedure. There are many advantages in performing a FNA. It is an easy, cheap, diagnostic test, and is associated with little to no risks. It is usually performed without sedation or anesthesia. Preliminary results can be obtained immediately, if in-house cytology is possible, or within 24-48 hours, if the slides are sent to a local clinical pathologist. Many tumors will exfoliate well making a diagnosis easy. Disadvantages include the fact that a definitive diagnosis may not always be obtained due to the small sample size and the lack of architectural information. Additionally, some sarcomas (mesenchymal tumors) will not readily shed cells, making a definitive diagnosis difficult to obtain. This test should be encouraged each time a mass is discovered on a patient. Once a diagnosis (definitive or preliminary) is obtained, then an appropriate therapeutic plan can be discussed.

Biopsy

This procedure involves obtaining a small piece of tumor. There are several different biopsy techniques that can be considered, including TruCut biopsies, punch biopsies, incisional and excisional biopsies. The decision of which technique to use will depend on the location, size, and shape of the mass. The advantages of performing a biopsy are the fact that there is the almost certainty of achieving a diagnosis, and more detailed information such as subtype and grade of a tumor can be obtained. The disadvantages of biopsy include the need for, at minimum, sedation and local anesthesia, but usually general anesthesia. Additionally, the cost of performing a biopsy is higher, the morbidity associated with the procedure is usually higher, and time to diagnosis is longer (usually 1 week). The decision to perform a biopsy usually comes when a FNA is non-diagnostic, or further information on the tumor is needed (such as subtype or grade) in order to make an optimal therapeutic plan.

COMMON CUTANEOUS/SUBCUTANEOUS TUMORS IN DOGS

Mast cell tumor

Mast cell tumors (MCTs) are the most common cutaneous tumor in dogs. They can present in all breeds, all sexes and ages. Common breeds include Boxers, Pugs, Boston Terriers. They can look like anything – so beware! MCT can be cutaneous or subcutaneous, and owners may describe that the mass is sometimes larger, and then gets smaller. Redness, swelling, and bruising of the mass can be reported as well. This is due to bioactive substances (histamine and heparin) released by the tumor cells. Fine needle aspirates are usually diagnostic, as the cells obtained are very characteristic (round cells containing many granules). After a MCT aspirate, it is not uncommon to see a small area of redness/bruising.{Henry, 2013 #482}

MCT are divided into 3 grades based on the Patnaik system (high (3), intermediate (2), or low (1)), or, into 2 grades based on the Kiupel system{Kiupel, 2011 #418} (Low/High). Information on the grade can only be obtained on histopathology. Most low-grade tumors/grade 1 and most grade 2 tumors carry an excellent prognosis with local control (surgical excision+/- radiation therapy) alone. Grade 3/some grade 2 tumors/high grade tumors carry a guarded prognosis. Their metastatic rate is high, with most common metastatic sites being the liver, spleen and lymph nodes. Median survival times are around 1 year for grade 3 MCT with local control and chemotherapy{Webster, 2008 #430}.

Soft tissue sarcoma

Soft tissue sarcomas (STSs) are a group of tumors which all present in a similar way, are treated the same and carry the same prognosis. They include histologies such as fibrosarcomas, peripheral nerve sheath tumors, hemangiopericytomas etc….STSs present as a subcutaneous mass, commonly on the limbs, although they can be found anywhere on the body. They are more common in older dogs, and no breed or sex predilection is reported. They are pseudo-encapsulated, meaning that the bulk of the tumor will easily shell out when surgery is performed. However STSs send out projections from the main tumor into the surrounding tissues, and the only way of achieving adequate local control is by removing the main tumor as well as 3cm of tissue around it.{Bacon, 2007 #484} STSs are divided into three grades, low (1), intermediate (2), or high (3). Local control remains the biggest challenge with these tumors, as even the highest grade tumors (grade 3), only have a 40% chance of metastasizing. The treatment of choice is surgical excision +/- radiation therapy{Burton, 2011 #483;Forrest, 2000 #486} to ensure optimal local disease control.

Lipoma

These are common subcutaneous tumors in older dogs. They are commonly soft, and can be very moveable. They are benign accumulations of adipose tissue. A lipoma will not undergo malignant transformation into a liposarcoma. Overtime lipomas may grow slightly, but any "lipoma" that suddenly starts to grow fast should be reevaluated with FNA or biopsy. Lipomas do not need to be surgically removed unless they become very large and bother the patient, or if they are in an area that causes mechanical obstruction. Some lipomas grow in between muscle masses (intermuscular lipomas){Case, 2012 #487} or grow into muscle, fascia etc (infiltrative lipoma){McEntee, 2000 #489}. These tumors are benign and will not metastasize, however they can cause mechanical trouble and discomfort to the patient. Recurrence is common after surgery, and radiation may be considered.

COMMON CUTANEOUS/SUBCUTANEOUS TUMORS IN CATS

Injection Site Sarcomas

Injection site sarcomas (ISSs) are also known as vaccine-associated sarcomas. They have most commonly been reported at the site of rabies or FeLV vaccinations, but are also seen after injections of long-acting steroids or other products. They can develop 4 weeks to 10 years post injection. The pathogenesis is not fully understood. The Vaccine Associated Sarcoma Task Force has rules for intervention (3-2-1 rules): if a mass is still growing after 1 month or if it measures more than 2cm, or if it is still present after 3 months, then further work-up needs to be performed. ISSs are poorly encapsulated and infiltrate along facial planes. The most aggressive approach to these tumors includes performing advanced imaging (CT scan) prior to wide surgical resection of the mass (with 3-5cm margins), as well as radiation therapy. Even with this aggressive approach, 30-40% tumors will still recur. Survival times range from a few months to several years, depending on the size and grade of the tumor, the extent of surgical resection, use or not of radiation therapy{Ladlow, 2013 #490}. They are a very frustrating disease to treat, and veterinarians are encouraged to follow rules for vaccinating cats on distal limbs, or on the distal tail.{Hendricks, 2013 #491}

Mast cell tumor

In cats, most mast cell tumors (MCTs) are cutaneous, do not invade into the subcutaneous tissues, and are readily treated with marginal surgical excision. They tend to be slightly raised, alopecic, pink to pale yellow, and usually measure less than 10mm. Cats may develop several cutaneous MCTs over time. Rarely, cutaneous lesions represent spread from abdominal disease (splenic or intestinal MCT). Feline MCTs are not graded[1].

CASE EXAMPLES

Several cases will be presented during the lecture.

CONCLUSION

It is vital to body map every patient that comes through the door. Early intervention is key if a cure wants to be achieved. Only work with facts- you can NEVER diagnose a mass based on its appearance, and it's not because a patient has had 10 lipomas in the past that the 11th mass will not be an aggressive neoplasia.

REFERENCES

1. Henry C, Herrera C. Mast cell tumors in cats: clinical update and possible new treatment avenues. J Feline Med Surg 2013;15:41-47.
2. Kiupel M, Webster JD, Bailey KL, et al. Proposal of a 2-tier histologic grading system for canine cutaneous mast cell tumors to more accurately predict biological behavior. Vet Pathol 2011;48:147-155.
3. Webster JD, Yuzbasiyan-Gurkan V, Thamm DH, et al. Evaluation of prognostic markers for canine mast cell tumors treated with vinblastine and prednisone. BMC veterinary research 2008;4:32.
4. Bacon NJ, Dernell WS, Ehrhart N, et al. Evaluation of primary re-excision after recent inadequate resection of soft tissue sarcomas in dogs: 41 cases (1999-2004). J Am Vet Med Assoc 2007;230:548-554.
5. Burton JH, Mitchell L, Thamm DH, et al. Low-dose cyclophosphamide selectively decreases regulatory T cells and inhibits angiogenesis in dogs with soft tissue sarcoma. J Vet Intern Med 2011;25:920-926.
6. Forrest LJ, Chun R, Adams WM, et al. Postoperative radiotherapy for canine soft tissue sarcoma. J Vet Intern Med 2000;14:578-582.
7. Case JB, MacPhail CM, Withrow SJ. Anatomic distribution and clinical findings of intermuscular lipomas in 17 dogs (2005-2010). J Am Anim Hosp Assoc 2012;48:245-249.

8. McEntee MC, Page RL, Mauldin GN, et al. Results of irradiation of infiltrative lipoma in 13 dogs. Vet Radiol Ultrasound 2000;41:554-556.
9. Ladlow J. Injection site-associated sarcoma in the cat: treatment recommendations and results to date. J Feline Med Surg 2013;15:409-418.
10. Hendricks CG, Levy JK, Tucker SJ, et al. Tail vaccination in cats: a pilot study. J Feline Med Surg 2013.

NEW TOOLS TO TREAT CANCER: VACCINES, METRONOMIC CHEMOTHERAPY AND TYROSINE KINASE INHIBITORS

Cecilia Robat, DVM DACVIM (Oncology)
Madison, WI

INTRODUCTION

Over the past few years, there have been many advances in the treatment of cancer in pets. In addition to surgery, conventional chemotherapy and radiation therapy, which have historically been the mainstays of cancer treatment, novel therapeutic approaches have been developed, providing exciting options to fight cancer. In this talk, we will discuss the top three new therapies for canine and feline cancer: the "melanoma vaccine", "metronomic chemotherapy" and tyrosine kinase inhibitors.

MELANOMA VACCINE

Melanomas are the most common oral tumor in dogs. They are highly metastatic, to the lungs, and local draining lymph nodes. They can present as a pigmented mass (2/3 cases), or, sometimes as an amelanotic (without pigment) melanoma (1/3 cases). Some breeds are more commonly affected, such as Cocker Spaniels, Golden Retrievers, Poodles. The treatment of choice is surgery (+/- radiation therapy) followed by medical therapy to slow down metastatic disease. Unfortunately, chemotherapy has not proven extremely effective in controlling melanoma metastasis, and many patients will die of pulmonary metastasis within a year of diagnosis.

Melanomas are very immunogenic tumors, meaning that they elicit a strong immune response. Using the immune system to fight melanomas is an interesting and logical approach to this disease. Several therapeutic vaccines have been evaluated, and one is available commercially. The Merial Melanoma vaccine received full USDA approval in 2011. It is labeled to extend median survival times in dogs with stage II or III oral malignant melanoma that have achieved local disease control via surgery or radiation therapy. It is a xenogeneic (from a different species) DNA vaccine, encoding human tyrosinase. Tyrosinase is an enzyme that is essential in the synthesis of melanin. It is commonly overexpressed in melanomas. When the vaccine is administered intradermally to canine patients, an immune response against the human tyrosinase occurs. Human tyrosinase is similar enough to canine tyrosinase that the immune response will also cross-target to canine tyrosinase, and therefore attack the canine melanoma cells[1,2].

The vaccine is administered every other week for 4 treatments, and a booster is given every 6 months. The cost is around $400 per dose. Results have been encouraging, and the vaccine is very safe. Further studies are necessary to fully understand the role and efficacy of the Merial melanoma vaccine.

METRONOMIC CHEMOTHERAPY

The term "metronomic chemotherapy" was first used 15 years ago, and is opposed to that of "conventional chemotherapy".

Conventional chemotherapy indicates the use of maximum tolerated dose therapy, ie the highest possible non-lethal dose. With this approach, chemotherapy treatments are administered every 1-3 weeks, and the goal is to kill the neoplastic cells. Eventually resistance develops.

Metronomic chemotherapy describes the administration of low-dose chemotherapy on a daily basis. The goal of metronomic chemotherapy is to kill endothelial cells of blood vessels- it is an antiangiogenic therapy. It is not a tumor specific therapy. It also appears to inhibit T-regulatory cells, which are cells that typically dampen the immune system to avoid auto-immune responses in healthy patients, but are recruited by tumor cells to decrease the body's anti-tumor immune response in patients with cancer[3].

Commonly used metronomic protocols include the use of low-dose cyclophosphamide +/- a non-steroidal anti0infammatory drug (NSAID)[4]. Other chemotherapy agents used in a metronomic fashion are CCNU and chlorambucil[5-7]. Side effects are very uncommon with metronomic chemotherapy, and the cost is low. It is important to monitor patients on long-term cyclophosphamide therapy for the development of sterile hemorrhagic cystitis. If this occurs, the drug should be discontinued immediately.

Limitations of the use of metronomic chemotherapy include the lack of complete understanding of the effects, as well as a relatively empiric dose and schedule. Further studies are necessary to evaluate this therapy as well as combination therapy with conventional-dose chemotherapy, NSAIDs and other drugs such as tyrosine kinase inhibitors.

TYROSINE KINASE INHIBITORS

Tyrosine kinase receptors are a family of cell-surface receptors that can either exist in an inactive state, or can be activated by binding of a ligand. Once turned-on, these receptors activate downstream pathways in the cell that will result in cell growth, division, migration, differentiation, etc. In certain cancers, one of these receptors is abnormal, as a result of gene mutation for example. The downstream pathways are erratically activated, and cells are unstable and prone to grow, divide, migrate when this is actually undesired. An abnormal proliferation of cells leads to cancer development[8].

Certain of these receptors have a very important role in angiogenesis (new blood vessel formation). Examples of these are VEGFR (vascular endothelial growth factor receptor) and PDGFR (platelet derived growth factor receptor). These are potentially dysregulated in many different cancer types such as hemangiosarcoma, osteosarcoma etc…

c-kit is a tyrosine kinase receptor expressed on certain normal cells (hematopoietic stem cells, melanocytes, mast cells). C-kit is activated by binding of the ligand called Stem Cell Factor (SCF). This receptor is abnormal in 30-50% mast cell tumors (MCT) in dogs. This is due to a mutation in the gene encoding the receptor (KIT). The mutation results in a constitutively activated c-kit receptor, meaning that the receptor does not need SCF to bind to it to be activated. Tumors with mutated KIT tend to act more aggressively, and carry a worse prognosis[9-11].

Palladia (toceranib) is a tyrosine kinase inhibitor (TKI) approved for the treatment of MCTs in dogs. It inhibits c-kit as well as PDGFR and VEGFR. It is given orally, every other day. It is efficacious at lower than label doses, and is typically used at a dose of 2.5-2.75mg/kg every other day (or Monday-Wednesday-Friday). Mutated MCTs respond better to toceranib than non-mutated tumors. Side effects include mild gastrointestinal upset, hypertension, increases in liver values. The cost for a 20kg dog is around $200-250 per month.

Kinavet (masitinib) is another TKI. It only has a conditional license in the USA, for recurrent or non-resectable grade 2 or 3 MCTs that have received no prior radiation or chemotherapy (except steroids). It is a federal offense to use it off label. It is given orally, every day, at a dose of 12.5mg/kg. Side effects include gastrointestinal upset, proteinuria, elevated liver values and myelosuppression. The cost for a 20kg dog is around $350 per month.

TKIs are an exciting new tool in veterinary oncology. Further studies are necessary to clarify indications for these drugs, as they seem to have positive effects on other tumors than MCTs as well. Future studies will evaluate their role alone or in combination with conventional therapy, and radiation therapy.

CONCLUSION

While we are still not curing most cancers in dogs and cats, we are making slow and steady progress. The past 20 years have provided oncologists and general practitioners with many new tools to treat cancer. As our knowledge of cancer pathogenesis advances, we will no doubt continue to see the advent of novel therapies, taking us a step closer to a cure!.

REFERENCES

1. Ottnod JM, Smedley RC, Walshaw R, et al. A retrospective analysis of the efficacy of Oncept vaccine for the adjunct treatment of canine oral malignant melanoma. Vet Comp Oncol 2013;11:219-229.
2. Grosenbaugh DA, Leard AT, Bergman PJ, et al. Safety and efficacy of a xenogeneic DNA vaccine encoding for human tyrosinase as adjunctive treatment for oral malignant melanoma in dogs following surgical excision of the primary tumor. Am J Vet Res 2011;72:1631-1638.
3. Burton JH, Mitchell L, Thamm DH, et al. Low-dose cyclophosphamide selectively decreases regulatory T cells and inhibits angiogenesis in dogs with soft tissue sarcoma. J Vet Intern Med 2011;25:920-926.
4. Elmslie RE, Glawe P, Dow SW. Metronomic therapy with cyclophosphamide and piroxicam effectively delays tumor recurrence in dogs with incompletely resected soft tissue sarcomas. J Vet Intern Med 2008;22:1373-1379.
5. Leach TN, Childress MO, Greene SN, et al. Prospective trial of metronomic chlorambucil chemotherapy in dogs with naturally occurring cancer. Vet Comp Oncol 2012;10:102-112.
6. Tripp CD, Fidel J, Anderson CL, et al. Tolerability of metronomic administration of lomustine in dogs with cancer. J Vet Intern Med 2011;25:278-284.
7. Schrempp DR, Childress MO, Stewart JC, et al. Metronomic administration of chlorambucil for treatment of dogs with urinary bladder transitional cell carcinoma. J Am Vet Med Assoc 2013;242:1534-1538.
8. London CA. Kinase dysfunction and kinase inhibitors. Vet Dermatol 2013;24:181-187 e139-140.
9. Takeuchi Y, Fujino Y, Watanabe M, et al. Validation of the prognostic value of histopathological grading or c-kit mutation in canine cutaneous mast cell tumours: a retrospective cohort study. Vet J 2013;196:492-498.
10. Thompson JJ, Yager JA, Best SJ, et al. Canine subcutaneous mast cell tumors: cellular proliferation and KIT expression as prognostic indices. Vet Pathol 2011;48:169-181.
11. London CA, Malpas PB, Wood-Follis SL, et al. Multi-center, placebo-controlled, double-blind, randomized study of oral toceranib phosphate (SU11654), a receptor tyrosine kinase inhibitor, for the treatment of dogs with recurrent (either local or distant) mast cell tumor following surgical excision. Clin Cancer Res 2009;15:3856-3865.

SEPSIS IN DOGS &CATS: PATHOPHYSIOLOGY AND RECOGNITION

Amy E. DeClue, MS, DVM, DACVIM (SAIM) and Dennis R. Spann, DVM, DACVIM (SAIM)
Columbia, MO and Roseville, CA

INTRODUCTION

Sepsis is derived from the Greek word sepo, a word meaning “decomposition of animal or vegetable organic matter in the presence of bacteria”. In the clinic, sepsis has come to mean infection with concurrent clinical evidence of the systemic inflammatory response (i.e., SIRS because of infection). At one time, sepsis and bacteremia were considered synonymous terms. However, as our knowledge of sepsis has grown, it has become evident that the clinical manifestations of sepsis relate to wide-spread inflammation and not necessarily wide-spread infection. Put simply, sepsis is an immunologic reaction to any type of infection: either localized or systemic in nature. Terminology used to describe the severity of sepsis and SIRS are listed in the box below.

In dogs and cats, bacterial infections are the most cause of sepsis. Gram-negative bacteria are the most common type of bacteria cultured, with *E. coli* being the most frequent isolate. However, any microbial organism (e.g., fungus, parasite, virus) may cause sepsis. For canine patients, sepsis most commonly originates from the gastrointestinal tract (septic peritonitis), followed by the respiratory tract. In cats, sepsis is commonly associated with septic peritonitis, pyothorax and hepatic abscessation.

> *Systemic Inflammatory Response Syndrome (SIRS)* – the complex clinical response to a nonspecific insult of infectious or noninfectious origin. In dogs, the diagnosis of SIRS is based on fulfillment of at least 2 of 4 criteria: tachycardia, tachypnea, hypo-or hyperthermia, and either leukocytosis, leukopenia or >3% bands. These criteria have been used to determine the severity of illness and prognosis in critically ill animals. In a study of 500 dogs presenting to a veterinary teaching hospital, mortality was significantly associated with the number SIRS criteria fulfilled
> *Sepsis* - the systemic inflammatory response to infection (i.e., SIRS because of infection)
> *Severe sepsis* – the systemic inflammatory response to infection associated with organ dysfunction and manifestations of hypoperfusion or hypotension.
> *Septic shock* – the systemic inflammatory response to infection with hypotension refractory to volume expansion.
> *Bacteremia* – the presence of bacteria in the blood stream. This is typically associated with sepsis but sepsis can develop without bacteremia.
> *Multiple organ dysfunction syndrome (MODS)* – altered function of 2 or more organs secondary to SIRS such that homeostasis cannot be maintained without intervention.
> *Acute respiratory distress syndrome (ARDS)* – a pulmonary inflammatory disorder characterized by noncardiogenic pulmonary edema, neutrophilic pulmonary inflammation and hypoxemia.

PATHOGENESIS

The sequence of events leading to sepsis is complex and incompletely understood. In the initial phases of infection, microbial products (e.g., endotoxin / lipopolysaccharide from Gram negative bacteria; exotoxins, peptidoglycans and superantigens from Gram positive bacteria; and fungal cell wall material) induce systemic inflammation through activation of immune cells. The induction of systemic inflammation may start at a local site like an abscess on the limb. For sepsis to develop from a local infection, inflammatory mediators and/or microbial products must enter into the systemic circulation and activate inflammatory cells throughout the body. Since Gram negative bacteria are the most common bacteria cultured from septic patients, the following discussion will focus on the interaction of Gram negative bacteria and the immune system.

During Gram negative sepsis, lipopolysaccharide (LPS), the glycolipid component of the cell wall of Gram negative bacteria, is released into circulation. Upon release, the lipid A portion of LPS binds to LPS binding protein. LPS is recognized via macrophage cell surface receptors like CD14. The main function of CD14, is to transfer LPS to toll like receptor (TLR)-4 and MD-2 for subsequent cellular activation. Once LPS binds to these cell surface receptors, the macrophage becomes activated. Activation of inflammatory cells results in the production of multiple inflammatory mediators that have been implicated in the induction and maintenance of sepsis.

There are many inflammatory mediators involved with sepsis and SIRS. Tumor necrosis factor alpha (TNF-α), interleukin- 6 (IL-6), nitric oxide and leukotrienes are examples of important mediators contributing to the pathophysiology of sepsis in dogs and cats. Sepsis is not simply the induction of inflammation, but rather it is induction of proinflammatory components of the immune system to the extent that they overwhelm normal anti-inflammatory counter regulatory mechanisms. This allows inflammation to go unchecked. However, the immune response during sepsis is a continuum. There can be vacillations between a pro-inflammatory response (i.e., SIRS) and an anti-inflammatory response which is often referred to as the compensatory anti-inflammatory response syndrome (CARS). Therefore, animals with sepsis may have a pro-inflammatory shift or an anti-inflammatory shift. Ultimately, the unchecked proinflammatory cascade leads to inflammatory cell infiltration, altered thermoregulation, vasodilation, vascular leakage, activation of coagulation, hemodynamic instability and multiple organ failure. Conversely, animals that have an anti-inflammatory shift develop of opportunistic infections that could also result in death.

Multiple organ dysfunction syndrome (MODS), is a devastating consequence of sepsis. Pathogenesis of organ failure (e.g., renal failure) during MODS is multifactorial but centers around the development of mitochondrial dysfunction. Circulatory collapse, microcirculatory changes, hypoxemia and inflammation, lead to tissue ischemia, reduced mitochondrial function and thus reduced cellular energy production. The resultant organ damage may be permanent, or may resolve when sepsis is cured. Some forms of organ dysfunction, like acute respiratory distress syndrome (ARDS), are the result of inflammation, not circulatory collapse and mitochondrial damage. During the initiation of ARDS, inflammatory mediators and microbial products activate pulmonary macrophages resulting in vasodilation, vascular leak, pulmonary edema and neutrophilic inflammation. ARDS is characterized by the development of noncardiogenic pulmonary edema, pulmonary inflammation and hypoxemia.

CLINICAL ASPECTS

Clinically, dogs can present in either a hyperdynamic or hypodynamic phase of sepsis. The hyperdynamic phase is a compensatory response, characterized by fever, brick red mucous membranes, tachycardia and bounding pulses. As the disease process progresses, a hypodynamic response characterized by hypotension, pale mucous membranes and hypothermia may be observed associated with decompensation. The hyperdynamic response is an early or relatively less severe systemic manifestation to infection while the hypodynamic response is evidence of an immediately life-threatening situation. Often dogs will have GI or respiratory signs associated with sepsis. Many dogs with sepsis will have myocardial depression and vasodilation leading to hypotension. Cardiovascular, gastrointestinal, hepatic, renal, endocrine and respiratory dysfunction/failure have all been associated with sepsis in dogs.

Cats with naturally acquired sepsis may develop signs of systemic inflammation but they typically do not have a hyperdynamic response during sepsis. Cats appear to develop septic shock (i.e., hypotension that is not responsive to volume expansion) more readily than dogs. Relative or absolute bradycardia (HR 120-150 bpm), hypothermia and abdominal pain (without an abdominal source of infection) are frequent, unique findings in cats with sepsis. Pulmonary inflammation and lesions consistent with ARDS were frequent findings at necropsy in one study of severe sepsis in cats and it is suspected that cats may more readily develop pulmonary failure than dogs. The mechanisms by which these feline specific manifestations of sepsis develop are unknown.

DIAGNOSIS OF SEPSIS IN DOGS AND CATS

The diagnosis of sepsis is accomplished by recognizing the systemic inflammatory response (SIRS) to infection and by identifying the inciting organism. There are several proposed schemes for identification of SIRS, but in general the criteria listed in the table 1 below are used. The more criteria that are met, the more likely it is that your patient has SIRS. However, it is important to note that these are non-specific criteria; any disease process could potentially result in SIRS.

Table 1: SIRS criteria in dogs and cats

Parameter	Dog	Cat
Temperature	> 102.6°F or < 99°F	>103.5°F or <100°F
Heart rate	>140	>225 or <140
Respiratory rate	>40	>40
White blood cell count	>19,500/ul or < 5,000/ul or >5% bands	>19,500/ul or < 5,000/ul or >5% bands

Identify the Infection

The next step is identifying the causative agent. It is important to remember that there are many noninfectious causes of SIRS. Don't forget to consider those differentials. Examples of noninfectious causes of SIRS include, but are not limited to, reaction to medications or vaccinations, envenomization, pancreatitis, heat stroke, trauma, burns, autoimmune disease and neoplasia. Ruling out any potential noninfectious cause of SIRS will help strengthen your clinical suspicion of infection. Some infections will be hard to recognize and simply a high level of suspicion will be enough to warrant therapy for sepsis.

Diagnostic investigation to identify the infectious organism(s) starts with a complete history and physical examination. For some cases, identifying the inciting cause will be simple (i.e., finding an abscess on a limb in a cat with a history of a recent cat bite), while other patients will require in-depth diagnostic testing. Based on historical and physical examination findings, an anatomic region can be usually be identified as the origin of sepsis (i.e., septic peritonitis in the vomiting dog with a history of consuming a foreign body, bacterial pneumonia in a dog with a cough and dyspnea). Thoracic and abdominal imaging are also helpful in identifying the potential source of infection and should be routinely performed in any septic patient. Use clinical clues to help guide sample collection (see table 2). When in doubt, collect samples from any possible source of infection. Samples should be evaluated via cytology/histopathology and with culture and sensitivity. Even if bacteria are not identified on cytology, culture and sensitivity should always be performed so that antibiotic selection can be tailored to the particular organism. Samples should be collected as quickly as possible and preferably prior to administration of antimicrobials. Aseptic technique should be used for sample collection to minimize sample contamination. For patients with suspected bacterial peritonitis, glucose concentration of the blood can be compared to the glucose concentration of the peritoneal fluid to achieve a rapid diagnosis. A difference of > 20 mg/dL between blood and peritoneal fluid glucose concentrations (blood > peritoneal fluid) is diagnostic for septic peritoneal effusion. Remember that infectious organisms that do not lend themselves well to culture may be the cause of sepsis so, depending upon the clinical presentation, serology and/or PCR for vector borne disease, viral infection and fungal infection etc. should be considered.

Table 2: Sample collection	Clinical clues to guide sampling	Evidence of infection (Note: a lack of these findings does not necessarily indicate a lack of infection in the sampled area)
Blood culture	Any (perhaps all) patient with suspected sepsis, newly discovered heart murmur	Positive culture
Bronchoalveolar lavage, endotracheal wash	Cough, dyspnea, interstitial or alveolar infiltrates on thoracic radiographs consistent with infectious pneumonia; (note that bacterial pneumonia is relatively rare in the cat)	Positive culture, degenerative neutrophils, intracellular bacteria or evidence of fungal organisms
Abdominocentesis	GI signs, abdominal distention, abdominal effusion, abdominal pain, lack of abdominal detail on radiographs or evidence of abdominal fluid on ultrasound	Positive culture, intracellular bacteria or evidence of fungal organisms, > 20 mg/dl difference between blood and fluid glucose concentrations
Thoracocentesis	Inspiratory or mixed inspiratory and expiratory dyspnea, muffled lung sounds in the dependent lung fields, evidence of pleural effusion on imaging	Positive culture, intracellular bacteria or evidence of fungal organisms
Arthrocentesis	Evidence of joint pain or effusion	Positive culture, intracellular bacteria or evidence of fungal organisms
Aspiration of subcutaneous masses	Any mass that has recently developed or changed in size, any draining or ulcerated mass, history of trauma in the region	Positive culture, intracellular bacteria or evidence of fungal organisms
Urine culture	Any patient with sepsis of unknown origin, azotemia, pyuria, bacteriuria, evidence of pyelonephritis on ultrasound	Positive culture, bacteriuria, pyuria. (Note: a lack of bacteriuria or pyuria does not rule out urinary tract infection or pyelonephritis)
Pericardiocentesis	Muffled heart sounds, evidence of a globoid heart on thoracic radiographs, evidence of pericardial effusion on ultrasound	Positive culture, intracellular bacteria or evidence of fungal organisms
Aspiration of abdominal organs	Identification of abnormal organ architecture or abscess on ultrasound	Positive culture, intracellular bacteria or evidence of fungal organisms
Diagnostic peritoneal lavage	Suspected abdominal source of infection but little or no abdominal effusion; any patient with sepsis of unknown origin	Positive culture, intracellular bacteria or evidence of fungal organisms

REFERENCES

Annane D, et al. **BMJ** 2004;doi:10.1136. Brady C, et al. **Vet Clin North Am Small Anim Pract** 2001;31:1147. DeClue et. al **JVIM**; 26:624-30. DeClue et. al **J Am Vet Med Assoc** 2011; 238:890-7. De Laforcade A, et al. **J Vet Intern Med** 2003;17:674. Hauptmann J, et al. **Vet Surg** 1997;26:393. Marik P, et al. **Intensive Care Med** 2004;30:748. Marik P, et al. **Chest** 1998;114:854.

SEPSIS: TEAM BUILDING FOR SUCCESSFUL OUTCOMES

Dennis R. Spann, DVM, DACVIM (SAIM) and Amy E. DeClue, MS, DVM, DACVIM (SAIM)
Roseville, CA and Columbia, MO

INTRODUCTION

Certified Veterinary Technicians serve vital roles in the care of animals with sepsis. These essential roles include: providing monitoring, nursing care, coordination of procedures, administration of medications, record keeping, providing feedback to the clinician and communication with pet owners. Since sepsis is a condition that results in dynamic, rapidly changing health status, aggressive monitoring, nursing care and early institution of therapy are particularly important. In this lecture, we will discuss mechanisms for building systems for a team oriented approach to the care of animals with sepsis.

Each veterinary practice's structure is different. The first step in developing a systematic approach to the care of animals with sepsis is to identify the sepsis team. Most teams will consist of a veterinarian and 1- 4 CVTs. The number of individuals involved in the team will begin to dictate how duties will be delegated. When a larger team is assembled, it is often advantageous to have a CVT team leader that will keep the team organized and on task. Often this individual will be responsible for maintaining the medical record for the patient. As you are considering how to build the team and develop a systematic approach to managing animals with sepsis, you should consider how your team will interact and what tasks that should be completed during the initial acute evaluation of the patient versus during longer-term hospitalization. Check lists insure that optimal management is performed and should be developed to prevent omission of critical steps in patient care. This is best accomplished by having a meeting of the sepsis team and discussing diagnostic and treatment goals. Based on those goals a management flow chart can be developed. Examples of such lists/flow charts are noted below.

Management flow charts should be made readily available, pre-printed and attached to a clipboard so that as an animal with suspected sepsis enters the ICU, immediate care can be provided based on your individual veterinary hospital guidelines. Hospital guidelines can be roughly divided into the following time frames: initial (i.e., first 1-6 hours) resuscitation, initial diagnostic and long-term (i.e., post-initial resuscitation) management procedures. By having protocols with specific guidelines and check lists in place, the team will function more efficiently and it is less likely that medical errors will occur.

The severity of bacterial sepsis varies from patient to patient. Some patients will require advanced life support measures to survive while others will require only basic supportive care. The key aspects of sepsis treatment include 1.) initial resuscitation, 2.) alleviating the inciting cause and 3.) aggressive supportive care. The treatment of sepsis can be roughly organized into initial resuscitation (i.e., first 1-6 hours) and long-term management (i.e., > 6 hours- discharge from hospital). Initial resuscitation includes not only restoration of hemodynamic homeostasis, but also alleviating the inciting cause through administration of antimicrobial drugs and/or debridement. Recently the concept of early goal directed therapy has become more prevalent in human medicine. Goal directed therapy is based on the idea of tailoring treatment to specific physiologic parameters thought to be associated with restoration of homeostasis and improved outcome. This approach is very amenable to check lists and flow charts to help direct therapy.

INITIAL RESUSCITATION PROTOCOL

Restoration of hemodynamic stability

Initial hemodynamic resuscitation of animals with sepsis is important for restoration of homeostasis. While there are no generally accepted, standardized initial resuscitation protocols for animals, guidelines have been established for people. These guidelines are outlined in the Surviving Sepsis Campaign: International Guidelines for Management of Severe Sepsis and Septic Shock. These guidelines were initially established in 2004 and then updated in 2008 and 2013. The Surviving Sepsis Campaign guidelines were developed by a group of physicians based on an extensive literature review. Recommendations were graded based on the quality of scientific evidence available. The authors recommended protocolized, quantitative resuscitation of patients with sepsis-induced tissue hypoperfusion. This approach is also termed early goal-directed therapy. Early goal directed therapy resulted in a 17.7% absolute reduction in 28-day mortality in a multicenter trial. The goals of early goal-directed therapy in people are to achieve each of the following during the first 6 hours of intervention:

- CVP 8-12 mmHg
- MAP ≥65 mmHg
- Urine output ≥ 0.5 ml/kg/h
- Superior vena cava oxygenation saturation of 70% or mixed venous oxygen saturation of 65%
- Additionally, resuscitation should be targeted to normalize lactate in patients with increased blood lactate concentrations.

While these goals have not been specifically tested in randomized, multicenter trials in animals, the general concept behind the goals could have value in veterinary patients. Resuscitation goals should be tailored to the unique physiologic parameters of the species. Parameters that could have value in developing a standardized treatment protocol in your practice include:

- Central venous pressure: Central venous pressure is an indicator of volume status. The most ideal CVP to achieve in dogs with sepsis has not been determined, but many veterinarians aim for a CVP of 7-10 cm. It is important to note that there are many non-volume related factors that can influence CVP and thus monitoring trends in CVP and developing patient specific goals could be more helpful than using a generalized guideline.
- Mean arterial blood pressure (MAP): A MAP of less than 60-65 mmHg is considered hypotension in the dog and cat. While the exact target blood pressure goals have not been determined, proposed target Doppler systolic blood

pressure post resuscitation is > 90 mmHg. Achievement of a Doppler systolic blood pressure > 90 mmHg within the first hour of fluid resuscitation was associated with increased survival in one study of dogs presenting to an emergency room with a variety of conditions. Induction of hypertension should be avoided.

- Urine output: In a dog or cat with normal hydration status, oliguria is defined as urine output less than 0.5-1.0 mL/kg/h and anuria is defined as urine output less than 0.3-0.5 mL/kg/h. Avoidance of oliguria or anuria is a reasonable goal for dogs and cats with sepsis.
- Additional parameters: Superior vena cava oxygen saturation (ScvO2), lactate and base deficit have been evaluated as prognostic markers for dogs with sepsis in a prospective clinical trial evaluating goal-directed therapy for sepsis. A higher ScvO2 (odds ratio – 0.76) and a smaller base deficit (odds ratio = 0.55) was associated with a lower probability of death (odds Ratio= 0.76). These data suggest that these parameters may have value in developing goal-directed therapeutic protocols in the dog although specific recommendations for target values have not been determined.

There is a continuum between early goal directed therapy and later management. While it is important to consider early resuscitation goals, maintenance of these goals during continued care beyond of the initial six hours of management is important. After initial resuscitation, it is not uncommon for animals to have recurrence of hemodynamic derangements.

Achieving hemodynamic support goals can be accomplished using boluses of isotonic crystalloids (Dogs 15-20 ml/kg, IV to effect; Cats 10-15 ml/kg, IV to effect) or colloids (Dogs, 5 ml/kg to effect; Cats 2 ml/kg to effect). Then, ongoing fluid therapy should be tailored to the needs of the patient to provide for maintenance requirements, correct interstitial hydration deficits, replace ongoing losses and continue to correct any hemodynamic derangements. Patients with sepsis will have a tendency to develop interstitial edema, including subcutaneous and pulmonary edema, because of increased vascular permeability and decreased blood colloid osmotic pressure. Administration of colloid fluids (e.g., hetastarch) may help prevent interstitial edema and should be considered for volume resuscitation and during maintenance of sepsis patients. Colloids are usually given in addition to crystalloid fluid support. While there is evidence in humans that hetastarch increases the risk for acute kidney injury, no evidence supporting or refuting this phenomenon is available in dogs and cats with sepsis at this time. The use of 0.9% NaCl has fallen out of favor in people with sepsis because of the amount of chloride in the solution. Administration of a chloride rich solution induces shock in animal models, promotes metabolic academia, induces a pro-inflammatory shift, reduces renal perfusion, increases the risk of acute kidney injury in people with sepsis and increases mortality in adult and pediatric people with sepsis. For these reasons, the use of a lower chloride solution like plasmalyte is recommended.

Despite aggressive volume resuscitation, some patients will require additional support to maintain normal blood pressure and tissue perfusion. The safest and most efficacious treatment for septic shock has not been determined for dogs or cats. Prior to administration of any sympathomimetic drug, care should be taken to ensure that the patient is not hypovolemic. Since the goal of treating septic shock is to maintain tissue perfusion, medications that cause vasoconstriction should be used only if absolutely necessary. Positive inotropic drugs (e.g., dobutamine) may be a good initial management choice for septic shock since they help combat decreased cardiac output caused by myocardial dysfunction without inducing peripheral vasoconstriction. However, care should be taken to avoid positive inotropic drugs in cats with evidence of HCM. If volume resuscitation and positive inotropic support has failed to restore adequate blood pressure, a vasopressor agent (e.g., dopamine, norepinephrine, epinephrine or vasopressin) could be added. Vasopressors should always be administered as a CRI and ideally, direct blood pressure evaluation should be used for monitoring. Although there are no clinical trials evaluating these drugs, epinephrine is generally considered the least ideal vasopressor. Critical illness related adrenal cortical insufficiency has been documented in dogs and cats with sepsis. It should be considered in any animal that has refractory hypotension after the patient is volume resuscitated. Judicious use of corticosteroids may address therapeutic failure of vasopressor therapy in these patients.

Alleviate the cause

The most important aspect of treating bacterial sepsis centers on the identification and eradication of the inciting cause. Although stringent effort should be made to identify the cause of sepsis, early antimicrobial treatment is critical for survival and should not be withheld pending culture results. In humans, early antibiotic administration has been shown to decrease mortality. Thus administration of appropriate antimicrobial drugs should be considered part of the initial resuscitation protocol. Broad spectrum, bactericidal antimicrobial agents (e.g., fluoroquinolone + penicillin) administered IV should be instituted as quickly as possible in patients with suspected sepsis. Antibiotic therapy should be selected based on the most likely type of organism given the site of infection. Depending upon the source of infection, surgical debridement may be necessary for infection control. Once culture specific antibiotic sensitivity is determined, the antibiotic with the narrowest spectrum of activity should be chosen and administered until there is complete clinical resolution.

INITIAL DIAGNOSTIC PROTOCOL

Factors that should be considered when developing initial diagnostic protocols have been reviewed in another set of proceedings for this conference (DeClue, Spann, Sepsis in Dogs & Cats: Pathophysiology and Recognition, ACVIM 2014). We refer you to those proceedings for additional detail. An example checklist for initial diagnostic testing for sepsis in dogs and cats has been included below:

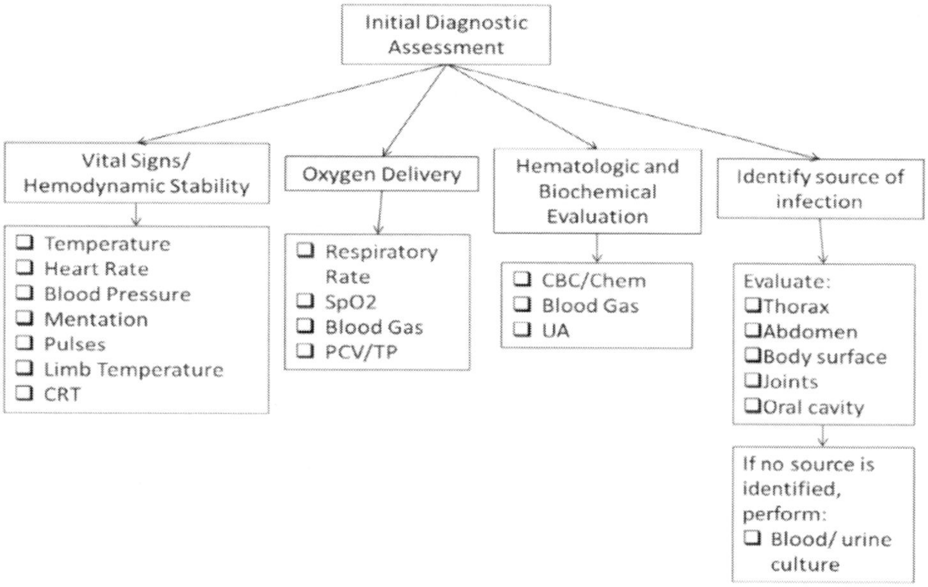

LONG TERM PROTOCOL

Development of a generalized long-term care protocol can be challenging as each protocol requires individual tailoring based on patient needs. However, general areas of focus should be developed to insure that important care aspects are not forgotten. A flow chart of management areas to consider along with more specific details are noted below:

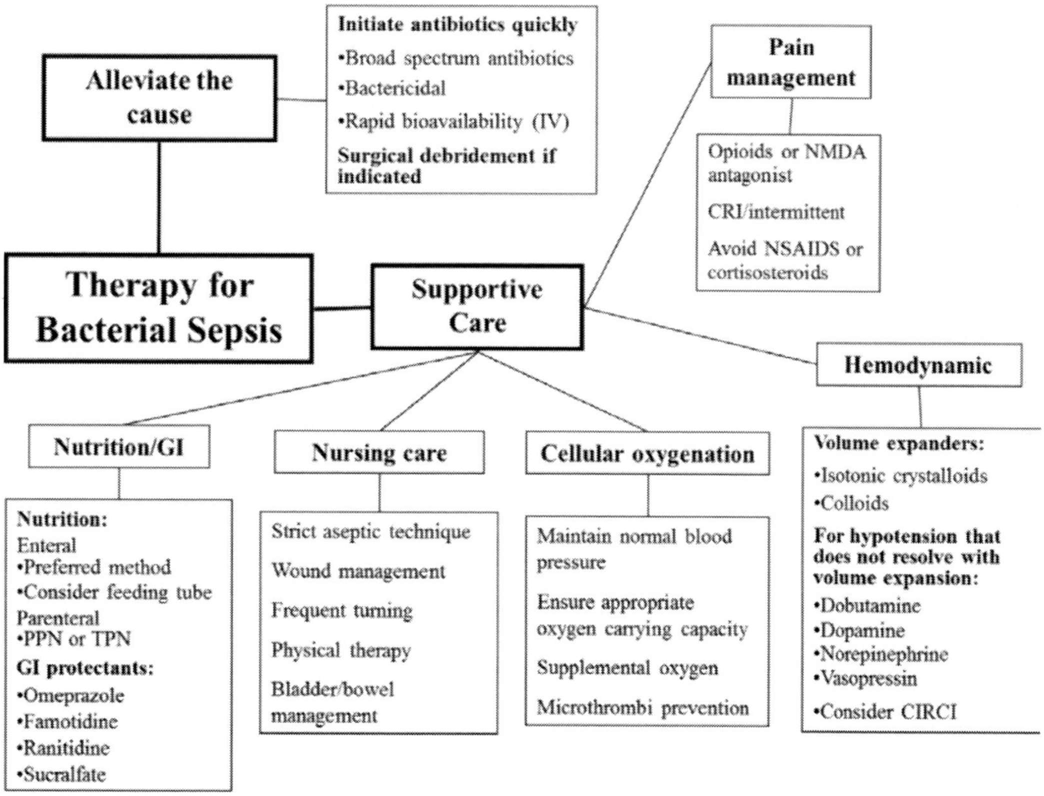

GI/Nutrition- Bacterial translocation from the GI tract can contribute to systemic inflammation during sepsis. Early placement of a feeding tube and initiation of enteral or, in patients with vomiting, parenteral nutrition will help maintain GI barrier function. Additionally, medication aimed at maintaining normal GI protective mechanisms (e.g., omeprazole, famotidine, sucralfate) could be considered. Hyperglycemia can be a complication of nutritional therapy, especially parenteral nutrition. Hyperglycemia has been associated with increased inflammation and a poorer prognosis in people with sepsis and SIRS. Although the importance of glucose homeostasis in dogs and cats with sepsis is unknown, iatrogenic hyperglycemia should be avoided.

Cellular oxygenation- Oxygenation, acid-base status, PCV or hemoglobin concentration and organ function should be closely monitored. Along with maintaining good tissue perfusion, maximizing cellular oxygenation will help maintain tissue viability and help avoid multiple organ dysfunction. Adequate cellular oxygenation may be accomplished by ensuring good oxygen carrying capacity (i.e., avoiding anemia), providing supplemental oxygen when indicated and preventing microthrombi formation. For the majority of patients, acid-base abnormalities are related to lactic acidosis secondary to poor tissue perfusion. Typically these abnormalities will resolve with resolution of hypotension and normalization of perfusion.

Nursing care/Pain management- Care should be taken to ensure adequate patient comfort including management of pain, careful catheter maintenance, bladder/bowel care and frequent patient turning/movement to prevent decubital ulcers. Almost all patients with sepsis will require analgesic administration for pain management.

REFERENCES

Conti-Patara et. al **J Vet Emerg Crit Care** 2012;22(4):409-18. Dellinger et. al **Crit Care Med** 2012; 41(2):580-637. Ertmer et. al **Crit Care Med** 2011; 39(4):783-92. Kellum et. al **Chest** 2006 Oct;130(4):962-7. Kellum et. al **Chest** 2004 Jan;125(1):243-8. Mutter et. al **Cochrane Collaboration** 2013; 1-27. Silverstein et. al J **Vet Emerg Crit Care** 2012; 22(6);666-73. Young et. al **Ann Surg** 2014;259(2):255-62.

MARIJUANA: THE HIGH POINTS

Dawn Merton Boothe, DVM, DACVIM, DACVCP
Auburn, AL

INTRODUCTION

Marijuana is formed from the dried leaves and tops of the hemp plant (*Cannibus sativa)* (Svienska 2008). Marijuana has been a part of recreational, religious and medical activities of a variety of cultures for over 5000 years (Krietzer 2009;Burns 2006). Indeed, it was among the most commonly prescribed medications in the United States Pharmacopeia until declared illegal in the 1930s. It subsequently has been declared a controlled substance (Class II: High potential for abuse, but significant clinical indication). The potential efficacy of marijuana for control of pain has led to a passage of laws allowing medical use in several states. This, in turn, has stimulated a flurry of scientific activity in an attempt to provide evidence for medical use of marijuana. Not surprisingly, pet owners have also been engaged in the conversation, with a potentially legitimate reason for administering marijuana. Not surprisingly, however, most of the scientific information generated as evidence is intended to support human, rather than veterinary use. To understand the allure of marijuana, and specifically if its promise of relief from what ails our pets is hype or hope, one must understand the plant.

Marijuana Ingredients

Marijuana is a pharmacologically (and toxicologically) diverse herb. *Cannibus* contains at least 480 unique compounds, with their presence varying with the plant product. Plant products include, in addition to marijuana, hashish and hashish oil, formed from the resin secreted by the plant. It is important to note that the amount of any one compound in the hemp plant can vary markedly, depending on the plant part.

The most well known of the compounds in the hemp plant are the cannabinoids, a term used to refer to a terpenephenolic compounds. The discovery of cannabinoids led to the recognition of the endocannabinoid system with endogenous cannabinoids. Since their discovery, both by pharmaceutical companies and substance abusers have synthesized synthetic compounds. "Phytocannabinoids" is thus used to refer to those occurring in the plant whereas "endocannabinoids" refers to endogenous and synthetic chemicals. Endocannabinoids also appear to be important as neuroprotectants (e.g, antioxidants, inhibition of calcium influx and excessive glutamate production), for example, that associated with CNS ischemia or hypoxia, or the presence of neurotoxicants. These effects appear to be mediated predominantly by CB1 (located particularly in the dorsal horn of the spinal cord) although CB2 also plays a role, depending on the tissue (Svizenska 2008). Cannibinoids also inhibit neuroinflammation (see therapeutic indications).

Close to 70 phytocannabinoids, divided into 10 classes, have been identified in the hemp plant, and particularly marijuana. Table 1 lists those associated with presumed therapeutic use (Brenneisen Ch. 2). Among the cannabinoid compounds found in marijuana, THC is the most pharmacologically and toxicologically relevant, and the most understood. It is THC that is responsible for most of the natural effects of the *Cannabis* plant. It acts by binding to the CB-1 receptor. CBD is the next "best" phytocannabinoid. In addition to its anxiolytic effects, it also reduces unpleasant side effects, primary due to potent inhibition of cytochrome P450 3A11 which otherwise would metabolize THC to much more potent psychoactive compounds.

In addition to the cannabinoids, marijuana contains approximately 140 different terpenoids which are responsible for its scent. The terpenoids yielded from a marijuana plant depend on the type of *Cannibis* (based on drug or fiber content), the part of the plant, its sex and age, whether or not it is cultivated in or outdoors, when it is harvested and the conditions at harvest, and how it is dried and stored. The serotinergic effects of marijuana (5-HT1A and 2A) may reflect the impact of these essential oils, contributing to analgesia and mood modification. Other components in the plant include nitrogen containing compounds (n = 70: alkaloids, amines); carbohydrates, including common monosacharides (n=13: fructose, glucose, mannose), selected disaccharides (sucrose, maltose), and several polysaccharides (eg, cellulose, pectin) as well as several sugar alcohols (n = 12; mannitol, sorbitol, glycerol). A number of flavonoids also are present (n=23); among them, apigenin has a wide variety of effects, including interaction with benzodiazepine receptors, resulting in an anxiolytic effect. Other ingredients include fatty acids (n=33) and others.

The Target: Cannabinoid Receptors

The endocannabinoid system is comprised of eicosanoid cannabinoid [CB; protein g coupled, negative to adenylyl cyclase] receptors [CBr], their endogenous ligands and the enzymes responsible for their synthesis and degradation. This system as a known contributor to physiology has been recognized for only about 25 years old. At least two CBr have been identified in many species, including the dog. CB1r occurs in the brain but also occurs in some peripheral tissues (cardiovascular, reproductive, gastrointestinal). They are responsible, in part, for central and peripheral regulation of food intake, fat accumulation and lipid and glucose metabolism. The dopaminergic reward pathway is stimulated by CB1 receptors, motivating eating, smoking and substance abuse. CB2r are located principally on immune cells, but this includes microglia. CB2 also is located on neurons where it may be associated with cell differentiation (Svizenska 2008). In the CNS, CBr are suggested to influence neurotransmitter release. At least 5 endogenous cannabinoids have been described, with anandamide (CB1 and 2 agonist, but higher affinity for CB1) being the most thoroughly studied. It is synthesized by post-synaptic neurons, acting as a retrograde messenger to influence neurotransmitter release. It is extremely unstable, being rapidly hydrolyzed to ethanolamine (an antimistamine) and arachidonic acid. Cannabinoids are able to disrupt short-term memory, impair cognition and time perception, alter mood while enhancing body awareness, discoordination, sleepiness, and reduce attention focus and the ability to "filter" irrelevant information.

As with many CNS active drugs, marijuana is associated with both tolerance (higher concentration needed to impart a similar pharmacologic effect) and withdrawal (a clinical syndrome of nervousness, tension, restlessness, sleep disturbance and anxiety). However, the long elimination half-life of the most active ingredient, THC (and others) appears to preclude a clear cut abstinence syndromw (Svizenska 2008). As with other addictive agents, laboratory rodents have been demonstrated to self medicate, suggesting an addictive component.

It is important to note that although interaction with cannaniboid receptors is unique among plants to hemp, other receptors are also targeted (as noted above: benzodiazepines, serotonin, others).

Table 1. Cannabanioids with potential therapeutic benefit

Compound	Proposed Therapeutic Use
CBGA (cannabigerolic acid)	Antibiotic
CBG (cannabigerol)	Antibiotic, antifungal, Antinflammatory, Analgesic
CBC (cannabichromene)	Antibiotic, antifungal, Antinflammatory, Analgesic
CBDA (Cannabidiolic acid)	Antibiotic
CBD (Cannabidiol)	Anxiolytic, antipsychotic, analgesic, anti-inflammatory, antioxicant, antispasmodic
CBN (Cannabinol)	Sedative, antibiotic, anticonvulsant, anti-inflammatory
THC (delta-9-tetrahydrocannabinol; delta-8 to a lesser degree)	Euphoriant, Analgesic, Anti-inflammatory, Antioxidant, Antiemetic
THCV	Analgesic, euphoriant

Medicinal Marijuana (?)

The proposed indications for medical marijuana have included, but are not limited to behavioral, sleep and gastrointestinal disorders, neuroprotection, antispasmodic but prokinetic, anorexia, nausea, glaucoma, diabetes, immunosuppression, malaria, anti-inflammatory and, of course, pain (Table 1, Izzo 2009). However, other potential indications have included A proposed advantage of medical marijuana compared to a single drug (e.g., dronabinol, a synthetic THC [Marinol®]), iis the multiple compounds contained in the plant. Two advantages are offered: 1. The compounds might act synergistically (a "synergistic" shotgun) to provide an enhanced desired pharmacologic effect while 2. at the same time, mitigating (one compound acting on another) undesirable effects. However, evidence for a synergistic benefit is lacking based on the lack of differences when THC is consumed as marijuana, versus Marinol® (humans). (Brenneisen 200X). Presumably, because marijuana contains so much THC, it may not be the most effective portion of the plant and it may contribute to more side effects (hence the question mark for this section; see also Marijuana and pets). Note that some plants may be designed to contain more or less of a specific target compound.

Pain and Inflammation/Immunomodulation: The capacity for cannabinoids to control pain is among the most studied responses. They are effective in both acute (phasic) and chronic (tonic) pain. They peripherally and centrally modulate processing of nociceptive signals. They act as antihyperalgesics. CBD has demonstrated efficacy in experimental models; the effects also appear to involve transient receptor potentials (Izzo 2009). CBD also influences T-cells, causing a generalized immunosuppressive effect. A number of mechanisms of immunomdulation have been proposed,including altered interleukin or tumor necrosis factor production or release, neutrophil migration, production of specific antibodies, etc. Arthritis and psoriasis are among the chronic inflammatory diseases in humans for which CBD has demonstrated or is suggested to have some potential efficacy.

Epilepsy: Experimentally, CBD attenuates experimentally-induced seizures in animals; this may reflect reduced calcium fluxes (Izzo 2009). THCV also has been associated with some anticonvulsant effects by virtue of its inhibitory effects on CB1.

Anxiolytic: These effects have been demonstrated in healthy human volunteers (Izzo 2009).CBD exerts benzodiazepine independent effects, possibly by activating post synaptic 5-HT1A receptors.

Neuroprotection: CBD is an antioxidant and as such has been proposed for treatment of Alzheimer's disease, Parkinson's disease and Huntington's disease. Restoration of calcium homeostasis may prevent apoptosis (Izzo 2009). In rodents, CBD reverses brain damage associated with ischemia.

Anti-Emesis/appetite suppression: Again, CBD has been demonstrated in animal models to be effective for the control of vomiting otherwise not responsive to 5-HT-3 antagonists. THCV and synthetic CB1 antagonists decrease food intake.

Diabetes mellitus: CBD inhibits development of diabetes in non-obese diabetic mice, including ameliorating clinical signs of disease. This appears to reflect, in part, control of pancreatic inflammation, but also reduction of oxidative stress in target tissues (eg, retina).

Bone formation: A number of cannabinoids (essentially all in Table 1) stimulate mesenchymal stem cells responsible for bone formation and fracture healing. CBD aldo controls bone resorption, reducing bone loss (Izzo, 2009).

Cancer: A number of the cannabinoids (all in Table 1) have antiprolierative-anti apotopic effects in a number of tumor cell lines.

Antimicrobial: CBC and CBG have demonstrated potent antibacterial effects towards selected microbes, including methicillin resistance staphylococci (MIC of 0.5 to 2 mcg/ml).

Marijuana and pets

Most of the information surrounding marijuana and pets is from a toxicologic standpoint. THC is among the compounds cited as a toxicologic hazard in detection (police) dogs (Llera 2008). It is the most common drug to which detection ogs are exposed. Both dogs and cats may become intoxicated with smoke inhalation as well as ingestion of food containing marijuana (or hashish). It is absorbed rapidly following either oral or inhalant administration with clinical signs evident within 30 to 60 minutes of ingestion, although one reference (Osweiler 2008) indicates onset as long as 12 hours after exposure. The drug is eliminated by hepatic metabolism and biliary excretion with elimination being complete in 5 days in dogs; duration of toxicity ranges from 30 minutes to 3 days, but 18-24 is the averag. Enterohepatic circulation contributes to the prolonged half-life. The most common signs of toxicity following ingestion in dogs include tachycardia, hypotension, depression, ataxia, vomiting (inducing emesis is not recommended in clinically depressed dogs because of the risk of aspiration), altered behavior, bradycardia, hypersalivation, weakness, hypothermia and seizures. Legalization of marijuana may be associated with an increased incidence of toxicity, with a 4 fold increase cited in one study (Meola 2012). Treatment is largely supportive, with sedation with benzodiazepines or phenothiazines as needed. Antiemetic therapy may be indicated.

No data exists regarding the efficacy of any portion of the hemp plant as an analgesic, or for other therapeutic indications. A commercially available dietary supplement purportedly consisting primarily of hemp stem is being marketed for dogs. Because the stem contains very little THC but a high proportion of CBD and other non-psycotropic cannabinoids, its claims is that it is a effective analgesic without psychotropic effects. While such a product is more appealing than marijuana as an analgesic, neither data supporting efficacy nor safety nor quality assurance data could be obtained.

REFERENCES

Brenneisen R. Chemistry and Analysis of Phytocannabinoids and other *Cannabis* Constituents. In: Forensic Science and Medicine: Marijuana and the Cannabinoids. Elshohly MA (ed), Humana Press Inc., NY, pp 17-49; Izzo AA: Non-psychotropic plant cannabinoids: new therapeutic opportunities from an ancient herb. Cell Press, Elsevier, 2009, 515-527;Kreitzer FR, Stella N. The therapeutic potential of novel cannabinoid receptors. Pharm Ther 2009; 122: 83-96;Llera RM, Vomer PA. toxicologic hazards for police dogs involved in drug detection. JAVMA 2006; 228: 1028-1032; Meola SD, Tearne CC, Haas SA, et al. Ealuation fo trends in marijuana toxicosis in dogs living in a state with legalized medical marijuana: 125 dogs (2005-2010). JVet Emer Crit Car 2012; 22: 690-696; Osweiler GD, Hovda LR, Brutlag AG, et al. Small Animal Toxicology, Wiley-Blackwell, 224-229; Svizenska I, Dubovy P, Sulcova A. Cannabinoid receptors 1 and 2 (CB1 and CB2), their distribution, ligands and functional involvement in nervous system structures: A short reviewe. Pham Biochem Beha 2008; 90: 501-511.

THE PLEASURE LAB

Charles H. Vite, DVM, PhD, Dipl ACVIM (Neurology)
Philadelphia, PA

INTRODUCTION

"Sleep my little baby-oh
Sleep until you waken
When you wake you'll see the world
If I'm not mistaken...

Kiss a lover
Dance a measure,
Find your name
And buried treasure...

Face your life
Its pain,
Its pleasure,
Leave no path untaken."

— Neil Gaiman, *The Graveyard Book*

Music. Sex. Social media. Sports. Drugs. Fast cars. Shopping.
These frequently give us a sense of pleasure and we seek them out for more.

Work. Attending classes. Laundry.
These less frequently give us pleasure and we postpone them if we can.

Death and taxes.
These we approach often passively as unchangeable.

Dentists. Zombies.
These we may avoid altogether.

Events evoking pleasure involve the medial forebrain, the ventral tegmental area, and the nucleus accumbens. Events we avoid involve the periventricular system. Passivity involves the septal nuclei, hippocampus, amygdala, and basal nuclei. What is going on inside our brain during times of pleasure, and why don't we feel pleasure all the time? What would happen to you and to the species if we felt "transcendent" all the time instead of just occasionally? How are the regions influencing pleasure determined? How can they be influenced? Do animals experience the same feelings of pleasure as we do?

fMRI AND PET

Functional MRI (fMRI) is a technique for non-invasively measuring hemodynamic changes following enhanced neural activity. Blood-oxygen-level-dependent (BOLD) contrast mechanisms are used and activation maps of the brain are generated. The interpretations of the activation maps are the key to understanding the involvement of specific brain regions in behavioral responses. Positron emission tomography (PET) uses specific positron-emitting isotope labeled biomolecules to examine the metabolic activity of specific brain regions. The methods and interpretation of fMRI and PET to evaluate the brain will be discussed.

INFLUENCING PLEASURE

Pleasurable experiences are associated with increased levels of dopamine in the limbic system. We will review the literature supporting this finding. We will discuss how dopamine is involved in both the expectation and the experience of pleasure. However, the administration of L-DOPA to people suggests that positive mood states are not specifically affected and that serotonin, norepinephrine, glutamate, GABA, endogenous cannabinoids and opioids, and others may play a substantial role.

DO ANIMALS EXPERIENCE PLEASURE THE WAY WE DO?

We will discuss the behavioral and imaging evidence that animals experience pleasure.

REFERENCES

Balcombe, J. Pleasureable Kingdom: Animals and the Nature of Feeling Good. Palgrave Macmillan 2007.
Bloom P. How Pleasure Works: The New Science of Why We Like What We Like. WW Norton and Co, 2010.
Liggins J, Pihi RO, Benkelfat C, Leyton M. The dopamine augmenter L-DOPA dos not affect positive mood in healthy human volunteers. PLOS ONE 7: e28370, 2012.
Linden DJ. The Compass of Pleasure: How Our Brains Make Fatty Foods, Orgasm, Exercise, Marijuana, Generosity, Vodka, Learning and Gambling Feel So Good. Viking, 2011.
Logothetis NK. What we can do and what we cannot do with fMRI. Nature 453: 869-878, 2008.
Oberlin BG, Dzemidzic M, Tran SM, Soeurt CM, Albrecht DS, Yoder KK, Kareken DA. Beer flavor provokes striatal dopamine relesease in male drinkers: mediation by family history of alcoholism. Neuropsychopharm 38: 1617-1624, 2013.
Sharot T, Guitart-Masip M, Korn CW, Chowdhury R, Dolan RJ. How dopamine enhances an optimism bias in humans. Curr Biol 22: 1477-1481, 2012.
Salimpoor VN, van den Bosch I, Kovacevic N, McIntosh AR, Dagher A, Zatorre RJ. Interactions between the nucleus accumbens and auditory cortices predict music reward value. Science 340: 216-219, 2013.

AUTHOR INDEX

AUTHOR INDEX

AUTHOR INDEX

American College of Veterinary Internal Medicine
1997 Wadsworth Boulevard Suite A
Lakewood CO 80214-5293

CURRAN ASSOCIATES INC
proceedings.com